Frontiers of Life

VOLUME IV

The Living World

ADVISORY BOARD
EDITORIAL DIRECTORS

Frontiers of Life

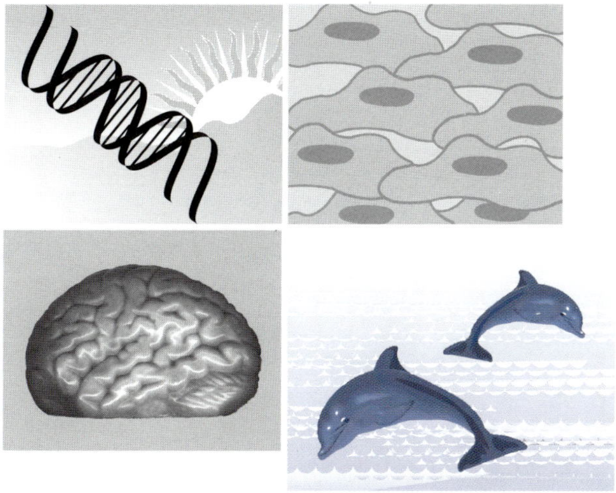

VOLUME IV

The Living World

ACADEMIC PRESS
A Harcourt Science and Technology Company
San Diego San Francisco New York Boston London Sydney Tokyo

TRECCANI
ISTITUTO DELLA ENCICLOPEDIA ITALIANA
ROMA

This book is printed on acid-free paper. ∞

Academic Press
A Harcourt Science and Technology Company
525 B Street, Suite 1900, San Diego, California 92101-4495, USA
http://www.academicpress.com

Academic Press
Harcourt Place, 32 Jamestown Road, London NW1 7BY, UK
http://www.academicpress.com

Library of Congress Catalog Card Number: 2001091214

International Standard Book Number: 0-12-077340-6 (set)
International Standard Book Number: 0-12-077341-4 (volume 1)
International Standard Book Number: 0-12-077342-2 (volume 2)
International Standard Book Number: 0-12-077343-0 (volume 3)
International Standard Book Number: 0-12-077344-9 (volume 4)

PRINTED IN THE UNITED STATES OF AMERICA
01 02 03 04 05 06 WP 9 8 7 6 5 4 3 2 1

FRONTIERS OF LIFE

EDITORS-IN-CHIEF:
David Baltimore, Renato Dulbecco, François Jacob, Rita Levi-Montalcini

VOLUME I

The Origins of Life

PART ONE
Origin and Evolution of Life
Edited by James E. Dahlberg and Glauco Tocchini-Valentini

PART TWO
Genetic Language
Edited by Fancesco Amaldi and Giuseppe Attardi

VOLUME II

Cells and Organisms

PART ONE
Cells and Cell Communities
Edited by Nica Borgese, Francesco Clementi, and Pietro De Camilli

PART TWO
The Immunological Systems
Edited by Antonio Lanzavecchia, Bernard Malissen, and Roberto Sitia

VOLUME III

Intelligent Systems

PART ONE
The Brain of *Homo sapiens*
Edited by Emilio Bizzi, Pietro Calissano, and Virginia Volterra

PART TWO
Building Blocks for Intelligence Systems
Edited by Daniel Amit and Giorgio Parisi

VOLUME IV

The Living World

PART ONE
Biology of Behavior
Edited by Enrico Alleva and Patrick Bateson

PART TWO
Discovery and Spoliation of the Biosphere
Edited by Elgene O. Box and Sandro Pignatti

PART THREE
Bioethics
Edited by Luigi De Carli and Stefano Rodotà

CONTENTS OF VOLUME IV

SECTION III / **CHANGES TO THE ENVIRONMENTAL CYCLE**

SECTION IV / **MAN IN A THREATENED ENVIRONMENT**

PART THREE

BIOETHICS

SECTION I / **BIOETHICS: HISTORY AND SIGNIFICANCE**

INTRODUCTION TO VOLUME IV

RENATO DULBECCO

This volume contains an overview of the relation between living beings and the world that surrounds them, as far as other living beings are concerned — with special attention paid to behavior and to ethical considerations — and as far as the environment in which all living beings abide, namely, the biosphere, is concerned.

The first part of the volume examines the biological bases of behavior. The ensemble of an individual's attitudes is the result of the interaction between two components: an innate component of genetic origin and one acquired through experience and upbringing. For example, weaving a web, for a spider, is an automatic form of behavior just as learning a language is for a human being. In both cases, however, the task is accomplished only if environmental conditions are suited to it.

The influence of the environment and of culture on the human species is evidenced by the differences one observes in the behavior of individuals belonging to different countries and regions of the world. Genetic influences, instead, have been studied using identical twins who share the same genes but have lived in different environments since birth. This line of research suggests that behavior may be explained in terms of the combined and interdependent effect of genes and social milieu, which represents personal experience and upbringing.

The observation of both humans and animals has shown that it is the experiences at the beginning of life that are especially important. This is particularly true for the human species, in which the characteristics of individuals' family ambience may exercise a substantial influence on their behavior once they become adults. One of the most debated problems is that of the origin of the aggressiveness that is manifest in many animal species. Many studies lead one to conclude that such behavior is strongly influenced by the environment (e.g., by conditions of stress), although a genetic component may also be hypothesized.

Of great importance is the capacity of individuals to learn from their peers. This capacity manifests itself in the songs of birds, which have regional components and in which the influence of imitation is evident. Among monkeys, the spreading of certain habits within a group, such as that of washing potatoes before eating them or the method used to capture ants, is interesting. The effects of imitation are even more evident among monkeys in captivity, where closer interactions between individuals take place. In the human species, language is a determinant factor, which amplifies possibilities of communication enormously.

On the whole, one may conclude that the genetic makeup of a species provides the frame within which culture develops; for culture to develop, however, the observation of the behavior of other individuals and learning are essential. Genes often fail to determine an individual's behavior but make its expression possible. How the various genetic or environmental factors act to determine behavior is not clear. The brain's plasticity, undoubtedly, is a predominant factor. Other factors are the physiological actions that serve as mediators between external stimuli and the brain, such as many hormones (e.g., corticotropin-releasing factor, which coordinates the hormonal and neurological mechanisms of stress).

Another example of complex behavior is the migration of animals, which are able to cover distances of thousands of miles by means of the most varied mechanisms. Long-distance migration, which takes place among many species that inhabit fluid environments such as air or water, is noteworthy because the organisms that begin their journey at a certain point in space can reach another point with extraordinary precision and regularity and then are able to return to their place of origin. This demonstrates that the individuals of such species follow directional information that is constantly with them. The nature of that information has been investigated especially in birds by studying their sense of orientation inside planetariums. In following the configuration of the stars or the position of the sun represented on the planetarium's vault, the birds tend to fly off in the same direction as they would in nature, and they are able to compensate for the sun's or stars' position in relation to the time of day or night. They possess precise instructions and, therefore, a correct map of the heavens which varies over time. Other directional mechanisms are based on magnetism, the polarization of light, or details in a landscape.

Biological rhythms constitute another form of interaction between individual behavior and environment. The activity of living organisms is synchronized by means of various kinds of cycles such as those of the sun, the moon, and the tides by way of a biological clock based on a molecular mechanism that employs the same components in different species.

The second part of the volume deals with the discovery and spoliation of the biosphere — that is, of the environment in which life exists and has developed; a complex view of life must therefore include the biosphere and its evolution in relation to the evolution of the various living species. The study of the biosphere is very difficult because of the extended interactions that exist among its multiple biological components (animals, plants, and microorganisms) and the influence of many nonbiological factors (e.g., the consequences of man's activity in terms of pollution).

Microorganisms in particular have had extremely important effects on the evolution of life for the past 2 or 3 billion years. Among them, the cyanobacteria, photosynthetic organisms that produce oxygen, have engendered the most important change in life on the earth: the appearance of eukaryotes. Also, the role of methanogenic microorganisms, which contributed to the production of the enormous quantities of methane that exist in the earth's crust, has been very important. In recent times, plants became the most important source of oxygen; they are also a dynamic component in the carbon cycle (another phenomenon of great significance to life), which is influenced both by human activity and by factors that are independent of man.

In the most recent period of the planet's history, man, in his eagerness to derive the greatest possible advantage from his rational and creative intellect, has introduced into the environment many substances that did not exist before. Some of them have turned out to be extremely toxic, to the extent that they threaten the life of other living beings. One such substance is DDT, which was widely used to combat malaria through the elimination of its parasite's vector, the mosquito. Many living beings have suffered from it, especially the birds, in which DDT causes the production of eggs with shells so fragile as to compromise their reproduction. After the abandonment of DDT, however, the incidence of malaria increased considerably. This creates a problem: Does man have the right to safeguard himself at the expense of other living beings by imitating other organisms (bacteria, fungi, and plants) which promote their survival in such a manner?

The last part of the volume has bioethics as its object. The progress of knowledge of man's biological nature has been the source of great benefits, which may be summed up in the fact that the average life span of human beings has doubled in the course of the past century. On the other hand, this progress has created new problems which touch upon the very heart of what life is. Knowledge of a genetic kind is at the center of the evolution; it has enabled us to establish that the development and physical functioning of individuals are strongly influenced by the genes they have received from their parents by means of a complicated causal process similar, in some sense, to a lottery. Advanced knowledge of man's genes (i.e., of his genome) opens new possibilities for the immediate future. By analyzing particular localized differences in DNA known as polymorphisms, for example, it will soon be possible to obtain the genetic profile of any individual. Such knowledge may be eminently beneficial because it can reveal, for example, the susceptibility or resistance of individuals to specific diseases and their response to chemical substances or drugs; it may also reveal, at least in part, individuals' physical or intellectual capacities and the presumable duration of their lives. There is the problem, however, that such discoveries render data on individuals' deepest secrets exploitable. The bad use of some scientific knowledge could have serious consequences. This type of consideration has given rise to a new discipline, bioethics, which attempts to define what uses of genetic information are legitimate and how to prevent abuses without compromising such uses as are acceptable from the moral and legal points of view.

One of the fields in the discipline concerns the prevention of hereditary diseases, which may be achieved by way of different types of intervention on individuals starting from the moment of their origin, during fertilization. The possibility of such interventions raises new problems, such as the need to define the moment when an individual's life begins and what the value of life is compared to that of the individual's health. Another strongly debated point is the role of industry in the development of genetic knowledge for practical and economic purposes — a quite delicate objective indeed when the need to patent human genes arises. It is difficult to achieve a consensus among the various representatives of society in regard to these problems because of the often profound differences that exist in the determination of what is good and what is evil.

PART ONE

Biology of Behavior

Edited by Enrico Alleva and Patrick Bateson

INTRODUCTION

PATRICK BATESON
ENRICO ALLEVA

Ethology is the biological approach to the study of behavior. Konrad Lorenz and Niko Tinbergen, who are usually regarded as its modern founders, treated behavior in the same way as any other aspect of an animal's anatomy or physiology in the sense that behavior patterns often have a regularity and consistency that relates to the obvious needs of the animal. Their approach brought the study of behavior into the synthesis of Charles Darwin's ideas with modern genetics. Although individuals are different from each other, closely related individuals are more like each other than unrelated individuals. If the family similarities are inherited genetically, then any survival or reproductive advantage afforded by a particular way of behaving means that, after sustained evolution, the behavioral characteristic will spread through the population until every member of the lineage behaves in the same way in this respect.

The interest in biological utility of behavior has led to many excellent studies of animals in natural conditions. A captive animal is usually too constrained by its artificial environment to provide a complete understanding of the functions of the great variety of activities that most animals are capable of performing. Studies in natural conditions have been an important part of ethology, played a major role in developing the distinctive and powerful methods for observing and measuring behavior, and led to the development of the sister discipline of behavioral ecology. The comparative approach continues to be an important part of the subject.

Mainstream ethology involves much more than mere observation. Even in the field, ethologists have been experimentalists and many have pursued their interests into the laboratory. Those who worked on the most efficient ways to find food in a natural environment have appreciated that their work raised important issues about how behavior is controlled. Some of the most striking ethological discoveries, such as imprinting and bird song learning in birds, have been made in artificial conditions. Although the ethological stance is emphatically that the understanding of mechanism should be set in the context of understanding the evolution and function of behavior, studies of the development and integration of behavior have markedly influenced how behavior in natural conditions has been interpreted. This has happened notably in the studies of perceptual factors and learning processes influencing mate choice and their implication for associated evolutionary theories of sexual selection.

It is also happening in areas of work generally lumped under the heading of "life-history strategies," which raise important issues regarding conditional responses to environmental conditions.

In recent years, enormous strides had been made in neuroethology and in understanding the hormonal basis of behavior. Links between the physiology of metabolism and behavior have been made as well as between behavioral state and the immune system. The relevance and value of molecular techniques has been realized. At the same time, experimentalists are less likely to hold all but one variable constant, and when the single independent variable is found to produce an effect it is the cause and everything else is deemed unimportant. A systems approach is essential in bringing together evidence that would otherwise have seemed unrelated. For instance, when a monitoring system signals that blood sugars have fallen to low levels, it is necessary to inhibit other activities that might interfere with the seeking out of food and raising the level of blood sugar. When an alarm signal warns that a dangerous predator is approaching, feeding better be suppressed in favor of running for safety lest the animal become the meal of another species. The various systems controlling different and incompatible behavior patterns must be in touch with each other. The need to express many different types of incompatible behavior in appropriate circumstances requires well-defined internal rules of organization.

The pressure to provide large computational needs has led to the evolution of ever more elaborate brains. An active area of research inquires into the extent that animals' cognitive abilities, perceptions, and self-awareness are similar to those of humans. Understanding animal perception and feelings has become especially important because of public concern about animal suffering. This can lead to naive projection of human intentions and emotions on animals that are extremely unlikely to have them. Seemingly purposeful behavior may be expressed by negative feedback from the environment just as it is in many human-made homeostatic devices such as those that regulate temperature in our houses. Complicated optimizing and time budgeting problems are often solved by rules of thumb that make time-saving short-cuts. Anticipation of the future can be generated by rules for associating neutral and biologically significant events. Nevertheless, every human makes assumptions that their fellow humans share their own capacity for emotion and rational decision making. Specifying what leads to such judgments makes exercise of projecting into other animals more transparent. This then leads us to ask whether an animal in question has anatomical, physiological, and biochemical mechanisms similar to those which in a human are known to be correlated with sensation, emotion, or intention. It also leads us to ask whether the animal performs in similar ways to humans who are believed to be thinking or experiencing sensations and whether its response is functionally similar to that expected of a human. In other words, the evidence must be set in the appropriate ecological and social context.

Although animals do many complicated things by employing simple rules in the organization of their behavior, the number of similarities between the behavior of fellow humans, to whom we readily attribute thought, intentions, and emotions, and that of complicated animals, such as mammals and birds, gives us serious grounds for reflecting on the fullness of their mental lives. This leads a great many ethologists to grant such animals the benefit of the doubt.

A classic ethological concern was with the inborn character of much behavior, and the subject was strongly associated with the development of a theory of instinct. However, instinct was defined in many different ways and evidence fulfilling one criterion, such as the presence at birth, did not necessarily imply existence of another, such as resistance to change by learning. Work by the developmentally minded ethologists has been important in illustrating how the processes of development involve an interplay between internal and external factors. Modern work has also eroded the belief that a good description of a species obtained from one habitat can be generalized to the same species in another set of environmental conditions. The variations in behavior within a species may, of course, reflect the pervasiveness of learning processes. However, some alternative modes of behavior are probably triggered rather than instructed by prevailing environmental conditions.

In understanding behavioral development it is important to appreciate that, in general, genes do not code for behavior in the sense that simple correspondence can be found between genes and behavior. The general point is that a nervous system generating observable behavior is a product of both gene expression and the conditions prevailing as that individual matured. Furthermore, which genes are expressed depends on the state of the nervous system. Therefore, external conditions may influence which genes are expressed. At least three different developmental processes lead to the adaptive complexity of behavior. Some do not depend on experience with the relevant environmental conditions, some are triggered by particular environmental conditions, and some involve specific rules for instruction by the environment. Any given behavior pattern may be affected by all three of these developmental processes.

Ethology and closely related subjects differ from many other fields of biology in that considerable time and effort are needed for discussion of conceptual issues. Molecular biology and the neurosciences, for instance, do not possess such an elaborate theoretical background and tend to rely largely on the power of sophisticated laboratory techniques to uncover and describe biological processes. The range of material found in the ethological literature is enormous. The strength is that the new material constantly surprises and opens up new lines of research. There is some tension between the view that variation in behavior is what matters most and the view that we should all be searching for universals. Of course, this reflects a difference in research goals. If they are pressed, most ethologists would agree that they were more interested in one goal rather than the other, even though they would disagree about which is the most important. Even so, many ethologists simply take delight in the study of behavior in all its diversity.

Modern ethology abuts so many different disciplines that it defies simple definition in terms of a common problem or a shared literature. The development and control of an individual's behavior can both be investigated at many levels — the whole animal, the systems physiology, and the molecular level. Behavior between individuals can be investigated in terms of interactions, relationships, and social structures. Those who call themselves ethologists now work alongside neurobiologists, social and developmental psychologists, anthropologists, psychiatrists, and many others. Ethologists are accustomed to think in ways that reflect their experience with free-running systems that both influence and are affected by many things around them. These skills have enabled them to understand the dynamics of behavioral processes and are prized by those people with whom they collaborate. For this reason, ethology will continue to play an important integrative role in the drive to understand what behavior patterns are for, how they evolved, how they developed, and how they are controlled.

General References

ALLEVA, E., FASOLO, A., LIPP, H. P., NADEL, L., and RICCERI, L. (Eds.). (1995). *Behavioural Brain Research in Naturalistic and Semi-Naturalistic Settings: Possibilities and Perspectives.* Kluwer, Dordrecht.

BATESON, P., and MARTIN, P. (1999). *Design for a Life: How Behaviour Develops.* Cape, London.

DAWKINS, M. S. (1995). *Unravelling Animal Behaviour,* 2nd ed. Longman, Harlow, UK.

MANNING, A., and DAWKINS, M. S. (1998). *An Introduction to Animal Behaviour,* 5th ed. Cambridge Univ. Press, Cambridge, UK.

MARTIN, P., and BATESON, P. (1993). *Measuring Behaviour.* Cambridge Univ. Press, Cambridge, UK.

John Durant

The National Museum of Science and Industry
London, United Kingdom

The History of Animal Behavior Study

Modern ethology has developed through formative, classical, and mature phases. Throughout this development, the field has been characterized by an insistence on the importance of naturalistic or seminaturalistic observation of behavior, an emphasis on the inborn ("innate") elements in behavior, an enthusiasm for the application of Darwinian evolutionary theory to behavioral analysis, and an interest in the comparative analysis of animal and human behavior. In the formative phase (ca. 1870–1920), many relatively isolated workers in Europe and North America brought together a centuries-old tradition of naturalistic observation of animals with the first fruits of Darwinian evolutionism in a series of scattered but nonetheless pioneering studies. During the classical phase (ca. 1920–1950), Lorenz and Tinbergen spearheaded the development of ethology, culminating in the formulation of a coherent intellectual and institutional basis for the subject. The mature phase (ca. 1950–1975) saw the rapid growth of ethology in both the German- and the English-speaking worlds. Lorenz consolidated a philosophically ambitious but scientifically conservative classicism in Germany, whereas Tinbergen inspired a philosophically more modest but scientifically more productive modernism in Britain and North America. Since 1975, the sheer success of functional and evolutionary approaches to behavior study has undermined the distinctiveness of the ethological approach. Today, the prospects for ethology as a distinctive field of inquiry appear to depend on the reintegration of evolutionary with causational and developmental analyses of behavior.

Prehistory

Curiosity about what we call animal behavior is as old as (in all probability, it is much older than) recorded thought. Aristotle's encyclopedia of animals, the *Historia Animalium,* contains observations not only of animal anatomy but also of animal habits; to these, many more were added by Pliny the Elder, whose *Historia Naturalis* was a source for most of the accounts of animal life that gained currency in the early centuries of the Christian era. The zoological writings that circulated during the Dark Ages and into the medieval period are commonly dismissed today as superstitious nonsense. In fact, many of these writings served purposes that we now find difficult even dimly to discern. Our remoteness from the medieval worldview makes it difficult for us to empathize with the way in which different animals were made to stand as tokens of good and evil in the richly illustrated medieval texts known as *bestiaries* (Fig. 1). Not recognizing the ways in which God, nature, man, and society were bound together, we may be inclined to dismiss as frankly absurd the medieval notion that the characters of different animals were intended to serve as moral and spiritual symbols to God-fearing men and women. However, the use of animal behavior study as a basis for moralizing about the human condition is a recurrent theme in the history of ethology. In this sense, we are not very far removed from the medieval desire to know humanity's proper place, behaviorally speaking, in the larger scheme of things.

FIGURE 1 The *Acerba* of Cecco d'Ascoli (*Acerba* Cod. Plut. 40.52 f. 36v. Biblioteca Medicea Laurenziana, Florence).

FIGURE 2 Portrait of John Ray. Oil painting attributed to Mary Beale (1633–1699) (Courtesy of the Natural History Museum, London).

Although it is important to recognize the true antiquity of our interest in animal behavior, it makes little sense to speak of even the earliest beginnings of the science of ethology before the birth of modern science in the sixteenth and seventeenth centuries. From about this time, it becomes possible to trace recognizable antecedents of modern behavior study. Ironically, however, these antecedents are mostly found in the work of clergymen–naturalists whose motives for the study of animals were every bit as pious as those of the monks who penned the bestiaries centuries earlier. For example, in the writings of Isaac Newton's Puritan friend, the theologian and naturalist John Ray (Fig. 2), we find a wealth of acutely observed detail concerning both the physical and the behavioral characteristics of animals, together with serious reflections on the nature and purpose of animal "instincts." Ray cast doubt on the medieval notion that swallows and swifts hibernate during the winter, and it was he who first saw clearly the relationship between bird song and defense of territory. The title of Ray's best known work reveals the motives for these and countless other contributions — it was called *The Wisdom of God Manifested in the Works of Creation* (1691).

A century after Ray, another English clergyman–naturalist made his mark by closely observing the animals all around him. In *The Natural History of Selborne* (1789), the Reverend Gilbert White simultaneously immortalized the life of an eighteenth-century English country parson and enlarged the fund of human knowledge concerning the habits and instincts of animals. White's observations, recorded at first in journals and letters to friends, reflected his love of the English countryside and his experience in one particular place. Whereas Ray had traveled extensively throughout Europe, White confined his attentions to the fields and hedgerows of his own Hampshire parish. Like Ray before him, however, he freely indulged his curiosity about the entire living world. He did not see himself as a student solely of animal behavior; among his observations are to be found many original behavioral insights. Thus, White provided the first close observations of the habits of swallows and swifts, and he is widely recognized as having been the first to use purely behavioral characteristics (in this case, songs) to distinguish among the different species of English "willow-wrens" or warblers.

In this brief account, Ray and White must stand for the long tradition of observational natural history in the centuries between the scientific revolution and the birth of modern ethology. A fuller account of this tradition would embrace not only clergymen–naturalists but also other gentlemen–naturalists, animal husbandrymen, gamekeepers, conservationists, wildlife artists, and many others whose work or pleasure it was to observe closely the ways of life of other animals. It is an interesting feature of this supposedly "amateur" tradition that it frequently provided criticisms of the views about animal behavior that were being expressed by the acknowledged leaders of scientific thought. Thus, in the eighteenth century the French gamekeeper Charles Leroy published sympathetic and at times frankly anthropomorphic accounts of the behavior of wolves that were explicitly opposed to the views of the renowned superintendent of the Royal Zoo at Versailles, the Compte Georges Louis Leclerc de Buffon. Whereas Buffon had drawn a sharp distinction between the simpler mechanistic habits and instincts of animals and the more complex, rational abilities of humans, Leroy insisted that animals had complex "moral characters" of their own which must be understood in relation to their distinctive ways of life in the wild. Leroy's criticisms were to be echoed frequently over the next two centuries by close observers who believed that armchair philosophizing in the name of biology was failing to do justice to the subtlety and the sophistication of animal behavior.

Darwinian Foundations

Field natural history was an important source of inspiration for early animal behavior study, but on its own it could never have created ethology. In addition to close observation of animals in natural or seminatural conditions, some sort of theoretical discipline was essential in order to bring coherence and meaning to empirical findings. The major ingredients of this theoretical discipline were provided in the mid-nineteenth century by Charles Darwin. Darwin was trained in the tradition of the clergymen–naturalists which flourished well into the nineteenth century. However, he was also heir to a more radical intellectual tradition of theorizing about organic and human origins. This tradition may be traced back to the work of the French biologist Jean-Baptiste Lamarck and the English doctor and polymath Erasmus Darwin (Charles' grandfather), both of whom had developed what would now be called evolutionary accounts of origins in which the direct effects of the environment and animal habits were important agencies of change. Charles Darwin was not overly impressed with the work of either Lamarck or his grandfather, but following his worldwide induction into field natural history aboard the H. M. S. *Beagle* in the early 1830s (Fig. 3), he became convinced that species change by a process that he termed *descent with modification.* Upon his return to England in 1836, he set out on a private voyage of intellectual discovery that culminated 2 years later in his formulation of the theory of evolution by natural selection.

a

b

FIGURE 3 (a) The brigantine H. M. S. Beagle. (b) Handwritten manuscripts by C. Darwin and examples of corals collected on the Cocos–Keeling atoll during the voyage of the Beagle in the Indian Ocean. (Photo Courtesy of Natural History Museum, London).

The theory of evolution by natural selection is important for the history of ethology for two reasons. First, through a long process of intellectual and ideological struggle that lasted well into the twentieth century, this theory gradually emerged as the central formulation of modern biology. Second, from the very beginning Darwin used his theory to explain not only physical but also behavioral and mental characteristics. In only a few months in 1837 and 1838, Darwin sketched in private the outline of an essentially materialistic theory according to which animal instincts evolved by natural selection in much the same way as physical characteristics, i.e., by the accumulation of many small variations, each of which conferred a distinct advantage in the struggle for existence. Not content with this, he developed an evolutionary psychology based partly on regular visits to the London Zoo to watch captive chimpanzees, partly on close observation of his own children, and partly on introspection of his own mental and emotional processes. Just how radical Darwin's thinking was in 1838 may be judged from the following entry in one of his private notebooks (Gruber, 1974, p. 281): "Origin of man now proved. Metaphysics must flourish. He who understand baboon would do more toward metaphysics than Locke." For a more complete account of Darwin's work on the evolution of instinct and intelligence and its influence on later generations, see Richards (1987).

None of these potentially explosive ideas saw the light of day until many years later. A chapter of the *Origin of Species* (Darwin, 1859) is devoted to instinct, and here Darwin deals with the examples of egg laying in cuckoos, slave making in ants, and hive making in bees in an attempt to show that his theory could account for the evolution of "the diversities of instinct and of other mental qualities of animals within the same class." On the question of human mental attributes, however, Darwin was much more restrained; it was only years later, first in *The Descent of Man* (1871) and then in *The Expression of the Emotions in Man and Animals* (1872), that he finally dared to return to the issue of the evolution of human behavioral and mental characteristics. Darwin's *Expression,* which Konrad Lorenz later heralded as a foundational work in the history of ethology, set out to trace the evolution of mental states through comparative analysis of expressions (Fig. 4). Darwin used postures and gestures (smiles, frowns, grimaces, etc.) as external physical signs of internal psychological processes. In this and other ways, his approach now seems similar to that which was used several generations later by classical ethologists such as Konrad Lorenz and Niko Tinbergen.

FIGURE 4 An animal expression from Darwin (1872).

Formative Phase (ca. 1870–1920)

With the wisdom of hindsight, it is easy to see in Darwin's writings most of the intellectual ingredients needed for ethology to flourish. As a matter of fact, this did not happen for approximately half a century. The reasons are complex and include the relative decline in the fortunes of Darwinism in the late nineteenth century and the preoccupation of many turn-of-the-century psychologists with very different approaches to behavior study. However, it is important to recognize that many people did follow Darwin's lead by undertaking biological studies designed to test and amplify some of his ideas. The two most important characteristics these people had in common were a belief in the importance of an extremely close acquaintance with the habits of animals in the wild and a conviction of the importance of a Darwinian approach to the explanation of these habits. Of particular significance were developments in Britain, the United States, and continental Europe (especially Germany and Holland).

In Britain, the most important contributions after Darwin were made by a long line of ornithologists, including E. A. Armstrong, H. Eliot Howard, J. S. Huxley, F. B. Kirkman, D. L. Lack, W. P. Pycraft, and E. Selous. Significantly, most of these men were not professional biologists; indeed, among the few who were it is worth noting that Huxley regarded his behavioral work as an occasional diversion from more serious professional activities, and Lack was only recruited into biology after having first undertaken a series of fine behavioral studies in his spare time as a schoolteacher. The case of Selous is illustrative of the nature of these men's contribution. A shy and solitary man, Selous pursued his passion for birds untiringly but alone. In a series of publications after 1900, he used observational diaries in an attempt to capture the finest nuances of bird behavior. However, he was no mindless empiricist. Selous was a committed Darwinian, and he regularly used his field observations to test evolutionary theories of behavior. For example, in the spring of 1906 he spent time in Holland observing the breeding habits of the ruff in order to test Darwin's claim that many female birds actively select their mates on the basis of subjective preference. His conclusion was decisively in Darwin's favor: "The selection on the part of the Reeves is most evident. They take the initiative throughout, and are the true masters of the situation. Quiet and unobtrusive as they are, compared to the Ruffs, their whole manner betrays conscious power" (Selous, 1906, 1907).

Selous's influence was considerable, both in England and on the European mainland. In particular, his work was cited extensively by the evolutionary biologist Huxley, who undertook very similar studies of his own. Huxley's interest in what he termed "scientific bird-watching" had first been aroused by one of Selous's books, which he read at about

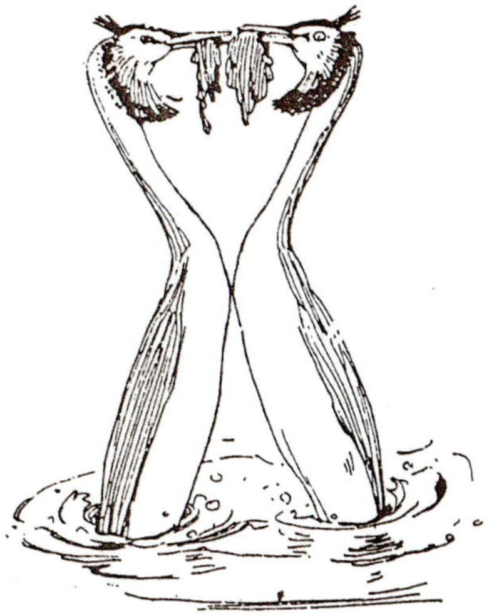

FIGURE 5 Out-of-text plate of the courtship of grebes from Huxley (1914).

the time that he left school. As his professional career prospered, Huxley regularly found time for ornithological fieldwork. In the spring of 1911, he first observed the "courtship" behavior of redshanks, and a year or two later he published an important account of what he termed the courtship habits of the great crested grebe (*Podiceps cristatus*) (Fig. 5). Huxley described the elaborate movements involved in the lengthy courtship of these birds as the gestures and postures of two performers caught up in some sort of evolutionary play. He suggested that they were part of a "curious set ritual" made up of movements which had once been of direct practical use to the birds but which had evolved into gestures of purely symbolic value (Huxley, 1914). Huxley appears to have been the first to use the term *ritualization* to describe this evolutionary process, and his approach to the study of courtship was to be widely cited by the leading ethologists of the 1920s and 1930s.

The antecedents of modern ethology in North America parallel those in England to a remarkable degree. Once again, important "amateur" contributions (such as those of George and Elizabeth Peckham on sexual selection in spiders) were influential on many professional biologists who sought to follow Darwin's lead by establishing a new way of studying animal behavior. Particularly important was a circle of colleagues and friends in the Midwest in the 1880s and 1990s, which included the biologists Charles Otis Whitman and his students Wallace Craig and William Morton Wheeler, the influential taxidermist and museologist Carl Akeley, and the Peckhams. Whitman undertook pioneering work on pigeon in which he used both morphological and behavioral characteristics to reconstruct the evolutionary relationships among nearly 40 wild species. His claim that "instinct and structure are to be studied from the common standpoint of phyletic descent" has been widely cited as evidence of his status as a true founder of modern ethology. In fact, this statement was embedded within a lengthy publication based on a series of lectures given at the Marine Biological Station at Woods Hole, Massachusetts, in 1897 and 1898 (of which Whitman was founder–director), and it is not clear how much immediate influence it had outside Whitman's personal circle (Riddle, 1919).

What is beyond dispute, however, is Whitman's considerable influence on his students Craig and Wheeler. A key here was Whitman's insistence on long and close observation of single species: His students are reputed to have greeted each other not with "What is your particular subject?" but rather with "What is your beast?" (see and Evans, 1970, p. 9). Using Whitman's large collection of living birds, Craig extended his mentor's behavioral work in a series of detailed studies that culminated in an important theoretical paper, "Appetites and Aversions as Constituents of Instincts" (1918). Opposed to the chain-reflect theory of instinct, Craig (1918, p. 106) invoked the analogy with human sexual behavior in a general appetitive theory of animal instinct in which, as he stated, "attention [using this and all the following terms in a strictly behavioristic sense], intelligence, memory, intensely emotional behavior, conflict of tendencies, hesitation, deliberation . . . rise, maintenance and decline of appetences" all played an important part. Although Craig appeared to lose touch with ethology in later life, there is no doubt that his early studies made an important contribution. In particular, his distinction between appetitive and consummatory behavior was fundamental to Lorenz's theory of instinct.

Finally, regarding developments in Germany, here again, significant behavioral work was well under way by 1900. Some of this work was undertaken by professional psychologists and zoologists, including Carl Groos, Otto Koehler, and Jacob von Uexküll, but here, too, the contribution of enthusiastic naturalists was of major importance. In 1896, a young man named Oskar Heinroth (Fig. 6) entered the

FIGURE 6 Portrait of Oskar Heinroth (1871–1945) (photo courtesy of A. Festetics).

Friedrich-Wilhelms-Universitat in Berlin to study zoology. At school Heinroth's passion for ornithology earned him the nickname "Duck Henry" (*Entenheinrich*), and part of his motivation in choosing Berlin for his university education appears to have been his wish to spend as long as possible observing bird behavior at the Berlin Zoological Gardens. Before his training was complete, Heinroth acted as an unpaid assistant to the director of the zoological gardens, and after graduating he continued in this role for several years until money could be found to make his position official (Heinroth, 1971).

Heinroth is probably the single most important figure in the early history of ethology. In the course of ornithological studies that culminated in the publication of a definitive, four-volume work titled *Die Vögel Mitteleuropas* (1924–1933), he laid the methodological and conceptual foundations of classical ethology. In many ways, he was the Whitman of northern Europe. As Whitman was convinced that good behavioral study must be based on the closest acquaintance with individual species, so too Heinroth was scornful of the poor standard of observation that underlay most existing work on animal behavior in Germany; as Whitman broke new ground in his comparative study of the pigeons, so too Heinroth paved the way for much that was to come later in his comparative study of the ducks, geese, and swans (*Anatidae*). Small wonder, perhaps, that Lorenz (as quoted in Evans, 1975, p. 7) would later say that "my scientific merit lies in discovering Whitman and Heinroth."

In his important paper on the *Anatidae* (1911), Heinroth began by contrasting his own interests with those of his contemporaries. Where others were concerned with the general biology of birds, he was interested in their "subtler life habits, manners, and customs." Heinroth criticized other German ornithologists for having sorely neglected these aspects of birds' lives. He pointed out that ornithological handbooks frequently failed to distinguish the quite different call-notes of male and female ducks, and that some even confused distinct species (e.g., whooper and mute swans) through ignorance of their distinct habits. In contrast, Heinroth used a combination of close description and photography to evoke the distinctive characters of the different species of *Anatidae*. In the course of this exercise, he described the "incitement behavior" (*Hetzen*) of female ducks, the "display postures" (*Imponierstellungen*) of male ducks, and the "triumph ceremonies" (*Triumphgeschreien*) which were performed by many geese and swans. He discussed how apparently trivial actions might serve as "greeting signals" (*Begrüssungszeichen*) or "intention movements" (*Intentionsbewegungen*), and he recounted the way in which the innate responses of hand-reared birds were often redirected to man (Heinroth, 1911).

If these ideas seem reminiscent of the early work of Konrad Lorenz, this is because Lorenz obtained them (and many others) directly from Heinroth. In the decade after 1926, when Lorenz's fiancée typed up his notes on the behavior of a hand-reared jackdaw and sent them to Heinroth (apparently without Lorenz's knowledge), the two men developed their ideas in such close collaboration that it is often impossible to distinguish their separate contributions. Although Lorenz brought with him from general biology (and particularly from comparative anatomy) many important ideas, he adopted the main elements of his early ethological research program from Heinroth. In a series of publications during the 1930s which are widely regarded as having established the boundaries of classical ethology, he took up one after another of the major themes in Heinroth's work. Using exactly the same techniques of naturalistic observation, hand-rearing, and hybridization, Lorenz extended his mentor's views concerning the role of instinct in the life of birds; he developed a more sophisticated analysis of the communicative and "releasing" functions of so-called "ritualized" behavior; and he refined the comparative behavioral analysis of the *Anatidae* (Fig. 7). In 1931, Lorenz wrote to Heinroth, "Are you . . . aware, Herr Doktor, that you are really the founder of a science, namely animal psychology as a branch of biology?" (Heinroth, 1971, p. 155). Five years later, the two men were discussing how best to introduce their views into the newly formed *Gesellschaft für Tierpsychologie*. In the end, their efforts were so successful that Lorenz was appointed joint editor with Koehler of the society's journal, the influential *Zeitschrift für Tierpsychologie* (Durant, 1981).

Classical Phase (ca. 1920–1950)

Lorenz's move into the field of animal behavior study in the late 1920s is generally regarded as the start of classical ethology. Clearly, there were many vitally important antecedents of Lorenz's work; however, it would be difficult to overemphasize Lorenz's importance in the development of classical ethology. What Lorenz took from Darwin, Huxley, Whitman, Craig, Heinroth, and others was a tool kit of theoretical ideas and practical techniques, but what he brought to this tool kit were considerable originality in the observation of animals (particularly captive or semidomesticated birds), great boldness in the articulation of ethological theories, and huge ambition for ethology as a scientific discipline. Combining all these qualities in what was undoubtedly a charismatic personality (few who met him ever forgot him), Lorenz was the one man above all others before World War II who appeared both able and willing to put the biological study of behavior firmly on the scientific map.

The most important ideas that Lorenz acquired from Heinroth were that all members of a species perform many invariant behavior patterns, and that these behavior patterns are similar to species-specific morphological characteristics in the sense that they appear to be inherited and,

FIGURE 7 Konrad Lorenz with the jackdaws (photo courtesy of Adelphi Edition).

by comparison between groups of related species, may be used for purposes of tracing evolutionary relationships. As he developed his view of these "innate" behavior patterns in the early 1930s, Lorenz came to view them as different in kind from all other forms of behavior, including reflexes and learned responses. Drawing Craig's distinction between appetitive and consummatory behavior and von Uexküll's concept of the intimate relationship between an animal's sensorimotor system and its perceived world (*Umwelt*), Lorenz developed a general theory of instinct according to which internal propensities ("drives") accumulate and provoke "appetitive behavior" until such time as either the animal encounters a specific "releasing stimulus" which triggers the performance of consummatory behavior via an "innate releasing mechanism" or else the drive is so great that it triggers consummatory behavior spontaneously in the form of so-called "vacuum activity" (Lorenz, 1937). This theory was the centerpiece of Lorenz's synthesis, not least because it held out the hope of a discoverable physiological basis for the innate behavior whose evolution he wanted to trace.

Looking back on the corpus of Lorenz's work, what is striking is its remarkably unchanging quality over time. Almost from the outset, we find present essentially all of the key ideas for which in later life Lorenz was to become both famous and (in some quarters, at least) notorious: the fundamental distinction between the innate and the learned; the existence of biological drives as the physiological foundations of instinctive behaviors; the analogy between instincts and morphological traits and hence the usefulness of comparative behavioral analysis in evolutionary studies; and the fundamental similarities between animal and human behavior. On this last point, it is interesting to observe that in his autobiography Lorenz's father Adolph remarked that he had not been keen on his son's choice of career, believing that "it was of no great importance to know whether herons were more or less stupid than they were thought to be." He continued, "Konrad contends that human psychology has much to learn from animal psychology and that there is no essential difference between these two branches of science" (Lorenz, 1936, p. 319). As early as 1931, Konrad had made this same point in a letter to Heinroth (1971):

> Who knows where today's human psychology will find itself, when once one knows of human behavior what is due to instinct, and what to intelligence: Who knows

how human morality would appear, with its drives and inhibitions, if one could analyze it in the same way as the social drives and inhibitions of a jackdaw? (p. 155)

Another unchanging quality in Lorenz's work was the generally discursive style of his writing. Throughout his career, he preferred to publish lengthy essays in which rich but essentially anecdotal behavioral observations were freely interspersed with theoretical and philosophical reflections, often of the most wide-ranging nature. Only once did he depart radically from this style, and this was in 1938 when he joined with the Dutch ethologist Nikolaas Tinbergen in coauthoring a publication (Fig. 8). Tinbergen had studied biology at Leiden University, and there the influence of his teachers (especially Jan Verwey) as well as of the English ornithologists led him toward the study of animal behavior (Baerends *et al.,* 1975). He had already completed a doctoral thesis on the orientation behavior of the digger wasp together with many studies of bird behavior by the time Lorenz paid him a visit in 1936. The following year, Tinbergen spent 3 months at Lorenz's home, and it was during this stay that the two men collaborated in the experimental study of egg-rolling behavior in the greylag goose (Schiller, 1957). The nature of this study (involving what Lorenz termed "unobtrusive experiments" designed to tease out the factors influencing undisturbed behavior) as well as the form in which it was eventually reported (which is conventionally scientific by the standards of the day) are far more characteristic of Tinbergen's than they are of Lorenz's later work.

FIGURE 8 Nikolaas Tinbergen with Konrad Lorenz in the 1960s (photo courtesy of Adelphi Edition).

Several factors served to unite Lorenz and Tinbergen in the 1930s. Both men aspired to make of animal behavior study something more than an occasional diversion for enthusiasts, both were dissatisfied with the conventional approach of the comparative psychologists to the understanding of innate behavior, both looked to make sense of innate behavior in broadly evolutionary terms, and both were aware of and admired the work of Selous and Huxley in England and Heinroth in Germany. Had the two men been able to continue their work uninterrupted, these similarities might conceivably have been the basis for a lengthier collaboration. As it was, however, the advent of World War II brought the whole of European ethology to a temporary halt; even more significant, it drove a wedge between the nascent ethological community in Germany and the equally fragile communities in England, Holland, and the United States. Lorenz and Tinbergen had dramatically different wartime experiences: The former put his medical training to use on the Russian front, whereas the latter was interned in a prison camp in which he helped to pass the time by writing stories for his children about the animals he had studied (Tinbergen, 1948, 1954).

It was Tinbergen who resumed work first at the end of the war. In particular, he led the move to establish classical ethology outside the German-speaking world. In 1938 he had spent several months at Yale University, and thereafter he systematically cultivated links with English-speaking biologists on both sides of the Atlantic. In 1940, he corresponded with Huxley and Lack in England about the need to reconstruct international cooperation among ethologists once peace had been restored, and in September 1945 he resumed the correspondence in order to urge the creation of a new European journal for the publication of ethological research. This move resulted in the creation of *Behaviour,* with W. H. Thorpe from Cambridge University as the first British editor. Thorpe had followed the progress of ethology in mainland Europe with keen interest, and he was determined to see the subject blossom in Britain. In the immediate postwar period, Thorpe masterminded the creation of a new center for ethological research in the form of the ornithological field station at Madingley, near Cambridge; he also conspired with Lack (who by this time was director of the Edward Grey Institute of Field Ornithology in Oxford) to lure Tinbergen to Britain. Thorpe also attempted to persuade Lorenz to move to Cambridge to be the first curator in charge of the new field station at Madingley. Although initially tempted by the suggestion, Lorenz preferred to stay in Germany with the support of the Max-Planck Gesellschaft (Thorpe, 1979). In the same year that Tinbergen moved to Oxford (1949), Thorpe hosted a hugely influential ethological symposium in Cambridge under the twin auspices of the Association for the Study of Animal Behavior and the Society for Experimental Biology (SEB) (Durant, 1986).

The SEB symposium served to reunite Lorenz and Tinbergen for the first time since the war. Lorenz's contribution to the symposium was a typically trenchant statement of his theory of instinct. On this occasion, he provided a graphic hydraulic model that purported to capture all of the main ingredients in instinctive behavior. Within this model, animals were presumed to possess as many distinct sources of motivational "drive" as they had specific instincts. Each drive was likened to a body of fluid accumulating continually in a reservoir for as long as the instinct remained unexpressed (Fig. 9). The fluid was prevented from escaping by a spring-loaded valve (the "innate releasing mechanism"), which could be opened by an appropriate pressure of fluid in the reservoir and/or an appropriate weight (the "releasing stimulus") in a scale pan attached to the reservoir. A variety of "innate" behaviors were thus released, depending in part on the pressure of fluid in the drive mechanism.

This was a general model applicable in principle to all animals (including humans) and rich in possibilities for empirical exploration and testing. The model was published in the proceedings of the SEB symposium in 1950, along with a parallel contribution from Tinbergen on the hierarchical organization of instinctive behavior (Lorenz, 1950). A year later, Tinbergen produced an even more systematic summary of classical ethology in his highly influential book, *The Study of Instinct* (1951). Summarizing the state of knowledge in all branches of ethology, this book also offered a theoretical synthesis around a hierarchical model of instinct that was broadly similar in conception to Lorenz's simpler drive model.

With the publication of the proceedings of the SEB symposium and *The Study of Instinct,* classical ethology became paradigmatic; that is, it came to be organized around a core of ideas (as exemplified in many key studies) that served as the organizing framework for further research (Kuhn, 1970). The key ideas, now widely available for the first time in English as well as German, embraced methodological and theoretical elements. Methodologically, close naturalistic observation remained central, but to this were added unobtrusive experiments of the sort favored by Tinbergen as well as hand-rearing, hybridization, and the conduct of systematic comparison for purposes of evolutionary analysis. Theoretically, the most important organizing principle of classical ethology was the notion of instinct. Both the SEB symposium and Tinbergen's *The Study of Instinct* were preoccupied with the causal mechanisms underlying behavior, and in the immediate postwar years a great deal of attention — and many arguments — centered on these mechanisms.

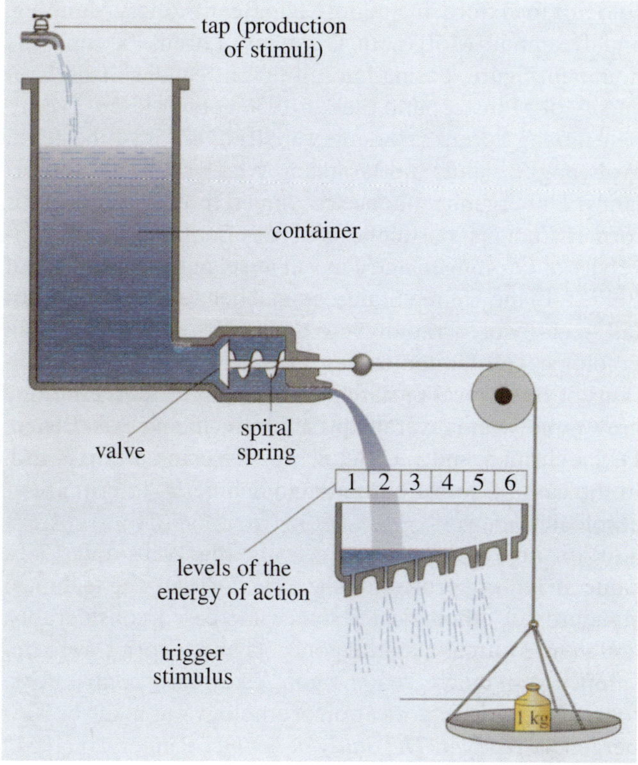

FIGURE 9 The psychohydraulic model of Lorenz, in which each motivational pulse is imagined as a liquid that, although the instinct remains inexpressed, accumulates in a reservoir. An innate trigger mechanism acts as a valve which opens at the onset of the trigger stimulus, seen as a weight placed on a balance, permitting the outflow of the liquid (adapted from Lorenz, (1950).

Mature Phase (ca. 1950–1975)

Ethology developed rapidly in the 1950s. In particular, substantial schools were created around key individuals in Britain, Germany, and The Netherlands, and links were rapidly created with animal behaviorists in the United States. In Britain, Bill Thorpe and Robert Hinde (who was appointed the first curator in charge of the Field Station at Madingley) were the crucial influences in Cambridge, whereas Tinbergen rapidly attracted a series of talented students to Oxford, including Colin Beer, Aubrey Manning, and Desmond Morris. In Germany, Lorenz became the dominant figure, basing himself first at Buldern (the University of Münster) and then in 1956 at Seewiesen. Prominent among Lorenz's students were Irenäus Eibl-Eibesfeldt, Wolfgang Schleidt, and Wolfgang Wickler. In The Netherlands, Tinbergen's influence continued through some of his former students, particularly Gerard Baerends at the University of Groningen and Jan van Iersel at the University of Leiden. In the United States, animal behavior study did not fall so easily or so rapidly into the pattern set by European ethology: The United States, after all, had its own traditions of behavioral research. Even so, interest in ethology grew rapidly, and several key people, including F. A. Beach, D. S. Lehrman, and T. C. Schnierla, began to take up — and, in the case of both Lehrman and Schnierla, to criticize — ethological ideas.

With international expansion in the 1950s and 1960s came increasing professionalization. Overall, the technical standards of ethological research increased considerably, and in many areas classical concepts and theories were developed with much greater rigor. A key early contribution to the conceptual clarification of ethology was made by Tinbergen. Already in *The Study of Instinct,* Tinbergen (1951) distinguished between different types of ethological investigation by devoting separate sections of the book to behavioral causation, development, and evolution. Later, these distinctions were captured more succinctly in what became known as the "four whys" of ethology. Faced with the question, Why does an animal behave in a particular way?, Tinbergen pointed out that the ethologist may answer in terms of *causation* (the physiological mechanisms by which the behavior arose), *ontogeny* (the developmental processes by which the behavior arose), *phylogeny* (the evolutionary history through which the behavior arose), or *function* (the survival value of the behavior). Tinbergen's main point was that these are not competing but rather complementary answers to distinct (albeit related) questions. He suggested that failure to distinguish clearly between these questions in ethological debate was the cause of much unnecessary confusion (Tinbergen, 1963).

A good example of the kind of problem that Tinbergen had in mind concerns the related terms "innate" and "instinctive." Among the classical ethologists, these terms were often applied both to behaviors that were believed to be adaptive and to behaviors that were believed to be produced without learning. The former usage is functional, whereas the latter is developmental; however, conflation between them can easily lead to the (in principle incorrect) conclusion that learning plays no part in adaptive behavior. This was one of the issues at stake in the debate which took place in the 1950s and 1960s regarding Lorenz's theory of instinct. The American animal behaviorist Daniel Lehrman was one of the first to explicitly criticize Lorenz's theory on the grounds that it appeared to rule out environmental factors in behavior development (Lehrman, 1953, 1970). Following a visit to the United States that included a friendly encounter with Lehrman, Lorenz published an account of his theory which avoided altogether the use of the distinction between the innate and the learned (Lorenz, 1965). In later publications, however, he tended to repeat his earlier formulations, and in any event, the issue did not disappear. It was one factor, along with the failure of empirical work to substantiate many of the more obvious predictions of the psychohydraulic model, that led to the general abandonment of Lorenzian instinct theory in the English-speaking world after 1960.

The difficulties encountered by instinct theory were part of a general shift of emphasis in ethology after 1950. In the prewar period, the focus of ethological interest was on causation and evolution. After 1950, many studies in these areas were of course conducted; increasingly, however, the center of attention moved to questions of development and function. In Cambridge, for example, Thorpe carried forward a program of research on behavior development. Thorpe's interest was in the development of bird song. Experimenting with male chaffinches reared in partial or total acoustic isolation, he was able to demonstrate the interaction of internal and external factors in the development of song. Thus, birds reared in isolation from a few days of age developed only a simple type of song. Learning was important but highly preferential: Male chaffinches learned chaffinch songs, not the songs of other species, and they learned these songs more readily during particular "sensitive periods" of behavioral development. Subsequent work by Peter Marler and others confirmed the complex integration of endogenous and experiential factors in the development of bird song. Work such as this played a part in undermining the classical Lorenzian distinction between the innate and the learned during ethology's mature phase (Thorpe, 1961; Marler and Mundinger, 1971).

While the Cambridge group was becoming increasingly involved in developmental studies, the interests of Tinbergen and his students were moving in different directions. Although Tinbergen had devoted over half of *The Study of Instinct* (1951) to causal questions, the difficulties encountered by both Lorenz's and his own models of instinct impeded rapid progress. In addition, the interest of purely historical, phylogenetic work [in which Tinbergen (1972) excelled in his extensive comparative studies of the behavior of gulls] was limited. In contrast, functional studies seemed much more attractive. Almost everywhere that Tinbergen and his students turned, they found fascinating stories to tell about the contribution of behavior to the survival and reproduction of animals in the wild. In this sense, it is no coincidence that the *Festschrift* produced by three of Tinbergen's students in 1975 (Baerends *et al.,* 1975) to mark his retirement from the Chair of Animal Behaviour at Oxford was titled *Function and Evolution in Behaviour.*

Despite its increasing sophistication in the 1950s and

1960s, ethology retained its status as a pioneering subject. Nowhere was this clearer than in the field of primatology, in which new field studies that were begun at this time achieved great scientific and social prominence. The key figure here, of course, was Jane Goodall. Introduced to the anthropologist Louis Leakey in 1957, Goodall started as an assistant secretary to him in Nairobi and then progressed through paleontological digs in the Olduvai Gorge to the place (the Gombe Stream Reserve) and the primates (a population of wild chimpanzees) that she was to make world famous. There are many ways of telling Goodall's story: as a powerfully mythic encounter between humans and their closest living relatives, as a romance involving a single woman steadily pushing back the frontiers of human knowledge, as a story of a white explorer at-large in postcolonial Africa; as a singular instance of the amateur making an important contribution to the study of animal behavior (Goodall had no formal training in ethology until Hinde agreed to supervise her for a doctorate in 1961), and so on. In its high-profile coverage of Goodall's work, *National Geographic* opted for a mixture of the mythic encounter and the heroic saga. It was this coverage, of course, which served to project Goodall and her chimpanzees as international celebrities. *National Geographic* coverage of Goodall's work started in 1963 and went on to include both feature articles and TV specials stretching well into the 1980s (Goodall, 1963, 1979).

The National Geographic Society's role in postwar primatology was not confined to reporting or to the work of Goodall. The society funded a considerable amount of primatological research, not only by Goodall and her team but also by many other women primatologists. The list is long, and it includes the late Dian Fossey and Birute Galdikas, both of whom also worked on higher primates (Fossey on the mountain gorilla and Galdikas on the orangutan) and achieved international fame. Overall, several things are striking about this tradition of ethological research: It was largely led by women; it was heavily dependent on precisely the kind of close, informal observation of animals in the wild that had been so important in ethology's formative period; it was consistently interested in the implications of nonhuman primate behavior for a better understanding of human evolution and contemporary human affairs; and it consistently attracted high-profile media coverage that served to make animal behavior study in general, and primatology in particular, internationally renowned (Haraway, 1989).

Popularization was an enduring feature of ethology throughout its mature phase. Once again, Lorenz led the way in this area—first with *King Solomon's Ring* (1952) and *Man Meets Dog* (1954) and then with a series of more overtly moralistic works, including *On Aggression* (1966) and *Civilized Man's Eight Deadly Sins* (1974). Tinbergen, too, produced popular books such as the semiautobiographical *Curious Naturalists* (1958) and popular films such as *Signals for Survival* (1970); in the case of the Oxford school, the most influential popularizations were produced not by the master himself but by one of his former students. Desmond Morris had been one of the first young biologists to be attracted to Oxford by Tinbergen's work. In the early years, he made significant scientific contributions to the study of communication in fish, birds, and primates, but a move into the presentation of TV wildlife programs in the early 1960s was followed by the publication of a best-selling book, *The Naked Ape* (1967), and from then on Morris became probably the best known popularizer of ethology in the world.

A great deal of popular ethology traded on the perceived relevance of animal behavior research for a better understanding of human behavior, and the moralistic tone of much of this literature excited considerable debate. Embracing what were also in part professional controversies about, for example, the nature of instinct and the relationship between the innate and the learned, the debate extended to include retrospective criticisms of the social and political uses to which Lorenz had put his own work in the late 1930s and 1940s. Close examination of many of Lorenz's wartime publications led some critics to claim that much of Lorenz's theoretical work was tainted with fascism. The argument focused in particular on Lorenz's attempt to explain certain types of supposed racial degeneration in humans by analogy with the degeneration of instincts in domesticated animals (Kalikow, 1983; Richards, 1987). This charge dogged Lorenz for many years, and it surfaced again in 1973 when (together with Tinbergen and the Austrian entomologist Karl von Frisch) he was awarded the Nobel Prize in physiology and medicine. The issue was raised during the press conference associated with the Nobel prize-giving ceremony to the acute embarrassment of Tinbergen, who was, of course, also present (Odelberg, 1974; Cloud, 1973). That this, the ultimate accolade for ethology as a mature scientific subject, should have been surrounded by controversy is perhaps the most graphic illustration of the extent to which intellectual and ideological tensions have continued to be created by ethologists' enthusiasm for drawing moral, social, and political inferences from their work.

Recent Trends

In 1975, the American entomologist Edward O. Wilson published a massive volume titled *Sociobiology: The New Synthesis*. A year later, the English ethologist Richard Dawkins produced a physically smaller but intellectually no less substantial book called *The Selfish Gene*. Although very different in style and even (in certain crucial areas) in substance, both of these books promoted a new kind of behavioral biology. Wilson did this by bringing together under a new label a vast amount of recent biological research on social behavior. In contrast, Dawkins was concerned with articulating as clearly as possible the wider implications of a "gene's-eye view" of behavior. In order to appreciate what has happened to ethology during the past 20 years, it is necessary to understand the nature and significance of these developments.

The intellectual inspiration for *The Selfish Gene* lies in the work of several evolutionary biologists who had been concerned with the application of the theory of natural selection to the explanation of social behavior. A particular view of natural selection had been implicit in the writings of Lorenz and other popular ethologists. In *On Aggression,* Lorenz had argued that the great restraint commonly seen in aggressive encounters between members of the same species had evolved for the benefit of the species. The theoretical implications of this position were articulated more clearly by the British biologist V. C. Wynne-Edwards (1962) in connection with the idea of population control. Briefly, traits can be selected for their benefits on the species only if natural selection operates at the level of (i.e., between) species. Such "group selection" was explicitly criticized, for example, by the British evolutionary biologist John Maynard-Smith and the American evolutionary biologist George C. Williams, both of whom reasserted the basic Darwinian idea of individual selection. These criticisms made all the more puzzling certain sorts of behavior (such as the Lorenzian restraint described previously) that gave every appearance of being "altruistic." In the attempt to resolve this puzzle, Maynard-Smith (1974) developed game theory models which showed how individual selection might favor restraint in a series of aggressive encounters, and Williams (1966) showed how individual selection might favor a certain sort of mutual back-scratching ("reciprocal altruism") in situations in which individuals could recognize and remember one another.

An even more radical approach to the evolution of apparently altruistic behavior was taken by the British evolutionary biologist W. D. Hamilton. In two key theoretical papers in 1964, Hamilton defined the circumstances in which selection would favor a gene that caused an individual to forego reproduction and, instead, to assist with the reproduction of biological relatives. Hamilton suggested that such a gene increased the "inclusive fitness" of its bearer, and he went on to show how this theory (which Maynard-Smith later termed *kin selection*) could help to explain perhaps the best known and most dramatic of all cases of apparent altruism among animals, namely, the existence of nonreproductive castes among the social insects (Hamilton, 1964), Darwin discussed the puzzle of the social insects in the *Origin of Species,* but now, for the first time, ethologists and evolutionary biologists had a precise explanation with which to work. In fact, the highly mathematical form of Hamilton's work resulted in its being widely ignored for several years. For many biologists, it was Dawkins' *Selfish Gene* which provided the first clear explanation of kin selection theory.

The evolutionary ideas of Hamilton, Maynard-Smith, Williams, and others were widely adopted by ethologists in the late 1970s, and the effects on their work (and thus on their subject) were radical. Where once ethologists had relied very heavily on the observation of supposedly typical species-specific behaviors, now they began to take an interest in the varying fortunes of individuals within social groups; where once they had asked questions about the survival value of innate behaviors or instincts, now they started to inquire about the reproductive benefits of particular behavioral strategies. Such questions demanded new methodologies. Above all, they introduced a new rigor into fieldwork which might have surprised even the most thorough among the field naturalists who contributed to the founding of ethology. In studies such as that by Clutton-Brock *et al.* (1982) on the red deer of the Scottish island of Rhum, the interpretation of behavior depends on the correlation of behavior with lifetime reproductive success for most or all individuals in a herd, and such correlations depend on a level of year-round, systematic observation that is unprecedented in the annals of classical ethology.

The sociobiological revolution has had a profound impact on ethology. It has completed the globalization of what was in its inception (but is no longer) a largely European discipline; it has transformed the ways in which ethologists think about and investigate the functions of behavior; and it has provided a set of theoretical models whose usefulness extends far beyond the study of animal behavior to include other aspects of organic evolution, such as the behavioral ecology of plants and microorganisms and the evolution of sex. At the same time, this latest phase in the history of ethology has been accompanied by intense argument about the wider implications of the subject for an understanding of human affairs. If nothing else, the sociobiology debate of the 1970s and 1980s has served to demonstrate that the development of more sophisticated evolutionary models of behavior has not resolved long-standing disputes about the extent to which human behavior can be explained in Darwinian terms (Kitcher, 1985).

Finally, it is interesting to note that the sociobiological revolution has reduced somewhat the distinctiveness of the ethological approach within biology. It has done this (i) by undermining particular aspects of the classical ethological view of animal behavior and (ii) by extensively decoupling evolutionary analyses of behavior from other types of behavioral analysis. The rapid progress of evolutionary ethology during the past 20 years has not been accompanied by equally rapid progress in other branches of ethological research concerned with, for example, developmental ethology or neuroethology. As a result, the classical ethologists' early vision of a genuinely integrated approach to the biological study of behavior has become a rather distant dream.

Today, the prospects for ethology as a coherent and distinctive field of inquiry appear to depend on the reintegration of evolutionary with causational and developmental analyses of behavior.

References

BAERENDS, G., BEER, C., and MANNING, A. (Eds.) (1975). *Function and Evolution in Behaviour. Essays in Honour of Niko Tinbergen.* Clarendon, Oxford.

CLOUD, W. (1993). Winners and sinners. *Sciences,* **13,** 16–21.

CLUTTON-BROCK, T. H., GUINNESS, F. E., and ALBON, S. D. (1982). *Red Deer: Behaviour and Ecology of Two Sexes.* Univ. of Chicago Press, Chicago.

CRAIG, W. (1918). Appetites and aversions as constituents of instincts. *Biol. Bull.* **34,** 91–107.

DARWIN, C. (1872). *The Expression of the Emotions in Man and Animals.* Murray, London.

DAWKINS, R. (1976). *The Selfish Gene.* Oxford Univ. Press, Oxford.

DURANT, J. (1981). Innate character in animals and man: A perspective on the origins of ethology. In *Biology, Medicine and Society, 1840–1940* (C. Webster, Ed.), pp. 157–192. Cambridge Univ. Press, Cambridge, UK.

DURANT, J. (1986). The making of ethology: The Association for the Study of Animal Behaviour, 1936–1986. *Anim Behav.* **34,** 12601–12616.

EVANS, R. (1975). *Konrad Lorenz: The Man and His Ideas.* Harcourt Brace Jovanovich, New York.

GOODALL, J. (1983). My life among the wild chimpanzees. *Natl. Geogr.* **124**(2), 272–308.

GOODALL, J. (1979). Life and death at Gombe. *Natl. Geogr.* **155,** 592–621.

GRUBER, H. (1974). *Darwin on Man. A Psychological Study of Creativity.* Wildwood House, London.

HAMILTON, W. D. (1964). The genetical evolution of social behaviour. *J. Theor. Biol.* **7,** 1–52.

HARAWAY, D. (1989). *Primate Visions. Gender, Race, and Nature in the World of Modern Science.* Routledge, London.

HEINROTH, O. (1911). Beitrage dur Biologie, namentlich Ethologie und Psychologie der Anatiden. In *Verhandlungen des V Internationalen Ornithologen-Kongresses.* German Ornithological Society, Berlin.

HEINROTH, K. (1971). *Oskar Heinroth, Vater der Verhaltensforschung, 1871–1945.* Stuttgart.

HUXLEY, J. S. (1914). The courtship habits of the great crested grebe (*Podiceps cristatus*), with an addition to the theory of sexual selection. *Proc. Zool. Soc.* **35,** 491–562.

KALIKOW, T. (1983). Konrad Lorenz's ethological theory; Explanation and ideology, 1938–1943. *J. History Biol.* **16,** 39–73.

KITCHER, P. (1985). *Vaulting Ambition. Sociobiology and the Quest for Human Nature.* MIT Press, Cambridge, MA.

KUHN, T. (1970). *The Structure of Scientific Revolutions,* 2nd ed. Univ. of Chicago Press, Chicago.

LEHRMAN, D. S. (1953). A critique of Konrad Lorenz's theory of instinctive behavior. *Q. Rev. Biol.* **28,** 424–434.

LEHRMAN, S. (1970). Semantic and conceptual issues in the nature–nurture problem. In *Evolution and Development of Behavior* (L. Aronson *et al.,* Eds.). Freeman, San Francisco.

LORENZ, A. (1936). *My Life and Work.* Scribner's, New York.

LORENZ, K. (1937). Über die Bildung des Instinktbegriffes. *Naturwissenschaften* **25,** 289–300, 307–318, 324–331.

LORENZ, K. (1950). The comparative method in the study of innate behaviour patterns. *Symp. Soc. Exp. Biol.* **4,** 221–268.

LORENZ, K. (1965). *Evolution and Modification of Behavior.* Univ. of Chicago Press, Chicago.

MARLER, P., and MUNDINGER, P. (1971). Vocal learning in birds. In *Ontogeny of Vertebrate Behavior* (H. Moltz, Ed.). New York.

MAYNARD-SMITH, J. (1974). The theory of games and the evolution of animal conflict. *J. Theor. Biol.* **47,** 209–221.

ODELBERG, W. A. (1974). *Les Prix Nobel en 1973,* pp. 11–146, 158–218. Norstedts Tryckeri, Stockholm.

RICHARDS, R. J. (1987). *Darwin and the Emergence of Evolutionary Theories of Mind and Behavior.* Univ. of Chicago Press, Chicago.

RIDDLE, O. (Ed.) (1919). *C. O. Whitman: Posthumous Works,* Vol. 3. The Carnegie Institute, Washington, D.C.

SCHILLER, C. (Ed.) (1957). *Instinctive Behaviour. The Development of a Modern Concept.* International Univ. Press, New York.

SEE, M. A., and EVANS, H. E. (1970). *William Morton Wheeler, Biologist.* Harvard Univ. Press, Cambridge, MA.

SELOUS, E. (1906). Observations tending to throw light on the question of sexual selection in birds, including a day-to-day diary on the breeding habits of the ruff (*Machetes pugnax*). *The Zoologist,* **10,** 201–219, 285–294, 419–428.

SELOUS, E. (1907). Observations tending to throw light on the question of sexual selection in birds, including a day-to-day diary on the breeding habits of the ruff (*Machetes pugnax*). *The Zoologist* **11,** 60–65, 161–182, 167–180.

THORPE, W. H. (1961). *Bird Song.* Cambridge Univ. Press, Cambridge, UK.

THORPE, W. H. (1979). *The Origins and Rise of Ethology.* Heinemann, London.

TINBERGEN, N. (1948). *Kleew.* Oxford Univ. Press, Oxford.

TINBERGEN, N. (1951). *The Study of Instinct.* Clarendon, Oxford.

TINBERGEN, N. (1954). *The Tale of John Stickle.* Methuen, London.

TINBERGEN, N. (1963). On aims and methods of ethology. *Z. Tierpsychol.* **20,** 410–433.

TINBERGEN, N. (1972). *The Animal in Its World. Explorations of an Ethologist, 1932–1972, Volume 1 Field Studies.* Harvard Univ. Press, Cambridge, MA.

WILLIAMS, G. C. (1966). *Adaptation and Natural Selection.* Princeton Univ. Press, Princeton, NJ.

WILSON, E. O. (1975). *Sociobiology: The New Synthesis.* Belknap Press of Harvard Univ. Press, Cambridge, MA.

WYNNE-EDWARDS, V. C. (1962). *Animal Dispersion in Relation to Social Behaviour.* Oliver & Boyd, Edinburgh, UK.

Robert A. Hinde
St. John's College
Cambridge, United Kingdom
Behaviour Department
Madingley
Cambridge, United Kingdom

Although ethology has been primarily concerned with animal behavior, it is now having an influence on the human social sciences. Examples include attachment theory in developmental psychology, the recognition of constraints on learning, the importance of coming to terms with dialectical relations between levels of social complexity, the development of a science of human relationships, and studies of human behavior in terms of their functional consequences.

Ethology, Psychology, and the Social Sciences

Introduction

The early ethologists were concerned with the behavior of animals but rejected both the black-box, stimulus—response approach of the operant psychologists and the undisciplined subjectivism then prevalent among many comparative psychologists in Europe. Its founders made attempts to apply ethological principles to human behavior (Lorenz, 1950), and some of their pupils called attention to trivial similarities between animal and human behavior in a manner which served only to bring ethology into disrepute. However, the initial thrust was directed almost entirely to animal behavior. The possibility that ethology could contribute to their disciplines was given little serious consideration by social scientists, in part because of their own prejudices and in part because of the brashness of some ethologists. However, in the decades that followed the 1950s ethological principles gradually penetrated the social sciences and are now beginning to have an important influence in a variety of areas.

What are those principles? Ethology had its theory, much of which is now discarded but, unlike the 1950s approaches to learning phenomena and unlike psychoanalysis, it was never characterized by theory. Rather, the enduring features of ethology consist of certain orienting attitudes. Among the most important of these was the view that the study of behavior must begin with careful observation and description under conditions that are natural to the species concerned, and that analysis should be accompanied by an awareness that the whole is more than the sum of its parts. In addition, ethologists held that the full understanding of an aspect of behavior required answers to four questions: What causes the behavior? (reference here being to factors both inside and outside the organism) How does it develop in the individual? How did it evolve? and What is or was its function?— or, to be more precise, Through what consequences is it maintained, or has it been maintained, in the repertoire of the species? These questions are logically distinct, although their study is often interfertile. The last two, of course, concern issues not then considered by psychologists but which, as discussed later, are now believed to provide important insights into aspects of human behavior.

During the second half of the twentieth century, as aspects of the ethological approach and ethological ideas infiltrated the human psychological and social sciences, orienting attitudes and techniques often merged, and in some areas it became difficult to distinguish "ethological" from more traditional studies of human behavior. The following sections attempt to illustrate the sorts of ways in which ethology is contributing to the understanding of human behavior: The examples chosen are only a small proportion of those that might have been chosen, and their selection has undoubtedly been influenced by my interests.

Questions of Development and Causation

Child Development and Psychiatry

Some of the early attempts to apply an ethological approach to child development involved the identification of distinct motor patterns comparable to the fixed action patterns which had been a focus of attention in animal studies. For instance, careful description of the sequence of reflex-like responses by which a neonate can find the nipple contributed to the solution of some problems experienced by breast-feeding mothers. Early on, ethologists started to describe human expressive movements, and their work linked with that initiated somewhat later by psychologists such as Ekman. In studies of nursery school-age children, Blurton Jones pioneered the use of ethological observational methodology and showed, for instance, that the seemingly aggressive rough-and-tumble play was not positively related to such items as frown, hit, and push, although it seemed to resemble them in many ways. However, the extensive description of motor patterns which he used led to an unnecessary and somewhat misleading mass of detail, and later applications of ethological observation have been adapted to determine the meanings behind actions.

The area in which an ethological approach has had far-reaching implications concerns the role of the parent–infant relationship in the development of personality. Bowlby (1969), a London psychoanalyst, saw that what had been called the "irrational fears of childhood"— fears of falling, of sudden loud noises, of being left alone, and so forth— would have been highly adaptive in the early evolution of hominids. This led to an integration of ethology, psychoanalysis, systems theory, and developmental psychology in attachment theory. Bowlby recognized that the parent–child relationship involves diverse patterns which have been adapted in evolution for the mutual benefit of both partners. Many of these, such as the infant's smile and the high-pitched voice that parents use in talking to their babies, have the function of cementing the bond between them. Bowlby postulated an "attachment behavior system" in the baby, involving interrelated patterns occurring in parent–child interaction, and a corresponding "parental behavior system" in the parent, with the concept of behavior system taken from studies of wasps and cichlid fish by Baerends. He further proposed that the child formed an internal working model of self, others, and relationships on the basis of his or her relationship with the principal caregiver, and that this influenced (and could be modified by) future behavior.

Bowlby emphasized the importance of the parent giving the child a feeling of security by means of measured and sensitive responsiveness to the child's needs and maintained that the nature and degree of the felt security provided by the parent could have extensive repercussions on personality development later in life. Ainsworth and colleagues (1978) developed a laboratory procedure for assessing relevant aspects of the parent–child relationship, thereby making possible the demonstration that security in early childhood is related to maternal sensitive responsiveness and has many correlates in later development. For example, in the absence of major life events children classified as secure at 1 year of age tend to get along better with peers in preschool. Interestingly, the gender differences seen in 4-year-olds seem to be primarily due to the insecure children.

Knowledge of the importance of security in childhood has led to the study of its role in adult relationships, and a variety of techniques have been devised to assess this. Thus, Hesse (1999) devised a sensitive interview to assess mothers' recollections of their relationships with their own mothers, and Bartholomew (1993) provided an elaboration of Bowlby's internal working model to distinguish two dimensions— one of positivity of the model of self and the other of positivity of expectations of others.

Such work has shown that the infant mind is not to be regarded as a "blank slate" (*tabula rasa*), but that infants are born with, or soon develop, a repertoire of behavior patterns, responsiveness to stimuli, and psychological capacities which form the basis of their subsequent development. This has been accompanied by an intensive study of infants' capacities and also of aspects of behavior that emerge early in development but seem independent of specific experience.

Of special importance here has been work on the nature

of learning. Associated with the tabula rasa view of the infant mind was the presumption that all human experiences could be assimilated with more or less equal ease and thus influence subsequent behavior. However, studies of animals had shown that learning abilities were often limited to particular contexts, with the animal adapting readily to circumstances in some situations but not in others (Tinbergen, 1951). For instance, Thorpe showed that learning played an essential role in the development of the song of the chaffinch (*Fringilla coelebs*). A chaffinch raised in auditory isolation from other chaffinches produced only a very simple song. However, chaffinches could not learn any song that they heard — only those with a note structure similar to that of chaffinch song. A related species, the bullfinch (*Pyrrhula pyrrhula*), could learn only the song of the male by whom it had been reared. Similar conclusions were reached by experimental psychologists. Thus, Garcia and associates showed that rats could learn to avoid poisoned food, even though the ill-effects could not have been felt until some time after they had eaten it. It has also been shown that rats associate the taste with the ill-effects much more readily than other stimuli present at the time. Such data led to the generalization that organisms, including humans, have predispositions to learn some things rather than others, and that the constraints are not simply a matter of task difficulty but depend on the subject matter and the context (Hinde and Stevenson-Hinde, 1973).

This has had widespread repercussions on conceptions of the nature of intelligence. Rozin, an experimental psychologist, contrasted the view held at the time by most of his colleagues that the evolution of intelligence involved the gradual improvement of a general learning capacity with the view that evolution has involved adaptive specializations for particular contexts. In recent years, the view that human intelligence depends on a more or less unitary general-purpose mechanism has fallen into disrepute. Fodor proposed that the mind consists of many discrete input systems feeding into a central cognitive mechanism responsible for "intelligence." Gardner, taking a different approach, distinguished seven types of intelligence — linguistic, musical, logical–mathematical, spatial, bodily kinesthetic, self-understanding, and other-understanding. Recently, as discussed later, evolutionary psychologists have proposed that the mind involves many special-purpose mechanisms.

The Social Sciences

As knowledge accumulates, it is inevitable that disciplines should demarcate their subject matter, become more distinct from each other, and divide into more specialist subdisciplines. This has certainly been the case in the human social sciences, in which individual psychology, social psychology, sociology, and anthropology occupy separate university departments and often have little contact with each other. However, ethologists studying animals in nature were confronted by the relations between the behavior of individuals, their relationships with each other, and the structure of the social group. For a long time attention was concentrated on the issue of dominance/subordinance, and the structure of animal groups was viewed principally in terms of hierarchies. However, more detailed studies, especially studies of nonhuman primates, revealed that dominance was not the only or even the most important issue, and that group structure could be very complicated. For instance, the structure of a rhesus monkey troop depends primarily on affiliative relations between relatives. There is often a series of matrilines each consisting of a matriarch, with her daughters ranking in inverse order of age and their offspring similarly arranged. The females are permanent troop members; the males form a hierarchy of their own, moving from troop to troop intermittently and forming occasional but usually not very durable bonds with females.

Such data led to the view that the several levels of social complexity could not be studied independently because they affect each other. For many purposes it is thus convenient to recognize successive levels: physiological/psychological processes within individuals, individual behavior, short-term interactions between individuals, and longer term relationships between individuals, groups, and societies (Fig. 1). An important distinction here is that between interactions and relationships. Regarding behavior, a relationship involves a succession of interactions between two

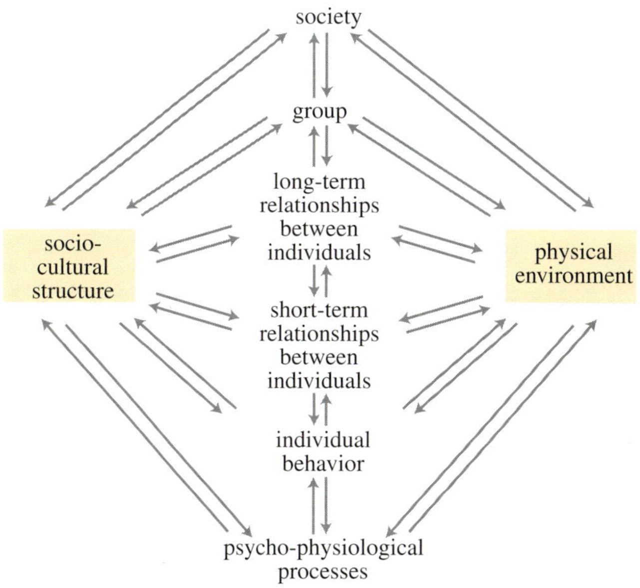

FIGURE 1 Simplified representation of the relationships between the different levels of social complexity (reproduced with permission from Hinde, 1997).

individuals who know each other, such that each interaction is influenced by previous interactions between them and often by the expectation of interactions in the future. Of course, there is more to it than that: interactions and relationships also involve cognitive and emotional aspects (Hinde, 1997).

Each of these levels involves properties not relevant to the level below. Thus, a relationship may involve one or many types of interaction: This property is not relevant to individual interactions. Also, in an interaction the behavior of one partner may or may not mesh with that of the other, meshing being a property irrelevant to the behavior of an individual in isolation. In addition, we use different explanatory concepts at each level. Thus, in explaining why two brothers are fighting (interaction level), we might focus on their desires for the same toy; however, if we wanted to explain why they were always fighting (relationship level), we might ascribe it to sibling rivalry.

Each of these levels affects and is affected by those adjacent to it. Thus, what occurs in an interaction is affected by the nature of the individuals involved and also by that of the relationship in which it is embedded. Also, the nature of a relationship is affected by the nature of the component interactions and, because A's relationship with B is affected by B's relationship with C, by the group to which the individuals belong. This implies that the several levels, including that of the individual, are not to be considered as static entities but as processes continuously being created, maintained, and degraded by the dialectical relations between levels.

All of this is equally applicable to our own species, but here there is an additional issue — the sociocultural structure. This term is used here to refer to the norms, beliefs, values, and institutions with their constituent roles which are more or less common to the individuals concerned, with the term structure indicating that the norms, values, etc. are at least loosely related to each other. This sociocultural structure also has dialectical relations with each of the levels of complexity; for instance, norms about marriage influence the frequency of divorce, and the frequency of divorce influences norms about marriage. In addition, although not considered here, each of these levels both affects and is affected by the physical environment.

It will be apparent that these several levels of complexity form the subject matter of different social science disciplines. However, the existence of dialectical relations between levels means that the development and causation of any aspect of human behavior cannot be fully understood at any one level of analysis — a view in direct opposition to that of Durkheim, who insisted that social phenomena could be understood only in terms of social variables.

An example will exemplify the importance of the dialectical relations between levels. Although one or two possible exceptions have been suggested, the special salience of snakes is fairly ubiquitous among human cultures, and the primary response to them is withdrawal and fear. Many hypotheses have been advanced to account for this. However, studies of children brought up in an institution who had never previously seen a snake showed that many started to show fear of snakes when 1 or 2 years old. How afraid a child became depended on the behavior of his or her caregiver: Strong fear responses by the latter could lead to a long-term fear of snakes. Parallel data are available from rhesus monkeys. Whereas wild-reared monkeys show fear of snakes, laboratory-reared monkeys do not. However, a laboratory-reared monkey shown a videotape of a wild-reared monkey being afraid of a snake is likely to become permanently afraid of snakes. A laboratory-reared monkey shown a similar videotape of a wild-reared monkey being afraid of a flower does not become afraid of either snakes or flowers (Mineka, 1987).

It thus seems that for both monkeys and humans, snakes are strongly salient as stimuli, and that there is a tendency to learn to be afraid of them from the behavior of another individual. However, in the human case there is an additional issue: Snakes play an important role in the mythology of most cultures. Usually, they symbolize evil or death. A wonderful example is provided by the Rubens paintings of the lost souls going down into hell with snakes gnawing at their genitals — an image likely enough to make anyone more than a little apprehensive about going to hell (Fig. 2). The role of snakes in mythology must be supported by the initial propensity to learn to fear snakes and must also play a part in maintaining the fear of snakes. However, in some cultures snakes have associations additional to or even instead of fear and evil. In India snakes are associated with fertility as well as with danger, and images of Shiva show the god with snakes twisted around him like ropes of jewelry, suggesting the ambiguity of his power as creator and destroyer. Full understanding of the role of snakes thus requires us to tease apart the relations between basic biologi-

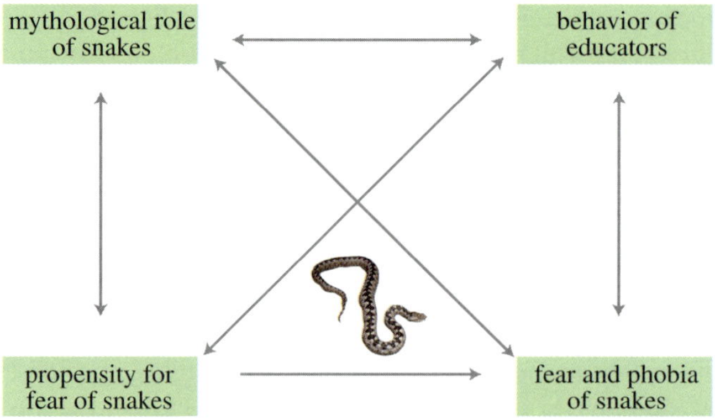

FIGURE 2 Schematic representation of the dialectical relationships giving rise to fear of snakes (reproduced with permission from Hinde, 1987).

cal propensities, relationships with other individuals, and the sociocultural structure.

Similar principles apply to many, perhaps all, aspects of human behavior. For example, the readiness of individuals to go to war is enhanced by propaganda which creates an image of the enemy as evil, dangerous, and subhuman. The success of this propaganda is due in large measure to the fact that it plays on certain basic human propensities — fear of strangers, aggressiveness, group loyalty (itself complexly based), and so on.

Regarding another issue, for reasons that are not immediately apparent, the study of personal relationships was neglected by psychologists until recently. Although clinicians had gathered extensive data, theirs was mostly a case-by-case approach which did not lend itself to the building of an organized body of knowledge. Recently, ethologists, social scientists, students of communication, and others have come together in the construction of a science of human relationships. In accordance with ethological principles, this involves the provision of a descriptive base, the specification of principles concerned with the dynamics of relationships, and statements about the limitations of those principles.

The descriptive base involves dimensions concerned with (i) the content of the interactions within the relationship (i.e., what do the participants do together?); (ii) the diversity of those interactions (do they interact in just one way or in many?); (iii) the quality of the interactions and of communication within the relationship; (iv) certain other qualities arising from the relative frequency and patterning of the interactions; (v) reciprocity versus complementarity of the interactions (i.e., do the partners show the same kind of behavior in their interactions, or does each show behavior different from, but complementary to, that of the other?); (vi) how often are they in conflict, and how do they attempt to resolve conflicts?; (vii) how is power distributed between the partners, and how do they believe it to be distributed? (viii) what is the balance between self-disclosure and privacy in the relationship?; (ix) interpersonal perception (e.g., does each see the other as he or she really is? Does each see the other as he or she sees himself or herself? Does he or she feel understood by the partner?); (x) satisfaction (Does he or she feel satisfied with the relationship in comparison with what he or she feels he or she deserves and in comparison with other possible relationships?); and (xi) commitment to the relationship.

The dynamics of relationships depend in part on the individual characteristics of the partners and on the relations between their characteristics, on social influences from outside the relationship, on attempts by the participants to maintain a coherent view of the world by seeking to reconcile apparently incompatible beliefs and values and by making appropriate attributions about the causation of events, on reciprocity in exchange, and on positive and negative feedback. Although the dynamics of interpersonal relationships are complex, a multidisciplinary science of human relationships is beginning to emerge (Hinde, 1997).

Questions of Evolution and Function

We have seen that ethologists attempt to answer questions not only about development and causation but also about the evolution and biological function of behavior. Can such questions usefully be applied to our own species?

Regarding the course of evolution, there is little to be said. Some human motor patterns and other aspects of human behavior closely resemble those of nonhuman primates; for example, the sequence of movements used by neonates to find the nipple. Some seem to be relics of patterns that were adaptive earlier in our evolutionary history. Thus, we have already seen how the "irrational fears of childhood" must once have been an important part of our ancestral repertoire, and the Moro reflex (a reflex response shown by infants and involving an initial spreading of the arms followed by clasping of the hands) seems to have been derived from the reflex by which young primates cling ventroventrally to their mothers. In other cases, possible links between human expressive movements and those of our nonhuman primate ancestry have been proposed: The case of smiling and laughter has been noted already. However, in general, the absence of evidence of their behavior reduces attempts to trace the detailed course of the evolution of human behavior to the level of speculation. However, careful examination of their artifacts is providing interesting evidence about the cognitive capacities of prehumans, and there have been some important attempts to assess the broad features of human evolution (Mithen, 1997).

With respect to functional aspects of behavior, the situation is very different, and a biological approach to human behavior is throwing light on some important issues. To convey the nature of this work, an historical approach seems best.

In 1964, W. D. Hamilton pointed out that natural selection could act to favor not only variants with greater individual reproductive success but also individuals who acted to increase that of related individuals even if it decreased their own. The occurrence and magnitude of the effect depend on the degree of relatedness between the actor and those who profit from the act. To take an oversimplified example, imagine a gene change that led to an individual being ready to give up its life for others. One copy of the gene would disappear if the bearer died, but if the act saved the life of more than two offspring (each of which has a 50% chance of carrying the gene in question), more than two siblings, more than eight cousins, and so on, then the gene would still increase in frequency. This provides an evolutionary explanation for many examples of behavior that appear to be detrimental to the actor, the self-sacrificial nature of parental behavior being a prime example. In 1975, E. O. Wilson applied Hamilton's ideas in a landmark volume, *Sociobiology,* a masterly integration of material from ethology, ecology, population genetics, and related disciplines. Its main aim was to interpret features of social organization from a knowledge of population parameters and the biological constraints imposed by the genetic constitution of the species. Although tarnished by unnecessary and bellicose claims that sociobiology would in due course engulf disciplines such as ethology, psychology, and the social sciences, Wilson's lead provided a base for many stimulating attempts to provide functional insights into aspects of human behavior. These are classified into the following overlapping categories.

The Sociobiological Approach

Although parental care may be understood in biological terms by the fact that the offspring are the carriers of the parents' genes, the dynamics of parental behavior are complicated by the fact that by devoting resources to one offspring, the parent may be reducing its ability to care for later ones. Thus, the offspring may demand more than what is appropriate for the parent to provide. This can give rise to a conflict of interests between parent and young. Although this does not necessarily imply that behavioral conflict between parent and child will occur (Bateson, personal communication), many aspects of the human parent–child relationship, from early parental protectiveness through weaning tantrums to adolescent conflict, fit this model. Anecdotal and other evidence indicates that parents are prone to commit more resources to a lastborn than to their older children, there being no subsequent offspring for whom resources should be conserved.

Behavior that helps others to the temporary detriment of the helper has also been ascribed to *reciprocal altruism,* the principle of "I'll help you if you are likely to help me later." This is in harmony with data showing that relationships between human adults are based in part on principles of social exchange in which reciprocity is expected, although the criteria of what is fair vary with the type of relationship and the circumstances. Some authors have suggested that relationships are guided by a social contract, and there is evidence that individuals not only resent being underbenefitted but also feel uncomfortable if they see themselves as overbenefitted (Hinde, 1997).

Human Behavioral Ecology

This aims to determine how ecological and social factors affect behavioral variability within and between populations (Borgerhoff Mulder, 1991) and often involves attempts to show that human behavior tends to maximize the reproductive success of the individual concerned or that of close relatives as a consequence of "evolved behavioral tendencies." For example, Kalahari hunter–gatherers, unlike women in many other non-Western societies, space their children widely, reproducing only about once every 4 years. Although this might suggest that they are not maximizing their reproductive success, Lee pointed out that the women gather much of their food and must carry their babies with them as they do so, thereby reducing the amount of food that they can carry. Blurton Jones (1987), making reasonable assumptions about the load that a woman must carry if her children are to be adequately nourished, showed that the evidence is in harmony with the views that the interbirth interval is set by the load that the mother can carry and that the interval of 4 years is optimal.

As another example, Borgerhoff Mulder studied reproductive success among the Kenyan Kipisigis. In this group, men alone own the land, sharing it equally among their wives. Borgerhoff Mulder found the size of a wife's share to be positively correlated with her reproductive success. Women tended to marry men who currently had fewer wives and more land, thereby maximizing their reproductive potential. Land is handed down to sons and not to daughters. The lifetime reproductive success of the men thus tended to be correlated with the size of the parental landholding. However, a similar relation was found for daughters, but here it was a consequence of the earlier menarche of girls brought up on large farms. A similar approach has been applied to many aspects of human behavior — foraging and diet, food sharing, sex differences in reproductive behavior, mating systems, parental care, and so on (Borgerhoff Mulder, 1991).

The results of such studies are of great interest, but these studies also have problems. Although many studies have shown a correlation between wealth and reproductive success in nonindustrialized societies, such a relation does not always hold. Thus, in Western societies the relation between wealth and reproductive success fluctuates with social conditions and may be an inverse one. There are in fact many situations in which humans do not behave in a biologically adaptive manner, and such instances require explanation. One possibility is that the tremendous changes in the human environment in recent millenia are such that behavioral propensities that were adaptive at an earlier stage in our evolution are no longer adaptive. Their current expression could be the result of the change in conditions; for instance, the "irrational fears" may not be adaptive in an urban environment. Also, maladaptive behavior may be an inappropriate or excessive expression of a propensity that is normally adaptive (e.g., gluttony). Again, individuals may be constrained by others or by the sociocultural structure to act in ways that do not favor their reproductive interests (e.g., religious celibates) (Hinde, 1987). Borgerhof Mulder (1991), who provides a careful discussion of these and other issues, is optimistic that studies of individually adaptive outcomes will shed light on the structure and evolution of human societies. She advocates the further development of games theory to deal with situations in which individual advantages are compromised by the strategies of other, more complex optimality models to deal with the diversity of the selection pressures operating, and she also advocates recognition of the importance of historical processes.

Evolutionary Psychology

This involves attempts to show that particular aspects of human behavior have been adapted for particular functions. It thus involves the assumption that the brain/mind comprises many specialized domain-specific mechanisms and depends in large measure on demonstrating that the "evolutionary design" of aspects of behavior provide efficiency, economy, and precision in operation. For example, the aspects of the opposite sex that humans find attractive appear to be such as are likely to maximize their reproductive success (Buss, 1994). Again, considerable evidence indicates that pregnancy sickness is a mechanism for protecting the fetus from toxins — a conclusion which puts quite a different perspective on the phenomenon. Barkow *et al.* (1992) have shown that the ability to detect violations of conditional rules of the form "if P then Q" is greater when they can be interpreted as involving cheating on a social contract. Thus, in one series of experiments subjects were asked whether a hypothesis had been violated in each of four instances shown by cards. In one case, the rule was "If a person was drinking beer, then he must be over 20 years old," and subjects were given four cards, each with information about an individual's age on one side and on what he was drinking on the other. The four exposed sides were "drinking beer," "drinking Coke," "25 years old," and "16 years old." The correct solution is that the first and last cards need to be checked. Although such problems, involving the possible violation of a social contract, produced the correct response in approximately 75% of subjects, problems in which the rule was merely descriptive and did not involve a social contract elicited a much lower success rate. The authors suggest that violations of a social contract elicit an evolutionarily adapted mechanism.

As with behavioral ecology, the evidence that the behavior in question is an adaptation is not always completely watertight; for instance, children have many more opportunities to learn about violations of social contracts, and to object to them, than violations of other conditional rules. However, the important issue in all these approaches is the focus on the influence of natural selection on human social behavior. Also, it is becoming apparent that this may have been, and is, much more complex than appears at first sight. For example, humans may change their environment, and this may affect the impact of natural selection (Odling-Smee *et al.*, 1996). The sociocultural structure influences mating patterns and mobility, with the same possible result. Thus, sociocultural change can influence genetic change (Boyd and Richerson, 1985; Durham, 1991).

Conclusion

Perhaps the most important lesson to be drawn from this discussion is that no one approach can lead us to a full understanding of the complexities of human behavior. No one discipline has the key. Rather, we must seek to integrate the findings of the behavioral sciences by teasing apart the dialectics between the successive levels of complexity. One route to this end involves the integration of causal/developmental and functional approaches. The first step is to identify "relatively stable characteristics" which, although individually variable, are present in all humans or in all members of an age/sex class. The stability of such characteristics across individuals might be a consequence of nature or common experience or, most likely, both. Behavioral examples are the salience of snakes and the propensity of the infant to bond with the mother. Then the more complex aspects of human behavior can be traced to interactions between such characteristics or between individuals, just as wartime propaganda plays on basic human propensities and on interactions between them. It is likely that the relatively stable characteristics, or most of their sequelae, would prove to be adaptive (Hinde, 1987). Such a course requires the integration of ethology with the social sciences.

References Cited

Ainsworth, M. D. S., Blehar, M. C., Waters, E., and Wall, S. (1978). *Patterns of Attachment.* Erlbaum, Hillsdale, NJ.

Barkow, J. H., Cosmides, L., and Tooby, J. (1992). *The Adapted Mind.* Oxford Univ. Press, New York.

Bartholomew, K. (1993). From childhood to adult relationships: Attachment theory and research. In *Learning about Relationships* (S. Duck, Ed.), pp. 30–62. Sage, Newbury Park, CA.

Blurton Jones, N. G. (1987). Bushman birth spacing: Direct tests of some simple predictions. *Ethol. Sociobiol.* **8,** 183–203.

Borgerhoff Mulder, M. (1991). Human behavioural ecology. In *Behavioural Ecology* (J. R. Krebs and N. B. Davies, Eds.), pp. 69–104. Blackwell, Oxford.

Bowlby, J. (1969). *Attachment and Loss. Vol. 1, Attachment.* Hogarth, London.

Boyd, R., and Richerson, P. J. (1985). *Culture and the Evolutionary Process.* Chicago Univ. Press, Chicago.

Bretherton, I., and Waters, E. (1985). Growing points of attachment theory and research. *Monogr. Soc. Res. Child Dev.* **50,** 1–2.

Buss, D. (1994). *The Evolution of Desire.* Basic Books, New York.

Durham, W. H. (1991). *Coevolution: Genes, Culture, and Human Diversity.* Stanford Univ. Press, Stanford, CA.

Hesse, E. (1999). The adult attachment interview. In *Handbook of Attachment: Theory, research and clinical application*

(J. Cassidy and P. R. Shaver, Eds.), pp. 395–433. Guilford Press, New York.
HINDE, R. A. (Ed.) (1983). *Primate Social Relationships.* Blackwell, Oxford.
HINDE, R. A. (1987). *Individuals, Relationships, and Culture.* Cambridge Univ. Press, Cambridge, UK.
HINDE, R. A. (1997). *Relationships: A Dialectical Perspective.* Psychology Press, Hove.
HINDE, R. A., and STEVENSON-HINDE, J. (1973). *Constraints on Learning: Limitations and Predispositions.* Academic Press, London.
LORENZ, K. (1971). Part and parcel in animal and human societies. In *Studies in Animal and Human Behaviour, Vol. 2* (K. Lorenz, Trans.), Methuen, London. (Original work published 1950)
MINEKA, S. (1987). A primate model of phobic fears. In *Theoretical Foundations of Behavior Therapy* (H. Eysenck and I. Martin, Eds.), Plenum, New York.
MITHEN, S. (1997). *The Prehistory of the Mind.* Thames & Hudson, London.
ODLING-SMEE, F. J., LALAND, K. N., and FELDMAN, M. W. (1996). Niche construction. *Am. Nat.* **147,** 141–148.
TINBERGEN, N. (1951). *The Study of Instinct.* Clarendon, Oxford.
WILSON, E. O. (1975). *Sociobiology.* Harvard Univ. Press, Cambridge, MA.

GEORGIA MASON
Department of Zoology
Oxford University
Oxford, United Kingdom

Animal Welfare

If we assume that animals have feelings of fear, pain, and frustration akin to our own, then many of the ways in which we keep animals seem likely to cause suffering. But identifying what needs to be improved, and how best to do it, requires considerable research. Animal welfare science uses a gamut of techniques, from measuring physiological signs of psychological stress to operant procedures designed to quantify preferences and aversions. As a young science, with the unenviable task of assessing inherently unmeasurable feelings, it still faces many problems and unknowns. However, through research, some specific problems faced by captive animals have been confirmed (e.g., that the lack of a nest is frustrating for battery-caged hens), and others have been revealed for the first time (e.g., that hard, slippery floors make lying down stressful and difficult for cattle). As the techniques used become increasingly reliable, the rate at which such problems are analyzed and understood should accelerate in the future. Future welfare research may also validate some general methods for improving the lifelong welfare of captive animals, such as ensuring that they are not maternally deprived when young and breeding selectively for low levels of frustration and fear.

Compassion and empathy were once the only tools available to those concerned about animal well-being. The occasional enlightened zoo curator might seek to reduce abnormal behavior or fear of people in his animal exhibits, but animal welfare issues were little addressed by research biologists until recently. The attention of ethologists, physiologists, and evolutionary scientists was first focused on the knotty problem of animal feelings by intensive farming (Table 1), which raised such questions as, do animals suffer in small cages, even though all their physiological needs for warmth, food, and water are met? (See Fig. 1.) Does it matter that they can only perform a limited range of activities? And when they perform abnormal behavior, does this indicate some kind of distress? (See Fig. 2.) These questions have gone some way to being answered in the past 20 years. Furthermore, the questions addressed by those investigating welfare issues have diversified enormously to include animal responses to human contact, the welfare significance of transportation, and many other issues. However, assessing animal welfare is still often a challenging task, dogged by unknowns and controversies. As a result, we still cannot always judge the merits of alternative housing systems, and we still cannot predict, from first principles, the responses that any given animal will show to a new treatment or housing system.

Is It Right to Treat Animals as if They Have Feelings?

Animal welfare means different things to different researchers. Some equate it with biological fitness, claiming

TABLE I

Percentage of Poultry Raised in Great Britain by Different Techniques[a]

YEAR	FREE-RANGE	NEST BOX	BATTERY
1948	88	4	8
1956	44	41	15
1961	31	50	19
1965	16	36	48
1977	3	4	93
1981	2	2	96

[a]This table shows how one intensive farming system, the battery farming of poultry, dramatically increased in the 1960s and 1970s. (From Fraser and Broom, 1990.)

FIGURE 2 The stereotypic behavior shown by this bear in captivity is disturbing for a human observer, but does that necessarily mean that the animal is suffering? (Courtesy of D. M. Broom)

that welfare is only reduced if an animal's ability to survive and reproduce is diminished. To others, emotions such as fear, anxiety, frustration, or boredom are of paramount importance: to them, an animal's welfare is only impaired if it is experiencing an unpleasant mental state (e.g., Dawkins, 1980, 1998; Mason and Mendl, 1993). Thus research into animal welfare is, for many, to assume animals have feelings, and it is certainly this assumption which gives the research its moral urgency. But is this valid?

FIGURE 1 In terms of food, water, and warmth, this mink, raised for its fur, has everything that it needs. Can we conclude that it has no welfare problems? (Photo courtesy of Georgia Mason).

Strictly logically, we simply cannot tell whether animals have feelings: the private, experiential aspects of subjective processes are inherently inaccessible to external observers. This is the "Other Minds" problem, much considered by philosophers such as T. Nagel (who famously asked the question, "What is it like to be a bat?"). However, two arguments, both appealing to the evolutionary continuity between us and other animals, support the idea that animals do indeed have feelings that are subjectively "pleasant" or "unpleasant."

The first argument is based on parsimony. Human feelings of pain and pleasure can be traced to the activation of anatomically distinct nocieptors and particular brain pathways such as the opioid- and serotonin-based systems responsible for the rewarding qualities of some psychoactive drugs (e.g., Rolls, 1999). Such neurological machinery is shared with other mammals, pain fibers are even more widespread among the vertebrates, and it seems unparsimonious to claim that the correlates of its activation somehow differ in us from all other animals.

The second argument is one of design. It is that we have emotions for an adaptive reason, to help ensure that we perform the behaviors that maximize fitness and avoid situations likely to put fitness at risk. Cabanac (1979), Dawkins (1990, 1998), and Rolls (1999) argue persuasively that this is why pleasurable feelings accompany mating or eating calorific food when hungry, whereas unpleasant feelings accompany being exposed to things that challenge survival (injury, the threat of predation, being very cold, and so on). According to this argument, an alien being with an anatomy and physiology quite different from our own might well have emotions, simply because this is a sensible way for a complex being in a complex world to be designed.

Such theses are not unanimously accepted. Some argue that where the problem faced by an animal can be addressed by simple rule-bound behavior ("find food—ingest it," "receive nociceptor stimulation—retreat," "detect predator—flee"), there is no reason to invoke feelings; after all, what additional function could they play? (e.g., Toates, 1995; Barnard and Hurst, 1996). Such authors suggest that feelings only play a role in a limited subset of behaviors (those with the most flexible or "declarative" control mechanisms, such as aversion learning); in cases where automatic responses cannot have evolved because the animal's situation is beyond that experienced by its ancestors (e.g., the "limbo" faced by a captive animal that has all its physiological needs met) and in situations where an animal's normal rules have not worked. There are, however, two problems for those not following the Cabanac/Dawkins/Rolls view. First, these arguments are contradicted by the way humans experience pleasant or unpleasant feelings in a range of situations which we might expect natural selection to have given us mere rules to deal with (from dealing with grit in one's eyes, to having sex). Second, to answer questions such as, "Is this behavior declaratively controlled?" "Is this situation beyond that which a species' normal rules can deal with?" or "Has this animal's rules failed it?" requires information that is difficult, if not impossible, to acquire. For a gazelle in the jaws of a lion, we could argue that it *is* suffering because its fleeing strategy has failed it, or that it is *not* suffering because it is merely following another rule—"if caught—struggle."

Faced with such difficulties, many adopt a "better safe than sorry" stance, based on the ethical view that it is better to mistakenly treat a nonsentient being as sentient than to do the obverse. I will follow this and assume that animals have feelings that accompany run-of-the-mill situations, such as being thirsty and then drinking, as well as the more extreme or evolutionarily unusual situations outlined earlier.

If Animals Have Feelings, How Can We Access Them?

If private experience is inherently inaccessible, how can we gauge the feelings of animals? One type of approach is based on the previous argument, that emotions have evolved to accompany strong motivations to reach or escape particular types of environment or to do particular activities. If we assume this, we can investigate what our subjects are highly motivated to reach or to avoid, and then use this information to judge the relative merits of different practices or housing systems. This approach is discussed in more detail later.

An alternative approach is to try and assess an animal's current welfare state, using aspects of its behavior or physiology believed to correlate with subjective feelings. There have been two main methods of selecting such measures. One method is to experimentally subject animals to "stressors," stimuli we feel certain must be unpleasant, such as hunger, electric shocks, or sensory deprivation. We can also observe animals known to be injured or ill. Data are collected on changes in these animals' behavior or physiology. Similar changes can then be used to infer poor welfare in animals in other circumstances. The second method is to observe the biological changes that occur in humans reporting feelings like fear, anxiety, or pain. This information is typically obtained from people who are sick, clinically depressed, bereaved, or faced by stressful events such as important examinations. Any changes we find that correlate with human subjects' negative emotions, can then be sought in nonhuman animals that we suspect of poor welfare. These two methods of seeking correlates for animal feelings have yielded a diverse, yet largely overlapping, range of measures; the following outlines are from the detailed reviews presented by Dawkins (1980), Broom and Johnson (1993), and Toates (1995).

Behavioral Changes

Behavioral signs of poor welfare include reduced levels of grooming, loss of appetite, loss of interest in mating, poor parental care, reduced interest in exploring the surroundings, increased timidity or aggression, and the performance of displacement activities (Fig. 3), and the subclass of repetitive stereotypies that derive from them, such as the leg-swinging of children doing difficult math homework (Fig. 3). Animals may also produce social signals of distress or need, such as particular odors or vocalizations (Fig. 4).

One particular type of situation assumed to cause poor welfare is frustration, in which an animal is highly motivated to do an activity (such as eat) but is unable to do so (because it is prevented from reaching food). The behavior animals show when frustrated typically includes aggression (see Table 2), attempts to escape, intention movements (brief repetitions of the start of the motivated behavioral sequence), and "vacuum" or redirected activities. These last are performances of the motivated behavior either directed to something quite inappropriate (e.g., a male kept in a cage without a female, mating with its food-dish) or "mimed" (e.g., hens in battery cages going through the complete movements of dust-bathing, even though on dust-free wire floors). All these behavior patterns may, if repeated often enough, develop into rhythmic, unvarying stereotypies, and are commonly used to diagnose frustration.

FIGURE 3 Primates show particular displacement activities, auto-grooming or yawning, when in situations of psycho-social stress, for example, being presented with food but not being dominant enough to monopolize it. (Photo courtesy of Dario Maestripieri).

Physiological Responses

In response to stressors, humans and other animals typically show some very rapid physiological changes, especially the release of adrenaline, an increase in heart rate, and the release of hormones by the adrenal cortex, as stimulated by adrenocorticotropin (ACTH). Prolonged activation of this last system (the hypothalamic–pituitary–adrenal, or HPA, axis) can sometimes result in chronically raised basal corticosteroid levels in the blood, and it can also sometimes change the sensitivity of this system. For example, chronically stressed animals often have a heightened adrenal cortex response to other challenges, or to exogenous ACTH. In addition, it may alter the negative feedback usually exerted by corticosteroids on ACTH release (Fig. 5). Because of this negative feedback, ACTH release can usually be inhibited by a synthetic corticosteroid called dexamethasone, but in humans suffering from endogenous depression, this does not happen. (For more details of the HPA axis, see Levine, this volume). A further physiological change in response to sustained exposure to stressors is an altered metabolism, particularly a reduced level of protein synthesis. Low growth rates, poor coats, slow healing rates, and a poor production of milk can therefore all reflect poor welfare.

TABLE 2

The Effect of Frustration Induced by Food Deprivation on the Number of Direct Attacks by Hens on Their Subordinates[a]

HOURS OF DEPRIVATION	MEAN NUMBER OF ATTACKS IN 20 MIN
2.5	3
5.0	11
7.5	18

[a] Increased frustration, elicited by increased food deprivation followed by the presentation of inaccessible food, results in increasing levels of aggression in hens. (From Broom and Johnson, 1993.)

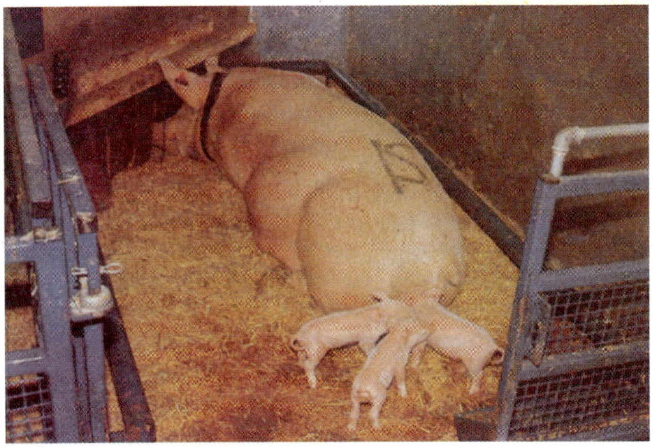

FIGURE 4 Underfed piglets call more and use more high-frequency calls than their well-fed littermates (courtesy of D. M. Broom).

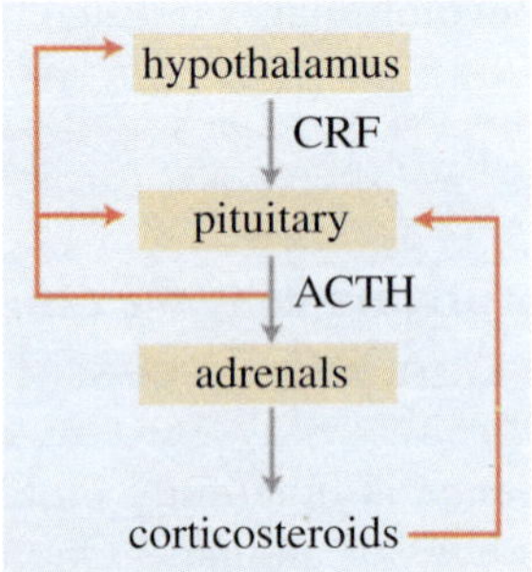

FIGURE 5 The negative feedback control loops (in red) of the hypothalamus–pituitary–adrenal system. The inhibitory effect of corticosteroids on the secretion of ACTH by the pituitary is reduced in humans suffering from depression and in animals exposed to inescapable electric shock. CRF, corticotropin-releasing factor; ACTH, adrenocorticotropic hormone (reproduced with permission from Toates, 1995).

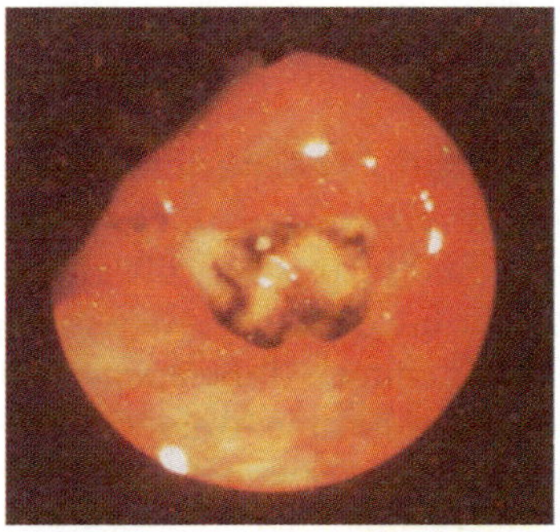

FIGURE 6 Gastric ulcers such as these are found in stressed humans, but they have also been found in rats and primates exposed to inescapable electric shock or social conflict (courtesy M. Beer-Gabel).

Pathologies and Prepathological States

Some indicators cause as well as reflect poor welfare; these changes include high blood pressure, arteriosclerosis, gastric ulceration (Fig. 6), and various forms of suppressed immunity. Suppressed immunity may be evident via *in vitro* tests of immune functioning (such as decreasing numbers of cytokine receptors on leukocytes involved in cell-mediated immunity), or via *in vivo* measures such as the antibody responses mounted against known antigens, or the propensity to develop clinical symptoms of disease. (Such issues are covered in more detail by Dantzer, this volume).

Why There Is No Such Thing as the Perfect Welfare Indicator

An important reservation applies to the use of all the indices above: with the exception of social signals, none of them have evolved to reveal an animal's welfare to observers. Each response is instead part of (or stems from, in the case of pathology) a functional suite of changes designed to minimize the specific effects of a particular challenge on the animal's fitness. This has three important implications for the welfare scientist.

First, it means that exactly which responses are elicited will depend on the precise nature of the challenge. For example, pain does not seem reliably to result in a corticosteroid response (e.g., Rushen, 1991), nor in displacement activities or stereotypy, animals tending instead to show reduced activity, presumably to reduce the risk of physical damage. Chronic cold similarly does not cause active behavioral signs of aversion or high motivation, animals tending instead to stay huddled to conserve energy, while severe dehydration likewise often does not result in a corticosteroid increase, presumably because this response causes increased body temperature, which would be harmful for an animal lacking water. The longevity of a challenge may also affect the responses it elicits. For example, repeated challenge may cause the habituation of corticosteroid responses, if these responses fail to assist the animal remove itself from the stressor. This is sometimes the case for subordinate monkeys, whose hypothalamic–pituitary–adrenal axes may become so unresponsive that they do not even react to electric shock.

The second interpretative problem arises because some events we might feel *sure* are unpleasant produce changes in our indices in an unexpected direction. Heart rate can *fall* as well as rise in response to increased stimulation, for instance; it falls in grouse that freeze as they detect the presence of a predator. The direction of the heart rate change thus seems to be adaptive, depending on the most appropriate behavioral response to the challenge. Similarly, immune responses can sometimes be enhanced rather than suppressed by exposure to stressors (see Dantzer, this volume), possibly because some challenges cause adaptive increases in the activity of particular branches of the immune system (Braude *et al.,* 1999). Lastly, restraint can sometimes decrease rather than increase the corticosteroid responses shown to acute challenge (reviewed Rushen, 1991). This may reflect the catabolic function of these hormones, which are of little use to an animal that is prevented from moving.

The third issue is that some of our "welfare" indices also change in circumstances we might assume *not* to be linked with negative emotions, but which place similar demands on the animal. For instance, corticosteroids rise in hungry calves given food; during coitus, in some male rats; just before laying, in hens; and while nursing, in female mammals (e.g., as reviewed by Broom and Johnson, 1993, and Toates, 1995). This reflects this hormone's metabolic role as fuel-mobilizer. Body weight is similarly affected by many factors other than psychological stress, and so can fall with changes in diet and activity levels. Immunological status may be influenced by circulating levels of testosterone, or by reproductive status, quite independently of stress. And as a final example, gastric ulceration can be due to dysfunctions that are nothing to do with stressors (e.g., the physical effects of hay and straw in the pyloric channels in calves, or local circulatory disturbance).

Overall, to make valid inferences about welfare using indices of current state, we therefore need a good understanding of their control mechanisms and evolutionary function.

Desires and Aversions: Measuring Animal Preferences

Instead of gauging its current state, we can measure the extent to which a subject animal would like to change its circumstances. Animals will learn very speedily to avoid food that makes them physically ill, to avoid areas in which they have been exposed to painful electric shocks, and to put considerable effort into reaching important resources such as food when they are hungry. These aspects of behavior can be exploited as a means of asking animals what they want (e.g., Dawkins, 1983, 1990), and this has a number of advantages — it is rather easier to interpret in welfare terms than some of the physiological and behavioral changes discussed in the preceding section; it avoids the need for invasive procedures; and it can also be used to pinpoint what specific aspects of an animal's environment are responsible for signs of poor welfare.

One powerful approach is to use measures derived from economic methods for assessing what human consumers value. One such measure is the elasticity of demand for a particular resource, which reflects its change in consumption or usage when made more costly (Lea, 1978; Dawkins, 1983). The area under the type of demand plot shown in Fig. 7 can also be used to calculate a value termed "consumer surplus," which is particularly favored by economists trying to assess human welfare. Another means of measuring strength of preference is to determine the maximum price an animal is prepared to pay to reach a resource (something known as the "reservation price" in human economics). For example, rats will cross a greater voltage electric grid to reach food if food deprived, whereas in mice,

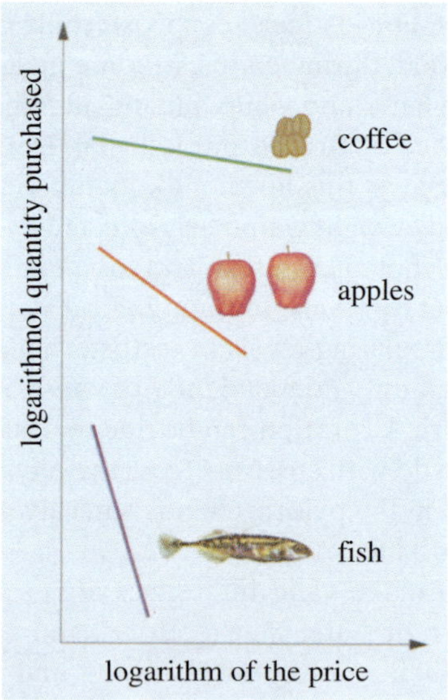

FIGURE 7 This plot of price against quantity purchased on a log-log plot is known as a demand curve. The slope of each line corresponds to the elasticity of demand for the commodity. Demand for coffee is quite inelastic: the amount of coffee bought is defended when its price increases. However, other goods, termed luxuries, are demanded far more elastically, showing a decline in consumption with price increases. Measuring the elasticity of demand for different resources, and the areas under their demand plots, can be used as a means of identifying what changes to an animal's environment will make the most difference to its welfare. From McFarland (1989).

responding for food simply breaks down, at high operant schedules, in animals that are not very food deprived.

Measuring Animal Preferences in Practice

Although much recommended, using these techniques in practice has proved surprisingly difficult. The first problem lies with producing data that give a valid picture of the priorities of the test animal. If the animal is being asked to perform a task, it must be able to associate the task equally easily with all the resources under offer and to perform it equally easily whatever its current motivational state. Hens, for example, sometimes appear to have difficulties associating pecking with access to litter, and so using this operant might artificially create the impression that litter is unimportant to them. To sidestep this problem, many researchers avoid artificial operants, using "natural barriers" such as heavy doors or narrow gaps that the animal must overcome to reach a resource. This last method potentially has another advantage — it can be adjusted so that the animal can pattern its behavior as it chooses, instead of having bouts interrupted as might happen if required to perform an operant for a set quantity of a resource. Another prerequisite for valid results is that the animal be tested in a closed economy, a setup in which the animal lives with all the resources under test for a realistically long period of time (Fig. 8). It would be invalid, for example, to use brief testing periods to investigate how important social contact is to an animal and between tests to re-house it in a home pen where it can interact with conspecifics.

Such experiments must also be carefully designed if they are to produce data relevant to the real world; if measuring motivation or aversion is aimed at improving a particular husbandry system, it must be done in such a way that applying the results is valid. Animals' preferences can be affected by their past experience, especially how they were reared when young; by how many activities they are able to do (a resource that might seem unimportant to an animal in an enriched enclosure might be highly valued by an animal in a small barren one); and also by the size of their time or energy budgets (think of how our own priorities shift if we are short of money, and shift again if we are short of time). This means, for example, that we must not use experimental subjects with large energy budgets if we are trying to assess the preferences of food-restricted farm animals (e.g., pregnant sows, farmed mink at some times of the year, and broiler breeders). A final, practical, difficulty is that animals do not always prefer what is good for them: they may work hard for brain stimulation or addictive psycho-active drugs, and most would probably be highly motivated to avoid veterinary attention. Applying the results of such tests must therefore be done with sensible appreciation of what is best for the animal in the long term. Because of these technical and interpretative issues, measuring preference has so far provided relatively few insights into animal

FIGURE 8 Mink in a closed economy will work hard for many resources and show a strong preference for swimming in water. From Mason, et al.(2001). (Photo courtesy of Claudia Vinke).

welfare. However, the potential of such techniques, as yet still in their infancy, remains enormous.

Putting These Procedures Together to Assess Animal Welfare

Evaluating state and measuring preference and aversion are thus the tools used for assessing welfare. But in practice, how should they best be used to make welfare judgments? The following questions need to be answered if we are to make informed inferences about welfare. And as will be seen, answering even one of these questions can involve a huge amount of research!

1. What Are the Exact Conditions in Which Our Subjects Are Kept? The answer to this question should include information on every possible aspect of long-term living conditions, including stocking density, the potential for performing normal behavior, social structure, feeding level, type of food, access to water, possible sources of discomfort, possible sources of pain, type of human contact, and so on. It should also consider rearing conditions, especially if they differ from how the animals are housed as adults. It should also detail treatments which may last only a short time, such as capture, medical treatment (e.g., the preventative medicine necessitated by many intensive farming practices), transport, mutilations such as tail-docking, experimental procedures, and so on. This analysis must be careful to appreciate how differences in sensory apparatus might make animals' perception of their environment very different from our own. (Fig. 9). All in all, this procedure catalogs the potential welfare issues for investigation.

2. How Does the Subjects' Behavior, Physiology, Appearance, and Health Compare with That of Genetically Similar Animals in Less Restricted Conditions? Comparing the subjects with a control group assumed to have good welfare will reveal if their behavior, health, and physiology differs from baseline values. This information is essential for revealing any behavioral or physiological differences that might indicate poor welfare. It will also suggest possible sources of poor welfare, by quantifying the impact of husbandry on our subjects' behavioral repertoire, social structure, etc. (Fig. 10).

3. Using the Information Gained in (1) and (2) to Identify the Most Likely Sources of Poor Welfare. Earlier we discussed how a full understanding of the mode of action and functional significance of welfare measures is

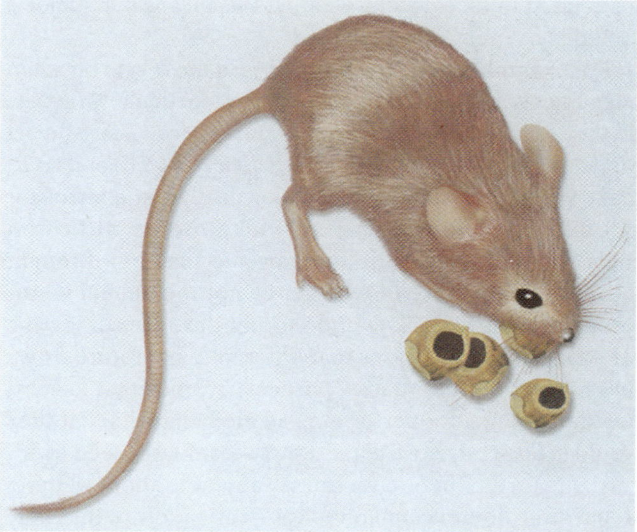

FIGURE 9 Rodents possess excellent ultrasonic hearing. Recently, it was realized that common laboratory equipment, such as oscilloscopes, emit sounds well within their range of hearing and these are probably emitted at an intensity which could be aversive. This illustrates the importance of understanding how animals' worlds may differ from our own.

FIGURE 10 Many features of junglefowl life are quite impossible in the crowded conditions of modern intensive farms. Wild junglefowl usually move around in flocks of one cock and a small harem of 2 to 5 hens; they spend a vast proportion of the day pecking and scratching; for safety, they can fly up to roosts in bamboo; and during the breeding season, the hens separate off to rear their chicks, choosing nesting sites well hidden from view (Dawkins, 1980). Whether the absence of such behavior causes frustration in caged hens is open to empirical investigation. (Photo courtesy of D. M. Broom).

essential before such measures can be interpreted. This understanding comes into play here, and we need it for two reasons. First, if there are differences between our subjects and those in supposedly "good" conditions, we need to judge whether these actually tell us our subjects are suffering. And second, if in contrast there are no differences, we need to think carefully whether this means our subjects are fine.

For example, if we find differences in heart rate between our subjects and those not exposed to a particular practice, do they reflect differences in emotional state? Clearly heart rate changes can reflect physical activity as well, and so to make inferences about welfare, our data must control for any differences in activity levels. Likewise, as corticosteroids have a homeostatic thermogenic function, brought into play irrespective of whether or not the animal is suffering, this must be taken into account if we are comparing HPA activity in situations that differ in temperature. If we find behavioral differences, particularly ones that suggest our subjects are not performing all the behaviors that they would in less restrictive conditions, we similarly need to investigate further before we can say anything about welfare. Behavioral differences may indicate frustration, or they may simply reflect how animals in different conditions have appropriately different behaviors. For instance, wing-flapping by hens occurs at very low levels in battery cages compared with birds in large enclosures. Wing-flapping is often described as stretching the wings, and it may be that the small size of the cage physically prevents this behavior. However, perhaps battery hens are simply not motivated to wing-flap: it could be a sexual signal, a social signal, or an intention movement to fly which simply is not elicited in a battery setup. Until we know this, it would clearly be premature to claim that battery hens are frustrated by their low levels of wing-flapping.

A full understanding of the measure used is vital, too, if we find no apparent differences between our subject and our "good welfare" control animal. If we find no differences in physiological indices, we need to consider whether factors such as habituation or even physiological exhaustion could explain this. For example, the corticosteroid levels of newly tethered bulls return to the level of untethered animals after a month. However, this would seem to be because the adrenal cortex has habituated, as it shows reduced responsiveness to ACTH; thus the central nervous system of long-tethered bulls may well be still maintaining a high output of ACTH. If we find no differences in stereotypy, again, we need to analyze where this might come from. The failure of some situations (e.g., cold) to elicit stereotypies has already been discussed. Thus the absence of stereotypies does not guarantee that subjects and controls have equally good welfare. And if we have used a within-subjects design, transferring our subjects to a less restrictive environment to see if this apparently improves their welfare, we have to be equally cautious if stereotypy levels do not fall. Stereotypies can become habit-like through repeated performance, and they might even become scars of previous welfare problems rather than indicators of current state. For example, the jumping stereotypies developed by voles in barren cages largely disappear in 2-month-old animals transferred to cages provided with vegetation, but they tend to persist in 14-month-old animals similarly moved to enriched cages. However, it would probably not be fair to conclude that the latter animals' welfare was unaffected by the move.

Having started to identify behavioral and physiological characteristics that seem to reflect poor welfare, we need to consider how to combine the different indicators. Severe behavioral or physiological abnormalities that might cause pain or illness are often given particular weight.

4. Is There Evidence of Severe Physiological or Behavioral Disturbance? Some changes in "welfare indices" do not just indicate poor welfare — they cause or anticipate states which are themselves deleterious to welfare. We often therefore give a good deal of weight to these measures. For example, sustained low immunity may result in an animal succumbing to disease, and prolonged high levels of corticosteroids may also cause or indicate metabolic, immunological, and reproductive problems. Included here might be some of the more extreme changes recorded in (1) and (2), such as persistent stereotypies that result in an animal injuring itself, abnormal behaviors that injure other animals, and gastric ulcers, highly likely to cause discomfort or pain. One should also record any behavior known

FIGURE 11 Outbreaks of bitten tails, as seen in this pig, are often preceded by restlessness and agitation. Tail-biting is obviously a cause of pain and distress to the recipient, and its behavioral precursors should be given high priority in welfare judgements. (Photo courtesy of Johnathan Cooper).

FIGURE 12 Primates will spend time and effort obtaining food from puzzle dispensers, even if they have food ad libitum, and even if the tasks it involves are far from "natural." From Gilloux, et al. (1992).

to be the precursor of pathologically abnormal behavior (Fig. 11).

5. What Conditions Do the Animals Themselves Prefer? This question is important because the recommendation that animals should be able to do much of their natural behavior, and the belief that high behavioral diversity is a sign of happiness, is prevalent in the welfare literature, yet as we saw earlier in (3) it is based on the often-untested assumption that absence is synonymous with frustration. Incidentally, in offering animals alternative environments one need not be constrained by what happens in nature. Animals may prove keen on very artificial analogs of what is provided in the wild (such as wheels to run on, in the case of rodents, and the food-dispensing games provided by some zoo curators; Fig. 12), or on social arrangements quite different from those in nature (for example, mink, which are highly solitary in the wild, can live in groups in captivity without aggression; indeed in such a system, they commonly sleep together and their levels of stereotypy are much reduced).

What Science Has Contributed to Animal Welfare

When the five questions above have been addressed, the final welfare evaluation will inevitably contain a subjective element. It can be hard to know how much weight to give to different indices, especially if some of them are nonindependent, or they suggest contradictory things about welfare. Even without this problem, as we saw earlier, the "Other Minds" issue means that it is logically impossible to relate such measures precisely to what the animal is actually feeling. Making welfare inferences is often made even harder in practice by the fact that most studies collect only a tiny subset of the data outlined earlier; they may measure preference, but not physiological responses to frustration; they may measure corticosteroid levels, but not adrenal responsiveness to ACTH or gastric ulceration — and so on. Yet despite these many problems, scientific research has already made some major contributions to animal welfare.

One important role of research has been to back, objectively, various "common sense" views, for example, that barren housing is suboptimal and that mutilating animals (e.g., clipping their teeth or trimming their beaks) causes pain. However, a second important role of research has been to identify some unanticipated problems faced by animals and to yield novel suggestions for improving welfare.

Long-suspected welfare problems confirmed by research, include that many current practices cause pain. For example, the behavior of male turkeys, which often have leg problems because of their rapid growth rates, can be quite changed if they are given analgesics. They eat and drink more, are more active, and, when presented with a female, can approach her with far more alacrity than untreated birds. All this suggests their leg conditions had previously caused sufficient pain to hamper movement. Mutilations, too, can be severe in their effects. Poultry are commonly debeaked to stop them pecking each other and causing injuries. Behavioral studies have shown that feeding, drinking, and preening all decrease after this mutilation, as if the beaks become too painful to use, while inactivity increases — birds spend more time standing inactive or sitting dozing. This last behavioral change can last as long as two weeks. Research has also confirmed the painful effects of "mulesing," an operation performed on sheep living in conditions where their ano-genital regions are prone to fly-strike. About 50 cm^2 of skin is cut away from around the tail and genital area, without anaesthetic. There is little behavioral response, or at least not immediately afterward — the sheep appear quite stoical. However, subsequently they show abnormal posture and locomotion and a very strong avoidance of the people involved in the procedure.

Some pain, fear, and discomfort can be avoided if animals are handled gently and considerately by the people that work with them. Research has confirmed the intuitively obvious importance of human behavior to animals, with stockmanship being shown to have profound effects on welfare. In experimental conditions, the plasma cortisol in battery hens after the rough handling typical on farms was three times the level seen in birds handled gently (Broom and Johnson, 1993). Data collection on working farms has also shown that pigs handled gently by their stockmen are less frightened of humans, have lower levels of circulating corticosteroids, and grow faster and produce more piglets yearly (reviewed by Hemsworth, 1997).

Another long-suspected welfare problem confirmed with data is that barren housing can cause frustration and other problems. For example, for battery hens, the stereotypic pacing performed the hour or so before they lay an egg has been shown to be due to frustration at not having a nest. Furthermore, hens will work hard to reach an enclosed nestbox, squeezing through narrow gaps that they would otherwise avoid, especially immediately before oviposition. Barren conditions seem suboptimal for pigs, too. Pigs kept in such conditions put on less weight than those whose environments were enriched with straw, and if also tethered, they show a reduced production of antibodies when challenged with antigens and may also become more aggressive (reviewed in Broom and Johnson, 1993). Tethered sows also show stereotypies, but research has identified another welfare problem as being at the root of these: underfeeding.

On farms, pregnant sows are fed some 60 percent of what they would eat given *ad libitum* food. This is to prevent them becoming overweight, and it also saves on feed costs. However, operant work has shown that pigs fed this amount are still highly motivated to feed, even directly after the meal, and furthermore, Terlouw *et al.* (1991) showed that this underfeeding is largely responsible for pigs' abnormal behavior. After experimentally altering both housing and feeding levels, behaviors such as sham-chewing, bar-biting, and chain-chewing were far higher in the pigs kept on low levels of food, as was excessive water-drinking, even if they were kept loose-housed in groups rather than being tethered alone in stalls (Fig. 14). Thus low feeding was confirmed by research to be a crucial welfare issue for pigs.

Animal welfare research has also played the valuable role of identifying problems not visible to the casual observer. For example, detailed ethological observations have identified slippery floors as a sources of frustration for cattle, because it makes lying down extremely difficult. Cattle on slippery floors repeatedly perform intention movements, sniffing the ground and performing the leg movements that are the usual preliminary to lying down, and the whole process of actually lying down can take as long as 20 min on some floors (Fig. 15). Physiological data have confirmed this as a source of stress: tethered heifers on a semi-slatted floor have a higher heart rate than animals housed on deep straw, especially when trying to lie down. This effect persisted for the 3 months of the study, even though differences in terms of corticosteroid secretion had disappeared by 1 month.

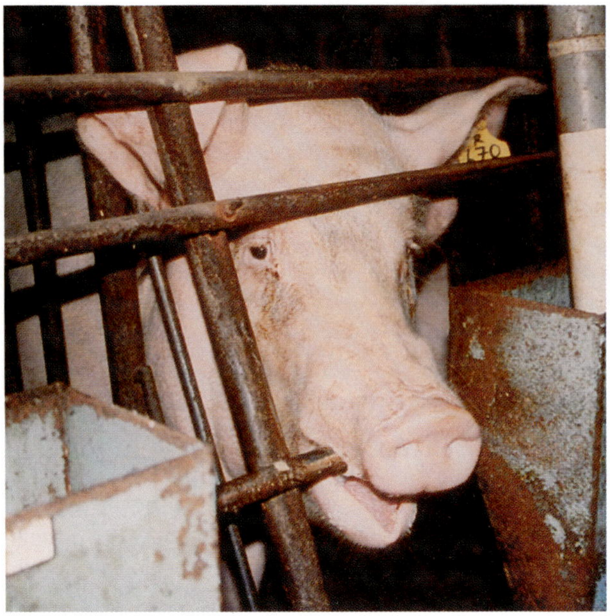

FIGURE 13 Behaviors such as bar-biting are far higher in pigs kept on commercially-widespread low levels of food, but are relatively unaffected by housing type. This suggests that bar-biting and similar behavior in pigs should cause us concern because they indicate sustained hunger. From Terlouw, *et al.* (1991).

Another previously unappreciated problem revealed by research is the stress caused by social interactions between conspecifics (e.g., as reviewed by Broom and Johnson, 1993). The discovery of elevated cortisol levels in tethered sows revealed the hitherto unsuspected problem of unresolved aggression between neighbors, thus indicating a need to redesign their housing; and there are yet more social welfare problems in animals housed in groups. Low- or middle-ranking individuals often experience chronic stress, and this can be true even of animals kept in fairly naturalistic settings. Chronically elevated corticosteroids are seen in low-ranking baboons, tree shrews, and mice; and subor-

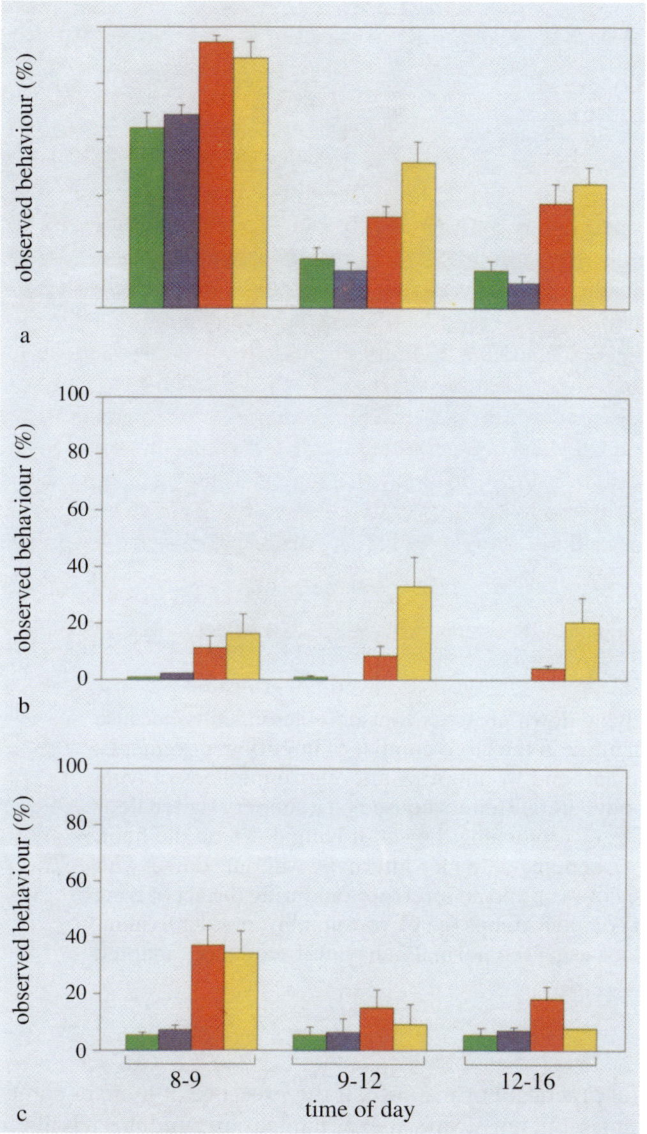

FIGURE 14 This graph shows the percentage of observations in which sows in different treatment groups are active (a), performing oral stereotypies like chain chewing (b), and drinking (c). Green: tethered but well-fed; Blue: loose-housed, well-fed; Red: tethered, underfed; Yellow: loose-housed, underfed. Adapted from Terlouw, *et al.* (1991).

dinate squirrel monkeys show a greater corticosteroid rise than dominants in response to catching and handling. Some of this stress may be because their low-rank means that they are forced to endure the worst conditions in the local environment; for example, subordinate cows are forced by the other members of their herd to stand in the wettest, dirtiest parts of the cow shed, and so they are much more likely to develop foot infections. Much stress can also result from the actual process of establishing a social hierarchy; piglets mixed at weaning have lower levels of play than an unmixed group, even 3 weeks after the event, whereas beef cattle that are repeatedly mixed show high levels of fighting and extreme physiological responses. However, the presence of social companions is still highly important to social animals. For example, sheep which are subjected to various routine procedures such as being put into a truck, dipped, or chased by a dog all show a rise in corticosteroids, but none of these has nearly as much effect as separating the sheep from the rest of the flock. Thus, shearing a sheep, when it is taken away from other sheep leads to higher corticosteroid levels even than slaughtering it.

Animal welfare science has also sometimes revealed that some practices are not the problems we might have predicted. Floors of fine-gauge hexagonal wire were thought to cause discomfort to battery-caged hens. However, when hens were given the opportunity to choose, they preferred to stand on the fine-gauge wire previously condemned as uncomfortable, instead of on the coarser type which had been recommended as an alternative. Another surprising result has emerged from another species of bird, this time in the wild. In a study of the possible impact of tourism on the Antarctic, human proximity was shown to elicit little or no heart rate response in penguins (Table 3).

Animal Welfare Research: Where Next?

Animal welfare research will continue to solve welfare problems on a case-by-case basis in the future. But it will do two further things as well. First, it will give us new means of improving animal welfare, such as altering animals to better cope with human husbandry. Second, it will give us a better understanding of how animals respond to challenge, both in terms of their adaptive, evolved mechanisms and how these mechanisms malfunction under duress. This will increase our fundamental understanding of animal biology as well as improve our abilities to choose and interpret welfare measures appropriately in applied situations.

Developing New Ways of Improving Animal Welfare

To date, most attempts to improve welfare have focused on changing current practices to suit the animal, and there remain several as yet unexplored techniques for doing this. However, another means of improving welfare is to adapt animals better to their environment, using science to produce individuals that are less frustrated by captivity. These could be selected for or produced in other ways; for example, there is growing evidence that the way animals are treated when young can have long-term effects on their responses to captivity.

One recurring theme in the stress literature is that predictability and control are important in determining responses to aversive stimuli. If a warning signal is always

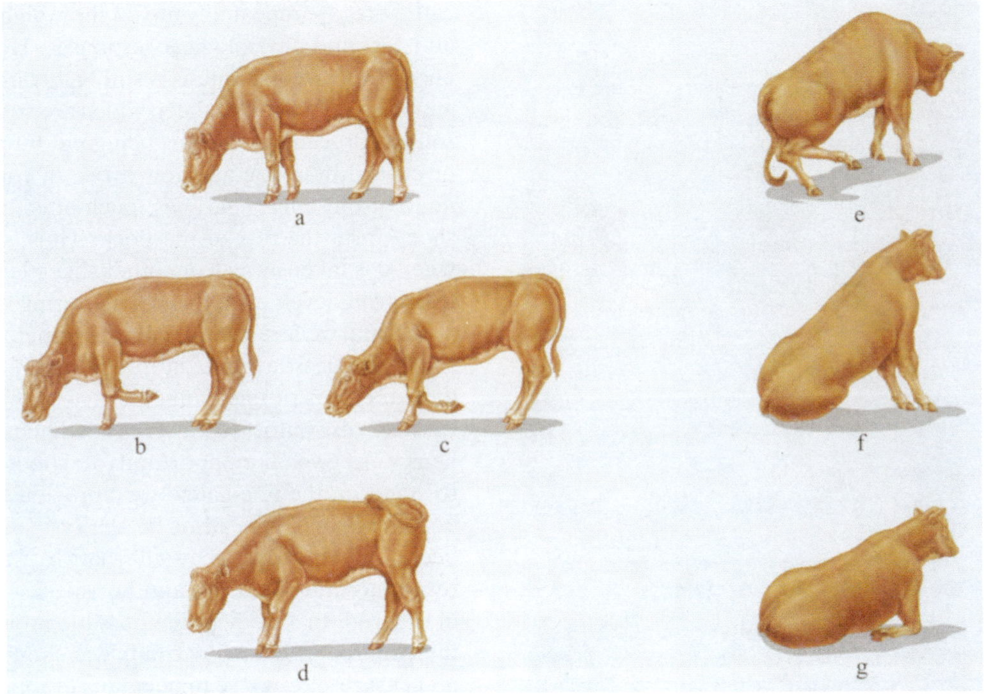

FIGURE 15 For cattle kept on slippery floors, the act of lying down presents considerable difficulty because their hooves are unable to find purchase. This difficulty is manifest in the large number of intention movements, by the increased cardiac rate, and by the fact that some animals end by adopting alternative methods of lying down, consisting of sitting initially with the rear part of the body. (Left) Heifers housed on a slippery slatted floor show behavioral alterations consisting of inhibiting, delaying, or prolonging the act of lying down on the floor, such as repeated sniffing of the ground without lying down (a), bending of the leg internally without contact with the floor (b), bending of the leg internally in contact with the floor (c), and an interruption during the act of lying down (d). (Right) Heifers lying on the floor by initially sitting on their rumps (e–g), presumably in order to minimize the pain associated with lying down on the slippery floor using the normal behavioral sequence (adapted from Andreae and Smidt, 1982).

sounded a few seconds before a shock, the corticosteroid response to it, and the levels of ulceration following repeated exposure, are far lower than if the shock is unsignaled. Animals will also choose environments with such warnings in preference to those without. In addition, in one seminal experiment, the gastric ulceration developed by rats given electric shocks that they could turn off, and hence partially control, was one third that of rats given exactly the same number and severity of shocks without being able to control their exposure. The former rats also lost less weight than rats given uncontrollable shock, and the corticosteroid response they showed could be suppressed by dexamethasone, in contrast to the corticosteroid responses of rats that could not terminate the shock themselves. Similar results have been obtained in a study of rhesus monkeys exposed to high-intensity noise (e.g., as reviewed by Toates, 1995). Such effects are thought to be because when shocks are signaled, or brought to an end with an operant, the animal can judge with confidence the periods for which its environment is safe. This principle has been little incorporated into husbandry regimes, but signaling could be a useful practice. For instance, if the presence of humans sometimes, but not always, means unpleasant handling, it is likely to be better for welfare if the former is signaled in some way (e.g., the stockmen wear particular clothes when about to inflict pain or discomfort). Making pleasant events more predictable, too, may also improve welfare. For example, if group-housed pigs are fed at variable intervals, with half given a reliable signal (a bell) that indicates that food is about to be delivered, and the other half given unreliable signals, high levels of aggression will be found in the latter group after a false signal; thus reliable feeding signals can reduce frustration-related aggression.

That animals should be able to control at least some aspects of their environment is more than just a means of improving adult animals' current welfare: if incorporated into young animals' rearing conditions, it can have lifelong effects on their ability to deal with stressors. The long-lasting effects of being reared in an uncontrollable environment have long been appreciated by primatologists and those studying child development. One way that young animals can control their exposure to frightening or stressful stim-

TABLE 3

Heart Rates (Beats/Min) of Penguins in the Absence or Presence of Humans[a]

Bird	HEART RATE IN THE ABSENCE OF HUMANS		HEART RATE IN THE PRESENCE OF HUMANS	
	Mean	Maximum	Mean	Maximum
1	91.3	114	91.7	96
2	88.4	108	90.6	108
3	89.2	105	91.8	103
4	73.3	78	72.7	96
5	84.1	96	89.4	132
6	82.4	108	78.5	84
7	82.6	90	79.9	90
8	81.6	96	80.4	90
9	85.1	102	82.3	96
10	88.1	96	89.6	108
11	83.3	96	87.2	120
12	83.6	102	83.5	90
13	74.0	90	73.6	84

[a]These data show that in contrast to previous studies, the heart rate response of Adélie penguins to human approach is negligible. One significant improvement of this study on previous work was the clever use of an infrared heart beat sensor concealed within an artificial egg. This avoided the need for capturing the birds to strap monitors to them, and it meant the experimental subjects were as naive and unhandled as the rest of the penguin population. (From Nimon *et al.,* 1995.)

uli is by retreating to their mother for comfort and reassurance. However, for most captive animals as they are currently housed, mother–infant interactions are routinely both impoverished (e.g., piglets born to sows in "crates" cannot interact normally with them) and prematurely curtailed (animals as diverse as pigs, cows, mink, rats, and mice are removed from their mothers before the age at which they would leave naturally). Several pieces of evidence suggest improving mother–infant interactions would have long-term effects on improving welfare.

First, in humans, the early loss of a parent, or an unstable social environment in childhood, can have deleterious effects on adults' abilities to cope with stressors (e.g., Rutter, 1991). Furthermore, the abnormal behavior, such as stereotypes, developed by caged mice and mink is greater the earlier they are removed from their mothers (Table 4). Finally, handling rodents as infants, for as little as 3 minutes a day has a dramatic effect on their later stress responsiveness: as adults, previously handled rodents show reduced corticosteroid responses to stressors such as electric shocks, particularly in terms of the rates at which their levels return to normal; fewer behavioral signs of stress in an Open Field Test; and increased levels of exploration. Early handled rodents also reach greater weights in adulthood. Several studies have suggested that these effects are mediated via the behavior of the mother, as the infants emit ultrasonic distress cries during handling and when returned to the nest, which elicit maternal care; it is this intense extra mothering which has the long-term effects on their development (Francis and Meaney 1999).

The early handling of rodents has another advantage — it reduces their resistance to being caught and held by humans as adults, highlighting another means of improving welfare. Fear of humans is a widespread problem for animals, yet gentle handling when very young will result in animals far less disturbed by human proximity. This has been effective, for example, in foxes farmed for their fur. Thus there are several ways in which the early experience of cap-

TABLE 4

Relationship between Abnormal Behavior and the Age of Weaning in Farmed Minks[a]

SIGNS OF EXCESSIVE GROOMING	PERCENTAGE OF ANIMALS WEANED PRECOCIOUSLY AT 7 WEEKS OF AGE			PERCENTAGE OF ANIMALS WEANED NORMALLY AT 11 WEEKS OF AGE		
	Female	Male	Total	Female	Male	Total
Clipped fur on tails at 6 months	26.0 (n = 50)[b]	17.6 (n = 51)	22.8 (n = 101)	7.8 (n = 51)	11.8 (n = 51)	9.8 (n = 102)
Clipped fur on tails at 10 months	26.1 (n = 46)	25.5 (n = 51)	25.8 (n = 97)	14.6 (n = 48)	14.0 (n = 50)	14.3 (n = 98)
Hairless tailtips at 10 months	19.6 (n = 46)	13.7 (n = 51)	16.5 (n = 97)	0.0 (n = 48)	0.0 (n = 50)	0.0 (n = 98)

[a]The level of tail-biting — an abnormal behavior in which animals suck and bite the fur from their own tails is lower in mink that are left with their mothers until the age at which they would naturally leave her in the wild. (From Mason, 1996.)

[b]n, number of animals examined.

tive animals could be manipulated to help minimize their welfare problems in later life.

An alternative strategy for improving animal welfare is to select for animals that are better adapted to captivity. Domesticated animals have been selected for growing and breeding well in captivity, but much unexplored potential exists for selecting, for example, animals that have low motivation to perform behaviors that are prevented by captivity. For instance, it might be possible to select hens which are not motivated to nest-build pre-egg-laying, and so would be less frustrated in battery cages. Further possibilities include breeding animals with higher pain thresholds or less fear of people. As our understanding of the genetic basis of brain mechanisms increases, and our skills with biotechnology grow, we may even be able to do this by genetically modifying animals to directly alter, say, aspects of amygdala functioning.

Refining Welfare Measurement by Understanding Animals' Responses to Challenge

The future of animal welfare research also lies in refining the indices we use by increasing the understanding of the complex processes involved in responding to challenge. These might include a better understanding of what types of factors elicit HPA activation (can this be predicted from the resource reallocation/mobilization that the challenge demands of the animal?); and what determines the rate at which such HPA responses wane with habituation if the challenge is repeated (the costs to the animal of not avoiding the challenge? the degree to which HPA activation actually helps the animal avoid it?). Likewise, future research should lead to an understanding of how and why the immune system responds in different ways to different scenarios. For example, it could well be that immune resources are redistributed, in a predictable way, according to whether the animal most risks physical injury to the skin or infection via the mucous membranes of the respiratory or digestive systems (Braude *et al.,* 1999). A functional understanding of behavior might also help us predict from first principles why some behaviors (such as foraging) seem particularly prone to persistence, and why some species seem more prone than others to persisting with functionless foraging to such an extent that it becomes stereotyped (Mason and Mendl, 1997).

Research into pathological processes will also help us develop or interpret indicators of poor welfare. For example, brain-imaging techniques that help elucidate the neurological bases of human depression and anxiety may one day help us diagnose similar states in animals. Cognitive tasks that help explain why humans with autism and schizophrenia behave as they do, may come to help us ascertain the extent to which captive animals are prone to psychological dysfunction. For example, aspects of forebrain functioning have been implicated in the stereotypies of autistics and schizophrenics and they might also affect how persistently animals perform similar behavior. The role such processes play in captive animals is an exciting area only starting to be explored.

Could future research also give us indicators of happiness? Several authors have suggested this as an important goal (e.g., Broom and Johnson, 1993; Toates, 1995). For example, although providing animals with resources that they have an inelastic demand for has been presented as a means of avoiding frustration, some have suggested that if we also give animals the time, energy, and resources to do elastically demanded, "luxury" activities, we might actually make them happy. Currently, we still know too little about how to assess pleasure to test this idea. Nevertheless, we could seek "pleasure indicators" in the same way that we have sought indicators of distress. First, we could look for behavioral and physiological changes only seen in animals with space, ample food and water, little predation, good climatic conditions, and a natural, strife-free social grouping. Such changes might include play, or vocalizations particular postures (Fig. 16). Second, we could see if contentment, happiness, or euphoria in humans has biological correlates we could search for in animals (there is some evidence, for example, that corticosteroid levels fall in humans when laughing). Thus welfare research in the future might help us maximize animal welfare as well as minimize its problems.

References

ANDREAE, U., and SMIDT, D. (1982). Behavioral alterations in young cattle on slatted floors. In *Disturbed Behaviors in Farm Animals.* (W. Bessei, Ed.). Eugen Ulmer, Stuttgart.

BARNARD, C. J., and HURST, J. L. (1996). Welfare by design: The natural selection of welfare criteria. *Anim. Welfare* **5,** 405–433.

BRAUDE, S., TANG-MARTINEZ, Z., and TAYLOR, G. T. (1999). Stress, testosterone and the immunoredistribution hypothesis. *Behav. Ecol.* **10,** 345–350.

BROOM, D., and JOHNSON, K. (1993). "Stress and Animal Welfare." Chapman & Hall, London.

FIGURE 16 A contented animal?

CABANAC, M. (1979). Sensory pleasure. *Quart. Rev. Biol.* **54,** 1–29.

DAWKINS, M. (1980). *Animal Suffering — The Science of Animal Welfare.* Chapman & Hall, London.

DAWKINS, M. S. (1983). Battery hens name their price: Consumer demand theory and the measurement of ethological "needs." *Anim. Behav.* **31,** 1195–1205.

DAWKINS, M. S. (1990). From an animal's point of view: Movitation, fitness and animal welfare. *Behav. Brain Sci.* **13,** 1–61.

DAWKINS, M. S. (1998). Evolution and animal welfare. *Quart. Rev. Biol.* **73,** 305–328.

FRANCIS, D. D., and MEANEY, M. (1999). Maternal care and the development of stress responses. *Curr. Opin. Neurobiol.* **9,** 128–134.

FRASER, D., and BROOM, D. (1990). *Farm Animal Behaviour and Welfare.* Bailliere-Tindall, London.

GILLOUX, I., GURNELL, J., and SHEPHERDSON, D. (1992). An enrichment device for great apes. *Anim. Welfare* **1,** 279–289.

HEDIGER, H. (1950). *Wild Animals in Captivity.* Butterworths, London.

HEMSWORTH, P. (1997). Human–animal interactions in agriculture and their impact on animal welfare and performance. In *Animal Choices.* British Society of Animal Science, Special Publication No. 20.

LEA, S. G. (1978). The psychology and economics of demand. *Psychol. Bull.* **85,** 441–466.

MASON, G. (1996). Tail-biting by mink is influenced by age at removal from the mother. *Anim. Welfare* **3,** 305–311.

MASON, G. J., and MENDL, M. (1993). Why is there no simple way of measuring animal welfare? *Anim. Welfare* **2,** 301–320.

MASON, G. J., and MENDL, M. (1997). Do the stereotypies of pigs, mink and chickens reflect adaptive species differences in the control of foraging? *Appl. Anim. Behav. Sci.* **53,** 45–58.

MASON, G., COUPER, J., and CLAREBROUGH, C. (2001). Frustrations of fur-farmed mink. *Nature.*

MCFARLAND, D. (1989). *Problems of Animal Behaviour.* Longman Scientific, Harlow.

NIMON, A. J., SCHROTER, R. C., and STONEHOUSE, B. (1995). Heart rate of disturbed penguins. *Nature* **374,** 415.

ROLLS, E. T. (1999). *The Brain and Emotion.* Oxford University Press, Oxford.

RUSHEN, J. (1991). Problems associated with the interpretation of physiological data in the assessment of animal welfare. *Appl. Anim. Behav. Sci.* **28,** 381–386.

RUTTER, M. (1991). A fresh look at maternal deprivation. In *The development and Integration of Behavior* (P. Bateson, Ed.), Cambridge University Press, Cambridge.

SALES, G. D., MILLIGAN, S. R., and KHIRNEKH, K. (1999). Sources of sound in the laboratory animal environment: A survey of the sounds produced by procedures and equipment. *Anim. Welfare* **8,** 97–115.

TERLOUW, E. M. C., LAWRENCE, A. B., and ILLIUS, A. W. (1991). Influences of feeding levels and physical restriction on development of stereotypies in sows. *Anim. Behav.* **42,** 981–991.

TOATES, F. (1995). *Stress — Conceptual and Biological Aspects.* Wiley, Chichester.

JOHAN J. BOLHUIS
Sub-Department of Animal Behaviour
University of Cambridge
Cambridge CB3 8AA, United Kingdom

Imprinting

Imprinting is the process through which social preferences of young animals become restricted to a particular stimulus or class of stimuli. A distinction is made between filial imprinting, in which a social bond is formed between the young animal (usually from a precocial avian species) and its parent, and sexual imprinting, which involves the formation of adult sexual preferences. The learning process during filial imprinting differs from that of Pavlovian conditioning, but some of the underlying mechanisms are similar to those found in associative learning. Apart from imprinting, under appropriate conditions filial preferences may also develop for stimuli without exposure to these stimuli. The system underlying such preferences is known as a predisposition. In both filial and sexual imprinting there is considerable flexibility with regard to the influence of experience on the formation of eventual social preferences.

Introduction

Imprinting is the name given to the development of a social preference of a young animal for its parent or for another individual as a result of exposure to that individual. It has been a prominent subject in the study of animal behavior since the vivid descriptions of the social attachments of birds by Konrad Lorenz (1937). Although Lorenz did not do so, in contemporary research into the formation of social preferences a distinction is made between filial and sexual imprinting. *Filial imprinting* is involved in the formation, in young animals, of an attachment to and a preference for the parent, parent surrogate, or siblings (Bolhuis, 1991). *Sexual imprinting* is involved in the formation of mating preferences that are expressed in later life (ten Cate, 1989; Bischof, 1994). In both cases most of the research has involved avian species. Until recently, research concentrated on filial imprinting. However, there have been important new developments in the field of sexual imprinting which will be discussed here. First, I briefly describe some general characteristics of the formation of filial preferences (Bolhuis, 1991), and then I discuss the underlying perceptual mechanisms. Two kinds of perceptual mechanisms play an important role in the development of filial behavior. The first is essentially a representation or memory of stimuli to which the animal is exposed. The formation of this memory is known as filial imprinting. The mechanisms of learning in imprinting have been the subject of considerable study, and some of the recent advances will be discussed here. Much is known about the neural mechanisms underlying filial imprinting, mainly from the work of Horn and collaborators. This neural evidence will not be discussed in this article; for further details the reader is referred to the general references.

The second mechanism involved in the development of filial behavior was already indicated by Lorenz (1937) when he pointed out that imprinting is not a universal phenomenon that applies to all species of birds. In Lorenz's terminology, some species of birds have an "innate perceptory pattern" (Lorenz, 1937, p. 247) of the appropriate "companion" (*Kumpan*), whereas in others the perceptory pattern is formed by experience with a particular object. The latter is equivalent to imprinting, whereas the former is approximately equivalent to a perceptual mechanism which has been called a *predisposition* (Horn and McCabe, 1984). In this context, predispositions refer to perceptual preferences that develop in young animals without experience with the particular stimuli involved. Recent evidence suggests that more generally predispositions play an important role in the development of early filial behavior in certain precocial bird species. In this article, the nature of these predispositions in the domestic chick (*Gallus gallus domesticus*) will be discussed.

Filial Imprinting in Precocial Birds

The formation of filial preferences has been studied mainly in precocial birds such as ducklings and domestic chicks. These birds can move about shortly after hatching and will approach and follow a conspicuous object to which they are exposed. In a natural situation this object is likely to be the bird's mother or sibling, but in the absence of these a wide range of inanimate mother surrogates can also elicit filial behavior (Bolhuis, 1991). For example, when a chick is close to an appropriate object it will approach it and attempt to snuggle up to it. When the object is removed the bird becomes restless and emits shrill calls. Initially, the young bird approaches a wide range of conspicuous objects to which it is exposed, but after it has had a certain amount of experience with a particular object the bird remains close to it and avoids novel ones. When given a choice between the familiar stimulus and a novel one, the bird preferentially approaches the familiar stimulus. Filial imprinting is defined as the process through which social behavior of the young animal becomes limited to a particular object or class of objects as a result of exposure to that object or class of objects (Bolhuis, 1991). This contemporary definition differs from Lorenz's original concepts, an issue to which I return later.

To study imprinting in the laboratory, chicks or ducklings may be hatched and reared in darkness and exposed to a conspicuous object (Fig. 1a) when they are about 24 h old. Chicks may be exposed to a stimulus (training) while they

FIGURE 1 (a) Examples of stimuli used in laboratory experiments to investigate the development of filial preferences in chicks. The box and the cylinder are made of Perspex, and they have colored filters and are lit from within. The stuffed fowl is a Burmese red jungle fowl hen (*Gallus gallus spadiceus*). All of the stimuli are mounted on motors with a central axle that can rotate. (b) The system used for investigating chick imprinting. The chick, placed in a running wheel of 30-cm diameter, can see a stimulus (the rotating box on the left) placed at about 50 cm, through the wire-mesh floor. When it attempts to approach or to move away from the stimulus, the chick rotates the wheel (adapted from Horn, 1985).

are in individual running wheels, as illustrated in Fig. 1b. The chick can see the imprinting object through the wire mesh floor of the running wheel. After imprinting training, the animals are returned to a holding incubator and kept in darkness until their preferences are measured in a choice test involving sequential or simultaneous exposure to the training stimulus and a novel stimulus. Although other measures have also been used, filial preferences are often expressed as the relative approach (or attempted approach) to the training stimulus compared to the combined approach to the training stimulus and to a novel stimulus. The resulting preference score is usually expressed as a percentage. A preference score of 100% indicates that during the test the animal approached the training stimulus only, whereas a 0% preference score means that only the novel stimulus was approached. A preference score of 50% indicates that the training object and the novel stimulus were approached equally. In general, the more conspicuous an object is to the human eye, the more effective it will be as an imprinting stimulus. Movement of the object increases its attractiveness, as does the presentation of an auditory stimulus with a visual stimulus. Imprinting is not an instantaneous process: It may take more than 1 h of exposure for a chick to achieve a significant preference for a simple artificial object, depending on the object and the training schedule.

Imprinting and Learning

Lorenz (1937) initially suggested that the process of acquisition of knowledge about the object to which social behavior becomes directed has nothing to do with learning, mainly because it was thought to occur within a sensitive period and to be irreversible. Subsequent research has shown that imprinting preferences can be reversed, and that there is not a strict sensitive period. Thus, these putative criteria do not distinguish between imprinting and other forms of learning. Lorenz stated, "Of course, it is a matter of personal opinion how much or how little importance one should attribute to these differences from associative learning" (p. 266). In fact, when the label "imprinting" is removed, Lorenz's views on the learning processes involved appear to be similar to current interpretations. Lorenz (p. 262) actually distinguished between imprinting — which he viewed as the acquisition of a "perceptual pattern" of the species — on the one hand, and "an acquiring process, that is . . . conditioning in a broad sense of the word," on the other hand. Lorenz went on to say that the latter process is "true associative learning which unites the different sets of releasing stimuli emanating from an individual companion" (p. 262). Thus, Lorenz clearly acknowledged that there was a process of acquisition of knowledge of the individual characteristics of stimuli. In fact, it is this process which is called filial imprinting by contemporary researchers. It has become clear that filial imprinting does not involve simple stimulus–response associations, but that a representation of the training stimulus is formed that can be used flexibly. For instance, following filial imprinting, chicks showed significantly faster rates of discrimination learning if one of the two stimuli to be discriminated was the imprinting stimulus, despite the fact that in the discrimination task approach to the novel stimulus was rewarded and approach to the imprinting stimulus was not. Thus, in this case the chicks could not make use of a simple association between the representation of the familiar imprinting stimulus and a representation of the approach response.

Bolhuis *et al.* (1990) reviewed some of the evidence for an associative learning account of imprinting and formulated the requirements for such an interpretation. Until recently, there was relatively little evidence pertaining to this issue. Some of the mechanisms underlying the learning process of imprinting were revealed when it was taken into account that in the natural situation the mother hen has a particular size, shape, and coloration and emits a specific call. Similarly, during imprinting training in the laboratory the young birds are typically exposed to a visual stimulus accompanied by a recording of the maternal call of a hen. In some of our studies chicks were exposed to either a visual stimulus together with a maternal call or to the visual stimulus alone. Subsequently, both groups received a preference test in which the visual training stimulus and a novel visual stimulus were presented without a maternal call. Chicks given the maternal call during training showed a significantly greater preference for the training stimulus than those that had not. Thus, the presence of a call during visual imprinting training can enhance the preference acquired for the visual stimulus. An interesting interpretation of these findings is that a similar mechanism might underlie the facilitation of imprinting and the potentiation of learning observed when subjects receive conditioning with a simultaneous compound. For instance, in rats, aversive conditioning of an odor is enhanced when that odor is paired with a taste during the conditioning session. It has been suggested that such potentiation may be a result of within-event learning. That is, during conditioning associations may form between the representations of the two elements of the compound stimulus. Similarly, during imprinting training with an audiovisual compound, representations of the auditory and visual components of the compound may become associated. The formation of such an association means that during a test in which only the visual component is presented, there will be two potential sources of preferential responding — that acquired by the visual component and that due to the associatively activated representation of the auditory component of the compound.

Bolhuis and Honey (1994) tested some of the predictions of such a within-event learning account of potentiation of imprinting. They exposed chicks to a compound stimulus containing an auditory element (a taped maternal call) and a visual element (a moving two-dimensional image). The potentiation effect was replicated in subsequent preference tests with the training visual stimulus and a novel visual stimulus. That is, chicks that had been exposed to the compound stimulus showed a significantly greater preference for the training (visual) stimulus than did chicks that had been exposed to the visual stimulus only. The authors found that additional imprinting training with the call reduced the preference for the visual stimulus, whether exposure to the call occurred after or before exposure to the audiovisual compound. In a subsequent study, Honey and Bolhuis (unpublished results) varied the temporal relationship between the visual and auditory components of an imprinting stimulus in a series of experiments involving an extended series of trials of brief exposure to the stimulus elements with relatively long interstimulus intervals. During imprinting training, presentations of a visual stimulus preceded, followed, or were separated by 30 s from presentations of a simultaneous compound comprising another visual stimulus and an auditory stimulus. In subsequent preference tests, chicks in each condition showed an equivalent preference to approach the visual stimulus that had been presented in compound with the auditory stimulus rather than the other visual stimulus. In contrast, in a separate study, Bolhuis and Honey (unpublished results) found that if the auditory stimulus was replaced with a heat reinforcer (and the chicks were trained in a cooled cabinet), imprinting proceeded better if presentation of the heat reinforcer followed presentation of the visual stimulus compared to when they were presented simultaneously. These results suggest that filial imprinting and Pavlovian conditioning proceed in different ways. In Pavlovian or classical conditioning, a neutral stimulus is paired with a motivationally significant stimulus. The neutral stimulus is called a conditioned stimulus (CS); in the famous example of Pavlov's dogs the CS was the sound of a bell. The motivationally significant stimulus is known as the unconditioned stimulus (US); for Pavlov's dogs this was the delivery of food. After several pairings of the bell with the delivery of food, the sound of the bell by itself was sufficient to induce salivation in the dogs, indicating that the animals had learned to associate the CS with the US. In the case of imprinting with an audiovisual compound stimulus, an alternative interpretation to that of within-event learning would assume that the auditory stimulus acts as a kind of US to the initially neutral visual stimulus (the CS). Imprinting would then be a kind of Pavlovian conditioning. Imprinting stimuli have been shown to have a certain motivational significance to the young animal: They can act as unconditioned stimuli in instrumental learning tasks. However, in conventional Pavlovian conditioning sequential presentation of the CS and the US is far superior to simultaneous presentation of these two stimuli. As has been shown, in imprinting with an audiovisual compound the opposite is true: Simultaneous presentation of the auditory and the visual stimulus is superior to sequential presentation.

The results with audiovisual compound imprinting are analogous to findings in similar (sensory preconditioning) paradigms in rats. They support the suggestion that the potentiation of imprinting is a product of within-event learning, which is undermined when one element of the compound is presented in isolation. Furthermore, the findings demonstrate that in filial imprinting, unlike Pavlovian conditioning, learning proceeds most readily when training involves the simultaneous presentation of stimuli.

Representations of Complex Stimuli

In an alternative interpretation of the findings in rats and chicks, the representation of a compound stimulus is thought to be configural. That is, in certain circumstances the elements of, for example, an audiovisual compound stimulus might come to address a single or unitary representation, and activity in this representation controls performance. Such an interpretation is accommodated by a neural net model of imprinting and object recognition that was developed by Bateson and Horn (1994). Figure 2 illustrates the simplest version of the model, consisting of three layers, an analysis, recognition, and an executive system, respectively. A basic feature of the network is that links between modules are strengthened when there is simultaneous activity in those modules, and that links are weakened when activity in a recognition module is not accompanied by activity in an analysis module.

Using this model, Bateson and Horn (1994) could simulate various characteristics of filial imprinting. Regarding the potentiation of imprinting, suppose that the auditory and the visual stimuli activate analysis modules A1 and A2, respectively. If one assumes that the dual input provided by the elements of the compound stimulus enhances the subsequent likelihood for either element to activate a mod-

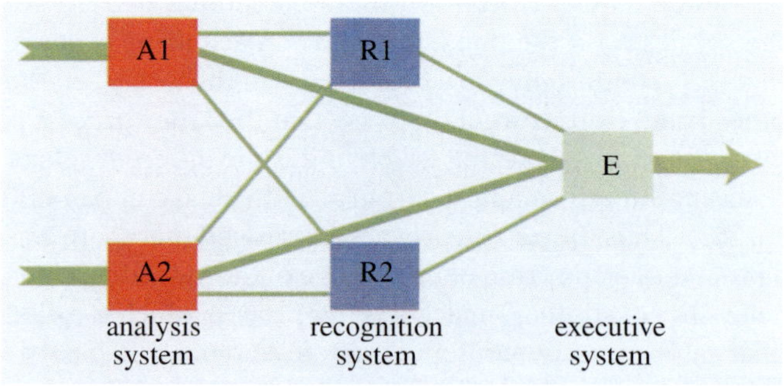

FIGURE 2 Flow diagram of the simplest version of the neural net model by Bateson and Horn (1994). The network consists of three systems (analysis, recognition, and execution), each containing multiple modules. For simplicity, only two analysis modules (A1 and A2) and two recognition modules (R1 and R2) are shown. All these modules are interconnected and are also linked directly to the module in the executive system (E) that controls filial approach behavior. The strength of the links connecting the modules is reflected in the thickness of the lines (reproduced with permission from Bateson and Horn, 1994).

ule in the recognition system (e.g., R1), then compound training would result in a potentiation effect. That is, the links between A1 and R1 and between A2 and R1 become stronger than they would have been if either of the stimuli were presented alone. To account for the observation that separate exposure to an element of the compound reduces the potentiation effect (Bolhuis and Honey, 1994), one would need to assume that if activity in R1 is not accompanied by the presentation of one of the elements, then its capacity to activate R1 will be reduced. Given that presentations of the auditory stimulus alone will produce activity in R1 that is unaccompanied by input from the visual stimulus, it is to be expected that input from the visual stimulus will become less able to activate R1. Finally, given that the model does not allow for residual activity in recognition modules after a stimulus has been removed and the appropriate analysis module is no longer active, it is clear that the model predicts simultaneous presentation of an auditory and a visual stimulus to be superior to sequential presentation.

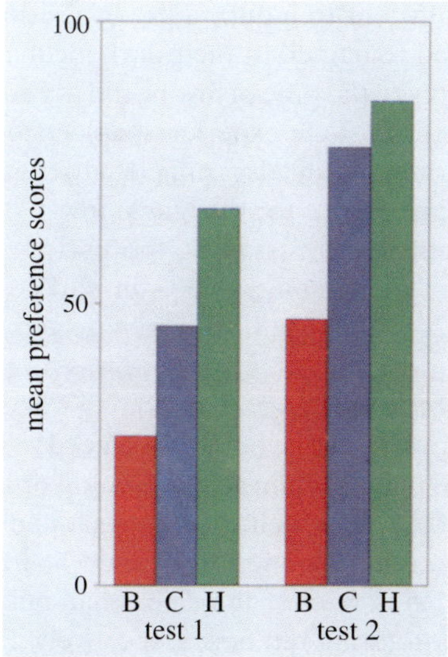

FIGURE 3 Mean preference scores, expressed as a preference for the stuffed fowl, of chicks previously trained by exposure to a rotating stuffed jungle fowl (H), a rotating red box (B), or white light (C). Preferences were measured in two tests in which the two stimuli were presented simultaneously. The first test was performed 2 h after the end of the imprinting session, whereas the second was after 24 h. Preference scores are defined by the number of attempts to approach the stuffed jungle fowl as a percentage of the total number of approach attempts (adapted from Horn, 1985).

Predispositions

As discussed in the Introduction, the development of filial behavior in the chick involves two perceptual mechanisms that are neurally and behaviorally dissociable (Horn and McCabe, 1984; Johnson *et al.,* 1985; Bolhuis *et al.,* 1989). I discussed some of the characteristics of the first mechanism, which underlies filial imprinting. Behavioral evidence for the second mechanism, that of a predisposition (Horn and McCabe, 1984), was provided by Johnson *et al.* These authors gave 1-day-old dark-reared chicks imprinting training by exposing them to either a rotating red box or a rotating stuffed jungle fowl hen (*G. gallus spadiceus;* Fig. 1). The approach preferences of the chicks were measured in a subsequent test in which the two training stimuli were presented simultaneously. Preferences were tested at either 2 h (test 1) or 24 h (test 2) after the end of training (Fig. 3). At test 1 the chicks preferred the object to which they had been exposed previously. At test 2 there was a significantly greater preference for the fowl in both experimental groups. Thus, the preference for the jungle fowl increased from the 2 h to the 24 h test and did so regardless of the stimulus with which the chicks had been trained. Chicks in a control group that had been placed in the training apparatus with no stimulus except for an overhead light also showed an increasing preference for the jungle fowl with time in preference tests conducted at the same time as those done for the experimental groups.

Johnson *et al.* (1985) proposed that the preferences of trained chicks are influenced by at least two different systems. On the one hand, there is an effect of experience with particular stimuli (reflected in the differences in Fig. 3)—that is, filial imprinting. On the other hand, there is an emerging predisposition to approach stimuli resembling conspecifics (Fig. 3). The results in Fig. 3 suggest that training with a particular stimulus is not necessary for the predisposition to emerge. The predisposition can in fact emerge in dark-reared chicks, provided that they receive a certain amount of nonspecific stimulation within a certain period in development.

Induction and Development of a Filial Predisposition

In subsequent studies (Bolhuis, 1991) it was found that visual experience was not necessary to "trigger" or induce the predisposition (Gottlieb, 1976). Chicks that were placed in wire-mesh running wheels (Fig. 1b) in total darkness for two periods of 1 h preferred the jungle fowl to the red box in a simultaneous choice test 24 h after the running wheel periods. Placement in the running wheels, in darkness, for two periods of 1 h (with 1 h in between during which the chicks were kept in a dark incubator) is a standard procedure that is used in many of the experiments discussed here and that is sufficient to induce a predisposition. However, chicks that had remained in their dark incubator from the time of hatching to the time of testing did not develop a preference. Therefore, some experience associated with being placed in the running wheels is sufficient to trigger the emergence of the preference for the jungle fowl.

In the predisposition studies, the period in which the chicks were in the running wheels in darkness began approximately 24 h after the birds had hatched. A significant preference for the stuffed jungle fowl was found 24 but not 2 h after the end of the period in running wheels. Hampton *et al.* (1995) investigated whether 24 h need to pass in order for the underlying mechanisms to develop or to exert their influence on the chicks' behavior. It was found that the predisposition becomes manifest at 10 and 5 h after placement in running wheels or after the chicks had been trained by exposure to a rotating red box, respectively. Furthermore, these authors showed that placement in running wheels was not necessary for the predisposition to emerge. Chicks that were taken out of their incubator and handled at the same times that the wheel-treated chicks would normally be handled showed a significant preference for the fowl over the red box at 5 but not at 2 h after the end of this experience. Similarly, Hampton *et al.* found that exposing 24-h-old dark-reared chicks to a maternal call of a chicken for 2 h was sufficient to induce the development of a visual predisposition 24 h later. Thus, evidence from the study by Hampton *et al.* suggests that different kinds of experience, in different modalities, are sufficient for the induction of the predisposition.

Interaction between Predispositions and Imprinting

On the basis of our results, we suggested that filial preferences develop as a result of an interaction between two perceptual mechanisms. We investigated whether the predisposition served as a filter or template, restricting the information that the chick can store in memory, or, alternatively, whether the predisposition acts independently of the effects of learning during development. We allowed dark-reared chicks to develop the predisposition, manifest as a significant preference for the stuffed fowl over the red box, by placing them in running wheels in darkness for 2 h when they were approximately 24 h old. When the chicks were approximately 50 h old, we trained them by exposing them to a rotating red box. In subsequent preference tests, it was found that these chicks, after the predisposition had developed, were still capable of learning the characteristics of conspicuous objects such as a red box. These results suggest that rather than the predisposition constraining the information that can be stored, the two mechanisms interact at the behavioral level. That is, the activated predisposition will bias the chick's approach response to certain stimuli, and the animal may learn the characteristics of those stimuli; however, the chick is still capable of learning the characteristics of other objects.

Stimulus Selection in Filial Predispositions

Stimulus selection is the phenomenon whereby different elements of a stimulus are differentially effective in eliciting behavior. The stimulus characteristics that are important for the filial predisposition to be expressed were first examined by Johnson and Horn (1988). These authors attempted to "titrate" the visual complexity of the stuffed jungle fowl against its "fowl-like" characteristics. To this end, the preferences of dark-reared chicks, which had been placed in running wheels for 2 h when they were approximately 24 h old, were tested either 2 or 24 h later. Stimuli used during the tests were an intact stuffed jungle fowl versus one in a series of increasingly degraded versions of a stuffed jungle fowl. The degraded versions ranged from one in which different parts of the model (wings, head, torso, and legs) were reassembled in an unnatural way to one in which the pelt of a jungle fowl had been cut into small pieces that were stuck onto a rotating box. At the second test, when the chicks were approximately 52 h old, the intact model was preferred only when the degraded object possessed no distinguishable jungle fowl features. In addition, Johnson and Horn demonstrated that chicks did not prefer an intact jungle fowl model over an alternative object that contained only the head and neck of a stuffed fowl. Thus, it seems likely that the head and neck region contains stimuli that are relevant for the predisposition. In subsequent experiments, Johnson and Horn found that the chicks did not prefer a stuffed jungle fowl hen over a stuffed Gadwall duck (*Anas strepera*) or even a stuffed pole cat (*Mustela putorius*). Thus, the predisposition is not species or even class specific.

Bolhuis and Hampton (unpublished results) provided dark-reared chicks with nonspecific stimulation (placement in running wheels) that would induce the emergence of the predisposition, and 24 h later the birds' preferences were tested in simultaneous tests involving different pairs of stimuli. One of the test stimuli used was a stuffed jungle fowl hen, of which the eyes had been covered with a patch of color-matched jungle fowl feathers attached to plasticine. Chicks showed an equally strong preference for a stuffed fowl with its eyes covered with patches of feathers and an intact stuffed fowl, compared to a red box. These findings indicate that areas outside the eyes provide stimulus elements that are sufficient for the predisposition to be expressed. However, in two separate experiments chicks showed a significant preference for an intact stuffed fowl over a stuffed fowl with covered eyes. These results suggest that eyes are an important stimulus, but that other aspects of the stimulus are also sufficient for the expression of a predisposition.

Thus, it appears that the eyes are not a sign stimulus that is uniquely important for the expression of the predisposition. It may be that stimuli outside the head also play a role in the expression of the predisposition. It is possible that the stimulus selection involved in the predisposition conforms to the rule of *heterogeneous summation,* whereby different stimulus characteristics contribute to the induction of the response independently, with the eyes forming an important stimulus, as well as other stimuli. Alternatively, it is conceivable that the chicks' response is governed by a combination of a configurational stimulus, which needs to have a head and neck as well as a body, and certain stimulus elements (such as the eyes) that are separately important, with the bird's response being governed by the rule of heterogeneous summation.

Sensitive Periods

As mentioned earlier, Lorenz (1937) suggested that one of the characteristics that made imprinting unique was that it occurred within a "critical period." Rather than the term critical period, which implies a fixed period, the term "sensitive period" is used in contemporary research. Reports of the length of sensitive periods vary widely, even within the same species studied; it has become clear that there is not a strict and narrowly defined sensitive period for filial imprinting. External experience plays an important role in changes in imprintibility; the end of an apparent sensitive period is largely a result of the filial imprinting experience and not so much a consequence of endogenous changes. Later, it will be shown that in sexual imprinting too, an apparent sensitive period proved to be less determining than had been assumed, and that sexual preferences may be influenced until the time at which sexual behavior is expressed.

In dark-reared domestic chicks there is a sensitive period for the induction of the filial predisposition. Dark-reared chicks which were placed in running wheels (for 2 h) at 12, 42, or 48 h after hatching did not show a significant preference in a preference test 24 h later, but chicks that received a 2-h period in the running wheels at 24 or 36 h after hatching significantly preferred the stuffed jungle fowl to the red box. Chicks in a control group were placed in the wheels at 24 h after hatching and given a preference test 72 h after hatching (i.e., at the same time as the chicks in the 48-h wheel placement group). These chicks showed a significant preference for the jungle fowl, demonstrating that the crucial factor is the time at which the chicks are placed in the wheels and not the time of testing. Thus, in dark-reared chicks there is a sensitive period between approximately 20 and 40 h after hatching during which placement in the wheels may induce the emergence of the predisposition.

Sexual Imprinting

Lorenz (1937) originally suggested that one of the consequences of imprinting was the formation of a sexual preference for the species with which the young bird had been reared. As outlined previously, today a distinction is made between filial imprinting and sexual imprinting (ten Cate, 1989; Bischof, 1994). Bolhuis (1991) listed some of the differences between filial and sexual imprinting; for example, the time course involved (early in life versus adulthood) and the period of time during which experience affects behavior (hours or days in the former and weeks in the latter). Furthermore, evidence by Bateson (1982) and others suggests that, in contrast to Lorenz's earlier suggestions, sexual preferences may be for individuals rather than for the species. In particular, Bateson demonstrated that the eventual sexual preference may be for an individual that is different from the ones with which the animal was reared.

In sexual imprinting studies birds are usually reared to a certain age with either parents of their own species or foster parents of a different species. Sexual preferences are tested when the animals are sexually mature, sometimes after a certain period of social isolation. Evidence for sexual imprinting is usually inferred from preferential responding (proximity to the test stimulus or directed courtship responses) to a member of the opposite sex of the rearing species. However, Bateson (1982) found that Japanese quail (*Coturnix coturnix japonica*) may develop a sexual preference for a certain individual—specifically a preference for an individual that was different, but not too different, from the ones with which it had been reared. Quail were reared in small groups. In simultaneous choice tests, male and female quail significantly preferred unfamiliar first cousins to familiar siblings, unfamiliar siblings, or unrelated individuals. Bateson suggested that this preferential responding is a mechanism to achieve optimal outbreeding, in which the deleterious effects of both extreme inbreeding and outbreeding are avoided.

For a long time, it was assumed that species-specific predispositions played a role in sexual imprinting. For instance, male zebra finches (*Taeniopygia guttata castanotis*) that were reared with Bengalese finch parents (*Lonchura striata*) and zebra finch parents in succession or with a mixed pair consisting of a parent of each of these two species preferred a female of their own species (ten Cate, 1989). However, after a detailed behavioral analysis, ten Cate discovered that zebra finch parents were involved in significantly more social interactions with the zebra finch young than were Bengalese finch parents. The strength of the sexual preference correlated with the amount of social interaction that the young males had received from the different parents. Furthermore, in a separate study it was found that in cross-fostered zebra finch males there was a correlation between the number of zebra finch siblings in the nest and the subsequent preference for zebra finch females.

Following Lorenz's original suggestions, it was long thought that once a sexual preference had been established it was stable and irreversible. However, recent evidence from two different laboratories has altered this view considerably and demonstrated that there is in fact a great deal of plasticity in the formation of sexual preferences in zebra finches (Bischof, 1994). When zebra finch males were reared with Bengalese finch foster parents for the first 40 days of life, they showed a strong courtship preference for Bengalese finch females in preference tests when they became adults. When such males were exposed to a zebra finch female for several months, they showed courtship behavior toward that female, but after separation from the zebra finch female the males once again preferred a female of the foster species in preference tests. However, if the males did not receive a preference test before exposure to the zebra finch female, most of them altered the direction of their courtship behavior toward a stable preference for zebra finch females. Direct physical interaction with the females was not necessary during the brief preference tests or during the longer exposure periods when the animals were sexually mature. Thus, sexual preferences could be altered by later exposure to a female of the other species, and brief exposure to a female of the rearing species was sufficient to "consolidate" or "stabilize" the animal's sexual preference. When zebra finch males that were reared by Bengalese finch parents were exposed to a Bengalese female first, they subsequently courted Bengalese finch females; however, when they were exposed to a zebra finch female first, a large proportion of the males changed their preference toward a zebra finch female. On the basis of these and other results, Bischof suggested that sexual imprinting is a two-stage process, with an acquisition phase during which the animal learns the characteristics of its parents and siblings and a consolidation phase during which the sexual preference is stabilized or modified according to the species of the individual to which the male is exposed.

Imprinting and the Development of Behavior

From a functional perspective, a filial predisposition may ensure that the young animal directs its attention to a particular class of objects (i.e., stimuli with a head and neck) or indeed to conspecifics. After the predisposition has fully developed, learning the visual and auditory characteristics of a particular object may lead to the recognition of individuals in filial and in sexual imprinting. It is important to understand how the development of filial preferences might proceed in the natural situation. The dynamics of the process as they have been reported may be peculiar to the laboratory conditions that were used (e.g., rearing in visual and tactile isolation and in the absence of maternal vocalizations) and do not necessarily reflect the normal course of development. In this article, I reviewed evidence that suggests that the complex visual and auditory experience that the young bird receives before and after hatching secures a rapid induction and development of a predisposition. For instance, Hampton *et al.* (1995) demonstrated that in the laboratory the predisposition in the chick is manifest between 2 and 5 h after the inducing experience. Furthermore, the predisposition is expressed as early as 2 h after complex visual experience (Bolhuis, 1991). Also, it is likely that in the natural situation, the necessary experience for the induction of the predisposition is available to the young bird at the appropriate time in the sensitive period, which is not a strict and absolute constraint.

The terms "maintenance," "facilitation," and "induction" were introduced by Gottlieb (1976) to distinguish some of the ways in which experience can affect the development of "species-typical perception." All these possibilities have been shown to occur in the development of filial preferences (Bolhuis, 1991). Hampton *et al.* (1995) concluded that, within the time course used in their own and related studies, the effects of experience on the development of the predisposition may be categorized as a case of induction; the predisposition is not expressed initially and only emerges after nonspecific experience. Gottlieb's terms, of course, are descriptive and do not elucidate the underlying mechanisms. However, Gottlieb's classification may be a useful way of categorizing effects of experience on behavioral development.

The finding that exposure to different kinds of stimuli may lead to the emergence of a predisposition (discussed previously) is an example of induction of a perceptual mechanism by nonspecific experience. Other work on filial imprinting suggests that the same principle applies to the facilitation of the development of perceptual mechanisms. In sexual imprinting in the zebra finch, evidence was found for maintenance (termed consolidation or stabilization) of sexual preferences through nonspecific experience (Bischof, 1994). The inducing, maintaining, or facilitating effect of nonspecific, seemingly unrelated, experience on specific perceptual mechanisms appears to be a fundamental principle of behavioral development. An important conclusion that can be derived from a review of recent research is that, in contrast to Lorenz's original suggestions, filial and sexual imprinting are not rigidly constrained by sensitive periods, irreversibility, or predispositions, although these factors do play an important role in some form or another.

Rather, there is considerable plasticity in the system that allows different kinds of experience to influence the formation of the animal's social preferences at different stages in development.

Acknowledgments

My research reported in this article was supported by a postdoctoral research fellowship from the Biotechnology and Biological Sciences Research Council (UK). I am grateful to Patrick Bateson for his comments on the manuscript.

References Cited

Bateson, P. P. G. (1982). *Nature* **295,** 236–237.
Bateson, P., and Horn, G. (1994). *Anim. Behav.* **48,** 695–715.
Bischof, H.-J. (1994). In *Causal Mechanisms of Behavioural Development* (J. A. Hogan and J. J. Bolhuis, Eds.), pp. 82–97. Cambridge Univ. Press, Cambridge, UK.
Bolhuis, J. J. (1991). *Biol. Rev.* **66,** 303–345.
Bolhuis, J. J., and Honey, R. C. (1994). *J. Exp. Psychol. Anim. Behav. Proc.* **20,** 240–248.
Bolhuis, J. J., De Vos, G. J., and Kruijt, J. P. (1990). *Q. J. Exp. Psychol.* **42B,** 313–329.
Gottlieb, G. (1976). In *Neural and Behavioral Specificity: Studies in the Development of Behavior and the Nervous System* (G. Gottlieb, Ed.), pp. 237–280. Academic Press, New York.
Hampton, N. G., Bolhuis, J. J., and Horn, G. (1995). *Behaviour* **132,** 451–477.
Horn, G. (1985). *Memory, Imprinting, and the Brain.* Clarendon, Oxford.
Horn, G., and McCabe, B. J. (1984). *Anim. Behav.* **32,** 288–292.
Johnson, M. H., and Horn, G. (1988). *Anim. Behav.* **36,** 675–683.
Johnson, M. H., Bolhuis, J. J., and Horn, G. (1985). *Anim. Behav.* **33,** 1000–1006.
Lorenz, K. (1937). *Auk* **54,** 245–273.
Ten Cate, C. (1989). In *Perspectives in Ethology* (P. P. G. Bateson and P. Klopfer, Eds.), Vol. 8, pp. 243–269. Plenum, New York.

General References

Gottlieb, G. (1971). *Development of Species Identification in Birds.* Univ. of Chicago Press, Chicago.
Hogan, J. A., and Bolhuis, J. J. (Eds.) (1994). *Causal Mechanisms of Behavioural Development.* Cambridge Univ. Press, Cambridge, UK.
Sluckin, W. (1972). *Imprinting and Early Learning.* Methuen, London.

GEMMA CALAMANDREI

Section of Comparative Psychology
Laboratory of Organ and System Pathophysiology
Istituto Superiore di Sanità
I-00161 Rome, Italy

The Ontogeny of Learning

For a young mammal, the ability to learn—that is, to establish causal relationships between events and to modify its behavior on the basis of that experience—is of great importance for adapting to possible and unpredictable changes in the environment. Starting in the early 1970s, the development of an ecological theory of animal learning recognizing the species-specific and age-dependent limits of associative processes gave great impulse to ontogenetic studies of these processes. The choice of conditioning procedures adapted to the motor and sensory characteristics of the immature animal has revealed unexpected capabilities of learning and memory in newborn animals, which appear to be able to learn and preserve memories of significant events before higher brain structures, such as the cortex and the hippocampus, have reached full functional maturity. In the human species as well, neonatal—even prenatal—learning seems to play an important part in the future organization of the central nervous system and of behavior.

Introduction

In many mammalian species, the young depend on parental care for a relatively long period of time before they are able to survive autonomously. Ontogenetic development is a gradual and continuous process lasting from birth to adulthood during which the behavior of an individual undergoes more or less rapid changes in a dynamic interaction between endogenous factors of maturation and environmental stimuli. In that process, the ability to learn or to establish causal relations between events and to alter one's behavior on the basis of that experience is of great adaptive importance in that it allows the young animal to respond with a relatively flexible behavior to possible, even if unpredictable, changes in the environment (Klopfer, 1988). The ability to learn is based on the extraordinary plasticity of mammals' central nervous system. Even in the earliest postnatal stages, it can carry out numerous complex functions, such as stimulus detection and sensory transduction, formation and storage of memories until recall, and performance of the appropriate motor response.

Early learning is of fundamental importance in laying the basis for the future organization of an individual's behavior. Nevertheless, until the 1970s, ontogenetic aspects received little attention from scholars of learning processes and memory. In the latter half of the twentieth century, the study of animal learning has been strongly influenced by controversies between the European school of ethology and the school of behaviorism, to which many US experimental psychologists referred. The behavioral psychologists were in favor of laboratory analysis of animal behavior in a limited number of species (rats, pigeons, and monkeys) in controlled environments (labyrinths, Skinner's cages, etc.). In this context, they tried to distinguish between learning processes depending on whether the change in behavior was induced by association between a stimulus and a reinforcement (as in classic or Pavlovian conditioning) or by association between a response and a reinforcement (as in the case of operant or instrumental conditioning). The objective of experimental psychology was to establish general laws explaining all associative learning processes, applicable to all animal species. The formulation of these laws could be guaranteed only by the use of experimental procedures that were as artificial as possible, minimizing the contribution of typical behaviors of the species being analyzed.

In the same period, ethologists were interested in functional descriptions of animal behavior in a natural context. The theoretical premise of the European school of ethology, of which Konrad Lorenz was the best known representative, was that evolution, through natural selection, favors the transmission of behavioral phenotypes in the same way in which it favors the transmission of the most appropriate anatomical features. Classic ethology, however, concentrated on the study of so-called *fixed action patterns.* These patterns are innate and stereotype behaviors that are activated at a certain stage in the animal's development as a result of exposure to a particular stimulus or series of stimuli. The ability of some species to learn particular responses represented an important area of study for ethologists; however, except for a few exceptions (imprinting in chicks, the development of song in birds, and habituation), examples of learning in nature were never subjected to rigorous experimental analysis. For ethologists, learning is a process that cannot be separated from the other biological characteristics of the species, viewed as a set of specialized abilities which have developed independently in each species to carry out a specific adaptive function. Therefore, the attempt to establish general laws for learning that were valid for all species was thought to be incongruous (Zolman, 1983).

For a long time, these two theoretical positions remained world's apart. Nevertheless, at the end of the 1960s, some experimental psychologists suggested adopting a more naturalistic approach to the study of animal behavior (Shettleworth, 1988). In fact, the results of some classic and instrumental conditioning experiments carried out in the laboratory indicated that there are biological limits to the associability of stimulus, reinforcement, and response, and that the phylogenetic history of a given species strongly influences that which an animal can learn and remember. In particular, the two theoretical bases of behavioral theory were thrown into question: the principle of equipotentiality, by which an animal can learn any association between events, and the principle of temporal contiguity, by which the association between events takes place only if they occur almost simultaneously.

In 1966, the results of an experiment by J. Garcia and R. A. Koelling on conditioned taste aversions in rats were published. The two authors demonstrated that the rats developed an aversion to a solution of saccharine and water if they drank it while being exposed to radiation or while receiving an injection of apomorphine or lithium chloride (treatments which induce the symptoms of food poisoning in animals), but that the aversion did not develop if the negative reinforcement was extroceptive (e.g., an electric shock). Furthermore, in the aversion to a taste stimulus, the association between the conditioned stimulus and the unconditioned one took place even if the two stimuli were not temporally contiguous. As a result, developing a conditioned aversion to a certain food or taste depended on the normal feeding behavior of the rat species, which is able in nature to learn to avoid foods that can be toxic for it even hours after ingestion. Thus, they demonstrated the importance of species-specific predispositions in learning processes, by which a given animal learns to associate two events more easily if they are likely to occur in a causal relationship in the real world. Subsequently, using an active and passive avoidance test requiring the animal to inhibit an escape response or to move from one compartment to another of a specific apparatus to avoid punishment, Bolles (1970) demonstrated that a rat's ability to learn was profoundly influenced by the animal's innate repertoire of defensive responses. Specifically, when in order to avoid punishment the experimental paradigm called on the emission of responses belonging to the rat's species-specific repertoire, such as running or jumping, learning of the response was successful. On the contrary, when the behavior required to avoid punishment involved lowering a lever, performance of the learning task was poor.

The existence of biological limits to learning represented one of the major challenges to the traditional theory of learning and forced many researchers to reconsider the methodology of laboratory tests so that they would take account of the ecological characteristics of the species studied. On the other hand, the ethological tradition had to acknowledge the usefulness of the classic and instrumental studies on conditioning carried out in the laboratory, which revealed an unexpected flexibility in the associative capabilities of a species with an apparently limited behavioral repertoire. Basically, in the 1960s and 1970s, the encounter between the psychological approach and the ethological approach to learning gave new impulse to the biology of learning and memory and stimulated interest in the ontogeny of those processes (Zolman, 1983).

The Ecological Approach to the Ontogeny of Learning

Recognition of the importance of species-specific predispositions to learning processes favored a substantially different approach to the study of behavioral development as well. The traditional point of view, which considered the young animal as a kind of incomplete and inferior version of the adult with a simplified behavioral repertoire, had relegated ontogenetic aspects to the margins of the behavioral sciences. In fact, until the beginning of the 1970s, the central problem of ontogenetic research was to determine at what stage of development an animal's learning and memory performance reaches that of the adult and whether or not early experiences can influence the cognitive abilities of the adult animal. To that end, the associative capabilities of the developing animal were tested with the paradigms normally used on adults, with the obvious result that the learning capacities of the neonate were found to be extremely limited or altogether lacking.

Instead, it appeared clear to some researchers that the immature animal has to deal with problems that are substantially different from those faced by adults. In particular, in altricial mammalian species the young are born blind with a nervous system that is still largely immature; they are unable to regulate their body temperature, to move efficiently, or to nourish themselves autonomously, and they require maternal care until weaning in order to survive. The sensorial universe of a newborn rat or kitten is quite different from that of the adult animal, and obviously its behavioral repertoire is functional to solving the specific problems of an immature organism, such as keeping in contact with the mother's body (the source of its food and heat) by means of thermotactile and olfactory stimuli. In a relatively short time, the developing animal's ecological relations change radically: Those changes are often accompanied by a complete reorganization of the organism and of its behavior as happens, to use the fitting metaphor of Bateson (1981), during the metamorphosis from caterpillar to butterfly. Just as it would be absurd to consider a caterpillar an incomplete version of a butterfly, Bateson claims it is misleading to search for the origin of an adult behavior in a corresponding behavior in the young. For example, the feeding behavior of a mammal changes radically after weaning, but sucking cannot be considered an "inferior" version of an adult feeding behavior.

Seen in this light, behavioral development is no longer a process through which the young animal become increasingly "capable" or "able" but rather a dynamic process during the course of which the individual sees the world from a perspective which progressively changes as its ecological relations change and as its sensory and motor systems mature (Spear, 1984). Neurobehavioral ontogeny is no longer based only on sequential processes of maturation, as maintained by preceding biodeterministic models, but also, and to the same extent, on the learning opportunities with which the neonate is faced in the nest's microenvironment from the moment of birth onward, if not during the prenatal period as well.

The studies carried out since the early 1970s, largely using altricial rodent species such as the rat or the mouse, revealed the richness and complexity of newborn mammals' behavioral repertoire. Very complex learning and retention capabilities emerged once procedures taking into account the ecology of the species and the motor and sensory characteristics of the immature animal began to be used. The ecological approach to the ontogeny of learning, highlighting the adaptive value of the neonate's behavioral repertoire, gave rise to a methodological revolution: Researchers prepared experimental paradigms suited to the specific capabilities of the young in which olfactory and thermotactile stimuli replaced the auditory or visual stimuli of classic conditioning procedures. Equal attention was given to the choice of both the behavioral responses with which to measure learning from among those normally expressed by the young in the initial postnatal stages, such as approach (generally toward the mother), sniffing (of the environment), and probing (for the nipple), and the reinforcements with which to induce instrumental conditioning (Zolman, 1983; Campbell, 1984; D'Udine and Alleva, 1984).

The Ontogeny of Learning in Rodents

Most studies of the ontogeny of learning and memory have been carried out on the two species of altricial rodents most frequently used in the behavioral sciences, the rat and the mouse. The young of these two species, as in other mammalian species, still have an immature nervous system at birth as well as a rather fetal appearance. The newborn rodent lives in an olfactory and tactile universe and uses these kinds of stimuli to locate the mother's belly and teats in order to suckle. During the first 2 weeks of postnatal life, most of the neonate's activity is related to feeding. With the maturation of its senses of vision and hearing and the development of its mechanisms of thermoregulation in the third week of life, the young's behavioral repertoire is remarkably enhanced to the point of resembling, upon weaning (21st day), a reduced version of that of the adult. Despite its fetal appearance, in the first hours after birth the newborn rodent learns and memorizes (mainly olfactory) information that is essential for its survival by means of processes that have a considerable degree of flexibility.

Numerous ontogenetic studies have shown the importance of early olfactory experiences in the neurobehavioral maturation of rodents. The olfactory cue that leads the neonate to its mother's nipple is the odor of amniotic fluid which the mother spreads onto her nipples while licking herself after delivery. This olfactory stimulus is recognized by the newborn animal as familiar from intrauterine life. However, Pedersen and colleagues (1983) showed that the preference for the odor of amniotic fluid is learned and not rigidly predetermined. In fact, if a small quantity of lemon essence is injected into the uterus during pregnancy, at birth the offspring seem to prefer the nipples marked with this scent rather than those marked with amniotic fluid. One-

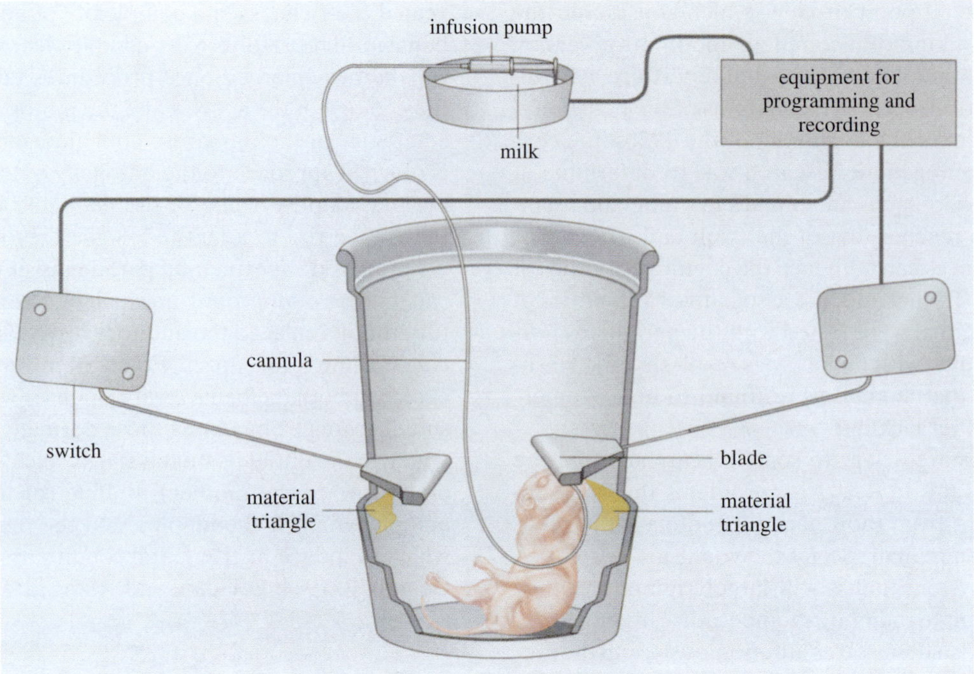

FIGURE 1 Diagram of the apparatus used for the experiment on instrumental conditioning of newborn rats. The triangular swatches of material attached to the paddles are impregnated with two different odors. When the neonate pushes upward against the paddles, from one of the two it receives a small quantity of milk through an intraoral tube. It soon learns to use the different odors to distinguish which paddle dispenses the milk and which does not (adapted from Johanson and Hall, 1979).

day-old rats that are administered milk through an implanted intraoral tube learn to move a paddle to receive milk (Fig. 1) and to distinguish successfully between two paddles marked with different odors (Johanson and Hall, 1979). At 1 week after birth, young rats learn to make their way through a Y labyrinth and to choose the right direction to reach the anesthetized mother at the other end of the labyrinth on the basis of olfactory cues.

Development and long-term retention of an olfactory preference for an artificial odor were obtained through the association of particular olfactory cues with the intraoral infusion of milk. A conditioned aversion to an odor can also be induced in 2-day-old rats (Rudy and Cheatle, 1977) if the exposure to that odor is associated with gastrointestinal malaise brought on by a toxic substance such as lithium chloride. A conditioned aversion to an artificial odor, which seems to last until adulthood, can also develop in intrauterine life, as shown by the early studies of Pedersen and colleagues (1983).

As previously discussed, the use of olfactory stimuli and behavioral responses typical of the neonate repertoire, such as sucking and orientation toward the mother and the nest, made it possible to detect associative capabilities even in the prenatal phase. More complex conditioning tests calling for the young rodent to use spatial information to orient itself in a Y, T, or radial labyrinth or to perform in an active or passive avoidance test show that the immature animal has an inferior learning capacity with respect to the adult, given its obvious motor limitations. However, if experimental procedures are changed to take the sensory and motor characteristics of the young animal into account, good learning capacities can be observed. For example, before weaning, newborn rats and mice find it more difficult than adult animals to learn an active or passive avoidance response. This kind of paradigm, generally used to test associative capacities in the adult, calls for the animals to inhibit an escape response or to move from one compartment to another of a specific apparatus to avoid punishment, generally a mild electric shock. The difficulties in learning the avoidance task are largely related to the expression of the conditioned response, hindered both by the high level of activation in the immature animal provoked by the electric shock and by the animal's poorer ability to orient itself to find the escape route. Eliminating the directional component of the escape response (the young animal only has to avoid touching the surrounding walls of the arena to avoid punishment) and replacing the electric shock with less stressful adverse stimuli, such as a cold draught, appreciably lowers the age at which learning appears (Stehouer and Campbell, 1980). The context in which the associative ca-

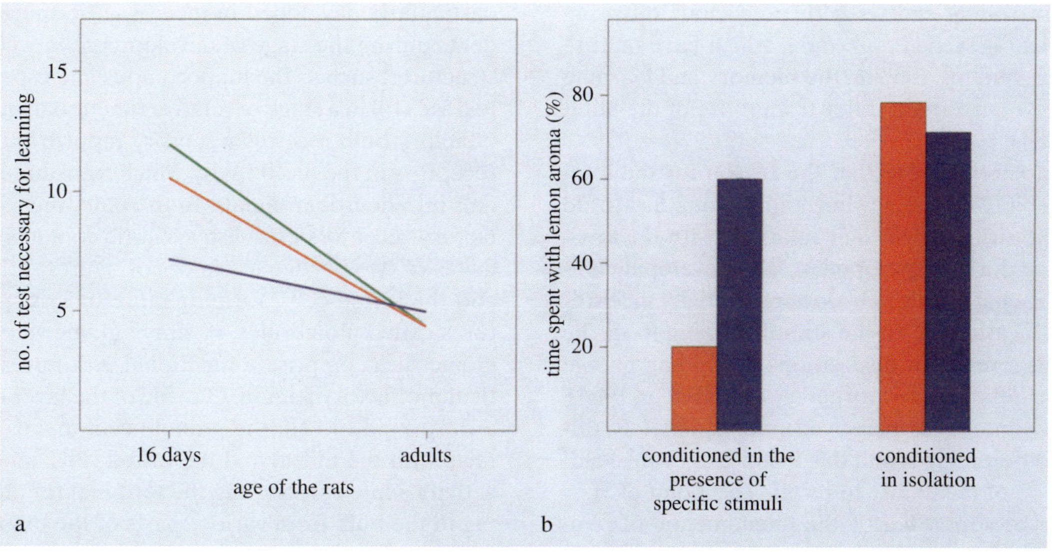

FIGURE 2 (a) Learning of a passive avoidance response by a 16-day-old rat in the presence and the absence of familiar olfactory stimuli constituted by bedding from the nest (in blue), clean bedding (in green), and the absence of bedding (in red). A more familiar environment facilitated acquisition of the response by the young but did not affect the performance of the adults. (b) Development of a conditioned aversion to a lemon scent following administration of lithium chloride (in red) in 2-day-old rats conditioned in the presence and the absence of other littermates. The aversion developed only if the association between odor and the toxic agent took place in the presence of conspecifics. The control group (in blue) was administered a simple saline solution (adapted from Smith and Spear, 1978).

pacities of the young are tested is also of decisive importance: It has been observed that a testing environment that closely resembles the young's ecological niche favors more precocious expression of associative capabilities. Smith and Spear (1978) demonstrated the incentive value of familiar stimuli, such as the odor of the nest or the presence of conspecifics, in modulating both the learning performance of 16-day-old rats in a passive avoidance test and the formation of a conditioned aversion to a scent (Fig. 2).

Infantile Amnesia

The experimental results of the past 20 years have shown that newborn mammals have an extraordinary capacity for learning due to a nervous system that is extremely plastic and ready to be shaped by experience. They have also revealed that during development, newborn and young animals generally have a lower memory capacity than adults and that their ability to retain and express a conditioned response improves progressively with age. This phenomenon has been called *infantile amnesia* (Fig. 3). As will be shown, infantile amnesia is still a strongly debated subject, even among psychologists interested in human cognitive development. Many authors have hypothesized that the lower memory capacity is due to the immaturity of the central nervous system, suggesting that the mechanisms of neuronal plasticity involved in the processes of learning and storing of information are not yet perfectly functional and that the development of adult-like memory capacities depends strongly on the later maturation of corticolimbic structures such as the hippocampus and the cortex (Nadel and Zola-Morgan, 1984). This hypothesis seems to be supported by

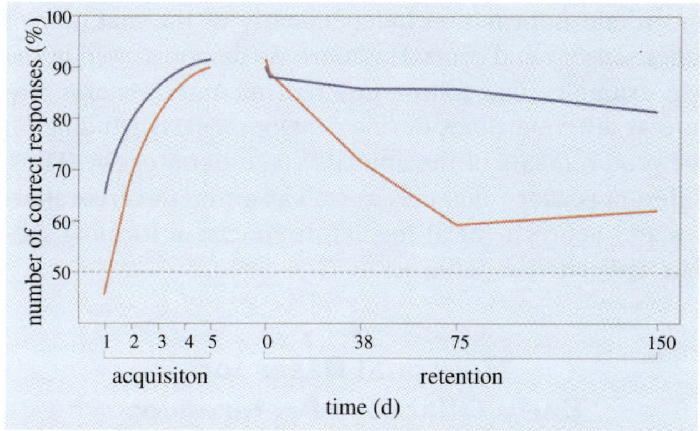

FIGURE 3 Differences between young (in red) and adult (in blue) rats in the acquisition and retention of a light/dark distinction. Although the learning performances between the two ages were not significantly different, the adults remembered the learned response for a longer period of time. The dotted line indicates the level of probability above which responses are significant (reproduced with permission from Campbell, 1984).

the fact that in rodent species with precocious offspring (e.g., the guinea pig, *Cavia porcellus*), which have mature central nervous systems at birth, the memory and learning abilities of the young do not differ from those of the adult (Campbell, 1984).

Other researchers believe that the reason for the inferior memory ability of the newborn and young has to do with the rapidity with which their relations with the environment change during development. The developing animal seems to be characterized by lower selectivity in learning and is not as attentive to the stimuli coming from the environment as a result of dispositions depending on age and the specificities of its sensory perceptions (Spear, 1984). The same environment is perceived in a completely different way at different ages and this factor can be misleading in evaluation of the ability to recall. This point of view would seem to be supported by the phenomenon of reinstatement, by which learned behaviors that would normally be quickly forgotten by the immature animal can be retained if the animal undergoes periodic repetition of some part of the initial experience, insufficient in itself to reestablish what was learned. Reinstatement procedures have also proven effective in favoring retention of learned experiences in human infants.

Probably all the mechanisms described contribute to the lower memory capacities of the young. In the past decade, developmental psychobiology has concentrated particularly on analysis of the neuronal mechanisms that regulate early learning with the aim of finding neurophysiological correlates explaining the differences in the long-term memory capacities of the young and of the adult. What seems to have emerged from these ontogenetic studies is that the development of long-term memory is not a unitary process related to the progressive maturation of the cortical structures. Rather, each sensory system has the ability to decode and retain information independently of the maturity of other sensory and central systems. As demonstrated in the two examples that follow, different memory systems mature at different times during development, depending on the requirements of the animal's stage of ontogeny. Thus, different sensory channels as well as different cerebral areas and neurochemical regulators concur in forming specific associative capabilities.

Neuronal Bases for Early Olfactory Preferences

As described previously, in rodents the memory of certain early olfactory experiences is retained until adulthood and plays an important role in establishing social, sexual, and family relations in the adult. The plastic mechanisms crucial for establishing olfactory preferences seem to take place in the olfactory bulb, a structure of the telencephalon particularly developed in macrosmatic species such as rodents during the stage of development in which the higher structures, such as the hippocampus and especially the cortex, are still in a stage of relative functional immaturity. The olfactory bulb receives a sensory input from the olfactory receptors in the nasal cavity which transmute chemical signals into neuronal signals. In the bulb, the axons of the olfactory receptors establish synaptic contacts with the dendrites of two particular classes of neurons, the mitral cells and the tufted cells, forming more or less spherical structures called glomerules. A single glomerule or a group of glomerules comprise a functional unit in response to particular olfactory stimuli. Outside of the layer of glomerules, a third kind of cell, the granule cell, makes synaptic contacts with the mitral and the tufted cells, modulating their activity while constituting the terminal for afferences leading to the bulb from various parts of the brain (Fig. 4).

When young rats are repeatedly exposed to an artificial

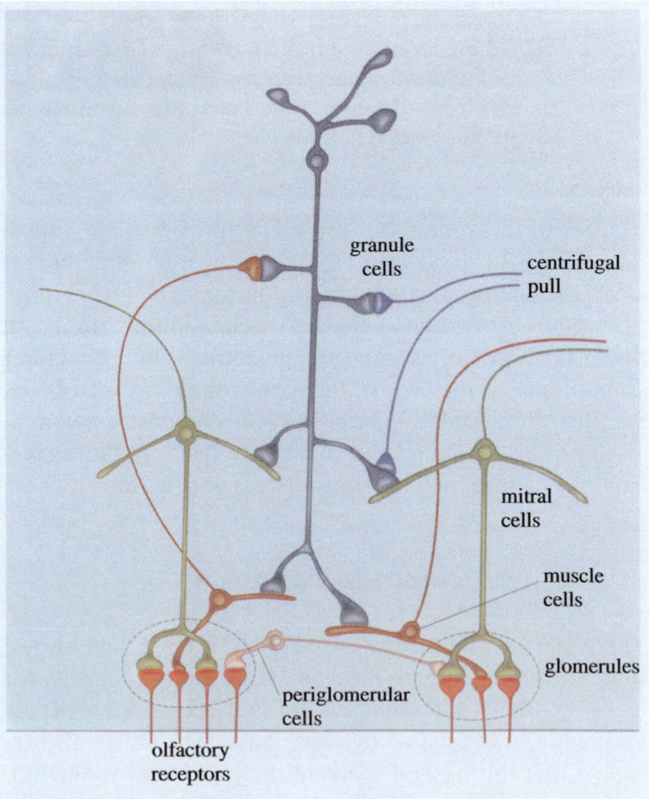

FIGURE 4 Simplified diagram of the neuronal circuits of the olfactory bulb. The axons of the olfactory receptors, which transmute chemical signals, reach the olfactory bulb and establish synaptic contacts with the mitral and tufted cells, forming almost spherical structures (the glomerules). The periglomerular cells and the granule cells are inhibitory interneurons, which receive the centrifugal afferences from other brain areas and modulate the electric signals emitted by the glomerules (reproduced with permission from Leon, 1987).

odor such as mint during their first week of life, they tend to develop a preference for that odor. Such a preference develops even more easily if the young simultaneously receive tactile stimulation similar to that dispensed by their mothers. At 20 days, the young prefer the odor previously associated with tactile stimulation over other familiar scents and show evident signs of behavioral activation when presented with the odor (Fig. 5). M. Leon and colleagues observed that the behavioral preference was associated with physiological and neuroanatomical changes in the glomerule layer of the bulb. In particular, following the olfactory experience associated with tactile stimulation, there was an increase in both the number and the metabolic activity of the glomerular neurons. More detailed analysis revealed that the synaptic relations between the different kinds of neurons of the olfactory bulb change so that subsequent presentation of the odor for which the animal has been conditioned produces a different electric output from a particular region of the bulb to other cerebral structures involved in the associative process, such as the amygdala and the pyriform or entorhynal cortex. These neurophysiological changes seem to be permanent and are modulated by the noradrenergic innervation of the olfactory bulb through the inhibiting action of the granule cells on other types of bulb cells, but they only take place if the animal is exposed to the odor/stimulation association in the first week of its postnatal life (Wilson and Sullivan, 1994).

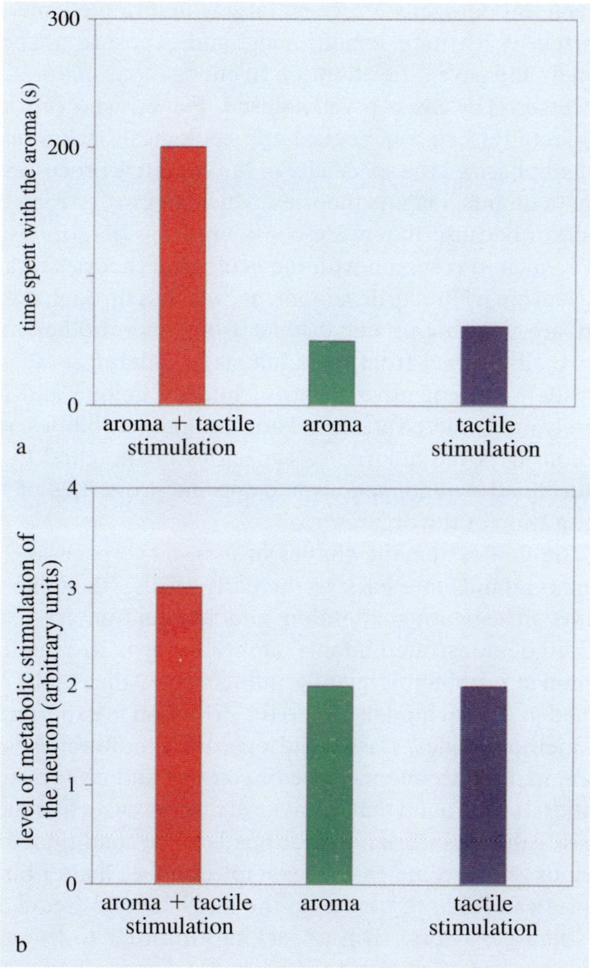

FIGURE 5 (a) Time spent on the mint scent by 20-day-old rats, exposed to the odor in the first 18 days of life, while receiving tactile stimulation similar to that normally dispensed by the mother (in red). The young controls receive only tactile stimulation (in blue) or only exposure to the odor (in green). Only the young that experience the association of odor and tactile stimulation develop a preference for the mint scent. (b) The metabolic activity of the neurons of specific areas of the olfactory bulb is significantly increased during exposure to the mint scent only in the young that experience the association of the odor to tactile stimulation. Learning of the olfactory preference is associated with neurophysiological changes in the bulbar neurons and their synaptic connections (reproduced with permission from Sullivan *et al.*, 1989).

During development, this model of neurobehavioral plasticity strongly resembles other plastic neurobehavioral phenomena, such as the development of ocular dominance in the visual system of the cat and some examples of olfactory learning in adult rodents, suggesting that the mechanisms of plasticity involved in olfactory learning are similar at different ages. A relatively simple structure such as the olfactory bulb can allow for the formation of long-term olfactory memory before the corticolimbic circuits involved in forming associations in the adult have become completely functional.

Maturation of Cholinergic Control of Behavior

Some kinds of learning which call for the integration of select information about the environment and the construction of a spatial representation of it mature relatively later in the course of development. One of the neuronal substrates that could potentially regulate these learning processes is the central cholinergic system. The central cholinergic neurons use the neurotransmitter acetylcholine and are grouped in a series of subcortical nuclei (basal, septal, and caudate nuclei) which account for most of the cholinergic innervation to the cortex and the hippocampus, two cerebral areas considered critical in learning and memory processes (O'Keefe and Nadel, 1978).

The central cholinergic system reaches functional maturity at approximately the 21st day (i.e., at weaning) in both the mouse and the rat, even though the system undergoes a considerable maturation from both the neurochemical and anatomical standpoints at approximately the end of the second week of life, marked by the establishment of septal–hippocampal connections and an increase in synthesis of the neurotransmitter acetylcholine. In parallel, during the second and third postnatal weeks, tests (such as passive avoidance) that call for response inhibition reveal a capacity for retention of at least 24 h, whereas the ability to learn tasks that require the animal to integrate the spatial information coming from the environment improve remarkably. In fact, it is only near weaning that, when placed in an unknown environment, the young animal has a habituation profile similar to that of the adult — a profile characterized by an intense exploratory activity at the beginning of the test which subsequently decreases. In general, maturation of the cholinergic system seems to coincide with the maturation of mechanisms of more selective attention, which allow the animal to integrate as much information as possible about test contingencies and to recall the information at a later time if and when a similar context appears. It is no coincidence that in the paradigms of classic conditioning, the phenomenon of latent inhibition usually only appears on the 18th to 20th days of life: If an animal is supposed to associate a stimulus with an adverse reinforcement (e.g., a sound with an electric shock) in a certain environmental context (e.g., an avoidance cage), preexposure to that test environment in the absence of the punishment will interfere with the subsequent learning of the association of the two events. Therefore, latent inhibition presumes that the characteristics of the environment have been recorded, preserved, and recalled. Basically, all the learning paradigms that call for information on "when" and "where" a specific event may take place only seem to become feasible after the maturation of cholinergic afferences to the cortex and the hippocampus, reaching adult levels with the maturation of the hippocampus (Nadel and Zola-Morgan, 1984).

Experimental manipulation of the cholinergic system involving the administration of drugs that block its activity or lesions in the septal–hippocampal connections or hippocampus makes it difficult for even an adult animal to learn and recall a spatial memory task. It was recently observed that the neurochemical and functional maturation of the central cholinergic neurons is regulated during critical stages of development by the trophic, tropic, and differentiating action of nerve growth factor (NGF) and other neurotrophines. These molecules play a key role in regulating the ontogenetic development of vertebrates by controlling cell growth and differentiation in various organs and tissues. Particularly in the central nervous system, the NGF synthesized during critical moments of the development of hippocampal or cortical neurons seems to direct the growth of the cholinergic afferences of the septum and the basal nuclei. Experimental manipulation of the endogenous levels of this neurotrophic factor slows or accelerates the maturation of the cholinergic function as well as that of retention capability in a passive avoidance test (Calamandrei and Alleva, 1996).

The Ontogeny of Learning in Human and Nonhuman Primates

The ecological theory of animal learning has also significantly influenced the study of human cognitive development. Developmental psychology tried to apply to children the principles of learning and memory derived from studies carried out on adults, without a real interest for development per se. Behavioral psychologists considered human behavioral ontogeny a process largely determined and organized by learning, conditioning, and exposure to social models and gave little attention to endogenous maturation processes. The Swiss psychologist J. Piaget, who (not coincidentally) had a biological and zoological background, first emphasized the specificity of the cognitive processes of human infants. Piaget's theories, which met with strong opposition because they were considered too deterministic, have much in common with the ecological theories of animal learning: Human development proceeds through stages that are continuous but distinct from one another and, above all, distinct from the adult stage. Therefore, Piaget considers the cognitive stages of infancy unique and not merely incomplete versions of adult cognitive abilities, and he considers the associative capacities of the child to be determined by endogenous and specific properties of the maturation of the organism.

The first systematic studies of associative processes in human infants date back to the early 1970s, when experiments investigating attention and habituation to visual stimuli demonstrated infants' ability to recognize familiar stimuli at birth. Subsequently, influenced by the results obtained in animal models due to the evolution in experimental methodologies, classic and operant conditioning tests were adapted to infants' specific sensory and motor capabilities. It was found that behavioral responses such as sucking, vocalizing, or orienting the head can be conditioned by various reinforcements. For example, only 24 h after birth, babies can learn to modulate the intensity and frequency of some behaviors (such as sucking) in order to listen to the sound of their mother's voice (DeCaspar and Fifer, 1980). Rovee-Collier and colleagues (Rovee-Collier, 1990) designed an operant conditioning test for human infants in which the baby determines the intensity of the reinforcement through his or her own behavior. In this test, which can be performed at 2 months of age, the infant learns to modulate the frequency and intensity of the movement of a colorful merry-go-round hanging above his or her crib by means of a ribbon tied to his or her ankle (Fig. 6). As the sessions progress, the infant learns to move his or her leg with increasing effectiveness in order to receive the reward — the toy's movement. The retention test, which involves measuring the infant's reaction in terms of frequency of leg

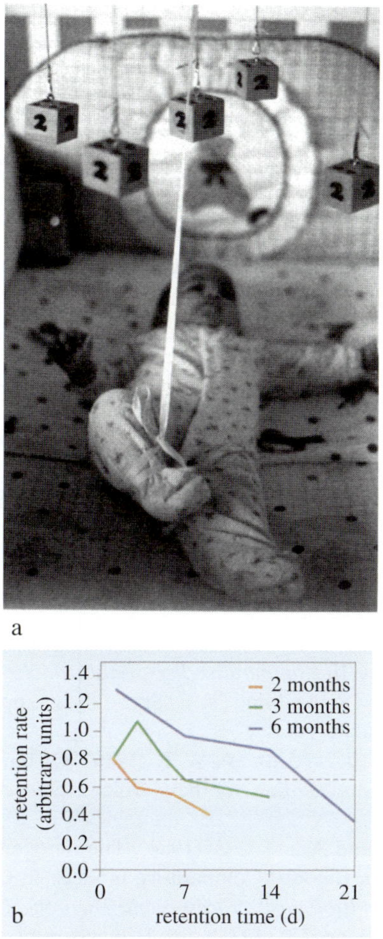

FIGURE 6 (a) Apparatus for the conditioning of newborn human infants. (b) Response retention in human infants 2 (in red), 3 (in green), and 6 (in blue) months of age. At 2 months, the infants forget the learned response 1–3 days after acquisition, whereas at 3 and 6 months they forget it after 7 and 14 days, respectively. The dashed line corresponds to the mean reference line above which the response is retained (adapted from Rovee-Collier, 1990).

movements when the merry-go-round is not moving and not connected to the infant's ankle, is carried out from 1 to 42 days after the learning session. Two-month-old infants forget the experience after 1–3 days; three-month-old infants remember the operating response for 6–8 days after learning. The capacity for long-term recall increases with the infant's age and reaches 21 days at 6 months of age.

As in young animals, the infants' mnemonic abilities also seem to improve with age. Analogously, the technique of reinstatement seems to work in human infants as well: Presenting some part of the test in the same context in which the conditioning took place (e.g., the same merry-go-round moved by the experimenter) for only 3 min 24 h before the retention test improves a 3-month-old infant's retention to 4 weeks after acquisition. Therefore, as in animals, memory of events learned does not seem to be definitively lost in the human infant and can be activated by renewed exposure to some component of the situational context.

In the human species, particular attention has been given to infantile amnesia — that is, adults' incapacity to remember events that took place in the first 2 or 3 years of their lives. As in the debate on the lower mnemonic capacities shown during development in animal models, two different positions have arisen concerning the origin of infantile amnesia. Some authors claim that the neurological structures (essentially those of the limbic system) designed for storage of long-term memories are not yet fully functioning in the first 2 years of life. The role played by the immaturity of these neuronal systems has been amply tested in nonhuman primates such as the macaque (*Macaca mulatta*). Classic cognitive tests (Bachevalier, 1990) which presuppose capacities for both visual and spatial discrimination (Fig. 7) have shown good correspondence between the cognitive development of monkeys and that of the human child. Although the ability to discriminate visually between objects is already present in both the macaque and the human infant in the first months of life, the ability to discriminate between objects on the basis of their familiarity only matures much later. In nonhuman primates the development of cognitive abilities is parallel to the synaptic maturation of the prefrontal cortex and the hippocampus.

On the basis of these results, some authors (Bachevalier, 1990) maintain that there are two ontogenetically separate memory systems in human and nonhuman primates: (i) a noncognitive habit system which is mature at birth and is mediated by corticostriatal structures and (ii) a real cognitive or representational system which uses a corticolimbic–

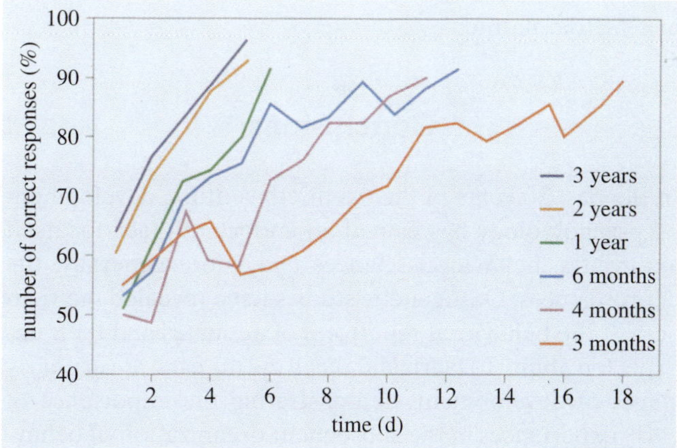

FIGURE 7 Learning curve of monkeys of different ages in an object discrimination test (DNMS). The learning performances gradually improve with age (reproduced with permission from Bachevalier, 1990).

diencephalic circuit that matures later and seems to be responsible for the implementation of conditioning paradigms that call for the subject's use of the spatiotemporal context.

The dichotomy between these two systems was criticized by Rovee-Collier (1990), who demonstrated that newborn human infants have excellent long-term memory capabilities, even in their first year of life. The neurological structures needed for perception, learning, and memory (e.g., those of the visual system) are already functional at birth, even if they continue to mature considerably during the first 2 years of life. The differences between the learning and memory performance of the infant and that of the adult therefore seem to be quantitative rather than qualitative. In the first 2 or 3 years of life, the perception, selection, and attention given to contextual stimuli seem to change. Furthermore, the development of language is one of the major changes that take place at this ontogenetic stage. Cognitive development could synthetically be characterized as the increase in the ability to generalize — that is, to extend the range of situations in which experiences learned at a very young age can be recalled and expressed. The debate on infantile amnesia is still open; in a thorough investigation, Howe and Courage (1993) hypothesize that maturation of the concept of self marks the emergence of autobiographic memory and, therefore, the ability to remember events that took place a long time ago.

Recent data indicate that some experiences undergone during the first year of life may have very important repercussions on the development of complex capacities such as the learning of language (Saffran *et al.,* 1996). Indeed, it seems that some characteristics of the mother tongue can actually be learned during intrauterine life. The study of fetal behavior seems to indicate that during prenatal life, in particular the last 3 months of pregnancy, the fetus is ready to receive and elaborate information, mainly of an auditory or olfactory nature.

Conclusions

In the past 30 years of the twentieth century, developmental psychobiology has gained an important place for itself among the behavioral sciences and, more generally, the neurosciences. Ontogenetic studies have revealed the richness of the behavioral repertoire of neonates and their unexpected ability to learn and recall events even at very early stages of development, demonstrating the importance of early experiences in the subsequent organization of behavior and the nervous system. The discovery of analogies between the plasticity mechanisms underlying early experiences of various kinds in different species of vertebrates has contributed to overcoming the old distinction between innate and learned, showing how a change in behavior on the basis of experience is always the result of the dialectic interaction between species-specific limits and the flexibility of the nervous system. The study of neonatal learning represents a unique opportunity to clarify how complex and competing nerve circuits and routes organize during brain ontogeny to give rise to the cognitive functions of the adult.

References Cited

Bachevalier, J. (1990). Ontogenetic development of habit and memory formation in primates. In *The Development and Neural Bases of Higher Cognitive Functions* (A. Diamond, Ed.), Vol. 608, pp. 457–485. N.Y. Acad. Sci., New York.

Bateson, P. P. G. (1981). Ontogeny of behaviour. *Br. Med. Bull.* **37,** 159–164.

Bolles, R. C. (1970). Species-specific defense reactions and avoidance learning. *Psychol. Rev.* **77,** 32–48.

Calamandrei, G., and Alleva, E. (1996). Neuronal growth factors, neurotrophins and memory deficiency. *Behav. Brain Res.* **66,** 129–132.

Campbell, B. A. (1984). Reflections on the ontogeny of learning and memory. In *Comparative Perspectives on the Development of Memory* (R. Kail and N. E. Spear, Eds.), pp. 23–35. Erlbaum, Hillsdale, NJ.

DeCasper, A. J., and Fifer, W. P. (1980). Of human bonding: Newborns prefer their mothers' voices. *Science* **208,** 1174–1176.

D'Udine, B., and Alleva, E. (1984). The ontogeny of learning capability in rodents: Comparative models. In *Contemporary Psychology: Biological Processes and Theoretical Issues* (J. McGaugh, Ed.), pp. 131–143. North Holland, Amsterdam.

Garcia, J., and Koelling, R. A. (1966). Relation of cue to consequence in avoidance learning. *Psychon. Sci.* **4,** 123–124.

Howe, M. L., and Courage, M. L. (1993). On resolving the enigma of infantile amnesia. *Psychol. Bull.* **113,** 305–326.

Johanson, I. B., and Hall, W. G. (1979). Appetitive learning in 1-day-old rat pups. *Science* **205,** 419–421.

Klopfer, P. (1988). Metaphors for development: How important are experiences early in life? *Dev. Psychobiol.* **21,** 671–678.

Leon, M. (1987). *Trends Neurosci.* **10.**

Nadel, L., and Zola-Morgan, S. (1984). Infantile amnesia: A neurobiological perspective. In *Infant Memory* (M. Moscovitch, Ed.), pp. 145–172. Plenum, New York.

O'Keefe, J., and Nadel, L. (1978). *The Hippocampus as a Cognitive Map.* Clarendon, New York.

Pedersen, P. E., Stewart, W. B., Greer, C. A., and Shepherd, G. M. (1983). Evidence for olfactory function *in utero. Science* **221,** 478–480.

Rovee-Collier, C. (1990). The "memory system" of prelinguistic infants. In *The Development and Neural Bases of Higher Cognitive Functions* (A. Diamond, Ed.), pp. 517–543. N.Y. Acad. Sci., New York.

Rudy, J. W., and Cheatle, M. D. (1977). Odor-aversion learning in neonatal rats. *Science* **198,** 845–846.

Saffran, J. R., Aslin, R. N., and Newport, E. L. (1996). Statistical learning by 8-month-old infants. *Science* **274,** 1926–1928.

SHETTLEWORTH, S. J. (1988). Varieties of learning and memory in animals. *J. Exp. Psychol.* **19,** 5–14.

SMITH, G. J., and SPEAR, N. E. (1978). Effects of the home environment on withholding behaviors and conditioning in infant and neonatal rats. *Science* **202,** 327–329.

SPEAR, N. E. (1984). Ecologically determined dispositions control the ontogeny of learning and memory. In *Comparative Perspectives on the Development of Memory* (R. Kail and N. E. Spear, Eds.), pp. 325–358. Erlbaum, Hillsdale, NJ.

STEHOUER, G. J., and CAMPBELL, B. A. (1980). Ontogeny of passive avoidance: Role of task demands and development of species-typical behaviours. *Dev. Psychobiol.* **13,** 385–398.

SULLIVAN, R. M., *et al.* (1989). *J. Neurosci.* **9,** 3998–4006.

WILSON, D. A., and SULLIVAN, R. M. (1994). Neurobiology of associative learning in the neonate: Early olfactory learning. *Behav. Neurol. Biol.* **61,** 1–18.

ZOLMAN, J. F. (1983). Ontogeny of learning. In *Perspectives in Ethology* (P. P. G. Bateson and P. H. Klopfer, Eds.), Vol. 5, pp. 275–323. Plenum, New York.

General References

HALLIDAY, T. R., and SLATER, P. J. B. (Eds.) (1983). *Genes, Development and Learning.* Blackwell, Oxford.

JOHNSON, M. H. (1993). *Brain Development and Cognition: A Reader.* Blackwell, Oxford.

KRASNEGOR, N. A., BLASS, E. M., HOFER, M. A., and SMOTHERMAN, W. P. (Eds.) (1987). *Perinatal Development: A Psychobiological Perspective.* Academic Press, Orlando, FL.

PIAGET, J. (1954). *The Construction of Reality in the Child.* Basic Books, New York.

SPEAR, N. E., and CAMPBELL, B. A. (Eds.) (1979). *Ontogeny of Learning and Memory.* Erlbaum, Hillsdale, NJ.

PETER MARLER
Section of Neurobiology, Physiology, and Behavior
University of California
Davis, California, USA

Song Learning

Birds communicate among themselves by means of a repertoire of vocalizations. Their brief calls are often monosyllabic, while their songs are extended patterns of sound. In nearly half of the 9000 species of birds, song is learned and, in the oscines, is characterized by complex vocal organs and distinctive forebrain circuitry evolved specifically to sustain song learning. Songs are usually androgen dependant and the associated brain circuitry is often sexually dimorphic with an extensive literation on both its extent and functional significance. Unlike calls, oscine songs do not develop normally if the young males do not have the opportunity to listen to the songs of the adults. The importance of learning is demonstrated by the presence of dialects whose geographical variations are comparable to those of humans. The development of learned song and innate foreknowledge of songs of their species can help in the process of between-species mimicry. However, the imitation of the other species' songs is rare in nature and if birds have a choice, they prefer to learn the song of their own species.

All birds have a repertoire of calls for communication about danger, food, sex, group movements, and many other purposes. Within this repertoire of a dozen or more sound signals, a distinction can be made between calls, which are usually brief and often monosyllabic, and song, which is a more extended pattern of sound, sometimes raucous, often tonal and melodic, and not infrequently a source of pleasure for human listeners (Catchpole and Slater, 1995). Defined in this way, virtually all birds have something that we can call a song, whether it be the crow of the cock or the music of nightingales. In approximately half of all 9000 species of birds, the song is learned. Most of the approximately 4000 probable learner species come from just 1 of the 27 orders into which taxonomists classify living birds. Excluded are waders, chicken-like birds, woodpeckers, cuckoos, doves, hawks, owls, and all marine and freshwater birds. The only groups that qualify are parrots, hummingbirds, and the very populous suborder of the Passeriforms known as the oscines, characterized by the complexity of their vocal organs and by their distinctive forebrain circuitry expressly evolved to sustain song learning. Parrots and perhaps hummingbirds seem to have independently achieved the same end with equivalent but neuroanatomically distinct brain circuitry (Striedter, 1994).

Songs are usually, although not always, a male, androgen-dependent prerogative (Brenowitz and Arnold, 1989; Brenowitz *et al.,* 1985). In species that are learners, the associated brain circuitry is often sexually dimorphic and there is an extensive literation on the extent and functional significance of this dimorphism and its developmental basis

(Nottebohm and Arnold, 1976; DeVoogd, 1991; Arnold, 1990; Arnold *et al.,* 1986; Brenowitz and Kroodsma, 1996). A larger song system in male brains clearly implies a greater commitment to song learning, and the correlation between the size of song system nuclei, the presence of song, and song repertoire size extends not only to comparisons between the sexes but also to comparisons between species, subspecies, and even between individuals (Nottebohm *et al.,* 1981; Canady *et al.,* 1984; DeVoogd *et al.,* 1993).

Unlike calls, which are occasionally learned but are typically innate, oscine songs always develop abnormally if a young male is reared out of hearing of adults. A common consequence of dependence on learning is the emergence of local song dialects, often varying on much the same geographic scale as dialects in human speech (Baker and Cunningham, 1985). Their distinctiveness depends on such factors as patterns of dispersal from the birthplace and the timing of song learning. When dialects are well-defined, it seems to be a general rule that songs, like a bird's own dialect, are especially potent as vocal signals, both to territorial males and to females ready to mate (Balaban, 1988; Baker, 1983; Baker *et al.,* 1981, 1987). A bird's dialect may or may not correspond to that which prevails in the natal area.

Many analogies can be struck between song learning and the acquisition of human speech. There is no nonhuman primate in which vocal patterns are culturally transmitted from generation to generation. Aside from cetaceans, which also engage in vocal learning, birds provide the most valuable animal model for studying the behavioral, hormonal, and neural bases of vocal plasticity (Konishi, 1985; Marler, 1991; Nottebohm, 1993).

Special Brain Mechanisms for Song Learning

Much has been learned during the past 20 years about the specialized neural mechanisms that mediate avian vocal plasticity since Nottebohm *et al.* (1976) discovered the song system in the canary brain, although many important issues remain unexplored. There are two main functional requirements for controlling song production. Appropriate configurations of the articulators must be set up, primarily the syrinx and the postsyringeal vocal tract. Also, the required pattern of airflow must be generated to set the articulators vibrating. These controls are required whether a song is learned or innate and, as might be predicted, some of the circuitry responsible is relatively conservative, located in the midbrain and the brain stem (Wild, 1994). The added requirements for songs to be learned are met by two major circuits in the forebrain — one to sustain the actual learning process, the so-called *anterior forebrain pathway,* and the other, the *motor pathway,* for control of the production of learned songs (Nottebohm, 1993) (Fig. 1). Disruption of the motor pathway always results in the distortion and even the elimination of learned songs, whereas the anterior forebrain pathway seems to be redundant except during actual song development (Bottjer *et al.,* 1984). As with innate songs, the motor pathway for learned songs must have access both to the vocal tract and the syringeal musculature (mapped in one part of the motor pathway) and to the respiratory system (mapped in another part), both in the robust nucleus of the archistriatum (Vicario, 1991). Demands on the motor pathway circuit are complicated by the existence of two semiindependent vibrators in the songbird syrinx, requiring precise coordination since both participate in the production of learned songs (Greenewalt, 1968; Suthers, 1990).

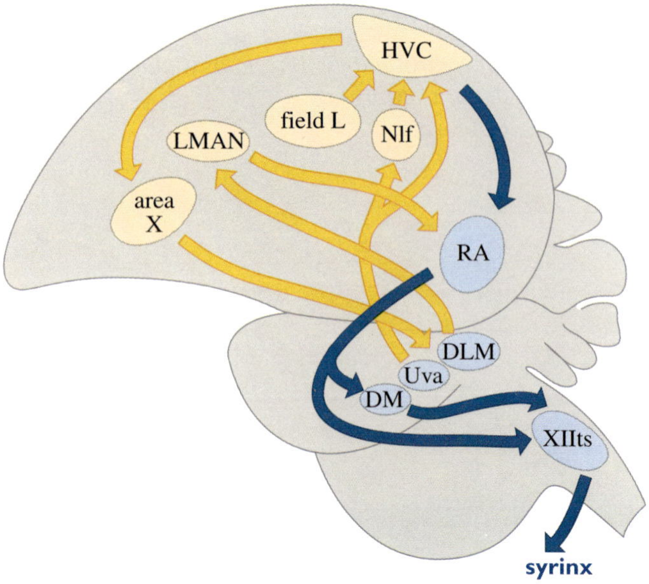

FIGURE 1 Diagram of the two major circuits in the song system of the oscine brain. The motor circuit (in blue) is for control of movements of the vocal apparatus in learned songs, whereas the anterior forebrain pathway (in yellow) is for song learning. There is a fork in the motor pathway from the robust nucleus of the archistriatum (RA) to the tracheosyringeal part of the hypoglossal nucleus (XIIts) and to the vocal apparatus. One branch goes directly to the hypoglossal nucleus, and the other passes through the dorsomedial part of the intercollicular nucleus (DM) in the midbrain, thought to be involved in the control of respiration. Inputs to the higher vocal center (HVC) in the basal forebrain pathway include diverse areas such as field L, the primary auditory forebrain area. The medial caudal neostriatum, another auditory area that also projects to the HVC, and the circuit for the production of innate songs, involving the intercollicular nucleus, are not shown. DLM, medial portion of the dorsolateral thalamic nucleus; LMAN, lateral magnocellular nucleus of the anterior neostriatum; NIf, nucleus interface; Uva, *nucleus uvaeformis* (adapted from Nottebohm, 1991).

The hearing of birds has been extensively studied in relation to their vocal behavior (Dooling, 1982). Many parts of the song system are responsive to song stimulation in addition to their motor functions. Strangely, one of the strongest stimuli, especially in the higher vocal center of adult males is the bird's own individually distinctive crystallized song (Margoliash, 1986). This finding has promoted the plausible speculation that adult males of some species use auditory feedback from their own song as a yardstick to calibrate the potency of songs of others when they hear them, especially those of conspecific neighbors and strangers.

Attunement of a male's song system to his own individual song, including all of its personal quirks, only emerges as songs crystallize (Doupe, 1993; Volman, 1993). Unfortunately, little attention has been paid to responsiveness of the brain of young birds to song stimuli prior to the production of plastic and crystallized song. Patterns of auditory responsiveness of the song in bird brain during the earliest memorization phase of song learning remain virtually unexplored (Whaling *et al.,* 1995). Behavioral tests at this age demonstrate innate responsiveness to many features of song, especially those of the bird's own species (Nelson and Marler, 1993; Whaling *et al.,* 1997). Details of the process of song memorization are consistent with involvement of the set of innate auditory predispositions that each species deploys, serving as a guide to the process of selective song learning. The neural mechanisms responsible await neurophysiological investigation, and until they are elucidated a full understanding of the physiological basis of song learning will continue to elude us. Currently, we do not know how songs are memorized by a young male, nor where they are stored in the brain in preparation for the guidance of song production.

One remarkable feature of the bird brain is the continuing development of new neurons in many parts of the brain, both in infancy and in adulthood, including the song system (Goldman and Nottebohm, 1983). Although the relevance of neurogenesis to song learning is unclear, there are many fascinating hints, including the demonstration — in something of a technical tour-de-force — that some new neurons are actually inserted into the song system and become active there (Paton and Nottebohm, 1984; Alvarez-Buylla *et al.,* 1990). It may be that neuronal turnover in the motor pathway prepares the way for the song plasticity that some birds such as the canary (Nottebohm and Nottebohm, 1978) display in adulthood.

Another place to look for song-related activity is in the forebrain auditory circuits concerned with the perceptual processing of song stimuli. One of these (the medial caudal neostriatum) has been identified as the site of gene activation in response to conspecific stimulation (Mello *et al.,* 1992; Mello and Clayton, 1994), and as a location where experience-dependent changes in electrophysiological responsiveness to song stimuli occur (Vicario *et al.,* 1996). Circuits extrinsic to the song system may prove to be more involved in song learning than has been thought.

Sensitive Periods for Learning

In song development, as in the ontogeny of other kinds of behavior, the rapidity and precision with which learning takes place varies during ontogeny from one stage of the life cycle to another. All songbirds seem to share a similar sequence of stages in the process of learning to sing. First is the acquisition phase, often restricted to a short period, in which a bird hears songs and commits some of them to memory. After storage for a period that ranges in duration from days to months, depending on the species, songs are recalled from memory and imitations begin to emerge. Renditions are relatively faithful to the original model in some species but in others depart from the tutor quite radically, with vocal inventiveness playing a significant role.

The separation in time between the acquisition or sensory phase and the production or sensorimotor phase ending with the production of crystallized, adult song is a significant one. In many birds the sequence occurs just once, often in early adolescence. In such cases, adult song patterns remain fixed thereafter, with their production waxing and waning with the passing of each breeding season. In other species, the sequence recurs repeatedly during adulthood so that song patterns may continue to change. Even close relatives, such as canaries and sparrows, can differ in this respect.

Several species of sparrows have a sensitive period for song acquisition beginning at about 2 weeks of age, soon after they leave the nest, and ending 6–8 weeks later (Nelson *et al.,* 1995). The precise duration varies, even in different populations of the same species (Fig. 2). Sensitive periods for learning are not fixed but changeable, within limits, depending on the strength of song stimulation, access to a live, interactive tutor (Baptista and Petrinovich 1986), and physiological factors such as hormonal states that vary with the season. Closure of the acquisition period can sometimes be delayed if songs of the bird's own species are withheld (Kroodsma and Pickert, 1980), as though termination of the sensory phase is delayed until adequate song stimulation has been experienced. If young are hatched late in the season and singing has already ceased for that year, the sensitive period may be extended until the following spring in some species, although this delay does not occur in all cases.

The Ontogeny of Learned Song

In contrast to song learners, nonlearners develop normal song when raised in isolation, even after deafening

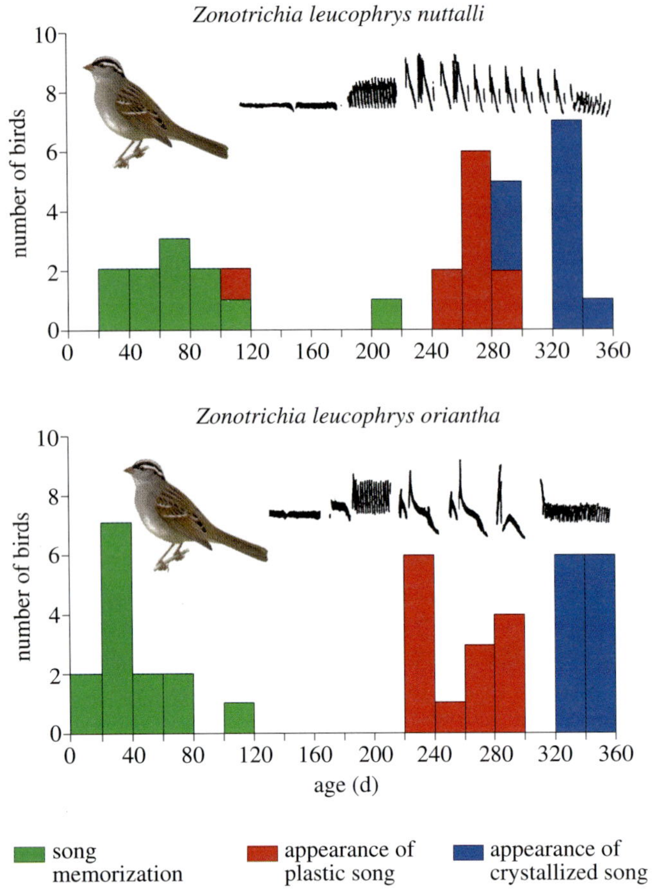

FIGURE 2 The timing of song production and acquisition based on tape recordings throughout the first year of life of male white-crowned sparrows (*Zonotrichia leucophrys*) raised in the laboratory. The birds analyzed represent populations of two different subspecies: one resident year-round, *Zonotrichia leucophrys nuttalli* (top), and the other migratory, *Zonotrichia leucophrys oriantha* (bottom). The migratory birds begin acquiring song earlier than the residents, peaking at 33 days instead of 52 days, and the acquisition period is significantly shorter. The sonograms above the histograms indicate a representative song of each population. Both are about 2 s in duration.

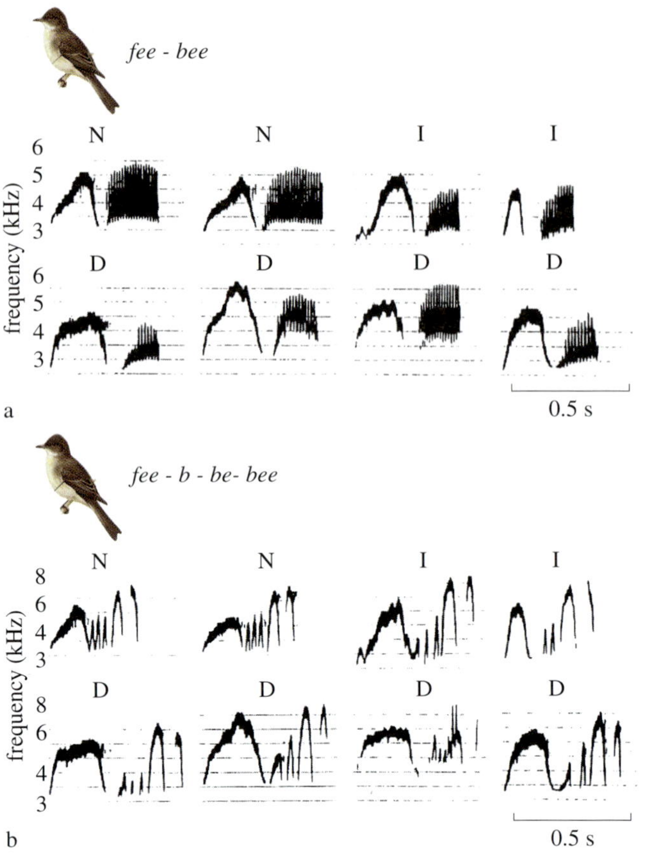

FIGURE 3 The oscine songbirds, among passeriform perching birds, all have learned songs, but the suboscines, which lack the special song system brain circuitry, do not. Here, examples are shown of the innate song of a suboscine passerine, the eastern phoebe (*Sayornis phoebe*). The different songs refer to birds which developed normally in the wild (N), in isolation (I), and after deafening (D). Two song types are represented, the *fee-bee* (a) and the *fee-b-be-bee* (b). Unlike oscines, whose songs are drastically modified by isolation or deafening, both isolated and deafened phoebes developed normal songs (adapted from Kroodsma and Konishi, 1991).

(Kroodsma, 1984; Kroodsma and Konishi, 1991). This is true of a male dove, a cockerel, and even of those passerines that are nonoscines, such as the New World flycatchers (Fig. 3). In such cases, when a young bird starts to sing, its very first efforts are clearly identifiable as immature versions of what will ultimately be normal adult song. Early attempts may be noisy and fragmented, but there is always a clear linear progression in the development of innate songs.

In birds that learn their songs, the developmental progression is quite different. They begin with the unique vocal pattern called *subsong,* which is a quiet, variable warbling that can begin as early as 3 or 4 weeks of age. It is reminiscent of the early babbling stage of speech development (Locke, 1993). Typically, it seems to bear no resemblance to mature song, although we know less about its structure and function than about any other aspect of song development. There are no imitations in subsong. In song learners there is a kind of vocal metamorphosis between subsong and later stages of song development. The amorphous structure and noisy spectral organization of subsong are consistent with a role in the acquisition of the general motor skills of singing, in coping with the complexities of the oscine syrinx, and also in honing the ability to guide the voice by the ear, a prerequisite for vocal imitation (Thorpe, 1961; Nottebohm, 1972; Marler and Peters, 1982b). Duration of the subsong period varies greatly, even within a species, depending on individual histories. In deaf birds and social isolates it may last for weeks, but a bird with adequate tutoring in youth may progress rapidly from subsong to the next phase (plastic song) in a few days.

In plastic song, signs of mature song structure first appear. Previously memorized song patterns emerge and begin to stabilize, first gradually and then rapidly, until adult, crystallized song is achieved and the young male is ready to launch a full season of mature singing (Fig. 4).

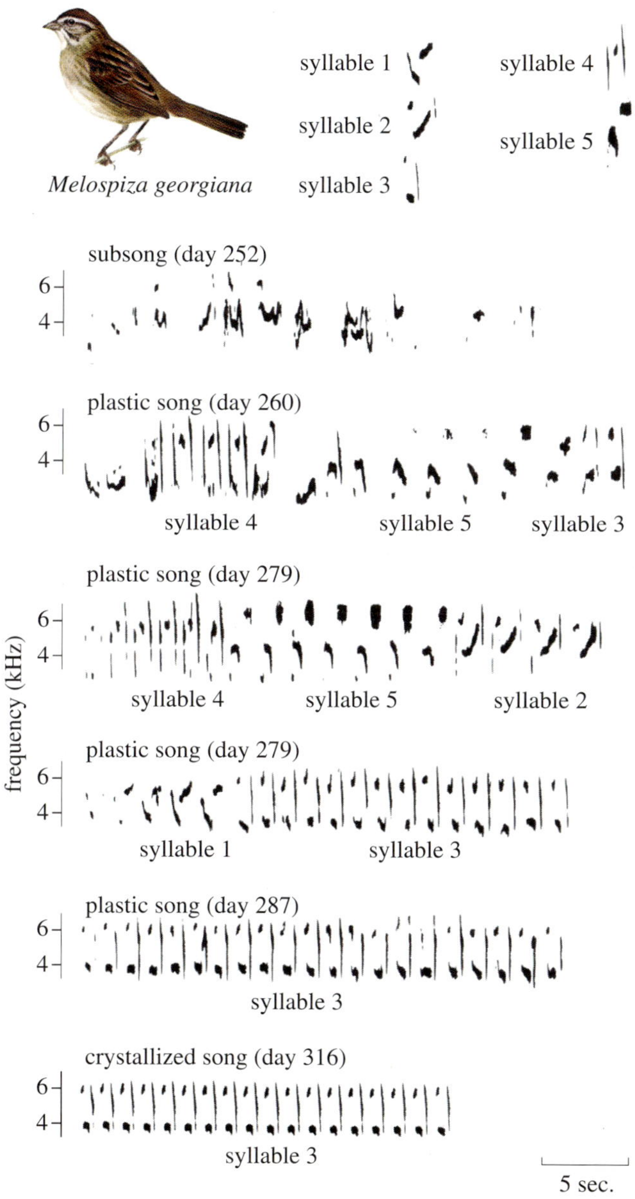

FIGURE 4 Development of the song of a male swamp sparrow (*Melospiza georgiana*) raised in the laboratory and tutored with tape recordings of song from 20 to 60 days of age. Some of his training song syllables are shown at the top right. The first emergence of imitations can be seen in his plastic song from Day 260, 200 days after the end of tutoring. His mature song crystallized at approximately 300 days of age (adapted from Marler and Peters, 1982a).

Overproduction, Attrition, and Action-Based Learning

Song learning has traditionally been viewed as "memory based." Songs are acquired and memorized and then used later to guide motor development by auditory feedback, with no practice at the time of memorization. Another kind of vocal plasticity, involving only working memory and with a key role for motor practice, occurs during the plastic song phase.

Several learner species overproduce during plastic song, generating more than they need for mature song and discarding excess plastic songs as the time for song crystallization approaches (Fig. 5). Some overproduced songs are invented, but many are imitated, and the richness of early song experience is one factor in determining the extent of the overproduced repertoire (Nelson *et al.*, 1996). However, overproduction and attrition also occur in birds tutored and raised in individual isolation in the laboratory (Marler and Peters, 1982a). In this case, the decision about which songs to discard seems to be random, whereas in nature it is guided by social experience.

The attrition phase in which overproduced songs are discarded from the repertoire coincides with the efforts of young male sparrows to establish their first territory. As it competes with adults, those songs of the young bird that match song types of older rivals, even if only approximately, are most effective in eliciting countersinging and perhaps in territorial defense (Beecher, 1996). Young males tend to retain and crystallize those plastic songs that are the closest match to those of their rivals and to discard the rest.

Effects on song attrition of male-to-male territorial interactions, either real or simulated, have been described in several species of sparrows (Nelson and Marler, 1994; Nel-

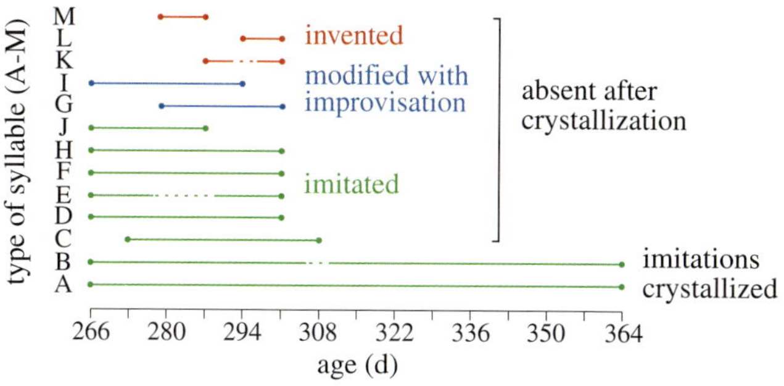

FIGURE 5 A diagram of song development in the same male swamp sparrow shown in Fig. 4. In addition to the eight tape-recorded songs imitated, this subject invented three more and modified two by improvisation. The excess of plastic song is discarded as the moment of song crystallization approaches. The adult repertoire, consisting of two song types, is typical for this species (adapted from Marler and Peters, 1982a).

son, 1992). In another species, the brown-headed cowbird, responses of females serve to shape male decisions about which overproduced songs to retain (West and King, 1988). The process by which a repertoire of varied motor patterns is selectively winnowed, with the selection process based on social feedback, has been termed *action-based learning* (Marler and Nelson, 1993). There are parallels with the development of other kinds of motor behavior, including play. Use-dependent processes of selective attrition may be widespread in the development of many kinds of behavioral activities. The actions that are discarded from the repertoire may have functional significance. For example, songs that are practiced in youth and then rejected may provide a potential memory bank for the future, perhaps consulted in assessing the songs of others. In species that retain vocal plasticity into adulthood, discarded songs may be redeployed as the basis for producing new songs later in life.

The life cycle time table for action-based learning is different from that for memory-based learning. Action-based learning has the potential to recur whenever, for whatever reason, an animal proceeds through a phase of variable action patterns which is then followed by use-dependent winnowing of the variants. A mechanism is required for selective reinforcement of certain motor variants over others. Unlike memory-based learning, highly specialized brain mechanisms such as the oscine song system are not required. As a consequence, action-based learning is likely to have a wide phylogenetic distribution. For example, although nonhuman primates seem to be incapable of memory-based vocal learning, they do have a limited degree of vocal plasticity, possibly mediated by action-based learning.

Effects of Isolation and Deafness

Regardless of the opportunity to learn, certain aspects of the normal singing behavior of the species will always develop in oscines. When sparrows are raised in isolation, for example, the note structure and tonal quality of their songs are aberrant, but aspects of basic song syntax develop normally (Fig. 6). Certain song features of the species are produced despite their absence from any of the tutoring songs a male has experienced — but with the important proviso that hearing must be intact.

The singing patterns of a hearing male oscine raised in isolation and of a male deafened early in life and raised in isolation are quite different. Typically, deaf male songs are highly degraded and variable, lacking most of the few normal features that an intact male can develop in isolation.

Songs of early deafened males are not completely amorphous. This is true even when the critical step has been taken to ensure that there has been no song practice prior

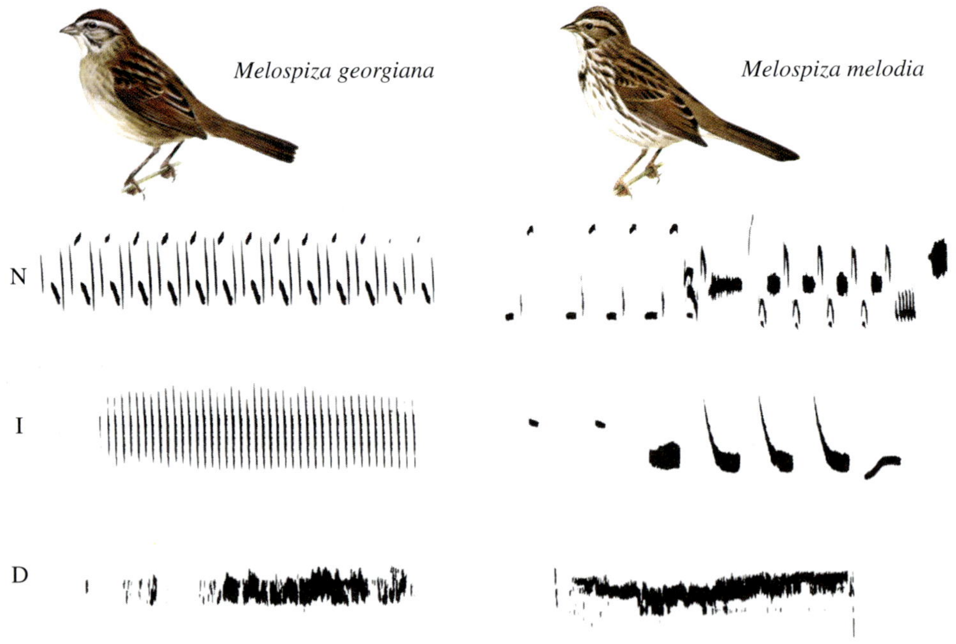

FIGURE 6 Normal songs (N) of swamp sparrows (*Melospiza georgiana*) and song sparrows (*Melospiza melodia*) compared with those developed by males raised in social isolation (I) and males deafened. The high degree of species specificity, which is very evident in normal song, is largely erased from the songs of deaf males (adapted from Marler and Sherman, 1985).

to deafening (Nottebohm, 1968). This step was lacking in some of the earliest studies. The emergence of some elements of species-specific song syntax in males deafened very early in life seems to implicate innate motor programs in song development (Marler and Sherman, 1983). The highly abnormal song patterns characteristic of deaf birds develop both if a male becomes deaf before tutoring and if he is deafened after the tutoring but before singing has developed (Konishi, 1965). There seems to be no internal brain circuitry that makes memorized songs directly available to guide vocal development. To transform a memorized song into a produced song, the bird must be able to hear its own voice. We can infer that some aspects of song production are guided by auditory templates for song, located in the brain, with specifications that vary from species to species (Konishi, 1965; Marler and Sherman, 1983). Neurobiologists have yet to demonstrate where these templates are located and how they operate (Whaling *et al.,* 1997).

Like ornithologists, wild birds rely heavily on song for species identification. Each bird has its own distinctive way of singing, and this is recognized by other species members. Songs of early deafened songbirds, on the other hand, are largely, though not entirely, lacking in species specificity and are often indistinguishable between species (Fig. 6). Songs of males raised in isolation but with hearing intact fall somewhere between. They retain some species-typical features, and they lack others.

When tape-recorded songs are played to wild territorial males or to females in hormonally induced estrus, experienced birds of both sexes respond strongly to normal song of their species but completely disregard deaf songs as social signals (Searcy *et al.,* 1985; Searcy and Marler, 1987). Similarly, songs of other species elicit none of the strong aggressive responses of territorial males, or the courtship of estradiol-primed females, that conspecific song evokes. However, those of social isolates do retain a minimal level of effectiveness. Thus, the efficacy of birdsong as a social signal is augmented during development both by learning and by ontogenetic involvement of innate auditory templates that encode certain species-specific features of normal song.

Learning Preferences

Learning theorists once believed that any stimulus could be learned, given appropriate conditions, and that differences in stimulus potency, although present, were unimportant enough to be ignored in theorizing about the mechanisms underlying learning. We now know that innate predispositions play crucial roles in the adaptiveness of natural learning processes, and song learning in birds provides many illustrations of this principle. Although birds are capable of learning a multiplicity of different sounds, including those of many other species (Fig. 7), other-species mimicry is relatively rare in the wild (Kroodsma and Baylis, 1982). Omitting mimics, which account for only about 15% of songbirds, it is adaptive for songbirds to avoid other-species songs as models for learning.

If males of two species, such as swamp and song sparrows, are brought into the laboratory and exposed to identical tape recordings that include both species songs, each displays a clear preference for song of its own kind, even if the song has never been heard before (Marler and Peters,

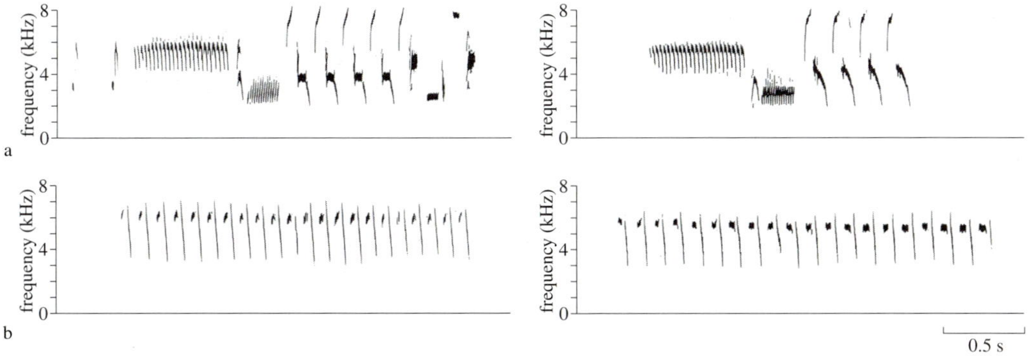

FIGURE 7 Heterospecific songs imitated by sparrows that have never heard conspecific song. In this example, male swamp sparrows (*Melospiza georgiana*) and song sparrows (*Melospiza melodia*) learned a tape-recorded song of the other species. Although sparrows typically reject the songs of other species as models for learning, they will sometimes accept them if given no other choice. Evidently, the vocal apparatus is able to articulate the sounds of the other species song. (a) (Top) Original song of the song sparrow; (bottom) imitation of the song of the song sparrow by a swamp sparrow. (b) (Top) Original song of the swamp sparrow; (bottom) imitation of the song of the swamp sparrow by a song sparrow.

1988, 1989). This preference does not arise from an inability to produce the other species' song. An alien song that a male rejects, when given a choice, may become acceptable if it is the only option (Fig. 7), especially when it is presented in a highly interactive situation by a live tutor caged nearby (Baptista and Petrinovich, 1986). However, given a choice, there is typically a preference for conspecific song.

The Role of Innate Knowledge in Song Development

The view of the brain as a *tabula rasa,* a blank slate — for too long a basis for the thinking of learning theorists — is patently absurd. Whatever the task, every brain brings to bear a set of neurally based predispositions, each with its own evolutionary and experiential history, about how best to proceed in dealing with a given set of learning problems. Some of these predispositions are generic and pan specific and others are highly singular and even species specific, as with the innate processes that the human infant brings to bear on the development of speech behavior and language. Such is the case with birdsong.

The young bird embarks on the process of developing song armed with innate predispositions, both sensory and motor, about how best to proceed. Some are generic, such as the widely shared tendency of songbirds to sing tonal sounds, evident even when they render imitations of sounds that were not originally strictly tonal (Nowicki *et al.,* 1992; Peters and Nowicki, 1996). Others are species specific. When sparrows learn to sing, they favor conspecific song. They do so on the basis not of a single ethological "sign stimulus" but a range of song features. Naive young birds, recently fledged, are highly discriminating with regard to the song stimuli they find most potent, as revealed by their tendency to give more begging calls in response to playback (Nelson and Marler, 1993). By playing experimentally modified songs to young birds, innate responsiveness has been demonstrated to a range of conspecific song attributes (Whaling *et al.,* 1995). It is clear from their behavior that young birds possess extensive foreknowledge about the song of their species before they embark on the process of learning to sing. The evidence indicates that they know much more than one would guess on the basis of the songs that a male produces when raised in social isolation. They behave as though auditory experience that matches their innate knowledge is required before that knowledge can be brought to bear on song development (Marler, 1984).

One way to view song learning is as a process of validation by use of a subset of innate knowledge, drawn from a more extensive library encoded in the brain, that details the rules for the singing behavior for each species (Marler and Nelson, 1992). This innate library is shared by all species members. The species-universal rules encoded there are sufficiently flexible to provide almost infinite opportunities for differences in individual behavior as a function of personal experience but are firm enough to restrict the development of excessive divergence between individuals.

There are analogies with the rules for orderly musical composition that each human culture possesses, encompassing the potential for an infinite number of different melodies (Jackendoff, 1994). The genetic underpinnings of the immune system provide another parallel, with the major histocompatibility complex providing enough alleles for a vast number of possible combinations. The imposition of limitations is important because it ensures that no matter how individualistic a bird's singing behavior may be, it still conforms closely enough to certain species-universal rules so that obstacles to communication with others are minimized.

In addition to predispositions evident in the sensory phase of song memorization, others are manifest during motor development. The emergence of some species-specific basics in the song syntax of early deafened males has already been noted (Marler and Sherman, 1983). A bird tricked into learning heterospecific songs may only reject them in late stages of plastic song as species-specific syntax emerges (Marler and Peters, 1989). Podos (1996) tutored young sparrows using experimentally modified songs with abnormally high syllable rates. They imitated the songs but reorganized them in invented patterns that conformed more closely to the temporal organization of their normal species song. Again, there is great flexibility in these motor predispositions; however, despite the fact that songs with abnormal temporal organization can be taught, there is an underlying tendency to conform to species-specific norms. The individual and population differences that characterize learned birdsongs are all-pervasive, but they rarely confuse experienced bird watchers when identifying the species responsible. With their genetic fitness at stake, the birds are hardly likely to be any less perceptive.

Other Aspects of the Song Learning Process

Some of the distinctive features of song learning recur in other cases of developmental plasticity. It is not uncommon for developing behavioral and neural systems to exhibit sensitive phases, when the potential for plasticity is unusually great. Completion of one stage is often a necessary precursor for embarking on the next (Rauschecker and Marler, 1987; Pinker, 1994).

Unlike some other widely studied forms of learning, no external reinforcement is required in song learning, even though social stimulation may augment the acquisition process. It suffices that the bird be exposed to an appropriate set of stimuli at the appointed stage of development for acquisition to occur. At the height of the sensitive period, learning can take place with great rapidity. Sometimes, remarkably few exposures to a bird's own species song are sufficient. In sparrows some acquisition can occur from 30 repetitions of a conspecific song. European blackbirds have learned 15–20 song presentations in a single day.

The virtuoso performer is the nightingale, which can learn a sequence of song types accurately after 20 presentations; some males can learn with as few as 5 presentations. The virtuosity of nightingales is such that they can learn strings of up to 60 different songs and reproduce them in the original order. The sequencing is only retained in subsets of 3–7 songs, however. After producing such a matched sequence, birds in one experiment then switched to another subset, beginning at a different point. These birds behaved as though they divide a string of many songs into manageable subsets of up to 7 songs using a strategy for memorizing long sequences reminiscent of that used by humans in memorizing strings of words (Hultsch and Todt, 1989, 1992). This is one of many avian accomplishments reminiscent of aspects of human speech behavior.

Song Development as a Creative Process

Perhaps most mysterious and intriguing of all is the ability of songbirds to invent new sounds. They achieve this in several ways. Some birds first imitate the components of songs they have heard and then, as plastic song progresses, recombine them to create new sequences (Marler and Peters, 1987). In this way, parts can be exchanged within a song, between different songs, and even between songs that have been acquired many months apart (Fig. 8). Rather than remaining faithful to the structure of imitated sounds, some birds find the temptation to improvise irresistible, especially during plastic song, so that songs originally imitated eventually become so changed as to be unrecognizable (Marler *et al.*, 1972).

Not uncommonly, birds invent new sound patterns that they have never heard. This is frequent in birds raised in isolation, with the result that although the songs of isolates are abnormally simple in some respects, they are also extremely diverse. The act of producing novel and distinctively patterned sound seems to be reinforcing, attesting to the strong and deep-rooted commitment of many songbirds to vocal inventiveness.

The drive to invention is stronger in some species than others. Certain birds remain rigidly faithful to their tutors, adding just sufficient individuality to personalize their song and no more. Adherence to their models makes them especially tractable as subjects for laboratory investigation of song learning. Perhaps for this reason, the more imaginative songsters tend to be underrepresented in the roster of favored subjects for scientific study. The huge individual repertoires that males of some birds possess, sometime with hundreds of song types built up by reiteration of the processes of imitation, segmentation, recombination, improvisation, and invention, are almost unmanageable as subjects for scientific study. However, for anyone interested in esthetic theory, the birds with large repertoires that give hu-

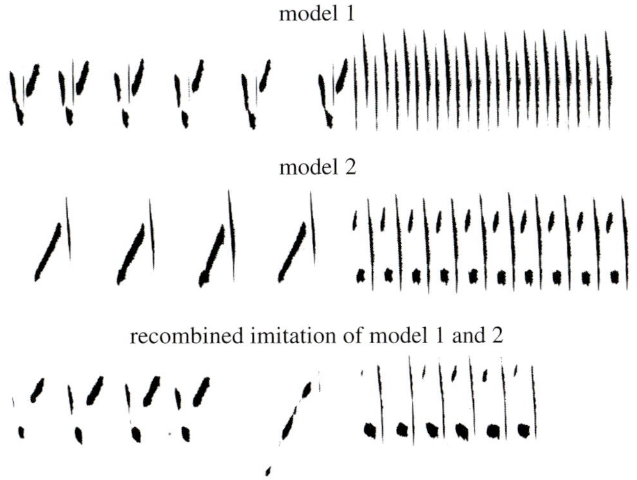

FIGURE 8 An example of a bird that created a new song by recombination. A captive male song sparrow (*Melospiza melodia*) was given only artificial songs to learn that were composed of two-parted songs of the swamp sparrow (*Melospiza georgiana*). He created a three-part song, approximating normal song sparrow syntax, by combining parts of the two different models (see Fig. 7).

man listeners so much pleasure, such as mockingbirds and their relatives, the Australian lyrebird, and certain thrushes, are unrivaled as subjects of choice. Utilitarian concerns are only one aspect of our fascination with birdsongs. If we are searching for animal models of the creative impulse, birdsong is one of the few that comes even remotely close to human standards.

Conclusions

As more bird species are studied, we find that each has achieved its own special solutions to the problems of developing a song. Songsters participate in a system of social communication that relies on signals learned from others. Some species learn readily from song stimulation alone, and often remarkably few song stimuli are necessary, as though the brain is primed to be especially responsive to songs of a bird's own kind. Other species take longer to learn and require social stimulation.

Despite the possession of extensive innate foreknowledge about songs of their species, and a strong inclination to favor them if given a choice, they can be persuaded in special circumstances to learn songs of others. Certain species naturally engage in the process of between-species mimicry and create a species-specific song by imposing distinctive temporal patterning on the imitations as they are rendered. Some birds are faithful to the song they imitate, and others use imitations as a basis for improvisation, retaining certain features of the tutor and changing others.

The diversity of these patterns of behavioral development is made possible because the brain of each bird species brings to bear a distinctive set of properties and predispositions to the task of learning to sing. The challenge for the neurobiologist is to understand not only the basic brain mechanisms that underlie the generic learning abilities of all songbirds but also the more subtle, singular means by which a given species adds its own unique flavor to the process of developing a song, adapted to its own particular set of social, ecological, and phylogenetic circumstances.

References Cited

Alvarez-Buylla, A., Kirn, J. R., and Nottebohm, F. (1990). Birth of projection neurons in adult avian brain may be related to perceptual or motor learning. *Science* **224,** 901–903.

Arnold, A. P. (1990). The passerine bird song system as a model in neuroendocrine research. *J. Exp. Zool. Suppl.* **4,** 22–30.

Arnold, A. P., Bottjer, S. W., Brenowitz, E. A., Nordeen, E. J., and Nordeen, K. W. (1986). Sexual dimorphisms in the neural vocal control system in song birds: Ontogeny and phylogeny. *Brain Behav. Evol.* **28,** 22–31.

Baker, M. C. (1983). The behavioral response of female Nuttall's white-crowned sparrows to male song of natal and alien dialects. *Behav. Ecol. Sociobiol.* **12,** 309–314.

Baker, M. C., and Cunningham, M. A. (1985). The biology of bird-song dialects. *Behav. Brain Sci.* **8,** 85–133.

Baker, M. C., Spitler-Nabors, K. J., and Bradley, D. C. (1981). Early experience determines song dialect responsiveness of female sparrows. *Science* **214,** 819–821.

Baker, M. C., Bjerke, T. K., Lampe, H., and Espmark, Y. O. (1987). Sexual response of female yellowhammers to differences in regional song dialects and repertoire sizes. *Anim. Behav.* **35,** 395–401.

Balaban, E. (1988). Bird song syntax: Learned intraspecific variation is meaningful. *Proc. Natl. Acad. Sci. USA* **85,** 3657–3660.

Baptista, L. F., and Petrinovich, L. (1986). Song development in the white-crowned sparrow: Social factors and sex differences. *Anim. Behav.* **34,** 1359–1371.

Beecher, M. D. (1996). Birdsong learning in the laboratory and field. In *Ecology and Evolution of Acoustic Communication in Birds* (D. E. Kroodsma and E. H. Miller, Eds.). Comstock, Ithaca, NY.

Bottjer, S., Miesner, E., and Arnold, A. P. (1984). Forebrain lesions disrupt development but not maintenance of song in passerine birds. *Science* **224,** 901–903.

Brenowitz, E. A., and Arnold, A. P. (1989). Accumulation of estrogen in a vocal control brain region of a duetting song bird. *Brain Res.* **480,** 119–125.

Brenowitz, E., and Kroodsma, D. E. (1996). The neuroethology of birdsong. In *Ecology and Evolution of Acoustic Communication in Birds* (D. E. Kroodsma and E. H. Miller, Eds.), Comstock, Ithaca, NY.

Brenowitz, E. A., Arnold, A. P., and Levin, R. (1985). Neural correlates of female song in tropical duetting birds. *Brain Res.* **343,** 104–112.

Canady, R. A., Kroodsma, D. E., and Nottebohm, F. (1984). Population differences in complexity of a learned skill are correlated with the brain space involved. *Proc. Natl. Acad. Sci. USA* **81,** 6232–6234.

Catchpole, C. K., and Slater, P. J. B. (1995). *Bird Song.* Cambridge Univ. Press, Cambridge, UK.

DeVoogd, T. J. (1991). Endocrine modulation of the development and adult function of the avian song system. *Psychoneuroendocrinology* **16,** 41–66.

DeVoogd, T. J., Krebs, J. R., Healy, S. D., and Purvis, A. (1993). Relations between song repertoire size and the volume of brain nuclei related to song: Comparative evolutionary analyses amongst oscine birds. *Proc. R. Soc. London Ser. B* **254,** 75–82.

Dooling, R. J. (1982). Auditory perception in birds. In *Acoustic Communication in Birds* (D. E. Kroodsma and E. H. Miller, Eds.), pp. 95–130. Academic Press, New York.

Doupe, A. J. (1993). A neural circuit specialized for vocal learning. *Curr. Opin. Neurobiol.* **3,** 104–111.

Goldman, S. A., and Nottebohm, F. (1983). Neuronal production, migration, and differentiation in the vocal control nucleus of the adult female canary brain. *Proc. Natl. Acad. Sci. USA* **80,** 2390–2394.

Greenewalt, C. H. (1968). *Bird Song: Acoustics and Physiology.* Smithsonian Inst. Press, Washington, DC.

Hultsch, H., and Todt, D. (1989). Memorization and reproduction of songs in nightingales: Evidence for package formation. *J. Comp. Physiol. A* **165,** 197–203.

Hultsch, H., and Todt, D. (1992). The serial order effect in the song acquisition of birds: Relevance of exposure frequency to song models. *Anim. Behav.* **44,** 590–592.

Jackendoff, R. (1994). *Patterns in the Mind: Language and Human Nature.* Basic Books, New York.

Konishi, M. (1965). The role of auditory feedback in the control of vocalization in the white-crowned sparrow. *Z. Tierpsychol.* **22,** 770–778.

Konishi, M. (1985). Birdsong: From behavior to neuron. *Annu. Rev. Neurosci.* **8,** 125–170.

Kroodsma, D. E. (1984). Songs of the alder flycatcher (*Empidonax alnorum*) and willow flycatcher (*Empidonax traillii*) are innate. *Auk* **101,** 13–24.

Kroodsma, D. E., and Baylis, J. R. (1982). A world survey of evidence for vocal learning in birds. In *Acoustic Communication in Birds* (D. E. Kroodsma and E. H. Miller, Eds.), pp. 311–337. Academic Press, New York.

Kroodsma, D. E., and Konishi, M. (1991). A suboscine bird (*Eastern phoebe, Sayornis phoebe*) develops normal song without auditory feedback. *Anim. Behav.* **42,** 477–487.

Kroodsma, D. E., and Pickert, R. (1980). Environmentally dependent sensitive periods for avian vocal learning. *Nature* **288,** 477–479.

Locke, J. L. (1993). *The Path to Spoken Language.* Harvard Univ. Press, Cambridge, MA.

Margoliash, D. (1986). Preference for autogenous song by auditory neurons in a song system nucleus of the white-crowned sparrow. *J. Neurosci.* **69,** 1643–1661.

Marler, P. (1984). Song learning: Innate species differences in the learning process. In *The Biology of Learning* (P. Marler and H. S. Terrace, Eds.). Springer-Verlag, Berlin.

Marler, P. (1991). Song learning behavior: The interface with neuroethology. *Trends Neurosci.* **14,** 199–206.

Marler, P., and Nelson, D. (1992). Neuroselection and song learning in birds: Species universals in a culturally transmitted behavior. In *Seminars in the Neurosciences* (P. Marler, Ed.), Vol. 4, pp. 415–423. Saunders, London.

Marler, P., and Nelson, D. A. (1993). Action-based learning: A new form of developmental plasticity in bird song. *Neth. J. Zool.* **43**(12), 91–103.

Marler, P., and Peters, S. (1982a). Developmental overproduction and selective attrition: New processes in the epigenesis of bird song. *Dev. Psychobiol.* **15,** 369–378.

Marler, P., and Peters, S. (1982b). Subsong and plastic song: Their role in the vocal learning process. In *Acoustic Communication in Birds* (D. E. Kroodsma and E. H. Miller, Eds.), Vol. 2, pp. 25–50. Academic Press, New York.

Marler, P., and Peters, S. (1987). A sensitive period for song acquisition in the song sparrow, *Melospiza melodia,* a case of age-limited learning. *Ethology* **76,** 89–100.

Marler, P., and Peters, S. (1988). The role of song phonology and syntax in vocal learning preferences in the song sparrow, *Melospiza melodia. Ethology* **77,** 125–149.

Marler, P., and Peters, S. (1989). Species differences in auditory responsiveness in early vocal learning. In *The Comparative Psychology of Audition: Perceiving Complex Sounds* (R. J. Dooling and S. H. Hulse, Eds.), pp. 243–273. Erlbaum, Hillsdale, NJ.

Marler, P., and Sherman, V. (1983). Song structure without auditory feedback: Emendations of the auditory template hypothesis. *J. Neurosci.* **3,** 517–531.

Marler, P., and Sherman, V. (1985). *Anim. Behav.* **33,** 57–71.

Marler, P., Mundinger, P., Waser, M. S., and Lutjen, A. (1972). Effects of acoustical stimulation and deprivation on song development in red-winged blackbirds (*Agelaius phoeniceus*). *Anim. Behav.* **20,** 586–606.

Mello, C. V., and Clayton, D. F. (1994). Song-induced ZENK gene expression in auditory pathways of songbird brain and its relation to the song control system. *J. Neurosci.* **14,** 6652–6666.

Mello, C. V., Vicario, D. S., and Clayton, D. F. (1992). Song presentation induces gene expression in the songbird's forebrain. *Proc. Natl. Acad. Sci. USA* **89,** 6818–6822.

Nelson, D. A. (1992). Song overproduction and selective attrition lead to song sharing in the field sparrow (*Spizella pusilla*). *Behav. Ecol. Sociobiol.* **30,** 415–424.

Nelson, D. A., and Marler, P. (1993). Innate recognition of song in white-crowned sparrows: A role in selective vocal learning? *Anim. Behav.* **46,** 806–808.

Nelson, D. A., and Marler, P. (1994). Selection-based learning in bird song development. *Proc. Natl. Acad. Sci. USA* **91,** 10498–10501.

Nelson, D. A., Marler, P., and Palleroni, A. (1995). A comparative approach to vocal learning: Intra-specific variation in the learning process. *Anim. Behav.* **50,** 83–97.

Nelson, D. A., Whaling, C., and Marler, P. (1996). The capacity for song memorization varies in populations of the same species. *Anim. Behav.* **52,** 379–387.

Nottebohm, F. (1968). Auditory experience and song development in the chaffinch, *Fringilla coelebs. Ibis* **110,** 549–568.

Nottebohm, F. (1972). Neural lateralization of vocal control in a passerine bird II. Subsong, calls and a theory of learning. *J. Exp. Zool.* **179,** 35–49.

Nottebohm, F. (1991). *Trends Neurosci.* **14,** 206–211.

Nottebohm, F. (1993). The search for neural mechanisms that define the sensitive period for song learning in birds. *Neth. J. Zool.* **43,** 193–234.

Nottebohm, F., and Arnold, A. P. (1976). Sexual dimorphism in vocal control areas of the songbird brain. *Science* **194,** 211–213.

Nottebohm, F., and Nottebohm, M. E. (1978). Relationship between song repertoire and age in the canary *Serinus canaria. Z. Tierpsychol.* **46,** 298–305.

Nottebohm, F., Stokes, T. M., and Leonard, C. M. (1976). Central control of song in the canary. *J. Comp. Neurol.* **165,** 457–486.

Nottebohm, F., Kasparian, S., and Pandazis, S. (1981). Brain space for a learned task. *Brain Res.* **213,** 99–109.

Nowicki, S., Marler, P., Maynard, A., and Peters, S. (1992). Is the tonal quality of birdsong learned? Evidence from song sparrows. *Ethology* **90,** 225–235.

Paton, G. A., and Nottebohm, F. (1984). Neurons generated in the adult brain are recruited into functional circuits. *Science* **225,** 1046–1048.

Peters, S., and Nowicki, S. (1996). Development of tonal quality in birdsong: Further evidence from song sparrows. *Ethology* **102,** 323–335.

Pinker, S. (1994). *The Language Instinct.* Morrow, New York.

Podos, J. (1996). Motor constraints on vocal development in a songbird. *Anim. Behav.* **51,** 1061–1070.

RAUSCHECKER, J. P., and MARLER, P. (Eds.) (1987). *Imprinting and Cortical Plasticity.* Wiley, New York.

SEARCY, W. A., and MARLER, P. (1987). Response of sparrows to songs of deaf and isolation-reared males: Further evidence for innate auditory templates. *Dev. Psychobiol.* **20,** 509–520.

SEARCY, W. A., MARLER, P., and PETERS, S. S. (1985). Song of isolation-reared sparrows function in communication, but are significantly less effective than learned songs. *Behav. Ecol. Sociobiol.* **17,** 223–229.

STRIEDTER, G. F. (1994). The vocal control pathways in budgerigars differ from those in songbirds. *J. Comp. Neurol.* **343,** 35–56.

SUTHERS, R. A. (1990). Contributions to birdsong from the left and right sides of the intact syrinx. *Nature* **347,** 473–477.

THORPE, W. H. (1961). *Bird Song.* Cambridge Univ. Press, Cambridge, MA.

VICARIO, D. S. (1991). Neural mechanisms of song production in songbirds. *Curr. Opin. Neurobiol.* **1,** 595–600.

VOLMAN, S. (1993). Development of neural selectivity for birdsong during vocal learning. *J. Neurosci.* **13,** 4737–4747.

WEST, M. J., and KING, A. P. (1988). Female visual displays affect the development of male song in the cowbird. *Nature* **334,** 244–246.

WHALING, C. S., SOLIS, M., CARRILLO, G., SOHA, J. A., MARLER, P., and DOUPE, A. J. (1995). Song learning in sparrows: From behavior to brain. *Soc. Neurosci.* **21**(2), 958.

WHALING, C. S., SOLIS, M. M., DOUPE, A. J., SOHA, J. A., and MARLER, P. (1997). Acoustic and neural bases for innate recognition of song. *Proc. Natl. Acad. Sci. USA* **94,** 12694–12698.

WILD, J. M. (1994). The auditory–vocal–respiratory axis in birds. *Brain Behav. Evol.* **44,** 192–209.

ROBERT DANTZER
INSERM U394
33077 Bordeaux Cedex, France

Emotional States and the Immune System

We all believe that intractable conflicts and the emotional turmoil that is elicited by these conflicts can impact on our health and be at the origin of such pathologies as cancer or autoimmune diseases. In addition, we all have the experience of minor ailments which tend to become unbearable in conditions of stress. The discovery that the immune system is closely connected to the brain and that psychological factors influence immune functions appears to give some credibility to these perceptions and beliefs. However, the relationship between stress, emotions, and disease is not a linear one, and much more needs to be achieved at the conceptual and experimental levels to understand how psychological factors can influence the onset, duration, and recurrence of disease.

We Tend to Believe That Emotions Impact on Health

In one of his first letters to Milena, the Czech writer Franz Kafka, who had already been suffering from tuberculosis for 3 years, wrote

> I only think of the explanation I once gave myself of my own case, and which must apply to many others. Things were such that my brain could not bear any longer all the troubles and torments which were inflicted upon it. It said: "I give up. But if there is still some organ which cares about the conservation of the whole, it should relieve part of my burden, and things will be able to last a little longer." It is then that my lung took over, it had apparently nothing to lose.

Like Franz Kafka, we all have the pervading tendency to believe that when the torments of our existence cannot be endured any longer, they are bound to impact with an unsurpassed vigor on our body, the soma, giving rise to the so-called psychosomatic disorders (Fig. 1). In the case of cancer, health psychologists have proposed that individuals who have a high social desirability and are unable to express their emotions are more at risk for developing tumoral processes than individuals who fully express their emotions (Temoshok, 1987). In a moving autobiographical account, Fritz Zorn, the son of a bourgeois family in

FIGURE 1 Scene from the opera *La Traviata* by G. Verdi. The protagonist, Violetta (the soprano Maria Corta), characterized by her beauty and extreme sensitivity, is dying of tuberculosis (photo by G. Neri).

Zurich, Switzerland, defined the neck tumor that was killing him as a physical expression of his mental state: "All the tears I had neither wept nor wished to weep during the course of my life had massed in my neck and created my tumor because their true function — to be shed — had not been attained."

The view that affective disruption impacts on health is the basic ingredient of many psychoanalytically oriented psychosomatic theories. The reasons why the psyche is unable to deal in an appropriate manner with the emotional turmoil which is evoked by our life torments vary from one theory to the other, as do the explanations for why a particular organ or part of the body (the lung in the case of Franz Kafka and the neck in the case of Fritz Zorn) cannot do anything else besides become the set-off of this failure (Dantzer, 1989).

This Relationship Is Difficult to Prove by the Methods of Epidemiology

Despite the apparent soundness of the relationship that Franz Kafka and Fritz Zorn drew between the disorders of their emotional life and their illness, there is still much disagreement on the real importance which can be ascribed to the influences of emotions on health. Not all individuals who experience emotional distress develop tuberculosis or cancer. Conversely, normal individuals without any previous history of depression and loss can develop these diseases. This means that psychological factors, when they exist, play the role of cofactors rather than that of the main causal factors. This is not necessarily what is perceived by patients and what psychosomatic theories tell us. These theories actually feed our craving for understanding why we become ill and why the disease process affects a particular organ. However, as pointed out by Susan Sonntag (1978) in an essay written after she discovered she had cancer, the explanations these theories provide are based on metaphors rather than on scientific evidence and they apply only to those ailments for which we do not know the causal factors, in the biomedical sense, and for which there is no effective therapy. In accordance with this opinion, many decades of literary elaboration on the exquisite sensitivity to tuberculosis of individuals with artistic temperament have been quickly wiped out by the discovery of the bacillus of Koch and the availability of effective antituberculosis chemotherapeutics. In contrast, the lack of proper understanding of the pathogenesis of tumoral processes and the limited value of many anticancer therapies can easily account for the still vivid belief in the role of psychic factors in cancer.

In the medical sciences, numbers are more important than beliefs and conclusions drawn from single case observations. If inappropriate emotional states can induce the onset of a disease or hasten its progression, it should theoretically be possible to evidence these negative influences of emotions on health by epidemiological approaches based on statistical analysis of a sufficient number of cases. There are some examples favoring such a possibility. In a study of 41 women with breast cancer, Greer and colleagues (1979) noted an association between patients' reactions to mastectomy and survival. On the basis of their answers to questions asked 3 months after the surgery and designed to reveal the manner in which they perceived their illness and the prospects for a successful outcome, patients were assigned to four categories: those who denied their condition and claimed that their breast was removed by precaution rather than by necessity, those who decided to become fully informed and to fight back, those who stoically accepted their condition, and those who were hopeless and helpless. Four years after, the mortality rate was 10% for the women of the first two categories and 38% for the women of the last two categories (Table 1). In other words, the risk factor associated with resignation and despair was about 4. How-

TABLE 1

Influence of Attitude toward the Disease on the Risk of Metastasis and Mortality in Women with Breast Cancer[a]

PSYCHOLOGICAL RESPONSE 3 MONTHS AFTER SURGERY	OUTCOME AT 5 YEARS: *N*	Recurrence free (%)	Death (%)
Denial or fighting spirit	20	75	10
Stoic acceptance or helpless/hopeless response	37	35	38

[a]From Greer *et al.* (1979).

ever, these results can hardly be generalized. In another epidemiological study on cancer patients with metastasis, Derogatis and colleagues (1979) found results opposite to those of Greer and colleagues. Mortality was less frequent in depressed subjects than in those who reacted to their condition more positively. David Spiegel reopened the debate by showing that patients with advanced breast cancer who participated in weekly group therapy and social support sessions had significantly greater survival times compared with patients who did not participate in such sessions (Spiegel *et al.,* 1989).

Such contradictory data are not necessarily surprising. There are many methodological problems with epidemiological studies of the influence of psychosocial factors on health, from the validity of the tools which are used by psychologists to assess personality and affective states to the power of the methods which can be used to determine the role of factors with relatively low risk values. The measurement of depressive mood provides a typical example of the first case. It is well-known that structured interviews such as the one used by Greer and colleagues do not necessarily provide the same results as self-answered questionnaires such as the one used by Derogatis and colleagues. Depressed patients do not easily recognize themselves as such when responding to a written questionnaire. In the second case, and as noted by epidemiologists, current methods of epidemiology are already at the limit of their sensitivity for revealing possible associations between nutritional factors and health, which typically yield risk factors of the same or even higher order of magnitude as those of psychosocial factors.

If Emotions Can Impact on Heath, It Must Be through the Changes in Physiology That They Induce

The difficulties that are encountered when trying to overcome subjectivity and assess in a quantitative way the influence of psychosocial factors on health are real and cannot be easily circumvented without drastic changes in strategy. As recognized by the science philosopher Georges Canguilhem in the 1960s, any elaboration on the influence of psychic factors on physiology is doomed to be of little help if it cannot be eventually backed up by a communication pathway which connects the brain to the organ under consideration. From a philosophical perspective, this is a manifesto for the substitution of materialism in place of mentalism.

In the case of psychosomatics, materialism finds its roots in the theory of stress which was elaborated by Hans Selye in the early twentieth century (Fig. 2). This was the first scientific attempt to translate psychogenic interpretations of illness in the language of biology and it was very much inspired by the then prevailing concept of homeostasis. Like

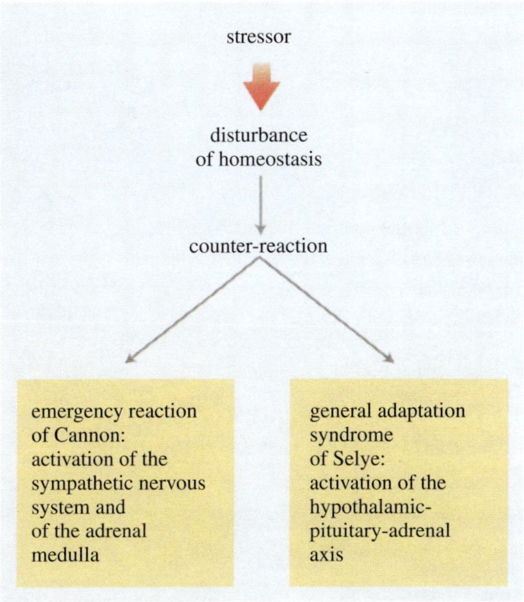

FIGURE 2 Relationship between stress and homeostasis. The stress response is the counterreaction to any disturbance of homeostasis and aims at reestablishing homeostasis. This type of response involves catecholamines that are released by the sympathetic nervous system and the adrenal medulla — the emergency reaction originally described by Walter Cannon — and glucocorticoids that are released by the adrenal cortex and play a pivotal role in the general adaptation syndrome described by Hans Selye.

other homeotherm animal species, human beings have physiological processes which enable them to keep their internal milieu constant despite irregularities in the external milieu. For example, blood glucose levels remain stable (approximately 1 g/liter) despite relatively large fluctuations in food intake. This is possible only because we are armed to deal with excess and insufficiency. If there is too large a deviation from homeostasis, our organism goes through a phase of alarm during which it mobilizes its resources to deal with the stressor. This phase is replaced by a phase of resistance when the stressor is overlasting. Exhaustion takes place when all the energy of adaptation has dissipated and it is followed by death (Fig. 3).

The main merit of the stress theory was not to replace the literary constructs of psychosomatics by the intuitive metaphors of equilibrium, tension, and energy but to propose a chain of events linking the brain to the peripheral organs and accounting for what Selye called adaptation diseases. This chain of events is represented by the sympathetic nervous system and the hypothalamic–pituitary–adrenal (HPA) axis. Activation of the sympathetic nervous system in response to a stressor takes place within a few seconds and results in the release of noradrenaline from nerve terminals and adrenaline from the adrenal medulla. These substances complement each other and have major effects

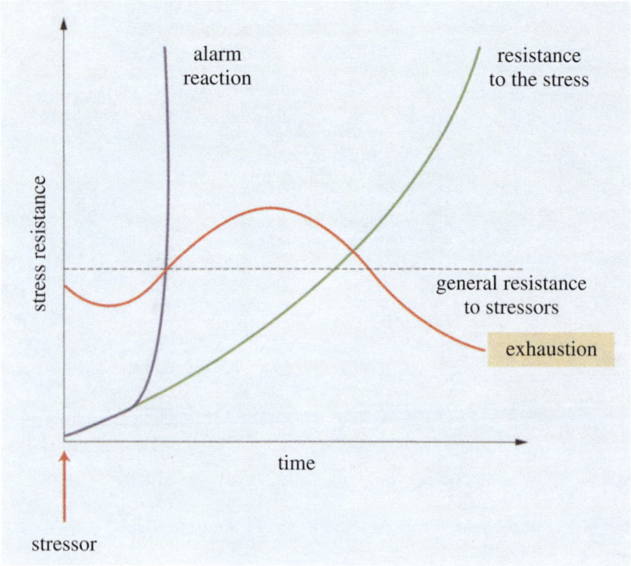

FIGURE 3 Selye's general adaptation syndrome. The figure illustrates how resistance to stress varies as a function of time. The red curve indicates the variations in the resistance state of a subject exposed to a stress agent. At the onset of the stress, there is an initial alarm reaction (blue curve) followed by a stage of resistance (green curve) during which resistance to the original stressor is increased but resistance to other forms of stressor (red curve) is lowered. If the stress continues long enough, the stage of resistance is replaced by a final stage of exhaustion which leads to death. Note that this final stage is reached much earlier if the initial resistance to stress is lower than normal. Although this representation is useful for educational purposes, it is unsupported by data about the time course of hormonal responses to stress. The dashed line indicates the normal level of stress resistance.

on cardiovascular function and metabolism. Their function is to redistribute blood to the brain and muscles and to provide to these organs more fuel in the form of glucose extracted from hepatic stores. Activation of the HPA axis takes longer since it is mediated by blood-borne mediators. Hypothalamic neurons release corticotropin-releasing hormone (CRH) in the blood of a specialized vascular system, the portal stalk. CRH is transported to the anterior pituitary, where it induces the synthesis and release of adrenocorticotropin hormone (ACTH). ACTH acts on the adrenal cortex to stimulate the synthesis and release of glucocorticoids (Fig. 4). The main glucocorticoid in man is cortisol. This hormone amplifies the effects of adrenaline and noradrenaline on their target organs and favors the production of glucose from lipids and proteins.

The stress response normally helps the organism to restore homeostasis. However, it can have deleterious effects if prolonged or excessive. This is particularly the case for glucocorticoids which have immunosuppressive properties. According to the theory of stress, inappropriate emotional states are accompanied by abnormal hormonal and autonomic responses, which in turn impact on physiology and metabolism and favor the induction of diseases. These diseases were called diseases of adaptation by Hans Selye, which means that they represent failure of the adaptive process which is normally triggered by exposure to stressors. The particular bodily organs which are affected during the general adaptation syndrome are those which are the weakest, either constitutively or because of previous pathological problems.

Stress-Induced Changes in Physiology Are Associated with Alterations in Immune Functions

To test the validity of the implications of the theory of stress, many experiments were carried out in the 1950s with the objective of assessing the influence of laboratory stressors on the course of immune-mediated pathologies or on immune functions. All these experiments confirmed that immune functions are very sensitive to the effect of stressors (Dantzer and Mormède, 1995). For instance, exposure to high-intensity noise increased mortality in virus-infected mice and decreased production of interferon, an antiviral agent manufactured by leukocytes. In the same manner, physical restraint decreased resistance of mice to herpes-like viruses and led to a reduction in endotoxin- and virus-induced synthesis of interferon. However, even at this early stage of investigation, it was apparent that the outcome of the stress response, in terms of progression of the disease process, was not always unfavorable and depended on the type of pathogen under investigation. For instance, the same physical restraint that decreased resistance to viruses in laboratory rodents attenuated somatic symptoms in a model of multiple sclerosis known as experimental allergic encephalomyelitis. Chickens housed in an environment with a high level of social interactions were more susceptible to viral infections but more resistant to bacterial and parasitic infections.

Stressors Influence Immune Functions Because the Brain Is Closely Connected to the Immune System

The main problem with all these studies was that the limited knowledge which was then available on mechanisms of immunity did not allow to go beyond phenomenology. The tendency was to consider the observed effects as an artifact due to the immunosuppressive effects of glucocorticoids. This field of research therefore drew little attention from other disciplines, including immunology, and quickly died

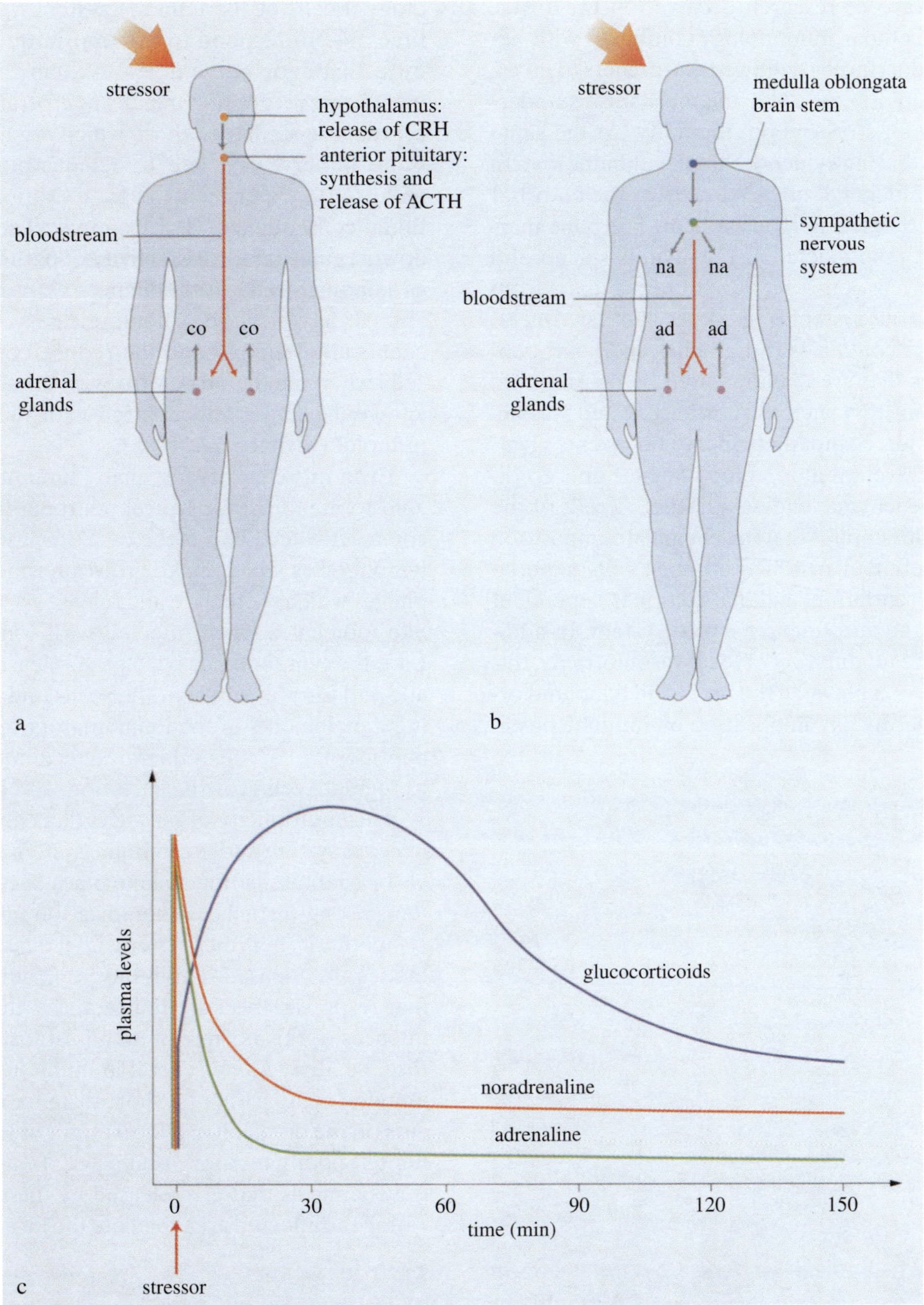

FIGURE 4 Neuroendocrine and autonomic mechanisms involved in the physiological response to stress. (a) Activation of the hypothalamic-pituitary-adrenal axis. The hypothalamic neurons induce release of corticotropin-releasing hormone (CRH) into the bloodstream. This hormone induces the synthesis and release of adrenocorticotropic hormone (ACTH) by the anterior pituitary. ACTH, lastly, acts on the cortex of the adrenal glands, stimulating the synthesis and release of glucocorticoids such as cortisol (co). (b) Activation of the sympathetic nervous system. In response to a stress event, activation of the sympathetic nervous system determines the release of noradrenalin (na) from nerve terminals and of adrenalin (ad) from the medulla of the adrenal glands. (c) The changes in plasma levels of adrenaline, noradrenaline, and glucocorticoids as a function of the time after the onset of a stressor (time 0). Note the short latency (a few seconds) of the catecholaminergic response (in green and in red) which depends on a "wired" neural system in contrast with the longer latency (up to 10–20 min) of the adrenocortical response which depends on an hormonal circuit

off. The second wave of research occurred in the 1980s, when progress in cellular immunology combined with advances in neuroendocrinology allowed researchers to investigate in a more sophisticated way the mechanisms underlying the influence of stressors on immunity. At the same time, there was increasing evidence that the immune system does not function independently of the rest of the body but is closely connected to the nervous system. The bone marrow and lymphoid organs, including the thymus, spleen, and lymph nodes, receive a dense network of nerve fibers from the autonomic nervous system (Fig. 5). As in other visceral organs, these fibers contain noradrenaline and neuropeptides (i.e., peptides that are manufactured by neurons and act as neurotransmitters) such as substance P and vasoactive intestinal peptide. Sympathetic nerve fibers enter lymphoid organs by traveling along blood vessels and go directly into the parenchyma, ending adjacent to cells of the immune system. This implies that the chemical composition of the microenvironment in which immune cells undergo proliferation, differentiation, and maturation is dependent on the activity of the autonomic nervous system. In addition to its role in the modulation of local vasomotricity, this anatomical connection between the brain and lymphoid organs would have hardly any implication on immune physiology if cells of the immune system were not at the same time able to respond to the transmitters that are released by sympathetic nerve fibers. Actually, immune cells carry membrane receptors for many neurotransmitters. These receptors are similar to those which are located in the membrane of nerve cells and, as demonstrated by many *in vivo* and *in vitro* experiments, their activation has marked functional consequences. For example, β-adrenergic receptors downregulate cytotoxic activity of natural killer (NK) cells, as demonstrated by the enhanced cytotoxicity of NK cells in animals in which the sympathetic nervous system has been chemically lesioned and the reduced cytotoxicity of these cells when incubated *in vitro* with noradrenaline, an effect antagonized by preincubation with the β-antagonist propranolol (Livnat *et al.,* 1985).

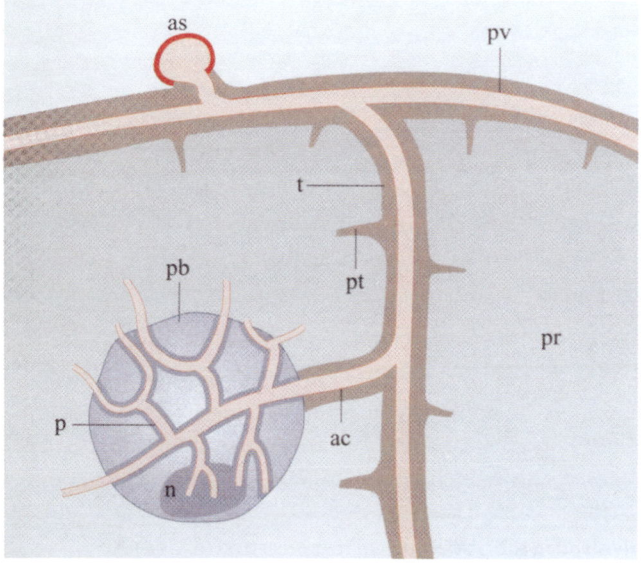

FIGURE 5 Schematic representation of the noradrenergic innervation of the spleen. Noradrenergic fibers enter the spleen around the splenic artery (sa), travel with the vasculature in plexuses (v), and continue into the spleen along the trabeculae (t) in trabecular plexuses (tp). Fibers from both vascular and trabecular plexuses enter the white pulp (wp), continuing mainly along the central artery (ca) and its branches. Noradrenergic varicosities radiate from these plexuses into the periarterial lymphatic sheath (p) but avoid the nodular areas (n). These parenchymal fibers end among fields of lymphocytes and other cell types. Rp, red pulp (adapted from Felten *et al.,* 1985).

Even more surprising, many immune cells are actually able to manufacture typical neuropeptides when they become activated. This is the case, for instance, for certain T lymphocytes which make growth hormone and for macrophages which synthesize and release ACTH and endorphin-like molecules when infected with some viruses (Fig. 6). Like the lymphokines which are released by immune cells and call into play many other cells and substances, including the elements of the inflammatory response, the neuropeptides that are manufactured by immune cells act locally to regulate cell growth, activation, and trafficking.

This multiplicity of biological links between the central nervous system and the immune system allows different levels of communication to take place between these two systems. Long-distance communication takes place when the sympathetic nervous system is activated and stress hormones are released. This is the case, for instance, during the fear response. In such circumstances, the functional consequences of this hormonal mobilization are likely to be diffuse, i.e., they affect several immune functions in different immune compartments. Short-distance communication occurs on the occasion of a local injury or trauma, when only a limited bodily territory is affected. In these circumstances, the chemicals that are released by the nerves fibers innervating the affected area regulate the local immune response.

Psychological Factors, Including Social Condition, Influence Immune Functions

Since different hormones have different effects on immune functions, the immune consequences of the physiological activation that occurs in response to environmental factors should depend on the patterning of the hormonal reaction. However, this patterning does not necessarily reflect the type of emotion that is experienced by the subject. Based on changes in circulating cortisol levels and activity of the autonomic nervous system, and in contrast with intuitive thinking, it has amply been demonstrated that sadness is

residues	1									10									
pituitary	NH_2-Ser	Tyr	Ser	Met	Glu	His	Phe	Arg	Trp	Gly	Lys	Pro	Val	Gly	Lys	Lys	Arg	Arg	Pro
	5'-TCC	TAC	TCC	ATG	GAG	CAC	TTC	CGC	TGG	GGC	AAG	CCG	GTG	GGC	AAG	AAA	CGG	CGC	CCG
lymphocyte	5'-TCC	TAC	TCC	ATG	GAG	CAC	TTC	CGC	TGG	GGC	AAG	CCG	GTG	GGC	AAG	AAA	CGG	CGC	CCG

	20										30									
pituitary	Val	Lys	Val	Tyr	Pro	Asn	Val	Ala	Glu	Asn	Glu	Ser	Ala	Glu	Ala	Phe	Pro	Leu	Glu	Phe-COOH
(continued)	GTG	TAC	GTG	TAC	CCC	AAC	GTT	GCT	GAG	AAC	GAG	TCG	GCG	GAG	GCC	TTT	CCC	CTA	GAG	TTC-3'
lymphocyte (continued)	GTG	TAC	GTG	TAC	CCC	AAC	GTT	GCT	GAG	AAC	GAG	TCG	GCG	GAG	GCC	TTT	CCC	CTA	GAG	TTC-3'

FIGURE 6 Comparison of the nucleotide sequences (with the corresponding amino acid coded in black) of the region coding for ACTH in the POMC gene. Lymphocytes from mice infected with Newcastle virus synthesize an ACTH that is indistinguishable from pituitary ACTH (adapted from Smith *et al.*, 1990).

not different from joy and elation. Appraisal of the situation and possibilities of action are actually more important than effects for determining the pattern of hormonal response that is elicited by the emotional stimulus (Dantzer, 1989). This means that the impact of stressors on the immune system should be critically dependent on the ability of the individual to cope in an appropriate way with the situation (Fig. 7). This was found to be the case since in mice exposed to a single session of inescapable electric shocks 24 h following tumor cell transplantation, the rate of tumor growth was accelerated, whereas mice exposed to escapable electric shock did not differ from nonshocked mice in time of tumor appearance and tumor size (Sklar and Anisman, 1979). In rats, a single session of inescapable but not escapable shock significantly reduced the incidence of rejection of transplanted tumor cells (Visintainer *et al.*, 1982). The detrimental effects of uncontrollability are likely to be mediated by changes in immune responses since inescapable but not escapable electric shock decreases cellular immune responses (Fig. 8). Rats submitted to inescapable electric shock on the first day and reexposed to a few "reminder" shocks on the second day displayed a lower proliferative response of blood lymphocytes to T cell mitogens than rats submitted to escapable electric shocks or apparatus control animals (Laudenslager *et al.*, 1983). The ability to predict the occurrence of a stressor is also important since rats submitted to unpredictable electric shock had lower proliferative responses of splenocytes to T cell mitogens than rats submitted to predictable shock or nonshocked animals (Mormède *et al.*, 1988).

Like many other animal species, human beings are gregarious and form enduring social bonds. These social variables have profound influences on physiology. Familiar and positive social relationships exert buffering effects on the stress response, whereas the formation of new relations and the disruption of established ones elicit considerable physiological activation. During the past 15 years, there has been a growing body of primate research demonstrating these two facets of the effects of psychosocial factors on immune responses (Fig. 9). Lymphocyte proliferative responses were reduced in young pigtail macaque monkeys following separation from their social companions (Reite *et al.*, 1981). The same effect was observed in both mother and infants following separation of mother–infant pairs of bonnet macaques (Laudenslager *et al.*, 1982). In squirrel monkeys injected with a benign virus to allow measurement of the primary antibody response, neutralizing antibody levels were reduced 10-fold in separated infants compared with control infants who remained with their mothers. However, this effect occurred only in infants which were separated from their mothers by being moved to a novel cage and not in infants that were permitted to remain in the home cage in which they had been reared. In the same manner, providing infants with familiar companions during the separation prevented the reduction in antibody responses (Coe, 1993). These observations illustrate the concept that the immune consequences of a stressful event are very much dependent on the cognitive and affective processes that are involved in the appraisal of this event. Individual differences are important in this respect. For instance, when lymphocyte proliferation responses of separated pigtail macaque and bonnet infant macaques were studied for each individual following separation, it appeared that only some infants displayed the decreased reactivity to mitogens (Laudenslager *et al.*, 1990). Immune changes occurred primarily in those infants that were the most reactive to the separation process, as assessed by agitation and distress vocalizations on the first day and slouched posture on subsequent days.

Psychological Factors Also Influence Immune Functions in Humans

The findings of altered immune responses in emotionally distressed experimental animals prompted a search for the possible occurrence of similar changes in human be-

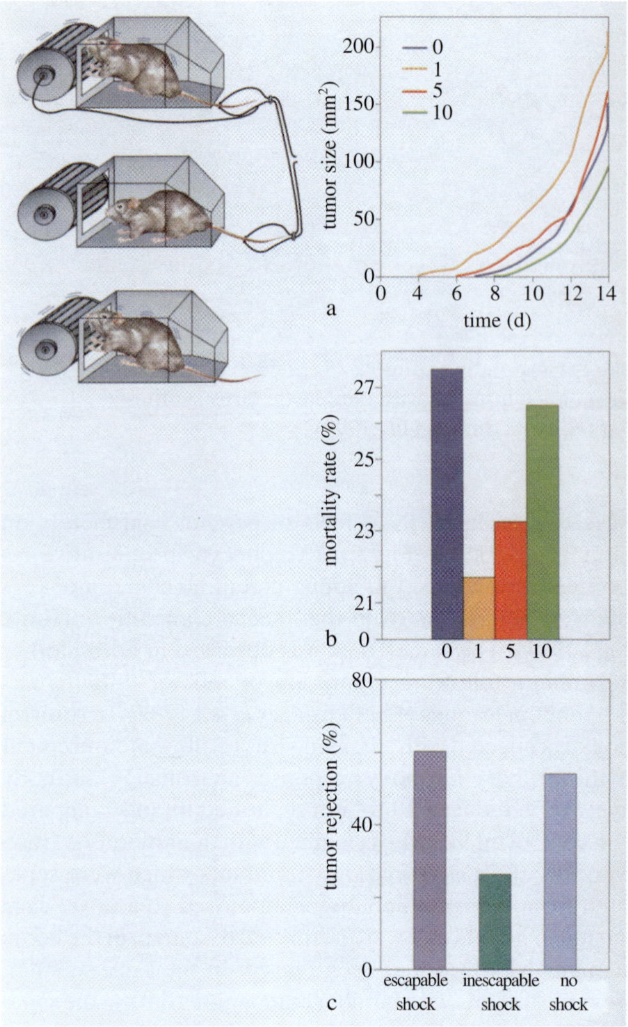

FIGURE 7 Differential effects of escapable versus inescapable electric shock on tumor growth (a) and tumor rejection (b). Rats or mice given the ability to terminate electric shocks by turning the wheel (left, top) are compared to animals which receive the same amount of shock but without any ability to influence them (left, center) and to nonshocked animals which are put into the apparatus only for the duration of the experiment (left, bottom). In the experiment illustrated in a and b, mice were exposed to 0, 1, 5, or 10 daily sessions of inescapable electric shock after tumor cell transplantation. Note that the size of the tumor (a) and mortality (b) were increased by one session of inescapable shock, but that there was a habituation of this effect upon repeated exposure to shock. In the experiment shown in c, rats were submitted to one session of escapable or inescapable electric shocks 24 h after implantation of the tumor, and tumor rejection was measured subsequently. Note that rats exposed to inescapable shock were deficient in their response to the tumor [reproduced with permission from Sklar and Anisman (1979) and Visintainer *et al.* (1982)].

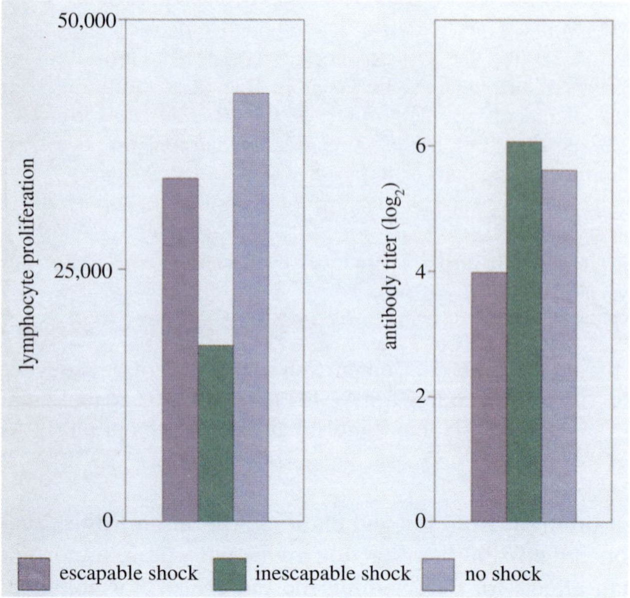

FIGURE 8 Influences of the possibility of control over electric shocks on cellular and humoral immune responses in rats. Rats placed in a shuttle box in which they could avoid impending electric foot shocks by shuttling from one compartment to the other were compared to animals placed in a second shuttle box coupled electrically to the first one so that their own behavior had no influence on the occurrence of shocks. A third control group of subjects were placed in a shuttle box in which no shock was delivered. Cellular immunity was measured after 10 daily sessions as lymphocyte proliferation in response to a mitogen (left; indicated by the amount of radioactivity incorporated into cultured splenocytes), whereas humoral immunity (right) was estimated by the titer in antibodies against sheep red blood cells injected on the fifth day of the experiment (adapted from Dantzer *et al.,* 1988).

ings. These studies are by necessity carried out on circulating populations of leukocytes, of which the composition and functions do not accurately reflect the ability of the organism's immune system to protect against pathogenic processes.

Short-term effects of emotions on immune functions can be observed in appropriate experimental circumstances. For example, Naliboff and colleagues (1991) reported that a 12-min mental arithmetic task induced rapid immune changes, in the form of a release of CD8 suppressor/cytotoxic T cells and NK cells in the circulation, and, in young subjects only, an increase in NK cell activity. The possibility of a differential effect of a controllable versus uncontrollable acute stress was investigated by Weisse and colleagues (1990). The stressor consisted of mild electric shock and loud white noise administered during a 30-min period in an unpredictable and intermittent manner. However, in contrast to the authors' prediction based on animal studies,

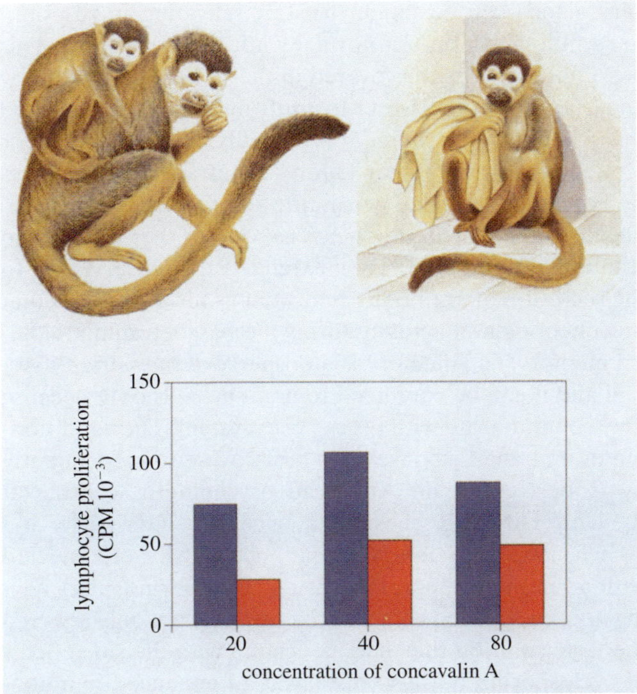

FIGURE 9 Squirrel monkeys, like other primates, are very social. When separated from their mother and put in groups with companions of the same age, they form strong attachments. The figure represents the lymphocyte proliferative response to different concentrations of the T lymphocyte mitogen concanavalin A before (in blue) and 2 days after separation (in red) from a peer companion (adapted from Friedman *et al.*, 1991).

subjects who could control the stressor displayed a lowered lymphocyte proliferation response to mitogens, whereas subjects who could not control the stressor were not affected despite increased anger and frustration.

A few studies have searched for a possible association between temperament traits and immune responsiveness. In view of the proposed importance of emotional repression in the pathogenesis of cancer, Esterling and colleagues (1990) asked 80 first-year undergraduates to complete a personality questionnaire and write an essay about a stressful event that had happened in their lives. The degree of emotional disclosure was scored according to the ratio of emotional-to-total words used in the writing. Antibody titers against Epstein–Barr virus were measured in a blood sample collected immediately after writing. High levels of antibody titers were observed in repressors independently of their degree of emotional disclosure but also in sensitizers who were low disclosers, which suggests that individuals who are aware of their heightened emotional response to environmental events but choose not to express it are more at risk than individuals who are not inhibited.

In clinically oriented studies, the emphasis has been mainly on negative life events and negative affective states. In the first case, Janice Kiecolt-Glaser and colleagues repeatedly observed altered immunocompetence — in the form of reduced NK cell activity, decreased lymphoproliferative responses, and reduced cellular immune control over latent herpesviruses resulting in increased antibody titers to Epstein–Barr virus — in adults experiencing difficulties in their marital life, in spouses of patients afflicted with Alzheimer's disease, and in medical students at the time of academic examination (Table 2).

In the second case, the possible link between depression and susceptibility to disease has been examined using immune responsiveness as a mediating factor. Depression can occur in many different contexts, including the context of loss. Bereavement is associated with altered immunocompetence, although the observed changes and the exact relationship between depressive symptomatology and immunity are not always clear. Schleifer and colleagues (1983) reported decreased lymphocyte proliferation in spouses of women who were terminally ill with breast cancer. Compared with immune responses tested before the death, lymphocyte responses were reduced 1 month after the death and tended to return toward baseline during the 4- to –14-month follow-up period. The relationship between depressed mood and immune responses was not tested in this study, which included only 15 subjects. In contrast, Irwin and colleagues (1987) reported that women who had been widowed for less than 6 months, women whose husbands were diagnosed with terminal lung cancer, and women whose husbands were healthy did not differ in their NK cell activity. However, NK cell activity was inversely related to symptoms of depression.

Findings of altered immune responsiveness in individuals diagnosed with depressive disorders are believed to

TABLE 2

The Influence of Life Events on Immune Functions[a]

	LEUKOCYTIC INTERFERON (UNITS/ML)	NATURAL KILLER CELL (% CIRCULATING CELLS)	CYTOTOXICITY (%)
Baseline	2000	16.3	45
Exam	80	9.1	30

[a]Forty second-year medical students were tested 6 weeks before end-of-year exams (baseline) and on the first day of exams. Note the extreme reduction in interferon production by activated blood white cells and the decrease in the number and cytotoxic activity of natural killer cells (from Glaser *et al.*, 1986).

be relatively common. Lower lymphocyte responsivity and sometimes decreased NK cell activity have indeed been reported in many studies of patients who were suffering from depressive illness (Weisse, 1992). However, a closer examination of the abundant literature on depression and immunocompetence reveals that reduced lymphocyte responsivity is mainly observed among the clinically depressed, particularly older patients diagnosed with a unipolar disorder severe enough to warrant hospitalization, whereas other measures of immunity do not provide any consistent picture. In addition, it is doubtful that dysphoric mood per se is responsible for the immune alterations observed in these studies.

Whether psychological distress can lead to adverse immunological changes providing a physiological pathway through which psychosocial factors might influence the course of disease is even more questionable. This possibility has been examined in HIV-infected patients since they are exposed to many psychosocial stressors, the impact of which on their immune status could account for the large interindividual variations they present in the duration of the asymptomatic phase of HIV infection and the course of the disease. In a study of 81 homosexual men, Schneiderman *et al.* (1994) observed that the anticipation and impact of being informed of their antibody status decreased lymphoproliferative response to mitogens and NK cell cytotoxicity in seronegative subjects. As expected, this decrease was followed by a gradual return to normal values following notification of the negative serostatus, which was associated with a reduction in distress and plasma cortisol levels. A reduced lymphoproliferative response to mitogens was also present in seropositive subjects but it was not further exaggerated by notification of HIV serostatus, in contrast to NK cell cytotoxicity. These immune changes were associated with increased anxiety and paradoxically decreased plasma cortisol levels. During the follow-up period, psychosocial factors predicted both immune changes and disease progression in HIV positive subjects. More precisely, high distress on diagnosis, denial increase surrounding this event, and low participation in behavioral intervention protocols were each strongly and positively related to disease progression. Although these findings indicate the existence of intriguing relationships between psychosocial factors, immune status, and HIV progression, there is obviously a need for assessing their generality.

Can the Immune System Be Submitted to Learning Influences?

There has been much excitement regarding the possibility of teaching the immune system to produce certain responses through behavioral conditioning, in the same way that salivary glands can be taught to start salivating in response to stimuli which do not normally affect salivary flow. This possibility was first discovered in the 1920s by Metalnikov and Chorine at the Pasteur Institute in Paris. They reported that when the experimenter repeatedly blows a trumpet or scratches the rabbit's ear before an intraperitoneal injection of heat attenuated germs, the afflux of leukocytes produced by the injection appears as soon as the conditioned stimulus is presented. They extended this observation to the secretion of antibodies and went as far as to claim that this phenomenon actually forms the basis of immunological memory, i.e., memory is a property of the nervous system and it can be conferred to immune cells only because they communicate with the nervous system. The conditioning of immune responses was rediscovered independently by Robert Ader, an American psychologist working at the University of Rochester, half a century later (Ader and Cohen, 1985). He observed that mice which were injected with an immunosuppressive drug, cyclosporine, after having drank a solution with a distinctive taste not only rejected the taste solution they had associated with the side effects of cyclosporine but also displayed an enhanced immunosuppression compared to mice which were not exposed to the taste solution. This evokes the anecdote of the woman who is sensitive to the smell of roses and starts sneezing when she sees a paper rose. Actually, allergic responses can be conditioned in experimental animals by pairing injection of an allergen with a nonimmune stimulus such as a distinctive smell or a blinking light. Although there is much experimental evidence in laboratory animals and even in human beings demonstrating the occurrence of alterations in immune response by behavioral conditioning, the exact nature of these alterations and particularly their specificity in comparison to the alterations in immune responses induced by stressors are unclear (Dantzer, 1989).

Multiple Mechanisms Account for the Influences of Psychological Factors on Immune Functions

In view of the multiplicity of the biological links between the central nervous system and the immune system, the exact mechanisms which mediate the influence of emotional states on immunity are not easy to determine. The possible role of increased levels of glucocorticoids is often the first to be examined. A clear example of the involvement of glucocorticoids in stress-induced changes in immune functions derives from a study carried out in aged rats (Sapolsky and Donnelly, 1985). In response to a stressor, aged rats secrete higher amount of corticosterone than do young rats and they do so because of a centrally mediated deficiency in the negative feedback regulation of the pituitary–adrenal axis.

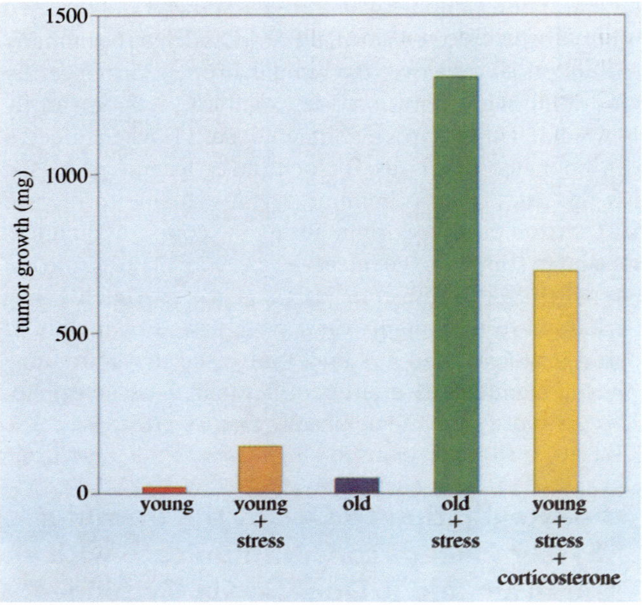

FIGURE 10 The role of glucocorticoids in the influence of age on the sensitivity to stress or tumor growth in rats. Tumor growth, measured by the weight of the tumor 15 days after inoculation, was significantly increased by stress in older but not in younger rats. When younger rats received injections of corticosteroid following stress, to mimic the increased adrenocortical response to stress of older rats, the effects of stress on tumor growth were increased (adapted from Sapolsky and Donnelly, 1985).

Aged rats that are inoculated with a syngeneic tumor are more sensitive to the tumor-enhancing effects of repeated stress, and this higher vulnerability can be replicated in young rats by enhancing their pituitary–adrenal response to mimic the response of old rats (Fig. 10).

There are many examples, however, of glucocorticoid-independent mechanisms in the alterations of immune responses that are caused by stressors, and it is clear that the negative influences of these hormones on lymphocyte functions have been overestimated. Many studies have shown that immune responses are normal or even enhanced under conditions of chronic elevations of circulating glucocorticoids and that the immunosuppressive effects of these hormones occur only when pharmacological doses of these compounds are used (Dantzer and Mormède, 1995). Actually, the major target cells for glucocorticoids are not lymphocytes but macrophages and monocytes, of which the ability to manufacture and release proinflammatory cytokines is tightly dependent on circulating levels of glucocorticoids.

Catecholamines, endogenous opioids, and certain pituitary hormones such as growth hormone and prolactin are likely candidates for mediating glucocorticoid-independent effects of stressors on immune responses. For example, in a correlative study of the hormonal and immune responses of 15 healthy subjects submitted to short-term physical and psychological stress, quantitative and qualitative changes in blood lymphocytes, monocytes, and granulocytes were negatively related to plasma adrenaline concentrations (Landmann *et al.*, 1984). Similar alterations in the composition of the circulating lymphocyte pool were caused by adrenaline injections and could account for significant reduction in lymphocyte proliferative response to mitogens (Eriksson and Hedfors, 1977; Crary *et al.*, 1983). The importance of catecholamines in mediating the effects of stress on immunity is not limited to the blood compartment. In rats exposed to intermittent electric shock, section of the splenic nerve abrogated the shock-induced decrease in antibody formation and splenocyte proliferative response to mitogens, which suggests that the splenic nerve transmits immunosuppressive signals to splenocytes (Wan *et al.*, 1993).

The Immune System Is Sensitive to Psychological Factors Because Its Functioning Is Regulated by the Nervous System

The increased sophistication in the methods used to study the impact of stress and emotional factors on immunity and the mechanisms that are involved in these effects leaves intact the question of the physiological significance of the observed changes. The mounting of an effective immune response requires a close cooperation between B lymphocytes, which are responsible for elaboration and secretion of antibodies, and T lymphocytes, which are responsible for cell-mediated immunity. The T lymphocytes which are involved in this cooperation are known as T helper cells. They exist in two subsets which differ by the type of lymphokines or cytokines they produce (Abbas *et al.*, 1996). Th1 lymphocytes produce mainly interleukin-2 (IL-2) and γ-interferon, which spur additional cell growth and macrophage activation. Th2 lymphocytes produce mainly IL-4 and IL-10. These cytokines deactivate macrophages and set B cells to multiplying and differentiating in plasma cells, which secrete antibodies. *In vivo*, many immune responses start predominantly as Th1 and then shift to Th2. Stress hormones have been proposed to be part of the mechanisms which drive the switch from a Th1 to a Th2 response (Rook *et al.*, 1994).

In terms of regulation, the possibility for the functioning of a visceral organ to be influenced by the central nervous system is an important adaptive advantage since it provides the organism with the possibility to adjust the level of functioning of the relevant organ according to the organism's actions, in real time and in anticipation of what can happen.

Removed from the organism and put in an organ culture system, the heart functions as a splendid self-regulatory machine since it is able to adjust its rate and strength of contraction to the amount of liquid it has to pump. However, when it is in the organism, it functions mainly under the influence of the sympathetic and parasympathetic nerves, which connect it to the brain, to adjust its work to what the organism is doing or is likely to do in the near future. The main problem with our views on the functioning and regulation of the immune system is that they have been shaped by studies which have considered immune cells in isolation, removed from the body and put into culture. It is certainly time to reintroduce the immune system in the body not just to study the influences of stress on its functioning but also to consider the way it is regulated *in vivo.* If the immune system is regulated by the brain, there must be a two-way flow of information between these two systems, i.e., the immune system must be able to inform the brain of its state of activation (Fig. 11). Since immune cells can manufacture the same neuropeptides which are synthesized and released by nerve cells, some authors have proposed that the immune neuropeptides take over the brain neuropeptides in conditions of immune activation. For example, it was previously shown that lymphocytes manufacture ACTH when infected with a noncytotoxic virus. Blalock and colleagues proposed that this ACTH from immune origin was able to replace ACTH from pituitary origin during the course of an infection and to stimulate the release of glucocorticoids from the adrenal cortex (Smith *et al.,* 1982). The experiments they carried out to prove their views were flawed by methodological problems, and it is now recognized that if immune cells can manufacture neuropeptides then these neuropeptides are more suited to act locally than at a distance.

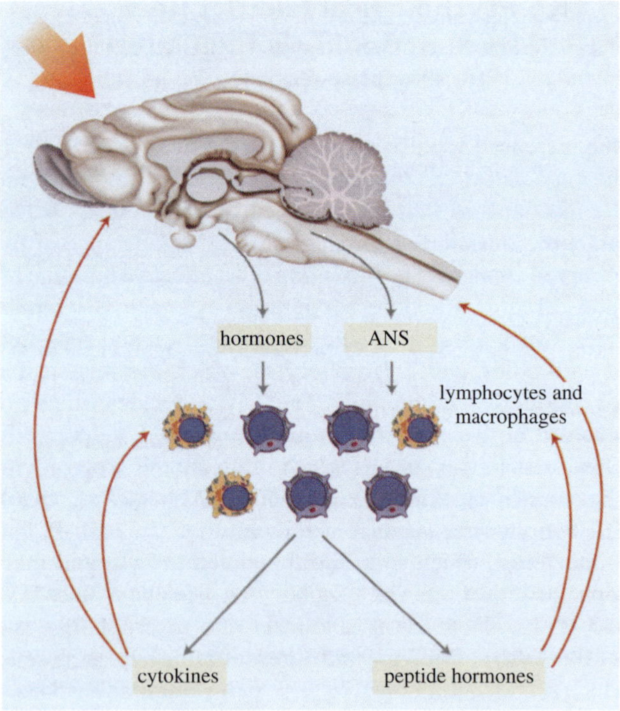

FIGURE 11 Communication pathways between the immune system and the nervous system. The influences of stress on the immune system are mediated by the autonomic nervous system (ANS) and the hormonal pathways that normally regulate functioning of the immune system, in response to communication signals from the immune system. These communication signals were initially thought to be neuropeptides manufactured by immune cells but they turned out to be cytokines, that is, substances synthesized and released by activated immune cells to regulate the mounting of the inflammation response (adapted from Dantzer and Kelley, 1989).

For Regulation to Occur, the Immune System Needs to Communicate with the Brain and It Does So via Cytokines

As already mentioned, immune cells communicate with each other primarily by secreting substances known as cytokines. These substances are diverse and potent chemical messengers. Binding to specific receptors on target cells, they are responsible for cell proliferation, activation, and differentiation. Proinflammatory cytokines which are manufactured by monocytes and macrophages play a pivotal role in the development of inflammation and the mounting of the immune response. The main proinflammatory cytokines are IL-1, tumor necrosis factor, and IL-6. These substances are the key communication signals from the immune system to the brain. By their actions in the brain, they are responsible for induction of fever and activation of the HPA axis.

Fever corresponds to a regulated increase in the set point for thermoregulation (Fig. 12) (Hart, 1988; Kluger, 1991). A feverish organism feels cold at normal environmental temperature. By increasing heat production and decreasing thermal losses, body temperature rises from 36.5 to 39 or 40°C and remains elevated for several hours. This increase in body temperature is important for fighting pathogens since it is accompanied by a cessation of cell proliferation except for immune cells, which grow and function better at higher than normal body temperatures. In addition, many pathogenic microorganisms stop proliferating at temperatures higher than 36°C.

Activation of the HPA axis is also an important component of the host systemic response to infection since the released glucocorticoids feed back on activated macrophages and monocytes to downregulate the synthesis of proinflammatory cytokines (Besedowsky *et al.,* 1986). As demonstrated by animal studies, blockade of this feedback action of glucocorticoids results in an excessive response of the host to infection, which can have deleterious consequences

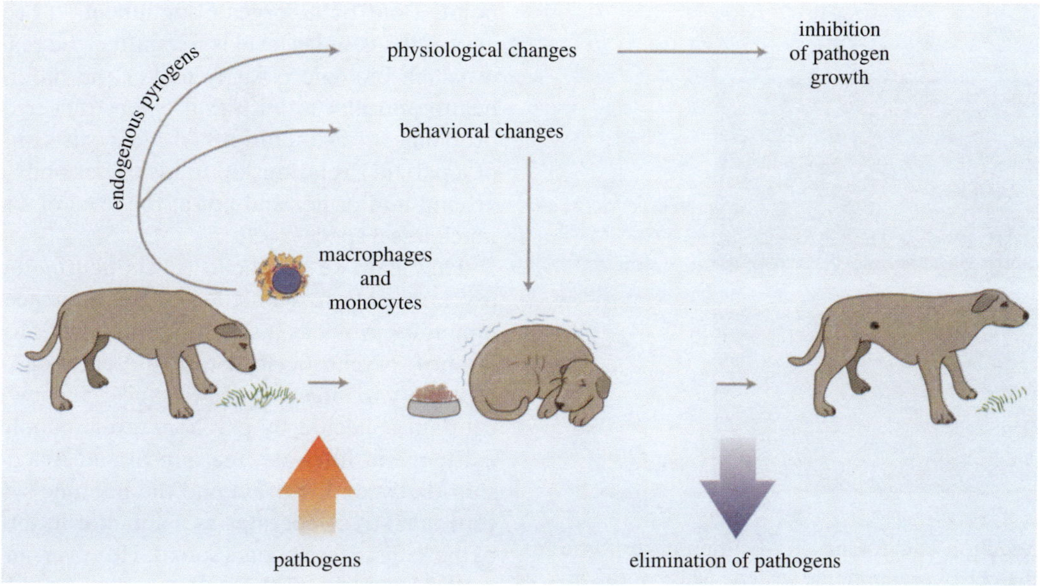

FIGURE 12 Illustration of the course of an acute infectious disease from the release of endogenous pyrogens or cytokines by activated macrophages and monocytes in response to exposure to pathogens. The course of the infection can be followed through the behavioral changes typical of sick animals to recovery with elimination of the pathogens. The behavioral changes of the sick animals include fever and the associated behaviors that potentiate it (adapted from Hart, 1988).

in terms of inflammation or sepsis. Lewis rats are very susceptible to inflammatory stimuli and, when challenged with an appropriate antigen, develop an inflammatory disease which affects limb joints and resembles rheumatoid arthritis. Fischer rats are similar to Lewis rats in terms of immunocompatibility, but they do not develop any inflammatory disease when exposed to the same antigen as Lewis rats. The reason for this difference is that proinflammatory cytokines are less effective in activating the HPA axis in Lewis than in Fischer rats so that the glucocorticoid feedback no longer works (Sternberg *et al.*, 1989). Correction of this deficit in Lewis rats by injecting glucocorticoids at the time of exposure to the inflammatory stimulus makes them similar to Fischer rats in terms of their inflammatory response. Conversely, blockade of the pituitary–adrenal response of Fischer rats to inflammatory stimuli makes them behave like Lewis rats.

Proinflammatory cytokines also have profound effects on behavior and mood since they are responsible for the malaise, fatigue, lethargy, decreased appetite, and disinterest for usual activities which characterize sick individuals (Kent *et al.*, 1992). The same symptoms can be induced in healthy individuals by systemic or central administration of recombinant cytokines, whereas administration of cytokine antagonists blocks the development of these symptoms in sick individuals.

How do proinflammatory cytokines act on the brain? These molecules cannot cross the blood–brain barrier and they actually do not need to do so since they are produced in the brain by resident macrophages known as microglial cells (Layé *et al.*, 1994) and act on brain cytokine receptors (Parnet *et al.*, 1994). Proinflammatory cytokines activate afferent nerve endings at the site of their release. This allows the transmission of information from the peripheral site of inflammation to the brain and results in the induction of expression of proinflammatory cytokines in the brain (Fig. 13) (Bluthé *et al.*, 1994; Layé *et al.*, 1995). The exact mechanisms leading from the expression of cytokines in the central nervous system to behavioral changes are not known. Binding sites for cytokines are present on glial cells and neurons, and mediators of the brain effects of proinflammatory cytokines include prostaglandins and nitric oxyde.

Biological Accounts of How Psychological Factors Influence Immune Functions Are Not Sufficient to Understand the Relationship between Stress, Emotion, and Disease

We have come full circle by showing not only that behavior and mental states affect immunity but also that immune activation has profound influences on behavior. This is so because the nervous system and the immune system are closely interlocked systems which can speak to each other

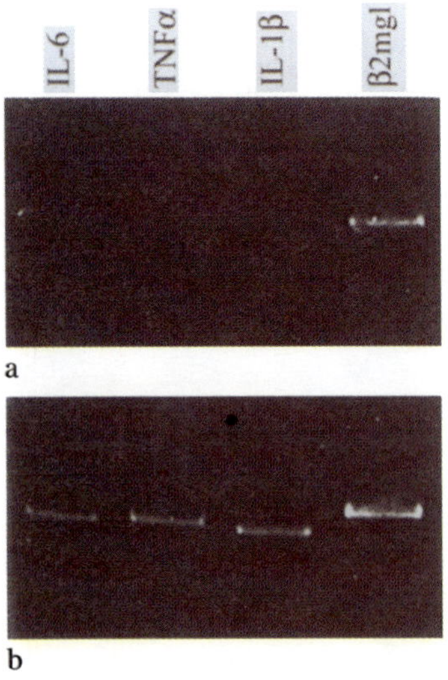

FIGURE 13 Induction of cytokines in the brain in response to signals from the peripheral immune system. Mice were sacrificed 2 h after an intraperitoneal injection of saline (a) or lipopolysaccharide (b), and total RNA was extracted from the hippocampus. The figure shows amplification products of the regions coding for IL-6, TNF-α, IL-1β, and β_2-microgobulin (β_2m). Note the increase in cytokine mRNA in response to lipopolysaccharide (adapted from Layé *et al.*, 1994).

primarily through the language of hormones. Immune cells can be seen as part of a diffuse sensory system which has the unique capacity of detecting the arrival of foreign invaders and to transmit this information to the brain. Conversely, the central nervous system has the dual ability to send signals via its sensory pathway to incite immune cells to invade the injured site of the body and to calm down activated immune cells via its neuroendocrine pathways.

Despite its achievements, psychoneuroimmunology has an ambiguous status both in the scientific community and in the general public since no effort has been made to clarify its theoretical positions. Psychoneuroimmunology willingly presents itself as an integrative scientific discipline which makes use of the concepts and methods of psychology, neurosciences, and immunology to better understand how the mind and the body interact to facilitate or impair good health. For example, this is the position which was adopted by psychoimmunologists when they were asked to present the latest advances in their discipline to the general public on the occasion of the Conference on Stress, Immunity, and Health which was organized by the American National Institute of Health in 1994. It is therefore not surprising that many of the discoveries made by this discipline form the scientific credentials of various holistic therapies aimed at reconciling the mind and the body.

This attitude is naive at best since it ignores the risk of confusing illness with cellular or organic dysfunction. As pointed out by Georges Canguilhem, "to seek out an illness at the cellular level is to confuse the realm of real life, in which biologic polarity makes the difference between health and illness, with that of abstract science, in which problems receive solutions." Illness exists only in the realm of a patient's relationship to his or her body and the world around him or her and not at the level of a specific organ, much less a specific cell.

How can we reconcile psychoneuroimmunology with psychosomatics? The study of the influences of stress on immunity provides a scientific framework for understanding how psychosocial factors and emotions can alter susceptibility to infection and to cancer. In view of the empirical data collected by psychoneuroimmunologists and the existence of intricate anatomical and functional relationships between the brain and the immune system, the concept that stress operates as a cofactor in the pathogenesis of these diseases appears sound. However, an exclusive biomedical reading of this relationship would be misleading. There are many routes by which stress can influence susceptibility to disease (Fig. 14) (Cohen and Williamson, 1994). Hygiene of life is often altered by stress: Stressed individuals sleep less well, smoke more, drink more alcohol, eat poorly, and are less prone to engage in health practices, including compliance with medical treatments.

Stress also influences how patients perceive and respond to their symptoms. Persons under stress may become more attentive to their internal physical state, more prone to attribute them to illness, and more likely to seek care. However, the reverse situation can also occur, with stress decreasing care seeking. These influences of stress on illness behavior are not minor since many studies of the relationship between stress and disease in humans rely on self-reported symptoms rather than on objective signs of disease. Last, but not least, the illness which afflicts the patient plays a very special role in his or her life, which is dependent on the existential meaning he or she gives to it.

The challenge facing us is not to go deeper into the molecular details of the connexions between the brain and the immune system, even if this knowledge is useful. It is to dare to confront the complexity of the relationship between stress, emotion, and disease and to seek comprehension of the different levels involved in this relationship, Friedman (1988):

> The overlap of the psychological and the spiritual, of physiology and psychology, their reconciliation in an acceptance of their respective specificities and interaction is beginning to promote a concept that will prove rich in promise. This is the idea of interdependence, of reasoning in terms of connections and not oppositions, of both "this" and "that"—without rejection, opposition, or exclusion.

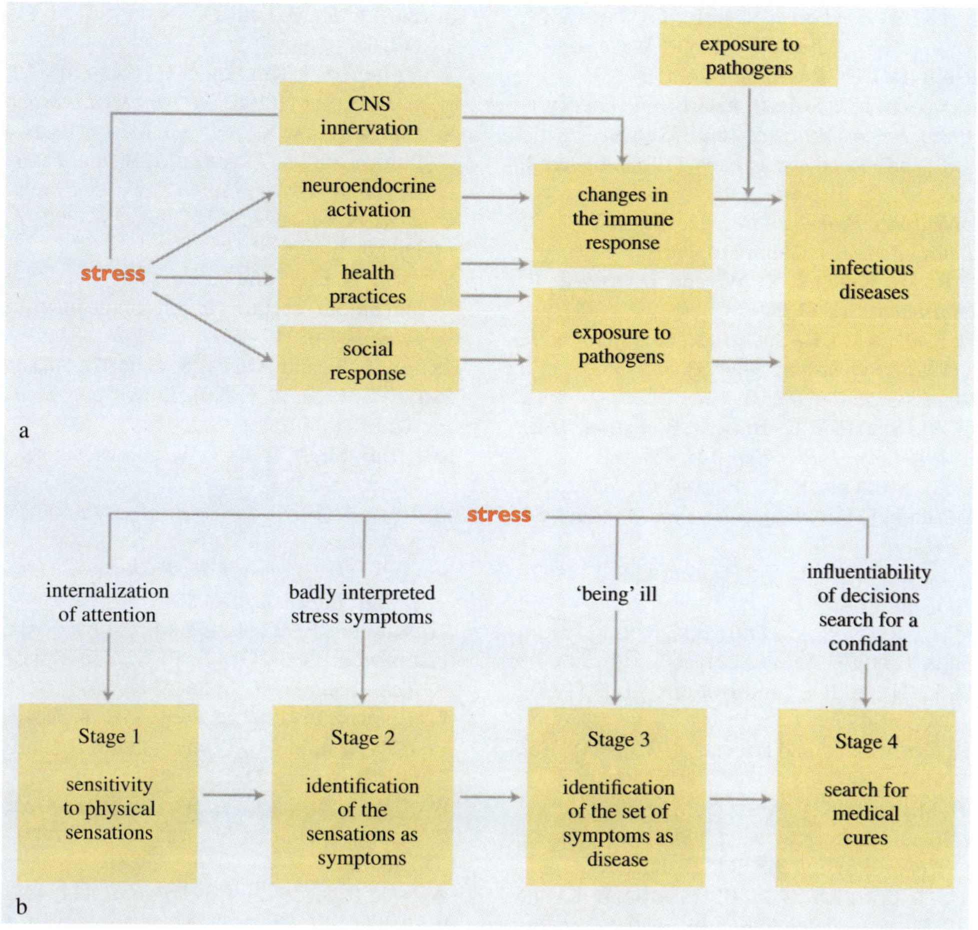

FIGURE 14 (a) Diagram illustrating the behavioral and biological pathways linking stress to the onset of infectious diseases. Note not only that the effects exerted by stress occur via autonomic innervation and neuroendocrine activation but also that changes in health practices and social responses are also important. (b) The psychological pathways linking stress to illness behavior. CNS, central nervous system (adapted from Cohen and Williamson, 1991).

References Cited

ABBAS, A. K., MURPHY, K. M., and SHER, A. (1996). *Science* **383,** 787–793.

ADER, R., and COHEN, S. (1985). *Behav. Brain Sci.* **8,** 379–426.

ANONYMOUS (1988). *Promesses et Menaces à l'aube du XXIe Siècel, Conférence of Nobel Laureates in Paris.* Odile Jacob, Paris.

BESEDOVSKY, H., DEL REY, A., SORKIN, E., and DINARELLO, C. (1986). *Science* **233,** 652–654.

BLUTHÉ, R. M., WALTER, V., PARNET, P., LAYÉ, S., LESTAGE, J., VERRIER, D., POOLE, S., STENNING, B. E., KELLEY, K. W., and DANTZER, R. (1994). *C. R. Acad. Sci. Paris Ser. III* **317,** 499–503.

CANGUILHEM, G. (1989). *The Normal and the Pathological.* Zone Books, New York.

COE, C. (1993). *Psychosom. Med.* **55,** 298–308.

COHEN, S., and WILLIAMSON, G. M. (1991). *Psychol. Bull.* **109,** 5–24.

CRARY, B., BORYSENKO, M., SUTHERLAND, D. C., KUTZ, L., BORYSENKO, J. Z., and BENSON, H. (1983). *J. Immunol.* **130,** 694–697.

DANTZER, R. (1989). *L'illusion Psychosomatique.* Odile Jacob, Paris.

DANTZER, R., and KELLEY, K. W. (1989). *Life Sci.* **44,** 1995–2008.

DANTZER, R., and MORMÈDE, P. (1995). In *Stress, the Immune System and Psychiatry* (B. Leonard and K. Miller, Eds.), pp. 47–67. Wiley, Chichester, UK.

DANTZER, R., *et al.* (1988). *Physiol. Behav.* **43,** 577–583.

DEROGATIS, L. R., ABELOFF, M. D., and MELISAROTOS, N. (1979). *J. Am. Med. Assoc.* **242,** 1504–1508.

ERIKSSON, B., and HEDFORS, E. (1977). *Scand. J. Haematol.* **11,** 275–286.

ESTERLING, B. A., ANTONI, M. H., KUMAR, M., and SCHNEIDERMAN, N. (1990). *Psychosom. Med.* **52,** 397–410.

Felten, D. L., Felten, S. Y., Carlson, S. L., Olschowska, J. A., and Livnat, S. (1985). *J. Immunol.* **135,** 755s–765s.

Friedman, E., *et al.* (1991). *Dev. Psychobiol.* **24,** 159–173.

Glaser, R., Rice, J., Stout, J. C., Speicher, C. E., and Kiecolt-Glaser, J. K. (1986). *Behav. Neurosci.* **100,** 675–678.

Greer, S., Morris, T., and Pettingale, K. W. (1979). *Lancet* **2,** 785–787.

Hart, B. L. (1988). *Neurosci. Biobehav. Rev.* **12,** 123–137.

Kafka, F. (1984). *Lettres à Miléna.* Gallimard, Paris.

Kent, S., Bluthé, R. M., Kelley, K. W., and Dantzer, R. (1992). *Trends Pharmacol. Sci.* **13,** 24–28.

Kiecolt-Glaser, J. K., Dura, J. R., Speicher, C. E., Trask, J., and Glaser, R. (1991). *Psychosom. Med.* **53,** 345–362.

Kluger, M. J. (1991). *Physiol. Rev.* **71,** 93–127.

Irwin, M., Daniels, M., Smith, T. L., Bloom, E., Smith, T. L., and Weiner, H. (1987). *Am J. Psychiatr.* **144,** 437–441.

Landmann, R. M. A., Müller, F. B., Perini, C., Wesp, M., Erne, M., and Bühler, F. R. (1984). *Clin. Exp. Immunol.* **58,** 127–135.

Laudenslager, M. L., Reite, M. R., and Harbeck, R. J. (1982). *Behav. Neurol. Biol.* **36,** 40–48.

Laudenslager, M. L., Ryan, S. M., Drugan, R. C., Hyson, R. L., and Maier, S. F. (1983). *Science* **221,** 568–570.

Laudenslager, M. L., Held, P. E., and Boccia, M. L. (1990). *Dev. Psychobiol.* **24,** 159–173.

Layé, S., Parnet, P., Goujon, E., and Dantzer, R. (1994). *Mol. Brain Res.* **27,** 157–162.

Layé, S., Bluthé, R. M., Kent, S., Combe, C., Médina, C., Parnet, P., Kelley, K., and Dantzer, R. (1995). *Am. J. Physiol. Regul. Integrative Comp. Physiol.* **268,** R1327–R1331.

Livnat, S., Felten, S. J., Carlton, S. L., Bellinger, D. L., and Felten, D. L. (1985). *Neuroimmunology* **10,** 5–30.

Metalnikov, S. (1934). *Rôle du Système Nerveux et des Facteurs Biologiques et Physiques dans l'Immunité.* Masson, Paris.

Mormède, P., Dantzer, R., Michaud, B., Kelley, K. W., and Le Moal, M. (1988). *Physiol. Behav.* **43,** 577–583.

Naliboff, B. D., Benton, D., Solomon, G. F., Morley, J. E., Fahey, J. L., Bloom, E. T., Makinodan, T., and Gilmore, S. (1991). *Psychosom. Med.* **43,** 331–342.

Parnet, P., Amindari, S., Wu, C., Brunke-Reese, D., Goujon, E., Weyhenmeyer, J. A., Dantzer, R., and Kelley, K. W. (1994). *Mol. Brain Res.* **27,** 63–70.

Reite, M. R., Harbeck, R., and Hoffman, A. (1981). *Life Sci.* **29,** 1133–1136.

Rook, G. A. W., Hernandez-Pando, R., and Lightman, S. L. (1994). *Immunol. Today* **15,** 301–303.

Sapolsky, R. M., and Donnelly, T. M. (1985). *Endocrinology* **117,** 662–666.

Schleifer, S. J., Keller, S. E., Camerino, M., Thornton, J. C., and Stein, M. (1983). *J. Am. Med. Assoc.* **250,** 374–377.

Schneiderman, N., Antoni, M., Irunson, G., Klimas, N., LaPerriere, M., Esterling, B., and Fletcher, M. A. (1994). In *Handbook of Human Stress and Immunity* (R. Glaser and J. Kiecolt-Glaser, Eds.), pp. 267–300. Academic Press, San Diego.

Sibinga, N. E. S., and Goldstein, E. A. (1988). Opioid peptides and opioid receptors in cells of the immune system. *Annu. Rev. Immunol.* **6,** 219–249.

Sklar, L. S., and Anisman, H. (1979). *Science* **205,** 513–515.

Smith, E. C., *et al.* (1990). *Proc. Natl. Acad. Sci. USA* **87,** 1057–1060.

Smith, E. M., Meyer, J. W., and Blalock, J. E. (1982). *Science* **218,** 1311–1313.

Sontag, S. (1978). *Illness as Metaphor.* Farrar, Straus & Giroux, New York.

Spiegel, D., Bloom, J. R., Kraemer, H. C., and Gottheil, E. (1989). *Lancet* **2,** 888–889.

Sternberg, E. M., Hill, J. M., Chrousos, G. P., Kamilaris, T., Listwak, S. J., Gold, P. W., and Wilder, R. (1989). *Proc. Natl. Acad. Sci. USA* **86,** 2374–2377.

Temoshok, L. (1987). *Cancer Surv.* **6,** 545–567.

Visintainer, M. A., Volpicelli, J. R., and Seligman, M. E. P. (1982). *Science* **216,** 437–439.

Wan, W., Vriend, C. Y., Wetmore, L., Gartner, J. G., Greenberg, A. H., and Nance, D. M. (1993). *Brain Res. Bull.* **30,** 101–105.

Weisse, C. S. (1992). *Psychol. Bull.* **111,** 475–489.

Weisse, C. S., Pato, C. N., McAllister, C. G., Littman, R., Breier, A., Paul, S. M., and Baum, A. (1990). *Brain Behav. Immun.* **4,** 339–351.

Zorn, F. (1988). *Mars.* Gallimard, Paris.

General References

Ader, R., Felten, D. L., and Cohen, N. (Eds.) (1991). *Psychoneuroimmunology,* 2nd ed. Academic Press, San Diego.

Glaser, R., and Kiecolt-Glaser, J. (Eds.) (1994). *Handbook of Human Stress and Immunity.* Academic Press, San Diego.

Leonard, B. E., and Miller, K. (Eds.) (1995). *Stress, the Immune System and Psychiatry.* Wiley, Chichester, UK.

Selye, H. (1956). *The Stress of Life.* McGraw-Hill, New York.

JACQUES BALTHAZART
LAUREN V. RITERS*

Center for Cellular and Molecular Neurobiology,
Research Group in Behavioral Neuroendocrinology
University of Liège
B-4020 Liège, Belgium

Hormones and Behavior

In adult animals and man, hormones affect a variety of behaviors either by acting as transcription factors in the nucleus of neurons (mostly steroids) or through binding with membrane receptors. These effects result in modifications of the intraneuronal physiology (peptidergic and peptide hormones). Steroids are often transformed into their target cells into metabolites that are behaviorally active, and this metabolism controls the behavioral action of the parent hormone. With relatively few exceptions, the steroid binding sites in the brain are remarkably stable in the different vertebrate classes indicating common mechanisms for controlling behavior, particularly reproduction. Binding of steroids to their receptors produces changes in transcription of messenger RNA and translation of proteins that results in a modified a neurotransmission and ultimately in behavioral activation. In addition, during ontogeny, steroids exert profound and irreversible actions on the brain that result in a permanent sexual differentiation of the behavioral phenotype. Specific examples of these processes are described; unanswered questions are pointed out and directions for future work are indicated.

Introduction

From a reductionist point of view, most behaviors can be viewed as a series of muscular contractions triggered by an organized series of nerve impulses. To activate a behavioral pattern, a hormone has to modify the electrical activity in specific areas or neurons in the brain. Hormones achieve this goal by modifying the concentration and/or the activity of neurotransmitters and neuropeptides. However, hormones do not induce behavior, they only change the probability of its appearance in a particular social situation. Additionally, the relationship between hormones and behavior is not unidirectional. Besides the influence of hormones on behavior, an organism's behavior (or other environmental stimuli) can also alter the endocrine state. For example, in seasonal breeders hormones secreted by the pituitary gland and gonads synchronize the onset of reproductive behavior with the time of year most conducive to reproductive success. In addition, secretion of these hormones is also influenced by environmental stimuli such as the behavior of conspecifics.

The brain is the major site of hormone action necessary for the expression of particular behaviors, but hormones also act on peripheral tissues in at least three ways that are relevant to behavior control (Hinde, 1970). First, they can change the nature of the sensory inputs to the brain. For example, androgens influence ejaculation in male rats by enhancing the sensitivity of the penis. Second, hormones influence the development of effector organs (e.g., certain muscles). Effector organs such as the muscles controlling

*Current address: Department of Zoology, University of Wisconsin, Madison, WI 53706, USA

the syrinx (a specialized structure in birds necessary for vocalization) are increased by androgens ultimately influencing the behavioral output triggered by central mechanisms. Finally, hormones are capable of altering external social signals provided by superficial structures. For example, steroid hormones induce the growth of external structures typical of male or female morphology, such as the comb and wattles of a chicken. A steroid-induced modification in behavior is therefore more than the consequence of the action of the hormone at the brain level.

Hormones Implicated in Behavioral Regulation

Specific neural circuits sensitive to hormones are known to participate in behaviors ranging from reproductive behavior to parental behavior and aggression. Steroids secreted by the adrenal glands and gonads have been most intensively investigated with respect to their influence on sexual behavior. The precursor for all steroids is cholesterol (Fig. 1). In response to signals from the anterior pituitary gland, progestins and corticoids are manufactured and these steroids are involved in maintenance of pregnancy and cessation of mating and in salt appetite, respectively. Additionally, signals from the pituitary induce the gonads to produce steroid hormones such as testosterone (T), androstenedione, and estradiol (E_2).

The genesis of behavioral endocrinology is often traced to early studies of Arnold Adolph Berthold, who in 1849 found that removal of the testes in young male chicks resulted in males whose appearance and behavior as adults did not become rooster typical. In contrast, castrated males whose testes were reimplanted or who received a testis from another bird developed both rooster-typical plumage and behavior suggesting that a blood-borne substance was responsible for induction of behavioral changes observed in adults. It is now established that androgenic steroids are necessary for the development of male-typical behavior and, for example, the injection of T into day-old chicks causes the development of adult morphological characteristics such as large comb and wattles as well as the activation of adult reproductive behaviors such as mounting and crowing (Fig. 2).

Investigations of steroids reveal several general principles of hormonal regulation of behavior and constitute the primary focus of this article. However, several other hormones implicated in behavioral regulation must also be mentioned briefly.

Hormones secreted by the anterior pituitary gland are known to modify a variety of physiological and behavioral processes. For example, prolactin seems to control the initiation or maintenance of behaviors such as egg incubation in birds, parental care in mammals, or food consumption. Additionally, anterior pituitary hormones are well-known for their participation in a neuroendocrine circuit involved in stress adaptation. During stressful circumstances (e.g., chronic subordination in a dominance hierarchy), adrenocorticotropic hormone (ACTH) is released from the anterior pituitary gland and stimulates corticoid release by the adrenals which normally feeds back to inhibit further release of anterior pituitary hormones. However, under persistent stressful conditions, this feedback mechanism no longer seems to operate so that high levels of ACTH are chronically present and animals display characteristic behaviors such as disturbances in feeding, aggression, learning, and memory.

The posterior pituitary peptides oxytocin and vasopressin have long been known for their involvement in mammalian milk ejection and water retention, respectively. Recently, a large body of evidence has implicated oxytocin in maternal and sexual behavior. Vasopressin has been found to be steroid dependent and involved in behaviors related to reproduction in males and in social memory.

Thyroid hormones, synthesized from the amino acid tyrosine, are necessary for physiological processes such as adaptation to changing temperatures and nutrient metabolism. Additionally, thyroid hormones regulate growth and differentiation via effects on growth hormone secreted from the anterior pituitary. These effects indirectly regulate behavior, but thyroid hormones also have more direct effects on behavior. In mammals hypothyroidism leads to severe mental retardation (cretinism) and delays sexual maturation. In humans, thyroid hormones have also been implicated in depression.

Several gut hormones are released from the gastrointestinal tract and function to aid in digestion. These hormones also regulate peripheral physiology necessary for coordinated behavior. Some of these gut hormones, such as cholecystokinin (CCK) and vasoactive intestinal polypeptide, are also found within the brain, in which they directly participate in behavioral regulation. For example, CCK plays a role in feeding and facilitates certain forms of learning and memory.

Cellular Mechanisms of Hormone Action

Steroids and other hormones act on the brain to control sexual and other behaviors by molecular mechanisms that are similar to those described at the level of peripheral organs (binding to specific receptors and modification of cellular physiology including protein synthesis). It cannot be excluded, however, that the central effects of steroids are also achieved through other brain-specific mechanisms involving, for example, direct effects on neuronal membranes.

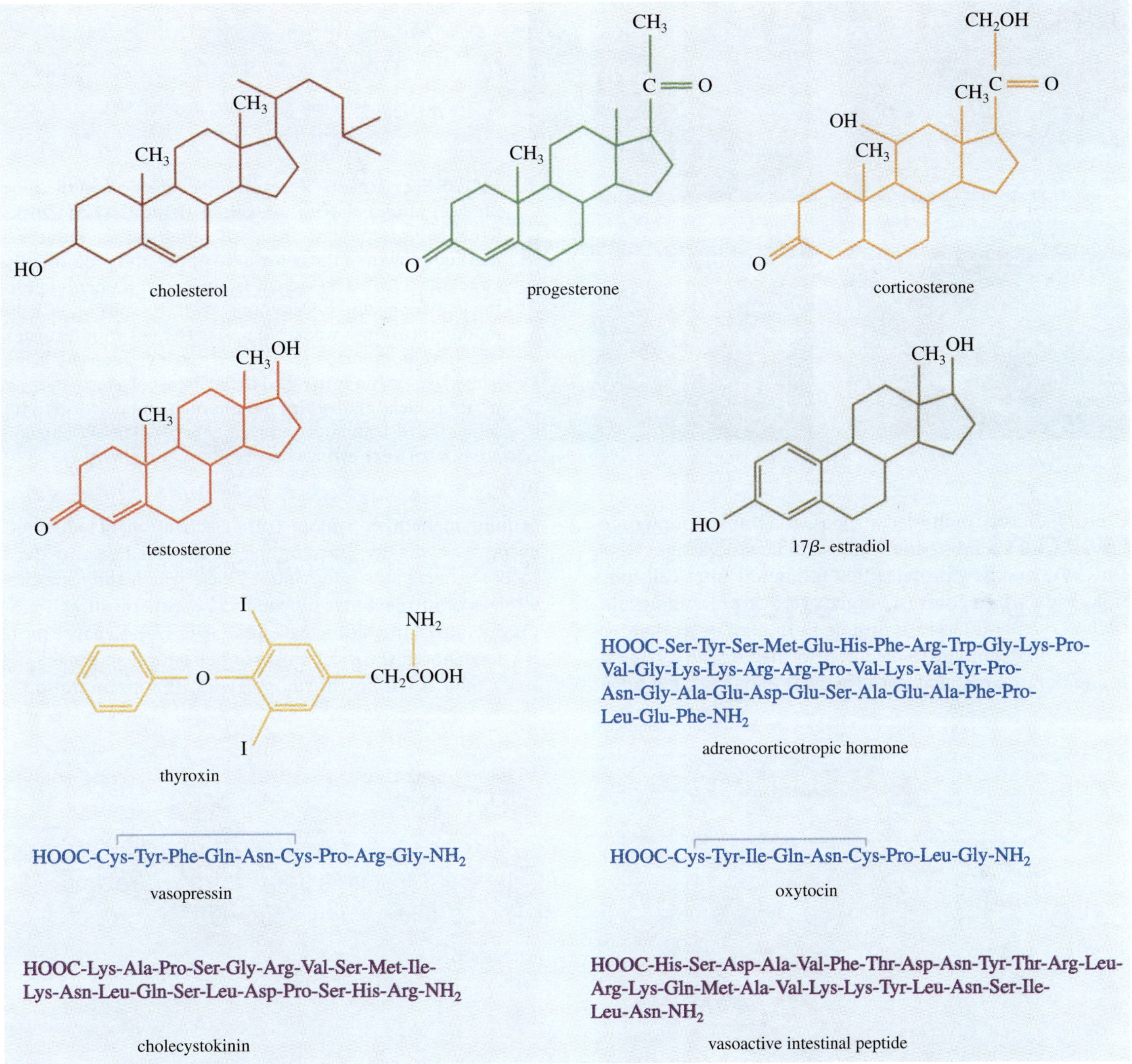

FIGURE 1 The chemical structure of some hormones that play important roles in the control of behavior. Four steroid hormones (progesterone, corticosterone, testosterone, and estradiol) are shown together with their precursor, cholesterol. The figure also shows one thyroid hormone (thyroxin) and several peptidergic hormones, adrenocorticotropic hormone, vasopressin, oxytocin, cholecystokinin, and vasoactive intestinal polypeptide. Many hormones implicated in behavior control are also proteins that are similar in chemical structure to the peptides shown here except that they are formed by much larger numbers of amino acids. Ala, alanine; Arg, arginine; Asn, asparagine; Asp, aspartic acid; Cys, cysteine; Gln, glutamine; Glu, glutamic acid; Gly, glycine; His, histidine; Ile, isoleucine; Leu, leucine; Lys, lysine; Met, methionine; Phe, phenylalanine; Pro, proline; Ser, serine; Thr, threonine; Trp, tryptophan; Tyr, tyrosine; Val, valine.

Two distinct types of interactions between hormones and their target cells have been described. Steroid hormones, and other similar lipophilic compounds such as the thyroid hormones, freely enter target cells and produce their biological effects by binding with specific intracellular receptors. The complex formed by the hormone and its re-

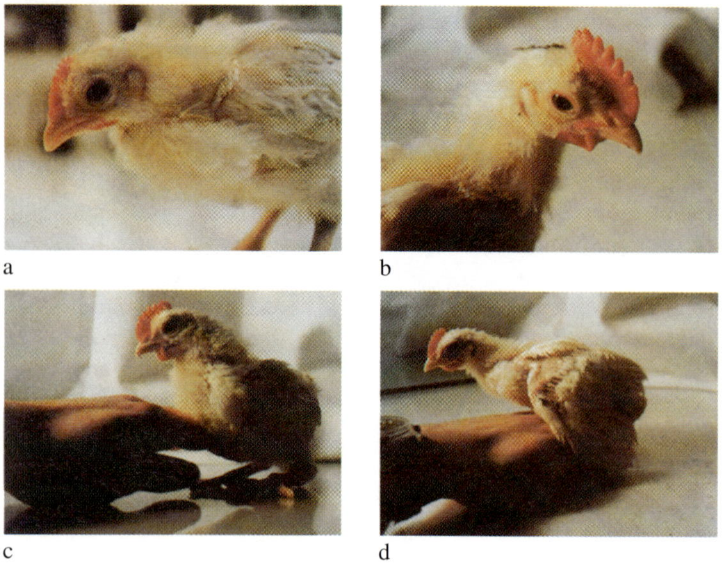

FIGURE 2 Effects of testosterone injection on the morphology and behavior of young domestic hen chicks. Treatment for a few days with this steroid induces a marked growth of the comb and wattles and promotes the appearance of early sexual behavior. The phenotype of the control group, injected only with the solvent, was not modified (a). In 2-week-old chicks treated with testosterone, development of the crest and wattles is seen (b). When a hand is gently approached to the breast of a testosterone-treated chick (c), the bird mounts on it and performs a full sequence of copulatory movements (d). These behaviors are not (or very rarely) observed in control birds.

ceptor then acts, in the cell nucleus, as a transcription regulation factor leading to the synthesis of new messenger RNA (mRNA) and new proteins that ultimately alter cell function (Fig. 3a). In contrast, larger or more polar molecules such as the peptidergic or protein hormones exert their actions through binding to receptors located on the cell membrane that are coupled to the activation of a second intracellular messenger system (often activation of adenylate cyclase). Typically, the binding of the hormone to these receptors activates adenylate cyclase which then initiates a series of intracellular biochemical events resulting in the production of the biological effects (Fig. 3b). Because more is known about the mechanism of behavioral action for steroids than for other hormone types, this presentation is

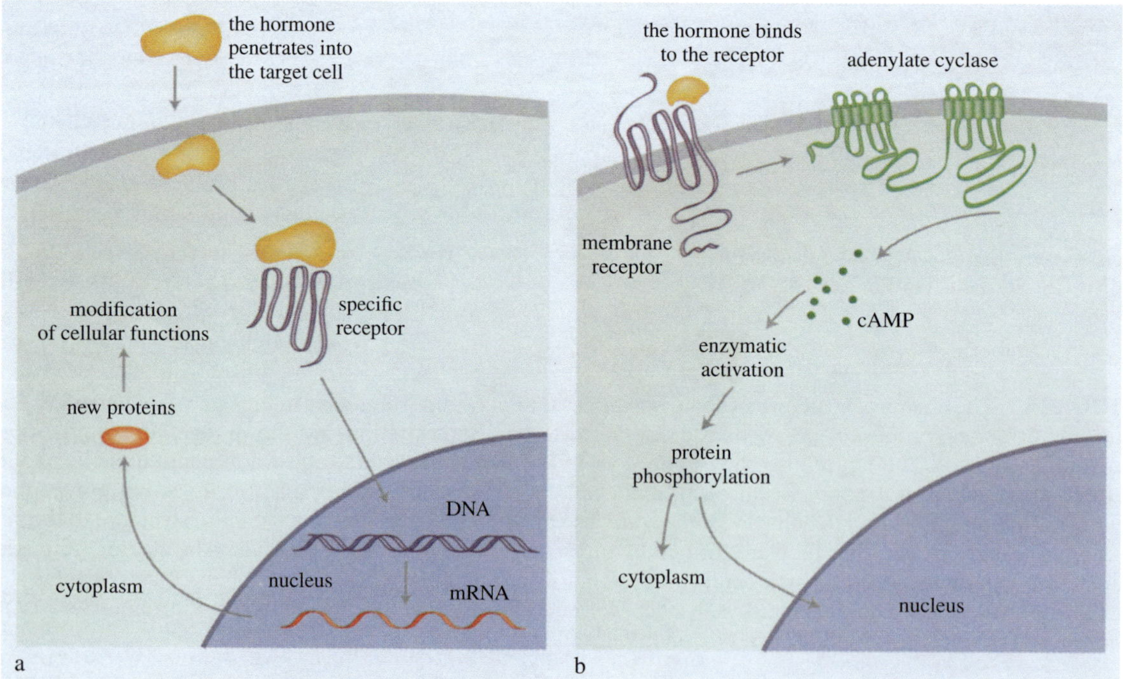

FIGURE 3 Schematic illustration of the two major modes of hormone action at the cellular level. (a) The steroid and thyroid hormones act as transcription factors at the nuclear level. (b) Peptide and protein hormones exert their effects through binding to membrane receptors and activation of a second intracellular messenger, such as cyclic AMP.

centered on these hormones to illustrate general principles of behavioral neuroendocrinology.

Intracellular Metabolism of Steroids

When they enter the target cells, steroids undergo important metabolic transformations (Celotti *et al.,* 1984; Balthazart, 1989).

Two enzymes (aromatase and 5α-reductase) catalyze the transformation of T into behaviorally relevant metabolites. Additional enzymes inactivate the steroid (5β-reductase) or produce reversible transformations of unknown behavioral significance (17β-hydroxysteroid dehydrogenase, 17β-HSDH; 3α- or 3β-hydroxysteroid dehydrogenase, 3α- or 3β-HSDH; Fig. 4).

The ratio of active versus inactive T metabolites that are produced in the brain and in peripheral structures can be affected by factors such as the sex, age, season, or hormonal condition of the subjects, providing a mechanism for the fine adjustment of behavior. Depending on the species, 5α-DHT and E_2, alone or in synergy, mimic most if not all effects of T. Copulatory behavior, for example, can be activated by 5α-DHT alone in the rabbit or in the guinea pig and by E_2 alone in the rat or in the Japanese quail. Estrogens (e.g., E_2) and progestagens (e.g., progesterone) are also metabolized in the brain and their metabolites play an important role in the activation of reproductive behavior.

Binding to Receptor

The neuroanatomical distribution of high-affinity binding sites for androgens, estrogens, and progestagens is remarkably constant in all vertebrate species (Kelley and Pfaff, 1978). This distribution was originally mapped in the

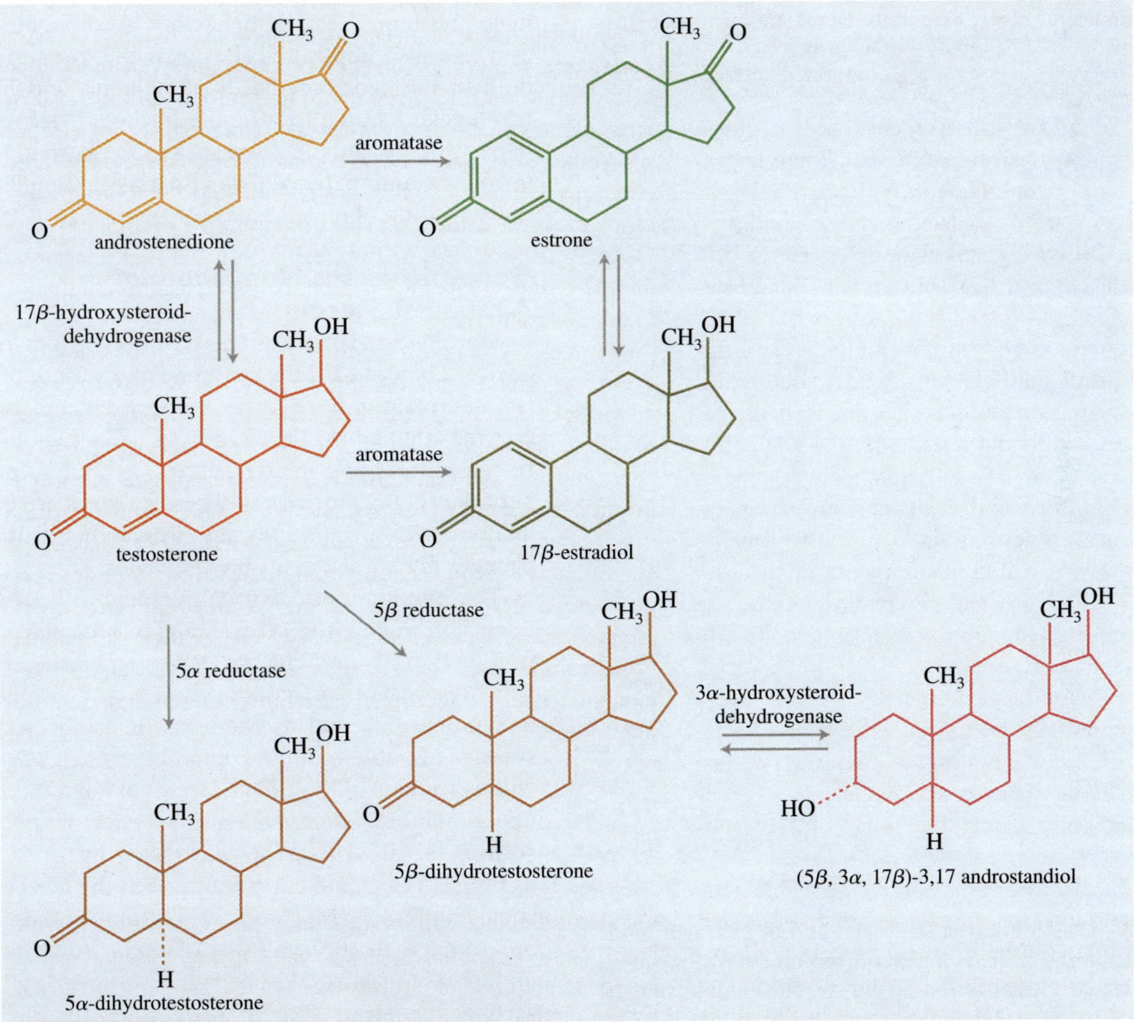

FIGURE 4 Metabolic pathways leading from testosterone and androstenedione to the formation of behaviorally active metabolites such as the estrogens estradiol and estrone, formed by aromatization, or the 5α-reduced androgens, including 5α-dihydrotestosterone. The formation of the behaviorally inactive 5β-reduced androgens is also illustrated.

1970s using *in vitro* autoradiographic techniques. Recently, immunocytochemical studies using mono- or polyclonal antibodies against the steroid receptors molecules have independently confirmed these distribution patterns.

Cells containing receptors for sexual steroids are localized in the medial preoptic area, the hypothalamus (anterior hypothalamic area, ventromedial nucleus, arcuate nucleus, and premammillary nucleus), telencephalic structures that are part of the limbic system (amygdala, lateral septum, and the bed nucleus of stria terminalis), and specific parts of the mesencephalon (optic tectum). A schematic presentation of binding sites for E_2 or androgen in the rat brain is shown in Fig. 5a. A similar distribution is observed throughout vertebrates from fishes to mammals.

The overall anatomical distribution of androgen-concentrating cells is similar to that of E_2-concentrating cells. However, differences have been observed in the intensity and the number of cells labeled as well as in their precise distribution within a given nucleus. In rat, the motoneurons of the spinal cord that control the bulbocavernosus muscle of the penis also contain androgen receptors. Some cases of specialization have been observed. In songbirds such as zebra finches and canaries, a network of neurochemically specialized brain nuclei controls both the learning and the production of song. Singing behavior is steroid sensitive and, accordingly, most nuclei in the so-called song control system contain androgen receptors (Fig. 5b). This presence of androgen receptors in telencephalic nuclei outside of the limbic system is unique in vertebrates and is functionally related to singing.

Only some of the binding sites for sexual steroids in the brain are involved in the control of sexual behavior. They have been identified by a combination of stereotaxic implantation of steroids in the brain and lesion experiments. In general, the medial part of the preoptic area and the ventromedial nucleus of the hypothalamus appear as critical and sufficient sites for the steroid activation of respectively male and female sexual behaviors. Additional sites, however, are also implicated. For example, progesterone acts both in the ventromedial hypothalamus and in the ventral tegmental area to promote lordosis. Androgen action in parts of the limbic system such as the nucleus of the stria terminalis and the amygdala also plays a significant role in the activation of male sexual behavior.

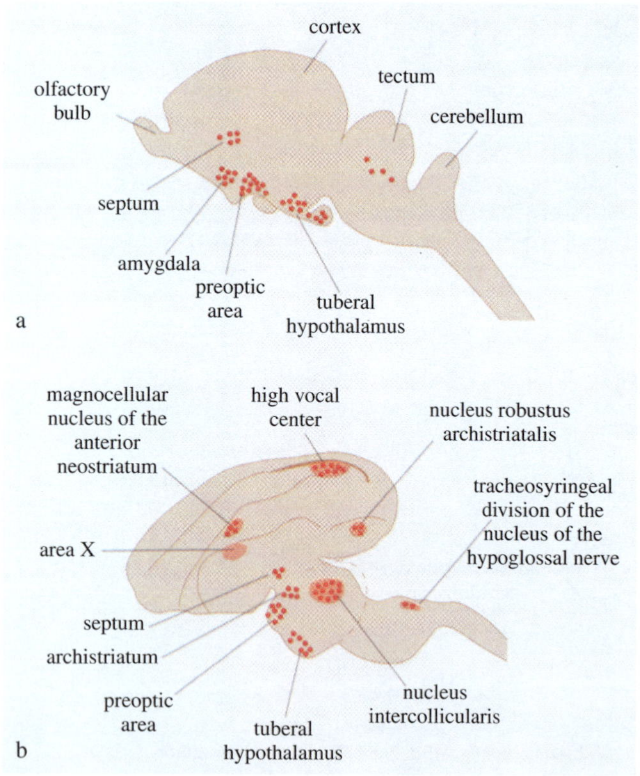

FIGURE 5 Distribution of androgen and/or estrogen binding sites in the rat brain as a generalized model of the vertebrate brain (a); specialization of the steroid binding sites as observed in the telencephalon of the zebra finch brain (b). In this species and in other songbirds, androgen receptors are found in several nuclei of the telencephalon implicated in the learning and production of song.

Genomic versus Nongenomic Actions of Steroids

Steroids acts largely as transcription factors to affect behavior. They control the synthesis of a variety of receptors (for neurotransmitters, for neuropeptides, and for steroids), enzymes that control the synthesis and catabolism of neurotransmitters and neuropeptides, as well as proteic and peptidic (neuro)hormones. These changes are confined to anatomically specific sites and ultimately result in specific changes in neural activity and behavior.

These genomic effects usually appear with relatively long latencies ranging from a few hours to a few days. Faster action of steroids on behavior, however, has been identified which suggests that these hormones may also act via fundamentally different mechanisms. It has been noted that steroids are able to alter excitability of neurons in culture within seconds of their application. In a limited number of experimental models, evidence has been obtained demonstrating the presence of specific receptors for steroids such as progesterone and corticosterone at the level of the neuronal membrane (McEwen, 1994). Some of these receptors are coupled to the synthesis of cyclic AMP or promote an increase in inositol phosphates turnover, reflecting the second messenger systems responsible for the biological effects of the hormone. Some steroids, such as 5α-reduced metabolites of progesterone, also modulate the ion flow through the GABA receptor. This type of cellular action may also contribute to the activation of some behaviors such as lordosis in the hamster.

Organizational Action Steroids

Many behaviors in animals, including humans, are sexually differentiated and are preferentially or exclusively exhibited in one sex (Goy and McEwen, 1980). It was initially believed that these sex differences result from the presence of a different hormonal milieu in males and females, such as high plasma T in the former and high plasma estrogen and progesterone levels in the latter. By the middle of the past century, however, this interpretation was found to be generally incorrect. Estrogens often cannot activate female-typical behaviors in adult males, whereas conversely T often fails to activate male-typical behaviors in adult females. An early and irreversible sexual differentiation of the brain was suspected based on a variety of experimental data, and in 1959 this work was conceptualized in a seminal paper by Phoenix and collaborators demonstrating in guinea pigs that, besides their effects on the activation of behavior in adulthood, steroids produce an irreversible sexual differentiation of brain and behavior during early life. This type of steroid action only takes place during a limited period of development (during embryonic or early postnatal life, depending on the species) that is usually referred to as the critical period.

In mammals, early exposure to steroids (T or its metabolite E_2) produces a masculine phenotype. Behavioral characteristics of the male are reinforced (masculinization), and more or less independently the capacity of males to display female typical behavior even in the presence of adequate levels of activating hormones (estrogens and progesterone) is diminished or lost (defeminization) (Fig. 6a). The development of the masculine and feminine phenotypes can be obtained irrespective of the genetic sex of the subjects by neonatal gonadectomy or injection of T. However, it is important to note that the development of the female phenotype is not a completely passive process. A growing body of evidence indicates that perinatal estrogen exposure is involved in the expression of normal adult female sexual behavior, at least in rats and mice.

The relative importance of the masculinization and defeminization processes vary from one species to another. For example, the sexual differentiation of male rats consists mostly of a defeminization associated with a limited masculinization of behavior, whereas in rhesus monkey a marked masculinization of behavior takes place that is not associated with defeminization.

More subtle differences in hormone exposure during development can also influence the degree to which an animal will display male- or female-typical behavior. A female rat positioned between two males *in utero,* and thereby exposed to slightly elevated concentrations of steroid hormones, will display higher levels of aggression and other male-typical behaviors as an adult compared to sisters who were not positioned between two males.

The masculinization and defeminization of the male in mammals are triggered by testicular androgens. However, it is often not T per se but its aromatized metabolite E_2 that is responsible for the differentiation of behavior at the brain level. Given that ovaries of pregnant females secrete high levels of E_2 that should easily permeate the placental barrier, one can wonder why these estrogens do not masculinize female embryos. However, the plasma of perinatal rats contains α-fetoprotein (AFP), a protein that binds to estrogens and could therefore make these hormones unavailable for brain differentiation. AFP does not bind to T, which, after being secreted from the testes, can reach the brain of males in which it exerts its masculinizing effects after being locally aromatized to E_2. Although widely accepted, this hypothesis remains somewhat controversial and may not apply to all species. In rats, it seems that AFP actually enters the brain and could serve more as a carrier for E_2 than as a means of sequestering the hormone in the plasma. Some species do not appear to have detectable levels of circulating AFP or possess an AFP homolog that does not appear to bind to estrogen, as found in primates for example. The relative importance of brain aromatization of T in the process of brain sexual differentiation could also be quite variable from species to species, and in monkeys, for example, androgens may be more important than estrogens for sexual differentiation.

In birds such as chicken, quail, and ducks that have been studied in detail, sexual differentiation exclusively affects the male-typical behaviors. The female reproductive behaviors (e.g., sexual receptivity postures such as squatting) are not sexually differentiated and can be activated in both sexes provided an adequate treatment with estrogens is administered in adulthood. In contrast, an extreme form of sexual dimorphism is observed for male-typical copulatory patterns (mounts and cloacal contact movements). These behaviors are exclusively observed in males and cannot be activated in females even after treatment with high doses of T. Contrary to what is observed in mammals, in the absence of gonadal steroids these avian species develop a male phenotype, whereas the female phenotype develops in the presence of estrogens. This physiological process can be experimentally manipulated so that the male or female behavioral phenotypes are induced in either sex by adequate embryonic endocrine manipulations. In Japanese quail, treatment of a male embryo with estrogens on Day 9 of incubation produces an adult that will never exhibit the male-typical copulatory patterns even after treatment with behaviorally active doses of T. Conversely, early treatment of

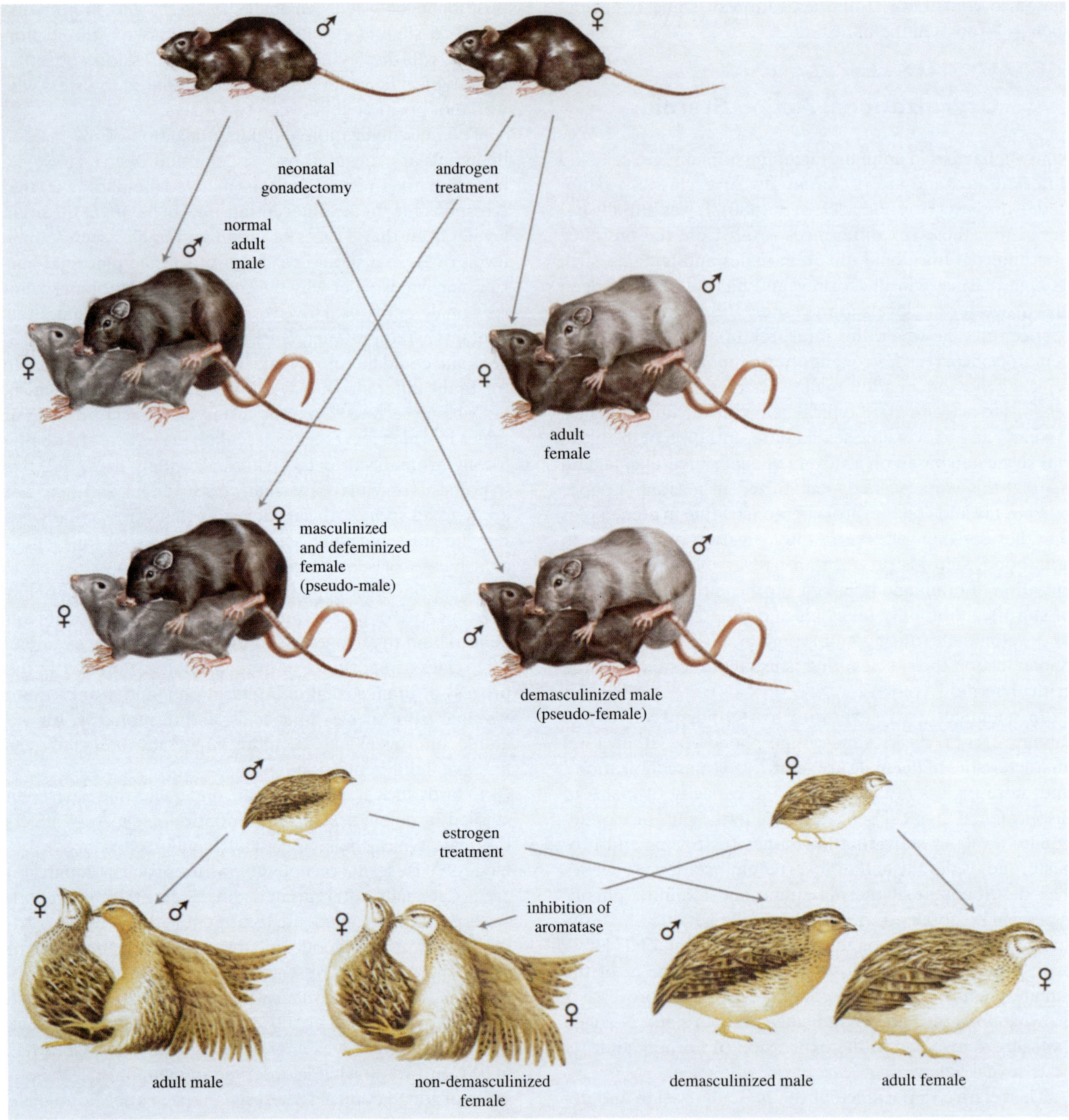

FIGURE 6 Schematic representation of the processes of sexual differentiation of reproductive behavior in rats (a) and in the Japanese quail (b). Neonatal or embryonic endocrine treatments that modify the adult behavioral phenotype of the treated subjects are illustrated in both species. The presence of steroids during a critical phase of development leads to a male-typical phenotype in rats but a female phenotype in quail.

female embryos with an aromatase inhibitor that blocks estrogen synthesis leads to the development of a fully masculine behavioral phenotype (Fig. 6b).

These contrasting patterns of differentiation in birds and mammals (only male-typical behaviors are affected in birds, whereas both male- and female-typical behaviors are differentiated in mammals; the relative absence of endocrine stimulation during early life results in male behavioral phenotype in birds and in female behavioral phenotype in mammals) may be related to the fact that females are the homogametic sex in mammals (XX), whereas males are homogametic in birds (ZZ). This observation could then highlight a more general rule according to which the phenotype of the heterogametic sex [male mammals (XY) and female birds (ZW)] would always develop in response to hormonal stimulation during ontogeny, whereas the behavioral phenotype of the homogametic sex would be the "default" observed in the absence of early endocrine stimulation (also referred to as the "neutral" sex). This principle must be accepted cautiously because the proximate mechanisms that might explain this connection between the nature of the sex chromosomes in males and females and the process of behavioral differentiation have not been identified and, furthermore, its limitations are unclear (only a few species have been studied).

An apparent exception to this rule has been noted in the case of zebra finches (*Taeniopygia guttata*), a species of songbirds that appears to conform to the avian pattern as far as the distribution of sex chromosomes is concerned (ZZ males and ZW females) but whose behavior differentiates by mechanisms that do not seem to follow the pattern identified in quail and chicken. In zebra finches, estrogens administered to males early in life demasculinize male-typical copulatory behavior as would be expected from studies performed on other avian species. However, the same early estrogen treatment masculinizes singing behavior in females as well as many aspects of the brain nuclei that control this behavior. It is therefore difficult to understand how the full behavioral repertoire of the male differentiates under physiological conditions in this species given that normal untreated males both copulate and sing in adulthood, whereas females never exhibit these two behaviors. All attempts to resolve this paradox based on the rat or quail model of sexual differentiation have been unsuccessful. It must be mentioned, however, that sexual behaviors are essentially controlled by the steroid-sensitive hypothalamic and limbic brain areas, whereas singing is essentially controlled by a circuit of telencephalic nuclei that represent a unique case of neurochemically specialized steroid-sensitive structures in the telencephalon (Fig. 5b). These steroid-sensitive structures have presumably appeared relatively late during evolution and it may not be totally unexpected that they differentiate by unique mechanisms that are still to be uncovered (Balthazart and Ball, 1995).

It must also be mentioned that in lower vertebrates many species do not appear to have sex chromosomes and that their brain and behavior differentiate by endocrine mechanisms that appear to be driven by the physical (e.g., temperature) or social environment (the presence of congeners of the other sex and position in a dominance–subordinate hierarchy).

The sharp distinction presented previously between activational (reversible, taking place throughout the life) and organizational (irreversible, limited to neonatal life) effects is somehow schematic, and it has in some cases been questioned. Its heuristic value has also been criticized (Arnold and Breedlove, 1985), but the distinction between organization and activation has, we believe, provided an exceptionally rich conceptual context for the development of behavioral neuroendocrinology.

Sex differences in the brain structures implicated in the control of behavior have been observed. These differences are morphological (groups of neurons in the preoptic area are larger in males, the size of the neurons, and the cell nucleus or dendritic arborization more developed in the ventromedial hypothalamus of females) but also biochemical. The concentration and turnover of various neurotransmitters or neuropeptides as well as the density of their receptors vary according to the sex (McEwen *et al.,* 1987). Some of these anatomical or biochemical differences persist in adult gonadectomized animals placed in similar hormonal environments. It is therefore likely that they are causally related to behavioral differences between sexes. Experimental evidence supporting this conclusion has been obtained in a limited number of cases. The organizational effects of steroids on behavior are the consequence of an action on the multiplication, migration, or death of neurons and also of a modulation of their functional differentiation (e.g., expression of neurotransmitters, neuropeptides, and their receptors). The detail of these effects, however, remains to be uncovered.

Specific Behavioral Systems

Male Sexual Behavior

Studies on male sexual behavior in rats have provided most of what is known about neuroendocrine regulation of male sexual behavior. Mating behavior has usefully been divided in two distinct components. The appetitive component consists of approaching and attempting to copulate with a female, whereas consummatory sexual behavior includes mounting, intromission, and ejaculation. Partly distinct neural mechanisms appear to mediate the two components of male sexual behavior but the experimental separation of these mechanisms is complicated because the different behaviors occur in sequence, and interference with early aspects of the sequence almost inevitably also affects subsequent steps. Some separation, however, has been observed and, for example, male rats with lesions of the preoptic area (POA) no longer copulate but will continue to work in an operant conditioning procedure for gaining access to a female.

The aromatization of T to E_2 within the POA and its 5α-reduction mediate both components of sexual behavior. These steroids influence the activity of neurotransmit-

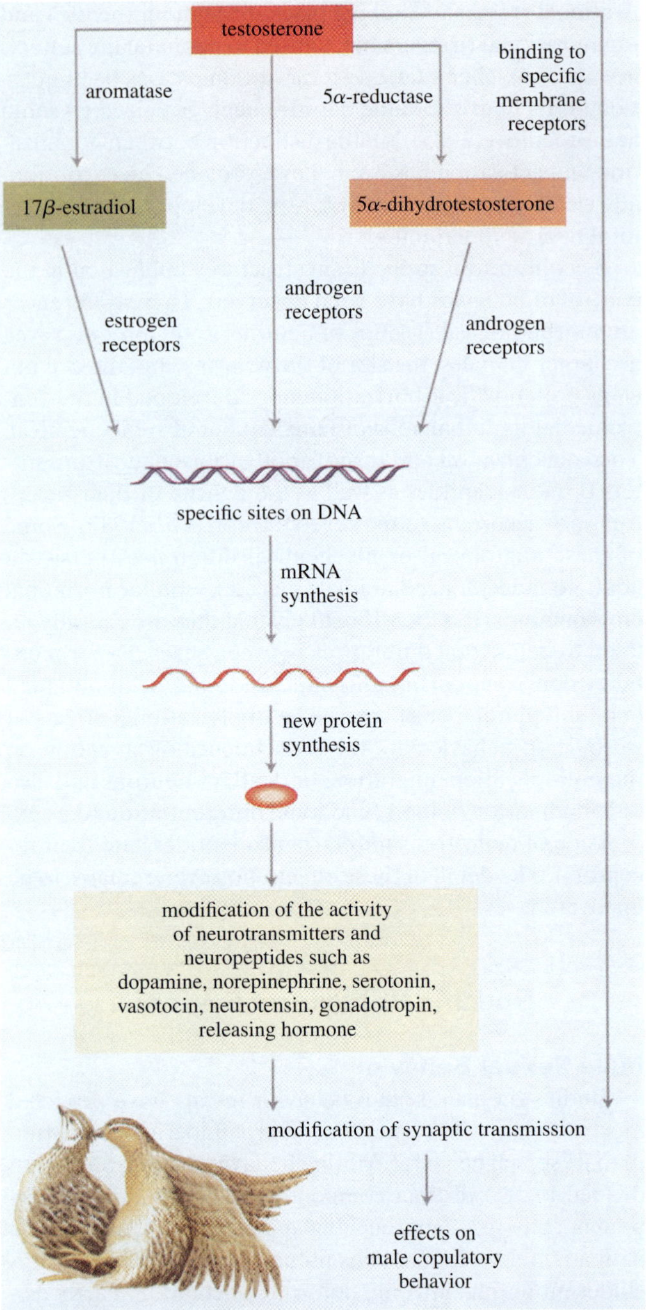

FIGURE 7 Schematic illustration of some neuroendocrine and neurochemical mechanisms involved in the activation of male sexual behavior.

ters either by modifying their synthesis or catabolism or by modulating the concentration of their receptors (Fig. 7). For example, androgens are known to modify catecholamine neurotransmission (Bitran and Hull, 1987). This modulation appears to represent one of the most important ways by which steroids affect behavior. Steroids also affect the concentration and release of specific neuropeptides [e.g., vasotocin, gonadotropin-releasing hormone (GnRH), and neurotensin] into anatomically specified brain structures. These neuropetides are often colocalized with neurotransmitters and they often modulate the action of classical neurotransmitters.

Although the relationships between steroids and neurotransmitters are complex, in general male sexual behavior is enhanced by postsynaptic dopamine but inhibited by serotonin or neuropeptide Y. The role of noradrenaline appears to be more complex, stimulating or inhibiting sexual behavior as a function of testing conditions, age, sex, and hormonal status of the animal. The types of receptors on which the transmitters act are also important and a transmitter can have opposite effects on sexual behavior depending on the type of receptor implicated.

Female Sexual Behavior

In most cases, the sexual activity of the female is restricted to a limited period of time when specific endocrine conditions are present. The hormonal determinants of female behavior differ according to the species. In rat, increased levels of estrogens are observed for a couple of days before a peak of progesterone that takes place immediately before the onset of sexual receptivity. In the ewe, however, the circulating level of progesterone is very low for more than 48 h before estrous but this behavioral state is associated with high circulating levels of estrogens.

Ovariectomy suppresses all aspects of female sexual behavior. To be effective in restoring sexual behavior in the ovariectomized female, a hormonal treatment has to mimic the sequence of the ovarian endocrine secretions (often a specific pattern of elevated levels of estrogens and progestagens) normally observed during the estrous cycle. Consequently, in the rat an injection of progesterone rapidly induces female sexual behavior only if given after 2 days of pretreatment with estrogens, whereas in the ewe several days of treatment with progesterone are required before an injection of estrogen can induce estrous behavior. In other species (e.g., prairie vole, ferret, and Japanese quail) estrogens alone are fully effective in restoring sexual receptivity in the ovariectomized female.

It has been established that E_2 ($\pm$ progesterone) affects the activity and probably the concentration of enzymes such as acetylcholine esterase (acetylcholine catabolism), tyroxine hydroxylase and dopamine β-hydroxylase (synthesis of dopamine and noradrenaline), catechol-*O*-methyl-transferase, and monoamine oxidases (dopamine and noradrenaline catabolism). Anatomically localized effects of steroids on the concentration or affinity of the receptors for most neurotransmitters and neuropeptides have been reported. Steroids similarly modulate the concentrations of mRNA for many neuropeptides (e.g., oxytocin, enkephalin, and GnRH).

If these neurochemical effects are relatively well-known, their significance in the control of behavior often remains unclear. Agonists and antagonists of specific neurotransmitters and neuropeptides have been directly injected into

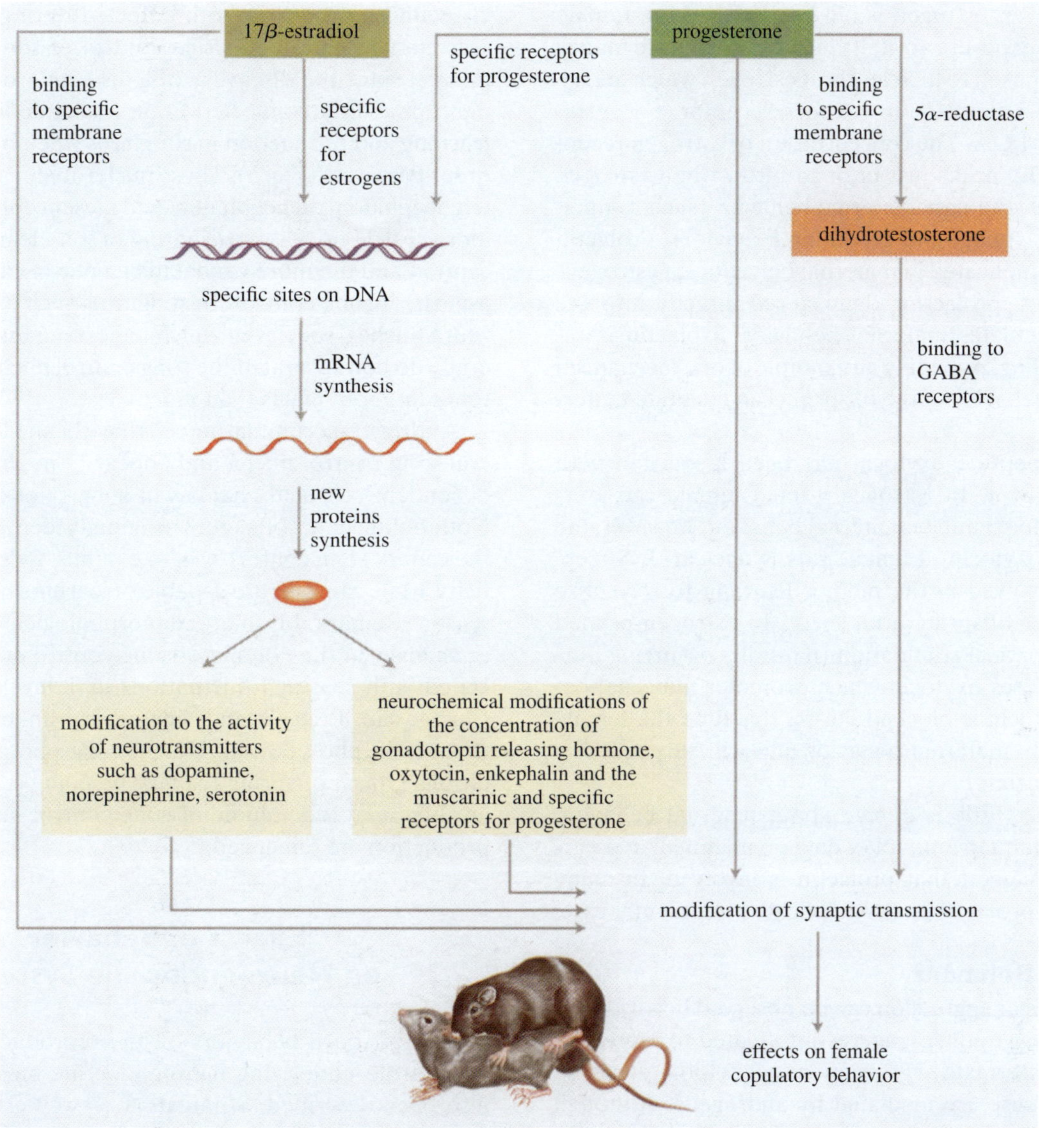

FIGURE 8 Schematic illustration of some neuroendocrine and neurochemical mechanisms involved in the activation of female sexual behavior.

specific brain structures, revealing their influence on reproductive behavior in females. The sexual receptivity of the female rat (lordosis), for example, can be facilitated by GnRH, oxytocin, vasopressin, substance P, CCK, and α-MSH, but it is inhibited by dopamine, serotonin, CRF, NPY, and β-endorphin (Fig. 8).

Through these neurochemical changes, steroids directly and indirectly modify synaptic transmission. E_2 and progesterone activate female sexual receptivity, at least in part, by modulating the noradrenergic transmission in the ventromedial hypothalamus at the pre- and postsynaptic levels. At the presynaptic level, the sexual steroids increase the noradrenergic transmission probably by controlling the liberation and the biosynthesis of the transmitter (modification of the turnover). This increased noradrenergic transmission contributes to the activation of lordosis in the female rat and guinea pig. Steroids can also affect the transduction of the noradrenergic signal at the postsynaptic level (Etgen *et al.,* 1992). E_2, for example, modifies the coupling with adenylate cyclase of some noradrenergic receptors (subtypes $\alpha 2$ and $\beta 2$).

Parental Behavior

After successful sexual behavior and reproduction, hormones play a critical role in parental behavior. In mammals, the female typically develops a hormonal substrate during pregnancy that prepares her to display parental behavior upon giving birth. This substrate can also be influenced by the behavior of the offspring and by prior maternal experience. Maternal behavior appears to be largely estrogen dependent. Estrogen acting within the POA initiates maternal behavior in the female of many mammalian species. However, estrogen is only effective if the female has had prior exposure to it. Typically, during pregnancy females are exposed to high levels of estrogen which are associated with an elevation in estrogen receptor concentration within the POA. The concentration of estrogen receptors within POA peaks just prior to birth, when estrogen can then act to stimulate maternal behavior (such as nursing, retrieving young, and grooming behavior). Prolactin has also been implicated in maternal behavior. In estrogen-primed females, prolactin administered directly into the POA also stimulates maternal behavior. Prolactin is released by suckling, providing an example of one mechanism by which the behavior of the offspring can facilitate maternal behavior.

The neuropeptide oxytocin has been associated with maternal behavior. In estrogen-primed female rats oxytocin appears to stimulate maternal behavior; however, the exact role of oxytocin in female rats is unclear. In sheep, oxytocin is involved in the mother learning to recognize and accept her offspring after birth. In estrogen-primed ewes, vaginocervical stimulation (naturally occurring during birth) releases oxytocin which promotes maternal behavior. If oxytocin is blocked during this time the female will not display maternal behavior and will subsequently reject her offspring.

In mammals, little is known about paternal behavior. However, the limited amount of data on mammalian paternal behavior suggest that prolactin is involved. In many avian species, prolactin is also implicated in parental care.

Aggressive Behavior

Many forms of aggression can be observed in a wide variety of animals. In males, aggression related to reproduction, such as intermale aggression, or aggression related to territorial defense are mediated by androgens. Although the relationship between androgen and aggression is complex, in general T levels are positively correlated with levels of aggression. Parallel fluctuations in T and levels of aggression are observed, for example, in animals that breed seasonally. This correlation is partly explained by the fact that administration of T produces heightened levels of aggression but, conversely, the aggressive interaction may increase plasma T levels.

In females, aggression is often expressed in the context of maternal behavior. Maternal aggression appears to be related to levels of progesterone rather than T. Studies in mice revealed high levels of maternal aggression to be related to high levels of progesterone. Administration of progesterone has also been found to elevate levels of aggression.

Song and Vocalization

Singing in birds has been extensively investigated in the context of reproduction. The males of many songbird species display seasonal fluctuations in both singing and T levels, with elevations in both detected during the breeding season. The neuroendocrine control of song has been of interest since the discovery of a discrete group of sexually dimorphic, interconnected brain nuclei dedicated to song learning and production in songbirds (members of the suborder Passeres) (Fig. 5). These nuclei are located within the telencephalon, diencephalon, and mesencephalon, with the more rostral nuclei participating in song learning and perception and the more caudal nuclei playing a role in motor aspects of song control. In songbirds such as canaries and zebra finches, species in which males sing but females generally do not, several of the song control nuclei are up to six times larger in males than in females.

Androgen-accumulating cells have been identified in several song control nuclei and appear to mediate androgen-dependent seasonal changes in song (Nottebohm, 1980). Nottebohm and colleagues originally demonstrated that the canary song control system exhibits T-dependent plasticity. Male canaries are capable of learning new songs each season. Remarkably, marked morphological changes, such as changes in the volume of song control nuclei, are associated with seasonal fluctuations in T levels and singing. Changes in T levels and singing are linked to seasonal changes in photoperiod. During long photoperiodic conditions typical of spring (canary breeding season) T levels are high and the volume of song control nuclei and song production are enhanced.

Effects of Behavior on Neuroendocrine Systems

Effects of social behaviors on the reproductive processes and on the underlying neuroendocrine mechanisms have also been described. Stimulatory as well as inhibitory effects of the sociosexual interactions have been demonstrated in both sexes. In laboratory mice, for example, the occurrence of puberty is accelerated in the presence of an adult male (Vandenbergh effect). In the anestrus adult female, the contact with the male can induce the resumption of ovulation and estrus cyclicity. This is the case in the laboratory mouse when anestrus is induced by a social stress (Whitten effect) as well as in the ewe or the goat during seasonal or postpartum anestrus. During the estrus cycle, the sensory stimulation among females can influence the length of the cycle as demonstrated in the rat or even result in synchronized ovulations (wild boar and human). In mice, keeping females in large groups can induce abnormalities in the estrus cycles (Lee-Boot effect). In colonies of African mole rat, the presence of a single sexually active female prevents sexual cyclicity in many other females. A similar phenomenon is observed in family groups of small rodents: The presence of the mother inhibits puberty in young females. In males and females of many species, the presence of a dominant subject can reduce or block the secretion of gonadotropins and androgens in a subordinate. In the male, the presence of females and sexual activity have a stimulatory effect on testicular size and T production (mouse, ram, boar, macaque, and ring dove). This effect appears to be due to a rapid stimulation of pituitary gonadotropin secretion. Many effects of social interactions on the endocrine system are mediated by pheromones perceived by the accessory olfactory (vomeronasal) system. For example, the odor of the ram's fleece induces in the anestrus ewe an immediate change in the rhythm of pulsatile secretion of luteinizing hormone, eventually resulting in the induction of ovulation.

Conclusion

Research carried out mostly during the past 50 years has identified endocrine processes that control the sexual differentiation and activation of several aspects of behavior in animals and man. However, many questions remain open, and in particular the ways in which hormones act at the neural level to promote the expression of specific behavior are largely unknown. It has been classically admitted that steroids act as transcription factors to promote behavior expression like they induce morphological and physiological effects but it is clear that other modes of actions, including binding to neuronal membrane and to subunits of the GABA receptor, will be documented in the near future.

It must also be stressed that the activational effects of steroids on behavior do not take place in a morphologically stable neuronal network. Steroids reorganize the neuronal circuits involved in the control of reproductive behavior. These morphological changes affect the number of synapses, the size and the ramifications of dendritic arborizations, the areas of contact between neurons and the neighboring glial cells, the size of the neuronal cell bodies, and even in some cases the number of neurons present in a given nucleus. In most cases, the exact function of these steroid-induced morphological modifications has not been determined.

Progress should also be made in refining these studies at the behavioral level. Although Frank Beach introduced the distinction between the appetitive and consummatory of behaviors in the field of behavioral endocrinology in the 1950s, most of the research analyzing the neuroendocrine basis of sexual behavior has been devoted during the past 40 years to the consummatory aspects of the behavior (intromission and ejaculation in male mammals and receptivity postures in females). This situation slightly changed during the past decade when it was realized that distinct brain regions and neurochemical systems may be implicated in the activation of these two aspects of behavior. Specific tests and experimental situations have been developed to quantify appetitive sexual behavior (motivation) in the male and in the female. Experiments carried out in rats and quail indicate a partial separation of the neurochemical processes that mediate the two aspects of sexual behavior. This topic of research should obviously be expanded.

Acknowledgments

The preparation of this article and experimental work from the authors' laboratory that is described here were supported by Grant R01 MH50388 from the National Institute of Mental Health, Grant CT94-0472 from the European Community, Grant 2.4555.01 from the Belgian FRFC, the University of Liège (Fonds Spéciaux pour la Recherche), and the Government of the French Community of Belgium (Action Concertée 99/04-241) to JB.

References Cited

ARNOLD, A. P., and BREEDLOVE, S. M. (1985). Organizational and activational effects of sex steroids on brain and behavior: A reanalysis. *Horm. Behav.* **19,** 469–498.

BALTHAZART, J. (1989). Steroid metabolism and the activation of social behavior. In *Advances in Comparative and Environmental Physiology, Vol 3: Molecular and Cellular Basis of Social Behavior in Vertebrates* (J. Balthazart, Ed.), pp. 105–159. Springer-Verlag, Berlin.

BALTHAZART, J., and BALL, G. F. (1995). Sexual differentiation of brain and behavior in birds. *Trends Endocrinol. Metab.* **6,** 21–29.

BITRAN, D., and HULL, E. M. (1987). Pharmacological analysis of male rat sexual behavior. *Neurosci. Biobehav. Rev.* **11,** 365–389.

CELOTTI, F., NAFTOLIN, F., and MARTINI, L. (1984). *Metabolism of Hormonal Steroids in the Neuroendocrine Structures.* Serono Symposia, Raven Press, New York.

ETGEN, A. M., UNGAR, S., and PETITTI, N. (1992). Estradiol and progesterone modulation of norepinephrine neurotransmission: Implications for the regulation of female reproductive behavior. *J. Neuroendocrinol.* **4,** 255–271.

GOY, R. W., and MCEWEN, B. S. (1980). *Sexual Differentiation of the Brain.* MIT Press, Cambridge, MA.

HINDE, R. A. (1970). *Animal Behaviour. A Synthesis of Ethology and Comparative Psychology.* McGraw-Hill/Kogakusha, Tokyo.

KELLEY, D. B., and PFAFF, D. W. (1978). Generalizations from comparative studies on neuroanatomical and endocrine mechanisms of sexual behaviour. In *Biological Determinants of Sexual Behaviour* (J. B. Hutchison, Ed.), 225–254. Wiley, Chichester, UK.

MCEWEN, B. S. (1994). Steroid hormone actions in the brain: When is the genome involved? *Horm. Behav.* **28,** 396–405.

MCEWEN, B. S., LUINE, V. N., and FISCHETTE, C. T. (1987). Developmental actions of hormones: From receptor to functions. In *From Message to Mind* (S. S. Easter, Jr., K. F. Barald, and B. M. Carlson, Eds.), pp. 272–287. Sinauer, Sunderland, MA.

NOTTEBOHM, F. (1980). Brain pathways for vocal learning in birds: A review of the first 10 years. In *Progress in Psychobiology and Physiological Psychology* (J. M. Sprague and A. N. Epstein, Eds.), Vol. 9, pp. 85–214. Academic Press, New York.

PHOENIX, C. H., GOY, R. W., GERALL, A. A., and YOUNG, W. C. (1959). Organizational action of prenatally administered tes-

testosterone propionate on the tissues mediating behavior in the female guinea pig. *Endocrinology* **65,** 369–382.

General References

BECKER, J. B., BREEDLOVE, S. M., and CREWS, D. (1992). *Behavioral Endocrinology.* MIT Press, Cambridge, MA.

MICEVYCH, P. E., and HAMMER, R. P. JR. (1995). *Neurobiological Effects of Sex Steroid Hormones.* Cambridge Univ. Press, Cambridge, UK.

NELSON, R. J. (1995). *An Introduction to Behavioral Endocrinology.* Sinauer, Sunderland, MA.

SEYMOUR LEVINE*

Department of Psychology
University of Delaware
Newark, Delaware 19716, USA

FRANCESCA CIRULLI

Behavioral Pathophysiology Section
Laboratorio di Fisiopatologia di Organo e di Sistema
Istituto Superiore di Sanità
I-00161 Rome, Italy

Stress and Coping

We address the perplexing issue of the definition of stress. In contrast to other more traditional definitions, which have not focused on either the stimulus or the response, we propose that stress is a state created by the brain. The hypothesis that we propose is that stress is a response of the central nervous system (CNS) to stimuli which represent uncertainty. The reactions to stress are regulated by the CNS and may vary as a consequence of experiential and genetic variables. The primary responses to stress involve changes in behavior and/or in the activity of the autonomic, endocrine, and immune systems. These outcomes, however, are interrelated and interdependent and can occur under many conditions. The responses, according to the view presented in this article, cannot define stress, whereas coping is defined as the reduction of uncertainty by any combination of control, predictability, or feedback.

Introduction

A precise, operationally defined, and universally accepted definition of stress does not exist. Although there is a massive literature on the biological and psychological consequences of a multitude of situations which have been categorized as stress, there is no consensus among workers in the field concerning the best definition. On numerous occasions this has resulted in the plea to abandon the concept and to focus on the specific psychobiological responses to a well-defined stimulus. The definitional problems become clear when the history of the concept is examined.

Historically, the two major figures responsible for the current use of the term "stress" are Walter B. Cannon and Hans Selye. Cannon emphasized the activation of the autonomic system in response to stress. In his famous description of the "flight–fight" response to stress, he demonstrated the importance of the secretion of hormones from the sympathetic nervous system, epinephrine and norepinephrine, from the adrenal medulla and the sympathetic nerves, respectively. For Cannon, stress was defined in terms of the stimuli required to elicit these responses. Selye (1956) was the first investigator to implicate the endocrine system within the framework of the stress concept. In his famous (or infamous) series of experiments, which led to one of the most robust theories in stress research (the general adaptation syndrome), he described a triad of responses that he hypothesized constituted stress: hypertrophy of the adrenal, stomach ulceration, and involution of the thymus gland. This triad of responses implicated (i) the involvement of

*Current address: Department of Psychiatry, Center for Neuroscience, University of California, Davis, CA 95616 USA

the endocrine system, with an increase in the secretion of both adrenocorticotropic hormone (ACTH) from the pituitary and corticosterone from the adrenal gland; (ii) increased autonomic activity, as evidenced by ulcerations of the stomach; and (iii) the activation of the immune system (involution of the thymus gland). In 1956, Selye was very specific in his statements that stress included all these manifestations and it was the presence of this complex of responses that constituted stress. Furthermore, Selye described these responses as occurring nonspecifically to any or all provocations, regardless of the stimulus. Thus, in one of the earliest attempts to provide an operational definition of stress, this was defined in terms of the response (Fig. 1).

In 1975, Mason presented a revisionist position which challenged the doctrine pronounced by Selye. Mason emphasized the crucial role for what he called "the psychological approaches involved in the emotional or arousal reactions to threatening or unpleasant factors in the life situation as a whole." He argued that the so-called nonspecificity of the endocrine responses to stress occurs because of the emotional component surrounding the experience associated with exposure to stress-inducing stimuli. Thus, the nonspecific response described by Selye is primarily behavioral or psychological in nature and "the interpretative processes underlying the nonspecific bodily responses probably involve a higher level of central nervous system function than was previously realized." In 1979, Hennessy and Levine hypothesized that the endocrine component of the stress response, the hypothalamic–pituitary–adrenal (HPA) axis, was a sensitive indicator of arousal and therefore its response was a reflection of heightened emotional arousal. This theory was in part an extension of the position presented by Mason. What both Mason and Levine accomplished was to make the brain an essential component of the stress concept.

It is important to note that in many ways Mason and Selye were attempting to understand very different aspects of stress. Selye's theory was based on an extensive series of studies that used massively invasive procedures over an extended period of time. It is unlikely that these studies could ever be done again, given the multitude of regulations that now govern animal research. In contrast, many of Mason's experimental paradigms used acute exposure to stress-inducing stimuli. There now exists a large literature which demonstrates that the neurobiology of stress is markedly different between acute and chronic conditions.

The problems surrounding the field are perhaps best stated by Engel (1985) in his severe critique on stress in which he noted that the term is used in very different ways: as a verb " to stress," as a noun "stress," and as an adjective "the stress response." Thus, it appears as though stress can and has been defined as the stimulus (input), the response (the output), or some intervening variable. In a somewhat different approach, Levine and Ursin (1991), in their attempt at a definition, described stress as consisting of three components: an input ("to stress"), processing of the input by the brain ("stress"), and the output, which results from processing ("the stress response"). For Levine and Ursin, any definition of stress must include all three components (Fig. 2). This approach is not entirely satisfactory because if we examine each component of this triad the final conclusion is that what is critical in dealing with the stress concept is the central component of this definition, the processor. In this article, we attempt to justify the definition of stress as a state created by the processor, the central nervous system (CNS). The central theme of this article is that stress is dependent on the response of the CNS to the input

FIGURE 1 Representation of stress used as the logo for the symposium on stress held in 1979 and dedicated to Hans Selye.

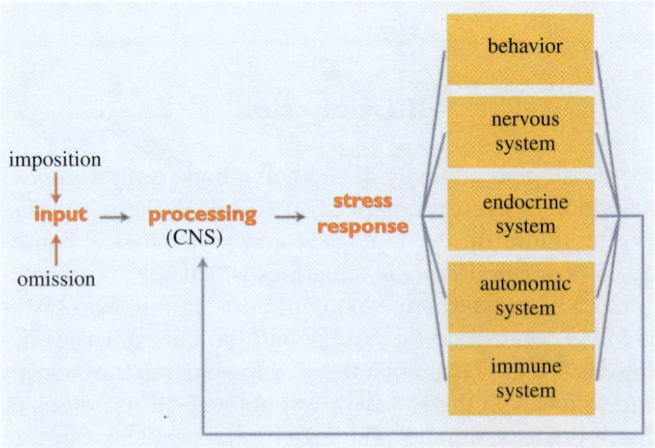

FIGURE 2 Levine and Ursin's (1991) model of stress. Stress is the result of three fundamental components: an input determined by the stress conditions, its processing by the central nervous system (CNS), and the response to the stress, which is expressed by different physiological and behavioral mechanisms.

or inputs, and that the specific response or complex of responses that are generally associated with it are regulated by the CNS.

In order to defend this hypothesis, it is necessary to discuss both the input variable(s) and the response components of stress. It is hoped that the critical relevance of the CNS will become apparent when we examine the relationship between all the components that we have come to associate with stress.

Input (to Stress)

Superficially, the diverse stimuli classified as stressful do not possess any obvious common physical characteristic. The list of stressful stimuli that have been used range from the so-called physical (e.g., sham surgery, hemorrhage, electric shock, intense noise, and injection of toxic substances) to stimuli that can be categorized as purely psychological (e.g., frustration, separation of mother and offspring, separation from peer groups, increased workload, and loss of relationships). Since all these stimuli are capable of activating one or several responses associated with stress, investigators have attempted to derive some common property of these events that would explain their capacity to elicit what appear to be very similar responses. Mason and Levine on different occasions have stated that the common element in all the stimuli that are considered stressful is psychological or emotional. Mason demonstrated that the acute presentation of a stimulus such as heat resulted in the activation of the HPA axis, whereas no response occurred when the animal was exposed to the same high temperature following a gradual change. On the basis of this and other studies, he concluded that the difference between these events was the emotional response associated with the acute change from ambient temperature to heat. Levine, in his attempt to find some common properties of the stimuli involved in eliciting a stress response, stated that the common element in all these events was the lack of information (uncertainty). Uncertainty could be a function of the absence of information or the failure to fulfill an expected outcome, such as receiving a positive reinforcement or avoiding a negative one. The absence or loss of control and/or lack of predictability also produces uncertainty which, when processed by the brain, activates the neurobiological events that result in a stress response.

There have been several attempts to discriminate between different types of stress. Some examples are neurogenic vs psychogenic and physical vs psychological. The problem with any of these classifications is that it is difficult to clearly separate the physical and psychological components from many so-called physical stress stimuli. There appears to be some anatomical basis for not assuming that all stimuli activate identical neural mechanisms. However, it is possible to identify purely psychological stimuli that contain no physical component. It may therefore be possible to identify some common elements that make these psychological situations stressful.

According to Levine and Ursin (1991), the contemporary view of stress can best be characterized as an impositional model. This view is derived primarily from the voluminous number of studies in which severely traumatic stimuli were used to elicit a stress response. There may be specific psychological situations which may fit the impositional model (e.g., defeat). However, from a psychological perspective, it has been hypothesized (Levine and Ursin, 1991) that stress-inducing stimuli have the property of representing the absence of critical features in the environment; by this is meant the lack of information (uncertainty or unpredictability) concerning the obtaining of positive outcomes or the avoidance of negative ones. Thus, stress stimuli, or signals representing these stimuli, indicate that something is missing and that whatever that something might be, it is highly relevant to the organism. Thus, we can categorize stress stimuli not only in terms of the occurrence of an event but also in terms of the omission or absence of critical features in the environment. The similarity between this view and the previously mentioned notion of uncertainty should be apparent. The are many examples of stress stimuli which are best described in this way (e.g., frustration, maternal loss, peer separation, and loss of employment). We are proposing that perhaps inherent to all psychological events that are viewed as stress is the omission or absence of critical information (uncertainty) that initiates the cascade of responses regulated by the brain.

This field in indebted to Jay Wiess (1972) for his pioneering work on coping. It is within the context of these studies that some of the crucial factors that are integral to the understanding of the regulation of the response to stress-inducing events have been elaborated. These studies demonstrated that a physiological response (stomach ulceration) to a well-defined physical stimulus (electric shock) can be modified if the animal is permitted to exert control in terms of regulating in some manner the length or intensity of the shock (control), is presented with information about when the shock is about to occur or terminate (predictability), or is given information concerning the efficacy of the response (feedback). In these experiments, Wiess employed what is now called a "triadic" design. One group of rats could terminate the shock by making an appropriate operant response, whereas the "yoked" animals received the identical shock experience as the rats that could control the shock but they were not allowed to make the operant response required to terminate the shock (Fig. 3). Subjects in the third group received no shock but were confined in the apparatus. The findings indicated that the most severe ulceration was observed in the yoked subjects, although they experienced the identical amount of shock as the rats

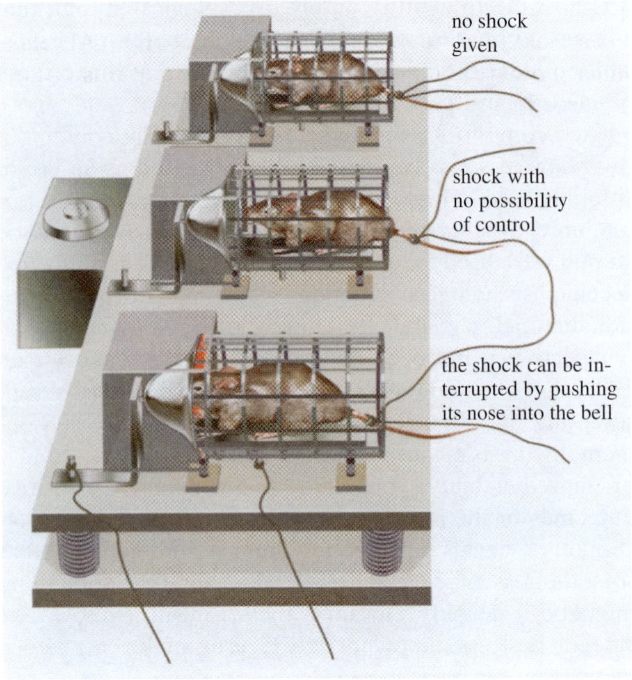

FIGURE 3 The experimental apparatus used by J. M. Weiss. The rat in the foreground receives an electric shock to the tail that can be interrupted by putting its nose into the bell and pressing the back panel. The rat in the center receives the same shock but has no possibility of control. The animal at the rear is the control and receives no electric shock (adapted from Wiess, 1972).

that could exert control. In other experiments by Weiss, animals were presented with a signal that preceded, and thus predicted, the onset of shock. Compared to rats which did not experience the warning signal (unpredictable), the animals which were presented with predictable shock exhibited significantly fewer ulcers. Furthermore, signals which followed a successful response also served to reduce ulceration (feedback).

These three major variables — control, feedback, and predictability — do have some common elements, namely, they can be viewed as having the capacity to reduce or eliminate uncertainty. Control provides the organism with the capacity to eliminate, or at least regulate, the duration and/or the intensity of the stimuli. Predictability, by definition, would serve to reduce uncertainty. Feedback can also be viewed in terms of reducing uncertainty since feedback provides information concerning the efficacy and the success of the response. We can speculate that cognitive and behavioral processes that reduce uncertainty would result in a reduction or elimination of one or more physiological responses to a stressful event. Such alterations in the response to stress as a consequence of one or a combination of the three variables would, by this approach, infer coping. Perhaps the most powerful of these proposed coping mechanisms is control. Although in the original experiments the dependent variable was represented by stomach ulcers, other physiological systems have been shown to be influenced by control, including the autonomic nervous system (ANS), neurochemical processes, and neuroendocrine and neural responses. Furthermore, the effects of control have been demonstrated in many diverse mammalian organisms, such as rats, dogs, monkeys, and humans.

It is important to note that the concept of coping as used by Weiss and others is very different from the approach of Lazarus and Folkman (1984). They defined coping as follows: "Coping involves problem-solving efforts made by an individual when the demands he faces are highly relevant to his welfare, that is, a situation of considerable jeopardy or promise, and when these demands tax his adaptive resources." Such a definition (i) emphasizes the importance of the emotional context in coping; (ii) allows inclusion of both the negative or stress side of emotions and their positive side of potential fulfillment or gratification; and (iii) emphasizes tasks that are not routine or automatized — that is, those in which the outcomes are uncertain and in which the limits of the individual's adaptive skills are approached.

What is clear from the definition of coping proposed by Lazarus is that some coping processes may increase the risk of maladaptive behavior, whereas others may improve adaptation and reduce the risk of negative outcomes. Thus, this definition of coping includes the possibility for both "good" and "bad" coping strategies. In contrast, in the theory of coping proposed by Weiss there can be no unsuccessful coping. Coping is defined by a positive outcome that for the most part is based on the reduction of pathological or physiological indices.

If stress is presumed to be a function of uncertainty created by the absence of relevant information or control, we are confronted immediately with the issue of how uncertainty is perceived and processed. We have no a priori reason to expect to be able to predict the effects of any given stimulus on the organism without knowledge of how the brain of any given individual is going to process a specific stimulus in a particular situation. This in part depends on the information stored in a specific brain for this stimulus situation. In other words, the processing of any event as uncertain depends in part on previous experience. For an organism to process a situation that represents uncertainty, some comparator functions are required. The only system capable of these functions is the CNS.

Processing Variable: The CNS (Stress)

Physiological activation in response to uncertainty is a central component in the model presented by Sokolov (1960). Although Sokolov's model was an attempt to account for the process of habituation, a broader interpretation of this model could also account for why uncertainty would result in the responses usually associated with stress. When a subject is presented with an unexpected stimulus, an orienting response occurs. Both behavioral and physiological responses are elicited by the stimulus. Physiological components of the orienting response include general activation of the brain, originally measured by increased electrical activity (EEG); changes in autonomic activity; and increased neuroendocrine activity. Recently, newer techniques such as the measurement of early immediate genes (e.g., *c-fos*) have demonstrated general activation of the brain following unexpected (uncertain) stimuli. The expression of these protooncogenes has been accepted as an indicator of neuronal functional activation in different situations, including stress. If the stimulus is repeated frequently, most of these reactions gradually diminish and eventually disappear. Sokolov's model is based on a matching system in which new stimuli or situations are compared with a CNS representation of prior events. This matching process results in the development of expectancies. Violation of these expectancies results in an arousal (stress) response. Uncertainty can therefore be viewed as a violation of expectancies. From this perspective, the evaluation of stimulus inputs by the individual is therefore crucial in determining whether a stimulus will elicit a stress response. This evaluation of the stimulus, rather than the stimulus per se, is under the control of the brain. It is not surprising, therefore, that what may appear to be identical events produce wide variations in both humans and animals in terms of the magnitude of the stress response and in the response system or systems that are activated. The fate of any input is determined by central factors or the state of the central processor, the brain. In most cases, therefore, a basic operation of the CNS is to perform some form of comparison. This appears to be true for simple responses such as the orienting reflex (Sokolov, 1960). Novelty can only be detected by signal evaluation taking place within the brain. This is also the case for uncertainty. This information has access to parts of the brain in which it is compared with what has been previously stored. These comparisons may be regarded as expectancy.

Bolles (1972) hypothesized that there are different types of expectancies—stimulus and response outcome expectancies. *Stimulus expectancies* refer to the storing of information concerning temporal relationships between stimuli (i.e., classical conditioning). The second type of expectancy results from learning about the consequences of specific acts. This is referred to as *response outcome expectancy* and is related to instrumental conditioning. Since coping can be viewed within the context of response outcome expectancies, it is the result of a learning process. An organism expects a positive outcome with a high degree of probability, and this expectation predicts a reduced stress response in the individual.

The development of positive response outcomes is dependent on some of the coping mechanisms previously described such as control, which is the capacity to make an active response in the presence of positive or aversive stimuli and to register the consequences of such acts. If the response results in a positive outcome, such as obtaining a positive reinforcement or being able to escape or avoid a noxious event, the brain will store this information as a positive outcome expectancy. This type of learning requires not only positive events to occur but also the brain to be able to code these results. The inability to cope, and therefore not register, positive response outcomes results in high levels of stress. The establishment of positive response outcomes reduces physiological activation both in humans (Ursin, 1988) and in animals. Thus, purely psychological events, such as the absence of information (novelty), failure to meet expectations, or the omission of a previously expected positive reinforcement (frustration), appear to be critical elements that elicit a stress response.

Individual Differences

Insofar as the crucial factor in this current attempt to define stress is the processing variable, it is important to elaborate that there are factors that can permanently alter the state of the brain so that what appear to be identical inputs can frequently result in different responses, both behaviorally and physiologically. It is clearly not possible to identify all the factors that may be responsible for individual differences. It has been proposed that some individual differences are genetically based (Suomi, 1987). There is abundant evidence, however, that some developmental experiences permanently alter what we have called processing. It has been well documented that early (pre- and postnatal) experiences can permanently affect the structure and function of the brain (Meaney *et al.,* 1993). During the past several years there has been a resurgence of interest in this area, and many new paradigms have been explored. Regarding ontogeny, the one manipulation that has been most extensively studied is stimulation in infancy, frequently called early handling. This term is in fact a misnomer. The experimental manipulation simply involves a brief period of maternal separation. Early handling has been shown to affect many of the outcome variables following exposure to stress in the adult. Behavioral, endocrinologic, neurochemical, and immunologic responses to identical stimulus inputs are very different in adult animals that have been "handled" as pups. In general, the effects of early handling appear to be an overall diminution of the stress response. Recently, other forms of early experience have resulted in an adult animal that shows an exaggerated response to stress. Prolonged periods of maternal separation (3 h daily from Day 2 to Day 14) are one example of an early manipulation that results in an accentuated behavioral and endocrinologic response to stress in the adult. Prolonged alterations of the processor can also be obtained by experiences in adulthood. These changes do appear to be reversible, in contrast to early experiences. Exposure to inescapable and unpredictable shock results in an organism that, in general, is hyperreactive to inputs and shows exaggerated outcome responses. These exaggerated responses do not appear to be task dependent and seem to generalize to numerous input stimuli. One important example of the effects of exposure to unpredictable, uncontrollable aversive stimuli is the phenomena of "learned helplessness."

Differences in the response to stress may also be a function of the context within which the stress-inducing stimuli occur. In humans, it is well established that the presence of social support increases tolerance to stressful life events (Isaacson and Janzon, 1986). Similar stress-reducing effects of social interactions have also been demonstrated in nonhuman primates that live in complex social structures (Levine, 1993). In summary, the view presented thus far focuses on the brain's capacity to process information concerning the obtaining of positive outcomes and/or the avoidance of negative outcomes. Stimuli which induce stress do so due to a lack of information (uncertainty). Coping mechanisms are effective insofar as they provide information and reduce uncertainty.

Outcome Variables (the Stress Response)

The general response to stimuli which appear to be stress eliciting is referred to as the stress response. The stress response is a function of the brain processing stimuli as stress and activating specific systems resulting in a cascade of responses presumed to be a consequence of the activation of one or several neural mechanisms which produce some measurable outcome. This outcome could be behavioral and/or any number of different physiological responses. It is a rare event that there is only one physiological system activated during stress. The most common situation is the expression of multiple outcomes that occur when the organism is experiencing the events we are designating as stress. The responses of different systems may have different dynamics (e.g., a different time course), but they will be emitted in some synchronous fashion. Thus, the responses that are observed are usually limited only by the capacity of the investigator in terms of the number of indices that one is capable of measuring. In other words, neural activation will rarely be selective for only one outcome. Usually, a scientist will focus on the system that he or she finds to be of interest and has proven to be reliable as an indicator of stress. The literature is full of the many faces of the stress response.

Molecular biology has provided the scientist with a wide variety of new methodologies to investigate the neurobiology of stress. As a result, our knowledge has increased enormously, but so has our confusion. It is abundantly clear that all the neural systems associated with regulating the outcome responses have an ongoing dialogue. For example, neuropeptides which play a direct role in regulating the HPA system also have numerous effects on the autonomic system and behavior. One such effect is that the major neurotransmitters of the catecholamine family all respond to stress stimuli, albeit idiosyncratically. Several of these neurotransmitters are directly involved in regulating the endocrine output. The reciprocal relationships between the endocrine, autonomic, and immune systems are well documented. The problem is that we do not yet understand all the vocabulary of this dialogue, let alone the grammar.

There is no a priori reason to pick a specific marker of the stress response as the one and only "stress marker." A good stress marker is one that works for a particular laboratory. It must be reliable, quantifiable, and sensitive. However, no particular physiological response that meets these criteria is more valid than any other. There are some important caveats. Although different outputs may indeed be regulated by psychological factors in a similar manner, it cannot be assumed that these distinct outputs bear a causative relationship. Control, for example, can reduce stomach ulceration and also reduce the magnitude of the glucocorticoid response. However, the levels of glucocorticoid are not responsible for the ulcerations which result from the lack of control.

The major physiological systems which have been used as indicators of stress are the ANS, the limbic–hypothalamic–pituitary–adrenal axis, and the immune system.

The Autonomic Nervous System

The ANS was one of the first physiological systems implicated in the response to stress. The ANS is responsible for controlling many of the peripheral responses that are associated with stress. The brain regulates the ANS through neural projections that branch from the brain down the spinal column and to the peripheral organs (Fig. 4). The ANS is divided into two apparently opposing systems: the sympathetic and the parasympathetic nervous systems. In the most simplistic description, the sympathetic nervous system is responsible for activating many of the responses to so-called emergency situations, including changes in heart rate, increased sweating, changes in skin conductance, increased blood pressure, increased gastric motility, and secretion of norepinephrine from the sympathetic nerves and epinephrine (commonly known as adrenaline) from the adrenal medulla. The responses of the ANS are largely involuntary, although there is evidence that biofeedback can produce some voluntary regulation of the ANS. The activation of the various components of the sympathetic nervous system is set in motion to prepare the body to respond appropriately to emergencies or to perceived emergencies. They are intimately involved in the fight–flight reaction. Since these reactions are regulated by direct neural connections, they occur rapidly. Some of the responses of this

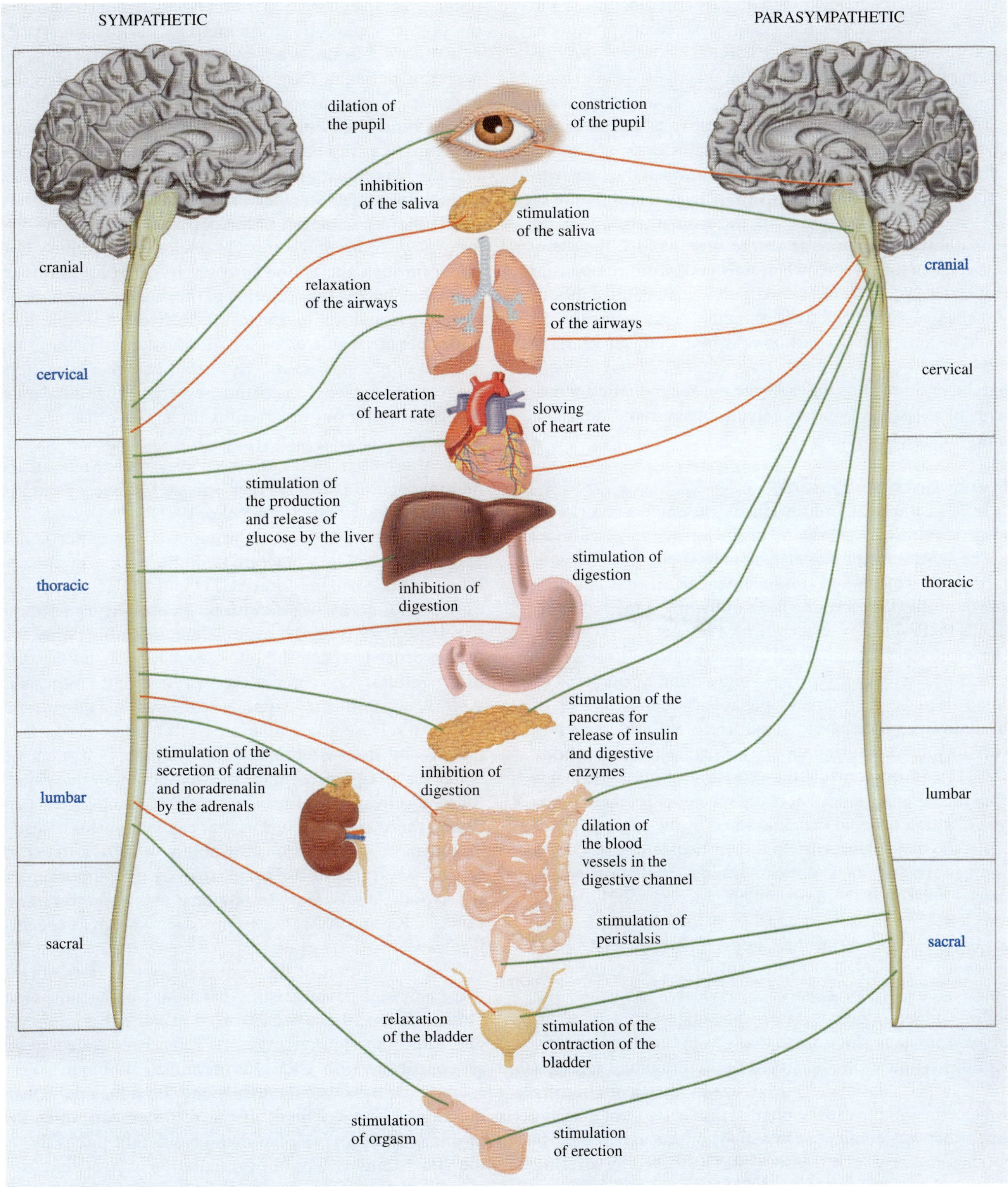

FIGURE 4 A simplified version of the stimulatory (in green) and inhibitory (in red) effects of the sympathetic and parasympathetic nervous systems on various target organs. The gangli of the sympathetic system are localized in the cervical, thoracic, and lumbar regions of the spinal column; those of the parasympathetic system are localized in the cranial and sacral regions.

system are also chemically mediated by norepinephrine and epinephrine. Since the release of these compounds from the sympathetic nerves and the adrenal medulla is also regulated by direct neural innervation, they are expected to be fast-acting processes.

In contrast, the parasympathetic nervous system generally plays an opposing role and mediates calm, vegetative activities. Following ingestion of a large meal the parasympathetic component of the ANS is activated and a state of calm and lethargy ensues. Thus, the sympathetic and parasympathetic projections from the brain course their way to specific organs where, when activated, produce opposite results. Clearly, it would be extremely maladaptive if both branches of the ANS were simultaneously activated. It would be analogous to putting your foot on the accelerator and brake at the same time. To prevent this from happening, the brain regions that activate the sympathetic component during stress typically inhibit, at the same time, the parasympathetic system.

The Endocrine System

In 1948, the late Geoffrey Harris presented a theory that was to revolutionize the world of endocrinology as we know it. The central theme was elegant in its simplicity. The pituitary gland, which was traditionally called the master gland, is responsible for secreting most of the major peptide hormones that regulate the secretion of the peripheral glands. These pituitary hormones are named for their target organs; for example, thyrotropic-stimulating hormone is so called because it stimulates the release and synthesis of thyroxin from the thyroid gland. Here, we are concerned with the pituitary hormone, adrenocortical tropic hormone (ACTH), which controls the secretion of glucocorticoids from the adrenal gland. There are species differences in the type of glucocorticoid that is secreted by the adrenal. In rodents, the major glucocorticoid is corticosterone (CORT), whereas in many other species — canines, primates, and humans — cortisol is the most potent glucocorticoid. ACTH and CORT, or cortisol, are considered to be the major stress hormones. Although in this discussion we are concerned only with the regulation of the stress hormones, the principles established by Harris are applicable to most of the body's endocrine glands. Harris postulated that the pituitary was in fact a slave organ that was in turn regulated by peptides synthesized and stored in the brain and released and transported via the portal vessels that anatomically connect the pituitary to the brain. If the synthesis of the neuropeptides was eliminated by lesions or damage to the hypothalamus, or if their transport was disrupted by severing the blood supply from the hypothalamus to the pituitary, this gland would no longer release its hormones and thus become nonfunctional. Thus was created the field of neuroendocrinology. Harris' theory was specific to the control of those hormones which are secreted by the anterior pituitary. Scharrer and Scharrer (1963) demonstrated that the posterior pituitary hormone [i.e., vasopressin (AVP)] is synthesized in the brain and transported directly to the posterior pituitary. Originally, Harris had postulated that for each pituitary hormone there was a corresponding releasing hormone synthesized in the hypothalamus, a small structure near the base of the brain. It is now recognized that the hypothalamus contains an array of both releasing and inhibiting factors which instruct the pituitary, in turn regulating the secretion of the peripheral glands. In some instances, the brain triggers the release of a pituitary hormone through the action of a single releasing hormone. Sometime it halts the secretion of the pituitary hormone by releasing a single inhibiting factor. More often, the pituitary hormones are regulated by the coordination of both releasing and inhibiting factors. To further complicate matters, the release of at least one of the most critical hypothalamic-releasing hormones involved in the stress response, corticotropin-releasing factor (CRF), is dependent on many afferent inputs from other neural structures. Some of the multiple neural pathways that activate the release of CRF have been described (Sawchenko, 1991).

We have discussed two of the important stress hormones, epinephrine and norepinephrine. In the periphery, the two other major hormones are ACTH and CORT. Activation of these hormones begins with the neural signals required to release CRF from the hypothalamus into the portal system in order to reach the pituitary. CRF acts on the cells of the pituitary to increase the secretion and synthesis of ACTH, which in turn stimulates the cells in the adrenal cortex to release and synthesize CORT (Fig. 5). Together, CORT and the secretions of the sympathetic system account for a large percentage of what occurs in one's body following stress. As is the case with most physiological processes, there are mechanisms that can inhibit this cascade. The primary mechanism is the ability of CORT to act on several structures in the brain, such as the hippocampus, the hypothalamus, and the pituitary, and to inhibit secretion of CRF and ACTH. This process is known as *negative feedback.*

This description of the endocrine system does not reflect many of the complexities that are still being unraveled. Other neuropeptides are involved in the stress response. One important neuropeptide, AVP, has been shown to act synergistically with CRF. Furthermore, although the responses to stress usually involve the hormones previously mentioned, the dynamics of the response can differ following different types of stimuli or different intensities of the same stimulus. Chronic presentation of stress-inducing stimuli changes the pattern of hormones released when compared to acute activation of these processes. In general, most stressors involve the hormonal changes we have outlined, with the glucocorticoid and sympathetic components being the most reliable. However, the rapidity and magni-

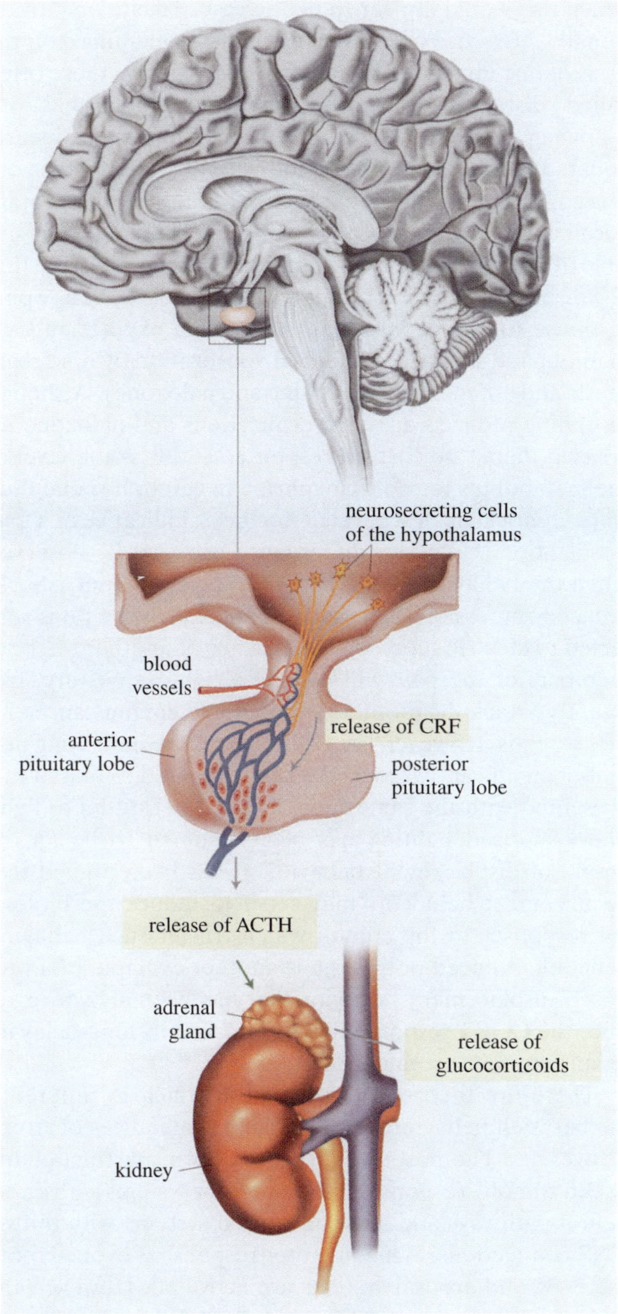

FIGURE 5 Scheme of the hypothalmus–pituitary–adrenal axis. Stimuli are processed by the brain and activate the hypothalamus to release corticotropin-releasing factor (CRF), which then activates the pituitary to release adrenocorticotropic hormone (ACTH). ACTH acts on the adrenal gland to stimulate the secretion of glucocorticoids.

tude with which the secretion of some particular hormone changes may vary according to the input. It appears that different stimuli produce a different hormonal signature that is specific to the stimulus and may also be idiosyncratic for a particular individual.

Immune System

Whereas it is possible to describe the essential aspects of the ANS and endocrine systems, albeit in a somewhat sketchy and simplistic manner, the immune system does not provide us with this luxury. This is a consequence of many problems unique to this system. In the past few years we have seen the emergence of a whole new discipline that has several names, including psychoimmunology, psychoneuroimmunology, psychoneuroendoimmunology, and neuroimmunomodulation. Part of the problem is that within the immune system there are diverse immune processes. Thus, there is cell-mediated and humoral-mediated immunity. Furthermore, the relationship between the immune system and the other systems (ANS and endocrine) appears to be very different. In both the endocrine and ANS the activation of the peripheral hormone is centrally mediated. In contrast, peripheral products from the immune system are capable of activating neural mechanisms. The most striking example of this process is the relationship between the cytokines and the release of CRF. This area of research has received considerable attention. It has been demonstrated that some of the cytokines produced by the macrophages, particularly interleukin-1, act directly on the hypothalamus to induce CRF release, which in turn leads to the increased secretion of glucocorticoids that have been shown to actively suppress many functions of the immune system. Thus, the homeostatic regulation of the immune system occurs in a very different manner than the homeostatic regulation of the endocrine activity. Much of the recent evidence suggests that the endocrine, ANS, immune, and central nervous systems interact and respond to pharmacological and physiological stimuli in a coordinated fashion. The presence of hormones, neurotransmitters, and receptors common to all systems supports the view that bidirectional communication must exist between all systems. Specifically, the brain–endocrine–immune responses to stress are an example of how these different systems respond in an integrated manner. In the case of the immune–endocrine relationships, the two primary candidates involved in coordinating such responses through the brain–endocrine and immune circuitry are CRF and interleukin-1.

However, although at one level the complexities of the multiple relationships between the various biological systems are somewhat daunting, the relationship between stress and illness is even more perplexing. A large clinical literature has documented that chronic stress may have profound health consequences. In one sense, stress has taken on the status of the twentieth-century plague. How often have we experienced the phenomena of our chosen medical practitioner attributing one or another of our unexplainable complaints to stress? Despite these intriguing results, it remains far from clear just how much stress, specifically chronic stress, makes one more vulnerable to diseases that in normal circumstances would be dealt with effectively by the immune system. The assumption that stress is related to health outcomes implies that these outcomes are based on some action of stress on the immune system. However, many of the immune parameters studied in the laboratory bear no known relationship to a specific disease entity. It is not possible to deny that there is some relationship between stress and health, and that chronic stress can produce many negative health outcomes. However, as Sapolsky (1994) so elegantly stated,

> Every child cannot grow up to be president, it turned out that merely by holding hands and singing folk songs we couldn't end all wars, and hunger does not disappear just by visualizing a world without it. Everything bad in human health is not caused by stress, nor is it in our power to cure ourselves of all our worst medical nightmares merely by reducing stress and thinking healthy thoughts full of spirit and love. Would that it were so. And shame on those who would sell this view.

Relationship between Behavior and Physiology

Previously, we stated that there was no specific reason to expect any one measure of the stress response to be any better than another, assuming a particular measure met certain criteria. However, when multiple measures of the stress response are obtained there often appears to be discordance between them. This lack of concordance is often the case when behavioral and physiological responses are obtained simultaneously. Thus, when feral-born squirrel monkeys living alone are exposed to a snake housed in a Plexiglas box placed above the cage, several behaviors indicative of fear are observed. The monkeys will emit alarm calls, display increased activity, and will avoid the snake by maintaining maximal distance. All these behaviors could be interpreted as a stress response. The monkeys will also show marked elevation of cortisol. If the snake is presented in the same way to animals living in a social group, all of the behaviors emitted are essentially the same but there are no changes in endocrine activity. Under the isolated living condition there is a concordance between the behavior and the physiological index of stress. In the social group condition there is a discordance.

When the behavioral and physiological indices of stress are in agreement, there is little problem with interpreting the outcome. The problem arises when there is a dissociation or discordance between two or more of the traditionally used outcome variables. Discordance implies that two variables that are assumed to be causally related change in different ways. Dissociation implies that neither variable is related. True dissociation would be apparent if the behavioral response varied and the physiological responses were constant and vise versa. There are several examples in which this would appear to be the case. To cite one, traditionally stress has been inferred in infant mammals on the basis of one important behavior — vocalization, more often called "distress" vocalization. An increase in vocalizations following separation from the mother has been linearly equated with stress. In a recent series of studies it was observed that following separation from the mother, infant vocalization in monkeys was inversely related to the magnitude of the cortisol response. Initially, we hypothesized that vocalization was an instrumental act that served as a coping response to reduce stress. In subsequent experiments we manipulated separation-induced vocalization by opioid agonists and antagonists (morphine and naloxone). Although morphine reduced distress vocalizations and naloxone increased them, the cortisol response as well as the level of the metabolites for norepinephrine in cerebral spinal fluid were identical in all separated monkeys, indicative of a true dissociation. However, discordance can also be observed when the behavior alters the physiological output. Shock-induced aggression is one such example. When a rat is subjected to electric shock in the presence of another rat, both members of the pair will assume a fighting posture, frequently attack each other, and in some circumstances inflict wounds. However, the ACTH response in fighting animals is significantly reduced when compared to that of rats presented with the same shock but not permitted to fight. Shock-induced fighting has been assumed to be an example of displacement behavior. It has been argued that displacement behaviors may serve to reduce the biological responses to the stimuli which trigger such behavior. Schedule-induced polydipsia is another example of a proposed displacement behavior that has been shown to reduce the CORT response that is produced by long-delay intermittent reinforcement.

There are theoretical grounds on which to differentiate between behavioral and physiological indices of stress. Groves and Thompson (1970) have drawn a distinction between stimuli–response (S–R) or motor systems with direct neural pathways and state or arousal systems with diffuse CNS connections. When a motor response is evoked, both the S–R and arousal systems are activated. However, the two systems appear to be independent. When a motor response is elicited repeatedly, the amplitude of the response may decrease (habituation) or increase (sensitization). Habituation is phasic and occurs in the motor system. Sensitization is tonic and believed to be regulated by the arousal system. Since S–R and arousal systems seem to be subject to different stimulation parameters (frequency affects habituation and intensity is more related to sensitization), dissociations between the two systems are likely, particularly in the case of repeated stimulation. Thus, in rodents repeated exposure to a novel environment results in a reduction of the motor response (e.g., reduction of freezing) as well as reduction of the autonomic response of defecation. The CORT response to repeated stimulation shows little if any reduction after many exposures.

Another example of dissociation is when the physiological indices respond differentially. One striking example occurred in an extensive study of men undergoing parachute training. Prior to and following the initial jump, all behavioral and physiological indices were elevated. Following repeated jumps from the training tower, the behavioral and most of the hormonal responses were no longer present. However, increases in testosterone levels, urinary epinephrine, and heart rate persisted throughout the training. Other instances of directional differences between endocrine and ANS systems have been reported in human and animals. One conclusion from these data is that different physiological outputs are responding to different psychological parameters or are regulated differentially by the brain. Henry and Stephens (1977) also suggested that there is an endocrine specificity associated with different behavioral modes in social stress situations. They believe that there exists an "amygdala system," possibly tied to catecholamines and the activation of the sympathetic nervous system, and a "hippocampus arousal mechanism" associated with the HPA system. That there are stress conditions which activate different components of the stress response seems to be well documented. How the brain organizes and determines which system is operating, and under what conditions, is one of the major questions for researchers studying the neurobiology of stress. When behavior can be presumed to be functioning as an instrumental or coping response, there are few problems since the behavioral response can be viewed as being functional and adaptive. In these instances, discordance is the expected outcome. When there is dissociation, we prefer to accept the physiological indicator as the one more closely reflecting what the brain is processing as stress.

Conclusions

In this article, we struggled with attempting to define stress and confronted many of the problems currently facing the field. The definition of stress as we have proposed it may be no more satisfactory than any other definition. It makes intuitive sense that stress is in the eye of the beholder. This statement implies that in order for any stimulus (input) to activate the brain to initiate the spectrum of stress responses, it must be processed by the brain as having specific properties (uncertainty). The implications of this definition are that unless the stimuli are processed by the brain and specific neural mechanisms are activated, the stimulus has no relevance for the organism. Furthermore, although the output to different stimuli may be the same (e.g., the release of adrenal hormones and sympathetic activation), the response does not define the stimulus as stress. This is contrary to popularly held beliefs. The hormones of the sympathetic nervous system, as well as the adrenal hormones, are a major component of the organism's adaptive machinery. Without these hormones, which perform vital metabolic functions, the body cannot effectively deal with the physiological insults it may encounter. Munck *et al.* (1984) hypothesized that the increased secretion of glucocorticoids acts as a suppressive mechanism to dampen some of the acute reactions to many physical challenges. Thus, the secretion is required to regulate and dampen the initial responses so that they do not damage the organism. The functions of these hormones, therefore, should be regarded as a homeostatic device to reestablish physiological balance. It is thus not surprising that different classes of stimuli may activate different or several neural pathways that respond to diverse inputs. Are we correct in assuming that any stimulus that activates the CRF neuron is to be considered as stress? The position we have taken is very akin to the original doctrine presented by Mason. It is not the output or the input that define stress but the emotional component that is registered and processed by the brain when it is confronted with informational discrepancy.

In both animals and man there is a consistent finding that uncertainty, lack of information, and the absence or loss of control produce outcomes that are consistent with an alarm state. Conversely, the presence of relevant information, behaviors that lead to positive outcomes, and behaviors that result in control reduce or ameliorate states of alarm (stress) with concomitant reduction of the physiological responses associated with this state. The views that have been presented depart from many of the traditional approaches to stress. However, they raise many issues that can and may be resolved as the methodology for studying the brain continues to become more sophisticated.

References Cited

Bolles, R. C. (1972). *Psychol. Rev.* **79,** 394–404.

Engel, B. T. (1985). In *Stress and Coping* (T. M. Field, P. M. MacCabe, and N. Schneiderian, Eds.), pp. 1–11. Erlbaum, Hillsdale, NJ.

Groves, P. J., and Thompson, R. F. (1970). *Psychol. Rev.* **77,** 419–450.

Harris, G. W. (1948). *Physiol. Rev.* **28,** 139–179.

Hennessy, J. W., and Levine, S. (1979). In *Progress in Psychobiology and Physiological Psychology* (J. M. Sprague and A. N. Epstein, Eds.), pp. 133–178. Academic Press, New York.

Henry, J. P., and Stephens, P. M. (1977). *Stress, Health and the Social Environment. A Sociobiologic Approach to Medicine.* Springer-Verlag, New York.

Isaacson, S. O., and Janzon, L. (Eds.) (1986). *Social Support—Health and Disease.* Almqvist & Wiksell, Stockholm.

Lazarus, R. S., and Folkman, S. (1984). *Stress, Appraisal, and Coping.* Springer-Verlag, New York.

Levine, S. (1993). *Psychother. Psychosom.* **60,** 33–38.

Levine, S., and Ursin, H. (1991). In *Stress, Neurobiology and Neuroendocrinology* (M. R. Brown, G. F. Koob, and C. Rivier, Eds.), pp. 3–23. Dekker, New York.

Mason, J. W. (1975). *J. Hum. Stress* **16,** 11–22.

Meaney, M. J., Bhatnagar, S., Larocque, S., Shanks, N., Sharma, S., Smythe, J., Viau, V., and Plotsky, P. M. (1993). *Ann. N. Y. Acad. Sci.* **697,** 70–85.

Munck, A., Guyre, P. M., and Holbrook, N. J. (1984). *Endocrinol. Rev.* **5,** 25–44.

Sapolsky, R. M. (1994). *Why Zebras Don't Get Ulcers.* Freeman, San Francisco.

Sawchenko, P. E. (1991). In *Stress: Neurobiology and Neuroendocrinology* (M. R. Brown, G. F. Koob, and C. Rivier, Eds.), pp. 55–72. Dekker, New York.

Scharrer, E., and Scharrer, B. (1963). *Neuroendocrinology.* Columbia Univ. Press, New York.

Selye, H. (1956). *The Stress of Life.* McGraw–Hill, New York.

Sokolov, E. N. (1960). In *The Central Nervous System and Behavior* (M. A. B. Brazier, Ed.), pp. 187–276. Josiah Macy Jr. Foundation, New York.

Suomi, S. J. (1987). In *Perinatal Development: A Psychobiological Perspective* (N. A. Krasnegor, E. M. Blass, M. A. Hofer, and W. P. Smotherman, Eds.), pp. 397–419. Academic Press, Orlando, FL.

Ursin, H. (1988). In *Neurobiological Approaches to Human Disease* (D. Hellhammer, I. Florin, and H. Wiener, Eds.), pp. 313–334. Huber, Toronto.

Weiss, J. M. (1972). In *Physiology, Emotion, and Psychosomatic Illness* (R. Porter and J. Knight, Eds.), pp. 253–265. Elsevier, Amsterdam.

HANS-PETER LIPP

Institute of Anatomy
University of Zurich
CH-8057 Zurich, Switzerland

Biological Variation of Cognition and Learning

Biological variation of cognition and learning is based on individual differences of the anatomy and the physiology of the brain within a given species, particularly in the limbic system, the higher order association cortex, and their related subcortical structures. The biological roots of such individuality are (i) differences in genes controlling development and maturation and (ii) developmental plasticity including lifetime changes of the brain. Molecular genetics methods permit one to selectively inactivate genes of putative importance for mechanisms of memory and learning and to understand how defined genetic differences result in differential phenotypes, whereas classical genetic methods and experimental natural selection can be used to study brain differences correlated with differential learning talents.

Introduction

No two animals or two humans behave alike; they differ in talents, temperaments, and spontaneous behavior. This individuality has four roots: the genome, the brain, life experience, and chance. However, only the two former fall into the realm of biology. Thus, biological variation of cognition and learning is defined here as predictable differences in learning abilities based on permanent or long-lasting differences in the brain which originate either from the genome or from developmental processes.

This article concentrates on nonhuman mammals, with some reference to birds and humans. Given the complexity of the issue, it first presents basic conceptual, behavioral, and neuroanatomical knowledge necessary to understand gene actions on development of brain and cognitive abilities. Then, examples of experimental studies in transgenic animals are given, and studies on hereditary covariations of brain structures and learning in the laboratory and their possible meaning for microevolutionary processes are discussed.

Prototype versus Individual: A Battle of Concepts

Historically, the concept of variability and individuality is based on two fundamentally different views. In the early nineteenth century, antievolutionary biologists such as Cuvier and Flourens advocated the principle that members of a species are invariant ("*l'espèce est fixe*") and that all variability — morphological, physiological, and behavioral — must be attributed to environmental influences. The opposing view has been developed by observational biologists such as Gall and Darwin: Here, individuality is the fundamental explanatory principle (Lipp and Wolfer, 1995).

This dichotomy is still present in modern biology. Fields with strongly reductionistic tendencies such as *in vitro* biology prefer the notion of a common prototype (a mouse is a mouse, and a cell is a cell) and interpret variability as measurement noise — a concept aptly expressed by the statistical terms "standard error" and "standard deviation." From this point of view, analyzing sources of errors is a necessary chore at best and has little heuristic value. On the other hand, evolutionary biologists, ecologists, behavioral biologists, and clinical scientists who work with less rigidly defined systems have difficulty viewing the major differences between individuals as an error of nature. In statistical terms, they prefer the notion of variance (i.e., the sum of the differences between individuals) in order to estimate similarities and dissimilarities in different situations or after different treatments. To them, analyzing the sources of variability is of fundamental importance for understanding the biology of complex systems.

This article does not attempt to reconcile the two views. However, one can safely assume that evolution has shaped the degree of determination (and also the degree of inherent variability) according to the system levels of the organism. Systems of fundamental importance for life processes (such as the structure of hemoglobin and its oxygen-binding capacities) must have been streamlined by evolution to display high predictability, whereas systems of less direct physiological relevance (such as neuronal nets subserving spatial memory) retain a genuine level of variability. From this perspective, variability is the biological norm, and cellular events are pseudomechanistic processes, an exception from the rule.

Forward versus Reverse Genetics

The dichotomy is also present in the actual trends to investigate the relations between genes, brain, and behavior. Forward genetics is based on a known alteration of the genome and investigates the behavioral changes associated with the mutation. The most straightforward approach is to change the genetic information by means of inserting additional genes (transgenes) or transgenic disruptions (knockouts) supposed to be of relevance for memory and learning and observing what happens. In reverse genetics, one observes behavior first and then seeks the causative genes (or environmental processes) either by examining the brain and its development or by so-called mapping techniques such as quantitative trait loci (QTL) analysis, which tries to link genetic differences to chromosomal markers with a known position in the genome map. These approaches are also known as genotype-to-phenotype and phenotype-to-genotype strategies, respectively. Another synonym in terms of system hierarchies is a bottom-up versus top-down approach.

Genetic Variation of Memory and Learning

Genetic differences in memory and learning are investigated by a special branch in genetics — behavior genetics. In humans, the most widely used method is twin research, but other multivariate statistical methods are used to partition the behavioral phenotypic variation into a genetic and an environmental component. In animals, genetic variation can be easily observed by studying inbred mouse and rat strains which are abundantly available for biomedical research, and there is a large database for strain differences in mice for a multitude of tasks. In many cases, the observed differences are polygenic; that is, they are caused by a large number of genes, each one contributing to the observed difference. At least in mammals, monogenic inheritance of behavioral performance that can be attributed to a single gene without other pleiotropic effects (a "psychogene") is not well documented.

There are good reasons to assume that many differences in learning and behavior are based on the action of only a few important (major) genes. First, selective breeding for learning differences in mice and rats has led to rapid differentiation, usually within three or four generations. If such differences were based on the actions of dozens or hundreds of genes, it would take many years to observe divergence. Also, QTL mapping has allowed the identification of putative chromosomal locations of a small number of unidentified genes underlying behavioral differences. Finally, molecular gene substitution techniques (gene knockouts) have produced mouse mutants in proteins believed to be important for memory and learning.

The following sections briefly review behavioral tests used to measure animal learning and cognition, examine the brain structures involved in such learning and where and how they are modifiable by genes, describe how developmental processes can mask or produce biological variation, and provide with some examples and emanating questions.

Animal Cognition and Its Measurement

In human psychology, *cognition* is defined as the capacity of perceiving a set of relations between objects or acts based on previous learning. In common language, it is called insight or understanding. For studying animal behavior, however, human insight is not always a useful guide. A pet owner watching his or her cats become nervous when the family prepares for a trip is led to believe that they anticipate the departure. This may be true, but the behavior of the cats could simply reflect the excitement and activity of the humans. Thus, inferring cognition for animals requires particular caution and experimental care. There are several lines of research in animal behavior which suggest that various forms of human cognition-like processes exist. However, not all of them are amenable to a systematic study of biological variation.

Imitation Behavior, Tool Use, and Counting

A classical cognitive behavior occurs when an animal learns the solution of a task by observing another animal. Imitative song learning of birds is usually not considered as cognitive, perhaps with the exception of imitation behavior by parrots. Imitative learning in mammals has been documented for great apes and, possibly, for other nonhuman primates, whereas claims of imitation learning in rats are still debated. However, the methodological difficulties in demonstrating imitation learning and related forms of cognitive social behavior (such as master–slave relations in rats) render this type of behavior unpractical for the study of biological variation, which requires large sample sizes and defined genotypes. The same conclusion applies to studies of animal counting abilities and tool use.

The Cognitive Map Concept

Probably the best studied class of cognitive processes is spatial orientation of rats in mazes. Natural behaviors of rodents (and most other animals), such as finding and retrieving food or exploring and defending territories, require the ability to memorize landmarks and to store them internally in the form of a "cognitive map"— a term introduced by the psychologist Tolman in the 1930s and 1940s. A cognitive map is thought to help animals find their way through the world by planning movements, setting goals, and making detours. In the 1970s, O'Keefe and Nadel (1978) refined the concept of a cognitive map. It was now thought to belong to a particular brain system which acquires information about its environment (exploration) and integrates this information into a neural representation by forming a constantly updated spatial map. The mental spatial map serves to compute trajectories to novel or familiar places and might also subserve nonspatial processes such as syllogistic reasoning in humans. The brain system holding the cognitive mapping system was thought to be the hippocampus — a hypothesis based on the observation that neurons in this brain structure were only active when rats visited a given space and on lesion studies in which the hippocampus was destroyed. O'Keefe and Nadel's book stimulated many debates and much research that led to the development of relatively simple and now widespread tasks measuring components of spatial memory in rodents. It is now generally recognized that such spatial memory tasks have a "cognitive" component. In most cases, the degree of cognition attributed to the animal depends on the researcher's expectations or hypotheses.

Spatial Memory Tests in the Laboratory

Radial Mazes. Spatial working memory is generally assessed in so-called radial mazes in which an animal faces eight or more arms, each one offering a food bait. After having consumed a bait, rats usually choose another arm and tend to avoid repeated visits of empty arms, which are counted as errors. After the first two or three choices, short-term spatial memory is required to remember the visited arms. Since previous experience is not useful for solving the task, this type of memory was termed working memory that applies only to a given trial. The problem of this and related tasks is that olfactory cues such as scent marks must be controlled, and that different species may have other foraging strategies. Mice, for example, have a strong tendency to enter arms in which they had found food before, which is probably meaningful from their point of view but noted as an error in the radial maze by the human observer.

Swimming Navigation. Long-term spatial memory of rodents (often referred to as *spatial reference memory*) is frequently measured in swimming navigation, an elegant task introduced and popularized by Morris (1984). Animals are released into a pool with opaque water, from which they can escape by climbing on a platform located immediately below the water surface (Fig. 1). Although the first contacts with the platform occur by chance, rats and mice quickly learn to find the invisible platform position, often after a few trials. They are able to do this by calculating their release position in the tank from visual landmarks surrounding the pool, which implies that they must have a mental representation of the target platform which enables them to calculate its position and to find it. Since the goal is not visible, the term navigation is justified here. Proof that spatial memory in the form of a map-like representation is responsible for the learning is derived from a procedure termed *probe trial.* After repeated training to a given position, the platform is removed or its location is changed (Fig. 1). When the rat or mouse continues to search around the old platform position (spatial retention), there is good reason to believe that the animal's behavior is actually based on a mental map; in other words, there is a cognitive component. Because there are no disturbing odor tracks, and because the swim paths of the animals can be recorded and analyzed thoroughly with the aid of computers, it is probably the most suitable laboratory task to assess animal cognition or parts thereof. On the other hand, it is also frequently misinterpreted. Many rodents learn poorly because smooth acquisition is disturbed by noncognitive responses such as thigmotaxis, the instinctive tendency of mice to swim along the walls. However, animals that do not learn the task well

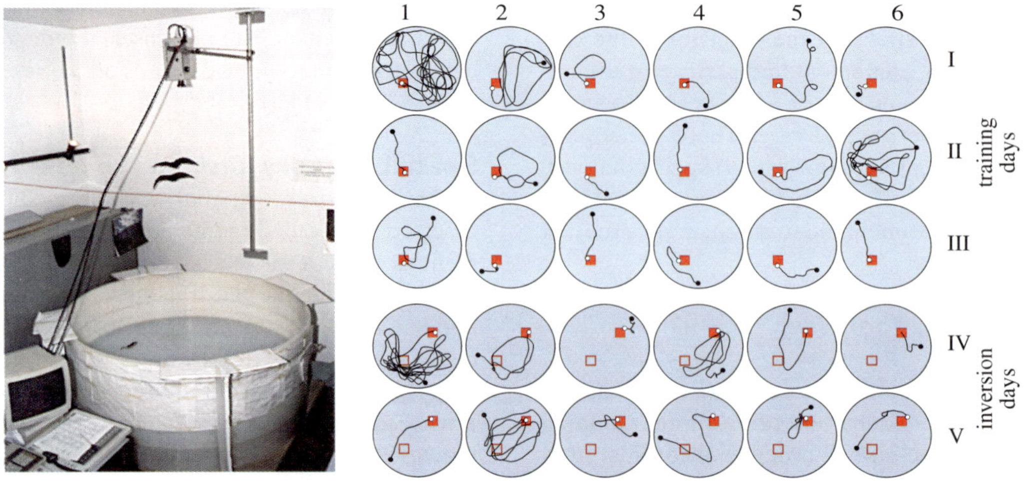

FIGURE 1 Swimming navigation in mice. (Left) The apparatus for recording the swim paths. Swim paths are stored in a computer and are analyzed off-line. (Right) The swim paths of one mouse tested for six trials a day (1–6) for 5 days (I–V), for a total of 30 trials. Each trial lasted until the mouse found the platform indicated by the red square or until 120 s had elapsed; the starting points are indicated by the black circles. Note the rapid learning on the first day. After the platform was rotated 180° (inversion), the animal searched in the original position (Days IV and V, white square). The proportion of time spent in the former target quadrant during the first 60 s of the first inversion trial (trial 1 on Day IV) can be a useful measure of spatial memory.

will show impaired scores of spatial retention in the probe trial, even if they have a perfect memory. In contrast to many other behavioral tasks, wrong conclusions can be avoided by careful visual inspection of swimming paths and by a statistical analysis which corrects the bias produced by noncognitive behavior.

To illustrate this point, Fig. 2 shows the results of a factor analysis of swimming navigation behavior in 1200 mice using several behavioral measures calculated by computer from the digitized swimming paths of mice. Most of the individual variability observed in swimming navigation (about 45%) can be explained by a factor correlated with

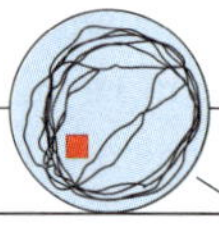

VARIABLE	factor 1	factor 2	factor 3
thigmotaxis	0.854	0.050	− 0.118
time (%) in the target quadrant	− 0.718	− 0.047	0.293
escape latency	0.889	− 0.219	− 0.111
swimming speed	− 0.071	0.969	− 0.047
time spent floating	0.708	− 0.428	− 0.016
spatial memory score	− 0.191	− 0.040	0.970
variability explained	45%	20%	15%

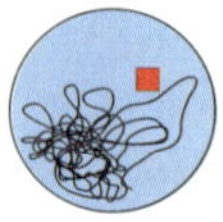

FIGURE 2 Results of a factor analysis carried out to investigate the variables strongly correlated with spatial learning in 1200 mice, whose swim paths were tested under standardized conditions (same number of trials and same period of platform reversal). The scores refer to the correlation coefficients between the factors and each of the original variables. The inset circles show examples of the behavior indicated. The experimental data for the single variables, with the exception of spatial memory, are the means of the scores calculated from 36 trials. The color scale indicates the correlation from the lowest absolute value coefficient (light blue) to the highest (dark blue).

long escape latencies, poor searching behavior in the target quadrant, and thigmotaxis. A second factor, unrelated to the first, is the swimming speed of mice and a tendency to float motionless for some time. It is also noncognitive and accounts for about 20% of the variability. A third factor, explaining about 15% of the variability, relates to the time that mice spend swimming over the old platform position — a measure of cognitive spatial memory. Note that this variable also has a negative loading on the first factor: Mice that show much thigmotaxis are unable to develop spatial memory. Overall, this analysis illustrates that most behavioral variation in this task can be explained by noncognitive factors, but it also reveals the presence of a memory factor completely unrelated to the interfering behavior. Thus, it is justified to infer spatial cognition in animals but only after comparing this behavior with other behaviors that are invariably present in this task but also in all other behavioral tasks.

Field Studies of Spatial Memory. Another, more ethological line of research stimulated by the cognitive map concept is the study of spatial memory and underlying neural structures across closely related members of a species. The best known paradigms are memory for food caches which show major differences within tit and corvid species, apparently dependent on the habitat used (Clayton and Krebs, 1995; Shettleworth, 1995), whereas mammals have been investigated less (Jacobs, 1995). Finally, ethological studies on long-distance navigation such as in homing pigeons can probably be classified in the category of cognition-based processes in animal behavior. As with the field studies in food-storing birds, analyzing the sources of individual variability is perhaps not impossible but impractical.

Other Forms of Learning

Biological variation is also studied in other forms of learning for which cognition is not essential, including Pavlovian conditioning which involves innate reflexes incorporated into a response toward a conditioning stimulus, imprinting to visual stimuli after hatching in precocial birds, and various forms of so-called instrumental conditioning in which an animal must perform a motor act in order to obtain a reward or to avoid a punishment. A special case is extrapolation. This term, introduced by the Russian animal psychologist Krushinski, denotes a nonlearned ability to extrapolate movements of objects — a concept of automatized cognition. Biological variation in some of these learning procedures will be described in the context of the experimental data presented later.

Brain Structures and Connections Subserving Memory and Cognition

Learning of the form of the stimulus–response type can be observed in very simple nervous systems. Cognition, however, requires complex brains which are able to extract relevant information that is assembled, blended into ongoing motor activities, transferred to memory, and stored in relational form as mental maps. A large body of evidence indicates that there are brain structures which are specifically involved in formation of mental maps, complex memories, and cognition of complex features and situations, although it is not always possible to draw sharp anatomical and functional boundaries. Many of these brain structures are anatomically and biochemically similar, with functional differences emerging largely from the way they are interconnected. Hence, many genes and developmental processes equally affect the cognitive and noncognitive parts of the brain. Since the main interest of this article is to understand how genes can affect memory and cognition, the following simplified description emphasizes those features of mammalian neuronal circuitry which permit such specific gene action.

Limbic Circuitry and Higher Order Association Cortex

Areal Organization of the Neocortex and the Hippocampal Formation. The cerebral cortex of mammals can be subdivided in the functional parts according to input and output organization. Many cortical areas in the posterior half of the cortex, the primary sensory cortices, receive sensory information conveyed from sense organs through so-called specific thalamic relay nuclei. The neurons of these nuclei send their axons in a precise point-to-point projection to the cortex, thereby creating visual, auditory, somatosensory, and kinesthetic maps of the body and the world. A narrow band of the cortex, the primary motor (or precentral) cortex, serves as the cortical output structure sending axons to the motoneurons of the brain stem and the spinal cord.

The remainder of the cortex has been called "association cortex" by the neuroanatomist Flechsig. The association cortex is organized as a patchwork of fields through which information flows in both directions. Cortical areas adjacent to the primary sensory cortex are known as unimodal association cortex and those adjacent to the primary motor cortex as premotor association cortex. These specialized association areas are reciprocally connected with the primary cortices but are also reciprocally connected to higher order multisensory cortex (or polymodal association cortex) in the parietal and temporal lobe. The premotor association cortex, in turn, is reciprocally connected to so-called higher order association cortex located mainly in the prefrontal cortex. The higher order association cortices of both sensory and motor type connect reciprocally with so-called paralimbic cortex (the cingular, retrosplenial, and entorhinal cortex), which finally connects reciprocally with the cortical portions of the limbic system — the amygdala and the hippocampus. The paralimbic cortex also receives olfactory afferents which bypass the sensory thalamus. A

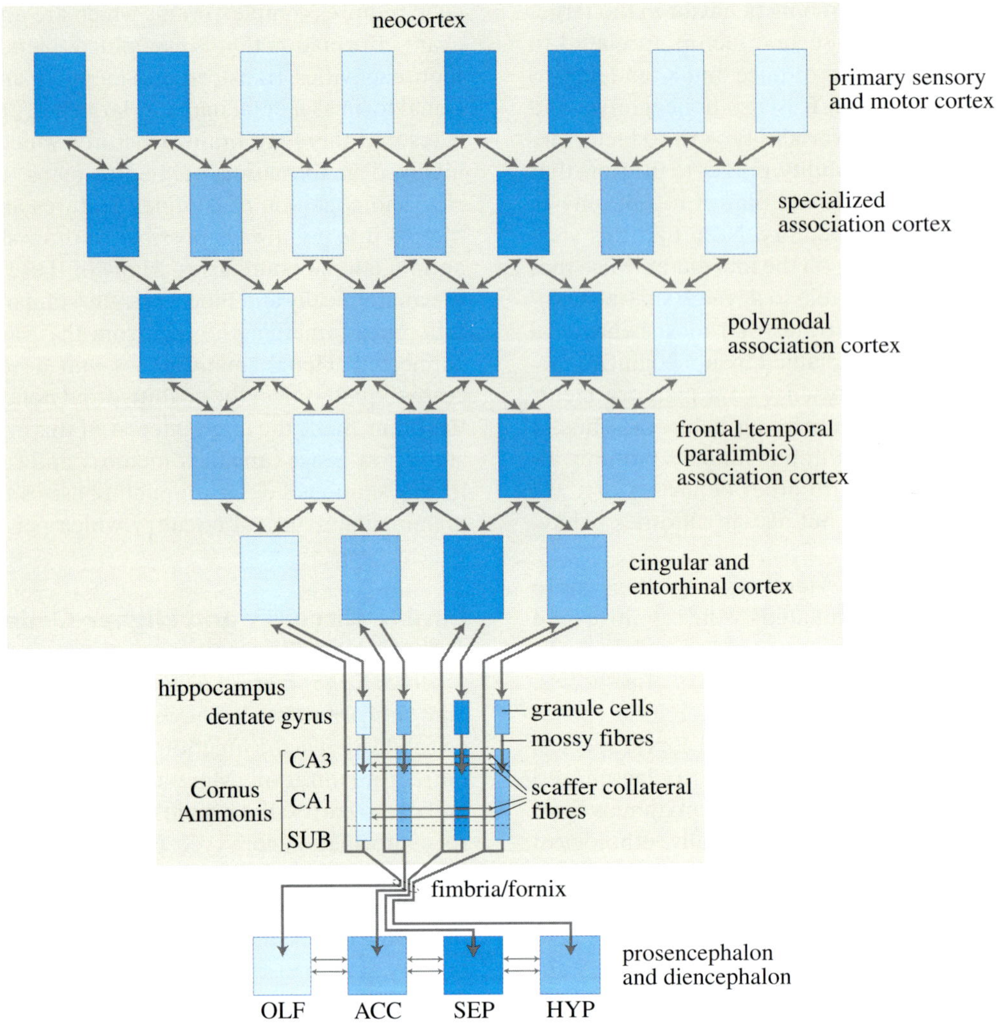

FIGURE 3 Reciprocal relations between the areal organization of the neocortex, the hippocampal formation, and the basal forebrain. The organization of information flow in the central nervous system is both convergent and divergent. The activation patterns produced at the cortex level (which represent thoughts, memories, emotions, etc.) converge toward the hippocampus and are synthesized as progressively more complex cognitive and motor activities. The hippocampus, in turn, returns the reelaborated input to the neocortex and, at the same time, coordinates activation of the prosencephalon and diencephalon. The different color intensities indicate the various patterns of activation. ACC, nucleus accumbens; CA1, CA3, subregions of the cornu ammonis of the hippocampus; HYP, hypothalamus; OLF, olfactory cortex; SEP, septum; SUB, subiculum.

scheme depicting the relations between cortical regions and the hippocampus is given in Fig. 3. A large body of evidence from humans and animals shows that higher order association cortex is responsible for cognitive processes, specialized memories, and their integration into complex and planned motor acts, whereas the paralimbic cortex, amygdala, and hippocampus act as general memory and retrieval structures.

Long Intrinsic Connections of the Cortex. Although the majority of fibers in the neocortex subserve reciprocal connections between adjacent cortical areas, long reciprocal fiber tracts connect primarily the physically separated higher order association cortices. These are divided into intrahemispheric (ipsilateral) and commissural bundles. For example, in humans the angular bundle connects ipsilaterally motor association cortex related to language control (area 44 or Broca's field) with higher order auditory association areas located in the temporal lobe, an arrangement of fundamental importance for speech. A similar fiber bundle, the cingulum, connects the paralimbic temporal and prefrontal cortex areas.

Commissures are fiber bundles which connect to the other half of the brain. In the forebrain, they include the corpus callosum, the anterior commissure, and the hippocampal commissure. Most of the fibers run to the corresponding cortical area (homotopic or reciprocal connections), and a lesser amount terminate in other contralateral regions (heterotopic connections). Callosal fibers preferentially connect association cortices, whereas primary motor and sensory cortices have relatively few commissural fibers. Functional specialization of the hemispheres (i.e., the localization of different talents in the right or left hemisphere) is mediated by these commissural pathways.

Subcortical Links between Cognition and Motor Systems

In discussing memory and cognition, it is often not realized that limbic and higher order association cortex is, as is the rest of the cortex, only a functional part of a much more extended circuitry consisting of recurrent loops which include the basal ganglia, the thalamus, and the cerebellum (Fig. 4). Fibers from all neocortical areas project to the basal ganglia, in which impulses are funneled and routed into thalamic nuclei projecting back to motor cortex and premotor association cortex (motor thalamus). Long, descending projections from the neurons in the motor cortex reach the spinal cord in the form of the corticopyramidal tract. The other target structures receiving input from the basal ganglia complex are the so-called supraspinal motor regions located in the formatio reticularis of the brain stem. This region orchestrates species-specific motor patterns and also sends descending fibers in the form of polysynaptic pathways to the spinal cord. Together, these two comparatively small pathways represent the final common pathway for all motor actions, that is, behavior (Fig. 4, right).

A parallel loop system contains projections from all cortical regions that end in the gray nuclei of the pons, from which an enormous number of fibers reach the contralateral cerebellum. The Purkinje cells of the cerebellar hemispheres, in turn, project to the deep cerebellar nuclei whose axons converge again on the contralateral motor thalamus, in which they blend with the input from the basal ganglia in order to reach the motor cortex.

Approximately parallel to the subcortical loops elaborating motor acts are fiber loops from the paralimbic cortex areas (Fig. 4, left). Collaterals descend to a ventral portion of the basal ganglia. From there, some fibers reach the motor thalamus, but others terminate in the nonspecific (mainly anterior) nuclei which project back to the limbic cortices. Other axons of the paralimbic cortex send signals to the pontine nuclei, from which they are relayed to the neocerebellum and then back to the motor and to the nonspecific thalamus. This latter limbic corticocerebellar loop is poorly investigated, however. The hippocampus sends fibers to the nucleus accumbens, the mammillary bodies, and other hypothalamic areas. Axons from the hypothalamus then reach the anterior thalamus, in which they converge with afferents from the limbic basal ganglia. The anterior thalamus (limbic thalamus) then projects back to the paralimbic cortex, closing the so-called Papez circuit. The amygdala, adjacent to the hippocampus, has very similar efferent projections, but one fiber tract descends to the reticular formation and reaches neuron groups regulating body functions.

In summary, activity patterns from all cortical regions are funneled into two parallel loops passing through the basal ganglia complex and the cerebellum but converge finally on the motor cortex. In this way, information from all cortex areas is blended into ongoing motor activity which remains inseparably intertwined with perception and cognition. Interaction between neighboring loops can take place in all the brain structures through which they pass. Direct interaction between physically separated loops, however, is restricted mainly to the cortex, with its long intrahemispheric and commissural fiber bundles.

Subcortical Links of Importance for Cognition and Memory

In addition to the corticosubcortical loops modulating ongoing motor activity, there is another set of subcortical connections relevant for memory and learning.

Reticular Formation. The phylogenetically old structure subserving learning and memory is the reticular formation of the brain stem, in particular its more rostral portions in the midbrain and the pons. It consists of an ill-defined neuronal network which embeds sensory and motor nuclei of the cranial nerves, passing ascending and descending fiber tracts from which it receives offshoots (collaterals), and various specialized subregions without clear anatomical boundaries that control respiration, blood pressure, and other vegetative functions. It is also connected to the cerebellum. In a simplified view, it may be perceived as the central processing unit of the brain, whereas the forebrain is comparable to a powerful graphic board which has been added to a small computer. This status can be derived from the presence of powerful efferent fiber systems leaving the reticular formation in order to control the activity of the entire forebrain but also of cerebellum, brain stem, and spinal cord.

Ascending Monoaminergic Projections. First, the formatio reticularis contains so-called monoaminergic cell clusters formed by noradrenergic, dopaminergic, serotoninergic and, in part, cholinergic neurons. These send ramifying axons in many regions of the brain, particularly the forebrain, in which they exert a powerful local control of neuronal activity, together with ascending cholinergic axons of the basal forebrain (which may be considered as a rostral extension of the reticular formation).

Reticular Formation and Nonspecific Thalamic Projections. The second way by which the reticular formation is able to control neocortical activity is through the thalamus. Apart from the motor and sensory relay nuclei

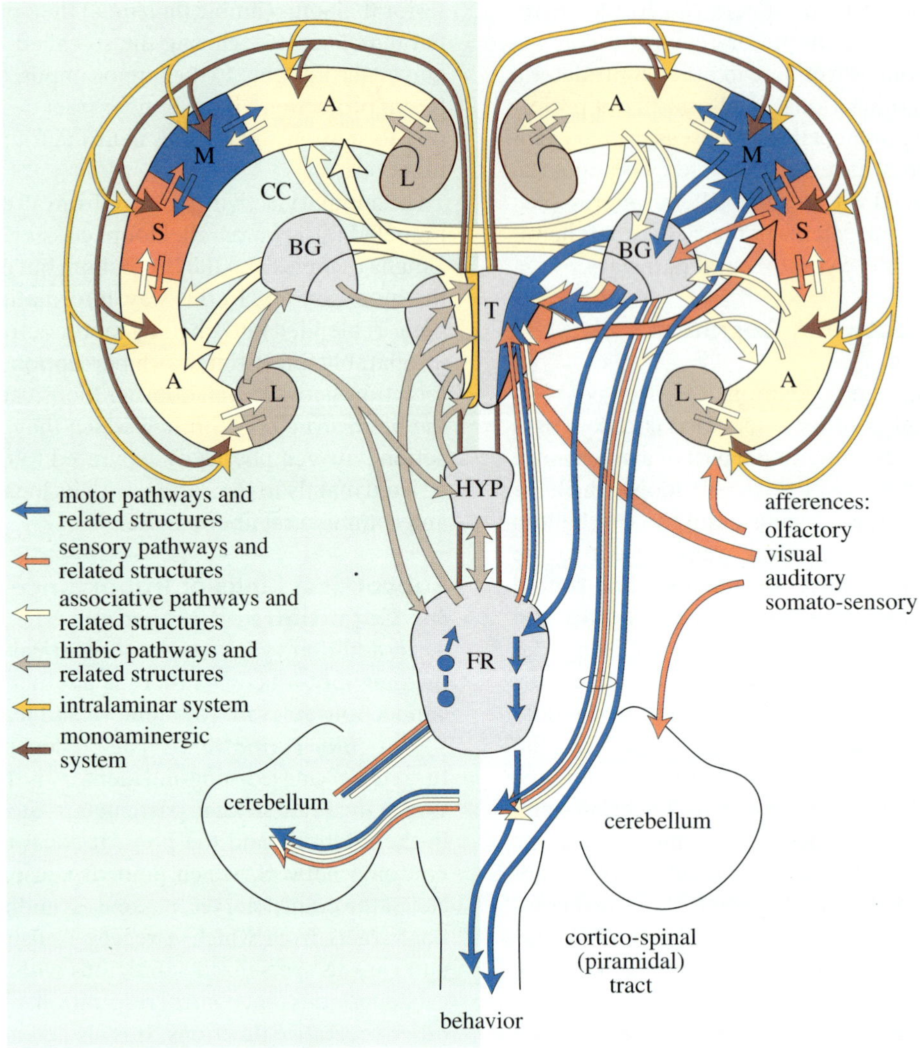

FIGURE 4 Schematic representation of the cortex and subcortical loops. The right side shows the main connections of the primary sensory (S) and motor (M) cortex areas: Sensory afferences (right, center) reach the thalamus (T) and from there are distributed in the sensory cortex fields. The information spreads into adjacent association cortex (A) areas and reaches the limbic cortex (L). The motor cortex gives origin to the fibers of the corticospinal tract (blue arrows) which mediates fine-tuned movements. The other motor output channel, which originates in the reticular formation (FR) and the brain stem, mediates species-specific motor behavior. The output activity of the motor cortex is constantly refined by two loop systems (band of multicolored arrows). Cortical fibers from all regions converge on the basal ganglia (BG) and send, via the motor thalamus, information back to the motor cortex. Another fine-tuning system is formed by cortical fibers from all regions reaching the contralateral cerebellum, which feeds signals back to the motor thalamus and then to the cortex. The left side shows the connective peculiarities of the cognitive (association and limbic) cortex. The principal input originates in the hypothalamus (HYP) and the reticular formation. The flow of information to and from the sensorimotor cortex is under subcortical control. Output goes to the other hemisphere via the corpus callosum (CC) and to the limbic cortex. The limbic cortex is connected to the basal ganglia, which in turn converge on the intralaminar system of the thalamus. Other fibers from the hippocampus and amygdala reach the hypothalamus and reticular formation, and from there the associative thalamus.

(sensorimotor thalamus), this structure contains another set of relay nuclei, the so-called nonspecific thalamic nuclei, which send topographically organized fibers either to the temporal and frontal association cortices or to the limbic cortex. These nonspecific relay nuclei serve to feed information from the reticular formation and the hypothalamus to the associative and limbic cortex.

Reticular Formation and the Intralaminar Thalamic Projections. The thalamus contains, in addition, a system of thalamocortical projection neurons that are interspersed in clusters, the intralaminar system. The axon endings of intralaminar neurons show much less topographic precision than the relay neurons of the sensorimotor or nonspecific nuclei since one ascending axon can ramify considerably within a cortex region. This intralaminar system has the ability of sensitizing or desensitizing extended cortical regions, exerting in this way the perhaps most powerful control of cortical activity. It is thus no surprise that the main afferents to the intralaminar nuclei originate in the reticular formation and the hypothalamus, with some intralaminar cell clusters also receiving input from fibers carrying pain signals.

Regions in which Genes Can Act to Change Memory and Cognition

Genetic influences specific for memory and learning can occur if a relevant brain region has unique structural, neurochemical, or developmental properties. The effects of a single gene on memory and cognition will be most distinct, however, when it affects some critical sites, such as anatomically recognizable circuitry bottlenecks, subcortical control systems, or physiological set points (e.g., for thermoregulation or circadian rhythms).

Hippocampus

This structure (Fig. 5) has many characteristics which make it an ideal target for biological variation without sensorimotor contamination. It holds a commanding position with respect to memory storage and retrieval (see Fig. 3); it contains clearly demarcated subregions, the CA fields, which show neurochemical specialization; and it is one of the three regions of the brain in which many developmentally active genes continue to be expressed (the others are the cerebellum and the olfactory bulb). The hippocampus also contains a projection system, the hippocampal mossy fibers, which is morphologically and biochemically unique in the brain, thus offering a target for highly site-specific genetic variation. Finally, parts of its circuitry differentiate very late, and it contains abundant receptors not only for many neuropeptides but also for circulating hormones.

In fact, studies in mice and rats have revealed intraspecific genetic variation in its size, the number of neurons, the topographical distribution of the mossy fibers, the quantity and numbers of several receptors for neurotransmitters, and the number of cholinergic septohippocampal neurons. Although the cytoarchitecture (the arrangement of neurons) of the hippocampus appears fairly similar across many species, other features such as the mossy fiber projection show considerable morphological variations between closely related species (see Fig. 13).

Amygdala

The amygdala, located adjacent to the hippocampus, likewise occupies a central position with respect to memory and learning but is more specialized for the coordination of emotional and visceral responses. A peculiarity of the amygdala is that it receives auditory input from the thalamus, a feature which may explain its particular role in fear

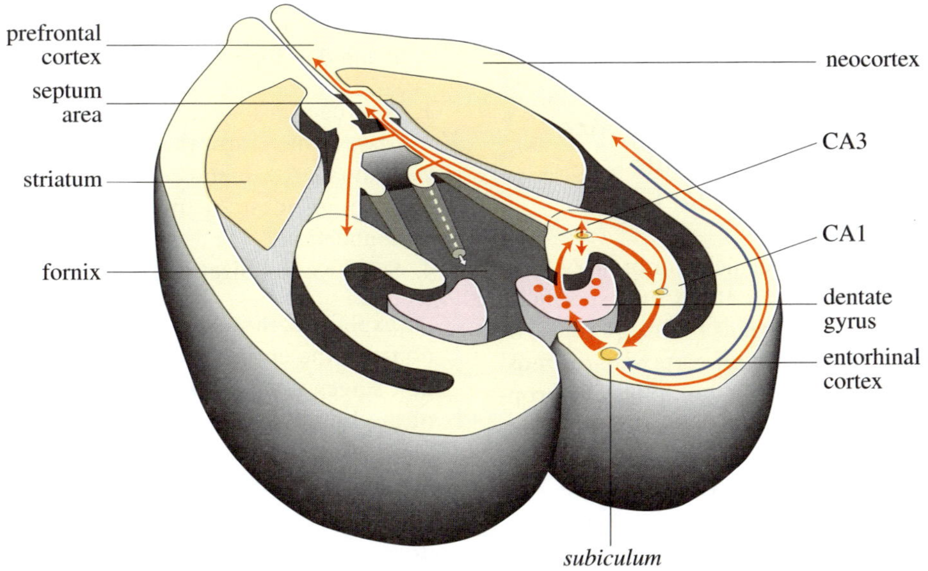

FIGURE 5 Scheme of the connections between the neocortex and the hippocampal formation in a horizontal section through a rodent brain. Information from the neocortex reaches the entorhinal cortex (blue arrow) and enters the lamellar loop (red arrows). In this loop the information proceeds via the dentate gyrus, the CA3 and CA1 hippocampal subfields, and the subiculum, returning to the entorhinal cortex. This loop is also (incorrectly) called the trisynaptic loop.

reactions to sounds. Given the presence of many neuropeptides and their receptors, one may expect to find genetic differences in this structure, particularly in individuals differing in emotionality. However, systematic investigation of genetic variation in the amygdala is rare.

Higher Order Association Cortex

These cortical fields appear to have less anatomical and neurochemical characteristics distinguishing them clearly from noncognitive parts of the cortex, except perhaps for the dopaminergic (mesolimbic) projection from the midbrain to the medial prefrontal cortex. Their most distinctive features are the subcortical afferents from the reticular formation and hypothalamus by way of the nonspecific thalamic relay nuclei and their extensive ipsilateral and commissural connections. Since the axons of these afferents compete developmentally for synaptic terminations, their proportion might be changed dramatically by genes controlling the formation and establishment of fiber tracts. One example of complete reorganization of connectivity by gene mutations is given in the section on developmental mechanisms.

Thalamus and Brain Stem

Although genetic variation of the thalamus has not been reported, genetic alterations of the thalamus have the potential to profoundly alter cortical activity, particularly if the intralaminar system is affected. Because of the difficulties in defining the reticular formation, genetic variation in this structure has not been observed. It occurs, however, within its embedded ascending control systems. In mice, there are many strain differences with respect to dopaminergic, noradrenergic, and serotoninergic systems.

When Genes Must Act to Influence Cognition: Flechsig's Rule (the Later the Better)

A particular gene without local specificity for brain systems governing memory or learning can attain such specificity through timing — a principle known as Flechsig's rule. For example, the higher order association cortex shares most features with the noncognitive cortex, but it has one property which makes it suitable to induce variation of cognition and learning: It shows protracted maturation. The best investigated marker of maturation is myelination, the continued ensheathing and thickening of axons by lamellae of myelin. As discovered by anatomist Flechsig in the late nineteenth century, some cortical fields in the human brain myelinate distinctly later than other fields, for example, coincident with the childhood period in which language develops and even later. Flechsig used such differential timing of myelination to define areas of "association" cortex presumably mediating the complex behavior appearing at those stages of behavioral development. The concept has remained valid and is widely although speculatively applied in neurology.

There probably are genes that act solely late in development such as those which govern puberty. On the other hand, many genes are chiefly active during brain development, but some of them continue to be expressed during adulthood. Since many brain-specific genes appear to have double roles during development and in adulthood, current efforts in molecular biology concentrate on the problem of producing targeted deletions which are activated in the adult brain, thereby avoiding the developmental consequences of the gene disruption. The concept that late-acting genes are most likely to cause variation in mental processes also implies that they often act jointly with environmental factors which are becoming increasingly more important during late brain development and maturation (Lipp, 1995).

How Genes Produce Variation in Memory and Learning

The main condition for a gene effect resulting in different cognitive or memory talents is that the effect remains constant over long periods. This sounds trivial, but unlike other organs, the brain has an immense capacity of plasticity and dynamic adaptation. The morphological plasticity is strongly reduced during adulthood, however, at least for distant connections and numbers of neurons. Thus, genetically induced structural variability is a potentially useful way of fixing mental traits.

Brain development can be divided into four stages:

1. Neurogenesis and neuronal migration — the two processes which form the shape of the brain and its parts
2. Axonal growth and establishment of connections
3. Synaptogenesis, neuronal death, and pruning of connections
4. Selective stabilization of connections and maturation

As explained, these developmental stages are not synchronous for the entire brain. There is heterochrony, with the first stages usually taking place in the more caudal parts of the brain. Genetic variation may occur during all four stages, but neurogenesis and formation of connections appear to be more suitable for establishing long-lasting changes.

Variation in Size

There is no doubt that species differences in brain morphology and function are caused by genes encoding neuronal numbers and other form-shaping factors, but the simple hypothesis that differential size of a brain part within a species should be related to differential functional capacity has been met with skepticism since first proposed by Gall. Nonetheless, there are many examples of such a relation. In canaries, the cell numbers and the volume of the hyperstriatum ventrale, a song control nucleus, show large differences between singing males and nonsinging females; the size of the nucleus correlates with the number of syllables recognizable in the song repertoire of a male bird and undergoes seasonal variation, being large during the song period and small during winter (Lipp and Wolfer, 1995). Similar correlations between structural size and special talents have been reported after comparison of hippocampal size in food-storing corvid species with different cache memory and in tit species storing food or not (Clayton and Krebs, 1995; Shettleworth, 1995). In mammals, hippocampal size in related species of kangaroo rats was found to correlate with habitat size, whereas sex differences in hippocampal size of mole species appear to be related to gender differences in home range (Jacobs, 1995).

In humans, such size differences appear most clearly in the form of left–right asymmetries. *In vivo* magnetic resonance in musicians with perfect pitch showed a stronger leftward anatomical asymmetry of the planum temporale, an auditory association cortex in the temporal lobe (Schlaug *et al.,* 1995). Thus, it appears reasonable to assume that bilateral and asymmetrical volume differences in brain regions subserving memory and cognition are responsible for other differential talents. They may be more difficult to detect, however, because the planum temporale is one of the rare cortex regions with measurable functions and defined anatomical boundaries.

Variations in Connectivity

The establishment of connections is another developmental process which offers many opportunities to produce genetic variation in mental functions. The formation of axonal connections is a highly dynamic process which includes pathfinding and establishing synaptic contacts. The process is competitive: Neurons usually grow exuberant numbers of axonal branches, with the axons competing with other afferents for contact sites on the target neurons. Neurons which fail to establish a sufficient number of synaptic contacts eventually die, leading to pruning or even complete elimination of connections. This developmental principle explains why even single genes can produce enormous quantitative and qualitative changes in long-range connectivity, particularly between different cortical regions in which every region receives thalamic, commissural, and ipsilateral cortical input. The most critical factor in producing differential connectivity is developmental timing. Even minute differences in the induction of axonal growth have the potential to produce remarkable shifts in the proportion of afferents to a neuronal population.

Antagonists of Genetic Variation: Brain Plasticity

Brain plasticity is an ill-defined term loosely applied to a variety of processes ranging from brain repair after injuries to learning. Whatever the particular mechanisms involved, it is clear that any single locus change at the genome level will never be reflected directly in memory and learning. It is masked or altered by various developmental and adult processes. In genetic terms, there is incomplete penetrance and pleiotropy. On the other hand, some mechanisms of brain plasticity have the capacity of inducing long-lasting changes as well. Thus, biological variation of cognition and learning can also result from developmental (epigenetical) processes based on chance or environmental factors.

Intrinsic mechanisms which mask, buffer, or transform genetic changes can be arbitrarily divided into two classes: (i) homeostatic regulation and genetic backup mechanisms and (ii) developmental reorganization, including adult plasticity and reorganization. Learning is another powerful factor in masking genetically induced variation, particularly in species with complex brains (Lipp, 1995). Since this article deals with sources of biological variation, learning will be omitted here.

Homeostatic Regulation: Possible Effects of Backup Genes

Homeostatic regulation is a property of systems based on sensor feedback mechanisms which allow to compensate for disturbances. This property is inherent in many neuronal systems and is thus an important buffer mechanism for many genetic alterations which do not have dramatic physiological consequences.

Homeostatic regulation of mutations may already be active at the gene level. Targeted disruption of genes can be achieved by inserting a foreign transgene. However, in many cases the results of such gene "knockouts" are surprisingly mild or missing. For example, disruption of the genes coding for fibrinolytic enzymes dissolving blood clots, namely, tissue plasminogen activator (tPA) and urokinase plasminogen activator (uPA), did not result in the expected severe pathology of multiple infarcts. However, breeding double-knockout tPA/uPA mutants lacking two enzymes from the fibrinolytic cycle resulted in sickness and early death in mice. This implies that the complete loss of an enzyme activates a backup system, perhaps an ancient silent gene coding for a similar enzyme.

We have also studied mice devoid of the PrP gene coding for the prion protein capable of inducing spongiform encephalopathy and expressed on many neurons. Extensive behavioral studies did not reveal any behavioral anomaly (Büeler *et al.,* 1992), and unpublished data from our laboratory showed that prionless mice proved to be resistant even to natural selection under harsh outdoor conditions described later in this article. These findings suggest either the presence of a backup gene with fairly decent compensatory capacities or that the physiological functions of the PrP gene are very subtle and thus nearly selection neutral.

Developmental Reorganization and Unpredictable Behavioral Effects in Knockout Mice

Reports of altered memory and learning following targeted gene disruption usually emphasize significant differences between wild-type and mutant mice. However, in our experience, even for mutations with severe effects on learning behavior there are some animals with decent performance, and there are regularly a few wild-type mice with poor learning behavior, even in samples without callosal dysgenesis. Careful analysis shows that the variability among the mutant mice is in most cases significantly larger. This is illustrated with data from swimming navigation learning in the previously mentioned knockout mice deficient for tPA. Tissue plasminogen activator not only is a clot-dissolving enzyme but also acts in the brain as an immediate early gene which is activated by seizures and plays an undefined role in brain development. The behavioral data showed a significant but moderate impairment of swimming navigation (Huang *et al.,* 1996). Why moderate?

Figure 6 shows the behavioral variation of 34 tPA knockout and 34 wild-type mice in the form of a box-and-whisker plot visualizing the variability of day-to-day learning scores including the extreme cases. It is evident that the control animals include some poor performers. The reason for this is probably the genetic background. Most wild-type controls in knockout mice represent a segregating Mendelian cross between C57BL/6 and 129/Sv. However, the strain 129/Sv shows poor learning abilities. It is thus not surprising to find a few individuals performing like those of the parental strain. Also, some mice try noncognitive strategies to find the platform, for example, a spiraling swim path which eventually leads to a hit of the platform.

The variability of the knockout mice is more disturbing, however. There is clearly an effect of the gene disruption since many mutant mice show prolonged swim paths. On the other hand, there are numerous knockout mice which show good or excellent learning. This implies that the effect of the gene disruption remains masked in many mice which somehow compensated the impairment due to the mutation. Hence, the mutant mice show significantly larger variability of their behavioral scores. Moreover, there is good evidence that individual mutant mice compensate their deficit in unpredictable ways. The same sample of mice was also tested for two-way avoidance learning, and this again revealed a significant impairment of the tPA mutants. However, there was no correlation between individual performance in swimming navigation and two-way avoidance. This suggests that each mouse compensated the mutation differently — one animal in a way that was compatible with avoidance learning and another animal in a way that was compatible with swimming navigation.

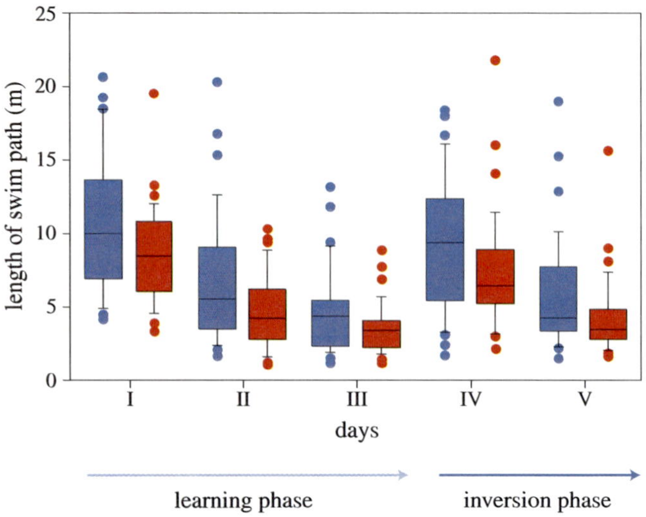

FIGURE 6 Variability of spatial learning in mice deficient for tissue plasminogen activator (in blue) and wild-type controls (in red) tested in the swimming navigation task of Morris. The behavioral measure is the mean length of the swim path. The blocks of trials are represented in box-and whisker form, in which the horizontal line represents the mean and the solid box the 25th and 75th percentiles (the range containing 50% of all scores). The two whiskers show the 10th and 90th percentiles (the range containing 80% of all scores). Above and below are the remaining extreme scores. During both the learning and the inversion phases, in general, the mice with the deletion show longer swim paths than the control mice, although even some of these show poor results in the test.

In conclusion, analysis of individual variability across different tasks indicates that the behavioral measures used to assess the effects of gene disruptions in mice are more or less remote reflections of the primary gene effect. In addition, the omnipresent developmental reorganization increases the risk of false-negative findings by completely masking mutation effects.

Developmental Reorganization of Cortical Connections Has Unpredictable Behavioral Consequences

It was previously described that genes have the capacity to dramatically alter thalamocortical and corticocortical connections, owing to the competition of axonal projections within a given target site. The same process also has the capacity to strongly mask any gene effect and can arise from purely epigenetical factors as will be shown in the following examples on callosal dysgenesis.

Targeted gene-deletion techniques in mice require embryonic stem (ES) cells in which gene disruptions are produced *in vitro.* The cells are then inserted into blastocytes of C57BL/6 mice and implanted into foster mothers. Some of the resulting chimeras will then transmit the inserted gene in the germline and are used to produce mutant mouse lines. Unfortunately, the standard donor strains for ES cells are 129 and BALB/c, both known for peculiarities in learning but also for dysgenesis and agenesis of the corpus callosum (Figs. 7 and 8). In the 129/J strain, about 30% of the animals have a reduced corpus callosum (CC), whereas about 10% of the animals show a complete absence of CC (Lipp and Wahlsten, 1992). The genetic architecture is not known exactly, but breeding studies have shown that callosal agenesis may result from the joint action of two or three recessive alleles. Mice which are recessive for all three loci such as I/LnJ show 100% penetrance of callosal agenesis. The strains BALB/c and 129 are probably recessive for only one or two loci, which results in many animals with reduced and some with completely missing CCs.

Such incomplete penetrance in an inbred strain can be explained by the developmental process of the formation of the CC. During the latest days of gestation, callosal fibers must pass to the other hemisphere using a narrow bridge, the lamina terminalis, formed by glial cells (Fig. 7). This sling permits passage of growing axons only during a short developmental period. Later arriving axons can no longer pass, and their growth cones wander in the ipsilateral hemisphere in which they form large aberrant longitudinal fiber tracts, the Probst bundles. If only a few axons succeed in making the passage, others may follow but the resulting CC remains small in size, a condition called *callosal dysgenesis*. Because of its exquisite sensitivity to timing, callosal agenesis and dysgenesis in humans is considered a teratological marker indicating delayed brain development.

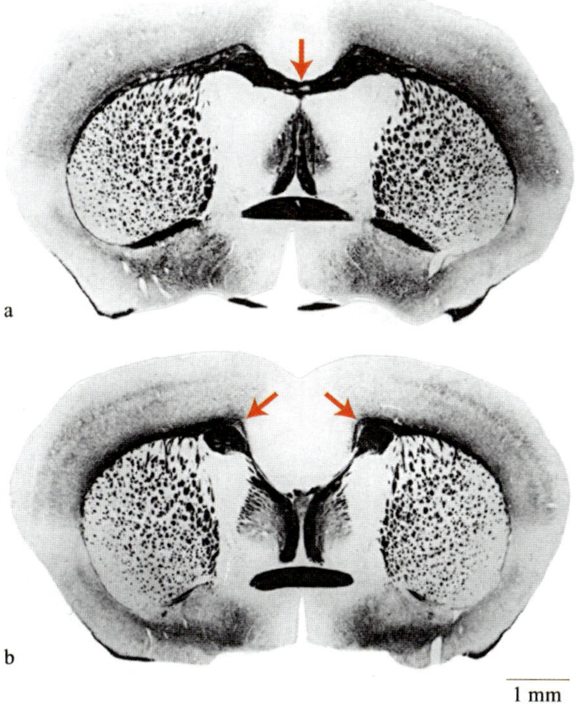

FIGURE 7 Agenesis of the corpus callosum in 129/Sv mice seen in coronal sections at the level of the anterior commissure. (a) The corpus callosum (arrow) in a normal mouse. (b) The compensatory Probst bundles (arrows) in an acallosal mouse.

At first glance, one might expect that acallosal mice might suffer from cognitive problems related to a missing connection between the hemispheres. The problem is much more complicated, however, because the fibers that could not reach the opposite hemisphere form new connections. These connections terminate in the appropriate laminae of the cortex, but because of the dynamics of competitive axonal growth it is not possible to predict which subregions become interconnected. The overall result appears as an unpredictable pattern of connections between ipsilateral associative cortex regions lacking callosal input. We have tried to analyze the consequences of such connective reorganization on cognitive and noncognitive behavior using mouse samples with congenital dysgenesis of the CC.

Swimming Navigation in 129 Substrains. Two substrains, 129/J and 129/Sv, both serving as donor strains for ES cells, were tested for swimming navigation learning. Afterwards, the cross-sectional area of the CC was measured in order to detect correlations with spatial learning (Fig. 9). In the 129/J mice, about 30% of the animals showed a complete agenesis of the CC, and the data revealed a moderate positive correlation: Mice with larger CC had shorter swim paths. However, it was also clear that some mice with complete CC agenesis showed very good learning. Only about 10% of the 129/Sv mice had complete agenesis, and their learning performance was inferior compared to that of the 129/J mice. Also, there was no correlation between size of the CC and swimming navigation learning. Again, mice with severely reduced CC showed extreme variability in swimming navigation learning. This suggests that reduced callosal connections, combined with development of additional ipsilateral fiber bundles, results in phenotypes showing extreme but unpredictable individual differences in spatial learning. Notice that the mice of the two samples were inbred (i.e., genetically identical). Thus, the variability of callosal size, altered connections, and learning performance is purely epigenetic.

Motor Coordination in Mouse Strains with Different Degrees of Callosal Agenesis. In order to test whether callosal agenesis is influencing sensorimotor coordination, several mouse samples were tested for their ability to step on small platforms, a task mastered with few errors by normal mice (Fig. 10a). This study revealed a significant genetic correlation: The smaller the average size of the CC, the higher the number of hindpaw slips (Fig. 10b). However, inspection of these scatterplots also reveals that the variability of the scores increased with reduction of the CC. To visualize this relation, a measure of individual variability, the standard error of the mean of the hindpaw slipping

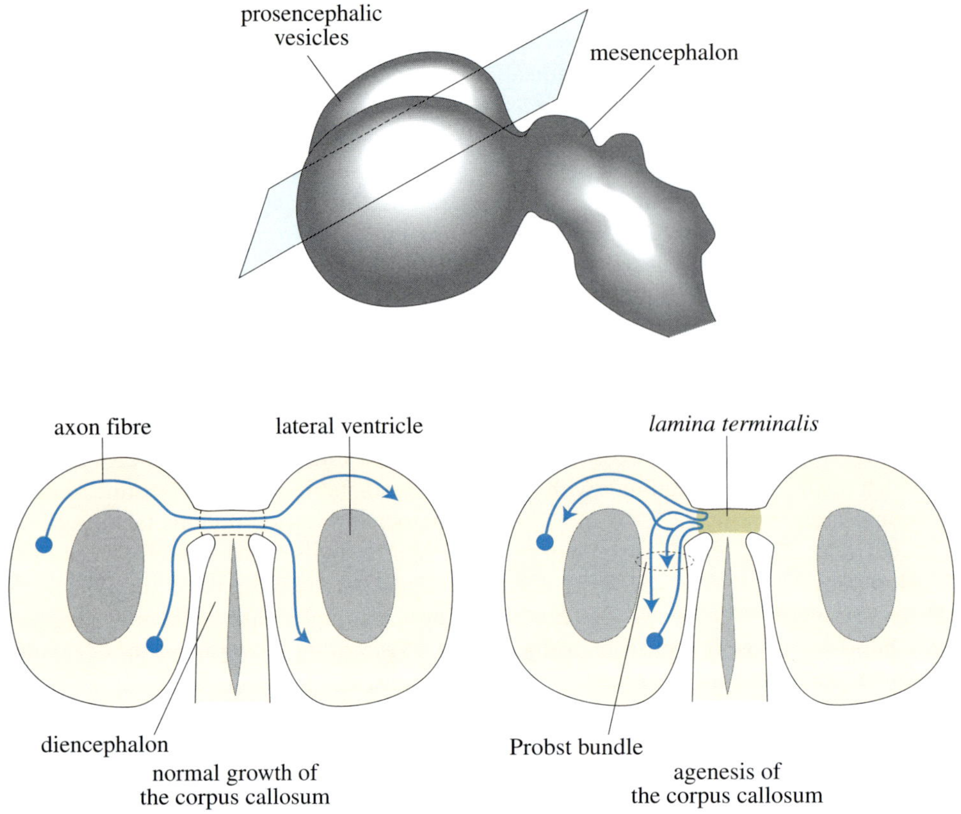

FIGURE 8 Growth of callosal axons in normal and acallosal mice in the embryonic mouse forebrain. (a) The frame indicates the plane of the sections. (b) Normally (left), the callosal fibers grow through a bridge of glial cells (with the lamina terminalis closing the rostral end of the diencephalon) and reach mirror sites in the other cortex. In acallosal mice (right), the glial sling is no longer permissive and the fibers form Probst bundles, establishing new synapses in the ipsilateral hemisphere.

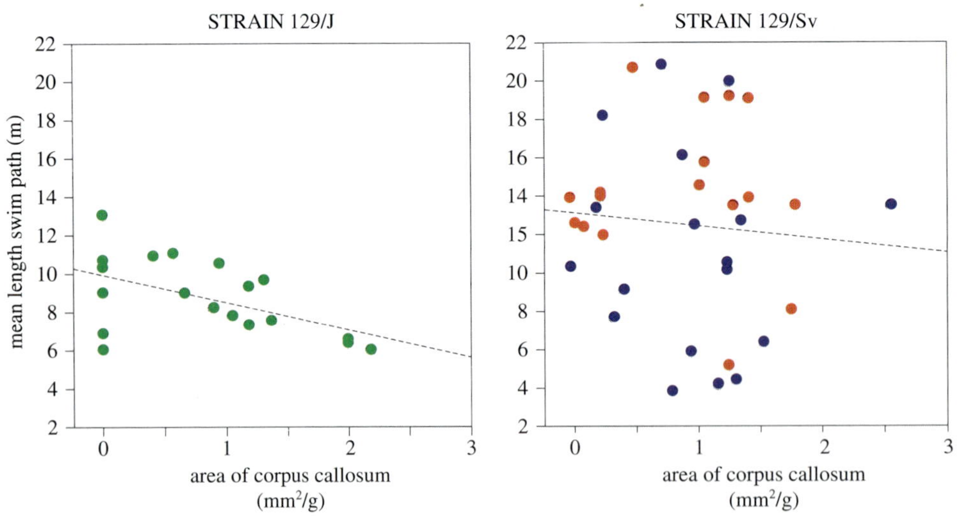

FIGURE 9 Correlation between the cross-sectional area of the corpus callosum (saggital median section) and the average length of the swim path in the acquisition period of swimming navigation in two substrains of mice (129/J and 129/sv) with partial or total agenesis of the corpus callosum. The mice were trained with six trials per day for 3 days. In the graph on the right, males are indicated in blue and females in red. The area of the corpus callosum is normalized with respect to the brain mass of the mouse. The dashed lines indicate the interpolated means of the values.

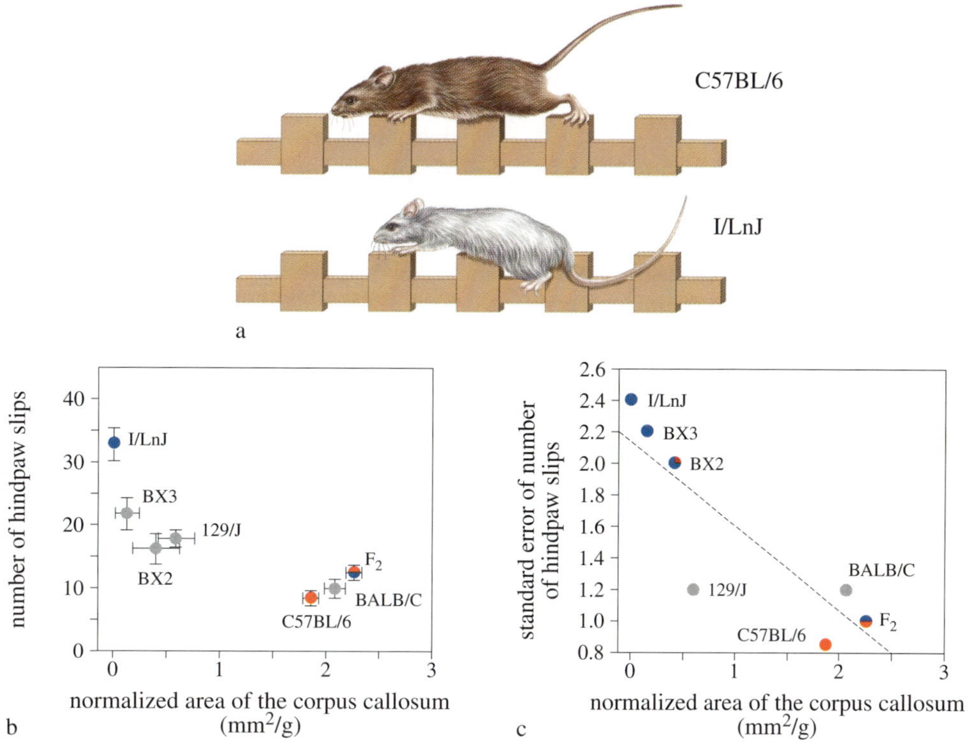

FIGURE 10 Hindpaw coordination in mouse strains with different sizes of corpus callosum. (a) The stepping bar apparatus for measuring motor coordination, showing the behavior of a C57BL/6 mouse with normal corpus callosum (top) and an I/LnJ mouse with complete agenesis of the corpus callosum (bottom). (b) The average number of hindpaw slips is negatively correlated with the average size of the corpus callosum in mouse pure strains or genetically defined crosses. (c) Even the individual variability (expressed as the magnitude of the standard error) is negatively correlated with the size of the corpus callosum. I/LnJ mice show very high, but also many very low, scores, whereas strains with a normally sized corpus callosum have much more homogeneous scores. The area of the corpus callosum is normalized with respect to the brain mass of the mouse. C57BL/6 and BALB/c, mouse strains with normal corpus callosum; I/LnJ and 129/J, mouse strains with total or partial agenesis of the corpus callosum; F_2, progeny from Mendelian crosses between I/LnJ and C57BL/6 mice (these mice show a normal corpus callosum); BX2 and BX3, second and third generation from backcrossing C57BL/6 × ILn/J hybrids to I/LnJ parents (backcrossing leads to a progressive elimination of the corpus callosum). Pure strains are indicated with single colors, and the hybrids are indicated with the colors of the parental strains.

scores, was plotted against callosal size (Fig. 10c). This revealed that reduction of the CC was also correlated with an increase in individual variability (Lipp and Wahlsten, 1992).

Minimizing the Masking Effects of Developmental Reorganization

Taken together, the results from both cognitive and noncognitive behavioral tasks showed that the altered cortical connectivity, induced by lack of callosal fibers and ipsilateral rerouting of callosal fibers, increased primarily individual variability. Such unpredictability is understandable because the basic reorganization of connections happened prenatally. In the mouse, this period is followed by a postnatal period of brain growth and development of about 20 days. During this time, many developmental processes can mask the circuitry changes through adjustment in synaptogenesis or by way of functional compensations at the synaptic level. The altered reorganization of cortical connections in acallosal mice is probably just one of many possibilities for developmental reorganization, exceptional only because the morphological changes can be easily detected.

The Importance of Timing. In order to minimize the interesting but confounding effects of developmental reorganization, it is thus necessary to study gene effects on memory and learning that remain confined to specialized brain structures which are formed late during development. There appears to be no targeted gene disruption specific for a brain region such as hippocampus or amygdala nor for a late maturing distinct cell population in the brain. Current efforts to produce targeted disruptions that can be triggered in adulthood may result in effects on memory and learning that are not marred by developmental reorganization. The alternative, the reverse genetics approach, is to concentrate on the natural variation of late-developing brain systems which are easily found in the hippocampus.

Size Variation of the Hippocampal Mossy Fiber System and Complex Learning

Mossy fibers are the axons of the granule cells in the dentate gyrus (see Fig. 3), have a unique neurochemical profile, and are known to develop postnatally. They can be easily visualized by Timm's stain because of their high content of zinc. A first set of mossy fiber collaterals returns to the dentate gyrus in the form of recurrent collaterals terminating on interneurons and granule cells (Fig. 11). In guinea pigs, these recurrent collaterals show physiological growth spurts at puberty and in midlife (Lipp and Wolfer, 1995), and they are also know to sprout into neighboring synaptic zones after these have lost afferent innervation because of pathological processes. Although the recurrent collaterals retain adult plasticity and the ability to reorganize, the topographical distribution of the other collaterals remains, once established, fairly invariant: A portion synapses on various cell types in the hilus of the dentate gyrus, a region known as CA4. The majority of the remaining collaterals terminate in a distinct layer above the pyramidal cell bodies in the hippocampal region CA3, forming the suprapyramidal mossy fiber layer (or stratum lucidum), whereas a smaller bundle of mossy fibers terminates on the basal dendrites of the pyramidal cells — the intra/infrapyramidal mossy fiber projection (IIP-MF). The latter shows enormous genetic variation, a feature characteristic of traits which remain constant over very long periods. Such stability is a prerequisite for behaviorally relevant structural changes. The reason for the stability of the mossy fiber pattern in CA3 is unknown. There seems to be a critical period during their postnatal proliferation from Days 5 to 15 in which they react to experimental manipulations such as lesions, hyperthyroidism, and alcohol intoxication, in particular the IIP-MF projection, but they appear to lose most of their growth capacity afterwards.

The stability of mossy fiber projections, their high-contrast staining pattern, and their lamination make them ideal for digital morphometry and correlative studies. Evidence for genetic variation of the mossy fiber projection was first provided by strain comparisons which showed that the IIP-MF projection shows remarkable size variation in comparison to the suprapyramidal and CA4 projection.

The Extent of the Infrapyramidal Mossy Fiber Correlates with Hippocampus-Dependent Learning Behavior

These studies prompted a search for correlations with learning behavior sensitive to hippocampal lesions. One such task is two-way avoidance (or shuttlebox) learning (O'Keefe and Nadel, 1978). The animals must learn to avoid a mild electrical foot shock (which is signaled by warning stimulus) by running to the other side of a two-compartment box, with the cycle being repeated many times (Fig. 12a). The task is difficult and confusing for rodents because there is no safe place in the shuttlebox. Studies of rats and inbred mouse strains with differential avoidance learning abilities showed that poor avoiders had larger IIP-MF projections (Fig. 12b). A critical role of size variation of the IIP-MF projection for avoidance learning was then demonstrated by increasing the size of the IIP-MF projections in strains known for small IIP-MF projections and avoidance learning. This was done by postnatal injections of thyroxine. The

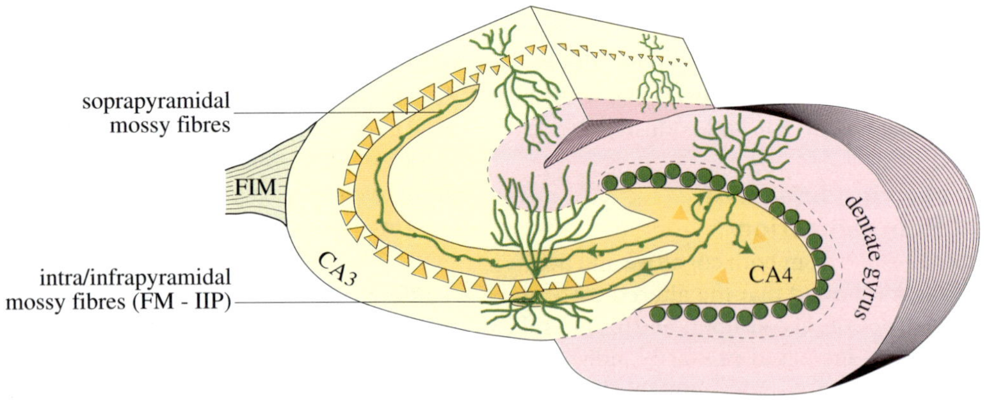

FIGURE 11 Diagram of the hippocampus and mossy fibers. The mossy fibers originate in the granule cells (green circles) of the dentate gyrus and project to hippocampal subregions CA4 and CA3. In CA3 they terminate above and below the pyramidal cell layer (triangles), respectively, as suprapyramidal and intra/infrapyramidal mossy fibers (IIP-MF). FIM, fimbria.

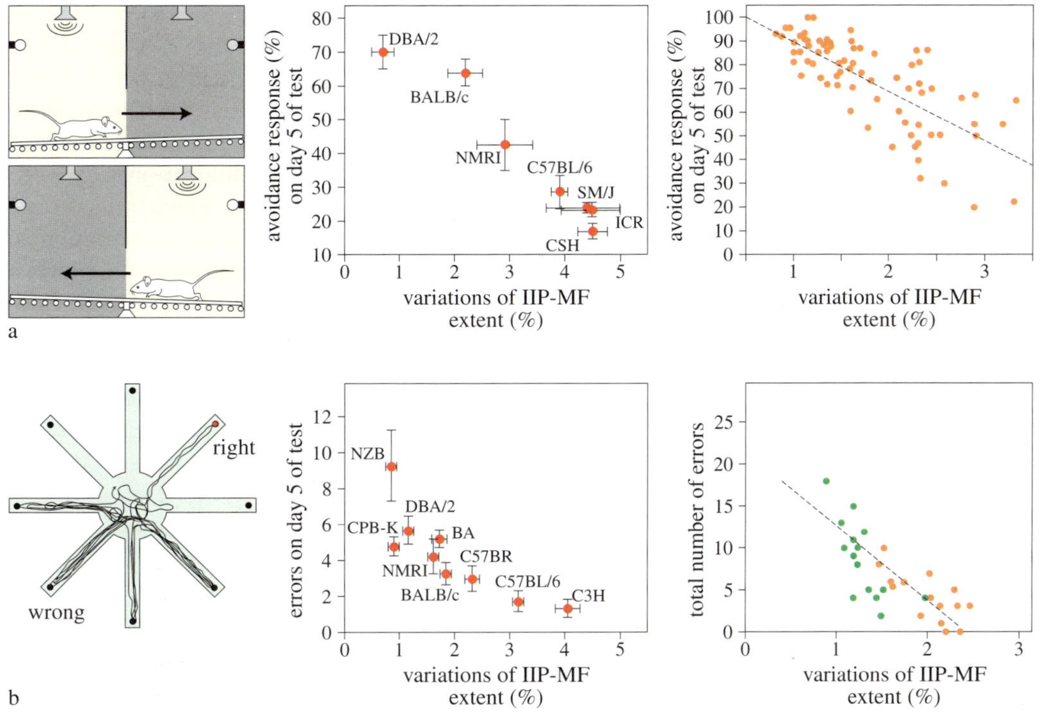

FIGURE 12 Variability of intra/infrapyramidal mossy fibers (IIP-MF) and covariation with learning scores in tasks sensitive to hippocampal lesions. (a) (Left) The shuttlebox apparatus for measuring two-way avoidance learning. (Center and right) Correlation between IIP-MF extent and avoidance responses: Inbred mouse strains with extended IIP-MF projections show poor avoidance learning. The error bars (standard error of the mean) give the variability within the isogenic animals. This variability is epigenetic, whereas the difference between the best (DBA/2) and the worst strain (CSH) gives the genetic variability. Heredity is clearly more important than environmental (ontogenic) factors in determining the level of two-way avoidance and the size of the mossy fibers. The same correlation is evident (right) in DBA/2 mice, in which postnatal thyroxine application altered the extent of the IIP-MF. (b) (Left) The radial maze task. (Center and right) The correlation between IIP-MF extent and errors: The number of errors is reduced in mouse strains with extended IIP-MF projections. Postnatal injections of saline (in green) and thyroxine (in orange) in the DBA/2 mice (right) produce subtle differences in the IIP-MF projection and correspondingly improve radial maze scores. The abbreviations indicate the different mouse strains.

adult animals showed a decrease in avoidance learning corresponding to the developmentally produced increase of the IIP-MF. Since two-way avoidance learning is often improved after hippocampal lesions, lower avoidance learning scores might actually correspond to better hippocampal function and learning.

To test this hypothesis, the relations between IIP-MF projection and spatial learning were investigated in the radial maze (Fig. 12) and in swimming navigation, both tasks being impaired after hippocampal lesions. The radial maze data studies yielded identical results to those of the studies in two-way avoidance: The best performing mouse strains had the largest IIP-MF projections (Fig. 12). Likewise, the thyroxine-induced hyperplasia of the IIP-MF projection during development remained correlated with an improvement of radial maze learning (Fig. 12). In swimming navigation, the extent of the IIP-MF projection correlated positively with the ability to learn an altered platform position (Lipp and Wolfer, 1995).

Moderate correlations between the extent of the IIP-MF were also found with open-field behavior, another behavior known to depend on the integrity of the hippocampus. The most clear-cut finding was that selective breeding for mice for increased locomotion as well as for rearing (an exploratory behavior in the open field) entailed an enlargement of the IIP-MF projections (Hausheer-Zarmakupi *et al.*, 1995).

These findings are all in line with the concept that the hippocampus is a structure mediating spatial learning and memory, and that variations of the infrapyramidal mossy fibers are related to a basic physiological process in the hippocampus directly involved in such learning.

The Extent of the IIP-MF Projection Correlates with Nonspatial and Nonlearned Behavior

The opportunity to have a morphological marker for subtle individual differences in hippocampal function prompted studies aimed at discovering correlations with behaviors thus far not tested by classical lesion studies, with some surprising results. Thus, it was found that male mice with short attack latencies toward an intruder in their home cage had reduced IIP-MF projections (Sluyter *et al.*, 1994). Another unexpected finding was that the left–right asymmetry of the IIP-MF in mice was significantly though moderately correlated with paw preference, the larger IIP-MF projection being on the side of the preferred paw. Also, mice with pronounced paw lateralization had generally bilaterally larger IIP-MF projection than those with missing paw preference (Lipp *et al.*, 1996).

Apparently, the hippocampus of rodents might also have some purely noncognitive functions since both paw preference and attacks against intruders are not learned and obviously do not have a cognitive component.

Cognitive and Noncognitive Functions of the Hippocampus as a Covariate of Mossy Fibers

Theoretically, these findings were difficult to reconcile with the view of the hippocampus as a specialized cognitive structure, and it became necessary to explain how structural variation of a relatively minor intrahippocampal trait could simultaneously influence cognition, memory, and noncognitive behaviors.

The explanatory hypothesis is that the hippocampal formation is orchestrating multiple processing in the cerebral hemispheres and in subcortical structures as visualized in Fig. 4. This implies that the hippocampus is constantly receiving a highly abstracted (coded) input depicting the activity states of the entire cortex as well as receiving input from subcortical structures governing nonlearned ("ethological") behavior patterns. This activity pattern is fed into reverberating loops by means of the mossy fiber projection and passes through the so-called lamellar (or trisynaptic) loop from dentate gyrus to CA3–CA1 and back to the paralimbic cortex. The primary function of multiple parallel feedback loops may not be cognitive at all. They could simply serve to temporarily stabilize the current activity patterns in the paralimbic and higher order association cortex as well as motivational states originating in the brain stem and hypothalamus. Such temporal stability is the prime condition for any organized behavior (which rarely changes abruptly) as well as for cognition and memory formation which require neuronal synchronization and coordination.

The neuroanatomical organization of the hippocampus, however, is such that afferent input is distributed at a right angle to the trisynaptic loop by means of Schaffer collaterals and axons of specialized cells in the CA4 region (see Fig. 11). Depending on the point of view, this arrangement can be perceived either as a grid (with the implication of a cognitive map) or, as preferred here, as an array of stabilizer loops connected by transverse fibers which allow cross-communication between neighboring and distant loops. This cross-communication serves to smoothen behavioral transitions and orchestrate coordinated changes in the connected higher-order association cortex.

This concept of hippocampal function implies that there must be an optimal balance between behavioral stabilization along and cross-communication between the lamellar loops. The structural variability of the mossy fiber projection is basically an altered balance between input and association systems in the hippocampus, with individuals with enlarged mossy fiber projections in the form of an additional infrapyramidal termination having potentially less, or less effective, intrahippocampal associational synapses. Structural variability of the IIP-MF projection then implies a shifted balance that affects the ratio of stability versus lability of hippocampal control. Larger IIP-MF projections may result in more stable intrahippocampal processing being resistant to alarming stimuli and, as a result, in improved cognitive abilities, superior memory formation, and behavioral predictability. Smaller IIP-MF projections, on the other hand, will be associated with increased behavioral irritability and impaired cognition and memory formation due to an inability to temporarily stabilize the activity pattern of the hemispheres. Obviously, optimal hippocampal function requires a compromise. An array of loops without interconnection will result in extreme behavioral stubbornness and inability to adapt and correct ongoing cortical processing, whereas uncontrolled communication between the loops entails hyperreactive behavior and the inability to form or recall complex memories. Both extremes might be expected to occur after destruction of the hippocampus.

This hypothesis explains most of the behavioral correlations of the IIP-MF projections. Mice with scanty IIP-MF projections show a form of mild hippocampal syndrome characterized by behavioral irritability which entails quick aggression toward intruders, an inability to perform continuous motor acts (a prerequisite of visible paw preference), superior two-way avoidance learning, reduced exploration behavior, and impaired performance in spatial learning and complex learning tasks. Mice with medium to large IIP-MF projections act consistently, are less irritable and more cautious, explore more efficiently, and do better in most hippocampus-dependent tasks except for two-way avoidance, which requires immediate reaction to external stimuli even in the presence of conflicting cues.

Why Are There Some Mice with Reduced Cognitive Abilities?

Having scarce IIP-MF projections equivalent to a mild hippocampal lesion syndrome would seem an undesirable property for mice, and one wonders why natural selection has not eliminated such an unhealthy feature, even if it were only of partial importance for animal cognitive behavior. Nonetheless, it occurs in a couple of mouse strains, and our investigations of wild house mice have consistently revealed individuals with fairly scanty IIP-MF projections.

One possibility is that mossy fiber variations might somehow influence behavior, but that the behavioral tests applied to mice are irrelevant and do not measure any biologically meaningful abilities of the animals — an argument often proposed by ethologists questioning animal psychologists. Another explanation is that laboratory mice are unrealistic models because of their domestication and behavioral degeneration, and one might dismiss the observed correlations as laboratory artifacts.

Natural Selection as Final Arbiter

Such arguments can only be answered by studying correlations between mossy fiber variation and behavior in an ecological context. If mossy fiber variations are of any relevance for natural behavior, they must respond to natural selection. If so, species with different lifestyles but comparable brains must show differences in their IIP-MF projections, and natural selection must occur within members of a given species.

The first condition, species differences, is documented in Fig. 13. Of the four species depicted, three have fairly large surface habitats, whereas one vole species (*Pitymys subterraneus*) lives in burrows. In the latter, an IIP-MF projection is barely visible, whereas it is well developed among the surface dwellers. This might be a chance result, but fur-

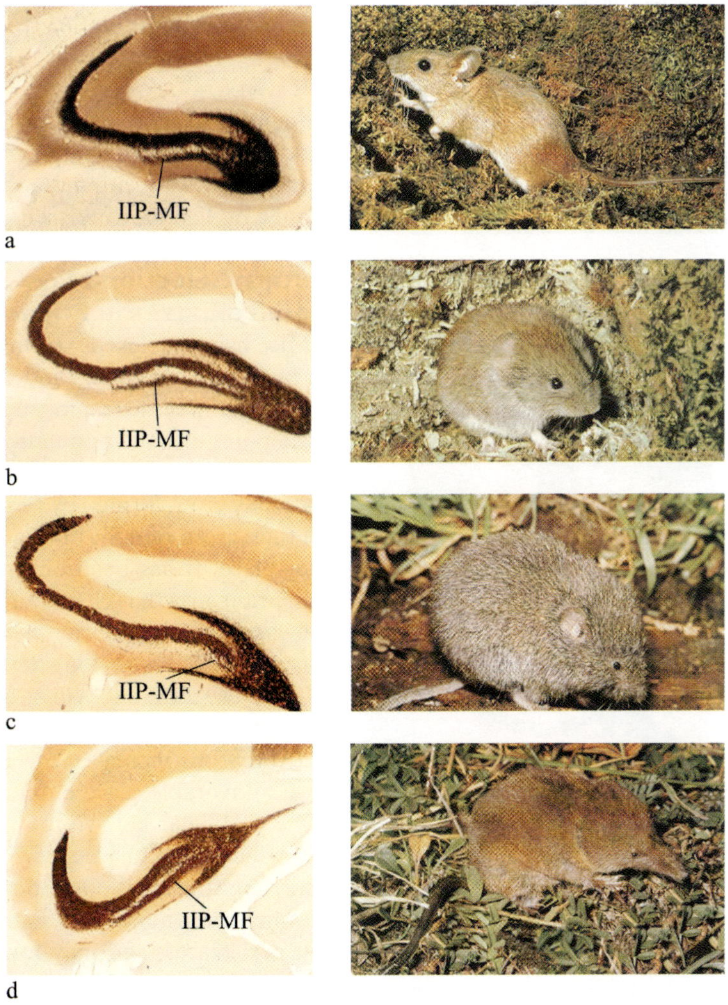

FIGURE 13 Mossy fiber patterns (left) in several species of small wild mammals (right): (a) *Apodemus sylvaticus;* (b) *Clethrionomys glareolus;* (c) *Pitimys subterraneus,* a subterraneously living mammal in which the IIP-MF are almost missing; and (d) *Sorex araneus* (reproduced with permission from Hauser, 1995).

ther studies that include other subterraneously living species such as moles will help to verify whether a complex surface habitat is associated with extended IIP-MF projections and superior spatial learning abilities.

In order to observe natural selection, laboratory mice with genetically different IIP-MF projections were crossed to obtain a sample of hybrid mice characterized by large genetic variability of this trait. These were then released into large outdoor pens in western Russia and have been kept there for several years (Fig. 14). Samples of mice are taken at yearly intervals and compared with control mice from the same crosses but which were kept under standard laboratory conditions and were randomly mated.

After 1 year outdoors, the sample analyzed still contained individuals with different IIP-MF projections (Fig. 14). However, the statistical analysis revealed that the feralized mice underwent a significant change toward smaller IIP-MF projections and increased suprapyramidal mossy fiber projections (Fig. 15). This finding is preliminary and does not prove natural selection since the altered mossy fiber pattern might also reflect an environmental factor. Nonetheless, it suggests that if natural selection is truly acting, it is not doing so, as expected, toward an intermediate size of the IIP-MF projection but toward a phenotype with inferior spatial abilities but shortened attack latencies. It is tempting to speculate that this reflects the ecological situation in which the mice had to live for most of the year in central covered shelters, a lifestyle characterized by

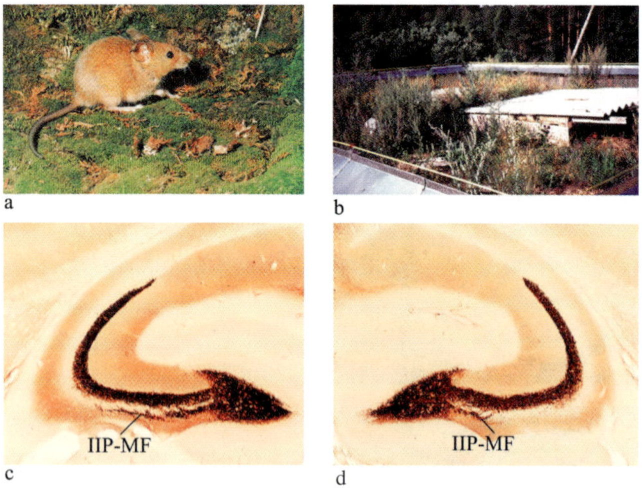

FIGURE 14 Natural selection of mossy fiber patterns. (a) House mouse (*Mus musculus*). (b) Outdoor pen used to study survival and natural selection of feralized mice. Both hybrids between wild and laboratory mice and crosses from laboratory strains survive under seasonal temperature variations from −30 to 30°C. (c) The largest IIP-MF projection found in a sample of 20 feralized mice. (d) The smallest IIP-MF projection found in the same sample (reproduced with permission from Hauser, 1995).

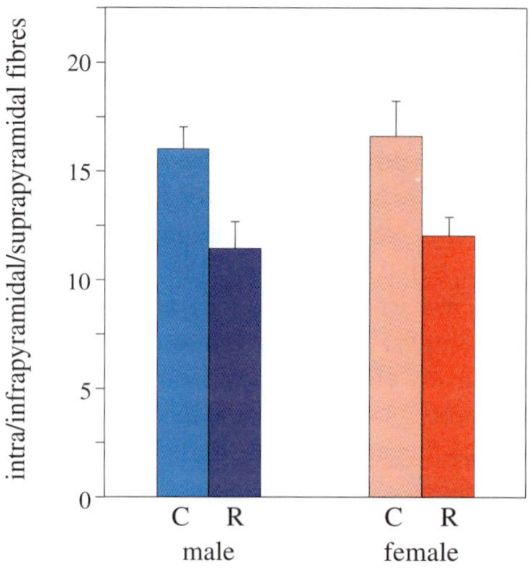

FIGURE 15 Variation of the ratio of intra/infrapyramidal fibers in laboratory mice returned to outdoor living (R) with respect to control mice (C). Outdoor life induces a reduction of the intra/infrapyramidal and an increase of the suprapyramidal mossy fiber projections.

frequent aggressive interactions but probably not very demanding in terms of spatial cognitive abilities. This interpretation fits well with a theory of disruptive selection developed by Van Oortmerssen.

Natural Selection May Cyclically Produce Extreme Phenotypes

Mice living in closed systems form a so-called *deme*. Under favorable conditions and with a large food supply, the population increases rapidly but then social stress increases to the point that the population breaks down, presumably because of stress-induced lack of maternal care. Then a new deme may grow, formed by remnant individuals, first slowly and then more rapidly, until a new population breakdown takes place. It has been observed that the peak phase of population growth is characterized by the presence of mice with short attack latencies, whereas the initial stages contain more mice with long attack latencies (Benus *et al.*, 1991). These differences in attack latency are hereditary, as shown by rapid selective breeding. Van Oortmerssen concluded that the increased social competition in a deme living in a favorable place entailed natural selection of mice with a particular behavioral profile termed "copers," animals which react aggressively or active (e.g., by flight) in conflicting situations. Genetically less active or more hesitant mice, the "noncopers," are driven away and emigrate, seek new places, and may found there new demes which grow undisturbed until the increase in the mouse population entails a new selection for aggressive individuals. Evidently, noncopers require superior exploratory and spatial abilities in order to survive emigration. Such disruptive selection explains why a normal rat or mouse population always contains a certain proportion of copers and noncopers (Cools *et al.,* 1993).

The Hippocampus as a Population Regulator?

The behavioral profile of copers and noncopers bears a surprising similarity with that of mice having differential mossy fiber projections: Placid noncopers are poor in two-way avoidance and superior in complex learning tasks, whereas the aggressive copers show largely the opposite pattern. Copers and noncopers differ in hormonal traits related to stress and aggression, but, as discussed previously, they are also different in mossy fiber distributions (Sluyter *et al.,* 1994). This suggests that the population dynamics in mouse demes is controlled by rapid natural selection of a few alleles controlling the development of brain systems underlying differential coping styles. Some of these genes may preferentially act on neuroendocrine systems, but others may influence the mossy fiber distribution because of their involvement in general irritability. The by-products of this selection are changes in cognitive abilities: less in the copers, which, in their crowded but simple environment, require primarily aggression for reproduction, and more in the emigrant noncopers, which may have better use for exploration and learning in new environments.

Time will determine whether the variability of the hippocampal mossy fiber circuitry reflects a participation of this structure in the short-term regulation of mouse populations. Ongoing studies must first convincingly demonstrate true natural selection effects in a given environment. A next step will then offer escape routes for emigrants leading into complex territories in order to test whether these animals will found demes with extended IIP-MF projections. Even if they do, the results can only be interpreted with a careful analysis of the ecological conditions and their demand for animal cognition. Clearly, the phenomenon of hereditary differential cognitive abilities is not restricted to rodents, but the putative hidden correlation with reproductive success is probably due to the rapid population dynamics of rodents and should not be extrapolated to humans.

Biological Variation of Cognition and Learning: Forward versus Reverse Genetics

This article discussed the two approaches to biological variation of learning. The forward (genotype-to-phenotype) strategy concerns gene deletions and their consequences, and the reverse strategy (phenotype-to-genotype) tries to understand the relations between structural and behavioral variability in the laboratory and the natural environment. The main problem of forward genetics is the developmental dynamics of the brain that mask straightforward relations between genes, brain, and behavior; its strength is the control over the beginning of the pathway from genes to behavior. The main problem of reverse genetics is to identify the genes and their contributions to the behavioral phenotype; its strength is a deeper understanding of the processes leading to biological variation of cognitive brain structures and its meaning for the individual and the population. However, it makes no sense to compare these approaches directly because forward genetics is in a stage of ongoing development. Nonetheless, it is clear that the genotype-to-phenotype pathway can only be understood if molecular approaches incorporate the knowledge of reverse genetics. The approaches must and will blend. This will take some time; the problems are complex and require interdisciplinary learning and patience. However, the reward is high because we will eventually solve a key problem of modern biology: We will determine the relations between genes, development, brain, learning, ecology, and evolution — for individuals and populations — in health and in disease.

Acknowledgments

The experimental studies presented in this article were supported by Grant No. 31-37497 from the Swiss National Science Foundation for Scientific Research. I appreciate the critical reading of the manuscript by David P. Wolfer.

References Cited

Benus, R. F., Bohus, B., Koolhaas, J. M., and Van Oortmerssen, G. A. (1991). Heritable variation for aggression as a reflection of individual coping strategies. *Experientia* **47,** 1008–1019.

Büeler, H., Fischer, M., Lang, Y., Blüthmann, H., Lipp, H.-P., DeArmond, S. J., Prusiner, S. B., Aguet, M., and Weissmann, C. (1992). The neuronal cell surface protein PrP is not essential for normal development and behavior of the mouse. *Nature* **356,** 577–582.

Clayton, N. S., and Krebs, J. R. (1995). Memory in food-storing birds: From behaviour to brain. *Curr. Opin. Neurobiol.* **5,** 149–154.

Cools, A. R., Rots, N. Y., Ellenbroek, B., and De Kloet, E. R. (1993). Bimodal shape of individual variation in behavior of Wistar rats: The overall outcome of a fundamentally different make-up and reactivity of the brain, the endocrinological and the immunological system. *Neuropsychobiology* **28,** 100–105.

Hauser, J. (Ed.) (1995). *Säugetiere der Schweiz: Verbreitung, Biologie, Oekologie.* Birkhauser, Basel.

Hausheer-Zarmakupi, Z., Wolfer, D. P., Leisinger-Trigona, M.-C., and Lipp, H.-P. (1995). Selective breeding for extremes in open-field activity of mice entails a differentiation of hippocampal mossy fibers. *Behav. Genet.* **26,** 167–176.

Huang, Y.-Y., Bach, M. E., Lipp, H.-P., Zhuo, M., Wolfer, D. P., Hawkins, R. D., Schoonjans, L., Kandel, E. R., Godfraind, J.-M., Mulligan, R. C., Collen, D., and Carmeliet, P. (1996). Mice lacking the gene encoding tissue-type plasminogen activator show a selective interference with late phase LTP in both Schaffer collateral and mossy fiber pathways. *Proc. Natl. Acad. Sci. USA* **93,** 8699–8704.

JACOBS, L. F. (1995). The ecology of spatial cognition. Adaptive patterns of space use and hippocampal size in wild rodents. In *Behavioural Brain Research in Naturalistic and Semi-Naturalistic Settings: Possibilities and Perspectives* (E. Alleva, A. Fasolo, H.-P. Lipp, L. Nadel, and L. Ricceri, Eds.), pp. 301–322. Kluwer, Dordrecht.

LIPP, H.-P. (1995). Genetic variability, individuality and the evolution of the mammalian brain. *Behav. Processes* **35,** 19–33.

LIPP, H.-P. and WAHLSTEN, D. (1992). Absence of the corpus callosum. In *Genetically-Defined Animal Models of Neuro-Behavioral Dysfunction* (P. Driscoll, Ed.), pp. 217–252. Birkhauser, Boston.

LIPP, H.-P., and WOLFER, D. P. (1995). New paths towards old dreams: Microphrenology or the study of intact brains in intact worlds. In *Behavioural Brain Research in Naturalistic and Semi-Naturalistic Settings: Possibilities and Perspectives* (E. Alleva, A. Fasolo, H.-P. Lipp, L. Nadel, and L. Ricceri, Eds.), pp. 1–39. Kluwer, Dordrecht.

LIPP, H.-P., COLLINS, R. L., HAUSHEER-ZARMAKUPI, Z., LEISINGER-TRIGONA, M.-C., CRUSIO, W. E., NOSTEN-BERTRAND, M., SIGNORE, P., SCHWEGLER, H., and WOLFER, D. P. (1996). Paw lateralization and intra/infrapyramidal mossy fibers in the hippocampus of the mouse. *Behav. Genet.* **26,** 167–176.

MORRIS, R. G. M. (1984). Development of a water-maze procedure for studying spatial learning in the rat. *J. Neurosci. Methods* **11,** 47–60.

O'KEEFE, J., and NADEL, L. (1978). *The Hippocampus as a Cognitive Map.* Clarendon, Oxford.

SCHLAUG, G., JANCKE, L., HUANG, Y., and STEINMETZ, H. (1995). In vivo evidence of structural brain asymmetry in musicians (see Comments). *Science* **267,** 699–701.

SHETTLEWORTH, S. J. (1995). Comparative studies of memory in food storing birds. From the field to the Skinner box. In *Behavioural Brain Research in Naturalistic and Semi-Naturalistic Settings: Possibilities and Perspectives* (E. Alleva, A. Fasolo, H.-P. Lipp, L. Nadel, and L. Ricceri, Eds.), pp. 159–192. Kluwer, Dordrecht.

SLUYTER, F., JAMOT, L., VAN OORTMERSSEN, G. A., and CRUSIO, W. E. (1994). Hippocampal mossy fiber distributions in mice selected for aggression. *Brain Res.* **646,** 145–148.

General References

COOK, N. D. (1986). *The Brain Code. Mechanisms of Information Transfer and the Role of the Corpus Callosum.* Methuen, London.

GALL, F. J. (1979). *Franz Josef Gall, Naturforscher und Anthropologe. Ausgewählte Texte* (E. Lesky, Ed.). Hans Huber, Bern.

GASIC, G. (1995). Systems and molecular genetic approaches converge to tackle learning and memory. *Neuron* **15,** 507–512.

KANDEL, E. R., and HAWKINS, R. D. (1992). The biological basis of learning and individuality. *Sci. Am.* **267,** 78–86.

PLOMIN, R., OWEN, M. J., and MCGUFFIN, P. (1994). The genetic basis of complex human behaviors. *Science* **264,** 1733–1739.

Leslie A. Real
Department of Biology
Indiana University
Bloomington, Indiana 47405, USA

Functional Rules for Decision Making

The cognitive information processing abilities of organisms often appear to be based on simple rules for efficiently and accurately representing the environment in which the organisms must live. Through evolution by natural selection, these rules should be structured so as to increase the ability of an organism to reproduce and survive within its environment. I illustrate the nature of such functional rules by examining the computational basis of finding and selecting among potential mates, choosing a set of essential resources from which to feed, and structuring social interaction leading to cooperation. Ultimately, we should be able to tie a specific computational rule to a neural mechanism that implements the rule. A comparative study of rules that emerge as dominant functional rules may also reveal some common features of evolved rule-based behavior.

Introduction

All organisms are faced with the difficult task of acquiring essential resources, such as food, shelter, and mates, in their struggle for survival and reproduction. Over the course of evolution, natural selection should favor those organisms that can act in an efficient manner to increase the likelihood of acquiring and utilizing these resources. Often, the behavior of an organism appears rule based, i.e., a pattern of responses to specific environmental conditions that can be described by a rule. What I call a *functional rule* is a rule which when applied consistently in the course of an organism's behavioral response to the environment increases its chances of survival and/or reproduction within that environment. Such rules may not be consciously applied but can manifest themselves as automatic responses encoded in the organism's neural machinery. For example, if by accident you were to step off the sidewalk and suddenly see an oncoming car, then you might follow the following rule: Jump back in a direction that maximizes the distance between your point of landing and the plane of motion of the car. Following such a rule would minimize the likelihood of collision and harm. There are competing rules that would not minimize harm, for example, jumping into the car's plane of motion. This rule maximizes the probability of harm and is clearly maladaptive (unless your intent is to collect a lot of insurance). Also, you could jump back along the trajectory in which you were walking. This rule only minimizes harm if you were originally walking at a right angle to the direction of the car. Any rule that does not minimizing the probability of being hit is not functional. My casual observation of people's responses (including my own) is that we indeed follow the functional rule and jump so as to maximize distance between the car's plane of motion and our landing point. We do not do this consciously nor do we perform a set of calculations. We simply respond. However, our response can be expressed as a set rule, and this rule appears to be implemented in the basic phenomenology of information processing in the nervous system.

Rule-governed behavior is characterized by two fundamental processes: encoding and computation (Gallistel, 1990). Information from the environment is translated through the senses into an encoded form that can be manipulated and stored according to specific computational rules. The application of a set of computational rules to a specific set of sensory data generates a specific set of responses or actions. By knowing the rules and the sensory perceptions of the organism, we can predict behavioral responses. The great task, of course, is to discover the underlying rules that govern specific kinds of behavioral responses.

The rules that are implemented within the organism's nervous system have evolved and are subject to natural selection in a manner analogous to any other aspect of the organism's phenotype. That there should be an evolutionary origin and differentiation in mental function has been noted since Darwin. However, there has been considerable recent interest in further exploring the implications of evolution and natural selection in interpreting an organism's cognitive abilities and limitations. Although many studies document varying degrees of adaptive specialization in cognitive function, few studies explore the adaptive nature of specific computational rules as aspects of the organism's functional cognitive architecture.

Different computational algorithms can generate selection differentials, at least in principle, either through affecting computational efficiency and accuracy or by producing different representations of the environment. As an example of algorithmic efficiency among functional rules, consider the following problem: Sum the numbers between 1 and 100. Most of us might attack this problem by simply sequentially summing numbers: $1 + 2 = 3, + 3 = 6, + 4 = 10$, etc. until arriving at 5050, which is the correct answer. Alternatively, one could note that $1 + 100 = 101, 2 + 99 = 101, 3 + 98 = 101$, and there are 50 such pairs of sums. Thus, the answer is $50 \times 101 = 5050$. Both methods lead to the same answer, but the second method is clearly more efficient and quicker than the first. If we try to carry out this calculation in our heads, the second method is probably more accurate as well. This example illustrates how different computational schemes may lead to different degrees of efficiency and accuracy in mental operation. Presumably, selection will operate to increase the efficiency and accuracy of essential computations involving behavior of immediate evolutionary consequence, such as food gathering or mate selection.

Different computational rules can also give rise to very different representations of the environment that may be crucial for leading to appropriate actions and decisions. For example, the central tendency in a series of events which may form the basis for the animal's expectation of rewards can be represented by a variety of computations. Consider the central tendency of a population of flowers in which nectar rewards conform to a uniform distribution between 1 μl and 20 ml. The three most common measures of central tendency — the arithmetic mean ($\overline{X}_A$), the geometric mean ($\overline{X}_G$), and the harmonic mean ($\overline{X}_H$) — generate three different measures and could potentially lead to very different choices and decisions:

$$\overline{X}_A = \frac{1}{20}\sum_{X=1}^{20} X = 10.5 \qquad [1]$$

$$\overline{X}_G = \left[\prod_{X=1}^{20} X\right]^{1/20} = 8.3 \qquad [2]$$

$$\overline{X}_H = \left[\frac{1}{20}\sum_{X=1}^{20}\frac{1}{X}\right]^{-1} = 5.6 \qquad [3]$$

Which of the three particular representations of central tendency best corresponds to observed patterns of behavior may depend on the specific task. For example, the arithmetic mean of rates of net energetic gain may characterize floral choice in bumblebees (Real, 1991), whereas the harmonic mean of intertrial delays may be employed by starlings when making foraging decisions in operant chambers (Reboreda and Kacelnik, 1991). As can be seen from the equations for central tendencies, in some cases currencies may be related. For example, minimizing delay (X) in the harmonic mean is equivalent to rate maximization in the arithmetic mean since $1/X$ is a rate. Which of the possible representations of central tendency is actually used may relate to general arguments for the applicability of each type of averaging method. For example, the geometric mean may best characterize systems with compounding effects, such as growth. Arithmetic averaging may best be employed in situation in which rewards are spatially distributed. Thus, natural selection may lead to one or another method of computational representation depending on the particular task or problem faced by the organism.

In the remainder of this article, I present examples of decision-making behavior that appear to be governed by, or at least can be explained by, functional rules.

Rules for Finding Mates

As a means of introducing the general method for formulating and exploring the evolutionary consequences of different functional rules, I examine the general problem of mate choice.

The number of offspring an individual produces over its lifetime is generally considered the most straightforward currency of evolutionary success. Consequently, natural selection may favor the formation of functional rules governing the choices that individuals make concerning their pattern of searching for and then selecting among potential mates. Bisexual animals, such as humans, must find and successively mate with members of the opposite sex to achieve genetic representation in future generations.

In a very insightful paper, Janetos (1980) pointed out that what may constitute the best functional rule for mate choice will depend on the behavioral repertoire of the individual. To illustrate this point, Janetos examined four alternative rules by which individual's might make decision about mates. The principal objective in the Janetos study was to determine whether a dominant rule exists that, when followed, generates a higher expected evolutionary success than the other possible rules of choice. Some rules for making choices about mates may be better than other rules for choice when they are acting on the same pattern of available information. From the postulated advantages associated with each decision rule, we can deduce the best kind of behavior that should be adopted by the organism.

First, imagine that in a given population one sex (usually males) actively seeks mates while the other sex (usually females) passively accepts overtures for mating. The passive sex, however, varies among individuals with respect to their quality. For example, in many fish species, males actively solicit mating from females and females are passive. However, females differ among themselves with respect to size, and size in fish is a fairly good indicator of how many offspring females will produce. Therefore, a male fish that mates with a very large fish will have many offspring, whereas that same male when mated with a smaller female will produce fewer offspring. Obviously, it is in the male's evolutionary interests to mate with as large a female as possible. Janetos (1980) proposed four different functional rules that an individual searching for mates might follow. The first rule assumed random selection of a mate. A searching individual encounters mates randomly and accepts the first encountered member of the opposite sex. The average benefit to the male is then determined by the average quality of available females. The second rule imagines that searching males possess a *fixed-threshold criterion* above which an encountered female is deemed acceptable and below which the female would be considered unacceptable. For example, a male fish might have a critical size limit for accepting a female. The third rule, the *best-of-n rule,* imagines that males sample a certain fixed number (n) of potential female mates and then select the females of highest quality from among the n sampled individuals. The males are free to return to any previously encountered female, and they can also remember the quality and location of all previously encountered females. In this rule, the decision variable is the number of females sampled (n) before a choice is made. The fourth rule Janetos examined, the *one-step decision rule,* assumes males accept or reject females as they are encountered. If the female is rejected, the male continues to search. Search terminates and mating occurs when the quality of the female exceeds the expected quality of females encountered through continued search. In this rule, the decision variable is an optimal threshold quality that terminates search.

Each of these four functional rules can be rendered into an explicit mathematical form, and for any given distribution of female quality the expected evolutionary fitness to the male can be computed. Each of the four functional rules shows an increase in evolutionary value to the male as the male increases the number of females sampled (Fig. 1). However, for any one sample size, there is a clear hierarchy in relative performance of the rules.

Random choice is the clear loser as a strategy. The male can always do better with one of the other three strategies. Under the assumptions of the model as articulated by Janetos, the best strategy, over all sample sizes, is the best-of-n rule. Since each of the four strategies also shows diminishing returns with each additional female sampled (i.e., the curve bends over with large sample sizes), Janetos concluded that the very best strategy is for a male to sample approximately five or six females and then choose the best out of that sample. His conclusion has been widely regarded as the most likely scenario for mate choice.

However, the performance of functional rules can drastically change if we modify some of the assumptions about the organism's behavior. Janetos neglected one important

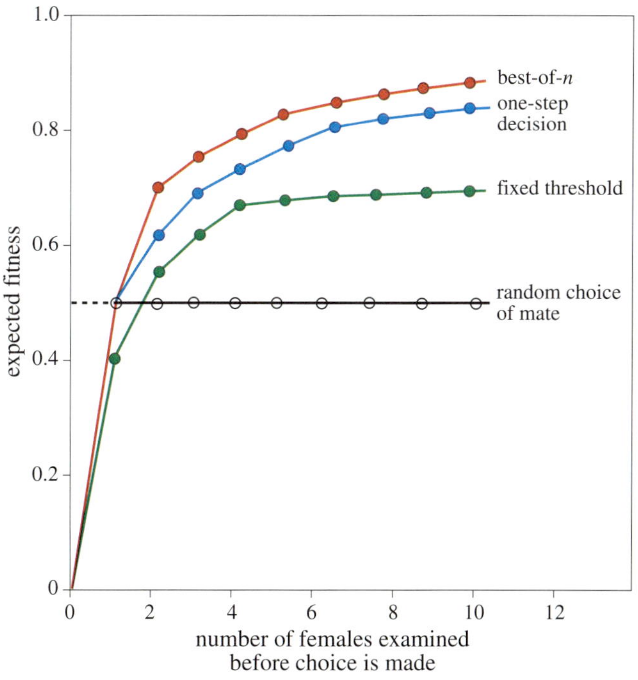

FIGURE 1 Comparative performance of the four functional rules examined by Janetos (1980) for female mate choice. All the strategies show decreasing returns for each female added to the sample. The best-of-n rule dominates all other functional rules. The rules are evaluated considering the expected evolutionary value (fitness; arbitrary units) to males as a function of the number of females sampled before making the choice (reproduced with permission from Real 1990).

feature of the mate-choice problem — the cost of search and sampling. It should not be a surprise that for costless search the best-of-n should dominate all other rules. In fact, since sampling has no cost, the best strategy is to sample all of the potential mates in the population and then choose the best of the population. However, the costs of sampling mates is often quite high, and rejected mates may not be available in the future. When search costs are added to the mate-choice problem, the performance of the different rules drastically changes.

In 1990, I presented modified models of mate choice which incorporate search costs. The mathematics of the models is complicated, but in essence the rules for choice are simple. For the best-of-n rule, the number of mates sampled increases until the sample size exceeds the cost of one additional search. For the sequential search rule, the critical threshold for acceptance is set to equate the cost of search with the expected fitness gain from one additional search. When these rules are compared, the sequential rule dominates the best-of-n rule over all search costs, a conclusion opposite from that reached by Janetos a decade earlier (Fig. 2).

When there are search costs (as there most likely always are) the superiority of sequential choice is easy to see. Under the best-of-n rule, searchers must sample a fixed number of potential mates before a choice is made, regardless of the quality of individuals encountered early in the sample

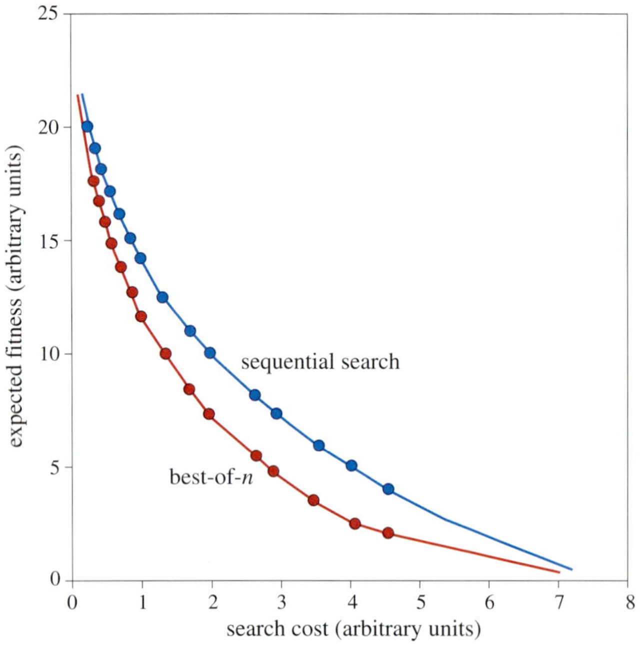

FIGURE 2 Comparative performance of the best-of-n rule and sequential search when there is a cost to sampling and search for mates. Over all ranges of search cost, sequential search dominates the best-of-n rule (reproduced with permission from Real, 1990).

sequence. If a very high-quality mate is encountered early, it must be bypassed as sampling proceeds, and this can be very costly. However, the sequential search rule is flexible and sensitive to the quality of the encountered mate. If the potential mate is of very high quality, search stops and no additional costs are accrued.

Experimental and laboratory investigations of the exact rules used in mate choice are under way. Most contemporary experiments are designed to evaluate the kind of functional rule used by searchers. It is still too early to determine if best-of-n or sequential choice is the rule followed by biological organisms. However, theoretical investigations of the sort carried out by Janetos are the first steps to articulating the most likely candidates for functional rules.

Perhaps the problem most successfully resolved by articulated functional rules is the question of how organisms make choices about what resources to use. Next, I discuss work on bumblebees as a model analysis of functional rules.

Rules for Choosing Resources

Every animal must collect food to provide the essential nutrients and energy required for growth and reproduction. In many species of insects that visit flowers for nectar, the link between resource gathering and reproduction is direct. In bumblebees, for example, single queens emerge in the spring, establish a nest, gather nectar and pollen, and lay eggs. The eggs are deposited within the nest and grow on food masses composed of a mixture of pollen and nectar collected by the queen. As workers emerge, they gather more nectar and pollen and the queen lays more eggs and the colony grows. The size and reproductive success of the colony are directly tied to the food-gathering abilities of the foragers. Since evolutionary success in these insects is tied to food gathering, presumably natural selection has promoted the formation of functional rules in workers that generate efficient foraging.

At any one time in nature, there are usually many flower species in bloom. What flower types should a foraging bee visit? Should the bee simply visit all the flower types as they are randomly encountered, or should it visit a subset of flower types based on some criterion of energetic efficiency? Naturalists in the nineteenth century observed that bees most often appear to specialize on only one flower type at a time and will restrict their visits to this one type. This kind of behavior has been termed *majoring*. Consequently, if bees appear to major on a single flower type, what is the basis for choosing one type over another type of simultaneously flowering plant species?

Since bees depend on energy and nutrients for reproduction, foragers may choose flowers that present the largest supplies of nectar and pollen in their flowers. If a bee is given a choice between flowers which present 1 μl of nectar in all flowers and a flower type that presents 5 μl of nectar in all flowers, it will almost certainly choose the flower type with the higher reward. However, flowers in nature are not uniform in the nectar rewards they provide but instead show considerable variation in the amount of nectar in each flower. When there is variation in the reward sizes in flowers, how do bees make choices?

This is a problem quite similar to how human financial investors choose among different stocks. Some stocks will be characterized by a high average rate of return but with high uncertainty, whereas other stocks will be characterized by low average returns but with low uncertainty. It has been recognized for many years that investors generally trade-off high yields for high uncertainty and that highly volatile stocks will only be chosen if they are accompanied by very high expected rates of return.

In 1980, I performed an experiment with bumblebees in which they were offered a choice between a constantly rewarding blue flower type and a variably rewarding yellow flower type with an average reward equal to that of the blue type. The bees consistently chose the constant over the variable type (Fig. 3).

This sort of behavior, in which animals choose constantly rewarding over variably rewarding alternatives (with equal expectation), is called *risk aversion* and appears to be widespread. Risk aversion has been documented in bees, wasps, birds, fish, small mammals, and humans. There are many functional models accounting for risk-averse responses (Bateson and Kacelnik, 1996). Next, I present one account based on the formation of functional rules for information processing in individual foraging bees and then discuss how these rules might be implemented in specific neural structures.

A Functional Rule for Floral Choice

When a bee visits a flower, it ingests a fixed quantity of nectar (μl) which translates into an energetic reward (Joules)

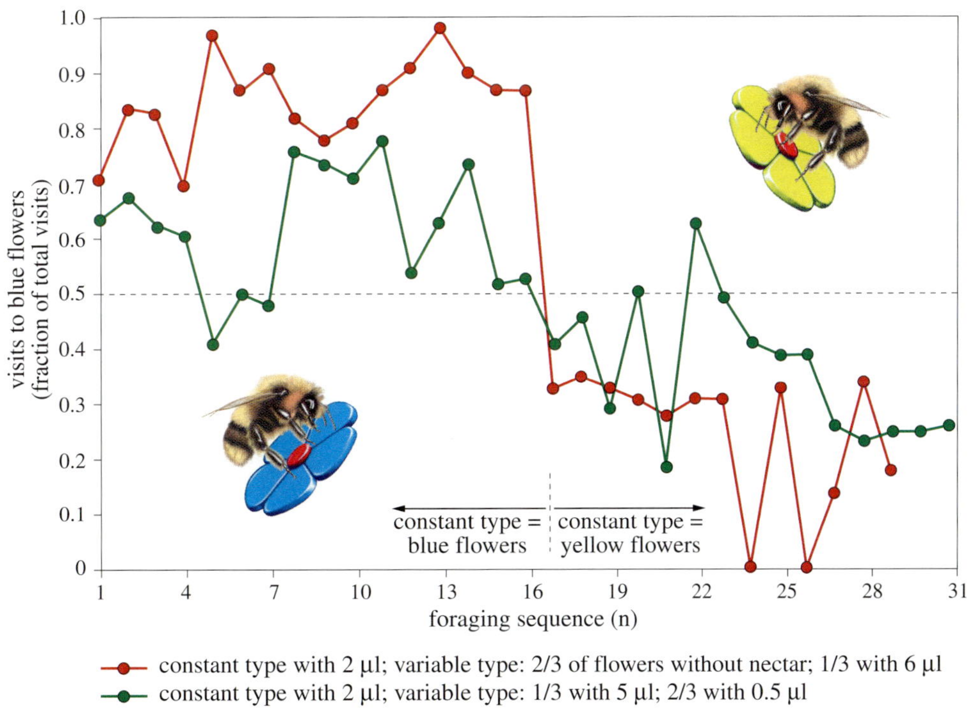

FIGURE 3 Preferences of bumblebees for flowers with constant rather than variable rewards in an artificial patch of blue and yellow flowers measured by the proportion of visits to blue flowers under different nectar distributions. Individual bees visited approximately 40 flowers during a single foraging sequence from a known distribution of nectar rewards. The green line corresponds to experimental trials in which the constant floral type contained 2 μl of nectar in every flower, whereas for the variable type two-thirds of its flowers had no nectar and the remaining one-third contained 6 μl per flower. The red response line corresponds to trials in which the constant floral type contained 2 μl in each flower, whereas the variable type contained 5 μl in one-third and 0.5 μl in the remaining two-thirds of the flowers. For trials 1–16, the constant floral type was blue. For trials 17–31, the constant floral type was yellow. If bees formed preferences based on expectation alone, then visits to blue flowers should always equal 0.5.

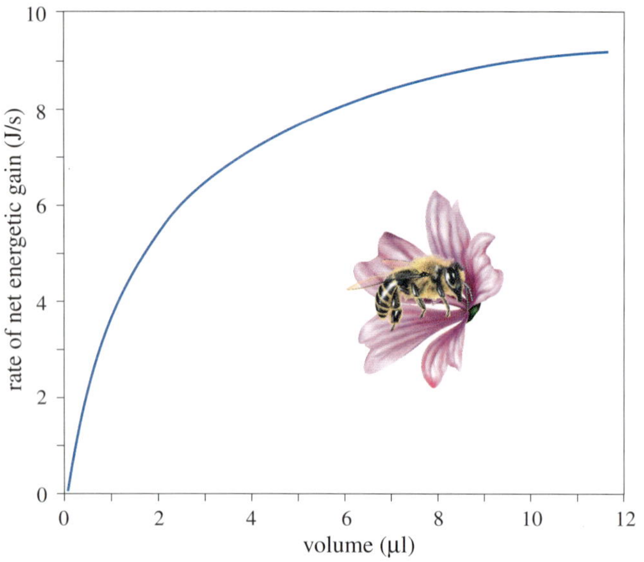

FIGURE 4 A plot of the rate of net energetic gain to a foraging worker bee as a function of nectar volume in a single flower. The relationship is based on the biomechanics of nectar extraction. When bees maximize the expected rate of net energetic gain, the nonlinear rate of gain accounts for the bees' sensitivity to variance in nectar rewards per flower.

based on the sugar content of the nectar. In 1987, Harder and I demonstrated that for an individual bee the rate of net energetic gain was a nonlinear function of nectar volume. That is, as the bee ingested more nectar the marginal rate of energy gain decreased (Fig. 4). This simple nonlinear relationship between nectar volume and rate of energy gain is sufficient to explain the pattern of risk aversion in the bees. Imagine that a bee can forage from two flower types, blue and yellow. In blue the bee receives a constant 2 μl of nectar, whereas in yellow it receives either 4 or 0 μl with equal likelihood. Harder and I suggested that bees choose the flower type that generates the higher average rate of energetic reward, and herein lies the interesting problem.

There are several alternative functional rules that could be constructed to characterize how bees calculate the average rate of energetic return. The structure of the averaging rule depends on how information is pooled as the animal experiences successive flowers during a foraging trip. For each flower in the sequence of n visits, let R_i represent the reward from the ith flower and T_i the time required to get the reward. The Harder–Real model assumes that bees maximize the expected net energy gain from calculations based on attributes of single flowers, i.e., the average of the set of R_i/T_i,:

$$E_1 = \left\{\frac{R_1}{T_1}, \frac{R_2}{T_2}, \ldots, \frac{R_n}{T_n}\right\} \quad [4]$$

Such an averaging scheme amounts to calculating the expectation of the ratio of two random variables, $E(R/T)$, which leads to the maximization of short-term energy gain. Several authors have argued, however, that organisms should maximize long-term gain rather than short-term gain.

Maximizing long-term gain amounts to calculating the ratio of expectations of the random variables, i.e., $E(R)/E(T)$. Actually, there is a range of possibilities between completely short-term and completely long-term gain corresponding to different ways of processing information from a string of flower visits. For example, rewards and times could be pooled across successive pairs of flowers in the string, i.e.,

$$E_2 = \left\{\frac{R_1 + R_2}{T_1 + T_2}, \frac{R_3 + R_4}{T_3 + T_4}, \ldots, \frac{R_{n-1} + R_n}{T_{n-1} + T_n}\right\} \quad [5]$$

or across triplets of flowers (E_3), quadruplets (E_4) of flower, and so on. In the theoretical limit, $n \to \infty$, for which $E_\infty = E(R)/E(T)$ which corresponds to the long-term average rate of energetic gain.

For each distribution of floral rewards used in our 1980 experiments and using the physiologically based expressions for R and T from the Harder–Real model (Fig. 4), we calculated the average rate of net energy gain E_k as a function of the frame size k over which samples are pooled. The frame size determines the fine structure of the processing of a particular string of information. In all our experiments, the correlation between average rate of energetic gain and observed preference was highest for short-term optimization (i.e., frame length $k = 1$ or 2) and lowest for long-term optimization (Real, 1992). These results strongly suggest that bumblebees form functional rules based on rates of return from individual flowers or on pairs of flowers, resulting in short-term energy maximization.

Implementation of Functional Rules

Martin Hammer and Randolf Menzel discovered an amazingly simple neural basis for reward learning in honeybees that may be the key to understanding the implementation of floral choice behavior. In a series of recent papers, Hammer and Menzel identified a specific neuron, VUMmx1, in the honeybee subesophageal ganglion that delivers information about reward in classical conditioning experiments (Hammer, 1993) (Fig. 5).

In many traditional learning experiments using honeybees, individual bees can learn to associate reward (sugar water) with specific odorants. Conditioning occurs when sugar water, placed directly on the bee's antennae, is paired with a puff of specific odor. Antennal stimulation by reward results in reflexive extension of the bee's proboscis and, following conditioning, presentation of odor in the absence of rewarding stimuli results in the same proboscis extension. Hammer discovered that direct stimulation of the VUMmx1 neuron can substitute for antennal stimulation using sugar rewards and that VUMmx1 stimulation can be

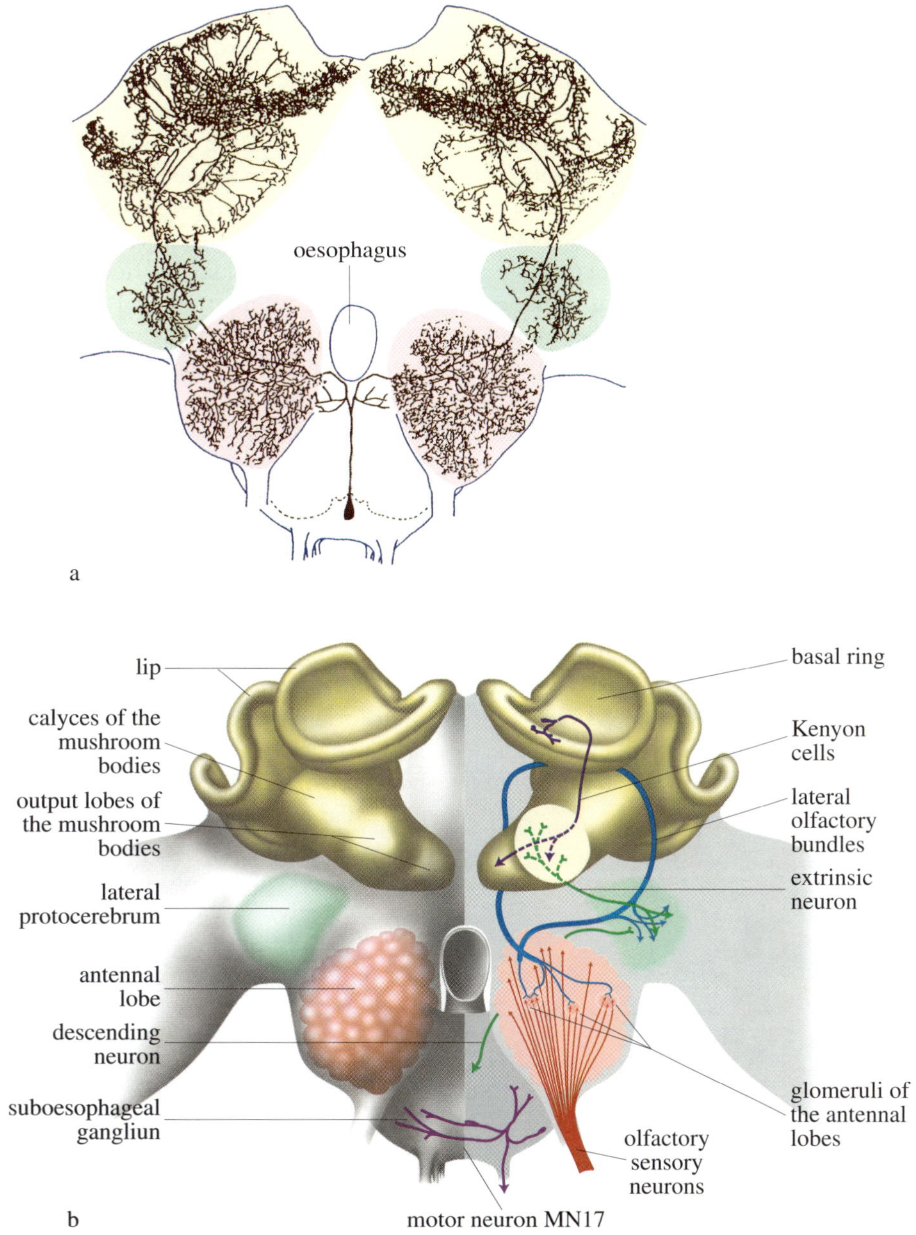

FIGURE 5 A schematic representation of the honeybee brain. (a) The diffuse branching pattern of the VUMmx1 neuron of a bee. (b) Diagram of the olfactory pathways and of the regions affected by VUMmx1. The olfactory sensory neurons (in red) from the antennae serve the glomeruli; two bundles of olfactory relays (in blue) connect the glomeruli to the calyces of the mushroom bodies and to the lateral protocerebrum. The Kenyon cells (in black) distribute the information to the output lobes of the mushroom bodies, whereas some extrinsic neurons (in green) connect the mushroom bodies with the lateral protocerebrum. Motor neurons such as MN17 (in violet), which generates the proboscis extension reflex, are localized in the subesophageal ganglion. It is thought that in insects descending neurons connect the lateral protocerebrum to the motor centers (adapted from Hammer, 1993).

paired with odors leading to the same conditioned response as reward learning. Hammer's results indicate the neural substrate involved in associative learning of rewards in the bee. Hammer and Menzel have also shown that stimulation of the VUMmx1 can be replaced by direct delivery of octopamine (the presumed neural transmitter) to the bee's brain regions corresponding to the antennal lobes and mushroom bodies.

Montague *et al.* (1995) constructed a simple neural network based on VUMmx1 that captures the essential features of floral choice under uncertainty and that mimics the results of our choice experiments in which bumblebees foraged on artificial fields of floral color types differing in reward distribution. In their simulation, the individual forager moves around in a three-dimensional arena. On the floor of the arena is a field of yellow and blue squares (160×160) representing flowers of two types. From a given location in the arena, the bee can see a portion of the field determined by a cone of vision that changes as the bee moves about the arena. The architecture of the bee's neural network is outlined in Fig. 6, where P is a simple linear neural unit that corresponds to VUMmx1. This unit pairs the reward stimulus $r(t)$ with color inputs and the output from P modifies motor behavior toward flowers of a given type or random tumbling into a new field of vision. The association between color inputs and reward stimulus follows a Hebbian learning rule which in their simulation becomes equivalent to the Rescorla–Wagner rule for classical conditioning (Rescorla and Wagner, 1972). The learning rate is characterized by a parameter λ, with $0 < \lambda < 1$ such that when λ is high the bee learns very quickly and alters behavior according to immediate short-term experiences. A low value for λ corresponds to a lack of short-term response but maintains the capability of responding over long-term average changes in the environment.

The link between our earlier experiments describing a functional rule for the individual bumble bee and the Montague neural network is established through the specific shape of the reward stimulus $r(t)$ as a function of nectar volume in encountered flowers in the arena. Rewards are scaled in value in the neural network according to the derived biomechanical function (Fig. 4) and Montague *et al.* (1995) use this relationship as the output from the reward neuron *R*.

Montague *et al.* (1995) modeled bumblebee foraging in the artificial arena in which the simulated floral field followed the nectar distributions in yellow and blue flowers that I used in my original 1981 experiments; that is, blue flowers contained 2 μl in each flower, whereas yellow flowers contained 6 μl in one-third of the flowers and 0 μl in the remaining two-thirds. The computer simulation produced a remarkably good fit to the observed pattern of choice by bees (Fig. 3), but only when the learning rate λ in the neural network was high ($\lambda = 0.9$). Thus, the neural network matched bee performance when the architecture of the learning rule corresponded to extreme short-term responsiveness. These neural network simulations based on very plausible neural mechanisms further support the general claim that bees use only short-term immediate experiences and functional rules based on single flower visits in shaping their foraging decisions.

The neural mechanisms involved in the decision-making behavior of bees may not be a biological curiosity of no particular relevance to human behavior. Rather, Montague and colleagues have convincingly argued that the bee's nervous system may be an ideal model for studying general nervous system organization for solving complex choices. The diffusely branching neuromodulatory system of the bee mediated through octopomine neurotransmitters has its analog in the diffusely branching dopamine-modulated neural systems ascending into the mammalian cerebral cortex. Consequently, we may be in a unique position to analyze the neural basis of complex decision making in humans using this simple invertebrate model.

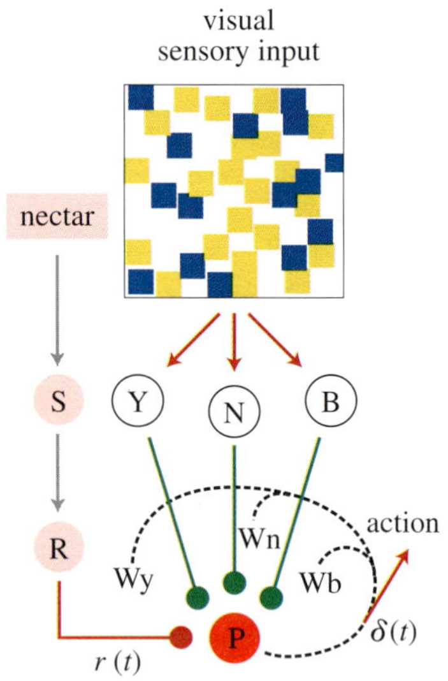

FIGURE 6 Neural network model for a foraging bumblebee. The virtual arena consisted of artificial arrays of blue and yellow flowers in a grid of squares. The output of neurons B, Y, and N represents changes in the percentage of blue, yellow, and neutral colors in the visual field of a bee. These neurons input to a diffusely projecting linear neuron, P, that affects motor selection in the bee [$\delta(t)$] and guides floral selection. Neuron P corresponds to the recently discovered VUMmx1 neuron in honeybees. R, reward neuron; S, olfactory sensory neuron; $r(t)$, reward stimulus (reproduced with permission from Montegue *et al.*, 1995).

Rules for Cooperation vs Deception

There are countless examples from nature of organisms engaged in what could only be called cooperative or helpful behavior. When startled, a white-tailed deer will flash its tail, signaling other neighboring deer to the presence of potential danger. A beaver will slap its tail to signal danger. Several species of birds postpone their reproduction when young in order to stay home and help their parents raise siblings.

However, cooperative behavior in both humans and nonhuman animals has been considered anomalous from an evolutionary perspective. Cooperation involves some form of helping others, often at some expense to the cooperator. However, from an evolutionary standpoint, organisms should always act to promote their individual interests rather than the interests of other organisms. In fact, Darwin noted that evolutionary success can be improved not only by reproducing more but also by causing others to reproduce less. Therefore, from the biological view, it even pays to harm rather than help other individuals, thereby reducing their reproductive success. How does cooperative behavior emerge?

Conditions promoting cooperation among interacting individuals have often been explored by using the prisoner's dilemma game (PDG). In this game, two individuals play against each other. Each player has two choices, either to cooperate or to defect. Each player acts independently of the other player and does not know how the other player will act on a current play of the game but can remember how the opponent has played in the past. The benefit or harm to each player depends on the combination of choices by both players. The payoff to each player is outlined in Table 1.

If both players choose to cooperate, each receives reward of $R = 3$ points. The reward could be dollars for humans or seeds for a foraging hamster. Mutual cooperation is a good strategy since both parties receive a reward. If one player cooperates while the other player defects, then the defector receives a large reward (e.g., $T = 5$), whereas the cooperator gets nothing ($S = 0$). If both players defect, then they each get $P = 1$ point. How should one play this game?

Suppose you are Player 1 and you believe that Player 2 will cooperate. If you cooperate as well, then you will get 3 points. However, if you defect you will get 5 points. What if you believe that Player 2 will defect? In this case, if you cooperate then you get 0, whereas if you defect you will get 1. Consequently, if your opponent chooses to cooperate, then it is best for you to defect, and if your opponent chooses to defect it is best for you to defect. Therefore, no matter what your opponent does, it is best for you to choose to defect even though both of you would do better by cooperating. Herein lies the dilemma. Self-interested behavior leads to outcomes which are of less benefit to both players had they managed to cooperate. Are there conditions in which cooperative behavior can be favored over entirely self-interested behavior?

TABLE I

Matrix of Possible Results for the Prisoner's Dilemma Game[a]

		PLAYER 2			
		Cooperation		Defection	
PLAYER 1	Cooperation	$R = 3$	$R = 3$	$S = 0$	$T = 5$
	Defection	$T = 5$	$S = 0$	$P = 1$	$P = 1$

[a]*R*, reward for mutual cooperation; *T*, temptation to defect; *S*, sucker; *P*, punishment for mutual defection.

In a brilliant and creative analysis of this problem, Axelrod (1984) challenged the world's leading game theorists to suggest strategies (functional rules) for winning the PDG when the game could be played repeatedly against the same player; this is called the iterated PDG. What rules, if any, consistently won against the other rules? Axelrod received strategies from 63 different theoreticians, and the rules ranged in complexity from very simple to those based on computing and updating complicated conditional probabilities based on patterns of previous play. Amazingly, however, one rule emerged as a clear dominant strategy—the tit-for-tat rule (TFT). This rule is not only the best strategy when played against other strategies but also was the simplest rule submitted. The rule is as follows: On the first choice, TFT always chooses to cooperate, and then this player chooses whatever his or her opponent chose in the last round of the game. If the opposing player defected, then TFT defects. If the player cooperates, then TFT continues to cooperate. This rule clearly performed better than all other rules when all were played against each other.

TFT's success as a functional rule relies on three characteristics. First, it begins by cooperating and thereby benefits from other cooperative strategies in the population. Second, it immediately retaliates by switching to defect when confronted by a player that fails to cooperate. Third, it is immediately forgiving and switches back to cooperate whenever an opponent reverts to cooperation.

This simple strategy appears unbeatable when well established in a population of players. However, imagine a population of players all of which only play defect. How could TFT invade this population? On the first play, TFT loses to everyone, whereas on all subsequent plays TFT does no better than defect. Therefore, TFT is a losing strategy in this population. How can it spread? Axelrod collaborated with a well-known evolutionary biologist, William D. Hamilton, to solve this problem. They noted that if small groups of like-playing agents could form associations, then the TFT strategy would at its origins be primarily playing against other TFT strategies and thereby experience an advantage. Hamilton and Axelrod proceeded to show that TFT requires only very small groups of interactors to achieve an initial advantage and spread through the population.

Conclusion

The best functional rules associated with the three behavioral problems discussed here (finding mates, choosing resources, and choosing to cooperate) point to some common features of behavioral organization. Each dominating functional rule is simple and therefore able to be implemented within simple neural structures. Also, each dominant functional rule appears to be short-sighted. That is, the best strategies consider only the immediate past or current conditions of the environment rather than incorporating information over long stretches of time. For bees collecting nectar, the rate of energy gain from single flowers is important. For cooperators, the only event that shapes behavior is the immediate past response of opponents.

I believe that this short-sightedness in decision making will generally be of some evolutionary advantage (Real 1992). Why? Organisms live in environments that are very structured in both time and space. The occurrences of specific phenomena are not generally random events but have some form of correlation either in time or in space. The reward size within flowers is usually spatially correlated such that high-rewarding flowers are in close proximity to other high-rewarding flowers, whereas low-rewarding flowers are in close proximity to low-rewarding flowers. Groups of genetically related individuals are often associated with each other in space and time, and these may form the necessary nuclei for the spread of cooperative strategies. Often, high-value encounters (food rewards or perhaps mates) are temporally correlated. In such cases, short-sightedness may be of value since it allows the organism to track changes in its environment and alter decisions based on immediate future consequences and conditions. It is this fundamentally simple structure within short-sighted functional rules that allows for the necessary flexibility to respond to rapidly changing environmental conditions.

References Cited

AXELROD, R. M. (1984). *The Evolution of Cooperation.* Basic Books, New York.

BATESON, M., and KACELNIK, A. (1996). Rate currencies and the foraging starling: The fallacy of the averages revisited. *Behav. Ecol.* **7,** 341–352..

GALLISTEL, C. R. (1990). *The Organization of Learning.* MIT Press, Cambridge, MA.

HAMMER, M. (1993). An identified neuron mediates the unconditioned stimulus in associative olfactory learning in honeybees. *Nature* **366,** 59–63.

HARDER, L., and REAL, L. A. (1987). Why are bumblebees risk-averse? *Ecology* **68,** 1104–1108.

JANETOS, A. C. (1980). Strategies of female choice: A theoretical analysis. *Behav. Ecol. Sociobiol.* **7,** 107–112.

MONTAGUE, P. R., DAYAN, P., PERSON, C., and SEJNOWSKI, T. J. (1995). Bee foraging in uncertain environments using predictive Hebbian learning. *Nature* **377,** 725–728.

REAL, L. A. (1981). Uncertainty and pollinator–plant interactions: The foraging behavior of bees and wasps on artificial flowers. *Ecology* **62,** 20–26.

REAL, L. A. (1990). Search theory and mate choice. I. Models of single-sex discrimination. *Am. Nat.* **136,** 376–404.

REAL, L. A. (1992). Information processing and the evolution of cognitive architecture. *Am. Nat.* **140,** S108–S145.

REBOREDA, J. C., and KACELNIK, A. (1991). Risk sensitivity in starlings: Variability in food amount and food delay. *Behav. Ecol.* **2,** 301–308.

RESCORLA, R. A., and WAGNER, A. R. (1972). In *Classical Conditioning* (A. H. Black and W. F. Prokasy, Eds.), Vol. 2, pp. 64–99. Appleton–Century–Crofts, New York.

General References

KREBS, J. R., and DAVIES, N. B. (1991). *Behavioural Ecology: An Evolutionary Approach.* Blackwell, Oxford.

REAL, L. A. (1994). *Behavioral Mechanisms in Evolutionary Ecology.* Univ. of Chicago Press, Chicago.

PIERRE L. ROUBERTOUX
MICHÈLE CARLIER

UPR CNRS 1294
Génétique, Neurogénétique and Comportement
Institut de Transgénose
45071 Orléans Cedex 2, France

Behavior Genetic Analysis

Behavior genetic analysis has yet to map and to identify genes involved in behaviors and their physiological correlates. Aggression has been selected to illustrate the advances in the field. Directional selections and Mendelian crosses have demonstrated the implication of the genetic factors in different measures of aggression in laboratory mice. Chromosomal linkage provides the means of discovering candidate regions. Invalidation of previously identified genes has shed light on physiological pathways between genes and this behavior, particularly regarding the role of neurotransmitters. Possible extensions of these approaches in ethology to develop "genetic ethology" are considered.

Introduction

Earl Green, who directed the Jackson Laboratory for more than 25 years, often said that the aim of genetics is "to find, to catch, to keep genes." We are far from the Nature vs nurture debate, heredity–environment controversy, or computation of the portions of variance due to genes or to milieu. Behavior genetic analysis is now engaged in transferring both concepts and techniques from genetics to neurobehavioral sciences. Therefore, the aim in the field is to find, to catch, and to keep genes involved either in behaviors or in physiological-associated mechanisms. To find genes is to map the genes on the chromosomes, to catch is to identify their molecular structure, and to keep is to preserve the discovered genes using techniques such as the development of congenic strains or molecular strategies that develop transgenic individuals. The task is perilous since it implies expertise both in genetics and in neurobehavioral sciences. The tools and concepts in genetics have advanced very quickly and, consequently, a geneticist trained in 1980 may not be fully competent in the current techniques. At that time, it was not possible to imagine that genes could be switched on and off by molecular tricks. The analysis of behaviors demonstrates an unexpected complexity for the behavioral traits under study several years ago. The refinement of ethological descriptions and concepts must be recognized. Psychiatric diagnoses are modified according to periodic publications from psychiatric societies or from the World

Health Organization. Cognitive psychology describes multiple facets of cognitive processes. The neurobiological technologies offer the opportunity to examine the functioning brain and this opportunity is now considered in human neurogenetics. Size or shape of brain structures can be measured, as can the functionality of particular regions during a well-defined task. Also, transmitter concentrations are currently assayed in our species. Moreover, the number, expression, and functional states of transmitter receptors can be assessed by the patch-clamp approach in other species. The further development of behavior genetic analysis needs the joint expertise of numerous fields.

What is the stake in the research of links between genetics and behavior or, more generally, between genes and integrated functions? Behavioral, physiological, and morphometric characteristics of the individuals are not transmitted from the parents to the offspring. The chromosomes included in the gametes carry the genetic material, and this material is transmitted. It is made of long sequences of chemical particles — the four bases (adenine, cytosine, guanine, and thymine) plus phosphodiester bonds and sugar phosphate backbones. Parents transmit atoms and molecules. The gene is a functional unit — a chain of several peptides or amino acids (a protein). Each amino acid is coded by a codon, and each codon is composed of three bases. The four bases arranged in various ways code for 20 amino acids plus stop codons and not for $(4 \times 4 \times 4)$ 64 codons. Most amino acids are represented by more than 1 codon. Thus, the genetic code is said to be degenerate. All the cells of an individual carry identical genetic information because of the mitotic process. All cells are genetic replicates of the original cell constituted by the fusion of the ovum and the spermatozoon. However, the cells have different shapes and functions, and the same chain of amino acids is differentially expressed in the cells of the liver or the brain or in the cells of two structures in one brain. This is due to the complex regulation of gene expression, controlled by other genes and by the environment. The incredible complexity of every living organism has roots in this original simplicity: a combination of four bases and 20 amino acids that leads to many proteins.

This has several consequences. First, it is impossible to recognize genes involved in functions or structures, but we can identify genes involved in individual differences, either in the normal or in the pathological range of variation. Hirsch (1982) stressed the differential aspect of genetic research. The genetic strategy consists of generating or observing meiotic recombinations and detecting cosegregations of identified genes with candidate genes for phenotypes. This strategy implies that individual differences were previously observed at both the genetic and the phenotypic levels.

Second, in these conditions the identified gene(s) is correlated with the observed difference in the population under study. Other genes should be discovered with another population exhibiting other genetic differences. This is not specific for behavior but is true for every level of functioning. This could explain different genetic findings with different populations. The numerous genes discovered for Alzheimer's disease, each explaining the observed variance in one population, illustrate this point. Parameters computed from quantitative genetic models provide different values for the same test, according to the genetic structure of the employed population. Consequently, the parameters often considered as describing the genetic architecture of the trait account for the characteristics of the population.

Third, specific and intraspecific (or individual) inheritance are two different things. *Specific inheritance* refers to differences between species that are due to different genomes. *Individual inheritance* refers to differences in information carried by the genome — the genotype. The basic genetic strategy that consists of observing the products of meiosis cannot be performed between species. In the closest species or subspecies parents might breed best, but the F_1 (i.e., hybrid) males are sterile. It is not possible to determine that a species-specific behavior is "genetically determined." Considerations about "innate" or "preprogrammed" traits refer to species-specific inheritance and are not falsifiable with genetic tools.

Finally, probably no gene is specific for behavioral differences and for every difference in structure or function far from the primary effect of the gene. One gene codes for a protein and most of the modifications occurring in its sequence (addition, deletion, and substitution of bases) modify the protein. This difference may induce modifications in the steps of cellular development and functioning. These modifications contribute to differences in shape, migration, and chemical communication within or between the cells, with consequences for behavioral differences. The genetic correlates of the behavioral differences that we observe are, at best, the indirect repercussions of the rearrangement observed in the base sequences. Therefore, the study of the genetic correlates of behavior cannot be dissociated from the analysis of the neuronal pathways between genes and behavior.

In 1976, we published *Génétique et Comportement* (*Genetics and Behavior*). Two hundred pages were necessary to present the state of the art. Since then, there has been an increase in the number of phenotypes analyzed, the species used, and the methods employed. In a recent review of the literature, more than 900 papers or chapters were published by specialists of behavioral sciences and genetics, biochemistry, and mathematics. It would not be possible to do similar work again and, of course, to present a state of the art covering normal and abnormal behavior in humans and neurobehavioral traits in other species. For these reasons, genetics with aggression in mice has been selected here. It is of interest for ethology and is considered a prime mover in the structure of societies. It fits with an illustration of the pitfalls and advances of genetics applied to behavior and to neuronal correlates of behavior. The conclusion will enable these considerations to be applied to other phenotypes and other species.

What Trait Should Be Used for a Genetic Analysis?

Aggression is a social behavior, and it must be ensured that performing genetic analysis with social behaviors is done properly. Genetic analysis generally deals with aptitudes (to learn, to explore, to build nest, and to prefer a nutriment) measured by reaction of the individual in a standardized situation. Social behavior is quite different. It is the succession of behavioral sequences occurring between two or more individuals: care to pups, courtship, copulation, and aggression. These sequences are the responses of the tested individual reacting to the cues (behavior, sensorial signals, odors, noises, and movements) displayed by the other subjects involved in the situation. One aptitude could remain unexpressed when it is not elicited — that is, when the corresponding cue is missing in the partner repertory. The succession of the sequences could differ in identical individuals with different partners. This has been shown several times since Ginsburg and Allee's pioneering work published in 1942. In the lab, we have extensively studied the strain NZB/BlN. Males from this strain display more attack behaviors when the opponent that has been selected for a dyadic encounter is from the C57BL/6 strain rather than from the BALB/c strain.

We can search for genes involved in aptitudes to initiate a social behavior since individuals exhibiting the aptitudes have genes, but we cannot search for genes of the product of the aptitudes. This point is often ignored (or forgotten) by researchers seeking genes for complex behavior involved in social relationships. A social structure has no genes. This is the pitfall of most genetic studies of aggression. Several researchers have employed the homogeneous set pair test in mice: One individual from one strain is tested with an opponent from the same strain. The cues emitted by the opponent are strain dependent and the measures performed with the male to be tested do not address the same mechanisms. Several strategies have been employed to circumvent this difficulty. The round-robin test developed by Ginsburg and Allee has some drawbacks because the same tested male is successively exposed to different opponents and defeat or victory in one round modifies the behavior in the following rounds. The standard opponent test has been developed as an alternative: A male is tested, with the identical individual (an opponent from another strain kept as much as possible in identical physiological conditions) providing the measure of the aptitude to initiate attack in reaction to the same stimulus. A stable phenotype resulting from a stringent definition is the prerequisite of any genetic or environmental analysis. This is the main problem of the genetic approach of psychiatric disorders in humans. Contradictory results and nonreplicated gene mapping are probably due to the variability of diagnoses. Findings with aggression in our species confirm the conclusions by Rutter (1996), who reviewed published data on twins. Concordance in criminality decreases from juveniles to adults for the two types of twins (monozygotic and dizygotic), and the contrast is higher in adults than in juveniles. Differences in age are associated with different types of crimes. This illustrates a very simple observation: Adult crimes are not the same as juvenile ones. The results from a genetic analysis are limited to a trait defined by its measure and the condition in which it was measured. The conclusion of any analysis, either genetic or environmental, will be limited to the conditions in which the trait was observed, including the stage of development, and cannot be generalized to generic behaviors observed at other ages.

There is a dilemma regarding the decision to select either a behavioral sequence corresponding to the previous criteria or a composite trait meaningful from a naturalistic standpoint. Several behavioral sequences occur during dyadic encounters: chasing, biting, fighting, rattling, and so on. Some authors have computed a composite index including all these measures on the basis of the following argument: Aggression is defined by the organization of these sequences, and it is thus difficult to select a single sequence for a genetic analysis since it would be nonrepresentative of a holistic phenotype. The problem is that the measures are not necessary related and it may be incorrect to consider them together. In Table 1 the only two variables highly related are the number of attacks and the number of chases.

The genetic or environmental analysis can be used for separated sequences, and a gene mapped for one behavioral study with one protocol cannot be generalized to other sequences or protocols. This point is illustrated by Michard and Roubertoux (1984). These authors performed a genetic analysis of the differences observed between females from two inbred strains of mice, NZB and CBA/H, for three pup-care behaviors: first retrieval latency of one pup, move away from one of the pups situated outside the nest, and staying in the nest after having transporting all the pups. These behaviors were measured during a retrieval test. The genetic analysis used 275 primiparous female mice from the two parental strains and their F_1, F_2, and backcross generations. The first retrieval latency was correlated with at least two segregating loci and a one-locus model was not rejected for the two other variables.

TABLE I

Correlations between Means (Kendall Tau) for the Six Aggression Variables Measured by Southwick and Clark (1968) in 14 Inbred Strains of Mice[a]

VARIABLE	NUMBER OF RATTLINGS	NUMBER OF CHASES	NUMBER OF ATTACKS	NUMBER OF FIGHTS
Number of chases	0.01			
Number of attacks	0.11	0.66		
Number of fights	0.04	0.00	−0.05	
Latency of the first attack	−0.09	−0.02	−0.23	−0.27

[a] Four males from the same strain are observed for 8 days. Data are reanalyzed by Michard and Carlier (1985).

Is There Evidence for Genetic Correlates of Aggression?

Population or inbred strain differences for a trait are often considered as providing evidence for genetic bases of the trait. An observed difference between two inbred strains of mice can be due to several factors, including the genes. Strain differences do not provide sufficient demonstration of a genetic contribution: Strains differ by their genes but also by the maternal environment they provide to the pups. This contribution is cytoplasmic, uterine, and postnatal, including care to pups and milk quality and quantity. It adds to or interacts with the effects of the genotype. This point has been documented for many behaviors, including sensorial and motor development in newborn mice (Roubertoux *et al.,* 1990, 1994) and aggression. We have shown that genes involved in initiation of attack behavior against a conspecific male that are carried by the Y chromosome have reverse effects according to the maternal postnatal environment that is provided by the male's mother (Fig. 1).

Carlier *et al.* (1992; Roubertoux *et al.,* 1994) present a set of methods for determining the effects of both maternal environment and genotype. The basic principle is to transfer the same genotype in different environments and to measure the phenotypic differential expression of the genotype according to these various environments. The appropriate use of embryo transfers or ovary transplantation with adoptions leads to the identification of the components of the environment acting in addition to or interacting with the genotype. Fortunately, these methods are restricted to infrahuman populations. In our species, it can be concluded using the risk method that genetics contributes to a trait when the similarity within a set of individuals is correlated with genetic relatedness. However, genetic relatedness in our species is associated with environmental proximity. Complementary situations are necessary to dissociate the two effects. Twin and adoption methods are widely used. Within-pair comparisons in monozygotic and dizygotic twins do not exhibit the significant difference that could be suggested by a large contribution of genetic factors for criminality in adulthood. A complementary approach using the adoption method has often been employed. The adoption method does not permit one to distinguish between genetics and environment because the adoptee has shared, at birth, for 9 months the same environment with the mother. We stressed this limitation of the adoption method 20 years ago. Data from different fields have demonstrated short- or long-term effects of the prenatal maternal environment. The effects of drugs, activity, food, and anxiety exposure during uterine life are known. Busnel *et al.* (1992) have shown genuine effects on perceptive development of uterine exposure to mother or father voice. However, considering jointly the findings from familial studies (including twin and adoption studies) one is led to conclude that genetic factors may be involved in adult criminality (Rutter, 1996). Anyway, strain comparisons in experimental populations or twin adoption methods do not provide information about the genes involved in the trait.

The first demonstration of the implication of genetic differences in differences for aggression was derived from breeding experiments. Breeding experiments consist of hybridizing individuals to induce recombinations and observing the products of the meiosis. The most direct evidence is provided by divergent selections. Several studies presenting positive responses to pressure of selection have been reported, the most recent of which is from van Oortmersen's team, which obtained high and low lines for attack latencies from a stock of wild house mice (Sluyter *et al.,* 1996). Mendelian crosses, including backcrosses, intercrosses, and di-

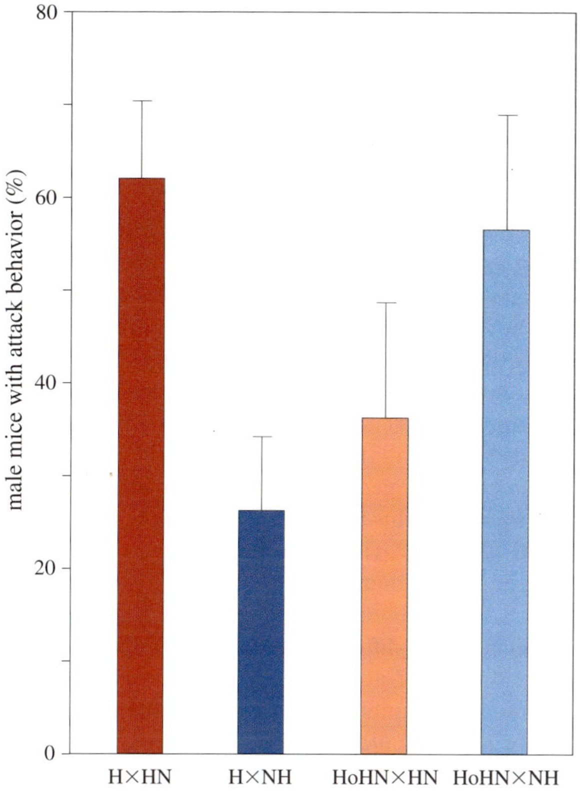

FIGURE 1 Interaction between genotype and maternal environment: The attack behavior of male mice can be modified by the maternal environment. (Left) H × HN males and H × NH males were born and reared by mothers of the H genotype and fathers of the HN or NH genotypes. The two genotypes differ from each other only in the origin of the Y chromosome (the first, N, from the NZB strain and the second, H, from the CBA/H strain). (Right) HoHN × HN and HoHN × NH male mice, again differing only in the origin of the Y chromosome, were born and raised by HoHN females, grafted with ovaries from females of the H strain. The results show that among the males reared in the H maternal environment, those that attacked more often had the Y chromosome from the N strain, whereas among those raised by the HN genotype females, those that attacked more often had the Y chromosome from the H strain (adapted from Roubertoux *et al.*, 1994).

allel crosses, imply quantitative genetic analysis to partition the components of either the mean differences or the variance. They also lead to consideration that genetic variability is correlated with variability in aggression.

These experiments do not provide information about the genes that are implicated in the phenotype. Indirect information is deduced about the number of genes. Several are at work when epistasis is present or when the response is improved during several generations as suggested by van Oortmersen team's data. Breeding experiments are analyzed according to complex mathematical models, with stringent conditions of application. When the criteria of this stringency are met, they inform about the structure of the genetic correlates [additivity, dominance, and epistatis (i.e., interactions between genes); Crusio as cited in Goldowitz *et al.,* 1992].

There have been many attempts to quantify the contributions of these components. The most famous of these estimates is the heritability coefficient. All have been written about extensively since Fisher (1951; Walhsten, 1990), who warned "the so-called coefficient of heritability, which I regard as one of those unfortunate shortcuts which have emerged in biometry for lack of a more thorough analysis of the data." However, we refer the reader to Wahlsten's paper and peer commentaries published in *Behavioral and Brain Sciences.* In short, heritability is part of the additive genetic contribution to the variability of the genotype. Its interpretation (and not its computation, unfortunately) implies several assumptions widely known: There is no dominance between allelic forms implicated in the phenotype, there is no epistasis, and the matings occur at random. At this stage it is easy to consider how the model is unrealistic for describing the phenomenon in a natural setting. Another important issue to bear in mind is that this coefficient is the proportion of variance accounted for in the population and not at the individual level. Considering a value for one population, no prediction can be made for one individual taken from this population: Heritability is not heredity. Moreover, because heritability is specific to the population in which it is estimated, the value cannot be used to describe another population in which environmental factors and gene frequencies differ. Most geneticists and even psychologists working in behavior genetic analysis are aware of these points but still continue to use heritability estimates. Why? If they are used to predict the success of a selective breeding experiment as illustrated by Roubertoux (1993) working on courtship behavior in the male guppy, they are used correctly. If they are used as only a first step to discover preliminary information, as argued by McGuffin *et al.* (as cited in Wahlsten, 1990), most researchers would abandon their computation since other techniques are available, even for human genetics. Moreover, behavior geneticists cannot ignore the fact that the misuse of heritability estimates may have catastrophic social consequences, especially in the field of intelligence (Roubertoux *et al.,* 1996).

There is experimental evidence for genetic correlates of aggression in mice. Have these genes been identified?

Genes Implicated in Aggression

There are two main strategies to recognize genes involved in individual differences for a function. The first begins with identified genes or chromosomal regions and employs methods either from recombinant DNA or from classical genetics (developing congenic strains). The second approach is the development of genetic mapping.

Recombinant DNA methods provide the means to integrate one gene from one species with the genotype of another species. The transfer is generally from human to rodents, most frequently the mouse, but the transfer to rat has recently begun. The gene is incorporated in the host's genotype and is transmitted according to Mendelian laws. Its expression induces proteins similar to those observed in human species. Homologies between the clinical or behavioral phenotypes in the two species remain to be discovered. It is also possible to transfer a large set of genes: A chromosomal region is cut and incorporated in yeast artificial chromosomes and then transferred to the host. With this method, it is possible to follow the effects of genes during development or to analyze their effects in the organism. The alternative consists of preventing the expression of a gene by molecular methods. In this case, the gene is said to be knocked out. The absence of the gene provides the opportunity to describe the function of the gene. This strategy is of interest because it identifies the physiological pathways between the gene and the behavioral trait.

Using recombinant DNA methods, several genes have been reported to be involved in the initiation or incrementation of attack behavior in mice. Serotonin or 5-hydroxitryptamine (5-HT) is released in the whole organism, particularly in the brain — cortex, hypothalamus, and spinal chord. The implication of 5-HT in aggression is well documented. Negative correlations between 5-HT and the number of attacks on a standard opponent were obtained by several authors. Other transmitters , such as metenkephaline, ACTH, β-endorphins, and GABA, exhibit correlations, suggesting their involvement in aggression. The picture is much more complicated. The effect of a transmitter is the consequence of a balance between the amount of transmitters and the functionality of the receptors. For 5-HT, seven families of receptors are known: 5-HT_{1A}, 5-HT_{1B}, 5-HT_{1C}, 5-HT_{1D}, 5-HT_3, 5-HT_4, and 5-HT_5. The problem is that there is no antagonist specific for each receptor. Consequently, the pharmacological tool is limited. René Hen's group (Saudou *et al.,* 1994) invalidated gene coding for receptor 5-HT_B. Receptors are lacking in homozygous mice and these mice present the highest frequency of attack behavior in different tests compared to both controls and heterozygous mice. Was the gene for aggression discovered? Chen *et al.* (1994) demonstrated that heterozygous mice knocked out for the α-calmodulin-kinase II gene also attacked more, whereas this behavior decreased in homozygous mice. Snyder's group has shown an augmentation of attack behavior in knockout mice for nitric oxide synthase (Nelson *et al.,* 1995). Many genes have been reported for their effects on aggression in the mouse when they are invalidated.

The gene-invalidation method has limits. A gene has pleitropic effects, as does the knockout gene. The effect of the knockouts are not tissue specific. We know that 5-HT_{1B} is distributed in the gut and involved in its motility. It might induce nutritive defects which are known to cause aggression in rodents. Also, the receptor is lacking from the early stages of development. Several knockout fetuses have die *in utero* or at birth due to the lack of receptor which is involved in crucial steps of development. It is useful to make three points. First, the methods are still evolving: Methods to produce time-dependent and tissue-specific knockout genes are now available. Second, we do not question the link of these genes with aggression. The caveat made only concerns the use of a pathological model (a null allele induces pathologies) to explain the observations made in the normal range of variation. Third, results with other knockout genes suggest that the effect of invalidation depends on the genetic background. This is well-known for learning; to our knowledge this has not been attempted with aggression. This could provide information about single versus several genetic mechanisms.

What about the genes carried by the Y chromosome? The association of differences in offense behavior, defined as the initiation of attack behavior against a conspecific male, with the variants of the Y chromosome in mice is the subject of a long-standing debate with regard to both humans and mice. Briefly, correlated response for male and female aggression in lines differentially selected only for female aggression has been reported by Cairns's group, suggesting that this behavior may not be transmitted by a locus linked to the Y chromosome (Hood *et al.,* 1988; Roubertoux *et al.,* 1994). Our research with the Y chromosome is of interest to show the difficulty of investigating a candidate gene even with laboratory animals. We used the test with standard opponent and we stress again our careful attempt to control the rearing conditions and to maintain constant the conditions over several years. Moreover, the prenatal and postnatal maternal environments were also maintained constant using the graft ovary method (Fig. 1). The frequency of males initiating attack behavior is higher in NZB than in CBA/H. We started a breeding experiment with reciprocal F_1s. It could be concluded that there was an effect of the Y since the HNF_1 group received the Y from the most frequently attacking strain and vice versa. However, several other factors contributed to this difference, including maternal effects and genomic imprinting. Only one solution was available to test for the Y effect: the development of congenic strains for the Y as Maxson *et al.* (1979; Roubertoux *et al.,* 1994) performed for other strains. The structure of the Y includes a part that is transmitted from father to son and a part that recombines with the X during the male meiosis and thus is partially transmitted from father to daughter. Performing repeated backcrosses between one parental female and the males from the cross, we obtained congenic strains for the specific part of the Y. The isogenicity of the background was also tested with molecular and histocompatibility methods. The mice were tested: Contrary to results obtained by Maxson's team, there was no effect of the specific part (Roubertoux *et al.*, 1994). Currently, we have reached the 45th generation (the strains are 10 years old), the results for which are identical. However, the story is much more complex; when the homogeneous set pair test is employed the effect of the specific part of the Y appears. The Y contribution is shown to interact with autosomal effects. These conclusions (Y interacting with autosomes) have also been reached by two independent research teams working in Storrs and Gröningen.

A complementary approach to identify genes involved in aggression consists of mapping. It follows alternative strategies of classical and reverse genetics. The prerequisite of classical genetics is the identification of the protein. Deficiency of enzymatic activity of monoamine oxidase (MAO) A was found in members of a human family in which impulsive aggression without conviction was observed (Brunner *et al.*, 1996). All the affected members of the family had a normal MAO B activity with a reduced activity for MAO A. The cause of the deficient activity was identified as a point mutation in the eighth exon and all the affected members carried the deficient allele. No additional kindred with MAO deficiency associated with impulsive aggression are known. It is highly probable that this abnormal MAO A activity explains very few cases of abnormal aggressive behavior.

Genetic mapping provides an exhaustive identification of the genetic correlates of a trait. It takes advantage of the results of human and murine genome sequencing projects. One issue of these projects was the mapping of short sequences of length polymorphisms (or microsatellites), which are small noncoding sequences that have a high level of polymorphism. Since they are noncoding, they are not subjected to selection. More than 100,000 are potentially available; 7000 have been mapped in the mouse and 5000 in humans. Every individual can be identified according to the allele that he or she carries for the typed locus (Fig. 2).

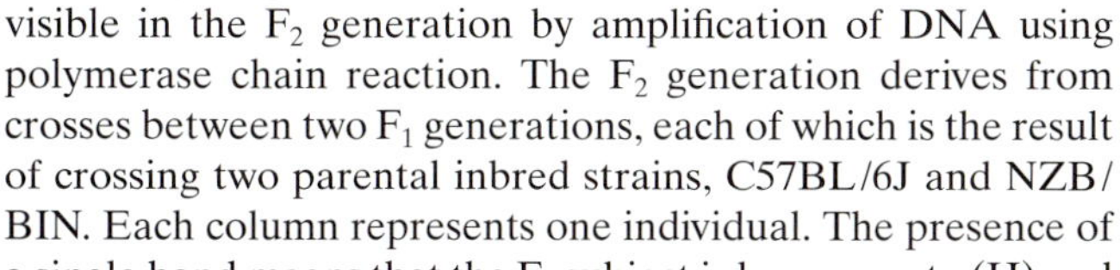

FIGURE 2 Gene mapping permits the identification of genes involved in a determined behavior. On the electrophoresis plate, polymorphic microsatellite bands for chromosome 5 are visible in the F_2 generation by amplification of DNA using polymerase chain reaction. The F_2 generation derives from crosses between two F_1 generations, each of which is the result of crossing two parental inbred strains, C57BL/6J and NZB/BIN. Each column represents one individual. The presence of a single band means that the F_2 subject is homozygote (H) and behaves as one of its parents. Two bands means that the F_2 is heterozygote (X) and behaves as an F_1 subject.

Many microsatellites are typed in a segregating population, and then the regression between the phenotypical values and the allelic forms is computed. This method provides the means to map genes with major effects but also genes with minor effects. It is possible to map genes explaining small percentages of variance and, consequently, fragments of polygenes can be identified. This technique, called quantitative trait loci (QTL) detection, is crucial since most of the behavioral or physiological phenotypes have polygenic correlates. Several genes have been mapped with this method for alcohol-related behaviors or exploratory activity in the mouse. It should provide solutions for complex phenotypes which have multigenic bases, particularly in the field of psychiatry, such as schizophrenia, manic depression, and autism. Regarding aggression in mice, in our lab Roubertoux and Le Roy employed QTL detection in an F_2 design between NZB (initiating attack) and C57BL/6 (which is pacifistic) with a standard opponent test, with A/jax male as the opponent. Four to six microsatellites per chromosome were selected; to date, approximately 40,000 typings have been performed. Analysis of the results is in progress, but they can conclude that microsatellites that are at the telomeric end of the X chromosome, close to the boundary of the pairing region, cosegregate with initiation of attack behavior. This finding concurs with our previous results which indicated that one of the genes involved in this behavioral difference in NZB and CBA/H was on the pairing X–Y region.

Other Species, Other Behaviors

What can one expect in this field during the next few years? Behavior genetic analysis has often been limited to human species (psychiatric disorders and cognition) and to studies of rodents, with a strong interest in the behaviors studied in the experimental psychology tradition (e.g., learning) or those providing models for human behaviors (drug dependence and reactions to pharmacological molecules). Little attention has been paid to species-specific behaviors resulting in a gap between genetics and ethology or evolution. These behaviors have only been considered as possible targets for selection. Their role in these processes as active forces to maintain or to increase genetic diversity should emerge as a consequence of genetic analyses.

Three goals may be considered in meeting this challenge. First, behavior genetic analysis should be extended to other behaviors well-known from previous ethological studies, such as care to pups, song production, mating behavior, and recognition of conspecifics. This is well illustrated by the results obtained by Yamazaki *et al.* (1988), who observed that male mice typically prefer to mate with females of different, non-self H-2 haplotype: That is, they prefer to mate with females with a histocompatibility system different from their own. An adoption study showed that this preference is mediated via maternal postnatal imprinting. Are these mechanisms at work in other behaviors?

Second, other species should be subjected to genetic analysis. This began several years ago when behavior genetic analysis became interested in *Drosophila* as is well illustrated by Jerry Hirsch's team work with taxies. In 1958, Hirsch started selective breeding for geotaxis (orientation and movement with respect to gravity) in *Drosophila melanogaster,* the fruit fly (Fig. 3). The high line (flies making up choices in a multiple-unit maze) had become stable in 1978 and the low line (flies making down choices) in about 1982 (Ricker *et al.,* 1985). Subsequently, they found evidence of partial reproductive isolation (mating preference). Three allozymes correlated with geotaxis expression: alcohol dehydrogenase (*Adh;* located on chromosome 2, region 2-50.1), amylase (*Amy;* located on chromosome 2-77.7), and 6-phosphogluconate dehydrogenase (*Pgd;* located on chromosome 1-0.5). However, by using hybrid subjects the authors observed a breaking up of the genotype–phenotype correlation for two of the markers (*Amy* and *Pgd*). After crossing reciprocally the high and low lines at the generation 699, they allowed free mating (i.e., the absence of selection) during 66 generations in each of the reciprocal sublines (HL and LH). An association between *Adh* and geotaxis score was detected in the LH subline but not in the HL subline. The authors concluded that an unknown gene correlate of geotaxis is situated close to the *Adh* gene, probably at 1 cm (Stoltenberg and Hirsch, 1996).

As illustrated by this time-consuming research, working with flies provides the opportunity to follow many generations of crossing. Another example can be drawn from Rose and his team, who focused on aging. They have bred 500 successive generations of *D. melanogaster* and in so doing have succeeded in doubling the fly's maximum life span from about 60 to about 120 days (Rose and Finch, 1994). Other species are available for genetic analysis. *Caenorhabditis elegans* (a microscopic worm) has proven its usefulness for chemotaxis, motor behavior, and aging. Few researches have used species such as birds or fishes of interest for ethology. We know the difficulties of performing genetics with social insects, but these difficulties could be overcome with bees considering the increasing knowledge of their genome.

Third, genetic analysis could take advantage of different genome projects performed for different species to end a speculative–polemic stage about the possible effects of genes on behavior. We can soon expect an exhaustive knowledge of the bases carried by the human and the murine genome; moreover, other genome projects with other species will soon provide extensive maps. Currently, mapping the candidate gene for a trait provides a candidate region. This region includes *N* previously known genes. The function of

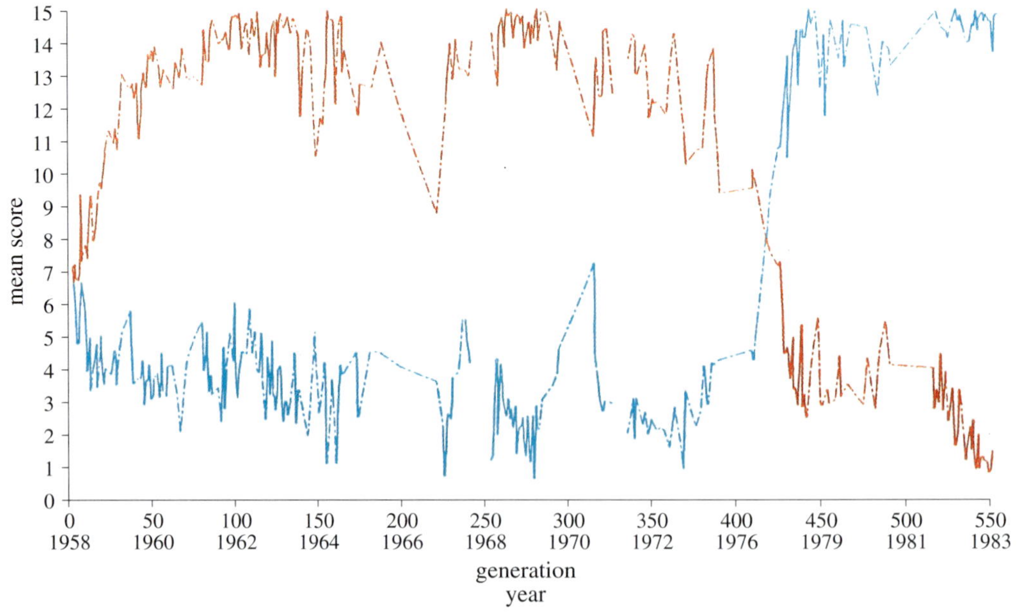

FIGURE 3 Measurement of the mean geotatic response of *Drosophila* females to selection in two lines (high and low). The response in the two lines gradually differentiated over time, more markedly in the high line. After 400 generations, a period of absence of selection led, in each of the two lines, to a regression to the mean response. The high line (which orient themselves in a complex maze; in red) became stable in 1978, whereas the low line (which do not orient themselves; in blue) became stable in about 1982. The solid and broken lines indicate forward and relaxed selection, respectively (reproduced with permission from Ricker and Hirsch, 1985).

every gene and thus its relevance for the trait will be investigated and tested according to species, either by transgenesis or by targeted mutation. The largest is the candidate region, which has the highest number of genes to test. Therefore, accurate mapping strategies (mapping within the smallest confidence interval) is the main challenge in genetics today.

References Cited

Busnel, M. C., Granier-Deferre, C., and Lecanuet, J.-P. (1992). *Ann. N. Y. Acad. Sci.* **662,** 118–136.

Chen, C., Rainnie, D. G., Greene, R. W., and Tonegawa, S. (1994). *Science* **265,** 291–294.

Figher, R. (1951). *Br. Agric. Bull.* **4,** 217–218.

Goldowitz, D., Wahlsten, W., and Wimer, R. E. (Eds.) (1992). *Techniques of the Genetic Analysis of Brain and Behavior: Focus on the Mouse.* Elsevier, Amsterdam.

Hirsch, J. (1982). In *Behavior–Genetic Analysis* (J. Hirsch and T. R. McGuire, Eds.), pp. 1–13. Penn, Hutchinson Ross, Stroudsburg.

Michard, C., and Roubertoux, P. L. (1984). *J. Comp. Psychol.* **98,** 154–164.

Nelson, R. J., Demas, G. E., Huang, P. L., Fishman, M. C., Dawson, V. L., Dawson, T. M., and Snyder, S. H. (1995). *Nature* **378,** 383–386.

Ricker, J. P., and Hirsch, J. (1985). *J. Comp. Psychol.* **99,** 380–390.

Rose, M. R., and Finch, C. E. (Eds.) (1994). In *Genetics and Evolution of Aging.* Kluwer, Dordrecht.

Roubertoux, P. (1993). *Int. J. Comp. Psychol.* **5,** 145–163.

Roubertoux, P. L., Carlier, M., Degrelle, H., Haas-Dupertuis, M.-C., Phillips, J., and Moutier, R. (1994). *Genetics* **135,** 225–230.

Roubertoux, P. L., Carlier, M., Spitz, E., and Hirsch, J. (1996). *Recherche* **283,** 70–79.

Rutter, M. (1996). In *Genetics of Criminal and Antisocial Behavior,* Ciba Foundation Symposium No. 14. Wiley, Chichester, UK.

Saudou, F., Amara, D. A., Dierich, A., LeMeur, M., Ramboz, S., Segu, L., Buhot, M.-C., and Hen, R. (1994). *Science* **265,** 1875–1878.

Sluyter, F., van Oortmerssen, G. A., de Ruiter, A. J. H., and Koolhaas, J. M. (1996). *Behav. Genet.* **26,** 489–496.

Stoltenberg, S. F., and Hirsch, J. (1996). *J. Comp. Psychol.* **110,** 252–259.

Wahlsten, D. (1990). *Behav. Brain Sci.* **13,** 109–161.

Yamazaki, K., Beauchamps, G. K., Kupniewski, D., Bard, J., Thomas, L., and Boyse, E. A. (1988). *Science* **240,** 1331–1332.

General References

Carlier, M., and Roubertoux, P. L. (1994). In *Traité de Psychopathologie* (D. Widlocher, Ed.), pp. 585–613. PUF, Paris.

Driscoll, P. (Ed.) (1992). *Genetically Defined Animal Models of Neurobehavioral Dysfunction.* Birkhauser, Boston.

Hirsch, J., and McGuire, R. R. (Eds.) (1982). *Behavior-Genetic Analysis.* Benchmark Papers in Behavior Vol. 16. Hutchinson Ross, Stroudsburg.

GIANLUCA TOSINI
MICHAEL MENAKER

*Department of Biology and National Science Foundation Center for Biological Timing
University of Virginia
Charlottesville, VA 22903, USA*

Biological Clocks

Daily variations in many organisms have been observed since the very beginning of time, but only since the second half of the twentieth century have biologists begun to show interest in this matter in a scientific manner. For many years it was believed that these rhythms were a simple response to the daily variations in the physical environment. The reawakening of interest in this type of study followed the demonstration that these variations in behavior or in physiology are controlled by accurate internal measuring systems: biological clocks. Many studies have shown that the majority of living organisms, whether animal or vegetable, possess an endogenic oscillator endowed with a period of oscillation similar to the duration of the astronomical day. In addition, many organisms possess clocks that oscillate with a period similar to that of other geophysical cycles that play an important role in their life.

Introduction

The most obvious biological rhythms are those that have fixed relationships to the astronomical day. Despite the fact that daily variations in the behavior of living things must have been observed for centuries, it is only since the 1950s that biologists have begun serious scientific investigation of these rhythms. For many years it was believed that the daily rhythms observed in behavior or physiology were imposed by rhythms in the physical environment to which organisms responded passively. It was the demonstration that plants and animals are not passive responders but rather that they have accurate endogenous time-measuring systems or biological clocks that generated new interest in the study of biological rhythms (Fig. 1). Thousands of studies have demonstrated that most organisms have internal clocks with periods close to the length of the astronomical day (i.e., 24 h); furthermore, many organisms have clocks the periods of which match other astronomically generated cycles that are significant in their lives (Table 1).

Biological Rhythms Have a Wide Range of Periods

The periods of biological rhythms range from milliseconds to much longer cycles with periodicities of several hours, days, or even years (Table 2). Some of these rhythms bear no direct relationship with cycles in the physical world; they have evolved in response to physiological demands and thus

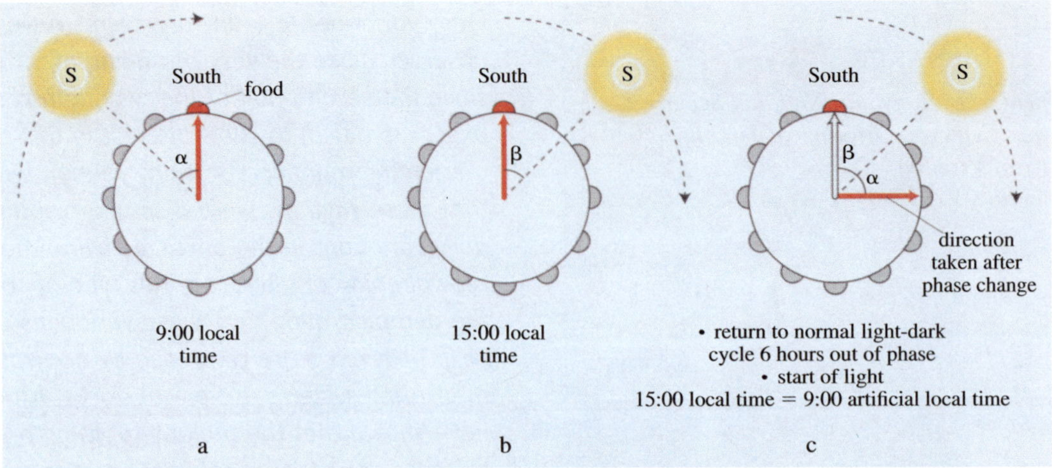

FIGURE 1 Klaus Hoffmann's classical experiments demonstrating, for the first time, the involvement of the biological clock in the orientation of starlings. The birds were initially trained to find food in a certain direction always at the same time of day (9:00) (a) and were able to choose the right direction even at a different time (15:00) (b) by estimating the change of position of the sun (S) during the day. They were then exposed to an artificial light/dark cycle 6 h out of phase with the natural day. When they were brought back to the normal light/dark cycle and tested at 15:00 for their ability to orientate toward the predetermined goal (c), the direction of their orientation was changed by 90°, corresponding to the direction they should have taken at 9:00 (their subjective time). α and β, orientation angles with respect to the sun at 9:00 and 15:00, respectively; red arrows, training and experimental directions; white arrow, expected direction (reproduced with permission from Hoffmann, 1954).

their periodicity depends on the dynamics of these demands (e.g., heart rate and spontaneous firing rhythms in many neurons). Other rhythms are tightly coupled to some external environmental cycle; they have evolved adaptive phase relationships to these cycles. Such rhythms have periods close to those of the astronomical day (*circadian*), the tides (*circatidal*), the phases of the moon (*circalunar*), or the alternation of the seasons (*circannual*). These rhythms persist when the organisms are isolated from the environmental cycles to which they are usually coupled, demonstrating that they are generated internally (*endogenous*). In the absence of environmental cues circadian rhythms have periods of approximately 24 hr (Fig. 2), whereas the periods of circatidal, circalunar, and circannual rhythms are approximately 12.4 h and 29.5 and 365 days, respectively (Table 2).

TABLE 1

Geophysical Cycles and Their Period Length

RHYTHM TYPE	GEOPHYSICAL CYCLE	PERIOD LENGTH
Circadian	Rotation on the earth on its axis	≈24 h
Circatidal	Gravitational attraction of the sun and the moon on the earth	≈12.4 h
Circalunar	Revolution of the moon around the earth	≈29.5 days
Circannual	Revolution of the earth around the sun	≈12 months

TABLE 2

Summary of Period Lengths of Various Biological Rhythms

TYPE OF CYCLES	PERIOD LENGTH	VARIABLES
Ultradian	milliseconds to hours	Firing rate of neurons Electrical activity of the brain Heartbeat Rate of breathing Hormone release
Tidal	≈12.4 h	Opening shells Locomotion/feeding
Circadian	≈24 h	Eclosion Feeding Temperature
Circalunar	≈29.5 days	Mating, egg laying Egg laying
Circannual	≈12 months	Hibernation Reproduction Migration Suicide (in humans)

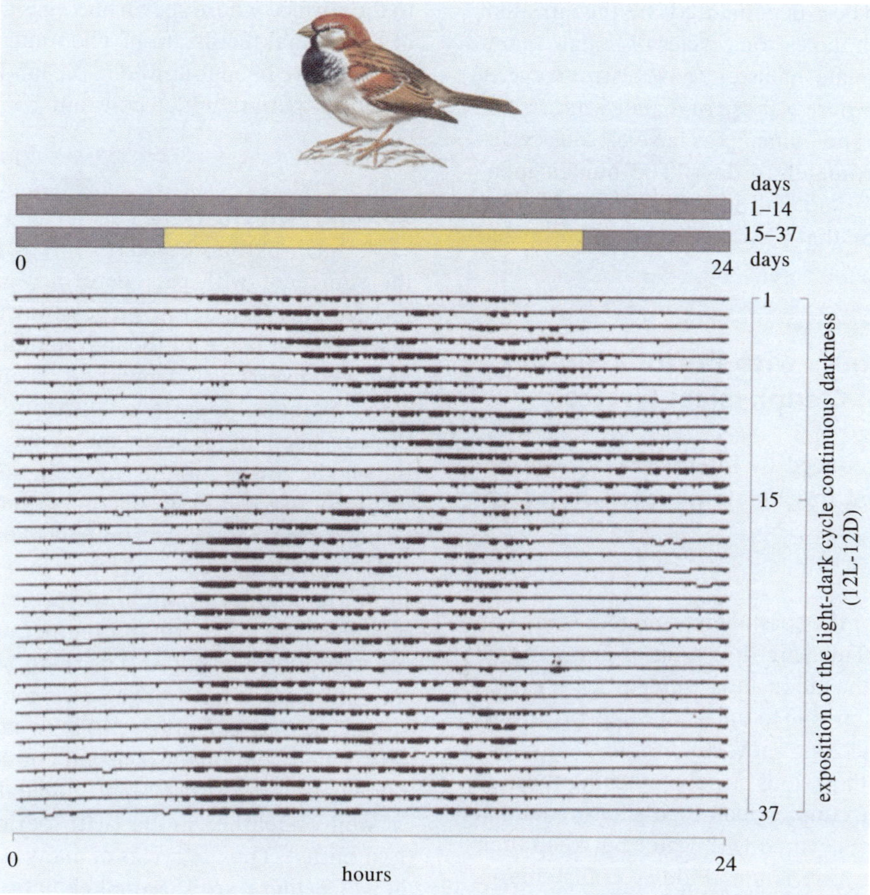

FIGURE 2 The circadian locomotor rhythm of the house sparrow in its free-running and light/dark entrained states. During the first 14 days of the record, the bird was maintained in continuous darkness and its rhythm free-ran with a period of about 25 h. The wider dark bars in each row indicate the activity periods during the 24-h period. Beginning on Day 15, the bird was presented with a daily light cycle of 12 h of light and 12 h of darkness (12L–12D), to which his rhythm synchronized (entrainment). This bird was blinded by removal of both eyes prior to the start of the experiment, demonstrating that synchronization with a light cycle can also occur through other photoreceptors (reproduced with permission from Menaker, 1968).

Biological Rhythms with No Relationship to Geophysical Cycles

Since most biological processes are regulated by control mechanisms involving feedback, which always generates oscillations, organisms have literally thousands of rhythms of which only four period classes are the products of biological clocks. Some scientists have found it convenient to lump the rest into two groups—those with periods shorter or longer than 1 day.

Ultradian Rhythms

Ultradian rhythms are endogenous rhythms with periods ranging from milliseconds to hours and with no obvious direct correlation with environmental cycles. Specific examples include the spontaneous firing rate of the cold receptors in the lingual nerve that has a periodicity of 100 ms and the pacemaker activity in the parietal ganglion of the snail (*Helix pomatia*) that has a period of about 5 s. Ultradian rhythms are also present at the behavioral level. The worm, *Arenicola marina,* which lives in burrows on sand flats in the intertidal zone, has a feeding rhythm with a period of 6–8 min. Similarly, some small mammals such as meadow voles, which are active during daylight hours, show bursts of activity followed by periods of rest in ultradian cycles with periods of about 2 h. Ultradian rhythms have also been demonstrated for the pulsed release of many hormones, including luteinizing hormone, growth hormone, and corticosterone.

Infradian Rhythms

Rhythms with periods longer than 24 h which are not related to geophysical cycles are called *infradian rhythms.*

The most common, and best-documented, are the infradian rhythms associated with the estrous cycles of female mammals. For example, female hamsters have estrous cycles with periods of 4 days, whereas those of female rats are 4 or 5 days in length; sheep and guinea pigs have estrous cycles with periods of approximately16 days. The human menstrual cycle may or may not fall into this category since there is reason to believe that it evolved to be synchronized by the lunar cycle.

Biological Rhythms with Periods Similar to Those of Geophysical Cycles

These rhythms are produced by internal clocks and are adaptively synchronized to geophysical cycles in the environment.

Tidal Rhythms

A primary feature of the coastal environment is the ebb and flow of the tides. Tides are the result of gravitational forces associated with the moon and exhibit a 12.4-h period from the phase of one low tide to the next. Many organisms inhabiting the intertidal zone show rhythms in their physiology and behavior with periods closely matching those of the tides. Oysters, for example, open their shells to feed at high tide when they are covered with water but keep their shells closed when the tide is low. Fiddler crabs emerge from their burrows at low tide to feed on the sand flats; as the tide returns they reenter their burrows. Several studies have shown that these rhythms persist with periods similar to those of the tides when the organisms are maintained in constant artificial environments with no tidal components; these results establish the remarkable fact that intertidal organisms have built-in tidal timers or, more accurately, circatidal clocks.

Lunar Rhythms

The approximately 29-day cycle of the intensity of nighttime illumination due to the waxing and waning of the moon is based on the moon's rotation around the earth and is clearly related to tidal rhythmicity. Because the lunar cycle predicts the spring (highest) and neep (lowest) tides, the marine midge (*Clunio marinus*) coordinates its eclosion, mating, and egg-laying activities with the lunar cycles. The famous grunion (*Leuresthes*), a marine fish, spawns during the spring tides on the sandy beaches of California using the phase of the moon as a timing cue. The spawning of some marine worms, most notably the palolo worm (*Eunice viridis*), is synchronized to the lunar cycle. These worms live in burrows in the coral reefs of the South Pacific. During the months of October–December, on the last day of the last quarter moon a segment of each worm breaks off and swims to the surface where sperm and eggs are released in an orgy of communal fertilization. The lunar rhythms of a related species have been followed in the laboratory and have been found to persist under constant conditions as circalunar cycles.

Annual Rhythms

Annual rhythms based on the earth's movement around the sun recur with periods of a year and are more pronounced the farther one moves away from the equator. Remarkably, in some birds and mammals seasonal rhythms have been shown to depend on an endogenous circannual clock. Some temperate zone mammals enter a condition of deep sleep and reduced metabolic activity (hibernation) during the winter months; by doing so, they avoid the harsh conditions of winter. Bouts of hibernation alternating with periods of activity occur annually in the field and persist for years with periods close to a year under constant conditions in the laboratory. Many bird species escape the rigors of winter in the northern and temperate climates by migrating to southern latitudes. The rhythms associated with such migration often have strong endogenous components. The stonechat (*Saxicola torquata*) is an excellent example. This bird, when held under constant conditions, still undergoes regular long-term cycles of gonadal growth and decline as well as feather molts (with periods of approximately 12 months). The life cycles of many insects that live in areas in which there are seasonal climatic shifts incorporate an annual period of diapause which is regulated by a circannual clock.

Daily Rhythms

Many, if not most, biochemical, physiological, and behavioral parameters exhibited by organisms show daily fluctuations; most of these persist in constant conditions, demonstrating that they are driven by an endogenous oscillator. Rhythms that persist in constant conditions with periods of approximately 24 h are called circadian (from *circa* = about, and *dies* = a day). Circadian rhythms have been described in almost all groups of organisms — from unicellular (and even some bacteria) to mammals — and we have now much information about their physiology, their organization, their genetics, and the molecular mechanisms that generate the periodic oscillations. Our knowledge of the mechanisms that underlie the other three circa rhythms is much less complete, so we devote the rest of this article to a discussion of what is known about circadian mechanisms, although some of the general discussion is applicable to biological clocks regardless of period.

General Properties of Circadian Oscillators

The study of biological clocks and the rhythms that they regulate is called *chronobiology.* Since some of the terminology is quite specific to this discipline, we begin with a chronobiological vocabulary. A *circadian oscillator* is a part (cell, tissue, or organ) of a living organism which is capable, when isolated and maintained in constant conditions, of generating a self-sustained rhythm with a period of approximately 24 h. A *rhythm* is defined as a recurrent event that is characterized by three features: period, amplitude, and phase (Fig. 3). The *period* is the length of time necessary to complete one full cycle (e.g., the time required to go from peak to peak). *Amplitude* is, roughly, the height of the peaks relative either to the troughs or to the zero crossings. *Phase* refers to the temporal relationship between a specific identifiable point on the cycle and a point on a reference cycle (e.g., the relationship between onset of the activity and the beginning of the light cycle in Fig. 2).

When an organism is held under constant conditions its circadian rhythm will persist with a period close to, but in most cases significantly different from, 24 h. This result establishes that the rhythms are self-generated or endogenous; the period of a rhythm under constant conditions is called its *free-running period.* The free-running period of a rhythm is directly related to the period of the biological clock which generates it, and it is therefore a useful measure of the state of the clock (e.g., how it is affected by particular genetic mutations).

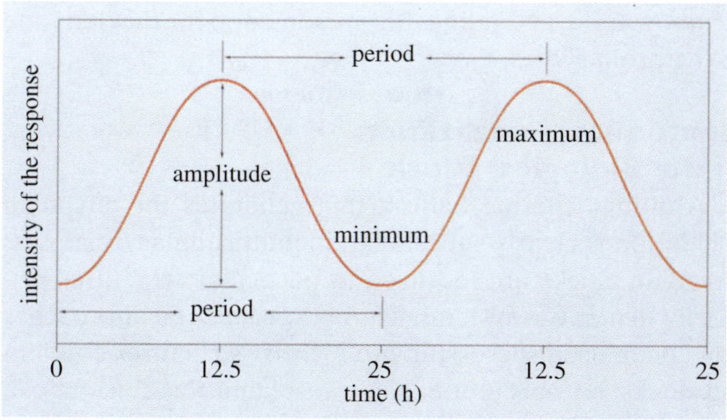

FIGURE 3 Diagram illustrating the formal properties of a rhythm. The period is the time required to complete one full cycle (25 h in this example). The amplitude is the difference between the maximum and the minimum values. The phase is the relationship that the rhythm assumed with respect to some other timescale; for example, in nocturnal rodents the phase of onset of the activity rhythm usually corresponds to the onset of the dark period of the light/dark cycle to which the animal is exposed.

Circadian rhythms are temperature compensated: Their free-running periods are changed only slightly by changes in the internal temperature of the organism (Fig. 4). The majority of biological processes have Q_{10}'s of 2 or 3 (i.e., the rate of the process doubles or triples with an increase in the temperature of 10°C), but circadian rhythms in all organisms tested have Q_{10}'s close to 1 (usually within the range of 0.85–1.3). Temperature compensation has important adaptive significance, especially in organisms that do not regulate their internal temperature physiologically (the majority). Clearly, a clock whose rate depends strongly on temperature would be of little use as a time-measuring device in the real world of fluctuating ambient temperature.

To be useful, circadian clocks, like watches, must be synchronized with the 24-h rotation of the earth on its axis. The process by which a circadian rhythm is synchronized with the environmental day/night cycle is called *entrainment* (Fig. 2). Much experimental effort has gone into the study of entrainment. Although there are many periodic factors present in the environment, very few have been shown to entrain circadian rhythms. The environmental light cycle is by far the most important; all organisms studied to date, including humans, have been shown to respond to this cue. Most organisms can be entrained by temperature cycles, and in a few birds and mammals timed food and drink availability and social interaction can act as entraining agents.

The general term for entraining agents is *zeitgeber* (German for "time giver"). A circadian rhythm is entrained when its period matches that of the entraining stimulus (Fig. 2). An entrained rhythm always assumes a fixed phase relationship with the zeitgeber. To produce entrainment the zeitgeber must alter the free-running period to match its own. If the free-running period is longer than that of the zeitgeber, the zeitgeber will act to shorten it; if the free-running period is shorter, the zeitgeber will lengthen it. This process has been analyzed as if the zeitgeber is resetting the clock in each cycle, and such analyses have produced models that account for entrainment quite well (Pittendrigh and Daan, 1976). Resetting is accomplished by instantaneously shifting the phase of the biological clock and phase shifts can be produced by single light pulses as well as by light cycles. The phase-shifting effect of single light pulses depends on the time in the circadian cycle at which they occur. A graphic representation of this effect is called a *phase response curve* (Fig. 5); the general shape of the phase response curve is similar in all organisms, but its details depend on factors such as the intensity and duration of the signal, the prior history of the individual, and species. If the phase response curve to single light signals and the free-running period are known, it is possible to predict how an organism will entrain to a particular light cycle. In the field, environmental zeitgebers, primarily light and temperature cycles, keep organisms synchronized with the environment

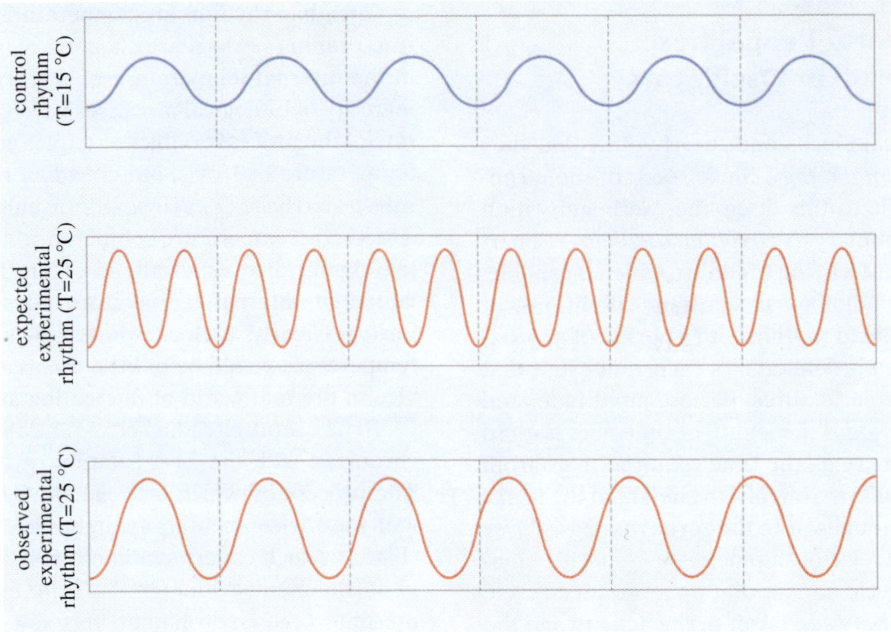

FIGURE 4 Expected and observed changes in the period length of a circadian rhythm when the organism is exposed to a 10°C increase in temperature. If the biological clock was not temperature compensated, after an increase of 10°C the free-running period would be halved. Instead, as indicated in the lower plot, the free-running period is only slightly affected by the change in temperature.

by means of their daily resetting action; the shape of the phase response curve and the free-running period largely determine the phase relationship between the organism and its environment (e.g., whether it is nocturnal or diurnal).

Circadian Clocks Are Ubiquitous

During the past 50 years the study of biological rhythms has been approached from many different perspectives: descriptive, formal (mathematical), physiological, and, recently, genetic and molecular. Originally, the major effort was concentrated on descriptive studies. These soon established that circadian rhythmicity is widespread in protozoa, fungi, plants, and animals. Among the protozoa circadian rhythms have been particularly well studied in *Euglena, Paramecium,* and *Gonyaulax.* The fungus *Neurospora* has clear rhythms of growth and spore formation, and it has been used extensively for biochemical and genetic studies of rhythms. Leaf movement, petal movement, and nectar secretion are just a few of the many rhythms demonstrated in higher plants. Among multicellular animals, circadian rhythmicity has been described in all the major groups, from coelenterates to mammals, including humans. Interestingly, such rhythms have also been described at the population level. Although any individual within a population may exhibit a certain kind of activity only once in its lifetime, the temporal distribution of these individual events may be rhythmic. One such rhythm that has been studied extensively is the rhythm of emergence of adult flies (*Drosophila*) from their pupal cases (Fig. 6). The timing of human births also shows this kind of rhythmicity, with birth being more likely to occur early in the morning than in the evening. Recently, robust circadian rhythms have been described in cyanobacteria, extending the phenomenon for the first time to the prokaryotes.

Anatomical Localization of the Biological Clock

Although the mechanism that generates the circadian oscillation is clearly subcellular, in multicellular organisms circadian clocks are localized to particular structures primarily in nervous systems. Three experimental approaches have been used successfully to identify structures containing clocks: lesion, culture, and transplantation. Lesions can be made in discrete areas of the brain or endocrine organs can be removed, and the effects on overt rhythms can be observed. Despite its drawbacks, this approach has been used for preliminary localization of biological clocks in several species. The clock controlling most of the circadian rhythms of mammals — the suprachiasmatic nuclei of the hypothalamus (SCN) — was first located using this method (Stephan and Zucker, 1972). However, the results of lesion experiments are difficult to interpret and this approach alone is

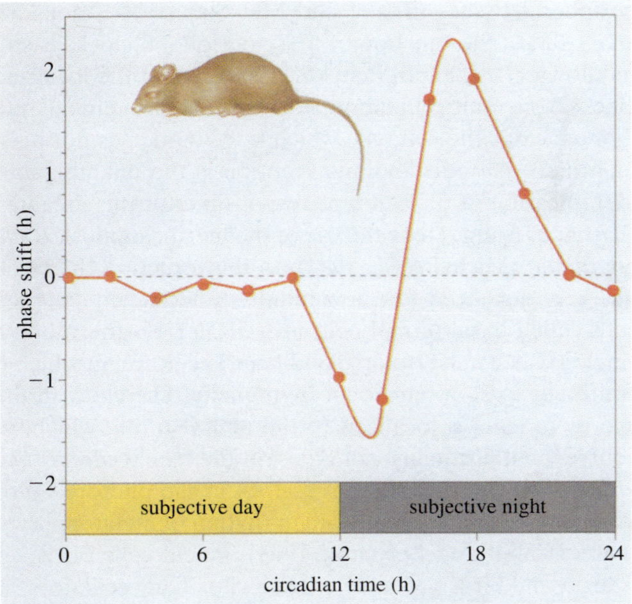

FIGURE 5 When light pulses are administered at different times during the circadian cycle to an animal kept in constant darkness, they produce a shift in the internal timing system. In nocturnal rodents, pulses given in the middle of the subjective day do not have any affect. A light pulse at the end of the subjective day, or early in the subjective night, delays the activity onset on the following day (negative phase shift); a pulse of light given in the late subjective night will advance the activity onset (positive phase shift). As shown, the magnitude of both delays and advances depends on the time at which the pulse of light is given.

not sufficient to demonstrate that a particular structure acts as a clock. When the results of lesion experiments are combined with data from culture and transplant experiments, unambiguous interpretations may be possible. It is sometimes possible to remove the structure from the organism and culture it under constant conditions for several days while recording some potentially rhythmic output (e.g., electrical activity, hormone release, and peptide synthesis). If in these circumstances the output persists rhythmically, this is a direct demonstration that the structure contains a biological clock. This approach has been used to demonstrate the presence of biological clocks in the retina and pineal glands of nonmammalian vertebrates and in the retina and SCN of mammals.

Although experiments of this kind confirm the presence of a circadian clock in the cultured tissue, they do not establish the clock's function in the intact organism. Transplantation experiments sometimes produce important information of this kind, and when they are successful, the results are often dramatic. For example, lesion of the SCN abolishes most circadian rhythms in rodents and transplantation of fetal SCN tissue into the lesion site restores the circadian rhythm of locomotor activity (but does not restore some other rhythms). Furthermore, using SCN tissue from mutant donors, it has been possible to show that the restored rhythm has characteristics of the donor's rhythm rather than those of the host. Similar experiments have confirmed the clock function of the pineal gland of sparrows, the optic lobe of cockroaches, and the brains of *Drosophila* and silkmoths.

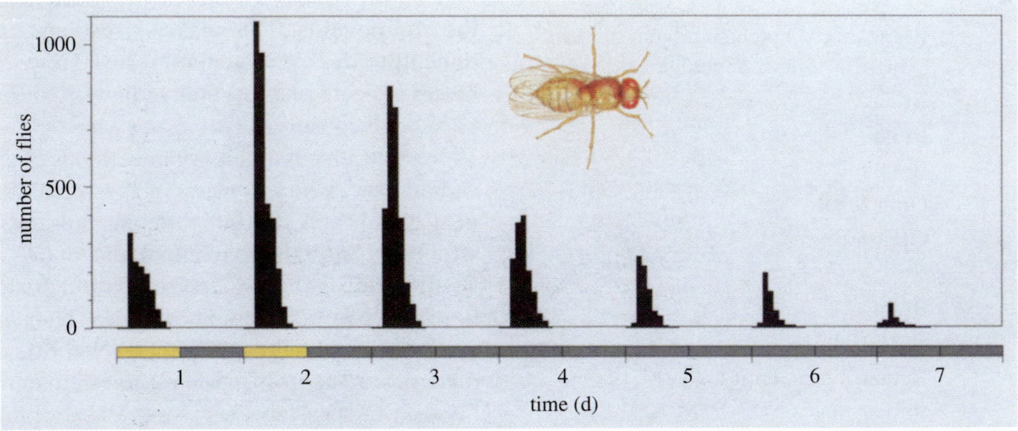

FIGURE 6 Rhythm of emergence from the pupal case (eclosion) in a population of *Drosophila pseudoobscura*. The first 2 days of data were obtained while flies were exposed to a light/dark cycle (12 h of light and 12 h of darkness); for the remaining part of the experiment (5 days), the flies were maintained in constant darkness. As the experiment progressed, the number of flies available for eclosion each day diminished (as clearly demonstrated by the decrease in the amplitude of the daily peaks), but the peaks continued to occur at approximately 24-h intervals (reproduced with permission from Pittendrigh, 1954).

Table 3 provides a summary of the sites for which direct or indirect evidence indicates the presence of a circadian clock although not necessarily its function. It is clear that multicellular organisms generally possess several anatomical sites that contain clocks, or at least circadian oscillators that have the potential to act as clocks. Usually, in animals these sites are located within the nervous system, but there are an increasing number of exceptions to this generalization. In *Drosophila,* for example, it has been shown that the Malpighian tubules — the insects' equivalent of kidneys — contain circadian oscillators (Giebultowicz and Hege, 1997). Surprisingly, there also appear to be circadian oscillators in the proboscis and on the edges of the wings of *Drosophila* (Plautz *et al.,* 1997). Among the vertebrates, the lizard *Iguana iguana* possesses at least five anatomically discrete circadian oscillators in the retinas, pineal, parietal eye, and another site unidentified in the brain (Tosini and Menaker, 1998). In general, nonmammalian vertebrates have several structures capable of self-sustained circadian oscillations (Table 3), whereas mammals seem to have a dominant biological clock in the SCN (Fig. 7) and a recently discovered circadian oscillator in the retina and in other peripheral tissues (Tosini and Menaker, 1996; Yamakazi *et al.,* 2000); the function of these peripheral clocks is still unknown (Tosini and Menaker, 1996). All of the localized clocks from multicellular organisms are heterogeneous and in most cases the cell type which is rhythmic has not been identified. An outstanding exception is the unambiguous identification of the cell type responsible for the circadian rhythm of firing rate in the eye of the marine mollusk *Bulla gouldiana* (Michel *et al.,* 1994). In this structure the clock has been localized to a few hundred electrically coupled cells called basal retinal neurons (BRNs). Furthermore, a single BRN can be isolated and kept in culture, in which it continues to show circadian rhythmicity. The clock in the eye of rodents is localized to the neural retina which, of course, contains many cell types; in the frog *Xenopus laevis,* it is known that there is a clock in the photoreceptor layer that regulates the circadian rhythm of melatonin synthesis (Cahill and Besharse, 1993). Pineal cells of fishes, lizards, and birds probably contain circadian oscillators, as do some mammalian SCN neurons (Welsh *et al.,* 1995).

TABLE 3

Anatomical Structures in Invertebrates and Vertebrates in which the Presence of Circadian Oscillators Has Been Demonstrated by *in Vitro* Experiments, by Lesion, by Transplantation, or by Other Techniques

ORGANISM	ANATOMICAL STRUCTURES
Invertebrates	
Fruit fly	Brain, eye, Malpighian tubules, sensory bristles, thorax, abdomen
Cricket	Brain
Spider	Brain
Silkmoth	Brain
Beetle	Optic lobe
Cockroach	Optic lobe
Mollusk	Eyes
Horseshoe crab	Brain
Vertebrates	
Agnatha	Retina, pineal, optic lobe
Fish	Retina, pineal
Amphibian	Retina, pineal
Reptile	Retina, pineal, SCN,[a] parietal eye
Bird	Retina, pineal, eyes, SCN
Mammal	Retina, SCN

[a]SCN, suprachiasmatic nucleus of the hypothalamus.

Organization of Circadian Systems

The several circadian oscillators in an individual organism interact with each other as well as with the environment and with the rest of the organism. What do we know about this "circadian organization"? It must contain three kinds of components: (i) oscillators that generate the rhythm, (ii) input pathways for the entrainment of those oscillators by the external environment, and (iii) output pathways which control many overt rhythms (Fig. 8). The ways in which these components are put together are likely to be much more variable from one organism to another than are the components. This makes circadian organization very difficult to discuss in general terms. However, in the vertebrates aspects of overall circadian organization appear to have been conserved.

Vertebrate circadian organization is built around a "circadian axis" which consists of the retina, the SCN, the pineal gland, and the links among these structures. These structures interact to control the many overt circadian rhythms through a variety of output pathways involving both nerve impulses and hormones. They are entrained by environmental light signals perceived by specialized circadian photoreceptors in the retina (in mammals and probably the other vertebrates as well), in the pineal gland (in all vertebrates except mammals), and in the brain (in all vertebrates except mammals) in unidentified areas. As can be seen from this brief description, the same structures often contain both circadian oscillators and the photoreceptors that link them to the environment. The same kind of organization is widespread among the invertebrates as well.

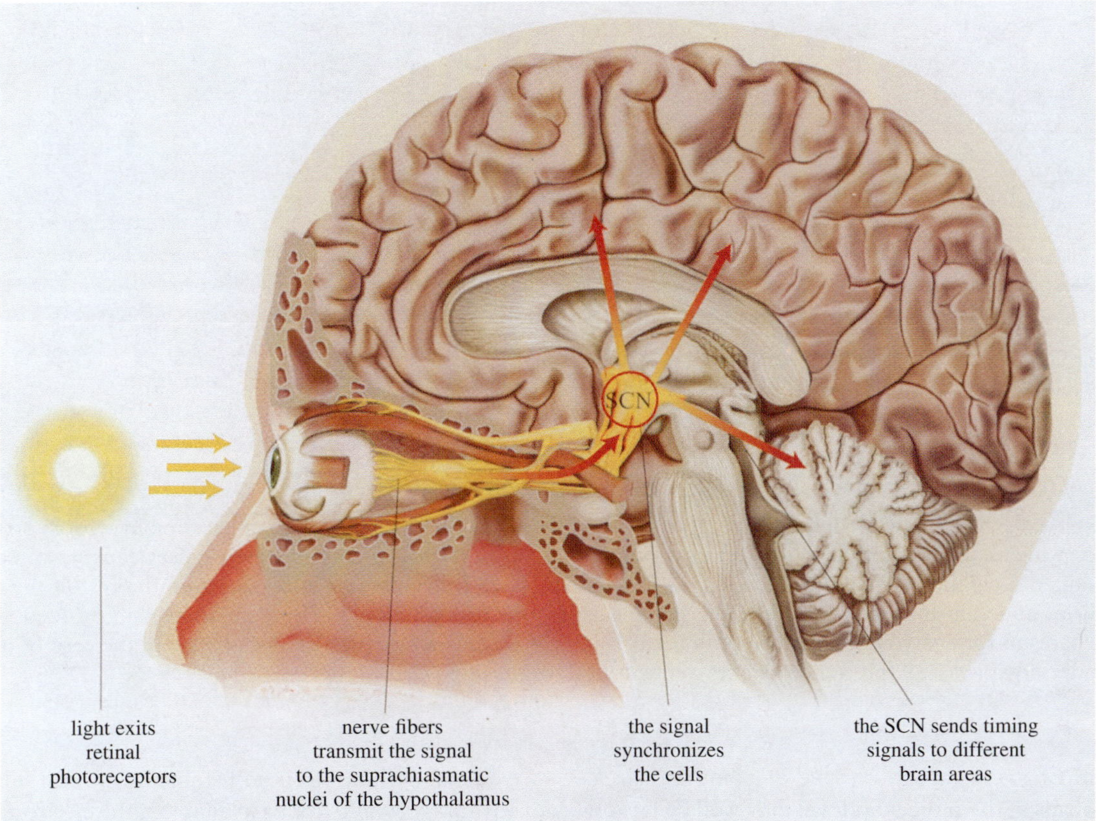

FIGURE 7 Mammalian circadian clocks are located in the retinas and in the suprachiasmatic nuclei of the hypothalamus. Environmental light information is transmitted to the suprachiasmatic nuclei of the hypothalamus (SCN) from photoreceptors located within the retina via a specific neural pathway (the retinohypothalamic tract). The suprachiasmatic nuclei of the hypothalamus control most of the behavioral, physiological, and hormonal circadian rhythms, whereas the clock located within the retina probably controls rhythmicity in many retinal processes. Interactions between the two clocks may also occur.

Generation of the Circadian Oscillation

Genetic and molecular analysis of circadian rhythmicity in *Neurospora* and *Drosophila* has provided insight into the operation of the oscillator at the subcellular level (Dunlap, 1996). In the 1970s, Konopka and Benzer (1971) initiated a genetic screen to identify mutations which affect the circadian system. They identified three mutant alleles of a single gene (*per* = "period"); per^1 flies had free-running periods longer than normal, per^s flies had short periods, and per^0 flies were arrhythmic. They mapped the *per* gene to the X chromosome. Later, the gene was cloned and now it has been intensively studied by several research groups. The levels of *per* mRNA and PER protein both exhibit circadian rhythms, and there is a 6- to 8-h lag between production of *per* mRNA and production of the protein. Recently, another circadian mutant of *Drosophila* was discovered (Sehgal *et al.,* 1994) and named *timeless* (*tim*) because mutant flies are arrhythmic. Homozygous *tim* animals not only lose behavioral rhythms but also lose circadian rhythmicity in the expression of *per* mRNA. In these flies, PER protein is not able to reenter the nucleus (Vosshall *et al.,* 1994) and therefore cannot regulate its own transcription as it is believed to do normally. These observations have led to the hypothesis that the dimer formed by the interaction of TIM and PER proteins regulates nuclear reentry and may be an important element in the generation of the circadian oscillation. Recent studies have revealed that other genes are involved in the molecular mechanism generating the circadian oscillation in Drosophila (Fig. 9, and Hardin, 2000).

Other organisms, including humans (Tei *et al.,* 1997), have been shown to possess similar genes, suggesting that the molecular mechanism that generates the circadian oscillation may have been at least partially conserved. In mammals, two circadian mutations have been identified (Ralph and Menaker, 1988; Vitaterna *et al.,* 1994) and cloned (King *et al.,* 1997; Lowery *et al.,* 2000). This aspect of

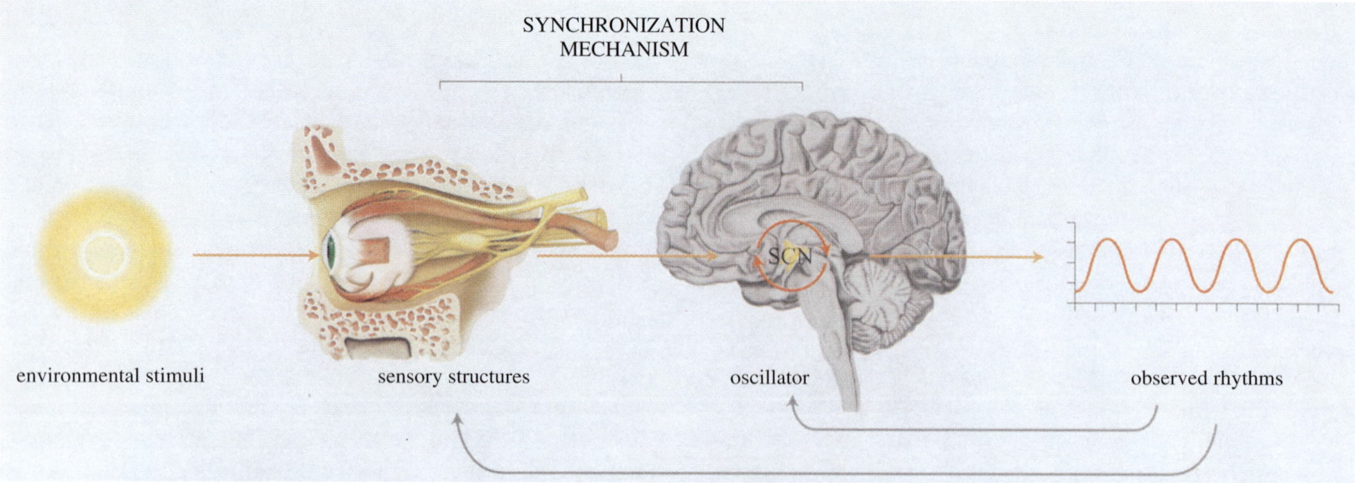

FIGURE 8 Schematic model of the components of a circadian system. The system must contain sensory structures which convey information about environmental stimuli (light), a cerebral mechanism capable of self-sustained oscillation (circular red arrows), and a system which transmits this information to the rest of the organism, in which the behavioral and physiological rhythms are generated. These rhythms can, by feedback mechanisms (gray arrows), influence both the oscillator mechanism and the sensory organs. In the case of unicellular organisms, all three components can be found within single cells.

the field is moving very rapidly on several fronts and circadian mechanisms at the molecular level may well be largely determined within 10 years.

Adaptive Significance

Circadian rhythms are such common features of organisms and are manifest at so many levels of biological organization that they must be of great adaptive significance. However, precisely because they are so widespread and the functions that they regulate are so diverse, it is difficult to assign a general role to them. Therefore, discussions of their adaptive value must be focused rather narrowly on specific cases and so may appear almost anecdotal. It is possible to impose some order on such a discussion by a somewhat artificial separation of the adaptive significance of circadian rhythms into two very broad categories.

Maintenance of Synchrony with the External World

The most obvious use to which circadian rhythms are put is the maintenance of synchrony between the organism's rhythmic functions and those of the external world — that is, the daily environmental cycles such as light, temperature, and humidity as well as the biological cycles of other organisms and of the ecological community. By using an environmental cycle to synchronize an internally generated rhythm, organisms gain the ability to "predict" future events because the phase relationship between the internal and the external oscillations is adjustable by natural selection. This advantage would not be available if the environment simply forced the biological events to occur. Entrained biological processes which take significant amounts of time to accomplish can be phased to environmental cycles in such a way that the end result of the process occurs at some later critical time. If, for example, the survival of an animal with a wide home range depends on its being back in its burrow by daylight, it must begin its return trip before perceiving the first light of dawn.

It is important to note that although organisms would be expected to use the most regular and reliable feature of the environment as an entrainment signal, the adaptive significance of a particular rhythm may well lie in its relationship to some other regular feature of the environment. For instance, the daily variation in humidity may not be sufficiently regular to serve as a precise entrainment signal, but organisms may achieve an adaptive phase relationship to it by entraining to the light cycle, which in turn bears a statistically reliable phase relationship to humidity.

There are several well-studied cases in which circadian rhythms are used by organisms as "continuously consulted clocks." Honeybees time their visits to different flowers at different times of day in such a way that they arrive while the flower is in its daily period of nectar secretion. Both the bees and the flowers are entrained to the environmental light/dark cycle and the bees consult their internal clocks throughout the day, presumably having learned which flowers are secreting nectar at which times. This synchronization is of obvious adaptive value for both organisms; it al-

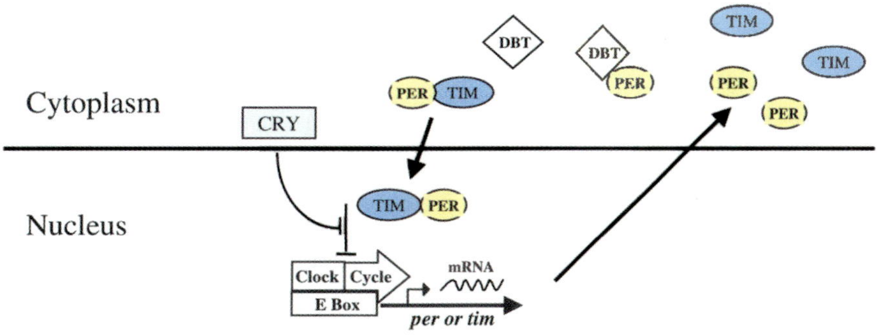

FIGURE 9 Simplified model of the molecular mechanism generating the circadian oscillation in Drosophila. In this organism transcription of the clock genes per and tim, is positively regulated by the heterodimer formed by the protein products of the clock and cycle genes. PER and TIM form heterodimers which can then translocate into the nucleus, thereby negatively regulating their own transcription by interfering with the positively acting CLOCK-CYCLE dimers. Moreover, the inhibiting action of the PER-TIM dimers is contrasted by the product of the cryptochrome (cry) gene. The action doubletime (dbt) gene, delays the formation of TIM-PER dimers by subtracting PER monomers which are degraded, following targeting via DBT-dependent phosphorylation.

lows the bees to gather the maximum amount of food with minimal effort and it ensures that the plant will be pollinated without the metabolic waste involved in continuous nectar secretion. Many species of birds perform complex feats of celestial navigation which depend on biological clocks that they consult throughout the day, much as human celestial navigation depends on an accurate chronometer. If the bird's clock is entrained in the laboratory to an artificial light cycle out of phase with the external day, the birds, when released in the field, will make a proportionate navigational error (Fig. 1).

The synchronization of annual reproductive cycles with events in the field which influence survival of the offspring is accomplished in many organisms, from plants to birds and mammals, by physiological response to seasonal changes in day length (*photoperiodism*). Photoperiodism depends on the ability of the organism to measure time. It must be able to distinguish between long and short days. Recent experimental evidence indicates that photoperiodic time measurement is accomplished by means of a circadian rhythm of sensitivity to the inductive effects of light. Light falling at some times in the organism's internal circadian cycle induces such photoperiodic responses as flowering in plants, diapause in insects, and gonadal growth in birds and some mammals. The same amount of light falling at other points in the cycle has no effect on these processes. The circadian sensitivity rhythm is entrained by the environmental light cycle with a phase relationship such that on long days light falls on the portion of the curve at which it is inductive, whereas on short days it falls only on the insensitive portion.

By appropriate differential phasing to environmental cycles, several species may reduce competition within the same spatial niche. For example, some species of both birds and bats are aerial and insectivorous but in most cases do not compete for food as a result of the different phase relationships which their predatory activities bear with the environmental light cycle. In the tropical rain forest, many frog species call for mates at night, but each species has a precisely defined temporal niche. By dividing the night in this way, inappropriate matings are reduced.

Maintenance of Synchrony among Rhythms within an Organism

Many physiological processes within the organism are rhythmic and maintain discreet phase relationships among themselves. Appropriate phase relationships among internal rhythmic processes may have great adaptive significance. Substrates for biochemical reactions, energy sources, and information-carrying molecules must be present at the right time as well as in the right place if organelles, cells, and tissues are to be integrated into an efficiently functioning organism. To accomplish this, processes which vary rhythmically must be controlled in such a way that they are appropriately timed and phased with respect to each other. Although we have just begun to document the adaptive value of internal temporal organization, there is ample evidence that such organization exists. It seems likely that although synchronization of particular behaviors with events in the external environment has a wide variety of specific adaptive values, the more general adaptive significance of circadian rhythms may lie in the control of internal biotemporal organization.

References Cited

Cahill, G. M., and Besharse, J. C. (1993). Circadian clock function localized in *Xenopus* retinal photoreceptors. *Neuron* **10,** 573–577.

DUNLAP, J. C. (1996). Genetic and molecular analysis of circadian rhythms. *Annu. Rev. Genet.* **30,** 579–601.

GIEBULTOWICZ, J. M., and HEGE, D. M. (1997). Circadian clock in Malpighian tubules. *Nature* **386,** 664.

HARDIN, P. E. (2000). From biological clock to biological rhythms. *Genome Biol.* **1:** 1023.

HOFFMANN, K. (1954). Versuche zu der im Richtungsfinden der Vogel enthaltenen Zeitschatzung. *Z. Tierpsychol.* **11,** 453–475.

KING, D. P., ZHAO, Y., SANGORAM, A. M., WILSBACHER, L. D., TANAKA, M., ANTOCH, M. P., STEEVES, T. D. L., VITATERNA, M. H., KORNHAUSER, J. M., LOWERY, P. L., TUREK, F. W., and TAKAHASHI, J. S. (1997). Positional cloning of the mouse circadian *Clock* gene. *Cell* **89,** 641–653.

KONOPKA, R. J., and BENZER, S. (1971). Clock mutants of *Drosophila melanogaster. Proc. Natl. Acad. Sci. USA* **68,** 2112–2216.

LOWREY, P. L., SHIMOMURA, K., ANTOCH, M. P., YAMAZAKI, S., ZEMENIDES, P. D., RALPH, M. R., MENAKER, M., TAKAHASHI, J. S. (2000). Positional syntenic cloning and functional characterization of the mammalian circadian mutation tau. *Science* **288,** 483–491.

MENAKER, M. (1968). Extraretinal light perception in the sparrow, I: Entrainment of the biological clock. *Proc. Natl. Acad. Sci. USA* **59,** 414–421.

MICHEL, S., GEUSZ, M. E., ZARITSKY, J. J., and BLOCK, G. D. (1994). Circadian rhythm in membrane conductance expressed in isolated neurons. *Science* **259,** 239–241.

PITTENDRIGH, C. S. (1954). On the temperature independence in the clock-system controlling emergence in *Drosophila. Proc. Natl. Acad. Sci. USA* **40,** 1118–1129.

PITTENDRIGH, C. S., and DAAN, S. (1976). A functional analysis of circadian pacemakers in nocturnal rodents. IV. Entrainment: Pacemaker as clock. *J. Comp. Physiol.* **106,** 291–331.

PLAUTZ, J. D., KANEKO, M., HALL, J. C., and KAY, S. A. (1997). Independent photoreceptive circadian clocks throughout *Drosophila. Science* **278,** 1632–1635.

RALPH, M. R., and MENAKER, M. (1988). A mutation of the circadian system in golden hamsters. *Science* **241,** 1225–1227.

SEHGAL, A., PRICE, J., MAN, B., and YOUNG, M. (1994). Loss of circadian behavioral rhythms and *per* oscillations in the *Drosophila* mutant *timeless. Science* **263,** 1603–1606.

STEPHAN, F. K., and ZUCKER, I. (1972). Circadian rhythms in drinking behavior and locomotor activity of rats are eliminated by hypothalamic lesions. *Proc. Natl. Acad. Sci. USA* **69,** 1583–1586.

TEI, H., OKAMURA, H., SHIGEYOSHI, Y., FUKUHARA, C., OZAWA, R., HIROSE, M., and SAKAKI, Y. (1997). Circadian oscillation of a mammalian homologue of the *Drosophila period* gene. *Nature* **389,** 512–516.

TOSINI, G., and MENAKER, M. (1996). Circadian rhythms in cultured mammalian retina. *Science* **272,** 419–421.

TOSINI, G., and MENAKER, M. (1998). Multioscillatory circadian organization in a vertebrate, *Iguana iguana. J. Neurosci.* **18,** 1105–1114.

VITATERNA, M. H., KING, D. P., CHANG A. M., KORNHAUSER, J. M., LOWREY, P. L., MCDONALD, J., DOVE, W. F., PINTO L. H., TUREK, F. W., and TAKAHASHI, J. S. (1994). Mutagenesis and mapping of a mouse gene, *Clock,* essential for circadian behavior. *Science* **264,** 719–725.

VOSSHALL, L., PRICE, J., SEHGAL, A., SAEZ L., and YOUNG, M. (1994). Specific block in nuclear localization of period protein by a second clock mutation, *timeless. Science* **263,** 1606–1609.

WELSH, D. K., LOGOTHETIS, D. E., MEISTER, M., and REPPERT, S. M. (1995). Individual neurons dissociated from rats suprachiasmatic nucleus express independently phased firing rhythms. *Neuron* **14,** 697–706.

YAMAZAKI, S., NUMANO, P., ABE, M., HIDA, A., TAKAHASHI, P., UEDA, M., BLOCK, G. D., MENAKER, M., and TEI, H. (2000). Resetting central and peripheral circadian oscillators in transgenic rats. *Science* **288:** 682–685.

General References

ASCHOFF, J. (1981). *Handbook of Behavioral Neurobiology. Biological Rhythms.* Plenum, New York.

KLEIN, D. C., MOORE, R. Y., and REPPERT, S. M. (1991). *Suprachiasmatic Nucleus: The Mind's Clock.* University Press, New York.

MOORE-EDE, M. C., SULZMAN, F. M., and FULLER, C. A. (1982). *The Clocks That Time Us.* Harvard Univ. Press, Cambridge, MA.

PITTENDRIGH, C. S. (1993). Temporal organization: Reflection of a Darwinian clock-watcher. *Annu. Rev. Physiol.* **55,** 16–54.

WETTERBERG, L. (1995). *Light and Biological Rhythms in Man.* Pergamon, New York.

FLORIANO PAPI

Department of Ethology, Ecology and Evolution
University of Pisa
Pisa, Italy

Animal Navigation

The spatial and temporal characteristics of migratory and other kinds of movements by which animals reach specific or localized goals, as well as the neural, sensory, and central processes that regulate such movements, continue to be the object of research efforts. Among the mechanisms of navigation, of particular importance are the "biological compasses," which elaborate astronomical information as well as information derived from the earth's magnetic field, and "mental maps," which represent the spatial relationship among environmental reference points such as landmarks. Reaching a goal can involve casual or systematic search strategies or can depend on information, whether genetic or learned, about the direction to take and the distance to cover. The ability to return to a location of departure can be based on information acquired during the outbound journey or this action can be done without such information. The latter case would be an example of true navigation if an animal knows how to calculate its current location relative to a goal location and then reaches the goal using one or more biological compasses. Calculating relative location can be achieved on the basis of a mental map of familiar locations, or it can extend beyond an animal's familiar area by a process of learning about distant, spatial information.

Introduction

During the course of evolution, the ability to move from one location to another, fundamental for nutrition and reproduction in the vast majority of animal species, has diversified and often become more complex as organisms have adapted to new environments. Many species began to range over large distances on the ground, in the air, or through water and at the same time acquired spatial orientation mechanisms that together are classified as animal navigation. *Animal navigation* is the capacity of an animal to reach a spatially defined and restricted goal that can be relatively distant. In general, in the vast literature on animal navigation, movements toward a goal that can be detected at the starting point by means of direct sensory contact are not considered examples of animal navigation. Also, human spatial experiences and terminology, such as navigation, pilotage, map, compass, and route reversal, are often used to describe the phenomena and mechanisms of animal navigation. In the majority of cases, navigational mechanisms are used to reach a location or area already familiar to an animal, and therefore the term *homing* or "return to home" is often used even when the goal is not a permanent residence.

The ability to navigate is common among arthropods and vertebrates that need to travel long distances to escape from periodic episodes of unfavorable climate, reach an area better suited for reproduction, search for sources of food over large areas, or even to return to nocturnal or diurnal resting sites.

Research Methods

To investigate navigational mechanisms in a particular animal species, it is first necessary to know as accurately as possible the spatial–temporal movement patterns of many individuals available for research. Historically, the primary method used is based on marking captured individuals with the use of bands or rings, tags, tattoos, or subcutaneously placed magnetic transponders and the subsequent recovery of a marked individual after it is released back to the environment. What is known about vertebrate migration, in particular bird migration, is very much a consequence of this system of capture–mark–recapture. However, this method is time costly and inefficient because the percentage of marked individuals eventually recaptured is usually very low. For migratory animals, the data obtained allow one to identify at best the exact location of departure and arrival, and occasionally resting or stopover locations, whereas other information about the route followed and the time taken remains unknown.

Other methods that provide partial data are direct observation, the capture and gathering of radar images of migrating birds, the hydrophonic recording of cetaceans, and, in general, the methods of telemetry (Priede and Swift, 1992). The use of emitted radio waves, or ultrasounds in the case of aquatic animals, allows one to follow animals for a more or less long period of time depending on whether the receiver is mobile or in a fixed location. Recently, in a method that has become increasingly used recording devices are attached to animals that place into electronic memory relevant data with respect to the direction and speed of movement as well as physical parameters of the environment (temperature, pressure, and light intensity) and physiological measures (activity, heart rate, and body temperature). If the recording device is coupled with a global positioning system (GPS), an animal's latitude, longitude, and altitude can be measured and recorded at relatively brief intervals. In this way, the movements of animals can be reconstructed with considerable precision and correlated with environmental and physiological conditions. However, this method requires the recovery of the data stored in the recording devices, except in cases in which the memorized data can be transmitted by radio.

In many situations, it is only possible to reconstruct the routes of animals relying on satellite telemetry. Satellites in polar orbit can receive low-power (1 W or less) signals from radio transmitters attached to animals. The satellites can calculate the geographic position of an animal relying on Doppler shifts in the radio transmitter signals and can transmit the relevant data back to Earth. Other, limited data from recording devices interfaced with the transmitter can also be included in the satellite signal that provides position information.

Satellite telemetry has been used with success in many vertebrate species capable of carrying a transmitter, which can weigh tens of grams, without difficulty. In aquatic animals, transmission is only possible when an animal surfaces. When the animal is under water, the transmitter remains silent to save the energy of the batteries. Among the main inconveniences of satellite telemetry is the temporal discontinuity in localizing the signal and, often, inaccuracies in the data provided. However, important technical improvements that can reduce or eliminate current deficiencies are expected. By taking advantage of satellite telephonics (cellular phone-like technology), it will soon be possible to transmit continually from animal to receiver any useful data, particularly exact geographic coordinates calculated by a GPS device.

Certain animals held in captivity actively display the tendency to escape in a particular direction. Exploiting this phenomenon, a large amount of data have been collected from subjects held mostly in arenas or circular cages, in which an animal's movements can be recorded using a variety of methods. In this experimental setting it is possible to manipulate sources of spatial information (e.g., information provided by the earth's magnetic field) with the goal of identifying environmental spatial cues used by animals for orientation.

General Characteristics of Long-Distance Movements

As previously mentioned, many journeys carried out by animals are seasonal migrations between an area of reproduction and another more favorable during the time between reproductive cycles. Departing from a permanent or temporary residence, or by following migratory prey, large distances can also be covered in the search for food.

Most tracks of long-distance journeys have been obtained from birds (Fig. 1), but similar data have been obtained from terrestrial and aquatic animals. The routes traveled are shaped by geomorphic factors and the characteristic physiology and ecology of a species. Some birds avoid crossing mountain chains or large areas of open water, whereas others fly considerable distances over the open ocean not only to reach oceanic islands but also to shorten the distance needed to reach their goal and to take advantage of favorable winds (Fig. 2).

Simply reconstructing the route taken by an animal can provide important clues that can be used to understand the navigational mechanisms involved. Many animal species carry out their movements by following landscape features such as rivers, valleys, and coastlines. Other species, or the same species in different situations, can maintain a constant direction even in a uniform environment, for example, while swimming in the ocean, crossing a desert, or flying

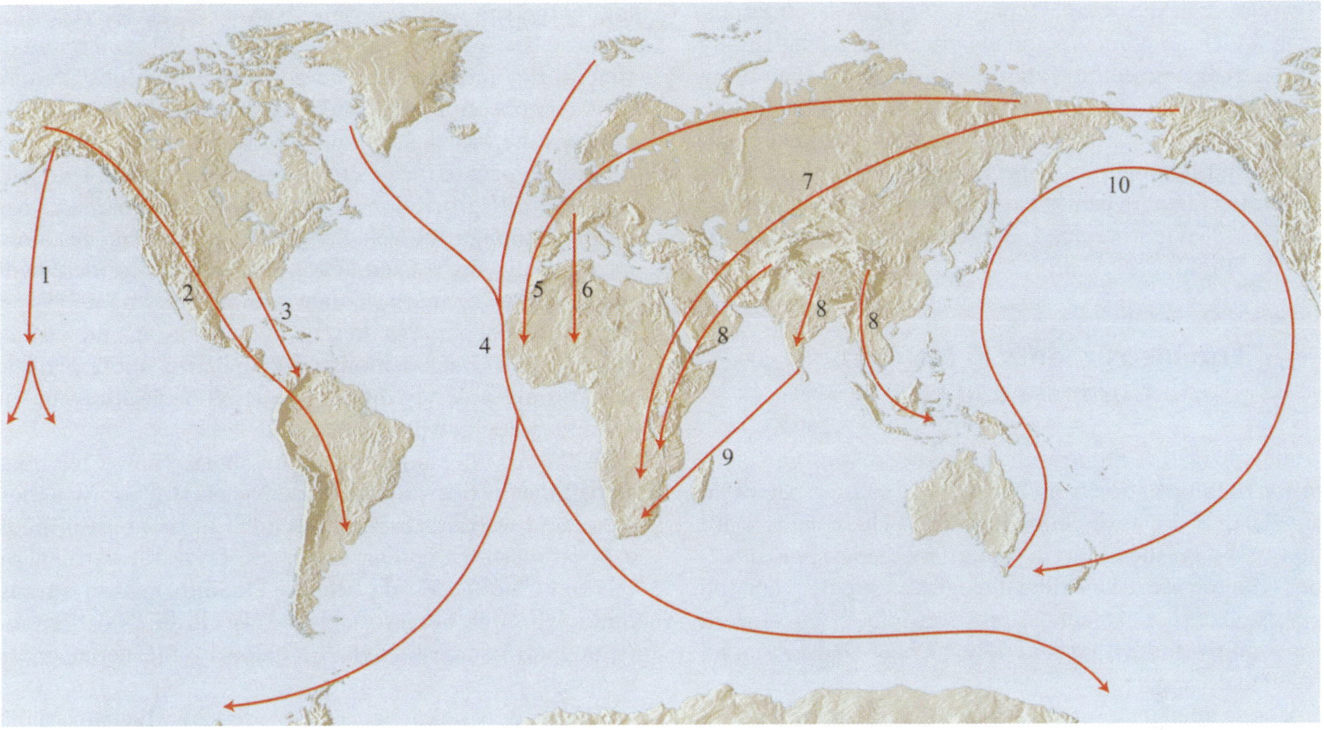

FIGURE 1 Schematic representation of the autumnal routes of transcontinental migration. 1, The Alaskan population of the Pacific golden plover *Pluvialis fulva* and other shorebirds headed for the Pacific islands; 2, Swainson's hawk, *Buteo swainsoni;* 3, migration over the Gulf of Mexico of many North American species; 4, Arctic tern, *Sterna paradisaea;* 5, ruff, *Philomacus pugnax;* 6, trans-Saharan migration of many European species; 7, an Alaskan population of the northern wheatear, *Oenanthe oenanthe;* 8, three populations of barn swallows, *Hirundo rustica,* directed toward different overwintering locations; 9, eastern red-footed falcon, *Falco amurensis.* Only the part of the migration over the Indian Ocean is shown; 10, complete round-trip, migratory route of the short-tailed shearwater, *Puffinus tenuirostris* (reproduced with permission from Berthold, 1993, p. 3).

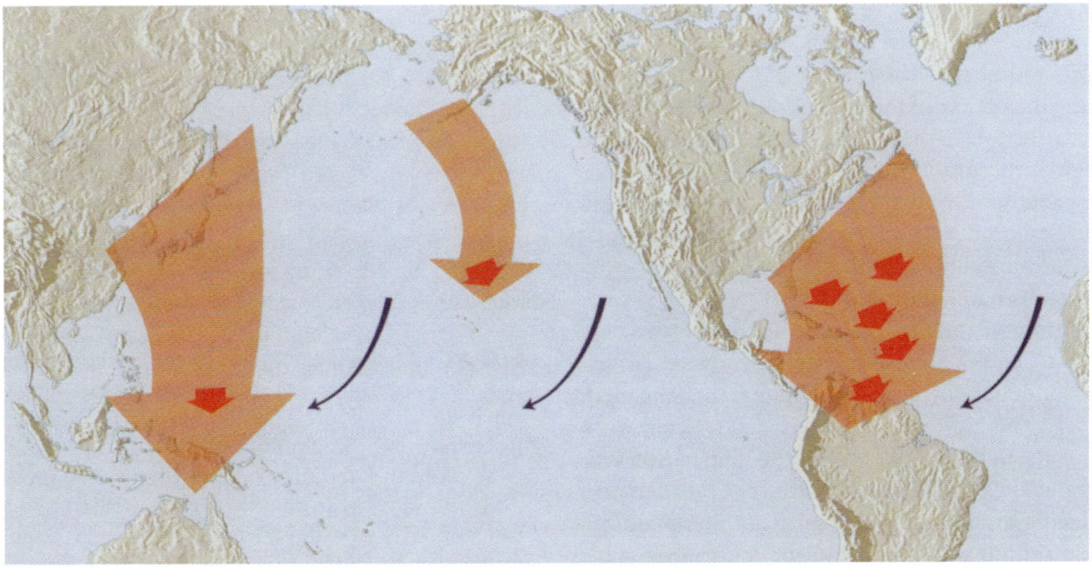

FIGURE 2 Migratory birds, even those that are strictly terrestrial, can fly over the ocean to reach their goal. The migration path (large shaded arrows) is the result of the directional heading of the animals (small red arrows) and drift caused by the winds (blue arrows) (reproduced with permission from Williams and Williams, 1990, p. 15).

above clouds or the ocean. Of particular interest is the manner in which an animal responds after being accidentally displaced off course, for example, by a storm that can bring an animal to a totally unfamiliar area. Also, the ability to compensate for experimental displacements, present in migratory and nonmigratory species (notable in this case is the homing pigeon), can be exploited as a research method (Baker, 1978; Gauthreaux, 1980; Alerstam, 1990; Gwinner, 1990; Berthold, 1991, 1993).

The Instruments of Navigation: Compass and Map

Animal navigation requires the active processing and elaboration of information from one or more sources before an animal can make its directional choices. These choices are often based on the ability to determine geographic directions with the use of so-called biological compasses and the possession of spatial mental representations, often referred to as cognitive maps, that generally can be used over a large, extended area.

Among animals, a very common orientation mechanism is based on the sun and is called the *sun compass* (Pardi, 1979). Despite its apparent movement through the sky, the sun can be used to recognize, identify, and maintain a specific geographic direction. To do so, it is naturally necessary that an animal determine in which position the sun is located at different times of day with respect to the cardinal directions. This acquisition is called *compass calibration.* Subsequently, an animal only needs to know the time of day to determine the correct angle of orientation with respect to the sun. For example, if an animal in the temperate region of the Northern Hemisphere wants to move in a southerly direction, it needs to know that the sun is found in the south at noon, and at that time it needs to move toward the sun. During the course of the afternoon as the sun gradually moves to the west, the animal moves with the sun more to its right until sunset, when the sun is 90° to its right. (The angle is exactly 90° only on the days of the equinoxes: March 21 and September 23.) To go south in the morning, the animal would instead have the sun to its left with the angle decreasing as noon approaches.

Information about time of day, and therefore the orientation angle to be taken with respect to the sun, is furnished by an endogenous circadian rhythm or so-called *biological clock.* The biological clock is synchronized with the rotation of the earth, in particular, the alternation between night and day. This phenomenon can be demonstrated by placing an animal under artificial illumination, which is phase shifted with respect to the naturally occurring alternation between night and day. Within a few days, all the endogenous rhythms would come into phase with the artificial light–dark cycle. Subsequently, an animal orienting by the sun would deviate from the expected direction because it would calculate the position of the sun based on time of day measured by its now shifted biological clock.

A complication of the mechanism of sun compass orientation is the fact that the angle of orientation is not taken up with respect to the position of the sun in the sky but with respect to its projection onto the horizontal plane or azimuth. Although the velocity of the sun as it moves along its arc in the sky is fixed at 15° per hour, the velocity with which the sun azimuth changes, although averaging 15° per hour, varies depending on time of day, season, and latitude (Fig. 3). It has been demonstrated that arthropods and vertebrates are able to compensate for the variations in sun azimuth velocity with considerable accuracy.

Although the sun compass has been known for quite some time, its operational mechanism is still poorly understood and warrants further research. In some arthropods, calibration of the compass with respect to the cardinal directions is not necessary because that information is transmitted through heredity. In contrast, in birds calibration of the sun compass might be aided by the geomagnetic compass.

Completely unknown is the process by which a variable endogenous rhythm is interfaced with the variations in the angle of orientation. It is only known that the process of adaptation to variations in sun azimuth velocity is, in at least a few cases, based on the direct experience of an ani-

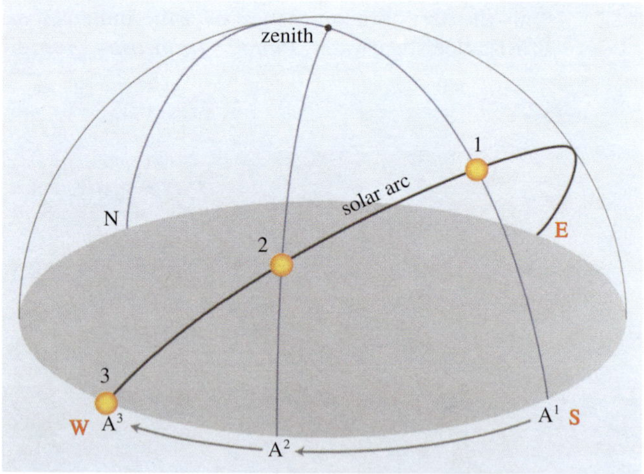

FIGURE 3 Solar arc at 45° north latitude at the time of the equinoxes of March 21 and September 23. The position of the sun is shown at noon (1), mid-afternoon (2), and sunset (3) and its corresponding azimuth (A^1–A^3), where A^1 is south and A^3 west. Note that the time taken by the sun to move from A^1 to A^2 is the same as from A^2 to A^3, demonstrating how velocity can vary. The celestial meridian that passes from A^1 to 1 and to the zenith and its projection beyond the zenith, or ante meridian, and the meridian that passes from A^2 to 2 and to the zenith are also marked. The arrow at the bottom indicates the direction of azimuth movement.

mal. Of considerable interest are the necessarily unique mechanisms of sun compass adaptation in tropical regions near the equator, where the sun appears to move clockwise through the south during part of the year and counterclockwise through the north in the other part. Particularly rapid is the shift from one kind of system to the other in migrant animals moving north–south or south–north through equatorial tropical regions.

The daytime sky permits the identification of geographic directions not only through the position of the sun but also through patterns of skylight polarization. Because the azimuth of the sun coincides with the location where the celestial meridian that passes through the sun reaches the horizontal plane (ground), for the purpose of a functional sun compass it is sufficient to identify the solar meridian or its continuation beyond the zenith, the solar ante meridian. To an eye capable of analyzing polarized light, the blue sky has a complex structure. The orientation of the plane of polarization and the intensity of polarization varies from one point to another in the sky, creating two symmetrical hemispheres separated by the solar meridian and ante meridian (Fig. 4). The pattern generated by skylight polarization moves as a single unit with the sun but nevertheless permits, even if the sky is only partially visible, determining the position of the solar meridian or ante meridian and therefore the azimuth of the sun.

The compound eye of ants and bees is provided with detectors for polarized light that are spatially arranged in a way to receive the most stimulation when the body axis is oriented parallel to the solar meridian. In this way, an insect can easily establish a reference direction from which it can determine a desired direction (Wehner, 1992). Birds are also able to perceive polarized light.

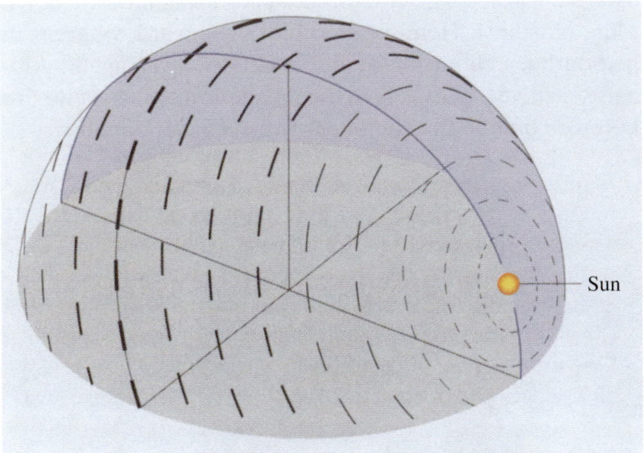

FIGURE 4 Schematic representation of polarized light superimposed on the vault of the sky. The plane of polarization is identified by the bars that are dispersed concentrically around the sun, located at an elevation of 24° above the horizon. The thickness of the bars indicates the percentage of polarized light with respect to the total light present. The line that passes through the sun represents the solar meridian and ante meridian (reproduced with permission from Wehner, 1987).

In the scientific literature one frequently finds separate treatments for orientation by the sun and orientation by skylight polarization. However, it is important to consider that many animals perceive the sky as a single complex or holistically. The sky serves as a source of information about the sun, polarized light, and differences in the intensity and spectral composition of the sky.

The existence of an orientation mechanism based on the stars (Sauer, 1957), called a *star compass,* has been demonstrated in some species of passerine birds and is probably common among birds that migrate at night. Demonstrating the ability to orient by the stars is relatively easy because on nights favorable for migration, birds, even if held in a circular cage, attempt to move or escape toward the direction of migration. If instead of being placed under the stars of the night sky birds are tested in a planetarium, the direction in which they attempt to escape can be changed at will by shifting the position of the stars projected on the ceiling of the planetarium.

Stellar orientation appears to be based on the recognition of constellations, particularly those found in the circumpolar area. The North American indigo bunting *Passerina cyanea* remains well oriented in a planetarium even when individual stars, including polar stars, are obscured. However, when the North Star and those stars found within 35° of it are eliminated, the birds are incapable of orientation. Planetarium experiments have also been used to demonstrate that the configuration of the constellations and their spatial relationship with respect to the celestial north pole is learned during the first weeks of life by observing the apparent rotation of the night sky. Birds held in a planetarium with a rotating night sky from the time of birth learn that north is identified by the star around which the constellations rotate, whatever star that might be. If, however, the planetarium stars are held immobile, the birds do not learn to orient (Emlen, 1975).

The possession of a lunar compass has been demonstrated in crustaceans that live in sandy littoral areas and carry out migrations perpendicular to the coast. However, a lunar compass used to orient toward specific objects has yet to be demonstrated.

The earth's magnetic field is a source of directional information that many animals use for orientation (Wiltschko and Wiltschko, 1995). When an experimental manipulation of the magnetic field leads to predictable variations in the orientation of an animal, one can deduce that the animal under study is able to determine the orientation of the magnetic field. Therefore, it would be legitimate to conclude that it possesses the biological equivalent of the magnetic compass. Evidence of this type has been obtained for approximately 40 different species of various types: mollusks,

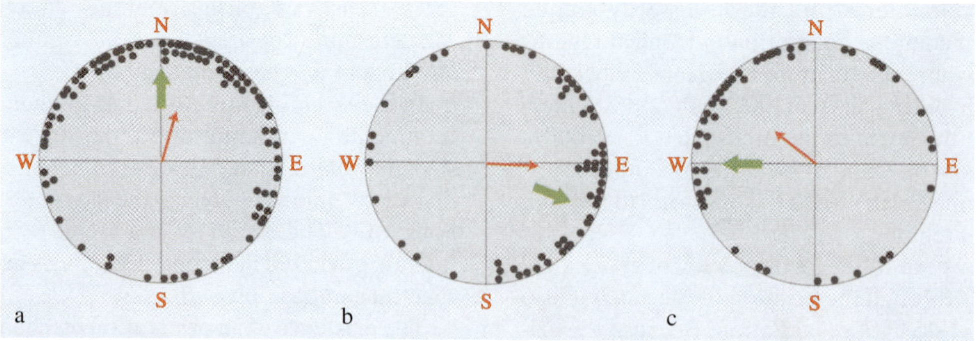

FIGURE 5 Orientation of European robins, *Erithacus rubecula,* held in a cage during the time of the spring migration when they show a tendency to move to the north. (a) The experiments were carried out in the natural magnetic field (the green arrow indicates magnetic north). (b and c) The magnetic north was rotated to east-southeast and west, respectively, causing a parallel change in the orientation of the robins. Each black point identifies the mean direction of one bird calculated on the basis of escape attempts that occurred on one night. The red arrows are the resultant mean vector of the distribution of the single points (reproduced with permission from Wiltschko, 1973, p. 29).

arthropods, and vertebrates. In many other cases, the ability to orient by the earth's magnetic field is only suspected as a consequence of the ability of the animals to maintain a constant route even in the apparent absence of other orientation cues; for example, nocturnal migratory birds that continue to fly in the correct direction even under a thick layer of clouds.

The most important experiments examining magnetic orientation have been carried out in the laboratory using passerine birds that migrate at night. Held in a circular cage in the absence of visual cues, such birds display imprecise and uncertain orientation. However, the extended observation of many subjects allows one to statistically determine an acceptable approximation of their preferred direction and to determine an effect induced by manipulating the magnetic field (Fig. 5). The birds are initially disoriented if the total intensity of the magnetic field is increased or decreased by 25%, but in a few days they can adapt to magnetic fields above or below this value. Some animals (e.g., salmon) appear to be able to recognize the polarity of the magnetic field, whereas in the majority of cases the biological compass functions like an inclination compass. In fact, the animals do not appear to be able to distinguish the polarity of the vector that represents the line of force of the magnetic field; rather, they appear to be able to distinguish only its inclination — the angle made by a line parallel to the line of magnetic force with a line tangent to the surface of the earth. The inclination of the magnetic vector is zero at the magnetic equator but increases in opposite directions toward the poles. The ends of the vectors that indicate north in the Northern Hemisphere point downward, whereas in the Southern Hemisphere they point upward (Fig. 6). Animals sensitive exclusively to the inclination of the vector are therefore unable to orient near the magnetic equator.

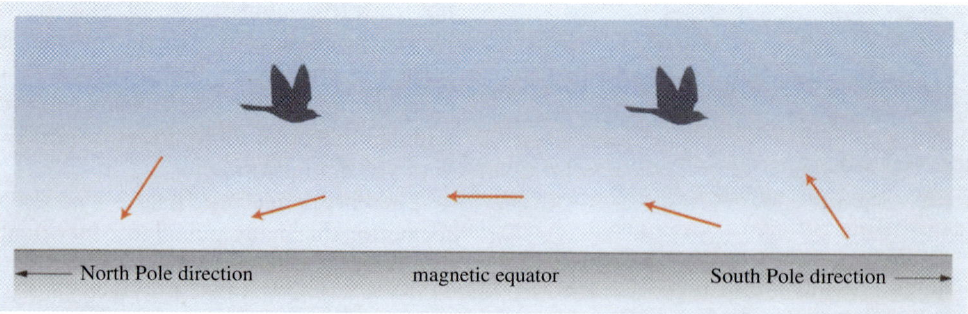

FIGURE 6 Representation of the situation that a bird must confront when migrating from the Northern Hemisphere to the Southern Hemisphere relying on its magnetic compass. The magnetic vector is identified by the arrow. If a bird could perceive polarity, it would be able to distinguish at any moment north (arrowhead) from south. However, because it is sensitive only to the inclination of the vector, it cannot orient where the inclination is zero. In other locations, it could deduce from the vector inclination only the direction of the nearest magnetic pole, toward the lowest end of the vector. The opposite direction is that of the magnetic equator.

The use of an inclination compass appears to create serious problems for transequatorial migrants. If we consider the case of a bird migrating in autumn from the Northern to the Southern Hemisphere, we can observe that initially the bird can orient south by moving in the direction in which the magnetic vector points up (Fig. 6). Reaching the zone of the magnetic equator, the bird would need to rely on a nonmagnetic compass to continue south until again finding a sufficient amount of magnetic inclination to orient. However, if the bird continues to fly in the direction in which the vector points upward, it would turn around and fly northward. The necessary reversal in orientation seems to be automatically released by passing across the magnetic equator. Laboratory experiments in garden warblers, *Sylvia borin,* have demonstrated a corresponding reversal in orientation after being held for 2 days in a horizontal field.

Many problems associated with magnetic orientation have yet to be resolved, especially the nature of the sensory receptors responsible for the perception of the magnetic field and how they work. Only in fish that possess electroreceptors is it known that these receptors can be used to perceive the magnetic field. In elasmobranch fish the electroreceptors are the ampullae of Lorenzini, which function like extremely sensitive measuring devices that can be used to record electrical potential differences among diverse points on the body surface. In addition to allowing orientation in a natural electric field based on electrical gradients, the electroreceptors, or ampullae of Lorenzini, can also measure potential differences induced by a fish's movement in the earth's magnetic field. These differences depend on the direction toward which a fish is swimming with respect to the earth's magnetic field. The induced potential is greatest with movement toward east or west and is zero with movement to the north or south. The level of excitation of the electroreceptors therefore informs the fish about the direction toward which it is moving. In other animal groups, research currently being carried out is guided by two principal hypotheses. The first assumes that magnetoreception depends on photoreceptors of the retina (Leask, 1977). The second hypothesis posits that magnetic perception is made possible by the presence of magnetite particles, which align themselves with respect to the ambient magnetic field, thus altering the state of unidentified receptor cells (Kirschvink *et al.,* 1985). In this case, magnetic perception would be light independent.

The coexistence of more than one type of compass in the same animal poses problems with respect to their relative importance and possible reciprocal calibration (Able, 1993). In migratory birds, navigation is based on the reciprocal interaction of more than one compass. The system is characterized by a sufficient degree of plasticity to allow an animal to adapt to varying situations during the course of its migratory journey relative to both magnetic (inclination, declination, and intensity) and astronomical (visible constellations and characteristics about the course of the sun at different latitudes, cloud cover, etc.) parameters. In relation to this last aspect, it is worth discussing as an example the Savannah sparrow, *Passerculus sandwichensis,* a nocturnal migrant that in autumn travels from Canada and the north-central United States to spend the winter in the southern United States, Mexico, and northern portions of Central America. At the time of migration, the direction of their movement can be recorded by placing individual birds in Emlen funnels, which are cages designed in the form of a funnel with walls lined with special paper that permits birds to leave tracks or records of their attempts to escape (Fig. 7). Birds captured in nature a few days after hatching and raised in the absence of important visual cues, but maintained in the normal light–dark cycle, are able to display a genetically determined orientation response in the appropriate migratory direction (south in autumn), relying on the earth's magnetic field. The direction taken using this primary magnetic compass can be modified if it is in conflict with celestial information. Indeed, when a young bird is exposed to a natural or artificial night sky, it can deduce from the rotation of the sky the direction of north, which can be used as a directional reference to recalibrate the magnetic compass if it is different. In addition, exposure to the daytime sky, even without being able to directly detect the sun, is sufficient for birds to recalibrate the magnetic compass by relying on polarized light. Viewing the sun in the absence of polarized light, however, is inadequate. The calibration of the magnetic compass occurs during the course of the entire migration and probably has the function of avoiding route errors due to geomagnetic declination, which can be considerable in certain regions. During the course of migration, directional information deduced from the crepuscular evening sky supercedes magnetic or stellar information (Able and Able, 1996).

In full development are studies investigating the routes taken by birds in Arctic regions, where oriented movements east–west or vice versa — movements characterized by a constant direction — can lead to a journey considerably longer than a great circle route, which would represent the shortest possible journey. A great circle route, however, would require that the flight direction be varied in continuation and it is unknown how an animal could do this.

Based on the behavior of many animals, one can conclude that they have a mental representation of environmental locations and points of spatial reference such as landmarks. One speaks of possession of a map. Depending on which characteristic one wishes to emphasize, the map can be defined as topographic, cognitive, or mental (Thinus-Blanc, 1987). Maps of this type are acquired through experience and generally have an extension that coincides with the area that an animal has had the opportunity to visit. For one to speak of a map, however, it is necessary that an animal is not limited to traveling by relying on a series of fa-

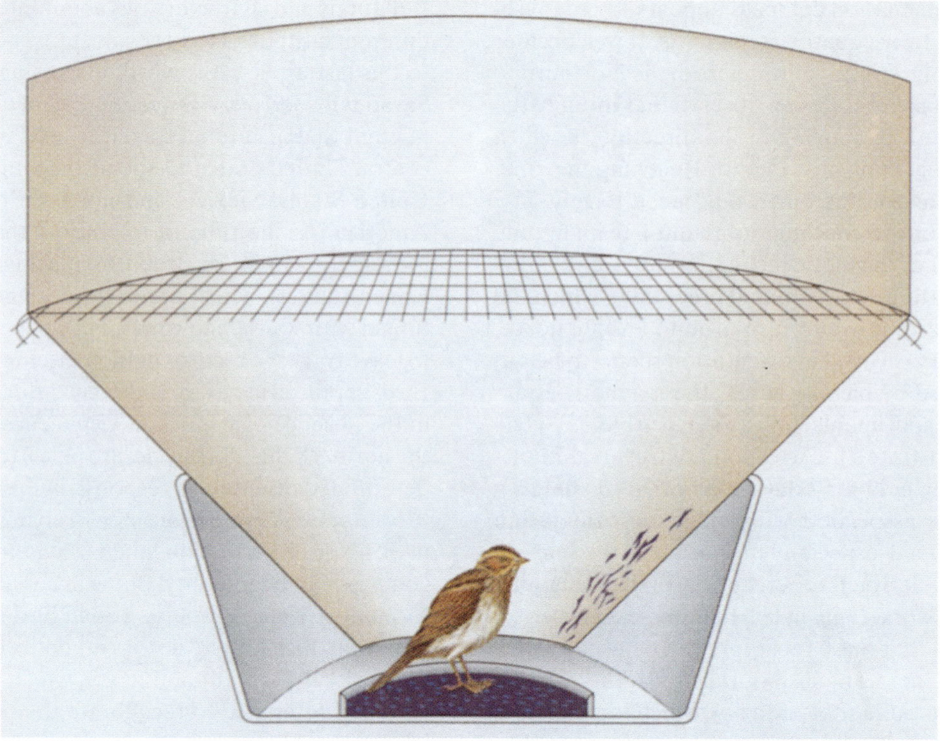

FIGURE 7 To examine the orientation of a small bird such as the Savannah sparrow, *Passerculus sandwichensis,* during the time of migration a bird can be placed in a funnel-shaped cage or arena, the Emlen funnel (here shown in half-section), whose walls are lined with absorbent paper. The feet of the sparrow rest on a sponge or pad bathed in ink. If oriented, during its attempts to escape, the bird will leave traces of ink with its feet in a particular sector of the funnel. The funnel can be screened in such a way as to block the view of any landmarks, as in the figure, or be located in a closed environment under controlled conditions (reproduced with permission from Emlen and Emlen, 1975, p. 105).

miliar reference points but is capable of determining novel spatial solutions or routes while moving among goal locations — for example, identifying with accuracy the most direct route between two locations that were previously visited separately. The ability to acquire a map has been clearly demonstrated in numerous vertebrates, whereas it is disputed whether some arthropods possess such a map.

The simplest way for an animal to use its map is to travel in the area covered by the map based solely on knowledge of the reference points, whether of a visible nature or of a chemical or acoustic type. More complicated is the use of a map in association with a compass. In this case, the map is mentally oriented with respect to directions in space, and the animal, recognizing its position with respect to its objective, can calculate the direction to maintain by relying on its compasses. The best demonstration of this phenomenon is provided by homing pigeons exposed to a shifted day–night rhythm. Even in a familiar area, such pigeons deviate from the expected direction as would be expected when relying on a mechanism of a functional sun compass.

There are cases in which an animal also displays the ability to orient with accuracy from unfamiliar locations. Therefore, one must assume that in such cases an animal is able to extend its map beyond the familiar area it has learned to recognize or that it has a nontopological map. The extension of a map can be accomplished by the acquisition of topological information about distant locations, as is the case in homing pigeons that use odors carried by the winds for this purpose. Nontopological maps, in contrast, represent only a working hypothesis that assumes an animal can build a map from two or more physical or chemical gradients whose isolines form a network comparable to the lines of latitude and longitude. The network can cover a large area, even the entire surface of the earth. An animal could learn to use it as a reference after having experienced, even during brief movements, how the gradients change in different directions. In a system based on two gradients, each point on the network would be characterized by two coordinates and represent an example of bicoordinate navigation.

Navigational Mechanisms

Given the structural and functional diversity of animals, and the different nature of their journeys, it is understandable that the navigational mechanisms put into action can

TABLE I

Classification of Navigational Phenomena

	TERMS USED IN THE TEXT (AND SYNONYMS FOUND IN THE SCIENTIFIC LITERATURE)	SOURCE OF INFORMATION (AND MECHANISM)
1	Random or systematic search	None
2	Genetically based orientation (vector orientation, vector navigation)	Genetic
3	Orientation based on trails (trail following)	Trails left during previous journeys
4	Route-based orientation (route-based navigation)	Information acquired during the outbound journey
4a	Route reversal (pilotage, Griffin's type I navigation)	Information acquired during the outbound journey (following in reverse order a series of reference points)
4b	Course reversal (Griffin's type II navigation)	Information acquired during the outbound journey (taking up the opposite geographic or magnetic direction)
4c	Path integration (dead reckoning, vector navigation, vector integration, integrative system)	Information acquired during the outbound journey (calculating current position with respect to goal using oneself as a reference)
5	Pilotage	An experience-dependent mental map (the mechanism does not require the determination of geographic directions)
6	True navigation (navigation, map-and-compass navigation, site localization)	External stimuli or signals to determine current location (the mechanism requires the determination of geographic directions)
6a	Map-based navigation (navigation, location-based navigation, pilotage)	External stimuli or signals to determine current location (location is determined using a mental map that represents the spatial relationship among topographical and geographical elements)
6b	Bicoordinate navigation (navigation, grid-based navigation comparative system, Griffin's type III navigation)	External stimuli or signals to determine current location (navigation is based on the recognition of physical or chemical gradients of considerable range)

be numerous and notably diverse, even if always based on a process of elaborating information (Table 1). The sources of the spatial information used can be usefully adopted as classification criteria (Papi, 1992). One must keep in mind, however, that many animals can rely on multiple mechanisms of navigation that they can use alternately or at the same time depending on the situation (Sinsch, 1992).

A strategy of random or systematic search can be adopted by animals in the absence of useful information. Ants that fail to find the entrance of their nest where they expect it carry out a systematic exploration of the environment by gradually expanding the search area. An American squirrel, *Tamasciurus hudsonicus,* when captured and released from an unfamiliar location, explores the territory in different directions in successive stages, returning each time to the point of departure (Bovet, 1992).

The genetic control of orientation is well-known from studies on passerine birds that migrate at night. Experiments have been performed primarily in autumn on subjects born the previous summer. At the time of their first migration, birds held in captivity become active at night and manifest a tendency to escape. This nocturnal activity is often referred to as migratory restlessness, and its seasonal duration approximately corresponds with the length of the journey that a bird would take if it were free. The direction chosen in captivity also corresponds with that of migration in nature and can change during the course of the migratory season in a manner similar to route changes that occur in nature. Given the fact that for migrant birds genetic information provides the direction, orientation, and length of the voyage, route information can be expressed by a vector. This kind of navigation is called *vector navigation.* Because a mechanism of this type does not seem sufficient to guide birds to a predetermined or known location or area, it would appear not to be an example of navigation as defined previously. It is assumed, however, that at the end of the voyage migrants would be attracted to ecologically suitable habitat, and therefore the choice of an overwintering area is ultimately shaped by conditioning.

Even more unusual is the case of the monarch butterfly, *Danaus plexippus,* an attractive milkweed lepidopteran whose larvae feed on plants of the genera *Asclepias.* Two populations exist in the United States: one west of the Rocky Mountains that carries out migrations of limited distance and overwinters on the California coast and the other east and central that is found in the middle of summer throughout the United States east of the Rocky Mountains north to the latitudes of the Great Lakes (Fig. 8). The adults of this population spend the winter in the high neovolcanic mountain belt of central Mexico and, in spring, migrate toward

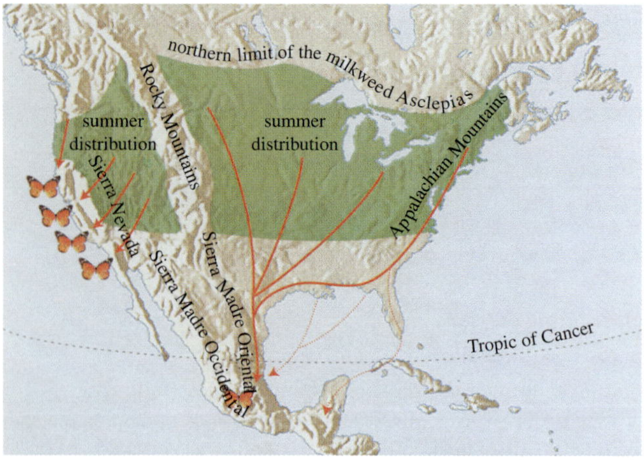

FIGURE 8 The summer distribution of two populations of the monarch butterfly, *Danaus plexippus,* in North America (colored areas) and the direction of autumn migration (continuous arrows). The dashed arrows identify hypothetical migratory routes (reproduced with permission from Brower, 1996).

FIGURE 9 This butterfly provides an exceptional example of the migration of the monarch butterfly. The individual was tagged by Dr. J. Fales in Maryland, USA, on September 2, 1992, and recovered on February 11, 1992, by Prof. L. Brower of the University of Florida in the mountains of the Sierra Campanario, State of Michoacan, Mexico. The straight line distance between the site of tagging and recovery is about 2800 km (courtesy of L. Brower).

the north and northeast. In about three to five generations, during which time the adults are continually moving farther north, the butterflies cover the entire area of their summer distribution (Fig. 8). In autumn, the adults of the last generation move to the south and southwest until reaching the location of their winter residence. The tagging of a large number of individuals, by fine tags applied to the wings (Fig. 9), has demonstrated that certain animals fly for at least 3000 km, with daily journeys of about 130 km. Little is known about their mechanisms of orientation. However, it is known that the butterflies are able to orient under complete cloud cover, try to exploit favorable winds, and compensate for wind drift when desirable. Information about the direction they need to maintain is genetic. The adults that migrate in autumn toward the location of their winter home have never been there before. In spring, the butterflies migrate to the north and northeast, and the successive generations gradually rotate their orientation from northeast to south.

The monarch butterfly represents a very promising subject for future research. Of particular interest is the presence in their tissue of toxic cardiac glycosides, cardenolides, that render the larvae and adults distasteful and cause birds that eat them to vomit. The larvae accumulate the cardenolides from various species of plants of the genera *Asclepias,* which have a limited geographic distribution. Analysis of the cardenolides allows one to determine from what species of *Asclepias* a larva has developed and, therefore, the area from which an adult has come (Brower, 1996).

With respect to vector navigation, one can also speak of bees, which orient toward a source of food signaled by their companions. As is well-known, a worker that has collected an abundance of food will, upon its return to the hive, execute on the honeycomb the so-called waggle dance or figure-eight dance from which his companions can deduce the distance and direction to a food source. In addition to the vector information, information about the quality of the food based on the odors that remained on the body of the returning forager bee is provided.

Orientation based on trails or tracks left behind, usually during the outbound journey, is used by many species of mollusks and ants (Chelazzi, 1992). This behavior is of limited interest as a method of navigation because it is not based on any phenomenon of memory, which characterizes other mechanisms of orientation based on information acquired during the outbound journey. In addition, these memory-based mechanisms are typically described in relation to homing, but we know that this kind of navigation is also used to return to a familiar location. The simplest example is when an animal reverses the route taken during the outbound journey, relying on memory for stimuli and reference points as in the reproductive migrations of salmon. At the time of reproduction, salmon carry out migrations called anadromous because they need to move against the prevailing water current. The extensively studied Pacific species of the genera *Oncorhynchus* serve as a useful paradigm or example of homing based on route reversal. The adults deposit eggs in the upper reaches of rivers, streams, and creeks. Only in a few species do the adults survive to reproduce again, whereas in others the adults die after egg deposition and fertilization. Shortly after birth, the young of some species begin a migration to their adult homes, whereas young of other species wait 1 or 2 years before beginning a journey that will bring them, depending on the situation, to a lake into which their natal river flows, to coastal ocean water near the mouth of the natal river, or even to the open ocean thousands of kilometers away. It requires 1–4 years to reach the age of reproduction, which is followed by the nuptial migration that brings a salmon back to the stream, creek, or lake where it was born. For salmon that migrate to the ocean, the first landmark is represented by the mouth of the river in which it descended. The mechanisms of orientation in the open ocean, however, remain poorly understood. It is suspected that they may be based on the use of a compass and may possibly consist of processes of true navigation. In the middle of the ocean individuals of the same species can congregate, even though they were born in and will return to rivers very distant from each other (Fig. 10).

In contrast, the freshwater return has been the subject of numerous observations and experiments. Salmon sometimes need to swim against the current for more than 3000 km and make a series of correct choices among rivers that flow into each other and among various rivers and streams that enter the same lake. These choices are the consequence of olfactory discrimination based on the memory of the odors the salmon perceived after birth and during the course of their outbound migration. This memory process appears to have the characteristics of imprinting and probably occurs in the presence of elevated levels of thyroxine in the blood. The evidence supporting the olfactory nature of this homing mechanism is broad and numerous. In an aquarium the salmon are capable of discriminating between water of different origins and show a preference for water that comes from locations to which they need to return. In addition, salmon deprived of the ability to smell show a reduced ability to home compared to intact fish. An additional and decisive piece of evidence demonstrating the role of olfaction was obtained by raising young salmon in the presence of artificial odorous substances. At the time of migration, these salmon are strongly attracted by the same artificial substances. However, it is likely that the return to the site of reproduction is not solely guided by olfaction but in some manner facilitated by auxiliary factors. This is indicated by the too high percentage of salmon deprived of olfaction that nonetheless succeed in reaching their goal. Furthermore, the site of reproduction can be reached by routes different from the routes followed during the juvenile migration (Dittman and Quinn, 1996).

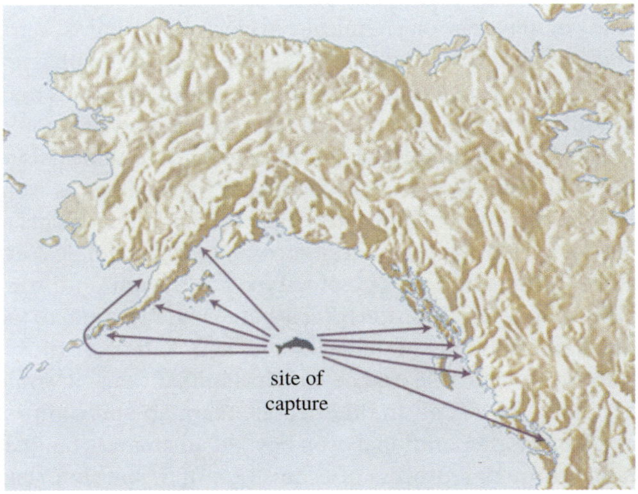

FIGURE 10 Individuals of one species of Pacific salmon, *Oncorhynchus nerka,* were captured together in the center of the Gulf of Alaska, tagged, and released. Some were later recovered at the mouths of very distant rivers as indicated by the arrows (reproduced with permission from Papi, 1992, p. 180).

A return carried out by means of route reversal, in contrast, implicates the use of a compass. In wasps that carry out excursions outside the nest to collect food, this mechanism is clearly operational. Surprisingly, wasps appear able to calculate the direction of their outbound journey even if they are transported passively inside a transparent container (Ugolini, 1987).

Among the mechanisms based on information acquired during the outbound journey, the most complex is called *path integration* (Wehner, 1992). In this case, an animal is able to take into account the direction and length of each part or phase of the outbound journey and at any moment calculate its position with respect to the point of departure. From this calculation it can deduce the length and direction of the return journey, another example of vector navigation. This mechanism is particularly useful in animals that carry out circuitous outbound journeys, for example, in search of local sources of food, and subsequently have or could have the need to return to their den via the shortest route. How animals determine the direction and length of the various phases of the outbound journey is mostly unknown, but based on the relatively few well-studied cases one can distinguish between one method that relies on external references and information (allothetic) and one that relies on internal information (ideothetic) derived from the central registration of movement or programmed movements.

Mammals rely on ideothetic information to update continually the calculation of their position with respect to the point of departure. However, as a journey progresses the calculation suffers from a progressive accumulation of errors as a function of both direction and distance. Consequently, even nocturnal mammals such as mice and hamsters prefer, when possible, the use of external references. The hymenoptera predominantly or exclusively rely on external information. Desert ants of the genus *Cataglyphis*

FIGURE 11 A worker ant of the species *Cataglyphis bicolor.* Workers from the *Cataglyphis* genera of desert ants are able to calculate at any time during an excursion away from the nest the direction and distance to home (courtesy of R. Wehner).

(Fig. 11) display an interesting spatial behavior that exploits the phenomenon of path integration. Workers carry out rapid, solitary excursions during the hottest hours in search of insects killed by the heat and subsequently return to the nest without the aid of an odor trail. Whereas the search for prey may involve a long and circuitous journey, the return home is by the shortest possible route (Fig. 12). At the moment an animal begins its return, it has already

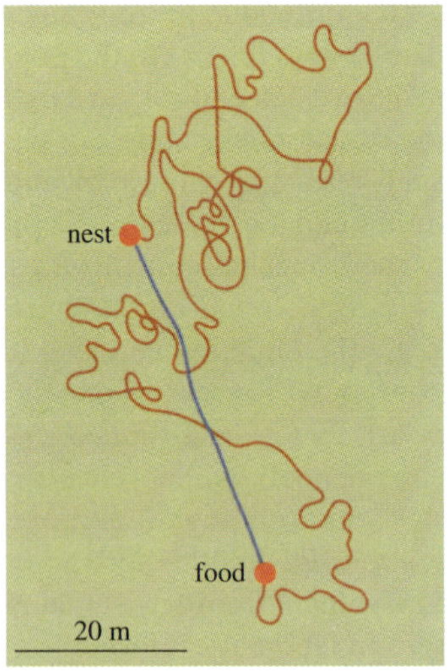

FIGURE 12 After having exited the nest, a worker of a species of desert ant, *Cataglyphis albicans,* explored the local environment along a circuitous route (red line) until it found food. Immediately it rapidly returned to the nest along a linear course (blue line). The route was reconstructed by observing the movement of the animal on a grid painted on the ground (reproduced with permission from Wehner, 1982, p. 19).

calculated the direction to take and the distance from the nest. In fact, if a worker that had just found food was displaced to another location, it would depart in the same direction it would have departed if it was not disturbed and, once it would have traveled the distance that would have returned it to its nest, the worker would begin a systematic search of the local environment in search of its nest. Information derived from the outbound journey — the direction and distance of the individual stages — is used to calculate the direction and distance of the return journey and is based, with respect to direction, on the position of the sun and skylight polarization. The method used to calculate the distance covered is uncertain. One of the distance elements is represented by the image that forms on their compound eye during locomotion. In fact, this is the principal calculation method for distance in bees, but *Cataglyphis* appears to use additional information. It is interesting to note that the mechanism of path integration is operational during the return home. If an ant marching toward its nest is forced to deviate from its route because of an obstacle, it will subsequently correct its route in such a way as to compensate for the deviation. This will occur even if the ant has been displaced to an area unfamiliar to it (Wehner, 1992).

A mechanism of *pilotage* is used when animals orient on the basis of familiar references, whether visible or of another nature, with only the aid of memory. Pilotage in its simplest form resembles route reversal but normally requires the possession of a topographic map.

The most complex forms of navigation, classified as true navigation, allow an animal to do without information accumulated during the outbound journey and, in many cases, permit reaching the objective even when departing from an unfamiliar location. True navigation enables an animal to calculate at any moment its position in space based on local cues, or physical or chemical environmental parameters, and to orient toward a goal using one or more compasses. The mechanism is called *map and compass.* With this definition one means that the animal may proceed like a person who, to orient toward his or her goal, determines his or her position in space on a topographic map, deduces from this the direction of his or her objective, and then orients in that direction with the aid of a compass. The simplest example is that of a topographic-type map acquired during visits to different locations. Such a map would extend to the limits of an animal's familiar range, and if the animal was transported and released beyond its familiar range it would be unable to orient. In other cases, the map and compass mechanism is operational even beyond an animal's familiar range through a process of acquisition of distant information. For example, one could imagine that the memory for landmarks never before visited but visible from very far away could be used for orientation should an animal find itself near them. However, the only well-described example of a map that extends beyond an animal's familiar range is that of the homing pigeon, which, in addition to a map of familiar locations, possesses an olfactory map of considerably larger range that is acquired using odors carried by the winds.

The homing pigeon is a domesticated race of the wild pigeon, *Columba livia,* that man has artificially selected for increased motivation, and possibly ability as well, to return home after being transported to a distant location and released (Fig. 13). Used for many centuries to carry messages, they are currently used almost exclusively for racing. The homing pigeon's capacity to find its home after being passively displaced to distant locations is legendary and is based in part on the training techniques employed by pigeon hobbyists or racers, who train their pigeons to return home from increasingly longer distances but always along the same directional route. The pigeons can therefore learn that home can always be found in a specific direction and are able to orient themselves with the sun compass and maybe magnetic compass when the sky is cloudy. Young, inexperienced pigeons often get lost — according to some authors about 50% become lost if they are transported 60 km or more — and their capacity to orient and return home can vary considerably depending on geographic area. Pigeons trained to return home from different directions acquire two maps, both of which function together with the sun compass. The first is a topographic map of familiar locations that extends over the familiar range of a pigeon. The second, so-called olfactory map is, like the topographic map, centered on the home loft but has a considerably larger range. The olfactory map, which enables a pigeon to return from unfamiliar locations, is acquired in the area of the home loft during the first months of life by associating odors carried by the wind with the direction from which the wind comes. In this way, a pigeon becomes familiar with the prevailing odors in the area that surrounds the home loft — odors that can be used to determine one's position with respect to home. For example, if wind from the north carries a specific odor — an odor found principally north of the home loft — a pigeon when released in this area would recognize the odor as being associated with north winds and would then need to simply fly in the opposite direction to return home.

FIGURE 13 Homing pigeons, *Columba livia,* can have different plumages but are easily distinguished by a trained eye from other domestic races.

The existence of two partially overlapping maps is demonstrated by the fact that pigeons deprived of the ability to smell are able to return home from familiar locations only. Also, pigeons that have undergone experimental damage to a region of the telencephalon of the brain, the hippocampus, lose their ability to organize familiar topographic references into a map-like representation or cognitive map of space. However, such pigeons can still orient home from both previously visited locations and unfamiliar locations by relying on their olfactory map, although they have difficulty returning home once near their loft because of an impaired ability to recognize the spatial position of their loft (Bingman *et al.,* 1995). To determine their position in space, pigeons rely not only on odors perceived at a release site but also on those they are exposed to during passive displacement. The latter informs a pigeon about the direction in which it is being transported and permits homeward orientation from very distant locations — locations even beyond the range of the olfactory map. The acquisition of the olfactory map by associating wind direction and different odor profiles has been demonstrated using developmental studies in young homing pigeons held in special aviaries that impeded, reduced, or altered this association. In other words, it produced predictable disturbances in navigation based on the treatment employed. More direct evidence is furnished by results obtained with pigeons exposed to artificial winds (Fig. 14). Some experiments carried out with wild birds indicate that the mechanism of olfactory navigation described previously may be broadly distributed. It is of considerable importance, however, to carry out further experiments, particularly ones examining the nature of the substances dispersed in the atmosphere that are the basis of the olfactory map. The nature, origin, and distribution of the odors used are completely unknown (Papi and Wallraff, 1992).

Our understanding of the mechanisms of true navigation is incomplete and many phenomenon are still unexplained. The well-known results of a displacement experiment carried out with 11,000 starlings during the 1950s demonstrated that adults, in contrast to young birds, are capable of true navigation (Fig. 15), a result that could be explained by the acquisition of an olfactory map similar to the one found in pigeons. Starlings deprived of olfaction have a reduced capacity to return in a familiar area. In other cases, however, the hypothesis of an olfactory map seems barely realistic and instead of an actual map, a network of

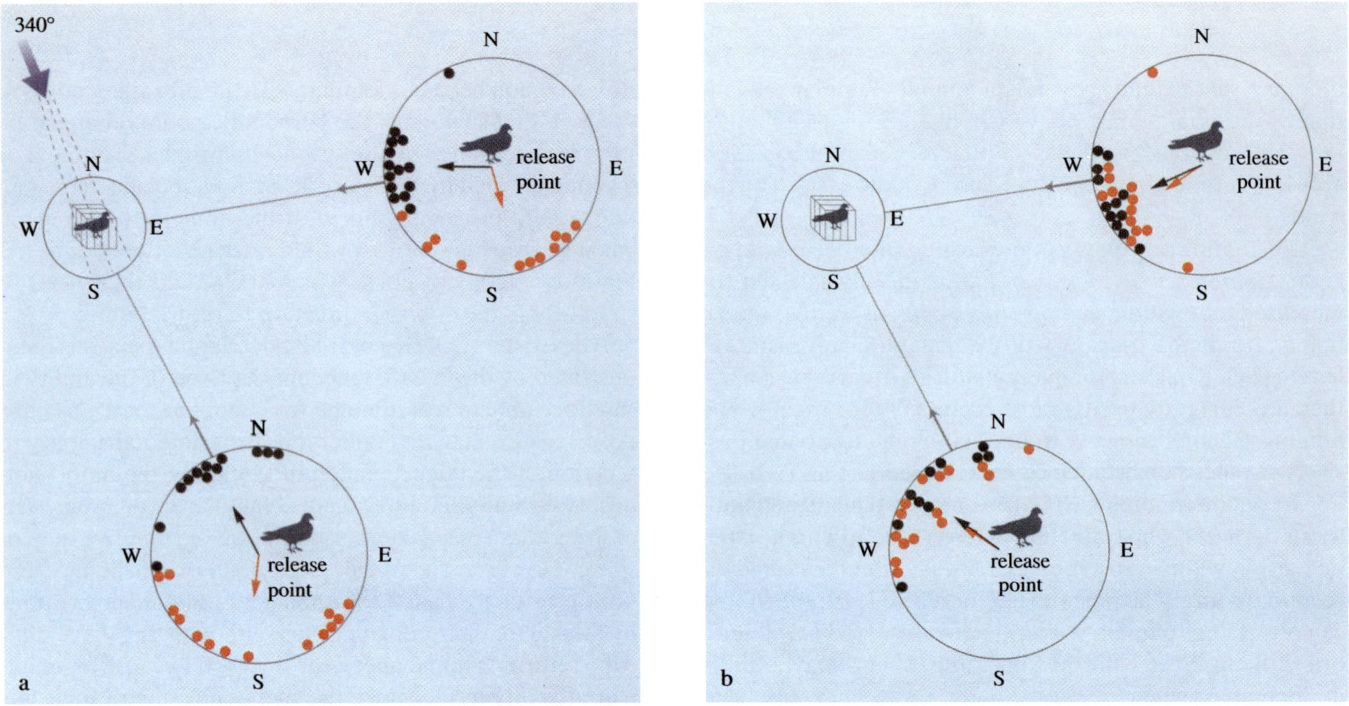

FIGURE 14 The olfactory map of homing pigeons is formed by the association between wind direction and the odors that the winds from different directions carry. This can be demonstrated experimentally: Two groups of pigeons were reared in separate cages exposed to the wind from all directions. (a) The experimental pigeons were also frequently exposed to an artificial wind produced by large fans, located about north-northwest (340°) of the cage, carrying an artificial odor of bitter almonds obtained from benzaldehyde. The pigeons were subsequently transported about 50 km to the east or east-southeast and released. The results are shown in the two diagrams (large circles) that display the direction taken by each pigeon after being released. Half of the birds were made to smell benzaldehyde before being released, and they oriented in the direction opposite the direction from which the artificial odor of bitter almond reached them (red points), whereas the other half, which were not exposed to benzaldehyde before being released, took off toward home (black points). (b) The control pigeons, which were not exposed to the artificial wind-borne odor, always oriented toward home whether they were forced to smell benzaldehyde prior to release (red points) or not (black points). In the diagrams each point corresponds to the direction taken by one pigeon, whereas the arrows represent the mean vector of the individual groups. The longer the arrow, the more similar were the directions taken by the individuals of a group (reproduced with permission from Ioalè *et al.*, 1990, p. 303).

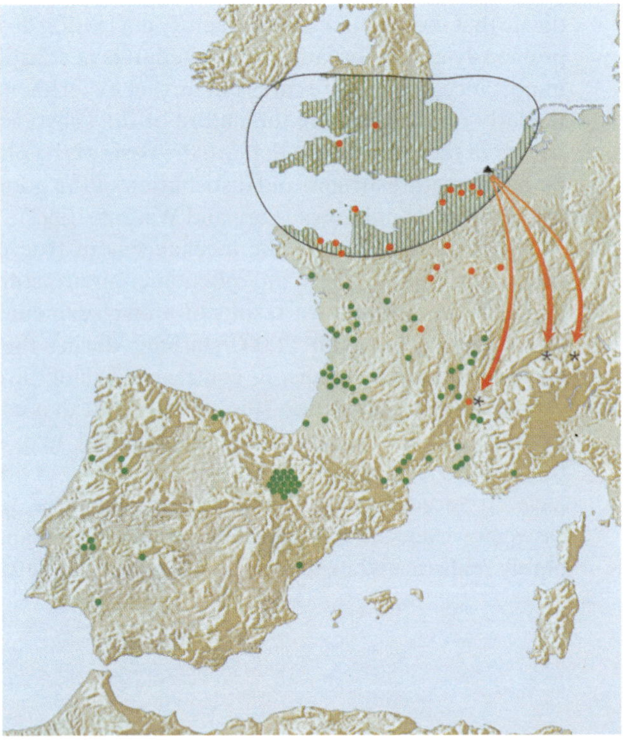

FIGURE 15 Starlings, captured in The Netherlands (black triangles) during their migration from the Baltic region to their winter quarters in northern France and the southern part of the British Isles (circled area), were released from three Swiss locations (asterisks). All the colored points identify the locations where an animal was later recaptured. Although young animals on their first migration, guided by genetic information, continued to move to the southeast (i.e., the direction of migration; green points), the adults, capable of true navigation, generally changed their route to reach their usual overwintering area (red points) (reproduced with permission from Perdeck, 1958).

FIGURE 16 This green turtle, *Chelonia mydas,* is carrying a radio transmitter on its carapace. The signal from the transmitter permits the satellites of the Argos system to identify its position.

physical or chemical coordinates as noted previously may be hypothesized.

Many marine vertebrates carry out long migrations during which, despite the uniformity of the environment and drifting created by the marine current or the winds in the case of birds, they succeed in reaching via a direct or at least economic route extremely distant and often point-like goals such as certain small oceanic islands. One of the most illuminating examples is represented by female marine turtles that deposit their eggs on the same beach every 2 or 3 years and then make a return journey to their nonbreeding residence. Depending on the species or population, the nonbreeding homes can be nearby or thousands of kilometers away. Satellite telemetry (Fig. 16) has permitted a complete reconstruction of some of the routes taken. Some take place along a coast and, consequently, could be based on pilotage. Others, in contrast, take place over open ocean and often in extremely deep water, and therefore in the absence of geomorphic references (Fig. 17). The turtles reveal a surprising capacity to directly head toward their objective, which can be very remote, and to appropriately correct their course during the journey (Papi and Luschi, 1996). The mechanism(s) that supports this behavior is unknown, but it has been proposed that the animals calculate their position on the basis of chemical stimuli or parameters of the earth's magnetic field.

Also, certain oceanic birds such as albatrosses, which nest on distant islands, are able to return to their nests after long flights in search of food. A wandering albatross, *Diomedea exulans,* leaving from the Island of Crozet in the Indian Ocean will travel 15,000 km, partly over open ocean and partly along the Antarctic coast, before returning safely to its nest (Jouventin and Weimerskirch, 1990). An analysis of the movements of these ocean navigators and the environmental conditions in which their voyages take place leads one to believe that they possess a mechanism that enables them to determine their position with respect to their goal; in other words, to locate themselves in geographic space. This appears to be confirmed by the fact that various

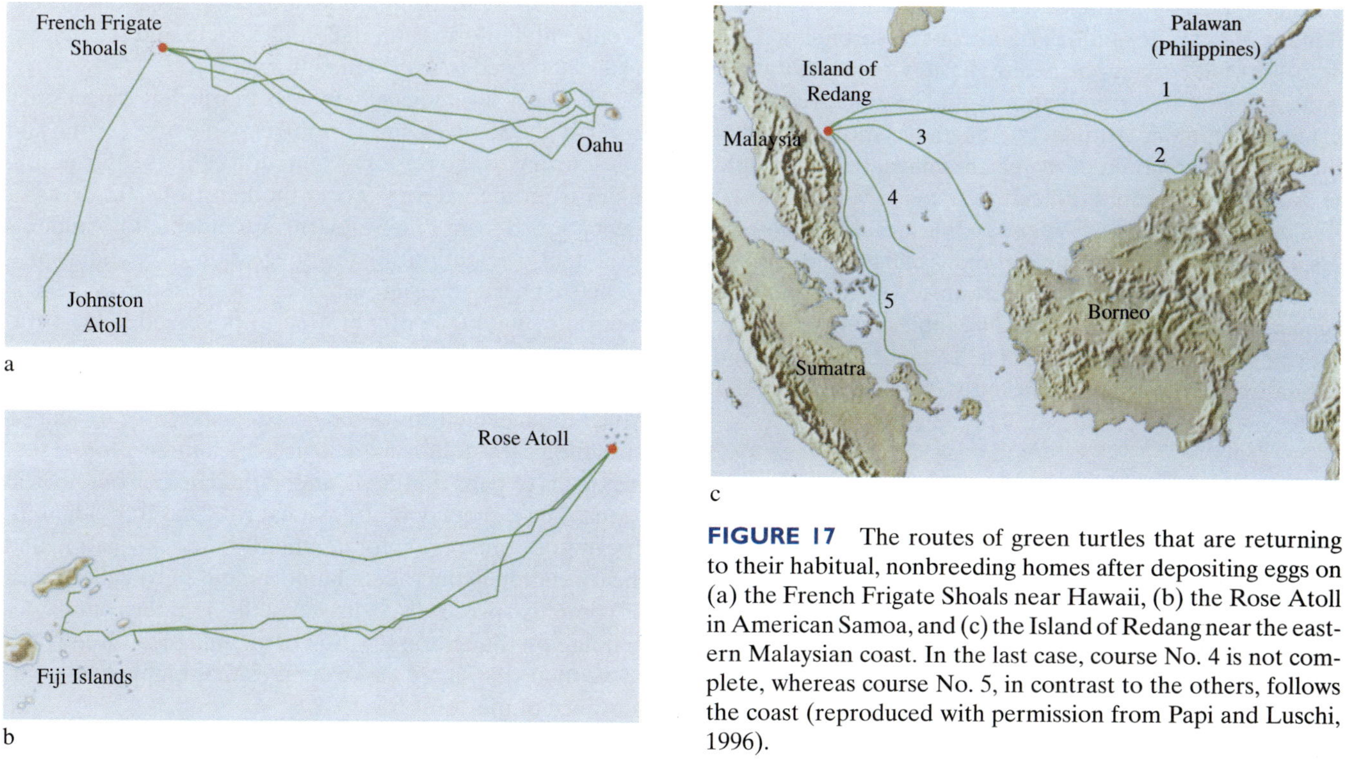

FIGURE 17 The routes of green turtles that are returning to their habitual, nonbreeding homes after depositing eggs on (a) the French Frigate Shoals near Hawaii, (b) the Rose Atoll in American Samoa, and (c) the Island of Redang near the eastern Malaysian coast. In the last case, course No. 4 is not complete, whereas course No. 5, in contrast to the others, follows the coast (reproduced with permission from Papi and Luschi, 1996).

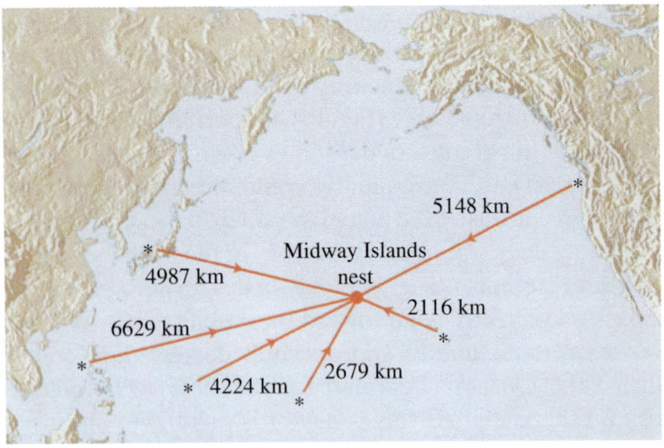

FIGURE 18 Schematic representation of the experiment in which Laysan albatrosses were transferred from their nest on Sand Island, Midway, to many distant locations in the area of the Pacific (asterisks). Of the 18 albatrosses displaced, 10 returned within 7–20 days and 4 within 2 months (reproduced with permission from Papi and Wallraff, 1992).

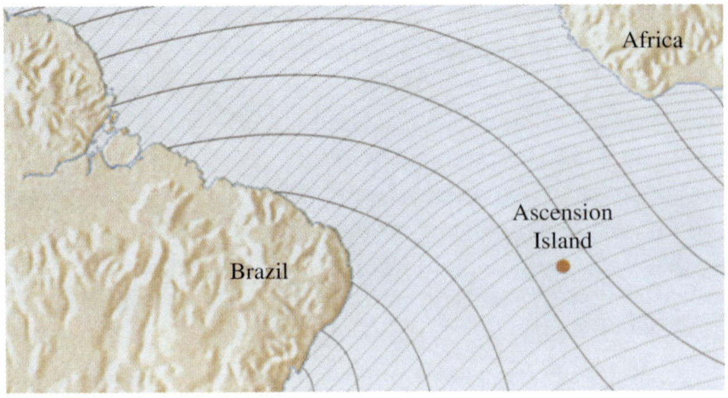

FIGURE 19 In the southern Atlantic the isobars for magnetic field inclination (dashed lines) and those for total intensity (continuous lines) form a network that could hypothetically be used by Brazilian green turtles during their migration to and from Ascension Island. The origin of the hypothesis is the discovery that in laboratory experiments, young turtles take up different directions depending on the amount of total intensity and inclination of the earth's magnetic field (reproduced with permission from Lohmann and Lohmann, 1996).

species of birds and pinnipeds captured and experimentally displaced are able to return from hundreds and even thousands of kilometers. Particularly demonstrative in this context is the example of the Laysan albatrosses transferred by plane and released at extremely distant locations (Fig. 18).

The mechanism used to determine geographic position is unknown, but plausible working hypotheses have been proposed. The determination of geographic or magnetic latitude appears accessible to the neural/sensory capacity of many animals. It would certainly not be surprising if the elevation of the circumpolar constellations or that of the sun at the moment of its culmination would be employed to determine geographic latitude. For magnetic latitude the total intensity or inclination of the earth's magnetic field would be available. It is more difficult to imagine what could be used as a second gradient whose isolines would be more or less parallel with geographic meridians and thereby furnish information on longitude. In certain areas, however, the same magnetic parameters could be sufficient to determine one's position in space because the isolines of intensity and inclination intersect at a sufficiently large angle (Fig. 19).

Final Considerations

As in any synthesis of a scientific field, the discussion of the available findings and discoveries summarized in this article has overwhelmed the presentation of gaps in our knowledge, as considerable as they are. In the field of animal navigation, the questions still unanswered are innumerable, but the prospects of answering those questions inspires the imagination with the exceptional and unexpected. There exist at least three levels of research on navigation that are accompanied by the hope of promising results. First is a descriptive level, which we can also define as morphological, designed to reconstruct long distance journeys. The spatial–temporal characteristics of animal movements, even when they can be exploited for economic value in some species or protect others from extinction, remain poorly studied. In addition, the accurate analysis of animal movements permits the formulation of hypotheses about the environmental information that guides them and the mechanisms employed to utilize the information.

A sensory level logically follows in which it is necessary to verify the sense organs that are capable of receiving and transmitting to the responsible neural centers information taken from the external world thought to be the basis of navigation. Sensory physiologists are indebted to students of animal orientation for the suggestion of research programs that have made important discoveries such as the perception of the plane of vibration of polarized light, ultrasounds, and infrared radiations.

Finally, there is a third level of central importance in which the mechanisms of navigation need to be explained physiologically following a logic of scientific progression that leads to behavior being understood in neurobiological terms. The discovery of the critical role played by the hippocampus brain region for landmark-based navigation and the revelation in the rodent hippocampus of so-called place or complex-spike cells, cells whose firing rates change depending on the spatial location of an animal (O'Keefe and Speakman, 1987), are two examples that highlight the importance of this level.

Whether a student of animal navigation works at one

level or another, or simultaneously at more than one level, he or she will be aware of confronting a field of research in which, more so than many others, the aphorism of Leonardo DaVinci "la natura è piena di infinite ragioni che non furon mai in isperientia" ("nature is filled with countless principles that have eluded experimentation") is strikingly applicable.

References Cited

ABLE, K. P. (1993). Orientation cues used by migratory birds: A review of cue-conflict experiments. *Trends Ecol. Evol.* **8,** 367–371.

ABLE, K. P., and ABLE, M. A. (1996). The flexible migratory orientation system of the Savannah sparrow (*Passerculus sandwichensis*). *J. Exp. Biol.* **199,** 3–8.

ALERSTAM, T. (1990). *Bird Migration.* Cambridge Univ. Press, Cambridge, MA.

BAKER, R. R. (1978). *The Evolutionary Ecology of Animal Migration.* Hodder & Stoughton, London.

BERTHOLD, P. (Ed.) (1991). *Orientation in Birds.* Birkhauser, Basel.

BERTHOLD, P. (1993). *Bird Migration.* Oxford Univ. Press, Oxford.

BOVET, J. (1992). Mammals. In *Animal Homing* (F. Papi, Ed.), pp. 321–361. Chapman & Hall, London.

BROWER, L. P. (1996). Monarch butterfly orientation: Missing pieces of a magnificent puzzle. *J. Exp. Biol.* **199,** 93–103.

CHELAZZI, G. (1992). Invertebrates (excluding Arthropods). In *Animal Homing* (F. Papi, Ed.), pp. 19–43. Chapman & Hall, London.

DITTMAN, A. H., and QUINN, T. P. (1996). Homing on Pacific salmon: Mechamisms and ecological basis. *J. Exp. Biol.* **199,** 83–91.

EMLEN, S. T. (1975). Migration: Orientation and navigation. In *Avian Biology* (D. S. Farner and J. R. King, Eds.), pp. 129–219. Academic Press, New York.

EMLEN, S. T., and EMLEN, J. T. (1975). *Sci. Am.* **213,** 105.

GAUTHREAUX, S. A., JR. (Ed.) (1980). *Animal Migration, Orientation and Homing.* Academic Press, New York.

GWINNER, E. (Ed.) (1990). *Bird Migration.* Springer, Berlin.

IOALÈ, P., *et al.* (1990). *Behav. Ecol. Sociobiol.* **26,** 303.

JOUVENTIN, P., and WEIMERSKIRCH, H. (1990). Satellite tracking of wandering albatrosses. *Nature* **343,** 746–748.

LOHMANN, K. J., and LOHMANN, C. M. F. (1996). Orientation and open-sea navigation in sea turtles. *J. Exp. Biol.* **199,** 73–81.

O'KEEFE, J., and SPEAKMAN, A. (1987). Single unit in the rat hippocampus during a spatial memory task. *Exp. Brain Res.* **68,** 1–27.

PAPI, F. (1992). General aspects. In *Animal Homing* (F. Papi, Ed.), pp. 1–18. Chapman & Hall, London.

PAPI, F., and LUSCHI, P. (1996). Pinpointing "Isla Meta": The case of sea turtles and albatrosses. *J. Exp. Biol.* **199,** 65–71.

PAPI, F., and WALLRAFF, H. G. (1992). Birds. In *Animal Homing* (F. Papi, Ed.), pp. 263–319. Chapman & Hall, London.

PARDI, L. (1979). Orientamento degli animali. In *Enciclopedia del Novecento,* Vol. IV, pp. 965–988. Istituto dell'Enciclopedia Italiana, Rome.

PERDECK, A. C. (1958). *Ardea* **46,** 12–14.

PRIEDE, I. G., and SWIFT, S. M. (Eds.) (1992). *Wildlife Telemetry: Remote Monitoring and Tracking of Animals.* Horwood, New York.

SAUER, F. (1957). Die Sternenorientierung nächtlich ziehender Grasmücken (*Sylvia atricapilla, borin* und *curruca*). *Z. Tierpsychol.* **14,** 29–70.

SINSCH, U. (1992). Amphibians. In *Animal Homing* (F. Papi, Ed.), pp. 213–233. Chapman & Hall, London.

THINUS-BLANC, C. (1987). The cognitive map concept and its consequences. In *Cognitive Processes and Spatial Orientation in Animal and Man* (P. Ellen and C. Thinus-Blanc, Eds.), Vol. 1, pp. 1–19. Nijhoff, Dordrecht.

WEHNER, R. (1982), *Neujahrsbl. Naturforsch. Ges. Zürich* **184,** 19.

WEHNER, R. (1987). *J. Comp. Physiol.* **161,** 520.

WEHNER, R. (1992). Arthropods. In *Animal Homing* (F. Papi, Ed.), pp. 45–144. Chapman & Hall, London.

WILLIAMS, T. C., and WILLIAMS, J. M. (1990). In *Bird Migration: Physiology and Ecophysiology* (E. Gwinner, Ed.), p. 15. Springer-Verlag, Berlin.

WILTSCHKO, W. (1973). *Inf. Org. Akad. Wiss. Lit. Mainz,* 29.

WILTSCHKO, R., and WILTSCHKO, W. (1995). *Magnetic Orientation in Animals.* Springer, Berlin.

EBERHARD GWINNER
Max-Planck Research Center for Ornithology
D-82346 Andechs, Germany

Bird Migration: Its Control by Endogenous Clocks

Every year, billions of birds leave their breeding grounds to spend the winter in more favorable areas. To reach the wintering sites, birds may have to travel more than 10 000 km and cross extremely inhospitable areas such as deserts or oceans. Migrations of this kind require a host of energetic adaptations and a well-developed capacity to orient in space and time. Recent studies indicate that for temporal orientation, migratory birds use sophisticated daily and annual clocks that are capable of running for many cycles even in the absence of external cues. Endogenous daily (circadian) clocks time the onset and end of migrations within the 24-h day. Endogenous annual (circannual) clocks determine the occurrence of migration at certain times of the year but are also responsible for seasonal changes in migratory direction. Moreover, endogenous circannual programs control the time course and to some extent even the distance of migratory flights. Under natural conditions, the endogenous circannual rhythms are synchronized by the annual cycle in day length (photoperiod) in such a way that an optimal temporal adjustment of migratory performance to season and latitude is guaranteed. In addition, a variety of supplementary environmental factors related to climatic and nutritional conditions modify the output of circannual clocks in an adaptive manner. Circannual rhythms also generate temporal frameworks for learning processes that enable birds to replace or supplement endogenous programs by mechanisms based on individual experience. Finally, the endogenous mechanisms controlling bird migration may be subject to rapid evolutionary change, as exemplified by the recent appearance of a new, genetically determined migratory pattern within a population of blackcaps.

Bird Migration: A Global Phenomenon

Every year, as autumn approaches, scores of birds leave their breeding grounds to migrate to regions in which the environment provides more favorable conditions for wintering (Fig. 1). Migratory behavior is particularly well developed in birds inhabiting high latitudes with inhospitable winter conditions. Thus, in many palearctic and nearctic species, northern populations are migratory, whereas southern populations are nonmigratory (Fig. 2).

The magnitude of the worldwide bird migration system is impressive. R. E. Moreau, outstanding British ornithologist and expert on African birds, estimated that the number of European birds that enter Africa each year from fall migration is at least 5 billion. Extrapolated from this estimation, there must be at least 50 billion birds worldwide that carry out regular seasonal movements. The guild of avian migrants comprises a mixture of travelers of many different kinds. Some stay in their home area literally to the last minute, escaping only when frost, snowfall, or ice make survival

FIGURE 1 A flock of white pelicans (*Pelecanus onocrotalus*) photographed in Turkey on their migration from the Black Sea to the Mediterranean Sea (photo courtesy of H. J. Fünfstück).

FIGURE 3 Corn buntings (*Miliaria calandra*) are territorial during the breeding season but gregarious during the rest of the year. Migratory performance varies from location to location. Whereas British birds are residents, most central European birds are short-distance migrants. The extent and timing of the migration vary greatly according to conditions (photo courtesy of A. Limbrunner).

impossible. In some parts of central Europe this holds true, for example, for the common buzzard (*Buteo buteo*), the mistle thrush (*Turdus viscivorus*), and the corn bunting (*Miliaria calandra;* Fig. 3) as well as for many species of waterfowl such as ducks and geese. Migrants of this type usually cover only short distances to slightly more favorable areas, and they return as soon as conditions improve. Because their migrations depend on acute weather conditions, migrants of this type have been called *weather birds* by the old ornithologists. At the other end of the scale are migrants that leave their breeding grounds each year at precisely the same time long before conditions deteriorate.

FIGURE 2 A flock of linnets (*Carduelis cannabina*) in migration. The linnet is a year-round resident in the Mediterranean part of its breeding distribution, but individuals from the northern part are migratory. Birds breeding in Scandinavia may travel as far south as northern Africa. Like many other songbirds that migrate in daytime, linnets form flocks during the migratory seasons (photo courtesy of A. Limbrunner).

An example is the common swift (*Apus apus*), which in many parts of Europe leaves during the first week of August, when insect food is still abundant. Birds of this type are usually long-distance migrants, covering distances of thousands of kilometers to remote wintering grounds. Long-distance migrants in this category include many species of waders (Fig. 4), the white stork (*Ciconia ciconia;* Fig. 5), and also many small songbirds such as the Scandinavian red-backed shrike (*Lanius collurio*), which spends the winter in southern Africa, almost 10,000 km south of its breeding grounds. Wheatears (*Oenanthe oenanthe*) from Alaska also migrate to South Africa. To reach it they cross the Bering Sea and then follow a route across the Asian conti nent and the Arabian Peninsula, a one-way distance of more than 16,000 km. Many North American migrants, such as the bobolink (*Dolichonyx oryzivorus*), travel to South American wintering grounds as remote as Argentina, 10,000 km from their breeding areas. Long-distance migrants of this type return to their breeding grounds late in spring. Their arrival dates are precise and depend only slightly on environmental conditions (Fig. 6).

Since migration times of birds in this category are highly predictable — as if timed by a calendar — such birds have been designated *calendar birds* by the old ornithologists. Indeed, in some regions of the world the peasants schedule their agricultural activities according to the passage of migratory birds. For instance, the Kelabits of Sarawak, Borneo, use a "bird calendar" based on the arrival and departure of migrants such as yellow wagtails (*Motacilla flava*) from northern Asia. In between the categories of extreme weather birds and extreme calendar birds are almost any

a

b

FIGURE 4 Waders are among the most spectacular long-distance migrants. Pairs defend territories during the breeding season, but during the rest of the year most species live in flocks. Golden plovers (*Pluvialis apricaria*) (a) and dunlins (*Calidris alpina*) (b) are migratory, breeding in the far northern parts of Europe and Asia (photo courtesy of A. Limbrunner).

type of intermediate migrants, ranging from birds that travel a few hundred kilometers to those that travel many thousands of kilometers.

In addition to the numerous birds that undertake regular seasonal migrations, every year at approximately the same time and over the same distances, some species, or populations of species, migrate only occasionally under certain environmental conditions. Again, these migrations can be restricted to a particular season in the years in which they do occur. These "eruption migrants," such as the waxwing (*Bombycilla garrulus*) or the nutcracker (*Nucifraga caryocatactes*), travel over long distances only in years following high reproductive success and/or shortage of food resources on their breeding grounds. In other instances, occasional migrations can be aseasonal. This is the case for Australian opportunistic breeders such as the zebra finch (*Poephila guttata*) or many species of Australian waterfowl, which follow the rainfalls that occur in irregular spatial and temporal patterns. Hence, the "nomadic" migrations of these birds

FIGURE 5 The white stork (*Ciconia ciconia*) is a long-distance migrant that breeds in Europe and winters in Africa. Western populations leave their breeding grounds in a southwesterly direction and winter in the Sahel zone of western Africa. Eastern populations migrate in a southeasterly direction to Turkey before they shift to a more southerly direction which leads them to their south African wintering grounds. As shown by early classical experiments, these migratory routes are partly the result of direct genetic factors and partly based on traditions resulting from individual learning (photo courtesy of A. Limbrunner).

FIGURE 6 Spectacular long-distance migrations are carried out by the dark-bellied brent goose (*Branta bernicla bernicla*). Birds of this subspecies breed in the Arctic part of Russia, mainly on the Taimyr Peninsula. From here they migrate to wintering sites in the southern North Sea area, southern England, and the Atlantic coast of France. During the homeward migration in spring, the birds first fly nonstop over approximately 2000 km to the tundra areas of the White Sea. Here, they rest and increase their fat depots before taking off for their final 3000-km flight (photo courtesy of A. Limbrunner).

are opportunistic with respect to timing as well as direction and distance.

Adaptations of Migratory Birds

The diversity of types of migrants is matched by a corresponding diversity of mechanisms by which migration is accomplished. The most sophisticated adaptations are found among long-distance migrants. Most conspicuously, these have evolved a variety of mechanisms to cope with the high energetic costs of migration and to enable the birds to fly in the correct direction at the correct time (Bairlein and Gwinner, 1994; Biebach 1996).

Metabolic Adjustments to Migration

Among the mechanisms that have evolved in response to the energetic demands of migration are adjustments of feeding, digestion, and metabolism that allow birds to deposit large amounts of fat before and during migratory seasons. These stores can subsequently be used as energy reserves for migratory flights. Some species, such as the red knot (*Calidris canutus*), the bar-tailed godwit (*Limosa lapponica*), and the ruby-throated hummingbird (*Archilochus colubris*), almost double their body mass by developing large subcutaneous and intestinal fat depots (Fig. 7). As a rule, long-distance migrants accumulate more fat than short-distance migrants, and among the former the largest amounts of fat are deposited by birds that are forced to cross ecological barriers such as deserts and oceans. Premigratory fattening results primarily from dramatic increases in food uptake (hyperphagia) but also from spontaneous shifts in diet (and hence nutrient selection) as well as an increased nutrient utilization. Garden warblers, for instance, eat up to twice as much while preparing for fall migration in August than during the preceding phase of their annual cycle. At the same time, their diet preference (as that of closely related blackcaps, *Sylvia atricapilla*) shifts from insects to fruits (Fig. 8). Fruits are amply available in autumn and, although their nutritional content is relatively low, feeding on fruits has almost no costs. During that time of the year, efficiency of nutrient uptake by the intestinal system is also increased. This is partly due to dramatic changes in gross gut anatomy. Bar-tailed godwits, for instance, increase their stomach mass by 30% during the first half of a 3-week refueling period that interrupts spring migration. By the end of this resting period, they have again reduced the stomach mass by 20%. Presumably, enlarging the stomach during refeeding at the resting site enhances food utilization, whereas reducing it just prior to migration decreases the costs of flight. For other species even more dramatic reductions in gut size during actual or simulated migration have been reported.

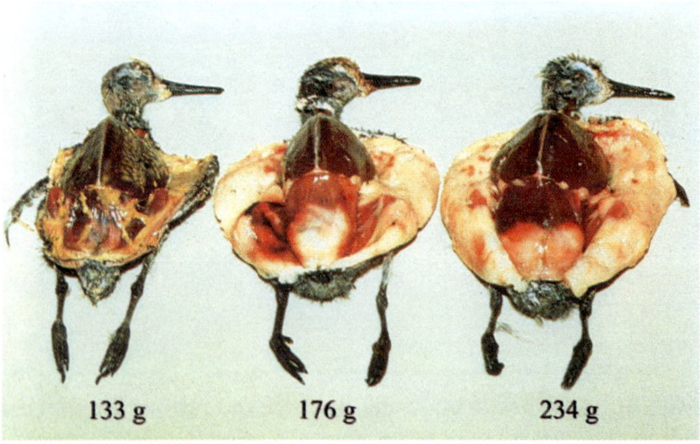

FIGURE 7 Changes in the size of the subcutaneous fat stores, the breast muscle, and the abdominal cavity of red knots (*Calidris canutus*) during their 3-week stopover in Iceland, where they refuel before migrating further north to their Arctic breeding grounds. These birds not only store considerable amounts of fat but also show drastic changes in fat-free mass of organs such as flight muscles, gut, and liver, as shown by the weight variations indicated (reproduced with permission from Piersma and Lindström, 1997).

FIGURE 8 A blackcap (*Sylvia atricapilla*) feeding on figs while resting in Israel. Many birds that normally eat insects prefer to eat fruits during migratory stopovers (photo courtesy of A. Limbrunner).

Although the intestinal tract often shrinks during migration, the flight muscles undergo drastic hypertrophy in preparation for migration flights. This is the case for long-distance migrating knots and other waders but also for songbirds such as the gray catbird (*Dumetella carolinensis*), in which flight muscle mass increases by about 35% prior to fall migration.

Orientation in Space and Time

Migratory birds not only have to cope with severe energetic challenges but also must appropriately orient their flights, both in space and time. The following sections concentrate on the way in which birds orient their migrations in time.

Migrations take up a considerable amount of time each year. As a consequence, migratory birds, particularly those covering long distances, are under severe pressure to organize their various seasonal activities appropriately and with the necessary accuracy, both relative to the natural year and relative to each other. Transequatorial migrants from high latitudes spend 6 months or more each year on their migratory journeys, leaving little time for other obligatory seasonal activities such as molt and especially reproduction. For tropical migrants such as the garden warbler (*Sylvia borin*) or the willow warbler (*Phylloscopus trochilus*), as for many arctic waders that arrive at their breeding grounds late in spring, the precise timing of homeward flight is extremely important: A premature takeoff could bring a bird to its breeding site too early. It might arrive there at a time in which conditions are still unfavorable and the bird might perish. A delayed homeward flight, on the other hand, may result in a late arrival at a time when territories are already occupied by competitors or the season is too advanced for the bird to successfully complete a breeding cycle. There is a corresponding need for fall migration to occur at the appropriate time, for instance, to ensure that there is sufficient time for the birds to molt in their winter quarters before spring takeoff. Migration and reproduction are mutually exclusive activities, and the same is usually also true for migration and molt. Hence, it is also necessary that these activities are appropriately timed relative to each other.

Temporal Control of Migration

In 1925, the Canadian ornithologist Rowan discovered that migratory activity, like many other seasonal functions of plants and animals, is under the control of photoperiod, the light fraction of the 24-h day. In one of his experiments, he kept American crows (*Corvus brachyrhynchos*) in winter in outdoor aviaries under the low-temperature conditions of the Canadian winter but with photoperiod artificially lengthened to simulate a spring day. After release in winter, the crows started to migrate toward their more northerly breeding grounds before the normal departure time (Rowan, 1926). Subsequent experiments with many other species confirmed the conclusion derived from Rowan's experiment — that increasing day length provides a stimulus for spring migration.

It is noteworthy that Rowan pointed out as early as 1926 that the photoperiodic control mechanism which he had discovered could not explain the timing of spring migration of many long-distance migrants.

Rowan was aware of the fact that such birds spend the winter close to the equator, where photoperiod is constant, or beyond it, where photoperiod increases during the temperate zone winter (Fig. 9). Rowan suggested that "those species that breed in the Northern Hemisphere and winter on the equator or cross it and winter in the Southern Hemisphere make necessary the assumption that there is another and internal factor involved, a physiological rhythm" (Rowan, 1926).

Internal Mechanisms Controlling the Onset of Migration

Although some anecdotal field data and preliminary experimental results supported Rowan's hypothesis of internal timing, it was not until the late 1960s that it was first rigorously tested. The study summarized in Fig. 10 was conducted with willow warblers.

Like many other small long-distance migrants, willow warblers are normally day-active but migrate exclusively at night. When kept in cages, they develop nocturnal activity during the migratory seasons, which can be measured, for instance, by means of microswitches mounted underneath the perches or by infrared or ultrasound motion detectors. Caged willow warblers born the previous spring were initially exposed to the natural photoperiod changes of their German breeding grounds and then, at the height of their fall migratory season in mid-September, transferred to experimental chambers in which photoperiod was held at a constant equatorial duration of 12 h, and temperature and nutritional conditions were also held constant (Fig. 10c). Two other groups of willow warblers served as controls: One remained in Germany and continued to be exposed to the photoperiodic variations of the breeding grounds (Fig. 10a), and the other was transported — at the time when the experimental group was transferred to constant conditions — to Zaire, an area within the species' wintering grounds where they experienced the natural photoperiodic conditions (and to some extent the changes in temperature and humidity) of their wintering area. Figure 10 shows that the birds of the experimental group kept under rigorously controlled constant conditions behaved essentially like the birds in the two control groups: They terminated fall migratory restlessness at about the normal time (December). Toward the end of the subsequent molting season they initiated spring migratory restlessness in February and March, the time of departure of free-living conspecifics. Obviously, the processes occurring in these equatorial migrants during the wintering season depend on endogenous timing factors and are primarily independent of environmental cues.

Experiments with other species of birds kept under constant environmental conditions have revealed that not only spring migration but also fall migration is triggered by endogenous timing mechanisms. Endogenous factors also control various processes associated with migration, such as fattening, shifts in preferred diet, and hyperphagia. Furthermore, the endogenous mechanisms determining

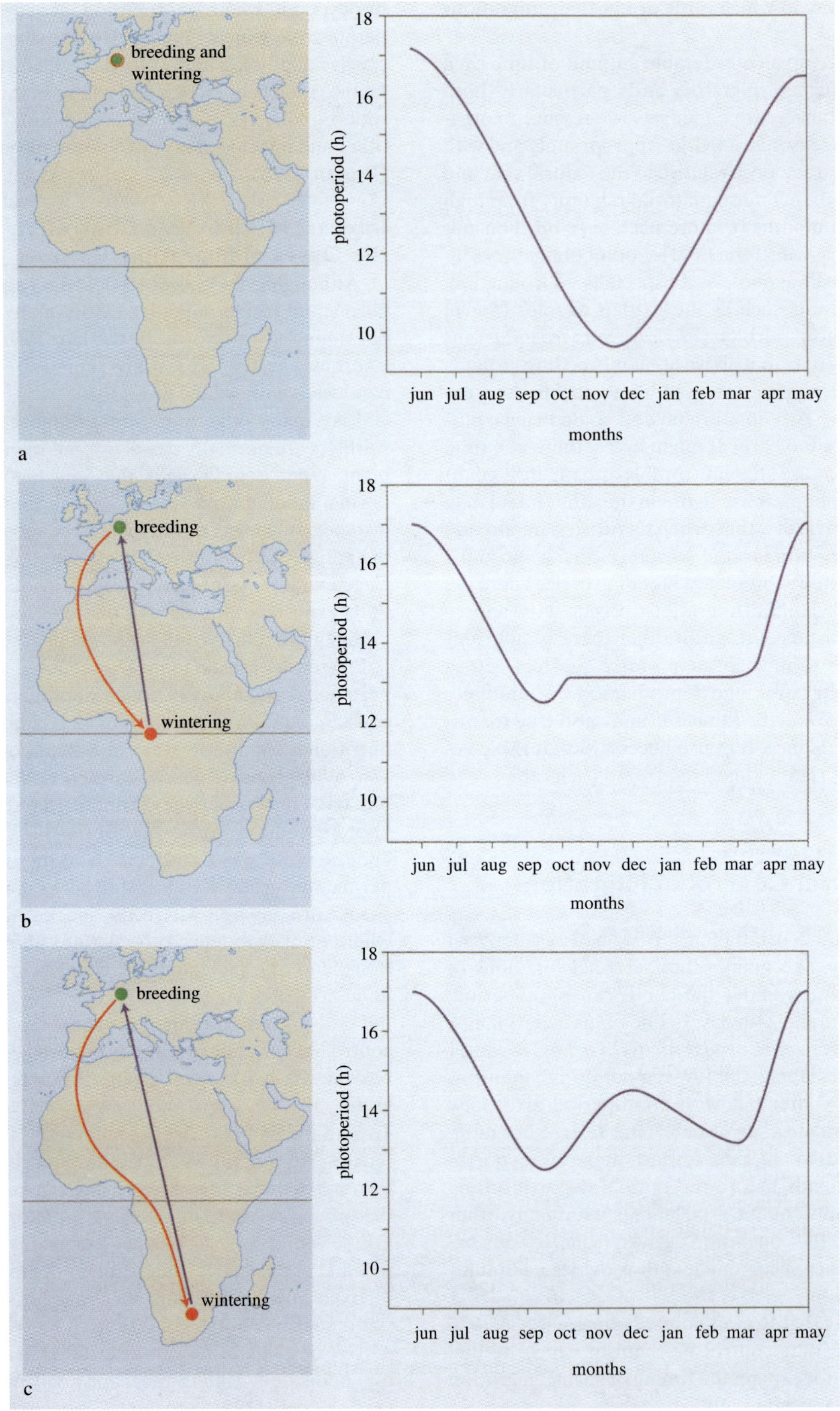

FIGURE 9 Photoperiodic changes experienced in the course of a year by temperate zone residents (a), equatorial migrants (b), and transequatorial migrants wintering at approximately 20° S (c).

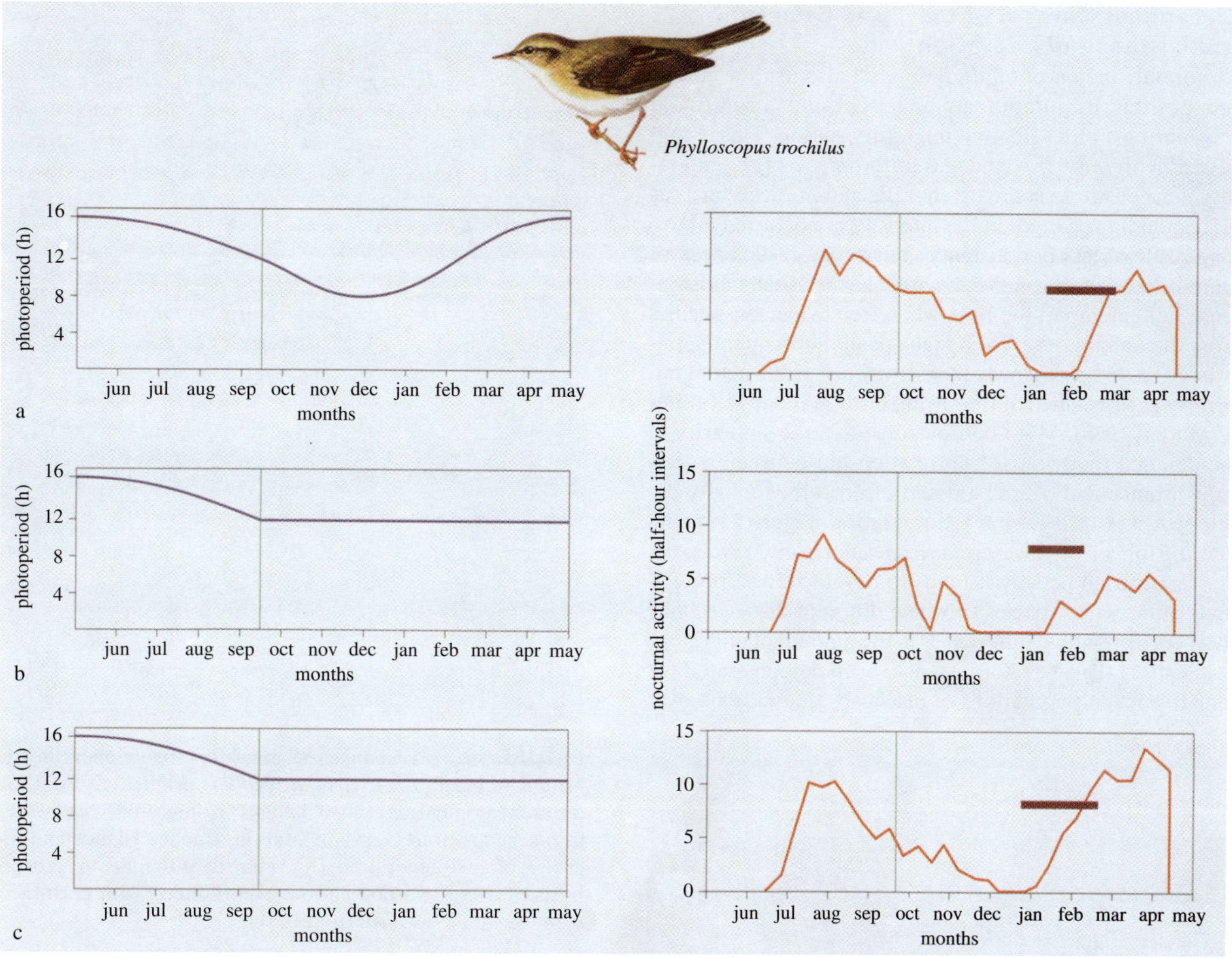

FIGURE 10 An experiment with willow warblers (*Phylloscopus trochilus*) to test whether the onset of spring migration in this equatorial migrant is endogenously controlled. Birds raised in captivity were initially kept under the natural photoperiodic conditions of their breeding grounds, but from September 20 to the end of the experiment in the subsequent May they were divided into three different groups: (a) one group remained in Germany and continued to experience the natural photoperiodic changes of the breeding grounds, (b) another was transferred to Zaire and exposed to the natural photoperiodic conditions of their wintering grounds, and (c) the third group was kept under a constant 12-h photoperiod and controlled-temperature conditions. On the left, the typical photoperiods of the specific areas are shown in blue, whereas the red curves on the right show changes in nocturnal migratory restlessness, expressed as the number of half-hour intervals during which birds were active each night. The horizontal bars indicate the occurrence of winter molt. The vertical lines (in green) correspond to the start of the experiment (adapted from Gwinner, 1968).

the onset of migration are organized in a species- or even population-specific manner (Berthold, 1990). That is, if kept in the same constant environment, early migrating species such as the willow warbler start migration earlier than late migrating species such as its close relative, the chiffchaff (*Phylloscopus collybita*). Cross-breeding experiments, for example, between a long-distance and a short-distance migrating population of the blackcap (*S. atricapilla*), have revealed that these differences in timing are genetically determined.

The mechanism underlying these endogenous timing functions is an oscillatory process with a period of about 12 months. It persists for many cycles in a constant environment, sometimes for a lifetime. Under constant conditions, the period of the rhythm is only approximately 1 year; hence, such rhythms are called *circannual* in analogy to circadian rhythms. This circa or "free-running" nature of the rhythmicity is of great significance because it excludes the possibility of an exogenous origin (Gwinner, 1996; Berthold, 1990).

Circannual Control of the Time Course and Distance of Migration

Not only the onset of migration but also its time course and, possibly, its duration are under circannual control, at least during a bird's first fall migratory season. This is suggested by the fact that changes in the speed of the migratory journey are to some extent reflected in caged birds as corresponding changes in the intensity of nocturnal migratory restlessness. For instance, migratory restlessness of long-distance migrating willow warblers is usually most intense in September, the time when free-living conspecifics cross the Sahara desert at a high speed. Subsequently, migratory restlessness slowly fades away just as the actual migratory performance of free-living birds gradually declines (Gwinner, 1968). More conspicuous than the similarity in the detailed time course is a relationship between migratory distance and overall amount of nocturnal activity developed by a bird during fall migration. Figure 11 shows that the longer the distance a species normally travels, the more nocturnal activity it tends to develop. Even populations of the same species traveling different distances may show corresponding differences in the amount of migratory restlessness. This has, for instance, been demonstrated for four European populations of blackcaps that migrate over different distances. The amount of migratory restlessness developed by caged birds of these four populations showed a trend paralleling that of migratory distance (Fig. 12a). Cross-breeding experiments revealed that these endogenously programmed differences in migratory pattern have a genetic basis: F_1 hybrids between German and Canary Island blackcaps exhibited a pattern intermediate between those of the two parental populations (Fig. 12b).

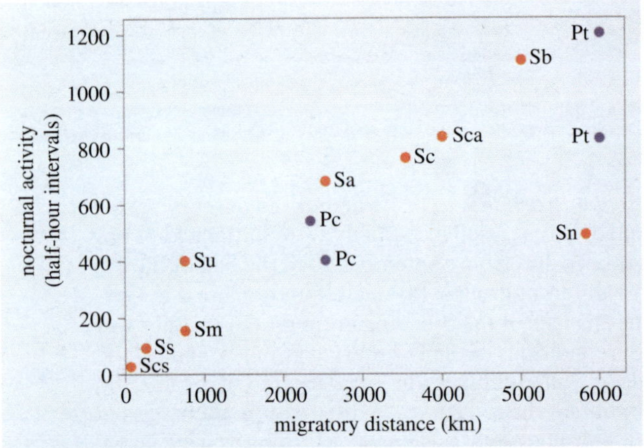

FIGURE 11 Relationship between the migratory distance normally covered by several species of *Sylvia* and *Phylloscopus* warblers and the total amount of migratory restlessness (expressed as the number of half-hour intervals with nocturnal activity) developed by groups of caged birds of these various species during their first fall migratory season. Sa, *Sylvia atricapilla,* blackcap; Sb, *S. borin,* garden warbler; Sc, *S. communis,* whitethroat; Sca, *S. cantillans,* subalpine warbler; Scs, *Sylvia couspiculata,* spectacled warbler; Sm, *S. melanocephala,* Sardinian warbler; Sn, *S. nisoria,* barred warbler; Ss, *S. sarda,* Marmora's warbler; Su, *S. undata,* Dartford warbler; Pc, *Phylloscopus collybita,* chiffchaff; Pt, *Ph. Trochilus,* willow warbler (reproduced with permission from Gwinner, 1986a).

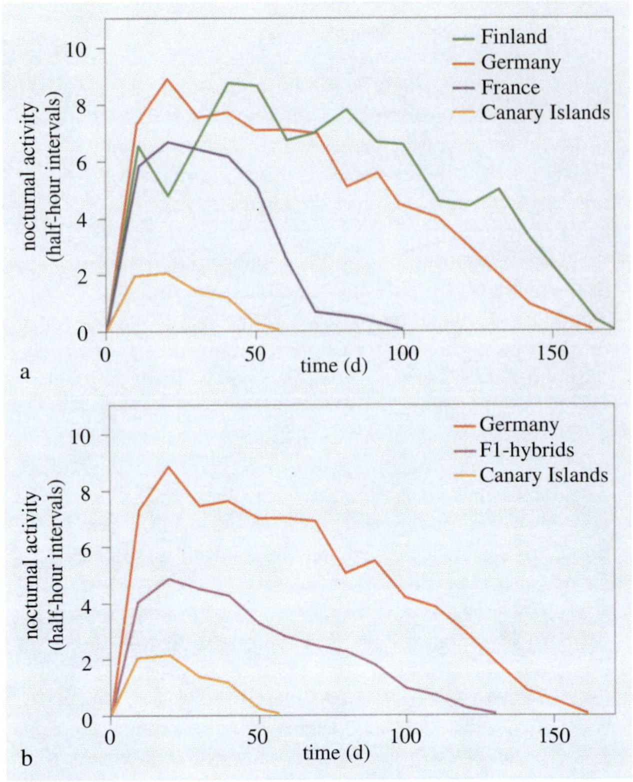

FIGURE 12 (a) Changes in migratory restlessness in four groups of blackcaps (*Sylvia atricapilla*) originating from four different populations. (b) Changes in migratory restlessness in blackcaps from Germany and the Canary Islands and in a group of first-generation (F_1) hybrids obtained by crossing birds of these two populations (reproduced with permission from Berthold and Querner, 1981).

The close correlation between migratory distance and amount of migratory restlessness suggests the possibility that the former may in fact be the direct consequence of the latter. Gwinner (1968) noted,

> The migratory restlessness of caged birds is the expression of an endogenous time program controlled by a circannual rhythmicity. It is organized in a species- or population-specific fashion in such a way as to produce just enough migratory restlessness (i.e., migratory time) to reach the specific winter quarters. In

other words, the birds would arrive in the vicinity of their wintering area when the program is terminated.

Such a mechanism might indeed be an essential part of the spatial orientation system of young birds migrating to their wintering grounds for the first time in their lives. As indicated by Perdeck's experiment discussed in the article by Papi on navigation (this volume), first-year starlings are apparently only capable of migrating in a particular compass direction. However, in order to arrive at their population-specific wintering area, it is not sufficient for the birds to migrate in a particular direction; they also must travel over a particular distance. The previous hypothesis proposes that this distance is determined by a circannual time program that, by specifying a certain number of migratory hours, establishes the length of a vector, which corresponds to migratory distance.

This vector-navigation hypothesis is supported by a variety of results. However, it is far from being proven. One of the problems it faces is that caged migrants usually appear to develop more nocturnal migratory activity than would be required for them to cover the distance between the breeding and wintering areas. However, although the beeline distances between breeding and wintering areas can be easily calculated, we usually do not know how straight a migrant flies. Depending, for instance, on wind conditions and geographic features of the landscape, birds "zigzag" along their average migratory route. How much this zigzagging adds to the beeline distance cannot currently be reasonably estimated. Further evaluation of the situation, and the previously discussed vector-navigation hypothesis, will critically depend on more detailed information about the actual flight paths of migrants.

Circannual Control of Changing Flight Directions

The migrations of many birds do not proceed along straight lines but rather show consistent seasonal changes. For instance, garden warblers (*S. borin*), like many other Western European birds, first leave their breeding grounds in a southwesterly direction but after they have reached the Iberian Peninsula or northwestern Africa, their migratory direction shifts to south or southeast. The subsequent homeward migration in spring proceeds along a more direct line from south to north (Fig. 13a).

This changing flight direction is determined by an endogenous circannual mechanism, as is indicated by the results of an experiment with garden warblers. Hand-raised birds were kept for 1 year in a constant 12-h photoperiod, but beginning with the first fall migratory season they were repeatedly tested in orientation cages. These cages were equipped with eight regularly spaced pairs of radial perches connected with eight microswitches (Fig. 14). The direction preferred by the restless bird could thus be monitored during each test night. Figure 13b shows that the warblers initially preferred a southwesterly direction but later in the fall they shifted to the southeast. In spring the birds headed northward. Hence, the garden warblers changed the direction in which their nocturnal activity was concentrated just as free-living birds change their actual flight direction. Since the experimental birds had been kept in a seasonally constant environment throughout, their directional shifts must have resulted from seasonally varying endogenous processes. Comparable results have recently been obtained from a few other species. Taken together, these results indicate that circannual rhythms, in addition to controlling the onset of migrations, their time course, and possibly their distance, are also basically involved in seasonal changes in migratory direction.

Environmental Modification

The results presented previously demonstrated the overwhelming significance of endogenous circannual mechanisms in the control of avian migrations. However, environmental factors also exert dramatic effects on the time course of migration, not only in typical weather birds but also in calendar birds which, as described previously, rely heavily on endogenous components.

Environmental factors can influence the temporal pattern of migration at two levels: (i) at the level of the circannual mechanism, affecting period and phase of the rhythm, and (ii) at the level of the output of the circannual system, modifying activities "downstream" without interfering with the basic clockwork.

Effects of Environmental Stimuli on the Circannual Mechanism. That environmental stimuli directly affect the circannual mechanism can be inferred from the fact that circannual rhythms free-run with a period different from 1 year under constant conditions while assuming a period of exactly 1 year under natural conditions. Hence, there must be periodic environmental stimuli, *zeitgebers,* that are capable of synchronizing the fast- or slow-running circannual oscillations with exactly 1 year. The major circannual zeitgeber is the annual rhythm in photoperiod. This has, for instance, been shown in experiments in which garden warblers were exposed to sinusoidal changes of photoperiod with a duration of either 12 or 6 months (Fig. 15). In each case, the rhythms of migratory restlessness and molt assumed the period of the respective photoperiodic cycle. As a consequence, the birds experiencing the 6-month photoperiodic cycle went through two phases of "fall" and "spring" migratory restlessness per calendar year and molted four times instead of twice.

Photoperiod thus corrects the period of the circannual rhythm in such a way that it coincides with the period of the natural 12-month environmental year. At the same time,

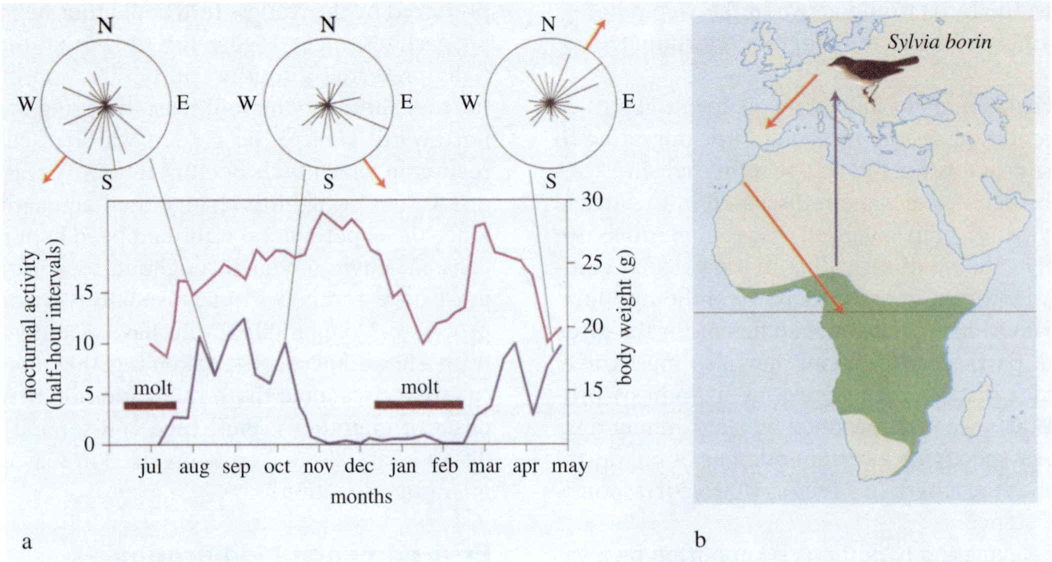

FIGURE 13 (a) Spontaneous seasonal changes of directional preferences in garden warblers (*Sylvia borin*) during nocturnal migratory restlessness. Garden warblers were kept throughout the experiment under a constant 12-h photoperiod and tested repeatedly in circular orientation cages (see Fig. 14). The birds had no view of the sky but were exposed to the natural magnetic field of the earth. The three circular diagrams summarize the results obtained in August and September (left), from October through December (middle), and from April through June (right) of the following year. The vectors are plotted on a relative scale such that the radius equals the greatest amount of activity in any one 15° sector. The red arrows show the directions of the mean vectors calculated for each test series. The graph shows the variations in nocturnal activity (in blue) and in body weight (in violet) as well as the occurrence of molt in a representative individual. (b) Direction changes of free-living garden warblers in their migrations between the nesting sites in Europe and the wintering grounds in central Africa (in green) (reproduced with permission from Gwinner and Wiltschko, 1978, 1980).

photoperiod exerts biologically meaningful effects on the angular velocity of the circannual oscillation, i.e., the rate at which the circannual system runs at certain phases. This can be illustrated with two examples. First, central European garden warblers normally spend the winter close to the equator, but some of the first-year birds may cross it (perhaps because they are carried too far south by their endogenous time program) and migrate into the Southern Hemisphere. This could cause serious problems since a bird that spends the winter too far south will have to travel a longer distance during the following homeward migration than a conspecific wintering further north. If not compensated for, this increase in distance and duration of fall migration would be highly disadvantageous: The bird would arrive in its breeding grounds too late — that is, when good territories are already occupied and when time has become too short for the successful completion of a breeding cycle. A compensation mechanism, however, does exist. It is provided by photoperiod and the way the bird responds to it. As the warbler crosses the equator it moves into the Southern Hemisphere summer with its long days, and the further south the bird gets the longer the photoperiod it perceives. These long days accelerate the bird's circannual cycle (i.e., the fall migratory season becomes truncated, winter molt is accelerated, and spring migratory activity becomes advanced) so that homeward migration starts earlier than in a conspecific wintering further north. In other words, the disadvantage of a southerly wintering bird having to migrate a longer distance than its northerly wintering conspecific is compensated for by an earlier departure time caused by long days.

Second, earlier in fall, the short photoperiods accelerate the system, and this has advantages too. We know that circannual rhythms start to run at hatching (or a date closely correlated with it), but within each population hatching is usually spread over a relatively long period of time. Since fall migration is initiated at a particular phase of the circannual cycle, the onset of fall migration would directly be determined by birthdate. Particularly for birds born very late, toward the end of the breeding season, this could cause problems since they may initiate migration too late (i.e., when conditions have already deteriorated). However, again photoperiod provides for compensation. Birds born late in the year grow up under decreasing photoperiods: The later they were born, the shorter the photoperiod to which they are exposed. Experiments have revealed that short photoperiods accelerate the system during this time of the year — that is, postjuvenile molt begins and ends ear-

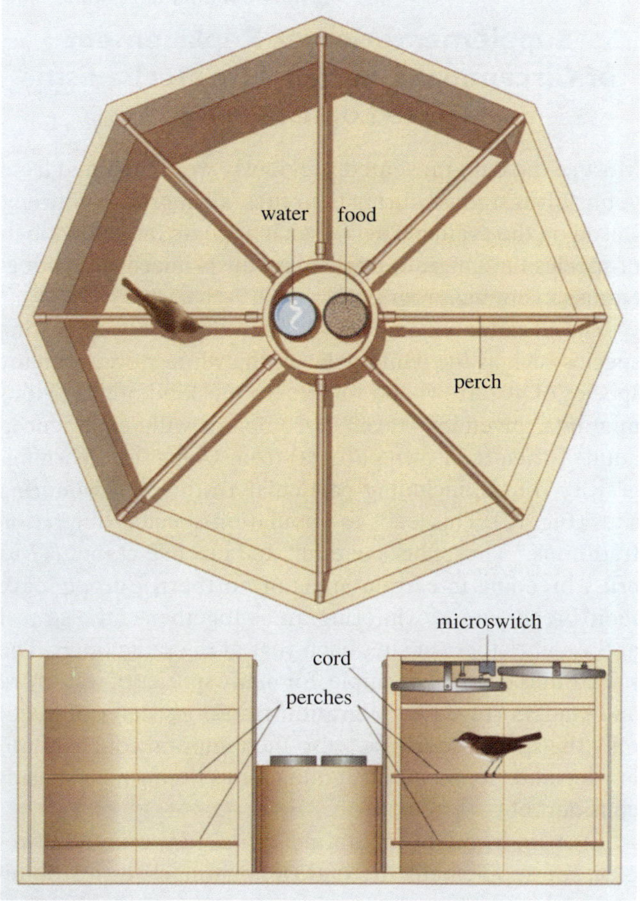

FIGURE 14 Orientation cage used to test directional preferences of small migratory birds at various times of the year. The cage has eight pairs of radial perches (top) and each pair of perches is connected by a string to a microswitch (bottom). When the bird hops on one of the perches, the microswitch closes an electrical circuit and the electrical impulse is stored on a computer. With the help of this apparatus it is possible to determine which part of the cage, and thus which compass direction, the bird prefers (reproduced with permission from Gwinner, 1986b).

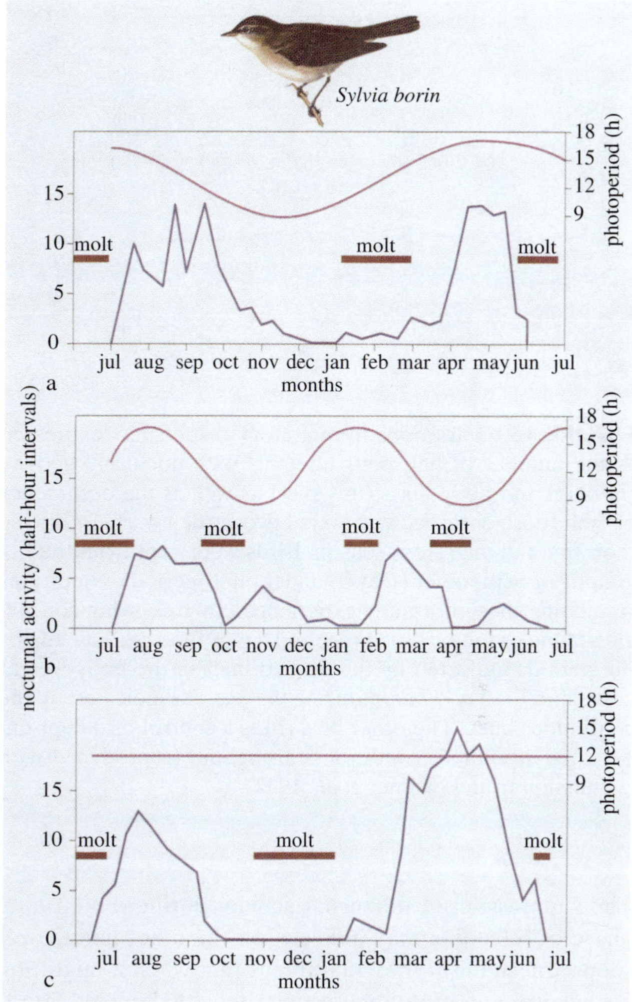

FIGURE 15 Changes in migratory restlessness (in blue) and occurrence of molt (horizontal brown bars) in garden warblers (*Sylvia borin*) kept under the photoperiodic conditions indicated (violet curves) (reproduced with permission from Gwinner, 1986a).

lier and fall migratory activity and fattening are initiated at an earlier age in young from late clutches than in young from early clutches. In other words, although in late fall and winter the circannual system is accelerated by long photoperiods, in late summer and early fall it is accelerated by short photoperiods. In either case, the acceleration serves the important biological function of keeping the birds in step with the season by correcting deviations of their endogenous calendar.

Modifying Effects of Environmental Cues on Clock Output. Although photoperiod influences the temporal pattern of migration at the level of the circannual oscillator by affecting its period, phase, and/or angular velocity, other stimuli only interfere with clock output without modifying the oscillator's pace. This obviously applies to extremely hostile environmental conditions such as headwinds, storms, or heavy rains, which may completely and instantaneously suppress migration. However, there are also more subtle kinds of interferences. A factor capable of either enhancing or reducing migratory activity is food availability. For example, Fig. 16 shows an experiment on a garden warbler that had accumulated large fat stores during its fall migratory season; when the bird was then deprived of food, it showed a slight increase in migratory restlessness. During the subsequent periods of refeeding, however, nocturnal activity was temporarily suppressed. The increased activity during food withdrawal may reflect a mechanism that enhances the migratory drive of a bird that has accidentally landed at a site providing no opportunity to refuel.

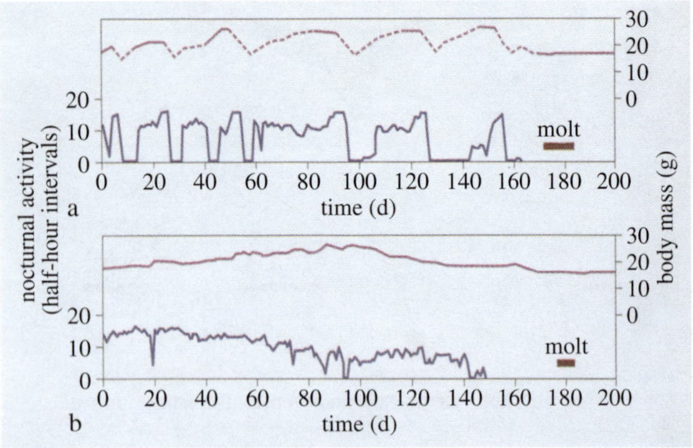

FIGURE 16 Variations in migratory restlessness expressed as the number of half-hour intervals with nocturnal activity (in blue) and body mass (in violet) as well as the occurrence of molt (horizontal brown bars) of two garden warblers during their first fall migratory season. Birds were kept from the beginning of September (Day 0) under photoperiodic conditions simulating those normally experienced by free-living conspecifics. One of the birds (a) received 8 g of food per day during the periods indicated by the dashed lines in the body weight curve and food *ad libitum* during the periods indicated by the continuous lines. The other bird (b) is a control bird kept under food *ad libitum* conditions throughout (reproduced with permission from Gwinner *et al.,* 1992).

The suppression of nocturnal activity during the ensuing phase of refeeding may indicate, on the other hand, a response mechanism that guarantees that when a migrating bird arrives at a profitable resting site with depleted fat reserves, it stays there for days to store up fat.

An experiment on North American red-eyed vireos (*Vireo olivaceus*) caught in southern Alabama in fall, just prior to crossing the Gulf of Mexico, suggests even more subtle effects of energetic conditions on migratory performance. Birds carrying a fat load sufficient for crossing the Gulf were quite likely to show migratory activity the following night and, if tested in orientation cages, preferred the expected southerly direction. Lean birds, in contrast, were less likely to embark on migration, and when they did they chose seasonally inappropriate directions (Sandberg and Moore, 1996).

Nutritional conditions and energetic state of the kind described previously almost certainly have no effect on the circannual mechanism. This is strongly suggested by the results of a study of garden warblers in which they were subjected to repeated food withdrawal and refeeding during the fall migratory season (Fig. 16). Although these treatments resulted in an overall reduction of migratory restlessness by about 20%, the birds so treated initiated winter molt, a reliable marker of circannual phase, at the same time as the untreated control birds.

Supplementation or Replacement of Circannual Programs by Mechanisms Based on Learning

Most of the data presented previously were obtained from young birds studied during their first fall migratory journey. Based on the available evidence, it appears that individuals of species that migrate singly may rely primarily on endogenous circannual programs.

The situation is somehow different in socially migrating species such as the white pelican, the white stork, or many species of waterfowl and waders. In this kind of gregarious migrants, juveniles travel in company with experienced adults (often their own parents) from which they learn migratory routes, including particular resting and wintering sites (Fig. 17). This leads to population-specific "migration traditions." Thus, flocks of adult and juvenile cranes (*Grus grus*) breeding in certain areas of northern Europe leave their breeding and wintering areas together and also rest at the same stopover sites each year at the same times. The same holds true, for example, for many species of waterfowl and waders (Fig. 17). Migration traditions determine not only the spatial course but also the temporal course of migration and, hence, modify endogenous timing mechanisms considerably.

Even individually migrating birds may eventually replace or supplement the endogenous mechanisms dominating migration control in first-year migrants with mechanisms that are based on learning. This is indicated, for example, by the fact that migrants may show a high site fidelity to both breeding and wintering grounds. Garden warblers, for instance, have been recaptured at the same locations in their breeding and wintering grounds during successive years. Individual birds of other species have even been found resting at the same sites along their migratory route more than once. Although this behavior suggests spatial learning, other studies indicate that the underlying learning process may not be restricted to memorizing landmarks. In his experiment with European starlings, Perdeck captured birds that were on their fall migration from their Baltic breeding area to their French and British wintering grounds and displaced them from The Netherlands to Switzerland. The first-year birds continued to travel in their original direction, a finding consistent with the hypothesis that they use a compass. In contrast, adult birds (i.e., birds that had traveled to their winter quarters at least once before) compensated for their displacement and headed toward the traditional wintering area of the population. Although these birds had never been in Switzerland before, they were able to determine, upon release, their position relative to their target area: They had somehow learned to navigate.

It is unknown what and how migratory birds learn, but

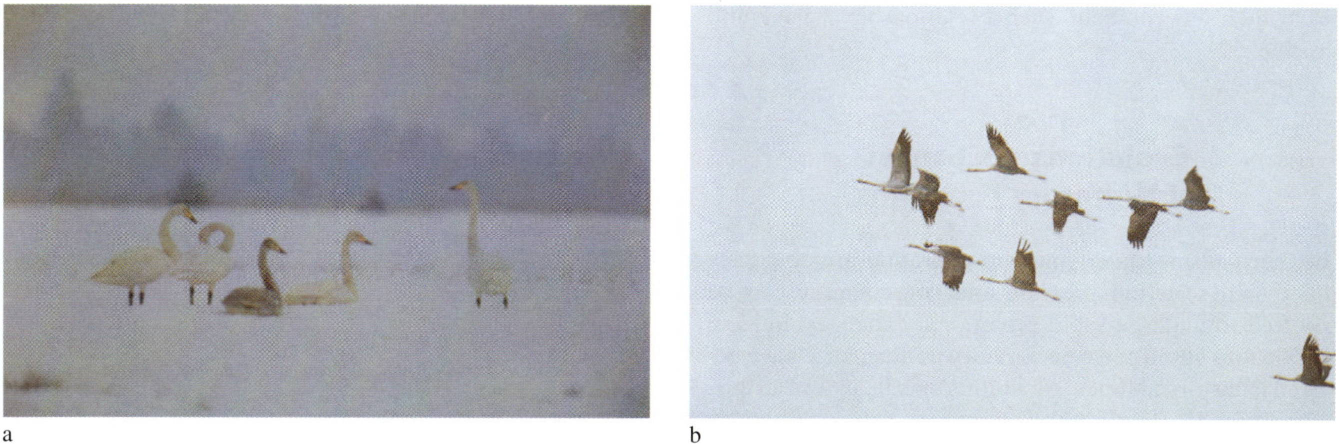

FIGURE 17 In the whooper swan, *Cygnus cygnus* (left), as well as in many other species of waterfowl and the crane *Grus grus* (right), young birds travel in the company of adults, thereby learning details of their migratory journey. This may result in "migration traditions," i.e., birds of a particular population travel along the same specific migratory route (photo courtesy of A. Limbrunner and T. Rödl).

it is likely that they learn much and that they learn fast. Recent data suggest that in connection with their spatial learning performance there may even be changes in brain anatomy. Experienced garden warblers have been found to have a larger hippocampus with more neurons than young conspecifics just prior to their first fall migration. In the nonmigratory Sardinian warbler (*Sylvia melanocephala*) such differences were not found (Healey *et al.,* 1996). It remains to be rigorously established whether the development of the hippocampus of the garden warbler is indeed related to its migratory behavior and, specifically, to learning processes associated with it. However, in view of the fact that hippocampal increases have been described for other birds — particularly food-storing species — that learn spatial proper-

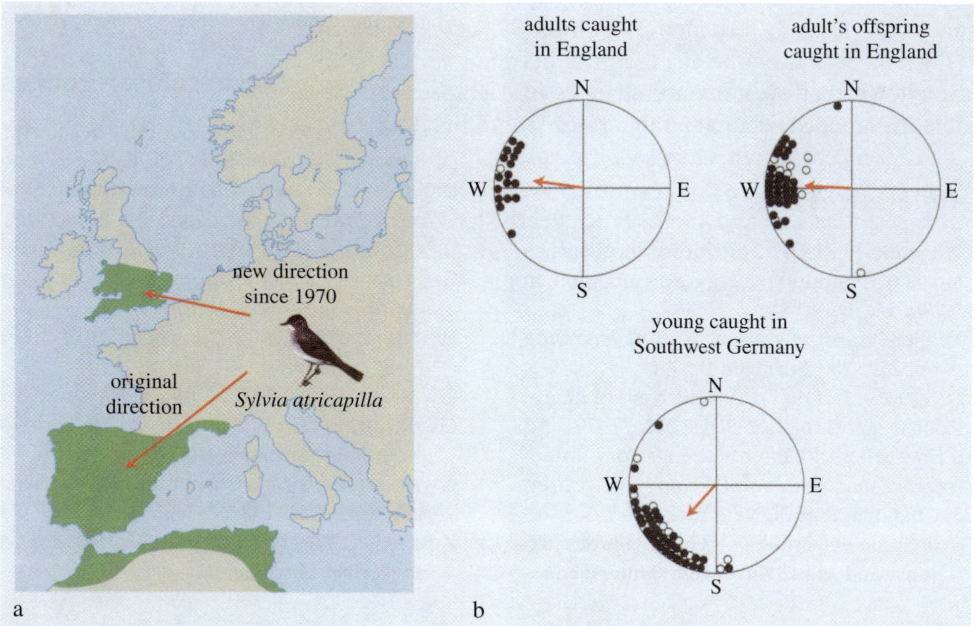

FIGURE 18 (a) Traditional and new migratory direction of central European blackcaps (*Sylvia atricapilla*) wintering (green areas) either in southwestern Europe or in England. (b) Orientation of blackcaps caught in winter in Britain (top left), their F_1 offspring (top right), and a control group from southwest Germany (bottom). Each symbol shows the mean direction of one individual during 15–20 tests. ●, Statistically significant mean vectors; ○, nonsignificant mean vectors. The red arrows show the mean vectors of each group (reproduced with permission from Helbig, 1994).

ties of their environment, such a relationship seems quite possible.

Evolutionary Changes of Migratory Programs

The mechanisms underlying bird migration are obviously under high selection pressure and consequently can be expected to undergo rapid evolutionary changes. Indeed, conspicuous alterations in migratory performance have occurred in historical times. As a rather spectacular case, during the past 30 years ornithologists have witnessed the establishment of a new wintering area in England by central European blackcaps. Originally, all birds of this population migrated in a southerly direction to wintering grounds in the western Mediterranean; however, starting in 1970, an increasing number of blackcaps ringed during the breeding season in Belgium, southern Germany, and eastern Austria have been recovered in a west to northwest direction, some in England and Ireland. This new migratory direction has a genetic basis. Blackcaps caught in England and tested in Germany during their subsequent fall migration preferred a northwesterly direction. The same was true for offspring of British winterers bred in aviaries, whereas simultaneously tested offspring of "normal" German birds preferred the traditional southwesterly direction. Finally, F_1 hybrids bred from parents with southwesterly and northwesterly directional preferences, respectively, oriented in an intermediate direction (Fig. 18). Selective breeding experiments with blackcaps have shown that evolutionary changes in migratory parameters may indeed occur at a rapid rate.

References Cited

Bairlein, F., and Gwinner, E. (1994). Nutritional mechanisms and temporal control of migratory energy accumulation in birds. *Annu. Rev. Nutr.* **14,** 187–215.

Berthold, P. (1990). Genetics of migration. In *Bird Migration* (E. Gwinner, Ed.), pp. 269–280. Springer, Berlin.

Berthold, P., and Querner, U. (1981). Genetic basis of migratory behaviour in European warblers. *Science* **212,** 77–79.

Biebach, H. (1996). Energetics of winter and migratory fattening. In *Avian Energetics and Nutritional Ecology* (C. Carey, Ed.), pp. 280–323. Chapman & Hall, New York.

Gwinner, E. (1968). Circannuale Periodik als Grundlage des jahreszeitlichen Funktionswandels bei Zugvögeln. Untersuchungen am Fitis (Phylloscopus trochilus) und am Waldlaubsänger (P. sibilatrix). *J. Ornithol.* **109,** 70–95.

Gwinner, E. (1986a). Circannual rhythms in the control of avian migrations. *Adv. Stud. Behav.* **16,** 191–228.

Gwinner, E. (1986b). Internal rhythms in bird migration. *Sci. Am.* **254,** 84–93.

Gwinner, E. (1996). Circannual clock in avian reproduction and migration. *Ibis* **138,** 47–63.

Gwinner, E., and Wiltschko, W. (1978). Endogenously controlled changes in migratory direction of the garden warbler, *Sylvia borin. J. Comp. Physiol. A* **125,** 267–273.

Gwinner, E., and Wiltschko, W. (1980). Circannual changes in migratory orientation of the garden warbler, *Sylvia borin. Behav. Ecol. Sociobiol.* **7,** 73–78.

Gwinner, E., Schwabl, H., and Schwabl-Benzinger, I. (1992). The migratory time program of the garden warbler: Is there compensation for interruptions. *Ornis Scand.* **23,** 264–270.

Healy, S. D., Gwinner, E., and Krebs, J. R. (1996). Hippocampal volume in migratory and non-migratory warblers: Effects of age and experience. *Behav. Brain Res.* **81,** 61–68.

Helbig, A. J. (1994). Genetic basis and evolutionary change of migratory directions in a European migrant *Sylvia atricapilla. Ostrich* **65,** 157–159.

Piersma, T., and Lindström, A. (1997). Rapid reversible changes in organ size as a component of adaptive behaviour. *Trends Ecol. Evol.* **4,** 134–138.

Rowan, W. (1926). On photoperiodism, reproductive periodicity, and the annual migrations of birds and certain fishes. *Proc. Boston Soc. Nat. Hist.* **38,** 147–189.

Sandberg, R., and Moore, R. (1996). Migratory orientation of red-eyed vireos, *Vireo olivaceus,* in relation to energetic condition and ecological context. *Behav. Ecol. Sociobiol.* **39,** 1–10.

General References

Aschoff, J. (1955). Jahresperiodik der Fortpflanzung bei Warmblütern. *Stud. Gen.* **8,** 742–776.

Berthold, P. (1988). The control of migration in European warblers. *Proc. Int. Orn. Congr.* **19,** 215–249.

Berthold, P. (1991). *Orientation in Birds.* Birkhäuser, Basel.

Berthold, P. (2000). *Bird Migration. A General Survey,* 2nd ed. Oxford Univ. Press, Oxford.

Berthold, P. (1996). *Control of Bird Migration.* Chapman & Hall, London.

Gwinner, E. (1986). *Circannual Rhythms.* Springer, Heidelberg.

Gwinner, E. (Ed.) (1990). *Bird Migration. Physiology and Ecophysiology.* Springer, Berlin.

Gwinner, E. (1996). Circannual clock in avian reproduction and migration. *Ibis* **138,** 47–63.

Zink, G. (1973–1985). *Der Zug Europäischer Singvögel.* Vogelzugverlag, Möggingen.

MONTSERRAT GOMENDIO

Department of Evolutionary Ecology
Museo Nacional de Ciencias Naturales
Consejo Superior de Investigaciones Científicas
28006 Madrid, Spain

Sperm Competition

When females mate with several males, rival sperm compete to fertilize the available ova and females may be able to exercise some choice within the reproductive tract. Differences in female reproductive tract morphology and physiology determine which ejaculate features are important for the success of males. Knowledge of the physiological mechanisms involved is essential in order to understand how male and female strategies translate into reproductive success. Sperm competition has selected for higher sperm production, more efficient ejaculates, higher copulatory rates, and mate guarding among males in polyandrous species. There may also be differences in ejaculate investment within populations, depending on the levels of sperm competition faced by males following different reproductive tactics. Females may mate with several males in order to avoid the costs of rejecting males or to obtain benefits which could be direct or indirect. Because males and females have different ways of maximizing their reproductive success, and what is beneficial for one sex may be detrimental to the other, there is sexual conflict. There may be conflict about whether there is sperm competition or not and also about the detrimental effects of male adaptations to sperm competition on females. Mating systems are the outcome of male and female mating strategies.

What Is Sperm Competition?

Darwin developed the concept of *sexual selection* to explain the evolution of conspicuous male traits, such as ornaments and weapons, which could not be explained by natural selection for improved survival. Conspicuous traits of this kind confer advantages to males in terms of competition over mates, either because they improve the ability to win in male–male agonistic encounters or because they make males attractive to females. Thus, sexual selection involves two different processes: male–male competition and female choice. Traditionally, behavioral ecologists have studied sexual selection under the assumption that mating success was equivalent to reproductive success. Thus, it was believed that males compete in order to copulate with as many females as possible, and females were perceived as being able to chose only at the behavioral level which males could copulate with them. Copulation was viewed as the end point for both processes. This view was dramatically altered after it became apparent that females of many species mate with more than one male during each sexual cycle. This fact has important implications because it opens up the possibility that both male–male competition and female choice continue beyond copulation at a physiological level, and it means that sexual access to females does not ensure fertilization. When females copulate with multiple males, ejaculates from rival males will compete to fertilize the available ova, and females may exercise some choice within the reproductive tract as to which sperm will achieve fertilization (Fig. 1). Strictly speaking, *sperm competition* refers

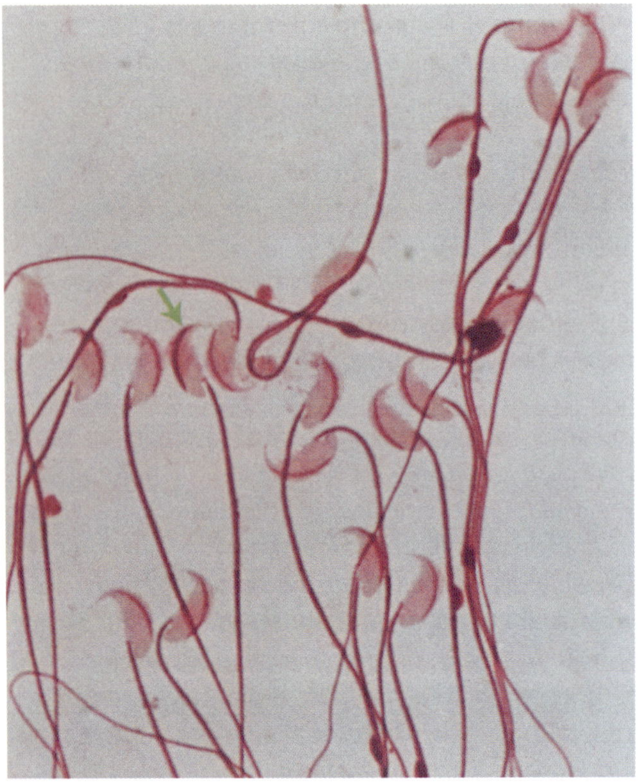

FIGURE 1 Rodent spermatozoa. In the spermatozoon the acrosome corresponds to the darker-colored zone (arrow). The spermatozoon at the far left of the photograph lacks an acrosome; thus, it cannot penetrate the ova vestments and is unable to fertilize (courtesy of E. R. S. Roldan).

only to competition between rival ejaculates, reflecting the emphasis on male adaptations. However, more attention is beginning to be given to the female perspective.

Sperm competition seems to be widespread among animals (Smith, 1984), as is "pollen competition" among plants. Female insects store sperm in specialized organs known as *spermathecae* in which spermatozoa remain viable for prolonged periods of time (up to years), making sperm competition possible even if the female remates at long intervals (Parker, 1970). For these reasons, insects were for some time the focus of most sperm competition studies. Birds were thought to be unlikely candidates for sperm competition since most species are socially monogamous, with males and females forming pairs which cooperate in rearing the chicks. However, this view has changed completely since the use of molecular techniques to determine paternity (such as DNA fingerprinting) has revealed that both males and females are often sexually unfaithful and engage in so-called extra-pair copulations (EPCs), which result in the production of multiply sired broods (Birkhead and Møller, 1992). These findings have revolutionized the study of avian mating systems. Among mammals the expectation was also that sperm competition should be rare because in most species females do not possess specialized sperm storage organs and, as a consequence, spermatozoa are short-lived. However, behavioral observations suggest that in many mammalian species copulations by rival males are close enough in time for rival sperm to compete, and even among some of the species in which observational evidence suggested monandry, this is certainly not the case since females produce multiply sired litters or, in species with singletons, offspring sired by males other than the male observed copulating with the female at approximately the time of ovulation (Gomendio *et al.,* 1998).

The remarkable mismatch between expectations about levels of sperm competition based on behavioral observations and patterns of genetic paternity is partly a consequence of our ignorance about underlying physiological mechanisms. We need to elucidate these mechanisms in order to understand how mating patterns translate into realized paternity.

Mechanisms of Sperm Competition

The physiological mechanisms underlying sperm competition differ greatly between taxa because the arena in which rival ejaculates must compete (i.e., the female tract) shows a striking degree of variation. A comparison of birds (Birkhead, 1996) and mammals (Gomendio *et al.,* 1998) will illustrate this diversity.

Female birds possess sperm storage tubules located at the uterovaginal junction (Fig. 2). Only a small fraction of the ejaculated sperm enter the tubules (approximately 0.1%); thus, there is a drastic reduction of sperm numbers in the vagina. By storing sperm in the tubules, females ensure an adequate reserve of sperm which lasts for a period of several days during which females ovulate, fertilize, and lay one egg after another. An interspecific comparison has shown that the duration of sperm storage is positively correlated with the spread of laying. Sperm storage tubules may also protect sperm from being pushed out of the female reproductive tract each time an egg is laid. Rival males seem to compete to have a greater proportion of their sperm filling the sperm storage tubules. For some time, it has been argued that sperm compete within the sperm storage tubules either by stratification (a last-in/first-out system) or by sperm displacement (the last ejaculate removing part of the previous ejaculate from the tubule). However, recent evidence suggests that after entering the sperm storage tubules, sperm is passively lost over time. Thus, as the period of time since the last copulation increases, fewer sperm are present in the sperm storage tubules. This explains why when two males copulate more than 4 h apart, the last male to copulate sires most of the offspring since by that time

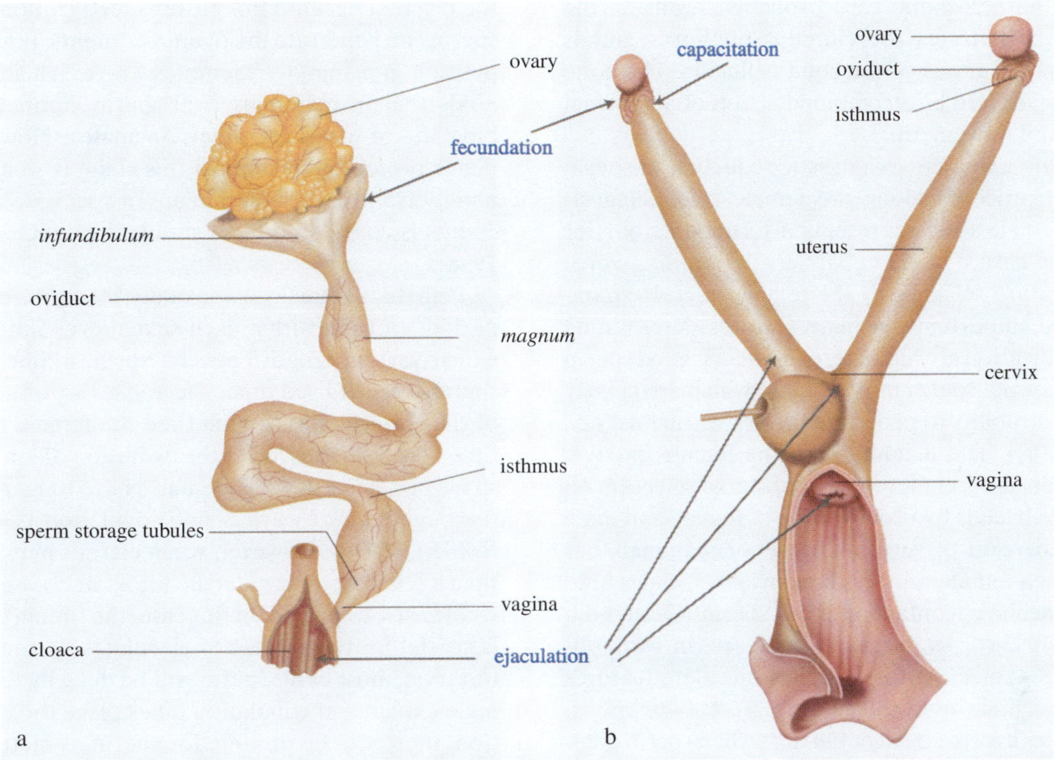

FIGURE 2 Schematic representation of the female reproductive tracts of birds (a) and eutherian mammals (b) (adapted from Gomendio and Roldan, 1993b).

enough sperm has been lost from the first male to ensure a predominance of sperm from the second male in the sperm storage tubules. In order to explain the high success of EPCs, it is also necessary to understand how ejaculates are produced and stored by the male. Because of the way sperm are stored in the male reproductive organs, a rested male will produce an ejaculate with more sperm and with faster-moving sperm than that of a recently mated male. The disproportionate success of EPCs can be explained by taking into account two factors. First, females cease to accept copulations from the pair male at approximately the time when egg laying starts; thus, as egg laying proceeds, sperm from the pair male are gradually lost from the sperm storage tubules. Second, males tend to engage in EPCs only after their own partner's fertile period has ended. Thus, a male performing an EPC will have experienced a period of low copulation frequency (during egg laying) and will thus produce a highly competitive ejaculate, which is likely to outcompete what remains of the partner's previous ejaculations.

In contrast, female mammals do not possess specialized sperm storage organs and as a consequence sperm remain fertile only for about 48 h once ejaculated in the female tract, whereas ova remain fertile for only about 24 h after ovulation. Female mammals may have no need to store sperm because, within each sexual cycle, all the ova are released and fertilized simultaneously shortly after ovulation. Ejaculated sperm are passively transported toward the upper sections of the female tract by muscular movements of the tract and by the activity of the cilia that line the lumina. At later stages, sperm must swim to go through physical barriers such as the uterotubal junction and to cover the last sections of the female tract. After going through the uterotubal junction, sperm in the lower isthmus of the oviduct attach to the mucosa of the oviductal wall, and those sperm that fail to attach die or lose their fertilizing capacity. At approximately the time of ovulation, a small fraction of the sperm present in the isthmus complete capacitation (a maturational process that enables sperm to fertilize) and detach from the wall. At this stage, spermatozoa acquire a peculiar type of motility known as *hyperactivation,* which is an efficient way of propulsion within the viscous fluids encountered in this section of the tract. The first spermatozoon to reach the vicinity of the ovum seems to be the one most likely to fertilize. Contrary to widespread perception, the ratio of sperm to ova *in vivo* is approximately 1:1. When the spermatozoon reaches the ovum, and after it has become capacitated, it starts to penetrate the cumulus oophorus. The acrosome reaction (i.e., the release of enzymes contained in the acrosomal cap) probably begins as the spermatozoon penetrates the cumulus oophorus and is completed on the surface of the zona pellucida. The zona pellucida is penetrated by a combined action of acrosomal enzymes and tail movement.

The ejaculate experiences a drastic reduction in sperm numbers as it proceeds along the female tract, suggesting that the female tract represents a formidable barrier for sperm. The nature of these barriers determines which sperm features are selected by the female tract. In mammals, the most important ejaculate features determining success at fertilization are high sperm numbers, good sperm motility and normal sperm morphology (which are closely related), and the ability to penetrate the ova vestments (i.e., acrosomal status). The magnitude of the female barriers likely represents different levels of intensity of selection on ejaculates, which leads to a coevolution of female and male traits. A comparison of several species of mammals has shown that when females possess long oviducts, males tend to produce ejaculates containing more sperm (Gomendio and Roldan, 1993a). As discussed later, sperm competition exacerbates selection for the same ejaculate features for which the female tract selects (Fig. 3). It is unknown whether females evolve stronger barriers when sperm competition prevails in order to enhance competition between ejaculates or to facilitate discrimination between them. For example, do females possess longer oviducts when there is sperm competition? There is no available evidence to test this hypothesis.

In mammals, rival sperm seem to compete to reach the ova first because this ensures fertilization provided the sperm can penetrate the ovum vestments; thus, sperm competition in mammals resembles a race. An alternative view holds that in mammals rival sperm compete directly by blocking or attacking other ejaculates. However, the evidence presented to support this claim is weak and derives mainly from studies on humans, a species which by all accounts is considered to be monandrous (Gomendio *et al.*, 1998).

Female mammals are sexually receptive during limited periods of time within each sexual cycle known as *estrus,* which vary in length between species. Since mammalian sperm is short-lived, males face the risk of not having any of their sperm alive by the time the female ovulates. This time constraint has led to the evolution of longer sperm life spans in species in which females have longer periods from the beginning of estrous until ovulation (Gomendio and Roldan, 1993a). However, when estrous periods are longer than a few days, sperm fertile life span does not last for the whole estrous period. In this case, the timing of copulations is crucial for the success of ejaculates. If a male copulates too soon, most of his sperm will be dead by the time the female ovulates. If copulation takes place too close to ovulation, there will be no time for sperm to mature within the female tract and they will be unable to fertilize when ovulation occurs. Finally, copulations taking place after ovulation have no chance of fertilization. Because synchronizing copulation and ovulation is very important, there are no order effects in mammals. The male that copulates at the right time will have a clear advantage, and if several males

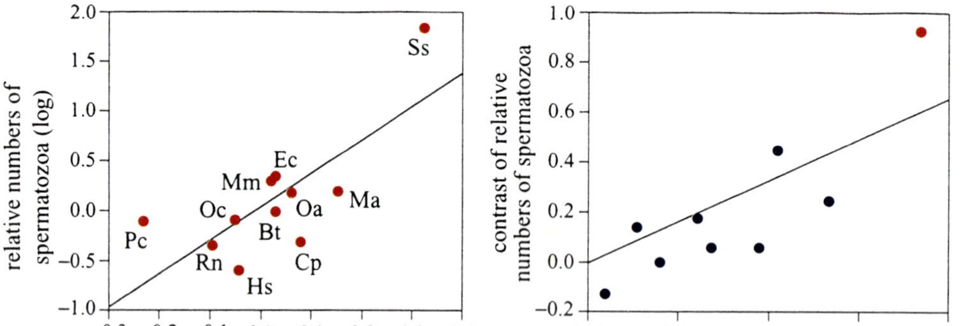

FIGURE 3 Relationship between relative oviductal length and the relative number of spermatozoa per ejaculate. Oviductal length and the number of spermatozoa were both corrected for body size. (a) Comparison of the values for various species: Pc, *Papyo cynocephalus;* Rn, *Rattus norvegicus;* Hs, *Homo sapiens;* Oc, *Oryctolagus cuniculus;* Bt, *Bos taurus;* Cp, *Cavia porcellus;* Mm, *Mus musculus;* Ec, *Equus caballus;* Oa, *Ovis aries;* Ma, *Mesocricetus auratus;* Ss, *Sus scrofa.* (b) Independent contrasts for each character. Contrasts produce comparative data statistically independent of phylogenesis for each branch of the phylogenetic tree characterizing the species. The regression line was constrained to pass through the origin (adapted from Gomendio and Roldan, 1993a).

copulate at the right time the one able to transfer a more competitive ejaculate is likely to achieve higher fertilization success.

The Male

Parker (1970, 1984) suggested that sperm competition resembles a raffle, with the male holding the greater number of tickets (i.e., sperm transferred to the female tract) having the greatest chance of winning. Selection for increased sperm production under sperm competition has been found in many taxa, including insects, amphibians, birds, and mammals. Thus, the increase in sperm numbers under sperm competition seems to have been favored in taxa with different underlying mechanisms, suggesting that its value is almost universal. Increased sperm production could result in ejaculates containing more sperm or in males being able to ejaculate more often. In both birds and mammals, males in species in which sperm competition is common have larger testes in relation to their body size, tend to copulate more often, and produce ejaculates with greater numbers of sperm (Birkhead and Møller, 1992; Gomendio *et al.*, 1998).

Sperm competition seems to have selected not only for greater quantities of sperm but also for an improvement of the quality of sperm. Thus, males in polyandrous primates (females mate with more than one male in each sexual cycle) produce more motile sperm per ejaculate than do males in monandrous primates (Møller, 1988). In addition, sperm competition should select for sperm which are able to outcompete rival sperm. Gomendio and Roldan (1991) suggested that, among mammals, sperm competition should select for longer sperm because these swim faster and will be more likely to win the race to fertilize the ova. Their comparative analysis shows that polyandrous species have longer sperm than monandrous species in both primates and rodents. Later, the same relationship between female mating patterns and sperm length was demonstrated in butterflies and birds. Whether the advantage of producing longer sperm under sperm competition is faster swimming speed for all these taxa is still unknown.

There are taxa, such as butterflies, in which males transfer nutrients to the female which are used for egg production and somatic maintenance and thus increase female fecundity and longevity. In these cases, polyandrous males also transfer greater amounts of nutrients than do monandrous males (Bissondath and Wiklund, 1995).

When the risk of sperm competition is high, males may try to improve their chances of fertilizing ova by mate guarding the female, and in this way either prevent mating by other males or at least delay any further matings. Mate guarding may take several forms, from aggressively excluding other males to prolonged copulations which physically prevent other males from copulating or the production of copulatory plugs. Depending on whether there is a first-male advantage, last-male advantage, or no order effects, it will pay males to guard females at different times. When there is a first-male advantage, males will compete to be able to locate females first. This has led to the evolution of more elaborate sensory organs in males than in females in many taxa. In 13-lined ground squirrels (*Spermophilus tridecemlineatus*) ovulation is induced by copulation and there is a first-male advantage, and consequently males do not guard after copulation. Therefore, the benefits of locating an estrus female before other males do are high, and this has led to very high mobility rates among males during the mating season and to complex cognitive abilities which allow males to anticipate the onset of estrus in females from one day to the next (Schwagmeyer, 1995). Whether or not males guard females will depend not only on the risk of sperm competition but also on the costs associated with the loss of opportunities to mate with other females. By comparing two species of squirrels, Sherman (1989) found that when other receptive females are difficult to find, males try to ensure fertilization of the females they have already mated with by mate guarding, whereas when other receptive females are readily available it is not beneficial for males to stay.

Recent theoretical models suggest that there are two contexts in which males within the same population should differ in the extent to which they invest in ejaculates (Parker, 1990a,b). First, if males are assigned favored and disfavored roles in a nonrandom fashion, the male in the disfavored role should invest more in ejaculates. Second, if there are sneakers and guarders, and sneak matings are rare, sneakers should invest more in ejaculates and, when guarders are able to detect sneak matings, they should increase the number of sperm they deliver into the female tract. In externally fertilizing fish the opportunities for sneak matings are obvious. In some species, such as the Atlantic salmon (*Salmo salar*), males follow two alternative strategies: They either mature as large anodromous males which dominate spawning sites and defend females or they mature as tiny precocious parr which gain fertilizations by sneaking in while anadromous males and females spawn. Supporting the previous predictions, parr males have a larger gonosomatic index, they produce a relatively greater number of sperm, a greater proportion of sperm are motile, and their sperm live longer (Gage *et al.*, 1995). In other species, males follow different strategies at different stages of their development. In coral reef fish (*Thalassoma bifasciatum*), adult males occupy spawning sites which they defend, whereas younger males either add their sperm to the matings of other males or take over mating sites and mate in groups. Thus, young males face higher levels of sperm competition than older territorial males, and as a consequence they have relatively larger testes. This difference is reversible since eventually young males grow older and become territorial, at which point their testes become reduced in size and total sperm output decreases. They shift from investing in ejaculates to investing in mate guarding (Warner *et al.,* 1995). In a population of shrews (*Sorex araneus*) two alternative strategies were observed: Some males established large, overlapping ranges in areas of high female density, inseminated more females, and fathered more offspring, whereas other males occupied small ranges away from females and had to make long-distance movements during the breeding season to visit female ranges; these males inseminated fewer females and produced fewer offspring. Examination of the testes revealed that, despite similar testes size, the latter type of males had more sperm reserves. This result has been interpreted as implying that males mating in disfavored roles are able to produce more sperm (Stockley *et al.,* 1994). However, this seems unlikely given the lack of differences in testes size. It seems more parsimonious to argue that the latter type of males had more sperm reserves because they had copulated less frequently through the breeding season. Nevertheless, this suggests an interesting mechanism: Males copulating in disfavored roles may have more sperm to spare if they also copulate less frequently; as a consequence, they may ejaculate more sperm when they have an opportunity to copulate. This could be the case for young or subordinate males, which rarely achieve copulations. As Parker predicts, there would be differences in the quality of the ejaculates, but these would not imply differences in sperm production by males occupying favored and disfavored roles.

A variation on Parker's hypothesis is that an individual male should be able to adjust ejaculate quality to the risk of sperm competition. Birds provide an interesting example since the same male may face a low risk of sperm competition when copulating with his pair female but a high risk of sperm competition when performing an EPC. To date, the available evidence suggests that males are not physiologically capable of modifying ejaculate contents (Birkehead 1996). However, as described previously, the fact that males tend to engage in EPCs after their partner's egg-laying period means that after a period of rest their ejaculates contain more and faster sperm. Whether the timing of EPCs by males is a result of priorities being given to mate guarding during egg laying or because otherwise EPCs would not be successful is not clear. Thus, even if males do not have the physiological ability to modify ejaculates, by manipulating the timing of copulations they may achieve the same goal.

The Female

Adopting the female's perspective entails, first, asking why should females copulate with more than one male. Mating is costly to females because they have to devote time and energy which could otherwise be spent on other activities; copulating females tend to be more easily detected by predators and are less mobile, so they find it more difficult to escape if detected; and reduced mobility may also lead to decreased foraging efficiency. Furthermore, close contact with males implies the risk of parasite transmission as well as the risk of contracting venereal diseases. Sexually receptive females may also be the target of sexual harassment by males and may be injured if they resist mating attempts or if males fight for sexual access to females.

When females are guarded by a male partner, mating with other males may have additional risks if the female is detected since the female's partner may respond by withdrawing or decreasing paternal care or by punishing the female. Polyandry may also have consequences for the female's offspring that may reduce her fitness, such as an increases in sib–sib conflict among multiply sired litters.

Given that the magnitude of the costs of mating with several males seems considerable, there are two scenarios which could explain the evolution of mating with multiple males by females. First, females do not benefit from mating with multiple males but incur high costs if they refuse to do so. These costs would be mainly the result of physical injury by harassing males. In many social primates that live in multimale groups, males attack females that reject their mating attempts. These attacks may cause serious injuries and may even lead to the death of the female. It is also common for males to respond to females' rejection by persistent courtship, which may also be dangerous for females, particularly in dimorphic species such as some marine mammals in which females may be crushed by courting males. In water striders, females seem to mate more frequently than is in their own interests because rejecting males in the face of repeated harassment may be even more costly than mating because it implies struggling to disengage males and because females have to move away from prime foraging areas to avoid harassing males (Rowe *et al.,* 1994). This female behavior has been termed *convenience polyandry.* Although there are contexts in which sexual harassment and punishment by males undoubtedly determine to a great extent the levels of female polyandry, they are unlikely to account for all cases.

Second, females do benefit from mating with multiple males and the benefits outweigh the costs. Although for some time the prevailing view was that females are "forced to accept" copulations by several males (hypothesis 1), accumulating evidence has revealed that females often solicit copulations by several males or actively search for other sexual partners. These observations suggest that in these cases, females must benefit from multiple mating. For females, there are direct and indirect benefits of mating with several males.

Direct Benefits

Male-Derived Nutrients. In butterflies males transfer a spermatophore which contains nutrients used by females for egg production and somatic maintenance. In these circumstances, females benefit from multiple mating because in this way they obtain more nutrients, resulting in increased reproductive output and increased longevity (Bissondath and Wiklund, 1995). Because females are in short supply of some of the male-derived nutrients, females are believed to actively forage for matings in order to obtain these limiting nutrients. The optimum number of matings for a female depends partly on the quality of the spermatophores she receives. To produce spermatophores is costly for males; thus, recently mated males can only deliver small ejaculates. Females compensate by increasing the number of matings after they copulate with a recently mated male.

Paternal Care. Females may mate with several males in order to increase the number of potential fathers and thus obtain more help in rearing the offspring. A comparison of two populations of alpine accentors and dunnocks illustrates this point (Davies *et al.,* 1996). Both species breed in groups in which there are two or more unrelated males and two or more females, and in which there is an alpha male that attempts to monopolize the females in order to sire most or all of the offspring. However, females resist alpha males' attempts to mate guard because it is in their interest to mate with other males in order to obtain their help in rearing the chicks. Other males will not help unless they are likely to have fathered at least some of the chicks. Females are very active at soliciting copulations from both the alpha and the beta males, and this may result in extremely high copulation rates, sometimes more than 1000 times per clutch (and on average approximately 100 to 200 times per clutch). These high rates of copulation are obviously far beyond what females would need to ensure fertilization of all the ova. The pattern of female solicitation in both species suggests that females are attempting to maximize male help with chick feeding. Females tend to solicit copulations from males with whom they have copulated fewer times. Among alpine accentors living in the French Pyrénées, alpha males reduce the amount of care with a decreased mating share, whereas beta males do not. Accordingly, in this population females prefer to give more matings to the alpha male. In contrast, alpha and beta male dunnocks living in the Cambridge Botanic Garden do not reduce their help, provided a critical share of matings is achieved. In this population, females prefer alpha and beta males equally and seem to attempt to give an equal share of matings to both males. It is likely that male alpine accentors demand greater paternity assurances in order to care for the offspring because this population lives in larger groups and therefore the alpha male is likely to have the chance of mating with other females, an opportunity not worth sacrificing unless paternity of the chicks is certain.

In other contexts, females may mate with several males not to obtain paternal care but to reduce the chances that their offspring will be harmed or killed by these males. In primates living in multimale groups, it has been suggested that females may copulate with several males to diminish the risk of infanticide.

Assurance of fertilization. Females copulate with more than one male to avoid the possibility of not being fertilized if they copulate with sterile males. Currently, we lack sufficient information on the prevalence of male sterility to test this hypothesis rigorously, but experimental evidence obtained with vasectomized zebra finches does not provide support for this hypothesis (Birkhead, 1996). There is also no evidence that more attractive males have better ejaculates, which could imply direct fertility benefits for females.

Female–Female Competition It has also been suggested that females may copulate with multiple males to monopolize sexual access to males (and their sperm) and prevent other females from being fertilized or from being fertilized by high-quality males. It is unclear whether this form of spiteful behavior could evolve.

Indirect Benefits

Increased Genetic Diversity of Offspring. In some eusocial insects, such as the desert leaf-cutter ant (*Acromyrmex versicolor*), polyandrous matings are common. In this species, females seem to have complete control over matings and mate on average with three males, and all males produce a similar number of offspring (Reichart and Wheeler, 1996). For social insect queens producing genetically diverse offspring could result in an enhanced division of labor, a colony sex allocation more in agreement with the queen's interests, increased resistance against pathogens and parasites, and reduced costs of diploid male production.

Genetic Bet Hedging. Females mate with several males to reduce the negative effects of occasional mating with inferior males. It has been suggested that mating behavior in female sierra dome spiders (*Linyphia litigiosa*) can be explained as genetic bet hedging (Watson, 1991). The female's first sexual partner tends to be a male which has fought other males off. Thus, females use male–male competition as the main filter for their first sexual partners. In contrast, secondary mates are apparently chosen by females and are not required to fight other males. Females could be remating to compensate the low uncertainty of having mated with a low-quality male, for example, if by chance the first male to find a female is a low-quality male. Because there is first-male sperm precedence, the first male (which has passed the stringent tests of intrasexual competition) is likely to sire most of the offspring.

Enhance the Genetic Quality of Offspring. Females may initially accept genetically inferior males if it is costly to look for males or because there is a risk of failing to mate altogether. However, if the opportunity arises to mate with genetically superior males later on, females may remate. In pair-bonded species such as birds, not all females have the opportunity to pair with high-quality males. Numerous studies on birds have shown that females paired to low-quality males are more likely to engage in EPCs, and that in most cases they involve males of higher quality than that of their partner. Obviously, females only obtain sperm from EPCs, so it is clear that females are gaining genetic benefits. A recent study obtained good evidence that this is the case. Female great reed warblers seek EPCs with males who have a larger song repertoire size than that of their partner, and there is a positive correlation between male repertoire size and postfledging survival of true genetical offspring (Hasselquist *et al.*, 1996). It is unclear why large repertoire size is a good cue to male genes for offspring viability. One possibility is that only old males can have large repertoire sizes and that by using male song as a cue females are choosing old males of proven viability.

In lizards it has been suggested that females cannot detect male genetic quality by assessing male phenotype. Under these constraints, females may mate with several males to enhance the chances of being fertilized by genetically superior males. This hypothesis assumes that genetically superior males perform better under sperm competition. Although in some populations of European adders and European sand lizards the number of sexual partners that females have is related to offspring viability, in others it is not. Thus, neither the relationship nor the underlying mechanisms are clear.

In populations with high levels of inbreeding, females may avoid the deleterious effects of inbreeding on offspring by multiply mating. High levels of inbreeding in males are related to poor-quality ejaculates (Roldan *et al.*, 1998). Thus, when females copulate with several males, the male with the lowest level of inbreeding may be more likely to fertilize the ova. If so, offspring would be less inbred and thus more viable, and sons would have more competitive ejaculates. This is one of the cases in which selection through sperm competition may be more efficient than selection at the phenotypic level, which may not show as clearly the effects of inbreeding. Sperm competition may also be a form of inbreeding avoidance in other contexts. In some mammals living in close family groups, it has been suggested that females seek matings with extragroup (or extrapair) males to avoid inbreeding. These females may also be constrained in their choice of partners. When females live in close social groups, mating with outsiders is probably the best way to reduce inbreeding in offspring, whereas avoidance of mating with intragroup males is likely to be costly.

Advantage of Solicitation of Sperm Competition. It has also been suggested that females would benefit from promoting sperm competition if by allowing to be fertilized by males producing the most competitive ejaculates they could pass on such traits to their sons. This hypothesis may explain how females could benefit from being fertilized by males with competitive ejaculates in a sperm competition context but not why multiple-male mating evolved in the first place. Modeling exercises have shown that multiple mating can spread rapidly provided there is a small fraction of females that copulate multiply originally (either because there are advantages to females or because the copulations are forced by males) and there are heritable differences among males in one or several of the following: number of sperm, success of sperm in fertilization, ability to displace sperm from previous copulations, and ability to prevent further copulations. In this situation, sons will inherit high fertilization efficiency and the trait will spread. Direct empirical evidence is lacking.

The Conflict

Males and females have different ways of maximizing their reproductive success. It is sometimes the case that what is advantageous for one sex is detrimental for the reproductive success of the other sex. Thus, each sex will try to manipulate the other in an attempt to maximize its own reproductive success, whereas the other sex will try to resist such manipulation. This situation may lead to an arms race between the sexes of interesting implications.

The conflict between the sexes can take a variety of forms. There may be conflict about whether there is copulation or not, the frequency of copulations, the duration of copulations, usage of sperm from different males for fertilization, the existence and duration of mate guarding, the timing of copulations, and the contents of the ejaculate.

In the first form, there may be conflict about whether there is sperm competition or not when it is in the female's interest to be choosy about her sexual partner(s) or to engage in a reduced number of matings, and males try to enhance their reproductive success by mating with as many females as possible. Water striders provide a well-studied example (Rowe *et al.*, 1994). Females store sperm for at least 10 days and they obtain no nutritional benefit from mating. The optimal strategy for females would be to copulate only a few times in order to obtain all the sperm they need to maximize their fertility potential. Further matings would not be advantageous for females because mating involves costs. When females copulate they have to carry the males, which slows down their movements and makes foraging less efficient; copulating females are also more visible and less capable of escaping from predators. On the other hand, sperm competition mechanisms are such that the last male to copulate obtains a greater share of paternity. Thus, for a male each new female encountered represents a good chance of producing more offspring. These divergent interests result in conflicting strategies: Males will attempt to copulate with the females they encounter, whereas already-mated females will be reluctant to mate. However, male persistence also imposes costs on females which may decrease their reluctance. To avoid copulation, females have to struggle to disengage males, which is costly energetically and increases predation risk, and there are also costs due to male harassment. Many factors influence the benefits and costs for males and females of each alternative: Male-biased sex ratios increase harassment rates, which decreases female reluctance and increases mating frequency. On the other hand, high predation risk causes a reduction in harassment rates, increases female reluctance, and reduces mating frequency.

In other contexts, nonpreferred males attempt to force copulations upon females and may develop morphological structures which make it difficult for females to avoid copulations. In sagebrush crickets the female feeds on the male's fleshy hindwings and the bodily fluids which leak from the wounds during copulation. Males allow this behavior because it keeps females busy during the time it takes them to complete spermatophore transfer. When males mate repeatedly, their hindwings become almost completely eaten and new females try to avoid these males. Males with reduced hindwings use a morphological device called the "gin trap" to hold rejecting females and force copulations upon them (Sakaluk *et al.,* 1995).

In some species, females counterevolve morphological structures which increase female control over copulation, such as the abdominal spines of female water striders.

The second form of conflict occurs when it is in the female's interest to mate with several males and males try to prevent females from doing so. This is clearly the case in pair-bonded species, such as birds, in which females paired with low-quality males benefit from copulating with better quality males. Males will attempt to mate guard their females, and females will attempt to deceive males. When females rely on male help to raise the offspring, the costs of being detected while engaged in EPCs are particularly high since males may respond to a decrease in paternity certainty by decreasing the amount of care they provide. However, when females are freed from their dependence on male assistance, they are able to express unrestricted mate choice and engage more often in EPCs. This is the case in cooperatively breeding superb fairy wrens (*Malurus cyaneus*), in which helpers cooperate in feeding and defending the young. When females breed in pairs they allow their partners a greater share of paternity to ensure their assistance in rearing the chicks. However, when females breed in groups, in which other helpers compensate for reductions in care by the dominant male, females produce broods containing more chicks sired by extragroup fathers (Mulder *et al.,* 1994).

In species living in one-male units, the harem holder will react aggressively to any attempt by his females to approach or copulate with extragroup males. In the hamadryas baboon, male controlling tactics involve a type of aggressive herding in which males vigorously bite females on the neck.

Specifically, when the risk of sperm competition is high, males may show adaptations to prevent or succeed at sperm competition which may be detrimental to females. Males may attempt to mate guard females, which may not only constrain their choice of mates but also decrease female foraging efficiency and increase their susceptibility to predators. Males may attempt to remove sperm from previous mates from the female tract or may attempt to copulate frequently or to engage in prolonged copulations in order to displace or dilute previous ejaculates. This may also be against female interests if they prefer previous males as sires of their offspring and because of the costs involved in prolonged copulations or repeated matings. Females may also suffer fertility costs when males respond to the risk of sperm competition by diverting energy into mate guarding and away from ejaculate expenditure. In the bluehead wrasee (*Thalassoma bifasciatum*), an externally fertilizing coral reef fish, territorial males defend spawning sites against smaller nonterritorial males. Territorial males with high spawning rates seem to partition ejaculates among successive females, which results in lower numbers of sperm per spawning and a lower fertilization rate for females (Warner *et al.,* 1995). There do not seem to be physiological constraints on sperm production by these males, but there seems to be a trade-off between gamete production and mate guarding.

A recent experiment with *Drosophila* dramatically illustrates how the sexes are involved in an arms race and how, when one sex is prevented from counterevolving, the other sex rapidly wins an advantage. In *Drosophila melanogaster* the seminal fluid reduces the competitive ability of sperm from other males, reduces the female's propensity to remate, and increases egg-laying rate. It also has substantial toxic effects on females because it decreases female survival. In an original experiment, males were allowed to evolve, whereas females were not allowed to retaliate. After 41 generations, adapting males did much better than control males: They produced more male progeny, were better at obtaining matings with already mated females, and were more capable of preventing females they had mated with from producing offspring by later males. However, this male adaptation led to a reduction in female survivorship which was mediated by an increased rate of remating and increased toxicity of seminal fluid (Rice, 1996).

It is often difficult to determine which sex has won the conflict or has gained an advantage. It has long been maintained that the intensity of selection on males to maximize the number of offspring produced is greater than selection on females to favor certain males in order to improve offspring quality (Parker, 1984). Thus, it has been assumed that male adaptations will prevail. However, in species with internal fertilization, no matter what behavioral and physiological strategies males follow, the ejaculate will eventually end up inside the female tract away from the male's control. This may allow females the possibility of developing physiological mechanisms to control paternity during the last stages at low cost. If so, intense selection on males to maximize fertilizations could be counteracted by the low costs of female control and a compromise would be reached. It has been suggested, for example, that female reproductive tract hostility may be a means of choosing the most competitive ejaculates, and that some of the complex interactions between sperm and the female tract could enable females to favor ejaculates from particular males (Roldan *et al.*, 1992). This exciting possibility has not been tested, and the evidence presented to support female control of paternity at a physiological level is still controversial. In a few studies it has been suggested that the fact that larger males win a greater share of paternity under controlled mating experiments indicates that females are selecting sperm from larger males. An alternative explanation is that larger males may perform better under sperm competition because they produce more competitive ejaculates and not because females favor them. More work is needed to clarify this issue.

The Outcome

Mating systems are the outcome of the reproductive strategies of all individuals living in a population. As previously shown, there will be conflict of interests between males and females, between competing males, and also between competing females.

Traditionally, mating systems were viewed as being the result of mate acquisition tactics by males which were aimed at copulating with as many females as possible. In this framework, male reproductive success was assumed to correlate with the number of females mated. Studies on sperm competition have revealed that this is only half of the story: In polyandrous species, sperm from rival males will compete within the female tract and females might exercise some choice at this level too. This new facet has many implications. First, male reproductive success cannot be equated to male mating success. Second, male and female reproductive strategies will also be molded to some extent by sperm competition mechanisms which will ultimately determine paternity. In monandrous species, some aspects of male mating strategies can be best understood as being aimed at avoiding sperm competition during the period in which ejaculates are most vulnerable.

In this context, male mating strategies may be viewed as a mixture of mate acquisition tactics and mate guarding tactics. Thus, there will be a trade-off between the number of females mated and how much effort is put into protecting ejaculates under the risk of sperm competition. This implies that in addition to the traditional factors considered to influence mating patterns (i.e., female spatial and temporal distribution), other factors related to the way in which sperm competition mechanisms work are also important. Gomendio and Roldan (1993b) suggested that differences in sperm competition mechanisms between birds and mammals may have influenced to a great extent the evolution of disparate mating systems.

In birds, females ovulate, fertilize, and lay their eggs in chronological sequence over a number of days (approximately at the rate of one egg produced per day) and may store sperm before and during egg laying in the sperm storage tubules (Fig. 4). Sperm competition mechanisms determine a last-male advantage. Females need male assistance to rear the chicks. The female accepts copulations from her partner for some days before the first egg is laid until egg laying starts, allowing the pair male a level of paternity certainty which will ensure his cooperation. In this way, she also ensures that enough sperm are stored to fertilize all the ova. The female benefits from securing a pair male's help, but she also benefits from producing offspring from high-quality males. Thus, the female ceases to accept copulations when egg laying starts to maximize the chances that, if a better male is found, an EPC will be successful. From the pair male's perspective, things look very different. A male would benefit from continuing to copulate during egg laying in order to maximize his chances of siring offspring. He is not allowed by the female to do this, but he will engage in retaliatory copulations if an EPC is detected. Potential EPCs are very costly to pair males due to the last-male advantage. Thus, males need to mate guard the female over the whole egg-laying sequence (i.e., days) to ensure paternity. The benefits of prolonged mate guarding are likely to have contributed to the evolution of the pair bond which is so common in birds. Finally, the point of view of potential extrapair partners should also be considered. Males are not likely to engage in EPCs during their partner's egg-laying period because at this time the priority is to mate guard. The imposed period of rest during egg laying implies that pair males have potentially highly competitive ejaculates toward the end of egg laying. This, together with the last-male advantage, means that it is relatively easy for males to obtain extrapair offspring since males only need to copulate once at the right time to outcompete the female's partner. The benefits associated with EPCs both for males and for females are likely to have favored the evolution of extrapair copulations in birds.

The mammalian scenario is radically different. Females are fertile for very limited periods of time because all ova are ovulated at the same time and remain viable for about 24 h, and females do not possess specialized sperm storage organs. As a consequence, mammalian spermatozoa are short-lived. Sperm competition mechanisms are such that there are no order effects. Instead, timing of mating in relation to ovulation, sperm numbers (in either one or several ejaculates), and ejaculate quality determine paternity

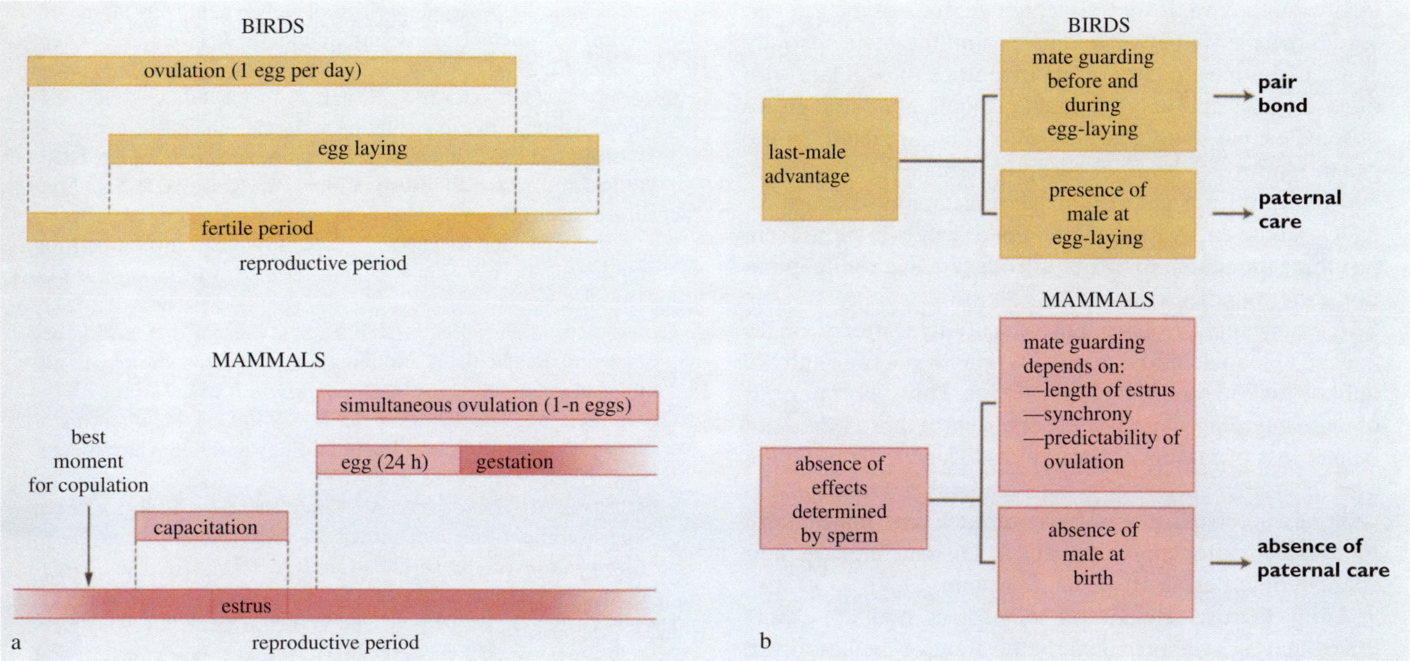

FIGURE 4 Characteristics of the reproductive cycle of birds and of mammals (a) that are postulated to have influenced the evolution of sexual behavior and mating systems (b) (adapted from Gomendio and Roldan, 1993b).

to a great extent. The best strategy for males is thus to copulate close to ovulation but allow time for capacitation to take place. There is far less need than in birds for males to defend females after ovulation, and since females undergo gestation and lactation males gain little from staying with a female. Since female mammals care for the offspring, there is little room for mammalian males to improve offspring survival and much to be gained from mating with other females. Taking these factors into consideration, it is understandable that polygyny is by far the most common mating system in mammals.

Among mammals, much variation in male mating behavior has been attributed to four characteristics of females: the extent to which female reproductive rate may be increased by male assistance in rearing offspring, the size of female ranges and core areas, the size and stability of female groups, and the density and distribution of females in space. It is likely that some aspects of female reproductive physiology in conjunction with sperm competition mechanisms will determine to an important extent mating strategies. These factors include the length of the estrus cycle, whether ovulation is spontaneous or induced, whether ovulation is predictable or not, and the degree of female synchrony in sexual receptivity.

Induced Ovulators

The first male to copulate triggers ovulation. It seems likely that males will have evolved sperm fertile life spans (normally ovulation takes place a few hours after copulation) and capacitation times that synchronize these events so that the sperm from the first male will be ready to fertilize at the time ovulation occurs. If this is the case, then the first male may have an advantage over males mating later on in the female cycle. If synchrony is so tight that males mating afterwards have very slim chances, then the first male to copulate should not guard the female. However, if males mating afterwards do have chances of fertilizing, then the first male should mate guard until ovulation occurs and fertilization by other males is no longer possible. Since ovulation tends to occur a few hours after copulation, mate guarding should in any case be short. Thirteen-lined ground squirrels (*S. tridecemlineatus*) are induced ovulators, ovulation takes place 8–36 h after copulation, the first male to copulate fathers most of the offspring, and males copulate several times with a female but do not mate guard (Schwagmeyer, 1995), suggesting that they fit into the first model proposed. In this species, males leave their mates after they have achieved a copulation longer than 9 min. It is possible that a copulation this long is necessary to induce ovulation. Laboratory studies have shown that, as expected, delays between copulations by two successive males favor the first one. In accordance with these findings, field data show that males that do not find an estrus female until at least 4 h after her first copulation tend not to copulate with her. A first-male advantage has also been reported in rabbit, which is also an induced ovulator.

Spontaneous Ovulators

Short Estrus. Regarding short-estrus ovulators, most males mating within an estrous period will have chances of fertilizing the ova. If there is a high degree of synchrony in sexual receptivity between females, and these are spatially close, males have plenty of opportunities to copulate with other females and may be unable to defend a large group of females. No mate guarding should be expected. In the ground squirrel *S. beldingi,* receptive females are clustered, and males do not guard females presumably because females can be easily located and males can gain more from resuming the search for other estrous females after copulation (Sherman, 1989).

If females are asynchronous, or spatially scattered, males may be able to defend one estrous female at a time and are unlikely to find other females in estrus. Thus, short periods of mate guarding should be observed since there would not be any loss of mating opportunities. This seems to be the case in another species of ground squirrel, *S. brunneus.* Females are in estrus for one afternoon each year; males guard for about 3 h after copulation and in this way improve their chances of fathering offspring (Sherman, 1989).

Long Estrus. Regarding long-estrus ovulators, only males mating at approximately the time of ovulation will have chances of fertilizing the ova. When females advertise ovulation, males should attempt to copulate with females at approximately the time of ovulation. In baboons, males compete to copulate when sexual swellings reach their maximum size (which pinpoints to a day or two the timing of ovulation). Outside this period, dominant males tend to ignore females, and subordinate or juvenile males take advantage of the opportunity to copulate with females. Whether there is mate guarding or not will depend on whether there is loss of opportunities to copulate with other females and thus on the degree of female synchrony. For this reason, consortships are more prevalent in baboons (nonseasonal breeders) than in macaques (seasonal breeders).

If females do not advertise ovulation, males will be unable to predict when ovulation will take place and thus will be unable to assess the value of copulations at different times throughout the cycle. In this case, females will always be synchronous to some extent because they will be receptive most of the time and males will not be able to pinpoint ovulation time. This situation will favor copulations at short intervals over long periods of time (to ensure fertilization) and possibly extended mate guarding. This seems to be the case in humans.

References Cited

BIRKHEAD, T. R. (1996). Sperm competition: Evolution and mechanisms. *Curr. Topics Dev. Biol.* **33,** 103–158.

BIRKHEAD, T. R., and MØLLER, A. P. (1992). *Sperm Competition in Birds.* Academic Press, London.

BISSONDATH, C. J., and WIKLUND, C. (1995). Protein content of spermatophores in relation to monandry/polyandry in butterflies. *Behav. Ecol. Sociobiol.* **37,** 365–371.

DAVIES, N. B., HARTLEY, I. R., HATCHWELL, B. J., and LANGMORE, N. E. (1996). Female control of copulations to maximize male help: A comparison of polygynandrous alpine accentors, *Prunella collaris,* and dunnocks, *P. modularis. Anim. Behav.* **51,** 27–47.

GAGE, M. J. G., STOCKLEY, P., and PARKER, G. A. (1995). Effects of alternative male mating strategies on characteristics of sperm production in the Atlantic salmon (*Salmo salar*): Theoretical and empirical investigations. *Philos. Trans. R. Soc. London B* **350,** 391–399.

GOMENDIO, M., and ROLDAN, E. R. S. (1991). Sperm competition influences sperm size in mammals. *Proc. R. Soc. London B* **243,** 181–185.

GOMENDIO, M., and ROLDAN, E. R. S. (1993a). Co-evolution between male ejaculates and female reproductive biology in eutherian mammals. *Proc. R. Soc. London B* **252,** 7–12.

GOMENDIO, M., and ROLDAN, E. R. S. (1993b). Mechanisms of sperm competition: Linking physiology and behavioural ecology. *Trends Ecol. Evol.* **8,** 95–100.

GOMENDIO, M., HARCOURT, A. H., and ROLDAN, E. R. S. (1998). Sperm competition in mammals. In *Sexual Selection and Sperm Competition* (T. R. Birkhead and A. P. Møller, Eds.). Academic Press, London.

HASSELQUIST, D., BENSCH, S., and VON SCHANTZ, T. (1996). Correlation between male song repertoire, extra-pair paternity and offspring survival in the great reed warbler. *Nature* **381,** 229–232.

MØLLER, A. P. (1988). Ejaculate quality, testes size and sperm competition in primates. *J. Hum. Evol.* **17,** 479–488.

MULDER, R. A., DUNN, P. O., COCKBURN, A., LAZENBY-COHEN, K. A., and HOWELL, M. J. (1994). Helpers liberate female fairy-wrens from constraints on extra-pair mate choice. *Proc. R. Soc. London B* **255,** 223–229.

PARKER, G. A. (1970). Sperm competition and its evolutionary consequences in insects. *Biol. Rev.* **45,** 525–567.

PARKER, G. A. (1984). Sperm competition and the evolution of animal mating strategies. In *Sperm Competition and the Evolution of Animal Mating Systems* (R. L. Smith, Ed.), pp. 1–60. Academic Press, San Diego.

PARKER, G. A. (1990a). Sperm competition games: Guards and extra-pair copulations. *Proc. R. Soc. London B* **242,** 127–133.

PARKER, G. A. (1990b). Sperm competition games: Raffles and roles. *Proc. R. Soc. London B* **242,** 120–126.

REICHART, A. K., and WHEELER, D. E. (1996). Multiple mating in the ant *Acromyrmex versicolor:* A case of female control. *Behav. Ecol. Sociobiol.* **38,** 219–225.

RICE, W. R. (1996). Sexually antagonistic male adaptation triggered by experimental arrest of female evolution. *Nature* **381,** 232–234.

ROLDAN, E. R. S., GOMENDIO, M., and VITULLO, A. D. (1992). The evolution of eutherian spermatozoa and underlying selective forces: Female selection and sperm competition. *Biol. Rev.* **67,** 551–593.

ROLDAN, E. R. S., CASSINELLO, J., ABAIGAR, T., and GOMENDIO, M. (1998). Inbreeding, fluctuating asymmetry, and ejaculate quality in an endangered ungulate. *Proc. R. Soc. London* **265,** 243–248.

ROWE, L., ARNQVIST, G., SIH, A., and KRUPA, J. (1994). Sexual conflict and the evolutionary ecology of mating patterns: Water striders as a model system. *Trends Ecol. Evol.* **9,** 289–293.

SAKALUK, S. K., BANGERT, P. J., EGGERT, A.-K., GACK, C., and SWANSON, L. V. (1995). The gin trap as a device facilitating coercive mating in sagebrush crickets. *Proc. R. Soc. London B* **261,** 65–71.

Schwagmeyer, P. L. (1995). Searching today for tomorrow's mates. *Anim. Behav.* **50,** 759–767.

Sherman, P. W. (1989). Mate guarding as paternity insurance in Idaho ground squirrels. *Nature* **338,** 418–420.

Smith, R. L. (Ed.) (1984). *Sperm Competition and the Evolution of Animal Mating Systems.* Academic Press, San Diego.

Stockley, P., Searle, J. B., MacDonald, D. W., and Jones, C. S. (1994). Alternative reproductive tactics in male common shrews: Relationships between male mate-searching behaviour, sperm production, and reproductive success as revealed by DNA fingerprinting. *Behav. Ecol. Sociobiol.* **34,** 71–78.

Warner, R. R., Shapiro, D. Y., Marcato, A., and Petersen, C. W. (1995). Sexual conflict: Males with the highest mating success convey the lowest fertilization benefits to females. *Proc. R. Soc. London B* **262,** 135–139.

Watson, P. J. (1991). Multiple paternity as genetic bet-hedging in female sierra dome spiders, *Linyphia litigosa (Linyphiidae). Anim. Behav.* **41,** 343–360.

ELISABETTA VISALBERGHI

Institute of Psychology
National Research Council
00197 Rome, Italy

DOROTHY FRAGASZY

Department of Psychology
University of Georgia
Athens 30602, Georgia, USA

Tool Use by Monkeys and Apes

The use of tools may enable or increase the exploitation of resources, such as foods that are difficult to obtain through direct action with hands or teeth. In the wild, the use of tools is widespread in chimpanzees but observed less often in other apes. However, great apes and several species of monkeys use tools readily in captivity. Appropriate anatomical and behavioral characteristics are necessary to support tool use. In most cases, an individual learns to use a tool through its own actions, although these often occur in a socially supportive context. Monkeys and apes do not learn to use tools merely from watching one another. The more an individual understands about the functioning of a tool, the more effectively the individual can use it or change it in ways that improve its effectiveness. When nonhuman primates use a tool, they do not necessarily understand how and why it functions.

Introduction

We all think that we know what we mean by the word "tool" and by the phrase "using a tool." However, as is often the case for words and phrases used in everyday speech, these terms are actually vague for scientific purposes. To determine whether and in what circumstances other species use tools, we need a precise definition. Does an archer fish use a tool when it spits water at an insect above the surface of the water, thereby knocking its prey to the surface, whereupon the fish captures it? Can one call water a tool? Does a Galapagos finch *Geospiza* use a tool when it probes into a small hole in a tree trunk with a needle from cactus bush? When it does this, it pierces a small grub, which is then extracted with the needle and eaten. Is the cactus needle a tool in this instance? Both of these cases involve the use of an object in a way that accomplishes something that would not be possible by the animal using its body alone. Is this sufficient to identify something as tool use? If we accept this position, what about the case in which a squirrel runs up a tree?

Benjamin Beck (1980) offers the following widely accepted definition of tool use in his book *Animal Tool Behaviour,* which is still the most complete catalog of tool use in animals:

> Tool use is the external employment of an unattached environmental object to alter more efficiently the form, position, or condition of another object, an-

> other organism, or the user itself when the user holds or carries the tool during or just prior to use and is responsible for the proper orientation of the tool. (p. 10)

Why Tool Use Is an Important Topic in Biology

Characteristics that seem to define our own species have always held fascination for students of human biology and evolution. Descartes formalized the notion during the middle of the seventeenth century that certain human characteristics reflect intellect and the presence of a divinely provided soul. According to Cartesian philosophy, such characteristics could not be present in other species without challenging our privileged status in the world.

Two hundred years after Descartes' works were published, evolutionary theories provided alternative views of human origins different from Descartes' philosophical view. Study of the evolutionary history of distinctly human characteristics became the province of an entire subdiscipline of anthropology (paleoanthropology). Discoveries in the past century by paleoanthopologists of objects identified as simple tools such as scrapers and cutting blades with early human remains led to prolonged discussions about the significance of tools in the evolution of our species. Using tools was still considered, as other characteristics were considered by Descartes, as a uniquely human characteristic. For example, the famous paleoanthropologist Louis Leakey (1961, p. 2) wrote that when an ape-like creature first made a tool to "a regular and set pattern" it became by definition Man.

Our contemporary fascination with tool using in nonhuman species reflects a profound appreciation of the importance of tool using to our own species. There is no doubt that the use of tools has empowered humans to diversify their way of life and to exploit resources not available to other primates. The paleontological and archeological records show that changes in tools throughout human history reflect accumulating mastery of physical relations and knowledge of natural processes. The tools provide a record of human workmanship– and, from the earliest periods of human history, one of the best records from which to infer the behavior of our ancestors. Moreover, using tools is linked in our minds to intelligence, and the emergence of tools in human history is thought to reflect the evolution of human intelligence.

Unfortunately for strict Cartesians, species other than humans display functionally similar behaviors with objects that, in humans, are identified as tool use. At least since Erasmus Darwin, the grandfather of Charles Darwin, in the late 1700s, observant scientists have published numerous accounts of tool using by nonhumans. Early reports of animals using tools were anecdotal in character. For example, Erasmus Darwin reported that he saw an old monkey in London cracking a nut with a stone. One hundred years later, the naturalist Thomas Belt, in his book *A Naturalist in Nicaragua,* described the behavior of his pet monkey as follows: "Sometimes, when there are broods of young ducks about, it would hold a piece of bread in one hand, and when it had tempted a duckling within reach, would seize it by the other, and kill it by a bite in the breast." As in our own species, observers assumed that these behaviors indicated intelligence and purpose. It will be shown that these assumptions may be unwarranted.

Apart from the issue of intelligence, tool using is of interest to biologists because it is a means by which an individual can extend what it can do or where it can live. For example, chimpanzees can open certain kinds of nuts only by cracking them with a stone. These nuts are a rich food source for the animals. Similarly, the finch using a cactus needle can obtain prey not otherwise accessible. Although it is often an assumption, it is generally thought that using a tool to solve an ecologically important problem (such as obtaining food or constructing shelter) confers an advantage over not using a tool.

Tool Use in the Animal Kingdom

Careful tabulation of Beck's (1980) book shows that tool using has been reported for invertebrates, fish, amphibians, reptiles, birds, and mammals. Animals use tools to crack open eggs, to extract larvae from woody substrates, during courtship behavior, and so on. Impressive examples of tool use can be found even among invertebrates. Weaver ants (e.g., *Oecophilla* and *Campanotus*) in tropical Africa, South America, and southern Asia hold their own larvae in their mandibles during nest construction and repair. When leaves are appropriately positioned by a team of workers, other workers move the larvae back and forth like shuttles across the leaf edges to "glue" the edges together with silk secretions from the larvae. Workers do not secrete silk; only larvae do this. In this example, the larvae are "tools" used by the workers. An example of tool use which also involves selection and some degree of preparation of the tool is given by Hunt (1996). Crows (*Corvus moneduloides*) from New Caledonia use two different types of tools to extract prey from the base of palm trees and the trunks of other trees. The tools are fashioned from the twigs and leaves of plants in this region. Twig tools require the removal of extra foliage and twigs and perhaps additional shaping of the tool. Leaf tools similar in shape to a blade of grass are cut from a larger leaf. The birds used a tool repeatedly, stored it in the feet during interruptions of use, and fetched it from a resting spot after the user changed trees. All these actions can be interpreted to indicate a plan on the part of the bird to prepare and use an object for a specific purpose. On the other hand, other explanations are also possible. The rich interpretation of planning, etc. is warranted only if experimental evidence can support rejection of a more parsimonious interpretation.

Tool Use in Chimpanzees in Natural Environments

There are more known examples of tool use in primates than in other animal groups, and for those primates in which tools are commonly used (humans, common chimpanzees, and capuchins in captivity), tools are used to reach a greater variety of goals, and a greater variety of objects are used as tools, than in other orders of animals. However, except for common chimpanzees (*Pan troglodytes*) and humans, it is not clear that primates are more frequent tool users in nature than are other animals. The clear star among tool-using primates (after humans) is the common chimpanzee (McGrew, 1992). These animals use tools everywhere in their natural range for feeding, obtaining water, cleaning the body, as aides during displays, and so on. The greatest variety of tool using occurs during feeding (Fig. 1).

The first well-documented cases of tool use in free-living chimpanzees were made by Jane Goodall in the late 1960s (McGrew 1992; Goodall, 1986). She observed chimpanzees using wads of leaves to sponge water from holes in trees during the dry season and probing for ants and termites inside their nests with grass, sticks, and other materials of appropriate shape and strength. Termites (*Macrotermes bellicosus*) construct mounds aboveground which have the consistency of concrete. To catch the termites from inside the mound, a predator must either destroy the mound or extract the insects through the small tunnels that are open to the surface. Chimpanzees do the latter. To extract this rich source of protein, the chimpanzee must delicately insert a slender probe down the tunnel (which can be curving) to reach the insects deep inside the mound. The termites "attack" the invading probe, grabbing with their mandibles. When the probe is withdrawn it may be coated with termites, which the chimpanzee can pick off with its lips and eat. Chimpanzees select probes from nearby vegetation or bring them from as far as 100 m away. Sometimes they modify the vegetation to suit the size of the tunnels. A human trying to copy the chimpanzees' actions reported that success at probing for termites required much practice, fine motor control, and knowledge about the locations and properties of termite mounds (Teleki, 1974). These observations were exciting because the way the chimpanzees used tools to fish for termites was sophisticated (in terms of the preparation of the tool and in the skill involved in its use) and seemingly indicative of a greater intelligence than had been granted to the great apes by most biologists (although not surprising to those who had worked with chimpanzees in captivity).

FIGURE 1 A young chimpanzee (*Pan troglodytes*) in the wild using a stick as a tool to catch termites (photo courtesy of P. Ryan/Wildlife/Panda Photo).

Since those early reports, several additional kinds of tool use have been reported for chimpanzees at many different sites, and reports of new forms of tool use appear regularly. For example, Yamakoshi and Sugiyama (1995) reported the use of tree branches to pound and soften the inner part of the crown of a palm tree ("pestle pounding") (Fig. 2). This report from Bossou was followed within a few months by another one from the same site concerning the use of sticks to collect floating mats of aquatic plants (a food source not previously seen to be exploited by chimpanzees). Table 1 lists the habitual forms of tool use in wild chimpanzees known by the early 1990s from five sites at which observations of chimpanzees have been carried out for many years. The significance of tool use in the lives of these animals can be appreciated from a recent report by the research team from the Primate Research Institute of Kyoto University on the chimpanzee community at Bossou that indicated that

FIGURE 2 A chimpanzee in the crown of a palm tree uses a branch to pound and soften the internal part in a behavioral sequence called pestle pounding: It spreads the mature leaves (a), exposes the petiole (b), pounds the exposed petiole with a branch (c), and extracts and eats the juicy pulp from the tree crown (d) (adapted from Yamakoshi and Sugiyama, 1995).

12% of the time spent foraging was devoted to foods that can be obtained only through the use of tools.

Nut Cracking in Wild Chimpanzees

One of the most interesting forms of tool use seen in wild chimpanzees is cracking nuts using stones as hammer tools. In western Africa, chimpanzees collect nuts of different species, bring them to a site with a natural "anvil" (a rock or exposed tree roots of an appropriate shape and hardness), place them in a flat or concave section of the anvil, and crack them open by pounding them with stones or large branches. The Boesches (Boesch and Boesch, 1983; Boesch, 1993) closely observed nut cracking in Taï Forest, Ivory Coast. The specific types of tools used vary across sites, but the general process is similar. Typically, females pound nuts more frequently and are more efficient nut crackers than males. The research team from Kyoto University at Bossou (Republic of Guinea) determined that young chimpanzees acquire nut-cracking skills over several years. They practice this activity at the same sites at which their mothers crack

TABLE I

Patterns of Tool Use Shown Repeatedly by Several Members of a Group of Wild Chimpanzees at Five Well-Known Sites[a]

	SITE				
PATTERN	Bossou, Guinea	Taï, Ivory Coast	Kibale, Uganda	Gombe, Tanzania	Kasoje, Tanzania
Termite, fish				x	x
Ant, dip	x	x		x	
Honey, dip		x		x	
Leaf, sponge	x			x	
Leaf, napkin			x	x	
Stick, flail				x	x
Stick, club	x?			x	x
Missile, throw	x			x	x
Self, tickle				x	
Play, start				x	x
Leaf, groom				x	x
Ant, fish					x
Leaf, clip	x				x
Gum, gouge	x				
Nut, hammer	x	x			
Marrow, pick		x			
Bee, probe		x			
Branch, haul	x				
Total	8	5	1	11	8

[a] Adapted from McGrew (1992).

nuts. They begin by pounding nuts on the ground directly or pounding the tool on the ground. It is only very gradually that they combine the tool and the nut in an appropriate manner by pounding the nut with the tool and by orienting the nut correctly. The same pattern of acquisition (with both effective and ineffective combinations of actions and objects) was observed in captive capuchins learning to pound open nuts with stones (Visalberghi, 1987; Fig. 3).

In all the years the Boesches observed the chimpanzees in the Taï Forest, they recorded only two cases in which the mother appeared to have intervened actively while her offspring was attempting (ineffectively) to crack nuts (Boesch, 1993). One mother repositioned the nut that her infant had incorrectly oriented on the anvil; another repositioned the tool in her hand in an obvious and deliberate manner while her infant watched closely. Boesch called these episodes *active teaching*. Despite the rarity of teaching of nut cracking, social circumstances undoubtedly fostered the infant's acquisition of this skill. Mothers stimulated the infants' activity with tools and nuts and facilitated the infants' acquisition of the skill once elements of nut cracking appeared. The mothers' behaviors varied in accord with the infants' ability to open the nuts. For example, mothers often left younger infants with hammers and unopened nuts at the site while they collected more nuts. Females without infants did not leave nuts or tools when they went off to search for more nuts. Older infants (which were able to use the tools, although inefficiently) were allowed to use tools and to crack nuts their mothers had brought to the site. These actions no doubt facilitated the youngsters' explorations with the appropriate objects and their opportunities to practice cracking nuts with appropriate tools. However, they are not "teaching" in the usual sense of the word (implying guidance).

Tool Use in Other Primates in Natural Environments

The incidence of tool use in other primates in natural conditions is decidedly less impressive than among chimpanzees. The other chimpanzee, the bonobo (*Pan paniscus*),

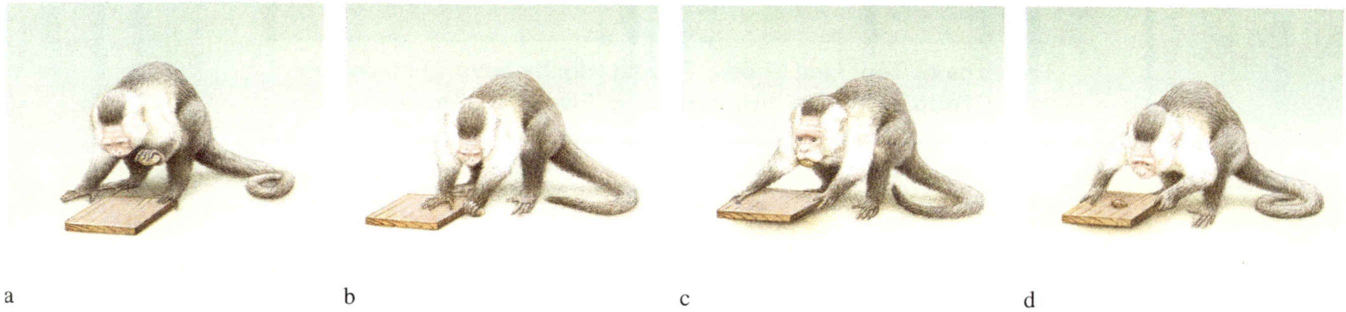

FIGURE 3 Capuchin monkeys (*Cebus apella*), while learning to use a piece of wood to crack open almonds, perform a variety of behaviors, some of which are patently ineffective. The capuchin pounds the nut on the tool (a) or next to the tool (b). With both hands, the capuchin pounds the tool on the ground, holding the nut in its mouth (c) or in its hands (behavior not shown), or it places the nut on the tool and then pounds tool and nut together on the ground (d). These behaviors became less frequent over time, whereas the correct use of the tool (i.e., the action of pounding the tool on the nut) became more frequent.

uses tools less often than common chimpanzees. Only recently has the use of tools in feeding been described for wild orangutans (*Pongo pygmaeus;* van Schaik *et al.,* 1996). The orangutans used sticks to extract social insects and to extract edible seeds from a large fruit protected by urticaceous hairs. Gorillas, the third genus of great apes, have never been observed to use tools in natural settings. All these genera use tools readily and effectively in captivity.

Among monkeys, capuchins stand out as masters at using tools (Visalberghi, 1990). There are instances of capuchins using tools in natural settings, in foraging, and in self-defense (Fernandes, 1991; Boinski, 1988). However, as is the case for the great apes other than common chimpanzees, the vast majority of observations of tool use come from captive animals. Captive capuchins, like the apes, readily and spontaneously use tools in many circumstances. Tool use in other species of monkeys in nature is limited to problematic forms such as dropping branches on visitors. These actions meet some of the criteria for tool use provided by Beck (1980) but not all. One reason they remain problematic is that these actions cannot be reliably provoked in captivity, and they are rare in nature as well. It is difficult to study the rare event systematically.

No tool use of any kind has been described for the other superfamily of nonhuman primates, the prosimians.

Tool Use in Captive Monkeys and Apes

Experimental studies of tool use began in the early part of the twentieth century by some of the historically most important comparative psychologists. Robert Yerkes (1927/1973, 1943), who established the first primate research center in the United States, included studies of tool use in great apes in his research program. Wolfgang Köhler (1925/1976), who had the good fortune to be on the island of Tenerife off the coast of Africa for the duration of the first World War, conducted famous experiments on tool use and other forms of problem solving in common chimpanzees. Köhler's apes stacked boxes to reach high objects, used sticks to retrieve a reward out of reach, and were given tasks for which both solutions were required to obtain, for example, a banana (Fig. 4). At about the same time, Heinrich Klüver (1933) was also exploring problem-solving abilities, including tool use, in capuchin monkeys. These researchers have been extremely influential in shaping and inspiring research on intelligence in nonhuman species. All of them had a profound appreciation of the richness and also the limits of their subjects' behavior.

Since the pioneering works of these and other authors, such as Ladygina-Kots (1935) in Russia, captive studies have been undertaken with the double aims of documenting which species among primates are able to use tools and, more interesting, the cognitive basis for tool-use behaviors. In captivity, unlike in nature, many species of primates use tools spontaneously if given the appropriate opportunities. Whereas virtually any mammal (or bird, for that matter) can be trained to use a tool (e.g., by rewarding approximations of the desired behavior), and some mammals (e.g., dolphins and elephants) exhibit occasional spontaneous use of tools in captivity, using tools in captivity is more characteristic of primates than other orders.

Undisputed cases of using tools in captivity have been reported for all species of great apes, many species of Old World monkeys, and several species of New World monkeys (Fig. 5). Most cases concern use of an object as an extension of the arm to reach something, as when a stick is used as a rake or a probe. Behavioral characteristics which support the spontaneous discovery and use of tools include manual dexterity, curiosity and responsiveness toward objects, and tendencies to explore objects and surfaces by means of other

FIGURE 4 A chimpanzee (*Pan troglodytes*) standing erect on stacked boxes and reaching with a pole for a banana placed too high to reach directly with the hand (adapted from Köhler, 1925/1976).

objects. There is a positive relation between the incidence of using tools and the degree to which these three behavioral tendencies are present.

How Is Tool Use Acquired?

Given that an individual belongs to a species which possesses the appropriate anatomy and behavioral characteristics to support tool use, how does the individual acquire a particular skill? Biologists and psychologists have speculated widely as to how nonhuman animals arrive at the efficient and sometimes elegant and skillful use of objects as tools. One common notion is that one individual learns to use a tool from observing another, as can often happen in humans. In nonhuman primates, research has shown that imitation as we typically think of it (watching and then reproducing novel actions) plays a null (monkeys) or a limited (great apes) role in learning a new tool-use behavior (Visalberghi and Fragaszy, 1990).

Apart from the special circumstances of apes reared in close contact with humans, the normal way of acquiring a new tool-use skill by nonhuman primates involves both social influences and individual discovery. For example, we presented to capuchin monkeys a food source inside a container (Fig. 6). The food could be obtained only through openings too small for the monkeys' hands but large enough to allow the insertion of sticks or straw, which were available to them. The monkeys showed immediate interest toward the food and tried every possible way to get it. For the capuchin monkeys, one possible way to get at the food was to poke into the container with any object at hand. This behavior is often seen during normal exploration, even when food is not involved. If the object happens to be the right size, shape, and strength and if it is inserted deeply enough, then the monkey will succeed in getting the food. Once a monkey has incidentally succeeded in this way, it is likely to repeat the actions. Other monkeys in the vicinity may observe closely and even obtain some food for themselves. Although social circumstances can increase an individual's interest in the tool, the container, and its environs, there is no evidence to date that capuchins can learn the specific details of how to use a tool from watching others. It is clear, however, that social influences can inhibit an individual's access to a site of particular interest to other group members and thus decrease the probability that the individual will discover how to solve the problem. Moreover, those individuals that reliably succeed in obtaining food from others may not become tool users themselves. Overall, what makes an individual a likely tool user is a mixture of factors ranging from exploratory and manipulative tendencies to immediate social context and individual social relationships. The notion that individual "intelligence" is the major re-

a b c

FIGURE 5 A young bonobo (*Pan paniscus*) that has been given a bundle of sticks held together by tape (a) carries out the task of extracting a reward container from a transparent tube. The diameter of the bundle is larger than the opening of the tube. The bonobo removes the tape from the bundle, sniffs the tape (b), and a moment later discards it. Finally, it uses one of the sticks obtained from the bundle to push the black reward container out of the tube. In the wild, bonobos, although able to do so, use tools less frequently than chimpanzees.

sponsible factor accounting for success has no experimental support.

Why is tool use more common in captivity than in natural settings? The possibility that captive animals are somehow "smarter" than their wild counterparts can be dismissed. Rather, it seems that the combination of abundant free time for play and exploration, small space, and few objects and surfaces all make discovery of tool use more likely than in nature. When a researcher places something in the cage which provides an interesting new focus for activity, the monkey's or ape's natural propensity to explore the interesting object is sufficient for the occasional incidental combination of actions and object which produces desirable consequences (Fragaszy and Adams-Curtis, 1991). Insight or sophisticated comprehension of physics are not required.

FIGURE 6 A female capuchin with her infant clinging to her side dips a stick into apple sauce kept in a vertical tube.

Analyzing Cognitive Elements in Tool Use

Although insight or comprehension are not required to use a tool, if they are present, they increase the effectiveness of tool use. Psychologists wish to determine if these processes, or others like them, are present in nonhuman primates and, if they are present, what role they play in tool-use behavior. The following statement from Reynolds (1982) illustrates a cognitive perspective on tool use:

> A tool is not an object but a mental program which can interrelate that object with others to implement anticipated external effects. Tools, in other words, are as much mental as material, and their description is not a photograph of the material object itself but an empirically verifiable characterization of the mental knowledge and behavioral programs which allow the object to be produced and used. (p. 377).

The more an individual understands about the functioning of a tool, the more effectively one can use it or change it in ways that improve its effectiveness. Furthermore, understanding allows an individual to deal with variations of the same problem without resorting to blind attempts to figure out a solution. The previously mentioned features are indeed part of what is special about the ways in which humans use and manufacture their tools. This view leads naturally to experimental investigation of how monkeys and apes cope with problems, especially when they are new. The novel features, for example, may be the form of the tool, the manner of its use, or various features of the task.

Both capuchins (Fig. 7) and chimpanzees (Fig. 8) can use

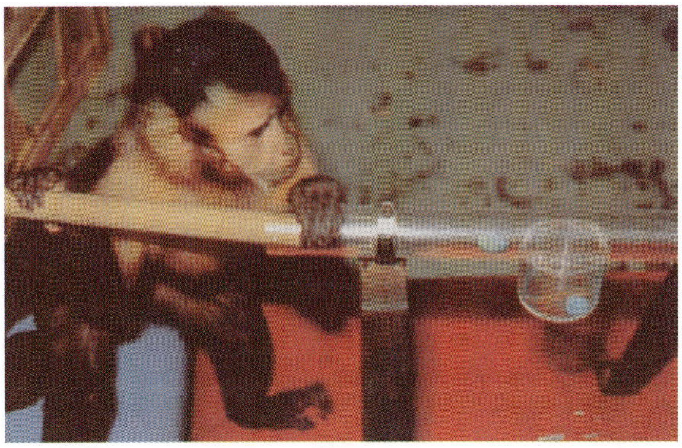

FIGURE 7 A capuchin monkey faces a difficult tool-using task — that of inserting a stick in one side of the tube to push food out of the other side. In this picture, the monkey is making an error — pushing the food (blue candies) into a trap in the center of the tube. There are candies visible in the trap which the animal has lost in previous attempts.

a stick to push an object out of a horizontal tube. One can ask if they have a similar understanding of the properties of the tool and the relations between their action and the outcome. This was tested by varying a property of the tube. The horizontal tube was modified so that it had a trap in its middle; thus, the tube was shaped like a T with a very short stem. To push something out of this tube one had to avoid moving the object in the direction of the trap. When faced with this task, capuchin monkeys that were otherwise proficient in pushing things out of tubes were suddenly performing at chance level (Visalberghi and Limongelli, 1996). One monkey eventually improved. Further testing showed that it was selecting the side of the tube in which to introduce the stick on the basis of its distance from the reward, without taking into account where the trap was located (Fig. 8).

FIGURE 8 Bobby, a 5-year-old chimpanzee (*Pan troglodytes*) confronted with the task shown in Fig. 7, is committing the same error as the monkey shown in Fig. 1. It opens its mouth at one end of the tube while pulling the stick and the food (red candy) toward itself from the other end. In this case, since Bobby inserted the stick in the same side of the tube as the food, without considering the trap in the center of the tube, the candy will be lost. Some older chimpanzees are able to choose the correct side in which to insert the stick to avoid this error.

When chimpanzees were presented with the same task, some of them also did not behave in ways suggesting a comprehension of the requirements of the task (Visalberghi and Limongelli, 1996). As can be seen in Fig. 8, a chimpanzee, despite repeated practice with the task and repeated experiences of losing a food item into the trap, did not figure out the effects of its action. It anticipated success, as indicated by its open mouth, while in fact it pushed the reward into the trap. Two older chimpanzees (out of four tested) achieved a type of performance which indicates that they were taking into account the presence of the trap and the position of the reward in relation to the trap. When they displaced the reward with the stick, they avoided pushing it into the trap.

The comparison between capuchins and chimpanzees in this task has shown that using a tool does not necessarily require understanding of its functioning. Moreover, we now appreciate that different species may have different understandings of the same problem.

Conclusion

There are many aspects of tool use in nonhuman primates that remain puzzling and that we expect will generate continued research. First, many primates use tools in captivity but they do not (except for chimpanzees) in natural environments. We do not fully understand why this pattern is evident. Second, we do not understand whether using tools is of importance in the lives of the animals that display it. In only a few instances, all in chimpanzees, has the value of using a tool been documented from the standpoint of solving a real-world problem for the animal. However, there is considerable variability across populations of chimpanzees in the frequency and forms of tool use. Does the variability reflect unimportance to the animals? Finally, we do not understand the psychological basis of tool use in any species, nor the extent to which the psychological basis varies across species. For the most part, our knowledge of tool use remains a catalog of interesting observations, the psychological basis and ecological value of which remain to be determined. Nevertheless, the parallels between tool-use behaviors in nonhuman species and our own species compel attention. Comparisons of psychological and ecological features of tool use in nonhuman primates and humans are likely to aid our reconstruction of milestone events in human evolution. In addition, these behaviors are strongly linked in our thinking with ideas about practical intelligence, and this is a sufficient reason for them to hold our abiding interest.

References Cited

BECK, B. B. (1980). *Animal Tool Behavior.* Garland, New York.

BOESCH, C. (1993). Aspects of transmission of tool use in wild chimpanzees. In *Tools, Language and Cognition in Human Evolution* (K. R. Gibson and T. Ingold, Eds.), pp. 171–183. Cambridge Univ. Press, Cambridge, UK.

BOESCH, C., and BOESCH, H. (1983). Optimisation of nut-cracking with natural hammers by wild chimpanzees. *Behaviour* **83,** 265–285.

BOINSKI, S. (1988). Use of a club by a wild white-faced capuchin (*Cebus capucinus*) to attack a venomous snake (*Bothrops asper*). *Am. J. Primatol.* **14,** 177–180.

FERNANDES, M. E. (1991). Tool use and predation of oysters (*Crassostrea rhizophorae*) by the tufted capuchin, *Cebus apella apella,* in brackish water mangrove swamp. *Primates* **32**(4), 529–531.

FRAGASZY, D. M., and ADAMS-CURTIS, L. (1991). Generative aspects of manipulation in tufted capuchin monkeys (*Cebus apella*). *J. Comp. Psychol.* **105,** 387–397.

GOODALL, J. (1986). *The Chimpanzees of Gombe.* Belknap Press of Harvard Univ. Press, Cambridge, MA.

HUNT, G. R. (1996). Manufacture and use of hook-tools by New Caledonian crows. *Nature* **379,** 249–251.

KLÜVER, H. (1933). *Behavior Mechanisms in Monkeys.* Univ. of Chicago Press, Chicago.

KÖHLER, W. (1976). *The Mentality of Apes.* Liveright, New York. (Original work published 1925)

LADYGINA-KOTS, N. (1935). Ditia schimpanze i ditia sheloveka (Infant ape and human child). *Sci. Mem. Museum Darwinianum Moscow,* 524–529. (Russian with English summary)

LEAKEY, L. S. B. (1961). *The Progress and Evolution of Man in Africa.* Oxford Univ. Press, London.

MCGREW, W. C. (1992). *Chimpanzee Material Culture.* Cambridge Univ. Press, Cambridge, UK.

REYNOLDS, P. C. (1982). The primate constructional system: The theory and description of instrumental tool use in humans and chimpanzees. In *The Analysis of Action* (M. van Cranach and R: Harré, Eds.), pp. 243–385. Cambridge Univ. Press, Cambridge, UK.

TELEKI, G. (1974). Chimpanzee subsistence technology: Materials and skills. *J. Hum. Evol.* **3,** 575–594.

TOMASELLO, M., and CALL, J. (1997). *Primate Cognition.* Oxford Univ. Press, New York.

VAN SCHAIK, C. P., FOX, E. A., and SITOMPUL, A. F. (1996). Manufacture and use of tools in wild Sumatran orangutans. Implications for human evolution. *Naturwissenschaften* **83,** 186–188.

VISALBERGHI, E. (1987). Acquisition of nut-cracking behavior by two capuchin monkeys (*Cebus apella*). *Folia Primatol.* **49**(3), 168–181.

VISALBERGHI, E. (1990). Tool use in *Cebus. Folia Primatol.* **54,** 146–154.

VISALBERGHI, E., and FRAGASZY, D. (1990). Do monkeys ape? In *"Language" and Intelligence in Monkeys and Apes* (S. Parker and K. Gibson, Eds.), pp. 247–273. Cambridge Univ. Press, Cambridge, UK.

VISALBERGHI, E., and LIMONGELLI, L. (1996). Acting and understanding: Tool use revisited through the minds of capuchin monkeys. In *Reaching into Thought. The Minds of the Great Apes* (A. Russon, K. Bard, and S. Parker, Eds.). Cambridge Univ. Press, Cambridge, UK.

YAMAKOSHI, G., and SUGIYAMA, Y. (1995). Pestle-pounding behavior of wild chimpanzees at Bossou, Guinea: A newly observed tool-using behavior. *Primates* **36**(4), 489–500.

YERKES, R. M. (1943). *Chimpanzees.* Yale Univ. Press, New Haven, CT.

YERKES, R. M. (1973). The mind of a gorilla. Parts 1 and 2. In *Classics in Psychology.* Arno Press, New York. (Originally published in *Genetic Psychol. Monogr.* **2,** 1–193, 375–551, 1927)

General References

Berthelet, A., and CHAVAILLON, J. (Eds.) (1993). *The Use of Tools by Human and Non-Human Primates.* Clarendon, Oxford.

BOESCH, C., and BOESCH, H. (1999). *The Chimpanzees of the Taï Forest. Behavioural Ecology and Evolution.* Oxford Univ. Press, Oxford.

KEN CHENG

Department of Psychology
Macquarie University
Sydney NSW 2109, Australia

Many animals have to solve the problem of finding a place. Two different and common methods for pinpointing a place are described. In path integration or inertial navigation, the traveling animal keeps track en route of the vector (direction and straight line distance) covered and returns to the starting point by reversing the computed vector. In landmark-based spatial memory, the target location is specified in terms of its spatial relations to surrounding landmarks. Research on ants, bees, birds, rodents, and humans is presented.

Spatial Representation

The Importance of Finding a Place

A frequent problem in the lives of many animals is finding a place. The survival and continuation of the species often depend on solving this problem. This article presents some examples of place finding in a range of creatures and focuses on two ubiquitous mechanisms that use route-based and landmark-based information, respectively.

The digger wasp is so called because the female digs holes in the ground in which to lay her eggs. The future larvae need food, so the wasp is faced with the problem of finding her nests again after she acquires the prey to provision her offspring. The information about the location of a nest is apparently acquired in a characteristic reconnaissance flight as she leaves the nest. The wasp circles her nest in larger and larger arcs before finally taking off. The digger wasp's continued progeny depends on finding the nests again.

A close relative of the digger wasp, the honeybee, finds her way time and again between her hive and a favorite clump of flowers. In temperate climates, successful foraging, and hence successful place finding, is crucial for the survival of the hive. Only an estimated 25% of newly established hives survive the winter. The honeybee's spatial prowess has been much studied, and its use of landmarks is discussed later.

Many birds store food in scattered locations in order to capitalize on short periods of food abundance. The most prolific food storer is the Clark's nutcracker, a robin-sized bird found on the eastern side of the Rocky Mountains in North America. Pine seeds form a large part of their diet, and they store many caches of seeds to survive the lean and cold winter months. They need to retrieve approximately 2200–3000 caches each winter. Field studies estimated that they can remember 70% of the caches they store, which is likely to be an underestimate because rodents often steal their stored hoards. In the lab, up to 90% of their attempted probes are at locations where they have stored food, even after a 30-day delay and after the food has been experimentally removed (Balda and Turek, 1984). The birds remember the locations of their caches for at least 9 months, making them perhaps the champions of spatial memory among all creatures.

Other food-storing birds include other members of the Clark's nutcracker's corvid family, such as jays, and members of the parid family, which include tits on the European continent and chickadees on the North American continent. Although none are as prolific as the Clark's nutcracker, memory for the location of stored food plays an important role in the lives of all these birds.

Place Finding as Servomechanism

To maintain life, many variables need to be kept within narrow ranges. Physiological servomechanisms help maintain such measures as blood sugar level and core temperature within acceptable ranges. Place-finding mechanisms can be conceived of as behavioral servomechanisms. A servomechanism tracks errors and reduces them. At its core is a comparator that compares a "desired" standard specification on a variable with a reading of the actual measure on that variable (Fig. 1). The difference between actual reading and standard specification constitutes an error signal. Error signals drive the servomechanism to generate actions that reduce the error, thus completing a negative feedback loop.

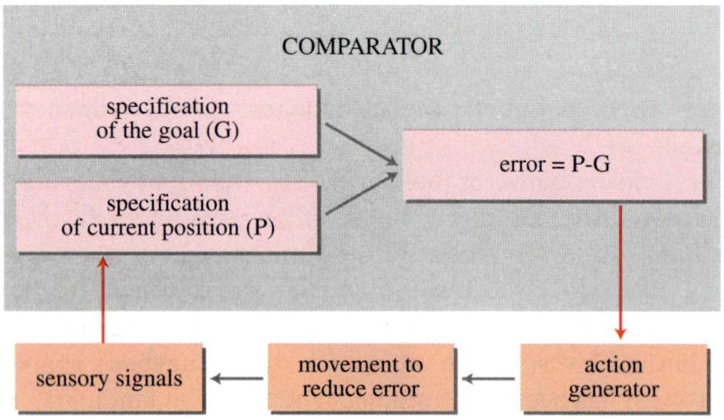

FIGURE 1 The mechanisms of place finding viewed as servomechanisms. The comparator encodes some specification of the goal to be reached, which is compared to a specification of the creature's current place. An action is then generated to reduce the discrepancy between current and target places, bringing the creature closer to the goal.

In the case of place finding, the standard specification is of the target place to which the animal needs to return. Different mechanisms specify the target in different terms. Two ubiquitous mechanisms are discussed here — path integration and landmark-based spatial memory. These mechanisms work in the last stage of navigation, when an organism has to zero in on an exact spot. For navigation over longer distances, see the articles on migration and navigation.

Path Integration

Path integration is also known as *inertial navigation* or *dead reckoning* (shortened from the term "deduced reckoning"). In path integration, the target place is the starting place, most typically the home of a creature, and places are defined in terms of the vector from the start. A vector is the classical algebraic vector consisting of a direction and a straight-line distance. What the creature has to do in using this mechanism is to keep track of the straight-line distance and direction from the start as it meanders on its outward journey. It literally has to add up, using vector addition, all the twists and turns it takes. The continuous addition of tiny pieces of the journey constitutes integration in the mathematical sense, which thus gives the mechanism its name. Another useful image is a tightly held thread pinned to the starting place (Wehner and Wehner, 1990). Wherever the creature wanders, that thread points straight to the start. Of course, the thread is computed in the brain rather than physical. The computed vector specifies where one is with respect to the start. The starting place is the 0 vector. The route vector thus constitutes the error. When it is time to head home, the creature moves in the opposite direction to the calculated route vector until it is reduced to 0. The process does not use landmarks to specify a place. It is limited in that unsystematic errors accumulate the longer one journeys, and without calculating one's place with respect to landmarks (also called taking a fix), no mechanism corrects the errors. When the errors are too large, the mechanism becomes useless. Nevertheless, path integration is ubiquitous and is extremely useful when landmarks are missing or when one is in new territory and relations between landmarks and goal are unknown.

Landmark-Based Spatial Memory

Landmark-based spatial memory is also known as *piloting* or *mnemotaxis* (mnemo for memory; in this case, memory of landmarks). The target place is defined with respect to its spatial relations to surrounding landmarks. Different creatures define different spatial relations, as discussed later. A creature's current position is also defined with respect to landmarks. The creature moves so as to reduce the discrepancy between the way landmarks currently appear and the way the landmarks should appear at the target place. This mechanism is not subject to cumulative errors, and taking a fix on landmarks often helps a creature update the calculations based on path integration. Piloting and path integration are thus often used in tandem.

Path Integration in Desert Ants

On the hot and barren deserts of Tunisia live ants of the genus *Cataglyphis,* a tiny 12-mg creature that is perhaps the champion of path integration. The home of the ant is a tiny inconspicuous hole in the ground on a mostly featureless plain. The unforgiving heat places a premium on timely returns home. It is not surprising that path integration is well developed in this insect.

Extensive studies of *Cataglyphis* have been carried out by Rüdiger Wehner and colleagues. On natural foraging trips, the worker may wander on tortuous curves in search of fodder. However, it returns home more or less in a beeline. Figure 2 shows one spectacular example in which a forager meandered a path length of 592 m before finding a morsel. Its return trip, shown by the red line, covered a homeward course of 140.2 m, only 1.06 times the length of the straight-line distance home.

The ant will rely on artificially provided landmarks near its home (Wehner *et al.,* 1996), but its primary mode is path integration. To experimentally ensure that path integration was being used, the ant was provided with an artificial feeding site at a particular distance and direction from home (Wehner and Srinivasan, 1981). Upon reaching the feeding site, the worker was captured in a dark box and displaced to another place. Should the ant be using landmarks to pilot its way home, it should head toward the actual nest. *Cataglyphis* did not do this. Instead, it headed in the direction that would have led it home had it not been displaced. It traveled for about the distance that would have taken it home and, not finding its home, made a sharp turn that indicated the start of a systematic search for home. Path integration continued during this search.

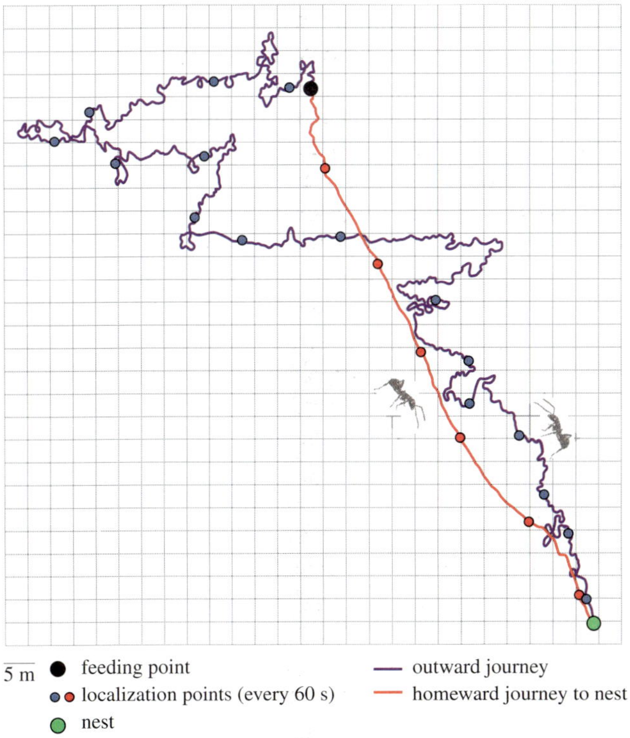

FIGURE 2 Path of a foraging ant *Cataglyphis fortis.* The ant meandered for 592 m (blue line) before finding food. The homeward journey to the nest (red line) is only 140.2 m, almost equivalent to the straight-line distance to the nest (reproduced with permission from Wehner and Wehner, 1990).

The search describes a pattern of a spiral outwards broken occasionally by a return to the start of the spiral (Müller and Wehner, 1994), with ever larger spirals as time passes. To return to the starting place of the search, the ant must keep track of its path. Again it was not using local landmarks. During its systematic search, the ant might be captured and displaced once again (Wehner and Srinivasan, 1981). The center of search was then displaced by the vector by which the ant was displaced. The insect could continue in this fashion until it succumbs to heat in about 2 h.

Determining Direction

To integrate a path, a creature must keep track of direction and distance. How hymenopterans such as ants and bees determine direction is well understood. They use what is called a *sun compass,* determining direction by two properties of the sun visible to them. One is the sun, and the other is polarized light created by the scattering of sunlight by air molecules (Wehner and Wehner, 1990; Wehner *et al.,* 1996). Humans are not sensitive to the pattern of polarized light, but many insects and birds are sensitive to the pattern. Hence, even when clouds cover the sun, a patch of blue sky is enough to let these latter creatures determine the direction of the sun.

The direction of the sun is not the only requisite information. Because the sun changes positions in the sky as time passes in a day, its movements must be accounted for. To do this, the creature must possess a clock that records the time of day, called a *circadian clock.* Most multicelled creatures possess circadian clocks.

Determining Distance

Originally, it was thought that insects measure distance by the amount of energy that they expend traveling. This hypothesis has recently fallen out of favor. In ants, Ronacher and Wehner (1995) found that the size of the morsel they were carrying home did not affect the distance that *Cataglyphis* traveled on the homeward trip, evidence against the energy hypothesis. Functionally, energy expended is an unreliable measure of distance precisely because the load carried might differ from one trip to another. In other experiments, a moving strip was placed under the ants en route home. The ants traveled on stationary Perspex, with a moving visual pattern underneath them. Compared with stationary patterns, a pattern moving in the same direction as an ant made it travel farther. A pattern moving in the opposite direction made the ant travel less. Thus, one measure of distance that insects employ is the visual flow. As the forager moves, the patterns on the ground appear to move backwards beneath it, and this visual flow can be used to calculate distance traveled. However, this factor did not account fully for the ant's performance, so Ronacher and Wehner suspected that kinesthetic cues generated by active muscle movements also contribute to the measure of distance.

In honeybees, Esch and Burns (1995) performed an ingenious experiment to rule out the energy hypothesis. The bees were trained to find food attached to a helium balloon. Gradually, the experimenters lifted the balloon up in the air. As it ascended, the bees had longer to travel to the balloon. Surprisingly, the foragers communicated to their hive mates by their dance that the distance was shorter. This paradoxical result can be explained if the honeybees are assumed to measure distance traveled by the amount of visual flow underneath them as they fly. The higher up they fly, the less the visual flow, and so the less the calculated distance.

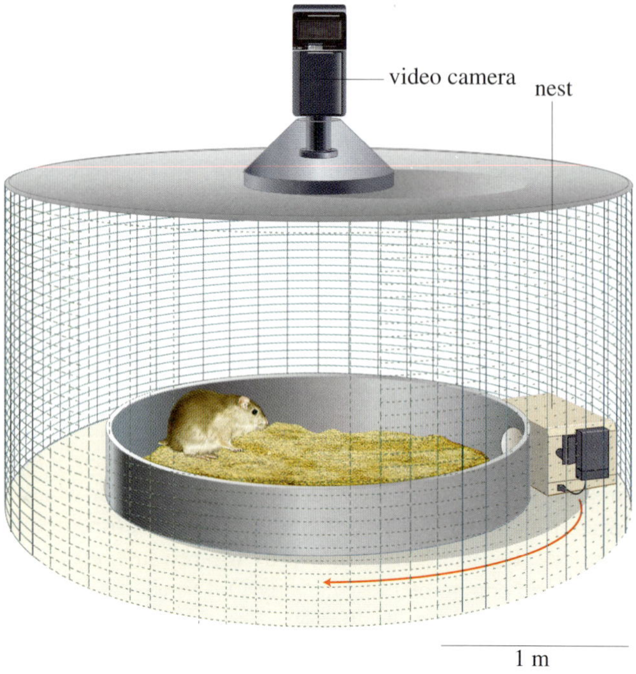

FIGURE 3 Apparatus for studying path integration in hamsters. The nest is just outside the arena and can be rotated. Landmarks may be provided on the enclosure. During experimentation, a thick layer of sawdust covers the arena. A video camera records the rodent's journeys (adapted from Etienne *et al.,* 1995).

Path Integration in Hamsters

The golden hamster hoards food. This motive was exploited in laboratory experiments on path integration carried out by Ariane Etienne. The experimental animal lived in a nest at the edge of a circular arena (Fig. 3). In the basic experimental task, it was lured into the arena to retrieve a bit of food for hoarding. The direction that the hamster traveled to return home was measured (Etienne *et al.,* 1998).

Every effort was expended in the experiments to prevent the hamsters from using any strategy other than path integration. Tests were done in the dark with an infrared camera recording the paths traveled by the creatures. The sand on the floor was swept to eliminate odor cues on the substrate. Sometimes, white noise was played to eliminate possible auditory cues. Also, sometimes the nest box was rotated while the hamster was in the arena to eliminate cues from the target. Even when all these possible sources of information are controlled for, the hamster that found food at the center of the arena headed more or less in the direction home (if the nest had not been moved) or in the direction of the nest before it had been moved (if the nest was moved in a trial).

Tracking Direction

Rodents moving in the dark are obviously not using the sun to determine direction. They use two sources of information. One source is their vestibular system in the inner ear. This system consists of semicircular canals filled with fluid. Rotation, whether active or passive, moves the fluid around in the canals, knocking them against sensitive sensory hairs. The other source of information comes from movement. Commands going out to muscles and kinesthetic information coming from the muscles can both be registered to keep track of the path being traveled.

Keeping track of one's turning is a limited enterprise in that errors accumulate without correction. When the errors are large enough, the path integration system is nonfunctional. Experiments from Etienne's laboratory show that after three passive rotations of 360° or five active rotations, the hamster is at chance at picking out the homeward direction. Experimentally, passive rotation was achieved by rotating the hamster on a platform as it was gathering food at the center of the arena for hoarding. Active rotation was achieved by pulling a bit of food on a string in front of the hamster. The food went round and round, inducing the hamster to turn to follow it.

Tracking Distance

Experiments from Etienne's lab show that the hamster can keep track of the distance covered by its active movement (Séguinot *et al.,* 1993). Visual cues are not necessary because they can do this in the dark.

Visual flow, however, has been shown to be one way of keeping track of distance in goslings (von St. Paul, 1982). These goslings were first imprinted on an experimenter.

They were then trained to find the experimenter at home whenever she disappeared, thus establishing a homing motive. Next, they were experimentally led away from home under active movement or passively in a cart. As long as the goslings saw objects passing by out the side of the cart, they could keep track of both how they were translated (requiring the estimation distance) and how they were rotated. Visual information is not necessary for keeping track of rotation: Even in a completely enclosed cart, the goslings kept track of rotation, presumably using their vestibular system.

Systematic Error in Path Integration

A systematic error is a bias in the average response. A nonsystematic error is equally likely to deviate from the correct value on one side or the other. A systematic error, on the other hand, is more likely to deviate in one direction rather than another from the correct value. When many nonsystematic errors are averaged, a value approximating the correct value is reached. A systematic error means that on average, errors are in one direction. I previously discussed nonsystematic errors, whose accumulation eventually renders path integration nonfunctional. A systematic error is also found in path integration in a wide range of creatures from ants to humans (Fig. 4).

The error occurs when the path-integrating animal is forced to make a sharp turn before striking out in the homeward direction. In such cases, the path integrator on average turns too much, past the homeward direction. Such an error means that creatures are not performing full trigonometric calculations in solving the task. Full trigonometric calculations on directions and distances would deliver, on average, the accurate direction; they would not produce any systematic errors. Some arithmetical shortcut to full trigonometric calculations is being employed (Etienne *et al.*, 1998; Wehner *et al.*, 1996).

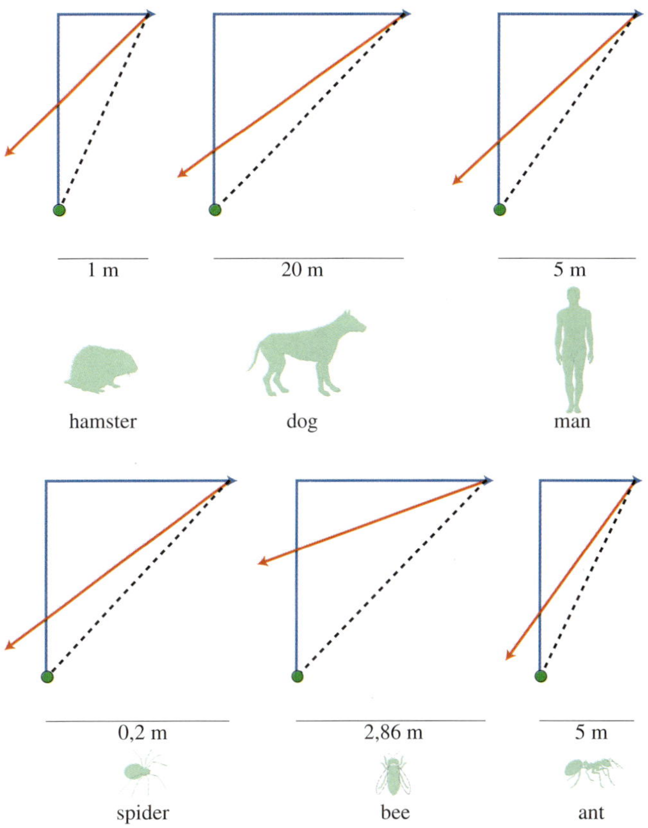

FIGURE 4 Systematic error in path integration. After making a sharp turn of 90° on the outward journey (in blue), species which use the path-integration strategy, from ants to humans, turn on average too sharply on the homeward journey (in red) (adapted from Etienne *et al.*, 1998).

Landmark-Based Spatial Memory in Honeybees

Nobel prize winner Karl von Frisch (1953) first demonstrated the use of landmarks, patterns of colors, in honeybees. His honeybee subjects lived in one of a row of rectanguloid hives. The front of each hive was distinctly colored. von Frisch's experiment consisted of shifting the colors of three hives — the one in which his subjects lived and two neighboring hives. The entire pattern of three colors was shifted one hive unit to the right or to the left. Under such conditions, every bee that came out of the experimental hive went into the neighboring hive whose color matched that of its usual home hive. Thus, the bees followed the pattern of colors on the front of the hives and ignored smells and sounds emanating from their home hive.

The most systematic studies of the use of landmarks by an invertebrate are from work on honeybees, mainly done by Thomas Collett (Cartwright and Collett, 1982, 1983; Collett and Baron, 1994). In these studies, free-flying honeybees were induced to enter a lab room to find strong sugar water. The experimental reward was much sweeter than anything they might find elsewhere, thus inducing them to return time and again to the lab. An experimental array of landmarks was set up in the lab, consisting typically of cylindrical tubes. The little dish of food to be found remained in a constant place with respect to the array. From one visit to another, however, the array was displaced, without rotation, in the lab. This means that the target did not stay in a constant location with respect to any landmark other than the experimentally provided tubes. Occasionally, individual bees were tested in the absence of the sugar water. Their search behavior was videotaped, and later frame-by-frame analysis extracted the amount of time the searcher spent at each small unit area of space.

Two such search distributions are presented in Fig. 5. In both cases, the array with which the bees were tested matched what they encountered during training (except that sugar water was absent). When trained and tested with a single cylinder as a landmark, the search distribution was

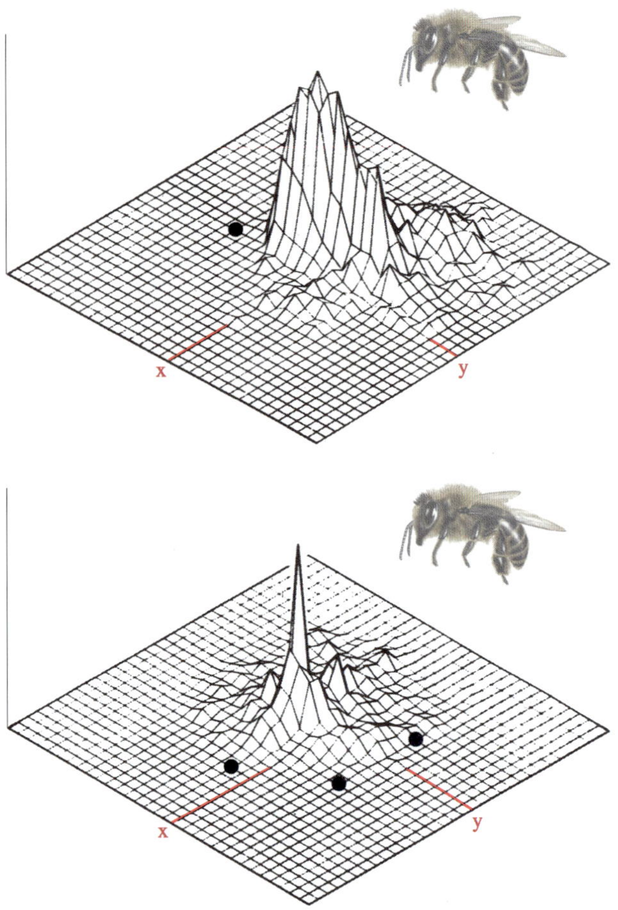

FIGURE 5 Two types of search distribution for honeybees: *x* and *y* indicate the coordinates of the food location. The blue circles indicate cylindrical tubes used as landmarks, with respect to which the food stayed in a constant position. During experimental tests, the food was absent (reproduced with permission from Cartwright and Collett, 1983).

scattered but nevertheless centered on where food would usually be found. This means that honeybees can determine compass directions from something other than the landmark array because a single cylinder gives no indication of direction. In fact, as noted later, further experiments show that the honeybee only uses external cues to determine direction and not the array of landmarks. With three cylindrical tubes as landmarks, the search distribution was more tightly clustered around the target, demonstrating the unsurprising fact that three landmarks pinpoint a location better than one.

Coding Distance

Interesting insights on the mechanisms of landmark-based memory can be obtained by transforming the landmark array on tests. With a single cylinder as an array, its size may be changed (Cartwright and Collett, 1983). Making the landmark larger makes the bees search farther away, whereas making it smaller makes the bees search closer. In part, then, honeybees judge the distance to a landmark by the size (angle) that an object projects on the eyes. Outside of experiments, this works well because objects do not change sizes capriciously. Honeybees can also use motion parallax to judge distance. Motion parallax arises from the movement of the perceiver. When we move our head from right to left, the visual scene appears to move from left to right. The amount and rate of movement, however, depend on how far an object is from our eyes. Objects that are near shift quickly and a large distance, whereas objects that are far shift slowly and little. This motion parallax thus provides a visual cue to distance. For the compound eye of the honeybee, these are the only cues to distance available.

Keeping Track of Compass Direction

Using compass direction poses some problems. Every time the insect turns, the direction in which a landmark should be with respect to itself changes. The mental "landmark map" must rotate with the bee, so to speak, and this seems a complicated process. In fact, no such rotation is necessary. The solution here is simple if not the most elegant: Bees fly facing the same direction when doing landmark-based search (Collett and Baron, 1994). Thus, they first approach the landmarks to be used. Somewhere near the goal, they switch to flying while facing the same direction and doing landmark-based search. Compass direction has now been reduced to the part of the eye on which a landmark should appear.

An Elemental System: No Coding of Landmark Configurations

When we code a location in terms of landmarks, we naturally think of configurations of landmarks. Suppose you were shown a dish of sugar water on a table. A blue cylinder is on the east side of the dish and a yellow cylinder on the west. From trial to trial, the whole array of dish and cylinders shifts about on the table but always in the same configuration. We naturally think of the dish as in between the cylinders. If the cylinders reversed positions on a "test," with the blue one west and the yellow one east, and the dish was absent, you would probably think that the dish should still be in between the cylinders. Human evidence in support of this claim is presented later; this is not the case for honeybees.

Recently, precisely this experiment was done on honeybees in my laboratory. They were trained to find food between two colored cylinders. For half of the bees, the yellow one was always on the east side and the blue one always on the west side, and for the other half the blue one was always on the east side and the yellow one always on the west side. On the crucial tests with the positions of the landmarks reversed, the honeybees hardly searched in the middle. Instead, they searched either west of the western cylinder or east of the eastern cylinder. At these positions, they maintained the correct compass direction and distance to one of the colored cylinders.

In general, when an array of experimental landmarks is rotated, honeybees do not "follow" the rotation. They try to match the compass directions and distances of individual landmarks in the following servomechanistic fashion (Cartwright and Collett, 1982, 1983). For each landmark, and landmarks are probably defined by their color, a compass direction and distance are defined as standards. The standards specify how the landmark should appear at the goal. As indicated previously, distance is specified by projected retinal size and motion parallax, and compass direction is equivalent to the part of the eye on which the landmark should appear. When doing landmark-based search, a landmark that appears left of where it should be induces the bee to turn to the left, whereas a landmark that appears right of where it should be induces the bee to turn to the right. A landmark that appears too near induces the bee to move away from it, and a landmark that appears too far induces the bee to move toward it. The bee averages the dictates of all these separate independent calculations. We can think of the system as using many separate servomechanisms, two for each landmark — one based on distance and one based on direction. Each tries to maintain just one standard: the right distance or direction (but not both) to one landmark. Also, the whole is simply the sum of the independent parts. Nothing in this system codes anything about two or more landmarks taken together, a configuration of landmarks. Such a system is called *elemental.*

Landmark-Based Spatial Memory in Vertebrates

Many vertebrates also use landmarks in searching for a place. The list includes food-storing birds (black-capped chickadees, European jays, and Clark's nutcrackers), nonstoring birds (chicks and pigeons), rodents (mice, gerbils, hamsters, and rats), and primates (monkeys, chimpanzees, and humans) (Cheng and Spetch, 1998). The list is surely much longer. Like honeybees, vertebrates use the metric properties of distances and directions. In contrast to honeybees, vertebrate species studied in detail — rats, gerbils, pigeons, and humans — sometimes use the configuration of landmarks.

Two Methods Yield Similar Results

Comparison of humans and pigeons stems largely from a collaboration between Marcia Spetch and me (Spetch *et al.,* 1996, 1997). We started with pigeons and later included humans for comparison. I had first pioneered experiments in which pigeons searched for buried food in an arena covered with wood chips. We called this the *field task.* Domestic pigeons were first trained to walk into the arena for a few visible grains or pellets. The food might be inside a bottle cap or in an exposed hole in the arena floor. Gradually, over trials, increasingly more of the food was covered until finally the wood chips covered the food uniformly. Landmarks were set about for the pigeon to use. They might be the walls of the arena, bottles, or blocks. After the bird had learned to find the buried food, occasional tests were given. As with the bees, on tests the food and its container (food cup or hole in the ground) were absent. In some cases, the entire arena was rotated so that the search area did not have any odors or other cues left from searching on rewarded training trials. Any landmarks in the arena were rearranged on such tests so that they appeared as they would on a training trial. A video camera recorded searching behavior, and frame-by-frame analysis later yielded search distributions.

On tests with landmarks arranged as in training conditions, the birds searched most where the food normally would be found. When landmarks were shifted, they typically shifted their peak place of searching as well, thus showing conclusively that they were using landmarks to guide their search.

Birds can also be trained to search on the surface of a computer monitor. Although pigeons had been tested on the computer before, the search task was pioneered in Spetch's lab. The monitor was covered with a Perspex panel to take the beating of pigeon pecks, and it was surrounded by a touch-sensitive frame. In our version of the frame, infrared beams crisscrossed vertically and horizontally. When something touched the screen, it broke the beams. The time and place of the touch were registered and stored on the computer. Such a setup is called a *touch-screen,* and the task is called the *touch-screen task.*

On the touch-screen task, the pigeon is first presented a target object on the screen, followed shortly by the delivery of food. A typical target is a 2-cm computer-drawn square. Food delivery may consist of a pellet of pigeon food rolling into a hopper or a hopper of grains being lifted up from below to give the bird a few seconds of access to food. Under such conditions, the bird learns to anticipate getting food when the object appears, gets excited, and starts pecking at the object before the food arrives — a process known as *autoshaping.* We can then put landmarks about the target and require the bird to peck at the target for food. The target stays in a constant location with respect to the landmarks. However, the entire array is moved about the screen from trial to trial, thus forcing the bird to localize by using the target–landmark array. Then, in a process called *fading,* the target is made dimmer and dimmer over trials until finally it is totally absent. The pigeon has now been trained to peck at an unmarked location for food, a location defined solely by its spatial relations to computer-drawn landmarks on the screen.

After the bird has been trained, unrewarded tests can be given. On such tests, a landmark array appears on the screen, but food delivery is withheld. The locations at which

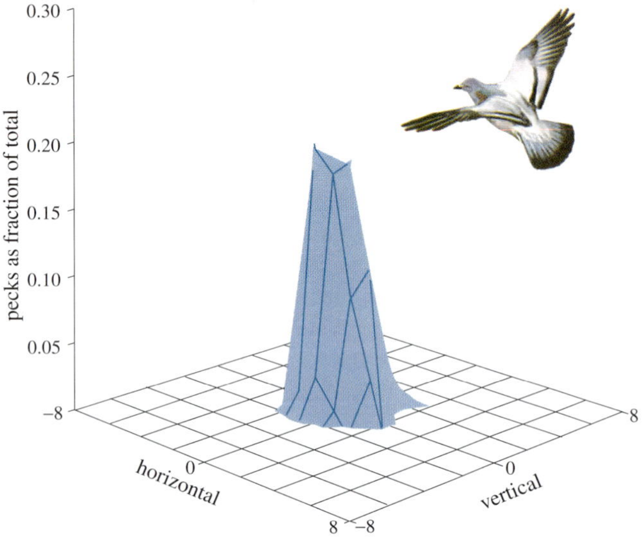

FIGURE 6 Search distribution of a pigeon on the touch-screen task. The number 0 indicates the coordinates of the target location. Landmarks provided for the bird are not shown. Each unit is approximately 1 cm. On experimental tests, no reward was dispensed (reproduced with permission from Spetch and Mondloch, 1993).

the pigeon pecks are registered on computer, saving the labor of video extraction. An example of a search distribution on the touch-screen task is shown in Fig. 6. The distribution was obtained with the landmark array matching what the birds encountered during training. Similar search distributions are obtained from the field task.

The field task and the touch-screen task differ in many ways. The touch-screen is vertical, whereas the floor is horizontal. The touch-screen is far smaller; on the floor, the bird can walk around in the space. Food is at the target location in the field task; it is delivered at some distance from the target location in the touch-screen task. Landmarks are much nearer (to the target and to each other) on the touch screen, and the bird can see the entire array in one glance. Landmarks are usually scattered and much farther apart on the floor. However, despite these differences, basically the same pattern of results is found on the two tasks.

We did analogous experiments on university students (Spetch *et al.*, 1996, 1997). Again, both field tasks and touch-screen tasks were presented. On the touch-screen task, the students poked at the screen with the eraser end of a pencil and received points for finding the target. The "field" for humans consisted of spaces on the table top and, in one case, a grassy plot outdoors. The students searched for a hidden object during these tasks. For humans, the two methods also produced the same pattern of results.

Humans and Pigeons Differ

Using the transformational strategy described previously, we modified the landmark array on some tests to examine what cues each species was using. One informative manipulation was to enlarge the landmark array by moving landmarks farther apart from one another. In one version of these experiments, four identical-looking landmarks were placed in a square array. The target was in the middle of the array. Touch-screen and field versions of this experiment were run on both species. When the array was enlarged, humans continued to search in the middle of the array. Students told us that they used some sort of an abstract "middle rule" in doing the task. They coded the target as being in the middle of the array.

The pigeons did not search in the middle on an expansion test. They searched at a location that was the right distance and direction from one of the landmarks. For example, if landmarks were 25 cm away northwest, northeast, southwest, and southeast, a bird might search 25 cm southeast of the northwest landmark. Different birds picked out different landmarks, but each bird was consistent from one test to another.

Both pigeons and humans used metric information (i.e., distances and directions) in landmark-based search, but they have different rules for using metric information. On expanded arrays, the human's search location matched the training situation in directions to landmarks but not in distance. The pigeon's search location matched both distance and direction, but only to one of the landmarks.

Vertebrates Use Configurations of Landmarks

Vertebrates sometimes use the configuration of landmarks. All species examined in sufficient detail do so, including gerbils, rats, pigeons, and humans.

The students who told us that they were using a middle rule were certainly indicating the use the configuration of landmarks: The middle is only defined with respect to a configuration of many landmarks. More convincing evidence than verbal reports came from a scrambling manipulation (Spetch *et al.,* 1996). Humans were trained to search in the middle of a square array of four distinct landmarks. On a scrambled test, the positions of the four landmarks were randomly interchanged, although a square array was maintained. Individually, each landmark pointed to some other target location than the center. For instance, if the top left landmark was switched to the top right position, it would point to a target to the right of the array. The students continued to search in the middle on a scrambled test, showing that they were using the configuration of the array in deciding where to search.

Consider again the pigeons' reaction to an expanded array of four identical landmarks with the goal in the middle. The birds searched at the correct location according to one particular landmark. To do so consistently, they must pick out that landmark. They could not do so by the characteristics of the landmark because all landmarks were identical in appearance. They could not pick out the landmark by its absolute position in space because the entire array was shifted about from trial to trial. They could only pick out a landmark by its position relative to the other landmarks in the array. That is, they identified a landmark by its position relative to the configuration of landmarks.

Experiments on gerbils were done by Collett *et al.* (1986). The gerbils searched in a circular arena covered with wood chips. The target was a hidden sunflower seed. Like the bee experiments, tubes served as landmarks. In one experiment, the gerbils were trained to find a target location defined in relation to two distinct landmarks. In one test, the array was rotated. The gerbils searched in the appropriate place according to the rotated array. They were thus defining directions with respect to the array. They could not define directions with respect to any individual landmark because it was cylindrical and gave no directional information. They must have been using the configuration of landmarks to define direction. Recall that honeybees, in contrast, do not "follow" the rotation of an array of landmarks in the fashion of the gerbils. The bees try to maintain the compass direction to a subset of landmarks.

In another experiment showing the use of configuration in gerbils, the training setup consisted of two identical-looking landmarks. The target location was in between the two landmarks but at some perpendicular distance from the line connecting the two landmarks. As in our pigeon experiments, the two landmarks were moved apart in some tests. Like the pigeons in the expansion tests, the gerbils searched at the correct location according to one of the landmarks. The same logic applies: They must pick out which landmark is which by their spatial relations to one another. Therefore, they must code the configuration of two landmarks.

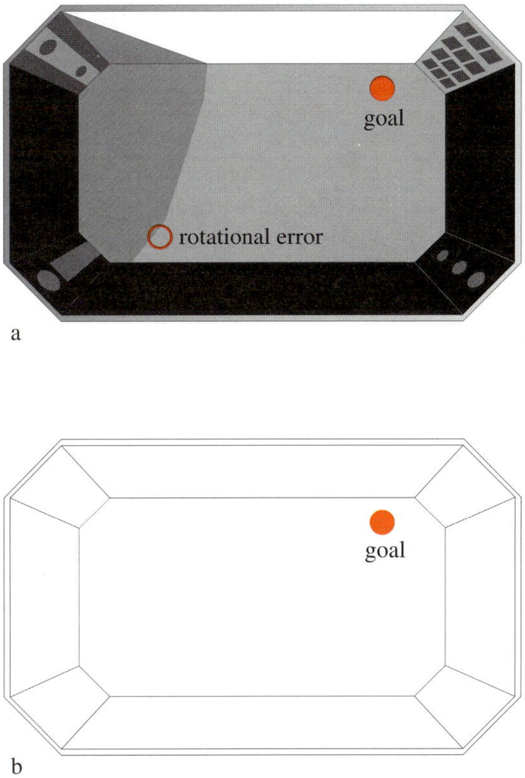

FIGURE 7 The geometric module in the rat. In this experiment, the rat has the task of remembering a site (goal) at which food is to be found. The search arena (a) is rectangular and contains many features that can be used as cues (one of the longer walls is white, whereas the other three are black; the corners have panels covered with material of different patterns, etc). The rat is then required to return to the same goal (where the food was buried). In most trials, the rat searched at the goal or at the location marked as rotational error, which is rotated by 180° from the goal through the center of the arena. To explain rotational error, it must be supposed that the rat uses a record containing only the broad shape of the environment (b). The record can be matched to the world in two ways, one of which produces the correct solution and the other of which produces the rotational error. In the first case, the top walls of the two pictures in the figure correspond, leading the animal to search in the correct place. In the second case, the top wall in a is made to correspond to the bottom wall in b; this leads the rodent to commit the rotational error (reproduced with permission from Cheng and Spetch, 1998).

Geometric Module in Rats and Children

In rats, evidence for the use of configuration derives from a systematic error that they make (Fig. 7a). They had the task of remembering a location at which they had been shown some sweet cereal, a favorite of rats. The search arena was rectangular and contained many features as cues. One long wall was white and three were black. In the corners were panels with different textures, different visual patterns, and different numbers of lights shining out of them. Smells of anise and peppermint emanated from two of the corners. The target location for each day differed so that the rat had to learn a location anew each day. In a trial, a rat was put into the arena with the food exposed, sitting in a dish. It was allowed to eat three pellets, after which it was removed. While the rat was out of the arena, the food was buried at the same location under the wood chips that covered the floor of the arena. Actually, this was effected by having a replica of the arena ready before the trial began. After a 90-s delay, the rat's task was to find the buried food again at the same place.

The rats sometimes dug at the correct location for food, but at other times they made a remarkable error (Cheng and Spetch, 1998). They dug at the location indicated in Fig. 7a as the rotational error. This location was rotated by 180° from the correct location through the center of the arena. They made this error almost as often as they made the correct choice. The error is not peculiar to rats. Two-year-old human children also make the same error. To explain this error, we have to suppose that the rat encodes some representation like that shown in Fig. 7b. This geometric module records the broad shape of the environment but not the discrete features on it, such as brightness of walls, visual patterns of panels, and smells. It is a record of pure configuration, shape abstracted from the features that make up the shape. On this record, the goal location is marked.

In the strangely designed experimental environment, which, unlike natural environments, is perfectly rectangular, using the geometric module presents an ambiguity. The rat has two ways of matching the map with the world. In one way of matching, the top of the record in Fig. 7b is matched to the top wall in Fig. 7a; this leads the animal to search at the correct location. However, in another match, the top of the record in Fig. 7b is matched to the bottom wall of Fig. 7a; this leads the rodent to make the rotational error. The existence of the rotational error forces this conclusion.

Other experiments ensured that the rats could actually discriminate the many features provided in the environment. In making the rotational error, the rats and children are (i) measuring the right distances and directions from walls, thus using metric information in landmark-based memory, and (ii) using a record of the pure configuration of shape — a strong sense of using a configuration.

Measuring Distance

How vertebrates measure the distances to landmarks has not been systematically studied. The visual systems of birds and mammals can register many cues for distance, including monocular cues, which require the use of only one eye, and binocular cues, which require the use of both eyes. Monocular cues include accommodation (how the eye muscles must be used to focus the eye on the object), motion parallax, texture and other gradients, and the size that a familiar object projects on the retina. Binocular cues include convergence (the angle that the two eyes have to be at for both eyes to focus on an object) and binocular disparity. The last named cue is based on the fact the two frontally placed eyes receive slightly different views of the world. Based on these differences, depth can be calculated. It is a good guess that a combination of all available cues are used.

Conclusions

This article examined path integration and landmark-based spatial memory, treating both as cases of servomechanisms. In path integration, a creature keeps track of the vector covered en route from a starting point. The vector points to the direction and distance to the starting point. The strategy requires keeping track of the distance and direction traveled simultaneously on a continuing basis. A large array of creatures, both vertebrate and invertebrate, use path integration. In landmark-based spatial memory, the goal location is defined in terms of its spatial relations to surrounding landmarks. Creatures examined in detail use the metric properties of distances and directions in using landmarks. Vertebrate creatures examined sometimes use the configuration of an array of landmarks. Honeybees, on the other hand, seem to only use landmarks elementally, each singly and independently, in specifying the target location. It is not known how the configural–elemental division is distributed across species.

References Cited

Balda, R. P., and Turek, R. J. (1984). In *Animal Cognition* (H. L. Roitblat, T. G. Bever, and H. S. Terrace, Eds.). Erlbaum, Hillsdale, NJ.

Cartwright, B. A., and Collett, T. S. (1982). *Nature* **295,** 560–564.

Cartwright, B. A., and Collett, T. S. (1983). *J. Comp. Physiol. A* **151,** 521–543.

Cheng, K., and Spetch, M. L. (1998). In *Spatial Representation in Animals* (S. D. Healy, Ed.). Oxford Univ. Press, Oxford.

Collett, T. S., and Baron, J. (1994). *Nature* **368,** 137–140.

Collett, T. S., Cartwright, B. A., and Smith, B. A. (1986). *J. Comp. Physiol. A* **158,** 835–851.

Esch, H. E., and Burns, J. E. (1995). *Naturwissenschaft* **82,** 38–40.

Etienne, A. S., Joris-Lambert, S., Maurer, R., Reverdin, B., and Sitbon, S. (1995). *Behav. Brain Res.* **68,** 103–116.

Etienne, A. S., Berlie, J., Georgakopoulos, J., and Mauer, R. (1998). In *Spatial Representation in Animals* (S. D. Healy, Ed.). Oxford Univ. Press, Oxford.

Müller, M., and Wehner, R. (1994). *J. Comp. Physiol. A* **175,** 525–530.

Ronacher, B., and Wehner, R. (1995). *J. Comp. Physiol. A* **177,** 21–27.

Séguinot, V., Maurer, R., and Etienne, A. S. (1993). *J. Comp. Physiol. A* **173,** 103–113.

Spetch, M. L., and Mondloch, M. V. (1993). *J. Exp. Psychol. Anim. Behav. Proc.* **19,** 353–372.

Spetch, M. L., Cheng, K., and MacDonald, S. E. (1996). *J. Comp. Psychol.* **110,** 55–68.

Spetch, M. L., Cheng, K., MacDonald, S. E., Linkenhoker, B., Kelly, D., and Doerkson, S. (1997). *J. Comp. Psychol.* **111,** 14–24.

von Frisch, K. (1953). *The Dancing Bees* (D. Ilse, Trans.). Harcourt Brace Jovanovich, San Diego.

von St. Paul, U. (1982). In *Avian Navigation* (F. Papi and H. G. Wallraff, Eds.). Springer, New York.

Wehner, R., and Srinivasan, M. V. (1981). *J. Comp. Physiol.* **142,** 315–338.

Wehner, R., and Wehner, S. (1990). *Ethol. Ecol. Evol.* **2,** 27–48.

Wehner, R., Michel, B., and Antonsen, P. (1996). *J. Exp. Biol.* **199,** 129–140.

General References

Cheng, K. (1995). In *The Psychology of Learning and Motivation* (D. L. Medin, Ed.), Vol. 33. Academic Press, New York.

Collett, T. S. (1992). Landmark learning and guidance in insects. *Philos. Trans. R. Soc. London, B* **337,** 295–303.

Gallistel, C. R. (1990). *The Organization of Learning.* MIT Press, Cambridge, MA.

Kamil, A. C., and Balda, R. P. (1991). In *The Psychology of Learning and Motivation* (G. Bower, Ed.), Vol. 26. Academic Press, New York.

MELISSA BATESON
Department of Psychology: Experimental
Duke University
Durham, North Carolina 27708, USA

Interval Timing

Interval timing refers to the ability to time intervals in the seconds to minutes range beginning at any arbitrary point in time. The interval timer is a flexible stopwatch-like clock that can be started, stopped, and reset by particular types of stimuli. A hallmark of interval timing is that the accuracy of a time estimate is proportional to the length of the interval being timed. What animals use the interval clock for has not yet been determined; however, it is likely to be involved both in optimal foraging and associative learning and may also be important in many other types of behavioral decisions. The neural mechanisms underlying interval timing have not been discovered; however, pharmacological and lesion studies suggest the involvement of different brain regions and neurotransmitter systems in the clock and memory stages of timing. Interval timing has been identified in a wide range of vertebrate species, and its characteristics are remarkably stable across species. It is therefore probably a very evolutionarily ancient ability.

Biological Clocks and Interval Timing

Time is a dimension that is ubiquitous in all environments, and many biologically significant events have predictable temporal characteristics associated with them. We should therefore not be surprised that many animals have evolved clocks that allow them to exploit efficiently the temporal regularities they encounter in the world.

To most ethologists, "biological clocks" are the endogenously generated oscillations that underlie the cycles of behavior that correlate with environmental periodicities such as day and night (the *circadian clock*), the tides (*circatidal clock*), the phases of the moon (*circaluna clock*) or the annual changing of the seasons (*circannual clock*). Indeed, these are the only types of clock explicitly discussed in most modern animal behavior or ethology texts.

However, psychologists have distinguished two distinct types of time sense in animals: the periodic or phase time sense and the interval time sense (Gallistel, 1990). The *phase time sense* relies on the various oscillations mentioned previously and refers to the ability of animals to predict events that occur at fixed points within the phase of a specific oscillator. For example, bees use their circadian time sense to visit a patch of flowers only during the period of the day when it is producing nectar, and oystercatchers probably use a circumtidal sense to fly to tidal mud flats for foraging only during the period when the tide is out.

Although phase clocks are useful for predicting events that occur at a fixed point within their phase, they are less appropriate for extracting information about other types of temporal regularity that animals may encounter. A biologically significant event may be predicted not by the state of a phase timer but by an external stimulus not locked onto any particular cycle. The type of time sense required to predict this event is usually referred to as an *interval* or *hourglass time sense* because in order to predict the significant event it is necessary to start a timer when a stimulus occurs and time the interval to the significant event. A foraging hummingbird provides an excellent natural example of the interval time sense in use. After the bird has emptied the nectar from a flower, the flower will take a fixed amount of time to refill. A bird that knows the length of this interval can return to the flower as soon as it has refilled.

In addition to allowing an animal to predict when an event will occur, the interval time sense also provides animals with the ability to extract temporal intervals. For example, a foraging starling might time the interval between each prey item it finds in a particular field. These intervals could be used to form an estimate of the rate of prey capture which is known to be an important currency in determining decision making in foraging starlings.

Although it is theoretically possible that the same basic clock could subserve both the phase and interval time senses, there is ample evidence that this is not the case and that different mechanisms underlie phase timing and interval timing. Henceforth, I refer to phase clocks and interval clocks. The term *interval timing* is usually restricted to the timing of intervals in the seconds to minutes range, and it is timing within this range that is the focus of this article. I will not consider the dedicated mechanisms that have evolved to time much shorter intervals such as the system used by bats to time the delay in the return of an echo.

Whereas phase clocks are described in most animal behavior and ethology texts, interval timing has received very little attention from ethologists. Historically, the study of interval timing has been the domain of operant psychologists who study how rats and pigeons respond to artificially generated schedules of reinforcement in the laboratory. As a consequence, work on interval timing has failed to attract the interest of ethologists and behavioral ecologists because of the lack of connection to the problems faced by animals in their natural environments. However, it is becoming increasingly clear that interval timing is likely to be vitally important in many aspects of natural behavior; hence, the aim in this article is to introduce the properties of interval timing and to discuss it from an ethological perspective.

Experimental Paradigms Used to Prove That Animals Have an Interval Time Sense

Killeen *et al.* (1997) provided the following operational definition of timing: "An organism is timing if our clock is a better predictor of its behavior than any other stimulus we can identify." Interval timing is not easy to see in animals in their natural environments because to provide proof of timing it is necessary to eliminate the possibility that the animal is in fact using some external cue. Hence, the evidence that animals have interval timers comes largely from the operant psychology literature in which animals are studied under extremely controlled laboratory conditions.

Operant psychologists have devised many paradigms to investigate timing in animals, which can be broadly divided into either estimation or reproduction procedures. In an estimation procedure an animal is required to attend to a time interval and then make some response based on the duration it estimates after the interval has elapsed, whereas in a production procedure the animal actually reproduces the time interval in real time with its own behavior.

Bisection: A Temporal Estimation Task

Probably the most widely used estimation procedure is *temporal bisection.* In the first stage of bisection the subject is trained to make a simple temporal discrimination between, for example, a tone presented for 2 s and an identical tone presented for 8 s. In each trial one of the two tones is presented. After the tone has finished, two levers are inserted into the box. If the tone was the 2-s one, then the animal is reinforced with food for pressing the left lever, whereas if it was the 8-s one it is reinforced for pressing the right lever (the assignment of lever to tone is arbitrary but always the same for a given subject). Animals such as rats and pigeons will readily acquire this temporal discrimination. Once the animals are performing reliably, probe trials are inserted in which tones of different durations to those on which the animal was trained are presented. In such probe trials the same two levers are still inserted into the box, but no reinforcement is given whichever the animal chooses. Data from the bisection procedure are commonly presented by plotting the lengths of the probe trials on the *x*-axis against the probability of choosing the 8-s lever on the *y*-axis. The function obtained is sigmoid in shape and yields two important pieces of information: The steepness of the curve gives an indication of the precision of timing, and the time corresponding to a 50% probability of choosing the 8-s lever indicates the subjective midpoint between 2 and 8 s.

The bisection procedure has many benefits. Unlike the peak procedure described later, it can be used to study a wide range of different intervals including very short ones. A second benefit is that it provides an estimate of the temporal middle, which many models of timing make predictions about.

The Peak Procedure: A Temporal Reproduction Task

The peak procedure is a temporal reproduction task that is a modification of the more basic fixed-interval (FI) schedule. In a FI schedule the subject is reinforced for the first response (key peck or lever press) it makes after a fixed interval of time has elapsed. FI schedules can either be free-operant procedures, in which the interfood interval is the only discriminative stimulus and the interval is reset each time food is obtained, or discrete trials procedures, which are the basis for the peak procedure. In a discrete trials FI procedure, a stimulus such as a tone or light comes on at the beginning of the interval and is extinguished when the animal meets the response requirement and receives food. Then there is an intertrial interval in which the stimulus is not present before the next trial begins. Although animals could theoretically respond at their maximum rate the moment the stimulus comes on, they do not do this. Instead, they commonly show a FI scallop or alternatively a break-and-run pattern: There is a pause in responding following the onset of the stimulus that lasts approximately two thirds of the interval followed by a rapid acceleration in responding toward the time at which reinforcement results (Fig. 1). The length of the initial pause can be analyzed to provide data about timing.

In the peak procedure 50% of the trials are standard FI trials, but the remaining 50% are "probe" trials in which the stimulus remains on for two or three times the normal FI. Probe trials terminate with an intertrial interval, and no reinforcement is given (Fig. 2). In these probe trials subjects behave as in a standard FI in the first part of the trial (since the two trial types are identical until this point), and then at some time after the criterion time they cease responding again. When the data from several probe trials are plotted with the time in the trial on the x-axis and the average response rate on the y-axis, a Gaussian "peak" function is usually observed (Fig. 3). This function yields interesting data, including the peak time, which is the time at which the rate of responding is at a maximum; the peak rate, which is the height of the function at the peak time; and the spread, which is the width of the function at half its maximum height. Peak time is commonly used as a measure of timing

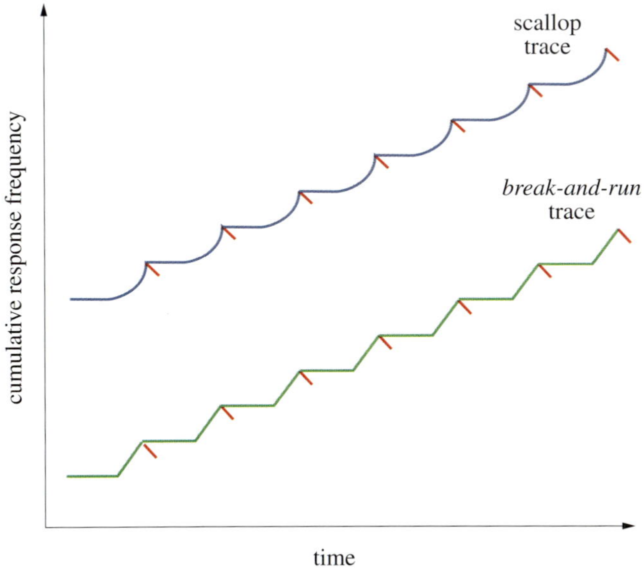

FIGURE 1 Traces produced by a cumulative recorder showing the typical behavior observed in a fixed-interval reinforcement task. The scallop (in blue) and break-and-run (in green) patterns are shown. In both cases, the response frequency (in arbitrary units) is plotted on the y-axis and time (in arbitrary units) on the x-axis. The diagonal ticks (in red) indicate reinforcement (reproduced with permission from Richelle and Lejeune, 1980).

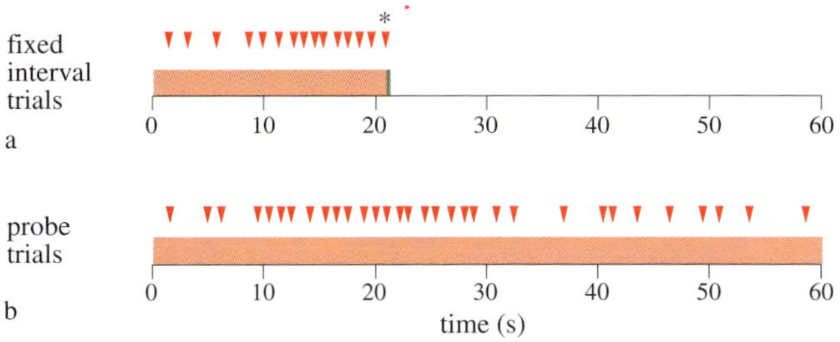

FIGURE 2 The peak procedure consists of mixed trials. (a) In the fixed-interval tasks, the light or tone signal appears at the beginning of the task and is turned off when the subject gives the first response (green vertical lines) immediately after the criterion interval (20 s). This response produces reinforcement (asterisk). The triangles represent the subject's responses. (b) In the probe trials the signal appears at the beginning of the task and is turned off after a period three times longer than the criterion interval (in this case, after 60 s), independently of the subject's behavior. In both cases, the red band corresponds to the time, within the trial, that the signal is active (adapted from Bradshaw and Szabadi, 1997).

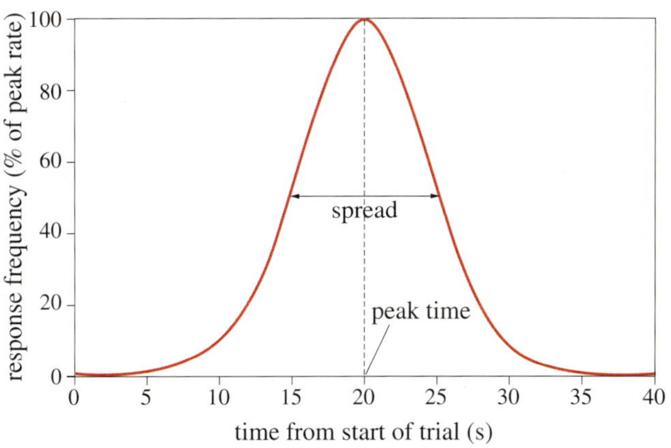

FIGURE 3 An ideal timing function obtained from probe trials of the peak procedure. The peak time is the time corresponding to the maximum height of the function and is used as a measure of timing accuracy. The spread is the width of the timing function at half the maximum height and is used as a measure of timing precision.

accuracy, whereas spread is used to reflect timing precision. One particular benefit of the peak procedure over the bisection procedure is that each probe trial can yield estimates of both these variables, unlike bisection in which many probe trials are needed to produce a timing function from which these measures can be extracted. This aspect of the peak procedure makes it particularly appropriate for studying transient effects on timing such as those produced by drugs.

An important variant of the peak procedure is the peak interval gap procedure. This is identical to the peak procedure with the exception that a gap, during which the stimulus is turned off, is inserted part way through a random 50% of the probe trials. By analyzing where the peak time occurs once the stimulus recommences in these gap trials, it is possible to determine how the animal handles the gap. There are three distinct possibilities. First, the animal may ignore the gap and run its clock through it, peaking at the normal criterion time after the beginning of the trial. Second, the animal may stop its clock during the gap, retain the value of the elapsed interval in memory, and resume timing the interval following the gap. In this scenario, peak times would be shifted to the right by the duration of the gap. Third, the animal may reset its clock in response to the gap. In this case, the peak times will be shifted to the right by the interval equivalent to the time before the gap plus the gap.

There are of course many other timing procedures in addition to those just described; however, the bisection and peak procedures have arguably yielded the most information about the nature of interval timing. One procedure that needs to be mentioned because of the problems associated with interpreting the data it produces is known as differential reinforcement of low rates of responding (DRL). In a DRL procedure, a response is reinforced if, and only if, it follows the preceding response by a specified temporal interval. Thus, DRL is similar to an FI in that a response is only reinforced after a criterion time, but it differs from an FI in that early responses are penalized. The data analyzed from DRL are the inter-response times (IRTs); in a subject that is timing well, the average IRT should be just longer than the criterion time. The problem with IRT as a measure of timing ability is that it will be influenced by an animal's ability to withhold responding as much as its ability to time. Thus, although good performance on DRL suggests good timing ability, failure on DRL should not necessarily be taken as evidence of poor timing ability.

Behavioral Features of the Interval Timing Clock

The procedures outlined previously have been used by experimental psychologists to elucidate many features of the interval timing clock. I will describe some of these features and compare them with those of the circadian clock, which is the best studied phase timer. In general, the interval clock can be compared to a stopwatch (Church, 1978; Roberts, 1983; Roberts and Church, 1978). It can be started when a stimulus is turned on, and it is reset either by the delivery of food or by the intertrial interval. Rats respond to the peak interval gap procedure by stopping their clock while the stimulus if off, retaining the interval that had elapsed before the gap in memory, and then resuming timing once the stimulus comes back on. The interval clock is not restricted to timing stimuli of a particular modality, and both rats and pigeons can time both tones and lights. Cross-modal transfer has also been observed in which a time interval learned with one stimulus is immediately transferred to a novel stimulus.

One of the most impressive features of interval timing is the broad range of durations that animals are able to time: The peak procedure has yielded timing functions with accurate peak times when the criterion time is as short as a few seconds to as long as 1.5 h (Bradshaw and Szabadi, 1997). In contrast, the circadian clock can only be entrained to time periods in the limited range of 19–27 h (Aschoff, 1980).

In interval timing the precision of timing is proportional to the interval being timed across the entire range of intervals that have been studied which spans three or four orders of magnitude (Gibbon *et al.,* 1997). Thus, the width of peak functions increases as the criterion time increases, but if peak functions are normalized by plotting the x-axis relative to the peak time and the y-axis relative to peak rate, then the timing functions obtained with different criterion times are superimposed (Fig. 4). This phenomenon is a hallmark of interval timing and has been dubbed the scalar property. It is not known whether or not phase timers also share the scalar property since there has been no comparison of the precision of timing when a phase clock is entrained to different intervals. It is usually found that the relative precision of phase timing is much higher than that of interval timing: The variation seen in the onset of daily activity in an individual golden hamster may be as low as 1% of the circadian phase, whereas interval timing typically shows a precision of between 15 and 35% of the interval being timed. Thus, the flexibility of interval timers in terms of the range that they can cover seems to be obtained at the cost of reduced precision.

Functions

Despite the large number of publications in the psychology literature on interval timing, the subject has been almost ignored by behavioral ecologists. Hence, we know very little about the selective pressures that shaped interval timing or the current uses that animals put it to. It seems obvious that having an interval clock should be beneficial, but for what do animals use it? That food seems to reset the clock and

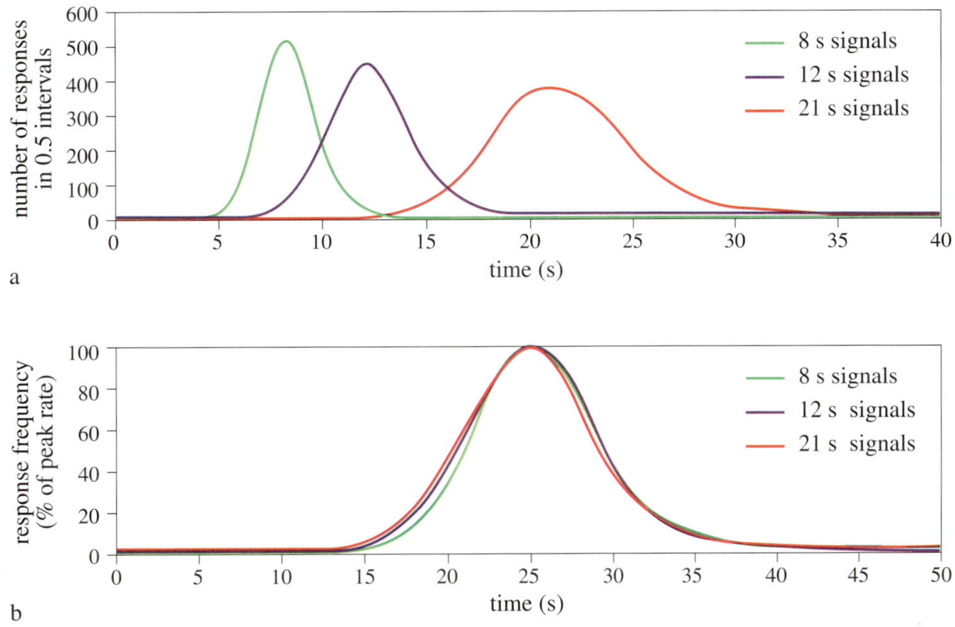

FIGURE 4 Timing functions obtained by calculating the pooled data for seven people tested on the peak procedure, with visual signals of 8-, 12-, or 21-s duration in different blocks of trials. (a) The actual data are shown. (b) The superimposition of the different curves is evident when the *x*-axis is normalized with respect to the peak time and the *y*-axis with respect to the peak rate.

that food-related stimuli are readily timed both suggest that interval timing is likely to be involved in foraging decisions. However, timing may also be involved in many other types of behavioral decisions, including mate choice, other forms of social behavior, and navigation.

In general, an interval timer provides animals with two types of information. First, it allows animals to predict the time at which an event will occur. Second, it allows them to estimate the duration of an elapsed time interval.

Prediction

The ability to predict when a significant will occur has many potential benefits. An animal that knows when events will occur can organize its time more efficiently and can prepare appropriately for upcoming situations. In the case of the hummingbird example mentioned earlier, a bird that can learn the amount of time it takes a particular flower species to refill after it has emptied it will have a clear fitness advantage over one that cannot. If the bird returns to a flower too early before it has completely refilled, the bird will not get the full volume of nectar it could have collected by waiting longer, whereas if the bird delays returning to the flower after it has refilled it becomes increasingly likely that another bird will get there first (Gill, 1988). Hummingbirds are experiencing a situation that is analogous to a free-operant FI schedule since the only discriminative stimulus available is the time between rewards.

For an natural example of the peak procedure we must consider the problem faced by many foraging animals of when to give up on the current patch and move on to a new one. If the interval between successive prey captures is relatively constant and the patch can suddenly deplete, then a forager should time the interval since the last prey capture and leave the patch as soon as it estimates that an interval longer than the normal interprey interval has elapsed. For example, flycatchers forage from a perch, making forays to catch insects. In one study it was observed that if a bird fails to catch an insect for 1.5 times the usual intercatch interval, it will leave its current perch in search of a new one (Davies, 1977). It thus appears that the birds are using interval timing to estimate patch depletion. This situation has been simulated in the lab using starlings foraging from a hopper on a FI schedule that could suddenly cease (Brunner *et al.*, 1992). The birds had the option at any point to leave the hopper and fly to a "new" one by flying back and forth between two perches. As suggested by the flycatchers, the starlings waited for a fixed proportion of the FI after failing to receive a reward before giving up and flying to a new patch. It is interesting to note that the giving up times were proportional to the FI rather than of a constant duration. If the bird waited a constant duration after timing the FI, then as the FIs increased the bird would have a higher probability of leaving the patch too early because of the decreasing precision of longer time estimates.

Timing is not only important for foraging behavior but also may play a role in many other types of behavioral decisions. Ring doves provide an example of how the ability to predict an event can be important in coordinating cooperative behavior. In this species, the male and female share incubation of the eggs, with the male taking the interval from approximately 10:00 to 17:00 h each day. If the onset of the male's incubation period is delayed 1 day, the offset will also be delayed, suggesting that the male uses an interval timer to predict the end of his bout of incubation (Gibbon *et al.,* 1984). It is not clear in this example why the male should time his bout rather than taking his cue to leave from the return of the female. It is also not clear why he uses an interval timer rather than his circadian clock since the female times her return to the nest according to the time of day.

Similarities exist between the peak procedure and classical conditioning experiments, suggesting that there could be connections between interval timing and associative learning. In both paradigms, an arbitrary stimulus (the conditioned stimulus; CS) is given at some interval before a reward (unconditioned stimulus; US) is available; the only difference between a classical conditioning experiment and a FI timing experiment is that the subject must make a response to collect the reward in the latter. Traditionally, learning theorists focused on the contingency between the CS and US in explaining learning; however, there is mounting evidence that the temporal contiguity of the CS and US is equally important in determining whether or not learning occurs and what types of conditioned responses are performed (Gallistel, 1994). Animals appear to learn the CS–US interval and the US–US interval, and it is the ratio of these remembered intervals, rather than just a simple CS–US association, that determines whether a conditioned response will occur (Gibbon *et al.,* 1977). It is also clear that the CS–US interval has an impact on the type of conditioned response performed. This makes good functional sense since the type of conditioned response that is appropriate will depend on the US (Timberlake *et al.,* 1982). For example, in the case of a stimulus-predicting food, if the CS–US interval is long, searching or hunting behavior may be adaptive, whereas if it is short handling and consummatory behavior are appropriate.

Estimation

The ability to discriminate time intervals is likely to be important in many different types of behavioral decisions. For example, a foraging animal may base future foraging decisions on how long it spent searching for food in two different patches or how long it took to handle two different prey types, and a female choosing a mate at a lek may base her decisions on how long a male displays. All these comparisons require the ability to estimate the duration of elapsed time intervals.

Estimating time intervals may provide a means of estimating many other variables of importance to animals. For example, the optimal foraging literature has identified rate of food intake as a currency that will often be highly correlated with fitness, and there is good evidence that many animals can make foraging decisions that optimize this currency (Kacelnik, 1984). Since this rate is computed by dividing amount by time, it is likely that interval timing is necessary for forming rate estimates. Estimates of time intervals could also be used to estimate amounts of food since there is often a strong correlation between the size of a food reward and the length of time it takes to handle and consume it. It is therefore extremely likely that interval timing is intimately involved in optimal foraging decisions. It is also possible that time intervals could be used to estimate distance traveled.

Finally, it is possible that interval timing could be involved in perceptual learning. For example, during the imprinting process a domestic chick forms a visual representation of its mother. The chick sees its mother from different directions and distances; however, these physically distinct views will be combined together in the same representation as long as they are seen within a short time of one another. The evidence for this is that following imprinting with two different objects, chicks will learn a discrimination between the two objects more quickly if the objects were presented 5 or more minutes apart during imprinting than if they were presented less than 30 s apart (Chantrey, 1974). These results suggest that a sense of time may be important in determining how animals form categories.

Comparative Studies

The comparative approach is often used to establish the adaptive significance of a behavior pattern. In the case of interval timing, this would involve comparing the timing abilities of related species and then attempting to correlate any differences found with differences in ecology.

The FI schedule has been used to study interval timing in a relatively wide range of vertebrate species and at least one invertebrate (Richelle and Lejeune, 1980). Mammals studied on FI schedules include mice, rats, cats, prairie dogs, horses, guinea pigs, Syrian hamsters, bats, beagle dogs, rabbits, hapalemurs, pottos, chimpanzees, baboons, squirrel monkeys, rhesus monkeys, and raccoons; birds studied include chickens, black vultures, crows, ravens, quail, European starlings, pigeons, and black-capped chickadees; fish studied include tilapia, gourami, and goldfish; and the only insect studied to date is the honeybee. In all species with the exception of one of the prosimians, the quail, the tilapia, and the honeybees, evidence for timing on FI schedules was observed. In the species that failed to show timing, it is not clear whether this failure reflects true deficits in timing ability or deficiencies in the experimental situations employed for testing.

However, attempts to draw any general conclusions from these data are beset with problems. The studies differed in many respects, including the response used, the reward used (e.g., food and water), the prior deprivation of the animals, and the intervals studied. Thus, it is possible that what appear to be species differences may actually result from differences in procedure. Even when two or more species are compared using identical methodology, differences in performance do not necessarily provide evidence for differences in timing ability because performance differences can also be produced by differences in motor abilities or motivation between species.

One of the best solutions to these problems is to compare only very closely related species that are expected to be similar in most respects apart from their timing abilities. Unfortunately, there have been surprisingly few careful comparisons of this type. One exception is a comparison of timing in pigeons and crows (Powell, 1974). The birds were trained on a DRL schedule that was gradually increased as long as the birds were still performing well. The pigeons were unable to time intervals beyond 10 or 30 s, whereas the crows could time intervals up to 120 s. This study is not subject to the usual criticism of DRL that the result could be attributable to a difference in the species' ability to withhold responding rather than their abilities to time because when the end of the interval was cued by a change in color of the response key, thus eliminating the need for timing, both species performed equally well at delays of 120 s. It thus appears that crows are capable of timing longer intervals than are pigeons. Although the conclusion of this study appears sound, it is difficult to speculate about the possible adaptive significance of this difference without data from more species.

Mechanisms

Perhaps one of the reasons that interval timing has not been more widely studied is that there is no obvious time measurement organ. However, many theoretical models have been produced to explain how timing could be achieved.

I briefly describe and contrast three timing models, all of which can produce a good quantitative fit to data obtained from the peak procedure (Bradshaw and Szabadi, 1997). Three different information processing components can be identified in all the models. First, measurement of time requires some type of clock. Second, all the models require some type of memory in which the time interval is represented. Third, all the models need a decision stage which uses output from the clock and the memory components to control behavior.

Scalar Timing

Scalar timing theory (Gibbon *et al.,* 1984) is a quantitative information processing model with explicit modules corresponding to the clock, memory, and decision processes (Fig. 5). The clock of scalar timing theory consists of a pacemaker that continually emits pulses. When the switch is closed these pulses are gated to an accumulator, such that elapsed time is a linear function of the number of pulses in the accumulator. The number of pulses in the accumulator is passed to working memory and finally to reference memory, in which a distribution of experienced intervals is stored. The length of a duration is recalled from reference memory by taking a random sample from the distribution stored there. The decision about when to start and stop responding on an FI is made by continually comparing a sample from reference memory with the current elapsed time in working memory. When the ratio of the current time to the memory sample is sufficiently close to 1, responding is initiated.

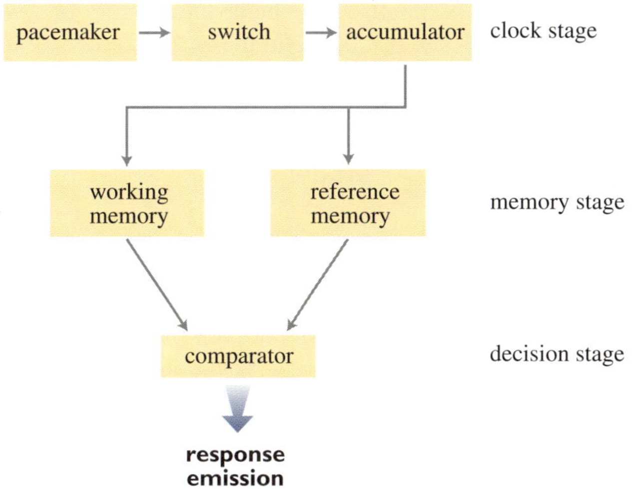

FIGURE 5 Model of information processing according to the scalar timing theory, showing the clock, memory, and decision stages.

Scalar timing theory is the most formally advanced of the existing timing models, and one of its great strengths is its ability to generate explicit quantitative predictions for many different timing procedures.

Behavioral Theory of Timing

As its name suggests, the behavioral theory of timing (Bradshaw and Szabadi, 1997) is a model based on observed behavior. Like scalar timing theory, the behavioral theory of timing has a pacemaker that emits pulses. Each pulse is assumed to be associated with a particular behavioral state, such as moving around a box, lever pressing, and eating. Thus, with each pulse from the pacemaker the animal progresses to the associated behavioral state. The speed of the pacemaker is assumed to be controlled primarily by the rate of reinforcement received by the animal. Thus, an animal is predicted to progress through sequential behavioral states more quickly on a 10-s FI than on a 30-s FI because the rate of reinforcement is higher on the 10-s FI. When reinforcement occurs, it strengthens the sequence of behaviors preceding it and this sequence is stored in memory. Thus, the representation of an interval corresponds to a sequence of behavioral states. When the animal is again placed in the presence of similar stimuli, the behavior sequence with the greatest strength is initiated.

The greatest strength of the behavioral theory of timing is its parsimony: It models timing without recourse to cognitive constructs such as switches, accumulators, and comparators.

Multiple-Oscillator Model

The clock of the multiple-oscillator model (Church and Broadbent, 1990) is composed of several oscillators with different periods spanning the range of intervals to be timed. The oscillators are completely coupled and are reset at the beginning of an interval. The elapsed time is determined by reading off the half-phase of each oscillator and thus yielding a vector in which each value is either +1 or −1. Working memory is assumed to be an autoassociation matrix of this vector, and reference memory is another matrix of the same dimensions as working memory. It is assumed that when reinforcement is received, reference memory is updated by computing a weighted average of the values in existing reference memory and the current contents of working memory. The decision about when to respond on a FI is made by comparing the vector representing current time with the reference memory matrix using a similarity measure.

The greatest strength of the multiple-oscillator model lies in its ability to predict detailed features of timing data for which the other models completely fail to account. Which, if any, of the three previously discussed models provides the best description of the timing mechanism is a matter of intense debate. Each of the models has strengths and weaknesses that reflect both the theoretical framework within which each model has been developed and the different data sets that they have been required to explain. It is likely that the choice of future timing models will need to be constrained not only by behavioral data but also by the growing body of neuropharmacological evidence regarding the internal clock.

Neuropharmacological Evidence

Compared with the circadian clock, little is known about the neural mechanisms that underlie interval timing in the seconds to minutes range. However, recent neruopharmacological research has provided evidence that the different stages of temporal information processing discussed previously may involve separate brain regions and be modified by different neurotransmitter systems (Meck, 1996).

There is substantial evidence that the speed of the clock is modulated by the functional level of dopamine in the basal ganglia. For example, if a rat is trained to reproduce a time interval using the peak procedure and is then tested on dopaminergic drugs, dopamine agonists such as methamphetamine and cocaine cause an immediate leftward shift in peak time that is proportional to the duration of the interval being timed. This is compatible with dopamine agonists speeding up the clock because the rat is reproducing the time interval earlier than the true criterion time. In contrast, dopamine antagonists such as haloperidol cause a proportional rightward shift in peak time that is compatible with a slowing of the clock. If the rats are trained on the drugs for several consecutive days, the timing functions gradually renormalize compatible with the criterion time being relearned with an internal clock of a different speed. If the drug is then suddenly withdrawn, the timing functions immediately rebound in the opposite direction to the initial shift observed, supporting this interpretation (Fig. 6). This pattern of effects is known as the *clock pattern.* Meck (1996) proposed that the neurons of the substantia nigra pars compacta may be the pacemaker cells used for interval timing because lesions of the substantia nigra cause a complete loss of temporal control of responding, but that systemic administration of L-DOPA (a precursor of dopamine) will restore timing.

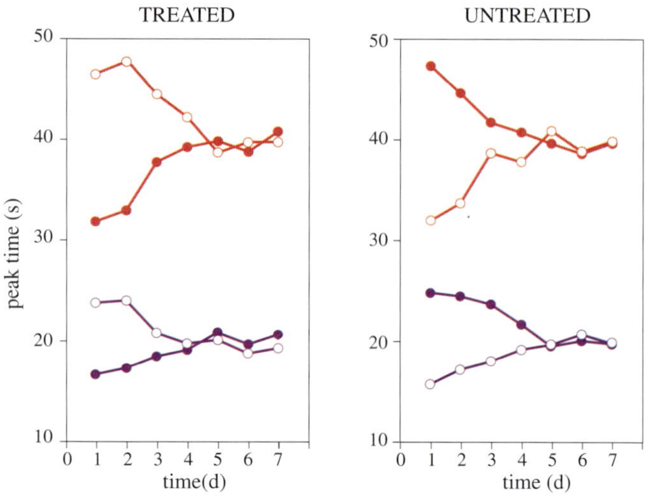

FIGURE 6 The clock pattern. Mean peak times of four different groups of rats initially trained on the peak procedure with criterion intervals of 40 s (in red) and 20 s (in blue) and tested under the influence of dopaminergic drugs which affect the animal's clock speed (left) or following interruption of the same drugs (right). ●, Rats treated with methamphetamine; ○, rats treated with haloperidol (reproduced with permission from Meck, 1996).

Temporal memory and attentional mechanisms appear to be correlated with aceltylcholine function in the frontal cortex. In rats, cholinergic drugs produce systematic distortions in temporal memory. For example, if a rat is trained to reproduce a time interval using the peak procedure and then tested with physostigmine, an acetylcholine esterase inhibitor, a proportional gradual leftward shift in peak time is observed. This shift is compatible with physostigmine causing shorter time intervals to be stored in reference memory, gradually replacing the veridical representation. In contrast, antagonists such as atropine cause a gradual proportional rightward shift in peak time compatible with the memories stored being longer. In both cases, when the

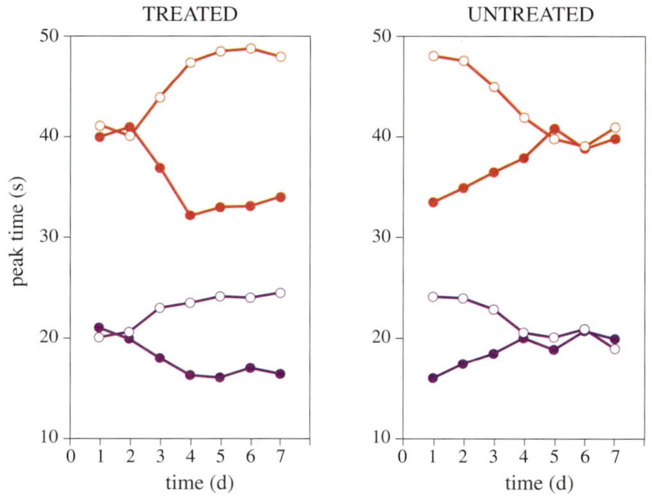

FIGURE 7 The memory pattern. Mean peak times of four different groups of rats initially trained on the peak procedure with criterion intervals of 40 s (in red) and 20 s (in blue) and tested under the influence of cholinergic drugs which affect memory (left) or following interruption of the same drugs (right). ●, Rats treated with physostigmine; ○, rats treated with atropine (reproduced with permission from Meck, 1996).

drugs are discontinued the functions gradually renormalize over a period of several sessions, compatible with the intervals being relearned with normal memory parameters. This pattern of drug effects is called the *memory pattern* (Fig. 7). In addition to the pharmacological manipulations, lesions of the frontal cortex are also found to produce shifts in peak time comparable to the memory pattern.

Meck (1996) suggested that the clock in the basal ganglia may be connected to the memory and attentional mechanisms in the frontal cortex by frontal–striatal loops. Evidence for the possible involvement of frontal–striatal loops in interval timing is emerging from functional magnetic resonance imaging studies of timing in humans.

Development

Very little is known about the ontogenetic development of interval timing. We do not know whether the capacity for temporal regulation is present from birth or hatching, or whether it appears at some later stage in development. We also do not know whether specific environmental experience is necessary for the normal development of interval timing.

Some technical obstacles have probably influenced the study of timing in young animals. First, because timing experiments involve training animals they tend to last many days, presenting a problem for study of timing in rapidly maturing species. Second, traditional techniques for studying timing depend on operant conditioning, which may be difficult in young animals with undeveloped motor systems. One solution to this latter problem is to study timing in a precocial species. A study on domestic chicks found good FI scallops in birds trained on 2- and 3-min FIs between 3 and 10 days posthatching, showing that chickens at least have the ability to time intervals from an early age (Marley and Morse, 1966). Given the evidence that interval timing is intimately involved in classical conditioning, we should perhaps not be surprised by this finding since there is ample evidence for classical conditioning being possible in fetuses prior to birth.

Phylogeny

When did interval timing evolve, and what is its evolutionary history? Given the broad taxonomic range of interval timing, it seems very likely that it is a primitive ability in the vertebrates, making it extremely evolutionarily ancient. More data on timing in invertebrates are desperately needed before we can make any conclusions about whether or not interval timing preceded the origin of the vertebrates. If there are strong connections between associative learning and interval timing as argued previously, then it is very likely that interval timing will be identified in the invertebrates since associative learning has been documented in many invertebrate species.

Understanding the mechanisms underlying timing may shed light on whether or not the clocks in different species are homologous. In this vein, pharmacological evidence is starting to support the idea of a common interval timing clock in vertebrates. Dopaminergic drugs, such as methamphetamine, that have been shown to increase clock speed in rats have also been shown to affect clock speed in both pigeons (Kraemer *et al.*, 1997) and European starlings. These data support the hypothesis that an interval timing clock similar to that present in modern-day birds and mammals was present in the common ancestors of these two classes. Comparable drug studies on fish will be necessary to demonstrate that this same clock is also responsible for interval timing in fish.

References Cited

Aschoff, J. (1980). Ranges of entrainment: A comparative analysis of circadian rhythm studies. In *Proceedings of the 13th Conference of the International Society of Chronobiology, Pavia* (F. Halberg, L. E. Scheving, and E. W. Powell, Eds.), pp. 105–112. Il Ponte, Milan.

Bradshaw, C. M., and Szabadi, E. (Eds.) (1997). *Time and behaviour: Psychological and neurobehavioral analyses.* In *Advances in Psychology,* Vol. 120. Elsevier, Amsterdam.

Brunner, D., Kacelnik, A., and Gibbon, J. (1992). Optimal foraging and timing processes in the starling, *Sturnus vulgaris:* Effect of inter-capture interval. *Anim. Behav.* **44,** 597–613.

CHANTREY, D. F. (1974). Stimulus preexposure and discrimination learning by domestic chicks: Effect of varying interstimulus time. *J. Comp. Physiol. Psychol.* **87,** 517–525.

CHURCH, R. M. (1978). The internal clock. In *Cognitive Processes in Animal Behavior* (S. H. Hulse, H. Fowler, and W. K. Honig, Eds.), pp. 277–310. Hillsdale, Erlbaum, NJ.

CHURCH, R. M., and BROADBENT, H. (1990). Alternative representations of time, number and rate. *Cognition* **37,** 55–81.

DAVIES, N. B. (1977). Prey selection and the search strategy of the spotted flycatcher (*Muscicapa striata*). *Anim. Behav.* **25,** 1016–1033.

GALLISTEL, C. R. (1990). *The Organisation of Learning.* MIT Press, Cambridge, MA.

GALLISTEL, C. R. (1994). Space and time. In *Animal Learning and Cognition* (N. J. Mackintosh, Ed.), pp. 221–253. Academic Press, San Diego.

GIBBON, J., and ALLAN, L. G. (Eds.) (1984). *Timing and time perception.* New York, New York Academy of Sciences.

GIBBON, J., BALDOCK, M. D., LOCURTO, C., GOLD, L., and TERRACE, H. S. (1977). Trial and intertrial durations and autoshaping. *J. Exp. Psychol. Anim. Behav. Process.* **3,** 264–284.

GIBBON, J., MALAPANI, C., DALE, C. L., and GALLISTEL, C. R. (1997). Toward a neurobiology of temporal cognition: Advances and challenges. *Curr. Opin. Neurobiol.* **7,** 170–184.

GIBBON, J., MORRELL, M., and SILVER, R. (1984). Two kinds of timing in circadian incubation rhythm of ring doves. *Am. J. Physiol.* **247,** R1083–R1087.

GILL, F. B. (1988). Trapline foraging by hermit hummingbirds: Competition for an undefended renewable resource. *Ecology* **69,** 1933–1942.

KACELNIK, A. (1984). Central place foraging in starlings (*Sturnus vulgaris*). I. Patch residence time. *J. Anim. Ecol.* **53,** 283–299.

KILLEEN, P. R., FETTERMAN, J. G., and BIZO, L. A. (1997). Time's causes. In *Advances in Psychology,* Vol. 120. Elsevier, Amsterdam.

KRAEMER, P. J., RANDALL, C. K., DOSE, J. M., and BROWN, R. W. (1997). Impact of *d*-amphetamine on temporal estimation in pigeons tested with a production procedure. *Pharm. Biochem. Behav.* **58,** 323–327.

MARLEY, E., and MORSE, W. H. (1966). Operant conditioning in the newly hatched chicken. *J. Exp. Anal. Behav.* **9,** 95–103.

MECK, W. H. (1996). Neuropharmacology of timing and time perception. *Brain Res. Cogn. Brain Res.* **3,** 227–242.

POWELL, R. W. (1974). Comparison of differential reinforcement of low rates (DRL) performance in pigeons (*Columba livia*) and crows (*Corvus brachyrhynchos*). *J. Comp. Physiol. Psychol.* **86,** 736–746.

RICHELLE, M., and LEJEUNE, H. (1980). *Time in Animal Behavior.* Pergamon, Oxford.

ROBERTS, S. (1983). Properties and function of an internal clock. In *Animal Cognition and Behavior* (R. L. Mellgren, Ed.). North-Holland, Amsterdam.

ROBERTS, S., and CHURCH, R. M. (1978). Control of an internal clock. *J. Exp. Psychol. Anim. Behav. Process.* **4,** 318–337.

TIMBERLAKE, W., WAHL, G., and KING, D. (1982). Stimulus and response contingencies in the misbehavior of rats. *J. Exp. Psychol. Anim. Behav. Process.* **8,** 62–85.

MARC BEKOFF

Department of Environmental, Population, and Organismic Biology
University of Colorado
Boulder, Colorado 80309-0334, USA

Social Play in Mammals

Play is an activity that has challenged students of behavior for a long time. Here, I concentrate on social play in mammals to make some general points about this fascinating behavior. After a general introduction, I mainly discuss some representative studies of how social play is initiated (solicited) and maintained in various canids (members of the dog family). I also discuss recent work on relationships between social play behavior and the study of animal cognition (cognitive ethology). The flexibility and versatility of play make it a good candidate for interdisciplinary discussion of comparative and evolutionary studies that will help us learn more about the nature of animal minds. Much exciting research is taking place in these areas of inquiry.

What Is Play?

"Play" is a very difficult word to define because it is a blanket term that covers a wide range of behavior patterns. From early serious efforts to come to terms with social play behavior, many definitions were offered (Bekoff and Byers, 1981, 1998; Fagen, 1981; Martin and Caro, 1985; Spinka, Newberry, and Bekoff, 2001), and these definitions provided the framework for numerous comparative and evolutionary studies. Today, it is difficult to pick up a journal that deals with the comparative study of animal behavior and not find at least one article that is concerned with the structure or function of social play (predominantly in mammals; Ortega and Bekoff, 1987). Although play is easy to recognize — people do agree about play when they consider possible examples — there does not seem to be any single definition that would apply to all instances of what is called play in the diverse species in which the activity has been observed. However, lacking a rigorous definition need not be an impediment to conducting solid research; requiring a rigorous definition prior to empirical research may unreasonably require possession of knowledge that must first be gained by empirical research (Allen and Bekoff, 1997).

To begin, I use a general definition that has weathered the years (Bekoff and Byers, 1981, pp. 300–301): "Play is all motor activity performed postnatally that appears to be purposeless, in which motor patterns from other contexts may often be used in modified forms and altered temporal sequencing." Martin and Caro (1985, p. 65) provide a slightly modified version of this definition. If the activity is directed

FIGURE 1 Social play among wolves. The wolf on the right is inviting his companion to play by rolling over in front of him.

toward another living being, it is called *social play* (Fig. 1); if it is directed toward an inanimate object(s), it is called *object play* (Hall, 1997); and if the activity carries the individual in a seemingly frantic flight about its environment, it is called *locomotor play*. Both social and object play (Fig. 2) can be locomotory in nature, and self-play can include object and locomotor play.

This working definition is a positive one in that play is defined in terms of what it is and not in terms of what it is not. For years, it was trendy to refer to social play as being not aggression, not predation, or not reproduction, but these sorts of definitions proved to be of little or no value to those who were interested in learning more about play. Note also that the definition centers on the structure of play sequences — what animals do when they play — and not on possible functions of play. Suffice it to say, functional accounts of play have resulted in unconsolidated mixes of speculations (R. Fagen, personal communication), and it is unlikely that there is one single reason why all animals play. There seem to be many possible and nonmutually exclusive functions of play (e.g., socialization, physical training, cognitive development, and energy regulation) depending on the species being studied and perhaps the ages and sexes of the participants and the location in which they are playing (Bekoff and Byers, 1981, 1997; Fagen, 1981; Martin and Caro, 1985; Burghardt, 1988, 1996; Byers and Walker, 1995). Furthermore, functional accounts of play that appeal to possible benefits and costs of playing are fraught with difficulties, as are most discussions of function in terms of potential benefits and costs. With respect to play, both short-term and long-term benefits and costs need to be considered, and it is extremely difficult to account for benefits and costs over long periods of time for individuals of long-lived species in which play has evolved.

FIGURE 2 Two hybrid Alsatian puppies playing together with a stick.

Despite its easy to recognize features, social play has been a very difficult behavioral phenotype with which to deal rigorously. Although there are very few people who would claim that animals do not engage in this easily recognizable activity, broad comparative analyses of play are still forthcoming, although the database has greatly increased during the past 5 years (Burghardt, 1996). Not only does play between species vary in structure (behavior patterns used and their temporal sequencing) but also there are marked individual differences within species that make generalizations difficult and tenuous. This variability can be the result of numerous factors, including the ages of the participants, their sexes, their relative social ranks, their experience with one another as play partners, their energy levels, their level of physical (aerobic and anaerobic) fitness, food abundance, the permissiveness of their caretakers, and where they play. Indeed, it is the flexibility and versatility of play that make it a good candidate for comparative and evolutionary cognitive studies. As Allen and Bekoff (1997) noted, behavioral flexibility is relevant to mentalistic attributions because it is connected to an organism's monitoring of its own performance. Indeed, behavioral flexibility and an individual's ability to adjust to variations in environmental situations are the main criteria to which Griffin (1992) appeals in his discussions of consciousness in many nonhuman animals.

Here, I concentrate on social play because much research is being conducted on this behavioral phenotype as a result of current interest in animal cognition (Bekoff and Jamieson, 1996; Allen and Bekoff, 1977) and also because social play seems to be important in the development of specific behavior patterns and for the development and maintenance of social bonds in various contexts, including reproductive activities. It is highly likely that much future research on play will concentrate on social play, and I believe that this is where the study of play may have its broadest appeal. I consider the behavioral flexibility that is apparent in the social play of many mammals to discuss different aspects of how animals initiate (solicit) and maintain social play and how these patterns of behavior are useful for learning about the cognitive abilities of animals and the evolution of cognition.

Social Play: Evolution, Cooperation, and Negotiation

Jethro (a dog) runs toward Sukie (another dog), stops immediately in front of her, crouches on his forelimbs (bows), wags his tail, barks, and immediately lunges at her, bites her scruff and shakes his head rapidly from side to side, works his way around to her backside and mounts her, jumps off, does a rapid bow, lunges at her side and slams her with his hips, leaps up and bites her neck, and runs away. Sukie takes wild pursuit of Jethro and leaps on his back and bites his muzzle and then his scruff, and then shakes her head rapidly from side to side. They then wrestle with one another and then part only for a few minutes. Jethro walks slowly over to Sukie, extends his paw toward her head and nips at her ears. Sukie gets up and jumps on Jethro's back and bites him and grasps him around his waist. They then fall to the ground and wrestle with their mouths.

This description of a play encounter between two dogs (it could be other canids, felids, nonhuman primates, or humans) shows that when they engage in social play they perform behavior patterns that are used in other contexts such as aggression, reproduction, and predation (Fig. 3). They, and other animals including humans, also use actions that are important for initiating and maintaining play, in this case "bows" (Fig. 4). Social play in nonhuman animals (hereafter animals) is usually a cooperative turn-taking venture, and an important question arises: How do animals negotiate cooperative agreements? I consider this and other questions in this article because the study of animal play provides access into animals' minds. Available data strongly suggest that play-soliciting actions seem to be used to communicate to others that actions such as biting, biting and shaking of the head from side to side, and mounting are to be taken as play and not as aggression, predation, or reproduction. Thus, bows are performed when the signaler wants to communicate a specific message about its desires or beliefs. Although we cannot be sure that two dogs, for example, have beliefs about the effects of their behavior on each other (Dennett, 1983), some data do suggest this possibility. For example, suppose we wanted to know why Sukie permitted Jethro to nip at her ears. One explanation may be that Sukie believes Jethro is playing. Also, perhaps Jethro believes that Sukie believes that Jethro is playing. Providing answers to questions such as these is one of the challenges of research in animal cognition.

FIGURE 3 Mounting as play of two domestic dogs.

FIGURE 4 Two dogs bow to each other to invite play.

The category of behavior called social play has challenged students of behavior for a long time (Bekoff and Byers, 1981; Fagen, 1981, 1993; Burghardt, 1988, 1996; Bekoff and Allen, 1992; Jamieson and Bekoff, 1993; Pellegrini, 1995; Allen and Bekoff, 1997). For some, social play was written off as a behavior that was not worth studying, whereas for others the difficulty of studying social play in a rigorous manner prompted them to claim that it was worth studying but that a sufficiently "scientific" study of social play could never be done. Others, recognizing that social play behavior was interesting and fun to study, embarked on a challenging journey that has yielded valuable data concerning questions such as the following: Why do animals perform behavior patterns that are unique to social play? Why do they perform sequences of behavior that incorporate actions from many different contexts? How do animals negotiate cooperation and agreements to participate in social play? How do they develop socially and learn rules of social interaction? and How do they ask permission to engage in social play? It also has become more permissible to speculate on the existence of a humorous or comic dimension in the natural lives of animals and the possible evolutionary implications of such a dimension (R. Fagen, personal communication; Burghardt, 1996). For example, we can ask how animals' obvious enjoyment of engaging in social play and their persistent efforts in luring others to play influences the development and evolution of this activity.

Understanding activities such as play may also be important for developing new research dealing with comparative approaches to cognition (Pellegrini, 1995). The broad comparative study of social play also lends itself to suggestions that ethology should be concerned with the evolution, adaptation, causation, and development of behavior (Jamieson and Bekoff, 1993). Studies of social play also cross disciplinary boundaries: Those interested in social play come from various academic areas (e.g., zoology, psychology, anthropology, computer science, cognitive science, medicine, philosophy, and sociology) and consider questions that deal with topics such as evolution, ecology, development, social communication, neurobiology, and cognition.

Social play has been described in most mammals in which its possible existence has been investigated (Fagen, 1981). Some of the most elaborate and complex forms of play occur in such mammals as the social carnivores (wild and domestic cats, wolves, coyotes, domestic dogs, and bears) and nonhuman primates, animals with highly developed cerebral cortices (Bekoff and Byers, 1981; Fagen, 1981). Perhaps it is a combination of these and possibly other factors that have played a significant role in the ability of individuals to engage in play — to have the cognitive skills to negotiate cooperative agreements, to ask for permission to engage in a specific activity, to perform behavior within a pretend context, and perhaps even to make mental attributions to others. With the goal of learning more about how studies of nonhuman animal cognition can inform studies of human cognition, social play is a good behavioral phenotype on which to concentrate because when animals play they typically perform behavior patterns that are used in other contexts (e.g., predation, aggression, or reproduction), and yet these behavior patterns take on a different meaning when performed in the context of play (Fig. 5). For example, when Jethro bit Sukie and shook his head rapidly from side to side, he did not continue to the point of eliciting submissive behavior from her, and she did not squeal and roll over submissively or run away. He also did not attempt to copulate with her when he mounted her from behind, and she did not try to shake him off of her back.

a

b

FIGURE 5 (a) Two dogs growl and hit each other with their paws while playing. Within play, growling does not lead to fighting. (b) Giving blows with the paws is a gesture inviting and leading to play.

It is very important to negotiate play and to agree that play is the name of the game. When animals play, behavior patterns for the most part are "borrowed" from other contexts, and individuals need to be able to tell one another that they do not want to eat, fight with, or mate with the other individual(s) but rather that they want to play with them. In most species in which play has been observed, specific actions have evolved that are used to initiate ("I want to play with you") or to maintain ("I still want to play with you regardless of what I just did to you or regardless of what I am going to do to you") play. These actions seem to function in negotiations between participants, the result of which is that they foster an agreement to engage in play rather than to partake in aggression or predation, for example. There is no solid evidence that animals invite others to play and then exploit them. Furthermore, self-handicapping (e.g., controlling the intensity of bites) and role reversals (dominant individuals assuming submissive roles only in play) have also been observed (Fig. 6).

In considering how play is initiated and maintained, I

FIGURE 6 Two huskies play at fighting.

briefly discuss issues such as the types of information that are shared during play and what play can tell us about the emergence of mind in animals. For example, pretend play, in which an individual creates an imaginary situation or distorts reality, has attracted the attention of many researchers interested in the development of human behavior and also seems to occur in nonhumans. Dogs and cats often chase their tails and bite them as if the tails were prey or a competitor, and they frequently attempt to kill their food bowl or mate with it. Some dominant animals also engage in self-handicapping or role reversing during play, in which they seem to pretend that they are not really dominant, perhaps to get others to play with them who would otherwise not do so (it is rare for the self-handicapping individual to then attempt to dominate their partner). Although there are few data concerned with the evolutionary precursors to social pretend play in human children, it may be the case that when a wolf is playing with another wolf by using behavior patterns that are typically used in sexual or predatory encounters, she is thinking about mating or eating her partner and pretending to do so, although the end result of the interaction — play and not killing or mating — does not allow this explanation to be made with any degree of certainty. However, requiring that the individual is thinking about the activity in which it is pretending to engage is not a criterion that is applied to human children, and there is no reason to require it in the case of animals (in which its occurrence could not be discovered by direct questioning: "Were you thinking about mating with your food bowl or with your play partner?"). Perhaps neurobiological analyses in nonhumans and humans would help us by showing that similar electroencephalograph (EEG) patterns are produced (multiply realized) in play and in what are deemed to be real instances of mating, preying, or fighting.

Before examining the data, one point needs to be addressed. Useful analyses of social play require words that some criticize as being anthropomorphic (animals enjoy playing, they gain pleasure from playing, they want to play, and they believe that others want to play with them). The point is that these critics hold that words such as "wants," "believes," or "intends" apply solely to humans. Although some categorically dismiss any analyses of animal behavior that use anthropomorphic terms, others believe that careful use of anthropomorphism can be useful for the study of animal behavior and animal minds (Allen and Bekoff, 1997). I believe that detailed comparative analyses of social play can be useful for informing questions about the evolution of animal minds: Do individuals have beliefs? Do they have desires? Do they make attributions of intentions to others? Do they represent internal states of others? and What is the content of their mental states? It seems clear that without a cognitive vocabulary, it would be very difficult to discuss social play in any useful way. When most animals play, they use specific signals to signal what seem to be their intentions — their desire to play and their belief that if they perform a particular behavior then play will occur — and also perform behavior patterns that are typically performed in other contexts but whose meaning is changed in the context of play. The richness and diversity of social play can only be captured by the use of mental terms.

Play Signals and Intentional Behavior: Do Animals Negotiate Agreements?

In order to discuss whether or not certain actions, hereafter called play signals, serve to communicate an individual's desire to engage in social play — "If I perform this action I believe that play will occur or will continue" — many other questions loom and beg to be studied. Do animals make plans for the future? What is represented in the heads of animals who desire to play or who are engaged in play? Do individuals desire to play with others? Do individuals believe that if they perform a particular behavior pattern then others will play with them? Do animals make attributions of mind to others? Do animals negotiate agreements when they want to play or want to continue to engage in play? Do beliefs lead to other beliefs, not just to behavioral responses? Asking whether or not animals are able to negotiate agreements seems to be the sort of question that could land someone in very hot water, especially when skeptics about animal minds are the audience. Some skeptics would ask, "How could an animal possibly negotiate anything in the absence of language?" Others might be very concerned with the use of anthropomorphic language, whereas some might be concerned about the use of unscientific folk psychological explanations. Some skeptics would also be concerned about the difficulty of making claims about the content of mental states of animals, and some would be concerned about the assumption that animals have any mental states at all. (The fact that there are many reasonable humans who seem unable to negotiate agreements does not seem to bother them in the least.)

Although some of the issues raised by skeptics deserve serious consideration, a full discussion is beyond the scope of this article (Allen and Bekoff, 1997). The point here seems to be a simple one. In some (perhaps most) instances of social play, I have argued (Bekoff, 1995), based on empirical data, that the participants need to cooperate and to negotiate agreements so that they can engage in ongoing social play. Also, after they agree to play there seem to be social regulations and social obligations to maintain the play mood. Fagen (1993, p, 192) also noted that "levels of cooperation in play of juvenile primates may exceed those predicted by simple evolutionary arguments."

Furthermore, there are data that suggest that animals often seek permission to play with another animal by performing behavior patterns that indicate that play is desired and not another activity (Fig. 7). For example, in the self-handicapping (e.g., inhibited biting or clawing) and role reversing (being chased rather than chasing) in which dominant animals engage to get more subordinate individuals to play, the dominant individuals are seeking permission to engage in play with animals that might otherwise avoid interacting with them. The end result is that the two individuals agree to cooperate, with self-handicapping and role reversing being a form of negotiating this agreement. These data are important to consider in discussions of the evolution of cognition because they indicate that certain cognitive abilities must be present even before an individual has the linguistic sophistication of a human.

FIGURE 7 A baby Siberian ibex invites his mother to play by neck twisting, a common play signal. The mother walks near to the rock face to avoid the interaction.

Initiating, Maintaining, and Agreeing to Play

Because play is composed of actions that are also used in other contexts, an individual needs to be able to communicate to potential play partners that it is not trying to dominate them, eat them, or mate with them. Rather, it is trying to play with them. Behavioral observations of many animals who engage in social play suggest that they desire to do so and believe that their thoughts of the future — how the individuals to whom their intentions are directed would be likely to behave — would be realized if they clearly communicated their desires to play using signals that in some cases seem to have evolved specifically to communicate play intention. In this view, play is seen as a cooperative enterprise, the result of negotiations that allow play to occur and decrease the likelihood that aggression or predation will ensue. Although other animals might also cooperate, including those for whom we are hard-pressed to grant the presence of sophisticated cognitive abilities (e.g., cleaning fish), the flexibility of play and the ease with which even two strange animals are able to play after only a brief introduction lead to the conclusion that interactions between players are not merely the result of (somewhat hard-wired and inflexible) coevolved systems but rather the result of ongoing negotiations that require careful assessments throughout an encounter. It seems to be the flexibility of play that is important in allowing animals to negotiate the desired outcome. Given the state of knowledge concerning animals for which we are reluctant to grant the presence of sophisticated cognitive abilities, it might be premature to dismiss them as not having well-developed cognitive skills that are used for different types of negotiations.

Do animals cooperate in play, and how do they agree or negotiate what it is they are going to do? To begin to answer these questions, we must consider in more detail the question of whether or not signals that appear to be used to communicate play intention (play-soliciting signals) to other individuals could foster the cooperation among participants that is necessary for play to occur. Such play-soliciting signals seem to transmit messages such as "what follows is play," "this is still play despite what just happened or is going to happen," or "let's play again, wasn't it fun" (Bekoff, 1995). Supporting evidence concerning the importance of play signals for allowing cooperative social play to occur derives from studies in which it has been shown that play-soliciting signals show little variability in form or temporal characteristics and that they are used almost solely in the context of play. For example, an action that is commonly observed in the context of social play in canids and some other mammals is the highly stereotyped bow. In certain canids, the bow seems to function to permit recipients to engage (or to continue to engage) in social play (Bekoff, 1977, 1995). Also, the first bows that very young canids have been observed to perform are highly stereotyped, and learning seems to be relatively unimportant in their development. These features of bows can be related to the fact that when engaging in social play, canids typically use action patterns that are also used in other contexts such as predatory behavior, agonistic encounters, or mating, in which misinterpretation of play intention could be injurious.

How are agreements reached in the context of social play? How is turn taking by individuals accomplished? In most species in which play has been described, play-soliciting signals appear to foster some sort of cooperation between players so that each responds to the other in a way that is consistent with play and different from the responses the same actions would elicit in other contexts; play-soliciting signals provide aid to the interpretation of other signals by the receiver. For example, in coyotes, the response to a threat gesture after a play signal had immediately preceded the threat or a play signal had been performed in the beginning of an interaction is different from the response to threat in the absence of any preceding play signal. The play signal somehow altered the meaning of a threat signal by establishing (or maintaining) a "play atmosphere." Unfortunately, there have been no other similar quantitative analyses, but observations of play in diverse species support the idea that play signals can and do serve to establish a social context that allows play to occur and alters the significance of behavior patterns that are borrowed from other contexts and used in social play.

Some other characteristics of bows and also some of the properties of social play support a cognitive explanation of play and can be used to inform questions about their use for initiating and maintaining social play. For example, bows occur throughout play sequences, but they usually occur at the beginning or toward the middle of playful encounters. Intentional explanations of the way in which bows are used in ongoing play are helpful for understanding how animals negotiate agreements, especially because of the apparent flexibility in how bows are used. For example, in a detailed analysis of the form and duration of bows (Bekoff, 1977), it was shown that bows are highly stereotyped and that bows were always less variable when performed at the beginning rather than in the middle of ongoing play sequences. These data were used to argue that bows were important in the initiation of play. Other observations also support this idea. For example, one dominant female coyote pup was successful in initiating chase play with her subordinate brother on only 1 of 40 (2.5%) occasions; her lone success occurred on the only occasion in which she had signaled previously with a bow. There may be more variability for bows performed during play bouts when compared with bows performed at the beginning of play sequences because of fatigue, the fact that animals are performing them from a wide variety of preceding postures, or because there is less of a need to communicate that "this is still play" than there is when trying to initiate a new interaction. Data for distinguishing among these three possible explanations do not currently exist.

Play signals are also used to maintain social play in situations in which the performance of a specific behavior during a play bout could be misinterpreted. This is of interest because it suggests that an individual knows what it is going to do next and possibly the effect that its behavior will have on the recipient. In a recent study, I (Bekoff, 1995) hypothesized that if bites accompanied by rapid side-to-side shaking of the head or other behavior patterns typically used in other contexts, such as aggression or predation, could be or were misread by the recipient and could result in a fight, for example, then the animal that performed the actions that could be misinterpreted might have to communicate to its partner that this action was performed in the context of play and was not meant to be taken as an aggressive or predatory move. In this view, bows would not occur randomly in play sequences; the play atmosphere would be reinforced and maintained by performing bows immediately before or after actions that could be misinterpreted. These ideas had not previously been analyzed empirically. I found that the distribution of bows during play that involves these actions is not random; instead, they tend to occur immediately before or after an action that can easily be misinterpreted, such as biting accompanied by rapid side-to-side shaking of the head. The bows seem to serve as a form of punctuation that clarifies the meaning of other actions that follow or precede them. In addition to sending the message "I want to play" when they are performed at the beginning of play, bows performed in a different context, namely, during social play, might also carry the message "I want to play despite what I am going to do or just did — I still want to play" when there might be a problem in the sharing of this information between the interacting animals.

Species differences were also found that can be interpreted from what is known about variations in the early social development of these canids. The interspecific differences are also related to the questions about how individuals monitor their own behavior — what they might know about themselves and others. For example, infant coyotes are much more aggressive and engage in significantly more rank-related dominance fights than either the infant (or adult) dogs or the infant wolves that were studied. During the course of my study, no consistent dominance relations were established in either the dogs or the wolves, and there were no large individual differences among the play patterns that were analyzed in this study. Social play in coyotes is typically observed only after dominance relationships have been established in paired interactions. Coyotes appear to need to make a greater attempt to maintain a play atmosphere, and indeed they also seem to need to communicate their intentions to play before play begins more clearly than do either dogs or wolves that have been studied (Bekoff, 1977). Indeed, as mentioned previously, one dominant female coyote pup was successful in initiating chase play with her subordinate brother on only 1 of 40 occasions, her lone success occurring on the only occasion in which she had signaled previously with a play bow. Furthermore, subordinate coyote infants are more solicitous and perform more play signals later in play bouts. These data suggest that bows are not nonrandomly repeated merely when individuals want to increase their range of movement or stretch their muscles. However, because the head of the bowing individual is usually below that of the recipient, bowing may place the individual in a nonthreatening (self-handicapping) posture. Self-handicapping may be explained by an individual's assessing a situation and deciding what it has to do to enable play to continue, and self-handicapping suggests that one is aware of how others will interpret its behavior. Tanner and Byrne (1993) observed a captive female lowland gorilla concealing her play face, and they concluded that the gorilla was aware of her spontaneous facial expression and the consequence it entailed, namely, that play would follow within a few seconds.

In summary, detailed data and careful observations of a wide variety of animals lead to the conclusion that play-soliciting signals appear to foster cooperation between players so that each responds to the other in a way consistent with play and different from the responses the same actions would elicit in other contexts. Thompson (1996) provides a comprehensive list of play-solicitation behaviors in primates, rodents, insectivores, carnivores, and ungulates. It is important to remember that cues other than those provided by visible actions may also be important in the solicitation of play (Fagen, 1981). The information that is shared by use of play signals or play markers concerns the readiness of an individual or individuals to engage in social play. Play can be a risky affair (Fagen, 1981) but also is an important activity for young and old animals alike, and by using play signals individuals can engage in and enjoy the activity. Perhaps it is the case that playing has benefits that are important for developing organisms as well as adults that make it unlikely that animals will use play signals to deceive others.

Social Play and the Comparative Analysis of Animal Minds: Evolutionary Continuity

Byrne (1995) wrote that

> Great apes are certainly "special" in some way to do with mentally representing the minds of others. It seems that the great apes, especially the common chimpanzee, can attribute mental states to other individuals; but no other group of animals can do so — apart from ourselves, and perhaps cetaceans." (p. 147)

Attempts to place humans apart from, and above, nonhumans have in a sense backfired because comparative research in animal cognition has demonstrated evolutionary continuity in many cognitive abilities and has shown how connected humans are to other animals. However, this is not to claim that the same mental states underlie behavior patterns that appear similar in members of different species. Claims such as Byrnes' (1995) in the previous quotation are premature and may well prove wrong when necessary comparative research is completed. (It is important to note that Byrne equivocates in this statement and in others in his otherwise excellent book, and he also realizes that little actually is known about the intellectual skills of, for example, carnivores.) It is a fact that very few nonprimate species have been studied concerning the possibility that they have theories of mind, and species-fair tests need to be developed and applied widely before primatocentric claims that can be used to argue against evolutionary continuity can be rigorously assessed. It is unlikely that methods used to study theories of mind in nonhuman primates can be directly applied to nonprimates — dogs and wolves may have theories of mind but it is highly unlikely that we will learn about them using methods that are used on primates — and even among primates it is clear that species-fair tests need to be developed to account for species differences in sensory and motor abilities, social organization, and habitat (Byrne, 1995).

There is little doubt that a broad comparative and evolutionary cognitive perspective will be very useful in future analyses of social play (Bekoff, 1995) and other behavioral phenotypes, and that these sorts of studies will be useful for those generally interested in the evolution of cognition. Evolutionary accounts of mental content are also needed. There simply are no substitutes for careful observational and descriptive studies and also for well-informed and well-executed experiments. They may be difficult to perform, but difficult does not mean impossible. Information about continuity of cognitive abilities will help to fill in the huge holes that exist in our current database. Individual variability also needs to be taken into account because it would be misleading to conclude on the basis of the behavior of a few members of a given species that all members of that species are unable to do something or do not perform specific behavior patterns that indicate well-developed cognitive abilities. Much research to date on animal cognition has been done on only a very limited number of species and on only a few members of those species in (often impoverished) laboratory environments. If research on animal cognition it to inform the study of human cognition, a broader database and broader evolutionary perspective are needed.

In this article, I used social play as a candidate for further research in comparative animal cognition. I asked, "What is it like to be a playing animal?" Because in some instances animals seem to know what actions are likely to violate the terms of play and they attempt to negotiate with their partner(s) so that play will continue, they seem to have the capacity to represent the mental states of others and make adjustments in their own behavior. This ability suggests an understanding of obligation or permission structures that may be crucial for understanding human performance on tasks that require more abstract reasoning.

Where do we go from here? Open-mindedness is the key. Double standards that demand that studies of animal play and animal cognition be more rigorous than studies in other areas of behavioral science are unlikely to be productive, especially when these sorts of biases lead to the demise of studies of nonhuman cognition. It is unlikely that all of humans' sophisticated cognitive abilities arose *de novo.* Interactions among supporters of different views are essential if we are to make headway in learning more about animal play and animal cognition and how studies of animal cognition are able to inform and to motivate the study of human cognition.

Acknowledgments

Portions of this article were excerpted with permission of Oxford University Press from Bekoff (1997). I thank Denise Cummins, Colin Allen, and Patrick Bateson for comments.

References Cited

ALLEN, C., and BEKOFF, M. (1997). *Species of Mind: The Philosophy and Biology of Cognitive Ethology.* MIT Press, Cambridge, MA.

BEKOFF, M. (1977). Social communication in canids: Evidence for the evolution of a stereotyped mammalian display. *Science* **197,** 1097–1099.

BEKOFF, M. (1995). Play signals as punctuation: The structure of social play in canids. *Behaviour* **132,** 419–429.

BEKOFF, M. (1997). Playing with play: What can we learn about cognition, negotiation, and evolution? In *The Evolution of Mind* (D. Cummins and C. Allen, Eds.). Oxford Univ. Press, New York.

BEKOFF, M., and ALLEN, C. (1992). Intentional icons: Towards an evolutionary cognitive ethology. *Ethology* **91,** 1–16.

BEKOFF, M., and BYERS, J. A. (1981). A critical reanalysis of the ontogeny of mammalian social and locomotor play: An ethological hornet's nest. In *Behavioral Development: The Bielefeld Interdisciplinary Project* (K. Immelmann, G. W. Barlow, L. Petrinovich, and M. Main, Eds.). Cambridge Univ. Press, New York.

BEKOFF, M., and BYERS, J. A. (Eds.) (1998). *Play Behavior: Comparative, Evolutionary, and Ecological Aspects.* Cambridge Univ. Press, New York.

BEKOFF, M., and JAMIESON, D. (Eds.) (1996). *Readings in Animal Cognition,* pp. 300–301. MIT Press, Cambridge, MA.

BURGHARDT, G. M. (1988). Precocity, play, and the ectotherm–endotherm transition: Profound reorganization or superficial adaptation. In *Developmental Psychobiology and Behavioral Ecology* (E. M. Blass, Ed.). Plenum, New York.

BURGHARDT, G. M. (1998). Play. In *Comparative Psychology: A Handbook* (G. Greenberg and M. Haraway, Eds.). Garland, New York.

BYERS, J. A., and WALKER, C. (1995). Refining the motor training hypothesis for the evolution of play. *Am. Nat.* **146,** 25–40.

BYRNE, R. (1995). *The Thinking Ape: Evolutionary Origins of Intelligence.* Oxford Univ. Press, New York.

DENNETT, D. C. (1983). Intentional systems in cognitive ethology: The "Panglossian paradigm" defended. *Behav. Brain Sci.* **6,** 343–390.

FAGEN, R. (1981). *Animal Play Behavior.* Oxford Univ. Press, New York.

FAGEN, R. (1993). Primate juveniles and primate play. In *Juvenile Primates: Life History, Development, and Behavior* (M. E. Pereira and L. A. Fairbanks, Eds.). Oxford Univ. Press, New York.

GRIFFIN, D. R. (1992). *Animal Minds.* Univ. of Chicago Press, Chicago.

HALL, S. L. (1998). Object play by adult animals. In *Animal Play: Evolutionary, Comparative and Ecological Perspectives* (M. Bekoff and J. A. Byers, Eds.). Cambridge Univ. Press, Cambridge, UK.

JAMIESON, D., and BEKOFF, M. (1993). On aims and methods of cognitive ethology. *Philos. Sci. Assoc.* **2,** 110–124.

MARTIN, P., and CARO, T. M. (1985). On aims and methods of cognitive ethology. *Adv. Study Behav.* **15,** 59–103.

ORTEGA, J. C., and BEKOFF, M. (1987). Avian play: Comparative evolutionary and developmental trends. *Auk* **104,** 338–341.

PELLEGRINI, A. D. (Ed.) (1995). *The Future of Play Theory: A Multidisciplinary Inquiry into the Contributions of Brian Sutton-Smith.* State Univ. of New York Press, Albany.

SPINKA, M., NEWBERRY, R. C., and BEKOFF, M. (2001). Mammalian play: Training for the unexpected. *Quart. Rev. Bio.* **76,** 141–168.

TANNER, J., and BYRNE, R. (1993). Concealing facial evidence of mood: Perspective-taking in a captive gorilla. *Primates* **34,** 451–457.

THOMPSON, K. V. (1996). Behavioral development and play. In *Wild Mammals in Captivity: Principles and Techniques* (D. G. Kleiman, M. E. Allen, K. V. Thompson, and S. Lumpkin, Eds.). Univ. of Chicago Press, Chicago.

General References

BEKOFF, M., and BYERS, J. A. (Eds.) (1998). *Animal Play: Evolutionary, Comparative and Ecological Perspectives.* Cambridge Univ. Press, Cambridge, UK.

BURGHARDT, G. M. (1998). Play. In *Comparative Psychology: A Handbook* (G. Greenberg and M. Haraway, Eds.). Garland, New York.

BENNETT G. GALEF, JR.

Department of Psychology
McMaster University
Hamilton, Ontario, Canada L8S 4K1

During the past decade, researchers working in a variety of subdisciplines within anthropology, psychology, and biology have made substantial progress in their study of imitative and nonimitative forms of social learning in animals. Nonimitative social learning is now known to play an important role in reproduction, food acquisition, and predator avoidance in a range of vertebrate species. Evidence is also available that not only great apes but also some birds and nonprimate mammals can learn by imitation.

Social Learning and Imitation

Why Study Social Learning?

Most scientists who study animal social learning are interested in one of two quite different issues. Primatologists and psychologists frequently want to know whether nonhuman animals are able to imitate behaviors that they have seen others perform. For these researchers, situations in which one animal can learn to do an act simply by watching another animal perform that act provide exciting opportunities to explore the cognitive abilities of species other than our own. Other researchers, those whose work on social learning in animals reflects a general interest in behavioral ecology, usually study social learning to understand how information acquired from others contributes to the development of adaptive patterns of behavior in free-living animals. These researchers are interested in social interactions that direct behavior in profitable directions regardless of whether imitation or some nonimitative social learning process that is less cognitively demanding than imitation is believed to support transmission of behavior from one animal to another. Both approaches to the study of social learning are useful, and both have venerable histories in the life sciences.

Since the end of the nineteenth century, there have been many attempts to define different types of social learning and to categorize instances of social learning in terms of the learning process on which each depends. Sometimes social learning seems to involve true imitation (learning to do a particular motor act from seeing it done), whereas other times it appears to rest on emulation (learning what is required for success in a task by watching a model perform, but not learning about the model's behavior per se; Tomasello, 1996), local enhancement ("apparent imitation resulting from directing the animal's attention to a particular object or part of the environment"; Thorpe, 1963, p. 134), or some other carefully defined behavioral process (Galef, 1988b; Whiten and Ham, 1992). Different types of social learning are believed to require different degrees of cognitive sophistication, so the question of what type of social learning is involved in the transmission of behavior from one individual to another has potentially important implications for our understanding of the cognitive abilities of animals.

Historically, the most important distinction, and the only one I make in this article, involves determining whether, in any particular case, an observer learns directly about the behavior of its model. In the various kinds of nonimitative social learning, an animal that watches another animal behave and sees the outcome of the behavior in which the actor engages learns something useful about either which parts of the environment are potentially important or what changes in the environment are caused by the actor's behavior (Heyes, 1996; Tomasello, 1996). Therefore, for example, if one chimpanzee were to watch another hunting grubs by turning over logs using its left hand to flip the logs end over end, the observing chimp might learn (i) that logs are interesting objects (learning about the environment), (ii) that food is revealed when the undersides of logs are exposed (learning about changes in the environment that the behavior causes), or (iii) that putting one's left hand under the end of a log and moving that hand rapidly upward is rewarded with food (learning about the actor's behavior). Only the last kind of learning, in which the observing chimpanzee learns directly about the behavior of the acting chimpanzee, would involve true imitation as imitation is currently defined (Heyes, 1996; Tomasello, 1996). Such "true" imitative learning seems to require that an animal or human first store a visual representation of the pattern of movement exhibited by another. Then, because the imitator rarely sees its own movements in a way that would allow it to match the sight of its limbs moving to the stored visual representation of the act to be imitated, the imitator must use its proprioceptive or kinesthetic senses to match its own movements to the stored visual representation. Such cross-modality matching is believed to be a cognitively demanding task that distinguishes true imitative learning from other possibly less complex, nonimitative forms of social learning.

Nonimitative Social Learning

Perhaps the simplest way to introduce outsiders to recent advances in the study of nonimitative social learning is to briefly describe a selection of recent analyses of socially learned behaviors with potential to contribute to survival or reproductive success. I chose the examples discussed here to indicate both the range of behaviors affected by nonimitative forms of social learning and the very different ways in which information acquired from others can facilitate acquisition of adaptive patterns of behavior.

Learning What to Eat: Food Selection and Poison Avoidance by Norway Rats

Decades ago, an applied ecologist trying to increase the efficiency with which rodent pests could be controlled discovered that when he repeatedly offered a single poison bait to any population of rats (*Rattus norvegicus*), his efforts at control failed dismally. Although there was a decline in rat numbers immediately after he introduced a poison bait into an area, target populations repeatedly exposed to the same bait soon recovered their original size. The reasons for failure when using a single bait repeatedly in the same area were fairly straightforward.

First, although most rats ate a lethal dose of a toxic bait shortly after it was first introduced into their colony's territory, a few members of most colonies ate only a small amount of bait the first time they tasted it, became ill, but did not perish. These surviving rats learned in a single trial to associate the taste of the bait with its ill effects, and they would eat no more of it.

Second, and even more discouraging for efforts to improve the economics of pest control, young rats that came of age in colonies whose members were survivors of a first encounter with a poison bait also totally rejected the bait that the members of their colony had learned to avoid. They would eat only foods that survivors were eating. Somehow, the poison avoidance learned by survivors was being transmitted to the young of their colony.

Such socially learned avoidance of poison bait by weaning rats is a robust phenomenon and is easy to capture in the laboratory (Galef, 1988a). Consequently, it has proven possible to explore in-depth the social learning processes that result in transmission of food choices from adult rats to juveniles of their species.

Potential Prenatal Influences on Flavor Preference Even before birth, a young rat can acquire information from its female parent about at least some of the foods that she is eating. Hepper (1988) fed garlic to pregnant rats late in gestation. An hour or less after litters were delivered by these garlic-fed mothers, Hepper gave the young to foster mothers that had never eaten garlic to rear. When the foster-reared pups were 12 days old and still ingesting only mother's milk, Hepper offered them a choice between two dishes, one containing garlic and the other onion. He found that pups that had been gestated by mothers that ate garlic while pregnant stayed near the dish containing garlic, whereas pups gestated by mothers that had not eaten any garlic during pregnancy did not prefer garlic to onion.

Flavor Cues in Mothers' Milk Flavors of foods that a rat eats while lactating can affect the flavor of her milk, and exposure to flavored milk affects the food preferences of rat pups as they wean. For example, weaning rat pups preferred the food eaten by a lactating female from whom they had suckled for several hours but did not prefer the same food after they interacted\ for the same length of time with a female that ate the same food and acted maternally toward the pups but did not give milk (Galef, 1977).

Effects during Weaning Galef and Clark used closed-circuit television and time-lapse video recordings to observe nine wild rat pups from three litters take their very first meals of solid food. All nine pups were observed to eat for the first time in exactly the same circumstances: Each ate at a site at which an adult was eating, and none ate at a nearby site at which no adult was present. Apparently, the physical presence of an adult rat at a feeding site made that site attractive to young rats and markedly increased the probability that they would wean to whatever food was to be found there (Galef, 1977). Indeed, simply anesthetizing an adult rat and placing it, while unconscious, near a feeding site made that site significantly more attractive to weaning rats than alternative locations that had no rat near them.

However, adult rats need not be physically present at a feeding site to guide their young to it. While eating, adult rats deposit residual olfactory cues in the vicinity of a food source, on any food they eat, and on the path they take when leaving a feeding site and returning home. All these odorants are attractive to young rats and cause them to approach and prefer feeding sites that adults have been using (Galef, 1977).

Effects after Weaning Galef and his students also found that after a young rat (an observer rat) interacts for a few minutes with a recently fed conspecific (a demonstrator rat), the observer exhibits a substantial increase in its preference for whatever food its demonstrator ate (Galef, 1988a, 1996). Exploration of the processes responsible for development of this socially induced preference in observer rats began with the finding that when observer rats were exposed to anesthetized demonstrator rats whose heads had been lightly dusted with particles of food (Fig. 1), the observer rats increased their preference for the food that they found on their respective demonstrators' heads. On the other hand, observer rats identically exposed to pieces of cotton wool dusted with food failed to develop a preference for that food (Galef, 1988a). Clearly, something about the presence of a demonstrator rat together with a food causes observer rats to alter their food preferences.

Experiments to discover the stimuli emitted by demonstrator rats that cause observer rats to increase their preference for foods associated with a demonstrator have shown that exposure to a constituent of rats' breath, carbon disulfide, and a food causes rats to increase their preference for the food, just as experience of a food together with a breathing rat increases preference for it (Galef, 1996).

FIGURE 1 An observer rat interacts with an anesthetized demonstrator rat whose head has been dusted with food (adapted from Galef and Stein, 1985).

Learning How to Eat: Exploitation of Pinecones by Roof Rats

The pine forests of Israel are inhabited by colonies of roof rats (*Rattus rattus*) that live on a diet consisting almost entirely of pine seeds and water (Terkel, 1995). Extraction of seeds from pinecones is a stable tradition in these rat colonies, permitting them to survive in areas in which pine seeds are the only food present in sufficient quantity to support a population of mammals.

Laboratory observations of the feeding behavior of rats taken from colonies living in pine forests have shown that the feeding method that allows rats to recover more energy from pine seeds than they spend in removing the tough, nonnutritious scales from pinecones requires that the rats take advantage of the structure of pinecones. The scales at the base of a cone must be removed first. Then the spiral of scales circling the cone's shaft to its apex must be removed in succession (Fig. 2).

Observation in captivity of rats taken from areas other than pine forests and offered pinecones to eat revealed that only 6 of 222 hungry, adult rats learned to use the spiral pattern of scale removal that permits a net energy gain from eating pine seeds. On the other hand, essentially all young rats reared by dams that efficiently stripped seeds from cones acquired the profitable technique. Apparently, some aspect of the interaction between mothers that strip seeds from pinecones and the young they rear is important in transmission of the efficient technique from one generation of rats to the next.

Further experiments demonstrated that experience of young rats in completing the stripping of scales from cones that had been started in the proper fashion by an adult rat (Fig. 3) allowed more than 70% of young rats to learn the efficient method of attacking cones (Terkel, 1995).

FIGURE 2 Sequence of the stripping of scales from a pinecone by a rat: from left to right, the scales are removed from the base to the top following their distribution on the cone (adapted from Heyes and Galef, 1996).

The tradition of pinecone exploitation seen in these forest-dwelling roof rats is of particular interest because it allows rats to thrive in a relatively sterile environment that would otherwise be closed to them. In this case, and perhaps in others as well, social learning has opened a new ecological niche to some members of a free-living species.

Learning What to Fear: Snake Avoidance by Rhesus Monkeys

How animals learn to avoid predators is difficult to understand because building an innate system to distinguish all harmless animals from all potentially dangerous animals seems almost impossible, given the wide range of predators to which many animals are vulnerable, and because learning to avoid predators by individual experience would not work very well either. Predators must be avoided the first time they are encountered. There is no time for trial-and-error learning.

It has been known for more than 40 years that although monkeys and apes that have been captured in the wild and brought into the laboratory will vigorously avoid contact with snakes or snake-like objects, monkeys born and reared in captivity are relatively indifferent to the same objects that elicit strong fear responses in their wild-born brethren.

Mineka and Cook (1988) studied effects on the responses of laboratory-reared monkeys to snakes after these monkeys watching wild-born monkeys exhibit their fear of snakes. They found that laboratory-reared monkeys, which initially showed little or no response to snakes, responded vigorously to the sight of a snake after seeing a wild-reared monkey respond to a snake. This effect on the responses of laboratory-reared monkeys to snakes did not diminish over 3 months.

It is interesting to note that monkeys would not learn to respond emotionally to any object to which they watched a conspecific react. When Mineka and Cook (1988) showed laboratory-reared monkeys videotapes of wild-reared monkeys exhibiting fear responses elicited by snakes but edited the videotapes so that the monkeys appeared sometimes to be exhibiting fear of snakes and other times to be exhibiting fear of flowers, the laboratory-reared monkeys learned to fear the snakes but not the flowers. Apparently, the mon-

FIGURE 3 A rat pup feeding near its mother on a pinecone she has started to strip.

keys learned selectively to associate snake-like objects with the fear responses of others of their species.

Learning with Whom to Mate: Choice of a Partner by Guppies

Although evidence of a role for social learning in the development of adaptive patterns of response to potential foods and potential predators is more than 20 years old, direct evidence of a role for social learning in selection of a sex partner was only recently obtained. Dugatkin and coworkers (Dugatkin, 1996) examined the role of social learning in mate choice by the wild relatives of the aquarium guppy (*Poecilia reticulata*). Guppies were selected as subjects for two reasons. First, the conditions under which guppies mate in the streams of Trinidad (where the guppies used in these experiments were captured) were such as to allow social influences on mate selection to occur. Second, guppies are relatively easy animals to breed, maintain, and observe in the laboratory.

In Dugatkin's experiments, an aquarium was placed between two water-tight end chambers, and at each end chamber was placed one of a pair of male guppies matched for size and overall brightness (qualities that are attractive to females of their species). A "subject" female was then placed in a clear container in the center of the aquarium and a second female, a "model" female, was confined by a glass barrier at one end of the aquarium near one of the two males.

Once the experimental situation had been set up, the subject female could watch as the male closer to the model female was allowed to court the model female for 10 min. At the end of the 10-min period of courtship, the model female and the barrier that had kept her at one end of the aquarium were both removed from the aquarium, and the subject female was released from confinement and was then allowed to choose between the two males in their end chambers. Of the 20 subject females that Dugatkin observed choose a male in this situation, 17 chose to stay near the male they had watched courting the model female.

Although the results of this first experiment were consistent with the hypothesis that female guppies copy one another's mate choices, there are several alternative explanations of the model females' preference that need to be examined before that hypothesis is accepted. For example, it is possible that female guppies, members of a species that in the wild tends to form shoals, might simply prefer an area in which they had previously seen two fish to an area in which they had previously seen only one fish. However, in an experiment identical to that described previously except that females rather than males were placed in the two end chambers of the apparatus, subject females chose the female in the end compartment near which the model female had been confined only 10 times out of 20, as would be expected by chance (Dugatkin, 1996).

Dugatkin's data demonstrate a strong social influence on mate choice by female guppies. His findings are also consistent with explanations that have been proposed for some intriguing behavioral phenomena observed in natural circumstances. For example, in species such as sage grouse, in which males compete directly for females on communal breeding grounds (leks), each day a few males enjoy access to almost all females. Details of the circumstances in which sage grouse mate in natural circumstances are consistent with the hypothesis that female grouse observe and copy the mate choices of others, leading to near unanimity in mate selection by hens on any day (Gibson *et al.,* 1991).

Even more puzzling is the observation that males of some fish species court and readily mate with females of other species, even though males mating with these females do not contribute any genetic material to the females' young. Schlupp *et al.* (1994) followed up on Dugatkin's observations of social influence on mate choice by female guppies and found that male sail-fin mollies, which frequently mate with females of another species (the Amazon molly), actually increase their access to females of their own species by such cross-species matings. Female sail-fin mollies tend to prefer males they have seen mate, even when the males mated with female Amazon mollies.

Learning How to Court: Song Learning in Brown-Headed Cowbirds

Brown-headed cowbirds (*Molothrus ater*) are particularly interesting to students of social influences on behavioral development because females of this species lay their eggs in nests constructed by members of other species (the brown-headed cowbird is "brood parasitic"). Consequently, young brown-headed cowbirds are always reared by foster parents that are members of some other species. A young cowbird has no contact with adults of its own species while growing up. Because of the isolation of juvenile from adult cowbirds, it was long thought that song learning by male cowbirds was not affected by social interaction. However, as a result of the insight and hard work of Meredith West and Andrew King, it is now clear that young male cowbirds, like the young of many other species of songbird, learn the fine details of their songs as a result of social interaction.

Male cowbirds from different geographical areas sing different variants of cowbird song, and these "song dialects" are maintained by differences in the responses of females from different areas to the songs that males produce. The same songs that, during the breeding season, are most likely to cause females to assume the "copulatory posture" that allows males to mate (Fig. 4b) are responded to by females at other times of the year with a "wing-stroke" display (Fig. 4a). After a female gives a wing stroke in response to a song, the male singing that song repeats it three or four times in succession, violating a basic rule of song producof a phase timer but by an external stimulus not locked onto any particular cycle. The type of time sense required to predict this event is usually referred to as an *interval* or *hourglass time sense* because in order to predict the significant event it is necessary to start a timer when a stimulus occurs and time the interval to the significant event. A foraging hummingbird provides an excellent natural example of the interval time sense in use. After the bird has emptied the nectar from a flower, the flower will take a fixed amount of time to refill. A bird that knows the length of this interval can return to the flower as soon as it has refilled.

In addition to allowing an animal to predict when an event will occur, the interval time sense also provides animals with the ability to extract temporal intervals. For example, a foraging starling might time the interval between each prey item it finds in a particular field. These intervals could be used to form an estimate of the rate of prey capture which is known to be an important currency in determining decision making in foraging starlings.

Although it is theoretically possible that the same basic clock could subserve both the phase and interval time senses, there is ample evidence that this is not the case and that different mechanisms underlie phase timing and interval timing. Henceforth, I refer to phase clocks and interval clocks. The term *interval timing* is usually restricted to the timing of intervals in the seconds to minutes range, and it is timing within this range that is the focus of this article. I will not consider the dedicated mechanisms that have evolved to time much shorter intervals such as the system used by bats to time the delay in the return of an echo.

Whereas phase clocks are described in most animal behavior and ethology texts, interval timing has received very little attention from ethologists. Historically, the study of interval timing has been the domain of operant psychologists who study how rats and pigeons respond to artificially generated schedules of reinforcement in the laboratory. As a consequence, work on interval timing has failed to attract the interest of ethologists and behavioral ecologists because of the lack of connection to the problems faced by animals in their natural environments. However, it is becoming increasingly clear that interval timing is likely to be vitally important in many aspects of natural behavior; hence, the aim in this article is to introduce the properties of interval timing and to discuss it from an ethological perspective.

Experimental Paradigms Used to Prove That Animals Have an Interval Time Sense

Killeen *et al.* (1997) provided the following operational definition of timing: "An organism is timing if our clock is a better predictor of its behavior than any other stimulus we can identify." Interval timing is not easy to see in animals in their natural environments because to provide proof of timing it is necessary to eliminate the possibility that the animal is in fact using some external cue. Hence, the evidence that animals have interval timers comes largely from the operant psychology literature in which animals are studied under extremely controlled laboratory conditions.

Operant psychologists have devised many paradigms to investigate timing in animals, which can be broadly divided into either estimation or reproduction procedures. In an estimation procedure an animal is required to attend to a time interval and then make some response based on the duration it estimates after the interval has elapsed, whereas in a production procedure the animal actually reproduces the time interval in real time with its own behavior.

Bisection: A Temporal Estimation Task

Probably the most widely used estimation procedure is *temporal bisection.* In the first stage of bisection the subject is trained to make a simple temporal discrimination between, for example, a tone presented for 2 s and an identical tone presented for 8 s. In each trial one of the two tones is presented. After the tone has finished, two levers are inserted into the box. If the tone was the 2-s one, then the animal is reinforced with food for pressing the left lever, whereas if it was the 8-s one it is reinforced for pressing the right lever (the assignment of lever to tone is arbitrary but always the same for a given subject). Animals such as rats and pigeons will readily acquire this temporal discrimination. Once the animals are performing reliably, probe trials are inserted in which tones of different durations to those on which the animal was trained are presented. In such probe trials the same two levers are still inserted into the box, but no reinforcement is given whichever the animal chooses. Data from the bisection procedure are commonly presented by plotting the lengths of the probe trials on the x-axis against the probability of choosing the 8-s lever on the y-axis. The function obtained is sigmoid in shape and yields two important pieces of information: The steepness of the curve gives an indication of the precision of timing, and the time corresponding to a 50% probability of choosing the 8-s lever indicates the subjective midpoint between 2 and 8 s.

The bisection procedure has many benefits. Unlike the peak procedure described later, it can be used to study a wide range of different intervals including very short ones. A second benefit is that it provides an estimate of the temporal middle, which many models of timing make predictions about.

The Peak Procedure: A Temporal Reproduction Task

The peak procedure is a temporal reproduction task that is a modification of the more basic fixed-interval (FI) schedule. In a FI schedule the subject is reinforced for the first response (key peck or lever press) it makes after a fixed interval of time has elapsed. FI schedules can either be free-operant procedures, in which the interfood interval is the only discriminative stimulus and the interval is reset each time food is obtained, or discrete trials procedures, which are the basis for the peak procedure. In a discrete trials FI procedure, a stimulus such as a tone or light comes on at the beginning of the interval and is extinguished when the animal meets the response requirement and receives food. Then there is an intertrial interval in which the stimulus is not present before the next trial begins. Although animals could theoretically respond at their maximum rate the moment the stimulus comes on, they do not do this. Instead, they commonly show a FI scallop or alternatively a break-and-run pattern: There is a pause in responding following the onset of the stimulus that lasts approximately two thirds of the interval followed by a rapid acceleration in responding toward the time at which reinforcement results (Fig. 1). The length of the initial pause can be analyzed to provide data about timing.

In the peak procedure 50% of the trials are standard FI trials, but the remaining 50% are "probe" trials in which the stimulus remains on for two or three times the normal FI. Probe trials terminate with an intertrial interval, and no reinforcement is given (Fig. 2). In these probe trials subjects behave as in a standard FI in the first part of the trial (since

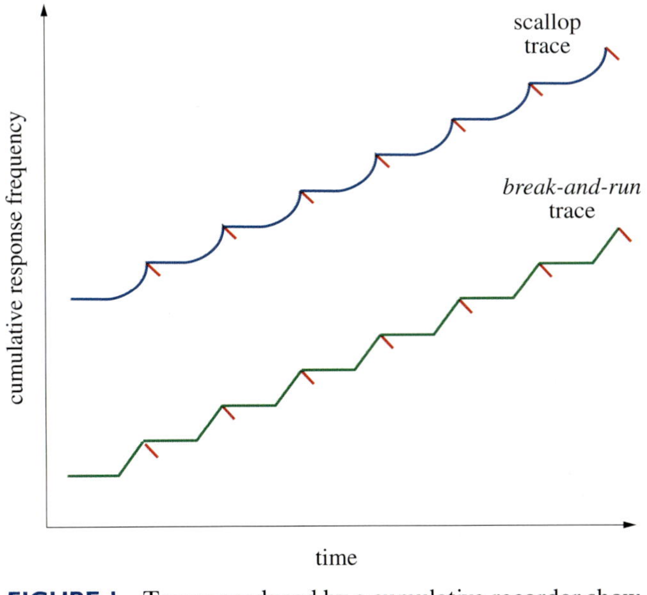

FIGURE 1 Traces produced by a cumulative recorder showing the typical behavior observed in a fixed-interval reinforcement task. The scallop (in blue) and break-and-run (in green) patterns are shown. In both cases, the response frequency (in arbitrary units) is plotted on the *y*-axis and time (in arbitrary units) on the *x*-axis. The diagonal ticks (in red) indicate reinforcement (reproduced with permission from Richelle and Lejeune, 1980).

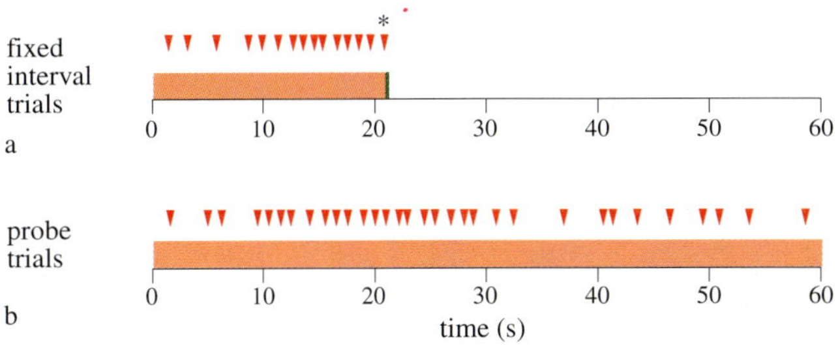

FIGURE 2 The peak procedure consists of mixed trials. (a) In the fixed-interval tasks, the light or tone signal appears at the beginning of the task and is turned off when the subject gives the first response (green vertical lines) immediately after the criterion interval (20 s). This response produces reinforcement (asterisk). The triangles represent the subject's responses. (b) In the probe trials the signal appears at the beginning of the task and is turned off after a period three times longer than the criterion interval (in this case, after 60 s), independently of the subject's behavior. In both cases, the red band corresponds to the time, within the trial, that the signal is active (adapted from Bradshaw and Szabadi, 1997).

the two trial types are identical until this point), and then at some time after the criterion time they cease responding again. When the data from several probe trials are plotted with the time in the trial on the *x*-axis and the average response rate on the *y*-axis, a Gaussian "peak" function is usually observed (Fig. 3). This function yields interesting data, including the peak time, which is the time at which the rate of responding is at a maximum; the peak rate, which is the height of the function at the peak time; and the spread, which is the width of the function at half its maximum height. Peak time is commonly used as a measure of timing

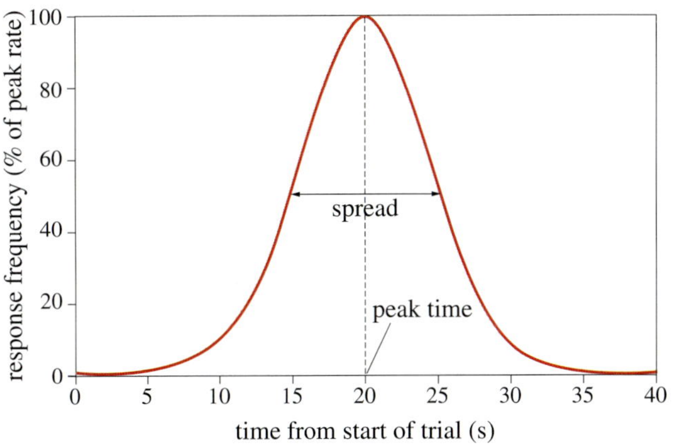

FIGURE 3 An ideal timing function obtained from probe trials of the peak procedure. The peak time is the time corresponding to the maximum height of the function and is used as a measure of timing accuracy. The spread is the width of the timing function at half the maximum height and is used as a measure of timing precision.

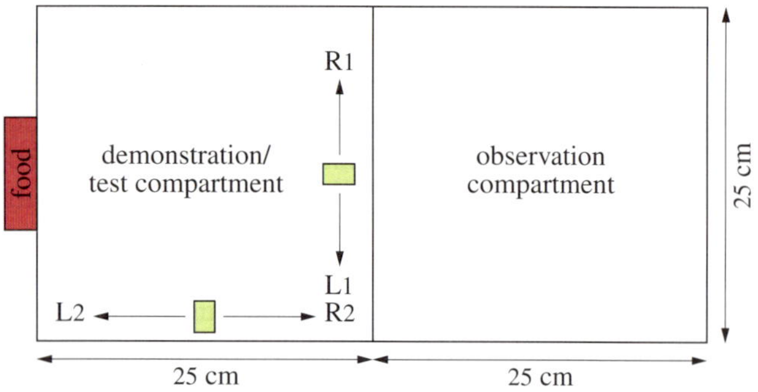

FIGURE 5 Schematic of the apparatus used by Heyes in the search for imitation learning in rats. The diagram shows the joystick in its starting position (between R1 and L1) and in the position to which it is moved for some observers during the test (at the front of the cage, between R2 and L2). An observer rat that saw a demonstrator move the joystick to R1 moved it toward R1 or R2 during testing — that is, in the direction used by the demonstrator. (adapted from Heyes *et al.,* 1992).

Moore (1992) provided startling evidence of spontaneous imitation of a human by an African gray parrot (*Psittacus erithacus*) that he named Okichoro. Moore housed Okichoro alone in a room that contained, along with the usual perches and toys, a microphone and video camera that allowed observation of Okichoro's behavior when alone and undisturbed.

Moore visited the aviary several times a day during a period of 5 years. During each visit, he repeatedly performed many distinct movements, each accompanied by a different spoken word or phrase. For example, as Moore left Okichoro's aviary each day he waved good-bye and said "ciao." Okichoro soon learned to say "ciao" and, by the end of the first year, was observed alone in his room saying "ciao" and at the same time waving his foot.

When in Okichoro's aviary, Moore also said "Look at my tongue," opened his mouth, and stuck out his tongue. Some time after Okichoro started to say "ciao" and wave, he was observed on the television monitor saying "Look at my tongue" and then opening his mouth and raising his tongue. Okichoro rarely opened his mouth and raised his tongue after saying "ciao" or waved his foot after saying "look at my tongue." In all, Okichoro was observed to copy many different actions involving six different body parts and to associate each action with its appropriate label (Fig. 6).

Some scientists accept such studies of social learning in nonprimates as providing examples of true imitation. Others reject them as examples of true imitation because it is difficult to know whether the motor patterns that the subjects used to produce a model's behavior were novel. It is possible that the acts that the animals copied were already in their behavioral repertoires and that they learned by observation only when to engage in acts they already knew how to perform. Whether production by an animal of an act already in its repertoire when it sees another animal perform that act should be considered true imitation is not clear. Difficult theoretical work remains to be done to reach consensus about exactly how imitation should be defined

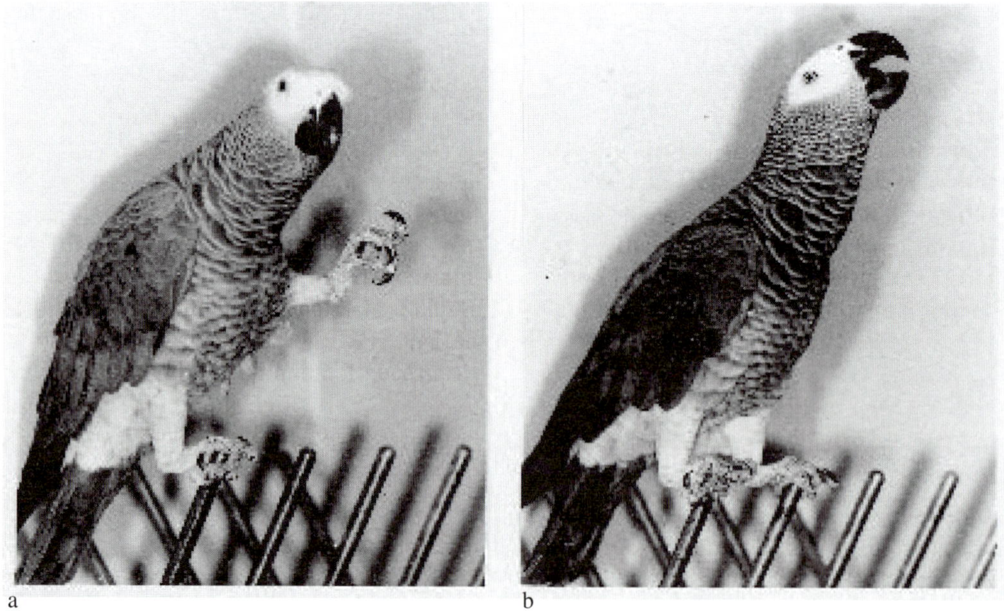

FIGURE 6 Examples of the imitation of human movements by the parrot Okichoro: (a) The parrot said "ciao" and (b) "look at my tongue" (adapted from Moore, 1992).

and what can be inferred regarding the cognitive abilities of animals from demonstrations of imitation of familiar or novel acts.

Conclusion

These are exciting times for students of imitative and non-imitative social learning in animals. After 100 years of effort, we may finally be in a position to start to explore the cognitive processes supporting imitation learning in animals and to compare such processes with those supporting imitation learning in humans. After all, even if both humans and apes imitate, they may not do so in the same way.

There is also strong evidence of an important role for social learning in the development of patterns of behavior that facilitate survival and reproduction by animals living in natural habitat. Such findings make important contributions to our understanding of the sources of information that animals use to respond adaptively to the challenges posed by their environments.

Two decades of hard work by scientists in many disciplines are starting to pay off. The next decade of research in social learning and imitation promises to be at least as productive as the past.

References Cited

DUGATKIN, L. A. (1996). Copying and mate choice. In *Social Learning in Animals: The Roots of Culture* (C. M. Heyes and B. G. Galef, Jr., Eds.), pp. 85–106. Academic Press, New York.

GALEF, B. G., JR. (1977). Mechanisms for the social transmission of acquired food preferences from adult to weanling rats. In *Learning Mechanisms in Food Selection* (L. M. Barker, M. R. Best, and M. Domjam, Eds.), pp. 123–150. Baylor Univ. Press, Waco, TX.

GALEF, B. G., JR. (1988a). Communication of information concerning distant diets in a social, central-place foraging species: *Rattus norvegicus*. In *Social Learning: Psychological and Biological Perspectives* (T. R. Zentall and B. G. Galef, Jr., Eds.), pp. 119–140. Erlbaum, Hillsdale, NJ.

GALEF, B. G., JR. (1988b). Imitation in animals: History, definition and interpretation of data from the psychological laboratory. In *Social Learning: Psychological and Biological Perspectives* (T. R. Zentall and B. G. Galef, Jr., Eds.), pp. 3–28. Erlbaum, Hillsdale, NJ.

GALEF, B. G., JR. (1996). Social enhancement of food preferences in Norway rats: A brief review. In *Social Learning in Animals: The Roots of Culture* (C. M. Heyes and B. G. Galef, Jr., Eds.), pp. 49–64. Academic Press, San Diego.

GALEF, B. G., JR., and STEIN, M. (1985). *Anim. Learning Behav.* **13,** 31–38.

GIBSON, R. M., BRADBURY, J. W., and VEHRENCAMP, S. L. (1991). Mate choice in lekking sagegrouse: The role of vocal display, female site fidelity and copying. *Behav. Ecol.* **2,** 165–180.

HEPPER, P. G. (1988). Adaptive fetal learning: Prenatal exposure to garlic affects postnatal preference. *Anim. Behav.* **36,** 935–936.

HEYES, C. M. (1996). Genuine imitation? In *Social Learning in Animals: The Roots of Culture* (C. M. Heyes and B. G. Galef, Jr., Eds.), pp. 371–390. Academic Press, San Diego.

HEYES, C. M., and GALEF, B. G., JR. (Eds.) (1996). *Social Learning in Animals: The Roots of Culture.* Academic Press, San Diego.

HEYES, C. M., *et al.* (1992). Imitation in rats: Initial responding and transfer evidence. *Q. J. Exp. Psychol.* **45B,** 81–92.

MINEKA, S., and COOK, M. (1988). Social learning and the acquisition of snake fear in monkeys. In *Social Learning: Psychological and Biological Perspectives* (T. R. Zentall and B. G. Galef, Jr., Eds.), pp. 3–28. Erlbaum, Hillsdale, NJ.

MOORE, B. R. (1992). Avian movement imitation and a new form of mimicry: Tracing the evolution of a complex form of learning. *Behaviour* **122,** 231–263.

RUSSON, A. E., and GALDIKAS, B. M. F. (1993). Imitation in free-ranging rehabilitant orangutans (*Pongo pygmaeus*). *Comp. Psychol.* **107,** 147–161.

SCHLUPP, I., MARLER, C., and RYAN, M. J. (1994). Benefit to male sail-fin mollies of mating with heterospecific females. *Science* **263,** 373–374.

TERKEL, J. (1995). Cultural transmission in the black rat: Cone feeding. *Adv. Stud. Behav.* **24,** 119–154.

THORPE, W. H. (1963). *Learning and Instinct in Animals,* 2nd ed. Methuen, London.

TOMASELLO, M. (1996). Do apes ape? In *Social Learning in Animals: The Roots of Culture* (C. M. Heyes and B. G. Galef, Jr., Eds.), pp. 319–344. Academic Press, San Diego.

VISALBERGHI, E., and FRAGASZY, D. M. (1990). Do monkeys ape? In *"Language" and Intelligence in Monkeys and Apes: Comparative Developmental Perspectives* (S. T. Parker and K. R. Gibson, Eds.), pp. 247–273. Cambridge Univ. Press, Cambridge, UK.

WEST, M. J., and KING, A. P. (1996). Social learning: Synergy and song birds. In *Social Learning in Animals: The Roots of Culture* (C. M. Heyes and B. G. Galef, Jr., Eds.), pp. 155–178. Academic Press, San Diego.

WHITEN, A., and CUSTANCE, D. (1996). Studies of imitation in chimpanzees and children. In *Social Learning in Animals: The Roots of Culture* (C. M. Heyes and B. G. Galef, Jr., Eds.), pp. 291–318. Academic Press, San Diego.

WHITEN, A., and HAM, R. (1992). On the nature and evolution of imitation in the animal kingdom: Reappraisal of a century of research. In *Advances in the Study of Behavior* (P. J. B. Slater, J. S. Rosenblatt, and C. Beer, Eds.), pp. 239–283. Academic Press, New York.

General References

GALEF, B. G., JR. (2000). Social influences on food preferences and feeding behaviours of vertebrates. In *Why We Eat What We Eat* (B. Capaldi, Ed.), American Psychological Association, Washington, DC.

ZENTALL, T. R., and GALEF, B. G., JR. (Eds.) (1988). *Social Learning: Psychological and Biological Perspectives.* Erlbaum, Hillsdale, NJ.

erative behavior. In this species, the male and female share incubation of the eggs, with the male taking the interval from approximately 10:00 to 17:00 h each day. If the onset of the male's incubation period is delayed 1 day, the offset will also be delayed, suggesting that the male uses an interval timer to predict the end of his bout of incubation (Gibbon *et al.,* 1984). It is not clear in this example why the male should time his bout rather than taking his cue to leave from the return of the female. It is also not clear why he uses an interval timer rather than his circadian clock since the female times her return to the nest according to the time of day.

Similarities exist between the peak procedure and classical conditioning experiments, suggesting that there could be connections between interval timing and associative learning. In both paradigms, an arbitrary stimulus (the conditioned stimulus; CS) is given at some interval before a reward (unconditioned stimulus; US) is available; the only difference between a classical conditioning experiment and a FI timing experiment is that the subject must make a response to collect the reward in the latter. Traditionally, learning theorists focused on the contingency between the CS and US in explaining learning; however, there is mounting evidence that the temporal contiguity of the CS and US is equally important in determining whether or not learning occurs and what types of conditioned responses are performed (Gallistel, 1994). Animals appear to learn the CS–US interval and the US–US interval, and it is the ratio of these remembered intervals, rather than just a simple CS–US association, that determines whether a conditioned response will occur (Gibbon *et al.,* 1977). It is also clear that the CS–US interval has an impact on the type of conditioned response performed. This makes good functional sense since the type of conditioned response that is appropriate will depend on the US (Timberlake *et al.,* 1982). For example, in the case of a stimulus-predicting food, if the CS–US interval is long, searching or hunting behavior may be adaptive, whereas if it is short handling and consummatory behavior are appropriate.

Estimation

The ability to discriminate time intervals is likely to be important in many different types of behavioral decisions. For example, a foraging animal may base future foraging decisions on how long it spent searching for food in two different patches or how long it took to handle two different prey types, and a female choosing a mate at a lek may base her decisions on how long a male displays. All these comparisons require the ability to estimate the duration of elapsed time intervals.

Estimating time intervals may provide a means of estimating many other variables of importance to animals. For example, the optimal foraging literature has identified rate of food intake as a currency that will often be highly correlated with fitness, and there is good evidence that many animals can make foraging decisions that optimize this currency (Kacelnik, 1984). Since this rate is computed by dividing amount by time, it is likely that interval timing is necessary for forming rate estimates. Estimates of time intervals could also be used to estimate amounts of food since there is often a strong correlation between the size of a food reward and the length of time it takes to handle and consume it. It is therefore extremely likely that interval timing is intimately involved in optimal foraging decisions. It is also possible that time intervals could be used to estimate distance traveled.

Finally, it is possible that interval timing could be involved in perceptual learning. For example, during the imprinting process a domestic chick forms a visual representation of its mother. The chick sees its mother from different directions and distances; however, these physically distinct views will be combined together in the same representation as long as they are seen within a short time of one another. The evidence for this is that following imprinting with two different objects, chicks will learn a discrimination between the two objects more quickly if the objects were presented 5 or more minutes apart during imprinting than if they were presented less than 30 s apart (Chantrey, 1974). These results suggest that a sense of time may be important in determining how animals form categories.

Comparative Studies

The comparative approach is often used to establish the adaptive significance of a behavior pattern. In the case of interval timing, this would involve comparing the timing abilities of related species and then attempting to correlate any differences found with differences in ecology.

The FI schedule has been used to study interval timing in a relatively wide range of vertebrate species and at least one invertebrate (Richelle and Lejeune, 1980). Mammals studied on FI schedules include mice, rats, cats, prairie dogs, horses, guinea pigs, Syrian hamsters, bats, beagle dogs, rabbits, hapalemurs, pottos, chimpanzees, baboons, squirrel monkeys, rhesus monkeys, and raccoons; birds studied include chickens, black vultures, crows, ravens, quail, European starlings, pigeons, and black-capped chickadees; fish studied include tilapia, gourami, and goldfish; and the only insect studied to date is the honeybee. In all species with the exception of one of the prosimians, the quail, the tilapia, and the honeybees, evidence for timing on FI schedules was observed. In the species that failed to show timing, it is not clear whether this failure reflects true deficits in timing ability or deficiencies in the experimental situations employed for testing.

However, attempts to draw any general conclusions from these data are beset with problems. The studies differed in many respects, including the response used, the reward used (e.g., food and water), the prior deprivation of the animals, and the intervals studied. Thus, it is possible that what appear to be species differences may actually result from differences in procedure. Even when two or more species are compared using identical methodology, differences in performance do not necessarily provide evidence for differences in timing ability because performance differences can also be produced by differences in motor abilities or motivation between species.

One of the best solutions to these problems is to compare only very closely related species that are expected to be similar in most respects apart from their timing abilities. Unfortunately, there have been surprisingly few careful comparisons of this type. One exception is a comparison of timing in pigeons and crows (Powell, 1974). The birds were trained on a DRL schedule that was gradually increased as long as the birds were still performing well. The pigeons were unable to time intervals beyond 10 or 30 s, whereas the crows could time intervals up to 120 s. This study is not subject to the usual criticism of DRL that the result could be attributable to a difference in the species' ability to withhold responding rather than their abilities to time because when the end of the interval was cued by a change in color of the response key, thus eliminating the need for timing, both species performed equally well at delays of 120 s. It thus appears that crows are capable of timing longer intervals than are pigeons. Although the conclusion of this study appears sound, it is difficult to speculate about the possible adaptive significance of this difference without data from more species.

Mechanisms

Perhaps one of the reasons that interval timing has not been more widely studied is that there is no obvious time measurement organ. However, many theoretical models have been produced to explain how timing could be achieved.

I briefly describe and contrast three timing models, all of which can produce a good quantitative fit to data obtained from the peak procedure (Bradshaw and Szabadi, 1997). Three different information processing components can be identified in all the models. First, measurement of time requires some type of clock. Second, all the models require some type of memory in which the time interval is represented. Third, all the models need a decision stage which uses output from the clock and the memory components to control behavior.

Scalar Timing

Scalar timing theory (Gibbon *et al.*, 1984) is a quantitative information processing model with explicit modules corresponding to the clock, memory, and decision processes (Fig. 5). The clock of scalar timing theory consists of a pacemaker that continually emits pulses. When the switch is closed these pulses are gated to an accumulator, such that elapsed time is a linear function of the number of pulses in the accumulator. The number of pulses in the accumulator is passed to working memory and finally to ref-

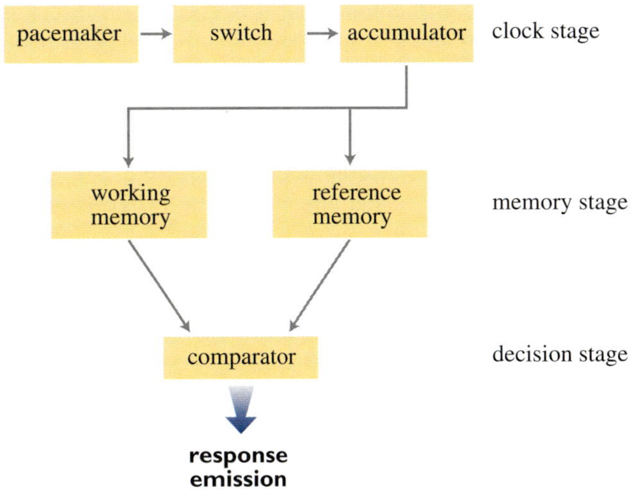

FIGURE 5 Model of information processing according to the scalar timing theory, showing the clock, memory, and decision stages.

erence memory, in which a distribution of experienced intervals is stored. The length of a duration is recalled from reference memory by taking a random sample from the distribution stored there. The decision about when to start and stop responding on an FI is made by continually comparing a sample from reference memory with the current elapsed time in working memory. When the ratio of the current time to the memory sample is sufficiently close to 1, responding is initiated.

Scalar timing theory is the most formally advanced of the existing timing models, and one of its great strengths is its ability to generate explicit quantitative predictions for many different timing procedures.

Behavioral Theory of Timing

As its name suggests, the behavioral theory of timing (Bradshaw and Szabadi, 1997) is a model based on observed behavior. Like scalar timing theory, the behavioral theory of timing has a pacemaker that emits pulses. Each pulse is assumed to be associated with a particular behavioral state, such as moving around a box, lever pressing, and eating. Thus, with each pulse from the pacemaker the animal progresses to the associated behavioral state. The speed of the pacemaker is assumed to be controlled primarily by the rate of reinforcement received by the animal. Thus, an animal is predicted to progress through sequential behavioral states more quickly on a 10-s FI than on a 30-s FI because the rate of reinforcement is higher on the 10-s FI. When reinforcement occurs, it strengthens the sequence of behaviors preceding it and this sequence is stored in memory. Thus, the representation of an interval corresponds to a sequence of behavioral states. When the animal is again placed in the presence of similar stimuli, the behavior sequence with the greatest strength is initiated.

nation example, this could occur if the pollinator discriminated among plants and failed to carry pollen from those flowers that did not produce adequate nectar. Finally, some level of cheating may actually be tolerated. This is expected only when the cooperating individuals are genetically related and partner choice is not possible. Relatedness enables the recipient to benefit by genetic proxy from the more selfish action of its partner.

The degree to which cheating occurs in any cooperative system depends on when partners benefit and whether mutual benefits require collective action. Mutualisms involving two individuals of either the same or different species are characterized by situations in which benefits are received immediately by all individuals as a direct consequence of their joint behavior. For example, African hunting dogs, *Lycaon pictus,* pursue and kill prey much larger than themselves by hunting in packs (Fig. 1). Foraging success, as measured by the amount of food killed per dog for each kilometer traveled, is highest for dogs living in large packs (Creel and Creel, 1995). As long as participation in cooperative hunts is required to capture and kill prey, cheating will not pay. Simultaneously received benefits almost certainly characterize other examples of cooperative hunting, including some tropical group-hunting ants that capture and consume prey much larger than themselves (Fig. 2).

If cooperation is defined solely as mutual benefit from any action, then a wide variety of selfish behaviors may also be considered cooperative when they inadvertently benefit other individuals. Any selfish acts by an individual that improve its ability to feed, warn, or defend itself or its young may simultaneously benefit other individuals by showing them where to feed, warning them of danger, or chasing off predators (West Eberhard, 1975). These types of situations have been termed *by-product mutualisms* and cannot be exploited by cheaters as long as there is no alternative behavior that provides greater reproductive success or survival to the selfish individual.

FIGURE 1 A pack of African hunting dogs (*Lycaon pictus*) disemboweling an antelope during a cooperative hunt. Each dog, by hunting in a pack, improves its success as measured in terms of the amount of meat obtained per kilometer traveled during hunting (photo courtesy of S. Creel).

FIGURE 2 An army of ponerine ants attacks and dismembers an earthworm.

Opportunities for cheating arise whenever beneficial acts require collective action or are separated in time. Exchanges involving some kind of donation, such as food, are referred to as *return benefit mutualism* when there is little or no cost to providing the donation and *reciprocal altruism* when the donation is costly (Trivers, 1971). Although reciprocal exchanges involving long time delays characterize a variety of human cultural practices, many evolutionary biologists believe reciprocity should be uncommon among animals because keeping cheaters out of such a system is likely to be difficult. For all individuals to profit from reciprocity, the value of past exchanges and identity of trading partners must be remembered. Cheaters which only infrequently repay their debts would be difficult to detect without considerable bookkeeping by the participants.

Although controversy remains over the extent and evolution of cooperation in nature, substantial theory and evidence indicate that faithful, repeated interactions as well as kinship among partners provide two important mechanisms for the evolution of cooperation. In the absence of partner fidelity or kinship, other social mechanisms are required to control the presence and behavior of individuals which attempt to garner more than their fair share of the benefits of cooperation. Several interesting possibilities for enforcing cooperation, including equal reproductive opportunity, partner choice, and mutual policing, have received attention recently. Although some evidence is consistent with each of these ideas, future empirical and theoretical studies will undoubtedly improve our understanding of how cooperation is maintained in different systems.

Many of the examples used in this article illustrate forms and mechanisms of cooperation among interacting animals.

Nevertheless, much of the theory has direct relevance to interactions between cells within organisms, interactions between different genetic elements within a cell, and even interactions between molecules in a primordial soup. Thus, the evolution of cooperation has implications for understanding parasite virulence, diseases such as cancer in which cell lineages compete instead of cooperate, and key transitions in the evolution of life from the origins of replicating molecules to multicelled organisms (Maynard-Smith and Szathmary, 1995).

Partner Fidelity Permits Cooperation

Reciprocity

Partner fidelity forms the basis for reciprocity. Trivers (1971) first pointed out that as long as two individuals help each other repeatedly, then over time cooperative pairs will survive and reproduce better than uncooperative pairs. Trivers termed this form of cooperation reciprocal altruism because he wished to explain cases in which aid is exchanged despite an immediate cost to reproduction or survival. If the benefit exceeds the cost, over time each participant profits from the interaction. However, individuals that fail to return the benefit do even better. Thus, for reciprocity to remain advantageous, prior association and memory are necessary for cooperative individuals to identify or retaliate against uncooperative individuals.

Food sharing in vampire bats, *Desmodus rotundus,* illustrates the potential advantage of reciprocity (Fig. 3). Female vampire bats regurgitate blood to their offspring and to each other when one has failed to obtain a blood meal on its own (Wilkinson, 1984). Each night, 7% of adults and 33% of juveniles fail to feed. Feeding failure occurs because the prey, which in most parts of Latin America is either cattle or horses, often detect and dislodge the bats. Young bats apparently do not avoid detection as readily as more experienced animals. Experiments with captive animals indicate that the bats not only identify and preferentially feed individuals from the same roosting group but also are more likely to give blood to a bat that has previously fed them. Such aid may be tremendously beneficial since a bat that fails to feed on three consecutive nights will die. In the absence of reciprocal blood sharing, annual mortality should exceed 80%, but female vampire bats are known to survive more than 20 years in the wild.

The mechanism by which vampire bats detect and exclude cheaters, those individuals which take blood without ever reciprocating, remains to be demonstrated. However, successful bats can ingest their body weight (more than 30 g) in blood. The stomach distention caused by such gorging sessions is likely to be noticed by another bat during episodes of mutual grooming. Mutual grooming frequently occurs between individuals just prior to blood regurgitation,

a

b

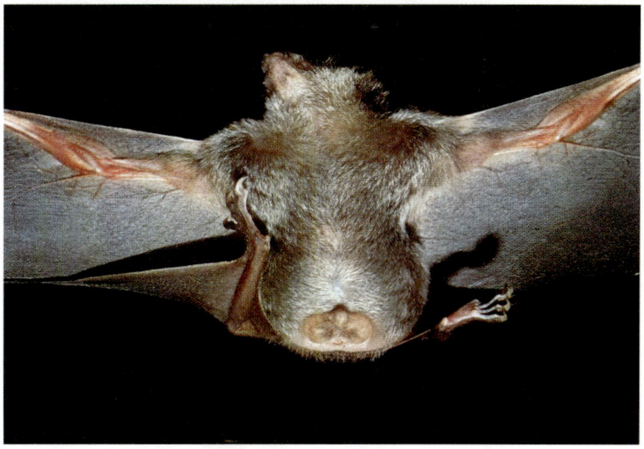

c

FIGURE 3 Vampire bats, of which the most common species is *Desmodus rotundus* (a), show various cooperative behaviors, such as blood sharing with other members of the colony and reciprocal grooming (b). During feeding, the bat, distending its stomach like a balloon, can consume its own body weight in blood (c).

as expected if grooming is involved in monitoring past feeding history. Since both mutual grooming and blood sharing in the wild only occur between individuals that have reliably roosted together in the same hollow trees for many months or years, partner fidelity appears essential for maintaining this cooperative system.

The Iterated Prisoner's Dilemma

The effect of partner fidelity on the evolution of cooperation has received considerable theoretical attention. Much of this theory involves analysis of a repeated two-person game known as the *iterated prisoner's dilemma.* Rather than barter money, biological players of this game win or lose points that reflect differences in their lifetime reproductive success or fitness — the currency of evolution. In one play of the game, when both players cooperate each receives a reward R compared with a punishment P if each fails to cooperate. The temptation T for a player to defect when it encounters a cooperative opponent provides the highest payoff, whereas a cooperating sucker receives the lowest score, S. In algebraic terms, a prisoner's dilemma is defined to exist when $T > R > P > S$ and $R > (T + S)/2$ (Fig. 4). In a single round of this game, the best strategy is to defect even though both players would receive higher payoffs if they both cooperated. This is because a player who defects either avoids receiving the sucker's payoff or obtains the temptation. Thus, defection by player A yields the best response to either possible action by player B for a single play of the prisoner's dilemma.

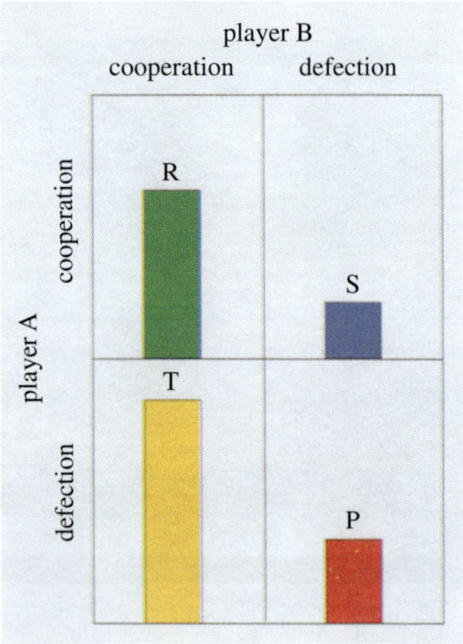

FIGURE 4 Matrix of the relative payoffs for player A in a prisoner's dilemma game. Players A and B gain points which reflect the difference in their success. In a round of the game, each player obtains a reward (R) if both cooperate and a punishment (P) if both fail to cooperate (defect). The temptation to defect if the other cooperates depends on the highest payoff (T), whereas cooperation with a defecting opponent gives the lowest score (S). To qualify as a prisoner's dilemma, the conditions must be respected that $T > R > P > S$ and $R > (T + S)/2$.

When this game is repeated over and over again, however, and the players base their decisions on previous interactions, then cooperation can emerge. In a series of computer tournaments, Axelrod (1984) discovered that the simple strategy tit-for-tat (TFT) outperformed all other competing strategies. TFT embodies the golden rule: It always begins by cooperating and on subsequent interactions simply copies whatever its partner did. Consequently, TFT is nice because it is never the first to defect and forgiving because it only defects once as long as its partner resumes cooperating. The success of TFT depends on there being a sufficiently large probability that two individuals will encounter each other again. In this way, TFT uses partner fidelity to exclude cheaters.

Recent theory indicates that TFT may represent only the first stage in the evolution of cooperation. Although TFT does as well or better than any alternative strategy in a constant environment, TFT is vulnerable to inadvertent errors. One accidental defection between two individuals playing TFT will lead to cycles of mutual defections until another mistake corrects the misunderstanding. Generous tit-for-tat (GTFT), a strategy which retaliates only two-thirds of the time in response to a defection, is more forgiving than TFT and therefore is immune to such occasional mistakes. Computer simulations by Nowak and Sigmund (1993) show that GTFT often gives rise to a strategy called PAVLOV. The terminology PAVLOV derives from the conditioned reflex response of this strategy to the payoff: It repeats its previous move when rewarded with either of the two more profitable payoffs, R or T, but changes behavior when punished with the least profitable payoffs, S or P. In other words, PAVLOV obeys the rule win-stay, lose-shift. PAVLOV outperforms TFT in an environment in which errors occur because it corrects mistakes and exploits an unconditional cooperator.

Several animal examples of cooperation appear to involve reciprocation and are consistent with strategies involving retaliation, such as TFT and PAVLOV (Dugatkin, 1997). For example, hooded warblers, *Wilsonia citrina,* exhibit increased aggressiveness in response to playbacks that simulate territorial incursions by neighbors rather than by strangers, as expected if there is retaliation against neighbors for not observing a territorial boundary. In some hermaphroditic serranid fish and polychaete worms, eggs and sperm are sequentially traded in small aliquots presumably to minimize the potential gain obtained from cheating by only transferring small, energetically inexpensive sperm rather than the larger, and therefore more energetically costly, eggs. Also, a series of experiments on sticklebacks (*Gasterosteus aculeatus*) and guppies (*Poecilia reticulata*) have shown that these fish often leave a school and approach a much larger predator in pairs. Predator inspection is assumed to benefit approaching pairs of fish because each learns what kind of predator is nearby and how likely it is to attack. Paired individuals interact repeatedly over time and take turns moving toward the predator, as predicted if each pair of moves represents one round in a prisoner's dilemma game.

The importance of the iterated prisoner's dilemma in explaining cooperation in animals remains controversial, however, because even though each of these examples is generally consistent with TFT, most have alternate explanations and some fail to mimic the payoff situation envisioned by the prisoner's dilemma. In particular, the iterated prisoner's dilemma assumes that both players make simultaneous decisions about their next move. If, instead, animals alternate moves, as is the case for food sharing, predator inspection, and territorial defense, then computer simulations show that yet another strategy, firm but fair (FBF), eventually outcompetes generous TFT and the Pavlovian win-stay, lose-shift (Frean, 1994). This new strategy is "firm" because it retaliates by defecting if it was a sucker. FBF is "fair" because it does not retaliate against a defector if it defected itself, and it cooperates with suckers rather than continue to exploit them. Thus, even though the order of play does not influence the outcome of a single round of the game, it will influence how to behave in situations in which partners exchange interactions over time.

Territorial defense in African lions, *Panthera leo,* provides an instructive exception to much of this theory because some individuals appear to exhibit unconditional cooperation, even when defection occurs. Male lions band together to form coalitions that compete for control of a pride of females. Small male coalitions are usually composed of unrelated individuals, whereas large coalitions invariably contain brothers. Although takeovers of prides by male coalitions occur infrequently — only every 2 or 3 years — territorial defense has been studied extensively by broadcasting roars through speakers to simulate intrusions by a competing coalition. All males in a coalition currently in possession of a pride approach playbacks independently of the behavior or relatedness of their companions (Fig. 5). The lack of any apparent cheating among males in a coalition makes sense because the reproductive success of each male hinges on retaining possession of the pride. Since all members of a coalition mate and share paternity, every male benefits by helping to drive away a competing coalition.

In contrast to the unfettered defense displayed by males, most female lions facing the same situation are less likely to rush out and defend the pride (Heinsohn and Packer, 1995). Females only approach speakers when the number of lions emitting roars is less than the size of the pride — in other words, when the odds are in their favor. Furthermore, individual females consistently differ in their willingness to lead the charge. Some females always act boldly and move toward playbacks, whereas others act fearful and lag behind. Although lead females often look back at their tentative pride mates, they continue to approach playbacks without exhibiting any form of retaliatory behavior. Female lions, therefore, appear to tolerate cheaters, at least in the context of territorial defense. Although such repeated cooperation in the face of apparent defection is inconsistent with even the most forgiving successful strategies for the iterated prisoner's dilemma, other examples of asymmetrical benefits to cooperation have been observed, but always among relatives. As discussed next, kinship provides an important reason for cooperation in these situations.

FIGURE 5 Male lions in a coalition exhibit cooperative defense. When a playback speaker broadcasts the roar of a simulated intruder, the male lion approaches the source of the noise (photo courtesy of C. Packer).

Kinship Promotes Cooperation

Kin Selection

Kinship refers to genetic relatedness. Reproduction in all animals involves duplicating and transmitting genes to offspring. The probability that an exact copy of a gene is located in another individual is called the degree of relatedness, symbolized by r. In humans and other diploid organisms (i.e., those that possess copies of genes on paired chromosomes), the degree of relatedness between a mother and her son is one-half because the son gets half of his genes from his mother and half from his father. Similarly, the degree of relatedness between a son and his mother is one-half for the same reason. Any pair of individuals with a common ancestor has some degree of relatedness, although r may be very small if the ancestor occurred many generations in the past. In all cases, r is measured by the probability that a typical gene in one individual has an exact copy in another due to direct descent from a common ancestor.

Kinship is important because cheating on a potentially cooperative interaction may not pay if a potential partner is related. Recall that natural selection favors traits that increase an individual's lifetime production of offspring and their survival. The reason for this is that parents are related to and share genes with their offspring. From the perspective of a gene, any behavior that increases its frequency in the next generation will be favored. Consequently, natural selection should also favor traits that increase the production and survival of other relatives, such as siblings or cousins. However, because siblings have a higher degree of relatedness than cousins, we expect animals to discriminate among and preferentially interact with their closest relatives.

If two animals are sufficiently related, even altruistic behavior may be favored. A gene influencing altruistic behavior will spread when the benefit to the recipient (B) times the degree of relatedness (r) minus the cost to the individual initiating the act (C) exceeds zero; in other words, when $rB - C > 0$ (Hamilton, 1964). This simple inequality, known as Hamilton's rule, forms the basis for a powerful theory called kin selection (Maynard-Smith and Szathmary, 1995). Kin selection influences the evolution of aid-giving and cooperative behavior, as well as many other forms of social behavior, in a variety of animals.

To illustrate how kin selection influences cooperative behavior, consider nesting and nursing behavior among house mice, *Mus domesticus*. House mice often give birth and nest with one or two other females and then indiscriminately nurse all the young in a communal nest (Fig. 6). An important advantage to forming a communal nest appears to be that two or more females are better at protecting their vulnerable pups from potentially infanticidal males. However, since lactation is usually the most energetically expensive period of any female mammal's life, why doesn't a female mouse try to avoid nursing pups in the communal nest? The answer appears to be that she is usually related to all the pups. House mice preferentially form communal nests with close relatives, such as sisters or daughters (Manning *et al.,* 1992). Consequently, females that provide milk to their own pups as well as their relative's pups leave more copies of their genes in future generations than females who are less cooperative.

FIGURE 6 A communal nest occupied by two female house mice (*Mus domesticus*). Females will nurse all the pups without discrimination. Experiments with congenic mice indicate that communal nesting females use the major histocompatibility complex (MHC) system to identify their relatives.

Kin Recognition

For kin selection to operate, female mice must be able to recognize and preferentially nest with close relatives. In other words, they need to be able to discriminate kin from nonkin. In some animals, kin recognition is based on past association. For example, a common smell acquired by two individuals that grew up together and fed on similar foods may impart a recognizable scent. Although a nest-specific odor might allow an animal to make simple categorizations, such as distinguishing siblings from unrelated individuals, it will not predict very well genetic relatedness in those animals that exhibit variation in relatedness within a brood, as will happen if two or more males share paternity. In these cases, animals would do better if they could somehow directly compare genes to assess relatedness.

Genetic kin recognition requires the presence of variable regions in the genome and the sensory ability to detect and discriminate individuals on the basis of these differences. Otherwise, genetic similarity caused by past ancestry may be confused with similarity due to chance. Studies using congenic house mice bred to be genetically identical except for alleles at the multiple histocompatibility loci (MHC) demonstrated that female mice form nests together depending on whether they have the same MHC haplotype (Manning *et al.,* 1992). Tremendous genetic variation at the MHC is maintained in all vertebrates to facilitate cell–cell recognition in the face of parasitic cell invasions from disease-producing organisms. Consequently, MHC loci are ideally suited to function as genetic indicators of relatedness and may do so in many vertebrates from tunicates to humans.

African lions resemble house mice in that females living together routinely nurse each other's pups (Fig. 7). A major cause of cub mortality is male infanticide following replacement of a coalition of males. Just as in house mice, the number of females living together determines how well cubs are protected from infanticidal males. However, lions go further than mice. Lactating females selectively permit more suckling by closely related cubs than by distantly related cubs (Pusey and Packer, 1994) (Fig. 8). How female lions recognize cub kinship levels is not known, but age, past association, and MHC similarity are all likely possibilities.

Kin Manipulation

When relatives differ in dominance and have restricted opportunities to live away from a social group, cooperation

FIGURE 7 A female lion snarls at a large cub while nursing her two small offspring and a third cub belonging to another female. Females give an average of 30% of their nursing to cubs other than their own (photo courtesy of C. Packer).

with unequal benefits among participants may occur. Communal breeding in groove-billed anis, *Crotophaga sulcirostris,* provides an example. These glossy black cuckoos live in social groups containing one to four pairs of birds and share a single nest. Each female lays all of her eggs in succession, with the most subordinate female laying first and the most dominant laying last. Before each bird begins laying, it removes several eggs already in the nest. As a consequence of this egg-tossing behavior, the dominant female contributes more eggs to the final clutch than any of the other females (Fig. 9). Once the eggs hatch, all members of the group participate in feeding and protecting the chicks. Why don't subordinate females leave and nest on their own? The reason appears to be that quality breeding habitat

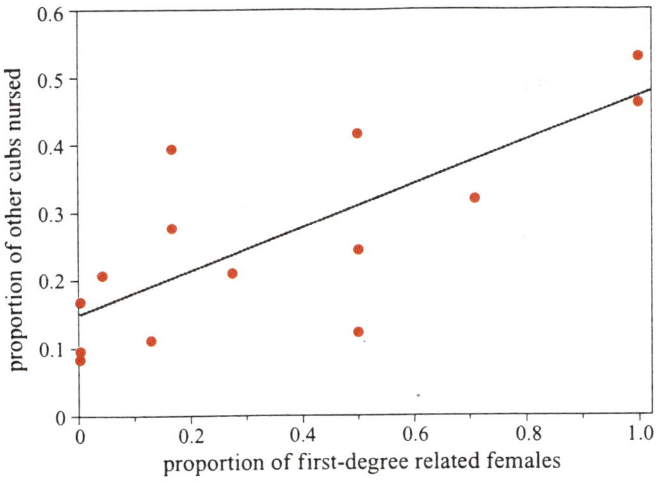

FIGURE 8 The proportion of cubs of other mothers nursed by female lions in a creche as a function of the proportion of all females which are first-degree relatives. The regression line implies that related cubs are preferentially allowed to nurse.

FIGURE 9 Two pairs of groove-billed anis (*Crotophaga sulcirostris*) at a communal nest. The dominant female lays in the nest only after removing several of the subordinate's eggs.

is typically unavailable in those parts of Central America where this bird is found. Consequently, a subordinate may produce more young in a cooperative group than she could on her own. If she is also related to the dominant, then she may gain further benefits by increasing the survival of related offspring (Vehrencamp, 1983).

Kin manipulation also occurs in mammals, including dwarf mongoose (*Helogale parvula*), African hunting dogs, cotton-top marmosets (*Saguinus oedipus*), and naked mole rats (*Heterocephalus glaber*). In each of these species, only a single female usually succeeds in rearing young (Fig. 10). Reproduction by other females in the social group is suppressed by yet to be determined pheromones or their young are killed at birth. Despite such treatment, these nonreproductive, but typically related, animals remain in the group, collect food, and help care for the young of the dominant pair. Naked mole rats represent an extreme example because they occur in large colonies in desert areas of eastern and southern Africa where they feed on large tubers of a few species of plants. Successful exploitation of this widely dispersed food source requires that a colony develop an extensive and ever-expanding network of tunnels. Once a tuber is found, workers carry pieces back to the communal nest and share them with other members of the colony.

FIGURE 10 A pregnant female naked mole rat (*Heterocephala glaber*) in the midst of a communal nesting area. This female gave birth to 27 pups the next day. The reproductive female in a colony is much larger than either males or females and is provisioned by scouts which collect tubers (photo courtesy of P. Sherman).

Workers advertise their find by vocalizations which recruit colony members to the food source. Even though workers may never reproduce, they indirectly benefit by helping the reproductive female because relatedness within a colony may be extremely high due to inbreeding, almost certainly from past incestuous mating. In contrast to most animals, naked mole rats rarely disperse.

Kin manipulation has also been reported in *Polistes* wasps (West Eberhard, 1975). In many species of paper wasps, sisters disperse together and found a new nest. Due to an unusual sex determination system known as *haplodiploidy* in which males carry one copy of chromosomes and females carry two copies, sisters from a singly inseminated mother have a 75% chance of sharing a gene which is identical by descent. Although sisters may initially cooperate to construct a nest, only one or perhaps two dominant individuals eventually develop ovaries and lay eggs. The remaining sisters often become workers which continue to build the nest and care for the brood. Once again, a subordinate wasp should, in an evolutionary sense, be willing to help as long as more copies of her genes will be carried by her sister's offspring than by her own offspring if she departed and attempted to nest on her own.

The Problem with Group Selection

Until the mid-1960s, cooperative behavior was commonly invoked as a mechanism for population regulation. This point of view is best captured by the phrase, "Cooperation occurs because it is good for the species." A detailed critique by Williams (1966) illustrated the flaw in this perspective. Unless there is little or no migration between groups of interbreeding organisms and frequent extinction of such populations, those individuals which selfishly reproduce as much as possible will outcompete those which limit reproduction to reduce competition. The primary exception to this conclusion occurs when animals are related to each other. The discovery that reproductive sacrifice is typically found in animals which form groups composed of close kin is therefore consistent with kin selection, rather than group selection, favoring cooperation.

During the past 15 years, however, an alternative form of selection at the group level has been advocated by some evolutionary biologists. Rather than depend on differential extinction of populations, this new type of group selection, *trait group selection,* depends on differential productivity of groups. Theoretical studies have shown that trait group selection may favor cooperative behavior among organisms if they mate at random prior to dispersal into groups on the basis of their predisposition to act cooperatively. In theory, such trait groups could be composed of unrelated animals, but the simplest mechanism producing such nonrandom assortment into groups is kinship association. When trait groups are composed of close relatives, kin selection, as outlined previously, is sufficient to explain the occurrence of cooperation. Although group selection has been demonstrated in the laboratory (Wade, 1977), few if any unequivocal examples have been identified in nature. Currently, perhaps the most plausible case of trait group selection involves cooperation among unrelated human groups (Wilson and Sober, 1994).

Enforcing Cooperation

Equal Reproductive Opportunity

The behavior of groove-billed anis, naked mole rats, and paper wasps might lead one to predict that all cases of cooperative breeding must involve closely related animals. However, in several species of ants and termites, nests may be founded by multiple reproductive females which may be distantly related if at all (Fig. 11). Where data are available, colonies founded by multiple queens tend to grow faster and are more likely to survive and produce reproductive individuals than colonies founded by single females. As a consequence, per capita reproduction is often higher in these multiple queen or polygynous colonies. Termites and ants differ from wasps, birds, and mole rats in that the decision to become reproductive is not reversible. Whereas workers in these other taxa are able to develop reproductive organs, most ants and termites do not have such physiological flexibility. Queens of the latter species typically must develop ovaries and extended abdomens before they can lay eggs. Also, although supernumerary wasp queens are invariably eliminated or end up helping but not reproducing within a few months after colony formation, some termite and ant queens may live together and continue to reproduce for

FIGURE 11 Thirty-three primary queens (with extended abdomens) and 17 primary kings (2 of which are visible at top right) of *Nasutitermes corniger* termites collected from a single colony in Panama (photo courtesy of B. Thorne).

much longer periods. Nevertheless, cooperative reproduction by multiple foundresses appears to represent a transient period typical only of the early phase of any social insect colony. In most cases, conflict eventually erupts and only a single queen survives. A plausible explanation for such temporary cooperation depends on foundresses being unable to determine who will eventually gain reproductive control. If every foundress has an equal chance of becoming the dominant reproductive in a colony, then sustained cooperation is expected.

Another intriguing potential example of equal reproductive opportunity acting to ensure cooperation involves the slime mold, *Dictyostelium discoideum.* The life cycle of this organism passes from a dispersed single-celled feeding stage to a multicellular aggregate or slug capable of coordinated movement (Fig. 12). When the slug reaches an ecologically appropriate location, a fruiting body containing spore cells is erected on a long stalk. Although kin selection may be part of the explanation for why stalk cells should sacrifice

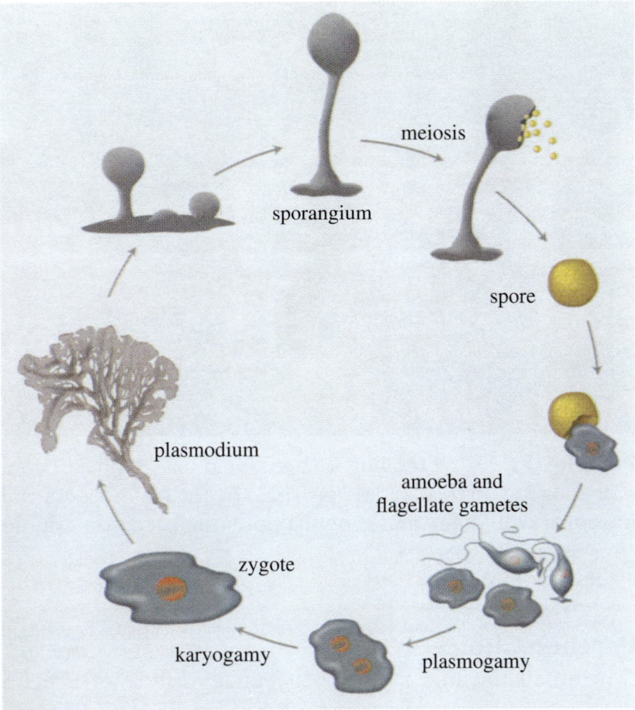

FIGURE 12 Life cycle of the slime mold *Dictyostelium discoideum.* The mature fruiting body (sporangium), held on a long stalk, produces the haploid spores. These hatch into amoebae, which multiply asexually. Amoebae aggregate, first by plasmogamy and then karyogamy, to form a migrating diploid slug (plasmodium), which eventually develops into a fruiting body.

themselves for their spore brethren, some slugs develop from up to 100,000 aggregated cells. Cooperation without kin selection may be favored if cell fate is determined at random in the developing slug. Recent work indicates, however, that cell lineages do not always cooperate to form spores (Strassman *et al.,* 2000).

Equal reproductive opportunity also typifies biological cooperation of a very different kind – the passage of chromosomes from a diploid cell into haploid gametes. Through the process of meiosis, each chromosome has a 50% chance of ending up in a gamete. Evidence that meiosis represents a bona fide example of cooperation derives from observations of cheaters. In a variety of organisms, meiotic drive occurs, i.e., one chromosome ends up in a disproportionate fraction of gametes. When a sex chromosome disobeys the laws of Mendelian segregation, distortion of the primary sex ratio results. Although such selfish behavior may be advantageous for genes on the cheating chromosome, the outcome of meiotic drive for a sex chromosome is eventual extinction when the limiting sex disappears unless some force intervenes to restore transmission frequencies. Thus, meiosis may be viewed as an elaborate mechanism for ensuring cooperation among the chromosomes (Maynard-Smith and Szathmary, 1995).

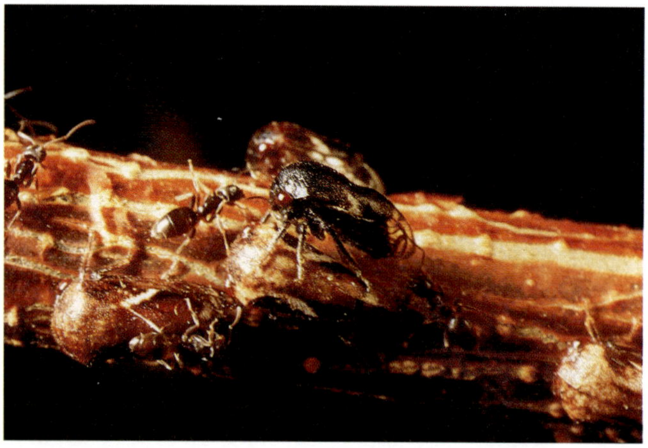

FIGURE 13 Ants tending a membracid leaf hopper. These ants will feed on the exudate provided by the leaf hoppers and in return will attack and repel its potential predators (photo courtesy of R. Cocroft).

Partner Choice

Mutualistic interactions between a variety of species provide good examples for how an alternative mechanism, partner choice (Bull and Rice, 1991), controls cheaters and maintains cooperation. Many species of homopterans, such as aphids and membracids, the larvae of many lycaenid and riodinid butterflies, and some plants such as ant acacias, provide sugar-rich solutions to ants in exchange for protection from predators or competitors (Fig. 13). Individual plants or insects could cheat by not producing rewards for the ants. Cooperation should persist as long as the ants are able to choose alternate plant or insect partners that provide higher rewards. At the same time, the insects or plants may withhold rewards if the attending ants do not provide adequate protection. Thus, cooperation may be regulated by partner choice as in any market economy (Noë and Hammerstein, 1994).

Evidence for partner choice derives from studies of mutualistic associations between some butterfly caterpillars and attending ants. Lycaenid and riodinid caterpillars produce sugar-rich solutions from a dorsal nectar organ, which functions only to produce rewards for ants (Fig. 14). Some caterpillars even attract ants by emitting vibratory acoustic calls and may emit chemical signals that cause the ants to attack approaching predators (De Vries, 1992). Isolated caterpillars produced more nectar when the number of attending ants was increased from one to two. Similarly, in those species of lycaenids which form large clusters, individual caterpillars failing to produce nectar are often killed by attending ants. Although the pupae remain attractive to ants by continuing to produce nectar, once eclosed, the emergent butterflies are vulnerable to ant predation, further demonstrating that cooperation in this system depends on the market value of nectar (Noë and Hammerstein, 1994).

Partner choice may also enforce mutualistic behavior within species. For example, in some primates and dolphins, males work together in alliances to gain access to sexually receptive females. Primate alliances appear to differ from dolphin alliances, however, in at least two ways. Primate alliances tend to be temporary and result in asymmetrical benefits to participants, much like in cases of kin manipu-

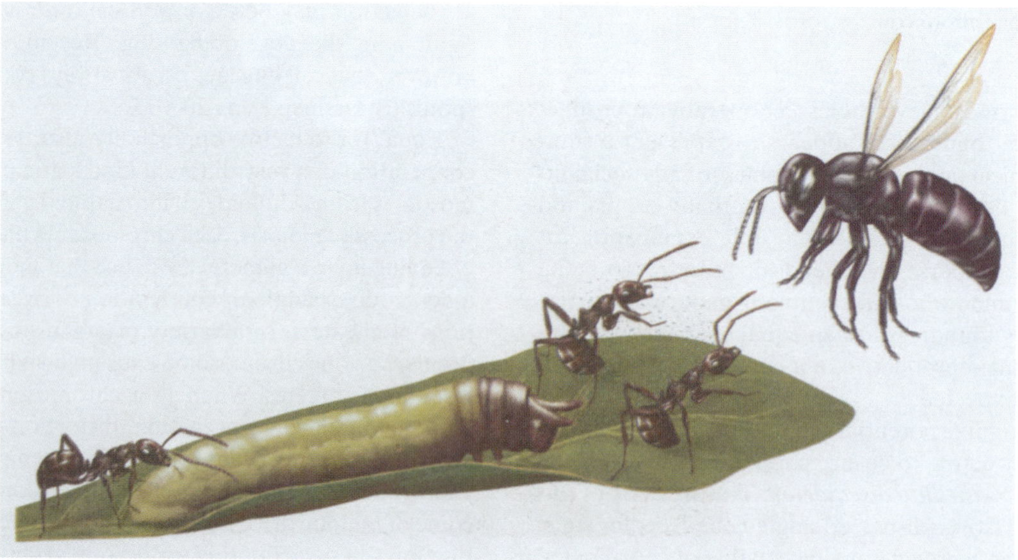

FIGURE 14 Riodinid caterpillars feed attending ants from their nectary organs (left). The caterpillars can elicit an attack response to predators from the ants by releasing chemicals that mimic the ants' alarm pheromone (right) (adapted from Vries, 1992).

FIGURE 15 Two bottlenosed dolphin (*Tursiops truncatus*) males, on opposite sides of a female, engaged in a simultaneous courtship display consisting of belly slapping. Males will form long-term alliances to gain access to receptive females. All the males in alliances appear to mate equally often with the escorted receptive females (photo courtesy of R. Connor).

lation. For example, a male baboon, *Papio anubis,* only enlists the support of another male in his troop when attempting to mate with a female currently in consort with a higher ranking male. If the alliance is successful, the dominant individual in the alliance has priority in mating. In contrast, some bottlenose dolphin, *Tursiops truncatus,* alliances persist for long periods with two or even three males working together on multiple occasions to display to and herd females. No evidence for asymmetrical mating success has been observed (Fig. 15). All males in an alliance often mate with the female being herded (Connor *et al.,* 1992). Greater cooperation among male dolphins than male baboons is therefore consistent with randomized reproduction, probably acting in addition to partner choice, as a mechanism to prevent cheating and foster cooperation.

Mutual Policing

An alternative to partner choice for enforcing cooperation is to repress competition among individuals in a potentially cooperating group. The best way to repress competition is to maintain a sufficiently high level of relatedness among all individuals in the group such that their reproductive interests coincide. For example, mutual policing in honeybees, *Apis mellifera,* depends on the level of relatedness. Female worker bees reproduce by laying unfertilized eggs which develop into haploid males. If, as usually happens, the queen has mated many times, then workers are less closely related to the haploid male offspring of their sisters than they are to their mother's offspring (Maynard-Smith and Szathmary, 1995). As expected, worker bees continuously inspect comb cells for the presence of eggs laid by other workers. If a worker-laid egg is discovered, it is promptly destroyed. Worker policing behavior (Fig. 16) breaks down if the queen dies (Ratnieks and Visscher, 1989).

An analogous situation occurs during the transmission of cytoplasmic genetic elements, including symbiotic organelles. In eukaryotes, genetic lineages of symbiotic cytoplasmic elements are prevented from mixing during

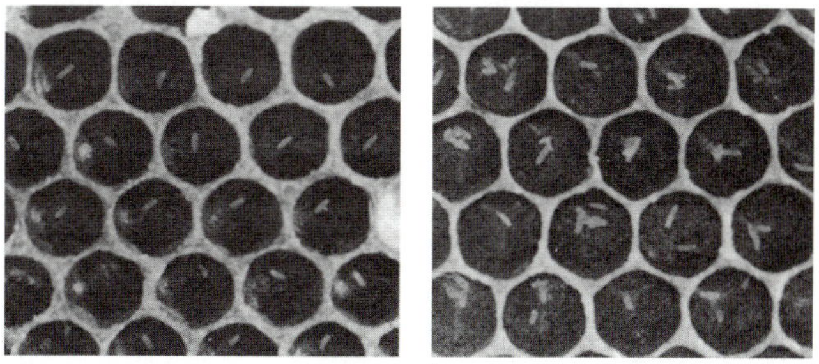

a b

FIGURE 16 Egg-laying patterns in honeybee (*Apis mellifera*) colonies in which a queen is present (a) or in queenless colonies (b). In the presence of a queen, the workers actively police the cells, destroying the eggs laid by other workers, whereas in the absence of a queen they actively compete among themselves for reproductive success.

transmission, thereby limiting the opportunity for genetic conflict. For example, considerable evidence supports the view that mitochondria and chloroplasts developed from bacteria that formed symbiotic associations with primitive eukaryotic cells. These organelles are now passed from mother to offspring. Such uniparental inheritance ensures that a zygote only contains organelles from a single genetic lineage. Biparental inheritance of organelles, in contrast, would lower relatedness among organelles within and between cells. Genetic variation among mitochondria could influence metabolic performance and lead to competition among cell lineages for transmission into the next generation. As any cancer victim will attest, intercellular competition is harmful to the parent organism. Uniparental inheritance therefore acts to maintain high levels of relatedness.

The most compelling evidence in support of uniparental vertical transmission-enforcing cooperation derives from examining variation in the degree to which parasitic organisms harm their hosts as a function of their mode of transmission. Virulence of both prokaryote and eukaryote parasites is associated with forms of transmission that permit infection by multiple genetic lineages (Maynard-Smith and Szathmary, 1995). Multiple infections are more prevalent with horizontal transmission than vertical transmission, especially when vertical transmission is uniparental. Experimental evidence in support of an association between virulence and horizontal transmission has been obtained for the bacterium *Escherichia coli* infected by a filamentous DNA phage (Bull *et al.,* 1991). When growth conditions permitted only vertical transmission, phage evolved to have minimal deleterious effects on bacterial growth rates. However, when the bacteria were grown to induce conjugation, thereby allowing horizontal transmission, phages with more harmful effects appeared. Comparative data on the effects of phage on bacteria also support an association between horizontal transmission and virulence (Maynard-Smith and Szathmary, 1995).

Several events in the history of life may also be interpreted as ways to ensure high levels of relatedness among potentially variable genetic entities that must cooperate to function properly (Maynard-Smith and Szathmary, 1995). For example, the ubiquity of reproduction from a single cell in multicellular organisms results in all the cells in a developing embryo being genetically identical. The early separation of germline cells from cells which give rise to the rest of the body may further control any possible cell lineage competition within the body. Early separation of germline cells also reduces the number of divisions that those cells must undergo and decreases the chance that somatic mutations will decrease genetic relatedness among the gamete-producing cells. A parallel case of separating potentially competitive cell lineages has been noted for many symbiotic organisms which are often housed in special organs within the body of the host (Frank, 1996).

Human Cooperation

Which, if any, of these mechanisms have been important for promoting cooperation among our ancestors? The answer is almost certainly all of them. Cooperative interactions in nonhuman vertebrate and human families are often consistent with expectations derived from kin selection and kin manipulation theories (Emlen, 1995). For example, assistance in rearing children is expressed to the greatest extent among the closest relatives. Kin selection, however, is less likely to explain interactions among families. To the extent that families banded together to form larger social units, repeated interactions must have created opportunities for temporary as well as more stable alliances based on partner choice and reciprocity. Ecological circumstances favoring long-term durable relationships between individuals in small groups would have laid the foundation for cooperation. Longevity coupled with an ability to remember individuals and their interactions provide all the necessary ingredients for stable reciprocity. As long as groups of individuals can anticipate many future opportunities for profitable exchange, cooperation in small groups seems inevitable.

Neither reciprocity nor kin selection adequately explain examples of ethnic cooperation in which very large groups of individuals develop common goals and will often go so far as to defend those goals by warfare. Examples of such large-scale cooperation do not exist among unrelated nonhuman animals. This suggests that ethnic cooperation may be the result of some cultural inheritance system in which the group, rather than the individual, is under selection. Cultural group selection has been proposed as an explanation for why humans differ from all other animals in cooperating, often against their own interests, with other unrelated individuals (Boyd and Richerson, 1991).

References Cited

AXELROD, R. (1984). *The Evolution of Cooperation.* Basic Books, New York.

BOYD, R., and RICHERSON, P. J. (1991). In *Cooperation and Prosocial Behaviour* (R. A. Hinde and J. Groebel, Eds.), pp. 27–48. Cambridge Univ. Press, New York.

BULL, J., and RICE, W. R. (1991). *J. Theor. Biol.* **149,** 63–74.

BULL, J. J., MOLINEAUX, I. L., and RICE, W. R. (1991). *Evolution* **45,** 875–882.

CONNOR, R. C., SMOLKER, R. A., and RICHARDS, A. F. (1992). *Proc. Natl. Acad. Sci. USA* **89,** 987–990.

CREEL, S., and CREEL, N. M. (1995). *Anim. Behav.* **50,** 1325–1339.

DE VRIES, P. J. (1992). *Sci. Am.* **267,** 76–82.

DUGATKIN, L. A. (1997). *Cooperation among Animals: A Modern Perspective.* Oxford Univ. Press, New York.

EMLEN, S. T. (1995). *Proc. Natl. Acad. Sci. USA* **92,** 8092–8099.

FRANK, S. A. (1996). *Am. Nat.* **148,** 1113–1124.

FREAN, M. R. (1994). *Proc. R. Soc. London B* **257,** 75–79.

Hamilton, W. D. (1964). *J. Theor. Biol.* **7,** 1–51.
Heinsohn, R., and Packer, C. (1995). *Science* **269,** 1260–1262.
Manning, C. J., Wakeland, E. K., and Potts, W. K. (1992). *Nature* **360,** 581–583.
Maynard-Smith, J., and Szathmary, E. (1995). *The Major Transitions in Evolution.* Freeman, San Francisco.
Noë, R., and Hammerstein, P. (1994). *Behav. Ecol. Sociobiol.* **35,** 1–11.
Nowak, M., and Sigmund, K. (1993). *Nature* **364,** 56–58.
Pusey, A. E., and Packer, C. (1994). *Behav. Ecol.* **5,** 362–374.
Ratnieks, F. L. W., and Visscher, P. K. (1989). *Nature* **342,** 796–797.
Strassman, J. E., Zhu, Y., and Queller, D. C. (2000). *Nature* **408,** 965–967.
Trivers, R. L. (1971). *Q. Rev. Biol.* **46,** 35–57.
Vehrencamp, S. L. (1983). *Anim. Behav.* **31,** 667–682.
Wade, M. J. (1977). *Evolution* **33,** 749–764.
West Eberhard, M. J. (1975). *Q. Rev. Biol.* **50,** 1–33.
Wilkinson, G. S. (1984). *Nature* **309,** 181–184.
Williams, G. C. (1966). *Adaptation and Natural Selection.* Princeton Univ. Press, Princeton, NJ.
Wilson, D. S., and Sober, E. (1994). *Behav. Brain Sci.* **17,** 585–654.

FELICITY HUNTINGFORD
Division of Environmental and Evolutionary Biology
Institute of Biomedical and Life Sciences
University of Glasgow
Glasgow G12 8QQ, United Kingdom

Aggression

Aggression (behavior that actually or potentially causes injury to another member of the same species) is widespread in the animal kingdom and typically results in the winner gaining access to limiting resources. It usually involves an initial exchange of displays from a distance, escalating to behavior involving direct physical contact and, finally, to injurious fighting. Conflicts may be resolved without full escalation, but serious, and sometimes lethal, injury is not uncommon. The degree of aggressiveness shown by different individuals of a given species is often highly variable and much research is aimed at explaining this variability. To illustrate frontiers in research into why animals fight as they do, this article describes studies of the endocrinology of aggression in lizards, the neurobiology of aggression in fish, the genetics of aggression in mice, and the behavioral ecology of aggression in fish and crabs. To show how an understanding of the biology of aggression can help to improve welfare and productivity of farmed animals, an example of the use of the cost–benefit approach of behavioral ecology to reduce fighting among farmed fish is presented.

Defining Aggression

The word "aggression" is frequently used in everyday language, where it refers to "an unprovoked attack . . . an assault" (*Oxford Shorter English Dictionary,* 1973). In some respects, this everyday usage is advantageous to students of animal aggression since it makes it easier for them to communicate their findings not only to other scientists but also to nonspecialists; in contrast, scientists studying seismology, for example, usually need to start any nonspecialist discussion by explaining what the word means. However, using this everyday word in a scientific context raises some problems because although most people know basically what aggression is, they often use the word to mean different things. In other words, there is an issue of definition that is not easily resolved and that has been the subject of extensive debate and controversy. Space, time, and the reader's interest preclude an extensive account of this debate, so this article simply highlights some of the controversial issues.

The most commonly used biological definition of aggression is "behavior that actually or potentially results in injury to another animal of the same species" (Huntingford and Turner, 1987). Inclusion of the word "potentially" deals with the question of whether an unsuccessful attack constitutes aggression (it does). The proviso that the injured animal must be a conspecific deals with the issue of whether the term should include attack by a predator on a prey species or the prey's counterattack, both of which may cause injury (it should not). This simple and useful definition does not solve the problem of whether the term aggression should apply to unintentional injury, to retaliation as opposed to unprovoked attack, to pain or damage caused in the best interests of the recipient, or to behavior causing psychological rather than physical injury. These are important issues and need to be resolved, but they cannot be dealt with using a definition that is generally applicable without making this definition so long and complex as to be useless. The best strategy, and the one adopted here, is to be flexible, using the word aggression as defined previously to indicate the broad category of behavior under discussion but specifying exactly what is involved where this is necessary for precision and clarity.

Patterns of Aggression in the Animal Kingdom

Behavior actually or potentially causing injury to a conspecific (i.e., aggression) is widespread in the animal kingdom; a few examples are shown in Fig. 1.

Thus, anemones grapple over space, clubbing unrelated neighbors with specialized tentacles until one or other moves; ragworms fight over burrows, battering each other with the proboscis that they also use for feeding and burrowing; male stag beetles fight over access to females, using their greatly enlarged mandibles to grasp and overthrow rivals; damselfish defend feeding territories on coral reefs, displaying with raised fins before charging at intruders and biting their fins; male frogs grapple over places from which to call for females (in the case of the gladiator frogs shown in Fig. 1, jabbing at each others eyes and eardrums with their sharp, curved thumbs); male lizards fight each other over breeding territories; cockerels and other birds fight for status; young eagles and other raptors regularly attack and kill their younger siblings; male rats arch their bodies and rattle their tails during squabbles over space; and male zebras lash at each other with their hooves during fights over females. Most animal fights involve just two or a few individuals, although there are cases in which large groups come into conflict, as when ants from neighboring colonies dispute feeding areas when they emerge from hibernation.

The physical details of fighting are as variable as the animals concerned, but there are some common features. First, animals usually have a repertoire of aggressive signals in addition to overt attack (Fig. 2). These include displays performed at a distance and noninjurious actions that involve direct contact as well as clearly harmful acts such as biting or kicking.

Fights usually (but not always) involve a process of gradual escalation, with displays performed at a distance tending to occur at the beginning, followed by direct, noninjurious contact and then overt attack. Animal fights can be, and often are, resolved by displays and harmless contact alone, but serious injury and death are not uncommon, especially when the animals concerned are fighting over something valuable (Table 1).

This relates to another general feature of animal fights — namely, that they are usually about resources, with the winner gaining exclusive or preferential access to something that both participants need. Access to disputed resources may be immediate, as when animals fight over a particular item of food or a specific mate, but it can also be delayed. For example, animals often fight over an area in which resources (food, mates, shelter, etc.) are likely to be found, in which case territoriality results; this is the case of the anemones, ants, damselfish, and lizards shown in Fig. 1 and the male cichlids in Fig. 2. In other circumstances, by winning fights animals gain dominance or social supremacy over rivals that, at some future time, translates into preferential access to resources (as happens for the cockerels shown in Fig. 1).

Within-Species Variability in Aggressiveness

The previous discussion may give the impression that whenever animals of a given species fight they do so in a species-typical manner; in general, this is the case. However, animals of the same species can be highly variable in how readily they fight. For example, three-spined sticklebacks (*Gasterosteus aculeatus*) are famous for the fact that breeding males attack red-chested rivals that enter their territory. Figure 3a shows data collected by counting the number of attacks given per minute by breeding male sticklebacks toward a red male placed into their territory in carefully standardized conditions in the laboratory. Clearly, the result is highly variable, with some males effectively ignoring the intruder and others biting it at rates of up to two per second.

Within-species variation in aggressive behavior can be even more extreme. For example, in a closely related species (the 10-spined stickleback, *Pygosteus pungitius*) some males do not fight but instead sneak into the territories of

FIGURE 1 Examples of animal aggression: (a) two beadlet anemones (*Actinia equina*) grapple over space; (b) ragworms (*Nereis* sp.) fight for possession of a burrow; (c) two male stag beetles (*Lucanus cervus*) fight with their mandibles over access to females; (d) male lizards (*Uta stansburiana*) fight each other over breeding territories; (e) cockerels (*Gallus domesticus*) fight for status; (f) an eagle chick (*Aquila chrysaetos*) attacks his own sibling to eliminate him from the nest; (g) a rat (*Rattus rattus*) in an aggressive position; and (h) a ritual fight between two male zebra (*Equus zebra*) (adapted from Huntingford and Turner, 1987).

a

b

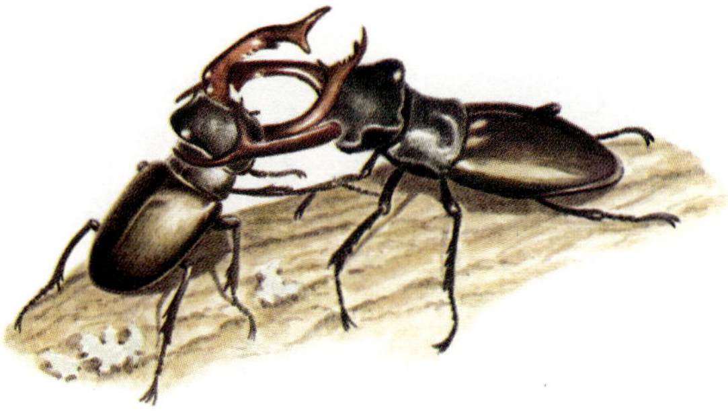

c

d

e

f

g

h

FIGURE 2 Consecutive phases of territorial fights between adult male cichlid fishes (*Tilapia zilii*). A preliminary display phase gives way to a period during which the fish engage in mouth wrestling, pushing, and pulling; after this the two fish vigorously follow each other, head-to-tail circling as the fish attempt to bite each other and pausing periodically to gasp for breath (photos courtesy of F. Neat).

another male, enter his nest after he has induced a female to lay eggs in it, and fertilize these before the territory owner has time to do so himself. This behavioral device for obtaining fertilization (which was first reported mistakenly as homosexual behavior; Morris, 1953) is one example of a common phenomenon — namely, the existence within a single population of the same species of alternative strategies for contesting limited resources. Other examples include various frogs and toads in which silent, nonterritorial males intercept and mate with females as they move toward calling males; more examples will be provided later.

Questions About Aggression

Variation in the extent to which members of a given species use aggression to win limiting resources is of interest in itself but can also be a very useful research tool. Studies designed to explain the variability in attack rates found in three-spined sticklebacks have shown that some of this can be accounted for in terms of the body condition of the fish concerned, with males with poor energy reserves being unenthusiastic fighters. This is a causal explanation, couched in terms of the mechanisms that underlie behavior. Some of the variability can be accounted for in terms of the pre-

TABLE I

Examples of Injury and Killing in Animal Fights[a]

SPECIES	OBSERVATION
Slugs	Conspecifics are injured (and eaten) when fighting for shelter.
Ants	Large numbers of ants are killed in battles over feeding territories.
Polistine wasps	Attempted nest takeovers are common and involve biting, grappling fights, and stinging. Corpses are often found under nests.
Fig wasps	78% of males in a single fig were injured, many seriously.
Spider crabs	Legs get broken off in fights over symbiotic anemones. In the wild, crabs of both sexes have missing legs and claws.
Fiddler crabs	25% of males have signs of injury probably inflicted by conspecifics.
Mantis shrimps	Species nesting high on shoreline where shelters are rare and essential fight fiercely and often fatally.
Blind gobbies	Fierce, injurious attacks on conspecifics approaching shelter.
Red-back salamanders	11% of animals in the wild have signs of injury to olfactory apparatus, probably sustained in fights over shelter.
Gladiator frogs	Extremely fierce, injurious fights over females: males often wounded or killed.
Meadow voles	82% of males and 57% of females in the wild have scars.
Musk oxen	5–10% of males are killed per year during the rut.
Mountain sheep	Many males have badly damaged nose and head, including broken bones. Rams have been found dead or dying from wounds gained during the rut.
Deer	84% of hides of male deer of more than 1.5 years have scars from fights. 23% of stags 5 years old and older show signs of injury (blinding and leg and antler breakage); in 6% this is permanent.
Cheetah	Males often injured and sometimes killed in territorial encounters.
Chimpanzees	Males and females of a small group were chased and killed by adult males from a larger group following troop fission.
Seals	Approximately 65% of seals identified as males in a single haul-out site have fresh neck wounds (some severe), probably inflicted during fights in the water.

[a] From Huntingford and Turner (1987).

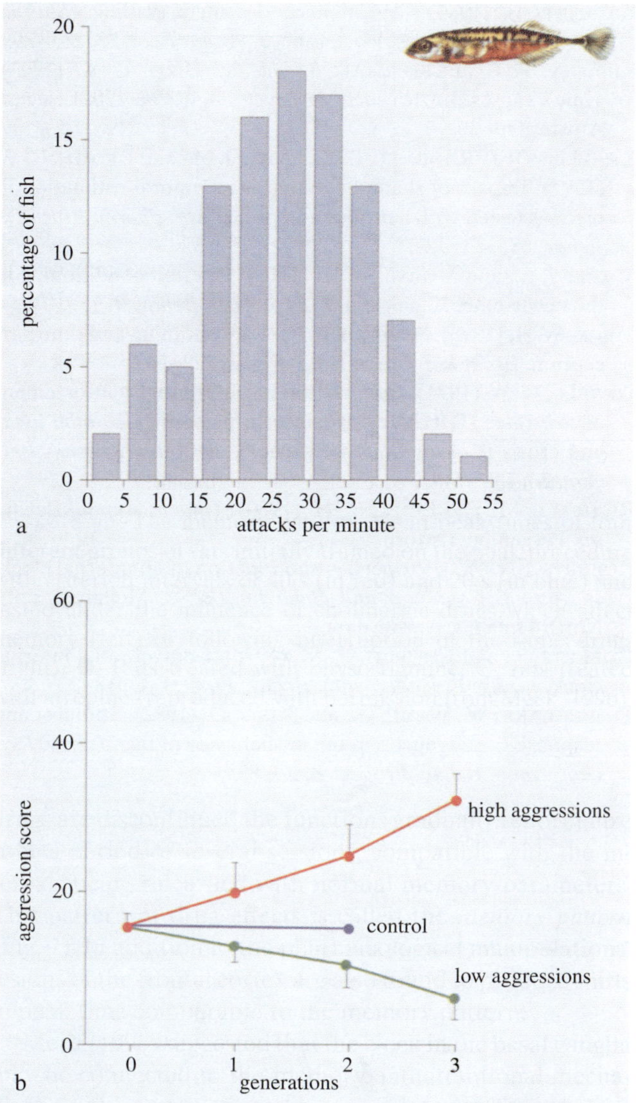

FIGURE 3 (a) Distribution of attack rates measured in a sample of breeding male three-spined sticklebacks (*Gasterosteus aculeatus*) with respect to rival males placed in their territory. (b) Average levels of aggressiveness in male sticklebacks selected for high and low aggression over three generations; only the most and the least aggressive males were allowed to breed (reproduced with permission from Bakker, 1986).

dation regimes prevailing at the sites from which the fish were captured; sticklebacks from sites with abundant predators are less aggressive than those from predator-free sites (Huntingford, 1982) since fighting makes them vulnerable to predatory attack. This is a functional explanation, accounting for low or high levels of aggression in terms of its consequences for the fitness. The behavioral differences between fish from different sites are probably based on genetic differences since selection experiments (in which lines were established in which only the least or the most aggressive males were allowed to breed) produced marked changes (decrease and increase, respectively) in aggressiveness in the selected lines (Bakker, 1986; Fig. 3b). This means that at least some of the natural variation in aggression among male sticklebacks is inherited. This is an explanation in terms of the developmental origin of behavioral variation.

Most current research on the biology of aggression seeks to explain why animals fight in a particular way, searching for causal, developmental, or functional explanations for this behavior. The next sections present in detail a series of examples that illustrate these different approaches. This article can only touch on the enormous literature on aggression; more detailed reviews can be found in Archer (1987) and Huntingford and Turner (1987).

Causes and Development of Aggression

Endocrinology of Alternative Competitive Strategies in Lizards

An extensive literature on the endocrinology of aggression has shown that, in species in which breeding males compete directly for access to females, the hormone testosterone (an androgen typically produced by male vertebrates) has a stimulatory effect on aggression, although this effect becomes weaker as the social organization of the species concerned becomes more complex. In addition to exerting effects on adult animals, testosterone may also act during development since in mammals, at least, exposure to this hormone at an early age sensitizes animals as adults to its aggression-promoting effects (Huntingford and Turner, 1987).

Lizards have proved a very useful group in which to study the physiological bases of aggression since many species are sufficiently large and robust to allow experimental manipulation in both the laboratory and the field. In addition, their behavior is simple enough to be easily recorded but complex enough to be interesting. It is well established for many species that (i) the increase in aggression seen in male lizards as the breeding season approaches and the bright color associated with reproduction are facilitated by rising androgen levels, (ii) castration of fully breeding males drastically reduces their aggressiveness, (iii) artificially administered androgens reverse this effect, and (iv) androgens are concentrated in regions of the brain that are known to be involved in the control of aggressive behavior (Becker *et al.*, 1992).

As described previously for sticklebacks, in several species of lizard, males compete for mates in one of two or more ways. Thus, it is not uncommon for a single population to include both males that are aggressive and defend territories and males that are nomadic, unaggressive, and obtain matings by sneaking; these behavioral differences are often associated with different coloration. For example,

CHANTREY, D. F. (1974). Stimulus preexposure and discrimination learning by domestic chicks: Effect of varying interstimulus time. *J. Comp. Physiol. Psychol.* **87,** 517–525.

CHURCH, R. M. (1978). The internal clock. In *Cognitive Processes in Animal Behavior* (S. H. Hulse, H. Fowler, and W. K. Honig, Eds.), pp. 277–310. Hillsdale, Erlbaum, NJ.

CHURCH, R. M., and BROADBENT, H. (1990). Alternative representations of time, number and rate. *Cognition* **37,** 55–81.

DAVIES, N. B. (1977). Prey selection and the search strategy of the spotted flycatcher (*Muscicapa striata*). *Anim. Behav.* **25,** 1016–1033.

GALLISTEL, C. R. (1990). *The Organisation of Learning.* MIT Press, Cambridge, MA.

GALLISTEL, C. R. (1994). Space and time. In *Animal Learning and Cognition* (N. J. Mackintosh, Ed.), pp. 221–253. Academic Press, San Diego.

GIBBON, J., and ALLAN, L. G. (Eds.) (1984). *Timing and time perception.* New York, New York Academy of Sciences.

GIBBON, J., BALDOCK, M. D., LOCURTO, C., GOLD, L., and TERRACE, H. S. (1977). Trial and intertrial durations and autoshaping. *J. Exp. Psychol. Anim. Behav. Process.* **3,** 264–284.

GIBBON, J., MALAPANI, C., DALE, C. L., and GALLISTEL, C. R. (1997). Toward a neurobiology of temporal cognition: Advances and challenges. *Curr. Opin. Neurobiol.* **7,** 170–184.

GIBBON, J., MORRELL, M., and SILVER, R. (1984). Two kinds of timing in circadian incubation rhythm of ring doves. *Am. J. Physiol.* **247,** R1083–R1087.

GILL, F. B. (1988). Trapline foraging by hermit hummingbirds: Competition for an undefended renewable resource. *Ecology* **69,** 1933–1942.

KACELNIK, A. (1984). Central place foraging in starlings (*Sturnus vulgaris*). I. Patch residence time. *J. Anim. Ecol.* **53,** 283–299.

KILLEEN, P. R., FETTERMAN, J. G., and BIZO, L. A. (1997). Time's causes. In *Advances in Psychology,* Vol. 120. Elsevier, Amsterdam.

KRAEMER, P. J., RANDALL, C. K., DOSE, J. M., and BROWN, R. W. (1997). Impact of *d*-amphetamine on temporal estimation in pigeons tested with a production procedure. *Pharm. Biochem. Behav.* **58,** 323–327.

MARLEY, E., and MORSE, W. H. (1966). Operant conditioning in the newly hatched chicken. *J. Exp. Anal. Behav.* **9,** 95–103.

MECK, W. H. (1996). Neuropharmacology of timing and time perception. *Brain Res. Cogn. Brain Res.* **3,** 227–242.

POWELL, R. W. (1974). Comparison of differential reinforcement of low rates (DRL) performance in pigeons (*Columba livia*) and crows (*Corvus brachyrhynchos*). *J. Comp. Physiol. Psychol.* **86,** 736–746.

RICHELLE, M., and LEJEUNE, H. (1980). *Time in Animal Behavior.* Pergamon, Oxford.

ROBERTS, S. (1983). Properties and function of an internal clock. In *Animal Cognition and Behavior* (R. L. Mellgren, Ed.). North-Holland, Amsterdam.

ROBERTS, S., and CHURCH, R. M. (1978). Control of an internal clock. *J. Exp. Psychol. Anim. Behav. Process.* **4,** 318–337.

TIMBERLAKE, W., WAHL, G., and KING, D. (1982). Stimulus and response contingencies in the misbehavior of rats. *J. Exp. Psychol. Anim. Behav. Process.* **8,** 62–85.

MARC BEKOFF
Department of Environmental, Population, and Organismic Biology
University of Colorado
Boulder, Colorado 80309-0334, USA

Play is an activity that has challenged students of behavior for a long time. Here, I concentrate on social play in mammals to make some general points about this fascinating behavior. After a general introduction, I mainly discuss some representative studies of how social play is initiated (solicited) and maintained in various canids (members of the dog family). I also discuss recent work on relationships between social play behavior and the study of animal cognition (cognitive ethology). The flexibility and versatility of play make it a good candidate for interdisciplinary discussion of comparative and evolutionary studies that will help us learn more about the nature of animal minds. Much exciting research is taking place in these areas of inquiry.

Social Play in Mammals

What Is Play?

"Play" is a very difficult word to define because it is a blanket term that covers a wide range of behavior patterns. From early serious efforts to come to terms with social play behavior, many definitions were offered (Bekoff and Byers, 1981, 1998; Fagen, 1981; Martin and Caro, 1985; Spinka, Newberry, and Bekoff, 2001), and these definitions provided the framework for numerous comparative and evolutionary studies. Today, it is difficult to pick up a journal that deals with the comparative study of animal behavior and not find at least one article that is concerned with the structure or function of social play (predominantly in mammals; Ortega and Bekoff, 1987). Although play is easy to recognize — people do agree about play when they consider possible examples — there does not seem to be any single definition that would apply to all instances of what is called play in the diverse species in which the activity has been observed. However, lacking a rigorous definition need not be an impediment to conducting solid research; requiring a rigorous definition prior to empirical research may unreasonably require possession of knowledge that must first be gained by empirical research (Allen and Bekoff, 1997).

To begin, I use a general definition that has weathered the years (Bekoff and Byers, 1981, pp. 300–301): "Play is all motor activity performed postnatally that appears to be purposeless, in which motor patterns from other contexts may often be used in modified forms and altered temporal sequencing." Martin and Caro (1985, p. 65) provide a slightly modified version of this definition. If the activity is directed

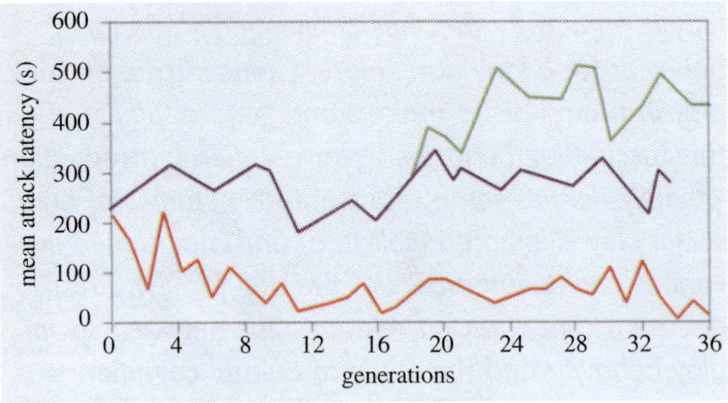

FIGURE 7 Attack latency in male mice selected over 35 generations into two lines characterized by a fast attack (low latency; in red) or slow attack (high latency; in green). The behavior of control mice is shown in blue (reproduced with permission from Benus, 1988).

dence that genetic differences are responsible for a significant component of the variation in aggressiveness among males from the original wild population. Interestingly, this difference in fighting behavior is associated with a suite of other behavioral differences: Overall, slow attackers are generally more flexible to changes in the environment than are fast attackers. How fiercely an individual male fights is therefore just one component of its inherited strategy for coping with its social environment (Benus *et al.,* 1991).

The inheritance of fast and slow attack in male mice was investigated in more detail by a series of crosses involving the two selected lines, designed mainly to examine the role of the Y chromosome. The main function of this chromosome is primary sex determination, but a variety of other roles have also been identified. The Y chromosome in mice, as in other mammals, is composed of two regions — one that pairs and recombines with part of the X chromosome at meiosis and one that does not and thus is paternally transmitted. Crosses were set up to produce a series of strains bearing various combinations of the pairing and nonpairing regions of the Y chromosome and the X chromosome from the fast- and slow-attacking lines. The genetic composition of these strains was related to attack rate in the relevant males (Fig. 8; Van Oortmerssen and Sluter, 1994).

The original difference in attack rate between the two parental lines is shown very clearly (compare 1 with 6 in Fig. 8), as is the fact that although all the hybrids are intermediate between the two parental lines, they are more similar to their fast-attacking parent. In effect, all but the pure slow line are relatively fast attackers. Among these relatively fast-attacking lines, the parental fast line, with a double dose of the "fast" pairing region of the Y chromosome, attacked significantly quicker than the line with one dose of the fast region (compare 1 with 2 in Fig. 8), which in turn attacked more quickly than the other hybrids which had received their pairing region from the slow strain (compare 2 with 3–5 in Fig. 8). Overall, these results suggest that rapid attack is largely dependent on the presence of the pairing region of the Y chromosome from the fast-attacking line. Elements on the nonpairing region may have a weaker, modulating effect since the hybrids that received the nonpairing region of the Y chromosome from the fast-attacking line were quicker to attack than the equivalent line receiving this section of the genome from the slow-attack line (compare 2 with 4 and 3 with 5 in Fig. 8).

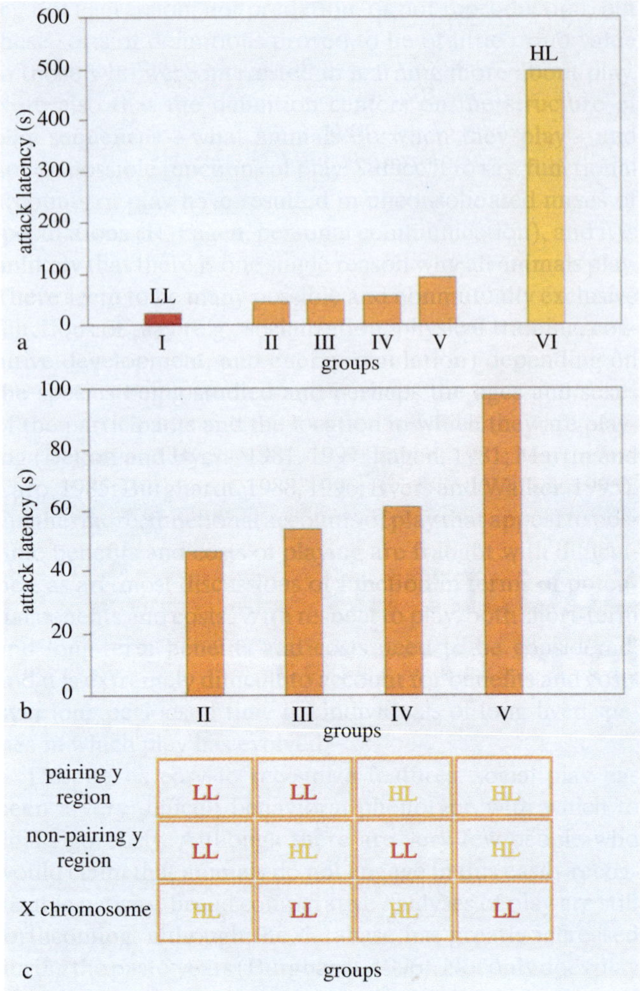

FIGURE 8 Hereditary transmission of low (LL) or high (HL) attack latency. (a) Crosses (in orange) between parents, one of which has a low attack latency (in red) and the other a high latency (in yellow). (b) Details of the attack latency of the crosses shown in a. All the F1 hybrids show a low attack latency. (c) Schematic diagram of the different combinations of the regions of the Y chromosome, pairing or nonpairing with the X chromosome, present in the F1 crossed lines. The regions of the Y chromosome pairing with the X chromosome are responsible for the attack speed. The nonpairing regions have only a modulatory effect (reproduced with permission from Van Oortmerssen and Sluyter, 1994).

Examination of plasma testosterone levels in the various strains described previously indicates that fast attack is unrelated to circulating testosterone levels in adult males; therefore, the behavioral difference is clearly not the result of a Y chromosome-induced difference in rates of androgen production in adult males. However, when the two selected lines and their reciprocal hybrids were castrated at 50 days of age (just after puberty) and injected with a standard dose of testosterone, strain-specific attack rates persisted. This indicates that the original behavioral difference was the result of greater sensitivity to the aggression-promoting properties of circulating androgens in mice that attack rapidly. It seems that a difference in circulating androgen during puberty, induced by elements in the pairing region of the Y chromosome, generates variation in sensitivity to circulating androgens, which in turn causes more rapid attack when male mice reach adulthood (De Ruiter *et al.,* 1992).

Overall, this series of studies demonstrated an important role for identified regions of the Y chromosome in generating different patterns of fighting in male mice. This does not rule out an additional role for autosomal elements in generating variation in aggressive behavior in this species. The Y chromosome effect is resistant to modification by pre- and postnatal maternal effects, although again this is not to say that postnatal experience cannot modify aggression in this species; indeed, the literature indicates that it almost certainly can do so (Archer, 1987; Huntingford and Turner, 1987).

TABLE 2

The Hawk–Dove Model

THE STRATEGIES

- Hawks launch straight into attack, win half their fights and are injured (and lose) in the other half.
- Doves only display and win half their fights.

THE PAYOFF MATRIX[a]

Payoff to	In a contest against Hawk	Dove
Hawk	$\frac{1}{2}(V - D)$: equal chance of winning or being injured	V: opponent always retreats
Dove	0: never wins but is not injured	$V/2 - T$: wins half of fights and both waste time

NET PAYOFFS

- In a population of doves, hawks gain a payoff of V while doves get a pay off of $V/2 - T$, so hawks do better.
- In a population of hawks, hawks gain a net payoff of $\frac{1}{2}(V - D)$ and doves gain a net payoff of 0. So if $D > V$, doves do better; otherwise, hawks do better.

[a] V is the value of winning, D is the cost of serious injury, S is the cost of slight injury, and T is the cost of spending time in display.

The Costs and Benefits of Aggression: Body Size, Resource Value, and Fighting in Lizards, Fish, and Crabs

The question of the adaptive significance of aggressive behavior (how and how much animals benefit in fitness terms from behaving in a particular way during fights) has attracted an enormous amount of attention in the past 20 years. The economic approach of behavioral ecology (in which the costs and benefits of behavioral options are identified and the behavior that maximizes net gains is deduced) has been applied to very good effect. In particular, the games theory approach, which recognizes the frequency-dependent nature of the costs and benefits of fighting, has been particularly influential (Maynard-Smith, 1974).

This approach involves defining the behavioral options open to a hypothetical population of animals, defined to highlight the question at issue. For example, in the famous hawk–dove model, which was designed to investigate why fights are often resolved by display alone even though the winner stands to gain a valuable resource, the two options or strategies differ in whether they use display alone (doves) or escalated fighting alone (hawks; Table 2). Hypotheses about how fights are resolved and about the costs and benefits of using these strategies are embodied in the payoff matrix, which defines the net gain acquired by individuals pursuing each of the defined strategies when paired against each category of opponent (Table 2; Fig. 9).

From the relative frequency of all strategies and the net payoffs, the overall fitness gains/losses of individuals adopting each strategy can be calculated. In particular, it is often possible to identify within the strategy set one or more that, if common in the hypothetical population, cannot be bettered by any other defined strategy. These are the so-called *evolutionarily stable strategies* and they represent the way that real animals should behave if the model depicts correctly the adaptive consequences of fighting.

For example, the hawk–dove model indicates that when the cost of escalated fighting is high, the hawk strategy (which involves launching straight into high-intensity fighting) is not evolutionarily stable but can be invaded by the dove strategy. This happens because at low frequencies, although they never win, doves never have to pay the cost of injury, whereas hawks incur serious injury in 50% of their fights. The model also shows that unless the benefit of winning is greater than the cost of injury, the evolutionary stable position is a mixture of hawks and doves (defined by the value of winning relative to the cost of injury) at which the two strategies gain equal fitness.

play? Why do they perform sequences of behavior that incorporate actions from many different contexts? How do animals negotiate cooperation and agreements to participate in social play? How do they develop socially and learn rules of social interaction? and How do they ask permission to engage in social play? It also has become more permissible to speculate on the existence of a humorous or comic dimension in the natural lives of animals and the possible evolutionary implications of such a dimension (R. Fagen, personal communication; Burghardt, 1996). For example, we can ask how animals' obvious enjoyment of engaging in social play and their persistent efforts in luring others to play influences the development and evolution of this activity.

Understanding activities such as play may also be important for developing new research dealing with comparative approaches to cognition (Pellegrini, 1995). The broad comparative study of social play also lends itself to suggestions that ethology should be concerned with the evolution, adaptation, causation, and development of behavior (Jamieson and Bekoff, 1993). Studies of social play also cross disciplinary boundaries: Those interested in social play come from various academic areas (e.g., zoology, psychology, anthropology, computer science, cognitive science, medicine, philosophy, and sociology) and consider questions that deal with topics such as evolution, ecology, development, social communication, neurobiology, and cognition.

Social play has been described in most mammals in which its possible existence has been investigated (Fagen, 1981). Some of the most elaborate and complex forms of play occur in such mammals as the social carnivores (wild and domestic cats, wolves, coyotes, domestic dogs, and bears) and nonhuman primates, animals with highly developed cerebral cortices (Bekoff and Byers, 1981; Fagen, 1981). Perhaps it is a combination of these and possibly other factors that have played a significant role in the ability of individuals to engage in play — to have the cognitive skills to negotiate cooperative agreements, to ask for permission to engage in a specific activity, to perform behavior within a pretend context, and perhaps even to make mental attributions to others. With the goal of learning more about how studies of nonhuman animal cognition can inform studies of human cognition, social play is a good behavioral phenotype on which to concentrate because when animals play they typically perform behavior patterns that are used in other contexts (e.g., predation, aggression, or reproduction), and yet these behavior patterns take on a different meaning when performed in the context of play (Fig. 5). For example, when Jethro bit Sukie and shook his head rapidly from side to side, he did not continue to the point of eliciting submissive behavior from her, and she did not squeal and roll over submissively or run away. He also did not attempt to copulate with her when he mounted her from behind, and she did not try to shake him off of her back.

a

b

FIGURE 5 (a) Two dogs growl and hit each other with their paws while playing. Within play, growling does not lead to fighting. (b) Giving blows with the paws is a gesture inviting and leading to play.

It is very important to negotiate play and to agree that play is the name of the game. When animals play, behavior patterns for the most part are "borrowed" from other contexts, and individuals need to be able to tell one another that they do not want to eat, fight with, or mate with the other individual(s) but rather that they want to play with them. In most species in which play has been observed, specific actions have evolved that are used to initiate ("I want to play with you") or to maintain ("I still want to play with you regardless of what I just did to you or regardless of what I am going to do to you") play. These actions seem to function in negotiations between participants, the result of which is that they foster an agreement to engage in play rather than to partake in aggression or predation, for example. There is no solid evidence that animals invite others to play and then exploit them. Furthermore, self-handicapping (e.g., controlling the intensity of bites) and role reversals (dominant individuals assuming submissive roles only in play) have also been observed (Fig. 6).

In considering how play is initiated and maintained, I

FIGURE 6 Two huskies play at fighting.

briefly discuss issues such as the types of information that are shared during play and what play can tell us about the emergence of mind in animals. For example, pretend play, in which an individual creates an imaginary situation or distorts reality, has attracted the attention of many researchers interested in the development of human behavior and also seems to occur in nonhumans. Dogs and cats often chase their tails and bite them as if the tails were prey or a competitor, and they frequently attempt to kill their food bowl or mate with it. Some dominant animals also engage in self-handicapping or role reversing during play, in which they seem to pretend that they are not really dominant, perhaps to get others to play with them who would otherwise not do so (it is rare for the self-handicapping individual to then attempt to dominate their partner). Although there are few data concerned with the evolutionary precursors to social pretend play in human children, it may be the case that when a wolf is playing with another wolf by using behavior patterns that are typically used in sexual or predatory encounters, she is thinking about mating or eating her partner and pretending to do so, although the end result of the interaction — play and not killing or mating — does not allow this explanation to be made with any degree of certainty. However, requiring that the individual is thinking about the activity in which it is pretending to engage is not a criterion that is applied to human children, and there is no reason to require it in the case of animals (in which its occurrence could not be discovered by direct questioning: "Were you thinking about mating with your food bowl or with your play partner?"). Perhaps neurobiological analyses in nonhumans and humans would help us by showing that similar electroencephalograph (EEG) patterns are produced (multiply realized) in play and in what are deemed to be real instances of mating, preying, or fighting.

Before examining the data, one point needs to be addressed. Useful analyses of social play require words that some criticize as being anthropomorphic (animals enjoy playing, they gain pleasure from playing, they want to play, and they believe that others want to play with them). The point is that these critics hold that words such as "wants," "believes," or "intends" apply solely to humans. Although some categorically dismiss any analyses of animal behavior that use anthropomorphic terms, others believe that careful use of anthropomorphism can be useful for the study of animal behavior and animal minds (Allen and Bekoff, 1997). I believe that detailed comparative analyses of social play can be useful for informing questions about the evolution of animal minds: Do individuals have beliefs? Do they have desires? Do they make attributions of intentions to others? Do they represent internal states of others? and What is the content of their mental states? It seems clear that without a cognitive vocabulary, it would be very difficult to discuss social play in any useful way. When most animals play, they use specific signals to signal what seem to be their intentions — their desire to play and their belief that if they perform a particular behavior then play will occur — and also perform behavior patterns that are typically performed in other contexts but whose meaning is changed in the context of play. The richness and diversity of social play can only be captured by the use of mental terms.

Play Signals and Intentional Behavior: Do Animals Negotiate Agreements?

In order to discuss whether or not certain actions, hereafter called play signals, serve to communicate an individual's desire to engage in social play — "If I perform this action I believe that play will occur or will continue" — many other questions loom and beg to be studied. Do animals make plans for the future? What is represented in the heads of animals who desire to play or who are engaged in play? Do individuals desire to play with others? Do individuals believe that if they perform a particular behavior pattern then others will play with them? Do animals make attributions of mind to others? Do animals negotiate agreements when they want to play or want to continue to engage in play? Do beliefs lead to other beliefs, not just to behavioral responses? Asking whether or not animals are able to negotiate agreements seems to be the sort of question that could land someone in very hot water, especially when skeptics about animal minds are the audience. Some skeptics would ask, "How could an animal possibly negotiate anything in the absence of language?" Others might be very concerned with the use of anthropomorphic language, whereas some might be concerned about the use of unscientific folk psychological explanations. Some skeptics would also be concerned about the difficulty of making claims about the content of mental states of animals, and some would be concerned about the assumption that animals have any mental states at all. (The fact that there are many reason-

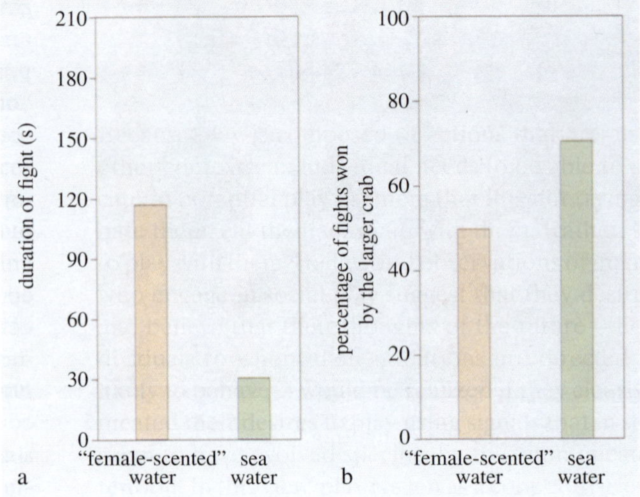

FIGURE 12 (a) Duration of fights between male swimming crabs (*Necora puber*) with and without the scent of a receptive female. (b) The proportion of fights won by the larger crab in the presence and the absence of female scent (reproduced with permission from Smith *et al.*, 1994).

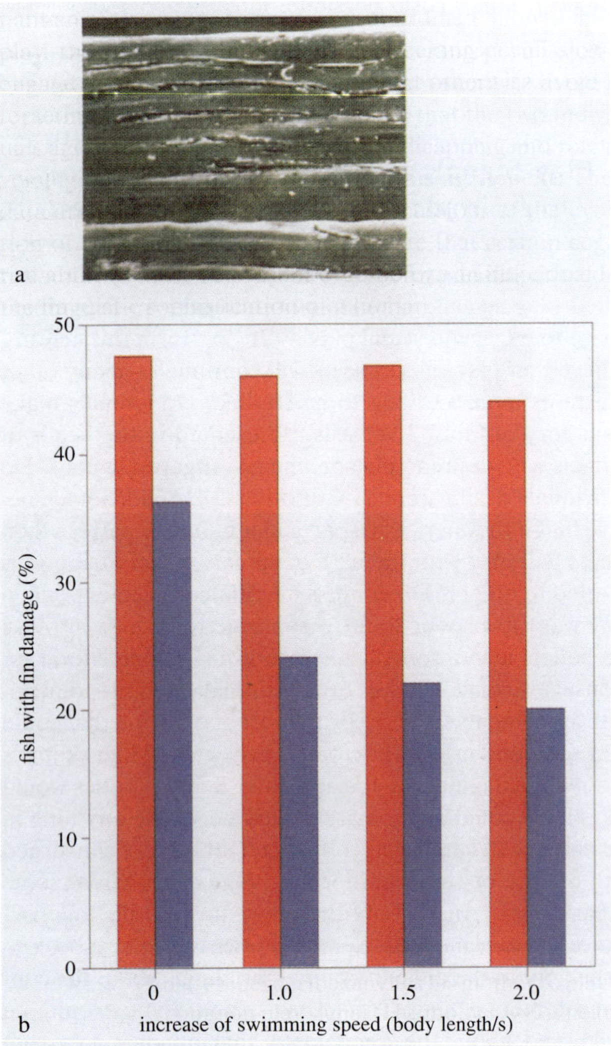

ticularly the smaller individuals, were improved (Christiansen and Jobling, 1990).

Thus, our knowledge of the consequences of fighting for individual fitness, resulting from pure, curiosity driven research, has been used to produce a simple and elegant solution to a fairly serious ethical and economic problem.

References Cited

Archer, J. (1987). *The Biology of Aggression.* Cambridge Univ. Press, Cambridge, UK.

Bakker, T. C. M. (1986). Aggressiveness in sticklebacks (*Gasterosteus aculeatus* L): A behaviour-genetic study. *Behaviour* **98,** 1–144.

Bass, A. (1992). Dimprohic male brains and alternative reproductive tactics in a vocalising fish. *Trends Neurosci.* **15,** 139–145.

Becker, J. B., Breedlove, S. M., and Crews, D. (1992). *Behavioural Endocrinology.* MIT Press, Cambridge, MA.

Benus, R. F. (1988). Aggression and coping. Ph.D. thesis, University of Groningen, Groningen, The Netherlands.

Benus, R. F., Bohus, B., Koolhaas, J. M., and Van Oortmerssen, G. A. (1991). Heritable variation for aggression as a reflection of individual coping strategies. *Experientia* **47,** 1008–1019.

FIGURE 13 (a) Scanning electron micrograph of the damage caused by biting to the dorsal fin of a farmed juvenile Atlantic salmon (reproduced with permission from Turnbull *et al.*, 1996). (b) The incidence of fin damage seen in a farm of Arctic charr (*Salvelinus alpinus*) before (in red) and after (in blue) a period of 63 days during which, by increasing the water flow, an increase in swimming speed was produced in the fish (reproduced with permission from Christiansen and Jobling, 1990).

Christiansen, J. S., and Jobling, M. (1990). The behaviour and the relationship between food intake and growth of juvenile Arctic charr *Salvelinus alpinus* L, subjected to sustained exercise. *Can. J. Zool.* **68,** 2185–2191.

De Ruiter, A. J. H., Koolhaas, J. M., Keijser, J. N., Van Oortmerssen, G. A., and Bohus, B. (1992). Differential testosterone secreting capacity of the testes of aggressive and nonaggressive mice during ontogeny. *Aggressive Behav.* **18,** 149–157.

Enquist, M., Leimar, O., Ljundberg, T., Mallner, Y., and Segerdahl, N. (1990). A test of the sequential assessment game: Fighting in the cichlid fish *Nannacara anomala. Anim. Behav.* **40,** 1–14.

Hews, D. K., Knapp, R., and Moore, M. C. (1994). Early exposure to androgens affects adult expression of alternative male types in tree lizards. *Horm. Behav.* **28,** 96–115.

Huntingford, F. A. (1982). Do inter- and intra-specific aggression vary in relation to predation pressure in sticklebacks? *Anim. Behav.* **30,** 909–916.

Huntingford, F. A., and Turner, A. K. (1987). *Animal Conflict.* Chapman & Hall, London.

Maynard-Smith, J. (1974). *Evolution and the Theory of Games.* Cambridge Univ. Press, Cambridge, UK.

Morris, D. (1953). The reproductive behaviour of the ten-spined stickleback (*Pygosteus pungitius*). *Behav. Suppl.* **6,** 1–154.

Oxford English Dictionary (1973). Clarendon, Oxford.

Sinervo, B., and Lively, C. M. (1996). The rock–paper–scissor game and the evolution of alternative mating strategies. *Nature* **380,** 240–243.

Smith, I. P., Huntingford, F. A., Atkinson, R. J. A., and Taylor, A. C. (1994). Strategic decisions during agonistic behaviour in the velvet swimming crab, *Necora puber. Anim. Behav.* **47,** 885–894.

Turnbull, J. F., Richards, R. H., and Robertson, D. A. (1996). Gross histological and scanning electron microscopical appearance of dorsal fin rot in Altantic salmon, *Salmo salar* L, parr. *J. Fish Dis.* **19,** 415–427.

Van Oortmerssen, G. A., and Sluter, F. (1994). Studies on wild house mice V. Aggression in lines selected for attack latency and their Y-chromosome congenics. *Behav. Genet.* **24,** 73–78.

JOHN D. REYNOLDS
School of Biological Sciences
University of East Anglia
Norwich NR4 7TJ, United Kingdom

Breeding systems are descriptions of mating behavior and parental care by individuals of both sexes. They include the number of mates obtained, the manner in which they are acquired, and the form and duration of parental care. Two key features of the environment have been linked to the evolution of various forms of breeding systems: temporal and spatial aggregation of resources. Recent studies have shown an additional role for direct choices of mates and distinguished between social partners and genetic partners. Theoretical advances have linked these aspects of sexual selection more clearly to parental care provided by mates, and phylogenetic studies are extending our understanding into the realm of historical transitions among forms of breeding system.

Breeding Systems

Introduction

Male northern elephant seals (*Mirounga angustirostris*), weighing 2300 kg, have something in common with 0.5-mm-long male water mites (*Unionicola formosa*). Males in both species fight vigorously with each other for mates, and the victors may end up with sole mating access to more than 50 females during a single breeding season. The study of animal breeding systems seeks to explain underlying reasons for the evolution of such extreme differences between the sexes in mating behavior. Perhaps these species share similarities in some key feature of their habitats and life histories. Indeed, differences among environments might explain the evolution of different mating behaviors in other species.

Breeding systems, also known as mating systems, can be defined as descriptions of mating behavior and parental care by individuals of both sexes (Reynolds, 1996). They describe variation among individuals of both sexes in aspects of reproductive behavior, including form and duration of parental care; form and duration of any pair bonds; number of mates acquired (both "genetic" and "social"); forms of courtship, coercion, and competition (including among sperm of different males); mating resources defended and offered; and the extent of mate choice (including among sperm).

These components include those originally given by Emlen and Oring (1977) in their classic formulation of "mating systems" as well as several new aspects of reproductive behavior which have come to light in the past two decades.

These include distinctions between genetic and social mates; studies of sperm competition, coercion, and mate choice; and a growing emphasis on variation among individuals in behavior combined with an equal emphasis on the behavior of both sexes, rather than an older tendency to focus primarily on males. The term breeding system is preferable to mating system because the latter implies a more restrictive focus on mating behavior alone (which is the domain of "sexual selection"), whereas the former seems more appropriate for including parental behavior — a key feature of any description and explanation of social aspects of reproduction. To understand the diversity of animal breeding systems, it is necessary to consider selection acting on each sex for providing parental care and/or competing for mates.

Differences between the Sexes

Darwin (1871) laid the foundations of our modern theory of breeding systems by emphasizing differences between the sexes in breeding competition. In most species one sex (usually males) attempts to mate with more of the opposite sex than vice versa. Variation among males in their success in such mating competition could exert a powerful influence on the evolution of weaponry such as the antlers of deer as well as gaudy display traits used to attract members of the opposite sex, such as the plumes of birds of paradise and pheasants (Fig. 1). Darwin termed mating competition and mate choice *sexual selection.*

A description of sexual selection, particularly variation in the numbers of mates obtained by each sex and how they acquire them, takes us part of the way toward describing a population's breeding system. The remaining task is to understand the evolution of parental care, which can be defined as any form of behavior that increases the fitness of an animal's offspring. In fact, parental behavior is tied closely to sexual selection, providing some hope to researchers concerned with constructing an integrated theory of breeding systems. If one sex provides little or no care for its young, it will maximize its lifetime reproductive success by mating with as many mates as possible. Thus, male black grouse (*Tetrao tetrix*), elephant seals, and lucanid beetles (Lucanidae), all of which provide no parental care, are larger than their female counterparts, probably as adaptations for fighting with one another for mates. In contrast, females in such species probably gain few material benefits from mating with multiple males and are free to concentrate on choosing the best mate, perhaps based on genetic quality or freedom from sexually transmissible diseases. Their fitness is maximized by focusing on parental care. In other words, females transfer their energy into the young from one or a few matings rather than into the creation of multiple mating opportunities for themselves (Andersson and Iwasa, 1996).

FIGURE 1 Male ring-necked pheasant (*Phasianus colchicus*) displaying (photo courtesy of G. Carrara/Panda Photo).

Relationship between Parental Care and Sexual Selection

Although parental care and sexual selection are clearly connected, a key challenge is to understand how they have coevolved and whether we can discern clear directions of cause-and-effect between them. Theories of sexual selection tend to take patterns of parental care as assumed starting points, whereas theories for parental care sometimes take sexual selection as a starting point. However, experiments have shown that each form of behavior may change in response to manipulation of the other. For example, in male cichlid fish (*Herotilapia multispinosa*) both parents guard the young. However, males will abandon females and their broods early if provided with an excess of unmated females, showing that paternal care is affected by opportunities for sexual selection. Thus, opportunities for sexual selection can dictate parental care. Conversely, in another species of fish, the sand goby (*Pomatoschistus minutus*), males provide all care without help from females. However, experimental reduction in the development time of embryos frees the males from their parental duties early, and males respond by increasing their competitiveness for new females (Kvarnemo, 1994). In this case, parental care dictates sexual selection. Thus, it would be wrong to assume a single direction of causality in the trade-off between sexual selection and parental care: Each aspect of the breeding system can respond to changes in the other.

It remains to be seen how many other species show such flexibility. It has been shown that in the barn swallow (*Hirundo rustica*), males that have particularly long tails provide less care for their nestlings than males with shorter tails (de Lope and Møller, 1993). However, experiments in which male tail length was elongated experimentally by gluing extra segments into the central feathers have shown that females still prefer males with the longest tails. This suggests that attractive males may maximize their reproductive success by reducing care in order to spend more time courting mates. This is another example of parental care being driven at least in part by sexual selection and not simply the other way around. It seems logical that evidence of such facultative responses in each direction can be extended to evolutionary timescales, such that parental care and sexual selection are expected to coevolve with feedback between them.

The fundamental trade-off between parental care and sexual selection just described does not mean that the sex which stays with the brood will necessarily be under the least sexual selection. Male bettas (*Betta splendens*), also known as Siamese fighting fish, are so pugnacious that fish hobbyists must keep them in separate aquaria. However, males provide all parental care, building and maintaining elaborate nests made out of bubbles of air gulped from the water's surface and also guarding embryos and young against predators. The key is that males can simultaneously guard the eggs provided by more than one female. Thus, they "process" young at a higher rate than females can produce batches of eggs. Therefore, mating with the first female does not impair a male's ability to mate with another one immediately afterward. This therefore explains why males experience a greater intensity of sexual selection than females in this species despite providing all the parental care. Similar explanations can be invoked for many other fish species such as most populations of sticklebacks (Gasterosteidae).

Potential Rates of Reproduction

The connections between mating competition and parental care have been formalized according to potential rates of reproduction of each sex, defined as the maximum rate at which individuals can produce offspring if given unlimited access to members of the opposite sex (Clutton-Brock and Parker, 1992). Other things being equal, the sex with the highest potential rate of reproduction will have the most to gain from competing for mates (Fig. 2).

This formulation accommodates species such as the betta fishes described previously because although males provide all postfertilization care, this activity is less time-consuming and energetically demanding per brood than in species such as birds which feed and/or brood their young and in female mammals which nurse their young. Males are therefore selected to compete more strongly for multiple mating opportunities than are females.

In addition to explaining sex differences in mate competition, sex differences in potential rates of reproduction have also been invoked to explain mate choice, the other major form of sexual selection (Johnstone *et al.*, 1996). The

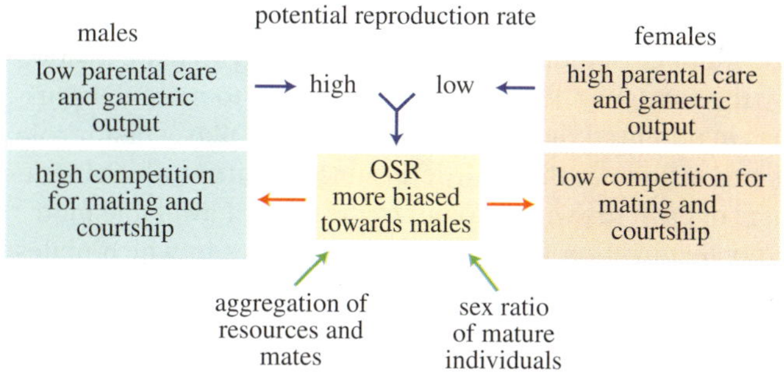

FIGURE 2 A model explaining selection in animal breeding systems. In most species, the males provide less care to the young, giving them a higher potential rate of reproduction than females. The operational sex ratio (OSR), the ratio of fertilizable females to males, is therefore male biased, which selects for greater competition among males for mates and courtship to females. Conversely, there is less selection on females to be competitive and they can afford to be more selective of mates. The OSR may be influenced by resource distribution (see Fig. 3) and by sex ratios of mature individuals.

sex which has the slower potential rate of reproduction (often females) can afford to be more selective in its choice of mate because the opposite sex will rarely be in short supply.

Consider a species such as the ruff (*Philomachus pugnax*), a shorebird in which females provide all parental care. If a female rejects the first three males she encounters on a display arena, she can mate with the fourth and she can still produce her summer's maximum of one brood of four young (perhaps with better genetic quality or perhaps sired by a male that was less likely to pass on a sexually transmitted disease to the female). However, if a male ruff passes up a chance to mate with a female, that will be a direct loss of one brood to his fitness because his potential rate of reproduction is limited only by his mating time and rate of sperm production rather than by 3 weeks of incubation and another week or more of parental care, as provided by the female.

Recent game theory models of this logic, based on differences between the sexes in the costs of lost mating opportunities, have shown that females should be much more likely to reject males than vice versa whenever they require more time to process broods, even if there are substantial differences between the sexes in variation in mate quality (Johnstone *et al.*, 1996). Thus, the costs of lost mating opportunities, set by differences in potential rates of reproduction of each sex, may govern the direction of sexual selection, other things being equal.

Operational Sex Ratios

A useful way of characterizing differences between the sexes in opportunities for mating is the *operational sex ratio* (OSR), first defined by Emlen and Oring (1977) as the ratio of available females to males and often calculated today as the number of individuals of one sex ready to mate divided by the total numbers of both sexes ready to mate. Disparities in potential rates of reproduction will play a major role in this ratio since they will determine the rate at which members of each sex will drop out of the pool of available mates after fertilization (Fig. 2). Thus, in species in which males provide no parental care, the OSR will be strongly male biased. Conversely, in species such as the red-necked phalarope, a shorebird in which males provide all the care during 40 days of incubation and brood rearing in contrast with 12 days for females to produce eggs, the OSR is strongly female biased. An OSR of approximately 0.5 would be expected in truly monogamous species, for example, in which both sexes rear the young and neither sex strays from the pair bond.

There are additional contributions to the OSR (Fig. 2), including differences between the sexes in sex ratios and mortality, behavioral responses to predation risk, emigration and immigration, age at maturity, reproductive life span, and distribution of each sex in space and time.

In species in which one sex fights for access to the other and performs conspicuous advertisements, such as pheasants, these factors may actually reduce the deviation in the OSR from 0.5 because the most competitive sex will be reduced numerically or its courtship and combat activities will be hampered by the risk of predation.

Role of the Environment

Temporal Aggregation of Resources and Mates

Predators are just one aspect of the environment that may shape a population's breeding system. Indeed, a traditional goal of research has been to match particular breeding systems with particular environments. For example, among anurans there is a tendency for species which have prolonged breeding seasons to have more intense sexual selection than explosive breeders in which breeding is confined to a few days. It has been suggested that prolonged breeders are more often territorial, whereas explosive breeders have direct scramble competition among males for females. Thus, short breeding seasons and temporal clustering of females reduce the ability of individual males to monopolize resources and mates (Emlen and Oring, 1977; Fig. 3).

Spatial Aggregation of Resources

In contrast to the effects of temporal aggregation, spatial clustering of breeding resources and members of the sex that is limiting to the other (i.e., having the lower potential rate of reproduction — often females) may promote sexual selection (Fig. 3). For example, among pinnipeds (seals and related species), the extent of polygyny is related to the degree to which females form aggregations during the breeding season. This, in turn, is related to the breeding habitat: In species breeding on isolated islands, one male can monopolize many females, whereas in species breeding on pack ice females are more dispersed and are more difficult for males to defend in groups. These differences in polygyny match differences in size dimorphism, with males being much larger than females in the highly polygynous species, in which there is a greater premium on fighting ability. For example, male northern elephant seals weigh 2300 kg, whereas females weigh only 900 kg. Defendability of breeding resources may also explain the previous example of the highly polygynous water mite *U. formosa.* These breed in freshwater mussels in North America (*Anodonta imbecillis*), which represent a discrete, defendable resource that can support many females and can be monopolized by a single male.

The previous examples from pinnipeds and water mites imply the following direction of causality: resource disper-

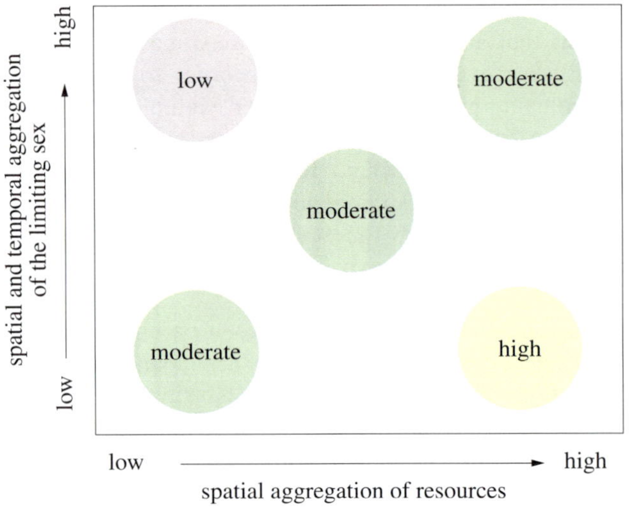

FIGURE 3 The importance of spatial and temporal aggregation of resources and the limiting sex (the sex with a lower reproductive potential) in determining the intensity of sexual selection. High spatial aggregation of nesting sites or territories leads to strong sexual selection when individuals of the sex with the highest potential rate of reproduction, often the males, can monopolize such resources, thereby gaining access to multiple mating partners that need them. This effect will be weakened by short breeding seasons and high reproductive synchrony of the limiting sex, often the females, because there is less time available for individual males to court and pair with females. Depending on the species and its ecology, intermediate levels of temporal and spatial aggregation of resources may lead to the most intense sexual selection if, for example, territories cease to be economically defendable and scramble competitions for mates ensue (reproduced with permission from Emlen and Oring, 1977).

sion → female dispersion → male dispersion. This scenario can be tested by altering the resource dispersion or removing members of one sex and monitoring the effect on the other sex. For example, in the gray-sided vole, *Clethrionomys rufocanus,* spacing of each sex was altered experimentally by moving caged individuals from one place to another (Ims, 1988). Free-ranging males tracked female dispersion, aggregating in experimentally established clusters of females, but free-ranging females did not alter their home ranges in response to manipulations of male dispersions.

This simple direction of causation from resource dispersion to females to males might be expected to evolve most often in species in which males do not provide any resources to females, as in most rodents, which lack paternal care. However, there is evidence from a study in Finland of another rodent in the same genus, *Clethrionomys glareolus,* showing active preferences for socially dominant males, even when male–male competition was excluded experimentally during choice tests (Horne and Ylonen, 1996). Studies of other rodent species have also shown female choice based on familiarity and relatedness. Thus, although dispersion of resources is undoubtedly important in setting the limits to numbers and quality of individuals encountered, and hence the potential for sexual selection, there is scope for an additional effect of females choosing among males directly.

The theory of female choice among resource-holding males has been formalized by the polygyny threshold model (Orians, 1969). This idea was inspired by marsh-nesting birds such as the North American marsh wren (*Cistothorus palustris*) and the red-winged blackbird (*Agelaius phoeniceus*). Imagine a female blackbird arriving at a marsh in the spring that has been divided into territories by males. Some males already have females on their territories, and others have yet to attract a mate. Should the newly arriving female settle with a male that already has a mate, therefore sharing his territorial resources with the other female, or should she choose an unmated male? The polygyny threshold model suggests that this decision will depend on the differences among territories in quality, such that the female may be able to rear more young by sharing a high-quality territory with another female than by choosing to be the only female on a low-quality territory (Fig. 4).

The polygyny threshold model is attractive because its simplicity gives it considerable generality about how variation in quality among members of one sex and the resources they hold should translate into a key aspect of breeding systems, namely, the number of mates animals should obtain. In particular, it predicts that a female choosing to mate in a socially polygynous situation should opt for a male offering sufficient resources for her to reproduce at least as successfully as if she had chosen to be alone with a male in a lower quality breeding situation. Unfortunately, nearly three decades of tests, mainly with birds, have largely failed

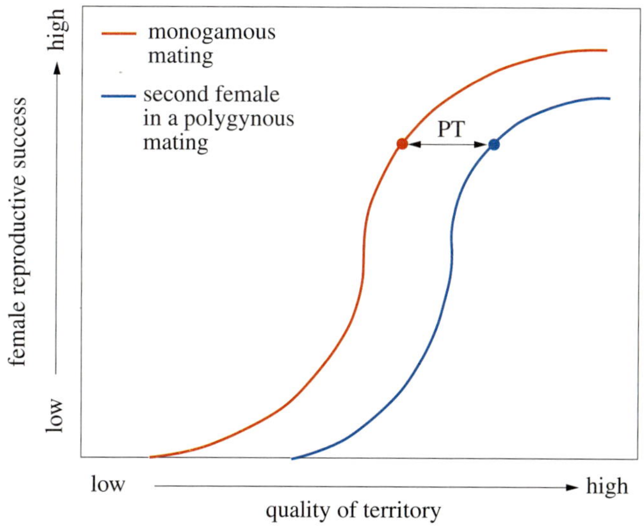

FIGURE 4 The polygyny threshold model. For a given level of territory quality, the reproductive success of a female that chooses a male which has already attracted a female will be lower than that for a socially monogamous female. The polygyny threshold (PT) is the difference in territory quality that would be needed to offset the cost to the second female of dividing the male's resources and/or parental care with an already settled female (reproduced with permission from Orians, 1969).

to support this model. It appears to fail for a variety of reasons, including the fact that females in some species do not necessarily have the free choice assumed by the model due to male harassment, aggression from already settled females, or imperfect knowledge about the quality of resources at the time of settlement. Thus, although the polygyny threshold model serves as a useful starting point for understanding mating patterns when resources are being offered, additional details of the biology of particular species must be added to understand which males attract multiple females.

Predictability of Resources and Mate Defense

If resources such as food for adults and young are spatially unpredictable, they will not be economically defendable, leading to reduced multiple mating and hence less intense sexual selection. For example, phalaropes (*Phalaropus lobatus, P. fulicaria, and Steganopus wilsoni*) and dotterels (*Charadrius morinellus*) are shorebirds which breed in the Arctic (*Phalaropus*), on treeless mountaintops (*Charadrius*), and in the prairies of North America (*Steganopus*). Each could be considered an opportunistic breeder, relying on unpredictable supplies of invertebrates as food and being subject to the vagaries of late snowmelts in the spring and, in the case of *Steganopus,* occasional droughts. These species also exhibit parental care by the male only, with females therefore having a higher potential rate of reproduction and hence under stronger sexual selection, as shown by stronger female–female aggression and brighter plumage than in males. However, the opportunity for sexual selection is restricted by unpredictability of food supplies, which is exacerbated by short breeding seasons. Therefore, females in these four species are not able to defend territories containing food or nesting sites required by males. Instead, their breeding system involves mate guarding, whereby females accompany one male at a time until the clutch has been laid, warding off rival females. The most successful females are thus only able to mate with two males sequentially during one breeding season.

This reduced sexual selection is in contrast with other shorebirds with male parental care, including temperate-breeding spotted sandpipers and tropical northern jacanas (*Jacana spinosa*). These birds breed along freshwater shorelines, where they have access to a more predictable source of food. Territoriality is thus an option, and competitive females attract multiple males which nest in their defended areas. The combination of simultaneous pair bonds with multiple males and longer breeding seasons leads to more intense sexual selection, reflected by greater size (but not plumage) dimorphism in these species.

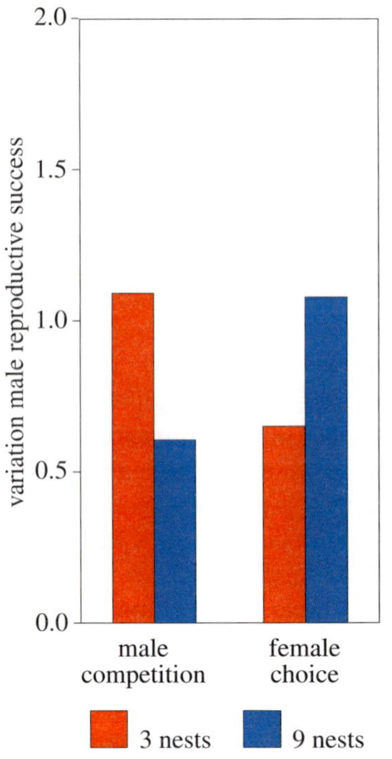

FIGURE 5 Variation in the components of male reproductive success in a population of sand gobies, *Pomatoschistus minutus,* from the southwest coast of Sweden. Six males and six females were introduced into aquaria which had been provisioned with either three nests (in red) or nine nests (in blue). The competition between males caused a higher variation of reproductive success in the group with three nests available. In contrast, female choice caused a higher variation in the group with nine nests (reproduced with permission from Forsgren *et al.,* 1996).

Resource Limitation and Variation within Species

Studies of variation in breeding systems within species have helped clarify the importance of resource limitation in determining mating patterns and the intensity of sexual selection.

The sand goby is a small benthic fish breeding around the coast of much of Europe. Males dig nests under mussel shells or stones and attract one or more females, which lay eggs on the ceiling of the nests. Males guard and ventilate the eggs until they hatch after 1–3 weeks, depending on water temperature. Forsgren *et al.* (1996) compared goby breeding behavior in two populations that differed in availability of shells for nest sites. A field experiment showed that when artificial nests were added to a Baltic Sea population, for which the natural nest density was only 0.2 sites/m^2, all of the 30 newly added sites were occupied by males within 54 h. In contrast, in a population on the Swedish west coast, for which the natural nest density was 9.6 sites/m^2, only 17 of 30 artificial sites were occupied after 96 h. At the low nest density site, males were significantly larger than females, consistent with greater predicted selection on their body size for more intense competition for nest sites. In the high nest density site there was no difference between the sexes in body size.

These results were supported by laboratory manipulations of numbers of nest sites. When six male and six female gobies were given either three or nine nest sites, variation among males in reproductive success stemming from mate competition was more intense in the three-site treatment. Conversely, variation in male success due to female choice tended to be less than that in the nine-nest treatment (Fig. 5).

The pattern was similar in fish from both populations, but there was less of an effect of nest site limitation on female choice in fish from the naturally site-limited population. Thus, limited breeding resources can lead to intensified sexual selection due to male–male competition while reducing the scope for female choice.

An elegant series of experiments have shown the importance of food limitation in the breeding systems of katydids (Tettigonidae). Males provide females of an Australian species with nutritious spermatophores, which females eat while being fertilized and incorporate into their ova. Gwynne and Simmons (1990) showed experimentally that when food was abundant, males were able to produce these spermatophores quickly and it was therefore unnecessary for females to compete for males. However, when food was limited, fewer males were able to offer spermatophores and females were therefore more competitive and mated more

FIGURE 6 Conflicts of interest in a trio of dunnocks, *Prunella modularis.* A beta male is pecking a female's cloaca before mating. This stimulates the female to eject sperm from previous matings, thus increasing the likelihood of paternity for the beta male and the probability that he will feed the nestlings. The dominant alpha male is about to interrupt the mating (adapted from Davies, 1992).

often than when food was abundant. Thus, the direction of sexual selection within a single population can switch according to the amount of food available.

One of the most striking examples of the role of resources in determining animal breeding systems is provided by the dunnock, *Prunella modularis* (Fig. 6). These birds may breed in pairs or in social groups ranging from polyandry to polygyny or both (polygynandry) (Davies, 1992). Davies and colleagues have shown that much of this variation can be attributed to the distribution of food. When food was limited, female home ranges were large, making it difficult for males to monopolize females and thus leading to polyandrous pair bonds with one female and two males helping rear the young. Conversely, when food was added, female home ranges shrank and males could monopolize females, leading to monogamous pair bonds. Variation in the dunnock's breeding system shows the importance of food limitation and also provides a case study for the role of temporal and spatial aggregation in resources (Fig. 3) since Davies (1992) has shown that polygynous pair bonds tend to occur when food is plentiful and two females in the male's home range are nonsynchronous in their timing of breeding.

Sperm Competition

Studies from a wide variety of taxa have shown that females often mate with more than one male during their fertile period. This leads to sperm competition and the potential for sperm choice by females in both internally and externally fertilizing species (Birkhead and Møller, 1998). In many species this form of sexual selection at the gametic level has had profound implications for the evolution of breeding systems. For example, species once thought to be monogamous have been shown to have greater scope for sexual selection than ever imagined due to extrapair copulations. Adaptations to sexual selection include increased ejaculate volume and sperm length, mate guarding, and the evolution of alternative male life histories and behavior.

Unusual Breeding Systems

Social Monogamy

Social monogamy is favored by several conditions:

1. Sexual selection is constrained by short breeding seasons relative to breeding synchrony of the sex with the lowest potential rate of reproduction (Fig. 3).
2. Sexual selection is constrained by resources widely dispersed and hence difficult to monopolize (Fig. 3.).
3. The efforts of both parents are required to rear the offspring.
4. The threat of sperm competition selects for mate guarding by males, thereby restricting opportunities for multiple pair bonds and leading to social monogamy.

Although approximately 90% of bird species are socially monogamous (i.e., one male forming a pair bond with one female), this is an exceptional breeding system in other taxa, including mammals, in which lactation by females gives males in most species more options for seeking extra matings. Monogamy is usually explained in birds by the need for both parents to feed the offspring, and indeed it has been shown that nonmonogamous pair bonds have evolved most often in avian taxa in which the young feed themselves, as in grouse and most shorebirds.

Monogamy in some carnivorous mammals, such as African wild dogs (*Lycaon pictus*) and gray wolves (*Canis lupus*), is thought to stem from males gaining more by bringing prey to the young and providing other forms of parental care than by abandoning the female after they have mated. Species such as these, which hunt large, dangerous, or difficult prey, are selected to hunt cooperatively, leading to complex social organizations characterized by strict dominance hierarchies and efforts by the dominant pair to monopolize mating within the pack. Social monogamy is thus an emergent property of foraging behavior, which is rooted in the environment.

Burying beetles (*Nicrophorus* spp.) provide another exceptional case of social monogamy (within insects) that can be explained by benefits of biparental care. They lay their eggs in vertebrate carcasses, which are an ephemeral resource in short supply and difficult to find. Thus, both members of the pair remain with the carcass as their larvae develop, defending it from other beetles. There is some flexibility in this system, with males of some species attempting to attract additional females to the carcass.

Monogamy does not preclude sexual selection because animals will still be selected to distinguish among mates according to their quality. Thus, many seabirds are monomorphic and brightly colored, and one would expect assortative mating to occur on the basis of such traits if they signal parental or genetic quality. Indeed, female birds may require males to provision them with nuptial gifts before allowing them to mate (Fig. 7).

Communal Breeding

If breeding resources are sufficiently scarce, competition may be so intense that subordinate animals may be selected to join a dominant pair in a communal group. This is one form of cooperative breeding, whereby young animals may help defend the territory and perhaps rear the young of a dominant pair in exchange for a chance of inheriting the territory in the future. Florida scrub jays (*Aphelocoma c. coerulescens*) are an example (Fitzpatrick and Woolfenden, 1988). These birds breed in stable populations in open oak scrub habitats in which all suitable breeding sites are monopolized in permanent territories. A breeding pair may be joined by up to six nonbreeding individuals, usually the breeding pair's offspring from previous years, which help the pair rear slightly more young than when alone and improve the adults' survivorship. Thus, the breeding pair benefits, and the helpers make slight gains in inclusive fitness while waiting for territorial vacancies.

FIGURE 7 A male roadrunner holding (*Geococcyx californianus*) a lizard as a nuptial gift while copulating (adapted from *BBC Wildlife Magazine,* 1992, p. 25).

The communal breeding previously described for the dunnock is different because the cooperating (and conflicting) members of the group are not close relatives. However, as with the scrub jay, they are thrown together by environmental circumstances according to the distribution and abundance of territorial resources.

Leks

Leks are communal display arenas in which males defend small token territories and offer nothing to females other than copulations. This lack of parental input from males causes an extreme asymmetry in the potential rates of reproduction of the sexes and a consequently strong male bias in the OSR, hence the potential for high variation among males in mating success and intense sexual selection. Lekking behavior is a rare breeding system with a patchy phylogenetic distribution, having evolved in birds at least 14 times (e.g., including black grouse, manakins, some hummingbirds, and one tropical flycatcher), mammals [e.g., Uganda kob antelope (*Kobus kob thomasi*) and hammerheaded bats (*Hypsignathus monstrosus*)], frogs and toads [e.g., hylid frog (*Ololygon rubra*) and natterjack toad (*Bufo calamita*)], and African cichlid fishes (e.g., *Cyrtocara eucinostomus*). Lekking species have received a great deal of attention because sexual selection is very intense. However, there is little consensus among researchers about why lekking has evolved, although there is no shortage of hypotheses, some of which focus on the role of female choice and others on male competition (Höglund and Alatalo, 1995). For example, a comparison among grouse species has suggested that in species in which females have large, overlapping home ranges (in response to dispersed food or nesting habitat), males have been selected to display in areas of overlap, where they will have a large rate of encounter with females. Leks may be located in "hot spots"—areas where there is high traffic from females visiting for other purposes. In some species, males may also have been selected to aggregate on leks to minimize the costs to females of sampling behavior and to facilitate comparisons.

In contrast to these female-based models of lek evolution, male-based models suggest that leks may form because of male–male dominance. Thus, the "hot-shot" model was produced as an alternative to the hot spot idea, suggesting that subordinate or less attractive males may cluster around dominant hot shots to intercept females. The "black hole" model suggests that leks form as a result of males taking advantage of a tendency of females to be attracted to groups, as in ungulates, with little active female choice among males within leks.

In considering lek evolution, it is important to bear in mind that although there could be a single underlying cause, it is perhaps more likely that lekking behavior has evolved for different reasons in different taxa. For example, female ungulates appear to have less freedom of choice among males within leks than do birds because male ungulates are capable of physically herding females. Thus, harassment by males in ungulates, as well as in certain birds such as the South American cock-of-the-rock (*Rupicola rupicola*) in which females are sometimes thrown to the ground by males, creates a different set of costs and benefits to each sex from joining aggregations than in species such as the ruff, in which once a male has attracted a female to his display court he bows his head forward and typically remains motionless. There is little direct interference with females, thereby giving females a greater scope for mate choice on the lek. It should also be noted that leks occur in habitats ranging from arid scrubland in the case of the sage grouse (*Centrocercus urophasianus*) to deep forest in many manakins, and that predation pressures on lekking individuals range from virtually none in the case of white-bearded manakins (*Manacus manacus trinitatis*) to many in the case of cock-of-the-rock. There may also be a difference between internally fertilizing species and external fertilizers if females in the latter have more control. As in "communal breeding," the term lek may therefore hide multiple phenomena within a single so-called breeding system.

Courtship Role Reversal

Courtship role reversal (also known as sex role reversal) refers to the direction of sexual selection, not to sex differences in parental care. Thus, male three-spine sticklebacks (*Gasterosteus aculeatus*) provide care for the young, but this species usually has "conventional" courtship roles because males are usually the most competitive sex, being more active than females in courtship and mating competition.

Examples of courtship role-reversed species were presented in the discussion of phalaropes, jacanas, dotterel, and spotted sandpipers. These species have in common the fact that males alone provide all the parental care. Because of the high energetic demands and long periods of care required by the offspring compared with those of fishes such as sticklebacks, males can only care for one brood at a time and the operational sex ratio is therefore skewed such that males limit the rate of reproduction of females. The result can be intense competition among females for mates.

Shorebirds illustrate the fact that, as with lekking, it seems unlikely that courtship role reversal can be attributed to any single environmental cause. Populations of jacanas face high rates of clutch loss due to predation, and it has been suggested that this could select for males to assume parental care while their mates replenish their energy reserves and prepare for a rapid clutch replacement. However, there is no evidence that temperate and Arctic-breeding shorebirds have higher rates of clutch loss than their biparental counterparts, and a recent comparative study of 96 shorebirds found no overall relationship between parental care and clutch predation (Reynolds and Székely, 1997). It has also been suggested that in species of shorebirds that lay a large mass of eggs relative to their body size, or which lay eggs quickly, males may be selected to reduce postlaying stress on their mates by assuming incubation. Again, comparative data do not support this hypothesis. It may be that jacanas have indeed evolved male incubation and consequent role reversal because of the need for females to be able to lay replacement clutches rapidly in high-predation tropical environments, whereas species such as phalaropes and dotterels have evolved male care because of selection on females to abandon clutches when capitalizing on unpredictable pulses of invertebrate food availability. Unfortunately, male care and role reversal in birds are so conservative phylogenetically that it is difficult to make more rigorous comparisons of the independent evolution of these traits.

Most of the cases of courtship role reversal in the animal kingdom occur in species in which males providing parental care are unable to care for the broods of more than one female in a short period of time. For example, female giant water bugs (Belostomatidae) attach their eggs to the backs of males. Males then clean, aerate, and protect the eggs for 3 weeks until they hatch. Males cannot carry the full clutches of two females at once, and in *Belostoma flumineum* females can produce a second clutch rapidly, such that male back space limits females early in the season. Courtship roles are not completely reversed, but males appear to be in control of the final mating decision.

A courtship role-reversed species that has been studied in detail is the pipefish *Nerophis ophidion* (Berglund *et al.*, 1989). Females lay eggs on the ventral surface of the male, and the male carries the embryos for an average of 4 weeks until they hatch. Females can produce eggs 1.8 times as quickly as males can care for them. Based on this difference in the potential rate of reproduction, one would expect courtship role reversal. Consistent with this prediction, females perform a courtship dance to males, displaying bright blue nuptial colors.

Hermaphrodites

In hermaphrodites an individual produces both male and female gametes but not necessarily at the same time. For example, sequential hermaphrodites may be either protogynous, starting as females and becoming males, or protandrous, starting as males and later becoming females. Sex change is favored when the relationship between reproductive success and body size or age differs between the sexes. Such differences are expected from the general breeding system framework which relates the form and intensity of sexual selection to ecology and parental effort.

Examples of protogynous species include many familiar coral reef fishes, such as wrasses (Labridae), parrotfishes (Scaridae), and groupers (Epinephilinae). In addition, in some species certain individuals will start as males and remain male throughout their lives. In the bluehead wrasse (*Thalassoma bifasciatum*), large terminal-phase males develop a bluish background color interrupted by a bold white band with black borders in the middle of the body. These males are strongly territorial and monopolize spawning sites over prominent coral outcrops. Females visit these males for spawning, and the pelagic embryos then drift away from the reef. The size-advantage model indicates that small males would have no chance of competing against such territory holders if they adopted the same behavioral tactics and bold sexual colors (Warner, 1988). Therefore, young fish either start life as females, which are able to spawn without competing for mates, or as primary males that adopt female coloration and behavior. These join mating assemblages and engage in intense sperm competition with other "sneakers" as well as the dominant territory holder. Typically, the largest females change sex and then bide their time as "floater" males, obtaining few or no matings but preparing to contest openings when the dominant male in a social group dies. At this point, reproductive success as a dominant male may greatly exceed that available as a female. Thus, the direction of sex change, as well as its timing and the identity of the individual that becomes male, can be explained by intense sexual selection due to different potential rates of reproduction in a breeding system in which males are able to monopolize resources critical for breeding.

Sex change from male to female (protandry) tends to occur in species in which male mating success is unrelated to male size or age. Individuals would maximize their reproductive success by beginning as males and then changing into females, thereby capitalizing on size-related fecundity benefits. Examples include some species of fishes such as anenomefish (*Amphiprion*) in which breeding resources are widely scattered and defended by both sexes, with less importance for direct male–male competition. Additional examples are pandalid shrimp, which are nonterritorial, and slipper limpets (*Crepidula fornicata*), which form stacks of up to 20 individuals, with the bottom-most individuals typically being large females and the smaller males on top (Fig. 8). It has been shown experimentally that removal of the female causes the next male in the stack to grow rapidly and change into a female. The size–fecundity advantage of this sex change may be augmented by freedom from sperm competition.

In simultaneous hermaphrodites such as hamlets (*Hypoplectrus*), small fishes which inhabit coral reefs, the potential rate of reproduction as a male is higher than as a female since eggs are larger than sperm and more costly to produce. Neither sex provides any care for the embryos. This causes a potential conflict between mating partners due to selection to offer only sperm. These fishes have evolved a resolution which fits remarkably well with expectations from the theory of cooperation. Individuals in a pair trade gametes with one another in an iterated "tit-for-tat" sequence, first donating eggs and then sperm. Indeed, this solution based on reciprocity has evolved independently in at least two other phyla — annelid worms and species of mollusks. Despite the apparent monogamy of this arrangement, one would still expect sexual selection to occur due to variation among individuals in quality as mates and in competitiveness. Furthermore, in several species of reef fishes that have been examined closely, there is evidence that the larger individual of the pair still obtains the highest reproductive success, in part by starting in the male role disproportionately often and sometimes by failing to reciprocate as a female.

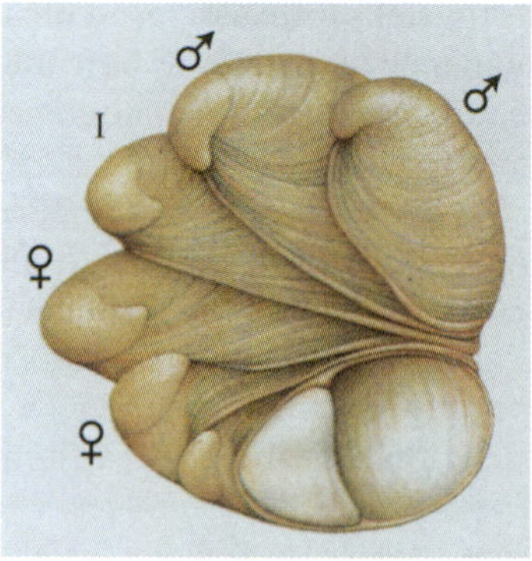

FIGURE 8 A stack of slipper limpets, *Crepidula fornicata.* The largest individuals, on the bottom, are females, and the smaller individuals on top are males (reproduced with permission from Hoagland, 1978).

Conflicts between the Sexes

Breeding systems are the outcome of conflicts. These conflicts occur among individuals of the same sex and between the sexes, even in species that form prolonged pair bonds. This is demonstrated dramatically by Davies' (1992) studies of dunnocks, in which some individuals gain a reproductive advantage at others' expense according to the number of birds in a breeding group (Fig. 6).

A female does best when paired with two males because her young are fed by both of her mates and worst when she must share a single male's efforts with another female. In contrast, an individual male has the highest reproductive success when paired with two females, except in poor habitats in which an extra male may be needed to help provision the nestlings from a single female. Otherwise, male reproductive success is worst when the female is shared with another male due to lost paternity. Monogamy yields intermediate success to both sexes. The resulting conflicts include a female soliciting a secondary male while its first mate attempts to prevent the other from joining the territory and mating with the female.

Whereas the classical theory of breeding systems overlooked much of this intersexual conflict, tending to focus on the point of view of the dominant or most polygamous sex (i.e., usually males), studies during the past decade have shown that in many species females play a much more active role than previously thought. This includes female–female competition for mates in species with otherwise "conventional" courtship roles, as has been observed in two lekking species — sage grouse and peacocks. Females may also incite males to compete, as in elephant seals, in which females are more likely to call loudly if mounted by a subordinate male than by a dominant male. Detailed investigations of female reproductive tracts in a variety of insects and other species have raised the possibility that females may be able to select among ejaculates of different males (sometimes referred to as *cryptic female choice*).

Extrapair matings in apparently monogamous species of birds have posed the biggest challenge to our understanding of breeding systems. These have come to light due to breakthroughs in DNA profiling and other molecular technology. For example, in the socially monogamous great tit, *Parus major,* an average of 17–20% of young are sired by a male other than the female's putative mate. Other examples of extrapair paternity include 22% in the barn swallow, 50–70% in the tree swallow (*Tachycineta bicolor*), and 35% in the indigo bunting (*Passerina cyanea*).

Equally startling is the finding that female red-winged blackbirds engage in so many extrapair copulations that apparent male mating success (based on the number of females on his territory) bears little resemblance to realized mating success (Fig. 9). It is ironic, or perhaps fitting, that these birds helped inspire one of the cornerstones of breeding system theory — the polygyny threshold model (Orians, 1969) — and have figured prominently in attempts to test it. Females control matings in this and most bird species, so we must focus on their mate-choice behavior in order to resolve this puzzle. Many potential solutions have been proposed (Table 1).

One possible explanation is that some males are particularly adept as social partners because of their ability to provide high-quality territories and provision the young, but these traits might have a weak genetic basis. Females might therefore choose such males as social partners but seek others for their genetic quality. In the case of red-winged blackbirds, however, males provide little care to the young, and despite repeated attempts by researchers to measure territory quality, few strong correlations have appeared. It remains to be seen how well this idea might apply to other taxa.

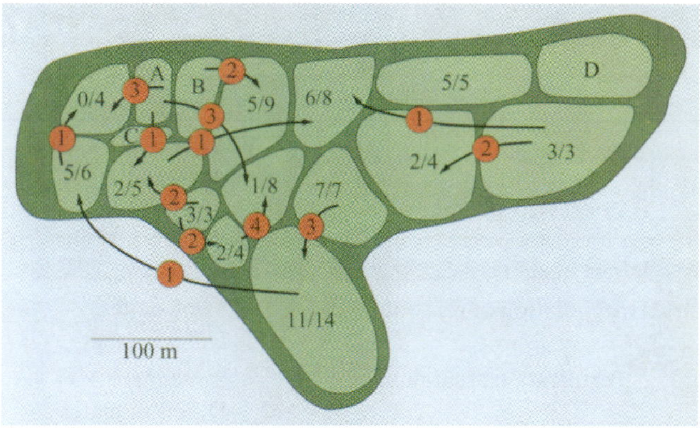

FIGURE 9 Effective mating success, based on parental DNA analyses of chicks, of territorial male red-winged blackbirds (*Agelaious phoeniceus*). The fractions in each territory indicate the number of chicks fathered by the resident male with respect to the total number of chicks found in nests within the territory. The arrows indicate extrapair fertilizations (the base of the arrow indicates the male, the head indicates the territory in which his progeny are found, and the number of progeny generated is indicated in the red circle). In the territories marked with letters, no data are available for the resident pair (reproduced with permission from Gibbs *et al.,* 1990).

Another possibility for females seeking extrapair fertilizations is that female–female competition may limit the options for individuals when settling on territories. This could give the females genetically inferior males. Females might also have imperfect information about the quality of their mates at the time of settlement but later discover their mates' shortcomings. In either case, females may then be selected to make the best of a bad social situation by improving the genetic quality of their offspring through extrapair copulations.

Phylogenetic Transitions

There have been recent attempts to reconstruct evolutionary transitions among key aspects of animal breeding systems. This "historical" approach involves tracing patterns of character change through time along phylogenies (or evolutionary trees). This complements direct observations and experiments with single taxa and provides the only explicit means of inferring coevolution and the evolutionary sequences of events. For example, Sillén-Tullberg and Temrin (1994) used a phylogeny of bird taxa to test the hypothesis that short-duration pair bonds (a prerequisite for social polygamy) were most likely to evolve in taxa in which the young feed themselves rather than requiring parents to

TABLE I

Hypotheses to Explain Multiple Mating by Females[a]

HYPOTHESIS	APPLIES TO SAME OR DIFFERENT MATES?	EXPLANATION
Direct benefits		
Stimulation of reproduction	Same mate	Copulation or ejaculate stimulate ovulation
	Different mate	Copulation or ejaculate stimulate ovulation
Fertilization assurance	Same mate	Ensure adequate sperm
	Different mates	Ensure adequate sperm
Sex ratio manipulation	Different mates	In haplodiploid insects, ensure adequate sperm; hence, females produced
Access to food, nuptial gifts	Same mate	Food or nutrients to female or ova during mating
	Different mates	Food or nutrients to female or ova during mating
Parental care	Same mate	Mate provides care if confident of paternity for current brood
	Different mates	Additional males provide care
Extra broods obtained	Different mates	In successive pairings, males provide care
Mate retention	Same mate	Multiple copulations reduce chance of mate loss
Male harassment	Same mate	Copulation less costly than resisting
	Different mates	Copulation less costly than resisting
Selfish females	Same mate	Keep other females from obtaining high-quality males
Mate assessment	Same mate	Male copulation may reflect his quality
	Different mates	Male copulation may reflect his quality
Genetic benefits		
Choice of paternity for benefits to offspring survival	Same mate	Devalue other males' sperm
	Different mates	Choose better male's sperm
Choice of paternity for benefits to offspring attractiveness	Same mate	Devalue other males' sperm
	Different mates	Choose better male's sperm
Avoidance of genetic defects from stored sperm	Same mate	Old sperm might be more prone to genetic abnormalities
	Different mates	Old sperm might be more prone to genetic abnormalities
Choice of mixed paternity for genetic benefits	Different mates	Reduce competition among genetically dissimilar siblings
Incite sperm competition	Same mate	Allow high-quality sperm to outcompete rivals
	Different mates	Allow high-quality sperm to outcompete rivals
Inbreeding avoidance	Different mates	Mating with multiple males reduces inbreeding
Genetic correlation with male multiple mating	Same mate	Females inherit tendency of males to mate often
	Different mates	Females inherit tendency of males to mate often

[a] See Reynolds (1996) for original references.

bring the food to them. The expectation was that if the young foraged for their own food, as in grouse, rather than having it brought to the nest, one of the parents would be sufficiently free from parental duties to desert the mate and perhaps obtain new mating opportunities. When Sillén-Tullberg and Temrin mapped pair-bond duration and the mode of feeding by the young onto their avian phylogeny, they found support for the hypothesis: Self-feeding young and short-duration pair bonds co-occurred in evolutionary lineages more often than expected by chance. This analysis thus links parental care and sexual selection to another key feature of avian life histories.

With sufficient information from a variety of taxa, it has also been possible to reconstruct directions of evolutionary change in aspects of animal breeding systems (Reynolds *et al.*, 2000). In another avian study, Székely and Reynolds (1995) showed that in shorebirds — a group of approximately 200 species of sandpipers, plovers, and allies — the ancestral form of parental care consisted of both sexes staying with the young until the time of fledging. However, at the base of one of the major lineages of shorebirds, the Scolopacida (which include sandpipers, snipes, phalaropes, jacanas, and allies), there was an early transition toward care provided predominately by the male. Since then, there has been a repeated tendency to evolve back toward increasing care by the female relative to the male. Increases in the relative amount of care by the female have been matched by reductions in the size of males compared with females. This is consistent with the idea that after the unexplained change toward predominately male care millions of years ago, sexual selection on males has caused different taxa to repeatedly revert to reduced male care. Further tests of evolutionary transitions in breeding systems in these and other taxa will become possible as more phylogenetic information becomes available.

References Cited

ANDERSSON, M., and IWASA, Y. (1996). Sexual selection. *Trends Ecol. Evol.* **11,** 53–58.

BBC Wildlife Magazine (1992, September 9). **10,** 25.

BERGLUND, A., ROSENQVIST, G., and SVENSSON, I. (1989). Reproductive success of females limited by males in two pipefish species. *Am. Nat.* **133,** 506–516.

BIRKHEAD, T. R., and MØLLER, A. P. (1992). *Sperm Competition and Sexual Selection.* Academic Press, San Diego.

CLUTTON-BROCK, T. H., and PARKER, G. A. (1992). Potential reproductive rates and the operation of sexual selection. *Q. Rev. Biol.* **67,** 437–455.

DARWIN, C. (1871). *The Descent of Man, and Selection in Relation to Sex.* Murray, London.

DAVIES, N. B. (1992). *Dunnock Behaviour and Social Evolution.* Oxford Univ. Press, Oxford.

DE LOPE, F., and MØLLER, A. P. (1993). Female reproductive effort depends on the degree of ornamentation of their mates. *Evolution.* **47,** 1152–1160.

EMLEN, S. T., and ORING, L. W. (1977). Ecology, sexual selection, and the evolution of animal mating systems. *Science* **197,** 215–223.

FITZPATRICK, J. W., and WOOLFENDEN, G. E. (1988). Components of lifetime reproductive success in the Florida scrub jay. In *Reproductive Success* (T. H. Clutton-Brock, Ed.). Univ. of Chicago Press, Chicago.

FORSGREN, E., KVARNEMO, C., and LINDSTRÖM, K. (1996). Mode of sexual selection determined by resource abundance in two sand goby populations. *Evolution* **50,** 646–654.

GIBBS, H. L., WEATHERHEAD, P. J., BOAG, P. T., WHITE, B. N., TABAK, L. M., and HOYSAK, D. J. (1990). Realised reproductive success of polygynous red-winged blackbirds revealed by DNA markers. *Science* **250,** 1394–1397.

GWYNNE, D. T., and SIMMONS, L. W. (1990). *Nature* **346,** 172–174.

HOAGLAND, K. E. (1978). *Malacologia* **17,** 353–420.

HÖGLUND, J., and ALATALO, R. (1995). *Leks.* Princeton Univ. Press, Princeton, NJ.

HORNE, T. J., and YLONEN, H. (1996). Female bank voles (*Clethrionomys glareolus*) prefer dominant males; But what if there is no choice? *Behav. Ecol. Sociobiol.* **38,** 401–405.

IMS, R. A. (1988). Spatial clumping of sexually receptive females induces space sharing among male voles. *Nature* **335,** 541–543.

JOHNSTONE, R. A., REYNOLDS, J. D., and DEUTSCH, J. C. (1996). Mutual mate choice and sex differences in choosiness. *Evolution* **50,** 1382–1391.

KVARNEMO, C. (1994). Temperature differentially affects male and female reproductive rates in the sand goby: Consequences for the operational sex ratio. *Proc. R. Soc. London B* **256,** 151–156.

ORIANS, G. H. (1969). On the evolution of mating systems in birds and mammals. *Am. Nat.* **103,** 589–603.

REYNOLDS, J. D. (1996). Animal breeding systems. *Trends Ecol. Evol.* **11,** 68–72.

REYNOLDS, J. D., and SZÉKELY, T. (1997). The evolution of parental care in shorebirds: Life-histories, ecology and sexual selection. *Behav. Ecol.* **8,** 126–134.

REYNOLDS, J. D., GOODWIN, N. B., and FRECKLETON, R. P. (2001). Evolutionary transitions in parental care and live-bearing in vertebrates. *Philosophical Proceedings of the Royal Society* (in press).

SILLÉN-TULLBERG, B., and TEMRIN, H. (1994). A phylogenetic analysis of the relationship between polygamy, pair bond length, and the characteristics of young in birds. In *Phylogenetics and Ecology* (D. Vane-Wright and P. Eggleton, Eds.). Academic Press, London.

SZÉKELY, T., and REYNOLDS, J. D. (1995). Evolutionary transitions in parental care in shorebirds. *Proc. R. Soc. London B* **262,** 57–64.

WARNER, R. R. (1988). Sex change and the size-advantage model. *Trends Ecol. Evol.* **3,** 133–136.

FRITZ TRILLMICH

Department of Animal Behavior
Bielefeld University
D-33501 Bielefeld, Germany

RUDOLF DIESEL

Max-Planck Institute of Behavior Physiology
D-82319 Seewiesen, Germany

Parental Care

Parental care has evolved many times independently in the animal kingdom among the invertebrates and the vertebrates. It consists of a great variety of behaviors beneficial to offspring by furthering their development and growth and increasing their chances of survival to maturity and eventual reproduction. Apparently, situations abetted the evolution of parental care in which early developmental stages face adverse conditions, such as harsh physicochemical environments, scarce food resources, intense predation, parasitism, and strong competition for rich food resources which young cannot defend. Benefits to offspring may accrue as incidental side effects of parental behavior but may also imply fitness costs to parents. The concept of parental investment explicitly includes only the latter case. Costs to parents may arise through reduced survival or fertility of caring (compared to noncaring) individuals. Due to this cost, selection on parents generally acts to ensure that care is restricted to their own young, leading to the evolution of mechanisms of parent–offspring recognition. Since parental investment decreases the future expected fitness of parents, conflicts regarding the duration and the extent of parental care are expected. Such conflicts may occur between pair partners for which both will be selected to let the other one work harder in offspring provisioning or defense or even to desert the mating partner, leaving it to rear the progeny alone. Conflicts similarly are predicted among siblings and between parents and offspring. Empirical evidence for conflict between pair partners and siblings is strong, but it is more equivocal for parent–offspring conflict. In the latter case, observed behavior that looks like conflict could also be interpreted as communication between parents and offspring about the state of each other as well as that of the environment from which the parent must garner food for the young. In some species (usually related) individuals help parents raise their offspring. In these cases, parental behavior is performed by animals which are clearly not the parents of the young. These helpers often gain some indirect fitness through increasing the reproductive success of close kin. The pinnacle of altruism in parental care is complete sterility of workers as observed in eusocial insects. The valuable resource "parental care" may

What Is Parental Care?

Reproduction is a necessary part of the life history of organisms. All animals of sexually reproducing species have had two parents which reproduced successfully since reproduction is the only way for the next generation to occur. However, successful reproduction does not automatically imply parental care. Many animals, especially marine animals, broadcast their reproductive products (ova and sperm) to the surrounding environment in which fertilized eggs may or may not successfully develop without ever coming into contact with the parents. This reproductive strategy is used, for example, by herring and characterizes one end of a continuum spanning widely varying reproductive strategies from release of many tiny sex products with no subsequent contact between parent and offspring to long-term *also be exploited by parasites that manage to trick other individuals of their own or other species into rearing their offspring. Parental care thus proves to be a valuable resource evolved for reproduction under difficult conditions that is contested between family members, intraspecifically and even interspecifically.*

contact between parents and offspring and extensive care of few large offspring. This latter kind of parental care is typical of mammals (including ourselves) and birds, but it is also common in lower vertebrates and invertebrates.

Care may also take another form that is used by many invertebrates: the provisioning of offspring that never come into contact with the parent(s). This reproductive strategy is observed in many solitary bees and wasps. In these species, females build cells in which they collect food. The female lays an egg on this food provision before closing the cell and leaves the developing offspring alone. The hatching larva feeds on this food store during its development. Due to the amount of work needed for such provisioning, mothers can only produce a few well-endowed offspring. We can thus distinguish parental care via *provisioning* of offspring without direct contact between parent and offspring and *direct care* in which parent and offspring interact for a protracted period.

It is not always easy to decide what constitutes parental care since offspring may sometimes benefit simply by the presence of an adult. For example, in many fish, males defend a territory with suitable spots for egg laying, thereby attracting females. This territory defense by the male also protects the eggs and the larvae hatching from them. To exclude such cases, we define *parental care* (Clutton-Brock, 1991):

in the broad sense as production of large, yolk-rich eggs, care of the fertilized egg inside and outside the parent's body, preparation of nests, provisioning of young (before and after birth), and support of offspring after nutritional independence.

in the narrow sense as care of eggs and young (including provisioning) after oviposition or birth and sometimes additional support of offspring after nutritional independence.

Parental care is essentially a descriptive term characterizing beneficial behavior of a parent toward offspring. However, there also exists a more life history-oriented concept, *parental investment,* introduced by Trivers (1972). This concept stresses fitness effects (usually measured in terms of reproductive success:

1. The behavior of the parent must demonstrably benefit the offspring, i.e., increase its chance to develop, grow, and become a reproducing adult.

2. Care behavior must impose a cost on the parent and should not be an incidental effect of its presence and of behavior performed for the parent's own benefit.

Benefit and cost in this definition are measured in fitness equivalents (e.g., lifetime reproductive success). Benefits to the offspring may show increased survival to maturity and enhanced fertility during its reproductive life span. Frequently considered costs to the parent are reduced fertility and increased mortality due to the effort expended on young. Parental care and parental investment are thus closely connected to one another and will be treated together here, although they are conceptually different since even lavish parental care may be almost cost-free to a parent in an environment with superabundant resources and lacking predators.

Parental Care Evolved as a Solution to Ecological Problems

Parental care evolved many times independently in various animal groups. In some, such as birds and mammals, it constitutes the characteristic, ancestral pattern of reproduction. In others, such as invertebrates, amphibians, fish, and reptiles, it is rather an exception and has evolved several times independently within each of these groups. This repeated, independent origin of parental care suggests that its evolution is an answer to ecological problems. The following examples illustrate such contrasting parental care strategies within a group and indicate the kinds of ecological problems that lead to the evolution of care.

The edible crab, *Cancer pagurus,* produces, like most of its fellow marine decapods, many small eggs (more than 1 million eggs of 0.3-mm diameter) that develop while carried on the mother's body. From the eggs hatch tiny larvae which are released into open water and spend considerable time feeding and growing in the marine plankton. Extremely few (<0.1%) of these larvae survive this critical part or their life history and ever see the benthos (bottom of the sea) to settle as juvenile crab, and even less live to reproduce. Except for the short period of retention of the developing eggs on the mother's body, this is an example of reproduction without parental care.

In contrast, the Jamaican bromeliad crab (*Metopaulias*

FIGURE 1 Parental care in the Jamaican bromeliad crab (*Metopaulias depressus*). A mother protects her young inside the enlarged leaf axil of a bromeliad plant.

depressus; Fig. 1) has evolved a completely different life history strategy involving extensive maternal care (Diesel and Schuh, 1993). It has colonized the mountain forests of central Jamaica and adapted to life in large epiphytic bromeliad plants. Bromeliads store small amounts of rainwater in their leaf axils. These pools are the essential resource for reproduction and survival in an otherwise dry habitat. The mother crab produces only few very large eggs (50 of 1.2-mm diameter), and the larvae are released into one leaf axil holding about 250 ml water. Owing to the large yolk resources provided by the mother, the nonfeeding larvae thrive entirely on yolk reserves (they are lecithotrophic). However, unfavorable acid conditions (pH 3 or 4), a low calcium and oxygen content of the axil water, and intense predation by voracious damselfly nymphs seriously threaten their survival. After metamorphosis, the juvenile crabs are preyed on by spiders and require more food than a leaf axil can offer for growth. These hostile abiotic and biotic conditions favored the evolution of an array of parental care. The evolution of such behaviors was made possible by the availability of the aggregated offspring within the mother's territory. The mother defends the larvae from predators and increases dissolved oxygen concentration by aerating the water and cleaning the axil of decaying debris. She fetches snail shells ($CaCO_3$) to neutralize the water and increase calcium content, both essential for successful molt of larvae and the buildup of their new carapace. The juvenile crabs remain in the nursery axil for up to 2 months. Too small to hunt for themselves, they are provisioned by their mother with prey, such as millipedes and beetles, captured elsewhere on the plant.

The benefit of such intricate and extended maternal care for the offspring was demonstrated in field experiments: Care substantially increases survival and growth of offspring. The cost of parental care is usually more subtle and less evident than its benefit; in the bromeliad crab, the extended care delays the molt and growth of the mother and, as fecundity increases with maternal size, intensive brood care reduces maternal lifetime fecundity.

As shown by this case, parental care can act as an evolutionary key innovation for the colonization of new environments, which offer less favorable conditions for early life cycle stages. What constitutes an inhospitable environment depends on the typical (and usually ancestral) natural history of a group. Decapod crustaceans typically live in the marine environment, and parental care evolved independently in crayfish and various freshwater crabs and proved instrumental in their successful colonization of freshwater and terrestrial habitats. For wood lice (Crustacea, Isopoda) with a long evolutionary history on land, parental care is essential in species such as the isopod *Hemilepistus reaumuri* (in this case even biparental care is essential) that colonizes hostile desert environments. Similarly, in teleost fish parental care evolved many times in response to the invasion of freshwater habitats (Baylis, 1981). Unfavorable environmental conditions (relative to ancestral environments), predation, and parasites are among the prime forces favoring parental care.

Fuel for the Ontogeny

One major problem for animals ranging from invertebrates to mammals is to provide the offspring with adequate nutritional resources for growth. This challenge led to manifold evolutionary solutions. A common and widespread pattern of preovipositional investment is the provisioning of eggs or embryos within the reproductive tract, for example, by furnishing massive amounts of yolk per egg, enabling subsequent lecithotrophic development of larvae, but also by providing external food resources or nutritional resources derived from a parent's soma.

Provisioning without Contact: Eggs on Rich Resource Provided by Parent

As mentioned previously, many species provision their offspring without direct interaction between parent and offpring. Female sphecid wasps of the genus *Philanthus,* for example, dig nest galleries with several brood cells into the ground and provision each cell with immobilized bees. An egg is then laid onto this food resource and the cell closed. The potter wasp, *Euodynerus foraminatus* (Eumenidae) even adjusts the amount of food according to the intensity of competition expected at the nest site, depositing more food with eggs when competition for mates and nest sites is expected to be high and large offspring size advantageous.

Provisioning with Contact: Resources Defended by the Attending Parent

Other species make use of rich nutritional resources, occurring unpredictably in time and space. An example is burying beetles of the genus *Nicrophorus:* When burying beetles encounter a fresh carcass of a mouse, for example, one or both parents bury, defend, and process the carcass, forming it into a ball. The female lays eggs near the carrion sphere and later when the larvae hatch, the mother (and in some species the father too) feeds them by regurgitation of food taken from the carcass (Trumbo as cited in Rosenblatt and Snowdon, 1996). Without the help of one or both parents, the offspring could not monopolize this resource, essential for survival and development. Too many intra- and interspecific competitors use the same resource, and some such as live-bearing flies even deposit larvae instead of eggs on carrion, thus giving their offspring a time and size advantage for which the larvae of burying beetles could never make up. Parental care by burying beetles is thus essential for safeguarding the rich but ephemeral resource for their offspring against competing organisms.

Provisioning with Contact: Resources Provided by Parent

In amphibians, a primarily aquatic group, a stunning array of parental care patterns independently evolved in various groups (Crump as cited in Rosenblatt and Snowdon, 1996). The typical anuran life history includes an egg and a tadpole stage, reflecting their evolutionary history in aquatic habitats. The tadpole stage is highly adapted to feed and grow in the aquatic environment. Many species of tropical anurans have evolved intricate pattern of parental care to enhance offspring survival under conditions unfavorable for early development.

One particularly adverse condition is the abundance and variety of predators in tropical aquatic systems. They threaten the survival of the highly nutritious and almost defenseless egg and tadpole stages. Two main solutions evolved in response to this threat of predation: (i) attending, guarding, and brooding early developmental stages in aquatic habitats, and (ii) a shift of early development away from open water into more terrestrial sites.

Leptodactylus ocellatus is a good example for the former group. Females produce floating foam nests containing the eggs. The mother attends the nest and guards it aggressively against potential predators. When the tadpoles hatch they remain together and move about in dense aggregation, which again is vigorously defended by the mother. In the aquatic *Pipa pipa,* the female carries her eggs embedded in her dorsum. Following early development protected in the dermal chamber, the fully metamorphosed froglets struggle out to start independent life.

Other lineages of tropical anurans evolved toward the second solution, adapting to a more terrestrial mode of reproduction. Since eggs and tadpoles cannot develop without water, dehydration needs to be reduced. *Dendrobates pumilio* parents achieve this by repeatedly moistening the eggs; the maki frog, *Phyllomedusa duellmani,* adds egg capsules that contain only water; and the African frog *Chiromantis petersi* produces an arboreal foam nest to keep the eggs from dehydrating. Parents may then transport the hatching tadpoles on their backs to water holes. In the terrestrial New Guinea microhylid frog *Cophixalus parkeri,* mostly males attend clutches for 85–100 days until hatching, thereby significantly reducing predation by arthropods, egg cannibalism, and fungal infection. The only species reported to care for froglets is the Jamaican cave frog, *Eleutherodactylus cundalli* (Leptodactylidae), in which the whole clutch of up to 70 fully developed froglets moves onto the back of the mother to be carried out of the cave in which breeding took place into the forest, in which the young froglets can find food (Fig. 2; Diesel *et al.,* 1995). The apex of parental care in frogs is represented by species that evolved provisioning of their offspring by transfer of materials from the soma, as in *Rhinodermata darwinii* (Rhinodermatidae), in which males swallow the freshly hatched tadpoles into the vocal sac. Here they develop, provisioned with materials from the paternal tissue, and they are released as froglets. Similarly, in some *Gastrotheca* species (Hylidae) eggs develop in dermal pouches that evolved a placenta-like structure that enables and regulates gas, water, and excreta exchange during a long developmental period in the safety of the maternal skin.

Other species that invaded terrestrial habitats oviposit in small amounts of open water in tree holes or water-storing plants. However, these nurseries lack sufficient nutritional resources to guarantee successful development of tadpoles. The Jamaican tree frog *Osteopilus brunneus* (Hylidae) solved the problem in an elegant way: Females deposit more than 100 eggs into a leaf axil and subsequently return to supply the tadpoles with in some cases more than 200 fertilized and later in the cycle unfertilized eggs (Crump as cited in Rosenblatt and Snowdon, 1996). An even more

FIGURE 2 Parental care in the Jamaican cave frog (*Eleutherodactylus cundalli*). A mother carries her froglets on her back from the cave into the forest, in which they can feed.

complex provisioning pattern evolved in dart-poison frog *Dendrobates pumilio.* Here, females transport and distribute single tadpoles into water-filled leaf axils. The mother remembers the location of each tadpole and subsequently returns repeatedly to all the different sites to provide a few unfertilized eggs at a time for each tadpole as food. In both species, tadpole survival depends entirely on maternal feeding.

The large variety of parental care that evolved in terrestrial breeding anurans shows that there is strong selection favoring care for eggs, tadpoles, and even froglets in environments in which these life history stages are highly endangered. Another evolutionary solution was to reduce the ill-adapted life history stages in terrestrial frogs. All species of the genus *Eleutherodactylus* have completely lost the tadpole stage and evolved direct development in which fully metamorphosed froglets hatch directly from the egg. At least one species (*E. jasperi*) even went a step further and evolved ovoviviparity.

Evolution of Parental Care by Male and Female

Despite the numerous descriptive and experimental studies on the significance of parental care in present-day species, we know little about its origin and why in some species males and in others females are the sex providing care. Given the wide variety in parental care and the many independent evolutionary lines leading to it, there are certainly many different ways by which it evolved. As an example, here we outline only one possible scenario for the origin of parental care.

For most oviparous ectotherms the origin of parental care probably was related to territorial defense of limited resources for survival or reproduction, such as feeding sites, shelter, oviposition sites, and courtship areas, by one parent (Baylis, 1981). If eggs and/or larvae occur within a parent's territory, one important precondition for parental care to evolve is met: the availability of the offspring to the parent in time and space (Clutton-Brock, 1991). Offspring may experience some protection purely by being located within a parent's territory. Such a situation may have occurred in the ancestor of the bromeliad crab *M. depressus,* in which female crabs probably defended a bromeliad plant as a hunting ground and sole water resource in an otherwise dry habitat. Larvae and juveniles occur in leaf axils and originally may have had increased survival chances simply due to the presence of the mother. In this situation, the mother may not have incurred any cost related to parental care, but such a situation could have provided the basis for the evolution of a more complex pattern of parental care as observed today. Similar scenarios may explain the origin of parental care in terrestrially breeding frogs and freshwater breeding teleost fish.

Three hypotheses have been suggested to explain the distribution of male versus female parental care in taxa with external fertilization: gamete order, certainty of paternity, and association of parent with embryonic stages. These hypotheses are not mutually exclusive and are based on the mate desertion model, which attempts to explain which parent will desert offspring first, leaving the other to either desert as well or care for the offspring alone (Trivers, 1972).

Trivers (1972) suggested that external fertilization such as in teleost fish and frogs should favor paternal care. In teleost fish paternal care probably evolved as a consequence of male territoriality. If males defend territories including oviposition sites, they may provide some care for the offspring and continue to court and mate with additional females. Hence, the cost of paternal care in terms of reduced mating opportunities is low (Trivers, 1972). In fish, male care is far more common than female care. One can imagine an evolutionary sequence of parental care in fish in which each state is ancestral to the subsequent state: no care → male-only care → biparental care → female-only care. However, it remains to be explained, for example, for frogs, why within a family or genus some species show male and others show female care, and the models are still too coarse to explain how such reversals of reproductive strategies evolved. Comparative studies based on much more knowledge of the natural history and the ecology of these species as well as a more precise knowledge of their evolutionary affinities may eventually help to explain these patterns.

Preconditions of Brood Care in Higher Vertebrates

Brood care is an expensive commodity, which usually is reserved by parents entirely for their own young, and animals without their own young usually react aggressively when approached by young. This can lead to the death of an unwary young animal that approaches a female that is not its mother. Such incidences can quite regularly be observed in the dense colonies of pinnipeds in which young are mauled to death by pregnant females or even by lactating females that are not their mothers (Trillmich, 1981). Mothers apparently can recognize their own young. This is a precondition for selective parental care: Adults must establish a specific bond with their own young and reject foreign young. Simultaneously, parents must develop responsiveness to young once they have their own offspring to ensure that parental care is provided.

Some kind of "recognition" is established in invertebrates in a very simple way. In the burying beetle mentioned previously, parents prepare a carcass of a small vertebrate (e.g., a mouse) into a ball, which they bury in the soil. The female lays her eggs near this rich food source and then male and female wait together for the juveniles to appear at the carcass. Eggs need approximately 2.5 days to develop into the first-instar larvae, which then approach the carcass. Females accept young that appear 8–12 h before this expected time and up to 24 h after this time. These larvae are fed by the parents irrespective of whether or not they are the female's own offspring. Larvae appearing outside this time window of acceptance are killed and eaten. The physiological nature of this window of acceptance is unknown. It correlates, however, with changes in juvenile hormone level in the female.

A quite extreme example of a motivational and physiological shift is provided by the gastric brooding frog *Rheobatrachus silus* of Australia (Tyler, 1983). In this species, females ingest their eggs immediately after fertilization and brood the young in their stomachs for a period of approximately 5 or 6 weeks before giving birth orally. Mothers fast during this period perhaps because the greatly distended stomach signals satiation. Young are apparently protected from digestion by the gastric secretions of the mother through the excretion of prostaglandines, which inhibit acid production and reduce the mobility of the gut.

Parent birds or mammals with altricial (i.e., immobile) young in a nest can initially rely on the location of young as a sufficient cue for their own offspring. It is therefore possible to exchange young in these species without any ill effect on parental care behavior until the young become mobile. In the cliff swallow, young develop individualized calls shortly before fledging which are learned by the parents. In these families, initial recognition of own offspring, therefore, depends on the location, but later parents learn to recognize their offspring individually, independent of location. Similarly, young learn to recognize their parents. Again, there is a time window within which recognition, in this case individualized recognition, becomes established. This time window is tuned finely to the development of young (e.g., the onset of locomotion).

In mammals the regulation of such a window of acceptance has been investigated in-depth for sheep (Levy *et al.* as cited in Rosenblatt and Snowdon, 1996). Here, the problem of offspring recognition is quite different because young are precocial; that is, they have fully developed sensory and locomotory systems. Young immediately move about and there is a danger that mothers might confuse young if they cannot recognize their own offspring. To ensure a safe learning period, females about to give birth separate from the group in which they are living. This enables the mother to learn individual features of her young immediately after birth without the possibility of confusing her own with alien young. Sheep without young and even pregnant ewes are quite aggressive against lambs. Shortly before and at parturition, mothers become very responsive toward lambs. This is induced by hormonal changes in steroid balance before and mechanical stimulation of the genital tract during parturition. These events trigger the release of oxytocin (the hormone that also stimulates milk ejection), which induces a sudden preference for amniotic fluid. This brings mother and newborn immediately into contact as the mother begins to lick amniotic fluid off her young and in this way begins to care for the newborn by drying its fur. This close contact offers an excellent opportunity to learn the individual odor of the lamb(s). Events during parturition lead to neurophysiological changes in the olfactory bulb that allow the learning of the offspring's specific olfactory characteristics. In this way, a selective bond between mother and young is established within a few hours after birth. Later, acoustic recognition also becomes established. Once the bond between mother and offspring has formed, alien young are rejected. Interestingly, alien young are not accepted even if smeared with the mother's own lamb's odor. However, mothers can be induced to learn an alien lamb's odor and thereafter accept it if given artificial cervicovaginal stimulation, which reinduces the initial responsiveness of the neural learning system. Mothers so treated will then accept young which they aggressively butted away minutes before. Once this imprinting-like process has occurred, the highly selective bond established between mother and lamb will guarantee the specificity of maternal care. This whole process occurs again when the ewe gives birth to her next lamb a year later. Interestingly, multiparous ewes seem to learn much faster and are less disturbed by distracting events around parturition than primiparous mothers, suggesting that the neuronal machinery of the former is primed by the first parturition experience.

Patterns of Brood Care in Higher Vertebrates

Birds

Birds are the only group in which social (although not necessarily genetic) monogamy is the rule. Birds, particularly songbirds such as great tits but also many large seabirds such as gulls, incubate their clutch together and feed the hatched young for considerable time until fledging. Male and female may contribute differently to the chores of family raising, but parenting is basically a joint venture. Frequently, there follows a long period of time after fledging of young when parents still care for the semi-independent young. Often, the male cares for fledged young during this period, thus enabling the female to gather resources for laying a second clutch. Postfledging parental care has until recently been little appreciated, but it plays an important rôle in easing the gradual shift of young to nutritional independence and may be vital for their survival. This pattern is typical of birds with altricial young.

In contrast, birds such as chickens, pheasants, and waders produce precocial young. These young can just as well be raised by one parent as by both. Only a few groups of birds with precocial young, notably geese and swans, rear young as a pair. In most other precocial species males contribute nothing to parental care and females raise young entirely on their own. Even though these young often forage independently immediately after hatching, this type of parental care can be quite demanding for the mother. The small young initially require frequent brooding to keep thermoregulatory expenditure of the small hatchlings within reasonable bounds in an environment in which ambient temperatures are far lower than body temperature (40.5°C in birds). Only exceptionally, roles are reversed and males rear the young alone, for example, in the Jacana and in some waders such as the spotted sandpiper.

Mammals

In addition to other features, mammals are characterized by lactation. The period of lactation is the most energy consuming in the life of a female mammal. It is not well understood why only females produce the taxon-specific milk. The physiological bias in ability to provide for young has often led to the evolution of exclusive female care of young and as a consequence to polygyny as the typical mammalian mating pattern. Monogamous social systems in which the male also contributes to parental care are the exception rather than the rule. Jackals, gibbons, and harvest mice are among the exceptional monogamous mammalian species. Even humans, for which in Western cultures monogamy is the prescribed norm, are mildly polygynous, and in fact the majority of societies that anthropologists have investigated allow polygyny.

In mammals as well as in birds, there are species with altricial and precocial young. Altricial young are typical for most rodents and carnivores, and precocial young are typical for ungulates and caviomorphs, the guinea pig and related species. The former are born naked, with eyes and ears closed, and depend largely on a nest or a burrow for protection from predation and on warming by the mother for thermoregulation. Litters of altricial young are larger than those of precocial young and at least initially grow much faster because the young do not need much metabolic power for activity and thermoregulation. In contrast, precocial young are born fully furred, with eyes and ears open, able to locomote and thermoregulate independently. They often follow the mother immediately after birth and, for example, in the antelopes of the African steppe, largely depend on flight and maternal defense for their safety. Primate young represent an intermediate strategy. Our young are well developed in their sensory abilities but take quite long to develop locomotor abilities to the point at which they can independently follow adults.

Among mammals, particularly the larger species, post-weaning care of young often lasts for long periods. This can take the form of sharing a territory, for example, in black bears, in which offspring may use the mother's territory for extended periods before dispersing. It may also take very elaborate forms in many group-living primates in which daughters inherit the maternal rank and are supported by her in competitive situations. This leads to the establishment of matrilinear support networks in which all the daughters of a mother and their children support each other and thus may come to dominate other females in a group. This structure is realized in many macaques, baboons, and vervet monkeys in which females stay in their natal group, whereas males emigrate shortly before or at maturity.

Parental Investment

Although it can easily be shown that young birds and mammals benefit from parental care since they will not develop properly when deprived of it, it is much more difficult to determine which mechanisms produce the costs of parental care mentioned previously. Of course, there are obvious energetic costs of feeding young by bringing food to the nest as in birds or producing milk as in mammals. However, once the period of parental care is over, parents should be able to recover condition and it seems that they could live just as before the reproductive event. However, it has been shown in many cases that parents caring for young considerably reduce their chances of further reproduction. One possible mechanism might be increased mortality incurred by parents which gather food for dependent young. They have to spend much more of their day in potentially dangerous sites in which they may encounter predators.

In addition to such direct mortality effects, recently a few physiological mechanisms causing a reduction in future reproductive success were demonstrated empirically. One such mechanism potentially diminishing future reproductive output may consist of reduced immune function (Sheldon and Verhulst, 1996). For example, the immune system of zebra finch parents was increasingly weakened the more young they had to feed, and this effect, of course, can make parents more susceptible to parasites or diseases. Reduced resistance to infection or parasites could increase parental mortality or reduce their future fertility or both. Lastly, caring for young now can reduce the fertility of parents in the next round of reproduction because, as shown for blue tits, parental care may interfere with proper molting. Parents caring for offspring late in the year (e.g., when they had a second brood) have to speed up the process of feather renewal and have to do so at a time when food abundance gradually diminishes. This may lead to growth of inferior feathers and in effect produce a poorly insulating plumage. Blue tit parents will then suffer adverse effects throughout the cold winter. They constantly have to spend more energy on thermoregulation than birds in a perfect plumage and this apparently reduced their condition in the next reproductive period.

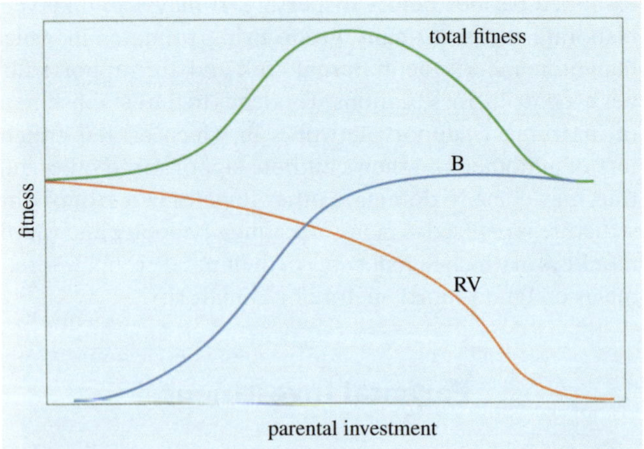

FIGURE 3 The concept of parental investment according to Trivers (1972). The blue curve indicates the benefits (B) and corresponds to the fitness gained by the parent from a given allocation of resources to the present brood. This benefit increases with parental expenditure up to a certain level, above which more expenditure is unlikely to produce more surviving young. The residual reproductive value of the parent (RV_P; red curve) shows how parental fitness through future reproductive episodes might decrease with increasing expenditure on the present brood. The green curve shows the expected total fitness expected for the parent as the sum of present and expected future reproduction for any given amount of parental expenditure now.

As mentioned previously, these trade-offs between reproductive effort now and a reduction in future reproduction have been called *parental investment* (Trivers, 1972). Figure 3 gives a graphical impression of this argument. Parents face the problem of producing a high gain in offspring fitness through parental care expended on the present brood if possible at a small loss to the parents' ability for future reproduction. The greatest net fitness gain is achieved when reproductive success through the present brood plus the expected future reproductive output is maximal. A cost of reproduction seems unavoidable given that resources to parents are not unlimited and that even when resources are unlimited, parents do not have an unlimited physiological capacity to process resources or to increase energy output for parental care. This will lead to trade-offs in which more allocation of resources to one function such as parental care behavior will necessarily lead to a reduction in another function such as growth and maintenance of the parental body or its immunological defense against parasites and diseases.

Conflict about Parental Care

Given this kind of trade-off involved in parental investment, one might expect conflict about the valuable resource "parental care." Theory indeed predicts such conflict between partners in a pair, between parent and offspring, and among offspring. How do such conflicts appear in real life and how do they influence parental care?

Conflict within Pairs

When we observe a pair of starlings feeding young, it is impressive to see how frequently both partners appear at the nest and pass whatever food they have collected to the gaping beak of a nestling. Both seem to work as hard as they can, particularly when the young are older and demand much food to maintain their metabolism and allow growth. However, theoretical considerations and experimental work demonstrate that parent birds feed their young less when working as a pair than when alone. When a starling male is slightly handicapped by gluing a piece of lead to its tail feathers, it reduces its contribution to the feeding of young. The female then increases her feeding frequency but not enough to make up for the lost contribution of the male. Apparently, she did not work as hard as possible when her mate was still fully able to work. This increase in work is even more obviously when a bird is widowed. The widowed bird quickly almost doubles its work to provide the young in the nest with sufficient food. Since even young in a nest in which both parents feed are generally not growing at the maximal rate, this suggests that in normal circumstances parents do not work as hard as possible to achieve the maximal growth rate of their young. Rather, every partner works only as much as necessary to keep the brood growing at an average rate. This is the outcome of an evolutionary stable strategy (EES) that protects a partner of a pair from exploitation by the other pair member. We can understand why parents limit their provisioning effort by remembering that parental care is costly: Both parents are under selection to restrain their investment in the present brood in order not to jeopardize future reproductive success. If one parent would respond to less work by the other by stepping up its own workload, it could be exploited by the partner, which in the extreme case could withdraw altogether from care for the current brood and attempt to find a partner elsewhere for an additional brood. The ESS, therefore, is simplistically said to work at a level slightly below the optimum for the growth of the brood and to replace only part of a reduced feeding effort of the partner to force him or her to keep working.

That birds indeed have the decision machinery to regulate their chick feeding effort depending on the fitness gains expected from a brood is perhaps best exemplified by the study of Davies (1992) on the dunnock. In dunnocks, males and females establish territories independent of the other sex in spring. Later, males try to adjust their territory boundaries to those of females in such a way that their territories encompass more than one female territory. Polygynous males realize the highest reproductive success. Females resist these male attempts because they achieve the highest reproductive success when paired with two males since broods fed by trios of two males and a female have highest fledging success. Females attempt to establish a situation with more than one male (polyandry), whereas males defend their females against other males and guard them aggressively, particularly during the egg-laying period. Males regulate their contribution to the feeding effort at the nest in relation to the mating access they had to a female during her fertile egg-laying period. If one of the males is kept away from the female during her fertile period, he withholds all parental care. He may even try to destroy the clutch to force a new breeding attempt. This exemplifies that birds indeed possess the mechanism to fine-tune the amount of parental care provided and shows how strong the conflict between the sexes about parental care may become within a family group.

Conflict within a Brood

Similar conflicts of interest occur in many species between siblings within a brood. The phenomenon has been described in many taxa and ranges from mild food competition among siblings to cannibalism as observed, for example, in some anuran tadpoles in which a large cannibalistic morph may develop and can even feed on less fortunate smaller siblings, although it prefers to feed on unrelated tadpoles.

An individual offspring that can monopolize parental care or garner most of this precious resource will grow best and have the greatest chances of successful recruitment to the breeding population. Normally, conflicts among close relatives are limited since individuals share many genes. Genes for behavioral strategies that reduce the survival chances of a close relative are therefore likely to reduce their own spread (Hamilton, 1964). However, despite this tempering influence of kinship, intense fighting of nestlings has been reported for many birds of prey, boobies, and egrets and also for mammals such as pigs, Galápagos fur seals, and hyenas (Mock and Parker, 1997). Piglets were actually shown to have evolved specialized weapons, functional canines and canine-like incisors, which erupt at birth, point sideways, and are used to fight for possession of the most milk-producing anterior teats of the mother sow.

Sibling conflict has been investigated most intensively in birds. In some eagle species, the clutch consists of two eggs, but the first hatched chick almost invariably kills its later hatched sibling by direct attack and by excluding it during feedings. Interestingly, this conflict is most intense shortly after hatching when food is not really limiting growth of the young eaglet. In these cases, the second egg in the clutch appears to serve exclusively as insurance against hatching failure of the first egg. Presumably, successful recruitment of one superquality young is more likely in these long-lived, slowly maturing species than would be recruitment of even one of two slight lower quality young. In species in which sibling aggression is not as intense, such as egrets, food abundance, as perceived by the young through the feeding frequency of the parents, greatly influences the outcome of sibling rivalry. Last hatched egrets are more resistant to attacks by older siblings when food is plentiful because the oldest sibling will be sufficiently satiated to allow more frequent uncontested feeding of the younger siblings. The feeding technique of the parents strongly influences the intensity of sibling rivalry. This can be seen in a comparison of the common egret and the great blue heron. In egrets, which feed boluses of small fish, larger siblings can reach up to the parent's bill and take the food directly, thus monopolizing the food and depriving younger siblings of access to it. In contrast, great blue herons feed on larger fish and regurgitate the prey into the bottom of the nest where young must try to pry off smaller pieces. This situation makes effective monopolization of a parental feeding impossible and food competition in great blue heron nests is much less intense than in egret nests (Mock and Parker, 1997.

Parent–Offspring Conflict

Strong sibling conflict seems to be counter to the interest of the parent. A parent would presumably benefit more from raising a large brood than the individual offspring in the brood and may therefore be expected to interfere with the fighting among the chicks. However, observations have failed to produce much evidence of parental interference in even lethal fights among offspring.

However, there is a much less dramatic context in everyday parental care for which parent–offspring conflict has been postulated. This concerns the interaction between feeding parents and begging young (Kilner and Johnstone, 1997). When starling parents feed young at the nest, young try to get the closest position to the parent, beg loudly, stretch their necks to gape closest to the parent, and widely open their mouths to make themselves maximally conspicuous. Special bright yellow markings around the edge of the mouth make this signaling additionally obvious. It is not entirely clear what is happening: Are the offspring signaling their need to the parent or is every individual in the nest trying to get more than its just share of the food brought by the parents? Theoretically, there could be conflict between parent and offspring about the amount of parental care expended on a given offspring (Fig. 4). This follows because parents are under selection to care equally for all their offspring — as long as the offspring are in similar condition — since they all are equally related to the parents. In contrast, a given offspring will value parental care expended on itself more highly than the same parental care expended on a full sibling since it shares only half its genes by common descent with the latter but of course can potentially pass on all of its own genes to the next generation. On the other hand, parent and offspring value the cost that a given amount of

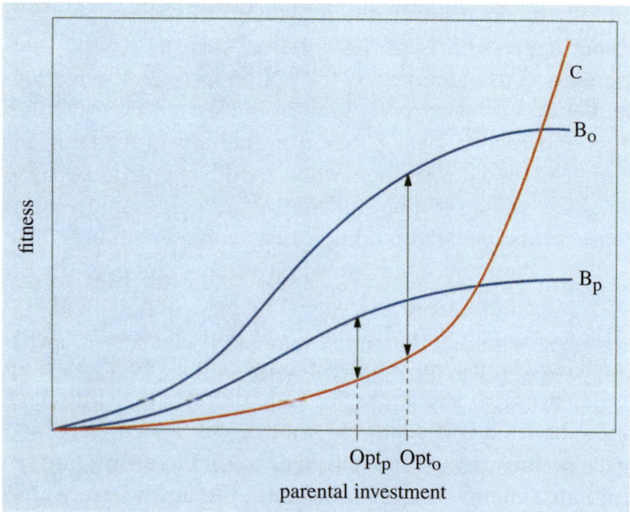

FIGURE 4 Parent–offspring conflict as modeled by Trivers (1974). Parents (B_p) and offspring (B_o) value the benefit to the offspring differently since the parent is equally related to all its offspring (with which it shares half its genes), whereas a single offspring has the opportunity to propagate all its genes. The cost to the parent (C) of a given parental expenditure can be measured in reduced future reproductive output. This has the same importance for parent and offspring since in a monogamous situation, both share half the genes of future siblings of the present offspring. As a consequence of this situation, the optimal parental expenditure (Opt_p), corresponding to the maximum difference between benefit and cost and hence to the maximum net fitness gain for the parent, is lower than that for the offspring (Opt_o).

parental investment inflicts on the parent equally because it implies reduced reproductive output of the parent, and both parent and offspring (in a monogamous situation) would value future offspring alike since both share half of their genes with future offspring of the parent. Due to this asymmetry in the selection on parent and offspring to value the benefits of parental care, Trivers (1974) postulated the existence of parent–offspring conflict (Fig. 4).

Alternatively, it has been argued that parents do not have all the necessary information about the state of the young. They can see and estimate an offspring's size and perhaps its condition, but it is much more difficult to determine the actual hunger state of a given young. Therefore, are we observing parent–offspring communication when observing feeding at a starling nest or are we witnessing parent–offspring conflict in which every offspring is trying to get more than its just share and parents have to try and counteract the offsprings' ploy? Many cases of presumed parent–offspring conflict have been investigated, but conclusions have been equivocal. There is good empirical support for the conclusion that begging indicates offspring need and that it influences parental provisioning. Apparently, the cost of begging to the offspring ensures that it does not pay to cheat this signal so that it is a reliable indicator of, for example, the hunger of the offspring. However, since young use multiple signals (acoustic, optic, proximity to the parent, how high it stretches its neck, etc.) to indicate their desire to be fed, it is not clear how actual behavior relates to the relatively simplistic theoretical models currently available (Mock and Parker, 1997).

It has been suggested that parent–offspring conflict is difficult to observe since parents can force their decisions on young simply because the parent is in a position of power relative to the young. However, detailed observations, for example, on egrets and on blue-footed boobies have provided no evidence that parents actually interfere in sibling conflict up to siblicide (Mock and Parker, 1997). In such cases, interference would be expected if parent and offspring conflicted about the threshold of food provisioning below which growth rate of young in the brood decreases so much that a younger sibling should be left to starve in order to allow successful raising of the older siblings. If parent and offspring agree on brood reduction, scramble competition could be allowed to weed out the weakest young.

Much of the observable behavioral conflict may perhaps better be interpreted as communication between parent and offspring about the state of the other partner in this interaction and the state of the environment (Bateson, 1994; Kilner and Johnstone, 1997). Consider, for example, time to weaning. When food is abundant, mammalian mothers in prime condition, for example, in vervet monkeys, may indeed rear their young quickly at near maximal growth rates and wean them early without much observable conflict between mother and offspring. When food is less abundant or maternal condition lower (e.g., due to old age or parasite infestation), mothers paradoxically wean their young much later. This appears largely to be caused by reduced rates of energy and nutrient transfer from mother to young so that young develop more slowly toward independence. Mothers in very poor condition or in environments with very little food wean their young earlier. They may actually terminate parental care at a stage in which young have hardly any chance of surviving without maternal care. These observations suggest that mothers and young adjust their interactions to the state of the mother and of the environment as long as this is possible within the environmental constraints. Nevertheless, there is an obvious conflict between the interest of the mother and her offspring, most blatantly if she terminates care of a still dependent offspring. In this case, a vervet monkey mother avoids excessive costs of parental investment, which may impose reduced fertility or even increased mortality risk on her, at the cost of losing an infant.

When the states of mother and offspring vary unpredictably with changes in the environment, communication about the states of parent and offspring should be honest but costly to avoid the risk of cheating. Otherwise, an offspring well able to support itself may cheat its parent by pretending to be very small and dependent — to sneak a little more milk or protection from its mother, thus enabling it to grow faster. Thus, mothers should force their offspring to signal clearly that they still need parental care, for instance, by rejecting them. This can be observed particularly obviously in the Galápagos fur seal, in which mothers often abandon a newborn if the previous young, by now a yearling, is still too small to survive on its own. In these cases, however, mothers usually behave very aggressively against their yearlings, threatening and biting them and chasing them away from the newborn (Fig. 5). Only yearlings less than approximately 14 kg (half the adult female weight) will endure this aggression and fight persistently for access to the mother's milk, thus communicating their urgent need for continued maternal care. Larger yearlings usually give up in the face of this maternal aggression and wean instantly. In these cases, the yearling apparently correctly signals its need for maternal support to the mother since small yearlings are not able to support themselves by independent foraging. At low body mass, they are poor divers and cannot forage efficiently enough to maintain themselves and ensure further growth (Horning and Trillmich, 1997).

Helping

Surprisingly, in some species nonparent individuals help rear offspring that is not their own. This phenomenon is called *helping* and includes behaviors such as feeding or protecting young by individuals that are not the parents of these young. If there is so much conflict about brood care, why do some individuals expend energy on raising another individual's young? Again, birds offer the best examples of this phenomenon (Brown, 1987). There are two main explanations. One that applies, for example, to the Florida scrub jay, has to do with very limited breeding habitat. Florida scrub jays breed exclusively in scrubland. All usable habitat is occupied by the population. Thus, young birds do not usually have the option to disperse, settle on their own territory, and breed there. Rather, they have to stay at home and wait for an opportunity to take over a neighboring territory or part of the home territory. While they are waiting for a vacancy, they can gain fitness by helping their parents. If, by feeding siblings or warning against predators, helpers enable parents to raise larger broods or more broods than the parents could rear if unaided, the helpers gain so-called "indirect fitness" (i.e., fitness that is gained not by own reproduction but through increasing the reproductive output of relatives — kin altruism; Hamilton, 1964). However, whenever these helpers are given a choice between own breeding and helping, they desert the parents and breed on their own. This became particularly obvious when a disease reduced the scrub jay population dramatically. After the population crash, no helping birds were observed and all previously helping individuals that survived were found settled and breeding on vacated territories. The major cause of helping of young Florida scrub jays is thus ecological and lies in the limitation of suitable breeding sites.

FIGURE 5 Parent–offspring and sibling conflict in the Galápagos fur seal (*Arctocephalus galapagoensis*). The mother defends her newborn pup against its yearling sibling, which also attempts to obtain maternal milk.

Another reason for helping is the lack of a suitable breeding partner. This occurs quite frequently in the pied kingfisher, in which females incubate the clutch longer than do males and do all the incubation during the night. They are therefore much more exposed to predation than males and thus become the rarer sex. Here, sons of a pair that cannot find a partner may act as helpers for the next breeding attempt of their parents, thus again gaining indirect fitness. Surprisingly, even totally unrelated males, called secondary helpers, may help in brood raising in this species. Because they have no genetic benefits from raising foreign young, this behavior is particularly difficult to understand. It was determined that these males are attempting to gain access to the female they are helping so that they can breed with her later. Apparently, the breeding male senses this threat to his status and he aggressively rejects unrelated helpers unless the brood is too large to be reared without additional help. On the other hand, they work much less in feeding young than sons of pairs. Obviously, the secondary helpers have a mechanism that keeps them from expending more than the absolutely necessary minimum of parental care on these young. They only feed enough to be tolerated by the female's mate but prefer to feed the female (a courtship behavior) rather than the young.

Thus, in helping, animals realize individual fitness advantages through contributing to the reproduction of relatives (kin selection) and/or through improving their chances for later own reproduction if current conditions (because of low chances of dispersal or lack of suitable breeding partners) are not permissive to breeding.

Parasitism of Brood Care

Given that brood care is such a valuable resource, it comes as no surprise that some animals exploit the brood care of others. In birds, such brood parasites depend on the ability to get their eggs accepted by hosts. This happens frequently within a species. For example, approximately 15% of all starling nests contain eggs that were not laid by the female that incubates and later feeds at that nest box but by other starling females that managed to place their eggs unnoticed into the nest. They can thus exploit the brood care of another pair to increase their fitness.

Truly parasitic species are found throughout the world. They include cuckoos, cowbirds, honey guides, and certain weaverbirds. These species, such as the European cuckoo, exploit the brood care of other species. They have to solve the problem of getting the egg accepted by the host species since eggs of parasite and host look different. This strategy either employs egg mimicry by the parasite or it can only function as long as the host species has not evolved egg discrimination which enables the host to reject parasitic eggs. If this hurdle is taken, the hatching cuckoo throws its step siblings out of the nest and can exploit the parental responses of its host parents to strong begging stimuli. Indeed, the young cuckoo begs so intensely that it appears to mimic a whole brood. This induces all forms of care, warming, feeding, and protection against predators in the fooled parents.

In other species parasitism has become tempered in its ill effects on the host. The parasite does not destroy the host's brood but participates in the care provided to the brood. However, this creates another problem: Parents can compare the parasitic young with own young and the parasite has to compete in this situation for parental care with the host's own young as, for example, in African widow birds that parasitize estrildid finches. Nestlings of these finches are noted for a wide variety of mouth markings and the parents will feed only nestlings that display the species-specific markings. Unlike the cuckoo, widow bird hatchlings do not remove the host's young from the nest. Instead, they copy the signals of the host's young so exactly that they are accepted as own young by the host parent. Selection for similarity (i.e., close mimicry) with the nestlings of the host has led to incredibly detailed copies of visual signals of the bill and gape. For example, the young of the paradise widow (*Steganura pardisea*) look so similar to the young of the melba finch (*Pytilia melba*) that it took ornithologists quite some time to notice that not all young in the nests of the latter species actually belonged to that species. Over evolutionary time, the parasitic species has closely copied the signal of the host to avoid parental discrimination during feeding of begging young. This again attests to the importance mechanisms of parent–offspring recognition. Indeed, in the widow birds (Viduinae), detailed mimicry of the respective host species has led to parallel evolution and speciation in brood parasite and host so that every host has one specific species of widow bird as parasite (Nicolai, 1974).

Eusociality

Eusociality in insects (particularly well-known from ants, wasps, and bees) exemplifies the pinnacle of altruism (Bourke, 1997). Here, sterile castes of females have evolved which help their mothers to raise siblings. In these species, evolution has thus led to the complete loss of own reproduction because the benefits reaped through helping are so major that individuals can afford to forsake production of own offspring. In essence, there is parental care which is never performed by the parent. The evolution of brood care in these societies is such a wide field that we can do no more than point the interested reader to, for example, the marvelous book of Wilson (1971).

References Cited

Bateson, P. (1994). The dynamics of parent–offspring relationships in mammals. *Trends Ecol. Evol.* **9,** 399–403

Baylis, J. R. (1981). The evolution of parental care in fishes, with reference to Darwin's rule of male sexual selection. *Environ. Biol. Fishes* **6,** 223–251

Bourke, A. F. G. (1997). Sociality and kin selection in insects. In *Behavioural Ecology: An Evolutionary Approach* (J. R. Krebs and N. B. Davies, Eds.), 4th ed., pp. 203–227. Blackwell, Oxford.

Brown, J. L. (1987). *Helping and Communal Breeding in Birds.* Princeton Univ. Press, Princeton, NJ.

Clutton-Brock, T. H. (1991). *The Evolution of Parental Care.* Princeton Univ. Press, Princeton, NJ.

Davies, N. B. (1992). *Dunnock Behaviour and Social Evolution.* Oxford Univ. Press, Oxford.

Diesel, R., and Schuh, M. (1993). Maternal care in the bromeliad crab *Metopaulias depressus* (Decapoda): Maintaining oxygen, pH and calcium levels optimal for the larvae. *Behav. Ecol. Sociobiol.* **32,** 11–15.

Diesel, R., Bäurle, G., and Vogel, P. (1995). Cave breeding and froglet transport: A novel pattern of anuran brood care in the Jamaican frog, *Eleutherodactylus cundalli. Copeia* **1995,** 354–360.

Hamilton, D. W. (1964). The genetical evolution of social behaviour I and II. *J. Theor. Biol.* **7,** 1–52.

Horning, M., and Trillmich, F. (1997). Ontogeny of diving behaviour in the Galápagos fur seal. *Behaviour* **134,** 1211–1257.

Kilner, R., and Johnstone, R. A. (1997). Begging the question: Are offspring solicitation behaviours signals of need? *Trends Ecol. Evol.* **12,** 11–15.

MOCK, D. W., and PARKER, G. A. (1997). *The Evolution of Sibling Rivalry.* Oxford Univ. Press, Oxford.

NICOLAI, J. (1974). Mimicry in parasitic birds. *Sci. Am.* **231,** 92–98.

ROSENBLATT, J. S., and SNOWDON, C. T. (Eds.) (1996). *Parental Care. Evolution, Mechanisms, and Adaptive Significance. Advances in the Study of Behavior,* Vol. 25. Academic Press, New York.

SHELDON, B. C., and VERHULST, S. (1996). Ecological immunology: Costly parasite defences and trade-offs in evolutionary ecology. *Trends Ecol. Evol.* **11,** 317–321.

TRILLMICH, F. (1981). Mutual mother–pup recognition in Galápagos fur seals and sea lions: Cues used and functional significance. *Behaviour* **78,** 21–42.

TRIVERS, R. L. (1972). Parental investment and sexual selection. In *Sexual Selection and the Descent of Man 1871–1971* (B. Campbell, Ed.), pp. 136–179. Aldine, Chicago.

TRIVERS, R. L. (1974). Parent offspring conflict. *Am. Zool.* **14,** 249–265.

TYLER, M. J. (Ed.) (1983). *The Gastric Brooding Frog.* Croom Helm, Kent, UK.

WILSON, E. O. (1971). *The Insect Societies.* Belknap, Cambridge, MA.

Frans B. M. de Waal
Filippo Aureli

*Yerkes Regional Primate Research Center and Department of Psychology
Emory University, Atlanta, Georgia 30322, USA*

Conflict Resolution in Primates

Research on nonhuman primates has produced compelling evidence for reconciliation and consolation —that is, postconflict contacts that serve to respectively repair social relationships and comfort distressed individuals, such as victims of attack. These discoveries have led to a theoretical framework within which conflict and conflict resolution form an integrated part of social relationships. Rather than view aggression as an antisocial or destructive "instinct," it is viewed as one of several tools available to negotiate social relationships. Implications of this new model of social conflict are discussed along with evidence for behavioral flexibility, the value of cooperation, and the possibility that distress alleviation rests on empathy, a capacity that may be present in chimpanzees but not in most other nonhuman animals.

Introduction

The social life of monkeys and apes is filled with activities from sex to play and from grooming to squabbling. Most conflicts are minor, but sometimes major fights break out in which individuals are injured or even killed. Instead of being attributable to some controversial and vague concept such as an aggression "instinct" or "drive," primatologists increasingly view aggression as a concomitant of the conflict that is inevitable when members of the same species, with their overlapping interests in mates and food, live together (Fig. 1). Groups would not exist if individuals did not gain from associating with each other, but this by no means eliminates the sources of competition. Individuals continue competing over resources while at the same time relying on each other for benefits such as antipredator defense and the finding of food.

For such an arrangement to work, conflict needs to be regulated in such a way that it does not undermine social relationships. Thus, aggressive behavior is subject to strong constraints imposed by the need to preserve cooperation. Despite the continuing popularity of the struggle-for-life metaphor, it is increasingly recognized that there are important costs associated with open competition and hence that there are sound evolutionary reasons for curbing competition. In the 1970s, social factors did not enter the equations of evolutionary biologists trying to explain the rarity of lethal violence. They thought in terms of the aggressor's risk of injury if the recipient were to defend itself. Currently, however, it is believed that the value of cooperation is at

FIGURE 1 Aggression in chimpanzees is usually inhibited and ritualized. Here, an adult male (left) attacks a female with hands and feet but does not use his dangerous canine teeth, which males sometimes use when fighting with other males.

FIGURE 2 A male bonobo (right) grooms a female. Grooming behavior in nonhuman primates originally served a hygienic function (i.e., for removal of dirt and ectoparasites) but has acquired an important social function as well. The act serves to establish and maintain relationships, hence fostering cooperation.

least as important a constraint on competition. The dependency of social animals on community life in general, and on the need for support and help from certain individuals in particular, makes aggression a socially costly strategy. The basic dilemma facing many animals, including humans, is that they sometimes cannot win a fight without losing a friend.

The Need to Preserve Relationships

From an evolutionary perspective, benefits of cooperation are most evident in relationships with a reproductive function, such as male–female and mother–offspring relationships. Relationships also produce benefits not immediately related to reproduction, however, such as when two individuals protect one another against attack, tolerate one another around resources, provide vigilance against predators, or cooperate during intra- or intergroup competition. It is of course assumed here that cooperative tendencies evolved because they paid off in the long run in terms of survival and reproduction.

Kummer (1978) defined the benefits that individual A provides to B as A's value to B. Any individual will try to improve this value: B will select the best available A, predict A's behavior, and modify A's behavior to its own advantage. In other words, B will invest in the relationship with A. Although most of B's investments may not lead to quick profits, such as immediately useful actions by A, they may help cultivate a relationship that is beneficial to both A and B over the long term. A good example of such investment is social grooming (Fig. 2). One primate may groom another for more than 1 h without any immediate return favor; after the session the two will simply part company, each going its own way. There are indications that grooming is altruistic in that it entails costs for the groomer, such as reduced time available for other activities and reduced attention to potential danger, whereas it provides benefits to the groomee in terms of hygiene and a calming effect. Why would one individual provide services to another (grooming is one of the most common activities in primate groups) if not to foster future beneficial exchanges?

What kind of exchange could this be? The most pervasive and effective cooperation within primate groups is the formation of alliances: Two or more individuals band together against a third. For example, a male attacks a juvenile and the mother rushes to the juvenile's aid at considerable risk to herself, or two males together overthrow the reigning alpha male, after which one of the victorious males becomes the new alpha (Fig. 3). These kinds of cooperation are critically important to primates: Their social rank, and sometimes their life, depends on it. Consequently, they need to get along with others, even with their competitors. Maintenance of valuable relationships despite occasional conflicts of interest is a critical requirement of group life.

The best known behavioral mechanisms that allows primates to repair social damage caused by hostilities is *reconciliation,* which is defined as a friendly reunion between former opponents: The reunion supposedly serves to return the relationship to normal levels of tolerance and cooperation. Human–animal relationships provide a helpful illustration. Köhler (1925) saw a need for "forgiveness" in the behavior of a juvenile chimpanzee, whose behavior he described as follows:

> The little creature, which I had punished for the first time, shrank back, uttered one or two heart-broken wails, as she stared at me horror-struck, while her lips

FIGURE 3 The most common form of cooperation within primate groups is the formation of alliances. The chimpanzee on the right is the alpha male of a large captive colony, but he would never have obtained this position without the support of the older male on the left. The alpha male has many reasons to groom this ally and repair their relationship as soon as conflict arises between them.

> were pouted more than ever. The next moment she had flung her arms round my neck, quite beside herself, and was only comforted by degrees, when I stroked her. (p. 261)

Apes seem to perceive a conflict with their caretakers as a threat to the relationship and try to control the damage with affectionate behavior. Although the environment of hand-reared apes is obviously unnatural, human influence does not appear to explain the phenomenon. In the next section, I review the current knowledge of conciliatory behavior, and other postconflict interactions, in chimpanzees and other nonhuman primates.

Reconciliation Behavior

Initially, the main aim of reconciliation research was to compare two alternative expectations concerning the effect of aggressive behavior on social relations:

Dispersal hypothesis: Losers of aggressive incidents are expected to avoid winners. The traditional notion of aggression as a spacing mechanism was based on experience with territorial species and on the idea that animals use aggression in order to maintain optimal "individual distances." This notion predicted reduced contact following aggression.

Reconciliation hypothesis: Individuals are expected to "undo" the damage that aggression inflicts on valuable social relationships. Such a tendency would be expressed in (i) increased contact following aggression and (ii) special reassuring and appeasing gestures during these contacts.

In support of the second hypothesis, de Waal and Van Roosmalen (1979) were the first to demonstrate that aggression leads to increased contact. Former opponents in the chimpanzee colony of Arnhem Zoo (The Netherlands) were found more often within 2 m of one another after rather than before a conflict. Moreover, the chimpanzees engaged in intensive body contacts following conflict and preferred such contact with the former opponent rather than with individuals that had not been involved in the conflict: Interopponent contacts constituted 30% of all postconflict contacts compared with a random expectation of 5.6%. These results demonstrated a pronounced conciliatory tendency in the chimpanzee, later confirmed by reports from the wild.

Although peacemaking strategies in the chimpanzee may not be surprising, few people would have predicted that an intolerant primate, such as the rhesus macaque, would follow similar strategies. Introducing a new, controlled methodology that compared contact rates following aggression with those during matched baseline periods, de Waal and Yoshihara's (1983) study of captive rhesus macaques produced evidence for reconciliation among these monkeys. Similar results were subsequently reported for a variety of other macaque species as well as for gorillas, bonobos, golden monkeys, vervets, mangabeys, baboons, and so on. Most of these studies were conducted in captivity, but reconciliation has also been demonstrated in wild primates. In most species studied to date, affiliative contact between former opponents is more likely during the first few minutes after a conflict, and postconflict attractiveness is selective — that is, the contact increase does not occur indiscriminately with all possible partners but specifically between recent opponents.

According to both the data on captive chimpanzees and descriptions of chimpanzees in the wild, reconciliation in this species involves distinct behavior patterns, such as embracing, gentle touching, and mouth-to-mouth kissing. In fact, chimpanzee opponents kiss one another 10 times more often during the first postconflict contact than during subsequent contacts. Not all primates show a similar behavioral distinction between first postconflict and other contacts, but stump-tailed macaques use another conspicuous behavior pattern rarely observed outside the reconciliation context — the "hold-bottom ritual" in which one individual, usually the recipient of aggression, presents the hindquarters, and the other clasps the presenter's haunches (Fig. 4). This ritual occurs in more than 30% of first postconflict contacts, which is a 20-fold increase compared to control contacts.

The kiss of chimpanzees and the hold-bottom ritual of stump-tailed macaques require high levels of intimacy and coordination. They assist the reconciliation process by making both the context and the meaning of the contact more explicit. This is in contrast to the implicit reconciliations of some other species, such as the rhesus macaque, which of-

FIGURE 4 Stump-tailed macaques (*Macaca arctoides*) reconcile with a typical hold-bottom ritual in which the subordinate (right) presents its hindquarters and the dominant holds its hips.

ten reestablish relationships through a brief, inconspicuous brushing contact that is, however, quite meaningful given the risks involved in mere proximity in this short-tempered species.

Reconciliation serves as a heuristic label for these contacts: It generates ideas about their function, such as that they repair social relationships or reduce social tension. The validity of this label has been confirmed by recent studies. Cords (1992), for example, carried out experiments indicating that reconciliation restores tolerance between former opponents. Following an aggressive incident between macaques, the monkeys were presented with two drinking nipples, side by side, from which they could obtain a sweet drink. The monkeys drank together more readily if reconciliation had taken place than without reconciliation. Observational studies of both macaques and chimpanzees confirm that reconciliation reduces the chances of further aggression: The frequency of renewed attack is reduced following reconciliation. These results support the hypothesis that reconciliation restores the relationship between former opponents.

One way of measuring whether reconciliation reduces social tension is to pay attention to self-scratching, a known correlate of stress and anxiety. Aureli and Van Schaik (1991) found a dramatically increased frequency of self-scratching in recent recipients of aggression. Following reconciliation, however, these signs of anxiety quickly returned to baseline level. Evidence for an anxiety-reducing function of reconciliation also derives from playback experiments in free-ranging baboons. Baboons responded less strongly to the sound of a threat scream by a former aggressor with which they had reconciled than by one with which they had failed to reconcile (Cheney *et al.,* 1995).

Both macaques and chimpanzees follow what seems a general rule among primates: Reconciliation aims to restore the most valuable relationships. In macaques, in which matrilineal kin relationships are particularly valuable, related individuals reconcile conflicts more often than unrelated individuals (Aureli *et al.,* 1997). In the chimpanzee, in which males form stronger intrasexual bonds than females, male conflicts are more often reconciled than female conflicts.

An experimental investigation of relationship value was carried out by Cords and Thurnheer (1993), who trained pairs of macaques to cooperate during feeding, thus enhancing their interdependency. Reconciliation following conflict increased dramatically once a partner had become useful for obtaining food, thus confirming the idea that reconciliation occurs in proportion to the value of the relationship.

Social Learning

Interspecific variability in aggressiveness and peacemaking tendency makes it possible to expose members of a given species to a social environment with dramatically different rates of these behaviors. This can be done by housing them with another species. If reconciliation is a learned social skill, one expects such a manipulation to affect postconflict behavior. This prediction was examined by de Waal and Johanowicz (1993), who exposed rhesus monkeys — a rather intolerant species with low levels of reconciliation — to a highly conciliatory species, the stump-tailed macaque. Both species belong to the genus *Macaca.*

Juveniles were housed in mixed-species groups of seven monkeys each for a period of 5 months. Following this period, they were observed for 6 weeks in groups of conspecifics only. Control rhesus monkeys, matched in age and sex to the experimental subjects, went through the same procedure without contact with another species (i.e., in all-rhesus groups). Initially, individuals of different species lived somewhat separate lives, sleeping and huddling in separate subgroups, but by the end of the 5-month period they were fully integrated and highly tolerant of each other.

Compared to the control experiment, the main result of the manipulation was a three- or fourfold increase in the proportion of fights followed by reconciliation. Rhesus monkeys that had lived with the kinder, gentler species apparently had learned peacemaking skills. This difference emerged gradually during the cohousing phase but was sustained following removal of the "tutor" species (Fig. 5). This result, which shows that reconciliation can be modified through environmental manipulation, has important implications. It shows that primate peacemaking is not some inflexible, instinctive pattern but subject to learning. This is probably even more true for our own species.

Complexity and Cognition

Like all macaques, pig-tailed macaques form matrilineal societies in which female kin associate with and support one

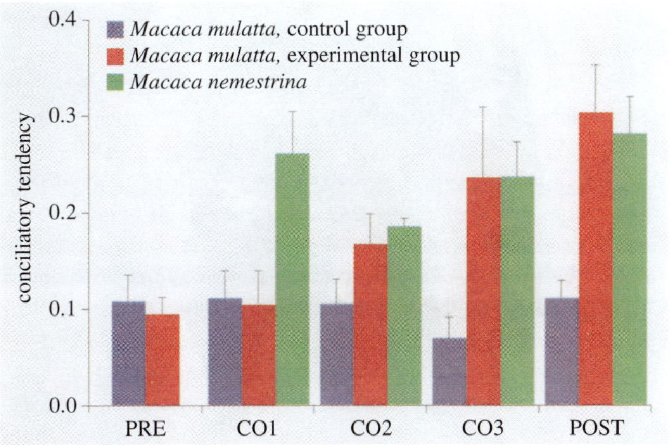

FIGURE 5 Mean proportion of aggressive encounters followed within 3 min by a reconciliation. The data are for a control group of rhesus monkeys (*Macaca mulatta*) housed with conspecifics only (in blue), a group of monkeys used as experimental subjects and housed with stump-tail monkeys (*Macaca arctoides*), another species of macaque (in red), and finally a third group in which the stump-tail monkeys acted as "tutors" for the experimental rhesus subjects (in green). The conciliatory tendency of the rhesus subjects was similar to that of the controls before cohousing (PRE) and during the first cohousing phase (CO1). It steadily increased during continued cohousing (CO2 and CO3) and remained above control levels even after the rhesus subjects had been separated from the tutor species (POST).

another. Apart from direct reconciliations between former combatants, Judge (1991) found that relatives of the victim tend to seek contact with the aggressor. For example, a mother may approach and groom the attacker of her daughter in what appears to be a reconciliation on her offspring's behalf. If such triadic reconciliation protects the victim's matriline against further hostilities, then all of its members benefit, including the individual that contacted the aggressor.

If monkeys sometimes reconcile for their matriline, the same mechanism could operate between groups. Intergroup relations tend to be hostile, but on many occasions adult females of different free-ranging groups have been observed to come together for grooming. Several such contacts took place shortly after intergroup fights and involved the alpha females of both groups. These contacts may serve to reestablish peace between groups.

The complexity of peacemaking interactions is even greater in chimpanzees. In this species, individuals act as if they know the meaning of reconciliation not only for themselves but also for others. Typical examples of the cognitive complexity of peacemaking in the Arnhem chimpanzee colony are presented in the following sections.

Third-Party Mediation

If, after a fight between them, two male rivals stayed in prolonged proximity without engaging in an actual physical reunion (an apparent deadlock situation), an adult female might initiate a grooming contact with one of the two. After several minutes of grooming, she would slowly walk to the other male, often followed by her grooming partner. If he failed to follow she might return to tug at his arm. After the three individuals had been together for awhile, with the female in the middle, she would then get up and stroll away leaving the males alone.

Deception

On six occasions a dominant female that had been unable to catch a fleeing opponent was observed to approach this individual some time afterwards with a friendly appearance, holding out her hand, only to change her behavior when the other came within reach. Reasons to regard the subsequent attack as the female's real intention are its timing (very sudden and without warning signals), the fact that all instances concerned victims capable of outrunning the aggressor, and the intensity of the punishment.

Opportunistic Reconciliation

Reconciliation may occur in a hurried fashion if continuation of the fight would harm the interests of both individuals. For example, in the years that the Arnhem colony was ruled by a coalition of Nikkie and Yeroen, the alpha male, Nikkie, could get in serious trouble during prolonged conflicts with his partner. The third male would begin an intimidation display, initially terrorizing the females and juveniles but later displaying increasingly closer to the two quarreling males. Nikkie was never observed to control the third male on his own. He would first approach his opponent, Yeroen, with a large grin, seeking an embrace. Only after reestablishment of contact with his partner would Nikkie go over to the third male to subdue him.

Control Role

In many primates high-ranking males break up fights. This so-called *control role* is perhaps best developed in chimpanzees. Males that adopt this role move from the usual support of winners in fights to loser support and dissociate their intervention tendencies from individual preferences (Fig. 6). That is, they are the only group members to intervene in fights impartially. They may move with all hair on end between two combatants until they stop screaming, scatter them with a charging display, or literally pry locked fighters apart with both hands. Their main objective seems to be to put an end to the hostilities rather than to support one party or the other. The following description illustrates how Luit, within weeks of attaining alpha status in the Arnhem colony, adopted the role (de Waal, 1982):

> On one occasion, a quarrel between Mama and Spin got out of hand and ended in biting and fighting. Numerous apes rushed up to the two warring females and joined in the fray. A huge knot of fighting, screaming

FIGURE 6 Mediation by an adult male chimpanzee (center), who came over as soon as he heard two females (extreme left and right) scream in a dispute over food. The male remained with stretched arms between them until the fight stopped.

FIGURE 7 Consolation in chimpanzees. A juvenile embraces a screaming adult male that has lost a confrontation with a rival.

> apes rolled around in the sand, until Luit leapt in and literally beat them apart. He did not choose sides in the conflict, like the others; instead anyone who continued to fight received a blow from him. I had never seen him act so impressively before. (p. 124)

Distress Alleviation and Empathy

Postconflict contact between former opponents should be distinguished from postconflict contact by one of the participants in a fight with a bystander, i.e., an individual that had not been involved in the fight. Such contacts, especially when initiated by the bystanders, cannot serve the same function as reconciliation (which is to repair a disturbed relationship). The most likely function of contact with bystanders is distress alleviation.

Sensitivity to the emotions of others emerges early in humans, such as when a nursery room with infants bursts out crying in response to the cries of one among them. This process, known as *empathic distress* or *emotional contagion,* provides the ontogenetic basis for cognitively more advanced responses to distress in which the actor understands the other's situation, distinguishes the other's distress from his or her own feelings, and acts out of genuine concern about the other's well-being. Hence, dependent on the precise mechanism involved, empathy, or the capacity to be emotionally affected by someone else's feelings, can be cognitively simple or complex.

An early study of chimpanzees found that affiliative contact between participants in a fight and bystanders occurred more often in the first minute following a conflict than in subsequent minutes. Moreover, first-minute contacts included more embracing and gentle touching than did contacts during subsequent minutes. Contacts with bystanders were labeled consolation (Fig. 7). Recent research confirmed these findings, demonstrating that it is particularly with victims of serious aggression that contact is made (de Waal and Aureli, 1996). Of course, these are also the individuals expected to be most in need of comfort.

In macaques, in contrast, none of the studies addressing this behavior have thus far produced evidence for consolation. Despite various measures and statistical methods, the finding has been the same in four different species: Affiliative contact between recipients of aggression and bystanders does not occur more often following conflict than during control periods. Despite this lack of evidence, qualitative observations suggest that macaques do engage in consolation behavior at least when very young. We have regularly seen infant rhesus monkeys attracted to the screams of one among them (e.g., after punishment by an adult or after a fall), approach the vocalizer, and establish contact. Occasionally, this resulted in a pile of infants clambering over each other. Given this strong empathic distress response, why is there so little of it left at later ages?

One possible explanation is that association with the recipient of aggression is fraught with risk. In macaques, recipients of aggression continue to attract aggression in the period immediately following the aggressive incident. If this elevated chance of further aggression extends to bystanders that approach recipients of aggression, this is a risk with which bystanders need to reckon. Perhaps these risks are less important in a chimpanzee society, with its looser and more tolerant hierarchy.

The alternative explanation is that chimpanzees achieve a higher level of cognition that allows a sharper distinction between self and other. It is generally recognized that the development of empathy in human children relates to this distinction and hence to the level of self-awareness of the child. Given that chimpanzees can recognize themselves in mirrors, whereas macaques cannot, it is possible that chimpanzees achieve a level of empathy that is simply not present in most other primates (Gallup, 1982).

How Is Aggression Avoided?

One function of dominance hierarchies, as found in many animals, is to regulate access to resources so that individuals do not need to contest priority every time there is potential for competition. Usually, the subordinate withdraws and the dominant obtains the resource. In this way, aggression is avoided and the relationship is not undermined. Loss of the resource by the subordinate may appear costly, but had this individual attempted to prevent the other from claiming the resource it most likely would not have gained much more than injuries and a damaged relationship. Conformity to the dominance hierarchy through avoidance of dominants is a most effective way of resolving routine disputes.

In addition to from avoidance, the approach of a dominant individual can also elicit special gestures or facial expressions in the subordinate. Primates have evolved ritualized status signals. Well-known examples are the silent bared-teeth display of rhesus macaques and the pant-grunt of chimpanzees. These submissive displays are completely unidirectional — that is, only one individual in a given pair shows these signals toward the other. The most likely function is appeasement in order to avoid harmful confrontation.

Social conflict can also occur in circumstances and for reasons that do not involve direct competition. As Mason (1993) notes, the most common source of conflict occurs when one individual expects some form of "satisfaction" from another but fails to achieve it. For example, one individual seeks to have sex with or groom another and is rebuffed. This type of conflict rarely leads to aggression but is resolved by a process of negotiation in which the participants exchange signals to increase the predictability of the other's future behavior and to facilitate the achievement of a common goal.

This process is more evident when individuals encounter one another for the very first time as arranged in some introduction experiments. At the beginning, the encounter is characterized by mistrust and/or hostility, after which various signals help to reduce the level of uncertainty and to establish a relationship. The type of relationship varies depending on the identity of the individuals and the interest that they have in one another. Two males can negotiate their dominance relationship, whereas one male and one female can enter a consort relationship.

This process of negotiation leads to the reduction of ambivalence and the achievement of common objectives or a compromise between the objectives of the two parties involved. In subsequent encounters the history of previous interaction strongly affects the process of negotiation. The fact that the two individuals already have an established relationship lowers the level of uncertainty and makes negotiation easier and the exchange of signals more subtle.

Sometimes, negotiation is not successful and one participant may attack the other. Aggression is also a possible outcome when group members want to follow different courses of actions (e.g., when there is the need for coordination of group activity but some individuals still forage while others are in the process of traveling). One of the most striking examples of overt aggression is when males compete over access to a mate. Among baboons, one male, alone or together with an ally, may attack another male who is in consort with an estrous female. However, nonaggressive tactics are also available to solve these competitive situations. Greeting rituals, in which one male gently touches or mounts another, are common during tensions among male baboons over sexual access to a female. When the number of estrous females is low compared to the number of males, the chance of mating for an individual male is reduced. Interestingly, in these circumstances the rate of male aggression remains the same but greeting rituals occur more often. This suggests that greetings decrease tension between males and lower the risk of intermale aggression through increased tolerance (Colmenares, 1991).

Another source of strong competition is food. In their natural habitat, primates spend most of their time searching for food. When they find a food source that can be monopolized by a few individuals, high competition is likely. This situation, however, does not lead automatically to high levels of aggression. Primates have developed ways to reduce social tension and the risk of potential aggression associated with food competition. For example, bonobos increase sexual behavior at or immediately after feeding time. Captive stump-tailed macaques and chimpanzees used to a regular feeding schedule anticipate conflict during feeding by grooming at higher rates before feeding time. The increased grooming is aimed at selected individuals probably to facilitate tolerance and feeding near them.

As soon as attractive food is in sight, chimpanzees become very excited and perform loud hooting, bluff displays, and submissive greetings. The occurrence of rituals in which dominant individuals display and subordinates submissively greet them in a prefeeding situation may function as status confirmation that allows the dissipation of tension and the relaxation of priority rights during feeding. At the same time, there is a dramatic increase of brief friendly interactions, such as kisses, embraces, and gentle touches. In experiments in which opportunities for such "celebration" were varied, it was demonstrated that this behavior reduces competitive interactions.

In captivity, primates have even more reasons to stop or prevent aggression. This situation, with its reduced interindividual distances, offers an excellent opportunity to study conflict management capacities. A traditional view predicts that aggression increases under crowded conditions. Recent research on nonhuman primates has suggested an alternative view. Primates, and perhaps other animals, are able to cope with high population density by conflict-avoidance strategies (de Waal, 1989). The way in which space restrictions affect social behavior varies among species and among situation, but the expected aggression increase under high-density conditions is often minimal and sometimes even reversed. In a recent study, chimpanzees were found to inhibit aggressive responses under short-term crowding, resulting in an actual reduction of aggression. Under long-term crowing, on the other hand, grooming levels increase in both macaques and chimpanzees. This affiliation strategy probably functions to increase social tolerance and limit the potentially damaging effects of aggression.

Conclusion

The main message from research on primate conflict resolution is that aggressive behavior is subject to strong constraints imposed by the need to preserve valuable social relationships. Realization of these constraints has resulted in a shift in attention from aggression as the expression of an internal state or drive to aggression as the product of social decision making. This framework is referred to as the *relational model* (Fig. 8) because it concerns the functioning of aggressive behavior within interindividual relationships. It examines aggressive behavior as only one of several methods to resolve conflict, but one that is problematic because it may undermine cooperation. In other words, winning a fight is one thing, but winning it without harming one's position in a complex support network is quite something else.

Results from studies in this domain contrast sharply with earlier views of aggression as a dispersive social force (i.e., causing increased interindividual distance). All monkey and ape species studied thus far, both in captivity and in the field, exhibit a significant tendency toward reunion following aggressive conflict, thus on average reducing interindividual distance. The observation that nonhuman primates cope with tensions, resolve conflicts, and reunite with their opponents has forced quite a different view of conflict. This shift in perspective has recently been tied to the evolution of morality in our own species. If it is agreed that empathy and a need to get along despite interpersonal conflict are essential ingredients of human moral systems — that moral systems are, in effect, systems of conflict resolution — the observed conciliatory and comforting tendencies of other primates hint at continuity with our behavior even in this domain (de Waal, 1996).

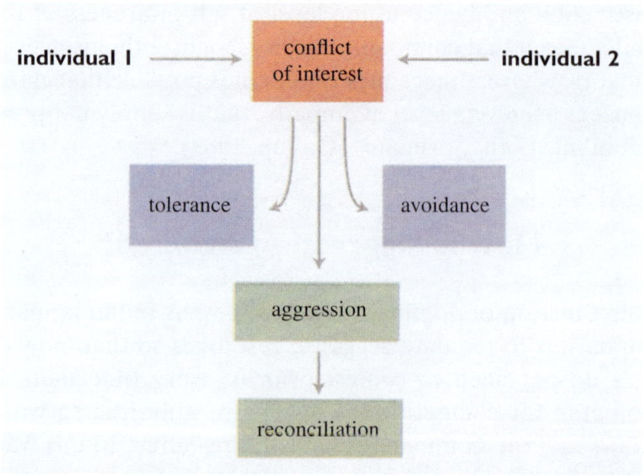

FIGURE 8 According to the relational model, aggressive behavior is one of several ways in which conflicts of interest are settled. Other possible ways are tolerance (e.g., sharing of a resource) or avoidance of confrontation. If aggression does occur, it depends on the nature of the relationship whether or not repair attempts will be made. If there is a strong mutual interest in maintenance of the relationship, reconciliation is likely. Parties negotiate the terms of their relationships by going through cycles of conflict and reconciliation.

This continuity, if confirmed by further research, will have far-reaching implication for how we view human existence. However, at a more concrete level there is ground for a thorough reevaluation of previous assumptions. For example, the reviewed findings indicate that most of the time aggressive behavior is a well-integrated component of primate social relationships. This view is radically different from the earlier view of aggression as a destructive force: Aggressive behavior appears to serve as a tool of negotiation within well-established relationships. Aggression can be used in this way precisely because its deleterious effects are buffered by postconflict interactions that alleviate anxiety and repair valuable relationships. The aftermath of conflict in primate groups thus provides a key to an understanding of the dynamic interplay between conflictuous and integrative tendencies.

References Cited

Aureli, F., and Van Schaik, C. P. (1991). Post-conflict behaviour in long-tailed macaques (*Macaca fascicularis*): I. The social events. II. Coping with the uncertainty. *Ethology* **89,** 89–114.

Aureli, F., Das, M., and Veenema, H. C. (1997). Differential kinship effect on reconciliation in three species of macaques (*Macaca fascicularis, M. fuscata,* and *M. sylvanus*). *J. Comp. Psychol.* **111,** 91–99.

Cheney, D. L., Seyfarth, R. M., and Silk, J. B. (1995). The use of grunts in facilitating interactions and reconciling opponents among adult female baboons. *Anim. Behav.* **50,** 249–257.

Colmenares, F. (1991). Greeting, aggression, and coalitions between male baboons: Demographic correlates. *Primates* **32,** 453–463.

CORDS, M. (1992). Post-conflict reunions and reconciliation in long-tailed macaques. *Anim. Behav.* **44,** 57–61.

CORDS, M., and THURNHEER, S. (1993). Reconciliation with valuable partners by long-tailed macaques. *Ethology* **93,** 315–325.

DE WAAL, F. B. M. (1982). *Chimpanzee Politics: Power and Sex among Apes.* Cape, London.

DE WAAL, F. B. M. (1989). The myth of a simple relation between space and aggression in captive primates. *Zoo Biol. Suppl.* **1,** 141–148.

DE WAAL, F. B. M. (1996). *Good Natured: The Origins of Right and Wrong in Humans and Other Animals.* Harvard Univ. Press, Cambridge, MA.

DE WAAL, F. B. M., and AURELI, F. (1996). Consolation, reconciliation, and a possible cognitive difference between macaques and chimpanzees. In *Reaching into Thought: The Minds of the Great Apes* (A. E. Russon, K. A. Bard, and S. T. Parker, Eds.), pp. 80–110. Cambridge Univ. Press, Cambridge, MA.

DE WAAL, F. B. M., and JOHANOWICZ, D. L. (1993). Modification of reconciliation behavior through social experience: An experiment with two macaque species. *Child Dev.* **64,** 897–908.

DE WAAL, F. B. M., and VAN ROOSMALEN, A. (1979). Reconciliation and consolation among chimpanzees. *Behav. Ecol. Sociobiol.* **5,** 55–66.

DE WAAL, F. B. M., and YOSHIHARA, D. (1983). Reconciliation and redirected affection in rhesus monkeys. *Behaviour* **85,** 224–241.

GALLUP, G. G. (1982). Self-awareness and the emergence of mind in primates. *Am. J. Primatol.* **2,** 237–248.

JUDGE, P. G. (1991). Dyadic and triadic reconciliation in pigtail macaques (*Macaca nemestrina*). *Am. J. Primatol.* **23,** 225–237.

KÖHLER, W. (1925). *The Mentality of Apes.* Vintage, New York.

KUMMER, H. (1978). On the value of social relationships to nonhuman primates: A heuristic scheme. *Social Sci. Inform.* **17,** 687–705.

MASON, W. A. (1993). The nature of social conflict: A pshychoethological perspective. In *Primate Social Conflict* (W. A. Mason and S. P. Mendoza, Eds.), pp. 13–47. State Univ. of New York Press, Albany.

General References

DE WAAL, F. B. M. (1989). *Peacemaking among Primates.* Harvard Univ. Press, Cambridge, MA.

MASON, W. A., and MENDOZA, S. P. (1993). *Primate Social Conflict.* State Univ. of New York Press, Albany.

John L. Gittleman

Department of Ecology and Evolutionary Biology
University of Tennessee
Knoxville, TN 37916, USA

The comparative method uses cross-taxonomic comparisons to investigate patterns of behavioral evolution. These comparisons explicitly involve taxa related by evolutionary descent, as reflected in phylogenetic trees. Spurred by a revolution of techniques and data collection for providing accurate phylogenies, precise tests of evolutionary hypotheses are now possible in comparative studies of behavior. Recent work has shown clear phylogenetic patterns of behavior with respect to historical origins, sequential changes, covariation with other traits, evolutionary lability, and rates of evolution.

Comparative Methods in the Study of Behavior

Every time a biologist seeks to know why an organism looks and acts as it does, he must resort to the comparative method.

Konrad Lorenz (1958)

The Importance of Phylogeny

Darwin only used one illustration in *The Origin of Species* (1859) (Fig. 1). A remarkably theoretical figure for the time, it showed a hypothetical "tree" of how organic diversity is hierarchically arranged: "All organic beings are found to resemble each other in descending degrees, so that they can be classed in groups under groups. This classification evidently is not arbitrary like the groupings of stars in constellations" (p. 411). The kinds of information conveyed in this tree lie at the heart of all comparative studies of behavior. By constructing a tree of evolutionary relationships, commonly called a *phylogeny,* it is possible to trace historical patterns of origin and change in a lineage. All comparative methods, whether in the context of behavioral study or in other areas of the life sciences [e.g., ecology (Wanntorp *et al.,* 1990) and neurobiology (Campbell, 1976)], rely on information from phylogenetic systematics. This is a new but healthy dependency. Early students of animal behavior well appreciated the importance of phylogenies, although few were available. Indeed, Konrad Lorenz and Niko Tinbergen were faced with not only hav-

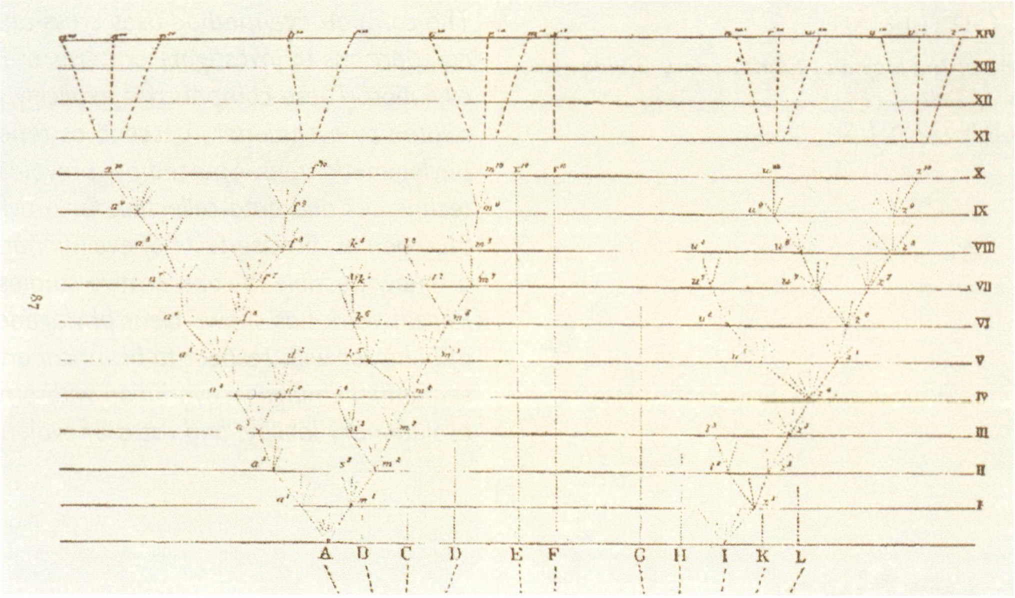

FIGURE 1 A hypothetical phylogeny from Darwin's *The Origin of Species* (1859). Used to illustrate the divergent pattern of evolution, it shows a tree-like picture of ancestor–descendent relationships and points of evolutionary change.

ing to collect original behavioral data but also working double-duty as systemacists, attempting to reconstruct phylogenies based on any type of information available (Burghardt and Gittleman, 1990) (Fig. 2). Today, the field of systematics is undergoing a revolution providing new techniques and data for drawing trees as well as using diverse information from morphometrics, molecular biology, and statistics, all leading to more precise pictures of the evolutionary relationships across taxa which can be used to test hypotheses of behavioral evolution.

The Nature of Phylogenies

Phylogenies contain information about common descent among organisms. This information is expressed in a branching diagram. It is important to recognize the distinction between a phylogenetic system of taxonomy and a Linnean system, until recently the more common. The Linnean system conveys information about the relative positions of taxa in a nested, hierarchical arrangement (i.e., taxonomic ranks such as kingdom, phylum, and class). Although Linnean categories can be used to deduce phylogenetic information, they do not necessarily do so. If used only to represent position in a hierarchy, Linnean categories contain no information or, at best, misleading information about common ancestry. For example, a tree based on a Linnean taxonomy would typically be symmetrical with equal branch lengths among taxa. This is clearly an unrealistic view of evolutionary history. This distinction is critical to comparative methods in the study of behavior (Fig. 3). Questions involving evolutionary origin, rate, and direction all involve the tenet of common descent and are directly affected by tree topology.

Phylogenetic trees provide specific types of information. *Nodes* represent putative ancestors. *Tips* are the ends of branches, typically represented by species in comparative study (Fig. 4). *Branches* of phylogenetic trees are lines connecting the nodes together or lines connecting a node to a tip. A *root* is the base, or most ancestral branch, of the tree. The subjects or taxa of a comparative study are technically called the *ingroup,* which must contain more than two taxa to be able to infer a phylogenetic relationship. The most closely related taxon to the ingroup is a *sister group.* Last, because it is critical to have a starting point to make an inference about evolutionary order, an *outgroup* is used to represent the ancestor of the ingroup, a group that must be independent of the study group.

When looking at a phylogenetic tree it is critical not to accept as fact a given view of evolutionary history: Phylogenetic trees are only estimates or hypotheses about phylogeny. Phylogeny is the result of a series of historical events that are typically unknown and unrecoverable. Comparative methods which necessarily rely on phylogenies will always be vulnerable to weaknesses of historical reconstruction. Fortunately, however, verification from the fossil record and analytical checks from computer simulation are resulting in increasingly reliable phylogenetic hy-

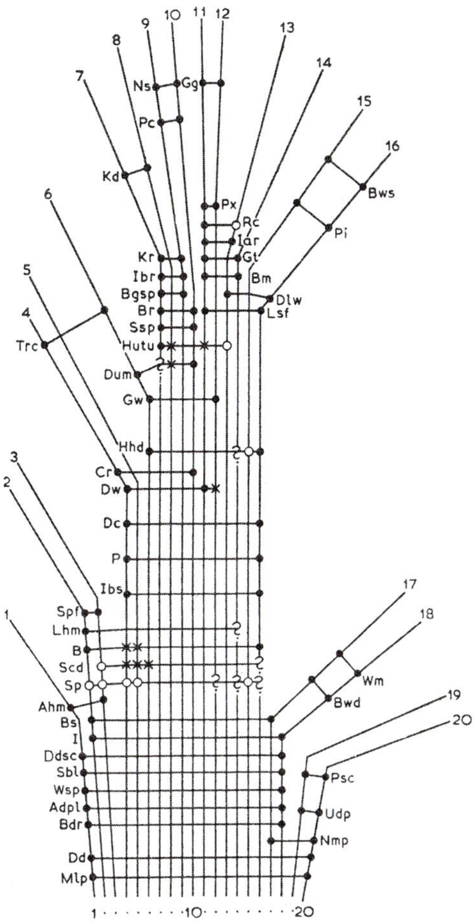

FIGURE 2 Phylogenetic tree of the Anatinae taken from K. Lorenz. The vertical lines represent species; the horizontal lines represent common characters. A cross indicates the absence of a character in a species. A circle indicates special emphasis and differentiation of the character. A question mark indicates the author's uncertainty. The species are shown as follows: 1, *Cairina moschata,* Muscovy duck; 2, *Lampronessa sponsa,* Carolina wood duck; 3, *Aix galericulata,* mandarin duck; 4, *Mareca sibilatrix,* Chiloe wigeon; 5, *Mareca penelope,* wigeon; 6, *Chaulelasmus streptera,* gadwall; 7, *Nettion crecca,* teal; 8, *Nettion flavirostre,* South American teal; 9, *Virago castanea,* chestnut-breasted teal; 10, *Anas* as genus including mallard, spot-billed duck, and Meller's duck; 11, *Dafila spinicauda,* South American pintail; 12, *Dafila acuta,* pintail; 13, *Poecilonetta erythrorhyncha,* red-billed duck; 14, *Poecilonetta erythrorhyncha,* red-billed duck; 15, *Querquedula querquedula,* garganey; 16, *Spatula clypeata,* shoveller; 17, *Tadorna tadorna,* shelduck; 18, *Casarca ferruginea,* ruddy shelduck; 19, *Anser* as genus; 20, *Branta* as genus. The characters are as follows: Mlp, monosyllabic "lost piping" emitted by an individual to maintain contact with its conspecifics; Dd, display drinking behavior; Bdr, bony drum on the drake's trachea; Adpl, Anatinae duckling plumage; Wsp, wing speculum; Sbl, sieve bill with horny lamellae; Ddsc, disyllabic duckling social contact call; I, incitement by the female; Bs, body shaking as a courtship or demonstrative gesture; Ahm, aiming head movements as a mating prelude; Sp, sham preening of the drake, performed behind the wings; Scd, social courtship of the drakes; B, burping; Lhm, lateral head movement of the inciting female; Spf, specific feather specializations serving sham preening; Ibs, introductory body shaking; P, pumping as prelude to mating; Dc, decrescendo call of the female; Br, bridling; Cr, chin raising; Hhd, hind-head display of the drake; Gw, grunt whistle; Dum, down–up movement; Hutu, head up/tail up; Ssp, speculum same in both sexes; Wm, black-and-white and red-brown wing markings of Casarcinae; Bgsp, black-gold-green teal speculum; Trc, chin raising reminiscent of the triumph ceremony; Ibr, isolated bridling not coupled to head up/tail up; Kr, krick whistle; Kd, koo-dick call of the true teals; Pc, postcopulatory play with bridling and nod swimming; Gg, "geeeeegeeeee" call of the true pintail drakes; Px, pintail-like extension of the median tail feathers; Rc, R calls of the female in incitement and as social contact call; Iar, incitement with anterior of body raised; Gt, graduated tail; Bm, bill markings with spot and light-colored sides; Dlw, drake lacks whistle; Lsf, lancet-shaped shoulder feathers; Bws, blue wing secondaries; Pi, pumping as incitement; Dw, drake whistle; Bwd, black-and-white duckling plumage; Psc, polysyllabic gosling social contact call of Anserinae; Udp, uniform duckling plumage; Nmp, neck dipping as mating prelude (reproduced with permission from Lorenz, 1941).

potheses (Purvis *et al.,* 1994). Moreover, the number of phylogenies currently being published is increasing at an astounding rate (Fig. 5).

Uses and Abuses of Comparative Data

Comparative studies live and die by breadth. On the one hand, data sets across broad taxonomic groups permit tests of the generality of a hypothesis and serve as the only solid basis for studying phylogenetic patterns of behavior. On the other hand, most comparative studies rely on data gathered by independent investigators using different methods and working under different conditions; bad data are the Achilles' heel of comparative research (Gittleman, 1989). Fortunately, with greater data accumulation, both of phylogenetic trees and of source behavioral data, information is now more verified and dependable than in earlier work. The fact that many behavioral studies are being replicated across populations adds to the validity of comparative data sets, particularly when comparisons are made by using single trait values for a given species. Furthermore, modern comparative studies involve greater *a priori* tests and findings which are statistically checked by computer simulation (Martins, 1996). Certainly, modern comparative studies now heed the warning of Tinbergen (1951) that the planning and execution of a comparative study should be as carefully fine-tuned as any experiment (Gittleman and Decker, 1994).

Another underlying problem leveled at comparative studies is that rarely do we know anything about what causal mechanism is driving a comparative result or whether a

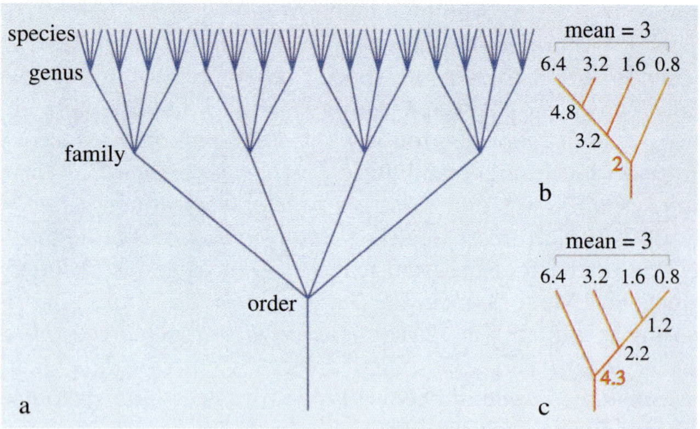

FIGURE 3 Linnean taxonomies can bias comparative studies. (a) A phylogeny implied by a Linnean classification in which all ancestral taxa contain four descendent taxa, all diverging at the same time. (b) A phylogenetic tree comprising four taxa with trait values known for the tips. The root value (in red) is 3.0 if taken as the mean of the tips, but it is 2.0 if topologies of the ancestors (bifurcation points or nodes) are given and these take values equal to the mean of their descendents [e.g., the root node: (0.8 + 3.2)/2 = 4.0/2 = 2.0]. (c) As in b but with a different topology subtending the terminals, and therefore the root value is 4.3 rather than 2.0. Topology can severely affect comparative studies in which ancestral values are estimated (reproduced with permission from Coddington, 1994).

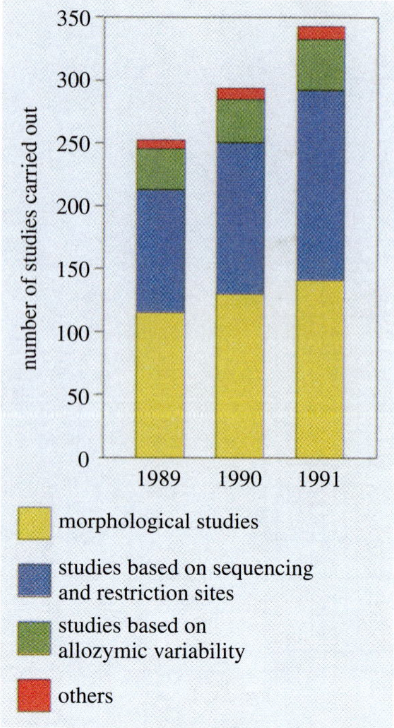

FIGURE 5 Phylogenetic studies during the period 1989–1991. Within each year, the type of data used is expressed as a proportion of the total for that year. The rate of publication of phylogenetic studies averaged 294 per year, increasing by 20% in 1990 and 16% in 1991. The rate increase held among all major taxonomic groups and types of data. Molecular studies involving sequencing and restriction site surveys increased especially fast, but the rate of accumulation of morphological studies also increased markedly (adapted from Sanderson *et al.,* 1993).

comparative test can distinguish between the effects of selection and those of drift, indirect selection, or uncontrolled environmental variation (Leroi *et al.,* 1994). To definitively test an adaptive hypothesis, it is necessary to directly measure genetic variation in the taxa involved, often requiring some type of experimental procedure. Comparative studies using a diversity of traits across vast phylogenetic lineages can rarely perform such work. Nevertheless, this criticism of the comparative method is really not a criticism of the method per se but rather a weakness resulting from limited availability of data and lack of understanding of the traits. As comparative data become more complete with information about causal mechanisms and heritability measures of traits, then the comparative method will work effectively.

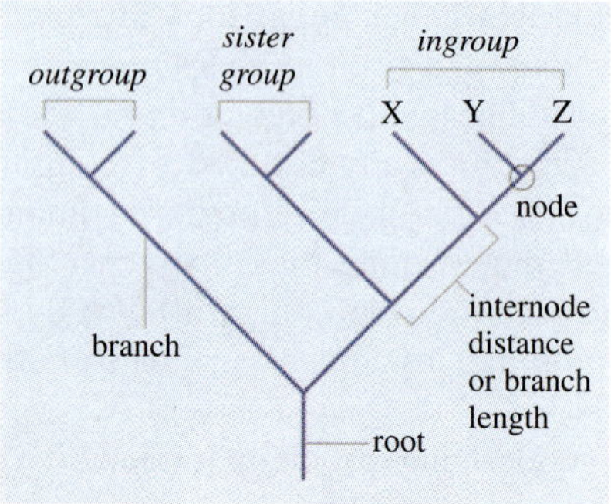

FIGURE 4 Components of a phylogenetic tree constructed from the reference group (ingroup) consisting of the taxa X, Y, Z (adapted from Brooks and McLennan, 1991).

New Comparative Methods

The comparative method, using cross-taxonomic comparisons to study patterns of behavioral evolution, dates back at least to Greek mythology. Aristotle used comparisons between carnivorous and herbivorous organisms to explain differences in sociality (Ridley, 1983). However, aside from the steady accumulation of data on the characteristics of organisms, is there anything really new about present-day comparative methods? The answer is yes. Two features of modern comparative studies vastly differ from methods used by early practitioners.

The first involves the method itself. Comparative methods are more systematic and quantitative. Generally, there

species	color	form	size
X	blue	square	small
A	blue	square	small
B	red	square	small
C	red	round	small
D	red	round	large

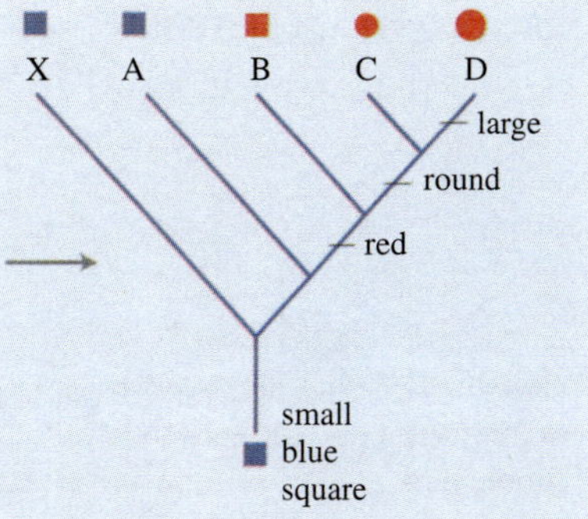

FIGURE 6 (Left) Distribution of three characters (color, form, and size) among four members of an ingroup (species A–D) and an outgroup (taxon X). (Right) A phylogeny of these taxa representing the distribution of the three characters. Species A is directly derived from the ancestral taxon, whereas the evolution of B is characterized by the change of color. C is derived successively from the same line, with a change of form in addition to the change in color, and in D a third character (size) also changes (adapted from Brooks and McLennan, 1991).

are two kinds of comparative methods, depending on the kind of data being analyzed: categorical and quantitative. Categorical methods essentially involve overlaying categorical or discrete traits (e.g., mating systems such as polygyny vs monogamy) onto a phylogeny. By pinpointing in what form and where traits occur at different nodes and branches, it is possible to see the evolutionary origin, maintenance, sequence of change, and covariation with other traits. For example, the distribution of categorical traits can show whether one trait is ancestral to another (Fig. 6). Potential difficulties with categorical methods are that it is difficult to assign statistical probabilities to comparative patterns and to investigate the degree of trait change in proportion to branch lengths. Quantitative methods entail statistical tests of covariation among quantitative traits (e.g., social group size, prey size, or mating frequency), given phylogenetic relationships among the taxa concerned. Essentially, quantitative comparative methods examine trait variation while controlling for phylogenetic structure in the data. Initially, the main reason for using a quantitative approach was purely statistical: Standard statistical tests such as regression assume independence of data points, but related species often share traits by common descent rather than through independent evolution (Fig. 7). Quantitative variables can be used to statistically test correlated evolution along a phylogeny using either the method of independent contrasts or autoregression (Fig. 8). Even though both quantitative methods are statistically sound under most conditions, each makes various analytical assumptions about how traits evolve (i.e., specific models of evolution) and the degree to which traits actually are correlated with phylogeny (Harvey and Pagel, 1991; Martins, 1996).

The other breakthrough in comparative methods, particularly for studying behavioral problems, is the availability of phylogenies which are independent of the traits being studied. Traditionally, phylogenetic trees are based on morphological traits such as skull shape and teeth size. Because behavior often varies with morphology, any comparative test of behavior using a morphological phylogeny runs the risk of being circular. The inception of molecular markers (e.g., allozymes and DNA sequences) to estimate phylogenetic relationships potentially breaks this circularity (Gittleman *et al.,* 1996). The field of molecular systematics

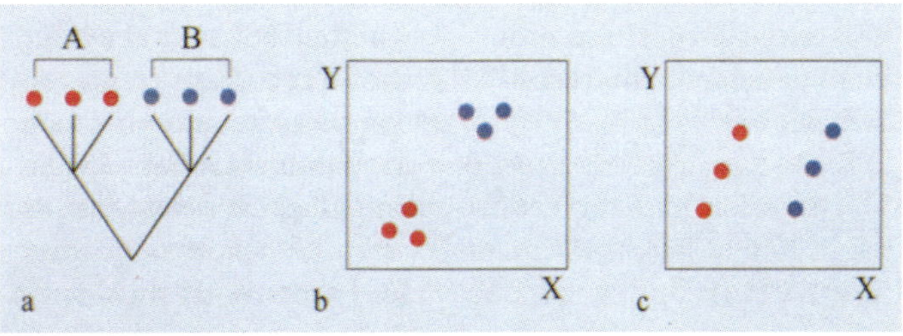

FIGURE 7 An illustration of the pitfalls posed by comparative data. Six species are related as shown in a. Comparative analyses treating species values as independent can find apparently significant correlations between Y and X even when there is no relationship within either clade (b) and can miss correlations that are strong within each clade (c) (adapted from Gittleman and Luh, 1992).

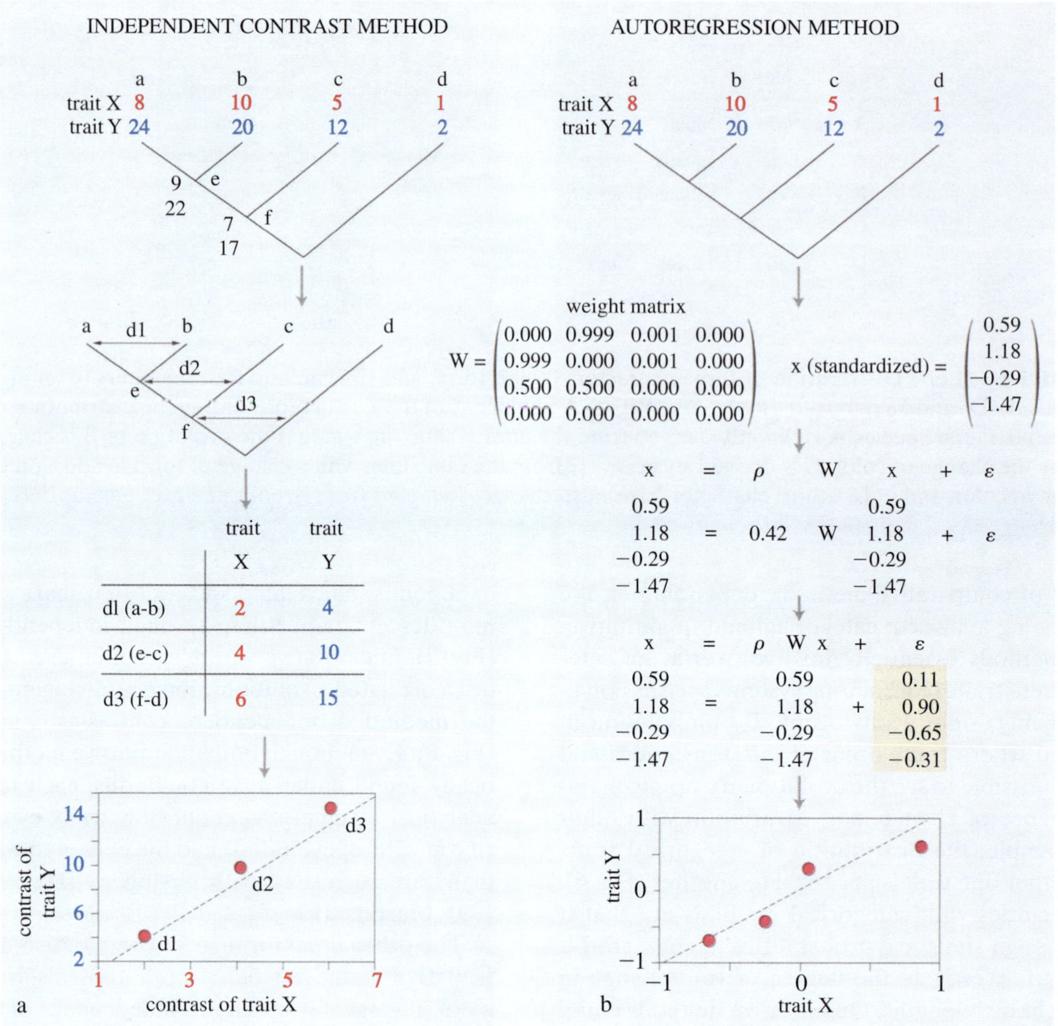

FIGURE 8 Illustration of the procedures used in the independent contrasts and the autoregressive methods. (a) The independent contrasts method, by calculating differences or contrasts between pairs of sister taxa, produces comparative data which are statistically independent of phylogeny. (b) Similar to regression, the autoregressive method estimates, for the taxa considered (a–d), trait variation (X and Y) as a function of phylogenetic relationships as reflected by a weighting matrix (W) relative to the single trait X. The residuals from the autoregressive analysis are considered independent of phylogenetic correlation and therefore can be used for comparative analyses. ρ, coefficient of autocorrelation which measures the correlation between the characters and the phylogenetic relationship (reproduced with permission from Gittleman and Decker, 1994).

provides tremendous promise for investigating a wide range of phylogenetic questions (Avise, 1994). These molecular trees are powerful "null hypotheses" for reconstructing historical pathways of behavior.

Comparative Methods Show Patterns of Behavioral Evolution

The following sections provide examples of how modern comparative methods have been applied to various behavioral problems. Emphasis is placed on the kinds of studies rather than on consistent patterns of behavioral phylogeny. These comparative methods are too new to determine whether behavioral evolution across independent lineages shows recurrent trends. Other reviews further illustrating comparative and phylogenetic approaches to behavior are available elsewhere (Miles and Dunham, 1993; Gittleman and Decker, 1994; Martins, 1996).

Origins of Behavior

One of the most vexing problems in the phylogenetic study of behavior is that the primary source(s) of data is collected by direct observation. This means that if behavioral data are not gathered as they occur or by some means of recording, they are forever lost to history. The availability of a phylogeny raises the exciting prospect of being able to reconstruct historical patterns or origins of behavior.

An example of how to use phylogenies and a comparative method to trace behavioral origins is a study of social behavior of manakins (Prum, 1994), a group of Neotropical bird species known for their sexually dimorphic plumage, lekking behavior, and elaborate male courtship displays. During breeding, two or more males exhibit an array of coordinated display patterns (Fig. 9). Based on detailed ethological description, these displays can be classified into three classes (Fig. 10): (i) simple display involving two or more males that perform synchronized versions of a display shown by a typical solitary male, (ii) *complex display* showing novel synchronized display patterns not performed by a solitary, and (iii) complex–coordinated display revealing a complex, coordinated display that is only shown as a form of mate attraction and results in differential mating success among coordinated males. The phylogenetic distribution of these display behaviors, when overlayed onto a phylogeny, indicates that no coordination in displays is the ancestral condition and that there are five evolutionarily independent origins of different displays within the manakins. Furthermore, the origins of complex displays are recurrent in the sense that all the species in a clade show them once they evolve. Such use of hanging comparative behavioral data onto a phylogenetic tree reveals that phylogenetic patterns of behavior can be consistently reconstructed. Also, there is the implication that, in contrast to

FIGURE 9 Illustrations of different display behaviors of manakins. (a) A pair of male white-fronted manakins (*Pira serena*) chasing one another from vertical perch to vertical perch in a uncoordinated manner. (b) One male golden-winged manakin (*Masius chrysopterus*) waits on the display log, whereas a second male flies to the log and performs the log-approach display (white arrows). (c) One male band-tailed manakin (*Pipra fasciicauda*) waits on the display perch (right) as the second male flies to the perch and performs the S-curve flight display (left). This display is primitive to the genus *Pipra,* whereas the coordinated drop from the perch is derived in the *Pipra aureola* clade. (d) Each male, from two to four, in a group of blue manakins (*Chiroxiphia caudata*) on the same perch flutters in succession from one end of the perch to the other as the remaining male(s) slide along the perch.

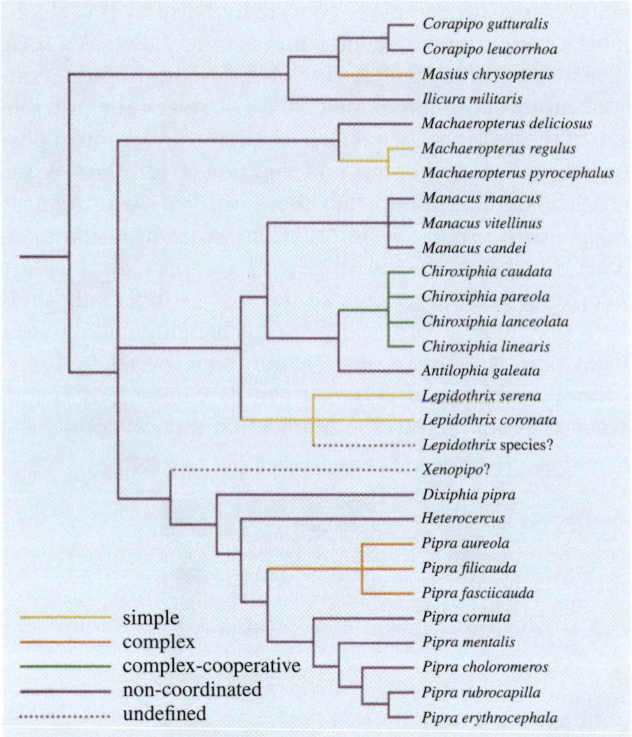

FIGURE 10 The phylogeny of coordinated and cooperative display behavior in the manakins (Pipridae). Three classes of behavior are recognized: simple, complex, and complex–cooperative. Coordinated behavior has had five independent origins in the family and has never been lost. All forms of coordinated behavior are phylogenetically independent of one another within the manakins (adapted from Prum, 1994).

the widely held belief of behavior being extremely labile, there are strong historical components to behavioral evolution.

Phylogenetic Sequences of Behavior

With a phylogeny in hand it is possible to determine the origin of a behavior pattern and also sequential changes during evolutionary history. Because behavior is closely intertwined with physiology and morphology, it is particularly insightful to examine sequential behavior patterns relative to other functional traits.

An elegant example of how phylogenetic analysis can be applied to sequential changes in traits is in a comparative study of salamanders (Lauder and Reilly, 1996) (Fig. 11). Detailed data were gathered on interrelated features of behavior (prey capture), morphology (jaw size), and motor patterns (head and jaw movements). Some of this information was painstakingly collected: Prey capture in salamanders is very rapid, taking between 30 and 120 ms and requiring frame-by-frame video recording and analysis. Mapping variation of these traits onto a phylogeny visualizes many trends. First, taxa such as *Ambystoma* and *Siren,* which are similar in morphology, show different behaviors. This result may be due to differences in motor patterns. Second, even though the closely related *Ambystoma* and *Necturus* are different in morphology, they are similar in behavior. Thus, counterintuitively, differences in morphology are not always matched by like differences in behavior. Third, and perhaps most striking, all of the four taxa show unique evolutionary features of behavior, morphology, and motor patterns. That is, there has been considerable phylogenetic divergence in all three classes of traits.

This type of comparative study clearly shows the complexity of identifying phylogenetic patterns of different traits. There is also a tangential message. If a phylogeny were constructed from one of these traits it would be very different from one based on another trait. This emphasizes how important phylogenies are to comparative study.

Correlated Change of Behavior

Correlation measures the intensity of relationship between variables. The difficulty in calculating correlation among comparative traits is to do so while accounting for the phylogenetic pattern also shown in the traits. The previously mentioned quantitative comparative methods (independent contrasts and autoregression) solve this analytical problem and, in essence, perform the same task of producing comparative trait values which are free of phylogenetic structure and therefore can be used to investigate correlation between traits.

Because these comparative methods are meant to account for phylogeny, it is necessary to first diagnose whether the traits in question are in fact related to phylogeny. There are many statistics which can be used. Essentially, some measure is needed of the relationship between phylogenetic distance (or taxonomic rank) and trait variation. In a study of social group size, life history patterns (e.g., gestation length and birth weight), and body size in canids, correlations were made between each of these variables and a molecular phylogeny (Geffen *et al.,* 1996) (Fig. 12). Not only do these traits correlate significantly with phylogeny but also they do so in an expected manner: Correlations are higher for more closely related species, thus reflecting greater trait similarity, whereas correlations precipitously decline among more distantly related species. Aside from checking whether quantitative traits are correlated with phylogeny, such descriptive statistics are also informative about the relative lability of trait evolution, as described later.

Once phylogenetic correlation is known, either the independent contrasts or autoregressive methods can be used to examine trait covariation. For example, in the study previously mentioned autoregression was employed to test whether the evolution of group size in canids is related to energetic investment of offspring. Even though neonatal

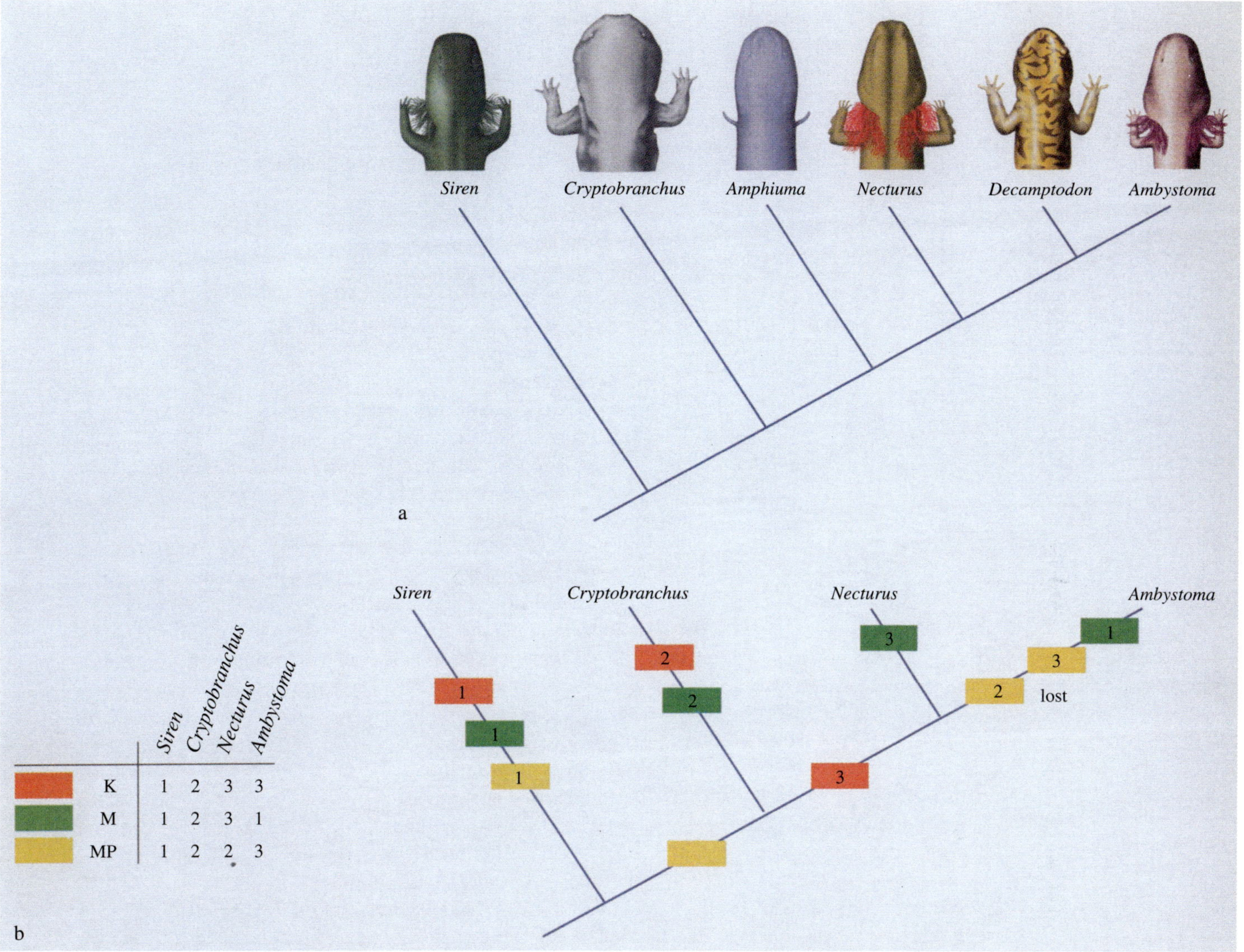

FIGURE 11 (a) Phylogenetic relationships of six taxa of salamanders. (b) Analysis of the phylogenetic changes in each of three traits: kinematic behavioral (K), morphological (M), and motor pattern (MP). The data matrix in the lower right codes the differences among taxa for each character and these are then mapped onto the phylogeny (adapted from Lauder and Reilly, 1996).

weight increases with female body weight, it does so proportionately, such that no support is given to the idea that group size evolves to adjust for greater demands imposed by having large offspring. The independent contrasts approach can be used in a similar manner to investigate correlation between quantitative variables. A fascinating example involves the use of contrasts in an unusually detailed comparative study of neuroanatomy and behavior (Fig. 13). The problem involves the relationship between bird song and brain anatomy. Independent contrasts were calculated across 41 oscine bird species (e.g., warblers, wrens, and sparrows) for song repertoire size, a measure of song complexity, and the relative volume of the high vocal center, an area housing nuclei directed connected to the syrinx (DeVoogd *et al.*, 1993). The analysis showed a strong relationship between song repertoire and the brain measure, even at different phylogenetic branching points, suggesting that the relationship has existed since the evolutionary origins of the oscines.

Quantitative comparative methods are quite effective for determining interrelationships between behavioral traits and other functional traits. Future developments with these methods will involve handling error variance of the trait values and multivariate statistical issues.

Relative Lability of Behavior

One of the reasons behavior has been used infrequently to construct phylogenies, or more generally studied from

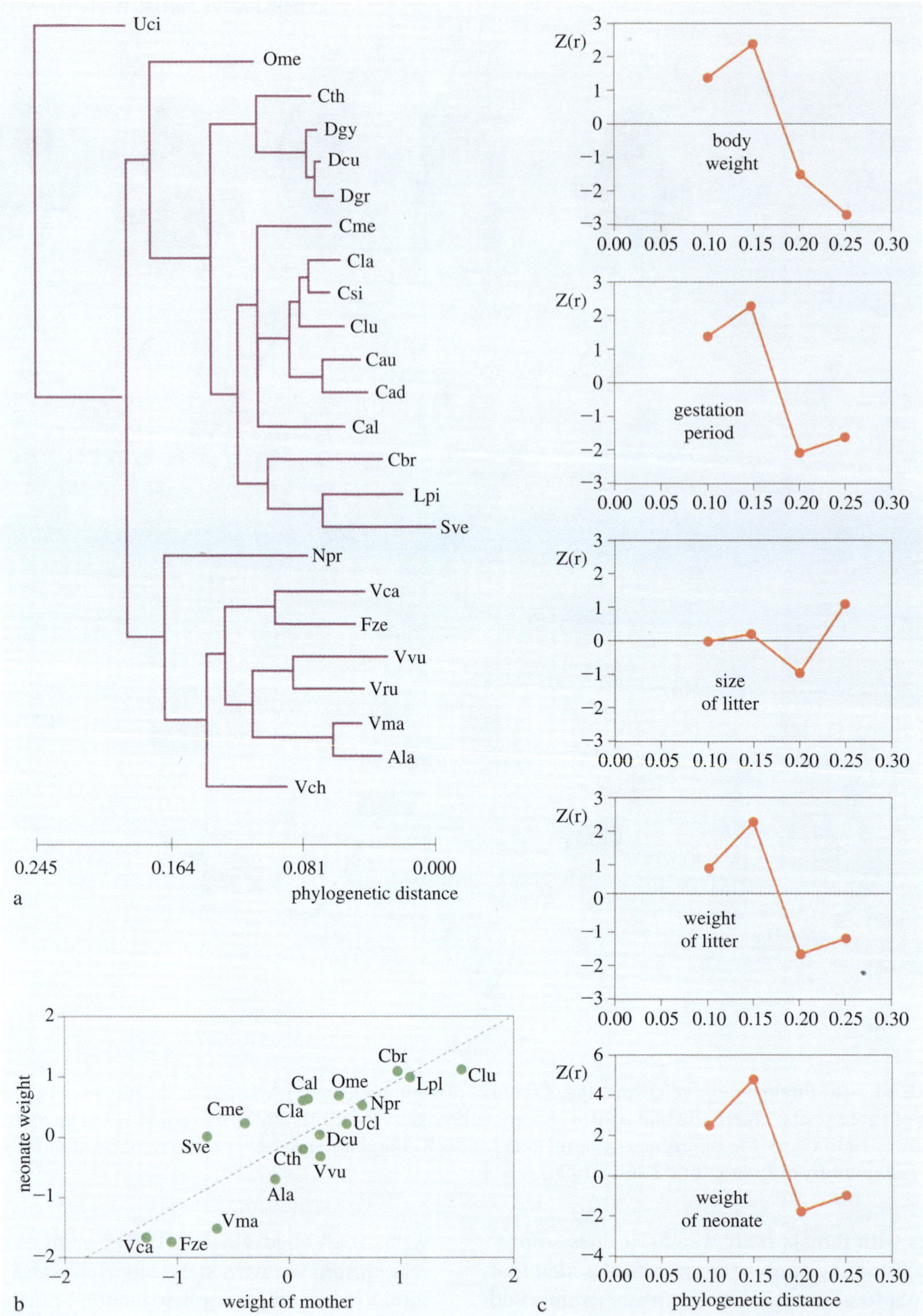

FIGURE 12 (a) A phylogenetic tree for 24 species of canids based on cytochrome b molecular sequence data. Uci, *Urocyon cinereoargenteus;* Npr, *Nyctereutes procyonoides;* Ome, *Otocyon megalotis;* Fze, *Fennecus zerda;* Vca, *Vulpes cana;* Vru, *Vulpes rueppelli;* Vvu, *Vulpes vulpes;* Vma, *Vulpes macrotis;* Ala, *Alopex lagopus;* Vch, *Vulpes chama;* Sve, *Speothos venaticus;* Cbr, *Chrysocyon brachyurus;* Lpi, *Lycaon pictus;* Clu, *Canis lupus;* Cla, *Canis latrans;* Csi, *Canis simensis;* Cau, *Canis aureus;* Cme, *Canis mesomelas;* Cal, *Cuon alpinus;* Cad, *Canis adustus;* Cth, *Cerdocyon thous;* Dgr, *Dusicyon griseus;* Dcu, *Dusicyon culpaeus;* Dgy, *Dusicyon gymnocercus.* (b) The relationship of mean female body weight to mean neonate weight. The slope is approximately 1.0 and does not support the argument that, in canids, larger females have relatively larger young. All trait values are adjusted for phylogeny using the autoregressive method. (c) Phylogenetic correlation (expressed as parameter Z) of different size and life history traits across canids, using the phylogenetic distances expressed in a. Except for litter size, all of the correlations are significant and in the predicted direction of high correlation among closely related taxa and low correlation among distantly related taxa (adapted from Geffen *et al.,* 1996).

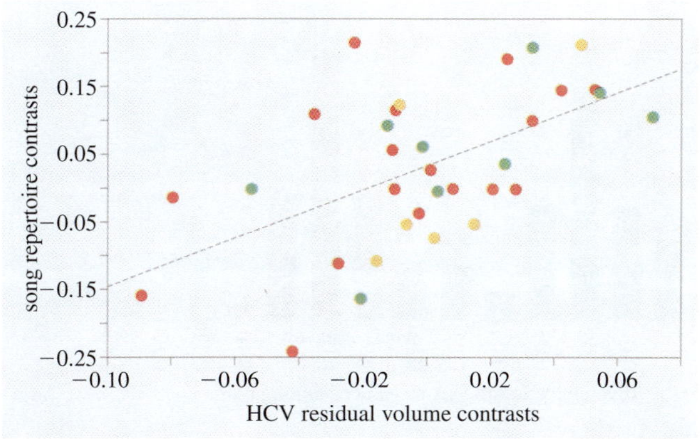

FIGURE 13 Correlation between size of song repertoire and residual volume of the high vocal center (HVC), analyzed using the independent contrasts method on 41 oscine bird species and their associated 34 contrasts. Contrasts between species (in green), genera and intermediate levels of phylogeny (in red), and families (in yellow) are distinguished. There is a strong association among these different phylogenetic levels, suggesting that this relation has existed since the early origins of the oscines (adapted from DeVoogd *et al.*, 1993).

historical perspectives (e.g., in paleontology), is because of evolutionary lability. That is, if behavior changes between generations or in response to environmental conditions, then behavior will leave poor traces of phylogenetic history. Even so, it is unclear whether lability in general is more characteristic of behavior than morphology or physiology.

Again, with some phylogenetic information and the same correlation statistics described previously in the canid example, it is possible to directly study the evolutionary lability of behavior. Recent work on gray-crowned babblers (*Pomatostomus temporalis*), a cooperatively breeding songbird from Australia and New Guinea, sheds light on this problem (Edwards and Kot, 1995) (Fig. 14). First, the phylogenetic information was amazingly precise, specifying the genetic distances from mitochrondial DNA (mtDNA) of 120 individuals belonging to 12 populations and representing distinct lineages. Measures were then taken of wing length for all individuals, which were assigned to one of 53 social groups. Correlations were calculated between phylogenetic distance among populations and each of these traits. As expected, significant positive correlations were found with wing length among individuals which differed in mtDNA by less than approximately 5%, whereas wing lengths from individuals differing by 7–11% in mtDNA were negatively correlated. In contrast, no relationship was found between phylogenetic distance and group size. An independent study of 8 higher level (family and above) mammal groups also showed that behaviors tend to reveal relatively little phylogenetic signal (Gittleman *et al.*, 1996). Thus, the presumed greater lability of behavior relative to morphology is supported both at lower and at higher taxonomic levels. The babbler study raises the further interesting issue regarding to what extent phylogenetic patterns are detectable below the species level.

These studies seem to support the view that behavior is labile. However, other work shows that the question of relative lability among traits may be more complex (de Queiroz and Wimberger, 1993). Across taxa extending from wasps to lizards, phylogenies were erected using a wide range of traits, including behavioral (e.g., feeding, nesting, and mating behaviors) and morphological (e.g., body weight and beak length) features. The behavioral and morphological phylogenies were statistically indistinguishable for any given taxon. Thus, for some taxa behavior patterns are equally "conservative" as morphological traits. Taken together, these studies highlight the fact that we know very little about why some behavioral traits in some taxa seem labile, whereas others are relatively fixed along phylogenetic lines. Now that the comparative tools are firmly in place, future work will reveal these general trends. This is of profound importance because we can determine whether different kinds of phenotypic traits, including behavioral ones, should be used in systematics. Furthermore, it addresses the question of the degree to which behavior is a driving force in evolution. As the influential evolutionary biologist Ernst Mayr (1963, p. 604) wrote, "A shift into a new niche or adaptive zone is, almost without exception, initiated by a change in behavior." Modern comparative research can finally assess this problem.

Rate of Behavioral Evolution

Behavior must evolve faster than other traits. Individuals change their behaviors from moment to moment, whereas other traits such as brain size or metabolic rate are relatively fixed. Thus, one would expect that the rate of behavioral evolution would be much greater than that of other traits, especially when measured across broad (phylogenetic) timescales as in cross-taxa comparisons.

Surprisingly, few studies have actually tested this idea, probably because any comparative test requires detailed and accurate information on time intervals between speciation events (i.e., nodes in trees). New phylogenies, especially molecular ones, have provided information about evolutionary history independent of phenotypic traits and have also included explicit time frames (Harvey *et al.*, 1996). Analytically, there are many ways to measure evolutionary rate (Gingerich, 1993). One of the first proposed is Haldane's (1949) measure of "darwins," which was originally used to measure rates of evolution in horses and recently in a many organisms, for example, to measure population change in marine snails and generation time in lab rats (Gingerich, 1993). Essentially, this measure calculates evolutionary rate as the amount of change between the initial value of a trait and the final value of a trait for an amount of time elapsed in a given time frame. Because rates of evolution change with temporal scale (i.e., evolu-

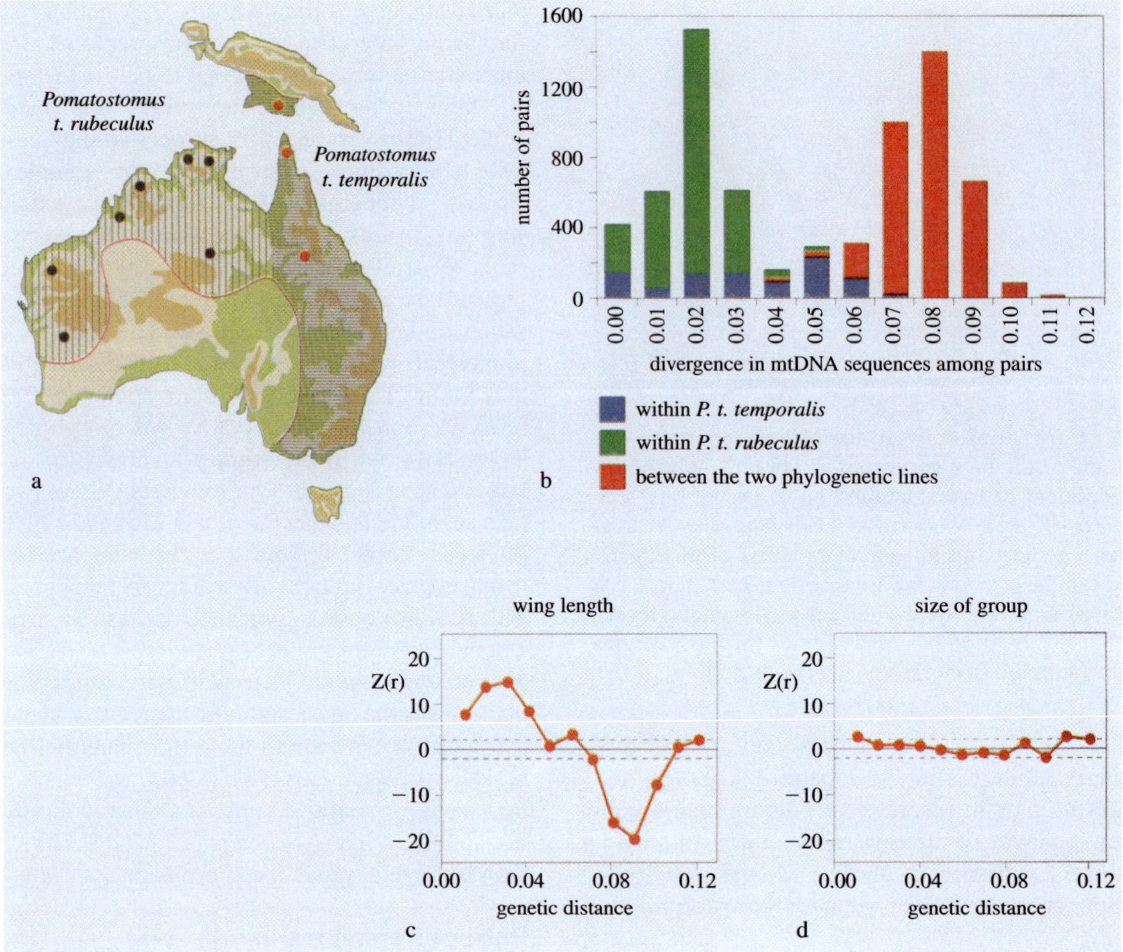

FIGURE 14 (a) Sampling localities (black and red circles) of seven populations of gray-crowned babblers (*Pomatostomus t. temporalis*) and three of the subspecies *rubeculus* (red circles) in Australia and New Guinea. (b) Divergences of mitochondrial DNA sequences, expressed as genetic distances, within each of the two phylogenetic lines of *P. temporalis* and between them. (c) Correlations (expressed as parameter Z) between wing length and phylogeny based on phylogenetic distances of the mitochondrial DNA. Values above or below the dashed lines are statistically significant. As predicted, high correlations (significant positive values) are found with closely related individuals and low correlations (negative values) with distantly related individuals. (d) Correlations between group size and phylogeny, as in c. With this behavioral variable no significant phylogenetic correlation is observed (adapted from Edwards and Kot, 1995).

tionary rate appears inversely proportional to timescale), darwins are typically plotted against time intervals as measured from phylogenetic distances and calibrated for each taxonomic group being studied. These so-called *darwin plots* (Fig. 15) reveal two statistical measures of rate: the net evolutionary rate by way of a slope and the instantaneous evolutionary rate for short time intervals by an intercept. A recent example of using darwins for comparing rates of behavioral versus morphological traits gives some support to the view that behavior evolves faster.

Across primates, darwin plots of social group size versus body weight — in this case, a molecular phylogeny was used to determine evolutionary relationships — show that slopes are steeper and intercepts are higher for group size, a finding that is consistent across many groups of mammals (Gittleman *et al.,* 1996). Therefore, by using a classic measure of rate together with modern molecular phylogenies, comparative analyses can examine questions which until recently were viewed as lost to history. This last example does not use the comparative method per se, as do previous examples. However, it rests on common bedrock: phylogeny. In this case, the phylogeny is used for more than inferring evolutionary relatedness; it also must provide accurate time frames in which evolution occurred. As temporal information becomes more reliable, comparative studies of tempo and rate will become more common.

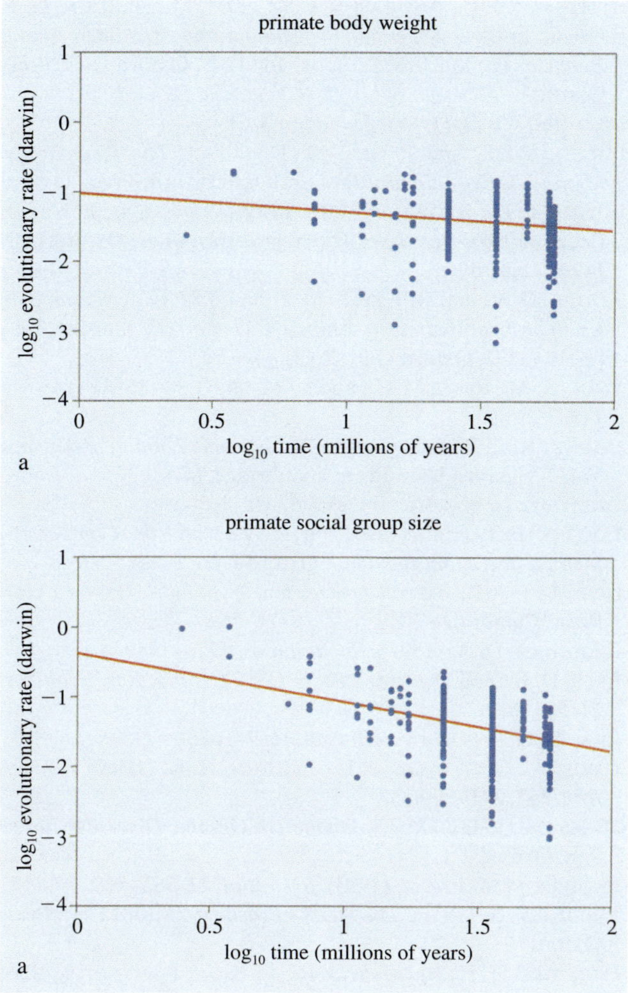

FIGURE 15 Relationship between evolutionary rate and time with respect to body weight (a) and the size of the social group (b). Based on the values of the slope and intercept, it is possible to deduce that group size is evolving at a faster rate than body weight (adapted from Gittleman *et al.*, 1996).

New Directions

Comparative studies are testing historical patterns of behavior with accuracy and rigor. A timely combination of comparative information on a variety of traits and organisms along with precise phylogenetic trees and statistical methods now allow detailed tests of hypotheses. Specifically, comparative approaches can effectively examine the evolutionary origin, rate, lability, and covariation of behavior patterns. This is an amazing development — one that would be a pleasant surprise to Darwin if he were to reappear and see the many applications of phylogenetic thinking. However, these are all problems of history, essentially reconstructions of the past. Can comparative methods also deal with issues of the present (experimentation) and possibly even predict the future (conservation)? Two lines of work are promising.

The first touches on a traditional weakness of comparative studies: lack of experimental tests. Can phylogenetic studies involve experimental controls and manipulation such that broad correlations can be pinned down to more specific causal mechanisms? A study of gasterosteid (stickleback) fishes shows one way of using phylogenies in an experimental setting (McLennan, 1991). Sticklebacks vary considerably in nuptial coloration, ranging from no coloration to gold, red, blue, and black. Phylogenies across six taxa of sticklebacks suggested that the origin and elaboration of nuptial coloration are related to (i) male–male interactions, (ii) parental care, and (iii) male–female interactions. Choice experiments were constructed in which female sticklebacks (*Gasterosteus aculeatus*) were allowed to select males based on color, breeding conditions, or behavioral affinities. Observations clearly indicated that females cue mainly on male body color rather than patterns of parental care or male–male interactions, suggesting that observed phylogenetic patterns relate to male–female interactions and very likely female choice of male body color intensity. Phylogenetic patterns can indeed be tested experimentally.

To preserve the earth's biodiversity, it is necessary to effectively identify species threatened by extinction and to protect those with characteristics which might make them vulnerable in the future. Comparative studies can directly help in two ways (Brooks *et al.*, 1992). First, phylogenetic trees may be used to identify which species are more unusual than others. Species such as the red (*Ailurus fulgens*) and giant pandas (*Ailuropoda melanoleuca*), which are the only species in a taxonomic subfamily (Ailurinae), are obviously more distinctive than some other carnivores such as the species-rich canids or felids. This alone does not translate into greater action to protect these species. However, a phylogeny used in this way can be an important pointer to which species should then receive immediate study and, in turn, be evaluated for potential problems. For example, detailed comparative behavioral studies have discovered some features of the giant panda that are significantly different from those of other carnivores which hamper the panda's ability to adapt to changing environmental conditions (Gittleman, 1994); specifically, slow neonatal development both in weight gain and in associated behavioral maturity. Other characteristics, such as types of mating systems, dietary preferences, or levels of aggression, may indicate future problems. Second, given that many characteristics are known to follow phylogenetic lines, if there is an endangered, relatively unknown species in a genus with many other, better known species, then the characteristics of these other species can be used as a short-term predictor for what characteristics might be threatening. Just as comparative methods are now readily available to reconstruct the past, even with characteristics such as complex behavior patterns, these same techniques may be instrumental in predicting the future.

References Cited

Avise, J. C. (1994). *Molecular Markers, Natural History and Evolution.* Chapman & Hall, New York.

Brooks, D. R., and McLennan, D. A. (1991). *Phylogeny, Ecology, and Behavior.* Univ. of Chicago Press, Chicago.

Brooks, D. R., Mayden, R. L., and McLennan, D. A. (1992). *Trends Ecol. Evol.* **7,** 55–59.

Burghardt, G. M., and Gittleman, J. L. (1990). In *Interpretation and Explanation in the Study of Animal Behavior* (M. Bekoff and D. Jamieson, Eds.), pp. 192–225. Westview, Boulder, CO.

Campbell, C. B. G. (1976). In *Evolution, Brain, and Behavior* (R. B. Masterton, W. Hodos, and H. Jerison, Eds.), pp. 107–113. Wiley, New York.

Coddington, J. A. (1994). In *Phylogenetics and Ecology* (P. Eggleton and R. Vane-Wright, Eds.), pp. 53–78. Academic Press, London.

Darwin, C. (1859). *On the Origin of Species by Means of Natural Selection; or, The Preservation of Favored Races in the Struggle for Life.* Murray, London.

Darwin, C. (1872). *The Origin of Species,* 2nd ed.

de Queiroz, A., and Wimberger, P. H. (1993). *Evolution* **47,** 46–60.

DeVoogd, T. J., Krebs, J. R., Healy, S. D., and Purvis, A. (1993). *Proc. R. Soc. London* **254,** 75–82.

Edwards, S. V., and Kot, M. (1995). *Evolution* **49,** 1134–1146.

Geffen, E., Gompper, M. E., Gittleman, J. L., Luh, H.-K., Macdonald, D. W., and Wayne, R. K. (1996). *Am. Nat.* **147,** 140–160.

Gingerich, P. D. (1993). *Am. J. Sci.* **293A,** 453–478.

Gittleman, J. L. (1989). In *Perspectives in Ethology* (P. P. G. Bateson and P. H. Klopfer, Eds.), Vol. 8, pp. 55–83. Plenum, New York.

Gittleman, J. L. (1994). *BioScience* **44,** 456–464.

Gittleman, J. L., and Decker, D. M. (1994). In *Behaviour and Evolution* (P. J. B. Slater and T. R. Halliday, Eds.), pp. 80–105. Cambridge Univ. Press, Cambridge, UK.

Gittleman, J. L., and Luh, H.-K. (1992). *Annu. Rev. Ecol. Syst.* **23,** 383–404.

Gittleman, J. L., Anderson, C. G., Kot, M., and Luh, H.-K. (1996). In *Phylogenies and the Comparative Method in Animal Behavior* (E. Martins, Ed.), pp. 166–205. Oxford Univ. Press, Oxford.

Haldane, J. B. S. (1949). *Evolution* **3,** 51–56.

Harvey, P. H., and Pagel, M. D. (1991). *The Comparative Method in Evolutionary Biology.* Oxford Univ. Press, Oxford.

Harvey, P. H., Brown, A. J. L., Maynard Smith, J., Nee, S. (Eds.) (1996). *New Uses For New Phylogenies.* Oxford Univ. Press, Oxford.

Lauder, G. V., and Reilly, S. M. (1996). In *Phylogenies and the Comparative Method in Animal Behavior* (E. Martins, Ed.), pp. 104–137. Oxford Univ. Press, Oxford.

Leroi, A. M., Rose, M. R., and Lauder, G. V. (1994). *Am. Nat.* **143,** 381–402.

Lorenz, K. (1941). *Studies in Animal and Human Behaviour,* Vol. 2. Harvard Univ. Press, Cambridge, MA.

Lorenz, K. (1958). *Sci. Am.* **199,** 67–78.

Martins, E. P. (Ed.) (1996). *Phylogenies and the Comparative Method in Animal Behavior.* Oxford Univ. Press, Oxford.

Mayr, E. (1963). *Animal Species and Evolution.* Harvard Univ. Press, Cambridge, MA.

McLennan, D. A. (1991). *Evolution* **45,** 1773–1789.

Miles, D. B., and Dunham, A. E. (1993). *Annu. Rev. Ecol. Syst.* **24,** 587–619.

Prum, R. O. (1994). *Evolution* **48,** 1657–1675.

Purvis, A., Gittleman, J. L., and Luh, H.-K. (1994). *J. Theor. Biol.* **167,** 293–300.

Ridley, M. (1983). *The Explanation of Organic Diversity.* Clarendon, Oxford.

Sanderson, M. J., *et al.* (1993). *Syst. Biol.* **42,** 562–568.

Tinbergen, N. (1951). *The Study of Instinct.* Oxford Univ. Press, Oxford.

Wanntorp, H.-E., Brooks, D. R., Nilsson, T., Nylin, S., Ronquist, F., Stearns, S. C., and Wedell, N. (1990). *Oikos* **57,** 119–132.

General References

Eggleton, P., and Vane-Wright, R. I. (Eds.) (1994). *Phylogenetics and Ecology.* Academic Press, London.

Harvey, P. H., Brown, A. J. L., Maynard Smith, J., and Nee, S. (Eds.) (1996). *New Uses for New Phylogenies.* Oxford Univ. Press, Oxford.

STEPHEN C. STEARNS
Zoological Institute
Basel University
Basel, Switzerland

The diversity of life on Earth is also a diversity of life histories. Higher plants and animals range in age at maturity from a few hours to about a century; in adult size from microscopic rotifers and nematodes to redwoods and blue whales; in fecundity from the handful of offspring born per lifetime to humans, albatrosses, and burying beetles to the billions born per season to oysters and orchids; and in life span from hours to centuries. What caused this diversity to evolve?

Life Histories

Life History Evolution Is Caused by the Interactions of Demography, Genetics, Trade-Offs, and Constraints

Why mature at a particular age and size, not earlier or later, larger or smaller? Why have many or few offspring? Why have a longer or shorter life span? Is aging inevitable? These questions define the scope of life history evolution, and this article presents an outline of how to answer them. Life history evolution shares with sex allocation and kin selection the credit for transforming evolution into a theory that makes successful quantitative predictions about phenotypes. The field owes its success to its concentration on the impact of age- and size-specific selection pressures on the evolution of age- and size-specific fecundity and mortality rates connected by trade-offs.

Effects from four sources — demography, quantitative genetics, trade-offs, and lineage-specific constraints — combine to cause life history variation. The causes flow in two directions: from variation in the genes through gene expression in development, morphology, and physiology to the phenotype and then back from variation among phenotypes in fecundity and survival (natural selection) to changes in gene frequencies (Fig. 1). A representation of the process that captures most of the essential elements shows life history evolution as an interplay between extrinsic mortality rates, intrinsic trade-offs, and lineage-specific constraints that limit trait variation.

Natural selection occurs when the individuals in a population differ heritably in their survival and fecundity. The first causal source, demography, is part of the explanation because survival and fecundity vary with age and size. De-

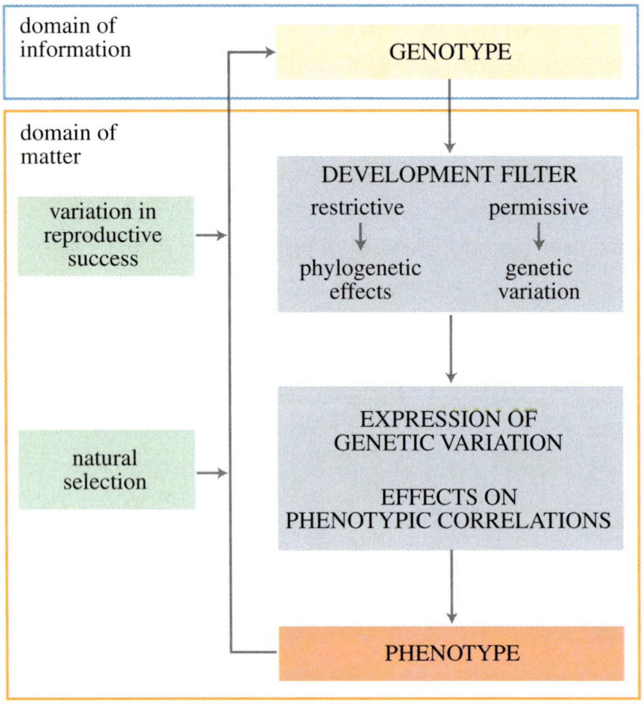

FIGURE 1 A schematic diagram of the causation relationships between the principal phenomena involved in organism evolution. Natural selection is expressed through reproductive success, where the variation is heritable. However, the connection between variation in genomes and variation in fitness is indirect, mediated by development and physiology, some of which is strongly influenced by phylogeny and by trade-offs connecting traits.

mography connects age- and size-specific variation in survival and fecundity to variation in fitness; it also indicates the strength of natural selection on different traits. However, it does not indicate how a population will respond to selection in the short term. For that we need to know how much genetic variation is present and how strongly the traits under selection are tied together by trade-offs.

Most life history traits are influenced by many genes of small effect. Their combined effect can be estimated through the methods of quantitative genetics, the second causal source. It is the appropriate tool for analyzing the genetics of most ecologically important traits. Developed for plant and animal breeding, quantitative genetics can be extended to natural populations by including the effects of phenotypic plasticity. Phenotypic plasticity has two important roles in evolution. First, by modifying the relationships of traits with fitness, it changes the selection pressures on traits across environments. Second, by changing the expression of genetic variation, it changes the genetic response to selection across environments.

Trade-offs are the third element of causation. A trade-off exists when an increase in fitness realized through a change in one trait is linked to a decrease in fitness in another trait. The direction that life history evolution will take in response to a novel selection pressure depends on the relative strength of the trade-offs that are present. Oft-encountered trade-offs include those between the number of offspring and their survival rate as juveniles and between reproductive investment and adult survival.

The fourth and final causal element derives from comparative biology. Comparisons among higher taxa raise striking questions. Do organisms that delay maturity live longer? Do mammals with larger litters have smaller offspring when both traits are adjusted for the body weight of the mothers? Such questions are answered by methods that weigh the effects of body size and taxonomic group on patterns of variation. They indicate how much of a pattern to attribute to history and lineage and how much to attribute to microevolutionary processes that operated within the local population in the recent past.

Next, I apply some of these methods to the major life history traits.

Age and Size at Maturity Are Pivotal Traits

Maturation divides a life into preparation and fulfillment. Age at maturity is a pivot for natural selection. Fitness is often more sensitive to changes in this trait than to changes in any other. With maturation, selection pressures change dramatically. Before maturation, aging cannot evolve; after maturation, aging begins.

Age and size at maturity evolve to the point at which the demographic pressure to mature early is balanced by the benefits of delayed maturity (Fig. 2). Kozlowski (1992) noted, "Small organisms are usually small not because smallness improves fecundity or lowers mortality. They are small because it takes time to grow large, and with heavy mortality the investment in growth would never be paid back as increased fecundity."

The principal benefit of early maturation is that rapid development increases the probability of surviving to reproduce. Organisms that mature earlier also have higher fitness because their offspring are born earlier and start reproducing sooner. The costs of early maturity are the benefits of delayed maturity, of which two are generally important. First, if delaying maturity permits further growth and if fecundity increases with size, then delayed maturity leads to higher fecundity. Maturity will be delayed for this reason until the fitness gained through increased fecundity is balanced by the fitness lost through longer generation time and lower survival to maturity. Second, if delaying maturity improves the quality of offspring or parental care, then delaying maturity improves the daily survival rates of the offspring. Maturity will be delayed for this reason until the

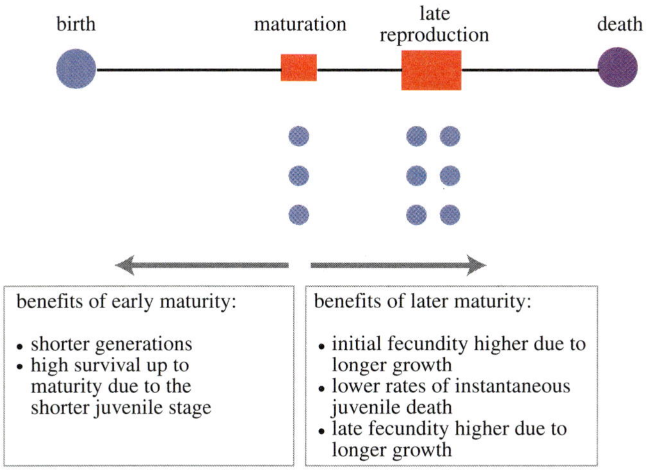

FIGURE 2 Principal benefits of maturation at different ages and sizes (indicated by the size of the red rectangles). The blue circles represent newborns.

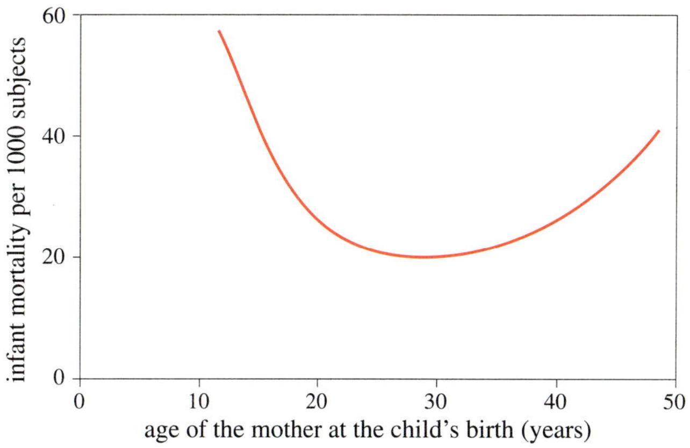

FIGURE 3 Variations in infant mortality rate as a function of the mother's age for humans. The data are based on 107,038 infant deaths documented by the U.S. National Center for Health Statistics in the 2-year period 1960–1961 (based on an unpublished analysis by R. Stafford).

fitness gained through the increase in juvenile survival rate due to the mother's older age at first birth balances the fitness lost through a longer generation time and that portion of survival to maturity that is attributed simply to the delay and not to any change in daily juvenile survival rate.

That age and size at maturity do vary among closely related species, among populations within species, and among individuals within populations suggests that they can respond rapidly to natural selection. Age at maturity responds to artificial selection in flour beetles and fruit flies, and genetic variation for age and size at maturity has been measured in many species. This suggests that age and size at maturity can be adjusted by natural selection to local conditions within populations. On the other hand, that body size and lineage strongly affect age and size at maturity suggests important constraints on age and size at maturity imposed by history and design.

Optimization implies that a balance of costs and benefits to fitness determines the value of a trait. We can be certain that one cost of early maturation, the instantaneous mortality rate of the offspring, does climb rapidly when females mature earlier. The capacity of younger and smaller females to produce offspring that can survive must decrease with age; at some point the female will be so small and poorly developed that she cannot produce any offspring (Fig. 3).

The major benefit of delayed maturity is that additional time for development allows the organism to grow larger and gain fecundity. Fecundity does increase with size in fish, amphibians, reptiles, many invertebrates, and most higher plants. Thus, earlier maturation results in shorter generation times and a shorter period of exposure to mortality before maturity; later maturation results in increased fecundity and lower juvenile mortality rates. At an intermediate age, an optimum should exist. The challenge is to estimate the relevant costs and benefits, model their interactions, and then predict the optimum and determine whether it matches the data (Fig. 4).

Models tested using data from flies, lizards, salamanders, and fish have performed well, accounting for 80–88% of the variation in age at maturity in natural populations (Bell, 1980; Roff, 1981; Stearns and Koella, 1986; Kozlowski, 1992). In general, age at maturity appears to be adjusted to an intermediate optimum.

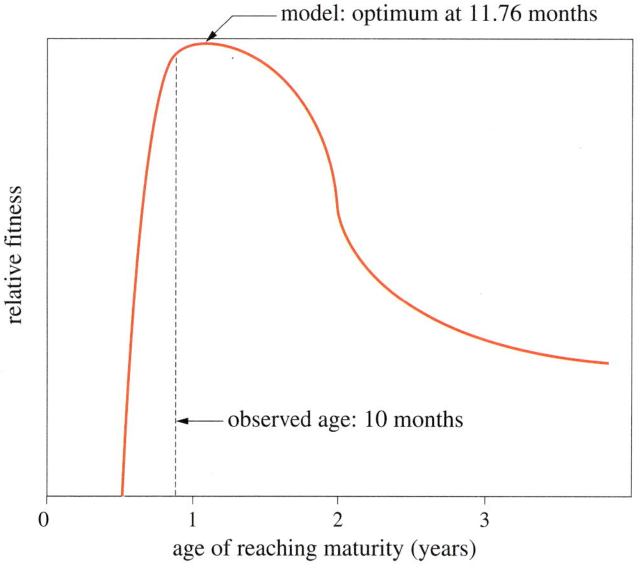

FIGURE 4 Comparison of predictions and observations for age at maturity in the eastern fence lizard in South Carolina. A model based on a linear increase in fecundity with size predicts a slightly greater age at maturity than is observed.

The Diversity of Reproductive Investment

The second major feature of the life cycle is the age-specific schedule of reproductive investment. How many offspring should be produced in each reproductive attempt? Many organisms share the record for the smallest clutch — one. The single offspring of a blue whale is as large at birth as an adult African elephant and grows to half its adult weight in the first 6 months of life. However, newborn whales are not especially large compared to their mothers. Bats have the largest offspring for their body size in the mammals, the record being held by *Pipestrellus pipestrellus*. It carries twins whose combined weight at birth is 50% of the weight of the mother after birth, and the mother must fly and feed while pregnant. The kiwi, the size of a large domestic chicken, lays the largest egg for its size of any bird — more than five times larger than the largest eggs from domestic chickens selected for large egg size. Its reproductive investment of 20%, however, is dwarfed by that of the caecilian *Dermophis mexicanus,* a tropical amphibian which gives birth to a clutch that weighs up to 65% of her post-birth weight. The species with the lowest reproductive potential per lifetime is probably not a large mammal or bird — elephants, whales, humans, and albatrosses all have a reproductive potential of 10–20 offspring — but a burying beetle, which may have only one reproductive season in which it lays less than 10 and perhaps as few as 5 eggs.

At the other end of the scale, many organisms have many very small offspring. The tiny seeds of orchids number in the hundreds of millions to billions. Large bivalves and some large fish (sturgeon, large catfish, and cod) release hundreds of millions of tiny eggs in each spawning. The world record for worst juvenile survival is probably held by an orchid, whose seeds weigh less than 1 μg and have a chance of about 1 in 1 billion of surviving to reproduce.

The analysis of the evolution of offspring number began with the British ornithologist David Lack, who suggested that altricial birds should lay a clutch with the number of eggs that fledges the most offspring. This is called the *Lack clutch* in his honor. Modern theory analyzes the effects that cause deviations from the Lack clutch. Trade-offs with other parental and offspring traits, temporal variation in optimal clutch size, and parent–offspring conflict will all reduce the Lack clutch. Opportunities for selective abortion and the potential to reduce broods flexibly to adjust the clutch size to unpredictable resource levels will both increase the Lack clutch (Fig. 5).

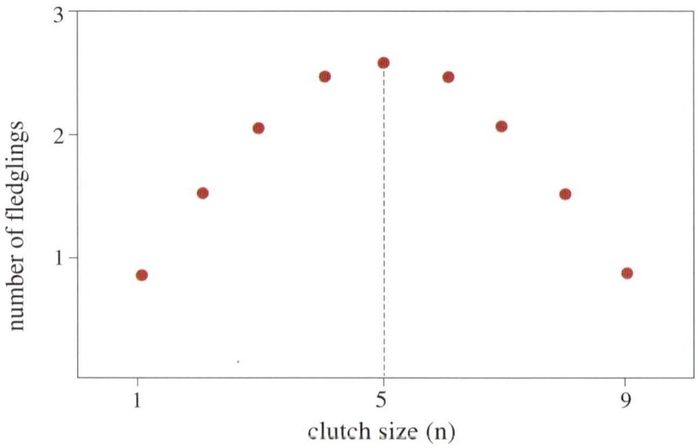

FIGURE 5 A model of the Lack clutch, based on the assumption that survival rates of fledglings decrease linearly as clutch size increases.

Clutch size may also be constrained within lineages so that no variation is possible. All birds in the order Procellariformes lay just one egg, in some species only every other year. These widely foraging seabirds vary in how far they travel to feed, in body size, and in age at maturity. The largest member of the order, the wandering albatross, takes up to 33 days and travels up to 15,000 km on a single foraging flight. Here, a clutch size of one is not surprising because the single chick, which has a special starvation physiology, must wait up to 1 month for a meal. These birds can live 50–60 years and lay a total of 20–25 eggs per lifetime. However, it is not clear that a clutch size of one is best for the smaller, less widely foraging species in the group. They appear to be constrained by phylogeny rather than adapted to their particular ecological conditions.

The Reproductive Effort Model Makes Quantitative Predictions That Withstand Experimental Test

The centerpiece of life history theory consists of models that explore the optimization of reproductive effort over the whole life span taking costs of reproduction into account. This approach, developed by, among others, Williams (1966), Schaffer (1974), Charlesworth (1994), Leon (1976), Michod (1979), and Kozlowski (Kozlowski and Uchmanski, 1987), is called the *reproductive effort model.* Its predictions are well confirmed by experiments carried out on kestrels in Holland (Daan *et al.,* 1990), on guppies in Trinidad (Reznick *et al.,* 1990), and on fruit flies in laboratory experiments (Stearns *et al.,* 1996).

The kestrel study yielded data of unusually good quality and applied the theory in great detail. The idea that framed the study was suggested by George Williams. He pointed out that the fitness of parents consists, at the moment of decision, of two components — the reproductive value of offspring that they are currently producing and their residual reproductive value (all their later contributions to fitness). For a bird, its present reproductive value is its clutch size multiplied by the reproductive value of one egg. The reproductive value of one egg is the sum of the probability that the egg becomes a bird that survives to reproduce once multiplied by its first clutch size, plus the probability that it survives to reproduce twice, multiplied by its second clutch size, and so forth for all clutches that it might have. The residual reproductive value of a parent is the probability that it will survive to reproduce once more multiplied by the expected number of offspring that it will then have, plus the probability that it will survive to reproduce a second time, multiplied by the expected number of offspring, and so forth until it dies.

Daan *et al.* (1990) calculated the reproductive value of the clutch and the residual reproductive value of the parents for experimentally reduced and enlarged clutches. If kestrels are correctly balancing the costs and benefits of reproductive investment, then the total reproductive value of the clutches actually laid should be larger than that of either the reduced or the enlarged clutches. These birds live near the North Sea in The Netherlands where survival is good and clutch sizes are large. However, the probability of surviving to be 1, 2, or 3 years old depends on the date the eggs are laid, as does the probability that a nest will succeed. When they accounted for these factors, the Dutch ornithologists found that the clutch size actually laid was the one that yielded the highest reproductive value. This is a remarkable result because the selection of effects to measure was made years before the calculation of the reproductive values, and most of the plausible costs of reproduction were measured, including those that are only realized in the following generation (Table 1).

Another impressive confirmation of reproductive effort models was achieved in field manipulation experiments by Reznick and colleagues on the guppy, *Poecilia reticulata.* Guppies are small, viviparous fish that live in shallow streams in Trinidad (as well as in home aquariums throughout the world). At some sites their main predator is a cichlid fish, *Crenicichla alta,* which can eat large, sexually mature guppies and causes high mortality rates in all size classes. At other sites, their main predator is a killifish, *Rivulus hartii,* which eats mostly small, juvenile guppies and causes lower mortality rates in all size classes. Before the experiments were begun, Reznick found that guppies from *Crenicichla* sites mature earlier, make a larger reproductive effort, and have more and smaller offspring than do guppies from *Rivulus* sites. The differences were heritable, and they corresponded with the predictions of reproductive effort models.

To demonstrate that predation caused the pattern, Reznick *et al.* (1990) manipulated natural populations by changing predators. This decreased the mortality rates in all age classes. After 11 years, or 30–60 generations, they observed significant life history evolution in the predicted direction (Table 2).

Before the manipulation began in 1976, under relatively high adult mortality, both males and females matured earlier and smaller, and the females had larger first litters, smaller young, a shorter interbrood interval, and made a larger reproductive effort. Eleven years later, guppies had evolved most of the differences seen in populations under similar selection pressures for a much longer time. Age and

TABLE 1

Computation of Reproductive Values in Experimentally Reduced, Unchanged, and Enlarged Kestrel Broods[a]

	REDUCED	CONTROL	ENLARGED
Brood			
Number of broods	28	54	20
Mean clutch size	5.25	5.19	5.40
Mean brood size (at 10 days)	4.53	4.20	4.83
Mean change in brood size	−1.74	0.00	+2.51
Mean number fledged	2.60	3.95	5.84
V_c, reproductive value of clutch	2.52	4.20	5.59
Parents			
Number of males and females	49	85	35
Local parental survival, p_1	0.653	0.588	0.429
V_p, residual reproductive value	9.88	8.89	6.49
Total			
V, total reproductive value	12.40	13.09	12.08

[a]From Daan *et al.* (1990). Brood manipulations were done at 10 days.

TABLE 2

Life History Patterns in Guppies after a Field Manipulation Experiment That Used Natural Predators to Increase Juvenile Mortality Rates and Decrease Adult Mortality Rates, Contrasted with Those Observed under the Same Two Natural Predators without Manipulation[a]

	REZNICK *et al.* (1990)		REZNICK (1982)	
	Control mortality		Introduction mortality	
LIFE HISTORY TRAIT	High	Low	High	Low
Male age at maturity (days)	48.5	58.2**	51.8	58.8**
Male size at maturity (mg, wet)	67.5	76.1**	87.7	99.7**
Female age at first birth (days)	85.7	92.3*	71.5	81.9**
Female size at first birth (mg, wet)	161.5	185.6**	218.0	270.0**
Size of litter 1	4.5	3.3*	5.2	3.2**
Size of litter 2	8.1	7.5	10.9	10.2
Size of litter 3	11.4	11.5	16.1	16.0
Offspring weight (mg, dry), litter 1	0.87	0.95	0.84	0.99**
Offspring weight, litter 2	0.90	1.02*	0.95	1.05*
Offspring weight, litter 3	1.10	1.17	1.03	1.17**
Interlitter interval (days)	24.5	25.2	22.8	25.0
Reproductive effort (%)[b]	22.0	18.5	25.1	19.2*

[a]From Reznick *et al.* (1990).
[b]The sum of four consecutive estimates of dry weight of brood divided by dry weight of maternal soma.
*$p < 0.05$; ** $p < 0.01$.

size at maturity, size of first litter, reproductive effort early in life, and size of offspring in the first two litters had changed and corresponded qualitatively and quantitatively to the differences found in the unmanipulated populations. Changes in the intensity of mortality rates caused by a manipulation of predation pressure led to rapid evolutionary change in life history traits in the direction predicted.

The rapidity of the change is striking. If the unmanipulated populations were at evolutionary equilibrium, then the reattainment of equilibrium following a major perturbation took just 30–60 generations, 11 years elapsed time, for most of the traits measured. The traits that changed most were those occurring early in life, confirming a basic assumption of the evolutionary theory of aging: Selection on reproductive performance earlier in life is stronger than selection on reproductive performance later in life.

The third confirmation of reproductive effort models was achieved with fruit flies in laboratory experiments on evolution (Stearns *et al.*, 1996). The first treatment consisted of constant high adult mortality and the second treatment of constant low adult mortality. To ensure that the effects of treatments were due to differences in mortality and not in density, adult and larval density were kept the same in all treatments. The evolution of reproductive effort and age and size at maturity was rapid and confirmed the reproductive effort model: Under high mortality, the flies eclosed earlier, increased fecundity early in life, decreased fecundity later in life, and evolved a shorter life span. The rate at which the traits changed corresponded with the intensity of the selection pressures predicted by the reproductive effort model. Thus, when an evolutionary change entails costs, further changes are made to reduce the costs. Because the only difference between the two treatments was the level of adult mortality, this experiment established differences in extrinsic mortality rates, interacting with intrinsic trade-offs, as a sufficient cause for the evolution of life history traits.

Life Span and Aging Are Explained by Both Life History Optimization and Nonadaptive Genetic Variation

The third major life history trait is reproductive life span, which varies widely among species. The longest lived invertebrates are sea anemones, lobsters, and bivalves; some clams can live 220 years. The shortest lived are rotifers, insects, and some crustaceans; some live less than 1 week. The record for the maximum captive life span among mammals in the world's zoos is held by the African elephant at 57 years, with the domestic horse and the spiny echidna tied for second at 50 years. Some humans live up to 120 years. The distribution of life spans in animals ranges from about 1 week in rotifers to about a century in some turtles, primates, bivalves, and large fish. Most animals reach their maximum life span at less than 30 years, but long-lived animals with life spans on the order of a century have evolved independently in the bivalves, fish, reptiles, and mammals. Plants vary in life span from weeks to many centuries. This needs explanation.

Life history theory views the evolution of the reproductive life span as a balance between selection to increase the number of reproductive events per lifetime and effects that increase the intrinsic sources of mortality with age. The first lengthen life, and the second shorten it. Age-specific selection pressures can adjust the length of life to an intermediate optimum determined by the interaction of selection with trade-offs intrinsic to the organism, whether there is any aging or not. The selection pressures that lengthen life decrease the reproductive value of offspring and increase the reproductive value of adults. These include lower adult mortality rates, higher juvenile mortality rates, and increased variation in juvenile mortality rates from one reproductive event to the next (Fig. 6).

The analysis of selection pressures that lengthen the reproductive life span starts with the distinction between *semelparity* (one reproductive event per lifetime) and *iteroparity* (more than one reproductive event per lifetime). All models agree that in a comparison of closely related semelparous and iteroparous forms with otherwise similar life histories (age and size at maturity and physiological condition), the semelparous forms should have higher fecundity. This appears to be the case. The reproductive allocations of semelparous plants are from 1.5 to 10.0 times as large as the reproductive allocations of their iteroparous relatives.

Because semelparity should evolve more easily in early than in late-maturing species, the late-maturing semelparous rosette plants of tropical alpine and subalpine regions and of deserts are of particular interest. Young (1990) did an exemplary long-term demographic study on one such group, the lobelias of Mount Kenya, where the semelparous *Lobelia telekii* lives near a close relative, the iteroparous *Lobelia keniensis.* Iteroparous *L. keniensis* flower only once every 7–14 years. To predict where the plants should be semelparous and where they should be iteroparous, Young applied a model developed by Charnov and Schaffer that told him under which mortality regime iteroparity was expected and under which regime semelparity was predicted. The long-term demographic data were consistent with the predictions of the model.

Young's (1990) study of lobelias on Mt. Kenya suggests that reproductive life span is adapted to local conditions.

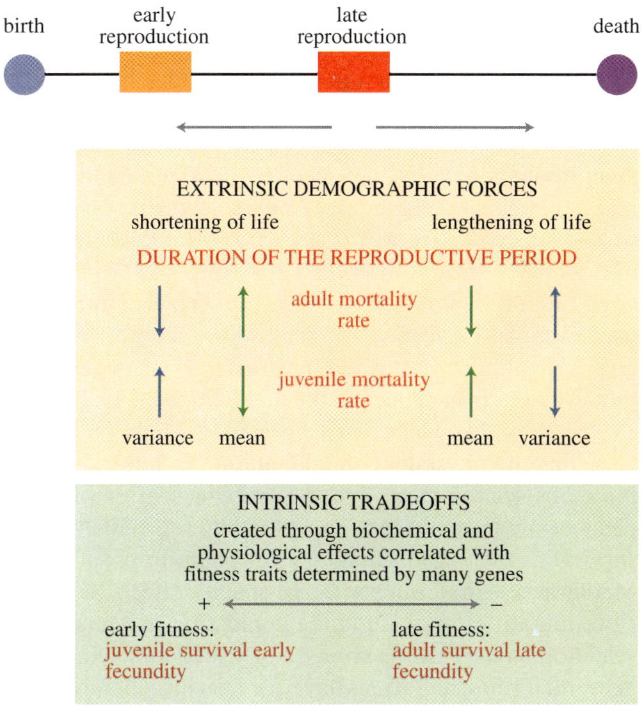

FIGURE 6 Diagram of evolutionary factors concerning life span and aging. The diagram depicts the balance of forces operating on the life span of a model organism. An increase in the mean or a decrease in the variance of juvenile mortality rates selects for increased life span by devaluing offspring relative to adult individuals. An increase in the mean or a decrease in the variance of adult mortality rates selects for decreased life span by devaluing adults relative to offspring. Once mortality and fecundity rates have established a point in the life cycle beyond which adults make little contribution to fitness, mutations with effects on repair mechanisms can invade the population. These produce pleiotropic effects that establish negative connections between fitness traits expressed early and late in life.

However, superimposed on the adaptive pattern determined by optimal allocation of resources to maintenance, growth, and reproduction are the maladaptive effects of aging. Why do many organisms become more susceptible to death as they age? Why do they not live forever if accidents and violence are avoided?

The effects of aging increase mortality rates and decrease fecundity rates in later life beyond the levels predicted from optimal allocation. The best measure of aging is one that combines both effects: In an aging population the product of age-specific survival and fecundity rates decreases with age. A necessary condition for the evolution of aging is that the organisms have a separation of germline and soma because the force of selection declines with the age of the soma, not with the age of the germline. However, the division of germline and soma is not necessarily developmental; it can also be achieved by a return to a single-cell stage in the life cycle, where the single cell in effect becomes the germline.

Weismann determined in the past century that aging is a property of the soma, and Fisher and Haldane knew 60 years ago that the force of selection declines with age. Hamilton (1966) and Charlesworth (1994) provided a clear mathematical understanding of how selection pressures on survival rates decline with age and do so at different rates in organisms with different life histories. That is the first key idea in the evolution of aging. In humans in industrialized countries, selection to increase the survival rate decreases to almost zero by the time one is 45 years old, a fact that does not bring cheer to the middle-aged (Fig. 7).

The second key idea was suggested by Medawar (1946). Genetics and physiology should interact with the declining force of selection to produce age-specific erosion of all aspects of organismal function, degrading reproduction and survival as the organism ages. A mechanism proposed by Medawar was mutations with age-specific effects. If deleterious mutations are held in the population by a mutation-selection balance, if the strength of selection declines with age, and if some mutations have age-specific deleterious effects, they should increase in frequency as the age class that they affect becomes older. This is now known as the *mutation accumulation hypothesis.*

Williams (1957) added another mechanism, one that appears to occur frequently: genes with conflicting effects on early and late age classes. Consider a gene with positive effects early in life, during the juvenile or early reproductive period, and deleterious effects late in life. Its net effect on fitness will be positive because the younger age classes make a larger contribution to fitness than do the older ones. Such mutations will spread, and if they are not fixed in the population they will maintain a negative genetic correlation between traits expressed early and late in life. This idea is known as the *antagonistic pleiotropy hypothesis*— antagonistic because the effects are positive in one age class and negative in another and pleiotropic because each such gene affects two or more traits.

FIGURE 7 The age-specific sensitivity of fitness to changes in survival rates for the population of the United States about 1940 (after B. Charlesworth, 1973).

Thus, the evolutionary answer to the question "Why must we age?" has three parts. First, reproduction must be asymmetrical in the sense that it is possible to clearly distinguish mother from offspring. Second, the force of selection declines with age; beyond a certain age organisms become irrelevant to evolution. Third, given these conditions two sorts of genetic effects become possible — the accumulation of mutations with stronger effects on older age classes than on younger ones and the accumulation of genes that benefit younger age classes at the expense of older ones. Maturation is the point in the life history at which these effects should start to occur; before maturation, they should not be seen. Thus, aging should follow the onset of reproduction with diffuse erosion of physiological and biochemical functions caused by many genes of small effect that produce aging as a by-product of selection for reproductive performance earlier in life, not as an adaptation. Aging caused by genes with large effects is not ruled out but is not expected to be the usual case. Genetically programmed cell death does occur, but there is no genetically programmed organismal death — only genetically programmed reproductive performance that may have death as a side effect.

It follows that fitness is maximized at a level of investment in repair that is less than would be required for indefinite survival. That is why we grow old and die. This insight, developed by Kirkwood (1987) and called the *disposable soma theory,* establishes a connection between the evolutionary theory of aging and gerontology. It predicts that aging should result from the accumulation of unrepaired somatic damage and that species with different longevities should exhibit corresponding differences in their levels of somatic repair. Repair is costly. More than 2% of the energy budget of cells is spent on DNA repair and proofreading, on processes that determine accuracy in protein synthesis, on protein turnover, and on the scavenging of oxygen radicals. Moreover, the genes that affect repair processes should have antagonistically pleiotropic effects. Decreases in extrinsic adult mortality will increase the value of investment in repair; coupled to that should be decreased investment in growth and reproduction, leading to longer life spans and lower reproductive efforts. Conversely, when extrinsic adult mortality increases, then investment in repair is devalued and there should be shifts to less repair and higher reproductive investment.

These ideas have been tested in experiments on fruit flies carried out by, among others, Rose, Partridge, Hillesheim, and Zwaan (Stearns and Kaiser, 1996) with consistent results: When the force of selection on older age classes is increased by only allowing older organisms to reproduce, aging is postponed. In some cases, increased life span is accompanied by reduced fecundity early in life, whereas in other cases it is accompanied by changes in larval growth rates, survival rates, and competitive ability. When selection is applied to the other end of the life, to size at maturity, the smaller flies that evolve live longer and have lower fecundity.

The evidence on the evolution of aging can be summarized as follows: Life span responds rapidly to selection in the laboratory in a manner consistent with evolutionary theory. Longer life must be paid for by reductions in performance early in life in lower fecundity, smaller body size, or reduced larval survival and competitive ability. These connections among traits expressed early and late in life confirm the mechanism suggested by Williams (1957).

Future Challenges

The analysis of life history evolution has revealed much about the selection pressures that shape life cycles and some of the genetics underlying that diversity. It has made the first steps toward explaining why the planet is occupied by organisms with such an enormous range of sizes, life spans, ages at maturity, and reproductive investments. The least understood and most challenging parts of the explanation are the mechanisms underlying trade-offs and the evolutionary origin of phylogenetic constraints on adaptation. Developing and applying the methods needed to unravel these two effects remain a challenge.

References Cited

BELL, G. (1980). *Am. Nat.* **116,** 45–76.

CHARLESWORTH, B. (1973). *Am. Nat.* **107,** 303–311.

CHARLESWORTH, B. (1994). *Evolution in Age-Structured Populations,* 2nd ed. Cambridge Univ. Press, Cambridge, UK.

DAAN, S., DIJKSTRA, C., and TINBERGEN, J. M. (1990). *Behavior* **114,** 83–116.

HAMILTON, W. D. (1966). *J. Theor. Biol.* **12,** 12–45.

KIRKWOOD, T. B. L. (1987). In *Evolution of Longevity in Animals* (A. D. H. Woodhead and K. H. Thompson, Eds.), pp. 209–218. Plenum, New York.

KOZLOWSKI, J. (1992). *Trends Ecol. Evol.* **7,** 15–19.

KOZLOWSKI, J., and UCHMANSKI, J. (1987). *Evol. Ecol.* **1,** 214–230.

LEON, J. A. (1976). *J. Theor. Biol.* **60,** 301–335.

MEDAWAR, P. B. (1946). *Modern Q.* **1,** 30–56.

MICHOD, R. E. (1979). *Am. Nat.* **113,** 531–550.

REZNICK, D. N., BRYGA, H., and ENDLER, J. A. (1990). *Nature* **346,** 357–359.

ROFF, D. A. (1981). *Am. Nat.* **118,** 405–422.

SCHAFFER, W. M. (1974). *Ecology* **5,** 291–303.

STEARNS, S. C., and KAISER, M. (1996). *Evolution* **50,** 795–806.

STEARNS, S. C., and KOELLA, J. (1986). *Evolution* **40,** 893–913.

STEARNS, S. C., KAISER, M., and BLARER, A. (1996). *Plant Species Biol.* **11,** 97–105.

WILLIAMS, G. C. (1957). *Evolution* **11,** 398–411.

WILLIAMS, G. C. (1966). *Am. Nat.* **100,** 687–90.

YOUNG, T. P. (1990). *Evol. Ecol.* **4,** 157–172.

General References

ROFF, D. A. (1992). *The Evolution of Life Histories.* Chapman & Hall, London.

ROSE, M. (1991). *The Evolution of Aging.* Oxford Univ. Press, Oxford.

STEARNS, S. C. (1992). *The Evolution of Life Histories.* Oxford Univ. Press, Oxford.

WILLIAMS, G. C. (1966). *Adaptation and Natural Selection.* Princeton Univ. Press, Princeton, NJ.

MALTE ANDERSSON

Department of Zoology
University of Göteborg
S-413 90 Göteborg, Sweden

Sexual Selection

Sexual selection, which is generated by competition over mates, has far-reaching evolutionary consequences for morphology, behavior, life history, and other characteristics of many organisms. Female choice favors conspicuous secondary sex ornaments and display in males competing to be chosen, but the evolution of mate choice is still debated. Ornaments can also be favored by male contests; other forms of sexual selection are scrambles, endurance rivalry, coercion, and sperm competition. Among the identified constraints on sexually selected traits are predation risk and energy costs, but quantitative demonstration of a balance between sexual and other selection pressures is still lacking. The great diversity of secondary sex traits among species also remains to be explained.

Introduction

Why do males make themselves more conspicuous than females by song, bright colors, large fins, feather plumes, or other gaudy ornaments? Such traits raise the risk of discovery by predators and may also have other costs. However, natural selection should favor traits that increase survival, not reduce it. Conspicuous male traits therefore presented a major problem for Darwin's theory of evolution by natural selection (Fig. 1).

Darwin (1871) proposed that such male traits are favored by sexual selection: competition over mates. He suggested that females prefer gaudy males as mates, and that this leads to evolution of conspicuous male traits even though they reduce survival. This cost is compensated by higher mating success.

Sexual selection became the most controversial of Darwin's ideas among his fellow scientists. It met much skepticism based on the mistaken assumption that female animals cannot make the necessary discrimination and choice among males. The theory has received wide support only in recent decades through mathematical modeling and empirical testing.

What Is Sexual Selection?

Why is there competition for mates? Sexual selection requires mating and sexual reproduction — the combination of genetic material from two parents in the progeny. The most obvious results of sexual selection also depend on there being two sexes with gametes that differ in size, with

FIGURE 1 (a) An adult male mandrill *Mandrillus sphinx* (photo courtesy of J. W. Warden/Overseas). (b) A male moth *Creatonotos gangis* attracts females with pheromones emitted from a specialized abdominal organ. (c) A male long-tailed widowbird *Euplectes progne* expands its tail, more than 50 cm in length, into a deep keel while performing a slow, conspicuous display flight when females visit his territory.

males making many small, mobile sperm and females making fewer large, nutritious eggs. This gametic difference underlies the evolution of sex differences in behavior and morphology.

Regardless of how sex and two gamete sizes evolved, which is still debated, once they exist the stage is set for sexual selection. A male can make many more gametes than a female, so males will compete to mate with as many or as fecund females as possible. Females, on the other hand, which invest their resources in fewer gametes, should be more careful in choosing mating partners.

Sexual selection of a trait is a shorthand for differences in reproductive success caused by competition over mates and fertilizations and related to the expression of the trait. Such differences can arise through many forms of direct or indirect mating competition. It may concern mate quality as well as numbers of mates and can be brought about by mate choice, scrambles, contests, competition based on endurance, or any other form of rivalry over mates (Andersson, 1994) (Table 1).

Note that mate choice by one sex suffices for there to be competition over mates in the other sex, even if the rivals

TABLE I

Competition Mechanisms of Sexual Selection and Traits Likely to Be Favored in the Competing Sex[a]

MECHANISM	CHARACTERS FAVORED IN THE COMPETING SEX
Scrambles	Early search and swift location of mates; well-developed sensory and locomotory organs
Endurance rivalry	Ability to remain reproductively active during a large part of the season
Contests	Traits that improve success in fights, such as large size, strength, weaponry, agility, or threat signals
	Alternative mating tactics of inferior competitors, avoiding contests with superior rivals
Mate choice	Behavioral and morphological traits that attract and stimulate mates
	Offering of nutrition, territories, nest sites, or other resources needed by the mate for breeding
	Alternative mating tactics, such as forced copulation
Coercion	Similar traits as for contests
	Morphological and other adaptations for forced copulation and other coercive behavior
Sperm competition	Mate guarding, sequestering, frequent copulation, production of mating plugs, or other means of preventing rivals from copulating with the mate
	Ability to displace rival sperm; production of abundant sperm to outcompete those of rivals
Infanticide	Similar traits as for contests

[a]Modified from Andersson (1994).

never meet. When a female accepts a male that fertilizes her eggs, they become unavailable to other males. Competition over mates and fertilizations, therefore, is the fundamental aspect of all sexual selection; also, that based on female choice also creates competition among males to be chosen. The distinction that is often made between sexual selection by female choice and by male competition is therefore misleading and so are the related terms "intersexual" and "intrasexual" selection. All forms of sexual selection are intrasexual in the sense of involving competition between members of the same sex. There is competition among males to be chosen (in female choice), to arrive first (in scrambles), to win fights (in contests), and so on (Table 1).

Pioneering work by G. Parker three decades ago made it clear that sexual selection in the form of sperm competition can also be important after copulation if the female mates with several males. Sperm competition might be combined with sperm choice, i.e., the female influences paternity through internal mechanisms after copulation with several males (Eberhard, 1996). Sperm competition has been abundantly demonstrated, and suggestive evidence for sperm choice is accumulating, but assessment of its relative importance requires more work.

The mechanisms of sexual selection that have attracted most interest are mate choice, contests, and sperm competition (Table 1). Much evidence shows that these three mechanisms are often important (Andersson, 1994). Mate choice and its evolution present one of the most complex problems in evolutionary biology, with challenges that have been broadly dealt with only recently.

Other forms of sexual selection have received less attention. One is endurance rivalry (Table 1), in which the ability to remain active at a lek or other mating site is crucial for the number of matings and offspring. Another mechanism is scrambles to find the mate before rivals do so. This can be achieved by (i) early emergence or maturation, with males often becoming active before females (protandry, occurring in some plants as well as animals), and (ii) swift localization of potential mates, which is favored by well-developed sensory organs, mobility, and spatial memory.

Sexual selection therefore may have far-reaching effects on the neurosensory system as well. For instance, it has probably contributed to the large eyes or antennae in males of many insects (Thornhill and Alcock, 1984). Search for females apparently favors better male spatial memory in some mammals. It is associated with larger male hippocampus, a part of the brain that is involved in spatial learning and is maximally developed during the breeding season (Jacobs, 1996). In plants, scrambles over pollination are likely to favor rapidly growing pollen tubes and pollinator-attracting traits such as conspicuous flowers, scent, and nectar (Andersson, 1994; Willson, 1994).

Yet another mechanism is coercion, in which a male uses threat or force to increase the chances that a female will mate with him and not with other males (Clutton-Brock and Parker, 1995). Sequestering, harassment, intimidation, and forced copulation are in this category. Other tactics used later in the reproductive cycle are induced abortion or infanticide that makes a female receptive to a new male.

Except for sperm competition, all mechanisms in Table 1 are also applicable to females. Sexual selection is usually stronger in males because they have a higher potential reproductive rate than females. Several mechanisms of sexual selection may often occur simultaneously, presenting the difficult problem of assessing their relative importance, which has rarely been done. Much more work is needed that measures several different mechanisms simultaneously.

Mate Choice and Mating Preferences

Despite much evidence that females (and also males) often choose their mates, how female choice favors male ornaments is still debated, as are the ways in which female preferences evolve. Several mechanisms have been suggested.

The Fisher Process

Darwin's theory of sexual selection was incomplete because it did not explain why females should be choosy. Early in the past century, Ronald Fisher suggested that "a sexual preference of a particular kind may confer a selective advantage, and therefore become established in the species." Suppose there arises genetic variation in a male trait such as tail length and that males with, for example, longer than average tails have a slight survival advantage owing to better agility. Assume also that females choose their mate, and that there is genetic variation in preferences among females. Those preferring males with a long tail tend to bear sons with high survival. Therefore, genes that code for longer tails in males will spread and so will genes that make females prefer long-tailed males because the two kinds of genes become associated in their offspring.

When this process continues, males with longer tails will be favored not only by better survival but also by higher mating success as the preference for long tail spreads among females. The higher mating success of long-tailed males helps carry the genes for the long tail and the associated female preference to yet higher frequency, and a feedback "runaway process" arises. Finally, the tail becomes so long that it reduces male survival, which balances the mating advantage. This brings the process to rest.

Fisher's idea assumes changes in gene frequencies, mathematical models of which have now been explored, verifying the logic of the theory (Pomiankowski *et al.,* 1991). Models of sexual selection by mate choice have become increasingly sophisticated, often using quantitative genetics or game theory for exploring various possibilities.

Indicator Mechanisms

One such possibility is that conspicuous, costly male traits are preferred by females because such traits indicate high heritable viability. If she mates with a highly ornamented male, a female is then likely to bear offspring that survive well.

Models show that genetic indicator mechanisms work most easily if ornaments are condition dependent and develop in proportion to the phenotypic condition and overall genotype of the male (Iwasa *et al.,* 1991). Many reasons have been suggested why some form of indicator mechanism might be important. One possibility is host–parasite interactions (Hamilton and Zuk, 1982). The degree of resistance to parasites or pathogens may be reflected in traits such as bright colors, which can then be used as cues in mate choice. Increasingly detailed evidence suggests that indicator mechanisms are sometimes important (Andersson, 1994; Møller, 1994; Johnstone, 1995).

Species Recognition

To evolve, it is not necessary that an ornament reflects higher survival. Improvement of species recognition that reduces the risk of hybrid matings is a plausible alternative. In several animals, there is evidence that selection for species recognition has played a role in the evolution of signals or mating preferences. The relation between sexual selection, species recognition, and speciation is currently an active research field. However, even if species recognition is involved, it does not fully explain the most conspicuous sex traits, such as the tail of the long-tailed widowbird (Fig. 1a). These tails are developed far beyond what is needed for species recognition; some additional process is likely involved.

Direct Phenotypic Benefits

Variation in parental ability might also lead to evolution of traits that advertise high parental quality, with immediate benefits to mate and offspring. For example, owing to a correlation with food abundance in the territory, male song rate might give a clue to his future performance as a parent and to the food situation for the female and her offspring (Kirkpatrick and Ryan, 1991; Andersson, 1994).

There are usually no obvious direct benefits of mate choice in lekking animals, in which females visit males only for copulation (Höglund and Alatalo, 1995). Genetic advantages of mate choice have therefore often been suggested for such species. However, some direct benefits of female choice, such as avoidance of disease or protection from coercive other males, are also possible in leks. Female preferences, however, have rarely been identified in lekking animals. Male contest competition may also contribute in the sexual selection of male secondary sex traits in many lekking species (Höglund and Alatalo, 1995).

Sensory Bias and the Origins of Preferences

A new male trait might also be favored by "sensory exploitation" because it happens to fit an already existing bias in the female sensory system, which arose in some other context (Ryan and Keddy-Hector, 1992). There is evidence for such bias in several species; later, one example is described for the Tungara frog.

Empirical Results

It took a century from Darwin's work until sexual selection became a lively research field in the 1970s, stimulated by theoretical advances. The study of sexual selection has since grown into a substantial part of evolutionary biology. Approaches range from field studies in even the most inaccessible parts of the world to white-coat laboratory work clarifying parentage with the latest methods of molecular biology and mathematical modeling of the mechanisms of sexual selection. In addition to the fascinating unsolved problems, these many faces of the subject contribute to its attractiveness. As a result, there are now hundreds of studies demonstrating sexual selection in organisms ranging from arthropods to birds and mammals (Andersson, 1994). Insects, birds, amphibians, and fishes dominate among the species studied. Among the male traits most often shown to be sexually selected are song and display, body size, visual ornaments, and territory or other resources.

Female choice has been found in more than 150 studies, so Darwin's once controversial idea has been abundantly corroborated. Also, male choice of mate is common, for example, in animals with indeterminate growth, in which males prefer large fecund females.

Male contests have also been found in many studies. Male scrambles or endurance rivalry have been documented much more rarely, perhaps partly because of a paucity of work on these mechanisms.

A major methodological breakthrough, DNA fingerprinting and related techniques, has greatly increased the accuracy of parentage determination, showing that paternity is often not what the social pairing pattern suggests. In some species, the social mates are almost always the genetic parents of the young. In other species, extrapair fertilizations are common. For example, a study of the socially monogamous reed bunting *Emberiza schoeniclus* showed that 55% of the young had fathers other than the male bonded to their mother and feeding the young. Male contribution of parental care increased with his degree of paternity (Dixon *et al.,* 1994).

Among the possible reasons why females seek extrapair copulations is heritable advantages to offspring fathered by high-quality males. Why species differ so much in the extent of extrapair paternity, however, remains a puzzling unsolved problem.

Genetics of Sexually Selected Traits and Preferences

Genetic correlations have been found between the preferred trait and the preference in sticklebacks *Gasterosteus aculeatus* (Fig. 2) (Bakker, 1993) and a few other species. These results suggest that sexual selection by mate choice has had substantial genetic consequences. A genetic correlation alone, however, does not reveal which mechanism is at work. A correlation is predicted by Fisherian runaway, indicator, and species recognition mechanisms. Developing critical tests for distinguishing among them, and estimating their relative importance in particular cases, remains a major challenge.

There is evidence for indicator mechanisms in several

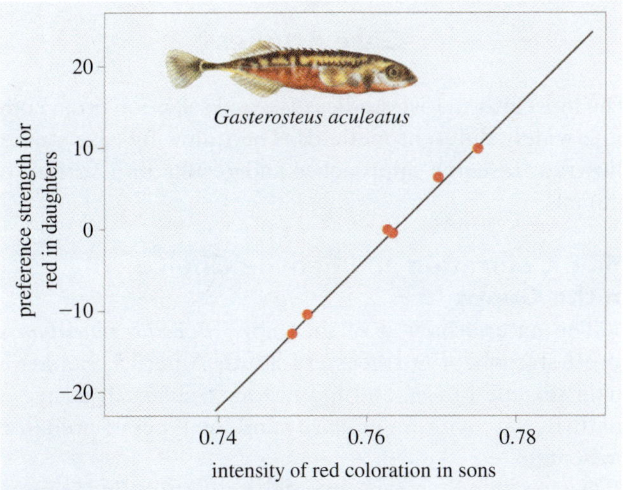

FIGURE 2 For stickleback *Gasterosteus aculeatus* fathers, there is a correlation between the intensity of red coloration in sons and the preference strength for red in the daughters. The red preference is expressed as the percentage of female movements toward the reddest male. The intensity of the red coloration is measured by the red index, given by the optical density of the red coloration and the optical density of the overall coloration (red, green, and blue) (adapted from Bakker, 1993).

species (Møller, 1994; Hasselquist *et al.,* 1996). On the other hand, rapid divergent evolution of male genital organs among insects and other animals seems best explained by Fisherian runaway evolution combined with sensory bias (Eberhard, 1996). Although it currently receives less attention than indicator mechanisms, Fisherian runaway selection remains a possibility in many cases.

Fluctuating Asymmetry

Much interest is focused on fluctuating asymmetry of secondary sex traits (sideways random deviations from perfect bilateral symmetry). Because both sides of the body are produced by the same genome, the degree of asymmetry will reflect its ability to counteract genetic and environmental stress and produce a harmonious phenotype. This opens the possibility for a genetic indicator process with a mating preference for symmetry. For example, fluctuating asymmetry in the long tail of the male swallow *Hirundo rustica* decreases with increasing tail length (Møller, 1994). A long tail should therefore reflect high developmental homeostasis, which might in part be heritable. Natural and experimentally induced asymmetry reduces flight performance in swallows. Long-tailed males tend to have lower asymmetry than males with shorter tails, and their offspring are less affected by parasitic mites. In an experiment manipulating tail length and asymmetry independently, female choice favored both symmetry and long tail (Møller, 1994). The role of fluctuating asymmetry in sexual selection is debated, however, and its analysis contains many pitfalls.

Case Studies

The most informative studies of sexual selection often combine widely different methods. The following case studies illustrate research approaches and results in a variety of animals.

Male Coloration and Female Choice in the Guppy

The natural habitat of the guppy *Poecilia reticulata* is forest streams of northeastern South America. Behavior, male sex coloration, and life history traits such as age at maturity vary with water clarity and the types of predators present.

An outstanding experimental field study in Trinidad showed that coloration and life history traits can evolve rapidly in response to changes in selection pressures (Endler, 1980; Reznik *et al.*, 1990). A cichlid, *Crenicichla alta,* is a major predator of adult guppies. Guppies from streams with this predator were moved to another stream, with predation mainly on juveniles by the killifish *Rivulus hartii.* As predicted by sexual selection theory, the guppies evolved more conspicuous male colors in the new stream. They also shifted their breeding to higher ages and produced fewer but larger young after 2 years (10–15 generations) in the new environment. After 30–60 generations (11 years) the life history changes clearly were genetic, as demonstrated by rearing guppies from the streams in laboratory tanks for 2 generations, eliminating environmental effects (Reznik *et al.,* 1990). This experiment in the wild convincingly shows that life history and secondary sex traits can undergo marked genetic evolution in surprisingly few generations.

Male guppies are highly polymorphic (Fig. 3), varying in number, size, position, and reflectivity of color patches. Endler (1980) tested the idea that the color pattern represents a balance between natural selection for crypsis by visually hunting predators and sexual selection for conspicuousness. The number and level of danger of predator species increase downstream in many guppy waters. A first test of predation effects is therefore to determine if the male spot pattern is related to the local fauna of predators. This is in fact the case: The size and number of color spots decrease with increasing risk of visual predation. Moreover, in the field transplantation test, the color patterns after 2 years had changed markedly in the predicted direction.

The evolution of color patterns was also examined over many generations in the laboratory using experimental populations derived from 18 sites in Trinidad and Venezuela. The spot patterns in the laboratory systems with different predators converged toward the pattern typical for the natural streams with the same predators in the wild.

Endler (1980) found that red and orange carotenoid color spots gave males an advantage in female choice of mate. Bright carotenoid spots may reflect high feeding success and general vigor, so the female preference may have evolved through a genetic indicator process, a possibility that remains to be tested in detail.

FIGURE 3 The male guppy *Poecilia reticulata* displays his bright color spots and large tail in courtship to a female (photo courtesy of R. Maier/Overseas).

Studies of different populations of guppies provide evidence for joint evolution of female preference and male ornament. The degree of preference for orange color spots in females from seven populations was correlated with the amount of orange in males from the females' population (Houde and Endler, 1990). The result suggests that male carotenoid color spots and female preferences have evolved together and diverged genetically among populations. Whether the joint evolution of trait and preference occurred by Fisherian, indicator, or other processes is not known.

In summary, experimental field and laboratory studies of guppies show that conspicuous male coloration is favored by female choice and disfavored by predation. There is evidence of joint genetic evolution of female preference and male orange color spots, as predicted by genetic models of sexual selection.

Advertisement Calls in the Tungara Frog

Some of the most thorough studies of sexual signaling have been done in frogs and toads. One example is the South American tungara frog *Physalaemus pustulosus* (Ryan, 1985). Males gather and call from water pools visited by females, which select their mates there. Males have two kinds of call, *whine* and *chuck* (Fig. 4a). Isolated males give only whines. In response to vocal competition from

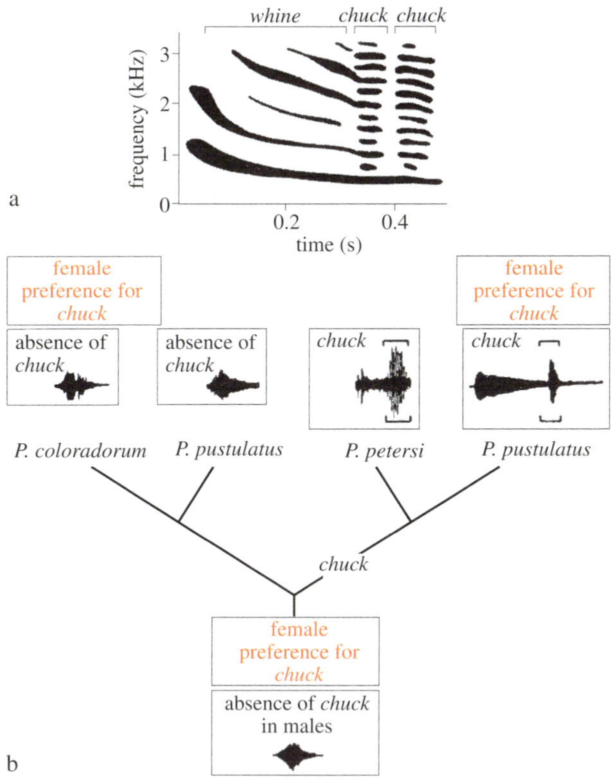

FIGURE 4 (a) The call of the tungara frog *Physalaemus pustulosus;* the sonogram shows a whine and two chucks. (b) Estimated phylogeny of the male chuck call and female preference in *Physalaemus* frogs. Only the closely related *P. pustulosus* and *P. petersi* emit the chucks, marked by square brackets in the oscillograms, which therefore likely evolved in their closest common ancestor. Female *P. pustulosus* and *P. coloradorum* prefer calls with chucks, suggesting that the common ancestor of all four species had the preference. If this is so, it probably evolved before the chuck itself (adapted from Ryan and Keddy-Hector, 1992).

other males they add chucks, which are more effective in attracting females.

There is evidence that the chuck call may have evolved by sensory exploitation, the response characteristics of the female ear having arisen for other reasons before the favored call. In support, there is a similar sensitivity to chuck calls in the related species *P. coloradorum,* although males of this species do not produce chucks. A possible reason is that both species inherited the sensitivity from a common ancestor, but only the ancestor of the tungara frog and its sister species *P. petersi* evolved a chuck call that takes advantage of the female sensory bias (Fig. 4b) (Ryan and Keddy-Hector, 1992).

The proportion of hatched eggs increases with the similarity in size between male and female (which is usually largest). It should therefore be adaptive for most females to mate with a male of larger than average size. Perhaps partly for this reason, females discriminate between chuck calls of different frequencies, which reflect differences in male size. In loudspeaker tests, females usually approached the speaker that gave the lower fundamental frequency, a response that should favor the larger of two males. In accordance, male mating success in the field increased with body size (Ryan, 1985).

Since chuck calls help attract females, one may ask why males do not use chucks more often. Part of the answer is the risk of predation by fringe-eared bats *Trachops cirrhosus,* which take many male tungara frogs at chorus pools (Fig. 5). Experiments show that bats are more attracted to calls that contain both chucks and whines than to those with whines only (Ryan, 1985). This probably explains why single males give no chucks and why chucks are not more common in tungara frog choruses. Bat predation therefore limits the use of the most effective mate attraction signal.

FIGURE 5 The call of male tungara frogs *Physalaemus pustulosus* attracts not only the females but also frog-eating bats *Trachops cirrhosus* (photo courtesy of M. D. Tuttle/Panda Photo).

Long Tails in Male Birds

Male long-tailed widowbirds *Euplectes progne* have one of the most remarkable ornaments among birds: an often more than 0.5-m-long tail (Fig. 1a), the function of which has been tested by changing its length experimentally (Andersson, 1982). Early in the breeding season, the territories of 36 males were mapped and their numbers of nesting females were counted. The color-ringed males were partitioned among nine groups with four males in each, as similar as possible in territory quality and tail length (Fig. 6). The tail was cut to about 14 cm in one of the randomly drawn males within each group. Each removed feather was glued to the corresponding feather of another male, elongating his tail by 20–30 cm. The two other males were controls. One of them was ringed only; the tail feathers of the fourth male were cut off at the midpoint and glued back again. The number of new nests that later appeared on the territory was then counted.

The result indicates that the long tail is favored by sexual selection through female choice of mate. Females preferred supernormal tails because the elongated males had highest success in attracting new females. Two alternative explanations were refuted: Shortened males did not become less active in courtship display, nor did they give up their territories more often than other males. The display of the tail (Fig. 1a) also suggests that it is used in attracting females and not in contests among males.

Similar experiments have since been done in several other birds. A particularly detailed study was done in the barn swallow *Hirundo rustica* by Møller (1994), who concluded that male tail length in this species is an honest indicator of viability and condition, reflecting heritable resistance to parasitic mites.

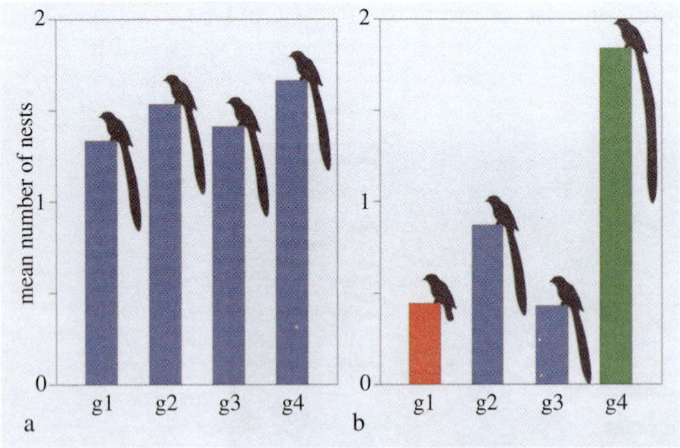

FIGURE 6 Experimental changes in the tail length of male long-tailed widowbirds *Euplectes progne* influence their capacity to attract females. The bars show the number of nests per territory in each of four categories of males (g1–g4) before the changes in tail length were made (a) and the number of new nests after the changes (b). After the tail treatment, the ability to attract females was changed according to the lengthened or shortened tail (adapted from Andersson, 1982).

Sexual Size Dimorphism in Elephant Seals

Elephant seals are among the most sex dimorphic of all mammals. Adult males of the northern elephant seal (*Mirounga angustirostris*) weigh more than three times as much as females (Deutsch *et al.*, 1990). Why is there such huge dimorphism in body size?

The answer is found in the mating system and the forms of competition over mates: male contests and endurance rivalry. Female elephant seals gather on beaches during the 3-month reproductive season to give birth and mate. Females form dense clusters, males fight over females (Fig. 7), and dominant males can defend large groups. In this highly polygynous system with much male aggression, only the largest bulls can achieve high rank and mating success. Less than one-third of the males copulate during a breeding season, and most matings are done by a few males. Large male body size, therefore, is favored by sexual contest competition (LeBoeuf, 1974; Deutsch *et al.*, 1990).

Another factor is the long breeding season in elephant seals, which favors male endurance. The capacity for energy storage increases more rapidly with body size than do metabolic costs. Large size and stored energy in the form of blubber therefore enable a male to remain for a long period on the beach, 91 days on average, and inseminate many females before he must go back to sea and feed (Deutsch *et al.*, 1990).

Females have some scope for mate choice. They usually protest against being mounted, especially by young males, doing so by loud calls and evasive movements of the hindquarters. The calls alert nearby males, and the dominant bull usually discovers, attacks, and chases away other males that try to mate. Females protest less against dominant males.

For these reasons, the great sexual size dimorphism in elephant seals is apparently a consequence of sexual selection favoring large male size in contests and endurance rivalry over females; female choice may also play a role. Contest competition over females favors not just large male body size. It is probably also the main reason why males in many vertebrates and arthropods have much larger weapons than females, such as spurs, horns, antlers, and tusks.

Male Song, Female Choice, and Offspring Survival in Great Reed Warblers

Male songbirds use song to attract females and defend the territory against rivals. Females in many species favor the males with the largest repertoires of song types. For example, the male sedge warblers *Acrocephalus schoenobaenus* that attract females first have the largest song repertoires. In the laboratory, females give more copulation solicitation displays to playback of large repertoires. Re-

FIGURE 7 Bull elephant seals *Mirounga angustirostris* fight violently over groups of females during the breeding season (photo courtesy of Hacana/Massart/Overseas).

sults are similar for the great reed warbler *A. arundinaceus* (Catchpole and Slater, 1995).

Why do females prefer males with large song repertoires? The results of a 10-year field study of great reed warblers go a long way toward answering this question. Hasselquist *et al.* (1996) individually marked and took blood samples from the members of the population, including the young, and recorded the song repertoires of the males. By accurate determination of paternity through DNA fingerprinting, male lifetime production of surviving offspring could be measured. There was a positive correlation between male song repertoire size and the postfledging survival of his offspring (Fig. 8). The study excluded variation among males in nongenetic effects, such as territory quality and male parental behavior, because the conclusion also held true for the offspring produced by extrapair fertilizations. Some females visit and mate with neighbor males that do not participate in raising their young. In all 10 cases identified, the extrapair male had a larger song repertoire than the female's social mate. It therefore seems that by seeking extrapair fertilizations from males with large song repertoires, females gain heritable survival benefits for their young. A genetic indicator process based on song therefore seems to be at work in the great reed warbler.

The reason why males with large repertoires have offspring with higher survival is not clear. As repertoire size increases with male age, by preferring large repertoires females may mate with males of higher than average and partly heritable survivorship, reflected in their age (Hasselquist *et al.,* 1996).

Perhaps only the males with the best genetic constitution can afford the cost of a large song repertoire. The higher vocal center of birds provides one of the few examples in which cell additions in the brain also occur in adults (Jacobs, 1996). The center grows each spring before the breeding season when males sing, and it shrinks when the season is over. Comparative studies show that songbird species with the largest song repertoires have the largest higher vo-

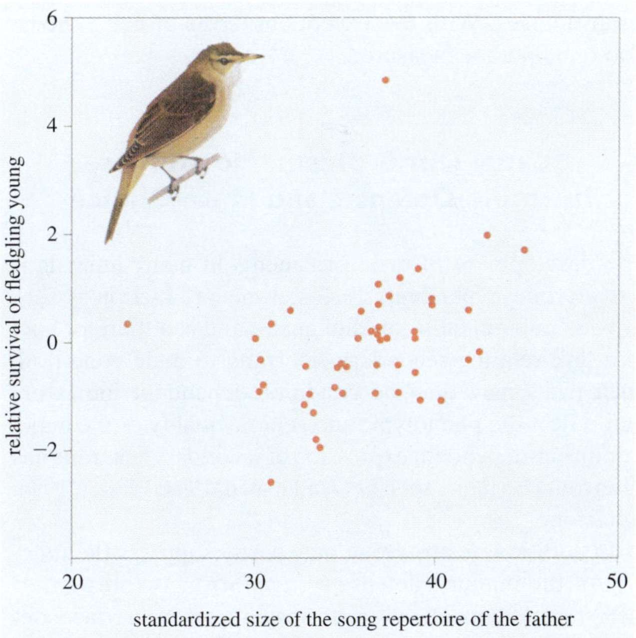

FIGURE 8 The relative survival of fledging young in the great reed warbler *Acrocephalus arundinaceus* (obtained from the correlation between the lifetime of fledglings and the lifetime of young born in the study area that have returned) is positively related to the size of the song repertoire of the father, given by the overall number of syllables used out of a total of at least 150 syllables possible (adapted from Hasselquist *et al.,* 1996).

cal centers. It therefore seems that sexual selection by female choice has affected not only song repertoire size but also a brain structure involved in song production.

Costs of Sexual Selection

Darwin pointed to predation risk, which is clearly greater in males than in females of many species, as a consequence of sexual selection (Fig. 5). However, several other costs can also limit the expression of secondary sex traits (Andersson, 1994). For example, contests over territories or females can cause injury or death. Sexually selected larger body size makes males more prone to starve than females during the juvenile growth period in birds and mammals.

Secondary sex traits can also impose energetic and foraging costs. Several forms of natural selection may therefore restrict secondary sex traits. Sexual selection can also be self-limiting. For example, large male body size in the redwing blackbird *Agelaius phoeniceus* is advantageous in male contests, but larger males spend more time foraging and have less time to defend their territory and attract females (Searcy and Yasukawa, 1995).

No study of secondary sex traits has yet produced a quantitative account showing the predicted balance between sexual selection advantages and natural (or sexual) selection disadvantages. With few exceptions, costs of mate choice also remain to be measured.

Sexual Dimorphism, Hormones, Immune Defense, and Mate Choice

The development of male ornaments in many animals is testosterone dependent. Testosterone production is sensitive to nutritional status and health and can therefore act as a link relating secondary sex traits to male condition. Such traits may then be condition-dependent indicators that reflect the phenotypic and genetic quality of the male. Condition-dependent expression of secondary sex traits has been found in many species (Andersson, 1994; Møller, 1994; Johnstone, 1995).

In addition, testosterone may partly suppress the function of the immune defense system. Strong expression of testosterone-dependent traits may therefore reduce defense ability against pathogens and parasites. Also for this reason, secondary sex traits may be condition dependent, with only individuals with low pathogen burden able to afford development of large sex traits. The importance of these mechanisms is debated, and the relationship between hormonal mechanisms, the immune defense system, and sexual selection by mate choice is the subject of much active research (Svensson and Skarstein, 1997). For example, von Schantz *et al.* (1996) found that the major histocompatibility complex genotype of male pheasants *Phasianus colchicus* is associated with variation in both male spur length and male viability. This suggests that females may influence the survival of their young by choosing males with long spurs.

Alternative Mating Tactics

More or less dimorphic variation in male structure and behavior is found in many species. The cause often appears to be sexual selection in the form of male contest competition. Some examples in fishes and arthropods are conspicuous aggressive or advertising behavior, large body size, weapons, or other secondary sex traits versus small size and sex traits and inconspicuous behavior. Such alternative mating tactics are related to variations in age, size, condition, or other phenotypic differences. They do not lead to equal reproductive success but arise from unequal situations and constraints for different individuals (Gross, 1996).

Other examples of alternative tactics favored by sexual selection are found in sex-changing organisms, for instance, some shrimps and other crustacea, mollusks, fishes, and plants. Whether individuals are males or females first depends in part on the nature and strength of competition over mates. Strong contest competition over females usually favors being female first and switching to the male sex as the individual grows larger, for example, in some fishes. When contest competition and the advantage of large size are weaker among males, fecundity advantages of large size in females may dominate, favoring being male first and changing to the female sex at large size (Charnov, 1982; Warner, 1988).

Sexual Selection in Plants

A main form of sexual selection in higher plants is scrambles among pollen to reach and fertilize ovules. Competition over pollination should favor efficient attraction and reward of pollinators and rapid germination and growth of pollen tubes in the style.

Selection through male function often favors more and larger flowers in hermaphrodites than needed from a female standpoint. The evolution of conspicuous floral display has traditionally been explained by advantages in cross-fertilization through attraction of pollinators. Sexual selection is also likely to be important, however (Andersson, 1994; Willson, 1994). Nectar production and flowering phenology may also be subject to sexual selection mainly through male function. In certain species (e.g., some orchids), there is pollen limitation of female reproduction. Sexual selection may then favor floral display through female and male function.

With important exceptions (Willson, 1994), botanists

have rarely considered in detail the parallels between plants and animals in sexual selection. New insights are likely to emerge as the similarities and differences between plants and animals in these respects become further clarified from an explicit sexual selection perspective.

References Cited

ANDERSSON, M. (1982). Female choice selects for extreme tail length in a widowbird. *Nature* **299,** 818–820.

ANDERSSON, M. (1994). *Sexual Selection.* Princeton Univ. Press, Princeton, NJ.

BAKKER, T. C. M. (1993). Positive genetic correlation between female preference and preferred male ornament in sticklebacks. *Nature* **363,** 255–257.

CATCHPOLE, C. K., and SLATER, P. J. B. (1995). *Bird Song.* Cambridge Univ. Press, Cambridge, UK.

CHARNOV, E. L. (1982). *The Theory of Sex Allocation.* Princeton Univ. Press, Princeton, NJ.

CLUTTON-BROCK, T. H., and PARKER, G. A. (1995). Sexual coercion in animal societies. *Anim. Behav.* **49,** 1345–1365.

DARWIN, C. (1871). *The Descent of Man, and Selection in Relation to Sex.* Murray, London.

DEUTSCH, C. J., HALEY, M. P., and LEBOEUF, B. J. (1990). Reproductive effort of male elephant seal: Estimates from mass loss. *Can. J. Zool.* **68,** 2580–2593.

DIXON, A., ROSS, D., O'MALLEY, S. L. C., and BURKE, T. (1994). Paternal investment inversely related to degree of extra-pair paternity in the reed bunting (*Emberiza schoeniclus*). *Nature* **371,** 698–700.

EBERHARD, W. G. (1996). *Female Control: Sexual Selection by Cryptic Female Choice.* Princeton Univ. Press, Princeton, NJ.

ENDLER, J. (1980). Natural selection on colour patterns in *Poecilia reticulata. Evolution* **34,** 76–91.

GROSS, M. R. (1996). Alternative reproductive strategies and tactics: Diversity within sexes. *Trends Ecol. Evol.* **11,** 92–98.

HAMILTON, W. D., and ZUK, M. (1982). Heritable true fitness and bright birds: A role for parasites? *Science* **218,** 384–387.

HASSELQUIST, D., BENSCH, S., and VON SCHANTZ, T. (1996). Correlation between male song repertoire, extra-pair paternity and offspring survival in the great reed warbler. *Nature* **381,** 229–232.

HÖGLUND, J., and ALATALO, R. V. (1995). *Leks.* Princeton Univ. Press, Princeton, NJ.

HOUDE, A. E., and ENDLER, J. A. (1990). Correlated evolution of female mating preferences and male color patterns in the guppy *Poecilia reticulata. Science* **248,** 1405–1408.

IWASA, Y., POMIANKOWSKI, A., and NEE, S. (1991). The evolution of costly mate preferences. II. The "handicap" principle. *Evolution* **45,** 1431–1442.

JACOBS, L. F. (1996). Sexual selection and the brain. *Trends Ecol. Evol.* **11,** 82–86.

JOHNSTONE, R. (1995). Sexual selection, honest advertisement and the handicap principle: Reviewing the evidence. *Biol. Rev.* **70,** 1–65.

KIRKPATRICK, M., and RYAN, M. J. (1991). The evolution of mating preferences and the paradox of the lek. *Nature* **350,** 33–38.

LEBOEUF, B. J. (1974). Male–male competition and reproductive success in elephant seals. *Am. Zool.* **14,** 163–176.

MØLLER, A. P. (1994). *Sexual Selection and the Barn Swallow.* Oxford Univ. Press, Oxford.

POMIANKOWSKI, A., IWASA, Y., and NEE, S. (1991). The evolution of costly mate preferences. I. Fisher and biased mutation. *Evolution* **45,** 1422–1430.

REZNIK, D. A., BYGA, H., and ENDLER, J. A. (1990). Experimentally induced life-history evolution in a natural population. *Nature* **346,** 357–359.

RYAN, M. J. (1985). *The Tungara Frog: A Study in Sexual Selection and Communication.* Chicago Univ. Press, Chicago.

RYAN, M. J., and KEDDY-HECTOR, A. (1992). Directional patterns of female mate choice and the role of sensory biases. *Am. Nat.* **139,** s4–s35.

SEARCY, W. M., and YASUKAWA, K. (1995). *Polygyny and Sexual Selection in Red-Winged Blackbirds.* Princeton Univ. Press, Princeton, NJ.

SVENSSON, E., and SKARSTEIN, F. (1997). The meeting of two cultures: Bridging the gap between ecology and immunology. *Trends Ecol. Evol.* **12,** 92–93.

THORNHILL, R., and ALCOCK, J. (1984). *The Evolution of Insect Mating Systems.* Harvard Univ. Press, Cambridge, MA.

VON SCHANTZ, T., WITTZELL, H., GÖRANSSON, G., GRAHN, M., and PERSSON, K. (1996). MHC genotype and male ornamentation: Genetic evidence for the Hamilton–Zuk model. *Proc. R. Soc. London B* **263,** 265–271.

WARNER, R. R. (1988). Sex change in fishes: Hypotheses, evidence, and objections. *Environ. Biol. Fishes* **22,** 81–90.

WILLSON, M. F. (1994). Sexual selection in plants: Perspectives and overview. *Am. Nat.* **144,** s13–s39.

Stefano Turillazzi

Dipartimento di Biologia Animale e Genetica "Leo Pardi"
Università degli Studi di Firenze
Florence, Italy

Centro di Studio per la Faunistica
Ed Ecologia Tropicali del CNR
Florence, Italy

Mary Jane West-Eberhard

Smithsonian Tropical Research Institute
Balboa, Republic of Panama
c/o Escuela de Biología
Universidad de Costa Rica
Costa Rica

Wasps as Models for the Study of the Evolution of Insect Societies

Social insects play an important role in the study of etiology and the evolution of social behavior. Currently the term eusociality, in fact, mainly applies to the insect group; the principle theories which explain the origin and evolution of eusociality have demonstrated that this concept has one general valence. Of the social insects, those pertaining to the order of Hymenoptera are particularly important. And of these, the wasp social models are considered for the study of social evolution and its mechanisms because several species have various degrees of sociality, from simple to complex societies. Some of the major themes of sociality and their study in wasps are discussed: communication and the mechanisms of social regulation in the colonies; behavioral flexibility and the origin of the queen-worker dualism; the control of colony organization; and social parasitism and nest mate recognition.

The Importance of Social Insects in the Study of the Mechanisms and Evolution of Social Behavior

In his introduction to his book, *Social Evolution*, Trivers (1985) remarks how life is intrinsically social and evolves through a process of natural selection which is itself social. All organisms interact with one another by means of direct contact or through the environment. The reproductive success of an organism, compared with others, is the main drive behind evolution. A social group, however, is not just a set of solitary individuals that find themselves together in particular circumstances. Social behavior, in the strict sense, is typical of those organisms which live in groups, within which relations between single individuals are characterized by well-defined and sometimes highly complex, organized patterns. Individuals belonging to the group can be seen to take on specific roles and tasks, which profoundly change their behavior compared with that of similar but solitary species.

Thus, social life is a characteristic of a limited number of species, obviously including our own. For this reason, man has long been interested in the study of social behavior in animals, in its proximal factors (i.e., its regulatory mechanisms), and its ultimate causes (i.e., evolution). The list of social species has considerably increased since the 1970s. This is mostly due to the huge amount of research inspired by the new evolutionary ideas forged by several researchers and in particular to the foundations of the theory known as kin selection, discussed briefly by Haldane and others and formalized by Hamilton (1964). In his book *Sociobiology,* Wilson (1975) proclaimed the birth of this new science which studies the evolution of social behavior, promoting a series of in-depth studies into the natural history, eco-ethology, and evolution of social species.

Social insects play an important role in these studies. All this fervor for study and research began in an attempt to solve one particular question, which Darwin asked himself in the *Origin of Species* — that is, the dilemma of the origin and evolution of sterile castes in social insects.

Eusociality and the Social Orders

Scientists adopt essentially three criteria to define eusociality: cooperative brood care, overlap of generations, and reproductive division of labor within the group. This last factor means that some individuals take on the role of reproductives, whereas others do not but instead specialize in tasks related to brood rearing and the maintenance or defense of the colony and its site. A similar, but temporary, type of group organization also occurs in many vertebrates. However, examples of lifelong sterility in one sector of the members of the group associated with a distinct morphology of the two classes of individuals have only been described in some invertebrates and in two species of vertebrates. One of the latter is the famous mole rat, *Heterocephalus glaber,* an east African rodent which builds large underground nests and whose colonies are very similar to those of termites.

Eusociality in invertebrates, other than insects, has recently been reported in crustaceans. In some species of shrimps of the genus *Synalpheus,* which live in large communities inside the sponges of the Caribbean coral reefs, some individuals reproduce and others defend the colony.

The phenomenon of sociality in spiders is thought to be favored by their use of silk, although the occurrence of truly eusocial species has not been well documented.

Insects are the group with the highest number of confirmed eusocial species. There are five orders in which eusocial species have been reported: Hymenoptera (bees, wasps, and ants), Isoptera (termites), homopterous Hemiptera (in aphids), Thysanoptera (thrips), and, perhaps, Coleoptera (beetles). The Hymenoptera is easily the order in which sociality has developed independently most times in the course of evolution — at least 11 times according to some authors. In some Hymenopteran families, such as the Formicidae (ants), all the species are social. Although all isopterans (termites) are eusocial, displaying various degrees of eusociality, eusociality is less frequent in the other three orders. In the Hemiptera, Aoki was the first to describe sterile soldiers in bamboo aphids, but recently such forms have also been found in *Pemphigus spirothecae,* the poplar aphids. In 1992, Crespi reported social characteristics in colonies in some genera of Australian gall-thrips (thysanopterans). In coleopterans, Kent and Simpson recently described a species of weevil, *Austroplatypus incompertus* (Curculionidae), believed to exhibit eusocial behavior.

The Principal Theories on the Origin of the Evolution of Eusociality

How can altruistic behavior — a behavior in which by lowering its own chances of survival and reproduction the performer (or worker) simultaneously increases that of the receiver (or queen) — be explained? Darwin also saw another facet to the problem: How did the sterile castes of so many insects evolve their different shapes and specializations? The explanations Darwin gave are similar to those accepted today. He hypothesized that selection acted at the group rather than the individual level. By favoring the group, or its family, an altruist, sterile individual also helped its blood relations, an explanation which in retrospect is very similar to the kin selection theory. The explanation Darwin offered for the second problem came later, in his 1868 edition of the *Origin of Species,* in which he hypothesized that heredity of characters involved transmission as well as development. He specified that the two forces were separate and that some traits of the discrete bee or ant worker morphology were still present, but not expressed, during development.

The Superorganism

Individuals in a social insect colony can be compared to the cells making up an organism. The idea is attractive and there are certainly some close similarities, but there are also substantial differences — the cells in an organism are all isogenic, but the individuals in an ant or bee colony are all genetically different. Despite this, the individuals in some societies (e.g., termites and *Apis*) are genetically remarkably similar, either as the result of long-term monogamy in the first case or through efficient reproductive control of the working castes in the second. The concept of *superorganism* applied to social insect colonies seemed to explain the phenomenon of altruism, but in fact it actually denied the possibility of altruism enhancing personal fitness and only favoring that of another individual. Group selection theories claimed that an altruistic individual simply derived some benefit from the increased success and survival rates of its own particular group. Today, classical group selection has few supporters for two main reasons. First, there is no need to postulate it if the phenomena it describes can be more simply explained by mechanisms behind individual selection. Second, it is practically impossible for it to work because the conditions it implied are so restrictive that they do not occur in reality (Starr, 1979)

Kin Selection

Hamilton (1964) pursued the ideas first proposed, but never completed, by Fisher and Haldane, and he examined the phenomenon of sterile castes in social insects, particularly in the Hymenoptera in which, as already mentioned, eusociality is thought to have developed at least 11 times. The English scientist related eusociality with the peculiar mechanism that governs reproduction and sex determination in this order of insects: haplodiploidy. This means that in hymenopterans the females develop from fertilized, diploid eggs, whereas males originate from parthenogenetic, haploid eggs, producing spermatozoa which are identical from the genetic standpoint. This results in asymmetrical kinship coefficients: On average, two full sisters will share three-fourths of their genes but they are only related to their mother by one-half. Hamilton suggested that the behavior of the workers, whereby they helped their mother queen to produce more sisters, was selected because in this manner more of their genes were passed down to other individuals than if they had reproduced directly. Although kin selection undoubtedly played an eminent role in the origin of the evolution of social insects in general, it seems that it was not the sole operating selective force. The condition of three-fourths kinship in fact rarely occurs, especially in primitively social species, because of the high frequency of multiple matings and the presence of multiple queens in the same colony (Wilson, 1975; West-Eberhard, 1978). Moreover, some social insects (e.g., termites) are not haplodiploid.

Other Theories

Other evolutionary mechanisms proposed to explain the origin and evolution of eusociality, based on the selection at the individual level, are parental manipulation and mutualism. Parental manipulation theory, proposed mainly by Alexander (1974), claims that sterile castes in social insects can be explained by selection at the level of reproducing individuals. If a queen managed to manipulate the phenotype of her own offspring by sacrificing some (the sterile workers) in favor of others (the reproducers), then she could pass down more copies of her own genes to the next generation. This would occur if she raised more or better reproducing offspring with the help of the sterile individuals than if she reproduced alone without their help.

Rather than explain the origins of sterile castes in social insects, the theory of mutualism proposed by Lin and Michener (1972) examines the conditions that may have favored the origin of social behavior. According to this theory, ecological pressures, such as the need for common defense against predators or the aggregation of huge numbers of not necessarily related individuals in particularly suitable nesting sites, could have led to the development of mutualistic interactions which later evolved into true and proper social behavior.

Today, eusociality in insects is believed to have developed and evolved both at the individual level and through kin selection, the latter particularly in family groups — not necessarily matrifilial groups but possibly groups containing a mixture of more or less closely related females, according to the polygynous family theory (West-Eberhard, 1978). This was the result of ecological pressures similar to those which modeled the same phenomenon in some vertebrates (Andersson, 1984).

Social Wasps as Models for the Study of Social Evolution and Its Mechanisms

Wasp Societies: From Simple to Complex

As previously shown, all species of ants and termites are eusocial. Their colonies show highly complex organizations, with communication systems based on pheromones and a worker caste, determined at the preimaginal stage and specialized for various types of tasks: defense, foraging, brood care, etc. On the contrary, in the wasps and bees, in addition to very socially evolved species such as honeybees, stingless bees, or several species of polistines and vespines, there are also solitary and other species which only meet the three criteria for eusociality during certain phases of their colonial cycle. In these species, the females can change caste during their adult life, which is why the bees and wasps are such important study subjects. Indeed, comparative methods applied to these groups allow us to reconstruct the probable evolutionary history of social behavior.

Evans hypothesized that the main types of possible socially primitive groups of insects such as wasps, in which individuals shared the same nest but did not exhibit a worker caste, were either matrifilial or aggregational. Michener devised a nomenclature to define the intermediate evolutionary stages. West-Eberhard (1978) and Pardi (1980) discussed the salient features (Roubaud, 1916; Wheeler, 1928; Cowan, 1991).

The family Vespidae comprises six subfamilies: Euparaginae, Masarinae, Eumeninae, Stenogastrinae, Polistinae, and Vespinae (Carpenter, 1991). The last three all exhibit various degrees of sociality. The distribution of the Stenogastrinae, or hover wasps (Fig. 1), is limited compared with that of the other subfamilies, stretching from southeast India to Southeast Asia and from Indonesia to as far as New Guinea. There are approximately 50 described species divided into 6 genera, 2 of which (*Stenogaster* and *Anischnogaster*) are only found in the Papua region and 4 (*Liostenogaster, Eustenogaster, Parischnogaster,* and *Metischnogaster*) only in the eastern region. The hover wasps live in primary forests and build their well-camouflaged nests on various kinds of substrates; nest architecture is highly varied but never displays a petiole. Their colonies are small and rarely exceed approximately 10 individuals. In some species, the majority of nests only have one female, and social interactions are limited to brief periods in which

FIGURE 1 A colony of *Liostenogaster flavolineata*, a stenogastrine wasp from Malaysia. The nest does not have a petiole.

FIGURE 2 *Polistes gallicus.* Several species of wasps belonging to the genus *Polistes* are among the animals most frequently employed in the experimental study of social behavior and its evolution. Note that the nest has a petiole.

FIGURE 3 Colony of swarming wasps of the genus *Polybia.*

the newly eclosed individuals remain on their nests with their mothers before leaving to found their own nests (Turillazzi, 1991).

The subfamily Polistinae has the most genera and species. According to a cladistic analysis proposed by Carpenter (1991), the subfamily comprises four tribes: the Polistini, with only 1 genus *Polistes;* the Mischocyttarini, with only 1 genus *Mischocyttarus;* the Epiponini, with 22 genera of Neotropical distribution; and the Ropalidini, with 4 genera of paleotropical distribution.

The genus *Polistes* comprises approximately 200 species; colonies are relatively small and their nests never have an envelope. It is important to note the fact that their geographical distribution covers the same area as most of the scientific laboratories concerned with the study of the evolution of behavior (Fig. 2). *Polistes* is a key group in modern sociobiology and a test organism for hypotheses of social behavior (Turillazzi and West-Eberhard, 1996). Nest foundation in this group is always performed by a solitary queen or associate foundresses, whereas swarming is the rule in the Epinonini, with numerous workers and a varying number of queens. Some species and genera of the Epiponini, which can sometimes form enormous colonies (Fig. 3), show remarkable morphological differences between the queen and the worker castes, but in *Polistes* the differences are only behavioral.

Vespine societies are extremely complex but in many aspects differ considerably from those of the polistines. Distributed mainly in the temperate region with the genera *Vespa* (hornets and yellow jackets), *Vespula,* and *Dolichovespula,* they exhibit a clear-cut caste dimorphism; their colonies, characterized by paper nests composed of several combs and a protective envelope (Fig. 4), are generally founded by a single queen. In these respects, the vespines are the most social wasps, although the swarming Polisti-

FIGURE 4 A nest of *Vespa crabro* (hornet) consisting of several overlapping combs, protected by an multilayered vegetable pulp envelope held together with a salivary matrix.

nae are undoubtedly more specialized in other aspects; indeed, although their colonies are seldom monogynic, sociality is obligatory since these wasps have no solitary founding phase. Moreover, they exhibit a sophisticated form of division of labor among the workers (Jeanne, 1991a).

Major Themes of Sociality and Their Study in Wasps

Nest Building: Society and the Environment

The nest is an important factor in the origin and evolution of group living in insects. According to Hansell (1984), the nest allows its builder to exert some sort of control over the environment. The creation of a more stable environment renders the foundation of new colonies more difficult and makes it more advantageous for the young to stay in the colony than to leave. As a result, the colony grows in size, and interactions between individuals become more complex. At the same time, the nest often serves as a protection against predators and parasites, plays a role in some particular forms of communication between individuals such as by bodily vibrations of the nest, and improves brood rearing opportunities with the possibility of food storage. However, it also limits the options of the builders as they become more dependent on the nest.

Nest architecture in social wasps is an extremely interesting field of study because animal architecture has undoubtedly reached it peak of variety of design and complexity in these insects. The paper by Jeanne (1975) on the evolution of vespid nests remains a classic in this sector. Jeanne argues that the architecture of a nest is the result of several selective pressures. The immobility of the social wasp brood makes them extremely vulnerable—they are legless larvae cramped into made-to-measure cells. Therefore, many larvae packed together make the nest a tempting source of food for both vertebrate and invertebrate predators. These characteristics are probably a secondary consequence of nest living. Originally, the nest was built to protect the brood, but in some respects it renders it more vulnerable; this in turn makes the nest even more important in a sort of self-reinforcing cycle.

The immature brood must necessarily rely on the adults to defend them if they are to survive. Defense may be direct, which explains why the adults are endowed with dreadful weapons such as stings and poisonous or irritating secretions as well as indirect means of protection in the form of certain features of the nest. One protective architectural device is the petiole, which limits access to the cells containing the brood or food to terrestrial predators such as ants. An envelope serves the same defensive purpose but is also effective against flying predators and parasites as well as pedestrian predators. Both structures require far more building material than is needed for simple brood cells; therefore, the amount of energy spent on building a complex nest must be taken into account when calculating the selective forces modeling its architecture. Sometimes, nest architecture is based on camouflage to blend in with the environment and thus give the larvae even better defense against predators. However, this type of design can entail a reduction in nest size and consequently a sort of involution at the social level of a given species. This seems to be the case in the genus *Metischnogaster,* in which its peculiar architecture, among the most mimetic of all the stenogastrines, has produced such a narrow nest that it can only reach a maximum of 13 or 14 cells: More cells would increase the risk of nest breakage (Fig. 5).

Hansell found that the choice and consistency of the building material could also influence nest design and indirectly the sociality of its builders. According to Hansell, the small stenogastrines are less social than the polistine wasps because of a fundamental difference in the quality of their building materials. The stenogastrines employ very short, wooden fibers held together with saliva but produce no matrix to efficiently bind the fibers together. The polistine wasps, on the other hand, can use longer plant fibers and bind them together with far more efficient substances. This allows the polistine wasps to build much larger nests and consequently opens the way to larger and more populous colonies.

FIGURE 5 The nest of the stenogastrine wasp *Metischnogaster drewseni* is highly mimetic but can never develop beyond a certain size.

Animal artifacts have often been considered as a sort of "frozen" behavior (Hansell, 1984). This gives an idea of the importance of nest architecture in the study of the behavior of a species. For this reason, nest architecture can sometimes be of taxonomic value. Nests can be extremely important in supplying phylogenetic information on social wasps and even help distinguish distinct but morphologically similar species. In some genera (e.g., *Polistes*), nest architecture appears to follow a constant design, but in others (e.g., *Mischocyttarus* or the stenogastrines) it can vary considerably. *Parischnogaster alternata* and *P. striatula,* for example, are very similar from a morphological standpoint but can be easily distinguished by the different shape of their nests (Fig. 6). Thus, wasps prove to be an excellent study subject for the evolution of building behavior.

Behavioral Flexibility: The Origins of Queen–Worker Dualism

As discussed previously, division of labor within colonies is one of the salient features of the phenomenon we define as eusociality. Division into castes is directly related to the problem of altruism. To paraphrase Darwin, Gadagkar observed that the evolution of eusociality follows two main phases: the origin of altruism and the origin of the traits that determine total and irreversible reproductive differentiation of the individuals. In this case, too, wasps have proved their worth as model study organisms. In some species, reproductive divergence only consists of different phases in the biological cycle of solitary females, with each female eventually becoming reproductively active following a period of ovarian stasis. In other species, such as the stenogastrines, the colonial cycle passes from a phase during which individuals follow a solitary life to others in which the colony shows all the characteristics of eusociality. In species such as *Polistes* and *Mischocyttarus,* there are again very simple forms of social organization but with a well-characterized worker caste, although in the majority of cases only at the behavioral and anatomic–physiological levels. In others such as *Belongaster,* these differences may also be morphological. Colonies of the genus *Ropalidia* range in size from very small (e.g., *Ropalidia formosa*) to very large (e.g., *Ropalidia montana*), accompanied by morphologically different queens and workers and where nest foundation occurs through swarming. This is the rule in Polybioides and the Neotropical genera of polistines, including the Epiponini tribe (Fig. 3). Their colonies can reach enormous sizes (*Agelaia vicina* colonies in Brazil can exceed 1 million individuals) and the worker caste performs an amazing variety of behaviors. Thus, although the Stenogastrinae are well suited for the study of the origins of altruism, wasps such as *Polistes* are model organisms for studying the characteristics of sterile individuals and evaluating the costs and benefits of their behavior. In fact, the origin and evolution of these traits are the principal problems of sociobiology.

How did queen–worker dualism originate? According to West-Eberhard (1996), the worker phenotype is based on properties of the ovarian ground plan of solitary wasps. The ovarian ground plan is the peculiar system whereby ovary size is related to hormonal activity (mainly that of the juvenile hormone which has a well-known gonadotropine effect in adult insects) and which affects behavior through neurohormonal coordination (West-Eberhard, 1996). The ovarian cycle in progressively provisioning solitary wasps is extremely precise. In *Zethus miniatus,* for example, the ovary produces one egg at a time, with several days between each deposition. During this interval, the mother cares for and nourishes the larva that developed from the previous egg. This reproductive cycle corresponds to a behavioral cycle. Cell building, in the period just before the egg is ready to be laid, is repeatedly followed by brood guarding and provisioning behavior, a stage characterized by the lack of developed eggs in the ovary.

Moreover, during the phase of ovarian development, the females of many solitary species are remarkably aggressive

a b

FIGURE 6 *Parischnogaster alternata* (a) and *Parischnogaster striatula* (b) are species which are practically identical morphologically, but they can be distinguished by the different architecture of their nests.

toward conspecifics approaching their nests. In the vespids, for example, the females which lay most frequently are the most aggressive, but this can also be found in various sphecids and in the pompilid *Auplopus semialatus.* We can therefore conclude that aggressive dominance, found in social species such as *Polistes* and one of the fundamental mechanisms of their social organization, is probably an ancestral trait which was already present in the solitary ancestors of the social wasps. If a female is unable to lay a developed egg because of lack of a suitable nesting place (the nest) or because she is undernourished or even prevented from doing so by a conspecific, then the egg can be reabsorbed, leaving traces in the ovary. West-Eberhard therefore suggests that reproductive division of labor in social wasps came about when aggressiveness related to ovarian state led to the splitting of ovarian cycle peculiar to solitary into two. The queen phase was characterized by developed eggs and the worker phase by nondeveloped eggs, with a concomitant division of behavioral roles.

The females of many solitary wasps, as in the case of *Zethus miniatus,* exhibit a wide range of possible alternative behaviors ranging from solitary nest foundation to usurpation of ready-built nests and the adoption of abandoned larvae by temporarily non-egg-laying females (Fig. 7). In stenogastrine wasps in which dominance hierarchies among the females of the colony have been reported, the females exhibit wide behavioral flexibility. They can follow several alternative behavioral strategies, from foundation and solitary reproduction to joining or usurping a colony of related (including the maternal nest) and nonrelated individuals. From this standpoint, the degree of sociality in the population of a species and the role an individual plays in the colony are clearly dictated by circumstances set by the social environment. The same is true for all primitively social wasps, such as *Polistes, Mischocyttarus,* and *Belonogaster,* in which the role of a female in a colony is determined by the colonial cycle and by the presence of an active queen, but inversion of social roles is nevertheless possible. One good example of this is afforded by workers or subordinate females which can reproduce if the queen disappears. In wasps in which caste differentiation is more clear-cut, the state of the ovary also regulates changes in the behavior of individuals over time, thus determining the complex social organization of the colonies of these in-

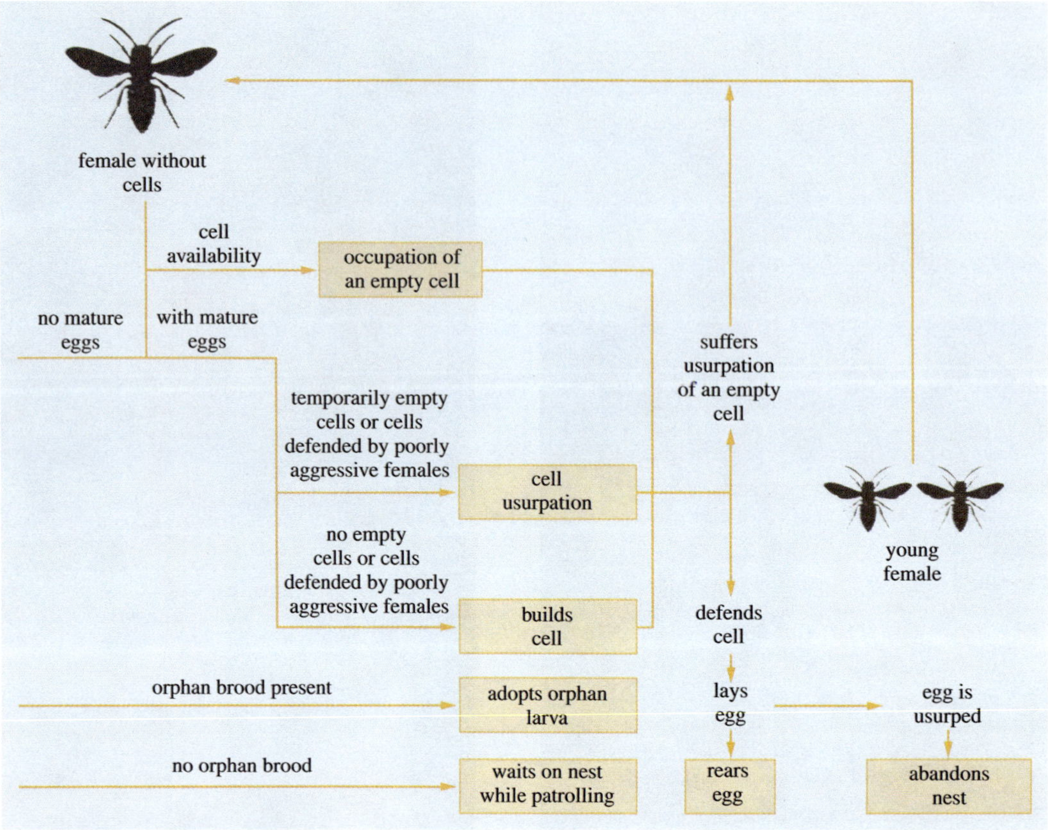

FIGURE 7 Diagram showing the possible behaviors in *Zethus miniatus* females still with brood and females which have already lost their brood.

sects. From the previous discussion, it can be seen how the three main phases of the ovarian cycle of solitary wasps may have served as the basis for the evolution of division of labor and its different characteristics in the sociality of wasps.

According to West-Eberhard, the events leading to the evolution of traits such as the initial expression of an ovarian cycle already occurred in preeusocial wasps. The juvenile hormone played an important role in this process by coordinating behavior and the physiological changes related to reproduction.

Communication and Social Regulatory Mechanisms in Vespid Colonies

According to Wilson (1975), communication exists when a given behavior or characteristic of one individual, a signal, causes a predictable change of behavior in another. Communication is therefore a characteristic of all animal societies. Hölldobbler and Wilson distinguish several functional categories of communication in insect societies, ranging from extremely simple phenomena, such as antennal contact between two ants meeting on the same foraging path, to highly spectacular manifestations, such as the mass movements seen in swarming, nest defense, and unified exploitation of a food source. All these categories employ the four main channels of transmission of information: optical, acoustic, chemical, and tactile. All the classes of communication mentioned by Hölldobbler and Wilson can be found in social wasp colonies. Nevertheless, for the most part, the study of communication in these insects is still at the descriptive stage.

Control of Colony Organization

Previously, the probable origins of castes were discussed. Here, the possible mechanisms controlling the phenomenon and which are therefore the basis of the social organization of a colony are discussed. It could be hypothesized that once we have resolved the nature of the fundamental mechanisms regulating the division of labor, then their application to increasingly larger colonies would finally reveal how their social organization actually works. As will be shown, this is not completely true, but the study of species with small colonies can greatly help us to unravel the problem. Many of the phenomena also occur in species with very large colonies, but less complex societies offer the opportunity to study them more easily from the experimental standpoint.

Experimental studies on the regulatory mechanisms in

social wasp colonies began in 1942 when Pardi first described dominance hierarchy in spring associations of *Polistes dominulus.* A system of ritualized behaviors replaces direct conflict and can be recognized by the establishment of a dominant individual and the akinesis of the subordinate (Fig. 8). Dominance relations not only determine reproductive roles but also organize the division of labor, with the subordinate females assuming the role of workers. The regulatory mechanisms controlling the colony in the foundation stage are therefore mainly behavioral and due to the aggressiveness of the dominant female toward her subordinates. Recognition of the roles of the different individuals in a colony is the basis of the ritualization of the behavioral modules. It substitutes combat and prevents its intensification, thus preventing disastrous confrontations once the hierarchy has been established.

However, what are the characteristics that guarantee a dominating rank and what are the mechanisms that inhibit reproductive potential in the auxiliary females? Initially, Pardi demonstrated that rank was directly correlated with the size of the ovaries of the different individuals. From his first studies, however, he also realized the importance of the activity of the endocrine glands, the corpora allata, in dominance (Fig. 8). From recent studies, especially those of Röseler (1991), it seems that several endocrine factors are involved in dominance behavior and that a deficit in one system can be compensated by greater activity in another. These endocrine factors are triggered by the activity of the corpora allata, which produces the juvenile hormone, and of the ovaries, responsible for the production of ecdysones; these two activities need not necessarily be synergic. In some species, prostaglandin, which is released after mating, may also be involved. Hormonal differences in *P. dominulus* females at the end of hibernation can therefore determine their rank in different founding groups. These physiological differences may be genetically influenced, induced through larval nourishment, or even settled by their order of emergence or other events that happened before hibernation. Other factors in the founding group, such as size and order of arrival in the nest, can also influence the organization of the hierarchy (Fig. 8). Conclusive results remain to be obtained.

The mechanisms behind the progressive loss of fertility in the subordinates, according to Pardi (1942), are of a "sensorial nature and directly depend on the presence of dominant females on the nests." He advanced the hypothesis that ovarian reduction evolved as a consequence of division of labor. Subordinates females, which primarily dedicated themselves to working away from their nest, underwent castration through their labor. The dominants were not only privileged because they did not perform any extranidal activities but also benefited trophically through the frequent regurgitation of the subordinates. Moreover, the fact that dominant females removed any eggs which the subordinates may have laid by eating them (a phenomenon

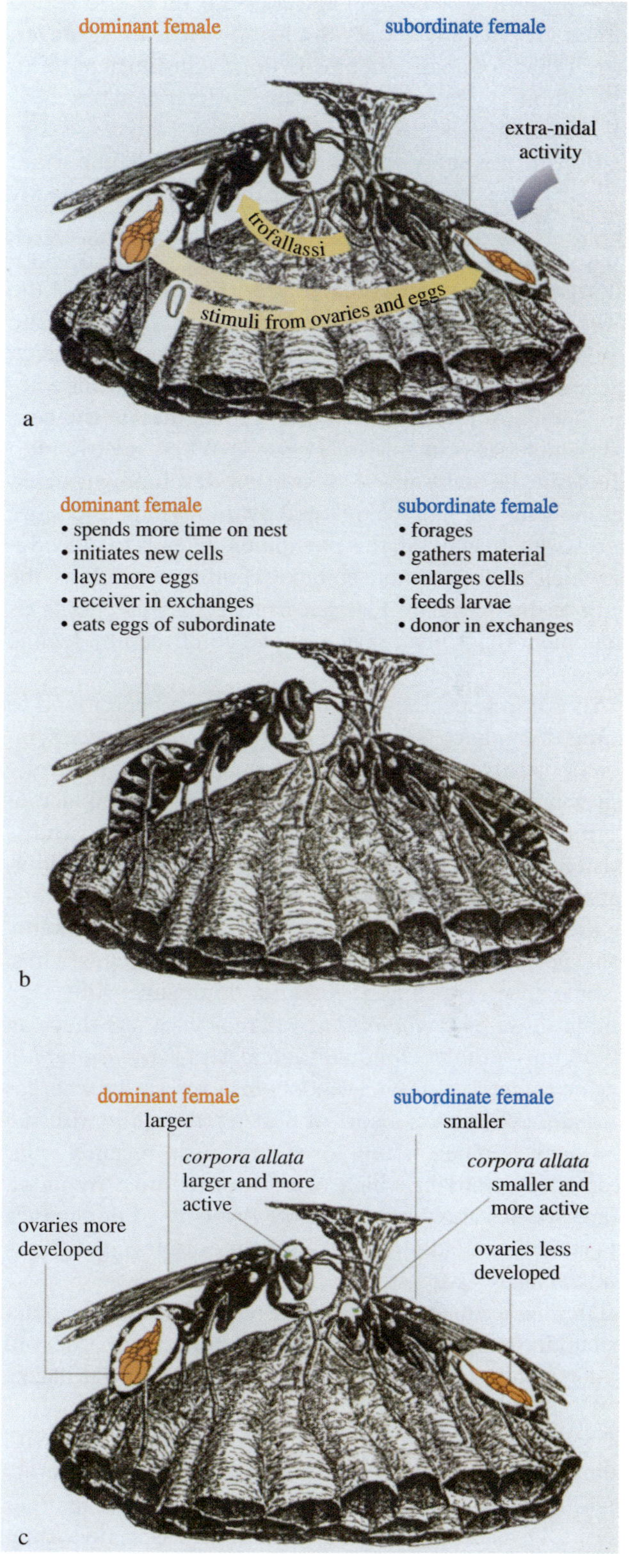

FIGURE 8 Characteristics of dominance and associate foundresses on a spring nest of *Polistes dominulus.* (a) Influence of dominance and extranidal activity on the ovaries of the subordinate females. (b) Behavioral characteristics of the two females. (c) Physical and anatomical characteristics of the two females.

called *differential oophagy*) and not vice versa also added to their trophic advantage. Studies by Röseler, however, suggest that dominance behavior alone is insufficient to totally inhibit egg-laying capacity in auxiliary females. Several factors may therefore act simultaneously to cause sterilization in the subordinates, one of the most important being that the dominant female, called alpha, is perfectly capable of laying eggs.

Dominance phenomena continue to be the basis of social organization in the colonies after the emergence of the workers. Pardi (1942) discovered that the workers undergo a preestablished cycle of ovarian development, which reaches its peak at approximately the 10th day after eclosion. This observation allowed him to formulate the general dominance principle for *Polistes:* "When any two individuals in the same phase of ovarian development meet, the one with the more developed ovaries will be the more aggressive." In general, this phenomenon leads to individuals which, under certain colonial conditions such as the death of the dominant females, can start laying eggs. This phenomenon has been reviewed and confirmed by Jeanne (1991a).

All *Polistes* species studied to date have been found to exhibit dominance behavior. Depending on the species, interactions between the females are characterized by strong aggressiveness and despotism until an almost total lack of hostility is reached, corresponding to a barely discernible division of labor. West-Eberhard (1996) showed that dominance rituals are characters that diverged rapidly during the course of evolution and which can differ markedly from species to species, even within the same genus. Thus, in some species such as *P. fuscatus,* dominance rituals resemble those of *P. dominulus,* whereas in others they can differ considerably. Dominant females of *P. erithrocephalus,* for example, wag their abdomens excessively, whereas *P. canadensis* installs a sort of nest territoriality, with the dominant females taking over the most recently constructed portions in which oviposition is most frequent. Other species show a remarkable diversity of dominance repertoires and differential oophagy, displaying various levels of aggressiveness.

Dominance hierarchies have also been reported and described in other wasps. For example, it has been observed in the small societies of the stenogastrines (Turillazzi, 1991). In *Parischnogaster nigricans,* the phenomenon is very marked and the rank of the females is correlated with their ovarian development. In contrast, dominance interactions are far more vague and difficult to determine in other species. Division of labor is evident in the associative foundations of *P. alternata,* with one of the females performing more extranidal activities and the dominant spending more time on the nest. However, there is no significant difference in the size of their ovaries, suggesting that both individuals offer a similar contribution to egg laying. Dominance hierarchies can also be seen, although with varying degrees of aggressiveness, in the polistines, which have colonies similar to those of *Polistes,* such as *Bellonogaster, Ropalidia, Mischocyttarus,* and *Parapolybia.*

Pardi (1942) noticed the similarity between dominance in *Polistes* and comparable phenomena described in vertebrates "even though on different psychic planes." He well understood how the recognition of individuals, which was possible in a small group of vertebrates, could pose a problem in wasp colonies consisting of hundreds or more individuals. Consequently, he proposed that rank could be recognized by olfactory or visual cues. This brings us to the problem of what kind of regulatory mechanisms the queen uses in the typically huge colonies of the swarming Neotropical polistines and vespines, which are far larger than those of *Polistes.* Dominance behavior has also been reported in the societies of swarming wasps. In *Metapolybia aztecoides* and *Synoeca surinama,* the workers perform a peculiar dance in front of the new queens. If the latter do not react aggressively, these workers will attack them increasingly more fiercely until they eventually dominate them and drive them out of their nest or force them to behave like workers. In the vespines, Montagner noticed a dominance hierarchy among the workers of the densely populated *Vespula germanica* colonies, similar to those described in *Polistes,* and how it determined division of labor within the worker caste.

Most authors believe that colonial control is first established and maintained through direct physical, aggressive contacts and oophagy and that only later are higher levels of aggressiveness related to individual recognition signals, easily recognized by nest mates. This allows subordinate individuals to avoid the costs of further combat. Finally, direct interactions become increasingly rarer, substituted increasingly more by signals of the queen's presence. The occurrence of recognition signals emitted by the queen has been confirmed in several swarming vespines and polistines. These signals, obviously chemical in nature, lead to the formation of "royal courts" in which the queen is constantly surrounded by many workers that follow, feed, and attend her. The classic and simple experiments in which a queen is removed from a colony inevitably cause a marked change in the behavior and physiology of the workers. As a colony grows in size, it can be predicted that control based on chemical signals will gradually take over from physical dominance (Fig. 9).

Spradbery suggests that in the small *Dolichovespula* colonies, which comprise approximately 100 workers, physical and pheromone factors work together. In species of the *Vespula rufa* group, there is still less physical dominance, although it still occurs in *V. atropilosa* and *V. acadica.* In *Vespa orientalis* a pheromone has been isolated from the heads of queens which controls some aspects of social organization, such as royal cell construction by the workers;

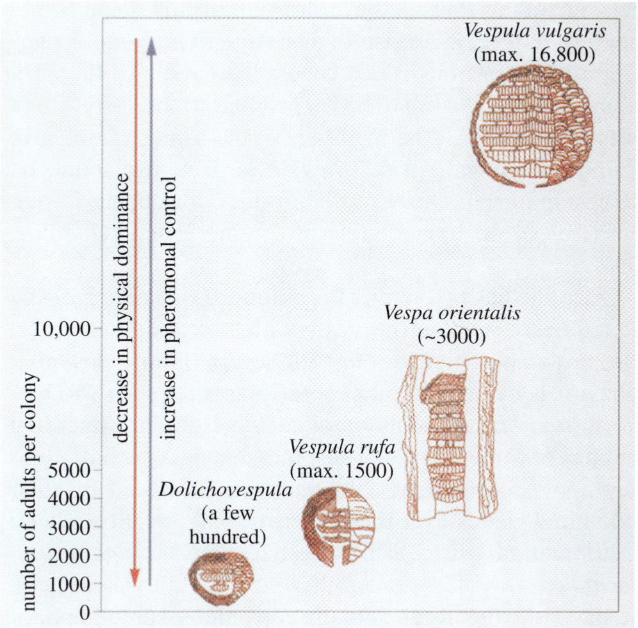

FIGURE 9 Schematic diagram of the behavioral and pheromonal control by the queen in various species of vespines, in relation to colony size.

there is still no proof, however, that this pheromone also suppresses ovarian development in the workers. Finally, in species of the *Vespula vulgaris* group, which has the largest colonies reaching tens of thousands of workers, experimental trials suggest that the queen can control ovarian development in the workers by means of chemical (probably nonvolatile) substances. These substances, which she applies on the nest, may be produced by her abdominal sternal glands. In the swarming polistines, findings suggest the presence of supplementary chemical control in several species belonging to genera such as *Protopolybia, Agelaia,* and *Polybia.*

With the advent of chemical recognition signals emitted by the queen, which work like a declaration of rank and status, it is a short step in evolution from recognition substances (releasers) to inhibiting substances (primers). This does not imply, however, the total disappearance of competitive phenomena or that the colonies are utopian, peaceful societies. According to Jeanne (1991b), the concept of competition should not be restricted to the concept of aggression and physical attack, but it should be extended to include even the most subtle and nonviolent interactions between workers even in the most evolved social wasps. This means that virtually all individuals of a colony are affected by the dominance system. In fact, it appears that all workers of all species of social wasps are able to reproduce, develop, and lay eggs at least at some time during their reproductive life. Moreover, the presence of an active queen is indispensable if ovarian development and queen-like behavior are to be inhibited in the workers. In other words, the colony is constantly under a competitive state of stress, which is upheld from the top of the reproductive hierarchy. Jeanne (1991b) again points out that even in large colonies, such as *Polybia occidentalis,* the workers stay inside their nests in the early stage of their lives, dedicating themselves to tasks which are relatively safe. Only toward the ends of their lives do they venture out to undertake more dangerous extranidal activities, as if they want to keep the possibility of reproducing themselves to the very last. Traces of individual discord therefore occur within even the most harmonious societies, and by studying these conflicts it is hoped that the mechanisms which guarantee such an apparently perfect organization will be discovered.

One of the questions Pardi raised regarding *Polistes* was whether rank recognition was based on olfactory cues, but this remains unanswered. Work by Maschwitch did not find any pheromones in these wasps. The question was recently readdressed by Downing and Jeanne, who, in an analysis on *P. fuscatus,* found evidence for a significant correlation between the rank of the individuals in a hierarchy and the size of some of their exocrine glands. Dominant females had more developed Dufour's gland and sternal glands than did subordinates, although other glands were similar in size in the two types of individuals. Therefore, some species of *Polistes* may be able to control the reproductive capacity of their competitors through the use of pheromones as well as by aggressive behavior. Furthermore, because some species show no evident signs of dominance behavior it cannot be ruled out that the dominant females may transfer inhibiting pheromones to the bodies of their subordinates.

Social Parasitism and Nest Mate Recognition

One phenomenon that could help us discover the true nature of colony control in *Polistes* is social parasitism. A social parasite is a species that has lost its worker caste and whose females must usurp the nest of another species to reproduce and exploit the workers to rear their own brood. Known in all the social Hymenoptera, these peculiar parasites are also found in the vespines. Three species occur in *Polistes,* all of which are prevalently distributed throughout the Mediterranean basin. In some cases, these parasites can have a catastrophic effect on the fitness of their hosts, with all the reproducing individuals emerging from the colony being the offspring of the usurper. In many colonies, however, the parasite does not totally take over the colony. As a result, the workers manage to complete ovarian development and lay their own eggs, and sometimes individuals leave their maternal nests to found new nests of their own.

Invasion of the host nest by the parasitic females always occurs at a particular moment of the colonial cycle, i.e., before the workers have emerged or when only very few are present (Fig. 10). Studies on nest invasion have revealed

FIGURE 10 Female of *Polistes sulcifer,* a social parasite of *Polistes dominulus,* on the nest of its host species.

that *Polistes* parasites employ two main strategies to take possession of the host nest — one aggressive and one nonaggressive. In the first method, typical of *Polistes sulcifer* and *Polistes semenowi,* the parasite fights the host foundress fiercely and eventually drives the dominant female from her nests, often mutilating her legs during the battle. However, the parasite does allow the subordinates to remain on the nest but subjects them to an intense treatment resembling dominance. In contrast, *Polistes atrimandibularis,* the third species of parasite, takes longer to arrive at its goal, bearing the attacks of the host foundress. These gradually become increasingly less convincing, until the parasite eventually takes over as the dominant female (Cervo and Dani, 1996).

Although conquering the nest may be risky for the parasitic queen, maintaining control over the colony is even more difficult. She must persuade totally alien workers to work for her and renounce not only their chances of reproduction but also the possibility of rearing a related brood. This suggests that the parasitic queen employs a far stronger means of colonial control than other *Polistes* and that she does not limit her methods to mere dominance behavior. There are two possible hypotheses. The first is the possibility that the parasites employ pheromones to inhibit aggressiveness, which they transfer to the host's body through careful licking operations during the process of dominance. This hypothesis is supported by the nonaggressive behavior that *P. atrimandibularis* displays when she first invades the host nest and by the fact that all three parasites often emit drops of clear liquid to the female hosts. It seems that the parasites are capable of manipulating the host's immature brood or of camouflaging themselves with the odor of the host colony, thus tricking the future host workers with regard to their true relationship with the usurpers. The "colony odor" is a particular signal that allows each member of the colony to distinguish nest mates from alien individuals. Recent studies on several social insects, including *Polistes,* have shown that this signal is of a chemical nature. It consists of a mixture of high-molecular-weight hydrocarbons which covers the cuticle of the individuals and has proved to be very similar to the secretion of Dufour's gland. All the members of the same colony have a similar chemical "print"; in *Polistes* it is also found on the nest material. The adults learn this odor during a brief, sensitive period immediately after emergence (Gamboa, 1996).

Once she has taken over the colony, the female parasite meticulously strokes the nests with her abdominal sternites, an activity that can last for several hours. A similar behavior is evident in other social parasites such as *Dolichovespula artica* and *Vespula austriaca* which attack the colonies of some species of vespines, during which it seems they apply some particular secretions produced by their abdominal glands over the nest surface. A recent study on *P. atrimandibularis,* which parasitizes *Polistes biglumis bimaculatus,* revealed yet another strategy. In this species, the parasitic queen can actually copy and assume the odor of the host colony, actively changing her own hydrocarbon print as the season proceeds by means of true and proper chemical mimicry. Immediately after nest invasion she starts to smell like her host, but her own brood keep the original odor of a parasite so that when the time comes for them to reproduce intraspecific recognition is possible. Some behaviors and the phenomenon of nest usurpation, which is very common in the genus *Polistes,* have been carefully analyzed and reconsidered. It has been determined that these behaviors, which are very conspicuous in the social parasites, also form part of the host repertoire, although in this case they are far more obscure (Cervo and Dani, 1996). Social parasites are therefore highly interesting subject matter since they can be considered an extension, and indeed amplification, of the system of colonial control in *Polistes* (Turillazzi, 1992).

Conclusions

In may respects, the importance of wasps as models and test organisms for the science of sociobiology can be compared to that of *Drosophila* in genetics and white rats in comparative psychology. We employ wasps to test various hypotheses on evolution, from the influence of demographic parameters on social life to the importance of nest mate and kin recognition and the role of kin selection in the origin and maintenance of social behavior. At the same time, they are used to study the mechanisms controlling the colony, communication between individuals, and the division of social roles. As study matter, they are also instruments for generating new ideas and hypotheses. Some of the fundamental discoveries of insect sociobiology (e.g., trophallaxis and nutritional caste control) were first discovered in wasps (Wilson, 1971). Likewise, observations on the biology and natural history of several species of social wasps led to generally convincing ideas which have marked modern studies on social behavior. Among these are the recognition of dominance hierarchy as a general characteristic in animal societies (Pardi, 1942) and Hamilton's findings on kin selection. Leaving aside the repulsion and fear these insects may arouse in the nonnaturalist, their societies can stimulate the fantasy and imagination of the attentive observer because they offer elements and ideas for some daring comparisons. As Hamilton observes, traits and emotions such as duty, rebellion, motherly love, violence, trickery, cowardice, and unification in the face of danger can all be found in a *Polistes* colony, so different and yet so similar to human society.

References Cited

ALEXANDER, R. D. (1974). The evolution of social behavior. *Annu. Rev. Ecol. Syst.* **4,** 325–383.

ANDERSSON, M. (1984). The evolution of eusociality. *Annu. Rev. Ecol. Syst.* **15,** 165–190.

CARPENTER, J. M. (1991). Phylogenetic relationships and the origin of social behavior in the Vespidae. In *The Social Biology of Wasps* (K. G. Ross and R. W. Matthews, Eds.), pp. 7–32. Comstock, Ithaca, NY.

CERVO, R., and DANI, F. R. (1996). Social parasitism and its evolution in *Polistes.* In *Natural History and Evolution of Paper-Wasps* (S. Turillazzi and M. J. West-Eberhard, Eds.), pp. 98–112. Oxford Univ. Press, Oxford.

COWAN, D. P. (1991). The solitary and presocial Vespidae. In *The Social Biology of Wasps* (K. Ross and R. Matthews, Eds.), pp. 33–73. Comstock, Ithaca, NY.

GAMBOA, G. J. (1996). Kin recognition in social wasps. In *Natural History and Evolution of Paper-Wasps* (S. Turillazzi and M. J. West-Eberhard, Eds.), pp. 161–177. Oxford Univ. Press, Oxford.

HAMILTON, W. D. (1964). The genetical evolution of social behaviour. I. *J. Theor. Biol.* **7,** 1–52.

HANSELL, M. H. (1984). *Animal Architecture and Building Behaviour.* Longman, New York.

JEANNE, R. L. (1975). The adaptiveness of social wasp nest architecture. *Q. Rev. Biol.* **50,** 267–287.

JEANNE, R. L. (1991a). The swarm-founding Polistinae. In *The Social Biology of Wasps* (K. G. Ross and R. W. Matthews, Eds.), pp. 191–231. Comstock, Ithaca, NY.

JEANNE, R. L. (1991b). Polyethism. In *The Social Biology of Wasps* (K. G. Ross and R. W. Matthews, Eds.), pp. 389–425. Comstock, Ithaca, NY.

PARDI, L. (1942). Ricerche sui Polistini. V. La poliginia iniziale in *Polistes gallicus* (L.). *Boll. Istituto Entomol. Univ. Bologna* **14,** 1–106.

PARDI, L. (1980). Le vespe sociali: Biologia ed evoluzione del comportamento. *Contrib. Centro Linceo Interdisciplinare Sci. Matematiche Loro Appl.* **51,** 161–221.

RÖSELER, P. F. (1991). Reproductive competition during colony establishment. In *The Social Biology of Wasps* (K. G. Ross and R. W. Matthews, Eds.), pp. 309–335. Comstock, Ithaca, NY.

ROUBAUD, E. (1916). Recherches biologiques sur le Guêpes solitaires et sociales d'Afrique. La genese de la vie sociale et l'evolution de l'instinct maternel chez le vespides. *Annal. Sci. Naturelles* **1,** 1–160.

STARR, C. K. (1979). Origin and evolution of insect sociality: A review of modern theory. In *Social Insects* (H. R. Hermann, Ed.), Vol. 1, pp. 35–79. Academic Press, New York.

TRIVERS, R. (1985). *Social Evolution.* Benjamin/Cummings, Menlo Park, CA.

TURILLAZZI, S. (1991). The Stenogastrinae. In *The Social Biology of Wasps* (K. G. Ross and R. W. Matthews, Eds.), pp. 74–98. Comstock, Ithaca, NY.

TURILLAZZI, S. (1992). Nest usurpation and social parasitism in *Polistes* wasps: New acquisitions and current problems. In *Biology and Evolution of Social Insects* (J. Billen, Ed.), pp. 263–272. Leuven Univ. Press, Louvain.

TURILLAZZI, S., and WEST-EBERHARD, M. J. (Eds.) (1996). *Natural History and Evolution of Paper-Wasps.* Oxford Univ. Press, Oxford.

WEST-EBERHARD, M. J. (1978). Polygyny and the evolution of social behavior in wasps. *J. Kansas Entomol. Soc.* **51,** 832–856.

WEST-EBERHARD, M. J. (1996). Wasp societies as microcosms for the study of development and evolution. In *Natural History and Evolution of Paper-Wasps* (S. Turillazzi and M. J. West-Eberhard, Eds.), pp. 290–317. Oxford Univ. Press, Oxford.

WHEELER, W. M. (1928). *The Social Insects: Their Origin and Evolution.* Kegan Paul Trubner, London.

WILSON, E. O. (1971). *The Insect Societies.* Harvard Univ. Press, Cambridge, MA.

WILSON, E. O. (1975). *Sociobiology: The New Synthesis.* Harvard Univ. Press, Cambridge, MA.

PART TWO

Discovery and Spoliation of the Biosphere

Edited by Elgene O. Box and Sandro Pignatti

INTRODUCTION

Elgene O. Box
Sandro Pignatti

The relationships that exist between living organisms and the environment define that extremely complex network which is the biosphere: In this part of the volume, we attempt to delineate that network's general characteristics through a description of individual aspects of particular significance. The environmental factors that regulate life, how they have interacted to form fauna and flora, the role of man, the negative consequences on the biosphere of man's actions, and future perspectives are all examined.

The guiding line connecting such disparate topics is provided by the notion of ecosystem, which allows us to interpret relationships between plants, animals, microorganisms, and environmental factors. This is an ambivalent relationship because if, on the one hand, living entities derive from the environment what they need to survive, on the other hand, the environment poses limits of tolerance that cannot be breached. Conditions outside of these extreme limits, in fact, are lethal.

The ecosystem is a system regulated by environmental factors. A *system* has been defined as "a set of parts that interact with one another," so the concept of system should be understood as referring to combinations of various elements that are coherent with each other. Regarding the biosphere, the set in question is that of the ensemble of living organisms made up of hundreds or thousands of species, if individual cases are considered, or millions of species if the entire biosphere is taken into consideration.

The study of systems is an object of thermodynamics, which first developed through the study of reversible processes. The second principle states that systems tend to reach states of equilibrium that correspond to conditions of greatest possible entropy. The process that leads to the equilibrium can be fully described by the thermodynamic variables. Equilibrium is a final condition of uniformity of such variables in which the system has no possibility of further transformation. All the physical and chemical systems of which we know tend toward this final condition of thermodynamic equilibrium, and this has led R. J. E. Clausius to imagine that, in the end, the entire universe will reach such a state, in the sense that it will encounter "entropic death."

A system that is truly isolated, however, does not exist insofar as any device devised for the purpose of shielding it would impair heat-transmission phenomena entirely: An isolated system, therefore, is an abstraction. In reality, it is closed systems (which exchange only energy with the outside) or else open systems (which exchange both energy and matter) that exist. These can import matter or energy from without in such a way as to lower the system's level of entropy. In this manner, it is possible that cyclical processes may come into being within the system. To understand how they work, consider a system made out of three components: The A component acts on the B component and the latter acts on the C component, which in turn acts on the A component. An increase in A can cause on C such a reaction as can further accelerate, or depress, the dynamics of A. Feedback circuits are thus established, and this is a very important aspect because, by way of retroactive rings, the system becomes structured, acquires order, and assumes complex forms of behavior. A particularly significant case is that in which the action of A and the reaction of C are close to balancing each other out and the system stabilizes in a condition far removed from equilibrium. In such a case, the system reaches a form of self-governance that allows it to remain active while avoiding the increase in entropy which would have taken place in an isolated system. In general, natural ecosystems are systems of this type, which through efficient self-governing and homeostatic systems find themselves in a stationary condition (i.e., they maintain the dynamics of vital cycles and do not tend to change over time). The biosphere is also a system of cycles that stabilizes in a stationary condition. Therefore, we may interpret these open systems as we would complex ones: They manage to maintain a low level of entropy, and this generally occurs by way of retroactive cyclical processes.

The biosphere is a system of the greatest inclusiveness on Earth and may therefore be considered to be an example of a complex system. In a highly simplified form, its functioning may be portrayed as a cyclical pattern (see figure). The source of energy is represented by the sun, which radiates energy in the form of light and heat; a large part of the energy that reaches the earth from the sun is then dispersed in space during nocturnal radiation in the form of low-heat thermal energy, and only a small part (1–2%) enters into the living system by way of photosynthesis. Light is a form of low-entropy energy that can activate further transformations. In plants, it is accumulated in the form of chemical bonds in reduced compounds, first as glucides and then, through subsequent metabolic cycles, as lipids, protides, and so on. The organic matter thus produced may feed the accumulation of biomass or contribute to increasing the biodiversity of the system, which continues to grow as a recursive process. In general, the two effects do not occur simultaneously. In the course of respiratory processes, yet another phase of dissipation occurs; part of the matter involved (carbon dioxide, nitrogen compounds, and nutrients), however, goes back into circulation and acts as a fundamental retroaction on photosynthesis.

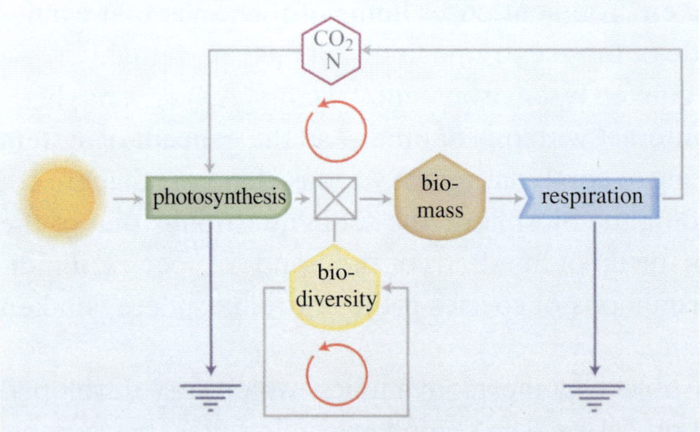

According to this representation, the biosphere appears to us as an open system that works in a cycle using energy from the sun. We may compare the process to the much more familiar one at work in internal combustion engines. Both systems require an energy source: The motor's energy source is gasoline, and that of the biosphere is the energy flow that originates in the sun. However, whereas in the biosphere the energy flow gives rise to metabolic cycles that lead to the formation of a certain quantity of organic matter in which part of the energy is stored in a useful form, in a motor the energy is not stored but degraded: In part it is used up as kinetic energy and in part it is dissipated in the form of heat.

In the ecosystem, the previously mentioned basic structure is articulated in a plurality of plant and animal components. The energy flow proceeds from the sun to the biological system through the process of photosynthesis, which leads to the production of organic living matter: biomass. It would be reductive, however, to consider living entities as an undifferentiated whole since they have functions that differ from one another and are integrated as structures of the complex system. A first category of living entities is made up of plants endowed with chlorophyll, (i.e., land plants and marine algae that accomplish primary production processes). Then the energy passes on to respiratory processes both in plants and, via the food chain, in herbivores, carnivores, and saprophytes. In the course of these transferals, there is a dissipation of matter and energy that may be immediate (respiration of single plants) or have a longer course (in which case the amount of energy that is dissipated is greater) by way of consumers (herbivorous and carnivorous animals or microorganisms such as molds and bacteria), whereas nutrients, nitrogenous substances, water, and carbon dioxide go back into circulation and are used up once again during photosynthesis. Another, no less relevant, aspect is the qualitative one: the tendency toward diversity that in the biosphere has led, through biological evolution, to the appearance of a multiplicity of organisms, each of which presents forms of adaptation that bind it to a particular niche in the grand cycles of planet Earth. The activity of the biosphere, therefore, consists of the production of biomass and of biodiversity.

Man worked his way into the dynamics of the biosphere, and his actions in general are finalized to the attainment of certain specific results which are thought to be useful or in any case desirable. In the first phases of hominization, man modified the biosphere through hunting, the use of fire, and traditional agriculture; the slowness of these processes, however, enabled the biosphere to progressively adapt. The situation changed radically with the onset of industrial development. Technology is based on experimental science, and therefore the phenomena that belong to it may be interpreted as chains of causes and effects. However, if we wish to have a view of the overall change of the biosphere, the single cause–effect link is not enough. In principle, it is always possible to find a relation between the size of a cereal crop and the quantity of fertilizer used to grow it; it is not possible, however, to obtain by extrapolation the rate of cereal growth starting from the formation of the kernel because the process governing such events is complex and the variables that influence it (such as the availability of water and quantity of heat) are interdependent. After vegetative growth, flowers will be produced and there will be fertilization, the ripening of fruit, and dissemination — phenomena that depend on a host of factors which differ from case to case — and in the end a new kernel will produce a new plant. The vital process is of the cyclical type, and this is one of its essential characteristics.

This may seem to be more than anything a philosophical question, but it is not so. In fact, the principle of causality is the basis of the scientific thought of the classical physics of Galileo Galilei and Isaac Newton. The motion of the celestial bodies, connected with Newtonian thought, is imagined as the perfect form of scientific knowledge which allows for an infinite predictability of phenomena. The movement of the planets in the solar system and eclipses can be predicted tens of thousands of years before they occur. When we consider complex phenomena instead, we are forced to abandon the beauty of the classical scheme (and also "grand unifying theories") and be content with a predictability that is increasingly limited. An instructive example is provided by meteorology: Today, it is possible to make almost exact predictions for the next 6 h and precise ones 24 – 48 h ahead of time; beyond this interval in time, predictability decreases and no one is capable of forecasting the weather 15 days hence.

Natural transformations entail continuous, small, and regular variations of the parameters that are involved; because of technology, however, man's action on the biosphere is massive and often produces sudden variations of individual parameters that break up the natural cycles. It is the assault on the planet which we have to confront each day. Man takes over the earth's resources and, as a result, new phenomena such as acid rain, eutrophication, pollution, a hole in the ozone layer, the greenhouse effect, and climactic changes loom menacingly over us. Well-being increases, but the biosphere is wrought asunder.

In this part of the volume, the relationship between environment and organisms is analyzed in terms of a set of fundamental points. In Section 1, the general conditions on the planet are discussed, beginning with the plurality of living entities that have resulted from biologic evolution, dealt with in the article by Anselm Kratochwil and Angelika Schwabe, and with the production of organic matter, dealt with in the article by Riccardo Valentini. These lead to the conception, advanced by Almo Farina in his contribution, of a system composed of the indivisible combination of environment and the living. The relation between the environment and the sphere of the living which, as described by Lukas Hottinger, went through some complex developments in biological times, today is threatened by global climactic change, a problem treated in the article by Elgene O. Box.

Section 2 deals with the role of man, who models the environment in its spatial structure and transforms nature into countryside, as Vittorio Ingegnoli illustrates. Largely in the wake of the discovery of food-producing plants in the Mediterranean basin, described by Sandro Pignatti, there was development of agricultural production, which created the new conditions on the landmass discussed in the article by Fabio Caporali. The opportunities for food consumption increased, thereby allowing populations to become concentrated in urban settlements, as Herbert Sukopp describes, and new habitats came into being in which organisms became adapted to a built-up environment, as discussed by Pierluigi Nimis.

Section 3 examines man's impact on the biosphere. Man's role in the biosphere gave rise not just to new possibilities: Human activities, in fact, produce substances that are foreign to the natural cycles and perceived as polluting, a problem discussed by Giuseppe Chiaudani. The imbalance that has resulted can be analyzed as a pathological phenomenon, as David B. Peakall maintains. A solution sought in the use of suitable biotechnologies, described by Walter Marconi, which can include the processes of biological purification that are described by Robert L. Irvine, Lisa I. Larson, David A. Irvine, Timothy J. Irvine, and Camille A. Irvine. These are techniques that require scientific capacities and the availability of capital.

Finally, Section 5, dedicated to ecology, deals with the question of the environment, and the problem shifts from the scientific and technical field to the economic and political one. We are facing a frightening imbalance between rich and poor, as Richard B. Primack demonstrates: What can we do about it? International institutions are deeply involved, as Bernd von Droste, Mechtild Rössler, and Anja von Rosentiel write in their article; however, they remain impotent and the formula of a sustainable development remains a myth, as Bruno Trezza shows. In the course of the past few centuries, politics has been dominated by the social question: Now, a confrontation regarding the no less dramatic environmental question is beginning.

At this point, it is clear that the human species, having attained the apex of its power, must pose itself a harrowing question: Within what limits is productive activity compatible with the survival of a world worthy to be lived in? We do not claim to have provided, in this part of the volume, an answer to this question: An answer which does not even exist, given the unpredictability of the complex system. The various contributions, however, can lead to a better understanding of the problems, and understanding is the indispensable premise to the organization of the search for solutions.

ANSELM KRATOCHWIL

Division of Ecology/Chemistry
University of Osnabrück
D-49069 Osnabrück, Germany

ANGELIKA SCHWABE

Institute of Botany/Vegetation Science
Darmstadt University of Technology
D-64287 Darmstadt, Germany

Evolution, Coevolution, and Biodiversity

The theory of biological evolution is of one of the most important, central theories of biology. Many evolution-theoretical theses can be proven, owing to the enormous scientific progress recently made in many biological fields (e.g., paleontology, biogeography, and molecular biology). The most important evolutionary factors are mutation, recombination, and selection. Of particular relevance is speciation, which is usually due to the geographic separation of populations and genetic isolation, in combination with an adaptation process induced by selection pressures, for example, of a special environment (allopatric speciation). However, speciation may also occur within an area and population, for example, by polyploidization (sympatric speciation).

Coevolution is defined as a process in the course of which two partners or entire partner systems (animals, plants, bacteria, or fungi) depend on one another in their evolution. Both acquire specific adaptations as a consequence of mutual selection pressure. Examples for a gene–gene coevolution, for a close coevolution between species (specific coevolution), as well as for a coevolution between species groups (diffuse or guild coevolution) will be presented. These examples concern flower–pollinator systems, the dispersal of plant species by animals, protection of plants by animal species, and coevolution between plant species and phytophagous insects.

Evolutive processes bring about biological diversity (biodiversity). The concept biodiversity will be defined, and the hierarchical levels of various biological organization forms will be differentiated. The different elements of diversity (taxonomic diversity, diversity of life-forms, diversity of spatial patterns, etc.), the diversity of organismic interactions, evolution-biological and ecological factors causing diversity, and functional processes of diversity will be discussed. Special attention will be given to intra- and interbiocoenotic diversity, the many and diverse phenomena occurring within and between biocoenoses. It will be shown that the preservation of biodiversity is of paramount importance; it is an essential prerequisite for the survival of man on Earth and has thus been incorporated into the concept of sustainable development.

Basic Principles

The theory of biological evolution influenced man's conception of the world — also on humanistic, social, and religious levels — in a radical and far-reaching way. Man's anthropocentric conception was first shaken in the sixteenth century by Nicolaus Copernicus, whose Copernican system moved the earth — and thus man — out of the center of the universe. The selection theory of the English naturalist Charles Darwin (1809–1882) upset it even further: The species *Homo sapiens* is related to all other organisms.

Biological evolution, coevolution between species, and the phenomenon of biodiversity (which is the result of evolutive processes) are causally closely connected. The factors discussed in the following sections are essential prerequisites for evolutive processes.

Energy

Without the sun as an extraterrestrial energy source, life would not be possible. The development of higher molecular chemical compounds due to photosynthesis, the primary production of autotrophic organisms (producers) — without which the consumers and decomposers (animals, bacteria, and fungi) could not have evolved — and temperatures favorable for the metabolism processes of living organisms are an essential prerequisite for organismic life. However, the ordered variety of different life-forms (biodiversity) within biocoenoses and ecosystems, reflected by numerous plant and animal species, is also based on a constant energy supply. The thermodynamic laws (basic principles of thermodynamics) apply both to inorganic matter and to organismic life. Order (neg entropy) within a system (in this case, a biosystem) is only possible when energy is available (supply of free enthalpy). Without energy supply, the components of a system (compartments) cannot ensure any transfer of matter or information. Such a transfer, however, is necessary for the functioning of the entire system, which is maintained by system-inherent characteristics — that is, the interactions of its compartments. Energy failure, on the other hand, leads to entropy (disorder) and breakdown in the system.

Organisms and Their Genetic Information

The genetic code (DNA and RNA) of organisms contains information regarding the preservation of the species- and organism-specific life processes and the organization of the constructive and the energy metabolism. Moreover, it ensures that the organism is able to respond to environmental influences and that the genetic information is passed on to the next generation.

Organisms and Their Structures and Functions

Organisms have structures: To survive in a specific abiotic and biotic environment, they have acquired morphological, physiological, and biochemical adaptations, and animals have also acquired ethological adaptations. The many and diverse environments (habitats as well as habitat complexes) bring about a diversity of structures. As a rule, structures of organisms (adaptations) are linked to certain functions and vice versa.

Mutation

Mutations are the basis of evolution. A mutation is a change in the genotype, either spontaneous and natural or induced by certain mutagenes (substances provoking mutations, high temperatures, and UV rays). The variation of genetic information within populations and between them (genetic variability) is due to mutations.

Selection

Mutation provides a wide variety of possible forms and phenomena; selection then "chooses" from the existing organisms those reaction types which are best suited, for example, to the prevailing environmental conditions. These types reproduce best ("high fitness") and thus pass on many of their genes and alleles to the next generation.

Time

The historicity of the taxonomic systems is one of the fundamental characteristics of biology. Biodiversity is the product of evolution and thus the result of phylogenesis effective over millions of years. Dollo's law on the irreversibility of evolution-historical processes implies that a species can only originate once. Intricate structures lost in the course of evolution can never be regained in their original form since combinations of random mutations and directional selection, as a rule, are not reproducible and evolution takes a linear course. However, there are examples showing, for instance, that some metabolic pathways which occur in different related groups have obviously been acquired by more simple mutation and selection steps (e.g., the Crassulaceae acid metabolism in plants).

In the following section, some basic principles of the biological evolution are explained. The evolution process may be considerably accelerated and the degree of coordinated adaptations may be especially pronounced when species or species groups exercise a significant selection pressure on each other (coevolution). Some particularly spectacular phenomena are due to coevolution. The forms of diversity will be described on the different hierarchy levels of biological systems (species, biocoenosis, ecosystem, and landscape).

Evolution

The Evolution Theory

The theory of the phylogenetic development of organisms, first outlined in Charles Darwin's epochal work *The Origin of Species by Means of Natural Selection* (1859), is essentially still valid today and has been substantiated by data from nearly all fields of biology.

The theory is based on two central principles: (i) All organisms are phylogenetically related to each other, and (ii) evolution is due to undirectional changes in the genetic material (mutation), leading to transformations of shape, function, and mode of life of organisms (species) and to directional selection by the abiotic and biotic environment.

Darwin's selection theory (with regard to phylogenesis

designated as *theory of evolution* or *theory of descent*) contradicts the thesis of the French zoologist Jean-Baptiste de Lamarck (1744–1829). According to Lamarck's thesis (developed in 1809), there is no genealogical relationship between the organisms. Evolution is caused by a change in the genotype, provoked by altered environmental conditions. The newly acquired characteristics are passed on to the next generations (Lamarckism) due to an organism-inherent impulse of perfection (psycho-Lamarckism). The selection theory, however, implies that only the phenotype, and not the genotype, can be changed, via modifications, by environmental influences, but environmental influences determine the selection processes and favor specific genotypes.

At the same time, but independent of Charles Darwin, the biogeographer Alfred Russel Wallace (1823–1913) ascertained the same main mechanisms of evolution. He did not work out a detailed theory, however.

Modern genetic (especially population-genetic) data have contributed to a better understanding of mutation processes and genetic variation and have thus confirmed and further substantiated the evolution theory. Disciplines such as ecology, ethology, developmental biology, systematics, special botany and special zoology, and paleontology have made considerable progress, and differentiated mathematical procedures have been developed. Consequently, evolutionary biology could be established as synthetic science (synthetic evolution theory; Huxley, 1974).

The course of phylogenesis and the factors which induced it are a topic of current research.

Evidence for Evolution

Important tasks of evolution research are the analysis of the change in species (anagenesis), reconstruction of the phylogenetic development ("phylogenetic tree" research), and demonstration of evolution factors (causal evolution research).

The following evidence for evolution has been gathered.

Data of Homology Research. Organisms can be distinguished from each other by certain organs and structures. Organs and structures of organisms whose similarities are due to the same hereditary information are homologous. As homologous organs change in the course of evolution, they have to be investigated with regard to certain criteria. The following homology criteria apply (Osche, 1966):

Criterion of position: Organs and structures are homologous when they occupy the same position within an organism (*homotopic organs*); for example, the position of labium, labrum, mandible, and maxilla of the mouthparts of insects (Fig. 1).

Criterion of continuity or steadiness: Organs and structures which are neither similar to each other nor occupy the same position within an organism (*heteromorphous* or *heterotopic organs*) can be recognized as homologous by transitions and intermediate forms. Examples can be given

(i) from the embryonic development — conversion of the serially and segmentally arranged, homologous appendices within the insect embryo into mouthparts and extremities (see Fig. 1);

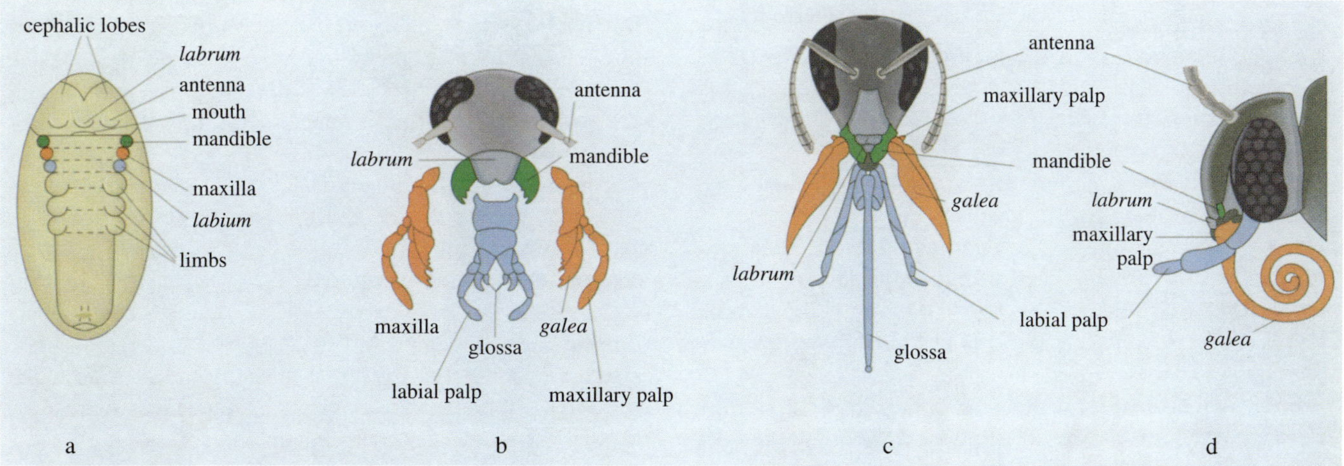

FIGURE 1 Homology of the mouthparts in different orders of insects. Dark blue, mandible; red, maxilla; light blue, labium. (a) Ventral view of the embryo of an insect, with the segmental rudiments of the extremity buds which in the head develop into mouthparts and in the thorax into the extremities. (b) Cockroach (genus *Blatta*); the mouthparts (shown separately) have the typical original chewing form. (c) Honeybee (genus *Apis*); maxilla and labium parts have been converted into a sucker and the mandibles are fully developed. (d) Butterfly (order Lepidoptera); parts of the galea are stretched into a proboscis, whereas the mandibles are reduced (adapted from Czihak *et al.*, 1976).

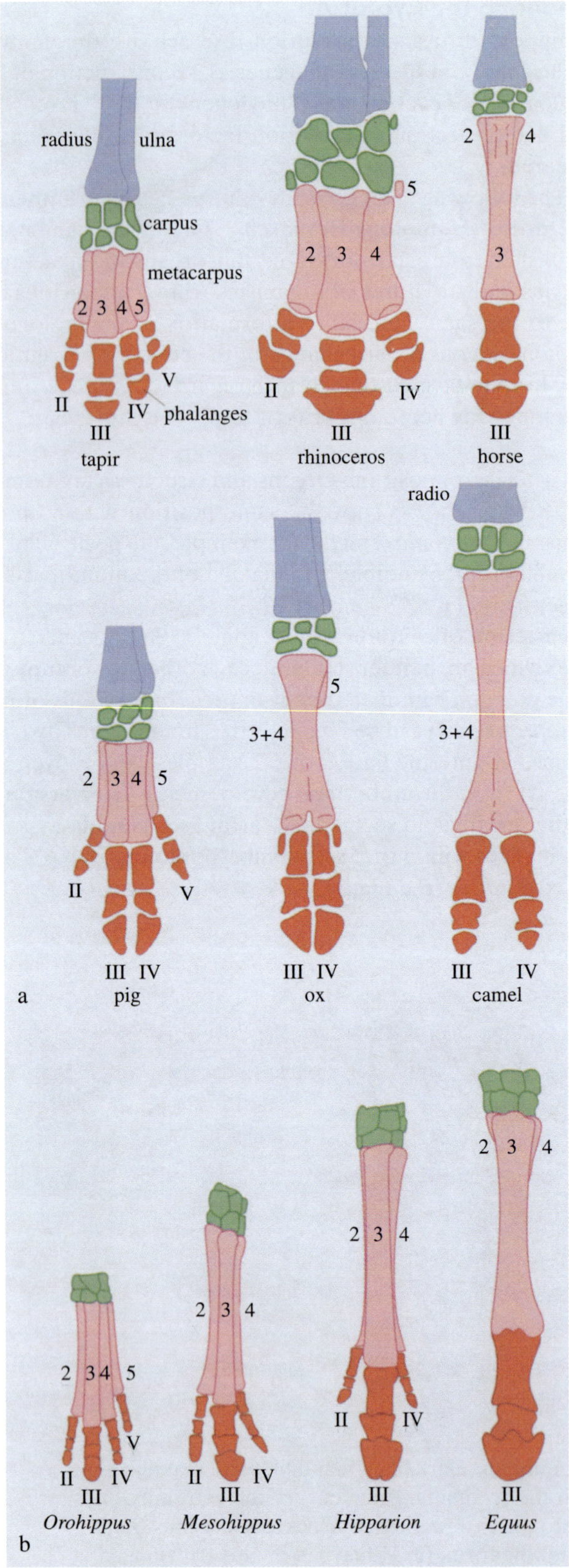

FIGURE 2 (a) Examples of the skeletal structure of the fore-extremities of extant hoofed animals. Perissodactyla: tapir (genus *Tapirus*), rhinoceros (genus *Rhinoceros*), and horse (genus *Equus*). Artiodactyla: pig (genus *Sus*), ox (genus *Bos*), and camel (genus *Camelus*). (b) Skeletal structure of the fore-extremities of a series of fossils showing the transformation of the horse from a slow woodland animal into a quickly running steppe animal. In *Orohippus* (Eocene), the first digit disappears; in *Mesohippus* (Oligocene), the fifth digit disappears and a more pronounced development of the third digit is observable, whereas *Hipparion* (Pliocene) is already a functional perissodactyl. Lastly, in the genus *Equus* (extant) the limb is typically one-toed. The numbers from 2 to 5 and from II to IV indicate respectively the metacarpals and the phalanges (adapted from Wehner and Gehring, 1995).

(ii) from the comparison of closely related living or fossil forms — reduction of the toes (phalanges) of hoofed animals (Fig. 2);

and (iii) from the comparison of serially recurring organs at the same individual of an organism — transition from normal leaves to sepals by progressive reduction of the upper leaf and transitions from petals to stamens (Fig. 3).

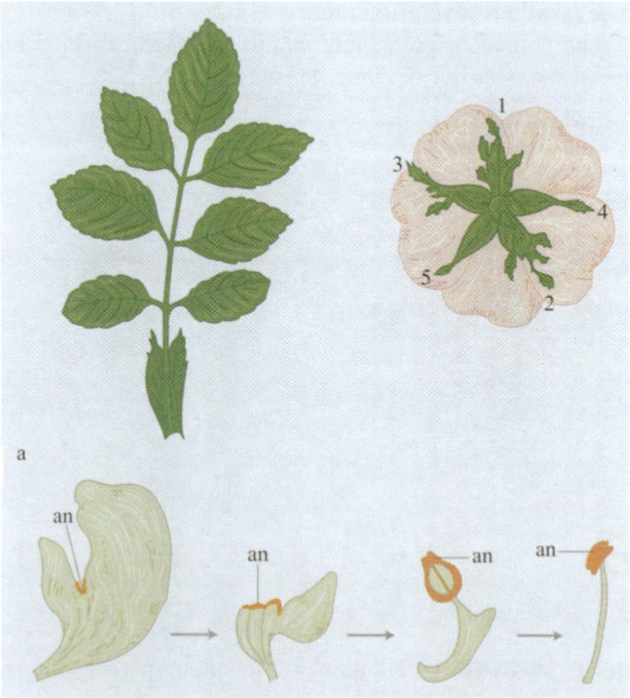

FIGURE 3 Metamorphosis of the leaves of a rose. (a) Normal leaf (left) and flower (right, seen from below), characterized by five sepals in a ⅖ divergence, with progressing reduction of the upper leaf. (b) Transitions from petals to stamens. An, anthers or their rudiments (adapted from Stocker, 1952).

Criterion of specific quality: Independent of their position within the organism, complex structures can be recognized as homologous when they have numerous similar individual structures (*homomorphous organs*); for example, teeth of vertebrates.

In addition to the determination of morphological characteristics, a homologization may also cover biochemical (comparison of enzymes, hormones, DNA, etc.) and ethological aspects (e.g., comparisons of inborn articulations of animal species).

Paleontological Data. Fossils constitute important evidence for the evolution of certain organism groups. In some cases, it is possible to precisely trace the sequence of adaptations and speciation — for example, for the evolution of horse-like mammals (Equidae) in the Tertiary and Quaternary Periods (Fig. 4).

Transition forms (connecting links) also play an especially important role between higher organization types. Examples include Ichthyostega from the Devonian, with characteristics of fish and amphibians (Fig. 5), and Archaeopteryx from the Jurassic, with characteristics of reptiles and birds (Fig. 6).

Biogeographic Data. Many organism groups are restricted to certain geographic areas. They probably originated in their current distribution area, and (natural) barriers (oceans, mountains, etc.) prevented a further dispersal. This would explain the fact that they are absent in other regions.

Species which only occur in a restricted area are called *endemics.* Continental islands are particularly rich in endemics; for example, Australia with its marsupials (Marsupialia). The same applies to old islands, frequently of volcanic origin; for example, the Galapagos Islands with their

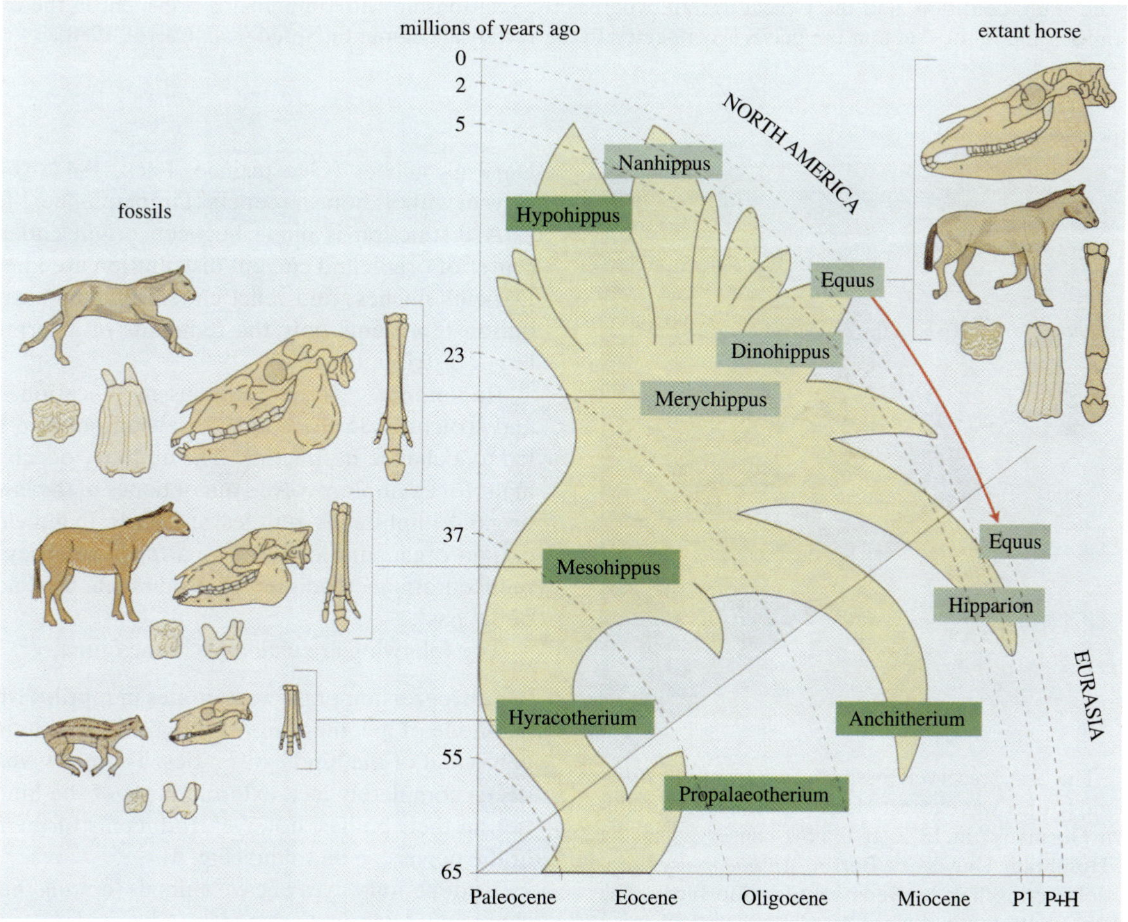

FIGURE 4 Phylogenetic tree of the horse (Equidae), with a selection of skeletal characteristics (skull, forearm, and molar teeth seen laterally and from above) of the important fossil genera. The species that ate leaves are shown in dark green; the grass eaters are shown in light green. The red arrow indicates the migration of *Equus,* which became extinct in northern America after the Pleistocene. Pl, Pliocene; P, Pleistocene; H, Holocene (adapted from Wehner and Gehring, 1995).

FIGURE 5 The fossil vertebrate *Ichthyostega* (in reconstruction) represents a transition form between a crossopterygian fish (Crossopterygii) and a typical amphibian. It is the oldest known quadruped land vertebrate and appeared more than 350 million years ago. All fossils of specimens of this genus were found in lacustrine water deposits of Mount Celsius on Ymer Island in Greenland. The relationship with fish is evident in the structure of the skull, dentition, and the typical fishtail, whereas the relationship with amphibians is evident in the extremities and in the fact that the pelvis is connected to the vertebral column (adapted from Carrol, 1988).

FIGURE 6 The best preserved fossil of *Archaeopteryx lithographica* found to date was excavated near Eichstätt (Altmühltal, southern Germany) in 1877. It is now conserved in the museum of Humboldt University, Berlin. *Archaeopteryx* represents the connecting link between reptiles and birds. The structure of the skull, toothed jaws, free fingers with claws, a missing carina, and the long reptile's tail composed of free vertebrae are all characteristic of reptiles, whereas the presence of feathers, a furcula, one toe directed backwards, and pneumatized bones are typical of birds. *Archaeopteryx lithographica* is a species that lived about 150 million years ago (photo courtesy of H. Jaeger).

Darwin's finches (Geospizinae; Lack, 1947) (Fig. 7) and Hawaii with its honeycreepers (Drepanididae; Mayr, 1943).

A distinction is made between origin endemism (the center of origin and current distribution are identical; e.g., Darwin's finches) and relict endemism (the current distribution represents only the remnants of a formerly much larger distribution area).

Rudiments. The continuous modifications of organs and structures in the course of evolution have frequently led to a change in function. The auditory ossicles of mammals, for example, evolved out of bones of the jaw hinges of original amphibians, reptiles, and birds. In individual cases, certain organs no longer have a function: They represent reduced organs (rudiments). Rudiments are "adaptations of the past."

The following are examples in the fauna:

Retrogression of the extremities in reptiles with a winding mode of locomotion: A python still has rudiments of a pelvis and of the hind extremities. Toothless whales (Mysticeti) completely lack external parts of the hind extremities; in their bodies, however, rudiments of the femur and of the pelvis can be found (Fig. 8).

Rudimentary eyes of cave animals (among the fishes the characine *Anoptichthys jordani;* Characidae) or of animals living in the soil (among the rodents the mole rat *Spalax typhlus;* Spalacidae).

Examples of rudiments in the flora include the retrogression of the five upper stamens in species of the genus *Scrophularia* (Scrophulariaceae), the occurrence of stamen remnants

FIGURE 7 A classic example for origin endemism is provided by Darwin's finches (*Geospizinae*) on the Galapagos Islands. At the end of the Tertiary Period, nearly 1000 km of land was colonized by a few individuals of the original *taxon.* Over the course of time, by food specialization the 13 current species could differentiate. They are shown grouped according to their alimentary preferences, which are strictly related to the morphology of the beak.

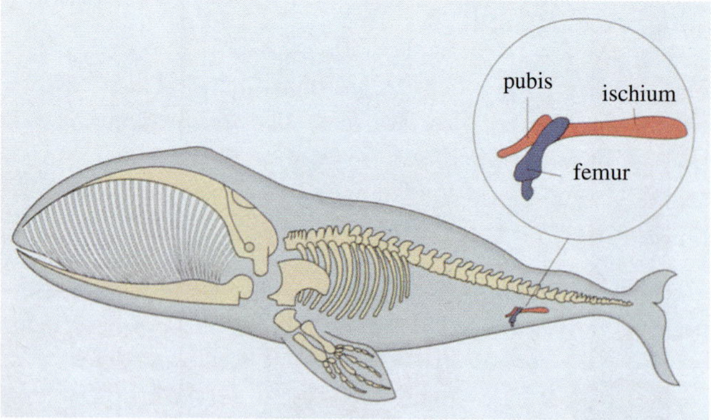

FIGURE 8 The toothless whales (Mysticeti family) completely lack external parts of the hindlimbs but, in their bodies, rudiments of the thigh bone (femur, in blue) and of the pelvis (pubis and ischium, in red) can be found (adapted from Czihak *et al.,* 1976).

in diclinous flowers (which evolved out of originally complete flowers), and the occurrence of no longer needed stomata in some aquatic plants.

Evidence Supplied by the Development of Individuals. In the course of their individual development (ontogenesis), numerous organisms form organs and structures which again regress completely (*circuitous development*). Many of these correspond to organs and structures typical of the organisms' phylogenetic ancestors. For example, as freshly hatched fish larvae, flat fish (Pleuronectifomes such as *Scophthalmus* or Pleuronectidae such as *Pleuronectes*) are bilaterally symmetrical; only at this stage do they develop an asymmetry (Fig. 9).

Toothless whales (also called whalebone whales; Mysticeti) absorb their food (tiny crustaceans) with a trap device, the baleen. In the embryonic stage, however, they develop rudimentary teeth, which then regress again. Their phylogenetic ancestors had teeth, like today's toothed whales (Odontoceti), such as the dolphins (Delphinidae). The Redcedar (Thuja, Cupressaceae) has short, scale-shaped leaves; in its adolescent stage, it first develops long needles typical of original specimens of the conifers (Coniferophytina).

FIGURE 9 Four stages of the larval development of the plaice (Pleuronectes family), showing the shift of eye and mouth opening from the original bilaterally symmetrical arrangement (top) to one side of the body (bottom).

Adaptations to the Environment

Evolution documents the variety of possible organismic adaptations to certain environmental factors in space and time. Adaptations are essential for an organism to survive and reproduce in a specific environment; in this way, the preservation of the species is ensured.

In this context, the environment is understood as a complex of external factors (ecofactors) which affect an organism directly and indirectly. This includes the factors essential for its survival (minimum environment) and the additional environmental influences (ecological environment). The environment of an organism thus comprises all abiotic and biotic factors affecting it in a positive or a negative way within the colonized habitat.

Abiotic (Physiographic) Environmental Factors. The abiotic (physiographic) environment is characterized by climatic factors (warmth, light, humidity, precipitation, wind, current, etc.), edaphic factors (physical and chemical properties of the soil), and orographic factors (geomorphology of a landscape), including exposition (position in relation to the compass point) and inclination (slope).

Biotic Environmental Factors. The biotic environment of an organism is determined by the interactions between the organisms: mutualism (symbiosis in the broader sense), predation, parasitism, intraspecific competition (competition within a species) and interspecific competition (competition between species), herbivory (absorption of plants), etc.

The Ecological Niche. The ecological niche of a species characterizes the relationship between a species and its specific environment. It is not understood as a spatial unit but as the dynamic relation system of a species' abilities and the habitat in which these abilities can develop (Hutchinson, 1965). The ecological niche is thus composed of an autophytic/autozooic dimension and an environmental dimension. The autophytic/autozooic dimension comprises the phylogenetically acquired morphological and physiological (for animals, also ethological) characteristics of the species, whereas the environmental dimension comprises the sum of all ecological factors effective within a specific habitat. Where both dimensions overlap, the ecological niche of a species is realized (Schmitt, 1987).

Evolutionary Factors: Mutation, Recombination, and Selection

The most important evolutionary factors are mutation, recombination, and selection.

Mutation. The genes located on the chromosomes determine the hereditary characteristics of organisms; their entirety is called *genome.* Depending on the degree of ploidy of its carrier, a gene may have several alleles. In a diploid set of chromosomes, two alleles determine a quality. A characteristic of an organism may be due to the combined action of several genes (polygeny), but it is also possible that one gene influences the development of several characteristics (polypheny).

Changes in the genotype can be caused by mutations, which may either be spontaneous (random mutations) or provoked by mutagenes (high temperatures, short-wave radiation, or certain chemical compounds such as colchicine). Mutations can cytologically be divided into four groups: genome, chromosome, gene, and point mutation. Point mutation is the mutation of one base. The phenomenon whereby genes are changeable by mutations is called *mutability.* Owing to the large number of all genes of an organism (100,000 to more than 1 million) there is a quite high mutation probability, even at a very low average spontaneous mutation rate of 10^{-4}–10^{-6} per gene and generation. Two or 3% of the individuals of every generation of the fruit fly *Drosophila* are mutated forms. In human beings, one mutation occurs per 100,000–200,000 gene replications. This means that every human being has on average one or two mutated alleles.

Mutations are responsible for the maintenance of a certain genetic variability within a population (*gene pool*). A population is defined as a group of individuals from the same species which form, at the same time (synchronously) and in the same spatial unit (syntopically), a potential reproductive community. Mutations provide the "raw material" for evolution (Fig. 10).

FIGURE 10 Variations in the color patterns of the elytra of the Asian lady beetle *Harmonia axyridis,* common in Siberia, Japan, Korea, and China (reproduced with permission from Ayala, 1978).

Recombination. By *recombination,* the recombination of alleles in the course of sexual reproduction (meiosis and fusion of gametes = syngamy) is understood. The developing new genotypes significantly extend the genetic variability of a population. Recombination is also a random process.

Selection. Selection "assesses" the genotype carriers: Those not adapted are eliminated. Only the adapted genotypes can pass on a high percentage of their alleles and genes to the next generation (high fitness). Selection stabilizes or alters the frequency of certain genes within a population (stabilizing and dynamic selection).

Selecting factors may be the weather (cold and dryness), competition for food, spatial competition, enemies and parasites, and for plants certain nutrients (e.g., nitrogen). Selection is always intraspecific (between the individuals of a species); it works opportunistically and not at random.

An example concerns industrial melanism. In its normal form, the light-colored peppered moth *Biston betularia* (Geometridae) can hardly be detected when it sits with spread wings on tree trunks overgrown with lichens. By mimesis (camouflage by similarity of the physiognomy to the underground) it largely escapes its enemies. Due to pollution of the air by sulfur compounds in industrial regions, lichen vegetation increasingly disappeared. In addition, the tree trunks were covered with soot. In 1848, a dark-colored (melanistic) form of the peppered moth (*B. betularius carbonarius*) was discovered for the first time in Manchester, England. The selection advantage (better protection on dark underground) led to a distinct increase in the number of these specimens: Fifty years later, 95% of the entire population, and from 1952 to 1956 even 98%, consisted of *carbonarius* forms (Kettlewell, 1972) (Fig. 11). Recent improvements in the field of technical environmental protection have resulted in a regeneration of the lichen vegetation and thus a selective advantage of the light-colored form.

Analogies Due to Adaptations. Organs and structures of organism groups not closely related may have to fulfill the same function and accordingly develop similar adaptations (analogies). The phenomenon in which these originally different structures and organs become increasingly more similar in the course of evolution is called *convergence.* An example of convergence in plants is the stem succulence (formation of a water reservoir to survive in dry habitats) in Asclepidiaceae, Compositae, Euphorbiaceae, Cactaceae, and Didiereaceae.

An example of convergence in animals is the streamlined shape of the body of different vertebrates: elasmobranch (shark), osseous fish (swordfish), fossil reptile (ichthyosaur), bird (penguin), and mammal (dolphin) (Fig. 12).

Especially remarkable are the convergences between the placental mammals and the marsupials (Marsupialia), the distribution of the latter is limited to Australia.

Speciation by Separation and Isolation

There are different definitions of a biological species:

Morphological–physiological species concept (morphospecies): A species is defined as all individuals (including their descendants) the essential characteristics of which (morphological, physiological, and also ethological, etc.) are identical. This definition is the only one applicable to fossil material.

Genetic species concept (biospecies): A species consists of actually or potentially crossing populations which are reproductively isolated from others, i.e., no genes are exchanged (species concept of the reproduction community).

Ecological species concept (ecospecies): Species make ecological demands on their environment; these demands are in part reflected by their ecological niche. Species living

a

b

c

FIGURE 11 Cryptic mimetization of the light and melanistic varieties of the moth *Biston betularia* in relationship to environmental pollution. (a) On the lichen-free bark (darkened by industrial fumes) of an oak tree near Liverpool, England, the melanistic black *carbonarius* form is better protected from its enemies than the light-colored one. (b) On a beech trunk covered with algae and the lichen *Lecanora conizaeoides,* which tolerates a slight pollution of the air, both forms of *B. betularia* are conspicuous. (c) On the lichen-overgrown bark of an oak tree in Wales the light-colored form can hardly be detected, giving it a considerable advantage with respect to predators (photo courtesy of L. Cook).

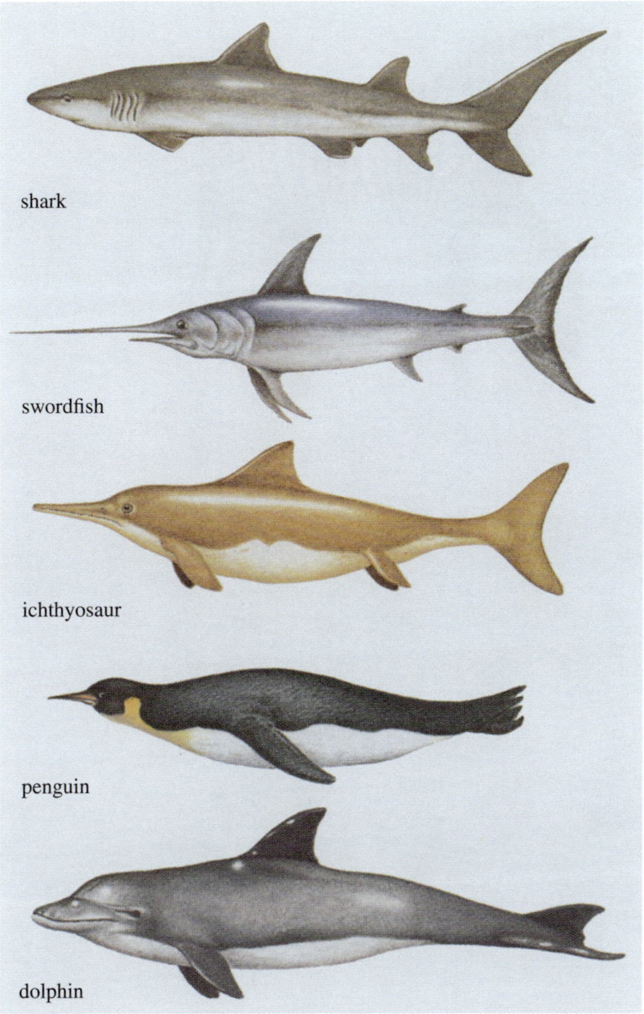

FIGURE 12 The streamlined bodies of elasmobranch (shark), osseous fish (swordfish), fossil reptile (ichthyosaur), bird (penguin), and mammal (dolphin) are perfect examples of convergence (as similarity of form) as a consequence of evolutionary adaptation to the same function (in this case, swimming).

syntopically and synchronously cannot form the same ecological niche.

Speciation. Apart from the allochronous speciation by historical species transformation over many generations (and consequently spanning long geological periods), there is a synchronous species formation, i.e., the segregation of a species into two sister species (cladogenesis).

Allopatric Speciation Step 1: Separation. The first step in an allopatric species formation is the (geographic) separation of originally genetically linked populations which may occur for several reasons: A few individuals overcome certain geographic barriers (mountains, deserts, and seas), macroclimatic changes in geological periods (ice ages) force the species to escape to different refugia, and increases in the sea level entail the separation of continental land parts (continental island formation). The lower variation in genotypes, new mutations, and differing environmental conditions (selective factors) may then bring about a speciation. The importance of separation is substantiated by the fact that species with a very large distribution area form geographic subspecies. Subspecies slightly differ with regard to their genetics and phenotype; each inhabits a certain geographic subregion within the area of the corresponding species (e.g., subspecies of the steppe zebra) (Fig. 13).

Allopatric Speciation Step 2: Development of Isolation Mechanisms. After a (geographic) separation of a continuously distributed population into two or more populations, mechanisms evolve against hybridization in a secondary contact zone, which would occur after a reintrogression. Possible mechanisms include metagamous isolation mechanisms (incompatibility of the genomes after mating) and progamous isolation mechanisms (prevention of mating).

Progamous isolation mechanisms include seasonal or cyclic isolation, i.e., different reproduction times (e.g., of butterflies) and different flowering times of plants; mechanical isolation, i.e., differences in the structure of copulation organs (e.g., of spiders); and ethological isolation, i.e., changes of optical, acoustic, and olfactory species characteristics by which animals recognize their mating partner (e.g., monkeys, birds, locusts, and butterflies) and by which plants "choose" their animal pollinators in the case of zoogamy.

Sympatric Speciation. Sympatric species formation occurs among the individuals of a species in the same distribution area. It is particularly frequent among plants. The chromosome set is duplicated (autopolyploidy), and the originating polyploid individuals are isolated from diploid ones (e.g., various ferns). About one-third of all plant species developed by polyploidy. Closely related polyploid species, however, may cross in certain cases and accordingly form new species by allopolyploidy that, in turn, are genetically isolated from the original species. Many of our cultivated plants (cotton, our cereals) were developed by allo- or autopolyploidy. Among animals, polyploidy is rare.

Macroevolution

Evolution processes in populations of species are called microevolution (changes of the allele frequencies in populations); macroevolution denotes the formation of new species, families, orders, classes, or phyla in flora and fauna.

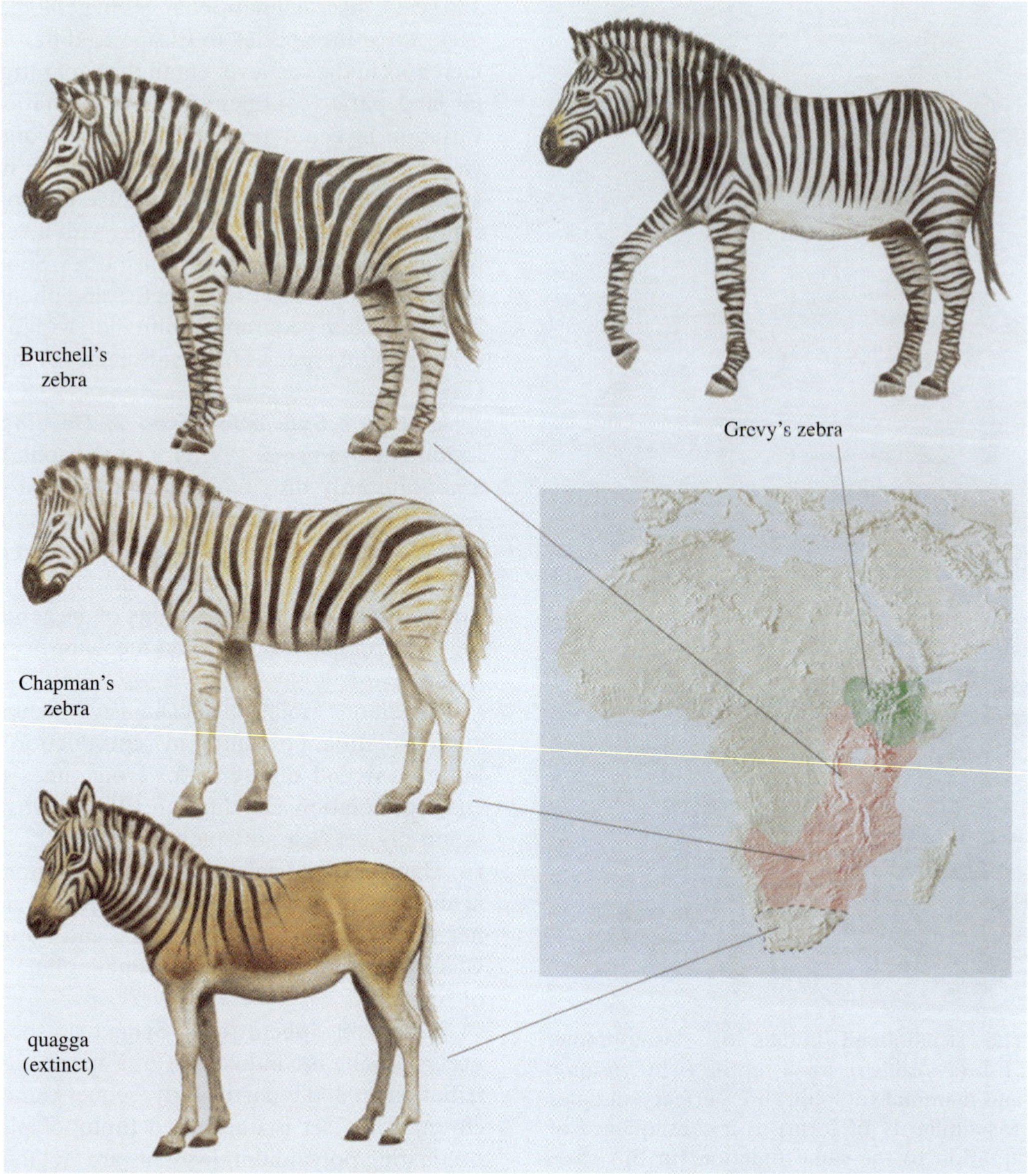

FIGURE 13 Several geographic breeds of the steppe zebra *Equus* can be distinguished by the varying stripe patterns of the coat and by the striping of the legs, which reduces increasingly toward the southern part of the territory. Grevy's zebra, *Equus (Dolicohippus) grevy,* lives in the northern regions of the territory (Kenya, Ethiopia, and Somalia); Burchell's zebra, *E. (Hippotigris) burchelli,* is diffuse in eastern and southern regions, together with the subspecies *chapmani* (Chapman's zebra) in the more southern part; and the quagga, *E. (H.) quagga,* now extinct, occupied the savanna of South Africa (adapted from *Wissen im Überblick. Das Leben. Die Welt der Modernen Wissenschaft: Zelle, Pflanze, Tier, Entwicklung, Evolution, Informationsverarbeitung, Verhalten,* Focus International/Verlag Herder, Stockholm/Freiburg, 1972).

The evolution of such higher organization forms does not occur rapidly but in small steps (additive typogenesis). Evolution processes (speciation) may be accelerated when organisms are able to realize a large ecological niche. Examples of this in phylogenesis are the conquest of the land by the fish group Crossopterygia, the conquest of airspace by birds, and the first terrestrial plants in the Devonian. Such a realization of a large ecological niche often entails a rapid splitting into numerous species (adaptive radiation); examples are Darwin's finches (Geospizinae) on the Gala-

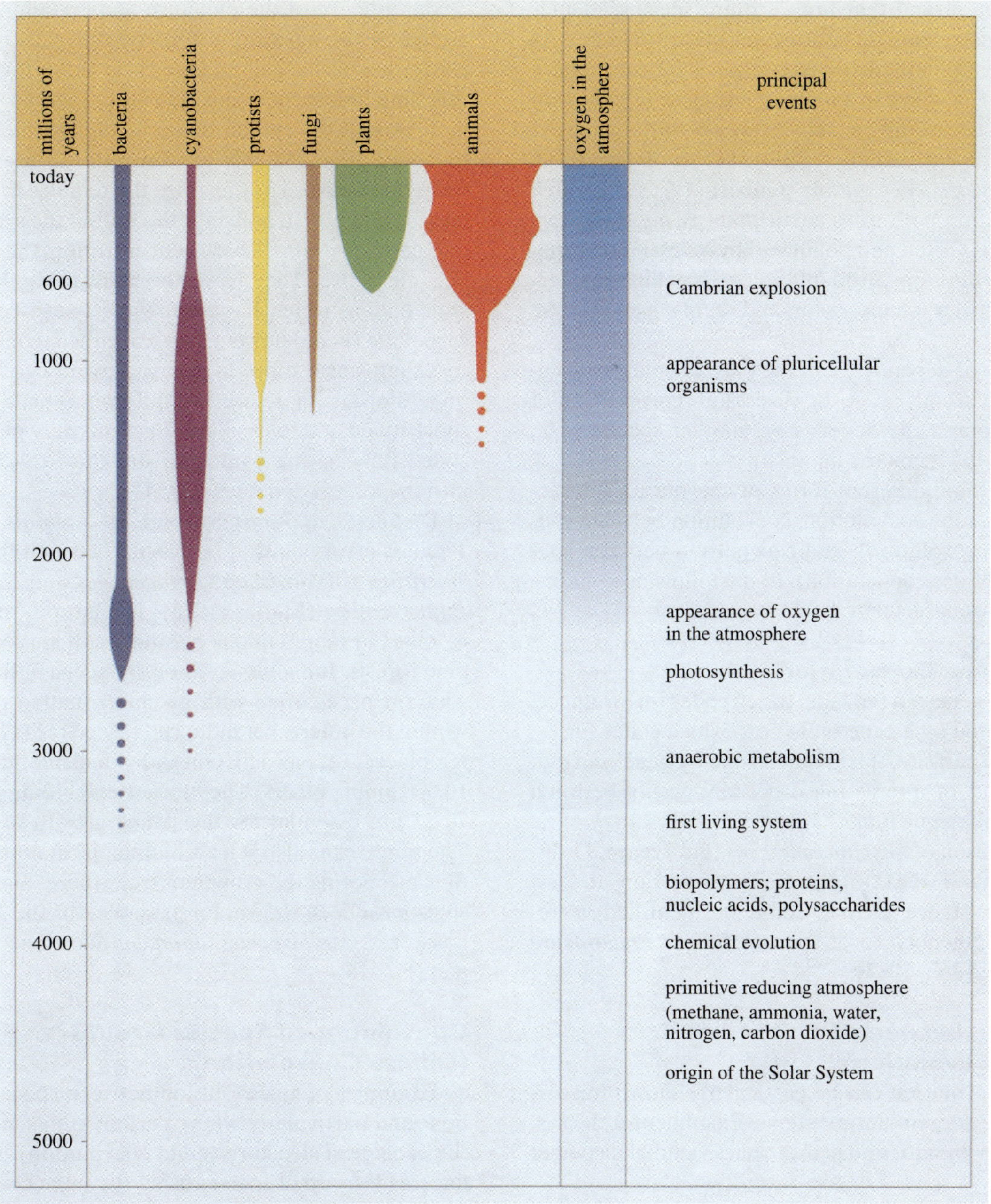

FIGURE 14 The geological timescale in relationship to the principal events which characterized the origin and evolution of life on Earth (adapted from Hickman *et al.,* 1997).

pagos Islands, honeycreepers (Drepanididae) on Hawaii, and marsupials (Marsupialia) in Australia.

The earth originated about 5 billion years ago and the first life-forms about 3.5–3.8 billion years ago. The development of prokaryotes (bacteria and cyanobacteria) began about 3.5 billion years ago. The first sponges occurred about 570 million years ago, fishes about 450 million years ago, reptiles about 280 million years ago, mammals about 200 million years ago, birds about 150 million years ago, and primates about 60–70 million years ago (Fig. 14).

Coevolution

Coevolution means that two partners or partner systems (plants, animals, fungi, or bacteria) depend on one another in their evolution and that both acquire specific adaptations as a consequence of mutual selection pressure. As a rule, it is distinguished between a *pairwise coevolution* (Janzen, 1980), in which in a more or less close relationship between two species the partners exert a continuous selection pressure on each other (*reciprocal coevolution*), and a *diffuse* or *network coevolution* (Gilbert, 1975), in which several species on both sides participate(d), e.g., a flowering plant that is visited and pollinated by several insect species and which develops, in adaptation to its pollinators, certain characteristics (shape, color, and scent) in the course of evolution.

The degree of dependence of the partners may vary significantly. In extreme cases, the successful reproduction of one species completely depends on another species (e.g., pollination of the fig by the fig wasp).

There are three different forms of coevolutive interactions: gene-by-gene coevolution, coevolution between species (specific coevolution), and coevolution between species groups (diffuse coevolution). In the following sections, these types are characterized.

Gene-by-Gene Coevolution

If a certain gene of a parasite, which codes for virulence, is complemented by a gene of its host, which codes for resistance to the parasite, this is called gene-by-gene coevolution. This kind of interaction especially occurs between plants and pathogenic fungi.

In a population of *Glycine canescens* (Fabaceae), 11 different phenotypic resistance patterns, based on at least 12 genetic resistance factors, could be identified as response to nine genotypes of the rust fungus *Phakopsora pachyrhizi* (Burdon, 1987).

Coevolution between Two Species (Specific Coevolution)

A close coevolution can be particularly shown for certain mutualistic or symbiotic systems. Examples for the systems flower–pollinator and plant species–animal disperser are given.

Flower–Pollinator Systems. The pollination of figs (Ficus and Moraceae) is very complicated (Wiebes, 1979). The urn-shaped inflorescences (syconia) of a fig tree, in which the extremely reduced flowers line the inner cavity, attract thousands of fig wasps of only a few millimeters in length (Agaonidae: genera *Ceratosoles, Blastophaga,* and *Sycophaga*). Each fig species (worldwide there are more than 1000) is pollinated by its "own" wasp species. The female wasp, transporting the pollen in two thorax bags, crawls through the very narrow, scale-covered opening of the inflorescence (ostiole) into the cavity. The stigmata are covered with pollen. Then the wasp jabs its ovipositor into the style of a flower and places an egg there. At this spot, as a consequence of the puncture and certain substances secreted by the fig wasp, a cell growth (gall) develops. The larva lives in this gall and feeds on the cell tissue. During this time, the carbon dioxide content within the syconium increases; at the end of the larval development it amounts to about 10%. At this concentration, only the male fig wasps are active. They fertilize the females in the gall, then leave it and bore a hole into the wall of the syconium. With decreasing carbon dioxide concentration, the female wasps become active. They leave the galls, fill both thorax bags with pollen, which has meanwhile been secreted by the carpellate (male) flowers, and leave the syconium. The process then starts anew. In order to avoid eggs being placed on all flowers, there are two different female flower types: short styled and long styled. Eggs are only placed in short-styled flowers; the ovipositor does not reach far enough into the long-styled ones (Fig. 15).

Dispersal of Plant Species by Animals (Zoochory). There is a very close relationship between the nutcracker *Nucifraga caryocatactes caryocatactes* and the arolla pine *Pinus cembra* (Mattes, 1978). In Europe, the nutcracker is found in mountainous habitats with spruce-larch-arolla pine forests. It mainly feeds on the seeds of the arolla cone which it pecks open with its chisel-shaped peak. For the winter, the nutcracker hides cedar seeds (2–11 nuts per hiding-place; average, 3.5). One individual may have more than 10,000 hiding places. The "forgotten" hiding places (about 20%) are essential for the young growth of arolla pines. The nutcracker also selects hiding places at the timberline, thus promoting the growth of trees there. Similar relationships have been shown, for example, for the Japanese nutcracker *N. caryocatactes japonicus* and *Pinus pumila* in Japan (Fig. 16).

Coevolution of Species Groups (Diffuse Coevolution)

Examples of a coevolution between species groups can be found particularly where certain guilds are involved in the ecological structure (guild coevolution). A *guild* is defined as a group of species using the same class of environmental resources in a similar manner — for instance, phloem feeders (greenfly: Aphidina), pollinators, and predators. In this context, it is important that several different species may be evolutionary vectors.

In mutualism, there are numerous examples of this form of coevolution: the many interactions between plants and their animal pollinators, between plants and their animal dispersers, and between plants and their animal "protectors."

Plants and Their Animal Pollinators. Many plant species have developed strategies to make use of the mobility of animals to attain a cross-pollination (*borrowed mobility*). By zoophily, the pollen is in a purposeful manner trans-

FIGURE 15 Symbiosis between the fig and the fig wasp (genus *Blastophaga,* family Agaonidae). (Left) The female fig wasp oviposits her eggs into the carpellate flowers, causing the formation of a gall in which the insect larvae develop. The male fig wasps hatch first within the gall, insert their abdomens into the gall flowers inhabited by the unhatched females, fertilize them, and leave the fig by a hole that they bore in the fruit. (Center) The fertilized female fig wasps leave the fig flower hours later. At the same time, the staminate flowers secrete pollen. The females fill their thorax bags with the pollen and then leave the fig through the holes bored by the males. (Right) After arriving at a young fig, in which the staminate flowers are still closed, the female squeezes through the ostiole. The pollen is put on the carpellate long-styled flowers, whereas the eggs are only oviposited into the short-styled flowers (adapted from Hickman *et al.,* 1997).

ferred in such a way that an exchange of genes is ensured. In the temperate zones, insects transport the pollen (entomophily); in the tropics and subtropics, this is additionally done by birds (ornithophily), bats (chiropterophily), and reptiles (saurophily), and in Australia it is also done by phalangers (Marsupialia; Faegri and van der Pijl, 1979). The flower visitors are "rewarded" by nectar and/or pollen, oil, and partly also by plant tissue.

The margin of flower visitors ranges from generalists (euryanthic flower visitors) such as bumblebees (*Bombus*) to specialists (stenanthic flower visitors) that can only use the flowers of certain plant species, genera, or families. Examples are the silker bee *Colletes cunicularius,* which exclusively visits willows (*Salix* and Salicaceae), and the andrenid *Andrena florea,* which is only found at the bryony species *Bryonia alba* and *B. dioica* (Cucurbitaceae).

Many and diverse coadaptations have evolved between flower visitors and plant species. Thus, the development of some primary attractants in plants (pollen grain types, nectar, and oil) correlates with the evolution of the structure of the mouthparts and of the pollen-gathering and -transport devices in animals (Table 1).

Pollen grains have specific surface structures which correlate with the structures of the pollen-gathering device of a bee. The bee *Lasioglossum lineare* (Halictidae), one of the main pollinators of the pasque flower *Pulsatilla vulgaris*

FIGURE 16 A nutcracker (*Nucifraga caryocatactes*) on an arolla pine (*Pinus cembra*) (photo courtesy of R. Oggiani/ Panda Photo).

(Ranunculaceae), has a specific gathering device at its hindlegs composed of particularly fine hairs. These hairs exactly fit into the sutures of the pollen grains of *P. vulgaris* (Kratochwil, 1988) (Fig. 17).

The entire flower morphology of zoophilous plant species is adapted to the morphology of the respective pollinator, its sensory physiological abilities (recognition of colors and scents, etc.), and its behavior ("handling" of the flower).

Certain adaptive syndromes have evolved between different animal groups (insects and birds) and the plant species pollinated by them. The syndrome of ornithophily (pollination by birds) is compared to that of chiropterophily (pollination by bats) in Figs. 18 and 19.

Plants and Their Animal Dispersers: Myrmecochory. The phenomenon by which ants disperse the seeds of plants and may in turn use the yellow to yellow-white lipoid appendage of the diaspore (elaiosome of the seed) for feeding is called *myrmecochory.* Myrmecochorous plant species include the ramson *Allium ursinum* (Liliaceae), violet species (*Viola* and Violaceae), the common celandine *Chelidonium majus,* and the holewort *Corydalis cava* (Papaveraceae and Fumariaceae). Worldwide, there are 70 plant families in which myrmecochory occurs (Beattie, 1983). Both in woodland regions and in dry open-land areas, myrmecochory may be important: Notably, many myrmecochores can be found in the fynbosch of Cape Province (South Africa), a sclerophyllous vegetation similar to the maquis in the Mediterranean region (Fig. 20).

Plants and Their Animal Protectors. The so-called ant plants (myrmecophytes) provide a comfortable habitat

TABLE I

FLOWERS	ANIMALS
Bird flowers	**Flower birds**
Flowering during the daytime	Active during the daytime
Garish colors, usually red	High spectral sensitivity to red
No landing surface	Too heavy for landing on a flower
Thick flower tissue	Hard beaks
Scentless	Poor ability to smell
Much nectar	Great need of nectar
Mechanism to retain nectar	Great need of nectar
Nectar hidden deep within the corollae	Long beaks and tongues
Hardly any flower characteristics	Great ability to learn
Bat flowers	**Bats**
Flowering during the nighttime	Active during the nighttime
White or creamy colors, often also greenish or purple	Color-blind, good sight in the immediate area, echolocation by ultrasound
Musty smell	Good ability to smell for distance orientation
Large bowl- or bell-shaped single flowers or larger inflorescences	Large pollinators with claws at the extremities to hold on to the flower
Much nectar	Great need of nectar
Large pollen amounts, large or many anthers	Pollen is the only protein source
The flowers/inflorescences are located outside the foliage, often directly at the stems (cauliflory); long corollae	Size, lower maneuverability in flight

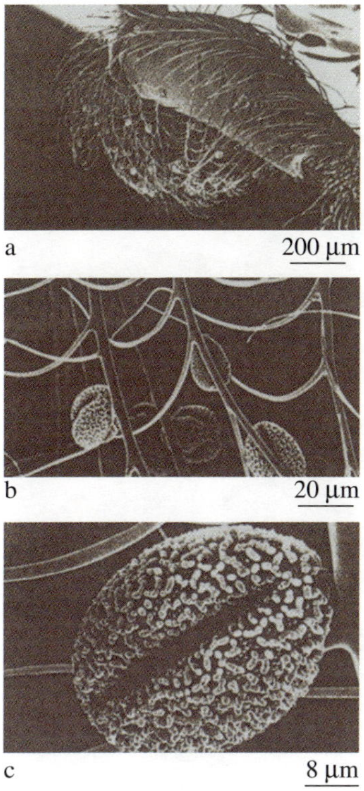

FIGURE 17 (a) Micrograph of the pollen-gathering device on the hind leg (femur) of a worker bee *Lasioglossum lineare.* (b) Micrograph of the pinnate hairs of the pollen-gathering device with pollen from the pasque flower (*Pulsatilla vulgaris*). (c) Micrograph of a pollen grain of *Pulsatilla vulgaris.*

for ants, which in turn protect the plant from phytophages and contribute to its feeding (myrmecotrophy).

Especially interesting are some ant plants of the subtropics and tropics, which also provide a living space for the ants (myrmecodomatia) and in turn use (via a transport of ions) the waste material of the ant colony as nutrients or are even freed from lianas by the ants (Fig. 21).

FIGURE 18 The flower visit of a bat of the *Glossophaga* genus (photo courtesy of M. D. Tuttle/Panda Photo).

FIGURE 19 The flower visit of a hummingbird (*Cynanthus latirostris*) to an ocotillo flower (*Fouquiera splendens,* Fouquieriaceae) (reproduced with permission from Reichholf and Weidensaul, 1990).

Coevolution between Plants and Phytophagous Insects. Closely related herbivorous animal species often eat closely related plant species. Within butterflies, the caterpillars of the whites (Pieridae) prefer crucifers. These possess as secondary plant matter mustard oil glycosides, by which the pierids recognize the larval plant. On bacteria, fungi, other insects, and mammals, these substances have a toxic or repulsive effect (Feeny, 1977).

Many plant species contain certain secondary substances (vegetoalkaloids, furanocumarine, and others) to protect themselves from being eaten (chemical defense).

Phenomena of coevolution also occur in predator–prey relationships and between hosts and parasites.

Biodiversity

In its original sense, diversity means variation, differentiation, and diversification, in contrast to uniformity. Diversity may be understood as something static: Heterogeneity,

FIGURE 20 A worker ant of the genus *Myrmica* transports a diaspore of *Corydalis cava* (Fumariaceae) into its nest. The white appendix is the elaiosome, whereas the seed is black (reproduced with permission from Zizka, 1990).

FIGURE 21 The ant plants of the genus *Myrmecodia* (madders, Rubiaceae) grow as epiphytes on other plant species. Therefore, they have to rely on ants to supply them with mineral salts. The thickened parts of their stems are traversed by passages and chambers which serve ants of the genus *Iridomyrmex* as living space (adapted from *Geheimnisse der Natur-Entdecken, Entschlüsseln, Erklären.* Bertelsmann Lexikon Verlag/Beazley, Munich/London, 1992).

then, denotes irregularities and variety denotes differences. Variability covers dynamic aspects. Diverse systems may be simple, but they also may be very complicated. As a rule, complexity is a sure sign of diverse systems: It is defined as something very intricate or complicated. Complexity covers the profundity of system structures and diversity their width. When assessing biological systems, diversity may also be seen as richness.

By biodiversity, biological diversity is understood: the total differentiation, variation, variability, complexity, and richness of life on Earth. Article 2 of the Convention on Biological Diversity of the IUCN, Rio de Janeiro, 1992 (as quoted in Bisby, 1995) states that

> 'Biological diversity' means diversity (according to differentiation, variation, variability, complexity, and richness) among living organisms from all sources, including, inter alia, terrestrial, marine, and other aquatic ecosystems and the ecological complexes of which they are part, this includes diversity within species, between species and of ecosystems.

Ranges of Validity of Biodiversity

Diversity is a fundamental quality, manifesting itself in the different organization levels of matter and energy. It is a characteristic feature of all levels of the nonbiological and the biological hierarchy (hierarchical diversities); there is diversity on every single level (Fig. 22).

The levels of life are particularly diverse; here, we generally distinguish between structural diversities and functional diversities. Data on diversity may be studied at each level of the hierarchical structure using two different approaches (Solbrig, 1991): a descriptive approach (e.g., identification, determination, description, and differentiation of elements and their components) and a functional approach (e.g., a causal analysis of the combination of the elements and their components as well as of absorption, transformation, and processing of energy and matter).

The subject of this analysis is the level of ecosystems in the broader sense: their biotic components (biocoenoses) and their habitats (biotopes). Also, the level of ecosystem complexes (landscape units) will be dealt with. Such complexes are formed by several ecosystems, the correlations of which follow certain rules. Since the Neolithic Period and increasingly in the past 150 years, man has considerably influenced ecosystems and ecosystem complexes in many parts of the world. A study of biodiversity therefore must include man–environment systems. An increase in and also a reduction of biodiversity may be anthropogenically caused.

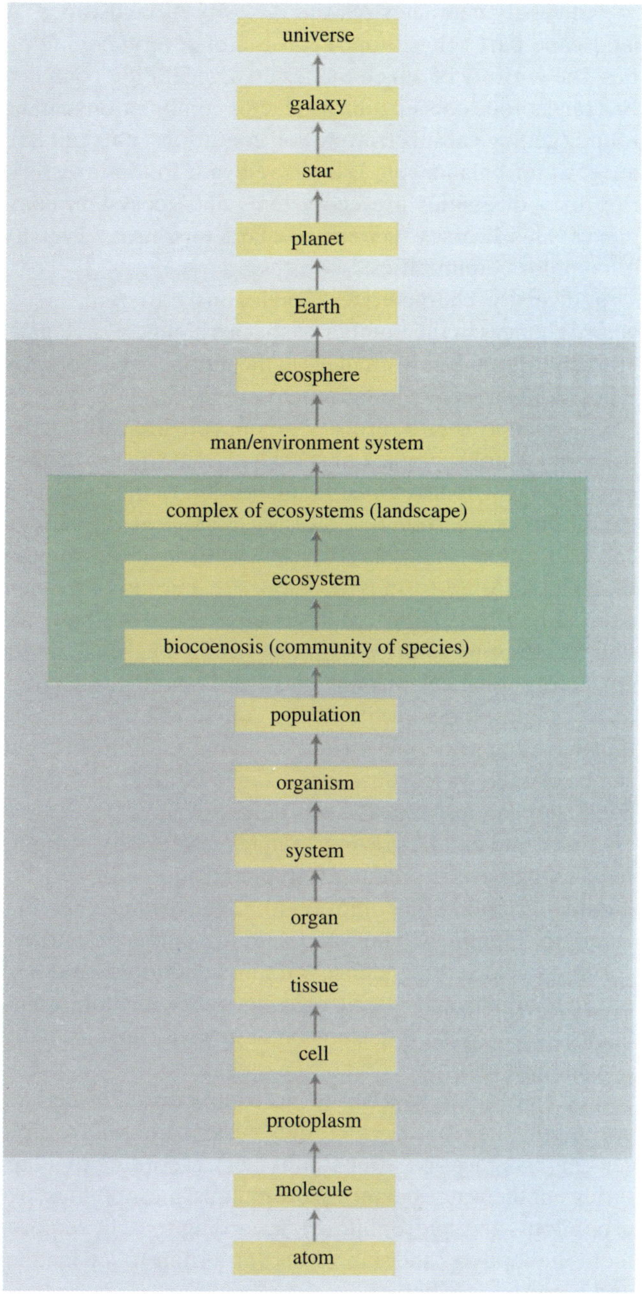

FIGURE 22 The hierarchical order of the different organization forms of organisms (gray square) and of matter and energy, each characterized by a structural and functional diversity. The range of biocoenology is indicated by the green square.

Forms of Diversity

The different forms of biodiversity may be assigned to four types:

Diversity of Elements (Element Pattern of Biodiversity).

1. Taxonomic and syntaxonomic diversity and species and coenosis diversity: Various species and coenosis diversity levels can be distinguished in different spatial units: α, β, γ, and δ diversity (Goetze and Schwabe, 1997).

α diversity refers to the species diversity of a certain area. It is described, for instance, by several calculation methods and the determination of indices. α diversity is also defined as "diversity within a community."

Gradients between different biotopes (habitats) can be analyzed by β diversity. This procedure is especially suitable for regions with ecological gradients (ecoclines), e.g., forest/open land areas and zonation complexes at water banks; however, it is less suitable for areas with pronounced discontinuities. This diversity type can also be described as "gradient diversity between communities."

γ diversity characterizes the diversity of landscapes. A landscape part (physiotope) consists of several communities, the entirety of which makes up a vegetation complex. In a physiotope certain uniform factor combinations can be found (geological substratum, soil conditions, nutrient balance, water balance, etc.). Units relevant to the investigation of γ diversities are ecosystems and ecosystem complexes. This diversity type can be characterized as "diversity of complex communities."

δ diversity characterizes (analogously to β diversity, where changes in the number of species along an ecological gradient are analyzed) changes in the number of vegetation types along an ecological gradient.

2. Diversity of life-forms: The concept "life-form" comprises the whole complex of species-specific qualities of an organism, which developed in adaptation to the particular conditions of a certain habitat (morphological, physiological, and ethological characteristics). Such life-forms can be typified. A life-form type belongs to a group of species, which often have different systematic ranks but have acquired, adapting to the conditions within a habitat, analogous morphological, physiological, and ethological characteristics and modes of life in the course of evolution and thus have the same life-form. For animals, life-form types can be classified according to feeding habit (e.g., phytophagous, zoophagous, parasitic, and detritophagous; filter feeders, substrate eaters, grazing animals, sap feeders, stinging suckers, gatherers, predators, trappers, and parasites), according to mode of locomotion (e.g., burrowing, crawling, climbing, jumping, flying, and running animals), and according to place of residence (edaphon, atmobios and herbicolous organisms = living on or in plants; phyllobios and lignicolous organisms = living on or in wood; epizoa, endozoa, and others).

For plants, different life-forms can be distinguished according to the way in which they survive the unfavorable season (classification after Raunkiaer), according to adaptations of the water balance (xerophilous, mesophilous, hygrophilous, and hydrophilous), according to light requirement (heliophytes and skiophytes), according to soil factors, and according to diet.

3. Diversity of spatial structures: A habitat can be divided into three different spatial structure types: stratotope, choriotope, and merotope. Such a differentiation is essential for the recording and analysis of synusia within a biocoenosis.

4. Trophic diversity: Classification into producer, consumer, and decomposer levels with further subtypes.

5. Phenological diversity: Characterization of time structures, diurnal and seasonal changes, periodic phenomena within a year, etc.

6. Genetic and population-specific diversity: Characterization of genetic variability and of the genotype spectrum, phenomena of homo- and heterozygosis and of gene drift, mutation rate of individual populations, and others.

7. Biochemical diversity: Characterization of different plant ingredients (e.g., alkaloids), partly important as biochemical defense against phytophages or scents as attractant for flower-visiting animals (Feeny, 1977).

Diversity of Interactions (Dynamic Pattern of Biodiversity). Among themselves, species create bi- and polysystems and thus form so-called biocoenotic links. These interactions between the organisms induce the emergence of characteristics which may contribute to stabilizing the system (quasi-stability in the species composition). Such interaction patterns can be divided into probioses (mutualism, symbiosis, and commensalism) and antibioses (predation, parasitism, etc.).

Mechanisms Causing Diversity (Causing Pattern of Biodiversity). Basically two different processes causing biodiversity can be distinguished: effects in evolutionary times (separation, speciation, and radiation) and effects in ecological times.

1. Effects in evolutionary times: In evolutionary time periods, biodiversity is attained by speciation (allopatric and sympatric). Of great importance in this case are the separation of originally linked populations, the subsequent differentiation of the separated populations, the development of isolation mechanisms, and the formation of different ecological niches. A decisive factor for high diversity rates is a slight extinction. An especially high species diversity is elicited by radiation. Examples include Darwin's finches (Geospizinae) on the Galapagos Islands (Lack, 1947) and the honeycreepers (Drepanididae) and fruit flies (Drosophilidae) of Hawaii (Mayr, 1943).

2. Effects in ecological times: In ecological time periods, a biocoenosis rich in species can only develop when communities immigrate and are newly formed. In this context, the number of ecological niches to be realized plays a decisive part. The ecological niche is not a spatial unit but rather the dynamic relation system of a species with its environment. It is composed of an autophytic/autozooic and an environmental dimension. The autophytic/autozooic dimension comprises the phylogenetically acquired morphological and physiological (and, for animals, also ethological) characteristics of the species, and the environmental dimension is the sum of all effective ecological factors. Where both dimensions overlap, the ecological niche of a species is realized (Schmitt, 1987). The breadth of the niche depends on the degree of specialization of the ecological niches which realize it. Niche overlaps can only be tolerated by species with a greater niche breadth.

Process of Functioning (Functional Pattern of Biodiversity). The question to what extent biodiversity contributes to the functioning of biocoenoses is controversial. There is no doubt that many organism species are constantly linked by certain interactions and that these relations may be obligatory. Such an interaction structure has only system character when it can be differentiated from other systems and when an independent matter flow is ascertainable. The differentiation of biocoenoses and ecosystems, however, has first a merely hypothetical character. Therefore, only theories can be developed in reply to the questions how much redundancy a biocoenosis or an ecosystem may tolerate without being impaired in the maintenance of their functional balance and whether there are upper and lower limits of biodiversity. The theory of biodiversity is closely linked with the ecosystem theory.

The more diverse the system, the more diverse must be its functional structure to stabilize the system. The element pattern and the diversity of interactions primarily contribute to this stabilization. Matter (nutrient) and energy flow are required to maintain the system and attain quasi-stability. The stabilization processes include matter and nutrient absorption, transformation, and transfer (as input–output reaction).

Intrabiocoenotic Diversity

A biocoenosis is composed of the plant community (phytocoenosis) colonizing a phytotope and the animal community (zoocoenosis) inhabiting a zootope. Owing to the physiognomically dominating higher plants, plant communities can be more easily analyzed and typified than zoocoenoses.

There are different pragmatic approaches to the study of biocoenoses and their diversity:

Investigation of taxonomic groups (zootaxocoenoses): classifying biodiversity

Investigation of functional groups or guilds, respectively (subsystems, smaller units, and functional groups of coexisting species which use the same resources in a similar manner): functional biodiversity

Investigation of certain relations (e.g., plant–insect complexes, food chains, and food webs): interaction biodiversity

Investigation of microhabitats (= synusia): classifying microhabitat biodiversity

More than 90% of all animal species living on land are bound to habitats characterized by their vegetation. The first step in the recording of an animal community may be a phytosociological characterization of the habitat because plant communities or vegetation complexes characterized by plant communities constitute typifiable units under ecological, structural, dynamic, chorological, and syngenetic aspects. Such a characterization of a habitat via its plant communities and plant community complexes is the starting point for a registration and analysis of biocoenological diversity.

The second step is a classification into microhabitats (= synusia); this classification should be based on three different spatial structure types: stratotope, choriotope, and merotope.

The different strata of a forest, for example, are designated as stratotopes; here, it can be distinguished between tree stratum, trunk stratum, herb stratum, etc., each colonized by its own stratocoenosis. Choriotopes, on the other hand, are independent vertical structures of the entire spatial unit or of parts of the stratotope (so-called choriocoenoses) such as the insect community of a tree or a shrub. Finally, in a habitat rich in structures, merotopes can be found (i.e., structure elements within a stratotope or a choriotope, such as organisms living on leaves or on bark or flower visitors) (Fig. 23).

Stratocoenoses. Analyses of taxonomic biodiversity demonstrate that each of these strata has its own animal species inventory, e.g., spider stratocoenoses in central European oak–birch forests. Comparisons of the strata of various plant communities, of the leaf and soil strata of a melick grass–beech forest (Melico–Fagetum), and of an oak–hornbeam forest (Querco–Carpinetum) show distinct differences in the species composition of earthworms in the stratocoenoses, especially in the leaf litter stratum (Rabeler, 1957) (Fig. 24).

Choriocoenoses. Other structural elements include special, clearly differentiable elements, so-called choriotopes; for example, a tree, a shrub, or a single plant, each with its community of phytophagous insects (phytophage complex). The diversity of a choriotope can be demon-

STRATOTOPE	STRATOCOENOSES	EXAMPLES
horizontal structures	sub-communities belonging to the stratotope	detritus stratum grassy stratum bush stratum foliage stratum
CHORIOTOPE	**CHORIOCOENOSES**	**EXAMPLES**
vertical structures of the entire spatial unit or of parts of the stratotope	sub-communities belonging to the choriotope	tree trunk bush corpse excrement anthill bird's nest
MEROTOPE	**MEROCOENOSES**	**EXAMPLES**
structure elements within a stratotope or a choriotope	sub-communities belonging to the merotope	leaf community wood community bark community flower community flower visitors

FIGURE 23 The three different spatial structure types (stratotope, choriotope, and merotope), the coenoses they comprise (stratocoenoses, choriocoenoses, and merocoenoses), and examples for each type.

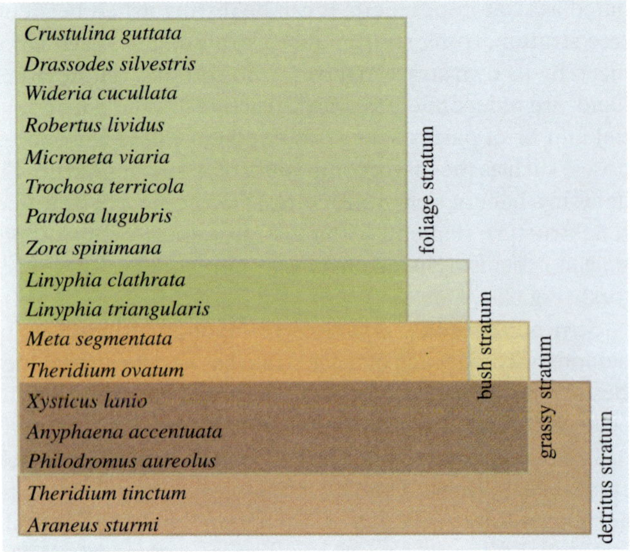

FIGURE 24 The dominant spider species from different strata in a central European oak–birch forest (data from Rabeler, 1957).

strated with the example of a bird's nest (Aßmann and Kratochwil, 1995). Bird species utilize very specific requisites to build their nests. The long-tailed titmouse (*Aegithalos caudatus*) builds highly characteristic nests in juniper (*Juniperus communis*) in northern Germany. An analysis of the nesting material shows that it consists of specific materials: certain moss species, lichen species, algae, etc. The composition depends on the plant community in which the nest is built. It is an orderly, habitat-typical structural diversity. The nest of a great tit (*Parus major*) is built in another way; also, this bird species is mainly found in quite different habitats. The diversity of species entails a diversity of the small structures created by them (Fig. 25).

Merocoenoses. The merotopes are parts of strato- and choriotopes. Strato-, chorio-, and merotopes combine to a special degree structural and functional diversity. Here, we discuss ecological niches, interaction levels, and relation structures between organisms.

The community of flower visitors corresponds to a merocoenosis, with the flowers representing merotopes. First, we find a systematic biodiversity of very different animal groups: Hymenoptera apoidea, Hymenoptera aculeata, Lepidoptera, Coleoptera, etc. Within this flower–flower visitor system, there is a functional diversity introduced by the visitor: for example, food relations (pollen, nectar, and oil) or certain other resource relations, such as the use of the flowers as warming-up places due to their parabolic mirror-like forms, as "rendezvous" places, as food source for predators and parasites, as overnight accommodation (e.g., for bees), or as provider of nesting materials. Flowers even supply scents used to mark swarming paths, as done by the neotropic, scent-gathering euglossine bees (Euglossinae). About 1400 plant species (belonging to 10 families) of oil-producing plants are known worldwide, and approximately 300 wild bee species specialized on them (Vogel, 1988).

MATERIAL USED FOR THE NEST	JUNIPER COMMUNITY (*Dicrano-Juniperetum*) long-tailed titmouse nests 1	2	JUNIPER COMMUNITY (*Roso-Juniperetum*) long-tailed titmouse nests 3	4	BIRCH FOREST (*Periclymeno-Fagetum*) great tit nests 5
LIVERWORTS					
Hypnum mamillatum	8.9%	23%	12.8%	38%	+
Dicratum scoparium	++	++	-	-	-
Campylopus introflexus	+	+	-	-	-
Lophocolea bidentata	+	-	-	-	-
cf. Dicranweisia	+	-	-	-	-
Hypnum mamillatum	-	-	+++	-	-
Eurynchium swartzii	-	-	++	-	-
Diranoweisia cirrata	-	-	-	+	-
Rhytidiadelphus squarrousus	-	-	-	-	50%
Scleropodium purum	-	-	-	-	40%
Brachythecium rutabulum	-	-	-	-	+++
Eurhynchium praelongum	-	-	-	-	+
Pleurozium schreberi	-	-	-	-	+
LICHENS	11.5%	13.1%	14.6%	9.8%	-
Hypogymnia physodes	+	+	+	+	-
Parmelia sulcata	+	+	+	+	-
Physcia tenella	-	-	+	+	-
Physcia adscendens	-	-	+	+	-
ALGAE					
Ulothrix spec.	+	-	-	++	-
Pleurococcus vulgaris	+	-	-	++	-
MISCELLANEOUS					
feathers	15.8%	19.4%	35.5%	11.6%	-
juniper bark	64%	12.6%	9.7%	4.7%	-
juniper berries	++	+	+	+++	-
juniper needles	++	+	+	+++	-
spider web	+	+++	+++	++	-
insect cocoons	-	++	-	-	-
heather	-	+	-	-	-
undefined	-	32%	32%	-	-
leaf remains	-	-	-	+	-

FIGURE 25 Structural diversities of four nests of long-tailed titmice (*Aegithalos caudatus*) and a comparison with the nest of a great tit (*Parus major*). Plus signs indicate the presence, more or less significant, of traces less than 1% (adapted from Aßmann and Kratochwil, 1995).

Moreover, plants also show different degrees of functionality (functional diversity). For the plant, the margin ranges from symbiotic relations, in which case the pollen-transferring insects are rewarded with food, to parasitism, which can be found in its most distinctive form in specimens of the genus *Ophrys:* The flowers imitate female bees and "sneak" by optical, olfactory, and tactile stimuli into the instinctive behavior of male bees to ensure a transfer of pollinia.

Species diversity and functional diversity always correlate with structural diversity. One example of this is the correlation between the structure of the pollen-gathering device of a bee and certain pollen grain structures.

The structural diversity of a flower–flower visitor merocoenosis is immense:

Optical diversity: the colors of the flowers in the visible and also in the ultraviolet wave range

Olfactory diversity: the multitude of different flower scents

Ethological diversity: the behavioral variety of flower visitors

Phenological diversity: the diurnal and seasonal variation of the occurrence of flowers and their pollinators.

Each plant community has its own animal communities on different levels (e.g., guilds). Often, animals need whole vegetation complexes. On the ecosystem level, the structural and functional diversity levels of different organism groups correlate with their specific abiotic environments. The biocoenoses or biocoenosis complexes are characterized by certain character species. However, each biocoenosis has its own range of diversity types and patterns. The greater the species diversity, the more varied are other diversity types: genetic diversity, space-structural and physiognomic diversity, biochemical diversity, phenological diversity, etc.

Interbiocoenotic Diversity

As a rule, landscapes are not composed of single biocoenoses but, rather, of biocoenosis complexes and a mosaic of different ecosystems. The development of individual vegetation units into associations, for example, is not arbitrary but follows certain rules. It is interesting that regularities on the species–biocoenoses (Thienemann's laws) and biocoenoses–biocoenosis complex levels follow the same natural laws.

In the first case, "The more variable the conditions for life of a site, the larger the species number of the respective community" and "The more the conditions for life of a biotope deviate from the normal and for most organisms optimum conditions, the poorer in species the biocoenosis will become, the more characteristic it gets, the more individuals of the single species will occur."

In the second case, "The more variable the environmental conditions of a habitat complex, the larger the number of its biocoenoses" and "The more the environmental conditions of a habitat complex deviate from the normal and for most biocoenoses optimum conditions, the poorer in biocoenoses the complex will become, the larger and the more characteristic are the occurring biocoenoses" (Schwabe and Kratochwil, 1994).

Does Each Species Have the Same Importance in a Diverse System?

It is undisputed that often the functional importance of species is not known and that there are frequently no indications as to how the species react to stress factors in a certain biocoenosis. There are species of central importance for an ecosystem, so-called keystone species, without which the entire system would break down. Such keystone species include *Phragmites communis,* which builds up large reeds (there are many links between reed and various animal species), and the beaver (*Castor canadensis*), which shapes entire river landscapes (e.g., the boreal regions).

The functions of many species in an ecosystem are not currently scientifically documented. Ehrlich and Ehrlich (1981) formulated the rivet hypothesis: Every species is comparable to a rivet joint in an airplane. Its importance cannot be predicted for any situation. There may be redundant subsystems. No one knows if missing rivet joints can be compensated or if just one missing joint may have severe consequences.

Applied Aspects of Biodiversity

Currently, about 1.5 million of the earth's animal and plant species have been described. Their actual number may vary between 5 and 30 million. For approximately the past 25 years, scientists have predicted the extinction of approximately 1 million species. An exponential tendency can be ascertained: From 1600 to 1900, every 4 years a species was eradicated by man and after 1900 a species was eradicated each year; currently, more than 1 species disappear per day. It is assumed that every hour 1 species is extinguished. By the end of the previous century, probably 20–50% of all species had disappeared (Fig. 26).

Under natural conditions, the net growth rate of species is 0.37% in 1 million years (i.e., 0.00000037%), which is an extremely low value. The natural extinction rate has thus been increased 10,000-fold by man; the decrease is at least 100 times higher than the loss of species in the past 65 million years. The rate of loss of genetic diversity on the level of populations is much higher.

The center of especially high biodiversity lies in the tropics, mainly in the tropical mountainous areas. On a few hectares of forest in Southeastern Asia or in the Amazon region, more tree species can be found than in the whole of

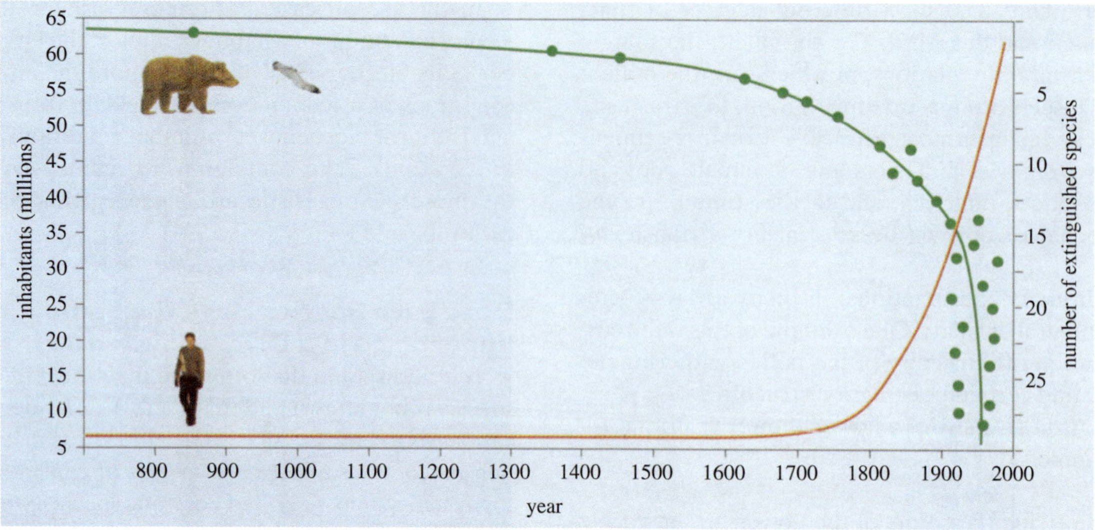

FIGURE 26 Demographic evolution in Germany and extinction rates of mammals and birds (depicted schematically).

Europe. In Venezuela's evergreen rain forest, there are at least 90 tree species per hectare. In some regions, the loss of biodiversity is significant: Worldwide, numerous ecosystem types are particularly endangered, including the tropical rain forests, certain marine ecosystems, islands in the sea, high mountain ranges, arctic and subarctic habitats, savannas, steppes, semideserts, large river systems, mangrove forests, many lakes, and also the landscapes in the countries in which we live.

A loss of biodiversity cannot be tolerated for ecological, ethical, religious, aesthetic, and cultural reasons, especially because the destruction of biodiversity is irreversible. To maintain biodiversity, developing theoretical principles and translating them into practical measures is one of the major tasks of the near future. The maintenance of biodiversity is closely linked to the survival of man on Earth and has thus been incorporated into the concept of sustainable development.

Acknowledgment

We thank Ms. Martina Lemme (University of Osnabrück) for translation of the article into English.

References Cited

Aßmann, T., and Kratochwil, A. (1995). Biozönotische Untersuchungen in Hudelandschaften Nordwestdeutschlands—Grundlagen und erste Ergebnisse. *Osnabrücker Naturwiss. Mitt.* **20/21,** 275–337.

Ayala, F. (1978). The mechanisms of evolution. *Sci. Am.* **239,** 48–61.

Beattie, A. (1983). Distribution of ant dispersal plants. *Sonderbd. Naturwiss. Ver. Hamburg* **7,** 249–270.

Bisby, F. A. (1995). Characterization of biodiversity. In *Global Biodiversity Assessment* (V. H. Heywood and R. T. Watson, Eds.), pp. 21–57. Cambridge Univ. Press, Cambridge, UK.

Burdon, J. J. (1987). Phenotypic and genetic pattern of resistance to the pathogen *Phakopsora pachyrhizi* in populations of *Glycine canescens. Oecologia (Berlin)* **73,** 257–267.

Carrol, R. L. (1988). *Vertebrate Paleontology and Evolution.* Freeman, New York.

Czihak, G., Langer, H., and Ziegler, H. (Eds.) (1976). *Biologie. Ein Lehrbuch für Studenten der Biologie.* Springer-Verlag, Berlin.

Darwin, C. (1859). *On the Origin of Species by Means of Natural Selection or the Preservation of Favored Races in the Struggle for Life.* Murray, London.

Ehrlich, P. R., and Ehrlich, A. H. (1981). *Extinction. The Causes and Consequences of the Disappearance of Species.* Random House, New York.

Faegri, K., and van der Pijl, L. (1979). *The Principles of Pollination Ecology,* 3rd ed. Pergamon, Oxford.

Feeny, P. (1977). Defense ecology of the Cruciferae. *Ann. Missouri Bot. Gard.* **64,** 221–234.

Gilbert, L. E. (1975). Ecological consequences of a coevolved mutualism between butterflies and plants. In *Coevolution of Animal and Plants* (L. E. Gilbert and P. H. Raven, Eds.), pp. 210–240. Univ. of Texas Press, Austin.

Goetze, D., and Schwabe, A. (1997). Linking ecosystems and biodiversity concepts by means of sigmasociology. *Coll. Phytosoc.* **27,** 798–812.

Hickman, C. P., Jr., Roberts, L. S., and Larson, A. (1997). *Integrated Principles of Zoology,* 10th ed. Brown, Dubuque, IA.

Hutchinson, G. E. (1965). The niche, an abstractly inhabitat hypervolume. In *The Ecological Theatre and the Evolutionary Play.* Yale Univ. Press, New Haven, CT.

Huxley, J. S. (1974). *Evolution. The Modern Synthesis,* 3rd ed. Allen & Unwin, London.

JANZEN, D. H. (1980). When is it coevolution? *Evolution* **34,** 611–612.

KETTLEWELL, H. B. D. (1972). *The Evolution of Melanism.* Clarendon, Oxford.

KRATOCHWIL, A. (1988). Zur Bestäubungsstrategie von Pulsatilla vulgaris Mill. *Flora* **181,** 261–325.

LACK, D. L. (1947). *Darwin's Finches.* Cambridge Univ. Press, Cambridge, UK.

MATTES, H. (1978). Der Tannenhäher im Engadin. *Münster Geogr. Arbeiten* **2,** 1–87.

MAYR, E. (1943). The zoogeographic position of the Hawaiian Island. *Condor* **45,** 45–48.

OSCHE, G. (1966). Grundzüge einer allgemeinen Phylogenetik. *Handbuch Biol.* **3**(2), 817–906.

RABELER, W. (1957). Die Tiergesellschaft eines Eichen-Birkenwaldes im nordwestdeutschen Altmoränengebiet. *Mitt. Flor. Soz. Arbeitsgem. N. F.* **6/7,** 297–319.

REICHHOLF, J. H., and WEIDENSAUL, S. (1990). *Kolibris. Fliegende Diamanten.* Verlag Karl Müller, Erlangen.

SCHMITT, M. (1987). Ecological niche sensu Günther and ecological licenses sensu Osche—Two valuable but poorly appreciated explanatory concepts. *Zool. Beitr. N. F.* **31,** 49–60.

SCHWABE, A., and KRATOCHWIL, A. (1994). Gelten die biozönologischen Grundprinzipien auch für die landschaftsökologische Dimension? Einige Überlegungen mit Beispielen aus den Inneralpen. *Phytocoenologia* **24,** 1–22.

SOLBRIG, O. T. (1991). *Biodiversity. Scientific Issues and Collaborative Research Proposals.* UNESCO, Paris.

STOCKER, O. (1952). *Grundriß der Botanik.* Springer-Verlag, Berlin.

VOGEL, S. (1988). Die Ölblumensymbiosen—Parallelismus und andere Aspekte ihrer Entwicklung in Raum und Zeit. *Z. Zool. Syst. Evolut.-Forsch.* **26,** 341–362.

WEHNER, R., and GEHRING, W. (1995). *Zoologie,* 23rd ed. Georg Thieme Verlag, Stuttgart.

WIEBES, J. T. (1979). Coevolution of figs and their insect pollinators. *Annu. Rev. Ecol. Syst.* **10,** 1–12.

ZIZKA, G. (1990). *Pflanzen und Ameisen. Partnerschaft fürs Überleben. Palmengarten Sonderheft 15.* Stadt, Frankfurt on Main.

General References

KRATOCHWIL, A., and SCHWABE, A. (2001). Ökologie der lebensgemeinschaften. *Biozönologie.* Ulmer Verlag, Stuttgart.

SCHULZE, E.-D., and MOONEY, H. A. (Eds.) (1993). *Biodiversity and Ecosystem Function.* Springer, Berlin.

Riccardo Valentini

Department of Science for Forest
Environment and Resources
University of Tuscia
Viterbo, Italy

This article analyzes how the resources of the biosphere and the energy transformation that takes place in it can affect our life conditions. I begin by describing the processes that regulate the biogeochemical cycle of water and the energy fluxes that take place between the biosphere and the atmosphere. In particular, I examine the factors controlling the energy dissipation mechanisms by vegetation. As an example of the influence that plant organisms have on climate, I discuss the impact of deforestation of the tropical rain forests of the Amazon and of the boreal forests of Siberia. I then analyze the importance of primary production as an essential element of the trophic organization of the planet, and I describe the carbon cycle on a planetary level, highlighting the role played by ecosystems in regulating the earth's climatic balance.

Energy Flux and Primary Production

Introduction

The terrestrial vegetation, 40% of which is composed of forests, covers approximately 23% of the total surface of the planet. If it were distributed in a single thin layer of photosynthesizing surface area, the foliar surface that forms the green mantle of vegetation could cover approximately 2.5 times the entire surface of the globe.

When we consider the importance of vegetation in our own lives or in the lives of other living organisms, we generally refer to the possibility of utilizing organic plant matter for the food consumption required to carry out our vital functions, or we think of extractable products that can be used to improve our living conditions, such as wood as a construction material and cellulose paste, oils, and plant fibers for various industrial applications. We rarely consider the important role that the earth's ecosystems have in regulating the conditions of our lives, for example, by influencing the characteristics of the atmosphere, the climate, the water cycle, erosion, and other processes that, in a less direct but extremely significant way, influence and make life possible on our planet (Valentini *et al.,* 1998a).

Through the processes of photosynthesis and transpiration, the world's ecosystems affect surface energy fluxes and alter the chemical composition of the atmosphere. About 4.5 billion years ago, when our planet first came into being, physical and chemical factors dominated the distribution of elements in the soil as well as the composition of both water and the atmosphere. When life first appeared on Earth about 3.5 billion years ago, biological processes became increasingly more important in determining the distribution of elements and influencing the planet's chemical and physical characteristics.

With the publication of *Gaia* in 1979, Lovelock focused the attention of the scientific community on the fact that the chemical composition of our current atmosphere is extremely "unusual" and that it is different than what one would expect for thermodynamic equilibrium. If life had not developed on Earth, the concentration of carbon dioxide in the atmosphere would have remained quite high and the surface temperature of the planet would have been very different — probably higher than it is now. In time, the concentration of oxygen would have increased slowly due to the dissociation of water caused by solar radiation, but it would only have risen to 1% of the current concentration. If life were to disappear from the planet today, denitrification processes would cease, leading to an increase in the level of nitrates and a decrease in the production of ammonia, which is almost exclusively biological in origin. Ammonia regulates the pH of rainfall, keeping it at a suitable value of about 6. Without ammonia, this value would decrease to 3 (for an acidity comparable to vinegar) and this would have a substantial impact on the oceans.

Table 1 lists the abundance levels of some of the gases that are currently present in the terrestrial atmosphere, which contains about 20% oxygen, together with the hypothetical concentrations at equilibrium. It can be seen that due to the abundance of oxygen, the current composition of the atmosphere could not remain constant if there were no biogenic fluxes capable of maintaining it. In fact, today's atmosphere has too much nitrogen, methane (more than 29 orders of magnitude compared to the equilibrium concentration), nitrous oxide, ammonia, and hydrogen. For at least 40 other biogenic gases, it is difficult to explain their enormous atmospheric concentration with respect to the amount of oxygen that is present (Levine, 1989). Given the fact that these substances are continuously being transformed by oxygen, their abundance can be explained solely by the existence of constant and considerable fluxes from the biosphere.

The amount of available energy on the earth's surface is the result of the balance between the energy coming from the sun and the energy that is dissipated through emitted radiation. The amount of incoming energy is a function of the reflectivity (*albedo*) of land surfaces. Ice reflects 80–95% of incident radiation, whereas a prairie reflects 30–40% and a coniferous forest reflects 10–15%. As a result, the seasonal variations in vegetation, like the impact of human activities, can trigger substantial changes in the amount of radiation that is absorbed by the earth's surface, appreciably altering the surface temperature. Many studies have demonstrated that the surface energy balance has important consequences with regard to the regional and global climatic regime.z

Today, there are numerous simulations of the global climate that take terrestrial vegetation and its properties into consideration in terms of surface energy fluxes (Sellers *et al.,* 1996). Several examples of the importance of surface energy fluxes have been reported in a study by Xue and Shukla (1993) that demonstrated how deforestation and changes in soil utilization in the Mediterranean and Sahel regions can lead to a decrease in precipitation, resulting in an increase in aridity and desertification processes.

In this article, I discuss how the earth's ecosystems can exert a significant impact on the climate of our planet at both a regional and a global level through the transformation of energy and the processes involved in the synthesis of organic matter.

TABLE I

Concentrations and Biogenic Fluxes of Atmospheric Gases

GAS	OBSERVED CONCENTRATION (MOLE FRACTION)	CONCENTRATION AT EQUILIBRIUM (MOLE FRACTION)	OBSERVED CONCENTRATION/ CONCENTRATION AT EQUILIBRIUM	BIOGENIC FLUX (10^6 t/YEAR)
Nitrogen	0.8	10^{-10}	8×10^9	1000
Methane	1.5×10^{-6}	$<10^{-35}$	$>10^{29}$	2000
Nitrous oxide	3.0×10^{-7}	10^{-20}	3×10^{13}	600
Ammonia	1.0×10^{-8}	$<10^{-35}$	$>10^{27}$	1500
Hydrogen	5.0×10^{-7}	$<10^{-30}$	$>5 \times 10^{23}$	20

A Working Definition of Biosphere

The biosphere is the part of our planet in which there exist the conditions that are fundamental for plant and animal life and in which living organisms, which are also part of the biosphere, exchange matter and energy. The biosphere includes the atmosphere, the land surfaces where plants live, the rhizosphere, which is dominated by roots and microorganisms, and also free bodies of water such as rivers, lakes, and oceans. Regarding aquatic ecosystems, the cycle of elements and the nature of the organisms that are involved are so different that the biosphere is generally subdivided into the geobiosphere, which includes terrestrial ecosystems, and the hydrobiosphere, which encompasses aquatic ecosystems (Walter, 1973).

Energy Transformations in Ecosystems

The circulation of water through plants and ecosystems represents the largest transfer of energy through a chemical substance in the biosphere. Plants can assimilate water through root absorption and they can convey it along the stem to bring it to the leaves. From here, through the stomata (tiny pores on the surface of the leaf), it is returned to the atmosphere in the form of water vapor. Plants can extract a considerable amount of water from the soil. A leaf can transpire a quantity of water equivalent to 200 ml per day, whereas a tree can transpire about 200 liters and a forest covering 1 ha can transpire 10,000 liters. Water transport and subsequent evaporation are thermodynamic processes that require a considerable amount of energy: It takes 2440 J to evaporate 1 g of water. Terrestrial ecosystems can be thought of as enormous and silent thermodynamic machines that, day after day, can transform significant amounts of energy.

The energy that sets all this in motion is solar energy, which is absorbed by plants and is utilized in various processes, including transpiration and photosynthesis. The way in which this energy is transformed is unique to each terrestrial ecosystem and it varies according to its structure, specific composition, and functionality.

Since most of the energy that is absorbed by ecosystems is used in transpiration and evaporation processes, I start by providing a general outline of the water cycle on a global scale, and then I note how the dissipation of energy can determine a set of functional properties at an ecosystem level. Lastly, I describe several cases in which the action of terrestrial vegetation affects climate on a regional scale.

The Global Water Cycle

Through transpiration, evaporation, and precipitation (as well as through ocean currents), water takes most of the energy it receives from the sun and transfers it from tropical regions to polar regions, somewhat like a pipeline that transfers heat from a boiler to the different rooms in a house.

The movements of water in the atmosphere determine the distribution of rainfall on the earth and the annual availability of water resources, thus determining plant growth and development and, more generally, the conditions suitable for biological life. If precipitation exceeds evaporation and transpiration, this can produce the surface flow (or runoff) that leads to erosion and the transport of solid matter toward the oceans.

The amounts of water involved in the global hydrological cycle (Fig. 1) are so large that they are conventionally expressed in units of kilometers^3. As one can easily discern, they are the largest water reservoirs on our planet: They hold 1,350,000,000 km^3 of water, or nearly 97% of all the water present on the earth's surface. Polar ice and continental glaciers furnish the second largest contribution to our wealth of water. The soil contains about 121,800 km^3 of water, about 58,100 km^3 of which can be found in the plant root area (Webb *et al.,* 1993). Man's needs are met by a relatively small amount of water contained in lakes and rivers. The amount of water present in underground water tables is difficult to assess, but recent estimates show that it can vary globally from 4,200,000 to 15,300,000 km^3; with the exception of the small amount extracted by man, this water is generally not available for the vital processes of the biosphere.

The atmosphere contains a relatively small amount of water – approximately 13,000 km^3. Nevertheless, large quantities of water move to and from the atmosphere as well as within it as a result of surface energy exchanges.

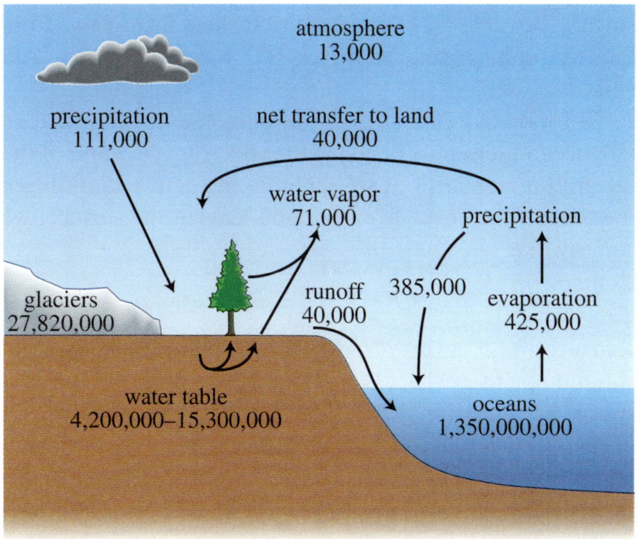

FIGURE 1 Diagram of the water cycle. The values of the compartments are expressed as kilometers^3 of water, whereas the flux values indicated by the arrows are expressed as km^3/year.

Evaporation takes about 425,000 km^3 of water from the ocean every year. Based on these numbers, it can be deduced that the mean time that water remains in the ocean is about 3200 years. Only part of this water, approximately 385,000 km^3/year, goes back into the oceans through rainfall. The remaining 40,000 km^3/year is transferred to the soil as water vapor. Plants can absorb water through their root systems and return it to the atmosphere as vapor: This process is known as *transpiration.* The combination of transpiration and evaporation that inevitably accompanies it in the presence of wet surfaces is defined as *evapotranspiration,* a term that is used to describe the flow of water from the surfaces of plants to the atmosphere.

The movement of water due to soil evapotranspiration amounts to about 71,000 km^3/year, and together with the 40,000 km^3/year that comes from the oceans this yields a precipitation of 111,000 km^3/year on land. These numbers show the great importance of terrestrial vegetation in recycling the water in the atmosphere. The water generated by the evapotranspiration process resides in the atmosphere for an average time of 1 year, far less than the amount of time that water resides in the ocean. These data highlight the fundamental role that terrestrial ecosystems play in the water cycle and in the energy transformation processes associated with it.

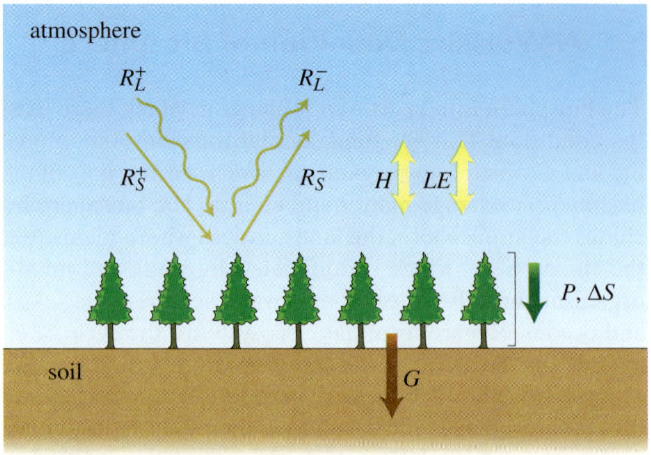

FIGURE 2 Surface energy balance of a plant canopy during the daytime. The energy fluxes for an ideal plant surface are sensible heat flux (H) and latent heat flux (LE) to and from the atmosphere, soil heat flux (G), and net radiative flux. The latter is the result of the radiative balance between radiation of the visible bands of the spectrum (R_S) and infrared radiation (R_L). The plus sign indicates fluxes directed from the atmosphere to the surface, whereas the minus sign indicates fluxes from the surface to the atmosphere. Other processes also take place in the space occupied by the plant canopy, such as heat storage (ΔS) and biochemical processes (P).

The Surface Energy Balance

If we consider a relatively uniform, even, and sufficiently widespread ideal surface that is separated from the atmosphere above it by a thin interface positioned directly over the soil, it is easy to calculate the energy balance since only vertical fluxes of energy need to be taken into consideration. There are essentially four types of energy fluxes for an ideal plant surface: net radiative flux (R_N) to and from the surface, sensible (H) and latent (LE) heat flux to and from the atmosphere, and heat flux (G) to and from the soil (Fig. 2).

Net radiative flux (R_N) is the outcome of the radiative balance in the visible bands of the spectrum (R_S) and of the infrared region (R_L). If we separate incident radiation going toward the cover from the one leaving the surface, we can say that

$$R_N = R_S^+ - R_S^- + R_L^+ - R_L^- \qquad [1]$$

By convention, fluxes directed toward the surface are considered positive, whereas the ones going toward the atmosphere are negative.

The visible solar radiation that reaches the surfaces is composed partly of direct solar radiation and partly of diffuse solar radiation. The component that leaves the surface is the fraction of incident radiation that is reflected by plant cover:

$$R_S^- = AR_S^+ \qquad [2]$$

in which A is the surface albedo, whose value will depend on the properties of the surface itself. High albedo values are observed with snow cover, bare soil without any vegetation, and so on, whereas for plant cover this magnitude can range from 0.1 to 0.3, indicating a significant level of absorption of visible radiation.

When there are no clouds, incident long-wave radiation belonging to the infrared region of the spectrum (R_L) depends mainly on temperature, moisture, and the concentration of greenhouse gases in the atmosphere, and it does not show any notable variation during the daytime. The presence of clouds significantly increases the amount of infrared radiation that reaches the surface. Outgoing infrared radiation depends on surface temperature, according to the Stefan–Boltzmann law:

$$R_L^- = \varepsilon\sigma T^4 \qquad [3]$$

where ε is surface emissivity (typically 0.90–0.98 for vegetation), σ is the Stefan–Boltzmann constant ($5.67\ 10^{-8} m^{-2}$/u\K^{-1}), and T is absolute temperature.

Since the infrared radiation leaving the surface is proportional to the fourth power of the temperature value, it shows significant daytime variations, with maximum values early in the afternoon and minimum values at dawn. Often, the plant surface temperature is still high during the early night hours, whereas the infrared radiation from the sky is less because only the blackbody of the air masses is present. As a result, the net radiative flux is negative.

Net radiation can be calculated by measuring its individual components or using special instruments called net radiometers, which are composed of two sensors that, with a rather broad spectral response (0.4–50 μm), simultaneously measure incident radiation and outgoing radiation, automatically providing the difference between the two (i.e., net radiative flux R_N).

Sensible heat flux (H) is the amount of heat that is conveyed via convection and turbulence between the surface and the atmosphere due to the temperature gradient that exists between them:

$$H = \rho c_p \frac{T_s - T_a}{R_h} \quad [4]$$

where ρ and c_p, respectively, are the density and specific heat of dry air at a constant pressure, T_s and T_a are the temperature of the plant surface and the surrounding air at a reference point over the cover. R_h is thermal resistance (i.e., the inverse of the heat transfer coefficient), which is dependent on wind speed and the aerodynamic characteristics of the surface. Sensible heat flux thus depends solely on the physical characteristics of the system, particularly surface heat and wind. The characteristics of the plant structure (branches, leaves, and height) also play an important role in the dissipation of energy. Small, acuate, or lanceolate leaves foster heat dissipation since they are aerodynamically efficient. This is the reason that in arid environments, most forms of plant life have developed this type of foliar structural adaptation to augment heat dissipation.

Likewise, the radiant flow of latent evaporation heat (LE), which represents the energy expended for evapotranspiration process, can be formulated as follows:

$$\mathrm{LE} = \frac{\rho c_p(\mathrm{VP}_s - \mathrm{VP}_a)}{\gamma(R_v + R_s)} \quad [5]$$

where γ is the psychrometric constant, VP_s and VP_a, respectively, are the partial water vapor pressures inside the leaves (specifically, in the substomatal cavities) and in a point in the atmosphere above the cover, R_v is the resistance to the eddy transfer of water vapor, and R_s is the resistance to transfer across the stomata (Monteith, 1975). It is interesting to note that R_v is a physical parameter, analogous to R_h, that is dependent on the aerodynamic characteristics of the surface and of wind speed, whereas R_s is a biological type of parameter correlated to the opening of the stomata, whose size is regulated by the plant's metabolism.

The radiant flow to and from the soil (G) can be estimated as

$$G = K\frac{dT}{dz} \quad [6]$$

where K is the thermal conductivity of the soil and dT/dz is the vertical thermal gradient of the soil. The flow of heat in the soil depends for the most part on the radiation conditions of its surface. If the soil has no plant cover, the flow of heat that enters the soil represents up to 50% of the overall net radiative flux; however, if the soil is protected by dense plant cover, as in a forest, this value decreases to about 2–5% of the net radiative flux.

During the day, the soil absorbs heat and as a result the flux is positive, whereas during the early night hours, the soil tends to give off heat rapidly, showing a negative flux.

An ideal surface, which is horizontally homogeneous and uniform, is rarely encountered in practice. In particular, the hypothetical thin interface between the atmosphere and the surface is purely an ideal concept since plant covers are composed of thick foliar layers. In particular, forests, whose trees can reach sizes of more than 50 m, encompass a large volume of space between the soil and the atmosphere above. In this space, other energy processes are also being carried out, such as heat storage and biochemical processes.

The energy fluxes associated with the presence of a finite volume ($\Delta S_B \Delta S_T \Delta S_E$) can be attributed to the following three processes that take place within the volume we are examining: heating of the biomass present in the volume, heating of the air, and the variation in the concentration of water vapor in the internal atmosphere of this volume. The entity of these energy fluxes is dependent on the density, specific heat, and temperature of the biomass present in the volume.

An additional term contributing to energy balance is the biochemical energy flux (P) that is used within this reference volume. Photosynthetic processes utilize solar radiation to produce organic matter, providing biological synthesis with energy flux. Inversely, respiratory processes represent a consumption of energy. Despite the fact that these energy fluxes are extremely important for the life of biological organisms and for the ecosystem as a whole, the amount of energy that is utilized in these processes is negligible with respect to the amount available for the ecosystem. In fact, this amounts to about 2% of all the available energy (with the exception of high-productivity crops, which can be as high as 8%).

In short, bearing the principle of the conservation of energy in mind, we can formulate the balance that refers to a plant surface (see Fig. 2) using the following expression:

$$R_N + H + \mathrm{LE} + G + \Delta S_T + \Delta S_B + \Delta S_E + P = 0 \quad [7]$$

Based on the considerations set forth previously, the terms that regard the energy fluxes associated with chemical reactions are negligible. Furthermore, as a first approximation, if we consider the terms regarding the energy stored inside the ecosystem volume as unimportant, the balance is reduced to the one for an ideal surface:

$$R_N + H + \mathrm{LE} + G = 0 \quad [8]$$

from which, by setting A equal to $RN + G$, we get

$$A = -(H + \mathrm{LE}) \qquad [9]$$

In this relation, A is the flow of energy that is effectively available to the ecosystem, which is involved in the two main processes of evapotranspiration and the transfer of sensible heat.

Energy Dissipation Mechanisms

By examining the simplified form of the energy balance equation, we can pose the problem of how terrestrial ecosystems use energy, how they dissipate it in various processes, and thus what their functional properties are. Traditionally, the balance of energy was studied to obtain practical information on water consumption in farming. In fact, by knowing the energy expended in the evapotranspiration process, it was possible to calculate the water required for crops and thus regulate irrigation more efficiently. Recently, numerous studies that have been extended to natural vegetation and forests have shown that the way in which terrestrial ecosystems dissipate absorbed solar energy is a characteristic functional property that varies according to the type of plant cover.

Jarvis and McNaughton (1986) were among the first to observe that the modes of energy dissipation are under the biological and structural control of vegetation. They thus introduced the concept of atmospheric coupling, expressed by what is referred to as the Omega factor. One way of expressing the evapotranspiration of a given plant cover as a function of the various biological and environmental parameters is the Penmann–Monteith equation (Monteith, 1975):

$$\mathrm{LE} = \frac{sA + \dfrac{\rho c_p \mathrm{VPD}}{R_a}}{s + \gamma\left(1 + \dfrac{R_s}{R_a}\right)} \qquad [10]$$

where A is the energy flux available in the ecosystem, ρ and c_p are the density of dry air and its specific heat, respectively, VPD is the vapor pressure deficit at a reference altitude over the cover, s is the slope of the saturated vapor pressure curve in relation to temperature, c is the psychrometric constant, and R_S and R_a are stomatal resistance and aerodynamic resistance, respectively. Due to the principle of the similarity of the eddy transport of mass and energy, the latter is the same for the transport of both sensible heat and of water vapor ($R_a = R_h = R_v$). Aerodynamic resistance is a function of wind speed, atmospheric stability, and the structural properties of vegetation, such as height, foliar area, and cover unevenness. In particular, for shorter plant covers that are relatively even and homogeneous (such as crops or prairies), R_a assumes a high value, thereby indicating low aerodynamic efficiency for the transport of energy. Instead, taller and more uneven vegetation (such as forests) show lower R_a values, denoting greater aerodynamic efficiency and thus easier transport of energy.

There are two limits in the Penmann–Monteith equation that represent the two situations I have described:

$$\lim_{R_a \to \infty} \mathrm{LE} = \frac{sA}{s + \gamma} \qquad [11]$$

$$\lim_{R_a \to 0} \mathrm{LE} = \frac{\rho c_p \mathrm{VPD}}{\gamma R_S} \qquad [12]$$

The first case involves a plant cover with a very high aerodynamic resistance. In these conditions, evapotranspiration depends solely on available energy, independently of the biological control carried out by the stomata, and the limit value of LE represents the evapotranspiration equilibrium, which is the maximum value in relation to available energy. This limit can be overcome only if there is an external input of energy.

In the second case, evapotranspiration is controlled more directly by the vapor pressure deficit and by the stomata. In this case, the vegetation can control evapotranspiration more effectively by regulating the opening of the stomata.

The first case represents short and homogeneous vegetation such as crops, whose evapotranspiration is more dependent on available energy. According to the terminology used by Jarvis and McNaughton (1986), these types of vegetation are "uncoupled" from the atmosphere and their Omega factor approaches 1. Instead, forest covers are more dependent on the vapor pressure deficits in the atmosphere and on stomatal opening conditions. In other words, they are "coupled" to the atmosphere and their Omega factor is close to 0. This proves that in the first case, energy is dissipated under the form of evapotranspiration (when there is sufficient water availability in the soil), whereas in the second case the biological control exerted by the plants lowers the dissipation of energy in the form of evapotranspiration to the benefit of sensible heat.

The sensitivity of the evapotranspiration process with regard to biological control can be analyzed by expressing the percentage variation of evapotranspiration in relation to the variation in stomatal resistance. The derivation of Eq. (10) yields

$$\frac{R_s}{\mathrm{LE}} \frac{\partial \mathrm{LE}}{\partial R_s} = -\frac{1}{\left(\dfrac{s}{\gamma} + 1\right)\dfrac{R_a}{R_s} + 1} \qquad [13]$$

The minus sign indicates a negative feedback, meaning that for each increment in stomatal resistance (closing of the stomata) there is a decrease in evapotranspiration. An increase in R_S increases the sensitivity of evapotranspiration to stomatal closing. Figure 3 compares three different types

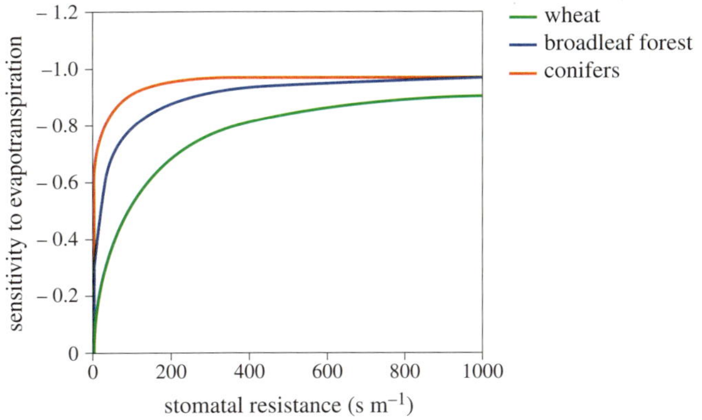

FIGURE 3 Trend in the sensitivity of evapotranspiration in relation to the stomatal resistance of various types of plant canopy. The more negative values indicate more effective stomatal control. Coniferous forests show greater evapotranspiration sensitivity to stomatal control, whereas stomatal control has a lesser effect in crops.

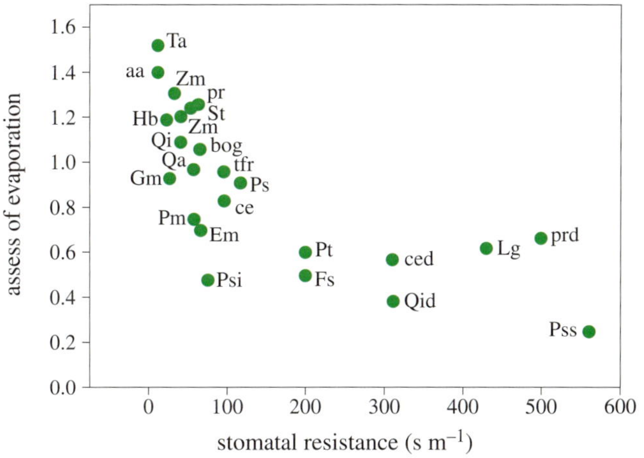

FIGURE 4 Relationship between the evapotranspiration/equilibrium evapotranspiration ratio and stomatal resistance for several terrestrial ecosystems. Ta, *Triticum aestivum* (wheat); Zm, *Zea mays* (maize); Pt, *Pinus taeda;* Qa, *Quercus alba* (white oak); pr, prairie; prd, dry prairie; bog, bog (central Siberia); Pss, *Pinus sylvestris* (Scots pine, central Siberia); Fs, *Fagus sylvatica* (beech); Qid, *Quercus ilex* (evergreen oak, dry); Qi, *Quercus ilex* (evergreen oak); Lg, *Larix gmelinii* (larch); Gm, *Glicine max* (soy); aa, alfalfa; St, *Solanum tuberosum* (potato); Hb, *Hevea brasiliensis* (rubber); Psi, *Picea sitkensis* (Sitka spruce); Ps, *Pinus sylvestris* (Scots pine); Pm, *Psuedotsuga menziesii* (Douglas fir); Em, *Eucalyptus maculata;* tfr, tropical rain forest (rondonian); ce, cerrado; ced, dry cerrado.

of plant cover (wheat, deciduous broadleaf forest, and coniferous forest), using R_a values of 80, 15, and 5 s m^{-1}, respectively.

The more negative sensitivity values indicate greater biological control over evapotranspiration. As would be expected, the sensitivity of evapotranspiration to changes in R_S is greater for forests and particularly for coniferous ones. It is thus clear that both the structure of the vegetation (expressed by the aerodynamic efficiency parameter) and biological control (expressed by stomatal resistance) control evapotranspiration and thus the dissipation of energy.

It is interesting to observe that most studies on the relationship between the biosphere and the atmosphere tend to indicate a functional convergence of various terrestrial ecosystems in amplifying their energy dissipation properties. For example, forests and natural vegetation in general, which have the highest aerodynamic efficiency and are thus better coupled to the atmosphere, also have the highest stomatal resistance values, which further increase the sensitivity of evapotranspiration to biological control. The opposite is seen with agricultural vegetation and prairies, in which the lower stomatal resistance values amplify their atmospheric uncoupling even further.

Shown in Figure 4 is the trend in the ratio between evapotranspiration and its equilibrium value (Eq. 11) as a function of stomatal resistance for a series of terrestrial ecosystems. As can be seen, the ecosystems with a strong stomatal control (a high R_S value) dissipate energy via evapotranspiration in a much lower quantity in a condition of equilibrium. This means that most of the absorbed energy is dissipated as sensible heat or, in other words, through a physical exchange with the surrounding atmosphere, with an ensuing impact on heating this atmosphere.

Inversely, as we move gradually to ecosystems with a less intense stomatal control in which the stomata are thus more open, the energy dissipated by evapotranspiration tends toward the equilibrium value, which depends solely on the amount of available energy provided by the sun. When most energy is dissipated in the form of evapotranspiration, the surface is colder and there is less of an impact on heating the surrounding atmosphere.

Before I examine the effect of the diversity in energy dissipation on regional climate, it is interesting to ask several questions: Why does the capacity to dissipate energy in the form of evapotranspiration vary so much in terrestrial ecosystems? Why do forests exert greater control over evapotranspiration than other types of vegetation such as prairies? In particular, with regard to forests, why are coniferous forests more sensitive to biological control than broadleaf forests?

One possible explanation is that forests and natural vegetation generally colonize harsher environments, in which water resources are scarce, variable for the most part, and highly unpredictable. In these conditions, strong biological control of evapotranspiration is required in order to control water loss during more adverse periods and to foster the recharging of reserves during more favorable periods.

In addition, evergreens must necessarily regulate the flux of water via stomatal control since they vegetate in conditions that vary greatly over the course of the year, whereas deciduous vegetation has a shorter period for its vegetative processes and thus cannot excessively limit stomatal opening. These differences are reflected in the different strategies that are observed in nature between water-spending species and water savers. It is interesting to note how the properties of energy dissipation, observed on the level of species and of individual foliar organs, are also reflected in the ecosystem as a whole, thus confirming the existence of functional properties on this level in biosphere–atmosphere relationships.

Climatic Implications on a Regional Scale

In light of the previous discussion, when we observe a given parcel of the terrestrial surface in terms of energy utilization, we find ourselves facing a mosaic of different territorial parts that reveal just as many ways of exchanging energy with the surrounding atmosphere. In this section, I analyze how energy dissipation at the level of terrestrial surface, and then at an ecosystem level, can affect regional and eventually global climate.

As mentioned previously, in the past there was a one-way view of the interactions between the biosphere and the atmosphere. In other words, the effect of climate on the life of biological organisms was taken into consideration, but no thought was given to the possibility that, in turn, the latter could also play an important role with regard to climate. It has recently been acknowledged that the effects of the characteristics of terrestrial surfaces and of the functional properties of the ecosystems comprising it are important in terms of climatic processes. Today, global circulation models that include surface characteristics and energy exchanges at an ecosystem level are becoming increasingly widespread in simulations of the climate of our planet (Sellers *et al.,* 1996). Additional studies have demonstrated that soil changes have an important effect on regional climate (Pielke *et al.,* 1991) and that there are systems in which the impact of vegetation on regional climate is so important that it determines the living conditions of the entire ecosystem.

An extensively studied case of the feedback between vegetation and regional climate is the Amazon basin (Schukla *et al.,* 1990), which is especially well-known because of the deforestation activities it undergoes every year and because of the impact that these activities have on the planetary carbon balance and on biodiversity. Nevertheless, a little known but extremely important aspect involving life in the Amazon ecosystem is the damage that deforestation is causing to the hydrological balance of the region. From a hydrological standpoint, the Amazon system is a closed system. Approximately 70% of its precipitation consists of water vapor that is evapotranspired from the vegetation. As a result, if the plant surface is reduced, precipitation will also decrease, thus altering the vital conditions of many plant and animal species adapted to the tropical wet climate of the forest. This trend has been predicted by climate simulation models when the effects of plant evapotranspiration are also incorporated. These results were subsequently confirmed through the observation of a decrease in precipitation over the Amazon basin during the past 20 years.

Another very interesting case with regard to ecological implications is the region of Siberia. The central regions of Siberia represent some of the most continental areas on the planet. In these conditions, vegetation can play a decisive role in terms of regional climate. This territory, which stretches for thousands of kilometers from the Ural Mountains to the Lena River, is dominated by alternating bogs and forests. Moreover, the latter are renewed as a result of frequent fires due to natural causes, such as lightning originating from the formation of convective clouds. In this mosaic of different ecosystems, the bogs and forests present different energy-dissipation behaviors. During the summer, which is characterized by high temperatures (often over 28°C) in these continental climate regions, the ratio between sensible heat flux and latent evapotranspiration heat is equivalent to about 1 in the bogs. This means that the available energy is equally distributed between the two processes. Instead, the forests have a ratio of about 4, which means that available energy is dissipated mainly in the form of sensible heat, whereas evapotranspiration is a very small component. Thus, the forest canopy generates a sensible heat flux that heats the atmosphere over it, triggering convective motion that transports considerable amounts of water evaporated from the bogs and thus gives rise to the formation of thunderclouds and precipitation (Fig. 5). Isotopic analysis of the rainwater shows that most of it is formed after the condensation of water vapor transpired by the plants. Therefore, this is a closed cycle.

Nevertheless, the most interesting aspect of the hydrological cycle in this region is the importance of the thunderbolts that are generated as a result of convective phenomena. The fires they cause make it possible to mineralize the organic matter present in the soil and enable renewal of these forests, which are normally poor in nutrients. This example shows how physical and biological phenomena are closely connected in a complex network of relationships. Each disturbance that modifies the distribution of the ecosystems throughout the territory, intervening in the alternation of bogs and forests, can have a significant impact on regional climate and thus also on the ecological succession phenomena that are the basis of the evolution of the vegetation in this region.

Primary Productivity and the Carbon Cycle

In the previous section, I showed that in terrestrial ecosystems most of the energy is utilized in the processes of evapotranspiration and sensible heat transport (sensible

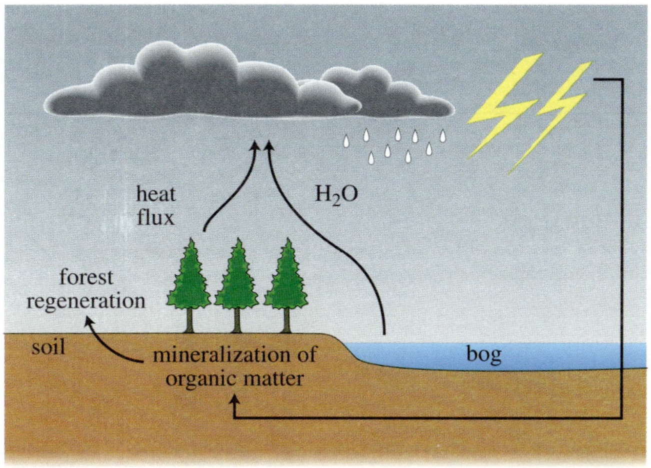

FIGURE 5 Diagram of the interactions between the forests and bogs and the atmosphere in the region of central Siberia and the implications on the formation of precipitation and the regeneration of boreal forests.

heat). It is through these energy fluxes that the terrestrial ecosystems play a fundamental role in the interactions between the biosphere and the atmosphere, affecting both regional and global climates.

Although only a small part of the available energy (about 2%) is expended in the synthesis processes of organic matter, these processes represent an essential mainstay for the life of all living beings since primary production constitutes the fundamental cornerstone for sustaining the trophic chains of the ecosystem. Nevertheless, the role of carbon metabolism by terrestrial ecosystems in the global carbon cycle and the impact of carbon dioxide concentrations in the atmosphere have not been studied extensively.

To describe these processes, first we need to recall some of the definitions of primary production. Gross primary production represents the amount of organic matter that is synthesized during photosynthesis, whereas net primary production is what remains of gross production less plant respiratory processes. Net primary production of the ecosystem is the part of the net primary production that remains once losses due to the respiration of heterotrophic organisms have been subtracted, including the decomposition of organic matter. These definitions can be applied to any level of the ecosystem and can be considered on an annual basis. However, many processes that affect primary production are carried out on a timescale that can range from decades to centuries. For example, fires, natural catastrophes such as flooding and windstorms, and biotic factors such as attacks by insects can disturb the ecosystem and have a considerable effect on primary production. These phenomena are particularly significant if they are considered over a vast spatial scale. If we consider the biosphere as a whole, for example, we will find very productive forests, other forests that have been exploited by man, and yet others that are subject to natural disturbances. As a result, a new concept has been introduced — the concept of net primary production of the biosphere, which represents the overall balance of biospheric carbon, including all the factors that comprise it.

If we observe the carbon cycle in all its components (Fig. 6), we can see that the major reserves of this element can be found in fossil sediments, which contain approximately 40,000 Gt (1 Gt = 10^9 t) of carbon, of which about 4000 Gt can be utilized for fossil fuels. The oceans contains about 38,000 Gt, equivalent to nearly 51 times the contents in the atmosphere. On land, the primary carbon reserve is composed of the soil, which contains 1500 Gt, whereas only slightly more than one-third of this amount can be found in the aboveground biomass. The atmosphere has the smallest reserve of carbon (about 750 Gt).

The most interesting aspect of the global carbon cycle is represented by the carbon fluxes between the various compartments. Between land and the oceans, there is a transport of dissolved organic carbon in river water of about 0.4 Gt/year. An equivalent flux is recorded in terms of inorganic carbon. However, the most important carbon fluxes are the ones occurring between the oceans and the atmosphere and between land and the atmosphere. The carbon exchange between the oceans and the atmosphere is approximately 92 Gt/year in the direction of atmosphere → oceans and about 90 Gt/year in the opposite direction. Thus, the oceans act as an element that can absorb atmo-

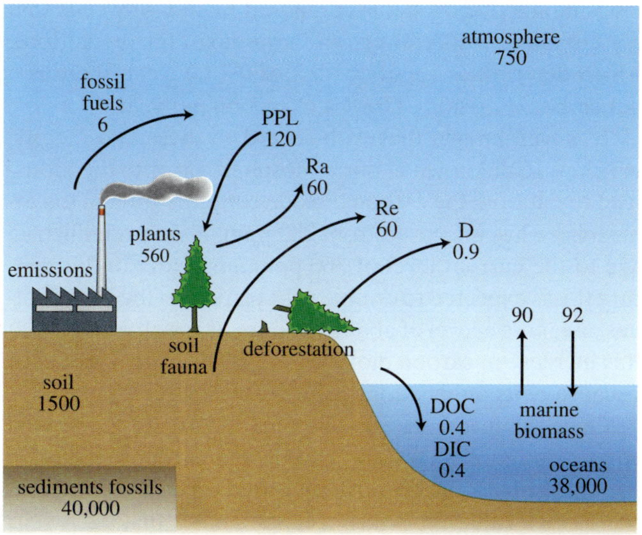

FIGURE 6 Diagram of the carbon cycle. The compartment values are expressed as Gt of carbon, whereas flux values (indicated with arrows) are in Gt/year. PPL, gross primary production; Ra, autotrophic respiration; Rh, heterotrophic respiration; D, deforestation; DOC, dissolved organic carbon; DIC, dissolved inorganic carbon.

spheric carbon, mostly due to the ability of carbon dioxide to dissolve into surface water and be deposited in the form of carbonates at the ocean depths.

The carbon exchanges between land and the atmosphere are characterized by several components, which are also the ones that are most influenced by human activities. Each year, the terrestrial ecosystems exchange 60 Gt of carbon with the atmosphere. This value represents the net primary production of the terrestrial ecosystems and it is the result of the difference between 120 Gt of gross primary production and 60 Gt released by plant respiration. The decomposition of organic matter releases approximately 60 Gt of carbon each year so that, in principle, the terrestrial ecosystems should be at equilibrium in terms of the carbon balance: Net primary production is consumed entirely by decomposers, and as a result the net primary production of the ecosystem is zero. However, it will soon be shown that the statement according to which the terrestrial component of the biosphere is at equilibrium must be reanalyzed based on new observations.

The planetary carbon balance also presents two unexpected factors that depend largely on human activities: the use of fossil fuels and tropical deforestation. Each year, the use of fossil fuels introduces about 6 Gt of carbon into the atmosphere, with a progressively growing trend. Of all the components connected with the carbon balance, the one tied to the exploitation of carbon fuels is the best documented and it continues to expand. Tropical deforestation is responsible for releasing about 1.6 Gt of carbon into the atmosphere each year. Considering the partial regrowth of the areas undergoing deforestation (which are nevertheless simplified ecosystems composed mainly of weeds that are completely different from the preexistent vegetation), which determines a carbon sink of 0.7 Gt/year, there is a net carbon emission of 0.9 Gt of carbon per year.

It is well-known that with regard to these two factors, which were unknown in our atmosphere prior to the industrial revolution, the carbon dioxide concentration in the atmosphere has increased from 280 ppm in the preindustrial age to the current level of 360 ppm, and early in this century this is expected to double with respect to the initial values, meaning a level of about 560 ppm. It is well-known that this increase in carbon dioxide is a matter of great concern because of its implications in terms of climatic changes.

Once the incoming and outgoing fluxes and the annual concentration changes are known, the carbon balance in the atmosphere can be verified. Based on data collected in the 1990s (Table 2), it has been estimated that each year approximately 6.9 Gt of carbon is released into the atmosphere, whereas the oceans can absorb about 2 Gt. Based on the concentration measurements, however, it has been ascertained that only 3.2 Gt of carbon enters the terrestrial atmosphere each year. Thus, we need to discover where the other 1.7 Gt/year of carbon goes that is unexplained by the observations and flux estimates. In other words, we need to identify the missing carbon sink, i.e., the specific factor that seems to oppose the increase in carbon dioxide concentrations in the atmosphere, slowing down the potential consequences of the greenhouse effect on climate.

TABLE 2

Carbon Flux Balance

PROCESS	CARBON FLUX (Gt/YEAR)
Emissions	
Fossil fuel	6.0
Deforestation (including regrowth)	0.9
Total emissions	6.9
Sink	
Atmosphere (increase in carbon dioxide)	–3.2
Oceans	–2.0
Total sink	–5.2
Missing sink	1.7

An initial hypothesis formulated to explain the unbalance that has been observed is the possibility that the oceans actually absorb more carbon than has been estimated. However, new measurement campaigns, and the study of ^{14}C distribution on the surface of the oceans, show that high carbon absorption rates are not possible under current conditions. In fact, although the carbon dioxide exchange with the oceans is not limited by surface diffusion, since carbon dioxide dissolves easily in water, the absorption is impeded by the difficulties with which surface waters mix with the deeper waters, in which carbon dioxide could effectively be stored. The other hypothesis, which has been confirmed by recent studies on the seasonal fluctuations of oxygen and nitrogen isotopes in the atmosphere, is that as opposed to what was predicted by classical theories, the terrestrial ecosystems are not in equilibrium with regard to the carbon balance and instead they represent a carbon sink. This would mean that the net primary production of the biosphere is not zero but positive.

Estimates of the net primary production of ecosystems are now available on a global scale through the data collected by the International Biosphere Program. These data, however, which have been limited mainly to the aboveground component, cannot explain the missing carbon sink in the biosphere since the net carbon balance is represented by the primary net production of the ecosystem, also including the decomposing processes of organic matter. Only recently has the development of new measuring methods made it possible to obtain direct measurements of the net carbon flux with the atmosphere and thus to derive estimates on the role of the terrestrial ecosystems on the carbon balance.

The eddy covariance measurement technique (Baldocchi *et al.,* 1996; Moncrieff *et al.,* 1997) is based on the correlation between the fluctuations in the vertical speed component of wind speed and the concentration of carbon dioxide in the atmosphere near the plant surface. The net flux (F) that crosses the surface unit in a given time unit is given by the time average of the product of wind speed (w) multiplied by gas concentration (c), in this case carbon dioxide:

$$F = \overline{wc} \quad [14]$$

where the bar represents the mean value over time.

Since air motion in the atmosphere is turbulent, both the vertical wind speed and the concentration can be broken down into a mean component and a fluctuating one ($\overline{w} + w', \overline{c} + c'$) from which

$$F = \overline{wc} + \overline{w'c'} \quad [15]$$

Moreover, since we are allowed to assume that there is no mean upward motion of the atmosphere, meaning that $\overline{w} = 0$, then eddy flux can be calculated using only the fluctuating components:

$$F = \overline{w'c'} \quad [16]$$

From an operating standpoint, it is necessary to have suitable instruments that can measure the vertical wind speed component and the carbon dioxide concentration in the atmosphere over the plant cover fast enough (about 20 times per second) to capture the rapid fluctuations comprising eddy flux. The vertical wind speed component is measured with a special anemometer that works using ultrasound pulses, and it can break wind speed down into its three spatial components. Carbon dioxide concentration is measured with an analyzer operating on the principle of the gas's absorption of infrared radiation. In fact, like all greenhouse gases, carbon dioxide shows a strong absorption band in the infrared region of the spectrum. These instruments are placed over the plant surface, and on an hourly basis they yield an ongoing measurement of the net carbon dioxide flux between the vegetation surface and the atmosphere. The advantage of this method is that the net flux value refers directly to the net primary production of the ecosystem integrated over the measurement area, which is approximately 100 ha.

Figure 7 offers an example of daily data for a beech forest in the central southern Apennines (Valentini *et al.,* 1996). It can be seen how the net carbon flux has a characteristic daily trend in response to light and thus to photosynthetic processes. The negative values during the day indicate a carbon dioxide flux toward the surface. Instead, at night the flux becomes positive: This means that it is directed from the surface to the atmosphere, indicating a loss of carbon dioxide during the respiratory processes.

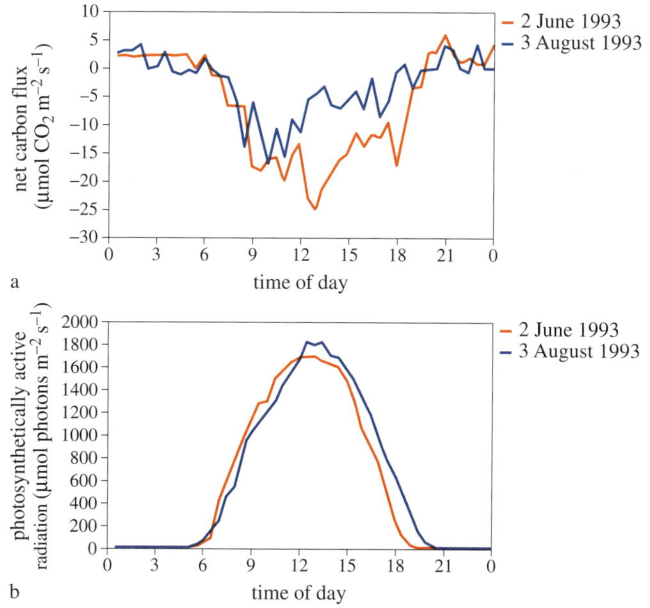

FIGURE 7 (a) Daily trend of the net carbon exchange between forests and the atmosphere over the course of 2 days. The negative values indicate absorption, and positive values indicate emission. (b) Trend of photosynthetically active radiation expressed in micromoles of photons per unit of surface and of time at different times of the day.

These observations can be repeated for several days, as shown in Fig. 8. This allows us to observe the dynamics of the photosynthetic and respiratory processes of the beech forest over the course of the season. During the winter, the forest, which has no leaves, is a source of carbon dioxide. In other words, the respiratory processes dominate over surface gaseous exchanges. As leaf development progresses, the carbon sink as a result of photosynthesis increases gradually and the daily balance of the ecosystem undergoes an overall shift in favor of carbon sink. During the summer, the rates are constant, with photosynthetic absorption that exceeds losses due to respiration. In autumn, with the decrease in foliar surface the losses due to respiration become more important and the ecosystem as a whole changes from a sink to a source of carbon dioxide going toward the atmosphere. By calculating the annual net carbon dioxide flux, we can assess if the ecosystem acts as a source or a sink of atmospheric carbon. This analysis (Valentini *et al.,* 1996) shows that, on an annual basis, the beech forest which was analyzed acts as a carbon sink element, with an accumulation capacity of approximately 432 g m^{-2} of carbon per year.

At the end of the 1990s, a international initiative called FLUXNET was initiated (Valentini *et al.,* 1998b). Various

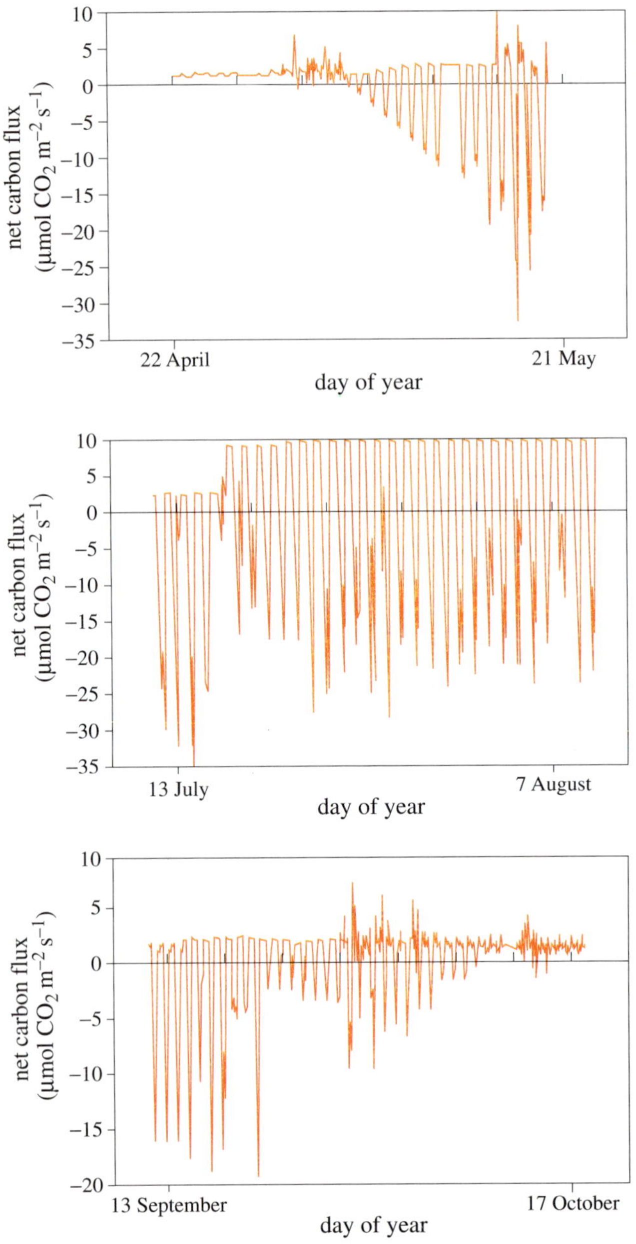

FIGURE 8 The trend of the net carbon flux for a beech forest at three times of the year (from top to bottom: approximately May, July, and mid-September to mid-October).

projects in different countries throughout the world are connected with this project, the goal of which is to study the capacity of terrestrial ecosystems to sink or release atmospheric carbon. Figure 9 presents a summary of some of the data on various European forests. As can be seen, some of the forests are sources of carbon dioxide (Scotch pine in Belgium and spruce in Sweden), whereas others, particularly in the Mediterranean area, show a significant atmospheric carbon sink.

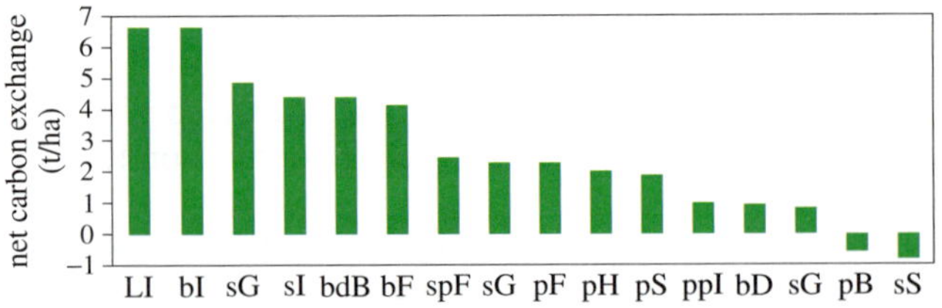

FIGURE 9 Distribution of ecosystem net carbon exchanges for various European forests, expressed in tons of carbon per hectare per year. The positive values indicate absorption of atmospheric carbon, whereas negative values indicate emission. LI, ilex forest, Italy; bI, beech forest, Italy; sI, spruce forest, Italy; sG, spruce forest, Germany; bdB, mixed beech and Douglas fir forest, Belgium; bF, beech forest, France; spF, sea pine forest, France; pF, Scotch pine forest, Finland; pH, Scotch pine forest, Holland; pS, Scotch pine forest, Sweden; ppI, poplar plantation, Iceland; bD, beech forest, Denmark; pB, Scotch pine forest, Belgium; sS, spruce forest, Sweden.

These data as a whole show a latitudinal distribution of carbon dioxide exchanges. In particular, it can be seen that the forests in the boreal regions have net annual carbon exchange values that are much lower with respect to those of forests in the temperate zones. This particular behavior can be explained by the importance of respiratory processes in the various ecosystems. In the boreal zones, the respiratory component of the carbon balance is relatively higher than that of temperate ecosystems. This is explained by the fact that in the boreal regions, the amount of organic carbon in the soil is much greater, especially in terms of the volatile component: Despite the fact that the temperatures are lower, the respiratory flux is relatively higher with respect to the photosynthetic processes. In addition, the respiratory component is substantially affected by the availability of water resources that determine a decrease in respiration in the ecosystems in temperate zones.

These observations nevertheless demonstrate that despite the fact that experimental evidence on a global scale suggests the presence of a significant carbon sink by terrestrial ecosystems, its geographical localization is still open to debate. In any event, the observations that have been made through new methods of studying carbon fluxes in the ecosystems confirm the importance of the biological component in the carbon cycle and, more generally, in regulating the chemical composition of the atmosphere and of the processes underlying our climatic system.

References Cited

Baldocchi, D., Valentini, R., Running, S., Oechel, W., and Dahlman, R. (1996). Strategies for measuring and modelling carbon dioxide and water vapour fluxes over terrestrial ecosystem. *Global Change Biol.* **2,** 159–168.

JARVIS, P. G., and MCNAUGHTON, K. G. (1986). Stomatal control of transpiration: Scaling up from leaf to region. *Adv. Ecol. Res.* **15,** 1–49.

LEVINE, J. S. (1989). Photochemistry of biogenic gases. In *Global Ecology: Towards a Science of the Biosphere* (M. B. Rambler, L. Margulis, and R. Fester, Eds.), pp. 51–74. Academic Press, Boston.

LOVELOCK, J. E. (1979). *Gaia: A New Look at Life on Earth.* Oxford Univ. Press, Oxford.

MONCRIEFF, J., VALENTINI, R., GRECO, S., SEUFERT, G., and CICCIOLI, P. (1997). Trace gas exchange over terrestrial ecosystems: Methods and perspectives in micrometeorology. *J. Exp. Bot.* **48,** 1133–1142.

MONTEITH, J. L. (Ed.) (1975). *Vegetation and the Atmosphere, 1.* Academic Press, London.

PIELKE, R. A., DALU, G., SNOOK, J. S., LEE, T. J., and KITTEL, T. G. F. (1991). Nonlinear influence of mesoscale land use on weather and climate. *J. Climate* **4,** 1053–1069.

SCHUKLA, J., NOBRE, C., and SELLERS, P. (1990). Amazon deforestation and climate change. *Science* **247,** 1322–1325.

SELLERS, P. J., *et al.* (1996). Modelling the exchanges of energy, water and carbon between continents and the atmosphere. *Science* **275,** 502–509.

VALENTINI, R., DE ANGELIS, P., MATTEUCCI, G., MONACO, R., DORE, S., and SCARASCIA MUGNOZZA, G. T. (1996). Seasonal net carbon dioxide exchange of a beech forest with the atmosphere. *Global Change Biol.* **2,** 199–208.

VALENTINI, R., BALDOCCHI, D. D., and TENHUNEN, J. (1998a). Ecological controls on land–surface atmospheric interactions. In *Integrating Hydrology, Ecosystem Dynamics and Biogeochemistry in Complex Landscapes* (J. Tenhunen and P. Kabat, Eds.). Wiley, Chichester, UK.

VALENTINI, R., BALDOCCHI, D., and RUNNING, S. (1998b). The IGBP–BAHC global flux network initiative (FLUXNET): Current status and perspectives. *Global Change Newslett.* **28,** 14–16.

WALTER, H. (1973). *Vegetation of the Earth.* Springer-Verlag, New York.

WEBB, R. S., ROSENZWEIG, C. E., and LEVINE, E. R. (1993). Specifying land surface characteristics in general circulation model. Soil profile data set and derived water holding capacities. *Global Biogeochem. Cycles* **7,** 97–108.

XUE, Y., and SHUKLA, J. (1993). The influence of land surface properties on Sahel climate, Part I: Desertification. *J. Climate* **4,** 345–363.

ALMO FARINA

Lunigiana Museum of Natural History
Aulla, Massa, Italy

Landscapes and Their Ecological Components

Landscape ecology is a discipline that studies environmental complexity, concentrating mainly on an analysis of the importance of spatial relationships between the various components (individuals, populations, communities, and land mosaics) of the real world. Defined as the spatial representation of an ecosystem, a landscape is composed of many superimposed environments as perceived by various species or modified by ecological processes such as flooding and fires. The way in which an organism perceives the environment that it inhabits is species specific, which allows us to talk of a geobotanical domain, an animal one, and an anthropic one. These complex landscapes are measured spatially by geographic information systems, and the resulting data are analyzed using a combination of geostatistics and fractal mathematics. Landscape ecology is used mainly for the management of human-scale landscapes and, more specifically, for the analysis, management, and conservation of nature; these offer the most fruitful application of the knowledge that landscape ecology provides about the workings of land mosaics.

Introduction

During the past few decades, it has become increasingly clear that in order to acquire a more thorough understanding of the complex processes that take place in the biosphere and, more generally, in the ecosphere, the spatial dimension of ecosystems must be broadened. An appreciation of the extent of the complexity of life on Earth has become widespread among both scientific researchers and the general public, and the globalization of environmental problems has made it necessary to broaden the scale of the analysis and of the interpretation of the phenomena involved. The broadening of the spatial scale has also led to the reassessment of the temporal one and, as a result, to the creation of a powerful and flexible analytical tool in which the spatial and temporal scales can be calibrated according to the system being considered. Ecological research, which until the beginning of the 1980s was rooted in concepts of environmental homogeneity, of equilibrium within the ecosystem, and of the separation of man from environmental processes, has undergone a profound change during the past decade. Indeed, when the spatial scale was broadened, ecological research began to view the environment as a heterogeneous system in which the variety of ecological processes are expressed as a state of dynamic equilibrium (*homeoretic flows*) and in which the processes governed by man become an intrinsic part of environmental mechanisms. Landscape ecology studies environmental complexity, concentrating mainly on an analysis of the importance of spatial relationships between the various components of the real world.

Heterogeneity appears to be the predominant pattern in

most landscapes, and increasing importance is attached to the role that it plays in determining ecological processes. A landscape contains heterogeneous characteristics that are expressed in discrete entities known as *patches,* which in turn make up a mosaic that is structurally and geographically distinct. The salient characteristics of these patches, such as dimensions, shape, type of vegetation, biological richness, abundance of organisms, and flow of nutrients, contribute to the organization and maintenance of the complexity of landscapes.

Landscapes

A *landscape* is defined as the spatial representation of an ecosystem. Thus, rather than an entity being described entirely in terms of its function (an ecosystem), it is defined in terms of its spatial characteristics (a landscape). The concept of landscapes represents a relatively new approach and one that is also rich with significance. Indeed, as the geographic dimensions are expanded, ever more processes, together with the increased biodiversity they bring, are included. A landscape, in fact, is a system of environments, each one of which is species specific and delimited by the species. Each species recognizes the boundaries of the environment it inhabits, identifies those characteristics that favor its existence, and reacts to the environment on a species-specific scale. A landscape is therefore the sum of many species-specific systems that can be superimposed to a varying degree and that contribute to the complexity of the environment. This vision of the environment is characterized by a large quantity of information that results in a set of retroactive mechanisms that help to regulate and calibrate the various processes and that make it possible for many organisms to survive and to perform multiple roles. All this tends to promote an autopoietic capacity, or the "creative" ability of the system to adapt to new conditions through homeoretic and homeostatic mechanisms.

When studying a landscape, it is important to be able to consider the components that are of immediate interest without losing sight of the whole. Such an approach is necessarily transdisciplinary, and analyzing a landscape does much to increase our understanding of the ecological complexity of a geographic area. A landscape can vary in size from a few centimeters to tens of kilometers. In fact, the term immediately brings to mind a large area, and in almost all Western cultures it is used to denote a scenographic or geographic reality. Nevertheless, every species perceives its own particular landscape in its own species-specific way, which can even change according to the physiological context in which that organism finds itself at any moment. The landscape is therefore perceived by man as his own environment. Consequently, the study of the way in which it functions is called *landscape ecology;* indeed, this dimension is the nearest to the complex processes that, through his own culture, man has applied to a natural substratum.

As a result of the variety of cultural and scientific approaches, there are many definitions of a landscape as perceived by man. It is the general character of a region (Humboldt, 1807); it is a heterogeneous area composed of a cluster of interacting ecosystems that reoccur in a similar manner throughout a region (Forman and Godron, 1986); it is the sum of physical, ecological, and geographic entities that embrace and bring together all natural and human processes and their patterns (Naveh, 1987); it is a particular configuration of topography, plant cover, use of the soil, and human settlements that are delimited by natural and cultural processes as well as by anthropic activities (Green *et al.,* 1996); and W. Haber (1996, personal communication) defines a landscape as an area of land that we perceive without paying particular attention to any one component that seems familiar to us.

Currently, most landscape ecology is concerned with landscapes that have been modified by man, even though its analytical principles and instruments can be validly applied to the study of landscapes that have not been subjected to the direct influence of man. Landscape ecology therefore studies complex systems and indeed often requires that special reference be made to a particular organism or group of organisms (Turner *et al.,* 1995). Consequently, a landscape perceived by human beings is quite different from a landscape perceived by coleoptera (Wiens and Milne, 1989) so that when one talks of a human landscape one must necessarily consider the cultural component that is associated with it.

The Contribution of Other Disciplines to Landscape Ecology

Precisely because landscape ecology is a complex discipline, its theoretical corpus is the result of bringing together and integrating a series of approaches and applications, among which the theory of island biogeography (MacArthur and Wilson, 1967) and geographical ecology (MacArthur, 1984) occupy preeminent positions. These theories were the first to emphasize the extent to which the degree of isolation and the dimensions of an island determine species richness and abundance. Islands that are small and distant from the mainland have a smaller number of species compared to those that are larger and nearer to the mainland. Furthermore, there is an inverse correlation between the risk of extinction and the probability of colonization. The smaller the island, the greater the risk of extinction, whereas the probability that another species will colonize it is correspondingly less. This theory can also be usefully applied to so-called *islands on land,* environments that due to the process of fragmentation have been reduced in size and have been surrounded by other environments that are hostile for given species.

Landscapes and Heterogeneity

A landscape consists of two fundamental components: the geographic or chorological component and the topological or functional one. In fact, these two components cannot be separated without drastically reducing the amount of information that is obtained when they are considered together. By definition, landscapes are heterogeneous environments: That is, they consist of various elements that are interrelated in different ways. This heterogeneity is a very important pattern since it in fact determines the ecological diversity of any region. The heterogeneity might be expressed as physically identifiable structures, such as woods and land under cultivation, or as processes such as variations in the thermal characteristics of slopes or the degree of the environment's species-specific suitability. At any rate, the degree of heterogeneity varies according to the spatial arrangement of the single component parts. Space is therefore an important component in determining the extent of the diversity of the various forms of life. It follows that the spatial arrangement of the various entities and their processes and relevant patterns can be considered as one of the main paradigms for introducing the spatial dimension as an element that is important per se. The presentation of landscapes as components in a "nest" hierarchy (one in which the highest level contains all the other levels) that incorporates the concept of scale has been a determining factor in linking the various paradigms (Allen and Starr, 1982; O'Neill *et al.,* 1986) and the various theories (Delcourt and Delcourt, 1988).

At the same time, a more satisfactory approach to dealing with complexity has been provided by non-Euclidean, fractal geometry (Mandelbrot, 1975), which is able to measure the complexity and the relative patterns of many natural systems and to establish relationships between elements that are only distantly connected. New theories of heterogeneity (Kolasa and Pickett, 1991) and the role of disturbance regimes in ecological processes (Pickett and White, 1985) represent significant steps forward and thoroughly incorporate the paradigm of ecotones (Hansen and Di Castri, 1992) and the processes associated with them, such as connectivity and connection (Merriam, 1984) and metapopulation models (Gilpin and Hanski, 1991), into landscape ecology. The recognition of the existence of a heterogeneous structure in landscapes has also made it possible to extend and adapt Pulliam's (1988) source and sink demographic model and to redefine the roles played by single patches in a landscape.

The Study of Landscape Ecology

The complexity of the issues they deal with, together with the fact that they make reference to a range of disciplines, has led to landscapes being viewed as entities that change character according to the approach that is adopted. The main texts on landscape ecology (Forman and Godron, 1986; Naveh and Lieberman, 1994; Forman, 1995; Zonneveld, 1995) have drawn ample attention to this characteristic while nevertheless expressing very different and often divergent visions of the discipline. Indeed, whereas Naveh and Lieberman have an anthropocentric, globalizing vision, that of Zonneveld is more geographical and concerned with typological classification and the approach of Forman and Godron is more geometric and views space on a large scale as the central factor that conditions processes.

It is time to reconcile and integrate the various concepts of what a landscape is, bearing in mind the different biological components (Farina, 1998). Historically, this discipline derived from the human geographical sciences, and the first description of families of patterns and processes was strictly linked to the anthropic sphere. Recent studies, mainly in North America, on the spatial arrangements of patterns and on processes that concern the soil, vegetation, and the animal component apart from the anthropic one, have resulted in a huge leap forward. There are indeed three perspectives to be considered in landscape ecology: the human one, the geobotanical one, and the animal one.

The *human perspective* allows us to dismantle and reassemble a landscape according to the functional entities that are relevant to human beings. The *geobotanical perspective* considers the spatial distribution of biotic and abiotic components, from the landscape of the soil to that "perceived" by plants and to the distribution of vegetal entities such as woods and grassland. Where plants are concerned, "perception" is taken to mean the ability to absorb information from the surrounding environment. This is directly related to the extent of adaptation, to the rate of colonization and extinction, and to the resistance to natural and man-made stresses.

The *animal perspective* considers the way in which each species perceives its own environment. It is obvious that a butterfly cannot use the same environmental parameters as a deer or a fish. Its life will therefore take place in and impact a precise part of a landscape. The combination of the interactions expressed by an animal community constitutes an important conditioning factor of the landscape seen as a complex whole. Man, performing the double role of biological species and intelligent organism, should also be considered in this context.

These perspectives are not in conflict because each one of them is concerned with exploring a domain of patterns and processes that are in fact components of the entire biological and ecological system. Furthermore, there are many more points in common, such as the perception of space and the spatial arrangement of patterns and processes, than there are differences. Thus, the common strategy adopted by landscape ecologists is to insert these patterns and processes into the most suitable spatial and temporal scale.

The human dimension of a landscape is probably the most complicated in that it superimposes man's animal and cultural components. This dimension relates to processes that have a very broad spatial and temporal scale. The biological dimension of man can be compared to that of animals, but the cultural component of humanity is unique, and it is this component that has the greatest interactive impact on a landscape, especially since man is inclined to use technologies that overcome the physical and biological limits of the natural systems in which he operates.

The three perspectives outlined previously cannot be considered separately, but their sectional nature is linked to the ability to adopt an approach that is interdisciplinary or even transdisciplinary. A landscape is so complex that even this approach does not provide explanations for all the processes; but it can shed new light on this complexity and define the limits of its own ability to provide explanations.

Landscape ecology is therefore one of the most promising ecological disciplines, and although it is highly differentiated and wide ranging, it is capable of expressing a single "spirit" linked to the finite dimension of the area of study. This discipline allows a more precise superimposition of information from the real world onto the virtual processes that today's postindustrial culture is promoting through globalization.

Currently, there is a high risk of landscape ecology being considered entirely from the anthropocentric point of view, and this might result in its being seen as a dogmatic discipline without either theoretical foundations or experimental backing. Furthermore, it seems unacceptably reductive to consider landscape ecology simply as a form of broad-scale ecology. In fact, one of the major strengths of this discipline is its ability to transfer information through various families of processes that take place on different spatial and temporal scales and to link knowledge from precise ecological research to environmental and landscape planning activities.

Theories and Models Embraced by Landscape Ecology

The hierarchical theory and the percolation theory combine with two models of population dynamics — the metapopulation model and the source and sink model — to create a powerful referential context for landscape ecology. Although they were formulated in various contexts, these theoretical elements share a common interpretation of the complexity of landscapes. They have played an undeniably important role in creating a homogeneous disciplinary corpus that offers a paradigmatic bridge over the complexities of landscapes.

The Hierarchical Theory and the Structure of Landscapes

The hierarchical theory (Allen and Starr, 1982; O'Neill *et al.*, 1986; Allen and Hoekstra, 1992) is a useful instrument for exploring numerous patterns and processes through various scales in space and time. Considering complexity as an attribute that is intrinsic to a landscape, the hierarchy paradigm explains how the various components located on certain scales enter into contact with other ones that are visible on different scales of resolution. The hierarchical theory views a system as a component in a larger system that consists of subsystems. As one moves from one system to another, the characteristics of the phenomena change — for example, the classification of a landscape as one goes from lower to increasingly higher levels in the hierarchy: *ecotope* (the basic unit in a landscape consisting of biotic and abiotic elements); *microchore* (the spatial distribution of ecotopes); *mesochore* (the environmental system composed of a group of microchores); *macrochore* (a mosaic of landscapes); and *megachore* (a group of geographical elements covering several kilometers).

A watershed is an example of a hierarchical system composed of subbasins that in turn are composed of even smaller basins. A system exists independently of its components and is generally able to organize itself and to transmit information; in other words, it is able to exist as a cybernetic system. A landscape exhibits its own type of complexity, and in order to understand it fully it is necessary to focus on a certain organizational level. There are innumerable hierarchical levels and thus an equal number of systems that are nested inside them in one way or another. The behavior of a given subsystem conditions nearby systems both above and below it. The speed with which the processes unfold and thus the scale in time are specific to each level. When going from one level to another, it is therefore necessary to adjust the resolution. The hierarchical theory contributes much, for example, to an understanding of the effects of various disturbance regimes. Thus, fire is a highly destructive element for the "tree system," but it is an element that, ecologically speaking, promotes "creativity" for the forest as a whole in that it establishes new conditions for many species of plants and animals.

The Percolation Theory

The percolation theory was formulated as a result of studies of the behavior of fluids in a medium (Stauffer, 1985) and has recently been used, with interesting results, in the creation of landscape models (Gardner *et al.*, 1987). Unlike diffusion processes in which every particle in a liquid moves in all directions, the percolation process takes place in finite regions of the medium, which the fluid does not leave. When the fluid occupies about 0.5928% of the entire surface of the matrix, it has reached what is known as the percolation threshold (Ziff, 1986); such a state guarantees, for example, that the cells in a matrix will be in contact with each other from one side of the matrix to the other. This theoretical limit is nevertheless an important concept with regard to studying real patterns, such as the expansion of an urban area or the spread of a wood as a result of secondary succession.

The relevance of this theory to the study of the behavior of landscapes is clear if one considers that around the critical threshold, contagion, forest fires, and demographic explosions of pathogenic agents exhibit their initial states of diffusion (Turner, 1987). Furthermore, this theory can be usefully applied to the study of animal behavior in a heterogeneous system. Indeed, the percolation process has been thoroughly tested with neutral models (i.e., square matrices with randomly filled elements). Neutral models assume a state of complete spatial independence between the various cells, and the habitats do not exhibit any autocorrelation. These models consider both the scale of resolution and the behavior of the single cells of the matrix and, for example, simulate the perception that the various species have of the environment that surrounds them. Since the perception of an environment is species specific, it is obvious that each species will perceive it as a percolating habitat (i.e., one that is more or less connected). These models can be used to predict when an environment that is in a state of fragmentation will begin to lose important characteristics, such as connectivity.

The Metapopulation Model

Increasing forest fragmentation is a general trend in natural environments that has created fragments of isolated woodland. For the remaining populations, fragmentation has increased the risk of extinction. It can be assumed that when a population lives in a fragmented and heterogeneous environment, it perceives the landscape in which it lives as a mosaic in which hospitable environments alternate with ones that are more or less hostile, and that contact between the various subpopulations is ensured only through processes of emigration or immigration. The risk of both local extinction and recolonization depends on the extent and existence of an exchange of individuals. The success of recolonization depends on many factors, including the ability to disperse, and these populations are therefore considered to be components of a metapopulation. The term *metapopulation* was first used by Levins (1970) to describe a population of populations (Gilpin and Hanski, 1991; Hanski and Gilpin, 1991). Metapopulations are systems in which the rates of extinction and colonization create flows of individuals that ensure a genetic connection between the subpopulations.

The metapopulation model is closely related to the theory of island biogeography (MacArthur and Wilson, 1967), but it is distinct from it in that the terrestrial systems in which the metapopulations develop are less hostile. In the theory of island biogeography, the distance between the various islands is the main barrier to dispersal, and the distance from the mainland is one of the factors that determine the rates of extinction and recolonization. In the metapopulation model, however, there is a constant relationship between the various subpopulations; as a result, the exchange of individuals and therefore of genetic inheritance is maintained at a high level. The metapopulation model can be usefully applied to fragmented areas in order to analyze their dynamics and predict the probability of extinction or recolonization of isolated populations.

The Source and Sink Model

The source and sink model was created by Pulliam (1988) in order to study the responses to environmental conditions among various species of birds. According to this model, which can be adapted very effectively to heterogeneous environments in which the resources are unevenly distributed, a source population is one in which the ratio between births and deaths is always above unity. In other words, in a source population there is always a surplus of individuals that will tend to leave it. On the contrary, in a sink population the ratio between births and deaths is below unity, and such a population can survive only if new arrivals from outside add to it. This model can also be used to classify patches in a heterogeneous environment according to their ability to sustain a source population or a sink population (Fig. 1). As a general rule, the larger patches in an environment will tend to act as source ones to the smaller patches. Whether a patch is a source one or a sink one will also depend on

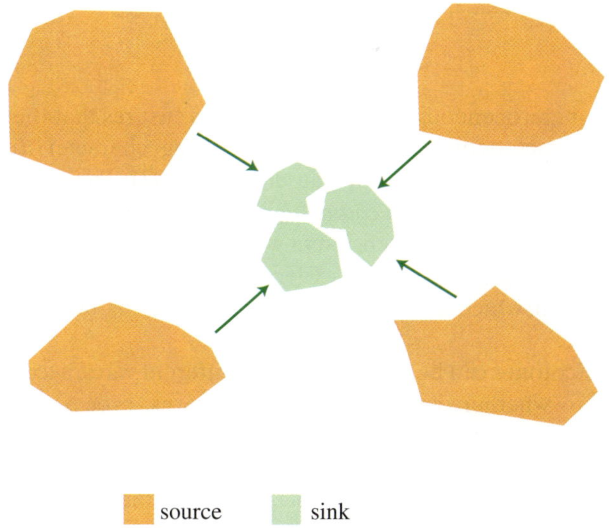

FIGURE 1 The source and sink model is used to analyze the distribution of the populations of a species in a heterogeneous environment not according to their density but to the ratio between births and deaths. A source area is defined as one in which there are more births than deaths, and a sink area is one in which the populations cannot survive because the number of births is less than the number of deaths. The continuing presence of a population in a sink area is guaranteed only by the arrival of individuals from source areas.

the seasonality of the resources and on the role that these resources play in the survival of a species at the various levels of aggregation. Indeed, this model can be usefully applied in fields outside population demography. Thus, if the resources in a given patch are such as to guarantee the survival of a species, that patch can be defined as a source patch.

There are variants of the source and sink model. For example, Watkinson and Sutherland (1995) coined the term *pseudo-sink* to describe an environment that appears to exhibit sink characteristics but in which, even if immigration comes to a halt, the population will stabilize at a minimum level, without disappearing. Furthermore, there are special cases in which environments with sink characteristics are so inviting to various species that they take on the appearance of source environments (Pulliam, 1996). This is the case with many environments that have been modified by man and that offer abundant resources but that do not, for example, guarantee adequate protection against predators or where, in the sensitive period for a given species, man produces a disturbance that interferes with successful reproduction or even makes it impossible. It must be pointed out that this model does not always allow one to deduce the qualities of an environment, especially if the demographic criterion is applied. There are times when a sink environment can be inhabited by more individuals than a source one, and there are populations in which most of the individuals inhabit sink patches.

Ecotones

The heterogeneous nature of landscapes ensures that there is contact between patches with varying characteristics (Pickett and White, 1985; Hansen *et al.,* 1992). These junction zones are called *ecotones.* Clements (1905) described ecotones as zones of tension in which the species of neighboring communities meet. Clements was the first to use the term, which is based on the two Greek words, οἶκος and τόνος.

Ecotones can be defined in many different ways, according to whether one is considering patterns, processes, or fully fledged habitats. They are therefore also habitats that serve as meeting points for species that are able to fully exploit the frontier zone characteristics of these environments (Ricklefs, 1973). In fact, even though the concept of ecotones is simple, their scalar characteristics complicate the task of studying them. For example, the closer one gets to an ecotone, the more difficult it becomes to distinguish what was clearly visible from a distance. Furthermore, the study of ecotones is complicated not only by these scalar characteristics but also by the fact that every biological species perceives and interacts with the environment in a species-specific way, with the result that the position of ecotones varies from species to species (Fig. 2). Thus, ecotones that can be identified by analyzing an aerial photograph or a geological map do not always exist as far as every species is concerned.

FIGURE 2 An ecotone (in red) is a transitional zone between different environments and is perceived in a species-specific way. Thus, there are species that do not perceive an ecotone (a), species that live entirely within it (b), species that see it as a hostile area and look on it as a barrier (c), and species for which the ecotone is part of their habitat (d).

Ecotones are important structures with regard to the functioning of landscapes. Indeed, an exchange of nutrients, water, spores, seeds, and animals takes place within these juncture zones between different patches. The fact that they are transitional zones also makes them excellent indicators of climatic changes linked to the increase in CO_2 and atmospheric pollution. Ecotones also have high rates of primary and secondary productivity since they generally benefit from the various characteristics of their constituent environments. In environments that have been modified by man, ecotones that are composed of copses, hedgerows, rows of trees, and the edges of fields are often refuge areas for many species that are sensitive to the disturbance regime created by man.

Along rivers and in lagoon areas, ecotones are junction zones between terrestrial and aquatic environments, intercepting the flow of nutrients moving from one to the other. The importance of ecotones has been recognized for a long time (Clements, 1897; Livingston, 1903; Griggs, 1914). Although authoritative ecologists have emphasized their importance at various times (Weaver and Clements, 1929; Odum, 1959; Daubenmire, 1968; Ricklefs, 1973), only recently have systematic studies on the subject been carried out (Di Castri *et al.,* 1988; Naiman and Decamps, 1990; Holland *et al.,* 1991; Hansen and Di Castri, 1992; Risser, 1995).

It now seems that the study of landscapes with reference to ecotones is capable of shedding much light on the way in which matter, energy, and individual organisms interact with a heterogeneous environment and thus with the very complexity of landscapes. It is no coincidence that the edges of mosaics have a great influence on the richness and the movement of organisms, on the flow and accumulation of matter and energy, and on the spread of disturbances. In order to understand ecological processes, it is therefore essential to study ecotones, especially when the object of study is landscapes, which by definition are a part of the real world that we have an interest in investigating. Ecosystem ecology, which focuses mainly on the mechanisms that operate within an ecosystem and which analyzes the internal parts of homogeneous areas, differs considerably from landscape ecology, which is concerned with the functioning of ecological systems operating within a geographical context.

It is possible to classify ecotones according to their origins. Holland (1988) suggested four types: ecotones that are created and maintained through human disturbance (hedgerows, plow land, and hay fields); those that are created and maintained by natural processes, such as ecological succession or the bridge between terrestrial and aquatic environments that results from beavers' dams; those that are created by natural processes and maintained by man, such as coastal lagoons or riparian woodland; and those created by man and maintained by natural processes, such as the marshy areas around artificial water reserves.

Generally, there are substantial differences between natural and man-made ecotones. In the former, the transition from one patch to another is gradual, whereas in the latter it is sudden and clearly defined.

Characteristics of Ecotones

Although not easy, it is nevertheless possible to identify the structure and functions of an ecotone. An ecotone covers the whole junction zone between one system and another. Its shape can be linear, as in man-made ecotones, circular, or convoluted, and it will determine the rate of transfer of environmental information, energy, and matter through the ecotone.

The biological structure is determined by the biomass or density of the dominant organisms. The structural contrast is determined by the difference between the structure of the ecotone and that of the neighboring environments. The internal heterogeneity is determined by the variability of the changes within the ecotone. The density of the ecotone is determined by the ecotonal surface area of a given area. There are also other important functional variables such as stability, which is a measure of the degree of the ecotone's susceptibility to change. Resilience, on the other hand, is a measure of the ecotone's ability to return to its previous state following a disturbance.

Both internal and external factors condition the organization of ecotones. The external factors are determined by processes that act on landscapes, and the internal factors represent elements of discontinuity. Many animal species, for example, create conditions that are hostile for others; furthermore, on an intraspecific level, territoriality can display ecotonal characteristics that are similar to a change in the vegetational conditions within a mosaic. Some species of plants modify the pH balance of the soil to such an extent that they create ecotonal conditions that are more or less hostile for other plants. The shade produced by a canopy can reduce the ability of other plants to grow. Many invasive species, such as the Jerusalem artichoke (*Helianthus tuberosus*), grow so densely that they prevent other species from establishing themselves.

Many animal species create ecotones as a result of their activities. An example is a small mountain field mouse, *Microtus nivalis,* that lives in grasslands above the treeline. Its incessant search for food, which does not even stop in winter, has a major impact on the distribution of the bilberry (various species of *Vaccinium*), whose evergreen stems it eats, especially in the cold season (Farina *et al.,* 1986). Similarly, as a result of digging holes and making molehills and nests above ground level, moles, ants, and termites modify the plant cover and even change the microtopography of the soil.

In summary, ecotones can be found both on a megascale, such as a bioclimatic region, and on a microscale, such as an area covering just a few square centimeters of ground. The role that they play in a landscape is fairly clear. They represent semipermeable membranes that allow processes of passive diffusion, such as the movement of air or water masses carrying matter or organisms, and also processes of active diffusion, by which organisms move biomass from one part of a landscape to another, thus modifying the environmental information as well.

There is no doubt that the presence of ecotones contributes to the stability of a system by guaranteeing a complexity that reduces the risk of sudden deterioration and the loss of the system's autopoietic capacity. Thus, a riparian ecotone (consisting of the vegetation along the bank of a river) reduces the effects of flooding by retaining, among other things, large quantities of organic substances that are then used in the food chain. Many organisms, including most amphibians, live only in ecotonal zones. The considerable species richness of the coral reefs, for example, is a function of the ecotonal characteristics of these structures. Furthermore, the greatest concentrations of plankton species are to be found in areas in which large masses of seawater meet.

Animal richness and abundance are greater along ecotones than in neighboring areas, but the linear nature of these environments also increases the risk of predation. In other words, these structures can play a variety of roles, even if by themselves they do not represent paradigms that are fully able to explain the complexity of landscapes or, therefore, the way in which they function. However, if this paradigm is coupled, for example, with the source and sink model or with the hierarchical theory, the combination is capable of yielding a great deal of information.

Cultural Landscapes

Most terrestrial landscapes have been profoundly modified by man, who has built towns, cut down forests, changed the vegetation, constructed roads and motorways, and brought about changes in the climate. In terms of the perspective adopted by landscape ecology, these modifications affect the spatial arrangement of patches, modify the fragmentation of large forest patches, and increase the number of transitional zones or ecotones.

Such modified systems exhibit an increase in both the fragmentation of the original forests and environmental heterogeneity. In these cases, landscape ecology is a discipline that offers an increasingly useful analytical tool in that it allows ecologists to study patterns and processes on different scales, including environments that have been modified by man (Fig. 3). Therefore, this approach has an increasingly important role to play in many forms of human activity that are directly related to the environment and its characteristics. Indeed, landscape ecology is able to study ecological processes over a broad range of spatial and temporal scales that reflect the complexity of the systems. By applying this discipline, it is possible to manage key species, residual forest zones, and edge networks and to influence or decide on urban development policies. Specifically, it can be applied to the problems of many landscapes that, having been modified by man, have acquired a high value in terms of complexity, biodiversity, and scenic beauty. These systems are called *cultural landscapes;* that is, landscapes that have been subjected to a man-made disturbance regime for a long time and in which the character and spatial arrangements of the patches are the result of complex retroactions between man and nature.

Cultural landscapes reflect the interactions between man and the natural environment and are complex phenomena with identities that are both tangible and intangible (Plackter and Rossler, 1995). In 1991, the secretary of the United Nations Educational, Scientific, and Cultural Organization set out a series of guidelines for identifying these landscapes. They generally exhibit a complex structure consisting of a fine-grained mosaic in which the physiotopes, or physical units of the landscape, are markedly localized and are used for specific and appropriate purposes (e.g., agriculture, forestry, and grazing). In upland and mountain areas, the slopes have often been terraced in order to facilitate the working of the land and, at the same time, to reduce erosion and retain nutrients (Fig. 4).

Cultural landscapes might represent a useful model for testing ways of ensuring that man's presence does not destroy resources or cause irreversible disruption to the environment. Currently, such an aim might seem utopian, granted the universal presence of technology, but the lesson that cultural landscapes teach us cannot be ignored and

FIGURE 3 A cultural landscape in Tuscany. Created and structured as a result of close interactions between agriculture, cultural heritage, and environmental dynamics, it requires a constant input of energy (in the form of work) from man.

FIGURE 4 Gullied hills in central Italy — a typical example of processes of erosion acting on a densely cultivated area. These processes, acting as refuge areas, ensure a high degree of biodiversity.

this model should be considered for many regions of the earth. Every cultural landscape is the product of the interaction between man and the environment over a very long period; thus, when human activity decreases, as happens increasingly often with the abandonment of farms in mountain lands, the fragility of this system becomes evident and cultural landscapes lose a part of their complexity. For example, once terraced land is abandoned, the terracing deteriorates rapidly.

Most cultural landscapes are different from each other, but they generally exhibit more linear structures and often have more open spaces compared to undisturbed landscapes. However, in some cases the opposite happens, such as in desert oases, in which the density of palm trees is greater than that in the surrounding desert. Cultural landscapes are generally the result of interaction between the resident population and the natural environment, but even nomadic populations such as the Mongols must have played a major role in shaping the character of the steppes. Entire regions, such as the Mediterranean, can be considered to be cultural landscapes, and the modifications brought about by man can have irreversible effects both on biodiversity and on ecological diversity. Therefore, many people are worried that these important phenomena, which also attract much tourism, will disappear. When evaluating what often appears to be natural and spontaneous, it is important to proceed with care; in regions such as the Mediterranean, biodiversity has been reduced over thousands of years, and there are no more new species to fill the ecological niches left empty by species that failed to adapt to man-made disturbances.

Cultural landscapes are fragile and need to be maintained by man; especially in dry climates, abandonment tends to result in unusually frequent fires, which reduce ecological complexity. Indeed, although most plants display a high tolerance to fire, it is precisely the greater frequency of this disturbance that prevents ecological succession from occurring and consequently increases the risk of erosion during periods of high rainfall.

Ecosystem Management

Currently, neither species management on a site-by-site basis nor ecosystem management using single species are considered to be effective. Landscape ecology does not consider single sites in isolation but rather the interactions between the various components. One of the most useful ways of studying ecosystems is to use a calibrated scale (Fig. 5). Indeed, a landscape can be considered as a geographically defined box containing not only spontaneous patterns and processes but also man-made ones. Watersheds can be considered as landscapes, but it is also possible to identify other functional units, such as ecotopes and micro-, meso-, macro-, and megachores. For example, it seems useful to study landscapes by applying the scale of watersheds, which can be considered as multifunctional units in which flows of water and the transfer of nutrients are distinctive processes.

Ecosystem management can also make effective use of umbrella species — that is, those that condition the functioning of the entire system. Thus, in the prairies of North America, the bison was a key species in that its grazing, trampling, and excrement modified the whole system. A similar role is performed by many ungulates in the savannas of Africa.

When a key species disappears, the whole system collapses and changes. Thus, it was probably the disappearance of the bison that caused the disappearance of many grasses and their subsequent replacement by shrubs.

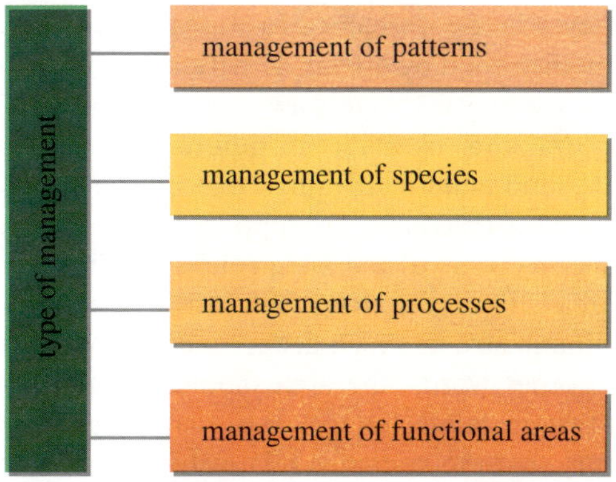

FIGURE 5 Ecosystem management can take place on at least four organizational levels. Each level requires instruments that are attuned to a particular spatial and temporal context.

Principles for the Conservation of Nature Reserves

Various principles deriving both from the theory of island biogeography (MacArthur and Wilson, 1967) and from the combination of paradigms in landscape ecology allow us to outline procedures for the conservation of areas of interest. First, it is important to estimate the minimum surface area involved. A tropical forest, for example, must be large enough to allow biodiversity. In temperate environments, even forests of a modest size support a high degree of biodiversity. Contiguous areas conserve more inner-forest species than disjointed ones, even when the total surface area is the same. In forest environments, patches that are separate but nevertheless near to each other have a greater number of species than patches that are far apart; also, patches that are disjointed but linked by corridors of protected areas are more conducive to biodiversity than patches that are completely isolated.

Granted similar environmental values, circular protected areas are preferable to elongated ones since the latter have a higher ratio between the perimeter and the surface area. These principles draw attention to the importance of surface area, the shapes of the patches, and the connection and the development of the ecotones.

Disturbance Regimes and Conservation

Increasingly, knowledge of the patterns of land mosaics appears to be an important element in the conservation of nature. In particular, large-scale disturbance regimes play a fundamental role in maintaining ecological processes (Baker, 1992). It is clear that disturbance regimes need to take place within large areas so as to affect only a part of the system. Indeed, in the case of fires, it is important that many patches should survive so as to create a shifting mosaic and enable disturbed patches to recover via secondary succession. Climax communities can thus be conserved if they are surrounded by a buffer of younger successional communities.

It is nevertheless unrealistic to hope that large areas will remain undisturbed by human activity, and thus the role of landscape ecology is to study a whole range of variables in order to evaluate the ability of different mosaics — forest remnants, land under cultivation, river systems, and urban areas and infrastructures — to support a reasonable degree of biodiversity. One of the main rules in delimiting a protected area is to ensure that it will include a zone of a certain minimum size in which disturbance dynamics can operate. In other words, the area should contain groups of patches that are subjected to a disturbance regime and that will ensure that the system is relatively stable in the short term. This should result in the creation of a shifting mosaic that has the greatest chance of guaranteeing ecological complexity. The boundaries of a nature reserve should coincide with the outer limits of disturbances, such as the line beyond which floods do not extend or at which fires burn themselves out naturally. It is advisable to avoid identifying a nature reserve with just one type of habitat; furthermore, it is desirable, where possible, to impose certain restrictions on activities beyond the boundaries of the reserve (Halladay and Gilmour, 1995) so as to enable the various processes and species to perform their functions fully.

Corridors and Conservation

Corridors, or narrow strips of a given environment that are wedged in between different environments, are very popular among planners but give rise to some perplexity among ecologists (Harrison, 1992). As often happens in a complex system, the function of a corridor does not always coincide with patterns such as hedgerows, rows of trees, and watercourses. In some cases, these structures are genuine corridors, but in others they do not function as corridors even though they exhibit patterns associated with them.

The more mobile species tend to move along preferential routes, but it is not always possible to identify these by applying human-scale criteria. Morphological structures such as canyons and mountain ridges often attract animals, for example, because they are easy to travel along. Hares tend to move along tracks made by domesticated animals, and the same is true of wolves and many other small mammals. Certain large carnivores, such as pumas, do not expect to find much prey in corridors, and therefore they move along them relatively quickly. For many animals, knowing their territory is essential for survival. This is true both of herbivores that move from one grazing site to another and of predators that systematically hunt for their prey in their territory.

The Conservation of Habitat Fragments and Relict Populations

In environments that are dominated by man, it is very common to find fragmented zones inhabited by more or less isolated populations whose survival depends on the degree of connectivity between the patches (Fig. 6). At the local level, populations can die out in certain patches; however, provided that this does not happen at the same time in all of them, the species is guaranteed to survive at the mosaic level. The size, shape, and quality of patches have a major influence on reproduction and the risk of predation. The qualitative character of a patch is not just a function of its shape; indeed, two patches of similar size but with different shapes have notably diverse ratios between surface area and volume.

Ecological Conservation and the Conservation of Processes

When using the paradigms of landscape ecology, it is easier and more realistic to pursue conservation policies directed at flows and processes rather than to concentrate on

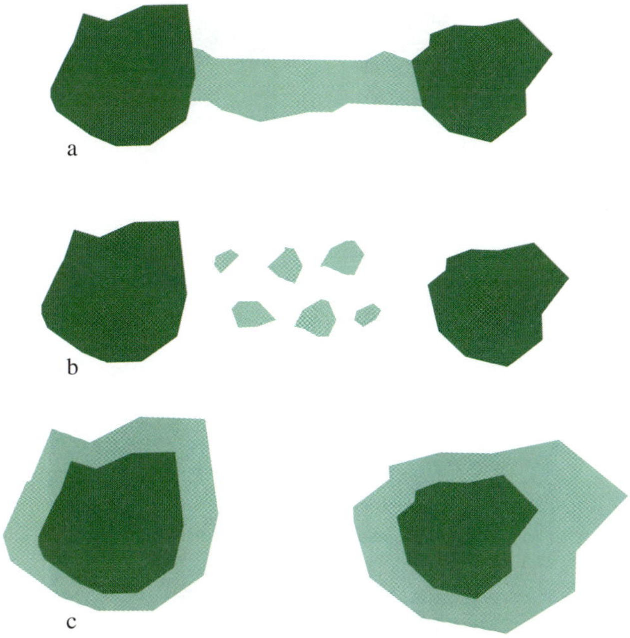

FIGURE 6 Various strategies for increasing the degree of connectivity between isolated areas. (a) Creation of a corridor, which permits the survival of many species in disjointed areas. (b) Improving the quality of the environment in intermediary areas. (c) Expanding the areas by creating buffer zones.

the conservation of a given species, at least when there are no particularly important species to conserve. It often happens that a species is present for short periods ranging from a few days to a few years; in this case, it will often be associated with particular stages of ecological succession. Furthermore, the rarity of a species is often a function of its demographic cycles, and it is possible for a species to be rare on the margins of its habitat and very common in the center. For example, the conservation of the migratory flows of birds from the western Palearctic cannot involve protecting just one single species but must involve identifying the areas in which the birds stop during migration. Most land migratory birds stop in environments other than those in which they winter and reproduce, and, as with warblers and thrushes, like open areas such as grassland or shrub-rich savanna, both of which offer ample opportunities for feeding. On the contrary, some birds are specialized during reproduction but become generalists during migration.

Hierarchical Patterns within Landscapes and the Conservation of Biodiversity

When a landscape is viewed as a hierarchical system, its various components exhibit a range of dynamics that vary according to the closeness of the relationship. It is therefore impossible to conserve these systems without taking this consideration into account, and in doing so one tends to conserve all elements of the biodiversity contained within a landscape. Once the widely held misconceptions concerning species conservation are cast aside, the conservation of both biological and ecological diversity becomes a strategic objective. Granted that this does in fact occur, modern economic models can coexist with biodiversity in such a way as to guarantee solutions that are ecologically valid. In other words, in a world that is dominated by man, it is essential to find disturbance regimes that mimic the natural disturbances that trigger environmental dynamics and that also promote a satisfactory level of economic activity. Indeed, the socioeconomic crises that have affected various societies at different periods in history and in different parts of the earth have generally been provoked by the sudden desynchronization of ecological and socioeconomic processes.

Methodological Approaches to Studying Landscapes

The spatial processing of data is a vital aspect of landscape ecology and requires a series of instruments, some highly sophisticated, that are able to collect and process geographical and ecological information. The instruments and techniques range from remote sensing to geographic information systems (GISs).

Analysis is carried out on at least two main levels. The first is concerned with the shape and size of the patches that make up a landscape, whereas the second is concerned with the characteristics of the land mosaic (Fig. 7). Consequently, it is important to consider the ratio between the length of the perimeter and the surface area of patches as well as their shape and size. An analysis of a land mosaic must first involve measuring the diversity between patches as well as their nearness and the extent of their aggregation. The distances separating the patches are particularly important with regard to estimating the degree of connectivity that any one system is able to guarantee for a given organism. Furthermore, greater distance necessarily implies an increase

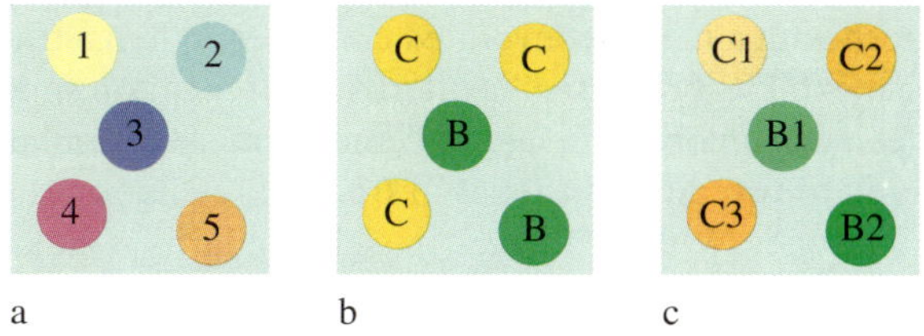

FIGURE 7 A quantitative analysis of a landscape can involve analyzing the single patches (a), analyzing the mosaic as a whole (b) and studying the various types of vegetation cover (in the example the A patches represent land under cultivation and the B patches represent woods), or a more sophisticated approach (c) in which the structural characteristics of each patch are studied, together with the various types of cover (C1–C3, B1, and B2).

in the risk of predation and a reduction in the likelihood of vector transport (the transport of matter or energy through the agency of physical forces of the movement of animals). Some techniques are based on an analysis of electronic images and are a function of the characteristics of single pixels. Such analyses can be carried out using data that are expressed in raster format (in which a process is represented by cells, pixels, or discrete spatial units).

The concept of *lacunarity* in a fractal distribution (Mandelbrot, 1975) provides the basis for an interesting approach to studying spatial patterns. An object with low lacunarity does not change when it is moved, whereas one whose gap distribution has a high degree of heterogeneity takes on a different spatial configuration when it is moved (Farina, 1998). It therefore seems useful to study environmental heterogeneity in terms of lacunarity. Since the analysis of edges is particularly important in this context, the complexity of a landscape can be studied not only with reference to the heterogeneity of the patches but also with reference to that of the edges (Fig. 8).

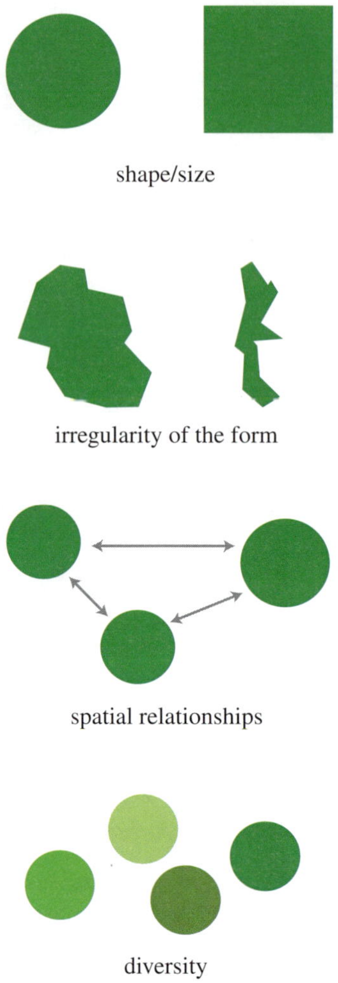

FIGURE 8 Information that is quantifiable at the patch level refers to shape and size, the irregularity of the shapes, the spatial relationships between the single patches, and the diversity between the types of cover.

Fractal Patterns in Landscapes

The spatial distribution of organisms and their aggregation reflects the heterogeneity of landscapes. Indeed, the distribution of organisms is determined by the unevenness of the distribution of resources and, therefore, of biomasses. Fractal geometry (Mandelbrot, 1975) offers a new and promising way of measuring this complexity.

In environments that have been modified by man, regular geometric or linear shapes predominate. In a natural system, however, the irregularity of the shapes precludes an Euclidean approach, which is unsuitable for describing environmental complexity. Fractal geometry is considered to have many promising applications, including the analysis not only of patterns but also of phenomena such as the unevenness of the distribution of organisms, their movements, and, more generally, their behavior.

The term *fractal* was coined by Mandelbrot in 1975 to refer to an irregular object whose irregularities are exhibited on every scale. A fractal can be taken to mean an object or a pattern with a highly convoluted character. When a fractal object exhibits broad-scale patterns that repeat on smaller scales, such an object is said to demonstrate self-similarity. There are two different types of fractals: regular ones and random ones. With the former, the self-similarity is the same on every level; in other words, the object is a copy of itself on every scale. The latter category, on the other hand, includes natural fractals such as clouds, coastlines, and organism abundance. However, self-similarity that is identifiable only statistically is a scale-dependent concept. For example, coasts are fractal objects, and their length depends on the scale of resolution that is used. Fractal geometry has a broad range of applications that include such areas of the natural sciences as geology, hydraulics, soil texture, dynamics, microbe transport, and plant structure. Fractal analytical techniques are particularly useful for studying objects which, due to their scalar properties, exhibit special complications. Thus, the abundance of an organism is a function of the scale with which environmental information is aggregated — that is, the degree of resolution that is used.

Geographical Information Systems

A GIS uses technology with which it is possible to process spatial data with computers and data input instruments such as digitizers and scanners and then express these data in graph form with printers and plotters. A good GIS generally makes use of digital cartography, remote sensing, and database management (Burrough, 1986).

GISs are considered indispensable for analyzing phenomena such as vegetation patterns, the distribution and movement of animals in a landscape, changes in the use of soil, and the modeling of ecological processes (Tomlin, 1990; Coulson *et al.,* 1991; Maguire *et al.,* 1991). There are two types of GISs: vectorial ones and raster ones. The former give a more accurate image of objects that reflects their natural contours, whereas the latter breaks the object into discrete parts or cells (pixels). The advantage of the raster system is mainly the ease with which information can be handled. Indeed, groups of cells that form part of a matrix can be aggregated, compared with their neighbors, and then reaggregated for different degrees of resolution. Furthermore, information from satellite images is in raster format and is relatively easy to handle with a raster GIS. The disadvantage of this system, however, is that it uses a great deal of memory space to store information.

Remote Sensing

The patterns in a landscape are generally easy to identify on satellite images, especially when 10 × 10-m sensors are used. The degree of resolution that is available for analyzing large geographical areas varies from 10 to 30 m per pixel and is sufficient to provide information on the patterns of the landscape under observation. It is indeed possible to gather a great deal of information from large areas, and remote sensing techniques can highlight transition zones between various environments. The complexity of the edges is a good indicator of the processes taking place in a given area. Remote sensing techniques make it possible to locate the position on the ground of patterns identified by sensors mounted on satellites. These global positioning systems (GPSs), which were developed for military purposes and for air and sea navigation, are now available at an affordable price (Leick, 1990; Hofmann-Wellenhof *et al.,* 1993) and offer great potential for application in the field of ecology (Trimble Navigation, 1994). Using triangulation, a group of satellites are able to identify the rover's whereabouts on the ground with a degree of accuracy ranging from 25 m to a few centimeters, depending on the configuration of the hardware. Thus, it is possible, due to GPSs, to designate areas, locate animals, and track their movements. Similarly, it is possible to scan an environment and take a census of birds by locating the sightings (Farina, 1998). The information is stored in real –time, and then at the end of the work session it is transferred to a GIS for processing. This technique makes it possible to carry out ground-level observation of patterns such as watercourses, coastlines, or ecotones.

References Cited

ALLEN, T. F. H., and HOEKSTRA, T. W. (1992). *Toward a Unified Ecology.* Columbia Univ. Press, New York.

ALLEN, T. F. H., and STARR, T. B. (1982). *Hierarchy. Perspectives for Ecological Complexity.* Univ. of Chicago Press, Chicago.

BAKER, W. L. (1992). The landscape ecology of large disturbances in the design and management of nature reserves. *Landscape Ecol.* **7,** 181–194.

BURROUGH, P. A. (1986). *Principles of Geographic Information Systems for Land Resources Assessment.* Clarendon/Oxford Univ. Press, Oxford/New York.

CLEMENTS, F. C. (1897). Peculiar zonal formations of the great plains. *Am. Nat.* **31,** 968.

CLEMENTS, F. C. (1905). *Research Methods in Ecology.* University Publishing, Lincoln, NE.

COULSON, R. N., LOVELADY, C. N., FLAMM, R. O., SPRADLING, S. L., and SAUNDERS, M. C. (1991). In *Quantitative Methods in Landscape Ecology* (M. G. Turner and R. H. Gardner, Eds.), pp. 153–172. Springer-Verlag, New York.

DAUBENMIRE, R. (1968). *Plant Communities. A Textbook of Plant Synecology.* Harper & Row, New York.

DELCOURT, H. R., and DELCOURT, P. A. (1988). Quaternary landscape ecology: Relevant scales in space and time. *Landscape Ecol.* **2,** 23–44.

DI CASTRI, F., HANSEN, A. J., and HOLLAND, M. M. (1988). A new look at ecotones: Emerging international projects on landscape boundaries. *Biol. Int.* **17** (Special issue), 1–163.

FARINA, A. (1998). *Principles and Methods in Landscape Ecology.* Chapman & Hall, London.

FARINA, A., TOMASELLI, M., and FERRARI, C. (1986). *Studi sui rapporti tra vegetazione e Microtus nivalis (Martins) nell'Appennino reggiano. Prime osservazioni* (Mammalia, Rodentia). *Boll. Zool.* **53,** 86.

FORMAN, R. T. T. (1995). *Land Mosaics. The Ecology of Landscapes and Regions.* Cambridge Univ. Press, Cambridge, UK.

FORMAN, R. T. T., and GODRON, M. (1986). *Landscape Ecology.* Wiley, New York.

GARDNER, R. H., MILNE, B. T., TURNER, M. G., and O'NEILL, R. V. (1987). Neutral models for the analysis of a broad-scale landscape patterns. *Landscape Ecol.* **1,** 19–28.

GILPIN, M., and HANSKI, I. (Eds.) (1991). *Metapopulation Dynamics: Empirical and Theoretical Investigations.* Academic Press for the Linnean Society of London, London.

GREEN, B. H., SIMMONS, E. A., and WOLTJER, I. (1996). *Landscape conservation. Some steps towards developing a new conservation dimension.* A draft report of the IUCN–CESP landscape Conservation Working Group. Department of Agriculture, Horticulture, and Environment, Wye College, Ashford, Kent, UK.

GRIGGS, R. F. (1914). Observations on the behaviour of some species on the edges of their ranges. *Bull. Torrey Bot. Club* **41,** 25–49.

HALLADAY, D., and GILMOUR, D. A. (Eds.) (1995). *Conservation Biodiversity outside Protected Areas. The Role of Traditional Agro-ecosystems.* IUCN, Cambridge, UK.

HANSEN, A. J., and DI CASTRI, F. (Eds.) (1992). *Landscape Boundaries. Consequences for Biotic Diversity and Ecological Flows.* Springer-Verlag, New York.

HANSEN, A. J., DI CASTRI, F., and NAIMAN, R. J. (1992). Ecotones: What and Why? In *Landscape Boundaries. Consequences for Biotic Diversity and Ecological Flows* (A. J. Hansen and F. Di Castri, Eds.), pp. 9–46. Springer-Verlag, New York.

HANSKI, I., and GILPIN, M. (1991). Metapopulation dynamics: Brief history and conceptual domain. *Biol. J. Linnean Soc.* **42,** 3–16.

HARRISON, R. L. (1992). Toward a theory of inter-refuge corridor design. *Conserv. Biol.* **6,** 293–295.

HOFMANN-WELLENHOF, B., LICHTENEGER, H., and COLLINS, J. (1993). *Global Positioning System, Theory and Practice,* 2nd ed. Springer-Verlag, New York.

HOLLAND, M. M. (1988). SCOPE/MAB technical consultations on landscape boundaries. Report of a SCOPE/MAP workshop ecotones. In *A New Look at Ecotones: Emerging International Projects on Landscape Boundaries* (F. Di Castri, A. J. Hansen, and M. M. Holland, Eds.). International Union of Biological Sciences, Paris.

HOLLAND, M. M., RISSER, P. G., and NAIMAN, R. J. (1991). *Ecotone. The Role of Landscape Boundaries in the Management and Restoration of Changing Environments.* Chapman & Hall, London.

HUMBOLDT, VON A. (1807). *Indeen zu Einer Geographie der Pflanzen nebst einem Gemälde der Tropenlander.* Tubingen.

KOLASA, J., and PICKETT, S. T. A. (Eds.) (1991). *Ecological Heterogenity.* Springer-Verlag, New York.

LEICK, A. (1990).*GPS Satellite Surveying.* Wiley, New York.

LEVINS, R. (1970). Extinction. In *Some Mathematical Questions in Biology. Lectures in Mathematics in the Life Sciences* (M. Gertenshaubert, Ed.), pp. 77–107. American Mathematical Society, Providence, RI.

LIVINGSTON, B. E. (1903). The distribution of the upland societies of Kent County. *Michigan Bot. Gazette* **35,** 36–55.

MACARTHUR, R. H. (1984). *Geographical Ecology, Patterns in the Distribution of Species.* Princeton Univ. Press, Princeton, NJ.

MACARTHUR, R. H., and WILSON, E. O. (1967). *The Theory of Island Biogeography.* Princeton Univ. Press, Princeton, NJ.

MAGUIRE, D. J., GOODCHILD, M. F., and RHIND, D. W. (Eds.) (1991). *Geographical Information Systems.* Longman, Harlow, UK.

MANDELBROT, B. B. (1975). *Les Objects Fractals: Forme, Hasard et Dimension.* Flammarion, Paris.

MERRIAM, G. (1984). Connectivity: A fundamental characteristic of landscape pattern. In *Methodology in Landscape Ecological Research and Planning: Proceedings of the First International Seminar of the International Association of Landscape Ecology (IALE),* pp. 5–15. Roskilde University Center, Roskilde, Denmark.

NAIMAN, R. J., and DECAMPS, H. (1990). *The Ecology and Management of Aquatic–Terrestrial Ecotones.* Parthenon, Park Ridge, IL.

NAVEH, Z. (1987). Biocybernetic and thermodynamic perspectives of landscape functions and land use patterns. *Landscape Ecol.* **1,** 75–83.

NAVEH, Z., and LIEBERMAN, A. S. (1994). *Landscape Ecology. Theory and Application,* 2nd ed. Springer-Verlag, New York.

ODUM, E. P. (1959). *Fundamentals of Ecology,* 2nd ed. Saunders, Philadelphia.

O'NEILL, R. V., DEANGELIS, D. L., WAIDE, J. B., and ALLEN, T. F. H. (1986). *A Hierarchical Concept of Ecosystems.* Princeton Univ. Press, Princeton, NJ.

PICKETT, S. T. A., and WHITE, P. S. (1985). *The Ecology of Natural Disturbance and Patch Dynamics.* Academic Press, San Diego, CA.

PLACKTER, H., and ROSSLER, M. (1995). Cultural landscapes: Reconnecting culture and nature. In *Cultural Landscapes of Universal Value* (B. van Droste, P. Plackter and H. Rossler, Eds.). Fischer in cooperation with UNESCO, New York.

PULLIAM, H. R. (1988). Source–sink, and population regulation. *Am. Nat.* **132,** 652–661.

PULLIAM, H. R. (1996). Sources and sinks: Empirical evidence and population consequences. In *Population Dynamics in Ecological Space and Time* (O. E. Rhodes, R. K. Chesser, and M. H. Smith, Eds.), pp. 45–69. Univ. of Chicago Press, Chicago.

RICKLEFS, R. E. (1973). *Ecology.* Chiron Press, Newton, MA.

RISSER, P. G. (Ed.) (1995). Understanding and managing ecotones. *Ecol. Int.* **22.**

STAUFFER, D. (1985). *Introduction of Percolation Theory.* Taylor & Francis, London.

TOMLIN, C. D. (1990). *Geographic Information Systems and Cartographic Modelling.* Prentice Hall, Englewood Cliffs, NJ.

Trimble Navigation (1994). *GPS Mapping Systems, General Reference.* Trimble Navigation, Sunnyvale, CA.

TURNER, M. G. (1987). Spatial simulation of landscape changes in Georgia: A comparison of 3 transition models. *Landscape Ecol.* **1,** 29–36.

TURNER, M. G., GARDNER, R. H., and O'NEILL, R. V. (1995). Ecological dynamics at broad scales. *BioScience* **29** (Suppl.).

WATKINSON, A. R., and SUTHERLAND, W. J. (1995). Source, sinks and pseudo-sinks. *J. Anim. Ecol.* **64,** 126–130.

WEAVER, J. E., and CLEMENTS, F. C. (1929). *Plant Ecology.* McGraw–Hill, New York.

WIENS, J. A., and MILNE, B. T. (1989). Scaling of landscape in landscape ecology, or landscape ecology from a beetle's perspective. *Landscape Ecol.* **3,** 87–96.

ZIFF, R. (1986). Test scaling exponents for percolation-cluster perimeters and ecology. *Phys. Rev. Lett.* **56,** 545–548.

ZONNEVELD, I. S. (1995). *Land Ecology: An Introduction to Landscape Ecology as a Base for Land Evaluation, Land Management and Conservation.* SPB Academic, Amsterdam.

LUKAS HOTTINGER
Museum of Natural History
Basel, Switzerland

Learning from the Past?

Learning from the past calls for understanding the complex processes that have governed nature during Earth's history and will shape its future. Analyzing the historical documents, we have to consider the heavy bias inherent in the biomineralization of carbonate hard parts with their comparatively high preservation potential. The latter preferentially belong to K-strategists from warm, stable, and oligotrophic environments. The analysis of the morphology of fossils is based on pattern recognition in four dimensions and functional morphology, including constructional and physiological constraints. From this, strategies of life and a hierarchical system of classification are conceived at a higher level of abstraction. Numerous parallel biozonations based on biotic evolutionary change in superposed strata of sedimentary rock are combined by complex correlations with abiotic signals in the rock record and calibrated by measuring decay rates of instable isotopes in order to construct a worldwide standard timescale. We admit causal relationships between synchronous historical events if the respective time and space dimensions match the resolution power of the timescale. The scaling of the biotic processes in space and time enforces a distinction of "short-term" events governed by migration of taxa and "long-term" events involving speciation and extinction. Learning from the past calls for focusing research on the short-term processes neglected for a long time and comprising the range of 100 to 100,000 years.

Introduction

Can we learn from history? If so, what can we learn? Among the historians, this utilitarian question was called to discussion in the second half of the nineteenth century when the German historian Leopold von Ranke in Berlin propagated history as an instrument to lead society into its future while the Swiss Jacob Burckhardt in Basel emphasized the impact of the genius of the time, different in every generation of historians, on the interpretation of historical documents. The dispute has never been resolved and continues in the present, in which in particular the rise and fall of economic performance is discussed and predicted in the light of the past in every daily paper.

In the life sciences, the historical perspective is produced by paleontology, the branch of science focusing on fossils. The latter documents the past 600 million years of history of life. However, life on Earth is much older, its early presence proven by direct, fossil documents and by extensive abiotic circumstantial evidence. Documents on the history of life registering more than its mere existence, however, are available only since the moment when hard parts appeared in the organisms as a sequel of the "agronomic revolution" (Seilacher, 1999). These, by their elaborate morphology, register functions and patterns permitting identification and are therefore of relevance to history. The corresponding time span of 600 million years since the appearance of organic hard parts up to the present time is called the Phanerozoic and constitutes the time frame of this article.

Benthic foraminifera comprise a group of eukaryotic

unicellulars with a biomineralized, calcium-carbonate cell envelope with a highly specific morphology permitting the identification of more than 50,000 species during the Phanerozoic. This group of microfossils is abundant, almost omnipresent, represented in all marine environments at almost all times since the beginning of the Mesozoic, 240 million years ago, and still living today in great abundance and diversity. The living foraminifera provide direct information on their way of life and of reproduction.

Foraminifera are unicellulars with a cell size beyond measure, playing an ecological role similar to meiobenthos on and in the sea bottom, to zooplankton, to sessile filter feeders on hard substrate, etc.—roles that are typical for lower invertebrates with tissues composed of differentiated cell types. However, the foraminifera play such roles as unicellulars, with their potential of particular performances such as rhizopodial activities including regeneration by molecular reactions, endocellular generation of biominerals for exocellular use, cloning of entire generations of autonomous individuals, and diversified endosymbiosis. These ambiguous characteristics of the foraminifera provide to the entire group a certain ecological autonomy leading to at least a partially separate ecosystem evolving in time and justifying its exclusive use to interpret the exceptionally numerous historical documents provided by the foraminiferal presence in the sedimentary record. Moreover, their economic interest due to their presence in the cuttings from boreholes has provided one of the most complete fossil records in existence, with an illustrated catalog on the species level and with an exceptionally extensive systematic literature. Therefore, many examples and also possibly some general views presented in this article are derived from the foraminiferal record, which may limit the validity of certain statements to this particular group.

What would we like to learn from Earth history and in particular from its fossil record? Learning means observing and understanding the past, drawing conclusions in the present about the mechanisms and processes governing history, and acting accordingly in the future. Triggered by the exploding population of the human species, the questions asked in our time are focused on the sustained availability of natural resources for food and energy supply, on the role of biotic diversity in our ambient environment, and on the functioning and coexistence of the individuals in the society and of different societies within the same area. Problems such as nuclear waste disposal or the increased carbon dioxide concentration in the atmosphere demonstrate the extension of the questions asked far beyond the secular time horizon of two or three human generations. This indeed justifies an attempt to search for answers in the current knowledge of Earth history.

In this article, no answers to specific questions can be given. Instead, I try to explain the reasoning and the different levels of abstraction on which Earth history is described and the extent to which mechanisms and processes are currently understood. I do not discuss what can be learned from the past but rather, if we can learn, what kind of answers we may expect.

The Biased Nature of the Documents of Earth History

The Fossil and Its Relation to Living Organisms

The fossil is the remains of an organism, not the living organism itself. Its survival through millions of years depends on severe selective processes which reduce the number of remains to insignificant amounts in relation to the number of organisms having lived on Earth through time. The sum of all processes involved to produce a fossil is called *fossilization.* The latter not only reduces the number of individuals but also selects them in a certain number of ways to be discussed later.

The fossil, within its taphocoenosis and together with the encasing sedimentary rock, constitutes a document of Earth history (Fig. 1). The original nature of the organism, its population within its community, and its habitat are altered by biotic and abiotic processes acting as selective filters: Scavenging and decay reduce most of the carcasses to their mineralized hard parts (shells, sklerites, and bones), which are often separated from one another by burrowing. Currents and turbulence often select the hard parts according to shape and size prior to final deposition. Diagenetic processes may alter the chemical composition, including the isotopic composition of the hard part and of the sediment which often provided a substrate for the living organism in its original habitat. Sometimes, only the activity of the organism may be preserved as traces in the sediment (trace fossils and footprints), documenting interesting features of behavior but rarely permitting the identification with precision of the organism responsible for the trace.

Diagenesis also governs the cementation of the grains, including the fossil hard parts in the sediment, and predetermines the conditions of reemergence of the fossil from its encasing rock by natural erosion and/or artificial preparation. Sampling of naturally eroded specimens in the field and preparation methods such as washing the fossils out of their encasement may also bias the record to a very large extent. Mostly geological methods (taphonomy, sedimentology, and sedimentary petrography) are used to assess the extent of alteration in a document of Earth history.

The succession of the historical documents in time constitutes the fossil record. The succession of taphocoenoses provides pictures of organismic evolution through time, and the succession of sedimentary rocks documents environmental change. The time relationships between the historical documents are described by correlating them to a standard timescale by stratigraphical methods.

The reduction of numbers by the processes of fossiliza-

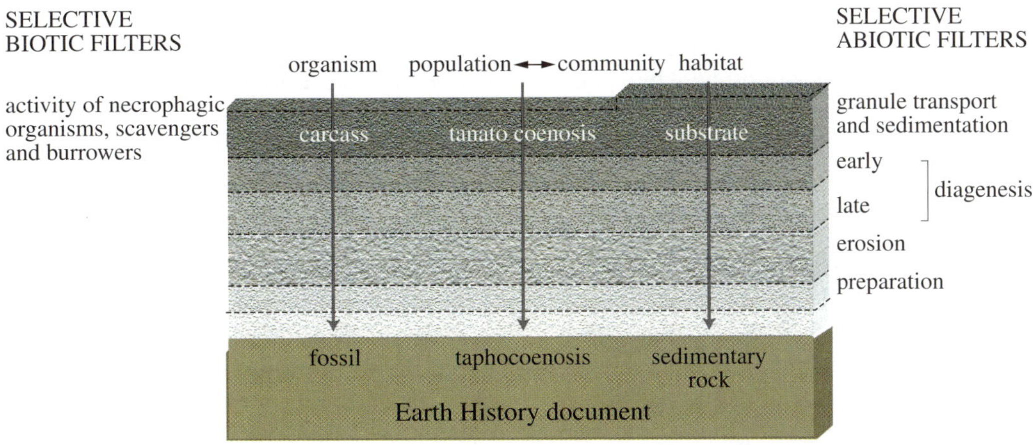

FIGURE 1 Schematic diagram of the taphonomic processes which lead to the formation of a fossil record, starting from organisms, populations, and communities. These processes act by biostratinomic mechanisms (from death to definitive burial) and diagenesis (from burial to the fossil discovery).

tion is indeed drastic when comparing the number of fossils with the number of living organisms. The chance of survival of information on extinct species is so small that only abundant, long-lasting, and widespread species may be recorded. Equivalents to extant species considered in ecology as "rare" have no chance of appearing in the fossil record, and when, eventually, we would find a single fossil specimen of a rare species, we would not be in a position to prove that it belongs to such a category. In particular, the chances to identify, locate, and document the allopatric origin of a species in the fossil record are practically nil.

Nevertheless, in many groups there are more fossils collected than we are able to study in a lifetime. The closer investigation of a complete vertebrate skeleton represents more or less the workload of a PhD thesis, (i.e., work for about 3 years), the structural analysis of a larger foraminifer perhaps 2 or 3 weeks, and the identification of a nannofossil maybe a few seconds. Thus, the accuracy of the historical record does not simply depend on the number of fossil specimens available for study but also on the quality and quantity of the information to be derived from every single fossil specimen.

As a rule, the number of specimens available from the fossil record is inversely related to the size of the remains. There are thousands of nannofossils in a spoonful of sediment, and there may be only one dinosaur in a mountain of encasing rock. On the other hand, the number and significance of data derived from the morphology of a fossil are often positively correlated with its size. In a coccolith, there is practically no information beyond its identity within the system of classification, whereas in the vertebrate skeleton there are numerous indications as to the lifestyle of the animal to be deducted from functional anatomy.

The significance of the fossil documents must therefore be weighted according to the groups of animals or plants investigated. The highest significance will be found in groups for which the fossil remains have the most significant morphology and for which many fossil specimens are available that match at least the human capacity of work. These relationships must be kept in mind when a particular sampling strategy is designed in order to answer specific questions about evolution or global environmental change.

The Systematic Bias Inherent in the Record of Hard Parts

The fossil remains of an organism usually consist of biomineralized hard parts preserving some of or all of their permanent morphology through fossilization. Often, the hard part is replaced by abiotic material during diagenesis or preserved only as cavity in the encasing rock. In these cases, the outer morphology of the hard part is preserved as mold, but eventual internal textures may be lost. The conditions of preservation thus depend very much on the chemical composition and on the physical properties of the biomineral used to produce the hard part in the living organism and on its chemical relation to the composition of the encasing sediment. Consequently, the chances of historical documentation of an organism depend to a very large extent on the organism's own process of biomineralization producing mechanically resistant parts of its body.

The most widespread and diagenetically most stable material for hard parts is calcium carbonate used preferentially in the aquatic realm, whereas on land cellulose is almost exclusively used by plants. The former material is much more resistant to oxidation than the latter (burning wood), and therefore its chances for preservation are much better. Both materials are very poor in proteins containing nitrates and phosphates. In the case of calcium carbonates, only very small amounts of proteins are necessary to govern hard part growth in order to give it the convenient shape needed for its mechanical or protective function.

Carbon in the form of carbon dioxide is available in the atmosphere without limitation and in the hydrosphere, particularly in shallow water, in large quantities; however, calcium is abundant only in the aquatic realm, in particular in shallow, warm water in which its solubility is comparatively high. The uptake of carbon from the atmosphere by a plant is linked to photosynthesis, and the uptake of calcium carbonate from the hydrosphere by an animal can be carried out at low net costs of energy, with or without the help of the photosynthesis of vegetal symbionts. Thus, the quantity of hard parts produced by an organism can be beyond measure with respect to the living biomass of the corresponding organism: The biomass of coral tissue is an infinitely small quantity in relation to the weight of the biomineralized structures supporting the living colony of polyps, and the biomass of living tissue in a large tree of the tropical forest is small compared to that of the wooden trunk supporting this kind of plant.

The second most important material used for biomineralized hard parts is silica (SiO_2). Because of its low solubility in seawater, its quantities available to life are seriously limited. In crystallized form (quartz) it would be much more stable than calcite and as such have the very high potential of preservation of a quartz grain. However, nature uses this material in a form which permits immediate recycling by producing amorphous hydrogels that dissolve immediately after use, i.e., after the loss of protection from dissolving by coats of organic matter. In these circumstances, siliceous hard parts are preserved only in environments in which the dissolved silica content is so high that dissolution of the hard part is seriously hampered. Local sources of dissolved silica are mostly of volcanic origin or fed by upwelling currents.

Larger quantities of silica from hard parts dissolving in the encasing sediment after deposition may migrate through the pores of the sediment and precipitate to form silica nodules or chert layers (Baumgartner *et al.,* 1995). Exceptionally, they leave behind cavities which may be filled subsequently with other minerals, thus preserving some of the morphology of the original organic, siliceous hard part. The quality of preservation depends in these cases in particular on the grain and pore sizes of the sediment.

Considering the present-day faunas and floras, including larval and juvenile growth stages, there are mechanically resistant organs or parts of organisms dominantly built as hydropneumatic structures (Gutmann, 1994) or as cuticular covers produced by organic material. Hydropneumatic structures need energy to maintain the pressure of the fluid in order to produce the mechanic stability needed. Both kinds of mechanical structures need proteins to create the recipient or cover. The proteins contain nitrogen in considerable quantities which is usually lost to recycling within the organism. This is particularly striking in all arthropods shedding their armor at each instar. In environments in which the availability of nitrogen is the limiting factor, mechanically resistant structures tend to be constructed using biomineralization as the most parsimonious way to keep nitrogen expenditure and losses as low as possible. Protein layers of only a few molecules of thickness are sufficient to control growth and shaping of the hard part. Therefore, the preferential preservation of biomineralized hard parts enhances the registration of oligotrophic environments and their corresponding inhabitants in the fossil record to a very large extent.

The vegetable mold covering today's soils as humus retains the major part of nitrates available in the biosystem. Humus is generated by the terrestrial vegetation cover developed in Devonian times by the "invention" of roots and a vascular fluid transport system in plants. The expansion of vegetation covers over large continental surfaces has hampered the fertilization of the coastal waters by nutrients, in particular by nitrates, from the continents since Devonian times. This major ecological event in Earth history is reflected by the change of frequency in the fossil record of protein-rich hard parts found in the pre-Devonian blackshales (graptolites and trilobites) in favor of calcium carbonate hard parts such as coral skeletons or brachiopod shells. On the other hand, the extant eutrophication of coastal waters by man stops biomineralization to such an extent that the community structure collapses due to lack of carbonate production providing the necessary substrate to the benthic organisms. This may illustrate the extent to which the record of calcium carbonate hard parts is selective in respect to the trophic nature of the environment.

The Encasing Sedimentary Rock and Its Relation to Extant Deposits

The interpretation of a sedimentary rock as a substrate on (or in) which the populations documented by fossil taphocoenoses have lived needs particular consideration. If the surface of the sediment is not cemented at the time of the living community under study, it is destroyed prior to the deposition of the next layer, either by erosion during storms or by burrowing. In the Gulf of Aqaba, throughout the photic zone, all moveable surfaces at the bottom are ploughed under by burrowing animals at least once a week (Reiss and Hottinger, 1984). This constitutes a fundamental process of nutrient recycling characteristic of oligotrophic seas with a vegetation cover of sea grasses (Fig. 2) in which oxygen is not a limiting factor. Therefore, we have to reconstruct the habitat of a particular fossil population by comparing it with extant, ecologically similar populations and support the analogies with data derived from functional morphology.

The comparison of fossil with recent habitats of benthic communities according to the principles of actualism is hampered by the fact that outcrops of sedimentary rocks reveal at best a more or less linear section of the area representing the investigated habitats for a particular time in Earth history, whereas on today's sea bottom the habitat can be studied in two dimensions as a surface which can be mapped. On the other hand, outcrops of sedimentary rock may document ecological gradients in time by the subsequent deposition of sedimentary layers, whereas the effects

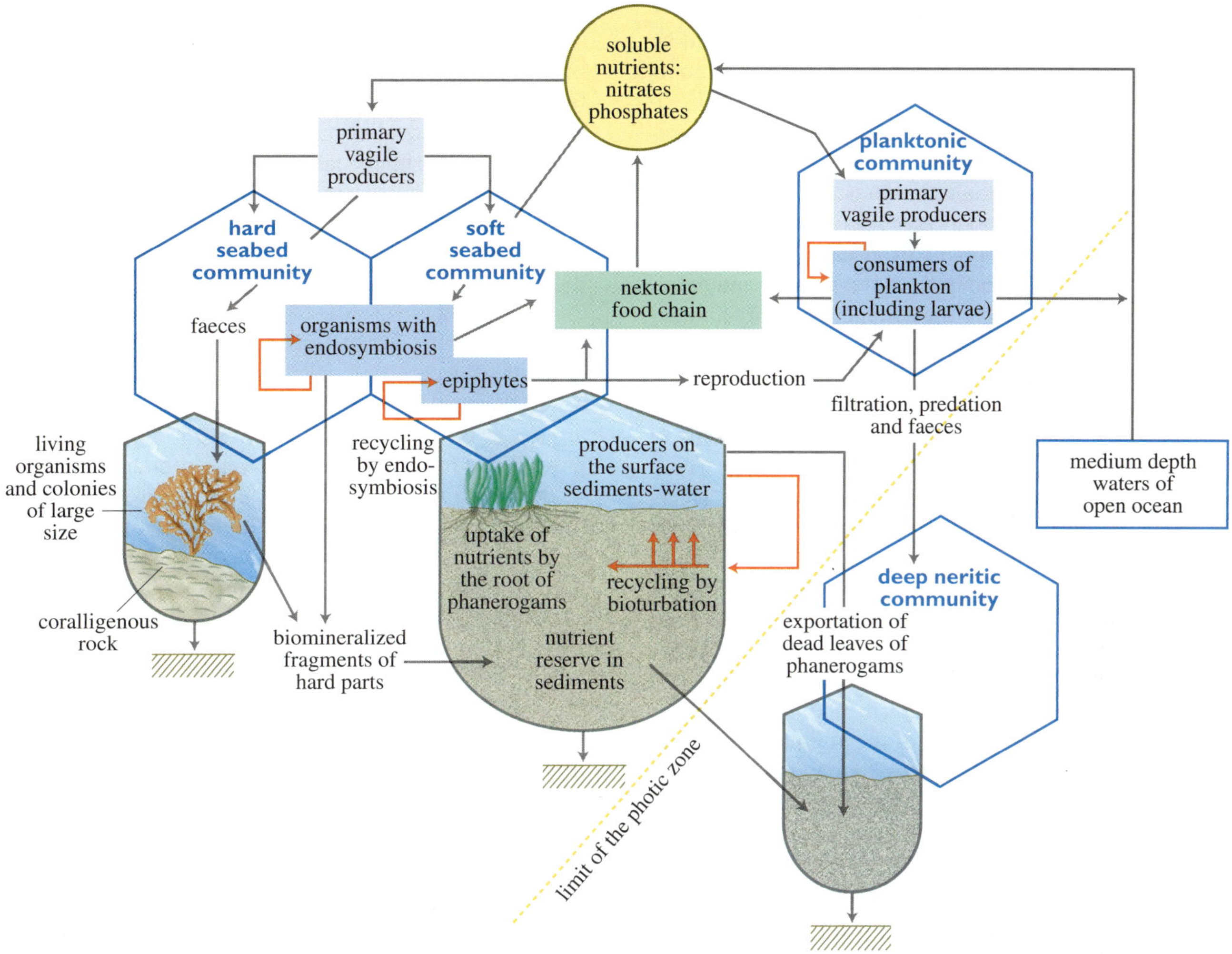

FIGURE 2 Nutrient circulation in carbonate-producing, oligotrophic, shallow environments. The recycling mechanisms are indicated by the red arrows. The role of the interplay of the relative concentrations of nitrate and phosphate is poorly understood.

of time gradients are difficult to investigate on the extant sea bottom: Only material from bore holes provides information on the recent history of the habitat and only at a single point of its areal extension (Fig. 3).

Mapping the area of extension of a population on the sea bottom may reveal, by correlation of the respective contours, dependencies of the organisms on mapable ecologic factors. In a two-dimensional system, the contours may be correlated with two independent factors such as depth and substrate: In the Gulf of Aqaba, the area of *Heterostegina depressa* populations is parallel to depth lines over the same substrate, parallel to the limits of substrate where hard bot-

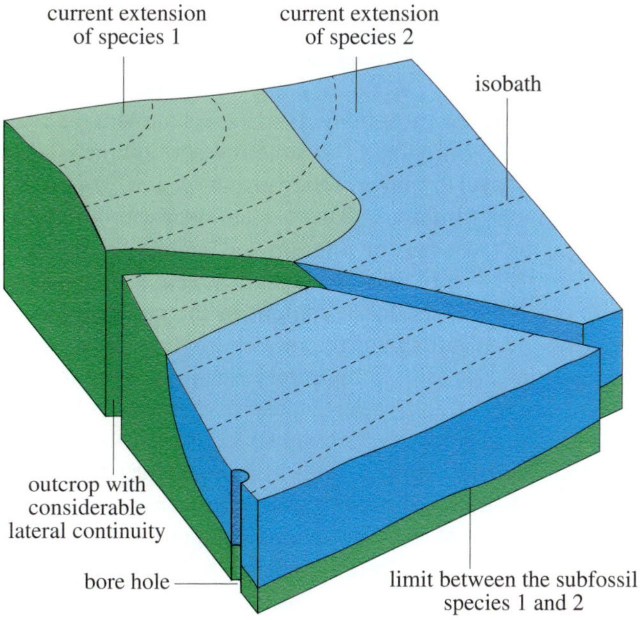

FIGURE 3 Relations between the extant surface of the seabed, an outcrop of sedimentary rock, and a bore hole in sedimentary rock relative to ecologically limited extensions of benthic populations. The model is derived from observation by scuba diving in the Gulf of Akaba, Red Sea.

toms are substituted by soft bottoms. The dependencies recognized by correlation of map contours are supported by biological understanding: Heterosteginids have symbionts and are therefore dependent on light. Experimental evidence shows them to be practically independent of particulate food (diatoms in particular) because of their endosymbiosis and therefore to colonize habitats with very low food density on hard substrates. The ecological factors depth and substrate are also independent from each other because light depends on absorption in the water column and therefore on depth, whereas substrate depends on local sedimentation rate governed by the exposition of the bottom relief to the dominating long-shore currents.

According to the structuring of the bottom topography, an area of a population may have dimensions of tens of square meters to several square kilometers. Imagine a continuous outcrop of fossil strata of similar dimension which, exceptionally, may be found in huge quarries or in desertic landscapes. In these circumstances, one would be able to follow the same bed along a single line crossing the original area in a random direction. A facies change within the same bed corresponding to a change in the composition of the taphocoenosis cannot be linked in these circumstances with a particular factor since the geometric relationship of one's limit in respect to the area limits of the population remains hidden. On the other hand, the fossil document, even if restricted to a single point of the original surface (e.g., in a bore hole), usually offers the possibility to examine the effects of ecological change with time on the facies and their respective taphocoenoses. Often, the changes observed in superposed beds are attributed to changes in depth of deposition ("shallowing" or "deepening-upward" sequences) without any discussion of other factors that may intervene as demonstrated by the closed contours of areas of a population in extant seas. Therefore, the identification of a depth gradient in geologic time should be supported by independent arguments preferentially derived from functional morphology of fossils used as depth indicators (Hottinger, 1997).

Neither the extant sediments nor the sedimentary rock record encasing the fossils document a continuum in time but rather, layer by layer, relatively short and irregularly discontinuous time units. There is an inverse relationship between the rate of sedimentation and the fossil richness in the historical documents. The higher the rate of deposition, the more diluted the fossil content in the sediment, neutralizing the advantage of the more complete documentation in time. Deposits yielding very rich fossil assemblages usually are very thin and particularly discontinuous in time. Communities living on hard substrate, for which the net sedimentation rate is practically nil, may produce under particular conditions taphocoenoses composed by fossils of significantly different age (condensed sedimentation). Where the net sedimentation rate is negative by erosion of previous sediments, fossilized organisms may be exhumed and redeposited elsewhere together with a taphocoenosis of the organisms living at the time of final deposition millions of years later. Often, the exhumed fossils are better preserved than the autochronous ones because their hard parts were stabilized by a first diagenesis providing, for instance, a protective coat of primary calcitic cement prior to their second fossilization. This process, called *reworking,* is relatively frequent when turbidites erode shelf edges and transport the eroded material by turbidity currents into deeper parts of the basin.

Diagenesis of Organismic Hard Parts

Today, the details of the physical and chemical processes called diagenesis, which transform the sediment into rock, are better understood than the transformation of the hard parts of the fossils. Dissolution, replacement, or alteration of the biomineralized hard parts are interesting phenomena which may produce indicators in analogy to the much more closely investigated (for economic reasons) diagenetic transformation of sedimented organic matter into coal, crude oil, or natural gas. The diagenetic effects on organic carbonate hard parts in particular depend on the crystallographic configuration of the original biocrystals (aragonite or calcite; the former is less stable than the latter), on the magnesium content in the calcite, and on the ultramicroscopic texture of the biomineralized hard part. These are the result of the different biological mechanisms of biomineralization genetically fixed and therefore characteristic of particular groups in the system of classification of the organisms. The combination of different diagenetic effects observed in different kinds of organisms within the same taphocoenosis may provide tools to differentiate the diagenetic history of a particular sediment at a particular site to an extent which may not be possible by considering only the diagenesis of the abiotic components of the sediment and their cements. An example is the different behavior in diagenesis of agglutinated, porcelaneous, and lamellar perforate cell envelopes of the foraminifera occurring together in the same limestone (Fig. 4). On the other hand, diagenetic behavior may be a welcome tool to identify particular submicroscopic textures such as the porcelaneous one tending to recrystallize first, prior to all other textures, in a gradient of increasing diagenetic activity.

Levels of Abstraction When Reading the Biotic Documents

The Significance of Morphology

The general focus of paleontology is the analysis and interpretation of the organisms preserved as fossils, i.e., their morphology, usually restricted to the one of their hard parts. The descriptive analysis of the morphology is never a goal in itself because it depends always on the questions asked. The latter depend on previous interpretations and conclu-

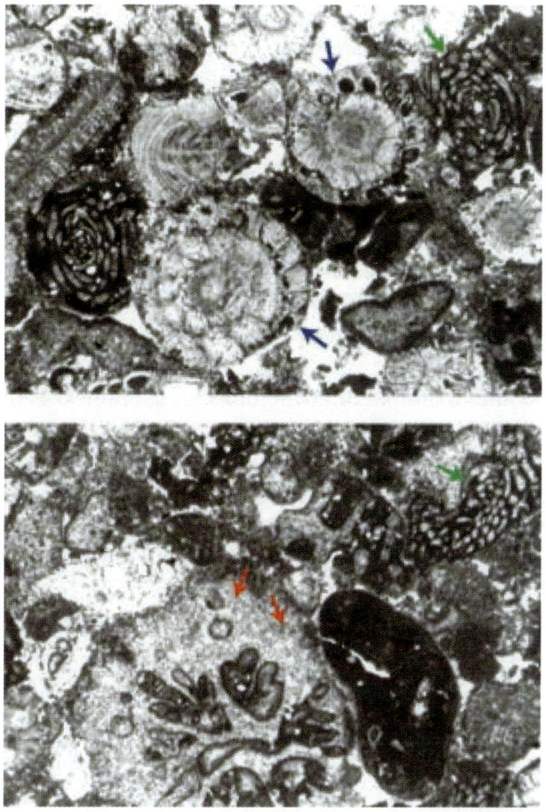

FIGURE 4 Differential preservation of foraminifera according to their mode of biomineralization in the same carbonaceous rock sample. The calcitic shell of the lamellar perforate rotaliid foraminifera, *Smoutina subsphaerica* (blue arrows), which contain little magnesium, show only mechanical damage due to transportation before final deposition but not spiral deformation. The calcitic shell of the porcelaneous alveolinid foraminifera *Glomalveolina primaeva* (green arrow), with a high magnesium content, is less stable and therefore appears deformed after the first diagenetic process which filled the cavities with sparitic calcite. This latter is more stable than the material of the shell and maintains the cavities open despite the spiral deformation of the shell. In the fragments of coral (red arrows), the hard aragonitic part was totally dissolved and substituted with sparitic calcite during the first diagenesis. The casing produced by this first generation cement is sufficiently stable to maintain the organic form during the successive diagenetic processes, including the compaction of the sediment. The sample is from Priolo near Siracusa, Paleocene (56 million years old; thin section in transparency, light microscope, ×20 enlargement). Note how it is possible, in random sections of hard rock, to identify the larger benthic foraminifera at the generic or even specific level due to their complex structure.

sions. Thus, paleontology is an iterative scientific process. At least once in a lifetime of scientists, the process will complete a full cycle.

Morphological analysis leads, on one hand, to (mostly partial) understanding of the biology of the organism and, on the other hand, to its identification. The organism's hard part preserved as fossil will have functions in the life of the organism which are in some way mechanical and therefore often linked to activities of the organism in its habitat, such as uptake of food with its teeth and mandibles or moving within its habitat with its limbs, fins, or wings. Functional relationships between the individual organism and its ambient environment are called *autecological* in contrast to "internal" functions supporting metabolism, reproduction, etc.

On the other hand, morphology is used to identify an individual as belonging to a certain category of biological classification. The classification is considered to reflect phylogenetic relationships between the categories. Phylogeny is a historical process in which geological time intervenes. Only the paleontological analysis of the fossil record can provide answers as to how to distinguish historically relevant inheritance from autecological functionality.

The almost infinite diversity of life on Earth and the complex interrelationships between its numerous protagonists call for abstractions in order to recognize processes and to understand their functioning and their coordination in hierarchical systems. We will not learn anything from the fossil record itself. It will be the insight into the functioning of nature through geological time that may teach us how to handle nature constituting our environment.

Typology: A Basic Concept of Cognition. Before introducing abstraction in reading historical documents, we have to first discuss a basic concept of cognition in science — so-called *typology.* Consider a large number of specimens or samples to be compared to each other. You can classify the specimens by measuring a single common character. In this case, each category created has the same "anonymous" value as its neighbor, like the units on a yardstick. The units can be defined or recognized as individuals only if counted by reference to a common reference point. If there are more than a single independent, common character to be measured, you will find a different classification according to which of the common characters you have chosen to measure.

Intuitively, one has learned to proceed in a fundamentally different way by choosing two or more specimens as so-called types to which the other specimens are compared with regard to their relative closeness. Each category or typological class created must be defined in a way to include its type specimen and to exclude all other types known. At least two types are necessary to define a typological unit because the definition is based on comparison (Fig. 5). Each typological unit is an individual and can be recognized as such without any common zero point of reference. For mnemotechnic and other reasons, in most cases it is meaningful to assign to each typological unit a name according to fixed rules. The name tag is tied to the type specimen, but this does not mean that the type specimen is more "typical" for this unit than for any other specimen. This simply means that its properties have to match the specific definition of the unit. This procedure ensures the stability of a formally named typologic unit when during subse-

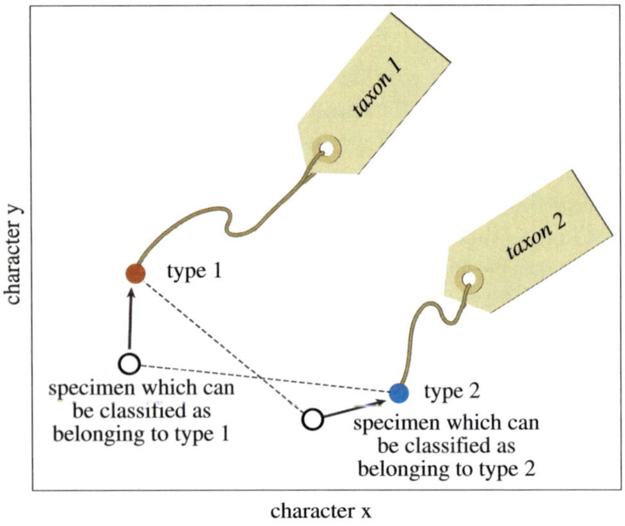

FIGURE 5 The classification of two specimens according to the typological method. On the basis of two quantitative characters *x* and *y,* the examples to be classified are assigned to the most similar reference *taxon*—that is, the one with more similar values for both characters.

quent research the definition of the unit has to be complemented or emended.

Larger typological systems of classification are hierarchical. Each typological unit on a particular level of the hierarchy is defined so as to include a single unit of the next lower level as type. In the Linnean system of classification of organisms, typology is explicitly used throughout its extended hierarchy: A particular species has a type specimen, a genus its type species, a family its type genus, and so forth.

Typology is not restricted to the classification of organisms; it is widely used in all sciences describing complex phenomena of nature in Earth history. In particular, I have to stress here that the term *facies,* widely used to classify rocks according to their genesis, also represents a typological system of classification, although it is much less formalized than the Linnean systematics. Facies units, low in numbers and without a clear hierarchy, usually lack a reference to types and are created ad hoc according to the specific questions asked. Thus, the same physical rock sample can represent a unit in different facies classifications, whereas the formal Linnean classification of organisms is unique and supposed to represent exclusively the history of descendence of life on Earth.

Functional Morphology. In order to understand the biological meaning of the morphology of organisms, and in particular of their hard parts, a comparison with extant, living organisms according to the principles of actualism is necessary. In this respect, the so-called living fossils (i.e., the representatives of almost completely extinct groups living today) play an outstanding role (Eldridge and Stanley, 1984). For example, the few species of *Nautilus* surviving as refugial populations today in a kind of submarine desert at the foot of large barrier reefs at 200- to 400-m depth are indispensable to understand the function of the tubular sipho, which is also visible in completely extinct ammonites and in the widely distributed nautiloids of the Paleozoic. This organ represents a device to pump the water from the gas chambers in the shell (phragmocone) serving as an instrument to control the buoyancy of the animal. On the other hand, the controversy about the interpretation of the ammonites as Carthesian divers (Seilacher and La Barbera, 1995; Jacobs, 1996) is difficult to resolve as long as no living fossil shows a mollusk to produce gas under considerable and variable pressures, as in the air bladder of fish.

Another example of the actualism's usefulness is the endosymbiosis of algal cells in the tissues of mollusks, corals, and larger foraminifera. This symbiosis can be directly recognized only in living animals because the symbionts are never preserved. However, symbiosis produces a certain number of morphological features in the hard parts which can be recognized in the fossil. In lamellar perforate foraminifera, the so-called eggholder structures, a kind of alveolar cavity below the pore chimney in the walls used for the symbiont's gas exchange, are such preservable structures indicating a dependence on sunlight of the carrier (Hottinger, 1998). Thus, the study of the biology of living fossils should be intensified in order to enhance the understanding of the fossil record. In this, the present indeed is the key to the past.

Functional morphology has to be based on morphogenesis, a subject which has been neglected for a long time—since Thompson's *On Growth and Form* (1917/1966) and Bonner's *Morphogenesis* (1952/1966). In order to support Earth history, the study of morphogenesis should be focused on groups of organisms with a good preservation potential presenting a comparatively complete fossil record, i.e., foraminifera, corals, brachiopods, bryozoans, mollusks, and echinoderms. Paleontology should also follow what present-day biology understands by morphogenesis on the molecular and cellular level (Rossomando and Alexander, 1992).

As Thompson showed almost a century ago, morphogenesis of the hard parts can be understood to a very large extent as the result of geometric constraints in the process of growth. Hard parts are "hard" (i.e., mechanically resistant) in order to meet the requirements of mechanical functions. These are in most cases autecological, concerning protection, stability in a moving medium (e.g., for filter feeding), motility, or food uptake. If these functions supported by the shape of the hard part are to be kept up during ontogenesis, the hard part must keep its shape during growth in order to remain functional.

Many ways of growth consist of adding new elements to previous ones without changing the latter's shape, particularly in shells. The only way to keep shell shape constant under these conditions of growth is to produce a spiral geometry of the hard part which maintains similarity (in

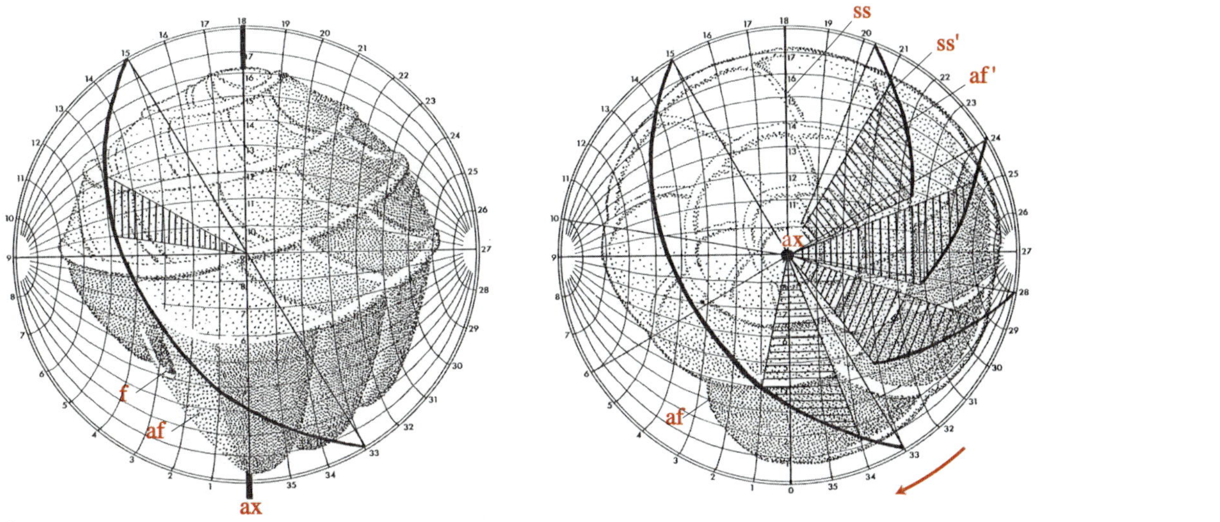

a

b

FIGURE 6 Spiral shells in a benthic rotaliid foraminifer. (a) Shell diagram in a Wulff net (projection of a hemisphere) in lateral (left) and dorsal or spiral (right) views. The inclination of the apertural face (af), with its aperture for the extrusion of the pseudopods (f), has a function for the motility of the animal. Note the constant inclination of the apertural face in successive growth steps as indicated by their projection on the Wulff net. The black delimitation of the hatched triangle (ss′) represents the projection of the septal suture (ss), and the pointed end of the triangle marks the position of the center of the umbilicus in the coiling axis (ax). The arrow indicates the direction of growth. Schematic diagram not to scale (reproduced with permission from Hottinger, 1985). (b) Camera lucida drawings of sections centered perpendicular to the coiling axis of the spiral shell of the rotaliid *Pseudorotalia indopacifica* (Indonesia; Pleistocene). (Bottom) Two megalosheric specimens; (top) a microspheric specimen illustrating the dimorphism of generations in the foraminiferan reproduction cycle (reproduced with permission from Billman *et al.,* 1980).

the geometric sense of the word) throughout growth in size by rotating the growing body around a spiral axis (Fig. 6). Therefore, spiral hard parts are widely used in foraminifera, mollusks, and brachiopods, all characterized by shells growing by linearly localized accretion.

Changes of shape in spiral shells during growth therefore indicate changes in the autecological functions of the shell and thus of the lifestyle of the organism. Of particular interest to the history of life are abrupt changes in the growth process marking the transition from larval to adult changes. Following the position of this mark within ontogeny through geologic time, we often will find a gradual reduction of juvenile stages in ontogeny within a phylum ("acceleration"). This may be used as a clock in biostratigraphy. Where ontogeny can be quantified by counting discrete growth steps, acceleration is a particularly useful, quantifiable instrument to measure geological time. However, groups having preserved discrete, countable growth steps in their hard parts are rare (foraminifera and cephalopods) or insufficiently investigated (tabulate coral and bryozoan colonies).

Quantifying shape in order to correctly describe all aspects of growth is a four-dimensional problem (in space and time) which I believe to be inadequately solved in today's paleontology. Where shells have to be sectioned to reveal early growth stages, as in all involute architectures, measurements in space are difficult or take a very long time to perform in a relevant number of specimens (serial sections as shown by Müller-Merz, 1980). Pécheux (1995) has shown for a larger foraminifer that measurements in two dimensions may be sufficient to answer relevant questions about growth and its variation (particularly depending on seasonality), but simple, linear measurements are of limited value.

In addition to the geometry of growth, the morphology of the organisms and their hard parts is heavily constrained by a multitude of factors determined by the specific history of the species and transmitted by heredity. For hard parts, the constraints are due to the mode of biomineralization linked to the devices of shaping the biomineralizate, the plan of organ construction linked to the mode of differentiation of tissues, and the specific plan of organism construction during ontogeny, including the coordination of growth of all organs within the organism. Thus, cascades of constraints are at work in addition to the autecological requirements to be fulfilled by the hard parts. All these morphogenetic factors are necessary for successful life of the organism and therefore are functional in the broadest sense. Their interplay with the hereditary conditions is responsible for the discrete, nonrandom morphology of the organisms. The morphological discreteness of the organisms inherent in their phylogenetic rise is the fundamental reason why organisms can be identified by their morphology.

Morphological Patterns. Observing, describing, and comparing different morphologies in different organisms is a matter of recognizing size, shape, and patterns. The latter are produced by morphological elements disposed in space according to rules of geometry. Patterns are by definition geometric features. They may or may not have functional or hereditary significance, and they may or may not be the consequence of morphogenetic constraints. Patterns exclusively describe a constant, mutual geometric relationship of morphological elements such as plates, bones, and teeth. Comparing different patterns, we have to identify each element in the systems considered in order to avoid comparing apples and pears. Comparable elements in different patterns are called homologous if they are generated during ontogeny by comparable rudiments (*anlagen*). The latter are interpreted to be of common hereditary origin. Homologies provide the main arguments in comparative anatomy with regard to when to interpret morphologic similitudes as genetic "closeness" and when to assign them overriding functional significance. The analysis of the fossil record may help to settle controversies arising from the discrepancy between the systematics of organisms based on their so-called homologous morphology and the analysis of their DNA in selected, conservative genes.

Patterns preserved in hard parts of fossils may be linear, developing along a line, two-dimensional, such as the arrangement of plates on a surface, or three-dimensional in space. Morphological elements may also emerge during ontogeny according to certain patterns. The latter may therefore also have a time dimension. Here, I give examples for each category.

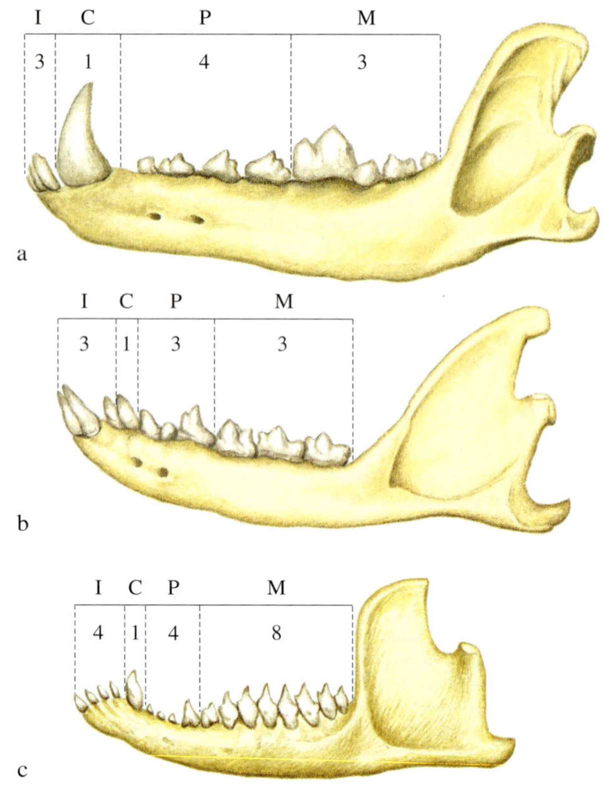

FIGURE 7 Example of a linear pattern: disposition of the teeth in the mandible of mammals. Labial view not to scale. Numbers and letters refer to the tooth formula: I, incisivas; C, canines; P, premolars; M, molars. (a) *Canis,* recent; (b) *Ptilocerus* (Scadentia), recent; (c) *Crusafontia* (Eupantotheria), Kimmeridgian, Late Jurassic (adapted from Carroll, 1987).

Linear patterns are observed in the linear arrangement of elements such as the teeth of most mammals (Fig. 7) or in the regular deformation of lines as in the cephalic suture of trilobites (sympar–opistopar), in the lobes along the sutures of gas chambers in the phragmocone of cephalopods, or in the folding (fluting) of septal sutures in foraminifera such as fusulinids or certain bolivinids and amphisteginids. These linear structures develop by intersection of two surfaces and are therefore part of a three-dimensional architecture. The identification of the homologous elements is usually given by following their genesis through ontogeny back to their rudiments. Bilateral symmetry of the structure as a whole may help to identify the elements by counts starting from the plane of symmetry. Linear patterns are often described by formulas, such as for mammal teeth or cephalopod lobes.

Two-dimensional patterns develop where surfaces of

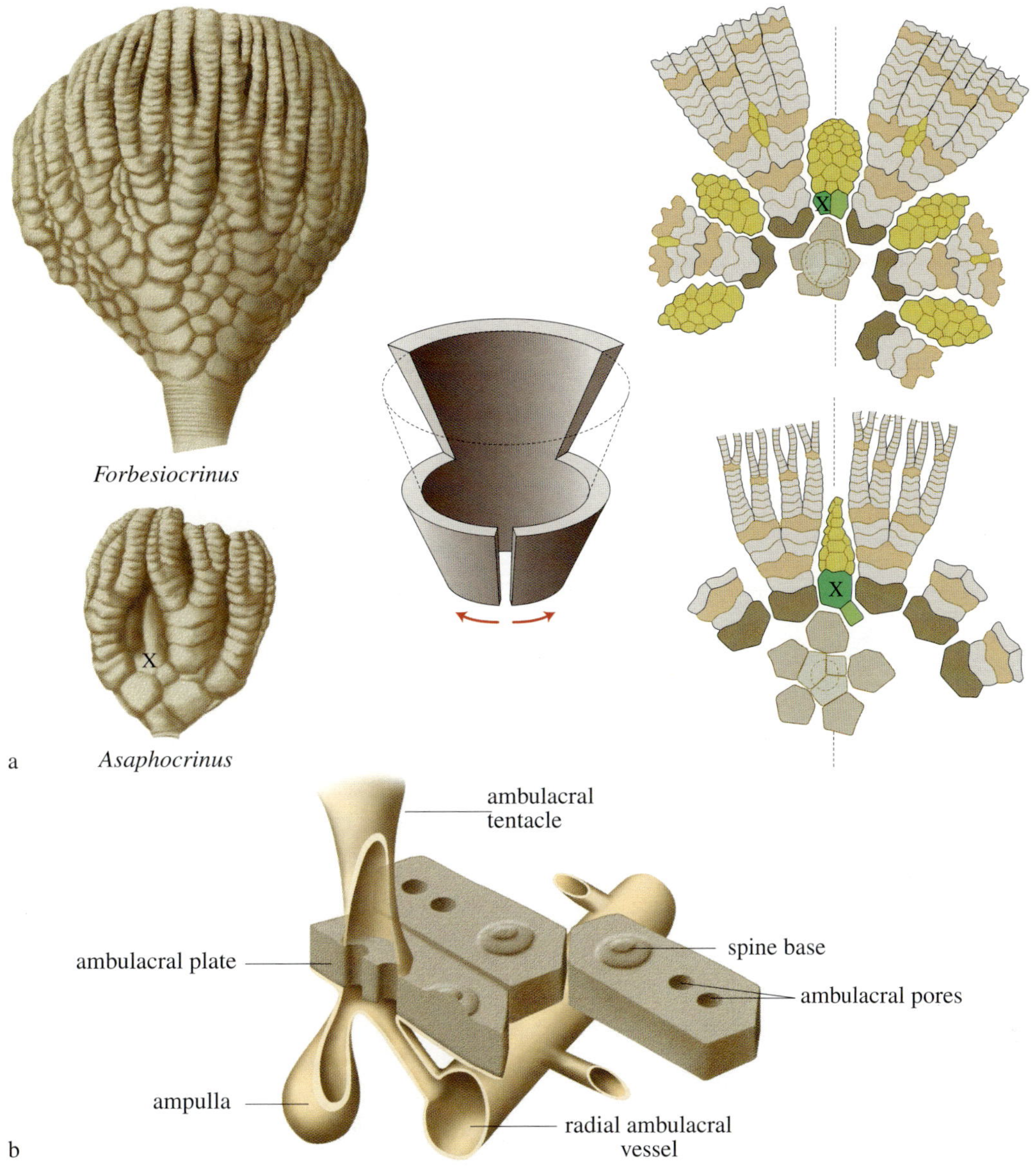

FIGURE 8 Example of a two-dimensional pattern: disposition of the plates in the calyx of two crinoids, *Forbesiocrinus* (Sagenocrinitidae) (Carboniferous) and *Asaphocrinus* (Homalocrinidae) (Silurian). The conical surface of the calyx (shown schematically on the left) is cut in the plane of symmetry (center) and extended (red arrows) on a plane surface (right; schematic stereograms not to scale) in order to represent the pattern of the disposition of the plates. The point of reference for comparing patterns is the so-called X plate at the base of the anus chimney which marks the plane of primary bilateral symmetry (dashed line) (adapted from Shimer and Shrock, 1955). (b) Example of a structural penetration in the echinoid ambulacral plate. A pair of ambulacral pores provide the communication of fluids between the radial ambulacral vessel, the ampulla, and the inflatable ambulacral tentacle through the biomineralized hard part also supporting a spine base and therefore having a mechanical function (in addition to the function of the corona as a whole).

an organism are covered by numerous hard parts forming shells, armor, or endoskeletons as hollow spheres (as in echinoids; Fig. 8). In the latter group in particular, the high specificity of the patterns permits an extensive and detailed system of classification. The identification of the elements is often based on the penetration of plates by other, linear elements such as the echinoderm tube feet perforating an ambulacral plate or their stone canal ending in the madreporan plate. Apparently, such penetrations need a particular complex package of genetic information distributed over many different genes in order to govern the coordinated growth necessary to realize the penetration during ontogenesis. The fossil record shows that the link between penetrating elements cannot be dissolved during phylogeny and therefore remains extremely "conservative." Only the structure as a whole can change its position in respect to unperforated elements or be suppressed altogether. Therefore, the controversy about the identification of the frontal scull bones in early amphibians and cross-opterygians has been solved by taking the foramen parietal penetrating the skull as a point of geometric reference (Kuhn-Schnyder, 1967). In a similar way, the innervation of bones in the splanchnocranium of the vertebrates is believed to be absolutely conservative and constitutes the basis of Reichert's theory on the middle ear (Goodrich, 1958).

Radial structures may produce two-dimensional patterns in corals or on the surface of discoidal diatom frustules. In corals, the identification of the pattern-forming radii may be helped by relating them to the pair of primary septa revealing the hidden primary bilateral symmetry of the polyp in hexacorallians, but there is no way to orient the calyx definitely up or upside-down. In diatoms, the radii cannot be identified because there is no point of reference.

Three-dimensional patterns can be demonstrated best in the disposition of stolon axes in larger benthic imperforate foraminifera (Fig. 9). Because the pattern has only minor functional significance, the distinction of different patterns leads to a classification which is often hampered by

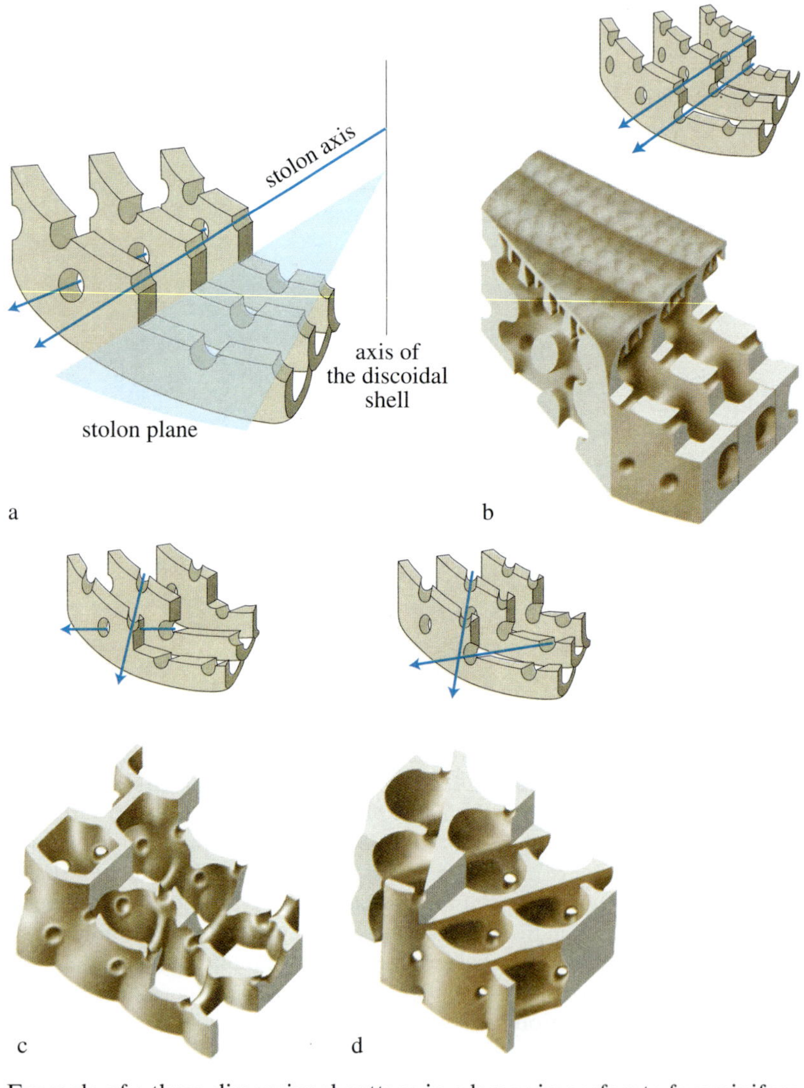

FIGURE 9 (a) Example of a three-dimensional pattern in a larger imperforate foraminifera, with the disposition of the stolon axes (arrows) and planes. In this example, the axes alternate in radial position on successive stolon planes. (b–d) Examples of different dispositions of the stolon axes (top) and of their effect on the architecture of the discoidal shells of the foraminifera (bottom). *Saudia* (Eocene) is characterized by stolon axes superimposed in radial position on successive stolon planes (b). *Amphisorus* (recent) shows crosswise oblique stolon axes, with apertures alternating in radial position on successive stolon planes (c). *Orbitolites* (Eocene) has crosswise oblique stolon axes, with apertures superposed in radial position on successive stolon planes [reproduced with permission from Hottinger (1979) and Drobne and Hottinger (1971)].

the overwhelming functionality of overall shell shape, producing many striking cases of homeomorphy.

Foraminifera also demonstrate very clearly the morphological effects when the same basic pattern is used for the same function in systems of different dimensions: So-called orbitoidal growth is defined by a two-dimensional, concentric, annular chamber arrangement in which the chamberlets occupy a position on shell radii alternating in each growth step. The same alternating positions in space are realized by a chessboard pattern in which one cycle of concentric chamberlets occupies the position of the white fields and the next layer the position of the black fields. In the first case, the overall shell shape will tend to be discoidal and in the second case it will tend to be spherical, both following a law of radial growth and both halving the volume of accretion per growth step necessary to fill the rapidly increasing chamber cavity volume prescribed by the geometry of radial concentric growth (Fig. 10).

The morphology of the adult always passes through earlier stages in ontogeny which, in the egg, is reduced mostly to a simple sphere. Whatever shapes are produced in between egg and adult during ontogeny must be functional at all times. Growth and differentiation are governed by the same constraints in ontogeny and phylogeny narrowly canalizing the morphology of the organism. This is the reason why ontogeny and phylogeny often show parallel trends of morphologic change and development with time.

Embryologists describe ontogenesis with timetables for the discrete differentiation and development of the organs within a growing organism. These timetables follow specific patterns in time (Fioroni, 1970, 1971). The fossil record produces a striking example of patterns in time in the ontogeny of the ammonite suture line as investigated by Schindewolf (1961, 1962, 1963, 1965, 1966, 1967, 1968). The sequence of the appearance of additional lobes during growth of the animal is different in different groups and follows a different pattern in time (Fig. 11). They are like music, in which the pattern is perceived as melody and/or as rhythm. The linear pattern of sutural deviation from a straight line in ammonites is comparable to the sound of a drum, whereas the ontogeny as a whole in higher organisms corresponds to a symphony performed by an orchestra, with its director acting according to the score written by genetic inheritance. Traces of this fundamental and general phenomenon of life are preserved in the fossil record only where the fossilized hard parts register discrete growth steps (the bars in the score). In the fossils of most other groups, only the final chord of the symphony is preserved. The septate shells of cephalopods and foraminifera are the most convenient fossils for analyzing patterns in the time dimension.

The similarity of patterns may be produced by their common genetic origin or by overriding functional requirements. The fossil record, if sufficiently accurate, helps to recognize functionality of a morphological feature if the latter is repeated in the stratigraphic column as part of parallel or subsequent phylogenetic trends and if these trends can be understood as having a function of selective value. In this respect, the fossil record constitutes an important key to the present.

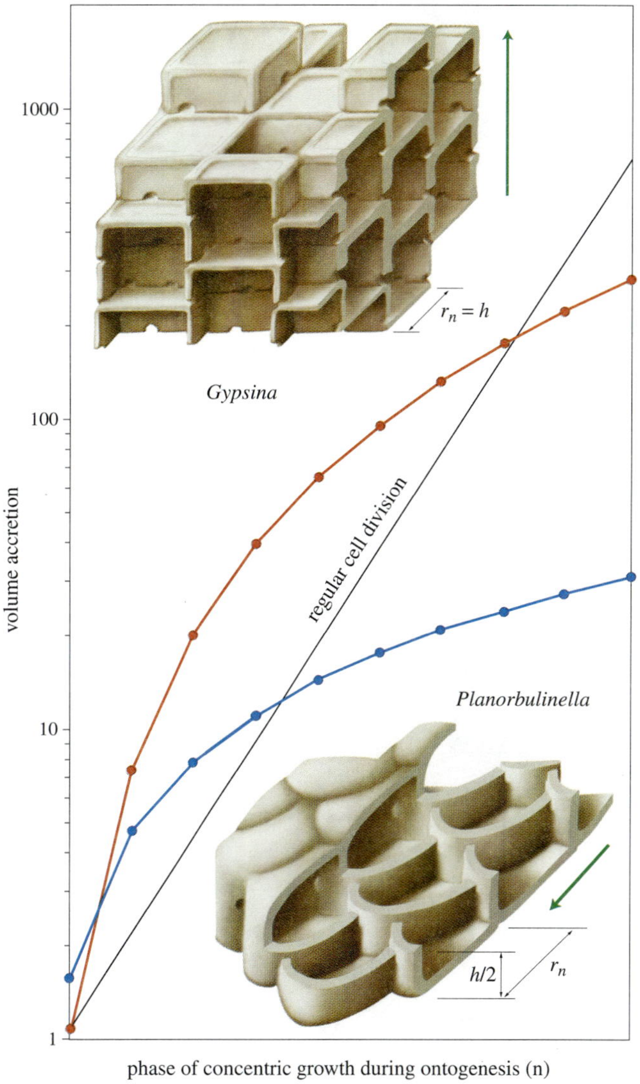

FIGURE 10 Volume accretion rates during discoidal (in blue) and spherical (in red) concentric growth of foraminifera having chambers with a alternating radial disposition ($\Delta V = V_n - V_{n-1}$) in comparison with a regular cell division ($V_n = 2V_{n-1}$), where V represents the volume at each growth stage (instar) or generation of cells in division, and n is the number of successive stages in the ontogenesis of the cell generations. The radial length (r_n) and the height (h) perpendicular to the radius of the chambers are normalized to the same dimensions, according to the essential function of each chamber, considered as an isotropic bioreaction chamber. Two-dimensional growth, as observed in *Planorbulinella,* produces discoidal shells; a three-dimensional accretion, as in *Gypsina,* generates spherical shells in which the chambers are spaced in the so-called chessboard model. The green arrows indicate the directions of growth (adapted from Hottinger, 1981).

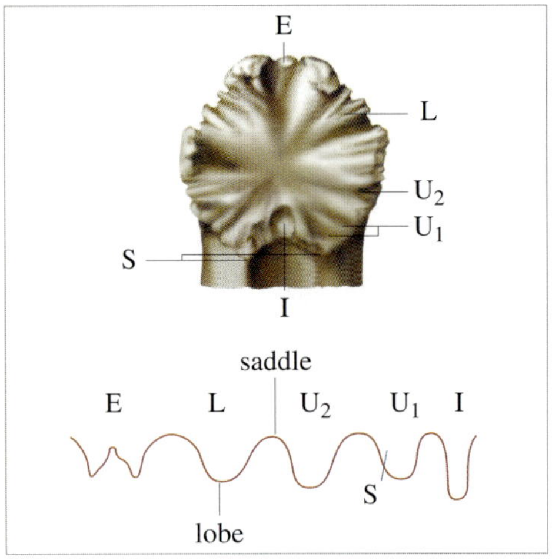
E
L
U_2
U_1
S
I
saddle
E L U_2 U_1 I
S
lobe

a

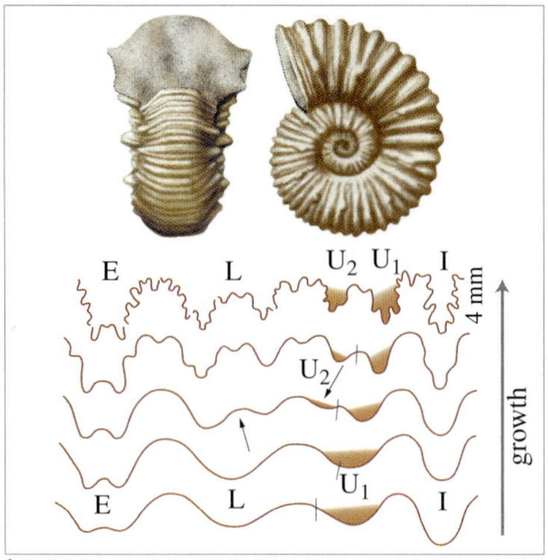
E L U_2 U_1 I
4 mm
U_2
growth
U_1
E L I

b

Cheloniceras

c

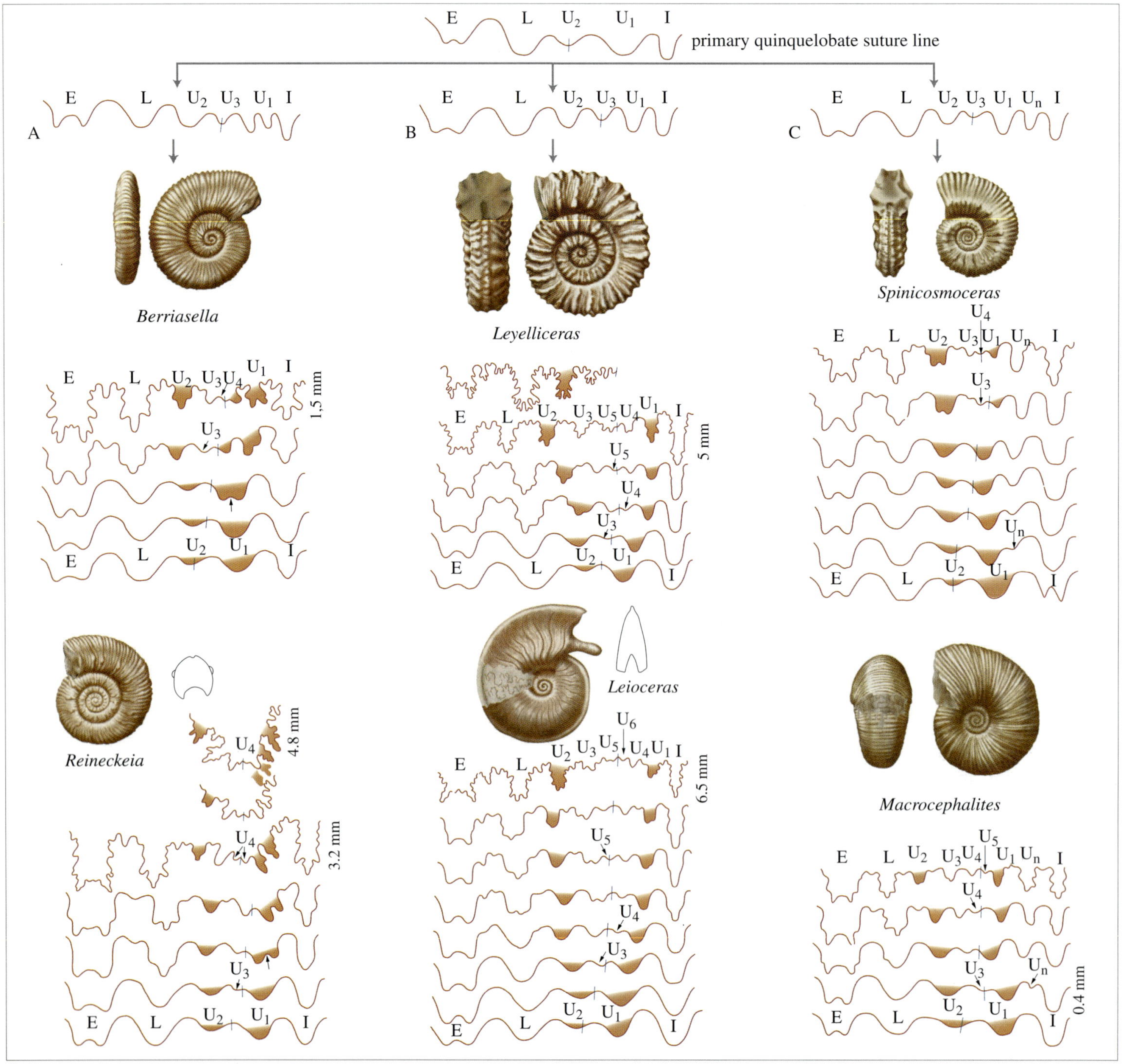
E L U_2 U_1 I
primary quinquelobate suture line
A
E L U_2 U_3 U_1 I
B
E L U_2 U_3 U_1 I
C
E L U_2 U_3 U_1 U_n I
Berriasella
Leyelliceras
Spinicosmoceras
E L U_2 U_3 U_4 U_1 I
1,5 mm
U_3
E L U_2 U_1 I
E L U_2 U_3 U_5 U_4 U_1 I
5 mm
U_5
U_4
U_3
E L U_2 U_1 I
U_4
E L U_2 U_3 U_1 U_n I
U_3
U_n
E L U_2 U_1 I
Reineckeia
4.8 mm
U_4
3.2 mm
U_4
U_3
E L U_2 U_1 I
Leioceras
U_6
E L U_2 U_3 U_5 U_4 U_1 I
6.5 mm
U_5
U_4
U_3
E L U_2 U_1 I
Macrocephalites
U_5
E L U_2 U_3 U_4 U_1 U_n I
U_4
U_3
U_n
0.4 mm
E L U_2 U_1 I

d

FIGURE 11 An example of patterns in the dimension of time: successive fluting of ammonite lobes with notations as lobe formula and diagrams according to H. Schindewolf (1961, 1962, 1963, 1965, 1966, 1967, 1968). (a) Front view of a lytoceratan chamber wall. The line E–I indicates the position of the plane of bilateral symmetry. The lobe diagram (bottom) represents only half of the chamber periphery. E, external lobe; L, lateral lobe; I, internal lobe; S, position of the whorl suture with respect to the lobes, indicated by a blue line in the lobe diagrams. (b) Ammonite phragmocone (chambered part of the shell) and lobe diagrams of early growth stages. The successive fluting diagrams are selected from the complete ontogenetic set when a lobe or a saddle is split—up to a given whorl diameter (indicated in millimeters). During ontogenesis, later chamber sutures may be further complicated by second- and third-order serrations. The lobe diagrams of the successive chambers are drawn at a comparable size and the umbilical lobes U_1 and U_2 are shaded in order to facilitate the comparison between the various stages. Note the splitting of the flutes in the distal direction (lobe splitting; not noted in the lobe formula) and in the proximal direction (saddle splitting) noted in the lobe formula as subscripts (U_1, U_2, etc.) according to the order of appearance in ontogeny. In this particular case (*Cheloniceras,* Douvilleiceratidae, Aptian), the pattern does not change with time, and the shape of the lobe diagram (ELU_2U_1I) remains constant. (c) Early growth stage, much enlarged, of a heteromorph ankyloceratan juvenile shell showing an early uncoiling covered by later regular coils, characteristic for many Late Cretaceous douvilleiceratids. The arrow indicates the approximate position of primary chamber suture. (d) (Top) Ammonoid basic patterns of septal fluting (ELUI). From the primary, quinquelobate suture present in all Mesozoic ammonites (except the quadrilobate phylloceratids, the small group of the tetragonites, the lytoceratids, and the ankyloceratids) derive the split primary umbilical lobe U_1 typical of psiloceratids, hammatoceratids, hoplitids, and perisphinctids (A); the undivided umbilical lobe, characteristic of the hildoceratids, the haploceratids, and the acanthoceratids (B); and the additional umbilical lobe U_n between U_1 and I produced by saddle splitting, typical of the stephanoceratids (C). In all groups, additional umbilical lobes $U_{3,4,5}$, etc. may be generated by saddle splitting in the umbilical region. (Bottom) Phragmocones and lobe diagrams in Jurassic–Cretaceous ammonites, derived from the patterns A–C. Modifications of pattern A are represented by *Beriasella* (latest Jurassic), in which the early splitting of U_1 is visible, and *Reineckeia* (Callovian, Middle Jurassic), in which the lobe diagram also shows an early splitting of U_1. Examples of modifications of pattern B are given by *Leyelliceras* (Albian) and *Leioceras* (Early Middle Jurassic), which has a complete shell with living chamber and its aperture including its auricles. Note the regular left–right addition of new umbilical lobes during ontogeny. Modifications of pattern C are evident in *Spinicosmoceras* (Callovian, Middle Jurassic) and *Macrocephalites* (Callovian), in which early saddle splitting produces U_3 and U_n on both sides of U_1 at an early diameter of 0.4 mm. Apparently, there is no connection between the tune of septal fluting, the shape, and the ornamentation (ribbing pattern and spines) of the shell.

The Significance of Size. The size of living organisms, independent of shape and body plan, is limited for multiple metabolic reasons to certain orders of magnitude. The macromolecular dimension determines the lower limit of size in an organism. No organism is known, either living today or preserved in the fossil record, to surpass several decades of meters in size. Consequently, the much larger surface of the earth harbors many independently functional, autonomous units of life (i.e., many organisms and not a huge, single one covering the total global surface).

Organisms interact within populations and communities with each other and with environmental conditions changing in space and time. Different environments demand different responses to be given by different kinds of organisms. The production of different kinds of organisms is considered to be due to evolutionary adaptation and, thus, to a historical process. There are large and small organisms of different kinds, and there are small, young and large, adult organisms of the same kind living together in the same community. Their size in relation to that of their partners in the community plays many important roles in the life of the community. This can be demonstrated by the fact that in many communities a very restricted number of populations coexist which have a very similar, often congeneric, morphology and usually a common history but very different adult sizes. These are called *odd partners* or, if there are only two of them, *odd pairs.* Extant odd partners in foraminifera live in the same habitat (e.g., on the same sea grass leaf) but reproduce at different periods in the seasonal cycle and have different growth rates and different reproduction periodicities. Odd partners are frequently observed in taphocoenoses of benthic foraminiferal shells (Hottinger, 1999) and can be found in other groups as well.

Adult size of an organism is related not only to the size of its habitat but also to the length of time during which habitat and food or nutrient resources are available. During this period of stable conditions, the organism has to grow and reproduce if the corresponding population is to survive. Large-sized organisms will need stable conditions over a comparatively long time in order to live through a long ontogeny up to their large size. For example, benthic foraminifera of the size of 0.5 to about 2 mm in diameter will live through about one or two dozen growth stages in a few weeks, whereas large-sized forms reaching 10–15 cm in diameter after several or many hundred growth steps may live several years. Thus, large-sized organisms are exposed for a longer time to the conditions of their environment and therefore will be under higher pressure to adapt their autecological functions to these particular conditions. In particular, the hard parts of the larger organisms will be highly adapted and their fossil remains will tell us more about their way of life than the fossils of smaller organisms. In this respect, the fossil record will also be biased by the relative size of the organism.

Sexual versus Clonal Reproduction. Modes of reproduction are rarely visible in the fossil record. An interesting exception are the benthic foraminifera exhibiting a reproduction cycle in which sexually and asexually produced generations are alternating, usually in relation to the seasonal cycle. The alternation in the mode of reproduction often produces a dimorphism in the first growth stages of the shell. The foraminiferal zygote, a product of sexual reproduction, is coated by a tiny biomineralized envelope of spherical shape, called a *microsphere,* which is preserved as the initial growth stage of the shell. The following generation is asexually produced by multiple mitosis. Each daughter cell will inherit, in addition to the double-chromosome set, a part of the mother's protoplasm and eventually some of the mother's symbionts. The volume of this initial stage in ontogeny (called *proloculus*) is large in respect to the zygote and therefore has a larger envelope, the *megalosphere.* Eventually, the megalospheric generation starts its ontogeny with a particularly voluminous "embryo" covered by a shell with its own complex architecture. Since there is a dimorphism in early stages of shell growth registering the mode of reproduction, we can investigate the relation of sexual versus asexual reproduction in the fossil record through geologic time.

Sexual reproduction will be favored by selection where a high genetic variability provides advantages (Bell, 1988; Law and Lewis, 1983). Genetic variability is called for under antagonistic conditions of life: In a food chain, for instance, the characters dominating the populations of the predator and the prey must be well equilibrated in order to maintain both populations over long periods of time. In this situation, minoritarian variants tend to be "overlooked" by the antagonistic mechanisms and may therefore have better chances to survive. Here, selection would work in favor of sexual reproduction enhancing variability.

Under mutualistic conditions (e.g., within symbiontic relationships), equilibrated characters dominating host and symbiont populations will enhance the symbiosis and favor both partners, whereas minority variants will be put under increased selection pressure as long as the perfectly matching genomes of host and symbiont remain stable. Thus, cloning will be a good solution to keep the genome perfectly identical. In order to avoid degeneration, mechanisms in addition to sexuality are frequently at work to guarantee the viability of the genome (e.g., by polyploidy).

In foraminiferal populations, the number of individuals produced by sexual reproduction (microspheric specimens) relative to the number of asexually produced, megalospheric specimens is about 1:20 or less in short-lived smaller benthics without symbiosis and about 1:100 to 1:10,000 in symbiont-bearing, large-sized groups. Planktic foraminifera reproduce exclusively by a sexual process despite the fact that some important species bear symbionts. This might be explained by the preponderant significance of predation on the zygotes produced in depth below the main predation pressure by the zooplankton in the photic zone. However, we are still just beginning investigations in this direction in order to support the assessment of life strategies in the fossil record.

The dimorphism of foraminifera may extend to the ultimate growth stages and thus also separate the generations by adult size. The microspheric generation always reaches a larger size than the megalospheric generation. Starting with a very small proloculus and reaching larger sizes, the microspheric generation must have a much longer ontogeny than the megalospheric, cloned generation. Keeping in mind the significance of odd partners discussed previously, we may interpret the size difference of the generations within a population in a similar way as reflecting different lifestyles but here within a genetically homogeneous population instead of an odd association of two closely related species. In some extant larger foraminiferal species, we know the megalospheric, short-lived generation with a voluminous initial volume, capable of rapidly initiated vigorous growth, to be linked with the seasonal input of nutrients in an oligotrophic environment, whereas the microspheric, long-lived generation has to survive at least one entire yearly cycle that includes at least two periods of starving in mid- to late summer and during winter. No wonder that the short-lived, opportunistic generation is constituted by clones, whereas the microspheric generation, exposed to heavy predation during the earliest phase of life as naked zygote and subsequently submitted for a long time to cyclically variable environmental requirements, is produced by sexual reproduction.

Life Strategy as a Level of Abstraction

Analysis and understanding of hard part morphology may lead to an interpretation of the lifestyle of organisms and populations within their communities and their ambient environments. In order to characterize their role in Earth history, it is useful to classify the lifestyles in categories of life strategies. Ecologists distinguish between strategies of life responding to constant or at least predictably cyclical availability of resources versus randomly distributed or erratically short-lived resources. In the first case, the carrying capacity of the space considered tends to be constant in time, and in the second case the carrying capacity will fluctuate. A constant carrying capacity is exploited best by keeping the density of the population in the space considered as constant as possible, whereas the more opportunistic strategy to exploit to their full capacity the limiting resources, available for a short time only, tends to produce populations irregularly fluctuating in numbers over time. Along this line of typological considerations, a linear gradient of life strategies is conceived. Parry (1981) produced a useful list of general criteria illustrating this concept. At its poles, a so-called *K* (for constant) *strategy* is opposed to an *r* (for high rates of growth) *strategy*. Adaptations to these strategies can be recognized in the morphology of the organism: K strategists will have long lifetimes, comparably large adult sizes, slow ontogenies, and highly developed and complex autecological adaptations and devices to stretch limiting resources by mechanisms of recycling (e.g., with symbiontic relationships). The growth of large, fixed colonies of filter feeders in the sea and of forests on land are another important expression of K strategy in which the space on solid substrate is the limiting ecological factor.

r Strategies can be recognized by the comparably small size of the organisms involved, constituting communities of low diversity. The organisms following this kind of strategy will exhibit rapid growth during ontogeny, short lifetimes, and high (usually exponential) rates of reproduction in order to profit from the temporal offer of resources.

Life histories, as characterized by rates of growth and reproduction, can be measured under experimental conditions in the laboratory and consequently modeled (Stearns, 1992). The fossil record provides only very limited possibilities to quantify growth and reproduction rates in relation to the individual life of an organism. In larger foraminifera, however, a regular deviation from the mean volume accretion rate per growth step must be interpreted as controlled by seasonality. The mean number of septa per cycle of deviation remaining constant during ontogeny permits the interpretation of the septa as regular markers of time during ontogenesis. Therefore, growth rates can be investigated in selected fossils preserving discrete growth stages in order to test the predictions of models for the evolution of life histories over geological time periods. Viable techniques to carry out the necessary measurements with affordable efforts still have to be worked out.

Mutualistic relationships between organisms will develop antagonistic relationships toward the r pole of the K–r gradient more often in the realm of K strategies. These relationships may be reflected in groups of fossils showing alternating sexual and asexual reproduction cycles as explained previously.

The linear K–r gradient of life strategies and its triangular elaboration by Grime (1977) are very simple and insufficient to classify ecosystems to a degree that every question can be answered. However, it provides a first step for understanding historical documents by explaining the correlation of certain types of morphology consistent with the notion of life strategy with certain types of sediments reflecting some general ecological situations in space or time, such as the trophic situation of an ecosystem.

Taxonomy as a Level of Abstraction

Beyond the concept of the organism as an autonomous functional unit of life, paleontology has to operate with units grouping the almost infinite number of individuals according to their morphological characters in some kind of system (Fig. 12). The ongoing debate about a meaningful biological system of classification cannot be reviewed here due to space constraints. However, I have to stress the fun-

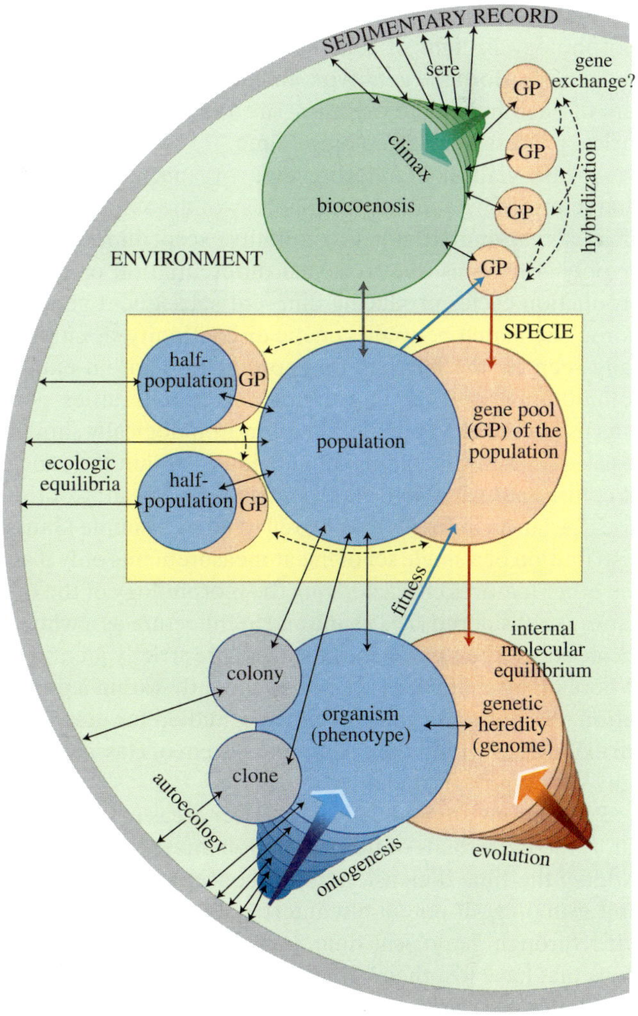

FIGURE 12 Organism, population, and community: their relation to time and to the environment. Note the concept of species on a higher level of abstraction.

damental difference between the paleontological and the neontological taxonomic systems despite their common formalisms in the international code of nomenclature following the Linnean tradition.

The neontological concept of classification is based on discontinuities in genetic variation; the paleontological concept must be based on discontinuities in the morphological variation. The latter is interpreted as the consequence of the former, justifying the use of common formalisms. The criteria for genetic discontinuity are given by the empirical observation that certain genomes are capable of successfully mating during the act of sexual reproduction, whereas others are not. The criteria for morphologic discontinuity related to genetic discontinuity are given by their empirical consistence in geological time as a phenotypic response to evolutionary processes. Neontology has developed an extensive discipline on the macromolecular level to understand variation in the genetic record present in each individual. Paleontology must develop its understanding of morphology in light of functional morphology and morphogenesis. Both the genotypic and the phenotypic characters of each individual organism are the result of the historical process during geological time.

An association of interbreeding organisms is usually characterized by a similar morphology of the associated individuals varying without discontinuity except dimorphisms or polymorphisms due to sexual differentiation or to reproduction cycles producing different subsequent generations. The match between genetic discontinuity, as empirically seen by the barriers for reproduction within extant communities, and the morphological discontinuities permits use of the term population for morphologically similar fossil specimens, including all individuals within the range of continuous variation. Here, I mention a caveat: Continuous variation in fossils may be reflected by a simple Gauss distribution of a single set of linear measurements only if all the other features characterizing the morphology of the organisms considered (i.e., shape and architecture as a whole, including all their ontogenetic stages) are strictly identical. Where many characters vary independently within a population, discontinuity can currently be studied for meaningful extensions within the taxonomic system of classification only with typological methods.

Populations are limited in space over areas extending as far as the individuals of the association interbreed actively. Adding the time dimension to this concept, we conceive a unit grouping all potentially interfertile organisms worldwide through geological time, including the metapopulations that have lost their genetic exchange with the mother population for some time. This unit is called *species* and has a historical dimension. A fossil species is delimited in time by its first appearance and by its extinction, whereas a recent species is delimited by its first appearance only.

Populations belonging to different species live together in a certain space or over a certain area in a unit, which is also delimited by discontinuities. This unit, the *community,* represents a state of equilibrium between its participants as well as between the participants and their ambient environment at a particular moment of its dynamic history. The latter is called a *sere,* which is easy to observe where cultivated land lies fallow long enough so that it returns to its natural state. Periodic destructive turnovers in the environment, such as seasonal fires in forests of dry areas, storms in shallow seas, or turbidity currents in deeper basins, as well as agricultural activities may keep the sere from developing into its full maturity, depending on the periodicity of the environmental dynamics. Historical documents usually reflect mature communities (climax) because only the mature stages of a sere exist long enough to have a chance of being preserved in the record. They are described under the "geological" notion "facies." Immature communities may be preserved only where the environmental periodicity is very regular for a very long time, such as the seasonal turnover of macroids in shallow seas (Hottinger, 1983).

The time dimension of seres is short with respect to evolutionary processes: The recovery of the vegetation cover of a Pacific island after the explosion of a hydrogen bomb has taken about half a century, and the recovery of the shallow benthic marine fauna under oligotrophic and warm conditions after the event at the Cretaceous–Tertiary boundary has taken 8–10 million years. The difference in speed of the recovery is due to the fact that in the first case the genetic information for all species involved in the community was preserved outside of the area of destruction. The recovery is therefore the result of immigration of populations. In the second case, the fossil record demonstrates that many species became extinct over a remarkably short period of time at or eventually beyond the resolution power of biostratigraphy. Where the genetic information was lost, the recovery of the communities was due to the evolution of new genomes. The corresponding phenotypes constitute communities which are different from the previous ones but equivalent with regard to their ecological significance. For instance, endosymbiosis, relative large size and long ontogenies, strong dimorphism of generations, and striking diversities characteristic of K strategist foraminifera in shallow oligotrophic tropical seas are restored in the phase of recovery.

Disparity of Morphological Units. The characters permitting the discrimination of morphological units and the definition of the discontinuities in their variation are of two kinds — qualitative and quantitative. Qualitative characters are present or absent. By definition, there can be no intermediates or variation. They are usually described as particular patterns, textures, structures, or architectures as previously explained. Quantitative characters of morphology vary, are usually measurable by linear measurements, and are circumscribed by notions of size, shape, or proportion. The separation of discrete morphological units is comparatively easy with qualitative, present–absent characters, and it is more difficult with the quantitative ones, which call for statistical treatment of measurements or counts and which may leave doubts about which unit to assign the specimens falling in an overlap of extreme variants.

In the exceptionally well-documented fossil foraminifera, the qualitative characters are constant over much longer periods of time than are the quantitative ones. The latter tend to change gradually in time in the same direction. If the direction of change can be understood as progressive adaptation by some insight into the functional meaning of the morphological character considered, we interpret this latter quantitative change as adaptation reflecting phylogeny and use it as biostratigraphic clockwork.

The succession of discrete morphological units in time as documented by the fossil record is the prominent objective argument to distinguish between grades and clades. At the same time, it permits the establishment of at least two levels in a hierarchy of characters, the specific and the generic levels, based on an objective criterion and in accordance with current evolutionary interpretations.

There is a consensus among workers on foraminifera to consider qualitative characters to be of generic rank and quantitative characters to be of specific rank in Mesozoic and Tertiary to recent larger benthics. However, there are some qualitative characters, such as the presence or absence of additional canaliform cavities in the polar basal layer of *Alveolina* (Hottinger, 1963), that depend on threshold values of quantitative characters (in this case, the elongation index). Complications of this type often remain hidden if the documentation of the group through time is not very accurate. Therefore, hierarchical levels in the traditional Linnean system may be more significant for answering historical questions than the numbers of branchings in cladistic systems of classification.

There is an additional argument to assign some reality to genera: Comparing the specific and the generic diversity in the populations over an ecological gradient, the relation changes in favor of specific diversity toward optimal conditions and in favor of generic diversity toward stressing conditions while the diversity of the community decreases (Fig. 13). This means that under conditions of stress, fewer discrete populations of more dissimilar nature characterize the community, whereas under optimal conditions the communities are composed of higher number of discrete populations of less dissimilar nature. Thus, disparity (i.e., the "morphological distance" between two discrete units) can be measured by the number of hierarchical levels of classification between them. Whether the higher levels of classification (families, orders, classes, etc.) have a significance beyond the needs of a complex formal classification depends on the accuracy of the fossil record in time and on the quality of the reconstruction of the history of the particular group of organisms concerned by the degree of morphological understanding of the characters used for classification. Current systematics of lower invertebrates should not be taken at face value for decendence, nor should they be used for quantification of diversity because the supraspecific levels of the hierarchy are poorly and arbitrarily defined in many cases. Therefore, units on the same level of

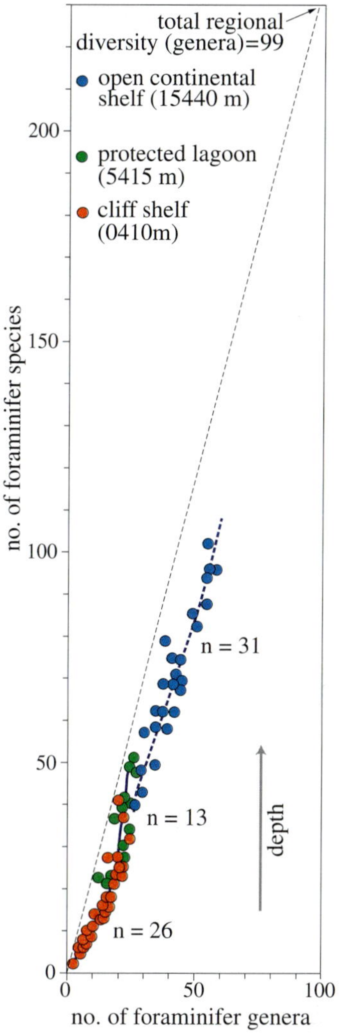

FIGURE 13 Relation of specific to generic foraminiferal diversity in tropical shallow waters. In shallow water, under the stress of photoinhibition and of disturbance by high turbulence, the diversity is low, with each genus represented almost exclusively by a single species. In greater depth, at which maximum diversity is reached under optimal conditions, many genera are represented by two or more species. Data collected at Lizard Island, Great Barrier Reef, Australia (reproduced with permission from Hottinger, 1988).

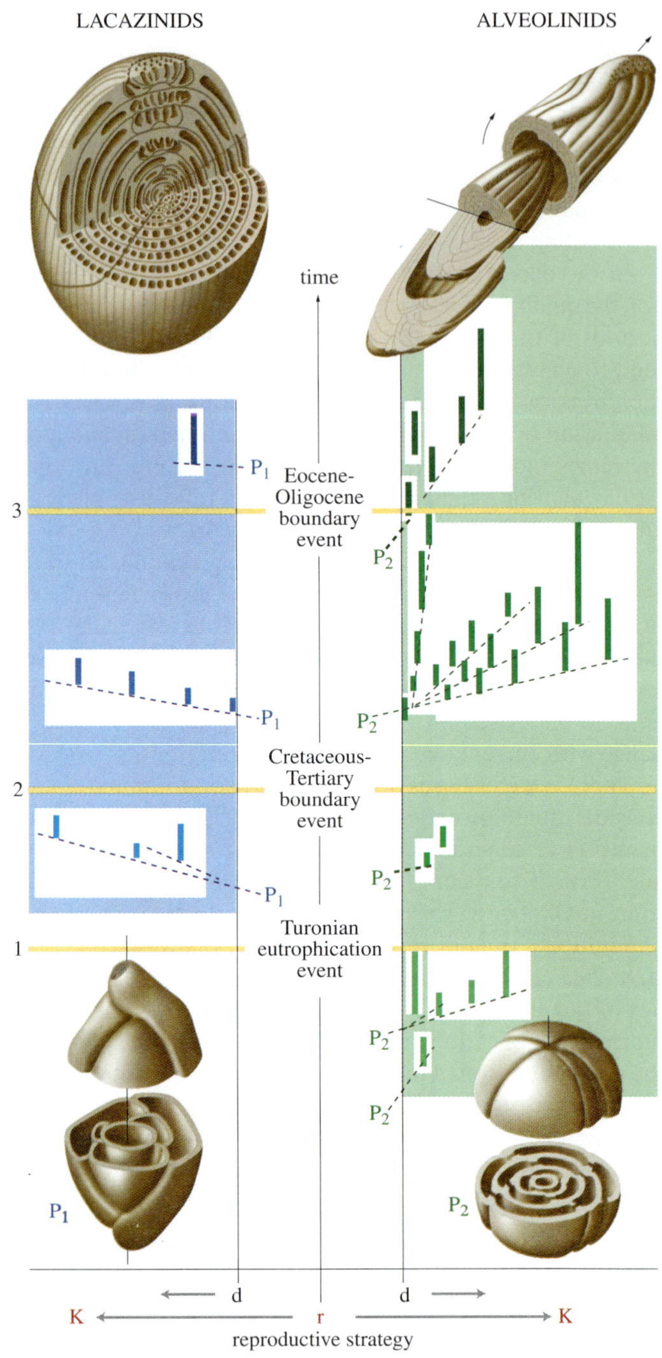

FIGURE 14 Significance of taxonomic hierarchy in the current systematics of Lacazinidae and Alveolinidae (Foraminiferida). Vertical axis, time; horizontal axis, life strategy gradient r and K, as reflected by morphology; d, limit between divided and undivided chambers; 1–3, discontinuities between global maturation cycles of the community). The colored surfaces indicate taxonomic units of family rank: in blue, the lacazanids, characterized by miliolid coiling with apertural axis, adult habit globular concentric; in green, the alveolinids with early streptospiral coiling, adult habit planispiral involute, tending to become fusiform. The dashed lines indicate the supposed descendence within a genus. P_1, progenitor species with miliolid habit and undivided chambers, characterized by an apertural axis and by the apertures in the polar position; P_2, progenitor species with pseudonummuloculid habit and undivided chambers, characterized by a stabilized coiling axis, at least in the adult stage, and by low slit apertures in the equatorial position. Note the very different weight of the genera in terms of species diversity.

the hierarchy may have a different meaning and may not be comparable (Fig. 14).

Timescale of the Phanerozoic

Biostratigraphy and Its Resolution Power in Time

Biostratigraphy is primarily an empirical science: One predicts the relative position of communities in a succession of sediments of comparable facies and verifies or falsifies the hypothesis by repetition of direct observations in the sedimentary record until the sequence of communities can be statistically proven (Hay and Southam, 1978; Guex, 1991). In many cases, particularly in the biostratigraphy of mammals and foraminifera, the assessment of sequences of communities in time is supported by evolutionary hypotheses defining the adaptational trends to perfection of functions, such as the progressive folding of enamel in the teeth of horses or the accelerated nepionic growth in radial foraminifera (Drooger, 1993). Many of these trends, if reduced to linear measurements, evolve with different speed in parallel, closely related phyla that nevertheless may control each other if present in the same taphocoenoses.

The smallest time unit in biostratigraphy is called a *biozone.* Since they are the smallest time units by definition, the limits between adjacent zones remain undefined and cannot be correlated because there is no adequate measuring method permitting the definition, with a significantly tighter scale, of the position of the zone limit. Therefore, the correlation of the biozone, according to a typological concept, concerns not its limits but its core (Pignatti, 1998).

Based on evolutionary processes, each biozone is a unique, individual unit bearing a name according to its respective index fossil. Recently, some widely used micropaleontological biozonations received a numerical designation in order to avoid frequent updating to keep up with the rapidly growing taxonomic knowledge: The number of taxa approximately doubles during each generation of scientists. However, this numerical designation does not abolish the individual identity of the biozone.

Biozones are defined by the appearance and disappearance of index species in relation to other species in the time gradient, as recorded in superposed sediments. Is the empirical notion of synchronized appearance of an index species over the extension of its eventually worldwide area compatible with the concept of allopatric speciation in evolutionary theory? In benthic foraminifera, the associations of many zonal index species are the same within their faunal province covering approximately a distance of more than 10,000 km. In the period of time corresponding to a biozone (i.e., approximately 0.5 to 1 million years), several new species would have to appear at a particular site anywhere in the faunal province under consideration and would have to spread all over the province in so short a time so as not to be noticed in the time frame of the biozone. The most elaborate example from extant situations is the spreading rate of the so-called lessepsian migrants (Por, 1978)—Red Sea species penetrating the Suez Canal, and from there the eastern Mediterranean, for a distance of 1000 km in 100 years. Lessepsian migration into and through the Suez Canal proceeds against the current in the canal and over seasonal salinity barriers. No migration in the opposite direction has been observed. There are 174 migrants known, belonging to many classes. The migration rate is very similar, independent of the mode of moving and of reproducing, for arthropods, mollusks, fish, and foraminifera—about 1000 km in 100 years. We may accept this migration rate as a crude model of migration velocity despite its possible dependence from something we might call "ecological attraction," an apparent disequilibrium in the diversity of the communities built up over millions of years by a land barrier separating the two marine basins involved. Admitting a spreading rate of the same or even of one lower order of magnitude for a new species of allopatric origin (i.e., 1000–10,000 years over the extension of the province), the appearance of the species would indeed be perceived as synchronous in the fossil record.

Biozonations are particular to facies representing the continental or the marine realms and, within the latter, the planktic, the shallow benthic, the open or restricted marine, or the deeper oceanic (below the calcium carbonate compensation depth) ecosystems. Each respective environment of deposition houses different groups of organisms evolving separately. Consequently, there are many parallel biozonations to be correlated. The correlation between the continental and the marine realms is the most difficult because the only direct link between the two realms, the pollen and spores blown out to sea, has a weak potential of conservation in the near-shore carbonate facies preserving the most important historical information in shallow water.

In order to produce a unified and worldwide timescale, the numerous parallel biozonations have to be integrated by a complex puzzle of correlations and intercalibrations according to a huge amount of data. The unified timescale uses as the smallest unit periods of time which are significantly longer than those of the biozonations. During some periods of Earth history, the difference in time between the most accurate biozonations and the standard time unit may reach one order of magnitude, and during others it may be less. The elementary units of the universal timescale are called *stages*. In contrast to zones, the limits between subsequent stages are measurable by the much narrower scale of biozones.

Stages are typological units with type sections at type localities. According to a Swiss stratigraphical commission (Trümpy, 1973), the time interval of a stage is defined as starting with the earliest biozone recognized in the type section and ending with the last biozone preceding the first biozone in the type locality of the subsequent stage. This is necessary because the type localities usually represent only part of the time corresponding to a stage (Fig. 15).

The stages, conceived with varying definitions throughout 150 years of scientific activity, originate from the direct observation that fossils in cyclical deposits on passive continental margins change significantly from one cycle to the next. Most type localities were designated in outcrops showing this kind of sediment because of their overwhelming richness in fossils. Therefore, the subdivisions of the timescale into stages are predetermined to some extent by the

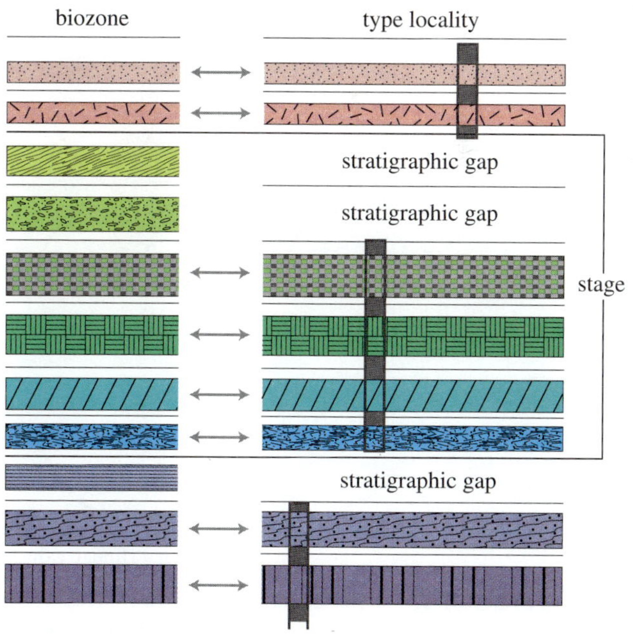

FIGURE 15 Time interval of a stage, a typological unit with type sections at type localities, measured by biozonation. A stage begins with the earliest biozone recognized in the type section and ends with the last biozone preceding the first biozone in the type locality of the subsequent stage. The arrows indicate biostratigraphic correlation.

periodicity of relative sea level rise and fall. The periodicity of the oceanic water level is said to have two independent sources: (i) the quantity of ice fixed in the polar regions of the earth and depending on the climatic temperature gradient from the equator to the poles and (ii) the amount of water displacement by the midoceanic ridges rising and falling according to the geothermal gradient created along the sources of ocean spreading. The interaction of the two main factors determines the relative level of the sea surface. Whether or not there are indirect relations between the two factors because of changes in large-scale geographic patterns (which may determine the global climate due to changes of the heat flow patterns in the oceans) is of no relevance since their periodicity is definitely unrelated.

The accuracy of the biozonal timescales depends to a large extent on the accurate perception of the index taxa and their morphology. The perception of the taxa depends on the accuracy of the sampling of the fossil record in the superposed sediments (Fig. 16). The resolution power of the biostratigraphical methods, however, is limited to the point at which the phenotypic morphology of two discrete units reaches the limits of discrimination. In many groups of fossils, the current accuracy of the biostratigraphical record is insufficient to exploit the morphology of the fossils to their very limits.

Because comparatively little is known about the intraspecific variation of the genome and the nature of discontinuities in its variability, we do not know whether the least significant difference of means in morphological characters corresponds to barriers for interbreeding or, in a more general sense, to relevant discontinuities in genetic variability. DNA analyses of conservative genes suggest that discontinuities in genetic variation may be even more closely spaced than the morphologic ones, as observed in planktonic foraminiferans (Vargas *et al.*, 1999). If this is also true for organisms with a more complex morphology, we will overcome one of the contradictions between the fossil record exhibiting slow, almost continuous "microevolutionary" change with time and the theory of punctuated equilibrium (Eldridge and Gould, 1972).

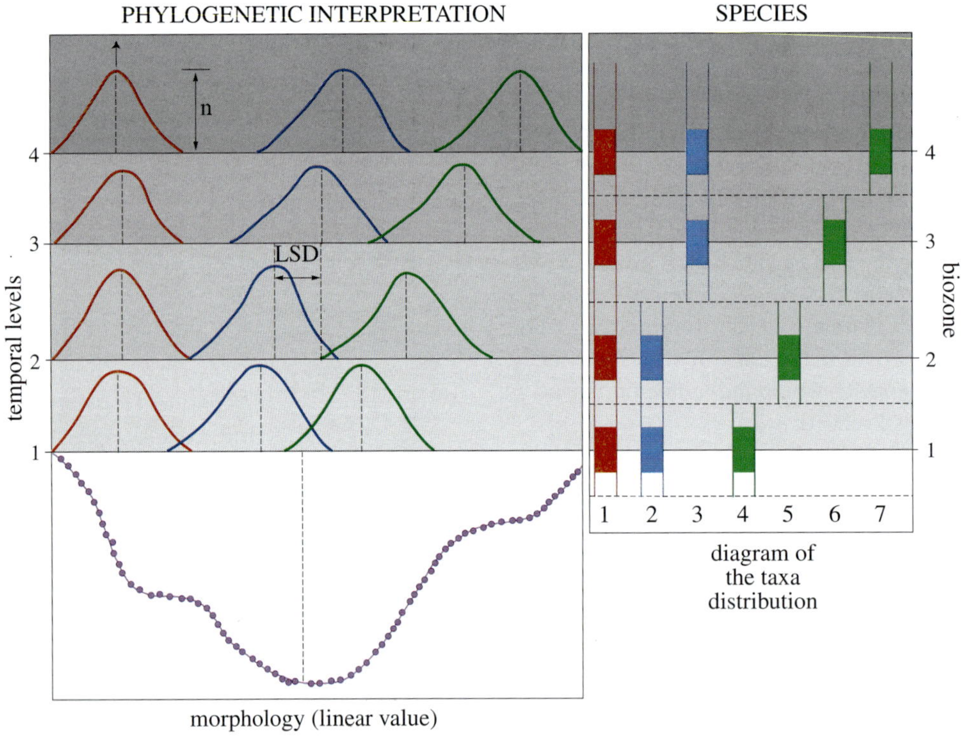

FIGURE 16 Resolution power of the biostratigraphic clock, discontinuity of the biozones, and their dependence on the accuracy of the stratigraphic record. The different colors indicate the phylogenetic interpretation of the fossils. The lower graph indicates the relative value of the morphology of the fossils. Under conditions of temporal aggregation (i.e., when the time differences between levels 1 and 4 are obscured by an inappropriate sampling strategy), the species which are phylogenetically closely correlated in evolutionary parallel lines must be aggregated into an artificial taxon, with an excessively wide variability. These taxa lose their value as fossil indicators. A similar view must be taken into consideration when the precision of the morphological analysis is inadequate. LSD, least significant difference of the mean values in successive levels; n, number of examples.

Abiotic Signals in Time

There are many methods to measure the abiotic variability of the sedimentary record in time. Sequence stratigraphy uses the signals produced by the basinwide geometry of sediment bodies combined with selected sedimentary features to reconstruct the relative rise and fall of the sea level. Because the oceans are in contact with each other, high or low stands of water level are synchronous throughout the globe. Among the methods to study the variation of chemical composition of the sediments in time, the analysis of stable isotopes is the most widely used in Earth history. Measures of physical features such as magnetic susceptibility have the advantage of detecting very minute differences in the sedimentary record and therefore produce a particularly narrow pattern of signals in time. All these methods produce anonymous signals needing a common point of reference to be identified and correlated. Since the sedimentary record is very incomplete and erratic, we need many more points of reference than the biozonation is able to provide except in the very youngest, Pleistocene to Holocene history, as far back as the decay of the isotope ^{14}C combined with dendrochronology provides absolute age control by making reference to the present day. Therefore, the significance of abiotic signals in the sedimentary record for time correlations is limited.

On the other hand, particular, widely occurring events affecting global or oceanwide environments by climatic change or, eventually, by the impact of extraterrestrial bodies, seem to produce abiotic signals of worldwide proportions in the sedimentary record. Some of these signals are correlated with a limited number of extinction events observed in the fossil record. For instance, at the Cretaceous–Tertiary boundary defined precisely by a mayor extinction event, we observe a significant excursion of the $^{12}C:^{13}C$ stable isotope ratio in many globally distributed localities. However, there is no proof of the event's duration by the fossil record beyond the mean duration of a biozone. In many investigated sections, even the zonal record is not complete, and the uncertainty about the event's duration is considerably higher.

The Meaning of Time Correlations

The complex puzzle of multiple, intercorrelated, biostratigraphical timescales is calibrated with methods of "absolute" dating based on the physical laws of decay of unstable isotopes in time at intervals corresponding approximately to stages and with margins of error corresponding approximately to the mean duration of a biozone. Along these lines, since the end of World War II a unified numerical timescale of worldwide significance has been worked out (Odin and Odin, 1990) and currently is corrected and updated with a steady flow of additional data. This is the backbone of Earth history. All processes and mechanisms of change and development in the history of life and of its ambient environment are described in respect to this scale of reference in time.

The mechanisms of evolutionary and environmental change formulated as hypothesis or known as fact are based on experimental evidence in the present-day world with perspectives of time corresponding to the investigator's professional lifetime or, often, to the funding duration of scientific projects (i.e., 3–5 years). However, mechanisms and processes determining Earth history occur for much longer time spans. Some of the abiotic processes, such as the decay of instable isotopes, may be extrapolated from short-term experimental experience; others, however, such as the opening of an ocean and the displacement of continents, although measurable today by centimeters per year, can be assessed with regard to their true extent over geological time only with geological methods. Likewise, biotic processes, such as the migration of species and the succession of communities forming a sere, occur over short evolutionary processes of historical impact (i.e., seen in the fossil record over long time spans). The causal effects between biotic and abiotic historical events have meaning for understanding processes only if the mode of operation of this process occurs over a similar long-term time span. Time correlations as a support for the causal relationship considered are only meaningful if the process occurs over geological time spans because the resolution power of the timescale is limited to about 1 million years.

Processes of Geological Dimensions

The reconstruction of the geography of the earth — that is, the distribution of continents and oceans on the surface of the globe, including the localization of the main river mouths and of high mountain reliefs for each period of history by reversing posterior displacements by tectonic movements (palinspastics) — is currently in full progress not only for economic reasons (petroleum exploration) but also because it constitutes the basis for understanding global change of climate and regional change of ecological conditions. In this respect, the palinspastic maps published by Dercourt *et al.* (1993) are of prime importance. With the support of understanding the physics of the extant hydrosphere and atmosphere we can, to a certain extent, reconstruct the climate and the current patterns in the sea and in the atmosphere. The state of knowledge in the reconstruction of the abiosphere during Mesozoic and Tertiary to recent times is advancing to a sufficient degree that data about the distribution of life are providing a clearer view of the historical evolution of the environment. However, it must be pointed out that the palinspastic data are aggregated over considerable time periods comprising several stages. Consequently, the processes to be investigated by correlations between worldwide distribution patterns of life and abiotic environment must have time dimensions far beyond the dimensions of discontinuities in Earth history and concern the steady flow of evolutionary change rather than biotic revolutions.

The response of life on Earth to environmental factors consists, beyond aut- and synecology, of a gradual differentiation of the communities in latitudinal direction. Threshold values (e.g., the 18°C minimum winter temperature in the sea limiting endosymbiosis in most corals) produce discontinuities in the variation of composition and diversity of the communities in the climatic gradient. If maintained during geological time periods, these continuities are establishing faunal provinces with their latitudinal limits. There are other unknown factors at work separating faunal provinces in the longitudinal direction. The most spectacular one is the limit between the Caribbean and Tethyan province in the Atlantic Ocean which has repeatedly separated faunal realms in shallow marine environments since mid-Cretaceous times. Perhaps, gyral surface currents or more complex current patterns are responsible for at least some longitudinal province boundaries by producing localized steep gradients of the fertility in the surface waters. Moreover, the distribution patterns of benthic foraminifera seem to indicate a possible differentiation of the faunal provinces in depth. It is suspected (Hollaus and Hottinger, 1998) that the thermocline in the water column produces at least one of the discontinuities in the depth gradient resulting in the horizontal facies units well-known by empirical depth zones such as "neritic" and "bathyal," also applicable for interpreting sequences of fossil taphocoenoses (Liebau, 1984). This is the marine counterpart of the differentiation of continental floras and faunas in the gradient of altitude.

Faunal provinces are the equilibrated steady state of the distribution of life on Earth. This steady state may be disrupted by events accelerating extinction events at particular moments in Earth history. This constitutes another kind of discontinuity, not in space but in geological dimensions of time.

Biotic Response to Discontinuities of Geological Dimensions in Environmental Change

Biotic discontinuity in geological dimensions as a response to discontinuity in global environmental change means accelerated rates of extinction (so-called *mass extinctions*), reducing drastically the diversity of the communities. Mass extinction means sudden and definitive loss of genetic information built up over long periods of steady diversification and adaptive specialization. The representatives of the polar ends of the K–r strategy gradient are most exposed to ecological revolutionary events. Keeping in mind the fundamental bias in the historical record, we will see at such a moment in Earth history the disappearance of the K strategists, (i.e., of the largest sized representatives of each group, often at the end of a food chain) and of the forms adapted to oligotrophic environments. Thus, at times of ecological revolution, the K strategists among the foraminifera, the so-called larger forams, disappear more or less simultaneously and are replaced by new ones after some amount of time. This is repeated several times during Earth history and apparently follows a fixed pattern.

How can we be sure that during (and by) ecological revolution, genetic information, relevant for processes of historical dimensions, is definitively lost and not hidden away in rare species reappearing after the recovery of environmental conditions (Lazarus species; Kauffman and Harries, 1996). In my view, historically relevant loss is indicated when a set of species playing a particular role in the ecosystem is substituted in this role by another set of new taxa. For the interpretation of Earth history, it does not matter if a certain taxon survives as a refugial or rare population at any isolated locality, such as the coelacanths that survived off Mayotte in the Indian Ocean without leaving any fossil record for 60 million years during the Tertiary Period. Apparently, their role during the Cretaceous was taken over by a modern teleost fish. "Not playing a role" means in this case that they did not succeed in reintroducing their genetic information corresponding to their status as coelacanths into the concert of teleost gene pools producing communitarian units in space and time sufficiently successful to be registered by history.

The recovery process after environmental revolution was studied at the Cretaceous–Tertiary boundary by Kauffman and Erwin (1995), taking into consideration mainly the mollusks in the mid-American shallow sea, and by Hottinger (1999), focusing on K strategist foraminifera throughout the western and central Tethys. Kauffman's model considers the substitution of taxa during time, including so-called disaster species, as a direct consequence of ecological revolution in the basin considered, in analogy to what might happen in the Adriatic Sea after a season of destructive algal bloom (Stachowitsch, 1984). In my view, this model mixes the concept of mass extinction, in which genetic information is lost, and mass mortality, a synecological concept within a restricted basin. Disaster species and pioneers belong to processes related to seres with a correspondingly short duration of about 10–1000 years. My model (Fig. 17) focuses on the recovery of the genetic complexity necessary to reinstall K strategists in the ecosystem over geological time spans, as reflected by the succession of phenotypes in geologic times and over oceanwide extension, within the frame of benthic foraminifera. Perhaps, the difference of approach and opinion about the question of recovery is at least in part due to the different significance of the mor-

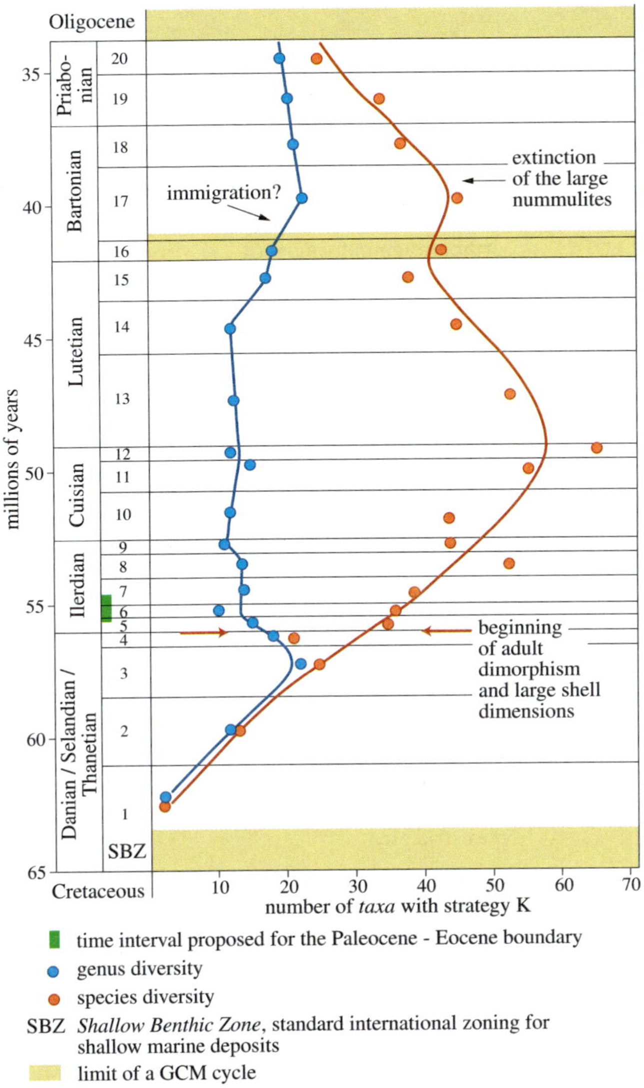

FIGURE 17 Generic and specific diversity per biozone (from 1 to 20) during two successive GCM cycles for K strategist tethyan foraminifera during the Paleocene and Eocene. Although the numbers of taxa may slightly change during current research, the trend of the two curves will be more differentiated than represented here (compare with Table 1) (reproduced with permission from Hottinger, 1998).

phology preserved in the fossil record: Foraminiferal morphology might be more expressive with regard to the life strategy in this group than a gastropod or lamellibranch shell in a much more complex molluskan organism.

The recovery of genetic complexity can be subdivided into a sequence of typical phases (Table 1). The sequence may be represented completely or partially according to the extent of losses of genetic information during the revolutionary event. After the Cretaceous–Tertiary boundary event, all larger foraminifera vanished. There are no K strategists for about 8–10 million years in the benthic record during the earliest biozone of the Paleocene, which, for precisely this reason, is comparatively long.

The following two biozones are characterized by the increase in a considerable number of new genera. This means that an increasing number of populations with a high degree of dissimilarity constitute the communities in shallow water. Several "strange" body plans appear for a short time ("failed crisis progenitors" according to Kauffman and Harries, 1996), as if nature would experiment with new morphologies. In this stage of the recovery process, all participant taxa produce shells of small adult size and have a weak or almost imperceptible dimorphism of generations. However, the structural complications in the body plan distinguish them clearly from the average benthic "smaller" foraminifer surviving at greater depths from the Cretaceous. So far, no taphocoenoses are known that would permit, by their preservation, a search for the traces of endosymbioses on the inner surface of their perforated walls.

During the third and fourth biozones, one or two representatives per basic body plan start to dominate the communities by number of specimens and by size: for porcelaneous foraminifera, planispiral involute shells tending to become fusiform (alveolinids), spherical concentric forms with an apertural axis (lacazinids), and annular concentric shells (orbitolitids); for lamellar-perforate forms, planispiral involute shells with a marginal cord (nummulitids), annular concentric shells with lateral chamberlets (orthophragminiforms), and trochospiral shells with umbilical plates (rotaliids). In these six morphological types, true progenitors in the sense of Kauffman and Harries (1996) representing six different families of the foraminifera, we start to recognize different assemblages in different depth zones within the photic realm. During this phase of recovery, the dimorphism of the adults is evident and reflects progressive differentiation of life strategies of the generations within the same species. So far, most of the genera (based on qualitative characters) are mono- or oligospecific.

During biozone 5, we observe a sudden diversification of the species belonging to the dominant, successful progenitor genera and an accelerated loss of many of the smaller ones. Adult size and extent of adult dimorphism rapidly gain importance in the species of the successful genera. The communities start to differentiate on the specific level in the ecological gradients according to depth of their respective habitat. At the same time, intraspecific variability may reflect the range of the species' habitat in the gradient. This stage in evolution marks the full recovery of K strategy in foraminifera which will last as long as environmental conditions remain undisturbed.

The previously discussed phases are reflected by the curves of taxonomic diversity per biozone when specific and generic diversity are compared (Fig. 17). Note that during

TABLE I

Global Community Maturation: Phases in a Cycle for K-strategist Foraminifera

PERIOD	PHASE	PRINCIPAL TENDENCIES	SALIENT ASPECTS[a]
Event at the Eocene–Oligocene boundary		—	Extinction of the planktonic foraminifera and of the orthophragminiforms Survival of the neritic benthos K strategist
Priabonian	3 (?)	Success of the competition	Appearance of new genera (*Spiroclypeus?*) together with the older ones
Late Bartonian	2 (?)	Progress of the competition	Insurgence of new genera; *Operculina alpina* dominates New competitors: *Pellatispira, Biplanispira, Heterostegina*
Lower Bartonian (Biarritzian)	5	Removal of competitors	Diminution of species diversity, increase of specific endemism; frequent monospecific endemism of species of large size
Lutetian, Cuisian, Upper Ilerdian	4	Size increase	Parallel evolution of diversified evolutionary lines; development of odd pairs; classes of over 10 cm size reached
Lower Ilerdian	3	Complete success	Diversification of the successful genera of different species; success of *Alveolina, Orbitolites, Assilina, Nummulites,* orthophragminiforms
Upper Paleocene	2	Experimentation of new life strategies	Increase in generic diversity and reduction of endemism; about 40 species of smaller size with complex structure
Lower Paleocene	1	Preparation of K strategy	No phenotypic response or no K strategy (?) No larger form No complex structure (?)
Event at the Cretaceous–Tertiary boundary		—	Extinction of the K strategist form; survival of the deep benthos and, in shallow waters, of *Laffitteina*

[a] Question marks indicate uncertain aspects.

the ecologically undisturbed phase there is a continuous increase in size [i.e., progressive adaptation, progressive differentiation in odd associations, and progressive geographical differentiation (endemism)] with time up to a point marking the end of Middle Eocene, when the largest sized taxa disappear, apparently substituted by a set of new genera. This minor revolution, possibly restricted to the shallow marine realm, seems to be unrelated to major environmental change and may be exclusively the consequence of biotic processes, either by failing to keep more successful immigrants out or by leaving open environmental niches after having genetic trouble by too many cloned generations in the symbionts that have to reproduce asexually within their host over its full reproduction cycle during many years.

A period of continuous, gradual biotic change in between two discontinuities is called a *global community maturation cycle* (GCM), in analogy to seres but involving major genetic revolutions at each cycle limit (Hottinger, 1998). Community maturation may occur in separate ecological realms as quasi-separate, parallel ecosystems. Revolutions in each system may be synchronous, totally unrelated in time, or delayed in respect to each other. Thus, we observe at the Cretaceous–Tertiary boundary biotic revolutions synchronously in the planktic and the shallow benthic realm, whereas the populations of the bathyal depth zone are substituted with considerable delay only at the Paleocene–Eocene boundary. The shallow benthic revolution at the limit Middle–Upper Eocene concerns only the most extreme K strategists in shallow water and not the smaller, less specialized ones, and it seems to have a very minor impact on the planktic microfauna. At the Eocene–Oligocene boundary, the plankton is substituted to a very large extent, whereas the shallow benthos is weakly affected and the deeper benthos more strongly by the substitution of the orthophragminiforms by lepidocyclinids, etc.

Therefore, the nature of the factors responsible for the discontinuities in the historical record must be different at each event in time. However, the current state of knowledge does not permit us to understand and define the exact nature of the different events producing the discontinuities.

Conclusions

The fossil record is determined by continuous, gradual change through time interrupted by environmental and/or biotic revolutions. The latter, if extended over geological dimensions, are characterized by accelerated genetic loss and recovery at the poles of the gradient of life strategy. In dimensions of shorter time and space, communities may be substituted by others with mechanisms involving migration of species under conservation of the genetic information.

The bias in the record favoring preservation of K strategists exaggerates the significance of the latter in disfavor of the still poorly understood, and poorly preserved, r strategists. Moreover, the emphasis on discontinuities in current research is obviously enhanced by the genius of our time characterized by fears of man-made catastrophes and by the experience in our lifetime of massively destructive events such as world wars and atomic explosions. The interpretation of Earth history, particularly the history of life during the Phanerozoic, should be reverted to a more equilibrated view. Concurrently, we have to consider the mechanisms of gradual change, often restricted to quantitative progress, as well as the mechanisms of revolution producing mostly qualitative change in the substitution of taxa with a very different genetic background in ecologically equivalent ecosystems. Such considerations must be supported by a better understanding of the functional and morphogenetic aspects of size, shape, and architecture of the fossil hard parts.

The processes involved in the historical change of life on Earth have dimensions not only in time but also in space. All dimensions involved have to match each other in order to establish meaningful relations of cause and effect in the working of the process. Many documents of Earth history have to be calibrated (i.e., reinvestigated concerning the match of the dimensions of the mechanisms believed to begin the process). According to their size class, there is a hierarchy of imbrication of processes in their dimensions of spatial range and duration determining the relations between life and environment, both under conditions of perpetual change driven by the relative instabilities of the individual genome, of the gene pool of populations, as well as of the earth's crust (Fig. 18). The processes at each level of imbrication are described at a particular level of abstraction in understanding, usually furnished with a particular terminology (Table 2).

The accuracy of interpretation of a fossil document depends on its respective place in the cycles of GCM. At mature stages, the accuracy is best, at least for K strategists, and at early stages it is worst. The number and kinds of phases present in the GCM cycle depend on the specific cycle considered because every cycle has different conditions for its start, depending on the "performance" of the previous one and on the extent and selectivity of the corresponding mass extinction. Therefore, any interpretation of a single historical document is relevant only within its historical context, which must include the respective GCM cycle and part of the previous one.

In order to attempt predictions of the future of mankind and its environment in dimensions relevant to man, it is insufficient to extrapolate short-term mechanisms understood and/or kept under control by experiment in the laboratory. Every situation in history, including our present time, is the result of historical processes which created the setting and the specific limiting conditions for shorter and spatially more limited processes. The calibration of already known processes in matching dimensions of space and time remains the most important task of current research, followed by the search for processes governing migration and acceptance or rejection of the immigrants in a new community.

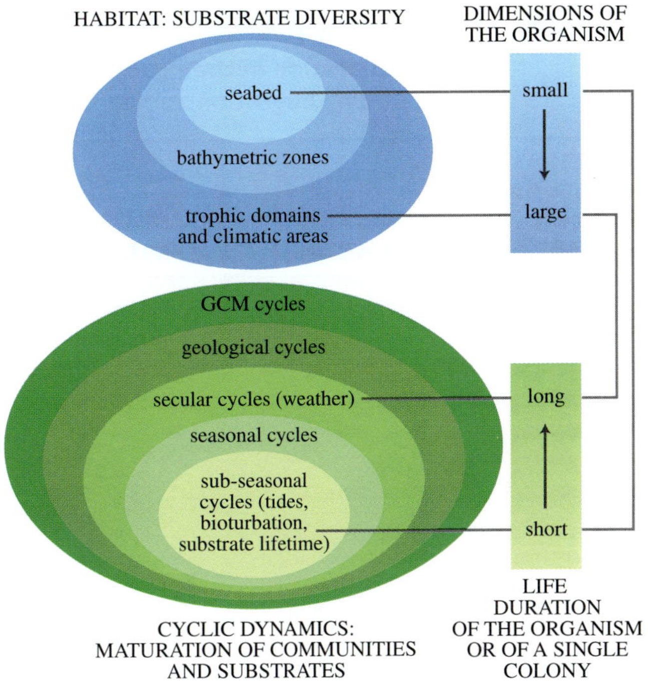

FIGURE 18 Imbrication of size categories in ecological factors in space (in blue) and time (in green) and their relation to the body size and life duration of the organism. An epiphytic foraminifer has a territory on a sea grass leaf of about 1 cm in diameter, whereas a blue whale has an oceanwide territory (compare with Table 2) (adapted from Hottinger, 1988).

TABLE 2

Calibration of the Processes of Biotic Change in Time Controlled Mainly by Migration and by Changes in Genomes or of the Gene Pools

CALIBRATION OF THE BIOTIC CHANGE IN TIME	BIOLOGICAL UNIT	SPATIAL EXTENSION	TEMPORAL EXTENSION (YEARS)	LITHOSTRATIGRAPHIC UNIT	CORRELATED PROCESSES
Single organism	Example	Territory	Mean lifespan (hours, years)	Example taken from a sample	Ontogenesis, autoecology, functional morphology, reproduction
Single taxonomic unit	Deme population	Habitat	($<1–10^5$)	One or more biostratigraphic samples	Intraspecific variation ecomorphology
				Surface of a stratum or a single stratum	Population dynamics, series, successions, mass mortality[a]
Multiple taxonomic units	Biotic associations, corporation	Biotope	($<1–10^5$)	Sequence of strata	Strategy for the common utilization of resources
			GENOMES OR GENE POOLS		
Single taxonomic unit	Species	Area	Biozone ($2.10^5–10^6$)	Single member Sequence of members	Evolution within a phylum
	Genus	Faunal province			Differentiation of ecological zones
Multiple taxonomic units	Community biocoenosis	Faunal province	Stages ($10^6–10^7$)	Succession, generally composite, of members or formations	Insurgence of endemism Mass extinction[a]
	Ecosystem	Global climatic band	Global community maturation cycle (10^7)		Macroevolution Creation of new somatic schemes Global community maturation cycle

[a] Accelerated change.

References Cited

Baumgartner, P. O., *et al.* (Eds.) (1995). Middle Jurassic to Lower Cretaceous radiolaria of Tethys: Occurrences, systematics, biochronology. *Mém. Géol. (Lausanne)* **23.**

Bell, G. (1988). *Sex and Death in Protozoa.* Cambridge Univ. Press, Cambridge, UK.

Billman, H., *et al.* (1980). *Schweiz. Paläontol. Abh.* **101,** 71–113.

Bonner, J. T. (1952). *Morphogenesis, an Essay on Development.* Princeton Univ. Press, Princeton, NJ. (Reprinted paperback, Atheneum, New York, 1963)

Carroll, R. L. (1987). *Vertebrate Paleontology and Evolution.* Freeman, New York.

Dercourt, J., Ricou, L. E., and Vrielynck, B. (Eds.) (1993). *Atlas Tethys Paleoenvironmental Maps.* Gautier-Villars, Paris.

Drobne, K., and Hottinger, L. (1971). *Slovenska Akad. Znan. Umetn.* **14,** 213–237.

Drooger, C. W. (1993). *Radial Foraminifera; Morphometrics and Evolution.* North-Holland, Amsterdam.

Eldridge, N., and Gould, S. J. (1972). Punctuated equilibria: An alternative to phyletic gradualism. In *Models in Paleobiology* (T. J. M. Schopf, Ed.), pp. 82–115. Freeman & Cooper, San Francisco..

Eldridge, N., and Stanley, St. M. (1984). *Living Fossils.* Springer, New York.

Fioroni, P. (1970). Die organogenetische und transitorische Rolle der Vitellophagen in der Darmentwicklung von Galathea (Crustacea, Anomura). *Z. Morphol. Tiere* **67,** 236–306.

Fioroni, P. (1971). Die Entwicklungstypen der Mollusken, eine vergleichend-embryologische Studie. *Z. Wissensch. Zool.* **182,** 263–349.

Goodrich, E. S. (1958). *Studies on the Structure and Development of Vertebrates.* Dover, New York.

Grime, J. P. (1977). Evidence for the existence of three primary strategies in plants and its relevance to ecological and evolutionary theory. *Am. Nat.* **111,** 1169–1194.

GUEX, J. (1991). *Biochronological Correlations.* Springer, Berlin.

GUTMANN, W. F. (1994). Evolution von Konstruktionen. Die Frankfurter Theorie. In *Morphologie und Evolution* (W. F. Gutmann, D. Mollenhauer, and D. St. Peters, Eds.), pp. 317–338. Kramer, Frankfurt on Main.

HAY, W. W., and SOUTHAM, J. R. (1978). Quantifying biostratigraphic correlation. *Annu. Rev. Earth Planetary Sci.* **6,** 353–375.

HOLLAUS, S., and HOTTINGER, L. (1998). Temperature dependence of endosymbiontic relationships: Evidence from the depth range of Mediterranean *Amphistegina lessonii* (Foraminiferida) truncated by the thermocline. *Eclogae Geol. Helvetiae* **90,** 591–597.

HOTTINGER, L. (1963). Les Alvéolines paléogènes, exemple d'un genre polyphylétique. In *Evolutionary Trends in Foraminifera* (von Koenigswald *et al.,* Eds.), pp. 298–314. Elsevier, Amsterdam.

HOTTINGER, L. (1979). In *Foraminifera 3* (R. H. Hedley and C. G. Adams, Eds.). Academic Press, London.

HOTTINGER, L. (1981). *Cahiers Micropaléont.* **2,** 45–55.

HOTTINGER, L. (1983). Neritic macroid genesis, an ecological approach. In *Coated Grains* (T. M. Peryt, Ed.), pp. 38–55. Springer, Berlin.

HOTTINGER, L. (1985). Spiralige Schalen. In *Die Spirale in Menschlichen Leben und in der Natur — Eine Interdisziplinäre Schau,* pp. 44–53. Gewerbemuseum Basel, Basel.

HOTTINGER, L. (1988). In *Atti Quarto Simposio Ecologia Paleoecologia Comunità Bentoniche, Sorrento.* Regional Natural Sciences Museum, Turin.

HOTTINGER, L. (1997). Shallow benthic foraminiferal assemblages as signals for depth of their deposition and their limitations. *Bull. Soc. Géol. France* **168**(4), 491–505.

HOTTINGER, L. (1998). Shallow benthic foraminifera at the Paleocene–Eocene boundary. *Strata Ser.* **1**(9), 61–64.

HOTTINGER, L. (1999). "Odd partnerships," a particular size relation between close species of larger foraminifera, with an emendation of an outstandingly odd partner, *Glomalveolina delicatissima* (Smout, 1954), Middle Eocene. *Eclogae Geol. Helvetiae* **92.**

JACOBS, D. K. (1996). Chambered cephalopod shells, buoyancy, structure and decoupling: History and red herrings. *Palaios* **11,** 610–614.

KAUFFMAN, E. G., and ERWIN, D. H. (1995). Surviving mass extinctions. *Geotimes* **14**(3), 14–17.

KAUFFMAN, E. G., and HARRIES, P. E. (1996). The importance of crisis progenitors in the recovery from mass extinctions. *Geol. Soc. London Special Publ.* **102,** 15–39.

KUHN-SCHNYDER, E. (1967). Paläontologie als stammesgeschichtliche Urkundenforschung. In *Die Evolution der Organismen* (G. Heberer, Ed.), Vol. 3, pp. 236–419. Fischer, Stuttgart.

LAW, R., and LEWIS, D. H. (1983). Biotic environments and the maintenance of sex. Some evidence from mutualistic symbioses. *Biol. J. Linnean Soc.* **20,** 249–276.

LIEBAU, A. (1984). Grundlagen der Oekobathymetrie. In *Paläobathymetrie. Paläontologische Kursbücher* (H.-P. Luterbacher, Ed.), Vol. 2, pp. 149–184. Paläont Gesellschaft, München.

MÜLLER-MERZ, E. (1980). Strukturanalyse ausgewählter rotaloider Foraminiferen. *Schweiz. Paleontol. Abh.* **101,** 5–6.

ODIN, G. S., and ODIN, C. (1990). Echelle numérique des temps mise à jour 1990. *Géochronique* **35,** 12–21.

PARRY, G. D. (1981). The meanings of r- and K-selection. *Oekologia* **48,** 260–264.

PÉCHEUX, M. J.-F. (1995). Ecomorphology of a recent, large foraminifer, *Operculina ammonoides. Geobios* **28**(5), 529–566.

PIGNATTI, J. S. (1998). The philosophy of larger foraminiferal biozonations, a discussion. *Dela-Opera SAZU (Ljubljana)* **34**(2), 15–20.

POR, F. D. (1978). Lessepsian migration. *Ecol. Stud.* **23.**

REISS, Z., and HOTTINGER, L. (1984). The Gulf of Aqaba, ecological micropaleontology. *Ecol. Stud.* **50.**

ROSSOMANDO, E. F., and ALEXANDER, S. (Eds.) (1992). *Morphogenesis, an Analysis of Development of Biological Form.* Dekker, New York.

SCHINDEWOLF, H. (1961). Studien zur Stammesgeschichte der Ammoniten. *Akad. Wiss. Lit. Mainz Abhandl. Math.-Naturwiss.* **10,** 635–743.

SCHINDEWOLF, H. (1962). Studien zur Stammesgeschichte der Ammoniten. *Akad. Wiss. Lit. Mainz Abhandl. Math.-Naturwiss.* **8,** 425–571.

SCHINDEWOLF, H. (1963). Studien zur Stammesgeschichte der Ammoniten. *Akad. Wiss. Lit. Mainz Abhandl. Math.-Naturwiss.* **6,** 259–432.

SCHINDEWOLF, H. (1965). Studien zur Stammesgeschichte der Ammoniten. *Akad. Wiss. Lit. Mainz Abhandl. Math.-Naturwiss.* **3,** 137–238.

SCHINDEWOLF, H. (1966). Studien zur Stammesgeschichte der Ammoniten. *Akad. Wiss. Lit. Mainz Abhandl. Math.-Naturwiss.* **3,** 324–454.

SCHINDEWOLF, H. (1967). Studien zur Stammesgeschichte der Ammoniten. *Akad. Wiss. Lit. Mainz Abhandl. Math.-Naturwiss.* **3,** 719–808.

SCHINDEWOLF, H. (1968). Studien zur Stammesgeschichte der Ammoniten. *Akad. Wiss. Lit. Mainz Abhandl. Math.-Naturwiss.* **8,** 139–188.

SEILACHER, A. (1999). Biomat-related lifestyles in the Precambrian. *Palaios* **14,** 86–93.

SEILACHER, A., and LA BARBERA, M. (1995). Ammonites as Cartesian divers. *Palaios* **10,** 193–506.

SHIMER, H. W., and SHROCK, R. R. (1955). *Index of Fossils of North America,* 5th ed. Wiley, New York.

STACHOWITSCH, M. (1984). Mass mortality in the Gulf of Trieste. The course of community destruction. *Marine Ecol.* **5,** 234–264.

STEARNS, S. C. (1992). *The Evolution of Life Histories.* Oxford Univ. Press, Oxford.

THOMPSON, D'A. W. (1966). *On Growth and Form* (Abridged ed.). Cambridge Univ. Press, Cambridge, UK. (Original work published 1917)

TRÜMPY, R. (1973). Arbeitsgruppe für stratigraphische Terminologie. Empfehlungen zur Handhabung der stratigraphischen, insbesondere lithostratigraphischen Nomenklatur in der Schweiz. *Eclogae Geol. Helvetiae* **66**(2), 479–492.

VARGAS, C. DE, NORRIS, R., ZANINETTI, L., GIBB, S. W., and PAWLOWSKI, J. (1999). Molecular evidence of cryptic speciation in planktonic foraminifers and their relation to oceanic provinces. *Proc. Natl. Acad. Sci. USA* **96,** 2864–2868.

ELGENE O. BOX
Department of Geography
University of Georgia
Athens, Georgia, USA

A Changing Planet

The changes in effect on our planet and the ones that will take place in the near future seem to be characterized in the first place by a drastic replacement of values based on a sense of community and the common good with those based on competition; in the second place by the acceleration of the modifications due to feedback in the Earth system and within the human community when the resources decrease rapidly in a certain environment. Another characteristic element is the imperceptibility of several of these changes, or the time delay with which ecological effects due to these causes become evident. In this manner a sudden and massive destruction of the natural world and diversity might occur. The possibility of sudden disaster would at least have the advantage that these ecologically aggressive activities would be blocked, while a slower degradation would allow a greater strain on resources and a continuous devastation of the Earth until we reach the point where life itself is completely degraded.

Introduction

The earth has always been changing. As we understand it, the earth is more than 4 billion years old. As recently as the past approximately 200 million years, it has developed geologically from a planet with a single supercontinent, Pangaea, to an expanding planet with continents Laurasia and Gondwanaland separated by the Tethys Sea and then to the current configuration of six main landmasses and four oceans. Some of the important geological events during this time (Table 1) and the order in which they occurred had profound and, from our viewpoint, perhaps unexpected implications for the earth's geographic development. For example, it is sobering to stand among the mountains of northern Sweden and realize that they represent the same mountain range as the Appalachian Mountains of eastern North America, and that these mountains are older than the Atlantic Ocean which now separates them.

With such profound changes in the earth's basic geography came changes in climate. The earth's biological development was shaped by both these geologic and climatic changes. For example, Australia was geographically isolated from other continents after about 80 million years before present (BP), and all of the bioclimatic and ecological niches which have developed in Australia since then were filled by the radiative speciation of its original, relatively small biota to produce its current diversity of marsupials, its unusual creatures such as the platypus, and its ecologically unique eucalyptus trees of current worldwide economic importance.

TABLE I

Some Main Events in Earth's Geological and Evolutionary History[a]

ERA/PERIOD	YEARS	EVENT	ORGANISM
Pre-Cambrian	From 3.5 b BP		One-cell oceanic organisms, photosynthesis (3.0 b BP)
	From 1.4 b		First eukaryotic organisms
	950–650 m BP?	Much of Earth glaciated	
Paleozoic Era (>225 m BP)			
Cambrian	600–500 m BP	Early Paleozoic orogenies begin	
Ordovician	500–440 m BP	Warm (but glaciated near end)	
Silurian	440–400 m BP	Climate similar to present?	First vascular plants appear, first land plants
Devonian	400–345 m BP	Climate similar to present?	
Carboniferous	345–280 m BP	Warm and humid	Forests of ferns, club mosses, and horsetails
Permian	From 280 m BP	Pangaea; glaciation at south Pole	First ginkgos and mosses
Mesozoic Era (225–65 m BP)			
Triassic	225–190 m BP	Breakup of Pangaea begins; climate cool and humid	Ginkgos and cycads dominant
Jurassic	190–135 m BP	Breakup of Pangaea: W Gondwana	Conifers and cycads dominant
	180 m BP	Africa and North America separate	
Cretaceous	135–65 m BP	"Greenhouse" climate, high sea level Major separations: South America, Africa, Madagascar, and India (130–100 m BP); North America and Europe (80 m BP); New Zealand and E. Gondwana (80 m BP); Siberia and Alaska join (70 m BP)	Angiosperms dominant, first marsupials, modern plant orders exist, some modern families and genera exist
Tertiary Period (of Cenozoic Era)			
Paleocene	65 m BP	Africa and Europe separate temporarily	
Eocene	50–40 m BP	Final separation of Europe and North America; Beringia becomes warmer; Australia separates from Antarctica; Himalaya forms as India collides with Asia	Most modern families, many general exist, Madro-Tertiary geoflora, first primates
Oligocene	38–23 m BP	Alps and Rocky Mountains form	
Miocene	23–5 m BP	Europe and Africa rejoined; islands form as Southeast Asia–Australian gap narrows; global climates become cooler and drier	
Pliocene	5–1.8 m BP	Land bridge joins North and South America; Cascade and Sierra Nevada mountains	Most modern plant taxa, Herbaceous plants abundant
Quaternary Period (of Cenozoic Era)			
Pleistocene	1.8–0.01 m BP	Glacial and interglacial periods: N Eurasia and N North America glaciated; sea level fluctuates	North–south migration of vegetation zones, with loss of taxa unable to cross barriers, tropical areas drier
	18 k	Last glacial maximum: sea 150 m lower, land bridges, and wider coastal plains	
	14–12 k	Warming and glacial retreat	
Holocene	From 10 k	Continued warming	Temperate and high-latitide soils and biomes redevelop
	8–5 k	Hypsithermal period (warmest)	
	5 k	Sea reaches present level, cooling and paludification	
	From 4 k		Destruction of primary land vegetation by man

[a]Times are in billions (b), millions (m), or thousands (k) of years before present (BP).

Sources: Pearson (1995), Hidore and Oliver (1993), Bridges (1990), Frakes (1979), Pielou (1979), Raven and Axelrod (1974), *Scientific American* (1976).

Current forces for change on Earth carry the potential for global climatic change, drastic losses of biodiversity and natural ecosystems, and complete transformation of the earth's soil and vegetation cover and thus its energy and hydrological balance, all driven by sweeping changes in human technology and patterns of land use. Feedbacks in these systems carry the potential for acceleration of current changes, which are already much faster than at any previous time in human history. Positive feedback in systems brings the potential for overshooting limits, especially when built-in time lags become longer in more complex systems. It is the extensive scale and accelerating pace of change that is unique and alarming. One recalls the parable of the lily pond (Brown, 1978), on which the water lilies are able to double their coverage every day and can cover the pond after 30 days. On the 29th day the pond is still only half covered.

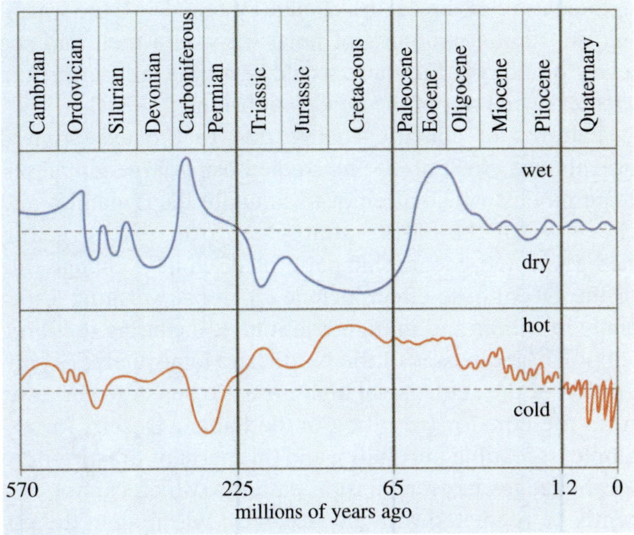

FIGURE 1 Variations in global mean precipitation (in blue) and temperature (in red) from the Cambrian to today, reconstructed on inferences based on fossil records and plotted as differences with respect to current values. The temperature scale is not linear (adapted from Frakes, 1979).

Climatic Change

In considering climatic change one must make an initial distinction between forces that cause drastic changes of short duration, which must be viewed as normal fluctuations in climate, and forces that cause true, relatively permanent change in the fundamental nature of Earth's climates. Sources of drastic but short-term fluctuations include forces such as El Niño events, volcanic eruptions which eject ash into the stratosphere (above the level of normal weather), and cycles which appear correlated with sunspot cycles (National Research Council, 1982). Sources of climatic change lasting many millennia or longer include such forces as plate tectonics (continental drift) and changes in the earth's orbital characteristics. The latter, articulated by Milutin Milankovic, involve cyclical changes in orbital eccentricity (a period of 100,000 years), tilt in axis of rotation relative to orbital plane (41,000 years), and wobble of the rotation axis (26,000 years). Some variations in global mean temperature and precipitation, as reconstructed from inferences based on fossil records, are shown in Fig. 1 (Frakes, 1979; Lamb, 1977). The decline into a period of alternating glacial and interglacial periods in the Pleistocene has been related to the Milankovic cycles (Hays *et al.,* 1976; National Research Council, 1982; Imbrie and Imbrie, 1980).

During the Pleistocene glaciations (1.8–0.1 million BP), climate and sea level changed rapidly as much of the earth's freshwater was alternately immobilized in ice caps and then released again into the oceans. Although the tropics and subtropics were not cooled as much as the higher latitudes, they became drier during glacial periods since there was less water in the global atmosphere. The last glacial maximum was about 18,000 years BP (Table 1). The current landscapes of especially northeastern North America and northwestern Eurasia, including their soils and vegetation covers, have developed entirely within approximately the past 14,000 years, after the ice retreated and climates warmed again.

Recently, several new anthropogenic drivers of potential climatic change have appeared: deposition of particulate pollution in the atmosphere (reducing solar radiation near the surface), depletion of the ozone layer (permitting greater influx of lethal ultraviolet radiation and perhaps important changes in atmospheric chemistry), and augmentation of the earth's natural "greenhouse effect" by injection of unnatural amounts of carbon dioxide (CO_2) and other gases into the atmosphere. These so-called "greenhouse gases" do not affect solar radiation but absorb outgoing Earth radiation, trapping more energy within the earth–atmosphere system, causing the atmosphere and the earth's surface to heat up ("global warming"). About two-thirds of the recent net injection of CO_2 into the atmosphere has come from burning fossil fuels (coal, oil, and natural gas), whereas the one-third has come from net deforestation in the tropics and subtropics (Houghton *et al.,* 1996). The level of CO_2 in the atmosphere had risen from its preindustrial level of about 280 parts per million (ppm) to about 360 ppm by the early 1990s, and it is expected to double to approximately 750 ppm (at current rates of burning) by the middle of the twenty-first century. Because of the magnitude of this change and the long residence time of this extra CO_2 in the atmosphere, the resulting global warming must be seen as a relatively permanent change (at least within the time frame of human experience and planning) in the nature of the earth's climate system. With the warming, substantial amounts of polar ice would melt and sea levels would rise, but there would be a time lag, permitting greater overshoot of the new equilibrium.

Patterns of climatic change due to the exacerbated greenhouse effect are being studied using large global climate models which attempt to simulate the dynamics and chemistry of the earth's atmosphere over several centuries. Expected geographic patterns of climate change due to the greenhouse effect include (i) global warming, especially in winter and in higher latitudes, including the large subpolar landmasses of the Northern Hemisphere; (ii) migration of the subtropical high-pressure belts further away from the equator (resulting in the Sahara Desert, for example, spreading into Italy); and (iii) perhaps drastic, more local changes in precipitation patterns, which cannot currently be predicted with any accuracy. Warming in the polar regions could be as much as 16–18°C in the Northern Hemisphere (Schlesinger, 1985), with warming of 3–6°C in midlatitudes and 2–5°C overall (Houghton *et al.,* 1996). Warming in the tropics would be slight, but drying could occur.

The difference between preindustrial and 1990s atmospheric CO_2 levels represents about 160 gigatons of carbon (gt C). If CO_2 levels do indeed double, this would represent an additional 800 gt C transferred into the atmosphere. It has been widely suggested (with supporting grants for scientific workshops, etc.) that much of this excess could be sequestered in the terrestrial biosphere in the form of biomass accumulation in growing forests (Henderson and Dixon, 1993). The difference between current storage of carbon in land vegetation (about 650 gt C; Houghton *et al.,* 1996) and the upper limit suggested by a climatic limitation model of terrestrial biomass accumulation (about 1600 gt C; Box, unpublished data) is also about 800 gt C. Thus, the additional 800 gt C could be stored in the terrestrial biosphere — but only if we abandon the planet and let it grow to its potential in forest and other vegetation. With current population increases, this — and the success of carbon sequestration schemes — seems unlikely.

Biotic Change

Evolution of recognizable life on land began perhaps one-half billion years ago, before the Paleozoic era. Some major biological developments in Earth's history are listed in Table 1, including dominance by ferns in the late Paleozoic (which produced our current fossil fuels), by conifers in the early Mesozoic, by angiosperms ("broad-leaved" seed plants) by the end of the Mesozoic Era, and the development of the modern plant families and genera during the Tertiary Period. It is thought that angiosperms evolved in Southeast Asia (Axelrod, 1952) in the Mesozoic, but this was also the time at which Pangaea was breaking up. Some taxa were carried to most parts of the world, but others were localized by the separations of what became the modern continents. Some plant families remain largely or only in the Northern Hemisphere, such as Betulaceae (birch and alder), or in the Southern Hemisphere, such as Proteaceae (especially southern Africa and Australia). In North America, a Madro-Tertiary geoflora may have existed across the middle and higher latitudes, with a more tropical flora in the southeast. With the cooling and drying of the continental interior during the mid-Tertiary, the eastern and western portions of this geoflora were separated and began to evolve separately, resulting in the large number of similar, vicariant but not identical species in the eastern and western United States, especially in the forest floras (e.g., *Quercus, Fraxinus, Acer,* and *Pinus*). Similar separations and parallel development occurred on other continents, especially Eurasia.

Biodiversity includes numbers of biological entities at various levels, not only species (Heywood *et al.,* 1996). For most parts of the world, biological surveys are still quite incomplete. Some current estimates of numbers of higher plant and animal species are summarized in Table 2 based on data in the global biodiversity report by the World Conservation Monitoring Centre (Groombridge, 1992). For plants, diversity is often correlated with warmer, wetter environments (Currie and Paquin, 1987) and is highest in the humid tropics. Reasons for this may include the longer uninterrupted biological history of the tropics, the higher activity (productivity) rates, and the relative stability of habitats over long periods, during which many very subtle biotic interrelationships have resulted in evolution of new species (Palmer, 1994). For animals, diversity varies more with phylogenetic group, with reptile diversity highest, for example, in many environments that are less hospitable for other taxa. Much of the earth's biodiversity, at the genus and species levels at least, is in the humid tropics, and it is disappearing rapidly as tropical forests are converted to more direct human land use.

Endemism is an important aspect of biodiversity and is often highest in relatively isolated regions. Data and maps describing endemism levels in different areas are also presented in the global biodiversity report. One very interesting area of high endemism is the Mediterranean-climate area of southern Africa, which various earlier classifications regarded as one of the world's main floristic kingdoms despite its quite small area (Schmithuesen, 1968). Loss of habitat in areas of high endemism is especially damaging since it results in losses of large numbers of species which occur nowhere else.

How many types of ecosystems and landscapes are there on Earth? Most global classifications include about 15–20

TABLE 2

Estimated Numbers of Higher Plants and Animals by Major Countries and Regions[a]

	PLANTS	MAMMALS	BIRDS	REPTILES	AMPHIBIANS
North America: USA and Canada*	17,000				
USA	18,956	346	650		
Mexico	20,000–30,000	439	961	717	284
Latin America, including Mexico*	85,000				
Brazil	55,000	394	1573	468	502
Europe, excluding USSR*	12,500				
Soviet Union	22,000	276		168	37
Africa, tropical/subtropical*	40,000–45,000				
North Africa*	10,000				
Tropical Africa*	21,000				
Southern Africa*	21,000				
South Africa	2,300	247	774	299	95
Asia, tropical/subtropical*	50,000				
Malesia*	30,000				
Indonesia	20,000	515	1519	511	270
Malaysia	12,000	264	501	268	158
China	30,000	394	1100	282	190
India	15,000	317	969	389	206
Turkey	8,472	116	284	102	18
Australia*	15,000	282	571	700	180
New Zealand	2,160		116	32	0

[a]Source: Global biodiversity report (Groombridge, 1992). All animal data are from pp. 139–141; plant data for individual countries are from pp. 80–82; *regional data are from p. 66 (see original sources therein).

main terrestrial biomes — that is, large floristic–faunistic units with relatively uniform potential natural vegetation (Archibold, 1995; Schmithuesen, 1968; Walter, 1985; Walter and Box, 1976). A recent attempt to represent the main types of potential natural terrestrial vegetation, for use with global-scale climatic and satellite-based models, derived about 50 types before concluding that no more types needed to be included (Box, 1995b). These types (some variants grouped) are listed in Table 3 and correspond largely with global classifications of ecoregions (Bailey, 1989, 1996) and biogeographic provinces (Udvardy, 1975). There is evidence that some of these types did not exist as recently as the last glacial maximum.

The main terrestrial vegetation types, as well as their constituent taxa (especially their larger, dominant plant growth forms), are constrained most generally by climatic factors. Their potential distributions are thus predictable over large areas using climatic data (Box, 1981, 1995a; Holdridge, 1947; Woodward, 1987). Global warming will shift the potential ranges of taxa and biomes, causing dislocations of distributions. The changes and loss of species, however, will be much faster than in the geologic past (Gates, 1993; Peters and Lovejoy, 1992). To some extent, it will be possible to postulate hypothetical patterns of climate change and predict the effects on natural vegetation types as well as species and assemblages. Warming patterns expected from the doubling of atmospheric CO_2 (without net drying) have suggested drastic dislocations of important forest trees of the eastern United States (Davis and Zabinski, 1992) and of woody plants in general in Florida (Box *et al.,* 1999). It is also possible to estimate the vulnerability (potential for breakup) of important current natural communities, as their constituent main species respond differently to changing climatic conditions (Lassiter *et al.,* 2001).

Using satellite data instead of climate data, it is also possible to estimate geographic occurrences of actual land cover types, including both natural vegetation and altered landscapes such as croplands. Both spatial and ecological

TABLE 3

Main Terrestrial Biome and Natural Landscape Types

Tropical rain forests	Tropical rain forest (lowland) Subtropical rain forest
Tropical seasonal forests	Tropical semi evergreen forest Raingreen forest
Tropical savanna, dry woods and scrub	Tropical dry evergreen forest Raingreen thorn-scrub Semievergreen dry scrub Tropical savanna
Tropical mountains	Tropical montane rain forest Tropical subalpine (cloud) forest Paramo alpine scrub Puna alpine grassland
Warm desert and semidesert	Warm shrub-steppe Mixed semidesert scrub Desert steppe
Mediterranean woodlands and shrublands	Mediterranean sclerophyll forest Sclerophyll scrub/shrubland Mediterranean dwarf-scrub
Evergreen broad-leaved forest	Evergreen broad-leaved forest Cool-temperate evergreen broad-leaved forest Subpolar evergreen broad-leaved forest
Temperate deciduous forest	Summergreen broad-leaved forest Summergreen broad-leaved woodland/parkland
Temperate rain forest	
Temperate grasslands	Tall-grass prairies Short and bunch-grass steppes Cool, maritime grassland
Cold-winter desert and semidesert	Continental shrub-steppe Continental mixed semidesert scrub Cushion-shrub semidesert Temperate desert grassland
Temperate mountains	Mesic montane conifer/mixed forest Dry conifer forest (including Mediterranean) Subalpine conifer forest and krummholz Alpine tundra
Boreal forest and woodland	Boreal dark-taiga conifer forest Boreal light taiga forest/woods Maritime broad-summergreenforest Forest–tundra conifer woodland
Polar tundra and cold desert	Polar tundra Maritime polar grassland and tundra Polar cold desert

resolution, as well as general accuracy of recognition, generally increase at finer spatial scales, where climatic constraints also become less important in determining geographic patterns.

One might try to imagine the response of a regional landscape to global warming by considering the effects of warming on important ecological processes, especially metabolism, near its northern and southern boundaries. For example, in much of east Asia and eastern North America, as well as in western to central Europe, summergreen deciduous forests are the natural vegetation and landscape cover type. These are bordered on the north by transition zones of mixed forest, followed by the boreal coniferous forest. On the south the deciduous forest is bordered by transitions (of various forms) to landscapes with evergreen broad-leaved forests (east Asia and eastern North America) or at least evergreen elements (Mediterranean Europe). With warming, respiration requirements will increase and may not be compensated by increases in photosynthesis, even if there is no net drying. Large trees in the forest canopy (as well as others) will eventually die, opening up the landscape for invasion by mobile secondary species. These will include both weeds and longer living but still aggressive colonizers, such as *Prunus serotina* and *Robinia pseudoacacia* in Europe (both from North America but already pest species). Although the warming may be greater toward the north, at least some of the invaders there may be other trees of the same general type, including more southerly acclimated variants of the same species. Toward the south, however, the invaders will mainly be faster growing weedy plants of different types, including weedy evergreens and subtropical species. Much of the landscape will thus be converted into unstable secondary vegetation, with broad tension zones along the southern boundaries of the natural vegetation regions (Soule, 1990).

Land Conversion: The Other Driver of Change

Man has been converting natural landscapes for his own purposes for at least 5000 years (Thomas, 1956). Although global warming is also caused by human activity, it may be no more important than land conversion as a driver of change in this century. Land conversion involves both expansion into currently seminatural but marginal lands and intensified use of already modified landscapes. Conversion of forested landscapes to agriculture, settlements, industrial areas, etc. reduces the vegetation cover and its buffering effect, thus destabilizing energy and hydrological balances and increasing local warming, drying, runoff, and flooding. In many cases, such as the humid tropics, conversion from forest to open land is relatively permanent due to soil modification. Even where this is not the case, fragmentation of forest cover starts a feedback dynamic which changes local and mesoscale climate, rapidly depletes wildlife habitats and populations, and fosters ever greater human use of the forest remnants.

Driving this ultimate transformation of the earth's surface are forces such as (i) continued rapid increase in the size of the human population; (ii) the unpredictable course of technological innovation, providing ever more effective tools for changing land covers; (iii) changes, often volatile, in economic conditions, which change the incentives for using the land and its resources in sustainable ways; and (iv) globalization of change, which immediately spreads the changes in economic conditions, in particular, across the whole world.

The 1960s term *popullution,* combining the effects of numbers and technological means, remains an accurate index of the pressure on the land (Ehrlich and Ehrlich, 1990) or "harrying capacity" (Adler-Karlsson, 1974). Globalization of communications, transportation, and economic systems now provides a powerful feedback mechanism for acceleration of change.

The demographic driver of landscape change has been understood for decades (Davis, 1967; Ehrlich, 1968; Borgstrom, 1969), but no effective response has emerged and none seems to be in sight. In their landmark book *The Limits to Growth,* Meadows *et al.* (1972) showed the urgency of the problem by demonstrating that they could stabilize their global simulation model of resources and human activity, without drastic destruction to Earth's productive capacity, only if stabilization of human population numbers were achieved immediately, in 1975, and not delayed until 2000. Thus, it is already too late for sustainable development over most of the globe (Meadows, 1995).

Projecting changes in land use, like anticipating the effects of new technologies (e.g., television and computer chips), is almost impossible. Some of the economic forces which drive land changes, however, are fairly obvious. As remaining areas of "undeveloped" land become smaller, they also become more valuable, both to the developer for his or her profit and to the poor subsistence-level farmer who must feed his or her family. We can thus expect rapid disappearance of the small remaining areas of unprotected natural landscape and enormous pressure from both the poor and the greedy on those areas which do have some protection status.

There is no need for more current, trendier, more academically correct terms and explanations for these phenomena — they were already well understood 30 years ago, but nothing was done because of economic and political forces. The most powerful of these forces comes perhaps from competition. Just as competition in an arms race wastes enormous amounts of resources, it also forces waste in commercial and other ventures. Even more important, where competition exists, one is not free not to compete since both resources and marketplace opportunities are limited (Hardin, 1968) or due to "social traps" (Cross and Guyer, 1980). Nor does the competitive market system of supply and demand regulate behavior in constructive ways (except near the middle of the spectrum) since scarcity fosters hoarding, irrational buying behavior (Ophuls, 1977), and windfall opportunities to make enormous profits quickly by deliberately exhausting the last of a dwindling resource and then reinvesting the profit (Fife, 1971; Hardin and Baden, 1977). Competition ensures that people and organizations are not free to think or act holistically for the good of the earth system and human society in general.

Changes in Human Values

Despite obvious depletion of economic resources, landscape destruction over larger areas, and clear if less visual downward spirals in standards of living, it is becoming ever more common for people and organizations to deny the situation. Sources of denial behavior include (i) vested interests of big business and other large organizations, (ii) agendas of ideological groups, (iii) dependence of careers on the "human dimensions" of change (as well as other human-only considerations, such as business, politics, religion, and most of academia), and (iv) simply an unwillingness to face the unpleasant. Ecological considerations threaten some established patterns of operation (Shepard and McKinley, 1969). The commercial and legal might of the vested interests, in particular, raises obstacles which effectively preclude holistic problem solving.

Overlooked but critical may also be threshold effects in human perceptions. What will happen, for example, if a new generation decides that the battle for conservation of remaining nature and for global sustainability is already lost? Societal values based on the common good, on right and wrong, may be rapidly replaced by values based on success in competition for what is left. This transformation (and acceleration of societal polarization) may already be under way in the United States in the form of the "conservative revolution" which began in the 1980s and is spread by the victory of aggressive capitalism and the forces of economic globalization. In such a case, "bad" becomes "good" because it is more competitively successful in getting what one wants. Some implications of such "business values" are suggested by the values which they replace (Table 4).

TABLE 4

Comparison of Values

BUSINESS VALUES	SOCIAL VALUES
Self-gain	Sense of the community, common good
Rapid exploitation	Conservation
Self-promotion, blatant advertising	Modesty, courtesy
Opportunism, cleverness	Respect, fairness
Aggressiveness	Civility
Freedom and absolute rights	Responsibility
Deception, "public relations," manipulation of information	Sincerity

Political and other forms of "correctness" are also fostered by this new value system since correctness promotes control.

Summary

Changes taking place on our planet now and in the near future may be characterized by the following features:

1. An overwhelming replacement of values based on community and the common good by values based on competition. (The word "aggressive" has rapidly become one of the most commonly used adjectives in American English, always promoted with "positive" connotations.)
2. Acceleration of change, due to feedbacks in Earth systems; to feedbacks in human systems when resources decrease rapidly in a finite environment; to time lags (imperceptibility of some changes) and consequent threshold effects; to polarizing changes from community-based values to values based on success in competing for vanishing resources; and to globalization, which quickly forces the economics of the strong and rich on the less strong and nonrich.
3. Massive, sudden loss of remaining naturalness and diversity once thresholds are passed.
4. The total unpredictability of many changes: Will the final exhaustion of land and other resources be a slow degradation or a sudden collapse? Collapse offers the advantage that it may stop some aggressive activities quickly, whereas slow degradation permits more hoarding, continued harrying of the land, and probably more complete degradation of the life-support capacity of the planet.

Any serious attempts to avoid the degradation and eventual desertification of the planet must label political correctness for what it is (self-serving ideology) and not be afraid to deal with questions which are considered "incorrect" or taboo. Foremost among these must be the question of how to reduce human population in areas which cannot support so many.

References Cited

ADLER-KARLSSON, G. (1974). *Harrying and Carrying Capacity.* Allmaenna Foerlaget, Stockholm.

ARCHIBOLD, O. W. (1995). *Ecology of World Vegetation.* Chapman & Hall, London.

AXELROD, D. I. (1952). A theory of angiosperm evolution. *Evolution* **4,** 29–60.

BAILEY, R. G. (1989). Explanatory supplement to Ecoregions Map of the Continents. Environ. *Conserv.* **16,** 307–309.

BAILEY, R. G. (1996). *Ecosystem Geography.* Springer-Verlag, New York.

BORGSTROM, G. (1969). *Too Many.* Collier, New York.

BOX, E. O. (1981). *Macroclimate and Plant Forms: An Introduction to Predictive Modeling in Phytogeography. Tasks for Vegetation Science, Vol. 1.* Junk, The Hague.

BOX, E. O. (1995a). Factors determining distributions of tree species and plant functional types. *Vegetatio* **121,** 101–116.

BOX, E. O. (1995b). Global potential natural vegetation: Dynamic benchmark in the era of disruption. In *Toward Global Planning of Sustainable Use of the Earth — Development of Global Eco-Engineering* (Sh. Murai, Ed.), pp. 77–95. Elsevier, Amsterdam.

BOX, E. O., CRUMPACKER, D. W., and HARDIN, E. D. (1999). Predicted effects of climatic change on distribution of ecologically important native tree and shrub species in Florida. *Climatic Change* **41,** 213–248.

BRIDGES, E. M. (1990). *World Geomorphology.* Cambridge Univ. Press, Cambridge, UK.

BROWN, L. R. (1978). *The Twenty-Ninth Day: Accommodating Human Needs and Numbers to the Earth's Resources.* Norton, New York.

CROSS, J. G., and GUYER, M. J. (1980). *Social Traps.* Univ. of Michigan Press, Ann Arbor.

CURRIE, D. J., and PAQUIN, V. (1987). Large-scale biogeographical patterns of species richness of trees. *Nature* **329,** 326–327.

DAVIS, K. (1967). Population policy: Will current programs succeed? *Science* **158,** 730–739.

DAVIS, M. B., and ZABINSKI, C. (1992). Changes in geographical range resulting from greenhouse warming. In *Global Warming and Biological Diversity* (R. L. Peters and T. E. Lovejoy, Eds.), pp. 297–308. Yale Univ. Press, New Haven, CT.

EHRLICH, P. R. (1968). *The Population Bomb.* Ballantine, New York.

EHRLICH, P. R., and EHRLICH, A. H. (1990). *The Population Explosion.* Simon & Schuster, New York.

FIFE, D. (1971). Killing the goose. *Environment* **13,** 20–27.

FRAKES, L. A. (1979). *Climates through Geologic Time.* Elsevier, Amsterdam.

GATES, D. M. (1993). *Climate Change and Its Biological Consequences.* Sinauer, Sunderland, MA.

GROOMBRIDGE, B. (Ed.) (1992). *Global Biodiversity: Status of the Earth's Living Resources. World Conservation Monitoring Center (et al.).* Chapman & Hall, London.

HARDIN, G. (1968). The tragedy of the commons. *Science* **162,** 1243–1248.

HARDIN, G., and BADEN, J. (Eds.) (1977). *Managing the Commons.* Freeman, San Francisco.

HAYS, J. D., IMBRIE, J., and SHACKELTON, N. J. (1976). Variations in the earth's orbit: Pacemaker of the Ice Ages. *Science* **194**(4270), 1121–1132.

HENDERSON, S., and DIXON, R. K. (1993). Management of the terrestrial biosphere to sequester atmospheric CO_2 [Special issue]. *Climate Res.* **3**(1/2), 1–140.

HEYWOOD, V. H. (Ed.) and Global Biodiversity Assessment Group (1996). *Global Biodiversity Assessment.* United Nationals Environment Program/Cambridge Univ. Press, Cambridge, UK.

HIDORE, J. J., and OLIVER, J. E. (1993). *Climatology, an Atmospheric Science.* Macmillan, New York.

HOLDRIDGE, L. R. (1947). Determination of world plant formations from simple climatic data. *Science* **105,** 367–368.

HOUGHTON, J. T., MEIRA FILHO, L. G., CALLANDER, B. A., HARRIS, N., KATTENBERG, A., and MASKELL, K. (Eds.) (1996). *Climate CHANGE 1995: The Science of Climate Change. Intergovernmental Panel on Climate Change.* Cambridge Univ. Press, Cambridge, UK.

IMBRIE, J., and IMBRIE, J. Z. (1980). Modeling the climatic response to orbital variations. *Science* **207,** 943.

LAMB, H. H. (1977). *Climate: Present, Past and Future. Vol. 2: Climatic History and the Future.* Methuen, London.

LASSITER, R. R., BOX, E. O., WIEGERT, R. G., JOHNSTON, J. M., BERGENGREN, J., and SUAREZ, L. A. (2001). Vulnerability of ecosystems of the mid-Atlantic region to climate change. *J. Soc. Environ. Toxicol. Chem.,* in press.

MEADOWS, D. H., MEADOWS, D. L., RANDERS, J., and BEHRENS, W. W. (1972). *The Limits to Growth.* Universe, New York.

MEADOWS, D. L. (1995). It is too late to achieve sustainable development, now let's strive for survivable development. In *Toward Global Planning of Sustainable Use of the Earth — Development of Global Eco-Engineering* (Sh. Murai, Ed.), pp. 359–374. Elsevier, Amsterdam.

NATIONAL RESEARCH COUNCIL (1982). *Solar Variability, Weather, and Climate.* National Acad. Press, Washington, DC.

OPHULS, W. (1977). *Ecology and the Politics of Scarcity.* Freeman, San Francisco.

PALMER, M. W. (1994). Variation in species richness: Towards a unification of hypotheses. *Folia Geobotan. Phytotax. (Praha)* **29,** 511–530.

PEARSON, L. C. (1995). *The Diversity and Evolution of Plants.* CRC Press, Boca Raton, FL.

PETERS, R. L., and LOVEJOY, T. E. (1992). *Global Warming and Biological Diversity.* Yale Univ. Press, New Haven, CT.

PIELOU, E. C. (1979). *Biogeography.* Wiley, New York.

RAVEN, P. H., and AXELROD, D. I. (1974). Angiosperm biogeography and past continental movements. *Ann. Missouri Botanical Garden* **61,** 539–673.

SCHLESINGER, M. E., and MITCHELL, J. F. B. (1985). Model projections of the equilibrium climatic response to increased carbon dioxide. In *Projecting the Climatic Effects of Increased Carbon Dioxide* (M. C. MacCracken *et al.,* Eds.), Report No. DOE/ER-0237, pp. 81–147.. U.S. Department of Energy, Office of Energy Research, Washington, DC.

SCHMITHUESEN, J. (1968). *Vegetationsgeographie,* 3rd ed. de Gruyter, Berlin.

Scientific American (1976). *Continents Adrift and Continents Aground. Readings from Scientific American (1963–75).* Freeman, San Francisco.

SHEPARD, P., and MCKINLEY, D. (Eds.) (1969). *The Subversive Science.* Houghton Mifflin, New York.

SOULE, M. E. (1990). The onslaught on alien species, and other challenges in the coming decades. *Conserv. Biol.* **4,** 233–239.

THOMAS, W. L. (Ed.) (1956). *Man's Role in Changing the Face of the Earth,* 2 Vols. Univ. of Chicago Press, Chicago.

UVARDY, M. D. F. (1975). *A Classification of the Biogeographical Provinces of the World. Occasional Paper No. 18.* International Union for Conservation of Nature and Natural Resources, Morges, Switzerland.

WALTER, H. (1985). *Vegetation of the Earth and Ecological Systems of the Geobiosphere,* 3rd ed. Spinger-Verlag, New York.
WALTER, H., and BOX, E. O. (1976). Global classification of natural terrestrial ecosystems. *Vegetatio* **32,** 72–81.
WOODWARD, I. (1987). *Climate and Plant Distribution.* Cambridge Univ. Press, Cambridge, UK.

General References

CATTON, W. R., JR. (1982). *Overshoot: The Ecological Basis of Revolutionary Change.* Univ. of Illinois Press, Urbana.
WALTER, H. (1968, 1973). Die vegetation der erde. In *Oeko-Physiologischer Betrachtung,* 2 Vols. Fischer-Verlag, Jena.

VITTORIO INGEGNOLI

Department of Biology
Faculty of Natural Science
University of Milan
Milan, Italy

Landscape Ecology

The concept of landscape was born in the protoagricultural period. Although its clear scientific formulation dates back to the beginning of the nineteenth century, it remained out of the ecological theory until the first half of the twentieth century. General ecology deals only with organisms, populations, and communities, and it reduces each ecological system in the ambiguous concept of ecosystem. Only recently, the new scientific paradigms shed light on the necessity of a reformulation of ecology and gave space to landscape ecology, which studies the structure, functions, and changes of interacting systems of ecosystems. This opens new theoretical perspectives in the field of the study of nature and allows further important applications. The theory of integrative levels of biological organization identifies the landscape as one of its specific levels and establishes that it is impossible to study a landscape using parameters derived from the ecosystemic level. Landscape ecology is a transdiscipline able to guide all the other applications such as ecological planning, environmental design and restoration, biological conservation, and environmental evaluation.

Introduction

Landscape ecology studies the structure, functions, and changes of systems of interacting ecosystems. This opens new theoretical perspectives in the field of the study of nature and allows further important applications. From a theoretical point of view, one can immediately observe that for the theory of integrative levels of biological organization, it is impossible to study a landscape using only the parameters derived from the ecosystemic level. On the contrary, one must recognize that the landscape, because it is a specific level of biological organization, has its own structure and its own dynamic. The landscape shows all the principal characters of any level of life organization. In fact, we can describe not only its specific structural characters but also its boundaries, its internal system of communications and biomass movement, its self-reproduction processes, its level of metastability, and its specific dynamic behavior. Therefore, landscape ecology opens the door to the study of unified ecology (Allen and Hoekstra, 1992), from the organism to the biosphere, and the anthropic component can also be studied.

From the standpoint of applications, landscape ecology is based on a transdisciplinary method. This method arises from the observation that there are several disciplines concerned with environmental studies but they do not all make the same contribution. The organizing principle of life remains essentially biological; therefore, only this principle can integrate all the others. For this reason, landscape ecology appears to be the transdiscipline able to guide all other similar application, especially in the field of ecological planning of land, environmental design and restoration, nature conservation, and environmental evaluation.

Today, much more than ever before, the solutions to critical problems such as habitat fragmentation and territorial planning, the control of the ecological web, and even the influence of the territory on organisms' behavior need the knowledge derived from landscape ecology.

Landscape and Ecology

Development of the Concept of Landscape in Science

The concept of landscape is very old. Its origin is linked with the first permanent settlements of human populations in the protoagricultural period (about 5000 BC), when the necessity of a functional exploration of the territory became a matter of survival. Man needed to gather information on the entire mosaic of biogeocoenosis forming his territory (i.e., the natural landscape) *conditio sine qua non* to be able not only to develop it in an agricultural and establishing way but also to manage and defend it. This approach was based on an action of conscious mutualism with nature and it shifted the religion toward the cult of the sun. It produced a first form of regionalization of culture, and major environmental changes on a large territorial scale began to occur.

There is various proof of this. Archeological findings dating back to the first protoagricultural villages (at the beginning of the Neolithic Period) show that men knew how to combine and transform the different elements of the natural landscape to create an anthropic landscape beneficial to them (Forman and Godron, 1986). The representations of maps dating back 4000 years in the territory of Camuni testify to the acquisition of the scientific concept of landscape (Ingegnoli, 1993). In the language of Jews, a more ancient civilization, the word "landscape" (*noff*) appeared in the Book of Psalms 3000 years ago (Naveh and Lieberman, 1990).

In Roman times, the landscape used to be called *regio* (region), indicating its geographical aspect, as noted by Columella, a famous agronomist of the first century BC. The landscape painter, who already had a good perception of prospective, used to be called "*pictor topiarius qui regiones formas pingit*" ("painter who paints the shape of the regions"), and the panorama used to be called "*prospectus.*" During the Renaissance in Italy, especially in the artistic and literary sphere (from Petrarca to Tiziano), people began to use the current meaning of the term, derived from the French *paysage* (from the Latin *pagus*), summarizing all the proper characters of a country. The meaning of these characters is evident in the papers of Leonardo da Vinci because his scientific interest and his studies on the form in all its numerous expressions occurred simultaneously. He considered "*sassi, piante e simili*" as "*le membra de'paesi*" ("stones, plants and similar things" as "the limbs of the countries"), comparing them both in a structural and in a functional sense to those of animals. Through his studies on alpine landscapes he started the analysis of geomorphology and proposed a first formulation of the concept of vegetation.

After Columella and Leonardo, a second step toward the use of the term landscape in a scientific way was taken by the English landscape architects of the eighteenth century. The first great transformations of the territory toward an industrial direction induced them to abandon the French formalism in shaping gardens and public parks and to favor a more natural kind of planning instead. Nevertheless, a perceptive concept of landscape still remained a prominent one.

Only at the beginning of the nineteenth century, with the German school of Alexander Von Humboldt, was the term landscape fully integrated into the scientific vocabulary. He defined the landscape as "*Der Totalcharakter einer Erdgegend*"—that is, the totality of all the proper characters of a specific territory. Nevertheless, from the first days of that century until the middle of the following century, the ecologists not only were not inclined to adopt the term landscape but also they did not deal with this argument.

After the introduction of the concept of ecology by E. H. Haeckel, a German naturalist and a painter, in the mid-nineteenth century, and after the formulation of the concept of biocoenosis by Moebius (1877), in 1921 the German geomorphologist S. Passarge (who was also a physician) began to speak on the physiology of the landscape following Von Humboldt's tradition, explaining modern concepts based on the integration of the climate geomorphology and the natural sciences. Ecologists did not follow all these attempts until 1939, when the German biogeographer K. Troll proposed the term "*landschaftsoekologie*"—landscape ecology (Troll, 1939).

Limits of Traditional Ecology

In reality, ecology was born as "autoecology" (i.e., as the study of the relationships between organisms and their environment) and evolved by separate and sometime even contrasting theories. Scientists know the famous dispute between population and ecosystemic ecologists: The concept of community (a complex of populations) and the concept of ecosystem (a system of ecological processes) were reworked by Odum (1971) through a compromise. On the other hand, O'Neill and his group note that these two different points of view have never been completely compatible. As a matter of fact, as Allen and Hoekstra wrote (1992), the conservation of matter and energy (which is a fundamental base for the concept of ecosystem) has no relevance for the concept of community; furthermore, a portion of an ecosystem (e.g., a pathway of species) may correspond to different portions of a community because each organism may be an autonomous member of a community. Even the ecology of large scales, such as the ecology of biomes and the biosphere, is divided into different categories: The first is especially related to biogeography, whereas the second one leads to "global ecology," starting from the famous "Gaia hypothesis."

Moreover, there is difficulty with traditional ecology when dealing with problems of applications on a territorial scale because the concept of ecosystem cannot be effective for all scales, especially when various spatial schemes interact. One must remember that the ecology of ecosystems considers the environmental heterogeneity as background noise (Blondel, 1986) with a secondary importance. However, the reality is very different: The heterogeneity and variability of the environmental mosaic represent a very important component in establishing the distribution of organisms and their interactions and also in an evolutionary sense. Moreover, the ecology of ecosystems did not try to obstruct the logic of the impact between man and nature, a deterministic reasoning that in reality derives from preconceived ideas which have no scientific meaning and which prevent people from managing the territory in an ecologically balanced way.

On the other hand, regarding attempts to overcome these difficulties by borrowing ecological methods from other disciplines to study the environment at a landscape scale, there are intrinsic limits in the theory of the integrative levels of biological organization. It does not make sense to analyze the ecological systems at the landscape level using the properties of the ecosystemic level because the ecosystemic level is a lower, simpler level of organization. It must not be forgotten that in 1971 Odum said that an evident principle of functional integration subsists inside the "biological spectrum," and therefore the results obtained at a certain level of organization help us in the study of a superior level but they can never completely explain the phenomena occurring at that level.

Today, much of general ecology is concerned with the ecology of organisms, populations, and communities and constricts each ecological system in the ambiguous concept of ecosystem. Finally, we have to consider that the theory of integrative levels highlighted by Odum is only a corollary of the most important principle of the "emerging properties" (Lorenz, 1978): An organic system is much more than the sum of its parts; therefore, it is sufficient to change the structural assemblage of the elements of a system to obtain a behavior not only different but also sometimes unpredictable. Therefore, it is absolutely necessary to analyze the properties of a specific level of a biological organization by studying the intrinsic behavior of the level because the information derived from lower levels is at the most partial. At the same time, it is absolutely necessary to refer to new scientific paradigms in light of which the ecological discipline is forced to renew.

New Scientific Paradigms

Adaptive Complex Systems

As Gell-Mann (1994) noted, the law of gravitation is a wonderful example of the simplicity of the natural principles which are at the basis of all phenomena we observe. This law is able to describe processes of aggregation of the matter which resulted, during the evolution of the universe, in the formation of galaxies and then of stars and planets, including the earth. Since the time of their constitutions, these celestial bodies have shown characters of complexity, diversity, and individuality. These properties assumed new meanings when complex adaptive systems appeared. On Earth, this development is linked with the origin of life and the process of biological evolution which produced the surprising variety of species, the behavior of organisms in the ecological systems, the ability of learning and thinking in animals and humans, the evolution of human society (included the economic–technologic aspect), and the formation and the evolution of landscapes.

Our species, at least in some respects, is the most complex on Earth and has succeeded in finding out much information on simple systems; however, studies on complex adaptive systems are just beginning. The common character of all evolutionary processes is that for each of them, a complex adaptive system acquires information on its environment and its interaction with it; the analysis of this information allows it to identify regularities, which are summarized in a sort of model able to provide a guide to act in the real world. In any case, different schemes exist in competition with others, and the results of the action in the real world are reintroduced in the system, influencing competition among schemes, at the moment in which the system has been transformed.

There is no doubt that the study of biological systems is a task for ecology because they are complex adaptive systems on a higher level than the level of organism–individual. However, it must be emphasized that the attempt of traditional ecology to actually remain in the sphere of Cartesian's determinism prevents us from entirely dealing with this problem. For example, the problem of the scale is misinterpreted because both a temporary pool and an entire geographic region may be defined as ecosystems; as a consequence, the landscape is relegated (in this case) to the geographic sphere and researchers think they will be able to study it only by using the knowledge of organisms, populations, and communities. In addition, the lack of a direction in time (time as an irreversible process) makes us consider nature as something without history, mechanistic, and under the human domain, with man who sees himself as its engineer.

Moreover, a description of nature according to deterministic laws, characterized by phenomena reversible in respect to time, is not capable of accepting novelty and creativity; thus, it is incapable of defining life outside an organism or a group of organisms. Therefore, it is simply unthinkable that a system of interacting ecosystems (the landscape) could be defined as a specific level of biological organization by traditional ecology. In addition, determinism not only obstructs contact with reality but also initiates a discussion of human freedom (Lorenz and Popper, 1985; Popper, 1996), whereas freedom presupposes an idea of creativity. Therefore, it is necessary to refer to new scientific paradigms.

New Paradigms of Reference in Ecology

Perhaps the major incentive for a new conception of nature comes from K. Popper, K. Lorenz, and I. Prigogine, who observed how nature creates its most fine, sensitive, and complex structures through the association of nonreversible processes with temporal dimension; these authors assert that indeterminism is compatible with realism — that an organic all is much complex than the sum of its parts and that life is only possible in a universe far from equilibrium.

The study of every branch of reality, which seems to be continuous in itself, begins with a separation from a systemic point of view. Due to the formulation of very important epistemological theories during the twentieth century, such as the theory of complementarity, the theorem of Goedel, the theory of information, and the theory of hierarchical systems, we know that mechanistic paradigms are no longer sufficient to determine the behavior of a system compatible with the laws of nature because the state of a system is not determined by its coordinates (Ingegnoli, 1971).

Thus, in order to know the state of an ecosystem or of a landscape, we cannot limit ourselves to a mechanistic paradigm. We cannot no longer suppose that the state of an object is independent from the way we observe it because the observation itself may change the sources of information on this object. Moreover, complex adaptive systems contain and include past disturbances in components of the lower level. All this suggests the importance of the history, the scale, and the contest as the bases of knowledge and integration.

Prigogine (Prigogine *et al.,* 1972; Prigogine and Nicolis, 1977; Prigogine, 1996) notes that we need a new objectivity different from both the one of the classic thermodynamic, for which what we are able to know coincides with what we are able to control (which identifies the knowable with the checkable), and the dynamic one, which tries to understand the evolution of a system starting from an instantaneous state. A structure created by a succession of extended fluctuations may only be understood in relation to its past: The past has been produced by unpredictable events and must be considered as unique and not repeatable. This assumption is in agreement with the principles of emerging properties and revalues without any doubt the importance of history in the ecological studies, even in the absence of man. Following the pioneers of this more complex and more complete meaning of the history, which dates back to C. Cattaneo and F. P. Braudel, it is interesting to note the concept of location of history, expressed by Zanzi (1995) as *avvenimentale* ("shaped by events"), in which different protagonists of history (such as trees, animals, men, configurations and structures of the ground or territory, etc.) evolve interacting with each others.

If minor changes in the initial conditions of the system produce conspicuous effects on it, effects which become large over time, the dynamic systems are considered unstable. For example, chaotic systems, in which trajectories almost coinciding at the initial stage diverge exponentially with the passing of time, are unstable. The difference between stable and unstable dynamic systems is therefore essential. Usually, the dynamic state of a system is represented by a point in space of the phases (i.e., the space of positions and momentum). Considering unstable systems, note — as declared by Prigogine (1996) — that the initial conditions are no more comparable to a point in space of the phases, but they correspond to a region of space described by a distribution of probability. Therefore, this is not a punctual (point for point) description but a landscape one. Moreover, as a consequence, there is a break in the temporal symmetry: Following it, the past and the future start to perform different roles in the statistic formulation as instability destroys the equivalence between the individual level and the statistic level and consequently probabilities assume an intrinsic meaning, not reducible to the case of individual trajectories. Then, the distribution of probabilities at the initial conditions permit us to include the complex microstructure of the space of the phases in the dynamic description. This information, which is additional in respect to the individual description, allows one to predict the evolution of the whole. Only in the description of stable systems can we return to a traditional dynamic. Usually in biology, however, stable systems are not interesting.

Each dynamic system may be characterized by a kinetic energy, which depends on the speed of its components, and by a potential energy, which depends on the interactions among the components. In a world isomorphous to a set of bodies, which are subject to no interaction (which are not liable to any interaction), there is no place for the arrow of time nor self-organization for life. Nevertheless, as observed by Poincaré (1906), the existence of resonance among the degrees of freedom of the system produces conditions of nonequilibrium. In these conditions, many different mutual relations occur among the components. A new sensitivity of matter to itself and to its environment takes place, associated with dissipation and not reversible processes.

The most famous example of this, noted by Prigogine, is the concentration of the intermediate product X in a chemical reaction: On the stable thermodynamic branch, we find a field of instability with the appearance of subsequent bifurcations (Fig. 1). When a system arrives in a fork point, disturbances become important, such as fluctuations which allow the system to choose one of the two branches of new relative stability. The evolution of this kind of system has a historic element. For example, if one observes the system in the position d_2 (Fig. 1), the system has passed through the states b_1 and c_1. Moreover, if we apply Prigogine's concept of self-organization of dissipative structures to the human society, it is possible to demonstrate that human activity, which is creative and innovative, is not foreign to nature. It can be considered as an extension and an intensification of some aspects already present in the physical world, which may now be more easily understood with the discovery of models for processes far from equilibrium.

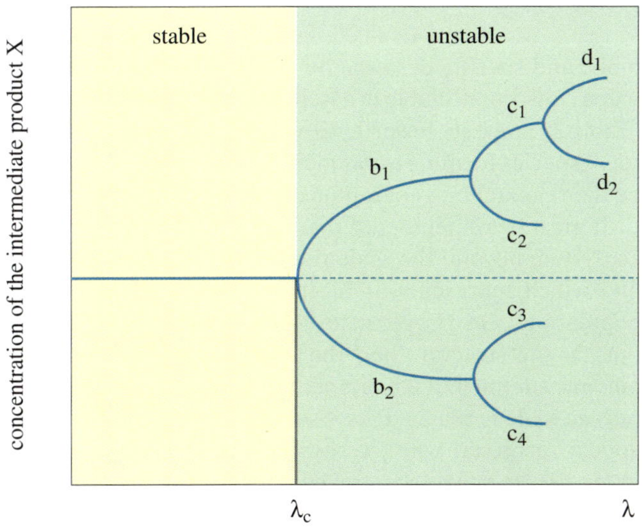

FIGURE 1 Successive bifurcations in the dynamics of a nonequilibrium system. Chemical reactions usually follow nonlinear dynamics; thus, for each value of the concentrations of initial and final substances, many possibilities of the concentration of the intermediate product X exist. Only one of these corresponds to the state of thermodynamic equilibrium. This solution may be extended in the nonequilibrium domain as a thermodynamic branch (dashed line). The concentration of X is represented as a function of the λ parameter, which measures the distance from equilibrium. Beginning with the critical point λ_c, the thermodynamic branch becomes unstable and successive bifurcations appear which correspond to alternative solutions (adapted from Prigogine, 1996).

Landscape Ecology: Toward the Frontier of Ecology

In light of the scientific theories discussed previously, the limitations of the approaches of traditional ecology may be easily demonstrated. The problems of scale, complexity, contest, relationship between man and nature, self-organization, and even definition of biological levels have received little consideration. In order to reach the frontiers of ecology, it is important to start from the point that biological organization depends on the dynamic of dissipative structures, associated with irreversible processes. The theory of dissipative structures and the concepts of metastability and incorporation of disturbances allow us to obtain the historicity of the biological systems (O'Neill *et al.,* 1986).

General ecology (Whittaker, 1975) views the landscape only as a geographical context for the ecological communities (or ecosystems)—that is, as a geographical gradient capable of affecting ecological structures and processes at the levels of organisms, populations, and communities. This definition is no doubt necessary, but it is not sufficient because it does not consider the landscape as a specific level of biological organization. As noted previously, to accept the "biological spectrum" (Fig. 2), as derived from the hierarchical theory, and to follow the principle of emerging properties should be enough to consider the landscape as an ecological system not describable through the characters of previous levels (populations, communities, or ecosystems). The landscape as a metaecosystem must own its specific characters and behaviors.

However, in order to define the landscape in the way we are sustaining, it is necessary to recall the concept of ecosystem, the biotic component of which is the community (Odum, 1971, 1993; Whittaker, 1975), and the theory of nonequilibrium thermodynamic (Prigogine *et al.,* 1972; Prigogine, 1996).

The Landscape as a System of Ecosystem

It must be remembered that when an intense energy flux passes through a system, some dissipative structures generally appear, characterized by high instability. In fact, the self-organized living systems capture energy, and this energy is used to produce new structures: Prigogine called this *order through fluctuation.* From this point of view, an ecosystem is an open system, far from thermodynamic equilibrium, and therefore dissipative. Consequently, the ecosystem is characterized by the same importance of internal and external exchanges, which can be described by the thermodynamic of systems far from equilibrium. In an open system, the inequality of Clausius–Carnot takes the following form:

$$dS = d_eS + d_iS \gtreqless 0$$

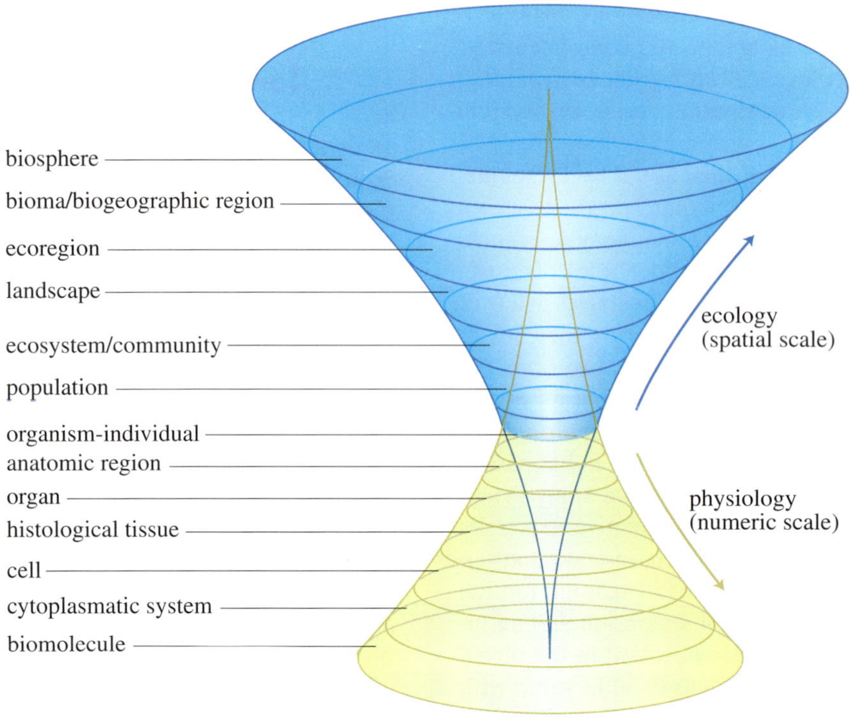

FIGURE 2 Biological spectrum showing the hierarchy of levels of biological organization including the landscape level. The spatial scale and the numeric scale are directed in opposite directions: Compare the level of biosphere (unique but with a big mass) with the biomolecular level (large numbers but small masses).

where d_eS is the variation of entropy due to exchanges with the environment, and d_iS is the production of entropy due to nonreversible processes within the system. Since d_eS can assume negative values, during its evolution a system can reach a state in which entropy is lower than that of the initial stage (i.e., a more ordinate state). This is applicable to both a forested area and an urbanized one, and it must have boundaries (i.e., "surfaces" or "membranes") for each ecosystemic unit and the existence of populations as "internal" structures. The external context, a specific system of ecosystems, is our landscape. This context is a system of ecosystems working directly with its elements exchanging species, energy, substance, and information and following the principle of emerging property, thus creating a proper structure and proper dynamics. A very important observation derives from the previous discussion: An element of the landscape assumes a specific functional role in its complex mosaic.

This system of interacting ecosystems is a very complex structure, difficult to focalize and to represent. Remember, however, that every biological system has a structure formed by well-defined functional sets in a context of variable substrates, from cells to ecosystems, landscapes, regions, and the biosphere. The landscape shows all the principal characters of any level of life organization because we may describe its specific structural characters and also its boundaries, internal system of communications and biomass movement, self-reproduction processes, level of metastability, and specific dynamic behavior. Moreover, the landscape is also identifiable in a well-determined range of scales (Table 1) since its lower level will be in proportion to the set of species forming a community and its higher limits will be the regional biogeographic district.

It must be remembered that it is not correct to study an ecosystem beyond the scale relevant to the entire community which represents its biotic part without knowing its populations. On a larger scale, when other elements emerge outside the structure and the functions of an ecosystem, one must refer to a landscape. For instance, when an ecomosaic scale is reached, new structures and new processes appear: ecotonal webs, ecosystems connectivity, porosity of landscape matrix, landscape apparatuses, landscape dynamics, new metastability strategies, and so on. Similarly, on a regional scale other features will appear that are impossible to study in a landscape, such as soil ordination, forest formation, faunal changes, biological gradients in rivers, and the climatic role in ecological changes. We must consider that with the widening of the scale in the ecological systems, not only many characters of the preceding level but also many dominant factors of control will change, as foreseen by the theory of hierarchic systems. Even the human factors of ecological control will change considerably as the scale increases, as may be seen in the structural differences between an aerial photographic survey (1:25,000) and a remote sensing satellite survey (1:1,000,000). In the first case, the human influence is usually relevant, whereas

TABLE I

Synthesis of Ecological Classification Systems of Hierarchical Character[a]

INTERNATIONAL NOMENCLATURE	ECOLOGICAL SCALE (QUALITATIVE)	APPROXIMATED SIZE (km^2)	MAIN MAP (COMMON SCALE)
Domain (zone, subcontinent)	mega–giga	1,000,000–10,000,000	1:2,500,000–1:10,000,000
Region (ecoregion, province)	macro–mega	10,000–100,000	1:500,000–1:2,000,000
Landscape (land mosaic, district)	meso–macro	100–1,000	1:25,000–1:200,000
Ecosystem (ecotope, site)	micro–meso	1–10	1:1,000–1:10,000

[a]Modified from Zonneveld (1995) and Bailey (1996).

in the second case geomorphic and topographic influences are often more relevant. The presence of human ecosystems may be a distinctive character and possibly an essential character of a landscape, even if analyzed on a very detailed scale.

The definition of landscape is now more clear; consequently, we can define landscape ecology as the portion of ecology which studies the landscape at a biological level, thus contributing to the development of integrated (unified) ecology and territorial planning. In this way, landscape ecology becomes necessary to nature conservation and must be considered one of the basic elements of unified ecology.

Principal Characters of a System of Ecosystems

Specific Structural Characters

There are mainly two descriptive models of the landscape: the fragmented model and the variegation model (Fig. 3). These models represent completely different outlooks, one static and one dynamic. However, it is not completely correct to set a mosaic of juxtaposed tesserae of landscape elements (fragmented model or mosaic model) against a variable geometry mosaic of species-specific tesserae (variegation model). In reality, each biological system has a structure made by well-defined functional groups in a context of substrates changeable in space and time (or variegated configurations over a fixed weft and whose elements are not only juxtaposed but also overlapped and intersected).

As a consequence, it is better to introduce the concept of ecological tissue (from the Latin *textus* or *textilis;* "textile") or ecotissue, which is a complex multidimensional mosaic formed by a main mosaic and a hierarchic succession of correlated mosaics:

$$\mathrm{Ect} = M_b \cup(\cap)\, M_1 \cup(\cap)\, M_2 \ldots \cup(\cap)\, M_k$$

where Ect is the ecotissue, M_b is the main mosaic, $M_1 \ldots M_k$ are the correlated mosaics, and the set-theoretical $\cup(\cap)$

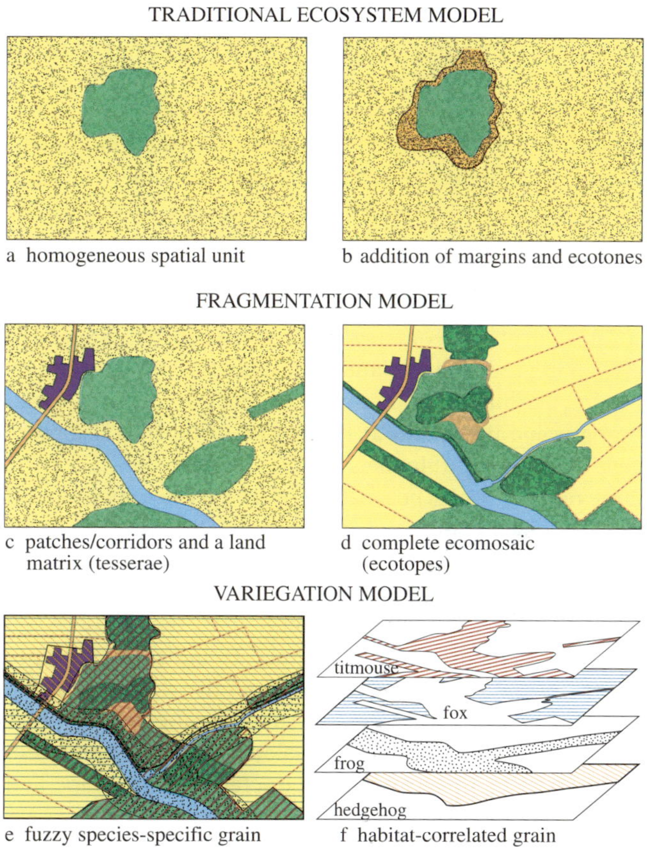

FIGURE 3 Spatial structural models in ecosystem ecology and landscape ecology. (Top) A traditional ecosystem model. A homogeneous spatial unit constitutes an ecosystem, but the context is not defined (a). Margin and ecotone belts bound the ecosystem unit, but the context remains undefined (b). (Middle) A fragmentation model. Patches and corridors emerge in a land matrix (c). The fragmentation model is completed by an ecological mosaic of juxtaposed tesserae (d). (Bottom) Variegated model. The ecological mosaic is composed of tesserae having a variable conformation according to the groups of animal species considered, producing a fuzzy grain (e). The same grain may be expressed by a group of species-specific superimposable mosaics (f).

symbols of union and intersection may be interchangeable or may even coexist, often being heterogeneous groups.

Note that the main mosaic is generally formed by the vegetational associations because they are the most important controller of the flux of energy and matter and can create the proper environment, interact with the soil, and form a microclimate. The element forming a landscape may be called a tessera or "ecotope. The ecotope has to be intended not only as a biogeocoenosis (following the German and Russian school) — that is, as the spatial representation of an ecosystem, even if in an integrated way — but also as a functional element because it acquires a role in the complex mosaic of which it is a part as a territorial niche.

The registration of a specific structural pattern and physiognomy of a landscape represents the earliest landscape studies. Since nature is chaotic and rarely can it be described using geometry, its structural heterogeneity is analyzed with the help of fractal geometry. According to Mandelbrot (1975), fractal geometry may be defined as the study of geometrical objects characterized by a fractional Euclidean dimension; Examples include a shoreline, the dimension of which is in between 1 and 2 since the length between two points tends to infinity if we change the scale, and the canopy of a tree, which presents a fractal dimension in between 2 and 3 and may be visually assimilated with the Euclidean concept of volume but may not be formalized as a volume through geometry. Two groups of forces seem to prevail in modeling the landscape, and they are revealed by the fractal dimension: human influences at microscale and topographic influences at macroscale. The fractal dimension of natural and human mosaics remains constant for a large range of scales because there are very few geometrical models which comprise the landscape structure (no more than five or six).

The distribution and the type of the grain of ecological mosaics and their fragmentation help in the study of the character of landscapes, for instance, by indicating the "porosity" of the landscape matrix. Even the structure of the landscape apparatuses (i.e., the set of ecotopes with a similar functional role in their mosaic) could be important in understanding the structural character of a landscape and in the classification.

Layer of Delimitation and Filter

In this context, ecosystem edges and ecotonal belts acquire new importance because in landscape ecology they constitute a system in some areas comparable with the cellular membranes in a histologic tissue (Forman and Moore, 1991), whereas in community ecology they represent only the interface between the homogeneity of an ecosystem and the heterogeneity of the surrounding environment. Note that the layers composed of vegetational biomass form the current boundary of an ecotope, whereas at the landscape level topographic and human barriers play a major role. In this sense, human actions have often had an evolutive function in defining many boundaries once undetermined in natural ecotopes, therefore giving a higher functionality to landscape mosaics. It is possible that one of the natural functions of this increase in human population could be the establishment of boundary web among different communities. Other animal population (e.g., beavers and, at a lower scale, ants and termites) transform undetermined boundaries into more defined ones, thus influencing heterogeneity and dynamics of ecological mosaics.

Internal System of Communications

Interecosystemic fluxes of energy and matter and chains of organisms permit internal webs of communication in an ecotissue. They allow the ecotissue to maintain a certain level of metastability. This informative system of a landscape is essential for the process of coevolution, which may be completed only at the landscape level; actually, an organism usually reacts in different ways depending on the scales of interaction with other levels of biological organization. The interactions among organisms lead to competition, the organism–environment interaction leads to mutualism, and the ecotope–ecotope interaction leads to coevolution.

Thus, the classic (Darwinian) concept of adaptation has to be expanded to include a new concept: In dealing with the (self)-organization of biological systems, the driving principle is not only the causal one but also the constraint one, and it forces work not only in terms of impact and conflict but also in terms of cooperation and integration. This is in agreement with Popper's (1996) statement: "Simply, it is not true we are shaped by the environment. We are the ones who look for the environment and who shape it. The nude gene searched for a protein environment and it built up its cover with proteins. . . . We try to modify and change our surrounding environment, then the one more distant, finally the entire Earth." If the relationship between man and nature is considered from this point of view, note it is not in contrast with the principles of evolution.

Thus, we overcome the preconception of the opposition between man and nature. As a consequence, the anthropic systems play a primary role in the system of communication and information at a landscape scale, and they seem to have evolved with a fundamental task of informative managing (Lorenz, 1978; Godron, 1984; Ingegnoli, 1993, 1997). The ability to plan the environment has an ecological and behavioral–ethological meaning of remarkable naturalistic importance and it expresses itself at the landscape level.

Internal Systems of Biomass, Matter, and Energy Displacement

The structure of an ecotissue affects different fluxes in a characteristic way: from the movements of organisms to the biogeochemic fluxes to the genetic ones. The movements of water and wind, or the movement of mineral particles, are related to the configuration of landscapes. Many studies have been conducted on this topic and many used neutral models, such as the percolation theory (derived from the modality of water infiltration in a soil), but many question have not been answered. The structure of an ecotissue is strictly linked with its ecotonal system because problems of filter, barrier, and corridors are important in both systems. In particular, the corridors seem to be the most interesting structures for the study of the movement of biomass.

In the displacement system, the concept of connectivity assumes great importance among the landscape elements: We may distinguish factors of connection and circuitation and differences among proper and improper corridors and ecotopes with connecting functions. Moreover, we can add the source–sink concept, which leads to the concept of metaclimax and metapopulation, which are interconnected systems of subunits in different evolutive phases. Also, the fluvial system represents a very important system of displacement because in the landscape it has the task of an excretory system. In fact, the ability to purify the catabolites from the surrounding ecomosaics is due to the structure and the functions of the complex corridor constituted by the river and its riparian vegetation (Ingegnoli, 1993).

Self-Reproduction

At the organizational level of ecotope and landscape, reproduction is different from the same processes pertaining to the organism or the cell, even if some basic factors remain valid. The information to be transmitted has to be circumscribed in time and space (e.g., in a propagule bank). An outbreak effect must happen—for example, a "zero event" (e.g., fire)—allowing the propagule to substitute previous organic structures (Oldeman,1990).

The concept of the zero event is strictly linked to a different valuation of disturbances in the ecological mosaics. Actually, in an ecological mosaic even high degrees of perturbation are structured because they are necessary for the self-reproduction of tesserae and ecotopes. Thus, without perturbations able to reset existing structures in an ecotope, its reproductive renewal would be impossible.

Also, recolonization plays an important role in self-reproduction. Pioneer phases tend to be stochastic, but pioneer dynamics are oriented by the same strain conditions which appear in a certain ecotissue.

Maintaining Metastability

The hierarchy theory indicates that interactions among the components at a lower level are controlled by slower interactions at higher levels. Therefore, the systems of ecosystems have greater homeostasis (i.e., capacity of incorporating disturbances) than their components; similarly, an ecotope has greater homeostasis than its populations. Only minor interactions may be cyclic or oscillating (such as the prey–predator relationship). Adding coevolution, self-reproduction, information exchange, and self-transcendence, even the landscapes are able to maintain a specific level of metastability. We know that the organization of biological levels may evolve only if the final state of a biological system is less unstable (more metastable) than the initial state. However, the maximum metastability of a system of an ecosystem does not correspond to the sum of maximum metastabilities of its components. Therefore, in a landscape in which each component is at its climax, the maximum metastability is not usually reached. Similar considerations are available for the biodiversity of a landscape (Blondel, 1986; Ingegnoli, 1993; Odum, 1993).

Consequently, the concept of climax must be reappraised, and this has great influence in theory and applications. Moreover, in theory, the ecological succession is no longer a linear (oriented) process since in a landscape it would result in a lower capacity to incorporate disturbances.

Specific Dynamic Behavior

It is possible to find a specific dynamic behavior of a landscape in various sectors.

General Landscape Processes. General landscape processes are those shaping the physiotopes, which are the geomorphologic substrates of ecotopes, and they concern climate, vegetation, soil, animals, and man. We may study the evolution of landscape types in relationship to their appearance on the biosphere, beginning with the first Palaeozoic colonization of the mainland. Relationships with morphologic processes are also analyzed: stable, intermediate, and unstable geodynamic environments and their spatial territorial distribution. To integrate morphogenetic actions into the landscape system, these actions must be defined through both a spatial criterion (separating localized from diffused phenomena) and a temporal criterion (as intermittent rather than discontinuous phenomena). Pedologic processes, especially catenal distributions, are very important, as Dokučaev noted in 1898 in the zonality law (Dokučaev, 1951). Finally, the most characterizing processes of ecotissues are the hydrologic ones, especially the different kinds of erosion and streaming runoff for types of ecotopes.

Functions of Element Configurations and Their Ecological Movements. Relationships among the characteristics of patches and the types of ecotissue are very useful in the valuation of the state of a landscape. Phenomena such as the interdigitation of a peninsula between two different types of landscape elements produce specific processes. Also, variation of the range of influence of the boundary effect in relationship with the dimension and the fragmentation of patches and corridors is a typical landscape function. Regarding movements depending on landscape structure, it is sufficient to remember that the source–sink phenomenon, the filter effect of boundaries and different evolutive states of the elements, and the geomorphic effects are responsible for them. These movements depend on a restricted number of vectors, such as wind; water; flying, terrestrial, and aquatic animals; and men and their means/equipment. Forces at stakes are diffusion, transportation, and locomotion.

Functions of Permeant Populations and Human Ecological Functions. Many functions are performed by populations. Among them is the process of dimorphism, which is maintained by antagonistic selective pressures that act at the landscape scale (metapopulation). Other processes include those of species in need of great heterogeneity of habitat for their lives: These species use the landscape differently and identify and transform resort of pasture, watering, refuge, meeting, ritualization, and so on. Also, the processes of extinction, isolation, and recolonization are included in the functional relationship among landscapes (or a portion of them) and permeant populations (which are populations regularly using different types of ecosystems).

Perhaps more specific are the concepts of *complementation* and *supplementation*. When a species needs different resources it may find them in ecotopes of different types. We deal with the complementation process if ecotissues with two types of habitat shelter a larger population than other ecotissues. On the other hand, we deal with the supplementation process if the population of an ecotope increases because its landscape unit is a portion of an ecotissue having in stock additional resources. The trend of population density of a species in a specific landscape is also a function of the sequential occupation of an optimal, then a suboptimal, and finally a marginal habitat.

Since the major permeant species is man, it is important to mention the landscape functions of man in accordance with ecology principles. Appropriate indicators (Turner and Gardner, 1990; Ingegnoli, 1991, 1993, 1997) allow us to note processes of formation of different types of human landscapes in relation to agricultural habitat per capita, residential and industrial needs, the ratio between human and natural habitat of a territory, biological territorial capacity, and so on and also to identify the limits of the structure of dense urban, sparse urban, suburban, rural, agricultural, and agricultural-protective landscapes for each ecological region. It is also possible to identify the minimum standard landscape ecological habitat, which is theoretically necessary to find the range for compensation actions in human transformations at local scale.

Disturbance and Transformation Dynamic. All landscapes have peculiar ways of transformation. They remain valid for a large range of scale. These methods may include homeorhetic processes (able to maintain a dynamic regime), such as the Mediterranean grasslands supported by cybernetic feedback controls of grazing pressure. Also, human landscapes show modality of transformations led by ecological laws, such as the law of complementarity of the phenomena of environmental transformations. These ecological laws in some way cause a change in culture and, consequently, in human behavior in order to maintain a metastatic equilibrium (i.e., an equilibrium around a stable attractor but capable of shifting to another attractor when the landscapes suffer heavy changing pressure).

Alteration and Pathology Processes. The degradation of a landscape begins when some alteration becomes evident at the ecotissue level. If altered ecotopes are substituted with other ecotopes, there is a transformation; if the ecotissue is not able to incorporate the disturbances of its complex mosaic, then degradation begins. Generally, out-of-scale disturbances disrupt the organization of ecological systems; therefore, the consequences are also out of scale. However, note that in some case (such as fragmentation; Fig. 4) a very small transformation may cause sensible alterations. Moreover, the degradation of landscapes may influence the physiology of many organisms living in it, including man. Finally, the study of landscapes allows one to create a picture of the most frequently recurring pathologies linked with these processes, such as structural and/or functional alteration of ecotissues, syndromes of unproper transformation, pollution degradation, and destruction from catastrophic disturbances.

New Perspectives in Ecology

The Unified Ecology

Here, we discuss the most significant aspects of landscape ecology contributing to the development of ecological theory. Determinism, linked to the concept of reversibility of time, has produced traditional ecology; as previously mentioned, new scientific paradigms have overcome the limits

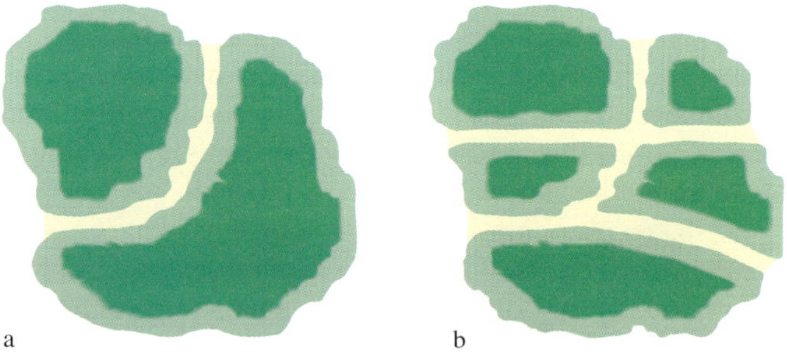

FIGURE 4 Examples of fragmentation of a forested ecomosaic. Two large forested patches (*Piceetum montanum*) are separated by a ski run (a). The surface of their internal part (dark green) measures about 170 ha, excluding the ecotonal belts (light green). The creation of two other ski runs crossing the first one (b) produces a fragmentation from two to five patches. The transformation causes a 9% decrease of the forested area. In contrast with this small change, note that the reduction of the inner part of the forest reaches 31.8%. This is an alteration 3.5 times larger than the felling area, impacting almost one-third of the entire area.

of traditional ecology, resulting in the concept of landscape ecology. Thus, the biological spectrum has been completed and the concept of ecosystem has been rescaled. Therefore, as some scientists have noted (Allen and Hoekstra, 1992), it is now possible to aim for a more integrated (unified) ecology that is more coherent and has more applications.

Today, ecologists claim the necessity of ecology as a predictive discipline and of scientists as people able to deal with complex systems of large scale. To reach this conclusion, one must consider both a consistent epistemological validation, as already noted, and a more correct interpretation of the hierarchy of levels of biological organization, which is the only way to achieve the integration of different branches of ecology.

The theory of hierarchic systems was introduced by O'Neill *et al.* (1986). Currently, we are overcoming the hierarchical model of traditional ecology (organism, population, community, ecosystem, biome, and biosphere) through the introduction of landscape ecology and regional ecology. Simultaneously, we are overcoming the deterministic concept of a limiting factor through the introduction of the more holistic concept of constraint. Nevertheless, the definitions of principal study criteria are still not correct.

It may be observed that a portion of the ambiguity of the ecosystem concept derives from its inclusion in a typically biotic hierarchy, even if it has a specific environmental (functional) meaning. The addition of the concept of the landscape level makes this hierarchy even more incoherent. In reality, moving from the more detailed scale to the global one, we can note (Table 2) the existence of three parallel hierarchies, respectively based on the biotic criterion, pertinent to organism or groups of organism; the environmental criterion, relative to geographical or ecological sectorial aspects; and the ecological (integrated) criterion, able to integrate the first two criteria.

Odum (1971, 1993) lists eight levels of scale. Nevertheless, the levels of scale in ecology may be reduced to six (Table 2) (seven if we consider the large unit of the biosphere), which correspond to six scales of observations linked to six levels of biological organization. The ambiguity still present in Fig. 2 derives from the adopted criteria. The biotic criterion may be acceptable by the ecological criterion only for the first two scales shown in Table 2 because in this case a complete integration between biotic and environmental criterion is impossible since these levels are not self-sufficient in an ecological sense. The integrated criterion, however, is obligatory at higher levels since it is the only one able to consider the entire biological organization without incompatibility between community and ecosystem, between biome and biogeographical region, and between biosphere and ecosphere and with a more explicit recognition of the landscape level.

Note that if we considered the landscape using the environmental criterion, it would only indicate the spatial contiguity of ecosystems or communities. Only through the integrated criterion may we define the concept of landscape as a specific level of biological organization, as already mentioned. As a consequence, ecology can only fully develop by following the integrated criterion.

TABLE 2

Hierarchical Levels of Biological Organization in Ecology: Comparison among the Scale of Analysis, the Biotic Criterion, the Environmental Criterion, and the Integrated Criterion (i.e., Properly Ecological)

SCALE OF ANALYSIS	BIOTIC CRITERION	ENVIRONMENTAL CRITERION	INTEGRATED CRITERION
Spot	*Organism*	Niche	Organism
Station	*Population*	Habitat	Population
Local	Community	Ecosystem	*Biogeocenosis*
Landscape	—	—	Landscape
Region	Biome	Biogeographic region	*Eco-region*
Global	Biosphere	Ecosphere	*Biogeosphere*

The Ecotissue Model as an Example of Integrated Ecology

The study of landscape remains incomplete without an integrated interpretation. Regarding the landscape structure, the fragmentation and the variegated models have already been mentioned, and the concept of ecotissue has been introduced. Here, it is useful to recall the development of interpretative models of ecological heterogeneity in territorial studies to demonstrate the necessity of the integrated criterion.

Both the concept of community and the concept of ecosystem are by themselves dependent on scale, but ecosystem ecology has never considered these levels in relationship to heterogeneity of an ecological contest at a higher scale: At most, they indicate (Fig. 3b) a boundary between two different ecosystems or an ecotonal belt if an environmental gradient separates the ecosystems (Odum, 1971; Whittaker, 1975). Realizing it would have been a mistake to disregard the level of landscape, scientists have simply emphasized the spatial configuration of elements (that are the ecosystems as biogeocoenosis, following the integrated criterion): patches and corridors (Fig. 3c) on a landscape matrix, often an indistinct matrix (Naveh and Lieberman, 1984; Forman and Godron, 1986). Through the concept of ecotope, it has been possible to employ the fragmentation model (Fig. 3d), which is able to represent a real and proper ecomosaic of juxtaposed fragments (tesserae) for a specific range of scale (Leser, 1997; Haber, 1990). Recently, many zoologists have observed that relatively few organisms perceive the landscape in similar biological ways because of the existence of stenotopics and euritopics taxa, which are taxa linked to limited local habitats (even one tessera) and taxa related to heterogeneous habitats (two or more ecotopes). Thus, the new landscape model is a variegation model (Figs. 3e and 3f) because the ecological mosaic is composed of tesserae having a variable conformation (e.g., fuzzy-edged boundaries), or it is an overlapping series of different patch-matrix mosaics (in proportion of main groups of steno- and euritopic habitats) (Wiens, 1998; Farina, 1997). Many researchers know that in reality both fragmentation and variegation models are needed (e.g., when studying a mosaic of forested landscape and its faunistic indexes).

The main limitation of these two models is seen when we exasperate the arguments which resulted in their creation, as pointed out by Allen and Hoekstra (1992). These authors noted that "if organisms of different species on the same spot of ground respond to different scaled patches of landscape, then the conception 'organism' becomes inconsistent as a scaled unit; alternatively, if we use the organism as the normalizing frame, then the landscape becomes a curved space." In fact, in a curved space such as the one mentioned in the theory of relativity, two observers reach different measures on the same objects. In summary, it seems that the two levels of biological organization, organism and landscape, might not be coexistent. Obviously, the reality is different, but paradoxes help to clarify ideas since it they are usual in mathematics. The problem is very difficult to solve but allows some interesting considerations. The organism undoubtedly remains a specific level of biological organization, referred to a proper group of scales, as demonstrated by the theory of hierarchical systems. On the other hand, landscape may not be interpreted as a curved space because we are not dealing with the conditions of the theory of relativity to distort the space.

The theory of complementarity states that the perception of reality may be dependent on our observation criteria, but the reality per se remains measurable. Therefore, when difficulties arise in the comprehension of a complex reality, it is necessary to derive a description able to consider its multiplicity, even through different but related ways. In reality, some observations in the variegation model are not true. Surely a euritopic species such as a locust will have a perception of the grain of ecological mosaic different from that of a stenotopic species, such as a Carabidae (*Coleoptera*). The euritopic species forages indifferently in a patch of meadow or in the closest patch of trees, but its physiologic responses will be not the same for each patch because the microclimate (e.g., light and humidity) changes, the quality of food (e.g., amino acids, lipids, and carbohydrates of leaves) changes, many predators change, and so on. Therefore, we may find diverse hierarchies of ecological factors, some of them deriving from the fragmentation model and other from the variegated model. Moreover, many animal species are definable as permeants (Odum, 1971), indicating their use of many types of ecosystems. These permeant species move with periodic seasonal rhythms, often creating real "paths," and they use different tesserae with a daily rhythm to drink, protect themselves from the sun, feed and hunt, and nurse. In addition, in nature we observe the coexistence of fixed and temporary species (not the extraneous ones) in a certain ecotope, with different but often interacting functions.

Therefore, it is not correct to compare a mosaic with juxtaposed tesserae (fragmented) to another composed of many overlapping mosaics (variegated). It would be better to talk about an ecological tissue or ecotissue (see Fig. 6) (i.e., a multidimensional mosaic formed by a main mosaic and a hierarchic succession of correlated mosaics), as mentioned previously.

After clarification of the integrated sense of the interpretative model of landscape, it is possible to resume discussion of the six levels of the biological organization according to integrated (unified) ecology: organism, population, biogeocoenosis, landscape, ecoregion, and biogeosphere. Observe that all these levels more or less interact with each level of the operative scale, even if the expression of each of them and the reciprocal influences change in relation to the range of different local scale. Undoubtedly, integration is a reigning law in ecology, but it does not proceed in an antihierarchic sense: Asymmetries are more than symmetries. That is, we cannot expect to find particular structures and particular functions proper to one or more different levels of organization in certain ranges of resolutions of local scale. However, it may be possible at other resolutions. In fact, nothing may be studied in full either at one level of scale only or at one level of organization only.

Applications of Landscape Ecology

Transdisciplinary Approach

In order to deal with landscape ecology applications and relevant related methods, it is necessary to take into account the managerial role of man in nature. This role derives from the observation that at the landscape level, human and natural ecotopes are integrated in complex ecological mosaics we can relieve as ecotissues. As a consequence, new scientific and cultural paradigms arise because landscape ecology is indispensable for a transdisciplinary approach and for all new methodological criteria of ecology applications: for restoration, environmental design, and (overall) biological conservation and ecological planning.

The transdisciplinary approach derives from the observation that disciplines interested in environmental studies are many but they do not have the same importance. The organizing principle of the different levels of expression of life-forms remains chiefly biological, and only the biological principle can act as integrator. Popper noted that nature has created a new dimension which reconciles nature and culture through the language and the possibility to formulate theories in linguistic form. Note that interdisciplinarity is no longer sufficient for the study of complex adaptive systems because it is linked to reductionistic and mechanistic bases acting through divisions in closed sectors and linear recomposition by sum. Instead, we have to follow a hierarchic methodology with an open spatial web, with prevailing holistic bases and integration principles that each candidate discipline can use — in other words, a transdisciplinary methodology.

As Naveh has often stated (Naveh and Liebermann, 1990), ecology could be a discipline possessing these characteristics, but only when it acquires landscape ecology principles — that is, integrated (unified) ecology. Note that landscape ecology is able to translate information from other disciplines into data which can be integrated with ecological principles (Ingegnoli, 1991, 1993). New ecological methodology must be analogous to medical methodology: Environmental evaluation has to depend on the comparison between the conditions of the examined ecological system and the conditions of a "normal" state. In other words, the relationship between pathology and ecology of the systems will allow a diagnosis, in a clinical sense, of the landscape to study or to plan. The clinical-diagnostic methodology needs proper ecological magnitudes. For landscape ecology, a quantity named BTC, or biological territorial capacity (Ingegnoli, 1991, 1999), could be useful. The

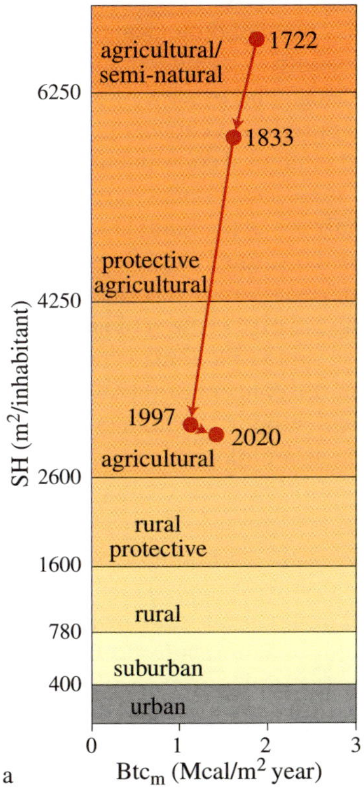

a

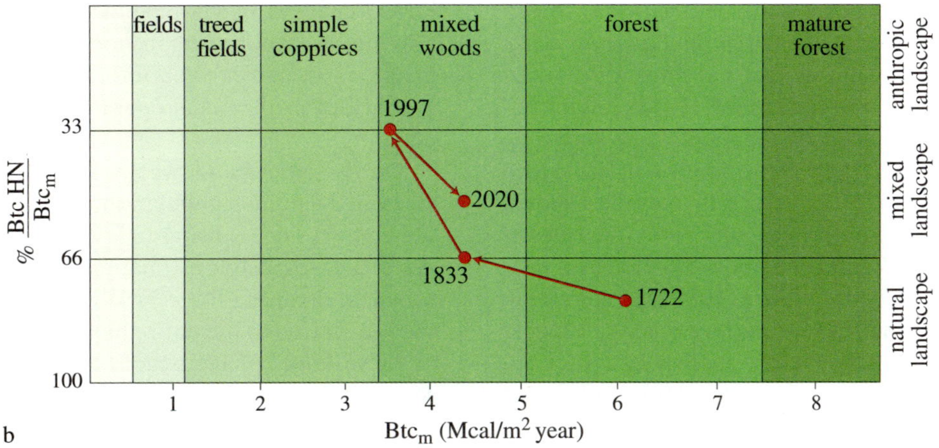

b

FIGURE 5 Model of control that allows one to follow the landscape transformations due to human actions. This kind of diagnostic model permits the verification of the changing of the landscape unit in both the human (a) and natural (b) habitats. It is based on the identification of thresholds of metastability for each type of landscape; in this case study, they have been evaluated for north Italy. SH, standard habitat per capita; Btc_m, medium biological territorial capacity for the entire territory. This model concerns the study of the Rivolta-Spino tract (about 6000 ha) in the Regional Park of the Adda River. Note that the renaturalization proposals try to invert the tendency of degradation of natural habitat before 2020 and to restrain the transformation of human habitat within the agricultural type of landscape. The dates for past centuries have been chosen according to the information available.

BTC represents the energy flux (Mcal/m^2/yr) that a vegetation system needs to dissipate to maintain its organizational level. It expresses the latent capacity of a landscape to return to its equalibrium state and permits the expression of an index (BTC) that can be estimated through a proper methodology related to the survey of vegetation (Fig. 5).

It is possible that the concept of environment will acquire a unique significance due to the integration between man and nature and the most serious environmental problems will assume new interpretations. The concepts of ecosystem and landscape will also be very different from the ones people have formally used in managing and controlling the environment. The concepts of health damage, environmental evaluation, territorial planning, and conservation biology will be completely modified.

Health Prevention

The control of air, water, and soil pollution is the completely dominant criterion of study and application. Alterations due to structural and spatial degradation of the ecological systems are hardly taken into consideration, even though they are more serious and difficult to correct (Ingegnoli, 1993). Remember that important behavioral functions of man are under genetic and physiologic control and may be influenced by the environment, mainly in the first years of life. The development of the synapses and neuronal webs into the telencephalic cortex is completed in humans in only about 2.5 years. If there is an environmental degradation, even in its structural and functional components (i.e., without any pollution), damages could be very serious and affect not only the psyche but also the hormonal system. Consequences become more serious in a domestic (and social) environment without stimuli for development. Dealing with systems of ecosystems, notwithstanding easier to treat syndromes deriving from pollution, even if they directly affect our health. Therefore, it is necessary to exercise more control over the state of systems of ecosystems to protect our health.

Environmental Evaluation

The environmental impact statement (EIS) is based mainly on the concept of impact (from the Latin *impactus,* which means "collision"). This fact indicates a prejudice and a contrast between man and nature. Hence, it is in contradiction with unified ecology. Consider the structuring role of disturbances and their incorporation in ecological mosaics, as mentioned previously. EIS methods should minimize out-of-scale disturbances. Moreover, the rules of EIS are limited to a few categories of works (environmental planning is not included); the term landscape has no scientific significance. Usually, there are no explicit references to the *erfolgskontrolle*— that is, the success rate after the completion of any action of environmental transformation; we note the lack of methodological references to a clinic diagnostic criterion — that is, unique valid criterion to evaluate the state of a biological system. Evidently, environmental evaluation is very important both for environmental design and planning and for controls. The influence of landscape ecological principles on environmental evaluation should produce a change in today's criteria.

Territorial Planning

Currently, the main problem of territorial planning is a heavy predominance of methods and concepts pertaining to urban planning. This is true even though experts in ecology are increasingly more frequently named as consultants in the analytical phases of plans. Nevertheless, political and economic choices continue to guide planning criteria, administrative divisions remain indifferent to the ecological divisions, landscape is considered only from an aesthetic and perceptive sense, and technologic infrastructures dominate every need of biological functionality of the territory. In many countries, architects and engineers completely control the design and process of planning, even if they do not have an appropriate knowledge of and a sufficient sensibility for the biological components of the landscape. Remember that these components are restricted not only to the natural habitats of the landscape but also to the human ones reinterpreted through a biological key. On the contrary, it may be necessary to invert the hierarchy of decisional values, giving priority to the ecological and the landscape components in all the planning processes. Reconsider the passage from the concept of territory simply as a geographic–economical support to human activity to the concept of landscape as a biological entity. Man has to conform with it. This will bring great changes both in planning methodology and in planning objectives. In fact, many scientists try to modify their methodology, using more modern mathematics instruments such as geographical information systems and multivariate statistical analysis. However, if they do not refer to a more advanced ecology, which is more appropriate for studies on territorial systems, any methodology will be not sufficient. For example, we could make some errors moving from the scale of a single ecosystem to the scale of an entire landscape if we do not consider the ecological changes in the type of dynamics behavior. This is also the case when using the more advanced instruments of mathematics and informatic formalization. Moreover, there is a lack of professional operators able to guide advanced planning.

On the other hand, we are at a turning point: The majority of the human population, which until recently lived in the countryside, is living today in towns and urban areas. According to reliable estimations, megacities and large suburban landscapes (Table 3) will absorb about two-thirds of the population of the entire biosphere by 2025. The growth of these kinds of landscapes may bring enormous environmental problems, such as pollution, epidemics, fragmentation of natural areas, an increase in temperature and

TABLE 3

Development of Principal Types of Human Landscape at the Biosphere Level[a]

TYPE OF LANDSCAPE	POPULATION: NUMBER (MILLIONS OF INHABITANS) AND PERCENT 1950		1975		2000		(2025)	
Urban residential	200	7.2	1100	25.6	2400	36.3	4600	51.1
Urban industrial	600	21.4	800	18.6	1200	18.2	1200	13.3
Rural agricultural	1600	57.1	2050	47.7	2700	40.9	2900	32.2
Rural industrial	400	14.3	350	8.1	300	4.6	300	3.4
Total population	2800		4300		6600		9000	

[a]Modified from Ermer *et al.* (1996).

rainfall, abandonment of agriculture, and destruction of ecological corridors. It must be remembered that the degradation of habitats, both human and natural, is the primary cause of extinction of species. Our society does not yet seem to be aware of this problem, as demonstrated by the metropolitan area of Milan, for which administration is not adequate for the urban landscape. It is sufficient to fly over the territory between the prealpine lakes (Lake Maggiore and Lake of Como) and the Po River to see that the city of Milan currently includes at least 50 municipalities extending for dozens of square kilometers. The loss of population of Milan that began in the 1980s seems to be a process going against the worldwide increase of urbanization, but this is not true. In reality, the population of Milan is increasing: It is sufficient to consider the urban phenomenon in its right boundaries to perceive it. As shown in Table 4, Milan has lost about 400,000 inhabitants, but the density of inhabitants per square kilometer of its province has increased by about 28.6%. The lower number of inhabitants derives from the detachment of the territory of Lodi. This demonstrates the completion of the process of urbanization of the entire province of Milan. More important, the province of Milan is less sprawling than the Milanese conurbation, which includes the southern part of the province of Varese and Como and surely exceeds 4 million people. Milan is currently one of the largest metropolises in Europe, but its framework and its services are not adequate because the urban phenomenon is not perceived as an ecological transformation of the territory.

TABLE 4

Changes in the Population of the Hinterland of Milan in the Period 1975–1995

DISTRICTS	1975	1995	% VARIATION
Number of inhabitants in Lombardy region	8,837,656	8,911,756	+0.8
Number of inhabitants in the province of Milan	4,031,110	3,720,789	−7.7
Inhabitants/km^2 in the province of Milan	1,459	1,876	+28.6
Number of inhabitants in the city of Milan	1,722,634	1,310,681	−23.9

Consequently, future urban planning should necessarily be preceded by ecological planning—that is, by an indispensable control on the ecological state of the territory, able to provide a reliable base to administrative choices. This planning must be elaborated at different space–temporal scales; referred to landscape units, not to single municipal areas; able to aggregate isolated patches; directed toward the creation of a web of natural corridors protecting watercourses and pledging connectivity for movements of key species among large patches; aimed at designing small patches and corridors to guarantee a high environmental heterogeneity; and able to maintain its historical trajectory since each transformation proceeding in a contradictory way with the history of landscape results in profound alterations in the ecotissue.

Through the analysis and diagnosis based on landscape ecology, naturalistic structures may be integrated with human structures: It must be an integration among ecological webs or, better, among landscape apparatuses with functional multiplicity, not a zoning of tesserae with rigorous constraints. Today, the conservation of nature acts exclusively with isolated residual patches; it does not matter if the patch is a local biotope or a large park—it is only a problem of scale. On the contrary, following the paradigms of landscape ecology, conservation consists of webs of ecotopes in a contest of complex mosaics and it considers the importance of the complementarity of human ecosystems

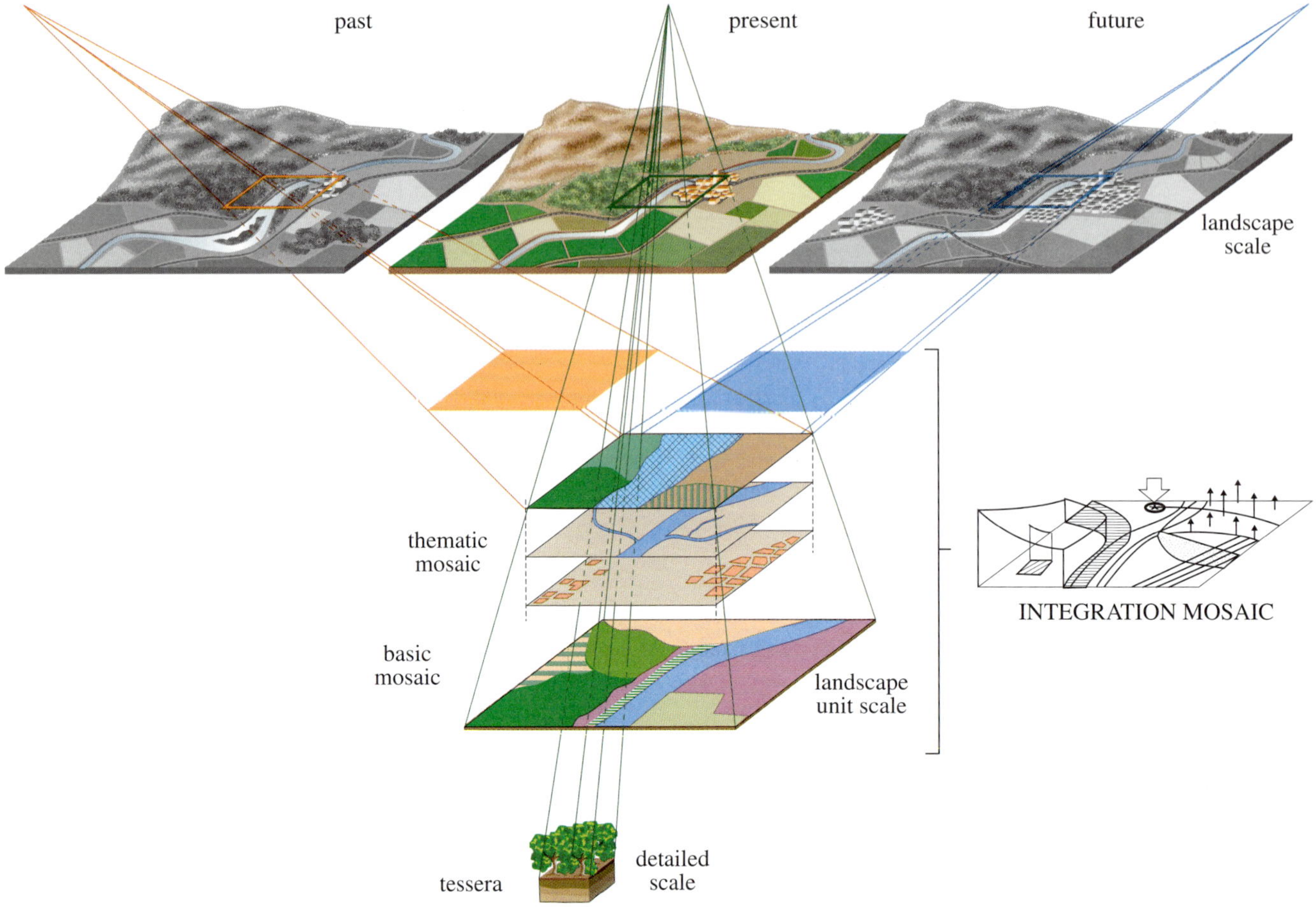

FIGURE 6 Schematic representation of the ecotissue structural model. It is a complex ecological mosaic that may be formalized as a hypervolume territorial niche and therefore multidimensional. This complex adaptive system may be studied only through sections and projections in order to gather information to be integrated with a hierarchical relationship into a basic mosaic, usually a vegetational one. It is necessary to consider constraints of different scales and intersections of thematic mosaics. In fact, the ecotissue results from the integration of three levels of subintegration: the set of spatial scales, from the landscape scale (top) to the detailed scale (bottom); the set of thematic mosaics; and the set of temporal scales (past, present, and future), which permit the reconstruction and forecasting of the evolutive dynamic. The result may be configured as an integrated mosaic following operative needs and criteria. It can be better defined as an operative chart of integration.

in highly populated countries, in which a process of coevolution between the two different types of ecosystems (natural and human) has been taking place for thousand of years. Therefore, the study of ecotissues becomes of central importance. It must be carried out through the integration of three types of subintegrations: The first one concerns the links among a group of scales from the regional to the local one, the second concerns the scale of a landscape unit (or landscape district) composed of a group of thematic mosaics, and the third concerns the temporal aspect (Fig. 6). In the analysis of landscape ecology, generally a preliminary interpretation of the ecotissue is necessary, such as that shown in Table 5.

Today, an ecologist usually works on different scales and the ecotissue concept obliges him or her to use new appropriate indicators. Therefore, at least four criteria must be added that are derived from landscape ecology: the definition of biological territorial capacity, the definition of landscape apparatuses, the definition of standard habitat, and the source and sink theory.

Biological Conservation

Biotopes and reserves are protected as sacred islands. Thus, the degradation of the entire surrounding landscape is allowed, even if biotopes and reserves are integral parts of this landscape. The destruction of habitats is undoubtedly the major problem of the loss of biological diversity, but it is considered less important than pollution.

On the other hand, disturbances and fragmentation due to human activity are seen only as negative factors, thus neglecting the structuring role of disturbances (even human ones) if they are not out of scale. As a consequence, we need

TABLE 5

Synthesis of the Principal Ecotissue Analysis

ECOLOGICAL SCALE	ECOTISSUE ANALYSIS	LEVEL OF INTEGRATION
Region	Main components Transformation dynamic Present ecological state	
Landscape	Landscape unit structure Transformation dynamic Present ecological state Comparison with regional data	
District, ecosystem mosaic	Geotopes analysis Biotopes analysis Human ecosystems analysis Mosaic context characters Change tendencies Operative objectives	Integration of the correlated mosaics in the main one Ecotopes diagnosis and operative criteria Ecological controls on forecasting structure
Site details	Species control Sample collecting Problematic points	

to develop a new concept of biodiversity and a new concept of naturalness. Similarly, principles of unified ecology are indispensable to control the dynamic of the ecological mosaic at different scales. In addition, we must consider all the different types of alterations of the levels of the biological spectrum and thus not focus on only extinct species. To emphasize how landscape ecology and conservation biology may change old kinds of thinking, we need to better specify the following focal points: biodiversity, naturalness, and conservation criteria (Massa and Ingegnoli, 1999).

Biodiversity. As a consequence of the conceptual importance of massive extinction episodes, the concept of biodiversity remains related to species. Effort to extend the concept of biodiversity to communities and habitats (α, β, and γ diversity of Whittaker, 1975) seems to be marginal. Moreover, people continue to relate the increase of biodiversity with the increase of stability of the ecological systems, independently from the spatiotemporal scale.

All this and its consequences represent a distortion of reality from a unified ecology point of view. Remember two basic assumptions: The concept of biological diversity exists and it is important at any level of biological organization, from genetic heritage to species and from population to landscape and biosphere; also, the evolution of complex adaptive systems requires phases with abundance in components and phases with a reduction of the components as a consequence of the increase in order and organization of living systems. For example, the evolution from Mediterranean maquis to an ilex/holm oak forest implies a reduction of biodiversity of almost 80% (Pignatti, 1994).

According to Prigogine, order through fluctuations (that are disturbances) seems to be one of the few general processes of our universe. The biological diversities must be emphasized with regard to specific biodiversity. The basic problem does not concern conservation in itself: It concerns the creative and evolutionary capability of life — that is, the process involving fluxes and phases of biodiversity. Death must be necessary, not at the individual level, but at the species, community, and landscape level: However, one thing is the death and another is the end of births, as noted by Soulé and Wilcox (1980).

Naturalness. The definition of *naturalness* in the Italian vocabulary of Italian Encyclopedia Institute is "quality, condition of something that is natural." Conservationists usually interpret it as "absence of the human activity," which is still a preconceived opposition of man and nature. Do not forget that man owns many attributes of the process that created him. The role of man is managing nature, as sustained by Lorenz and Popper (1985): "Human sensibility to values corresponds to an evolutionary event since a play of all against all takes place in our brain as in external world. . . . The model of rational solution of problems remains the same in all fields: the combination of creativity and critic. It's essentially an evolutionary model."

Among all the revolutions during the history of life, some of the most important are the origin of prokaryotes, the origin of eukaryotes, the development of pluricellular organisms, the development of superior vertebrates, and the origin of a species (human species) potentially able to understand the evolutionary process.

At the present evolutionary state, man should be conscious of all the aspects mentioned previously. A new concept of widespread naturalness is necessary: According to Giacomini (1965), it implies a global readjustment according to natural laws, including human habitats. However, this is not sufficient. The current concept of naturalness must abandon any Cartesian reference to mechanistic nature since nature has a historical perspective and it cannot be dominated by man, who is its engineer in conflict with it. Natural reality is an evolutionary and historical reality. Order and disorder play the same role: Nature cannot be understood because it is open to unpredictable creative perspectives.

Therefore, naturalness may be defined as the prevailing consistence of nature. It derives from the concept of wilderness of Thoreau (1849): "In naturalness is the preservation of world." It also derives from the words of J. Muir: "Going to the mountains is like coming back home; naturalness is a necessity; parks and reserves located on the mountains are useful not only as sources of wood and water, but also as source of life." Valleys with great predators and rare species, large forests and prairies, and the absence of consumer development on the Italian Alps are among the examples of a naturalness including man.

Conservation Criteria. In conservation planning, it is much more important to preserve the possibility of the spreading of species than to focus on particular rare species, even if they are endemic. Remember the paradox of management in conservation of nature (Forman and Collinge, 1996): "The probability to reach an immediate goal is major in small areas, but the probability to reach a long-term success is bigger in the large areas." Therefore, our conservation efforts must move to the landscape and the regional scales. In addition, studies must be elaborated at different spatiotemporal scales, and they have to be referred to landscape units, not to administrative boundaries. As the first goal, methodology must verify if areas preserved as "natural parks" are really playing this role in the regional landscape system (Fig. 7). Then, the process of ecological analysis comprises at least three major phases:

1. Diagnostical studies on disturbances, state of vegetation, human habitat/natural habitat ratio, the method of transformation of ecotissue, changes in fragmentation, ecotonal web, source and sink ecotopes, key species, genetic problems, etc.
2. Therapeutics proposal: few large patches, vegetated corridors along the rivers, arrangements of biotopes in the matrix, connectivity, strategic points, areas of new development, restoration, and bioengineering
3. Ecological controls and final design: controls on transformation of the ecotissue, control of regeneration niches, disturbances incorporation, areas of inference of natural and human web, monitoring, etc.

In conclusion, remember that this methodology must be supported by more general criteria of territorial planning. Thus, objectives of conservation must be integrated with the natural one, such as conservation of historic–architectonic evidences and their environment or protection and socioeconomic sustainability of local human populations. Ecological integrity and human basic needs must develop in a mutual consonance. Compensating readjustments, bioengineering works, constructional equalizations of infrastructures, and settlement make this accord possible if unified ecological principles are followed.

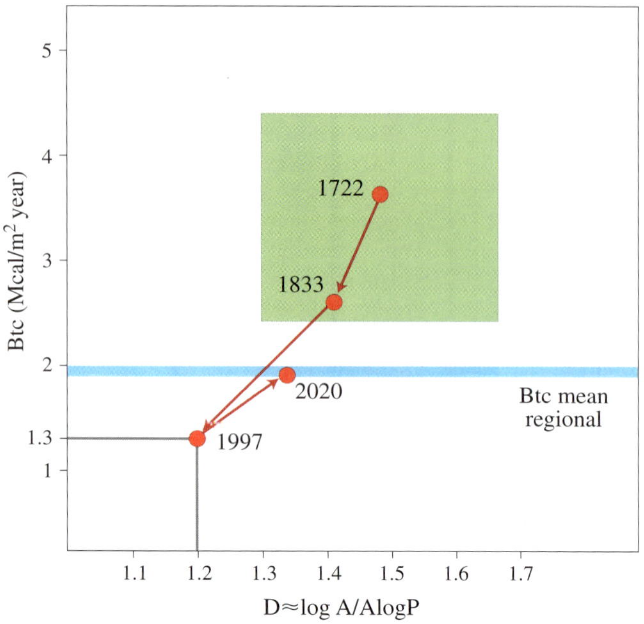

FIGURE 7 Model to verify the ecological functionality of a natural park. This synthetic model is necessary to verify the present state of the ecological system (landscape unit) of a natural park, comparing it with the medium regional Btc (in blue) and the optimal surroundings (in green) for a certain kind of park; *D*, fractal dimension, where *A* is the area and *P* the perimeter, calculated as the total set of forested patches in the studied area. This model refers to the Rivolta-Spino tract (about 6000 ha) in the Regional Park of the Adda River. Note that in 1997, the studied portion of the park did not have the ecological landscape of a park because the Btc was definitely lower than the regional one and the D value was 1.2 (corresponding to the shape strictly limited to the river corridor). Consequently, it is necessary to direct the transformations of renaturalization toward the lower boundary of the optimal field. This model permits a quantification and control of the principal project measures for an adequate action on the landscape unit.

Acknowledgments

I thank Zev Naveh and Richard Forman for their scientific encouragement, Jurgen Ott for the critical observations to an earlier draft of the manuscript, and Elena Giglio for her aid in the correction of the text.

References Cited

ALLEN, T. F. H., and HOEKSTRA, T. W. (1992). *Toward a Unified Ecology.* Columbia Univ. Press, New York.

BAILEY, R. G. (1996). *Ecosystem Geography.* Springer, New York.

BLONDEL, J. (1986). *Biogéographie Évolutive.* Masson, Paris.

DOKUČAEV, V. V. (1951). *Writings,* vol. 6. Akad. Nauk, Moscow.

ERMER, K., HOFF, R., and MOHRMANN, R. (1996). *Landschaftsplanung in der Stadt.* Ulmer, Stuttgart.

FARINA, A. (1997). *Principles and Methods in Landscape Ecology.* Chapnam & Hall, London.

FORMAN, R. T. T., and COLLINGE, S. K. (1996). The spatial solution to conserving biodiversity in landscapes and regions. In *Conservation of Faunal Diversity in Forested Landscapes.* (R. M. DeGraaf, Ed.). Chapman & Hall, London.

FORMAN, R. T. T., and GODRON, M. (1986). *Landscape Ecology.* Wiley, New York.

FORMAN, R. T. T., and MOORE, P. N. (1991). Theoretical foundations for understanding boundaries in landscapes mosaics. In *Landscapes Boundaries.* (A. J. Hansen and F. di Castri, Eds.). Springer-Verlag, New York.

GELL-MANN, M. (1994). *The Quark and the Jaguar. Adventures in the Simple and the Complex.* Freeman, New York.

GIACOMINI, V. (1965). Significato e funzione dei parchi nazionali. In *I Parchi Nazionali in Italia.* Istituto di Tecnica e Propaganda, Rome.

GODRON, M. (1984). *Ecologie de la Vegetation Terrestre.* Masson, Paris.

HABER, W. (1990). Basic concept of landscape ecology in planning and management. *Phys. Ecol. Jpn.* **27,** 131–146.

INGEGNOLI, V. (1971). Ecologia territoriale e progettazione: Significati e metodologia, In *L'ingegnere di Fronte alla Sopravvivenza Umana,* Coll. Ing. Vol. 1, pp. 398–400.

INGEGNOLI, V. (1991). Human influences in landscape change: Thresholds of metastability. In *Terrestrial and Aquatic Ecosystems: Perturbation and Recovery* (O. Ravera, Ed.), pp. 303–309. Horwood, New York.

INGEGNOLI, V. (1993). *Fondamenti di Ecologia del Paesaggio.* Milano, CittàStudi (Utet), Milan.

INGEGNOLI, V. (Ed.) (1997). *Esercizi di Ecologia del Paesaggio.* Utet-CittàStudi, Milan.

INGEGNOLI, V. (1999). Definition and evaluation of the BTC (biological territorial capacity) as an indicator for landscape ecological studies on vegetation. In Windhorst, W., and Enckell, P. H. (Eds.) Sustainable landuse management: The challenge of ecosystem protection. *EcoSys: Beitrage zur Oekosystemforschung,* Suppl. Bd. **28:** 109–118.

LESER, H. (1997). *Landschaftsoekologie.* Ulmer, Stuttgart.

LORENZ, K. (1978). *Vergleichende Verhaltensforschung: Grundlagen der Ethologie.* Springer, Vienna.

LORENZ, K., and POPPER, K. (1985). *Il Futuro è Aperto.* Rusconi, Milan.

MANDELBROT, B. B. (1975). *Les Objects Fractals: Forme, Hasard et Dimension.* Flammarion, Paris.

MOEBIUS, K. (1877). *Die Auster un die Austernwirtschaft.* Berlin.

NAVEH, Z., and LIEBERMAN, A. (1990). *Landscape Ecology: Theory and Application.* Springer, New York.

ODUM, E. P. (1971). *Fundamentals of Ecology.* Saunders, Philadelphia.

ODUM, E. P. (1993). *Ecology and Our Endangered Life-Support Systems.* Sinauer, Sunderland, MA.

OLDEMAN, R. A. A. (1990). *Forests: Elements of Sylvology.* Springer, New York.

O'NEILL, R. V., DE ANGELIS, D. L., WAIDE, J. B., and ALLEN, T. F. H. (1986). *A Hierarchical Concept of Ecosystems.* Princeton Univ. Press, Princeton, NJ.

PIGNATTI, S. (1994). *Ecologia del Paesaggio.* Utet, Torino.

POINCARÉ, H. J. (1906). *La Science et l'Hypothèse.* Flammarion, Paris.

POPPER, K. (1996). *Tutta la Vita è Risolvere Problemi: Scritti sulla Conoscenza, la Storia e la Politica.* Rusconi, Milan.

PRIGOGINE, I. (1996). *La Fin des Certitudes. Temps, Chaos et Lois de la Nature.* Odile Jacob, Paris.

PRIGOGINE, I., and NICOLIS, G. (1977). *Self-Organization in Nonequilibrium Systems.* Wiley, New York.

PRIGOGINE, I., NICOLIS, G., and BABLOYATZ, A. (1972). Thermodynamics of evolution. *Phys. Today* **25,** 23–28.

SOULÉ, M. E., and WILCOX, B. A. (1980). Conservation biology. Its scope and its challenge. In *Conservation Biology. An Evolutionary–Ecological Perspective* (M. E. Soulé and B. A. Wilcox, Eds.), Sinauer, Sunderland, MA.

THOREAU, H. D. (1849). *A Week on the Concord and Merrimack Rivers.* Franklin Library, Boston.

TROLL, K. (1939). *Luftbildenplan und Ökologische Bodenforschung.* Zeitschrift & Gesellschft Erdkunde, Berlin.

TURNER, M. G., and GARDNER, R. H. (1990). *Quantitative Methods in Landscape Ecology.* Springer, New York.

WHITTAKER, R. H. (1975). *Communities and Ecosystems.* Macmillan, New York.

WIENS (1998). Spatial scaling in ecology. *Funct. Ecol.* **3,** 385–397.

ZANZI, L. (1995). Viatico per una avventura nella storia della Val Grande. In *Val Grande, Storia di una Foresta* (L. Alberti, G. Rainaldi, and E. Rizzi, Eds.). Anzola d'Ossola, Fondazione Arch. Enrico Monti.

ZONNEVELD, I. S. (1995). *Land Ecology.* SPB Academic, Amsterdam.

General References

COLUMELLA, L. J. (1977). *De re Rustica.* Harvard Univ. Press, Cambridge, MA.

SANDRO PIGNATTI

Department of Plant Biology
University of Rome "La Sapienza"
Rome, Italy

Human–Vegetation Coevolution in the Mediterranean Ecosystem

The first domestication of edible plants for agricultural purposes took place in about 9000 BC in the eastern Mediterranean basin. There, the natural habitat was composed of relatively species-poor evergreen forest vegetation, whereas herbaceous flora were confined to extreme environments inside marginal habitats (cliffs, tall mountains, and coastal areas). Man began to interfere with this situation, especially by means of fires, in a very remote era. Where the forest was burnt, large open spaces were created which represented an opportunity for the species growing in the extreme environments to spread. The intense speciation that developed in this manner gave rise to an entirely new kind of flora, typical of environments that have been modified by human intervention. Numerous edible species form a part of this flora, including the precursors of many cultivated plants: wheat, barley, rye, oats, vetch, and lentils. A synergism between the disruptive action of the human population and the spreading of vegetation that contained many edible plants thus began to develop. In this way, a process of plant–human coevolution began which then became essential to the development of Mediterranean culture.

Domestication of Agrarian Plants

The process of domestication of agrarian plants began in many areas of the earth in an entirely independent manner: in the Mediterranean, India, Iran, China, sub-Saharan Africa, Central America, and the Andes. The most ancient traces of agriculture, dating back to 9000 BC, have been uncovered at Jericho in the eastern Mediterranean and in other, slightly more recent sites located in the Fertile Crescent (the area within the territories of Syria, Iraq, and Jordan) and in Egypt.

The domestication of agrarian plants in the eastern Mediterranean (Renfrew, 1973) undoubtedly has been of primary importance not only because it was the first such domestication but also because certain of the plants derived from that area are to this day the most widespread on a global scale and provide the greater part of the foodstuffs used on the planet: wheat, barley, rye, and oats (Diamond, 1997). It was here that, at about the same time, the domestication of some of the most important breeding animals, such as the cow, sheep, goat, dog, and pig (the latter was also domesticated in other areas), also took place. No matter how much all this is undoubtedly tied to the particular climatic conditions of the area, it is well-known that Mediterranean-type climates also characterize four other parts of the world in which no such important domestication events took place: California, central Chile, South Africa, and western Australia (Di Castri and Mooney, 1977). It is therefore interesting to ask why domestication first developed in a Mediterranean environment rather than, for example, in the tropics and why, out of the five possible such areas, it did so in the Mediterranean and not in the others where the climate was almost identical. The answer is to be sought in the evolutionary history of Mediterranean flora, which is profoundly different from the floras of other Mediterranean-type ecosystems, and in the history of human presence in the various areas. Mediterranean flora, therefore, constitutes an interesting example of how the biodiversity of a group of plants can condition sociocultural development and be conditioned by it at the same time.

A Rapidly Changing Environment and Biodiversity

The Mediterranean basin possesses a flora composed of about 20,000 species (Greuter *et al.,* 1984–1989) and is therefore one of the territories on the globe with the highest floristic density (Malyshev, 1975). We still know little about the evolution of this flora, mostly because the scientific approach that is most widespread (not only in evolutionary biology) is the reductive one, which leads one to consider evolution as a phenomenon that only concerns what is alive, neglecting its retroactive aspects that become established at the level of population, species, community, and ecosystem. In order to delve deeper into the problem of the domestication of agrarian plants, one must first examine the climatic characteristics of the Mediterranean environment and the evolution of flora in that area. It is also possible to identify some events which have had consequences of a general character (Pignatti, 1978, 1979, 1995).

The Mediterranean Bioclimate

The Mediterranean bioclimate is a temperate, warm, and at times almost subtropical climate with hot and arid summers and relatively mild winters (Di Castri and Mooney, 1977; Pignatti, 1986). The average annual temperature is between 14 and 18 or 19°C; annual precipitation is between 600 and 1200 mm, but it can reach levels as low as 350–400 mm in the subdesert areas. Freezing spells and snow are rare and, in general, last only a few days.

For the plant's physiological conditions, this climate results in a period of stress during the summer months, when the high temperatures cause a high level of transpiration and the soil dries up on the surface at least. These negative conditions for plant life are compounded by a decisive reduction in precipitation, sometimes drought, which can last from 2 to 4 months. In general, such conditions force plants to completely suspend their vital activities; therefore, the summer represents a period of almost total vegetative pause. The ligneous species of Mediterranean flora are for the most part evergreen plants (i.e., they maintain their leaves during winter), since the freezing spells which might otherwise damage them are absent; during the summer they keep their leaves, but photosynthesis in general falls to very low levels or stops altogether. The herbaceous species, instead, usually lose their leaves entirely during the summer period, maintaining only subterranean organs such as bulbs or rhizomes, or they pass the arid period in the form of seeds. The climate characterized by summer aridity is proper both to the areas that surround the Mediterranean Sea and, with insignificant differences, to other Mediterranean-type ecosystems.

Derivation from Two Floristic Strains

Two components may be distinguished in the vascular flora of the Mediterranean area: groups with predominantly western affinities (Pignatti and Pignatti, 1989), which include especially species of the acidophilic maquis and the impoverished soils, and groups with predominantly eastern affinities, which include the majority of the deciduous arboreal species and the vegetation of salty and coastal environments. Along with these two components, which usually predominate, there are other floristic currents, especially groups with paleotropical, Atlantic–temperate, and nordic affinities, the latter probably as a result of the glaciations.

A Desert Environment during the High Miocene

About 7 million years ago, a general rising of the Gibraltar threshold took place and the connection between the Mediterranean Sea and the Atlantic Ocean was interrupted. The Mediterranean Sea became an enclosed basin and strong evaporation determined the formation of salted deserts. The spreading of groups of halophytes, such as *Limonium, Salicornia, Arthrocnemum, Salsola,* and *Artemisia gr. maritima,* largely of Mesogaean affinity, probably occurred during this period. The emergence of the sea depths also allowed for the spreading of groups of mountain species such as the thorny *Astragalus* that from their Caucasian–Iranian center spread across the Mediterranean mountains all the way to the Sierra Nevada. The high Miocene had a briefer duration, lasting less than 1 million years, during which time the connection with the Atlantic was reestablished and the basin filled anew. This phase was important for the appearance of species in marginal environments and extreme living conditions, such as peaks, crests, cliffs, and ledges. These species constitute an important component of the biodiversity of Mediterranean flora. They are plants that are small in size and which may be defined as opportunistic insofar as they have greater capacities of adaptation to particularly severe conditions. They spread in diversified environments (euryplastic species), and many of them evolved further. At the end of the Pliocene, the climate became subtropical, with abundant rainfall from autumn to spring and aridity in the summer: What was to become the most important climatic characteristic of the Mediterranean region thus became established. Later, the glaciations took place, which in the Mediterranean basin resulted in a lowering of forest boundaries and some arid phases in the coastal regions.

Variations in Sea Level

Repeated changes in the sea level took place during the last phases of the Pliocene and in the Pleistocene that profoundly changed the shape of the coasts of the Italian peninsula and of the islands. This led to a submersion of the earth's surface to 270–280 m below sea level during the "Selinuntian" and, on the contrary, to a lowering of the sea level to almost 100 m below the current level during the glaciations. The submersions affected flora especially with regard to the distribution of endemic species or in any case those species specialized in coastal and insular environments, which were eliminated by the flora of less elevated islands. The islands which were completely submerged in the Selinuntian (Vulcano, Stromboli, Lampedusa, Zannone, and Linosa) present a significantly smaller number of species (on average 260) compared to the others (Pantelleria, Favignana, Ponza, Marettimo, Levanzo), which have 460 species on average (Pignatti, 1995). Regarding the distribution of the *Limonium* species, which are typical of coastal environments, only one endemic species is found on the islands belonging to the first group, whereas seven endemic species are present on those belonging to the second group. Here again, we are dealing with a highly biodiversified group of plants, whose evolution is centered on a marginal environment with extreme characteristics such as cliffs and rocks along the coast.

Polyploidy

Speciation phenomena are often accompanied by an increase in ploidy since the mutations responsible for the formation of new species may be followed by a doubling in the number of chromosomes. Alternatively, the new genotype may be the result of the hybridization of two preexisting genotypes which bear affinities with one another and become doubled only later. It is believed that this takes place in our flora about half of the time with groups that manifest polyploidy frequently (pteridophyte, *Rosa, Leucanthemum, Festuca,* and *Carex*) and others in which polyploidy is practically absent (gymnosperms and fagaceae).

From this standpoint, the Mediterranean environment presents entirely particular conditions; climax vegetation offers only limited occasions for speciation since it is an environment in which almost all the niches are saturated, whereas vegetation that is exposed to disruption resides in highly unstable conditions. New ecological niches are therefore created, to which new genotypes may possibly adapt.

These conditions are reflected in the frequency of polyploid species, whose ecological division is extremely diversified (Pignatti, 1978): In fact, they are rare in stabile vegetation, such as the climax cenospecies of a forest character, and more frequent in types of vegetation that are exposed

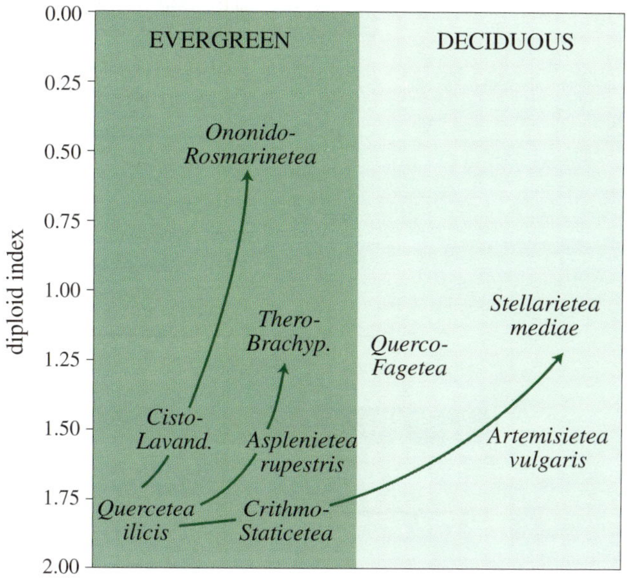

FIGURE 1 Diploid index of the main classes of Mediterranean vegetation. The index is given by the ratio between the weighted sum of the diploid species and that of the polyploid species.

to disruptive influences (Fig. 1). It is therefore possible to affirm that a disruptive influence renders possible the evolution and stabilization of new species, possibly accompanied by polyploid-forming processes. Polyploid species are therefore more frequent in disrupted environments.

The Evergreen Woods of Mediterranean Type

At the end of the Pliocene, the Mediterranean environment was characterized by a climatogenous vegetation in a warm, Mediterranean type of climate: It was either evergreen forest or a deciduous sort of vegetation, or it was linked to the savanna. Climatogenous associations are floristically very stable and sparse (in the ilex groves, there are currently only 10–25 species in a standard area of 100 m^2). They represent a conservative environment which does not favor the evolution of plant species. Plants belonging to the fagaceae, oleaceous, and ericaceous families, and elements with tropical affinities such as the *Laurus, Myrtus, Smilax,* and *Rubia* genera, prevail, whereas the families which supply the largest contingent of species to Mediterranean flora (*Caryophyllaceae,* cruciferous, papilionaceous, composite, and even the gramineous) are poorly represented among climatogenous vegetation.

Forest flora, as already mentioned, is rather poor: According to the study conducted by Braun-Blanquet (1937), which embraces the whole of Mediterranean France, there are only about 60 species connected with this type of envi-

TABLE 1

Relation between Phytomass and Biodiversity in the Mediterranean Environment

	FOREST	MAQUIS	GARIGUE	NITRO-PHILOUS GRASSES
Phytomass (t/ha)	150	60	10	3
No. of species	60	60	400	150

TABLE 2

Biodiversity of Mediterranean Vegetation

	SOUTHERN FRANCE, (BRAUN-BLANQUET, 1952)			BARCELONA, (BOLOS, 1962)		
	Genera	Species	Associations	Genera	Species	Associations
Ilex grove	58	69	2	58	66	2
Maquis	65	78	2	60	68	2
Garigue	235	435	28	171	286	10

ronment (Table 1). The situation is the same in the maquis, which differs primarily as a result of the ilex, with its bushy rather than arboreal carriage, but otherwise has a similar specific makeup. The garigue (Table 2), instead, which is attained by means of a further degradation as a result of fire and pasture (Fig. 2), presents a flora which is far richer (Naveh and Whittaker, 1979).

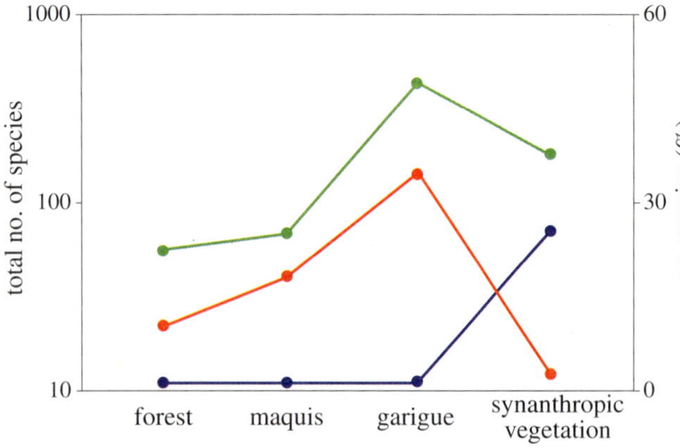

FIGURE 2 Floral diversity in different Mediterranean environments ordered according to the degree of human impact: forest, maquis, garigue, and synanthropic vegetation. The total number of species (on a logarithmic scale) is indicated in green, whereas the sum of leguminous and graminaceous plants is in red and the sum of the *Chenopodiaceae* and other wild grasses is in blue; both these groups are shown as percentages with respect to the total number of species.

According to a uniformitarian principle, it might be maintained that in such a situation, the greatest diversification of flora took place outside of the realm of climatogenous vegetation in environments outside the forest, such as peaks, crests, cliffs, breccia formations, and on the rocks along the coast.

Evolution by Reduction

Composed of about 6000 species belonging to more than 1100 genera, Italian flora reveals noncausal relations between morphology, biology, ecology, and the distribution of individual species; it appears that the results may be generalized to include all Mediterranean flora. One may distinguish, among the most varied groups, species which we deem to be either ancestral or derivative (Fig. 3), which in extreme cases present the various traits. Ancestral species have a well-developed vegetative apparatus; are either ligneous shrubby or fruticose plants, or in any case perennial, diploid, with a restricted and often punctiform area of distribution; and their natural habitats are marginal environments such as peaks, cliffs, breccia formations, and coastal rocks. Derivative species, on the contrary, have a reduced vegetative apparatus; are annual herbaceous species, polyploid, with a very ample and often pan-Mediterranean area of distribution; and their habitats are man-made environments.

Between the two extremes, there are a myriad of inter-

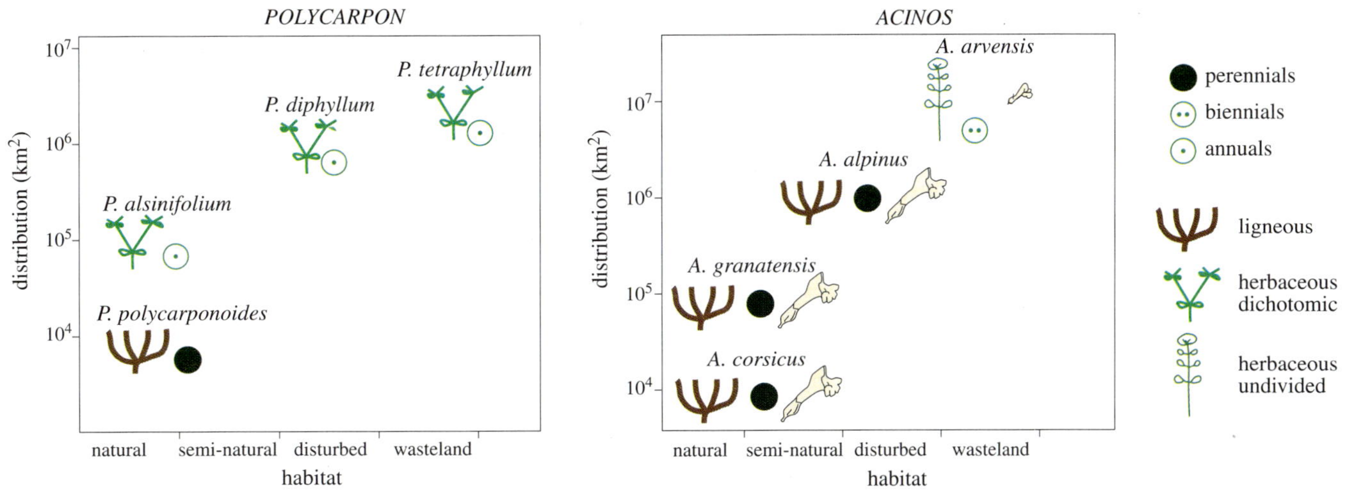

FIGURE 3 Relationships between the type of habitat, the wideness of geographic distribution, and the length of life of the *Polycarpon* and *Acinos* groups of Mediterranean flora. In the *Acinos,* the proportional length of the corolla is indicated in yellow.

mediate cases. A model has been described (Pignatti, 1979) in which it is hypothesized that at the end of the Pliocene, the ancestral species grew only in marginal habitats because forest climatogenous formations occupied the most favored stations with low-diversity cenospecies (Fig. 4). A genetic reservoir was thus created in the marginal habitats, whence an intense speciation process began during the Pliocene. This process depended on the formation of new open niches capable of being colonized by entirely new genotypes and led to a frequent passage to polyploid states.

The new niches were formed as a result of certain mechanisms. The climatic variations connected with the glaciations determined a contraction of forest vegetation, the expansion of steppe-like strips, variations in sea level and glaciation phenomena, albeit restricted to a few groups of tall mountains. Starting with the Riss–Würm interglacial, however, the consequences of man's action on natural vegetation also began to appear. An imposing speciation process was thus initiated (Fig. 4b), probably in a relatively short time, which has undoubtedly had very important consequences on the whole of Mediterranean flora. Once again, the invasion of environments lacking competition took place. The genetic reserves were situated in the marginal environments in which, as previously mentioned, a shrubby and herbaceous flora had specialized. Plants belonging to peak, crest, cliff, coastal rock, and breccia environments gave rise to an entirely new flora that colonized the environments that had been disturbed by human intervention (Table 3). The opportunistic species which, by exploiting a brief cycle (a few months) and the intense production of seeds, were able to spread very efficiently prevailed in this flora. Today, annual species constitute more than 30% of all Mediterranean flora and amount to about 5000–7000 different species, and a significant number of them may well have originated by means of the mechanism described pre-

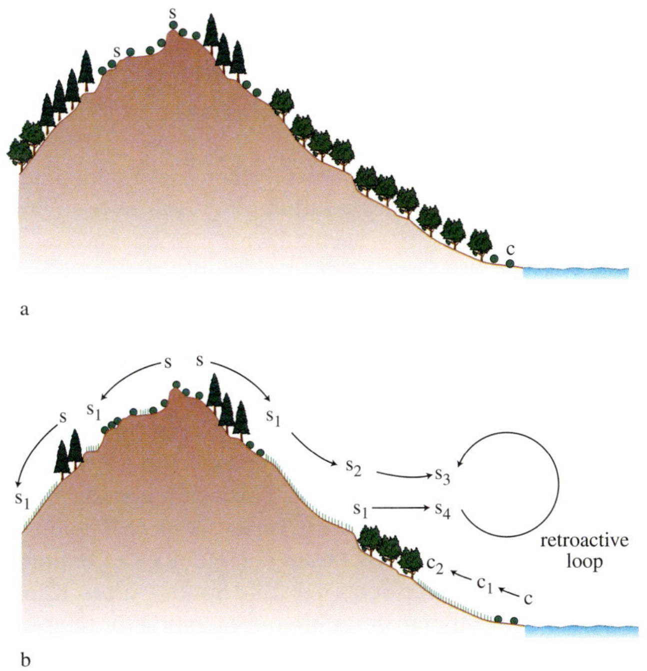

FIGURE 4 (a) Hypothetic reconstruction of the distribution of vegetation in the Mediterranean area at the end of the Pliocene. The Mediterranean environment was occupied by forest vegetation — in lower areas by evergreen broadleafs (*Quercus*) and by conifers (*Abies, Cedrus,* and *Pinus*), together with *Ilex* and *Taxus* in the higher regions. The bush flora was concentrated in the marginal summit environments (s), where the growth of trees was prevented by the wind, and the coast (c), where the limiting factor was the salinity. (b) Schematic representation of the formation in marginal environments, at the beginning of the Pleistocene, of a new flora composed of herbaceous species which occupies the surface that has remained free. s_1–s_4 represent the species derived from the flora of the summit environments (s), whereas c_1 and c_2 are the species derived from the coast environment (c). Here, the retroaction (see Fig. 5) between the action of man and speciation is inserted.

TABLE 3

Diversity of Mediterranean Environments

	MARGINAL HABITATS	MAQUIS AND GARIGUE	EVERGREEN FOREST
Diversity	Low	High	Low
Endemism	Very high	High	Low
Polyploidy	Very low	High	Low
Annual species	Absent	Frequent	Scarce

viously. The forest environment, on the other hand, in which all the niches were already saturated, remained deprived of these opportunistic species and did not give rise to particular speciation processes.

The Ecology of Fire and the Origins of the Garigue

Under natural conditions, the evergreen forest dominated by ilexes is subject to degradation as a result of fire (Trabaud and Lepart, 1980). Such a factor may be considered natural in the Mediterranean ecosystem, and it had its effects even before man settled in the region. In fact, even today it is possible to observe the effects of spontaneous fires in areas with Mediterranean-type vegetation in western Australia and South Africa. Since man settled in the Mediterranean basin, however, the relatively rare fires due to natural causes have been outnumbered by those willfully or unwillingly provoked, which have undoubtedly been more frequent. Man used fires in various ways in relation to hunting: Study of the last extant communities of hunter–harvesters (Australian aborigines and South African Khoi-San) has shown that man set fire to vegetation for the purpose of driving wildlife out into the open and thus set fires that could propagate for many kilometers, or he burned brushwood in order to produce, within the span of a few weeks, an ephemeral herbaceous vegetation capable of attracting herbivorous animals and thus concentrating them and making them more easy to hunt. One may therefore accept that ample areas of forest in the Mediterranean basin were set on fire and thereby transformed into empty habitats; the latter thus became available to new genotypes which formed by chance as a result of mutation, hybridization, and polyploid formation. With the advent of neolithic culture, which in the Mediterranean probably developed around 9000–6000 BC, animal grazing began to become widespread, which in turn caused further damage to the woodlands, especially as a result of the spreading of goats. With the iron age, about 3000 years ago, man became capable of destroying woods by cropping the oldest trees and cutting down the rest. In all these cases, the result was substantially the same: The evergreen forest was replaced by various kinds of shrubby or herbaceous vegetation (Fig. 4).

The garigue is the sort of vegetation that follows a fire, has the appearance of a low-rising maquis, with individual small-sized shrub species (such as cistus, heather, and rosemary) growing between small grassy areas, and it often covers a surface of only a few square meters. The term "garigue" is to be found throughout the entire area of the Provençal languages, from Catalonia to the south of France to the alpine valleys in the environs of Cuneo in the Piedmont. Ecological conditions vary greatly as a result of the greater or lesser depth of the soil, the presence of stones, the greater or lesser acidity of the substrate, and the possibility of the formation of pools or quagmires in the wet season; as a result, there is an enormous variety of different associations and specific combination. The garigue environment, therefore, is a mosaic composed of multiple associations and this explains its high floristic density. There are similar formations in the eastern Mediterranean which are just as rich, such as the *Phrygana* in Greece and the *Batha* in Israel, although both are characterized by the prevalence of thorny shrubbery. As a result of its intermediate position, Italy is endowed with examples both of garigue (especially in Liguria, Tuscany, and Sardinia) and of *Phrygana* (in Puglia, Sicily, and Sardinia near Cagliari). In the garigue, there are ephemeral grassy fields with gramineous plants (*Avena, Briza, Catapodium, Cynosurus, Dactylis, Lolium,* and *Vulpia*), leguminous plants (*Lathyrus, Lupinus, Medicago, Pisum, Trifolium,* and *Vicia*), composite plants (*Crepis, Geropogon,* and *Hypochoeris*), and other genera such as *Capsella, Draba, Erophila, Echium, Erodium, Geranium, Hutchinsia,* and *Sherardia.* Where there is often fire, the ground is progressively smoothened until it becomes a sort of siliceous sand on which plants of the cistus family prevail (*Tuberaria* and *Helianthemum*). Among the shrubs, thyme and rosemary grow on soil which is rich in limestone, and sometimes the *Phrygana* has *Sarcopoterium spinosum.*

Where man's action on his environment and on vegetation is even more intense, groups of herbaceous species may be observed growing close to settlements, in pens and folds, and in the animals' places of gathering and repose — all environments in which organic matter accumulates in the form of waste, refuse, and human and animal excrements which enrich the soil with nitrates. A nitrophilous herbaceous flora which includes *Atriplex, Borago, Chenopodium, Cichorium, Diplotaxis, Parietaria, Raphanus, Rumex, Scorzonera, Sonchus, Tanacetum, Taraxacum,* and *Urtica,* many of which are edible at least after cooking, is the result.

To better understand the relations between flora and man in his capacity as author of environmental impact and unconscious generator of new niches, one must examine those plants which may be considered as having been useful to him during Paleolithic times. At that stage of human culture, the vegetable world represented man's greatest resource (Hehn, 1911): From it, man derived tools, medicines, food, and many other things, even if uses for purposes of nutrition were quantitatively predominant.

In the evergreen forest environment, there are few edible plants: strawberry trees and, in the more arid places, olive and carob trees; one can also find mushrooms, which are relatively scarce, however, given that it is an arid environment. Acorns from evergreen oak trees such as ilexes and cork oaks are rich in tannin and therefore not edible. On the other hand, the acorns of *Quercus tenoreana,* which grows on hill slopes in sub-Mediterranean climates, are sweet and in Roman times constituted a basic foodstuff for the poorest stratum of the population; they were still harvested and sold in the early twentieth century. The chestnut tree has also been of great nutritional importance. These arboreal species, however, live in deciduous woodlands and are not present in the Mediterranean environment.

The flora of the garigue includes numerous nutritious species, none of which by itself is capable of maintaining a whole population; however, in their ensemble, they are quite capable of providing a varied and balanced diet. In addition, in the majority of cases these species provide dried fruits, which can be preserved easily. The flora also provides starchy plants (Pignatti, 1995) — the gramineous ones such as barley, oats, rye, and primitive forms of wheat, such as *Aegylops* and *Dasypyrum;* starchy bulbs and tubers, such as asphodel, "lampagione," wild onions, and *Orchis;* protein-rich seeds in leguminous plants, such as vetch, lentil, chickpea, lupine, and broad bean; oleaginous seeds, such as linseed and rapeseed; and vegetables such as corn salad, goat's beard, and chicory. To the list of these basic foodstuffs one may add a relative abundance of aromatic herbs, such as garlic, rosemary, oregano, mint, pennyroyal, sage, thyme, lavender, myrtle, juniper, fennel, and parsley, many of which also have medicinal effects.

Regarding nitrophilous flora (Pignatti, 1983, 1987), which grows near human settlements, there is an almost limitless series of vegetables, such as *Urtica, Atriplex, Chenopodium, Silene, Crepis,* and *Traxacum,* whence come salads, chicory, wild spinach, succory, oyster plant, and many others; however, basic foods that are rich in protein or digestible carbohydrates are missing.

A retroactive loop (Fig. 5) is thus established to the extent that the spreading of the garigue causes an increase in the availability of foodstuff, which in turn causes an increase in population with accompanied increases in the number of fires that destroy vegetation and the further extension of the garigue. The system functions as a cycle and the process acquires an autocatalytic character.

This vegetation, which played a fundamental role in sustaining the human population before and during the passage from the condition of harvester–hunter to the domestication of edible plants, is an ephemeral flora, just as the environment that follows fires and in which it vegetates is ephemeral. The species that it comprises have a brief life span and often complete their vital cycle in only 1–3 months. Nevertheless, this is an imposing presence: There are thousands of species, some of which are quite diffused. The phenomenon has been compared (Pignatti, 1978) to the appearance of structures of dissipation according to the thermodynamics of irreversible processes developed by Prigogine (Fig. 6).

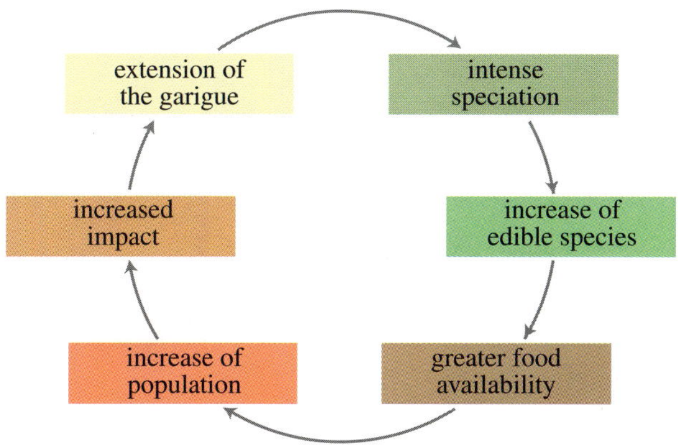

FIGURE 5 Diagram of the retroaction between the action of man on the vegetation (especially by means of fire) and the diffusion of edible vegetable species responsible for the formation of the Mediterranean plant landscape.

Agriculture represents the first method that man developed to concentrate energy and manage it in such as way as to satisfy the necessities of the system. In fact, throughout the entire course of humanity's development up to the industrial revolution, man has been entirely dependent for his needs, directly or indirectly, on solar energy. The sun, however, releases energy in a diffused form which can be neither conserved nor transferred. The cultivation of useful plants, instead, allows one to obtain products that are rich in energy (such as grains and cereals) that may be employed as energy sources whenever and wherever they are most needed. In this way, the creation of cities and complex societies is made possible.

The Mediterranean Orchard

During the first phases of agriculture, land was considered a common good and it is likely that, with the aid of rudimentary tools in cut stone or wood, a sort of itinerant agriculture developed. With the advent of metallurgy (3000–5000 BC), it became possible to build more efficient instruments (hoe, spade, and sickle) and especially plows, which enabled man to start to exploit the strength of draft animals (especially

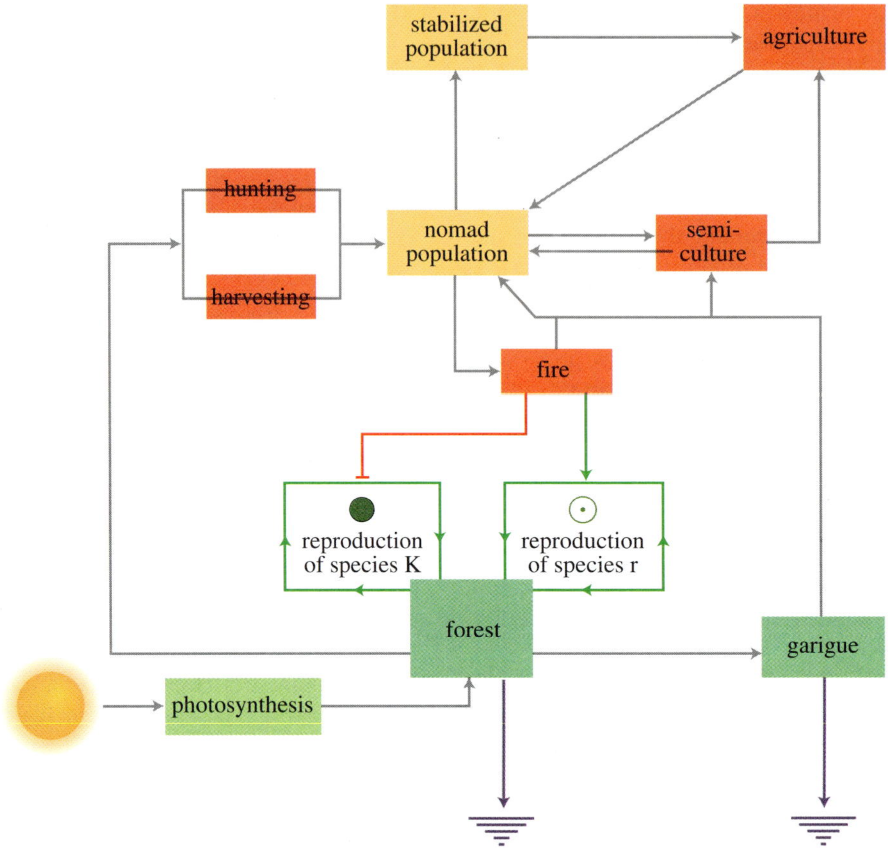

FIGURE 6 Diagram of the energy flow for the man–vegetation interaction during the periods of transition of man from hunter–harvester to farmer; in the natural environments a forest vegetation develops with scarce production of edible substances. At this stage, man intervenes with fire, moving the dominance from species K (trees) to species r (annual herbaceous). A positive retroaction is thus established that leads to a condition of semiculture, to the stabilization of the population, and, finally, to agriculture. The gray arrows represent transfer of energy, blue arrows represent dissipation of energy, green arrows represent inducing actions, and red arrows represent inhibiting actions.

bovines, but also horses, donkeys, mules, and camels) and allowed an even greater concentration of energy. The result of this greater efficiency was the sedentary life of populations; for agricultural production, this led to a further evolution, especially by means of irrigation and fertilization. In other words, man began to model the countryside (Naveh and Lieberman, 1984).

Production was concentrated in areas that were increasingly well equipped, and this was achieved by means of the introduction of ligneous plants. Farmers confronted the problem in a creative manner by means of an imitation of the ilex grove using useful plants (Hehn, 1911) such as olive trees, caper trees, strawberry trees, vines, and fig trees (all of which grow spontaneously among Mediterranean forest vegetation). In this way, the model of the Mediterranean orchard was achieved (Sereni, 1961), which developed especially in Greece and on the islands.

At this point, it is fundamental to consider the possibility that the system, which was becoming increasingly differentiated, might have received an ordering input from without (Pignatti, 1991). The nature of the latter may have varied, but it would have had to consist of a constraint which made the population accustomed to certain forms of behavior while not being too intensive; otherwise, it would have blocked the development of society. For the fluvial cultures (Egypt and Mesopotamia), such a constraint was the reliance on the regular flooding of rivers; for Greek coastal cities it was the scarcity of farmable alluvial land which imposed the creation of colonies overseas; and for the Italic peoples it was the settlement of steep hills, which were almost natural fortresses. Mediterranean civilization arose and consolidated itself in these conditions, at first in connection with small city-states (the culture of the *polis;* Pignatti, 1994).

Transcontinental Diffusion

The history that followed is well-known (Cavalli-Sforza and Cavalli-Sforza, 1993; Diamond, 1997). In ancient times, the crops which had begun in the Mediterranean spread toward the north (northern Europe), the south (subtropical Africa), and especially the east (all the way to the valleys of the Indus and Ganges). After the great geographical discoveries at the end of the Middle Ages, they were also brought to the Americas and, later, to South Africa and Australia. Today, wheat (in general along with rye, barley, and oats) is grown in all the temperate zones of the globe and has also spread to the temperate cold areas (in Europe up to a northern latitude of 65°) and to the subtropical and tropical ones; there are authentic granaries in the United States, Canada, Argentina, and Australia. Among the plants which did not originate in the Mediterranean, only two (rice and maize) have attained a comparable diffusion.

The main reason for the success of crops that originated in the Mediterranean climate is probably connected with the ecophysiological characteristics of the Mediterranean species, which are for the most part annual herbaceous species with brief cycles that germinate in autumn (or at the beginning of spring), blossom in the spring, and whose fruits ripen in May–July. In this climate, they adapt egregiously to the living conditions in the fields, which host the cultivated species only for periods of a few months, during which time they must complete their entire life cycles; after harvesting, the soil is left to rest.

At this point, we may attempt to answer the question posed at the beginning of this article: Why did domestication first develop precisely in the Mediterranean basin and not in the other ecosystems with a Mediterranean climate? The answer must take two factors into account: the evolutionary time spans and biological characteristics of the various floristic stocks.

Regarding the time spans, it should be borne in mind that the domestication process is the result of a close interaction between humans and plants: The longer the period in which such an interaction has taken place, the greater the chances that it will have been successful. The Mediterranean, from this viewpoint, is in a unique position compared to Mediterranean-type ecosystems: Man had certainly been present here in the form of *Homo erectus* in quite ancient times, as demonstrated by various sites (Monte Carmelo, Ternifine, Atapuerca, and Isernia) which have been established as being 0.5–1 million years old.

Since then, man's presence in the Mediterranean has been constant, first as *H. erectus* and later as Neanderthal man and as *Homo sapiens,* up to the appearance of our own species. The use of fire is documented as having taken place at least 0.5 million years ago; at the beginning, perhaps, man relied on the spontaneous ignition of fires, but at the site at Terra Amata (200,000 years ago) he appears to have been able to light it for his own needs. By comparison, human presence in areas with a Mediterranean-type climate is far more recent: 40,000 years in Australia, 12,000 years in California, and 11,000 years in South America. In Africa, man has been present since the last phases of the Pliocene but always in climatically non-Mediterranean areas, such as the savanna in Tanzania and Ethiopia and the high-altitude prairies in the Transvaal; the settlement of the province of the Cape of Good Hope, which does have a Mediterranean climate, is much more recent, probably dating back to no more than 10,000 years ago (Diamond, 1997). Basically, in the Mediterranean basin man has been able to share his living environment with edible plants for a period 25–100 times longer than the period in which his brothers have done so in their Mediterranean-type climates, and this undoubtedly represented a decisive advantage with regard to domestication.

The intrinsic characteristics of Mediterranean flora are of no lesser significance. It descends from an ancient stock (Atlantic or Mesogeaen), but successive events changed it and enabled opportunist, euryplastic types capable of rapidly evolving by way of hybridization and polyploidy to emerge. After having come into contact with man, especially by means of fire, the flora reacted with an imposing process of speciation. Nothing of this kind took place in the other ecosystems with a Mediterranean-type climate: The flora of western Australia is almost entirely wanting in herbaceous species, and the annual species are limited to the ephemeral blossoming that takes place in desert environments after periods of rain; in South Africa, there is a specialization in bulbous plants; and in California and Chile, there is no like abundance of species capable of colonizing the environments man has created there. Speciation by reduction and human–plant coevolution are phenomena that are entirely peculiar to the Mediterranean environment.

Conclusion

The flora and vegetation of the Mediterranean region present a particularly marked biodiversity. This is the result of ancient and recent events which have deeply influenced the biography of the area; Italy, situated in a central position, has also been affected by them. In particular, there has been an intense speciation process connected with the spreading human presence.

The appearance of new species caused by increases in population led not only to numerical increases in Mediterranean flora but also to its diversification through the selection of annual species or ones which, in any case, have short cycles and represent a significant increase in polyploidy. A retroactive connection was established between the development of the human population and the evolution of flora which has impressed a cyclical character on the system. Human–plant coevolution has been a decisive factor in the development of civilization in the Mediterranean environment and represents what is probably a unique event in the history of humanity.

References Cited

Braun-Blanquet, J. (1937). *Les Groupements Végétaux de la France Méditerranéenne.* CNRS, Paris.

Cavalli-Sforza, L., and Cavalli-Sforza, F. (1993). *Chi Siamo. La Storia della Diversità Umana.* Mondadori, Milan.

Diamond, J. (1997). *Guns, Germs and Steel: The Fates of Human Society.* Norton, New York.

Di Castri, F., and Mooney, H. (1977). *Mediterranean Type Ecosystems, Origin and Structure.* Springer-Verlag, New York.

Greuter, W., Burdet, H. M., and Long, G. (1984–1989). *Med-Checklist,* Vols. 1, 3, and 4. Conserv. Jard. Bot., Geneva.

Hehn, V. (1911). *Kulturpflanzen und Haustiere in Ihrem Übergang aus Asien Nach Griechenland und Italien Sowie in das Übrige Europa.* Borntraeger, Berlin.

Malyshev, L. I. (1975). The quantitative analysis of flora: Spatial diversity, level of specific richness and representativity of sample areas. *Bot. Zhurn.* **7,** 1537–1550.

Naveh, Z., and Lieberman, A. S. (1984). *Landscape Ecology, Theory and Application.* Springer-Verlag, New York.

Naveh, Z., and Whittaker, R. (1979). Structural and floristical diversity of shrublands and woodlands in northern Israel and other Mediterranean areas. *Vegetatio* **41,** 171–190.

Pignatti, E., and Pignatti, S. (1989). Life forms distribution and phytogeographical affinities of the Canarian flora. *Flora* **183,** 87–95.

Pignatti, S. (1978). Evolutionary trends in Mediterranean flora and vegetation. *Vegetatio* **37,** 175–185.

Pignatti, S. (1979). Plant geographical and morphological evidences in the evolution of the Mediterranean flora (with particular reference to the Italian representatives). *Webbia* **34,** 243–255.

Pignatti, S. (1983). Human impact in the vegetation of the Mediterranean basin. In *Man's Impact on Vegetation* (W. Holzner, M. J. A. Werger, and I. Ikusima, Eds.), pp. 151–161. Junk, The Hague.

Pignatti, S. (1986). The consequence of climate on the Mediterranean vegetation. *Annali Bot.* **42,** 123–130.

Pignatti, S. (1987). The relationships between natural vegetation and social system in the Mediterranean basin. In *Vegetation Ecology and Creation of New Environments* (A. Miyawaki *et al.,* Eds.), pp. 35–45. Tokyo Univ. Press, Tokyo.

Pignatti, S. (1991). Man–environment relationships in the earliest phases of civilization. *Biol. Forum* **84,** 83–90.

Pignatti, S. (1994). *Ecologia del Paesaggio. Il Paesaggio Vegetale Italiano.* UTET, Torino.

Pignatti, S. (1995). Biodiversità della vegetazione mediterranea. *Atti Convegni Lincei* **115,** 7–31.

Renfrew, J. M. (1973). *Palaeoethnobotany. The Prehistoric Food Plants of the Near East and Europe.* Columbia Univ. Press, New York.

Sereni, E. (1961). *Storia del Paesaggio Italiano.* Laterza, Bari.

Trabaud, L., and Lepart, J. (1980). Diversity and stability in garrigue ecosystems after fire. *Vegetatio* **43,** 49–57.

Fabio Caporali

Department of Plant Production
University of Tuscia
Viterbo, Italy

Ecosystems Controlled by Man

Demographic and technological pressures have appreciably influenced natural ecosystems, resulting in changes in the distribution of the earth's surface and the intensity of use. Urban and industrial systems represent the extreme cases of the substitution of the natural components of native systems with anthropogenic components. Agricultural systems, on the contrary, are cases in which natural components undergo maximum change. Historically, the main type of change in land use was the increase in the area destined for agriculture. The problem of ecosystem regulation addresses both the conservation of the remaining natural ecosystems and the rationalization of exploitation of ecosystems at different levels of anthropization. Because sustainable development means, above all, the development of equitable relations between man and man and between man and his environment, agroecosystem regularization must be managed in a way that will guarantee four fundamental characteristics: productivity, stability, equitable sharing of benefits, and sustainability.

Regulation or Control of Ecosystems

The level of man's interference in the ecosystems has increased with the growth of demographic and technological pressures. Needs related to subsistence and the development of human activity such as agriculture, industrialization, urbanization, and transport, have all contributed radically to changing the original ecosystems, resulting in structural and functional changes which testify to the degree of human intervention.

However, human interference is not only purposive, like that of other species, but also purposeful (i.e., it is purposefully conducted according to its aims). By regulation or control of ecosystems, we therefore mean all the intentional interference exerted on the ecosystems aimed at reaching a preestablished objective. The type of control will depend on these preestablished targets. Although the problem of the objectives is of a cultural nature and depends on the level of maturity of the society, the problem of planning control will involve methodological aspects, such as the links and coordination between institutions, and technical problems relating to the control mechanisms to be adopted to change the structure and function of the ecosystems. Regulation is a level of interference programmed by the controller. The latter cannot foresee all the possible responses of an ecosystem influenced by climatic factors, individual organisms, populations, and communities and therefore characterized by its own independent development strategy which is only partially predictable. Thus, every action of regulation is inevitably accompanied by a set of nonintentional or unpredictable effects. Consequently, it cannot be denied that the more we know about the range of reactions of the ecosystem, the more responsible and reliable will be the level of regulation. Recent developments in various branches of ecology are helping to change both the goals and the mechanisms of ecosystem regulation. Unfortunately, the unpredictable consequences of past management have led to a global environmental crisis characterized by pollution, impoverishment of resources, and loss of biodiversity and also a social crisis with imbalance between the rich and the poor, migratory flows, crime, and drugs. These consequences urgently require new principles of ecosystem control which can redirect human development toward sustainability.

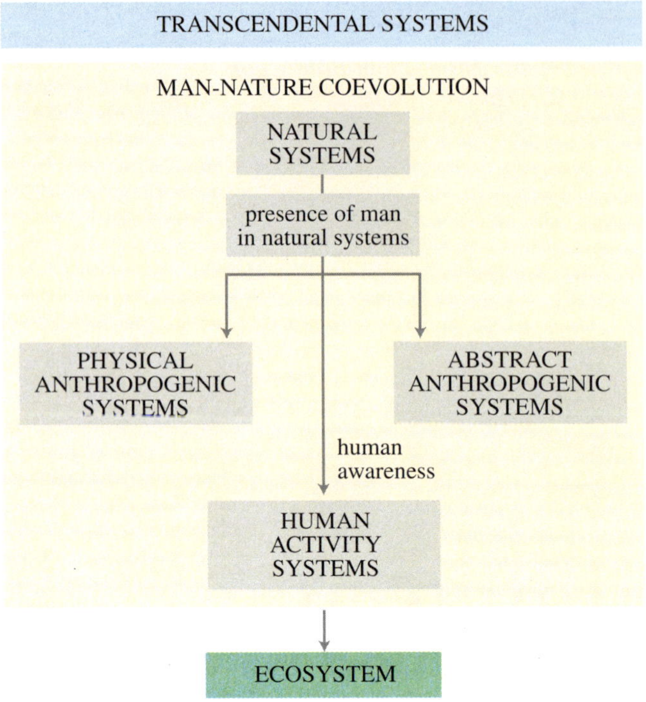

FIGURE 1 Organization of the five classes of systems constituting a map of the universe. The natural systems comprise man, who through his creativity enriches them with abstract (thoughts) and physical (manufactured goods) components, organizing human activity systems. The resulting coevolutionary global system, the ecosystem, is the true scenario in which we operate, and the transcendental systems operate beyond current human knowledge (reproduced with permission from Checkland, 1981).

Human Interference with the Ecosystems

To explain the structure of reality, the processes involved, and the role of man, Checkland (1981) suggested the use of a holistic type of instrument called *systems thinking,* which simultaneously embraces epistemological and scientific values. This analytical tool of reality allows the construction of a map of the universe based on five different class systems (Fig. 1): natural systems, physical anthropogenic systems, abstract anthropogenic systems, human activity systems, and the transcendental systems which go beyond empirical knowledge. The first four classes are grouped together in a supersystem which we can define as the coevolution of man and nature, whereas the transcendental systems lie outside, in the impossibility of objectively pinpointing their relation with the other classes. In effect, the coevolutionary nature–man system is the process of becoming that which defines reality, resulting from the confluence between natural and social history.

It is no surprise that man and his activities are fully part of the concept of an ecosystem. Dating back to the first formulations of this concept (Tansley, 1935), human activity finds its space in ecology because it is considered an extremely powerful biotic factor which increasingly disturbs the balance of preexisting ecosystems and in the end destroys them, creating new, different ones in their place. The ecosystemic unit therefore includes man–nature dualism.

The state of coevolution is seen in different gradualisms, depending on the extent that anthropogenic systems have changed the natural systems. Human activity systems, such as agriculture, industry, and urbanization, are the result of organizational activity created by the abstract anthropogenic systems. This organization is achieved by replacing or changing the natural components of the ecosystems and adding physical anthropogenic systems (manufactured goods, machines, etc.). At the present moment of coevolution between man and nature, anthropogenic processes are so preeminent that they have become the most important element in ecosystem evolution, on both the local and the planetary scale. Urban and industrial systems are cases of maximum substitution of natural components of the native ecosystems with anthropogenic components. Agricultural systems, on the other hand, are cases in which the natural components of the native ecosystems have undergone the most change. In any case, the purpose in changing ecosystem use is to increase the capacity for sustaining human activity. Changes in ecosystem management are made by changing the area and intensity of its use. Currently, the problem concerns the control of the spatiotemporal distribution of ecosystems managed by man as well as their internal organization.

Since ecosystems are open and hierarchical systems, control or regulation can only occur at ranks below the scale of intervention. The higher hierarchical ranks, on the contrary, impose limits and exercise a sort of "control of the control." Absolute ecosystem regulation by man is therefore impossible since it is impossible to control the entire planetary system, the final controller. However, nonintentional effects do unfortunately happen and escape human control. These occur at the higher hierarchical levels, above those of anthropogenic intervention (e.g., the greenhouse effect, acid rain, and the hole in the ozone layer), and can hinder the development of future strategies.

Controlling the Spatiotemporal Distribution of Ecosystems

Since the eighteenth century, and especially in the twentieth century from the 1960s onwards, the effects of changes in ecosystem use have become global. Widespread changes, practically over the whole world, have produced global variations in characteristics shared by the whole planet (e.g., the climate). It must be emphasized that global changes affect all regions, regardless of whether they have contributed or not to their production (Houghton, 1994).

Although they do not cover such vast areas as do the other types of ecosystems, the urban and industrial systems use so many resources (energy and materials) and emit so much output (gas and liquid effluents, solid waste, etc.) that they are held to be a major source of global change in substrate quality, particularly air and water. The effects of this type of system have raised doubts about the current tendency in all developed and developing countries to concentrate their human populations in urban rather than rural areas (Table 1). If the developing countries, in which population growth is typically high (Fig. 2), were to follow the same type of technological development based on the combustion of fossil fuels which to date has been typical of the industrialized nations, the environmental consequences could be apocalyptic (Bresso, 1993).

Considering the three most important areas of the world for size and intensity of ecosystem use, Table 2 shows that CO_2 flow from fossil fuels in the two industrialized areas is well over half (United States) and four times (former Soviet Union) the net fixation capacity of the relative terrestrial ecosystems. This emphasizes how the catabolism of modern human society far exceeds the anabolism of the natural ecosystems.

TABLE I

Urban Population Growth during the Period 1950–2025[a]

	URBAN POPULATION (%)		
	1950	1990	2025
Developing countries	17	34	57
Developed countries	54	73	84
World	29	43	61

[a]From Bongaarts (1995).

Historically, the greatest change in land use has been the increase in areas destined for agriculture. Ten thousand years ago, the earth's ecosystems were occupied by natural or seminatural vegetation. Today, approximately one-third of the world's surface is covered by agricultural lands subject to cultivation or grazing (Houghton, 1994). Directly or indirectly, man uses about 40% of the sun's energy, which is fixed annually through photosynthesis by terrestrial producers (Vitousek *et al.,* 1986). Currently, the area destined for agriculture amounts to 4.7 billion ha, 30% of which is arable (arboreous and grassland) and 70% permanent pastures. Most of the changes in land use have taken place since the seventeenth century (Asner *et al.,* 1997) and have

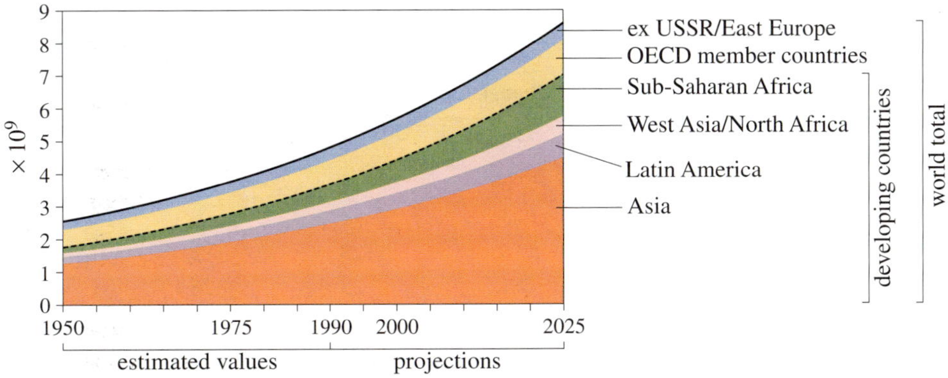

FIGURE 2 Estimated population values (1959–1990) and projections (1990–2025) for different areas. Population growth varies considerably in the different geographical areas, remaining stable in developed countries and tending toward an increase in developing countries. The area with maximum population growth is Asia. The increase of approximately 3 billion inhabitants on the planet by the Year 2025 will result in extra food, energy, and material requirements as well as further pressure on the environment. OECD, Organization for Economic Cooperation and Development (reproduced with permission from Bongaarts, 1995).

TABLE 2

Estimated Carbon Absorption and Release in the Form of CO_2[a]

	FORMER SOVIET UNION	UNITED STATES	BRAZIL
Net production of the ecosystems	568	293	−16
Wood harvesting	−148	−202	−18
Change in soil use	0	−4	−132
Farming	−76	−2	−85
Total for the area	344	85	−251
Logged and stored wood	33	36	3
Fossil fuels	−1.139	−1.306	−53
National total	−762	−1.185	−301

[a] All data are given in Tg/years ($T = 10^{12}$). Positive values indicate net carbon absorption, and negative values indicate carbon transfer to the atmosphere or outside the national area. Adapted from Turner *et al.* (1997).

occurred quite diffusely in all areas of the world (Table 3). With regard to arable land, half of this increase occurred in the twentieth century (Houghton, 1994). Consequently, natural terrestrial ecosystems (forests and meadows) have been drastically reduced for conversion into agroecosystems.

The tendency of deforestation for agricultural purposes has practically ended in the temperate belt, but it is continuing at a fast rate in developing countries in the tropics. The Food and Agriculture Organization estimates that the rate of deforestation from 1980 to 1990 in the belt of tropical countries (Houghton, 1994) was 15.4 million ha per year, an increase of more than 36% compared to the late 1970s. This increase in deforestation rate, however, differs from continent to continent: The largest increase is in Asia (+93%), America is intermediate (+32%), and the smallest increase is in Africa (+12%). It has been estimated that at the current rate of deforestation there will be no tropical forests left in about 100 years (Houghton, 1994).

The global effects of changing native tropical ecosystems into agroecosystems are beginning to emerge. One is the change in climate related to the imbalance of the water cycle due to the lower levels of transpiration of trees, the emission of greenhouse gasses, and the reduction of the ozone layer in the stratosphere (Schimel, 1995; Turner *et al.*, 1997). Other effects are loss in soil fertility and the associated decrease in agricultural potential leading to abandonment of the land to uncertain destiny (Houghton, 1994; Asner *et al.*, 1997) and also the loss of irreplaceable biodiversity at both the community and species levels (Wright, 1990).

The data in Table 2 show the transfer of CO_2 to the atmosphere in Brazil following changes in soil use from forest to agriculture. This country has the highest deforestation rate, estimated at 1–2 million ha per year in the 1980s. CO_2 emission derives from combustion of phytomass and degradation of the carbon surplus left on the ground. CO_2 production from this source in Brazil is more than twice the

TABLE 3

Variations in Percentage of Forests and Woods, Pastures and Meadows, and Farming Land from 1700 to 1980[a]

	FOREST AND WOODS		GRAZING PASTURES		FARMING LAND	
LOCATION	Absolute Change (× 10^6 ha)	% Change	Absolute Change (× 10^6 ha)	% Change	Absolute Change (× 10^6 ha)	% Change
Europe	−18	−7.8	−52	−27.4	70	104.5
North America	−74	−7.3	−125	−13.7	200	6666.7
Russia and north Asia	−197	−17.3	−3	−0.3	200	606.1
China	−77	−57.0	−28	−2.9	105	372.1
Southeast Asia	−18	−7.1	−33	−26.4	41	1275.0
South Asia	−155	−46.3	−2	−1.1	157	296.2
North Africa	−24	−63.2	−63	−5.6	87	435.0
Tropical Africa	−284	−20.9	106	10.1	178	404.5
Latin America	−294	−20.3	159	26.2	135	1928.6
Developed countries in the Pacific	−21	−7.9	−31	−4.9	53	1060.0
World total	−1.162	−18.7	−72	−1.0	1236	466.4

[a] Adapted from Asner *et al.* (1997).

amount from fossil fuels. This situation is antithetic to the United States and the former Soviet Union, the other two industrialized nations considered. In these countries, the only appreciable CO_2 flow comes from fossil fuels and practically none derives from changes in land use. Agricultural activity leads to even more CO_2 reaching the atmosphere because it favors an increase in the oxidative processes of the organic substances during tillage; this effect is particularly accentuated in hot and humid climates as found in the tropics. Overall, 86% of Brazil's total net CO_2 flow toward the atmosphere derives from deforestation for agriculture and agricultural purposes. Nevertheless, the total CO_2 transfer to the atmosphere from fossil fuels in the United States is about four times that of Brazil. Considering the difference in population — 263 million in the United States compared to 150 million in Brazil — the net global and pro capita CO_2 production is much lower in the latter than in the former.

These data demonstrate the major differences in the development models for countries at different levels of industrialization and at the same time invite reflection on the reasons behind them. The "hunger" for farming land in the tropics also points to the unsatisfied appetites in other areas of the planet.

Agriculture as a Human Activity System

Agriculture is a human activity system derived from the integration of natural and anthropogenic systems, both abstract and physical (Fig. 3). The system is the product of a network of social connections linking all the important nodal points (Fig. 3, boxes) which elaborate and distribute information. It converges in a central, operative nodal point consisting of the farmers operating on the natural ecosystems and who transform them into agroecosytems.

In turn, the network of social connections revolving around the farmers consists of interactions between important institutions (politics and administration, research and professional training, industry and commerce, and consumers); this is the preliminary informative context which eventually determines the choice of the farmers. The activity of the farmers is determined by comparing the patterns gained from individual and institutional information. The activity of the farmers concerns well-defined, elementary sectors of the land (farms); together these comprise a mosaic which characterizes the whole of the rural environment. In Italy, about 3 million farms cover a total area of approximately three-fourths of the national territory.

The native territory on which agriculture is practiced is the natural ecosystem subject to anthropogenic change; the resulting territorial unity of coordination is called an agroecosystem. Like all ecosystems, agroecosystems are built on a continuous hierarchical gradient, ranging from the microhabitat near the cultivated plant to the macrohabitat of the farming area. It is helpful to consider the farm as the elementary operative unit in agriculture and as an agrosystem representing the actual meeting point between the interests of man, oriented toward directing natural ecosystem

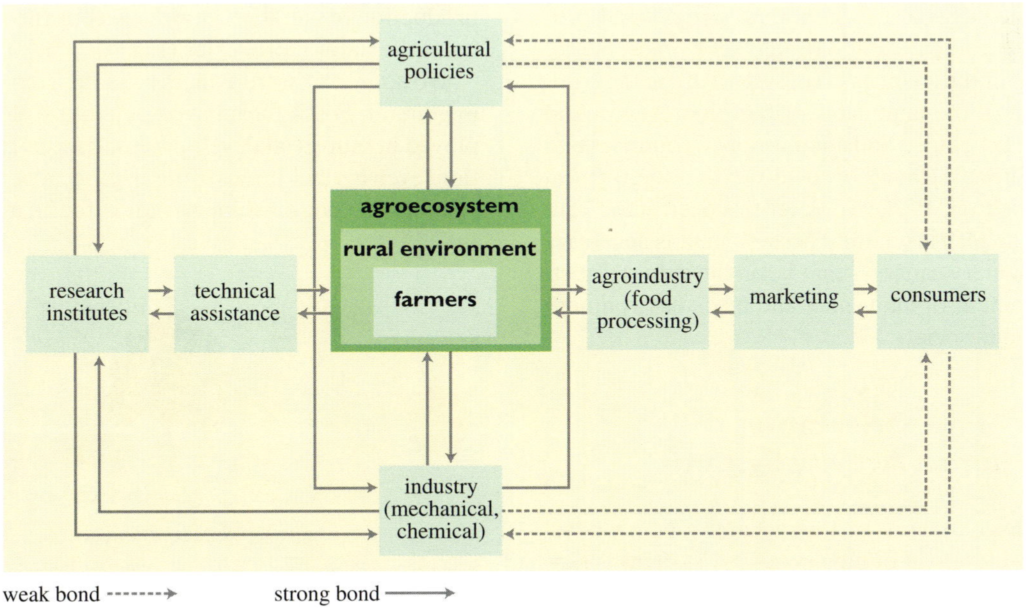

FIGURE 3 Farming as a human activity system. Farmers, who are the hub of the system, plan their activity as a function of the interactions with the other components of the farming system. These interactions first emerge through the exchange of information and then in the form of exchange of energy and matter between the agroecosystem and the context in which it is inserted. The context is now holistic and is the interaction of national economies within the international scenario.

productivity to his own advantage, and the strategy of the natural systems, striving to maintain the climax community (i.e., the stability of the system, conservation of resources, and biological diversification) (Caporali, 1991).

Through the increasing globalization of communications, economics, and politics, the environmental informative context is practically permeating the entire planet through international agricultural and commercial policies between states or aggregations of states, the activity of multinationals in strategic sectors of biotechnology, mechanical and chemical supports to agriculture, etc. In addition to the global conditioning of human activity by nature through the climate, an equally global type of conditioning — a "cultural climate" by man — is occurring. Thus, national farming methods, for which traditionally their main element of control was a country's government, are today further conditioned by a wider system of international relations.

In the industrialized nations, in which distribution of the different types of ecosystems has remained constant, the problem of controlling changes in ecosystem use has apparently been solved. This is not so, however, in tropical countries, in which local and international mechanisms both favor deforestation. The former stem from constant demographic increase, related to poverty and the pressure of multitudes of landless people who are excluded from any property rights in their own countries, in which most of the land is in the hands of a small minority of landowners. The latter, on the contrary, are linked to international trade requirements aimed at supplying the rich countries with low-priced agricultural products and also to the needs of the poor countries which must expand their agricultural production for export to pay their national debt. Thus, one of the main causes of the changes in ecosystem use in the tropics is correlated to the unequal distribution of wealth and power between developed and nondeveloped countries and between the different classes of society within the tropical countries (Houghton, 1994). To deter this destructive tendency toward the planet's natural ecosystem, it is necessary to act on the primary causes — that is, on the values directing the development of man–man and man–environment relations in human society.

Past Orientation of Human Activity Systems

Recognizing the constraints of abstract anthropogenic systems on the development of human activity systems is fundamental if we are to interpret the past and set the direction for the future. In practice, the type of environmental management depends strictly on the cultural pattern it interprets — that is, the ruling paradigms (or system of values) that were established during the cultural development of human society.

The paradigm still in force today is based on the roots of the Cartesian dualism between man, who thinks (subject), and nature, which man conceives (object). This has led to the prevailing tendency of viewing man as dominating over nature, which with her wealth of resources must be tamed and exploited. Consequently, a theory of environmental management (economics) was established which considered nature as an unlimited supply of resources and a dump for disposing waste. Production service activity aimed at maintaining the planet's living conditions (Westman, 1977) was not recognized and so had not been protected. Only the recent ecological outlook is righting this false interpretation and creating a new paradigm — that of coevolutionary (Norgaard, 1984) or sustainable development of human activity systems. Currently, there are two types of paradigms with conflicting ways of thinking and underlying systems of values: egoassertion and integration (Capra, 1996). The first is based on the position of man's alienation and dominion over nature and the second on the recognition of the sense of belonging to everything that systematic ecology embraces, which leads to the recognition of the principle of man's responsibility toward nature and himself (Jonas, 1993).

The acceptance of a system based on the egoassertion paradigm has resulted in the history of the development of human society during the past centuries being characterized by the progressive orientation of society toward producing material goods to the point of excessive consumerism. This process has inevitably led to the progressive exploitation of the planet's resources, especially all types of energy, and the progressive pollution of the environment because the production of metabolic waste exceeds the recycling capacity of the natural processes (Fig. 4).

No source of stored energy is safe from human plunder in general. Fossil fuels have been and still are largely employed in industrial development, transport, and urbanization; even agriculture uses them, especially for building and running farming machinery and synthesizing chemical com-

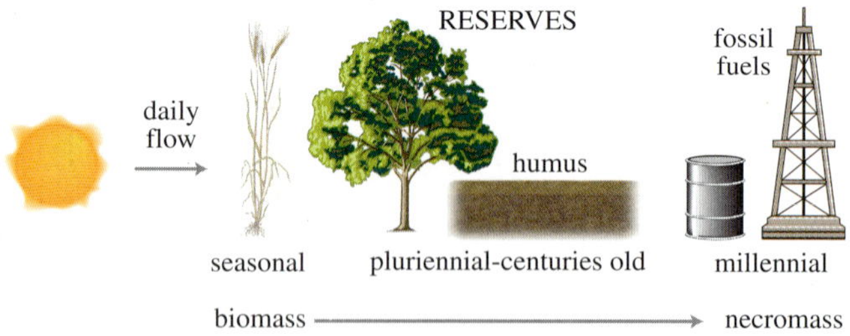

FIGURE 4 Methods of storing (reserves) of solar-derived energy sources (flow). Through photosynthesis, plants accumulate solar energy in the form of carbon bonds or plant biomass. The amount of accumulated biomass is in relation to its structure and the time required for degradation. The surplus is accumulated in the soil by biochemical transformation and in time produces various forms of energy.

pounds requiring high energy input, such as nitrogen-based fertilizers. The forest biomass (which needs tens to hundreds of years to accumulate) not only has been overexploited for timber and firewood but also has been continually replaced by new agroecosystems (which is still occurring). The biomass which has accumulated in the soils of the meadows or in the soils of ancient arable lands as slowly decomposing organic matter (humus) has been, and still is, under assault to make room for new agroecosystems or to intensify their use. The attack on the organic carbon bonds in nature is so widespread that humanity can be said to be suffering from a kind of pathological disorder, definable as "carbon hunger," which risks undermining the very foundations of human survival.

To invert this tendency, a cultural change is necessary toward a paradigm of integration, which identifies the environment as a comprehensive and common asset. It is comprehensive because interdependence rules and because the environment cannot be divided into separate parts, and it is common because as an asset it belongs to everybody, and everybody has the right to enjoy it and the responsibility of conserving it and, if possible, improving it. This cultural view is beginning to make its mark on the dominant paradigm of egoassertion, as seen, for example, in the recent evolution of the key concept of welfare, which has been the primary goal in the theory of human activity management systems (Bernetti and Casini, 1995). From the original concept of welfare related to the level of utility (opportunist theory) and subsequently to the possession of goods and services (opulence approach), today we conceive welfare as a cultural value that stresses what we can or cannot do or what we can or cannot be (the theory of freedom). Goods are thus relegated to simple means aimed at other ends, and the environmental scenario becomes an integral part of the concept of welfare, including the beauty of natural values, improvement in environmental quality, lower pollution levels, and intergenerational equality.

Humanity tends to convert natural ecosystems into agroecosystems (i.e., into systems organized to channel solar energy to its own benefit). An organization of agrosystems based on environmentally and socially compatible criteria would probably also be able to extract useful elements for organizing the whole of society along the lines of sustainability. In this case, agriculture could also act as an "ecology teacher" (Caporali, 1991).

Agroecosystem Regulation

The organization of an agroecosystem must fit its goals. Specifying the goals of an agricultural activity system must precede all types of planning and therefore all types of agroecosytem control. The choice of goals depends on the dominant type of culture, which determines the practical solutions to be adopted for implementing the agroecosystems. In the industrialized era, the main objective was maximization of the profits of an agricultural enterprise by maximizing agricultural production, both plant and animal. The entire agricultural system (political and financial policies, the development of research and industry, agricultural professionalism, and the preferences of the consumers) was oriented in this direction. Currently, this direction has been under discussion because of its effects in the industrialized nations and its implications at the world level. These effects were noted in production planning which produced a surplus of agricultural products with no market as well as in their environmental impact. The latter included the loss of soil fertility, loss of biodiversity, eutrophication, and water pollution, which all result in extra social costs (external effects). Finally, there were social consequences such as the rural exodus from the mountain and hill regions with increased migratory fluxes (Ammassari *et al.,* 1997). In light of these unexpected consequences and the systematic outlook gained from ecological knowledge, the aim of the objectives has widened to include social and environmental aspects and is oriented more toward the quality than quantity of agricultural development.

In more developed nations, the legislation and the curricula of the most important educational institutions, such as universities, are showing tangible signs of a change aimed at directing the cultural and professional formation, and therefore behavior, of all the components of the agricultural system (from the farmers to the consumers) along ecocompatible criteria. Since 1985, for example, the legislation of the European Community has introduced laws (culminating in regulation 2078 in 1992) and financial measures in favor of ecocompatible farming—that is, the type of farming which, in the opinion of the American Society of Agronomy, improves the quality of the environment and the natural resources on which it depends. It supplies food and fiber for human needs, it is economically valid, and it improves the standard of living of the farmers and of society as a whole. On the educational front, plenty of consortiums of European universities are offering parallel curricula at different levels of specialization within the ambit of community programs. These also foresee the mobility of students and academics among the associated seats.

To reach the expected social and environmental goals, agrosystems must be organized around four fundamental characteristics (Conway, 1987): productivity, stability, equitable sharing of benefits, and sustainability. These properties come to light when an agroecosystem is considered as a territorial entity, organized by a methodical mind which joins physical and biological resources together, native and imported, to satisfy personal and social needs. The structural model of the agroecological system in Fig. 5, which can for study purposes can be applied to any scale, is particularly useful for farms—socially relevant, elementary parcels of land influenced by the flux of ideas, energy, and matter.

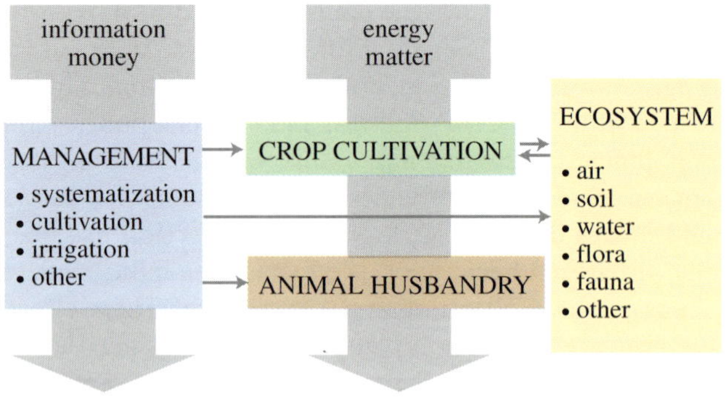

FIGURE 5 An agroecosystem model at the farm level. The farm is the most important agroecosystem for social purposes. Its structure is organized in relation to how the farmer elaborates the flow of information (including economical) which comes from the society. The farmer accordingly introduces organisms (crops and animals), energy and matter from the outside, shaping the preexisting ecosystem to farming needs (adapted from Caporali and Onnis, 1992).

The environmental context and agroecosystems are linked together through exchanges in the input and output stemming from cultural information, including the values of the society and the farmer, socioeconomic information (i.e., the flow of money), technical information for crop farming and livestock management, genetic information derived from biological evolution and related to crop and animal use, and the flow of energy and matter. These last two are in the form of natural inputs (e.g., solar radiation and rainfall) or artificial inputs (e.g., machines, fertilizers, and phytochemicals) or natural outputs (e.g., water and air currents) and seminatural outputs (e.g., farm products exported for the market). The natural ecosystem is radically changed both structurally and functionally with the introduction of new components chosen by man and by their spatiotemporal organization, which depends on the criteria derived from all the inputs of the previously mentioned information.

Productivity

Productivity concerns the capacity of the agroecosystems to store solar energy in cultivated fields under the form of crop biomass, which can later be exploited and transformed into animal products (e.g., meat, milk, and wool) to the advantage of man. Every cultivated field can be considered a solar power station, driven by the photosynthetic activity of the crops. The set of cultivated fields in a farm determines its capacity to store solar energy. The main characteristic of agroecosystems is that part of the energy and matter accumulated as biomass leaves the system as harvest for export to the market. Figure 6 illustrates the basic working model (regarding the exploitation of solar radiation flow) of a mixed farm agroecosystem (with plant and animal products). The flux of radiant energy activates the pasture chain (crops–livestock–man) and waste chain (surplus from crops and plant and animal breeding) by integrating the three main components of an agroecosystem (crops, livestock, and the soil). Soil fertility, which is the key element for expressing crop productivity, is strongly influenced by the level of integration between cultivation and animal breeding as well as by the typology of each production.

The productivity of an agroecosystem such as a farm depends on the interaction between all the natural and anthropic factors involved in its management: the climate, the soil and plants, the disposition of the crops in space (monocultures or combinations) and in time (free or fixed rotations), and how the crops interact with animal breeding (e.g., through the recycling of excrement). Moreover, if the biomass is to accumulate on the fields, it must be defended from attacks by natural enemies, predators, and parasites, particularly phytophagous and phytopathogenous ones. All interventions regarding crop management, from sowing to harvesting, are performed with machinery or material which require energy for their building and function. *Auxiliary energy* is the complex of the energetic support needed to run a farm which relies on the use of fossil fuels. Actually, agroecosystem productivity is the process of transforming input into output (goods and services). Consequently, it can be measured in terms of transformation efficiency of input used and output produced. In addition to the usual way of expressing productivity as dry product, energy, proteins, etc. per unit of area and time (which indicates the importance of the soil as a natural resource), productivity can be expressed as the intensity of use of each resource employed, for example, by calculating the input–output ratio for energy, capital, manpower, etc. This allows the system's transformation efficiency to be measured, and extremely useful numerical indices can be obtained to evaluate sustainability, which is the other fundamental property of agroecosystems.

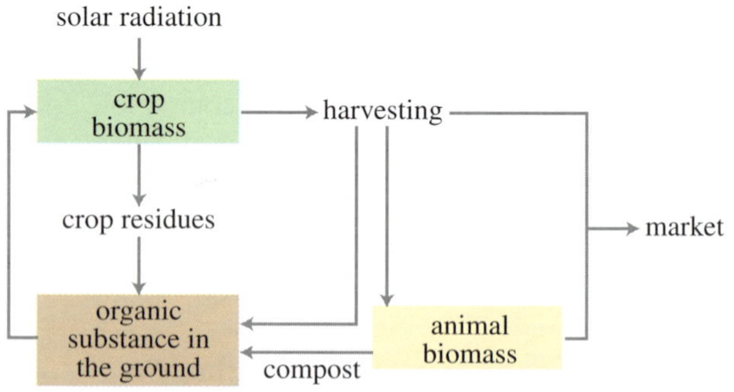

FIGURE 6 Basic agroecosystem operation. This working model illustrates the path of solar radiation which activates the grazing and waste chains within a mixed agroecosystem (i.e., one with plant and animal production). Integration between grazing and waste gives the system greater autonomy and sustainability (from Caporali and Onnis, 1992).

Direct intervention on the fundamental components of an agroecosystem (crops, livestock, and soil) and their level of integration and type and entity of input can drastically influence productivity. Recent agricultural history has shown that the main target in the industrialized era has been maximum yield for crops and animal breeding. Attention was concentrated more on cultivation and breeding than on the system embracing them; consequently, the farm was interpreted more as an aggregation of crops and livestock than as an integrated system. In this perspective, even a single crop or breed of animal could constitute a single agroecosystem under the form of monocultural orchards, cereal farms, or monospecific animal rearing. These targeted maximum specialization and aimed at what was assumed to be the maximum control of the system and maximum economical profit for the enterprise. All this was possible due to high external input of cheap energy and low-cost material (they did not have to pay the consequences of the external effects they produced) which was transformed into high output, in part expected (high farming production) but in part unexpected (negative environmental effects which have caused external effects on the entire living community). The working of an agroecosystem was blatantly constructed along a linear plan, organizing agricultural productive processes using the same criteria operating in factories and totally disregarding the role of the natural ecosystem, its physical and biological components, and its cyclic processes which integrate and sustain life. The exaltation of productivity, understood as an increase in yield per acre or per head of cattle and celebrated as a great social achievement (the green revolution), could only happen to the detriment of energy resources and matter and to the detriment of environmental quality, simply because productivity had never been considered as the ratio between input employed and output generated during the production process. Even genetic patrimony of the plants and domestic animals, the fruit of biological evolution and human selection under specific environmental conditions, has been profoundly altered and standardized to meet the cultural conditions typical of high input, both for sustaining and for defending agricultural production. Several ecotypes and local races of crops and farm animals, selected by generations of farmers for their resistance and their capacity to enrich habitats in which climatic and pedological conditions are difficult, have been substituted by new genotypes. However, their performance is only valid if accompanied by a package deal of technical interventions necessary for their support but which are at risk if the inputs are inadequate or under conditions of strong environmental stress. The loss of biological diversification, which resulted from the industrial control of farming within the agroecosystems as well as within its components, endangers the other three fundamental properties of agroecoystems — stability, sustainability, and equitable sharing of benefits.

Currently, a new and important tool for altering the productive capacity of farm species, both plants and animals, is being implemented: biotechnology, which is a method of genetic engineering that can transfer fragments of DNA between totally different types of organisms. The large-scale introduction of biotechnology in farming in high-input agrosystems, for example, with genetically modified crops to resist the use of herbicides, can only lead to negative impact on the environmental, public health, and society. On the other hand, if the same crops are modified so that they are more resistant to environmental stress (cold, hot, and drought) or attack by noxious phytophagous and phytophathogenous agents with a goal of reducing auxiliary energy input (irrigation, nitrogen-based fertilizers, pesticides, etc.), then biotechnology can offer an important contribution to compatible farming. This type of farming, with a strong local influence on the quality of its products and production processes, can prove to be fundamental in rural development and in the promotion and defense of human and land resources.

Stability

Maintaining high and constant productivity levels is an important objective of agricultural activity throughout the world. In its most accepted sense, stability, or constant productivity, is the level of fluctuation in yield in response to short-term changes in climate and farming and marketing trends. However, the concept of short-term stability can be extended to include fluctuation levels of other output related to farming — for example, loss of the agroecosystem water supply, the degree of erosion, and the availability of nutrients or pesticides, all of which can influence decision making and regulation activity of crop production levels. Losses can be calculated by applying models simulating scenarios which become increasingly more realistic as better and more accurate data on the environment and farming are inserted.

To some extent, the farmer can control the degree of output fluctuation through the management of the most important factors which limit productivity (i.e., water supply, soil fertility, and interventions against weeds and phytophagous and phytopathogenous competitors). Climatic trend is variable and can only be predicted on a statistical basis. Consequently, climatic limitations are more imposed than controlled, although direct intervention such as irrigation can help increase the water supply for crops. Irrigation, however, implies an independent water supply from outside the farm and a financial position which can meet the costs of the irrigation plant and its use. Often, these conditions do not exist. To guarantee production stability in the face of changes in the climate and the consequential changes in soil fertility and competitor aggressiveness (weeds and phytophagous and phytopathogenous agents), farming systems must be planned to allow the choice of species whose spatiotemporal organization responds best to environmental stress. Some general principles related to this strategy can be applied at any intracompany scale of intervention (i.e., in planning the whole farm) to its farming systems and to each crop or cultivated field. The common basis of these planning principles is to guarantee diversity in the system it intends to implement.

At the level of agroecosystem farming, the highest level of diversification can be ensured by mixed farming (i.e., simultaneous plant and animal production). In this case, interactions that develop between the crops and the domestic animals, due to the virtuous complementarity between pastoral and waste chains, favor the return of excrement to the soil as compost. This integration increases the organic content of the soil, favoring water retention and all the microbiological processes that govern plant nutrition and their good state of health. In the case of mixed farming, crop rotation must include extensive pluriennial fodder crops, usually leguminous plants such as alfalfa and pluriennial trefoils, which are not found in the simpler crop systems in exclusively vegetable farming. If the land is allowed to lie in fallow for their life span (2–4 years), these plants enhance the humification rather than the mineralization phase of the soil, thus increasing its organic content and the associated advantages.

In fact, leguminous plants can restore the nutritive properties of soils through nitrogen fixation, thus making it available to the next crops. The advantages of nitrogenized substances released in this manner last for several years and act as a strong regulatory element within the system for maintaining yield stability. Table 4 gives the relevant significant data obtained from a comparison between crop systems in the Mediterranean area with and without alfalfa in crop rotation. Stability in crop yield for the expected value was automatically achieved for all crops which followed alfalfa and which where not treated with synthetic compounds (herbicides and nitrogenous fertilizers). Similar results without alfalfa were only obtained by resorting to nitrogenous fertilizers (importation into the system). The latter, however, have not proved efficient under unfavorable climatic conditions, as shown by data for hard wheat in 1989, which was a year of drought. Data suggest that yield stability in the farming system can be maintained either through informative input (planning appropriate crop rotation) or through external energy and matter input (nitrogenous fertilizers and herbicides); however, external energy and matter input result in additional financial costs, uncertain outcome, and an added environmental impact caused by a decrease in subsequent crop production due to the use of herbicides.

Pluriennial fodder crops in the farming systems guarantee additional stabilizing elements for yield stability (at high values) and other outputs (at low values). For example, their curbing effect on weeds and phytophagous and phytopa-

TABLE 4

Grain Yield (kg/ha) for Sunflower Seeds, Hard Wheat, and Oats in Relation to the Previous Crop, Nitrogen Level, and Treatment with Chemical Herbicides or Mechanical Weed Killers in the Mediterranean Areas[a]

			CROP PRECESSION					
YEAR			Annual crops (6 years) + alfalfa (4 years)			Annual crops (10 years)		
			Added nitrogen (kg/ha)					
1988								
			0	65	130	0	65	130
	Sunflower seeds	Chemical herbicides	2476	2770	3406	1637	1944	2996
		Mechanical weeding	3150	2995	3068	1574	2272	3553
1989								
			0	72	144	0	72	144
	Hard wheat	Chemical herbicides	1537	1095	1057	1013	373	285
		Mechanical weeding[b]	2806	1510	1584	1253	716	1460
1990								
			0	0	0	0	0	0
	Oats	Chemical herbicides[b]	3500	3333	3330	1708	2100	2567
		Mechanical weeding[b]	3625	3708	3458	2125	2000	2625

[a]From Caporali and Onnis (1992).
[b]Not carried out.

thogenous elements (which prosper in monocultures and sequential plants in crop rotations) is fundamental in farming. They also reduce runoff, erosion, the loss of nutritional elements, and the spread of pesticides in the environment.

Finally, some advantages of yield stability can be achieved even at the level of a single crop or one cultivated field if certain criteria are adopted which favor some degree of interspecific (combination of different species) or intraspecific diversification (combination of varieties or ecotypes). In the first case, the ecological complementarity of species with different nutritional requirements is exploited [e.g., combining graminaceous and leguminous plants, different posture (colonization of aerial space), or soil exploration (colonization of different strata by the roots)]. In the second case, complementarity is achieved by means of different genotypes which respond differently to climatic variations and parasitic and predatory attack.

Equal Opportunity

Equal opportunity refers to the social relations in force under a working agroecosytem. Equal opportunity means equal access to resources for farming and to the benefits derived from farming.

With regard to resources, the first to be considered is the earth, on which the possibility of practicing agriculture depends. As previously mentioned, agriculture is an interactive relationship between man and nature that is performed on elementary territorial units usually called farms. The first requisite for this relationship to be profitable, at the individual level (for the farmer) and the social level (for the rest of the community), is precisely the type of relationship between the farmer and the land in terms of quality and duration. The history of agriculture teaches us that the best situation is when the farmer owns the land (i.e., when the relationship is long-lasting and motivated). In this context, Haussmann (1964) emphasized the concept of symbiosis between man and land to define the tendency of the human being to best express his or her organizational creativity and predisposition to work when he or she can own and freely use his or her own life scenario. Under these conditions, he or she shapes the land physically and operationally, following his or her own aesthetic values too, as if it were an appendage of his or her own house. This social status of farmer–landowner was only gained through effort, struggle, and perseverance in western Europe, as testified by the history of land reforms in several countries in the past century; moreover, it was the fundamental characteristic of the colonization of North America (but with the cost of eliminating the native populations). Unfortunately, it is not a prerogative of all peoples of all continents but only a social conquest of the landless peasants who fought for it. Where farmer–landowners lack this condition, and therefore there is no bond of affection between man and his scenario, there is a loss of interest, both aesthetic and functional, which is reflected negatively in the operation of the farm. What is left is a predominantly economical interest of the nonfarmer owner (an individual, a society, or even a government), who views the land only as a profit-making tool. Consequently, "symbiosis with the land is replaced with the criterion of self-interest and speculation" (Hausmann, 1964).

A study in China from 1987 to 1996 (Prosterman *et al.,* 1996) clearly revealed that the lack of the right to own arable land, or at least to have a sufficiently long-term concession for land use, was the main reason preventing the country from reaching alimentary self-sufficiency. In fact, farmers who do not own their land or are not granted use of the land are not interested in carrying out improvements which could increase crop yield. It has been suggested that China should adopt regulatory measures to implement two major objectives aimed at land ownership: long-term land investment and the cultivation of nonused lands. An additional goal which has been suggested is to prevent urban and industrial expansion encroaching arable land.

Regarding resources, some of the most important are seeds, fertilizers, and mechanical means. The success of the crops, the ease of burden from the most arduous farming tasks (tilling the land and harvesting), conquering more land per unit of human labor, and the speed of farming operations all greatly depend on employment of these resources. Purchasing power for these resources requires a favorable economic situation which will allow access to credit and financial incentives. The majority of farmers in developing countries can rarely meet these conditions; therefore, the poverty level limits any possibility of farm planning which can guarantee adequate productive levels (Murdoch, 1990). Currently, equal opportunities do not exist between farmers in rich and poor countries or between rich and poor farmers of the same country. This inequality in the access to resources is a source of social and environmental imbalance. The poorer farmers are forced to leave their livelihood and their lands, increasing local, regional, and continental migratory flows, which in the long term generate unsustainable conditions of urban migration and growing social, environmental, and sanitary tensions due to unemployment, crime, and a predisposition to disease.

Regarding the trade and use of seeds, a critical debate is currently raging over patenting genetically improved seeds (mostly the product of biotechnology) that seed-producing companies are promoting and trying to legally impose on farmers in all areas of the world. Since the advent of genetic improvement (protected by patent) by the seed-producing companies, farmers who were once themselves promoters of genetically improved production through selection and free to use and exchange their own improved seeds find that they must purchase their seed stock from the market. In practice, farmers are being increasingly deprived of their capacity of supplying for themselves the most important resource of all in farming—the seeds that germinate their crops and that they are now forced to buy.

The problem of access to agricultural resources is similar to that of marketing outlets for farm produce. States or groups of states have always had protectionist criteria to guarantee advantages within areas of protected interest. In general, trade barriers favor rich countries, which buy low-priced farm produce from poor nations and simultaneously sell their own technology at high prices (Murdoch, 1990). As a result, the economical diversity between rich and poor countries tends to increase rather than decrease.

Faced with this unfair social situation, it is easy to understand what a great regulatory effort must be made if we are to achieve equal opportunity, which would allow more balanced farming methods and agroecosystems in all areas of the world, in a spirit of solidarity and harmony among the peoples.

Sustainability

Sustainability is the most complex property of agroecosystems and concerns the capacity to maintain long-term productivity in relation to both internal and external planning conditions. Long-term productivity can in fact be maintained if the natural mechanisms which support it are not compromised, if the production processes do not pollute, and if the effects of the external cultural and socioeconomic forces do not damage the structure and functionality of the agroecosystems.

Because it embraces the different aspects of the multifunctionality of human activity systems, the concept of sustainability is used in the everyday sense to indicate the ideal development model that should be followed (i.e., sustainable or coevolutionary development). The ecosystem model, as a representation of working reality, identifies the development strategy found in nature (i.e., a strategy based on auto-organization, autocontrol, and autoevolution as a long-lasting development strategy or ecodevelopment) (Caporali, 1991). The fundamental principles on which ecodevelopment is based are the integral use of solar radiation to guarantee productivity and habitability, the recycling of matter, and progressive biological diversification. These three principles should also be considered when planning and implementing agroecosystems and, therefore, to achieve ecocompatible agriculture. Since the creative force of nature is based on the integrated use of natural resources, on the total recycling of matter, and on the promotion of biodiversity, these factors should also be applied to agroecosystems to improve efficiency, the degree of autonomy, and stability while simultaneously reducing the energy input responsible for negative environmental impact. Fundamentally, at the company level the goal is to construct agroecosystems which stress biophysical complexity, both within and outside the cultivated fields, more than do today's specialized farming systems, which tend to separate animal and plant production and favor monocultures and monosuccessions.

In practice, the planning of environmentally compatible agroecosystems should seek to achieve a unit of management (farm) in which the components are integrated and balanced so that all the main functions of an ecosystem are guaranteed. Therefore, an ecosystem should work like a solar energy plant capable of converting the sun's energy into biomass with the maximum continuity possible during the year; like a basin for collecting rainfall, which would favor infiltration rather than surface runoff and erosion; like a basin for biological fixation of atmospheric nitrogen through frequent and extensive growing of leguminous species in rotation and in association; like mixed farming systems, in which animal breeding and plant growing permit the maximum integration possible between the pasture and waste chains; and like an integrated biological community to promote prevention of infestation through biological control based on the diversity in and outside the cultivated fields.

Evaluation of Agroecosystem Performance in Terms of Sustainability

The sustainability of an agroecosystem refers to its capacity to achieve the main object for which it was designed (i.e., to support the trophic chain aimed at human nutrition). Therefore, production must reach sufficient quantitative and qualitative levels, which must be maintained on a temporal scale with a degree of self-sufficiency, to keep dependence on external sources of energy and matter as low as possible, to reduce management costs, and to avoid negative environmental impact. Production levels not only concern the output which can be extracted from the system but also the degree of input–output transformation (i.e., transformation efficiency). Sustainability of an agroecosystem is best when it manages to achieve a high yield but keep the intensity of factor use low (e.g., low external energy input per output unit). At the same time, the production mechanisms, such as soil fertility or the use of biological control against weeds and phytophagous and phytopathogenous agents, should remain intact. Some key aspects of production efficiency and autonomy (Caporali *et al.,* 1989; Azar *et al.,* 1996) can furnish adequate agroenvironmental performance indicators (APIs) to assess the properties of the agroecosystem, especially at the farm level. In this case, it is possible to combine input and output analysis at progressive levels of company organization to quantify the exchange of energy and matter as well as the monetary value between the internal components of the agroecosystem and between the agrosystem and the external world. In this manner, a series of APIs (indicators of dependence, reemployment, removal, autonomy, input yield, etc.) can be obtained, which can serve to evaluate the sustainability of the farm.

The acceptance of what is or what is not an agroecosystem today renders sustainability extremely precarious.

"Landless" livestock farms are a case in point: In these farms, many animals are reared in the minimum space, representing the maximum degree of agroecosystem simplification. Landless livestock rearing works without the support of farm-produced fodder (grazing chain) and without excretion recycling (waste chain) *in loco* but relies on the import of food from the market and placing excretion on the market for disposal. These agroecosystems function on "import" sustainability in the first case and "export" in the second. Their sustainability in practice revolves around the external cultural and economic conditions which allow their acceptance by society. Today, for example, many ethic concerns regarding the conditions of constraint and stress that the farm animals suffer and also sanitary concerns due to the risk of food chain contamination through the use of chemicals and animal feed, which can also be dangerous to humans, raise the issue of whether these breeding systems should be maintained.

Finally, sustainability of animal breeding in industrialized countries actually depends on the general tendency to include large amounts of cereal grain and legumes in animal feed for ruminants to improve transformation into milk and meat. Whereas ruminants can do without the grain due to their peculiar digestive system, which is capable of digesting cellulose contained in leaves, this is not true for monogastric organisms such as man, for whom grain-derived products are indispensable. In view of the increase in human population, this massive alimentary flow of grain cannot continue to be directed toward the ruminants, whose authentic alimentary compatibility is linked to the leaves in the fodder.

Orientation of Human Activity Systems toward Sustainable Development

For sustainable development of humanity, there must be a greater effort toward decentralized types of human activity systems that mostly exploit the direct flow of solar energy and rely least on the auxiliary flow of fossil fuels.

Regarding the employment of the ecosystems, it is important to define a point of equilibrium between the quota of the existing natural ecosystems that we should save from a change in use and the degree of exploitation intensity of the ecosystems undergoing the highest degree of change in their natural components (agroecosystems). Environmental compatibility must be defined in physical terms as derived from a state of substantial equilibrium of numerous local ecosystems which produce a state of equilibrium of the overall ecosystem of a country and the planet (Bresso, 1993). Since information input concerning better internal planning of ecosystems can always substitute external energy and matter input, institutions concerned with cultural formation and professional training (universities and schools) and with legal constitution (political and administrative bodies) are most responsible for the regulation of the use of ecosystems. The development of ecologically oriented cultural formation and legislation can constitute a powerful mechanism for informative feedback able to repair human activity systems and direct them toward ecocompatibility.

At the catchment basin level, the current ecosystem approach seems to be the best tool available to sustain planning activity, management, and environmental restoration aimed at sustainable development (Van Haveren *et al.*, 1997). If we are to achieve this ideal, the high-integrity habitats still remaining must be conserved, populations and habitats must be managed within the context of entire catchment areas, and the primary causes of habitat degradation must be identified and corrected. Management must work in harmony with natural processes so that in the future conditions will reflect the past distribution of the natural conditions. A network of natural refuges must be restored and protected throughout the territory, and future planning must take into account long-term benefits. These main principles should be enforced through a circle of monitoring activity and appropriate management in which data obtained from monitoring can furnish the necessary feedback to redirect adaptive action. Because the most important catchment basins in the world are transnational, any intervention on international catchment areas should be managed and coordinated through newly constituted international bodies (Bresso, 1993).

Conclusions

If constructed along the lines recommended to guarantee productivity, stability, equal opportunities, and sustainability, agroecosystems are the most suitable forms of land use to satisfy man's needs and at the same time ensure conservation of the environment. Consequently, farmers should be entrusted with not only the task of providing the basic foods and raw materials for humanity but also that of safeguarding the land as a common heritage of resources for future generations. This upgrading of their professional profile must benefit not only the farmers but also society as a whole. The farmer will gain greater social prestige (also through compensation) for his services as guardian of the land. At the same time, society will benefit from a more widespread and vital agricultural system over the territory, from a strong motivation for demographic decentralization, and also from the advantages derived from the use of solar energy and the recycling of material, land diversification, and creation of new jobs in agriculture, industry, and tourism. Finally, society will gain in social and environmental stability.

In an industrialized and densely populated nation such

TABLE 5

Requirements and Values of a Park

REQUIREMENT	VALUE
Landscape significance (results of a coevolutionary process between natural and human forces)	Political
Naturalistic significance (natural physical and biological characteristics)	Economical
Anthropological significance (archeological and historical–monumental characteristics)	Educational and recreational

as Italy, the areas most suitable for testing integrated forms of sustainable development are the national parks, whose natural characters merit conservation and exploitation. Their institution depends on their needs and values (Tables 5 and 6) and, through the active participation of the local population, should encourage the convergence of cultural socioeconomic forces toward a common objective of conservation and exploitation of resources for promoting man–habitat synergy.

TABLE 6

Principal Values of a Park

VALUE	DESCRIPTION
Political (social promotion)	Search for public consensus Decentralization of power of the territory Active participation by the population
Political (environmental promotion)	Internal management directed toward natural productivity Concern regarding interactions with the environment
Economical	Increase in employment Qualitative orientation of farming production Increased cultural and naturalistic tourism Development of local crafts Acquisition of financial resources (European Community, state, county, etc.)
Educational	Promotion of ecological culture oriented professional training
Recreational (environmental promotion)	Promotion of physical and psychological well-being

Today, national parks are as much an efficient legislative tool for governing a limited area as they are convincing test beds to verify the organizational capacity of the richer societies to take the direction of ecocompatibility. The results at the social and environmental level in the areas suitable for parks can be useful in providing the necessary experience for managing entire nations and indeed the whole world along similar lines.

References Cited

Ammassari, L., Palleschi, M. T., Caporali, F., Donnhauser, C., Quaglia, G., Nori, L., Bassoli, M., and Serino, G. (1997). *Il Sistema di Qualità nei Processi Produttivi Agroalimentari Ecocompatibili. Itinerari Formativi.* Franco Angeli, Rome.

Asner, G. P., Seastedt, T. R., and Towsend, A. R. (1997). The decoupling of terrestrial carbon and nitrogen cycles. *Bioscience* **47,** 226–234.

Azar, C., Holmberg, J., and Lindgren, K. (1996). Socio-ecological indicators for sustainability. *Ecol. Econ.* **18,** 89–112.

Bernetti, I., and Casini, L. (1995). Ambiente e benessere: Un approccio secondo le teorie delle scelte sociali di Sen. *Riv. Econ. Agraria* **50**(1), 3–40.

Bongaarts, J. (1995). Global and regional population projections to 2025. In *Population and Food in the Early Twenty-First Century: Meeting Future Food Demands of an Increasing Population* (N. Islam, Ed.), pp. 7–16. International Food Policy Research Institute, Washington, DC.

Bresso, M. (1993). *Per un' Economia Ecologica.* La Nuova Italia Scientifica, Rome.

Caporali, F. (1991). *Ecologia per l'Agricoltura: Teoria e Pratica.* Utet, Torino.

Caporali, F., and Onnis, A. (1992). Validity of rotation as an effective agroecological principle for a sustainable agriculture. *Agric. Ecosyst. Environ.* **41,** 101–113.

Caporali, F., Nannipieri, P., Paoletti, M. G., Onnis, A., and Tomei, P. E. (1989). Concepts to sustain a change in farm performance evaluation. *Agric. Ecosyst. Environ.* **27,** 579–595.

Capra, F. (1996). *The Web of Life: A New Synthesis of Mind and Matter.* Harper Collins, London.

Checkland, P. (1981). *Systems Thinking, Systems Practice.* Wiley, New York.

Conway, G. R. (1987). The properties of agroecosystems. *Agric. Syst.* **24,** 95–117.

Haussmann, G. (1964). *La Terra e l'Uomo: Saggio sui Principi di Agricoltura Generale.* Boringhieri, Torino.

Houghton, R. A. (1994). The worldwide extent of land-use change. *Bioscience* **44**(5), 305–313.

Jonas, H. (1993). *Il Principio Responsabilità: Un'Etica per la Civiltà Tecnologica.* Einaudi, Torino.

Murdoch, W. (1990). World hunger and population. In *Agroecology* (C. R. Carrol, H. Vandermeer, and P. Rosset, Eds.), pp. 3–20. McGraw–Hill, New York.

Norgaard, R. B. (1984). Coevolutionary development potential. *Land Econ.* **60,** 160–173.

Prosterman, R. L., Hanstad, T., and Ping, L. (1996). Can China feed itself? *Sci. Am.* **275,** 70–76.

SCHIMEL, D. S. (1995). Terrestrial ecosystems and the carbon cycle. *Global Change Biol.* **1,** 77–91.

TANSLEY, A. G. (1935). The use and abuse of vegetational concepts and terms. *Ecology* **16,** 284–307.

TURNER, D. P., WINJUM, J. K., KOLHUGINA, T. P., and CAIRNS, M. A. (1997). Accounting for biological and anthropogenic factors in national land-base carbon budget. *Ambio* **26**(4), 220–226.

VAN HAVEREN, B. P., WILLIAMS, J. E., PATTISON, M. L., and HAUGH, J. R. (1997). Restoring the ecological integrity of public lands. *J. Soil Water Conserv.* 226–231.

VITOUSEK, P. M., EHRLICH, P. R., EHRLICH, A. H., and MATSON, P. A. (1986). Human appropriation of the products of photosynthesis. *BioScience* **36,** 368–373.

WESTMAN, W. (1977). How much are nature's services worth. *Science* **197,** 960–963.

WRIGHT, D. H. (1990). Human impacts on energy flow through natural ecosystems, and implications for species endangerment. *Ambio* **19,** 189–194.

General References

BAWDEN, R. J., and ISON, R. L. (1992). The purposes of field-crop ecosystems: Social and economic aspects. In *Ecosystems of the World* (C. J. Pearson, Ed.), pp. 11–35. Elsevier, Amsterdam.

BRETH, S. A. (Ed.) (1996). *Integration of Sustainable Agriculture and Rural Development Issues in Agricultural Policy: Proceedings of the FAO.* Winrock International Institute for Agricultural Development, Arkansas.

CAPRA, F., and PAULI, G. (Eds.) (1995). *Steering Business toward Sustainability.* United Nations Univ. Press, New York.

CHAMBERS, R. (1983). *Rural Development. Putting the Last First.* Longman/Wiley, Harlow, UK/New York.

MATSON, P. A., PARTON, W. J., POWER, A. G., and SWIFT, M. G. (1997). Agricultural intensification and ecosystem properties. *Science* **277,** 504–509.

PIMENTEL, D., HUNTER, M. S., LAGRO, J. A., EFROYSOM, R. A., LANDERS, J. C., MERVIS, F. T., MCCARTHY, C. A., and BOYD, A. E. (1989). Benefits and risks of genetic engineering in agriculture. *BioScience* **39,** 606–613.

Herbert Sukopp
Institute of Ecology
Technische University Berlin
D 12165 Berlin, Germany

The Urban Environment

The study of the urban environment or urban ecology means investigating the living organisms in relation to their environment in towns and cities using scientific methods. In planning, urban ecology means "urban design" in the normative sense. Early studies of the urban environment shared the astonishing discovery that environments created by man provide habitats for characteristic species and that these species reoccur under similar conditions. Analyses have shown that urban areas include a wide variety of habitats, organisms, and communities. The variety of species found in urban areas often exceeds that in the urban hinterland. The altered soil, climate, and water conditions in cities have a corresponding impact on the composition of species occurring in urban areas. First, urban areas are heterogeneous, consisting of a variety of settlement structures, land uses, and small-scale habitats. This creates many specific and even unusual ecological conditions. Second, historically, species which have been introduced into an area through human activity, directly or indirectly, have begun their dispersal in urban areas and therefore occur there most frequently. With increasing settlement size, trade and traffic in and out of the city increase, and thus the proportion of nonnative species in the flora increases. Typically, cities show a mosaic of habitats with increasing degrees of human impact on a gradient from the outskirts to city centers representing different ages or time spans of human impacts. The close relationship between certain species and the urban environment means that these can be used as bioindicators for specific environmental factors. Many European states develop programs for nature and nature conservancy in towns and cities. The main principles of urban nature conservation are described.

Introduction

An increasing proportion of the world's human population is living in urban areas. Since 1700 the urban proportion of the world population has been increasing steadily. The United Nations forecast that by the Year 2025, 60% of the world's population will be living in urban areas, compared to 29% in 1950. The 50% mark will be reached between the years 2000 and 2010. In 2025, more than a dozen cities will have more than 20 million inhabitants each, and some will have more than 30 million. Twenty-three of the 25 largest urban conglomerations on the planet will be in Africa, Asia, and Latin America. Fig. 1 shows demographers projections for the evolution of the world and urban populations from 1950 to 2025. Although world population during this period will increase by a factor of 3.26, the number

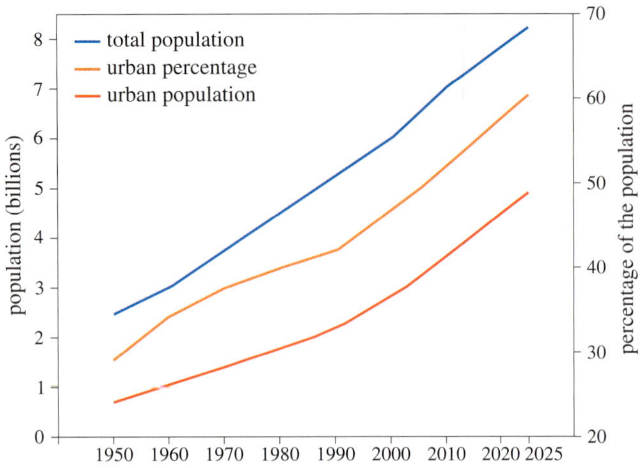

FIGURE 1 The growth of world total population and urban population from 1950 to 2025. In 1950, the urban population was 29.18% of the total population; between 2000 and 2010 it will pass the 50% level and, according to precise estimates, in 2025 will reach 60.1%.

of inhabitants in the world's cities will increase by a factor of 6.72.

Urban activities, such as building, traffic, or industrial production have strong impacts on the quality of air, water, and soil and influence the living conditions of humans, plants, and animals. For human well-being, contact with living parts of the environment is essential. This is the main reason why in urban areas a minimum number of plant and animal communities must be conserved.

Whereas a few decades ago the general public did not take much notice of the state of urban nature, today there is growing concern. This is illustrated, for example, by the "Green Paper on the Urban Environment" published by the European Community (1990) which stresses the importance of conserving urban nature. In Germany, the Federal Nature Conservation Law (Bundesnaturschutzgesetz) demands that nature must be protected, managed, and developed in populated as well as in unpopulated areas.

Historically, urban settlements are classified as preindustrial (with walls and gates; traders and crafts), industrial (extensive in rings and sectors), or postindustrial (with employment concentrated in the tertiary and quaternary sectors).

History of Human Influence on the Urban Environment

To understand current urban biota and their ecosystems, it is necessary to view them as a result of historic development. In south and central Europe, the process of redevelopment of a forest vegetation after the last ice age was not completed when human influence began to cause disturbances on a local scale. Large-scale disturbance, however, only began with clear-cutting of extensive areas for agriculture. At the same time, human impact on the hydrology of the landscape increased.

Historically, cities have fought nature back to create a cultural, artificial environment as opposed to the more natural environments prevailing outside towns and cities. Nevertheless, in many cities fields and gardens were included for many centuries (Fig. 2). Today, however, cities usually consist of a mixture of densely settled areas in the historic centers, remnants of agroecosystems ("encapsulated countryside"), and even near-natural areas in urban forests, parks, and nature reserves.

Open spaces in cities represent modifications of older sites. The similarity between former and present habitat conditions decreases with time and along a gradient from the periphery to the center. Historic phases of urban development, like the preindustrial, industrial, and postindus-

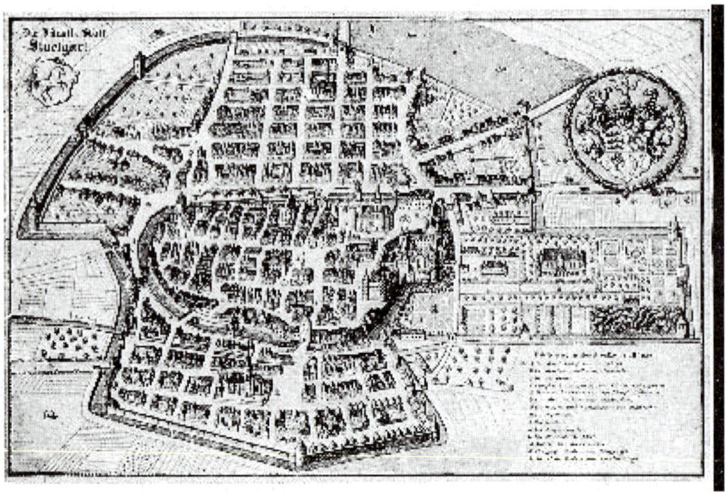

FIGURE 2 The city of Stuttgart, Germany, in the seventeenth century, in an engraving by M. Merian. The densely built area of the old town is surrounded by new districts with fields and gardens. (Right) At the fringe of the urban area is the prince's large pleasure garden (reproduced with permission from Hennesso, 1970).

trial phases, created specific site conditions and favored different plant and animal communities. Even today, parts of cities with different ages show different plant species and communities. In these, typical species of rural biotopes can often still be found. Also, living quarters of different ages differ in their species composition, as Aey (1990) found for medieval, turn-of-the-century, and post-World War II housing estates in Lubeck. Combined, urban ecology and landscape history may contribute to an enlarged basis for ecosystem research and its application in settlement areas.

Urban Ecology

In policy and planning, the term *urban ecology* is used for programs to develop sustainable cities. In science, however, urban ecology means investigating the living organisms in relation to their environment in towns and cities using scientific methods, in the same way as other branches of ecology study forests or the sea. In the scientific sense, ecology only describes analyses, without forming value judgments. In planning, the same word means "urban design" in the normative sense.

Once, it was thought that the ecology of urban areas was not worth studying. Cities were viewed as antagonistic to life. It was assumed that few plants or animals could survive in an urban setting and that urban animal and plant communities were random assemblages of species. Attempts to discover patterns or the reasons for such patterns were thought to be futile. This view began to change more than 25 years ago. Since then, the ecological aspects of cities have been intensively researched. It became obvious that urban communities were not merely coincidental.

The beginnings of urban ecology can be found in investigations of the flora (Deakin, 1855; Nylander, 1866) and fauna of urban areas and their occurrence and as part of the "natural history" tradition. In 1823, Schouw used the expression *plantae urbanae* for plants which occurred near villages and towns and stated, "In most cases, these plants are of foreign origin, which explains why they are only found near towns and villages." Early studies of ecology in cities shared the astonishing discovery that environments created by humans provide habitats for characteristic species and that these species reoccur under similar conditions. Analyses have shown that urban areas include a wide variety of habitats, organisms, and communities. The variety of species found in urban areas often exceeds that in the urban hinterland.

With increasing suburbanization, it has become increasingly difficult to separate analysis of urban areas from that of the hinterland. The word "hinterland" is used here to refer to areas which are functionally connected with the city but they cannot be precisely defined in space. Relationships between towns and their hinterlands involve economic, cultural, and ecological aspects. The immediate hinterland, or suburban area, must also be differentiated from the distant hinterland.

We are living a global lifestyle, and our cities are connected, in an ecological sense, to ecological systems worldwide, not just to systems in the immediate hinterland. Today, urban areas are provided with food, raw materials, and natural resources from a distant hinterland which stretches across the globe. A dense transport network involving land, water, and air makes it possible to transport huge quantities of goods and resources (Figs. 3 and 4).

Research on urban areas has always meant research on large cities. This applies to all the disciplines which investigate specifically urban topics. Social problems and health concerns are the main reason for research to be carried out.

The City as a New Type of Environment

Urban ecosystems differ from nonurban ones in many ways. Although most of the factors which affect urban ecosystems also operate in nonurban areas, the combination of these factors means that unique ecosystems develop with species combinations peculiar to urban areas.

Construction and business require a division of the city into zones of densely built-up and loosely built-up areas. Small gardens and rubbish dumps, heaps of ruins and rubble, as well as vast fields are typical of the inner border zone, whereas forest areas and large parks are typical of the outer border zone (Table 1). Air pollution and warming, changes in the groundwater level, and backfills of great expanse are the consequences of urban construction and business (Fig. 5). Because the volume of import of construction materials, raw materials for finished products, and foodstuffs is greater than that of the waste materials which have been taken away, in the course of time the ground level of the city has risen. A eutrophication of many locations and compaction or sealing of the soil are linked with the extent of the depth of the cultivated layer. Eutrophication through urban waste concerns not only the waste dumps and sewage fields (which are irrigated with wastewaters in order to clarify the latter biologically) but also almost all expanses of water, and it influences the composition of species of plant animal communities.

Urban Climate

Cities have particular climatic characteristics (Landsberg, 1981; Kuttler, 1993; Horbert *et al.,* 1983), such as the following:

Higher air pollution: Gaseous pollution is 5–25 times more, and condensation nuclei are about 10 times higher.

Altered radiation: There is 5–15% less hours of sunshine, 20–25% less direct solar radiation, approximately 10% less surface albedo, and 12% more atmosphere back-

FIGURE 3 Import of food resources into the European Union in millions of tons (adapted from Bundesministerium für Ernährung, Landwirtschaft und Forsten 1984; data of the Statistics Office of the European Community and of the Statistischen Bundesamtes).

radiation leading to an increased net radiation of 11% at noon or 47% in the evening.

The relative humidity is between 2% (winter) and 10% (summer) lower; on clear days this difference can be 30%.

Wind speeds are reduced by 10–20% due to the roughness of the city surface; calm days occur 5–20% more often.

Annual mean precipitation is increased by up to 20%.

The ecologically most important result of these effects is increased temperature. Cities are "heat islands" or "hot spots" on the surface of the earth. The difference depends on the size of the city: It can reach 9°C on clear days; there can be a difference of 0.5–1.5°C in the yearly mean temperature.

Climatic conditions within a city can vary considerably, depending on type of construction, paving, location in the city, and, especially, the distance to large vegetated areas. According to these changes, different climatic zones, usually more or less concentric, can be distinguished. The city heat island usually covers the same area as the built-up area, but changes in wind direction can result in other areas also being temporarily heated. Air pollution is caused by traffic, heating of buildings, power stations, and industry. Trace gases, excepting low-level ozone, occur at concentrations 5–50 times higher than usual.

Urban Soils and Waters

Two soil types are characteristic of urban areas: deeply cultivated garden soils with high nutrient content and high water-holding capacity and soils formed from rubble (e.g., from buildings demolished during war). The latter are alkaline, dry, and well aerated in the beginning but develop with humus accumulation. They are usually low in nitrogen but have moderate to high contents of phosphorus, calcium, potassium, and other nutrients (Blume, 1993).

In urban soils, extremely high amounts of phosphate, borate, and sulfate are found (5, 10, and 500 times the amounts in natural soils, respectively). Phosphate and borate especially occur in highly eutrophic soils developed

FIGURE 4 Importation of mineral resources into Germany, where they undergo physical and chemical processes and are transformed in large part into waste material. The figure does not mirror the actual situation, which changes according to economic interests, but simply shows the transfer of raw materials, in some cases for decades, which continue to make their effects felt (adapted from Deutsche Forschungsgemeinschaft, Bonn).

from sewage sludge and mud. Sulfate is found in substratums in which technogenic processes are in progress (e.g., coking plant deposits, ashes, sewage sludge, municipal waste, and mud).

The total contents of heavy metals Cd, Cu, Ni, Pb, and Zn in urban soils cover a wide range. Brick-and-mortar debris, municipal waste, and sewage sludge are especially rich in Cu, Pb, and Zn, whereas the Cd and Ni loads are low in all soils.

Regarding organic matter contents, soils developed from anthropogenic substratums differ widely from those developed from natural parent material. Developed soils have low contents of litter compounds, especially of easily decomposable litter components (proteins, hemicellulose, and cellulose). Organic chemicals in urban soils include atracene, 2,4-D, and pentachlorphenole. Urban soils often contain amounts of polycyclic aromatic polycarbons (PAHs) that exceed the critical loads (intervention values). Brick-and-mortar debris and municipal waste are potentially highly hazardous, whereas other soils developed from ashes and mud are less hazardous. Coking plant deposits are characterized by high contents of PAHs with a predominance of the more rapidly soluble and transferable congeners, high contents of cyanide, and extremely high contents of sulfuric acid; thus, they are classified as potentially extremely hazardous.

The properties of urban soils, their distribution patterns and assemblage, as well as the intensity of pedogenesis differ significantly from those of soils in natural landscapes. Because the exploration of urban ecosystems is a comparably young branch of science, the discoveries made so far cannot be regarded as valid knowledge (Blume and Schleuss, 1997). Material inputs into cities have exceeded outputs for centuries (Mumford, 1961). Therefore, the ground level in older districts has risen by several meters. Surface sealing and the resulting reduced infiltration capacity, as well as the extraction of groundwater for human usage, further reduced the groundwater level. Therefore,

TABLE I

Characteristics of the Flora and Vegetation in Zones 1–4 of Berlin

	CITY CENTER		OUTSKIRTS	
	Zone 1 (built-up)	Zone 2 (partially built-up)	Zone 3 (more central)	Zone 4 (more external)
Vegetation-covered surface (%)	32	55	75	95
Vascular plant species/km^2	380	424	415	357
Rare species/km^2	17	23	35	58
Hemerochores (%)	49.8	46.9	43.4	28.5
Archeophytes (%)	15.2	14.1	14.5	10.2
Neophytes (%)	23.7	23.0	21.5	15.6
Woods (%)	17.0	21.5	17.9	35.2
Dry prairie species (%)	7.5	8.2	8.9	11.6
Fertile grassland and pasture species (%)	15.2	15.1	15.2	13.7
Marshy ground and bog species (%)	3.1	1.8	5.5	6.9
Ruin and wall species (%)	24.1	19.4	22.2	13.4
Herbaceous weed species (%)	15.3	14.8	14.5	7.6

urban soils have become considerably dryer over the centuries, except in gardens and parks, which have become moister through extensive watering. It is worth mentioning that the rise in ground level means that some urban soils are totally dissociated from the underlying substrates. Of course, this affects the plants and animals. Urban soils are generally eutrophicated and overcompacted. Soils in industrial areas, under waste deposits, on filtration fields (land onto which wastewater was directed as a form of purification), and on roadsides are in some cases heavily polluted. The extent of the "cultural layer" is related to the eutrophication of many areas and the compacting or loosening of soils within the settlement. Eutrophication through urban waste affects not only rubbish deposits and irrigation fields but also all the water courses in the areas and the composition of terrestrial and aquatic communities.

The temperature of rivers that flow through heavily industrialized and urbanized areas can be increased 5°C or more (Detwyler and Marcus, 1972), and these rivers acquire large amounts of sewage. Lester (1975) found a proportion of river flow to effluents of 5:1, which during dry periods decreased to as low as 2.5:1.

The most common steps in purification of sewage are mechanical and biological treatment, but only a third step of chemical purification can sufficiently reduce the concentration of organic substances to prevent eutrophication. In many cities, the general drainage system is a combined system, carrying sewage as well as run-off. If this general drainage system is connected to the water purification plants, this system has the advantage that both types of drainage are purified, whereas with separate systems only the sewage is treated. The combined system, however, has practical disadvantages. During times of heavy rainfall, which are even more severe in city regions, the flood locks regularly open and sewage and rainwater are channeled untreated into rivers and lakes. Especially in summer, this may cause breakdowns (die-off of fishes) in subsections of the freshwater ecosystems.

Flora, Fauna, and Biotic Communities

Urban Flora and Fauna

The altered climate, soil, and water conditions in cities have a corresponding impact on the composition of species occurring in urban areas (Wittig, 1991; Klausnitzer, 1993). These changes are greater the larger the urban area and the closer one gets to the city center. Comparisons between urban and rural areas and between different cities and city edge/inner-city gradient studies have been established as methods of analyzing these differences.

The number of ferns and flowering plant species per unit area is higher in cities with more than 50,000 inhabitants than in the surrounding area. In central Europe, the number of ferns and flowering plant species correlates closely with population size and/or density. In small and medium-sized towns, between 530 and 560 species are usually found; 650–730 species are found in cities with 100,000–200,000 inhabitants, 900–1000 species are found in older cities with populations of between 250,000 and 400,000, and in cities

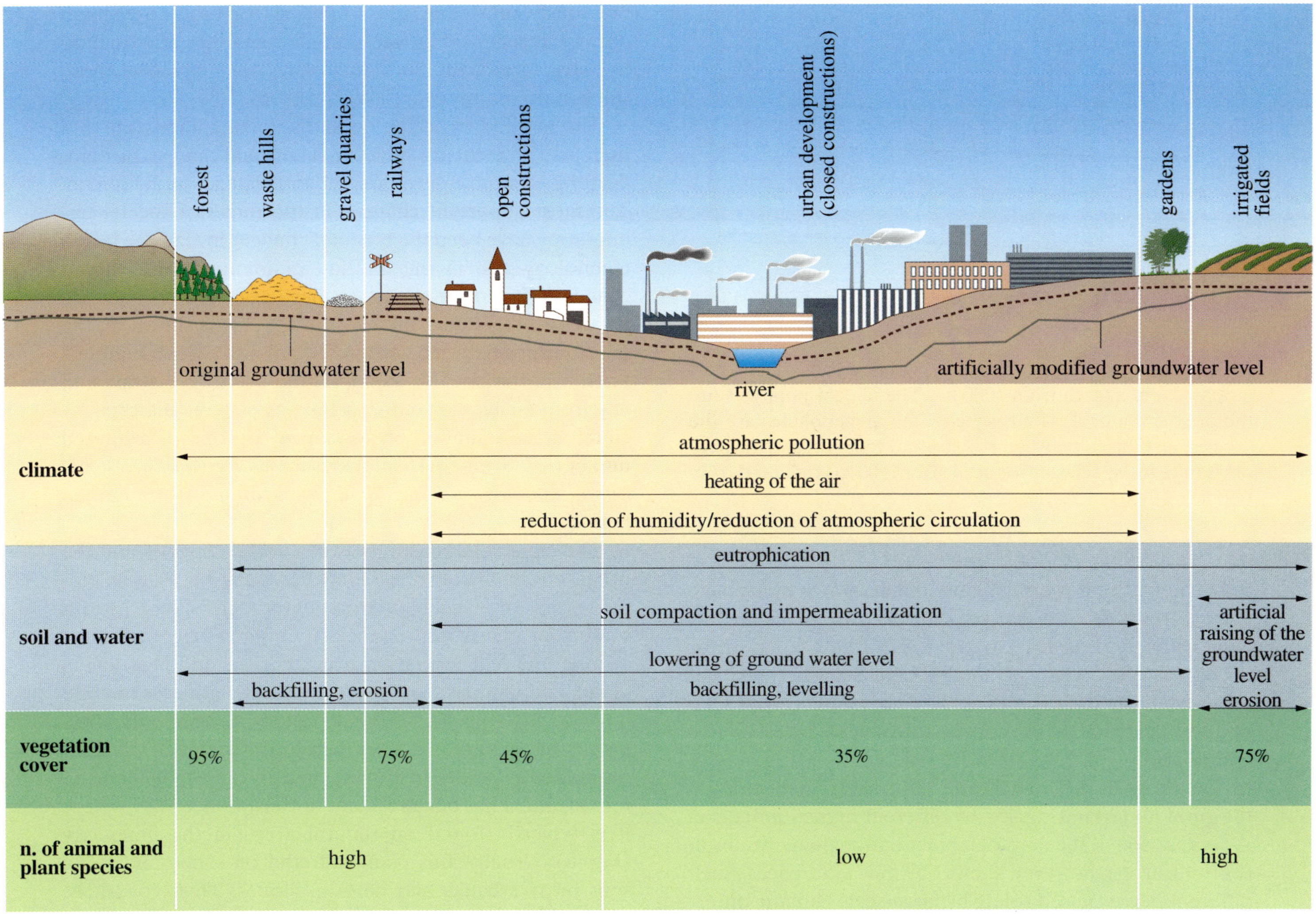

FIGURE 5 Changes in the biosphere caused by a large city. The arrows indicate the radius of action of the single changes. Heating of the air, reduction of humidity, and compaction and impermeabilization of the soil affect mainly the urban areas of open and closed construction. Atmospheric pollution extends from the forests to the cultivated fields; eutrophication extends from the waste dumps to the irrigated fields; and lowering of the groundwater level affects a band between the forests and the peripheral gardens, whereas backfilling and soil erosion affect only the extraurban areas. (Bottom) The percentages of vegetation cover and the richness of the animal and plant species are shown (adapted from Sukopp, 1990).

with more than 1 million inhabitants the number of species usually exceeds 1300. There are various reasons for this. First, urban areas are heterogeneous, consisting of a variety of settlement structures, land uses, and small-scale habitats. This creates many specific and even unusual ecological conditions. Second, historically, species which have been introduced into an area through human activity have begun their dispersal in urban areas and therefore occur there most frequently. With increasing settlement size, trade and traffic in and out of the city increases, and thus the proportion of nonnative species in the flora increases (Fig. 6, Table 2). The increase in the number of species by immigration is mainly due to man's activity either directly, as in the case of ornamental plants, or indirectly (e.g., when impurities get into the transported materials or seeds). These species are called *hemerochories* or *anthropophytes.*

Today, the number of naturalized plants in cities by far outnumbers that of extinct species (Landolt, 1992), resulting in a high number of species in urban areas, as mentioned previously. A main question is whether the process of man-made species migrations has passed beyond its zenith. If not, an enormous increase in species richness in some areas, in particular urban ones, seems to be possible in the future. However, we do not know enough about the mechanisms of invasion processes to make even approximate prognoses (Trepl and Sukopp, 1993). The num-

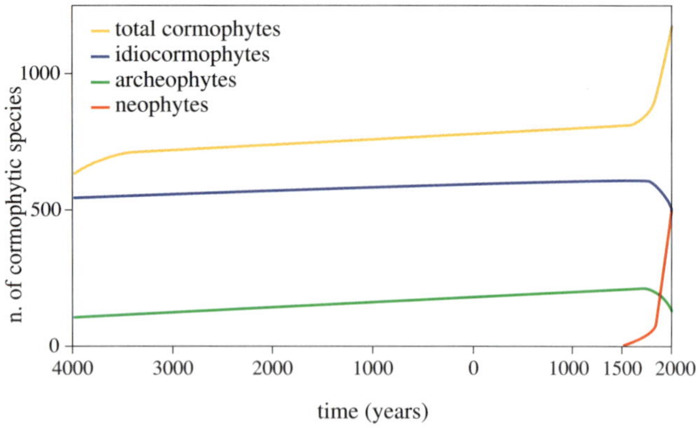

FIGURE 6 Changes in the numbers of flowering plants (cormophytes) in Zurich from 4000 BC. The idiocormophytes are the original autochthonous species, the archeophytes are the species introduced between 5000 BC and 1500 AD, and the neophytes are the species introduced after 1500 AD (adapted from Landolt, 1992).

ber of nonnative flowering plants and diaspores which have been introduced into central European urban areas has been estimated. At least 10,000 species have been introduced. However, when the number of plants which have become established (385) is taken into account, it can be seen that less than 5% of the species introduced have become naturalized.

The urban flora is remarkably different from the flora of the rural hinterland caused by different human influence. A comparison of the flora of large urban agglomerations in central Europe, however, shows only few differences. Such differences are caused solely by large-scale climatic differences. They increase with the distance of urban agglomerations from each other. However, the differences are still astonishingly minor even between cities as distant as Warsaw and Brussels. A connection of the type of flora to the previously prevailing soil conditions in urban areas can only be recognized by very detailed analyses; the qualitative structure of the flora, however, does not reflect those soil conditions any more (Wittig, 1996).

The method by which nonnative plants immigrate and their actual extent are subject to continual change. Both reflect the economic development and changes in the region. The most important changes in the range of species immigrating have been the result of changes in transportation technology and in the method of immigration. Habitat changes affecting immigration may have been of lesser importance (Trepl and Sukopp, 1993). This is shown by the relationships between certain species and certain methods of transportation. For example, the use of rye straw and hay from Sicily and southern Italy as packing material facilitated the immigration to central Europe of a limited number of "fruit companions" such as *Reseda lutea, Chenopodium vulvaria,* and *Plantago coronopus* for a limited period.

The close relationship between certain species and the urban environment means that these can be used as bioindicators for specific environmental factors. The growth season for plants in the inner-city area is extended, blossoming and leaf formation occur earlier, and heat-loving plants and animals from more southerly latitudes migrate into the city (Fig. 7). Sticky goosefoot, *Chenopodium botrys,* is an example of a temperature-dependent plant. It originated in southern Eurasia and in the Mediterranean region, and it has been resident in Berlin since 1889, where it is typically found on ruderal areas in the inner city (Fig. 8). Usually, this plant is found on sandy and stony soils on riverbanks and on talus slopes. Thus, roadsides, wine-growing areas, and derelict sites became secondary habitats. Human activity in central Europe has increased the number of these open, chalky, sandy to gravelly locations. However, major long-term occurrence of this species can only be found north of the Alps in areas with warm summers, such as in the upper Rhine floodplain, in Berlin and the Ruhr area, and on the burning spoil heaps in Lille.

In addition to numerous nonnative species, many native organisms also profit from urban habitats (Fig. 9), especially if the conditions are similar to those of their original habitats. In the natural landscape, these plants and animals may originally have occurred on tree-free sites such as on gravel or mud banks along rivers and streams, in open areas within forests, on the alpine pastures of large mammals, or on ground disturbed by fire, earthquakes, landslides and avalanches, volcanic activity, and coastal erosion accretion. Because most large cities are located on large rivers, erosion and sedimentation areas along the rivers act as important initial habitats for urban apophytes (Fig. 10).

However, environmental changes and species transport cannot fully explain the changes in the composition of

TABLE 2

Number of Cormophytes and Higher Plants in the City of Zurich

	TOTAL SPECIES	DIS-APPEARED SPECIES	NEW SPECIES
5000 BC	650	—	—
Late Neolithic	750	—	100
Medieval	850	20	120
1839	950	20	120
1905	1050	30	150
1990	1100	70	150

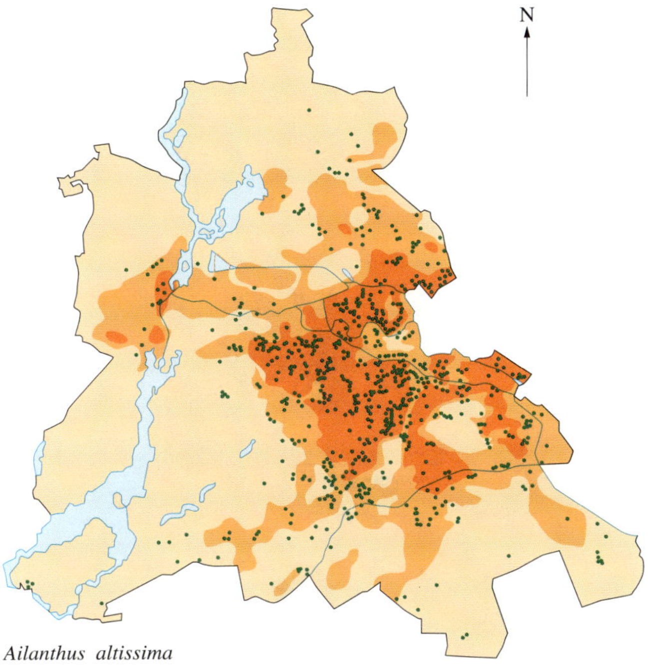

FIGURE 7 Distribution of Tree of Heaven, *Ailanthus altissima,* in West Berlin (green dots) in relation to the different temperature zones. The areas colored from yellow to red correspond to zones with increasing temperature. This plant, originally from Asia, tends to grow in the warmer zones (adapted from Kowarik and Böcker, 1984).

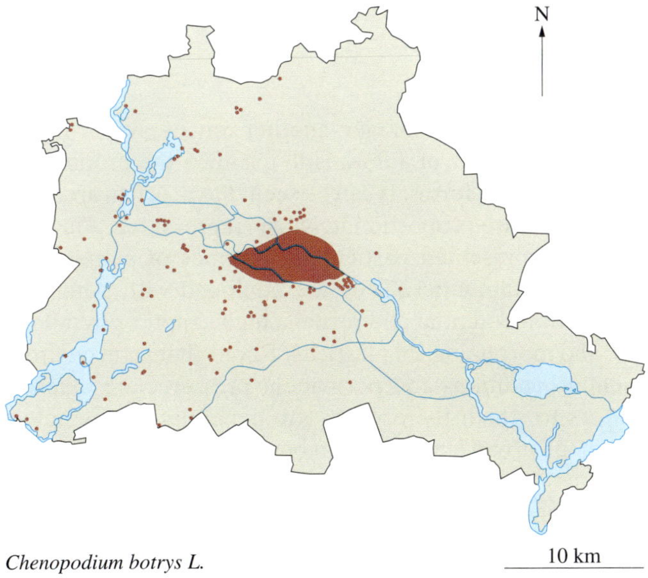

FIGURE 8 Distribution of *Chenopodium botrys* in Berlin. The red areas indicate the zones in which the plant is diffuse, whereas the single red dots show the single locations in which it was found (adapted from Sukopp, 1971).

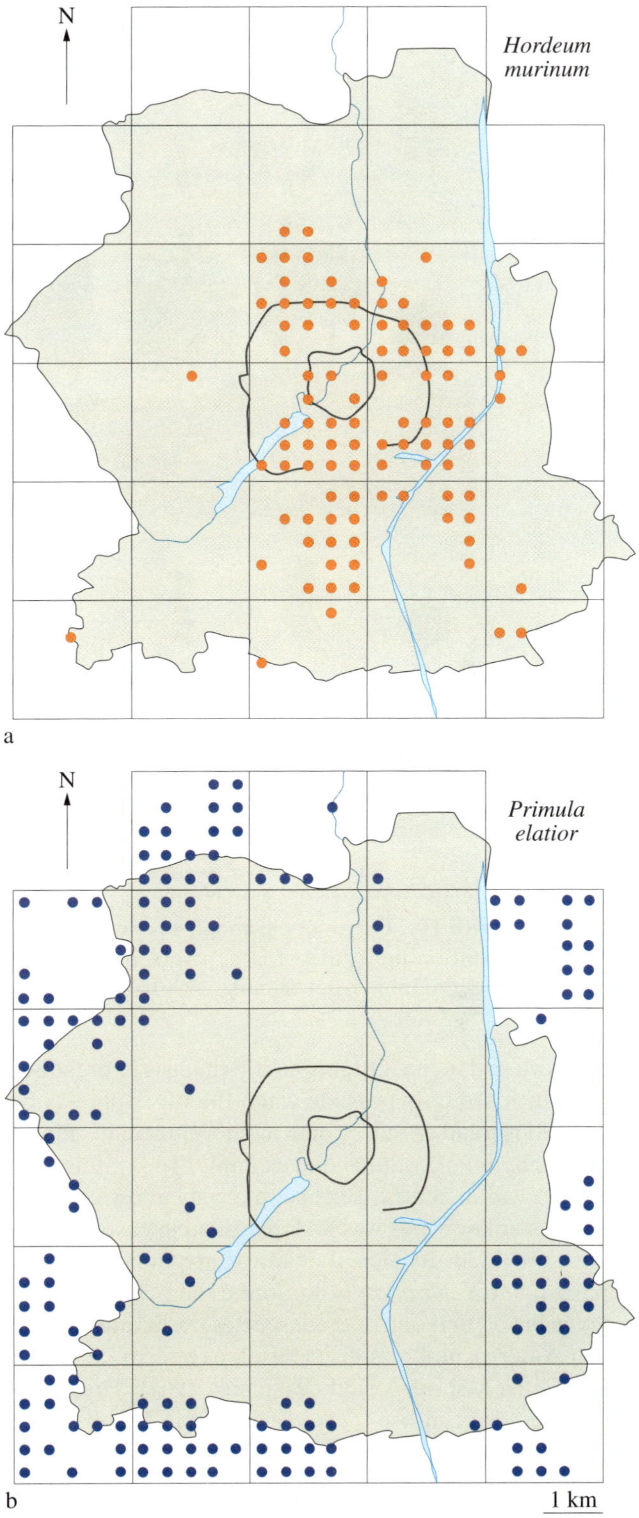

FIGURE 9 Distribution pattern in the city of Münster of *Hordeum murinum,* an esotic species confined to the inner city (a), and of *Primula elatior,* a native species confined to the outskirts (b) (adapted from Wittig *et al.,* 1985).

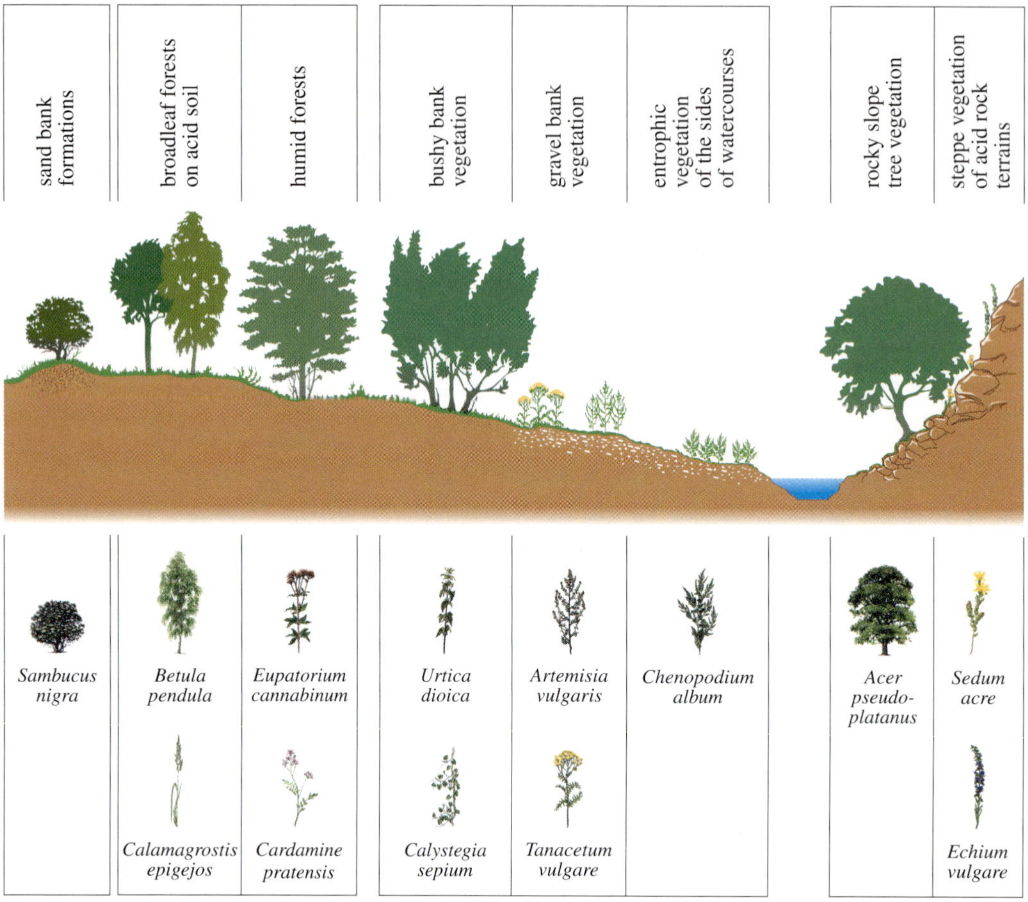

FIGURE 10 Origin of the apophytes (indicated below) present in urban flora. Since the majority of large cities are built on the banks of large rivers, many urban apophytes come from natural environments of the watercourses (adapted from Schmitz, 1996).

urban flora and fauna. Often, genetic changes in organisms occur which are only possible when the organism has become widespread in cities and in the immediate hinterland. *Xanthium albinum* is one example. Today, this plant is widespread in central Europe, but only a few centuries ago it was unknown. It was first identified in eastern Brandenburg in 1830. Its closest relatives are found in North America, but *X. albinum* is not found there. This leads to the conclusion that one of these species was imported from North America and established itself as a separate species in an relatively short period (Wagenitz, 1964). The spread of the evening primrose (*Oenothera* spp.) in Europe seems to have followed a similar pattern. Today, more than 15 species have been identified, few of which are identical to their North American ancestors. These European species developed over a period of 350 years after the first importation of the American parent species to Europe. Both of these examples show that when conditions permit, new species can develop over a short period of a few centuries. New species and subspecies have developed over historical timescales, not just over geological ones.

According to Raunkiaer, another way of measuring the ecological quality of a flora is to examine the proportions of different life-forms. It can be seen that in urban areas annual plants are proportionately overrepresented. This life-form is represented particularly in types of pioneer and ruderal communities. They are equipped with a high reproduction potential and reveal a high genetic, physiological, and morphological plasticity. Rapid change in environmental conditions and permanent pressures are external factors to which therophytes are generally well adapted. Thus, the urban landscape can be distinguished by its rapid changes which favor the spread of r strategists, a category that includes most neophytes. K strategists, which depend on a steady system of environmental relations, subsequently decline under urbanization. This affects the indigenous flora in particular. Regarding the success of the ruderal flora in cities, it should be pointed out that the former natural habitats for ruderal flora, such as riverbanks that were periodically destroyed under natural conditions, also experienced rapid changes. Thus, ruderal species are well adapted to such conditions.

A additional essential aspect is the propagation of pioneer and ruderal species. The majority of them are spread by wind and therefore can reach isolated open areas, such as inner-urban green areas. The low constancy in urban biotopes is well illustrated by the high fluctuation in the floristic composition. This fluctuation is also brought about by the large proportion of ephemerophytes, which are species that appear for a short time.

The dependence on human settlements varies among animals. Some species have not been found outside city areas. Others are found in such high concentrations that they can be described as typical urban species, including street pigeons, house sparrows, crested larks, and swifts. Species which usually live on rocks and boulders in mountain areas often occur in urban areas, where buildings act as "rock faces." Species which usually live in caves and species of hollow trees are also found in cellars and other rooms and in roof spaces.

Total numbers of animal species in urban habitats show similar trends to those of plants. At least in many groups of invertebrates, birds, and mammals (except large carnivores), species numbers in cities are higher than in adjacent nonurban areas. The highest species numbers are found in areas of the outer and inner suburbs; in city centers and new housing developments they are considerably lower (Gilbert, 1991; Klausnitzer, 1993). Many of the nonnative species of animals and fungi which prefer urban habitats originate from warmer regions. Behavioral changes take place as a response to the altered conditions in urban areas; for example, the mating season of house sparrows takes place earlier as a result of artificial lighting.

Since different plant and animal species show varied adaptability to the human living space, methods were developed to evaluate this adaptability. The most widespread is probably the concept of a classification according to synanthropy (Povolny, 1971), which indicates the relative dependence of various organisms on humans. Animals which are dependent on the immediate human living space are designated as *eusynanthropic* or *obligatorily synanthropic,* whereas species which appear both in the city and in the surrounding country have a low or facultative synanthropy.

A summary of the ecological requirements of the urban wild flora can be attained by carrying out an analysis of indicator indices such as those of Ellenberg (1979), in which an average of all the indices is taken for all species. Kunick (1982) compared many cities. The analysis shows that ecologically large quantities of light and high soil reaction values, as well as a certain aridity tolerance, seem most important. There is also a slight tendency toward a relatively high demand for nitrogen (Fig. 11).

Specific urban–industrial environmental effects are often investigated using lichens on tree bark (epiphytes). The distribution pattern of various lichen species is examined because they differ in their sensitivity to air pollution

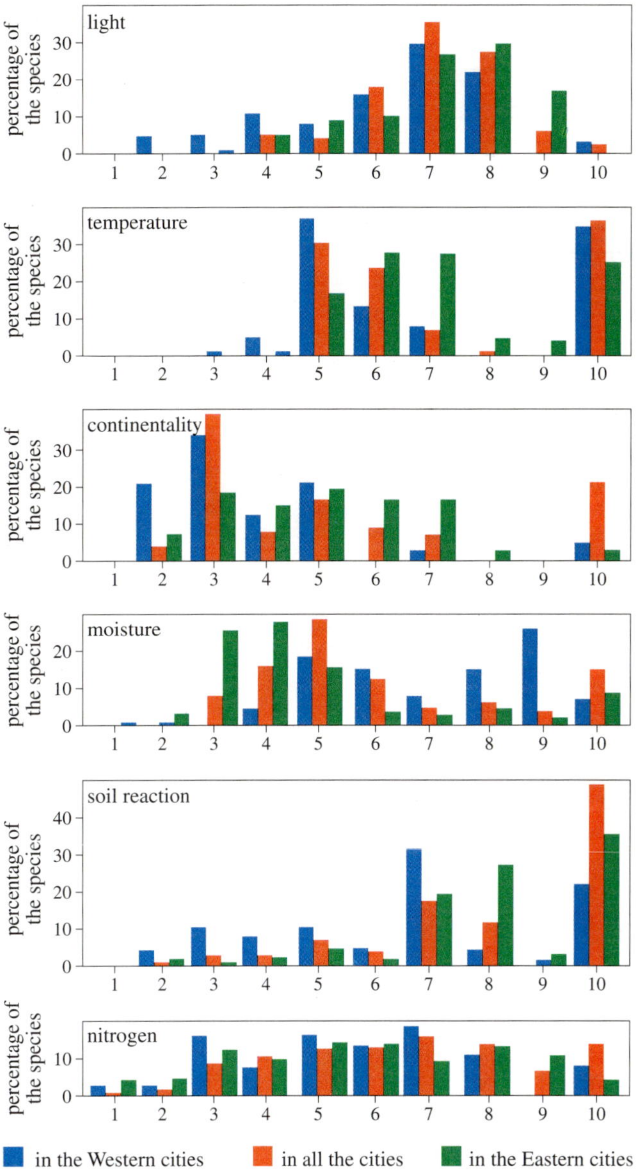

FIGURE 11 Ecological demands of the urban wild flora: 2061 species from nine European cities were analyzed using the indicator indices of Ellenberg, which allow the quantification (with values from 0 to 10) of the optimum of the response of the different species to six environmental parameters. High values of light, soil reaction, and nitrogen are shown to be particularly important (adapted from Kunick, 1982).

(Fig. 12). A lichen-free-zone in the city center is usually followed by a zone of transition, in which lichen growth markedly increases toward the city edge. The "normal zone" begins where the number of species, their combination, and their extent correspond with the conditions in the hinterland. In addition to lichens, grasses and herbaceous plants are also used as bioindicators (Steubing and Schwantes, 1992). For trees, early ecophysiological measurements of

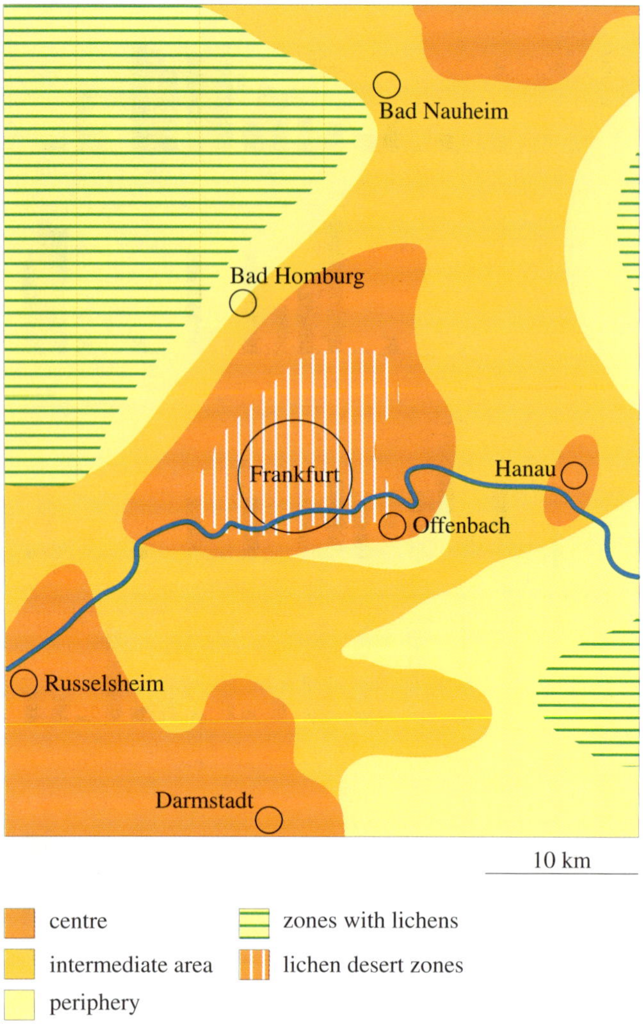

FIGURE 12 Distribution of various lichen species around Frankfurt on Main. The lichens, almost absent in the urban areas, increase progressively toward the periphery (adapted from Stuebing, 1978).

gas exchange and pigments indicate stress and damage (Lichtenthaler, 1996), showing the state of health and thus allowing help for ill trees.

Biotic Communities

Urban ecosystems differ from those of their hinterlands with regard to resource and energy budgets. In the middle latitudes, urban energy use in oil, coal, gas, electricity, and food usually amounts to 25–50% of the natural energy use. In addition, the use of solar energy for the biosynthesis of organic compounds is very limited because the biomass of the primary producers is low in urban areas. Green plants do not form the "energy base" that they do in natural ecosystems. The consumers — humans and many individuals of a small number of domestic and wild animal species — must be provided for by the import of energy in the form of resources from the hinterlands. Decomposer organisms play a very limited role. Flows of materials rarely form complete circles; therefore, both urban ecosystems and those in the hinterlands are burdened with high levels of waste. Despite the rapid changes in flora and fauna, habitat-related combinations of organisms are formed comprising species which originated in differing habitats. An outline of urban plant communities is given by Wittig (1991).

The most important characteristic of urban ecosystems seems to be their low degree of integration. They are nonequilibrium systems, and stochastic processes predominate over deterministic. Successions in urban biocoenoses — which are, compared with nonurban ones, subject to strong and extremely variable anthropogenic influences — are strongly linked to site history. Hence, these successions are not deterministically directed toward a predetermined state; they are dominated by chance, unpredictable, and not repeatable. No reasonable approximation to climax conditions exists. The initial species composition remains relevant for further development.

A major reason for the (relative) unpredictability of successions in urban ecosystems is the high degree to which they are subject to invasions of "aliens"; the biogeographical spectrum of species composition of cities is very different from that of the surrounding countryside. The main cause may be (i) the conditions of naturalization (i.e., in the high invasibility of the ecosystem's biocoenoses) or (ii) the conditions of dispersal (introduction and transportation). Disturbances generally increase invasibility and urban ecosystems are disturbed ones. Towns are open to invasions of alien species, and their number is unpredictable (Trepl, 1994).

Urban biocoenoses are an extreme example of communities produced by successive invasions and not by coevolutionary development. In principle, the historic uniqueness of urban situations (i.e., of combinations of environmental factors and organisms) differentiates urban ecosystems from most natural ones, even those subject to strong disturbance.

The number of animal species depends to a high degree on the number of plant species. However, the real number of plant species remains far below the number which would be possible under the climatic conditions. This is due not only to the dispersal barriers but also to interactions, especially competition. For cities, less dispersion barriers exist (cities are centers of the transportation system) and the number of plant species is hardly limited by competition: Wild-growing aliens and especially cultivated crops and ornamental plants considerably increase the number of plant species which are (potentially) relevant for animals. Therefore, there is reason to assume that there will be a considerable future increase in the number of species in cities. It will presumably exceed by far the numbers expected outside settlements as a consequence of the worldwide mixing of species (Trepl, 1994).

The Distribution of Urban Habitats

Although whole cities can be viewed as large ecosystems, especially with regard to energy and matter fluxes, structurally and functionally they form complexes of various interconnected ecosystems (Sukopp and Wittig, 1993; Rebele, 1994). Urban ecosystems and the composition of urban plant and animal communities are greatly dependent on human activities. The extent of these impacts varies in time and space. Typically, cities show a mosaic of habitats with increasing degrees of human impact on a gradient from the outskirts to city centers representing different ages or time spans of human impacts. A common way to study the specific urban biota and their habitats is by investigation of their history (Aey, 1990) and by comparison along rural–urban gradients (McDonnell and Pickett, 1990). Some features of habitats vary clinally with distance from city centers to the suburbs or with time. The organisms and communities in these habitats react to human influences in various ways and are consequently different for each structural unit of the city. The distribution of habitat types in cities is an important prerequisite for nature protection and town planning. This information is widely available for many European cities (Starfinger and Sukopp, 1994).

Urban Forests

The trees growing on streets first come to mind when one reflects on the question of trees in the city and reads the corresponding literature (Meyer, 1982). They are subject to extreme stress and their often sickly appearance is noticed by many city residents. Major disturbances include earth movements during the construction of streets, soil compaction due to trampling and vehicles, eutrophication and rise of pH values, as well as mowing and herbicide application in existing streets. Salinity is a major factor in the soils of urban streets due to de-icing salt, which is widely used in city streets in climates with cold winters. Today, its use is restricted in many European countries.

Trees are exerting positive functions for climate and air hygiene in urban environments. However, long-living trees have to be adapted to urban conditions for many years (Meyer, 1982; Brod, 1991): emissions from traffic, industry, and heating, water stress caused by sealing and runoff, urban soil conditions, high temperature, and wind. Trees are sinks for harmful substances by filtering and uptake via roots and leaves. These strains influence the functions of trees, leading to diminishing of growth, susceptibility to climatic changes, disease, and even death.

Disasters such as the epidemic of Dutch elm disease and the death of trees from the use of de-icing salt have made an awareness of urban forests particularly pertinent. Inventories of tree stocks have been drawn up in many cities and lists of tree species developed which classify plants according to their resistance to various pollutants (e.g., salt resistance and SO_2 resistance) and their general suitability for the city. Recently, studies of tree health have been made with the help of aerial photographs using infrared false color.

The influences of inner-city areas of vegetation on the urban climate were investigated by Stülpnagel (1987, 1990). He found a reduction in temperature not only in a green area but also up to 1.5 km away from it. This climatic the dust load of the air. Jonas *et al.* (1985) studied this ability of different tree species and compared it to that of other vegetation. If trees are selected for planting, however, it has to be taken into account that they are harmed by deposits of harmful substances.

Even in areas adjacent to cities, and more so in cities, forests are disturbed by urban activities, such as building of houses and roads, recreation, and emissions. In a comparison of urban and rural forest stands in southeastern Wisconsin, Sharpe *et al.* (1986) found urban sites disturbed by houses and yards, dumping, footpaths, and other factors that were not evident in the rural sites.

In central Europe, the overriding influence of the urban environment has led to a situation in which a homogeneous basic stock of trees predominates in all large cities (maple, oak, chestnut, plane tree, and beech) despite the very different original natural conditions in those areas.

Parks and Gardens

The habitats provided by parks vary considerably depending on the age, size, function, topography, and especially the philosophy of the designer. The Royal Parks of London, due to their age, are outstanding habitats for plants and animals. This constancy of care and use over time has resulted in a high diversity of niches and a wide range of species. In every case, mown grass containing scattered trees is the leading habitat often well supported by areas of woodland. Formal beds of roses, annuals, herbaceous plants, and rockeries usually occupy a relatively small area because they are expensive to maintain. The value of these habitats to wildlife depends on factors such as their size, plant species diversity, structure, availability of microhabitats, and shelter, but management is also crucial (Gilbert, 1989). The flora of parks and other green spaces is characterized by a high percentage of alien species selected for their form or color.

The gardens of the suburbs represent favorable living

spaces in that they lie in the transition zone between countryside and city and the environmental stresses imposed by cities are still relatively slight. The wealth of species in these gardens is very appreciable (Owen and Owen, 1975). The word "garden" derives from Hebrew meaning "a pleasant place," and many of the plants grown in domestic gardens emphasize this character by being highly ornamental. Originally, they were chosen from the more useful or showy of the native species, but these were soon supplemented by large numbers of exotics as plant collectors scoured the world for horticultural material. Later, plant breeders working on both native and exotic elements produced a wide variety of hybrids and cultivars so that the range of spectacular plant material available is now very large. Also, the weed flora of gardens shows great variety, including bird seed aliens. Owen (1983) found at least 80 weeds in her 0.6-ha garden.

All towns contain cemeteries. These can vary in size and in age. In general, the earlier a cemetery was established, the better it is for wildlife today. Relic grassland, heathland, and woodland often survive within their boundaries, contrasting with the secondary communities which invade neglected areas. Lichens on stonework, herbicide-resistant communities around kerbs, mown grassland on the verges, gravel in the grave spaces, and mature structure planting together produce a varied but integrated vegetation.

Aquatic Systems

Rivers form important mixed-habitat corridors as they run through towns, and where canalization is not too severe or water pollution too oppressive, they can be the richest of all urban wildlife sites. Canals are among the oldest structures in towns. They provide a wildlife resource of outstanding value. The associated habitats are also of considerable interest (Fig. 13). Seven ecological zones can be recognized across a canal (Gilbert, 1989).

The shore area, the littoral, is the habitat of an invertebrate fauna rich in species and of a fish fauna which lives on these invertebrates and therefore is especially worthy of protection. Characteristic of this zone are the benthic macrophytes such as rush, sedge, and floating plant communities which are steadily declining in urban waters. In many cities, the decline of reed stands has been carefully followed for years. First, the changes of the hydrological conditions and then, in a second stage, boat traffic, eutrophication, and attacks by parasites caused the destruction (Sukopp and Markstein, 1978).

Housing Areas

The vegetation in housing areas is subjected to catastrophic disturbances every time buildings are demolished and rebuilt. The disturbance regime and, consequently, the species composition of housing areas are closely linked to the age of the housing area. Aey (1990) found typical properties of soils and flora in housing estates of different ages in Lubeck, Germany. The soils of the oldest parts were richer in humus and in nutrients (nitrogen and phosphorus). The plants of the oldest parts were typically native forest plants, which had high requirements for nutrients and humidity. On the more recently disturbed soils of the

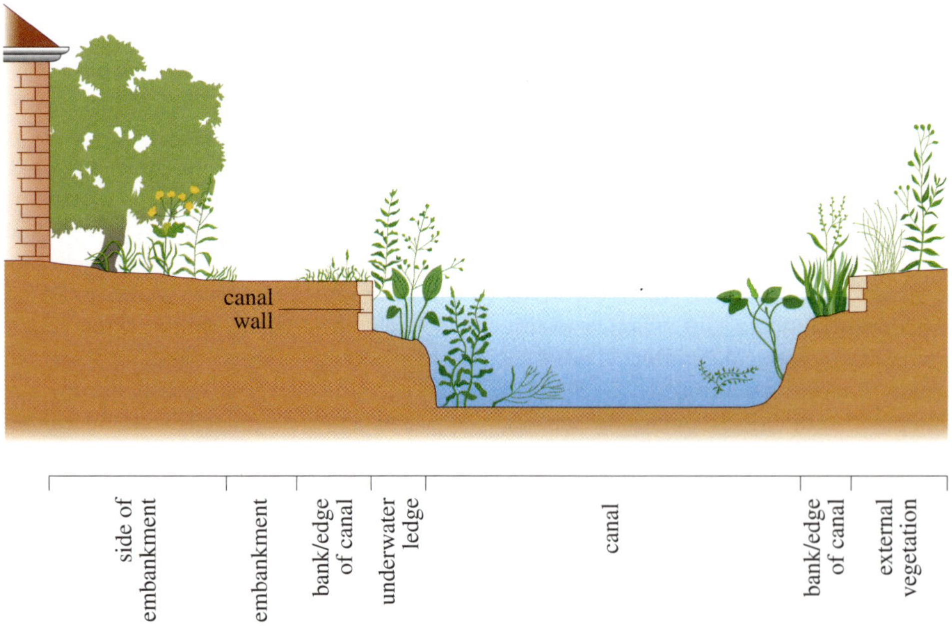

FIGURE 13 Diagrammatic cross section of an inner-city canal, one of the natural habitats in the city that is most rich in species (adapted from Gilbert, 1989).

newest (ca. 25 years old) housing area, the flora contained a high proportion of nonnative annuals.

Plants growing on buildings positively influence the climate in the buildings. Extensive investigations confirm the considerable benefits to microclimate and air hygiene of climbing plants on walls (Köhler, 1993). A cover of plants on roofs can reduce the surface temperature, filter the air, fix harmful substances, and reduce heat losses during the winter (Darius and Drepper, 1984).

Archeological sites are characterized by the increased presence of indicators for elevated temperature and light; the availability of water and nutrients is correspondingly low. Archeological areas in the urban framework are subsystems with elevated floristic diversity (Fig. 14). In addition to their significance as a cultural heritage, they offer important ecotopes for the urban flora (Celesti *et al.,* 1993).

Urban Wasteland

Once land is urbanized, it usually remains in urban use. However, some time may pass between the demolition of old buildings and the construction of new ones. In recent decades, an increasing number of inner-city sites have been left vacant after either industrial or traffic (railway) uses were discontinued due to economic or political reasons. The soils, climate, and water regime of these sites are generally strongly altered by human influence, but the vegetation can often develop with relatively little disturbance for some time.

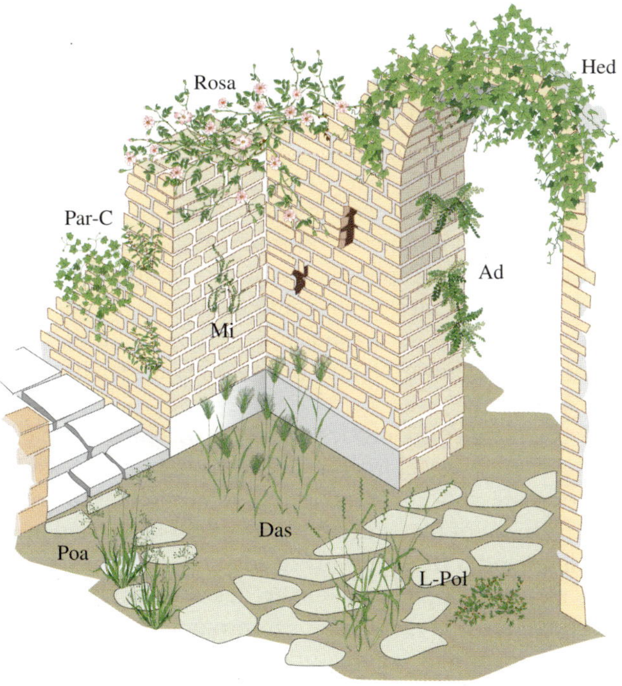

FIGURE 14 Main vegetation types within the archeological area of Ostia Antica. Ad, *Adiantetum;* Par-C, *Parietario-Cymbalarietum muralis;* Mi, *microphylla;* L-Pol, *Lolio-Polygonetum arenastri;* Poa, *Poa-Brachypodion phoenicoidis* with pioneer colonizers such as *Poa bulbosa, Barbula convoluta, Tortula muralis,* and successively xerophytes, principally *Vulpia ligustica* and *Brachypodium distachyum;* Das, thick grassy formations, mainly *Dasypyrum villosum;* Hed and Rosa, natural climbing plants such as *Hedera helix* and wild or naturalized roses (adapted from Pignatti and Giomi Visentin, 1989).

Large railway sites and stations represent a special ecological situation. The relation between soil and vegetation is mainly influenced by man-made substrate (banking material of sand, brick stones, mortar, industrial cinder, and gravel) (Fig. 15). Alien plant species comprise up to 40% of the flora due to more than 100 years of support of species immigration with transport goods, enrichment by dispersal of naturalized ornamental plants, and urban climate with higher temperature, especially on dark-colored gravel and coal slag. The variety of parent soil substrate, character and intensity of human impact, and the time available for vegetation development are responsible for a high diversity of vegetation structures from open, poor, sandy sites to nutrient-rich mixed deciduous forests (Figs. 16 and 17).

In the North American and northern European regions, an increase in southern species has been observed as a consequence of the heat island characteristic of a city. Their dependence on the urban area increases with the distance from their geographical optimum (Tischler, 1980) as well as their ecological demands (Saarisalo-Taubert, 1963). This close connection can be used as a positive indicator. A good example of a positive indicator of urban environments in central Europe is wall barley (*Hordeum murinum*). In principle, the climatic conditions bring about an increase in xerothermic species which tolerate well the increase in temperature and the reduced humidity.

The variety of species on inner-city wasteland is surprisingly large. Thus, a site at Berlin's Lützowplatz in Berlin-Tiergarten accommodates 140 seed plants and at least 200 types of insects. The carefully maintained lawns and bushes of the nearby Tiergarten Park have at the most one-fourth as many insect species on the same area.

The Future of Urban Nature

The urban way of life is viewed ambivalently: With man's growing demands for space, the protection, preservation, and development of natural landscape, adequate protection of groundwater and surface water, as well as soil protection are only possible in a settlement system which concentrates the essential, environmentally damaging functions. This system is the city (Deutscher Städtetag, 1988, p. 600). On the other hand, the urban way of life is said to destroy "its own bases. The nature on which the city as the most artificial of all worlds is erected cannot stand it any more" (Häussermann and Siebel, 1988).

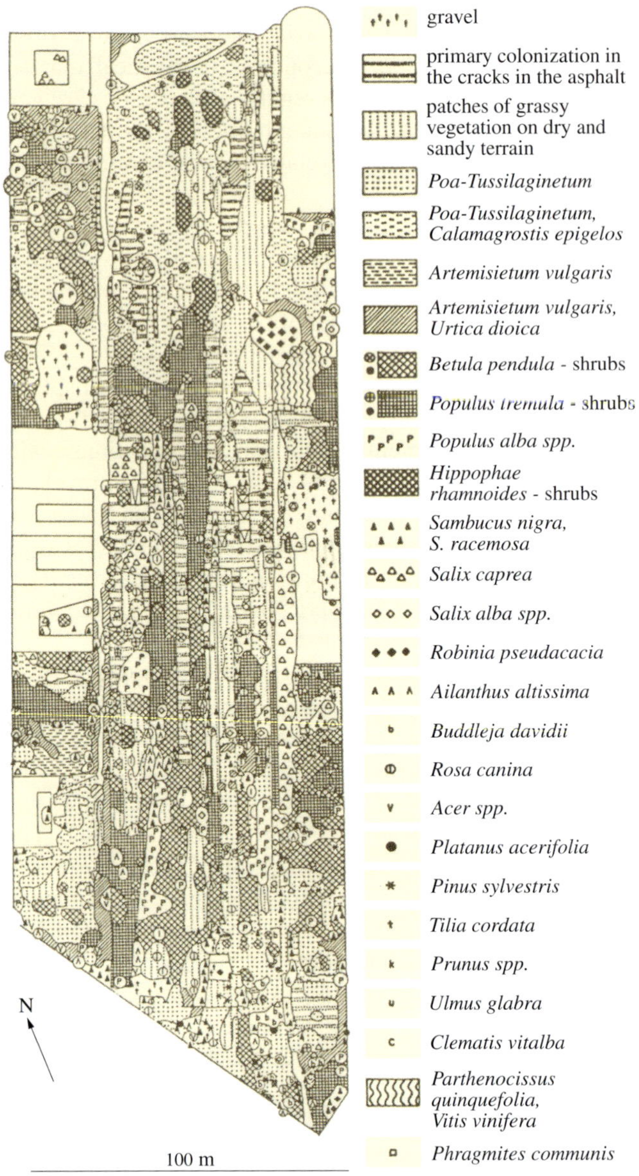

FIGURE 15 Vegetation map of a 25-year-old urban wasteland, formerly a railway station, in the center of Berlin. Trees and shrubs prevail on the remains of the railway lines with grasses mainly on the platforms.

Ecological or, perhaps better, "environmentally sustainable" urban development must be viewed as a permanent process in which the ecological, spatial, and social objectives are coordinated with one another on an equal footing. Urban planning only deserves the attribute "ecological" if the following basic principles of ecological planning are taken into account (Fig. 18):

1. Historical continuity: Primary ecotopes or those with a long history of the same kind of management are particularly valuable for nature conservation. They should be identified and protected.

2. Urban ecological zonation: Remnants of natural ecosystems and ecosystems which depend on agricultural or forestry management (encapsulated countryside) are most easily maintained in the outer districts or on the fringes of cities. The inner city, in which the effects of urbanization are more intense, can more easily support the specialized urban plant and animal communities.

3. Maintenance of local variety: All places which hold ecotopes or species uncommon in the area should be identified and protected (Fig. 19).

4. Maintenance of diversity: Notwithstanding the value of large blocks of a single ecotope, diversity is important in urban areas. Every effort should be made when creating new open space to avoid uniformity and to encourage diversity.

5. Value of corridors: In order to reduce the effect of isolation on plant and animal populations, islands of green should be linked by green corridors. The provision of small "stepping stone" open spaces is also valuable. The scientific evidence for the effectiveness of green corridors, however, is poor.

6. Value of size: A single large area of green space is more valuable ecologically than an equivalent area of separate open spaces.

7. Prevention of disturbance to existing green space: To further the conservation of ecotopes and species, there must be a general presumption against disturbance of well-established green spaces with a seminatural flora and fauna.

8. Use of native species in landscaping: In the outer districts of cities, it is preferable to use trees, shrubs, and herbs native to that geographical area of the country rather than alien species or cultivars in planting schemes. Ground conditions and/or the constraints imposed by pollution on the urban climate, however, may restrict the range of useful native species.

9. General consideration for the environment in development: Built development does not necessarily lead to drastic losses in wildlife habitat. In some circumstances, redevelopment of old sites/areas can result in an increase in wildlife habitat. Every effort needs to be made to harmonize built development with the needs of wildlife and to make the most of any chances to protect features of value and to create new seminatural habitats in the execution of work.

10. Functional incorporation of buildings into the ecosystem: Buildings need not act as complete ecological blockages in the development of corridor systems. In heavily built-up areas an increase in wildlife habitat can be achieved by the use of roof gardens, climbing plants on walls, the use of window boxes, and so on. Bird nesting cavities or bat roosting sites are better designed into structures than simply left to chance. In this way, possible nuisance will be avoided.

11. Avoidance of pollution: Every effort should be

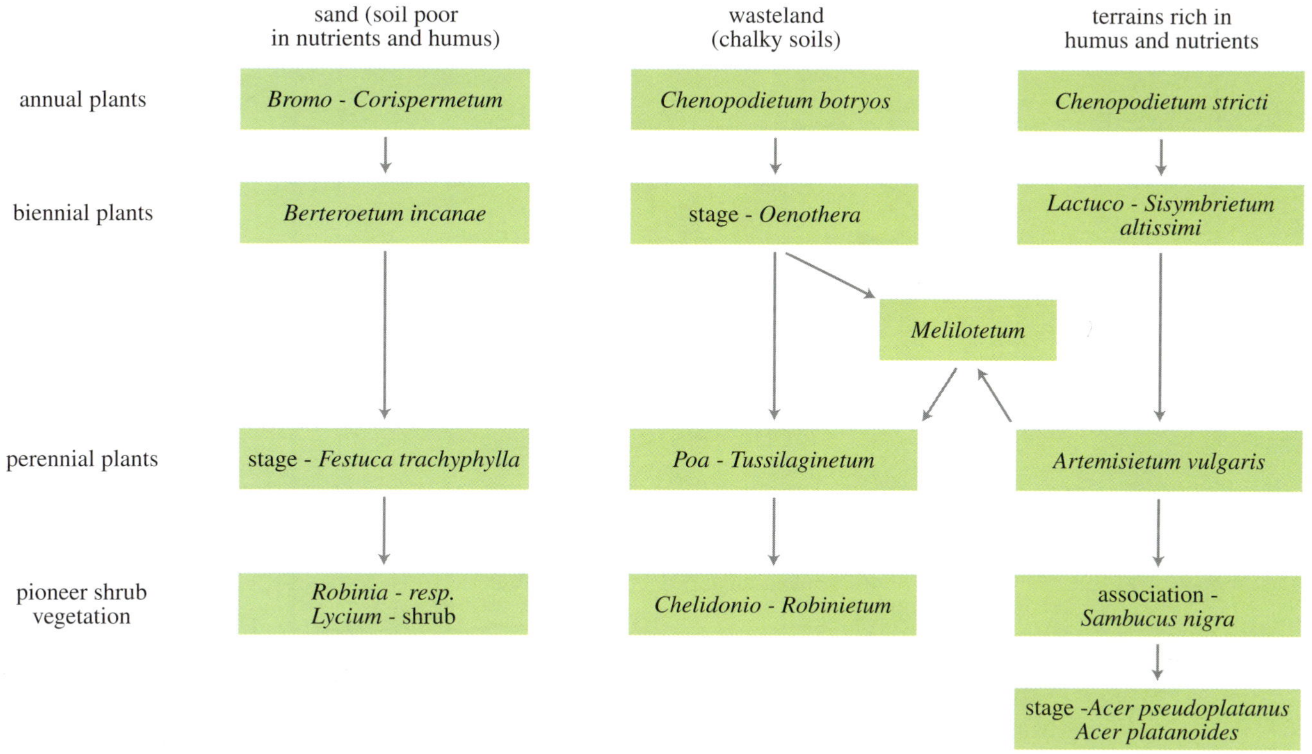

FIGURE 16 Succession stages of ruderal vegetation in the urban area of Berlin. Different types of vegetation develop in the different types of soil.

made to reduce existing levels of chemical pollution of the soil, air, and water. Levels considered acceptable in terms of human health are still highly damaging to many species of plants and animals.

12. Zonation of use of open spaces: The use of open spaces should be zoned and channeled to ensure that only light use is given to areas susceptible to disturbance and trampling. The natural carrying capacity of different ecotopes and the influence of topography and soil type on erosion need to be taken into account at the planning stage. Wetlands, woodlands, steep slopes, and light acid soils are particularly vulnerable.

13. Education of the people: In order to safeguard wildlife within a city, it is necessary to inform the public about it, to interest people in wildlife, and to involve local residents in active measures to protect it.

14. Availability of study areas for schools: It should be an objective to ensure that every school has access to at least one area on which ecology can be studied, either within the school grounds or within a 5- to 10-min walk of the school.

Many European states develop programs for nature and nature conservancy in towns and cities. The chief aspects are contained in the World Conservation Strategy.

The traditionally developed system of parks and green areas can be expanded by a system of existing, ecologically functioning areas which have not yet been "planned," such as areas resulting from war damage or demolition which are spontaneously settled by animals and plants.

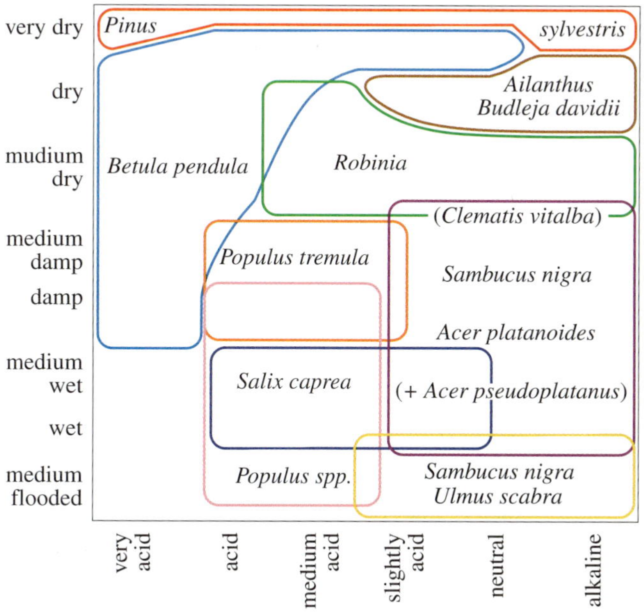

FIGURE 17 Range of tolerance of soil moisture and acidity for the most important trees and shrubs present in the city of Berlin (adapted from Sukopp, 1990).

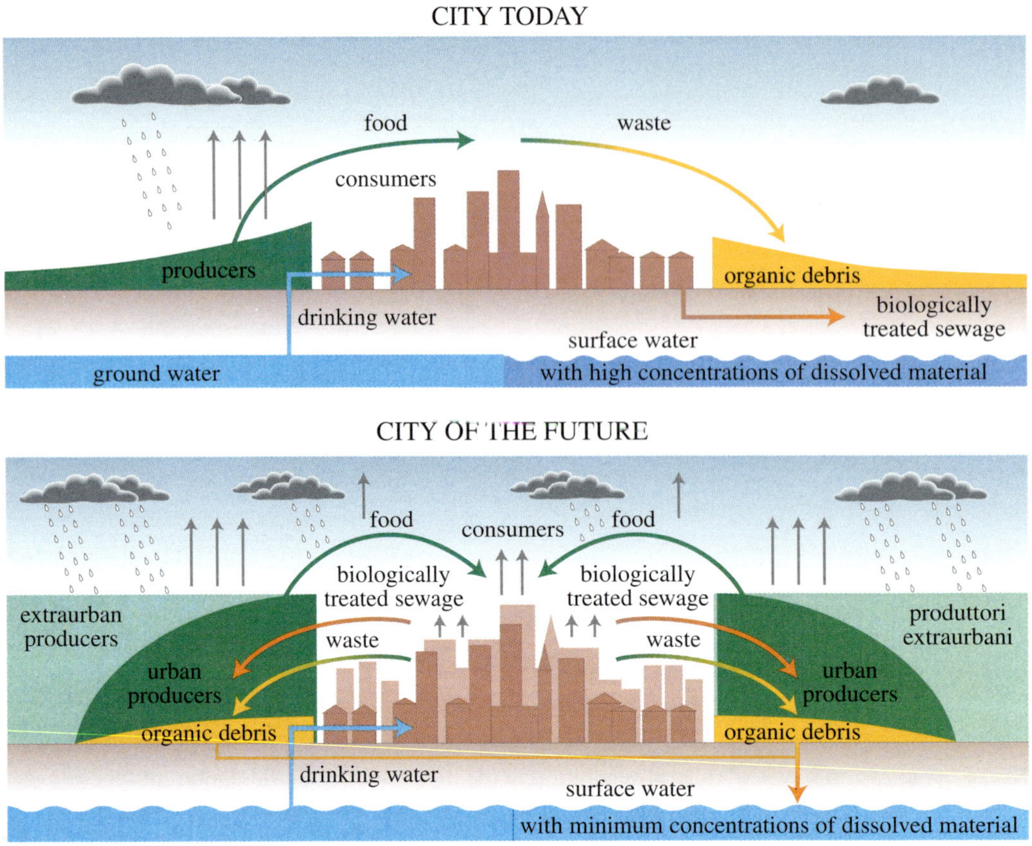

FIGURE 18 Water and organic material cycles in a city of today (top) and of the future (bottom). Although cities today have open cycles and cause a high contamination of the surrounding environment, in an ecological city of the future the cycles will be closed in a short-circuit, limiting contamination of the external environment as much as possible (adapted from Ripl and Hildmann, 1995).

In 1977, the first ecological park was set up on inner-city wasteland. The William Curtis Ecological Park in London was created on an old lorry park. The aim was to establish "typical rural plant communities," especially to give city children access to nature.

Methods of Urban Habitat Mapping

Planning requires an areal inventory, comparisons, and evaluations. In order to develop and carry out systematic conservation and development measures, habitat mapping has been found to be an appropriate and necessary measure.

Habitat mapping in rural landscapes was initially directed at registering valuable biological–ecological areas worthy of protection. The entire urban landscape, including areas put to extreme uses, consists of a series of habitats of different types, all of which (even those apparently unworthy of protection) fulfill a function. This is a point which is especially significant for built-up areas.

A biotope map has uses in planning, programs, and other measures at various levels, according to the amount of detail it contains. For example, it can be used via landscape and urban planning as an ecological contribution in the development planning, in expert planning of nature and landscape conservation (e.g., demarcation of protected areas and evaluating intervention),and in communal programs and measures (communal nature conservancy programs, laying out and caring for biotopes, and supporting private measures).

The aim of guidelines for mapping applicable for the whole of West Germany (work group "Methods for biotope mapping in built-up areas") is to present a largely uniform methodological basis for an overall relevant assessment of the situation and for an ecological evaluation of the urban and village ecosystems. Results will then be more comparable, can be expanded on (e.g., by institutes at the state level or by the regions and communes), and more emphasis can be placed on the aims of nature conservation in the settlement areas.

The application of such mapping guidelines should not serve purely scientific ends but should also relate to measures to increase environmental conservation, protection,

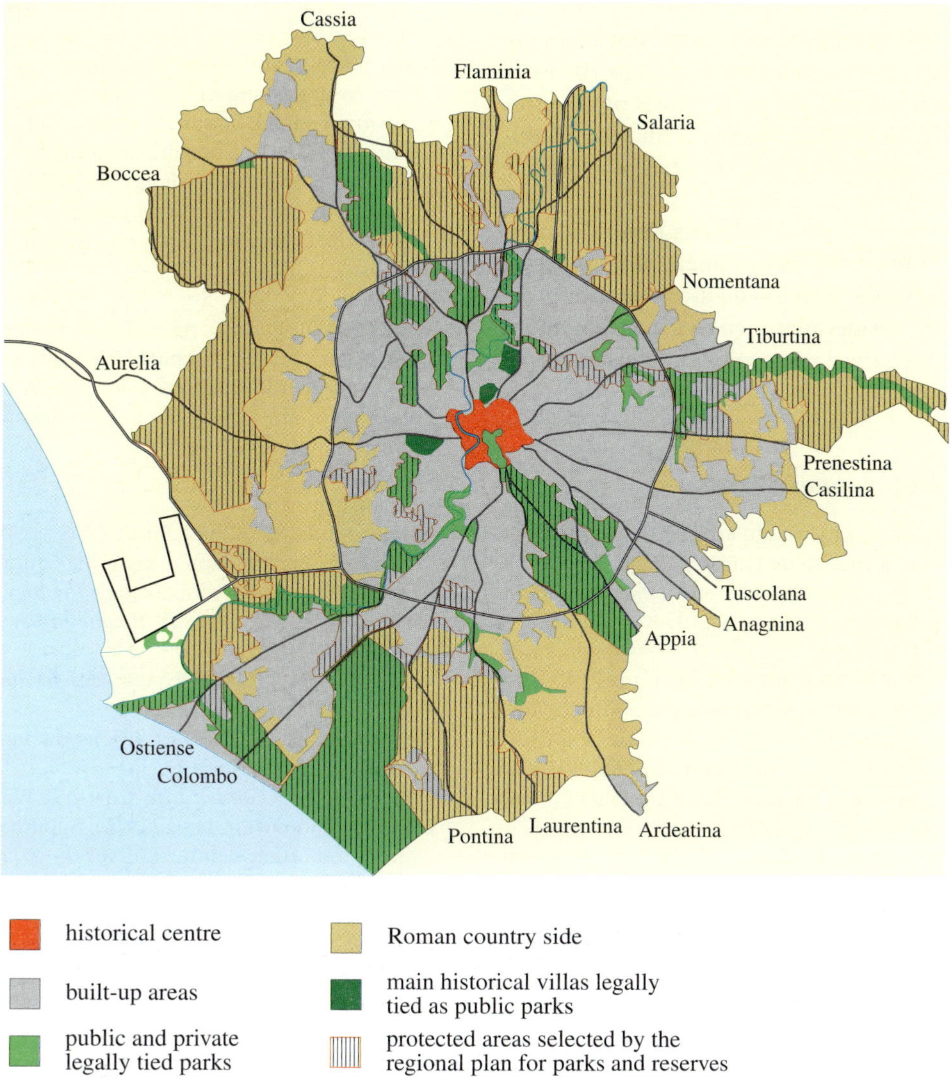

FIGURE 19 Distribution of the green areas in Rome. Although the vegetation in this city is considerably degraded, its wide distribution represents a positive example (adapted from 1995).

and development. This includes developing programs to protect species and biotopes at the community level (Arbeitsgruppe Methodik der Biotopkartierung im Besiedelten Bereich, 1993) and providing a basis for the investigation of the environmental compatibility of land use plans and local plans.

The numerous methods used for biotope mapping in settlement areas can be classified into three categories (Sukopp and Weiler, 1986):

Selective mapping covers only habitats worthy of protection. This presupposes a framework of evaluation for biotopes worthy of protection and thus of mapping.

Representative mapping maps examples of all types of use and the results are transferred to all areas with the same use structure. In this way, biologically and ecologically characterized habitat types or habitat type complexes are obtained. The evaluation of the biotope is not directly coupled to the registration.

Overall mapping covers the biological and ecological features of all actual habitats (i.e., the entire area). Registration is initially independent of evaluation.

A common feature of representative and overall mapping is the fact that the results relate to the entire area under investigation and that the evaluation of the habitats is a separate step. There are no fixed boundaries between the mapping methods. For the basic investigations in a built-up area the working group recommends the representative method.

Selective mapping has the important advantage of being relatively quick, with less demands on personnel and money. It can provide quick results for building and nature conservancy planning. However, there are various reasons why it is not adequate to use the selective method alone. It is usually only possible to judge whether a habitat is worthy of protection in the context of its surroundings. If areas worthy of protection are selected on the basis of selective mapping, then it is possible that the importance of a habitat which plays a vital role as a link between other larger refuge habitats will not be recognized.

The restriction to rare types of habitats raises the danger of losing sight of the direct environment of most urban inhabitants. Parallel to the development of selective mapping methods, these considerations led to the development of methods for the biological and ecological registration of the entire built-up area.

References Cited

AEY, W. (1990). Historical approaches to urban ecology: Methods and first results from a case study (Lübeck, West Germany). In *Urban Ecology: Plants and Plant Communities in Urban Environments* (H. Sukopp, S. Hejny, and I. Kowarik, Eds.), pp. 113–129. SPB, The Hague.

ARBEITSGRUPPE METHODIK DER BIOTOPKARTIERUNG IM BESIEDELTEN BEREICH (1993). Flächendeckende Biotopkartierung im besiedelten Bereich als Grundlage einer am Naturschutz orientierten Planung. *Natur Landschaft* **68,** 491–526.

BLUME, H.-P. (1993). Böden. In *Stadtökologie* (H. Sukopp and R. Wittig, Eds.), pp. 154–171. Fischer, Stuttgart.

BLUME, H.-P., and SCHLEUSS, U. (Eds.) (1997). Bewertung anthropogener Stadtböden. *Schr. R. Inst. Pflanzenernährung Bodenkunde Univ. Kiel* **38,** 1–346.

BROD, H.-G. (Ed.) (1991). *Straßßenbaumschäden — Ursachen und Wirkungen.* Ecomed, Landsberg.

CELESTI GRAPOW, L. (1995). *Atlante della Flora di Roma. La Distruzione delle Piante Spontanee come Indicatore Ambientale.* Argos, Rome.

DARIUS, F., and DREPPER, J. (1984). Rasendächer in West Berlin. *Gartenamt* **33,** 309–315.

DEAKIN, R. (1855). *Flora of the Colosseum of Rome.* Grombridge, London.

DETWYLER, T., and MARCUS, M. G. (Eds.) (1972). *Urbanization and Environment. The Physical Geography of the City.* Duxbury, Belmont.

DEUTSCHER STÄDTETAG (1988). Zum Begriff "Stadtökologie." *Der Städtetag* **9,** 600–601.

GILBERT, O. L. (1989). *The Ecology of Urban Habitats.* Chapman & Hall, London.

HÄUSSERMANN, H., and SIEBEL, W. (1988). Die Stadt war immer auch eine Maschine. *Die Zeit* **3**(6).

HENNESSO, D. (1970). *Geschicte des Stadtgrüns I.* Patzer-Verlag, Berlin.

HORBERT, M., KIRCHGEORG, A., and STÜLPNAGEL, A. VON (1983). Ergebnisse stadtklimatischer Untersuchungen als Beitrag zur Freiraumplanung. *Umweltbundesamt Texte* **18,** 83.

JONAS, R., HORBERT, M., and PFLUG, W. (1985). Die Filterwirkung von Wäldern gegenüber staaubbelasterter Luft. *Fortwiss. Centralbl.* **104,** 289–299.

KLAUSNITZER, B. (1993). *Kologie der Groástadtfauna,* 2nd ed. Fischer, Stuttgart.

KÖHLER, M. (Ed.) (1993). *Fassanden- und Dachbegrünung.* Ulmer Verlag, Stuttgart.

KOWARIK, I., and BÖCKER, R. (1984). *Tuexenia* **4,** 9–29.

KUNICK, W. (1982). Comparison of the flora of some cities of the Central European lowlands. In *Urban Ecology, the Second European Ecological Symposium, Berlin 8–12 September 1980* (R. Bornkemm, J. A. Lee, and M. R. D. Seaward, Eds.), pp. 13–22. Blackwell, Oxford.

KUTTLER, W. (1993). Stadtklima. In *Stadtökologie* (H. Sukopp and R. Wittig, Eds.), pp. 113–153. Fischer, Stuttgart.

LANDOLT, E. (1992). Veränderungen der Flora der Stadt Zürich in den letzten 150 Jahren. *Bauhinia* **10,** 149–164.

LANDSBERG, H. (1981). *The Urban Climate,* International Geophysics Series 28. New York.

LICHTENTHALER, H. K. (Ed.) (1996). *Vegetation Stress.* Fischer, Stuttgart.

MCDONNELL, M. J., and PICKETT, S. T. A. (1990). Ecosystem structure and function along urban–rural gradients: An unexploited opportunity for ecology. *Ecology* **71,** 1232–1237.

MEYER, F. H. (Ed.) (1982). *Bäume in der Stadt.* Ulmer Verlag, Stuttgart.

MUMFORD, L. (1961). *Die Stadt. Geschichte und Ausblick.* Colonia, Berlin.

NYLANDER, W. (1866). Les lichens du Jardin du Luxembourg. *Bull. Soc. Bot. France* **13,** 364–372.

OWEN, J. (1983). *Garden Life.* Chatto & Windus, London.

OWEN, J., and OWEN, D. F. (1975). Suburban gardens: England's most important nature reserve? *Environ. Conserv.* **2,** 53–59.

PIGNATTI, E., and GIOMI VISENTIN, G. (1989). *Braun-Blanquetia* **3,** 271–278.

POVOLNY, D. (1971). Synanthropy. In *Flies and Disease* (B. Greenberg, Ed.), pp. 17–54. Princeton Univ. Press, Princeton, NJ.

REBELE, F. (1994). Urban ecology and special features of urban ecosystems. *Glob. Ecol. Biogeogr. Lett.* **4,** 173–187.

RIPL, W., and HILDMANN, C. (1995). *Schr.-R. Vegetationskd. Sukopp-Festschrift* **27,** 201–207.

SAARISALO-TAUBERT, A. (1963). Die Flora in ihrer Beziehung zur Siedlung und Siedlungsgeschichte in den südfinnischen Städten Porvoo, Loviisa und Hamina. *Ann. Bot. Soc. Vanamo* **35,** 1–190.

SCHOUW, J. F. (1823). *Grundzüge einer Allgemeinen Pflanzengeographie.* Reimer, Berlin.

SCHMITZ, G. (1996). Phyophagen-Komplexe von *Artemisia vulgaris* L. und *Tanacetum vulgare* L. (Asteraceae) und deren Beeinflussung durch zunehmende Urbanität der Standorte. Dissertation, University of Bonn.

STARFINGER, U., and SUKOPP, H. (1994). Assessment of urban biotopes for nature conservation. In *Landscape Planning and Ecological Networks* (E. A. Cook and H. N. Van Lier, Eds.), pp. 89–115. Elsevier, Amsterdam.

STUEBING, L. (1978). In *Handbuch fur Planung, Gestaltung und Schutz der Umvelt* (K. Buchwald and W. Engelhard, Eds.), Vol. 2. Bayern Landwirtschaft Verlag, Munich.

STEUBING, L., and SCHWANTES, H. O. (1992). *Ökologische Botanik Einführung in die angewandte Botanik.* Quelle & Meyer, Heidelberg.

Stülpnagel, A. von (1987). Klimatische Veränderungen in Ballungsgegieten. Dissertation, Technische Universität Berlin.

Sukopp, H. (1971). Beiträge zur kologie von Chenopodium botrys L. *Verh. Bot. Ver. Prov. Brandenburg* **108,** 3–74.

Sukopp, H. (Ed.) (1990). *Stadtökologie.* Das Beispiel Berlin, Berlin.

Sukopp, H., and Markstein, B. (1978). Die Ufervegetation der Berliner Havel, Veränderungen 1962–1977, Schutz, Pflege und Entwicklung. In *Bedeuntung der Ufervegetation in Binnengewässern, Schutz, und Regulierung. Arbeiten des Deutschen Fischer,* Vol. 25, pp. 16–29.

Sukopp, H., and Weiler, S. (1986). Biotopkartierung im besiedelten Beriech der Bundesrepublik Deutschland. *Landschaft Stadt* **18,** 25–38.

Sukopp, H., and Wittig, R. (Eds.) (1993). *Stadtökologie.* Fischer, Stuttgart.

Tischler, W. (1980). *Biologie der Kulturlandschaft.* Fischer, Stuttgart.

Trepl, L. (1994). Towards a theory of urban biocenoses: Some hypotheses and research questions. *Memorabilia Zool.* **49,** 15–19.

Trepl, L., and Sukopp, H. (1993). Zur Bedeutung der Introduktion und Naturalisation von Pflanzen und Tieren für die Zukunft der Artenvielfalt. *Rundgespräche Kommission Kologie* **6,** 127–142.

Wagenitz, G. (1964). Compositen (Korbblütler) II. Teil. In *Hegi, Illustrierte Flora von Mitteleuropa.* München.

Wittig, R. (1991). *Kologie der Groástadtflora.* Fischer, Stuttgart.

Wittig, R. (1996). Die mitteleuropäische Großstadtflora. *Geogr. Rundshau* **48,** 640–646.

Wittig, R., *et al.* (1985). *Flora* **177,** 265–282.

Pier Luigi Nimis

Department of Biology
University of Trieste
Trieste, Italy

Artistic and Historical Monuments: Threatened Ecosystems

Italy has the largest heritage worldwide in historical monuments, ancient books, parchments, paintings, sculptures, ancient tapestries, and textiles. Some are kept outdoors, some indoors, and others even under water; their preservation is a responsibility of all Italians to mankind. These works of art are attacked by many organisms, and the biologist regards them as outright ecosystems. Open-air monuments mostly host photosynthetic organisms, such as cyanobacteria, algae, lichens, mosses, and higher plants, whereas those stored indoors are attacked by heterotrophic organisms, such as bacteria, fungi, and insects. Biology can help their conservation and restoration in that it can identify such organisms and the ecological conditions that allow their growth. This knowledge allows the biologist to judge the effectiveness of restoring treatments. Much has been done and much more still remains to be done.

Works of Art as Ecosystems

The richness and variety of Italian artistic heritage is unequalled worldwide. All Italians are responsible for its conservation. However, it is not an easy task: Physical, chemical, and biological agents attack any item. Often, changes thus produced are irreversible. Living beings play an important part in this process; an ancient illumination in a library, a Roman statue in an archeological site, and a fresco or a painting in a church are all under aggression. Many organisms may destroy them or change them enough to mask the artist's idea. Thus, the biologist's knowledge can prevent these works of art from disappearing.

Illuminations, statues, and paintings can be viewed in a biological light: They are ecosystems, with primary producers, decomposers, and predators, like any African plain seen in many television documentaries. An imaginary documentary titled "The Statue Ecosystem" would unveil a new and remote world in which no acacias grow and in which no elephants, hyenas, lions, or hunting dogs dwell. Instead, this ecosystem includes primordial cyanobacteria; weird colonies of algae (the sea is not the only place in which they live), attacked by squadrons of sneaky fungi; unearthly lichen landscapes; and tiny moss forests, inhabited by horrible creatures that human eyes have rarely seen — everything from sleepy herbivorous tardigrades to ferocious predator carabids.

Therefore, a statue is in itself an environment in which all ecological phenomena can be observed: competition, host–parasite and prey–predator relationships, and successions. It is also the substratum for the complex reactions of biogeochemical cycles, such as nitrogen, carbon, and sulfur cycles.

Restoration ecology is a recent field of study in which much still remains to be done. Italy holds a prominent position in this field, if not in others. The available knowledge is based on the work of many biologists, who work for prestigious institutions that study the decay and conservation of works of art. Examples of such institutions are the Istituto Centrale per il Restauro (Central Institute of Restoration), the Istituto per la Patologia del Libro (Institute for Book Pathologies), the Istituto per la Ricerca sul Legno (Institute for Wood Research), various centers of the Consiglio Nazionale delle Ricerche (National Council for Scientific Research), the Monuments and Fine Arts Office research laboratories in Bologna and Venice, university departments, and other research centers. This article aims to observe statues, churches, illuminations, and parchments from the biologist's point of view; thus, the material traces of our cultural roots will become tiny, complex, and fragile ecosystems, threatened with destruction by the same laws of nature that generated them through human intelligence.

Biodeterioration of Works of Art

Biodeterioration is the name biologists give to the damage that biological agents produce on various materials. Two types of organisms are involved: autotrophic and heterotrophic. The former — cyanobacteria, algae, lichens, mosses, and vascular plants — draw energy from sunlight through photosynthesis; thus, they do not use the substratum as an energy source but as a mere support or, at most, as a source of micronutrients or water or both. Therefore, these organisms can grow on inorganic substrata, such as monuments, stained-glass windows, or metal objects, when these are exposed to sunlight. On the other hand, heterotrophic organisms classified as decomposers (such as bacteria and fungi), herbivores, and predators (usually snails and insects) draw energy by decomposing organic molecules. These organisms live on organic matter (other organisms, paper, wood, parchment, and some types of paint) used as a direct source of energy irrespective of the lighting conditions; this makes them the primary biodeterioration agents for works of art kept indoors.

Heterotrophic organisms often produce irreversible damage when they feed directly on artwork made of organic matter, whereas they are less involved in the biodeterioration of outdoor monuments. Autotrophic organisms, on the other hand, can cause both reversible and irreversible damage. For example, the color changes that algal patinas cause on church façades are easily restored — when not associated with physical or chemical damage — by removing the algae.

Both hetero- and autotrophic organisms mainly bring about two types of biodeterioration: (i) physical actions (*disintegration*), such as the effect of tree roots on marble monuments, physical disruption of rock by lichens, and holes dug by insects in wood, and (ii) chemical actions (*decomposition*), such as the irreversible alterations that fungi cause to fabrics, paper, and wood, the chromatic changes due to fungal and bacterial pigments, and the solution of calcareous rocks by endolithic cyanobacteria and lichens.

Biodeterioration can be both direct and indirect. The former is caused by organisms that permanently use a particular substratum for support or nourishment, such as the destruction of books by fungi and bacteria or the perforation of submerged columns by mollusks. Indirect biodeterioration is caused by organisms whose ecology is not tied to a particular substratum; examples are guano deposits on statues and monuments that, incidentally, can give rise to direct biodeterioration by other organisms or chemical and physical alterations. At the extreme, even human vandalism can be considered indirect biodeterioration.

There is a marked ecological difference between works of art kept outdoors and those kept indoors. The former are attacked primarily by autotrophic organisms, which grow on inorganic materials (stone, glass, and metal). The latter, made entirely or in part of organic material (paper, parchment, wood, glue, and paint), host heterotrophic organisms. Outdoor monuments are the substratum for complete miniature ecosystems, inclusive of primary producers (photosynthetic organisms); on the contrary, indoor artwork is a "halved ecosystem" in which decomposers prevail.

Ecological Study of Outdoor Monuments

The use of inorganic constructing materials, such as stone, bricks, and mortar, throughout its history has left Italy with an amazing architectural heritage that endures through time. Other cultures, such as those of the Far East, having built their monuments in wood, are left with a much more deteriorated or recent memory of the past, mainly because of the action of biological agents. Nevertheless, Italian monuments are still prone to physical, chemical, and biological deterioration because they are mostly made of calcareous rock such as marble. Glass, of siliceous nature, is another lithic substratum, but it is not subject to colonization by microecosystems, except when air humidity levels are high. This condition is rare in Italy, but the stained-glass windows of French churches, for example, exposed to humid Atlantic winds, are attacked by fungi, lichens, and algae that discolor them or make them irreversibly opaque.

Monuments built in stone are mainly colonized by photosynthetic organisms — cyanobacteria, algae, lichens, mosses, and vascular plants — that use the substratum uniquely as a support to reach light, their primary source of energy. Sometimes, though, the colonization of stone surfaces starts with chemoautotrophic organisms, such as particular types of bacteria, which are able to draw energy from inorganic substances without light. These organisms prepare the substratum for the development of various series of biological successions. Heterotrophic organisms arrive later because they need to feed on the organic substances that autotrophic organisms have produced. These organisms are less important in the biodeterioration of lithic substrata and they will be discussed only with regard to organic substrata preserved indoors.

As in any ecosystem, biological colonization of monuments depends on a variety of factors, such as exposure, substratum, and the availability of nitrogen compounds. Studying the ecology of the communities that invade monuments can help identify the factors that favored their growth, thus contributing to the planning of strategies for restoration and preservation.

Bacteria

Bacteria that colonize monuments are mainly autotrophic, such as sulfur and nitro- and cyanobacteria; the alterations that they cause are of a chemical nature. Nonphotosynthetic sulfur bacteria convert limestone into gypsum (especially in sulfur dioxide-polluted areas), irreversibly covering the surfaces of statues and capitals with horrible black crusts. Nitrobacteria cause less damage, although they can alter lithic substrata by producing sulfuric and nitric acids. Nitrifying bacteria are the first to colonize stone surfaces; they mainly produce nitrates, which often foster the growth of other microorganisms including lichens. Cyanobacteria were among the first prokaryotic organisms to colonize land; they have chlorophyll and play an important role in the deterioration of stone surfaces, especially calcareous ones. Some even penetrate beneath the surface, producing typical erosion phenomena; others become encrusted in thick calcium carbonate secretions, deforming their substratum. Free from competitors, many coccal cyanobacteria cover in dark patinas large vertical areas exposed to sunlight, such as walls facing south. In addition to the esthetic damage, these dark patinas, in the Mediterranean area, cause the temperature to increase by up to 8°C higher than that of the noncolonized, lighter-colored surfaces, facilitating further physical deterioration (Garty, 1990). Cyanobacteria need water in a liquid state to carry out photosynthesis. In their search for the long-lost and nearly lunar world they conquered alone, cyanobacteria colonize hot, arid, and sunny surfaces, with water flowing over them for brief periods; all these conditions can be found on any vertical surface facing south and washed down by rain. Cyanobacteria cells, enclosed in gelatinous secretions, can retain water for long periods, thus delaying desiccation of the stone surface and favoring chemical alterations, such as the solution of calcium carbonate. Nitrifying bacteria, like cyanobacteria, are able to fix atmospheric nitrogen, enriching the surface of statues and bas-relieves with nutrients for further harmful colonizing organisms.

Algae

The algae living on works of art are green algae, ancestors of higher plants. They are simple, unicellular, or threadlike organisms with few defenses against desiccation. Algal patinas, frequent on monuments, are among the first to colonize stone and glass surfaces after bacteria, but only damp, shady surfaces allow them to grow and cause damage. Typical places are usually the base of monuments or surfaces facing north. However, algae produce only chromatic alterations, coloring the substratum green or red-orange (if the algae are of the genus *Trentepohlia*). These patinas can be easily removed by using biocides or by mechanical means. Nevertheless, algae produce carbon dioxide, retain water, and often secrete chelating agents, such as aspartic acid, citric acid, and oxalic acid, that alter — if only slightly — the substratum. On stone surfaces, damage is limited to the most superficial layer and is usually repaired easily; on the contrary, on stained-glass windows, small chemical alterations can result in irreversible esthetic damage.

Lichens

Lichens are a symbiotic association between algae or cyanobacteria and fungi. These organisms are certainly the most conspicuous and important colonizers of outdoor monuments, including both glass and lead surfaces of ancient stained-glass windows; only heavy pollution can prevent their growth. The most noticeable effects of lichen colonization are of a chromatic nature: Italian archeological sites and parks are populated by mythological heroes, venuses, nymphs, and horses whose noses are *Caloplaca* red, whose heads are *Candelariella* yellow, and whose bodies are *Aspicilia* gray. The north façade of the Orvieto duomo is an example of the esthetic damage produced by lichens. The duomo was built with dark basalt and light-colored limestone in alternate bands. The dark bands are colonized by light-colored species, whereas the light-colored bands host orange or dark species (Nimis and Monte, 1988). The result is bizarre and may not be that unpleasant, but it is certainly far from what the artist had in mind (Fig. 1).

However, threat due to lichens may be much more severe: They can produce much worse damage, such as major alterations of stone surfaces through biogeophysical and biogeochemical processes. Biogeophysical alterations are caused by the penetration of fungal hyphae (the thin filaments that make up the fungus) beneath the stone surface and by the contraction and expansion of the lichen subsequent to desiccation and rehydration (Seaward, 1988). These contraction–expansion cycles are due to the lichen content in gelatinous and mucilaginous substances, whose volume strongly depends on water content. These move-

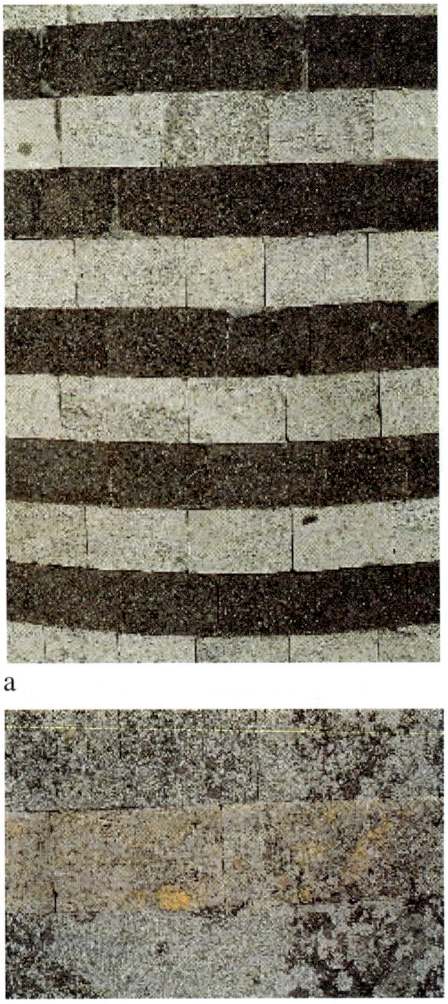

a

b

FIGURE 1 (a) The south façade of the Orvieto duomo not colonized by lichens, with the typical alternate bands of dark basalt and light-colored limestone. (b) The north façade of the Orvieto duomo: three bands, two made of basalt and the central one of limestone, extensively colonized by lichens.

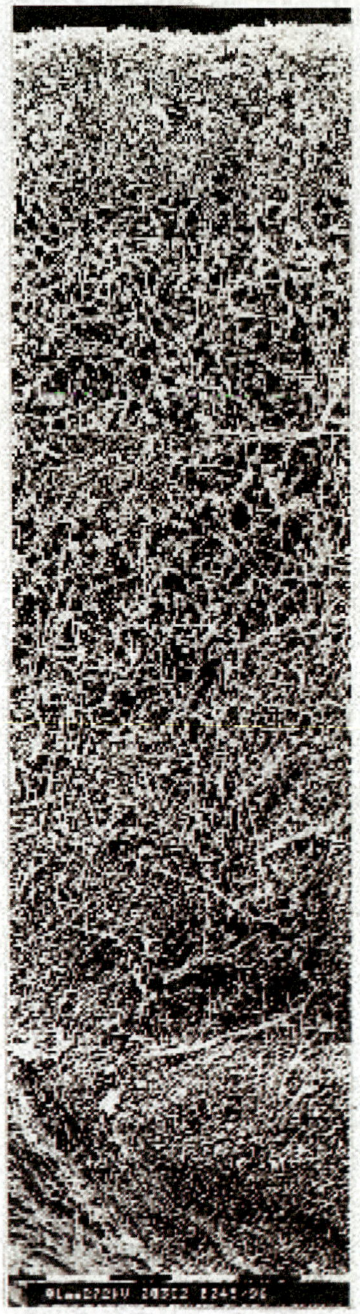

FIGURE 2 Scanning electron micrograph of the endolithic thallus of the lichen *Verrucaria baldensis* in calcareous rock, partially decalcified in acetic acid. The penetration of the fungal hyphae in the rock is evident (reproduced with permission from Pinna *et al.,* 1998).

ments result in the lifting of the marginal part of the lichen and in a "peeling" effect on the stone surface, with the detachment of superficial layers. The depth of penetration of the hyphae depends on the composition of the substratum, the type of lichen (foliose or crustose, epilithic or endolithic), and the species. The symbiotic algae rarely penetrate deeper than 2 mm and can usually be found less than 1 mm from the stone surface; the fungal hyphae may penetrate up to 15 mm in very porous carbonate rocks, although these too usually penetrate no more than 2 or 3 mm. The hyphae can at times penetrate through the mineral crystals, but more often, especially in siliceous rocks, it is much easier for them to dig their way between the crystals.

Endolithic lichens are a particular case: Their growth takes place entirely inside rocks, usually calcareous ones, and they are often invisible to the naked eye because their color is identical to that of the rock (Fig. 2). Their fruiting bodies, however, protrude on the surface, leaving small hollows as they fall; this effect is called *pitting.* The small holes, measuring 0.2–2 mm across, are a typical weak spot for chemical aggressions, such as the solution of calcium carbonate in water or the penetration of pollutants. Little was known of these organisms before the studies carried out by Tretiach (1998). Some of the findings are truly unexpected: The amount of chlorophyll per unit area enclosed in rock may be the same as that found in tree leaves, but the primary productivity of an endolithic lichen is definitely smaller. The growth of these lichens is often very slow—not more than a fraction of a millimeter per year. Strange as it may seem, many statues, bas-relieves, or cathedral walls live, breath, and carry out photosynthesis; they contain as much chlorophyll as the trees in the parks in which they are situated.

The chemical alterations lichens produce are due to three substances they secrete: carbon dioxide, lichen compounds with complex properties, and oxalic acid. Carbon dioxide, produced through respiration, when in an aqueous environment produces an acid solution that, although weak, can solubilize relatively insoluble salts such as calcium and magnesium carbonates in limestone, dolomite, a variety of marbles, and plasters that contain carbonates. All these are converted into much more soluble bicarbonates. This process is common to all organisms, and in the case of lichens it is not particularly relevant to biodeterioration.

Many of the organic compounds secreted by lichens, commonly called *lichen acids,* have complex properties due to polar chemical groups that can chelate metallic cations by donating electron pairs. Lichens produce a variety of chelating substances in large amounts, and they often take on rusty colors due to iron accumulation, especially on siliceous rocks. Oxalic acid is one of the most active acids in metal deterioration and in cation exchange. Different lichen species have different oxalate production capacities; moreover, the type of oxalate produced depends on the cations available in the environment—that is, on the mineral composition of the substratum. Calcium oxalate is found in lichens in two different crystals: whewellite and weddellite. When secreted, calcium oxalate forms an extracellular insoluble deposit. Species obliged by their physiology to live on calcareous substrata have much higher calcium oxalate levels than species living on siliceous substrata. Lichen production of calcium oxalate has given rise to heated debate among restoration ecology experts. The difference between a "new" stone surface and the "age patinas" that cover ancient artwork is evident: Many restorations have been criticized for removing these patinas. When thus restored, churches, bas-relieves, and statues take on an almost modern appearance, as if they were made of plaster. Many perceive this type of restoration as artificial compared to the passing of time. The vast yellowish or brownish films (*time patinas*) are almost totally made of calcium oxalate, but their origin is still open to debate. There have been two hypotheses put forward. First, calcium oxalate derives from organic substances used in the past as protective treatments or for esthetic purposes or both; Second, calcium oxalate is secreted by lichens and, incidentally, by bacteria and fungi. Both hypotheses need to be studied further, but it is possible that both are correct, depending on the case in question (Lazzarini and Salvadori, 1989).

Bryophytes and Higher Plants

In average humidity conditions, the colonization of outdoor monuments usually starts with nitrobacteria, followed by various stages of lichen vegetation or, if humidity conditions allow it, by algal patinas. The growth of mosses, ferns, and higher plants is also delayed because of the need for soil accumulation. In certain conditions, though, this accumulation is fast, such as when dust or stone weathering products are deposited on horizontal surfaces or between stones. Ancient brick structures are typical examples: As soon as roots find their way between bricks, the resulting mechanical effects can be devastating (Caneva and Roccardi, 1991).

Many circles of walls in Rome are emblematic in this respect. Typical higher plants found here are pellitories, common throughout Italy, or capers, typical of the Mediterranean part of Italy. Fig trees, ailanthus, ivy, and other trees or bushes make their way between bricks, taking advantage of weak spots in the structure or of the poor resistance of mortar. Their roots also produce chemical alterations by secreting acid substances, but mechanical effects are by far the most dangerous. In some archeological sites, tree roots penetrate in underground cavities, causing irreversible damage to buildings and frescos. In addition to roots, ivy clings to the substratum with its aerial rootlets, producing the detachment of vast portions of superficial layers of construction stones (Caneva and Salvadori, 1989). Occasionally, the presence of higher plants in archeological sites may be useful. In areas where the water table is close to the surface, the flooding of underground buildings is frequent. Planting trees with high evaporation rates is an effective way to lower the water level. Eucalyptuses, for instance, are among the most efficient "biological pumps" known to man. Moreover, a line of trees can be a windbreak, modifying the microclimate, reducing evaporation or direct solar radiation, hindering the wind's weathering effects, and filtering nitrogen compounds from possible nearby cultivated land.

Humble bryophytes (mosses and liverworts), having no roots, are less dangerous. Nevertheless, because they are able to retain large amounts of water (12 liters/m^2 or more), they foster biogeochemical deterioration (Fig. 3). Bryophytes also tend to accumulate soil in the form of dust and organic residues. As in any typical primary succession, higher plants thus colonize the substratum.

FIGURE 3 Top of the head of a limestone statue in the Villa Manin Park in Passariano (Udine) covered by the nitrophilous lichen *Xanthoria calcicola* (yellow) and by the moss *Grimmia pulvinata* (gray) (photo courtesy of S. Del Bianco).

Ecological Study of Indoor Artwork

It may seem that works of art kept in museums, churches, and houses are more protected from biological aggression than outdoor monuments. However, this is true of artwork made of inorganic matter only: The best way to preserve a statue is to keep it in a museum. Any item made of organic matter, if kept outdoors, has a grim and short life compared to a statue; many physical, chemical, and biological agents act synergistically and rapidly destroy it. This is why books, parchments, fabrics, and wooden objects — artwork all or in part made of organic matter — are kept in closed and sheltered places. Thus, some of the ecological factors that allow the growth of damaging organisms may be kept in control. Simply screening off sunlight eliminates all autotrophic organisms, control over temperature and humidity helps limit the growth of bacteria and fungi, and insect aggressions are easily reduced by disinfestation. Nevertheless, even indoors artwork is not perfectly safe: Composite armies of heterotrophic organisms are ready to attack every time control over environmental conditions fails perfection, and a perfect control is far from easy on a broad scale.

Before providing a review of the troops responsible for biodeterioration, one of the worst examples of indirect deterioration deserves to be mentioned. It is before anyone's eyes, in most churches; it may not attract much attention and, paradoxically, it is caused by intelligent and well-meaning men. It is striking that today most churches of Venice — famous throughout the world for its masterpieces of painting — have endless lines of pitiless candles piously shedding horrible patinas of lampblack on defenseless blackened paintings or, even worse, on newly restored paintings. The tradition of lighting candles is deeply rooted in worship practices and religious culture. However, there is very little awareness that in 1 year, one single candle may cause more damage than whole generations of fungi, bacteria, and algae. The gloomy colors of many ancient paintings are not the work of depressed painters but are due to candles lit with the best of intentions; they are certainly not meant to disfigure the object of worship. The stubborn tradition of lighting candles in Italian churches is totally unacceptable today. This problem, complex in many ways, should be a priority for the curie and the Monuments and Fine Arts Offices.

In addition to man and his ominous candles, indoor environments host swarming battalions of heterotrophic organisms ready to attack any item made of even minimal quantities of organic matter. Wood, paper, parchments, leather, and cloths are the refectory of legions of bacteria, fungi, and arthropods. Not even paintings or frescos are safe, being composed of delicious specialties such as amid, gums, sugars, glycerin, various types of gelatins, linseed oil, and egg yolk.

Wood

Lignin, cellulose, and hemicellulose are among wood's principal components. Lignin is the most resistant to biological aggression; its worst enemies are fungi, especially ascomycetes. Also, newly synthesized cellulose is not easily attacked by biological agents, but a long exposure to physical and chemical agents engenders amorphous spots that facilitate aggression by microorganisms. Among the most important are the basidiomycetes *Serpula lacrymans, Poria* spp., and *Coniophora puteana;* all have enzymes that convert cellulose in glucose, leaving a brownish residue of non-decomposed lignin (Allsopp and Seal, 1986). Hemicellulose is the least resistant of the three components: It is easily hydrolyzed by bacteria and fungi.

Conservation of wood artwork outdoors is a lost battle: The proliferation of bacteria, fungi, and insects is difficult to control effectively and for a long period of time. Indoors, the intensity of microbiological aggressions mostly depends on air humidity. Asco- and deuteromycetes attack wood items kept in humid underground rooms; wood becomes soft if damp and extremely fragile if dry. The humidity of dry wood is in equilibrium with the relative humidity of the surrounding environment; microbiological attacks are relevant only if water content is higher than 20%. Therefore, adequate air-conditioning is the best defense against bacteria and fungi. Insects, however, are still free to attack. Coleopterans, lepidopterans, termites, hymenopterans, and many others can produce much damage to dry wood, not easily attacked by bacteria and fungi. These attacks are carried out with an ingenious biological trick. Most insects, because they are unable to synthesize the enzymes necessary for breaking up lignin and cellulose, host bacteria and fungi (which do produce these enzymes) in their intestine, in which the humidity conditions necessary for the growth of such microorganisms is provided.

Paper

Cellulose is the principal component of paper, together with lignin, hemicellulose, pectines, various types of wax, tannins, and proteins in different proportions, depending on the type of paper and on the time of fabrication. It is possible that paper fabricated in the Middle Ages was much richer in cellulose, and therefore less prone to biological attacks, than most paper fabricated in recent times (Kowalik, 1980).

The principal agents of biodeterioration of paper are bacteria and fungi, and again the level of aggression depends on the water content of substratum. Books kept in dry environments survive longer than those kept in high-humidity conditions. The attacks of fungi and bacteria appear at first as stains or as a discoloration of ink. Stains are differently colored depending on the type of organism and on the type of paper. The discoloration of ink, on the other hand, is due to tannase, an enzyme — synthesized by some fungi of the genera *Aspergillus* and *Penicillum* — that catalyzes the hydrolysis of gallotannate. Bindings, which may contain hygroscopic substances, are among the first parts of a book to be damaged. Prolonged growth of fungi and bacteria causes paper to become felt-like and fragile. Its characteristics are altered (e.g., its acidity), fostering the growth of other species of fungi and bacteria more suitable to the new ecological conditions. Even in books kept for centuries in libraries, biologists can study ecological successions similar to those that, in "normal" ecosystems, turn meadows into forests. Books that undergo floods, such as those damaged in the Florence flood in 1966, are a particular case. Pages become cemented in solid blocks and are not easily separated. This is due to the rapid growth of bacteria and fungi that occurs during imbibition. These organisms rapidly degrade cellulose, producing oligosaccharides (with mucus-like properties) which glue the pages together, making the book totally useless.

Today in Italy, too many ancient books and papers are kept in ecological conditions that turn them into ecosystems at risk. Even in the rare cases in which rooms are adequately air-conditioned, insects are still dangerous: It is exactly in these conditions that these organisms have their ecological optimum. Among the most dangerous is a devilish little insect that goes by the graceful name of silverfish (*Lepisma* spp.), commonly found in houses. This small, flat animal can penetrate virtually anywhere; it easily slips inside books, merrily feasting on pages that no one will ever be able to read again. Other insects, instead of eating paper, feed on the fungi and bacteria that live in books, especially close to bindings; their excrements are a rich source of food for saprophagous bacteria and fungi. Similar processes may be observed in the biodeterioration of cloths, although there are important differences between fabrics of vegetable derivation (linen and cotton) and fabrics of animal derivation (wool).

Leather and Parchment

Although paper is of vegetable derivation, leather and parchment are of animal derivation and therefore have a high protein content. Organisms that decompose such materials must synthesize specific proteolytic enzymes (proteases and peptidases). Italian culture owes some of its most ancient papers to parchment, which is made from sheep or goat skin (the best were made using fetus skins). Parchment is mainly composed of collagen, which is hydrolyzed by specific enzymes (collagenases) synthesized by bacteria of the genus *Clostridium* (Kowalik, 1980). Collagen, however, is often depolymerized during the processing of parchment; thus, it may be decomposed by nonspecific proteolytic enzymes, synthesized by many bacteria and microfungi. Leather, whose chemical properties are similar to those of collagen, undergoes similar decomposition processes. During processing, though, leather is often treated with tannins, which inhibit biological attacks, especially those of bacteria. As with other materials, insects may still strike; especially dermestids and tineids can banquet on tannin-treated leather.

Fabrics

Ancient fabrics consist of fibers of plant or animal origin, which have different chemical composition. Vegetable fibers mainly consist of cellulose, whereas animal fibers mostly consist of protein compounds such as keratin in wool or sericin in silk. The former are mainly attacked by cellulose-hydrolyzing fungi such as deuteromycetes; the latter can be attacked by many bacteria and fungi, actually more effective in the deterioration of wool than of silk. Again, the control of relative air humidity can prevent part of the damage. The situation is quite different with insects: Everyone is familiar with the damage moths may produce in wool. Therefore, it is not surprising that the preservation of huge arras kept in Italian churches and museums should be quite difficult; obviously, hanging bags full of mothballs is not a solution.

Paintings

Wood, paper, leather, parchment, and cloth have relatively homogeneous compositions. On the contrary, paintings are composed of a wide variety of materials — so wide, in fact, that their biodeterioration is quite fascinating to biologists. Chemical composition of paintings varies in layers. The undermost layer is often wood — and its biodeterioration is that typical of wood items — but it may also be made of lime or plaster, covered with animal or vegetable glues. Whatever the type of material or biodeterioration process, damage to the base of the painting can be quite serious, resulting in cracks in the painting. Usually, the undermost layer of paintings is a canvas of vegetable nature or, as in the case of watercolors, paper. In these cases, the biodeterioration agents are those typical of fabrics and paper. Canvas is usually covered with colors of mineral derivation, often diluted with animal or vegetable oils, glues, gelatins, etc. Finally, the protective varnishes, which cover paintings and are often renewed throughout the years, are mostly made of organic matter. Therefore, paintings offer a rich and varied menu to many heterotrophic organisms — bacteria, fungi, and, fortunately, few arthropods. A painting is therefore one of the most interesting "artistic ecosystems" among those found indoors. Fungi of the genera *Aspergillus, Penicillus,* and *Trichoderma* feed on tempera and color binders; *Phoma* and *Aureobasidium* destroy oil colors; *Geotrichum* feeds on casein binders; and *Mucor* and *Rhizopus* attack various types of glues (Ionita, 1971). Therefore, the type of biological aggression depends on the substratum. Also, biodeterioration may concern all the different layers of paintings or be limited to the color layer, depending on the type of colors and diluents used. Usually, the undermost layers of paintings are those at risk of biodeterioration, although many species of organisms also attack colors, damaging them irreversibly (Fig. 4). Bacteria and fungi are mainly active in conditions of high air humidity levels, whereas insects prefer dry environments and tend to limit their attacks to the parts in wood and paper of paintings (the base or the frame).

Frescos and other inorganic types of murals are a special case because they are attacked by actinomycetes (Giacobini *et al.,* 1988). Many murals, situated in damp churches, crypts, caves, and tombs, are disfigured by whitish or grayish patinas of these fungi, which metabolize nitrites and nitrates and reduce sulfates (Fig. 5).

FIGURE 4 Detail of fungal attack on a painting on canvas (photo courtesy of O. Salvadori).

FIGURE 5 White patina on the frescos of an Etruscan tomb in Tarquinia due to the growth of actinomycetes (reproduced with permission from Caneva *et al.,* 1991).

Ecological Study of Submerged Artwork

Italian coasts conceal a huge historic and artistic patrimony. In the course of centuries, hundreds of ships have sunk with their precious loads, and only recently has Italy started to discover this treasure of historical heritage. To study biodeterioration processes in an aquatic environment, a close examination of marine ecology is necesary, and this is beyond the scope of this article. Therefore, what follows is only a brief account of such processes.

It is obvious that almost all organic matter left for long periods of time under water has a short life. Books, parchments, and cloths, once under water, are lost forever. The only exception is wood, especially that of ship hulls. In a terrestrial environment, lignin is mainly attacked by bacteria, fungi, and insects. Millions of years of evolution have brought about organisms able to decompose wood in the environment in which this is frequent — the terrestrial environment. In an aquatic environment, especially at great depths, wood is rare; in natural conditions, wood at sea mostly floats. Even under water, some bacteria and fungi are able to decompose wood, although less effectively than in a terrestrial environment. Aquatic bacteria and fungi are both aerobic and anaerobic; various species produce different types of damage, such as superficial erosion, cavitation, and the formation of small holes. Occasionally, especially in shallow water, submerged wood may become covered in algal colonies, which do not much alter their substratum. Greater damage is caused by animals. The sea may not be populated by insects, but their absence is amply compensated by the abundant presence of mollusks and crustaceans, which may cause serious damage to submerged artwork. Mollusks involved in biodeterioration are mainly of the genera *Teredo, Bankia,* and *Martesia;* they dig deep holes in wood structures. Crustaceans are mostly isopod genera, and they cause wood to become extremely fragile. Most of these organisms, however, are active only when wood is close to surface. In the past, these animals were sailors' nightmares because the hulls of ships were made of wood. Wrecks, kept at greater depths by the weight of their loads, are infrequent items in marine ecosystems and therefore less prone to biological aggression.

Even artwork made of stone is subject to biological attacks in marine environments. The biodeterioration produced by mollusks is the most devastating. These organisms can perforate rock, digging holes as deep as 10 cm, through mechanical action or by secreting acid substances. Roman columns near Pozzuoli are a well-known example of this phenomenon. Because the area is subject to bradyseism, these columns have been alternatively above and under water through the centuries. Through the holes left in the columns by marine organisms, it is possible to reconstruct the raising and lowering movements of the ground through time. Other organisms, such as some species of calcareous algae, thickly encrust with calcium carbonate the surfaces of submerged objects.

FIGURE 6 A limestone statue in the Villa Manin Park in Passariano (Udine) covered with various communities of nitrophilous lichens (photo courtesy of E. Rui).

What Can Be Done?

Works of art and historical papers are small but nonetheless complicated ecosystems. Control and prevention of biodeterioration must consider a wide variety of organisms, which attack different parts of works of art, and each of these organisms has a different ecological optimum. Restoration and preservation must not only aim at eliminating potentially harmful organisms but also at controlling the ecological factors that may foster recolonization.

Outdoor Restoration

Eliminating organisms from statues, walls, and basrelieves should satisfy esthetic criteria, but a close preliminary ecological and biological study of the object of restoration should also be undertaken. Three factors should be attentively considered: species, the causes of their growth, and the damage they produced. Many monuments, for instance, are covered with fungi, lichens, and other nitrophilous organisms. Simply eliminating them without identifying the source of nitrogen compounds is useless: The same organisms will grow back in just a few years (Fig. 6).

Simple measures such as coverings or canalization and deviation of rainwater can be taken if a monument is exposed to rain (or to rainwater from gutters) and if the growth of biodeterioration agents arises from water availability. When nitrogen compounds are brought by the wind from nearby cultivated land, a line of trees may be planted as a windbreak. If eutrophication is due to guano deposits, the bird population should be brought under control or structures that prevent birds from alighting on monuments should be constructed. In some cases, however, nothing can be done to prevent damage. In many parks of Italian villas, statues are covered with colonies of nitrophilous lichens because of guano deposited by birds. Usually, birds are attractions in parks, so it is obviously impossible to eliminate them, and fair venuses and nymphs cannot be burdened with disturbing crowns of thorns. In these cases, removing the lichens would only be a temporary remedy; moreover, biological patinas may only be removed through mechanical action, per se damaging to stone surfaces. It is best to accept that the idea of contrasting nature with candid marble statues was queer from the start. Once placed in a park, statues become part of that ecosystem, and biological patinas should be accepted as a natural consequence of it. This may not please art critics, but it may teach man that he cannot always bend Nature to his will, and it may teach art critics the importance of ecology, an almost ignored discipline in the eighteenth century. Should a statue have an extremely high artistic value, the best thing to do is put the original in a museum and leave the park with a copy.

Organisms are usually eliminated with the use of biocides. These compounds are a heterogeneous class of chemical agents, and none are specifically meant or sold for restoration purposes; the restoration market is so poor that the cost of producing them would be much too high. Biocides in use mainly come from two fields: medical science (disinfectants such as quaternary ammonium salts) and agronomy (herbicides). When in contact with the target organism, these substances either inhibit photosynthesis (as urea derivatives do) or interfere with other metabolic processes. In choosing between these substances, characteristics such as efficacy, interactions with the substratum, and toxicological properties must be considered. Efficacy (i.e., the level of action against biodeterioration agents) should be relevant at minimal doses, for an ample range of susceptible organisms, and of long duration. Biocides must not interfere chemically or physically with the substratum; they must not react with certain components of stone (the worst case possible) or alter the appearance of monuments by changing their color (yellowing or whitening) or brightness. All biocides, in different measure, are toxic. Therefore, both restoration operators and the environment are at risk. Washing off of water-soluble biocides can result in unwanted toxic effects on insects, plants, and animals. Moreover, biocides can modify the ecosystem, fostering colonization by more aggressive or nonsusceptible organisms.

In my opinion, biocides are often misused, especially for eliminating microorganisms on outdoor monuments. This misuse is the result of transferring to autotrophic organisms in open-air ecosystems the restoration techniques correctly used for heterotrophic organisms in sheltered environments. Many algal, lichen, and bryophyte patinas are easily removed by mechanical means without the help of potentially toxic substances, whose use should be avoided in areas visited by tourists. Moreover, mechanical removal is necessary even after the use of biocides. In some cases, the use of biocides is extremely damaging, as in the case of endolithic lichens (Fig. 7) living on limestone monuments (Tretiach, 1998). When killed, the lichen, deeply rooted in

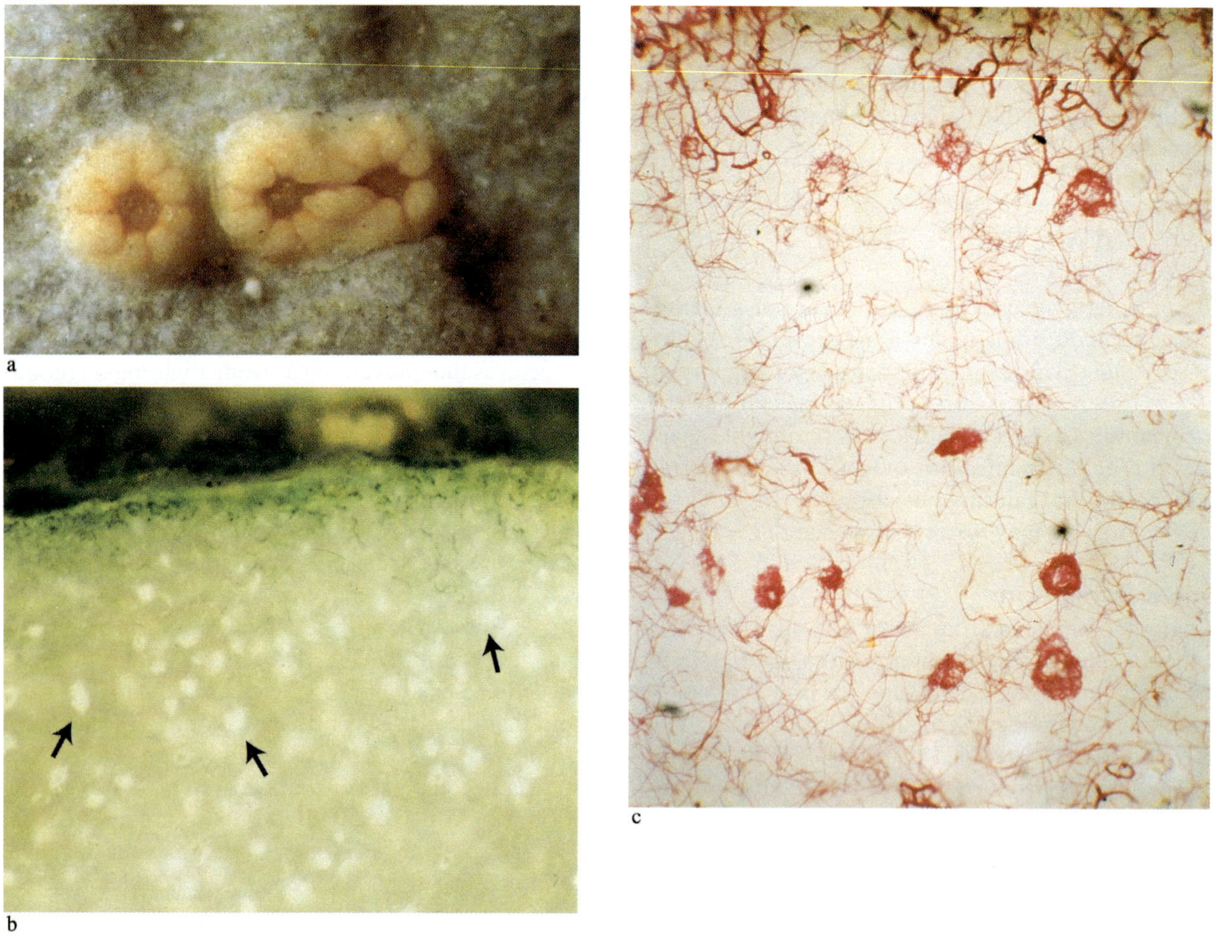

FIGURE 7 (a) Thallus and fruiting bodies (orange) of the endolithic lichen *Petractis clausa.* (b) A polished section of the same lichen in which the symbiotic cyanobacteria colonies (green) and the areas of solution of the calcareous substratum (indicated by arrows) are visible. (c) Detail of the polished section, stained with Schiff reagent to reveal the fungal hyphae, aggregated in balls in the areas of substratum solution (photos courtesy of O. Salvadori and M. Tretiach).

the stone substratum, causes the exfoliation of the outermost layer (*lithocortex*) that detaches in small scales, leaving the underlying stone surface exposed. This surface, however, is extremely porous because of the penetration of the fungal hyphae, leaving the monument defenseless against the aggression of physical, chemical, and biological agents (such as rain, wind, atmospheric pollutants, and organisms).

It is worth asking whether it is always appropriate to remove organisms from monuments. Until a few years ago, the answer would have been affirmative in any situation or condition. Originally, any monument was certainly free of biological colonization of any kind. Bringing it back to its original state through restoration was seen as a priority by restoration experts. Today, the growing importance of the biologist's point of view has partially changed this state of things. Columns, statues, and historical buildings can offer allochthonous substrata for fungi, plants, and animals that would otherwise be absent from the area. Many archeological sites in Latium, the Italian region with Rome, have monuments made from exotic materials from distant regions of the Roman Empire. Today, such monuments host a remarkably rich and diverse lichenic flora. Moreover, many of the more than 500 lichen species known to live in Latium are found only on ancient monuments (Nimis *et al.,* 1987). Sardinia's nuraghi, with their subcylindrical shape and the consequent great variety of microclimatic conditions, host a great number of rare lichen species.

Italy is a signatory of the Rio Biological Diversity Treaty, and this involves severe restrictions on a global scale. The high biodiversity of many archeological and monumental areas is per se a richness worthy of preservation, which indicates the artistic and historical value of a site. The cultural importance of biological diversity is a relatively new concept and not always a priority in artistic and monumental site management. A preliminary study of the importance of organisms from a biological point of view is absolutely necessary when planning interventions aimed at the removal of these organisms from such sites. This, of course, may engender conflicts between biologists and the Monuments and Fine Arts Office. Guidelines for the resolution of such conflicts are still a debated and urgent matter. However, collaboration between biologists and art historians may also produce otherwise unthinkable historical valorization of the biological component of archeological sites. For instance, many of Sardinia's nuraghi, especially those built with basalt, are covered with a red-orange lichen, *Xanthoria calcicola,* whose color inspires even the names of some nuraghi, for example, *nuraghe ruju* ("red nuraghe"). This lichen causes no relevant damage to the stone substratum so that its removal would be absurd even from a cultural point of view, especially if the nuraghe is named after the lichen. Another example is the north wall of a Roman temple in Ostia Antica. This brick wall is completely covered with a thick carpet of a hanging gray lichen, *Roccella phycopsis.* This lichen does not produce relevant esthetic or mechanical damage — the wall is gray rather than reddish — but someone may have the brilliant idea of removing the "mold" from the wall. However, *Roccella* is far from being a mold: It was the product Romans mainly used for extracting purple dye. This dye is of an inferior quality to "royal purple" extracted from *Murex* shells (it is less resistant to light), but it was also much less expensive. Ships loaded with *Roccella* sailed the Mediterranean from Sardinia, the Balearic Islands, and African coasts to Ostia. Many islands depended economically on this lichen and numerous traces of this commerce are still found in toponymy. The use and commerce of *Roccella* was so much revived in the Renaissance that the lichen was named after a Florentine family of merchants. Therefore, instead of getting rid of the mold, it would be far more interesting to inform the visitor on the routes of Roman *Roccella* trade and on the dying properties of the lichen. This is only one of many cases in which scientific culture and arts and humanities should, for the good of Italian heritage, go hand in hand.

Indoor Restoration

As discussed previously, indoor ecosystems usually have no primary producers: The main biodeterioration agents are bacteria, fungi, and insects. The ideal growth conditions for bacteria and fungi are high relative humidity, high temperatures, and scarce ventilation. Some species are favored by strong light, whereas others need substrata rich in nitrogen compounds (dirt, dust, substances used in restoration, etc.). Bacteria and fungi propagate by spores or conidia. Biologists studying the organic matter content of air also study the propagule content of air in indoor environments, such as libraries and archive collections. These studies are important for evaluating potential risks both for books and papers and for the public or people working at these institutions. In fact, the propagules of many species are a hazard for health.

High humidity of indoor environments is one of the most damaging biodeterioration factors because it favors the growth of fungi and bacteria. Some of the most ancient and important documents of human culture, such as Egyptian papyri or the famous Qumran manuscripts, were preserved by the dry climate of the subdesert areas where they rested for millennia. Italy's climate is very different from those of Palestine or Egypt. Being a narrow peninsula cooking in a bain-marie — the Mediterranean — the control of air humidity is a priority even indoors. Paper, parchment, and wood are all hygroscopic to some degree, and as such their conservation requires a careful control on temperature and relative humidity through air-conditioning. Relative humidity can also be controlled through the use of hygroscopic substances. In addition to natural air humidity, another problem must be considered. Italian archeological sites and museums are among the most visited worldwide, and the presence of visitors may increase both temperature and relative humidity indoors. Moreover, the "microclimatic needs" of visitors often differ from those necessary to prevent biodeterioration. The microclimate of cathedrals, crypts, museums, libraries, and art galleries deserves further study from the ecological standpoint. These environments host a variety of microclimates. In some churches, "nave breezes" or "cupola breezes" could be measured, just like sea and coast breezes are measured. Buildings could be studied to determine the areas with more frequent temperature inversions in which, therefore, water condenses more easily. This would avoid the conservation of paintings in places in which the microclimatic conditions allow water condensation, which is particularly dangerous for the back of a painting. Studying these factors is essential to planning careful biological control. Usually, acceptable temperatures are expected to vary between 18 and 20°C, whereas relative humidity should be between 50 and 65% (Gallo, 1985).

To date, the efforts that Italy has devoted to the preservation of a heritage that belongs to humanity are not sufficient. Nearly the totality of churches in Italy are still not monitored for biological deterioration agents, and the same can be said of proper air-conditioning. Moreover, candles still burn undeterred. In some environments, such as caves, crypts, catacombs, and other underground rooms, microclimatic factors are almost impossible to keep under control, especially when these environments host visitors daily. Lighting conditions have a major influence on biodeterioration, both directly and indirectly. Organic matter is degraded by light and this makes environments more susceptible to biological attacks. In particularly humid environments, artificial light may foster the growth of algal patinas (Fig. 8).

To reduce damage caused by light, the time of exposure can be shortened and the intensity of light diminished by the use of filters to eliminate ultraviolet radiation and reduce red and infrared wavelengths. On the contrary, algal patinas that grow by artificial light may be controlled by ultraviolet irradiation without the use of biocides, but only when substrates are not damaged by it (stone, mortar, and bricks). Often, however, the needs of visitors are the same of biodeterioration agents.

FIGURE 8 Proliferation of green algae in patinas under artificial light in an underground room of the *Domus Aurea* in Rome (reproduced with permission from Caneva *et al.*, 1991).

Even when lighting problems are solved, most insects and many fungi are still active biodeterioration agents, able to grow in complete obscurity. The use of fungicides or insecticides cannot be avoided, but, apart from being toxic to man, the efficacy of these substances is limited in time. Control over some insects, especially those that feed on wood, is particularly difficult. Contrarily to what many think, termites are not at all rare in some parts of Italy. These animals grow and feed inside wood, rarely eating their way to the surface like other insects do. This makes damage difficult to discover at an early stage: Roof beams and other wooden objects can be turned to crumbs before anyone is aware of what's going on inside the wood. Sometimes, the conservation of the immense patrimony that Italy has inherited may seem a Sisyphean toil. Nevertheless, Italians must devote all efforts to this task: This patrimony belongs to the whole of humanity.

Acknowledgments

I thank Prof. G. Caneva, Dr. M. Castello, Dr. E. Rui, Dr. O. Salvatori, Dr. N. Skert and Dr. M. Tretiach, who have kindly helped with constructive criticism of the paper or have supplied material for it.

References Cited

Allsopp, D., and Seal, K. J. (1986). *Introduction to Biodeterioration.* Arnold, London.

Caneva, G., and Roccardi, A. (1991). Harmful flora in the conservation of Roman monuments. In *Int. Congr. Biodet. Cultural Property, Lucknow,* 212–218.

Caneva, G., and Salvadori, O. (1989). Sistematica e sinsistematica delle comunità vegetali nella pianificazione di interventi di restauri. In *Il Cantiere della Conoscenza, il Cantiere del Restauro: Atti del Convegno di Studi. Bressanone, 27–30 Giugno 1989* (G. Biscontin, M. Dal Colle, and S. Volpin, Eds.), pp. 325–335. Libreria Progetto, Padova.

Caneva, G., *et al.* (1991). *Biology in the Conservation of Works of Art.* ICCROM, Rome.

Gallo, F. (1985). *Biological Factors in the Deterioration of Paper.* ICCROM, Rome.

Garty, J. (1990). Influence of epilithic microorganisms on the surface temperature of building walls. *Can. J. Bot.* **68,** 1349–1353.

Giacobini, C., De Cicco, M. A., Tigliè, I., and Accardo, G. (1988). Actynomycetes and biodeterioration in the field of fine art. In *Biodeterioration 7: Selected Papers Presented at the Seventh International Biodeterioration Symposium. Cambridge, UK, 6–11 September 1987* (D. R. Houghton, R. N. Smith, and H. O. W. Eggins, Eds.), pp. 418–423. Elsevier, London.

Ionita, I. (1971). Contribution to the study of the biodeterioration of the works of art and of historic monuments. II. Species of fungi isolated from oil tempera and paintings. *Rev. Roum. Biol. Botanique* **16,** 377–381.

Kowalik, R. (1980). Microbiodeterioration of library materials. Part. I. *Restaurator* **4,** 99–114.

LAZZARINI, L., and SALVADORI, O. (1989). A reassessment of the formation of the patina called "scialbatura." *Stud. Conserv.* **34,** 20–26.

NIMIS, P. L., and MONTE, M. (1988). Lichens and monuments. *Stud. Geobot.* **8,** 1–133.

NIMIS, P. L., MONTE, M., and TRETIACH, M. (1987). Flora e vegetazione lichenica di aree archeologiche del Lazio. *Stud. Geobot.* **7,** 3–161.

PINNA, D., *et al.* (1998). An anatomical investigation of calcicolous endolithic lichens from the Trieste Karst. *Plant Biosyst.* **132,** 183–195.

SEAWARD, M. R. D. (1988). Lichen damage to ancient monuments: A case study. *Lichenologist* **10,** 291–295.

TRETIACH, M. (1998). Ecophysiology of calcicolous endolithic lichens. *Giorn. Bot. Ital.* **129,** 159–184.

General References

CANEVA, G., NUGARI, M. P., and SALVADORI, O. (1994). *La Biologia nel Restauro.* Nardini, Firenze.

CANEVA, G., NUGARI, M. P., PINNA D., and SALVADORI, O. (1996). *Il Controllo del Degrado Biologico, i Biocidi nel Restauro dei Materiali Lapidei.* Nardini, Firenze.

GALLO, F. (1992). *Il Biodeterioramento di Libri e Documenti,* 5th ed. Centro di Studi per la Conservazione della Carta, Rome.

LIOTTA, G. (1991). *Gli Insetti e i Danni del Legno: Problemi di Restauro.* Nardini, Firenze.

MANDRIOLI, P., and CANEVA, G. (Eds.) (1998). *Aerobiologia e Beni Culturali. Metodologia e Tecniche di Misura.* Nardini, Firenze.

NIMIS, P. L., PINNA, D., and SALVADORI, O. (1992). *Licheni e Conservazione dei Monumenti.* CLUEB, Bologna.

GIUSEPPE CHIAUDANI

Department of Biology, Ecology Section
University of Milan
Milan, Italy

The Scientific Basis for Regulatory Actions on Water Resources Quality Control

Since the 1960s, the perception of harm caused to natural environments has led the political and legislative authorities of industrially developed countries to introduce, or renew, regulations to protect the environment. Now it is necessary to reflect on results achieved with the control policies to determine whether regulations and acts concerning aquatic ecosystems have been effective without modifying their delicate natural equilibria but using their great capacity of transformation and fitting within the natural processes. I define some aspects of the problem of micropollutants and note what regulatory actions are necessary for effective control and prevention of water pollution. An analysis of scientific knowledge will be carried out with the intention of verifying how further development may help to improve the current methods for pollution control, and I try to define the scientific basis for future strategies.

Definition and Application of Water Quality Criteria

First, it is necessary to consider what *micropollutants* are; this term is in fact ambiguous and may assume different meanings according to the context in which it is used. Water pollutants may be classified in many ways; for instance, according to chemical characteristics, physical state, environmental compartments in which they are discharged or found, the sources of pollution from which they originate, the types of effects they cause, and the target organisms that can be affected.

In this context, we consider as micropollutants all those elements or substances that can be discharged as a result of human, domestic, or industrial activities and that can be found in water bodies in limited quantities. The concentration increase caused by human activities, with respect to the quantities or concentrations present in nonpolluted environments, can cause deleterious effects on organisms or ecosystems and/or modify biogeochemical cycles.

This definition is in agreement with the concept of *specific pollutant* elaborated by the Water Management Group of the Organization for Economic Cooperation and Development that states that a substance, introduced into the environment as a result of human activities, under given conditions reduces the quality of water due to its toxic effects on human beings and on aquatic ecology. Such aspects distinguish specific pollutants from the classic global parameters, such as biochemical oxygen demand, chemical oxygen demand, and suspended solids. Specific pollutants are often defined by other organizations as trace pollutants, micropollutants, or recalcitrant pollutants.

About 30 elements, molecules, or classes of substances are included in the list of emission limits in the Italian law for the prevention of water pollution (Law No. 319, also called Merli law). About 1500 molecules are commonly detected and classified by the Index of Solubility, Toxicity, and Biodegradability of the Water Research Centre in England. Furthermore, the U.S. Environmental Protection Agency (EPA) has established quality criteria for water for about 40 elements or substances. Currently, there is a trend in many countries to increase the number of substances to be kept under control or to include them in new regulations.

The most commonly accepted definition of water pollution is given by the Joint Group of Experts on the Scientific Aspects of Marine Pollution, which in 1970 defined marine pollution as "the introduction by man directly or indirectly, of substances or energy into the marine environment (including estuaries) resulting in such deleterious effects as harm to living resources, hazard to human health, hindrance to marine activities; including fishing, impairment quality for use of seawater, and reduction of environmental amenities." This definition mainly considers water pollution as a limitation to possible uses, and therefore quality criteria have been formulated for the various uses.

At the end of 1990s, however, the definitions of quality criteria were rejected for those uses not requiring high water quality and only four main uses were taken into consideration: those of drinking, agricultural, bathing and recreation, and aquatic life. In the 1990s, the opinion which considers the quality criteria for aquatic life as the most important when planning environmental policy was developed and consolidated. An aquatic ecosystem, in which structures and functions are not disturbed, certainly presents a water quality such that it is immediately suitable for a variety of uses or requires only very simple treatment. Moreover, in those years, it was accepted that every aquatic environment, for each pollutant, has a certain receptive capacity defined by the range between the natural concentration and that concentration which causes observable and undesirable effects.

These concepts have been the basis for the definition of quality criteria for aquatic life by different international and national organizations, such as the European Union, the European Inland Fisheries Advisory Commission (EIFAC), the EPA, and the Italian Water Research Institute.

In general, water quality criteria are defined by reviewing the available scientific information and defining limits of environmental concentrations not to be exceeded. According to EIFAC, for instance, quality criteria for aquatic life and in particular for freshwater fish should satisfy the following needs: They should permit all stages in the life cycles to be successfully completed and should not produce conditions in river water which would either taint the taste of the flesh of the fishes or cause them to avoid portions of rivers where they would otherwise be present or cause the accumulation of deleterious substances in fish in such a degree that they are potentially harmful when consumed. Indirect factors, such as those affecting fish food organisms, must also be considered in the definition of quality criteria.

Here, it is advisable to mention the concepts of *standard, criterion,* and *objective* because these terms are frequently used in the context of water protection. Their meanings are certainly not unique, although certain definitions have been widely used by the majority of researchers and water quality managers and they deserve to be mentioned here.

The quality standard applies to any definite rule, principle, or measure established by authority. The fact that it has been established by authority makes a standard somewhat rigid, official, or quasi-legal; this does not necessarily mean that the standard is fair, equitable, or based on sound scientific knowledge. It may have been established arbitrarily on the basis of inadequate technical data corrected by a proper factor of safety. When scientific data are not consistent, such arbitrary standards may be justified.

The quality objective represents an aim or a goal, and it may designate an ideal condition. However, it does not imply strict adherence or rigid enforcement by authorities. The objective is to obtain approval among engineers on boards and commissions that intend to achieve water pollution control by persuasive methods and cooperative actions.

A quality criterion designates a condition defined by means of a critical review on scientific information and that is suitable to maintain the structural and functional characteristics of the ecosystem. Unlike a standard, it has no connotation of authority other than that of fairness and equity; nor does it imply an ideal condition.

Once quality standards, objectives, and criteria for the aquatic life are fixed, there are different procedures for obtaining the desired quality criterion (Fig. 1). Two different approaches have been followed by different countries: the approach that does not take into consideration the type and use of the receiving water body (rigid effluent standards) or does not adapt the quantity of the pollutant to the natural characteristics of the receiving water body (rivers, lakes, and coastal waters) and that which takes into account the characteristics of the pollutants, namely, toxicity, persistence, and bioaccumulation (flexible effluent standards).

The first approach, which does not consider the characteristics of the receiving water body, must necessarily be very restrictive in order to be effective even in the most critical situations. As a result, in many cases the application of this criterion can lead to the overestimation of the need

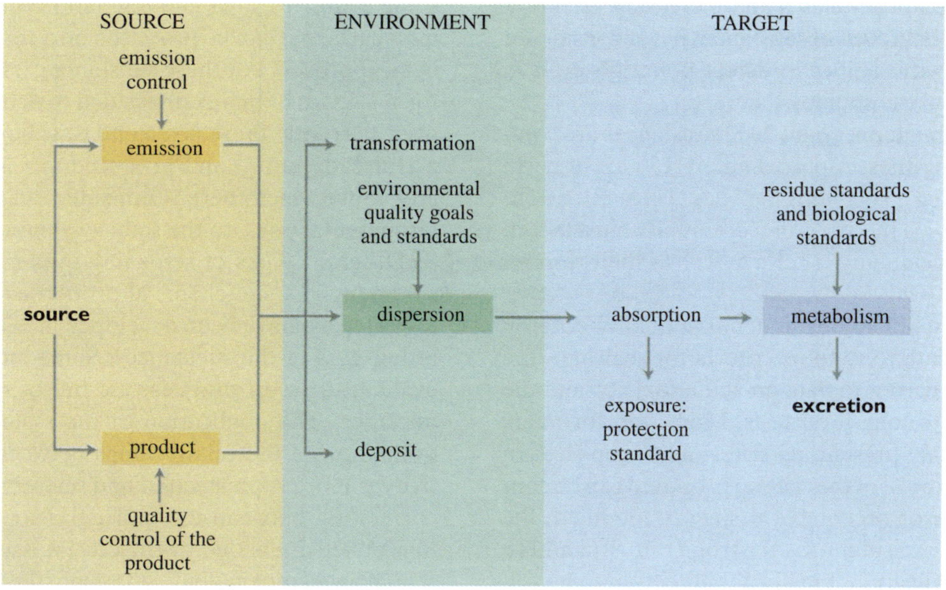

FIGURE 1 Schematic representation of transport, system control, and the processes of transformation of a toxic substance through the various environmental sectors.

of restoration and hence to excessive and unnecessary cost. More permissible regulation may be more economical, but it may lead to a reduction in the level of protection. In fact, by definition this procedure does not include the quantity of pollutant discharged over a period of time or the number of discharges in the same receiving water body, and it also does not take into account its receptive capacity. The main advantage is the facility of administrative management.

The second approach, on the other hand, appears to be excellent from many standpoints, especially because it requires a case-by-case treatment level which considers the receptive capacity and the water use. Therefore, this procedure implies law application costs, and it offers reliability in protecting the environment. This procedure is certainly preferred when setting up regulatory action for effluents emission limits; despite being theoretically valid, from the practical standpoint it is difficult to apply because it requires the knowledge of many basic elements.

First, knowledge of the receptive capacity of the water body is necessary. Such knowledge implies a case-by-case study of the typology of the water body, its hydrological balance with particular reference to the critical flow, its actual dilutant capacity, its oxidizing capacity, its chemistry, its biological structure, its thermic variations, its relationship with groundwater, the preexistence of other sources of pollution, and all other factors that can influence the actual maintenance of a safe concentration of the discharged pollutant. Second, knowledge of the evolution of the sources of pollution is necessary. This analysis requires detailed information on population characteristics and on urban and industrial development, etc. for many local situations and, obviously, within the framework of general economic planning.

Therefore, in many countries, particularly in the European Community, an intermediate approach has been followed. This procedure allows an adaptation of the effluents standards to the typology of the receiving water body and to groups of pollutants. It is also possible to set limits in other parts of the ecosystem rather than in the water body (organisms or sediments).

In summary, during the 1990s there was a tendency to consider the subjects on a case-by-case basis, to separate the ways of operating with different kinds of pollutants, and to establish different criteria for receiving water bodies with different characteristics. An example of this procedure concerns the different regulations based on the different scientific background for mercury and phosphorus, or even the differences between regulations of flowing water and lakes and marine waters. It must also be remembered that the classification of certain substances in a "black list," which includes substances not to be discharged and not to be used, and in a "gray list," which includes those substances which may be discharged within certain limits and under controlled conditions, is commonly accepted.

The Function of Scientific Research in Defining Water Quality Criteria

As a necessary premise, it should be remembered that in recent years there has been relevant progress in equipment, development tools, and analytical methodology in response to increasing pollution problems and increasing attention toward substances or groups of substances never considered before and also to the ability to detect pollution even in those areas far from the sources.

In the European Union, about 9000 substances are commonly used; among these, approximately 150 are used in quantities exceeding 50,000 tons per year. Thus, the extent of the problems which the scientific community must face in attempting to define quality criteria as well as their responsibility is evident.

There are many difficulties in defining quality criteria: the lack of basic data to support criteria for seawater, the excessive emphasis given to data on acute toxicity, and the lack of reference to long-term tests. Moreover, for many pollutants commonly present, no criterion is proposed. In other words, about 50% of the criteria proposed can be considered acceptable or acceptable if slightly modified; the remaining criteria are submitted to strong criticism and to proposals for modification. Finally, the direction to be followed for future developments needs to be clarified, and information needs to be obtained on the chemical behavior of the pollutant on environmental fate, on the toxicity of its principal chemical forms, and on possible transformations and bioaccumulation.

An even more controversial subject that needs further experimental research concerns the problem of mixtures of toxicants; for example, it is sufficient to consider the divergent positions taken by two different organizations on the matter. The EIFAC (1980), after an extensive literature review on the effects of mixtures of toxic compounds on fishes, concluded that experimental evidence suggests that the proposed quality criteria for single toxins can be applied even in cases in which several pollutants are present. On the other hand, the EPA (Stephan, 1985) states that the concentration of a single pollutant in the mixture is noneffective if four or fewer of the five proposed quality criterion for the case in which the single pollutant are present; thus, it is necessary to reduce the quality criterion of each pollutant present in the mixture.

In summary, one can infer that "traditional" approaches serve to cover large gaps in knowledge or at least to clarify very controversial points. Concerning the "nontraditional" research, there are two main currents: The first regards the forecasting of the effect and the fate of pollutants in ecosystems, and the second regards the rebuilding of the ecological cycle of the molecules even in quantitative terms. These two branches of research are supposed to produce suitable information for pollution control before pollutant emission in the ecosystems occurs. For instance, in the 1990s several substances (DDT, Aldrin, and Dieldrin) were banned due to their relevant effects on environment and human health. This fact demonstrated the need to forecast the risk related to the use of many chemical substances. In some countries, laws and regulations on the new synthesis molecules that can be produced on a commercial scale have been approved. For these substances, it is prescribed to perform a series of highly diversified toxicological tests. Moreover, currently there is a trend to submit molecules which are already in use and produced in very large quantities, and about which there is not sufficient information on environmental risks, to the same evaluation.

Different types of screening tests (integration of measurement of some chemical properties and use of toxicological tests) have been developed in order to rank the hazardousness of the substances. Some integrated systems of evaluation are of practical use but of scarce scientific importance. The application of the systems of quantitative evaluation of the relationships between structure and activity is more sophisticated and scientifically valuable. The correlation between the molecular structure and the biological activity is one of the crucial issues of advanced research in ecotoxicology, although there are still few experimental results. Concerning these topics in toxicology, quantitative approaches with aquatic organisms were attempted at the end of the 1990s.

Currently, data on toxicology do not permit the development of sufficiently precise predictive methods. On the other hand, there are good results, as already documented, from studies about the accumulation of toxins in the muscle of fish, for which excellent correlation between the bioconcentration factor and the partition coefficient *n*-octanol/water has been found.

The widescale application of these methods and the further development of research in this sector will allow the definition of more accurate and scientifically reliable lists of dangerous substances. Problems similar to those previously described are also important in the nontraditional research field — that is, with regard to the definition in quantitative terms of the biogeochemical cycle of the various molecules or elements (Fig. 2). The statement made by Odum (1959) that "the new generation of ecologists will have to know the cycle of each substance" (Fig. 3) seems more prophetic than ever. In the 1990s, system analysis was increasingly applied in ecology and some researchers tried to explain the functioning of ecosystems by describing the environment in all its complexity by means of different models. This may be a way to overcome certain difficulties that intuition alone cannot explain and to fully understand the complex and dynamic functioning of an ecosystem and the related systems of feedback control. Simulation models, or tests performed in partially controlled ecosystems, concerning the definition of the cycles were developed in a much more accurate way with respect to the first works (Metcalf *et al.*, 1971) on pesticides and certainly are more practical than the complex detailed models developed for a few pollutants, such as that related to DDT.

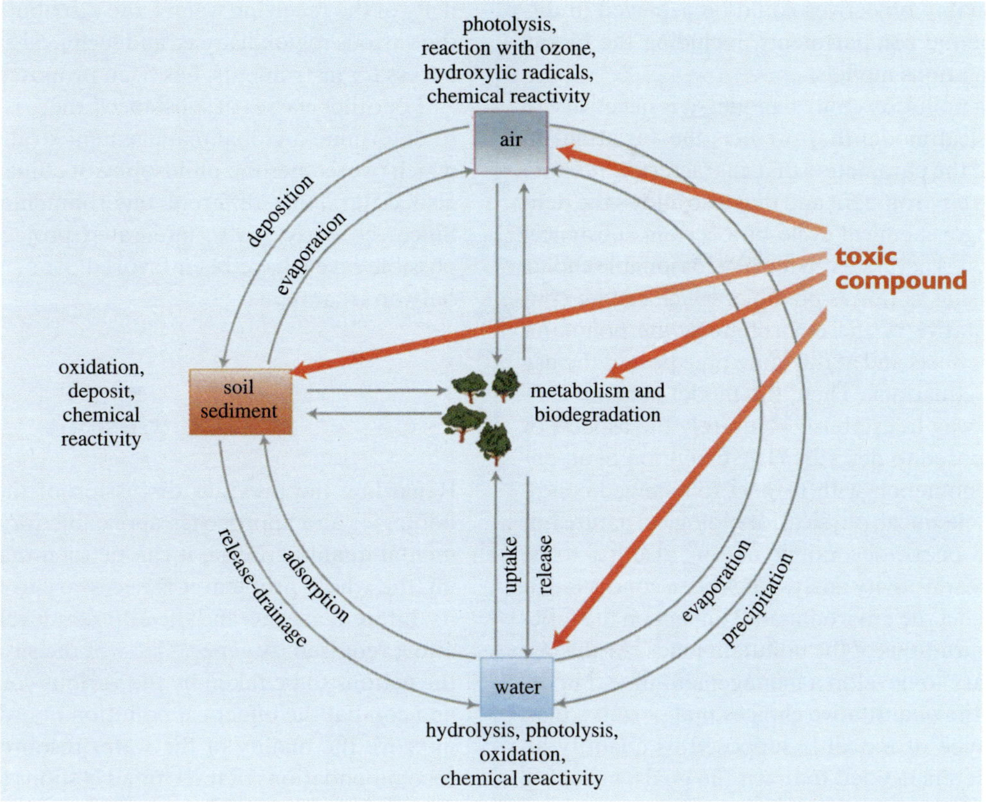

FIGURE 2 Origin, transformation, and transfer of a pollutant in the environment.

The problem related to the definition of an expected environmental concentration has been the subject of much theoretical research (Baughman and Lassiter, 1978; Branson, 1978; Haque, 1980) but has already led to the first applications. Therefore, it seems possible, on the basis of the current scientific trends, to implement a type of environmental management able to fully using the assimilative capacity (absorption, transformation, and storing) of each environmental compartment.

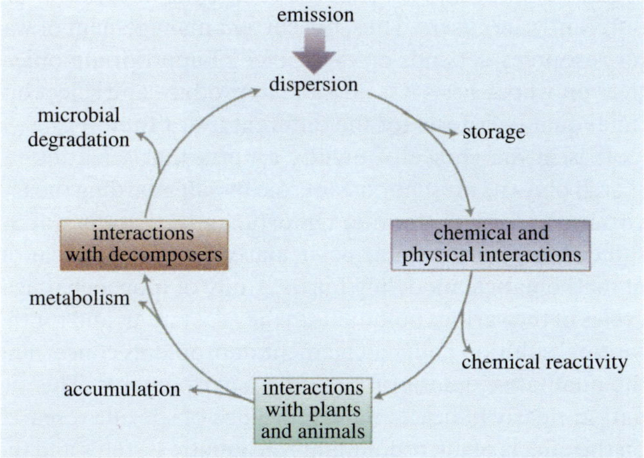

FIGURE 3 Schematic representation of the environmental dynamics of a chemical substance and its interactions with the environment and living organisms.

Remarks on the Global Ecosystems Management Approach

For the control of polluting substances, the following have been implemented: effluents emission limits (rigid effluent standards), criteria for the water bodies based on uses, and, in certain cases, classification of the water bodies. These limits or criteria were sometimes expressed in order to take into account their annual distribution, admitting their normal distribution. For example, 50% of the cases should be lower than a certain value and 90% lower than another value. Following the results of scientific research, there is a trend toward the possibility of global management of the pollutants, the levels of which would be regulated in all steps (see Fig. 1). In the simplest case, a substance could be banned (black list) or strictly limited in use; in more complex cases, limits should be imposed on the total quantities to be introduced into the environment. In other circumstances, many quality objectives could be achieved in different environmental compartments, including the biotic, by applying simulations models.

To construct a pollution control model, it is necessary to construct a physical model that predicts the variations in space and time of the parameters that characterize the state of pollution of an environment and that also allows the definition of the biogeochemical cycle of a certain substance. For this purpose, it is necessary to make a reasonable choice of parameters, using as few as possible, which can be considered representative of the environment, the pollutant, and the possible effects and at the same time permit the use of mass balance equations. Then, the model must be calibrated in such a way to establish accurately the aspects of interest it is supposed to describe. The evolution of an environmental phenomenon with respect to changed conditions, not only of chemical, physical, or biological nature but also of industrial or socioeconomic nature, also has to be established. Once sensitivity analysis has been conducted, it is possible to predict the environmental impact on the function of possible variations of the pollutant loads. As the last step, it is necessary to develop a management model in order to carry out the quantitative choices that permit a plan of action. This type of model is supposed to quantify all those key parameters needed to reach the goals, including socioeconomic parameters.

Relatively simple computer programs have been developed that allow an evaluation of the maximum predictable concentration of an organic substance in a simple ecosystem. By applying the model to 10 substances, this type of analysis proved to be very useful. By combining scenarios, laboratory data, and system analysis, it is possible to define the critical points on which to carry out research and obtain better information on the fate and the distribution of pollutants. In the early 1980s, a procedure for assessing the environmental concentration of consumer products such as detergents and additives contained in the washing powers was developed. The evaluation includes pro-capite consumption of the product, the distribution and efficiency of water treatment plants, the dispersion of sludge, and the dilution factor in the receiving waters. It is evident that there has been a shift from purely theoretical models to operative models on a more or less wide scale and of more practice type.

An example of a global strategy is the control of phosphorus, the main cause of the eutrophication of inland and coastal waters. For this element, data available for the choice of strategies to apply countrywide have been evaluated, possible lines of action for the control of phosphorus in relation to the sources (metabolic by-products, detergents, agriculture, husbandry, and industries) have been determined, and an integration of the various types of intervention, also taking into account the geographic specificity of the receiving waters, the distribution of the loads in the various regional areas, and technical and administrative necessary instruments, has been proposed.

For other classes of substances, there is ongoing research to determine a global management strategy for pollutants which overcomes the philosophy of effluents standards and also criteria for different environmental compartments. Since the early 1990s, integrated projects concerning all possible levels have been favored over strategies centered only on treatments.

Conclusions

Regarding the previous discussion of the different methodologies and approaches applicable for defining environmental quality criteria, it can be seen that when considering the whole problem it is necessary to take into account its different aspects and the different levels of investigation effort required. A general view of the situation concerning the actions to be taken by the various countries to prevent and combat the effects of pollution in order to protect and improve the quality of the water resources is given in the recommendations of the United Nations (1977) conference on water that are known as the "Mar del Plata Action Plan," which is still of great interest today.

The control and management of water quality, in relation to the general interest of the society, implies the choice of priorities and objectives, the selection of criteria through which quality must be verified in relation to fixed objectives, and the definition of standards for the criteria chosen in order to pursue and achieve the goals. The establishment of the objectives is a political decision which must be made on the basis of social, economic, and technological considerations, taking into account the conflicting interests of the different water users. Thus, the current management of water resources depends on the choice of appropriate objectives on whose basis it is possible to produce and select optimal quality criteria for the different uses (Table 1).

It is in the choice of quality criteria that scientific research plays its most important role by safeguarding the environment for a better life. Unfortunately, there is still insufficient scientific data, particularly for the application of mathematical modeling for the study of biogeochemical cycles of the various pollutants in the ecosystems and for the rational solution and management of problems concerning the qualitative deterioration of the environments. This situation clearly indicates the importance of specific research on the effects of micropollutants on aquatic systems and the necessity of reducing feedback time between the production of scientific data and their application in government environmental policy and management.

TABLE I

Actions Needed to Protect and Improve the Water Quality Resources

ACTION	DEFINITION
Locating of settlements	Devote careful attention to the availability of water and the effects of environmental pollution when deciding on the location of new settlements
Planning land use	Apply appropriate land-use planning as a tool for preventing water pollution, especially in the case of groundwater
Use of low environmental impact substances	Stimulate the development and use of substances which minimize hazards to human health and the environment, taking into account toxicity, biodegradability, bioaccumulation, and eutrophication
Non introduction of toxic substances in aquatic systems	Counteract with all appropriate measures the introduction into water of hazardous substances, such as DDT, polychlorinated biphenyl, mercury, and cadmium, taking into account the special requirements of developing countries
Survey of pollution levels	Conduct surveys on current levels of pollution in surface water and groundwater resources, and establish monitoring networks for the detection of pollution
Monitoring agricultural pollution	Conduct research on the measurement of surface and groundwater pollution caused by agricultural fertilizers and biocides, particularly regarding their adverse environmental impact
Establishment of analysis laboratories	Establish, when and where necessary, laboratories for the systematic and routine analysis of water samples, including physical, bacteriological, and biological analysis
Monitoring air pollution	Increase efforts to monitor and assess the effects of the deposition of airborne pollutants in water from distant sources and reduce the total emission of such pollutants, i.e., by applying the best available technology that is economically feasible
Biological controls	Conduct biological control where chemicals are used in the control of water-related organisms
Quality control standardized criteria	Use, where possible, uniform criteria, methods, and standards for assessing and monitoring water quality, compiling data and classifying waters with regard to their use
Defining quality standards	Establish quality standards for the various beneficial uses of water, whenever possible, taking into account the degree of development and the social and economical conditions of each situation
Updating pollutant lists	Prepare and continuously update a list of water pollutants and a harmonizing terminology in the field of water pollution control, in collaboration with existing international organizations engaged in similar work
Developing simulation models	Recognize that techniques such as simulation, parametric modeling, and computerized analysis need to be developed to facilitate solutions to problems in the field of pollution control
Establishing institutions against pollution	Set up adequate institutions, where necessary with appropriate coordinating machinery, or strengthen those that already exist to enable them to be more effective in the fight against pollution
Law application	Apply legislation, regulatory measures, and systems of incentive charges as to discharge of pollutants that guaranteed quality goals will be reached within certain periods of time; the discharge into the aquatic environment of dangerous substances that are toxic, persistent, and bioaccumulative should be gradually eliminated
Regulating industrial waste	Regulate the discharge of industrial, urban, and mining wastes into water bodies by establishing the necessary control measures in the context of an overall water management policy, taking into account qualitative and quantitative aspects
Industrial pollution control	In combating industrial pollution, undertake a careful analysis of the cost-effectiveness ratio of the different measures proposed for pollution control
Evaluation of measures against pollution	Recognize that it is necessary to undertaken studies to evaluate the best possible approach to control pollution on an industry-by-industry or pollutant-by-pollutant basis, in accordance with national requirements and with the nature and level of industrial development
Costs at the polluter expense	Adopt the general principle that, as far as possible, direct or indirect costs related to loss of amenity attributable to pollution should be paid by the polluter
Improvement of treatment systems	Increase the number and improve the operation and establish comprehensive monitoring of facilities and technologies for the treatment of wastewater, giving greater attention to alternative (especially low-energy) methods of waste treatment and land application or other economic use of wastes
Decontamination measures	Ensure fast decontamination of water pollution during natural and man-made hazards

References Cited

Baughman, G. L., and Lassitfr, R. R. (1978). Prediction of environmental pollutant concentration. In *Estimating the Hazard of Chemical Substances to Aquatic Life* (J. Cairns, Jr., K. L. Dickson, and A. W. Maki, Eds.), STP No. 657, pp. 35–54. ASTM, Philadelphia.

Branson, D. R. (1978). Predicting the fate of chemicals in the aquatic environment from laboratory data. In *Estimating the Hazard of Chemical Substances to Aquatic Life* (J. Cairns, Jr., K. L. Dickson, and A. W. Maki, Eds.), STP No. 657, pp. 55–70. ASTM, Philadelphia.

European Inland Fisheries Advisory Commission Working Party on Water Quality Criteria for European Freshwater Fish (1980). Water quality criteria for European freshwater fish; Report on combined effects on freshwater fish and other aquatic life of mixtures of toxicants in water. Food and Agriculture Organization, Rome.

Haque, R. (Ed.) (1980). *Dynamics, Exposure and Hazard Assessment of Toxic Chemicals.* Ann Arbor Pub., Novato, CA.

IMCO/FAO/UNESCO/WMO/IAEA/UN Joint Group of Experts on Scientific Aspects of Marine Pollution (1970). Reports. IMCO/FAO/UNESCO/WMO/IAEA/UN Joint Group of Experts on Scientific Aspects of Marine Pollution.

Metcalf, R. L., Sangha, G. K, and Kapoor, I. P. (1971). Model ecosystem for the evaluation of pesticide degradability and ecological magnification. *Environ. Sci. Technol.* **5,** 709–713.

Odum, E. P. (1959). *Fundamentals of Ecology.* Saunders, Philadelphia.

Stephan, C. E. (Ed.) (1985). *Guidelines for Deriving Numerical National Water Quality Criteria for the Protection of Aquatic Organisms and Their Uses.* U.S. Environmental Protection Agency, Washington, DC.

United Nations (1977). *Mar del Plata Action Plan. Reports of the United Nations Water Conference, Mar del Plata 14–25 March 1977.* United Nations, New York.

David B. Peakall

Monitoring and Assessment Research Center
The Old Coach House, King's College
London W8 7AD, United Kingdom

Ecotoxicology is the field of study that integrates the ecological and toxicological effects of chemical pollutants. Although the greatest interest of ecotoxicologists is effects on higher organizational levels, such as population, community structure, and ecosystems, it becomes increasingly difficult to relate effect to any specific chemical as one moves up the organizational levels. This article is organized into three main topics: the movement of pollutants in air, water, soils, and through the food web, together with biotransformation; the effects of pollutants on living organisms; and the effects of pollutants at the level of population, community structure, and, as far as possible, on the whole ecosystem.

Ecotoxicology

Introduction

Ecotoxicology merges the fields of ecology and toxicology, although the definitions and the emphasis that various workers put on certain aspects of the subject very much depend on whether they came from the ecological or toxicological fields. Since 1977, when the term *ecotoxicology* was first coined by Trehaut, ecotoxicology has become a field of study it is own right and it should be taught as such. Nevertheless, the field is still influenced by the biases of the approaches of various practitioners.

The original definition of Trehaut was "the branch of toxicology concerned with the study of toxic effects, caused by natural or synthetic pollutants, to the constituents of ecosystems (including human), vegetable and microbial in an integral context." This is clearly the definition of a toxicologist. Moriarty, in his pioneering book *Ecotoxicology: The Study of Pollutants in Ecosystems* (1983), defines ecotoxicology as "the study of the effects of pollutants on ecosystems." This is equally clearly the definition of an ecologist.

Although no one would dispute the importance of knowing the effects at an ecosystem level, it is only rarely that we can measure effects of pollutants on the ecosystem as a whole. Thus, ecotoxicology is forced to concentrate on lower organization levels than the ecosystem. Forbes and Forbes (1994), after discussing many other definitions, defined ecotoxicology as "the field of study which integrates the ecological and toxicological effects of chemical pollutants on populations, communities, and ecosystems with the fate of such pollutants in the environment."

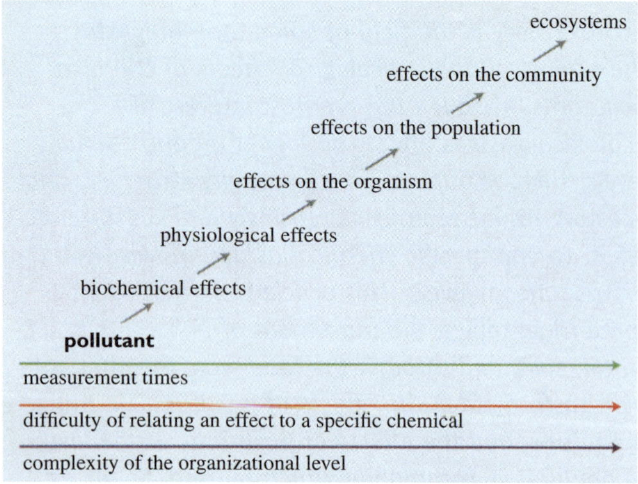

FIGURE 1 The possibility to discriminate the effects of a chemical compound in relationship to the time frame of measurements and increasing levels of organizational complexity of the ecosystems. With increases in the organizational level it is more difficult to relate any effects seen to a specific chemical and the observations need longer measurement times (reproduced with permission from Walker *et al.,* 1996).

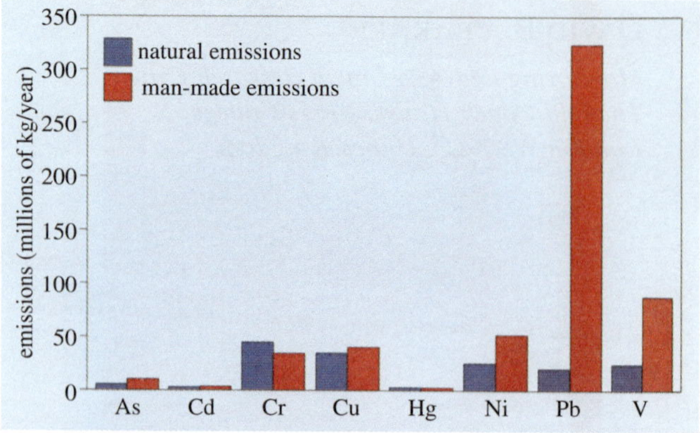

FIGURE 2 Natural and anthropogenic emissions of metals (reproduced with permission from van Emden and Peakall, 1996).

As one moves up the organizational levels it become more difficult to relate any effects seen to a specific chemical and the time frame of measurements also increases. These concepts are shown diagrammatically in Fig. 1.

This article is organized into three main topics: the movement of pollutants in air, water, soils, and through the food web together with biotransformation; the effects of pollutants on living organisms; and the effects of pollutants at the level of population, community structure, and, as far as possible, on the whole ecosystem.

Pollutants and Their Fate in Ecosystems

The major classes of pollutants and their movement and fate in the environment are considered here. Some of the pollutants are considered by their chemical class (i.e., metals and organochlorines) and others by the route by which they enter the environment (i.e., other pesticides and the by-products of energy production).

Metals

The metals, of course, are naturally occurring, so we must deal with the effects of the excess toxic metals put into the environment due to human activity. The relative importance of natural and anthropogenic inputs of metals is given in Fig. 2. The metal with the highest ratio of man-made to natural emissions is lead. This is largely due to the use of lead in gasoline; it is calculated that 80–90% of lead in ambient air in the 1980s came from this source. In Western countries, this source has decreased considerably, but it still remains a problem in many areas of the world.

Although the metallic elements cannot be created or destroyed, organometalic compounds can be made and broken down. In addition to tetraethyl lead produced for use in petrol, another man-made organometal which has leached into the environment is tributyltin, which is used as an antifouling agent for boats. The effects of this on the marine environment, especially on mussels, is discussed later in a case study.

A few organometals are made naturally, the most important of which is methyl mercury. Inorganic mercury can be methylated by microorganisms, and this form of mercury, dimethyl mercury, is much more readily taken up by animals. Flooding of land can liberate mercury from the soil; mercury can then become methylated and accumulate in the biota in the course of hydroelectric projects. One such project was the Churchill–Nelson River Hydroelectric Development Project in Manitoba, Canada. Levels of mercury in fish increased soon after impoundment and in predatory fish increased from approximately 0.5 to approximately 2.5 ppm. Mercury levels in fish showed no decline 5 years after impoundment. Elevated levels were noted in fish-eating mammals, although no toxic effects were noted. A substantial proportion of the native population had blood levels above 20 ppb, but no adverse effects were detected. Another Canadian project that is causing considerable concern and controversy is the James Bay Project. This project involves the diversion of three rivers and the flooding of 14,000 km^2 of land in which the soil is rich in mercury, and the project affects an area the size of France. There is considerable scientific debate about the timescale for the rise and subsequent decline of mercury levels in fish; it has been claimed that the aggregate lake would be one of the most mercury-polluted water bodies in the world and would remain a significant problem for 80–100 years.

Organochlorines

Although some organochlorines (OCs) occur naturally, those of ecotoxicological importance are man-made compounds created either as pesticides or for industrial purposes. Among the former are such well-known compounds as DDT and dieldrin, and the latter include polychlorinated biphenyls (PCBs) and polychlorinated dioxins.

The OCs are characterized by their stability and low water/high lipid solubility. These two factors cause these compounds to be bioaccumulated up the food chain. This phenomena, based on data from the North American Great Lakes, is shown in Fig. 3.

One of the earliest examples of the problems caused by bioaccumulation regarded grebes at Clear Lake in California. This lake was spraying with DDD in 1949 to control a gnat which in the summer months multiplies to such numbers as to cause a serious nuisance although it is not a health hazard. However, bioaccumulation of this pesticide up the food chain caused the loss of the entire population of western grebes, which had numbered 1000 pairs before spraying was started.

One of the most widespread effects of organochlorine pesticides was a reduction of many species of fish-eating birds and birds of prey throughout the Northern Hemisphere. This is considered in more detail in case studies presented later.

A group of OCs used for a wide variety of industrial purposes that subsequently caused serious environmental problems are the PCBs. These highly stable compounds were first used industrially in the early 1930s but were not detected in the environment until the mid-1960s. The toxicological assessment of this complex group of chemicals is considered later.

FIGURE 3 Accumulation of DDT up the Lake Michigan food chain. Total values of DDT are expressed in ppm (reproduced with permission from Walker *et al.,* 1996).

The discovery of organochlorine pesticides in the fauna of Antarctica in the late 1960s was the first dramatic demonstration of the global nature of the contamination of the environment by these compounds, since none of them had been used on that continent. Although the levels there were low, the same is not true of the Arctic. Here, substantial levels have been recorded in wildlife and the levels of PCBs in the native population of arctic Canada are approximately four times higher than those in the population of the cities in the south of the country.

Plants are an important means of monitoring global pollution by organochlorines since there is a linear relationship between the concentration of chemicals in foliage and in air. The global patterns of the various organochlorines are quite different. Hexachlorobenzene (HCB), which on the basis of its high vapor pressure would be expected to have a long atmospheric residence time, has a fairly symmetrical pattern with low values in the tropics and much higher values at high latitudes. This trend can be explained by the cold condenser effect which is particularly important for a volatile compound such as HCB. This view is supported by the fact that levels on Mount Everest and Mount Kenya are higher than at other sites in the tropics (Calamari *et al.,* 1995). It is this condensation of organochlorines from the atmosphere that accounts for the high levels in the arctic fauna. The amount transported depends on the volatility (Fig. 4). In contrast to HCB, DDT plant concentrations are highest in the tropics and subtropics. Here, the usage is higher than in countries at higher latitudes and the high temperature increases the vapor pressure of this nonvolatile compound.

Other Pesticides

Compared to the organochlorine pesticides, the organophosphates (OPs) and carbamates are often said to be more toxic and less persistent, and their effects are more transient. Although there is truth in these points, they are an oversimplification. The range of toxicity of OPs is considerable and some are less toxic than the OCs. Recently, there have been concerns, at least regarding human health, about the long-term effects of these compounds, particularly in the Gulf War syndrome. However, the most obvious effect on wildlife has been direct mortality, which has often been on a large scale.

Many instances of mortality of birds caused by OPs are known. In North America, three-fourths of the incidents reported involve diazinon, fenthion, parathion, or phosphamidon, whereas in Europe carbophenothion and chlo-

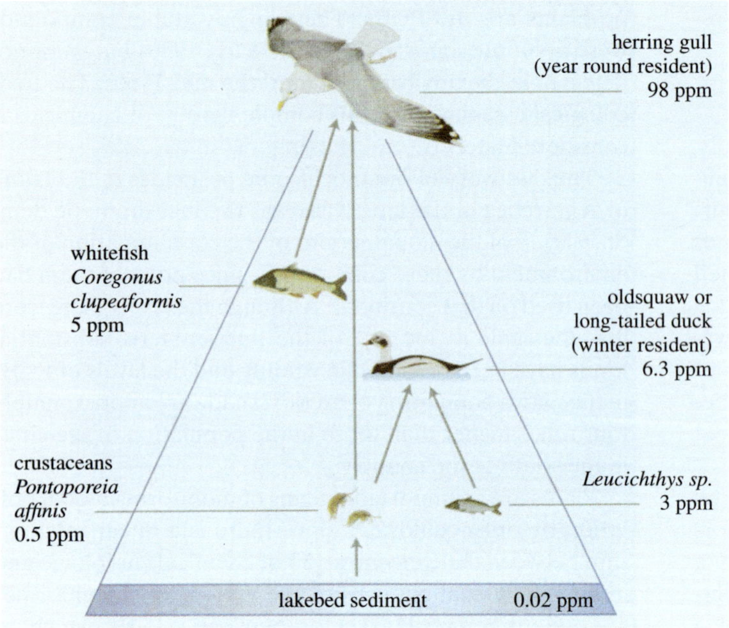

FIGURE 4 Transport of pesticides from the tropics to the Arctic. The degree of transport depends on the volatility.

rofenvinphos are the principal pesticides causing problems. Among the carbamates, carbofuran has caused the most problems. Some of the instances of mortality involve considerable numbers — for example, 10,000 American robins (*Turdus migratorius*) on berry fields in Florida and 500 greylag geese (*Anser anser*) in Scotland.

In these cases, the mortality was confirmed by direct counts, but potentially more serious are much larger losses that have been estimated on the basis of losses in small areas and then extrapolated to the total area sprayed. Based on the reduction of singing males following spraying of the forests of New Brunswick with phosphamidon, it was estimated that 2.9 million songbirds were killed. Using the number of carcasses found at specific sites and multiplying this by the total area treated, the U.S. Environmental Protection Agency estimated that mortality caused in the United States from the use of carbofuran on corn was 1.3–1.6 million birds per year. Pimentel *et al.* (1993) estimated that 67 million birds are killed annually in the United States. Of course, there is considerable uncertainty, but the data do suggest that pesticide usage could be a significant mortality factor.

Another group of compounds that are used as pesticides are the pyrethroids. Natural pyrethroids have been in use since the 1820s, but widescale use of synthetic pyrethroids dates only from the 1980s. The pyrethroids are an example of the artificial nature of the separation between natural and synthetic chemicals. Pyrethroids are rapidly metabolized and have low toxicity to warm-blooded animals. However, they have high toxicity to aquatic invertebrates.

Herbicides are widely used in large quantities and their usage is increasing at a faster rate than that of other categories of pesticides. Many herbicides, such as phenoxyacetic acid and urea-based herbicides, have low toxicity to warm-blooded animals, although some, such as bipyridinium (e.g., paraquat), can be highly toxic. However, the greatest concern from a wildlife point of view about the usage of herbicides is the alteration of habitat. The impact of these habitat changes on avian population in the United Kingdom is considered as a case study.

Petroleum Hydrocarbons

These are naturally occurring compounds, but man's extraction and subsequent transportation of them throughout the globe has greatly increased the environmental exposure of this group of chemicals. The annual input of petroleum hydrocarbons to the oceans of the world has been estimated at 3.2 million tons. Figure 5 gives a diagrammatic representation of the relative importance of the routes of entry into the environment. Waste disposal and transportation are the two largest sources. Spills, despite the attention that they receive, account for only about 12% of the total.

Chemicals Associated with Energy Production

Energy usage is fundamental to human comfort but cannot be produced without environmental costs. Carbon dioxide emissions from fossil fuels are an important by-product of energy production and the largest contributor to the "greenhouse" effect. Countries and regions with the most carbon dioxide emissions are shown in Fig. 6. The United States is the largest single contributor, in which 4% of the world's population produces more than 20% of the pollution. Currently, the production of carbon dioxide per person is only 0.7 tons/year in China compared to 5.3 tons/

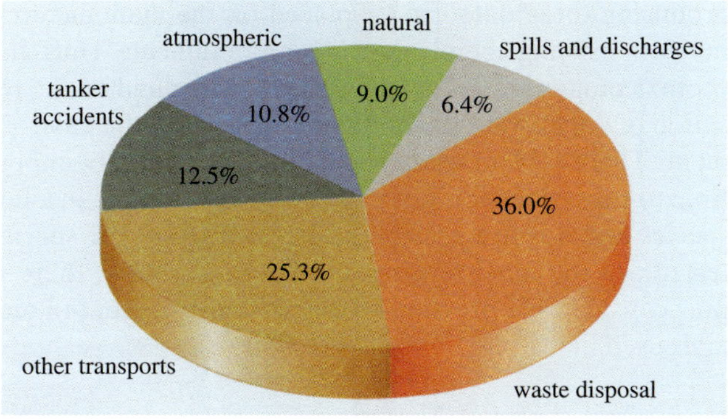

FIGURE 5 Sources of oil emission into the marine environment (reproduced with permission from van Emden and Peakall, 1996).

year in the United States. Nevertheless, the rate of increase in China is rapid, and China is projected to become the largest single contributor. The changes expected due to global warming provide an interesting possibility of applying ecotoxicology on a global scale and are considered later.

The oxides of sulfur and nitrogen are also produced in large amounts; for the former, combustion of coal is the major source, and transportation is the major source for the latter. The oxides of nitrogen and sulfur are carried for hundreds of miles; the acidification of lakes in Norway has been blamed on coal burning in the United Kingdom. The consequences of "acid rain" are considered later.

Chlorofluorocarbons

Like the PCBs, the chlorofluorocarbons (CFCs) were first synthesized in the early 1930s and initially appeared to be harmless because they are inert and nontoxic. There are several different CFCs and a shorthand numbering system is commonly used. The two most commonly used compounds are F-11 and F-12. The first digit is the number of fluorine atoms and the second is the number of hydrogen atoms plus one. Thus, F-12 contains one fluorine and one hydrogen atom.

Not until the 1970s were studies undertaken that found that CFCs were accumulating in the upper atmosphere. The problem that occurs from this accumulation is that the ozone layer, which protects the world from harmful ultraviolet (UV) radiation, is destroyed. The global nature of the problem is shown by the fact that although most of the CFCs were released in the Northern Hemisphere, the greatest effect has been seen over Antarctica. The effects of the "ozone hole" are considered later.

Other Chemicals

The previous categories do not cover all classes of chemicals found in the environment but are those of greatest concern. The total volume of chemicals entering the environment every year is immense: more than 6000 million tons of carbon dioxide, 100 million tons of sulfur dioxide, and 60 million tons of nitrogen oxides. Two and half million tons of pesticides and 300,000 tons of lead enter the environment annually.

Are there new problem in store for ecotoxicologists? This is a difficult and, in an absolute sense, impossible question to answer. Two examples of "sleeper" chemicals—PCBs and CFCs—already discussed do not give confidence in our ability to predict environmental effects of chemicals produced on a large scale. There is a strong case for an international program to discover the extent to which high-volume chemicals are actually present in the environment. For these chemicals, calculations should be made to determine which part of the environment they will most likely reach, and then that part of the environment should be examined to determine if the compounds are present.

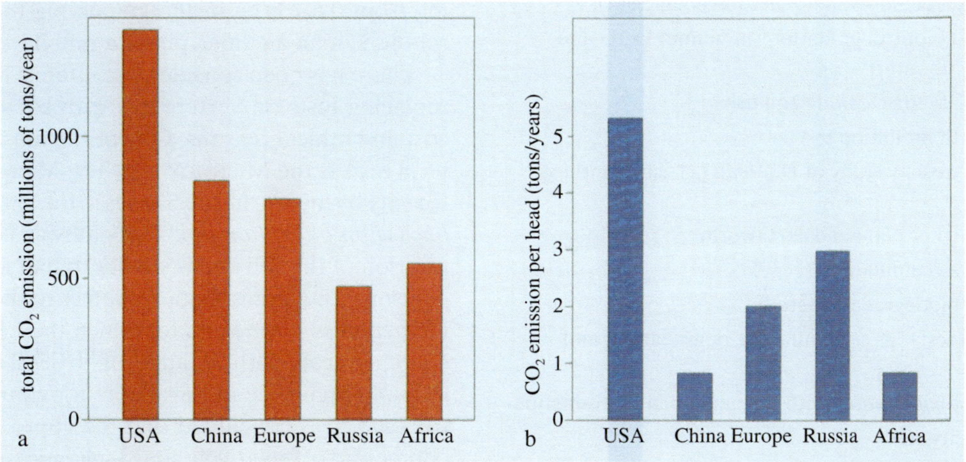

FIGURE 6 Total (a) and per head of population (b) emissions of carbon dioxide (reproduced with permission from Flavin and Tunali, 1996).

Effects of Pollutants on Organisms

Although in ecology we are more concerned with populations and communities than the individual, most tests, especially those officially adopted, are single-species tests. Another fundamental difference between human toxicology and ecotoxicology is that for human studies one is concerned with the extrapolation from many species to one species (man), whereas in ecotoxicology we are concerned with extrapolation from a few species to a vast number of diverse species.

In the following sections, the official ecotoxicology tests are briefly described, followed by a discussion of indicator species and the biomarker concept.

Official Ecotoxicology Tests

The testing of industrial chemicals within the European Economic Community (EEC) is based on a tiered approach, with the following levels: level 0 for all new industrial chemicals, level 1 when production exceeds 100 tons/year, and level 2 when production exceeds 1000 tons/year. The ecotoxicological data requirements for these various levels are summarized in Table 1.

The EEC level 0 tests are often referred to as the minimum premarket data tests, and it is an important point that these tests apply to new chemicals. The responsibility for producing these data can be placed on the manufacturer, but with existing chemicals this is more difficult. Thus, the ecotoxicological tests for a new chemical depend on the reaction of only two of the millions of species in the environment. This problem of extrapolation is essentially impossible to solve since it is not possible to test more than a few species and it is not possible to know whether the species tested are the most sensitive ones. Furthermore, there is now considerable pressure to decrease the number of animals used in testing.

TABLE I

Ecotoxicological Data Required by European Economic Community[a]

LEVEL	DATA
0	Physicochemical properties (vapor pressure, water solubility, octanol–water partition coefficient).
	Abiotic and biotic degradation (minimum BOD and BOD/COD ratio)
	Acute toxicity for *Daphnia* and fish
1	Algal growth inhibition test
	Prolonged toxicity study of *Daphnia* (21 days) and fish (14 days)
	Test on a higher plant and earthworm
	Test for bioaccumulation
	Prolonged biodegradation study
2	Additional tests for accumulation, degradation, and mobility
	Prolonged toxicity study with fish, including reproduction
	Avian toxicity study
	Other tests may be required based on likely movement within the environment

[a] From Directive 79/831/EEC.

Reducing Animal Testing

Twenty years ago, when the current testing systems were designed, there was little concern about the numbers of animals used in testing. Although this is not the place to debate the ethics of the issue of animal rights, there is no doubt that it has become a political force throughout the Western world and there is considerable pressure to alter testing procedures in the near future. There is a paradox of growing public demand for greater safety regarding the effects of chemicals on human health and the environment and at the same time a demand for fewer or even no animal experiments.

The two main avenues available, short of banning all testing, are to reduce the number of animals used in a specific test and to use alternative tests.

Procedures are now available to reduce the number of animals used in the basic LD_{50} test, which determines the dose of a chemical that kills half of the test animals. New methods, based on small numbers of animals, have been put forward and reasonable agreement between the results from these procedures and literature values has been found. Furthermore, high precision in the LD_{50} value is not important due to the high degree of uncertainty regarding the extrapolation from test animals to humans or other animal species. For example, the LD_{50} for dioxin varies by more than two orders of magnitude for mammalian species, ranging from 0.6 μg/kg in the guinea pig to 115 μg/kg in the rabbit. Is man a guinea pig or a rabbit?

There has been considerable interest in the possibility of replacing tests on vertebrates with bacteria tests for both cost and ethical reasons. One of the most widely used bacteria tests is the Microtox test. The Microtox test evaluates toxicity by measuring decreases in the light output of *Photobacterium phosphoreum.* This pathway is linked to the respiration of the cell and is similar to that found in higher organisms. The largest data set refers to the fathead minnow (*Pimephales promelas*), for which the LD_{50}s of more than 400 chemicals with a range of toxicities spanning 10 orders of magnitude are available for comparison to Microtox data. It is concluded that the approach works well for compounds of relatively simple chemical structure with one reactive group, but it does not work as well with complex structures, especially those whose toxicity is greatly affected by pH changes. Nevertheless, the Microtox test is becoming widely accepted as an important part of the testing procedure of chemicals in many countries.

The Biomarker Approach

An approach that is being increasingly used to quantify exposure and its impact is to monitor biological endpoints (biomarkers) in animals and plants.

There have been several attempts to define the term biomarker, but the definition of biomarker used here is "a biological response to a chemical or chemicals that gives a measure of exposure and sometimes also of toxic effect in an organism." It is proposed that biological responses measured above the organism level be considered as "ecological indicators." Although ecological indicators are obviously important, they are less well developed than biomarkers and their current status is considered later.

There have been many attempts to subdivide biomarkers. The journal *Biomarkers* makes the following division: biomarkers of exposure, biomarkers of effect, and biomarkers of susceptibility. The latter category is considered to include genetic factors and changes in receptors which alter the susceptibility of an organism to exposure to a chemical substance.

However, these subdivisions seem artificial. By definition, all biomarkers are biomarkers of exposure. Whether they are also biomarkers of effect depends on the state of knowledge. For example, the induction of mixed function oxidases is well-known as a biomarker of exposure to a variety of chemicals, and there are many examples that show that chemicals in the environment cause induction, although the consequences of the induction is unknown. In a few cases, such as for the fish-eating birds in the North American Great Lakes, enough data are available to relate this induction, expressed as dioxin equivalents, to embryonic death. This concept is considered in more detail later.

Specific and Nonspecific Biomarkers. The specificity of biomarkers to chemicals varies greatly. Both specific and nonspecific biomarkers have their place in environmental assessment. A nonspecific biomarker can indicate that a pollutant is present in a meaningful concentration but does not indicate which chemical is present. Based on this information, detailed chemical investigation can be justified. In contrast, a specific biomarker indicates which chemical is present but gives no information on the presence of other chemicals.

Some of the best studied nonspecific biomarkers are the heme-containing enzymes known as the mixed function oxidases (MFOs). This large family of enzymes is a major component of the defenses of organisms against toxic chemicals in their environment. Originally evolved, perhaps as long as 2000 million years ago, to handle naturally occurring toxic compounds, they now play an important role in the detoxification of man-made chemicals.

The concept of using induction of MFO activity in fish as a monitor of pollution of the marine environment by oil was put forward in the mid-1970s. Since then, a wide variety of studies on fish have been published, including the study of Los Angeles sewage outlets and pulp mill effluent in Sweden. Induction of MFOs by paper mill effluent has proved to one of the most sensitive biomarkers for tracking this particular pollution.

The usefulness of MFOs as a biological monitor has been clearly demonstrated for both polynuclear aromatic hydrocarbon and organochlorine pollution in a wide range of organisms. Since the response is caused by a variety of chemicals, the system is capable of showing exposure sufficient to cause a biological response to many xenobiotics. Conversely, it indicates little about the causative agent(s) but can be used to delimit an area for which it is worth the time and expense of more detailed investigations.

In contrast, the inhibition of *d*-aminolevulinic acid dehydratase (ALAD) is a sensitive and specific indicator of lead exposure. ALAD is an enzyme in the heme biosynthetic pathway and its inhibition reduces the oxygen carrying capacity of blood. Although blood lead levels greater than 30–40 mg/dl can be considered diagnostic of lead poisoning in wildlife species, determination of the ALAD activity ratio can detect lead exposure prior to the onset of other toxicological responses and therefore act as a sensitive early warning of lead exposure. Two major sources of lead into the environment are lead shot used in waterfowl hunting and the use of leaded gasoline. In many countries, there has been a strong move toward unleaded gasoline, and in some countries lead shot has been banned. Nevertheless, on a worldwide basis, both are still serious sources of this toxic metal into the environment.

Concept of Meaningful Exposure. Biomarkers have the advantage over chemical analysis in that they can demonstrate whether or not an organism is meaningfully exposed. For some classes of persistent organic chemicals, especially the organochlorines, detection limits have been reduced to parts per trillion. Although these man-made chemicals can be detected in almost all samples, the physiological significance is rarely known.

With biomarkers it is possible to determine if the physiology of the organism is significantly different from normal. If so, then the organism can be considered to be meaningfully exposed; equally important, if the physiology is not significantly different, then the organism can be considered not to be meaningfully exposed even though the chemical(s) can be detected. The ability to determine whether or not an organism is meaningfully exposed is important in the decision-making process with regard to whether action should be taken and also in making the decision whether or not remedial action has been successful.

The following criteria need to be met before the concept of meaningful exposure can be used:

1. Data must be available on the normality for each biomarker. This is complex because of the diversity of species

involved: It is impossible to have data on all species. The question of indicator species will be discussed later.

2. Biomarkers are needed to indicate the health of the major functions of the organism and to be able to assess the impact of the major classes of chemicals of concern. Although we have not reached this point yet, the rate of progress toward the goal is encouraging.

A list of criteria for evaluating biomarkers in addition to those on specificity modified from Huggett *et al.* (1992) is given in Table 2.

Selection of Indicator or Sentinel Species. One of the fundamental problems in ecotoxicology is the extrapolation of results found on a few species to many species. Obviously, the number of species that can be studied is extremely small compared to the many millions that exist, and thus great care should be taken in their selection. Two concepts which have invoked considerable support and controversy are that of *keystone species* and the *most sensitive species.* The concept of keystone species — that is, species that, if removed, cause the collapse of many other species — was put forward approximately 30 years ago and attracted considerable interest by ecologists. Despite some successes, the concept has not become widely established and has not been taken up by ecotoxicologists. The concept of the most sensitive species was put forward by toxicologists on the grounds that if the most sensitive species was protected then the remaining species would also be protected. Regrettably, the search for the most sensitive species has not been successful. For example, tributyl tin caused reproductive failure of mollusks at concentrations of a few nanograms per liter, several orders of magnitude lower than it causes effects in other classes of organism, but for most pollutants mussels are quite resistant and on this basis have been used for a major monitoring program, the "mussel watch."

The criteria for choice of indicator species for ecotoxicological studies are given in Table 3. Although the most important consideration is to select the right species for a

TABLE 2

Criteria for Evaluating Biomarkers[a]

CRITERIA	VALUATION
Biological specificity	It is important to know which classes of organisms the biomarker may be used on. The inhibition of AChE can be applied throughout the animal kingdom, whereas the induction of vitellogenin is confined to those vertebrates that lay eggs.
Clarity of interpretation	How clear-cut is the end point as an indicator of exposure to anthropogenic stress? Can the end points be clearly distinguished from natural stresses? It is valuable to know the mechanism of response to the chemical in assessing this point.
Time of responses	The response time of different biomarkers can vary widely from nearly instantaneous to years. Depending on the type of study, slow or rapid manifestation may be desirable.
Permanence of response	It is important to know how long the response lasts. If it is transient it can readily be missed. The inhibition of AcHe, especially in blood, is transient and thus it is necessary to know when the exposure occurred to assess the importance of the degree of inhibition. In contrast, the inhibition of ALAD is only slowly reversed.
Reliability	This can be considered under two headings: environmental influences that modulate the organism's response to a chemical and inherent variation in the biological response to a given exposure. It is important to know the extent of all variation in order to have a reliable biomarker.
Methodological considerations	Important considerations are precision (analytical reproducibility of the method) and cost and ease of the assay. Although many reliable assays have been developed, there is a need for standardization, along the lines used in analytical chemistry, so results from different laboratories are comparable.
Relative sensitivity	It is important that the biomarker be sensitive compared to other end points, such as mortality or reproductive impairment, and it is important to know the relative sensitivity compared to other biomarkers.
Validation in the field	For a biomarker to be useful in environmental assessment it must be validated in the field. Organisms in the field are subjected to a much wider range of variables than occurs in the laboratory.
Linkage to higher level effects	A biomarker is more useful if there is clear linkage to effects at higher levels of organization. Studies on invertebrates have been particularly fruitful because population changes occur more rapidly than in higher species.

[a] Based on Huggett *et al.* (1992).

TABLE 3

Criteria for Choice of Indicator Species

The species should be widely distributed, relatively abundant, and easy to collect. With increasing concern regarding collecting, the first two considerations are particularly important. There is an increasing tendency to use "lower" organisms since our concerns over animals are, in general, phylogenetic in nature.
It is valuable to be able to ascertain the age of the animal by inspection. To indicate contamination occurring just before collection, young or short-lived animals should be used. Where one needs information on accumulative buildup, long-lived animals are required.
If measurements of local changes are required, indicator species must be sedentary in their habitats. When contamination of a large area is to be studied, species with extensive ranges are more appropriate. In either case, the range of the indicator species must be known.
The diet and the position of the indicator species within the food web should be known.

specific study, it is useful, as far as possible, to use species for which a good deal of background data are available.

Microcosms and Mesocosms

The original distinction between microcosms and mesocosm has become blurred. The former term is generally used for small-scale models of natural ecosystems in artificial containers maintained in the laboratory, whereas mesocosms can be defined as bounded and partially enclosed experimental units outside the laboratory. However, the difference between laboratory and outside experiments is not universally accepted and some workers use size as their criteria. Any cutoff based on size is obviously arbitrary and original definition based on laboratory or outside experiments seems more logical. If this distinction is not made, then the best course is to treat them as a continuum. Nevertheless, both terms remain in frequent use.

The ecological objections to basing regulations on single-species testing are strong (Cairns *et al.,* 1996). No species lives in isolation but rather is part of a living, dynamic community. Nevertheless, the extent to which a system consisting of a few algae, bacteria, and invertebrates can be considered to mimic the natural environment is hotly debated. The difficulty with the microcosm/mesocosm approach is that although as systems become larger they become more realistic, the experiments become more expensive and less reproducible.

A vast number of different systems have been used and the account that follows necessarily lists only a few. Studies of specific natural systems have been made by placing blocks of polyurethane foam in the area to be studied and allowing colonization by indigenous organisms to occur. The foam is then removed to the laboratory in which it can be treated with chemicals at various doses or, alternatively, samples from different areas along a gradient of pollution can be collected. The taxonomic richness and composition have been found to be most reliable indicators of the effects of chemicals. From a regulatory viewpoint, there is the difficulty of extrapolation of effects on lower organisms to those on higher organisms which are of more interest when human health is to be considered. There is also the difficulty of finding sufficient expertise in the taxonomic identification of groups such as the protozoan.

Artificial ponds have been widely used for the determination of effects of chemicals on aquatic systems. By scaling up to this extent, there are difficulties in obtaining a large enough sample size to study a dose–response and the degree of reproducibility. Typically, there is an initial period when the ponds are open to each other to allow exchange of water and microorganisms. Then, for the experiment the ponds are closed off from each other, and the chemical and fish (often in cages which allow free movement of microorganisms in and out) and added to the system.

Artificial streams have also been widely used, starting with the work of Tidy more than 100 years ago; Tidy stated that he considered that "it was absolutely necessary that the water experimented upon should be running, not merely exposed to light and air in a bottle." Designs of stream systems vary considerably in size and complexity. Some small fiberglass models of streams which are easy to use have been used to study effects on microbial communities. A limitation is that the substrate bears no resemblances to that found in nature. More complex systems, in which a portion of the water of a stream is diverted, have been used to study macroinvertebrate communities. At the top end of the size scale, limno corrals (enclosure systems in open water) and littoral corrals (enclosure systems at the edge of a lake) have been used to study the fate and effects of chemicals on water quality parameters, plankton, macroinvertebrates, and fish.

The critical factors in the design of microcosms and mesocosms are to increase their realism and reduce their variability. The difficulty is that as the complexity of systems is increased to handle the first concern, the ability to replicate and reduce variability is decreased.

Monitoring Biological Systems

Many approaches have been used to monitor biological systems. The following are the most important: measuring concentrations of pollutants in indicator or sentinel species, assessing the effects of pollutants on organisms and relating them to the concentrations of the pollutants in these organisms or their surroundings, monitoring the effects of pollution on the presence or absence of species from a site or changes in species abundance, and detecting genetically different strains of species which have evolved resistance in response to a pollutant.

Monitoring of Chemical Residue Levels

The first approach has been widely used and there are many long-term national programs and even some international programs such as mussel watch. Although these programs give useful data on the temporal trends and geographical variations, they have a tendency to be an end in themselves. The question "What are we going to do with the information when obtained?" is too rarely asked.

One of the longest running programs is that of the Canadian Wildlife Service based on the eggs of herring gulls (*Larus argentatus*). The program was started in 1974; eggs were collected from 13 colonies throughout the Canadian portion of the Great Lakes. The data obtained from one of the sites for DDE is shown in Fig. 7a. It should be noted that initially the levels dropped rapidly after restrictions of the use of this pesticide were put in place. Then in the 1980s a plateau was reached, and little change has been reported during the past 15 years. Starting in 1987, analysis of pooled samples, rather than of individual eggs, was carried out. This meant a considerable saving in analytical costs, although it did result in less information since the variations in the data were no longer known. However, since the individual eggs were stored in a tissue bank, it is still possible to reanalyze the data on an individual egg basis. The value of storing samples was shown when dioxin was first identified in herring gull eggs in 1981. It was then possible to analyze the levels in eggs collected for 10 years prior and to show that a marked decrease had occurred during this period (Fig. 7b).

A critical question is which species should be selected for a given program, as already discussed. In the case cited previously, it was clear that fish-eating birds were being affected and the herring gull, which nests colonially in considerable numbers on the ground, was the most accessible species.

Assessing Effects on Organisms

There are two ways of assessing effects on organisms. The first is to use the knowledge from monitoring programs and search for effects when the concentration of a pollutant (or pollutants) is high enough for one to suspect, based on toxicological knowledge, that an effect is likely. The second approach is to search for effects and, if found, relate them to pollutant levels.

The difficulty with the first approach is the limited amount of toxicological data that are available, especially on target species. The difficulty is compounded by the fact that only rarely do we deal with single compounds. Once one is faced with complex mixtures which vary both temporally and geographically, the problem becomes virtually unsolvable.

The usual assumption is that the effects of mixtures of chemicals are additive. Although this is generally true in a relatively small but important number of cases, the toxicity is substantially greater than additive (i.e., there is a potentiation of toxicity). The term *synergism* is also used to describe this phenomenon. There is also the reverse case in which there is antagonism and the total effect is less than additive. The concept is shown diagrammatically in Fig. 8.

In order to deal with the problem of potentiation in a

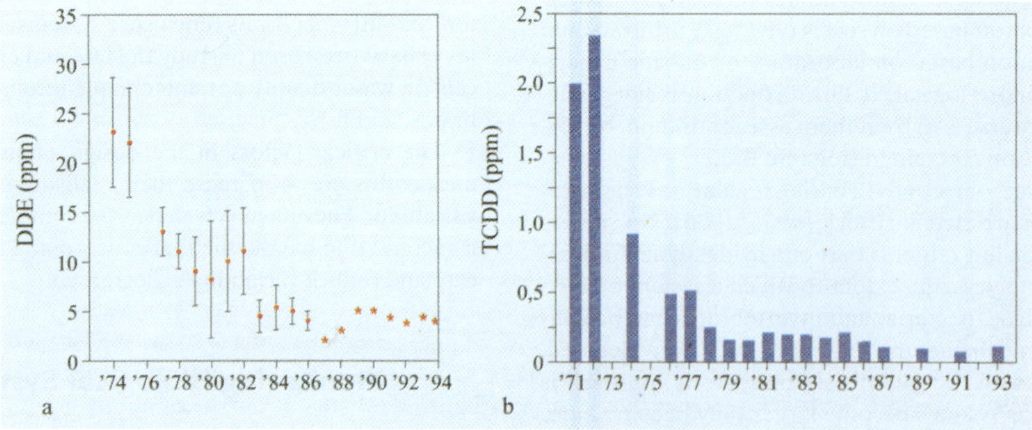

FIGURE 7 (a) Levels of DDE (1,1-dichloro-2,2-bis-*p*-chlorophenylethylene) in herring gull eggs from Lake Ontario during the period 1974–1994. The data until 1986 represent the mean and standard deviation of samples of 10 eggs. Data from 1987 onwards (asterisks) refer to pooled samples of 10 eggs each. (b) Levels of TCDD (2,3,7,8-tetrachlorodibenzo-*p*-dioxin) in herring gull eggs from Lake Ontario during the period 1971–1993. The data shown are based on pooled samples of 10 eggs each.

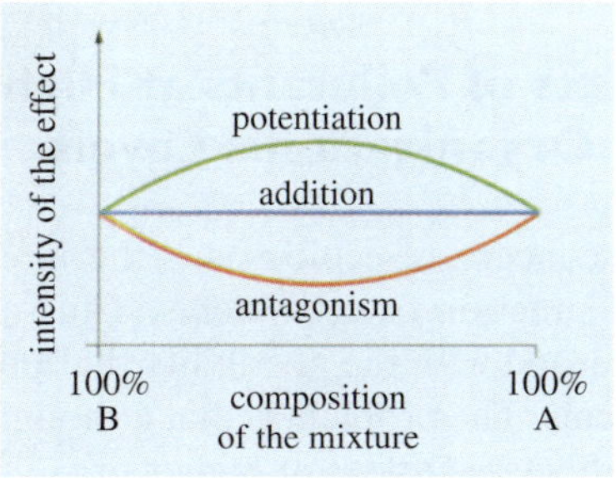

FIGURE 8 Possible effects on the toxicity of the mixture of two chemicals (reproduced with permission from: Walker *et al.*, 1996).

scientific manner — as opposed to trying innumerable combinations at random — one needs to know the mechanism of action of the pollutants. There are two possible mechanisms for causing potentiation. First, compound A inhibits an enzyme that detoxifies compound B. The rate of detoxification of compound is B is thus reduced. Second, compound A induces an enzyme which activates compound B and thus increases the activation of compound B.

An example of the first system is the combination of pyrethroid insecticides and certain fungicides that inhibit ergosterol biosynthesis. These fungicides can inhibit the MFO system of bees that normally detoxifies the pyrethroids. Potentiation of up to 20-fold has been reported in bees.

An example of the second system is the activation of polycyclic aromatic hydrocarbons to carcinogenic metabolites after the MFO system has been induced by compounds such as dioxins. The high incidence of tumors in the beluga whale (*Delphinapterus leucas*) in the St. Lawrence River in Canada is believed to be caused by this mechanism.

The second approach has been discussed previously and here only the use to which it has been put in hazard assessment will be considered. The advantage of the biomarker approach is that if a change is seen, it demonstrates that the pollutant is present in sufficient concentration to cause an effect.

Changes in Species Composition

In general, ecotoxicologists are more interested in effects at the population level or above than effects on individuals. Currently, one of the major concerns is the loss of biodiversity, although the effects of pollution are often less than those caused by destruction of habitat. However, there is increasing evidence that phenomena such as acid rain and global warming are having effects at the ecosystem level. These problems are considered in subsequent sections.

Many formal programs have been devised to examine species composition. In the United Kingdom, the River Invertebrate Prediction and Classification System has been set up to assess the biological quality of rivers. The basic information collected is used to generate site-specific predictions of the macroinvertebrate fauna expected in the absence of major environmental stresses. The fauna observed in an area suspected of pollution can then be compared with the fauna predicted in the absence of pollution. Differences between the two indicate the presence of environmental stresses which need investigation. There are currently approximately 9000 sites in the United Kingdom used to assess biological quality in this way.

In the United States, the Index of Biotic Integrity is another measure of the effects of contaminants on aquatic communities. This index uses fishes and the physical characteristics of the water to derive a quantitative index of aquatic community integrity. This index can be used to assess the impact of contaminants on the health of the aquatic environment following spills. The system is used in several states as a regulatory mechanism, but the problem with adopting it for wider regulatory use is the geographical variability in various regions of the country.

Evolution of Resistance Caused by Pollutants

Resistance is usually, but not always, controlled by a single gene. Most studies of resistance have been performed on the effects of pesticides, although there are many cases of resistance to heavy metals.

Resistance of insects to pesticides has become a major problem and several hundred species are currently resistant. Insects may be resistant to more than one insecticide and often to insecticides of more than one class. When resistance to more than one insecticide is achieved by a single mechanism, this is called *cross-resistance,* but when several resistance mechanisms are involved this is called *multiple resistance.*

The most important type of mechanism of resistance is the occurrence of metabolic pathways that increase the rate of breakdown of the insecticide. Other mechanisms include behavioral responses that allow avoidance of the insecticide, decreased sensitivity to the insecticide, or morphological changes such as decreased permeability of the cuticle.

In most cases, the mechanisms that allow individual insects to resist the insecticide are present in a small number of individuals before exposure starts. These individuals are able to survive and breed, whereas their companions die. Thus, the genetic makeup that was rare becomes common and the pesticide loses its effectiveness. This change is not without cost to the insect because those individuals that are resistant to insecticides are less fit from other standpoints. For example, energy used on detoxification is not available for growth. In other words, resistant alleles are favored at polluted sites but selected against at nonpolluted sites.

Attempts to control the pests affecting cotton in the southern United States are a good example of the changes to community structure that can occur with resistance. From 1945 to the mid-1950s, there was almost complete control of the main pests (the boll weevil and the cotton fleahopper) by organochlorine pesticides DDT and endrin, and there was a spectacular increase in production. In the mid-1950s, the boll weevil was becoming resistant to organochlorine pesticides and it was necessary to switch to organophosphorus compounds, which increased the costs but cotton was still profitable. Initially, control was maintained by increasing the dosage and frequency of application of pesticides. Although the boll weevil was controlled in this manner, the former minor pests, the bollworm and tobacco budworm, increased in numbers and were becoming major pests. By the mid- to late 1960s, all the pests were highly resistance to all insecticides and cotton production dropped dramatically. The introduction of integrated pest management based on new short-season cotton varieties and traditional cultural practices were needed to solve this problem.

Although the majority of the cases of resistance have been due to pesticides, other environmental pollutants have caused similar changes. One of the most interesting is the occurrence of industrial melanism. The blackening of industrialized parts of the United Kingdom was a result of the industrial revolution started in the late eighteenth century. One result was that the melanistic forms, which are better camouflaged in such environments and thus better protected against predators, increased. Although this has occurred in many species of arthropod, the best examples are moths. The incidence of melanism has risen, in some cases dramatically, since about 1850. In the case of the peppered moth (*Biston betularia*), the first melanist specimen was found in Manchester in 1848, and by 1895 98% in that area were melanic. Recently, legislation such as the Clean Air Act has decreased atmospheric pollution, and the frequency of melanism has begun to drop and the occurrence is now about 50%. The clear relationship between the frequency of melanic form and air pollution is shown in Fig. 9.

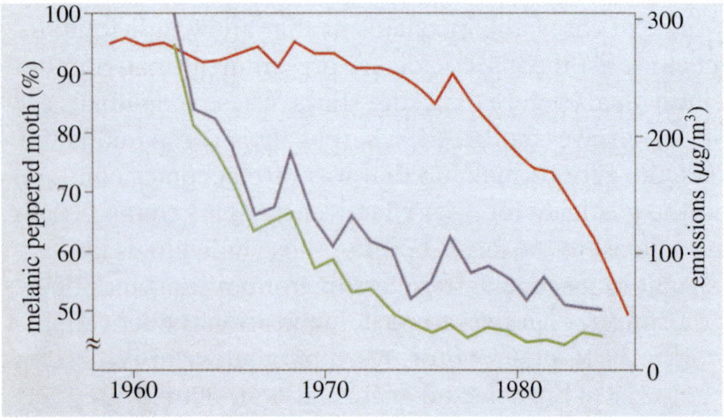

FIGURE 9 Decline in the frequency of the melanic form of the peppered moth, *Biston betularia,* in Manchester (in red, ordinate scale on left) in relation to emissions of smoke (in green, ordinate scale on right) and sulfur dioxide (in blue, ordinate scale on right) (reproduced with permission from Walker *et al.,* 1996).

Effects of Pollutants at Higher Organizational Levels

The ultimate concern of ecologists is the integrity of the structure and function of ecosystems. Thus, ecotoxicologists are concerned with the possibility that pollutants can affect ecosystems. The problem is that although the importance of the effects increases at higher organizational levels, so does the time of response and the difficulty of linking specific effects with specific chemicals (Fig. 1). Chemical effects on ecosystems can be difficult to distinguish from natural fluctuations.

The nature of pollution has changed over the years. Acute, localized problems have given way to chronic, widespread problems. In the 1960s and 1970s, there was considerable concern regarding the regional effects of acid rain on aquatic and terrestrial ecosystems. Recently, the effects of chlorofluorocarbons on the ozone layer and global warming have received much attention.

Acid Rain

Although the term *acid rain* was coined by Angus Smith in 1852 to describe problems around Manchester, it was another 100 years before widespread regional effects of acid rain were noted. At first, the effects of acid rain were localized around factories, but the building of high stacks to disperse the pollution and increasing amounts discharged by vehicles have converted a local problem into a widespread problem. The global input of the oxides of sulfur is approximately 100 million tons per year and the oxides of nitrogen about 60 million tons per year. The main source of the former is coal used in power stations and that of the latter is vehicular emissions.

The effect of the acid rain that is caused by these pollutants is better understood for the aquatic than the terrestrial environment because changes in pH and in the concentrations of ions are more readily measured than in soil.

In Norway a survey of 1000 lakes revealed that the number of barren lakes had doubled between 1974 and 1986. Much of the pollution which affected Norwegian lakes came from the United Kingdom, emphasizing the widespread nature of the problem. One of the most important aspects of acidification is the changes caused by the bioavailability of metals. The increase of aluminum concentrations can cause reproductive failure in fish and is considered to be the main cause of the loss of fish populations in acidified lakes. The increase of mercury does not have such a direct effect on fish but does lead to accumulation of this toxic metal in the food chain.

Detailed studies have been made in Canada to determine the ecosystem stresses caused by acidification in an experimental lake situation. In one study, the pH of a poorly buffered lake in northwestern Ontario was reduced from 6.8 to 5.0 during a period of 8 years (Schindler *et al.,* 1985).

The main findings in addition to the expected effects on fish populations were changes in phytoplankton species composition, the disappearance of benthic crustaceans, and the appearance of filamentous algae. Despite these changes in species composition, acidification did not cause any decease in primary production or in nutrient concentrations. These experimental studies allowed a detailed assessment of many lakes acidified "naturally" by human activities.

The decline of forests in Europe has become a serious political issue, especially in Germany, in which the term *waldsterben* (forest death) has become a household word. Damage to the silver fir was found in many areas in the 1970s, and since then it has occurred in other species, most notably several species of pines. In some areas of Germany, it was estimated that 80% of the trees were affected. Recently, doubt has been cast on the conclusion that acid rain is the primary cause of waldsterben. A detailed review (Kandler, 1992) concluded that the suggestion that large-scale declines of European forests due to acid rain have taken place since the late 1970s is not corroborated by 10 years of research. Kandler points out that there is no good correlation between dieback and the spatial and temporal distribution of air pollutants. It seems likely that a variety of factors, both natural and unnatural, are at work on forest dynamics.

The Hole in the Ozone Layer

In 1984, using land-based instruments in Antarctica, it was found that the concentration of the stratosphere ozone layer (which protects the earth from harmful UV radiation) was only 40% of the values found in the 1950s.

Initially, this finding was not confirmed by data collected by satellite because the default value for the ozone concentration had been set above the new low figure: The computer had been told to ignore values below 180 Dobson units, whereas the values were down to 125 units. Once this problem was cleared up, it was shown by satellite monitoring that ozone depletion was occurring through the stratosphere except in the tropics. The major concern has been the possible increase of cancers, especially in high northern latitudes because although the depletion is not as severe as in Antarctica there are many more persons living in this area.

On the environmental front, concern has been expressed about the effects of increased UV radiation on the productivity of plankton. However, the estimated effects (10–20%) are small in comparison to annual fluctuations, and the critical question of whether this sort of reduction, if it occurred at the low point of natural fluctuation, is important has not been answered. Studies using mesocosms indicate differential effects on algae and insect larvae such that although the algae decreased initially, this loss was more than compensated for by a subsequent decline in the insect population which reduced predation on the algae. Whether such effects occur in nature remain to be demonstrated.

The depletion of ozone in the stratosphere is caused by chlorofluorocarbons (CFCs). These compounds are chemically inert, hence their value as "safe" substances to use in refrigerators and as propellants. The international aspects of the problem are clearly demonstrated by the fact that although 95% of the CFCs are released in the Northern Hemisphere, the redistribution is rapid enough for the concentrations in both hemispheres to be similar. Indeed, the greatest effect has been in Antarctica. Rowland and Molina (who were subsequently awarded the Nobel prize for their work) were able to demonstrate that CFCs (the rate of release was about 800,000 tons/year) did not break down but accumulated in the atmosphere. Thus, the cause for the hole in the ozone layer was found, and the likely chemical culprit was identified.

The control of CFCs is a case in which international action has been taken. A series of protocols to reduce the use of these compounds has been implemented. Despite some failures of enforcement and some black market usage of the compounds, the regulation of CFCs ranks as a success for the international community (Tolba, 1992).

Climate Change

One of the most contentious issues in ecotoxicology today is climate change. The one fact that is agreed is that the levels of carbon dioxide in the earth's atmosphere are rising steadily. In contrast, the impact of the increase of carbon dioxide (and other greenhouse gases) has been vigorously debated. This is hardly surprising since reducing the 6000 million tons of carbon dioxide emitted annually by human activity will be costly (Wigley *et al.,* 1996). Although some modest reductions have been agreed on by the European Union, it has not yet been possible to get action from the largest single producer, the United States, in which the emissions per person are five times the global average and they are more than twice those of western Europe. Although it should be possible to stabilize and even slowly reduce the emissions of carbon dioxide in the industrialized world, increases from the developing world cannot be avoided. The industrialization of China, based largely on coal, is rapidly increasing the emissions from that country and emissions will likely overtake those of the United States early in this century (Flavin and Tunali, 1996).

Despite the fact that the subject is controversial, consensus is emerging from the Intergovernmental Panel on Climate Change that global warming is occurring and that a doubling of the carbon dioxide levels (which if the current rates continue will occur by 2030) will cause significant increases in global temperatures. Estimates of the actual increase in temperature vary, as do opinions on the seriousness of the effects. One of the major effects will be alterations to the rainfall patterns of the world, and another is a rise in sea levels.

The possibility that there are combined effects of global warming, acid rain, and ozone depletion has been put forward based on studies in northern Canada. The key to these changes is the loss of dissolved organic carbon (DOC) in the surface layers of lakes. The DOC of lakes is reduced by a decrease of rainfall and increase of evaporation caused by global warming. This allows greater penetration of the water column by UV radiation, which has increased due to the hole in the ozone layer. The impact of acidification, which reduces the productivity of some plants, increases the degree of UV penetration. Changes among the algal communities have been found, with the greatest effects in clear, shallow lakes.

Case Studies

Effects of Persistent Pollutants in the North American Great Lakes

Collectively, the Great Lakes are the largest body of freshwater in the world. The very vastness of the Great Lakes was a factor in their pollution. They are so large that "dilution seemed the solution to pollution." The waterway of the lakes was a major factor in the opening up and development of this part of the New World. The harnessing of Niagara Falls to generate electricity led to massive industrial development along the Niagara River. Now, approximately 36 million people live within the Great Lakes basin and more than 13,000 manufacturing and industrial plants are located there. The result was that by the 1960s serious wildlife problems were occurring on these inland seas. The residue levels of contaminants found and their effects have been comprehensively reviewed (Environment Canada, 1991a,b).

Initial observations in the early 1970s showed poor reproduction and deformities of chicks in a colony of common terns (*Sterna hirundo*). Subsequent investigations showed severe reproductive impairment of the herring gull and disappearance of the double-crested cormorant (*Phalacrocorax auritus*) in the lower Great Lakes. Poor reproduction of the cormorant was also found in Lake Huron.

The hypothesis that these severe effects over a wide area were caused by pollutants, notably OCs, was immediately put forward. It was based on a correlation between total OC levels and reproductive effects. The problem that was much more difficult to deal with was the relating of these effects to a specific chemical or chemicals. Apart from the difficulties inherent in dealing with the effects of complex mixtures, there were two additional problems. First, the levels of most OCs cross-correlated strongly with each other. Second, the effects seen — embryotoxicity, edema, structural abnormalities, and behavioral changes — were known from laboratory studies to be caused by a wide range of OCs. Only in the case of eggshell thinning in cormorants was it possible to assign a specific cause with a high degree of certainty.

Detailed field studies, which included egg injection, egg exchange experiments, and nest attentiveness studies, were undertaken. These investigations were made more difficult by the then (late 1970s) rapidly decreasing levels of most OCs following bans and restrictions on their usage. By the late 1970s, only a generalized correlation with pollutant levels could be established.

By the early 1980s, the reproductive problems of fish-eating birds were confined to a few specific areas rather than occurring on a lakewide basis as they had a decade before. However, the decline of levels of organochlorines which had been a feature of 1970s halted, and levels remained essentially constant during the late 1980s and early 1990s so that there is now a chronic pollution problem which can be expected to continue well into this century.

In hindsight, the problem was that neither the analytical chemistry nor the experimental toxicology were sophisticated enough. The confirmation that the chlorinated dioxins (PCDDs) and chlorinated dibenzofurans (PCDFs) were present in the Great Lakes was not made until 1980. Isomer-specific analysis was soon developed and now analytical chemists routinely report on the levels of 80 OCs compared to a dozen or so a decade ago. Similarly, although it had long been known that the number of isomers of PCBs was large, almost all the early toxicological work was on the commercial mixtures. The work on specific isomers soon revealed large differences in their activity, and studies on structure–activity relationships were undertaken.

Meanwhile, molecular biologists were using dioxin (specifically 2,3,7,8-tetrachloro-*p*-dioxin) for their work on the isolation and characterization of receptors. The identification of the Ah receptor in the mid-1970s by means of its strong specific binding to dioxin was a key finding in bringing molecular biology into the realm of toxicology. The Ah receptor is responsible for the control of the MFO systems, which are responsible for an important defense mechanism against foreign compounds.

The application of this complex biochemistry into field investigations has been by means of expressing the complex mixtures of organochlorines as "dioxin equivalents." This concept is based on the correlation found between the concentration required to induce a specific monooxygenase (alkyl hydrocarbon hydroxylase) and the concentration required for toxic effects for a large number of PCBs, PCDFs, and PCDDs. In dioxin equivalents, the activity of the most powerful compound, 2,3,7,8-TCDD, is set at 1 and the potency of the other compounds is calculated from the correlation. This number can then be multiplied by the concentration and the equivalence in terms of dioxin calculated. Although the potency of other compounds, such as individual PCBs, is much lower than that of dioxin, the concentrations are often much higher and therefore they frequently contribute more to the total dioxin equivalent than does dioxin. The approach is now used in reverse: The degree of induction of the enzyme is measured and then converted into dioxin equivalents. This bioassay approach is rapid and inexpensive compared to the conventional chemical analysis by gas chromatography–mass spectrometry.

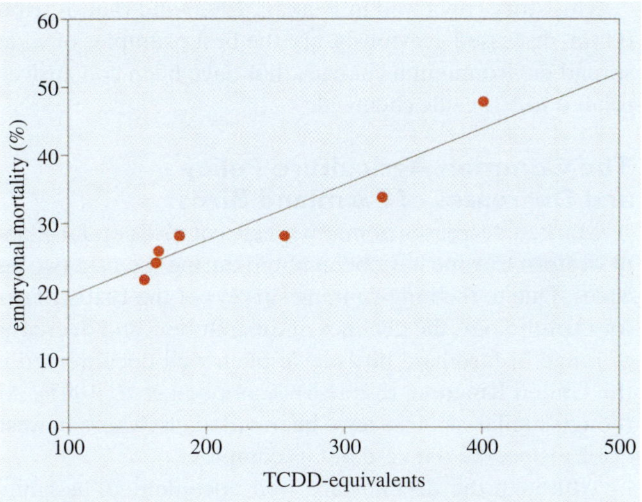

FIGURE 10 Relationship between embryonic mortality in colonies of Caspian terns, *Hydroprogne caspia,* and the dioxin (TCDD) equivalent in their eggs (reproduced with permission from Peakall, 1992).

The dioxin equivalents of egg samples from fish-eating birds collected from colonies of cormorants and terns in Michigan and Ontario have been determined. When the reproductive success of double-crested cormorant and Caspian tern (*Hydroprogne caspia*) colonies was plotted against dioxin equivalents of eggs from each colony, a high degree of correlation was found (Fig. 10).

The reproductive failure of fish-eating birds in the North American Great Lakes is an example in which two initially different lines of research — molecular biology and investigations by field biologists — eventually blended to provide a comprehensive answer to the problem.

Reproductive Failure of Mollusks Caused by Tributyltin

A well-studied example of the effects of pollutants in the marine environment is that of tributyl tin (TBT) leached from antifouling paints. TBT has been used as a molluskicide on boats, quays, and other marine structure since the mid-1960s, and for the next decade its use increased as it became widely recognized as being extremely effective. Problems soon emerged, with population declines of oysters and whelks in northern France and southern England.

The most detailed studies have been carried out by the Plymouth Marine Laboratory in the United Kingdom following the first finding of imposex in the dog whelk (*Nucella lapillus*) in 1969. Imposex (females developing male characteristics) is expressed by the development in females of a small penis. Skin can grow between the genital papilla and the penis, and in the most extreme cases this occludes the papilla, preventing egg liberation and thus causing reproductive failure (Bryan *et al.,* 1986).

Following the first report of imposex in Plymouth Sound, a broader survey around the southwestern peninsula of England was made which revealed that imposex was widespread, with the most marked effects along the Channel coast. A significant finding was the relationship between the degree of imposex and the proximity of the affected population to harbors. This suggested that imposex was caused by a pollutant associated with the boating industry. Another important finding was that the degree of imposex was increasing markedly after its first discovery, suggesting that the level of pollution was increasing. These findings suggested that TBT could be the causative agent.

This hypothesis was confirmed by laboratory studies which allowed a cause–effect relationship to be established. In these studies, levels of TBT as low as a few nanograms per liter were enough to cause sterilization in dog whelks. The no-observable-effect-level has been given as less than 1.5 ng/liter by the environmental criteria document published by the World Health Organization. Sterilization of some females occurs in locations with 1 or 2 ng/liter and is total in areas with 6–8 ng/liter.

Subsequent to the first findings in France and England, imposex has been widely reported in the United States, Southeast Asia, and Japan. Even more disturbing is a recent report of imposex in whelks in the open areas of the North Sea and its correlation with shipping traffic densities.

In France, initial restrictions on the use of TBT were imposed in 1982, followed subsequently by a complete ban. In the United Kingdom, the use of TBT on small boats and aquaculture cages was banned in 1987. In the United States, many states have imposed restrictions, some of which are based on leaching rates, rather than outright bans on the use of TBT. The leaching rate of TBT from paints has been considerably decreased in recent years by incorporating the TBT in a copolymer matrix.

Following the bans and restrictions on the use of TBT, recovery of the whelk populations has been noted in some areas. However, the use of TBT is reported to be still to be high in some areas, especially the Far East.

DDE-Induced Eggshell Thinning in Raptorial Birds

In 1965, a meeting was held in Madison, Wisconsin, which was attended by persons interested in peregrine falcons (*Falco peregrinus*) from Europe and North America. Decreases in population were reported throughout the Holarctic, and in eastern North America the peregrine had disappeared completely. Ratcliffe, in the United Kingdom, noted increased egg breakage of this species in the mid-1950s but did not, at that time, follow-up on the observation.

When the population decline was highlighted, Ratcliffe remembered the egg breakage and measured the eggshell thickness of many eggs collected in the United Kingdom that were in museums and private collections and demonstrated that a significant decrease had occurred starting in the mid-1940s. Subsequently, it was demonstrated that egg-

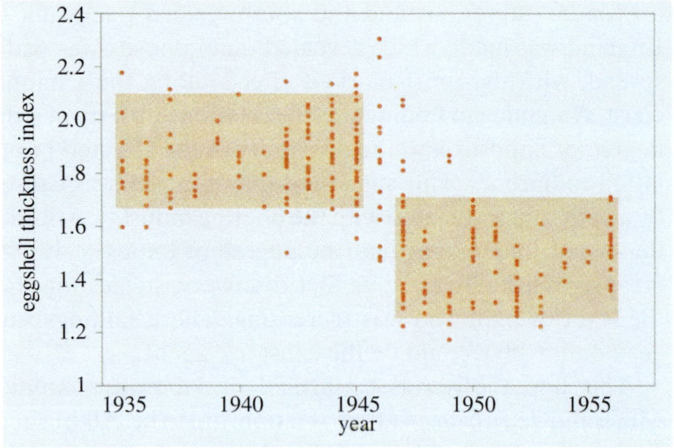

FIGURE 11 Temporal changes in the eggshell thickness index for the peregrine falcon in Great Britain. Shaded areas represent the 90% confidence limits calculated between 1935 and 1945 and between 1947 and 1956 (reproduced with permission from Peakall, 1993).

shell thickness had decreased in many areas of the world (Fig. 11).

Experimental work was carried out on the American kestrel (*Falco sparverius*) that established a cause-and-effect relationship to DDE. It was also possible to demonstrate that the relationship between the degree of eggshell thinning and DDE residue levels in the laboratory was the same as that found in the field. Two interesting aspects of the phenomena of eggshell thinning are the wide difference in the sensitivity of different species and the finding that eggshell thinning is caused, at environmentally reasonable doses, only by DDE and very closely related compounds. Several of the most common avian test species (quail, pheasant, and chicken) are almost completely insensitive, and others, such as the duck, only moderately so. However, many species of fish-eating bird and raptors are extremely sensitive. Thus, it is unlikely that preregistration studies would, even today, have found this particular adverse effect of DDT.

The mechanism is considered to be via the inhibition of Ca-ATPase in the oviduct reducing the transport of calcium to the site of eggshell formation, but details of the pharmacodynamics involved in interspecies variation remain unclear. It was soon put forward that an 18–20% reduction of eggshell thinning was enough to cause reproductive failure and population decline. This has been confirmed by data collected on a worldwide basis for the peregrine falcon (Fig. 12). In many parts of the Northern Hemisphere there has been a marked decrease in the residue levels of DDE and an increase in the thickness of eggshells. This had led to a recovery of peregrine populations in almost all parts of its range. In eastern North America, since the species had completely disappeared, it was necessary to set up a reintroduction program. This was highly successful and now there is a self-sustaining population.

This story (reviewed in Peakall, 1993) and that of tributyl tin, discussed previously, are the best examples of widespread environmental changes that have been conclusively related to a specific chemical.

The Common Agriculture Policy and Decreases of Farmland Birds

Marked decreases of many species of birds on farmland in western Europe have been noted during the past two decades. Due to the long-running surveys of the British Trust for Ornithology, the declines in the numbers and decreases of range of farmland birds have been well documented in the United Kingdom (Table 4) (Campbell *et al.,* 1997). Although similar changes have been noted elsewhere in western Europe, the data are not as complete.

Although the declines are well documented, assigning a cause or causes is more difficult. Given the scale of the changes, both temporally and geographically, an epidemiological approach has to be used. The principal criteria for epidemiology are a time–order relationship of the association, the specificity of the association, and the coherence of the association. These epidemiological criteria do not provide the cause and effect proof of the experimentalist but form a process and framework on which a balanced judgment can be developed.

Time–Order Relationship. It is essential that the time of the onset of the effect relates to the time of occurrence of the proposed cause. In this case, the timing of the decline (Table 4) relates to the onset of the most recent phase in the intensification of farming, which includes the increased use of pesticides. However, other management practices have also changed over the same period, such as an increase in crop specialization in arable areas and a change from spring to autumn sowing of cereals.

Consistency of the Association. The declines have been seen in farmland throughout the United Kingdom and over much of western Europe. Populations of farmland birds appear to be faring better in eastern Europe, where farming is still more traditional. However, as stated previously, the data from other European countries are much less detailed than those available in the United Kingdom.

Coherence of Association. The idea that a decrease in food supply for birds due to increased insecticide and herbicide use and decreased habitat due to herbicides (i.e., loss of noncrop plants in fields) could affect the survival of birds is biologically reasonable. For both plants and invertebrates, the Joint Nature Conservancy Council report states that "the overall, albeit far from complete, picture is one of general decline in both numbers and diversity, particularly in the last 20–30 years." The hypothesis that intensive farming is to blame is strengthened by the fact that bird populations are higher on organic farms, although it was not possible to determine which factor (pesticide usage, habitat quality, rotation, etc.) was primarily responsible for the effect.

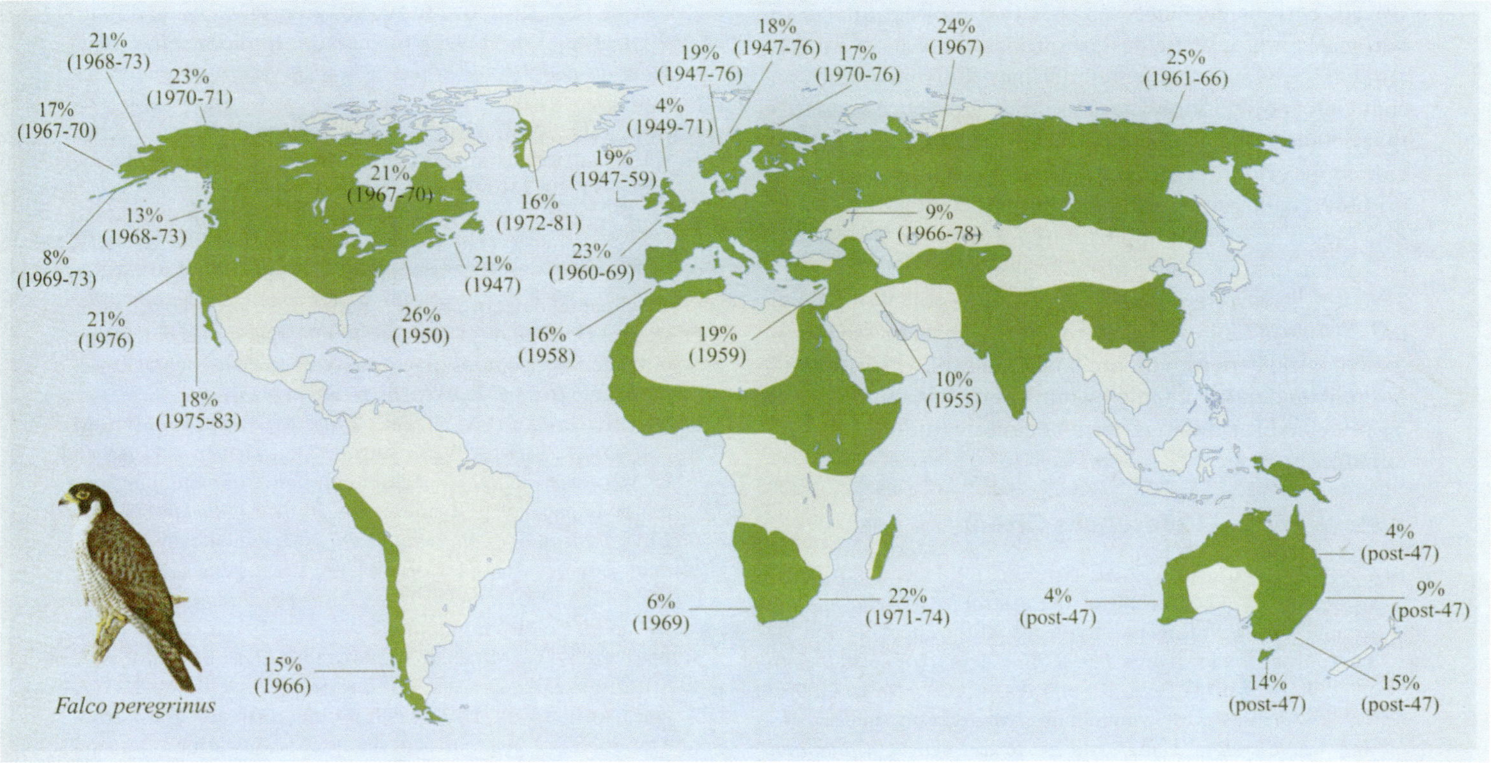

FIGURE 12 Global distribution of eggshell thinning in the peregrine falcon (*Falco peregrinus*) in different areas of its world distribution (in green) (reproduced with permission from Peakall, 1993).

The most detailed research to date is that carried out on the gray partridge by the Game Conservancy (Potts, 1986). This study, spanning 30 years, included long-term field monitoring in support of modeling studies and experimental manipulations of pesticide use. It produced convincing evidence that partridge chick survival has been reduced by the indirect effects of herbicides and broad-spectrum insecticides. Reduction of pesticide usage alone was shown to lead to significant improvement in chick survival.

Although it is to be expected that several factors are in-

TABLE 4

Changes in Population Numbers and Ranges of Selected Species during the Past 20–30 Years in the United Kingdom, Timing of Decline, and Current Total Population Estimate[a]

SPECIES	POPULATION TREND ON FARMLAND, 1969–1994 (%)	YEAR DECLINE STARTED	POPULATION (PAIRS), 1988–1991	RANGE CHANGE, 1970–1990 (%)
Tree sparrow (*Passer montanus*)	−89	1978	110,000	−20
Gray partridge (*Perdix perdix*)	−82	Uncertain	145,000	−19
Turtledove (*Streptopelia turtur*)	−77	1979	75,000	−25
Bullfinch (*Pyrrhula pyrrhula*)	−76	1977	190,000	−7
Song thrush (*Turdus philomelos*)	−73	1975	990,000	−2
Spotted flycatcher (*Muscicapa striata*)	−73	Before 1969	120,000	−2
Lapwing (*Vanellus vanellus*)	−62	1985	210,000	−9
Reed bunting (*Emberiza schoeniclus*)	−61	1977	220,000	−12
Skylark (*Alauda arvensis*)	−58	1981	2,000,000	−2
Linnet (*Carduelis cannabina*)	−52	1977	520,000	−5

[a] From Peakall and Carter (2001).

volved, current evidence suggests that indirect effects of herbicides and insecticides are probably the most important cause of the documented declines. Current studies on changes of populations where set-aside is practiced and further detailed studies of organic farming should help to provide more evidence. If more intensive farming follows the proposed enlargement of the European Union, this would provide an additional opportunity for detailed studies, albeit at the further cost to farmland birds.

With the current common agricultural policy (CAP) of the European Union, we have embarked on a massive ecotoxicological experiment, the final results of which are still not fully understood. Only by unraveling this can the CAP be reformed for the long-term benefit of wildlife and sustainable farming.

References Cited

BRYAN, G. W., GIBBS, P. E., HUMMERSTONE, L. G., and BURT, G. R. (1986). The decline of the gastropod *Nucella lapillus* around southwest England: Evidence for the effect of tributyltin from antifouling paints. *J. Mar. Biol. Assoc. UK* **66,** 611–640.

CAIRNS, J., JR., BIDWELL, J. R., and ARNEGARD, M. E. (1996). Toxicity testing with communities: Microcosms, mesocosms and whole-system manipulations. *Rev. Environ. Contamin. Toxicol.* **147,** 45–69.

CALAMARI, D., TREMOLADA, P., and NOTARIANNI, V. (1995). Relationships between chlorinated hydrocarbons in vegetation and socioeconomic indices on a global scale. *Environ. Sci. Technol.* **29,** 2267–2272.

CAMPBELL, L. H., AVERY, M. I., DONALD, P., EVANS, A. D., GREEN, R. E., and WILSON, J. D. (1997). A review of the indirect effects of pesticides on birds, JNCC Report No. 227.

ENVIRONMENT CANADA (1991a). *Toxic Chemicals in Great Lakes and Associated Effects. Vol. I. Contaminant Levels and Trends.* Environment Canada, Toronto.

ENVIRONMENT CANADA (1991b). *Toxic Chemicals in Great Lakes and Associated Effects. Vol. II. Effects.* Environment Canada, Toronto.

FLAVIN, C., and TUNALI, O. (1996). New strategies for stabilizing the world's atmosphere. *WorldWatch Paper* **130,** 84.

FORBES, V. E., and FORBES, T. L. (1994). *Ecotoxicology in Theory and Practice.* Chapman & Hall, London.

HUGGETT, R. J., KIMERLE, R. A., MEHRLE, P. M., JR., and BERGMAN, H. L. (Eds.) (1992). Biomarkers. In *Biochemical, Physiological, and Histological Markers of Anthropogenic Stress.* Lewis, Boca Roca, FL.

KANDLER, O. (1992). Historical declines and diebacks of central European forests and present conditions. *Environ. Toxicol. Chem.* **11,** 1077–1093.

PEAKALL, D. B. (1992). *Animal Biomarkers as Pollution Indicators.* Chapman & Hall, London.

PEAKALL, D. B. (1993). DDE-induced eggshell thinning: An environmental detective story. *Environ. Rev.* **1,** 13–20.

PEAKALL, D. B., and CARTER, N. (2001). Decreases in farmland birds and agricultural practices: A huge ecotoxicological experiment. *Toxicol. Ecotoxicol. News,* in press.

PIMENTEL, D., ACQUAY, H., BILTONEN, M., RICE, P., SILVA, M., NELSON, J., LIPNER, V., GIORDANO, S., HOROWITZ, A., and D'AMORE, M. (1993). Assessment of environmental and economic impacts of pesticide usage. In *The Pesticide Question. Environment, Economics and Ethics* (D. Pimentel and H. Lehman, Eds.), pp. 47–84. Cornell Univ. Press, Ithaca, NY.

POTTS, G. R. (1986). *The Partridge. Pesticides, Predation and Conservation.* Collins, London.

SCHINDLER, D. W., MILLS, K. H., MALLEY, D. F., FINDLAY, D. L., SHEARER, J. A., DAVIES, I. J., TURNER, M. A., LINSEY, G. A., and CRUIKSHANK, D. R. (1985). Long-term ecosystem stress: The effects of years of experimental acidification on a small lake. *Science* **228,** 1395–1401.

TOLBA, M. K. (1992). *Saving Our Planet.* Chapman & Hall, London.

VAN EMDEN, H. F., and PEAKALL, D. B. (1996). *Beyond Silent Spring.* Chapman & Hall, London.

WALKER, C. H., HOPKIN, S. P., SIBLY, R. M., and PEAKALL,. D. B. (1996). *Principles of Ecotoxicology.* Taylor & Francis, London.

WIGLEY, T. M. L., RICHELS, R., and EDMONDS, J. A. (1996). Economic and environmental choices in the stabilization of atmospheric CO_2 concentrations. *Nature* **379,** 240–242.

General References

CALOW, P. (Ed.) (1993). *Handbook of Ecotoxicology,* Vols. 1 and 2. Blackwell, London.

HOFFMAN, D. J., RATTNER, B. A., BURTON, G. A., JR., and CAIRNS, J., JR. (1995). *Handbook of Ecotoxicology.* Lewis, Boca Raton, FL.

WALTER MARCONI
Department of Chemistry
University of Rome "La Sapienza"
Rome, Italy

The Development of Biotechnologies for the Environment

In this article, I discuss the contribution of biotechnology to novel techniques for the synthesis of chemicoindustrial, agroindustrial, and pharmaceutical products and to the pretreatment of broad-consumption goods, through which a reduction of polluting potential, compared to traditional processes or products, is achieved. After a brief overview of the improvements which, due to biotechnology, have been achieved in the plant species, I describe processes based on the use of enzymes (in either a free or an immobilized state) and of cells, whether microbial or of plant origin. I then discuss the possible contributions of biotechnology to mineral and fossil-fuel processing and to the production of broad-consumption materials (fertilizers, energy vectors, and polymers) with low environmental impact. Finally, I outline the sector's most promising possible future developments.

Introduction

A unique and peculiar characteristic of biotechnologies is that they are based on the close interaction between the natural sciences and engineering technologies. This interaction leads to the employment of entire organisms, cells, and cellular constituents in the creation of novel products and development of new production or "service" technologies. The latter category includes the most varied methodologies, from the destruction of polluting agents and reinstatement of ecosystems to the host of analytic or therapeutic methodologies which employ components of a biological type. The development of these technologies requires the cooperative application of biological disciplines (biochemistry, microbiology, genetics, molecular biology, etc.) and physical ones (chemistry, physics, electronics, etc.).

Biotechnology contributes to the protection of the environment not only by making the disposal of wastes and the reinstatement of polluted environments easier to accomplish but also by allowing for the production of broad-consumption goods by means of processes that are less polluting in terms of both reagents and wastes. The sectors in which they are employed, which are many and of great importance, are related to chemical and agroindustrial production. Of great importance, for example, is the use of cheaper or renewable raw materials not only for chemical productions but also as fuels that are less polluting than petroleum by-products (e.g., ethanol, biodiesel, and the hydrogen that is obtained from the fermentation of agricultural waste).

Another sector in which biotechnology is applied is in the introduction of processes that are less polluting than their corresponding chemical processes, which do not entail the use of organic solvents or heavy-metal-based catalysts, or in the selective introduction in chemical compounds of specific functional groups in bland processing conditions (room temperature and pressure conditions, water used as solvent, close to neutral pH, etc.). Biotechnology may also be used in the exploitation of biotransformations that give rise to intermediate or finished products which cannot be properly synthesized by chemical means, such as chiral intermediates for the production of drugs or phytopharmaceutic products. In addition, biotechnology is used in the removal of toxic substances during fossil-raw material treatment processes, such as the desulfurization of petroleum or carbon, or those that use microorganisms in the course of the procedures for the extraction of metals. The continuous monitoring of pollutants by means of biosensors in areas that may be far apart from each other and very large is another application of biotechnology.

In the near future, the list of biotechnology's uses will also include the production, by means of extensive crops of plant species that have been modified using genetic engineering, of high added-value chemical products for specialized purposes, such as new proteins, carbohydrates, plant enzymes, particular oils and fats, biodegradable plastic materials, and new drugs. In the future, the nonalimentary use of agriculture and crops may become extremely important in the industrialized countries, in which food production is currently more than sufficient, and have positive effects also in terms of environmental impact.

The manipulation of development, metabolism, and resistance to diseases and parasites has always been the goals of the genetic improvement of cultivated plants. Biotechnologies provide extraordinary new tools for genetic improvement. The innovative aspect of the biotechnological approach is the possibility of identifying and manipulating the genes responsible for plants' vital functions and the possibility of creating transgenic plants in which one or more of these genes are modified or expressed in ways unlike those in which they are expressed in nature. Biotechnologies also allow one to transfer into plants genes that are entirely foreign to them, derived from other organisms, in order to improve them, protect them, or make them produce useful substances.

A better understanding of the metabolism of carbon, carbohydrates, and nitrogen compounds, and of the molecular mechanisms at work in plants' adaptations and interactions with microorganisms, will allow us to create varieties capable of growing in extreme conditions and on marginal terrain, thereby limiting the use of chemical fertilizers which are polluting and costly. Also, the growth of flowers and the ripening of fruit can be modified if genetically controlled.

More nutritious plants can be created by altering the composition of seeds' reserve proteins, which represent the edible part of some of the most important crops (cereals and leguminous plants). The genes that code for seeds' reserve proteins, in fact, were among the first to be isolated and studied in detail. It is also possible to intervene by modifying the composition of extractable oils used either as foodstuffs or in the production of detergents, for example, as in the case of rape oil with increased lauric acid content.

Particularly spectacular have been the results biotechnology has already achieved in the protection of crops from insects and diseases. The economic aspect of the problem, in this case, is represented by the $3–5 billion that is spent throughout the world each year to keep insects in check by chemical means (with enormous environmental damage). The first examples of transgenic plants (tobacco and tomato) that are resistant to insects due to a gene derived from a bacterium date back to no later than 1987 or 1988, when the first biotechnological plants resistant to viruses, due to a derivative of viral genes, were also developed. Since 1996, commerce in numerous plants that are insect resistant due to the presence of bacterial genes (maize, potato, and cotton) and others resistant to viruses (marrow and papaya) has been authorized in the United States. The first generation of biotechnological products already available on the market in the United States includes plants that are resistant to the various herbicides (maize, soy, and cotton) and allows one to dose more carefully and to circumscribe the use of these highly polluting chemical substances.

Regarding the use of biotechnology in chemical production processes, it must first be explained that the vast majority of chemicoindustrial technologies still use, and will continue to use for a long time, petroleum or gasses associated with petroleum (such as methane) as raw materials. This is because of their abundance and low cost and because of the existence of a great many petrochemical technologies which have been optimized.

Given the high levels of environmental pollution encountered in every phase of the petroleum extraction and refining cycle (including transportation and deposit), however, a greater ecological awareness has led to its replacement, wherever possible and economically viable, with methods of synthesis based on the fermentation of natural substrates using yeast or bacteria (or enzymatic systems derived from them). The most notable example is ethyl alcohol, which in Brazil is no longer produced starting from ethylene but through the fermentation of sugar-containing substances (especially sugarcane).

Almost 5 million automobiles in Brazil today use pure ethanol and 9 million use a mixture of bioalcohol (20%) and gasoline. A government program (known as "Proalcol"), for which $6.97 billion has been set aside, has made such a "miracle" possible. The savings that have resulted from reductions in the importation of petroleum have amounted to $12.48 billion. Moreover, sugarcane residues are used as natural fertilizers in plantations at São Paolo and are responsible for a 30% increase in productivity. Ethanol-based fuel is also produced in the United States; currently, production volume is 1 billion U.S. gallons (1 gallon is equal to about 3.8 liters) or 4 million cubic meters. It is obtained through the fermentation of maize starch and sold for $1.20 to $1.30 a gallon.

The production of organic acids by way of fermentation has been very important as well. The production of citric acid (acidulent) via the fermentation of by-products of the processing of grain with *Aspergillus niger* has almost completely replaced its extraction from citrus fruit, which is less economical and much more polluting. Analogously, lactic acid is derived from the fermentation of molasses produced during the processing of saccharose, of serums obtained during the processing of dairy products, or of residues of the extraction of starch from maize. The advantages inherent in the use of enzymatic instead of chemical catalysts to transform, modify, degrade, or synthesize complex chemical compounds are many. Enzymes catalyze almost every type of organic reaction: oxidation, hydrolysis, elimination, double-bond addition, and isomerization. They are characterized by high catalytic specificity, allow one to work in conditions that are very bland with regard to temperature and pressure, and have a very low environmental impact because of their reduced output of secondary products and impurities.

The development of new biocatalysts requires the assaying of many microorganisms for the purpose of identifying those endowed with the desired activity in order to then isolate the enzyme or enzymes responsible for it. The techniques for producing microbial cultures suited for the development of mutant strains for specific reactions are very important, as are the techniques for the isolation and removal of enzymes that catalyze undesired reactions. All this requires, on the part of the industries engaged in the sector, considerable investments in terms of time and money.

Noteworthy applications are expected to also derive from the production of enzymes that have been modified in order to increase their catalytic activity and make them suitable for working for long periods under adverse processing conditions, such as high temperatures, extreme conditions in terms of pH or salinity, and the presence of organic solvents. These structural modifications of enzymatic molecules will be carried out by combining, on the basis of information provided by modern techniques for structural enquiry, the techniques for protein engineering with those for the chemical transformation of polypeptides.

The development of recombinant DNA techniques has had a notable impact on the present-day production and application of biocatalysts, especially through increases in the quantities of enzymes produced by specific microorganisms and the modification of the configuration of native enzymes for the purpose of altering their specificity or stability.

The first of these factors is largely pursued by the main producers of enzymes since the enzymatic activity of microorganisms that exist in nature is usually too low to be of commercial interest. A widespread technique consists in isolating the gene that codes for the desired enzyme and cloning it in another organism (host organism) using a plasmid as gene vector; the plasmid consists of sections of extranuclear DNA that are present in microorganisms and capable of incorporating, by means of recombinant DNA techniques, the gene in question and of transferring it to the host microorganism. The latter, once it undergoes the intense reproductive activity typical of fermentative processes, will express substantially higher levels of the enzymatic protein coded for by the gene that has been introduced. The isolation of the gene that codes for the enzyme and its transfer are complex molecular engineering operations, which today are possible to accomplish due to the rapid increase in knowledge of the nature of genes and enzymes that occurred in the 1980s and 1990s. The enzymes are employed to isolate the DNA fragment that codes for the protein one wants to produce (restriction enzymes) and also to insert the fragment into the new host organism which will synthesize the protein. The insertion into the plasmid of the extraneous DNA fragment to be introduced in the host microorganism is carried out by ligases. In a well-accomplished system of expression, the protein produced by the cloned gene may make up to 70% of the overall amount of cellular proteins, with the added advantage that the operations for isolating the enzyme after fermentation are simplified.

The modification of an enzyme's configuration, instead, has yet to find widespread application as a way to increase biocatalysis; it requires techniques such as site-specific mutagenesis, which consists of changing particular amino acidic residues involved in the structure of the enzyme's active site in such a manner as to alter, for example, the latter's shape or size in order to enable it to transform substrates that differ from those typical of the native enzyme (Fig. 1). Changes in the amino acid sequence are obtained by applying recombinant DNA techniques to alter, in the plasmids of the host organism, the DNA segments that code for the sequence.

Enzymatic Technologies

Enzymes used in bioconversion processes may be either in a free state in aqueous solution or immobilized in matrices of various nature; alternatively, the whole cell may be employed as an enzymatic catalyst. The use of enzymes in solution is very widespread. Proteases and lipases, for example, are used in this form in washing machine detergent

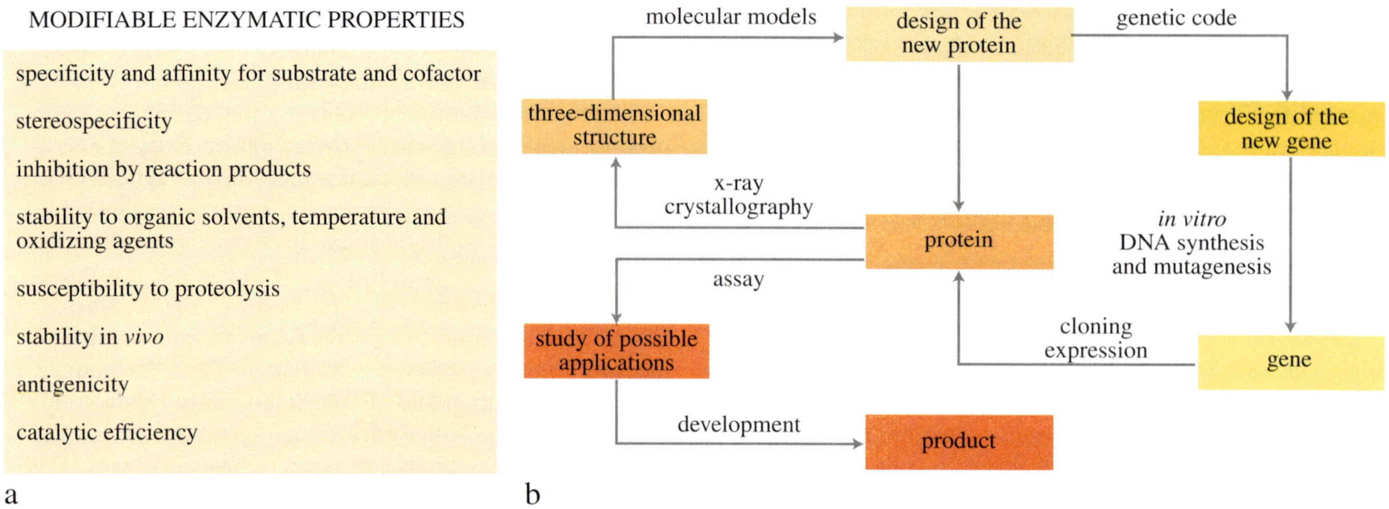

FIGURE 1 Modifications of enzymatic properties possible using genetic engineering (a) and the repeated passages involved in enzymatic engineering (b).

powders, in which they help to eliminate tough stains from fabrics at notable savings in detergent and, consequently, with lesser negative environmental impact.

The cost of producing extractive enzymes and enzymes of microbial origin is often high, however, just as is the cost of purification processes. Therefore, different techniques for their immobilization and for the employment of immobilized enzymatic preparations in industrial reactors, especially in the synthesis of particular-use, pharmaceutical, and fine-chemical products, have been developed.

Use of Enzymes in Solution

An important chapter in the industrial use of biotechnology concerns the production of derivatives of maize. These procedures, which led to the production of highly purified sorbitol and maltol, make ample use of hydrolytic enzymes that gradually degrade starch into low-molecular-mass materials such as glucose and maltose and, after the latter's conversion by means of catalytic hydrogenation, yield sorbitol and maltol, respectively. These substances are used mainly as sweeteners and have replaced synthetic products such as saccharine and cyclamates, which are unsafe from the toxicological standpoint and synthesized through processes with high environmental impact. Their importance as sweeteners is the result of the strong demand on the part of consumers who suffer from disorders that are quite widespread in highly industrialized countries, such as diabetes and obesity; moreover, they have been used as additives to replace other, less innocuous synthetic chemical products in a wide variety of productive sectors, from the food industry to the pharmaceutical one and from the cosmetic industry to that of personal hygiene products. In addition, sorbitol is employed, after having been etherized with epoxides (ethyl, and especially propylene, alcohol), as a basic constituent in the production of polyurethanes (foam rubber and linings) with special properties such as excellent dimensional stability, a high level of resistance to aging, and a high reticulation density. It is also used in the production of thermosetting compositions (melamine resins).

Starch is a highly polymerized polysaccharide composed of glucose molecules (D-glucose or dextrose) linked to each other by way of α-glucosidic bonds. The general formula for starch $(C_6H_{10}0_5)_n$, where n is about 1000, reflects its intimate structure only vaguely since starch contains a linear polymer (amylose) in which all the bonds are in position 1.4 and a ramified polymer (amylopectin) in which 20–30 units of glucose form chains with 1.4 bonds and the ramification takes place through 1.6 bonds (Fig. 2).

Starch milk undergoes certain processes to yield polyols, or polyfunctional aliphatic alcohols (Fig. 3). Hydrolysis, carried out by means of bacterial α-amylase by *Bacillus licheniformis* or *B. subtilis,* leads to the formation of maltodextrines. In order to degrade the amylopectins as well, and thereby obtain increased yields of glucose and maltose in more purified forms, enzymes such as isoamylase or pullulanase are used in this phase that are capable of acting on the bonds that in amylopectins ramify the glucosidic units in position 1.6 (Warren, 1996).

Then, depending on the product one wishes to obtain, one proceeds with saccharification by means of glucoamylase to obtain glucose syrups or with incubation with amylase to obtain maltose solutions (Fig. 4). The glucose and maltose syrups are then transformed by treatment with hydrogen at high temperature and pressure into sorbitol and maltol, respectively (Fig. 5). After having been purified in continuously operating demineralization and purification

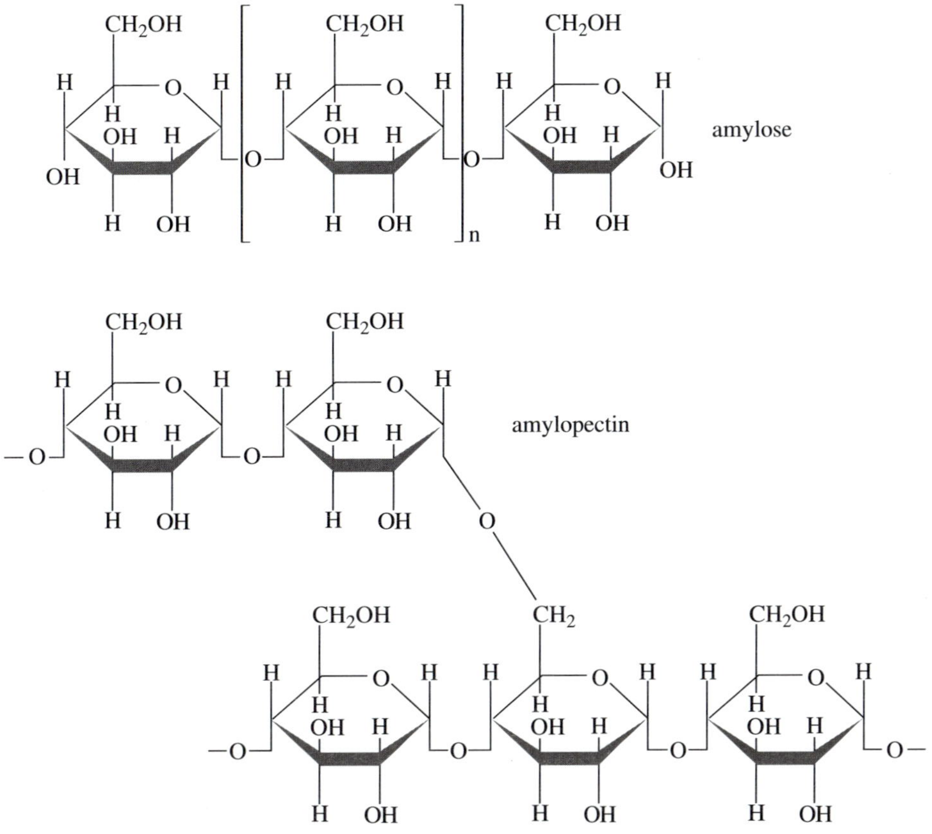

FIGURE 2 Chemical formulas of amylose and amylopectin.

columns filled with ionic exchange resins (both cationic and anionic), these syrups are dried (to a solids value of at least 99.7%) by means of air-tight concentrators.

The dairy-cheese industry presents a problem which is extremely significant from economic and environmental standpoints: the utilization of the lactose contained in dairy serums. Lactose (β-D-galactopyranosyl-D-glucopyranose), which is composed of one molecule of glucose and one of galactose, is a low-value sugar that makes up more than 75% of the dry weight of the serum and accumulates in the world in the amount of about 1.2 million tons per year. For many reasons, it currently has very few market outlets, es-

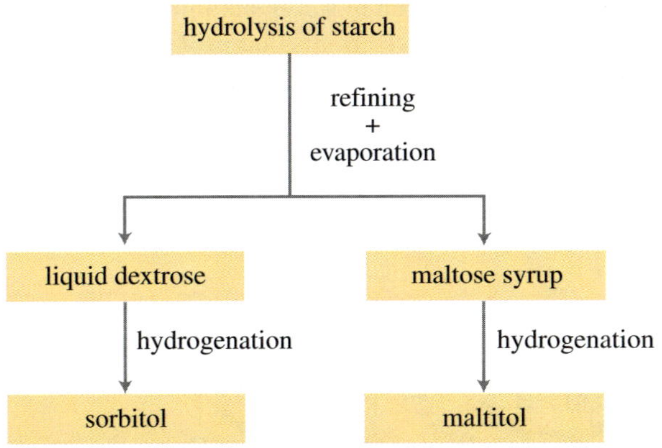

FIGURE 3 Block diagram of the processes for obtaining polyoils from starch.

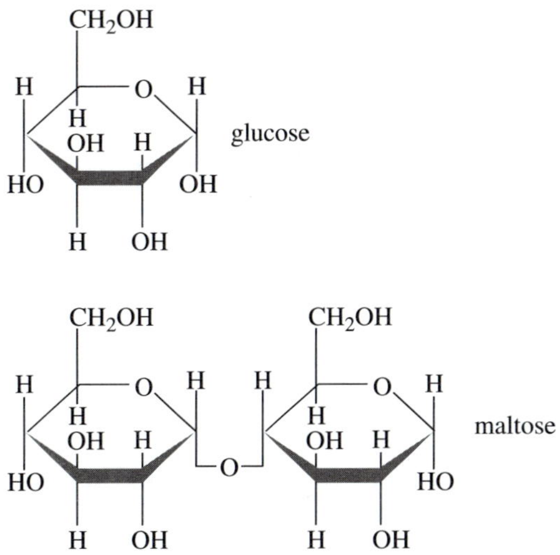

FIGURE 4 Chemical formulas of glucose and maltose.

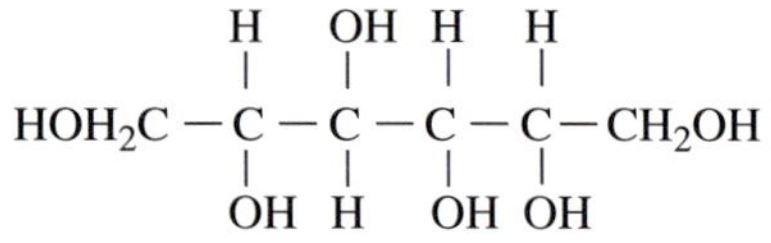

sorbitol

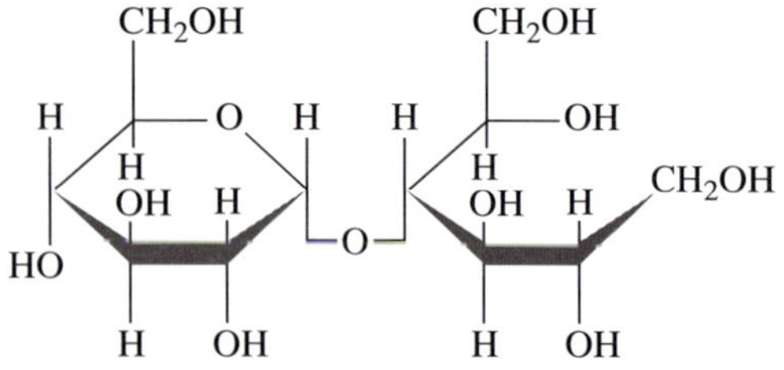

maltitol

FIGURE 5 Chemical formulas of sorbitol and maltitol.

pecially because of its scarce digestibility due to an intestinal intolerance on the part not only of numerous humans but also of important species of breeding animals.

Numerous enzymatic transformations are currently being studied to transform lactose into final products of a certain interest, such as lactitol, lacttulose, or lactobionic acid; by far the most important transformation of lactose, however, consists of its division by means of the enzyme β-galactosidase into its two constituent monosaccharides (glucose and galactose), both of which may also be used as foodstuffs for humans (with some limits regarding galactose) (Fig. 6). The reaction is carried out continuously, making the serum flow into a reactor filled with immobilized fungic lactase (β-galactosidase) at a flow velocity such that an 85–90% hydrolysis of the lactose takes place in the outflowing serum. When this simple and inexpensive technology is applied to lactase-containing yeast or fungic cells instead of to the immobilized enzyme, it yields biocatalysts endowed with 70–80% of the activity of the free enzyme in aqueous solution — an activity comparable to that of preparations containing the immobilized enzyme. Also, the stability over time of the activity of the immobilized cells has been shown to be excellent.

A relatively recent and environmentally highly significant application of biotechnology concerns, as already mentioned, the use of enzymes in the detergents industry, in which proteases and lipases capable of removing organic contaminants from fabrics have replaced toxic (perborate and hypochlorite) or polluting (polyphosphates) substances. The proteases and lipases used for this purpose are always of microbial origin since their synthesis through the fermentation of microorganisms allows them to be produced at costs that are considerably reduced compared to those of other animal or plant sources. Moreover, it has also been possible to improve the productivity of these enzymes in thermophile microorganisms, which are capable of surviving and reproducing at temperatures higher than 80–90°C, and enzymes that are active not only at room temperature but also at high washing temperatures have thus been obtained. Due to their catalytic efficiency, the proteases and lipases demolish stains and deposits of a complex nature which would resist the action of the detergents alone.

Lipases are widely used in industrial processes for the hydrolysis of oils and animal and vegetable fats and also in the production of glycerine and fatty acids. Here too, since enzymes operate in conditions that are much more bland than those that exist in traditional processing, their employment guarantees the greater purity of products, less pollution, and lower levels of energy consumption.

An important sector in which protease, but also lipase, enzymes are applied is that of the degreasing and tanning of animal skins. The advantages are better product quality and, most important, a drastic reduction of the serious environmental pollution associated with these processes. Since the skins' different layers are composed of protein such as collagen, α-keratin, and elastin (in addition to globular proteins such as albumin, globulin, and glycoproteins), the use of proteases of different specificity has allowed the elimination, by way of selective hydrolysis, of all the skins' constituents other than collagen. In this way, after the first processing phase (soaking) in which the salted and dried untanned skins are soaked in water, alkaline proteases and trypsin are used to remove nonfibrillar protein (prevalently, albumins and globulins); in the next phase, which consists of the depilation and bating of the skins in view of their subsequent coloring with chromate, trypsin and smaller quan-

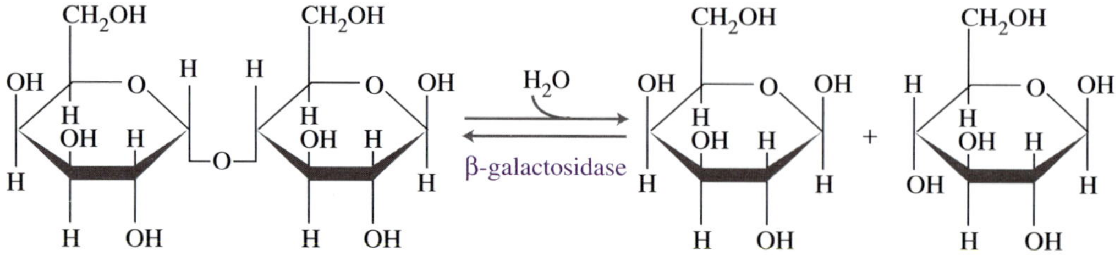

FIGURE 6 Hydrolysis of lactose by means of β-galactosidase.

tities of alkaline and neutral proteins are used. Depending on the kind of skin to be produced, a different pH, temperature, processing time, and type and relative quantities of enzymes in solution are used in the second phase. It is foreseen that in the future processing will take place under pressure to accelerate the spreading of the enzymes into the various strata of the skin.

Enzyme Immobilization

The advantages in using enzymes that are in an immobilized state are illustrated in Table 1. There are also some disadvantages, of course, such as the cost of immobilization, the loss in activity which it often entails, the extreme difficulty of making high-molecular-weight substrates (such as peptides, polysaccharides, or polynucleotides) react with insoluble enzymatic samples, and the greater purification that is often required of enzymes that are to be immobilized.

Since such advantages and disadvantages often depend on the immobilization techniques used, numerous methods of immobilization have been tested which are based on the formation of chemical bonds between the enzyme and the solid matter that acts as its support or on the entrapment of the enzyme within the latter. Immobilized molecules are generally more stable than ones in solution. This property is the result of the lesser conformational elasticity of the protein molecule's three-dimensional structure.

Whatever type of immobilization is used, the characteristics of the support material may strongly condition the success of the immobilization and of the subsequent application of the immobilized enzymatic sample. First, the cost and availability of materials may affect the entire cost-effectiveness of the process, especially in cases in which transformation costs must be kept very low. Other characteristics of materials that may affect the system are their physical form, chemical and mechanical stability, hydrophily, permeability, electrostatic charges, and resistance to microbial attack. The principal characteristics of a support for industrial use are summarized in Table 2.

In general, there is no ideal and universal method of immobilization, nor is there an ideal and universal support: Both the fixing procedure and the support must be chosen on the basis of the structural characteristics of the biological molecule and its applications.

TABLE 1

Advantages of the Use of Immobilized Enzymes

Ease in developing continuous processes
High stability and resistance to mechanical stress and microbial contamination
High reaction speed as a result of the high catalyst concentration
Possibility of reusing the catalyst for repeated reaction cycles
Less output of pollutants

TABLE 2

Main Characteristics of Support for Industrial Applications

Low cost
Mechanical stability
Thermal stability
Extended surface area
Chemical stability
Resistance to microbial attack

Cell Immobilization

The immobilization of whole cells of microbial and plant origin represents an evolution in techniques for the immobilization of enzymes and for their use in heterogeneous catalysis. It entails lower production costs compared to enzymes that have been immobilized separately and, most important, provides the opportunity to obtain molecules with complex structures and high added value starting from simple substrates by way of a sequence of chemical reactions catalyzed by the various enzymes present within the cells. Due to the cell's well-organized structure, this technique allows one to obtain biosynthetic reactions which require the participation of coenzymes. The immobilized cells finally find a valid application in the production of microbial metabolites of interest to industry (ethanol, organic acids, vitamins, etc.), in a way that is alternative to fermentative processes. The immobilized cells may be either dead or live. In the former case, the cells can be assimilated into a small envelope which contains the active enzyme or a combination of enzymes. If the cells are live, all their metabolic components are intact and therefore they are capable of reproducing.

The cells that are used may be either growing or in a stationary state (resting cells). There are various methods with which to immobilize them, homologous to those used with enzymes. The three main procedures are incorporation in a polymeric material, attachment to a support by adsorption, and reticulation with bifunctional reagents. Incorporation in synthetic or natural materials is the most commonly used method. The materials which are generally used in this way are polysaccharides (natural or chemically modified) or synthetic polymers such as polyacrylammide. This system is an alternative to the use of enzymes; therefore, the immobilized cells are usually live. The advantages and disadvantages of using cells are summarized in Table 3. Cells may also be used in media composed of organic solvents when the solubility of substrates and products requires it.

TABLE 3

Advantages and Disadvantages of the Use of Immobilized Cells

ADVANTAGES	DISADVANTAGES
Elimination of enzyme extraction and purification costs	Limitations of a diffusive character due to the presence of cell walls and membranes
Use of the entire enzymatic activity present in the cells	Possibility of secondary reactions because of the presence of numerous active cellular components
Greater stability of the enzyme, which remains in its native state and natural environment	Less enzymatic activity per unit of support surface
Ease of immobilization and fewer denaturing effects	Impossibility of using exocellular enzymes

TABLE 4

Biotransformations of Considerable Commercial Interest Carried Out by Means of Immobilized Enzymatic Catalysts

PRODUCT	CATALYST	PRODUCTIVE CAPACITY (TONS/YEAR)
Acrylamide	Nitrilohydrolase	10,000
6-Aminopenicillinic acid	Penicillinacylase	1,000
Aspartate	Thermolysine	1,000
L-Aspartate	Aspartase	100
L-Malate	Fumarase	100
Isopropylmiristate	Lipase	100
p-Hydroxyphenylglycine	Hydantoinase carbamylase	100

Live immobilized cells have an intact genetic inheritance and are therefore capable of reproducing after they have been put into some nutritive medium, the so-called culture medium for free cells. If free and immobilized cells are similar from a metabolic standpoint, they are not so from a technological one. Immobilized cells can be separated easily from the reaction medium, are easily reusable, and are particularly suited for continuous-flow processes.

This technique constitutes an alternative to the normal fermentative processes that are used to produce alcohols, amino acids, organic acids, vitamins, solvents, and other primary and secondary metabolites of industrial interest. Immobilization enables the microorganism to be reused by transforming, in the specific case of fermentative processes, the operative system from a discontinuous one (of the batch type) to a continuous one by maintaining a high cell population and, ultimately, a high fermentative velocity. This greater metabolic productivity which results from a high microbial density allows the reactor that contains the immobilized cells to operate at hydraulic dilution velocities, which are notably greater than those allowed by the maximum specific speed of growth of the microorgansim in a free state (Marconi, 1989). Plant cell cultures and their utilization for productive purposes, both in a free and in an immobilized state, will be discussed later.

The Use of Microbial Enzymes in Heterogeneous Systems

Some typical biotransformations of considerable commercial interest that may be obtained using immobilized enzymatic catalysts are listed in Table 4. The activity and the stability of the enzymes, which are indispensable prerequisites for the economic optimization of the various processes, have in every case been improved using protein engineering techniques.

In the future, chiral chemical intermediates (i.e., those endowed with a given spatial configuration among the two or more possible ones) will continue to grow in importance for the pharmaceutical industry. In the medium term, biotransformations will undoubtedly play an determinant role in the preparation of such compounds and replace traditional techniques (Roberts, 1990) such as the selective crystallization of a single chiral intermediate out of a mixture of them, collection from natural sources (especially plant species), and asymmetrical synthesis by means of special catalytic systems (stereoselective) capable of directly synthesizing only one of the statistically possible enantiomers. In general, these technologies are highly polluting because they employ solvents to extract active principles, complexing agents (e.g., ammines and asymmetrical phosphines), and organic solvents used in chemical processes that are stereoselective (asymmetrical synthetic) of the desired enantiomer.

The greatest development of biotechnology has taken place in the pharmaceutical sector: In the United States, for example, the annual revenues of the 10 largest companies that applied biotechnology to health during the 1986–1996 period increased from \$6.2 to \$26.5 billion. The powerful governmental organization which in the United States approves the introduction of new drugs and food products on the market, the Food and Drug Administration (FDA), established by decree in May 1992 that, in the case of drugs composed of racemic compounds, approval depended on the petitioning industry's provision of extensive pharmacotoxicological documentation on each of the enantiomers present in the mixture. The cost and complexity of these procedures were such that at the end of the 1990s more than 20% of the drug market in the United States (the estimated value of which is more than $250 billion per year) was comprised active principles made up of a single enantiomer. As a result of the increasing demand for chiral active principles, the market of chiral intermediates, which are necessary for their production, has experienced at the world level an abrupt increase to $1 billion per year and is growing rapidly.

Production of Drugs and Phytopharmaceuticals. It is in the pharmaceutical sector that biotechnology has achieved its greatest commercial success and that the greatest efforts toward development and marketing have been accomplished. Antibiotics are the most important category of chiral drugs in this context. Therefore, they constitute an ideal category of compounds for enzymatic or genetic engineering treatments, given these techniques' characteristic action on mixtures of compounds endowed with the same chemical composition but different spatial configurations.

Despite the fact that penicillin was the first antibiotic discovered, it has retained great therapeutic (and therefore commercial) importance due to the enormous amount of research that has been carried out on it, which has engendered derivatives with broader antibacterial spectrums of action and resistance to inactivation on the part of the pathological microorganisms on which the penicillin acts. These new penicillins are based on the introduction in the molecule of substituents that differ from those naturally introduced by the microorganisms that synthesize penicillin in the course of fermentation and especially on the variation of the chemical substituents in penicillin's basic "skeleton" [i.e., 6-aminopenicillanic acid (6-APA)].

The industrial production of 6-APA via the enzymatic hydrolysis of penicillin (produced by fermentation) that was introduced in the 1980s, and that of 7-aminodeoxyacetylcephalosporanic acid (7-ADCA) from cephalosporine G, at the end of the 1990s reached levels of many thousands of tons per year.

Penicillinacylase (PGA) is specific for penicillin's lateral chain (phenylacetic acid). PGA produced by *Escherichia coli* (Fig. 7a), which hydrolyzes penicillin G (benzylpenicillin), is used in industry. Another enzyme which is used is PGA synthesized by *Bovista plumbea,* which instead hydrolyzes penicillin V (phenoxymethylpenicillin), obtained by adding 2-phenoxyethanol instead of phenylacetic acid as a precursor during fermentation. These two enzymes also hydrolyze the cephalosporines obtained as a result of the chemical broadening of the respective penicillins, as shown in Fig. 7b for cephalosporine G. Since it was first introduced, the economic efficiency of the process has been constantly improved through the genetic manipulation of microorganisms so as to increase their enzymatic production and through the fine-tuning of immobilization techniques.

6-APA is the product one uses in order to synthesize semisynthetic penicillins, compounds in which phenylacetic acid is replaced by other groups that endow the molecule with greater stability and a broader antibacterial spectrum and which are therefore of greater therapeutic utility. For a long time, the only way to produce 6-APA was by chemical hydrolysis at low temperature (−40°C) and the use, for the hydrolysis of the amidic bond, of chloroderivatives of phosphorus and silicon and aromatic ammines in organic solvents. Therefore, the reasons for the success of the enzymatic process are obvious: Not only does the use of di-

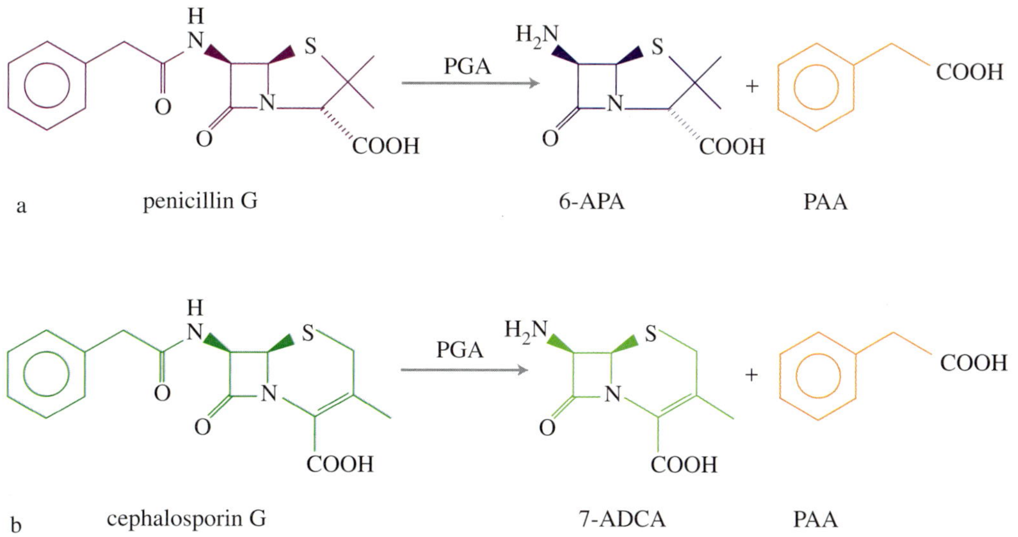

FIGURE 7 (a) Hydrolysis of penicillin G by means of penicillin G acylase (PGA) to obtain 6-aminopenicillinic acid (6-APA). (b) Hydrolysis of cephalosporin G by means of PGA to obtain 7-amino-deoxyacetylcephalosporanic acid (7-ADCA). PAA, phenylacetic acid.

methylaniline in the chemical process eliminate worries about the safety of the product but also it solves important environmental problems which, as a result of the carcinogenic nature of aromatic ammines, would otherwise still plague the process. Moreover, phosphorus chloride is a difficult compound to handle and to store, and it requires special precautions in the utilization and treatment of effluents. Finally, the enzymatic process yields great quantities of 6-APA, at least equal to those of the chemical process (Johann, 1993).

A Swiss company that is a leader in the production of nicotinic acid (an important intermediate for pharmaceutical and fine-chemical products) and nicotinamide (vitamin PP or B_3) recently developed at the industrial level a biotransformation of 3-cyanopyridine in nicotinamide which has been shown to be competitive with the chemical synthesis of methylethylpyridine (Fig. 8). To partially hydrolyze the cyanopyridine, the immobilized enzyme nitrilohydrolase (from *Rhodotorula rhodoctous*) is used, which is similar to that employed to hydrolyze acrylonitrile into acrylamide.

An important field of application of chiral synthesis is that of intermediates for phytopharmaceuticals. One of the most important classes of the latter is that of the derivatives of 2-phenoxypropionic acid, herbicides sold as racemic mixtures whose biologically active form is always composed of a single isomer. Although the racemic mixture obtained by means of chemical synthesis can be separated in its enantiomers using traditional crystallization techniques, there are notable economic and environmental advantages in producing the herbicide directly from intermediates with a high optical purity. These advantages include both the size of the market for these products and the fact that, in principle, the techniques for the synthesis of chiral drugs that have already been developed may be adapted to produce chiral intermediates for phytopharmaceuticals.

In a procedure that was developed with this aim in mind (the Zeneca process), the key intermediate is racemic α-chloropropionic acid, an equimolar mixture of the two enantiomers that contain the chiral carbon atom in an absolute configuration (R and S, respectively). The goal is to selectively dehalogenate (R)-α-chloropropionic acid so as to leave the S isomer unaltered; the latter will be used subsequently in pure form to synthesize the herbicide (Fig. 9). The dehalogenation, which transforms the (R)-α-chloropropionic acid into (S)-lactic acid, is obtained using microbial cells that have been selected to produce the specific enzyme R-dehalogenase in abundant quantity. The cells are produced separately by fermentation and then added to the substrate in a reactor. The reactor, when adequately stirred, works in a discontinuous manner. At the end of the reaction, the cells are removed by filtration. The capacity of the installation is 2000 tons/year. It is worth noting that the Zeneca process is not the only technology which may be used industrially for this sort of production: At the Massachusetts Institute of Technology, a process has been developed which is based on the selective esterification with butanol, catalyzed by the enzyme lipase, of the sole isomer S, which (given the notable difference in physical characteristics) may be separated from the free R acid using standard techniques.

Biotransformations. The hydrolysis of animal and plant fats underlies some very important traditional processes used to produce long-chained fat acids (whose alkaline salts constitute the soap) and glycerol. These processes are carried out industrially by means of alkaline solutions or powerful mineral acids, and therefore they are highly polluting. The generalized use of enzymes for this type of process is currently too expensive; instead, they are being increasingly used for the biotransformation of fat acids and their derivatives. In these reactions, which are of very notable interest to industry and often impossible to reproduce by chemical means, lipases are prevalently employed. The latter are of various type and specificity: Some lipases can reduce to soap the ester bond localized on glycerol's secondary hydroxyl but cannot bond to the two primary alcohol hydroxyls. Usually, lipases are not specific with regard to the type (the presence or absence of insaturations) and length of the chain of the fat acid, a characteristic which allows them to be applied with great versatility. The importance of the market for them has induced enzyme producers to develop not only microorganisms capable of producing larger quantities of enzymes but also enzymes that are

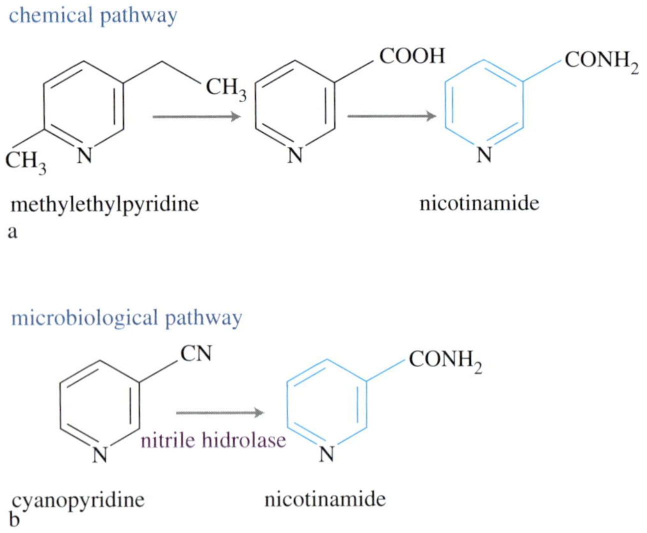

FIGURE 8 (a) Chemical synthesis of nicotinamide starting from methylethylpyridine. (b) Lonza process for the microbiological synthesis of nicotinamide starting from 3-cyanopyridine. The use of the enzyme nitrile hydrolase, obtained from *Rhodotorula rhodoctus*, permits the partial hydrolysis of cyanopyridine.

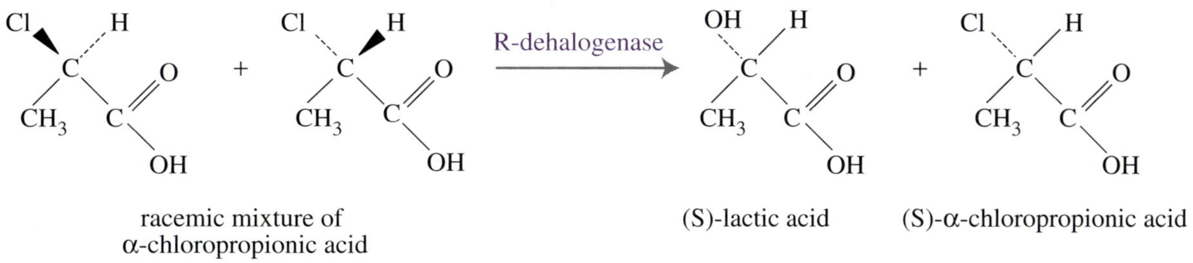

FIGURE 9 Reaction diagram of the Zeneca process for (S)-α-cloropropionic acid.

more stable and able to work in broad ranges of temperature and pH and in the presence of organic solvents.

The lipases catalyze the hydrolytic scission of the triglycerides present in many oils and fats into their constituents (glycerol and long-chained aliphatic acids). In the absence of water, the same enzymes also catalyze the reaction of synthesis of ester, or the transesterification reaction, which allows one to replace one or two acid residues in the lipid molecule with others of different type (with an unsaturated chain, a lower number of carbon atoms, etc.), according to the diagram shown in Fig. 10 in which R‴ is the new substituent one wishes to introduce. This reaction allows one to obtain a new oil or fat with improved physical properties or structural or nutritional characteristics.

A peculiar characteristic of this technology is that its products are very pure. Moreover, the specificity and bland reaction conditions of the enzymatic transesterifications (or of the esterifications) may be exploited to produce esters which would be too expensive or impossible to synthesize by conventional means, such as certain unsaturated acids or alcohols. For example, a multinational corporation has developed a new triglyceride based on plant fats to be used in artificial milk for newborn babies because its structure is very similar to that of the fats present in human milk. In general, "structured" triglycerides which present advantages from the nutritional standpoint comprise a market that is developing rapidly, and numerous forms are already available in Japan and the United States.

Hundreds of tons of esters are currently produced for the cosmetics industry by means of immobilized lipase which, instead of the usual hydrolysis of the esters, causes them to be synthesized starting from acid and alcohol, in the absence of water:

$$\mathrm{RCOOH} + \mathrm{R'OH} \xrightarrow{\text{lipase}} \mathrm{RCOOR'} + \mathrm{H_2O}$$

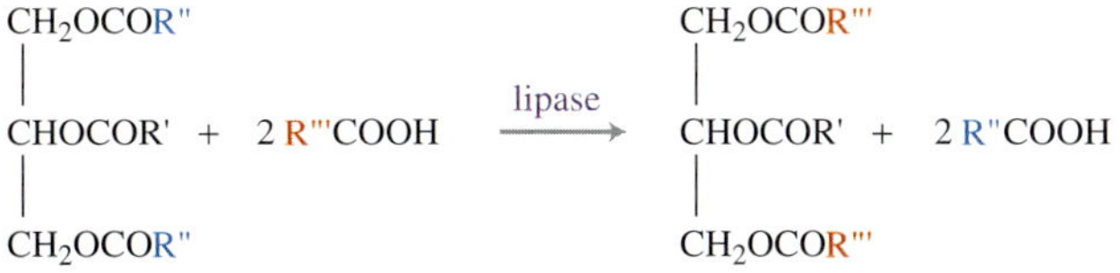

FIGURE 10 The transesterification reaction catalyzed by a lipase. Substitution of one or two acid residues in the lipid molecule allows one to obtain a new oil or fat with improved physical properties or structural and nutritional characteristics.

Microbial lipases combine an elevated stereoselectivity (i.e., the capacity to act on a single enantiomer in a racemic mixture) with broad substrate specificity, in the sense that they exert their catalytic action on molecules that may contain, on the optically active carbon atom, substituents of quite varied character. Furthermore, they are active (something which is rare for an enzyme) in organic solvents in the almost total absence of water. Therefore, they are being actively developed for the purpose of resolving racemic mixtures of alcohols, acids, or esters. A very important application concerns nonsteroidal antiinflammatory drugs, often composed of derivatives of propionic acid that has been replaced with aryl groups at position β (Naproxen, Ibuprofen, etc.). Since only one of the two isomers present in the specialty is endowed with therapeutic activity, and since these drugs are not without collateral effects, there is an interest in resolving the racemic mixture so as to market specialties that contain only the active isomer. Lipases (especially those of yeasts and molds) have been shown to be quite effective in hydrolyzing the esters (methyl, ethyl, etc.) of the two stereoisomers at different rates and in synthesizing, starting from racemic acid and alcohol, the ester of only one of the two forms before the synthesis (chemical and not enzymatic) of the ester of the other stereoisomer begins (Servi, 1996). This allows for a relatively easy separation of the desired product.

Amino Acids and Their Derivatives. Amino acids are the fundamental constituents of proteins and their industrial production is aimed primarily at the pharmaceutical and food industries.

Pharmaceutical companies require amino acids because peptides and proteins are used for therapeutic purposes, and this has led, especially since the end of the 1980s, to the development of drugs composed of particular amino acids and their derivatives. Their typology is quite varied and is not limited to amino acids that are present in nature; there is great demand, for example, for some D-amino acids, which are obviously not present in nature (except in very rare instances which are not relevant from this point of view). Various synthetic sweeteners and flavor enhancers (such as aspartate and monosodium glutamate) are composed of amino acids, and a great deal of research and development is currently taking place in this area given the importance of the sector in the world market which, at the end of the 1990s, was estimated to retail $3 billion annually. The commercial success of aspartic acid has proven the success of the industrial synthesis of peptides and their use as additives in foodstuffs.

Some amino acids are produced on a large scale directly by fermentation, whereas others must be synthesized enzymatically or in part chemically and in part enzymatically. The progress that has been made in the application of genetic engineering techniques to improve the efficiency of fermentative processes has had a decisive impact on the production of some of them. During the 1990s, for example, numerous processes for the synthesis of L-phenylalanine from benzaldehyde were developed; the fermentative method, however, appears to be economically preferable as a result of the productive improvements obtained due to genetically modified microorganisms.

Japanese industries are on the front line with regard to the production of amino acids and their derivatives, and they have proven to be quite able, not only in identifying and optimizing suitable biocatalysts but also in developing the relative industrial processes on a large scale. A Japanese company, for example, is the market leader in the production of L-aspartic acid for pharmaceutical purposes, a market which produces 6000 tons a year. It is synthesized from fumaric acid and ammonia by means of the enzyme aspartase, according to the formula

$$\text{HOOCCH}{=}\text{CHCOOH} + \text{NH}_3 \rightarrow$$
$$\text{HOOCCH}_2\text{CH}\,(\text{NH}_2)\,\text{COOH}$$

The biocatalyst is neither the enzyme nor an enzymatic preparation but rather *E. coli* cells with high aspartase activity that have been modified using recombinant DNA techniques and immobilized in a polysaccharide gel (carrageenin). Thus immobilized, the cells are packed into a continuously operating, fixed-bed reactor. The fact that the cells are immobilized enables one to employ continuously operating reactors and thereby avoid the costs of extracting the enzyme. Moreover, the catalyst is extremely stable; its activity is halved after 3 years of operation.

The use of aminoacidacylases to obtain optically active and highly purified amino acids starting from the racemic amino acid acetylated to the aminic group represents an important chapter in industrial enzymology and may be accomplished using the enzyme either immobilized on polymeric matrices or in solution in reactors with ultra-filtering membranes. A company on the forefront in the development of membrane reactors on an industrial scale has developed many enzymatic techniques to produce L-amino acids and D-amino acids. The first application was to use acylase in the production of L-valine and L-methionine, neither of which can be obtained inexpensively through fermentation. The membrane reactor allows for the resolution of the racemic mixture of the chemically synthesized *N*-acetyl amino acid. Several hundreds tons of L-amino acids of very high steric purity are produced each year by means of this technology (Fig. 11).

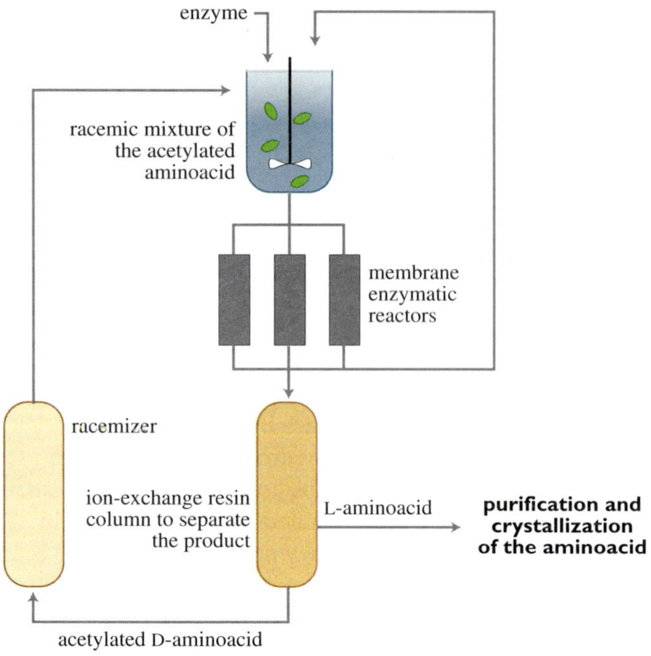

FIGURE 11 Diagram of the process to obtain stereochemically pure amino acids using membrane enzymatic reactors. The racemic mixture of acetylated amino acid is put into the enzymatic reactor. The enzyme, dissolved in water, is retained by the membrane because of its large molecular size, which prevents it from passing through pores, and selectively hydrolyzes the L-enantiomer. The L-amino acid and the unaltered acetylated amino acid in the D- form cross the membrane to the separation zone, made up of a bed of ion exchange resin. Here, the L-amino acid is absorbed, while the acetylated D-amino acid then undergoes racemization. The racemic mixture is recycled to the reactor. The L-amino acid, instead, is purified and crystallized.

The commercial interest of amino acids such as L-aspartic acid and L-phenylanine is largely the result of the fact that they are the fundamental raw materials used to synthesize aspartase, a low-calorie sweetener which is not present in nature, endowed with much higher edulcorative power than saccharose (150–200 times greater), and completely metabolized and innocuous, as evidenced by long-term toxicological studies and the FDA's approval of its free commerce in the United States. From the chemical standpoint, it is a dipeptide derived from the condensation of L-aspartic

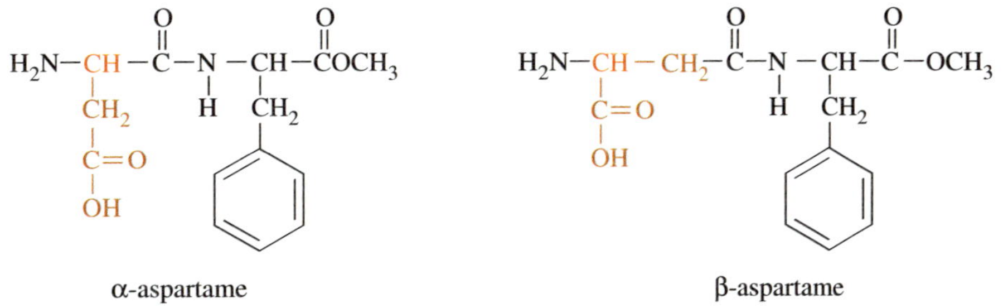

FIGURE 12 Chemical formulas of α- and β-aspartame.

acid with the methyl ester of L-phenylanine. Its sweet taste depends on the L-shaped conformation of the amino acids, the presence of the methyl ester group on the phenylanine, and on the latter's selective condensation with the carboxyl in the aspartic acid at position α with respect to its own amine group. This last characteristic is essential to prevent the production of β-aspartase as well, which has a bitter taste (Fig. 12). Via traditional chemical synthesis, instead, the formation of β-aspartase is inevitable (20–40% depending on the method used); given the extreme similarity, from the chemical standpoint, of the two compounds, its removal requires complicated technologies. Furthermore, chemical synthesis requires complex protection stages and the subsequent reestablishment of functional groups, in addition to the use of L-shaped amino acids, which are much more costly than the racemic form. A Japanese company has introduced enzymatic synthesis, which avoids all these drawbacks and uses the enzyme thermolysine, a neutral protease synthesized from *Bacillus thermoproteoliticus.* The reaction is carried out in water-saturated ethyl acetate at 50–60°C using the enzyme that has been immobilized in a hydrophilic polymeric matrix and, as reagents, L-aspartic acid (whose amine group is protected with a carbon benzoxylic group which is later eliminated by hydrogenation) and the methyl ester of D,L-phenylanine. Because of the regioselectivity and stereoselectivity introduced by the enzyme, the reaction only takes place between the carboxylic group of the aspartic acid at position *a* and the L isomer of the phenylanine's ester (Fig. 13). In 1999, the enzymatic production of aspartase on the part of the largest companies in the sector amounted to 2000 tons per year, for a turnover of about $100 million.

Also of notable industrial importance is the process for the enzymatic synthesis of *para*-hydroxyphenylglycine, which is used instead of phenylacetic acid in modern-day penicillins (amoxycillin) and semisynthetic cephalosporines (Fig. 14).

Terpenic Derivatives. An interesting application of the resolution of racemic mixtures with immobilized cells operating in a nonaqueous system, albeit in an organic solvent, is the stereospecific hydrolysis of the racemic esters of menthol on the part of immobilized cells of *Rhodotorula minuta,* in which only one of the two constituents of the mixture of menthol esters (dextrorotatory and levorotatory) is hydrolyzed and L-menthol is produced (Fig. 15). This substance is used as an aroma in many broad-consumption products such as chewing gum. The fact that it works in an organic solvent is the result of the products' insolubility in water.

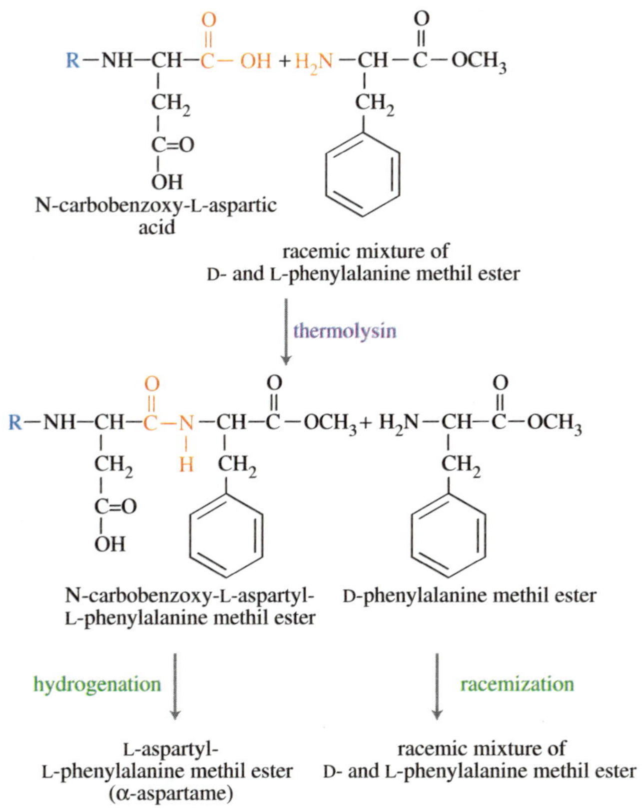

FIGURE 13 The enzymatic production of aspartame. L-Aspartic acid, whose amino group is protected by a carbobenzoxylic (R) group, reacts with the methyl ester of phenylalanine in the presence of the enzyme thermolysin. The reaction takes place only between the carboxylic group of the aspartic acid at position α and the L-isomer of the phenylalanine ester. The carbobenzoxylic group is eliminated by hydrogenation to obtain α-aspartame.

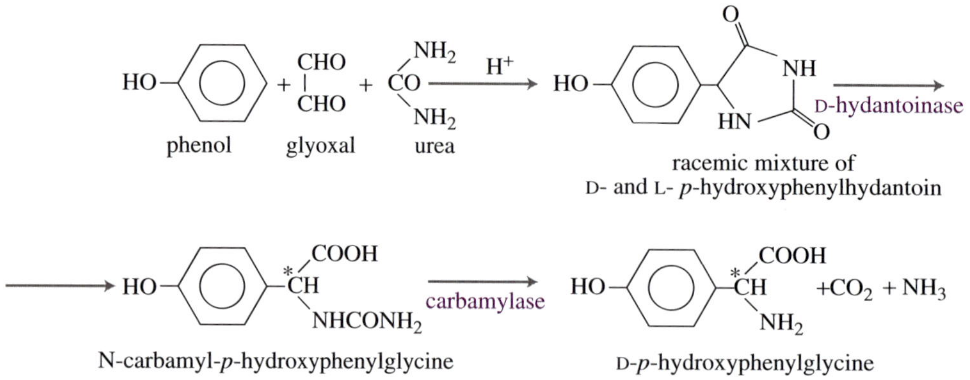

FIGURE 14 Synthetic process of D-*p*-hydroxyphenylglycine. The process is based on the chemical synthesis of a racemic hydantoin. The D-hydantoin is hydrolyzed by the enzyme D-hydantoinase into *N*-carbamyl-*p*-hydroxyphenylglycine. This, in turn, is transformed into *p*-hydroxyphenylglycine by the enzyme carbamylase or carbamyl hydrolase. The L-isomer of the hydantoin, which is not transformed by the D-hydantoinase, undergoes racemization on heating and is converted into an equimolar mixture of D- and L-isomers which are recycled to react with the enzyme. Both enzymes are obtained from cells of *Agrobacterium radiobacter* grown in a cultivation medium which maximizes the activities of both enzymes. The process is carried out using the high-activity *A. radiobacter* cells, without isolating the two enzymes. In the discontinuous process, the two stages of hydrolysis are carried out in 48 h and the D-*p*-hydroxyphenylglycine is separated by crystallization from the unaltered solution of L-hydroxyphenylhydantoin.

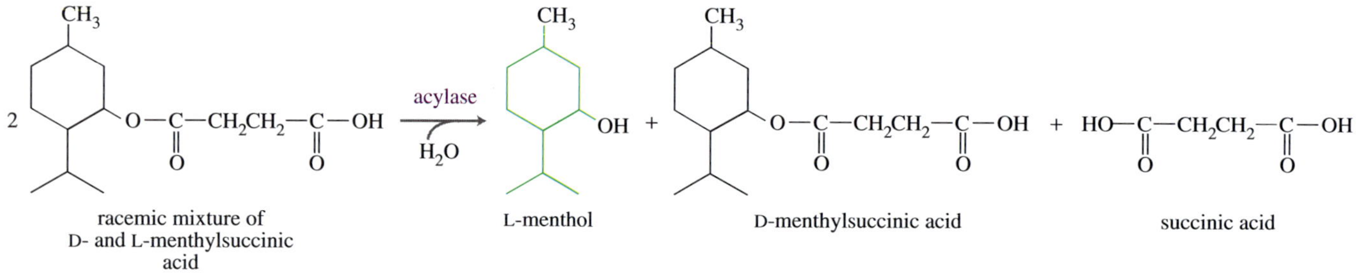

FIGURE 15 Stereospecific hydrolysis of D- and L-menthyl-succinate by immobilized cells of *Rhodotorula minuta,* with formation of L-menthol, a flavor enhancer in widespread use.

Productive Technologies That Employ Plant Enzymes and Cells

Plant cells and, in general, plants as a whole make considerable contributions to many industrial sectors, especially to the pharmaceutical and cosmetic ones. In pharmaceutics, for example, 25% of drugs contain products that have been isolated from plants; the most common of these substances are steroids, cardiac glycosides, alkaloids, and acetogens (Table 5). Note that almost all product categories are listed in Table 5. It should also be noted that improvements in the use of primary agricultural productions, often achieved by means of an intelligent synergic combination of the genetic improvement of plants and innovative enzymological processes, have not only yielded innovative foodstuffs and industrial products but also, due to notable reductions in production costs, led to an optimal exploitation of existing agricultural resources.

TABLE 5

Substances Obtained from Plant Cell Cultures

Nucleic acids	Enzymes	Peptides
Organic acids	Calcons	Proteins
Alkaloids	Flavonoids	Perfumes
Amino acids	Cardiac glycosides	Pigments
Antileucines	Lipids	Flavor enhancers
Antimicrobials	Naphthoquinone	Steroids
Antihumorals	Nucleotides	Tannins
Antivirals	Oils	Terpenes
Anthraquinones	Insecticides	Polysaccharides
Fragrances	Opiates	Vitamins
Carbohydrates	Hormones	Sugars

A fundamental concern that will continue to drive the evolution of these technologies in the near future is the need to design productive systems that are able to perceptibly diminish current levels of pollution. Moreover, it is predictable that the introduction of new biotechnologies will cause, in fundamentally important sectors such as foodstuffs, major changes in the behavior of consumers. In industrialized countries, in fact, in addition to an overproduction of traditional agricultural products, consumers are always better informed and more demanding with regard to dietary characteristics, nutritional content, the nondangerousness of foods, and also in relation to transformation, conservation, and packaging techniques. Basic products have thus been modified so as to improve their nutritional and functional characteristics, and much has been undertaken, by way of a great variety of techniques based on the use of microorganisms and tissue cultures, in sectors related to the production of important metabolites such as aromas, pigments, vitamins, antioxidants of natural origin, enzymes, and lipids. The production of natural substances by means of *in vitro* cultures of plant cells no doubt has its advantages compared to the isolation of the same products from botanical materials from plantations of the traditional type. The main advantages are independence from environmental factors (climate, crop infections, seasonal geographic, and geopolitical factors); the possibility of obtaining products of higher and more consistent quality through the control of the variables in the process; a greater elasticity in decision making in regard to the possibility of determining, on the basis of market demands, where, when, and in what quantity to produce; and higher yields compared to those obtainable using traditional extractive technologies.

Of greatest interest in plant cell cultures is the production of drugs that are difficult to synthesize chemically or that can replace molecules that are synthetically complex and polluting (Van der Esch *et al.,* 1993). The main drugs that can be obtained from higher plant organisms and their respective pharmacological activities are listed in Table 6.

Other sectors can also profit considerably from the high added-value products obtainable via plant cell cultures, including the food industry, the cosmetics industry in the production of essential oils (rose and jasmine oils from the respective species), and the phytopharmaceutical industry in the production of pyrethroids from *Chrysanthemum* spp. This kind of culture, although capable of providing an increasing number of substances, has a low level of cellular production. Research efforts in this sector, therefore, are aimed at obtaining cell lines that are more productive, at making them genetically stable, and, at the identification of the biophysical parameters that will allow processes to grow to industrial scale (Miele, 1997).

Plant cell cultures are carried out using the callus of the plant as a raw material, a mass of undifferentiated tissue that is formed as a cicatricial outgrowth each time a plant suffers a wound. After it has been explanted and placed in a culture medium enriched with the right nutrient and growth-regulating substances (auxins, gibberellins, etc.), this cellular mass is capable of developing differentiated cells which, by means of continuous or semicontinuous fermentative techniques similar to those used with microbial-type organisms, can be made to produce the desired substances.

TABLE 6

Main Drugs of Plant Origin

PLANT SPECIES	DRUG	ACTIVITY
Atropa belladona	Atropin	Anticholinergic
Catharanthus roseus	Vinblastine	Antileukemiac
Chondodendron	Tubocurarine	Muscle relaxant
Cinchona ledgeriana	Quinine	Antimalariac
Colchicum autumnale	Colchicine	Tubocurarine
Datura metel	Scopolamine	Anticholinergic
Digitalis purpurea	Digitoxin	Cardiotonic
Digitalis lanata	Digoxin	Cardiotonic
Dioscorea deltoidea	Diosgenin	Contraceptive
Papaver somniferum	Codeine	Analgesic
Uragoga ipecacugha	Emetine	Amebicide

The immobilization of plant cells was first tested as a means to study the behavior of these cells in a medium that simulated the one present in plant tissues in order to examine cellular redifferentiation and, subsequently, to verify the advantages of immobilization in the production of some very costly natural substances. The slow growth of plant cells compared to microbial ones makes it advisable to carry out the immobilization in sterile conditions so as to avoid the risk of bacterial contamination.

The immobilization of plant cells no doubt provides greater advantages than the immobilization of cells of bacterial origin. Because of the higher investment and management costs of producing a biomass from them, plant cells grow very slowly, attain very low cellular densities, their growth is affected by the synthesis of inhibitors, and their culture media are very expensive. The possibility of using this biomass more than once, therefore, provides significant economic benefits.

Transformation and Synthesis of Complex Substances by Way of Plant Cells

Plant cells are very large (100 μm), sensitive to mechanical stress, and also very sensitive to chemical reactants; they are therefore best immobilized by inclusion in polymeric matrices. The capacity of plant cells to transform and synthesize substances with very complicated molecular structures and of great commercial interest may be exploited us-

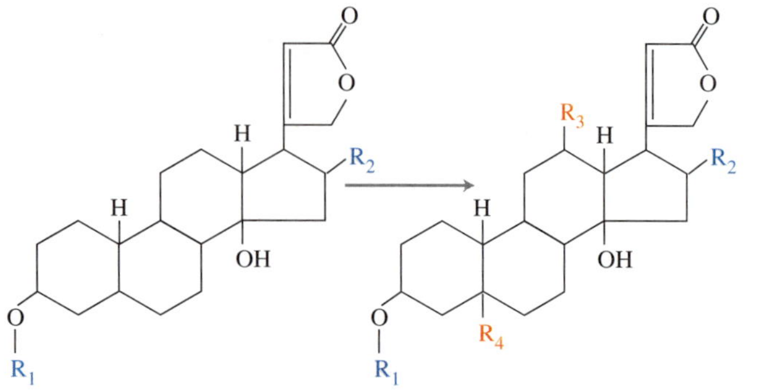

FIGURE 16 Biotransformation of glycosides by means of immobilized plant cells. The introduction by chemical means of the functional groups R_3 and R_4, usually hydroxyls, is extremely difficult.

ing three distinct biosynthetic activities: simple biotransformations, synthesis from precursors, and synthesis *ex novo.*

The first category includes such reactions as hydroxylation, glycosidation, methylation, and the reduction or oxidation of very complex molecules. An example of biotransformation of glycosides with immobilized cells of *Digitalis causata* or of *Daucus carota* is shown in Fig. 16); the introduction of the functional groups R_3 and R_4 (generally oxydrils) by chemical means would be extremely difficult.

The synthesis of precursors is a complicated process which leads to a series of biotransformations starting from certain precursors. A practical example is provided by immobilized cells of *Catharanthus roseus* which, starting from tryptamine and secologanina, yield ajmalcine, a substance effective against hypertension and with tranquilizing effects (Fig. 17).

The synthesis of complex secondary metabolites starting from common sources of carbon such as saccharose (synthesis *ex novo*) is the most ambitious application of plant cell cultures, based on the coordinated exploitation of the enzymatic patrimony the cell has developed to carry out its metabolic functions.

The difficulties with these processes are of two types: The cells contain enzymes that catalyze undesired reactions, and the cell's wall and membrane can prevent or delay the spreading of substrates and products. These drawbacks are being obviated by means of agents capable of inducing both an appropriate permeabilization of the cell and selective enzymatic inactivation.

From a certain point of view, biotransformations may be less interesting than biosyntheses since other, more simple, catalysts (enzymes or microbial cells) are able to carry out the same reactions. Biosynthesis, on the other hand, requires a greater control of cellular metabolism and it is difficult for a simple substrate to be transformed into a com-

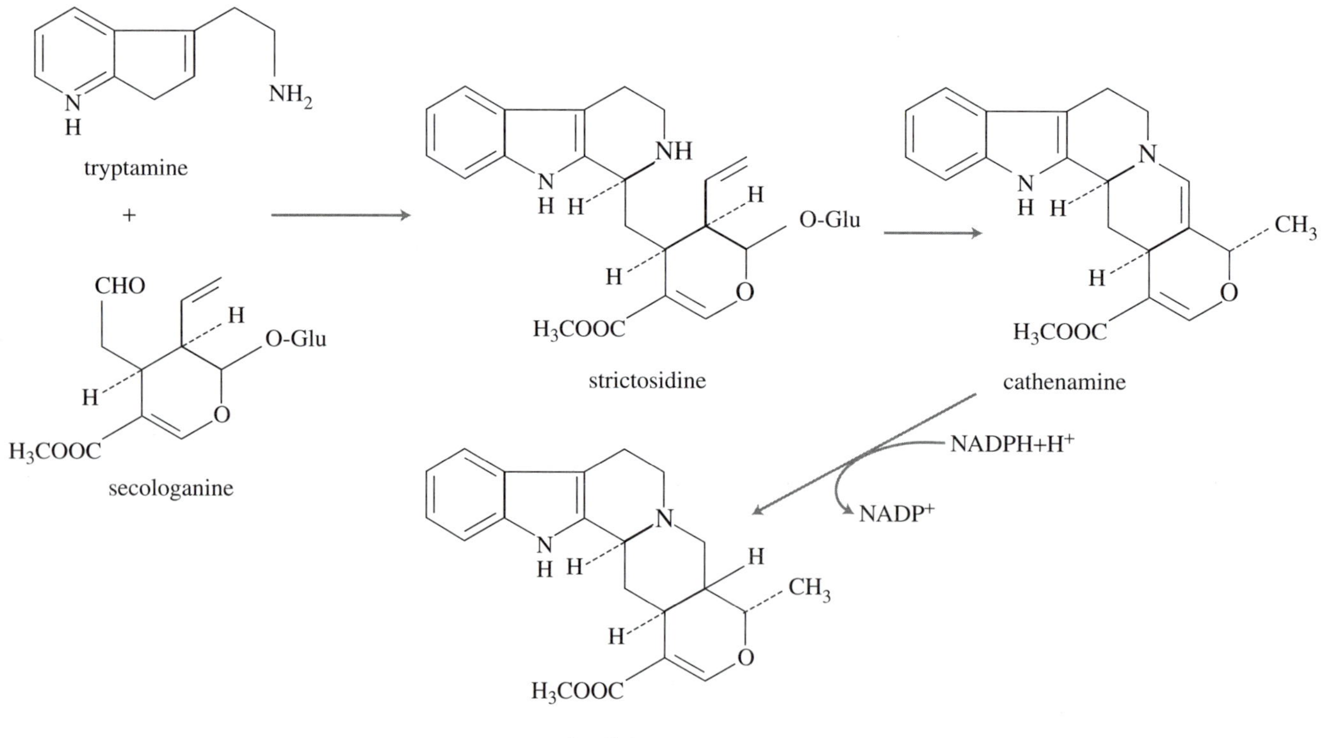

FIGURE 17 Biosynthesis of cathenamine and ajmalicine starting from tryptamine and secologanine by way of a complex series of biotransformations carried out using immobilized cells of *Catharantus roseus.*

plicated structure with high bioconversion efficiency. The complexity of cellular metabolism and the limited knowledge of control factors impose an empirical use of hormones and of other cellular mediators in the promotion of the formation of the desired metabolite, keeping collateral reactions at a minimum.

An interesting example of how the combination of sophisticated techniques can lead to spectacular improvements in productive processes based on advanced biotechnologies is provided by recent developments in the synthesis of shikonin. A naphthoquinonic derivative of dark red color, sciconin is used as a coloring substance, an antibacterial substance, and an antiinflammatory agent, and it is produced industrially by means of cell cultures of *Lithospermum erythrorbizon* (Fig. 18). In a recent study, these plants cells were placed in bioreactors and immobilized in calcium alginate, and a two-phase system was used that employed *n*-hexadecane in the organic phase, which was capable of dissolving, by removing it from the aqueous phase, most of the shikonin that had been produced. In this way, 25-fold increases in productivity were achieved with the same reaction volumes compared to the traditional technology based on the use of nonimmobilized cell cultures.

It is certain that the most important contribution to the improvement of these catalytic processes will come from genetic engineering. By means of recombinant DNA technology, in fact, one can increase the production of the enzymes that regulate the activity of the catalytic complex; one can also modify the mechanisms that regulate metabolism in order to maximize the metabolic flow toward the desired products, blocking the secondary pathways. These research efforts, together with those aimed at improving the operative stability of biocatalysts and the efficiency of reactors, will decisively condition the success of the industrial application of biotechnology also in the sector of plant products.

Biotechnology and Refining Processes

A sector of biotechnology which is of increasing interest in terms of its applications concerns the replacement in the extractive industries of highly polluting chemical technologies which often require high levels of energy consumption. Particular interest is being paid to the introduction of these

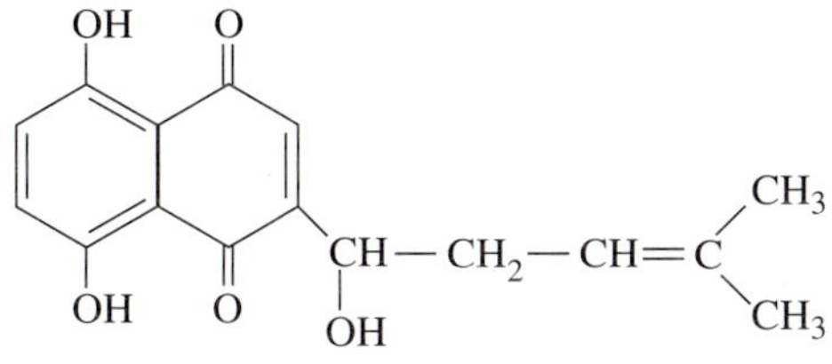

FIGURE 18 Chemical formula of shikonin.

innovative methodologies in the mining and metallurgic sector, and in that of fossil fuels such as petroleum and coal, given the size in volume of their production and the resulting environmental pollution that traditional technologies can cause. These new biotechnologies must compete against economic factors that are quite penalizing, such as the low added value that characterizes raw material transformation processes and the fact that decades (and sometimes centuries) of use of traditional technologies have led to their technical and economic optimization. It is therefore likely that an extended application of the new biotechnologies will result not only from an improvement of the microbial strains employed (in terms of productivity, selectivity, and resistance to toxic products) or from the optimization of the processing cycle but also from an increased awareness among the general public of environmental problems and, as a result, from the enactment of suitable legislative norms. The main fields of application of these biotechnologies are essentially the microbiological treatment of metallic minerals and the microbiological desulfurization of fossil fuels.

The Microbiological Treatment of Metallic Minerals

For some time, chemical industrial processes for treating minerals have been called into question because of their negative environmental impact; in particular, the problem concerns sectors dealing with extraction, the treatment of minerals, and the disposal of mining reflux and of wastes resulting from metallurgic processing.

Given the increasing attention that legislation is paying to the environment, it is likely that the tendency will become stronger in the future. These processes require oxidative phases carried out at high temperature that are very costly (pyrometallurgy) or acid leaching with oxidative and reducing (hydrometallurgy) that entail the emission or use of toxic substances, such as sulfur dioxide, hydrogen sulfide, and carbon monoxide. This situation has encouraged research in alternative technologies that are potentially less polluting and costly and that are also suited for treating materials with low metal contents that cannot be treated using conventional techniques.

In this context, biotechnologies that can effect high-speed oxyreductions at room temperature have elicited notable interest. Biohydrometallurgic processes which utilize microorganisms to treat minerals and to purify reflux contaminated by heavy metals represent a potential technological alternative to conventional processes for treating minerals, such as pyrometallurgic and hydrometallurgic processes. For this reason, we are witnessing all over the world the development of integrated industrial processes of the chemicobiological type which open new frontiers in the biohydrometallurgy sector.

The biological oxidation of sulfurous minerals takes advantage of a particular group of microorganisms which can oxidize the mineral at standard temperature and pressure. Such microorganisms are chemoautotrophic (i.e., they get their energy from chemical substances and use CO_2 as their main source of carbon). The compounds from which these microorganisms derive energy are usually inorganic reduced compounds (reduced sulfurous compounds, reduced nitrogenous compounds, ferrous compounds, etc.). When bioleaching is used, the energy source is the oxidation reaction of the sulfurous mineral or the oxidation reaction of the ferrous ion in solution, depending on the type of microorganism. The bacteria most commonly used in bioleaching processes are those that belong to the genus *Thiobacillus,* especially *T. thiooxidans* and *T. ferrooxidans.*

The bacterial leaching process takes place through various physical and chemical interactions among the bacteria, the sulfurous minerals, the ions, and the precipitated compounds in the culture medium. As a result, the system is very complex and an exhaustive definition has yet to be given. On the basis of current knowledge of the process, the dissolution of the sulfurous minerals is considered as taking place by way of two main mechanisms. In the first (direct mechanism), the dissolution of the mineral is caused by the microorganisms that adhere to the surface and use the mineral as a source of energy (Fig. 19). In this case, direct biological oxidation occurs. In the second mechanism (indirect mechanism), oxidation of ferrous ions on the part of bacteria that are in a free state in solution occurs, followed by the chemical oxidation of the mineral on the part of the ferric ion. After the chemical oxidation, ferrous ion and elementary sulfur are formed, which are then oxidized biologically (Fig. 19b).

The main industrial applications of bioleaching involve such minerals as uranium, copper, and the auriferous substances. In these processes, natural populations of sulfobacteria such as *T. acidophilus* and *T. ferrooxidans,* which can be isolated from the effluents of uranium mines, are normally employed. *Thiobacillus acidophilus* (or *T. organoparus*), in particular, is a bacterium that oxidized iron and sulfur and uses sugars such as glucose and fructose, organic acids such as citrate and malate, and amino acids such as aspartate glutamate. *Thiobacillus ferrooxidans* is a typical lithoautotrophic species which derives the energy required to render the carbon in CO_2 organic from the oxidation of various sulfides. Both these bacteria grow at very low pH values (1.5–2.0). The solubilization of uranium is catalyzed by these microorganisms, whereas the oxidation of reduced uranium is obtained by means of a ferric ion:

$$UO_2 + 2Fe^{3+} \rightarrow UO_2^{2+} + 2Fe^{2+}$$

The reoxidation of the ferrous ion at low pH values is very slow in the absence of a chemical oxidant or a biological catalyst. *Thiobacillus ferroooxidans* allows for the rapid replacement of the ferric ion, which is used up by means of the following reaction:

$$4Fe^{2+} + O_2 + 4H^+ \rightarrow 4FE^{3+} + 2H_2O$$

Thiobacillus ferrooxidans, on the other hand, is able to directly oxidize the reduced uranium compounds (such as the sulfate of uranyl and dioxide UO_2), without directly involving the couple Fe^{3+}/Fe^{2+} but acting as an external carrier of electrons according to the following reactions:

$$2U^{4+} + O_2 + 4H^+ \rightarrow 2U^{6+} + 2H_2O$$

$$2UO_2 + O_2 + 4H^+ \rightarrow 2UO_2^{2+} + 2H_2O$$

The production of uranium in Ontario, Canada, obtained via chemical leaching followed by liquid–solid separation and ion exchange has been increased to 210 tons annually through the insertion of a bioleaching phase in the conventional grinding process.

The solubilization of copper and zinc follows the microbial attack on the part of chalcopyrites or blendes

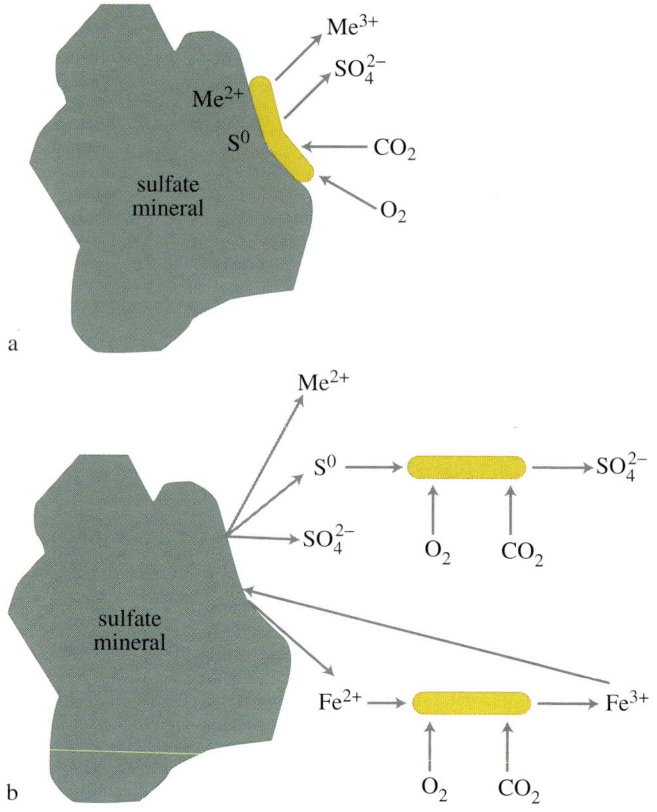

FIGURE 19 (a) Direct mechanism of the bioleaching process. The bacteria adhere to the mineral. (b) Indirect mechanism of the bioleaching process by free bacteria in solution, followed by chemical oxidation of the mineral by ferric ion. Me represents iron, copper, or zinc, depending on the mineral being treated.

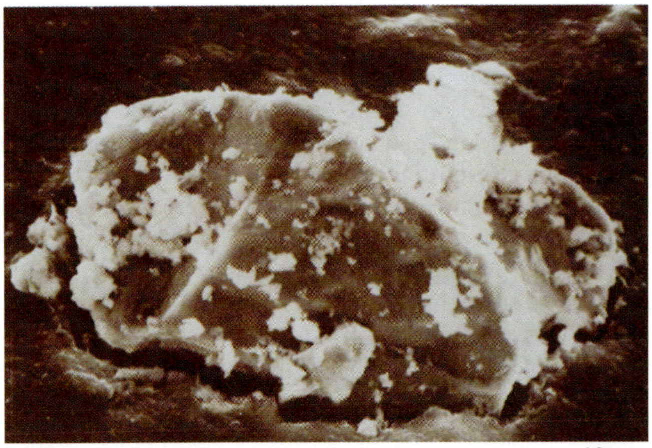

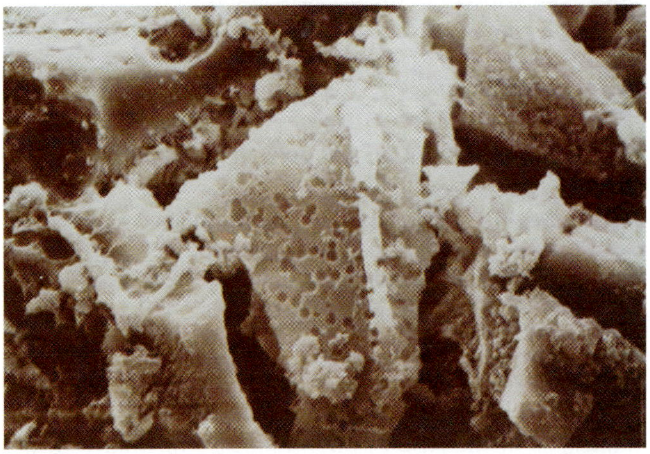

FIGURE 20 Microphotograph of the mineral chalcopyrite before (a) and after (b) treatment with microorganisms.

(Fig. 20). In the United States, the copper obtained in this manner represents more than 10% of the overall industrial production.

Gold can be recovered from certain minerals by way of gravimetric methods or direct leaching with cyanide, with extraction yields of up to 90%. Cyanidization is a relatively simple and inexpensive process which may be applied to a variety of gold-containing minerals and residues. Sulfurous minerals often contain gold and other precious metals, which are finely distributed inside the sulfide crystals; the fact that the particles are encapsulated gives the material an inert character so that the fractions are often lost as refuse. To obtain a satisfactory recovery, one must break the sulfide crystals, thereby freeing the auriferous particles, and then apply a conventional extraction method.

The choice of process with which to treat the mineral or the refractory concentrate is based on technical, economic, and environmental considerations. Various methods cause the oxidative degradation of a mineral matrix composed of sulfides: roasting oxidation, pressure oxidation, bacterial oxidation, and chemical oxidation.

The biological leaching process has only recently been taken into consideration for the treatment of auriferous minerals and concentrates. In certain cases, biological treatment has produced 30–90% increases in the amount of gold and 45–95% increases in the amount of silver that can be extracted. The bioleaching process dissolves the sulfide's matrix, thus exposing the precious metals to chemical attack.

To conclude, I mention the biological treatment of mine effluents. Mine effluents are often rich in metals and very acidic; for obvious reasons, a neutralizing and purifying treatment is required. To obtain a successful precipitation of the iron, however, the ferrous ion derived from simple sulfides such as blendes and complex sulfides such as chalcopyrites must be oxidized before it is neutralized. Oxidation by means of *T. ferrooxidans,* which has already been accomplished at the pyrite mines at Yanahara, Japan, allows for savings of 70% compared to oxidation by means of nitrogen oxide (Imaizumi, 1986). On the basis of these results, the Japanese government developed an installation for the neutralization of the acid discharges of the Matsuo mines. This installation uses *T. ferrooxidans* which, by oxidizing the ferrous ion to ferric ion, allows it to precipitate.

Microbiologic Desulfuration of Fossil Fuels

It is well-known that the presence of sulfur in fossil fuels poses problems of an environmental and technological order. For example, SO_x emissions (mainly SO_2) contribute to the creation of acid rain, the formation of particulate, the phenomena of pipeline and instrument corrosion, and the poisoning of catalysts used in refining and in combustion. Regulations tend to block SO_2 (and H_2S) emissions close to the petroleum refineries, directly at the mouth of gas-producing wells, and within electrical and thermal energy plants. At the same time, tolerance levels of the compound in products of petroleum origin are constantly being lowered. In some European states, such as Finland and Sweden, fiscal policies have been enacted which favor the marketing of gas oil with very limited sulfur contents (0.001–0.005%), and in the United States the low-emission vehicles program foresees a sulfur content in gasoline of 30–40 parts per million.

The sulfur contained in petroleum fractions cannot be eliminated easily, and the difficulties increase as the fuel's boiling point rises; thus, it is more difficult to desulfurize gas oil than gasoline as a result of the greater variety and refractoriness to chemical treatment of the sulfured compounds that are present in the former. Moreover, dehydrosulfuration treatments carried out in refineries at high temperatures with pressurized hydrogen and the necessary catalysts not only produce highly polluting solid residues but also transform the sulfur contained in the hydrocarbon charge into hydrogen sulfide (H_2S). The latter smells extremely bad, is notably toxic, and requires further oxidative treatments which transform it into sulfur dioxide.

Biotechnology is advancing specifically in the sector of the desulfuration of petroleum products (Klein *et al.,* 1994). It is showing possibilities of undoubted interest because of the favorable conditions under which it may be employed, in terms of temperature, pressure, pH, and the resulting simplicity of installations and reduced energy requirements; because of the capacity of numerous bacterial strains to desulfurize aromatic organic molecules (e.g., dibenzothiophene, one of the compounds that are least reactive in conventional hydrodesulfuration installations); and because of biotechnology's limited environmental impact.

In petroleum, sulfur is bound to the aliphatic chains by way of sulfur, disulfur, and polysulfur bridges. In particular, it may be found in organosulfured molecules such as mercaptans, sulfides, thiophenes, benzothiophenes, and dibenzothiophenes (Oshiro *et al.,* 1995). The sulfur can be removed by means of oxidative attacks of the C–C bond, by means of either reductive or oxidative attacks of the C–S bond, and by means of photosynthetic mechanisms. The oxidative breaking of the C–S bond is the most interesting from the technological standpoint; it takes place via biochemical pathways that can lead either to the total degradation of the sulfured molecules or to the detachment of the sole sulfur (Fig. 21).

An American company, in collaboration with some petroleum companies, is involved in a pilot project in which a procedure for the desulfuration of intermediate products distilled from petroleum by means of *Rhodococcus rhodocrous* is being tested. The pilot installation, built in St. Louis, Missouri, has a capacity of 0.5–5 barrels per day. The same company plans to extend the process to gasoline and low-acidity petroleum, and to this end has signed agreements with other petroleum companies. Feasibility studies carried out using crude petroleum and fractions of intermediate products of various type have shown that 40–80% of the sulfur originally present can be removed, and it is believed that these results can be further optimized. The economic analysis that has been performed has led to the conclusion that, on an industrial scale, the total cost of desulfuration would be about $3 per barrel (one barrel being equal to about 160 kg) for crude petroleum and $2 per barrel for gas oil.

Figure 22 shows the diagrammatic representation of another process, developed by ENI (Italian national hydrocarbons company), which works continuously using a strain of *Anthrobacter* that is grown separately and added to the desulfurer after having been put in a centrifuge.

Regarding the microbial desulfuration of carbon, its interest is the result of stringent environmental necessities connected with the increasing presence in the atmosphere of sulfur oxides and acids due to the combustion of fossils with high carbon contents. Moreover, the elimination of the sulfurs from carbon increases its heat-producing properties. Sulfur is present in carbon either in inorganic form (pyrite and sulfate) or as finely dispersed organic sulfur (dibenzothiphene and benzothiophene), which resists conventional purification methods. Various microorganisms are employed to remove pyritic sulfur. The most common are *T. ferrooxidans* and *Leptospirillum ferrooxidans,* whereas the most promising results in the release of organic sulfur have been obtained using thermophilic bacterial species (capable of acting at temperatures of 70–80°C), such as *Sulfolobus acidocaldarius.* The advantages of biological treatment are the selectivity of attack (small losses of useful carbon), the absence of ashes, the recycling of concomitant by-products (acid solutions and iron and aluminum sulfides), the low cost of the process, and the possibility of carrying out the desulfuration in the proximity of mines, far from inhabited areas.

Bioproductions of Environmental Interest

An interesting sector for the experimentation and productive implementation of biotechnology is that of the production of broad-consumption goods such as polymers, energy vectors, or products for agriculture, whose characteristics are "clean" from the standpoint of their environmental impact. Sometimes, it is a matter of employing, in a closely interconnected fashion, both chemical and biological techniques, as in the case of pheromones, which have turned out to be of fundamental importance in the fight against insects. Each of these substances, depending on its chemical formula, has an extremely selective effect on each individual species and allows one to attract only those insects that belong to a specific class into special sites where they can then be eliminated. The biological sciences have provided the keys to the mechanism of action of these substances, which are now being produced by chemicosynthetic methods.

Another interesting possibility, in the sector of energy vectors, is the production of gas oils through the esterification or transesterfication of oily fractions obtained from genetically modified versions of plants such as rape, sunflower, and soy. The environmental advantages would be enormous because fuels of this type, unlike the equivalent petroleum-based products, would be entirely sulfur free and because botanical sources (unlike fossil fuels) are renewable.

Having established that the limiting factor of these fascinating possibilities is the high cost of the technologies and products concerned, it is still interesting to analyze them in greater detail.

Microbial Insecticides

The use of chemical insecticides to control harmful insects presents certain negative aspects, by far the most no-

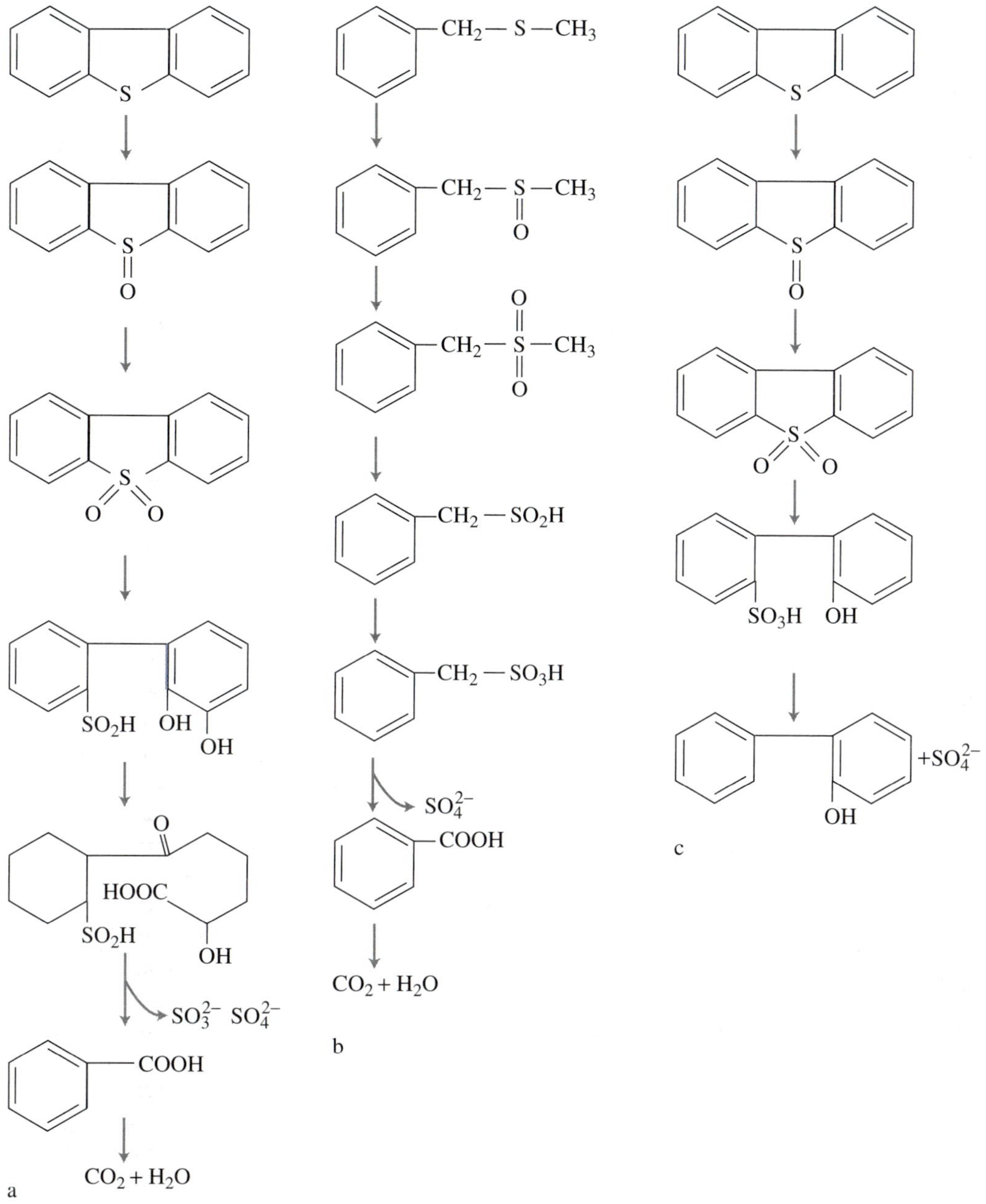

FIGURE 21 Desulfuration by means of oxidative breaking of the C–S bond: destructive (a, b) and non-destructive (c) pathways.

table of which are their cost, persistence in the environment, and the need to use ever-stronger doses as a result of the development of mechanisms of resistance or of losses in efficacy. These drawbacks have fueled efforts to develop biological forms of control based on the use of predators, parasites, pathogens, and attractive or repulsive natural substances. In the sphere of microbiological control, *Bacillus thuringiensis* is the most studied entomopathogenic organism. It is a bacterium that is quite widespread, found normally in soil and other common habitats in which insects abide, and can generate spores under particular environmental conditions. The majority of endomopathogenic bacteria form endospores inside the mother cell (sporangium). This species, in particular the aerobial and sporigenous types, produces during the sporation phase a protein crystal known as a *parasporal body,* which is mainly responsible for the entomopathogenicity of the bacterium. However, the proteins that make up the crystal may also be present in other parts of the bacterial cell, such as the endosporium or the spore wall, which may therefore have an

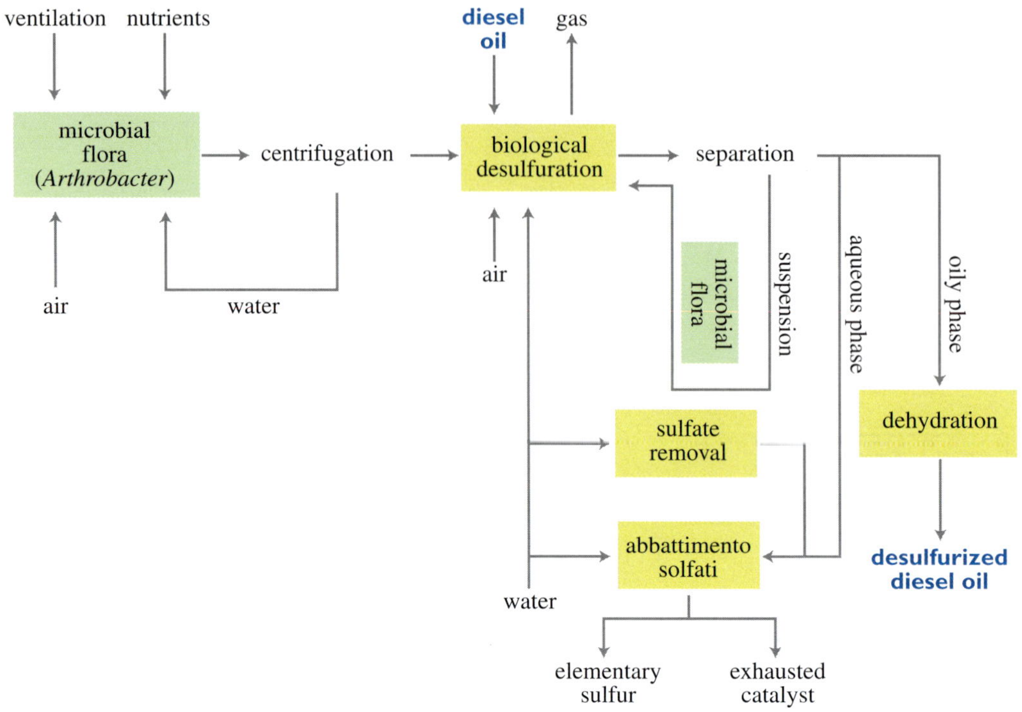

FIGURE 22 Block diagram of the procedure developed by the ENI (Ente Nazionale Idrocarburi, Italian national hydrocarbons organization) research laboratories for the biological desulfuration of petroleum products.

insecticidal effect of their own. These toxins are not dangerous for birds, mammals (including man), and many species of useful insects (bees, predators, and parasites of harmful insects).

Bacillus thuringiensis has a circular chromosome about 1.5 μm in length and a molecular mass of 5 kDa. The species' vital cycle is composed of two clearly distinguished phases: In the first, there is an exponential growth due to the multiplication of the vegetative cells; in the second, the latter mature into sporangiums (Fig. 23). Each sporangium contains a spore and one or more parasporal bodies. Sporulation is induced by the scarcity of certain nutritional elements. In order to have an insecticidal effect, *B. thuringiensis* must be ingested. It produces different toxins, generally in the course of the vegetative phase (exotoxins), except for the toxin that is contained in the protein crystal (endotoxin). The composition and concentration of these substances vary greatly and are influenced both by the genetic characteristics of the bacterial strain and by environmental factors (Rowe and Margalitis, 1987). The classification of the different biotypes of *B. thuringiensis* may be carried out on the basis of their immunological properties by subdividing the species into serotypes or, more common, categorizing them on the basis of their pathological specificity, distinguishing pathotypes that infect different groups of insects. For example, many strains have an effect on the larvae of lepidoptera, whereas others only infect larvae of diptera or coleoptera.

Since the 1960s, *B. thuringiensis* has been employed to produce biological insecticides to be applied on the leaves of plants. They are usually produced from the fermentation of individual strains of the bacterium on culture media that are poor and undergo centrifugation. The end product is usually a powder which is either soluble in water or may be scattered on the terrain in the form of granules. One of the limitations of this type of product stems from its mechanism of action, which is based on its ingestion on the part of larvae so that it is unsuited for controlling insects that feed inside their host plant.

Therefore, research has been directed toward the production of the proteins of the bacillus inside plants through the creation of transgenic plants. In the 1990s, this approach led to the production of plants such as maize, cotton, tomato, and potato that resist the attack of various phytophages because they can produce the toxins of *B. thuringiensis* (Wilson *et al.*, 1992).

Finally, it should be borne in mind that the ever more widespread use of substances produced by *B. thuringiensis* has led to the selection of insect strains that are resistant to its toxins. In particular, insects with very broad genetic pools can develop mechanisms to resist the bacterium more easily and more quickly. Therefore, it should be emphasized that it is very important that the microbiological tack be part of an integrated pest management policy that allows one to influence the population dynamics of the insects to be kept in check, intervening at different levels and thus

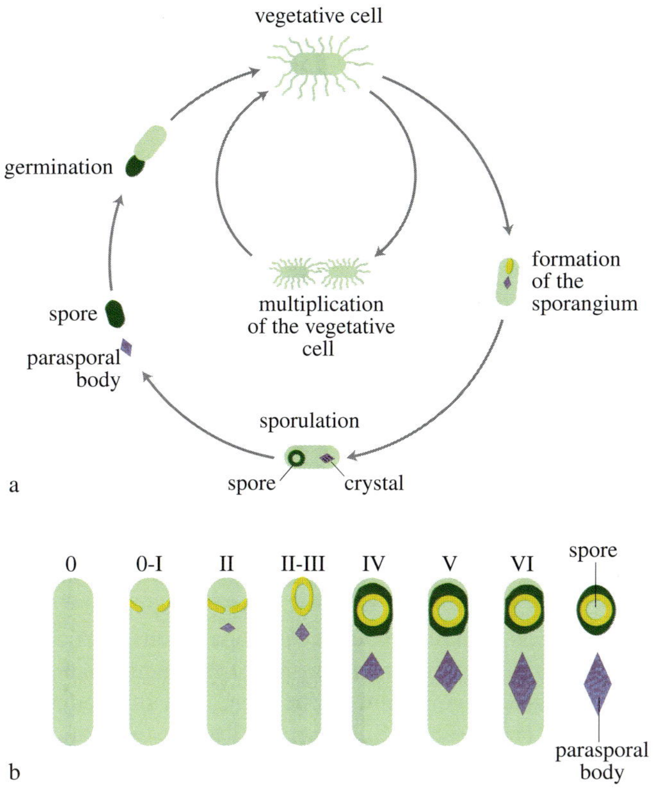

FIGURE 23 Schematic representation of the biological cycle of *Bacillus thurigiensis,* subdivided into two distinct phases: germination of the spores and multiplication of vegetative cells (a) and their maturation into sporangia (b). Each sporangium contains a spore and one or more parasporal bodies. The parasporal body is produced by the bacillus during sporulation and is made of a protein crystal which is responsible for the entomopathogenicity of the bacterium.

preventing the selection of characters that are resistant to limiting factors characterized by a small number of genes.

Biotechnologies for the Production of Polymers

An extremely interesting sector of application of biotechnological productions is that of polymers of biological origin. Their sources may be plants, animals, or microbes and their applications range across fields from mining (e.g., in petroleum extraction) to the most common consumption goods (foils for packaging and agricultural uses, biodegradable containers, etc.) and to a great and increasing number of therapeutic products. It is interesting to note how products that have existed for centuries and are of enormous economic interest for vast parts of the planet, such as the natural rubber extracted from *Hevea brasiliensis,* have been able to beat the competition of alternative technologies of a chemical nature. Natural rubber has done so due to the optimization of methods of cultivation and the amount of rubber lattice produced by the plant. Natural rubber is made out of a polymer of 2-methylbutadiene (isoprene) with a particular structure (1,4-*cis*):

$$\left(\begin{array}{cc} CH_3 & H \\ | & | \\ C & = C \\ | & | \\ -CH_2 & CH_2- \end{array} \right)_n$$

After the vulcanization of sulfur, this structure confers upon the natural polymer its well-known elastic and extraordinary dynamic properties (i.e., the quality of developing very little heat even after repeated solicitations). As a result, it constitutes a prime polymer for huge productive outputs such as tires for heavy vehicles (trucks, busses, etc.).

This example, however, is an exception (albeit an economically relevant one) since the interest of applying biotechnologies in the field of natural polymers, although it does not regard polydiolefins, is great with regard to polysaccharides and their combinations with synthetic polymers. The latter, just like aliphatic polyesters, have the characteristic of being biodegradable (i.e., decomposable on the part of the microorganisms present in the natural ecosystem). Since the microorganisms effect the degradation by means of their enzymatic system, it is important to establish the types of bonds and structures that the enzymes are capable of degrading. It has been shown that, as one might well have expected, peptide-, polysaccharide-, and ester-type bonds are subject to hydrolysis. Moreover, characteristics such as crystal structure, high molecular weight, the reticulations between polymeric chains, or the elevated hydrophobicity of the polymer constitute drawbacks with regard to its biodegradability. This has caused the field of investigation to become restricted to only a few categories of polymers (polysaccharides or aliphatic polyesters) (Hocking, 1992).

Chitin, for example, which derives from the 1,4-β linkage of acetylglycosaminic residues and exists in great quantities in nature, is a constituent of the exoskeleton of crustaceans and insects and of the cell wall of filamentous mushrooms and certain algae (Fig. 24). Its deacetylation by means of chemical hydrolysis yields chitosan. As a result of their excellent mechanical properties, both chitin and chitosan have numerous potential applications; chitosan is currently produced on a semiindustrial scale and sold by various producers for uses that range from flocculents to additives for the food industry, coverings for seeds, and films. Due to their biodegradability, the use of chitin and chitosan in the biomedical field for the production of cicatrizing substances and artificial skin appears to be very promising since the molecules of the polysaccharide can be progressively replaced by the epithelium in its organizational phase.

An important category of polysaccharides comprises those of microbial origin—that is, products of the fermentation of microorganisms such as dextran, which is pro-

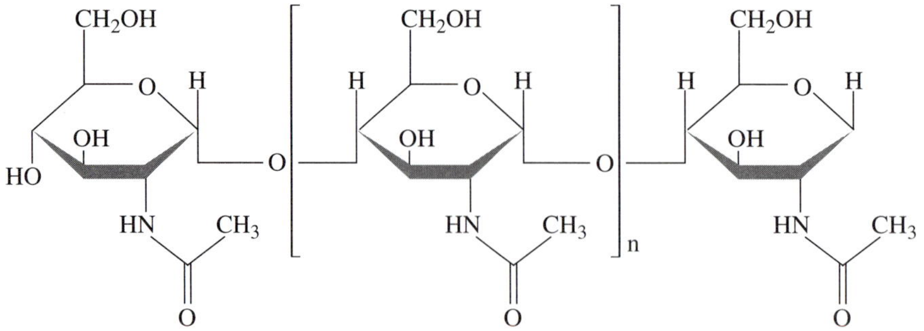

FIGURE 24 Chemical formula of chitin.

duced by *Leuconostoc mesenteroides* grown on saccharose and used as a fundamental constituent of artificial human blood plasma (*plasma expander*). In addition to dextran, xanthan, another polysaccharide of microbial origin obtained through the fermentation of *Xantomonas campestris,* is also very important. It is made of cellulosic-type skeleton bonded to lateral units containing mannose and glucuronic acid and is used to stabilize suspensions in food products, as a thickener of the drilling muds in oil fields or natural gas reservoirs, and especially in the "tertiary" recovery of petroleum. The spontaneous extraction of petroleum (i.e., the natural discharging of crude oil as a result of gaseous pressure in deposits) on average accounts for no more than 35–40% of the petroleum contained in any single deposit. In order to "displace" the drops of petroleum that are withheld as a result of the rock's porous characteristics, varied techniques are used which range from the simple repressurizing of the oil well to sophisticated methods such as those based on the use of surface-active agents. One of the most commonly used techniques is that of pumping into the deposit water that has been "thickened" with polymers such as xanthan, which are capable of providing highly viscose solutions at very low concentrations. Other microbial polysaccharides of increasing interest are hyaluronic acid, which is polysaccharidic in nature (Fig. 25), and aliphatic polyesters such as poly-β-hydroxybutyrate and polylactic acid and their copolymers.

Hyaluronic acid, which used to be extracted from rooster crests via complex, costly, and polluting procedures, is now obtainable in industrial quantities from the fermentation of bacteria of the genus *Streptococcus.* This technique is extremely competitive on the economic level and the product it yields, once purified, may be used in cosmetic, and especially in pharmaceutical preparations, particularly for ophthalmic surgery (the curing of cataracts), as an anti-inflammatory agent, and in the treatment of rheumatoid arthritis. Hyaluronic acid can be used in the form of microspheres of a few dozen micrometers in diameter to convey drugs: As the microspheres are gradually biodegraded *in vivo,* the drug which they contain or which is covalently bound to the polymer is released at a constant and programmed rate. Aliphatic polyesters of microbial origin, given their total biodegradability and lack of toxicity, are used especially to make containers for foods or hygienic products; however, they also have sophisticated biomedical applications in orthopedics or in the controlled releasing of drugs. These products still cost too much to be used on a massive scale; in order to reduce their cost and to simplify the operation of purifying products, studies are under way that are aimed at cloning the genes responsible for the biosynthesis of the polymers in bacteria such as *E. coli* and in plants. Polyesters derived from lactic acid, instead, are prepared for polymerization with ionic catalysts of their cyclic dimer (lactide), and biotechnology is used essentially in

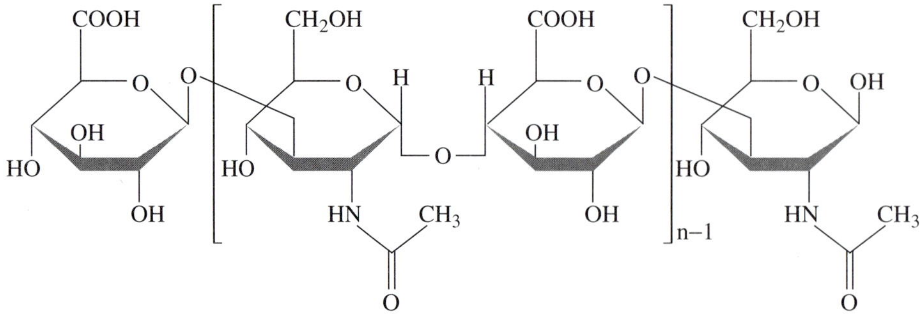

FIGURE 25 Chemical formula of hyaluronic acid.

the fermentative process which yields very sterically pure L-lactic acid. These polymers have uses analogous to those of other aliphatic polyesters.

Microbiologic Production of Hydrogen

Sources of fossil energy are present in nature in limited quantities and in the course of combustion produce carbon dioxide, one of the major causes of the greenhouse effect. Moreover, their combustion also produces NO_x and SO_2, which are the major causes of acid rain. The discovery of alternative, renewable sources that do not produce substances that are harmful to the environment and dangerous to man is therefore of fundamental importance.

The recent attention that has been paid to the use of methane does not solve the problem. Although more abundant than petroleum, methane is not a biologic product and cannot, as a result, be thought of as a renewable source. In fact, although methane is purer than petroleum and whatever impurities it does contain (such as CO_2 and H_2S) can easily be eliminated, its combustion still releases CO_2, which only worsens the greenhouse effect that is warming up the planet. The case of biogas can be different: It is made of a mixture of methane and carbon dioxide and is a by-product of the microbiologic treatment of effluents in anaerobic-type fermenters. The effluents may differ: urban, zootechnical, or agroindustrial residues (such a residues from pig sties or the milk and cheese industries). The biogas one obtains has little heating power because its methane content is only slightly greater than 50% and is burned in specially made burners. This type of fuel's contribution, in any case, remains marginal in relation to global energy supplies. Analogously, alternative sources such as of those of a ligneous cellulose sort (pruning residues, corn cobs, multicellular algae hydrocultures, etc.) are expensive to collect, transport, stock, and sometimes dry (algae), and they do not lighten the greenhouse effect since their combustion produces CO_2.

From the standpoint of energy, hydrogen represents an ideal source since it produces water when it burns. Currently, it is produced industrially from fossil fuels (natural gas or coal) by way of steam reforming. In these procedures, the emission of pollutants takes place at the production site, and therefore they do not solve the environmental problem. Moreover, they require nonrenewable raw materials and it is therefore reasonable to suppose that production costs are destined to increase in the future. An alternative is to produce hydrogen from water; this is the most enticing way from the environmental standpoint, and therefore it is in this direction that a great deal of research is being directed. Not only electrolysis but also plasmolysis, magnetolysis, and the use of sunlight are being assayed. The latter technique is the most interesting since it employs a renewable source of energy which is also the most abundant on Earth. The production of hydrogen from water and solar energy is closely connected with the possibility of converting luminous energy into chemical energy by way of photochemical processes. The study of these processes is still in an explorative stage. Among the systems being studied are the combined photovoltaic cells–electrolyzer systems, photoelectrochemical systems, and photocatalytic systems.

Biological systems have always attracted attention as potential producers of hydrogen, but concrete application have yet to be found. The cost of the energy to be supplied to the biological systems, in fact, is only slightly inferior, and at times even superior, to the value of the energy that is produced. Therefore, the possibility of applying biotechnology to the production of hydrogen depends on the availability of inexpensive sources of energy. From this standpoint, photosynthetic bacteria, which can use solar energy as an energy source with an efficiency of 7 or 8% and gather the protons required to produce hydrogen from the degradation of organic waste, are undoubtedly very promising (Morimoto and Myake, 1994). In general, it should be noted that the biological production of hydrogen can be accomplished using both *in vivo* and *in vitro* systems. In the first instance, it takes place due to either phototrophic organisms (which use light as an energy source) or chemotrophic organisms (which rely on the metabolism of chemical compounds). In the latter instance, hydrogen is produced by enzymatic systems that have been extracted from microorganisms and coupled to semiconductors and to chemical compounds capable of carrying electrons. Often, hydrogen is obtained from the degradation of organic substrates that takes place alongside the scission of water (Miura *et al.,* 1994). The yield may therefore be greater than the amount of stoichiometric hydrogen present in the organic substances at the beginning. *In vivo* systems of the phototrophic type include algae, cyanobacteria, and phototrophic anoxygenic bacteria. Those of the chemotrophic type include aerobic and anaerobic bacteria and higher organisms (Fig. 26). In all cases, the production of hydrogen is catalyzed by enzymatic systems that are complex and almost always work in combination with each other.

In the current economic–energetic scenarios in Western countries, the production of hydrogen to be used as fuel does not appear to be a competitive option and more promising middle-term perspectives may instead be found in the use of hydrogen as a raw material in chemical industrial productions, especially in view of its production from materials whose cost is zero or negative and that are present in liquid effluents or solid wastes. However, there is no doubt that biological pathways to the production of hydrogen present notable environmental advantages. The biological systems (see Fig. 25), in fact, allow one to fix CO_2 (algae and cyanobacteria) and NO_x (especially cyanobacteria) and therefore produce organic substances of a certain value (starches, glycerol, formate, etc.) starting from saline substances alone (at least with regard to cyanobacteria).

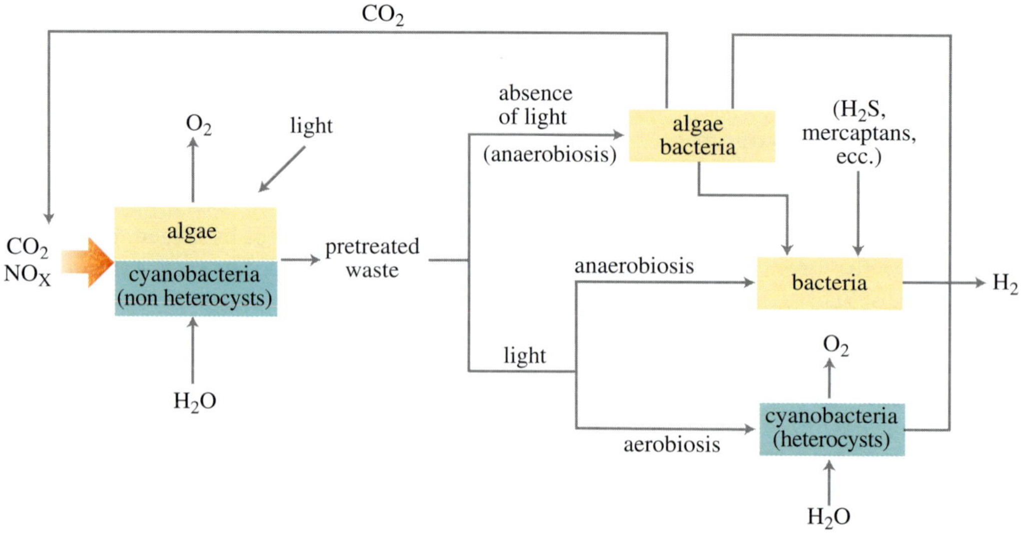

FIGURE 26 Alternative biological pathways for the production of hydrogen.

Moreover, they make it possible to photolyse water with less energy than is required by physical processes: In biological processes, electrons come from organic substrates and water, and the energy of light is used essentially to activate the metabolic processes. The physical lysis of water, for example, requires 242.25 kJ/mol of select energy, whereas cyanobacteria and algae allow one to recover hydrogen from substances that have been completely oxidized (CO_2 and H_2O) at the sole cost of luminous energy. Finally, biological systems allow one to treat waste effluents, whether solid, liquid, or gaseous.

Conclusions

Because the development, improvement, and extension of the applications of a new technology depend largely on the concomitant development of the scientific disciplines that constitute its theoretical and experiment basis, it is comforting to note that the fundamental sciences that are upstream of biotechnology (cellular and molecular biology, recombinant DNA techniques, computerized projections of molecular structures, molecular modeling, etc.) are experiencing a period of tumultuous development, capable of sustaining a rapid applicative extension. The latter will continue to be conditioned by economic factors, but one of its most important aspects will always be the decreased environmental impact that it implies. Therefore, an increased sensibility of public opinion and legislators in regard to these aspects can notably influence its success.

Regarding the agricultural sector, biotechnologies are the last, in chronological terms, among the many technologies that have led to an amazing development in the productivity of foodstuffs in the world through further increases in crop yields, cost decreases, and the development of new products with high added values that are capable of satisfying the growing needs of consumers and the agroalimentary industry. These products include new seeds and new pesticides, products for veterinary diagnostics and therapy, food additives, enzymes for the food industry, foods with higher nutritional values (e.g., with higher essential amino acid contents), and agricultural products with improved processing characteristics. The manipulation of varieties of maize, cotton, and tomato is aimed at introducing, in their genetic makeup, genes derived from bacteria (e.g., *B. thurigiensis*) which regulate the synthesis of numerous toxins capable of acting, through various mechanisms, on a variety of insects. Due to this novel approach, the plants that have been genetically combined with the heritage of microorganisms are capable of defending themselves in an autonomous manner from numerous agents that would otherwise plague them. A further advantage of this technique is that the lethal effects of these toxins are limited to the insect species which they affect, which is not the case with traditional chemical insecticides that can exert an undifferentiated toxic action on all living beings, including useful insects or higher organisms.

The predictable and notable environmental advantages of biotechnology will consist of a decreased production of waste products, the simplification of industrial cycles, and a more limited use of fertilizers. Regarding this last fundamental point, important research efforts are being made both in order to induce the production of ammonia in the soil on the part of nitrogen-fixing microorganisms put there for this purpose and to discover bacterial species capable of providing nitrogenous fertilizers to plant species other than the leguminous ones. The latter, in fact, do not require the use of fertilizers in order to grow since bacteria of the genus *Rhizobium* (which are capable of transforming nitrogen in the atmosphere into ammonia via the complex enzymatic–coenzymatic system known as nitrogenase) abide symbiotically in their radicular nodules.

The technologies that produce chemicals, pharmaceuticals, and food products using enzymes (preferably immobilized) appear rich in possible applications as a result of the important progress achieved in recent years in the field of recombinant DNA technology, which promises that it will be possible to produce virtually any enzyme one wishes on a large scale and at reasonable costs. In fact, a new phase of enzymatic biochemistry based on molecular genetics is under way in which it is possible to design and produce enzymes with particular properties based on the specific needs of industrial catalysis, such as specificity and affinity of substrates, stereoselectivity of reactions, stability in the presence of nonaqueous solvents, heat, and oxidation, resistance to proteolytic degradation, and catalytic efficiency.

Acknowledgments

I thank Paolo Costantino of the "La Sapienza" University of Rome and Massimo Cristofaro of the ENEA Casaccia Research Center for their important suggestions and useful discussions on some of the topics discussed in this article.

References Cited

Hocking, P. J. (1992). The classification, preparation, and utility of degradable polymers. *J.M.S. Rev. Macromol. Chem. Phys.* **C32**(1), 35–54.

Imaizumi, T. (1986). Some industrial applications of inorganic microbial oxidation. *Jpn Biotech Bioeng. Symp.* **16,** 363–371.

Johann, P. (1993). The application of hydrolases in organic synthesis. *Chemspec Europe–93 BACS Symposium.*

Johnson, D. B., Ghauri, M. A., and Said, M. F. (1992). Isolation and characterization of an acidophilic heterotrophic bacterium capable of oxidizing ferrous iron. *Appl. Environ. Microbiol.* **58**(5), 1423–1428.

Klein, J., Van Asserden Nipfeifer, F., and Schatch, S. (1994). Microbial desulfurization of coal and oil. *Fuel Processing Technol.* **40,** 297–310.

Marconi, W. (1989). Processi enzimatici industriali. In *Enciclopedia del Novecento,* Vol. 8, pp. 106–120. Istituto della Enciclopedia Italiana, Rome.

Miele, L. (1997). Plants as bioreactors for biopharmaceuticals: Regulatory considerations. *Trends Biotechnol.* **15,** 45–50.

Miura, Y., *et al.* (1994). Hydrogen production by a combination of marine green alga and photosynthesic bacteria. *Workshop on Biological Hydrogen Production,* Tokyo, November 24–25.

Morimoto, M., and Myake, J. (1994). Development of environmentally friendly technology for the production of hydrogen. Hydrogen energy progress X. In *Proceedings of the 10th world hydrogen energy conference, Cocoa Beach, FL, 20–24, 1994,* p. 951.

Oshiro, T., Hirata, T., and Izumi, Y. (1995). Microbial desulfurization of benzotiophene in the presence of hydrocarbons. *Appl. Microbial Biotechnol.* **44,** 249–252.

Roberts, S. M. (1990). *Recent Developments in the Use of Enzyme-Catalyzed Reactions in Organic Synthesis.* Academic Press, San Diego.

Roberts, S. M., and Turner, N. J. (1992). Recent developments in the use of enzyme-catalysed reactions in organic synthesis. *J. Biotechnol.* **22,** 227–232.

Rowe, G. E., and Margalitis, A. (1987). Bioprocess developments in the production of bioinsecticides by *Bacillus thuringiensis. CRC Crit. Rev. Biotechnol.* **6,** 87–127.

Servi, S. (1996). Biocatalysis in the preparation of homochiral pharmaceuticals. *Chim. Industria* **78,** 959.

Vander Esch, S. A., Maccioni, O., and Vitali, F. (1993). Biomolecole d'interesse industriale da fermentazioni vegetali. *Biotec* **5,** 20.

Warren, R. A. (1996). Microbial hydrolysis of polysaccharides. *Annu. Rev. Microbiol.* **50,** 183–212.

Wilson, F. D., Flint, H. M., Deaton, W. R., Fischoff, D. A., Perlak, F. J., Armstrong, T. A., Fuchs, R. L., Berberich, S. A., Parks, N. J., and Stapp, R. R. (1992). Resistance of cotton lines containing *Bacillus thuringiensis* toxin to pink bollworm and other insects. *J. Econ. Entomol.* **4,** 1516–1521.

General References

Katchalski-Katziz, E. (1993). Immobilized enzymes, learning from past successes and failures. *Trends Biotechnol.* **11,** 471–478.

Various authors (1995). *Trends Biotechnol.* **13,** 312–409.

Wiseman, A. (1996). Novel biocatalysts will work even better for industry. *J. Chem. Ed.* **73,** 55.

Robert L. Irvine[1]
Lisa I. Larson[2]
David A. Irvine[2]
Timothy J. Irvine[3]
Camille A. Irvine[2]

[1]*Department of Civil Engineering and Geological Sciences*
University of Notre Dame
Notre Dame, Indiana 46556, USA

[2]*SBR Technologies, Inc.*
South Bend, Indiana 46628, USA

[3]*Warrington College of Business*
University of Florida
Gainesville, Florida 32611, USA

Biological Depuration

Biological depuration, the cleansing or purification of unwanted contaminants from liquids, solids, or gases, occurs in the natural environment and in engineered systems for substances that are biodegradable. By utilizing carbon and/or energy sources present in the environment, microorganisms such as bacteria and fungi convert both contaminants and noncontaminants into carbon dioxide, water, cell mass, and a plethora of other compounds that become part of nature. When natural biological processes do not keep pace with the rate of contaminant addition to the environment, the water we use, the air we breathe, and the land on which we live become polluted. This article focuses on how microbial-based biological systems are engineered in a systematic and controlled manner to rid the environment of unwanted contaminants at an accelerated rate and how technology can be used to assist nature in its quest to serve all life-forms in the earth's biosphere.

Historical Background

Pollution in the form of contaminated water, soil, and air has plagued the health and progress of civilizations throughout the ages. From the hunter and gatherer societies of 30,000 years ago to the modern industrialized nations of today, human populations have had to suffer and overcome the adverse effects of environmental contamination. The earliest contaminants were found in the water supplies of prehistoric man.

Much early sickness was caused undoubtedly by what we would today call pollution. The presence of human gut bacteria such as *Escherichia coli* and deposits of lead in the drinking water enervated early man. Recent analysis of the 200,000-year-old Broken Hill hominid from Zambia has produced evidence that he suffered from lead poisoning due to an ore lode underlying the water supply of the cave dwelling (Janssens, 1970).

The transition from hunter–gatherer groups to nomadic herding systems and eventually to settled agriculture farming communities during the Neolithic Period has been described as "the most fundamental change in human history" (Ponting, 1991). By allowing output of food to increase, the concept of "property" to develop, and surplus food production to grow, the agricultural transition became the basis of a human revolution.

Out of agriculture grew the community. The development of towns and cities ushered in an era that underscored nature's limited ability to use biological processes to cleanse itself whenever the rate at which pollutants enter the environment is excessive. Pollution was defined by the British biologist Kenneth Mellanby in 1972 as the "presence of toxic materials introduced into our environment by man" (Mellanby, 1982). However, it is also the disruption of natural soil and water regimes by the displacement or mobilization of natural substances. A classic modern example is the pollution of rivers and coastal ecosystems by soil and silt washed off the land. Salinization is a phenomenon in this category, and this destroyed Sumarian civilization. During a period of about 1700 years, from 3500 to 1800 BC, Sumerian agriculture in southern Mesopotamia declined and wheat productivity fell because of salt pollution. When flat land is irrigated, lack of proper drainage causes the water to seep into the underlying groundwater and raise the water level. As the soil becomes waterlogged, salts rise and high levels of evaporation from the soil surface leave salt crusted on the fields. The Sumerians, still unaware of the folly of harvesting on flat land, could only lament the crippling effect on crop output and record that the "earth turned white" (Ponting, 1991). As recent as 1989, according to the World Health Organization and the United Nations Environment Program, "increasing salinity is [still] one of the most significant and certainly the most widespread forms of groundwater pollution." Salinity seriously affects 7% of the world's irrigated cropland, mainly in India (24% of the total irrigated area), the United States, Pakistan, Iran, Iraq, and Egypt (Meybeck, 1989).

Although the Sumerians were overwhelmed by salinization, the Romans were more successful in implementing systems to avoid and solve the hazardous effects of pollution. The first sewerage system was the Roman Cloaca Maxima, built in the sixth century BC during the Etruscan dynasty of the Tarquins. The initial purpose of the massive structure was to drain the swamp between the Palatine and Capitoline hills, leading eventually to the establishment of the Roman Forum, which became the hub of the republic and later the empire. The hydraulic pioneers of the ancient world, the Romans constructed a whole network of *cloacae,* or sewers, as well as a maze of aqueducts bringing water into the city.

The rapid pace of advancements in water management set by the Romans did not continue into the Middle Ages. The majority of the towns and villages of medieval Europe were hampered by unsophisticated waste management techniques and challenged nature's ability to purify itself. Most sewage flowed directly into open gullies and trenches that ran down the streets. Sewers and cesspools had been developed, but their efficacy was doubtful. Sewers were often blocked or simply emptied into the nearest river or stream, whereas cesspools overflowed and tended to leak into people's wells. Many towns and cities had only rudimentary regulations for the disposal of waste, and teams of "rakers" or "scavengers" were often employed to remove garbage from cities.

Water pollution and deplorable urban conditions intensified with the onset of the industrial revolution. The challenge of the urban environment was felt first and most acutely in Great Britain. In the 1820s and 1830s the populations of many British cities were increasing by 40–70%. Refuse from urban living increased the pollution of rivers and other streams. Frederick Engel gave voice to the frustrations of many when he described the Manchester river, Irk, as a "narrow, coal-black, foul-smelling stream, full of debris and refuse . . . the most disgusting, blackish-green, slime pools . . . from the depth of which bubbles of miasmatic gas constantly arise and give forth a stench unendurable" to anyone standing on the bridge 40 or 50 ft above the water (Engels, 1975). As the public health steadily declined, the push for environmental regulation and developments in sanitary engineering grew not only in Great Britain but also throughout the Western world.

Although sanitary reformers of the early Victorian era accomplished much in regard to water pollution, they were not able to remove the limitations placed on nature's thirst for self-purification. Building on a growing number of medical and sociological studies, Edwin Chadwick, the most famous of Britain's reformers, published in 1842 a groundbreaking report that proved that disease was related to filthy environmental conditions, which in turn were caused by lack of drainage, sewers, and garbage collection. His report recommended an improved kind of sewer, the connection of water closets to the sewer, and a constant stream of water under high pressure (as opposed to the more prevalent intermittent system) to carry the waste away before it had a chance to stagnate and putrefy. Chadwick acknowledged that discharging sewage into a river caused pollution and injury. Chadwick sought for and thought he had found a solution by diluting the sewage with water, carrying it in covered sewers to land some distance away, and distributing it as liquid manure by irrigation. By this means, he concluded, the health problems by the pollution of natural streams could be avoided. Although Chadwick's recommendations did little to halt the discharge of pollutants into natural waterways, his report became the basis of Great Britain's first public health law, which created a national health board and gave cities broad authority to build modern sanitary systems.

Urban industrial pollution first emerged as a significant public policy problem in the United States in the second half of the nineteenth century, largely as a result of the enormous amounts of waste produced by U.S. industrial cities. Early water pollution studies sponsored by state and municipal authorities in the United States often pointed to industrial wastes as the most important cause of declining water quality and led to a massive program of sewerage construction that began in the 1870s. City officials viewed the waste-carriage system of human waste removal as a great improvement over the traditional methods of individual storage and private scavenger service removal. Unfortunately, the sanitary reformers who pushed for the adoption of this new technology did not foresee that removal of sewage in large volumes to rivers and lakes would lead to frequent outbreaks of typhoid and other waterborne diseases. Sanitary experts understood the link between bacteria and disease; this was one of the main reasons they had advocated the waste-carriage system. The flaw in their thinking concerned the ability of flowing water to purify itself in a time period short enough to protect potential users from harm. According to this theory of dilution, flowing water in a sizable lake or river dispersed and purified sewage through natural processes. Thus, sewage could be discharged into large bodies of water that also served as a source of drinking water without threatening public health. The experts were correct to a point, but they exaggerated a stream's capacity to assimilate unlimited amounts of waste and to do so quickly. Students of waste disposal eventually understood that a stream purifies itself through the actions of naturally occurring bacteria that utilize free oxygen to consume organic waste in a process sometimes referred to as *oxidation.* A stream's capacity to purify itself is thus a function of the dissolved oxygen present, and this supply of oxygen can be exhausted before it can be replenished. In other words, there was a finite limit to how much organic matter a lake or river could assimilate at one time.

In order to understand how the natural environment can be manipulated or how systems are engineered to destroy contaminants, a pragmatic description of microbiology, biochemistry, and contaminants is presented in the initial sections of this article. These sections are followed by an in-depth discussion of biological treatment processes.

Biodegradation

Although environmental contaminants are generated by all forms of life and by geological and meteorological phenomena (e.g., volcanic activity), attention is usually focused on contaminants produced by human activity at the industrial, agricultural, mining, and domestic levels. Independent of their origin, contaminants eligible for destruction are either organic or inorganic food sources that provide energy or nutrition to an existing microbial consortium or to one that has been established in a waste treatment facility. In either case, virtually all contaminants that can be consumed by microorganisms are said to be *biodegradable* or soft. Specifically, this means that the microbial consortium contains the genetic information necessary to produce catabolic enzymes which extract energy from the contaminants and convert them into intermediate biochemical products. Enzymes for anabolic pathways use the extracted energy and intermediates to synthesize the compounds necessary for the microbe to reproduce. The microbial consortium present must therefore possess the pool of genetic information needed for the appropriate biochemical reactions to take place or the contaminant will persist in the environment (i.e., the compound is nonbiodegradable and is said to be *recalcitrant* or hard).

Contaminants are judged hazardous or nonhazardous independent of their biodegradability. A substance that is hazardous for one group of microorganisms may be a valuable food source for another group. Accordingly, many hazardous compounds can be and are biodegraded in the environment. Given sufficient time, nature can rid itself of all biodegradable compounds, hazardous or not. For many reasons, including recreation, conveyance, and public health, the public often cannot wait for nature's biological processes to depurate contaminants present in water, air, or land. As a consequence, these compounds must be removed in biological waste treatment facilities.

Minimum Essentials of Microbiology

Microbiology is the branch of biology in which the physiology, reproduction, symbiosis, impact on ecosystems, occurrence in nature, and significance in science and industry of living minute organisms not capable of being seen by the naked eye are investigated. Microbiology includes the study of bacteria (bacteriology), fungi (mycology), protozoa (protozoology), and viruses (virology). Because of their biochemical and environmental versatility, bacteria are the most commonly used microorganisms in systems designed to destroy biodegradable contaminants. Bacteria are also frequently used in biological research and biochemical processing.

Multicellular organisms such as plants and animals have eukaryotic cells. Cells found in lower life-forms, such as bacteria and blue-green algae, are prokaryotic. Unlike prokaryotic cells which contain no membrane-bound organelles, the cytoplasm of eukaryotic cells includes intracellular organelles, such as vacuoles, mitochondria, endoplasmic reticulum, and nuclei, which are separated from the cytoplasm by individual membranes. A semipermeable plasma membrane separates the cytoplasm from the extracellular environment.

Prokaryotes are 1000–10,000 times smaller than eukaryotes. Many are motile. Bacteria are spherical, rod-like, or spiral and are the smallest of the single-celled microbes, with typical dimensions of 0.5–3 μm. They can be as "wide" as 0.1 μm and as "long" as 50 μm. Their shape is defined by a structural cell wall. Bacteria have one double-stranded, circular chromosome. Many environmentally important bacteria contain additional genetic information on a plasmid. Because of their small size, bacteria reproduce at extremely high rates, with division times between 1 and 4 h not uncommon. This makes bacteria extremely valuable to virtually all biological treatment processes. Not all environmentally important microorganism are prokaryotes. Fungi and algae are eukaryotes and have been used to a limited extent in systems engineered to destroy contaminants. Both fungi and algae, of course, are ubiquitous in the environment and make significant contributions to natural biological depurative processes.

Some bacteria grow best when attached as biofilms to surfaces; others grow best when unattached (i.e., in suspension as dispersed organisms or in clusters or flocks). Strict aerobes only grow in the presence of oxygen, whereas strict anaerobes are destroyed by oxygen and facultative anaerobes grow in the presence or absence of oxygen. Microorganisms convert contaminants to new cell mass, carbon dioxide, water, and other end products which depend on the nature of the contaminants and the specific microorganisms present.

Microorganisms require an energy source, a carbon source, and nutrients for growth. Photoautotrophs such as algae use sunlight as the energy source and carbon dioxide as the carbon source; chemoautotrophs use a reduced inorganic such as ammonia or iron (+2 oxidation state) as an energy source and carbon dioxide as the carbon source; heterotrophs use organics for both the energy and the carbon source. Nutrients, elements not found in either the carbon and energy source, must also be supplied because materials produced by living cells (e.g., proteins, carbohydrates, fats, and nucleic acids) collectively contain a wide variety of elements, including carbon, hydrogen, oxygen, nitrogen, phosphorus, and sulfur. In addition, proteins, when used as enzymes, require trace concentrations of metals, such as sodium, potassium, iron, manganese, magnesium, and calcium. Nutrients required at relatively high levels are referred to as *macronutrients* (e.g., nitrogen and phosphorus); those required at low levels are called *micronutrients* (e.g., trace metals). Some cells do not contain the genetic information needed to synthesize selected required intracellular organics, such as amino acids and vitamins. Along with inorganic nutrients, these "essential" preformed organic intermediates must be either present in the environment or provided for organisms to grow.

A microbial population will establish in the environment or in an engineered system designed for pollutant destruction if all nutrients and the carbon and energy sources are available and the microbes present contain the genetic information necessary to produce the catabolic enzymes which extract energy from the contaminants. Genetic information for the bulk of the catabolic enzymes and all of the anabolic enzymes is present on the chromosome. Additional genetic information, especially for the production of enzymes that catalyze the breakdown of xenobiotics, is typically found on plasmids. As a result, the plasmid genes often provide a microbial consortium with the information needed to rid the environment of unwanted contaminants.

Facilities designed for the treatment of domestic wastewaters or sewage are virtually always biologically based. The wastewater flow contains all the essential nutrients for treatment. These systems typically employ a mixed culture (vs a pure culture) of bacteria. Although other microorganisms are present (e.g., single-celled protozoa), bacteria are primarily responsible for the aerobic destruction of organics and the removal of nitrogen and phosphorus. Both fixed film (e.g., trickling filters) and suspended growth (e.g., activated sludge) systems are employed in the biological treatment of domestic wastewaters. Most systems installed today are of the suspended growth variety. Biological solids produced in such systems contain living and dead biomass and suspended solids that are partially decomposed by the microbes present in the treatment facility. These biosolids are usually processed further in aerobic or anaerobic digesters. The aerobic digesters primarily stabilize the waste biosolids by mineralizing (i.e., producing carbon dioxide and water) the soft organics and thus reducing the mass of biosolids that requires disposal in incinerators or on land or as a soil conditioner or fertilizer. Anaerobic digesters meet similar goals while producing methane gas and hydrogen. These gasses are often used to generate electricity.

In the past, industrial wastewaters were treated using physical and chemical facilities more often than biologically based systems. Because biological systems are usually less expensive than their physical and chemical counterparts and because of the high cost associated with meeting today's more stringent environmental requirements, most industries have intensive programs designed to develop biological treatment facilities whenever possible. Contaminants present in industrial wastewaters often create a greater challenge than those in domestic waste. A greater variety of xenobiotics at higher concentrations, nutrient deficiencies, more hostile temperatures, pH, salinity, and heavy metal levels, and slower bacterial consumption rates demand more careful attention in treatability studies designed to determine parameters needed for design. Once selected, however, aerobic (and sometimes anaerobic) bioreactors can be designed and operated to meet the permit limitations required by existing regulations.

Minimum Essentials of Biochemistry

One specific enzyme is required to catalyze a given biochemical reaction. Each enzyme is composed of one or more proteins. A protein requires at least one dedicated gene to direct its synthesis and coenzymes (nonprotein organic molecules) and/or cofactors (inorganic ions) to bind to or associate with the protein to allow the desired catalytic activity. The specific genetic information or deoxyribonucleic acid (DNA) content of a microorganism or cell determines what enzymes are produced. In bacteria, genetic information is located on the chromosome or on a plasmid.

In order for growth to occur, the food source or substrate (or contaminant) must be able to actively pass through the cell membrane. For heterotrophic bacteria, the substrate may be a carbohydrate, protein, lipid, or some other organic compound (either biologically produced or a xenobiotic). Substrates that are large molecules may have to be partially degraded (e.g., enzymatically) in the extracellular environment so that the products can be actively transported across the cell membrane. Once transported, the compounds must be substrates for intracellular enzymes. Such compounds are considered biodegradable when a succession of reactions that result in the production of innocuous final products take place. If fully mineralized, simple inorganic ions (e.g., Cl^- or NH_4^+) and carbon dioxide and water are the final products.

In general, substrates or energy sources are reduced organics for heterotrophs and reduced inorganics for autotrophs and are commonly referred to as *electron donors.* Energy or electrons are "extracted" from the substrates and transferred to extracellular and intracellular electron acceptors. Common extracellular electron acceptors include oxygen, oxidized forms of nitrogen (e.g., nitrite and nitrate), and sulfate. During catabolism some fraction of the electrons stored originally in the substrate or contaminant are transferred enzymatically to oxidized form of flavin and nicotinamide nucleotides to produce the reduced forms of these nucleotides, specifically one reduced flavin adenine dinucleotide ($FADH_2$) and two reduced nicotinamide nucleotides ($NADH + H^+$ and $NADPH + H^+$). Other important products of catabolism are stored chemical energy [e.g., as adenosine triphosphate (ATP)] and low-molecular-weight intermediates. The reduced nucleotides are used along with the chemically energy to transform the intermediates to macromolecules during anabolism. During anabolism the reduced nucleotides help reduce intracellular precursors and are oxidized to FAD, NAD^+, and $NADP^+$, which are once again used during catabolism. Reduced nucleotides are also oxidized using extracellular or exogenous electron acceptors. These reactions produces large quantities of ATP and markedly increase the rate at which electron donors or contaminants are converted to biomass, carbon dioxide, water, and other innocuous by-products.

The basic understanding of biologically based treatment facilities as electron transfer engines provides the foundation necessary to design and operate these facilities. In the parlance of environmental engineers, this is seen best by conducting a mass balance on chemical oxygen demand (COD) around the system. The COD of a sample is the mass of oxygen needed to convert all the organics present in the sample to carbon dioxide and water. This measurement is usually obtained by using a strong oxidizing agent, such as dichromate or permanganate. In an aerobic biological system or in anaerobic systems that have alternative exogenous electron acceptors such as nitrite and nitrate (these systems are referred to as *anoxic*), the total mass of oxygen or alternative exogenous electron acceptor (expressed as oxygen equivalents) consumed by the biomass must be equal to the mass of COD that enters the system less the mass of COD that leaves the system (i.e., in the treated effluent and in the waste biosolids). In anaerobic systems with no alternative exogenous electron acceptors, the total mass of COD that enters must be equal to the mass of COD that leaves the system. Thus, the mass of hydrogen plus methane that is generated (expressed as COD) must be equal to the mass of COD that enters the system less the mass of COD that leaves the system in the treated effluent and in the waste biosolids.

Stoichiometry and Kinetics

The rate at which electron donors (i.e., contaminants) accumulate in the environment depends on the nature of the contaminants; genetic information in the microbial consortium existing at the point of contamination; nutrient availability; the presence of inorganic and organic toxins; the suitability of environmental conditions, such as pH, salinity, and temperature; and the rate at which oxygen is supplied as an exogenous electron acceptor.

In order to integrate these factors into a tool that can assess the rate at which biological depuration takes place in nature or in engineered systems, data must be collected and translated into stoichiometric and kinetic expressions. The reaction schemes thus developed can then be used by conservation principle-based mathematical models (e.g., mass balances) to assess the fate of pollutants in the environment. A brief summary of the stoichiometric and kinetic schemes currently used in such models follows.

Any single reaction can be written simply as

$$\nu_1 A_1 + \nu_2 A_2 + \cdots + \nu_i A_i + \cdots + \nu_m A_m = 0 \qquad [1]$$

$$\text{or} \quad \sum_{i=1}^{m} \nu_i A_i = 0 \quad \text{for } i = 1, \ldots, m \qquad [2]$$

where $A_1, A_2, \ldots, A_m$ are the "m" components for this reaction; $\nu_1, \nu_2, \ldots, \nu_m$ are the stoichiometric coefficients for components $A_1, A_2, \ldots, A_m$, respectively, with $\nu_i < 0$ for reactants and $\nu_i > 0$ for products; and r is the rate of the reaction (moles/volume − time). The rate of reaction, r, is a constant function defined by the rate of formation of each component (e.g., r_{fi} for the ith component) and is given by the following relationship:

$$\frac{r_{f1}}{\nu_1} = \frac{r_{f2}}{\nu_2} = \cdots = \frac{r_{fi}}{\nu_i} = \cdots = \frac{r_{fm}}{\nu_m} = r \qquad [3]$$

This notation can be easily extend to multiple-reaction systems as follows:

$$\begin{cases} \nu_{1,1}A_1 + \nu_{2,1}A_2 + \cdots + \nu_{i,1}A_i + \cdots + \nu_{m,1}A_m = 0 \\ \cdots \\ \nu_{1,j}A_1 + \nu_{2,j}A_2 + \cdots + \nu_{i,j}A_i + \cdots + \nu_{m,j}A_m = 0 \\ \cdots \\ \nu_{1,n}A_1 + \nu_{2,m}A_2 + \cdots + \nu_{i,n}A_i + \cdots + \nu_{m,n}A_m = 0 \end{cases} \quad [4]$$

$$\text{or} \quad \sum_{i=1}^{m} \nu_{i,j}A_i = 0 \quad \text{and } j = 1, \ldots, n \quad [5]$$

where $\nu_{i,j}$ is the stoichiometric coefficient for component A_j in the ith reaction ($\nu_{i,j} < 0$ for reactants and $\nu_{i,j} > 0$ for products), n is number of reactions, m is number of components, and r_i is the rate of the ith reaction. As for the single-reaction system, the rate of each reaction in a multiple-reaction system, r_i, is a constant function defined by the rate of formation of each component in a each specific reaction and is given by the following relationships:

$$\frac{(r_{f1})_j}{\nu_{1,j}} = \frac{(r_{f2})_j}{\nu_{2,j}} = \cdots = \frac{(r_{fi})_j}{\nu_{i,j}} = \cdots = \frac{(r_{fm})_j}{\nu_{m,j}} = r_j \quad [6]$$

$$(r_{fi})_j = \nu_{i,j}r_j \quad [7]$$

$$r_{fi} = (r_{fi})_1 + (r_{fi})_2 + \cdots + (r_{fi})_n \quad [8]$$

where $(r_{fj})_i$ is the rate of formation of the jth component in the ith reaction, and r_{fj} is the total rate of formation of the jth component in an n reaction system and is simply equal to the sum of the rates of formation of the jth component in each individual reaction.

An example of stoichiometric and kinetic relationships commonly used in the design of biological wastewater treatment systems is shown here. In this example, two reactions — a growth reaction and a death reaction — are presented on a mass basis. The first reaction describes the growth of a mixed culture of aerobic bacteria. As written, the reaction includes two reactants, substrate (S) and oxygen (O_2), and, except for biomass (B), does not include other products (e.g., CO_2 and H_2O). The stoichiometric coefficient for biomass in the first reaction ($\nu_{1,B}$) represents the yield of biomass for each unit mass of substrate or contaminant consumed. In this reaction, 0.3 kg of biomass would be produced for each kilogram of substrate utilized. The second reaction, or death reaction, depicts endogenous respiration: 0.9 kg of oxygen is consumed and 0.11 kg of cellular debris (CD) or "dead" biomass is produced for each kilogram of biomass consumed.

$$\text{Growth:} \quad (-1)S + (-0.58)O_2 + (\pm 0.3)B + \cdots = 0$$

$$r_1 = \left(\frac{\mu}{\nu_{1,B}}\right)C_B = \left(\frac{\mu_{\max}}{\nu_{1,B}}\right)\frac{C_S C_B}{(k_S + C_S)}$$

$$\text{Death:} \quad (-1)B + (-0.9)O_2 + (0.11)\text{CD} + \cdots = 0$$

$$r_2 = K_2 C_B$$

The rate function for the growth reaction models the classical Monod expression, where μ represents the specific

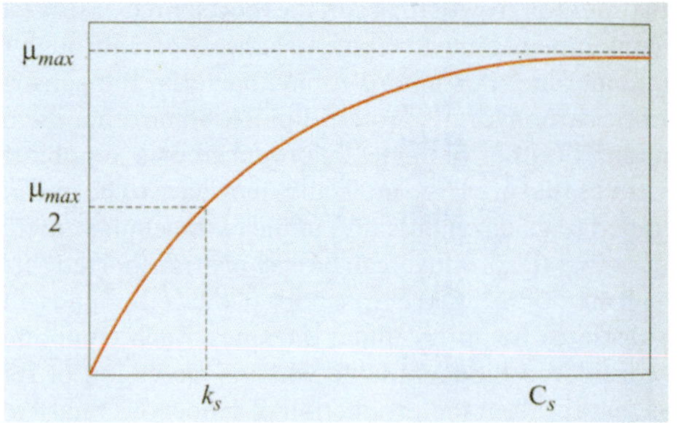

FIGURE 1 The rate function of the growth reaction following the Monod expression. *m* is the specific growth rate of biomass, C_S is the substrate concentration, and k_S is the semisaturation constant.

growth rate of biomass in kilogram biomass produced per kilogram of biomass present per unit period of time. The functional form of the Monod expression is shown graphically in Fig. 1. Inspection of the curve and its mathematical form ($\mu = \mu_{\max}C_S/(k_S + C_S)$) results in the understanding that this rate expression is approximately first order with respect to substrate at low substrate concentrations (e.g., $C_S < k_S$), zero order (i.e., independent of substrate) at high substrate concentrations (e.g., $C_S > k_S$), has a maximum value of $\mu_{\max}$, and is equal to $\mu_{\max}/2$ when the substrate concentration is numerically equal to k_S. Typical average values for μ in biological wastewater treatment systems range from 0.05 to 0.3 per day.

The rate function for the death equation is usually taken as first order with respect to the biomass concentration as shown. A typical value for K_2 is 0.05 per day. This suggest that approximately 5% of the biomass will be lost to death or endogenous respiration each day. The r_{fB} for this two-reaction system is given by

$$r_{fB} = (\nu_{1,B})\left(\frac{\mu}{\nu_{1,B}}\right)C_B + (-1)K_2C_B = (\mu - K_2)C_B \quad [9]$$

Therefore, the specific rate of formation of biomass for both reactions, r_{fB}/C_B, is equal to $\mu - K_2$. As a result, the net mass of biomass produced in the growth reaction is mediated by the destruction of biomass and endogenous respiration defined by the second reaction. This is an important consideration in design because the operating and capital costs associated with the handling of biosolids are appreciable fractions of their respective costs.

Because of the high cost of energy and because of the critical impact that inadequate oxygen supply has on overall performance, careful attention must also be given to the rate of formation of oxygen, r_{fO} ($= (\nu_{1,O}\mu - \nu_{2,O}K_2)C_B$). The specific rate of formation of oxygen for both reactions, r_{fO}/C_B, is normally referred to as the *specific oxygen uptake rate* and, except for the endogenous oxygen uptake rate that takes place at zero substrate concentration, looks like the Monod rate form shown in Fig. 1.

Process of Biological Destruction

Biological treatment systems use microbial communities to remove or eliminate hazardous and nonhazardous organics and inorganics present in domestic wastewaters, industrial wastewaters, leachates, air, and soils. These engineered systems replicate the environment and mitigate nature's limitations, especially those that result from local imbalances that slow the rate of self-purification.

Microbial communities selected and enriched in engineered systems depend on many factors, including the contaminant and how it varies with time; the mass flow rate of individual waste constituents; the presence or absence of electron acceptors (i.e., aerobic, anoxic, or anaerobic conditions); the frequency with which aerobic, anoxic, or anaerobic conditions change; the intensity of feast and famine conditions (e.g., the frequency of change from high to low growth rate); the nature of any by-products formed; and the tendency for attachment (i.e., growth in suspension vs fixed to surfaces).

Properly operated, engineered biological treatment systems can select and enrich for a robust microbial consortium and then manipulate its physiological state to handle normal variations in contaminant load. This level of control is very difficult to achieve in the natural environment. The following sections focus on the design of biological systems for the treatment of wastewater. Although engineered biological systems are used to remove contaminants from gas streams and soils, their aggressive application has occurred only recently and will be discussed briefly (Irvine and Wilderer, 1988).

Engineered Systems for Biological Depuration of Wastewater

Ideal Reactor Configurations

Conventional design of biological reactors uses "ideal" reactor configurations to model or represent "real" systems. Two ideal configurations are most commonly employed: completely mixed (batch, semibatch, or continuous flow) and tubular (continuous plug flow or continuous flow with backmix). Continuous flow reactors can be either recycle or no recycle. Mass balances or "mathematical models" are easily developed from these ideal reactor representations. Five reactor types are discussed here: the completely mixed batch reactor (CMBR), the completely mixed flow reactor (CMFR), the sequencing batch reactor (SBR) (Irvine and Ketchum, 1988), the plug flow reactor (PFR), and the plug flow reactor with dispersion (PFDR) (Weber and DiGiano, 1995).

Mass Balances

Mass balances resulting from application of conservation principles are sets of algebraic, ordinary, and/or partial differential equations. The resulting sets of equations or models are used for reactor design and/or analysis.

A general form of the law of conservation of mass is stated here. Each term has the unit mass per unit time. As can be seen from the following expression, a component (e.g., the *i*th component, A_i) will accumulate in the system when its net rate of entry into the system exceeds its net rate of exit from the system. At steady state, there is no accumulation in the system and the net rates of entry and exit are the same. The relationship developed previously for the rate of formation is independent of the system modeled. Because the stoichiometric coefficients for all reactants are negative, the rate of formation is negative for reactants. In a multiple-reaction system, some components are both reactants and products depending on the reaction considered. In this case, the net rate of formation can be both positive and negative during the course of the reaction. *Nitrification,* the sequential oxidation of ammonia to nitrite and nitrite to nitrate by two different autotrophic bacteria, provides an excellent example of such an occurrence. In a batch reactor, nitrite is often observed increasing during the early stages of nitrification as nitrite is being oxidized to nitrite and then decreasing after all the ammonia is oxidized (and thus the production of nitrite ceases):

Rate at which A_i accumulates in system = rate at which A_i enters system − rate at which A_i leaves system + rate at which A_i formed in system ± rate at which A_i enters or leaves the system by other sources or sinks

Completely Mixed Batch Reactor. A simple schematic representation of a batch reactor is shown in Fig. 2, where the symbol V represents the liquid volume and C_i is

FIGURE 2 Schematic diagram of a completely mixed batch reactor. V represents the volume of liquid and C_i the concentration of the *i*th component.

the concentration of the ith component, A_i. Since the reactor is completely mixed, C_i represents the concentration of the ith component at any location in the reactor. In the mass balance the initial concentration of the ith component is $C_i(0)$, and any source (e.g., supply of oxygen through mixing or aeration) or sink (e.g., volatilization of contaminants with a high vapor pressure) is represented by the symbol S_i. The plus sign is used for sources and the negative sign for sinks. Of course, all sources and sinks must be accounted for in this expression.

Mass balances such as the one depicted in Fig. 2 are often used to develop stoichiometric and kinetic relationships. For the two-reaction system described earlier, a mass balance on substrate for a constant volume CMBR and no other sources or sinks can be used to evaluate $\mu_{\max}$ and k_S. This is done by plotting the reciprocal of the negative of the specific substrate removal rate ($[(-dC_S/dt)/C_B]^{-1}$) against the reciprocal of the substrate concentration on arithmetic paper. This double-reciprocal plot is virtually identical to the one used to evaluate the same coefficients for Michaelis–Menton enzyme kinetics.

CMBRs are used extensively in the fermentation industry. Their use in full-scale biological wastewater treatment systems is less common. The SBR system, however, does incorporate a batch treatment period during its normal operation.

Completely Mixed Flow Reactor. CMFRs are commonly used in a suspended growth, mixed culture biological wastewater treatment in systems referred to as *activated sludge.* These systems are mostly operated with recycle but sometimes with no recycle. No-recycle systems include a general class of biological wastewater treatment reactors used in rural communities. These systems have been called oxidation ditches, stabilization ponds, lagoons, etc. A schematic for a no-recycle CMFR and an associated mass balance is shown in Fig. 3. The volumetric flow rate and the concentration of the ith component in the feed stream are given as Q and C_{i0}, respectively. The hydraulic residence time (HRT), τ, is equal to the reactor volume divided by the volumetric flow rate. The HRT for these systems may be as short as 1 day and as long as 10 days. These ponds or lagoons are often designed to use algae as a source of oxygen. Because algae grow to elevated levels during summer months, if the HRT is more than 2 days, effluents from these systems are considered to be a source of pollution.

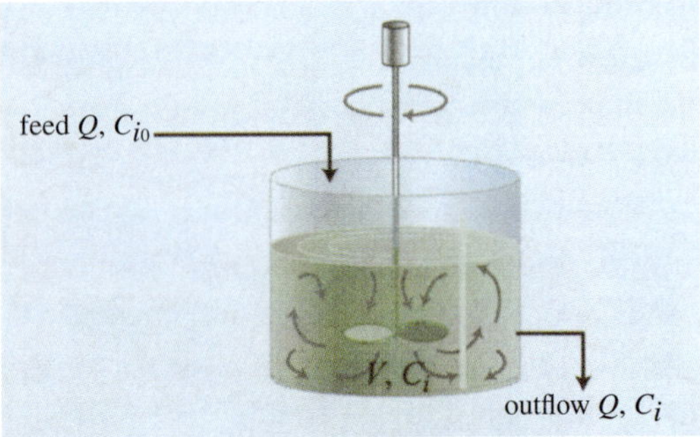

FIGURE 3 Schematic diagram of a no-recycle completely mixed flow reactor. Q represents the volumetric flow rate, C_{i0} is the concentration of the ith component in the feed stream, and C_i is the concentration of the ith component in the outflow.

Microbiologists and biochemists who operate no-recycle bench-scale CMFRs refer to them as *chemostats* and typically use pure rather than mixed cultures. No-recycle CMFRs have also been used to model natural bodies of water, such as ponds and lakes.

Designers often use CMFRs with recycle in the bench-scale and pilot-scale treatability studies. A simple schematic of such a system is shown in Fig. 4, in which the tank on the left represents the biological reactor or CMFR and the tank on the right represents the gravity separator (clarifier or sedimentation tank). Because the biological reactor is completely mixed, the concentration of the ith component in the reactor is identical to the concentration of the same component in the reactor's effluent. The subscripts A, e, w, and r designate concentrations or flows leaving point A, leaving the second tank as the liquid effluent, leaving the second tank as the biosolids effluent, and returning or recycling from the gravity separator to point A, respectively.

The biomass concentration in CMFRs with recycle is much higher than that of a CMFR without recycle and typically ranges from a low of about 1.5 g/liter to a high of about 4 g/liter. The resulting HRT is correspondingly lower, with that for domestic treatment facilities ranging from 0.25 to 1 days and that for industrial wastewater treatment systems

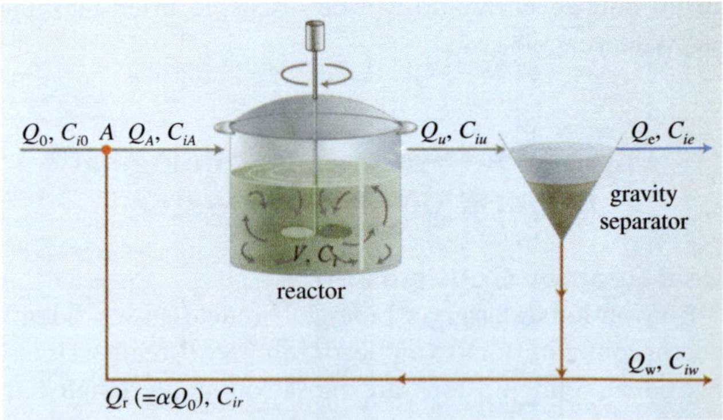

FIGURE 4 Schematic diagram of a completely mixed flow reactor with recycle. Q represents the flow, C_i the concentration, and α the recycle ratio. The subscripts indicate the paths: feed (0), the sum of feed and recycle (A), the outflow from the reactor (u), the liquid effluent (e), and the biosolids in excess (w) and recycled (r). A is the confluence point of the recycle and feed flows.

ranging from 0.25 to more than 10 days. CMFRs with recycle are the systems of choice in large cities, in which the cost of land is high.

Sequencing Batch Reactor. The SBR is a periodically operated mixed culture, suspended-growth system. Although clearly an activated sludge system, the SBR differs from a conventional continuous flow system. The SBR is a time-oriented system, with flow, energy input, and tank volume varying according to some predetermined, periodic, operating strategy. Because it is a controlled, unsteady-state-activated sludge system, the operator can employ a wide variety of operating strategies to select and enrich the desired microbial consortium.

As can be seen from Fig. 5, each reactor in an SBR system has five basic operating modes or periods, each of which is named according to its primary function. The periods are *fill, react, settle, draw,* and *idle.* The periods fill (contaminated wastewater is received and reactions are initiated) and draw (treated effluent is discharged) must occur in each complete cycle for a given reactor. React (desired reactions are completed), settle (biosolids are separated from the treated effluent), and idle (a period of inactivity between draw and fill) are added or eliminated as needed. For example, simple adjustments in the time allowed for aerobic react and mixed react controls the effectiveness of the biological removal of inorganic forms of nitrogen and phosphorus, the extent of sludge bulking (i.e., the settling characteristics of biomass), and the removal of hazardous organic compounds found in industrial wastes and landfill leachates. The SBR models mathematically as a variable volume CMFR followed in series by a CMBR (Irvine and Ketchum, 1988).

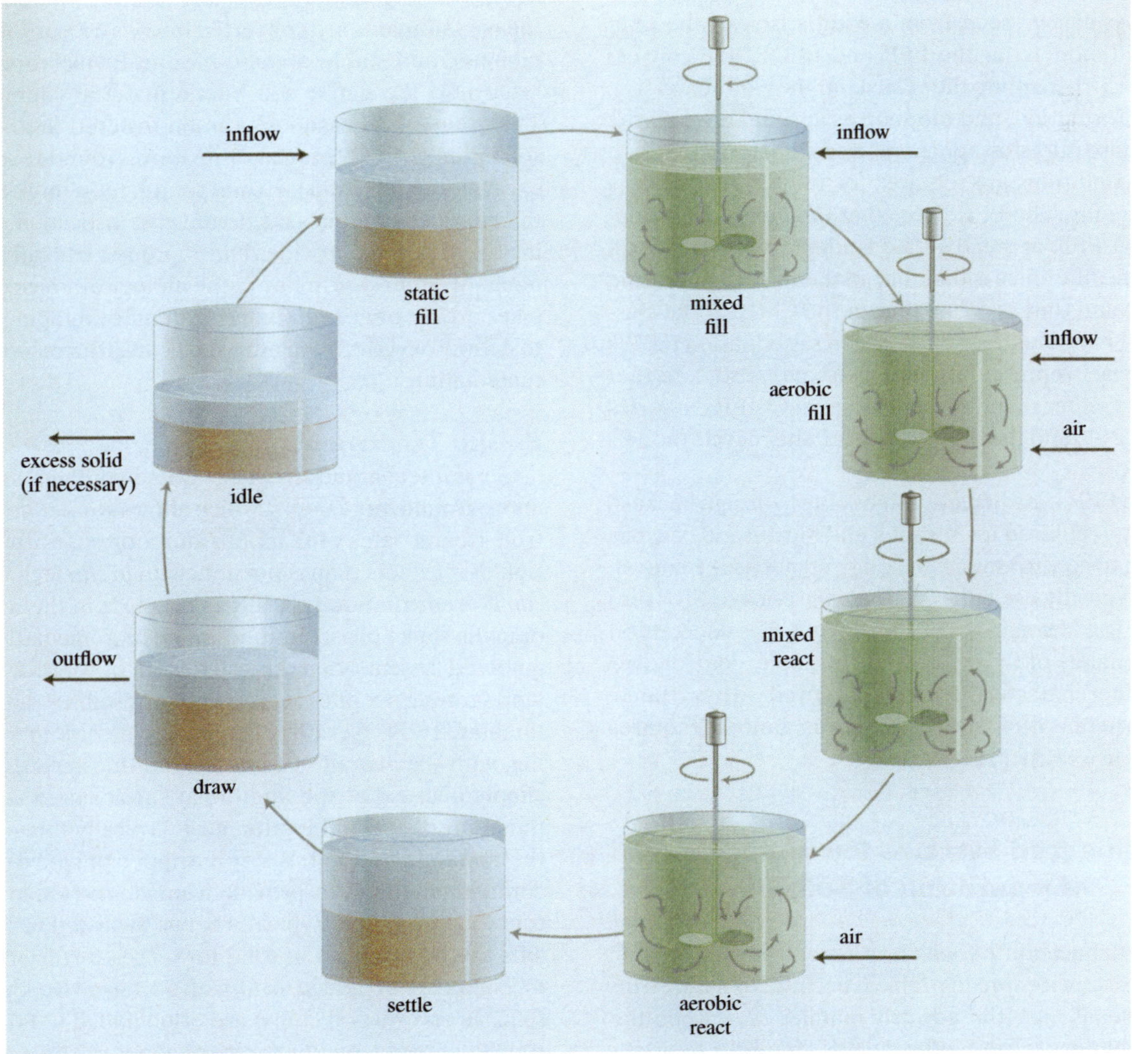

FIGURE 5 Operative scheme of a sequential batch reactor with five periods: fill, react, settle, draw, and idle.

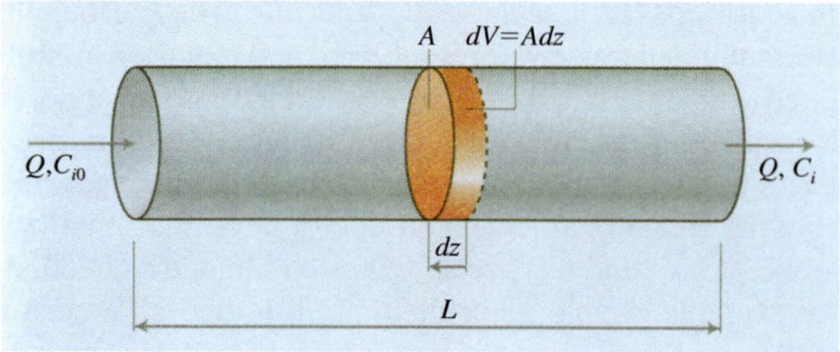

FIGURE 6 Schematic diagram of a tubular plug flow reactor or plug flow reactor with dispersion. dV represents the infinitesimal volume, A the cross section, z the axial coordinate, and L the length of the reactor.

Tubular Reactors. The PFR and the PFDR can be represented simply by the schematic depicted in Fig. 6. In these reactors the HRT is defined in the same way as the CMFR but represents the travel time in the reactor. The difference between the two models is the diffusion coefficient, D_i. As can be seen from a comparison of the one-dimensional models for the PFR and PFDR, the convective term (which involves the velocity in the z direction, v_z), the rate of formation, and the sources and sinks are all the same. Because diffusion appears as a second derivative, two boundary conditions are required for the PFDR and only one is needed for the PFR. Note that the model for a one-dimensional PFR at steady state is identical to that for a CMBR. The difference is that time in the PFR is space time (i.e., HRT) and time is actual time in the CMBR. The classical Streeter–Phelps dissolved oxygen sag equation results from this exact representation of a stream. Also note that an infinite number of infinitesimally small CMFRs in series model as a PFR and that a PFR with infinite recycle models as a CMFR.

Because PFRs are virtually impossible to design because of the energy required for aeration and mixing and because of the need to control sludge bulking and nutrient removal, designers typically use either CMFRs in series or SBRs instead of PFRs. The number of CMFRs in series depends on the requirements of the system. For example, control over sludge settling characteristic can be achieved with two tanks in series, whereas biological phosphorus removal requires as many as five tanks in series.

Engineered Systems for Biological Depuration of Soils

Biological depuration of soils contaminated with organic compounds is a widespread problem because of nature's inability to cleanse itself in a timely manner even when the target contaminants are biodegradable. In order to accelerate the rate of cleanup, biologically friendly conditions, such as nutrient and oxygen availability, pH, and moisture content, must be improved. When contaminants introduced into soils through human activity (e.g., improper disposal methods, accidental surface spills, and leaking underground storage tanks) are met with aggressive action, the time required for remediation is shortened and the costs are minimize, especially when biologically based solutions are implemented.

Common physical/chemical methods for removing contaminants from soil include incineration, chemical oxidation, vapor extraction, low-temperature thermal stripping, and soil washing. However, these methods can be relatively expensive because the extracted contaminant or incinerated soil must often be further treated or disposed of, adding to the overall treatment costs. Alternatively, many organic contaminants are readily biodegraded by indigenous soil microbes and are potentially amenable to bioremediation. Biological soil treatment is typically the least expensive treatment option for contaminated soils (Environmental Protection Agency, 1992, 1993). During bioremediation, the contaminants are converted into water, carbon dioxide, biomass, and stable organic products which are incorporated into the native soil humic material (humification). These biological reactions can be fostered in undisturbed soil in the ground (*in situ*) or in aboveground reactors with excavated soil (*ex situ*). *Ex situ* treatment is more expensive than *in situ* techniques due to the cost associated with moving the soil, but it provides better process control and treatment. In either case, many of the biologically based technologies are aerobic and revolve around developing methods to deliver oxygen to the microbial consortia responsible for remediation.

Ex Situ Treatment

Ex situ techniques treat excavated contaminated soil in aboveground bioreactors, which allows greater process control, greater rates of reaction, and more uniform distribution of reactants than is possible with *in situ* techniques. *Ex situ* bioremediation typically takes place in the solid phase or in the slurry phase. In solid-phase bioremediation, an engineered system often takes the form of windrows (e.g., in land farming) or piles (e.g., biopiles or soil heaping) of contaminated soil. Aeration is typically passive in land farming, with the overall mass of oxygen transferred increased simply because of the additional surface area exposed to the atmosphere. In biopiles, air is typically forced through the contaminated soil. Water is dripped or sprayed over the contaminated soil to provide a moisture content suitable for biodegradation (typically between 45 and 65%). Nutrients can be provided in solid form (i.e., fertilizer, manure, or compost) or added in dissolved form. Bulking agents (e.g., straw or wood chips) are often added to fine-grained soils to enhance pneumatic permeability. The energy consumed in land farms or by forced aeration is small, making solid-phase bioremediation one of the least expensive *ex situ* treatment methods for contaminated soil. However, the effectiveness of solid-phase bioremediation is often limited by heterogeneous contaminant removal and low rates and extent of biodegradation. Slurry-phase bioremediation (usually called bioslurry treatment) refers to the decontamination of contaminated soil or sludge by mixing in water (typically 5–40%) to form an aqueous slurry, which is then mixed and aerated in a tank or a lagoon to achieve contaminant biodegradation.

Slurry-phase bioremediation is more expensive than solid-phase treatment because of the energy required to keep the soil particles in an aqueous suspension. However, mixing in bioslurry reactors lowers mass transfer resistances by dispersing isolated pockets of contaminant and disrupting aggregates of clay and silt particles. Homogenization provides intimate contact between added nutrients, contaminants, microorganisms, and oxygen. The result is a greater rate, extent, and uniformity of contaminant biodegradation than is typically realized in solid-phase systems.

Slurrying followed by solid-phase bioremediation combines the advantages and minimizes the drawbacks of each treatment method when used alone. Such systems use a short slurrying time to achieve homogenization and maximum rates of contaminant destruction and microorganism production, followed by completion of biological treatment in a solid-phase (energy nonintensive) system (Cassidy, 1995).

In Situ Treatment

In situ techniques treat contaminated soil in place by removing contaminated groundwater and/or gas and replacing it with uncontaminated water and/or gas charged with oxygen, nutrients, and other amendments required for contaminant biodegradation in the subsurface. Because of health and safety problems associated with the movement of contaminated soils, especially soils contaminated with hazardous materials, the many regulatory authorities have shown a preference for *in situ* technologies. In such systems, the ground serves as the bioreactor. Delivery of oxygen to the subsurface is typically the rate-limiting step for *in situ* bioremediation systems. Four different types of *in situ* techniques have received most attention: *in situ bioremediation,* in which contaminant biodegradation is stimulated in the saturated zone; *bioventing,* in which contaminant biodegradation is stimulated in the vadose (or unsaturated) zone; *biosparging,* in which contaminant biodegradation is stimulated in both the saturated and the vadose zones; and *intrinsic bioremediation,* in which the contaminants are allowed to biodegrade in either the saturated zone or the vadose zone without the active engineering input (except for monitoring wells).

In situ bioremediation systems typically consist of a combination of groundwater injection wells (or infiltration galleries or trenches) and one or more groundwater recovery wells. In most cases, the recovered groundwater is treated to remove contaminants prior to amendment with nutrients and/or an oxygen source before reinjection. Treatment of recovered groundwater is typically done with air stripping units, adsorption onto activated carbon, oil–water separators, bioreactors, advanced oxidation units (e.g., ozone), or a combination thereof. Nitrogen and phosphorus are typically provided in the form of salts of ammonia and orthophosphate. Oxygen is typically provided by sparging the water with air or pure oxygen before reinjection or by adding hydrogen peroxide.

Bioventing systems consist of a series of air extraction or injection wells with screens that extend through the zone of contamination. Extraction wells are more commonly used than injection wells in order to better control the volatile loss contaminants to the atmosphere. Nutrients are commonly supplied by dissolving in water, which is leached over the contaminated area. Water is required in the vadose zone to avoid desiccation caused by venting. Interestingly, this technology developed from soil vapor extraction (SVE) systems in which the primary removal method was thought to be physical. Users of SVE systems were unable to account for all the remediation by considering the extracted contaminants only. The difference in many cases resulted from biological consumption.

Biosparging (sometimes called air sparging) is relatively new technology developed to overcome the two major rate-limiting processes associated with *in situ* bioremediation in the saturated zone — oxygen availability and dissolution of nonaqueous-phase liquids (NAPLs). Biosparging injects air under pressure into the saturated zone to deliver oxygen for contaminant biodegradation in the saturated zone. Sparging not only provides enhanced oxygenation relative to *in situ* bioremediation but also encourages the dissolution of NAPLs through the agitation and mixing produced by the sparged air bubbles. Increased dissolution of NAPLs enhances the general bioavailability of the NAPLs in the saturated zone. The bubbles produced from sparging result in increased volatilization of organic contaminants from the saturated zone into the vadose zone. Therefore, biosparging usually includes air extraction wells (i.e., bioventing) to provide oxygen for contaminant biodegradation in the vadose zone and for control of fugitive volatile organic compounds (VOCs).

Intrinsic bioremediation (sometimes called natural bioremediation or natural attenuation) is a remediation process in which the contaminant concentration in contaminated aquifers is reduced without engineering assistance or manipulation of the ecosystem. Intrinsic bioremediation is a passive technique which simply exploits the natural ability of site-adapted microbes to biodegrade contaminants. The underlying principle is that a biologically active zone is present in a contaminated aquifer, in which a stable indigenous consortium of microorganisms develops and biodegrades the contaminants. The biologically active zone can be aerobic or anaerobic, depending on the conditions required for contaminant biodegradation. The presence of a biologically active zone depends on the continual supply of materials essential for growth (i.e., nutrients and electron acceptors) and occurs naturally at the site. Contaminants passing through the biologically active zone under the natural flow conditions at the site are consumed by the microorganisms, which allows the contaminant plume to be managed so that migration of the plume off site does not occur.

Engineered Systems for Biological Depuration of Air

Biofiltration is an established air pollution control technique in Europe, in which more than 500 full-scale installations were constructed before 1991 (Leson and Winer, 1991). Intense interest in biofiltration in the United States began during the early 1990s, even though true concern for air pollution began in the 1960s. A growing environmental awareness in the United States was accompanied by passage of increasingly stringent environmental regulations, such as the Air Quality Act of 1967, the 1970 Clean Air Act, and the Clean Air Act Amendments of 1977 and 1990 (Hunton and Williams, 1998).

The treatment of waste gasses using biofiltration reduces treatment costs, eliminates secondary waste streams, and is as effective or more effective at removal of biodegradable contaminants than other technologies. Biofiltration is significantly less expensive than any other treatment technology (Bohn, 1992). Even though biofiltration is the lowest cost alternative for most air pollution control problems, several technical challenges have limited its use. The following problems are commonly reported:

1. Excessive microbial growth: Biomass produced in overloaded systems or in systems operated for long periods of time often clog biofilters, causing channeling, short circuiting, and excessive head loss across the filter bed.
2. Inadequate moisture control: Excess moisture reduces VOC and oxygen transport rates and insufficient moisture limits microbial activity.
3. Inadequate pH and nutrient control: High or low pH levels and inadequate nutrient supply decrease or eliminate microbial activity.
4. Short-term increases in contaminant loading: Severe variations in waste gas characteristics (e.g., VOC and volatile inorganic compound concentration, relative humidity, and temperature) can result in dynamic "shock" loads that exceed biological reaction capacities.

Biofiltration

Biofilters originated more than 40 years ago for use in controlling odors from the agricultural, food processing, and wastewater treatment industries. Early biofilters used in Europe consisted mostly of venting odorous air through soil beds or compost piles. In general, biofiltration involves the passage of contaminated gasses through a porous solid medium which supports a biologically active layer of microorganisms. As gasses flow through the medium and past the biofilm, contaminants partition from the gas phase into the aqueous and solid phases (e.g., support medium and biofilm), in which microorganisms consume the contaminants just as they do in biological treatment facilities for wastewaters and/or soils.

The basic physical components of a biofilter are a gas distribution system, a support medium for the microbial consortium, and a humidification and drain system to control moisture content. Typically, a blower forces contaminated gas through a network of perforated pipes. Humidification may be provided to such a system by saturating the air with moisture before it enters the biofilter, by sprinkling water on the surface of the packing material, by adding water to the bed using a soaker hose, or by some combination of these techniques. A drain system is usually provided as a means of removing excess water.

Although packing media used in conventional biofilter beds consist mostly of peat or compost, a wide variety of other materials have been used, including soil, sand, wood chips, bark, sawdust, activated carbon, ceramic monoliths, ceramic pellets, sintered glass, polystyrene beads, ground tires, and polyurethane foam (Moe *et al.,* 1997; Baltzis, 1998; Kinney *et al.,* 1998). Typical additives to the packing media include bulking agents, buffering agents, nutrients, and microbes. Bulking agents, which prevent bed compaction, increase bed permeability for improved gas distribution, and decrease pressure drop across the bed. These have included Styrofoam spheres, perlite, vermiculite, bark, wood chips, and ground tires. Buffers adjust the pH by countering any hydrogen ion production (or consumption) resulting from microbial activity. When an inorganic support medium is used, nutrients are usually added either with the packing material before biofilter assembly or in a nutrient solution sprayed on or mixed with the packing material after construction. Whatever the packing material used, several desirable properties should be incorporated, including high porosity, appropriate pore size, low density, and an ability to sorb water (Moe, 1999).

Alternative Systems

Although biofiltration was the first system for biological waste gas treatment, several alternate process configurations have been developed to biologically treat gas-phase contaminants. The systems are known as biotrickling filters, bioscrubbers, sparged gas reactors, and porous membrane reactors.

A biotrickling filter has a free-flowing liquid phase. In this system, a nutrient-containing stream is continuously recirculated over the packing medium. Packing media are usually inert and rigid and may consist of gravel, plastic, ceramic, or a variety of other materials. A bioscrubber uses a tower or column to transfer contaminants from the gas to the liquid phase, which undergoes biodegradation in a separate reactor. Some towers used for contaminant transfer use flat plates, others are filled with inert packing, and some use a spray for contaminants to partition into the droplets that fall through the contaminated gas stream. A sparged gas reactor is a liquid reactor (e.g., a CMFR) that is sparged with contaminated gas. Contaminants partition from the gas phase into the liquid phase, in which they are biodegraded by a suspended culture of microorganisms. The porous membrane reactor uses a microporous hydrophobic membrane made of material such as polypropylene, polyethylene, or silicone. Contaminants travel from the gas phase, through the membrane, to the liquid phase, in which they are degraded by a suspended growth (e.g., activated sludge) or fixed-film microbial population (Moe, 1999).

References Cited

BALTZIS, B. C. (1998). Biofiltration of VOC vapors. In *Biological Treatment of Hazardous Wastes* (G. A. Lewandowski and L. J. DeFilippi, Eds.). Wiley, New York.

BOHN, H. (1992). Consider biofiltration for decontaminating gases. *Chem. Eng. Prog.* **88**(4), 34–40.

CASSIDY, D. P. (1995). Bioremediation of soils contaminated with hydrophobic compounds using slurry and solid phase techniques. Ph.D. dissertation, University of Notre Dame, Notre Dame, IN.

ENGELS, F. (1975). The condition of the working-class in England, 1844–45. In *Collected Works* (K. Marx and F. Engels, Eds.), Vol. 5. New York.

ENVIRONMENTAL PROTECTION AGENCY (1992, November 5). TheSuperfund innovative technology evaluation program: Technology profiles, EPA/540/R-92/077. Environmental Protection Agency, Washington, DC.

ENVIRONMENTAL PROTECTION AGENCY (1993, May 8). Bioremediation in the field, EPA/540/N-93/001. Environmental Protection Agency, Washington, DC.

HUNTON and WILLIAMS (1998). *Clean Air Handbook* (F. W. Brownell, Ed.), 3rd ed. Government Institutes, Rockville, MD.

IRVINE, R. L., and KETCHUM, L. H., JR. (1988). Sequencing batch reactors for biological wastewater treatment. *Crit. Rev. Environ. Control* **18**(4), 255.

IRVINE, R. L., and WILDERER, P. A. (1988). Aerobic processes. In *Standard Handbook of Hazardous Waste Treatment and Disposal* (H. Freeman, Ed.). McGraw-Hill, New York.

JANSSENS, P. (1970). *Palaeopathology: Diseases and Injuries of Prehistoric Man.* Baker, London.

KINNEY, K. A., WRIGHT, W., CHANG, D. P., and SCHROEDER, E. D. (1998). Biodegradation of vapor phase contaminants. In *Bioremediation: Principles and Practice* (S. K. Sikdar and R. L. Irvine, Eds.), Vol. 1, pp. 601–632. Technomics, Lancaster, PA.

LESON, G., and WINER, A. M. (1991). Biofiltration: An innovative air pollution control technology for VOC emissions. *J. Air Waste Management Assoc.* **41**(8), 1045–1054.

MELLANBY, K. (1982). *The Biology of Pollution.* Arnold, London.

MEYBECK, M., *et al.* (1989). *Global Freshwater Quality: A First Assessment.* Basil Blackwell, Oxford.

MOE, W. M. (1999). Operation and performance of a periodically operated gas-phase biofilter using a novel polyurethane foam medium. Ph.D. dissertation, University of Notre Dame, Notre Dame, IN.

MOE, W. M., IRVINE, R. L., and MONTEMAGNO, C. D. (1997). Preliminary investigation of polyurethane foam medium for use in vapor phase biofiltration. In *Proceedings of the 1997 CSCE–ASCE Environmental Engineering Conference: Protecting People and the Environment, Edmonton, Alberta, Canada* (S. J. Stanley and C. J. W. Ward, Eds.), pp. 1755–1766. CSCE, Montreal.

PONTING, C. (1991). *A Green History of the World.* Penguin, London.

WEBER, W. J., JR., and DIGIANO, F. A. (1995). *Process Dynamics in Environmental Systems.* Wiley, New York.

General References

BAILEY, J. E., and OLLIS, D. F. (1986). *Biochemical Engineering Fundamentals,* 2nd ed. McGraw-Hill, New York.

BREEZE, L. E. (1993). *The British Experience with River Pollution, 1865–1876.* Lang, New York.

BRIMBLECOMBE, P. (1988). *The Big Smoke.* Routledge, London.

COLTEN, C. E., and SKINNER, P. N. (1996). *The Road to the Love Canal: Managing Industrial Waste before EPA.* Univ. of Texas Press, Austin.

ERGAS, S. J., SCHROEDER, E. D., and CHANG, D. P. (1992). Biodegradation technology for volatile organic compound removal from airstreams; Phase I: Performance verification. Prepared for Research Division, California Air Resources Board, final report under Contract No. AO-32–137.

FITTER, R. S. R. (1946). *London's Natural History.* Collins, London.

GOTTLIEB, R. (1995). *Forcing the Spring: The Transformation of the American Environmental Movement.* Island Press, Washington, DC.

HAMLIN, C. (1987). *What Becomes of Pollution: Adversary Science and the Controversy on the Self-Purification of Rivers in Britain, 1850–1900.* Garland, New York.

HAYS, S. P. (1987). *Beauty, Health, and Permanence: Environmental Politics in the United States, 1955–1985.* Cambridge Univ. Press, Cambridge, UK.

KEHOE, T. (1997). *Cleaning up the Great Lakes: From Cooperation to Confrontation.* Northern Illinois Press, Dekalb.

LERNER, S. (1997). *Eco-Pioneers: Practical Visionaries Solving Today's Environmental Problems.* MIT Press, Cambridge, MA.

MCNEILL, W. (1979). *Plagues and People.* Penguin, London.

PAEHLKE, R. C. (1989). *Environmentalism and the Future of Progressive Politics.* Yale Univ. Press, New Haven, CT.

SCHEFFER, V. B. (1991). *The Shaping of Environmentalism in America.* Univ. of Washington Press, Seattle.

TARR, J. A., MCCURLEY, J., and YOSIE, T. F. (1980). The development and impact of urban wastewater technology: Changing concepts of water quality control, 1850–1930. In *Pollution and*

Reform in American Cities, 1870–1930 (M. V. Melosi, Ed.). Univ. of Texas Press, Austin.

TCHOBANOGLOUS, G., and SCHROEDER, E. D. (1985). *Water Quality.* Addison-Wesley, Reading, MA.

WOHL, A. S. (1983). *Endangered Lives: Public Health in Victorian Britain.* London.

ZHU, H., KEENER, T. C., ORTON, T. L., BISHOP, P. L., KHANG, S. J., and SIDDIQUI, K. (1997). The effectiveness of aeration recirculation in controlling VOC emissions from publicly owned treatment works. *J. Air Waste Management Assoc.* **47,** 1259–1267.

RICHARD B. PRIMACK
Biology Department
Boston University
Boston, Massachusetts 02215, USA

The Gap between Rich and Poor: How Economic Inequities Threaten Biodiversity

Loss of diversity in many biological communities, particularly the tropics, is directly related to poverty in local human communities. In many areas of the world, poor families that lack access or title to farmland subsist by converting forests or wetlands into agricultural plots, destroying fragile habitats in the process. Although such plots usually are short term, the long-term damage to biological diversity is incalculable. However, for many poor families, wild plants and animals may be the only affordable option for medicines, fuel, and building materials. As biological communities become increasingly degraded, the poor consume wild resources faster than they can regenerate, creating an inexorable downward spiral of impoverishment for both the human and the biological communities. This cycle of poverty is exacerbated by the excessive and wasteful consumption practiced by developed nations, which habitually consume disproportionate quantities of the world's natural resources. As natural resources, including energy, food, and even genetic diversity, are claimed by industries in the developed world, fewer resources are available for the growing ranks of poor people in developing nations. Ultimately, the fate of biological diversity depends on reducing the gap between rich and poor people in developed and developing nations. Wealthy nations must be convinced to limit their consumption of natural resources; some advances have been made toward this goal through recycling initiatives, research supporting renewable energy sources, and other programs, but developed nations still far outstrip developing nations in the proportion of resources they consume. Likewise, poor nations must be assisted in reaching a higher standard of living to reduce reliance on threatened resources. Programs that improve literacy, promote lower birth rates, and, most important, provide alternative sources of income to poor families have made great strides in improving local living standards to the benefit of local ecosystems. Programs which provide direct economic benefits to local communities that preserve biological diversity offer some of the greatest promise. If the rich variety of biological life is to be preserved, then the extremes of human existence—extreme wealth with its wasteful excesses as well as extreme poverty—must be eliminated.

Introduction

There is no mystery as to why biological diversity is showing a rapid, worldwide decline. Biological communities are destroyed and species driven to extinction either because someone is making money by doing it or because people are simply using up resources in order to survive. However, this truth oversimplifies a tremendously complicated problem: Underlying threats to biodiversity lie not so much in why resources are extracted but in how they are extracted and consumed—how much is extracted, how quickly it is extracted, and how efficiently it is used. It has long been acknowledged that people living in wealthy industrialized nations consume excessive amounts of natural resources, particularly nonrenewable energy resources such as petroleum, as well as contribute disproportionately to global pollution [Lacayo, 1997; World Resources Institute/United Nations Environment Program/United Nations Development Program (WRI/UNEP/UNDP), 1994]. At the same time, developing nations, many of which have large numbers of people living at or below poverty level, are attempting to acquire the prosperity enjoyed by the industrial nations, and their principal method of achieving this goal is to consume biological resources, such as tropical forests and fisheries, for large-scale consumer and industrial use. Unfortunately, industrial resource extraction in many parts of the world tends to be rapid, indiscriminate, and destructive, whereas use of the extracted resources usually is inefficient at best and blatantly wasteful at worst. However, the economic benefits of industrial extraction are often limited to a small percentage of the population, leaving the remainder to rely on low-wage jobs and small-scale extraction of biotic resources for their day-to-day sustenance. Such small-scale activities, such as subsistence hunting, collection of wild plants, and even shifting agriculture, generally do not harm biological diversity when the practitioners are adequately spaced across the landscape. However, as intensive industrial extraction leaves fewer natural areas available to local people, an ever-increasing human population relies on a rapidly shrinking land area. Biological communities rapidly become degraded as the poor consume wild resources faster than they can regenerate, creating an inexorable downward spiral of impoverishment for both the human and the biological communities.

How is this problem to be addressed on a global scale? The first step is to recognize that biological resources are limited, and that biological communities cannot be expected to survive if they are continually mined for resources. The second step is to use resources more efficiently, limiting the amount of waste, pollution, and other potential threats associated with resource exploitation. The third step is to curb human population growth, a factor that is beginning to strain even common resources. Finally, and most important, the underlying social and economic factors that lead to uncontrolled exploitation of biological communities must be addressed: In blunt terms, the gap between the rich and poor of the world must be closed at both ends by eliminating the excesses of the wealthy as well as the desperation of the poor. These admittedly formidable goals require efforts at all levels — personal, institutional, and international — to educate individuals on ways of reducing household waste and to mandate cleaner, more efficient methods of manufacturing in both the developed and the developing world.

The Limits of Biological Resources

Rates of Habitat Destruction

If there is one principal threat to biological diversity, it is habitat loss. In many countries of the world, particularly where human population density is high, most original habitat has been destroyed. Worldwide, more than 50% of vegetated areas have been disturbed or destroyed by human activities; even where protected areas have been established to safeguard biodiversity, nearly half of the protected land area has been subjected to medium to high levels of disturbance (WRI/UNEP/UNDP, 1994). In tropical Asia, approximately 62% of the wildlife habitat has been lost, with particularly high levels of destruction reported for China (99% of primary forests converted), Bangladesh (96%), Sri Lanka (86%), India (78%), and Vietnam (76%). Sub-Saharan Africa has lost approximately 84% of its forests, with habitat loss higher than 75% in countries such as Mali (78%), Uganda (79%), Rwanda (80%), Nigeria (80%), and Senegal (82%). Some smaller countries, such as Burundi and Sierra Leone, have lost more than 90% of their forests (WRI/UNEP/UNDP, 1994). Recent rates of deforestation are highest in the tropical forests of Africa, Asia, and the Americas, where many countries have lost 10–33% of their forests within the past 15–20 years (Brown *et al.*, 1997).

The destruction of tropical rain forests has come to be synonymous with the loss of species. Tropical moist forests occupy 7% of the earth's land surface — mostly located, by accident of history, in developing nations — but they are estimated to contain more than 50% of the planet's species (Myers, 1986), with a complexity of species interaction and specialization unparalleled in any other community. The original extent of tropical rain forests and related moist forests has been estimated at 16 million km^2, based on current patterns of rainfall and temperature (Myers, 1984, 1986, 1991; Sayer and Whitmore, 1991). However, tropical deforestation rates are alarmingly high and are still growing (Fig. 1). A combination of ground surveys, airplane photos, and remote sensing data from satellites showed that in 1982 only 9.5 million km^2 remained, an area about equal in size to the continental United States. By 1991, another 2.8 million km^2 had been destroyed, and the subsequent 6 years have witnessed the loss of nearly 12.6% of the forests present in 1991 — a total loss of nearly 3.65 million km^2 in only 15 years (Brown *et al.*, 1997). Where forest are not completely destroyed, much of what remains is so badly degraded that the species composition and ecosystem processes of the community are greatly altered.

Habitat loss is also taking place in aquatic ecosystems. Lakes, rivers, wetlands, mangroves, and coral reefs have all been heavily damaged by human activities, either directly by exploitation of aquatic species or diversion of water or indirectly through pollution. Particularly in regions of limited or unreliable rainfall, access to adequate, clean supplies of fresh water can have a pronounced effect on plant, animal, and invertebrate populations. When fresh water is diverted for (or polluted by) human use, the impact on biological diversity may be profound.

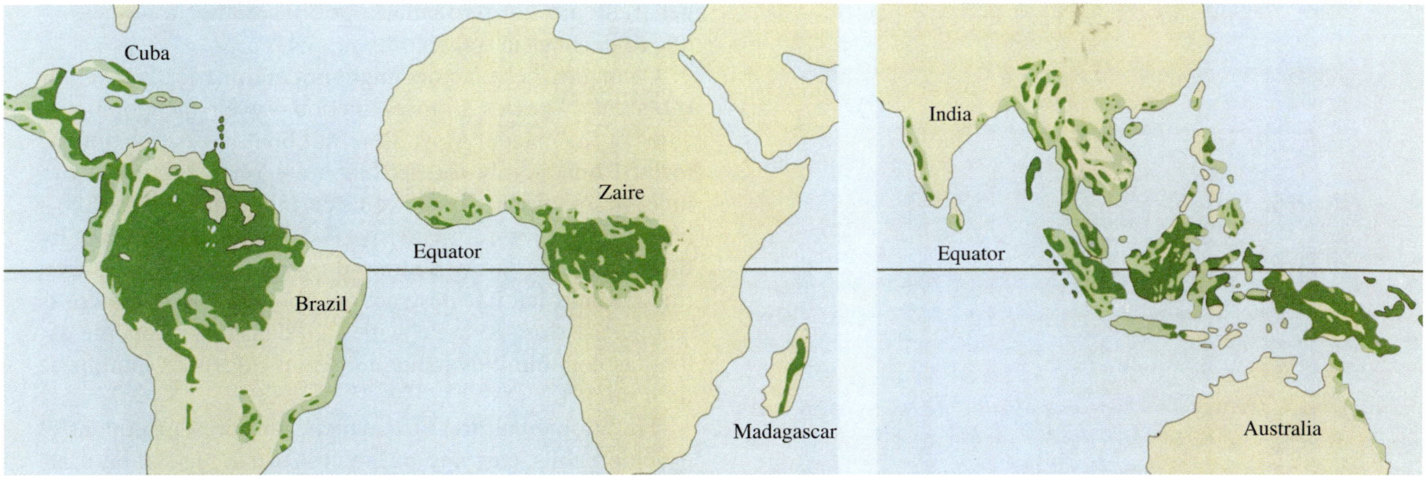

FIGURE 1 Tropical rain forests are found predominantly in wet, equatorial areas of the Americas, Africa, and Asia. Tropical forests once covered a wider area (indicated in green), but human activities have resulted in the loss of a great deal of rain forest, reducing the limits to the areas shown in dark green. In the light green areas the forests are still present, but these are secondary forests in regrowth after cutting, plantations of rubber or teak, or forests used for harvesting wood or for fuel production. Only the areas in dark green indicate tropical rain forests still intact and large enough to support their biodiversity (adapted from Primack, 1998).

Population, Poverty, and Biological Diversity

The major threats to biological diversity — habitat destruction, habitat fragmentation, habitat degradation (including pollution), overexploitation of species for human use, introduction of exotic species, and increased spread of disease — are all related, directly or otherwise, to an ever-increasing use of the world's natural resources by the expanding human population. The greatest destruction of biological communities has occurred during the past 150 years, as human population has increased from 1 billion in 1850 to 5.7 billion in 1996 (Brown *et al.,* 1997; Fig. 2), with almost one-fourth of that increase occurring in just the past decade. Although the rate of population growth worldwide has slowed slightly in recent years, it is still high in many areas of tropical Africa, Latin America, and Asia, where the greatest biological diversity is also found.

All else being equal, more people equals less biodiversity. People use natural resources such as fuelwood, wild meat, and wild plants, and people convert vast amounts of natural habitat to agricultural and residential purposes. Because some degree of resource use is inevitable, population growth is partially responsible for the loss of biological diversity, and some scientists believe attempts to protect biological diversity will only be successful if the size of the human population is controlled (Hardin, 1993). However, the point also must be made that global resources are still sufficient to support the current human population, if only they could be distributed more equitably. Human activities are not automatically incompatible with biological diversity; people have lived in nearly every terrestrial ecosystem of the world for thousands of years as hunters, fishermen, farmers, and gatherers. Traditional societies utilizing innovative irrigation methods and a mixture of crops were often able to support relatively high human population densities without destroying the environment or the surrounding biological communities. For example, at the height of Maya civilization approximately 1200 years ago, the tropical forests in Mexico, Belize, Guatemala, and Honduras had densities of up to 500 people per square kilometer, yet the range of species apparently was still maintained (Gomez-Pompa and Kaus, 1992; Fig. 3). Today, these same areas have less than 10% of their precontact levels as a result of poverty, enslavement, disease, and warfare that became widespread after the arrival of Europeans. Despite the low levels of human population, however, many parts of this region have experienced unprecedented declines in animal and plant populations, loss of forest cover, and tremendous destruction and degradation of habitat. Although this may partly be a product of technological capabilities — the ancient Maya lacked chainsaws, tractors, and even draft animals, which have been used by modern inhabitants to clear vast tracts of forest — it is also likely that a major difference between the ancient and modern residents of the Central American tropical forests was one of mind-set. The ancient civilizations had grown up in these forests, relying on them over the centuries for their sustenance and livelihood. As a consequence, they developed many techniques to safeguard the long-term viability of the land and maintained viable forest ecosystems. Modern residents, on the whole, seek quick profits and short-term benefits when they clear forests for plantation agriculture or pasture, and indeed such ventures

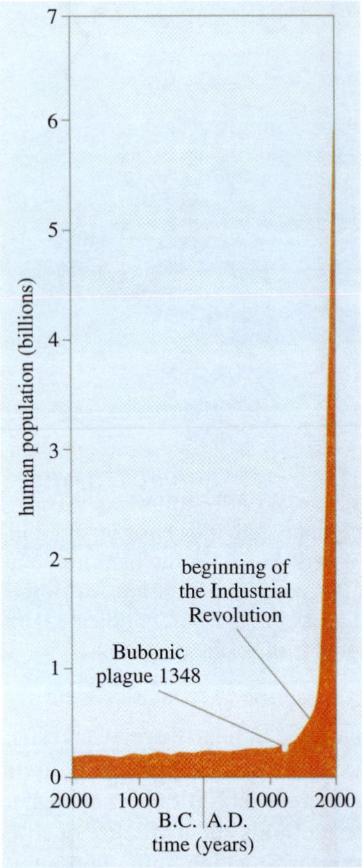

FIGURE 2 The demographic growth of the human population in the past 4000 years. The growth has increased spectacularly since the sixteenth century; in the past 150 years it increased from 1.0 billion in 1850 to 5.6 billion in 1996. If the current growth rate is maintained, the world population will double in less than 40 years (adapted from Primack, 1998).

generally are not particularly profitable after a few years due to declines in soil fertility.

Changing these factors might not automatically save all threatened species from extinction — even the populous cities of the ancient Maya were not built in pristine, intact forests, but merely forests that were relatively intact by modern standards — but it would certainly improve the likelihood of survival for many species. The important point is that population growth alone does not cause species extinction and habitat destruction. Biological communities can persist close to areas with high densities of people as long as communities make conscious efforts to minimize their impact.

Traditional cultures that depend on forests frequently develop habits, customs, or laws to regulate use of natural resources. Sometimes, this regulation breaks down during war, political unrest, and social instability. When this happens, insecurity or fear may trigger a scramble to consume resources that for generations had been used sustainably. The higher the density of people, the more closely their activities must be regulated, and the greater the destruction that can result from a breakdown in authority. The rise of industrial capitalism and materialistic modern societies has greatly accelerated demands for natural resources, particularly in the developed countries. Inefficient and wasteful use of natural resources is also a major cause of the decline in biological diversity. Whether motivated by subsistence needs, individual greed, or corporate policy, once a mentality of short-term exploitation of resources is established in a community or a nation, it is difficult to restrain. A revealing example is the devastation that occurred in China's forests during the Cultural Revolution: Strict regulations against cutting trees were no longer enforced, so farmers

FIGURE 3 The tropical forest has returned to cover the ruins of the Mayan cities. A thousand years ago, Mayan farms and cities occupied a wide area of the Central American lowlands, with no apparent loss of species.

cut down trees at a tremendous rate, stockpiling wood for fuel, construction, and furniture making (Primack, 1988).

Impact of Subsistence Activities

Much of the destruction of the world's biological diversity is caused by people who are desperately poor and are simply trying to survive. In many instances, government regulations created with the intent of alleviating poverty encourage people to establish title to unclaimed, wild lands by "improving" them — that is, by clearing the natural vegetation to establish permanent agriculture or pasture, even when lands are not suitable for these uses. Some of this land conversion is permanent, but most converted lands are placed under shifting cultivation so that they return to secondary forest during the fallow period (Fig. 4). On a global scale, about 86,000 km^2 per year (61%) of rain forest destruction results from small-scale cultivation of crops by poor farmers, most of whom have been shifted to the agricultural frontier to practice temporary farming out of desperation and poverty (Myers, 1991). Included here is land degraded each year for fuelwood production, mostly to supply local villagers with wood for cooking fires. More than 2 billion people cook their food with firewood, so their impact is significant. Another 29,000 km^2 per year (21%) is destroyed through small-scale commercial logging in clear-cutting and selective logging operations. Approximately 15,000 km^2 per year (11%) is cleared for cattle pasture. The relative importance of these activities varies by geographical region: Logging is a more significant activity in tropical Asia and America, cattle ranching is more prominent in tropical America, and farming and fuelwood gathering are more important in tropical Africa with its rapidly expanding population (Kummer and Turner, 1994; WRI/UNEP/UNDP, 1994; Rudel and Roper, 1996). Extending the projection forward reveals that at the current rate of loss, there will be virtually no tropical forest left after the Year 2040 except in the relatively small areas under protection. The situation is actually even more grim than these projections indicate because the world's population is still increasing and poverty is on the rise, putting ever greater demands on the dwindling supply of rain forest.

FIGURE 4 River fish are the main source of protein for the Tukano people of the Amazon basin. The Tukano people protect the forest along the river margins and prohibit fishing in certain stretches of river (photo courtesy of P. Patmore).

Traditional Societies and Biological Diversity. People in rural societies have always hunted and harvested the food and other resources that they need to survive. As long as human populations were small and the methods of collection unsophisticated, people could sustainably harvest and hunt the plants and animals in their environment without driving species to extinction. However, even in preindustrial societies, intense exploitation has led to the decline and extinction of local species. Excessive hunting by humans has been blamed for a series of large mammal extinctions occurring 40,000 years ago in Australia and 12,000 years ago in the New World, coinciding with the presumed timing of human migration into these areas (Eldredge, 1997). However, as human populations have increased, their use of the environment has escalated. Methods of harvesting have also become dramatically more efficient. Guns are used instead of blowpipes, spears, or arrows for hunting in the tropical rain forests and savannas. Powerful motorized fishing boats and efficient "factory ships," some equipped with sophisticated radar and sonar equipment for locating schools of fish, harvest fish from all the world's oceans. Even small-scale local fishermen now have outboard motors on their canoes and boats, allowing them to harvest more rapidly and over a wider area. Wasteful practices such as "blast fishing" (use of explosives to kill masses of fish within a given area, usually about 5 m) not only kills more fish than the fishermen actually collect but also can severely damage fragile reef habitat (Jennings and Polunin, 1996). Overexploitation by humans has been estimated to currently threaten about one-third of the endangered, vulnerable, and rare species of vertebrates (World Conservation Monitoring Center, 1992), placing it behind only habitat loss and degradation in importance.

In traditional societies, restrictions often existed to prevent the overexploitation of natural resources. For example, the rights to specific harvesting territories were rigidly con-

FIGURE 5 One method used to protect the howler (*Alouatta pigra*) habitat is to create aerial bridges over man-made gaps in the forest such as roads. These bridges not only allow the monkeys to move between forest patches in relative safety but also allow visitors to see the monkeys in the open (photo courtesy of R. Horwich).

trolled; hunting in certain areas was banned; there were often prohibitions against taking females, juveniles, and undersized individuals; certain seasons of the year and times of the day were closed for harvesting; and certain efficient types of harvesting were not allowed (Fig. 5). These restrictions, which allowed traditional societies to harvest communal resources on a long-term, sustainable basis, are almost identical to the rigid fishing restrictions developed and proposed for many fisheries in industrialized nations (McGoodwin, 1990). Among the most highly developed restrictions were those of the traditional or "artisanal" societies of Micronesia and Polynesia (Johannes, 1978). In these societies the resources of the reef and lagoon were clearly defined, and the possibility and consequences of overharvesting were readily apparent.

Rather than being a threat to the "pristine" environment of tropical forests, traditional peoples have been an integral part of these forests for thousands of years. The current mixture and relative densities of plants and animals in many biological communities may reflect the historic activities of people in the area, such as selective hunting of certain game animals, fishing, and planting of useful plant species. The commonly practiced agricultural system, known variously as swidden agriculture, shifting cultivation, or slash-and-burn agriculture, also affects forest structure and species composition by creating a mosaic of forest patches of different ages. In this system, the trees in an area are cut down, the fallen plant material is burned, and crops are planted in the nutrient-rich ash. After one or several harvests, the nutrients are washed out of the soil by the rain; the farmer then abandons the field and cuts down a new patch of forest for planting. This system works well and does not degrade the environment as long as human population density is low and there is abundant forest land.

Local Action to Protect Biological Diversity. In many cases, local people already protect biological communities such as forests, wildlife, rivers, and coastal waters in the vicinity of their homes. Such protection is often enforced by village elders with justification based on religious and traditional beliefs. Governments and conservation organizations can assist local conservation initiatives by providing legal title to traditional lands, access to scientific expertise, and financial assistance to develop needed infrastructure. One example is the Baboon Sanctuary in eastern Belize, created by a collective agreement among a group of villages to maintain the forest habitat required by the local population of black howler monkeys (*Alouatta pigra*) (Horwich and Lyon, 1997). Ecotourists visiting the sanctuary must pay a fee to the village organization and additional payments if they stay overnight and eat meals with a local family (Fig. 6). Conservation biologists working at the site have provided training for local nature guides, a body of scientific information on the wildlife, funds for a local natural history museum, and business training for the village leaders.

In the developing world, a rigid separation between lands used by local people to obtain natural resources and strictly protected national parks is often not possible (McNeely, 1993a,b). Local people may live in protected areas or traditionally use the resources of the protected areas. Also, considerable biological diversity often occurs on land owned by local people. Many examples exist in which people are allowed to enter protected areas periodically to obtain natural products. In biosphere reserves, local people are al-

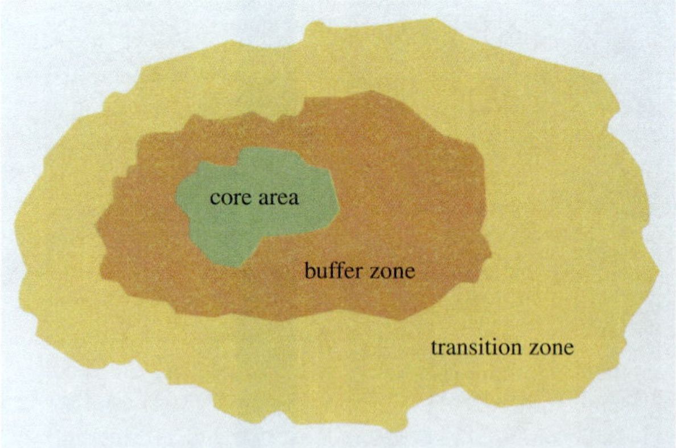

FIGURE 6 The general pattern of a reserve organized by UNESCO in the Man and the Biosphere program. The reserve includes a severely protected core area surrounded by a buffer zone in which regulated traditional human activities are permitted together with monitoring and nondestructive research. In turn, this is surrounded by a transition zone of sustainable development, experimental research, and agriculture in the villages (adapted from Primack, 1998).

FIGURE 7 In the Chitwan National Park in Nepal, local residents can collect cane grass and thatching materials. Park officials weigh the bundles in order to keep the harvest at a sustainable level (photo courtesy of J. E. Lehmkuhl).

FIGURE 8 A Huastec Indian woman in an indigenous managed forest in northeastern Mexico. The photograph shows the harvesting of sapote fruit (*Manikara achras*) and cuttings of a frangipani tree (*Plumeria rubra*) for successive planting (photo courtesy of J. Alcorn).

lowed to use resources from designated buffer zones (Fig. 7). Large game animals are harvested for meat in many African game reserves (Lewis *et al.,* 1990). Through such compromises, the economic needs of local people are included in conservation management plans to the benefit of both the people and the reserve (Fig. 8). Such compromises, known as integrated conservation–development projects, are increasingly being regarded as one of the best conservation strategies (Wells and Brandon, 1992). Failure to make such compromises, however, may aggravate conflicts between park managers and local people, usually to the detriment of the park; even where local people are in favor of conservation measures, restricting access to resources needed for daily living can create resentments and, occasionally, violent opposition. Poaching, sabotage, and even physical attacks on government employees have occurred in parks such as the Chitwan National Park in Nepal (Nepal and Weber, 1995) and the Maya Biosphere Reserve (Ponciano, 1997). Even where parks are not the issue, government restrictions on hard-pressed resources may stimulate similar reactions. For example, threats and attacks against Forest Service and Parks Service employees are becoming increasingly common in the western United States, where ranchers and loggers believe that their livelihood is threatened by government regulations regarding public lands.

Local people can manage the environment to maintain biological diversity, as shown by the traditional agroecosystems and forests of the Huastec Indians of northeastern Mexico (Alcorn, 1984). In addition to their permanent agricultural fields and swidden agriculture, the Huastec maintain managed forests (known as *te'lom*) on slopes, along watercourses, and in other areas that are either fragile or unsuitable for intensive agriculture. These forests contain more than 300 species of plants from which the people obtain food, wood, and other products. Species composition in the forest is altered in favor of useful species by planting and periodic selective weeding. These forest resources provide Huastec families with the means to survive the failure of their cultivated crops. Comparable examples of intensively managed village forests exist in traditional societies throughout the world (Oldfield and Alcorn, 1991; Redford and Padoch, 1992; Nepstad and Schwartzman, 1992).

Conservation biologists need to work with charitable and humanitarian organizations to assist local people in organizing and developing sustainable economic activities that do not damage biological diversity. Integrated conservation–development projects represent one such approach. Foreign assistance programs need to be carefully planned to help alleviate rural poverty rather than primarily benefiting urban elites, as they often currently do. Because the funds of conservationists are limited and problems are urgent, conservation efforts should be concentrated on people who are affecting areas of major biological importance. Conservation biologists and organizations are increasing their participation in programs promoting smaller families. Reducing human population growth should be closely linked to efforts aimed at improving economic opportunities and halting environmental degradation (Dasgupta, 1995). Laws need to be changed so that people can obtain titles and leases to harvest selected trees and use rangelands only as long as the health of the biological community is maintained.

Links between Industry and Destruction of Biodiversity

Although the subsistence activities of local people in the developing world have an important impact on habitats, a major cause of the destruction of biological diversity is the unequal use of natural resources worldwide. People in industrialized countries (and the wealthy minority in the developing countries) consume a disproportionate share of the world's energy, minerals, wood products, and food (Parikh and Parikh, 1991; WRI/IUCN/UNEP, 1992). Each year, the average U.S. citizen uses 43 times more petroleum products, 34 times more aluminum, and 58 times more phosphate fertilizer than the average citizen of India (WRI/UNEP/UNDP, 1993). This excessive consumption of resources is not sustainable in the long run. If this pattern is adopted by the expanding middle class in the developing world, it will cause massive environmental disruption. The affluent citizens of developed countries must confront their excessive consumption of resources and reevaluate their lifestyles as they make suggestions and offer aid to curb population growth and protect biological diversity in the developing world (Fig. 9).

In much of the world today, resources are exploited as rapidly as possible. If a market exists for a product, local people will search their environment to find and sell it. Whether people are poor and hungry or rich and greedy, they will use whatever methods are available to secure that product. Sometimes traditional groups will even decide to sell the rights to a resource, such as a forest or mining area, in order to use the money to buy desired goods. In rural areas the traditional controls that regulated the extraction of natural products have generally weakened; in many areas into which there has been substantial human migration, or where civil unrest and war have occurred, such controls may no longer exist at all. In countries beset with civil conflict, such as Somalia, the former Yugoslavia, Zaire, and Rwanda, there has been a proliferation of firearms among rural people and a breakdown of food distribution networks. In such situations the resources of the natural en-

FIGURE 9 The wealthy, developed countries often criticize the poorer, developing nations for a lack of sound environmental policies but seem unwilling to acknowledge that their own excessive consumption of resources is a major part of the problem (cartoon by S. Willis; reproduced with permission from Primack, 1998).

vironment are available to whoever can exploit them. The most efficient hunter can kill the most animals, sell the most meat, and make the most money for himself and his family. Animals are also sometimes killed for target practice or simply out of spite toward the government.

Overexploitation of resources often occurs rapidly when a commercial market develops for a previously unexploited or locally used species. One of the most pervasive examples of this is the international trade in furs, in which species such as the chinchilla (*Chinchilla* spp.), vicuña (*Vicugna vicugna*), giant otter (*Pteronura brasiliensis*), and numerous cat species have been reduced to low numbers because of hunting for furs. Overharvesting of butterflies by insect collectors, that of orchids, cacti, and other plants by horticulturists, that of marine mollusks by shell collectors, and that of tropical fish for aquarium hobbyists are further examples in which whole biological communities have been targeted to supply an enormous international demand. It has been estimated that 150–300 million tropical fish are sold worldwide for the aquarium market, with many times that number killed during collection or shipping. The legal and illegal trade in wildlife is responsible for the decline of many additional species. The world market in wildlife is surprisingly large: an estimated $5 billion per year, with annual trade involving approximately 40,000 primates, 4 million live birds, and 10 million reptile skins (Table 1). Major exporters are primarily in the developing world, often in the tropics, and most major importers are developed countries, including Canada, China, the European Economic Community, Hong Kong, Japan, Singapore, Taiwan, and the United States (Hemley, 1994).

Billions of dollars are also involved in the illegal trade in a black market linking poor local people, corrupt customs officials, rogue dealers who know they are breaking the law, and wealthy buyers who do not question the sources from which they buy. This trade has many of the same characteristics, the same practices, and sometimes even the same players as the illegal trade in drugs and weapons.

The pattern of overexploitation of plants and animals in many cases is distressingly similar. A resource is identified, a commercial market is developed for that resource, and the local human populace mobilizes to extract and sell the resource. The resource is extracted so thoroughly that it becomes rare or even extinct, and the market identifies another species or another region to exploit in place of the first. Commercial fishing fits this pattern well, with the industry working one species after another to the point of diminishing returns. Commercial forestry companies often behave similarly, with loggers extracting less desirable tree species or trees of smaller size in successive cutting cycles until there is little marketable timber left in the forest. As a result of timber extraction based only on a short-term perspective, almost all West Indies mahogany (*Swientenia mahoganii*) has been removed from the Caribbean islands, and the extensive cedar forests of Lebanon, which previously covered 500,000 ha, have now been reduced to a few isolated fragments. In Indonesia, the Borneo ironwood tree (*Eusideroxylon zwageri*) has always been highly valued for its durable timber and its extraction carefully regulated by village customs. However, traditional sustainable patterns of extraction are giving way to a government-sanctioned industry that is rapidly removing all the available trees. Also, in the process of harvesting a few of the most valuable trees, careless logging operations using mechanized equipment may destroy most of the surrounding trees.

Governments and industries often claim that they can avoid the overharvesting of wild species by applying mod-

TABLE I

Major Targeted Groups of Worldwide Trade in Wildlife[a]

SPECIES	NUMBER TRADED EACH YEAR	COMMENTS
Primates	25,000–30,000	Mostly for biomedical research; also for pets, circuses, zoos, and private collectors
Birds	2–5 million	Mainly perching birds, but also legal and illegal trade in large parrots
Reptiles	2–3 million	Also 10–15 million raw skins and 50 million manufactured articles, mainly from the wild but increasingly from farms
Ornamental fish	500–600 million	Most saltwater tropical fish come from the wild, sometimes using illegal methods which damage other wildlife and the surrounding coral reef
Coral reef	1000–2000 tons	Reefs are destructively mined to provide aquarium decor and coral jewelry
Orchids	9–10 million	Approximately 10% of the international trade comes from the wild, sometimes deliberately mislabeled to avoid regulations
Cacti	7–8 million	Approximately 15% of traded cacti come from the wild, with smuggling a major problem

[a] Worldwide trade in wildlife is valued at more than $10 billion per year, not including timber and edible fish (adapted from Hemley, 1994).

ern scientific management. As part of this approach, an extensive body of literature has developed in wildlife management, fisheries, and forestry to describe the maximum sustainable yield — the greatest amount of a resource that can be harvested each year and replaced by population growth without causing harm to the population. Calculations using the population growth rate and the carrying capacity (the largest population that a given area can support) are used to estimate the maximum sustainable yield, which typically occurs when the population size is at half the carrying capacity. In highly controlled situations in which the resource can be quantified, such as plantations of timber trees, it may be possible to achieve maximum sustainable yield. However, in many real-world situations, harvesting a species at the theoretical maximum sustainable yield is not possible, and attempts to do so lead to an abrupt species decline.

In order to satisfy local business interests and protect jobs, governments often set harvesting levels too high, resulting in damage to the resource base. Species that migrate across national boundaries and through international waters are particularly difficult to harvest sustainably due to the problems of coordinating international agreements and monitoring compliance. Illegal harvesting may also result in additional resource removal not seen in official records, and a considerable proportion of the remaining stock may be damaged during the harvesting operation. There is an additional difficulty if harvest levels are kept fairly constant even though the resource base fluctuates; a normal harvest of a fish species during a year when fish stocks are low due to unusual weather conditions may severely injure or destroy the species. However, governments are beginning to recognize the need for drastic measures to prevent complete destruction of species. A recent example of the lengths to which some governments will go occurred in Canada. During the 1980s, the Canadian fishing fleet harvested large amounts of cod off Newfoundland, despite declines in the species' population. As a result, cod stocks dropped to 1% of their original numbers, and the government was forced to close the fishery in 1992, eliminating 35,000 jobs. The magnitude of this crisis was underscored by an unusual episode of international conflict: A Spanish fishing boat, searching for commercial fish in international waters off the coast of Canada, was accosted, boarded, and confiscated by the Canadian coast guard. The Spanish labeled this incident an act of piracy because the boat had been captured outside the limits of Canadian waters, but the Canadians retorted that, given the hardship their fishing industry had suffered because of the crash in fish populations, they were justified in using any means necessary to protect the fisheries from continued overexploitation. Not surprisingly, this position was supported by the United States, which had also been forced to curtail fishing as commercial species declined. Such examples clearly demonstrate that models of maximum sustainable yield are often inappropriate and invalid for the real world.

In many cases, the factors causing habitat destruction are the large industrial and commercial activities associated with a global economy, such as mining, cattle ranching, commercial fishing, forestry, plantation agriculture, manufacturing, and dam construction, initiated with the goal of making a profit. Many of these projects are funded by national governments and international development banks and are touted as providing jobs, commodities, and tax revenues. Others are initiated and funded by large multinational corporations. However, this exploitation of natural resources is often neither efficient nor cost-effective because the emphasis in these industries is on short-term gains, often at the expense of the long-term sustainability of the natural resources and generally with little regard for the local people who depend on the resources. Timber companies that lease forests, mining companies that lease the rights to mineral resources, and ranchers that rent rangeland from the government often damage the land and reduce its productive capacity in pursuit of short-term profits. In so doing, they not only leave behind a degraded ecosystem that may be unfit for recolonization by many primary forest species but also often expand the frontier of human activities by building roads and trails that migrants use to gain access after the industries have moved on.

The destruction of tropical rain forests is frequently caused by demand in industrialized countries for cheap agricultural products, such as rubber, palm oil, cocoa, and beef, and for low-cost wood products (Fig. 10). During the 1980s, Costa Rica had one of the world's highest rates of deforestation as a result of conversion of rain forests into

FIGURE 10 Tropical forests in the Brazilian Amazon are being cut down to make way for cattle ranches. Similar scenes are common in Southeast Asia, where logging by timber companies is followed by shifting cultivation (photo courtesy of T. Stone).

cattle ranches (Downing *et al.,* 1992). Much of the beef produced on these ranches was sold to the United States and other developed countries to produce inexpensive hamburgers. Adverse publicity resulting from this "hamburger connection," followed by consumer boycotts, led major restaurant chains in the United States to stop buying tropical beef from these ranches. Even though deforestation continued in Costa Rica after the boycott, the boycott was important in making people aware of the international connections promoting deforestation.

Another well-documented linkage involves the forests of Thailand and the livestock industry of The Netherlands. The Dutch import large quantities of tapioca (made from cassava), palm-kernel cake, and other tropical products to use as cattle feed. In order to supply this market, farmers in Thailand had increased their cultivation of cassava and other fodder plants from 100,000 ha in 1965 to 1 million ha by the mid-1980s. Fully 25% of the deforestation in northeastern Thailand can be related to the clearing of land by the 500,000 families involved in cassava cultivation. What seems on the surface to be a minor Dutch import to supply a minor domestic industry in fact has major environmental effects (Primack, 1993). A priority for conservation efforts is to help provide the information, programs, and public awareness that will allow the greatest amount of rain forest to persist once the current cycle of destruction ends.

Efforts to preserve biological diversity, however, sometimes come into conflict with actual or perceived human needs. Increasingly, conservation biologists, policy makers, and land managers are recognizing the need for sustainable development — economic development that satisfies both present and future needs for resources and employment while minimizing its impact on biological diversity. The concept of sustainable development has been applied in a variety of ways. As defined by some ecological economists, economic development implies improvements in organization but not necessarily increases in resource consumption and should be clearly distinguished from economic growth — that is, material increases in the amount of resources used. If sustainable development is to be a useful concept in the conservation of biological diversity, it must emphasize improving current development and limiting growth. Investing in national park infrastructure to improve protection of biological diversity and provide revenue opportunities would be an example of movement toward sustainable development along these lines, as would implementation of less destructive logging and fishing practices.

The economic implications of protecting endangered species can be staggering. Many large corporations, and the policy organizations that they fund, have misused the concept of sustainable development to "green-wash" their ongoing industrial activities without any change in practice (Fig. 11). Probusiness groups have been formed that use environmental-sounding rhetoric to argue against the U.S.

"Sir, would you take this latest warning of ecological disaster and pooh-pooh it for me?"

FIGURE 11 Business interests and governments are often unwilling to acknowledge and deal with environmental problems. Citizen activism is sometimes needed to convince them that concern for biological diversity often makes economic as well as ecological sense (drawing by D. Fradon; reproduced with permission from Primack, 1995).

Endangered Species Act, the most important law protecting species in the United States, promoting instead "wise use" of natural resources — in most cases, uses that prioritize human needs for jobs, raw materials, and economic development over the requirements of species or ecosystems. Limitations on logging in sensitive areas such as the Pacific Northwest, strongly advocated by environmental organizations, have been fiercely resisted by business and citizen groups in the region as well as by many politicians. After years of negotiations, legal maneuvering, and political lobbying, a solution to this ongoing controversy still has not been found. Recognition that intact watersheds are needed to provide abundant clean water and to maintain salmon populations and the valuable commercial and sport fisheries they support may eventually tip the balance toward the preservation of these forests.

The Gap between Rich and Poor

The needs of poor people and industries often come into direct conflict, to the detriment of biological communities. In many countries there is extreme inequality in the distribution of wealth, with the majority of the wealth (money, good farmland, livestock, timber resources, etc.) owned by a small percentage of the population. In many cases, "consumption" of forest resources may be accomplished simply by installing (and enforcing) a "No Trespassing" sign around a large tract of forest, designating it a protected area and excluding local inhabitants from hunting or collecting materials for food, medicine, and fuel in this area. Such actions lead to a phenomenon loosely classified as the *crisis of the commons:* With no land or resources of their own, rural people are forced to exhaust the few remaining common lands, destroying biological communities and hunting endangered species to extinction (Kummer and Turner, 1994). When the No Trespassing sign belongs to a forest reserve or park, open conflict may develop between conservationists and local people. Often, however, the land belongs to private, wealthy landowners, and the forest is quickly converted to pasture or plantations. An equally common pattern in many countries of the developing world is for local farmers to be forced off their land by large landowners and business interests, often backed by the government, the police, and the army. These local farmers often have no choice but to move to remote, undeveloped areas and attempt to eke out a living through shifting cultivation. As noted earlier, shifting cultivation has little impact where human population density is low, but in these cases the farmers are often confined to an area too small to sustain them without damaging the environment. Political instability, lawlessness, and war also force farmers off their land and into remote, undeveloped areas where they feel more safe. Rather than being called "shifting cultivators," however, these newly arrived people would be more accurately described as "shifted cultivators" in order to distinguish them from traditional farmers who have long inhabited rain forest areas. Such "shifted" farmers may lack familiarity with practices, such as terracing or mulching, that could improve soil and nutrient retention. They also may be unwilling or unable to spend the time and money necessary to develop more sustainable forms of agriculture on land that they do not own and may not occupy for very long.

International Markets and Biological Diversity

Biological diversity is concentrated in the tropical countries of the developing world, most of which are relatively poor and are experiencing rapid rates of population growth, development, and habitat destruction. The developed countries of the world (the United States, Europe, Japan, Canada, etc.) require the biological diversity of the tropics to supply genetic material and natural products for agriculture and industry. The general public of the developed world is also anxious to preserve the unusual species, such as elephants, lions, and birds, which are so interesting to see in zoos and nature reserves, read about, and watch on television. Developing countries are often willing to preserve biological diversity, but they may be unable to pay for the habitat preservation, research, and management required for the task. How can countries work together to preserve biological diversity?

Protection of biological diversity is a topic that must be addressed at multiple levels of government. Although the major control mechanisms that currently exist in the world are based within individual countries, agreements at international levels increasingly are being used to protect species and habitats. International cooperation is an absolute requirement for several reasons. First, species migrate across international borders. Conservation efforts must protect species at all points in their ranges; such efforts in one country will be ineffective if critical habitats are destroyed in a second country to which an animal migrates. For example, efforts to protect migratory bird species in northern Europe will not work if the birds' overwintering habitat in Africa is destroyed. Efforts to protect whales in U.S. coastal waters will not be effective if these species are killed or harmed in international waters. Species are particularly vulnerable when they are migrating because they may be more conspicuous, more tired, or more desperately in need of food and water.

Second, international trade in biological products is commonplace. A strong demand for a product in a wealthy country can result in the overexploitation of the species by a poor country to supply this demand. When wealthy people are willing to pay high prices for exotic pets and luxury items, such as tiger bones, rhino horn, and bear gall bladders, poachers or poor, desperate people will take or kill even the very last animal to obtain this income. To prevent overexploitation, education, economic alternatives for the people involved, and control and management of the trade are required at both the points of export and import.

Third, the benefits of biological diversity are of international importance. The community of nations is helped by the species and varieties that can be used in agriculture, medicine, and industry; the ecosystems that help regulate climate; and the national parks and biosphere reserves of international scientific and tourist value. Wealthy countries of the temperate zone that benefit from tropical biological diversity need to be willing to help the less wealthy countries of the world that preserve it.

Finally, many problems of environmental pollution that threaten ecosystems are international in scope and require international cooperation. Such threats include overfishing and hunting; atmospheric pollution and acid rain; pollution of lakes, rivers, and oceans; global climate change; and ozone depletion. Consider a river such as the Danube, which flows through Germany, Austria, Slovakia, Hungary, the former Yugoslavia, Bulgaria, Rumania, and Ukraine before emptying into the Black Sea, itself an international body of water. Only countries working together can solve these problems.

Foreign Debt and Foreign Aid

Governments in the developing world sometimes justify their development policies, such as clear-cutting of forests and the lack of pollution controls on industry, as necessary actions intended to raise revenue to pay off their foreign debts. It is an unfortunate paradox of the international banking system that poor people in the developing world are driven deeper into poverty to pay the banks of wealthier countries, and that the biological riches of the tropical world are destroyed to pay national debts rather than used to supply the needs of their inhabitants. A better goal for foreign aid would be to provide resources — expertise as well as money — to ward off social unrest, prevent epidemic disease, protect biological diversity, and address other problems stemming from the increasing poverty.

Increasingly, groups in the developed countries are realizing that if they want to preserve biological diversity in species-rich but cash-poor countries, they cannot simply provide advice: A financial commitment is also required. Institutions within the United States represent some of the largest sources of financial assistance. The aid these organizations provide is substantial: In 1991, a total of 1410 conservation projects receiving aid from U.S. institutions were identified in 102 developing countries, accounting for a total investment of $105 million (Abramovitz, 1994). The predominant sources of funds were U.S. government agencies ($70 million), such as the Agency for International Development and the National Science Foundation; charitable foundations ($20 million), such as the Mellon Foundation, W. Alton Jones Foundation, and Pew Charitable Trusts;

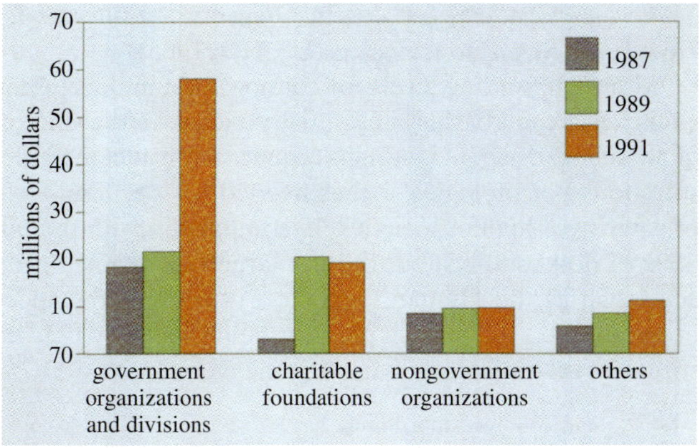

FIGURE 12 Funding from U.S. sources for biodiversity research and conservation efforts in developing countries (by type of funding organization): government organizations and divisions; charitable foundations; nongovernment organizations; and miscellaneous institutions, such as universities, zoos, and museums (adapted from Abramovitz, 1994).

and nongovernment organizations ($10 million), such as the World Wildlife Fund, Conservation International, and the Nature Conservancy (Fig. 12). Conservation investments by large foundations increased sevenfold between 1987 and 1991, and government funding tripled during this period, demonstrating that tropical conservation has clearly been targeted as a funding priority. These projects were carried out by government agencies (e.g., U.S. Fish and Wildlife Service and the Peace Corps), nongovernment organizations (e.g., the Wildlife Conservation Society and the World Wildlife Fund), museums (e.g., the Field Museum of Natural History and the Smithsonian Institution), botanical gardens (e.g., the Missouri Botanical Garden and the New York Botanical Garden), zoos, and universities.

Less than 1% of the $11 billion spent each year by the U.S. government in foreign assistance funds biodiversity protection. Within the United States, the leading government agency providing economic and humanitarian aid is the U.S. Agency for International Development (USAID), with a budget of more than $5 billion per year. With programs and missions in about 60 developing countries, USAID is a dominant presence in foreign assistance activities. To reflect growing environmental concerns, the U.S. Congress amended the Foreign Assistance Act in 1983 to make the conservation of biological diversity and the protection of endangered species an integral part of USAID foreign assistance activities. The amendment requires USAID to spend money each year on biodiversity projects, particularly supporting local conservation efforts, education, and ecosystem protection. To provide guidance for this effort, a panel of government and nongovernment representatives was convened to formulate an official U.S. policy on biodiversity that was released as the *U.S. Strategy on the Conservation of Biological Diversity: An Interagency Task Force Report to Congress* (USAID, 1985).

Although funding levels for conservation in developing countries are increasing substantially each year, the amount of money being spent is still inadequate to protect the great storehouse of biological riches needed for the long-term prosperity of human societies. In comparison with the billions of dollars allocated to other large U.S. science projects, such as the Human Genome Project and the space program, the $105 million per year being spent by U.S. institutions on biological diversity is meager indeed.

International Development Banks and Ecosystem Damage

Ironically, current methods of funding development frequently encourage habitat destruction. Tropical deforestation has been greatly accelerated by poorly conceived, large-scale projects financed by the major multilateral development banks (MDBs): the World Bank, which lends to all regions of the globe, and the regional MDBs, including the Inter-American Development Bank, the Asian Development Bank, and the African Development Bank. The MDBs loan more than $25 billion per year to 151 countries to finance development projects (Rich, 1990). The impact of the MDBs is actually even greater since their funding is often linked to financing from donor countries, private banks, and other government agencies; the $25 billion funding of the MDBs attracts about another $50 billion in loans, which makes the MDBs major players in the developing world. Although the goal of the MDBs ostensibly is economic development, the effect of many of the projects supported is to exploit natural resources to create exports for international markets. In many cases, these MDB-funded projects have resulted in the destruction of ecosystems over a wide area, involving soil erosion, flooding, water pollution, health problems and loss of income for local people, and loss of biological diversity.

Another source of international funding, the International Monetary Fund (IMF), also can profoundly affect biological diversity through its policies of economic reform. Created in the wake of World War II, the IMF is a "bank of last resort" for troubled economies (Lacayo, 1997). When countries find themselves facing economic collapse, as occurred in Mexico in 1994 and in Thailand, Indonesia, and South Korea in 1997, the IMF can offer a package of loans predicated on economic reform. The idea behind such bailouts is to prevent the kinds of severe economic instability that ultimately led to the rise of fascism in Europe; indeed, such packages generally accomplish their goals largely because the countries asking for funds must submit to stringent rules for restructuring their economies. Unfortunately, the austerity measures imposed by the IMF can seriously impact biodiversity because governments are required by the bank to simultaneously increase taxes and cut back "luxury" programs such as social spending and conservation programs. These measures create increased hardships on the poor at the same time that they reduce efforts to maintain wild habitat. Moreover, one of the ways that countries free themselves of the IMF's control is by establishing self-sufficient industries, some of which may rely on increased exploitation of natural resources.

International funding agencies have begun to recognize their roles in damaging global biodiversity. In recent packages for Argentina and Thailand, for example, the IMF included provisions to support health care and education and reduce unemployment (Lacayo, 1997) — measures that may alleviate social factors promoting habitat loss. The World Bank also has begun to more closely examine its role in damaging ecosystems: A study of its own projects found that 37% could not be considered satisfactory if environmental criteria were included in the project evaluation. For example, major projects financed by MDBs are dam and irrigation systems that provide water to agricultural activities and generate hydroelectric power. Although dams provide important benefits, these projects destroy free, wild rivers and often damage large aquatic ecosystems by changing water depth and current patterns, increasing sedimentation, and creating barriers to dispersal. As a result of these changes, many species are no longer able to survive in the altered environment. Ironically, the destruction of biological diversity may be a key to the failures of some of the large international dam projects that have devastated tropical rain forests. The loss of plant cover on the slopes above water projects often results in soil erosion and siltation, with resulting loss of efficiency, higher maintenance costs, and damage to irrigation systems and dams. Protecting the forests and other natural vegetation in the watersheds is now widely recognized as an important and relatively inexpensive way to ensure the efficiency and longevity of these water projects while preserving large areas of natural habitat. In one study of irrigation projects in Indonesia, it was found that the cost of protecting watersheds ranged from only 1 to 10% of the total cost of the project, in contrast to an estimated 30–40% decrease in efficiency due to siltation if the forests were not protected. One of the most successful examples of an effective environmental investment was the $1.2 million irrigation sector loan by the World Bank to assist in the development and protection of Dumoga-Bone National Park in northern Sulawesi, Indonesia (Wells and Brandon, 1992). A 278,700 ha primary rain forest, which included the catchment area on the slopes above a $60 million irrigation project financed by the World Bank, was converted into a national park (Fig. 13). In this case, the World Bank was able to protect its original investment us-

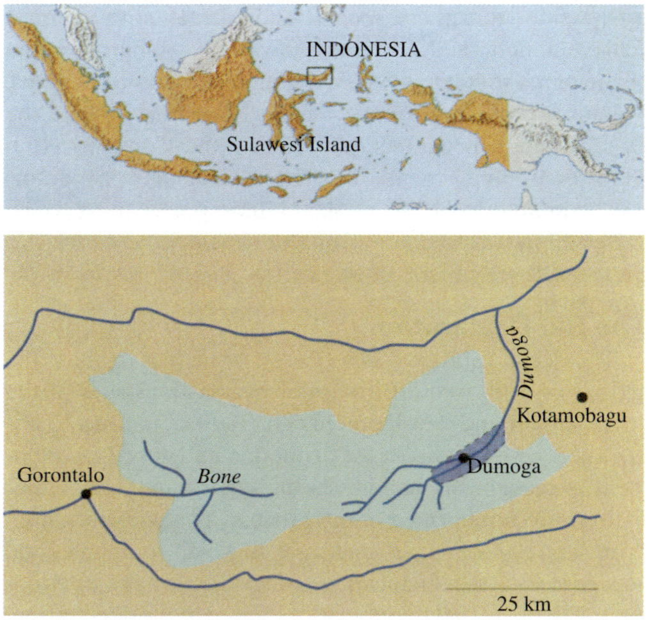

FIGURE 13 (Bottom) The Dumoga-Bone National Park (in green) on the northern arm of Sulawesi Island, Indonesia (top inset). The park protects the watershed above the Dumoga Irrigation Project (in blue) (adapted from Wells and Brandon, 1992).

ing less than 2% of the project's cost and create a significant new national park in the process.

Changing the Funding Process

Why are big projects approved? If many of the large international development projects are so economically unsuccessful and environmentally harmful, why do host countries want them and why do the MDBs agree to finance them? Projects are often funded because economists make overly optimistic predictions on production schedules and prices of commodities and minimize potential problems. Surveys and pilot studies are not undertaken or their results are minimized; comparable projects elsewhere are not evaluated. The environmental costs of projects are often ignored or minimized since these variables are considered external to the economic analysis. However, an accurate model of a project would include all of its costs and benefits, including the effects of soil erosion, the loss of biological diversity, the impact of water pollution on the health and diet of local people, and the loss of income associated with the destruction of renewable resource.

Host governments often want large projects to proceed regardless of unfavorable reviews since the projects provide temporary jobs, economic prosperity, and, as a result, some release from social tensions for the duration of the project. Local business leaders, especially those with close links to the government, endorse the projects because they can make large profits on project contracts. Industrialized countries which support the banks encourage these loans to stabilize governments in the host countries that are friendly to their interests but which may lack popular support.

How could the MDBs operate more responsibly? First, they should stop environmentally destructive projects. This step would require the banks to develop economic cost–benefit models for development projects that include the environmental and ecological effects of projects. Also, banks need to encourage open public discussions from all groups in a country before projects are implemented. In particular, the banks should allow public examination, independent evaluations, and discussions of environmental impact reports before a project is approved for funding.

Because the MDBs are primarily financed by the governments of the major developed countries, such as the United States, Japan, Germany, Italy, the United Kingdom, France, and Canada, the policies of the MDBs can be scrutinized by the elected representatives of the MDB member countries, the national media, and conservation organizations. In particular, because some of the ill-conceived projects of the World Bank have been publicly criticized in the U.S. Congress and elsewhere, the World Bank has reacted by requiring new projects to be more environmentally responsible, hiring ecological and environmental staff to review new and ongoing projects, and adopting a management policy called "new environmentalism" that recognizes the linkages between economic development and environmental sustainability (Fig. 14). This policy recognizes that there is a high environmental cost of inappropriate economic policies, reducing poverty is often crucial to protecting the environment, and economic growth must incorporate environmental values.

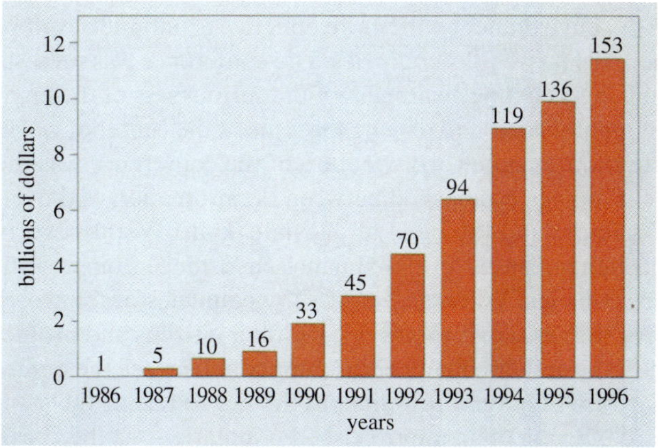

FIGURE 14 World Bank financing for environmental conservation has increased dramatically during the past decade. The numbers over the bars represent the numbers of projects financed (adapted from Primack, 1998).

Global Efforts to Protect Biological Diversity

The need for increased funding for biodiversity remains great at local, national, and international levels. Currently, about $4 billion is spent each year in the budgets of protected areas, but $17 billion would be required for adequately managing the world's biological diversity (Heywood, 1995). Although $17 billion sounds like an enormous amount of money, it is trivial compared to the $245 billion spent each year on agricultural subsidies and the $1 trillion spent on military defense. Certainly, the world's priorities could be modestly adjusted, giving more resources to the protection of biological diversity. Instead of the United States rushing forward in a race against itself to produce the next generation of fighter aircraft, missiles, and other weapons systems, what about spending what it takes to protect biological diversity? Instead of the world's affluent consumers buying the latest electronic gadgets to replace the ones that still work, what about contributing more money to conservation organizations and causes?

Protecting the environment is ultimately a global task. Despite continued destruction of key resources and ecosystems, significant strides have been made in adopting a global approach to sound environmental management. Two recent markers of progress are the Earth Summit, held for 12 days in June 1992 in Rio de Janeiro, Brazil, and the Climate Change Summit, held in Kyoto, Japan, in December 1997. The first of these meetings, the UN Conference on Environment and Development (also known unofficially as Eco '92 and the Rio Summit), included representatives from 178 countries, with more than 100 heads of state, plus leaders of the UN and major nongovernment and conservation organizations in attendance. The purpose of the conference was to discuss ways of combining increased protection of the environment with more effective economic development in less wealthy countries. The conference was successful in heightening awareness of the seriousness of the environmental crisis and placing the issue at the center of world attention. A noteworthy feature of the conference was the clear linkage established between the protection of the environment and the need to alleviate Third World poverty through increased levels of financial assistance from wealthier countries. Whereas the wealthy countries of the world have the resources to provide for their citizens and protect the environment, many of the poor countries see the immediate use of natural resources as the key to raising the standard of living of an impoverished population. At the Earth Summit, the wealthy countries were collectively told that they would have to assist the poor countries of the world to protect the global environment and tropical biodiversity.

The conference participants discussed and eventually signed five major documents and initiated many new projects. Aside from these specific achievements, the central achievement of the Earth Summit was the willingness of the participants to keep working together on long-term goals. The most contentious issue was deciding how to fund the Earth Summit programs, particularly the Convention on Biodiversity and Agenda 21. The cost of these programs was estimated to be about $600 billion per year, of which $125 billion was to come from the developed countries as overseas development assistance.

The Rio Declaration

The declaration provides general principles to guide the actions of both wealthy and poor nations on issues of the environment and development. The right of nations to utilize their own resources for economic and social development is recognized, as long as the environments of other nations are not harmed in the process. The declaration affirms the "polluter pays" principle, in which companies and governments take financial responsibility for the environmental damage that they cause. As stated in the declaration, "States shall cooperate in a spirit of global partnership to conserve, protect, and restore the health and integrity of the earth's ecosystem. In view of the different contributions to global environmental degradation, states have common, but differentiated responsibilities."

Convention on Biodiversity

The Convention on Biodiversity has three objectives: protecting biological diversity, using it sustainably, and sharing the benefits of new products made with wild and domestic species. Although the first two objectives are straightforward, the last point recognizes that developing countries should receive fair compensation for the use made of species collected within their borders. Indigenous people should also share in the benefits derived from biological diversity, particularly when they have contributed their own local knowledge of species. In the past, industrialized countries have developed new crops, medicines, and other biotechnology products based on tropical species and have not returned any of the technology, new products, or profits to the developing country in which the wild species was originally found. Developing provisions and compromises in international intellectual property rights laws to fairly share the financial benefits of biological diversity is proving to be a major challenge, with some progress and some roadblocks still to overcome. The treaty, ratified by almost 170 nations, affirms that countries have certain rights over species occurring within their borders and responsibilities to use them sustainably. The U.S. Congress has still not ratified the treaty because of perceived potential restrictions on its enormous biotechnology industry.

Agenda 21

This 800-page document shows the linkages between the environment and other issues that are often considered separately, such as child welfare, poverty, women's issues, technology transfer, and unequal divisions of wealth. Plans of action are described to address problems of the atmosphere, land degradation and desertification, mountain development, agriculture and rural development, deforestation, aquatic environments, and pollution. Financial, institutional, technological, and legal mechanisms for implementing these action plans are also described. The United States has responded to Agenda 21 by issuing a report that outlines how the goal of sustainability could be applied to the United States; the report is titled "America: A New Consensus for Prosperity, Opportunity, and a Healthy Environment for the Future."

Although all three of these documents clearly state the link between environmental health and human well-being, developed nations unfortunately did not fully accept the sacrifices that would be required to promote these values. Because existing levels of overseas development assistance amount to $60 billion per year for all activities, implementing these international conventions would require a tripling of the current foreign aid commitment. This increase in funding was not agreed to by the developed countries, struggling to balance budgets and pay for domestic programs. As an alternative proposal, the Group of 77, a group of developing countries, suggested that industrialized countries increase their level of foreign assistance to 0.7% of their gross national product by the Year 2000, roughly doubling their level of assistance. Although the richer countries agreed in principle to this figure, no schedule was set to meet the target date. Of 21 donor countries, only 5 countries have reached the specified target levels in terms of percentage gross national product donated: Norway (1.4%), Denmark (0.96%), Sweden (0.92%), The Netherlands (0.88%), and Finland (0.76%), with the United States notably lagging behind at 0.2%.

The frustration of the developing countries was eloquently summarized by Dr. Mahathir bin Mohamed, Prime Minister of Malaysia:

> The poor countries of the world have been told to preserve their forests and other genetic resources on the off chance that at some future date something is discovered which might prove useful to humanity. But now they are told that the rich will not agree to compensate the poor for their sacrifices, arguing that the diversity of genes stored and safeguarded by the poor are of no value until the rich, through their superior intelligence, release the potential within.

One major new source of funds for conservation and environmental activities does exist: the Global Environment Facility (GEF), created in 1991 by the World Bank along with the UN Development Program and the UN Environmental Program. Most of the funds for the program were authorized and the projects approved by the time of the Earth Summit in June 1992. The GEF was established as a 3-year pilot program with a budget of $862 million to be used for funding projects relating to biodiversity, international waters, global warming, and ozone depletion.

The latter two topics were of principal concern in the recent Climate Change Summit in Kyoto. Delegates from 170 nations met to work out a treaty to limit the emissions of greenhouse gases that are damaging the ozone layer and promoting global warming. Industrial nations, particularly the United States, lead the world in the quantity of emissions of CO_2 and other gases; reducing these emissions would require strict, expensive measures to reduce waste from automobiles and industrial processes. When the European Union proposed measures that would return emissions to 85% of 1990 levels, the United States initially balked, fearing that such restrictions would hamper its economic prosperity and pointing out that developing nations would not be required to limit pollutants under the European Union plan. The conference seemed deadlocked until U.S. Vice President Albert Gore — a well-known supporter of environmental causes — unexpectedly attended the conference and announced that the U.S. delegation had been ordered to negotiate with greater flexibility, paving the way for an agreement. Although the willingness of the United States to negotiate these issues is an important breakthrough, it must be noted that there are many in the U.S. Congress who will not accept the agreement either because they are unwilling to sacrifice current prosperity for a nebulous future benefit or because they do not believe that such reductions are truly necessary. Even among scientists, a vocal minority continues to argue that global warming is nonexistent or is a product of natural forces, not human activities. As long as such opposition and doubt remain, it will be difficult or impossible to convince the leaders of many industrial nations that action is necessary.

Considering that the Earth Summit and Climate Change Summit were the moments when the world's leaders and public attention were focused on environmental issues and there was maximum pressure to get something accomplished, the inability of the major industrial countries to perceive the depth of the problems is a severe disappointment. Raising additional funds for the implementation of the Convention on Biodiversity and Agenda 21 will be an ongoing process. Also, decisions on how this money is to be allocated have not been satisfactorily resolved, with a major division remaining between developing and developed countries regarding control of the programs.

Conclusions

Developed countries of the world often place a greater emphasis on the preservation of biological diversity than do the poorer Third World countries that have the most biological diversity. Developed countries, conservation organizations, zoos, and botanical gardens should provide financial support to developing countries to establish and maintain national parks and other protected areas. This seems fair and reasonable when the developed countries have the funds to pay for these parks and are able to utilize the protected biological resources in their agriculture, industry, research programs, zoos, aquaria, botanical gardens, and educational systems. Related economic and social problems must be resolved at the same time, particularly those relating to poverty and war. Reducing or forgiving foreign debt payments, debt-for-nature swaps, and environmental trust funds may be additional mechanisms to achieve these goals. Individual citizens can donate money and participate in organizations that further advance these conservation goals.

One of the most serious challenges facing conservation biology is reconciling the needs of local people and the need to preserve biological diversity. How can poor people, particularly in the developing world but also in rural areas of developed countries, be convinced to respect nature reserves and biological diversity when they are desperate to obtain the food, wood, and other natural products that they need for their daily survival? Park managers need to find compromises, such as the concepts of biosphere reserves and integrated conservation–development projects, that allow people to obtain the natural resources that they need to support their families and yet not damage the natural communities of the park. At national and international levels, the world's resources must be distributed more fairly to end the inequalities which exist today. Effective programs must be established to stabilize the world's human population. At the same time, the destruction of natural resources by industries must be halted so that the short-term quest for profits does not lead to a long-term ecological catastrophe. Management strategies to preserve biological diversity also need to be developed for the 95% of the terrestrial environment that will remain outside of protected areas as well as for the vast, largely unexplored marine environment.

Citizens and governments of developed countries must also become aware that they bear a direct responsibility for the destruction of biological diversity through their overconsumption of the world's resources. People need to realize how their own actions and lifestyle on the local level can have an influence far beyond their immediate community. Widespread changes in lifestyle and alternative markets for environmentally friendly and recycled products can have a positive effect on the environment. Equally important may be a simple change in perspective. Consider, for example, the euphemisms used to distinguish the wealthy, industrial nations from the less wealthy, nonindustrial nations: Industrial nations of North America and Europe were labeled "developed" or "First World" countries, whereas nonindustrial nations of Africa, Asia, and Latin America were referred to as "underdeveloped," "developing," or "Third World." "Underdeveloped" or "developing" implies that nonindustrial nations are somehow falling behind the "proper" course of human society, represented by the "developed" nations, which are implicitly viewed as having attained the benchmark level of human use of natural resources. An alternative perspective might consider the possibility that the so-called developed nations are, in fact, overdeveloped nations — that the amount of mechanization, consumption, and sheer waste in such countries has gone well above the appropriate levels, no longer merely meeting the needs of normal day-to-day living but rising to the point of self-indulgent decadence. Although this statement is true within societies as well as between them — certainly in the United States, which typifies the overdeveloped nation, there are people that live in moderate to desperate poverty in stark contrast to the excessive and luxuriant lifestyles of many wealthy people — as a whole, a majority of the people considered poor by U.S. standards have far greater access to goods and services than the poor in most developing nations. If people in the developed nations can realize that many of the things they consume are unnecessary, they may be encouraged to "downsize" their lives and consume less. As Peter Raven, director of the Missouri Botanical Garden, remarked regarding the accelerated loss of biological diversity, "You can think about it on a worldwide basis, and then it becomes discouraging and insoluble, or you can think about it in terms of specific opportunities, seize those opportunities, and reduce the problem to a more manageable size" (quoted in Tangley, 1986).

Acknowledgments

Portions of this article are derived from *Essentials of Conservation Biology* (Primack, 1993) and *A Primer of Conservation Biology* (Primack, 1995). Elizabeth Platt assisted with the editing of the manuscript.

References Cited

Abramovitz, J. N. (1994). *Trends in Biodiversity Investments: U.S.-Based Funding for Research and Conservation in Developing Countries, 1987–1991.* World Resource Institute, Washington, DC.

Alcorn, J. B. (1984). Development policy, forests, and peasant farms: Reflections on Huastec-managed forests' contributions to commercial production and resource conservation. *Econ. Bot.* **38,** 389–406.

Brown, L. R., Renner, M., and Flavin, C. (1997). *Vital Signs 1997: The Environmental Trends That Are Shaping Our Future.* Norton, New York.

Dasgupta, P. S. (1995). Population, poverty and the local environment. *Sci. Am.* **272,** 40–45.

DUFOUR, D. L. (1990). Use of tropical rainforests by native Amazonians. *BioScience* **40,** 652–659.

ELDREDGE, N. (1997). *Dominion.* Univ. of California Press, Los Angeles.

GOMEZ-POMPA, A., and KAUS, A. (1992). Taming the wilderness myth. *BioScience* **42,** 271–279.

HARDIN, G. (1993). *Living within Limits: Ecology, Economics and Population Taboos.* Oxford Univ. Press, New York.

HEMLEY, G. (1994). *International Wildlife Trade: A CITES Sourcebook.* Island Press, Washington, DC.

HEYWOOD, V. H. (1995). *Global Biodiversity Assessment.* Cambridge Univ. Press, Cambridge, UK.

HORWICH, R. H., and LYON, J. (1997). Community development as a conservation strategy: The Community Baboon Sanctuary and the Gales' Point Manatee Reserve. In *Timber, Temples, and Tourists: Conservation and Development in the Maya Forest of Belize, Guatemala, and Mexico* (R. B. Primack, D. B. Bray, H. A. Galletti, and I. Ponciano, Eds.). Island Press, Washington, DC.

JENNINGS, S., and POLUNIN, N. V. C. (1996). Impacts of fishing on tropical reef ecosystems. *Ambio* **25,** 44–49.

JOHANNES, R. E. (1978). Traditional marine conservation methods in Oceania and their demise. *Annu. Rev. Ecol. Syst.* **9,** 49–64.

KUMMER, D. M., and TURNER, B. L., II (1994). The human causes of deforestation in Southeast Asia. *BioScience* **44,** 323.

LACAYO, R. (1997, December 8). IMF to the rescue. *Time.*

LEWIS, D., KAWECHE, G. B., and MWENYA, A. (1990). Wildlife conservation outside protected areas: Lessons from an experiment in Zambia. *Conserv. Biol.* **4,** 171–180.

MCGOODWIN, J. R. (1990). *Crisis in the World's Fisheries: People, Problems, and Politics.* Stanford Univ. Press, Stanford, CA.

MCNEELY, J. A. (Ed.). (1993a). *Protected Areas and Modern Societies: Regional Reviews of Conservation Issues.* IUCN, Gland, Switzerland.

MCNEELY, J. A. (Ed.). (1993b). *Building Partnerships for Conservation.* IUCN, Gland, Switzerland.

MYERS, N. (1984). *The Primary Source: Tropical Forests and Our Future.* Norton, New York.

MYERS, N. (1986). Tropical deforestation and a mega-extinction spasm. In *Conservation Biology: The Science of Scarcity and Diversity* (M. E. Soule, Ed.). Sinauer, Sunderland, MA.

MYERS, N. (1991). Tropical deforestation: The latest situation. *BioScience* **41,** 282.

NEPAL, S. K. and WEBER, K. E. (1995). Prospects for coexistence, wildlife and local people. *Ambio,* 238–245.

NEPSTAD, D. C., and SCHWARTZMAN, S. (Eds.) (1992). *Non-Timber Products from Tropical Forests: Evaluation of a Conservation and Development Strategy.* New York Botanical Gardens, Bronx.

OLDFIELD, M. L., and ALCORN, J. B. (Eds.) (1991). *Biodiversity: Culture, Conservation and Ecodevelopment.* Westview, Boulder, CO.

PARIKH, J., and PARIKH, K. (1991). *Consumption Patterns: The Driving Force of Environmental Stress.* UNCED, Geneva.

PONCIANO, I. (1997). Forestry policy and protected areas in northern Guatemala. In *Timber, Temples, and Tourists: Conservation and Development in the Maya Forest of Belize, Guatemala, and Mexico* (R. B. Primack, D. B. Bray, H. A. Galletti, and I. Ponciano, Eds.). Island Press, Washington, DC.

PRIMACK, R. (1988). Forestry in Fujian Province (People's Republic of China) during the Cultural Revolution. *Arnoldia* **48,** 26–29.

PRIMACK, R. B. (1993). *Essentials of Conservation Biology.* Sinauer, Sunderland, MA.

PRIMACK, R. B. (1995). *A Primer of Conservation Biology.* Sinauer, Sunderland, MA.

PRIMACK. R. B. (1998). *Essentials of Conservation Biology I,* 2nd ed. Sinauer, Sunderland, MA.

REDFORD, K. H., and PADOCH, C. (Eds.) (1992). *Conservation of Neotropical Forests: Working from Traditional Resource Use.* Columbia Univ. Press, New York.

RICH, B. (1990). Multilateral development banks and tropical deforestation. In *Lessons from the Rainforest* (S. Head and R. Heinzmann, Eds.). Sierra Club, San Francisco.

RUDEL, T., and ROPER, J. (1996). Regional patterns and historical trends in tropical deforestation, 1976–1990: A qualitative comparative analysis. *Ambio* **25,** 166–170.

SAYER, J. A., and WHITMORE, T. C. (1991). Tropical moist forests: Destruction and species extinctions. *Biol. Conserv.* **55,** 199–213.

TANGLEY, L. (1986). Saving tropical forests. *BioScience* **36,** 4–15.

WELLS, M., and BRANDON, K. (1992). *People and Parks: Linking Protected Areas Management with Local Communities.* The World Bank/WWF/USAID, Washington, DC.

WORLD CONSERVATION MONITORING CENTER (1992). *Global Biodiversity: Status of the Earth's Living Resources,* compiled by the World Conservation Monitoring Center. Chapman & Hall, London.

WORLD RESOURCES INSTITUTE/UNITED NATIONS ENVIRONMENT PROGRAM/UNITED NATIONS DEVELOPMENT PROGRAM (1992). *World Resources 1992–1993.* Oxford Univ. Press, New York.

WORLD RESOURCES INSTITUTE/UNITED NATIONS ENVIRONMENT PROGRAM/UNITED NATIONS DEVELOPMENT PROGRAM (1994). *World Resources 1994–95: People and the Environment.* Oxford Univ. Press, New York.

General References

NATIONS, J., PRIMACK, R. B., and BRAY, D. (1997). Introduction to the Maya forest. In *Timber, Temples, and Tourists: Conservation and Development in the Maya Forest of Belize, Guatemala, and Mexico* (R. B. Primack, D. B. Bray, H. A. Galletti, and I. Ponciano, Eds.). Island Press, Washington, DC.

SNOOK, L. K. (1997). Sustaining harvests of mahogany (*Swietenia macrophylla* King) from Mexico's Yucatán forests: Past, present and future. In *Timber, Temples, and Tourists: Conservation and Development in the Maya Forest of Belize, Guatemala, and Mexico* (R. B. Primack, D. B. Bray, H. A. Galletti, and I. Ponciano, Eds.). Island Press, Washington, DC.

Bernd von Droste
Mechtild Rössler
Anja von Rosenstiel
UNESCO World Heritage Center
75007 Paris, France

"Environment" is a complex field of action because it involves a wide range of different actors in the scientific, political, and general public domain. We highlight the process of building international institutions with the growing concerns about environmental destruction since 1972. The main focus is the establishment of international institutions and programs with the implementation of international environmental treaties, mainly in the framework of the United Nations system. These institutions have become critical in the implementation of environmental policies and provide effective mechanisms in environmental protection activities on a worldwide scale.

The Environment and International Institutions

The Earth is one but the world is not. We all depend on one biosphere for sustaining our lives. Yet each community, each country, strives for survival and prosperity with little regard for its impact on others.

Our Common Future, the report of the World Commission on Development and Environment (1987)

Introduction

Human impact on natural processes has increased to the extent that an evolutionary responsibility is involved, such as atmospheric change, which is believed to be caused not only by natural factors but also by human activities. The pollution of water and the depletion of the ozone layer have produced global effects that concern humanity as a whole. However, humankind has powerful technologies, institutions, and legal instruments to guide and determine its future to be environmentally sound.

The fact that nongovernmental actors are key players in the field of environment and sustainable development implies that in order to be effective, the United Nations (UN) system increasingly has to work with and influence these actors. In line with a global trend of stronger reliance on markets, trade liberalization, and transboundary investments, in the near future the UN system will likely have to rely more on market forces and economic instruments rather than on traditional regulation tools. Examples include UN involvement in the design and operations of global greenhouse gas emission trading schemes (as foreseen under the Kyoto

TABLE I

List of Abbreviations

ACC	Administrative Committee on Coordination
ACSD	Inter-Agency Committee on Sustainable Development
AGBM	Ad Hoc Group on Berlin Mandate
AG 13	Ad Hoc Group on article 13
AIJ	Activities implemented jointly
CBD	Convention on Biological Diversity
CCC	Convention on Climate Change
CCD	Convention to Combat Desertification
CFCs	Chloroflourocarbons
CHM	Clearinghouse mechanism
CITES	Convention on International Trade in Endangered Species of Wild Fauna and Flora
CMS	Convention on the Conservation of Migratory Species of Wild Animals
COP	Conference of the Parties
CSD	Commission on Sustainable Development
ECOSOC	Economic and Social Council
FAO	Food and Agricultural Organization
GEF	Global Environmental Facility
GEMS	Global Environmental Monitoring System
GNP	Gross national product
IACSD	Inter-Agency Committee on Sustainable Development
IAEA	International Atomic Energy Agency
ICAO	International Civil Aviation Organization
ICCROM	International Center for the Study of the Preservation and Restoration of Cultural Property
ICOMOS	International Council on Monuments and Sites
IFF	Intergovernmental Forum on Forests
ILO	International Labor Organization
IMO	International Maritime Organization
INFOTERRA	International Referral System for Sources of Environmental Information
IOC	International Oceanographic Commission
IOMC	Inter-Organization Program for the Sound Management of Chemicals
IPCC	Intergovernmental Panel on Climate Change
IPF	Intergovernmental Panel on Forests
ITTA	International Tropical Timber Agreement
ITTO	International Tropical Timber Organization
IUCN	International Union for the Conservation of Nature (now the World Conservation Union)
MARPOL	International Convention for the Prevention of Pollution from Ships
MOU	Memorandum of understanding
NGO	Nongovernmental organization
ODA	Official development aid
OECD	Organization for Economic Cooperation and Development
Ramsar Convention	Convention on wetlands of international importance, especially, waterfowl habitat
SBI	Subsidiary Body on Implementation
SBSTA	Subsidiary Body for Scientific and Technological Advise
SBSTTA	Subsidiary Body for Scientific, Technical, and Technological Advice
UNCED	United Nations Conference on Environment and Development
UNCLOS	United Nations Convention on the Law of the Sea
UNDP	United Nations Development Program
UNDPCSD	United Nations Department for Policy Coordination and Sustainable Development
UNEP	United Nations Environmental Program
UNESCO	United Nations Educational, Scientific, and Cultural Organization
UNIDO	United Nations Industrial Development Organization
UNGA	United Nations General Assembly
WCED	World Commission on Environment and Development
WHO	World Health Organization
WMO	World Meteorological Organization
World Heritage Convention	Convention concerning the protection of the world cultural and natural heritage
WRI	World Resources Institute
WWF	World Wildlife Fund

protocol), transferable resource use rights (e.g., fish stocks and ocean mineral resources), benefit transfer schemes (e.g., related to bioprospecting issues under the Convention on Biological Diversity), and debt-for-environment and sustainable development operations. Therefore, it is possible to conceive that innovative institutional UN structures designed for such purposes will be implemented in the near future (Table 1). These considerations fall largely outside the scope of this article.

Environment as a field of action is a complex, cross-sectoral, dynamic, and fragmented process that involves a wide range of different actors and institutions. After a period of emancipation, there are now strong efforts to reintegrate the environment into general international policy making. Thus, most international institutions have "greened" their activities by establishing green programs and subbodies. Therefore, international financial and development institutions are also increasingly involved in implementation of environmental treaties. Nongovernmental actors, including private industry, the scientific community, academic institutions, and environment and development organizations, "are all seen as critical players in the development of an environmental policy" (UNEP/Coord. 7/15, July 4, 1997). In this article, however, we concentrate on the environment in the institutional context of the UN system and the international legal regime it has established (Fig. 1).

The UN has convened many international conferences including the two main international events in this domain, the Stockholm Conference in 1972 and the Rio Conference in 1992, which changed humankind's thinking about the environment. These conferences led to a chain reaction in terms of institutional arrangements from the international to the local level and also to an increase of environmental concern across different disciplines, such as development studies, economics, and sciences. The core values and basic principles that were adopted at these conferences constitute a major guideline for environmental activities. Agenda 21, the action plan adopted at Rio 1992, is an important point of reference for environmental decision makers for how to translate into practice the overall goal of environmentally sound development. Thus, UN activities are in the center of a rapidly evolving environmental policy.

The Way to a Coherent Environmental Policy

A structure of international environmental cooperation has been emerging and developing and may provide a bridge to more inclusive international relationships in the future.

Caldwell (1984, p. 5)

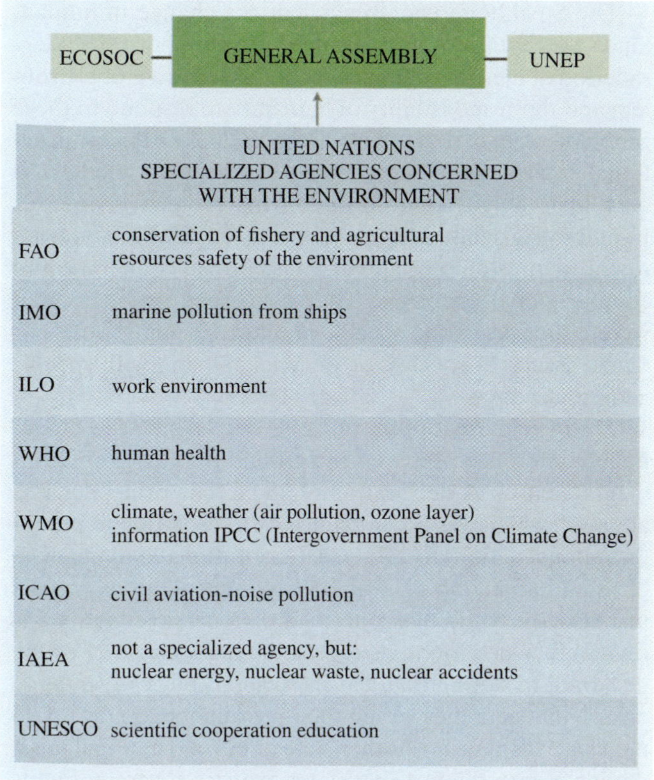

FIGURE 1 United Nations specialized agencies and programs for the management of natural resources. For acronyms, see Table 1.

The United Nations Conference on the Human Environment, Stockholm, 1972

The Stockholm conference was convened by the UN General Assembly as a response to the growing concern of the international community for the protection of the environment. Despite the fact that global changes had been taking place for a long time, their impact on the environment had reached a threshold at the time of the Stockholm conference that made the public aware of these changes. The lack of adequate- environmental management and policy to deal with stress that economic growth puts on the earth's life-support system was increasingly perceived by the international community (in 1968, the Intergovernmental Conference of Experts on the Scientific Basis for Rational Use and Conservation of the Resources of the Biosphere under the auspices of UNESCO had already called on humanity to take up its responsibility) and finally led to the UN Conference on the Human Environment, the first global conference on the environment, held in Stockholm on July 5–16, 1972, which assembled representatives from more than 114 states and a large number of international institutions and organizations and nongovernmental observers.

The Stockholm conference led to a change in humankind's perception of its relation to the biosphere. The degradation of the environment was perceived as a global problem and the responsibility of humanity to respond to these problems was seen as urgent. Despite the fact that international environmental agreements had been adopted as far back as 1902 (the Convention for the Protection of Birds Useful to Agriculture signed in Paris), most of the international treaties did not aim at international cooperation to combat global problems. They concerned problems that were either discipline specific in nature, dealing with particular issues or species, or regional, dealing with specific geographic areas.

Through the adoption of three documents, the Stockholm conference prepared the ground for global environmental action: A declaration of 26 guiding principles was adopted, which was accompanied by a resolution to put international action on a different institutional and financial basis and an action plan of 109 recommendations was set up. Also, the restrictive concept of national sovereignty was reviewed, which found its expression in Principle 21 of the declaration, stating that it is the responsibility of states to ensure that activities within their jurisdiction or control do not cause damage in another state or beyond national jurisdiction. Given the fact that states exercise sovereign rights over all natural resources on their territory and that these rights can only be limited by voluntary means, the recognition by the international community that these rights were not absolute was a very important step for further international cooperation.

In the spirit of the Stockholm conference, the UN General Assembly created the United Nations Environmental Program (UNEP), headquartered in Nairobi, Kenya. It is operated by a governing council of 58 members elected by the UN General Assembly that meets once per year and has to report to the General Assembly through the Economic and Social Council (ECOSOC). The governing council is assisted by a secretariat, headed by an executive director. Despite the fact that UNEP does not have the status of a UN specialized agency and only has limited financial resources, it is the advocate for the protection of the global environment within the UN system. Its main functions are to facilitate and promote the adoption and implementation of international agreements and guidelines and to review and assist with the further development of international environmental law (Fig. 2). In addition, UNEP administers secretariats for six global environmental agreements (the Convention of International Trade in Endangered Species, the Convention of Migratory Species, the Basel Convention, the Convention on the Protection of the Ozone Layer, the Montreal Protocol on Substances That Deplete the Ozone Layer, and the Convention on Biological Diversity) and has a coordinating function among the secretariats of the other international environmental treaties.

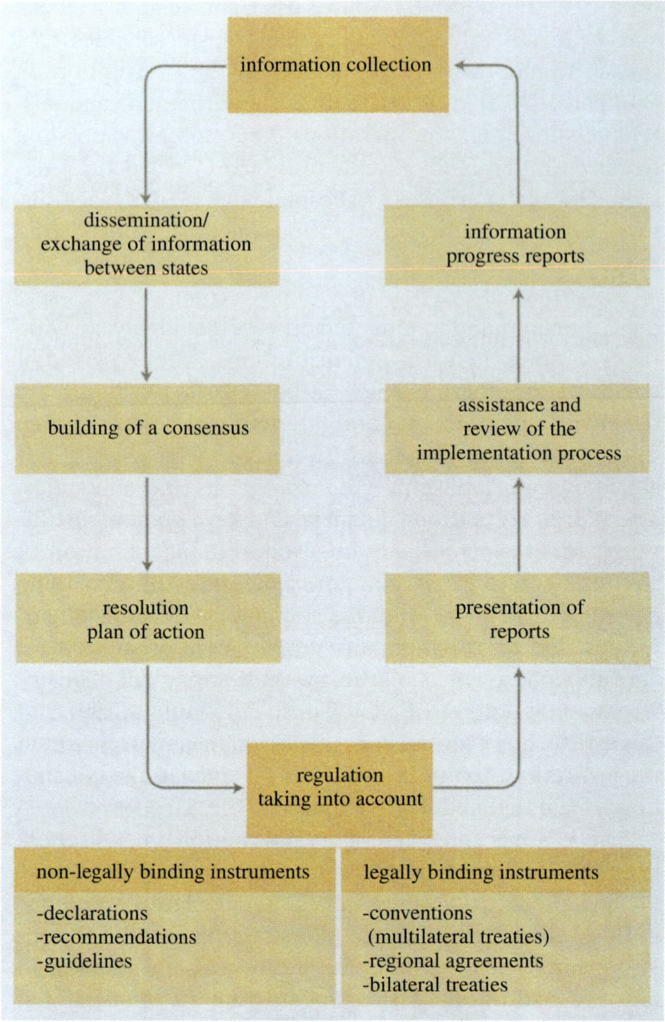

FIGURE 2 Functions of UNEP for interstate cooperation in favor of a global environment monitoring system and for the management of environmental data.

The Stockholm conference was also a catalyst for environmental action on the national level. As a follow-up to the conference, many countries set up an institutional framework to become active in the environmental field which in many cases did not exist at that time.

In reaction to the advancement of information technology and the increased capacity of communication, the Action Plan, adopted at the Stockholm conference, provides for a global environmental assessment program, which was translated into action by the creation of Earthwatch (the Inter-Agency Committee on Sustainable Development named UNEP as task manager for Earthwatch, reinforcing the coordinating and catalytic role that UNEP has played in Earthwatch since its founding). Earthwatch is composed of the Global Environmental Monitoring System, the International Referral System (Infoterra), the International

Register of Potentially Toxic Chemicals, and the state of the environment report and environmental data reports since 1987 with the purpose to collect the necessary information to guide and determine adequate environmental management. The other two main components of the Action Plan are environmental management, including proposals for natural resources management, and supporting measures relating to training, education, and information.

1972–1992

The period between 1972 and 1992 is characterized by the joint effort of UN institutions to build international consensus as a basis for an international response to the pressing global problems that were later addressed by the Earth Summit. The first step in this direction was the adoption of the World Charter for Nature by the UN General Assembly in 1982. This charter expressed the reformulated perception of the international community in its relation to the biosphere. It called for an ecocentric approach in the sense that every life-form should be regarded as unique and should be respected regardless of its worth to humans.

One year later, the World Commission on Environment and Development (WCED, often referred to as the Brundtland Report) was established by the UN General Assembly as an independent "think tank" to review environment and development in order to develop adequate proposals for new forms of international action and cooperation. The report, "Summary of Proposed Legal Principles for Environmental Protection and Sustainable Development Adopted by the WCED Experts Group on Environmental Law, Our Common Future" (1987), called on states to use living natural resources and ecosystems sustainably and to that end proposed a UN program on sustainable development. In order to achieve sustainable development, cross-disciplinary and integrated action was proposed since "global crises . . . are not crises: An environmental crisis, a development crisis, an energy crisis. They are all one" (p. 4). The report therefore can be seen as the "catalyst" for the Rio Summit 5 years later, with sustainable development as a leitmotif of the conference.

United Nations Conference on Environment and Development

The UN Conference on Environment and Development (UNCED), often referred to as the Rio conference or "Earth Summit," was held in Rio de Janeiro, Brazil, in June 1992. It assembled more than 100 heads of state and government and more than 20,000 nongovernmental organizations (NGOs) representatives.

At that conference, a new global partnership for sustainable development was launched. This new course of sustainable development implies for each nation and the world community at large that the social, economic, cultural, and environmental dimensions of development must be treated in an integrated and balanced way. The conference therefore constituted the turning point in the international community's thinking about development. Although the indivisibility of environment and development has been criticized in the sense that the environmental movement would lose its autonomy to be only supplementary to the development process, this connection can be said to be a dramatic step forward in the evolution of international environmental concern since it puts environmental issues at the center of international attention. After long negations, unprecedented consensus was reached on three "soft law" documents, the Rio Declaration, Agenda 21, and the Forest Principles, and two "hard law" conventions about climate change and biological diversity (Fig. 3).

Although all three soft law documents are nonlegally binding instruments, which mean that there is no ratification procedure and in case of noncompliance their provisions cannot be enforced, they serve as a point of reference for international decision-making bodies and have a high moral value since they are the expression of global consensus and political commitment at the highest level.

The Rio Declaration. The Rio Declaration, a declaration of 27 principles, reaffirmed the sovereign right of states to exploit their own resources pursuant to their own environmental and developmental policies. Principle 2 of the declaration enlarged Principle 21 of the Stockholm Declaration through expanding the sovereign right of states to their development policies. The heart of the declaration is its explicit acknowledgment of the right to development and the interconnection drawn between sustainable development and environmental protection. To this end, it calls upon states to cooperate, including the transfer of scientific and technological knowledge. Special priority should be given to the needs of developing countries. In this respect, the principle of common but differentiated responsibilities of states is stated, which entails that different standards of implementation should be applied and financial incentives should be given to developing countries. To this end, the responsibility of developed states in the pursuit of sustainable development has to be acknowledged, given the pressures the societies of the developed states place on the global environment and in view of their command of technologies and financial resources (Principle 7). The declaration even deals with different techniques for the implementation of international standards. In comparison with the Stockholm Declaration 20 years earlier, the declaration is a more specific instrument of reference and reflects the dominating theme of the summit, which is the indivisibility of development and environment. Although the Rio Declaration contains many fundamental principles, it did not fulfill the expectations of a new charter as a basis, setting out the sovereign rights and reciprocal responsibilities of all states on environmental protection and sustainable development, especially since the secretariat of UNCED had originally pro-

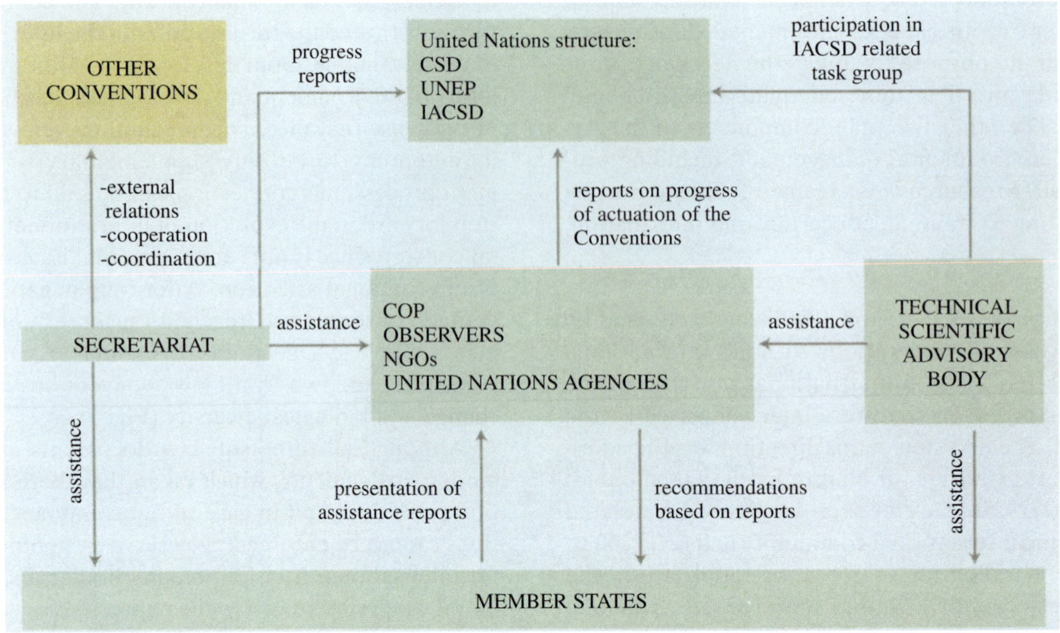

FIGURE 3 The institutional framework of the Rio Conventions. For acronyms, see Table 1.

posed the creation of an earth charter at an early meeting of the UN Preparatory Committee for the 1992 Earth Summit. Therefore, UNCED Secretary General Maurice F. Strong, Green Cross International with the support of the government of The Netherlands, and other independent experts formed the Earth Council. Its priority is to support national entities in the formation of national policies set out in Agenda 21.

Agenda 21. Agenda 21 (Fig. 4) is a comprehensive action plan to lead the international community toward sustainable development in the twenty-first century. It is suggested that the overall implementation of Agenda 21 over the period 1993–2000 will cost about $600 billion annually. In 40 chapters, Agenda 21 concentrates on a thematic and comprehensive approach to hundreds of environmental and development program areas. It includes a preamble and four sections: Section 1 deals with social and economic dimensions, Section 2 with conservation and management of resources for development, Section 3 with strengthening the role of major groups, and Section 4 with means of implementation. The seven chapters of Section 1 comprise the issues of combating poverty, changing consumption patterns, demographic dynamics, and sustainability and also protecting and promoting human health conditions, promoting sustainable human settlement development, and integrating environment and development in decision making. Thus, Section 1 concentrates on an integrated approach to economical, social, and development problems that the Brundtland Report had suggested earlier.

Section 2 concentrates on a thematic approach to the following: atmosphere, planning and management of land resources, deforestation, desertification and drought, sustainable mountain development, sustainable agriculture and rural development, biological diversity and sound management of biotechnology, oceans and seas, freshwater, toxic chemicals, hazardous wastes, and sound management of solid wastes and radioactive waste. By doing so, it suggests a systematic, thematic, and ecosystematic approach to all environmental decision makers, replacing the ad hoc and piecemeal nature of environmental policy.

Section 3 recognizes that "one of the fundamental prerequisites for the achievements of sustainable development is broad public participation in decision making, including new forms of participation" (para. 23.1–2). In addition, key groups for the implementation of Agenda 21 are addressed. Section 4 deals with financial resources and mechanisms, transfer of environmentally sound technology, cooperation and capacity building, science for sustainable development, promotion of education, public awareness and training, national mechanisms for national capacity building in developing countries, international institutional arrangements, international legal instruments, and information for decision making. In the context of institutional arrangements, the key role of NGOs, including the scientific community, the private sector, and women's groups, in the implementation of Agenda 21 has been recognized. The cross-sectoral and comprehensive character of Agenda 21 made this document an "instruction plan" and overall point of orientation for the international "machinery" for how to work in the future. Thus, Agenda 21 constitutes the basis for the

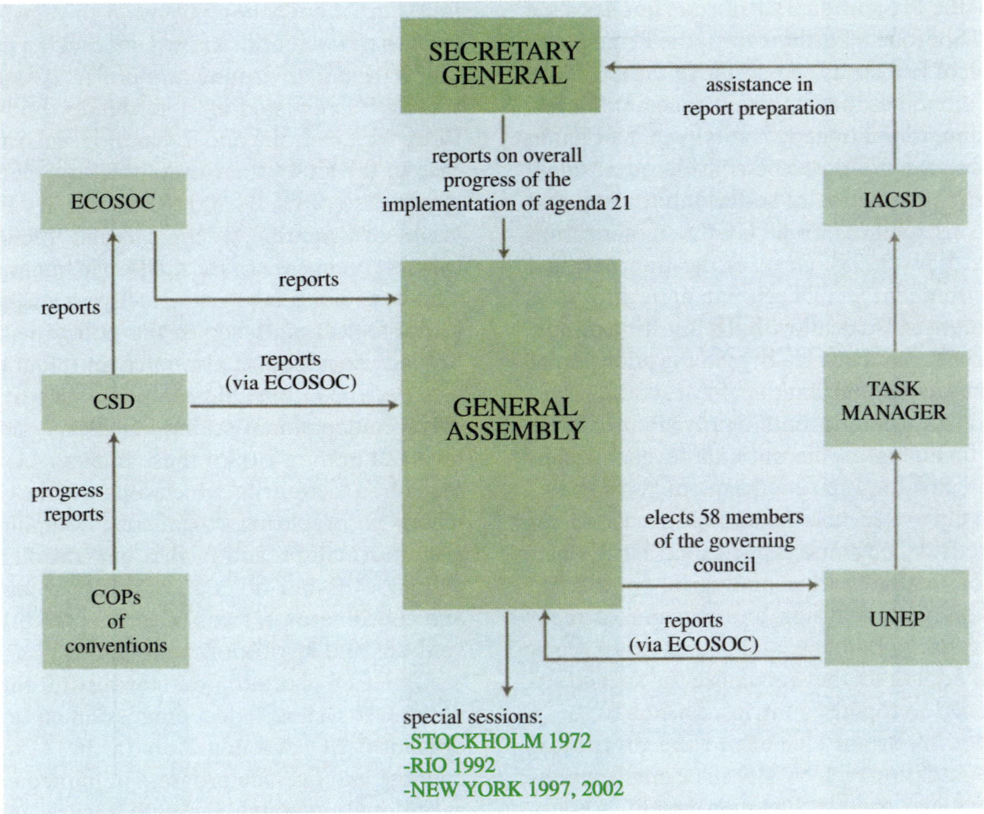

FIGURE 4 The institutional framework set out in Agenda 21.

overall goal of a coherent global environmental strategy. Moreover, it has become the basis for many national action plans, and more than 1800 cities and towns worldwide have created their own "local Agenda 21."

Forest Principles. The statement of principles for a global consensus on the management, conservation, and sustainable development of all types of forests reflects the first global consensus reached in regard to forests. It states that all forests should be sustainably managed for their social, economic, and ecological services and benefits. The widest participation of people in forestry should be actively promoted. The declaration is composed of general principles and does not provide for a specific action program since opinions of developed and developing states on the issue of forests were too divided.

Hard Law Instruments. At the occasion of UNCED, two legally binding treaties were also signed: the Convention on Biodiversity (CBD) and the Convention on Climate Change (CCC). In addition, the UNCED recommended that the UN General Assembly establish an Intergovernmental Negotiating Committee, which began work in 1993 to prepare a convention to combat desertification in countries experiencing serious drought and/or desertification. Since these agreements were either adopted or initiated as a negotiating process, they are further referred to as "Rio Conventions." In contrast to the Agenda 21, the Rio Declaration, or the Forest Principles, the conventions form part of the so-called hard law since they constitute enforceable commitments of states to fulfill their obligations. The "convention" as one form of international legal regulation was seen as the effective response to the growing global problems. The legally binding character, however, often has the disadvantage of leading to very general obligations which are difficult to translate into practice instead of specific commitments, to which states are generally very reluctant to submit. Thus, these obligations usually represent the smallest denominator of common interests and opinions.

The conventions incorporate certain principles set out by the Rio Declaration and Agenda 21 as guiding principles of international environmental policy: The Precautionary Principle implies that in cases in which there are threats of serious or irreversible damage, lack of full scientific certainty shall not be used as a reason for postponing cost-effective measures to prevent environmental degradation (Principle 15 of Rio Declaration, para. 35.3 of Agenda 21, Article 3, para. 3 of the CCC, and the preamble of the CBD). Thus, in contrast to the so-called "polluter pays" principle, which deals with compensation for environmental harm that has already occurred and which states that "the polluter should bear the costs of pollution" (Principle 16 of

Rio Declaration), the Precautionary Principle implies a so-called "no regret" approach. Furthermore , the Principle of Common Concern of Humanity (preamble of the CCC and the CBD), which embodies the important concept of international partnership, is mentioned. It entails that a certain environmental issue can no longer be considered as solely within the domestic jurisdiction of states due to its global importance and consequences for all [IUCN, Commission on Environmental Law (1995), Draft of the International Covenant on Environment and Development, p. 33.].

The Rio Conference also dealt with the question of how to implement its decisions: The idea behind the adoption of global conventions was "global thinking, local action." Consequently, the Rio Conventions contain provisions on their implementation, including on the national level. In addition, the indivisible link between environment and sustainable development led to a certain thematic approach to implementation. The Rio Conventions put a special emphasis on the exchange of scientific and technological knowledge, development, adaptation, diffusion and transfer of technologies, human capacity building, and education to reach their objectives. By doing so, they recognize the special situation and needs of developing countries, which have been suggested to be an important element to the reversal of environmental degradation by UNCED since environmentally sound technologies protect the environment, are less polluting, and use resources sustainably.

The centerpiece to reviewing and assessing progress is a reporting or monitoring procedure which can be defined as the measurement of a situation in a time. Any scientific assessment also depends heavily on regular reporting of the status of conservation, as does any efficient environmental management. Especially in view of the advancement of information technology, the collection and dissemination of information, dependent on the communication of detailed information related to implementation by the contracting parties, plays a key role in effective implementation on the regional and national levels.

In terms of institutional arrangements, the Rio Conventions originally set out the same system: The supreme decision-making body is the Conference of the Parties (COP). The COP comprises all ratifying parties and gives the observer's status to national and international organizations which have relevant expertise to provide it with information relevant to each convention and to the United Nations, its specialized agencies, as well as any state member thereof or observers thereto not party to the convention. The COP must promote and review the implementation of each convention. To this end, it examines and comments on the reports handed in by the states member to the convention. On the basis of these reports the COP makes recommendations and publishes reports on the convention's implementation. Since the COP has the right to add amendments and protocols to the convention, it has the very important function to "keep the convention alive," which means adapting it to new scientific knowledge and the evolving system of international environmental policy. In achieving this task, the COP is assisted by a subsidiary body, the Subsidiary Body for Scientific and Technological Advice (SBSTA) in case of the CCC, the Committee on Science and Technology in case of CCD, and the Subsidiary Body for the Provision of Scientific, Technical, and Technological Advise (SBSTTA) in case of the CBD. The main function of these advisory bodies is to give effective scientific, technical, and technological guidance to the policy makers in the COP. The advisory bodies also have an integrating factor since other actors in the field can turn to it for reference. In addition, independent scientific experts and institutions are involved in the work of these bodies. Also, the COP is assisted by a secretariat, which supports the COP by arranging meetings, preparing documents, compiling and transmitting information, and facilitating consultations and others actions. Another function of this secretariat is to establish and coordinate "external relations" to other relevant conventions and institutions.

UNCED also initiated new institutional developments in the UN system. The Commission on Sustainable Development (CSD; Agenda 21, para. 38.11) was created to encourage and oversee progress in implementing Agenda 21 at national, regional, and global levels and to promote integrated, rationalized decision making on environmental and development issues. It consists of representatives of 53 states elected by ECOSOC with due regard to equitable geographical distribution. Due to the growing importance of NGOs in the field of the environment, they are granted an observer status at the CSD. The CSD is supposed to be a forum for the further implementation of Agenda 21 since it is not provided with any legal or budgetary authority.

For additional support to ensure effective monitoring, coordination, and supervision of the involvement of the UN system in the follow-up to the Rio Conference, the Inter-Agency Committee on Sustainable Development (IACSD; Agenda 21, para. 38.16) was created. Rio also strengthened the catalytic role of UNEP to stimulate, promote, and coordinate environmental activities in the UN system (Agenda 21, para. 38.21).

1992–1997

Despite the initial success of Earth Summit, the 5 years since Rio were characterized by difficulties in translating the Earth Summit commitments into reality. Many countries established environmental ministries and presidential-level councils for sustainable development, formulated new national environmental policy statements and strategies and reviewed existing legislation to ensure national compliance with the Rio Conventions. However, most developed countries have still not reached the UN's target of committing 0.7% of their gross national product (GNP) to official development assistance. On average, official development aid (ODA) as a percentage of the GNP of developed countries has drastically declined in the post-UNCED period from 0.34% in 1992 to 0.27% in 1995. Similarly, technology transfer and technology-related investment from public and private sources have not been realized as outlined in Agenda 21.

Among the achievements since UNCED are the entry into force of the UN Framework Convention on Climate Change, the Convention on Biological Diversity, and the UN Convention to Combat Desertification; the conclusion of an agreement on straddling and migratory fish stocks; the adoption of the Program of Action for the Sustainable Development of Small Island Developing States and the elaboration of the Global Program of Action for the Protection of the Marine Environment from Land-Based Activities; and the entry into force of the UN Convention on the Law of the Sea.

A positive step forward was the establishment, restructuring, funding, and replenishment of the Global Environment Facility (GEF). The creation of the GEF was followed by a 3-year pilot phase, which ended in June 1994. After this pilot phase, GEF was replenished with $2 billion. GEF is striving for universal participation, and currently 156 countries are participants. Countries must be party to the CCC or CBD to receive funds from the GEF in the relevant focal area. The GEF's main function is to finance agreed incremental costs of activities which generate global environmental benefits but would not be included in national development funding. Only a certain number of subject areas are eligible for funding: climate change, biological diversity, international waters, ozone depletion, and, since 1992, desertification and deforestation in relation to these environmental issues. The World Bank functions as secretariat, trustee, and administrator for the GEF program. UNDP provides the technical assistance, capacity building, and project preparation. UNEP is responsible for ensuring the GEF program is consistent with existing and emerging conventions and protocols, provides a secretariat for the Scientific and Technical Advisory Panel, and is also responsible for strategic planning. GEF is currently the operating entity of the financial mechanism of the CCC and the CBD.

Rio Plus 5: "Moving Sustainable Development from Agenda to Action"

Fifty-three heads of state or government and 65 ministers of environment and related areas attended the General Assembly's special session "Rio Plus 5," held June 23–27, 1997. The primary purpose of Rio Plus 5 was to review progress on implementation of commitments achieved since the UNCED 5 years ago and to reenergize commitment to further action on goals and objectives set out by the Rio Earth Summit. However, to the disappointment of many delegates, NGOs, and the media, only few new commitments were made by the international community and the final text adopted by consensus among delegates from more than 165 countries, titled "a Programme for the Future Implementation of Agenda 21," contains carefully worded compromise language. Delegates were unable to agree on a self-standing political declaration that was to summarize the outcome of the special session. Instead, a statement of commitment was included as a preamble to the final document, in which governments reaffirmed Agenda 21 and the Rio principles and recommitted themselves to the global partnership established there. Despite the fact that momentum was gained in preventing climate change, forest loss, and freshwater scarcity, states did not reach an agreement on definitive policies in relation to quantified emission reduction targets or about a convention on the protection of forests.

The Rio Conference initiated a comprehensive approach to sustainable development and set out a thematic outline of action. Rio Plus 5 put an emphasis on implementation and regarded cooperation and collaboration of all actors involved in the process as a key element to the success of Agenda 21's implementation. Given that the 5 years since the Rio Convention were characterized by an accelerated globalization of interactions among countries in the areas of world trade, foreign direct investment and capital markets, and rapid economic and social changes, the institutional arrangement set out at UNCED must also respond to these changes. Another challenge will be to achieve greater coherence in various intergovernmental organizations and processes since the number of decision-making and UN bodies has increased.

In this context, the General Assembly called on the conferences of parties of the Rio Conventions and governing bodies of conventions relevant to sustainable development to cooperate in exploring ways and means of collaborating in their work to advance the effective implementation of the conventions. The General Assembly suggested that the motto should be "synergy versus duplication," which implies that already existing actions should be coordinated and become more efficient instead of losing precious time and resources through manifold action. To that end, it has been proposed to collocate the secretariats of these conventions, to improve schedules of meetings, and to integrate national reporting requirements. In terms of institutional arrangements, the convention secretariats should be provided with effective support and services to overcome the difficulty that these activities demand with regard to a notable investment of resources and a contribution of extra staff time of secretariats involved. One of the responsible institutions to steer this process is UNEP because it is the leading global environmental authority that sets the global environmental dimension of sustainable development and advocates for the global environment within the UN system. The General Assembly has appointed UNEP to establish coherent linkages between relevant environmental conventions.

Second, the IACSD and its system of task managers are mandated to enhance systemwide intersectoral cooperation and coordination for the implementation of Agenda 21 and for the promotion of coordinated follow-up to the major UN conferences in the area of sustainable development. In terms of cooperation, country profiles prepared for the IACSD could become first-access documents containing general institutional and macroeconomic information with cross-references or links to other databases holding more specific information collected by the conventions and organizations of the UN system (UNEP/CBD/COP/3/29).

The General Assembly considers the CSD as the central forum to review the implementation of Agenda 21. The CSD was addressed to make an effort to attract greater involvement in its work of ministers and high-level national policy makers and also the international scientific community, the major stakeholders affected by the implementation process of Agenda 21, and NGOs through the establishment of roundtables and dialogue sessions. In the view of the General Assembly, the CSD should also act as a forum to exchange national experiences. It has been suggested that states may provide the commission with relevant information regarding their incorporation efforts on a voluntary basis. Also, on the regional and subregional level, it is called to act as such a forum, including voluntary regional exchange of information regarding initiatives and collaboration in the context of sustainable development.

In addition to the critical overview of the implementation process of Agenda 21, it has been criticized that "the current UN system is not able to enforce compliance with international accords" (A /s-19/1, "Overall Review and Appraisal of the Implementation of Agenda 21," para. 25) since sanctions within the framework of the conventions are only of political nature. In the context of the reporting procedure about the implementation process, states can only be asked to make a greater effort, but no independent body exists to enforce their compliance to the convention obligations.

Although the final document of the special session states that all the subject areas enumerated in Agenda 21 should be treated "on equal footing," Rio Plus 5 has de facto established a new priority list of environmental issues to be addressed in the near future, with freshwater being particularly high on the list. Energy, transport, sustainable tourism, and natural disasters were recognized for the first time as autonomous areas that need to be dealt with in the context of sustainable development. Energy has been noted to be vital to social and economic development, but it was emphasized that sustainable patterns of production, distribution, and use of energy are crucial. To achieve sustainability, states reaffirmed the necessity to ensure international cooperation for promoting energy conservation and improvement of energy efficiency, the use of renewable energy and research, and the development and dissemination of innovative energy-related technology. The CSD will discuss the energy issue at its ninth session.

It was discussed at Rio Plus 5 that the transport sector, as the largest end user of energy in developed countries and the fastest growing one in developing countries, should be submitted to an integrated transport policy which particularly encourages international cooperation in the transfer of environmentally sound technologies to accelerate the phasing out of the use of leaded gasoline. Voluntary guidelines for environmentally sound transport need to be promoted, as do partnerships at the national level.

The future role of tourism, as one of the world's largest industries and one of the fastest growing economic sectors, has been analyzed by the General Assembly as crucial in regard to environmental conservation and protection. In this context, the efforts of developing countries to broaden the concept of tourism to include cultural and ecotourism need to be merited by the international community and its financial institutions. It was proposed to integrate tourism in the implementation of Agenda 21 since it uses resources and creates environmental costs and benefits in the process. Moreover, national policy development and capacity in areas of physical planning and the use of economics on the conservation of biodiversity and fragile ecosystems, such as coral reefs, mountains, coastal areas, and wetlands, have also been considered. The CSD discussed tourism at its 1999 session.

Thematic Approach to the Institutional Framework Made by Treaty Arrangements

Biological Diversity

Convention on Biological Diversity. The current decline of biodiversity is mostly due to human activities. The loss of biodiversity stems mainly from habitat destruction, overharvesting, pollution, and the inappropriate introduction of foreign plants and animals. In response, there has been growing concern of the world community to conserve biodiversity: Its importance was emphasized at the Stockholm conference in 1972 (in Principle 2, "natural resources of the earth . . . must be safeguarded"; in Principle 3, "Man has a special responsibility to safeguard and wisely manage the heritage of wildlife and its habitat"). Also, in many legally nonbinding declarations the conservation of biodiversity has been seen as vital (e.g., World Charter for Nature, 1982, principles 2–4; WCED Expert Group on Environmental Law, principle 3; and Caring for the Earth, 1991, principles of a sustainable society, p. 9.). As late as

May 1989, however, an Ad Hoc Working Group of Technical and Legal Experts was established by UNEP to prepare an international legal instrument for the conservation and the sustainable use of biodiversity. The group was renamed the Intergovernmental Negotiating Committee when the formal negation process started in February 1991. The convention was adopted on May 22, 1992, in Nairobi, Kenya, and was finally opened for signature during the UNCED conference held in June 1992 in Rio de Janeiro. It entered into force on December 29, 1993. As of May 27, 1997, the convention had been ratified by 169 parties.

The convention was an important step toward the conservation of biodiversity since it was the first treaty to deal with genetic resources on a global level in a comprehensive perspective.

The convention had three objectives: the conservation of biological diversity, where *biological diversity* means the variability among living organisms from all sources including, inter alia, terrestrial, marine, and other aquatic ecosystems and the ecological complexes of which they are part (this includes diversity within species, between species, and ecosystems) (Art. 2); the sustainable use of its components; and the fair and equitable sharing of benefits arising from the utilization of genetic resources (Art. 1). In order to achieve these objectives, states have to develop national biodiversity strategies, plans, or programs and are obliged to integrate the conservation of biological diversity and the sustainable use of its components into relevant sectoral and cross-sectoral plans, programs, and policies (Art. 6). In terms of conservation, the convention follows the "global thinking, local action" approach. Thus, the most significant actions under way are taking place at the national and local levels. By 1995, more than 20 nations had developed biodiversity strategies or actions plans (*http://www.wri.org./wri/rio-5/rio5biod.html#overview*). The convention also foresees provisions about research and training (Art. 12), public education and awareness (Art. 13), exchange of information (Art. 17), and technical and scientific cooperation (Art. 18) since the information gap between north and south relating to environmental problems and solutions still persists. With regard to the conservation of biodiversity, *in situ* conservation—"the conservation of ecosystems and natural habitats and the maintenance and recovery of viable populations of species in their natural surroundings and, in the case of domesticated or cultivated species, in the surroundings in which they have developed their distinctive properties" (Art. 2)—is the primary approach and *ex situ* conservation—"the conservation of components of biological diversity outside their natural habitats" (Art. 2)—is only complementary to the former. The scope of conservation includes not only wild species and genetic diversity but also the *in situ* conservation of human-created plant varieties and animal breeds. According to its objectives, the convention's conservation measures address protected areas and nonprotected areas. The convention also provides for access to genetic resources (Art. 15), access to and transfer of technology (Art. 16), and the handling of biotechnology and distribution of its benefits (Art. 19). The regulation on fair and equitable sharing of benefits arising from the use of biotechnology was one of the most controversial since some states interpreted this provision as a threat to their intellectual property system and other states feared the costs of such a regulation. At Rio Plus 5, the need for international cooperation to enable developing states to compete in the emerging market of biological resources was emphasized again. The convention does not include detailed rules on genetically modified organisms. At Rio Plus 5, the international community agreed on the need to quickly complete the CBD by a biosafety protocol, complemented by UNEP biosafety guidelines.

The main body to review and steer implementation of the convention is the COP. The second meeting of the COP in Jakarta, Indonesia, gives an example of the central and dynamic role of the COP. The scope of the convention was expanded to include a program for action for implementing the convention with respect to the marine and coastal biological diversity. The adoption of such a program also shows the important role of the technical advisory body (SBSTTA) in the framework of the CBD since the COP had heavily built on the recommendations made by the SBSTTA. The third meeting of the SBSTTA in Montreal will focus on the assessment of the status and trends of biological diversity in inland waters. To facilitate and promote the scientific and technical cooperation and cooperation in general, the convention called for the establishment of a clearinghouse mechanism (CHM), which has now been established. The notion of clearinghouse derives from the English clearinghouses which in former times were institutions in the financial world that cleared credits and balances between banks (Art. 23). Its main function is to provide for information exchange. The work of collecting, structuring, and reviewing the data will be executed by the contracting parties through the provision of national reports and the voluntary provision of information sources by specialized agencies experienced in the field of biodiversity. The secretariat of the CBD has the responsibility to disseminate the information to all partners and to avoid the duplication of information. In addition, the CHM is charged with matching the demand and support side in the context of scientific and technological cooperation. The CHM therefore does not play the role of a custodian, but it provides the urgently needed basis for cooperation and cross-sectoral actions in the field of biodiversity. At Rio Plus 5, the importance of the establishment of the CHM was explicitly stressed again. In this context, the Internet constitutes a promising tool for making the CHM successful.

The GEF serves as the interim financial mechanism for the CBD (*http://www.biodiv.org/chm/chm-bac1.html*).

During the GEF's pilot phase, (1991–1993), about $300 million was spent on about 57 biodiversity projects (Art. 39).

The CBD is one of a family of international environmental agreements related to biodiversity. The original idea to have a convention that would absorb all the existing treaties relevant to biodiversity was dropped in favor of a framework or umbrella agreement that would built on existing conventions. Since the COP of the CBD has adopted an ecosystem approach over time, it depends on a close working relationship with the other conventions. This approach is consistent with the trend toward coordination and cooperation on the international level. Thus, close cooperation between the secretariat of the CBD and the other biodiversity-related conventions is maintained in the form of memoranda of understanding (MOU) (*http://www.wri.org/wri/rio-5/rio5biod.html #overview*). Also at the sixth meeting on coordination of the secretariats of environmental conventions in Nairobi, participates agreed to perform a feasibility study on harmonization of information for biodiversity-related treaties. Since any convergent action is a very complex and long process, the COP of CBD proposed a step-by step method, concentrating on specific practical activities, such as reporting, among a small group of conventions that have as central concerns the conservation and sustainable use of biological diversity and its components.

Convention on International Trade in Endangered Species of Wild Fauna and Flora. The Convention on International Trade in Endangered Species of Wild Fauna and Flora (CITES) was adopted on March 3, 1973, in Washington, DC, and entered into force on July 1, 1975. It has established a worldwide system of controls on international trade in endangered wild fauna and flora and specimens derived from them. Trade in these species must be authorized and is restricted by government-issued permits or certificates. The objective is to ensure that trade in such species is regulated and monitored to ensure their sustainability. As of August 26, 1996, 132 states had ratified the convention.

The convention lists all species that are endangered or could become endangered in three separate appendices, according to the severity of the extinction threat. Appendix 1 includes all species threatened with extinction that are or might be affected by trade. Currently, 830 species are listed in Appendix 1. Appendix 2, which lists 30,000 species, comprises all species that may become threatened with extinction unless trade is strictly regulated. It also includes species that are similar in appearance to endangered species and must therefore be regulated to ensure that efforts to protect endangered species are effective. Appendix 3, listing 229 species, includes all species that any party identifies as being subject to regulation within its jurisdiction for the purpose of preventing or restricting exploitation. Amendments to the appendices are made on a regular basis, whereas a two-third majority is necessary to add or subtract a listing to Appendixes 1 and 2. Every state party to the convention, however, can add or subtract a listing from Appendix 3, and this is also true for reservations or exceptions to Appendix 3.

To control trade in endangered species, any species listed in Appendixes 1 or 2 has to be accompanied by an export and import permit when it enters the importing country. Species listed in Appendix 3 must be accompanied by an export permit alone. A species that is discovered without a permit in the country of import is considered illegal and must be send back to the country of export. For species listed in Appendix 1, a permit is only given if the responsible scientific and management authorities of the state of export advise "that export will not be detrimental to the survival of that species" and that the species was not taken illegally. For species in Appendix 2, a permit is not granted in cases in which the designated scientific authority determines that trade should be controlled to maintain the species "well above the level at which that species might become eligible for inclusion in Appendix 1." For species in Appendix 3, permits are generally granted without special considerations. These enforcement measures are accompanied by corresponding national measures. According to the convention, states should "penalize trade in, or possession of, such specimens, or both; and to provide for the confiscation or return to the state of export such specimens."

The CITES secretariat cooperates with the CBD in the effort to harmonize national legislation and strategies, to explore opportunities for access to GEF funding, and to coordinate research, training, and public awareness activities. Moreover, the COP of CITES reviewed the implementation and effectiveness of CITES with regard to the threat that international trade poses for biodiversity. It concluded that the key factor to a close working relationship is national-level cooperation, including cooperation among trade ministers, customs authorities, police forces, and other enforcement agencies (Decision 10/4 of the COP of CITES).

Convention on the Conservation of Migratory Species of Wild Animals. The Convention on the Conservation of Migratory Species of Wild Animals (CMS) was adopted on June 23, 1979, in Bonn, Germany, entered into force on November 1, 1983, and currently has 51 parties. The convention provides a framework for the conservation of migratory species and their habitats by adopting strict protection measures for migratory species or their separate populations that have been categorized as endangered (listed in Appendix 1 of the convention). The convention also provides for the conclusion of international agreements for the conservation and management of migratory species or their separate populations that have an unfavorable conservation status or would significantly benefit from international cooperation (Appendix 2 of the convention). The convention includes provisions for joint research and coordinated management activities. It also has a COP, a standing committee, which provides policy and administrative guidance between the regular meetings of the COP. A scientific council, composed by experts appointed by member states and the COP, gives technical and scientific advise. The secretariat, under the auspices of UNEP, gives administrative support to the convention. The CMS and the CBD secretariat work closely together.

The Convention on Wetlands of International Importance Especially as Waterfowl Habitat. The convention (often referred to as the Ramsar Convention), which was adopted in 1971 in Ramsar, Iran, currently has 103 contracting parties. The convention is the only one of its kind to deal with a specific habitat type. Until the Brisbane Conference of the parties in 1996 in Australia, the convention did not have a general strategy for its implementation. The convention's mission has been defined as being "the conservation and wise use of wetlands by national action and international cooperation as a means to achieve sustainable development throughout the world." The concept of "wise use" has to be interpreted as sustainable use, defined by the COP as "sustainable utilization for the benefit of humankind in a way compatible with the maintenance of the natural properties of the ecosystem."

Through this important review, the convention through its COP has incorporated the "post-Rio frame of mind," a comprehensive approach to wetlands. The focus of the convention can be enlarged from its original emphasis on waterfowl to a much broader range of conservation and sustainable use issues at the ecosystem level. Also, a resolution on Ramsar and water and a recommendation on toxic chemicals in wetlands were adopted by the Brisbane Conference. In addition, the Ramsar secretariat has signed a MOU with the secretariats of the CBD and CMS and will establish close working relations with the CCD and CITES. In return, the COP of the CBD discussed biodiversity with regard to wetlands at its third meeting and invited the Ramsar Convention to act as a lead partner of the CBD on wetlands and inland water ecosystems. The MOU between the convention and CBD set out four areas of action: institutional cooperation, especially between the scientific and technological advisory bodies; the exchange of information and experience; the coordination of programs of work; and joint conservation action. In addition, consultation and reporting measures between the secretariats on the implementation of the MOU are planned (UNEP/CBD/COP/3/Inf.38).

Convention Concerning the Protection of the World Cultural and Natural Heritage. The Convention Concerning the Protection of the World Cultural and Natural Heritage (World Heritage Convention) was adopted on November 16, 1972, by the General Conference of UNESCO in Paris and entered into force on December 17, 1975. Its objectives are to promote international cooperation for the protection and conservation of sites of outstanding universal value. The convention is unique since it encompasses both natural and cultural heritage. In retrospect, those who drafted the convention were very clear-sighted 20 years before the "Rio frame of mind" because they recognized the indivisibility of nature and culture. In the convention, *world heritage* is defined as natural or cultural heritage or mixed cultural and natural heritage (Art. 1 and 2). The criteria to determine the "outstanding universal value" of cultural heritage, including cultural landscapes and natural heritage, are set out in the operational guidelines of the convention. Cultural properties represent a masterpiece of human creative genius; exhibit an important interchange of human values; bear a unique or at least exceptional testimony to a cultural tradition or civilization; are an outstanding example of a type of building or architectural or technological ensemble or landscape, traditional human settlement, or land use; and/or are directly or tangibly associated with events or living traditions, ideas or beliefs, or artistic and literary works. Cultural landscapes can be considered as "the combined works of nature and man, embracing the diversity of manifestations of the interaction between humankind and its natural environment." Natural heritage includes sites that "contain the most important and significant natural habitats for *in situ* conservation of biological diversity, including those containing threatened species of outstanding universal value from the point of view of science or conservation" and also fulfill the condition of being "the most important sites for the conservation of biological diversity."

The substantive connection between nature and culture has been reinforced by efforts in the framework of a so-called "global strategy" to make the world heritage list more balanced and representative in terms of completing the identification of world heritage sites of outstanding universal value for all regions of the world, on the one hand, and to correct the imbalance between natural and cultural sites, on the other hand.

Each state party to this convention is obliged to identify and delineate the different properties situated on its territory (Art. 3) as defined by the convention and to submit, insofar as is possible, to the World Heritage Committee an inventory of the property forming part of the natural and cultural heritage situated on its territory and suitable for inclusion in the World Heritage List provided for by the convention (Art. 11, para. 2). By signing the convention, states pledge not only to conserve the sites of outstanding universal value situated on their territory but also to protect their national heritage. To this end, state parties to the convention will do all, to the utmost of their own resources, to identify, protect, conserve, present, and transmit their natural and cultural or national heritage to future generations. To ensure that effective and active measures are taken, states endeavor, insofar as is possible, and appropriate for each country to adopt a general policy to integrate the protection of that heritage into comprehensive planning programs and to take appropriate legal, scientific, technical, administrative, and financial countermeasures. Furthermore, states are obliged to promote scientific and technical research and training of personnel (Art. 5).

The convention provides for international financial, scientific, and technical assistance (Art. 22), but the responsibilities to fulfill the duties described previously reside with the state parties (Art. 4). To achieve the objectives set out by the convention, a World Heritage Fund is provided for, which receives its income from obligatory and voluntary contributions. The total amount available under this fund for international and emergency assistance is $4.5 million per year. Given the fact that it is a modest sum, priority is given to the most threatened sites, such as those on the List of World Heritage in Danger.

The World Heritage Committee is entrusted to take decision in the following areas: First, it makes the decision to include a property in the World Heritage List that has been nominated by a state party to the convention, and it has the right to delete a site from that list. Second, in case a site is threatened by "serious and specific dangers" the committee has the right to put the site on the List of World Heritage in Danger. Third, the committee makes decisions regarding the use of the World Heritage Fund and the allocation of the different types of international assistance. Finally, it examines the state of conservation of the properties forming part of the World Heritage List.

The World Heritage Committee, which meets in December each year, consists of representatives of 21 state parties to the convention, of which 7 members are elected for the World Heritage Bureau, a smaller executive body that prepares the work of the committee. The bureau meets twice a year in June and in December. The UNESCO World Heritage Center was set up in 1992 by the director-general of UNESCO to ensure the day-to-day management of the convention. It organizes the sessions of the World Heritage Bureau and Committee; provides advice to state parties in the preparation of nominations; organizes technical assistance in case of approval by the chairperson, the bureau, or the committee; and coordinates both the reporting on the state of conservation of properties and the emergency action to be undertaken when world heritage properties are threatened. It is also responsible for the administration of the World Heritage Fund. It cooperates closely with the advisory bodies to the conventions IUCN, ICOMOS, and ICCROM and with the state parties to the convention, funding agencies, and other convention secretariats.

In terms of cooperation and collaboration among the different environmental conventions, the CBD seeks relevant information on the implementation of its articles dealing with general measures for conservation and sustainable use and *in situ* conservation, including experience of relevant conventions (decision II/7 of the COP of CBD). The World Heritage Convention prepared the ground for future reporting by state parties. It has conducted numerous activities in relation to the assessment of the state of conservation of properties. The convention's committee introduced a system of monitoring and reporting on the state of conservation of world heritage properties by the state parties in 1994 that was the result of a process started in 1982. In 1992, the World Heritage Committee revised the selection criteria for natural sites to better acknowledge biodiverse sites of outstanding universal value (see No. 44, para. 4 of the operational guidelines of the World Heritage Convention). As a consequence, the CBD parties now have the opportunity to collaborate with the state parties in fulfilling their obligation under the CBD to identify components for biodiversity important for conservation. To that end, the World Heritage Committee and the CBD could work together to systematically identify potential world heritage sites of high biodiversity value for inclusion in the tentative list and the World Heritage List of the convention.

The idea to create a so-called "global list" of "hot spots" of biodiversity was originally suggested for the CBD in the negotiation process but was not included in the CBD, which has adopted a national approach to biodiversity conservation (see Art. 3 of CBD, which leaves it to the states to identify protective areas for biodiversity conservation). Thus, the World Heritage Convention's effort to include the most biodiverse sites of the world in the World Heritage List could be complementary to the obligation of states to identify biodiverse sites on the national level under the CBD (Art. 3). Moreover, such a list of hot spots could be helpful as a point of reference for a systematic approach to biodiversity protection on the international level. To achieve effective conservation and protection of biodiversity, the conventions could also work together in cases in which, under the World Heritage Convention, action such as mitigation and rehabilitation measures needs to be taken to remove a world heritage site from the List of World Heritage in Danger.

The CBD recognizes that, among many other values, sites of high biological diversity may also have cultural value. and it acknowledges the interactions between culture and biological resources in many of its provisions. Special attention is also given to the rights of indigenous peoples. The World Heritage Convention became the first international legal instrument to identify and protect cultural landscapes in 1992. Through the inclusion of "cultural landscapes" in the World Heritage List, the Convention recognizes the close links between people and the environment, and therefore also of indigenous peoples and their natural environment. The protection of traditional cultural landscapes is also helpful in maintaining biological diversity with regard to the conservation of genetic diversity among agricultural crops: "A wide range of plant diversity has been lost this century as farmers have abandoned their traditional cultivars and 'landraces' for genetically uniform, high yielding varieties" (*http://www.fao.org/waicent/faoinfo/sustdev*). Moreover, traditional forms of land use can contribute to modern techniques of sustainable land use since these traditional forms often "have helped enable societies to avoid overexploitation and to live within the limits imposed by the availability of resources" (von Droste *et al.*, 1995, p. 26).

Forests

The subject of forests is related to the entire range of environmental and development issues and opportunities, including the right to socioeconomic development on a sustainable basis. Deforestation, conservation, and use of genetic resources and biological diversity and also the contribution of forests to reduce global warming, the fate of indigenous forest dwellers, and international trade in forest products have become a matter of international concern in the past few years.

The program on the further implementation of Agenda 21, paragraph 37, states that "since the adoption of the Forest Principles at the Rio Conference, tangible progress has been made in sustainable forest management at the national, subregional, regional, and international levels and in the promotion of international cooperation on forests." A decisive step forward was the establishment of the open-ended ad hoc Intergovernmental Panel on Forests (IPF) by the CSD at its third session. The panel was mandated as an intergovernmental forum to pursue a consensus and formulate options for further actions in order to combat deforestation and forest degradation and to promote the management, conservation, and sustainable development of all types of forests. The panel was requested to promote multidisciplinary action at the international level consistent with the Forest Principles, taking into account the Rio Declaration on Environment and Development and Agenda 21. To this end, the IPF drew heavily on the resources of international organizations (ITTO, FAO, UNDP, DPCSD, and UNEP). The panel had a small secretariat and was funded by voluntary extrabudgetary contributions. In addition, an International Task Force on Forests was formed (by FAO, UNDP, UNEP, ITTO, World Bank, UNDP, CSD, and the secretariat of CBD). Its strategy for effective implementation of CSD–IPF proposals for action considers actions at global, regional, and national levels and existing strategies, mandates, policies, and programs of work of International Tropical Timber Agreement members and other relevant organizations.

In the last of its four sessions, the IPF adopted the "Report of the Ad Hoc Intergovernmental Panel on Forests," which completed 18 months of political dialogue that helped define the world's "forest agenda." It stressed the need to implement the UNCED forest-related decisions on international and national levels, including the examination of sectoral and cross-sectoral linkages. To this end, international organizations and multilateral institutions are called on to intensify their cooperation and collaboration. In addition, appropriate legal mechanisms are called for. The panel adopted proposals for action and reached consensus on key issues, such as improved national forest programs and forest resources assessments. Also, the Rio Plus 5 summit agreed that international cooperation and collaboration are vital for further progress on the issue. The panel was an intergovernmental body and not an expert group; however, NGOs heavily influenced the IPF process in taking advantage of their right to participate as observers. During IPF-3, NGOs were permitted not only to make interventions on the floor during official negotiations but also to make direct comments on the texts and on other delegations' proposals. NGOs in the IPF-3 therefore pushed the limits of official UN rules on NGO participation.

At Rio Plus 5, it was decided to continue the intergovernmental policy dialogue through the establishment of an ad hoc, open-ended Intergovernmental Forum on Forests under the aegis of the CSD to promote the implementation of the IPF proposals for action; to review the progress in the management, conservation, and sustainable development of all types of forests; and to consider matters pending on the program elements of the IPF.

At the Rio Plus 5, states again failed to adopt an international legally binding instrument related to the protection of the world's forests. There are various difficulties in reaching an agreement on a forest convention. The major timber-producing countries still do not support the idea of a legally binding instrument. In addition, given the complexity and vastness of the subject, it is very difficult to consolidate the wide range of opinions involved. The reestablished forum, however, is also mandated to identify the possible elements of and work toward consensus for international arrangements and mechanisms, such as a legally binding instrument. The CSD will conduct a comprehensive review at its eighth session in 2002.

Oceans and Seas

Oceans and seas have long been regulated by international agreements. Most of these treaties, however, regulate the free access to, or the free use of, the sea rather than the protection of the marine environment.

United Nations Conference on the Law of Oceans and Seas. The most comprehensive international agreement on the marine environment, which is regulated by global and regional rules, is the United Nations Conference on the Law of Oceans and Seas (UNCLOS). UNCLOS has the objective to create "a legal order for the seas and oceans which will facilitate international communication, and will promote peaceful uses of the sea and oceans, the equitable and efficient utilization of their resources, the conservation of their living resources, and the study, protection and preservation of the marine environment." The adoption of UNCLOS was a step forward in marine environmental protection since a global authority was established over oceans and seas, and it was also a step forward in terms of environmental protection. Thus, UNCLOS has the potential to end the "tragedy of the commons" because the maritime zone outside a state's national jurisdiction often serves as an area to discharge its environmental wastes instead of disposing of its wastes on its own territory.

In the context of marine pollution, UNCLOS does not provide for specific provisions but often makes reference to international law outside the convention. In the case of ocean dumping by ships, for example, it requires states to establish a legal regime that at least meets the standards set by the International Maritime Organization and the International Convention for the Prevention of Pollution from Ships. Such an obligation, however, is not provided for in terms of land-based maritime pollution, which encompasses all resources of maritime pollution from the land or coast, such as pollution by agricultural effluents, and which is the most significant source of the marine pollution. In this context, UNCLOS only calls on states to take into consideration international rules and standards related to land-based pollution. In terms of marine environment, it calls on states to conserve living resources and to cooperate in view of the conservation of marine mammals.

Progress in fighting marine pollution has also been made through the adoption of the Global Programme of Action for the Protection of the Marine Environment from Land-Based Activities Including through the Atmosphere at the 18th session by the UNEP governing council (pollution by land-based activities is also dealt with by UNCLOS, the Paris Convention, the OSPAR Convention, three UNEP regional seas protocols, and the Baltic conventions). UNEP was mandated to host the secretariat. The global program of action is expected to contribute to the implementation of UNCLOS provisions concerning the protection and preservation of the marine environment (Part XII), particularly pollution from land-based sources. This global action program was preceded by the Montreal Guidelines. Despite the fact that these guidelines are not mandatory, they could provide the basis for a future global instrument since they provide for basic obligations by states to prevent the pollution of the marine environment of other states and areas beyond the limits of national jurisdiction and foresee measures to prevent, reduce, and control pollution from land-based sources.

Since UNCED, progress has been made in "negotiation of agreements and voluntary instruments for improving the conservation and management of fishery resources and for the protection of the marine environment." Also, the "conservation and management of specific fishery stocks for securing the sustainable utilization of these resources, has been achieved." However, "the decline of many fish stocks, high levels of discards, and rising marine pollution continue" (program on the further implementation of Agenda 21, para. 36).

At Rio Plus 5, oceans and seas were one of the major issues because the protection of the marine environment was considered to be especially relevant to coastal zones and small island states. States were asked to implement the CSD decision (Decision 4/15) that called for periodic intergovernmental reviews by the CSD of all aspects of the marine environment and related issues, as described in Chapter 17 of Agenda 21. Governments were also asked to prevent or eliminate overfishing through the adoption of management measures and mechanisms to ensure the sustainable management and utilization of fishery resources. At its fourth session, the CSD also stressed the importance of effective conservation and management of fish stocks.

Freshwater

Water resources are essential for satisfying basic human needs, health, and food production and the preservation of ecosystems as well as for economic and social development in general. In view of the growing demands for water as a limited resource, and its unsustainable use, Rio Plus 5 gave the highest priority to the serious problems in relation to freshwater.

Water in the sense of a natural resource was mentioned in the Stockholm Declaration (Principle 2: "Natural resources including water . . . must be safeguarded for the benefit of present and future generations"). Agenda 21 provides for seven program areas to protect the quality and supply of freshwater resources which present a comprehensive approach to the protection of water resources. The general objective of these program areas is that adequate supplies of water of good quality are maintained for the entire population of this planet while preserving the hydrological, biological, and chemical functions of ecosystems (para. 18.2). In order to achieve this objective, innovative technologies, including the improvement of indigenous ones, are needed. Due to the widespread scarcity, gradual destruction, and aggravated pollution of freshwater resources in many world regions, Agenda 21 calls for water resource planning and management (para. 18.3).

On the occasion of Rio Plus 5, freshwater was one of the major issues. The conference called for an integrated approach to watershed management, including issues related to pollution and waste and the interrelationship between water and land, including mountains, forests, upstream and downstream users, estuarine environments, biodiversity and the preservation of aquatic ecosystems, wetlands, climate, and land degradation and desertification. The General Assembly stated that international cooperation for technological transfer and the financing of integrated water resources programs and projects have to be strengthened, and the participation of major stakeholders has to be ensured. It was recognized that developing states have limited resources that need to be shifted to higher value, less water-intensive modes of agricultural and industrial production. Moreover, sustainable production has to be accompanied by training and education measures. Global consensus was reached on the fact that water constitutes a social and economic good with a vital role in the satisfaction of basic human needs, food security, poverty alleviation, and the protection of ecosystems. Finally the urgent need for action in the field of freshwater was emphasized. An intergovernmental dialogue under the aegis of CSD was recommended to built a consensus on the necessary actions and in particular on the means of implementation and on tangible results and to initiate a strategic approach relating to all aspects of the sustainable use of freshwater for social and economic purposes.

Moreover, the COP of the Ramsar Convention recognized that wetlands also have important hydrological functions, including groundwater recharge, water quality improvement, and flood alleviation, and that there is an inextricable link between wetlands and water resources.

Atmosphere

Convention for the Protection of the Ozone Layer and Montreal Protocol on Substances That Deplete the Ozone Layer. The objectives of the convention are the protection of human health and environment against adverse effects resulting from or likely to result from human activities which deplete or are likely to deplete the ozone layer. Negotiated under the aegis of the UNEP, the convention was adopted in 1985 and entered into force in 1989. In order to create a real incentive for developing countries to join the agreement, the convention was the first global convention to introduce differentiated obligations for developed and developing states and to set up a financial mechanism for the fulfillment of their obligations. In addition, it provides for cooperation for the protection of the stratospheric ozone layer through the exchange of scientific, technological, and other relevant information.

Several protocols have been amended to phase out the production and consumption of specified ozone-depleting substances in accordance with an agreed schedule since the convention did not foresee any targets or timetables for emission reductions of those substances. The Montreal Protocol was adopted in 1987 and was celebrated as a landmark agreement in the context of the environment since states could agree on specific legally binding obligations. It was amended to phase out the ozone-depleting substances chlorofluorocarbons (CFCs) and halons by the Year 2000. In 1992, it was again revised, moving forward the phase-out date to 1996 for CFCs and to 1994 for halons, except for essential uses (a different schedule has been set for developing countries; Art. 5). The Copenhagen Amendment, adopted by the Fourth Meeting of the Parties to the Montreal Protocol and entering into force on September 22, 1993, advanced the phaseout of all the controlled substances of Annex A and B to the Protocol to January 1, 1996, except for halons, which were phased out by January 1, 1994.

The United Nations Framework Convention on Climate Change. The convention was opened for signature at the June 1992 Earth Summit in Rio de Janeiro. It was signed during the Earth Summit by heads of state and other senior officials from 154 countries (and the European Community) and entered into force on March 21, 1994. By November 5, 1997, 171 states were party to the convention. Its ultimate objective is to stabilize greenhouse gas concentrations in the atmosphere at a level that prevents dangerous anthropogenic interference with the climate system. Such a level should be achieved within a time frame sufficient to allow ecosystems to adapt naturally to climate change, to ensure that food production is not threatened, and to enable economic development to proceed in a sustainable manner (Art. 2). The driving element behind the adoption of a global instrument was not the clear-cut evidence of existing environmental harm but rather scientific uncertainty about the impact of human activities. This precautionary approach shows the international community's increasing sensitivity to the environment.

The convention nevertheless has fewer binding commitments than other environmental agreements and therefore is still a work in progress. In contrast to the specific legally binding obligations set out in the Montreal Protocol and its amendments, agreement between states on specific commitments in the framework of CCC is much more difficult since greenhouse gases cannot be substituted by other means and therefore entail severe economic consequences. The convention comprises a reporting mechanism, including inventories of greenhouse gas emissions by source and greenhouse gas removals by "sink," and reports on national climate change strategies. Furthermore, parties have to adopt national programs for mitigating climate change and develop strategies for adapting to its impacts. They also agreed to promote technology transfer and the sustainable management, conservation, and enhancement of greenhouse gas sinks and "reservoirs" (such as forests and oceans). In addition, the parties promised to take climate change into account in their relevant social, economic, and environmental policies; to cooperate in scientific, technical, and educational matters; and to promote education, public awareness, and the exchange of information related to climate change.

The principle of differentiated responsibilities, already referred to in the preamble of the convention, is applied in the context of the adoption of policies and measures aimed at returning individually or jointly their greenhouse gas emissions to 1990 levels by the Year 2000. Members of the Organization for Economic Cooperation and Development should "take the lead in modifying long-term trends in anthropogenic emissions." Countries with economies in transition are allowed a certain degree of flexibility.

The convention established two permanent subsidiary bodies: the SBSTA and the Subsidiary Body for Implementation (SBI). The SBSTA provides the link between the scientific, technical, and technological assessments and the information provided by competent international bodies and also the policy-oriented needs of the COP. The SBI develops recommendations to assist the COP in its review and assessment of the implementation of the convention and in the preparation and implementation of its decisions. Furthermore, the COP created another subsidiary body at its first session in Berlin, known as the Ad Hoc Group on Berlin Mandate for "a protocol or another legal instrument," which would inter alia set quantified objectives for limiting and reducing emissions of greenhouse gases within specified time frames for developed country parties to the convention. The Ad Hoc Group on Article 13 (AG 13), which deals with the resolution of questions regarding implementation, was set up to consider the establishment of a multilateral consultative process and its design. The permanent secretariat of the CCC is located in Bonn, Germany, and is functionally independent. Its head is appointed by the UN secretary general in consultation with the parties.

The convention defines a mechanism for providing financial resources for projects which address climate change. This financial mechanism is operated, on an interim basis and under the guidance of the COP, by the Washington, DC-based GEF (*http://www.unep.ch/iucc*). Projects supported by the GEF are implemented through three agencies: the UN Development Program, UNEP, and the World Bank. The GEF provides resources for investment projects having global environmental benefits, such as projects that reduce emissions of greenhouse gases by increasing energy efficiency and the use of renewable energies. The GEF also supports the increasing of the capacity of developing countries to implement the convention and to prepare national communications of the COP. It promotes bilateral and multilateral cofinancing and the leveraging of private sector participation and resources.

The COP also instituted in 1995 a pilot phase for activities implemented jointly (AIJs) among Annex 1 parties and, on a voluntary basis, with non-Annex 1 parties at their request. The financing of AIJ shall be additional to resources provided through the financial mechanism and ODA flows. AIJs are to yield real, measurable, and long-term environmental benefits related to the mitigation of climate change which would have occurred without these projects.

United Nations Convention to Combat Desertification in Countries Experiencing Serious Drought and/or Desertification, Particularly in Africa

Drylands cover about 30% of the world's terrestrial surface and are home to 900 million people. Defined as arid, semiarid, and dry subhumid areas, they are among the world's most fragile ecosystems. Various factors contribute to widespread natural resource degradation in dry areas: climatic cariation, inappropriate land use and agricultural practices, increasing population density, economic pressures, and changes in land tenure patterns.

In June 1992, the UNCED recommended that the UN General Assembly establish an intergovernmental negotiating committee, which began work in 1993 to prepare a convention. The convention was adopted in Paris on June 17, 1994, and entered into force on December 26, 1996. As of June 30, 1996, 159 countries were parties of the convention. Long before the adoption of the convention, a wide range of actions were put forward to respond to the problems related to desertification. The convention therefore concentrates on providing guidelines for action rather than recruiting support for the need for action. Its objective to combat desertification and mitigate the effects of drought will be primarily implemented by national action programs. In this context, strong emphasis is placed on the participation of populations and local communities in the decision-making process of the convention since they are often the most knowledgeable. This "bottom-up" approach also aims at establishing a better understanding of the nature and value of land and scarce water resources in affected areas and to work toward their sustainable use. The CCD does not provide for a new financial mechanism to support these activities. Instead, a global mechanism is foreseen to increase the effectiveness of the existing financial mechanisms. However, the convention does call for "new and additional funding." In institutional terms, the convention follows the Rio pattern. At the first COP meeting in Rome in October 1997, it was decided that the permanent secretariat of the CCD should be located in Bonn, Germany.

In terms of coordinating activities, the IPF has called for a report on experiences in afforestation and rehabilitation of forest ecosystems, especially in countries with fragile ecosystems affected by desertification and/or drought.

Basel Convention on the Control of Transboundary Movements of Hazardous Wastes and their Disposal

The convention was adopted on March 22, 1989, and entered into force on May 5, 1992. As of July 1996, the depository of the convention had received 100 instruments of ratification, accession, acceptance, or approval. The secretariat of the convention is administered by the UNEP.

International concern over hazardous wastes begin in the industrialized countries. As a response, stricter domestic rules for an environmentally sound management of these wastes were passed, which led to increasing exports of hazardous wastes from industrialized to developing countries.

In view of this situation, the convention aims at the strict regulation of transboundary movements of hazardous wastes by submitting its parties to the obligation to manage and dispose wastes in an environmentally sound manner. The main principles of the convention include the reduction of hazardous wastes to a minimum consistent with their environmentally sound management. The treatment and disposal of the wastes should take place as close as possible to their source of generation, which again indicates the problem of exports. The generation should be reduced and minimized at the source. To control the implementation of these principles, the convention provides for a monitoring mechanism, assistance for environmentally sound management, and promotes cooperation between parties. Moreover, technical guidelines for the management of hazardous wastes are developed.

At the first conference of the parties in 1992, an ad hoc working group of legal and technical experts was established to prepare a draft protocol on liability and compensation for damage resulting from the transboundary movement of hazardous wastes and their disposal. Given the fact that legally binding obligations on liability and compensation could entail tremendous costs, state parties had difficulty reaching an agreement on the matter. At its second meeting, the COP decided to establish an immediate prohibition of all transboundary movements of hazardous wastes, which are destined for final disposal from OECD to non-OECD countries, that was later included in an amendment to the convention. In terms of implementation, a consultative subgroup of legal and technical experts to study all issues related to the establishment of a mechanism for monitoring implementation of and compliance with the convention was established by the COP's third session.

Hazardous Substances and Activities

UNEP set out the program Hazardous Substances and Activities to ensure that international trade in all types of potentially harmful chemicals is conducted in a safe and environmentally sound manner. To this end, it will undertake to review, update, and strengthen the London Guidelines for the Exchange of Information on Chemicals in International Trade, promote their wide and effective implementation, and consider the development of legally binding instruments and other appropriate programs. In addition, UNEP provided an international forum to private sector parties (industry and NGOs) to prepare a code of ethics on the international trade in chemicals to complement the London guidelines. The CSD, at its second session in 1994, called for the code's application on as wide a scale as possible and reinforced industry's role as a major player in furthering the objectives of Agenda 21.

UNEP, ILO, and WHO have also established and Intergovernmental Forum on Chemical Safety, whose need was identified in Agenda 21 (Chapter 19). UNEP, ILO, FAO, WHO, UNIDO, and OECD agreed on a memorandum of understanding concerning the establishment of an Inter-Organization Program for the Sound Management of Chemicals (IOMC), which entered into force on March 13, 1995. IOMC is composed of representatives of the six participating organizations and meets regularly to discuss matters of common interest in the field of chemicals. UNEP also participates in an intergovernmental negotiating committee, aiming at the preparation of an international legally binding instrument for the application of the "prior informed consent" procedure.

Finally, UNEP started an assessment process concerning the need for international action to develop a global, legally binding instrument for the reduction and/or elimination of emissions and discharges and, where necessary, the elimination of manufacture and use and illegal traffic of persistent organic polluters.

Conclusion

Following the 1972 UN Conference on the Human Environment at Stockholm, efforts by intergovernmental, nongovernmental, national, and regional bodies in the service of the environment have been joined. The crucial institutional development in this process has been described in this article, with a focus on the institutional development within and related to the UN system. Coordinated actions within the UN on the problems of the environment have led to many environmental treaties and to institutions to implement these. The 1992 UNCED promoted a concept of environmentally sound development, outlined in the sustainable development concept of the Brundtland Commission Report and the UN perspective document for the Year 2000 and beyond. This concept manifested as an action program, Agenda 21, which was adopted at the Rio conference in 1992.

The period from 1972 to 1992 was characterized by the idea that all countries face common environmental problems and by the global determination to harmonize policies relating to the environment. This determination also had impacts on the development of scientific disciplines, both in the social and in the natural sciences analyzing the relationships between socioeconomic driving forces, environmental changes, and impacts on humankind. The advancement of scientific knowledge about the environment was crucial in the international effort to respond to threats to the environment, taking into account increasing complexity, fragmentation, and specificity of the issues. Worldwide scientific cooperation and research programs and immense technological progress assisted in the monitoring and assessment of the state of the environment.

The UN system faces several challenges in regard to the environment. The global nature of environmental problems calls for concerted international action. To some extent, this is made easier through recent regionalization of political decision making (i.e., EU, ASEAN, and NAFTA). On the other hand, national sovereignty concerns are still effectively blocking strong binding commitments to address environmental problems. The fact that governments increasingly rely on markets and economic instruments, in addition to classical regulation tools, implies that the UN system will also have to adopt a more market-based approach, internally and externally. The ongoing globalization of the economy and the environment will lead to new challenges in this century that the existing institutions will have to face.

Nongovernmental actors or nongovernmental organizations are associations not between states exclusively on the basis of an international treaty setting out the specific purpose of the organization but between nonstate actors.

Acknowledgments

The authors thank Mr. Peter Dogse and Ms. Sarah Titchen for a critical review of the article and all convention secretariats for their comments and corrections to an earlier version of the article.

Reference Cited

von Droste, B., Plachter, H., and Rössler, M. (Eds.) (1995). *Cultural Landscapes of Universal Value. Components of a Global Strategy.* Fischer, Jena.

General References

Caldwell, L. K. (1984). *International Environmental Policy. Emergence and Dimensions.* Duke Univ. Press, Durham, NC.

Glowka, L., *et al.* (1994). A guide to the Convention on Biological Diversity, Environmental Policy and Law Paper No. 30. IUCN, Gland.

McCormick, J. (1997). *The Global Environmental Movement.* Wiley, New York.

Sand, P. (Ed.) (1992). *The Effectiveness of International Environmental Agreements — A Survey of Existing International Instruments.* Cambridge Univ. Press, Cambridge, UK.

Sand, P. (1992). UNCED and the development of international environmental law, Yearbook Vol. 3.

Sand, P. (1990). *Lessons Learned in Global Environmental Governance.* World Resources Institute, Washington, DC.

Internet Sites

Basel Convention: *http://www.unep.ch/sbc.html*
Biodiversity Forum: *http://www.worldcorp.com/biodiversity*
CITES: *http://www.unep.ch./cites*
Climate Change: *http://www.unep.ch/iucc.html*
Convention on Biological Diversity: *http://www.biodiv.org*
Convention on Migratory Species: *http://www.wcmc.org.uk/~cms/*
Desertification Convention: *http://www.unep.ch/incd/*
Earth Council: *http://www.ecouncil.ac.cr*
Food and Agricultural Organization: *http://www.fao.org*
Global Environmental Facility: *http://www.worldbank.org/html/geg/geg.html*
IMO: http://www.imo.org
International Panel on Climate Change: *http://www.unep.ch/ipcc*
IUCN: *http://www.iucn.org/themes/law/elp*
Montreal Protocol: *http://www.unep.org/unep/secretar/ozone/home.htm*
Ramsar: http://iucn.org./themes/ramsar
United Nations: *http://www.un.org*
United Nations Environmental Program: *http://www.unep.org*
World Resources Institute: *http://www.wri.org*

BRUNO TREZZA

Department of Public Economics
Faculty of Economics and Commerce
University of Rome, "La Sapienza"
Rome, Italy

The Myth of Sustainable Development

Can the idea of sustainable development be put into practice? In this article, I illustrate an approach to the problem of the defense of the environment which differs from the traditional tack. The main idea is that the environment should be considered a complex, self-organizing system in which human activities take place; the important element is how these activities interact with the cycles that constitute the environment. The problem is that of reconciling the needs of the economic system with the need not to disrupt the cycles of the biosphere and thereby to preserve the environment. Using certain indicators relative to the conversion of energy by photosynthesis and to the overall quantity of energy that man uses, I demonstrate the speed at which the biosphere's cycles are being disrupted and the urgency of the need to develop a new way of dealing with the problem. Once the relation between economic system and environment is put into the right perspective, it is clear that the only way to confront the problem is to consider the environment not as a resource to exploit but rather as an asset to preserve.

Introduction

The degree to which the environment has been destroyed is well-known and it is perfectly clear that the current level of impact of human activities on the environment cannot be maintained without irreversible damage being caused to life on Earth. In its report to the United Nations Conference on Environment and Development (UNCED), the Business Council for Sustainable Development stated that "we cannot continue with our current methods of employing energy, managing forests and agriculture, protecting plants and animal species, handling urban development and production of industrial goods" (Schmidheiny, 1992). The World Resource Institute (1992) has declared that "the world today is not projected toward a sustainable future but, rather, toward a variety of human and environmental disasters." The World Bank, in its examination of predicted increases in production by the Year 2035, has affirmed that "if environmental degradation and pollution were to increase in parallel to the increase in production, the result would be a terrible environmental disaster" (World Bank, 1992). The Worldwatch Institute has concluded that "if we do not succeed in converting our self-destructive economy into an economy that is sustainable from the environmental point of view, future generations will be overcome by environmental degradation and social disintegration" (Brown, 1993).

Even at a quick glance, the list of the damages that have been caused and the level of environmental degradation that has been reached are staggering: overall increase in

FIGURE 1 Deforestation in the Amazon (photo courtesy of Leroux/Explorer/Overseas).

temperature, reduction of ozone in the stratosphere, deposition of acid substances, toxic wastes, extinction of species, deforestation, degradation of soils, and depletion of hydraulic, ichthyic, and renewable resources. This list, which is already worrisome, must be extended as a result of the interrelations that exist between different phenomena, insofar as different kinds of environmental damage tend to reinforce their reciprocal negative effects: Deforestation (Fig. 1) is the cause of the degradation of soils, the extinction of living species, and the increase in temperature; the depositing of acid substances kills the forests (Fig. 2) and pollutes groundwater tables; UVB extraradiation, due to the reduction of the ozone shield, damages fish larvae and phytoplankton; and so on. In brief, all the problems become more serious and environmental degradation is simultaneously increased and accelerated.

Between 1952 and 1991, the population of the world increased by a factor of 2.17; annual production of many traditional products such as steel, zinc, and aluminum increased by a similar factor, between 1.6 and 3.5 depending on the products, but all products connected with petroleum and transportation have increased at much higher rates, reaching factors of increment of 5.7 times. Since these productions and the consumption that is connected with them are among the major causes of pollution, not only have the economic system and human activities developed at a very high rate but also the highest rates of growth have taken place in sectors where pollution is greatest.

FIGURE 2 The effects of acid rain on a fir tree (photo courtesy of Overseas).

To complete the picture, one must take into account the predicted increase of the world population: The average projection among those computed by the United Nations (UN) assumes that by the Year 2030 the world population will be 8670 million, as long as the rate of growth decreases from the rate of 1.6% in 1990 to 0.9% in 2030 (UN, 1995). Moreover, by 2025, 85% of the world population will be concentrated in the less wealthy countries, with an age distribution in favor of young persons who, plainly, require higher growth rates in order for jobs to be created. Since the rich countries will tend to maintain, if not to further increase, their standard of living, the burden on the environment in the next 30 years cannot but increase considerably.

Equivalent Energy from Global Photosynthesis and from Petroleum, Coal, and Gas

A simple but efficient way to evaluate the speed and force with which human activities affect the environment consists of comparing the equivalent energy derived from the production of crude oil, coal, and gas with the solar energy that was converted by global photosynthesis, for example, in the years 1952 and 1991 (Table 1). The data show that the equivalent energy derived from fossil fuels has increased 3.33 times since the beginning of the 1950s, whereas the energy converted by photosynthesis (vegetation plus marine plankton) may be taken as having been constant, even though it has probably diminished by at least 10% as a result of the destruction of the tropical forests.

TABLE I

Comparison between the Equivalent Energy Obtained from Crude Petroleum, Coal, and Natural Gas and Energy Converted by Global Photosynthesis[a]

	PRODUCTION (10^9 GJ/YEAR)	
ENERGY SOURCE	1952	1991
Crude petroleum	24.10	124.68
Coal	35.48	101.09
Natural gas	10.33	72.80
Total	69.91	233.05
Global photosynthesis	108.78	108.78
Total	178.70	341.83

[a] Calculations were carried out by S. Pignatti and W. Bocola of the "La Sapienza" University of Rome.

Until about 1952, the part relative to photosynthesis was considerably greater than that used for industrial processes in the overall flow of energy that passed through the biosphere; 40 years later, the amount of energy used for purposes of production and consumption was more than twice that employed for photosynthesis. In terms of energy, the relation between the energy used by man and that used for photosynthesis, which is the basis of the life cycle, has increased from 64 to 214%. At a certain point, in the course of the 1960s, man became the most important factor in the evolution of the ecosystem, but the scientific community and humanity as a whole seem not to have become sufficiently aware of this epochal event. Whereas the process of photosynthesis produces biological molecules and oxygen, which are entirely compatible with the life process and wholly recyclable, the products that derive from the industrial use of energy are almost always difficult to recycle and, therefore, polluting. The amount of damage is immense and there is not much time left in which to intervene.

As discussed later, the comparison between the energy used for photosynthesis and that used by man is far more significant than any comparison that takes into account the total amount of energy derived from the sun. The reason for this is that the energy used for photosynthesis is at the base of the life cycle on Earth, and it is precisely this cycle that is being compromised by pollution and environmental degradation. Only such a comparison gives one a significant idea of the amount of impact that human activity has on the ecosystem, intended as a system which is evolving.

The gravity of the situation should have led to an immediate and drastic modification of the relation between human activity and the environment, thereby also changing the relation between the economic system and the biosphere; to accomplish as much, it would have been necessary to profoundly change man's overall attitude toward the ecosystem and the cultural model that underlies it. What took place instead was only a modest attempt to intervene in the process of environmental degradation, limiting the impact of the new situation on the economic system and its requirements. This was accomplished by drawing attention to sustainable development—that is, the opportunity to create a situation in which the demands of development and functioning of the economic system might remain unaltered and at the same time allow the environment to be preserved.

The conditions for sustainable development have not been found and environmental degradation has continued almost entirely undisturbed, whereas theoretical discussion has been animated by many contributions and empirical analysis extended by many new instruments, none of which are really suited to deal with the problem.

In this article, I present an overview of the theoretical foundations of the economic analysis of the environment, examine the tools that doctrine recommends and that have been adopted in practice, and suggest an alternative and, I hope, a more correct method of dealing with the extremely serious theme of the relation between human activity and environment.

The State of Theory

From the Classics to the Neoclassics

The first considerations of the relation between economic activities and environment date back to classical thought, especially to Malthus (1826/1946) and Ricardo (1821/1947), in relation to a study on the impossibility of sustainable development as a result of the depletion of resources occurring simultaneously with increases in population. Whereas the former grew according to a linear progression, in fact, it was believed that population grew at a geometric rate.

Through these authors' analysis, the hypothesis of an unlimited growth rendered possible by the strengthening of markets and continuous increase in the division of labor in the manufacturing industries, already contemplated in the work of Smith (1789/1927), was turned upside down. Whereas Smith, in fact, failed to take any factor that was external to the economic system into account, Malthus and Ricardo considered it in terms of a dynamic relation with environmental resources and the possibility of their increase; as a result, the only way out was believed to be a stationary state in which a population of finite dimensions gave rise to a production and consumption that were constant and permanent over time. The renewal of resources would guarantee that this stationary state could be perpetuated.

Malthus's and Ricardo's thesis was also fundamentally shared by Mill (1871/1952), who in this regard introduced the important consideration of the relative scarcity of natural resources so as to consider the economic effect of price increases before the attainment of the ultimate limit of the stationary state. Moreover, Mill also introduced another important consideration — that of nature as a supplier of services that are connected to the quality of life and that may be compromised by continuous economic growth. In this sense, he did not consider the stationary state as the sad result of a development at the limits of sustainability but rather as a condition in which the freely chosen limitation of population, production, and consumption might give rise to an adequate quality of life. Mill wrote, in fact, "I sincerely hope, for the good of posterity, that they accept the stationary state well before things inevitably force them to suffer it."

The developments that followed classical theory, the neoclassical theory born in approximately the middle of the eighteenth century, stemmed from the law of diminishing returns and proceeded with the simple generalization according to which an increase in the quantity of a product, whether in agriculture or industry, may be the result of an increase in any one of a number of productive factors, although in decreasing increments, and that equivalent proportional increases of all the factors should in any case give rise to increases in the quantity of product of similar proportion (linear homogeneity of the productive function). This notion of the homogeneity and divisibility of productive factors enabled neoclassical theory to abstract the productive process from the physical and historical conditions of production and to view it solely in relation to the system of consumers' choices. A closed universe thus came to be created — the realm of economics — in which the activities of production of goods and services had no connection with the environment and appeared to be related only to each other by relations of functionality.

In this context, the environment was considered only in terms of nonrenewable resources, such as petroleum or mining deposits, which posed the problem of how best to manage them: a relation between immediate and future consumption or, in other words, between the well-being of present generations and future ones. A pertinent analysis in this regard is that of Hotelling (1931), who developed a rule for the management of such resources which could bring the problem back within the confines of economic categories. According to the rule, prices are indexes of relative scarcity — scarcity in relation to uses, either direct or indirect, for purposes of consumption — so that the value of an exhaustible resource should increase over time as its depletion approaches or as its relative scarcity increases. In such a situation, a correct use of the resource implies that it may be employed every time the rate of growth of its value is less than the profitability of an alternative financial activity, represented by the current rate of interest on capital.

In effect, the rule is only the application of the fundamental rule of neoclassical economics in regard to decisions with respect to time: A reduction in the direct or indirect use of goods for purposes of present consumption in favor of future consumption is only possible when the increase in value thus attained is less than the interest rate. A traditional tale reports that George Washington cut the cherry tree on his property when the rate of growth of the tree turned out to be less than the interest rate.

Subsequently, the theory introduced an opportune distinction between renewable and nonrenewable resources, stock resources, and flow resources. Renewable resources may be employed up to the point at which their renewability is at risk; beyond that point, a process of deterioration would begin which may sometimes even be irreversible. It was on these bases that the concept of a minimum level of conservation was introduced (Ciriacy-Wantrup, 1968).

Neoclassical Theory and the Environment

The reduction of the problem of the environment to that of resources and their management enabled neoclassical theory to face the emergence of the environment issue by means of the conceptual tools that are proper to economic analysis and, in particular, to the theory of affluence, which were perfected in the 1960s.

The pertinent conceptual scheme, in fact, presents the environment as a series of goods (the environment's resources) whose use is valued, as with any other good, by the members of society. As with every other good, the valuation is accomplished in relation to the good's capacity to contribute to the satisfaction of the consumers' utility. Since environmental goods are usually freely available, there is no price which need be paid in order to use them, and therefore they cannot be subject to efficient valuation and management.

To attain efficiency, various tools have been identified, just as others have been set up to keep track of environmental impact from various standpoints. The latter, and all the other tools that economic theory takes into consideration, will be critically examined later.

The Debate on Exhaustible Resources and Sustainable Development

An efficient use of resources does not, plainly, prevent them from being depleted. As a result, a great deal of attention has been paid to the problem of the depletion of natural resources. An empirical study carried out in 1963 by Barnett and Morse on behalf of the American foundation Resources for the Future showed that in the period between 1870 and 1957, the real cost of productive factors provided by nature had decreased relative to other production costs. The cause of this phenomenon, it was maintained, was the compensatory benefits resulting from technological progress (Smith and Krutilla, 1979).

At the beginning of the 1970s, this optimistic position was put into question both by the oil embargo of 1973, which gave rise to an enormous abrupt increase in the prices of raw materials, and by growing environmental pollution. Also, the publication of the Massachusetts Institute of Technology (MIT) report for the Club of Rome on the limits of development (Meadows *et al.,* 1972; Mesarevic and Pestel, 1974) contributed to casting doubt on those premises. Until then, it had been a common opinion among economists that natural and environmental resources did not pose important obstacles to the continuous growth of the economy. The MIT report raised three problems — the environmental, food, and energy crises — referring them to the system Earth as to a unitary system. The report employed simulation models that, on the basis of fixed proportions in pro capita consumption of nonrenewable resources and of a constant growth of the economy and of population, predicted the total depletion of the main mining reserves and the attainment of a point of no return in terms of global pollution.

In this way, the environmental issue was again being posed in classical terms — in terms of economic growth and depletion of nonrenewable resources. The reactions to this neo-Malthusian approach were quite strong; both the methods and the data that had been used in the report were questioned, and models based on a more optimistic vision in regard to the scarcity of natural resources were advanced. In support of the critics, the hypotheses underlying the neoclassical view of the environment problem were reaffirmed, and it was claimed that technological innovation could compensate in an unlimited fashion the restrictions associated with the depletion of resources, that the energy and materials that can be used in the economic process are available in far greater quantities than those present in known reserves (insofar as there is a broad margin of replacement of different types of resources with one another), and that the market price mechanism is capable of signaling the growing scarcity of resources in such a way as to allow them to be rationed and conserved over time (Bekermann and Oates, 1974; Nordhaus, 1974; Kahn *et al.,* 1976).

The question of the relation between economic growth and environment could not but become tied with the problem of the development of the countries of the Third World, thus giving rise to that which later became known as sustainable development. At the start of the 1980s, it was shown how the worsening of the crisis of the Third World countries should be viewed both as a crisis of development and as an environmental crisis (World Bank, 1984). The environmental crisis was perceived for the most part as desertification, depletion of traditional energy sources, and destruction of tropical forests. At the United Nations Conference on the Human Environment held at Stockholm in 1972, sustainable development was identified as a general topic. In fact, the conference established that development and environmental protection had to be pursued together by means of rational planning. On that occasion, the United Nations Environment Program (UNEP) was also instituted as part of the UN.

UNEP played a considerable role in preparing the *World Conservation Strategy,* a text in which the contents of sustainable development were laid down for the first time (IUCN, 1980) and identified in the preservation of the ecological processes that are essential to the production of food, health, and other aspects of human survival on the planet; in the preservation of genetic diversity in the animal and plant worlds; and in the sustainable development of ecosystems, especially regarding the species exploited by man. The most authoritative exposition of sustainable development, however, was achieved with the publication of *Our Common Future* (World Commission on Environment and Development, 1987), the Bruntland report, in which such development was defined as "the development that satisfies the primary needs of the present without compromising the capacity of future generations to satisfy their own needs."

The vagueness of the definition elicited many subsequent stipulations; sustainable development has been defined by the World Commission on Environment and Development (WCED; 1987) as follows:

> Sustainable development requires that the fundamental needs of everyone be satisfied and that everybody have the opportunity to satisfy the aspiration to a better life.... Sustainable development must not deteriorate the natural systems that sustain life on the earth: the atmosphere, waters, soil, and living beings.

In the UNESCO report, economic development was distinguished from economic growth, given that the latter requires the use of greater quantities of materials and energy, whereas the former allows for improvements without an increase in aggregate consumption beyond levels sustainable by the environment (or qualitative evolution as opposed to quantitative evolution). However, in the final document of the International Conference of the United Nations on Population and Development that was held at Cairo in September 1994, it was stated once again that "the aims and ends (pursued by the international community) are economic growth within the limits of sustainable development, etc." (UN, 1994).

In any case, given the current conditions of the world and forecasts about the growth of the world population, independently of how the issues pertaining to growth, development, equity between generations, etc. have been dealt with, it is perfectly clear how such a definition implies not only the preservation but also the growth of productive potential and consumption on a world scale.

Defined in this manner, the theory of sustainable development paints the picture of the best of all possible worlds, given that it combines the defense of the ecosystem with the aspiration to a higher standard of living, which implies a consistent rate of growth on a world level. It does not ask the wealthy countries to make any sacrifices or the Third World countries to forsake their hopes of developing, but only that it all take place in respect of the environment without even posing the problem of the actual existence of such a possibility. In effect, in this manner the problem is not solved but merely removed.

Sustainable Development and Natural Capital

Sustainable development necessarily implies the nonexhaustibility of natural resources; some authors believe this concept to be the same as that of the maintenance of natural capital. *Natural capital* may be defined as "everything from which one obtains a flow of productive services over time and which is subject to control in its production processes" (Herffindal and Kneese, 1974). In this manner, the accent is placed on those things which natural resources and manufactured goods have in common and one assumes the possibility of replacing one with the other.

In this regard, economists at the London Centre for Environmental Economics have put much effort into sustaining that the maintenance of natural capital stock is the necessary condition for sustainable development, whereas one cannot instead count on the automatic evolution of technology in the direction of the protection of the environment. Sustainable development "can mean either that wealth pro capita grows over time thanks to the replacement of natural capital with capital that is reproducible by man; or that the growth over time of pro capita wealth is subject to the condition that natural capital not diminish" (Holmberg, 1992).

Of the two possibilities, the London school believes only the second to be practicable, insofar as it holds some environmental resources to be irreplaceable, that knowledge about ecosystems is limited (so that they must be preserved and not altered), that many phenomena are irreversible (such as the disappearance of biological diversity), and that the poorest and weakest are also those who suffer most from environmental degradation. It has also been noted how the principle of the conservation of natural capital is coherent with the view of the world which recognizes the right to existence of other species, independently of their utility to man (Pearce and Turner, 1990).

The value of a theory of this kind depends on how one identifies and measures natural capital. With regard to this aspect, the solution which has been advanced is far from satisfactory since, having abandoned the idea of measuring it in physical terms because of the heterogeneity of the resources to be taken into consideration, the quantification of environmental capital is viewed in monetary terms – in terms of the current prices of the various quantities of each resource. Aside from the fact that many resources do not have a price (air, water, natural habitat, etc.) and that, in any case, the prices ascribed to them are meaningless (Hueting, 1980; Barbier, 1989), such an approach falls back within the confines of traditional neoclassical theory insofar as it identifies natural capital with resources and tries to measure the latter using parameters that belong to economics. In addition to the question of principle that it raises, the theory in question represents neither a particularly significant nor an innovative variation of the neoclassical approach to the environmental problem.

On the whole, economic theory in its different expressions maintains that every economic activity requires the following of the environment: raw materials and energy as productive factors, the recycling of wastes, and the maintenance of a life-support system. Any increase in production places these elements under increased pressure. The concept of environmental impact coefficient (EIC) was developed to attempt a quantification of this pressure (Jacobs, 1991). Reductions in EIC are believed to be possible by means of changes in the composition of production toward forms that are less harmful, the replacement of the most harmful productive factors with others that are less polluting, and an increase in the efficiency of resources as a result of technological progress. The World Bank explicitly affirms that "whether environmental limitations will or will not place constraints upon the growth of human activity depends on the scale of replacements, technological progress, and structural changes" (World Bank, 1992).

To get a better idea of the effort which is required, the impact of human activity on the environment has been expressed by means of the equation $I = PCT$ (Ehrlich's equation), where I is environmental impact, P is population, C is pro capita consumption, and T is the impact coefficient per unit of consumption (Ehrlich and Holdren, 1974). In the equation, T is unknown and C and T cannot be considered to be independent of each other. Nonetheless, the equation allows one to perform some simple calculations to determine the effects of different scenarios. For example, if one assumes that the population will double in the next 50 years and that pro capita consumption does not decrease, T must be cut almost by half if one wishes environmental impact to remain constant. It is precisely by following this line of thought that Anderson (1992) maintains that, on the basis of the World Bank's projections in regard to population growth, it should be possible to reduce the EIC to the extent required by Ehrlich's equation in all the most important areas of environmental impact. Anderson, however, makes no attempt to integrate the results obtained in the various fields so that it is unclear whether the extra energy and industrial production which would be required to improve agriculture are compatible with the hypotheses of industrial pollution he advances elsewhere.

In principle, there will be situations in which there are inefficiencies (as a result of inadequate government policies or market failures) which may be emended both at the level of environmental impact and at the level of costs. Naturally, there will also be situations in which improvements in terms of environmental impact will imply increased production costs and, therefore, decreased economic growth. Opinions differ with regard to the relative importance of the two situations so that some authors write that "saving the environment will no doubt limit productive growth and probably imply lower income levels" (Tinbergen and Hueting, 1992), whereas the World Bank (1992) maintains that "the advantages of environmental protection are often great and its costs in terms of income reduction modest if the appropriate policies are applied."

The majority of effort has been directed at establishing the characteristics of sustainable development and finding efficient techniques to manage renewable and nonrenewable resources so as to reach a situation of sustainable development founded exclusively on the use of renewable resources. This problem has been approached primarily in the form of adequate mechanisms based on systems of prices, taxation, and rights on the environment that can limit productive processes and protect the environment. If we assume that there exists a coherent relation between an efficiently managed economy and the safeguarding of the environment, the main question appears to be that of passing from the current system, which is harmful to the environment, to a way of life that is compatible with sustainable development.

Ecological Economics

Toward the end of the 1960s, J. S. Mill's approach was resumed and the problem of the desirability of continuous economic growth in the industrialized countries was explicitly posed (Mishan, 1967); at the same time, the limits of such growth were analyzed for the first time in light of physical theory (Boulding, 1966; Georgescu-Roegen, 1971). The question of the convenience and opportunity of unlimited growth became matter for discussion, whereas such an opportunity had been taken for granted by neoclassical theory since it represented the result of the increased wealth of consumers, who were the ultimate beneficiaries of economic activity. According to Mishan, in advanced economies it is not individuals' absolute income that is important but their relative income so that the aim of continuous growth is not relevant to all of society since not everyone can improve on their relative condition; moreover, he emphasized the disquieting elements connected with continuous growth and advanced the idea of "zero growth" (Mishan, 1977). In this way, Mishan, on the one hand, dealt a blow at the fundamental axiom of neoclassical theory without criticizing it in sufficient depth and, on the other hand, simply did not realize the nature of the deep reasons that lead advanced industrial systems, which are based on profit, to tend toward a limitless growth.

The trend of thought that has dealt with the limits of growth in connection with the laws of physics in general, and of thermodynamics in particular, has been more interesting. The first law of thermodynamics and the principle of the conservation of mass clarify that matter and energy cannot be either created or destroyed but only transformed; as a result, economic activity, which consists of production and consumption, is nothing other than a particular modality of such transformations. Every form of matter and energy that is consumed must therefore return to the environment in the form of residue (w) and in a different physical state. Therefore, a linear process connecting resources and production does not exist; what exists instead is a circular process that also involves the environment and its capacity to recycle wastes. The process is not described by the relation $(R) \rightarrow (P, C)$, where R represents employed resources and P and C represent production and consumption, but rather by the relation $(R) \rightarrow (P, C) \rightarrow (w) \rightarrow (R')$, where R', or the resources left at the end of the process, will be equal to or less than R depending on the capacity of the environment to recycle the wastes, w; this function is represented by the symbol $\rightarrow$. It is only natural to think that by increasing w, a continuous increase in the use of resources will lead to a situation in which wastes can no longer be recycled or resources renewed.

Georgescu-Roegen (1971, 1977, 1979), while working within this framework, drew attention to the second law of thermodynamics, according to which matter and energy in isolated systems must of necessity proceed from a state of order to a state of disorder and, in any case, become increasingly less usable (increase in entropy). Every physical and biological process, and therefore every economic one, contributes to the increase in the level of entropy on the earth, insofar as it utilizes ordered materials and energy flows and transforms them into disordered forms of mater and energy. Moreover, it is impossible to recycle the totality of materials without employing a greater amount of energy than that which is obtained as a result. In terms of the previous formula, it should always be the case that $R' < R$.

The scarce relevance in practical terms of Georgescu-Roegen's position based on the necessary increase in entropy is due to the fact that the earth is not an isolated but an open system which receives energy from the sun so that the entropy on Earth is not necessarily bound to increase. Life, in fact, may be considered an increase in the amount of order connected to the continuous flow of energy that feeds it. Moreover, the energy that man uses — and, in any case, is contained in all fossil fuels — is extremely modest compared to the amount of energy coming from the sun. As a result, Georgescu-Roegen's assertion in reality states nothing more than the fact that fossil fuels, such as petroleum and mineral deposits, cannot be reproduced.

I believe, however, that a whole analytical trend has been born, which is still in a developmental phase, that goes by the appellative *ecological economics.* According to its approach, basic elements are provided (Klaassen and Opschoor, 1991) by the circular interdependence that incorporates the main processes taking place in the environment and takes into account the fundamental laws of ecology, by the independence of values and needs from individual preferences (e.g., society may attribute a higher value to the preservation of the environment than do single individuals), and by the existence of a hierarchy of values (as ultimate models, sustainability and environmental compatibility are adopted: the protection of species and ecologically practicable models for the employment of resources).

In carrying out these general principles, some authors have attempted to identify principles of coherence between economic activity and environment. In particular, it has been affirmed (Daly, 1991) that

1. One must limit the scale of the economy to the capacities of the biosphere, defining the former as the population level multiplied by pro capita consumption of resources and the latter as net primary product — that is, the solar energy that is captured by organisms capable of photosynthesis.
2. Technological progress must be directed toward a more efficient exploitation of scarce resources.
3. The exploitation rate of renewable resources must be less than the latter's rate of renewal, and the rate of residue emission must be less than the environment's rate of residue assimilation.
4. All investment aimed at exploiting nonrenewable resources must be compensated by equal investments in renewable ones.

According to this approach, the general norm contained in the first principle, which subordinates the scale of economic activity to the capacity of the biosphere to contain it, is qualified and specified by the definition of economic activity as the employment of resources and further specified by the subsequent principles which govern precisely the use that is made of natural resources. In this manner, the entire theoretical construct again falls back into the traditional approach that tends to reduce the problem of the environment to that of an adequate management of natural resources. The reference to the capacity of the biosphere considered as energy that flows due to photosynthesis and the problem of the environmental recycling of wastes lack a proper context and are expressed in terms that are not homogeneous with respect to the other affirmations. Moreover, it is naive to suppose that there might actually be a way of replacing nonrenewable resources with other, renewable ones; in effect, this amounts to nothing more than the idea, albeit in a different form, of replacing natural capital with man-made capital.

The Institutionistic School

A new line of thought based on the contributions of Myrdal (1972, 1978) and known as the *institutionistic school* (Bromley, 1989; Hodgson, 1991; Kapp, 1991) has dealt with the problem from the standpoint of systems in such a way as to recognize the complexity of natural systems and of the relation between economic systems and environmental ones.

According to this school (Swaney, 1972), there are two extreme manners of dealing with nature which one may define as *quiescent earth mentality* and *frail earth mentality,* with possible intermediate conceptions. In the first case, one is faced with an inert Earth on which humans can act as they please without putting biological life-support functions at risk. In the second case, one is faced with a very fragile ecological equilibrium which can be easily broken as a result of human activity. The institutionistic view falls in between these two extremes, in the sense that it considers the biosphere to be composed of a variety of ecosystems which are integrated with and interdependent on each other and which have different degrees of stability; some are highly resilient, whereas others are quite fragile and, if disturbed, can easily break down. As a result, the management of environmental problems must take more than one aspect into consideration. In the first place, the interrelations between ecosystems must be taken into consideration because they give rise to ecological chains of causes and effects and define systems of resources. The elimination of negative results is all the more difficult when the ecological chain underlying the phenomenon is long and complex so that an understanding of these relationships must necessarily accompany any action of impact on the environment. One must also bear in mind that man's capacity to influence the environment has reached unimaginable levels so that it is no longer possible to base oneself on the prejudicial hypothesis of an inert planet Earth.

According to the institutionistic theory, human activity causes some of the links in the ecological chain to break — links which cannot be easily reconstituted because of the irreversibility of evolutionary phenomena. There is also the possibility that as a result of the length and complexity of the ecological chain, it becomes difficult to trace the causes of undesired effects on the environment, just as it is difficult to evaluate the overall effects of human activities. Furthermore, the evaluation of these aspects must take into account the consideration that social systems, in turn, are complex systems so that the needs of society may differ from the needs of the individual subjects that comprise society. In particular, if resources were to become private property, even their efficient management on the economic level would give rise to results that are unacceptable from the point of view of society as a whole, if only because future generations would not be present on the markets that currently handle the environmental resources.

Some authors have examined the interrelations between values and social structures insofar as social values are affected by the structures that make up society which, in turn, are influenced by dominant social values: An interactive process thus comes into being, which is the basis for the evolution of human society. According to this theory, the development of the modern and contemporary era has taken place on the basis of the hypothesis that ecological systems were immutable and that there was no relation between human activity and modifications of the environment. An important consequence of such an approach has been the replacement of solar energy, which provided renewable energy, with fossil fuels (i.e., nonrenewable resources).

The institutionistic approach has the advantage of not wanting to develop a theory or methodologies aimed at making environmental parameters part of the economic process and of emphasizing the need for a pluralistic and interdisciplinary approach. It also highlights the importance of the ethical norms on which decision-making processes must be based, even if they reaffirm the assumption of man's basic needs and their a priori knowledge as the foundation of the policies to enact (Kapp, 1991).

Current Tools for Change and the Defense of the Environment

Analysis of the Cost–Benefit Relation

According to neoclassical theory, goods and services have a value only insofar as they contribute to satisfying the final consumption needs of the members of society. In a perfect market, such as the competitive one, prices vary according to the capacity of goods to contribute, whether directly or indirectly, to the satisfaction of final consumers. In this manner, the price system becomes an adequate indicator for the allocation of resources to different uses so as to maximize the satisfaction resulting from consumption. Often, however, either because the structure of the market is inadequate or because of other reasons, the price system fails to reflect these capacities adequately and therefore does not allow for an efficient allocation of resources: This is the case with environmental resources or goods. The main tool developed by economic theory to deal with such situations is cost–benefit analysis. The purpose of this analysis, in fact, is to identify, for society as a whole, the ensemble of costs and benefits — always understood in terms of final consumption — that are connected with any decision of an economic nature. In other words, it is a question of evaluating the ensemble of direct and indirect increments and reductions, present and future, of consumption levels according to the estimation of all the subjects involved.

This sort of analysis maintains that there is only one criterion for choice since the aim of any economic action is given by the level of consumption and by the degree of consumer satisfaction. Cost–benefit analysis has been and is still today at the root of the evaluations carried out by all international organizations (World Bank, Bank of Asia, etc.) and by almost all national governments in regard to projects of investment that are funded with international or public monies. Even recent methodologies, such as multicriterion analysis, which are not in line with such an approach are considered, in effect (although wrongly, as discussed later), as extensions of or improvements on this general criterion.

The importance of the criterion on which cost–benefit analysis is based is due to the need to assign a univocal aim to economic activity in such a way as to allow the value of any good, service, or resource to be determined in terms of a quantifiable unit of reference: consumption. If consumption and the satisfaction that the various subjects derive from it are, in fact, the sole aim of economic activity — in other words, if those activities which are aimed at satisfying consumers define economic space — then all value judgments are out of place in the evaluation of economic actions since, by interfering with the latter's ultimate aim, they situate themselves, by definition, outside of economic space. It is true that within the confines of such an analysis there is also room for value judgments in the form of the relative importance that is attributed to the various individual evaluations so that elements that are considered politically significant may be taken into account. Such value judgments, however, are always situated outside of economic space and do not deal with the evaluation of particular activities and decisions; rather, they only define contour conditions such as evaluations relative to the distribution of income between individuals or generations or over time — evaluations which, from the standpoint of economic analysis, present themselves as data that are a priori with respect to the process governing the efficient allocation of resources.

The idea of bringing the environmental problem back within the confines of an efficient management of resources, and therefore back into the heart of economic activity, explains why cost–benefit analysis has given rise to almost all the tools and methodologies for the evaluation of environmental problems.

Externalities, Pollution Taxes, and Other Tools

All the tools drawn from current economic theory and employed to monitor environmental impact are based on the idea that environmental damage can be handled by means of payments or economic evaluations. All evaluations of environmental damages caused by economic activity must be carried out by means of a functional relation capable of connecting the amount of pollution with the level of economic activity; in this manner, the optimal level of pollution can be established and efficiency can be attained.

The theoretical hypothesis relevant to the analysis of the interactions that exist between production and environment foresees that ecological factors not taken into account by the market can be internalized by making the producers of pollution pay a penalty, which is considered a form of negative externality. In reality, economic theory introduced the concept of *externality* in relation to phenomena that had nothing to do with environmental impact and only later gave rise to this extended form of applicability. Externalities arise when the activities of one subject interfere with those of another so that correctives must be introduced to avoid the situation in which decisions that subjects make independently of one another and that do not take account of interference may give rise to decisions that are inefficient at the systemic level. The problem arises from the incompatibility of the division of decisional capacity among the various subjects with respect to a system that cannot be subdivided into parts or isolated subsystems; when decisional authority belongs to a single subject, externalities disappear.

A specific application of this point of view consists of creating rights over environmental damage that can be bought and sold like any other good (Coase's theorem). In this way, a market is generated and efficient solutions may be achieved. Other tools are based on the taxation of environmental damage: Taxes on pollution may be introduced in such a way as to modify cost structures and induce individual producers to establish levels of activity that are compatible with an optimal level of pollution (Pigou's taxes).

These ways of establishing and evaluating the effects of human activity on the environment are rendered ineffective by the fact that the question is not correlated with an inefficient allocation of resources which may be corrected by means of an adequate system of penalties and taxes but, rather, with the far more complex problem of the interaction between human activity and environment, as the institutionistic school has always maintained. Environmental degradation is independent of both the problem of the distribution of decision-making power among different individuals and the difficulty of establishing adequate prices for environmental resources. Rather, it is connected with the overall impact that productive and consumption activities have on the cycles that constitute the life of the biosphere. Tools such as externalities and taxes on pollution, aimed at evaluating and eliminating environmental damage, are intrinsically misleading and unable to achieve the effects for which they were devised: Environmental damage and pollution considered from an economic standpoint and with an eye to the attainment of economic efficiency (Pareto's efficiency) are always different from environmental damage and pollution considered from the biophysical standpoint. In this way, when subjects act as economic agents they are induced to pay a price for the destruction of the bases of their own existence as living beings. In the same way, the most industrialized countries are authorized to pay the less developed countries for their acceptance of the pollution generated by the needs of the richest consumers.

Indicators

Indicators were devised to control factors that are crucial to the maintenance of the ecological bases of our lives and to draw attention to the damaging of the ecosystem. Indicators of sustainable development were introduced in environmental management to also establish whether a particular developmental trend is actually sustainable. These indicators refer to economic, social, and environmental elements and are based on indices of a monetary and energetic type.

The simplest economic indicators introduce prices for the purpose of modifying the gross national product (GNP) and enact compensations for environmental damage connected to the depletion of mining resources, erosion of the soil, pollution of waters, etc. In addition to the difficulty of finding acceptable definitions and measures of the quantities to include when revising the GNP, the greatest problem consists, on the one hand, of the impossibility of evaluating environmental goods that cannot be traced back to the market in monetary terms (thereby ignoring essential ecological functions) and, on the other hand, in the failed understanding of the interactions that exist between economy and environment. Other attempts are based on the identification and measurement of natural capital in such a way that a loss of value of the latter is considered to be an indicator incompatible with sustainable development.

Socioeconomic indicators are similar insofar as they attempt merely to take into account social magnitudes that are not considered in GNP calculations, such as levels of nutrition, health, and schooling. With the Index of Sustainable

Economic Welfare, a measure of long-term environmental damage associated with climatic changes was introduced. This index leaves the problem of the measurement and valuation of natural resources, such as forests and mines, unsolved and neglects, as do all economic and socioeconomic indicators, the interconnections (which are crucial to the understanding of the process of dynamic interaction) between economy and the environment with which economic activities clash.

Ecological indicators deserve greater attention, but their use is limited by our inadequate knowledge of the way in which the ecosystem works. Many indices use net primary production — that is, the energy that is left after having subtracted the energy used for breathing purposes by primary producers from solar energy. Such indices, however, only provide quantitative information about the ecosystem, whereas qualitative evaluations can only be achieved through the study of biodiversity, which is still in a pioneering stage.

Environmental Data and Economic Accounting

The UN (1992) integrated environmental and economic accounting method known as the System for Environment Accounting (SEA) is another tool for monitoring the effects of the economy on the environment. SEA was conceived as ancillary to the System of National Accounts of the United Nations (SNA) so that the need to describe the relation between environment and economic system would not generate difficulties within the SNA. For this reason, satellite systems have been created to deal with problems such as the production of housing, social security, and education outside of the accounting mainframe that is SNA.

Therefore, SEA had to be composed of separate blocks which contained the description of the interaction between environment and economic system in physical terms and imputed costs for the utilization of natural resources and extension of the boundaries of SNA by way of the consideration of internal and external environmental protection as a productive activity. This type of analysis, however, like the proposal Leontief advanced in 1974, does not allow one, for example, to follow the full course of wastes after they have become a part of the environment since a definite boundary to productive processes needs always to be introduced.

The identification of important activities in the environmental field has been put in the form of a national patrimony account which, however, requires a synthesis of physical data that will never be attained for a variety of reasons. The most important of which results from the difficulty in defining natural capital and in determining the difference between stock and flow. Moreover, the environment varies over time and the components of natural capital change considerably in relation to the timing and manner in which pollution manifests itself. For example, rain can clean the atmosphere, the percolation of chemical substances can pollute waters for years, and the carbon dioxide in the atmosphere can stay there for centuries and create a permanent greenhouse effect. The need to make SEA compatible with SNA necessarily implies that the environment is treated as a given resource, like any other productive installation, and not as a system in evolution of which human beings are part.

Problems of Measurement

There exists great uncertainty in regard to the measurement in biophysical terms of the impact of human activities on the environment, insofar as the "background noise" of natural changes often disguises the effects of human actions. More serious is the inadequacy of our understanding of the chain of causes and effects (Carpenter, 1994). Environmental response to human activity cannot be easily predicted because it is not linear and therefore not analyzable by means of the usual models. In these conditions, current efforts to understand the functioning of systems that involve many resources, multiple local ecosystems, and large-scale systems prove inadequate. The lack of valid statistical indicators has also been emphasized in a report by the Asian Development Bank (1990) in which it is stated that "statistical information on the environment is scarce, often inaccurate, rarely comparable from country to country, and rarely available in a temporal series covering a sufficient number of years."

In general, measurement difficulties are the result of ignorance about the ecosystem on a regional scale; the nonlinearity of the ecosystem's responses to stress; natural variability; errors in sampling and in the analysis of assays carried out in the field; inadequate monitoring in space and time; the deliberate counterfeiting of data for political, social, or cultural reasons; and the inadequacy of research on the relations between natural and controlled ecosystems (Carpenter, 1994).

Economic Activity and Ecosystem

The most important characteristic of economic systems according to neoclassical economic theory consists of the possibility of achieving positive results from a general standpoint by means of individual choices and decisions. All attempts to allow the market to monitor environmental effects are based on this characteristic.

The basic assumption underlying the theory is that productive process can be adequately defined. To this end, productive factors, products, and the relations that bind them must be clearly determined, which requires the existence of a specific limit within which the productive process is defined and outside of which no effect is taken into consideration. Beyond that limit, the productive process must not have any effects on the environment or, if it does, such an effect must be negligible for all practical purposes; if this is not the case, the limit would have to be indefinitely extended. In this manner, decisions can be made by individual operators once each has a complete picture of the entire process under his or her control and of all its consequences. Externalities, indicators, etc. are tools by means of which single operators can take specific and well-defined environmental damage into account. This allows one to maintain the decentralized decisional structure which is characteristic of the current economic system.

The problem of the relation between economic activity and ecosystem in this context is formulated in terms of the allocation of limited resources and of compensation for well-defined and circumscribed effects on the environment by means of specific tools connected with individual productive processes. New boundaries to each productive process are established in this manner and the structure of the entire system is left unchanged. The shadow price system is precisely intended to bring back into the confines of the productive process, and therefore to the attention of every individual operator, the effects on the environment that do not have an explicit economic valence and therefore cannot be taken into consideration by individual agents operating in a market economy. Such effects should be examined, decisions should be modified, and the environment should be protected.

In accordance with such an approach, the ecosystem must be divided into separate and independent subsystems so that production and consumption processes may also be separated and attributed to individual operators. In reality, there exists only one limited resource which is represented by the ecosystem taken as a whole, insofar as production and consumption on a world level are so great that, for every practical purpose, significant subdivision of the ecosystem is impossible, just as it is impossible to limit the effects on the environment within the confines of a specific dimension or a specific space. Whenever certain determinate effects are taken into consideration, new ones emerge which must also be specifically considered, and this results in an endless series of new events that reflect the impossibility of dividing the ecosystem into separate unconnected parts.

It is therefore not surprising that all the approaches that have been suggested and which have been briefly described lack an adequate theoretical basis for integrating nature and human society in a coherent picture and are incapable of developing the empirical tools which would be required to make truly effective decisions. There is no valid proof of the existence of a possible sustainable development, however one may define it, just as there is no certain knowledge of the interaction between human society and ecosystem: The existence of sustainable development was assumed as given. The contradiction inherent in the hypothesis is also revealed by the fact that the current unanimous consensus in favor of moving from the current unsustainable process of development to one that is sustainable is not matched, nor could it be, by any consensus regarding to how to achieve such a passage.

Research Institutions and Problems of a Cultural Nature

Two further problems deserve attention. The first pertains to the poor results attained by research institutions. For example, in 1969 the U.S. Congress approved a National Environment Policy Act (NEPA) which requested an appraisal of environmental impact for all major federal projects. These appraisals, however, were carried out only on a procedural level and no increase in knowledge was ushered in as a result of NEPA activities. Also, after more than 10 years of activity, no significant result has been produced by the National Acid Precipitation Program (NAPAP), an institution financed by the U.S. Congress. Instead, accusations of manipulation of data to protect financial or electoral interests have been voiced. NAPAP has been described as "an example of how appraisals on a broad scale should not be carried out" (Schindler, 1992). Research institutions in other countries do not appear to yield results that are very different regarding the environment.

The second, and perhaps more important, element concerns the fact that ecological problems are viewed not only from the scientific point of view but also from different religious and cultural perspectives. Currently, these problems are being increasingly viewed from an ethical point of view in many areas of the world, especially in the less industrialized areas, as evidenced by the return to teaching from religious texts. The current growth of population and consumption is creating an unsustainable situation which will give rise to a harsh clash over political and ecological questions without the mediation of a set of common scientific and economic values, to which the Western world is accustomed.

Human Activities and the Ecosystem

A Definition of Environment

Everything that takes place does so in a context, and the context in which man lives and operates is the ecosystem. The ecosystem, however, is not static and immutable; rather, it too is the result of a process – the slow evolutionary process that allows life to exist on Earth (Fig. 3).

The ecosystem is a self-organized system that has evolved for a long period of time and is composed of subsystems that are connected and interact with one another. These subsystems are also self-organized since this is life's principal characteristic. Insofar as they are complex systems, both the ecosystem and its subsystems have properties that are the result of the ensemble of functional interrelations, which cannot be adequately analyzed through the study of one or more elements isolated from the whole (Capra, 1996).

The dynamics of the ecosystem may be briefly illustrated in the following manner. The energy radiated by the sun is for the most part dispersed; only a part of it is used for

a

b

c

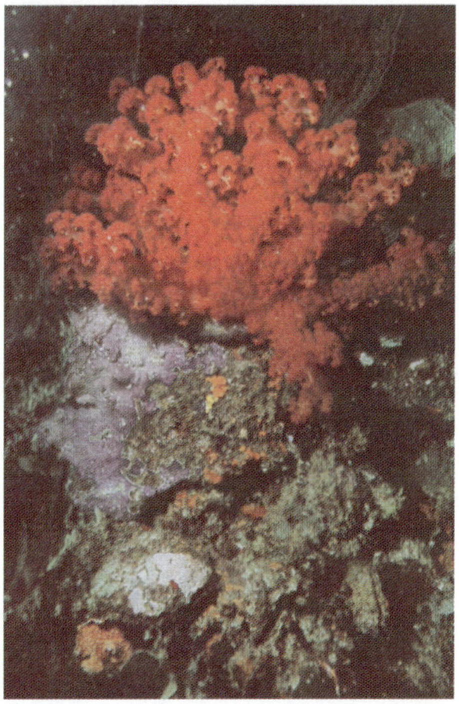

d

e

FIGURE 3 Examples of ecosystems. (a) Rain forest in French Guyana (photo courtesy of F. Hoogervorst/Panda Photo). (b) Woods of the Cansiglio in Italy (photo courtesy of Pigazzini/Overseas). (c) Woodland savanna in the Masai Mara National Park in Kenya (photo courtesy of A. Petretti/Panda Photo). (d) Australian coral reef (photo courtesy of Fontes/Overseas). (e) Tundra in central Alaska (photo courtesy of Ferri/Overseas).

photosynthesis and thereby transferred to producers and consumers. The dissipation, imposed by the second law of thermodynamics, is then completed through the process of breathing. A certain number of material components (carbon, nutrients, etc.) are fed back into the environment and may be used again in another cycle. A retroactive effect is thus created so that the ecosystem may be represented as a recursive system that operates cyclically and uses solar energy as a low-entropy source of energy. The intensity of the sun's emission of energy has been the same since the most recent geological eras; also, the retroaction based on the carbon and nutrient cycles has not shown any significant changes.

The ecosystem includes a broad spectrum of living organisms from producers, which turn inorganic material into organic one, to consumers, which demolish the organic material into its elementary components (Fig. 4). In between these two extremes are all the living beings that are capable of transforming organic matter into some other form of organic matter. Life on Earth changes continually, albeit slowly, but living beings perceive the system as unaltered since their life is short compared to the speed of change. Species, on the contrary, adapt and thus become a part of the evolutionary process. The shorter the life of individual living beings, the faster and easier species adapt to the transformations of the ecosystem: It is for this reason that individual living beings are born and die.

As a result of the ecosystem's complex nature, the understanding of its functioning most probably presents at least the same level of difficulty that one finds in the study of Ashby's nonbanal machines and in that of systems that are sensible to initial conditions (SIC).

A banal machine is a structure characterized by inputs and outputs related to one another in specific and unalterable ways. A simple example is that of the typewriter, in which the letters are connected to the keys: If the relation is unknown, it may be discovered by means of trial and error. An Ashby nonbanal machine is a structure in which inputs and outputs are related in ways that may change, and the changes are brought about by the impulses the machine receives. An example is a typewriter in which the relation between letters and keys changes depending on which key is struck. In such a situation, the structure of the overall relation connecting letters and keys is very difficult to discover, even if it is strictly deterministic. Ashby's nonbanal machines may be seen as black boxes represented by mathematical structures characterized by a finite input vector, a finite output vector, the internal state function, and the output function.

Even in the favorable cases in which input, output, and internal states are known, the number of transition schemes that may be generated is enormous. Even without taking into account the computational complexity that is a function of the length of the identification sequence, a system with only three binary inputs, three binary outputs, and four internal states has 1048 possible transition schemes (Laise, 1994).

SIC systems, otherwise known as chaotic systems, are systems of differential equations in which the variables of state, x, depend on initial conditions in such a way that the difference between $x = f(t, x_0)$ and $x = f(t, x_1)$, where x_0 and x_1 are two different initial conditions, increases extremely rapidly with the passing of time t (Devaney, 1989). In order to be able to predict the evolution over time of a system of this type, initial conditions must be known to a high degree of precision. The quantity of information relative to initial conditions increases very rapidly along with the temporal amplitude over which the system's behavior is analyzed so that it becomes computationally impossible to open the observational window beyond certain limits; after a certain span of time, the values of the effective variables may differ a great deal from those that have been computed so that any calculation, even if carried out with the help of powerful computers, presents itself as distorted and untrustworthy.

Moreover, it should be noted that since it is a system composed of complex and mutually interacting subsystems, the biosphere cannot be considered the sum of separate parts without falling into a serious problem of efficiency with regard to information theory. This may be shown synthetically in the following way. If one takes two proper subsystems, X and Y, the algorithmic content, $H(X, Y)$, required to describe X and Y at the same time (i.e., as parts of the same unit) is given by the expression $H(X) + H(Y) + H(X{:}Y)$, where $H(X)$ and $H(Y)$ represent the informational content necessary to describe each individual part and $H(X{:}Y)$ represents the degree of interdependence between X and Y. In information theory, this yields the following equation:

$$H(X{:}Y) = -\log\left[\frac{P(X)P(Y)}{P(\langle X, Y\rangle)}\right] + 0(1) \geq 0$$

where $P(X)$ and $P(Y)$ represent the probability of the event in parentheses, $P(\langle X, Y\rangle)$ represents the probability of the combination $\langle X, Y\rangle$, and 0(1) indicates that there exists at most a difference of order 1 connected to the machine that carries out the calculation. As a result, if there is interdependence between X and Y — that is, if they are parts of the same set (or system) Z — the number of bits, or the amount of information, required to know X and Y separately is much greater than that needed to know Z in its entirety (i.e., X and Y are considered as a unique system) (Chaitin, 1979).

The Perception of the Environment

For purposes of understanding, man tends to divide the system in which he lives into two parts: A part that is observed and a part that surrounds it, which includes everything else and is considered as an environment or context.

FIGURE 4 The cycle of energy and nutritive substances in an ecosystem. Solar energy (yellow arrows) is captured and transformed by the plants (i.e., the producers). Herbivorous animals, or primary consumers, assume the energy contained in the plants (green arrows) and, in turn, form the nutriment (red arrows) of carnivorous and insectivorous animals, or secondary consumers. The superpredators, or tertiary consumers, feed on the secondary consumers. All waste and remains of dead organisms are decomposed by way of the detritus chain which ends with the intervention of unicellular algae and bacteria (brown arrows). Dissolved organic substances are returned to the plants (blue arrows), thus becoming once again a part of the cycle and making the ecosystem a self-sufficient whole.

The relation between cause and effect is confined to the system that is observed, and a clear line of demarcation is traced between the experiments that have been carried out, actions that have been taken, rules that have been applied, and the rest of the world considered as environment, or context, within which all this takes place. Although scientists are aware of the limitations and the conjectural character of such a division, technical applications such as those developed for industrial production require that the dividing line be clearly traced and not susceptible to change over time; only in this way can economic activities be evaluated and carried out in practice.

The basic hypothesis underlying any economic activity is that the production process is well-defined; this implies the existence of a clear boundary beyond which it is assumed that there will be no effect on the environment or, if there is one, that it may be entirely neglected for all practical purposes. In this way, decisions may be made by individual operators to the extent that each operator, within the established boundaries, has a complete view of the whole productive process he or she controls and of its consequences.

A simple representation of the biosphere as recursive system may be expressed as $S \rightarrow (A \rightarrow B \rightarrow \cdots \rightarrow Z \rightarrow A)$, where S indicates the energy coming in from the sun and the arrows indicate the transformation from one subsystem $(A, B, C, \ldots)$ to another, all forming a part of the biosphere. In this representation, the sequence $B \rightarrow C$ may be considered as a process that leads from B to C, but at the level of system the opposite, albeit in a more indirect way, is also or may well be the case. The only recourse for the biosphere is the sun's energy that sustains the entire life cycle, in which impulses and results, causes, and effects may be distinguished from one another only in relation to specific cases and subsystems but not for the system as a whole.

Production processes, instead, are conceived as one-way paths that start with the inputs, or causes, and end with the outputs, or effects: They may be represented as sequences of the type $(I \rightarrow O)$, which are equivalent to the $B \rightarrow C$ sequence. Energy, materials, installations, etc. are the resources which activate the inputs, whereas nature is considered as a natural resource. The economic process requires that I and O be attributed values — for example, P and V — in such a way as to be able to calculate the difference $OV - IP$, which must be positive for the process to be possible. Human activity is composed of an enormous number of processes of this type, all taking place within the biosphere and generally sustained by sources of energy other than the sun. These processes, which may be conceived of only if abstracted from the recursive system as a whole, have an impact on the environment in many unpredictable ways.

To identify production and consumption processes and attribute them to individual operators, the ecosystem must be subdivided into separate subsystems which do not interact with one another. The ecosystem as a whole, however, is a fully integrated process and should be viewed as a complex self-organized system. Not only does such a conception come naturally to mind but also it is advanced by many experts (Capra, 1996). Since the entire structure of production and consumption has reached such a vast scale as to give rise to a series of retroactions which change the workings of the ecosystem, the attempt to contain the effects within the confines of a specific dimension or determinate space and to confront the consequences of this interdependence in terms of partial analyses (such as those underlying the current economic approach) for the purpose of evaluating environmental impact in terms of individual actions or projects is entirely inadequate. The sole purpose of such an approach is to maintain the separation between observed system and environment, the functional characteristics of the economic process, and the ways it is perceived.

The Nature of Economic Activities and Economic Theory

In light of the previous discussion, it is extremely important to define what the real end of economic activity. Current economic theory maintains that the end of economic activity is the satisfaction, by means of consumption, of human needs; in this sense, the search for profit is only a means toward this end. On the basis of this hypothesis, economic theory has also established which conditions must exist for an economic system to be considered efficient and therefore neutral with respect to the ends that society wishes to pursue.

This formulation did not belong to the classics, nor to Karl Marx, obviously (Trezza, 1993), and in recent times Keynes worked out an important distinction between an economic system aimed at generating profit and one aimed at maximizing the utility derived from consumption. Keynes defined the former as *entrepreneurial economy* and the latter as *cooperative economy;* the latter system is at the base of neoclassical theory. As a result, the purpose of economic policy is to make an entrepreneurial economy behave like a cooperative one; Keynes called such a state of affairs *neutral economy* (Keynes, 1987; Rymes, 1989). Keynes clearly states that "in an entrepreneurial economy, *the only aim is to make money*" (as quoted in Rymes, 1989); in *The General Theory of Employment, Interest and Money,* Keynes (1936) also connects aggregate investment to profit alone. Of course, Marx also thought that profit was the sole aim of economic activity. The validity of the definition developed by Keynes and of the resulting distinction between two types of systems may be easily appreciated by noting the way things have actually evolved and the way in which firms are managed and valued on the capital markets.

The accounting system proper to the current economic system, since it is a system that has profit as its end, uses monetary criteria to establish the validity of the results it obtains. In fact, all the balances of firms, like all economic determinations, wind up as a number in the general accounting of prices in monetary terms, which rids one of all judgments of value. In this way, the economic system is separate from every other part of human society, and the results that are sought and achieved by economic operators within the economic sphere are evaluated using the same measuring rod that defines the sphere. No value judgment is allowed since the only accepted measure is expressed in terms of the one-dimensional metrics of the system based on monetary prices. In this way, the simple rule whereby everything that gives rise to greater profits is to be preferred to what engenders lesser ones is established.

In order to be able somehow to represent the functioning of this system by means of the principle of consumer satisfaction, economic theory has been modeled in such a way as to pose economic determinations using a system of accounting which in some manner could conform to that effectively used in the real economic system. In this way, theory has been able to provide a plausible description of the functioning of the economic system. Value judgments have been expunged from economic theory as well, and the economic system is considered as separate from the rest of society. Here, too, economic activities are to be evaluated on their own terms using the accounting system of an economic space, presented here in terms of utility of consumption: A greater consumption, with the same distribution among individuals, is preferable to a lesser consumption, and comparisons can be carried out in scalar or one-dimensional terms.

To attain such a result, however, one had to introduce hypotheses which not only are not substantiated by reality but also lack a sufficient logical basis: in particular, the hypothesis according to which the consumer has a "rational" ordering within the space of present and future goods and the hypothesis that the consumer never reaches a condition of satiety within the gamut of his or her possible choices (Mas-Colell *et al.*, 1995). The weakness of these hypotheses and of the theory based on them has been pointed out many times both within the framework of a traditional approach (Kahnemann and Tversky, 1984; Laise and Valentino, 1995; Mas-Colell *et al.*, 1995) and within that of globally critical positions (Geogescu-Roegen, 1984; Simon, 1985). Furthermore, the hypotheses on which the theory of consumer choice is based are entirely in disagreement with many recent studies on the concept of human rationality (Lackoff, 1987; Damaso, 1995; Hofstadter, 1995) and also with recent and fortunate developments in logic, such as "fuzzy logic."

It should also be noted that the majority of firms and institutions employ multicriteria analyses to formulate their choices, but such a kind of analysis is incompatible with the theory of choice based on the maximization of utility insofar as it contradicts the very hypothesis on which the latter is based. Multicriteria analysis, in fact, implies or gives rise to non-"rational" orderings, contrary to what the theory requires, and therefore prevents the construction of supply and demand functions with the desired properties. Only in cases in which all the criteria may be combined into a single one are the results that have been obtained compatible with the economic theory (Laise and Valentino, 1995). In reality, therefore, the majority of economic operators act using methodologies that are incompatible with the hypotheses on which neoclassical economic theory is based.

By means of the negation of the true nature of the economic system, the theory has attempted to create a system that justifies the existence and functioning of the real economic system on different and more acceptable, but nevertheless misleading, bases; it has failed, however, to generate a theoretical corpus capable of representing such a functioning in a way that is entirely coherent.

The Relation between Economic Activity and Ecosystem

The economic system may be considered a growing subsystem that interacts with the ecosystem within the context in which economic activities take place. In this sense, the economic system is comparable to any other system of living beings that forms a part of and interacts with the ecosystem.

All living beings produce waste with their metabolism; this waste becomes a part of the chain of vital and material cycles on which life on Earth is based. Production and consumption activities interact with the ecosystem and introduce into it many additional processes of a chemical and physical nature. Both because of the high rate of waste production and because of the creation of chemical compounds that are difficult to break down, this gives rise to a process of accumulation that progressively reduces the ecosystem's capacity to transform waste into substances useful to life and gives rise instead to an ever faster process of environmental degradation. The problem of the environment, therefore, is not connected with the exploitation of natural resources but with the way in which human activities interfere with the very functioning of the ecosystem. Alterations that destabilize vital and material cycles, and accelerate the evolution of the ecosystem toward new and unknown conditions, thus arise.

Pollution and environmental degradation are the effects of the retroaction of the overall system on the subsystem that man perceives. Naturally, these effects are apperceived by people directly through their own direct experience, scientific measurements, and public statements — within the normal sphere of human life in which judgments of value have their place and in which all the diverse human actions are judged. The causes of environmental degradation, on the other hand, are part of the autonomous economic system which is governed by the rules of the market and in which everything is appraised in terms of economic accounting and where judgments of value have no role to play.

Economic theory presents the free market system as one that can optimize the use of scarce resources so that instead of submitting economic activities and decisions that concern the environment to the value judgments expressed by society, one tries to make the effects of environmental degradation somehow relevant to economic decisions. Many attempts have been made in this direction, using all the tools that the theory was able to devise, in order to make environmental damage perceivable at the economic level by way of measurements and evaluations effected in terms of the accounting system of the economic sphere. In so doing, every time new damage was taken into consideration, new dividing lines between economic activities and the environment (i.e., the context in which the activities take place) have been traced. These dividing lines have to be, and indeed have been, continually modified as time goes by.

The sun, air, water, noise, quality of life, and the environment as a whole are elements essential to human life, but the economic process cannot take them into consideration unless there is a price associated with them. Whenever this takes place, all these elements are transformed into utilizable resources (i.e., goods that may be bought and sold). In this way, the current organization of society is maintained and environmental degradation becomes a technical problem — that of establishing the right system of prices — whose solution is entrusted to experts and economists.

Aside from the fact that one may attribute prices and values only to some of the environment's components, such an attribution fails to adequately safeguard the environment given that the problem lies in the upsetting of the cycles on which the ecosystem is based. To make this point clearer, consider any consequence of human activity that exercises an impact on the environment, such as the emission of CO_2. In general, there are two possibilities: The first consists of establishing the level of CO_2 emission per unit of time and distributing that amount among the various activities ordered according to their degree of social importance; the second foresees that an efficient price, tax, or other tool developed by economic theory to quantify the emission be established.

In the first case, the extent of the CO_2 emission phenomenon is decided outside of the economic system on the basis of a judgment of value relative to the preservation of the atmosphere. Production and consumption activities must adapt so as to remain within the limits that have been established, whatever the relative impact on the economic system may or may not be; society as a whole must then act so as to make such adjustments socially bearable.

In the second case, the price is determined within the economic system in such a way as to create an exchange between the harmful phenomenon (in this case the emission of CO_2) and all other economic goods; the emission may be considered as a negative resource and evaluated like all other resources in such a way as to allow the economic system to function efficiently and without hindrance. The level of environmental impact, if calculated in terms of efficient prices or public taxation, is determined in a flexible manner so that the final effect on the environment is the result of the balance of forces and economic interests; in this way, the defense of the environment is sold and bought just like any other economic good.

Moreover, in the economic determination of environmental damage, the object of the evaluation is not the environment and its deterioration but rather the economic perception of the damage or of the drawbacks suffered by a given number of subjects who are more or less directly involved in the productive and consumption activities under investigation.

Finally, the valuation, taxation, and definition in accounting terms of environmental characteristics are carried out separately and in connection with well-defined environmental damages and specific production and consumption activities. The environment, on the other hand, is a well-integrated process, and what counts is the overall impact of all human activities so that the overall effect of all barters between environmental damages and economic interests becomes even more unpredictable.

In this situation, the concept of sustainable development tacitly implies the capacity to dominate and govern the interaction between man and environment in such a way as to protect the latter and accomplish that which man desires. According to widespread opinion, the current situation should not give rise to special worries since every damage caused by technology may be compensated for and solved using other technologies that are extant or will be discovered in the near future. In reality, in order to make sustainable development possible, one would first have to fully understand the overall extent of the impact that human activity has on the environment; to do this, a level of knowledge is required that is currently not available and probably cannot be attained, among other reasons, because of the computational complexity indicated in regard to Ashby's nonbanal machines and SIC systems. The adaptation of living species to the environment is never complete and always attained by means of a series of successive attempts, often with dramatic changes in the levels of extant populations. The condition of man is the same as that of any other species, and a great deal of caution is required when one thinks oneself capable of controlling man's interaction with the ecosystem. It is not even possible to subordinate the relation between man and ecosystem to the needs of the economic system because it is autonomously defined in its own sphere and subject to the need to continuously expand.

Conclusion

The environment is the context in which man lives and in which human society (also a system that is evolving) must find its place. Like any living being, man has always caused changes to occur in his environment and, therefore, in the biosphere; the changes in the preindustrial world, however, were limited in space and scope, and they did not unbalance the biosphere as a whole. Moreover, the relative slowness of environmental change has never set man face to face with the need to safeguard it in overall terms and on a planetary scale but only in relation to local conditions that were far more directly accessible and understandable. By changing the nature, the scope, and the rate of growth of human activities, the industrial revolution has engendered the problem of the compatibility between the ensemble of human activities and the safeguarding of the biosphere. Human society has become the foremost cause of environmental evolution and must therefore accept responsibility for that evolution. Even if the final effects on the environment cannot be identified, the greatest possible effort must be made to understand the nature and intensity of the impact of human activity on the environment so as to limit the damage it causes. It seems that man, instead, wants the best of both worlds: the sustainability of ancient times and the continuous development of modern ones. It is difficult to say when he will awake from this dream, but it is not difficult to foresee that it will be a horrifying awakening.

The use of tools developed by economic theory and aimed at the defense of the environment gives rise to an unstable and unsustainable relation with the biosphere. All those procedures, in fact, pursue the aim of quantifying environmental damage by attributing a price or an economic value to it; in so doing, they lead environmental characteristics into the economic sphere and evaluate them together with all the other elements that are factored into establishing costs and earnings. To consider environmental damage and the other conditions that are necessary to life as resources to be viewed economically and used like any other resource enables the problem of the environment to be seen as part of the economic sphere and as subject to the particular accounting proper to that sphere. The ecosystem is thus allowed to become destabilized by interfering with its cycles. Since destabilizing the ecosystem means destroying the bases of life on Earth, this means putting the suppression of life on the same level as any other activity taking place within the sphere of production and consumption: the price paid for the acceptance of environmental degradation thus becomes the price of homicide, on a planetary level, of future generations if not of those that are already alive today.

Even if it will never be possible to reach a complete understanding and full capacity to predict the results of the impact of human activities on the biosphere, in order to take some steps toward the preservation of the environment the greatest effort must be made to study the ecosystem and the subsystems that comprise it. Global evaluations should be carried out so that guidelines for all human activities may be devised. Since it is not possible to carry out a separate evaluation of the environmental impact of each of them, all relevant processes of production and consumption should be reviewed on the basis of these global evaluations and on that of the relative guidelines in order to ascertain in the best possible manner how they interact with environmental processes and how these interactions may be lead back to levels sustainable from the environmental standpoint. One would need to identify levels and forms of consumption other than the current ones an sustainable on the basis of forecasts of increases in the world population. At this point, objectives might be indicated, ways of proceeding might be suggested, and reasonable choices might be enacted.

In other words, it is necessary for the awareness of environmental damage, so often denounced, to give rise to a plan for changing the overall ensemble of human activities, not just to limited and specific modes of intervention, and therefore the bankrupt attempt to put the defense of the environment on the level of economic decision making should be abandoned. This would allow for a step forward to be taken both with respect to the declarations of intent to intervene in the defense of the environment that have been repeated at international meetings (Rio de Janeiro Conference, 1992) and with respect to insufficient and partial commitments taken in this regard (Treaty of Montreal, 1978).

We must become aware that it is impossible to confront the problem of the defense of the environment through decentralized decision making, in a context of continuous development, based on the economic evaluation of environmental damage in the form of scarce resources in such a way as to preserve the main characteristics of the current economic system. Value judgments should be reintroduced so that decisions might be of an explicitly political nature since they concern questions of choice of living standards in terms of forms and levels of consumption, preservation of the ecosystem, and the protection of the environment on both the local and global scale. Particular attention should be paid to the effects of this state of affairs on the economic system and on the constraints that must be placed in order to maintain production and consumption profiles within the preestablished limits. A new form of the economic sphere would have to be conceived and a new relation between economic and political decisions identified.

The choice that humanity is called on to make requires more than a dramatic change in the way in which the economy is usually considered. Rather, what is required is an overturning in the economic mechanism and Western point of view of the environment. What is required is a veritable cultural revolution, given that one must replace the concept of exploitation and control of the environment with that of coexistence with it. In order to be effective, ethical principles must be perceived by every individual as necessary and legitimate, given that they can never be made operative by any set of rules or laws to which a deep and full respect has not been granted. A great cultural effort is required so that the necessary social and behavioral changes may take place.

The enormous difficulty of setting out along this path is obvious and greatly increased by the fact that although the impact on the environment is global, and also the economic system is proceeding toward a planetary global dimension, the seat of political decisions, and therefore of judgments of value, remains scattered among the many extant political realities born in the course of the human race's historical evolution. The various political realities often answer to different ideologies and conceptions and are connected with particular and not infrequently contrasting interests; nor is there on a world level such an authority as might be capable to put itself above the nation-states and confront the environmental problem on its planetary global scale, where it must necessarily abide.

The simple displacement — which many have suggested — from dominion of the environment to its custody is not sufficient since man cannot consider himself the custodian or guardian of nature but, rather, as one of its inhabitants — as a creature in the mother's womb. As such, man lacks, and for a long time to come (if not forever) will continue to lack, the knowledge and the capacity to understand, and much less to govern, a system of such complexity. To this day, the only proof man has had that the environment in which he lives is suited to him is his presence on the earth; hence the supreme folly of any major modification of the ecosystem.

References Cited

ANDERSON, D. (1992). Economic growth and the environment. In *World Development Report 1992.* World Bank, Washington, DC.

ASIAN DEVELOPMENT BANK (1990). *Economic Policies for Sustainable Development.* Asian Development Bank, Manila.

BARBIER, E. (1989). Economics, natural resource scarcity and development. *Environ. Conserv.* **14**(3), 457–466.

BEKERMANN, W., and OATES, W. (1974). *In Defense of Economic Growth.* Cape, London.

BOULDING, K. E. (1966). *The Economics of the Coming Spaceship Earth* (H. Jarret, Ed.). Johns Hopkins Univ. Press, Baltimore.

BROMLEY, D. W. (1989). *Economic Interest and Institutions.* Basil Blackwell, New York.

BROWN, L. (1993). *State of the World.* Norton, New York.

CAPRA, F. (1996). *The Web of Life.* Anchor, New York.

CARPENTER, R. A. (1994). Can sustainability be measured? *Ecol. Int.* **2,** 11–25.

CHAITIN, G. J. (1979). Toward a mathematical definition of "life." In *The Maximum Entropy Formalism* (R. Levine and M. Tribus, Eds.), pp. 477–498. MIT Press, Cambridge, MA.

CIRIACY-WANTRUP, S. V. (1968). *Resource Conservation: Economics and Policies.* Univ. of California Press, Berkeley.

DALY, H. E. (1991). *Steady State Economics,* 2nd ed. Island Press, Washington, DC.

DAMASO, I. (1995). *L'errore di Cartesio. Emozione, Ragione e Cervello Umano.* Adelphi, Milan.

DEVANEY, R. L. (1989). *An Introduction to Chaotic Dynamic System.* Addison-Wesley, Redwood City, CA.

EHRLICH, P., and HOLDREN, M.-J. (1974). Human population and the global environment. *Am. Sci.* **62,** 282–292.

GEORGESCU-ROEGEN, N. (1971). *The Entropy Law and the Economic Process.* Harvard Univ. Press, Cambridge, MA.

GEORGESCU-ROEGEN, N. (1977). The steady state and ecological salvation: A thermodynamic analysis. *Bioscience* **27,** 266–270.

GEORGESCU-ROEGEN, N. (1979). Comments on the papers by Daly and Stiglitz. In *Scarcity and Growth Reconsidered* (V. K. Smith, Ed.), Johns Hopkins Univ. Press, Baltimore.

GEORGESCU-ROEGEN, N. (1984). Analytical representation of economic decisions under multiple criteria. In *MCDM: Past Decade and Future Trends, a Source Book of Multiple Criteria Decision Making* (M. Zeleny, Ed.). JAI Press, Greenwich, CT.

HERFFINDAL, O., and KNEESE, A. V. (1974). *Natural Theory of Natural Resources.* Merrill, Columbus, OH.

HODGSON, G. (1991). *Economia e Istituzioni.* Otium, Ancona, Italy.

HOFSTADTER, D. R. (1995). *Concetti Fluidi e Analogie Creative.* Adelphi, Milan.

HOLMBERG, J. (1992). *Policies for a small planet,* from the International Institute for Environment and Development. Earthscan, London.

HOTELLING, H. (1931). The economics of exhaustible resources. *J. Political Econ.* **39,** 137–175.

HUETING, R. (1980). *New Scarcity and Economic Growth: More Welfare through Less Production?* North-Holland, Amsterdam.

INTERNATIONAL UNION FOR THE CONSERVATION OF NATURE (1980). *World Conservation Strategy: Living Resource Conservation for Sustainable Development.* United Nations Environment Program/World Wildlife Fund, Geneva.

JACOBS, M. (1991). *The Green Economy: Environment, Sustainable Development, and the Politics for the Future.* Pluto Press, London.

KAHN, H., BROWN, W., and MARTEL, L. (1976). *The Next 200 Years: A Scenario for America and the World.* Morrow, New York.

KAHNEMANN, D., and TVERSKY, A. (1984). Choices, values and frames. *Am. Psychologist* **39,** 341–350.

KAPP, K. W. (1991). *Economia ed Ambiente.* Otium, Ancona, Italy.

KEYNES, M. (1936). *The General Theory of Employment, Interest and Money.* Macmillan, London.

KEYNES, M. (1987). *Collected Writings.* Vol. 13, pp. 14, 29. Macmillan, London.

KLAASSEN, G. A. J., and OPSCHOOR, J. B. (1991). Economics of sustainability or sustainability of economics: Different paradigms. *Ecol. Econ.* **4.**

LACKOFF, G. (1987). *Women, Fire and Dangerous Things. What Categories Reveal about the Mind.* Univ. of Chicago Press, Chicago.

LAISE, D. (1994). *Teoria dei Sistemi e Teoria della Politica Economica.* Università La Sapienza, Dipartimento di Informatica e Sistemistica, Rome.

LAISE, D., and VALENTINO, P. (1995). *Teoria Economica Neoclassica e Scelte Multicriterio.* Università La Sapienza, Dipartimento di Economia Pubblica, Rome.

LEONTIEF, W. (1974). *Les Répercussions de l'Environnement sur la Structure Économique.* Essais d'Économique, Paris.

MALTHUS, T. R. (1946). *An Essay on the Principle of Population,* 6th ed. Murray, London. (Original work published 1826)

MAS-COLELL, A., WHINSTON, M. D., and GREEN, J. R. (1995). *Microeconomic Theory.* Oxford Univ. Press, New York.

MEADOWS, D., *et al.* (1972). *The Limits to Growth. A Report for the Club of Rome's Project on the Predicament of Mankind.* Universe, New York.

MESAREVIC, M., and PESTEL, E. (1974). *Mankind at the Turning Point. The Second Report to the Club of Rome.* Dutton, New York.

MILL, J. S. (1952). *Principles of Political Economy,* 7th ed. Longman/Green, Reader and Oyer, London. (Original work published 1871)

MISHAN, E. J. (1967). *The Costs of Economic Growth.* Staples, London.

MISHAN, E. J. (1977). *The Economic Growth Debate: An Assessment.* Allen & Unwin, London.

MYRDAL, G. (1972). *Against the Stream.* Pantheon, New York.

MYRDAL, G. (1978). Institutional economics. *J. Econ. Issues* **12.**

NORDHAUS, W. (1974). Resources as a constraint on economic growth. *Am. Econ. Rev. Papers Proc.*

PEARCE, D. W., and TURNER, K. (1990). *Economics of Natural Resources and the Environment.* Harvester Wheatsheaf, London.

RICARDO, D. (1947). *Principles of Political Economy and Taxation,* 3rd ed. Murray, London. (Original work published 1821)

RYMES, T. K. (Ed.) (1989). *Keynes's Lectures, 1932–35.* Univ. of Michigan Press, Ann Arbor.

SCHINDLER, D. W. (1992). A view of NAPAP from North the border. *Environ. Appl.* **2.**

SCHMIDHEINY, S. (1992). *Changing Course: A Global Business Perspective on Development and the Environment* (with the Business Council for Sustainable Development). MIT Press, Cambridge, MA.

SIMON, H. A. (1985). *Causalità, Razionalità, Organizzazione.* Mulino, Bologna.

SMITH, A. (1927). *An Inquiry into the Nature and Causes of the Wealth of Nations,* 5th ed. Strahan & Cadell, London. (Original work published 1789).

SMITH, K. V., and KRUTILLA, J. V. (1979). The economics of natural resource scarcity: An interpretative introduction. In *Scarcity and Growth Reconsidered* (K. V. Smith, Ed.). Johns Hopkins Univ. Press, Baltimore.

SWANEY, J. A. (1972). Elements of a neo-institutional environmental economics. *J. Econ. Issues* **21** (4).

TINBERGEN, J., and HUETING, R. (1992). GNP and market prices. In *Population, Technology and Lifestyle: The Transition to Sustainability* (R. Goodland, H. R. Daly, and S. EI Serafy, Eds.). Island Press, Washington, DC.

TREZZA, B. (1993). *Origine e Sviluppo delle Teorie Economiche.* CISU, Rome.

UNITED NATIONS (1992). *System for Integrated Environmental and Economic Accounting (SEEA).* United Nations, New York.

UNITED NATIONS (1994). *UNO international conference on population and development.* Il Cairo, documento finale, par. 1.18.

UNITED NATIONS (1995). *Demographic Year Book.* United Nations, New York.

WORLD BANK (1984). *World Development Report 1984.* Oxford Univ. Press, New York.

WORLD BANK (1992). *World development report 1992.* World Bank, Washington, DC.

WORLD COMMISSION ON ENVIRONMENT AND DEVELOPMENT (1987). *Our Common Future.* Oxford Univ. Press, New York.

WORLD RESOURCE INSTITUTE (1992). *World Resources.* Oxford Univ. Press, New York.

PART THREE

Bioethics

Edited by Luigi De Carli and Stefano Rodotà

INTRODUCTION

Luigi De Carli
Stefano Rodotà

The profound transformations that have taken place in biology and biomedicine, not only in terms of technology but also of methods and concepts, have created new perspectives for research, and their influence on living systems, especially on the life and health of human beings, has been greater than anyone had foreseen. The relation between science and technology has become an increasingly close one and the boundaries between basic and applied research have been reduced. These changes have provoked a change in the role of the scientist, who has been entrusted with new tasks and responsibilities as a result of the novel demands resulting from biotechnological progress. The bioethical question follows from these premises. More than a codified and systematic discipline, bioethics is a continuously expanding area of enquiry into often interrelated problems that concern both the experimental and the human sciences and result from the possibility to intervene upon living systems. This explains the apparent lack of a guiding line to connect the essays of the part of the volume dedicated to bioethics, as well as a certain redundance in the way some topics have been dealt with. The essays have been grouped into sections according to a criterion of homogeneity that considers genetics and medicine, biotechnology and experimentation and, finally, the relation between science, technology and society to be unifying themes.

The first section contains three essays which may be viewed as pertaining to fundamental bioethics insofar as they deal with the problem of the sources, methods and objectives of bioethics and with the psychological context, at the individual and collective level, in which bioethics operate. A recurrent and paradigmatic example is that of medically-assisted human procreation, an endless source of new and unprecedented cases. Through the accounts of protagonists, Remo Bodei analyzes the existential problems that the dissociation between generation and fecundity has created. On a more general level, he notes that biotechnology has provoked profound changes in concepts, convictions, habits and feelings, and confronted man with dilemmas and paradoxes that bioethics attempts to solve through a comparison between differing world views. Carlo Augusto Viano emphasizes the need for innovation in moral rules so that the contribution that biotechnology makes toward the improvement of the quality of life may be appreciated. The possibility of intervening so as to improve the human genome should not be ruled out, along with an increased understanding of genetic heritage and the improvement of techniques of genetic manipulation. According to Henri Atlan, the general principles contained in religious traditions and moral philosophies cannot provide practical answers to specific ethical questions. The author distinguishes three successive levels of ethical reflection which have as their object the individual's experience of pleasure and suffering, his projection in time and in the collectivity, and the communication of different normative criteria among different cultures and religious traditions. This sequence of levels would lead to a universal ethics built upon a pragmatic base.

The Human Genome Project is achieving an evermore important place in the bioethical debate over questions that concern biomedicine. Alberto Piazza, in his essay, illustrates the Project in all the phases that will lead to the complete reconstruction of human genetic heritage. The possibilities and limitations of genetic screening are described in the essay by Antonio Cao and Maria Cristina Rosatelli, who present an analytic study of the principal screening programs that deal with the prenatal and neonatal periods and with adult life. A reading of their essay leads one to conclude that a critical bioethical evaluation of screening for new diseases will serve to optimize its positive effects on the genetic endowment of the populations involved. A correct use of genetic screening both at the individual level and at that of whole populations is guaranteed by the genetic counseling that must precede and follow upon every diagnostic action. Angus Clarke, in his essay, identifies nondirectivity, respect of autonomy and informed consent as being the ethical principles which must inspire genetic counseling. The question of informed consent is taken up also by Christian Hick, who further develops it in relation to applied therapy and to medical research. Access to genetic tests, which is vitally important for the safeguarding of the health of individuals, families and the collectivity, is becoming of growing importance in the context of labor relations and insurance practices. In this regard, the essay by Carlos M. Romeo-Casabona focuses on the measures to adopt in order to avoid the potential encroachments on the freedom of individuals and their privacy that would clash with the principle of equality and with that of solidarity. Parts of the human body at different levels of organization such as tissues, cells and macromolecules can be isolated, replaced or exchanged for essentially therapeutic purposes. Giovanni Berlinguer's essay examines the complex moral and juridical problems inherent in the most recent developments of such a practice. In the context of the so-called "technological marketplace" of the human body, which includes also DNA sequences, the problem of rights of patent over the whole range of biotechnological products and inventions arises. Gilbert Hottois, instead, presents a synthesis of the problems dealt with in the preceding essay from a broader perspective, in which the European and the American models in regard to the relation between body and marketplace are compared also from the philosophical and sociopolitical point of view. The former model is based on the fundamental principle of the dignity of man, the latter on individual liberty and the right to dispose of oneself as one pleases.

The therapeutic need that justifies a replacement or corrective operation underlies gene therapy as well. The methods and strategies that are employed to correct hereditary defects by means of exogenous DNA are described in the essay by

Mauro Giacca and Arturo Falaschi, who also report on the current situation of clinical experimentation. In the same essay, the authors conclude that somatic gene therapy is ethically acceptable, whereas they consider germinal gene therapy as impracticable even for exclusively therapeutic purposes. In his essay on the end of life, Maurice A. M. De Wachter calls medical technology into cause as an important resource for controlling the conditions and modalities of dying. Concern over the safeguarding of the integrity of human genetic material and the need to respect the dignity of the individual are at the root of the norms, directives or recommendations made by national and international organizations for the purpose of regulating assisted procreation. Bartha M. Knoppers and Lori Luther present an exhaustive picture by comparing the official positions of different countries in regard to the use of reproductive technologies for clinical applications and in research. An in-depth account of the technical and scientific aspects of research carried out on embryos is presented in the essay by Matteo Adinolfi and Joyce Harper, who show how important the biomolecular and genetic investigations carried out during the early phases of development are for the improvement of reproductive technologies and the progress of knowledge of the mechanisms underlying hereditary diseases.

Bruno Silvestrini tackles the problem of animal experimentation as an essential step in the process of both basic and applied research. In his essay, the relation between experimentation and scientific progress is examined in a strictly biological key in order to understand if and to what extent man can make use of other living beings to satisfy his own needs. The same theme is taken up and developed, especially from the philosophical and moral point of view, by Dietmar von der Pfordten, who poses the question of the recognition of moral status to non-human living beings inasmuch as they are endowed with their own legal rights. The traditional anthropocentric model is matched against the non-anthropocentric one which, in its strongest version, extends the ethical valuation of animals also to plants and microorganisms. Dietmar Mieth ponders the environment and the changes that it has witnessed as a result of biotechnology; by defining the domain of ethics, the author clarifies the distinction between ethics and morality and identifies the principle characteristics of ethics in scientific autonomy, the interdisciplinary approach and the integration of knowledge. In light of philosophical considerations open also to theological interpretations, the concept of natural order of which biological equilibriums are an expression is developed. The essay by Alessandro Palmieri and Roberto Pardolesi, which raises the question of the limits within which the right to patent should be admitted, examines the advantages for basic and applied scientific research on the one hand and concerns of an ethical nature on the other.

The massive application of biomedical technologies in therapy has raised the problem of the gathering and dividing of resources not only from the economic point of view, but also from the ethical and philosophical one, since it is the very definition of health that is at issue. The main ethical questions that Ruth Chadwick discusses in her essay concern the different interests of individuals and of the community in regard to the use of the means provided by health care assistance, the problem of public involvement, the interpretation of the concepts of justice and equity, the relation between rationing and the dictates of traditional medical ethics. Finally, the public dimension of bioethics, with special attention paid to assisted procreation, is dealt with in the essay by Wayland Kennet and Joseph Thornton, which provides a broad and detailed overview of laws, norms and opinions, documented with extracts of original documents released by various countries and international and supranational organizations. In addition to the effect of philosophical and religious traditions, the essay also analyzes the role of public opinion.

HENRI ATLAN

Ecole des Hautes Etudes en Sciences Sociales
75006, Paris, France

Human Biology Research Center
Hadassah University Hospital
91120 Jerusalem, Israel

Genealogy and Levels of Ethics

Problems of biomedical ethics are used as privileged instances to investigate the sources of practical answers given to ethical questions and to evaluate the role of religious traditions and moral philosophies in the process of decision making. General principles do not appear to be sufficient to deduce solutions to particular problems. Searching for what may be universal in practice, several levels of ethics are distinguished, from the immediate experience of pleasure and pain to the different systems of standards of existing moral orders inherited from cultures and religious traditions in different societies. The latter experience is associated with various experiences of the sacred and is modulated by the language, memory, imagination, geography, and history of each society. The possibility of building a pragmatic universality from local to general, from the existing cultures and systems of norms, is compared with what is done in moral philosophy in which universality is still that of the individual in his or her singularity.

The Sources of Ethics: Religion and Moral Philosophy

Twenty to 30 years ago it would have been very difficult to predict the current revival of ethics as a field of interest. One of the main reasons for this revived interest is related to the new possibilities offered by the explosion of biotechnologies which has taken place during this short period of time. However, it is not only a matter of quantity. The nature of our relations to science and technology has also changed. Not only have science and technology dramatically improved the quality of our lives but also they have created new realities, with new ethical problems unheard of before, without providing ways to solve them.

The following is a very simple but striking and paradigmatic example: Being a mother has always implied two different biological functions. These two functions, ovarian and uterine, have always been performed by the same woman, the mother being at the same time the provider of the ovum and of the uterus for pregnancy (Fig. 1). Technology has made it possible to dissociate these two functions. For the first time in the history of mankind, the woman who is pregnant and gives birth can be different from the one who gives the ovum with its genetic material. The following is the obvious question: Who is the mother? This question did not exist before the progress of biotechnology. On the other hand, science and technology, which have created the problem, are unable to solve it. When one asks people this question, some answer that it is self-evident that the mother is the genetic one; others say that it is self-evident that the

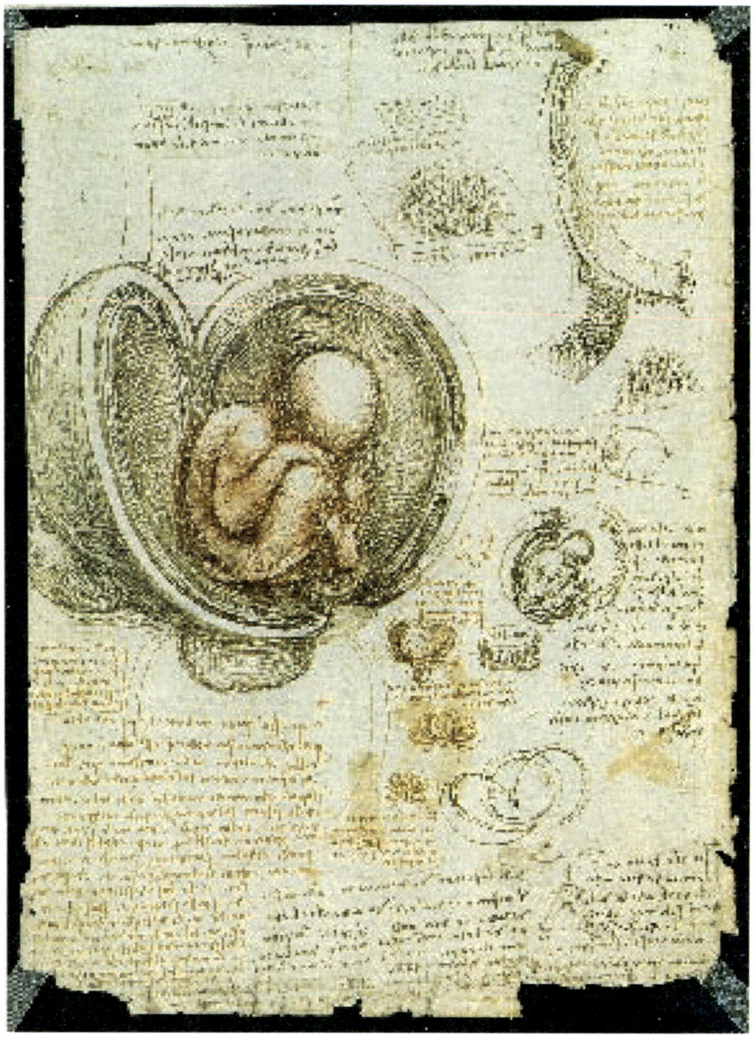

FIGURE 1 Drawing by Leonardo da Vinci: cleavage of a human uterus containing a fetus. The majority of the other designs by Leonardo illustrate the passage of maternal blood from the placenta to the fetus (1510–1512) (reproduced from Leonardo da Vinci, *Anatomical Notebooks;* Photo courtesy of Windsor Castle, Royal Library, f 19102 r).

mother is the one who is pregnant and gives birth. It is clear that biology alone does not give an answer: Both women are biological mothers.

This example is paradigmatic because it outlines a new relationship between scientific knowledge and ethical problems. It is not just a matter of naturalistic fallacy (i.e., of not deducing ethics from scientific knowledge). It was recognized long ago that what should be cannot simply be deduced from what is. Science is necessary to state and analyze the technical aspects of a problem, but it is unable to provide and justify value judgments about good and evil. However, in the example of the two mothers, there is something even more puzzling: Science and technology have created a problem not only of values, concerning what should be done, but also of identity concerning the relation between mother and child: "Who is the mother?" is not only a question about what should be but also a question of identity about what it is. Again, the answer to the question created by science and technology does not come from science and technology Where does it come from? More generally, where are we going to look for answers to ethical problems that science and technology create without giving solutions?

This is a new way of asking the question of the source of ethics. Where does ethics come from if not from natural sciences? In answering this question, two sources of ethical judgments are obvious: religion and philosophy.

Religion in all its forms has provided not just mythical

and traditional visions of the world, with which science has competed successfully. Religion has always been the source of norms and rules of behavior, indicating what is sacred and what is profane, what is good and what is evil, and what should be done and what should be forbidden. Therefore, it is natural that many people confronted with difficult ethical problems turn to religious teaching for guidelines, even if they do not believe in the articles of faith of this or that religion and even if the traditional views of the world seem obsolete and contradicted by scientific teachings.

This is the case, for example, when people are frightened by the endless possibilities of new technologies. There is a feeling that technology develops itself by itself, with no one in control knowing what will be the outcome, since it is enough that some procedure is technically feasible to be actually achieved, however strange and far-fetched it might be when applied to humans, such as human reproductive cloning and other breeding techniques. For this reason, restrictive rules enunciated by a church sometimes play the role of barriers, even for nonbelievers, to prevent this extreme scientistic attitude which justifies everything in the name of technical progress. However, religion as the sole source of ethics for us today suffers from several pitfalls. Although it provides believers with meanings and values for experiences of life and death, it suffers from a lack of accessibility. The foundations of its teachings, dogmas, and articles of faith are accessible only to a restricted community of believers. This is due, among other reasons, to the vocabulary and language, mostly theological, which is used to justify this or that position. In addition to this lack of accessibility, these general ideas or articles of faith serve to justify contradictory attitudes with regard to practical applications. For example, God's will can be used to justify the desire of procreation at any price as well as a general prohibition against medically assisted techniques of procreation. Sometimes, religious teachings seem to be consistent with a kind of natural theology prescribing the following of natural courses of behavior and noninterference with natural processes. However, this principle does not hold very long when confronted with practical questions since plain nature with no intervention is not always desirable. Nature is responsible for terrible diseases and catastrophes, and plain medicine has always been a fight against natural diseases by means of human interventions.

Apart from religion, philosophy has always set as its goal the search for universal ethics which would be acceptable by all humans because of the universal value of reason. Indeed, philosophical thought is very important to try to understand critically what science tells us about nature and not to believe blindly what science seems to tell us through undue generalizations and metaphors leading to fantasies and science fiction. For example, the Human Genome Project, the goal of which is to determine the structures of all our genes, was often presented as a research project which will tell us everything we need to know about human nature and at least everything about the mechanisms of all diseases. This is not quite true, and this misconception is based on a widespread metaphor among biologists which compares the structures and functions of the genes to a computer program. This is the usual notion that the genetic program is supposed to be written in our DNA molecules. Based on this notion taken literally, we have read again and again that knowing the structures of our DNA would be equivalent to having deciphered the "book of man." We have been told that because of this achievement we shall be able not only to cure all diseases, as if all diseases could be reduced to some genetic abnormality, but also to cure social diseases such as aggressiveness and criminality since we would no doubt discover the gene of criminality and genetic therapy would allow us to eradicate it (Koshland,1989). This was written as an editorial in the journal *Science* by its chief editor in the 1980s. It is therefore important that such statements are subject to philosophical criticism in order to distinguish what is real (i.e., the existence of genetic determinations associated with some DNA structures) from what is extrapolation and metaphorical anthropomorphism, such as the idea of a computer program with instructions directed to the achievement of some purpose of life supposedly written in our genes.

In addition, there is much to be learned from the great philosophers of the past and from the confrontation with their systems, not so much at the level of their general principles as at the level of their methods, in their use of reason to go back and forth from knowledge of nature to subjective experiences of human passions, imagination, and dreams associated with reason and language.

However, not all the different moral philosophies are equivalent and equally adapted to our current state of knowledge and to the relationship which has been developed between objective natural sciences and our subjective experiences of life and death. For example, I argue that ancient stoic philosophy and the ethics of Spinoza, in which freedom and happiness are searched for in a world of absolute determinism, are more adapted to our present-day physicochemical mechanistic representations of living organisms, including ourselves. Specifically, it would be possible to show that such philosophies are better adapted to the scientific representations of our times than modern idealist philosophies, Kantian and post-Kantian, based on the idea of a human free will outside, or parallel to, the realm of natural laws, as if man was in nature like "an empire within an empire," according to the ironic wording of Spinoza. However, like the great principles derived from theology, the great moral philosophical systems fail to provide solutions that everybody would be able to integrate to solve ethical problems practically, in real life.

Particular Situations versus General Principles

Like religious articles of faith, great philosophical principles break down when we are confronted with real, complex situations. Consider, for example, the well-known Kantian principle never to use a human being as a means only, but always as an end. When organs are taken from the body of a human being who just died and transplanted into the body of a patient, it is difficult not to see that the former is used as a means for the benefit of the latter. Nevertheless, this attitude now not only is generally accepted but also is even praised as a sign of solidarity and brotherhood between humans. Of course, one can try to reword and reinterpret Kant's principle in order to accommodate it with our practice by saying, for example, that a dead body is not a person. However, in other circumstances, the dignity of a (dead) person is recognized to the dead. It is clear that our stance on this issue, whatever it is, is a rationalization after the fact and that the practice of organ transplants with all its difficult ethical and emotional problems is actually accepted or rejected on the basis of other more practical arguments.

Similarly, the idea that life is sacred cannot be applied in all circumstances. It is too general and too vague since it does not take into consideration the existence of many levels of organization in living organisms. This idea amounts to ignoring the differences between the life of an amoeba, a fly, a fish, a bird, a dog, or a man. Put to the extreme in the name of the sacredness of life, one could equally glorify the life of parasites, bacteria, and viruses and also that of cancer cells, which are very much alive.

Even if restricted to human life, we would not put at the same level a whole human being and its organs or limbs, or cells, taken from the blood or the skin, for example, and cultured in the laboratory. Most countries forbid by law the selling and buying of the human body and its organs, considered as parts of the human body. However, different countries have conflicting laws regarding the commercialization of blood and blood products. The same principle of respect for human life does not lead to the same conclusions concerning the status of living cells of human origin. To which level in the parts of the human body are we going to extend this principle — to that of the tissues, the cells, or even the molecules?

Another example is very instructive. The question of personhood is certainly one of the most difficult to deal with in the context of modern biology. We learn from biology about mechanisms, physical and chemical, whereby our organisms are built and do what they do. All this has nothing to do with ancient questions about the soul or the essence of life and what makes a human person. Nevertheless, human social life, the legal system, and the notion of responsibility are based on the idea that we are persons and not things, and that together with personhood goes dignity and respect.

Because of technical possibilities of interfering with the beginning (and the end) of the existence of human beings, the question has been raised regarding when a human embryo becomes a person. Biology by itself cannot give an answer to this question since the notion of person is not a biological one but rather a legal, social, and moral one. The Catholic Church gives the apparently simple answer that a person begins with the fertilization of the egg. However, fertilization is a continuous process which takes a few hours. Moreover, this answer is hardly convincing for many people who view the egg as a single cell and the early embryo as a mere shapeless mass of cells and not a subject or any kind of object or being that one could easily recognize as a person. Embryonic development is a continuous, progressive process producing at the end a newborn baby whom almost everyone would recognize as a beginning of a person. Therefore, different times between fertilization and birth have been proposed as critical, like thresholds such that we should consider the embryo as a person endowed with its dignity and its rights after that time and as a mere mass of cells (i.e., an object) before that time. However, all these thresholds rely on different presuppositions about what is a human person. One should not think that one of them is scientifically better founded than the other. Again, biology by itself has nothing to say about what is a person.

The definition of a threshold, whatever it is — time of fertilization or a few days or a few weeks — is always arbitrary and can be justified only by examining its consequences when applied to practical situations. This is the reason why, in France, the notion of preembryo has not been accepted, nor has any other threshold which would define in general when an embryo begins to be a person. Rather, particular procedures to be performed on embryos are allowed or forbidden depending on their purpose and the circumstances in which the embryo happens to be in the hands of a biological medical team.

Along these lines, I mention a philosophical standpoint which dates back to Aristotle, was taken up by the Catholic Church until the time of Saint Thomas and then abandoned, and was accepted by the Jewish masters of the Talmud and is still valid today for Jewish rabbinical law. According to this tradition, the criterion for our recognition of a human person is nothing else than our immediate recognition of the human body and a human face, which makes us immediately recognize a human being and distinguish him or her from another species. Only when the embryo has the form of a human body and a human face with the eyes on the face and not on the sides do we consider him or her a human being (Fig. 2). This stage was estimated at 41 days after fertilization. Before 41 days an embryo is considered as a plain liquid object and, for example, abortion

FIGURE 2 Edvard Munch, *Heredity I* (1897–1899) (courtesy of Edvard Munch Museum, Oslo; Photo by Scala).

until then, either spontaneous or provoked, is not considered to be the death of a human child.

Of course, this does not imply, for example, that abortion before 41 days should always be permitted or that one could be allowed to do with the embryo whatever one wanted. Since a child can eventually grow out of the embryo, we are responsible for possible effects on the future child. In other words, even if some threshold is set up, the early embryo is important in relation to the future person who might develop, even though its present status might be that of a raw material.

Again, this position, no more than others, cannot be justified on *a priori* grounds. It can be judged only by examining what is implied in terms of practical behavior in various circumstances. First, it means that our definition of a human being does not have to be mediated by some theoretical analysis but is the result of our instant perception of a human body and face and this is part of our own sensibility as human beings in communication with one another. In addition, this definition has several practical implications. One of them is a reluctance to relate personhood to some theoretical function by a given property — scientific, philosophical, or theological — such as the existence of a normal nervous system, a consciousness, or soul, such that some humans could nevertheless be deprived of the quality and dignity of human beings if they lack these given properties. Another implication has to do with our relationship to other living species and our attitude about animal experimentation (Fig. 3). This criterion for personhood draws a clear demarcation between human and animal species.

Reduction and control of animal experimentation, especially when it is likely to be painful, is certainly an ethical progress for human beings. However, its motivations should not be searched for in a kind of extension to animals of our relation to human persons. Our moral relation to animal species, similar to that to nonhuman nature in general, should not be dictated by the dubious idea of animal rights or rights of Nature but rather by our notion of human duties and rejection of human cruelty. As a matter of fact, human morality toward animals is more easily seen when the appearance of the animal is closer to the human form. We are more reluctant to experiment on monkeys and dogs, even if we do it for the sake of human lives, than on mice, frogs, or flies because the form of the body and face of the

FIGURE 3 Monkey undergoing a laboratory experiment (courtesy of Antivivisectionist Scientific Committee).

former is closer to that of our own. We can more easily interpret their behavior as similar to a human one.

We are much less inhibited with respect to experimentation involving insects, plants, bacteria, or cells in culture, even if they are of human origin. This closes the circle and shows us that what is important regarding this difficult question of animal experimentation is our projection of ourselves on the object of experimentation, more than a problematic theoretical knowledge about the essence of living beings which we are dealing with. Assessment of likely human benefits and likely animal suffering, and decisions on a case-by-case basis, seem to involve commonsense ethics rather than general statements supposed to be general principles.

In fact, this last observation about general principles versus case-by-case analysis applies not only to the problem of the person but also to most situations and ethical problems created by new sciences and technologies. This is indeed one of the main lessons I have learned from my work with the French National Ethics Committee for Health and Life Sciences, and this is the main point I stress now.

In general, real-life situations are too complicated to be dealt with by means of general principles from which the appropriate solution could be deduced. Often, as in the case of the two mothers, we begin to react almost instinctively and others react in opposite ways with apparently equal justification.

In a second step we think about and try to rationalize and understand the reasons for each stand. Then, we may eventually reach an agreement on what should be done in a given case without necessarily agreeing on more general issues. This observation does not mean, of course, that thinking about general principles is not useful. As a matter of fact, case-by-case analysis enables us to extract from particular situations some general guidelines which may be applied, under some conditions, to other cases. However, these guidelines are more helpful in stating the problems than in providing ready-made solutions.

For example, physicians are often faced with their most difficult ethical problems when they have to take care of critically ill or disabled patients with whom they necessarily develop a complex relationship. On the one hand, the patient is alienated by the disease and the suffering and depends to a large extent on the physician; on the other hand, caring of the patient implies for the physician not only good technical practice but also respect for the dignity of the patient as a person. Moral philosophers and bioethicists have theorized this situation by stating that the physician must submit to two general principles, the *principle of beneficence* and the *principle of autonomy.*

According to the principle of beneficence, the good of the patient in terms of pain relief and well-being must be sought. According to the principle of autonomy, the patient must be treated as an autonomous person and his or her values and choices must be respected. However, these two principles often contradict one another when one has to apply them in concrete situations. According to the principle of beneficence, the whole truth about a critical condition should not be told without precaution since it may aggravate the suffering and even the medical state of the patient. However, the principle of autonomy implies that the patient be told the whole truth in order to make medical choices and decisions as a person treated with respect and dignity. Practically, the physician will try to find a compromise, different in each case and adapted to each particular situation, in order to satisfy as much as possible these conflicting demands.

Thus, it is true that general principles can help to analyze and rationalize the situations. However, the solutions must be invented in each particular situation by carefully examining what makes a particular case not only similar but also different from other cases.

This attitude is often depicted and condemned as "casuistic" and as "moral relativism," meaning that there is no absolute good — no universal way to decide what is good and what is evil. Under each particular situation, different people would react in different ways and what is good for one could be viewed as evil for the other and vice versa. Philosophers have tried to establish universal ethics on the grounds of reason. Therefore, moral relativism is viewed as a failure of these attempts to define universal values.

However, we can realize that some sort of relativism is not necessarily a failure by searching for the source of ethics in practice as it is lived by most individuals and not only as a moral system or doctrine elaborated by philosophers. Unfortunately, the philosophical path is not followed by most people. Only a few individuals, if any — apart from some philosophers and their followers — are led in their lives by such a philosophical ethics. The kind of universality reached by philosophy is a singular one, valid for these few individuals. Moreover, different philosophers disagree on what should be considered the universal good.

Humanization through Society

In practice, most people behave according to sets of norms and values which have been inherited by their society from history and ancient traditions. In other words, ethics everywhere is a collective reality not grounded in objective rational knowledge but in the culture and history of a given society. When we think about the possible universality of ethics we must take into account, in addition to the individuals and the whole human species, an intermediate reality — that of the group or the society. This reality, intermediate between each individual and the whole of mankind, is well-known to anthropologists but often forgotten by idealist philosophers and theologians. In fact, human individuals become human; they acquire their humanity and their moral values not directly by instant exposure to the whole human species but through a family, a group, and a society with its language, culture, tradition, and history. Also, for each society a given religious tradition composed of mythology and rituals has always played a central role in shaping the collective set of values and norms.

This reality of the group, intermediate between the individual and the human species, is difficult to deal with in classical moral philosophy because it seems to imply the negation of universal ethics so that there is nothing else but moral relativism. However, this idea is based on an all-or-none view, in which ethics would either be universal, in the sense of being based on universal values from which one can deduce what is good and what is bad, or would not exist, in which case anything goes since what is good for one is bad for the other. In fact, there is an intermediate situation — the real one — in which explicit and implicit normative judgments on good and evil are inherited by different societies from their traditions, which can always be traced to ancient religions even if through crises, reactions, and revolutions. Even secularization and humanist traditions are not neutral and universal. Secularization is always a separation and departure from a given religion; therefore, it does not have the same meaning and the same content everywhere.

Based on this intermediate reality, another kind of universality can be conceived and actually constructed. Not an abstract and theoretical universality deduced from principles and values, but a possible common existence despite different value systems. Such a common existence is even more necessary now than before because of the pluralist character of increasingly more societies in increasingly more countries, where people belonging to different religions and value systems are forced to live together.

To understand how such a practical universality is possible and is being constructed piece by piece, ethical demands must be analyzed at different levels.

Pleasure and Pain: The First Level of Ethics

The most difficult question we confront when reflecting on the nature of moral judgment or on the genealogy of ethics (there are various ways of doing this) concerns the relation between the normative and the descriptive. My general point of view is that there exists a first level of ethics that is first in the sense that the question concerning the relation between the normative and the descriptive is resolved in terms of it or at least it seems to be. At this level of ethics the main concern is avoiding pain and seeking pleasure. This level, at which pleasure is sought and pain avoided, is also first in the sense that it is probably the most universal level of ethics. However, things become more complicated because various interrelated levels of ethics actually exist, and as we pass through them the way of understanding pleasure and pain changes. It is precisely this superposition and integration of different levels of ethics that produces the richness and also the difficulty of moral problems. To be sure, these problems cannot be reduced to considerations of pleasure and pain; however, pleasure and pain cannot be eliminated from them.

The relation between the normative and the descriptive is a difficult problem because we must avoid falling into the naturalistic fallacy — that is, the confusion of what is with what should be. The objectivity of knowledge cannot be preserved if what is the case is confused with what we would like to be the case. Conversely, what we decide should be the case is not necessarily what is the case. If we discover that we have aggressive tendencies, this does not necessarily means that we cannot, and indeed must not, decide to oppose them. Similarly, we decide that it is good to treat and heal diseases, knowing that diseases are an integral part of natural realities. Moreover, our knowledge of what is (e.g., of the mechanisms of disease and of our drives) can help us to oppose this and therefore to change it so that it conforms more with what we judge should be (within, of course, the limits of certain constraints).

However, having recognized that what is the case is different from what should be the case, the question of how they are related remains and it presents itself in a different manner at every level. The question is asked first at the level of moral judgment, and in this respect we say "this is good and it must be sought; this is bad and it must be avoided." Where does the normative character of moral judgment come from? (This is what distinguishes it from other types of judgments; for example, from judgments of fact on what exists or does not exist or from judgments of truth on what is true or false). The question is found perhaps more in the forefront at other levels of ethics and notably at the level that some call *meta-ethics,* in which moral judgments are themselves judged in order to evaluate on what terms a moral judgment is a good moral judgment and also if one particular system of values and norms is better than another. At this level, the question meets with that of moral relativism for those who wrongly believe (as mentioned previously) that the only alternative is between an absolute relativism — according to which everything has equal value and within which the observation of different systems of values and norms present in different cultures can yield nothing more than a simple list of mores and customs — and, on the contrary, a universal ethics that would enable us to judge in what terms this or that system is more "advanced" or "enlightened" than another.

The thesis I defend here is that the origin of the normative character of ethics on all its different levels is to be found in our experience of pleasure and pain. I do not intend by this, of course, the raw, immediate sensory experience of pain or well-being that accompanies the fulfillment of a need or a desire, but rather this same experience after it has been modified, transformed, interpreted, and sometimes inversed — in short, after it has passed through the prisms of our cognitive capacities. When these same cognitive capacities are not processing sensory input related to pleasure or pain, they are at the origin of other types of judgments. Nevertheless, it is by acting on a certain aspect of pain or pleasure that our cognitive capacities come to play a role in moral judgment.

Cognitive Functions and Ethical Competence versus Ethical Performance

Here, I can only briefly evoke questions still widely open to debate concerning the nature and workings of our cognitive activities, even if these same workings are also at the origin of this property particular to human beings that allows us to make judgments about good and evil. These cognitive workings or mechanisms are tied with what we could call, in analogy with our linguistic competence, the ethical competence of *Homo sapiens.* In question are the mechanisms of memory, language, our capacity of representation and symbolization, and our intentionality, which is understood as a source of signification and meaning and also, more simply, as a source of projects and well-defined activities. To be sure, an understanding of the workings of the activities of what is called in French *l'esprit* and in English the *mind* or, more precisely, the mind–brain unity is surely pertinent for our purposes. In the absence of these activities, most notably in the case of animals, there is no ethical competence at all. All of these activities, even those that appear the most simple, involve complex processes that are the object of the cognitive sciences: neurophysiology, psychology, linguistics, artificial intelligence, etc. I am working on the introduction to a physical theory of intentionality (Atlan, 1991, 1992, 1998) with the help of a model of the emergence of projects within automata networks in which no intentionality is postulated *a priori.* We must understand how a purely causal sequence of brain states producing a certain effect can transform itself into a finalized procedure in which it is the representation of this same effect that will subsequently trigger the repetition of the very same sequence. Such a sequence then becomes the completion of a project. What is in question, then, is a relatively simple and limited form of the emergence of intentionality. This is nevertheless far from being a trivial issue because, apart from human beings, this state of affairs can only be observed in apes capable of using objects as tools. However, I do not discuss this issue, even though it is pertinent to the current topic. At issue here are only elements of ethical competence and not of ethical performance, or performances, to sustain the analogy with linguistics. Our capacity to make moral judgments tells us very little about the content of these judgments. Also, it is as a function of the content of our moral judgments that the question concerning the relation between the descriptive and the normative is raised.

The idea that the origin of the normative character of ethics is found in the experience of pleasure and pain is not new. It was already present in the philosophers of ancient Greece, not only in the Epicureans but also in particular in Plato and Aristotle. The experience of pleasure and pain is distinctive in the sense that it is simultaneously cognitive and normative. Sensation or perception are immediately and automatically joined by a mode of behavior that aims to avoid pain or seek pleasure. Here, we encounter complex phenomena — and this is why I hesitate to speak of "primitive" facts — because there is no symmetry between pleasure and pain. Moreover, pain understood as a sensation is different from suffering, which is already on a neuropsychological level. Neurosurgery as well as studies based on the data acquired from positron emission tomography cameras have demonstrated the existence of nerve pathways for suffering (i.e., for a state illustrating a disagreeable characteristic) different from that of pain understood as a discriminating sensation (Willis, 1995; Craig *et al.,* 1994; Gybels and Sweet, 1989). The fact nevertheless remains that, at least in the case of the individual and his or her immediate sensory perception, there exists an experience in which the knowledge of a personal bodily state, or of a state of affairs in the world, immediately determines a flight reaction, which is a type of protojudgment about what must be done. Next, once states of this sort have been memorized, they are taken up by reason, imagination, and language, which make of them the object of a type of knowledge that is less immediate but which retains the trace of a normative character of pain that must be avoided and of pleasure that must be sought. It is this point that the ancients understood well, notably the Epicureans, certain of whom Cicero qualifies as "subtle" because they are not content "to judge by sensation what is good and what is bad, but instead pretend that one can also understand by means of the mind and reason . . . that pleasure must be sought for its own sake and pain avoided for its own sake"(as cited in Voelke, 1993, p. 40). However, these philosophers also saw that this task was not so simple, and we find in the work of Epicurus (1983a) a description of the moral problem that is, I believe, exhaustive, because in a matter of a few sentences he sums up all the difficulties at the origin of the problem's almost infinite complexity:

> And it is just because pleasure is the first innate good that we do not choose every pleasure; but sometimes we pass up many pleasures when we believe a greater pain follows. And we believe many pains to be better than pleasures when a greater pleasure follows. So every pleasure is a good thing, inasmuch as it is congenial to our nature, but not every one is to be chosen. Just as every pain too is a bad thing, but not every one is such as to be avoided. What remains is that every pleasure must be appreciated by a wise examination of the advantages and disadvantages that follow. For at some times we treat the good thing as bad and, conversely, the bad thing as good.

Everything is present in this quotation if we highlight two expressions that resume two important problems with which generation upon generation of philosophers have ceaselessly grappled. The first of these problems concerns the question of human nature: "So every pleasure is a good thing, inasmuch as it is congenial to our nature." The other problem concerns the "wise examination of the advantages and disadvantages." What, then, is this wise examination that sometimes causes us to "treat the good thing as bad and, conversely, the bad thing as good"? Who, indeed, does this "we" refer to? A "we" who sometimes takes good for evil and evil for good?

It could be the "we" of the philosopher, who decides by reason, or it could also be all of us, each one of us who expresses in our individual behavior an existing morality. Therefore, I think it is important to discuss this distinction between the morality that is found in a philosophical system and that, or those, actually practiced by existing societies.

The Philosophical Path

I first examine the case of the philosophers, noting that in one way or another they inevitably address the question of being congenial to nature. Indeed, we learn that for the Epicureans "pleasure and pain are a gift common to all living beings. The one is congenial to their nature, the other is foreign to it. It is based on this idea that we decide between what to choose and what to avoid" (Epicurus, 1983b). Also, it is not only the Epicureans who have expounded this point. For all naturalistic philosophers, the good is in harmony with nature in general, with the nature of all living beings, or with the nature of man, and it is for this reason that we must choose it. In other words, the normative (what we must choose) is identified with the cognitive (what nature is). Therefore, knowledge of nature is at the same time knowledge of good and evil. It is obviously this Socratic identity that critical philosophy has destroyed and rejected under the name of the "naturalist fallacy." However, this makes it no less necessary, although more difficult, to search for a possible relationship between the two different orders.

At this point, when the question of a possible knowledge of good and evil is treated within the context of a study of the relationship between ethics and cognition, it is difficult not to evoke the biblical myth of the first man and the tree of knowledge (Fig. 4). This is especially appropriate since later I discuss the role of myths in the genealogy of ethics. Amid the multitude of interpretations that this myth has given rise to, it is interesting to recall those of two philosophers, both rationalists though nevertheless very different from each other: Maimonides and Spinoza. Although at odds on many essential points, they agree that a knowledge of good and evil distinct from a science of the true and the false is merely makeshift, accepted in the absence of something better. In the paradise before the fall, rational knowledge of the true and the false rendered knowledge of good and evil useless and even nonexistent. For Maimonides, "man distinguishes the true from the false by reason, and so it is with all intelligible things." On the contrary, the good and the bad, like the beautiful and the ugly, are not of the intelligible or necessary order but are of the order of opinion and probability. Insofar as Adam possessed a perfect and complete knowledge of all knowable, intelligible things, "he possessed no faculty that applied to the realm of opinion, and he did not even understand this realm." Good and bad as such did not exist for him; all that did exist for him was what reason had him understand as necessary. It is only the loss of this knowledge of all things as intelligible that brings him to acquire a new state, in which things become available to him in a manner other than through reason — that is, in a contingent manner following the opinions he makes of them as beautiful or ugly, good or bad.

FIGURE 4 Tree of the knowledge of good and evil (miniature by Berthold Furtmeyer from the missal of the archbishop of Salzburg; photo courtesy of Bayerische Staatsbibliothek, Munich).

In Part 4 of *Ethics,* Spinoza (1972) also refers to this same biblical myth: "If men were born free they would form no conception of good or evil as long as they remained free" because "he is free who is guided by reason alone. He, therefore, who is born free and remains free has only adequate ideas." Moreover, he has "no conception of evil" because any such conception would be the product of incomplete, confused, and inadequate ideas; "consequently (since good and evil are correlative) [he has] no conception of good." However, Spinoza continues this line of reasoning by observing that this hypothesis (namely, that men are born free) is impossible because they are only part of nature and thus are influenced and determined by external factors. It is this impossible hypothesis, Spinoza states, that "Moses seems to be alluding to in the famous story of the first man." According to Spinoza, it is from within the perspective of this hypothetical but impossible original type of man that Moses "related that God prohibited free man from eating of the tree of knowledge of good and evil, and that as soon as he ate of it, at once he began to fear death rather than to desire life."

In any case, it is knowledge of good and evil that characterizes the human condition after the fall. This approximate and conflicting knowledge replaces the knowledge obtained by the clear and distinct ideas of reason, the only possible source of freedom. Thus, for these philosophers, knowledge of good and evil constitutes a state of alienation for humans, as they are, born into the real world, in opposition to the intuition of a world of freedom and happiness conceived through rational knowledge of the true.

Thus, the role of science and philosophy would be to give us access, with the help of reason, to a knowledge of good and evil that would be true since it would be derived from the "true" nature of man. It is in this sense that true knowledge is conceivable. Such true knowledge is characterized by Spinoza as that which reason teaches us is either useful or harmful to the nature of man. Here is not the place to discuss the criticisms of this special kind of utilitarianism of which the ethical dimension would be contentious. It is useful simply to remember that, in the context of the ethics, what reason shows to be most useful to man is other men (Spinoza, 1972). Whatever the case may in fact be, we have here the possibility, at least theoretical, of a natural foundation for ethics. However, this evidently implies that we have adequate knowledge (i.e., perfect and complete knowledge) of the nature of man.

Obviously, this is far from being the case. In place of such knowledge we find instead religious and nonreligious ideologies that have pretended and continue to pretend to represent true knowledge of nature in general and of human nature in particular. In addition, even if we admitted we were able to acquire such a knowledge of good and evil, the knowledge would have no force on its own. As purely intellectual, disembodied knowledge, whatever truth it might possess would not allow it to impose itself on the ideas of good and evil we normally formulate from our experiences and our passions. This is another way to highlight the difficulty involved in accounting for the relation between the descriptive and the normative. Spinoza (1972) notes, "True knowledge of good and evil cannot restrain any affect insofar as this knowledge is true, but only insofar as it is itself considered as an affect"—that is, only to the extent that this knowledge becomes incorporated and internalized to the point of acquiring the power of a passion. In other words, the true does not force anything on us. It can only be a norm in reference to itself and not to any mode of behavior. This point is of capital importance. The content of knowledge can only be transformed into a norm if it is invested in a passion that produces a sense of well-being or of suffering—that is, a sense of something agreeable to be sought and of something disagreeable to be avoided.

For our life in society, we have only inadequate knowledge of good and evil. Such knowledge can in practice, at least for the philosopher, only be that of a provisional ethics inherited from the society within which the philosopher lives and in relation to which, for that matter, his personal ethics may impose on him some difficulties. In other words, the aspiration to obtain true knowledge of good and evil in accordance with which evil might disappear is perhaps a universal aspiration. It is in any case the aspiration of certain thinkers and philosophers who would like to find within our affective life the same certainties found in logic and mathematics or, at most, the same type of evidence found in empirical science. Instead, in reality we are confronted with existing moralities.

The Practical Sociocultural Path

Leaving behind the philosopher's theoretical reflection, then, we find ourselves facing a multiplicity of opinions concerning systems of values and norms, inherited from our social and individual histories. It is based on these norms and opinions that we make value judgments, deciding what is good and what is evil. However, we can still go behind the multiplicity of these value systems and ask what is common to them and thus universal. First, there is the fact of their very existence. The existence of systems of values and norms is a universal fact in the sense that it characterizes all societies, even if the content of these systems is not the same. Also, it is here that we find again the first level of ethics—a level common to all these different systems because each one is conditioned by the physiology of *H. sapiens* through the experience of pleasure and pain, of what makes us feel good and what makes us feel bad, whatever may be our way of naming and describing this experience or, of course, of integrating it into a socially constructed and inherited system.

We can freely call this first level of ethics, in which good and evil merge with pleasure and pain and with good and bad, the fundamental level or level zero. For those who prefer to assign to it a positive connotation, we call it the fundamental level; for those who prefer a negative connotation, resulting from its obvious poverty in relation to the richness and complexity of moral problems, we call it level zero. In the case of most modern philosophers, especially the successors of Kant, it seems that the experience of pleasure and pain is too poor and too limited to play a role in the intricacies of what we call the problems of consciousness, freedom, and responsibility, which are the daily bread of ethical thought.

At the turning point of these considerations, we again encounter Spinoza (1972) because he takes up this identification between good and evil and between good and bad and makes of it the foundation of his theory of the passions: "Insofar as we perceive anything to affect us with Joy or Sadness, we call it good or evil; and therefore the knowledge of good or evil is nothing other than the idea of Joy or Sadness . . . idea which follows necessarily from the very affect of Joy or Sadness"—that is, the affect in itself, insofar as we are conscious of it, insofar as we form a conception of it, or insofar as we have "the idea of the idea of it." As Proposition 8 states, "The knowledge of good and evil is nothing other than the affect of Joy or Sadness, insofar as we are conscious of it." However, in this case the good and the bad are no longer reduced simply to pleasure and pain but to what he calls the affects of joy and sadness, which he defines in reference to an increase or decrease in our power of acting. This definition allows us to broaden the notion of pleasure and pain to include the notions of well-being and suffering. These latter notions are less restrictive and encompass greater stimuli at a level of organization that would be a "common gift" not, as for Epicurus, to each living being but to each being endowed with reflexive consciousness, and in this case to all human beings.

After this first level, things become complicated very quickly because of other characteristics specific to human beings, notably our cognitive capacities of representation, memory, symbolization, intentionality, etc. The experiences of pleasure and pain resulting, individually and collectively, from the contingencies of our personal histories are locked into memory, and it is the content of these experiences that subsequently feeds our more or less rationalized imagination. In other words, due to the effect of memory and imagination what we call good and bad is no longer reduced to the immediate perception of what makes us feel good and what makes us feel bad. The meaning of the words "good" and "bad" is split by the imagination, and this happens on two accounts. First, in terms of time: What is good or bad within the present sensation is projected out in time. Second, in terms of space: Our personal experience of what makes us feel good and what makes us feel bad is projected on others who resemble us and to whom we attribute the same experiences, and the meaning is thus broadened into a collective good or evil. It is this twofold splitting in two of the things we call good and bad that is conveyed by the ambiguity of this expression. In the case of the word good, this is the ambiguity between, on the one hand, something understood to be good because it is a source of pleasure, well-being, or happiness, and, on the other hand, something understood to be good in terms of a reflexive or conditioned judgment (i.e., a moral judgment). The same ambiguity exists in the case of badness, whether it be understood as something causing pain or as a wrong act. This ambiguity is even more obvious in languages making use of a word derived from the Latin *malum,* whereas the two words "bad" and "evil" are used in English in opposition to the two meanings of "good." Thus, on the one hand, there is evil as pain and suffering, which the search for well-being and good(ness) tend to eliminate; on the other hand, there is evil as badness — that of wrong acts, vice and perversion, and hate and injustice — from which the teachings of ethics are supposed to deliver us. This second evil sometimes joins the first when the wrongdoing is a suffering inflicted on someone else or even on oneself in a more or less distant future.

It is the realization of the gulf between good and bad, perceived directly in sensation and the results of their more abstract perception by the imagination when projected onto others or onto oneself at some time in the future, which constitutes what we call the "knowledge" of good and evil. Since this good or bad is imagined and projected out toward others or in time, it is not perceived directly as good or painful; it can only be the object of a reflexive judgment (i.e., a moral judgment). This judgment is marked by the ambiguity of what we call good and evil, which bring together their two aspects within consciousness or within the representation we have of them.

This ambiguity is evident in the notions of joy and sadness and of happiness and unhappiness. Inextricably bound with these notions are those of immediate individual sensation and its twofold splitting in two, foisted out onto others and projected in time by the imagination. It is precisely here that we find our second level of ethics.

The Second Level: Strategies to Defer

This second level of ethics can be characterized generally by individual or collective strategies in which the search for good and the avoidance of bad are deferred. We believe, or think we believe, that something currently believed to be good could be the source of suffering for others or even for ourselves at some time in the future and, conversely, that something currently believed to be bad could produce feelings of goodness for others or for ourselves in the future. The second level of ethics is that at which the search for pleasure and the avoidance of pain are tied to an imagined projection in space and time, which in practice postpones the satisfaction of desire.

Indeed, the desire we all share to avoid suffering and immediately enjoy pleasure is necessarily deferred more or less in accordance with the circumstances and especially with our value systems, our philosophies, and our traditions. Each one of these is expressed, after all, in norms of behavior that indicate for different types of behavior when, how, and how much their satisfaction should be deferred. It is not possible to remain in the present, and both memory and imagination suggest possible ways to defer satisfaction. The most difficult question, then, concerns how much one can and should put such satisfaction off. If satisfying all desire immediately in an animalistic or child-like fashion is impossible, conversely, deferring this satisfaction indefinitely, even by repression or sublimation, is also impossible. As King Solomon said in *Proverbs* (13, 12), "A deferred wish is an ailment of the heart; a growing desire is like a tree of life."

This second level of ethics is constituted by a set of judgments about when, how much, and how to defer the satisfaction of our desires. This second level is also universal with regard to its existence but diversified with regard to its content. The law of satisfying or deferring desire is experienced differently in accordance with the varieties of ways of satisfaction and deferment and the alternations between them. These alternations create a rhythm that characterizes each system of ethics as a particular instance of the law of desire and its sublimation. Each instantiation of this type constitutes a particular definition of good and evil, in general produced by the history of each society and that of its collective imagination, in relation with that of its individual members.

Thus, the second level of ethics, understood as a strategy according to which the desire to benefit from pleasure and the desire to avoid pain are deferred, is the level of reflexive judgment at which good and evil are no longer directly perceived as what is good and what is painful. The instinctive urges of the animal or of the infant are transformed in the adult human by knowledge, at once rational and imaginative, which accompanies the development of memory and new cognitive faculties.

The Third Level: Moral Judgments on Moral Norms

It is at this stage that we encounter a third level of ethics. Indeed, it is at this level that we make judgments about moral judgments, about the criteria we use to decide between what is good and what is evil. It is here also that the question of relativism arises. Are all criteria for distinguishing between good and evil equivalent? Do all different points of view amount to the same thing in the end — the moral standards of the gang, the nazis, human rights advocates, the master and the slave, those who hold that "thou shalt not kill," those who think that "might is right," etc.? Are there criteria that would allow us to judge that one system of morality is better than another, or even that one moral system is actually immoral? The totality of these criteria would thus constitute a universal ethics. We therefore find ourselves at a third level, at which we formulate judgments about the value systems found at the second level.

Here again, it is useful to make the distinction between philosophy and existing moral systems. It is possible to outline two general ways of thinking about this. On the one hand, there is the great tradition of occidental moral philosophy that culminates with the work of Kant and the Age of the Enlightenment, according to which reason would enable the creation of a universal ethics. On the other hand, one can think about a universal ethics to be established not deductively from principles of reason but, rather, pragmatically from the observation of existing moralities.

The Singular Universal of the Philosopher

Let us envisage first the postulate of the universality of a rational ethics that would teach to each person what is good and what is evil. As we have seen in the case of the ancient Greek philosophers, this type of ethical system is derived from a certain understanding of human nature, itself taken as universal. It is for this reason that this philosophical system has always maintained a close relationship to the sciences, the natural sciences as well as the human sciences, which are intended to teach us the nature of man. However, in relation to ancient and classical philosophy, a decisive turn was taken with critical philosophy. This latter recognized the impossibility of founding morality on pure reason and the categories of the sciences of nature but did not, for all that, renounce the universality of practical reason as a possible source of universal ethics. This change in perspective was conditioned by the birth of modern science in the seventeenth and eighteenth centuries — by what we call the scientific revolution which concerned physics and chemistry especially and, to a much lesser degree, the life sciences. This is why it could be shown that with the present extension of this mechanistic revolution to biology, the very continuation of critical philosophy's project leads today to a calling into question of this humanist philosophy, founded on the hypothetical existence of a suprasensible domain of human freedom, since this humanist philosophy could benefit from the vitalistic and finalistic climate of the age (Atlan, 1991).

In a very schematic fashion, within the context of this philosophy, a necessary link is taken for granted between the use of morality and the reality of our free will. In other words, our moral judgment (at the second level) would only be a good moral judgment (from the point of view of the third level) if it were made by a subject endowed by a voluntary consciousness that decides freely about its future actions. Free will and the power of individual determination clearly seem to be conditions of access to the second level of ethics. However, this judgment about what a good moral judgment should be is itself far from being universal. It arises because we are used to the idea that our moral judgments are the result of an internal debate we hold with ourselves, which precedes the act of our reflexive choosing between something that is good and something that is evil. However, the inevitable character of these deliberations and of their eventual causal effect is not evident, neither from the anthropological standpoint nor from the philosophical one. Moreover, as an outcome of contemporary biology, which is physicochemical and mechanistic in nature, we uncover most of the implications for our behavior of an absolute, mechanical determinism of nature, presupposed already in the scientific rationality of the classical age but from which, at that time, life and human beings appeared to escape. It is thus perhaps our task to revive a certain way of viewing the world, difficult for us to imagine and long since forgotten, which has been hidden behind the too simple idea of a responsibility based on accountability which would itself be only conceivable on the assumption of free will.

On the contrary, it seems that we will have to return to an ancient, Homeric view of the world, according to which men were possessed by the gods, who used them and acted through them. However, if we are possessed and determined in our behavior nowadays, it is no longer by a mythical pantheon of gods, themselves personalized in man's image, but instead by impersonal mechanisms — those of the chemical reactions of the molecules in our genes and in the metabolism of our cells. One could attempt to tabulate the respective advantages and disadvantages of these two different ways of being possessed: on the one hand, by personal, mythical gods, human in a certain sense and projections of ourselves in whose image we see ourselves, and, on the other hand, by impersonal mechanisms, over which, as a result of technology, we nevertheless have some hold.

In any case, our contemporary world is one of impersonal laws in which what is problematic and difficult to understand is not the possible knowledge of a timeless, absolute determinism. On the contrary, each discovery of a new type of physical or biological mechanism, especially in the form of a mathematical law or a formal, timeless structure, even if their field of application is local and limited, puts us face to face with this world of impersonal laws. What is actually problematic in this mechanistic view of the things that rule us, then, is our social and subjective experience of responsibility, which is held to be dependent on the free bidding of our will.

Contrary to what people often say about modern science, which attempts to harness uncertainty and chance by confronting the complexity of reality, it does not abandon this presupposition of causal determinism or the classical ideal of knowledge by causes. The indeterminacies of contemporary probabilistic science no more make way for free will than do the holes present in classical determinism. This is the reason why, from an individual and speculative point of view (i.e., philosophical as used by philosophers and their public), this criticism of Kantianism Alexis Philomenko (1969) talks of the "agony of Kantianism"— could well give rise to something like a modern version of Spinozism. Although he was pushed aside and marginalized for three centuries, it seems as if, all along, Spinoza had really just been waiting for scientific reason, in which he invested so much importance, to guide enough men along part of the way. Indeed, only this experience of an understanding of things can allow the Spinozistic solution to the problem of ethics. Only experience of understanding can allow intellectual knowledge itself to serve as a criterion of judgment between good and evil. It does so not in terms of the content of its knowledge but in itself by making us experience that "what is good for me" is the very fact of understanding, and "what is bad for me" is the obstacle to understanding.

In the "battle of pantheism" of eighteenth-century Germany, Kant had to take sides against Spinoza. Apparently, like Kant, Spinoza drew out the consequences of a rationalism in which algebraic geometry and the new physics served as a paradigm. However, in doing so, and contrary to Kant, he marginalized himself in relation to the theologicophilosophical tradition of free will without giving up on an ethics of freedom gained through knowledge, which is much more difficult to conceive and even more so to accept. Thus, in this type of philosophy responsibility is not based on free will. On the contrary, it is responsibility that is primary — not necessarily accompanied by guilt — within a world given to us at once determined and intelligible.

This ethics with overtones of Stoicism, which Spinoza nevertheless attempted to differentiate from Stoicism, is ill adapted to techniques of mass communication, at least for the moment. It requires a very special type of learning and teaching skill to get its message across and actually implemented. Although it is rational and universal in its principles, it is only realized, if this is possible, in the existence of the individual. It can thus only be a matter of a singular universal that would belong only to the philosopher or the wise man.

The Constructed Universality of Intercultural Argumentation

Therefore, at this third level of ethics we must again return to existing moralities and why we cannot avoid observing, between the individual and the universal, the reality of the diverse social groups in their concrete, living forms of behavior and existence. We cannot ignore the place where the systems of values and norms that shape individual behavior actually develop. As mentioned previously, this place always takes the form of a particular social group with its own culture, language, and history. Also, if we do not wish to fall into a relativism in which everything is of equal value, we can search for, and indeed find, common elements that might not only bring these existing moralities in contact with one another but also start a dialogue between them. For example, in a genealogical study of the source of a specific ethical system as it is actually lived by a certain group, we always come across the trace of religion — that is, of a group of beliefs, myths, and rituals that the individuals belonging to the group have internalized through their education. This continues to hold true even if, especially in the case of modern secular societies, the individuals are able to exercise to varying degrees their power of critical reflection and contribute to the modification of the accepted values and norms.

The interest of such a genealogical study lies in avoiding a simple acknowledgment of the existence of a plurality of cultures, their moral standards, and their religions — an acknowledgment that would amount to nothing more than a simple list of mores and customs. Instead, one may attempt to identify, behind this diversity, common elements that could facilitate the meeting of and dialogue between different cultures and traditions. Such encounters and such reciprocal respect are also part of our reality. This is not just a pious wish. It becomes a reality when individuals nourished by different traditions live together in pluralistic societies and also when different societies, peoples, and nations are condemned to coexist peacefully after renouncing war as a habitual way of resolving their conflicts.

In other words, to state that a universal ethics that imposes itself on everyone by the very force of reason does not exist is not necessarily to resign oneself to moral relativism. Another sort of relativism is possible in which reason does not legislate like a tribunal but rather functions as a tool enabling dialogue. It does not serve as an ultimate foundation based on a blank slate but instead as a means of dialogue based on different moral experiences. In following this path, we can construct, with reason's help, a system of ethics that is concretely universal, under the condition that we begin from existing moralities by means of rhetorical argumentation and rationalization.

Contrary to the way things might appear, this point of view constitutes the only real barrier preventing the relativism of knowledge from sliding into a nihilism in which "everything is equal" yet all the while refusing to recognize any abstract absolute. Whether this absolute masquerades as rational and scientific or as revealed truth, it is clear to us today that the result is at risk of being the same: a totalitarian regime complete with (holy) wars of extermination, in which belief goes hand in hand with intolerance. It is only by means of a dialogue between different cultures and traditions, a dialogue that involves a search for compromise, that we can hope to arrive at, progressively and empirically with no imposed revealed truth nor excommunication, a universality founded on the recognition of the other. Evidently, no one nor any thing can guarantee either the truth or the success of this approach. The risk is always that dialogue and negotiation will be fruitless and that conflict will be settled once again, as it always has been, by war.

Nevertheless, as will be discussed later, such a pragmatic approach already has some modest and partial success to its credit. It is possible to believe in its relative efficiency when we note the nature of the framework within which the dialogue can be undertaken. I do not mean a theoretical framework that would bring us back to laying down principles acceptable to all — principles other than the simple desire to or the necessity of coming together and talking — because this would again bring us face to face with the impasses of the abstract universal already discussed. It is perhaps this very point we can bring against the philosophies of argumentation, such as those of Habermas or Apel, which are nevertheless far advanced in this direction.

Instead, I intend the framework necessarily created in the very practice of such dialogue. This practice can only exist because the moral experiences produced and conveyed by the mores of different societies, although different and sometimes irreconcilable, do not end up completely isolated. They can only be opposed in the first place because they are not incommensurable in the sense that they result from the same way of seeing good and evil, which itself remains universal as long as it can be reduced to the first level of "what is good for me" and "what is bad for me." All agree that something painful feels bad and that something pleasurable feels good. This is why argumentative and even rhetorical reason can take hold of this first level and seek reconciliation based on this common ground of experiences. In this search for a common denominator, a common final end can be admitted more easily when it is limited to the suppression of suffering and the search for well-being, without of course forgetting that different cultures are divided on the strategies of such a search, given the necessity of deferring the satisfaction of desire and extending it to the other. In other words, the worry to remain rooted somehow in this first level of ethics — the only one that is truly universal in the sense that it concerns the suffering and the happiness of each individual in his or her singularity and in the present instant — at least encourages the different systems of norms to meet on this common ground. Later, one can come to understand, without necessarily justifying, in what sense the normative character of the second level of ethics of a given foreign social group results from the individual history of the sensibilities of the group. These sensibilities are expressed first in the language and in the words that designate good and evil. According to different languages, the good and the evil, the just and the unjust, the generous and the cruel, the pure and the impure, the clean and the dirty, the fragrant and the putrid all have kept a close proximity with what designates well-being and suffering. The pure and simple identity in the romance languages of words based on *bonum* and *malum* which designate our first two levels is not always found in other languages. For example, in English "good" and "bad" are transposed into "good" and "evil" or into "right" and "wrong," which seem to have nothing to do with "pleasure" and "well-being" or "pain" and "suffering." The differences in these words express, in the wake of Freud, what we might call varying degrees of the sublimation of the physiological realm into a sociocultural one. However, I insist that behind these differences there is a common element that seems to be important in the history of moral sensibilities — a common element that is simultaneously, like religions, a source of rapprochement and opposition.

Modified States of Consciousness and the Sacred: A Source of Unity and Diversity

At the second level of ethics things are complicated not only by imagination and memory, which project pleasure and pain outward in time and space, but also by another factor which is worth consideration. This factor is a certain understanding of the realm of the sacred according to which this realm is taken to be an objective reality. Through a genealogy of ethics, understood as a rediscovering of the entirety of sources covered over and forgotten by new interpretations, we find within religions a very different type of experience of pleasure and suffering linked to what is known as altered states of consciousness. The role of hallucinogenic plants in contemporary shamanistic religions and their probable role in ancient religions is one of the elements providing support to this observation. However, these altered states of consciousness, characterized in diverse fashions as ecstatic, mystical, etc., are not produced simply by the ingestion of plants. Although they afford access to another reality, these states of consciousness also characterize moments of "normal" life, which some people have not hesitated to call a separate reality: dreams, sexuality, a particular relationship to danger and death in extreme situations lived out in a heroic fashion, including in war, and also in certain aspects of the aesthetic experience, in which all this is transfigured in the work of art. Although such experiences as these can today be lived out in a nonreligious context and can be integrated into a positivistic and even materialistic view of the universe, these experiences, in almost all religious traditions, are at the origin of what is known as the realm of the sacred.

I defended the hypothesis that religions in general and the meaning of the sacred have their origin in these types of experiences (1986). According to this hypothesis, the basic role of myths and rituals, underlying whatever symbolic rationalizations we may give to them, is to bridge the gap between these two separate realities that correspond to normal and modified states of consciousness. Indeed, behind their diversity, these experiences share the fact that, unlike other experiences, they are pleasurable or painful insofar as they afford access to what appears to be another reality, a world different from the one in which we usually live. Thus, if we allow that the systems of values and norms actually existing in societies are inherited from their mythical iconography and representations as well as religious rituals — a point of view that is difficult to contest — it is then easy to find the traces of these experiences of the sacred in the genealogy of ethics. Without neglecting the classical interpretations of the role of myth and ritual in the organization of traditional societies, when one takes seriously both the psychical and the physiological reality of these rituals they appear as modes of behavior allowing the individual to bridge the gaps between two realms and piece together his or her fragmented existence, in which the individual finds himself or herself participating in two separate realities, two different temporalities, two distinct worlds — that of the everyday and that of the sacred. The role of shamanistic rituals in the treatment of disease presents one of the most blatant examples. The hidden cause of the evil from which the patient is suffering is found in a world to which only ecstatic experience gives access, and the ritual acts on this hidden cause. This interpretation does not negate others that, for example, highlight the integrative function of ritual and myth, as a result of which social cohesion and politicoreligious power are assured. However, as discussed later, this interpretation does make it possible to understand the phenomenon of *internalization of the social* — that is, how individuals discover the power of this other reality, existing at the very heart of their being and at the same time indissociable from the collective representations conveyed by the imagination of the group and from the rules of behavior that accompany it.

I know that many people who are willing to recognize that the origins of actual moralities exist in religions are nevertheless unwilling to concede that experiences of the sacred assimilated with dreams and hallucinations are at the origins of religions. Not only in their histories but also at their sources, we must distinguish between those religions that have conserved a relation to the sacred that is lived out in the same fashion as the shamanistic religions (which maintain only the most tenuous of relations with the written word) and those religions that are considered to be more "advanced," notably those known as the great religions, which have pushed back and covered up the reality of these experiences with literature and philosophy of great richness. In the West, this tendency has given rise not only to the intellectual sophistication of theology but also, after numerous detours by direct influences and by countereffects, to similar sophistication in philosophy and the sciences. On the contrary, shamanistic religions have in most cases remained oral and, perhaps as a result, have been considered as primitive in the pejorative sense of the term. These distinctions are certainly pertinent from the standpoint of the history of ideas and of civilizations. However, the fact nevertheless remains that this dimension of ethics rooted in the experience of the sacred as objective reality, unusual for moral philosophy, cannot be ignored under the pretext that it presents an irrational dimension tied up with illusion and mystification.

The dream is not an illusion because an illusion is an error. Even the waking dream, for the man who knows he is dreaming, is a lived experience. To be sure, it is difficult to think of its content as objective, and the only way to do so is through interpretation, although it is illusory to seek an objective truth independent from the subjectivity of the interpreter. However, it is an empirical experience all the same, despite the changes brought to the perception of space and time that characterize it. To the extent that dream-like and hallucinatory experiences, induced or not, have played — and indeed continue to play — a role in the institution of the realm of the sacred and of the norms of behavior that result from it, it would be mad to ignore them in our strive for universality, under the pretext that the content of these experiences, not objective, are often mutually exclusive and difficult to communicate. The reality of these experiences seems to be a characteristic of the species just as universal as our upright posture and reason itself.

It is important to recognize this level of ethical experience because it enriches and complicates considerably our second level. In addition to imagination and memory, which project pleasure and pain out into time and space, these experiences of the sacred through hallucinatory, dream-like or other means divide good and evil into a double register or into two separate realities, if we wish to call them thus, corresponding to these two modes of perception. We see the effects of this in the moralities revealed and ritualized in traditional societies. However, these effects are also present in certain types of behavior valued and ritualized in our own societies, even if the meaning of the sacred is here greatly diminished and has long since been separated from these dream-like and hallucinatory origins. An extreme example, pathological and tragic although banal, is that of drug addiction, in which pleasure is sought although it is also a form of suffering and can even be fatal. A particular feature of these experiences is that they simultaneously act on the individual in his or her singularity and also on collective representations, myths, and rituals. This is why these experiences suggest a possible mechanism of an internalization (within the individual) of a reality that belongs first to the group. This includes a means of internalization of the law and norms of the group.

It is in this manner that the first level of ethics, individual and immediate, becomes modulated to the second level by the norms of the search for good and of the avoidance of evil, where good and evil are understood as occurring simultaneously individually and collectively and as having both a sensory and a collectively imagined nature based on experiences of multiple realities. It is these norms, internalized by the individuals of a group, which make them say "this is good" in the sense that it "increases happiness" and "this is bad" in the sense that it "causes pain." This is the case even if the judgments are sometimes opposed to a certain individual's immediate sensory perception of what is good or bad, either for himself or herself or for another. Between cultures, this internalization of norms by the individuals occurs differently according to the society to which the individuals belong or in which they grew up. To facilitate the coming together of different cultures and value systems, it is thus important to understand how the relation between these norms and the individual and immediate level of pleasure and pain is the result of the history and the memory of particular collective experiences. For this purpose, genealogical and ethnological studies — even applied to our own societies — can prove fruitful in discovering the elements of these experiences that underlie language, social structure, organization of time and space, and the myths and rites that express our memory of all these.

In this fashion, the third level of ethics, in which different systems of values and norms coexist without being normative in the same sense as a universal moral philosophy, can avoid being the ground of a moral relativism which would permit nothing more than the acknowledgment of a list of mores and customs. Instead of being the place of a meta-ethics, as is often stated, this third level could be that of an ethics of argumentation in which different systems of values and norms could enter into dialogue. The only rule of dialogue would be the necessity to resolve practical problems of coexistence while making the greatest effort to avoid war as much as possible. Success is not guaranteed, but things are not as utopian as one might think. There are various factors that can influence eventual success, and the relatively recent appearance of ethics committees in pluralistic societies affords us a view of some of these factors. One of them, as noted previously, is the presence within all individuals of the common denominator of the experience of pleasure and pain, even if this experience is covered over with the deposits of the group's history and culture.

Another factor, also very important, is a particularity of our ways of communication and argumentation by which we attempt to answer the question "why?" in terms of causes, reasons, or motivations. What is at issue, then, is a most remarkable property that appears as a theoretical weakness present in many of our ways of explanation but a property which can nevertheless be turned into a strength, an advantage that helps us to live our intersubjectivity.

Underdetermination of Theories by the Facts: A Help in Reaching Agreement

This is a phenomenon that has been analyzed in different contexts under the heading of the underdetermination of theories, first in physics by Duhem and then by Quine in epistemology and within his work on translation. It is present today in our way of designing models of dynamic systems by means of networks of automata (Atlan, 1989, 1991b, 1992).

In a very schematic sense, every dynamic system (e.g., a network of neurons, cells in the immune system, and metabolic reactions within the cell) demonstrates the property of converging toward one of its stable states (or, more generally, one of its attractors whether it be stable, oscillating, or chaotic) through a multitude of ways. These ways differ not only with respect to different initial conditions, which might arise from historical contingencies or prior determination, but also with respect to the different structures of the networks. A network's structure depends on the way in which its constituent elements are interconnected and different interconnections express different models or different theories. However, although each network structure produces one or several attractors which can thus be computed and predicted, different network structures can produce identical attractors and thus predict the same observations.

This fact can be demonstrated very easily and by relatively simple examples. It can be shown that in the majority of systems made up of elements of which we can observe only relatively few discrete states (e.g., the states of a neuron's activity or the states of a given population of immune system cells), the number of observable states of the network is much smaller than the number of possible structures of interconnections and therefore of theories or models constructed from these same elements. Indeed, the first varies as a power N (where N represents the number of connected elements) and the second as a power N^2. The number of possible states in a network containing N elements is $2^N, 3^N, \ldots, q^N$ if each element can be found respectively in 2, 3, . . . , or q states. On the other hand, the number of connecting structures is $2^{N2}, 3^{N2}, \ldots, p^{N2}$ if the possible number of values representing the force of each connection can be, respectively, 2 (e.g., 1 or 0; that is, existence or absence of a connection), 3 (e.g., 1, 0, −1, if a connection could be inhibiting), or p (in the case of connections for which the "weight" could vary and take p possible different values). For these numbers to be of a similar order of magnitude, and for there to be no underdetermination, the following condition must be met: $q^N \geq p^{N2}$ (i.e., $q \geq p^N$). This condition is met only in controlled experiments, when systems can be divided into subsystems in which the value of N is small and in which the elements can be observed in an almost continuous fashion, such that the value of q approaches infinity. Here, I have assumed that the only possible observed states are the attractors, the number of which cannot be larger than the number q^N of states of the network. However, even if we assume that we can observe the transients, together with the attractors, this simply multiplies the number of possible observations by q^N (i.e., the number of initial states) since each of them leads to one of the attractors through a different transient states. The underdetermination is not reduced very much in this case since q^{2N} is still, in general, much smaller than p^{N2}.

The result is that in most cases hundreds or even hundreds of thousands of different models (their number increases in an exponential fashion with N) predict the same observed facts. Therefore, when we are faced with natural systems, such as biological ones that we study *in vivo,* after having discovered the structure and the function of the elements that constitute them (i.e., cells, populations of cells or of molecules, etc.) the difficulty of theorizing does not come from a lack of predictive theory but, on the contrary, from a surfeit of valid theories that predict the same observations, without our being able to determine if one would be more "true" than the others.

This fact, of course, indicates a weakness of our theorizing capacity. However, when the same phenomenon is applied not to theoretical models but to natural systems (e.g., to networks of neurons that actually make up our brain and not to those formal neuron networks we use to simulate them), it appears on the contrary as a property of structural stability (i.e., robustness) in their dynamics. When applied to the working of our brain, this characteristic of converging toward a common final state through different ways and despite different structures perhaps helps to explain a remarkable characteristic of our intersubjectivity: namely, that it is much easier to agree on actual conclusions than it is to agree on the way in which they should be achieved. This point perhaps appears curious and even shocking to our deductive, rationalizing thought. However, we often unwittingly experience it. The deliberations heard in ethics committees are just one case among many others that provide striking examples. When confronted with a particular question concerning whether a certain practice is morally acceptable or not, it often happens that people react spontaneously in the same fashion before actually formulating the reasons for their position. On the contrary, difficulties arise when it comes time to analyze and justify the declared moral character of the practice in question, whereas its being acceptable or not seemed evident from the outset. For example, all members of the French National Ethics Committee for Life Sciences were in agreement that the use of aborted human embryos in the cosmetics industry was unacceptable. However, among individuals with opposing conceptions of the nature of personhood and the status of the embryo, it proved very difficult to agree on a minimum of reasons for justifying this position. In other words, when confronted with a situation, especially a well-defined situation, we first react emotionally following our sense of what we feel to be good or evil. However, if it should turn out that our conclusions are actually the same (even if only because there is a limited number of them), we now see that it is perhaps better not to ask why and not to analyze our motivations too deeply. This questioning gives rise to often irresolvable disagreement concerning beliefs and principles.

Fortunately, we do not give up this questioning and we try to argue about the different motivations which have led to our common conclusion. Thus, relative misunderstanding, or what Wittgenstein called "the vague of natural languages" (in opposition to the precision of logicomathematical languages), is sometimes what allows dialogue to begin. Also in this fashion, although general agreement on *a priori* principles and concepts is out of reach, a real, existing universal ethics can be constructed, little by little, by means of searching heuristic procedures, and this can be done from the bottom up rather than deductively from top to bottom. In this process, finding ourselves in agreement about a certain practical conclusion, even though our interlocutor bases claims on a value system different from our own, we may be moved to relativize our own motivations, which we believe to be those of our own ethical system, without devaluing them. In other words, the initial hidden misunderstanding concerning our motivations sometimes enables us to develop a dialogue. Moreover, the process of developing the dialogue may make us more aware of the misunderstanding and thus sometimes relativize the oppositions between our basic beliefs.

Return to the First Level: The Minimal Moral of Indignation

It is here that the importance of the first level of ethics (which had never disappeared) can now reappear and that its weight in the course of the argument can push us along quickly toward new agreements.

We increasingly witness these practices when we observe, case by case, how problems of cultural coexistence are treated when dialogue and negotiation are admitted as the only ways of avoiding conflict. I mentioned the existence of ethical committees in pluralistic societies. Another example, at a different level, is that of a wide-ranging scope of public agreement on a humanitarian basis. What remains universal and allows us to agree on moral grounds — outside, of course, of shared interests which may force us to avoid war — is the invocation of good and evil at the first level through the immediate sensory experience, even the spectacle, of suffering and well-being. We witness the suffering of the other through the media, and this sets off the movement of public opinion of which we want to be a part. These movements of public opinion are of course ambiguous because, among other reasons, there exists an evident asymmetry between good and bad, and this limits the wide-ranging consensus.

It is much easier to come to agreement about what we call bad or suffering than what we might call happiness. This is the reason why these wide-ranging agreements are based on a minimal ethics that could be classified as a morality of indignation, which is in fact the most immediately and the most universally accessible. We are more spontaneously determined to combat an evil we sense or witness than to search for a happiness we find difficult to imagine because the search for well-being is much more individual and it stretches out over time. It is easier to come to agreement about this minimal ethics of prevention of suffering and visible evil than about more elaborate ethical standards that would help to bring about good and well-being. This is the reason why the morality of indignation witnessed in the face of suffering is the most immediately and the most easily consensual.

However, this type of morality is not composed only of advantages. It is a slightly naive morality of the present moment, which relies on a feeling produced spontaneously by an image in the very instant of our perception of this image. It is a morality in which one image follows on the heels of another, a morality of the underdog in which the person suffering is necessarily right even if he or she is wrong in the long term. As such, this type of morality of indignation is susceptible to all types of manipulations of information, playing, for example, on a childish type of emotional blackmail. However, it is better than nothing. It is a first level that we would be wrong to neglect. The avoidance of suffering, whether it be the imagined, future suffering of oneself or the suffering of the other, constitutes at the very least a point of contact at which dialogue can begin.

It is remarkable that this attitude leads us to a new type of universalism, in which a pragmatic agreement can be realized between peoples whose traditional ideas of good and evil are nonetheless different. The Universal Declaration of Human Rights is the best, and perhaps the only, example. It is becoming increasingly evident that different cultures understand the notion of human rights in a different manner than the successive writers of the declaration during the French Revolution and after World War II. This means that almost all peoples of the planet have come to agreements despite or as a result of these misunderstandings. The condemnation of torture, slavery, and racism as well as the notions of crimes against humanity and protection of the environment on a global level are milestones on the way to a universal morality. They have been obtained more from the reactions of horror produced by evil actually inflicted, to the eyes of the greatest of numbers, than by the force of conviction of a so-called natural morality founded on an image of the supreme good and imposed by universal reason. This procedure does not necessarily negate the philosophical idea of the universal, such as that of the Kantian imperative of never using man only as a means. However, it radically modifies the origin of the universal. It is no longer necessarily a question of the maxims of practical reason arising from a certain understanding of human free will that is imposed on that reason; instead, at issue are rules of behavior accepted as minimal means to preserve concrete individuals against bodily and emotional pain.

Conclusion

In conclusion, I resume my presentation with a diagram (Fig. 5). Going from the bottom up, there are three levels of ethics that I have attempted to distinguish, along with a fourth level, still hypothetical and still to be constructed,

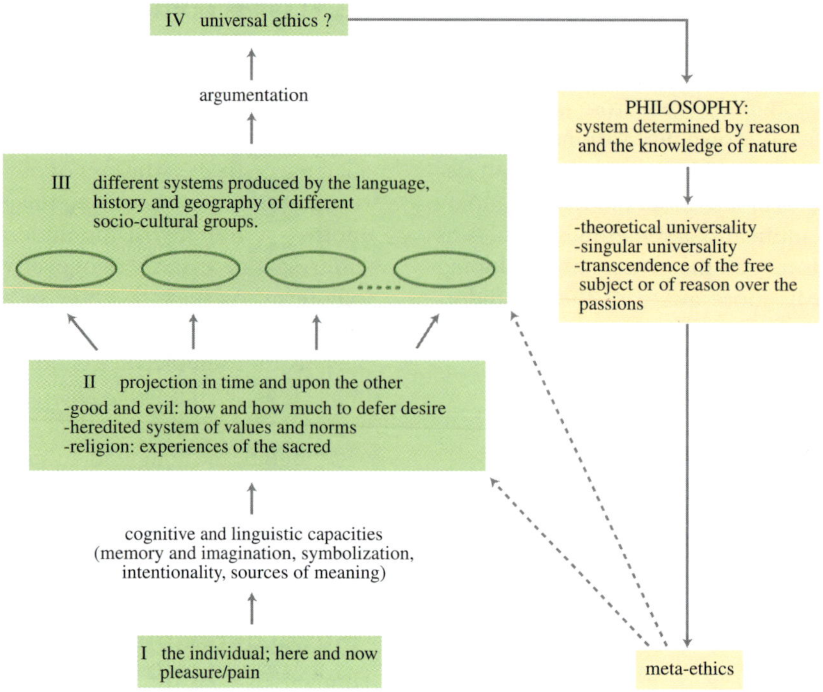

FIGURE 5 Diagram illustrating the levels of ethical reflection described in the text (levels I–III) and a fourth level (IV), still hypothetical and to be constructed.

which is that of a real, existing universal ethics that would not be formulated by moralists alone. This does not strip of its practical interest the solitary work of the philosopher, who tries, by means of a parallel although nevertheless singular and individual path, to rise first of all to a metaethical level of which the theoretical universality would be that of reason. However, in Fig. 5 the arrows that descend from this level into the second and third levels deductively, from principles, are depicted by broken lines because their effect on the real content of these levels is still quite limited.

The previously mentioned observation on the underdetermination of theories and its implication — that it is easier to agree on a conclusion than on the motivations to reach that conclusion — may not be agreeable to proponents of a theoretical rational thinking that is unique and universal. However, it gives us some optimism about our chances of success in a search for a kind of pragmatic, constructed universalism.

Of course, there is no guarantee that this dialogue between cultures, or even opposite stands within the same culture, as in the case of abortion, will not fail and occasionally result in violence or even war. However, this is the only way to reach some kind of universality in practice and not just in the minds of some individual thinkers.

Acknowledgment

This work was supported by an Ishaiah Horowitz Scholarship in Residence in Ethics and Philosophy of Biology.

References Cited

Atlan, H. (1986). *A Tort et à Raison, Intercritique de la Science et du Mythe.* Seuil, Paris. [English translation: L. Schramm (Ed.), *Enlightenment to Enlightenment, Intercritique of Science and Myth.* State Univ. of New York Press, Albany, 1993]

Atlan, H. (1989). Automata networks Theories in Immunology: Their utility and their underdetermination. *Bull. Math. Biol.* **51**(2), 247–253.

Atlan, H. (1991). *Tout, non, Peut-être. Education et Vérité.* Seuil, Paris.

Atlan, H. (1992). Self-organizing networks: Weak, strong and intentional. The role of their underdetermination. *Nuova Critica,* 19–20, 51–70.

Atlan, H. (1998). Intentional self-organization. Towards a physical theory of intentionality. Thesis 11.

Craig, A. D., Bushnel, M. C., Zhang, E. T., and Blourquist, A. (1994). A thalamic nucleus specific for pain and temperature sensation. *Nature* **372,** 770–773.

Duhem, P. (1981). *La Théorie Physique. Son Objet. Sa Structure.* Marcel Rivière & Cie/Vrin, Paris. (Original work published 1914)

Epicurus (1983a). Letter to Menoeceus. *Lives and Sayings of Famous Philosophers* **10,** 121–135.

Epicurus (1983b). Letter to Herodotus. *Lives and Sayings of Famous Philosophers,* **10.**

Gybels, J. M., and Sweet, W. H. (1989). *Neurosurgical Treatment of Persistent Pain.* Karger, Basel.

Koshland, D. (1989). Sequences and consequences of the human genome. *Science* **146,** 189.

Maimonides (1963). *The Guide of the Perplexed* (S. Pines, Ed.), Part 1, Chap. 2. Univ. of Chicago Press, Chicago.

Philonenko, A. (1969). *L'Oeuvre de Kant,* Vol. 1, pp. 335–336. Vrin, Paris.

Quine, W. V. O. (1960). *Word and Object.* MIT Press, Cambridge, MA.

Voelke, A. J. (1993). *La Philosophie comme Thérapie de l'ame.* Editions Universitaire, Fribourg, Switzerland.

Willis, W. D., Jr. (1995). Cold, pain and the brain. *Nature* **373,** 19–20.

Spinoza, B. (1972). *Ethics* (R. Cantoni and F. Fergnani, Eds.). UTET, Torino.

CARLO AUGUSTO VIANO
Department of Philosophical Disciplines
University of Turin, Italy

Bioethical Problems and Ethical Alternatives

Bioethics is the attempt to elaborate guidelines and standards to address the problems that have emerged from contemporary medicine. In fact, the transformations of medical technology have suggested new questions concerning choices relating, for instance, to procreation and to how to consider death. Many of these questions were not recognized in the ethical codes of the medical profession. Bioethics thus took on the task of reformulating moral indications to maintain the traditional ethical obligations in the face of the new situations or of finding new moral orientations capable of giving positive answers to the new questions. Other problems stem from the developments of genetics which enable us to glimpse the possibility of intervening on the structure of the human DNA. Bioethics, understood as ethics of the quality of life, does not exclude that medicine may in the future, also by means of genetic measures, improve human life, and it therefore maintains that the problems posed by the need to innovate moral rules have to be dealt with in order to permit the enjoyment of the good things that genetic measures could make available to us.

The Crisis of Medicine

It was Potter who, in 1971, proposed the name *bioethics* in his book *Bioethics: Bridge to the Future.* With this term he referred to a new ethic capable of guaranteeing respect for human values, reorienting not only medicine but also man's whole activity which, since the scientific and industrial revolution, had become a threat, "a sort of canker" for nature.

The first alarm of this type had sounded with the end of World War II when, after the dropping of the atom bomb, such a philosopher as K. Jaspers announced the start of the atomic age, characterized by the fact that science and technology had eluded man's control. Jaspers took up the very reasons of the movements that had seen a danger in the sciences of nature, especially in the physical and chemical ones, closely linked with technology. The alternative to those forms of knowledge was not always conceived in a clear, even way, but one of the constants remained opposition to industrial society, obsessed by the increase of material well-being, and one of the most popular and repeated formulas was the call for human values, which industrial society had lost or betrayed.

As opposed to many prophets of anti-industrialism, Potter held that new morals had to be elaborated by science because the instinct, which had been sufficient to guide human behavior and to keep it within the bounds of compatibility with nature, was now no longer sufficient. However, a new ethic could not be expected from the physical and mathematical sciences and so it was necessary for a "biological revolution" to start from the science of life. Traditional biology, however, had to be transformed to undertake the new task, bringing its knowledge into line with human values. Not only would the preservation of humanity derive therefrom but also the quality of life would be enhanced: In this sense, to Potter bioethics had to be an ethics of life which would spontaneously bring back to morality a medicine no longer dominated by traditional scientific knowledge.

Potter also expressed apprehension about the fate of industrialized Western society, which had reemerged around the end of the first postwar period. During the 1960s, people had begun to talk again about the population explosion as an imminent danger because the attempts to control the growth of population in the Third World (i.e., where the ratio between population and resources was most disadvantageous) seemed to have failed. However, an awareness of the unequal distribution of resources had also appeared: The industrialized world was viewed as absorbing the major part thereof, although it had a minority of the world population. It was the type of life of Western society and the level of its consumption that constituted a world threat. Therefore, it was necessary to change the way of life, especially that of the population that lived in the industrialized countries, rather than the quantitative relation between resources, particularly food, and population.

In the beginning of the 1970s, however, the trend appeared of generalizing the problem. The intensive technology of world industry tended not only to make resources scarce in relation to the population but also risked exhausting them and making the survival of technologies untenable. The ratio between population and resources was replaced by that between scientific and technological development and the environment. In 1972, B. Ward and R. Dubos presented an alarming report at the conference on the environment organized by the United Nations that claimed that within a generation the technological civilization would have reached its breaking point. In the same year, The Club of Rome was formed and launched the formula of "zero development," considering it urgent to halt the growth of the industrialized countries. The fuel crisis of 1973 and 1974 appeared to many as the first sign of the scarcity of energy resources.

The same year that Potter put forward bioethics as a new ethical theory, A. Hellegers joined the Center of Human Reproduction and Bioethics of the Kennedy Institute at Georgetown University. Hellegers, who was an obstetrician, agreed in part with Potter's general view but assigned to bioethics the task of addressing clinical cases, in which doctor–patient relations are essential but in which multidisciplinary competences (sociological, theological, juridical, and philosophical) are also needed, and a purely biological outlook is not sufficient.

Paying attention to the patient caused another line to emerge, which had determined the prehistory of bioethics and accounted at least in part for different contents present in the projects of Potter and Hellegers. Just as the advent of the atomic age had helped to raise doubts regarding the compatibility between technological progress and natural balances, so too the revelations of the Nuremberg trials on the experiments carried out in German concentration camps (Fig. 1) cast a sinister light over a seemingly positive aspect of modern medicine, which brought it closer to the other fields of scientific knowledge — that is, regarding its concern about searching for experimental controls.

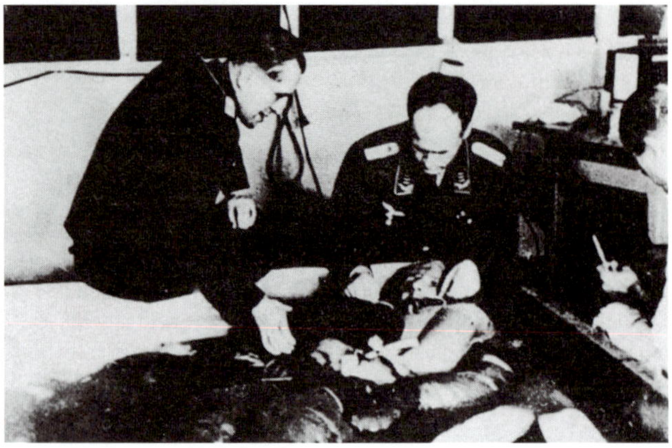

FIGURE 1 A prisoner in the Dachau concentration camp subjected to a hypothermia experiment (photo courtesy of Publifoto-Olympia/AGF).

In the German lagers the experiments had been conducted on imprisoned persons with no legal guarantees. It subsequently emerged that elsewhere, too, ordinary prisoners had been used for experimentation or had been induced to accept it. The experiments revealed at Nuremberg could have been based on groundless scientific hypotheses or on projects distorted by ideologies, but it was the introduction into medicine of more rigorous scientific controls that had induced those in power to test scientific hypotheses and drugs on significant numbers of people. Nevertheless, the interest in the advancement of knowledge and in the benefits that could have stemmed therefrom mingled with the competition among medical schools and the search of the pharmaceutical industries for profit.

In this situation, one might be induced to sacrifice the freedom and the interest of the people used in experiments to the progress of knowledge and of the collective benefit, if not to industrial profit. It seemed that medical experimentation could not fail to involve people, thus running the risk of instrumentalizing them. Moreover, it was clear that the laws, in force also in Germany, had not been enough to protect people against the abuses of experimentation, and that this might be repeated if the possibility were admitted of depriving classes of citizens of the protection of the laws. This showed the need to have recourse to different normative instruments than the positive laws that could be used to judge political systems and were capable of inspiring lines of conduct for doctors and health organizations. The idea thus gained ground that rights existed that were over and above those sanctioned by the positive laws, which no positive system could have suspended and no particular condition of the subject could have canceled.

The Nuremberg Code of 1946, the Declaration of Geneva of 1948, and the international code of medical ethics of 1949 (documents which were later absorbed in the Helsinki Declaration of 1964 and were taken up in the directives for biomedical research of the Council of International Organizations of Medical Science in 1993) outlined the conditions that medical experimentation had to meet. It was necessary to have consensus expressed by free persons who were able to give their effective refusal and who were provided with sufficient information to be able to decide. In no case, however, could experimentation be proposed that did not bring some benefit to the persons involved and it would have been necessary to exclude all and any discrimination in the choice of the persons on whom to conduct this research, protecting those belonging to the weaker groups of the population. Some important consequences for bioethics derived from this approach: On the one hand, it found its own rules in norms and directives contained in international documents, which were not backed by the coercive and sanctionary capacity of state laws, and, on the other hand, these were rules that were difficult to apply directly not only because they could lead to conflicts with the positive laws but also because it was not easy to refer them to the concrete cases presented by practical medicine. To deal with the problems that would have emerged from this new approach, the health institutions would have created special organs, ethical committees, which could have taken into account the norms contained in international documents that were not always taken up into national laws. These ethical committees would have introduced a certain anomaly into traditional systems because they would have been organs without any legal power, whose action would have taken the form of advice and possibly of only moral sanctions, or they would have exercised indirect action (e.g., by approving or disapproving research programs), thereby conditioning the acceptance or the financing thereof by the institutions at which they could have been carried out.

The Revolt against the Doctor

Between 1969 and 1974, a code of patients' rights took shape in the United States. Promoted by a private association, this was later accepted by the organization of hospitals and recommended by the Federal Department of Health to all hospital organizations. The regulations were to govern not relatively rare procedures, such as clinical experiments, but ordinary medical practice, and thus they limited the doctor's autonomy in his or her daily routine. In the United States, this also occurred because public health services were not available as in Europe, and health assistance was entrusted to private insurance. Hence, the hospital authorities, which increasingly had to answer legal claims from insurance companies, sought to involve the patients with the request for explicit consensus for examinations and therapies, which could also be used in court cases.

This process, together with the disciplining of experiments, transformed the figure of the doctor such as it had taken shape in the Western tradition, threatening his or her power to decide what treatment should be applied. The final choice lay with the patient, who was well informed, and informed consensus became the minimum condition that any medical practice had to satisfy. In addition to the patient, an organ such as the ethical committee could act, formed as it was not only by doctors but also by jurists, philosophers, sociologists, psychologists, and representatives of the religious communities; the weight assigned to the medical component differed from country to country.

With the passing of the idea that the doctor, with all his or her knowledge, should unconditionally have the patient's trust, and the center of decision being shifted to the patient, new problems arose regarding not who had to decide (doctor or patient) but on what a decision could be made. While the doctor had had the power to make the final decisions, it had seemed that he or she was bound by moral rules capable of delimiting his or her powers, whereas it was far less easy to identify the bounds within which the patient's claims should come and the criteria on the basis of which the doctor could possibly refuse to satisfy them. With the recognition of the patient's rights, the doctor could no longer claim his or her superiority based on his or her professional competence and the presumption of knowing what was good for the patient, and it was even difficult for the doctor to apply the moral limits within which he or she should continue exercising his or her profession.

The impression was therefore that a particular way of regarding the doctor's position, a conception that was indicated as "medical paternalism," was passing away and that in its place there was a contractualistic interpretation of the doctor–patient relationship. As a matter of fact, from the historical standpoint, the idea that between doctor and patient there was a form of contract, possibly an implicit one, was a return to old times because in ancient society the doctor was not a public figure, protected by the institutions or on whom some authority was bestowed. In general, in traditional societies the doctor was one who provided his or her work in the service of the customer, who assessed the doctor's capacity to restore his or her health and the doctor's professional sincerity, and who was supposed to guarantee the patient against the danger of seeing his or her life jeopardized without his or her consensus or, even worse, to favor relations or enemies. The customer could defend himself or herself against the doctor's incompetence or fraud only by taking private action against him or her.

It was not until the end of the eighteenth century that academic medicine, on the strength of its learned tradition, succeeded in gaining what has been regarded as the "epistemological mirage" of the mechanical and mathematical sciences and in repudiating its relation with popular and traditional practices, which could have represented an alternative to that epistemological approach. To pursue that mirage, medicine had sought to locate the various diseases (i.e., to associate them with anatomical lesions) and to reclassify them, without taking account of the human reality of the patient, enclosed in the preordained space of the hospital. In this way, it could be considered that, far sooner than becoming a reliable form of knowledge of human pathologies and of useful intervention against them, medicine had been an attempt to frame the pathological human reality by separating it from the normal reality and making the hospitals no longer places of succor and charity but veritable barriers between sick humanity and that considered normal. Stemming from the direct application of "scientific concepts" to human ways of behaving, modern medicine had from the very outset been the most exasperated form of technological domination of society. Anatomy, an indispensable instrument in that program, had given the doctor complete control of the patient's body until death and after it, when an autopsy would enable the lesion that had caused the disease to be sought. Medicine had become an important component of modern individual and collective life, serving the judicial organs and the military apparatus, the inspirer of measures for the fight against endemic and epidemic diseases and of reforms to improve public hygiene — all things that had shared in turning the doctor into a public figure and in causing medicine to be regarded as a science of public utility. Thus, the doctor had become one of the policemen of social life, guarantor of the well-being and the life of the members of society. The doctor alone, with his or her publicly recognized professional status, could judge what were the most favorable conditions of life and who kept watch over the duties imposed on the population to avoid diseases. In this perspective, the treatment of diseases ceased being a private matter for which the individual made provision himself or herself, turning to the doctor at his or her own initiative and following the doctor's indications according to his or her own judgment.

It could be claimed that the population had duties to be respected to prevent socially dangerous diseases and that care of the health was a task for the state: Thus, the doctor no longer answered directly to the patient, the doctor's customer, but became responsible to the political authority. Successively, the planning of health assistance and the industrialization of medicine would have contributed increasingly more to making the doctor a public figure and a sort of alliance would become sanctioned between medicine and morality, which would have formed the basis of codes of conduct and of professional ethics. The transformation of medicine into a science of public utility could fall in with the idea that Christianity had inherited from Platonism regarding life as a place of suffering but also as a divine gift and a road along which the soul is purified from the body and lays the foundations for a life consisting only of noncorporeal satisfactions. In this perspective, the preservation of life, rather than the restoration of health, became the doctor's fundamental duty.

By refusing the primacy of the doctor and claiming the patient's rights, the natural alliance between medicine and morals was repudiated, to the point of a real revolt against the doctor, eventually casting doubt on the reliability of experimentation in general and the legitimacy of subjecting human beings to experiments. If experimentation without the consensus of the persons involved presupposes that the medical institution is capable of identifying the general interest and is able to pursue it independently of the choices of the individuals, the demand for consensus, too, seems somehow to presuppose that persons may be regarded as instruments. This would be a sign that medicine, and modern science in general, has lost its sense of the correct order of things and, instead of placing itself in the service of man and his sufferings, regards man as a means for affirming itself. In this way, the same results were reached as those achieved in other ways in the revolt against science, starting with the rejection of atomic energy and of industrial society in general.

Although in the classical phase of the modern age medicine, having made itself a science of public utility, seemed to have incorporated the ethics of life, contemporary medicine — wholly dependent on the mechanical view of nature, on the pretext of pursuing progress or collective well-being — presumed to utilize persons and became the most radical form of industrialization applied to human beings. This was all the more serious because in today's society, medicine, intensely technological, had permeated broader sectors of individual and collective life, giving rise to what was described as a "medicalization" process: Babies are born, the diet is chosen, and people die with the assistance of a doctor. Having reached the end of the process of dehumanization and mechanization, medicine, while having become more persuasive, had broken its relations with morality which it had incorporated, and the traditional rules of ethics (professional ethics, codes of conduct, and medical ethics) were no longer sufficient: A new instrument was needed to bring medicine back to morality, understood as a patrimony of principles common to all men. At this point, it was unthinkable to go back to the epistemological turning that had been the origin of modern medicine because, despite the success of cultural propaganda against the horrors of contemporary science and technology, the model represented by physics and chemistry was becoming fully affirmed in Western medicine. The task of bringing present-day medicine back to morality therefore had to be entrusted to a special normative discipline, namely, bioethics, which could certainly not be the new scientific ethics spoken about by Potter nor even a mere updating of the codes of conduct and of professional ethics.

The Start of Life

Once it had been recognized that medicine does not in itself tend to achieve morally good ends, it could be hoped that moral rules and beliefs common to all men could be imposed on the technicians of medicine by means of the acknowledged right for the patient to assert his or her own personal choices. However, the defense of the patient's rights revealed that this was not the case. It was the beneficiaries of medicine who, in the name of their rights, were in favor of medicalization, claiming that doctors should assist them with contraceptives, with the interruption of pregnancy, with treatment of sterility, and with the application of euthanasia — all "services" once prohibited or left for popular medicine and in any case placed beyond what had seemed the ethical bounds of modern medicine. The ethical committees, which were born from a realization of the insufficient nature of the traditional ethical institutions, could no longer be regarded as instruments to bring medicine back under the control of morality but rather had to address new situations generated by the internal transformations of medicine and by the attitudes of its users.

The question of abortion was one of the first matters in which the awareness emerged that a new situation had come about. In much of modern Western legislative systems abortion was illegal; apart from the professional and social discredit that would have attached to one indulging therein, severe penalties were also envisaged. These, however, were relatively recent prescriptions and connected with the acquisition of an academic status by doctors. Abortion, in fact, had often been left to unofficial medical practitioners, especially to women, and it continued to be widely practiced by women or doctors often on the fringes of official medicine, even when it was excluded from the medical measures permitted. In the ancient world abortion was a widespread practice and was not condemned. Only abortion procured following aggression was punished because it was considered an injury caused to a women or to her family. From this point of view, Greek and Roman practice was no different than that of the Oriental, Jewish, or Christian civilizations. It was only in the Hippocratic oath that the doctor was required not to provide a woman with any means of causing an abortion, but that oath was probably a body of rules for a small group of ancient doctors, perhaps inspired by Pythagoras.

The condemnation of abortion, which was introduced into Christianity, aimed more at defending the inviolability of the soul, the work of God, than life. In fact, it did not lead to an absolute prohibition of interrupting pregnancy since it was held that the embryo was not "animate" in the first stage of its life. St. Thomas Aquinas supposed that only autonomous movements could prove the presence of the soul and believed that it was only after 40 days for males and 80 for females that the fetus was able to move by itself. This viewpoint depended on the assumption that all spontaneous movements derive from the soul and that it is the soul that gives human form to the body. Once this conception of movement was abandoned and it was recognized that the embryo has human characteristics before spontaneous movements can be attributed to it, Christian doctrine excluded the lawfulness of early abortions and stated that abortion was willful murder right from conception. In the twentieth century, the absolute condemnation of abortion was rigorously sanctioned by official Catholic doctrine and has recently been confirmed in the Pastoral Constitution *Gaudium et spes* of Vatican II, in Pope Paul VI's encyclical *Humanae Vitae,* and in the 1974 declaration on procured abortion by the Sacred Congregation for the Propagation of the Faith. This position has been confirmed several times by Pope John-Paul II.

In the nineteenth century, the banning of abortion before the supposed capacity of the fetus to move of its own accord was also introduced into civil legislation. Lay doctors, in fact, confident of the scientific basis of their competence, considered Thomist animism to be impracticable and boasted that they were the heirs of Hippocrates and of his oath, which seemed to guarantee the reliability and correctness of medicine. Again in 1948, the World Medical Association resumed use of the Hippocratic oath and sanctioned the obligation to respect human life "from the moment of its conception." This attitude was based not only on the results of modern embryology but also on the conviction that medicine was a service in favor of life.

Abortion started to be regarded in a completely different way when it was viewed not so much as a way of getting rid of an undesired pregnancy or birth as a measure to address a precise health danger. The religious prohibition of abortion did not take into account the dangers to which women could be exposed by pregnancy and childbirth: If necessary, women should be prepared to sacrifice themselves to favor the birth of a child. However, the opposite could happen — namely, that the mother might wish to sacrifice her maternity so as not to bring a sick child into the world. This circumstance occurred in a uniform and generalized way when the use of thalidomide, a tranquilizer prescribed for pregnant women, became widespread in many countries, and it was found that it could be the cause of serious malformations (Fig. 2).

In 1967, Great Britain legalized abortion if carried out by a doctor for the purpose of preventing harm to the physi-

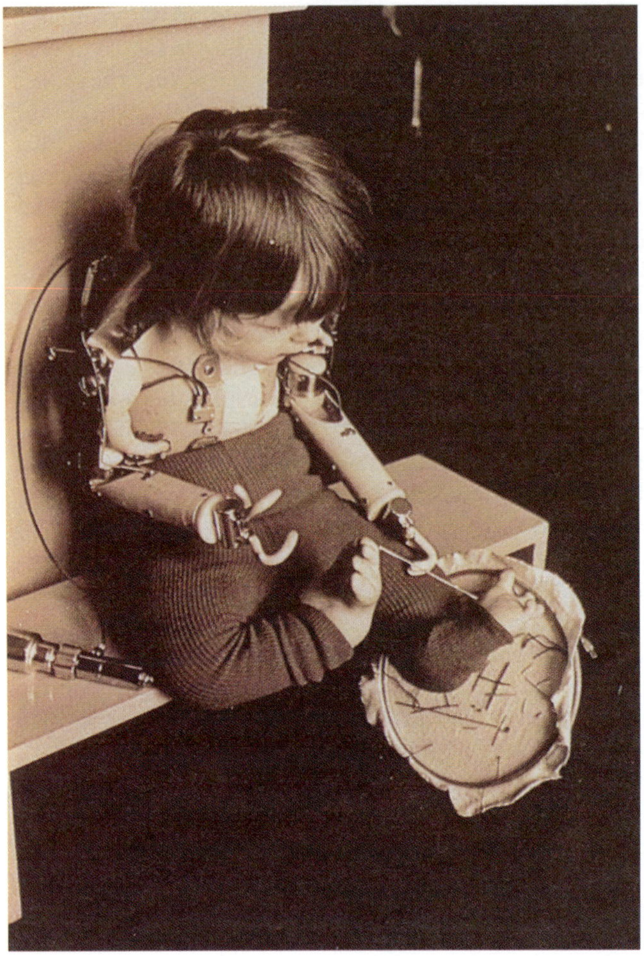

FIGURE 2 A phocomelic child, the outcome of administration of thalidomide to the pregnant mother (photo courtesy of Keystone-Sygma/Neri).

cal or mental health of the mother or a child already present in the family or if there was a real danger of the child being seriously disabled if born. The sole condition was that the abortion should not involve more serious danger than the possible advantages. In the United States, several states started to permit abortion to safeguard a woman's health, to avoid the birth of handicapped children, or to permit the interruption of a pregnancy due to incest or rape. Finally, the Supreme Court, with its decision in the case of *Roe v. Wade,* stated that abortion was a constitutional right. In fact, recourse to this practice increasingly spread as medicine, with prenatal diagnosis, became increasingly more capable of foreseeing anomalies in the future baby.

With the cited decision, the United States Supreme Court admitted that the state is interested in the potential life of the fetus but established that until such time as the fetus becomes "vital" the woman's right to privacy prevails over the interest of the state. This decision aroused lively reactions, especially from the "movement for life," which saw in it a substantial lack of interest in protecting the fetus and maternity and the conferment of an undue preference for the right of the woman, but again raised the problem of the beginning of a life that could be considered as properly human. There was thus the risk of a return to Thomist theology and of opening a discussion on the moment when the fetus actually becomes vital.

Artificial insemination complicated the problem of abortion. In the attempt to treat sterility, the point was reached at which a woman could be fertilized not only by artificially introducing sperm into her body but also (in 1978) by fertilizing an ovum *in vitro* and then placing it in a woman's uterus (Fig. 3). This provided new medical prospects but also posed new problems. Embryos could be obtained even outside of the female body, and rules were needed to establish what procedure this would entail. Difficulties were increased by the fact that to have a certain percentage of success, it was necessary to produce more embryos than those to be implanted (*supernumerary embryos*) and criteria had to be established to decide what to do with them. They could be cryopreserved for further implants in other women, or experiments could be conducted on them to obtain better knowledge of the twin embryo intended for implantation or even for other general scientific purposes. However, embryos could also have been produced not to be implanted but to be subjected to experiments. This presupposed that things could be done to an embryo that would not have been legal to do to a person. This made the problem of the status of the embryo more acute, requiring an answer to the question "What is an embryo?," which seemed to raise biological, ethical, and legal problems.

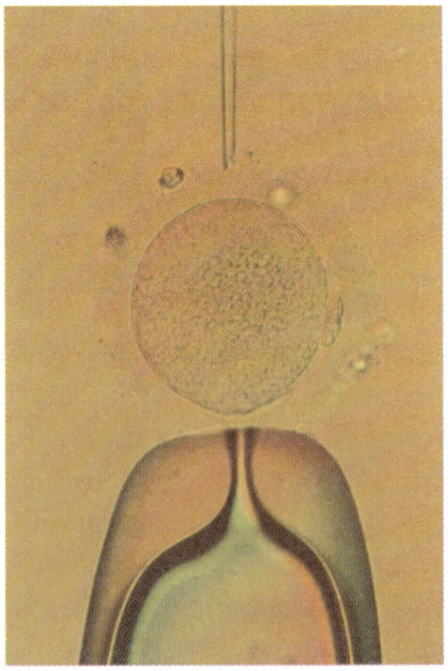

FIGURE 3 *In vitro* fertilization technique in a human: DNA of a single sperm cell is injected with a micropipette (on right) into an ovocyte (photo courtesy of CC Studio-Science Photo Library/Neri).

It could be claimed that the embryo, resulting from the fusion of gametes, has a human character and so must have the same protection as human beings, a protection that must be identical from fecundation until death because the genetic patrimony remains constant. Against this position it was objected that the appearance of the embryo does not coincide with fecundation because from the fusion of the gametes to the penetration of the sperm into the ovocyte requires more than 20 h. Moreover, the probability that a human being will be born from an embryo is not much greater than the probability that an embryo and then a human being will be born from ovocytes and sperm. Hence, if the equality of the protection to be afforded to an adult individual and to an embryo is based on the continuity of the process that leads from one to the other, the problem arises of extending that protection to the preembryonal human genetic material. Finally, the embryo possesses in the initial stage properties that clearly distinguish it from a human being already born: Its cells are totipotent (i.e., from each of them a cell of any type could develop), and two or more genetically equal twins could derive from one and the same embryo — all things that cannot happen to individuals already born. The last possibility becomes invalid at approximately the 14th day after fecundation, after which the development of the embryo seems to proceed in continuous form toward the formation of an adult individual, incapable of separating into two or more equal individuals. The possibility thus appeared of attributing individuality to the embryo after the 14th day, possibly calling the entity that has developed until that moment the preembryo.

Full moral and legal protection is normally given only to persons, and supposing that only human beings may be persons, it is difficult to attribute personality to genetic material just because it is human or to an entity considered a human individual just because it possesses a patrimony of genetic material destined to remain constant in time. In juridical and philosophical culture there already existed interpretations of the concept of personality that were profoundly different from one another, but this difference became actual opposition when used to resolve the problems posed by the need to define the status of the embryo. The solutions based on religious beliefs have considered the notion of personality as a substitute for the notion of soul and have made of this an inherent property of the embryo, as the seat of the soul (Fig. 4). Those who have assumed such a position have sometimes admitted the possibility of discussing whether the personality is formed immediately or when the embryo has lost the possibility of giving rise to monozygotic twins. Catholic theology has willingly returned to Boethius and has considered the personality as a "rational substance" which actually becomes manifest in the fully formed individual but which may also potentially exist in the embryo immediately after fecundation. Opposed to the substantialistic interpretations of the personality are the functionalistic or relational interpretations, according to which the personality is not a permanent and constant property in a person but rather a complex of relations or the exercise of functions. It is not always present,

FIGURE 4 The descent of the soul into the embryo, depicted in a miniature of the *Liber Scivias* (114q-1151) of Hildegard of Bingen (Germany), Abbey of St. Hildegard (photo courtesy of Benediktinerinnenabtei St. Hildegard).

and its content and its degree may vary, even in the life of one person.

The appearance of the embryo as an object of medical practices raised the problem of safeguarding objects not capable of choosing, a problem encountered in the practice of informed consensus which appeared as a form of safeguarding the patient against medical authoritarianism and paternalism and, in general, of weak subjects against dominant ones. With the embryo fully corresponding to the human person, the mother or both parents — persons perfectly able to formulate this consensus — were deprived of the possibility of so doing. In reality, the absolute or quasi-absolute prohibitions of abortion were aimed at protecting not weak subjects but human life in general, which, considered sacred, should to the doctor be a supreme good and to the believer a gift of God. Consideration of the embryo and of the unborn baby receded so much into the background that it was attributed a constant preference for life, whatever the ways in which it might have lived it.

The End of Life

Many judicial decisions, solicited by what could be called the "successes" of contemporary medicine — that is, the possibility of maintaining in life persons irreversibly deprived of the capacity to breathe by themselves or persons in possession of automatic mechanisms such as for breathing but irreversibly without consciousness and no longer capable of eating or drinking — were to have important consequences on the concept of life.

In 1976, the Supreme Court of New Jersey ruled that, on the basis of the constitutional right of privacy, the family of Karen Quinlan, a patient no longer able to breathe by herself, could request the "breathing apparatus," considered an artificial means of sustaining life, to be switched off without being guilty of willful murder. Instead, in 1983 the Supreme Court of Missouri denied the parents of Nancy Cruzan authorization to remove the feeding tube because Nancy was in possession of her vegetative functions but was unable to express her preferences, and the state attributed to itself an interest in preserving its citizens' life. Suspension of medical assistance was not admitted until 8 years later because someone claimed to remember that Cruzan would have wanted that way. In Britain, judges listened to the unanimous opinion of doctors that Tony Bland, whose cerebral cortex had been destroyed in a sports accident in 1989 and who was in a persistent vegetative state, no longer had any type of consciousness and that there was no possibility of any improvement of his condition. The judges stated that, when a patient is incapable of expressing any consensus of his or her treatment, the doctors are under no legal obligation to persist with initiatives which they know will be of no advantage to him or her and can only ensure the continuation of his or her biological life without any other benefit. The American judges based the medical procedure on the decisions of the patient who is in full possession of his or her good health and must always be able to express his or her consensus to health measures. The British judges instead based their decision on the opinion of the doctor, who has to judge whether the life of a patient no longer able to decide has acceptable prospects or constitutes a threat to his or her dignity. In the first case, the possibility emerged of legalizing suicide, recognizing the patient's complete disposal of his or her own life. In the second case, the road to euthanasia could be opened up, authorizing doctors to hasten the death of incurable patients. To avoid these results the concept of *therapeutic vehemence* was taken up, which means the unjustified continuation of treatment without any prospect of cure and with the risk of worsening the patients' sufferings. Avoiding therapeutic vehemence would not mean artificially shortening life but simply letting nature take its own course; for this reason, the churches and religious organizations or movements also took up this concept.

Since life maintenance treatments are almost always carried out when the patient is no longer able to decide, documents have been excogitated in which citizens can give, at any moment and before entering the terminal stage of life, instructions as to the cures to which they wish to be, but especially not to be, subjected. These initiatives are intended to guarantee the citizen's right of self-determination and to leave to him or her, and not to the doctor or the judge, the choice of the way to die. They have been especially successful in the United States, in which in 1991 a federal law recognized the possibility of signing an advance directive, called *testament of life* or *living will* in which a person can indicate the ways in which he or she would like to be treated if he or she should enter the terminal phase of an incurable illness and were to lose the capacity to exercise his or her own choices. In 1992, a similar measure was introduced in Denmark, and the British Medical Council invited its members to take the indications of patients into account.

In the United States, these advance directives are required increasingly more often in health institutions. Also, in countries in which they have not received any legal recognition a conspicuous bioethical literature has developed on the subject and forms have been proposed in which it is explicitly requested not only that no extraordinary measures be applied but also that not even forced hydration and feeding be applied, except to the extent required to avoid suffering, that secondary illnesses should not be treated, and that palliative measures able to shorten survival should also be carried out. To counteract what they regard as the euthanasia component of lay-inspired advance directives and to confirm their simple refusal of therapeutic perseverance, understood as the adoption of disproportionate therapies, Catholic movements have submitted their own forms of advance directives in which it is requested that life should not be prolonged in an unnatural and unreasonable way, and euthanasia is explicitly refused.

When in 1967 C. Barnard performed the first heart transplant, discussions of the definition of death had already begun at Harvard University, but after said operation a commission of that university proposed that persons in an irreversible coma should be considered dead. Authorizing the taking of hearts that were still beating from persons made it more likely that such operations would succeed and also led to an increased number of organs available. The Harvard commission presented what was termed *cerebral death* as the result of the application of the instruments of ascertainment available to modern medicine which no longer needed to observe the ceasing of cardiorespiratory and nerve functions with the traditional means, but which could use such apparatuses as the electroencephalograph (Fig. 5) to ascertain the irreversible ending of the cerebral functions necessary for maintaining the vital functions. In 1981, the conclusions of the Harvard commission were taken up by a U.S. presidential commission, when the new definition of death had already been entered in the law books of many countries and of many of the states in the United States, whereas in Great Britain it had been adopted directly by the doctors without much formality.

Ethics of Life and Ethics of Dignity

Even those who consider conception as the precise starting point of human life, accepting the concepts of cerebral death and therapeutic vehemence and the legality of palliative treatments, give a less univocal meaning to the phrase "end of life." In fact, it has become increasingly difficult to distinguish between artificial death and natural death, especially since the terminal stage of the illness occurs almost entirely under the control of the physician and reanimation techniques have made the borderline between life and death uncertain. There are already some who consider the concept of cerebral death the starting point for a further transformation of the idea of death. Because it is possible to diagnose the irreversible loss of the functions of the cerebral cortex, a loss that permits the automatic vital functions to continue but prevents the exercise of the consciousness, we could speak of cortical death because keeping patients with no cortex alive would make no more sense than keeping patients alive with their hearts beating with the help of equipment.

Even those who deny the disposability of life at its initial moment, considered a rigid term, tend to admit that those concerned may decide, perhaps with a document compiled before they lose their capacities, the end of their lives. Acceptance of the concept of therapeutic vehemence and recognition of palliative treatments have introduced, even in conservative bioethical positions of Christian inspiration, the idea of the possibility of choosing in some way how one's life will end. The most radical development of this issue has led to the assertion of one's full powers to dispose of one's own life, an idea that raises the problem of suicide. This action has been depenalized in modern legislations, which, however, consider it a crime to help a person to take his or her own life. However, on many sides it is beginning to be urged that medical assistance to a suicide is no different than the action whereby a doctor puts an end to the sufferings of a terminal patient by stopping therapeutic treatments, and there are doctors who have declared that they

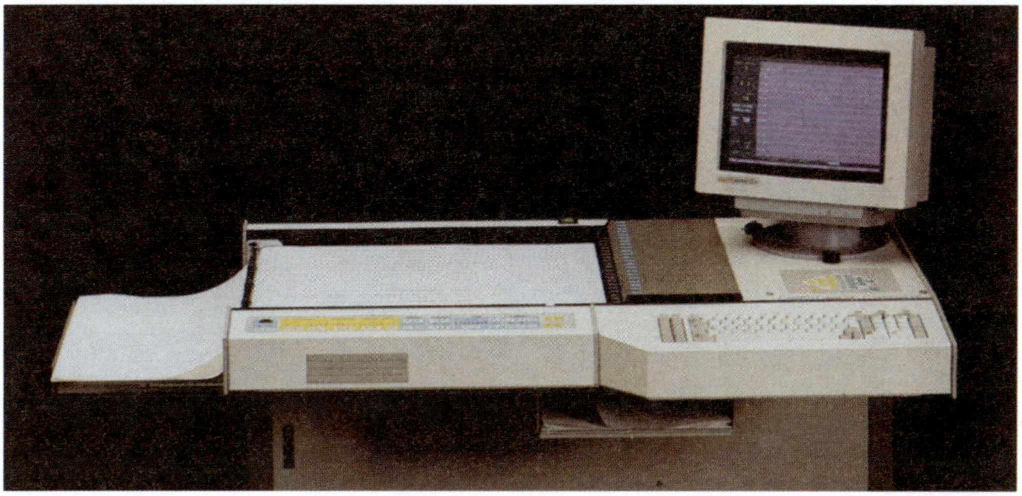

FIGURE 5 Electroencephalograph with monitor for display of bioelectrical signals (photo courtesy of Esaote Biomedica).

have provided their services in such cases or that they are disposed to do so. Experiments of this sort have been carried out, or it has been sought to carry them out, to different extents in the United States, Australia, Colombia, and especially Holland. Nevertheless, in 1997 the United States Supreme Court declared that assisted suicide is not unconditionally constitutional, as district courts and appeal courts had ruled and as physicians, patients, and philosophers claimed. The Supreme Court recognized that the state is interested in safeguarding the life of its citizens and may limit the right of medical assistance to suicides, but it did not exclude returning to the question if concrete cases were to raise the doubt of unconstitutionality of possible state laws. Even the philosophers who defended the legitimacy of assisted suicide before the U.S. Supreme Court claimed that it should have functions of euthanasia, whereas it cannot be admitted that a doctor may help just anyone to take his or her own life.

The relative uncertainty about the initial and the final moments of life has made it difficult to support the principle of the sacredness of life formulated in purely biological terms. Indeed, it has been sought to reformulate this principle without referring it to precise biological realities or to interpretations of the person as rational matter. The concept of sacred is an instrument with which one can assert a protection and which can be preserved even if its object changes. Not every aspect of life is sacred, but those that derive from profound choices, which lend dignity to life, are sacred, and dignity may be the object precisely protected by the sacred character recognized for life. The mentally deranged person who expresses satisfaction with his or her existence does not, however, lead a dignified life and, if before the obfuscation of his or her faculties he or she had chosen not to live the terminal phase of his or her life under such conditions, these are the person's real, authentic decisions. Likewise, one who turns to the doctor for assistance in a "passionate suicide" is not acting on the basis of those capacities that are the groundwork of his or her human dignity.

The same could also be said for the start of life: Abortion or assisted fecundation are not mere manipulations of possibly living biological material, nor acts with which to dispose arbitrarily of the lives of others. Rather, they are choices based on conceptions, moral principles, or ways of life of those assuming them. It is inevitable that in such situations somebody will intervene in the existence of the others and at times make decisions in their place, but to assess these interventions it is necessary to examine whether they are respectful of the dignity of others and especially of the persons for whom the choice is made.

The ethics of dignity are much older than the ethics of life; they pertain to many religious beliefs and are incorporated in modern medicine. These ethics suppose that men behave on the basis of a conception of what is good and a global vision of their lives — a legacy of the philosophical tradition according to which man is a rational being (i.e., a being heedful of different motivations than his immediate stimuli). However, although classical philosophical ethics assigned a fixed content to "good," to the correct vision of life and to rationality, the ethics of dignity — in its modern, contemporary version — considers rationality a function with a variable content, able to generate various moral plans. It therefore refuses one of the postulates of the ethics of the sacredness of life, according to which life takes place fully in the traditional forms of family and social organization and should be accepted with the sufferings it contains. Precisely because it considers rationality to have a flexible structure, the ethics of dignity admit that different choices may be made (of acceptance, refusal, or correction) with regard to the biological aspects of life, of the sufferings to which it exposes us, and of the family and social structures in which it takes place. To the ethics of life, autonomy is above all a way of subtracting everything that is human, from conception to death, from the naturalistic dominion of the world and from the action of technology, whereas to the ethics of dignity it is man's possibility of self-determination using science and technology that makes them instruments to broaden the range of the possibilities of determination.

Genetics

The notion that health depended in part on the natural, unchangeable structure of individuals, inherited from their parents, was widespread even in popular culture and in part guided matrimonial choices, especially in traditional societies, in which family membership was an important element for evaluating the procreative partner. Contemporary science has related many serious diseases which have afflicted persons from birth or which develop at a certain moment in life independent of the environment to genes (i.e., the fundamental components of the genetic stock). In sick persons many genes appear different than those of healthy persons; these genes may also be present in persons who, though not affected themselves, could transmit them to their descendants. The genes responsible for diseases can therefore be traced in family groups, in which some members contract these diseases.

Direct knowledge of the possibility of transmitting genes responsible for diseases has posed new problems. On the one hand, it has opened the possibility of making informed procreative choices, such as abstaining from procreating or turning to abortion or assisted fecundation with an ovum or sperm from a donor whose genetic stock entails no risks. On the other hand, it has stimulated research for a new type of therapy. In fact, in principle it is possible for the genes responsible for diseases or those making the person prone to diseases to be replaced or corrected in the cells of a person's body, in which they function as regulators of biochemical reactions. An intervention of this sort is local (i.e., at the level of the soma) and does not replace or correct the genes responsible for diseases in all the cells of an individual, especially the germ cells. Hence, the individual subjected to somatic gene therapy can transmit to his or her offspring the genes that the treatment has altered in many cells in his or her body.

Operations on genes could be regarded as a form of eugenics — that is, as an attempt to select individuals in possession of given qualities such as the absence of certain diseases or resistance to others. Even the "normal" traditional procreative strategies, which do not assume a knowledge of the genetic stock of the generative partners but use their phenotypic characteristics, are implicit eugenic choices. The mechanisms of these choices are complicated and have not been explored in-depth. In the long run they aim at adapting to the environment, but this condition is very indefinite because the environment is a complex reality. The social environment acts strongly on reproductive choices, both by introducing filters in the choices of the reproductive partners and by guaranteeing the sustainability of choices that are disadvantageous from the purely biological standpoint. In this sense, social organization, technological progress, and traditional medicine have enabled genetic diseases or diseases having a strong genetic component to be maintained.

Independently of direct corrective measures on the genetic stock, the genetic knowledge already available, and that which will become available in the future, could enable the negative repercussions of social selection over natural selection to be corrected in part. Knowledge of the reproductive partners' genetic stock and prenatal diagnoses in fact permit procreative choices targeted on excluding the birth of sick persons or ones destined to fall ill, and assisted fecundation can become an important instrument of negative eugenics. This also includes purely therapeutic measures on the somatic cells performed to exclude genes responsible for diseases, such measures generally being considered morally acceptable. Similar operations on the germ cells could exclude given genes not from the cells of individuals but from the human genetic stock. Such a project, applied on a wide scale, could lead to irreversible results for the human species. Currently, although somatic gene therapy is difficult, germ therapy is technically impossible. However, in itself the exclusion, as widespread and possibly definitive as is feasible, of disease-bearing genes from the human genetic stock should be viewed as a wholly positive fact, the culmination of the attempt that has animated the genome project of exploring the human genetic stock first and foremost to reconstruct a framework of the genetic factors of diseases. Germinal gene therapy, on the other hand, has aroused ethical reservations; it seems risky to eliminate on a definitive basis certain genes and interferences between different genes. Under these conditions, there might be the risk of eliminating genes that are important for the exercising of important functions or which could turn out to be important in situations which are currently unforeseeable.

Even greater reservations are obviously aroused by what could constitute a form of positive eugenics — that is, the application of gene therapy not to treat genetic diseases certainly recognized but for purposes of "improvement" to upgrade normal performances which, as intelligence, memory, physical strength, beauty, and so forth, are found to a different degree in everyone. These are functions that are not easy to describe accurately, and it is even more difficult to establish any connection between them and genes. It is probable that there is a connection, but it is also probable that the genes determine parts or aspects of complex forms of behavior, in which many genes operate. In this case, the nexus between the various functions of a gene and between the various functions of different genes is even less known than in the case of diseases. Apart from the technical feasibility of measures of this sort, it is likely that for a long time it will be difficult to know exactly the consequences of an "improving" measure and even to judge certain measures rather than others as definitely of an improving nature. There would be even greater problems if said measures were to be carried out on the germ cells.

The principle of responsibility, which insists on abstaining from operations whose consequences are not known, has been invoked against the prospect (albeit only theoretical) of designing eugenic projects based on knowledge and alteration of the genome. Attempts to modify the genetic stock are typical undertakings of this type since neither the complete structure of the genome nor the repercussions that its modification, even only partial, could have on the interactions between individuals and the environment are known. This principle presupposes there to be a phase displacement between technically possible measures and knowledge of their consequences: This is a situation that is characteristic of certain moments in the development of genetics, which is precisely of the present moment and could become accentuated in the future. In these conditions, this principle can be a useful rule for regulating attempts to use genetic techniques in measures whose consequences are not sufficiently known. This phase displacement could be destined to continue in such a way as to exclude the admissibility of a certain number of applications of genetic technology. However, at times the principle of responsibility has been used to suggest that the balance of nature is based on an edifice that is too complex to be modified by man, and from this standpoint genetics has been considered a threat to man's actual survival, similar to how physics was regarded when it was employed in the production of the atomic bomb.

New Problems

Recent developments in genetics have shifted attention to problems other than those posed by birth, death, and the perfecting of traditional medical techniques. These are problems connected with the fact that genetics has determined a considerable phase displacement, which will perhaps increase in the near future, not only between operations technically possible and their consequences but also between accessible knowledge and possible therapeutic measures. The prospects opened up by genetics indicate that knowledge of the genetic basis of particular nonpathological but important aspects of the personality could increase. Persons could thus know particulars of the trends of their behavior or of their ways of responding to external stimuli, environmental situations, and so on and in general know far more about themselves than in the past. This prospect further attenuates the link between medicine and illness, in accordance with a tendency that has already taken shape with the progressive medicalization of many nonpathological aspects of life, such as sexuality, procreation, birth, and death, but it can influence the way in which human persons are usually conceived, regarded as centers of self-projection and of decision. These aspects of the personality will continue to be important, but the choice among available moral projects could take far more account of biological factors than in the past. Furthermore, cultural and moral experiences which have so far appeared as primary constituents of the personality could appear as responses to biological factors hitherto unknown.

This brings up the problem of the availability of biological information, primarily to the individuals concerned. There could be persons who would not like to know the biological information regarding them for two reasons: either because they do not want to know their prospects of catching incurable diseases or because they do not want to be conditioned by biological information in what they consider their autonomous moral choices. It would be an easy matter to renounce biological information if the collectivity decided globally not to seek this, whereas it is more difficult to make a selective renunciation possible which presupposes the choice by each citizen of the type of information he or she intends to have or not have. This presumes a complex health organization in which the information that a person may perhaps not wish to have could be available to someone else.

Biological information could also have a social or collective use. If it were available, society could use it so as to make decisions about its members, excluding them from given social positions or directing them to others. If placed in circulation, said information could produce discriminations, excluding certain persons from benefits or depriving them of the freedom to choose. These problems indicate that measures should be studied to safeguard the right of individuals to privacy of biological information regarding them and which defends them against social discriminations based on the construction of biological maps. However, these issues are framed in more profound ethical prospects, which are barely starting to take shape.

The first reaction, faced by the possible prospects opened by genetics, is usually one of defense and induces people to consider genetic information as private property, the privacy of which must be defended. Nevertheless, from the biological standpoint, the genetic stock is anything but private since it represents that which every individual has received from his or her parents, transmits in part to his or her offspring, and possesses in common with his or her relations. Knowledge of the section of family genetic stock that an individual possesses can be important to those who share it with him or her. Combining the necessities of making available what can be of use to others, the exercising of one's choices, and the safeguarding of privacy requires the development of a system in which new freedoms and new social relations can be created. In fact, the appearance of genetics suggests the possibility of rethinking the way in which culture has hitherto represented the system of identities and differences which make up societies. In a complicated society, the various groups forming it make images of their cultural and historical differences. Contemporary liberal societies have sought not to attach too much importance to these images, denying that cultural differences have a natural basis and refusing to take them into consideration when important resources have to be distributed fairly. On the other hand, natural differences have not been considered a good basis for social differentiation because it has been considered that they are altered by the social differences through which they emerge: The results of the "natural lottery," from which every individual extracts his or her own fate, are thus largely masked by the "social lottery." The development of genetic knowledge could, in the future, lead to reconstructing the map of natural differences between individuals in a manner relatively independent of the system of representations with which groups and individuals imagine themselves. At least in part, the natural lottery could be perceived beyond the social lottery, and knowledge of a certain number of differences could improve the utilization of the differences among individuals.

In the scenarios that it allows us to glimpse, genetics is often considered a threat, as though it could provide mankind with the instrument to shape definitively a large part of the living world. In fact, by manipulating the genetic stock of animals and even of men, individuals in the future could be obtained responding to models chosen in advance. The first cases of cloning of mammals indicate that this has already started. To avoid prospects of this sort, the principle of responsibility as mentioned previously has been invoked, but the catastrophic forecasts of genetic manipulations presuppose technologies that are not currently available; even if they were, it is not certain that they would be applied to such an extent as to endanger natural balances.

Natural balances are also the result of mankind's diffuse cultural choices, in themselves not without dangers. To date, every generation has partially chosen successive generations in a broadly indirect way and through cultural schemes, but the procreative conducts suggested by genetic screening or assisted fertilization enable part of the set of genes of unborn babies to be determined directly, at least negatively, excluding genes responsible for diseases. In the future, theoretically, part of the set of genes of one's offspring could be determined positively so as to guarantee that they will possess given qualities or it could even be determined wholly, reproducing exactly the set of genes of the parents by a cloning process. In this way, we could do without sexual reproduction, at least in cases in which difficulties are encountered; a person unable to have a child, or a healthy child, with his or her companion could theoretically have one by himself or herself.

A use of genetics of this type, although not eliciting the catastrophic scenarios evoked by the total control of mankind with biological instruments, would introduce grave perturbations into the structure of society and the moral beliefs on which this stands. In the case of cloning, a person would be twin to his or her own parent, but assisted fertilization had already aroused objections of this type. A person born by this technique could have a legal father different than his genetic father, who may well remain unknown. A person, donating his sperm or her ova, could generate after his or her death, a woman alone could generate a child with the sperm of a man whom she never met or could give birth to a child of whom she is not the mother from the genetic standpoint, and so forth. From all this, an upsetting of moral rules has often been perceived. It has been feared in particular that procreative technology, made even more powerful by genetic knowledge, would make children of the species artificial products, modeled according to the desires of the parents or of just one parent — a decidedly disquieting aspect of modern technological society, in which every means is sought to realize the desires of persons without considering the value thereof.

This diagnosis of contemporary society and of the position occupied in it by medicine has underpinned the demand that it be brought back under the control of traditional morality. This requirement has caused many conservative moralists and ones of Christian inspiration to become harsher in their beliefs to the point of viewing the embryo as a person and the genetic stock as the real guarantor of personal identity, and it has caused many liberal moralists to spiritualize their beliefs, recognizing dignity to human life only if it is modeled on ethical conceptions and dominated by motives of a higher character. In reality, the demand for medicine has increased, often in a manner wholly independent of religious belief and, as new medical techniques became (and become) available, has extended recourse to doctors in situations (especially regarding procreation, sexuality, and death) in which it would be difficult to consider pathological. In this situation, an inverse process is taking shape to that which is usually depicted as medicine coming back under the aegis of morality because it is medical technology itself that is not only asking for new rules but also suggesting new types of behavior, both to avoid the negative consequences of medical technology and to change the way that people think about this technology. What has already taken place with the change in the concept of death with the advent of transplant surgery could occur in other fields, with the transformation of beliefs and rules concerning birth, death, suicide, kinship, and so on. Also, perhaps occasions might eventually be found for changing the very representation of ourselves and the structure of our moral behavior.

Two Ethical Prospects

From the previous discussion, two different interpretations of morality and the ethical theory taking it as its object are outlined. The first interpretation considers morality as the intrinsically correct structure of human behavior and ethical theory as the study of the rules that guarantee that it will be correct. From the practical standpoint, ethical theory could promote and orient the restoration of morality whenever it appears compromised, a possibility which, in today's world, appears connected with the primacy of the technological civilization. The second interpretation considers ethical theory as an instrument for promoting and orienting the correction of current morality, which in itself is not necessarily good. In the contemporary world a prospect of this last sort has been proposed by utilitarianism, a movement which, aimed at freeing current morality from its prejudices, regards moral principles not as absolute, evident truths but as rules to be judged on the basis of the well-being that they generate. Therefore, a moral principle, when in contrast with a technique that enables well-being to be increased, should be corrected. The essential constituent of well-being is pleasure, and therefore whatever contributes toward diminishing pain and suffering is considered a positive element. In this perspective, medicine has been viewed by utilitarians as an important instrument in constructing collective well-being and, indeed, utilitarianism has been one of the basic cultural elements in the consideration of medicine as a science of public utility. As previously discussed, this interpretation of medicine has become an ingredient of those ethics based on the intangibility of life, which have made life a good in itself, a divine gift to be accepted whatever its content. Utilitarianism, instead, has rejected the idea that suffering has a meaning in itself: The individual is fully master of his or her life, of which he or she may freely dispose, to the point of suicide, and can attempt to diminish sufferings, provided this does not lead to greater sufferings for other individuals.

Although utilitarianism has made the diminishing of suffering the central point of its program and has subordinated the generalized safeguarding of life to the pursuit of the quality of life, it has broadened the consideration of life worthy of protection. Since total well-being is diminished by any suffering, everything that can feel suffering must be taken into account, and among the beings that have to be considered are animals different from man. This position seems to form part of the prospects that emerge from genetics. Although the ethics of life have often taken the form of the set of genes as an absolute starting point and have considered the set of genes as the guarantor of the permanence of human individuality, or of the human personality from fertilization until death, scientists of utilitarian persuasion have observed that the separation between human life and animal life is not absolute because the genetic stock of the human species is related to that of other animal species. Recognition of an affinity between man and the other animal species has induced the attribution of rights to animals and the doubting of the legitimacy of inflicting suffering on them, using them for experiments or for food. On the other hand, this affinity has induced people to consider animals as deposits of biological resources, all the more in demand since medicine adopted the biological perspective.

The presence of two opposed tendencies is a feature of ethics of utilitarian derivation. In fact, when it interprets moral rules as instruments to achieve the maximum collective well-being, utilitarianism tends to authorize transfers of well-being from one individual to another because the sum of small decreases in individual well-being can mean a significant increase in total well-being. However, there exists a limit to transfers of this sort because a suffering of any sort cannot be inflicted on an individual if it produces a greater pleasure for another individual. Medical practice presents many cases of transfer of benefits from one person to another: The clearest case is that of medical experimentation, from which, as previously discussed, one of the branches of bioethics started. The instrumental use of persons for experimental purposes is unanimously rejected but, both in experimentation considered legitimate and in ordinary medical practice, transfers of benefits considered acceptable take place. The large-scale organization of medicine, with research conducted on populations by way of screening, has given even more evidence to transfers of this sort.

Health screening, the use of animals and the safeguarding of their well-being, and the relation between the privacy of biological information (especially of a genetic nature) of an individual and the eventuality that this may prove useful to other individuals are all cases in which it is necessary to plan behaviors and rules that make the well-being of one individual compatible with that of the others. The original utilitarian idea that benefits may freely be transferred from one individual to another or the benefits of every individual can be increased by increasing those of the collectivity is now regarded as optimistic. The distribution of benefits requires mature social instruments which enable the various preferences of individuals to be matched.

Bioethics addresses issues generated by the meeting of traditional arrangements of preferences with relatively rapid changes in medical techniques and practices which interfere with the beliefs that underpin these arrangements. The recognition that sufferings exist that are generated by medicine has interfered with the idea that life is a good to be safeguarded in any case. On the other hand, knowledge of the condition of a human being that is derived from the set of genes that his or her parents can transmit to him or her has made procreative choices very different acts than those that traditional morality considered. Lastly, genetics has shown that deep-seated relations between the human species and other animal species or among individuals of the human species exist which are not always clearly reflected by explicit cultural representations. Genetics can offer techniques and perspectives capable of acting as turning points to the transformation of moral beliefs, made possible or required by contemporary medical practice. This is the possibility of acting on the profound structures of life, not only with targeted corrective modifications but also with actual genetic manipulations. These have appeared to be violations of the structure of life and have almost always been looked on with horror. However, genetic manipulations may appear as essential instruments of "biological" medicine — that is, of the medicine that exploits biological resources rather than the traditional chemical and physical resources. With these, biological products, and theoretically even parts of organisms, can be obtained from animals or from human embryonal materials; in theory, again, genetic manipulations can also produce animals to be used for purposes for which animals are already used today, such as experimentation or food.

Radical doubts have been mooted regarding the legitimacy of these uses of animals. Those voicing them have claimed that human beings should change their ways of life to avoid using animals for experiments and for food, if necessary paying the prices that such a transformation would entail even if these prices would relate to a transition period, compensated by greater advantages in a longer period. Nor have voices been lacking claiming that it would be possible to give up instrumentalizing animals precisely on the strength of our confidence in science. Stimulated by this renunciation, man could substitute experimentation on animals with theoretical simulations and could find new forms of nourishment not involving the consumption of meat.

Genetic manipulations could have another function, which would provide an alternative outlook to the preceding one. They could in fact select animals on which to carry out experiments which would not have the negative effects currently considered unacceptable, or animals could be raised without having to undergo the physical and psychological sufferings to which they are currently subjected in intensive-type units. These manipulations may not be extensive enough to compromise the survival of the natural species but could concern only the individuals to be subjected to experiments or to be raised. This practice would be more disturbing if it were also practiced with human beings, and genetics has often been pinpointed as the instrument with which human society could be manipulated in a total manner. Obviously, there is no technique that cannot be used badly and the possibility of genetic manipulations on men would indeed raise the issue of their regulation. Meanwhile, genetic knowledge could have important consequences on the selection made naturally through procreative choices, following prenatal diagnosis and genetic screening.

In the future, greater knowledge of the genetic stock could permit a better distribution of natural differences, and techniques of manipulation could make it possible to plan the natural characteristics of human beings. This would be an important and an absolutely new challenge for mankind, which would involve the maturing of social rules and institutes suitable to regulate it. It is probable that there is no principle or set of principles from which to deduce the appropriate rules and that to find these rules mistakes will be made and sufferings caused. In the maturation phase minimum guidelines could be formulated which will allow new normative forms to be tested, avoiding inflicting foreseeable sufferings and guaranteeing, in the forms required by the new technical possibilities, conditions of freedom and of fair access to the resources made available by said techniques.

General References

Beauchamp, T. L., and Childress, J. F. (1983). *Principles of Biomedical Ethics,* 2nd ed. Oxford Univ. Press, New York.

Bompiani, A. (Ed.) (1996). *Bioetica in Medicina.* CIC Edizioni Internazionali, Rome.

Callahan, D. (1970). *Abortion: Law, Choice and Morality.* Macmillan, New York.

Cavalieri, P., and Singer, P. (Eds.) (1993). *The Great Ape Project: Equality Beyond Humanity.* Fourth Estate, London.

Dworkin, R. (1993). *Life's Dominion, an Argument about Abortion, Euthanasia and Individual Freedom.* HarperCollins, London.

Engelhardt, H. T., Jr. (1986). *The Foundations of Bioethics.* Oxford Univ. Press, New York.

Ford, N. (1988). *When Did I Begin? Conception of the Human Individual in History, Philosophy and Science.* Cambridge Univ. Press, Cambridge, UK.

Foucault, M. (1963). *Naissance de la Clinique: Une Archeologie du Regard Medical.* PUF, Paris.

Glendon, M. A. (1987). *Abortion and Divorce in Western Law.* Harvard Univ. Press, Cambridge, MA.

Harris, J. (1985). *The Value of Life.* Routledge & Kegan Paul, London.

Harris, J. (1992). *Wonderwoman and Superman: The Ethics of Human Biotechnology.* Oxford Univ. Press, Oxford.

Jonas, H. (1985). *Technik, Medizin und Ethik. Zur Praxis des Prinzips Verantwortung.* Insel Verlag, Francoforte sul Meno.

Lamb, D. (1990). *Organ Transplants and Ethics.* Routledge, New York.

Mori, M. (1995). *La Fecondazione Artificiale: Una Nuova Forma di Riproduzione Umana.* Laterza, Rome.

Mori, M. (1996). *Aborto e Morale: Un Manuale per Capire, un Saggio per Riflettere.* Saggiatore, Milan.

Neri, D. (1995). *Eutanasia. Valori, Scelte Morali, Dignità delle Persone.* Laterza, Rome.

Nuland, S. B. (1994). *How We Die: Reflections on Life's Final Chapter.* Random House Large Print in association with Knopf, New York.

Potter, V. R. (1971). *Bioethics: Bridge to the Future.* Prentice Hall, Englewood Cliffs, NJ.

Ramsey, P. (1989). *The Patient as Person: Exploration of Biomedical Ethics.* Yale Univ. Press, New Haven, CT.

Reich, W. T. (Ed.) (1978). *Encyclopaedia of Bioethics,* Rev. ed., Vol. 5. Free Press, New York.

Rodotà, S. (Ed.). (1993). *Questioni di Bioetica.* Laterza, Rome.

Sgreccia, E. (1991). *Aspetti Medico-Sociali. Manuale di Bioetica.* Vita & Pensiero, Milan.

Singer, P. (1994). *Rethinking Life and Death. The Collapse of Our Traditional Ethics.* Text Publishing, Melbourne.

Spinsanti, S. (1987). *Etica Bio-Medica.* Edizioni Paoline, Cinisello Balsamo.

Thompson, J. J., *et al.* (1992). In *Introduzione alla Bioetica* (G. Ferranti and S. Maffettone, Eds.). Liguori, Naples.

Viafora, C. (1989). *Fondamenti di Bioetica.* Ambrosiana, Milan.

Ward, B., and Dubos, R. (1972). *Only One Earth: The Care and Maintenance of a Small Planet: An Official Report Commissioned by the Secretary-General of the United Nations Conference on the Human Environement.* Penguin, London.

REMO BODEI
Department of Philosophy
University of Pisa
Pisa, Italy

Bioethics and Biotechnology

Biotechnologies have altered the notion of family based on blood ties, the concept of personal identity, the feelings associated with the most solemn phases of existence (conception, birth, and death), and the rhythm of life cycles. Involuntary blood ties are increasingly being replaced with elective ones. In this way, the awareness of an increased dominion of the human species on itself and, at the same time, the anguish of whoever is called on to make decisions the effects of which may at times be unforeseeable, is becoming more accentuated. Bioethics has become a battlefield for opposing views and brought attention back to corporeality both at the individual level and at the level of future generations by way of interventions on genetic makeup. In this article, the existential problems that arise from the separation of generation and fecundity, particularly in cases of artificial heterologous insemination and donation of egg cells, are analyzed. In this way, the increasing complexity of the "elementary forms of family ties" and of the feelings both of persons born as a result of such procedures and of their biological and social parents, when their roles are divided, are highlighted.

The Impact of Biotechnologies

For a long time, medicine was considered a type of knowledge that supported nature, which limited itself to helping nature in the healing processes that organisms carry out. In the modern era, however, the habit of "letting nature take its course" has been qualitatively and quantitatively undermined by means of sciences and technologies aimed at the total control of spontaneous natural processes. Until recently, it was simply external nature and the present which were controlled, whereas today such a project of control has been extended to include the internal nature of man and concerns future generations as well. This means that although in the past technology treated nature as inanimate physical matter, today it is man and, more generally, living matter in its hereditary makeup which have become the specific objects of biotechnologies. The latter are putting into question convictions, habits, and ideas that are thousands of years old that until recently were believed to have been founded on solid evidence or even on the authority of divine revelation. Nothing to date has seemed less worthy of being put into doubt, for example, than the fact that an individual is born into the world, according to the old and proven methods of natural sexual reproduction, with a body and mind that are subject to congenital disorders and malformations, suffering, enjoyment, and dying along with the lot of his or her organs. However, biotechnologies force us to rapidly reformulate, even at the level of common sense, many of the parameters due to which daily life has unfolded in the succession of generations, such as the notion of personal identity, the network of affective family relations, the role of sexuality, the ethical and juridical norms which regulate the rights of individuals and families, life cycles, the texture variety, and intensity of specific passions. From the standpoint of the latter, for example, the system of feelings which mark the most solemn moments of human existence — conception, birth, paternity and maternity, and death — has been altered. Even the configuration of the imaginative sphere, insofar as it is conditioned by our former biological or mental limits and by the complementary desire to avoid them, has been modified.

That which seemed to be imposed by the hard laws of necessity or the inscrutable will of God is transformed into an object of choice, enabling one to be a social father when one is sterile, a mother at a menopausal age, or the a child of unknown parents (because one is born of a woman who received a donation of sperm or in a "rented uterus" or from a widow years after the death of her husband). Moreover, that which appeared morally and naturally indivisible — sexuality and procreation or procreation and the parent figure — due to contraceptives, especially of the pharmacological kind, and to modern fertilization techniques has begun to fall apart, transforming what was once the energy of affective ties into a fluctuating and disquieting energy which has yet to determine how to distribute itself and causes disconcertment and pain. In the same way, the body — intended as an organism composed of indissoluble parts — becomes divided and the separate organs can be exchanged, passing from one organism to another.

The individual thus becomes the bearer of entirely new and heavy responsibilities; he or she is entrusted with choices which were once the most jealous prerogative of states or of churches. As a result, everyone is faced with metaphysical scenarios which involve them even in the remote future: life and death, the scansion of one's own and others' life cycles, the transplantation of organs, and the management of genetic makeup. In fact, it is plausible that one may be able to intervene both on the somatic line, for therapeutic purposes, by means of the insertion of pieces of DNA in the corrected staminal cell and also through the manipulation of an individual's germinal line, in such a way that the inserted modifications become a constitutive part of his or her genome and are transmitted hereditarily (Stich and Jackson, 1979; Glover, 1984; Capron *et al.,* 1993; Di Mauro and Mafettone, 1997). The biological sciences change from substantially pure forms of knowledge to ones capable of displacing the frontiers of life and death and of comprising a *bricolage* of organs.

Biotechnologies do not end with the human world, however, even though it is there that their ends lie. In fact, they change our ways of eating (by means of genetically modified foods such as seeds, plants, and hybrids which were previously unobtainable) and of curing ourselves (by means of drugs produced using genetic engineering techniques or of the construction of models of diseases induced in transgenic animals). Genetically modified organisms, moreover, are now a largely diffused industrial product on which converge the efforts of powerful multinationals that currently benefit from the fact that the European Parliament, on May 12, 1998, approved a directive which allows biological materials to be patented. Vegetable organisms which are immune from attack by certain larvae have thus been created. By inserting a gene from the bacterium *Bacillus thuriginsis,* which secretes a specific toxin, in the DNA of certain vegetables, the larvae of lepidopters such as *Heliotis zea* are warded off (Fig. 1). On the other hand, life has been manipulated for a long time in the agricultural and veterinarian fields, for example, through the artificial selection of seeds or animal species.

It becomes possible to transfer living matter from one body to another; it is rendered compatible by modifying genes using molecular biology. Different existences and human stories are thus set in relation to one another, which will meet even beyond death. The barriers between species virtually disappear. The heart of a pig can be used for a man, to the point where the philosopher H. Jonas (1985), while projecting his fears, was able to affirm

> I repress the metaphysical shudder which overcomes me when I think of the horror of animal–human hybrids, which, in a perfectly consequential manner, have already appeared among the practical perspectives of molecular biology under the guise of research into recombinant DNA.

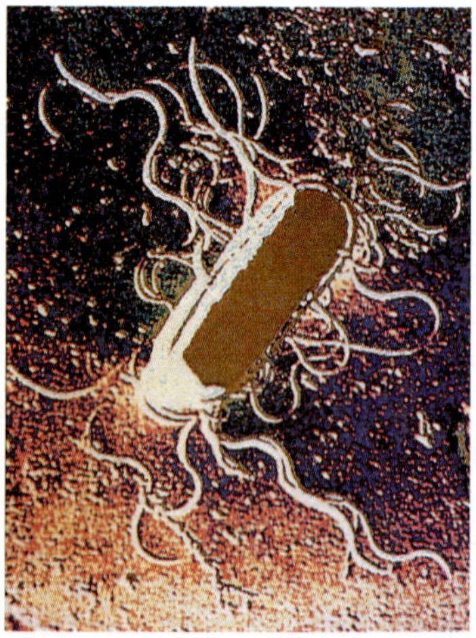

FIGURE 1 Microphotograph of the bacterium *Bacillus thurigensis* (photo courtesy of Pasieka/SPL/Neri).

Bioethics: A Branch of Philosophical Ethics

Bioethics, a term perhaps coined by the cancerologist V. R. Potter (1971), was born with the declared intent of founding a bridge between science and ethics, two shores which seemed destined to remain separate (Fig. 2). It represents a form of frontier knowledge both in the local sense, insofar as it lies in between two disciplines which were once separate, and in the temporal sense because it is young and continuously growing due to the previously unimagined effects of biotechnologies and, in particular, to genetics and molecular biology. According to Reich (1978), it is "the systematic study of human conduct in the realm of the life sciences and of health care, when such human conduct is evaluated in the light of moral values and principles."

Bioethics has become a part of philosophical ethics, revitalizing public ethics which had become, on the one hand, absorbed by positive legislation and, on the other hand, relegated to the sphere of private morality. It also contributes to shifting interest from the sphere of the soul and the invisible to that of corporeality, bearing out the conviction that not only do we have a body but also we are a body. The problem of corporeality — which might have been secondary in the Western tradition which insisted on the body prison of the soul and on the Christian or Cartesian dualism of body and soul — is brought back to the forefront when biotechnologies render the in-depth modification of the bodily dimension possible (Melchiorre and Cascetta, 1983; Nancy, 1992; Huisman and Ribes, 1992).

FIGURE 2 Cover of the book *Bioethics: Bridge to the Future* by V. R. Potter published in 1971 (reproduced with permission).

In this way, ethics ceases to be a specialized topic for philosophers and becomes, with fewer mediations than in the past, a part of the spheres of common sense, medicine, and law. Even if the disproportion between principles and norms, on the one hand, and their actual application, on the other hand, is evident, one should not underestimate the importance of ethics in today's world and think that its rules are useless sermons. On them and on the effort made to observe them, in fact, depend a sound life in the present and in the future of coming generations.

Given its increasing importance in society, therefore, today bioethics can claim with good reason its autonomy with respect to traditional ethics. It is nevertheless true that the historical origin of its concerns dates back to a period that long precedes the development of biotechnologies. Bioethical questions are born in the Western world in the background of a moral crisis which is tightly bound with the faltering of a series of convictions both ethical and ontological. In fact, already with the breaking up of religious unity at the time of Luther, "it was no longer possible to hope to live in a society which aspired to a single moral conception, dictated by a single supreme moral authority" (Engelhardt, 1986). Morality thus was pushed back into the inner world of individual conscience or had to become a part of positive right or of the commandments of churches and states. Bioethics, with its links to the body, has given back to morality its public, visible character; it has made it into a shared patrimony (of worries, if nothing else). It is, in fact, "developing like the common language of a world which is interested in health care but lacks a common ethical conception" (Engelhardt, 1986). In a regime of "polytheism" of values, since we cannot agree on ultimate moral principles, even though we are faced with ultimate questions, we are led to confront others with proposals whose claims are based on consensus and on sound argumentation. Engelhardt, in an interview published in a daily newspaper in 1998, defined lay morality as "the default condition of society, a condition of lack, which may not be ideal but is the most we are capable of attaining."

Antidestiny

The shifting of the frontiers of life as knowledge, genesis, quality, duration, and result modifies the individual's horizon of expectations and, therefore, the understanding that everyone has of themselves and of others. Many conditions which once appeared to be bound to the hard and impenetrable necessity of the world are changed into matters of choice, into an "antidestiny" (Rodotà, 1997), posing the question whether what is technically possible is also legitimate (an obvious point should be remembered; namely, that everything that biotechnologies introduce as innovations in the realm of nature follow the laws of nature). What is implied by the possibility of knowing ahead of time the probable duration of our existence or the nature of the disorders (such as Huntington's disease, which manifests itself at approximately age 40) by which we and our children are likely to be affected? This is information which employers and insurance companies would certainly be happy to be privy to. However, should it be imparted to those who are directly involved or not?

The fact that the spectrum of possibilities becomes broader implies that there is a decreased trust in the power and normative character of the presumed laws of nature as we have understood them until now. This also implies a complementary tendency to favor, especially in the future, a greater degree of artificiality in one's own body or the prevalence of acquired feelings, for example, of affective paternity or maternity instead of blood ties. This does not take place in a simple or painless way, however. In fact, one asks oneself: If life and death were programmed and bureaucratized — organized with sperm banks and possibly with centers for the conservation of "state cadavers," which would provide organs to those who need them — would they not become banal and deprived of intrinsic value?

By dealing with "ultimate" questions, bioethics renders problematic what previously was considered normal and indistinctly located in the inert background of our direct moral concerns. The solution of our difficulties, in fact, was usually provided by our separate faiths or beliefs or left to the realm of individual conscience. Now, instead, individuals are being charged with unprecedented and wearisome responsibilities, insofar as they are being called on to make decisions which in the past had been delegated to the great emitters of norms and moral directives — to "ethical banks" such as churches, states, or parties.

Given the stakes, it is almost inevitable that conflicts and fanatical reactions be unleashed which, in addition to lacerating the conscience of the individual, set different cultures conceptions and religious faiths in a virtual collision course, amplifying their differences. These differences are already great as a result of globalization phenomena, which bring distant parts of the planet in touch with one another, and of the widespread cohabitation of different ethnic groups on the same territories.

As a result of biotechnologies, the gap between the possibilities for innovation and their acceptance on the social cultural and religious level becomes wider. Often, reactions of rejection or strong differentiation take place, and the furrow between consolidated ethical and religious norms and attitudes that are more open to and aware of the opportunities created by scientific research becomes deeper. Sometimes, such levels of radical behavior are reached in the blind defense of one's own reasons and dogmas that a citizen may act in open contrast to the laws of his or her country or a believer may either oppose the teachings of his or her own religious denomination or accept the latter's directives aimed at fighting whoever puts its credo into question. To realize the size of the problem, one need only think of the polemics with respect to artificial insemination, the state of embryos, or the genetic therapies that are already being envisaged for intervening on the human genome. In the face of the common feeling of "vertigo" which is caused by the (real or virtual) successes of biotechnologies, confronted with the implications (many of which are unknown) that the latter might well entail, moral conscience is perturbed and people's feelings are wrought asunder.

One may say that, on the whole, the fault runs in between two conceptions, between the defense of the sacrality and gratuity of life (which is upheld by believers in revealed religions but not only by them) and the view of those who emphasize the quality of life. Those who maintain the former point of view, who sometimes proclaim their positions to be simply rational in character since they recognize absolute moral norms "to be respected always and in any case, without exceptions" (Melina, 1988; Sgreccia, 1988; Cattorini, 1990), center their arguments on the idea of the person or on the uniqueness of every human individual and on his or her difference from other animals (Sgreccia, 1988):

> The self cannot be reduced to digit, number, atom, cells, neurons. . . . Inside every man, inside every human being, the world is entirely summed up and acquires meaning, but the cosmos is at the same time surpassed and transcended. Every man harbors within him the sense of the universe and the whole value of humanity: A person is a unity, a whole and not a part of a whole. . . . From the time he is conceived to the moment he dies, in every situation of suffering or good health, the human being is the point of reference and the measure of what is legitimate and what is not.

Although the opposition between religious and laypeople hides and distorts the plurality of positions that exist as far as bioethics is concerned, in the lay camp — using the term in the sense of a way of thinking which is antidogmatic, tolerant, and open to doubt and correction of acquired stances — there is much discussion today about the person. In this context, however, the concept of person represents an attempt to find a minimum common multiple which will put the accent on the dignity of every individual man. Having avoided the Thomist notion of person as *individua substantia rationalis naturae* (indivisible substance of rational nature), it is difficult to bind the person to the sole psychological dimension of self-consciousness, which would exclude, for example, fetuses at a certain moment in their development. Whatever definition of person one chooses, something essential is lost, and the least agreement between the various conflicting sides, at present, dissipates.

From the lay point of view, the theme of the quality of life is seen in the following terms: "The Catholic world may be 'eternal ethical values in new situations'; the lay motto 'new ethical forms for new situations,' inspired by the fear of the evil that old ethical forms may produce in new situations" (Scarpelli, 1996). Scarpelli (1987) successfully summed up the sense of laypeople's intentions and positions:

> I will therefore proceed in my reasoning *etsi Deus non daretur,* that is, as a layperson. The layperson, in fact, is not someone who denies God (the negation of God is itself a proposition about God, a speaking of the ineffable) but someone who reasons outside of the hypothesis of God by accepting the unbreachable limits of human knowledge and existence. . . . But there is no truth in ethics. The values of truth and falsehood are suited to propositions which belong to a descriptive–explicative–predictive discourse, not suited to those which belong to a prescriptive–evaluative one; the very historical variety of principles leads one to believe that they are the fruit of cultural, social, and personal processes, and that they cannot be traced back to an abstract and meta-historical zone of truth which can be immediately intuited by every intellect. In its principles, therefore, ethics is logically arbitrary, but that does not mean that it is humanly arbitrary. Everyone is engaged in ethical experience and problems with his own organic and cultural conditioning, with his baggage of ties and affections, in the individual and concrete situation into which he was born and which has changed along with him in the course of his existence. . . . An ethics is therefore always and radically individual; it is precisely that which ensures human nonarbitrariness.

We are faced here with two positions which appear to be antithetical or, at least, difficult to reconcile and integrate with one another and which lead to veritable ethical aporias. The matter of contention is both physical and metaphysical: substantially, it is life itself, intended not only in the biological sense but also in the spiritual and cultural one. For believers, life and the body should in general be conceived as a divine gift, as something gratuitous, of which we are not free to dispose of, almost like livery of the Lord's servants, received when one is born and to be given back at one's death (Flores d'Arcais, 1997). In addition, we are not the proprietors of our own bodies: Cells multiply and die, blood flows, endocrine glands secrete their products, white blood cells attack infections, and our lungs expand. There is something that lives in us which is not ours and does not depend directly on our will, hence the sense of the mystery and sacred nature of existence and the fulcrum of arguments against the idea of individual autonomy that is a part of lay positions.

If for those who believe, therefore, we are not lords in our own homes, and divinity or the Nietzschian "great reason of the body" seem to know better than we do and better than our "small reason" what is best for us, why then aspire in Promethean fashion to become the lords and even the creators of ourselves by means of science and technology? Why dare what is impossible or impious by escaping the constrictions of chance, the genetic lottery? Why demand control over the whole of life, from its source to its results, passing through all its critical phases. Why separate and scatter what God, nature, or society have always (since time immemorial) united, namely, sex and procreation and procreation and parental responsibility?

Today, the idea of the sacred nature of life clashes with the will of many to regulate their own existence in such a way as to improve its quality and dignity as well as with the tendency of scientific knowledge and of biotechnologies to go beyond the limits that are set anew at every development. Also, it will certainly not be the phantoms which hover about many discussions (about new Frankensteins or new *Lebensborn* divisions) that block these processes.

The New Dilemmas Created by Biotechnologies

The challenge raised by medical progress and by biotechnologies poses numerous and difficult dilemmas for bioethics and introduces paradoxes that are far from easy to solve. In the form of "exercises in perplexity," we now try to determine what consequences have already ensued and those which presumably will manifest themselves on the terrain of feelings and of personal identity.

First, the perception is changed — initially in a thoughtless manner and then in ways that are increasingly aware — both of the broadening of the horizon of one's own life and of the risks that inevitably are connected with any attempt to explore the unknown. With the shifting of this horizon line, expectations and fears, projects, and renunciations change. Along with it, the extent and "longitude" of desires grow as well. The happiness and unhappiness of individuals and of their possible descendants, forms of love and respect for offspring or parents, a sense of responsibility, family and social hierarchies, rights and duties are all deeply affected. All the more so given the fact that for the first time in its approximately 170,000–190,000/years of existence, the exponents of the species *Homo sapiens* are beginning to think like the only beings on the planet who are capable not only of altering the environment but also of changing themselves and the generations of future centuries through a direct manipulation of the "code of life." These billions of mammals who inhabit every corner of the earth and live in all sorts of climates, however, have become (because of their number and overall technical development) a potentially harmful species, one which is virtually dangerous for the survival of all the forms of life that are gathered inside the thin layer of the biosphere, which is only 30 km thick.

The massive introduction of biotechnologies, however, above all generates disconcerting paradoxes in the sense of theories and situations which run against the δόξα, the opinion that is understood and consecrated by tradition. Alongside them grow false or simply premature expectations from the results of those that have been called "the laboratories of happiness" (Flamigni, 1994): One expects too much and too soon. The paradoxes especially affect the intimacy of our relations. Today, one can ask oneself quite brutally, according to the current expressions, what it feels like to be the offspring of a test tube or of "science" and what it feels like to have an unknown biological father who, in normal circumstances, one will never be allowed to meet (except in Sweden). The forms of family relations to which we have been accustomed thus become displaced, the architecture of family roles is changed, and its presumptively immutable principles are replaced in such a way that even the proverbial *mater semper certa, pater incertus* is no longer necessarily true.

The family, that "ubiquitous institution," no doubt varies in space and in time, but it can also change in depth not only as a result of aided procreation in general but also especially because of heterologous insemination (also known as artificial insemination by donors; AID) and the donation of egg cells or the possibility of choosing the sex of offspring. The family institution would thus be rocked in its foundations from the juridical and moral standpoints as well as from a psychological perspective.

The procreative act, the most secret and intimate, is thus reduced to the level of an artificial and programmed scientific experiment, something which Goethe poetically prefigured (Jonas, 1985) in *Faust* in the scene in the laboratory in Act 2 (Fig. 3), in which Wagner produces the *homunculus* by way of retorts and alembics (translation by Philip Wayne, 1959):

That old style we declare
A poor begetting in a foolish fashion.
The tender core from which life used to surge,
The gracious force that came from inward urge,
Which took and gave, for self-delineation,
Blending near traits with far in new mutation,
To this we now deny its lordly height;
What if the beasts still find it their delight,
In future man, as fits his lofty mind,
Must have a source more noble and refined.
(He turns to his hearth)

FIGURE 3 Wagner creates the *homunculus:* an 1858 engraving by Engelbert Seibertz for Goethe's *Faust* (photo courtesy of Archiv für Kunst und Geschichte).

Look! There's a gleam!—Now hope may be fulfilled,
That hundreds of ingredients, mixed, distilled—
And mixing is the secret—give us power
The stuff of human nature to compound;
If in a limbeck we now seal it round
And cohobate with final care profound,
The finished work may crown this silent hour.

Whereas contraception separates sexuality (and its accompanying pleasure) from reproduction, the techniques of induced fertilization, which were first used in the veterinary sector, separate generation from fecundity. In a Western world in which (unlike in the developing countries) infertility is increasing (in the United States, one out of every six couples is sterile), it has become possible for a couple to have children when one of the two parties is sterile or the woman is of menopausal age. However, is it right to want to have children "at all cost"? Vegetti Finzi (1997) stated,

> For the time being, a system of contradictions which it is very difficult to defuse is at work. On the one hand, to refuse medically assisted procreation in the name of an inviolable natural character of the generative process amounts to reducing man to the biolog-

ical and physiological identity of sexual relations; on the other, by opening the transmission of life to social exchanges, one risks handing filiation over to the laws of the market and of speculation.

Were these technologies to become widespread, it would be the family itself — as in the great philosophical utopias from Plato to Campanella — which would observe the reasons of its own existence vanish. In fact, it derives its legitimacy from blood ties, from the forbiddance of incest, and from patrilinear transmission. By relegating biological reproduction to the margins, the family structure "is left an empty shell, a purposeless construct" (Vegetti Finzi, 1997). As noted previously, ascribed, unwilled, blood ties tend to be replaced by conscious, elective, and affective social ones.

The problem is the use to which we will be able to put this greater freedom. The fear, not difficult to understand, is that we will not be able to manage it — that we will not be on par with the knowledge and wisdom that nature shows, despite all its mistakes. The bond (which has been in existence since "tombs, courts, and altars made the human beast take pity on itself and others") between biological paternity and maternity, and social and juridical paternity and maternity in marriage, is being rescinded.

The possible decline in relative importance of the monogamic family as we, to this day, have known or imagined it to be — although it allows for "new and as yet untried aggregations" — threatens those factors of individual identity which are known to us, dividing, for example, the single maternal figure into biological mother, gestating mother, and social or legal mother. Harris (1992) notes,

> Progress in *in vitro* embryology has on the contrary divided the concept of maternity in two, making it possible to distinguish between a *genetic mother,* or she whose gamete has been used and whose DNA has provided the child with 23 chromosomes, and a *gestational mother,* or she who has carried it in the womb and given birth to it. And, naturally, there remains the possibility of a third, distinct mother, the "social mother," who raises and educates the child. In an analogous manner, birth represents a traditional watershed. In law it marks the boundary between two crimes, abortion and infanticide; and almost all celebrate it as the arrival of the new human being. Except that we are entering an epoch in which it will be possible to practice ecto-genesis, that is, to bring the fetus to completion in an artificial uterus. When that happens, the notion of birth will lose a large part of its juridical and social significance.

In this last case, a very deep and visceral relationship will be broken — that between mother and child carried in the womb for 9 months. In fact, "whereas sperm functions as a system of signs which is transmitted through the bodies and is to that extent impersonal, the 'container uterus' constitutes the link between biology and biography" (Vegetti Finzi, 1997).

Stories of *in Vitro* Fertilizations and Artificial Inseminations

When, on July 25, 1978, at Oldham, northern England, Louise J. Brown came into the world (Fig. 4), the first human being to be born through *in vitro* fertilization and embryo transfer, a new era had begun whose protagonist recalled 20 years later, underlining the emotional side of her experience (Polito, 1998),

> Dad and mom told me everything when I was 5 years old. They turned on the television and showed me this cassette, yes, this very one. On that glass dish are one of mom's eggs and dad's sperm. A little stirring and that's it. It was the morning of November 10, 1977. That same evening I became the fertilized hope of human life. Shortly after midnight I was already an embryo placed in the warmth of my mother's uterus. But I don't think I was born then. I was born in this other video cassette, which was shot 9 months later. It was July 25, 1978, exactly 20 years ago. That pink spot among the green smocks is me. Since then I have never been free of that label: the test-tube baby. The first in the world. . . . At the age of 5 I understood almost nothing of what my parents were explaining to me. But they were forced to do it then. I was about to start school. And children were terrible, mom told

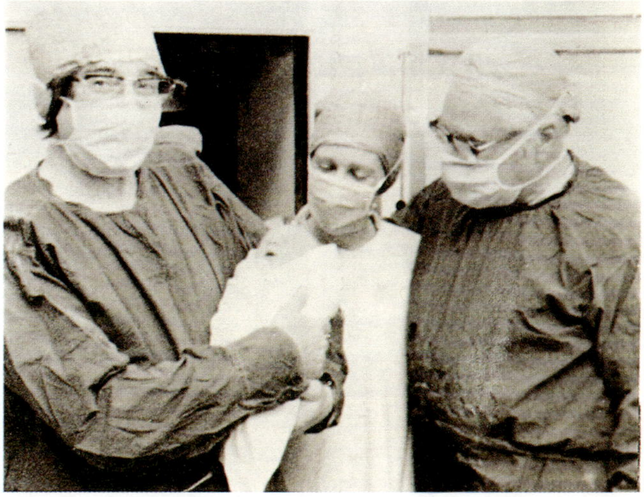

FIGURE 4 Doctor Robert Edwars with, in his arms, Louise J. Brown, the first child to have been conceived *in vitro,* at Oldham General Hospital, Lancashire, July 25, 1978 (photo courtesy of Hulton-Getty/Ronchi).

me. The most cruel. Every time I passed by on the street they would sing: the test-tube baby, the test-tube baby. The problem was that everybody thought that I had actually grown up in a test tube, and that that's where I had come out from. They looked into the stroller to see if I was different, who knows, a bit longer and thinner than a normal child, with an egg-shaped head, a sort of newly invented extraterrestrial.

An attempt at artificial human fertilization had been made in Italy in 1961 (Fig. 5). Mori (1995) notes,

> The sixties began with the news released through the daily press on January 13, 1961, that Daniele Petrucci of the University of Bologna had succeeded in producing an embryo *in vitro* and in keeping it alive for 29 days, carefully filming the entire experiment and interrupting it for moral reasons. The scandal was enormous and Petrucci (who appears never to have produced any data) was expelled from the university and found refuge in the Soviet Union.

It is true that, especially as a result of artificial insemination, there is the risk of disembodied procreation and of a therapy which might be classified as "a sort of mechanism for quality control in the productive process" (Lauritzen, 1993), of treating the egg and sperm cells as anonymous, depersonalized factors; however, it is also true that such procedures lead to natural processes (nothing, in effect, occurs in a way that is contrary to nature) which will result in a child who perhaps will be more loved than others.

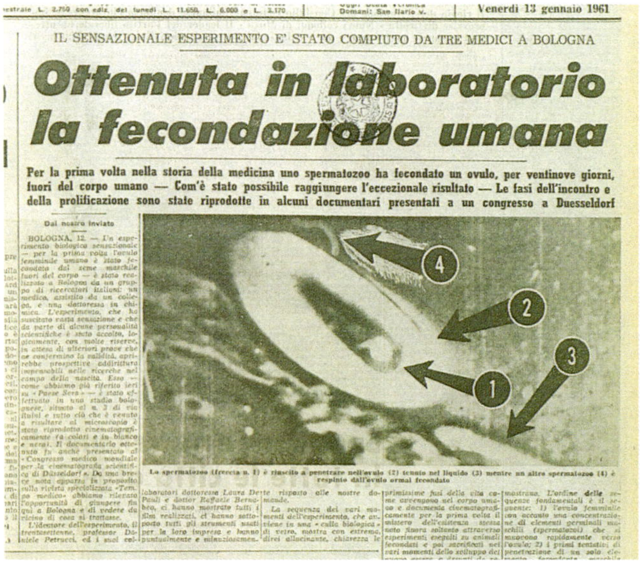
Venerdì 13 gennaio 1961

IL SENSAZIONALE ESPERIMENTO E' STATO COMPIUTO DA TRE MEDICI A BOLOGNA

Ottenuta in laboratorio la fecondazione umana

Per la prima volta nella storia della medicina uno spermatozoo ha fecondato un ovulo, per ventinove giorni, fuori del corpo umano — Com'è stato possibile raggiungere l'eccezionale risultato — Le fasi dell'incontro e della prolificazione sono state riprodotte in alcuni documentari presentati a un congresso a Duesseldorf

Dal nostro inviato

BOLOGNA, 12 — Un esperimento biologico sensazionale — per la prima volta l'ovulo femminile umano è stato fecondato dal seme maschile fuori del corpo — è stato realizzato a Bologna da un gruppo di ricercatori italiani: un medico, assistito da un collega, e una dottoressa in chimica. L'esperimento, che ha suscitato vasta sensazione e che da parte di alcune personalità ...

FIGURE 5 The daily newspaper *Il Paese* (January 13, 1961) publicly announces the experiment, carried out at the University of Bologna, in which a human embryo was produced *in vitro* (photo reproduced with permission).

AID families or so-called artificial families, instead, pose difficulties of a different order which heighten the problem of the relationship between the freedom of individuals and the well-being of society. The first case of AID appears to have occurred in 1884 in Philadelphia with the wife of a Quaker merchant who was sterile. The case, which was enacted and discussed at the Jefferson Medical College, was described in 1909 in the review *Medical World* by a person who at the time was a student and perhaps the donor, given that he was the "most handsome of the class." For a few years, the news caused a sensation. The ethical questions extant today concern the problem of "how to satisfy the needs of individuals while at the same time safeguarding the integrity of social organization" (Snowden and Mitchell, 1983).

The practice of artificial insemination was condemned by the archbishop of Canterbury in 1945. Aside from the reasons of principle which lend support to the disfavor with which it is also seen by the Catholic Church, I believe that its psychological and existential consequences can indeed be serious. AID produces instability in couples and may encourage dissimulation (more or less honest) on the origins of the child in the fabric of interpersonal relations. The child, in fact, is "hers" but not "his" so that both prefer to pretend, when dealing with relatives (especially his) and with society, that the child is the offspring of his or her social or putative father. The tendency toward secrecy (which is not always in the best interest of the child) can poison the couple's relationship and cause it to become vulnerable to other people's opinions. There is the concrete possibility that the secret will be betrayed, perhaps even in front of the child, by either the husband or the wife out of resentment for one another or as a result of not being psychologically strong enough to keep the secret from everyone else, to bear the whole weight of a suffered decision. It is also my opinion that artificial insemination puts the figure of the father in doubt, both in the field of the real and in that of the imagination (which in this case may even be more important), and that all this may be at the root of an unsolvable dispute. On the other hand, insemination procedures have become relatively simple: All one needs is a donor so that outlawing the practice would only lead to an increased rate of secret and clandestine events.

There are several reasons for choosing AID. For the woman, there are basically two forms of psychological pressure that get absorbed during her upbringing: that she must have a child and that she must have it in a marital context. Many women, moreover, suffer from "baby hunger." This, in effect, is one of the reasons (not always corresponding to the truth) why some husbands say they accepted to become putative fathers (Snowden and Mitchell, 1983). The legal father may also be afraid that he will become psychologically dependent on the donor and that mother and child will form an alliance which excludes him. A conflict of roles takes him over (Snowden and Mitchell, 1983):

The *genitor* is the biological connection, the male who provides the sperm which fertilizes the egg cell in the female. In this case the donor is the parent. The *pater* is the father, in other words, the male who is legally responsible for the child's well-being, the person to whom the child owes filial obedience and who usually plays the role of head of the family.

Not knowing one's father's identity can be the cause of a tormented, lifelong search. Some letters reveal it, like the one written by a reader of the *Guardian* (July 5, 1979) who advised against AID, telling how he, the son of an unknown father, spent a miserable childhood trying to identify his father among strangers: "I seek him on the basis of his likeness [to me] in the crowd, outside the hostels of the Salvation Army, on television, everywhere. This is my identity crisis and it is very hard to bear." In another letter, a young man born by way of AID confesses, "I never worried about being a bastard. No, what upset my whole being was that no one knew my 'real' father: as if half of myself didn't exist, doesn't exist" (Snowden and Mitchell, 1983). The solution represented by mixing semens so that one can never know who fathered an AID child is, according to some experts, morally reproachable, especially if it is a doctor who uses such a method (Jonas, 1985):

> He has been assigned an unheard of role in the history of medicine. Unless the couple brings the donor of the semen with them, he must find him or, rather, he must find the semen. This, vulgarly speaking, turns his role into that of a pimp. . . . Here, however, there is a diabolical move which allows one to entirely circumvent the juridical problem, by making the donor's identity *de facto* impossible to determine: the so-called "semen cocktail" made out of semen of various origins which is provided for the insemination.

Those who defend AID insist on three points. First, the historically determined character of filial and paternal relations, as shown by the fact that the children of adulterous parents (in Italy up until 1975) or of incestuous ones even today, although biological paternity is undeniable, were not or are still not considered such, whereas adopted children, although they lack any genetic relation with the couple who takes them in, were and are recognized as children as far as the law is concerned. Such a historical change in the context of filiation and paternity should be viewed in the broader context of the transformation of the idea of the family. The latter, in fact, currently includes some new typologies, which are also the result of a more widespread recourse to divorce, such as "one-person families, de facto families of various types, families without children, families with their own and adopted children, and so on" (Mori, 1995).

The second point is the observation that AID families (2% of all French families) are far more stable than others, insofar as they are less involved in divorces (Mori, 1995). The third point is that the psychophysical equilibrium of children born according to the procedure (except as far as the effects of the secrecy surrounding the modalities of their birth are concerned) is generally not affected. However, Mori (1995) notes,

> But in the absence of litigious tendencies though in the presence of a high number of AID-born children, one is led to conclude that AID is a cohesive factor as far as the family is concerned, since the latter is tempered by the experience of sterility and by the joy of the new birth and discovers a newfound unity. . . . Today objectors insist on the damage of a psychological nature resulting from the fact that the child is deprived of "normal" parental figures. But such an objection seems to be based on the traditional concept of filiation (tied to the old concept of marriage) and immediately disappears as soon as one adopts the new concept in light of empirical facts: In practice, AID-offspring are (on average) taken care of more and better than others and do not present any particular psychological disturbance. . . . One must admit that whoever was born thanks to AID would never have been born without such a technique, so that since we are considering the possible damage suffered by the newborn child, one should say that the alternative for him is between being born thanks to AID and not being born at all, and if he is born surrounded by tender loving care it appears that the fact of having put him in such a condition may hardly be defined as harmful.

These arguments are complemented by the fact that the "new reproductive rights" view the employment of assisted insemination techniques not as a therapeutic remedy against infertility but rather as an unalienable prerogative of individuals *ut singuli,* even before they are considered as members of a family. This is the principle of the autonomy of individual choices which is establishing itself despite traditional family-centered visions. This does not rule out that tensions may be inevitable and even predictable, as in the case of a social father who has disclaimed the child (now 11 years old) who was born by means of AID. The legal case, begun in 1990, had yet to be resolved in 1998 (Bandera, 1998):

> In the meantime, the child's family has fallen apart. He now lives with his mother, a nurse, in a small town about a dozen kilometers from Cremona, amid unimaginable hardships, exposed to the psychological burden of the curiosity the case has created in his schoolmates and people in general. The father runs a bar a few kilometers away, but he has reiterated that he does not want to have anything to do with the two

of them, justified by the fact that the child was not a product of his own semen.

Egg Cell Donation

Symmetrical to AID (on the female side) is egg cell donation. Without taking into consideration the more complex situations (such as the combination of two egg cells from two different women — an older one who provides the nucleus with her own genetic patrimony and a younger one who instead provides the cytoplasm), even this standardized technique can cause psychological reactions of rejection and strong perplexity.

Such is the case of a 40-year-old woman who decided to resort to egg cell donation (Bonaccorso, 1998):

> It took me 10 years to have a child. . . . First the usual routine of doctors and gynecologists. Everyone had something to say, and you feel like something I wouldn't know how to define, knocked around left and right. And you keep wanting the child and you can't have one unless you decide to take a radical decision. Which breaks you up. Either you accept or no child. And your husband is putting pressure on you. And you start thinking that he might leave you and find someone who does have egg cells, nice and ready, by the dozen. A woman who is waiting for nothing better than to find someone like him, like your husband, desperately in search of a child of his own, or maybe two or three. You have these almost crazy fantasies, you only notice people who have children, it seems to you that everybody can and you can't. A tragedy . . . perhaps that is what it was for me and it still is anyway, despite the child, who is wonderful and whom I adore. But I know he has nothing that belongs to me. He has some of his father's traits but not mine. Not even a single detail. Nothing. And he looks like a woman whom I never met and know nothing about.

Note the feelings of fear and jealousy directed at a specific person who has never been seen or encountered which appear, in a different form, also in another woman, of 38 years, to whom egg cell donation was suggested (Bonaccorso, 1998):

> You will never be able to have children, the gynecologist told me. I felt as if I was dying. The only alternative was egg cell donation. Nothing else. And I didn't feel up to it. And so everything fell apart, marriage included. How can you tell your child: I'm sorry, you know, but in order to have you I used another woman's egg cell, and I don't even know who she is. To keep it a secret? Even worse. Egg cell donation splits you into two. How do they manage to live with a choice like that?

The Mother–Grandmother and the Aunt–Mother

Many other paradoxes capable of causing disconcertment and scandal arise from other applications of medicine and biotechnologies. One of the paradoxes that has troubled consciences is the case of the "mother–grandmother"—not so much that of Mary Shearing, a 53-year-old Californian woman, who in November 1992 gave birth to twins using the egg cells that had been donated by a 20-year-old woman and fertilized with her husband's semen as that of an Italian woman of menopausal age who convinced her daughter to allow herself to be artificially inseminated with the semen of her mother's partner. However, there is also the case of the aunt–mother: In January 1995, Elizabeth was born in Rome from one of four embryos that were obtained by means of *in vitro* insemination and frozen, which her mother (who had died in a car accident 2 years earlier) had left. The woman who carried out the pregnancy was the sister of the husband of the deceased woman, or the child's paternal aunt. In 1997, the Italian gynecologist P. Billotta announced the case of a woman who was pregnant with two fetuses formed by embryos belonging to two different couples of parents. In fact, one of the natural mothers lacked a uterus, whereas the other could not face the trauma of childbirth because she suffered from heart disease.

Recently, there has been great disconcertment at the news of the insemination of a woman by means of the sperm that had been extracted from her dead husband. There was talk of "fertilization from the tomb" or of a bridge between the mystery of birth and that of death. Here are the facts: The July 1998 issue of the *New Scientist* reported the case, then taken up by all the major dailies, of a 29-year-old English widow already in the first month of pregnancy who had had her uterus implanted with an egg cell fertilized by the sperm of her dead husband. It happened at the Century City Hospital in Los Angeles. Professor Cappy Rothmann, who performed the operation, is certain to have acted ethically: "This operation is ethical because it gives people hope and reduces the pain for the sudden loss of someone dear."

The birth of a child from the other world, of a child as the continuation of a dead parent, caused diverse reactions. Whereas from the lay camp it was observed that, in this manner, a blow had been dealt to "that nucleus of feelings which underlies the distinction between dead and alive, in which our religious, moral, and civil traditions flourish" (Giorello, 1998), on the Catholic side the conviction was reaffirmed to the effect that "artificial insemination is always wrong." The following was added (Sgreccia, 1998):

> But here there is the aggravating circumstance of the total lack of willingness and consciousness of the man, there is the macabre aspect of the intervention on a dead body. These are aberrant zootechnical experiments. . . . A fundamental principle has been violated in the deepest and most awful manner: Human procreation must be based on a personal act on the part of the married couple, a free and responsible act of love. It is evident that extracting the semen of a dead man is as far as one can possible get from such a requirement.

The Determination of Sex

Among the possibilities which present themselves today in the context of assisted procreation, there is also that of selecting the characteristics and the sex of the newborn child and — in what appears to be a science-fictitious perspective — that of cloning and parthenogenesis.

The anticipated choice of the child's characteristic traits already appears to be controversial, insofar as it reminds one of Nazi eugenics, even though, in effect, it is the least harmful of all. Some have argued that it makes no sense to prevent parents from having such an option. (Harris, (1992) noted,

> To me it seems as if the problem can be put in the following terms: either these traits — hair or eye color, sex and the like — are important or they are not. If they are not important, why not allow people the chance of choosing them? And if they are important, can it be right that they should be left to chance?

More discussed is the case of the choice of the sex of the newborn child (cited previously as being equivalent to the color of eyes or of hair), both in the way it has already been applied in England and using the more sophisticated methods for selecting spermatozoa (in either case, however, the patrilinear or matrilinear system at work in a given society would, over the generations, be altered).

In certain private clinics in London it has been possible, in fact, to choose the sex of the child since British law ruled that it is discriminatory to deny women of Indian origin the right to prefer male children, given that in their culture the birth of baby girls is considered highly undesirable. In India it is sometimes avoided by resorting to abortion after the identification of the fetus's sex so that out of 8000 abortions on the part of couples who had performed a prenatal diagnosis for such a purpose, 7,997 fetuses of female sex were found and the remaining 3 cases were the result of a diagnostic error (Mathieu, 1990; Rodotà, 1995).

Even those who are in favor of sex selection, without resorting to late abortion, have some perplexities on the subject. Harris (1992)states,

> Naturally, sex selection is not a new phenomenon of our era. Selective infanticide has existed for thousands of years in various societies and is still practiced, at times even in our own society. Amniocentesis also reveals sex and a mother might decide to interrupt a pregnancy precisely on the basis of such information. In the future, moreover, with all probability it will become relatively easy to determine sex by selecting the spermatozoon to be used for fertilization. In the animal field, we already have techniques for the identification and separation of the spermatozoon in relation to the sex that it would determine. . . . The most widespread worries are connected with the possibility that some societies should favor the production of male children in such a marked fashion that the discrimination against females would attain epidemic proportions. Even though it would appear that such patterns of behavior should be destined to correct themselves on their own, no doubt in the short term they can give rise to much unhappiness. It is worth remembering that, in any case, since a certain number of genetic diseases are connected with gender, in the absence of other therapies the choice of a particular sex for one's offspring may be suggested by reasons which are quite strong.

In general terms, one can see looming on the horizon yet another danger, hinted at since 1932 in *Brave New World* by A. L. Huxley, a writer who came from a family of biologists. In the book it is hypothesized that a rigid social hierarchy composed of individuals (alpha, beta, and gamma) is created in which some are genetically programmed to give orders and others to follow them (Fig. 6). What is perhaps less well remembered is the fact that in the novel there is a part of the globe that has not been subjected to this very particular gene therapy, and that there are so-called "savages" on whom the author seems to place his hopes for the liberation of future humanity and the parallel destruction of such hierarchies. Although the style in which it was envisaged in 1932 is science fictitious, today a scenario such as Huxley's could become plausible. It could be, in fact, that advanced biotechnologies will come to concern the inhabitants of the most economically rich and developed nations, favoring the growth not of races but certainly of generations of artificially treated individuals, always less prone to diseases and malformations, programmed in their numbers and sex, leaving the greater part of mankind, which does not have access to these technologies, abandoned to the lottery of nature and history.

Underlining his own "principle of responsibility," Jonas

FIGURE 6 Title page of the first edition of *Brave New World* by A. L. Huxley published in 1932 (reproduced with permission).

(1985) warned against programming ahead of time the life of unborn human beings by determining their sex or individual characteristics, by cloning them, or by intervening in a risky manner — albeit with the best intentions — on their genetic patrimony at the germinal level. He therefore defends the "right of every human life to find its own way and to be a surprise for itself." If this does not happen, "the gamble in life" is

> defrauded of its attractive and also anguished opening. The past has been allowed to precede the future by way of an inauthentic knowledge of it and this has been done in the most intimate sphere: in the sphere of the question "who am I?" This question must come from the secret. Yes, the secret, which is already a condition of asking and searching, is, for whoever is looking for an answer, even the condition of the possibility of *becoming* perhaps precisely *that* which *then* will be the answer. The artificial being known from the start, the subjective absence of the secret, destroys the condition of authentic growth.

Chance, which is the enemy of the antidestiny propounded by biotechnologies, is thus revalued and defended against its critics, insofar as human life has the right of ignorance of beginning and end, has the right to be — according to the lesson of Augustine taken up by H. Arendt — an absolute *initium:* "Chance, the productive source of the evolution of the species. Chance: in every sexual reproduction the guarantee that every individual who is born is unique and that no one is entirely identical to anyone else. Chance looks after the existence of what is always new, of what has never existed" (Jonas, 1985).

Parthenogenesis and Cloning

The unrepeatable uniqueness of the human individual and the value of his or her authenticity are prospectively put into question by two techniques — one more unthinkable and remote and the other more likely or more close to becoming reality. The former foresees that, by means of artificial parthenogenesis, only female individuals are born by way of a process which induces cellular multiplication inside the egg cell. Harris (1992) writes,

> A curious but far from insignificant by-product of the massive amount of work being performed in the field of human embryology is represented by the possibility of artificial parthenogenesis. Parthenogenesis is a process in which the unfertilized human egg cell (like those of other animals and mammals) is stimulated to develop without fertilization. The result of the process is an individual of the female sex who is the equivalent of a clone of her mother. It is believed that parthenogenesis takes place by chance, though rarely, in many species; with egg cells collected for *in vitro* fertilization, cellular multiplication can be induced through the simple expedient of dipping the unfertilized egg cells into alcohol. . . . If they wanted to, a group of women with the necessary technological resources could give rise to an exclusively female yet vital society, i.e., one which is capable of reproducing itself without recourse to men, not even in the humble role of sperm donors or males for reproductive purposes.

The other technique instead aims at cloning single individuals. Cloning is nothing other than a type of asexual reproduction, common among plants, and which we are now able to also cause among animals by introducing the nucleus of a cell into the cytoplasm of an enucleated egg cell. Dolly, the famous sheep, has recently shown the feasibility of the project which seemed science fictitious only a few years ago and has opened the way to an already possible cloning of a human. This has unleashed fears and fantasies, which render plausible what the writers of the nineteenth and twentieth centuries had simply imagined as regards doubles, token persons, and photocopied individuals. It is a characteristic of clones to have the same genetic patrimony and this no doubt represents an advantage for the possible transplanting of organs without possibility of rejection. Many people today are already thinking with horror about the possibility of duplicates of themselves being kept alive as containers of organic spare parts.

Were human cloning to take place (and the chance of it occurring in the immediate future, despite the moral condemnation and juridical sanctions against it, appears high), we would be faced with further paradoxes, such as those mentioned by Prodi (1987):

> In this way, a woman can carry within her or construct a copy of her dead husband, rebuilding him out of one of his cells that has been preserved in liquid hydrogen, and grow him up and find him there in front of her as young as he once had been. If she married him again, would it be incest? Entirely novel juridical relations. And the right to property? Endless disputes over inheritances, perhaps spanning the centuries (the cell which donates the nucleus can stay in cold storage indefinitely, and permit the perfect reconstruction of the couple centuries later).

Philosophers, especially Anglo-Saxon ones, have already described many of these scenarios, according to some of which one ends up not distinguishing any more between the original and the copy (Williams, 1973; Nagel, 1979; Nozick, 1981; Parfit; 1984).

In other fields as well, biotechnologies alter the apparently simple notions and representations of one's own body and of personal identity which until today have been considered rather obvious, albeit difficult to explain or justify rationally. Through them, new and unhoped for possibilities of improving and curing one's body arise, and the body thus acquires an intrinsic dignity: It is no longer the prison of the soul nor corruptible flesh or skeleton. It becomes the subject and object of new rights which are no longer only economic or political, but which concern life and health: In this way, the body becomes "juridified" (Rodotà, 1995).

Organ Transplanting

Among other things, biotechnologies allow one to receive in one's own body the organs of another person, alive or dead, or of genetically modified animals. Also, the status of the dead has changed, and therefore also our imagination's relationship to them, since many cadavers are no longer buried whole according to rituals or abandoned to the dissolving power of the elements, but some of their parts continue to live in another body, in a novel contamination of the live and the dead. One might even go as far as to say that some organs live twice — live two lives in different bodies. Some have problems accepting such a contamination, but necessity, in this sector, creates the rules (Degos, 1994).

The problem of organ transplants touches the nucleus of personality in another manner, i.e., at the economic and cultural level. It is no secret that there is a scarcity of organs for transplanting in the world and that there is a vast and flourishing market that deals in them, which in India and Turkey is legal. In the same way, it is also known that it is always those who are most poor who sell organs to those who more wealthy. In this sense, it has been rightly said that one should not aim solely at frontier bioethics, at the mapping of the genome, at gene therapy, or at the invention of advanced drugs because there is also an everyday bioethics which concerns the majority of mankind and which touches on the prevention and care of individuals who have been destroyed, in the course of a brief existence, by the hardships of the environment in which they live (Berlinguer, 1989; Berlinguer and Garaffa, 1996). Hence, the importance of the distribution of investments in so-called "big science" (i.e., in the large scientific enterprises), insofar as investing in one field, given that resources are limited, means neglecting others.

Personal identity is touched on also in relation to the problem of when the self begins to exist with respect to the development of the fertilized egg cell or embryo. The relevance of the question may be explained, in juridical and existential terms, in relation to the lawfulness of abortion. Is the egg cell that gets fertilized during conception already a person, someone who must be protected by the law so that willful abortion can already be equated with a sort of infanticide, Or — as Ford (1988), a Salesian priest, sustains — do human beings begin to exist as individuals only after the 14th day with the formation of the primitive stria and only after that moment are embryos entitled to protection by the law? The Warnock commission has spoken in terms of preembryo when referring to the product of conception before the 14th day (Warnock, 1985) and, within this time frame, the question of lawful protection of the embryo does not arise. In broader terms, this position is also shared by Mori (1995): "We are certain that the preembryo is not a person because we know that only a few days after conception we have neither individuality nor rationality, and this is more than enough to solve our problem. It is obviously difficult to speak of self where prenatal life is concerned. Does it depend on when the fetus is capable of experiencing pain or pleasure? On when a first inkling of consciousness is born?

In Italy, the question has raised a discussion on the revision of Article 1 of the Civil Code in such a way as to move the attribution of juridical capacity from the newborn infant to the human egg cell immediately upon fertilization or to the fourth week of its prenatal development. Many persons, however, have noted that in this way one undermines the freedom and responsibility of men and, especially, of women in determining the most opportune moment to ensure the zygote, embryo, fetus, or conceived entity a harmonious situation of growth and satisfactory integration in society (Celotti, 1997; Massarenti, 1997).

Conclusions

As a result of the many changes that biotechnologies have introduced, the very idea of life as spontaneity, autopoiesis, is weakened. At the same time, the aura of mystery which surrounds all that takes place in us in order to keep us healthy and alive without our consensus tends to become thinner. If we have never been completely the rulers of our own bodies and their functions, nor have ever tried to be (as opposed to oriental thought, in the West since the time of Aristotle, it has been believed that one cannot interfere with the so-called vegetative soul), we are now in a position to intervene in various ways and with efficacy on living matter. As a result, not only is maternity assisted, as we say, but also the whole of our body and, in part, also our sense organs and even our mind are assisted due to many interventions, drugs, and prostheses. From this point of view, what matters most are not changes introduced at the somatic level, but the possible transformations — with uncertain results — at the level of germinal cells. The controlling of genetic patrimony, in fact, makes it vulnerable not only to manipulation but also to impoverishment — to the possible loss of richness and variety and to the reduction of everyone to common models.

In any case, the problem does not appear to be that of condemning the right of individuals to a more reasonable life by means of biotechnologies. It is a question of posing limits and of creating a culture of limits, of avoiding a privatization or a private interest in scientific discoveries, and of not posing obstacles, dictated only by fear, to scientific research; it is not through fear alone — or, to use Jonas' (1985) expression, through "the heuristics of fear" — that one can control the processes that are at work. Fear remains a source of bad advice. That which instead is useful is to focalize together the constraints, possibilities, consequences, and expectations of biotechnologies. It takes humility not to take sides immediately in referendum-style fashion either in favor or against things one knows nothing about because ignorance does not give anyone the right to either believe or disbelieve, and ethics will also serve better as a guide than as a brake.

Acknowledgments

I thank Armando Massarenti and Maurizio Mori for their help in finding some of the bibliographical materials.

References Cited

Bandera, S. (1998). Disconosce il figlio concepito in provetta: Deciderà la consulta. *Il corriere della sera* **2.**

Berlinguer, G. (1989). In *Bioetica Quotidiana e Bioetica di Frontiera* (A. Di Meo and C. Mancina, Eds.). Laterza, Rome.

Berlinguer, G., and Garrafa, V. (1996). *La Merce Finale. Saggio sulla Compravendita di Parti del Corpo Umano.* Baldini & Castoldi, Milan.

Bonaccorso, M. (1998). Noi madri vere di figli in provetta. *Corriere della sera* **1.**

Capron, A., *et al.* (1993). *Costruire la Vita.* Elidir, Rome.

Cattorini, P. (1990). Qualità della vita negli ultimi istanti. *Med. Morale* **2,** 273–293.

Degos, L. (1994). *I Trapianti di Organi.* Saggiatore, Milan.

Di Mauro, E., and Mafettone, S. (1997). Il futuro non è più quello di una volta. Contributi per un'etica dell'ingegneria genetica. *Filos. Questioni Pubbliche* **1,** 83–110.

Engelhardt, H. T., Jr. (1986). *The Foundations of Bioethics.* Oxford Univ. Press, Oxford.

Flamigni, C. (1994). *I Laboratori della Felicità. La Cultura della Sterilità tra Successi e Delusioni.* Bompiani, Milan.

Flores d'Arcais, P. (1997). *Il debito e la dignità. Micromega* **2,** 25–40.

Ford, N. M. (1988). *When Did I Begin? Conception of the Human Individual in History, Philosophy and Science.* Cambridge Univ. Press, Cambridge, UK.

Giorello, G. (1998). *Corriere della Sera* **16.**

Glover, E. J. (1984). *What Sort of People Should There Be?* Penguin, Harmondsworth, UK.

Harris, J. (1992). *Wonderwoman & Superman.* Oxford Univ. Press, Oxford.

Huisman, B., and Ribes, F. (1992). *Les Philosophes et le Corps.* Dunod, Paris.

Jonas, H. (1985). *Technik, Medizin und Ethik. Zur Praxisi der Prinzip Verantwortung.* Insel, Frankfurt.

Lauritzen, P. (1993). *Pursuing Parenthood Ethical Issues in Assisted Reproduction.* Indiana Univ. Press, Bloomington.

Mathieu, V. (1990). Bioetica in cammino. In *La Dimensione Etica nelle Società Contemporanee.* Edizioni/Fondazione Giovanni Agnelli, Torino.

Melchiorre, V., and Cascetta, A. (Eds.) (1983). *Il Corpo in Scena. La Rappresentazione del Corpo nella Filosofia e nelle Arti.* Vita & Pensiero, Milan.

Melina, L. (1988). In *Quale Vita? La Bioetica in Questione* (A. Scola, Ed.). Leonardo Mondadori, Milan.

Mori, M. (1995). *La Fecondazione Artificiale.* Laterza, Rome.

Nagel, T. (1979). *Mortal Questions.* Cambridge Univ. Press, Cambridge, UK.

Nancy, J.-L. (1991). *Corpus.* Cronopio, Naples.

Nozick, R. (1981). *Philosophical Explanations.* Cambridge Univ. Press, Cambridge, UK.

Parfit, D. (1984). *Reasons and Persons.* Oxford Univ. Press, Oxford.

Polito, A. (1998). Intervista a Louise Brown. *Il Venerdì di Repubblica* **31.**

Potter, V. R. (1971). *Bioethics: Bridge to the Future.* Prentice Hall, Englewood Cliffs, NJ.

Prodi, G. (1987). Prefazione al libro di D. Lamb. In *Il Confine della Vita.* Mulino, Bologna.

Reich, T. W. (1978). *Encyclopedia of Bioethics.* Free Press, New York.

Rodotà, S. (1995). *Tecnologie e Diritti.* Mulino, Bologna.

Rodotà, S. (1997). La fine del destino. *Micromega* **2.**

Scarpelli, U. (1987). Bioetica: Alla ricerca dei principi. *Biblioteca Libertà* **99.**

SCARPELLI, U. (1996). Quale base comune per la riflessione bioetica in Italia? Dibattito sul Manifesto di bioetica laica. *Politeia* **1,** 41–42.

SGRECCIA, E. (1988). *Manuale di Bioetica.* Vita & Pensiero, Milan.

SGRECCIA, E. (1998). *Corriere della Sera* **17.**

SNOWDEN R., and MITCHELL, G. D. (1983). *The Artificial Family. A Consideration of Artificial Insemination by Donors.* Unwin, London.

STICH, S. P. (1978). *The Recombinant DNA Debate,* pp. 188–203. Prentice Hall, Englewood Cliffs, NJ.

VEGETTI FINZI, S. (1997). *Volere un Figlio. La Nuova Maternità tra Natura e Scienza.* Mondadori, Milan.

WARNOCK, M. (1985). *Introduction.* In *A Question of Life.* Blackwell, Oxford.

WILLIAMS, B. (1973). *Problems of the Self.* Cambridge Univ. Press, Cambridge, UK.

General References

AA. VV. (1988). *La fecondazione artificiale. Questioni morali nell' esperienza giuridica.* Giuffré, Milan.

AGAZZI, E. (Ed.). (1993). *Bioetica e Persona.* Angeli, Milano.

BARAN, A., and REUBEN, P. (1989). *Lethal Secrets: The Shocking Consequences and Unresolved Problems of Artificial Insemination.* Warner Books, New York.

BEAUCHAMP, T. L., and CHILDRESS, J. F. (1989). *Principles of Biomedical Ethics.* Oxford Univ. Press, Oxford.

BLANK, R. H. (1984). *Redefining Human Life. Reproductive Technology and Social Policy.* Westview, Boulder, CO.

BROWN, N. (1983). *The Worth of Persons. A Study in Christian Ethics.* Catholic Institute of Sydney, Sydney.

COREA, G., *et al.* (1985). *Man-Made Women: How New Reproductive Technologies Affect Women.* Indiana Univ. Press, Bloomington.

GALIMBERTI, U. (1983). *Il Corpo.* Feltrinelli, Milan.

GOETHE, J. W. (1970). *Faust.* Mondadori, Milan.

HARE, R. M. (1993). *Essays on Bioethics.* Clarendon, Oxford.

HARING, B. (1971). *Morality Is for Persons.* Farrar, Straus & Giroux, New York.

HARRIS, J. (1989). *The Value of Life.* Routledge & Kegan Paul, Boston.

JONAS, H. (1984). *Philosophical Essays. From Ancient Creed to Technological Man.* Chicago Univ. Press, Chicago. [In Italian: *Dalla fede Antica all'Uomo Tecnologico,* Mulino, Bologna, 1991]

KEANE, N. P., and BREO, D. L. (1981). *The Surrogate Mothers.* Everest House, New York.

LADRIERE, P. (1991). Le concept de personne, héritier d'une longe tradition. In *Biomédicine et Devenir de la Personne* (S. Novaes, Ed.). Sevil, Paris.

LECALDANO, E. (1995). Etica. In *La Filosofia* (P. Rossi, Ed.), Vol. 3. UTET, Torino.

MORI, M. (Ed.) (1988). *Questioni di bioetica.* Editori Riuniti, Rome.

NOVAES, S. (Ed.) (1991). *Biomédicine et Devenir de la Personne.* Seuil, Paris.

ONFRAY, M. (1989). *Le Ventre des Philosophes.* Grasset, Paris.

RAMSAY, P. (1970). *Fabricated Man. The Ethics of Genetic Control.* Yale Univ. Press, New Haven, CT.

SCARRY, E. (1990). *La Sofferenza del Corpo.* Mulino, Bologna.

TRISELIOTIS, J. (1973). *In Search of Origins: The Experience of Adopted People.* Routledge & Kegan Paul, London.

VIANO, C. A. (Ed.) (1990). *Teorie Etiche Contemporanee.* Bollati-Boringhieri, Torino.

WALTERS, W., and SINGER, P. (Eds.) (1983). *Test Tube Babies: A Guide to Moral Questions, Present Techniques and Future Possibilities.* OUP, Melbourne.

ALBERTO PIAZZA
Department of Genetics, Biology, and Biochemistry
University of Torino
Torino, Italy

The Human Genome Project

The ambitious project of sequencing the entire human genome, mapping, and analyzing every single DNA gene, derives from the assumption that only a complete knowledge of its structure and organization can allow us to fully understand its specific functions. This knowledge, which is extremely important for both scientific and practical reasons, employs the techniques of molecular biology which are summarized in the first part of this article. The second part analyzes the remarkable bioethical implications deriving from a progressively increasing knowledge of the human genome and from the prospects of its practical application. In fact, one might affirm that the knowledge of the human genome with its wide spectrum of technological implications has had an essential role in promoting the development of the bioethical theme in its most modern expression.

The Human Genome Project

The Human Genome Project is an international research initiative in which the United States, Japan, and a few Asian and European countries take part. Its objective is the mapping of the human genome; in other words, localizing all the 60,000–70,000 genes (Fields *et al.*, 1994) on the chromosomes and reconstructing their DNA sequence (of about 3×10^9 nucleotides). The cost of this operation was originally estimated at approximately $1 per nucleotide distributed over a period of 15 years with an expense of approximately $200 million dollars per year. The possibility of programming such exacting research started taking shape in approximately 1985, when geneticists began to exploit the resources of molecular biology to isolate many disease-causing genes. It was obvious then that a systematic study of all human genes, of their precise localization on the chromosomes, and of the molecular structure of their products would not only enable extremely refined and timely diagnostics but also, more generally, result in a profound progress in biological knowledge, especially in regard to obtaining answers to medical problems which were believed insoluble, such as gene therapy. The project became operative in 1988 in the United States when the American Congress recommended to the National Institutes of Health and to the Department of Energy to redistribute a significant part of their funds to grant the project. In the United States, but perhaps more in European countries, the project caused alarm both for scientific strategy reasons (e.g., the fear that more funds would be distributed to the project to the detriment of others) and for the social, ethical, and legal implications of a project which was considered too advanced from certain points of view in regard to the social and political context of the communities for which it is intended. To this should be added the complication of the genomic variability: An individual's genome is different from another's, and since it would be pointless to identify the whole DNA sequence of each individual it is important to understand within what limits the analysis of a limited number of genomes or their fragments can be generalized.

As we know from experience, however, hoping for a deferment of the knowledge deriving even by a partial pursuit of this project would be a useless if not damaging attitude. It would seem more constructive to be aware of the problems and to deal with them before they occur rather than deal with them, unprepared, postfactum. The Human Genome Project will probably be carried through the beginning of the twenty-first century: As a result of scientific progress, together with the laws of the market in the biotechnological field, the mapping phase of the human genome on chromosomes has been completed (Fig. 1), and identifying their content in terms of DNA sequence should occur in the near future. The knowledge of one's own genes will become a normal acquisition in our everyday life: Genetic information will be increasingly more accessible.

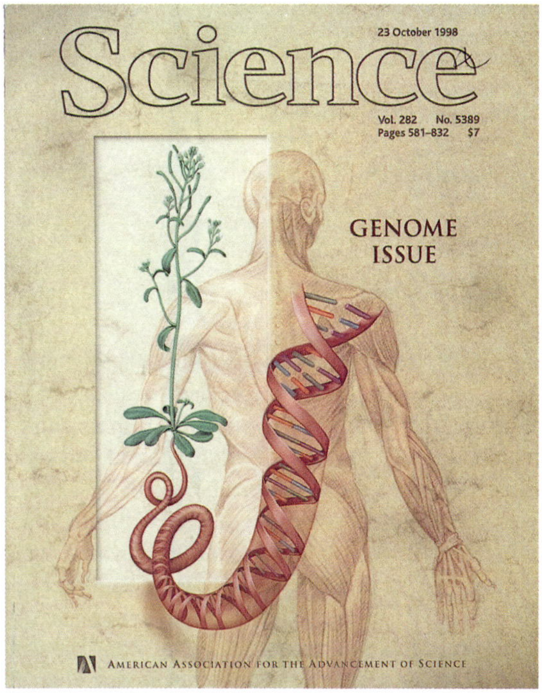

FIGURE 1 Cover of the journal *Science* devoted to the most recent research on the human genome and its prospects. The illustration symbolizes the differences and the similarities in various organisms — all related and all different.

Also, genetic counseling services will have to become integrated to the other diagnosis and care health services, and somehow their access will have to be ensured.

Regarding current knowledge, it may be interesting to give a brief summary of the research and analysis strategies used for their pursuit.

The State of the Art

High-Resolution Genetic Maps

The possibility of identifying the sequence of many nucleotides essentially depends on the preliminary identification, along the DNA of each chromosome, of reference points for the relative localization of the genes constituting the genome. These reference points constitute the chromosome map, and a map is obviously all the more important the closer these points are to one another and the more they are homogeneously distributed. Mapping human genes therefore means defining the 24 types of chromosomes on which they are located and fixing their relative position on the chromosome: As on a highway the location of a car can be calculated from its distance from the nearest tollgate, the location of a gene can be calculated from its distance from the nearest map point, and in the metaphor the geographic position of the tollgates network constitutes the reference map. In the case of human chromosomes, there are different types of maps, each of which has a different resolution level, being more or less thick. A genetic map shows the location within a chromosome of DNA markers — genes or segments of DNA without any special function but polymorphic (i.e., with different sequences — which can be identified by their family transmission. The distance between markers is deduced from the frequency with which pairs of markers are transmitted from one generation to the next on the same chromosome or on different chromosomes; the idea is that the closer the markers, the more seldom they recombine onto different chromosomes.

The first human genetic map was published in 1987 (Donis-Keller *et al.,* 1987), but its resolution — namely, the average distance between a marker and the next one in terms of number of nucleotides — is not only approximate if referred to the number of nucleotides but also too large (more than 10 millions).

Short tandem repeat (STR) microsatellite polymorphic markers — very short and repeated DNA segments composed, for example, of two nucleotides, citosine and adenine, which are homogeneously and frequently distributed along the entire human genome probably without any functional significance but whose number of repeats is constantly inherited — have been identified. This identification has made it possible to construct a much finer, high-resolution map. Its most recent version, published by a French research laboratory (Dib *et al.,* 1996), has reached an average density equivalent to a microsatellite marker every 200,000–500,000 nucleotides in which the order of magnitude varies from one chromosome to the other. This result is considered sufficient for a genetic mapping of the human genome and the efforts of researchers are now directed at the construction of its physical high-resolution mapping.

Physical High-Resolution Maps

Unlike genetic maps, whose measure of the distance between markers is expressed in terms of the frequency with which pairs of markers segregate together in the meiosis, physical maps specify distances measured directly in number of nucleotide bases between reference points along each chromosome. The resolution that can be obtained for these maps is also at different levels. Cytogenetic maps describe the localization of DNA markers (including genes) in relation to reference points (bands or breaking points) visible on the microscope or physically hybridized with DNA segments marked with a fluorescent molecule so as to visualize the position of the marker to be mapped on the chromosome through a fluorescence microscope. The resolution of the map obtained through this procedure (called fluorescent *in situ* hybridization; FISH) varies from 1 million bases (1 Mb) in chromosomes in normal metaphase to a few thousand bases if one manipulates the chromosomes by diminishing their condensation.

By digesting the genomic DNA or the DNA of a specific chromosome with restriction enzymes, cutting sites locate a map (restriction map) which will be more or less thick according to the frequency of the cutting sites and therefore according to the battery of selected enzymes. As in the case of genomic libraries, the most difficult technical problem is not so much that of cutting the cloned fragment [e.g., into artificial yeast (YAC) or bacterial (BAC) chromosomes with inserts of the megabase order] into ever smaller fragments (to be read by a sequencer which cannot read more than 1 kb of DNA at a time) as, once the DNA has been fragmented and sequenced, to reconstruct the order of the nucleotides on the longer fragments from which one started (assembling process) so as to obtain the sequence of the DNA contained in the entire chromosome. For this purpose, the libraries of chromosome clones are arranged in blocks of DNA *contig;* that is, a group of DNA clones with DNA segments in common so as to cover redundantly and without gaps a DNA region too vast to be cloned in only one piece. The arrangement of contigs is essentially based on the possibility of identifying a common sequence of approximately 100 nucleotides for each couple; this sequence should be specific (i.e., it does not exist anywhere else in the genome) and identifiable through the polymerase chain reaction (PCR). These DNA sharing regions are called STSs and, being unique, can be reference points on physical maps (STS maps). The construction of contigs is constantly verified with the FISH technique, which visually associates the clones to specific regions of chromosomal bands.

The first high-resolution human genome map was published by the Centre d'Etudes du Polymorphisme Humain (CEPH) using a library of 33,000 YAC clones with inserts of about 0.9 Mb mapped by about 2100 STSs (Cohen *et al.,* 1993). This map, although still incomplete particularly in some chromosomes, has provided a main reference basis for the scientific community and for future research, especially because it was made immediately available on the web (*http://www.cephb.fr*). The same researchers published a map of YAC contigs covering 75% of the human genome consisting of 225 contigs 10 Mb long on average (Chumakov *et al.,* 1995).

Expressed Sequence Tag Maps

From the beginning of the Human Genome Project, it was debated whether one should carry out the task of sequencing indiscriminately all the nucleotide (3 Gb) constituting the human genome or whether it would not be preferable to start analyzing only that DNA fraction which, translated into RNA, codes for the synthesis of gene products expressed in specific tissues or organs. The length of this codifying DNA in man is about 1.5 kb on average, but, unlike other organisms, human genes are to be found only once (on average) every 40–50 kb of DNA; therefore, their DNA represents about 3% of the entire human genome, thus saving much effort by not sequencing the noncoding DNA. Those who want to pursue the entire project, to the simple assertion that the genetic information relevant for medical purposes is that which is expressed, object that it is not easy to find this information in all the tissues, in all the cell types, and during the various development stages. Also, with this selection one does not take into consideration the DNA sequences which do not synthesize specific molecules but somehow control their synthesis, the so-called regulating genes flanking the codifying genes in other organisms. They can be identified and mapped on the entire genome transforming the messenger RNA expressed in the various tissues and organs into complementary DNA (cDNA) through the enzyme inverse transcriptase. From human cDNA libraries, constructed in a similar way to the DNA libraries, one can generate specific STSs for every coding DNA sequence: These specific STS are called expressed sequence tags (ESTs), which serve as markers. Their collection, characterization, and relative mapping on the individual chromosomes have been one of the most successful cooperative research efforts of the past few years. Many projects grouped under the Integrated Molecular Analysis of Gene Expression consortium made it possible to sequence tens of thousands of cDNA fragments (each 200–300 nucleotide bases long) from clones derived from messenger RNA found in the brain, the body

TABLE I

The Human Genome in Numbers (1998)[a]

CHROMOSOME	MAPPED GENES	IDENTIFIED GENES STILL TO BE MAPPED	TOTAL GENES	TOTAL EST	TOTAL STS	GENE MAPPING EST (= STS)	MICROSATELLITES	OTHER POLYMORPHISMS	TOTAL POLYMORPHISMS
1	593	1,201	1,794	41,992	5,820	3,114	29	426	1,255
2	327	969	1,296	28,009	4,378	2,257	779	280	1,059
3	273	765	1,038	22,129	4,307	2,015	728	649	1,377
4	213	512	725	17,069	5,339	1,478	605	378	983
5	232	592	824	19,656	3,731	1,529	661	344	1,005
6	332	648	980	22,760	3,608	1,893	646	384	1,030
7	304	647	951	21,636	4,973	1,594	574	428	1,002
8	178	491	669	18,678	3,021	1,206	613	327	940
9	223	502	725	19,991	2,382	1,248	471	154	625
10	192	532	724	14,446	2,720	1,371	519	253	772
11	372	666	1,038	37,260	5,075	1,755	560	643	1,203
12	313	576	889	22,947	3,269	1,585	553	201	754
13	93	244	337	5,707	2,053	703	364	150	514
14	178	415	593	20,014	2,027	1,047	358	185	543
15	150	407	557	15,713	1,955	1,029	312	207	519
16	207	355	562	13,926	2,077	849	374	285	659
17	351	567	918	22,427	2,981	1,263	405	478	883
18	83	236	319	6,116	1,420	523	291	77	368
19	354	428	782	14,408	1,686	1,114	305	241	546
20	123	349	472	11,008	1,545	758	310	105	415
21	92	89	181	3,441	1,348	305	221	178	399
22	144	232	376	13,515	1,476	565	162	216	378
X	417	362	779	15,125	4,719	874	491	577	1,068
Y	30	23	53	926	421	106	32	125	157
Total	5181	10,607	15,788	428,900	66,511	30,181	10,334	6,865	17,199

[a] Source: *Genome database,* Johns Hopkins University, *http://gdbwww.gdb.org/;* further updating on *http://www.ncbi.nlm.nih.gov/genemap.*

muscles, the liver, the heart, and the lymphocytes. A very large number of ESTs have thus been classified; they identify and map thousands of genes on the individual chromosomes because generally several ESTs correspond to a single gene.

In 1998, the map of ESTs available for the 24 human chromosomes was integrated by the previously mentioned microsatellite high-resolution genetic map; each EST, being a specific STS, can start a PCR reaction and its association with other genetic markers can be easily identified. One of the most important goals was the physical mapping of 30,181 human genes, corresponding to about half the expected ones (Deloukas *et al.*,1998), which closes the first phase of the Human Genome Project — namely, of having available a sufficiently detailed physical map of each chromosome.

All the information dealt with so far and updated in 1998 for each human chromosome is summarized in Table 1. In Table 2, information is given on the sequencing of the genomes of some nonhuman organisms whose organization can give useful insights for the sequencing of the human genome. The number of microorganisms for which the sequencing of the genome has been planned is increasing particularly because of the great interest that industry has shown in this sector.

Short-Term Objectives

Under the title *New Goals for the U.S. Human Genome Project, 1998–2003,* the heads of the two major American research institutions which financed the Human Genome

TABLE 2

Genome Sequencing in Nonhuman Organisms[a]

ORGANISM	GENOME SIZE (Mb)	SEQUENCED GENOME (%)	YEAR	GENES IDENTIFIED (ESTIMATED)
SV40	0.005	100	1978	
Mitocondrial DNA	0.0166	100	1981	37
Saccharomyces cerevisiae	12.1	100	1996	6,220
Escherichia coli	4.6	100	1997	4,288
Bacillus subtilis	4.2	100	1997	4,000
Synechocystis sp.	3.6	100	1996	3,168
Archaeoglobus fulgidus	2.2	100	1997	2,471
Haemophilus influenzae	1.8	100	1995	1,740
Helycobacter pylori	1.7	100	1997	1,590
Methanococcus jannaschii	1.7	100	1996	1,692
Borrelia burgorferi	1.3	100	1997	863
Mycoplasma penumoniae	0.8	100	1996	677
Mycoplasma genitalium	0.6	100	1995	470
Methanobacterium thermoautotrophicum	1.8	100	1987	1,855
Caenorhabditis elegans	97	100	1998	19,100
Drosophila melanogaster	165	9		(100,000)
Mus musculus	3000	0.2		(60,000–70,000)
Homo sapiens sapiens	3000	6		

[a] Updates at *http://www.ncbi.nlm.nih.gov/Entrez/Genome/org.html.*

Project—the National Institutes of Health and the Department of Energy—present the ambitious schedule for the completion of human genome sequencing by 2003, 2 years ahead of the previous projections (Collins *et al.*, 1998). In the following sections, I describe some of the main objectives, which are relevant because of the effort of international cooperation (not only American but also at least 30–40% European) to which they necessarily appeal. Because of the special context of this contribution the ethic, legal, and social implications of the whole project will be discussed later.

Sequencing the DNA of the Human Genome

In 1998, the genetic and physical genome maps were completed at a sufficient resolution (one genetic marker on average about every 500,000–700,000 DNA nucleotides with more than 50,000 mapped STSs) to be able to move on to the following sequencing phase with an error frequency not higher than 1 nucleotide out of 10,000. The aim is to sequence one-third of the genome by 2001 and to complete the work by the end of 2003. The objective seems very ambitious: In effect, at the end of 1998 180 human Mb and 111 nonhuman Mb were sequenced, at a rate of about 90 Mb per year. The projection of an exponential growth of the working capacity is certainly favored by the policy adopted by the whole scientific community taking part in the project of making the results immediately and freely available on the web. In other words, it was agreed that each even partially "assembled" sequence more than 1 kb long should be released in a data bank accessible to the public within 24 h of its generation.

The quantity of sequences produced each year has an exponential growth. From a rate of about 90 Mb per year at an approximate cost of 0.50¢ per nucleotide (1998), one forecasts to be able to develop innovative automation, miniaturization, and robot systems which can analyze 500 Mb per year at half the cost, namely, 0.25¢ per nucleotide.

Analysis of Genome Variability

An individual's genome is different from that of another, and this diversity is at the basis of the biological process of evolution. The most common sequence variations are those for which one nucleotide in the DNA is replaced by another nucleotide; these variations, which have an average frequency of about 1 every 1000 nucleotides, are called single nucleotide polymorphisms (SNPs). An analysis of the types, frequencies, and distribution of these polymorphisms both at the individual level and at the population level is particularly important in the study of not only human evolution but also diseases whose origin or predisposition are genetically controlled. SNPs, being stable, abundant, and more or less homogeneously distributed along the genome, lend themselves to automatic analysis on a large scale and will certainly be precious markers for the study of complex diseases, such as cancer, diabetes, and cardiovascular and mental diseases. Several genes, each one on a reduced scale, could contribute to their risk. Very sophisticated technologies are being elaborated for a fast and large-scale identification of SNPs on the entire genome. The objective has also been established to produce a map of at least 100,000 SNPs, 1 every 30,000 nucleotides, by the end of 2003. The April 15, 1999, issue of *Nature* published the news that 10 pharmaceutical companies, together with the Cambridge Sanger Center (Great Britain) funded by the Wellcome Trust (a British philanthropic association), set up a consortium in order to fund a $45 million project for the mapping of the genomic variation associated with the most common diseases through the identification of 300,000 SNPs. An interesting condition set by the Wellcome Trust for the realization of this cooperative effort is free access without any conditions to the map during its construction; the idea is that in these precompetitive projects the quick and immediate diffusion of information acts as a catalyst for multiplying inventions and resulting applications, which are much more profitable in the long term.

Functional Genomics

The knowledge of the structure of a gene or of its components is a fundamental though not exclusive aspect of the project of sequencing the entire human genome. The next step is to learn its function deriving from the interaction of each individual's genome with its environment, first inside the organism under normal conditions and then in contact with other organisms, such as those pathogenic microorganisms to which our genome is appointed to react more or less successfully. As a consequence, the topics related to the study of functional genomics are a further bet of human genome analysis. These studies mainly concern the comparison of differently structured DNA sequences so as to define their relevant aspects for the functions of genes, large-scale messenger RNA analysis in the different tissues and of expressed proteins, the different experimental strategies to deprive genes of their function or to alter their function in animal models, and the strategies for creating libraries of noncoding DNA sequences responsible for other functions such as the control of the gene function, keeping the chromosomal structure, and replication and recombination.

Unfortunately the technology for cloning the full-length human cDNA (the ESTs previously described represent fragments impossible to arrange in contigs to obtain the respective full-length cDNA) or for the identification of rare transcripts is not beyond the experimental phase. So far, the only available full-length cDNA bank is limited to the organism *Saccaromyces cerevisiae* with approximately 6200 transcripts composing its genome. It is in the development of these technologies that the genome project is planning to use its resources in the next few years. Fortunately, the necessary technology for the simultaneous parallel interrogation of the expression profile of a complete repertory with high sensitivity and recognition fidelity (in any case a very high number of transcribed genes) was given a boost in 1996–1998 by the possibility of miniaturizing and reading automatically thousands of hybridizations in parallel. The method developed by P. Brown and collaborators since 1996 (Brown and Bostein, 1999) is based on the robot-guided microdeposition and on the consequent cDNA chemical immobilization in microarrays arranged on 2 × 2 microscope slides which can contain 10,000 cDNAs. The probes for the interrogation of these microarrays consist of fluorescent cDNA with different colors. The hybridization is read on the fluorescence focal microscope. Figure 2 was obtained by reading on the microscope the microarray containing the complete 6200 cDNA of *S. cerevisiae* yeast genome interrogated with 2 cDNA probes derived from mRNA extracted from two different growth conditions.

Another method developed by the American company

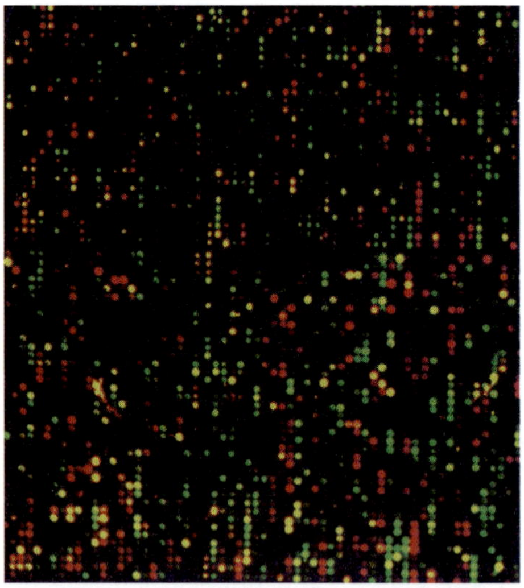

FIGURE 2 Illustration of the fluorescence obtained by reading with the focal microscope the microarray containing the complete 6200 cDNAs of the *Saccaromyces cerevisiae* yeast genome. The cDNAs have been studied with two fluorescent cDNA probes derived from mRNA extracted from two different growth conditions (in green, transcripts induced by the growth in the presence of glucose for 12 h; in red, transcripts induced by the growth in the absence of glucose; in yellow, transcripts both in the presence and in the absence of glucose) (reproduced with permission from Schneider, 1998).

Affymetrix differs from the previous one in that the microarray does not contain cDNA sequences but synthetic oligonucleotides. These microarrays are called DNA chips because the technology used for their construction is based on photolithography, commonly used to draw semiconductor microcircuits. The technique for "printing" oligonucleotides with a known sequence on a glass support makes it possible to arrange in a 1.28×1.28 chip up to 400,000 oligonucleotides with different sequences, each 25 nucleotides long. Also for this technique, the probe is marked with fluofores, enabling one to visualize and quantify the hybridization through a fluorescence focal microscope. In Fig. 3 the microarrays from the two technologies are shown, as is the interrogating procedure used to obtain the gene expression profile of many genes. Figure 4 illustrates a biochip used to identify DNA mutations in two genes of the HIV-1 virus (Lipshutz *et al.*, 1999) and the strategy for identifying mutations and polymorphisms in a DNA or cDNA sequence.

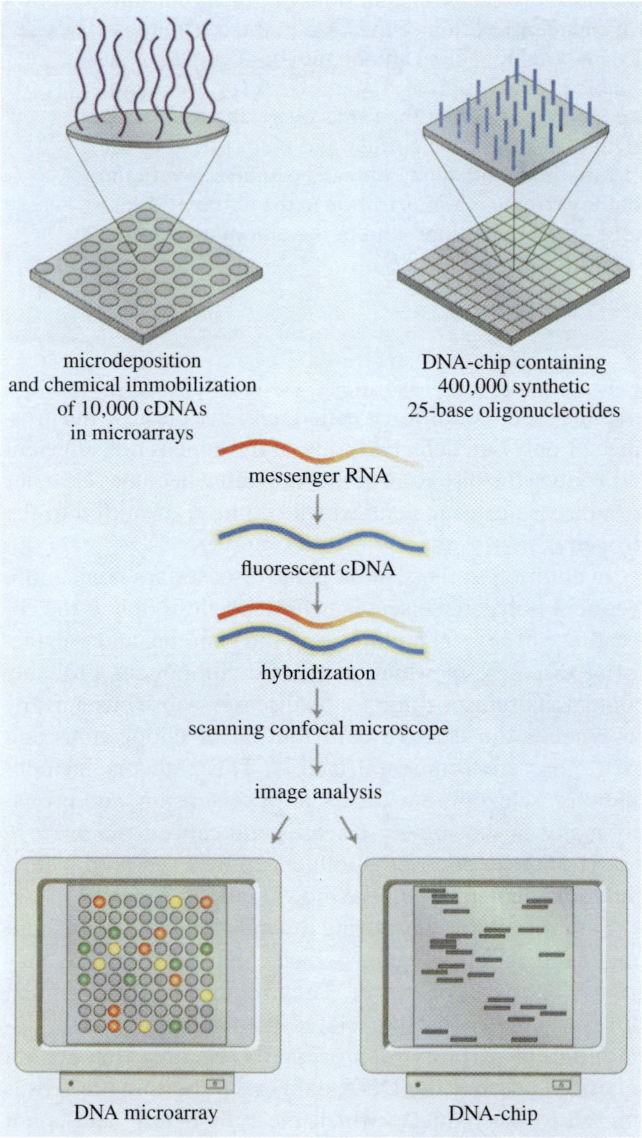

FIGURE 3 Simplified scheme of the two procedures followed to obtain cDNAs and DNA chip microarrays (adapted from Schneider, 1998).

Comparative Genomics

All organisms are connected by a common phylogeny; therefore, an important part of the project has been devoted to the study of nonhuman organism genomes (Table 2). The comparative analysis of genomes could provide useful patterns for the study of universal biological mechanisms and useful information for the analysis of more complex information. At the same time, the analysis of less complex genomes than those of the human could be the basis for pilot experiments aimed at probing on a reduced scale the validity of technological innovations to be subsequently and systematically applied on scales of a slightly larger order of magnitude. The intention is to complete the sequence of the *Drosophila* genome by 2003, whereas the completion of the murine genome sequence is set for the Year 2005.

Bioinformatics

Informatics is an essential support for the elaboration of this project. There are at least five subobjectives which the scientific community deems necessary to submit to the analysis of its information specialists: improving the contents and the usefulness of databases, making their structure more flexible; developing faster algorithms to generate, interrogate, and annotate the sequence data produced in ever increasing amounts; adapting the existing databases to the functional studies; developing statistically more "intelligent" algorithms to represent, analyze, and foresee similarities and differences between sequences; and creating support mechanisms promoting robust and flexible operative systems for producing easily transferable software potentially shared by users in different laboratories.

Education and Training

This project has made it necessary to train new professionals on the verge of scientific disciplines proper (biology, medicine, chemistry, mathematics, physics, informatics, and engineering—to which the prefix "bio" can be associated) and of those disciplines which are improperly (because they indeed all are) called social (law, ethics, communication, and sociology). The integration between disciplines, and as a consequence between languages with such different histories, is far from easy. Therefore, a new term, *genomics,* was created; included in this word is the necessity of a complex interdisciplinary competence covering skills which meet and cross in different technologies from bioinformatics, of which there are very few experts, to patent jurisprudence, which must be considered in an international

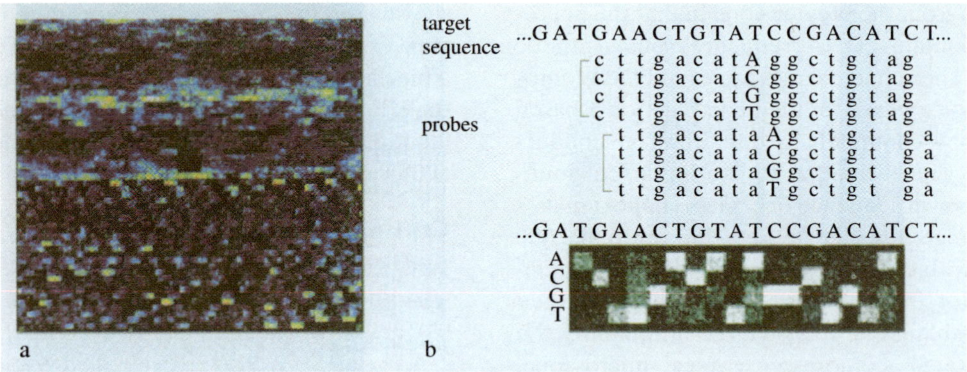

FIGURE 4 Microarray for the analysis of DNA or cDNA sequences. (a) Image of a microarray for typing the HIV-1 virus hybridized with fluorescent copies of the viral PR (protease) and RT (reverse transcriptase) genes. The different colors identify the mutations and their localization along the sequence. The microarray measures 0.8 × 0.8 cm. (b) Interrogation strategy for the identification of DNA (or cDNA) mutations or polymorphisms. Four oligonucleotides of 17 bases each are used as probes for the identification of the base in the central position of the probes. The probe giving the stronger fluorescent signal (lighter color) among the four devoted to interrogating the target sequence is the one forming the most stable hybrid (duplex), T (target)–A (probe) in this specific case. The following nucleotide in the target sequence is interrogated in the same way using four other probes, which have the same length in this example. The length of the oligonucleotides and the position of the target base can change. In order to increase the precision of the method, both filaments can be analyzed with the same procedure. The effectiveness of the technique consists of the extreme miniaturization in the automatic deposition of oligonucleotides, in the parallel analysis of the probe–target reactions, and in the automation of the confocal microscope reading (reproduced with permission from Lipshutz *et al.,* 1999).

frame. The globalization process involving the economy of developed countries cannot but influence education and training processes which must overcome the traditional borders between disciplines. The academic world must restructure its teaching and research programs; it will have to offer more incentives to the students to cultivate this new discipline, which currently is monopolized by a few universities often cofinanced by private companies as well as by many small industries which are often the only resource for immediate employment for the new professions but whose aim is obviously profit and not training. Finally, one should not forget the problem concerning the best way to inform public opinion about this new discipline: Social and political impact in this field often produce emotional reactions both because of incorrect information and because, especially in Italy, scientific training is not yet considered an important economic and cultural investment factor.

Problems and Perspectives

Human Genome and Hereditary Diseases

Currently, more than 6500 genetic diseases are known that are caused by the alteration of individual genes (monogenic), many of which are lethal. Some of them only appear when both copies of the gene inherited from the parents are defective (i.e., homozygous with regard to the mutated gene), including cystic fibrosis, sickle-cell anemia, and thalassemias. The condition is called *recessive* because the presence of only one defective copy of the gene is not sufficient to produce the disease, but the individual becomes a carrier of a disease-causing gene which can be transmitted to the progeny.

In addition to these monogenic diseases are much more frequent polygenic diseases, whose origin is due to the coexistence of several mutations in the genome, and polyfactorial diseases, in which a genetic component probably controls a predisposition to the disease, whose onset in any case needs the decisive contribution of endogenous and exogenous environmental factors. This category includes diabetes, schizophrenia, arterial hypertension, and probably many cardiovascular diseases and cancer; the onset in adults of these diseases is influenced by a series of factors generally dependent on the individual's lifestyle.

Somatic alterations during the individual's development under the action of agents which can alter the DNA may trigger the onset of cancer. They require the involvement of several genes, mostly related with cellular growth, differentiation, and survival processes. Generally, they are not inherited because the DNA damage in the somatic cells is limited to the tissue to which the cells belong and is not transmitted to the following generations. However, in addition to cancers due to somatic mutations, cases of hereditary predisposition to the onset of cancer have been observed, although more rarely; these are due to the transmission of defective copies of certain genes (oncogenes) responsible for the neoplastic transformation or involved in the mechanisms of DNA repairing.

From the Diagnosis of the Disease to the Diagnosis of Disease Predisposition

The progress of knowledge of the human genome will broaden and accelerate genetic research, thus enabling us to extend genetic analysis not only to the diagnosis of an increasing number of genetic diseases but also to the detection of genetic predisposition to polygenic or polyfactorial pathologies. Although not properly genetic diseases because they are influenced by the presence of environmental factors, these have a hereditary transmission; in fact, they preferably affect individuals with a specific genotype. A typical example of a disorder whose onset can be predicted through a genetic test is Huntington chorea, which has a 1:20,000 frequency in European populations and appears at an age of onset between 30 and 50 years with a progressive loss of motor control and with neurological (dementia) and psychical (depression) problems until death usually 15 years after the diagnosis, often by suicide. It is a terrible disease because of its late onset and its slow and fatal progression. It is caused by a loss of neurons in the brain, and it was among the first to be mapped on chromosome 4p through linkage analysis using DNA restriction sites (Gusella *et al.,* 1983). The gene was cloned in 1993 (Huntington's Disease Collaborative Research Group, 1993) and the DNA sequence allowed the observation in the affected people of the presence of CAG trinucleotide sequences repeated several times inside the gene. From the number of triplets it is possible to predict exactly whether and when (the onset age is inversely proportional to the number of triplets) a carrier will develop the disease.

The analysis of a complex gene such as that responsible for cystic fibrosis, another lethal disease (people survive on average until the age of 30) mapped on chromosome 7q in 1985 and cloned 4 years later (Kerem *et al.,* 1989), was much more difficult. In fact, although 70% of patients present a specific mutation in the gene coding the protein responsible for the disease, a negative test for this mutation does not completely exclude cystic fibrosis because this disorder is associated with more than 600 different mutations within the same gene.

Early diagnosis by genetic tests is also possible for other monogenic diseases, whether dominant autosomic (myotonic dystrophy, Marfan syndrome, type 1 neurofibromatosis, and osteogenesis imperfecta), recessive autosomic (phenilchetonuria, β-talassemia, α-1-antitripsin deficiency, and sickle-cell anemia), or linked to the X chromosome (Duchenne and Becker muscular dystrophy and hemophilia A and B).

It seems much more difficult to predict the evolution of common diseases, such as certain neoplastic processes and cardiovascular pathologies, given that mutations in several genes contribute to their onset. It is known, for instance, that the risk of developing a colon cancer increases if there is a mutation in five different genes, whereas a mutation in at least two separate genes determines a predisposition to the cancer of the breast. One woman out of 10 in the Western world is liable to be affected by this disease by the age of 85, with a mortality rate of 25%. However, only 5% of cancers of the breast are hereditary (Lynch *et al.,* 1984), and only 80–85% of them can be associated with mutations in the two previously mentioned genes, BRCA1 and BRCA2 (Ford *et al.,*1998). As a consequence, specific genetic tests for BRCA1 and BRCA2 will not be entirely predictive for breast cancer, both in its hereditary forms and especially in the nonhereditary ones.

The most immediate applications from the analysis of the human genome mainly concern the possibility of a prenatal diagnosis to identify genetic mutations responsible for specific hereditary diseases detectable at birth or in the neonatal period. It is possible to obtain a diagnosis through amniocentesis only after the 15th or 16th week; however at Weeks 8–10 this can be done through a biopsy of the chorionic villi, but this presents more risks of complications than the previous technique.

Much more complex are problems due pre- or postnatal genetic diseases with a late onset — namely, those which will express themselves in adults. In fact, the prescription of a genetic test would be correct and necessary if there was an adequate therapy or if it was at least possible to modify the evolution of the disease with a timely medical treatment, reducing its possible complications. On the other hand, it seems questionable for all those cases for which no therapy is available; it could cause the individual, when still healthy, more damage than real benefit. Therefore, the forthcoming of a molecular predictive medicine requires a reconsideration of benefits and damages of medical science and consequently of the ethical principle of benefit.

If the right to know one's own genetic type cannot be renounced, it should also be possible to admit the right not to know it, particularly in those cases in which a preventive knowledge of the disease would only cause an anticipation of psychological suffering without any concrete therapeutic advantage. The case of subjects affected by the aforementioned Huntington chorea is exemplary. In fact, in these cases family analysis can meet with severe obstacles, for instance, if it is impossible to obtain one or more samples from the progeny or if the mother wants to know the risk of disease for her fetus but not her own (principle of autonomy).

Molecular medicine introduces a new kind of doctor–patient approach in the prognostic phase. Genetic tests do not always identify the presence of a given disease, even at an early stage of its evolution, but rather the presence of the mutation of a gene which can determine the onset of the disease. This condition could be described as predisposition, susceptibility to the disease, and potential or probable risk. The ability to predict a genetic susceptibility to develop specific pathologies will create a new figure in the medical world: the unpatient (Jonsen *et al.,* 1996), who will not be considered a patient in the traditional sense of the term, not being under medical treatment, but will not even be considered a completely healthy person free from any pathological condition. The possibility of predicting with certainty that a subject is liable to develop certain pathologies, when he or she is still perfectly healthy, may also imply a discrimination of the individual in various fields of his or her everyday life. The need therefore arises to protect single individuals from a bad use of genetic information, avoiding their leading to social behavior characterized by discriminating attitudes which limit individual freedom and rights.

Prediction of Work-Related Diseases

The identification of an increasing number of genetic markers also makes it possible to apply it in working environments, particularly in terms of secondary prevention of diseases related to a working activity. Actual cases are exemplified in the rich literature on the subject (U.S. Congress, 1990a): An example is the ethical problems arising from the genetic screening recommended for selecting workers assigned to work with potentially cancerogenous chemicals. It is possible that some individuals have a different susceptibility to the risk of cancer when exposed to these substances because they are provided with metabolic enzymes which can reduce their possible toxicity more or less rapidly. In order to test the validity of this hypothesis, it is necessary to ascertain the relation of causality between metabolic phenotype and cancer in different populations, with an acceptable percentage of assignment errors in carrying out the correspondence between genetic mutation and metabolic phenotype of the subjects (test sensitivity and specificity). An important preliminary condition for a correct interpretation of these correlation data is the proof that the metabolic phenotype is independent from the clinical state or from other prognostic factors of cancer.

If the causality relation between phenotype and cancer is statistically proved, a genetic screening aimed at identifying the susceptibility to become ill in the working environment is ethically advisable only if the objective is the protection of the worker's health and if it respects the following principles: autonomy of decision, namely, being free to choose whether or not to be tested and freedom in choosing the job, once one has obtained full information on the nature of the potentially cancerogenous exposition and on the limits of the test; benefit, because obviously the worker would benefit much more from nonexposition to the risk before even finding out his or her state of genetic susceptibility — as a consequence, the employer has the moral obligation to avoid the use of cancerogenous substances in the worksite before activating any genetic screening program; and justice, because the distribution of genetic polymorphisms can vary from one population to the next and the less susceptible groups could be preferred to the others, thus creating a discrimination in the work access.

Lappe (1983) suggested a list of ethical criteria which should enable researchers to identify genetic tests safeguarding the workers' rights. They include an attainable goal, the active participation of workers undergoing the tests, equal access opportunities, effective execution protocols, no obligation, informed consent, protection of the individuals' health, access to information, counseling and follow-up services, proven connection between tests and therapy, and protection of the right to privacy regarding the results from the tests.

Genetic Predisposition in Human Behavior

Until the 1980s, the methodology for the study of behavioral genetics was based on formalism elaborated by quantitative characters genetics (Falconer, 1981) applied to a wide range of behavioral characters: cognition skill such as intelligence, aspects of the personality such as emotionality, and diseases such as schizophrenia and depression. The objects of the analysis were families, twins, and adopted children and the analysis consisted of studying the differences between individuals in an attempt to separate environmental and genetic factors. The interactive or additive effect of the various genes contributing to the variation of the behavioral character in a population was taken into consideration without distinguishing the contribution of each gene and was associated with a "heritability" index in which it was very difficult to identify the part of variability contributing to the genetic transmission alone without confusing it with the cultural transmission prompted by the environment.

During the following 10 years, the integration of molecular genetics methods with animal models made it possible to obtain a finer resolution of genetic loci which can be involved in the expression of an abnormal behavioral trait. The identification of loci controlling quantitative traits (quantitative trait loci) was developing very rapidly, thus enabling the planning of research projects with the goal of mapping genes of susceptibility to complex diseases, such as schizophrenia (Moises *et al.,* 1995) and emotional lability (Flint *et al.,* 1995). The increasingly refined identification of susceptibility genes will make the specific analysis of the genetic component of complex traits more realistic. Collaborations on national and international scale have been activated for the study of some pathological behavioral traits such as schizophrenia, manic depressive psychosis, alcoholism, and autism; there is now a rich literature documenting research progress in this field (Plomin *et al.,* 1994; Rose, 1995; Cloninger *et al.,* 1996; Sherman *et al.,* 1997).

Genetic Tests, Personality, and Behavior

Geneticists studying behavior are well aware that the object of their research can raise social and ethical concerns, and most of them do not conceal the serious responsibilities of the scientific community for not having prevented — on the contrary, for having sometimes favored — in the past an extreme eugenic trend in Great Britain and in the United States, which culminated in the infamous attempt of German Nazism to pursue an impossible racial purity also with regard to personality and behavior. Some geneticists even question whether research in this field should not be stopped until the ethical problems associated with behavioral genetics are settled. This attitude, however, does not allow for the fact that ethical problems have a continuous evolution simultaneously with innovation in knowledge and technology: The possible answers to today's ethical problems do not necessarily solve or avoid tomorrow's problems.

One could more radically wonder why one should study the genetic influence of behavioral traits, given that the results could raise delicate social and political questions and could be used to give a misleading rationale for discriminations and abrogation of social programs. The answer could be that the analysis of behavioral genetic and nongenetic traits helps us better understand the appearance of certain individual behaviors. The knowledge of data by public opinion and the formulation of programs and of political projects should not refer to science-fiction myths but to those results which serious scientists are gradually accumulating. As is often the case in Italy, it is a question of communication between those who carry out this kind of research, those who render it public, and the people who can use its results, all problems which can acquire especially important ethical aspects.

There are at least two aspects with regard to which the analysis of genetic influence on personality and on human behavior can raise ethical and social concerns and therefore require special attention: genetic counseling and the study of populations or groups of individuals, defined or otherwise selected, within the same population.

With the increasing knowledge of genetic components of complex behavioral traits it is to be expected that many parents will inquire about the possibility of influencing the genotypes of their future children so as to ensure the genotype they prefer. The task of geneticists will become even more complex than it is with regard to the traits determined by a single gene. On the one hand, they will have to respect the choice of the people consulting them; in making decisions concerning their family, people exercise their right of autonomy. Geneticists will have to give them full and updated information, suggesting at the same time that this information might change within a short period of time. On the other hand, they will have to make it clear that even if one or more genes contributing to the determination of a complex character have been identified, the knowledge of a single gene has a limited predictive validity with regard to the relevant phenotype. Also, it is not easy to explain, to people who are anxiously waiting for firm answers to complex problems, the concept that each gene, if individually considered, can have a limited effect on the trait and also the subtle statistical reasoning suggesting that it may be useless to probe the presence of a single gene unless the interaction with the other relevant genes is sufficiently known.

The genetic differences between individuals within the same population are the condition without which Darwin's evolution cannot exist. These differences include variations in physical characteristics, personality, intellectual and artistic skills, and the inclinations which are often the object of a strong social interest. The average values of these characters often differ among people depending on sex, age, interests, work, geographic area, and many other criteria. These differences often cause well-rooted ideological distortions; there is a common tendency, in fact, to exaggerate their significance. A slight average difference between two groups with regard to a certain character is interpreted as if almost all the individuals of one group surpass almost all the individuals of the other group with regard to that character. This seldom happens with simple or complex genetic characters, behavioral and nonbehavioral, that have been measured in man. The tendency of many people to exalt the differences between groups and the injustices which can derive from this attitude often induce, even on the part of the scientific community, the extreme opposite tendency — that of asserting that these differences do not exist. It is preferable to adopt a different strategy: an accurate value analysis of the differences between the interested groups and the predictive strength of these differences (normally modest) for the individuals and also the education of public opinion and media to a balanced interpretation of these differences. Consider, for example, the case of a personality trait such as emotional stability in men and women, a behavioral aspect which is often analyzed even by nonspecialists. The assumption that all men are emotionally more stable than women is a mere stereotype. With regard to this character, as for almost all behavioral characters, the variability within each group is much greater than the variability indicated by the average difference between the two groups divided by sex which would suggest a slightly greater stability in men. This means that the knowledge of the sex has an almost null value in determining an individual's emotional stability. However, the stereotype resists:

One of the reasons is that even a slight difference between the averages of a character in two selected groups can become a disproportion between the two groups, particularly if one considers only the extreme values of the character. As is known, stereotypes are based on the observation of extreme values.

The Knowledge of the Human Genome and the Insurance System

Progress in the knowledge of the human genome risks greatly distorting the insurance system, which is notoriously based on the principle of risk sharing among the people insured. The possibility of ascertaining, through genetic tests, the presence of an individual's predisposition to get certain diseases, and generally to know an individual's genetic profile and an indication of his or her future state of health, has raised the problem of the possible use of these tests for a more balanced evaluation of the individual risk and a more adequate reformulation of actuarial calculations on life expectancy. However, this would mean that people who are to be insured would have to subject themselves to genetic tests and the results would be communicated to the insurance companies; this would be a clear violation of individual privacy, which would only be carried out with the consent of the person to be insured.

The question regarding the right of companies to have access to the results of screening tests was dramatically raised by the diffusion of HIV infection among the population. There followed an inflamed debate between the insurance world, which claims access to genetic information for a more balanced calculation of premiums, and those who stress the danger of an uncontrolled access to these data and the necessity of preserving the right of each individual to privacy with regard to his or her genetic condition.

In the United States, the National Institutes of Health and the Department of Energy Working Group on Ethical, Legal, and Social Implications (ELSI) of the Human Genome Project and some political groups maintain that insurance companies should not have access to genetic information (Pokorski, 1997). Other institutions, including an organization protecting the interests of individuals affected by genetic diseases in Great Britain, believe that this request is unrealistic, particularly if one takes into account the progressive diffusion of genetic tests, but they ask that their use in the insurance field be controlled by state laws (Masood, 1996).

The European Parliament, when dealing with the problem of genetic privacy, suggested that companies should be forbidden, through ad hoc laws, the use of the results of genetic tests. However, the juridical scenario is currently rather heterogeneous. In Italy, Germany, Japan, Spain, Portugal, and Great Britain, there are no laws governing the use of genetic tests and of their results by insurance companies. In Great Britain, the association of British insurers does not consider genetic tests a necessary condition for an insurance policy, but it holds the right of access to previous genetic analysis (Wilkie, 1998). In France, genetic information can be used for purely medical or research purposes, but the norms in force have never specifically dealt with the question of genetic tests in the insurance field, although insurance companies have actually accepted a deferment in the use of genetic information.

In the United States, 10 states have forbidden the use of genetic information in health insurance policies, whereas there seem to be no restrictions in the field of life insurance. On the contrary, the possibility of determining the premiums through a more accurate evaluation of the individual risk is a strong incentive toward the acquisition of genetic analysis results. This is why the American Council of Life Insurance dissociated itself from the final document of a work group on genetics and insurance policies, organized by ELSI in 1993, urging companies to adopt a deferment on the use of genetic information (Masood, 1996).

The problem of genetic tests in the insurance field had a particular impact in the United States because of the peculiarities of the health system in force. As is known, it is based on the system of private insurance; in other words, citizens sign, of their own initiative, accident policies, insurance policies, and the so-called "health care" policies covering medical fees deriving from accidents or disease. These policies are made through cumulative contracts generally signed by employers, which pay a large part of the premiums, in favor of their employees. It should be noted, however, that the Ohio bill, voted for by the majority of representatives of that state in July 1993 and later approved by the U.S. Senate, explicitly forbids American insurance companies providing health care policies to use genetic information for this kind of policy. The Committee on Assessing Genetic Risk: Implications for Health and Social Policy of the National Academy of Sciences' Institute of Medicine adopted the same policy and it has also asked for a postponement in the use of genetic tests in insurance policies, except for life policies.

Identification through DNA for Forensic Purposes

The use of DNA polymorphisms in the legal field was initially welcomed with obvious enthusiasm; however, it became the object of criticism with regard to the limitations and causes of error of the techniques which were employed and because of the fact that such delicate analyses were carried out by laboratories without the necessary experience. An insufficient or inadequate DNA analysis in the course of investigations whose consequences are often decisive can lead to serious legal errors both in the civil field (e.g., erroneous attributions of paternity) and in penal proceedings; an effective control on the quality and standard of methods was obviously necessary. The debate on the validity of DNA analysis was particularly inflamed in the United States, in which the Office of Technological Assessment, which was charged with evaluating the disputed aspects of personal identification through DNA fingerprint (digital DNA print), confirmed the scientific foundation and reliability of the use of the DNA test in the forensic field (U.S. Congress, 1990b). Much attention was also given to the problem of identification through probability calculus and to the correct use of the frequencies of genetic traits with regard to both the reference population and the number of samples (Evett and Weir, 1998). It is known, in fact, that under favorable conditions it is possible to obtain results of exclusion of identity. These have the characteristics of an adequate scientific certainty if they are obtained with the help of several DNA polymorphisms. In cases in which the exclusion does not occur, it is impossible to state a full proved identity but only its probability through adequate statistical calculus on a database reliable in terms of the number of random individuals and of which specific ethnic group is sampled.

It is also necessary to consider the problems related to the setting up and management of national and international correlated databases which receive a regular input of information on the genetic profile of individuals subjected to DNA test. The access to these databases must be controlled in such a way as to prevent the use of information for purposes other than institutional ones, such as by industries or companies for the selection of personnel or by banks and insurance companies in connection with loans or life insurance policies.

In certain countries, particularly in the Anglo-Saxon world, criminal archives already exist in which the DNA profiles of people who have been accused or condemned for serious offences, particularly habitual ones, are maintained in a computerized and centralized system which can provide useful information for comparative checks in the case of investigation of biological traces. Some people believe that the storing of genetic profiles and their continuous comparison with those obtained at crime scenes are an abuse of individual rights; they therefore oppose the setting up of these archives. It is obvious, in fact, that faced with the prospect of individual DNA typing and their record, the needs of justice inevitably conflict with the right of any citizen to be adequately protected not only against possible legal errors or excesses but also against the unjustified violation of his or her personal freedom (Fiori, 1995).

Bioethical Implications

Research on the human genome cannot avoid raising ethical problems. The notion of responsibility may change: The partial or total genetic definition of a character, particularly if related to behavior (e.g., alcoholism), can render the boundary between pathologic and illicit behavior uncertain, with a range of intermediate shades to which the interventions of social protection and the legal tools must be able to conform.

However, the main problem connected with increasing knowledge of our genes is genetic discrimination. The origin of this danger is in the long interval between the prediction or the diagnosis of a genetic (or gene-related) disease and the practical possibility of adopting a therapy. If all diseases which have been diagnosed on the basis of genome knowledge could be cured immediately, the problem would not differ from the general one of a health policy which must evaluate costs, benefits, and effectiveness. The specificity of a disease caused or predicted by our genetic code but incurable can be discriminating for the healthy individual for whom the prognosis of the disease is made if this information is accessible to third parties (e.g., the employer or the insurance company). It can also be discriminating for the individual affected by the disease if access to health and social services is differentiated according to the therapeutic possibilities. Finally, it can be discriminating because there are few genetic counseling services in national health systems due to the high cost of training qualified personnel.

An attitude which throughout history has always led to genetic discrimination in its worse form — racism — is that by which it is believed that in the interest of society its members must be induced to preserve "beneficial" genes (positive eugenics) or to eliminate "damaging" genes (negative eugenics). This attitude, which leads to restrictions of reproduction behavior or to the adoption of selective abortion, has different justifications when it is the result of a responsible choice of a couple or when it reflects the safeguard of a public interest — something which is considered common good by the state. In this second case, the eugenic justification — an original sign of the scientific community which found fertile ground in political ideology — is based on a blunder: the belief of being able to change significantly the frequency of the "damaging" genes. In effect, even if one considers sterilizing all individuals affected by genetic diseases, it is possible to work out that the number of generations necessary to eliminate them is very high, on the order of centuries if not millennia. Omitting ethical considerations, it is very difficult to find a serious justification for a measure of social policy when its effects are projected into such a distant future.

The justification adopted by some jurisprudence, that the state has the duty to intervene to protects interests which individuals are unable to protect in matters of reproductive choices, is very delicate: It would make sense only if one wanted to accept that the state can decide better than the parents whether to have children and which ones. Such proxy would be so dangerous as to endanger the most elementary rights to freedom, but it meets with constraints and dramatic choices between conflicting values in matters of demographic policy (e.g., in China).

The Human Genome Project also raised other preoccupations: the fear that its results could lead to discrimination to the detriment (and stigmatization) of individuals and groups of individuals; the possibility of the scientific community not having free access to information regarding new discoveries for commercial reasons or because of the request of patents; the reduction of the human being to its gene sequences and the attribution of human social problems and other specific problems to genetic causes; the lack of respect for the values, traditions, and integrity of populations, families, and individuals; and an inadequate commitment of the scientific community to plan and carry out genetic research according to protocols and strategies open to the public.

The Human Genome Organization, a nonprofit international community to which belong scientists who carry out genomic research, has provided guidelines and procedures so as to render these preoccupations unjustified and to ensure certain ethical standards. It set four main principles as the basis of its recommendations (Knoppers *et al.,* 1996): The statement that the human genome is part of a common inheritance for humanity as a whole; the recognition of the international rules on the rights of man; the respect for the values, traditions, culture, and integrity of those who are the object of genetic research; and the acceptance and the claim for the principles of human dignity and of freedom. Such principles are an integral part of the Universal Declaration on Human Genome and Rights of Man adopted by UNESCO in November 1997.

Genetic Tests and Multifactorial Diseases

Here, I do not analyze in detail the ethical and social implications that the work, function, and behavior of the medical geneticist might have vis-à-vis an increasingly detailed knowledge of the human genome. However, one specific, though exemplary, case is illustrated: the responsibility of the medical geneticist at a time in the future when he or she will be able to prescribe routine genetic tests to evaluate the patient's risk of being affected by multifactorial diseases.

One of the reasons why genetics manipulating DNA and related genetic tests are raising hopes and preoccupations is the recent discovery of DNA mutations which, if associated with others or concomitant with unfavorable environmental factors, predispose to diseases of great social impact (at least in the Western world), such as tumors, cardiovascular diseases, and the various syndromes of senile dementia. They are very serious and frequent diseases; therefore, genetic tests for their prevention, diagnosis, and therapy interest many people. Also, with regard to the nonmedical but psychological, sociological, and cultural aspects of these diseases, the potential of genetic tests in a preventive phase is going to raise echoes and hopes of the public which impose a great sense of responsibility.

Since the 1980s, due to various experimental tests it has been possible to identify human genes (or their products) specifically altered in tumor cells: Among more than 6600 classified hereditary diseases, at least 340 hereditary traits predispose the onset of cancer as the main expression of the disease or as its complication. For more than 20% of patients affected by cancer, there is an oncological history in their relatives of first or second degree; between 1 and 5% of all tumors are associated with syndromes of a genetic nature. The tumor seems to be the consequence of the accumulation of mutations taking place in the DNA of somatic cells (but also, rarely, of germinal cells) in specific genes whose function is the control of proliferation, differentiation, and cellular death or in genes involved in DNA-repairing mechanisms. Currently, genetic tests enable us to obtain clinically useful information (for an alternative medical decision of proven effectiveness) in certain hereditary syndromes such as familial retinoblastoma (*Rb* gene), familial colon polyposis (*APC* gene), von Hippel–Lindau syndrome (*VHL* gene), and type 2 multiple endocrine tumors (*RET* gene). Genetic tests for the identification of nonsymptomatic carriers in families in which one of these pathologies is present are considered an essential part of a correct praxis of medical assistance.

There is a second group of hereditary syndromes predisposing to cancer for which some of the responsible genes have been identified (e.g., nonpolyposis-associated familial colorectal cancer, genes *hMSH2, hMLH1, hPMS1,* and *hPMS2;* familial breast and ovarian carcinoma, genes *BRCA1* and *BRCA2;* Li–Fraumeni syndrome, gene *tp53;* ataxia–telangiectasia, gene *ATM;* xeroderma pigmentosum, gene *XP;* and neurofibromatosi type 1, gene *NF1*). For this group, genetic tests should only be offered in a frame of genetic and clinical research; experimentation protocols are not sufficiently defined and have not shown a proven diagnostic effectiveness. In particular, the case of *BRCA1* and *BRCA2* genes associated with family breast cancer (Blackwood and Weber, 1998) must be emphasized because of the great social, psychological, and welfare impact associated with this form of tumor. There are many cases in the United States of asymptomatic women, although positive on the genetic test, who voluntarily subjected themselves to total preventive mastectomy; this is an exemplary indication of the ethical and social responsibility of genetic counseling as well as of the structures and the people involved in it.

If the presence of cancer genetic risk is ascertained, genetic counselors have the responsibility of discussing with the individuals at risk the advantages and limitations of all available options in terms of secondary prevention as well as the importance and the potentiality of a possible search of germinal mutations in genes controlling susceptibility to cancer development. The consent to carry out genetic tests must be informed; in other words, there must be consent deriving from a discussion with the geneticist that is adequate in terms of the capacity of understanding and the perception of the disease risk of the subject, in such a way that his or her choice among the available options (including the refusal of any test) is free and aware. The results of the tests must be communicated to the subject by discussing with him or her the meaning of the genetic analysis which has been carried out as well as advantages and limitations of the accessible options of medical care in light of the specific results of the test and also by suggesting further meetings so as to evaluate the degree of understanding of the supplied information and its psychological impact (Vineis, 1997).

There is a reductive tendency by some researchers, often encouraged by wealthy grants from pharmaceutical or insurance companies (especially in those countries in which they are charged for health services), to examine their equals essentially as "gene carriers." An example is provided by developments following the recent identification of genetic tests for the prediction of susceptibility to mental diseases, specifically those with a late onset such as Alzheimer's disease, which has become a major disease due to the increase in life expectancy of Western populations. Many genes have been associated with this invalidating disease; three (*APP, PS1,* and *PS2*) can contribute to the diagnosis associated with a family form with early onset, and one (*APOE*) is potentially predictive because one of its mutations is associated with a higher susceptibility. In fact (according to data updated to 1998), less than 2% of patients with Alzheimer's disease carry one of the diagnostic mutations (McConnell *et al.,*1998): Patients who have been clinically diagnosed find their diagnosis confirmed by the test; the only beneficiaries from a psychological standpoint are the relatives for whom the test is negative, whereas the relatives with a positive test are in the condition of knowing their future without having available effective preventive measures or possible therapies at the onset of the disease, apart from some measures to postpone the first symptoms.

For genetic tests of a predictive kind, such as that for the mutations of the *APOE* gene, there are still too many false-positive and false-negative reactions introduced in the laboratory routine. The ethical responsibility of those who want to use these tests is further complicated by the fact patients suspected to have Alzheimer's disease may have lost the capacity of consenting to the genetic test. Any action taken by the genetic counselor on the basis of the information from a genetic test on the person affected could have implications for his or her relatives. As a consequence, a responsible approach to the elaboration of a decision on behalf of a person affected by dementia should be to appoint a single person to represent those who give their consent but also include in the decisional process the consent of the relatives who could claim their right not to know the results of the genetic test. The most balanced attitude with regard to genetic tests, particularly if aimed at predicting mental diseases, is to discourage their widespread use until their medical benefit is apparent for individual patients. This does not mean eliminate it from genetic experimentation because the identification of genes associated with these diseases could produce more effective pharmacological treatment (Nuffield Council on Bioethics, 1998).

References Cited

BLACKWOOD, M. A., and WEBER, B. L. (1998). BRCA1 and BRCA2: From molecular genetics to clinical medicine. *J. Clin. Oncol.* **16,** 1969–1977.

BROWN, P. O., and BOTSTEIN, D. (1999). Exploring the new world of the genome with DNA microarrays. *Nature Genet.* **21,** 33–37.

CHUMAKOV, I., RIGAULT, P., LE GALL, I., and COHEN, D. (1995). A YAC contig map of the human genome. *Nature* **377,** 174–297.

CLONINGER, C. R., ADOLFSSON, R., and SVRAKIC, N. M. (1996). Mapping genes for human personality. *Nature Genet.* **12,** 3–4.

COHEN, D., CHUMAKOV, L. I., and WEISSENBACH, J. (1993). A first-generation physical map of the human genome. *Nature* **366,** 698–701.

COLLINS, F. S., PATRINOS, A., JORDAN, E., CHAKRAVARTI, A., GESTELAND, R., WALTERS, L., and the members of the DOE and NIH planning groups (1998). New goals for the U.S. Human Genome Project, 1998–2003. *Science* **282,** 682–689.

DELOUKAS, P., *et al.* (1998). A physical map of 30,000 human genes. *Science* **282,** 744–746.

DIB, C., *et al.* (1996). A comprehensive genetic map of the human genome based on 5.264 microsatellites. *Nature* **380,** 152–154.

DONIS-KELLER, H., *et al.* (1987). A genetic linkage map of the human genome. *Cell* **51,** 319–337.

EVETT, I. W., and WEIR, B. S. (1998). *Interpreting DNA Evidence. Statistical Genetics for Forensic Scientists.* Sinauer, Sunderland, MA.

FALCONER, D. S. (1981). *Introduction to Quantitative Genetics.* Ronald Press, New York.

FIELDS, C., ADAMS, M. D., WHITE, O., and VENTER, J. C. (1994). How many genes in the human genome? *Nature Genet.* **7,** 345–346.

FIORI, A. (1995). Le analisi del DNA a fini giudiziari: Aspetti etici e deontologici. *Riv. It. Med. Leg.,* 209–238.

FLINT, J., CORLEY, R., DEFRIES, J. C., FULKER, D. W., GRAY, J. A., MILLER, S., and COLLINS, A. C. (1995). A simple genetic basis for a complex psychological trait in laboratory mice. *Science* **269,** 1432–1435.

FORD, D., *et al.* (Breast Cancer Linkage Consortium) (1998). Genetic heterogeneity and penetrance analysis of the BRCA1 and BRCA2 genes in breast cancer families. *Am. J. Hum. Genet.* **62,** 676–689.

GUSELLA, J. F., *et al.* (1983). A polymorphic DNA marker genetically linked to Huntington's disease. *Nature* **306,** 234–238.

HUNGTINGTON'S DISEASE COLLABORATIVE RESEARCH GROUP

(1993). A novel gene containing a trinucleotide repeat that is expanded and unstable on Hungtington's disease chromosomes. *Cell* **72,** 971–983.

JONSEN, A. R., DURFY, S. J., BURKE, W., and MOTULSKY, A. G. (1996). The advent of the "unpatients." *Nat. Med.* **2,** 622–624.

KEREM, B., ROMMENS, J. M., BUCHANAN, J. A., MARKIEWICZ, D., COX, T. K., CHAKRAVARTI, A., BUCHWALD, M., and TSUI, L. C. (1989). Identification of the cystic fibrosis gene: Genetic analysis. *Science* **245,** 1073–1080.

KNOPPERS, B. M., HIRTLE, M., and LORMEAU, S. (1996). Ethical issue in international collaborative research on the human genome. *Genomics* **34,** 272–282.

LAPPE, M. (1983). Ethical issues in testing for differential sensitivity to occupational hazards. *J. Occup. Med.* **25,** 797–808.

LIPSHUTZ, R. J., FODOR, S. P. A., GINGERAS, T. R., and LOCKHART, D. J. (1999). High density synthetic oligonucleotide arrays. *Nature Genet.* **21,** 20–24.

LYNCH, H. T., *et al.* (1984). Genetic predisposition to breast cancer. *Cancer* **53,** 612–622.

MASOOD, E. (1996). Gene tests: Who benefits from risk? *Nature* **379,** 389–392.

MCCONNELL, L. M., KOENIG, B. A., GREELY, H. T., RAFFIN, T. A., and the Alzheimer Disease Working Group of the Stanford Program in Genomics, Ethics & Society (1998). Genetic testing and Alzheimer disease: Has the time come? *Nat. Med.* **4,** 757–759.

MOISES, H. W., *et al.* (1995). An international two-stage genome-wide search for schizophenia susceptibility genes. *Nature Genet.* **11**(3), 321–324.

NUFFIELD COUNCIL ON BIOETHICS (1998). *Mental disorders and genetics: The ethical context,* report of the Bioethical Working Party on Mental Disorders and Genetics. Nuffield Council on Bioethics, London.

PLOMIN, R., OWEN, M. J., and MCGUFFIN, P. (1994). The genetic basis of complex human behaviors. *Science* **264,** 1733–1739.

POKORSKI, R. J. (1997). Insurance underwriting in the genetic era. *Am. J. Hum. Genet.* **60,** 205–216.

ROSE, R. J. (1995). Genes and human behavior. *Annu. Rev. Psychol.* **46,** 625–654.

SCHNEIDER, C. (1998). *Quademi Sci.* **100D,** 34–43.

SHERMAN, S. L., DEFRIES, J. C., GOTTESMAN, I. I., LOEHLIN, J. C., MEYER, J. M., PELIAS, M. Z., RICE, J., and WALDMAN, I. (1997). *Behavioral genetics '97: ASHG statement.* Recent developments in human behavioural genetics: Past accomplishments and future directions. *Am. J. Hum. Genet.* **60,** 1265–1275.

U.S. CONGRESS, OFFICE OF TECHNOLOGY ASSESSMENT (1990a). *Genetic Monitoring and Screening in the Workplace.* U.S. Government Printing Office, Washington, DC.

U.S. CONGRESS, OFFICE OF TECHNOLOGY ASSESSMENT (1990b). *Genetic Witness: Forensic Use of DNA Tests.* U.S. Government Printing Office, Washington, DC.

VINEIS, P. (1997). Ethical issues in genetic screening for cancer. *Ann. Oncol.* **8,** 945–949.

WILKIE, T. (1998). Genetics and insurance in Britain: Why more than just the Atlantic divides the English-speaking nations. *Nature Genet.* **20,** 119–121.

General References

CAVALLI-SFORZA, L. L. (1995). The Human Genome Diversity Project. In *Actes 1995,* Vol. 2, pp. 71–80. CIB de l'UNESCO, Paris.

DE CARLI, L., BONCINELLI, E., DANIELI, D. A., and LARIZZA, L. (1998). *Genetica Generale e Umana,* Vol. 2. Piccin, Padova.

DULBECCO, R., and VEZZONI, P. (Eds.) (1998). *Il Progetto Genoma. Quaderni Sci.* **100D.**

STRACHAN, T., and READ, A. P. (1996). *Human Molecular Genetics.* Bios, Oxford.

CARLOS M. ROMEO-CASABONA

Universities of Deusto and the Basque Country
48.080 Bilbao, Spain

Genetic Information: Collection, Access, and Use

Genetic analysis is part and parcel of modern health care (preconception, preimplantation, and prenatal diagnosis as part of genetic counseling, and postnatal diagnosis) and provides valuable information. Advances in research are continually widening the range of information obtained and enhancing its accuracy and predictive ability. This information is vital not just for the individual to whom it refers but also for others, especially genetic family members but also other parties or entities to which a potentially healthy body is deemed necessary for the performance of certain activities (i.e., labor relations and health insurance). There arises the issue of access to the results. Personal freedom and privacy, the individual right to self-determination about personal information, are elements which can be affected. Effective protection of these interests has to be ensured or justification be given if they are to be restricted in favor of other interests. Discrimination might prevail over the principles of equality and solidarity. To prevent this, one has to reinforce the aforementioned principles with informed consent and the so-called right not to know and ensure that information is kept in strict confidence. Thus, the right to a private life protected from outside intrusion serves today as a guarantee of protection of the individual decision made.

Genetic Information: A New and Peculiar Subject of Legal Protection

One of the short- and medium-term goals of human genome research is to gain knowledge of the characteristics of human DNA and its components, particularly genes. There is particular interest in learning how genes function and what their role is in the transmission of biological heredity. One consequence of such information is that we will be in a position to know which genes are responsible for certain conditions (illness) and the mechanisms by which these are manifested and transmitted. More remote for the time being is the possibility that we will be even able to know which genes influence the appearance of certain tendencies, skills, or abilities in persons.

Genetic testing of individuals can bring to light important personal and family information, such as biological information on a person's current and future health, including mental health, even though this may be limited to giving advance warning of a propensity or predisposition to certain disorders, or information on reproductive capacity and the future health of offspring. Noncodificant DNA (DNA profiles or the so-called "genetic fingerprints") can also serve to reveal ties with other persons (e.g., paternity or the commission of a crime), and even genetic information should help us clarify certain intellectual capacities, behavioral trends, aptitudes, etc., although this is a basis for great discussion.

The potential information obtained from genetic tests conducted on a person is different from any other information. Its origin and characteristics do not depend on the individual will of the subject; its source is indestructible since it is present in almost all the cells of the body while alive and usually even when dead. Lastly, it is permanent and unalterable, save for spontaneous genetic mutations or ones triggered through genetic engineering or as the result of other external agents (e.g., radioactive ones). In any case, such mutations will always be partial and limited.

The information which can or could be obtained from genetic testing raises a series of problems associated with the information itself, access thereto, and the uses of such data, given that the interests of the data subject may be in conflict with those of other persons (including the biological family), with collective health and safety interests, and even with interests of an entirely different kind (e.g., economic). Access to such information provides knowledge of highly important aspects of the tested individual and directly affects his or her innermost sphere. However, the information is also highly valuable for protecting his or her health or that of any offspring. Uncontrolled dissemination of the information would pose a serious danger, especially because of the risk that humans would be turned into "transparent" or "crystal" citizens.

As stated by Malem-Seña (1995),

> A completely transparent society is one without hope. All hope feeds on a portion of opaqueness . . . deliberation and doubt are possible only when we know that things are possible, but not when they are necessary or impossible. The greater the transparency, the fewer the doubts and the less deliberation. (p. 136)

Furthermore, there is the danger that it might lead to discrimination against individuals, against families, or in the workplace and also discrimination when taking out life, sickness, or retirement insurance or when seeking official permission or authorization, loans, etc. There is also the added risk that certain populations or groups will be stigmatized as "defective" or "unfit" to participate in social relations. Such discrimination may even take on racial tinges, given that certain genetic conditions are more frequent in specific ethnic groups. In addition, it should also be recalled that because of the wealth of information that must be collected in order to be usable, automated processing is required, with all the benefits this entails. However, at the same time the (highly personal) information becomes more vulnerable, as occurs with all data which is subject to automated processing and communicated by means of informatic and telematic procedures (Romeo-Casabona, 1988, 1993).

The problems previously mentioned can have a direct bearing on (i) personal freedom (if genetic testing is conducted without consent), (ii) privacy (if the results are accessed, even legitimately or for a lawful purpose), and (iii) other fundamental rights, depending on the use which is made of the information (if misused or used for discrimination or a purpose other than that originally authorized). Hence, a careful weighing is required in each case and the appropriate safeguards need to be put in place to achieve the correct solution (Martínez-Bullé Goyri, 1995).

The first individual right which must be guaranteed is the principle of the freedom and autonomy of the individual, which also implies initial freedom of choice and consent before undergoing testing and even before providing biological samples for the test. Another legal consequence is the protection of genetic data as confidential information and the corresponding duty of secrecy on the part of the health care professionals and other persons (or entities) who, for whatever reason, have access to the data. From this perspective, priority has to be given to guaranteeing the confidential nature of information concerning each individual as a means of protecting personal private life (a crucial part of which is privacy) and other rights and particularly as a way of preventing discrimination, which must be prohibited absolutely in this (highly propitious) sphere.

This danger has already been appreciated on the international level, and firm stances have been taken with a view to prevention. Some international rules ban the discrimination of individuals on the basis of their genetic characteristics. The Council of Europe's *Convention for the Protection of Human Rights with Regard to the Application of Biology and Medicine: Convention on Human Rights and Biomedicine* (1997) expressly prohibits "any form of discrimination against a person on grounds of his or her genetic heritage" (art. 11). The UNESCO *Universal Declaration on the Human Genome and Human Rights,* adopted by its General Conference in 1997, states that "everyone has a right to respect for their dignity and for their human rights regardless of their genetic characteristics" (art. 2.a) and, specifically, that "no one shall be subjected to discrimination based on genetic characteristics that is intended or has the effect of infringing human rights, fundamental freedoms, and human dignity" (art. 6).

In summary, the concerns focus particularly on deciding who, in what circumstances and for which predetermined objectives, can perform genetic tests and obtain the corresponding biological samples; who will be given access to the information resulting from the testing of the samples, and to whom the information will be passed on and what use can be made of it; and what measures shall be taken to protect the information and, where appropriate, the biological samples (The Privacy Commissioner of Canada, 1992, p. 28).

The Protection of Privacy and Genetic Data

Concern with the legal protection of privacy has increased in recent decades as a result of the multiplication and increased potential of procedures capable of violating privacy, including technical means of collecting and transmitting images and sound and ways of compiling and processing information in general and personal data. Perhaps because the phenomenon has only recently become intense that until now it has not been viewed as necessary to clearly define what must be protected, in order then to put in place the most adequate procedures for protection.

Furthermore, current social notions and relations have gradually eroded the bastion of the sphere of privacy. Requirements derived from greater state intervention and the satisfactory enjoyment of the public freedoms proper to a social and democratic state based on the rule of law have, often to the detriment of the right to privacy, made it legitimate for authorities to gather an ever-increasing amount of information on private or confidential aspects of the lives of citizens in order to be able to carry out their duties and provide services more efficiently (e.g., in the field of genetics, information on reproductive health or the hereditary criminal or aggressive tendencies in criminals), as with the right to information and freedom of expression.

In the private sector also, awareness is growing that information or possible access thereto are vital to the success of certain activities, even if in so doing the private sphere of persons is affected (e.g., employers having genetic information on the future health of applicants for a job or insurers having such information on its customers) (Romeo-Casabona, 1993). Therefore, the urgent need to establish precisely what constitutes privacy is clearly seen. Thus, the patrimony or autonomy-based notion (the famous nineteenth-century American notion of "the right to be left alone") has been replaced by the explicit attribution of privacy as an inherent personality right, reinforced as a human right (in the *Universal Declaration of Human Rights* of 1948, art. 12) or fundamental right (as in the Spanish Constitution of 1978, art. 18), one which transcends even the exercise of other public or private rights.

This evolved perspective has been adopted by the German Constitutional Court in its formulation of the idea of the right to self-determination with regard to information when it refers to the "power of the individual, derived from the notion of self-determination as regards information, to decide essentially for himself when and to what extent details of situations concerning his personal life may be disclosed" (ruling of December 15, 1983, *BVerGE,* t.65, p. 1ff). To this characteristic one must also add its contextual relativity, diffuse and dynamic nature, and expansive potential as well as the fact that it may be shared occasionally since access by others to one's private realm does not necessarily mean that it ceases to become private, particularly when the person concerned is obliged to share (Romeo-Casabona, 1993).

In summary, one might understand privacy as meaning all manifestations of individual (or family) personality, knowledge of which is reserved to said person or over which he or she exercises control of some kind when others are involved (both private individuals and the authorities). Consequently, the right to privacy also involves recognition of this reserved knowledge or control over others. It should be noted that I am referring here not just to knowledge but also the person's own development, which extends the scope of the right and in so doing may open a debate which I will not enter into here. Clearly, when the person does not have a legitimate mechanism of control over others, the manifestation of personality no longer remains within the personal realm. In such cases protection is possible only to the extent that the person's reputation or honor are affected since these are closely bound with privacy and have always been acknowledged as sharing a common zone (Morales-Prats, 1984; Lucas-Murillo de la Cueva, 1992).

The concept of privacy is a relatively new one in European comparative legislation, which views it as a wider and more global (than "intimacy") set of aspects of personality which, considered individually, may lack intrinsic significance but when taken together provide a picture of the personality of the individual which he or she is entitled to reserve for himself or herself. It is worth noting that the aforementioned notion is but one of the facets which may by violated by misuse of personal information, whether held on computer or not. In addition to privacy, other personal interests of the individual may also be violated (which are in fact aspects of the right to a private life), such as the exercise of certain fundamental rights and public freedoms, without transfer of data even being necessary.

In the light of the previous discussion — that is, the importance which genetic information holds for the individual concerned and the fact that it can be protected via the privacy notion — the need can be seen for an overall framework to organize said protection.

Protection of Manifestations of Privacy

In the case of this first profile of privacy, there are three main avenues of protection, each with different legal instruments in terms of their nature and scope. It should be noted, however, that since the privacy is inevitably relative and diffuse in nature, protection focuses on certain (the most important) manifestations of personality and occasionally, particularly in connection with data protection, extends beyond privacy to take in other legal interests of a related, but at the same time different, nature. These facets are as follows (Romeo-Casabona, 1993):

1. The protection of privacy, as a manifestation of the redoubt of personality, refers to the realm of privacy which is reserved directly and exclusively for the individual in question (or his or her family), with whom rests the decision as to the extent of the protected area, albeit with certain limitations. Protection of this facet of privacy can be secured by protecting certain manifestations of secrecy and privacy, in both the civil and the penal spheres.

2. The protection of privacy as a manifestation of shared confidentiality: this includes aspects of privacy to which, by law or due to the nature of social relations or relations among individuals, others have access although they are legally bound to keep secret. From this is derived the duty of professional secrecy, or official secrecy in the case of public servants, the observance of which can be demanded by the individual, who may even take action for any liability which might arise. Breaches of professional secrecy are also generally covered in both civil and penal law.

3. The protection of privacy with regard to data processing using new information and communication technologies: Because of the special implications, this aspect deserves separate mention. Here, the scope of privacy centers particularly but not solely (given that some of the aspects mentioned previously are also applicable) on the fact that the individual has to furnish private and even confidential data concerning himself or herself, even though he or she is acknowledged as having a certain amount of control over the information. This control merges here with so-called "informatic freedom and identity," denoting the control by the individual of his or her own personal data (i.e., whether it is accurate and appropriate, control over the purpose and use of such data and its passing on to others), regardless of whether the information is private. It should be noted that we are dealing here with two different protected interests which overlap and merge largely as a result of the powers of control which tend to be acknowledged as existing (Lucas-Murillo de la Cueva, 1992; Muñoz de Alba Medrano, 1995; Romeo-Casabona, 1988; Sola Reche, 1991). It has already been noted that mere reference to privacy cannot cover satisfactorily other elements of the protection of the personal realm present in informatic freedom or the right of self-determination with regard to information, unless the scope of privacy were extended considerably, with the corresponding danger that it would be even more imprecise and would be less effective as a legally protected interest.

Also applicable is the Council of Europe convention of January 28, 1981, on the protection of persons with regard to the automated processing of personal data. The European Union's directive on the subject (Directive 95/46/EC of the European Parliament and Council, October 24, 1995, on the protection of physical persons with regard to the processing of personal data and the free circulation of such data) must also be borne in mind. The directive prohibits "the processing of personal data which reveal racial or ethnic origin . . . and the processing of data relating to health or sexuality" (art. 8.1). However, it does make an exception

> when the processing is necessary for prevention or for medical diagnosis, health care and assistance or medical treatment or management of health services, provided that it is conducted by a health professional who is subject to professional secrecy, in accordance with national legislation or the regulations established by the competent national authorities, or by a person who is bound by an obligation which is equivalent to secrecy. (art. 8.3)

In summary, the directive exceptionally permits the processing of health data for the aforementioned medical or health care reasons as long as the duty of professional secrecy on the part of the person who performs the processing is guaranteed. It should also be understood that the consent of the individual concerned is required (art. 8.2.a). Here, too, penal law may intervene to protect the previously mentioned aspects of personal data, whether or not they deserve to be viewed as private, provided that the principle of minimum intervention is not exceeded.

The role of nonpenal legislation on data protection is essentially to prevent abuse in the form of unauthorized or inappropriate use of the information. It thus implies the adoption of measures governing the gathering, processing, use, and passing on of the data. Current data protection laws tend to incorporate the following legal interests: (informatic) privacy, informatic freedom and identity, and free and full enjoyment of personal rights. The first aspect, protection of privacy with regard to the information, is grounded chiefly on the recognition that the data are confidential, which leads specifically to the confirmation of the duty of secrecy and the establishment of offenses for breaches thereof. The second aspect, as previously shown, stems from the recognition that data subjects have a right to information and a right of access to the data as well as the right of amendment and deletion (of erroneous data) and also recognition of principles concerning the quality of the data, information on data collection, consent, security, and the passing on of the data. This entire arsenal of guarantees and rights is designed to serve as a framework to ensure that the automated processing of personal data, when considered necessary, does not undermine the full and free exercise of the other civil and political rights normally recognized in constitutional texts (Fig. 1).

FIGURE 1 Control room of the Health Care International hospital (Scotland), without paper documents. This hospital does not have a traditional archive: The information concerning each patient is stored in a centralized computer system, which the doctors may access from any terminal in the building (photo courtesy of Fraser-SPL/Neri).

Data relating to health and racial origin are generally acknowledged as lying at the core of privacy. Along with others, they make up so-called "sensitive" or "highly sensitive" data (together with information on ethnic origin, political opinions, trade union affiliation, religious or other beliefs, and sex life), and here the connotations of genetic testing are clearly seen. The category of such data (highly sensitive) implies the establishment of stronger protection guarantee, such as prohibition of access and automated processing without the subject's consent or, in other cases, restrictions (generally set by a law) governing such access and use. In this sense, the European convention of 1981 (art. 6) and the EU directive of 1995 establish the exception that "processing must be necessary to safeguard the vital interest of the data subject or other person, in the case where said subject is not in a position to give his consent" (art. 8.2.c, 8.3).

Extension of the General Normative Framework on Protection of Privacy and Personal Data to the Specific Protection of Privacy and Genetic Data

The previously mentioned general criteria on protection of privacy or of personal information and data (whether confidential or not) comprise the legal framework in which one would have to situate the specific protection of individual genetic information. However, it must also be kept in mind that, in addition to privacy, other fundamental rights and public freedoms may also be involved. However, the special nature of individual genetic data requires special protection and one has to ask whether existing instruments satisfy this need.

As has been correctly noted, the first criteria lies in affording greater protection to medical records in general and in reducing the relevance of those records to coverage for health care (Yesley, 1997). Another line of thinking which is finding increasing favor is that each person is also entitled to decide to whom, when, and to what extent such information can be disclosed. Thus, the passing on of information obtained by means of genetic testing should be prohibited unless consent is given by the data subject or legal representative in the case of minors or legally incapacitated persons.

Genetic information also affects the unborn, who carry their exclusive information as of the moment of conception, whether this be natural or assisted. This raises a new problem but still requires legal instruments of protection similar in scope to those for persons already born. The rights and duties of the unborn have to be exercised by their future legal parents (also if conception has been achieved using donor gametes) just as parents exercise the rights and duties of their minor children in discharge of their parental authority.

Genetic data must also be protected vis-à-vis other persons even in the case of a biological relative who seeks information concerning the possible presence in him or her of a pathological gene similar to that discovered in the data subject and also inherited from the parents (Danish Council of Ethics, 1993) or when the doctor considers it necessary for family members to know the information because they too might be affected by the results of the genetic test. The Group of Advisers on the Ethical Implications of Biotechnology (1996) also insists on the need to maintain confidentiality in these cases.

One frequently finds, however, that the subject consents to the disclosure of the information to members of the family or may even provide the information himself or herself (De Sola, 1994/1995). However, it is important in the genetic counseling process to stress to the subject the importance for others (kin) to partially relieve the doctor of his or her professional duty of secrecy or for the subject to give the information (Annas *et al.,* 1995).

The application of the right to self-determination with regard to information — discussed in 1983 by Germany's Constitutional Court in connection with information stored in databases — to individual genetic data poses many problems. As previously mentioned, it proves difficult to maintain intact when set against other general or private interests as well as the need for modifications (Simitis, 1994).

It is worth recalling that a genetic data subject, from the biological perspective (the legal standpoint may view things differently), is not the individual himself or herself but rather his or her entire biological family. A separate issue is that of the legal duties which fall on family members to whom information affecting them is disclosed (De Sola, 1994/1995).

The undisputed importance of an adequate legal instrument has led the Council of Europe to pass a new recommendation on the protection of medical data [Recommendation No. R (97) 5], which includes specific regulations on health-related genetic data. For its part, UNESCO's *Universal Declaration on the Human Genome and Human Rights* (1997) proclaims the protection of confidentiality with regard to genetic data: "Genetic data associated with an identifiable person and stored or processed for the purposes of research or any other purpose must be held confidential in the conditions foreseen by law" (art. 7). Also, in order to protect human rights and fundamental freedoms, limitations to the principles of consent and confidentiality may only be prescribed by law, for compelling reasons within the bounds of public international law and the international law of human rights (art. 9). There should be no overlooking the need for new regulation of (or a completion of existing norms) some aspects of the protection of individual genetic information.

Comparative law provides many solid solutions to the protection of genetic data, with recognition given to the primacy of the subject's wishes. The following are examples of specific provisions:

Switzerland: The Constitution of the Swiss Confederation (following the reform of 1992), art. 24, states, "The genetic material of a person may not be tested, recorded, or disclosed without the subject's prior consent, except where expressly permitted or imposed by Law".

Austria: Law on genetic techniques of 1994 (*Gentechnikgesetz*), in a very comprehensive way, provides "(1) any person who conducts genetic testing or orders it to be carried out shall keep secret the data obtained, in accordance with the present federal law, and having due regard for the following rules: 1. If he so requests, the data subject shall be given access to all the information concerning him. 2. Any unexpected results of immediate clinical relevance shall be made available to the data subject, as well as any other information he expressly requests. Particularly in cases where the subject did not make a specific request, the notification shall be made in such a way as to not cause undue concern to him; in exceptional cases, the notification may be omitted entirely. 3. Nonanonymous data may only be used for a purpose other than the original one for which it was collected with the express, written consent of the data subject. 4. Data may be passed on only to (a) persons engaged directly in the gathering, processing, and evaluation of the data in the center in which they were obtained, (b) the data subject, (c) persons listed in art. 65, paragraphs 3 and 4, (d) the doctor on whose instructions the test was conducted, and the doctor in charge of the treatment or responsible for diagnosis, (e) other persons, only to the extent that the data subject has given his express and written consent, which can be withdrawn in writing. 5. Data must be protected adequately to avoid access by unauthorized persons. 6. Data which have not been made anonymous may only undergo automated processing in the center in which they were collected, and only within the medical sphere for which the genetic test was ordered; they must be stored separately from other data and may only be retrieved by persons authorized in accordance with the present federal law, and only by means of a specific mode of access. 7. The obligations set out in Sections 3–6 above shall apply to all persons intervening in the performance of genetic tests or the administration or storage of the data obtained therefrom. (2) To the extent that the present federal law does not provide otherwise, the Data Protection Law BGBI, No. 565/1978, shall remain in force, together with other provisions establishing special duties regarding secrecy and communication" (art. 71).

France: Its parliament has passed some provisions to protect the processing of personal data with the purpose of medical research (Law No. 94-548 of July 1, 1994, *relative au traitement de données nominatives ayant pour fin la recherche dans le domaine de la santé*).

Spain: The law regulating the automated processing of personal data (Organic Law 5/1992, of October 29, *de regulación del tratamiento automatizado de los datos de carácter personal*) provides for special measures of protection for personal health data and lays down the conditions for automated processing by health professionals and public and private health institutions. Briefly, this provision is limited to the remit made to health legislation on access to information and confidentiality (art. 7.3 and 8) (Heredero Higueras, 1994), which time will show to be insufficient, given the nature of such information and the persons who might be given access. One of the laws mentioned (unnecessarily) is Spain's General Health Law of April 25, 1986, which recognizes confidentiality as a patient and user right (art. 10.3 and 61). This rule is also applicable to genetic data.

Italy: The law on the protection of individuals and other subjects with regard to the processing of personal data (Law No. 675 of December 31, 1996) provides "1. Health professionals and public health institutions may, even without being authorized by the Garante, process personal data allowing the disclosure of health conditions, exclusively with regard to the data and operations required in order to safeguard the data subject's bodily integrity and health. Where the self-same purposes concern a third party or the public as a whole and the data subject fails to give his consent, that data may be processed upon authorization by the Garante. 2. Personal data allowing the disclosure of health conditions may be communicated to the data subject only by a physician who must have been appointed either by the data subject or by the controller. 3. The authorization as per paragraph 1 shall be granted, except in cases of special urgency, after consulting the Higher Council for Health Care [Consiglio Superiore di Sanità]. It shall be prohibited to communicate data obtained in breach of the limitations laid down in the said authorization. 4. Dissemination of data allowing the disclosure of health conditions shall be prohibited, except where it is necessary for the prevention, detection, or control of offenses, subject to compliance with the provisions applying to this sector" (art. 23).

The Right Not to Know

In addition to the right to information, or right to know, which also covers individual genetic information, reference tends to also be made to the reverse, i.e., the right "not to know." This is a right which has arisen essentially in connection with genetic information, although it has also occurred with similar controversy in other health areas (Abbing, 1995; De Sola, 1994/1995; Rodotà, 1992; Romeo-Casabona, 1994).

The right not to know means that by not undergoing genetic testing the subject avoids knowing whether he or she is carrying a genetic condition or whether he or she might do so in the future. This attitude might be criticized as selfish if it serves to block scientific progress or prevent family members or future offspring from knowing their condition. Hence, arguments could be advanced in favor of neutralizing this right with legal coercive measures. However, it is important to recall that if someone invokes the right not to know (a "right" of which he or she is most probably unaware), it generally means that he or she "already knows." In other words, the express desire not to undergo testing means that the subject already knows that he or she belongs to a genetic disorder risk group or is at least aware of similar circumstances in a relative. He or she is also aware that, as things stand today, science cannot provide the means to prevent, treat, alleviate, or palliate the possible genetic condition. Therefore, he or she opts not to receive further information (i.e., on whether he or she is carrying a deleterious gene or will suffer the disease), particularly if the condition is serious or deadly and manifests itself late in life. In this way, his or her development and social life are not affected by the knowledge that neither he or she nor science can do anything to prevent the future onset of the condition. Even if this individual process of reflection were not sufficient grounds, the objective knowledge that at the time in question there are no scientific means or know-how to prevent or palliate the condition also leads to the conclusion that such diagnostic tests should not be allowed in such circumstances (Abbing, 1995). Consequently, the right to protection of a private life free from outside intrusion reemerges today as the guarantee of an individually made decision based on the understanding that the right not to know is merely a manifestation of the right to privacy and to respect for one's private life (Martínez-Bullé Goyri, 1995).

As is known, the right to respect for private and family life is expressly recognized in the *Universal Declaration of Human Rights* and the 1950 European Convention for the protection of human rights and fundamental freedoms (art. 8.1). For its part, the Council of Europe's Convention on Human Rights and Biomedicine also underscores the importance of this right in the area of the biomedical sciences (art. 10.2). Nonetheless, this convention (art. 26.1) and the 1950 convention (art. 8.2) set out a limitation, namely, when the interference is in accordance with the law and is necessary in a democratic society for the protection of public health or for the protection of the rights and freedoms of others. This restriction may be relevant for our purposes here, although one first has to determine in which cases it can be viewed as being necessary for a democratic society. Moreover, although certain interferences in the private lives of others may be warranted to prevent the risk of an infectious disease (the 1950 convention accepts in this case that a person may be detained lawfully; art. 5.1.e), such interference may be questionable when there are as yet no specific subjects whose health, rights, and freedoms have to be protected (e.g., someone who has not yet been conceived).

Violation of this right would have to be considered in penal law as an attack on individual freedom (if the subject's wishes are ignored when the sample is taken) or on another legally protected interest derived from a fundamental right (e.g., privacy or, in the broader sense, private life). It would also be necessary to verify if the means used can be matched (exceptionally, I think) to the corresponding penal offenses and, in civil law, whether they can be considered to have caused injury or harm for which compensation may be sought.

Exceptions to Confidentiality

Having established that a person has rights over his or her own genetic information, that penal protection is merited because of the legal interests at stake, and that said rights must prevail over the interests of others, it is nevertheless possible to consider the rights of the latter when their health or lives are seriously affected. Specific regulation is not required for this. Rather, it can be achieved on grounds of the conflict of duties (Abbing, 1995; Malem-Seña, 1995) or, as the case may be, the state of necessity, techniques which exist in the majority of penal legislations. The Nuffield Council on Bioethics (1993) also acknowledges a possible exception to the duty of confidentiality in certain cases. To resolve the problem, all the interests at stake in the conflict have to be weighed, not just the legally protected interests. In order to use the conflict of duties solution, it is necessary to first determine the obligation of information toward others on the part of the person who performed or proposed the test (e.g., a doctor in relation to his or her patient), a duty which generally speaking cannot be substantiated in law (Annas *et al.,* 1995). However, in weighing the factors one could accept the principle of freedom and autonomy as a prima facie precedent, which would place the burden of proof in his or her favor in all cases in which such a conflict arises (Malem-Seña, 1995).

Specific Cases

The Use of Genetic Analysis in Labor Relations

It has been mentioned several times that the research being carried out on the human genome implies an incalculable benefit for the treatment of numerous illnesses linked to genes; nevertheless, we do not know much about most of them, even about their treatment. However, another advance of importance is that it should be possible to predict with enough precedence if a person will develop an illness or at least presents a genetic predisposition to suffer it in the function of the exogenous agents in its manifestation (polygenic and multifactorial genetic illnesses). Also, when the research becomes more advanced, it will be possible to take measures of genetic engineering (i.e., intervene in the genes that present some defects) or use other procedures with the purpose of avoiding the illness or, if that is not possible, to reduce its most harmful effects. This is known as *predictive–preventive medicine.* Nevertheless, the information that could be obtained about any person could also be used for not so desirable ends by a measure which will be of great discrimination for the person from which that information derives. Regarding the workplace, it is argued that firms may be interested in submitting their workers or candidates to genetic analyses, and if the results were positive (i.e., they reveal the future suffering of an illness or merely a predisposition to it) they could decide to fire workers or not hire candidates despite being qualified for the job and not presenting any symptom of any illness and, in some cases, despite no existing link between the future development of the illness and the job. In effect, this would be a discriminatory practice (Ammon, 1995; Catalano, 1995; Natowicz *et al.,* 1992; Slovak, 1993; Sprumont and Guillod, 1995; Wiese, 1994; Zeitz, 1991). To proceed to the juridical value of this matter, several aspects must be considered.

Objectives of the Proofs and Their Validity

We should work with several hypotheses that could explain the objectives pursued with the performance of these tests, even although some might seem unlikely.

First, the pursued objective by the employer could consist of obtaining the genetic analysis in a general way with a double goal, accumulated or alternative: (i) to proceed to a negative selection with the purpose of not contracting the candidates to whom any anomaly of a secure appearance in the future will be diagnosed (illness of monogenic mendeliane hereditary origin) or a simple predisposition of multifactorial origin (e.g., polygenic illnesses linked with the cardiorespiratory system and cancer), with known exogenous factors implicated or otherwise with work conditions (i.e., a given workplace, radiation emission, liberation of fumes or toxic solids, and other pollutant factors of the work environment), or (ii) with positive selection criteria to select the most suitable workers for a job based on their genetic characteristics (more physical resistance to sleep or to some toxic agents that could provoke cancer, and so on). In this case, testing is justified because the tests would serve the exclusive interests of the employer, to the detriment of the candidates for the job, not only because in this way the most suitable candidates would be selected from the biological standpoint but also by acting in this way it would not be necessary or urgent to adopt preventive measures and reduce health risks from the productive activity and the labor environment. These possibilities were expressly rejected by the European Parliament (Resolution of 1989).

However, we must remember the correction in the pointed observation that the major tendency to require genetic aptitude tests in relation to a job is linked to the social security systems that are partially controlled by the company that hires, auto insurance companies, or intermediary private firms (i.e., in the United States), and that this tendency decreases in systems of universal security or in which it is assumed by the public powers by several procedures (as occurs in the European Union).

It seems more likely that these tests could be intended for use in a more limited and selective way, with the purpose of detecting possible predispositions of the candidate to develop an illness as an effect of the risk factors that the concrete labor environment, linked with the job, presents. Also, it should be more desirable that they will be obtained to adopt the opportune measures to prevent risks and labor accidents and to protect the health of the workers in relation with a given workplace, which is one of the duties of the employer, in addition to providing periodical controls, and to reduce in this way the costs to the firm or the community.

One should also take into account that it is usual to distinguish to this effect two types (International Labour Office, 1993): (i) *genetic screening,* which refers to the identification of individuals with characteristics or particular inherited illnesses, even those that present a major susceptibility from the environment; and (ii) *genetic monitoring,* which refers to periodic examinations with the purpose of identifying mutations in the genetic material of a person that might be induced by the environment, allowing the establishment and prevention of damage because examination could occur at the time of identifying the risks of a given labor environment, with the purpose of eliminating or reducing them.

Potential Benefits Derived from Genetic Tests

On the whole, the knowledge of the genetic risks in the workplace presents several points of interest:

1. For the workers, with the purpose of preventing illness or treating it or changing jobs, which is compatible with an information system and advice to the margin of the labor regime.
2. For the employers, in a way that it may reduce the costs derived from the future potential illnesses of their workers (with respect to both the productive activity and to the responsibilities for accidents originated by their workers and the contributory repercussions by labor of anticipated retirements because of illness), to exclude the ones that present these higher risks or reduce them.
3. For other workers, if their security is guaranteed and accidents that could injure them can be prevented (not the contagion of transmissible illnesses, however) originated by the partners with predisposition.
4. To third persons concerned with the firm (i.e., customers) because the accidents derived from the appearance of the illness of the employee that presents predisposition could be avoided (e.g., drivers of passengers transportation, and airplane pilots).
5. For the state, with the purpose of satisfying in a more efficient way its functions of preventing accidents and protecting the health of workers through the stabilization and imposition of prevention risks measures and labor hygiene.

Nevertheless, the apparent benefit of obtaining genetic analysis when hiring workers is nullified because this is not the most appropriate moment for the compliment of the several pointed objectives, unless the candidates present some reason for unsuitability. On the other hand, it can result in the abusive practice of obtaining information against the job candidates which could give rise to populations excluded access to the job market (De Sola, 1995), a new social class of rejected people including the families of the analyzed subjects.

However, it is acceptable to avoid risks to third persons, but the procedure with the least costs to the persons must be determined. No doubt that this enumeration of persons and bodies concerned with the genetic information in the workplace reveals a plurality of interests that are sometimes concurrent but on other occasions antagonistic and contrasting.

On the other hand, the genetic research does not have to be contemplated from a maniquea perspective — that is, absolutely in favor of its presumed goodness or absolutely against its perverse effects — because it also presents several nuances. It must, consequently, deal with the contrasting interests or at least find the adequate balance between them (Falçao de Oliveira, 1995), which ensures the equality of positions and that the beneficial aspects of the genetic analysis in the labor world may be developed, eliminating the negative ones. This reflection of relativization of these matters has to take into account that probably the great majority of citizens could be carrying deleterious genes that could cause illnesses by exogenous or environmental mechanisms so that it would be necessary to renounce to the possible decisive value of the genetic analysis for the selection of candidates or the promotion of the most suitable workers.

In this context, to the jurist and to the legislator arise some doubts or problems that have to be solved before adopting a resolution. These include the diagnostic liability, in the actual state of the science, of the genetic analysis in relation to the predisposition or genetic susceptibility that one person will develop a multifactorial illness in the future; predictive or prognostic liability in relation to given environmental factors; and the capacity of premature detection of the symptomatology of the corresponding illness because it may be accepted as an irrefutable starting point in these discussions that the manifestation of these illnesses would always have to be violent and intense and therefore harmful for the security of third persons.

On the whole, we want to establish the best way to determine the possible predictive value that has to be recognized in these analyses (Wiese, 1994). To this, we have to add the disposability (scientific and material) of preventive ways concerning the predisposition of the concrete illness and the capacity to reduce the exogenous factors of the workplace surroundings that have a major potential to cause illness. It is true that most of these doubts will disappear in the future as the research on these kinds of illnesses progresses at the same time that more exact knowledge of the causal incidence of the risk factors for susceptible and unsusceptible persons is obtained.

On the other hand, it is necessary to take into account the concept of illness in the workplace (which is different from the other sectors of the legal system such as the civil and the penal, which deal with any modality of handicap for health, for instance, a carrier of HIV, or even preventive medicine, as long as a risk situation requires a preventive treatment). In effect, we have to take a functional approach as a beginning point, understanding it to be that which invalidates or decreases in a appreciable way the capacity to perform the concrete job. Consequently, the detection of a predisposition could not be qualified as an illness in this sense because it does not invalidate or decrease the capacity of working, and it does not present actual symptomatology. We still do not know if in the future it will arise; the relation between the predisposition and the workplace as a factor that caused it must be verified. This must be done without the prejudice that the preventive medical measures could be adopted and the control to avoid or decrease the risks of its appearance (e.g., in a similar way as it occurs with the persons that present positive HIV), independently of the opportune measures that could be taken in the workplace.

Legislative Provisions

From the actual legislations it cannot be deduced, in general, that it is compulsory to submit to genetic analysis. However, the position of nonequality in which the demands of the job or the worker instead of the requirements of the employer to be submitted to them cannot be forgotten. Currently, there are very few specific regulations in national legislation (the European ones, except Sweden, Working Environment Act of 1991; Austria, Law 1994, art. 67; and also those of the United States) (Yesley, 1997), but very soon there will be more state provisions because there are several bills under discussion and parliamentary approbation.

These first legislative initiatives are significant in that (i) some way of legislative intervention is necessary with our interests, on the one hand, and the novel, extraordinary, and transcendence of the matter, on the other hand, that impose a specific regulation; and (ii) we can use other legislative experiences as a model for countries of similar social background or, on the contrary, as examples of what should be avoided if the experience is not favorable.

In this matter, it is also very important not to forget other initiatives of other supranational bodies, such as the International Labour Office (1993), the Council of Europe (1996 Convention on Human Rights and Biomedicine, art. 12), and the European Union (Directive 90/679/CEE), that can contribute to reach a normative uniformatization and homogeneity. Another very significant observation is that in the framework of the European Union, the freedom of the occupational contraction enables candidates to direct their job demands to the employers of those member states that do not have regulations concerning health analysis (in particular, the genetic tests) or, on the contrary, those with very restrictive provisions, as long as the union does not establish the opportune legal criteria that should be applied to all member states.

Genetic Analysis and Insurance Contracts

It has been pointed out that one of the paradoxes of health care is that more biomedical information could result in the deterioration of a desirable situation (Murray, 1995). Insurance companies quickly realized that genetic analyses of future clients could be of capital importance for the so-called private personal insurances (i.e., life, health, and accidental) and can establish based on the results of these analyses some conditions that are more or less severe (e.g., the fixing of the premiums that are applied to the client) or even reject contracting in some cases (Yanes-Yanes, 1994/1995).

The insurance sector is most interested in the research on the human genome. The legitimate interests of these companies, ruled by the principle of economic benefit, are confronted with those of the clients, who intend to insure their futures or those of their families without losing their privacy and without suffering discrimination (Menéndez-Menéndez, 1995). However, it is equally likely that a person knowing his or her predisposition to a serious illness with no cure, as determined from a genetic analysis, might cover up this eventuality when applying for a policy.

Therefore, the prevention of prejudices for both parties will require a revision of the corresponding legislation, with the purpose of concealing the different interests and ensuring the necessary balance while avoiding any discrimination (Capron, 1995). In any case, the mere predisposition to develop an illness as opposed to actually having an illness will not be enough to change the conditions of a contract. In case of doubt, the prevailing interest of the person who obtains insurance should be guaranteed, particular his or her right to informative self-determination, taking measures to avoid having the genetic analysis generalized by the insurance company (Yanes-Yanes, 1995).

Regarding obligatory health insurance for workers, in the health systems in which they have to contract it through private companies, insurance could be provided by the state in cases in which the genetic diagnostic does not reveal a predisposition but a high probability of illness.

In the United States, a special task force on genetic information and insurances made the following recommendations on the matter (Task Force on Genetic Information and Insurance, 1993):

1. Information about past, present, or future health status, including genetic information, should not be used to deny health care coverage or services to anyone.
2. The U.S. health care system should ensure universal access to and participation by all in a program of basic health services that encompasses a continuum of services appropriate for the healthy to the seriously ill.
3. The program of basic health services should treat genetic services comparably to nongenetic services and should encompass appropriate genetic counseling, testing, and treatment within a program of primary, preventive, and specialty health care services for individuals and families with genetic disorders and those at risk of genetic disease.
4. The cost of health care coverage borne by individuals and families for the program of basic health services should not be affected by information, including genetic information, about an individual's past, present, or future health status.
5. Participation in and access to the program of basic health services should not depend on employment.
6. Participation in and access to the program of basic health services should not be conditioned on disclosure by individuals and families of information, including genetic information, about past, present, or future health status.
7. Until participation in a program of basic health services is universal, alternative means of reducing the risk of genetic discrimination should be developed. As one step, health insurers should consider a moratorium on the use of genetic tests in underwriting. In addition, insurers could undertake vigorous educational efforts within the industry to improve the understanding of genetic information.

Genetic Analysis in Other Social Fields

Regarding the authorizations or administrative licenses that request a previous medical examination (e.g., driver's license for motor vehicles and a license to have weapons), these must also not discriminate, and the genetic tests will not have to be compulsory. Nevertheless, if in the future they become compulsory, and in the case that it is a prerequisite, obtaining these kinds of tests with a shorter periodicity to check for the appearance of an illness that would invalid the license or authorization could become legally binding, without the possibility to reject or retire it.

Biological paternity tests and genetic fingerprinting in criminal investigations, instruments available to the justice system, raise other considerations which must be taken into account, given that they involve a conflict between the interests of the affected person and the general collective interest (Romeo-Casabona, 1996).

Some Criteria for Regulation

Although by no means exhaustive, the following criteria might be advanced for the protection of genetic information (without going into details of specific aspects relating to health or labor relations, etc.) (Romeo-Casabona, 1996):

1. The protection of genetic information gained from genetic tests should be ensured by the requirement of a greater protection given to medical records in general and by the reduced relevance of those records to coverage for health care.
2. The free and informed consent of the data subject must be guaranteed for all genetic testing, save where for urgent reasons and for his or her own benefit this cannot be obtained beforehand. Consequently, prohibition must be established to prevent the compulsory and particularly the coercive imposition of such tests for the sake of public or private interests. Some exceptions are admissible with regard to the obligatory giving of samples for genetic fingerprinting, under the terms outlined later.
3. Information given to the subject before he or she consents to any form of genetic testing must be complete and include details of the nature of the test and the results (foreseen or otherwise) which might be obtained. Similarly, the refusal of the subject to be given information regarding his or her own genetic data must be respected (the right not to know).
4. Genetic testing is admissible when other diagnostic health tests are imposed by law, to the extent that they are more reliable for the legally established objective and that they provide no other kind of information, particularly that of a predictive nature. The law may set exceptions to this, but only if there exist proven general interests (e.g., important reasons of public health and security) or clearly prevalent interests of other persons.
5. In all cases, the secrecy of the results of genetic testing must be guaranteed and restrictions of access provided, even with regard to members of the family of the data subject(s), although solutions may exist in law should a conflict of interests arise which can be resolved via the conflict of duties approach. Genetic information must be protected using guiding principles and similar (albeit reinforced) measures to those established for automated personal data (quality of the data: relevance, accuracy, purpose, deletion, right of access and amendment by the data subject, etc.;

physical and logical security, separation of data to permit partial or selective access, etc.), regardless of whether it is obtained by force of law (if such provisions exist) or with the consent of the subject. The corresponding duty of secrecy must extend to all those who, in the course of their work, have access to the genetic information.

6. The information obtained must be used only for the originally intended purpose(s). More specific and restrictive measures should be provided for the transfer of genetic information to other files and the transnational flow of such data, given that current data protection laws tend not to afford sufficient protection.

7. The results of genetic tests conducted for health purposes must not give rise to discriminatory practices nor to interference in the private life of the subject. In particular, they should not serve to condition in any way access to health care or social benefits or to curtail reproduction (e.g., through abortion or sterilization), save for any legally established conditions of access to assisted reproduction techniques.

References Cited

ABBING, R. (1995). Genetic information and third party interests: How to find the right balance? *Law Hum. Genome Rev.* **2,** 38, 40, 53–65.

AMMON, U. (1995). Genome analysis in the world of work. Contribution to the discussion on the need for regulation in the Federal Republic of Germany. In *The Human Genome Project: Legal Aspects,* Vol. 4. BBV Foundation, Bilbao, Spain.

ANNAS, G. J., GLANTZ, L. H., and ROCHE, P. A. (1995). The *Genetic Privacy Act and Commentary.* University School of Public Health, Health Law Department, Boston.

CAPRON, A. M. (1995). Insurance and genetics: A purposive analysis. In *The Human Genome Project: Legal Aspects,* Vol. 3. BBV Foundation, Bilbao, Spain.

CATALANO, G. (1995). Genetic analysis of workers in Italy: A legal approach. In *The Human Genome Project: Legal Aspects,* Vol. 4. BBV Foundation, Bilbao, Spain.

CAVOUKIAN, A. (1995). Confidentiality issues in genetics: The need for privacy and the right not to know. *Law Hum. Genome Rev.* **2.**

COUNCIL OF EUROPE (1996). Convention for the protection of human rights with regard to the application of biology and medicine: Convention on human rights and biomedicine. Council of Europe, Strasbourg.

COUNCIL OF EUROPE (1997). Recommendation No. R (97) 5, on the protection of medical data. Council of Europe, Strasbourg.

DANISH COUNCIL OF ETHICS (1993). Protection of sensitive personal information. With special reference to genetic data. In *Ethics and Mapping of the Human Genome. Fifth Annual Report.* Danish Council of Ethics, Copenhagen.

DE SOLA, C. (1994/1995). Privacy and genetic data: Situations of conflict (I, II). *Law Hum. Genome Rev.* **1/2.**

EUROPEAN COMMISSION (1995). Directive 95/46/EC of the European Parliament and Council, on the protection of physical persons with regard to the processing of personal data and the free circulation of such data. European Commission, Brussels.

EUROPEAN PARLIAMENT (1989). Resolution on ethical and legal problems derived of the genetic manipulation, Document A 2–327/88.

FALÇAO DE OLIVEIRA, G. (1995). Genetic knowledge and labour law. In *The Human Genome Project: Legal Aspects,* Vol. 4. BBV Foundation, Bilbao, Spain.

GROUP OF ADVISERS ON THE ETHICAL IMPLICATIONS OF BIOTECHNOLOGY, EUROPEAN COMMISSION (1996). Opinion on the ethical aspects of prenatal diagnosis, No. 6.

HEREDERO HIGUERAS, M. (1994). La protección de datos de salud informatizados en la Ley Orgánica 5/1992, de 29 de octubre. *Derecho Salud* **2**(1), 180.

INTERNATIONAL LABOUR OFFICE (1993). *Worker's Privacy, Part III, Testing in the Work Place,* Vol. 12, No. 2. International Labour Office, Geneva.

LUCAS-MURILLO DE LA CUEVA, P. (1992). La protección de los datos personales ante el uso de la informática en el derecho español. *Estudios Jurisprudencia* **3,** 15–17.

MALEM-SEÑA, J. F. (1995). Privacy and genetic mapping. *Law Hum. Genome Rev.* **2.**

MARTÍNEZ-BULLÉ GOYRI, V. M. (1995). Genética humana y derecho a la vida privada. In *Genética Humana y Derecho a la Intimidad.* Univ. Autónoma de Mexico, Cuaderno del Núcleo de Estudios Interdisciplinarios en Salud y Derechos Humanos, Mexico.

MENÉNDEZ-MENÉDEZ, A. (1995). The genetics code and insurance contracts. In *The Human Genome Project: Legal Aspects,* Vol. 3. BBV Foundation, Bilbao, Spain.

MUÑOZ DE ALBA MEDRANO, M. (1995). La informática frente al derecho a la intimidad. El caso de la información genética. In *Genética Humana y Derecho a la Intimidad.* Univ. Autónoma de México, Cuaderno del Núcleo de Estudios Interdisciplinarios en Salud y Derechos Humanos, México.

MURRAY, T. H. (1995). Genetic information and access to health insurance. In *The Human Genome Project: Legal Aspects,* Vol. 3. BBV Foundation, Bilbao, Spain.

NATOWICZ, M. R., ALPER, J. K., and ALPER, J. S. (1992). Genetic discrimination and the law. *Am. J. Hum. Genet.* **50,** 465–475.

THE PRIVACY COMMISSIONER OF CANADA (1992). *Genetic Testing and Privacy.* The Privacy Commissioner of Canada, Ottawa, Canada.

RODOTÀ, S. (1992). Le droit face aux dilèmmes moraux de la vie et de la mort. In *XXth Colloquy on European Law, Glasgow, 1990.* Council of Europe, Strasbourg.

ROMEO-CASABONA, C. M. (1988). *Poder Informático y Seguridad Jurídica.* Fundesco, Madrid.

ROMEO-CASABONA, C. M. (1993). Tendencias actuales sobre las formas de protección jurídica antes las nuevas tecnologías. *Poder Judicial* **31,** 163–164.

ROMEO-CASABONA, C. M. (1994). Questions de droits de l'homme dans la recherche en génétique médicale. In *Ethique et Génétique Humaine, 2nd Symposium du Conseil de l'Europe sur la Bioéthique.* Council of Europe , Strasbourg.

ROMEO-CASABONA, C. M. (1996). Genetics and law. In *Law in Motion, First World Law Congress.* Kluwer, Deventer.

SIMITIS, S. (1994). Allgemeine Aspekte des Schutzes genetischer

Daten. In *Genanalyse und Persönlichkeitsschutz.* Polygraphischer Verlag, Zurich.

SLOVAK, A. J. M. (1993). Should atopic employees be excluded from specific occupations? *Occup. Med.* **43,** 51–52.

SOLA RECHE, E. (1991). La protección penal de la intimidad informática. *Rev. Facultad Derecho Univ. Laguna* **11,** 186.

SPRUMONT, D., and GUILLOD, O. (1995). Implications of genetic knowledge in work relations. In *The Human Genome Project: Legal Aspects,* Vol. 4. BBV Foundation, Bilbao, Spain.

TASK FORCE ON GENETIC INFORMATION AND INSURANCE (1993). *Report.* National Institutes of Health, Bethesda, MD.

UNESCO (1997). *Universal Declaration on the Human Genome and Human Rights.* UNESCO, Paris.

WIESE, G. (1994). *Genetische Analysen und Rechtsordnung.* Luchterhand Verlag, Neuwied.

YANES-YANES, P. (1994/1995), Personal insurance and genetic information. *Law Hum. Genome Law* **1/2** 186–258.

YESLEY, M. S. (1997). Genetic privacy, discrimination, and social policy: Challenges and dilemmas. *Microbial. Comp. Genom.* **2,** 21–33.

ZEITZ, K. (1991). Employer genetic testing: A legitimate screening device or another method of discrimination? *Labor Law J.*

General References

KNOPPERS, B.-M., CADIET, L., LABERGE, C. M., *et al.* (1992). *La Génétique Humaine: De la Information à la Informasitation.* Litec, Paris.

ROMEO-CASABONA, C. M. (1996). *Del Gen al Derecho.* Univ. Externado de Colombia, Bogotá.

Antonio Cao
Maria Cristina Rosatelli

Department of Biomedical Science and Biotechnology
University of Cagliari
Cagliari, Italy

Genetic Screening

Genetic screening is the search procedure for particular genotypes that are determinant or predisposing causes of diseases in the detected individuals and their offspring. According to the target population, three types of genetic screening can be distinguished as follows: newborn screening, heterozygote screening, and prenatal screening. In most of the programs newborn screening is focused on the detection of newborns affected by phenylketonuria and hypothyroidism, two diseases whose missed detection in the neonatal period implies development of mental insufficiency. The second is mainly concerned with Tay–Sachs disease, β-thalassemia, and in prospective cystic fibrosis. Finally, prenatal screening deals with prevention of the hemolytic disease of newborns from Rh incompatibility and detection of at-risk pregnancies for chromosomal aberrations, especially Down's syndrome, and neural tube defects. Improvement in the knowledge of genetic diseases due to the Human Genome Project leads us to presume that the number of genetic diseases that can be targeted for screening will increase remarkably in the near future.

Introduction

Genetic screening refers to a research procedure aimed at identifying, in a specific population, those individuals whose genotype determines or confers the predisposition to develop a specific disease either in the individuals being screened or in their offspring. Another type of screening, carried out for research purposes only, consists of a search for the incidence of a disease-associated polymorphic trait in a specific population.

The main types of genetic screenings, categorized according to the target population, are newborn screening, screening for heterozygotes for autosomal recessive diseases in adults, and prenatal screening. Recently, a screening for common genetic diseases, transmitted according to an autosomal-dominant pattern, in relatives at risk has also been added to this consolidate group of screening. The aim of the latter is to detect in the preclinical phase diseases that appear in adulthood and to timely provide an adequate therapy or prophylaxis to limit the disease-associated consequences. This article will not deal with screening for polymorphisms and the identification in the preclinical phase of diseases manifesting in adulthood. Newborn screening is specifically aimed at identifying genetic diseases in which a timely proper therapy or prophylaxis could allow the prevention of the disease-determined organ damage. Heterozygote screening allows the identification of couples who are carriers for autosomal recessive disorders, as well as carriers for X-linked recessive disorders, at risk to transmit to their offspring a severe genetic disorder. Prenatal screening allows the detection of genetic diseases in the prenatal stage in women not defined as being carriers for genetic disorders.

The following are general recommendations for genetic screening:

1. Genetic screening is an appropriate form of medical intervention provided that the following criteria are met: confirmed benefit for target population and agreement from both target population and general practitioners, feasibility, prevalence of benefits over costs, proper education of the population, availability of adequately equipped laboratories, and setting up of pilot programs.

2. Organization of a screening program should be the responsibility of a health service (preferably a public one), acting in concert with relevant organizations.

3. Before undertaking a genetic screening, rationale and goals must be defined. The latter might require a medical intervention or family planning. Both these aspects should be discussed with the general public.

4. Population screening, aimed at timely medical treatment, should be included in a program, as should planning the timely recontact of those people determined to be positive. The program should include the capability to perform accurate diagnosis, carry out genetic counseling, offer effective therapy, and provide accurate long-term follow-up.

5. Under the technical profile the chosen screening procedure will ensure the maximum specificity (limiting the false positives) and sensitivity (limiting the false negatives) and should have a high predictive efficiency (ratio of real positives/false positive). Obviously, the procedure should be constantly evaluated to prove its efficacy.

6. Every program will be designed taking into consideration both the ethnical and the geographic variants for either genotype or phenotype characterizing each disease.

7. Accurate understanding of its contents is a required condition for participation in the program.

8. Explanation of the meaning and outcome of clinical testing will be provided to patients and physicians involved in the screening program.

9. The medical/legal impact and results of the program should be constantly evaluated, which will allow the addition of proper amendments to the original project.

10. Screening programs should be associated with educational programs on general genetics and related to the specific aspects of genetic diseases for the screening subjects, including teaching of genetics and probability law and the training of general practitioners on preventive genetics.

11. In general, participation in a screening program should be voluntary.

12. Absolute confidentiality should be kept for results that should be disclosed only to the subject (National Academy of Sciences, 1975; American Academy of Pediatrics, 1976). Developments in the molecular diagnosis of genetic diseases will result in the identification of an increasing number of genetic diseases, potentially subjects for screening. For every amendment to the current screening guidelines, workable on the basis of recent research, preliminary, accurate evaluation of the risks and benefits needs to be performed.

Newborn Screening

General Aspects

Newborn genetic screening essentially deals with some inborn errors of metabolism and congenital hypothyroidism. Inborn errors of metabolism constitute a group of genetic diseases caused by a specific enzyme deficiency producing disease because of the accumulation of metabolites from the enzymatic reaction involved and/or because of a lack of reaction product. The rationale for this screening is based on the evidence that a large proportion of inborn errors of metabolism and congenital hypothyroidism are not recognizable at birth, but present in infancy, childhood, or even adulthood, when irreversible organ damage has already occurred. Detection of these defects at birth allows proper treatment, thus permitting the prevention or at least the attenuation of the evolution of the disorder. Phenyketonuria (PKU), a disorder caused by a defect in the metabolism of the amino acid phenylalanine, was the first metabolic disorder subjected, in many populations since the 1960s, to newborn screening. In the 1970s, congenital hypothyroidism screening was associated with screening for PKU. Guidelines for newborn screening vary in different Western countries. In Europe, the minimal screening program focuses on PKU and hypothyroidism. Most German and Swiss people are also screened for other metabolic defects. In the United States, screening programs vary from state to state. The metabolic diseases most commonly screened, besides PKU, are galactosemia, maple syrup urine disease, homocystinuria, tyrosinemia, and byotinidase deficiency, which are due, respectively, to defective metabolism of galactose, branched-chain amino acids, homocystine, tyrosine, and organic acids. The inclusion of screening for sickle-cell anemia, congenital adrenal hyperplasia, and cystic fibrosis (CF) is still controversial. In developing countries, newborn genetic screening is not operative or there are only limited pilot programs.

Newborn screening has nevertheless allowed the collection of extremely well-defined incidence data for different genetic diseases. In general, the values obtained indicate a higher incidence than those previously obtained (Table 1). Due to screening, the spectrum of the phenotypic variations of different genetic disorders has often been defined.

Ethical Aspects

Recently, improvement in molecular genetics and its potential applicability in newborn screening has settled an open debate on the ethical aspects of this kind of screening,

TABLE I

Incidence of Metabolic Genetic Disorders Detected by Newborn Screening

DISORDER	INCIDENCE/ LIVE BIRTHS[a]
Cystic fibrosis	1:2,500
Congenital hypothyroidism	1:4,000
Cystinuria[b]	1:7,000
α_1-Antitrypsin[c] deficiency	1:8,000
Phenylketonuria	1:12,000
Histidinemia[d]	1:17,000
Imminoglycinuria	1:20,000
Hartnup's disease[b]	1:26,000
Wilson's disease[e]	1:35,000–40,000
Hyperprolinemia[d]	1:40,000
MCAD deficiency[f]	1:40,000
Congenital adrenal hyperplasia from 21-hydroxylase deficiency	1:40,000
Galactosemia	1:57,000
Biotinidase deficiency	1:60,000
Adenosine deaminase deficiency	1:100,000
Maple syrup urine disease	1:200,000
Homocystinuria	1:200,000

[a] Average incidences are reported. There are relevant differences in the incidence data among different ethnic groups.

[b] Disease resulting from specific aminoacid transport defect in kidney and, in some subtypes, also in intestine.

[c] Disease that may produce pulmonary emphysema and chronic hepatic disease ending in cirrhosis.

[d] Disease due to inborn error of metabolism.

[e] Disease resulting from liver and brain copper accumulation.

[f] Disease resulting from defective medium-chain acyl-CoA dehydrogenase deficiency.

especially on the employment of DNA specimens for purposes that differ from those of the original screening. From this debate, some principles on which newborn screening programs should be based have been developed: Newborn genetic screening is to be considered a medical act in the context of preventive medicine and should be offered to all newborns in the target populations; parents and the general public should be informed about the purpose of the program, the diseases subject for screening, and the procedures employed for detection. The methodologies to be employed are those whose sensitivity, specificity, and reliability have been tested in pilot studies on the target population. The programs must provide information relative to the results of the screening to the parents and should also be associated with a proper control of the screen-positive newborns in order to confirm diagnosis and/or consent for other appropriate medical intervention or prophylaxis.

Moreover, the program may consent to the use of specimens for research provided that confidentiality is kept and that this is done only on well-defined terms. If the screening program is compulsory or based on automatic participation, in the absence of specific refusal, information on the specific studies to be performed on DNA should be given to the population involved. For epidemiological investigation of diseases for which any medical intervention or any benefit to the newborn cannot be achieved, confidentiality should obviously be kept, and in order to avoid individual or public stigmatization or discrimination the results should have no connection with the program and free and voluntary access to individual outcomes should be made available. When the screening program is based on individual informed consent, individual agreement for research respecting confidentiality, as well as for epidemiological studies, is required. Finally, complete confidentiality should be kept unless specific consent is given. DNA specimens should be kept in such way to ensure the same confidentiality as that for other medical acts. Genetic newborn screening can employ DNA analysis provided that molecular pathology and genetic heterogeneity, for the specific disease and population, have been completely defined (American Academy of Pediatrics, 1992).

In many European countries, as in several states in the United States, newborn screening is mandatory. In other U.S. states, informed consent is needed instead. Concerning this issue, it is almost unanimously agreed that informed consent is not required, provided that the screening focuses on diseases whose diagnosis and treatment could ensure benefit to the newborn who is a target for screening (Fletcher *et al.,* 1985; Laberge, 1994; Knoppers *et al.,* 1996).

During the newborn screening for curable diseases, the presence of the carrier state for a recessive autosomal disease not related to the disease being screening for might be detected, and its presence should be adequately disclosed to the parents. Moreover, free choice should be given to parents to decide whether to continue investigation to establish whether only one partner of the couple or both are carriers for the specific genetic disease detected. As previously stated, a basic prerequisite for newborn genetic screening is its utility in the treatment of the newborn. There no doubt about the existence of critical requisites for screening for some diseases, such as PKU and hypothyroidism. For other metabolic diseases subject to screening, including galactosemia, maple syrup urine disease, and homocystinuria, the benefits of screening are certainly less evident because of the sudden appearance, complexity of treatment, and rare occurrence of these diseases (Rowley, 1984). Currently, other diseases proposed for screening, such as adenosine deaminase defect, a congenital immunological defect, and tyrosinemia, due to their rarity are not included in screening programs except for some specific populations. Inclusion of newborn screening for cystic fibrosis and sickle-cell disease (Table 2) is still controversial but is recommended

TABLE 2

Diseases for which Newborn Screening Is Recommended or Controversial

SCREENING RECOMMENDED	SCREENING CONTROVERSIAL
Phenylketonuria	Adenosine deaminase deficiency
Hypothyroidism	Tyrosinemia
Sickle-cell anemia	Galactosemia
	Homocystinuria
	Maple syrup urine deficiency
	Homocystinuria
	Biotinidase defect
	Cystic fibrosis
	Congenital adrenal hyperplasia
	X-linked mental retardation with fragile X
	G6PD deficiency

because timely diagnosis would permit a prompt therapeutic approach to reduce the risk of premature death. For Duchenne progressive muscular dystrophy, the most common progressive muscular disease, screening is suggested for appropriate genetic counseling that could prevent a second child being born that would be affected by the disease. For this purpose, it must be noted that the efficacy of the incidence of the disease obtained with this kind of preventive approach is minimal. On the other hand, it should be considered that for newborn lacking any symptoms, there is not an adequate ethical requirement for screening of Duchenne muscular dystrophy. If the screening is eventually performed, informed consent is nevertheless needed.

Principal Diseases Subject to Newborn Screening

Hyperphenylalaninemias

Hyperphenylalaninemias are congenital defects of the hydroxylation of phenylalanine (Phe) to tyrosine (Tyr). The mean incidence is 1 every 12,000 live births. Hyperphenylalaninemias present extremely heterogeneous genotypes and are due to mutations of different genes coding proteins involved in the hydroxylase reaction. The most common defects concern the enzyme phenylalanine hydroxylase (hyperphenylalaninemia with or without phenylketonuria). Hyperphenylalanemias due to phenylalanine hydroxylase defect are genetically very heterogeneous; there are more than 100 different mutations that show a population-specific distribution. Newborn screening for hyperphenylalaninemias assesses Phe through a semiquantitative assay based on the inhibition of bacterial growth, determined by increased Phe levels. The maximum normal level for Phe is 2 mg/dl.

Untreated PKU results in severe mental retardation sometime associated with autism, hyperactivity, convulsions, aggressive behavior, eczema, and light skin and hair. Mental defect is related to the negative effect that Phe excess has on myelination and on brain protein synthesis. The presence of pathologically elevated Phe levels leads to initiation of a Phe diet. The diet should be carefully followed and critically reevaluated in the light of periodical serum Phe assessment. Good results are achieved when diet regimen is introduced before the third week of the infant's life. Although timely therapy allows prevention of mental retardation and neurological alterations, some modest difficulties in learning may still occur even in patients properly treated. General consensus on a lifelong diet has been reached. In fact, interruption of diet in children or adolescents has always been followed by mental deterioration. An additional advantage of this protocol is the prevention of organ damage (mycrocephaly, congenital cardiac defects, and intrauterine growth retardation) caused in a non-PKU fetus from a non-PKU mother not on a low Phe diet. These alterations are the consequence of the negative effect that exceeding maternal Phe exerts on organ development.

Due to the extreme molecular heterogeneity of hyperphenilalaninemia in most populations, it is currently not possible to substitute the procedure usually employed with direct mutations analysis; nevertheless, for each case detected through the standard procedure, the diagnosis should be completed with a search for the specific mutation. This information, together with the study of haplotypes in the case of unknown mutations, is of practical utility in the detection of carriers among families and for prenatal diagnosis. Moreover, even if the correlation between specific mutation and phenotype is not precise, the detection of a specific mutation may allow more precise diagnosis, optimize therapy, indicate the severity of the case, and determine the long-term prognosis.

Congenital Hypothyroidism

Congenital hypothyroidism has an incidence of 1 in 4000 live births. This disorder is produced from embryogenic defects resulting in thyroid agenesis (50–55%) or dysgenesis (30–35%) (mostly not inherited), congenital metabolic defects of biosynthesis of thyroid hormones (10–15%), and congenital defects or metabolic errors in the hypophysis or the hypothalamus (4%) with resultant reduced production of the thyrotopic hormone.

Since thyroid hormone is needed for the development of the central nervous system, every diagnostic delay of congenital hypothyroidism, even a short delay, may result in permanent damage to the nervous system, with consequent mental retardation. Therefore, screening programs for congenital hypothyroidism have operated in Europe, Japan, and the United States for many years (American Academy of Pediatrics, 1993; Irons, 1993). The testing material is represented by blood specimens collected on filter paper by the fifth day of life to avoid interference due to hypophysis and thyroid postpartum stimulation. Two different strategies are currently used. In the United States serum tyroxine (i.e., the thyroid hormone T4) is measured, which should allow the diagnosis of all forms of hypothyroidism except the rarest forms due to peripheral resistance to thyroid hormones. In Europe, thyroid-stimulating hormone (TSH) is measured. An increase in TSH enables identification of the most common forms of hypothyroidism due to thyroid defect; however, with this approach rare cases due to hypophysis and hypothalamus defects are missed. TSH examination, however, is more sensible than T4 measurement. Both procedures may give false-negative or false-positive results, and obviously the best results are achieved by the concurrent evaluation of both T4 and TSH. Positive cases may be confirmed by a complete evaluation of thyroid function. After the eventual diagnostic evidence has been obtained, T4 treatment should be started immediately. When the therapy is started within the first month of life, the outcomes are excellent, with the exception of the association of fetal–maternal hypothyroidism, which may produce cerebral damage in the prenatal stage. Therapy should be monitored by determining TSH and T4 levels. T4 should result in high normal values, whereas TSH levels normalize very slowly. After several years, the treatment can be interrupted for a brief period for diagnostic reevaluation.

Sickle-Cell Anemia

Sickle-cell anemia is one of the most frequent autosomal recessive genetic diseases. Its incidence is particularly elevated in equatorial Africa, Arabic mainland, and India. Certainly, due to the slave trade, an elevated incidence also occurs in North and South America and in the Caribbean. The most common genotype of sickle-cell disease is homozygosity for the *HbS* gene, resulting from a single nucleotide substitution at codon 6 of the β-globin gene, which determines glutamic acid substitution with valine. Less commonly, sickle-cell anemia is due to compound heterozygosity for *HbS* and β-thalassemia (a disease caused by the defective production of the β-chains of the Hb molecule) or some Hb variants, such as *HbC* or *HbO* (Table 3).

The most relevant clinical features of sickle-cell disease are anemia and crisis of vascular occlusion. From early infancy the disease may result in several complications which may be fatal, including hyperacute sepsis from *Streptococcus pneumoniae* and splenic sequestration crises. Detection of this disorder in newborns enables a control program to be set up, which includes penicillin prophylaxis and pneumococcus vaccination, that may result in reduced mortality of affected children. The previous considerations are the rationale for newborn screening. Nevertheless, it has not been definitively established whether screening for sickle-cell anemia has a positive impact on morbidity and mortality.

TABLE 3

Different Genotypes Causing Sickle-Cell Anemia

FREQUENT	RARE
Sickle-cell anemia (HbS/HbS)	HbS/HbD disease
HbS/β-thalassemia (β^0 or β^+)[a]	HbS/HbO Arab diseasc[b]
HbS/HbC	$\delta\beta$-Thalassemia[c]

[a]Depending on whether β-thalassemia is due to reduced (β^+) or absent (β^0) β chain production.

[b]HbC, -D, and -O are different abnormal hemoglobins due to single-nucleotide substitutions.

[c]$\delta\beta$-Thalassemia is a thalassemia disorder characterized by the absence of δ and β chain production.

Screening is carried out through Hb electrophoresis on cellulose acetate and agar citrate. A valid alternative is electrofocusing of hemoglobin. These procedures enable the identification and quantification of various types of hemoglobins (S, C, A, A_2, and F). Sickle-cell anemia screening is ongoing for all newborns in the United States, Cuba, Great Britain, and France for minorities at risk. Obviously, this screening implies crucial adverse effects, especially for populations of ethnic black minorities. Among these effects are stigmatization, loss of self-esteem, and discrimination. Regarding discrimination, especially in workplaces, preventive measures should be instituted considering that under this profile many abuses have occurred in the past in the United States. For this purpose, absolute confidentiality must be effectively ensured. Screening for sickle-cell anemia is seemingly justified only in the presence of a health care system organized for properly following up these hemoglobinopathies. During the screening program sickle-cell carriers, as well as carriers for other hemoglobinopathies, should be identified. Carrier identification, however, requires an adequate genetic counseling program (Cao, 1987, 1994; Angastiniotis *et al.,* 1995; Petrou and Modell, 1995).

Cystic Fibrosis

Cystic fibrosis (FC) is the most common autosomal recessive inherited disorder among people of Caucasian ancestry, with an incidence of 1 out of 2000–3000 live births. Principal clinical features are chronic obstructive pulmonary disease (due to the peculiar density of mucus); persistent pulmonary infection, especially from *Pseudonomas aeruginosa;* meconium ileus (intestinal obstruction due to elevated meconium density); pancreatic failure; and defective growth. Diagnosis, clinically suspected, is confirmed by the finding of increased chloride values in sweat (> 60 meq/

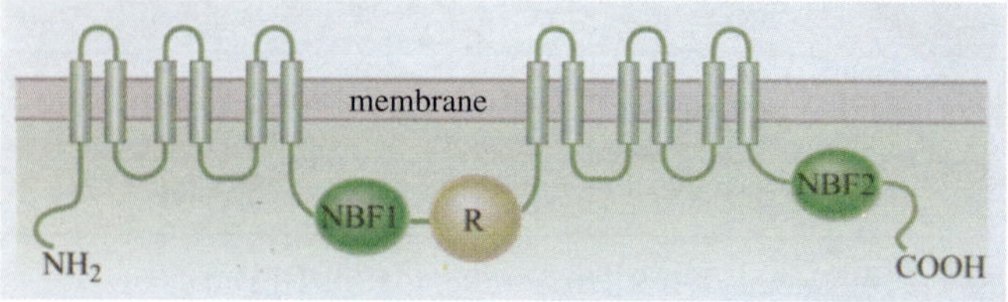

FIGURE 1 The membrane chloride channel. Mutations in the gene for this protein are responsible for cystic fibrosis. The NBF1 and NBF2 domains are sites for ATP binding; R is the regulatory domain.

liter). CF is due to a molecular defect of a membrane chloride channel. The most frequent mutation (about 70% in northern Europe and 50% in Italy) causing CF is a three base-pair deletion coding phenylalanine at codon 508. In addition to this mutation (called ΔF508), there are more than 500 mutations, some quite frequent (with incidence of 1–3%) and others detected only in individual families (Fig. 1). Distribution of the different mutations is population specific. Therapy is based on postural drainage, physiotherapy, antibiotic cyclic administration to limit damage resulting from lung infections, and substitutive pancreatic enzyme therapy. Prognosis has considerably improved, and average survival is 30 years. Neonatal diagnosis might result in early therapy and therefore prevent or delay complications. The previous considerations are the rationale for genetic screening. Nevertheless, definitive evidence that an early diagnosis might improve the course and prognosis of the disease is lacking. Detection of at-risk couples before the beginning of pregnancy, after informed consent has been obtained, certainly is a considerable additional advantage. The procedure employed is quantitative evaluation of trypsin with an immunological methodology for blood specimens collected on filter paper. This technique has a high frequency of false-negative results (5–10%), especially among cases of meconium ileus. However, there is also an elevated number of false positives. In those cases determined to be positive (increased trypsin values), genotype determination through direct gene analysis for the most common mutations in the target population is suggested.

X-Linked Mental Retardation with Fragile X

X-linked mental retardation with fragile X is the most frequent cause of inherited mental retardation. This syndrome is caused by the expansion of a trinucleotide repeat cytosine-guanine-guanine (CGG) located in the untranslated region of a gene of unknown function dubbed fragile mental retardation 1 (*FMR-1*). The incidence of fragile X syndrome is 1 in 1500 males. Normal individuals have less than 5 copies of the repeat. Two types of mutations can be distinguished, i.e., premutation and full mutation (Fig. 2). The premutation which consists of a modest expansion of the repeat (from 60 to less than 200 copies) is not associated with mental retardation. The full mutation is characterized by a greater expansion of the repeat (> 200 copies) and is associated with mental retardation. Premutation may expand to a complete mutation only through maternal transmission. Pre- and postnatal diagnoses, as well as premutation carrier detection, are based on direct DNA analysis.

Currently, the only available screening for this syndrome

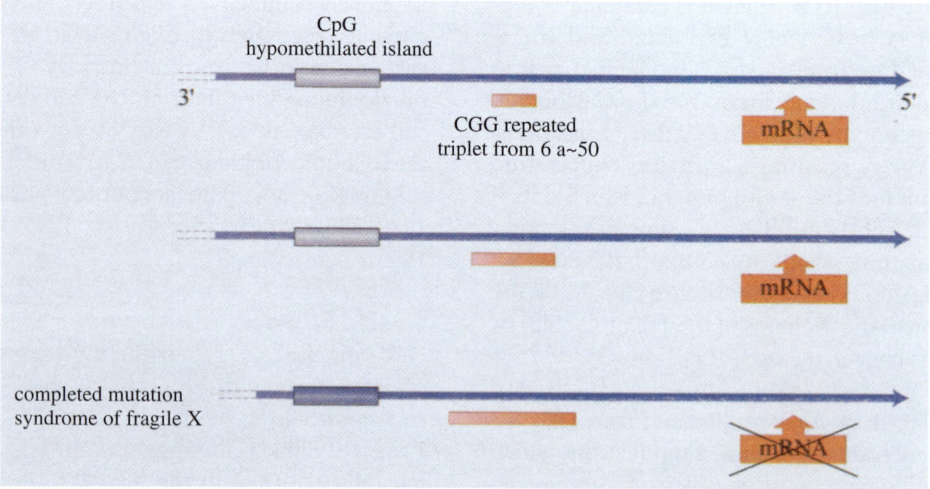

FIGURE 2 Schematic representation of the 5′ untranslated region of the *FMR1* gene, associated with fragile X syndrome. The gene contains CGG triple repeats in the region adjacent to the CpG island flanking the coding region. Normal subjects may have 6–50 CGG triplets. In the premutation case, there are 50–200 repeats. In the full mutation, there are more than 200 triplets and the CpG island is hypermethylated. The premutation may expand to a complete mutation only through maternal transmission.

is for subjects with mental retardation since it allows detection of at-risk families that have appropriate genetic counseling. For newborn screening lacking one of the basic requisites (i.e., benefit for the newborn) there is no rationale for application. In fact, a unique advantage for newborn screening related to mental retardation might be the early detection and consequent genetic counseling. Concerning women carriers for the full mutation, it must be noted that prediction of mental retardation level is not possible; therefore, detection might involve an elevated anxiety level among family members. Screening for childbearing women, aimed at detecting premutation carriers, would be justified by the frequency and the severity of the syndrome. The limitation of the use of this procedure is essentially a technical issue since a simple approach able to define both premutation and complete mutation is not available. Improving DNA analysis procedures may result in the development of a procedure with the requested characteristics in the near future. Lack of knowledge regarding the frequency of carriers of *FMR-1* point mutations and the risk of carriers with borderline repeats is another issue (Mandell *et al.,* 1994).

Glucose-6-Phosphate Dehydrogenase Deficiency

Glucose-6-phosphate dehydrogenase (G6PD) is an ubiquitous enzyme, catalyzing the first step of the hexose monophosphate shunt which is the main source of NADPH production. NADPH, crucial for preservation and regeneration of reduced glutathione (GSH), has among its various rules the essential one of detoxification of hydrogen peroxide. This activity is particularly crucial for red blood cells, which are very susceptible to oxidative damage and lack other enzymes that produce NADPH. The G6PD defect is the most common among the inherited enzymopathies. The highest incidence is observed in tropical and subtropical Africa and in several Mediterranean regions. The most common clinical features are neonatal jaundice and acute hemolytic anemia. The latter is induced from ingestion of fave beans, use of oxidant drugs, and from several infections, particularly typhoid fever and acute hepatitis. G6PD deficiency is genetically extremely heterogeneous. More than 400 different variants of this defect are known, most of which are molecularly defined. Some of these variants are associated with chronic hemolytic anemia.

Newborn screening for G6PD deficiency is controversial since many of the requisites, previously mentioned, are lacking. Nevertheless, newborn screening occurs in several high-incidence areas, such as Sardinia, Thailand, and Malaysia. The utility of screening is limited to the information on the risk of exposure to agents inducing hemolysis, whereas prevention of neonatal jaundice is not possible. Due, at least partially, to screening, the incidence of acute hemolysis from G6PD deficiency has been drastically reduced in these countries. Screening, however, should be inexpensive, limited to homozygotes and hemizygotes, since simple procedures are not apt to detect heterozygotes.

Heterozygote Screening

A second type of screening consists of the screening the adult population to detect heterozygotes for autosomal recessive disorder or X-linked disorders. For subjects determined to be positive, it is suggested that they undergo genetic counseling. Programs for identification of heterozygotes can identify couples at risk for producing children affected by genetic disorders before a proband birth (i.e., a subject affected by the disease); thus, these programs have outstanding preventive efficacy.

The following are unanimously agreed on requisites necessary for undergoing heterozygote genetic screening: elevated incidence of the disorder in the target population for screening; availability of inexpensive and simple methodologies for detection of heterozygotes, with limited outcome of false positives and possibly false negatives; and alternative reproductive choice available both for at-risk couples for autosomal recessive disorders and for couples whose female partner is a carrier for an X-linked disorder. Disorders that have indication for heterozygote screening are β-thalassemia, Tay–Sachs disease, CF, and sickle-cell disease. The main experience in preventive genetics concerns Tay–Sachs disease and β-thalassemia. Programs relative to these two disorders and general principles deriving from such expertise will be discussed briefly; other disorders, whose rationale for heterozygote screening is still debated (especially CF), will not be discussed.

Tay–Sachs Disease

Tay–Sachs disease (TSD) is one of the GM_2 gangliosidoses. GM_2 gangliosidoses are autosomal recessive disorders characterized by intralysosomal GM_2 ganglioside overload and are mostly due to a defect of the β-hexosaminidase enzyme. Two different molecular forms (isoenzymes) of β-hexosaminidase are known: β-hexosaminidase A, composed by one α subunit and one β subunit, and β-hexosaminidase B, composed of two β subunits. α and β subunits are codified by the distinct A and B genes. TSD and its variants are caused by A subunit gene mutation. Isoenzyme A is defective, whereas form B is normal. Gene B defect induces a similar, even more rare, disease known as Sandhoff disease, which is associated with defective activity of both A and B isoenzymes. Rare forms of this group are caused by a defective function of an activator. Under the clinical profile they can be considered neurogenerative progressive disorders, whose most severe form is TSD, leading to death within the fourth year of life. All the GM_2 gangliosidosis forms are genetically heterogeneous due to an elevated number of different mutations. Among the Ashkenazy Jews population, TSD is more homogeneous since 90–95% of the molecular defects are composed of only two mutations. No cure is available for GM_2 gangliosidoses. By employing an artificial fluorescent substrate, healthy carriers can be identified because they have reduced β-hexosaminadase enzymatic activity in serum (Kaback, 1993). The same procedure applied to amniotic cells or chorionic villi can be used to perform prenatal diagnosis of the disease. Currently, both healthy carrier identification and prenatal diagnoses are confirmed by direct mutations analysis.

In 1970, a screening program, including genetic counseling, began in the United States. This program was directed to childbearing-age subjects belonging to the Jewish population, in whom TSD incidence is of 1 out of 4000 live births and the carrier rate is 1:30. Later, this type of program was applied to several Jewish communities throughout the world. The numbers of heterozygotes and at-risk couples are reported in Table 4. Couples at risk were given nondirective genetic counseling, in which various reproductive options were discussed, including prenatal diagnosis. Most of the couples counseled chose prenatal diagnosis and pregnancy termination if there was evidence that the fetus was affected (Table 5).

Only in a very limited number of cases was the prenatal diagnosis of affected fetus not followed by therapeutic interruption of the pregnancy. The reasons were mainly diagnostic delay in ongoing pregnancy or religious or ethical beliefs against abortion. These programs led to a dramatic reduction, on the order of 90%, of the birth rate of TSD in the populations targeted for screening.

β-Thalassemia

β-Thalassemia, one of the most common autosomal recessive disorders, is caused by a reduced (β^+) or absent (β^0) production of the β-chains of the hemoglobin molecule.

TABLE 4

Tay–Sachs Disease Heterozygote Screening

COUNTRY	NO. TESTED	CARRIERS IDENTIFIED	AT-RISK COUPLES IDENTIFIED
United States	712,818	27,750	657
Israel	159,544	4,229	263
Canada	55,922	2,922	57
South Africa	11,638	1,286	36
Europe	10,927	725	21
Brazil/Mexico	1,682	96	20
Australia	473	10	2
Total	953,004	36,418	1056

TABLE 5

Worldwide Prenatal Diagnosis of Tay–Sachs Disease (1969–1992)

	COUPLE IDENTIFIED AT RISK BY		
	Prior Affected Offspring	Carrier Screening	Total
Pregnancies monitored	1118	1298	2416
Affected fetuses	268	201	469
Elective abortions	250	201	4510[a]
Tay–Sachs disease fetuses missed	2	1	3
Unaffected offspring born	827	1054	1881

[a] Eighteen infants affected with infantile Tay–Sachs disease as predicted.

This disease is very frequent in Mediterranean basin populations, particularly in Sardinia and Cyprus, and also in the Middle East, India, and far Western countries. Due to emigration and slave trade, β-thalassemia has spread throughout the world, (i.e., the United States, northern Europe, Canada, South Africa, and Australia). Homozygotes for β-thalassemia present with a severe anemia caused principally by red blood cell precursor destruction in bone marrow (ineffective erythropoiesis) and by a peripheral hemolysis. Clinical features are characterized by an early anemia occurring at approximately the sixth month of life, spleen and liver enlargement, and bone alterations due to the expansion of bone marrow as it attempts to compensate for anemia. If not treated, β-thalassemia patients die in early childhood. Instead of producing hemoglobin (Hb) A and small quantities of Hb A_2, typical in normal adults, homozygotes for β-thalassemia produce large quantities of Hb F and traces of Hb A_2; Hb A is markedly reduced (β^+ thalassemia) or completely absent (β° thalassemia). The health carrier state, or heterozygote for β-thalassemia, is characterized by the presence of a reduced volume of red blood cells, reduced corpuscular mean hemoglobin concentration, and increased Hb A_2 percentage (>3.5%). Treatment of the homozygous state implies a continuous transfusional regimen and chelation therapy (desferroxamine B) to eliminate iron overload due to transfusions. This treatment increases survival to 30 years of age. A valid alternative to this therapy is bone marrow transplantation from HLA identical siblings or parents, which provides a cure in approximately 80–90% of cases, especially if performed before the occurrence of complications. Currently, there are only a limited number of appropriate donors. β-Thalassemia is genetically markedly heterogeneous; approximately 200 mutations are known. The number of mutations is nevertheless limited in each population. A β-thalassemia carrier detected through red cell indices evaluation and quantitative determination of Hb A_2 by chromatographic or electrophoresis procedures is studied with molecular methodologies to assess the type of mutation. The same molecular methodology is employed for prenatal diagnosis on chorionic villi or amniotic cell DNA (Cao and Rosatelli, 1993; Petrou and Modell, 1995).

In the mid-1970s, screening programs for β-thalassemia, associated with genetic counseling and prenatal diagnosis, were introduced in Sardinia, Cyprus, continental Greece, and among ethnic minorities in Great Britain (i.e., Cypriots and Pakistani) (Cao, 1987, 1994; Angastiniotis *et al.*, 1995; Petrou and Modell, 1995). Since then, screening has been extended to continental Italy, Sicily, and to at-risk ethnic minorities in other European countries and the United States. Table 6 and Figure 3 show results on prenatal diagnosis and the relative decrease of β-thalassemia homozygotes in the Sardinian population. On the whole, 95% prevention has been achieved. The remaining cases are mainly the result of lack of information and less frequently, due to the choice of maintaining pregnancies of homozygote fetuses for ethical issues, to carrier state, or to prenatal misdiagnosis or nonpaternity.

Characteristics of Heterozygote Screening Programs

Screening programs for TSD and β-thalassemia have many characteristics in common. On the whole, programs concerning β-thalassemia in many aspects have followed the Tay–Sachs screening program model. The target population has been always adults of childbearing age. In the Tay–Sachs screening program and in many programs for β-thalassemia, the subjects for screening are mainly married

TABLE 6

Prenatal Diagnosis for β-Thalassemia in Sardinia (Overall Results)

	FETAL DNA[a]	FETAL BLOOD[b]
Pregnancies monitored	3842	1131
Homozygous fetuses	996	286
Failures	7	10
Misdiagnosis	1	2

[a]Carried out by analysis of chorionic villi DNA.
[b]Carried out by fetal blood analysis before DNA analysis was available.

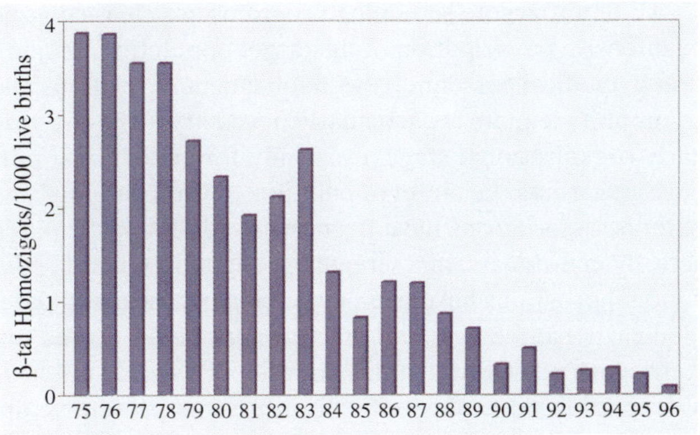

FIGURE 3 Reduction of the number of homozygotes for β-thalassemia from 1975 to 1996 as a result of the prevention program introduced in Sardinia.

or premarried couples. Otherwise, the target population is represented by pregnant women. Premarriage screening is certainly more appropriate, providing different alternatives for couples in whom both partners are carriers. A screening program at the conclusion of compulsory school, as carried out in Cyprus and Lazio, is also effective. It must be noted, however, that heterozygote screening of reproduction directed at children in primary schools or even to younger infants is not necessary since counseling cannot be effective. Some programs carried out in Greece attempted to submit subjects to compulsory screening, but the results on information issues were weak. On the other hand, in a field such as heterozygote screening it is impossible to find any ethical rationale for compulsory screening. Every individual undergoing screening has the right to know its purpose, the benefits that might derive from the knowledge, and the possible adverse outcomes, and individuals should be provided with this information so that they can make the choice of whether or not to participate in the program. Therefore, preliminary informed consent should be obtained from each subject to be screened. Nevertheless, even for those programs that do not require informed consent (all programs concerning β-thalassemia), subjects should be informed in an easy and comprehensible way of the meaning of being carrier, of what carrier state implies, and of the reproductive options available for a carrier. Whether every subject should be properly informed of the possible psychological consequences arising from the knowledge that he or she is a carrier is debatable. Absolute voluntary choice has been completely respected in all these programs. It should be noted that the Orthodox Church in Cyprus, to discourage marriage among carriers, requires before marriage a certificate attesting that both partners have undergone screening. Unquestionably, if there is evidence that they are both carriers, the church does not refuse to celebrate marriage.

All heterozygote screening programs are characterized by intensive participation of the target population. In addition to medical personnel, the involvement of religious and community leaders is particularly necessary, from the very early organizational stage, especially for educational purposes regarding the target population. Several patients' and parents' associations have been involved and participated actively in β-thalassemia screening programs. Overall, educational programs have been directed at medical and paramedical personnel (general practitioners, pediatricians, obstetricians, nurses, and family counselors) who do not have a deep knowledge of genetic disorders and preventive approaches. Education on genetic disorders subject to screening is given to the general public through various communicative channels (mass media, conferences, etc.). In Cyprus, each year 1 week is devoted to thalassemia, during which pamphlets and reviews on the issues are distributed. Educational programs on β-thalassemia have been introduced in secondary school in Greece, Sardinia, and Cyprus. Informative material on how and where to perform testing is available in key locations, such as medical consulting rooms (especially those of pediatricians and obstetricians), familial planning centers, and matrimonial agencies. The message is comprehensive for the following issues: possible identification of the healthy carrier; homozygote state inducing a severe, potentially fatal, disease for which no cure is definitive but only palliative, except for bone marrow transplantation which is feasible only in a limited number of cases, depending on the availability of HLA identical donors; and possible identification of the disease in the fetal stage.

Under the technical profile, experience derived from the programs previously mentioned suggests that it is of primary concern to properly conform screening services and laboratories to the demands of genetic counseling and prenatal diagnosis. A negative aspect is certainly failure to meet the demands of the educational policy. The procedures employed for screening should be simple, precise, and applicable on a large scale; special care should be given so that the incidence of false negatives is minimized. False positive should also be limited to avoid unnecessary anxiety. Therefore, screening procedures should be carried out in the fastest possible way. Population screening implies identification of a broad number of subjects to be addressed later by genetic counseling that should be offered to individuals as well as couples. Particular attention should be paid to those subjects whose test results are unreliable or require analyses to be extended to other family members. The subjects in this situation obviously have to cope with a state of considerable anxiety. Genetic counseling relative to these programs has been mostly carried out in conformity with internationally agreed directives. Particularly, counseling should avoid coercion and should be based on a confidential meeting with the single carriers or with the couple. Any attitude that might, in some way, restrain the couple's freedom should be avoided. Information disclosure is aimed at providing the counselee with those elements that are helpful for making an informed choice. First, unbiased information on the natural history of the diseases and treatment possibilities should to be disclosed as well as information on the existence of attenuate forms of the disease and on the ongoing research in the field. Additional information concerning the available reproductive choices for a couple composed of two carriers should be provided, including the possibilities of ending the relationship, choosing a childless marriage, or having a child and undergoing prenatal diagnosis. Preimplantation and preconceptional diagnoses, besides artificial insemination from a healthy donor, should also be discussed in this circumstance. Generally, details are also given on the procedures employed to obtain fetal specimens for analysis (amniocentesis, chorionic villi, and fetal blood) as well as on the various risks that such procedures imply for pregnancy and for the fetus, on the probability of failure to obtain adequate specimens for diagnosis, and on the risk of misdiagnosis.

A critical aspect of genetic counseling is the disclosure of genetic risk to the carrier's relatives. Disclosure is provided to the subject identified as the carrier in the form of a definite message for relatives. Moreover, educational material is provided for the relatives, who after reading the information can contact health centers for further information or undergo screening. This praxis has been very effective; in Sardinia, for instance, as a result of this so-called "inductive screening," through counseling of a limited number of subjects most of the predictable carriers on the basis of the genetic frequency of β-thalassemia were detected. Direct disclosure to carriers' relatives from the involved centers is otherwise ethically unacceptable since this is considered a violation of confidentiality. Absolute confidentiality should be maintained regarding the screening outcome, which should only be accessible to the subject concerned. The same confidentiality should be observed for DNA specimens, which obviously should not be used, without explicit consent, for purposes other than the primary purposes for which they were collected.

During screening for recessive autosomal genetic disease carriers, a finding of nonpaternity might be incidentally detected. In this situation, it must be remembered that the most important priority of the geneticist is to protect the well-being of the entire family. On the other hand, it should be considered that in many societies the woman is extremely vulnerable to physical, social, psychological, and economic abuse. Every moral judgment on nonpaternity is therefore inappropriate. Concerted consent unquestionably states that disclosure of nonpaternity to the husband should be avoided. Nevertheless, especially for purposes of providing information relevant for future childbearing, it is necessary to tell the truth to the woman, without the husband or partner present. The geneticist should discuss the issue with the mother, suggesting the best time to notify the husband of the situation and also emphasizing the potential psychological benefits of the disclosure, including release from the burden of keeping a secret and honesty in the family relationship. If the mother decides to tell her husband or partner, the counselor should provide adequate psychological and social support. Nevertheless, an honest and open attitude is not possible in numerous social/cultural contests. If it is not possible to discuss the situation with the mother alone, it is better not to provide the information to anyone than to risk harm to the mother and the child. If a husband or partner asks directly about his paternity, the geneticist should follow the principle of preventing in any way harm to the mother.

Heterozygote screening program should be submitted to constant critical evaluation of the organizational and technical issues and also the social impact and psychological implications on individuals detected as carriers. In general, negative effects, such as stigmatization and loss of self-esteem in the carriers, are limited. When undertaking new programs, possible negative effects should be carefully evaluated through pilot studies.

Heterozygote screening programs have achieved a considerable reduction in disease incidence and in general have enabled education of carriers on the disease and the affordable approaches for prevention. Otherwise, screening programs have a modest impact on heterozygous frequency since the majority of heterozygotes derive from marriage between a normal subject and a heterozygote rather than between two heterozygotes (Milunsky, 1992; Cao and Rosatelli, 1993; Cao, 1994).

Cystic Fibrosis Carrier Screening

As already noted, CF is the most common genetic disease transmitted in a recessive autosomal manner in Caucasians. The mutation indicated as ΔF508 is present in 70% of carriers of north European origin; minor frequencies are observed in southern Europe. As formerly reported, there are other (~700), although less frequent, mutations. The CF spectrum of mutations is population specific. In populations of European origin, approximately 85–90% of carriers can be identified through studies of 6–12 population-specific mutations. The number of mutations detectable with limited testing is greater than 700 in several relatively homogeneous populations, such as Ashkenazy Jews and the British. Due to a marked genetic heterogeneity, it is difficult for a test based on DNA analysis to identify more than 95% of the mutations. To achieve this objective requires the capability to perform the test on 20–40 different mutations. The test with a sensibility of 85% has a rate of false-negative outcomes >15% (Table 7). In these conditions testing can detect only 72% of couples at risk to have an affected child. In the case of a carrier–noncarrier couple, and a test with 85% sensibility, the risk of having an affected child increases from 1 in 2000 to 1 in 660. This unquestionably could cause marked anxiety. Only if the percentage of the detection of mutations can be increased to 95% will the risk for a carrier–noncarrier couple be 1 in 1964, which is equal to the risk in the absence of screening. The major technical problems concerning the screening for CF carriers are due to the marked genetic heterogeneity and the lack of simple inexpensive testing for carrier identification.

TABLE 7

Proportion of Detectable Mutations and Residual Risk of Having a Cystic Fibrosis-Affected Child for a Carrier/Noncarrier Couple

DETECTABLE MUTATIONS (%)	RESIDUAL RISK TO HAVE A CYSTIC FIBROSIS-AFFECTED CHILD (%)[a]
70	0.32
75	0.25
80	0.20
85	0.15
90	0.10
95	0.05
96	0.04
97	0.03

[a] General risk for couple before testing (0.04%).

Severity of the disease in each patient is only partially correlated with the type of mutation. Therefore, it is impossible to accurately determine the clinical phenotype on the basis of the mutation detected in parents. Moreover, for most genotypes there are too few known cases to allow an accurate prediction of clinical phenotype.

The product of the gene responsible for CF is a chloride channel. Concerning the mutations only partially inactivating the channel, much research on increasing the residual activity by pharmacological manipulation is ongoing. Studies to evaluate a potential gene therapy by the introduction of the normal copy of the gene responsible for CF in respiratory epithelium cells, through retroviral adenoviral or liposomal vectors, are also ongoing. This information is relevant for possible public consent for CF prevention programs and should be provided during genetic counseling.

Significant scientific improvements in this field have resulted in the possibility of CF control through carrier screening and prenatal diagnosis, and this has also been achieved for β-thalassemia and TSD. Concerning screening for subjects with CF family history, consent has been reached. Screening should be preceded by appropriate educational and genetic counseling policies. The most favorable age for screening is adulthood, preceding conception, so as to provide the largest possible variety of reproductive options. Identification of carriers, in this situation, is almost 100%. It must be considered that these subjects have direct CF experience that enables them to make choices on concrete bases.

Regarding population screening, an important debate on the utility of performing it and the target populations to be examined is ongoing. It is commonly thought, a view also shared by the American Society of Human Genetics, that before introducing CF screening in the whole population, the effectiveness of educational material and laboratory capacity should be assessed, and genetic counseling characteristics and positive and negative aspects of screening should be accurately analyzed. A particular aspect is the effect that the screening has on a couple in which one partner has been identified as a CF carrier and the other, for whom test results are negative, might be a potential carrier of a rare mutation not included in the panel tested. Currently, almost 1 couple in 15 (90% of mutations identified) could be included in this category, who have a risk of 1 in 1000 to have affected offspring. Several pilot programs concerning both the childbearing adult population and women with ongoing pregnancies have been undertaken.

Adult population screening through general practitioners or family planning centers had shown contrasting results with regard to acceptability (Watson *et al.*, 1991; Bekker *et al.*, 1993; Tambor *et al.*, 1994). The best results had an acceptance of 70% with the screening carried out by the general practitioners and 90% with the screening performed by family planning centers. It is worthwhile to note that in some programs a high number of relatives of carriers, detected in the screening, have requested to be examined, thus indicating the absence of stigmatization following carrier detection. Furthermore, the majority of detected carriers still had the information after 6 months and were glad to have acquired this information, even though in some cases it caused anxiety. Subjects determined to be CF negative were also glad to have had the tests performed and kept the information long term, especially crucial information revealing that testing might not detect a significant proportion of carriers and does not detect all carriers (Watson *et al.*, 1991).

The second group of pilot studies concerned pregnant women, who have certainly shown a greater willingness to accept testing. For this purpose, both sequential and couple approaches have been employed. In the sequential approach, testing on the woman is first performed, and only if there is a positive result is testing extended to the partner. Because of a long period of uncertainty, this procedure involves elevated anxiety. In order to avoid an adverse outcome in this approach to couples in whom both partners are tested, testing is defined as positive only if both partners results are positive; otherwise, it is negative even if the results of just one partner indicate that he or she is not a carrier for the mutations analyzed. Screening acceptance has proved to be very high, almost 70% with both the approaches; furthermore, for couples determined to be at risk, after genetic counseling 90% chose to perform prenatal diagnosis. Similarly, these studies have also shown that the participants in the screening, both those determined to be positive and those determined to be negative, developed nothing more than a transitory state of anxiety; in fact, they have never felt stigmatized and were glad to have been tested. A critical prerequisite for obtaining this outcome was certainly obtaining appropriate knowledge of CF, in addition to accurate genetic counseling.

Another aspect to be considered is the cost–benefit analysis, which is strongly influenced by the percentage of couples who choose to interrupt pregnancy of the affected fetus. Analysis carried out in Israel, with 95% testing sensibility and with a high percentage (97%) of decisions in favor of prenatal diagnosis and option versus pregnancy interruption (92%) in the event of an affected fetus, has shown that costs for prevention of an affected child are lower than costs for therapy. Of course, in analyzing costs and benefits other factors should be considered, including the possibility that carrier couples could cope with other pregnancies without repeating the testing; the necessity of including among the costs for patient treatment other induced costs, such as if one parent loses his or her job because of the necessity of assisting an affected child; the travel expenses of commuting from the home to health centers; and costs for other relatives requiring screening. It should also be considered that screening costs are considerably reduced, almost 13 times, in the second generation since testing might be offered only to children of known carriers.

There are still many uncertainties on the utility of suggesting a population preventive screening for CF. The attitude toward screening in Europe generally seems to be more positive than that in the United States. It is likely that in the near future CF screening will be universally adopted in Western countries. Once adopted, some guidelines, on the whole in conformity with the directives of the American Genetic Society, should be followed in population screening. Screening should be voluntary, confidentiality should be ensured, and informed consent should be requested. Moreover, before undertaking screening, benefits and risks, especially regarding possible stigmatization, should be discussed; screening should thus be followed by appropriate genetic counseling; laboratories performing testing should be submitted to accurate qualitative control; screening should be available to the entire population; the target population should be composed of adult couples who have not yet conceived so that the couple can take advantage of all the reproductive options available; and screening should be preceded and followed by an educational campaign addressed to the population.

CF screening certainly has a great social and medical value since it represents an extraordinary information vehicle for screened subjects to obtain knowledge of the disease and the methodology of its prevention. This information will allow the couples that choose pregnancy interruption to avoid the birth of an affected child and those who choose not to interrupt pregnancy of an affected child to cope emotionally and practically with the cure for a CF affected child.

Genetic Screening in Pregnant Women

Genetic screening for apparently healthy pregnant women is apt to identify congenital defects in couples without any a priori risk. This screening is essentially based on three diagnostic procedures: determination of Rh compatibility, evaluation of the fetus through ecography, and biochemical analysis of maternal blood aimed at identifying at-risk women for chromosomal aberrations and neural tube defects.

Determination of Rh Compatibility

Determination of Rh compatibility between mother and fetus is aimed at preventing hemolytic disease of the newborn due to blood group incompatibility. Hemolytic disease from isoantibodies presents when the mother shows Rh negativity and the fetus is Rh positive. In this condition, fetal erythrocytes, which in limited number cross the placenta and reach maternal circulation, stimulate anti-Rh antibody production. Anti-Rh antibodies through the placenta reach the fetus and induce red blood cell hemolysis, causing severe anemia. The incidence of this disease varies depending on the prevalence of Rh-negative subjects in the population; however, it varies remarkably in different populations (Table 8).

TABLE 8

Negative Rh Genotype Incidence in Different Populations

POPULATION	%
Caucasian	11–21
American	14.4
Australian	14.7
Indian	8
Black American	5.5
Japanese	0
Chinese	0
North American Indian	0

The Rh system is composed of three membrane proteins that are codified by two linked genes mapping on the chromosome 1 short arm — one codifying antigen D and the other codifying Cc/Ee antigens, which are produced by a process of alternative splicing. The three antigens groups are cDE(R^2), Cde(R^1), and cde(r), which in the white population have a frequency of 0.14, 0.41, and 0.39, respectively.

Rh-positive subjects include heterozygotes who inherited the D gene from only one parent and homozygotes who inherited the D gene from both parents. Rh-negative D genotype is due to total or partial absence of the Rh D gene. Immunization of the Rh-negative mother may occur through transfusion of Rh-positive blood or through transfusion of fetal red blood cells in maternal circulation during pregnancy or at term. This latter is the most common, which also explains why hemolytic disease from Rh incompatibility is exceptional in the first pregnancy. Rh immunization occurs more frequently when pregnancy or delivery are complicated, as in toxemia, delivery by cesarean section, or manual placental removal — all conditions that facilitate transplacental hemorrhage. Fetus–placenta hemorrhage mainly occurs during delivery or abortion or is consequent to diagnostic manipulations, such as amniocentesis or fetoscopy. Of course, the greater transplacental hemorrhage, the more frequent and severe the immunization. Prevention of hemolytic disease from Rh incompatibility in newborns, with Rh-positive fathers and Rh-negative mothers, is based on the treatment of the Rh-negative mother with anti-Rh antibodies soon after delivery, miscarriage, or diagnostic manipulation. No other important ethical involvement is implied besides the impartiality requested in providing this type of prevention to whoever is at risk for hemolytic disease due to Rh incompatibility in the newborns. In coping with this problem the couple, in fact, does not have any possible alternative to prevent disease either to save the fetus or the newborn if the mother has already been immunized. Rarely is hemolytic disease in the newborn due to anti-c, anti-E or anti-Kell (another membrane antigen) antibodies.

Echography

Fetal echography is currently the best procedure for prenatal detection of congenital anomalies. This methodology discloses major defects but allows the detection of minor anomalies indicative for chromosomal abnormalities or other genetic defects. Currently, two fetal ultrasound images are suggested for all pregnant women, the first performed by the 12th week and the second by the 20th week of pregnancy, to evaluate the growth of the fetus and to regularly assess major and minor defects. Among the most frequent genetic defects detectable by fetal ecography are neural tube defects (anencephaly and spina bifida); hydrocephaly holoprosencephaly; macrocephaly; cystic adenomatous lung malformation; diaphragmatic hernia; fetal hydrops; defects of the abdominal wall; gastrointestinal, cardiac, and renal defects; and many skeletal dysplasias and oligo- or polyhydramnios that may suggest the presence of fetal abnormalities.

In addition to these major abnormalities there are a series of minimal anomalies indicative for the possible presence of chromosomal aberrations. Among these, particularly relevant are the findings of cystic hygroma of the neck, the simple nuchal translucency, shortness of the femur, or intrauterine growth retardation. In fetuses showing congenital anomalies at echographic examination, a high incidence of chromosomal aberrations has been detected—approximately 2% in the presence of a single defect and 22% when multiple anomalies are observed. By considering the problem in another perspective, it can be stated that 90% of the fetuses with trisomy 13 and trisomy 18, and those with triploides or with Turner's syndrome, present malformations detectable through an accurate ultrasound analysis. In trisomy 21, fetal anomalies are instead more subtle so that only 40% of the fetuses can be detected or suspected by ultrasound. For all these cases, analysis of fetal karyotype is suggested. Chromosomal analysis is indicated both in the second trimester, when pregnancy interruption may be carried out, and in the third trimester since the presence of chromosomal abnormalities may be useful for the treatment of the newborn and for genetic counseling for further pregnancies. The most reliable procedure is fetal blood analysis for which results can be obtained within 3–5 days, which surely reduces parents' anxiety.

TABLE 9

Risks for Down's Syndrome as a Function of Maternal Age

MATERNAL AGE AT EDD[a] (YEARS)	RISK	MATERNAL AGE	RISK
<25	1:1500	38	1:190
25	1:1350	39	1:150
26	1:1300	40	1:110
27	1:1200	41	1:85
28	1:1100	42	1:65
29	1:1000	43	1:50
30	1:950	44	1:35
31	1:800	45	1:30
32	1:680	46	1:20
33	1:570	47	1:15
34	1:470	48	1:11
35	1:380	49	1:8
36	1:310	50	1:6
37	1:240		

[a]EDD, expected date of delivery.

Maternal Serum Screening Analysis for Chromosomal Aberrations and Neural Tube Defects

Down's syndrome (trisomy 21) is the most common cause of severe mental deficiency associated with multiple major and minor anomalies. The most common minor abnormalities are oblique palpebral fissures, epicanthic folds, excess skin at the nape of the neck, macroglossia, flat occiput, depressed nasal bridge, short fifth finger, single palm or flexion crease, and marked hypotonia The most common major malformations are congenital hearth defects. The incidence of this disorder is 1 case in 700 live births. People with Down's syndrome have varying degrees of mental disability; however, often the disability is severe. Some individuals with trisomy 21 lead satisfied, independent lives, whereas other are completely dependent on their parents or relatives. Almost 50% of Down's syndrome pregnancies will miscarry. Approximately one-third of Down's syndrome cases have a serious heart defect. The average life expectancy of a person with Down's syndrome is about 60 years. Most develop Alzheimer's disease-like symptoms (i.e., progressive dementia) after the age of 40. Advanced maternal age represents the unique a priori risk of having a pregnancy with a Down's syndrome fetus. Table 9 indicates that the risk increases from 1 in 1500 for those under 25 years of age to 1 in 6 for those over the age of 50.

Neural tube defects are due to a defective closure of the primitive neural plate and comprehensive anencephaly and varying types of spina bifida (myelomeningocele, meningocele, and spina bifida occulta). These are among the most common and severe inherited defects. Their origin, including environmental and genetic factors, is multifactorial. In the absence of prenatal screening, the incidence is 1 in 400 live births. Among the neural tube defects anencephaly is fatal in the neonatal period. The natural history of spina bifida is variable. Most children with spina bifida have paralysis of the legs, fecal and urinary incontinence, hydrocephaly, and, less often, mental retardation. Neural tube defects result from a defective primary neurolation by which the primitive neural plate produces the neural tube.

Screening of maternal serum allows identification of women with an increased risk of bearing a child with Down's syndrome or neural tube defect. The most complete test employed is the so-called quadruple test, which consists of analysis of the serum markers α-fetoprotein (AFP), non-conjugated estriol, and free α and β subunits of the human chorionic gonadotropin (α-hCG and β-hCG). Testing is performed in the second trimester (Week 16). Variants of the approach that are employed less and less reliable also exist. Other serum markers are employed in the first trimester, but such an approach is not widespread. Four serum markers together with maternal age and nuchal translucency evaluation allow a quantitative evaluation of the relative risk for Down's syndrome. In pregnancies with Down's syndrome, α-fetoprotein and conjugated estriol are reduced, whereas free α-hCG and β-hCG are increased. On the basis of this analysis, together with maternal age, an probability estimation can be made: Tests showing a 1 in 300 risk are considered positive and options for further analysis are discussed. Approximately 1 woman in 20 has a screen-positive result; 1 in 60 screen-positive women will have a Down's syndrome pregnancy. Table 10 shows the probabilities for screen-positive women to develop Down's syndrome as a function of the maternal age. Regarding neural tube defects, only AFP is analyzed, which for the previous condition is increased. Women with values 2.5 higher than normal are considered positive. Genetic counseling follows, in which options for further assessment are discussed. For 1 woman in 50 with a screen-positive result, 1 in 20 has a pregnancy with an affected fetus. On the whole, 2 out of 3 pregnant women with a fetus affected by a neural tube defect and certainly all pregnancies with anencephaly fetus are identified. Both ultrasound analysis and quadruple test are preceded by adequate counseling. Every couple, especially the woman, has the right to know the purpose of the analysis and its implications. For instance, both ultrasound analysis and quadruple test may lead to detection of a condition for which no definitive therapy exists and miscarriage is the only alternative. It must be assumed that other conditions that can be treated at birth or thereafter may be detected. Knowledge of Down's syndrome and neural tube defects should be assessed during a couple's preliminary counseling. It is also necessary to establish that these analyses may report at-risk values but not definitive results, and that both false positives (screen-positive and unaffected fetus) and false negatives (screen-negative and affected fetus) may occur. It must be clearly stated that screening is a procedure aimed at identifying people at risk and therefore does not lead to complete certainty. All information which derives from screening (e.g., risk of sex chromosome anomalies and of nonrelevant disorders) should always be disclosed. Also, the sex of the fetus, whether the results are normal and uncertain, should be communicated. The information should be directed at both parents, timely, and discussed personally, and the couple should be given the opportunity to comment the outcome. If the mother wishes to have the information, it should be given to her.

A clear and precise description of Down's syndrome and the neural tube defects, including all emotional, educational, and financial aspects, should be provided. It must be reaffirmed that the disease is not to be imputed to the couple's or single partner's specific behavior or attitude. Information on the actual and future prospects for therapy should always be given. It must frequently be mentioned that there are many other pathologies that, although not subject for screening, may induce anomalies in the fetus. Before offering screening to the couple (i.e., ultrasound examination and quadruple test), it must be ensured that, if required, other procedures for further evaluation are affordable. Opportunity to perform fetal ultrasound examination and quadruple test should be offered even to those couples showing negative attitudes toward abortion. Eventual use of blood specimen for purposes different from chromosomal aberration and neural tube defects screening requires specific consent after explicative information. Exhaustive, appropriate information on the risks and advantages of these additional procedures should be given to the couple, thus allowing them free choice whether or not to undergo the screening. In the case that the two partners disagree, the woman should make the ultimate decision. In the occurrence of screen-positive outcome, a second nondirective counseling is requested to establish the affordable options and specifically whether the couple is willing to perform prenatal diagnosis.

TABLE 10

Probability of a Screen-Positive Result and Percentage of Down's Syndrome Related to Maternal Age

MATERNAL AGE (YEARS)	PROBABILITY OF SCREEN-POSITIVE RESULT	PREGNANCIES WITH DOWN'S SYNDROME DETECTED BY SCREENING (%)
<25	1:45	45
25–29	1:30	50
30–34	1:15	60
35–39	1:5	80
40–44	1:2	95
>45	1:2	>99
Median	1:20	65

Conclusion

As previously discussed, genetic screening is divided into three categories according to the target population: newborn screening, heterozygote screening, and prenatal screening. Principal requisites for a genetic disorder to be rationally considered as subject for newborn screening are prevention, through timely diagnosis, of negative effects of the disease (benefits for the newborn); high incidence; availability of simple screening procedures with low risk of false-positive/false-negative outcome; and the possibility to treat and follow appropriately the affected newborn.

Newborn genetic screening for phenyketonuria and hypothyroidism has proved to be of extreme individual and social medical relevance since it allows, through preclinical diagnosis, the avoidance of neurological damages that late diagnosis might imply. This type of screening has proved not to have adverse events. As for all other medical acts, especially those concerning genetics, absolute confidentiality should be maintained for the results. Seen its certain preventive efficacy, the most relevant ethical problem is the equal opportunity to access to this type of screening for populations throughout the world. Unfortunately, currently this preventive measure is only affordable for populations of the most industrialized countries. A second problem that has recently emerged is the possible employment of the specimens collected for screening for purposes other than the original purpose, essentially aimed at a search for DNA variation in peripheral blood specimens. Most investigators believe that informed consent from a newborn's parents is needed. An acceptable, although debated, alternative is the use of anonymous, unidentifiable specimens. Among the other newborn screenings proposed, CF and sickle-cell disease have been included in the newborn screening process in many populations (limited to at-risk populations). The rationale for these types of screenings is prevention of morbidity and severe complications through early diagnosis. Advantages derived from screening for these disorders, however, are still unproven. The effectiveness of G6PD-deficiency screening is also debatable, considering the lack of diagnostic requisites mentioned previously. Nevertheless, G6PD screening is operative in several at-risk populations, in which it has proved to be effective, especially under an educational profile, leading to a decrease in the incidence of adverse events due to the assumption of medications and fave beans with oxidative potential in subjects with G6PD deficiency.

Heterozygote screening for recessive autosomal disorders in adulthood is aimed at informing couples with two heterozygotes of the risk of producing affected offspring, allowing them to avoid this occurrence, by diagnostic procedures (prenatal diagnosis and preconception or preimplantation diagnoses) and eventual interruption of pregnancies of affected fetuses. Heterozygote screening also allows couples at risk to be prepared at the birth of a child affected from severe genetic disease, thus enabling them to cure and give assistance to the baby. This type of screening has been largely employed for TSD among the Ashkenazi Jews and among Mediterranean populations for β-thalassemia. In these cases, screening proved to be extremely effective since the accurate information provided to at-risk populations allowed them rational reproductive choices based on full knowledge. The number of subjects born with these disorders has decreased considerably. Notably, stigmatization and loss of self-esteem have not occurred among individuals defined to be at risk. Basic requirements for this type of screening are voluntary participation and nondirective genetic counseling. It is also fair that screening should be preceded by informed consent, which is currently only employed for TSD. For β-thalassemia, in order to maintain equality in distribution of the facilities, screening should be provided for all at-risk populations in developing countries, in which the disease has a very high incidence; however, economic and organizational conditions currently preclude the introduction of any preventive genetics service. Likewise, the previous conditions should also be applied to sickle-cell disease. In the near future, once the current technical obstacles connected to the incomplete capability to detect the spectrum of mutations in at-risk populations are overcome, CF will also more likely be subject for heterozygote adult screening. Another potentially useful heterozygote screening, once technical difficulties are resolved, is that for fragile X mental retardation. In the near future, by improving our knowledge on molecular genetics and as a result of the Human Genome Project, the molecular basis for many genetic diseases will be defined, and for most of them newborn or heterozygote screenings or both will be offered.

Prenatal screening for congenital defects, especially those associated with neural tube defect and the most frequent chromosomal aberrations, has allowed diagnosis of many fetal malformations in ongoing pregnancy and given couples important information for making decisions on eventual pregnancy termination or obtaining adequate training to care for an affected child. Due to the screening for chromosomal aberrations by quadruple testing, many at-risk categories have been more precisely defined, thereby reducing the number of cases undergoing prenatal diagnosis in the absence of specific indications.

Geneticists should be aware of advantages and disadvantages of each new screening proposal, always considering the ethical aspects and especially the fact that the screening should primarily be advantageous to the individuals and families subject to the medical act. Finally, one the most relevant problems to be considered is the possible discrimination regarding the workplace and insurance of subjects with genetic diseases.

References Cited

American Academy of Pediatrics (1976). The task force on genetic screening: The pediatrician and genetic screening (every pediatrician a geneticist). *Pediatrics* **58,** 757–764.

American Academy of Pediatrics (1992). Committee on Genetics: Issues in newborn screening. *Pediatrics* **89,** 345–349.

American Academy of Pediatrics (1993). AAP Section on Endocrinology and Committee on Genetics and American

Thyroid Association Committee on Public Health: Newborn screening for congenital hypothyroidism: Recommended guidelines. *Pediatrics* **91,** 1203–1209.

Angastiniotis, M., Modell, B., Englezos, P., and Boulyenkov, V. (1995). Prevention and control of haemoglobinopathies, *Bull. World Health Org.* **73,** 375–386.

Bekker, H., Modell, M., Denniss, G., Silver, A., Mathew, C., Bobrow, N., and Marteau, T. (1993). Uptake of cystic fibrosis testing in primary care: Supply push or demand pull? *Br. Med. J.* **306,** 1584–1586.

Cao, A. (1987). Result of programmes for antenatal detection of thalassemia in reducing the incidence of the disorder. *Blood Rev.* **1,** 169–176.

Cao, A. (1994). 1993 William Allan award address. *Am. J. Hum. Genet.* **54,** 397–402.

Cao, A., and Rosatelli, M. C. (1993). Screening and prenatal diagnosis of haemoglobinopathies. *Baillières Clin. Haematol.* **6,** 263–286.

Fletcher, J. C., Berg, K., and Trnoy, K. E. (1985). Ethical aspects of medical genetics. A proposal for guidelines in genetic counselling, prenatal diagnosis and screening. *Clin. Genet.* **27,** 199–205.

Irons, M. (1993). Screening for metabolic disorders. How are we doing? *Pediatr. Clin. North Am.* **40,** 1073–1085.

Kaback, M. (1193). Tay–Sachs disease-carrier screening, prenatal diagnosis, and the molecular era. An international prospective, 1970 to 1993. The international TSD data collation network. *J. Am. Med. Assoc.* **270,** 2307–2315.

Knoppers, B. M., Caulfield, T., and Kinsella, T. D. (Eds.) (1996). *Legal Rights and Human Genetic Material.* Montgomery, Toronto.

Laberge, C. (1994). Public health rational for newborn screening and civic values. In *New Horizons in Neonatal Screening* (J. P. Farriaux and J. L. Dhondt, Eds.), pp. 25–38. Excerpta Medica, Amsterdam.

Levy, H. L., and Cornier, A. S. (1994). Current approaches to genetic metabolic screening in newborns. *Curr. Opin. Pediatr.* **6,** 707–711.

Mandel, J. L., Biancalana, V., Boiux, J. C., and Cossèe, M. (1994). Screening for the fragile X-mental retardation syndrome. In *New Horizons in Neonatal Screening* (J. P. Farriaux and J. L. Dhondt, Eds.), pp. 261–272. Excerpta Medica, Amsterdam.

Milunsky, A. (Eds.) (1992). *Genetic Disorders and the Fetus: Diagnosis, Prevention and Treatment.* Johns Hopkins Univ. Press, Baltimore.

National Academy of Sciences (1975). *Genetic Screening: Programs, Principles and Research.* National Academy of Sciences, Washington, DC.

Petrou, M., and Modell, B. (1995). Prenatal screening for haemoglobin disorders. *Prenatal Diagn.* **15,** 1275–1295.

Rowley, P. T. (1984). Genetic screening: Marvel or menace? *Science* **225,** 138–144.

Tambor, E. S., Bernhardt, B. A., Chase, G. A., Faden, R. R., Geller, G., Hofmab, K. J., and Holtznan, N. A. (1994). Offering cystic fibrosis carrier screening to an HOM population: Factors associated with utilization. *Am. J. Hum. Genet.* **55,** 626–637.

Watson, E. K., Mayall, E., Chapple, J., Dalziel, M., Harrington, K., and Williamsson, R. (1991). Screening for carriers of cystic fibrosis through primary health care services. *Br. Med. J.* **303,** 504–507.

General References

Fernandes, J., Saudubray, J. M., and Van Den Gerghe, G. (1995). *Inborn Metabolic Diseases: Diagnosis and Treatment,* 2nd ed. Springer-Verlag, New York.

Kennet, A., and Wald, N. J. (1994). Antenatal screening for Down's syndrome. In *The Yearbook of the Royal College of Obstetricians and Gynaecologist* (J. Studd, Ed.), pp. 229–243. Parthenon, London.

Knoppers, B. M., Caulfield, T., and Kinsella, T. D. (Eds.) (1996). *Legal Rights and Human Genetic Material.* Montgomery, Toronto.

Milunsky, A. (Ed.) (1992). *Genetic Disorders and the Fetus: Diagnosis, Prevention and Treatment.* Johns Hopkins Univ. Press, Baltimore.

Scriver, C. R., Beaudet, A. L., Sly, W. S., and Valle, D. (Eds.) (1995). *Metabolic and Molecular Bases of Inherited Disease,* 7th ed. McGraw–Hill, New York.

World Health Organization (1995). *Guidelines on Ethical Issues in Medical Genetics and the Provision of Genetic Services.* World Health Organization, Geneva.

ANGUS CLARKE
Institute of Medical Genetics
University Hospital of Wales
Cardiff, United Kingdom

Genetic counseling is a process of communication between a health professional and a client. The scope of genetic counseling is broad and it may take place in many settings. This article describes the process of genetic counseling as it occurs in specialist genetic counseling clinics; discusses some of the ethical issues raised in genetic counseling, including issues relating to genetic testing; and examines related issues in the context of population genetic screening.

Genetic Counseling

Process of Genetic Counseling

Referral

The process of genetic counseling begins at the point of referral of the client, patient, or family to the clinical genetics unit. The letter of referral from the family's general practitioner or the hospital specialist will usually explain the client's concern and what the diagnosis is thought to be. It may also make clear whether the referral has been initiated by the client/family because of their concerns or by the referring clinician.

The client or family may wish to find out more about the condition that affects one or more family members. They may wish to obtain a precise diagnosis for the condition, if that has not already been established; be given an explanation for what has happened in the family; learn what the future holds for the affected individual(s), including the risk of specific complications and the possibilities for prevention and treatment; be given an explanation of the pattern of inheritance of the condition; learn the likelihood that someone else in the family will also be affected, whether someone currently without the illness may subsequently develop it, or whether a future child may be born with the condition; learn how they can avoid the birth of another affected child; and learn about any applicable prevention or surveillance for complications of the disorder.

The genetic counseling client may also wish to consider genetic testing. It may be possible to identify those individuals who are likely to develop an illness in the future (predictive genetic testing). It may be possible to identify a fetus that will have, or would be likely to have, a specific genetic disorder if the pregnancy continued and the fetus became a child (prenatal genetic testing). In these situations, clients will often want to know the risks of the problem arising, what the tests can offer (how sensitive and specific they are), and what the process of testing would entail in practice for them.

Referring physicians may have their own reasons for referring a family for genetic assessment and counseling. There may be some diagnostic uncertainty that the physician would like to be resolved that is not in his or her professional interest, even if the client does not attach much importance to the precise label that is used to describe the condition in his or her family. The referring physician may also want members of the family to comply with his or her recommendations about predictive or prenatal genetic testing or about family planning; he or she may want the family to take the necessary steps to ensure that no other affected child is born. The family may or may not share this perspective.

Listening

Once a referral to the genetic counseling service has been received, the first meeting of the genetic counseling team with the client will be arranged. The principal professional goal at this stage is to listen to the clients and understand what they are seeking from genetic counseling (Fig. 1). What do they already know, or suspect, about the family condition? Do they want more information about the condition and how to cope with it? Do they want a more precise diagnosis? Are they considering predictive or prenatal genetic tests?

It is also helpful at this stage for information to be gathered about the family and about the genetic problem in the family. If the individual referred to the genetics service is not affected by the particular condition, then it may be necessary to review the medical records of affected relatives; thus, the permission of these individuals must be sought and their records must be reviewed before it is possible for the client's questions to be properly answered.

In some areas, including Wales, the initial meeting with the client usually takes the form of a visit by a genetics nurse, coworker, or associate (a specialist nurse or other fieldworker) to the family home. Such preclinic home visits by genetic associates are not possible for all centers but have many advantages. The genetic counselor knows before the clinic visit the issues that are likely to be most important for the client and can prepare accordingly. If other family members have to be contacted, and if medical records need to be reviewed, then this can be achieved in the interval between the home visit and the genetic counseling clinic. Without a previous home visit, additional clinic visits may be needed.

Because the communication process during the home visit is two-way, the client learns what to expect from the

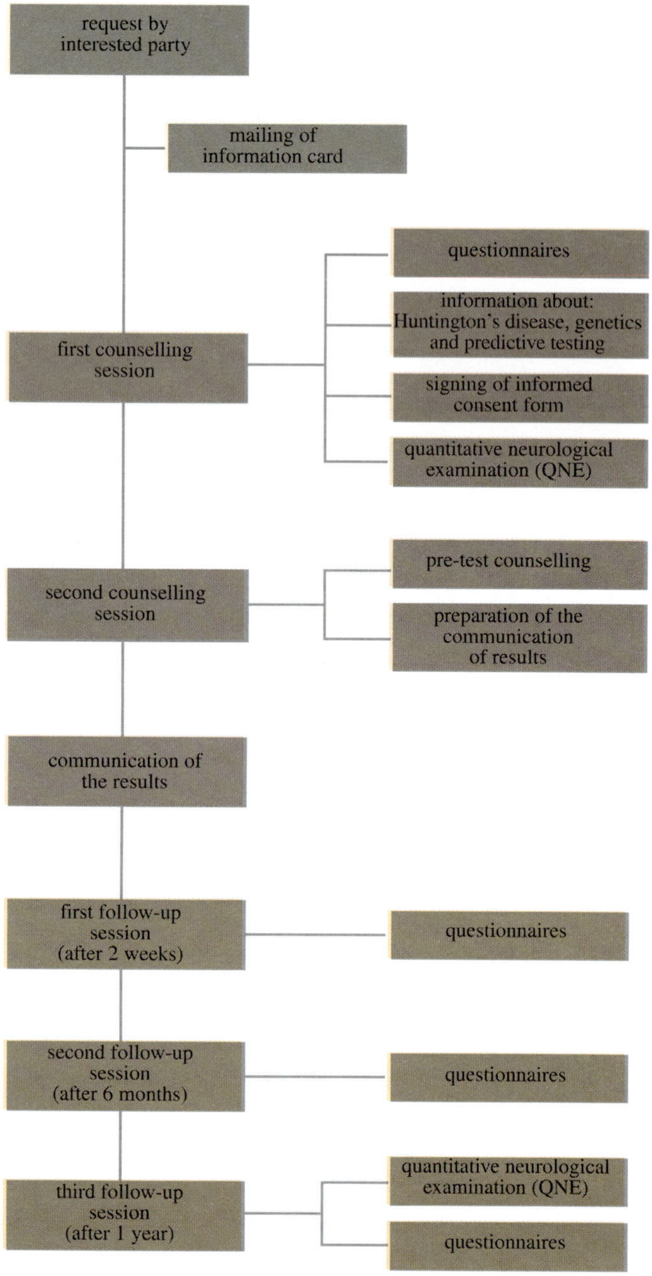

FIGURE 1 Guidelines for predictive genetic testing (reproduced with permission from Benjamin *et al.*, 1994).

genetic counseling clinic. The expectations of the professionals and the client can be aligned in advance, and this helps to maximize the value of the consultation. Even when there has been a preparatory home visit, it is helpful to ask the client what he or she is seeking from genetic counseling when he or she comes to the clinic; the client's expectations of genetic counseling may have changed as a result of the home visit, and his or her revised expectations must be ascertained.

Listening is an active process, perhaps the most important core component of good communication, and genetic counselors must draw heavily on the necessary skills. Some clients find it difficult to formulate what they are expecting from genetic counseling, and great sensitivity may be needed to help them to explain their questions or concerns. Others are not looking for anything from genetic counseling and have been referred to the service because the referring physician thought it would be useful. In such cases, the counselor must consider if he or she can respond constructively to the referring clinician without creating or reinforcing difficulties for the client. Working toward a diagnostic label has a high priority in most systems of medical thought but may be inappropriate if the labeling process could damage a family's fragile emotional equilibrium and if the label would be unlikely to lead to any therapeutic benefit.

An important question for the genetic counselor is why the referral process has been initiated now rather than last year or next year. Sometimes the reason will be obvious, as in the event of a high-risk pregnancy or the recent recognition of developmental problems in a young child, but often the trigger for the referral will not be immediately apparent. Factors that may have triggered the referral will often emerge in the discussion between the family and the genetics coworker when he or she visits their home, and such information can be very helpful in managing the subsequent clinic consultation in the most helpful and appropriate manner. This is particularly true if different members of the family adopt different approaches to "their" genetic condition (or their condition of being at risk). They may have different expectations of the referral or different degrees of interest in it, and an appreciation of this allows the genetic counseling team to remain neutral in relation to potential family disagreements.

The Medical Formulation

Before the genetic counselor can respond to the client's questions and concerns, it is necessary for the medical basis of the concerns to be assessed. This process may be brief and straightforward or it may be complex and time-consuming. Essentially, the diagnosis of the condition in the affected family members needs to be confirmed; if no diagnosis has been made, then it may be necessary to examine the affected individuals and to consider appropriate investigations. In many cases, there will be no firm diagnosis even after a thorough assessment, and then it will be more difficult to answer the family's questions with confidence and authority.

Diagnostic uncertainty is frequent in the context of a child who has multiple physical anomalies and developmental delay. About half of children with such problems have a diagnosis — an explanation of the underlying cause for the various problems — and the other half do not. Have the unusual physical features and developmental difficulties arisen by chance in the child, or do they have a single, common cause that amounts to a distinct condition?

If the family's concerns include questions about risk — perhaps of the client developing a cancer or a degenerative illness or of a future child being affected by a particular condition — then the genetic counselor may need to carry out complex calculations to answer these questions. The counselor must be aware of the limited validity of some risk calculations based on restricted sets of published data. Even if there is no issue of mathematical complexity, the counselor may need to review the medical literature or genetic databases in order to respond appropriately to the client's questions. Details of the family condition may need to be discussed with colleagues at national or international meetings so that the counselor will be confident that he or she is giving the best available information to the client.

The work entailed in the medical formulation of the family's problem can vary from the trivial and transparent — readily understood and appreciated by clients — to the heroic and opaque. If research work is required, it may be invisible to the family members or they may be closely involved throughout the whole process. Difficulties (dissatisfaction) can arise when the diagnostic formulation remains imprecise despite intensive investigations. Dissatisfaction can also arise if the full medical formulation of the family's problem (e.g., the DNA sequencing of one or more genes in the search for a mutation) is technically possible but in practice cannot be arranged. If such molecular genetic testing is only possible in a few research laboratories with an interest in the family's condition, then testing may become unavailable once the research focus of those laboratories changes. Diagnostic testing laboratories may be unlikely to introduce testing for a rare disease, even though it would be technically feasible, because it may be financially nonviable if the demand for such testing would be limited. Families that have contributed to an international research effort may then be denied services that are technically feasible only as a result of their previous support.

Communication of Information to Clients

Once the medical diagnostic formulation has been undertaken, it is possible to respond to the client's questions. The way in which this is done depends greatly on the assessment made by the genetic counseling team (the clinician and coworker) of the level of understanding of the client and other family members and of their emotional preparedness. The information that may be imparted to the clients may be received by them as "good" news or as "bad" news; different individuals could even interpret the same information in quite different ways. The counselor must present information in the way most appropriate to the individual concerned — appropriate to both his or her level of understanding and his or her emotional state. The way in which information is imparted to the client will depend on judgments made by the genetic counselor in light of the preclinic home visit (where applicable) and of the counselor's assessment of the client in the "listening" phase of the counseling process. Very different accounts would be given to a research biochemist and to a client with learning difficulties.

If information about risk is to be given, then the way in which this is presented will have to reflect the general educational level of the client and, specifically, the client's numeracy and arithmetical confidence. It may be helpful to present both the probability of an event, X, and the probability of not-X; furthermore, probabilities may need to be presented in several different forms (as fractions, decimals, percentages, and odds ratios) (Shiloh and Sagi, 1989). Thus, the probability of X happening may be given as $\frac{1}{4}$, 0.25, 25%, and 1:3 against; the probability of not-X would be $\frac{3}{4}$, 0.75, 75%, and 3:1 in favor. If the probability of an event is small, then graphic, visual representations may be helpful. The accompanying verbal descriptions of degrees of risk may be powerful factors in shaping a client's response to risk information, and thus counselors often try to avoid value-laden descriptions such as "high risk" or "low risk."

When the information to be imparted is likely to cause distress, particular care needs to be taken in the way in which it is presented (Standing Conference of Voluntary Organisations for People with a Mental Handicap in Wales, 1989; Parsons *et al.*, 1996). The necessary planning includes deciding who should be present at the consultation: the client with his or her partner or parent or friend and without an audience of students or superfluous staff. The physical setting should be a private place not subject to interruptions. The unwelcome information should be imparted "honestly and directly by someone who conveys warmth and understanding." The client should have access to the counselor on subsequent occasions to ask questions and should be given time and space to consider his or her feelings.

Particular experience has been gained in "breaking bad news" in the context of malignant disease, and Buckman (1992) describes the process in six steps: (i) getting started, (ii) finding out how much the patient knows, (iii) finding out how much the patient wants to know, (iv) sharing the information (aligning and educating), (v) responding to the patient's feelings, and (vi) planning and follow-through. Many of the lessons are applicable to the field of clinical genetics and genetic counseling. One especially important lesson is that clients may find it most helpful to be given information in stages. By determining what a client already knows and what he or she wants to know now and providing that bolus of information, the counselor can allow the client to return for more information in a few days, weeks, or months, digesting the unwelcome facts in stages. Giving clients control over the rate at which they are given information is crucial and requires considerable sensitivity.

The central role of information giving in genetic counseling emphasizes an important way in which genetic counseling differs from "counseling" in the more usual, generic sense. Counseling has been defined in many ways: the British Association of Counselling (1984) states that "the task of counseling is to give the 'client' an opportunity to explore, discover, and clarify ways of living more satisfyingly and resourcefully." The ethos of counseling is that of a facilitative, enabling, and empowering process in which the client is supported in his or her personal growth and development. Counseling is therefore usually a client-centered process of supportive and reflective communication, as developed by Carl Rogers and others. Genetic counseling differs in that the counselor has a body of technical knowledge about an area of biomedicine (human genetics) on which he or she has to draw to speak with authority to clients, who have presented their questions to the counselor-as-expert. The genetic counselor, therefore, often speaks with an authority that would not be appropriate in other types of counseling; genetic counseling is inevitably authoritative although, it is hoped, not authoritarian. Several ways in which genetic counseling has incorporated some of the concerns of generic counseling to guide good practice and to promote effective communication between client and counselor (listening to the client and aligning expectations early in the genetic counseling process) have already been considered. In the next section, other ways in which genetic counseling distances itself from the more prescriptive, authoritarian practices found in other branches of conventional medicine are discussed.

Responses to Information

Providing information to clients about their genetic situation can have far-reaching implications for them and other members of their families. Whether reflectively and deliberately or tacitly and by default, we all shape our own lives. We may construct an explicit life plan with set targets and goals, we may allow ourselves to drift through life without a sense of direction, or we may allow a coherent but implicit, unformulated life plan to emerge spontaneously, perhaps unconsciously. Many of the expectations we have of our future lives are based on assumptions about our likely future health and lifespan and our likely future obligations to care for parents, partners, or children; whether these assumptions are implicit or explicit, it can still be very distressing to have them challenged by an exponent of contemporary biomedicine.

Those at risk of an adult-onset genetic disease may construct their life plan on the assumption that they will develop the family illness and become disabled or die at a relatively early age; they may therefore decide not to marry or not to have children. These are the "cautious planners." Other individuals in a similar position may deny and repress any thought of their risk of illness and even fail to comply with recommended precautions to minimize their possible health problems — the "ostrich" position of denial. If these people find out that the assumption on which they have build their lives is false, they can suffer distress and feel regrets about the practical life decisions they have made which now seem inappropriate. The cautious planner who decides not to marry and subsequently finds that he or she is not at risk of the disease may regret having thrown away his or her chance of happiness with a former lover. The "ostrich" who finds that he or she is developing the family's illness may be full of regret for perhaps having transmitted the condition to his or her children or for not taking steps to minimize the impact of the condition on his or her physical health.

Those identified as at risk of having a child with a genetic disorder can also respond in different ways to this information. They can decide either to take their chances (perhaps a 1 in 2 or 1 in 4 chance in each pregnancy), to have no children, or to seek prenatal diagnosis in a pregnancy with the option of terminating the pregnancy if the fetus is determined to be affected. The decision that individuals and couples are likely to make will vary with their experiences and also with their sense of values and their beliefs. If a couple has had a child affected by a serious disease, or if one member of the couple had an affected brother or sister, that will certainly shape their future plans. The type of care required by their affected child, if he or she is still alive, may also influence them; they may feel able to cope with one affected child but not with two or three. Experiences of prenatal diagnosis in a previous pregnancy, especially of a termination of pregnancy, may influence the decision that a woman will make on a subsequent occasion.

Where state legislation is independent of any religion so that prenatal diagnosis and the termination of affected pregnancies are at least permitted, the formal religious affiliation of individuals seems to make little difference to their reproductive decisions. This is not to say that moral considerations are not extremely important but rather that the moral considerations taken into account in the practical reasoning and decision making of individuals are not shaped so much by the codified beliefs of religious institutions as by the individuals' personal and collective social experiences of disease and disability and of responses to disease and disability. Both their own responses and the anticipated responses of others are important, as is the available family networks of support, the health and social care systems as they operate in practice, and people's confidence in the wider social provision of support.

Decisions

When faced by a decision, some individuals struggle to itemize and weigh the possible reasons for making each particular decision as "objectively" as possible. Others make an instantaneous and intuitive judgment. Yet others think through a range of possible outcomes and then make a decision with which they believe they can live, perhaps the least bad option in difficult circumstances. One factor that influences these processes is the individual's understanding of risk information they may be given; many individuals respond to risk information dichotomously, believing either that "it will happen" or "it will not happen". They then make their decisions on the basis of this belief about what will happen rather than incorporating the risk information per se.

The process of decision making in these contexts is not well understood. A rationalistic, utilitarian calculus is sometimes assumed to be the best type of decision making, but such a calculus has to exclude all but a narrow range of types of evidence from consideration and will therefore always be unsatisfactory.

The genetic counselor will usually present the full range of options in relation to any decision confronting his or her client and will usually frame them in as neutral a fashion as possible, trying not to press the client to choose one option rather than another. This may entail leading the client through a consideration of each possible course of action to give him or her a chance to reflect on the implications (for himself or herself and for others) of each option. This is termed *scenario decision counseling* and is one approach to the counselor's task of "helping clients reach a decision wisely, rather than reach a wise decision" (Shiloh, 1996).

Support

Genetic counseling, if carried out empathetically by a caring professional, will inevitably entail the provision of emotional support to the client. The experience of being listened and attended to will often be supportive and can be therapeutic. There are limits to the support that can be provided in genetic counseling, however, although how these limits are set will differ according to circumstances.

The genetic counselor may operate within constraints of time or expertise that do not permit him or her to embark on psychotherapy even for a problem related closely to the genetic counseling. Although a counselor with sufficient expertise may be able to provide brief therapy, any more prolonged course of treatment would probably necessitate referral to a generic (not genetic) counselor or psychotherapist. Many genetic counselors regard it as their role to recognize the need for such treatment and to make the necessary referral without themselves initiating therapy, although this situation may change as the professionalization of genetic counselors develops. To open unresolved issues, such as an unresolved bereavement response to the death many years ago of a sibling or a child, would not be appropriate unless the counselor had both the time and the skill to offer ongoing support to his or her client.

Genetic counselors will often meet patients whose ongoing medical, social, or educational support is inadequate. In such cases, there is an obligation to help the patients and their families to obtain the appropriate support within the constraints of what is realistically attainable in practice. The usual approach to this is to refer the patient to another agency or professional who can provide the support required or to put the individual or family in contact with a lay support group — a self-help group of involved families. Some clinical geneticists do become involved in the practical provision of medical support, but this is usually restricted to the coordination of surveillance for complications of a specific disease (e.g., a familial cancer syndrome). Other genetic counselors may become heavily involved in campaigning on behalf of a particular group of patients, but this is a personal matter for them rather than intrinsic to their professional role.

Genetic Testing

The term *genetic testing* covers many different activities and must be used with caution. It can be used to for any molecular genetic tests, but then it would include the use of DNA-based diagnostic tests for infectious diseases or prognostic tests in certain malignancies even though the tumor might not be familial. We restrict the term to testing of an individual's genetic constitution rather than testing for pathogenic organisms or testing for molecular changes within tumors. We also extend the term to encompass tests of an individual's genetic constitution that do not depend on molecular genetic methods but instead use cytogenetic or conventional biochemical or pathological methods so that a chromosome-based test, an X-ray, a muscle biopsy, or a biochemical measurement might all function as genetic tests if they provide information about the individual's genetic constitution.

It is also essential to distinguish between genetic testing of the individual and testing of a population or large group, when the term *genetic screening* may be used. Because the word "screening" has so many possible applications, it is best to avoid its use in relation to genetic testing except when the test is offered to members of a large population or group rather than to specific individuals on the basis of their personal characteristics or family history. Such genetic screening tests applied to a population are discussed later and must not be confused with genetic screening (i.e., DNA-based searching) of a specific gene for mutations or with screening for complications of a genetic disease, such as screening for tumors in a family cancer syndrome. Using the term genetic screening too freely will lead to much imprecision of thought and consequent confusion.

Genetic testing should not be considered in isolation from genetic counseling and should only be made available in association with the provision of appropriate genetic information and counseling, both before a test is performed and afterwards. The counseling provided in association with genetic tests follows the lines already discussed, and in particular the use of scenario decision counseling may be especially helpful in this context to enable clients to think about the possible personal implications of testing. What might be the impact on them and on their family of the four principal possible outcomes — being tested and having a favorable result, being tested and having an unfavorable result, being tested and having an unanticipated result of uncertain significance, or not being tested?

Diagnostic Genetic Testing

When a patient presents with signs or symptoms of clinical disease, it is necessary to establish a diagnosis as accurately as possible both to give a prognosis and to plan treatment. If the signs or symptoms are the result of a genetic disorder, then investigations are likely to establish a genetic diagnosis. These investigations may entail conventional laboratory tests or molecular genetic methods, but the significance of the test results — establishing that the disease is genetic in origin — will be similar. There are likely to be repercussions both for the patient and for the family when a genetic diagnosis is made. These are unavoidable, given the nature of the disease, but the distress and disruption caused in this way can sometimes be ameliorated by careful consideration of the way in which the testing is carried out.

It is currently recognized as good practice to ensure that written informed consent is obtained from the patient, when valid consent is possible, before certain genetic diagnostic investigations are performed such as tests for Huntington's disease (Fig. 2) and similar neurodegenerative disorders with profound implications for other members of the family. This is not to obstruct the making of such diagnoses but to ensure that the potential implications for others are considered at an early stage rather than not being appreciated until afterwards. It is also necessary to ensure that genetic counseling is available to concerned relatives once the diagnosis of a genetic disorder has been established.

Predictive Genetic Tests

In contrast to diagnostic genetic testing, which is carried out to establish the diagnosis of overt disease, is predictive genetic testing. Because genetic tests can be applied at any stage of life and irrespective of the individual's state of health, they can be used to identify those who will (or are likely to) develop disease in the future. This extension of the scope of diagnostic possibilities has many consequences. This was recognized at the advent of molecular diagnostics for Huntington's disease (HD), when procedures for pretest genetic counseling of at-risk but currently unaffected individuals were developed (Craufurd and Harris, 1986). These pretest counseling procedures include careful explanations of the limitations of testing, especially the residual uncertainty about the likely age of onset of symptoms in those who will be affected, and an exploration of the likely outcomes of the testing process for the indi-

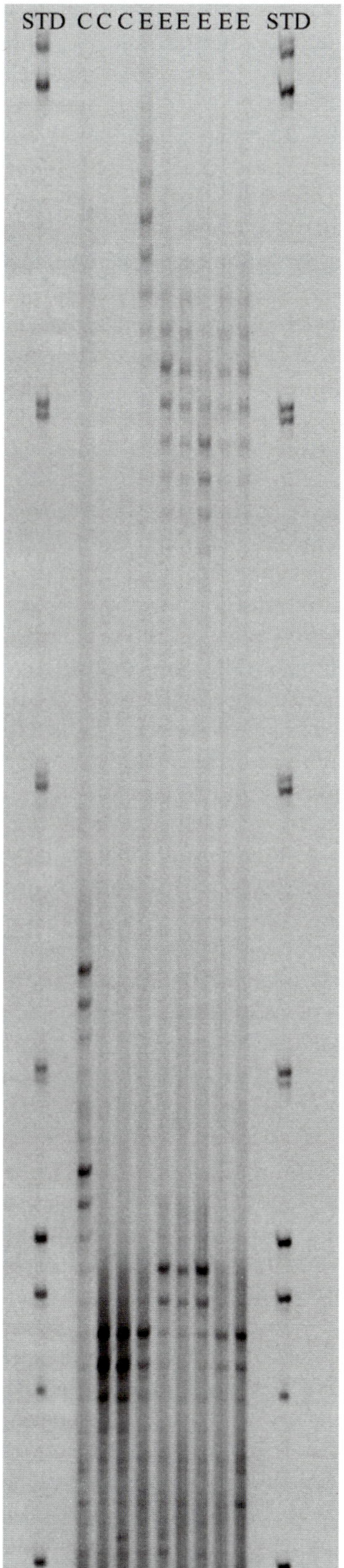

vidual and other members of the family. This amounts to "scenario decision counseling" as discussed previously and is carried out over at least two sessions weeks apart.

If the protocols developed in the early stages are employed, it is clear that predictive testing for HD can be provided without causing undue distress and that the process of testing is experienced, on balance, as being helpful by those given unfavorable results as well as by those given favorable results. Adverse consequences of testing do arise, however, in those given low-risk and those given high-risk results (Bloch *et al.,* 1992; Huggins *et al.,* 1992; Codori and Brandt, 1994; Lawson *et al.,* 1996), as might be anticipated from the previous discussion. Furthermore, it has become apparent that many individuals at risk for HD, who come forward for predictive testing, decide during the pretest counseling process not to proceed with the test. Only a small minority of those eligible for testing actually do it, so the favorable evaluation of predictive testing for HD only applies to those who were willing to comply with the pretest counseling protocols and who did not change their minds about testing during this process. It would not be valid to argue that the experiences of predictive testing so far have been favorable so that testing could be offered without similar pretest counseling and support (Ball *et al.,* 1994; Codori *et al.,* 1994; van der Steenstraten *et al.,* 1994).

This caution is reinforced by the problems that arise in practice during and after the predictive testing process (Morris *et al.,* 1989); at least one important problem arose in 46% of cases in one international series (European Community Huntington's Disease Collaborative Study Group, 1993). These problems can be grouped as problems of inappropriate referral (e.g., referral without consent or under pressure from a third party), problems involving relatives (e.g., when an individual wishes to be tested but an at-risk parent does not wish to know his or her own mutation status, which would be revealed as affected if the concerned individual was tested and was given an unfavorable result), and problems relating to disclosure of results (to other members of the family or to third parties such as employers or insurance companies).

In contrast to HD, for which there is currently no effective treatment, predictive testing is possible for a range of other disorders in which therapeutic interventions are available. For many familial cancer syndromes, the identi-

FIGURE 2 Electrophoretic plate used for the molecular test for Huntington's disease: expansion of the $(CAG)_n$ region in the *IT15* gene, responsible for the disease, in affected and at-risk individuals (E). In contrast to normal subjects (C), a series of bands are seen at the top which correspond to alleles with more than 40 CAG triplets. Each vertical band corresponds to one individual. STD, DNA standard used to determine the molecular mass of the fragments in test (courtesy of Marina Frontali).

fication by molecular genetic means of those who are susceptible to the relevant cancers is very helpful for management. One-half of those at risk are spared the (sometimes invasive) procedures necessary to diagnose cancers at an early, treatable stage, and the health services are spared the cost of the unnecessary surveillance for tumors in those not at risk. Prophylactic surgery can be provided to those genuinely at high risk of specific malignancies. Conditions for which genetic testing is certainly helpful include familial adenomatous polyposis coli, retinoblastoma, and multiple endocrine neoplasia. The benefits of predictive genetic testing are less certain for other conditions such as familial breast cancer, for which there are currently no interventions of proven worth for those at high risk and further evaluations are required to clarify the situation. The model of pretest counseling and posttest support that might be most appropriate for predictive testing across a range of diseases in a regular service setting (not in a specialist research clinic) was discussed by Harper (1997).

Prenatal Genetic Testing

When a couple or a woman know that they are at risk of having a child affected by a serious genetic disease or congenital malformation, they may be very apprehensive about embarking on a pregnancy at all. Other couples in the same situation may decide to take their chances. In addition to the information given in genetic counseling, many factors (social, moral, economic, and others) influence the decisions made by such families. One factor that has had a considerable impact on the decisions made by such families during the past 20 years is the increasing availability and scope of prenatal genetic diagnosis and the consequent possibility of the selective termination of affected pregnancies. This topic is therefore inseparable from the wider debate about abortion in general, but in this article I address only the specific, additional issues raised by the termination of pregnancies for genetic disease or fetal malformation.

The availability of accurate prenatal diagnosis for certain disorders has altered the pattern of family planning, and of family life in general, in many contexts. For many pediatric genetic diseases, it has meant that couples at risk of affected children may have one affected child but are less likely to have a second affected child as long as the first has been diagnosed before the second child is conceived. In families with sex-linked disorders, in which carrier females can be identified before they have even a single affected child, the effect can be greater. In the context of Duchenne muscular dystrophy, for example, the sisters of affected males may well be carriers. Those girls who grow up with affected boys, and watch them over the years become unable to walk and then become progressively weaker until they die, often have strong views about taking risks with their own children. Some will not choose prenatal testing out of a sense of loyalty to their brothers, but many are unwilling to take even a very small chance of having an affected son. Before the advent of accurate prenatal diagnosis, these women remained childless, had fetal sexing followed by the termination of pregnancies with male fetuses of whom at least 50% would have been unaffected, or took their chances. Now they can usually be told with certainty whether or not they are carriers. If they are carriers, and if they choose to proceed with prenatal diagnosis, they will be able to continue pregnancies in which the fetus is male but unaffected. Prenatal diagnosis enables these families to achieve the type of family to which most couples reasonably aspire, and this can be achieved at the cost of fewer terminations of pregnancy.

In families that have firsthand experience and an accurate understanding of the disease from which they or their children are at risk, genetic counseling and prenatal diagnosis raise relatively few specific ethical difficulties apart from the general questions regarding abortion. Of course, the family may need to have the formal facts of inheritance explained to them and they may require information about the prenatal diagnostic procedures under consideration, but they are in a good position to weigh the consequences of each possible course of action and to choose between them. This situation is quite different from the context of antenatal population screening programs discussed later.

The central ethical question is whether the genetic disorder in the family is serious enough to warrant termination of an otherwise wanted pregnancy. The usual response to this dilemma from clinical geneticists and genetic counselors is to insist on their professional neutrality and to leave the decision to those who are intimately concerned and who will have to live with whatever decision is made. This response — a passively nondirective complicity with the client — is not entirely adequate because it ignores the responsibility of the counselor to ensure that their clients have taken account of the full range of likely consequences of their decisions. It also ignores the broader social consequences of genetic counseling and testing services and the collective consequences of the sum of individual choices made as a result.

Professional respect for the policy of nondirectiveness in genetic counseling is especially important in the context of prenatal genetic diagnosis, in which any other approach would be abhorrent, but the meaning of "nondirectiveness" is complex and contentious and need not simply amount to "laisser faire" or to a shrugging off of responsibility onto the client. It may be thoroughly appropriate for a genetic counselor to explore with his or her clients how they have made their decision; he or she can thereby ensure that the decision has been made in light of the true facts about the disease and that his or her clients have considered the full extent of the implications of their decision.

Despite the value of adherence to nondirectiveness in prenatal genetic counseling, it can be difficult for professionals to conceal their disagreement with some families, especially where prenatal diagnosis and the termination of an affected pregnancy are sought for a very treatable disorder (e.g., phenylketonuria; PKU), for an apparently trivial or largely cosmetic condition (such as a simple cleft lip), or because the fetus is the "wrong" (socially undesirable, usually female) sex or has the wrong (socially inconvenient) father. These contexts slide from the realm of medicine and medical genetics into the social abuse of genetic technology. Professionals' responses to such requests will vary, and prescriptive guidelines are not going to be helpful. Possible responses include an immediate refusal even to discuss the questions, an in-depth exploration of the couple's reasons for their request, or referral for a more expert course of counseling or psychotherapy.

Finally, mention should be made of technical developments that will confront us and our clients with a different set of issues. Preimplantation genetic diagnosis has become possible for some disorders when a few (often just two) cells are removed from a very early embryo following *in vitro* fertilization; only unaffected embryos will then be implanted *in utero*. There remain numerous questions of safety and reliability to be explored here, but if these are resolved then such prenatal diagnosis may be regarded as preferable to conventional prenatal diagnosis because the question of terminating an established pregnancy does not arise; the onerous nature of *in vitro* fertilization means that this is unlikely to be used widely except by those couples with fertility problems. A more general issue is raised here, however, if preimplantation testing for nondisease, but socially important, traits becomes technically possible: Would it be acceptable to implant the early embryos (the fertilized ova) that were most likely to develop into intelligent, tall, and athletic people?

The development of noninvasive methods of prenatal diagnosis will also alter the ethical issues raised by prenatal diagnosis because currently genetic testing of a fetus entails hazard to the pregnancy—a risk of miscarriage of about 2 or 3% for chorionic villus sampling at 11 weeks' gestation and of about 1% for amniocentesis at 16 weeks' gestation. For those seeking prenatal diagnosis even with the current hazards, such techniques as the testing of nucleated fetal red blood cells sorted from a maternal blood sample will be a great benefit. On the other hand, prenatal diagnosis may be sought for less serious conditions and women who are ambivalent about prenatal diagnosis may be more easily pressured into tests and terminations of pregnancies that they do not want. We need to remain alert to these potential problems in the future.

Population Genetic Screening

Population screening for genetic disease raises many more ethical difficulties than does genetic counseling and testing within families known to be affected by genetic conditions. First, screening programs are proactive, in contrast to standard genetic counseling, which is reactive; instead of responding to the needs of individual families, screening entails offering a test to many individuals at low risk of the specific problem who may never have had any concerns about it before the offer of the test was made. Before raising anxieties and concerns in many people about a problem that is unlikely to affect them, and before burdening them with the need to make decisions with which they would often prefer not to have been confronted, it is necessary to ensure that a screening program will do more good than harm.

Second, families which have already been affected by genetic disease will often know much about the condition and will be in a good position to decide whether or not they wish to proceed with further genetic testing. Most members of the population, however, will be relatively ignorant about the condition, and they will not find it easy to weigh the potential benefits of testing against the problems it may cause. They may find that the offer of genetic screening by respected health professionals convinces them that it is clearly worthwhile and beneficial for them to participate; they may feel persuaded by the offer of the test without really appreciating the possible disadvantages and without weighing the issues. Indeed, it is very difficult to offer a screening test in a nondirective manner (Clarke, 1991, 1997a).

For a population genetic screening program to be offered properly, therefore, it is necessary for a careful evaluation to be made of the full range of its likely consequences before it is introduced on a wide scale. It is also necessary to ensure that each member of the population who is offered the test experiences the offer as a genuine choice and is given sufficient information in advance so that he or she can make a decision without feeling pressured. These standards may be difficult to meet.

Newborn Screening

The screening of blood spots taken from newborn infants in the first week of life for evidence of PKU became possible in 1963, when an appropriate method was first devised. Programs of screening for this and other eminently treatable metabolic disorders, such as congenital hypothyroidism, were subsequently introduced in most developed countries. PKU affects 1 in 10,000 infants in Britain and varies substantially in frequency between populations; congenital hypothyroidism affects about 1 in 4000 infants in Britain and is more uniform between populations (except where iodine deficiency is endemic). These screening programs raise few ethical concerns because they clearly meet the World Health Organization criteria for being worthwhile (Wilson and Jungner, 1968). Because of the clear benefits of early diagnosis to affected infants, allowing prompt dietary or hormone replacement therapy, such screening is effectively mandatory in many countries and states.

The question of the need for informed parental consent to screening has been raised, but this is not widely regarded as a serious problem because the medical case for universal screening is very strong. Parental knowledge about newborn screening is certainly very deficient in Britain (Statham *et al.,* 1993). In one study in the United States, the level of parental knowledge was improved by an educational program introduced to accompany a state requirement for parental consent before screening could be performed, and there was no decrease in the proportion of infants successfully screened (Faden *et al.,* 1982; Holtzman *et al.,* 1983). Whether the extra effort is worthwhile can still be debated, and similar efforts have not been made in many other countries.

Other disorders for which the benefits of newborn screening programs are probably sufficient to justify their introduction include sickle-cell disease and cystic fibrosis in the relevant population groups. More difficult issues arise in the context of newborn screening for other, less treatable genetic disorders. Where the medical benefits to the child of an early, presymptomatic diagnosis are less certain or more marginal, the possible disadvantages of an early diagnosis need to be considered and weighed against the potential benefits. There is a wide range of such disorders in which newborn screening is technically feasible but of limited or uncertain overall benefit; such disorders include galactosemia and maple syrup urine disease. The reason why screening may be judged of unproven benefit varies with the different conditions: Some infants will present before the newborn screening result can be obtained, early diagnosis may make little difference to the natural history of the disease, or the available laboratory methods may be insufficiently sensitive or specific.

Another important disease for which newborn screening is technically possible is Duchenne muscular dystrophy (DMD). This affects about 1 in 4000 infant boys. Affected boys often present at 2–4 years with progressive weakness and wasting of the muscles, although the diagnosis may not be made until 4–6 years; affected boys become unable to walk by 8–12 years and often die in their teenage years or early twenties. No effective treatment is available, and the principal benefits of an early diagnosis are the avoidance of distress from a prolonged diagnostic process and the availability of genetic counseling before a second affected child is conceived. These benefits, such as they are, accrue to the family rather than to the child. Are they sufficient to shatter the family's early happiness with an apparently healthy son, altering irrevocably the relationship between the affected child and his family?

A social evaluation of newborn screening for DMD has been available in Wales since 1990, and evidence suggests that families do find the early diagnosis to be helpful. There are two caveats, however. First, newborn screening for untreatable disorders such as DMD must be offered as an optional test on the basis of the parents' informed decision. This requires a very different ethos from the other newborn screening tests, which in turn necessitates an educational program for the midwives who offer the tests to the parents and collect the blood samples. Second, the presymptomatic nature of the test confers an advantage in that there is no clinical urgency in pursuing the diagnosis. It is therefore possible to plan the process of disclosing the test result to the family in conjunction with the primary health care team and to employ a protocol developed to optimize this difficult process (Bradley *et al.,* 1993; Parsons and Bradley, 1994; Parsons *et al.,* 1996). With these safeguards, families seem to prefer the diagnosis of DMD in their child to be made earlier rather than later.

The planned introduction of new technologies such as tandem mass spectrometry into newborn screening will provide further challenges because many infants will be identified as having abnormal metabolites in their serum but the clinical significance of these will not always be apparent until years later. For many of the disorders diagnosed in this way, there may also be no effective therapy. There are potentially serious adverse consequences for a family that is told that their child has "something" but that it is difficult to identify more precisely, probably cannot be treated, and may never cause the child any problems. Technological progress does not always translate readily into human progress.

Reproductive Genetic Screening

There are two types of reproductive genetic screening tests: antenatal screening tests offered to all pregnant women that examine the fetus and screening tests to identify healthy carriers of autosomal recessive diseases, such as cystic fibrosis, the hemoglobinopathies, or Tay–Sachs disease, and potentially also to identify female carriers of X-linked conditions such as the fragile X syndrome.

Antenatal screening programs are intended to identify, and usually to terminate, pregnancies affected by Down's syndrome (Fig. 3) and neural tube defects (anencephaly and spina bifida).The methods available range from asking a simple question ("How old are you?"; the probability that a woman will give birth to an infant with Down's syndrome increases with her age) to biochemical measures of fetal or placental polypeptides in the mother's blood and ultrasound scans of the fetus (Fig. 4). These tests are not hazardous, except when the results lead a woman to undergo amniocentesis to enable a more precise, diagnostic test to be performed, nor are they completely sensitive or specific. These tests are constantly being refined, however, so I focus on the issues and principles at stake rather than details of the methods employed.

Carrier screening programs identify healthy carriers of genetic conditions, whose children may be at risk of being affected. In the case of autosomal recessive diseases, a child will only be affected if both parents are carriers. The frequency of carriers of autosomal recessive diseases varies between different populations, with cystic fibrosis (Fig. 5) be-

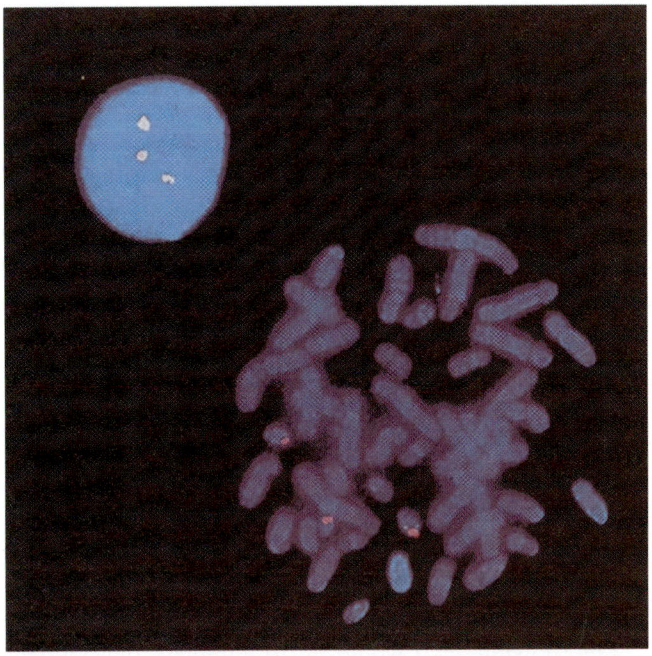

FIGURE 3 Photomicrograph of chromosomes (bottom) from a fetus with Down's syndrome. The chromosomes from the nucleus of the cell (top) have been treated by a fluorescence *in situ* hybridization technique which highlights (pink dots) the trisomy (three copies) of chromosome 21 (photo courtesy of King Holmes/SPL/Neri).

ing especially common in those from northwest Europe, sickle-cell anemia being more common in those from West Africa, and the other hemoglobin disorders being common throughout the Mediterranean, the Middle East, Southeast Asia, and much of Africa. A couple identified as both

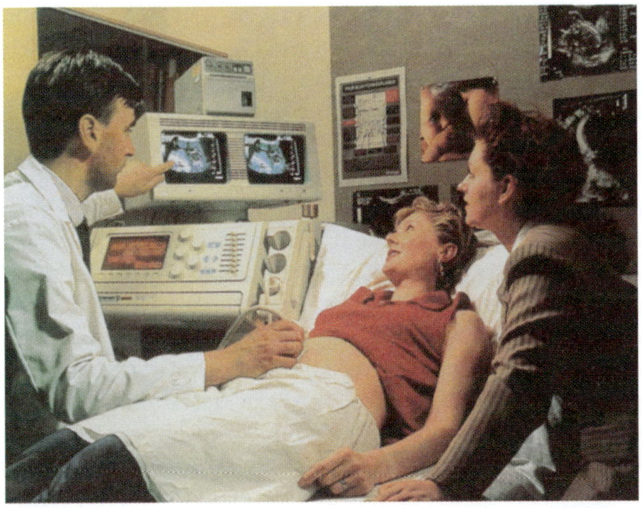

FIGURE 4 Ecographic scan of the fetus in a specialized clinic (photo courtesy of SPL/Neri).

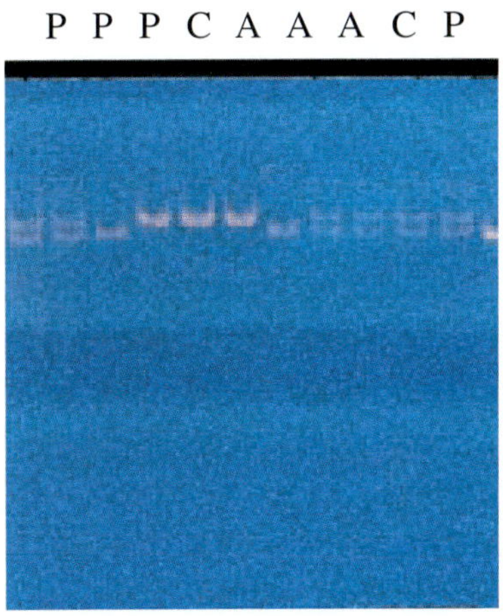

FIGURE 5 Electrophoretic plate used for the molecular test for cystic fibrosis. When illuminated with ultraviolet light, the DNA fragments appear as pink fluorescent bands. In affected patients (A) the lower band is missing and only the intermediate band is seen. Carriers (C) show a normal upper band but lack the band immediately below it. Normal noncarrier subjects (N) have only the lower band (photo courtesy of Molecular Genetics, East Anglian Regional Health Authority/SPL/Neri).

being carriers of an autosomal recessive disorder will have a 1 in 4 (25%) chance of having an affected child in any pregnancy and would therefore be eligible for prenatal diagnosis.

The ethical issues raised by reproductive genetic screening include those raised by prenatal diagnosis but also additional issues because reproductive genetic screening entails the proactive application of tests to a population that may otherwise have had few concerns.

The first issue is the offer of testing. The way in which the offer of prenatal screening tests is made to women is often highly directive (manipulative), with compliance being portrayed as the "responsible" course of action and also being routinized so that clients are not encouraged to think through the issues (Marteau *et al.,* 1992a,b; Rothman, 1988; Sjogren and Uddenberg, 1988). Fetal ultrasound scans are often portrayed ("offered") as being social occasions, with the prime goal seeming to be the taking of the first picture of the baby for the family album; in the collusion between anxious mother and anxious ultrasonographer, the search for fetal anomalies is often not given much emphasis.

In the context of population carrier screening tests, it is clear that the way in which the test is offered is the major factor influencing whether testing is undertaken (Bekker *et al.,* 1993; Tambor *et al.,* 1994; Payne *et al.,* 1997). Where strong social pressures are applied, very high rates of carrier screening can be achieved, as in premarriage carrier screening for β-thalassemia in Cyprus and in antenatal carrier screening for cystic fibrosis in Edinburgh and Manchester, United Kingdom. It may be difficult to determine whether a high rate of uptake of a test occurs because the test is actively desired as being relevant and helpful or whether high uptake indicates a degree of (subtle, well-meaning) coercion.

Second, women whose pregnancies are found to be affected or at high risk are often unprepared for this possibility and are given little support and less time to make their decisions. The information available antenatally to women and couples about the disorder for which screening is offered often gives a distorted and negative view which is different from the positive, optimistic impression conveyed by the literature given to the parents of children with the same condition (Lipmann and Wilfond, 1992). Equivocal or unanticipated results are especially likely to cause problems because the immediate response of a pregnant woman or couple may be to terminate the pregnancy; only later is it appreciated that the fetus may not have had a serious disorder at all but perhaps a condition such as Turner's syndrome. Those who do terminate their pregnancies are likely to be given little support and to suffer sustained psychological distress (White-van Mourik, 1992).

Third, individuals identified as healthy carriers of autosomal recessive disorders gain from this knowledge in that they can make informed reproductive decisions. They may also experience disadvantages from being identified as carriers of such diseases. They may experience social stigmatization, being seen as less desirable marriage partners (Stamatoyannopoulos, 1974; Evers-Kiebooms *et al.,* 1994), or they may have residual anxieties about their own health despite explanations from health professionals (Marteau *et al.,* 1992c; Zeesman *et al.,* 1984; Loader *et al.,* 1991). In some contexts, as in the United States in the 1970s, carrier testing can be abused to reinforce racial discrimination because of differences in gene frequencies between ethnic groups. In a society with private health insurance, carriers may also experience discrimination in employment and insurance, even when there is no rational basis for the discrimination (Billings *et al.,* 1992; Geller *et al.,* 1996).

Fourth, there are also wider social consequences of reproductive genetic screening. There are the far-reaching effects on the experience of pregnancy, which becomes provisional ("tentative") rather than wholeheartedly welcomed until the fetus has passed its genetic tests (Rothman, 1988). Professionals collude with pregnant women in not discussing the reason for genetic tests in a pregnancy, or they use the bland and misleading discourse of reassurance ("to make sure the baby is healthy") when the tests may only serve to allay anxieties that were raised by the offer of testing in the first place (Press and Browner, 1994; Browner and Press, 1995), and of course they do not reassure those women in whom screening generates an increased risk or an abnormal result.

Antenatal screening programs may also have consequences for those members of society who are affected by the conditions for which screening is performed. Although some affected individuals do believe that screening is helpful (wishing perhaps that they had not been born), there are many others who feel hurt and wounded by the effort of medical scientists to eliminate individuals like them from society. There are few studies of this sensitive area, but there are grounds for thinking that the parents of children with Down's syndrome and spina bifida are more likely to be viewed as irresponsible and blameworthy because they did not avoid the birth of their affected child as they perhaps could have done (Marteau and Drake, 1995; Julian-Reynier *et al.,* 1995). This type of social intolerance has wider implications for the very nature and quality of our society and our acceptance of difference and disability.

Finally, in making decisions about whether to promote antenatal screening programs, policymakers will want to identify the goals of the program and then to assess how effectively those goals are met. What then are the goals of antenatal screening? What is its purpose?

One possible goal would be to eliminate the physical suffering of affected individuals. This could plausibly apply in some cases of spina bifida but would be less relevant to anencephaly and only sometimes relevant to Down's syndrome, from which many affected individuals do not suffer physically and can attain a good quality of life. Given the difficulty of predicting which fetus with spina bifida or Down's syndrome would suffer physically as a child if the pregnancy continued, and given that many affected individuals regard their lives as worthwhile, it is difficult to sustain this justification as the reason why society promotes prenatal screening as it does.

A second possible goal is to avoid the costs of providing health and social care for affected individuals, whose needs are certainly greater than those of other individuals. This seems to be the dominant justification for prenatal screening employed in medical publications written from a public health perspective (Hagard and Carter, 1976; Wald *et al.,* 1992; Cuckle *et al.,* 1995) and can be portrayed as the "genetic cleansing" paradigm of prenatal screening (Clarke, 1997b).

One variation of this model is the claim that fetal anomaly ultrasound scanning can significantly reduce the perinatal mortality rate and hence the burden of disease (Saari-Kemppainen *et al.,* 1990). Unfortunately, the reduction in perinatal mortality is largely accounted for by an increase in the rate of terminations of pregnancy for lethal disorders and thus represents a shift of inevitable mortality from the perinatal period to before 24 weeks' gestation. This need not indicate an improved outcome, however, because families may experience less distress and be given more support after a perinatal death than after a second-trimester termination of pregnancy.

A third plausible goal of prenatal screening, the one

most espoused in genetic counseling circles, would be to maximize informed reproductive decision making (Modell and Kuliev, 1993). This is the "informed choice" paradigm of prenatal screening, but it is not entirely clear that it is a coherent policy because it begs such questions as "Which reproductive decisions are to count?" and "How are these context-setting decisions to be made, and by whom, before the clients ever have the chance to make their 'informed decisions'?" (Chadwick, 1993; Clarke, 1993). There are likely to be powerful interests that operate to determine these more fundamental decisions (Lippman, 1991).

An alternative response to the fact of genetic disease and to the existence of prenatal screening programs is to acknowledge the flaws in most such programs — the difficulty of achieving fully informed and considered decisions in the context of mass antenatal screening — and to work toward improving the context within which screening operates. Improvements in the training of midwifery, ultrasonography, and obstetric staff may lead to an improved quality of pretest information and counseling and an improved quality of postresult counseling and support. Through promoting health and social services for those with genetic conditions, the general confidence of prospective parents in the willingness of society to care for those with disability, whether or not it is genetic in origin, will be increased and this may reduce the sense of coercion to participate in prenatal screening that is now a common experience. This places more emphasis on the nurture of those affected by genetic conditions and less on their prevention.

Susceptibility Screening

Population screening for susceptibility to disease is currently restricted to screening of blood pressure and serum cholesterol, and the case for screening healthy adults for hypertension or hypercholesterolemia, in the absence of specific indications, is not firmly established. There are important disadvantages to being identified as having an increased blood pressure or an increased serum cholesterol level, however, including psychological morbidity from being labeled as sick — in effect, being found to have a silent disease — and potential practical problems obtaining insurance or employment (Lefebvre *et al.,* 1988; Tijmstra, 1990). With the rapid genetic dissection of the common, complex diseases of the developed nations, it may soon be technically feasible for population screening for genetic susceptibility to such diseases as common cancers, heart disease, diabetes mellitus, hypertension, and Alzheimer's disease.

In the absence of specific interventions known to be of benefit to those screened, however, it would be difficult to justify the introduction of such screening. Not only would such programs be a drain on limited resources but also they could have negative consequences for those screened. They would be likely to generate much anxiety in those given increased risk results, and there could also be paradoxical health consequences as a result of inappropriate health behaviors: Those given a low risk of heart disease could lead an unhealthy lifestyle and hence worsen their chances, and those given a higher risk could believe that disease is inevitable and hence have no motivation to comply with health-promotion advice (Tijmstra, 1990; Clarke, 1995).

In the context of genetic counseling, there would be concern that the introduction and promotion of such susceptibility screening could be driven by the need of biotechnology corporations to recoup their major investment in research. Pretest "counseling" could then function as commercial promotion rather than a neutral provision of information and the facilitation of the client's rational decision-making process. Tests could be sold as "sexy," "responsible," or "rational" to different client groups. Commercial pressure could easily undermine the nondirective stance of pretest genetic counseling — it has been shown how the mode of offer of carrier testing for cystic fibrosis is the major determinant of uptake.

Although some professional groups would challenge or distance themselves from such practices, others would be willing to profit from them. The alliance of enthusiastic health professionals with industrial and commercial interests can be a powerful force, as in the promotion of fetal ultrasound scanning, maternal serum screening for Down's syndrome, and carrier screening for cystic fibrosis (Koch and Stemerding, 1994). It may become necessary for governments to introduce legislation tightly controlling the offer of such tests directly to the public (by post or over-the-counter) and even their marketing through health professionals (Clarke, 1995).

Goals of Genetic Counseling

It is difficult to define the goals of genetic counseling in a form that can be measured. Instead, qualitative definitions that apply to the ethos and process of genetic counseling have been developed. This is a weakness in the struggle to secure adequate funding for clinical genetic services when in direct competition for resources with other specialties for which outcomes are more readily quantified. The definition of genetic counseling given by Fraser (1974) is still widely used: genetic counseling is

> a communication process which deals with the human problems associated with the occurrence, or risk of occurrence, of a genetic disorder in a family. This process involves an attempt by one or more appropriately trained persons to help the individual or the family to
>
> (i) comprehend the medical facts, including the diagnosis, the probable course of the disorder, and the available management;

(ii) appreciate the way heredity contributes to the disorder and the risk of recurrence in specified relatives;

(iii) understand the options for dealing with the risk of recurrence;

(iv) choose the course of action which seems appropriate to them in view of their risk and their family goals and act in accordance with that decision; and

(v) make the best possible adjustment to the disorder in an affected family member and/or to the risk of recurrence of that disorder.

This definition makes abundantly clear the absurdity of trying to assess the quality or effectiveness of genetic counseling in terms of some simplistic, population-based measure, such as the birth incidence of a genetic condition or the number of terminations of pregnancy following prenatal diagnostic procedures. To apply such measures would be deeply disturbing not only to many genetic counseling professionals but also to individuals with genetic disease and to all those interested in civil liberties; it would lead, in effect, to a eugenics program masquerading as genetic counseling.

Given that our goals are essentially client set, defined afresh in each consultation, we need some way of assessing how useful our services have been to each client. Recording client expectations before genetic counseling and then assessing how completely these expectations have been met is one possible approach. The client's initial expectations, however, may have been inappropriate because he or she may not have had enough information to know what to expect from genetic counseling. Accordingly, it has been proposed that a retrospective assessment by each client of the extent to which his or her genetic counseling met the expectations he or she should initially have had, if he or she had been better informed at that stage, might be the most appropriate measure of success (Clarke *et al.,* 1996).

The pressure on genetic counseling providers from the purchasers or commissioners of health services to demonstrate their effectiveness and worth in simple, numerical terms may still be intense, however, and will need to be resisted by convincing arguments that such predetermined goals simply do not apply in this field. If this argument fails to persuade, then genetic counselors may be pressured to achieve certain population targets (e.g., a certain number of pregnancy terminations for "costly" disorders) and will then feel pressure to lead ("direct") their clients to make the appropriate decisions. One great advantage of the collective professional adherence to nondirectiveness in genetic counseling, even if it is inadequate in certain contexts, is the strength it gives professionals to resist this sort of coercion by those who pay for health services (Clarke, 1990, 1991; Harper and Clarke, 1997).

Recurrent Issues in Genetic Counseling Practice

Autonomy, Informed Consent, and Nondirectiveness

The insistence by genetic counselors on the importance of nondirectiveness is understandable on many counts. Genetic counseling emerged as a discipline during the decades after World War II, when the complicity of German biological and medical scientists in the promotion of the Nazi doctrines of racial superiority and especially of race hygiene became abundantly clear. Genetic counselors wished (and currently wish) to distance themselves from Nazi-associated "eugenics." During the same period, the field of medical ethics, at least in the Anglo-Saxon world, became dominated by the principle of autonomy, very much in tune with the trend toward individualism apparent in political, social, and economic affairs in the second half of the century. Furthermore, counseling has emerged as a profession and, indeed, almost a way of life. "Nondirective counseling" is associated with one of the pioneers in "counseling-as-a-route-to-self-fulfillment," Carl Rogers, who founded one of the dominant schools of humanistic psychology. Adherence to nondirectiveness is therefore a very appropriate guide to thought and practice in the young disciplines of clinical genetics and genetic counseling.

Genetic counselors wish to make it clear that they do not tell their clients what to do. They want their clients to retain responsibility for their decisions and actions. They will go to great lengths to avoid prescribing the appropriate course of action for their clients. Even when it is perfectly clear from the context what the counselor thinks is the best course of action, he or she will package the information given to his or her client in such a way that he or she can deny any suggestion that he or she has given advice. The good (i.e., nondirective) counselor would never tell his or her client what to do.

Respect for autonomy and privacy, a strict policy of prior informed consent for any diagnostic procedure or disclosure of information, and adherence to nondirectiveness together form a troika that gives genetic counseling much of its current character. This is quite distinct from the ethos in many other medical disciplines so that doctors who refer patients to clinical genetics may have a very different understanding of what genetic counseling will entail. They may regard it as obvious that their patient should be sterilized, or have a termination of pregnancy, because that is (to them) the obvious course of action; the function of the genetic counselor is to explain this obvious necessity to the patient and, if necessary, persuade the patient to comply. Such conflicting agendas are not as frequent now as they were in the past, but they do still occur. When these events are witnessed in such an extreme form, it is easy to see that the emphasis on nondirectiveness is necessary.

In areas of genetic counseling practice other than prenatal diagnosis and the predictive testing for untreatable disorders, in which absolute respect for client autonomy is very appropriate, there may be occasions when it would be legitimate for genetic counselors to recommend a particular course of action. They may recommend compliance with protocols of surveillance for complications of a genetic disease. They may recommend that a client passes on information about a genetic disease to other members of the family. Such recommendations do go beyond the letter of nondirectiveness but do not necessarily conflict with its spirit. We must take a broader view of the counselor's responsibilities than strict adherence to nondirectiveness might permit, acknowledging that the exclusive focus on client autonomy is inadequate. At the same time, it is essential that the genetic counseling focus on nondirectiveness should be understood by other health professionals and should be incorporated into aspects of their work, particularly in routine programs of antenatal screening in which a greater emphasis on fully informed consent and respect for client autonomy may be overdue.

Children and Genetic Testing

Genetic testing of children will be appropriate in many health care contexts, such as when a child presents with an illness that may be genetic in origin or when a child is at risk of a genetic disorder that would be likely to cause symptoms in childhood. In the first case, genetic testing is essential to make a diagnosis (standard medical practice). In the second case, there is little to lose by testing the child at whatever stage the parents would find it most helpful. It is worth discussing this issue carefully with the parents so that the question of timing is thought out, but there is no virtue in declining to test a child in such a situation in which the parents will otherwise be liable to overinterpret any minor problem or illness as evidence of the genetic disease beginning to manifest even though the child may be totally unaffected. In addition, it will be important to carry out genetic testing in childhood if treatment or surveillance may improve the long-term clinical outcomes for the child as a future adult, as perhaps is the case in some familial hyperlipidemias.

There are two contexts, however, in which the genetic testing of children is more contentious: predictive testing for late-onset (i.e., usually adult-onset) disorders and testing for genetic carrier status when it will only be of significance to the child when he or she is older and is considering his or her reproductive plans. In these contexts, there are reasons for deciding not to carry out genetic testing in childhood. At this point, I must make it clear that the word "child" is not defined simply by chronological age, for example, as being less than 16 or 18 years of age. If a younger child, perhaps 13 or 14 years old, requests genetic testing on his or her own behalf, and if the child demonstrates a sufficient understanding to appreciate what is at stake in such testing, then testing may be appropriate (Alderson, 1993; Binedell *et al.,* 1996).

There are three reasons for suggesting that predictive or carrier testing of children, at the suggestion of their parents or a third party, may be unwise. The first is that the child's future capacity to make up his or her mind (the child's future autonomy as an independent adult) has been abrogated. Second, the child will have lost control of the test result; the confidentiality that an adult would expect as a right will have been lost. Third, the parents' knowledge of their child's test results may distort the child's rearing. For example, parental expectations of a child's school performance may be lowered if they know he or she is destined to develop HD. If a child carries a genetic condition that could affect her future children, then the child's parents may try to channel her away from an interest in babies and boyfriends. We do not know that such problems would arise, but they are plausible; the fact that only a small minority of at-risk adults choose to undergo predictive testing for HD reinforces these concerns. Furthermore, the abrogation of the child's future rights of autonomy and confidentiality definitely would result from such testing, although we do not know how seriously this would be viewed in retrospect by the child as a future adult.

Attitudes regarding these concerns vary enormously, both within the medical and genetic counseling circles and among families affected by genetic disorders. Requests for predictive and carrier tests have often been received by genetic units, but a consensus among genetic professionals and lay support groups has developed that would not support predictive testing of a child for an adult-onset condition (Craufurd and Harris, 1986; Bloch and Hayden, 1990; Harper and Clarke, 1990; Clinical Genetics Society, 1994; Wertz *et al.,* 1994; Dalby, 1995; American Society of Human Genetics/American College of Medical Genetics, 1995; Clarke and Flinter, 1996). There is more diversity of views in relation to testing for carrier status in childhood, and the issues are more finely balanced. Most agree, however, that the appropriate decision is that which will be in the child's long-term best interests, if that can be determined. The argument that a child is the possession of his or her parents, who can therefore exert property rights over him or her and have any test performed that they choose, receives little support (Montgomery, 1994) except perhaps in North America (Pelias, 1991; Sharpe, 1993). There is a clear need for further research in this area (Michie and Marteau, 1996).

Confidentiality within Families

Respect for confidentiality has been a cornerstone of medical ethics ever since Hippocrates, but it has been challenged in two influential British reports (Royal College of Physicians, 1991; Nuffield Council on Bioethics, 1993) and in an American report (American Society of Human Genetics, 1998). These documents raise the issue of what a genetic counselor is to do when a client refuses to pass on to other members of his or her family some information that the counselor thinks could be relevant and important to those other individuals. All three documents reached a similar conclusion: If the risk to other members of the family is real and substantial, and if the counselor has failed to persuade the client to pass on the information, then the counselor may overrule the client's decision and inform the relative(s) himself or herself. The arguments brought forward in these documents are crudely utilitarian, maintaining that breaking one client's confidentiality is more than balanced by the harm prevented by informing other relatives of some potential genetic problem. They fail to give proper consideration to three important points.

First, the consequences of breaking medical confidentiality in the context of genetic disease, if this becomes generally acknowledged within the profession as a proper course of action, may have serious consequences for the whole practice of clinical genetics and genetic counseling. Clients will come to know that genetic information can be passed on to other members of their families with impunity, and those who feel a need for privacy will simply not give correct information about their families. This could have a damaging effect on the general conduct of genetic counseling. It could also result, paradoxically, in fewer family members being informed about genetic risks because those clients who are initially cautious and give an "edited" version of their family tree may find it difficult subsequently to admit to this deception. Without the professional threat to confidentiality, such a client would probably have been persuaded to pass on the necessary items of information to his or her relatives but this may now not happen.

Second, the harm to be prevented may never happen and it will be difficult to know when the need for other members of the family to know becomes urgent. If the counselor is already in contact with them, then discussing the family genetic condition will hardly be a breach of confidence (even if some detail is withheld). If the counselor is not in contact with them, then how does he or she know when the need to make contact is sufficiently urgent to break confidentiality? Finally, there will often be practical difficulties in contacting other members of the client's family unless they are already known to the counselor.

Given these considerations, it can be maintained that respect for privacy of medical and genetic information should be especially strong in genetic counseling. The development of guidelines to define the "exceptional" circumstances in which confidentiality can (or should) be breached would result in these circumstances no longer being exceptional and would likely cause many more problems in practice than it is intended to remedy. That these proposals have emerged at all demonstrates the inadequate conceptual framework (too narrowly utilitarian) that is employed by some ethicists and which is unable to take full account of the likely consequences of their proposals.

Confidentiality and Third Parties

There has also been discussion about genetic privacy in relation to third parties, usually institutions such as insurance companies, employers, or the state. Two questions arise: What limits should society place on the uses to which genetic information about individuals may be applied? and How should citizens be protected from illegitimate uses of their genetic information? I consider only the first question here.

Concern has been expressed that insurance companies may use genetic test results to identify those who are more or less likely to suffer ill health or to die prematurely. This may not be relevant to many families now because the only available tests are for the uncommon, single-gene (Mendelian) disorders; the situation may change rapidly, however, when genetic tests for susceptibility to the common, complex diseases become available. If some companies used these tests, then they could offer cheaper insurance to those with better health prospects; other companies would then be forced to adopt the same strategy. This could result in the undermining of the whole basis of insurance — the collective pooling of resources to meet unpredictable risks faced by all. This would be a particular problem in countries such as the United States in which health care is funded through private health insurance. In Europe, the impact would be felt predominantly in the life insurance industry because health services are less subject to market forces.

One interim solution to this, adopted in The Netherlands and being considered elsewhere, is to restrict the right of insurance companies to use genetic test results in decisions about life insurance policies linked to house purchase when the policy is of an average size. Companies would have the right to full details about genetic test results and family history information when considering an application for an unusually large policy. Where health care is linked to private insurance schemes, it would be especially important to apply a similar restriction (Harper, 1993; Harper and Clarke, 1997).

Other applications of personal genetic information by private corporations can be envisaged, although they may not be as immediate as the applications in insurance. Employers might wish to use genetic information in recruiting so as to select the healthiest, the most hard working, or the most intelligent. They may also wish to identify those most likely to suffer from occupational diseases, and this could be helpful if it was not used to the disadvantage of the susceptible individual or as an excuse for not improving safety standards for all. Educational institutions may wish to select students on the basis of intelligence and motivation. The use of genetic information in these contexts might be unwise because there are many nongenetic factors that influence health, motivation, and intelligence, but this overemphasis on genetic factors could become important if encouraged by prominent academics and powerful biotechnology interests. Such widespread misuse of genetic testing could conceivably lead to a dystopia in which an individual's social role and social worth would be fixed by his or her genetic profile even more firmly than by the current system of social class and wealth. This system would gain legitimacy from its base in the (pseudo)rationality of (misapplied) human genetics.

Another application of genetic information is the compilation of genetic databases by national governments that contain information on an increasing proportion of the population. A DNA data bank on all military personnel is under development in the United States, and in Britain a police database with genetic information on convicted criminals and some suspects is also being established. The justification for the police database in Britain is to help fight crime — to identify suspects who have left traces at the scene of a crime. However, might such data banks find other, less commendable, and less widely debated applications?

Our Disabling Society

Everyone is "disabled" at some point in their lives. We all start out as infants who require total care, and many of us end our lives in a similarly dependent state. At any time in between, our abilities may abruptly change. We may run a marathon one day, suffer injury in a road accident the next, and be tetraplegic thereafter. We may be afflicted by a genetic condition that limits our abilities in one way or another. Then we may find that our capacities are limited still further by the society around us. This social dimension of disability includes both physical facts, such as architectural designs that may limit wheelchair access, and social facts such as the (un)willingness of individuals to behave with courtesy and consideration to others.

Clinical genetics and genetic counseling may be thought to encourage our striving for perfection, especially for the "perfect baby." We help families to avoid having children affected by serious physical and mental problems. Does this necessarily entail a lack of respect for those individuals who have such problems? In confronting genetic disease and disability, and in talking with families about how they can respond to their situations, are we denigrating those who are affected by these conditions?

My personal answer is that we may well promote a shallow and dehumanized image of disease and disability, and therefore of humanity, unless we are very careful. This may not be our aim, but it may be the perceived consequence of our clinical activities. There is a tension between a proper respect for those with disease and disability and a respect for those who wish to avoid having other children with these problems. How can respect for affected individuals then be reconciled with respect for the wishes of parents to avoid such problems in their children?

There is no single, clear answer except to hold an awareness of the tension in one's mind and to maintain a sense of human solidarity with both poles of the tension. This may be easier to suggest on paper than to achieve in practice. One essential step is for genetic counselors to resist any attempt to assess the worth of their services by reference to their impact on population measures such as birth incidences or rates of pregnancy termination. Adopting such goals and measures would totally discredit the profession's attempts to show human solidarity with those affected by genetic conditions.

Cross-Cultural Issues

The final topic for discussion is that of cross-cultural barriers in genetic counseling. This is an area that has scarcely been broached in any systematic way, but I indicate some obstacles to transcultural genetic counseling and a few studies that have begun to explore these issues.

Language barriers and the difficulty of communication through interpreters — whether members or friends of the family or professional interpreters — are clear, especially in a discipline in which precise communication is much more important and more subtle than with questions such as "Does it hurt here?"

There are also the beliefs that each cultural group holds about the other, with all the dangers of stereotyping. Professionals may think, "Oh, she's a XXX so she won't want to know about prenatal diagnosis" or "I won't mention predictive testing because they all believe in fate." The thoughts of the clients may equally suffer from stereotyping: "I can't admit to feeling XXX because it'll make him/her think that I'm too primitive" (Karbani *et al.*, 1997).

Where health behaviors and uptake of services differ according to ethnic or cultural group, it is important that material conditions are not ignored because poverty is associated with membership of a minority ethnic group in many countries. The contribution of deprivation to the perinatal mortality among a Muslim community in Britain may too easily be ignored (Proctor and Smith, 1997). The complex of factors underlying the lower uptake of antenatal services by black women in the West Midlands may also be ignored if too superficial a view is taken (Green and France-Dawson, 1996). The communication problems in antenatal counseling of women from a Hispanic population in the United States have also been described (Rapp, 1988).

Two very specific contexts that arise regularly in discussions about the cross-cultural aspects of genetic counseling are sex selection and consanguinity. Is a couple's preference for children of one sex, especially in a community in which social discrimination against females can be strong, any less justifiable as grounds for pregnancy termination than a woman's preference not to continue a pregnancy because it would be socially inconvenient, perhaps disrupting her education or career? Most western European genetic counselors would affirm the difference between these situations because to practice fetal sex selection would be to reinforce the prejudice that provides the grounds for the termination. However, they are usually unwilling to extend this logic to the case of prenatal screening for (i.e., against) Down's syndrome.

The practice of consanguineous marriages, especially close consanguinity, which may be difficult to avoid in some immigrant communities, will increase the incidence of autosomal recessive diseases for any particular frequency of the recessive disease gene in that population. There are distinct social advantages to consanguinity, however, which more than counter this concern in the minds of many people throughout the world. These include a strengthening of the social position and security of women in the family. Although it may be reasonable to encourage marriage between more distant relatives rather than very close relatives, there is no medical justification for the antipathy to consanguinity that is manifest in some circles (Modell, 1991; Clarke and Parsons, 1997).

The culture-bound and instrumentalist perspective of some studies of cross-cultural genetic practice demonstrates the tight control that majority groups have on those from disadvantaged ethnic minority groups (Nidorf and Ngo, 1993). It is hoped that true cross-cultural communication about genetic issues can be developed, but this will require the investment of time, effort, and resources and a high degree of personal commitment from genetic counselors.

Concluding Remarks

Genetic counseling is a complex area of communication that has not been adequately described. Many ethical and practical moral questions are raised in genetic counseling practice, both for the counselor and for the client. Some of these have been discussed in this article. I do not expect readers to agree with the representation of genetic counseling that is given here or with the judgments that are made in the text, whether implicitly or explicitly, but I will be more than pleased if this article is helpful to those who wish to consider the social and ethical issues raised in the application of the new genetic technologies to society.

References Cited

ALDERSON, P. (1993). *Children's Consent to Surgery.* Open Univ. Press, Buckingham, UK.

AMERICAN SOCIETY OF HUMAN GENETICS/AMERICAN COLLEGE OF MEDICAL GENETICS (1995). Points to consider: Ethical, legal and psycholosocial implications of genetic testing in children and adolescents. *Am. J. Hum. Genet.* **57,** 1233–1241.

AMERICAN SOCIETY OF HUMAN GENETICS SOCIAL ISSUES SUBCOMMITTEE ON FAMILIAL DISCLOSURE (1998). Professional disclosure of familial genetic information. *Am. J. Hum. Genet.* **62,** 474–483.

BALL, D., TYLER, A., and HARPER, P. S. (1994). Predictive testing of adults and children: Lessons from Huntington's disease. In *Genetic Counselling: Practice and Principles* (A. Clarke, Ed.), pp. 63–86. Routledge, London.

BEKKER, H., MODELL, M., DENNISS, G., SILVER, A., MATHEW, C., BOBROW, M., and MARTEAU, T. (1993). Uptake of cystic fibrosis testing in primary care: Supply push or demand pull? *Br. Med. J.* **306,** 1584–1586.

BENJAMIN, C. M., *et al.* (1994). *Am. J. Hum. Genet.* **55,** 606–617.

BILLINGS, P. R., KOHN, M. A., DE CUEVAS, M., BECKWITH, J., ALPER, J. S., and NATOWICZ, M. R. (1992). Discrimination as a consequence of genetic testing. *Am. J. Hum. Genet.* **50,** 476–482.

BINEDELL, J., SOLDAN, J. R., SCOURFIELD, J., and HARPER, P. S. (1996). Huntington's disease predictive testing: The case for an assessment approach to requests from adolescents. *J. Med. Genet.* **33,** 912–918.

BLOCH, M., and HAYDEN, M. R. (1990). Opinion: Predictive testing for Huntington disease in childhood: Challenges and implications. *Am. J. Hum. Genet.* **46,** 1–4.

BLOCH, M., ADAM, S., WIGGINS, S., HUGGINS, M., and HAYDEN, M. R. (1992). Predictive testing for Huntington's disease in Canada: The experience of those receiving an increased risk. *Am. J. Med. Genet.* **42,** 499–507.

BODDINGTON, P. (1994). Confidentiality in genetic counselling. In *Genetic Counselling: Practice and Principles* (A. Clarke, Ed.), pp. 223–240. Routledge, London.

BRADLEY, D. M., PARSONS, E. P., and CLARKE, A. (1993). Experience with screening newborns for Duchenne muscular dystrophy in Wales. *Br. Med. J.* **306,** 357–360.

BRITISH ASSOCIATION FOR COUNSELLING (1984). *Code of Ethics and Practice for Counsellors.* British Association for Counselling, Rugby.

BROWNER, C. H., and PRESS, N. A. (1995). The normalization of prenatal diagnostic screening. In *Conceiving the New World Order: The Global Politics of Reproduction* (F. D. Ginsburg and R. Rapp, Eds.), pp. 307–322. Univ. of California Press, Berkeley.

BUCKMAN, R. (1992). *How to Break Bad News.* Macmillan, London.

CHADWICK, R. (1993). What counts as success in genetic counselling? *J. Med. Ethics* **19,** 43–46.

CLARKE, A. (1990). Genetics, ethics and audit. *Lancet* **335,** 1145–1147.

CLARKE, A. (1991). Is non-directive genetic counselling possible? *Lancet* **338,** 998–1001.

CLARKE, A. (1993). Response to: "What counts as success in genetic counselling?" *J. Med. Ethics* **19,** 47–49.

CLARKE, A. (1995). Population screening for genetic susceptibility to disease. *Br. Med. J.* **311,** 35–38.

CLARKE, A. (1997a). The process of genetic counselling. Beyond non-directiveness. In *Genetics, Society and Clinical Practice* (P. S. Harper and A. J. Clarke, Eds.), pp. 179–200. Bios, Oxford.

CLARKE, A. (1997b). Prenatal genetic screening. In *Genetics, Society and Clinical Practice* (P. S. Harper and A. J. Clarke, Eds.), pp. 119–140. Bios, Oxford.

CLARKE, A., and FLINTER, F. (1996). The genetic testing of children: A clinical perspective. In *The Troubled Helix: Social and Psychological Implications of the New Genetics* (T. Marteau and M. Richards, Eds.), pp. 164–176. Cambridge Univ. Press, Cambridge, UK.

Clarke, A., and Parsons, E. (Eds.) (1997). *Culture, Kinship and Genes: Towards Cross-Cultural Genetics.* Macmillan, Basingstoke, UK.

Clarke, A., Parsons, E., and Williams, A. (1996). Outcomes and process in genetic counselling. *Clin. Genet.* **50,** 462–469.

Clinical Genetics Society Working Party (1994). The genetic testing of children. *J. Med. Genet.* **31,** 785–797.

Codori, A.-M., and Brandt, J. (1994). Psychological costs and benefits of predictive testing for Huntington's disease. *Am. J. Med. Genet. (Neuropsychiatr. Genet.)* **54,** 174–184.

Codori, A.-M., Hanson, R., and Brandt, J. (1994). Self-selection in predictive testing for Huntington's disease. *Am. J. Med. Genet. (Neuropsychiatr. Genet.)* **54,** 167–173.

Craufurd, D. I. O., and Harris, R. (1986). Ethics of predictive testing for Huntington's disease: The need for more information. *Br. Med. J.* **293,** 249–251.

Cuckle, H. S., Richardson, G. A., Sheldon, T. A., and Quirke, P. (1995). Cost effectiveness of antenatal screening for cystic fibrosis. *Br. Med. J.* **311,** 1460–1464.

Dalby, S. (1995). GIG response to the UK Clinical Genetics Society report "The genetic testing of children." *J. Med. Genet.* **32,** 490–494.

European Community Huntington's Disease Collaborative Study Group (1993). Ethical and social issues in presymptomatic testing for Huntington's disease: A European Community collaborative study. *J. Med. Genet.* **30,** 1028–1035.

Evers-Kiebooms, G., Denayer, L., Welkenhuysen, M., Cassiman, J. J., van den Berghe, H. (1994). A stigmatizing effect of the carrier status for cystic fibrosis? *Clin. Genet.* **46,** 336–343.

Faden, R., Chwalow, A. J., Holtzman, N. A., and Horn, S. A. (1982). A survey to evaluate parental consent as public policy for neonatal screening. *Am. J. Public Health* **72,** 1347–1352.

Fraser, F. C. (1974). Genetic counseling. *Am. J. Hum. Genet.* **26,** 636–659.

Geller, L. N., Alper, J. S., Billings, P. R., *et al.* (1996). Individual, family and societal dimensions of genetic discrimination: A case study analysis. *Sci. Eng. Ethics* **2,** 71–88.

Green, J., and France-Dawson, M. (1997). Women's experiences of screening in pregnancy: Ethnic differences in the West Midlands. In *Culture, Kinship and Genes: Towards Cross-Cultural Genetics* (A. Clarke and E. Parsons, Eds.), pp. 120–132. Macmillan, Basingstoke, UK.

Hagard, S., and Carter, F. A. (1976). Preventing the birth of infants with Down's syndrome: A cost–benefit analysis. *Br. Med. J.* **1,** 753–756.

Harper, P. S. (1993). Insurance and genetic testing. *Lancet* **341,** 224–227.

Harper, P. S. (1997). Presymptomatic testing for late-onset genetic disorders. Lessons from Huntington's disease. In *Genetics, Society and Clinical Practice* (P. S. Harper and A. J. Clarke, Eds.), pp. 31–48. Bios, Oxford.

Harper, P. S., and Clarke, A. (1990). Should we test children for "adult" genetic disease? *Lancet* **335,** 1205–1206.

Harper, P. S., and Clarke, A. J. (1997). *Genetics, Society and Clinical Practice.* Bios, Oxford.

Holtzman, N. A., Faden, R., Chwalow, A. J., and Horn, S. D. (1983). Effect of informed parental consent on mothers' knowledge of newborn screening. *Pediatrics* **72,** 807–812.

Huggins, M., Bloch, M., Wiggins, S., Adam, S., Suchowersky, O., Trew, M., Klimek, L., Greenberg, C. R., Eleff, M., Thompson, L. P., Knight, J., MacLeod, P., Girard, K., Theilmann, J., Hedrick, A., and Hayden, M. R. (1992). Predictive testing for Huntington's disease in Canada: Adverse effects and unexpected results in those receiving a decreased risk. *Am. J. Med. Genet.* **42,** 508–515.

Julian-Reynier, C., Aurran, Y., Dumaret, A., *et al.* (1995). Attitudes towards Down's syndrome: Follow-up of a cohort of 280 cases. *J. Med. Genet.* **32,** 597–599.

Karbani, G., Godsil, S., and Mueller, R. (1997). The role of unconscious fantasy in the giving and receiving of genetic counselling. In *Culture, Kinship and Genes: Towards Cross-Cultural Genetics* (A. Clarke and E. Parsons, Eds.), pp. 158–164. Macmillan, Basingstoke, UK.

Koch, L., and Stemerding, D. (1994). The sociology of entrenchment: A cystic fibrosis test for everyone? *Soc. Sci. Med.* **39,** 1211–1220.

Lawson, K., Wiggins, S., Green, T., *et al.* (1996). Adverse psychological events occurring in the first year after predictive testing for Huntington's disease. *J. Med. Genet.* **33,** 856–862.

Lefebvre, R. C., Hursey, K. G., and Carleton, R. A. (1988). Labeling of participants in high blood pressure screening programs: Implications for blood cholesterol screenings. *Arch. Int. Med.* **148,** 1993–1997.

Lippman, A. (1991). Prenatal genetic testing: Constructing needs and reinforcing inequities. *Am. J. Law Med.* **17,** 15–50. [Reprinted in *Genetic Counselling: Practice and Principles* (A. Clarke, Ed.), pp. 142–186. Routledge, London, 1994]

Lippman, A., and Wilfond, B. S. (1992). Twice-told tales: Stories about genetic disorders. *Am. J. Hum. Genet.* **51,** 936–937.

Loader, S., Sutera, C. J., Segelman, S. G., Kozyra, A., and Rowley, P. T. (1991). Prenatal hemoglobinopathy screening. IV. Follow-up of women at risk for a child with a clinically significant hemoglobinopathy. *Am. J. Hum. Genet.* **49,** 1292–1299.

Marteau, T. M., and Drake, H. (1995). Attributions for disability: The influence of genetic screening. *Soc. Sci. Med.* **40,** 1127–1132.

Marteau, T. M., Slack, J., Kidd, J., and Shaw, R. (1992a). Presenting a routine screening test in antenatal care: Practice observed. *Public Health* **106,** 131–141.

Marteau, T. M., Plenicar, M., and Kidd, J. (1992b). Obstetricians presenting amniocentesis to pregnant women: Practice observed. *J. Reprod. Infant Psychol.* **11,** 3–10.

Marteau, T. M., van Duijn, M., and Ellis, I. (1992c). Effects of genetic screening on perceptions of health: A pilot study. *J. Med. Genet.* **29,** 24–26.

Michie, S., and Marteau, T. M. (1996). Predictive genetic testing in children: The need for psychological research. *Br. J. Health Psychol.* **1,** 3–14.

Modell, B. (1991). Social and genetic implications of customary consanguineous marriage among British Pakistanis. Report of a meeting held at the CIBA Foundation on 15 January 1991. *J. Med. Genet.* **28,** 720–723.

Modell, B., and Kuliev, A. M. (1993). A scientific basis for cost–benefit analysis of genetics services. *Trends Genet.* **9,** 46–52.

Montgomery, J. (1994). The rights and interests of children and those with a mental handicap. In *Genetic Counselling: Practice and Principles* (A. Clarke, Ed.), pp. 208–222. Routledge, London.

MORRIS, M., TYLER, A., LAZEROU, L., MEREDITH, L., and HARPER, P. S. (1989). Problems in genetic prediction for Huntington's disease. *Lancet* **2,** 601–603.

NIDORF, J. F., and NGO, K.-Y. (1993). Cultural and psychosocial considerations in screening for thalassemia in the Southeast Asian refugee population. *Am. J. Med. Genet.* **46,** 398–402.

NUFFIELD COUNCIL ON BIOETHICS (1993). *Genetic Screening—Ethical Issues.* Nuffield Council on Bioethics, London.

PARSONS, E. P., and BRADLEY, D. M. (1994). Ethical issues in newborn screening for Duchenne muscular dystrophy: The question of informed consent. In *Genetic Counselling: Practice and Principles* (A. Clarke, Ed.), pp. 95–112. Routledge, London.

PARSONS, E., BRADLEY, D., and CLARKE, A. (1996). Disclosure of Duchenne muscular dystrophy after newborn screening. *Arch. Dis. Child.* **74,** 550–553.

PAYNE, Y., WILLIAMS, M., CHEADLE, J., *et al.* (1997). Carrier screening for cystic fibrosis in primary care: Evaluation of a project in South Wales. *Clin. Genet.* **51,** 153–163.

PELIAS, M. Z. (1991). Duty to disclose in medical genetics: A legal perspective. *Am. J. Med. Genet.* **39,** 347–354.

PRESS, N. A., and BROWNER, C. H. (1994). Collective silences, collective fictions: How prenatal diagnostic testing became part of routine prenatal care. In *Women and Prenatal Testing: Facing the Challenges of Genetic Technology* (K. H. Rothenberg and E. J. Thomson, Eds.), pp. 201–218. Ohio State Univ. Press, Columbus.

PROCTOR, S., and SMITH, I. (1997). Factors associated with birth outcome in Bradford Pakistanis. In *Culture, Kinship and Genes: Towards Cross-Cultural Genetics* (A. Clarke and E. Parsons, Eds.), pp. 97–110. Macmillan, Basingstoke, UK.

RAPP, R. (1988). Chromosomes and communication: The discourse of genetic counselling. *Med. Anthropol. Q.* **2,** 143–157.

ROTHMAN, B. K. (1988). *The Tentative Pregnancy: Prenatal Diagnosis and the Future of Motherhood.* Pandora, London.

ROYAL COLLEGE OF PHYSICIANS COMMITTEES ON ETHICAL ISSUES IN MEDICINE AND CLINICAL GENETICS (1991). *Ethical Issues in Clinical Genetics.* Royal College of Physicians of London, London.

SAARI-KEMPPAINEN, A., KARJALAINEN, O., YLOSTALO, P., and HEINONEN, O. P. (1990). Ultrasound screening and perinatal mortality: Controlled trial of systematic one-stage screening in pregnancy. *Lancet* **336,** 387–391.

STANDING CONFERENCE OF VOLUNTARY ORGANISATIONS FOR PEOPLE WITH A MENTAL HANDICAP IN WALES (SCOVO) (1989). *Parents Deserve Better. A Review and Report on Early Counselling in Wales.* SCOVO, Cardiff, UK.

SHARPE, N. F. (1993). Presymptomatic testing for Huntington disease: Is there a duty to test those under the age of eighteen years? *Am. J. Med. Genet.* **46,** 250–253.

SHILOH, S. (1996). Decision-making in the context of genetic risk. In *The Troubled Helix: Social and Psychological Implications of the New Human Genetics* (T. Marteau and M. P. M. Richards, Eds.), pp. 82–103. Cambridge Univ. Press, Cambridge, UK.

SHILOH, S., and SAGI, M. (1989). Framing effects in the presentation of genetic recurrence risks. *Am. J. Med. Genet.* **33,** 130–135.

SJOGREN, B., and UDDENBERG, N. (1988). Decision making during the prenatal diagnostic procedure. A questionnaire and interview study of 211 women participating in prenatal diagnosis. *Prenatal Diag.* **8,** 263–273.

STAMATOYANNOPOULOS, G. (1974). Problems of screening and counselling in the hemoglobinopathies. In *Proceedings of the 4th International Conference on Birth Defects* (A. G. Motulsky and F. J. B. Ebling, Eds.), pp. 268–276. Excerpta Medica, Amsterdam.

STATHAM, H., GREEN, J., and SNOWDON, C. (1993). Mothers' consent to screening newborn babies for disease. *Br. Med. J.* **306,** 858–859.

TAMBOR, E. S., BERNHARDT, B. A., CHASE, G. A., FADEN, R. R., GELLER, G., HOFMAN, K. J., and HOLTZMAN, N. A. (1994). Offering cystic fibrosis carrier screening to an HMO population: Factors associated with utilization. *Am. J. Hum. Genet.* **55,** 626–637.

TIJMSTRA, T. (1990). The psychological and social implications of serum cholesterol screening. *Int. J. Risk Safety Med.* **1,** 29–44.

VAN DER STEENSTRATEN, I. M., TIBBEN, A., ROOS, R. A. C., *et al.* (1994). Predictive testing for Huntington's disease: Nonparticipants compared with participants in the Dutch program. *Am. J. Hum. Genet.* **55,** 618–625.

WALD, N. J., KENNARD, A., DENSEM, J. W., CUCKLE, H. S., CHARD, T., and BUTLER, L. (1992). Antenatal maternal serum screening for Down's syndrome: Results of a demonstration project. *Br. Med. J.* **305,** 391–394.

WERTZ, D. C., FANOS, J. H., and REILLY, P. R. (1994). Genetic testing for children and adolescents: Who decides? *J. Am. Med. Assoc.* **272,** 875–881.

WHITE-VAN MOURIK, M. C. A., CONNOR, J. M., and FERGUSON-SMITH, M. A. (1992). The psychosocial sequelae of a second-trimester termination of pregnancy for fetal abnormality. *Prenatal Diagn.* **12,** 189–204.

WILSON, J. M. G., and JUNGNER, G. (1968). *Principles and Practices of Screening for Disease.* World Health Organization, Geneva.

ZEESMAN, S., CLOW, C. L., CARTIER, L., and SCRIVER, C. R. (1984). A private view of heterozygosity: Eight-year follow-up study on carriers of the Tay–Sachs gene detected by high school screening in Montreal. *Am. J. Med. Genet.* **18,** 769–778.

General References

ABRAMSKY, L., and CHAPPLE, J. (Eds.) (1994). *Prenatal Diagnosis: The Human Side.* Chapman & Hall, London.

BOSK, C. L. (1992). *All God's Mistakes: Genetic Counselling in a Pediatric Hospital.* Univ. of Chicago Press, Chicago.

CLARKE, A. (Ed.) (1994). *Genetic Counselling: Practice and Principles.* Routledge, London.

DUSTER, T. (1990). *Backdoor to Eugenics.* Routledge, London.

LEWONTIN, R. C. (1991, 1993). *The Doctrine of DNA: Biology as Ideology.* Penguin, Harmondsworth, UK.

MARTEAU, T., and RICHARDS, M. P. M. (Eds.) (1996). *The Troubled Helix: Social and Psychological Implications of the New Human Genetics.* Cambridge Univ. Press, Cambridge, UK.

GILBERT HOTTOIS

Brussels Free University — CRIB
Brussels, Belgium

The Body and the Market

Contrary to the view that the human body is a marketable good, bioethics respects the dignity thereof. I discuss this opposition by analyzing the main topics of the debate: the removal and grafting of organs, banks of human tissues, and patents for inventions that use components of human origin. There are two main views: the more European model of the body as not an object of commerce as energetically argued by France, according to which any commercial use of the human body should be regulated and controlled by the state, and a more open free market trend, considered in Europe as typically American, which stresses individual freedoms and considers the individual as the owner of his or her body. A fundamental part of the debate turns on the aim of the donations and exchanges that have the human body as their object. This aim, which should be therapeutic, is in reality problematic because technology, money, and those biomedical practices that respond to individual requirements or desires make individuals free to regard their body as something that can plastically be remodeled.

Dignity of and Market for the Human Body

European, particularly French, bioethics indissolubly links the principle of the nonmarketability of the body with the fundamental principle of the dignity of man. A section of a report of the National Advisory Committee on Ethics (1994) states, "Money reifies everything it buys and impresses a mark of equality on that which it has reified, not by chance, but by virtue of its very nature. This is the secret of its power." This observation centers the essential features of the problem in question. In fact, although the notion of dignity attributes an intrinsic value to the thing to which it is referred, that which has the character of dignity becomes in some way sacred, inaccessible and impossible to be disposed of. The value in itself of something excludes that it can be exchanged or changed, without being irremediably harmed.

The dynamic today designated with the term *market,* within the context of the globalization of the neoliberal economy, uses mainly two universal operators of unlimited power of transformation and exchange: money and technology. This dynamic is furthermore increased by individual and collective desires and illusions.

In some well-known pages of *Das Kapital,* Marx describes money as the universal medium, the general equivalent, which enables anything to be exchanged. Due to the invention of money and to its application to all aspects of human existence, there is no longer anything that does not have its price, including that which in some contexts (i.e., culturally and historically) has been regarded ontologically and essentially as sacred — that is, inviolable and priceless. In itself, money is not worth anything, it cannot permit anything, and it does not make possible the physical transformation of what is real. For its power to become effective, the intervention of the other exchange operator, technology, is essential. The market, as the space in which the two operators of transformation and power — money and technology — are manifest and converge, is the universal that dissolves all intrinsic value, all ontological axiology. Moreover, this space of transformation of the financial and technical economy, within which technical, scientific, and biomedical research and development (which I will indicate by the acronym TSRD; technical and scientific research and development), occupies a very important position, and the neoliberal ideology is strongly complementary. In a way, the neoliberal doctrine is the symbolic representation that should permit growth and participation in the desymbolizing dynamics of unlimited, universal trade. The neoliberal doctrine is hostile to state intervention which organizes society and the economy from the outside and from above, starting from a plan, from a project, an ethical notion, or a representation considered true and good. The neoliberal state has no function other than that of permitting the free play of interests, autonomies, and individual or private desires, safeguarding and guaranteeing ownership, the value of money, conventions, and freedoms. The basic assumption is that the individual who rationally pursues the maximization of his or her interests, and it is supposed that *Homo oeconomicus* always effectively behaves in this manner, necessarily contributes to the general interest. The market is the space in which this calculation of interests is constantly carried out through the play, with money as the medium, of supply and demand. The prices on which the various interests reach agreement express balances which, albeit temporary, are the best possible ones between divergent interests which interact in a nonviolent manner. These balances optimize individual and general interests more than could the state. This is the essence of the doctrine of free trade and of *laissez faire,* according to which, once the effects of some of its transitory and apparent benefits have worn off, state regulation of nonviolent trade (social protectionism) leads not to a reduction but to an increase in human suffering.

Here, I do not examine in detail the general and basic criticisms leveled at the neoliberal monetary techno-economy and relating to two basic idealizations: the abstraction of *H. oeconomicus,* an informed, rational calculator, and the abstraction of the market as a space for egalitarian negotiation without any constrictions or gray areas. These criticisms tend to demonstrate that the neoliberal techno-economy is first and foremost in the service of the interests of those who, in given circumstances which may be very ephemeral or last for many generations, prove to be the strongest and the most astute.

However, it is not my intention to discuss the vices and virtues of the market in general. I intend, instead, to consider whether it is opportune to allow the transformations and the trading relations that in one way or another have human bodies or parts of human bodies as their object, to be regulated by the market, particularly in relation to TSRD. This field of inquiry is extremely extensive and varied (Rodotà, 1996). I intend to deal with three aspects: the transfer of organs, human tissue banks, and requests for patents.

Transplants of Organs

Depending on which organ is transplanted (Fig. 1), the donor may be alive or dead, but the general rule is that organs (liver, kidneys, eyes, heart, etc.) are off the market and not commercially obtainable. This is why we speak of donation and thus of a gratuitous act whose aim must be therapeutic. Regarding postmortem donation, two criteria are envisaged. The first is that of presumed consensus (*opting out*), according to which every person is considered as consent-

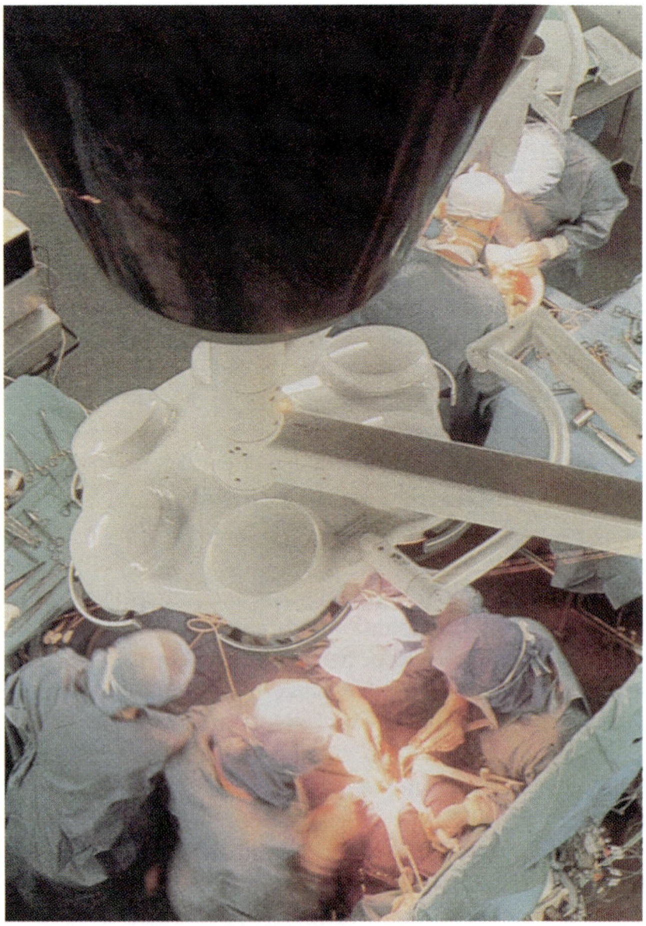

FIGURE 1 Surgical team during a liver transplant operation (photo courtesy of Plailly-Eurelios-SPL/Neri).

ing unless he or she has made a declaration of dissent when alive; in this case, *de facto* or *de jure,* the person's family and relations are almost always also consulted. The second one is that of explicit consensus (*opting in*), on the basis of which the donation requires a preliminary assent by the donor recorded in a document. The first system is more widespread in Europe and the second in the United States.

Whatever criterion is adopted, the organization "off the market" of the transplanting of organs is far from having resolved all the problems, which are represented by a persisting penury and scarcity of resources and waiting lists, by a parallel illegal market characterized by clandestine trafficking, and by favoritism. Due to these problems, and also on the basis of fundamental considerations concerning the status of the human body, some have proposed opting for a market not wholly without any specific regulation. I therefore think it advisable to set out the arguments used in favor of this choice (Lemennicier, 1991; Andrews, 1995; Fagot, 1995, 1996; Radcliff Richards, 1996):

1. The need for greater respect of the freedom of individuals and of their rights over their bodies (alive or dead): This personal right to one's own body is not clearly understandable except as a right of ownership and of full availability. If a poor man decides to sell one of his kidneys, this choice must be considered, from his point of view and on the basis of the situation in which he lives, the freest choice possible. If he is forbidden to make this decision with the pretext that he is not really free, his situation is not improved and his freedom is not increased; quite the contrary. Furthermore, the opting-out system would favor the trend toward the dissimulated appropriation of human bodies by the state, which disposes thereof in the name of the public good and without the due respect for individual autonomy.
2. By adopting the solution presented previously, everyone would see his or her interests recognized: The rich sufferer could be treated and the poor one could improve his or her situation. A mutually advantageous exchange cannot be harmful to society. By forbidding it, the ones concerned are abandoned to their respective form of misery for the sake of the consciences of those who do not have to address this problem since they are neither poor nor ill. In this way, in fact, we limit ourselves to protecting the good moral conscience, represented by certain principles and values, of a fraction of society: Prohibition in no way modifies the injustice and the real disparity of conditions.
3. The intrinsic value or the dignity of the human body, which justifies prohibition on the basis of a certain status of the human body, is a very confused notion from the philosophical and ethical standpoint. If the body is considered in terms of possession, of personal ownership, there is no confusion. The foundations on which prohibition is based are in reality obscure and irrational; they pertain to the domain of emotion, tradition (beliefs), or intuition (ontological proof of value). This sort of justification is limited and cannot be universalized, and it eludes rational discussion.
4. The fact that we are able to dispose of our bodies as if we are the owners thereof does not harm others' rights (on the contrary in this case). For that matter, many illicit uses of the body are tolerated, such as in certain sports and high-risk jobs, prostitution, or in more general ways of behaving that are harmful to health.
5. One's relation to one's own body is a private issue that is in the sphere of personal morals and not of public law since it concerns exclusively the individual. This prohibition stems from an undue confusion between morals and law. Through the law, a social trend is to impose on all its concept of the body, its value, and its morality. The idea that the individual is the owner of his or her own body, on the contrary, is morally neutral since it does not impose using the body for one purpose or another. This idea, raised to legal status, does not condition the free moral choices of individuals. It does not forbid, for example, the free donation of organs by persons inclined toward this sort of charity and philanthropy. The two criteria, that of donation and that of sale, can perfectly well coexist.
6. The scarcity of organs, although not overcome, is at least reduced, and with it the temptation to turn to clandestine traffic and to the parallel black market. The quality of the organs and of assistance is greater because any inefficient operators would quickly be ousted from the market.
7. Recognizing a right of individual ownership to the body and permitting its access to the market does not mean renouncing all forms of public and specific regulation of such a market, in the sense of safeguarding the health and social justice. Control of prices and of the suitability of operators and intermediaries and a system of intervention and reimbursement for the most needy are certainly feasible and certainly necessary.

These arguments have aroused certain objections which are simultaneously justifications of the principle of noncommercialization. I briefly mention them because they can easily be inferred from the previous discussion:

1. The status and the ontological dignity of the body forbid it from being treated in whole or in part as an impersonal commercial good. The commercialization of human bodies is the most unworthy form of moral degradation.
2. The freedom offered by the market is a caricature of personal autonomy: The living individual who sells an organ of his or her body is anything but free.
3. The commercial disintegration of individual bodies will produce an effect of dissociation in the social corpus because it destroys community symbols and sentiments at a fundamental level—that of basic solidarity, of altruism and generosity, and of consideration for others' bodies and therefore for relations with other persons. No human society can be based on such a splitting up of its members. The

system of symbols relating to the gift as a transparent, disinterested relationship is therefore socially indispensable to offset the dissociating effects of the market, which takes human relationships into consideration only through calculation of interests.

4. In addition to these important symbolic risks, there are physical risks for both donors and receivers. There is reason to believe that it will not be the individuals enjoying the best state of health who will decide to sell their organs. Poor quality will therefore prevail over good quality.

5. There is nothing to guarantee that the market will effectively resolve its problems left partially unresolved by the systems of nonpayment (e.g., the problem of scarcity of organs), especially because mercantile logic is liable to discourage any altruistic gift.

In conclusion, although the principle of the nonmercantile gift remains almost universally dominant at ethical and juridical levels, the discussion, fed by realistic, utilitarian, and consequentialistic arguments, has continued to develop in recent years.

Tissue Banks

What are known as "tissue banks" occupy a central position in the general problem of relations between human body and market. However, as for all bioethical questions, the real situation indicated by this expression, examined more closely, is varied and complex. It also includes activities currently being developed, closely associated with TSRD, and therefore still inadequately defined and scarcely regulated.

In the United States, the definition of tissue is very extensive. In fact, it includes every functional grouping of cells. In the Old World, instead, on the basis of the directives of the Council of Europe, it is preferred to use a more differentiated definition which takes the various tissues into consideration separately and in which ethical and juridical considerations immediately emerge. The tissues include every part of the body, including surgical samples, placentas, and embryo tissues, but the blood and its derivatives, the gametes, and the embryos (embryo bank) form a separate category, as do the hair and nails. These are not only elements of a very heterogeneous nature but also, for the most part, they have formed the object of well-defined regulation. Although excluded from the problem of tissue banks, blood is nonetheless a point of reference since the majority of questions are concerned with it, particularly those relating to marketing. Moreover, the question relating to the organs, which concern a complex of tissues, is still much debated. However, the problems relating to the tissues and to the organs are clearly distinguished.

The tissues are characterized by the possibility of conservation and continuous growth (cell line), generally by the lesser urgency of operations, by the greater ease of being obtained, by the possibility of multiple transformations, and by far more important actual and potential economic and financial interests. The possible uses of the tissues are therapeutic (transplants, reconstructions, and transfusions), diagnostic, and didactic; they concern research, archives, and collections; and they also concern the industrial production of medicines, medicaments, or cosmetics. This variety determines the diversity of the more or less specialized banks both as sources of materials for therapy or for industry and as sources of research, for archives or for collections. Among the functions carried out by the banks, the main ones are collection, which implies the selection and screening of donors; characterization, purification, and treatment necessary for conservation; conservation; and distribution. These basic functions are not exclusively passive and can extend to particular treatments, such as the preparation and transformation of tissues with a view to their industrial use. Processing is a complex ensemble of transformation processes which involve more or less substantial modifications. The Nuffield Council on Bioethics (1995) developed a list (not an exhaustive one) of approximately 100 therapeutic or research products derived from human tissues.

A type of bank in full development and with a certain future is the DNA bank (Fig. 2), sometimes referred to as a *biobank*. Such banks are practically analogous to the tissue banks insofar as they effectively conserve tissues. However, their particular purpose is the preservation and use of the genetic information contained in these tissues and not their direct use. The biobanks are the object of specific questions relating to the protection of personal data, in view of the immense importance of genetic information for the individual, and to the exploitation of genetically interesting persons and groups. Their scientific, technical, and economic importance will certainly become strategic in relation to the inventory of human and nonhuman genetic biodiversity currently being compiled.

The separation between the respective spheres of competence of the bank and industry is fundamental. This distinction derives from the definition of the operations to which the tissues are subjected. The operations of collection, purification, conservation, preparation, and distribution are functions that pertain to the banks. That which instead concerns the transformation of the tissues for purposes of industrial production is outside of their province. However, the operations carried out by the banks presuppose many transformation and elaboration activities, and an industrial group can legitimately possess its own banks. Furthermore, research can lead to a patentable and marketable invention. It is very difficult to draw a clear distinction between exclusively conservative activities and industrial activities, particularly between conservative treatment

FIGURE 2 DNA bank in the Genset Center at Evry (photo courtesy of Raguet-Eurelios/Neri).

and substantially modifying treatment. The essence of the problem is in the distinction between activities and functions that cannot in any way be economically advantageous and others which, at least to a certain extent, could prove to be advantageous.

The difficulty consists in articulating activities that should be kept outside the sphere of the profit-generating market and other activities conditioned by the preceding ones and able to produce considerable profits. At the two extremes there is, on the one hand, the gratuitous personal donation and, on the other hand, the transnational bioindustry. The tissue banks are between these two extremes.

The case of *Moore v. Regents* of the University of California in 1990 (argued before the Supreme Court of California) made history. During the course of medical treatment, samples of tissue from J. Moore's body unexpectedly proved to be of considerable scientific and commercial interest, which led to a patent and an industrial application. Moore then claimed a share of the profits. The court turned down his claim on the basis of the principle that the individual does not possess a right of ownership of his or her cells, whereas it censured the circumstance that Moore had not on any occasion been informed of the use to which the samples repeatedly taken of his body tissues had been put.

This illustrates the "out-of-the-market" status of the human body in the case of tissues. The main arguments for and against this status do not substantially differ for organs or tissues. A more idealistic and deontological position, which unreservedly supports the principle of gratuitousness, opposes a more variegated and divided utilitarian trend which emphasizes the unresolved problems and suggests that a certain commercialization could perhaps help to resolve them. Both these trends appeal to the rights of man: The former to stress the dignity of the human being and the latter to place the emphasis on individual rights and freedoms.

Compared with the question of organs, the debate on the tissues is more moderate and more lively. It is more moderate in that the perception of what is sacred, of dignity, and of intrinsic value is weaker for the majority of tissues than for the organs. It is more lively because the commercial and financial interests at stake, both actual and potential, are far more considerable. The debate is also much more complex because of the diversity of the uses and treatment for which tissues are intended and due to the many different actors that come into these processes. The predominant position is that of the gratuitous nature of the donation, which does not exclude a recompense for the costs and inconveniences incurred by the donor, and of the nonprofit character of the operations performed by the banks (collection, preparation, conservation, and distribution). This means both that they cannot earn a profit on them and that the gains possibly obtained have to be reinvested in research and development activities able to improve the services provided by the bank, which may not, for instance, distribute dividends or invest on the stock exchange or in other economic activities. The justification most frequently adopted is that the bank does not transform and does not produce: It is said to be a substantially passive intermediary. On the contrary, bioindustry does transform and produce, and this productive and basically creative activity can produce legitimate profits. Furthermore, it would be improper for a bank that receives tissues free of charge to make a profit on donations that instead should limit itself to administrating in the best possible way, that is, in the interests of therapeutic medicine. The general therapeutic aim of the ensemble of the operations in fact plays a decisive role also with respect to lucrative bioindustrial activity, which ideally should lead exclusively to the production of medicines or medical devices.

If the products that derive from human tissues originally received as gifts enter a market that no longer has any relation with therapeutic aims, then there are problems with the ethical logic on which the gift is based — solidarity, altruism, and helping those who suffer. Various authors have proposed the imposition on all bioindustries using human tissues of a public system of special taxation or the obligation to plow back part of their earnings into medical research so as to make their activity morally acceptable. The proceeds from the taxes would be paid into the coffers of sociohealth assistance to be used in favor of the most needy. The scenario thus outlined with its requirements and its postulates assuredly raises many questions. It takes on significance only insofar as the banks' activities are effectively quite distinct from other activities and if the goal of therapy is rigidly respected. In reality, things are not always like this, and the need — now an impelling one — to regulate the activity of the banks of human tissues is a sign of the existing fracture between the ideal and the real. This need is undoubtedly imposed to an even greater extent by questions of quality, safety, and health. In other words, it means explicitly making the tissue banks responsible for two complexes of ethical needs: respect for the person and safeguarding health.

Regarding respect for the person, the first imperative principle is the donor's consensus. It is necessary for the donor to be informed as fully as possible of the uses to which his or her gratuitous donation will be put. If subsequently the prospect arises of an unforeseen use, it is advisable for the donor to again be informed and consulted. A second group of problems concerns respect for private life, privacy, the donor's anonymity, and not divulging the actual or potential information (particularly genetic information) implied by the possession of a person's tissues. This question can certainly come into conflict with another requirement — that of quality and safety.

Concerning health, it is fundamental not to use on individuals any tissues or products derived from tissues that are not absolutely safe. This is a very important problem, if one considers the extraordinary diversity of the possible sources of tissues and the contexts in which they could be taken or collected. In this connection, the identification of the source and the possibility of tracing it and thus the obligation to keep all the information relating to said source (whether a deceased or a live person) are of fundamental importance. The clinical history of the donors should be known, and all the tests necessary to specify and assess the innocuous nature of the sample and to file the results thereof should be carried out.

The free circulation of things and persons, internationalization (i.e., the globalization of trade), and the role played by the multinational corporations make this sort of control increasingly more difficult and more indispensable. According to many observers, abandoning goods and products based on tissues to the space of free trade and laissez-faire involves physical risks for individuals and for the collectivity that cause more concern than merely symbolic and moral risks. The worldwide growth of the biomedical economy, within a very heterogeneous (especially with regard to health) and changing planetary environment, causes one to suppose that these risks will not diminish in the near future.

In conclusion, the necessity for a clear definition of the functions, rights, and responsibilities of the tissue banks can no longer be postponed. Whether public or private, the banks should require the approval of a control body. In this way, their relationship with commercial activities could be defined.

Patentability of Inventions That Presuppose the Use of Elements Coming from Human Bodies

Undoubtedly, the question of the "patentability" of living human beings has caused the largest conflicts between concepts of a moral character (as well as political) and technoeconomic logic (likewise political). This debate, which is ongoing in animated fashion, reached its apex on the occasion of many applications for patents with inventions as their object presupposing the use of elements of human origin, and in particular the nucleotide sequences. The agreement reached in the spring of 1998, after approximately 12 years of discussion, between the European Commission, Council, and Parliament regarding the *Directive on the Legal Protection of Biotechnological Inventions* marked a decisive turn of events.

The controversy appears to have started in 1980 with the *Diamond v. Chakrabarty* case in which the question of the patentability of a recombined genome microorganism able to break down a number of hydrocarbons was debated (Fig. 3). The United States Supreme Court ruled that "everything that has been made by man" must be consid-

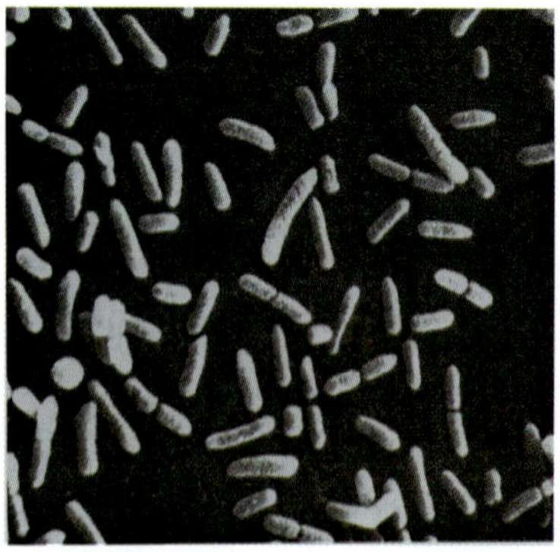

FIGURE 3 The microorganism obtained by biochemist A. Chakrabarty in the early 1970s from the recombined genomes of four different species of the *Pseudomonas* bacterium (photo courtesy of General Electric Research and Development Center).

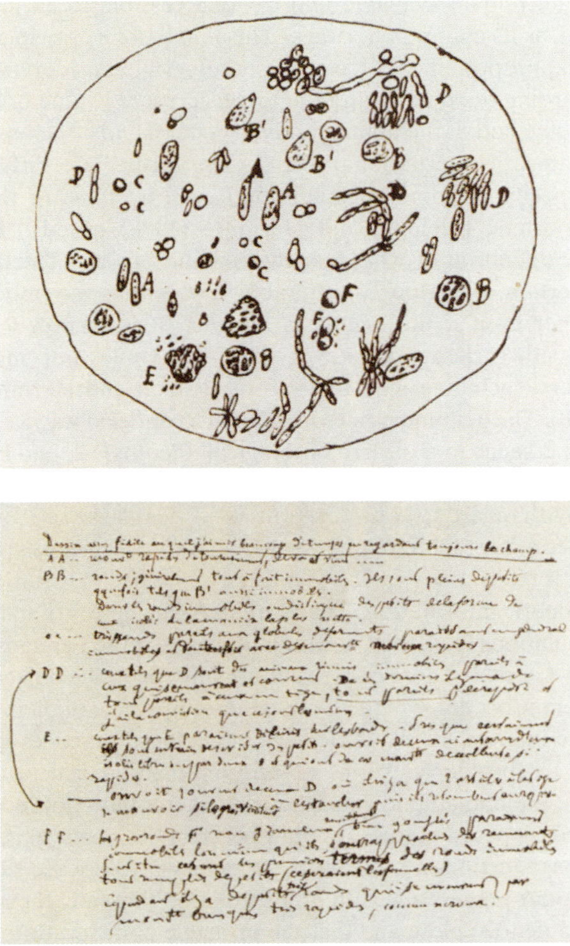

FIGURE 4 Manuscripts written by Louis Pasteur regarding his research on active organisms in the fermentative process (photos courtesy of Bibliothèque Nationale, Paris).

ered as patentable, and that the living/nonliving opposition is irrelevant. This is certainly not a revolutionary position: As long ago as 1873 Louis Pasteur obtained a patent for a yeast (Fig. 4). The naturalization of the human being, regarded as one of the living species, and the development of the genetic technosciences, reinforcing this naturalization, led within a short time to the question being raised of the patentability of elements of human origin. In this case, too, the answer favored continuity since patents for many elements of human origin (e.g., blood coagulation factor VIII) had been granted long before. Despite this, the debate on ethical questions of great importance concerning not only the rights of the individual over his or her body but also the principles of contemporary Western society based on economic competition and on TSRD was started and is still ongoing. This debate has been fed by the fact that legislation regarding patents contains what is sometimes referred to as the "moral clause," which states that an invention contrary to law and order and to morality cannot be patented. The need to respect the just moral order could not fail to encourage the notion of the dignity of the human body from being raised.

The essence of the debate can be understood only on the fundamental basis of the definition of the technical and juridical criteria of patentability. This presupposes novelty or newness in relation to the legislation on patents (i.e., something that was not available until that moment), inventive activity (research work), and industrial applicability or the possibility of useful and economically adequate production. In other words, the application for a patent introduces new information that can lead to other inventions available to society. This is possible because the presumed invention starts to be legally protected from the moment the application is made. The purpose of the patent, which is generally granted for a period of 20 years, is therefore twofold: to encourage the divulging and the circulation of socially and scientifically useful information and to protect the work performed by the inventor by means of guaranteeing that his or her invention may not be exploited by anyone without his or her consent. The patent in itself is not identified with the right to exploit the invention, because industrial production and placing the product on the market require that specific conditions be respected, but only the right to prevent others from exploiting it. It offers the inventor the possibility of entering the market with the advantage of possessing a monopoly.

The inventor is seldom an isolated individual because as such he or she would have great difficulties in imposing himself or herself on the market. Interests in biotechnological inventions concern the big bioindustries (pharmaceuticals, biomedical technologies, and the agrofood industry). These interests are international and thus also political. Bioindustry plays a crucial role in economic competition between the countries and the major regions of the world, such as Japan, Southeast Asia, the United States, and the European Union. During the 1990s, 75% of the patents in the biomedical field were granted in the United States and Japan and less than 25% in Europe. The link between the development of the bioindustries and technoscientific research is very tight and entails considerable and constant investments in the field of research, which have to be offset and compensated through commercialization. It is ridiculous to limit oneself to opposing economic and technoscientific competition, on the one hand, and ethical demands, on the other. The progress achieved by biomedical TSRD is enthusiastically followed and expected by an immense number of persons in a poor state of health, and almost all the associations of sufferers support whatever contributes toward such progress. For this reason, the system of patents proves essential: Where this system is not applied or operates badly, TSRD declines or disappears. In Europe, in the name of the dignity of the human being, no clear, common legislation for the authorization of patents which have what is proper to living beings, particularly human beings, as their object has been introduced, resulting in the poor functioning of this system. This situation has given rise to contradictory decisions and to appeals against applications for patents. The legal uncertainty thus created and maintained is favorable neither to the economy nor to the TSRD. The European Directive of 1998 attempts to regulate this problem.

What, according to the directive, are the patentable elements of human origin? Article 5 of the directive answers this question, stating,

> An element isolated from the human body or otherwise produced by means of a technical process, including the complete or partial sequence of a gene, can be regarded as a patentable invention, even if the structure of this element is identical to the natural element.

It must be pointed out that these provisions refer to gene sequences. However, their formulation is much broader and may be applied, in principle, to any part of the human body (e.g., the tissues) which matches the conditions of patentability and is not excluded by specific legal provisions. What is assuredly not patentable is the body of an individual in its entirety, nor any part of it that is effectively a constituent component of the individual's body.

An important aspect of the debate which might appear to be overlooked in the previous citation turns on the distinction between discovery and invention because the patented element must not be structurally different than the natural element. This is a very complex problem because it concerns general epistemological considerations relating to the nature of contemporary science which, as technoscience, invents or produces its objects rather than discovers them; it also depends on highly technical statements regarding what is actually patented and is not identical to the original element or in an identical context to that of the original element, even if the information is the same in both cases.

The concrete issue, in its economic scope, is the following: The further upstream the patent is granted (i.e., to an isolated gene whose codifying function is known and which is indicated as susceptible, for example, to a therapeutic use), even though the applicant for the patent may still be far from finalizing the therapeutic product, the broader and more indeterminate will the future monopoly be that is granted to the patent holder, who thereby becomes the master of the TSRD sector relating to the gene sequence patented. It is for this reason that some have repeatedly stated that it is necessary to grant patents as far downstream as possible, ideally only to the production processes and to the medicines destined to be placed on the market. One of the ethical justifications of this position is that the information contained in the human genome and acquired by it should not be the object of appropriation and exploitation by vested interests (national or private) but should remain available and freely accessible to all mankind, as its common heritage. This mandatory principle of nonappropriability is based on ontological considerations regarding the dignity and the intrinsic, sacred value of the genome and of the human body in general and also considerations of international justice. The former are difficult to specify, as sometimes admitted even by some of their supporters: Edelman and Hermitte (1988) noted, "The very definition of what it is that troubles us is subjective, uncertain, or rather unutterable." It is thus a symbolic danger even if the enemy is known: "It is market logic that slowly desacralizes everything that comes into man's sphere, including man himself" (Edelman and Hermitte, 1988). The genome is here likened in a confused way to human essence in a bizarre ontological, theological, and biological amalgam.

With regard to considerations of justice, these refer first and foremost to the Third World, where a conspicuous part of our biodiversity can be identified and thus the variations in genetic information. There are many instances of native populations (e.g., the Guaymi of Panama or the Papuans of New Guinea) whose cells in culture have been the object of patent procedures in the United States by virtue of their genetic interest. It is not clear whether these practices give rise to specific questions of economic and social justice since the human body and the parts thereof are not recognized a special status. It could be claimed, however, that poverty justifies a substantial pecuniary recompense since the poor possess a wealth of information that only the rich are able to exploit, and that the principle of the gratuitousness of the donation appears in the specific case as particularly unjust. It cannot be denied that these practices also pose serious questions relating to the information and to the consensus of those concerned.

The sentence in the Moore case cited previously ruled against Moore, insofar as he claimed a share in the profits obtained by a patent which presupposed at its origin the use of his cells, since he had not participated in any way in the work that had led up to the invention and to the marketable products. Nevertheless, the sentence found Moore to be in the right in that he had not been informed of the destination reserved for the cells taken from his body.

Respect for informed consensus is an absolutely essential ethical condition for the applications for patents for biotechnological inventions which presuppose elements taken from the human body. This information should be as exhaustive as possible and therefore include, if necessary, any industrial development of the researches conducted on the samples taken. If this development and the application for the patent are not foreseeable initially and therefore not contemplated in the original consensus, it is as important to contact the person concerned and to inform him or her of the unexpected developments, taking note of his or her opinion. Within the context of possible economic development of samples that reveal exceptional properties and, for example, great therapeutic interest, the principle commonly accepted of the gratuitous nature of the donation seems at least problematic. It is understandable that individuals and populations, all the more so when theirs is a miserable lot, should claim to negotiate the price of these samples.

Situated at the point of intersection between ethical, legal, and technical and scientific, economic, and political requirements, the problem of the patentability of elements taken from the human body proves particularly complex. However, it cannot be expected that legal provisions governing patents will resolve all problems and, in particular, the ethical problems habitually posed upstream and downstream of the applications for patents. Regarding upstream applications, it is clear that the research that leads to a patent application must be conducted respecting the ethical rules concerning, for example, experimentation on human beings, whether this experimentation or the treatment lead to a patent or otherwise. Regarding those downstream, the granting of a patent does not authorize the exploitation, production, or marketing of the patentable discovery. These are successive stages for which specific ethical conditions and legal rules exist. Before receiving the authorization necessary to be placed on the market, a medicine has to be subjected to a lengthy and complex process, which in turn includes many ethical imperatives. In the final analysis, it is society that is responsible for the use to which the invention and the new products toward whose discovery it has contributed will be put. It is obvious that any invention can be used well or badly and on the basis of many different aims.

The Euro-French Model and the Global American "Threat"

The opposition indicated in the title of this section is plainly schematic because the positions referred to are neither simple nor homogeneous. Nevertheless, it corresponds approximately to the real trends and to the widespread representations underpinned by the actors. To be convinced of this, it is sufficient to examine French bioethical literature relating to the body and its parts or to its products. In this connection it is possible to consult the provisions of the National Ethical Advisory Committee (CCNE) and the studies by Fagot (1995, 1996), Sève (1994), Rameix (1998), and Lemennicier (1991). France, which was also very active in drawing up ethical provisions for the Council of Europe, in this field plays a policing role and does not hesitate to report any mercantilist deviations of the European Union. Consider in this context the CCNE's denunciation of the 1989 European directive concerning medicines derived from human blood and plasmas:

> Our disapproval is first and foremost focused on the use of such expressions as "raw material" to designate blood and plasma, and "medicinal preparation" to designate the products prepared on the basis of blood and plasma. Such terminology, in itself distorted, paves the way to a deformation of concepts, principles, and practices.

France, the first country to set up a national ethical committee, promulgated in 1994 a series of laws directly concerning the human body and a statute in which the principle that this is not on the market and not in commerce plays a decisive role. A statement of symbolic identity, whether French or European, is necessarily taken as a "counterdefinition" of what is depicted as a bugbear and more or less demonized, without considering the nuances. Furthermore, an identity, especially when it directly concerns just laws and true values, naturally claims to be universalist. Thus, and there are profound historical motivations urging in this direction, the French model claims to be based directly on the universalist tradition of human rights, very keenly felt in the Council of Europe which has promulgated the Convention on Human Rights and Biomedicine. This reference to the philosophy of human rights could be contested because for many bioethical problems, human rights have no relevance or are open to different, often opposed interpretations (Hottois, 1996).

Although the notion of dignity may be invoked in support of the principle of the body outside of the market, individual freedoms and rights to dispose freely of oneself and not to suffer impositions regarding private life indicate contrary directions. If human rights were originally dictated by the need to protect the individual against illicit interventions by the state, then the laws aimed at establishing what the individual may or may not do with his or her body seem to be at least debatable and not to descend necessarily from the human rights tradition.

Before examining in greater detail the two trends opposing each other in this field, note that they are only trends of thought and that the bioethical, biojuridical, and biopolitical reality on both sides of the Atlantic is far more composite.

The Euro-French Model and Its Critics

The philosophical and ethical basis of the model which I call Euro-French is derived from Rousseau and Kant. From Rousseau, this model has drawn the idea that the individual accedes to freedom (i.e., becomes a subject) only by virtue of belonging to the social corpus, of his or her participation in the formation of the general will of the nation. It is the state (public law) that constitutes the individual–subject and makes him or her a citizen. No individual freedom can thus prevail over the general will or the public good of which the nation-state is the expression. The individual body is considered a member of the social corpus and individuals are usufructuaries rather than owners of their bodies. Individuals, as citizens, are children of the nation-state for which they must be prepared to sacrifice even their lives. We are therefore in the presence of a paternalistic political philosophy in which the principle of the common good tends to prevail clearly over the principle of autonomy, and this is perfectly in keeping with a certain paternalistic tradition of medicine. The nation-state, the general will, plainly knows better than the individual what has to be done in the interest of everyone's good, which is identified with the good of each person. Insofar as this state has a universalist claim, the good thus defined also has to be understood as rational and universal.

From Kant, instead, comes the prohibition to objectivize and instrumentalize the individual in the name of the dignity of the person: The human person is a free, rational spiritual entity. According to Kant, the exercise of individual freedom, correctly understood, cannot come into conflict with the exercise of the reason which is universal: The principle of the autonomy of reason consists of this view. To Kant, the person is not identified with his or her body but, through the mediation of various traditions and philosophical–religious arguments (Roman Catholicism and the dogma of incarnation, Aristotelianism and the concept of substance, and the phenomenology of the body), his model stresses the inseparability of the person from the body and automatically transfers to the latter the value and the dignity associated with the person. The indivisibility of the body and the person is also the recurrent theme of the preparatory studies to the formulation of the French laws on bioethics (Braibant, 1988). This does not prevent a philosopher such as Sève (1994), even if a convinced supporter of these positions, to express his perplexity—"The body is and is not, at one and the same time, the person"—and to add that the problem consists of the fact that "starting from a matter of ethical concern, that of avoiding the dehumanization of biomedicine, refusing the depersonalization of the body, it is thought opportune to support it with an ontological argument."

Introducing the body and the parts of the body onto the market dominated by money, technology, and individual desires (or pseudo-freedoms) is tantamount to destroying these symbolic syntheses: The body is separate from the person who owns it, its integrity is atomized into an infinite number of parts, and it becomes a modifiable entity able to be transformed and exchanged according to irrational personal whim. These deviations are perceived as literally diabolical (i.e., antisymbolic) destroyers of the dignity of the person and solidarity and of the links that constitute the basis of society, whose members are said to perceive themselves and others as arbitrarily liable to dismemberment. The splitting up of the individual body causes the splitting up of the social corpus and therefore the annihilation of the nation that might allow itself to be dragged into this impasse.

Introducing individual bodies onto the free trade market represents the gravest symbolic risk. Therefore, it is necessary at all costs, through ethical argument and through the law, to preserve society's constituent symbols, even if the reality of practices and the situations is far removed from the ideal of solidarity, brotherhood, equality, and gratuitousness. In this connection, Fagot (1996) states, "It is clear that to the national committee, the individual risk for hemophiliacs of contracting AIDS . . . is nothing with respect to the social risk implicit in recognizing that antihemophiliac components extracted from human blood are a marketable object."

In summary, the Euro-French model of the body outside of the market ideally sanctions that the individual body is inviolable, indisposable, and inaccessible—that it is not the heritage or the property of the individual and thus not in any way marketable and convertible into money, not even the tiniest parts of it. Some of these prohibitions can be broken only with the individual's explicit consensus and for a therapeutic purpose (possibly a cognitive one) or with his or her presumed consent in the case of impediment or after his or her death. At the origin of any utilization of the human body or a part of the human body there is always a donation, a gratuitous act of solidarity, to be strictly controlled by the public power, especially when the donation is the first stage of a process that leads to a marketable end product. This control, which guarantees its ethical character, must verify that the final use of the donations is good (i.e., essentially medical) and that any profits are properly used (e.g., reinvested in biomedical TSRD) or in any case limited (e.g., through the levying of taxes). In summary, with reference to the human body, the state, and not primarily technology or money via the market, should absolutely act as mediator in every transformation or exchange because only public mediation can guarantee the morality of the transformations (i.e., their character of not being not injurious to the *socius*), of the constituent solidarity of society, and of the national but ideally universal public good. Any other way of proceeding ends in the destructive irrationality of society and, presumably, of mankind.

Among the possible criticisms is assuredly the confused nature of the philosophical bases of this model and other observations briefly listed here:

1. The fact (e.g., with regard to the matter of contaminated blood) that the public organization does not provide a guarantee of quality and physical safety of transformations and exchanges: Voluntarism, the morality of good intentions, can lead to a lessening of the sense of responsibility for a variety of reasons.

2. Reservations regarding the moral quality of the state and of those who serve it: The absence of control is more

serious than in the case of private operators who can be under the surveillance of the public authority and be sanctioned by the market.

3. The hardly transparent and not very effective management of public bodies responsible, for example, for the fact that the companies that collect human tissues by paying for them manage to introduce products onto the market at a price lower than that imposed by the public organisms that collect blood and tissues without having to pay for them.

4. The feeling of not belonging, of not being the owner of one's own body, whereas the state is the real owner receiving the donations of those who are alive and the bodies of the dead: This is a profound feeling of an unacceptable affront to the freedoms and the fundamental rights of the person.

5. The conviction that individuals, in general, are in a more suitable position to defend the interests of their body than that of the state: In the Euro-French model, the state negotiates with researchers and businessmen for the exploitation of the body and its products, without any control of the persons who are without any power of negotiation on the subject (Fagot, 1996).

The opposition of the two trends dealt with here has at times been ironically commented on by stating that although the American citizen is the owner of his or her body, the French citizen is merely the occupant of a body that in reality belongs to the state.

The American Threat

It must not be forgotten that whatever is said about the body and the market must be considered in relation to the economic war in which the countries and the major regions of the world confront each other materially and ideologically. In the context in which Europe and the United States usually appear in competition and in which every country plays its own cards, the question of the body, its parts, and its products takes on a particular symbolic value. It is usually considered that the United States is the advocate of the introduction of everything onto the globalized free market, including the last traditional stronghold of natural humanity — the body (Kuttner, 1997). Saving the body from the market and from money thus becomes an ideological objective of capital importance. Examining American bioethical literature, one immediately realizes that reality is instead much more complex and variegated, and that the debate is not going on only between Europe and the United States but also, and in lively fashion, within the United States. For that matter, as already stated, Europe is also far from having a homogeneous position, and the various bodies (e.g., the Council of Europe and the European Commission) or the various countries of Europe are addressing these problems in ways that at times prove not to be convergent. As already emphasized, in this field there are in fact two main trends opposed to each other — one mainly characterized by the support of Europe and the other by that of the United States. Simplification in outlining the two opposed positions has first and foremost a didactic purpose and serves to clarify the terms of the debate and to enable them to be understood.

The philosophical basis of the American model is to be sought in the work of Thomas Hobbes and especially John Locke. It assumes that individuals enjoy certain inalienable natural rights, including the right of ownership, which is more precisely a right of appropriation. Natural things belong to all and to no one. However, a thing becomes the legitimate property of a person the moment it has been made the object of the activity exercised by that person. As a general rule, it is by way of labor, which has manifold forms, that the thing of no one finds an owner. The thing that has been appropriated thereby becomes a good able to be freely exchanged and sold. This approach makes it possible to resolve fairly easily many questions that concern research, patentability, and production, at the origin of which there is an element of the human body. Instead, it proves less resolutive concerning question relating to the starting point, although it can be claimed that a person's body is also, and in case in part, the product of the activity, of the work of said person, and that he or she may freely dispose of it, just as he or she disposes of other more exterior goods which he or she has acquired, kept, and modified with his or her labor.

Many other currents of the English-speaking philosophical–religious tradition are more or less decidedly in favor of recognizing a relationship of ownership and of full disposability between the individual and his or her body. Among them I must cite the great tradition of individualist, economic, and political liberalism, the nominalist empiricism extraneous to ontological values, or Protestantism and a more contractualist practice of medicine and human relations in general. It should also be pointed out that there is a more resolute will, with respect to Europe, to keep the state and the churches, as well as the public space (governed by the law and by the state) and the private or community sphere, which comes wholly within the domain of morals and religions, separate. It is also necessary to consider a concept of medicine less linked to the idea of natural integrity which tends to justify only measures of a rigorously therapeutic character. American medicine, which aims at contributing toward comfort and well-being, seems more open to the demands that originate from individual desires and which, if compared with the needs defined by nature, might appear very artificial. However, it is the very image — evolutionist and neo-Darwinian — of nature that is different.

Critics of this model denounce in particular the fundamental confusion generated by it between individual freedom (i.e., autonomy, in the rigorous meaning of the term) and irrational individual desires profoundly conditioned (limited and orientated) by the situation (economic, cultural, social, and psychological) of individuals. Moreover, criticism of the model stresses the neo-Darwinian dominion of the reason of the strongest said to be manifest in the neoliberalist ideological system and the breaking up of society through the destruction of all forms of solidarity and all symbolic social links, with the subsequent assigning of social cohesion to technostructures manipulated by private interests. Lastly, the model would lead to the global increase of inequality and suffering, which are diametrically opposed to the illusions of utilitarian ideology. In the end, there would be the dissolution of the welfare state of universal vocation, the only one able to curb the destructive effects of the war between particular and self-centered interests.

Beyond the Therapeutic Horizon

Dignity, invoked as the basis for respecting the gratuitousness and the liberal character of every transfer of an element of the human body, is compatible on principle only with the therapeutic aim of the immediate or derived (i.e., after industrial transformation) use of parts sampled. When this aim is not therapeutic but, for example, cosmetic or when it is the origin of commerce for a profit, a sort of moralization of the process is required through the obligation to reinvest a substantial part of the profits in TSRD or by means of a taxing system to the benefit of medical and social assistance. This approach raises numerous objections.

Medical practices of a voluptuary character or aimed at satisfying requirements of well-being have long existed, and such practices, of which the archetype is plastic surgery, have continued and will continue in the future to develop and become diversified. Money and the most advanced techniques are fundamental for these practices. Since they are used in a great variety of situations, it is very difficult to draw the dividing line between the therapeutic operation and one that cannot be defined as such. This brings us to consider the second objection: What does therapeutic mean? What is the criterion that enables us to distinguish with certainty a therapeutic practice from a non-therapeutic use of biomedicine?

According to the traditional medical ideal, which is also an ideology, a measure may be defined as therapeutic if it is aimed at restoring natural integrity that has been accidentally harmed. However, what is natural integrity in a world in which nature is conceived as an ensemble of unforeseeable, contingent evolutionary processes and in which culture, which defines what is natural and what has to be respected in nature, is inflexibly multiform? In this world — our world — there is no longer a "right" form that can be universalized and applied to every individual, as an expression of nature and a sign of health. The borderline between the normal and the pathological has become relative, contextual, or multiform, according to the communities and the individuals concerned.

There are, Nietzsche claimed, "countless states of health" because man is "by far, among all sick beasts, the one who is sick for longest and most profoundly" and because man is "the beast not yet fixed" and we have to be "animals of experimentation." The definition of the concept of health, established by the World Health Organization, which designates a state of complete physical and mental well-being, is so sweeping and utopian as not to exclude *a priori* that the individual may have recourse to medicine for any sort of intervention. Also, when French legislation on bioethics (laws No. 653 and No. 654 of 1994), which is very sensitive regarding the respecting of the dignity of the human body, imprudently states in Article 3 that "the integrity of the human body can be harmed only in case of the person's therapeutic necessity," it seems to be admitting a disassociation between the state of the body and the needs or desires of the person, which can in principle justify any operation and calls to mind the more freely postmodern expressions of Engelhardt (1996): "As persons we can make our bodies objects of our judgment and manipulation. We can find ways in which we could have been more attractive."

Who more than the person is fitted to assess the therapeutic necessity (which does not mean the most appropriate means) justifying an operation on his or her body? The example of transsexuality (Salas, 1994), in which the medical indication is particularly uncertain and in which the desire component is intense, is very instructive in this regard. At the basis of the requests for the satisfaction of the desires and the needs of individual well-being, there are always a lack and a form of suffering. Undoubtedly, the question to answer is whether a given medical response is the most adequate one and, without any doubt, the doctor is, in the majority of cases, the one in the best position to clarify this by means of a dialogue with the patient. However, this help has certain limits and varies considerably according to the case, especially when it is a question of requirements and desires that transcend the individual and concern the human condition itself and the way in which each individual puts himself or herself in relation thereto. Is it necessary to accept the fact of growing old or to be prepared to do anything to hinder it and delay it, particularly if one has the necessary money available to undertake these attempts? Must one resign oneself to sterility or seek to remedy it? Must inherited handicaps be accepted, or should one try to annul them? Must the technology of prostheses that enable so many physical handicaps to be overcome or remedied (Fig. 5) be developed, although it is known perfectly well that this type of technoscientific creativity will

FIGURE 5 Anthony Volpentest, winner of the 200-m flat race at the Para-Olympics at Atlanta, August 1996 (photo courtesy of Squire-Allsport/Neri).

inevitably lead to the creation of the cybernetic organism and sensorial and other sorts of prostheses that are curative and not just palliative in character? The same problem happens with regard to the research for gene therapy aimed at eugenics. Suffering is a laboratory in which the faces of the future are invented. The doctor should not be a priest; he or she does not have the right answer. The doctor is simply an individual who, due to the position he or she occupies professionally in society, has the possibility of acting on other individuals at their request. Certainly, society can and must delimit the use of such powers. However, society is far from being philosophically, morally, and religiously homogeneous and stable.

Where runs the dividing line between natural suffering, changeable desire, and rebellion against human limitations? Where is the borderline between fruitful rebellion, which is a condition of progress, and self-destructive rebellion, and who can establish it? In the majority of societies considered as traditional and primitive, regarded as close to nature, the body has been an object to be shaped according to individual and collective fantasy, and the vestments of the religious cults have no cause to envy the present-day excesses of fashion with regard to the metamorphoses of the physical appearance.

Past, present, or future, natural or cultural, individual or specific, the body cannot be considered as intangible and unchangeable. It is fundamentally plastic, evolutive, polymorphic, and inhabited by the desire to be abandoned. The body of the human species has been slowly modeled on the basis of that of the hominids to permit the development of the brain and the use of the hand. The great French paleoanthropologist A. Leroi-Gourhan (1964), who showed the decisive importance of this anatomical evolution for the expansion of the brain and thus for the full development of what we call intelligence, does not hesitate to project this evolution into the future, predicting a progressive desuetude of the body in favor of the brain connected to its technical extensions. After an initial reaction of emotional and almost physical repulsion, reflection induces us to ask ourselves, Is this unworthy for man? What is man?

The contemporary imagination, fed by the development of communication, virtuality, and simulation technologies, is densely populated by fantasies of "disembodiment" or "disincarnation." Thus, in a well-known book written in the 1980s, Gibson (1984), the inventor of the term "cyberspace," contrasts those who are linked to corporeity and who go on living mainly, if not exclusively, in the material, corporeal world—eating, drinking, and especially moving about physically ("traveling was a trick of those who were linked to corporeity")—with those who live, think, and act almost exclusively in cyberspace and are at the point of becoming a sort of total prosthesis–body–universe for the brain. Independently of the future of the species, is not the body of each person perhaps first and foremost a process and experience of metamorphosis, from the zygote to senescence, by way of adolescence? The program of this marvelous and terrible plasticity is imposed on us, with its numerous risks and with the limits and obligations that it involves. In the name of what would it be necessary to renounce changing this program or diversifying it insofar as we acquire the capacity to do so?

The postmodern technocultures and cybercultures (Aronowitz *et al.*, 1996) that develop in the meshes of the conurbations and in the world communications networks combine sciences, techniques, markets, and symbols in spaces in which incessant processes of transformation and exchange are taking place. These are no longer situated in any way in a material world. It is for this that the dualist body–person concept, still widespread and often serving to justify in a philosophically clumsy way the full availability the person enjoys vis-à-vis his or her body, is inappropriate and outdated. The person is not a substance (spirit) but a fiction of freedom, a way of signifying that the members of the human species and this same species are nonfinite, open, indeterminate processes and that they can, to create this future, use materials and products of the past. The transcendence associated with metaphysical–religious dualisms is vertical and static; it exceeds the corporeal and limited condition of the human being only on the symbolic plane, sublimating it. The technosciences instead open the possibility to an operative transcendence of the species: They enable the natural limits associated with the human condition and body to be effectively surpassed. They also permit individuals and the collectivity to go beyond these natural and traditional configurations and limits in a different way.

In an extraordinary novel titled *A Miracle of Rare Design,* with the subtitle *A Tragedy of Transcendence,* Resnick (1994) tells a story of research and metamorphosis in which a writer, due to the biomedical technosciences, successively takes on the morphology and the sensitivity of many different forms of life, variously intelligent and sentient, that live on planets very different than Earth. Since these experiences can absolutely not be expressed in the language of men, he decides to abandon writing and to devote himself totally to this technobiological alchemy that threatens to destroy him slowly. He therefore ends up choosing one final metamorphosis: a form of life that does not have an existence of its own but which combines in itself the particular features of all those he has known.

We are therefore very far from the ontological or phenomenological body and from its natural dignity, hostile to change and to exchange. Offsetting this, we are approaching the body without organs and the machines with desires of Deleuze and Guattari (1972): the body of desire, change, and exchange *a priori* unlimited that capitalist society, its techniques, and its markets would release and repress at the same time.

Finding Good Fictions

The following description by Kroker (1996) brutally but effectively represents the problem: "The organic body knows that it will die before it can be metamorphosized into a virtual state, and so it desperately seeks enslaved bodies, and those of young people in particular, in search of an elixir of life: kidneys, pancreas, eyes, and hearts."

We are living in a world in which ontology, metaphysics, fundamentalism, and all the guiding ideas deriving from these, such as that of God, truth, being, nature, essence, and value in itself, are in crisis, and we consider that this crisis is not evil. The nihilism associated with this crisis presents many positive, emancipatory, diversifying aspects of promising creativity, possibility, and hope. We are living in a world of fictions, in the broadest meaning of the term, which includes symbolic (ideological, juridical, moral, and institutional) and technophysical fictions (e.g., the technoscientific discoveries and inventions and their applications), a world of inherited fictions (natural and traditional) and new fictions (TSRD and cultural forms). Fictions are never pure: They combine the past and the future, the material and the immaterial, nature and artifice, law and morality, and the symbolic and the technical. Within the living species, the markets, currencies, prostheses, the liberal donation, patents, definitions of death, cell machines, and the biobanks we find fictions that function in synergy or in opposition and of which nature itself is part as product or producer, variously symbolized according to the epochs and the cultures. In this world of fictions, it is the new technophysical fictions of the TSRD that play the destructive or creative role of driving force. It is very important to accompany this tirelessly inventive technoscientific dynamic with symbolic (moral, legal, political, and institutional) fictions so that the ensemble may offer living creatures and ourselves the greatest opportunities of realization. The sufferings of living creatures are not fictions, even if they may be aggravated and alleviated by symbolic fictions and technical fictions.

What are the good symbolic fictions? I do not believe they are to be sought among the antifictions of the past, such as essences, values in themselves, necessary truths beyond discussion. The dignity of the human body as an intrinsic value descends from these. I think, therefore, that it cannot be invoked without arousing criticisms. In this sense, I approve of the effects of deontologization, emancipating mobilization, and free exchange and free change, performed by the markets and by techniques and technologies. Nevertheless, insofar as these markets and these techniques are excessively misled to the advantage of certain fractions of mankind, they cause greater sufferings and offer hardly any supplementary possibility of realization to the rest of mankind, perhaps the majority. The markets must be regulated so that they are no longer exclusively subordinated to the dominant vested interests. In this sense, the ideology of radical neoliberal economism, which is today a particularly active fiction, must be criticized, especially because it claims to sustain the truth and to be the instrument that will necessarily lead in the medium term to the maximization of everyone's interests — to a more egalitarian and more just world. There is no fiction that is definitively and absolutely good. Such a fiction would be identified with truth. Appealing to truth is a particularly poorly enlightened, immodest, and risky way of alleging the goodness of one's own fiction.

References Cited

Andrews, L. B. (1995). My body, my property. In *Life Choices: A Hasting Center Introduction to Bioethics* (J. H. Howell and W. F. Sale, Eds.). Georgetown Univ. Press, Washington, DC.
Aronowitz, S., Martinson, B. E, and Menser, M. (Eds.) (1996). *Technoscience and Cyberculture.* Routledge, London.

BRAIBANT, G. (1988). *De l'Éthique au Droit.* La Documentation française.
COUNCIL OF EUROPE (1997). *Convention sur les Droits de l'Homme et la Biomédecine.* Council of Europe.
DELEUZE, G., and GUATTARI, F. (1972). *L'Anti-Oedipe. Capitalisme et Schizophrénie.* Minuit, Paris.
EDELMAN, B., and HERMITTE, M.-A. (Eds.) (1988). *l'Homme, la Nature et le Droit.* Bourgois, Paris.
ENGELHARDT, H. T. (1996). *The Foundations of Bioethics.* Oxford Univ. Press, Oxford.
EUROPEAN COMMISSION (1998). *Directive sur la Protection Juridique des Inventions Biotechnologiques.* European Commission, Brussels.
FAGOT, A. (1995). Does non-commercialisation of the human body increase graft security? In *Organ and Tissue Transplantation in the European Union* (Y. Englert, Ed.). Nijhoff, Dordrecht.
FAGOT, A. (1996). Problèmes d'éthique médicale posés par de nouvelles techniques thérapeutiques: Greffes d'organes, de tissus et de cellules. In *l'Éthique à la Croisée des Savoirs* (P. Livet, Ed.), Vrin, Paris.
FRANCE (1994). Lois n°94–653 et 654 relatives au respect du corps humain, au don et à l'utilisation des éléments du corps humain, à l'assistance médicale à la procréation et au diagnostic prénatal. et au diagnostic prénatal. et au diagnostic prénatal.
GIBSON, W. (Ed.) (1984). *Neuromancer.* HarperCollins, London.
HOTTOIS, G. (1996). *Entre Symboles et Technosciences.* Champ Vallon/PUF, Seyssel/Paris.
KROKER, A. (1996). Virtual capitalism. In *Technoscience and Cyberculture* (S. Aronowitz, B. Martinson, and M. Menser, Eds.), Routledge, New York.
KUTTNER, R. (1997). *Everything for Sale.* Knopf, New York.
LEMENNICIER, B. (1991). Le corps humain: Propriété de l'Etat ou propriété de soi? Droits **13.**
LEROI-GOURHAN, A. (1964). *Le Geste et la Parole.* Michel, Paris.
NATIONAL ADVISORY COMMITTEE ON ETHICS (1994). *Les avis de 1983 à 1993.* National Advisory Committee on Ethics, Paris.
NUFFIELD COUNCIL ON BIOETHICS (1995). *Human Tissue. Ethical and Legal Issues.* Nuffield Council on Bioethics, London.
RADCLIFFE RICHARDS, J. (1996). Nefarious goings on (kidney sales and moral arguments). *J. Philos. Med.*
RAMEIX, S. (1998). Corps humain et corps politique en France. In *Laval. Théologique et Philosophique.*
RESNICK, M. (1994). *A Miracle of Rare Design.* Tor Books, New York.
RODOTÀ, S. (1996). Entre droits et marchés. Réflexion sur le corps 'juridifié'. In *Le Génome et son Double* (G. Huber, Ed.). Hermès, Paris.
SALAS, D. (1994). *Sujet de Chair et Sujet de Droit: La Justice Face au Transsexualisme.* PUF, Paris.
SÈVE, L. (1994). *Pour une Critique de la Raison Bioéthique.* Jacob, Paris.

General References

ARNOUX, I. (1994). *Les Droits de l'être Humain sur son Corps.* Presses Univ. of Bordeaux, Bordeaux, France.
CHARLESWORTH, M. (1993). *Bioethics in a Liberal Society.* Cambridge Univ. Press, Cambridge, UK.
ELSHTAIN, J. B. (Ed.) (1995). *Politics and the Human Body: Assaults on Dignity.* Vanderbilt Univ. Press, Nashville, TN.
ENGELHARDT, H. T. (1991). *Bioethics and Secular Humanism.* SCM, London.
ENGLERT, Y. (Ed.) (1995). *Organ and Tissue Transplantation in the European Union.* Nijhoff, Dordrecht.
HOTTOIS, G. (1996). *Il Paradigma Bioetico.* Rubbettino, Messine.
HOTTOIS, G., and PARIZEAU, M.-H. (1993). *Les Mots de la Bioéthique. Un Vocabulaire Encyclopédique.* De Boeck-Université, Paris.
HOWELL, J. H., and SALE, W. F. (Eds.) (1995). *Life Choices: A Hasting Center Introduction to Bioethics.* Georgetown Univ. Press, Washington, DC.
RENSBERGER, B. (1996). *Life Itself.* Oxford Univ. Press, Oxford.

GIOVANNI BERLINGUER

Department of Animal and Human Biology
University of Rome "La Sapienza"
Rome, Italy

Organ and Tissue Transplants

The present is characterized by growing successes in transplant techniques and by processes which lead in opposite directions: an utmost emphasis on the value of the human body and widespread commercialization. The first ethical problem related to transplants is the definition and ascertainment of death. The concept of "cerebral death" has become broadly accepted in law and in practice, and it constitutes a sure guarantee against abuses. The second problem concerns the property of the body and decisions about the removal of organs from cadavers. Although a proprietary idea of the body may be subject to confutation, the will of the individual in regard to the uses of his or her body must prevail over other wills in the name of the principle of autonomy. Many laws envisage that, after adequate information and consultation, the principle of "tacit consent" in the donation of organs be adopted. Other ethical problems, less discussed but just as important, concern the organization of services, the allotment of tissues and organs, and the distribution of resources among the various fields of medicine and prevention. In the last section of this article, the ethical and practical perspectives of transplants are examined.

The History and Meaning of Transplants

Tissue and organ transplants represent one of therapy's most extraordinary forms of progress, which has been achieved due to surgical techniques combined with basic biological knowledge such as the discovery of blood groups, the immune nature of organ rejection, tissue types, and immunosuppressive drugs. In the wake of the first attempts carried out at the end of the nineteenth century and the actual start of blood transfusions at the beginning of the twentieth century, transplants began to develop rapidly with the end of World War II. Currently, the list of the tissues involved is vast and constantly growing (blood, cornea, marrow, skin, cartilage, bone, and heart valves), as is that of the organs (kidneys, heart, lungs, liver, intestine, pancreatic islets of Langerhans, the small bones of the middle ear, larynx, trachea, and thyroid) and of laboratory-grown human cells and tissues which can be transferred from one individual to another for therapeutic purposes. The spectrum of possibilities has broadened to include nervous tissues; furthermore, the number of failures has decreased and the survival time of recipients of transplants has lengthened.

With rare exceptions resulting from religious beliefs or traditions, the humanitarian significance of transplants has been greeted with widespread approval by public opinion since the first blood transfusions were carried out "from arm to arm" as salvaging acts of generosity. Techniques for the conservation and transplantation of tissues, and later of organs, have deprived donations of their immediacy but increased their benefits in time and space. Since the shift from reproducible tissues such as blood to nonreproducible organs such as kidneys and the heart was accomplished and the possibility of extraction from cadavers was added to that of donations from living beings, and since the development of widespread use of transplants, complex moral and juridical problems have arisen. Among the latter, the most prominent are the definition and ascertainment of death, the regulation of explants, the criteria and priorities in the assigning of tissues and organs, consensus about donations, the property and marketability of the human body, and the distribution of resources between transplants and other branches of medicine.

The difficulty and intrinsic interest of these themes must be combined with the idea that the exchanging of body parts has its roots in myth, literature, and religion. In tradition and in art there is the miracle of Cosma and Damiano (Fig. 1), who allegedly amputated a sick leg and replaced it with that of an Ethiopian (from a black man to a white one, like the first heart transplanted by C. Barnard—an unquestionable demonstration of the unity of the human species).

Blood, which in saints, heroes, and the young is considered the distilled essence of their qualities, has often assumed a special symbolic value. In Ovid's *Metamorphoses,* the witch Medea convinces the daughters of old Peleus to bleed him, promising to fill his veins with young blood: "*ut repleas vacuas iuvenali sanguine venas.*" In medieval pharmacopeia, blood was considered the remedy to various ills such as gout and tertian malaria, and since the fifteenth century the news (or legend) has been passed down of the extreme attempt to save a dying Innocent VII by transferring to him the blood of three children. An analogous practical and symbolic interest has sometimes surrounded male genital organs. This explains the popularity of the surgeon

FIGURE 1 Saints Cosmos and Damian, considered the patron saints of pharmacists, combine their talents and transplant the leg of an Ethiopian onto Deacon Justinian (courtesy of Beato Angelico, Convent of Saint Mark, Florence).

S. Voronoff, who, at the beginning of the twentieth century, injected extracts or grafted the testicles of monkeys to give new vigor to his patients. One must go return to these roots, and to irrepressible desires to rejuvenate, to perpetuate oneself in one's descendants, to reintegrate sick organs and, at the limit, to become immortal in order to understand the expectations, which transcend therapeutic necessities, generated by the growing, although certainly not infinite, possibilities of transplants. The history of medicine has shown, in fact, that the greatest advancements in the health of the population have depended on *ex ante* interventions, on overall improvements in living and working conditions, nutrition, and education, and on individual and collective behavior.

Transplants, therefore, may well be considered one of the most important advancements and at the same time one of the most debated fields in modern medicine. Trohler (1993) noted,

> Transplants mark the tip of the iceberg of the type of intricacies which the progress of the whole of medicine faces: Their therapeutic and social functions are seen by those who work in the field, by the general public, and by historians as problematic, everything but obvious. . . . The history of transplants illustrates the age-old psychological or social reasons why doctors insist on concentrating on problems that are rare in reality, but interesting on the conceptual level.

Rarely dealt with in this context is the fact that when a patient who is seriously ill is saved by an organ which has been extracted from a cadaver, the transplant represents an extraordinary success; however, before the latter can take place, there must almost always have been, unfortunately, two failures. That one resorts to such an operation only as a last possible therapy, in fact, means that other preventive or curative measures have been lacking or have failed. Also, the fact that there are organs available, usually belonging to young persons, shows that there has been a premature, often avoidable, death. This has been proved, for example, by the decreasing trend in lethal consequences as a result of motorcycle accidents since the introduction of the obligation to use protective helmets on the basis of a law which in Italy was opposed by much prejudice and by powerful industrial interests.

At conferences and in reviews which deal with transplants, such as *Transplantation Proceedings,* it is rare that one finds references to an organic vision of individual or collective health. Also, only rarely are those who carry out the operation on the two subjects (explant and transplant) interested in reconstructing their anamnesis and signaling, on the basis of the latter, the timely modalities of prevention and cure which might be useful to others. This does not in any way diminish the meaning of the transformation of the cruel principle *mors tua, vita mea* into a positive act, which renders health and life to an individual who is maimed or soon bound to die. Such a fact, however, causes one to wonder about unilateral enthusiasms with respect to such interventions and shows that we must act in various ways to limit our reliance on them.

On the general level, the ethical and juridical problems connected with transplants go well beyond the sphere of medicine, sensitive as it may be. The fact that it has been proven that many parts of the human body (even nervous tissues) can be replaced, and that the possibility exists of extracting them not only from living beings and cadavers but also from embryos and anencephalic newborns, has put traditional philosophical and juridical conceptions of the body and of personal identity into question.

The Human Body: Emphasis on Its Value and Commercialization

A question such as "What is a man who has had a brain transplant?" and the equally hypothetical answer "It is a brain which has been endowed with another body" could not even have been posed before the era of transplants. The forbidding of brain transplants (currently impossible) and of genital glands (already possible) by legal enactment is the result of the need to protect peoples' identities and descendants. There are also extreme cases in which identity is subject to doubts which did not arise before and which today may often have controversial practical solutions. For example, the positive attitude assumed by the American Medical Association (Council for Ethical and Juridical Affairs, 1995) toward the explantation of organs from anencephalic newborns who are still alive has caused some discussion because anencephalics cannot be considered "donors" (this is what the document states) since donation implies an act of will which for them is impossible, because of the risk that rushed and arbitrary explants will be extended to other deficient subjects, and because of the differences and analogies that exist between the absence of a brain (anencephaly) and cerebral death. These are questions which imply a redefinition of the concepts of body and personal identity.

These examples, among the many that are possible, give rise to stimulating scientific/philosophical considerations which belong to the so-called frontier bioethics, whose field of research is made up of those phenomena which had been impossible before the most advanced developments in biomedicine and which impose a retrospective analysis of concepts developed in the past. In the field of everyday bioethics (Berlinguer, 1999a) lies the task of examining more deeply the context and relations which affect everyone's lives.

Said context appears to be dominated by two opposed tendencies. One is toward the utmost emphasis on the value of human corporeality, a phenomenon without precedent in history except perhaps in ancient Greece (only as far as free male subjects were concerned). The most relevant fact is the extension of the duration of life which now tends, for many individuals, to approximate the natural bounds of the human species. This result has been achieved not least because of medicine, which has removed age-old plagues from our bodies, invented efficient cures for many diseases, solved many motor sensorial and sometimes mental deficiencies with prostheses and other technical apparatuses, reduced or eliminated pain, and made possible the use of body parts donated by one man to the advantage of another (Porter, 1997). The control of sexuality has contributed to increasing the possibilities of self-determination, especially for women, and to transforming procreation into an act which can be freely chosen. The care of one's body, physical and athletic activity, and the striving toward fitness and beauty have become widespread needs. Many other items might be added to this list; more important, each of the positive points I have listed should be accompanied by a description of its distortions (such as violence and pedophilia in the sexual field) and of enduring or growing inequalities (e.g., in the length of life spans).

The other tendency, which runs counter to the emphasis on the value of human corporeality as an intrinsic value, is the latter's commercialization. Humanity has already experienced the phenomenon: Slavery, in fact, implied the buying and selling of bodies *in toto*. Today, the possibility of using their separate parts has given rise to a "technological market" whose catalog can include DNA sequences (object of patents), male and female gametes, cells and tissues, blood, marrow, corneas, uteruses (which are objects of "gestation contracts"), and organs; it currently excludes the brain, and the outlawing of the cloning of human beings prevents them, for now, from being put on sale.

Over the intrinsic value of the human body looms — and is sometimes even legally used — its commercial value. Prestigious reviews of an economic character, although less precise than official publications, have impudently calculated its overall worth. On the basis of a macabre calculation carried out in 1969, the market price of the components of a human body (5 lbs of calcium, 1.5 lbs of phosphorus, a bit of potassium, sulfur, and sodium, and a pinch of magnesium, iron, copper, and iodine) was 98¢, which had increased to $8.84 in 1994 as a result of inflation. The current price of the separate parts, were they to be obtainable and sellable, has been estimated to be about $300,000 (Nash, 1994). It has been calculated differently according to general and special parameters, as can be seen in Fig. 2.

The existence or even the mere possibility of such a technological market has raised numerous moral and juridical controversies which will be examined later. The two premises enunciated so far, the first on the moral significance of

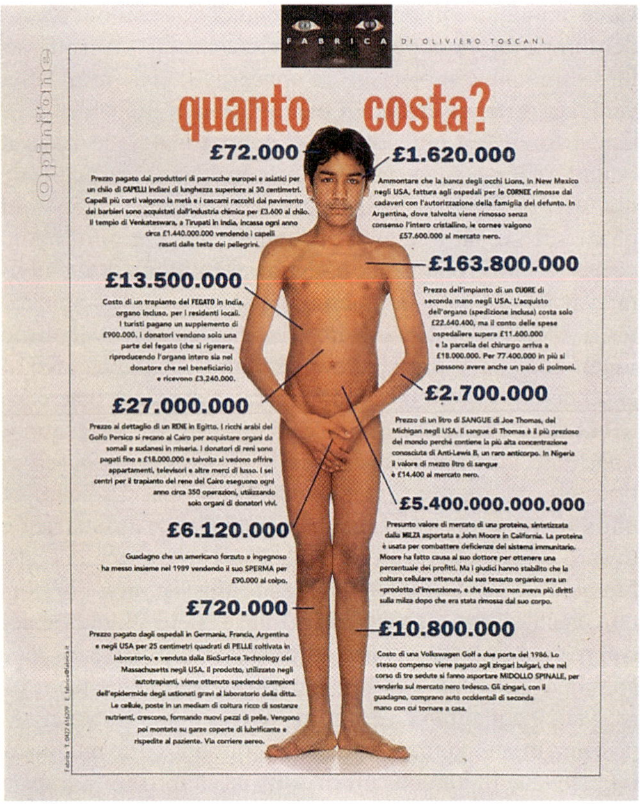

FIGURE 2 The price in Italian lire of the various parts of the human body, in relation to the state of the market, in a composition from the "Fabrica" series by Oliviero Toscani (courtesy of O. Toscani).

transplants and the second on the relation between emphasis on the value and commercialization of the body, make it possible to frame the most systematic treatment of the various ethical problems pertaining to transplants.

The Commerce and Donation of Blood

Different ethical and practical problems are posed depending on whether the source is a cadaver, which is almost always the case, or a live human being. However, the source may also be a laboratory product or some other living species, and depending on the typology of the "materials" to be used, which may be either renewable (such as blood and marrow) or not (organs), it may come in pairs (kidneys) or be unique, and so on. In general, renewable tissues such as blood are taken from a live human being, and vital organs are taken from a cadaver.

The most important ethical question in regard to the transfusion of blood (which foreshadowed for several decades the debate on the relation between the donation and commercialization of organs) has been whether it should be considered a marketable item or a social resource. Transactions can be of two types: One consists in payments in currency and the other in granting those who give blood (donors) a blood credit to be cashed in should they be in need of transfusions. Both have been widely applied in the United States and have raised both objections of principle and accusations of violation of the principle of justice; it has been observed, in fact, that poorer populations have a greater need to sell and, on the other hand, less opportunities to buy blood in case of necessity.

In Great Britain, where donation had been widely established, the sociologist R. Titmuss published in 1970 the classic text *The Gift Relationship,* in which he maintained that donation represents a personal decision aimed at solidarity and is an act which fulfills the desire of individuals to act in favor of their community, thereby strengthening social cohesion (Titmuss, 1970, 1997). Titmuss's criticism of the system at work in the United States was also based on safety considerations on the evidence that the purchasing of blood makes the transmission of type B hepatitis more probable, albeit by a small percentage; this later occurred in the case of AIDS. This criticism was accompanied by scorn for "the pathetic idea, present in some circles in the United States, that computers allied with plasmapheresis will solve the moral dilemmas of the blood market." In 1970, Titmuss emphasized the external effects of the marketing of medical assistance and of systems for the transfusion of blood:

> They should now be redefined and evaluated in international terms. They go beyond the consequences of the importation of blood and plasma; they include the incalculable effects of exporting, in the form of models for the poorer countries, the values and methods of the blood market just like the effects of exporting health care systems which raise profits to the utmost.

In summary, one might concur with Titmuss that the gift relationship expressed by the donation of blood represents an example in which the virtue of compassion becomes linked with an interest in cohabitation in a spirit of solidarity, and that its antithesis, the buying/selling process, "represses the expression of altruism, erodes the sense of community, lowers scientific standards, limits personal and professional liberties, legitimates the pursuit of profit in hospitals, subjects crucial sectors of medicine to the laws of the marketplace" (Titmuss, 1970). Thirty years after the publication of the book, integral blood in the United States is collected almost entirely through donations, although it is still commonly purchased for the purpose of extracting its components industrially. In its rules, the European Union has forbidden all form of remuneration for blood collection and entrusted the latter exclusively to voluntary donations.

Definition and Ascertainment of Death

In order to explant organs from a cadaver, the first condition is that death be ascertained not only beyond all reasonable doubt but also beyond all possibility of mistake and risk of abuse. The second is that there be some consensus.

The first condition requires the existence of a definition of death which is shared and respectful of civil rights. For a long while this was difficult because death is simultaneously an event and an idea, and both elements have their history. The ascertainment of the event depends on instruments: In the past, one employed a mirror which ceased to be clouded by the breath of the dying person, and today one makes sure the electroencephalogram is flat and checks other neurological signs. The idea, too, changes with the times and with cultures. Since antiquity, the idea of life has been associated with the presence of breathing and that of death (expiration) with the abandonment of the body on the part of the soul or of the vital breath (pneuma). According to a more materialistic conception, the two ideas are associated with the unitary nature of the body: "Alive, I react as a mass; dead, I react as molecules" is a famous definition of D. Diderot's. Having established that the body's unitary nature substantially depends on the functions of its central nervous system, increasing reference has been made to the cessation, either partial or total, of the latter's activity.

Starting in the 1960s, the refining of instruments with which to investigate cerebral functioning has been complemented by two novel developments: the possibility of maintaining some cardiorespiratory activity by means of machines in the absence of cerebral activity and the need to have organs that are still intact for transplantation purposes. Diagnostic instruments, artificial survival, and demand for cures are the elements underlying the definition of cerebral death: The latter consists in the irreversible cessation of all the functions of the entire brain — that is, of both hemispheres (cortex) and the trunk, which orders and controls fundamental organic functions. This definition was formulated in 1968 by a Harvard Medical School special committee and, in a relatively brief time, was almost universally accepted both by doctors and by the legislators of the various nations. The latter have also established rules to guarantee death's ascertainment and to distinguish it from other phenomena, such as hypothermia and pharmacologically induced states which might simulate it. The fundamental distinction, which is full of ethical values, is between cerebral death and coma: The latter, in fact, ranges from the more or less serious impairment to the total loss of consciousness, motility, and sensibility, with total or partial preservation of vegetative functions.

Although commonly accepted, the definition of cerebral death is the object of methodological and substantial objections. Among the first kind, doubt has been cast on the statement first enounced with great clarity by Pope Pius XII, according to which it is entirely up to medicine to define what death is on the practical plane (certainly not on the philosophical one). Moreover, it has been suggested that the Harvard definition is characterized by a double "original sin." Its original motivation, later replaced with more diplomatic but analogous formulations (Singer, 1996), was based on two requirements which, in some sense, are extrinsic to the subject: the existence "of a strong demand for tissues and organs, also for the tissues and organs of patients whose brain has been irremediably damaged, so as to allow salvageable patients the chance to live" and the enormous burden which the maintenance of patients suffering from "total loss of intellect" represents for families and hospitals. These motivations, although founded on humanitarian considerations (the second, however, shows the influence of the tendency to limit the right to assistance), might open the way to discriminatory ethical criteria, a competitive notion of life, and violation of the principle of the equality of the rights of every individual. These reservations mix with the perplexities of public opinion with respect to the new definition of death. These perplexities may be the result not only of old ways of thinking and the disconcerting novelty that the heart–lung system can be kept functioning by means of machines but also of the idea that doctors, in addition to defining technical criteria, can arrogate unto themselves the right to decree the death of an individual. The question "But are they really dead, the bodies from which organs are explanted?" may be answered by saying that more than a matter of decision, it is a matter of observation. The death of the organism, in fact, may be deduced from the total inactivity of the brain since it has been identified as the principal element of human individuality and fulcrum of all systems and since, on the practical level, no one has ever come back to life after ascertained cerebral death.

These objections, which we might define as pertaining to methodological ethics, have been matched by others of a substantial nature which lead to opposite conclusions. One is based on the idea that "a person is not defined by the mere persistence of biological functions," if self-consciousness and rationality are lacking (Engelhard, 1991). This reasoning leads one to want to replace cerebral death (hemispheres + trunk) with cortical death; that is, with the cessation of the superior mental activities — cognitive capacities, consciousness, and thought, which are located in the cortex. Another objection is based on the hypothesis that there might be residual neurological and hormonal functions in cerebral death as it is defined according to current instrumentation, and that further tests, along with the improvement of instruments, might make them visible, as has happened in the past with those signs of the brain's and other organs' activities which were not previously perceivable.

Despite these objections, the concept of cerebral death is becoming increasingly more broadly accepted. It may even be that it is "a convenient fiction," as P. Singer maintains, but to replace it with cortical death implies a notion of the individual as a purely rational being and opens the way to many abuses. To state that the truly dead cannot provide organs, and to maintain that the infallible sign of death is the putrefaction that follows in the wake of rigor mortis, may often imply the will to consider as alive bodies that organically are not so at all and, most of all, it implies foregoing making use of their parts, which are objects of residual life, in order to give life and health to other suffering subjects. In practice, this would imply renouncing altruism altogether.

Who Decides about the Body

Among the two conditions for explantation from a cadaver, the first, a shared definition of death which is respectful of civil rights, may be considered as having been achieved more than the second, which pertains to consensus. The problem has as its context one of the basic principles of bioethics, autonomy, which, on the one hand, implies the protection of the body's integrity and, on the other hand, the right of its owner to dispose of it as he or she wishes. Is this principle still valid when the boundaries of life have been crossed? On the juridical plane, a living body enjoys rights to the extent that it is a physical person, whereas a dead body cannot be considered in the same manner even though specific norms for its protection may still exist. On these grounds, it is a fact that, also in the absence of permission from relatives, the state may dispose of a cadaver in order to ascertain the causes and circumstances of death for public purposes (i.e., in the case of diseases or crimes). This has been adopted as a possible justification for the use of the body's parts for more noble ends such as saving a human life.

The laws, however, are not inclined toward considering the dead body as *res nullius* or as *res omnium* which should be taken care of not by the individual or the family but by the state. Even if respect for the body may be distinguished from its lack of tangibility, and even if the carrying out of transplants corresponds to the common good, the latter cannot be pursued by mere power of authority, against everyone's good; most of all, the problem cannot be seen only from the juridical perspective. The cult of the dead, not only as spirit and memory but also as conservation of their integrity, is ancient and almost universal given that bodies are destined, according to many beliefs and almost all religions, either to survive or to be reborn. Moreover, on the anthropological plane, almost everybody feels a spontaneous sense of distaste or repulsion at the thought of their cadaver being plundered, and for relatives the explantation may make the death of their dear one appear even more definitive and painful and interfere with the "natural course" of their mourning. However, the opposite idea may also affirm itself, the awareness that one's own body or that of one's relative may be useful to, and life saving for, others. The sense of donation may prevail, whether it be accompanied by a sort of "altruistic egotism" or not, based on an awareness that in a different form from that of the continuity promised by religions, something of oneself will survive in others, in unknown persons who may be able to benefit. For this to take place, however, the act must be spontaneous. A donation which took place out of compliance would be, both in word and in substance, an obvious contradiction; the voluntary nature of the act would also be undermined if a refusal to effect it should generate, on the part of others, accusations of insensibility or senses of guilt. The spontaneity of the donation, instead, entails psychological benefits for the donors (it has been found that among relatives who have agreed to explants, only 8% have later regretted it), stimulates the sense of responsibility of operators, and contributes (Titmuss, 1970) to the cohesion of micro- and macrocommunities.

In reality, however, requests for transplants are growing at a faster rate than the number of immediately available organs. Borrowing a term from economics, as is fashionable these days, one says that there is an increasing imbalance between supply and demand. This situation is particularly serious in Italy, where the number of donors has increased considerably in recent years (Fig. 3) but is still far lower than the European average, and there are strong inequalities between the regions of northern, central, and southern Italy (Figs. 3b and 3c). This practical reason has induced many countries, especially European ones, to introduce by means of legal enactments, in addition to affirmative consent, presumed consent (known also as tacit assent). Italy, too, has followed this path, also on the basis of an opinion of the National Bioethics Committee (1991) which has deemed it advisable to effect a

> transition towards a juridical régime in which the personal choice to donate one's organs and the qualified assumption of consent on the part of whoever has not pronounced themselves may be recognized, even in the presence of a norm which prescribes that one should render one's will explicit.

That is, will to accept or to refuse the explantation. This approach was approved in the disposition on the subject of explants and transplants of organs and tissues in the Italian law No. 91 of April 1, 1999.

Despite the fact that the motivation which gave rise to this approach is primarily utilitarian in nature, the norm

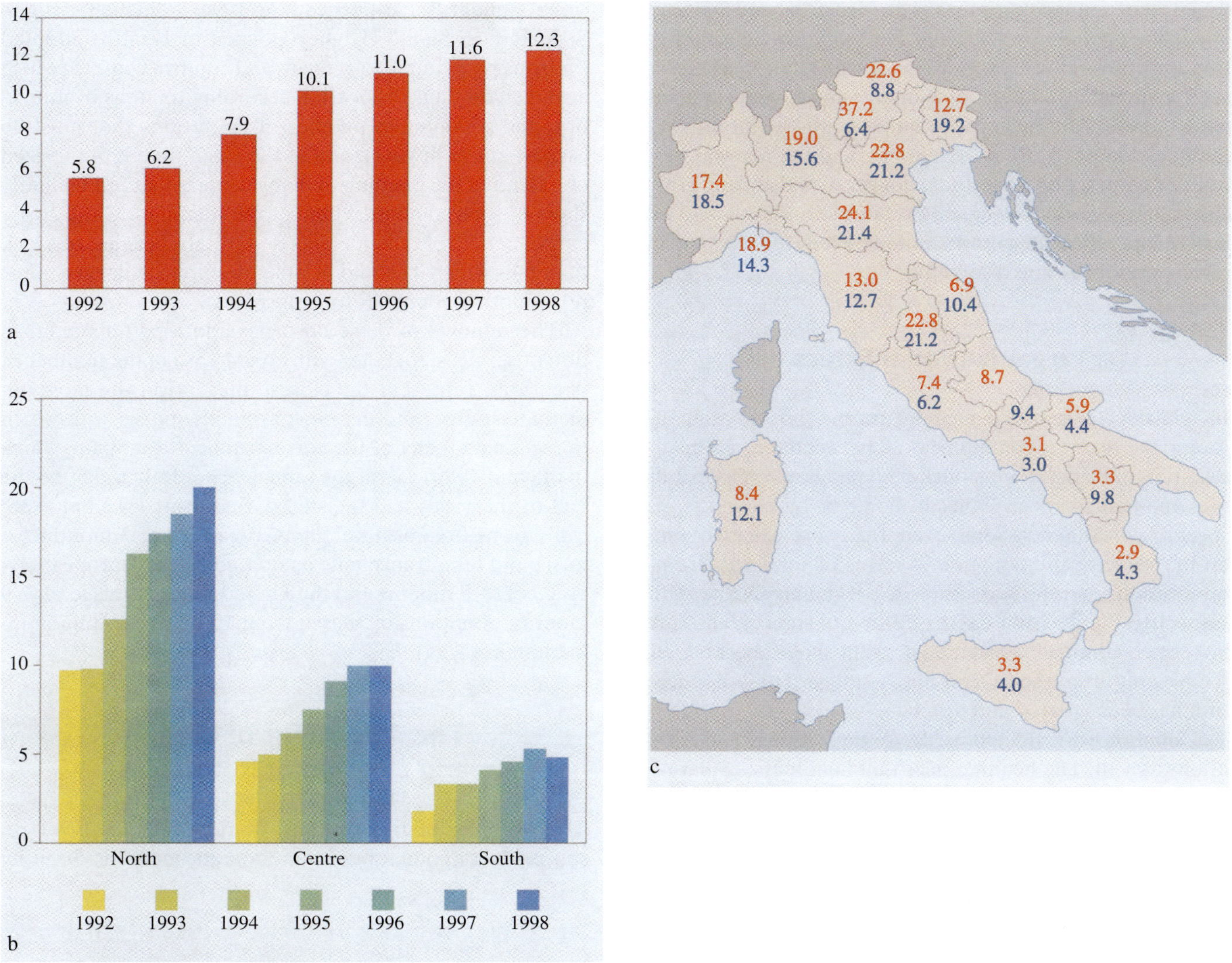

FIGURE 3 (a) The number of donors per million inhabitants in Italy between 1992 and 1998. In 7 years, the number of donors has more than doubled, but it is still the lowest in Europe. The Italian Association of Organ Donors (AIDO) insists on the fact that "the real problem is not one of consent but of organization." (b) The number of donors per million inhabitants in northern, central, and southern Italy between 1992 and 1998. The imbalance between the three areas has increased every year. In the south, many transplant centers have been created, but there is not the same sort of commitment toward promoting and organizing donations. (c) The number of donors per million inhabitants in the various Italian regions in 1997 (in red) and 1998 (in blue) (redrawn with permission from Quintieri and Pugliese 1999).

which entrusts persons with the task of expressing their own inclinations presents, on the ethical plane, the advantage of privileging the autonomy of subjects in relation to the will of their relatives, the former being called on to express, while they are still alive, an anticipated decision with respect to their body,. This autonomy is also becoming established in other fields, as is evidenced by the introduction of informed consent in doctor–patient relationships, which might be better defined as "conscious participation" and whose meaning should not be turned upside-down and become only a signature that the patient gives for the doctor's legal protection; it is also evidenced by the so-called "living will," in which subjects declare ahead of time, in case they should be incapacitated to do so later on, when life-preserving systems which have become forms of therapeutic relentlessness with respect to the dying person should be interrupted.

If this norm plainly favors the autonomy of subjects over that of their relatives, its relation to the public good is not quite as automatically discernable. Its application implies a widespread awareness of the possibility and motivations of donation and requires that each citizen be asked in a reasoned and unimposing manner to decide ahead of time about the possible use of his or her body for transplantation purposes. However, if a law of this kind, marked by such a pregnant content of human values, were applied with an authoritarian spirit or bureaucratic methods, as if the question posed to citizens concerned the renewal of a utilities contract or a fiscal fee, doubts would accrue, criticism of a state which decreed the fates of bodies would be justified, practical advantages would be reduced, and moral effects would become disquieting.

Organizational Ethics

In addition to knowledge about norms, information or, rather, the two-way communication between the health authorities and citizens must include data about the possibilities and limits of transplants; it should be based on the respect of personal decisions (even if they are negative) and on the value of the donation in terms of solidarity. Communication, therefore, acquires a preliminary value with respect to the decision expressed by the subject, and any law which admits of presumed consent should become effective only after a social consensus in regard to its application has been created and tested.

Donations are influenced by organizational considerations as well. The health offices that handle the explantation, conservation, and distribution of organs; the liaisons between the places where explants takes place (often reanimation centers) and those where transplants are carried out; the programming of services; collaborative spirit; and efforts to overcome territorial and social imbalances are all obvious practical necessities, but they imply a specific organizational ethic. The latter includes the professional deontology of doctors and nurses, the humanization of relationships with patients and their families, and the duties of those who program, coordinate, and direct the services. In these fields, when ethical and communicative requirements are neglected, the value of the transplant is inevitably discredited and, almost as reprisal, the thesis that the lacunae in the system depend essentially on the fact that there exists but a scarce propensity of citizens to make donations gains in credibility.

When these ethical and communicative bases are missing, the majority of available organs (which, unfortunately, are not lacking given the frequency of deaths, often avoidable, that occur among the young) are neither collected nor utilized because the separate services and their interactions are poor, and disinterest, autarchy, and jealousy prevail. In Italy, positive steps have been made in the organization of and connection between services, and both the number of transplants and their rate of success have been increasing.

However, too many centers are authorized to carry out transplants, many of which perform only a few operations with a resulting waste of resources and lowering of professional standards. Paradoxically, in 1994–1996 the Consiglio Superiore di Sanità (Higher Council of Health) adopted rigorous procedures to activate and confirm centers (which also include their dislocation according to areas of utilization and a minimum number of operations that must be carried out each year), and at the same time it authorized or extended the opening of many centers that fell blatantly short of such standards. One of the humiliating and costly consequences of these defects is that Italy, because of a lack of organs which it should be able to supply on its own, must rely on other countries for transplants.

The removal of these obstacles and a better organization of services, together with an increase in the number of donations, represent the closest alternative, the most immediate source, and the most available means with which to overcome many of the current difficulties. Spain, which in the mid-1980s was at the same level of Italy today, by the end of the 1990s had registered, due to its laws but especially to its organization, an average wait of 3 months for heart and lung transplants compared to the European average of 6–8 months and the United States average of 7–9 months; donations increased from 10 to 29 per million inhabitants (Fig. 4).

The Allotment of Organs and Division of Resources

The problem of the absolute or relative scarcity of resources is common for many therapeutic methods. Scrutiny

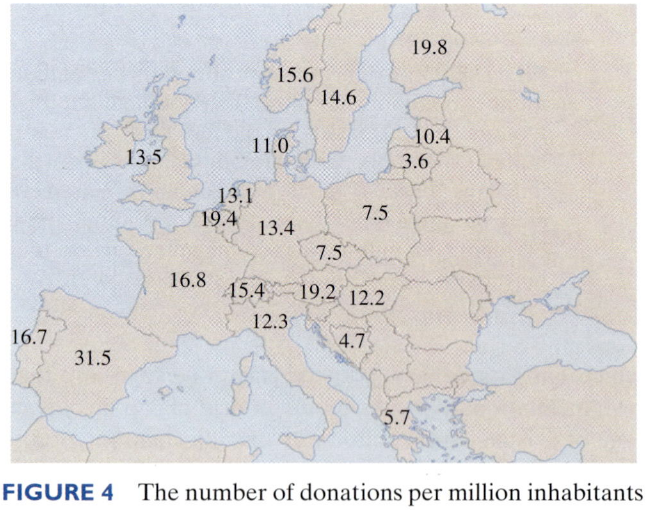

FIGURE 4 The number of donations per million inhabitants in Europe in 1998. The country with the greatest number of donors relative to its population is Spain. The principle of tacit assent has been put into practice in that country with a widespread public information campaign and participation; however, most of all, an efficient organization has been created that covers the entire national territory, with regional headquarters and branches in the most important hospitals (reproduced with permission from *Organs and Tissues,* 1999).

of this problem is sometimes distorted by insisting exclusively on financial resources and neglecting that which cannot be traced back to goods (drugs) and professional instruments or activities, such as solidarity (essential in the case of organs and tissues) and minimum-cost therapies (e.g., rehydration in the case of serious childhood gastroenteritis). The problem, however, has always existed. After every discovery of new drugs such as antibiotics, there has always been a period in which they were accessible only to a minority, after which time mass production commenced. With blood transfusions, the problem was dealt with by increasing blood drawings, with conservation and fractioning, which enables one to increase the utility of a single bottle by distributing the individual fractions to different patients on the basis of their specific requirements. With organs, this has been a problem since the beginning and has become progressively more acute as the successes of transplants have progressed at a faster rate than increases in the number of donations. Criteria for the allotment of organs, therefore, often imply tragic choices and have given rise to animated discussions, especially in relation to the ethical question of equity.

The question "to whom?" has different answers, depending on whether the organs come from a live individual or from a cadaver. In the first case, which is more rare, the destination is almost always preestablished in favor of a person who has close relations to the donor. Laws and practice are oriented, in fact, toward admitting donation on the part of a live individual only when the organ is indispensable for the survival of close blood relatives, in some cases spouses and on rare occasions friends: Also, the World Health Organization has upheld this restrictive orientation. Two arguments speak in its favor. The first, which is of a practical and moral nature, is that although the collection of blood and sperm has no negative effects and the removal of egg cells and marrow entails a more invasive operation and begs anesthesia, where organs are concerned, even if they be double (such as kidneys) or subject to regeneration (such as the liver), the donor will still suffer certain damage and be at considerable risk. The second argument, which is of a consequential nature, is that by enlarging the circle of donors beyond the immediate family it would be easier (and difficult to control, given the sometimes impelling needs of whoever "supplies" and the weight of the pressure that may be brought to bear by whoever demands) for monetary compensation for explants or other contractual types of agreement to become common practice.

In the allotment of organs derived from cadavers, biological criteria are deemed fundamental: the typology of tissues, blood types, body sizes, urgency, and probability of success. This list of criteria is extended by others which, on the moral plane, fall on the dividing line between objectivity and arbitrariness, such as the date patients were put on the waiting list, their age and presumable life expectancy, and plainly selective considerations. There was considerable dismay when, in the state of Oregon, a first attempt was made to choose the recipients of organs on the basis of the origin of their diseases and to distinguish, among cases of cirrhosis of the liver, between those due to hepatitis and those that were the result of alcohol and to put the latter (in what was a typical case of "victim blaming") at the bottom of the list; even a comparative evaluation of the possible contribution that each potential recipient might have made to society had been suggested.

The explicit codification of selective criteria is not frequent, however. Just as for the "rationing" of all other types of therapy, which is being more frequently invoked in the context of restrictions to public assistance (Berlinguer, 1999b), the greatest risk is that they be decided without any explicit motivation. Such a practice inevitably consolidates an asymmetrical system based on the dislocation of services, information, public and private relations and influences, the census, and the cultural level of patients. A study carried out in Rome on the length of time kidney recipients had to wait, for example, showed that subjects with only a basic education waited 2.5 times longer than those with university degrees.

In the United States, in particular, Caplan (1995) notes,

> There are reasons to be concerned about the global impartiality of the present system. First of all, for the role of money in determining not who obtains a transplant in any given hospital, but who is considered and admitted to a transplant center. A transplant is a costly procedure. . . . In order to be taken into consideration for a transplant a person must be able to face these costs. . . . In the United States and in many other countries, insurance is the key which determines who has access to transplant centers and, if they are qualified, their placing on the waiting list.

Also, the period after the transplant operation is subject to the same rule. In 1996, there was much dismay at the case of a 25-year-old woman, L. Cunningham from Philadelphia, who 11 years before had received a new heart and, after the operation, had enjoyed federal subsidies for the poor. At 21 years of age, the subsidies were interrupted. Ms Cunningham still continued for some time to enjoy the benefits of free antirejection drugs, but eventually even they were denied her and she let herself die. The doctors who had operated on her certified that the death had been caused by the interruption in the administration of the drugs. In addition to problems of microallotment, which answer the question "To whom should an organ be allotted when clinical conditions are equivalent?," there are obviously also problems of macroallotment, which concern the health care systems: Unlike those services that are based on the principle of universality, in fact, the insurance system causes the selective filter to be situated upstream.

Relation between Body and Market

The Council of Europe's (1996) Convention on Human Rights and Biomedicine established, in Article 21, that "the human body and its parts shall not, as such, give rise to financial gain." The explanation for the specification "as such" which is part of the convention makes it clear that economic gain is legitimate for professional activities based on the utilization of organs, tissues, or cells, and that the body and its parts cannot be bought and sold. An analogous principle is accepted in the United States as per the National Organ Transplant Act of 1984, which declares it to be "illegal for anyone to buy, receive, or in any way transfer any human organ [identified as kidney, liver, heart, lung, pancreas, bone marrow, cornea, eye, bone, or skin or others on the basis of rules established by the Department of Health, Education, and Welfare; the buying and selling of sperm, ovules, and blood are instead allowed] as a good to be used in human transplants, if the transfer implies interstate commerce." In 1992, the United Nations and the World Health Organization pronounced themselves in the same sense.

Shortly after the first experiments by the physiologist and surgeon A. Carrel, who in 1912 received the Nobel prize for medicine due to his research on the suturing of blood vessels and experiments on organ transplants, the possibility that in addition the advantages of such practices, a new road might also have been opened to the buying and selling of individual body parts was soon hypothesized (Gramsci, 1918):

> Life itself will become a form of merchandise. . . . Life, all of life, not only the mechanical activity of limbs, but the very physiological source of such activity, becomes detached from the soul and becomes merchandise to be bartered. It is the fate of Midas, the symbol of modern capitalism.

That this be the symbol of a system is an opinion, that this has become in part reality is a fact, and that such a tendency could increase is a possibility.

Facts, especially when they are dismaying, often become associated with legends, such as that in various parts of the world there have often been abductions and kidnappings, especially of children, so as to take their organs for the purpose of transplants. In the drafting of the volume *La Merce Finale. Saggio sulla Compravendita di Parti del Corpo Umano* (*The Ultimate Good. Essay on the Buying and Selling of Human Body Parts;* Berlinguer and Garrafa, 1996), a detailed analysis of all cases of this type that had been signaled in the press was carried out without finding that any of them had any bearing in fact. Also, the denunciation of the disappearance — attributed to the same cause — of 3000 among the 4000 Brazilian children that were adopted in Italy between 1988 and 1992, irresponsibly presented by the French deputy L. Schwartzenberg before the authoritative tribune of the European Parliament, has been carefully invalidated by an accurate census of the same children, who were all found to be alive and almost all happy in their adoptive Italian families. Finally, the international investigation carried out by Rothman (1998) has also produced a negative result. He added the following caustic and realistic consideration to his investigations:

> However rapacious medical workers may appear to be, it is not plausible that teams of surgeons, specialized nurses, anesthesiologists, technicians, and blood transfusion experts, highly trained and sophisticated on the medical plane, should conspire to kill for the organs or accept them from off the street. Had they done so, at least one case would have come to light.

In addition to the practical and moral harm caused by this legend in discrediting the practice of transplantation and adding a new horror to those already attributed to the applications of science, there is another negative effect: that of hiding, in the fog of invented facts, two phenomena which instead are a part of reality.

The first, which has been documented by us and by others in many countries (Kimbrell, 1993), is that since the start of the 1990s, when the surgeon R. Crockett was condemned in London for having transplanted in four British citizens kidneys that had been purchased from four Turkish citizens, similar abuses have taken place and continue to take place in many countries, in ways that are either legal or tolerated by the law. For example, in India kidney transplants prospered for years by means of organs that had been sold and explanted from desperately poor subjects so as to be used by patients capable of paying for them. When, in 1994, Italian laws intervened in order to forbid transplants from a living being, at first there was a notable decrease but subsequently the phenomenon returned to former levels because the norms allowed for donations not only among relatives but also in other, albeit authorized, cases; abuses soon flourished. Another example, different in reality but similar in inspiration, is provided by the large business that takes place in many hospitals in China, where a 1984 law which allows the use of organs taken from individuals condemned to the death sentence is being used to provide a fast transplant (even preestablished in time since the sentence can be carried out earlier or later according to the request of clients) to patients from other parts of Asia.

The second phenomenon, complementary to the first, is the tendency of various surgeons (Sells, 1992) and philosophers (Engelhardt, 1991) to justify commerce in this area. According to the former, there are fundamentally practical reasons, such as the scarcity of kidneys and the immorality of prohibitions which "take away all hope from the very poor and the dying" who had been waiting to sell so as to be able to live or to buy so as to survive their physical impairment. The latter adduce reasons of principle: the right to property and freedom of allocation of private resources, starting with oneself. Some (Radcliffe-Richards *et al.,* 1998) maintain that the ideal solution would be to "reduce poverty to the extent that the option of selling one's organs is no longer the best option," neglecting the fact that allowing such purchases might make it convenient to maintain a part of the world in a perpetual state of poverty so as to induce it to sell the only good which it would produce in abundant quantity (i.e., oneself). Others affirm the right of disposing of one's own body as one wishes, without making any distinction between disposability and marketability.

One starts, fortunately, with the fact that "when the practice of buying kidneys from live sellers came to light, it elicited such horror that all the professional associations denounced the practice and almost all countries have declared it illegal" (Radcliffe-Richards *et al.,* 1998). These tendencies, however, should not be underestimated. Arguments in favor should be discussed in-depth, but the fact that such a market continues to exist and the tendency to justify it to expand, perhaps by regulating it with a "right" price and a sort of centralized agency, constitutes an obstacle to donations. Spinsanti (1994) noted, "Diffidence with respect to all forms of transplant is growing. The doctors who carry them out feel the malaise grow around them, almost as if they were involved in a reproachable activity. The donation of organs is decreasing." For it to increase, in addition to the necessary appeals to solidarity, there are three guarantees which can act as a stimulus or, if they are missing, as a brake: the certainty of death, which the criterion of cerebral death should enable one to feel sure about; equity in allotment, which is sometimes posed in questioning terms; and the nonexistence of a market parallel to the flow of donations. In order to be effective, impediments should have a universal character, as the laws against the marketing of bodies *in toto* (i.e., slavery) had in the eighteenth and especially the nineteenth centuries. However, even more important are the growth of moral conscience and the deontological commitment of professionals, without whose specialized labor the transplantation of organs that have been bought and sold would not be possible.

Ethical and Practical Perspectives

In the field of organ and tissue transplants, in which scientific progress and biotechnologies are ever more rapidly intertwined, it is difficult to make forecasts. A. Sauvy used to say that the only sure thing, as far as long-term demographic projections were concerned, is how much future generations will laugh when they read them. A similar certainty could be advanced in the field of transplants, with an eye, however, to short-term forecasts and to one's contemporaries. What may be attempted, however, is to reason, albeit in a summary fashion, on the ethical aspects of tendencies that are already at work, which foreshadow the future. Pig cardiac valves are already a type of xenotransplant, as are the iron lung, the kidney substitute used in dialysis, and the synthetic limb prostheses, which in some manner are also like artificial organs. I hope that from the following discussion, two aspects will clearly stand out: (i) that ethical controversies will continue and probably become more complex and (ii) that many technical and scientific innovations are quite promising, so much so that it is possible to envisage that in the future "traditional" transplants will be considered only a transitional phase in this type of therapy. There is no reason to engender illusions, but there is an important reason — prudence — to avoid introducing in the laws and in manners such moral shortcircuitry as the buying and selling of organs.

Can Everything Be Replaced or Duplicated?

The fact that it is possible to replace an increasing variety of parts of the human body may lead one "to imagine a future society in which people consider their own body (except the brain) in the same way that today they consider machines" (Lockwood, 1993). When the body starts to tire, one will proceed with the periodical purchasing of second- or thirdhand organs, as in the case of automobiles; also, in the future perspective, "the body with which one is born will be considered just like the first house in which one has lived, no doubt full of sentimental ties but not the only house in which one will spend one's life" (Lockwood, 1993). Moral objections to this hypothesis may also be the result of practical considerations. It presupposes, in fact, that there be subjects in our species whose role is to produce spare parts, or that the average life span should become so short that a sufficient number of persons die young, or at least at a convenient age, and that the demand for organs guaranteed for sufficiently long-term use be satisfied. However, there is also a scientific objection pertaining to personal identity. Current knowledge has gone beyond the notion that the latter resides in a single organ, the brain. Together with, and in relation to, the latter, and to the nervous and neurovegetative system as a whole, there are two other regulating systems — the hormonal and the immune systems; also, body structure, voice, gestures, and the system of external relations (to wit, the house and feelings) contribute toward the formation of personal identity. The thesis of M. Lockwood, which appears to be hypermaterialistic, substantially repeats the old Cartesian idea of the body as a "machine made out of bones and flesh," separate from feelings and thoughts, which instead would be the product of the soul — an idea which has been countered on the philosophical plane and is now considered invalidated by scientific anthropology.

Another possible source of "spare parts" could be cloning, whether it be the result of the insertion of the nucleus of a differentiated cell in an enucleated egg cell or of the precocious division of the embryo when its cells have yet to differentiate. The idea has been suggested, for example, in the case of parents with a child who requires a transplant, that with the birth of a clone, he or she would have available a biological duplicate of his or her organs. Then the idea was carried further, advancing the possibility that every individual who is born be accompanied by an alter ego, possibly without a brain, who would serve the function of supplier of whatever organ or tissue should be required. The fact that one thinks about the creation of human duplicates in the interest, which in some cases may be sentimentally understandable, of other subjects casts light on and renders objections to human cloning even more valid. The latter are based on the idea that it is chance, and not predetermination, which forms the basis of individual freedom. Moreover, as H. Jonas observed, being someone's constructed copy (which is not a natural one, as in the case of twins) upsets one's image and perception of oneself, one's self-discovery, and manifesting of oneself through one's choices. It is on these bases that, in 1998, the first supplementary protocol of the Council of Europe's Convention on Human Rights and Biomedicine, which prohibits all "cloning of human beings," was approved.

Cultivated Tissues and Cells

The previously mentioned protocol refers to human beings. Different ethical criteria apply in cases in which cloning or other forms of reproduction, increasingly connected with biotechnologies, concern individual body parts, which today range from DNA sequences to cells and tissues and tomorrow will involve the production of biological structures that function as organs, aimed at the therapeutic treatment of lesions or deficiencies of the human organism. The benefits which may be derived from such practices are very broad on the practical level and, on the moral level, provide an alternative to the dilemmas that arise in the collection and allotment of the "materials" available to date. As a result, there is widespread consensus in favor of their introduction, and ethical appraisal shifts to three questions that are all but accessory: source, property, and use.

Regarding sources, the most complicated issue is whether turning to human embryos is acceptable or not. The Council of Europe's convention (Article 18) establishes that "the creation of human embryos for research purposes is prohibited," and it may be presumed that such a perspective will also include purposes of production. Even the use of aborted embryos and fetuses is controversial, both because of the fear that it might open the way toward commissioned child bearings and abortions and for reasons of principle shared by those in favor of protecting embryos insofar as they constitute potential persons and by others out of respect for a project of human life. A new problem has come to light in the wake of the recently accomplished isolation of so-called staminal embryo cells, whence the various human tissues spring after successive processes of differentiation: Starting with such cells, tissues might even be produced in the laboratory (Gearhart, 1998). The paradoxical aspect is that the law which forbids experimentation on embryos in the United States does so in a way which, on the one hand, is total and which goes well beyond the limitations foreseen by the Council of Europe's convention and, on the other hand, it is limited to research carried out in federally funded institutions. Many American scientists believe this to be absurd and also that the regulation "has not put an end to privately funded research, which therefore continues without the tight controls that affect research carried out with public funds" (Wade, 1998). It is not easy to understand the logic of the two opposed ethical criteria, distinguished on the basis of the source of financing.

The question of ownership of the body *in toto* (which is connected with its use) has always been the object of debate on the moral level. Today, the discussion also involves isolated sequences of DNA, cells, tissues, and organs reproduced outside of the body, and everything is rendered more complicated by the fact that the law and jurisprudence, whose formulations have been conditioned by the competitive stimuli of an industrial economy, have gone beyond the traditional distinction between invention and discovery in the application of patents (Donati and Tallacchini, 1997).

Regarding use, today there is an increasing tendency to consider as mere objects the products obtained by means of genetic engineering techniques in which human cells are involved. Also, the directive of the European Union on patents recognizes a broad right to patent that which already exists in nature, including sequences of the human genome, as long as they have been identified using technological means (but how might it be otherwise?) and recognized in their functions. Nelkin and Andrews (1998) commented on this tendency: "In this climate, developments in biotechnology are increasingly binding the biomedical sciences to aggressive marketing, which is invading practically every sector of human life"; such marketing produces a growing sense of disappointment and suspicion because "the invasion of the human body on the part of commercial practices increasingly challenges individual and cultural values, encourages exploitation through the collection and use of tissues, and transforms the latter (and potentially people as well) into sellable products." It is likely that this will create incentives for industrial interests to invest in research (which many persons believe should be encouraged especially by other sources and with other stimuli), but beyond the moral judgment is the fact that, in this manner, highly selective criteria in regard to access and to the use of biotechnological advancements are introduced from the start.

Artificial Organs and Tissues

Since the end of the 1960s, the invention of artificial "external" life-support apparatuses for respiration, circulation, kidney function, thermoregulation, etc. has led to many benefits for the prematurely born, nephropaths, or patients requiring reanimation. The ethical problems which have arisen are more complementary than intrinsic to their use, which is quite widespread. They concern themes which, in many respects, are common to other technologies as well: distributive equity; the high costs and, as a result, "the justification for spending vast sums of money to provide life-support systems to a small number of individuals, with scarce potential benefits, while many do not receive even the basic medical assistance which notoriously provides ample benefits which, however, are less spectacular" (Bunker, 1995); the selection of patients; therapeutic relentlessness; and the choice of the moment in which to interrupt the treatment.

Promising steps forward have recently been achieved in the internal replacement of tissues, including fractions and elements of blood; in the recovery of functions such as movement, speech, and perception; and in the construction of organs. The artificial heart, for example, is one of the goals which is considered possible. In these fields, ethical problems analogous to those mentioned previously may arise, but others (and this is proof of the fact that biotechnologies create but can also solve moral problems) may instead be removed. For example, the production using genetic engineering techniques of factor VII in blood (which hemophiliacs are lacking) can ensure widespread therapies at a low cost and help balance the relation between donations and transfusions. The problematic area at the border between ethics, anthropology, and psychology, however, may become broader, especially if the internal integration of biological and artificial processes also involves mental processes, and even other functions in which organs and tissues are involved.

Xenotransplants

This same problematic area might also concern, albeit in a marginal fashion, xenotransplants — that is, the use of live cells, tissues, and organs taken from a nonhuman animal and transplanted or implanted in a human being. The presence in humans of natural xenoreactive antibodies has rendered this practice impossible to date, but in the near future, by means of genetically modified animals, it might become possible and widespread. In addition to bringing myth back to mind (Fig. 5) and exciting the fantasy of illustrators (Fig. 6), the practice has raised two objections of an ethical character.

FIGURE 5 Transformation of Odysseus's companions into pigs. Circe offered Odysseus's men a potion mixed with cheese, honey, and wine. After they had filled their stomachs with it, they grew pig's heads and bristles and grunted like swine, although their minds remained as human as ever (reproduced with permission from Weiss, 1998).

One objection, which has been voiced since the start, concerns the right of making use of animals. The opposition has been strongest for the animals closest to man, the primates: Not only are their species protected because they are in danger of becoming extinct but also some of them, such as chimpanzees, are on the border of self-consciousness and communication which defines human beings. Since one speaks mostly of pigs in regard to xenotransplants, however, answers to criticism have come primarily not from the ethics of animal rights activists but from everyday common sense: It was immediately pointed out, in fact, that there is no moral difference between eating them and using their organs for another purpose. Some differences actually do exist (Tallacchini, 1997): The practice of xenotransplants requires a substantial increase in animal experimentation, which in other fields is declining. The genetic modifications which are introduced may render the animals more frail and vulnerable to suffering; their lives, in conditions of isolation and environmental sterility, are the obvious negation of their existence in nature. One may, however, become convinced that although legitimate problems of principle do subsist, the saving of human lives (which is a duty and not a free choice, as instead is the decision whether to eat meat or not) would undoubtedly lead to their solution.

In the meantime, another objection has become prevalent, one based on scientific considerations and strength-

FIGURE 6 Cover of the January 22, 1998, issue of *Nature,* dedicated to the fears and hopes surrounding xenotransplants (reproduced with permission from *Nature,* 1998, 391).

ened both by research on the origins of AIDS, a disease whose cause is attributed to a modified virus of monkey origin, and by the transmission from ovines to bovines and then to the human species of the prions that are the agents of bovine spongiform encephalopathy or "mad cow" disease: The risk of a spreading of xenosis, as the transmission of infective agents from one species to another, has been raised.

On the historicobiological plane, this process is exactly the opposite of the immunity which served to fight and finally to overcome human smallpox. The latter, in fact, was induced by means of the agent of an animal disorder (i.e., cowpox); by freeing some individuals from the risks of the disease, at the same time it freed others from the risk of contagion. A xenotransplant, on the other hand, may save the life of one individual and at the same time spread biological agents that are lethal or pathogenic for others.

On the ethical level, four considerations may be taken into account (Bach *et al.,* 1998): A public risk requires a public mechanism to determine its acceptability and a means to express acceptance of the risk; the public risk is not a unique event, and its evaluation and regulation should be repeated; the standard model of informed consensus to medical intervention must be modified because the risk involves other subjects and requires the monitoring of the persons with whom the transplant recipient has contacts; and there exists the possibility that the infective agent could become even more pathogenic in humans and thus represent a risk for other animals (and via the latter for other humans). To this one should add the problem of the freedom of the patients, who have to accept the need to live in isolation and to remain under observation for the rest of their lives, to make their private data available to public health agencies, and to be bound by the permanent preclusion to giving away any of their own body parts. Whereas these limitations might be individually acceptable as the price which must be paid in exchange for health, the fundamental fact remains that this type of transplant "is born in competition and possible conflict with borderline situations" (Tallacchini, 1997). One might define a typical case of conflict between individual and collective interests, with the difference that the situations do not exactly border one another: The diseases may, in fact, involve other living species and affect human beings after some time and from a distance. We are therefore referring to the global risks and to future generations.

For these reasons, a moratorium has been proposed which should allow the risk to be evaluated and the preparation of some methods for controlling its consequences, which in order to be effective must be global indeed. Powerful professional and scientific–industrial interests, however, have been opposed to it, especially in the field of the production of immune-depressive substances and transgenic animals (Butler, 1998). It is therefore possible that the ethical principle of caution, which should prevail in the face of scientific uncertainty and was expressed by the ancients with the motto "*in dubium abstine,*" shall be neglected. On the perspectives of transplants, one may therefore embrace the thesis that "prevention, a better system of supply and distribution of organs and the tacit consent laws should be our primary aims, with the new surgical techniques and the development of synthetic organs as potential long-term solutions" (Berger, 1996).

References Cited

BACH, F. H., FISHMAN, J. A., DANIELS, N., PROIMOS, J., ANDERSON, B., CARPENTER, C. B., FORROW, L., ROBSON, S. C., and FINEBERG, H. V. (1998). Uncertainty in xenotransplantations: Individual benefits versus collective risks. *Nat. Med.* **4**(2), 141–144.

BERGER, A. H. (1996). Animal-to-human transplants: The ethics of transplantation. *Sci. Eng. Ethics* **2,** 482.

BERLINGUER, G. (1999a). *Bioetica Quotidiana.* Firenze, Giunti,

BERLINGUER, G. (1999b). Realities of and alternatives to rationing in international perspective. Paper presented at the Symposium on Rights and Rationing in Health Care, University of Bristol, Bristol, UK.

BERLINGUER, G., and GARRAFA, V. (1996). *La Merce Finale. Saggio sulla Compravendita di Parti del Corpo Umano.* Baldini & Castoldi, Milan.

BUNKER, J. P. (1995). Artificial organs and life support systems. In *Encyclopedia of Bioethics* (W. T. Reich, Ed.), Vol. 1, pp. 198–206. Macmillan, New York.

BUTLER, D. (1998). Last chance to stop and to think on risks of xenotransplants. *Nature* **391,** 320–324.

CAPLAN, A. L. (1995). Organ and tissue transplants. Ethical and legal issues. In *Encyclopedia of Bioethics* (W. T. Reich, Ed.), Vol. 4, pp. 1888–1894. Macmillan, New York.

COUNCIL OF EUROPE (1996). Convenzione per i diritti dell'uomo e la dignità dell'essere umano riguardo alle applicazioni della biologia e della medicina, approvata dal Consiglio dei ministri, Strasbourg, November 19.

COUNCIL FOR ETHICAL AND JUDICIAL AFFAIRS (1995). Anencephalic infants as organ donors, Code of medical ethics. *Am. Med. Assoc. Rep.* **1,** 49–52.

DONATI, L., and TALLACCHINI, M. (1997). Ingegneria tissutale bioetica e prodotti artificiali. *Med. Morale* **2,** 267–285.

ENGELHARD, H. T., JR. (1991). *Manuale di Bioetica.* Saggiatore, Milan.

GEARHART, J. (1998). New potential for human embryonic stem cells. *Science* **282,** 1061–1062.

GRAMSCI, A. (1918). La merce. In *Avanti!*

KIMBRELL, A. (1993). *The Human Body Shop. The Engineering and Marketing of Life.* Harper Collins, London.

LOCKWOOD, M. (1993). La donazione non altruistica di organi in vita. In *Questioni di Bioetica* (S. Rodotà, Ed.), pp. 139–147. Laterza, Rome.

NASH, E. P. (1994, August 14). What's life worth? *The New York Times Magazine,* 34–36.

NATIONAL BIOETHICS COMMITTEE (1991). Donazione d'organo ai fini di trapianto, 7 ottobre, Presidenza del Consiglio dei ministri, Dipartimento per l'informazione e l'editoria, p. 11.

NELKIN, D., and ANDREWS, L. (1998, September/October). Homo economicus. Commercialization of body tissue in the age of technology. *Hasting Centers Rep.,* 30–39.

Organs and Tissues (1999). **1,** 7.

PORTER, R. (1997). *The Greatest Benefit to Mankind. A Medical History of Humanity.* Norton, New York.

QUINTIERI, F., and PUGLIESE, O. (1999). *Organ Extraction and Transplantation Activity in Italy.* Higher Institute of Health.

RADCLIFFE-RICHARDS, J., DAAR, A. S., GUTTMANN, R. D., HOFFENBERG, R., KENNEDY, I., LOCK, M., SELLS, R. A., and TILNEY, N. (1998). The case for allowing kidney sales. International forum for transplant ethics. *Lancet* **351,** 1950–1952.

ROTHMAN, D. J. (1998, March 6). The international organ traffic. *New York Review of Books,* 14–17.

SELLS, R. A. (1992). The case against buying organs and a future market in transplantation. *Transplant. Proc.* **34,** 2198–2202.

SINGER, P. (1996). *Ripensare la Vita. La Vecchia Morale non Serve Più.* Saggiatore, Milan.

SPINSANTI, S. (1994). Per i trapianti d'organo, un circuito della solidarietà. *Clin. Ospedaliera* **5,** 73–76.

TALLACCHINI, M. (1997). Etica e xenotrapianti. Un primo approccio. *Vita Pensiero* **9,** 577–593.

TITMUSS, R. (1970). *The Gift Relationship. From Human Body to Social Policy.* Allen & Unwin, London.

TITMUSS, R. (1997). *The Gift Relationship. From Human Body to Social Policy, Original Edition and New Chapters* (A. Oakley and J. Ashton, Eds.). LSE, London.

TROHLER, U. (1993). Surgery (modern). In *Companion Encyclopedia of History of Medicine* (W. F. Bynum and R. Porter, Eds.), pp. 1022–1023. Routledge, New York.

WADE, N. (1998, November 20). Cells fuel debate on ethics. *Science Times* (Suppl. of *The New York Times*).

WEISS, R. A. (1998). *Nature* **391,** 327–328.

CHRISTIAN HICK*

Department of Medical History and Medical Ethics
Cologne University
D-50931 Köln, Germany

Informed Consent

According to a concise definition proposed in 1993 by Finkelstein et al., *informed consent is an "autonomous action by a patient authorizing a physician to initiate a medical plan." To make this very condensed statement more meaningful, we first outline the most important steps in the history of informed consent, both in clinical practice and in medical research. We then discuss the ethical meaning of informed consent along its five core elements of disclosure, competency, understanding, voluntary choice, and autonomous authorization before addressing recognized exceptions, the philosophical background, and possible criticism of the informed consent doctrine. Subsequently, problems with informed consent in special situations, such as with psychiatric or elderly patients or with children and adolescents, in emergency medicine research and in genetic testing will be considered. Finally, a brief outlook will be given on what could be expanding obligations for physicians originating from the informed consent doctrine in the near future.*

Introduction

> Of all the elements ingredient in the reorientation that has occurred in medical ethics over the past couple [of] decades, the informed consent doctrine is potentially the most profound and far-reaching. . . . Informed consent is the cutting edge of the patient autonomy movement.
>
> —Wear (1993)

This is one of the many statements that recognize informed consent as being "the foundational concept of modern biomedical ethics" (Meisel and Kuczewski, 1996). Finkelstein *et al.* (1993) defined informed consent as an "autonomous action by a patient authorizing a physician to initiate a medical plan." This definition comes at the end of substantial historical changes concerning the relation between a patient and his or her physician. For an adequate understanding of the doctrine of informed consent, it is therefore essential to review how this change came about and how the concept of informed consent was historically shaped in clinical medicine and in medical research.

*christian.hick@uni-koeln.de

History of Informed Consent

The doctrine of informed consent is a modern one. The term was first introduced in 1957; a more detailed discussion did not start until approximately 1972. Katz (1984) even went as far as to say that "the doctrine of informed consent surfaced, seemingly out of nowhere" and that "disclosure and consent . . . have no historical roots in medical practice." This is obviously true if one refers, as Katz does, to the concept of informed consent as we know it today and tries to find it in history. From a less anachronistic point of view, it has been said that "truth telling and consent-seeking have long been part of an indigenous medical tradition" (Pernick, 1982). In an effort to reconcile these two positions, Faden and Beauchamp (1986) argued that although disclosure and consent-seeking have always been part of the medical practice, before 1957 this practice was governed by a *beneficence model* rather than by an *autonomy model* of the patient–physician relationship. This background model fundamentally changed the character of disclosure. In the beneficence model the physician's primary obligation was to provide benefit for the patient, whereas in the autonomy model the emphasis lies on respecting the free and voluntary, "self-legislating" decisions of the patient as an autonomous agent.

Informed Consent in Clinical Medicine

Ancient and Medieval Medicine. In the *Corpus Hippocraticum* (Fig. 1) there is no explicit discussion of consent by patients or respect for patients' decisions. Moreover, it is even advised "[to conceal] most things from the patient, [to turn] his attention away from what is being done to him, [to reveal] nothing of the patient's future or present condition. For many patients through this cause have taken a turn for the worse" (*Decorum XVI*).

However, there is other evidence in the Hippocratic writings that the importance of a therapeutic understanding with the patient was well acknowledged: "The art consists in three things: the disease, the patient, and the physician. The physician is the servant of the art, and the patient must combat the disease along with the physician" (*Epidemics* I, 11). This statement, recognizing the importance of the patient's involvement in the medical treatment, comes immediately after the famous dictum that the physician's aim should be above all "to do good or to do no harm." Therefore, this embryonic form if not of shared-decision making then at least of concerted action seems to be governed by the principle of beneficence.

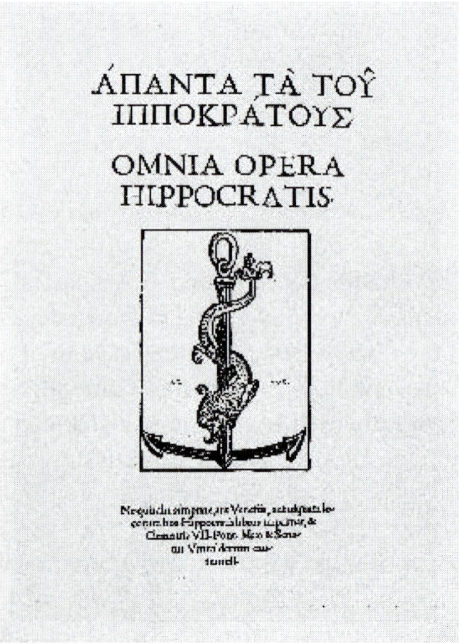

FIGURE 1 Frontispiece of the Aldine edition of the *Hippocratic Corpus* (Venice, 1526).

In Plato's *Laws* the description of how a free physician treats his or her patients (those being free men as well) contains some elements of what later became part of the legal and moral doctrine of informed consent. A free physician does not merely treat a patient; he or she investigates the causes and the nature of the patient's disease, considers the social context in which the patient lives, and even learns from the patient (*Nomoi,* 720d). Before prescribing anything, the physician tries to "convince" the patient (*sympeíthein*) by discussing the details of the prescription and the best way of restoring health. Unfortunately, the Greek verb *sympeíthein* has a rather broad meaning, with possible translations being "to discuss something with someone," "to convince," "to persuade," or "to talk someone into something." However, Plato seems to argue that the best way to achieve medical success is to come to an agreement with the patient concerning his or her medical treatment.

In medieval medicine the Hippocratic tradition of benevolently veiling the patient the truth of his or her disease was continued by the surgeon Henri de Mondeville (1260–1325). Mondeville recommended that a surgeon "ought to promise a cure to every sick person" (Welborn, 1938) while discussing the patient's real medical condition exclusively with the patient's relatives. Only if the primary goal of treating the patient according to the physician's art and knowledge was not in danger could the doctor respect particular requests of the patient. Truthfully informing the patient or engaging in a shared decision-making process did not seem to be at the heart of the perceived moral problems for physicians in this period. An analysis of confession manuals from the early fourteenth to the early sixteenth century, guiding priests in what to ask confessing physicians, shows that at this time other areas were of concern: professional competence and diligence, a reasonable fee for treatment, and charity treatment for the poor or an obligation to care, even for desperate cases (Amundsen, 1981). The modern concept of informed consent thus seems to be lacking a foundation in medieval medical practice, maybe simply because the transfer of information was something that neither seemed possible nor desirable, medicine being an art in the context of an increasingly corporate medieval urban society. To share and discuss information with a patient would have been in contradiction to the self-conception of the physician as the privileged member of a medical guild, holding a monopoly on medical knowledge and medical practice.

Enlightenment Medicine. With Enlightenment philosophy, the context of medical practice changed dramatically. Personal freedom and freedom to choose became the cornerstones of human dignity. Freedom to choose together with "enlightenment" (i.e., the information of the public) were believed to be the way leading to personal happiness for ever more individuals. For this period, I focus my review mainly on the extensively studied prehistory of informed consent in Anglo-American medical texts (Faden and Beauchamp, 1986) because these constitute the historical background for the later development of the informed consent doctrine in the United States.

John Gregory (1724–1773) (Fig. 2), with his *Lectures on the Duties and Qualifications of a Physician,* has been shown to be one of the most important figures at the foundation of modern medical ethics (Strätling, 1998). Influenced by Hume's moral philosophy, Gregory bases his medical ethics on the principle of "sympathy" with the patients and believes that it demands an "exercise of humanity." On this ground, he advocates a policy of informing the patient, limited only by the desire not to harm him or her by this information (Gregory, 1772):

> A physician is often at a loss when speaking to his patients of their real situation when it is dangerous. A deviation from truth is sometimes in this case both justifiable and necessary. It often happens that a person is extremely ill; but yet may recover, if he be not informed of his danger. It sometimes happens, on the other hand, that a man is seized with a dangerous illness, who has made no settlement of his affairs, and yet perhaps the future happiness of his family may depend on his making such a settlement. In this and other similar cases it may be proper for a physician, in the most prudent and gentle manner, to give a hint to the patient of his real danger, and even solicit him to set about this necessary duty.

FIGURE 2 Portrait of John Gregory. Engraving by R. Earlom from the picture of G. Chalmers (photo courtesy of the Library of Medicine, Bethesda, MD).

In any case, it is necessary for the physician to tell the entire truth to the relatives of the patients, enabling them to prepare themselves for the inevitable, should the patient die. This obligation of truth telling, according to Gregory, is a "painful office": "To a man of a compassionate and feeling heart, this is one of the most disagreeable duties in the profession: but it is indispensable."

Shared decision making, however, was a concept that had not yet found its time: The responsibility for any therapeutic action remained with the doctor. However, a certain "gentleness and flexibility" of the physician was thought to be necessary, not for the sake of autonomy as such, but to ensure better compliance with the physician's directives: "The government of a physician over his patient should undoubtedly be great, but an absolute government very few patients will submit to" (Gregory, 1772).

Notwithstanding this still paternalistic conception of medical practice, Gregory argued for a larger participation of the patient in the therapeutic process. The doctor has to carefully investigate the adequateness of any remedy proposed by the patient and not just refuse this proposal "from a pretended regard to the dignity of the profession." Here, an important conceptual element for the further development of informed consent is formulated in what could be said to be the credo of enlightenment medical ethics (Gregory, 1772):

> Every man has a right to speak where his life or his health is concerned, and every man may suggest what he thinks tends to save the life of his friend. . . . It becomes them to interpose with politeness, and a deference to the judgment of the physician; it becomes him to hear what they have to say with attention, and to examine it with candour.

Gregory's enhanced recognition of the patient's preferences prepares the ground for a less asymmetrical patient–physician relationship as the precondition to any further step on the road to informed consent.

In the same line, Benjamin Rush (1745–1813), a disciple of Gregory, pleaded for a transfer of information from the physician to the patients and for a general medical enlightenment of the public with the aim of achieving better medical outcome. Rush advocated truthful, comprehensive information and patient education (Faden and Beauchamp, 1986). His aim, however, was not a process of shared decision making but a better patient's compliance to the physician's recommendations. Although stressing the value of liberty and reason above tradition, the enlightenment did not envision a pluralistic, open situation in which a patient would be allowed to choose between different medical options or no medical treatment at all. There remained the conviction that for every given situation there was one medically adequate treatment, which only had to be explained before this treatment, according to what medical science had indicated as the patient's best, could proceed. This strict and exclusive adherence to the beneficence principle was rooted, on a more fundamental level, in the belief that there can only be one truth, to be found by reason and science, and therefore only one adequate medical treatment, known exclusively to the scientifically educated doctor. Therefore, conclusions that strike us today came rather natural, even for "enlightened" physicians: "The obedience of a patient, to the prescriptions of his physician, should be prompt, strict, and universal. He should never oppose his own inclinations nor judgment to the advice of his physician" (Rush, 1811).

Thomas Percival (1740–1804), who might have met Gregory in Edinburgh during his clinical studies and almost certainly knew Gregory's *Lectures* (Strätling, 1998), summarized his convictions on the moral obligations and the professional etiquette for physicians in his *Medical Ethics; or a Code of Institutes and Precepts Adapted to the Professional Conduct of Physicians and Surgeons* (1803). Although Percival does not mention consent solicitation or respect for patient decisions, he examines the problem of truthfulness, trying to establish a balance between truth telling and beneficence. In severe illness, the physician should not make "gloomy prognostications . . . but . . . give to the friends of the patients timely notice of danger . . . and even to the patient himself if absolutely necessary" (Percival, 1803). If, however, telling the truth could possibly harm a patient, the principle of beneficence prevails:

> To a patient . . . which makes inquiries which, if faithfully answered, might prove fatal to him, it would be a gross and unfeeling wrong to reveal the truth. His right to it is suspended, and even annihilated; because, its beneficial nature being reversed, it would be deeply injurious to himself, to his family, and to the public.

Nineteenth- and Twentieth-Century Medicine. The precepts in Percival's *Medical Ethics* were to become the basis for the American Medical Association's (AMA) first *Code of Medical Ethics* (1847). The AMA code accepted Percival's justification of beneficent deception while stressing that in any exchange between physicians, veracity is "a jewel of inestimable value . . . the lustre of which ought never to be tainted for a moment, by even the breath of suspicion."

It was not until the publishing of *Physician and Patient* by Worthington Hooker (1849) that strict veracity between patients and physicians was praised as a value in itself. Hooker strongly advocated truth telling in any circumstance and overtly rejected the concept of beneficent deception:

> The good, which may be done by deception in a *few* cases, is almost as nothing, compared with the evil which it does in *many* cases. . . . The evil which would result from a *general* adoption of a system of deception [shows] the importance of a strict adherence to truth in our intercourse with the sick.

Hookers attacks on the AMA's moral views on veracity, however, did not have much impact: All subsequent revisions of the AMA code in 1903, 1912, 1947, and 1957 retained the concept of beneficent deception and continued to adhere to the paternalistic understanding of the patient–physician relationship found in Percival's *Medical Ethics* (Faden and Beauchamp, 1986). The 1980 revision was the first that referred to a respect for the emerging legal requirements on the subject of informed consent, to be balanced with the traditional, beneficence-guided professional standards.

This indicates that the doctrine of informed consent, as we know it today, did not primarily emerge from ethical reflections inside the medical community. It finally prevailed only at the end of a series of legal decisions which led to a subsequent shift in medical codes and behavior: "The law was beginning to weave its garment of requirements of consent — ultimately of informed consent — but there was stillness in medicine and moral philosophy" (Faden and Beauchamp, 1986).

Law acknowledged patient's right to information as early as the eighteenth century. One of the first explicit formulation in the Anglo-American tradition is found in the 1767 court rule, *Slater v. Baker and Stapleton,* in which it is stated that "it is reasonable that a patient should be told what is about to be done to him, that he may take courage and put himself in such a situation as to enable him to undergo the operation." This right to information, however, was still based on the beneficence assumption that the information was needed for the patient's best interest, and not due to a respect for his or her autonomy. Moreover, the Slater case, in which the plaintiff claimed that Drs. Baker and Stapleton had "ignorantly and unskillfully" rebroken his partially healed leg fracture and injured him, does seem to be classified under what would later be called malpractice law.

In the United States, the legal history of informed consent in its current understanding started with the *Schloendorff v. New York Hospital* (1914) case. The court ruled that a patient's right to "self-determination" was the primary justification for a physician's duty to obtain the patient's explicit consent before performing any medical procedure:

> Every human being of adult years and sound mind has a right to determine what shall be done with his own body; and a surgeon who performs an operation without his patient's consent commits an assault for which he is liable in damages.

The rationale of medical decision making by patients was thus legally rooted in the right of freedom from bodily invasion. A violation of this right was regarded as battery, even if the treatment was performed *lege artis* and even if it was ultimately successful.

Further development of the concept was brought about by another landmark court ruling (*Salgo v. Leland Stanford, Jr. University Board of Trustees, 1957).* Martin Salgo, a 55-year-old patient with leg pain and intermittent limping, suffered from permanent paralysis of his legs after translumbar aortography. Salgo claimed that the physician had not warned him about this possible side effect of the diagnostic procedure. The court acknowledged this claim and stated that any consent given by the patient demanded prior information, by his or her physician, about any risks and benefits and possible alternative procedures to be legally valid:

> A physician violates his duty to his patient and subjects himself to liability if he withholds any facts which are necessary to form the basis of an intelligent consent by the patient to the proposed treatment. . . . Discussing the element of risk a certain amount of discretion must be employed consistent with the full disclosure of facts necessary to an informed consent.

Mere consent was no longer deemed sufficient; any consent must be given after truthful information has been provided on the consequences of consent. The term *informed consent* with its core meaning was born. However, the amount of information to be given (e.g., which particular risks should be disclosed) was left to the physician's discretion.

A later court ruling (*Natanson v. Kline,* 1960) strengthened the emphasis that had been put on patient self-determination by *Salgo.* It stated that

> if [the patient is] of sound mind [he may] expressly prohibit the performance of life-saving surgery, or other medical treatment. A doctor might well believe that an operation or form of treatment is desirable or necessary but the law does not permit him to substitute his own judgment for that of the patient by any form of artifice or deception.

To enable the patient to make a reasonable judgment on the proposed treatment the physician has the duty to disclose and to explain to the patient "in a language as simple as necessary the nature of the ailment, the nature of the proposed treatment, the probability of success or of alternatives, and perhaps the risks of unfortunate results and unforeseen conditions within the body." The *Natanson* court still limited the scope of this informed consent to those disclosures "a reasonable medical practitioner would make under the same or similar circumstances," explicitly formulating what has been called the "professional standard doctrine" of informed consent.

Possibly the most influential court decision in the history of informed consent was the 1972 trial *Canterbury v. Spence,* which dealt with the question whether a 1% risk of paralysis after laminectomy would have needed to be disclosed. The *Canterbury* court ruled that the patient is only able to exercise his or her right of self-determination if he or she "possesses enough information to enable an intelligent choice." The duty of informed consent was made independent of a patient's request for disclosure. Moreover, the court stated, against the professional standard doctrine that prevailed in the *Natanson* ruling, that an expert is not needed to assess the adequateness of the information provided. The information would have to be given in general nontechnical terms. The physician must communicate the inherent hazards of the proposed treatment, the alternatives to this treatment, and the consequences if no treatment is performed. Independently of any established professional standards of information, those risks must be communicated which are "material" to the patients decision — "material risk" or "lay standard" doctrine. The standard of informed consent cannot be conceived as one which physicians may decide on by themselves but rather as a standard set by law along the lines of a lay's informational needs in any particular situation: "The scope of the physician's communications to the patient . . . must be measured by the patient's need, and that need is the information material to the decision. . . . All risks potentially affecting the decision must be unmasked."

A material risk is viewed as a risk that a reasonable person in the patient's position would hold significant in deciding whether or not to forego the proposed therapy. More precisely, the significance of a risk must not only be determined by its incidence but also by the degree of harm which might be involved. Even a small chance of serious harm must be considered significant to the patient's decision and thus subject to the duty of disclosure in informed consent. This change from the professional standard doctrine to the material risk doctrine had important implications. Now, it became the jury, (i.e., nonprofessionals) deliberating on the potential materiality of risks for the patient, not the medical experts, who decided if adequate informed consent had been provided. The *Canterbury* court ruled that "respect for the patient's right of self-determination on particular therapy demands a standard set by law for physicians rather than one which physicians may or may not impose upon themselves." For the first time, the quality of a physician's interaction with his or her patient was, at least insofar as information on and consent to treatment were concerned, open to public revision and possible sanctions by nonprofessionals.

At the same time the *Canterbury* court defined two legitimate exceptions from the "basic social policy" of informed consent: (i) unconscious or incapable patients, when instant treatment is mandatory to avoid harm and (ii) when risk disclosure would entail psychological reactions possibly aggravating the patient's medical condition. Explicitly rejected, however, was the conception that a physician might refrain from giving all pertinent information because divulgence could result in the patient refusing a treatment the physician considers medically necessary.

After *Canterbury,* the legal doctrine of informed consent underwent only minor changes and adjustments (Faden and Beauchamp, 1986). Starting in 1975, however, there was a strong tendency to replace the common-law doctrine of informed consent with statutory laws: Most states enacted informed consent legislation that mostly adhered to the professional standard doctrine, falling behind what had been reached with *Canterbury.*

Until the beginning of the 1970s, informed consent was discussed in a strictly legal sense along court rules and subsequently introduced statutes (Murphy, 1981). The legal reasoning behind this discussion was that since the patient bears the risk of injury in any medical procedure, it must be his or her right to choose whether to accept this risk or not, even if he or she might make, in the opinion of his or her physician, a bad choice. The patient, as every other member of society, explicitly has a right to be left alone. In the medical literature the concept of informed consent started to be discussed and mostly rejected as impossible to fulfill in the 1970s. Kaufmann (1983) noted that "medicine's reaction to informed consent was both critical and defensive. There was a noticeable attempt to debunk the concept as a myth and a fiction." As a reaction to this criticism, the statutory laws on informed consent in many U.S. states tried to contain the potentially infinite duty of disclosure implicit in the notion of informed consent, within the limits of a specific formal procedure: A written and signed consent form was accepted as sufficient evidence of adequate information disclosure in case of a litigation. Only as the discussion of the concept shifted from mere legal considerations to an ethical analysis did the medical community adopt the concept and finally make it a mandatory requirement of any medical procedure. This shift from a legal doctrine to an ethical concept was implicitly contained in the very notion of informed consent. Once the (legal) obligation to communicate with the patient on the proposed intervention was established, this obligation of enhanced communication made the old paternalistic paradigm of medical practice difficult to maintain. One expression of this change, outside the legal sphere, was the *Patient's Bill of Rights* published in 1973 by the American Hospital Association. Concerning the subject of informed consent, it said:

> The patient has the right to obtain from his physician complete current information concerning his diagnosis, treatment, and prognosis in terms the patient can be reasonably expected to understand.
>
> The patient has the right to receive from his physician information necessary to give informed consent prior to the start of any procedure and/or treatment.
>
> The patient has the right to refuse treatment to the extent permitted by law and to be informed of the medical consequences of his action.

Therefore, informed consent was no longer only seen as a *legally* necessary procedure, to avoid liability, but as a *moral* right of patients, that doctors have a moral obligation to respect (Faden and Beauchamp, 1986):

> For perhaps the first time in any influential document of medical ethics, the physician was compelled, by claim of right, to incorporate patients into the decision-making process and to recognize their right to make the final authoritative decision.

Later, the concept of informed consent was addressed by the Judicial Council of the American Medical Association. This AMA statement can be seen as a "testament to the impact of the law of informed consent on medical ethics" (Faden and Beauchamp, 1986). It reads (AMA Judicial Council, 1984) as follows:

> The patient's right of self-decision can be effectively exercised only if the patient possesses enough information to enable an intelligent choice. The patient should make his own determination on treatment. Informed consent is a basic social policy. . . . Social policy does not accept the paternalistic view that the physician may remain silent because divulgence might prompt the patient to forego needed therapy. Rational, informed patients should not be expected to act uniformly, even under similar circumstances, in agreeing to or refusing treatment.

Medicine had finally recognized the value plurality of modern societies and freed itself, at least on a conceptual level, from a utopian understanding of medical benefit that regarded this benefit as being identical for all reasonable beings. This new understanding of patient autonomy as being in the center of any health care decision — the "philosophical turn" in the history of informed consent — was canonized by the President's Commission for the Study of

Ethical Problems in Medicine and Biomedical and Behavioral Research in its report *Making Health Care Decisions: The Ethical and Legal Implications of Informed Consent in the Patient–Practitioner Relationship* (1982). It was stressed that consent to treatment has to be made within the framework of a person's own goals and values, and that in the case of competent persons there can be no substitution to autonomous self-determination. In case of conflict, autonomy had to prevail over beneficence (The President's Commission, 1983):

> The primary goal of health care in general is to maximize each patient's well-being. However, merely acting in a patient's best interest without recognizing the individual as the pivotal decisionmaker would fail to respect each person's interest in self-determination. . . . When the conflicts that arise between a competent patient's self-determination and his or her apparent well-being remain unresolved after adequate deliberation, a competent patient's self-determination is and usually should be given greater weight than other people's views on that individual's well-being. . . . The patient [should have] the final authority to decide.

Informed Consent in Medical Research

Informed consent in medical research with human subjects developed only in the second half of the twentieth century. This may be related to the fact that large-scale scientific research in medicine did not develop before this time. Additionally, there seems to have been "no broad interest in consent to research prior to the Second World War" (Faden and Beauchamp, 1986). The starting point for the introduction of informed consent into the mainstream of medical research was set at the Nuremberg Trial (Fig. 3). Considering the atrocities of "medical" experimentation on prisoners in German concentration camps, the Nuremberg judges wanted to go further than merely condemn these acts as "crimes against humanity." The opening statement of a list of 10 principles, constituting the Nuremberg Code, stipulated what should be permissible in medical experimentation with human subjects (Military Tribunal, 1949):

> The person involved should have legal capacity to give consent; should be so situated as to be able to exercise free power of choice, without the intervention of any element of force, fraud, deceit, duress, overreaching, or other ulterior form of constraint or coercion; and should have sufficient knowledge and comprehension of the elements of the subject matter involved.

Some consider that the Nuremberg Code's first principle, that a subject's voluntary and informed consent be obtained whenever any research is performed on human subjects — be it prisoners, healthy volunteers, or patients in a therapeutic or in a nontherapeutic setting — is in its generality, even today, an "unfilled legacy" (Katz, 1996).

FIGURE 3 Nazi doctors, defendants in a hearing of the Nuremberg Trials, July 19th, 1947 (photo courtesy of AKG).

The principle of consent to medical research was, by 1946, far from being accepted by the medical community in the United States (Annas, 1992). In the years after Nuremberg, there were numerous reported incidents involving research without adequate consent. For example, until the early 1950s it was generally accepted practice to test new polio vaccines without any form of consent on institutionalized, mentally retarded children (Koprowski *et al.,* 1953). Other violations of the rule of informed consent in research followed, the most discussed being the injection of living cancer cells into patients of the Jewish Chronic Disease Hospital in Brooklyn, New York, in 1963 and the Tuskegee syphilis study, initiated in the 1930s, in which black men with a syphilitic infection were involuntarily and without information or consent watched without any treatment (until 1973) to gain insight into the "natural course" of untreated syphilis (Fig. 4). The case of the Jewish Chronic Disease Hospital, however, provoked at least a reaction by the Board of Regents of the State University of New York, which censured the physicians involved in the study: "A physician has no right to withhold from a prospective volunteer any fact which he knows may influence the decision."

There was a growing awareness developing that informed consent must be obtained in medical research to respect a subject's right to self-determination. In 1964, the World Medical Assembly in Helsinki adopted the *Recommendations Guiding Medical Doctors in Biomedical Research Involving Human Subjects* (Declaration of Helsinki). Three of its 12 basic principles explained how informed consent in research should be obtained:

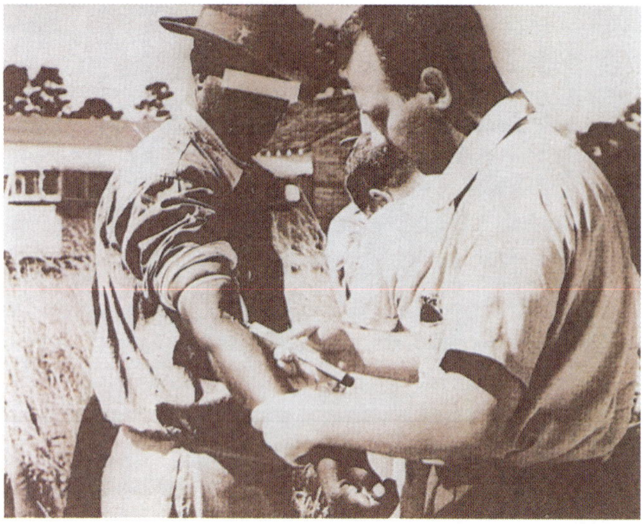

FIGURE 4 Early 1950s. Blood drawn from a subject during the Tuskegee study experiment carried out in Tuskegee, Alabama (photo courtesy of the Centers for Disease Control, Atlanta).

In any research on human beings, each potential subject must be adequately informed of the aims, methods, anticipated benefits, and potential hazards of the study and the discomfort it may entail. He or she should be informed that he or she is at liberty to abstain from participation in the study and that he or she is free to withdraw his or her consent to participation at any time. The doctor should then obtain the subject's freely given informed consent, preferably in writing.

When obtaining informed consent for the research project the doctor should be particularly cautious if the subject is in a dependent relationship to him or her or may consent under duress. In that case the informed consent should be obtained by a doctor who is not engaged in the investigation and who is completely independent of this official relationship.

In case of legal incompetence, informed consent should be obtained form the legal guardian in accordance with national legislation. Where physical or mental incapacity makes it impossible to obtain informed consent, or when the subject is a minor, permission from the responsible relative replaces that of the subject in accordance with national legislation. Whenever the minor child is in fact able to give a consent, the minor's consent must be obtained in addition to the consent of the minor's legal guardian.

These demands of the Declaration of Helsinki, the first medical code on informed consent in research by a medical body, have been largely incorporated into national laws and regulations. However, the Helsinki declaration has been criticized because for therapeutic research (i.e., research that has immediate potential benefits for the patient), informed consent was not required if giving the information was believed not to be "consistent with patient psychology." This exception has been regarded as a loophole, leaving the act of obtaining informed consent up to the physician's discretion.

In the United States, a paramount step in regulating research on human subjects was the development of a explicitly philosophical foundation for informed consent in the report of the U.S. National Commission for the Protection of Human Subjects in Biomedical and Behavioral Research (1978). Consent was now seen not as a mere procedural protection of research subjects against the risk of medical actions but as grounded in the principle of "personal dignity" or of "respect for persons," thus linking the notion of informed consent to autonomy, a fundamental concept of traditional moral philosophy. From this philosophical basis the commission developed several applications for research ethics on institutionalized mentally handicapped persons, children, or prisoners. Much of these recommendations subsequently became federal policy.

Faden and Beauchamp (1986), at the end of their historical analysis on informed consent in clinical medicine and biomedical research, note an important difference between these two fields: "In clinical medicine physicians have been *exhorted* to solicit consent, by appeal either to medical ethics or legal self-interest; in research, scientists have in many cases been *compelled by regulation* to obtain informed consent."

The Ethical Meaning of Informed Consent

Ethical vs Legal Doctrine of Informed Consent

The historical development of the notion of informed consent showed a shift from the merely legal to an ethical understanding of the term. To clarify the meaning and the elements of informed consent, this ethical meaning of informed consent must be separated from the legal one.

In the first, ethical sense, informed consent occurs "if and only if a patient or subject, with substantial understanding and in substantial absence of control by others, intentionally authorizes a professional to do something" (Beauchamp and Childress, 1994). The ethical meaning of informed consent must thus be conceived as the authorization by the patient as an autonomous agent of a medical intervention to be performed on him or her. The aim of informed consent, understood in this ethical sense, would be to enable patients or research subjects to autonomously decide about authorization or refusal of medical or research interventions (Katz, 1984).

In the second, legal sense, informed consent is constituted by adherence to certain social rules that any institution must follow, defining how to obtain consent before any specific therapy or research may be undertaken. To date, the most advanced legal doctrine of informed consent has been developed by common law and statutory enactments in the United States. It must be stressed, however, that this legal doctrine of informed consent was created in the context of an atypical group of patient–physician encounters (Sprung and Winick, 1989), in which patients, contrary to the vast majority of cases, were dissatisfied with or even injured by medical procedures. The following are the legal core requirements of this informed consent doctrine that physicians must fulfill:

1. Information on the medical problem
2. Information on the recommended intervention and its associated risks and benefits
3. Information on risks and benefits if no intervention is performed
4. Information on alternative interventions with their associated risks and benefits

To give informed consent, the patient must be competent to do so; competence is always presumed unless there are sufficient reasons to the contrary. In the United States' legal doctrine, a physician does not have to disclose the risks of which an average patient will already be aware, nor does he or she need to force information on unwilling patients. The legal doctrine of informed consent is thus primarily a doctrine of adequate disclosure of information.

Ethical content and the legal form of informed consent, which have historically developed in close interaction, must be separated not only on a conceptual level but also in medical practice. As Little and Leeder (1996) pointed out, confusion between the ethical meaning of informed consent (a good in itself that promotes patient autonomy) and the legal perspective on informed consent (risk of being sued if no signed informed consent paper can be produced) can lead to overemphasizing the legal understanding of informed consent, neglecting the (ethically) due respect for patient self-determination: "The need to act defensively in the matter of informed consent can readily override any spontaneous or learned sensitivity to the vulnerability of sick people, and their need to understand their illnesses and the potential for healing."

Ethical Elements of Informed Consent

In its canonical understanding, the concept of informed consent consists of five conditions that must be met in a given clinical situation (Meisel and Roth, 1981; Beauchamp and Childress, 1994; Little and Leeder, 1996):

1. Full disclosure of all relevant information by the doctor
2. Competence of the patient to appreciate what the information signifies
3. Understanding of the facts and issues by the patient
4. A voluntary choice by the patient
5. An autonomous authorization by the patient for treatment or entry into a trial

Disclosure. Disclosure of information is a complex process and requires much more than a simple transmission of data from the physician to a patient. Most patients and relatives will remember only 25% of the information they receive in any single interview (Cramond *et al.,* 1967), so care must be taken to reiterate any serious and relevant information when informed consent is to be obtained. The scope of adequate disclosure must contain the diagnosis, the nature of the proposed intervention with its associated risks and benefits, and the consequences of not performing this intervention. It is important to note that a significant historical shift regarding how far disclosure should go could be observed in the past few decades: Whereas in the 1960s most physicians maintained that cancer patients should not be told their diagnosis, at the end of the 1970s virtually all physicians agreed that the diagnosis must be told (Novack *et al.,* 1979). Moreover, as indicated in the history of informed consent, the standard on how to judge whether adequate disclosure of relevant information was given has shifted in the United States from the professional standard to the "reasonable person" standard, whereas the professional standard doctrine remains prevalent in the United Kingdom. Under the reasonable person doctrine court rules in the United States have, for example, required disclosure for a 3% chance of death or paralysis or a 1% chance of loss of hearing. Disclosure was not required for a 1.5% chance of loosing an eye or a 1/800,000 chance of aplastic anemia (Sprung and Winick, 1989).

In Germany, and for some recent cases in the United States, the criterion for adequate disclosure has been what could be called the "subjective standard" doctrine (Goldworth, 1996). Information that must be disclosed, according to this standard, includes that which the person to whom the disclosure is being made would consider as relevant for his or her specific decision. The application of this subjective standard of disclosure, however, seems to require a comprehension of the patient's biographical situation that goes further than what is usually known in the clinical setting (Little and Leeder, 1996). Nevertheless, the subjective standard doctrine of disclosure in informed consent is acknowledged as being particularly able to ensure that patients receive the information they need. Moreover, disclosure according to the subjective standard may also facilitate the subsequent understanding of this information by the patient and generate trust in that the physician truly respects the interests of his or her patient and is not just seeking to comply with a legal obligation (Goldworth, 1996).

Competency. Any patient is presumed to be competent to make decisions for himself or herself unless he or she has been formally judged to be incompetent. A competent patient has the right to refuse any treatment proposed, even if this would entail a serious risk for his or her health. Competency thus is the necessary condition of any informed consent decision.

The U.S. President's Commission report (1983) indicated three abilities necessary for a patient to be competent: (i) the ability to reason and deliberate, (ii) the ability to understand and communicate information, and (iii) the possession of values and goals. Competence accordingly would consist in weighing all relevant information and making an informed decision. This understanding of competence as the ability to make a reasonable decision, however, has its limits because it tends to link competence to decisions that are regarded by others as "rational." A richer conception of competence holds that competence requires that the person can be hold accountable for the decision made. A competent decision thus would be a decision that truly can be said as being one's own decision, however rational or irrational it may appear to an outside observer (Elliot, 1991). Additionally, this focus on accountability, not on merely intellectual ability to decide, stresses the fact that not only intellectual but also emotional disturbances (e.g., major depression) can reduce a patient's competence to consent to treatment (Elliot, 1997).

Competence is not merely present or absent, but must be acknowledged along a continuous scale. Beauchamp and Childress (1994) proposed a ranking of incompetences that should permit an evaluation of the patient's state of competence (Table 1). However, even given this scale, the clinical evaluation of competence is a difficult issue, often requiring structured tests or special expertise to assess the precise level of competence relevant to informed decision making (Fitten *et al.*, 1990).

Understanding. Informed consent is to be assimilated more to an educational and understanding process than to a mere transfer of informational bits. Adequate understanding, however, is only difficult to obtain in clinical reality. Cassileth *et al.* (1980) found that 1 day after disclosure only 60% understood the purpose and the nature of the intervention proposed and only 55% could list the major risks involved. It must also be acknowledged that given the complexity of medical procedures, there may remain difficulties in understanding no matter how carefully and extensively information is disclosed (Ingelfinger, 1972).

Faden and Beauchamp (1986) identified two levels of understanding in the informed consent process. First, the patient must understand that he or she is actually authorizing a medical procedure. This may sound trivial; however, according to a survey of physicians and members of the general public, the majority of both groups did not associate the meaning of informed consent with "giving permission." "authorizing," or "agreeing" to treatment. Informed consent was most of the time understood as a mere explanation of the medical treatment proposed (Harris *et al.*, 1982). Second, the patient must understand all information that is material to his or her decision. The materiality of the information given for a particular patient is determined by his or her personal history and his or her social context. What is material to understand may thus differ from one individual to another — a fact that physicians engaged in an informed consent procedure must be aware of.

There are numerous factors that may compromise a patient's understanding of the medical information given. The most important factors that should be kept in mind are the following (Faden and Beauchamp, 1986):

1. Linguistic difficulties
2. Different cultural background
3. Inferential errors involving risk information: All statistical risk information, alien by its nature to the common information dealt with in ordinary life, is subject to strong "framing effects"; understanding is heavily influenced by the context in which particular information is given, e.g., if outcomes are expressed as probabilities of survival or probabilities of death.
4. Information overload
5. Stress and anxiety experienced by the ill patient in a clinical setting
6. Pain
7. Mental disturbances related to underlying medical conditions

Voluntary Choice. Early in the history of transplant programs, there were concerns whether free and voluntary choice is possible, given the strong pressures inherent in many social situations (e.g., in the families of potential live organ donors) (Louisell, 1966) Moreover, most patients seem to believe, after the disclosure of medical facts that permits choosing between different options, that they do not have any choice but to accept or reject the proposed intervention en bloc (McCullum and Schwartz, 1969). There are three ways to impair true voluntariness: coercion, manipulation, or persuasion. Although coercion and manipulation obviously are always in contradiction with authentic

TABLE I

Evaluation of Competence[a]

1. Inability to express preference or choice
2. Inability to understand one's situation
3. Inability to give a reason
4. Inability to give a rational reason though having some supporting reasons
5. Inability to give a risk/benefit related reason though able to give some supporting reasons
6. Inability to give a reasonable decision, judged by the "reasonable person" standard

[a]From Beauchamp and Childress (1994).

informed consent, persuasion may, if the argument is valid and governed by a desire to benefit the patient, still permit ethically valid informed consent to be obtained. This might be the case, for example, if a doctor persuades a patient with polyposis coli to have a colectomy performed, thus sparing the patient the development of colon carcinoma (Little and Leeder, 1996).

However, the basic principle governing the introduction of informed consent in clinical practice must be remembered: A person's voluntary choice has a moral value in itself and must thus be respected independent of the content of that choice (Goldworth, 1996).

Autonomous Authorization. By evaluating the risks and benefits as disclosed by the physician, patients should come to a rational autonomous decision concerning the specific medical intervention proposed. In reality, however, such a rational decision-making process only seldom seems to be operating. Decisions of patients are more often based on previous experiences and personal feelings (Faden and Beauchamp, 1980). One study on the impact of informed consent in kidney donors showed that none of the donors weighted alternatives in a way that could be assimilated to rational decision making (Fellner and Marchall, 1968). However, even if a decision authorizing (or refusing to authorize) a medical treatment cannot be said to be fully autonomous, the informed consent given can in many cases still be said to be valid because autonomy, as well as competence, comes in degrees.

The five standard criteria of disclosure, competency, understanding, voluntary choice, and autonomous authorization are analytically clear, but they fail to entirely catch the specificity of a real interaction between a patient and his or her doctor in a clinical setting. In reality, the medical recommendations of a doctor (i.e., his or her value judgments on a specific medical option) count more than would a mere disclosure of information (e.g., of the results of empirical studies). Therefore, Beauchamp and Childress (1994) propose a revised model of informed consent, comporting seven main elements within three categories (Table 2).

TABLE 2

Elements of Informed Consent[a]

CATEGORY	ELEMENT OF CONSENT
Preconditions	*Competence* to understand and decide *Voluntariness* in deciding
Informational elements	*Disclosure* of material information *Recommendation* of a plan *Understanding* of the two preceding elements
Consent elements	*Decision* in favor of a plan *Authorization* of the chosen plan

[a]From Beauchamp and Childress (1994).

An even more flexible reconceptualization of informed consent has been suggested by Sprung and Winick (1989), aimed specifically at a reconciliation of the legal concept and the medical reality of informed consent. They offer a sliding-scale approach to informed consent, in which the greater or lesser need of informed consent is determined by a set of clinical variables (Table 3). Sprung and Winick argue that the doctrine of informed consent "does not apply uniformly in every situation, but that its application does and should vary considerably based on the presence or absence of these variables."

Exceptions from Informed Consent

There are four exceptions recognized from the informed consent doctrine. These exceptions are granted either because informed consent is impossible to obtain or because a balance is needed between the value of individual autonomy and what is medically believed to be in the best interest of the patient.

Emergency Exception. If the patient is threatened with serious harm if informed consent must be obtained prior to an intervention, thus delaying medical action, the intervention may be performed without informed consent. In this case the consent of the patient is assumed. For the emergency exception to be valid there must be an immediate, not just a future, danger for the preservation of life or limb.

Incompetency Exception. With incompetent patients there are three cases that can be distinguished. First, a doctor may treat incompetent patients without their permission. Second, a consent from an incompetent patient does not automatically allow the doctor to treat the patient. Third, not treating an incompetent patient who has refused treatment may make the doctor liable. Also, general incompetency, where patients are incapable of making any rational decision, and specific incompetencies, where patients may be able to make only certain types of decisions, must be discriminated. Moreover, competency and incompetency may change with time and often need reassessment.

Therapeutic Privilege Exception. If a physician considers that fully informing a patient on a certain medical condition would seriously harm the patient physiologically or psychologically, such information may be withheld by the doctor. This exception is difficult to define precisely because legal controversy exists regarding whether a professional standard (general medical opinion) or a lay standard (a reasonable person) should be applied. The therapeutic privilege exception does not legally cover a physician not disclosing information because he or she believes this would make the patient refuse the proposed intervention: In this case, the harm is said not to come from the disclosure as such but from the "wrong" decision made by the patient. Such a wrong decision, however, must be accepted if the patient is competent.

In considering the therapeutic privilege exception, the

TABLE 3

Sliding Scale Evaluation on the "Need" of Informed Consent[a]

VARIABLE NO.	GREATER (NEED FOR) INFORMED CONSENT	LESS (NEED FOR) INFORMED CONSENT
1	No emergency	Emergency
2	No waiver by patient	Waiver by patient
3	Competent patient	Incompetent patient
4	Disclosure of information not potentially harmful	Disclosure of information may be harmful
5	Invasive procedure	Noninvasive procedure
6	Experimental procedure	Routine procedure
7	High risk/benefit ratio	Low risk/benefit ratio
8	Patient objects	Patient assents
9	Acceptable medical alternatives available	Acceptable medical alternatives not available
10	Physician approaches patient	Patient approaches physician
11	Known physician	Strange physician
12	Demanding patient	Nondemanding patient

[a] Variables to be considered in determining the required extent of informed consent: Nos. 1–9 may be more important than Nos. 10–12 (From Sprung and Winick, 1989).

doctor must be aware that several studies have not found the disclosure of risk information to be associated with worse compliance or outcome. In most circumstances a lack of information has been found to be more harmful than a tactful and truthful disclosure (President's Commission, 1982).

Waiver Exception. If the patient wishes not to be informed on a specific intervention but consents in general to whatever action his or her physician might consider appropriate, the intervention may proceed without informing the patient. In this case, the physician needs to obtain a written and signed waiver of consent from his or her patient.

If the emergency, incompetency, or therapeutic privilege exception applies, it is legally recommended to obtain surrogate consent from spouse, parent, relative, or a legally appointed guardian. Hospital ethics committees or courts may also act as a surrogate forum for obtaining consent. The surrogate's rationale for consenting to a patient's treatment should be based on a "substituted judgment," a "best interest," or a "reasonable person" standard.

The Philosophical Background of Informed Consent

Informed consent has been described as being one of the most important consequences drawn from the concept of patient autonomy. However, the autonomy concept is far from clear and unambiguous. It is therefore necessary to further analyze this concept to gain a better understanding of this core component of informed consent.

Faden and Beauchamp (1986) introduced a first distinction that can be made between autonomous persons and autonomous actions. The so-called autonomous person, the "competent chooser," is a too demanding and misleading ideal in the context of informed consent. For a satisfactory understanding of the nature of informed consent the emphasis must be on the autonomy of specific actions (of refusal or consent). Therefore, even a person who would be said to be nonautonomous (e.g., a child) can, in certain circumstances, make an autonomous decision; an autonomous agent can also fail to do so, for example, because he or she is deceived by a misrepresentation of important data.

Additionally, it must be recognized that there are degrees of autonomy. This scalability of autonomy can be explained by a model that identifies three basic conditions of autonomous actions (Faden and Beauchamp, 1986). An action may be called autonomous if the subject acts intentionally, with understanding, and is not controlled by the influence of others. Because both understanding and noncontrol can be present to a greater or lesser extent, the resulting autonomous actions can subsequently be said to be more or less autonomous. This pragmatic understanding of autonomy seems to be better adapted to seize the nature of informed consent in medicine than the more "idealistic" and global conceptions of autonomy as "freedom" or "authenticity." Specifically, the concept of degrees in autonomy resolves a common problem with informed consent in clinical medicine, in which patients are said to be unable to fully understand the nature of the procedure to which they are asked to consent. Such a complete understanding that would result in a fully autonomous decision is an ideal too strong to be reached. Informed consent presupposes nothing but a sufficient degree of autonomy — what Beauchamp and Faden (1986) called a *substantially* autonomous consent. Obviously, the threshold of what should still count as "substantially autonomous" is difficult to establish. However, it is only by this imprecision that the concept of autonomy is made applicable to clinical reality. The degree of autonomy (i.e., the degree of understanding and the degree of freedom from other's control) is not absolutely given; there can be different degrees in any specific social and medical context.

Based on this model of autonomy, Faden and Beaucham (1986) put forward a concise definition of informed consent, understood as a specific instance of an autonomous action. Informed consent in this view is given if a patient with substantial understanding and in substantial absence of control by others intentionally authorizes a professional to perform a medical procedure.

Criticism: The "Myth" of Informed Consent

"Mention informed consent to a clinician and my experience is that you are more likely to get a groan as well as talk about the myth of informed consent" (Wear, 1993). Informed consent is often regarded as an "unattainable goal that has been transformed into a legal requirement" (Sprung and Winick, 1989). Most of the critics of informed consent see a profound incompatibility between the principle of patient autonomy at the core of informed consent and the principle of beneficence at the heart of the physicians' care for their patients. Here, I review the most frequent objections.

One of the principal objections of physicians against giving the patient all information necessary is that on this basis, most of them would decline any intervention, fearing the risks of which they have been informed. This common assumption, however, is empirically wrong, as investigations on this subject have shown since the beginning years of the informed consent debate. As early as 1971, a study of informed consent for angiographic procedures showed that only 4 of 232 patients refused to have angiograms after being carefully informed of all relevant risks, and most of the patients desired to have this information (Alfidi, 1971). Denney and coworkers (1976) found that in 40 female patients recommended for hysterectomy and given extensive information on the nature, risks, benefits, and alternatives of the intervention in the form of a booklet, preoperative anxiety levels were not different from those of a matched control group not given written information. Moreover, postoperative anxiety was less pronounced in the informed consent group and only a small percentage of these patients stated that they would have preferred not to have had the information.

Another objection against providing informed consent is the conception that the patient has no means to adequately understand the medical information given to him or her. This, however, can at least partly be overcome by adapting the disclosure of information to the patient's capacity of understanding. Moreover, according to the concept of substantial autonomy, "full" understanding is not required; an understanding of the material facts is sufficient. Finally, in some cases, physicians may wrongly conclude that a patient did not understand medical information just because he or she made a choice different from the one they would have made in the patient's situation.

A third line of criticism argues that patients "don't want to know." Many patients, so the argument goes, would simply prefer to put themselves in the hands of their doctors and be dependent on their judgment without any further information, even on material risks. Although this category of patients certainly exists, empirical data seem to indicate that these patients are not in the majority. Moreover, the waiver exception permits the handling of any wishes to not be informed in the informed consent doctrine.

Informed consent, when given in its legal sense, is said to destroy the fiduciary character of the relationship between the physician and the patient. If a doctor is forced to discuss possible alternative treatment options with his or her patient (e.g., radical mastectomy, lumpectomy, or radiation therapy for breast cancer) the patient could think that the physician does not have enough professional competence to choose between these different alternative treatment options by virtue of his or her medical knowledge. In this view, "the integrity of his or her physician" is seen as the best guarantee for the rights of the patients (Laforet, 1976).

A related criticism argues that the legal duty of the doctor, as put forward by the *Canterbury* case, to inform the patient on any aspects a "reasonable person" would consider as "material" for his or her decision, independent of any professional standards of disclosure defined by the medical community, must be seen as an assault on the autonomy of the medical profession, unprecedented in any other licensed profession.

The disclosure of likely outcomes, of statistic probabilities of survival, does not help the patient to make a decision regarding his or her particular situation: Stochastic information, as given in the informed consent process, often obfuscates the individual situation. Moreover, the sheer amount of information to be given for truly informed consent is impossible to transmit in any real-world clinical situation. This criticism is based on the misunderstanding that patients would want to "micromanage" their care (Meisel and Kuczewski, 1996). On the contrary, patients should not compete with a physician's understanding of every medical detail, but they are, under the informed consent doctrine, entitled to information on the principal goals of treatment, the number of therapeutic options, and the basic risks associated with them — information that may well be given in a reasonable amount of time.

Finally, the concept of informed consent has been denounced as relying too heavily on the patient's autonomy, which by the very fact of being ill is often compromised in the medical situation. This criticism seems to presuppose a (too strong) definition of autonomy as full authenticity. A scaleable concept of autonomy, as introduced earlier, is not problematic with regard to coping with different degrees of autonomous actions.

Even if some of the criticism of the doctrine of informed consent has its merits, and actually led to a refinement of the concept, it must be acknowledged that the practice of informed consent has produced a major shift toward a richer patient–physician relationship by installing reasonable discourse into the silent world of the doctor and the patient (Katz, 1984). The criticism, however, rightly emphasizes that the process of informed consent must be seen less as the rigid application of legally valid and complete consent forms and more as the hermeneutic "skill in understanding what the patient is seeking" (Little and Leeder, 1996). If informed consent is conceived as a process of shared decision making, in which physician and patient have a mutual exchange of facts and values, the patient's autonomy and the physician's concern for the patient's well-being can finally be reconciled, thus resolving the presupposed antagonism between the principle of patient autonomy and the physician's strive to beneficence (Meisel and Kuczewski, 1996).

Special Issues

Psychiatric Patients

It seems obvious that informed consent in psychiatric practice faces difficulties not encountered in other fields of medicine. However, there has been a tendency to extend the right of consenting to treatment to psychiatric patients since the 1980s (Gendreau, 1997). Increasingly, the idea that a person diagnosed as mentally ill would be incapable of any autonomous form of consent and that in this case a beneficent or paternalistic intervention even against the person's wishes would be justified is being abandoned. This tendency is reflected by the United Nations "Principles for the Protection of Persons with Mental Illness and for the Improvement of Mental Health Care," adopted in 1991, which specifically addresses the issue of informed consent in psychiatry in its Principle 11. Although this principle clearly states that "no treatment shall be given to a patient without his or her informed consent," it provides exceptions in cases of involuntary admission, incapacity, unreasonable refusal, urgent necessity, prevention of harm to the patient or to other persons, major medical or surgical procedures, and psychosurgery. The exception concerning "unreasonable refusal" has been especially criticized as being a return to "a vague 'best-interests' standard, ceding all decision-making power concerning medication to professionals" (Rosenthal and Rubenstein, 1993). If this interpretation is correct, this would be regrettable because a consistent practice of obtaining informed consent from psychiatric patients may confer an additional therapeutic benefit of its own by enhancing the self-esteem and reducing the helplessness of the patient. It must be stressed, however, that treatment refusal in a manifest psychotic state is often not in any way autonomous or consistent with prior beliefs but influenced by the underlying pathological process (Schwartz *et al.*, 1988). Therefore, there is a danger of undue respect for a nonautonomous refusal in patients who do not yet meet the criteria of involuntary treatment but would nevertheless be spared suffering if treatment could be promptly instituted.

As a possible solution to the dilemma of informed consent in psychiatric patients, who at a certain point of time lack the capacity to autonomously consent to or to refuse medical treatment, the concept of "advance consent" (Ulysses contract) has been proposed (Dresser, 1984). Under this concept, stabilized psychiatric patients, whose symptoms are in remission, consent in advance to specific treatment options (e.g., hospitalization or electroconvulsive therapy) should they become necessary in the case of a relapse.

Special provisions must be found for safeguarding informed consent procedures in psychiatric research. Empirical investigations have shown that in this field, investigators frequently neglect addressing issues related to the subject's capacity to consent (Applebaum and Roth, 1983). A study on decisional capacities found that 52% of schizophrenic and 24% of depressed subjects versus 12% of seriously medically ill subjects demonstrated substantial impairments in an inpatient setting (Grisso and Appelbaum, 1995).

To address the issue of informed consent in psychiatric research, a distinction must be made between research that has the potential to generate long-term or immediate benefit for the individual patient (*therapeutic research*) and research that has no such potential and mainly aims at enhancing medical knowledge of the disease process (*nontherapeutic research*). According to the Helsinki declaration, therapeutic research with patients who cannot give informed consent is permitted if consent by a legal guardian can be obtained. The U.S. National Commission for the Protection of Human Subjects in Biomedical and Behavioral Research (1978) came to a similar conclusion, permitting therapeutic research in incompetent patients in case of proxy consent, and it even authorized some low-risk forms of nontherapeutic research if relevant to the subject's disease. Nontherapeutic research with patients incapable of giving informed consent is a major subject of controversy in Europe, especially regarding the European Bioethics Convention. There seems to be a basic consensus, however, that this type of research may only be allowed under carefully defined conditions. The following criteria, which must be fulfilled before such research can be considered, are currently under discussion (Helmchen, 1996):

• The research must be related to the medical problem of the patient.

• There must be no alternative to the proposed research.

• The knowledge to be expected for diagnosis or treatment of the patient's illness must be considerable.

• Risks and inconveniences must be minimal.

• Refusal behavior of the patient must be accepted.

• The legal guardian or proxy must consent to the research.

• The study design has to be approved by an institutional review board.

Analogous to the Ulysses contract in the clinical treatment of psychiatric patients, or to advance directives in the general medical setting, advance consent for nontherapeutic research on incompetent patients has been advocated, for example, in the field of dementia research (Sachs, 1994).

Elderly Patients

Older patients are at risk for impairments in decision-making capacity due to cognitive impairments or to manifest psychiatric illness. A review of 12 studies addressing the decision-making capacity and thus the feasibility of informed consent in the elderly (Christensen *et al.,* 1995) reached the following conclusions:

• Comprehension, not only information recall, was found impaired in all eight studies that addressed the question.

• Decision-making capacity was reduced in nursing home residents and acutely medically ill hospitalized patients. Age was found to be related to a reduction in decision-making capacity in the majority of the studies.

• Consent forms in particular were less readily understood with increasing age.

A broader review of the literature on informed consent in elderly patients (Sugarman *et al.,* 1998) obtained similar results. To cope with these difficulties in obtaining informed consent, the authors recommend novel formats (e.g., video and storybooks) and procedures (use of health educators and multiple disclosure sessions).

Children and Adolescents

Children and adolescents cannot give informed consent in the classical sense. However, progress in medical treatment leaves families and their children with ever more complex treatment decisions, for which the right choice is not easy to make (Fig. 5). McCabe (1996) stressed that "we need to support minors' involvement in decision making, particularly for treatment decisions where the clarity of the 'right choice' fades, where treatment preferences are based upon personal values and 'quality of life' issues." McCabe (1996) stresses the following advantages of integrating children and adolescents into something equivalent to an informed consent process:

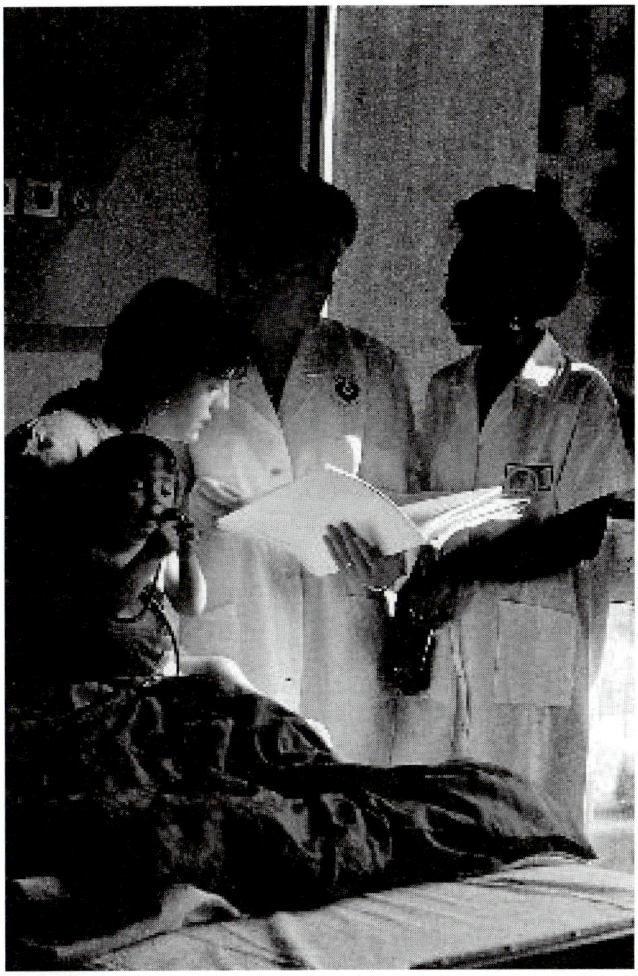

FIGURE 5 Medical consultation in the pediatric department of a hospital (photo courtesy of Grison-Rapho/Neri).

1. Respect for the ethical principle of self-determination
2. Fostering of open communication, thus enhancing satisfaction with medical care
3. Better cooperation with treatment
4. Enhanced sense of control
5. Respect for children's capacities, providing opportunities for further development

Therefore, with increasing age and increasing capacity to comprehend the medical situation, some form of "mature minor's" consent must be considered. Studies have found that by the age of 9 children may have sufficient capacity to at least participate in the decision-making process (Leikin, 1993). In a study specifically designed to address and compare the competency to consent in four age groups (9, 14, 18, and 21 years), it was shown that the 14-year-old group demonstrated the same level of competency as the two adult groups (Weithorn and Campbell, 1982). Therefore, McCabe (1996) advocates different levels of involvement for children in medical decision making ranging from information about illness and treatment over shared decision making with parents or guardians to fully autonomous decision making. However, care must be taken not to have children or adolescents making decisions beyond their capacity or, worse, serving a tie-breaker role in cases of dissent on medical treatment between the other parties involved.

In case of research on children and adolescents, because of their higher vulnerability special safeguards have been established in many countries. In the United States, for example, both child assent and parent or guardian permission are necessary for any entry into clinical studies. Additionally, the research may only be performed if it has minimal risk or, if the risk is greater than minimal, "presents the prospect of direct benefit to the subjects great enough to justify the risk." In case the research is of greater than minimal risk, but still has a minor risk and does not provide direct benefits to the child or adolescent concerned, there must at least be a likelihood that the research provides for a better understanding of the subject's disorder, and the intervention must be commensurate with the patient's general medical situation.

However, these strict requirements tend to limit necessary research in this age group so that there has been a proposal for a system of age-graded informed consent to moderate risk protocols that would consider informed consent to research acceptable in children aged 12–14 years and older (Arnold *et al.,* 1995).

Emergency Medicine Research

In emergency medicine research there is a balance that must be found between "protecting patients and putting excessive obstacles in the way of needed and legitimate research activities" (Ellenberg, 1996). In patients that are temporarily incompetent due to the emergency situation, it has been commonly accepted practice to obtain informed consent by a relative before inclusion into any clinical study. The patient representative necessary for this proxy consent, however, is often not available as quickly as necessary. In this context, the concept of *deferred consent* was developed. Patients would be included in the study immediately, and consent was obtained as soon as possible from a relative or the patient. In the United States, this concept of deferred consent was rejected in 1993 by reviews of both the Food and Drug Administration (FDA) and the Department of Health and Human Services (DHHS). New regulations demanded that research without consent in emergency patients must be "necessary to preserve life" (FDA) or impart only "minimal risk" to the patient (DHHS). Both demands have been very difficult to meet so that research on emergency patients has been virtually shut down at least in the United States. A revision to these FDA and DHHS regulations that permits for waiving of the informed consent requirements in specific circumstances was made in 1996. A waiver of informed consent, permitting research on emergency patients, may be granted if an institutional review board can document that the following conditions are met (FDA, 1996):

1. The human subjects are in a life-threatening situation for which available treatments are unproved or unsatisfactory.
2. The participation in the research holds out the prospect of direct benefit to the subjects.
3. The research protocol defines a therapeutic window within which the investigator will commit to contact a legally authorized representative to obtain informed consent.
4. Additional protections of the rights and welfare of the research subjects will be provided. This includes consultation with representatives of the communities in which the research will be conducted and from which the subjects will be drawn.

Genetic Testing

Genetic screening is increasingly able to detect inherited mutations that put the concerned subjects at a substantially higher risk for future development of disease, such as different forms of cancer. Problems arise because once the increased risk for cancer is detected, there is still little that can be done to prevent the disease. Moreover, genetic testing for cancer is increasingly performed outside of research protocols in clinical practice. In a consensus statement the task force of the Cancer Genetics Studies Consortium (Geller *et al.,* 1997) came to the conclusion that genetic information differs in at least three ways from normal clinical information, warranting special consideration of informed consent in this context: (i) It affects an entire family; (ii) it is laden with symbolic meaning, with genes being regarded as the unique determinators of future health and behavior; and (iii) its probabilistic nature is (even for physicians) difficult to handle — healthy individuals become reclassified as being potentially ill.

The consensus paper of the task force identified the following special requirements for informed consent in genetic testing:

1. The purpose of the test must be made clear, being performed either for research or for the identification of personal risk.
2. The possible ambiguous nature of the results must be stressed in the context of multiple genes with numerous mutations possibly being associated with a particular type of cancer. Testing positive for a particular gene does not mean that cancer will certainly develop; it may only "replace one type of uncertainty with another" (Geller *et al.,* 1997).
3. The participants of the tests must be warned of the possible psychological responses to a positive test, such as anxiety, depression, or anger.
4. Patient's must be informed that, if tested positively,

they may have difficulties in obtaining health, life, or disability insurance or may have to pay higher premiums.

5. Alternatives to genetic testing should be discussed (e.g., closer clinical follow-up).

Expanding Physicians' Obligations

Informed consent has been about providing the patient with all material information necessary to make a knowing and intelligent decision regarding medical treatment. Recent developments in the United States common law and statutory regulations indicate that physicians' obligations concerning what information they must disclose to their patients are being expanded (LeBlang, 1995). A Maryland court ruling (*Faya v. Almaraz*) considered that a HIV-infected surgeon was under the doctrine of informed consent and obligated to disclose his HIV infection to his patients, even though the risk of a transmission of HIV infection was considered by experts — given proper barrier techniques — as being only a theoretical possibility. Other similar court rulings indicate that a physician's physical condition (e.g., substance dependence) may well become part of what is material for a patient's decision and thus needs to be disclosed when obtaining informed consent. Going further, some commentators suggest that a physician might equally be under an obligation by the informed consent doctrine to disclose his or her personal experience or his or her record of success with a certain medical procedure (e.g., laparoscopic cholecystectomy).

Physicians have expanded obligations of information to any research or economic interests that might influence the physician's treatment recommendations. In the landmark case of *Moore v. Regents of the University of California* (1990), the court ruled that a physician who performed a splenectomy on a patient with hairy-cell leukemia to slow disease progression would have been under an obligation to disclose to the patient that he was also planning on establishing a cell line from the patient's T lymphocytes for research and economic purposes. The reason for this obligation to disclose any additional research or commercial interest linked to the procedure to be performed was seen to consist in the possibility of conflicting loyalties (LeBlang, 1995):

> Given the fiduciary nature of the physician–patient relationship, disclosure of such a conflict is essential to ensure that the patient is capable of making an intelligent informed decision regarding not only whether to go forward with a proposed treatment course but even whether the physician–patient relationship should continue.

Analogous disclosure obligations are increasingly stated to exist in situations in which physicians refer patients to facilities in which they have an investment interest. LeBlang (1995), in his review on the expansion of physicians' obligations of disclosure based on the fiduciary nature of the physician–patient relationship, came to the conclusion that in the future these disclosure standards based on the doctrine of informed consent may give heightened potential for legal accountability.

References Cited

Alfidi, R. (1971). Informed consent: A study of patient reaction. *J. Am. Med. Assoc.* **216**(8), 1325–1329.

American Hospitals Association (1973). *Statement on a patient's bill of rights.* Hospitals **47,** 41.

American Medical Association Judicial Council (1984). *Curr. Opin.* **1984,** 29–30.

Amundsen, D. W. (1981). Casuistry and professional obligations: The regulations of physicians by the court of conscience in the late Middle Ages. *Trans. Stud. College Phys. Philadelphia* **3,** 22–39, 93–112.

Annas, G. J. (1992). The Nuremberg Code in U.S. courts: Ethics versus expediency. In *The Nazi Doctors and the Nuremberg Code* (G. J. Annas and M. A. Grodin, Eds.), pp. 201–222. New York. Oxford University Press, Oxford.

Applebaum, P. S., and Roth, L. H. (1983). The structure of informed consent in psychiatric research. *Behav. Sci. Law* **1,** 9–19.

Arnold, L. E., Stoff, D. M., Cook, E., Cohen, D. J., Kruesi, J., Wright, C., Hattab, J., Graham, P., Zametkin, A., Castellanos, F. X., McMahon, W., and Leckman, J. F. (1995). Ethical issues in biological psychiatric research with children and adolescents. *J. Am. Acad. Child Adolescence Psychiatr.* **34**(7), 929–939.

Beauchamp, T. L., and Childress, J. F. (1994). *Principles of Biomedical Ethics.* Oxford Univ. Press, Oxford.

Cassileth, B. R., Zupkis, R. V., Sutton-Smith, K., and March, V. (1980). Informed consent — Why are its goals so imperfectly realized? *N. Engl. J. Med.* **302**(16), 896–900.

Christensen, K., Haroun, A., Schneiderman, L. J., and Jeste, D. V. (1995). Decision-making capacity for informed consent in the older population. *Bull. Am. Acad. Psychiatr. Law* **23**(3), 353–365.

Cramond, W. A., Court, H. J., Higgins, B. A.., Knight, P. R., and Lawrence, J. R. (1967). Psychological screening of potential donors in a renal homotransplantation program. *Br. J. Psychiatr.* **113,** 1213–1221.

Denney, M. K., Williams, D., and Penn, R. (1976). Community medicine – informed consent – emotional responses of patients. *Postgraduate Med.* **60,** 205–209.

Dresser, R. (1984). Bound to treatment: The Ulysses contract. *Hastings Center Report* **14,** 13–16.

Ellenberg, S. S. (1996). Informed consent: Protection or obstacle? Some emerging issues. *Controlled Clin. Trials* **18,** 628–636.

Elliot, J. (1991). Competence as accountability. *J. Clin. Ethics* **2,** 167–171.

Elliot, J. (1997). Caring about risks. Are severely depressed patients competent to consent to research? *Arch. Gen. Psychiatr.* **54,** 113–116.

FADEN, B. R., and BEAUCHAMP, T. L. (1980). Decisionmaking and informed consent: A study of the impact of disclosed information. *Social Indicators Res.* **7,** 313.

FADEN, R., and BEAUCHAMP, T. (1986). *A History and Theory of Informed Consent.* Oxford Univ. Press, Oxford.

FELLNER, C. H., and MARCHALL, J. R. (1968). Twelve kidney donors. *J. Am. Med. Assoc.* **206,** 2703.

FINKELSTEIN, D., SMITH, M. K., and FADEN, R. (1993). Informed consent and medical ethics. *Arch. Opthalmol.* **111,** 324–326.

FITTEN, L. J., LUSKY, R., and HAMANN, C. (1990). Assessing treatment decision-making capacity in elderly nursing home residents. *J. Am. Geriatric Soc.* **38,** 1097–1104.

FOOD AND DRUG ADMINISTRATION, DEPARTMENT OF HEALTH AND HUMAN SERVICES (1996). Protection of human subjects: Informed consent, No. 45 CFR 46. Food and Drug Administration, Department of Health and Human Services, Washington, DC.

GELLER, G., BOTKIN, J. R., GREEN, M. J., PRESS, N., BIESECKER, B. B., WILFOND, B., GRANA, G., DALY, M. B., SCHNEIDER, K., and KAHN, M. J. E. (1997). Genetic testing for susceptibility to adult-onset cancer. The process and content of informed consent. *J. Am. Med. Assoc.* **277**(18), 1467–1474.

GENDREAU, C. (1997). The rights of psychiatric patients in the light of the principles announced by the United Nations. *Int. J. Law Psychiatr.* **20**(2), 259–279.

GOLDWORTH, A. (1996). Informed consent revisited. *Cambridge Q. Healthcare Ethics* **5,** 214–220.

GREGORY, J. (1772). *Lectures on the Duties and Qualifications of a Physician. A New Edition, Corrected and Enlarged.* Strahan & Cadell, London.

GRISSO, T., and APPELBAUM, P. S. (1995). The MacArthur treatment competence Study. III. Abilities of patients to consent to psychiatric and medical treatments. *Law Hum. Behav.* **19,** 149–174.

HARRIS, L., *et al.* (1982). Views of informed consent and decisionmaking: Parallel surveys of physicians and the public. In *Making Health Care Decisions,* Vol. 2, pp. 17–316. President's Commission, Washington, DC.

HELMCHEN, H. (1996). Summary and proposals. Common European standards and differences, problems and recommendations. In *Informed Consent in Psychiatry. European Perspectives of Ethics, Law and Clinical Practice* (H. Koch, H. Helmchen, and St. Reiter-Theil, Eds.), pp. 380–411.

HOOKER, W. (1849). *Physician and Patient; or a Practical View of the Mutual Duties, Relations and Interests of the Medical Profession and the Community.* Baker & Scribner, New York.

INGELFINGER, F. J. (1972). Informed (but uneducated) consent. *N. Engl. J. Med.* **287,** 465–466.

KATZ, J. (1984). *The Silent World of Doctor and Patient.* New York.

KATZ, J. (1996). The Nuremberg Code and the Nuremberg Trial. A reappraisal. *J. Am. Med. Assoc.* **276**(20), 1662–1666.

KAUFMANN, C. L. (1983). Informed consent and patient decision making: Two decades of research. *Soc. Sci. Med.* **17**(21), 1657–1664.

KOPROWSKI, H., JERVIS, G. A., NORTON, T. W., and NELSON, D. J. (1953). Further studies on oral administration of living polio virus to human subjects. *Proc. Soc. Exp. Biol. Med.* **82,** 277–280.

LAFORET, E. G. (1976). The fiction of informed consent. *J. Am. Med. Assoc.* **235,** 1579–1585.

LEBLANG, T. R. (1995). Informed consent and disclosure in the physician–patient relationship: Expanding obligations for physicians in the United States. *Med. Law* **14,** 429–444.

LEIKIN, S. (1993). Minors' assent, consent, or dissent to medical research. *IRB Rev. Hum. Subject Res.* **15,** 1–7.

LITTLE, J. M., and LEEDER, S. (1996). Logic, hermeneutics, and informed consent. *Eur. J. Surg.* **162**(1), 3–10.

LOUISELL, D. W. (1996). Transplantation: Existing legal constraints. In *Ethics in Medical Progress* (G. E. W. Wolstenhome and M. O'Connor, Eds.), pp. 78–103. Little, Brown, Boston.

MCCABE, M. A. (1996). Involving children and adolescents in medical decision making: Developmental and clinical considerations. *J. Pediatr. Psychiatr.* **21**(4), 505–516.

MCCULLUM, A. T., and SCHWARTZ, A. H. (1969). Pediatric research hospitalization: Its meaning to parents. *Pediatr. Res.* **3,** 199.

MEISEL, A. M., and KUCZEWSKI, M. (1996). Legal and ethical myths about informed consent. *Arch. Int. Med.* **156,** 2521–2526.

MEISEL, A. M., and ROTH, L. (1981). What we do and do not know about informed consent. *J. Am. Med. Assoc.* **246,** 2473–2477.

MILITARY TRIBUNAL (1949). *Trials of war criminals before the Nuremberg military tribunals under Control Council Law 10,* 2 Vols. U.S. Government Printing Office, Washington, DC.

MURPHY, W. J. (1981). Development of the concept of informed consent. *Dent. Clin. North Am.* **26**(2), 287–305.

NOVACK, D. H., PLUMER, R., SMITH, R. L., *et al.* (1979). Changes in physicians' study. Attitudes towards telling the cancer patient. *J. Am. Med. Assoc.* **241**(9), 897–900.

PERCIVAL, TH. (1803). *Medical Ethics; or a Code of Institutes and Precepts Adapted to the Professional Conduct of Physicians and Surgeons.* Russell, Manchester, UK.

PERNICK, M. S. (1982). The patient's role in medical decisionmaking: A social history of informed consent in medical therapy. In *President's Commission for the Study of Ethical Problems in Medicine and Biomedical and Behavioral Research (1982). Making Health Care Decisions,* Vol. 3. U.S. Government Printing Office, Washington, DC.

PRESIDENT'S COMMISSION FOR THE STUDY OF ETHICAL PROBLEMS IN MEDICINE AND BIOMEDICAL AND BEHAVIORAL RESEARCH (1982). *Making Health Care Decisions: The Ethical and Legal Implications of Informed Consent in the Patient–Practitioner Relationship.* U.S. Government Printing Office, Washington, DC.

PRESIDENT'S COMMISSION FOR THE STUDY OF ETHICAL PROBLEMS IN MEDICINE AND BIOMEDICAL AND BEHAVIORAL RESEARCH (1983). *Deciding to Forego Life-Sustaining Treatment.* U.S. Government Printing Office, Washington, DC.

ROSENTHAL, E., and RUBENSTEIN, L. S. (1993). International human rights advocacy under the principles for the protection of persons with mental illness. *Int. J. Law Psychiatr.* **16**(3/4), 257–300.

RUSH, B. (1811). *Sixteen Introductory Lectures.* Bradford & Innskeep, Philadelphia.

SACHS, G. A. (1994). Advance consent for dementia research. *Alzheimer Dis. Associated Disorders* **8**(4), 19–27.

SCHWARTZ, H. I., VINGIAN, W., and PEREZ, C. B. (1988). Autonomy and the right to refuse treatment: Patients' attitudes after involuntary medication. *Hospital Community Psychiatr.* **39,** 1049–1054.

SPRUNG, C. L., and WINICK, B. J. (1989). Informed consent in theory and practice: Legal and medical perspectives on the informed consent doctrine and a proposed reconceptualization. *Crit. Care Med.* **17**(12), 1346–1354.

STRÄTLING, M. (1998). *Die Begründung der neuzeitlichen Medizinethik in Praxis, Lehre und Forschung. John Gregory (1724–1773) und seine. Lectures on the Duties and Qualifications of a Physician.* Lang, Frankfurt.

SUGARMAN, J., MCCRORY, D. C., and HUBAL, R. C. (1998). Getting meaningful informed consent from older adults: A structured literature review of empirical research. *JAGS* **46,** 517–524.

WEAR, S. (1993). *Informed Consent. Patient Autonomy and Physician Beneficence within Clinical Medicine.* Kluwer, Dordrecht.

WEITHORN, L. A., and CAMPBELL, S. B. (1982). The competency of children and adolescents to make informed treatment decisions. *Child Dev.* **53,** 1589–1599.

WELBORN, M. C. (1938). The long tradition: A study in fourteenth-century medical deontology. In *Medieval and Historiographical Essays in Honor of James Wesfall Thompson* (J. Lea and E. N. Anderson, Eds.), pp. 344–357. Univ. of Chicago Press, Chicago.

General References

KOCH, H., HELMCHEN, H., and REITER-THEIL, ST. (Eds.) (1996). *Informed Consent in Psychiatry. European Perspectives of Ethics, Law and Clinical Practice.* Nomos Verlagsgesellschaft, Baden.

REICH, W. T. (Ed.) (1991). *Encyclopedia of Bioethics,* pp. 1233–1270. Macmillan/Prentice Hall, New York/London.

ROZOVSKY, F. A. (1995). *Consent to Treatment. A Practical Guide,* 2nd ed. Little, Brown, Boston.

BARTHA M. KNOPPERS
LORI LUTHER

Center for Public Law Research
University of Montreal
Montreal, Canada

A Comparative Overview of Medically Assisted Procreation

Taking part in a comparative discussion on assisted fertility means to take into account how the same ethical-legal principles may be translated into different cultural models. This does not deny the validity of these principles, but instead confirms their universality. The positions assumed on an international, regional, or domestic level, related to both the clinical and the experimental aspects of reproductive technologies, reveal similarities and differences in several key points. These involve the actions to be included in these technologies, the status, access, donation, and preservation, in addition to the production and possible uses of human genetic material. Even though, in definitive terms, what counts most is the political courage to intervene on this question, the range of options—which go from prohibition to legislative protection to professional self-governance and a free market approach—demonstrates how difficult it is for a Government to intervene in a matter that is so private and personal.

Introduction

Although many countries have adopted legislation to regulate medically assisted procreation, others, more reticent to do so, have produced recommendations or guidelines in this regard. Despite social, economic, cultural, and political differences between countries, what consistently emerges is an increasing preoccupation with, and respect for, human genetic material and for the long-term implications of its use for our concept of human life, procreation, and family. This has a direct effect on the clinical and research aspects of medically assisted procreation.

This article focuses on official national positions. A preliminary section briefly considers medically assisted procreation from general international and regional perspectives, including both governmental and nongovernmental positions. We then compare and contrast national positions (Australia, Austria, Canada, Denmark, France, Germany, The Netherlands, Norway, Spain, Switzerland, the United States, and the United Kingdom) on the regulation of reproductive technologies with regard to the clinical aspects. We specifically consider issues of definition, access, donation, donation after death, storage, and anonymity. Finally, we discuss research particularly with regard to available zygotes and embryos, embryos created solely for research, time limitations, prohibited practices, and the status of preimplantation diagnosis.

International and Regional Positions

Given rapid advances and discoveries made with respect to assisted human reproduction, international approaches are necessarily set within the framework of safeguarding human dignity. Human rights, which derive from human dignity, are irretrievably linked to respect for the human body and its constituent parts, including, but of course not limited to, reproductive tissues and cells.

Although not directly related to medically assisted procreation, respect for human genetic material is evident in the preamble of UNESCO's proposed "Universal Declaration on the Human Genome and the Protection of Human Rights," which notes that human genome research must fully respect human dignity and individual rights. Accordingly, in relation to scientific genetic research the declaration provides that

> [b]enefits from advances in biology and genetics concerning the human genome, shall be made available to all, with due regard to the dignity and human rights of each individual. . . . [Research] shall seek to offer relief from suffering and improve the health of individuals and humankind as a whole. (art. 12 (a), (b) *in fine*)

It also maintains that "[n]o research or its applications . . . should prevail over the respect for human rights, fundamental freedoms, and human dignity" (art. 10).

Among international nongovernmental organizations, the Ethics Committee of the Human Genome Organization (HUGO) also addresses the issue of research under the Human Genome Project and Human Genome Diversity Project in its "Statement on the Principled Conduct of Genetic Research" (1996). In its background principles, the statement refers to "the acceptance and upholding of human dignity and freedom." Although both the UNESCO declaration and the HUGO statement aim to protect human rights with regard to human genome research and are not concerned with reproductive technologies per se, these instruments are illustrative of the respect that human genetic material has recently garnered. As such, these guidelines also have indirect effect on the regulation of medically assisted reproduction.

Several other statements of international nongovernmental organizations bear mention and are more specifically related to the issue of assisted reproduction and research. Beginning with the most recent, the International Planned Parenthood Federation's "Charter on Sexual and Reproductive Rights" (1996) seeks to provide a framework for reproductive rights and access to reproductive health services within the context of human rights provisions already recognized in national laws, international human rights documents, and other consensus documents. It maintains that, within the established "right to decide whether and when to have children" (art. 8), comprehensive reproductive health care services, which inter alia include prevention and appropriate treatment of infertility, should be made accessible, affordable, acceptable, and convenient to all users. Moreover, "the right to health care and health protection" (art. 9) and "the right to the benefits of scientific progress" (art. 10) should also include within legal limits the right to assist couples and individuals in achieving their reproductive goals as well as to provide access to all methods of fertility regulation, treatment for infertility, and reproductive technologies that are safe and acceptable.

The International Law Association in its 1988 "Resolutions on Reproductive Technologies and the Protection of the Human Person" proposed to extend the general provision of the "inherent dignity in all human beings" since it provided that "irrespective of any legal qualification," the human embryo is entitled to the same "inherent dignity as a member of the human family" (art. 2). Accordingly, its provision limiting research or manipulation of human genetic material to therapeutic purposes subject to the approval of an ethics committee (on pain of penal sanctions) applied equally to the embryo since nontherapeutic experimentation on the human embryo is prohibited.

The World Medical Association's 1987 declaration on in vitro fertilization and embryo transfer also refers to human dignity inasmuch as it notes that the principles of the Helsinki Declaration must be respected. The declaration particularly comments on aspects of research and donation and provides that physicians must consider their moral, legal, and ethical implications as well as the moral and ethical principles held by patients (pp. 2–3). Thus, a physician should not violate his or her personally held moral principles. In addition, the physician must refrain from any act which would conflict with legal or ethical norms or the moral principles of patients. This means that physicians have a "greater responsibility to communicate fully with the patients who will participate" in research and that the informed consent of those patients must meet the requirements of law as well as the special level of professional responsibility dictated by ethical standards. Moreover, physicians are required to respect the health of both the mother and the embryo.

Regionally, on November 19, 1996, the Council of Europe adopted the "Convention for the Protection of Human Rights and Dignity of the Human Being with Regard to the Application of Biology and Medicine: Convention on Human Rights and Biomedicine" that, upon signature by member states is binding on those states. The convention provides that it "shall protect the dignity and identity of human beings and guarantee everyone, without discrimination, respect for their integrity and other rights and fundamental freedoms with regard to the application of biology and medicine" (art. 1). "The interests and welfare of the human being shall prevail over the sole interest of society or science" (art. 2). It specifically refers to the "use of techniques of medically assisted procreation" and in regard with choosing a future child's sex, that it "shall not be allowed . . . except where serious hereditary sex related disease is to be avoided" (art. 14). Furthermore, even though the convention does not take a specific position on the legitimacy of research with embryos, it nonetheless provides that where embryo research *in vitro* is permissible, such research must "ensure the adequate protection of the embryo" and that "[t]he creation of human embryos for research purposes is prohibited" (art. 18). An additional protocol relating to the latter topic is actually in progress.

Interestingly, an information document developed by the Council of Europe in 1989 titled "Principles in the Field of Human Artificial Reproduction" which considered in far greater detail all issues regarding assisted reproduction and related permissible research endeavors, was never adopted by the Council of Ministers. Although originally intended to be a recommendation, it failed because two member states disapproved of assisted conception techniques. These principles explicitly approved of, inter alia, assisted reproduction for heterosexual couples, the storage of one's own gametes for future personal use, and exceptional donation of gametes and embryos.

The European Parliament, in its 1989 resolution on artificial insemination *in vivo* and *in vitro,* again along the lines of human dignity, recognized the value of life and especially the right of the human being to protection. The resolution further expressed concern at the potential waste of embryos during *in vitro* fertilization techniques. Accordingly, the resolution stipulated that techniques for assisted reproduction must not be considered abusive to the embryo, that only the number of ova that can be implanted may be retrieved, and that the selection of *in vitro* embryos should not be permitted. Moreover, the resolution considered that medically assisted procreation should only be undertaken for therapeutic purposes — that is, to overcome infertility.

Although these international and regional instruments provide a normative backdrop for medically assisted reproductive techniques and related research issues, an in-depth analysis at the national level will more completely reveal points of consensus and, as will be shown, issues of divergence on both the clinical aspects and the research issues.

National Positions: Clinical Aspects

Definitions

The disparity of legislation and recommendations in the field of reproductive technologies begins at the definitional level. Although the terminology used, such as zygote, pre-embryo, embryo, and fetus, is linked to varying stages of development from the moment of fertilization, different definitions are often dependent on the goals sought.

Two countries have attempted to define human life from implantation onwards. In Australia, the 1983 report of the National Health and Medical Research Council stated in its guidelines that the terms "fetus" and "fetal tissue" include, respectively, all or part of what is called the embryo, fetus, or "neonate" from implantation to complete gestation. The Infertility Treatment Act of 1995 of the state of Victoria defines both zygote and embryo. The zygote covers the developmental stage from the beginning of penetration of the egg by the sperm (Fig. 1), up to but not including syngamy (defined as the union of two gametes in fertilization to form a zygote, set at 24 h after fertilization), whereas the embryo covers development from syngamy onwards (§3).

The 1993 report of the Canadian Royal Commission on New Reproductive Technologies explicitly described the biological process. The stages of development begin with the zygote, the fertilized egg until approximately 14 days after fertilization, and then proceed to an embryo until the eighth week after fertilization, which finally develops into a fetus from the ninth week of fertilization until birth (p. 148). Likewise, Bill C-47 in Canada (the Human Reproductive and Genetic Technologies Act of 1996) similarly defined the aforementioned three stages of development but referred to each as a human organism (§2). Following the call for elections in 1997, this bill died and must be reintroduced.

Other countries under consideration have only provided definitions for the term embryo. In Germany, the Embryo Protection Law (1990) defined the embryo as "the

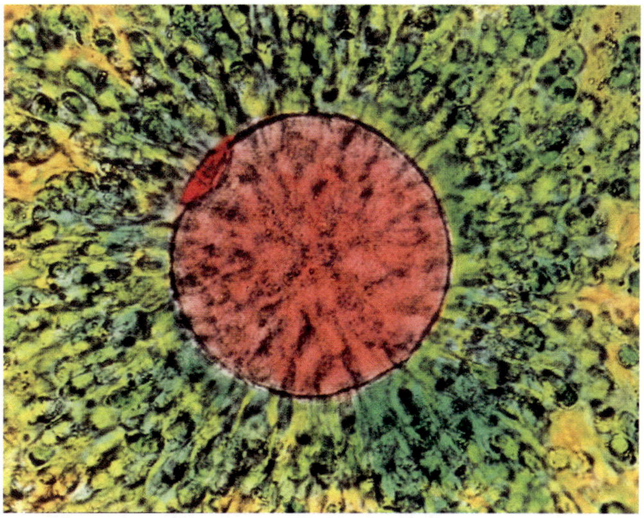

FIGURE 1 Scanning electron microphotograph of a human egg surrounded by spermatozoa competing for fecundation (photo courtesy of Kjeldsen/SPL/Neri).

human egg cell, fertilized and capable of development, from the time of fusion with the nuclei, as well as each totipotent cell removed from an embryo that is capable, in the presence of other necessary conditions, of dividing and developing into an individual" (§8.1). "Capable of development" refers to the "first 24 hours after fusion of the nuclei during which time it is established that the embryo is capable of developing beyond the single-cell stage" (§8(2)) (Fig. 2). In Switzerland, both the 1996 Bill Relating to the Federal Law Concerning Medically Assisted Procreation and the Medicoethical Guidelines on Medically Assisted Procreation (1990) of the Swiss Academy defined the embryo as the product of fertilization until the end of organogenesis; the guidelines refer to the earliest stages preceding implantation (No. 1.5). Finally, the United Kingdom stipulated in the Human Fertilisation and Embryology Act of 1990 that an embryo means "a live human embryo where fertilization is complete." The act, which only applies to embryos *ex utero,* further specified that fertilization is not complete until the appearance of a two-cell zygote (§1(1)).

Finally, in the United States, the National Institutes of Health (NIH) Human Embryo Research Panel (1994) defined the preimplantation human embryo as a fertilized ovum *in vitro* that has never been transferred or implanted in a uterus. Although not considered to have the same moral status as infants and children, the panel was of the opinion that preimplantation human embryos warranted serious moral consideration as a developing form of human life. The 1994 guidelines of the Ethics Committee of the American Fertility Society, however, preferred to use the term preembryo to mean the stage that ends 14 days after fertilization. The preembryo was considered to be in a unique position since medically assisted procreation isolates it and makes it accessible to observation or intervention in a manner distinct from more advanced embryos and fetuses. Although not a person, the committee considered that the preembryo should be accorded special status since "it is a genetically unique living human entity that might become a person if implanted and exposed to a receptive uterus" (p. 32S).

FIGURE 2 Microphotograph of a human egg 24 h after fertilization. The two haploid (maternal and paternal) pronuclei (in orange), the zona pellucida (in yellow), a few spermatozoa (gray, bottom right), and the polar globules (in red) are visible (photo courtesy of Motta/SPL/Neri).

In summary, terminological problems clearly exist since the term embryo often refers to the zygote stage, whereas the term preembryo could be viewed by some as diminishing the humanity of the developing entity from conception onwards. Nevertheless, this enumeration of definitions demonstrates national divergence as well as terminological "manipulation" for moral and political purposes and further sets the stage for disparate views on access to reproductive technologies.

Access

In addition to financial and counseling issues, access to assisted reproduction raises the question of whether every individual has a right to "found a family" regardless of, for example, age, sexual orientation, or medical indication. Access to reproductive technologies therefore refers to the civil status of the applicants, their ages, and medical indications.

Regarding civil status, some legislation and recommendations only allow heterosexual couples in marital or stable marital-like relationships access to assisted conception. Often, the best interests of the child are explicitly taken into consideration, with psychological, social, and familial factors of paramount concern. Under this approach, medical indications, as well as the risk of transmission of serious or grave genetic disease, might also permit access but only once the previous prerequisite conditions are satisfied.

The Human Fertilisation and Embryology Act of the United Kingdom provides that regard shall be taken of the welfare of the child "including the need of that child for a father" (§13(5)) without further legal requirements such as marriage. The Code of Practice of the Human Fertilisation and Embryology Authority is more specific. It provides explicit information about factors to be considered in regard to the welfare of the child (§3.12–3.32). Nevertheless, this consideration is not as evident in the legislation of France (art. L.-152-2 Code of Public Health), Australia (state of Victoria) (§8 (1)), Austria (§3), and Switzerland, which are more direct in their approach and require that those seeking reproductive assistance be couples who are married or who are living together as though they were married without further qualification concerning the interests of an eventual child.

Couples who are not married or are of the same sex may be faced with restricted or more stringent access to reproductive technologies. Thus, in Austria, for example, under the federal Reproductive Medicine Law of 1992 (one of the most restrictive reproductive technology laws in the world), although married (i.e., heterosexual) couples must obtain a reproductive "license" to use donor gametes *in vivo* (i.e., not frozen) (§7), the procedure for nonmarried common-law couples is even more restrained. They must also obtain a license indicating the permissible duration of treatment and additionally submit to an advisory procedure before a court or notary public prior to each treatment as well as pay a fee (§8). Moreover, although the status of nonmarried couples living together may arguably allow same-sex couples to avail themselves of reproductive technologies, France's Law No. 94-654 of July 30, 1994, on the donation and use of elements and products of the human body, medically assisted procreation, and prenatal diagnosis, explicitly provides that the requisite couple be a man and a woman (art. L.152-2 Code of Public Health).

In contrast, the Royal Commission on New Reproductive Technologies in Canada would not limit access to reproductive technologies to married or cohabiting couples but rather recommended that such services be made widely available irrespective of sexual orientation, civil status, or economic situation (recs. 94(f), 99(d), 121, 145). Similarly, Spain's Law No. 35/1988 on assisted reproduction procedures permits access to all women who are of the age of majority (18) and fully competent (art. 6.1). Arguably, this would allow same-sex female couples and single women to proceed with assisted reproductive procedures. In this vein, The Netherlands, in a 1997 report on *in vitro* fertilization, mentioned in the summary of its recommendations that "[e]xcluding all lesbians and single women where IVF is medically indicated, may be unlawful" (p. 14).

Regarding age, the Canadian Royal Commission imposed a menopausal age prerequisite and recommended that women availing themselves of reproductive technologies be required to be of childbearing age (rec. 141, 162, 172). Likewise, the Ethics Committee of the American Fertility Society expressed medical and social concerns with regard to pregnancy beyond the ordinary childbearing age. However, in special circumstances, in which the medical status of the woman would permit childbearing and extensive counseling and evaluation of the social situation had occurred, the committee was of the opinion that egg donation not be excluded on the basis of age alone. Generally, however, the statistical likelihood of age associated with mental or physical disabilities led the committee to believe that medically assisted reproductive technologies for women and men beyond their forties was contraindicated. In this regard, the committee believed that any age limits on reproduction should apply equally to males and females (p. 49S). In The Netherlands, the 1997 report on *in vitro* fertilization (IVF) mentioned that women older than 45 should only receive IVF in the context of research (p.99). The 1997 report also restricts participation in such research to women aged 48 or less at the start of the procedure. For women using frozen donated ova or donated embryos, a maximum age of 44 at the start of treatment is prescribed. Finally, Spain's Law No. 35/1988 on assisted reproduction procedures provides only that a woman must be at least 18 years of age to undergo assisted reproductive procedures (art. 2(1) (b), 6(1)). The advanced age of a woman ostensibly does not bar access to alternate reproductive techniques. The law stipulates that a woman "shall be informed of the possible risks for her descendants and during her pregnancy resulting from her unsuitable age" (art. 6(2)). However, a provision of general principle indicates that assisted reproductive procedures may only be employed where there is no serious health hazard to the mother or any resultant descendants (art. 2(1)). This latter restriction may therefore temper the seeming absence of an age limit.

Although the majority of countries agree that there should be recourse to medically assisted procreation for medical indications, such as fallopian tube blockage, impaired sperm production, or complete gonadal failure, other countries view infertility as a larger health problem in accordance with the WHO definition of health. Thus, whereas Canada, the United States, and Spain favor a more social rather than functional definition of infertility, France is explicit in requiring that the infertility be of a pathological nature and medically diagnosed (art. L.152-2-2 Code of Public Health). It is interesting to note the position of The Netherlands, which provides in its 1997 report on IVF that, like the United Kingdom, "when deciding whether to help a couple, a doctor should consider the (future) interests of any child that might be born following IVF" (p. 17).

Finally, several countries provide that reproductive assistance should also be made available in order to avoid the transmission of a serious hereditary disease to a fetus. In this regard, France refers to avoiding the transmission of a particularly grave disease (art. L.152-2 Code of Public Health), whereas the Infertility Treatment Act of 1995 of the state of Victoria in Australia more liberally mentions risk of transmission of disease or genetic anomaly (§8.30 b) and Switzerland's Bill Relating to a Federal Law Concerning Medically Assisted Procreation provides that in addition to being grave, the hereditary disease must also be incurable (art. 5.1 a.b.). Again, in The Netherlands, although no specific mention is made concerning genetic indications, interests of any child that might be born following IVF must be considered by a doctor deciding whether to help a couple (p. 17). Conversely, in keeping with its restrictive nature, Austria's Reproductive Medicine Law of 1992 explicitly excludes grave genetic diseases as an appropriate indication for access to reproductive technologies (art. 2). The issue of access gives rise to the question of which countries permit sperm, egg, zygote, or embryo donation.

Donation

Reproductive technologies may also involve the donation of unfertilized or fertilized gametes. Although sperm donation does not engender the same concern as egg donation, the psychosocial implications for the donor and recipient couple as well as the best interests of the child who results are nevertheless considered in both instances.

Donor sperm would allow a heterosexual couple (in which the male has impaired sperm quality or quantity), a same-sex female couple, or a single woman (where legislatively permitted) the possibility of having a partially genetically related child. Although third-party sperm donation for purposes of artificial insemination is now generally thought of as ethically unproblematic (sexual orientation issues aside) and typically routine (subject to consent), limitations have been imposed by some countries. Austria's Reproductive Medicine Law of 1992, as mentioned previously, only exceptionally allows access to third-party sperm where the sperm of a husband or partner is incapable of reproduction (§3). Even genetic indications are insufficient to permit access to such a treatment. Likewise, restrictions have also been set by Switzerland's medicoethical guidelines on medically assisted reproduction and its bill relating to the federal law concerning medically assisted reproduction inasmuch as mentioned earlier only married couples can benefit from third-party sperm donation (No. 3.2, art. 3(3)). In contrast, the Royal Commission in Canada permits access to third-party sperm based on medical criteria and social factors. Likewise, countries such as Australia (the state of Victoria), Britain, The Netherlands, Norway, Spain, and the United States provide no limitations on the access to third-party sperm except, of course, appropriate consent.

The donation of unfertilized or fertilized eggs, for IVF procedures, was also found to be subject to varied restriction in the countries under study. In Austria (§3), Germany (§1(1) 1 & 2), and Norway (§2-10), the donation of both fertilized and unfertilized eggs is forbidden, with the latter country only permitting IVF with the couple's own egg and sperm cells. The donation of fertilized eggs is prohibited under the Danish Act on a Scientific Ethical Committee System and the Handling of Biomedical Research Projects (§14(5). The majority of the Danish council, however, in its "Ethics Report on Assisted Reproduction" (1995), exemplified its general opposition to IVF by recommending that the donation of both unfertilized and fertilized eggs be forbidden. The majority of the council was opposed to fractioning motherhood, and indeed parenthood, into genetic and social segments.

France, on the contrary, by virtue of its Law No. 94-654 of July 30 1994 on the donation and use of elements and products of the human body, medically assisted procreation, and prenatal diagnosis permits unfertilized eggs to be donated provided the donors are married or cohabiting and already have had a child (now incorporated into art. L.673-2 Code of Public Health). Donation is allowed only when assisted reproduction has not been successful in the requesting couple using their own gametes (art. 152.6 Code of Public Health). Indeed, "[s]pouses or partners who, in order to procreate, have recourse to medical assistance requiring the intervention of a third-party donor, must give their prior consent to a judge or notary" (art. L.152-10 Code of Public Health). Although in France as a general rule the donation of fertilized eggs is not permitted, in exceptional cases a couple can give their consent for a fertilized egg to be received by another couple (art. L.152-3, 152-5 Code of Public Health). Donation requires the written consent of the couple who are the genetic originators of the fertilized egg and the approval of the judicial authority (art. L.152-5 Code of Public Health). The approving authority (typically a judge) must ensure that the recipient couple will safeguard the child's familial, educational, and psychological interests.

More liberally, the Ethics Committee of the American Fertility Society in the United States considered the donation of unfertilized and fertilized eggs "ethically acceptable" and would further permit designated donation "since there seems to be no compelling reason" to exclude such donation "from among parties known to each, such as relatives or friends" (p. 49s, 51s). In The Netherlands, the 1997 report on IVF also authorizes the donation of unfertilized and fertilized eggs. Moreover, it mentions that "[t]he alternative of IVF with 'double' donor gametes, may be chosen if the couple want to use sperm or ova (or both) from a family member. This method at least ensures a certain genetic relationship with one or both of the prospective parents" (p. 89). Although the Royal Commission on New Reproductive Technologies in Canada would have allowed the donation of unfertilized and fertilized eggs, such donation would be restricted to women already undergoing surgical procedures or egg retrieval as part of their own treatment (rec. 174). Egg retrieval solely for the purpose of donation would be impermissible, as would designated donation (rec. 172). Bill C-47, it should be noted, was more expansive in its donation approach. Egg donation for the purpose of insemination of a woman would be permitted as long as the donor who produced it consented to this purpose. Finally, the Infertility Treatment Act of 1995 of the state of Victoria in Australia (§12) and the 1997 report on IVF of The Netherlands (pp. 89–91) provides similar consent requirements and thus maintain the possibility of donation. Although consent is always essential, it is of particular importance in those countries that permit donation after death.

Donation after Death

Transferring stored zygotes/embryos after the death of a partner raises questions not only with regard to succession and inheritance rights but also with respect to the best interests of the child, particularly in relation to the repercussions of coming into existence from a father or mother known to be dead. The legislation or recommendations of Australia, Austria, Canada, Denmark, France, Germany, and Switzerland all forbid insemination and/or donation after the death of the genitor. In Australia, the state of Victoria's Infertility Treatment Act of 1995 is explicit in its prohibition and stipulates that a person must not (i) inseminate a woman with sperm from a man known to be dead, (ii) transfer to a woman a gamete from a person known to be dead, (iii) transfer to a woman a zygote or embryo formed from a gamete from a person known to be dead, (iv) form a zygote with sperm from a man known to be dead, or (v) form a zygote if the woman who produced the oocyte used to form the zygote is known to be dead (§43).

The Royal Commission on New Reproductive Technologies in Canada recommended that zygotes only be stored for a couple's own use up to the death of either partner (rec. 180; see also rec. 171), despite its recommendation that single women and same-sex (female) partners be allowed access to donor insemination. Bill C-47 does not refer to donation after death. It did, however, provide that a sperm donor must provide consent to the intended use of the sperm, thus perhaps permitting postmortem insemination with consent. Germany's Embryo Protection Act of 1990 provides a maximum 3-year penalty or fine for anyone who "knowingly fertilizes artificially an egg cell with the sperm of a man after his death" (§4(1)3). Finally, Switzerland's constitution, federal bill, and medicoethical guidelines also prohibit postmortem insemination. The guidelines suggest that both implantation and insemination after the death of one the genitors would be disallowed and that, in the case of embryo conservation, it be destroyed "at the latest, on the death of one of the members of the couple."

Both the United States and the United Kingdom allow for donation and/or insemination after the death of the donor. The Ethics Committee of the American Fertility Society found the insemination of a deceased husband's sperm permissible upon his wife's request. The committee was of the opinion that the existence of a spousal relationship prior to both the preservation of sperm and the husband's death was sufficient to make such an arrangement acceptable (p. 42S). In the United Kingdom, the Human Fertilisation and Embryology Act of 1990 allows the individual to decide what should be done with his gametes in the event of death, provided written consent is obtained prior to death. Thus, if one provides consent to storage and specifies that the stored gametes may be used following death, such direction would not be contrary to the law (schedule 3, ¶2.2b). Unlike the previously mentioned countries which may be concerned with the administration of estates, succession, and inheritance rights, the UK act provides that where an embryo is created with a man's sperm following his death, he is not posthumously considered the father of the resulting child regardless of whether the woman is using her deceased husband's sperm or that of an unknown donor (§28 (6) (b)). The resulting legal status of the child may possibly serve to ultimately discourage postmortem fertilization. In any event, donation after death requires, at the least, conservation of sperm and hence raises concern with regard to permissible duration of storage.

Storage

Although sperm cells and fertilized eggs can be frozen and subsequently thawed in order to enhance assisted reproduction techniques, maintaining the viability of cryopreserved unfertilized eggs is often more difficult. Legal and ethical issues regarding rights of ownership and decision-making authority, as well as practical issues related to the regulation and to the consequences of long-term storage, add to the complexity of storing sperm, surplus eggs, and embryos. In any event, the ultimate goal of the storage of gametes and/or embryos is intended to increase the chances of pregnancy.

The freezing or storage of sperm for the purpose of donor insemination programs, or where medical treatment threatens to destroy sperm cells (e.g., cases of vasectomy, orchiectomy, or chemotherapy), is generally not forbidden. Moreover, sperm cryopreservation is often not governed by any time limitation (Fig. 3). Norway's 1994 Act Relating to the Application of Biotechnology in Medicine allows

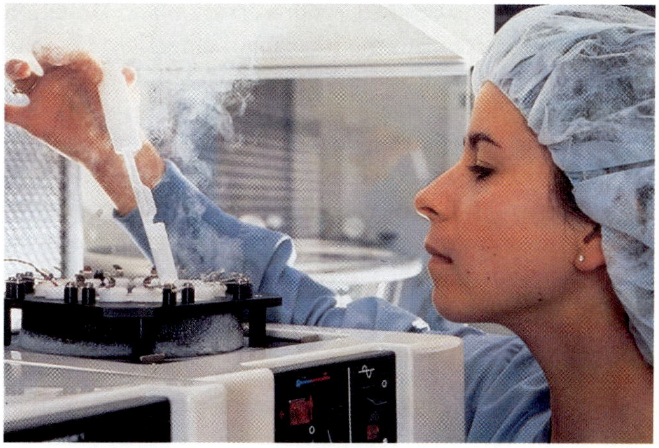

FIGURE 3 Freezing of oocytes and sperm for *in vitro* fertilization (photo courtesy of Morgan/Science Source/Neri).

sperm to be frozen or stored by establishments that are authorized to carry out assisted reproduction techniques (§2.5). The act does not set a time restriction for preservation nor stipulate when, if ever, the sperm must be destroyed. Likewise, the Ethics Committee of the American Fertility Society did not recommend a limitation period for sperm storage. The committee was of the opinion that sperm cryopreservation was ethically and medically acceptable within the confines of appropriate protective procedures and full disclosure regarding long-term storage (p. 42S). Accordingly, men seeking sperm preservation need to be informed of the reduced survival characteristics of frozen–thawed sperm in terms of fertilizing capacity. The Royal Commission in Canada allows sperm to be frozen in centers with a license for such purposes.

Despite the aforementioned absence of restricted storage and/or preservation periods for sperm, some countries do impose time limitations. Spain's Law No. 35/1988 on assisted reproduction imposes a maximum 5-year conservation period on semen (art. 11(1)), whereas the Infertility Treatment Act of 1995 of the state of Victoria in Australia (§51(1)) provides for a 10-year period, after which time the sperm must be destroyed. The Human Fertilisation and Embryology Act of 1990 (§14(3)) authorizes a storage period not exceeding 10 years. The act also foresees the possibility of a shorter period or, in circumstances as may be specified in the regulations, a longer period (§14(5)). Indeed, it also provides an exception for patients struck by cancer. In this case, sperm will be frozen until the patient reaches 55 years of age or, at his request, for a shorter period. In fact, any time limitation provided by law provides the possibility for patients to request a shorter storage period.

Stored and cryopreserved unfertilized eggs may be required to be returned to the woman from whom they originated, donated for research purposes, or donated for the purpose of inducing pregnancy in another woman. Denmark's 1994 Order on the Cryopreservation and Donation of Human Eggs permits the freezing of unfertilized human eggs for a period of no longer than 1 year, after which time they must be destroyed (§1(1), 3(1)). Before being thawed, the order requires the consent of the genitor/donor (§2(2)). Other countries permit the storage of unfertilized eggs under the general rubric of gametes. Accordingly, as noted for sperm preservation, Australia allows a 10-year preservation period, whereas the United Kingdom provides a statutory storage period not exceeding 5 years as the license may specify. As already mentioned, it also provides the possibility to substitute this storage period for a shorter one or a longer one in circumstances as specified in the regulations. Moreover, as for sperm, it foresees the possibility for patients to request a shorter storage period or a longer one for donors struck by cancer, until the age of 55.

In contrast, Norway's 1994 Act Relating to the Application of Biotechnology in Medicine statutorily bans the cryopreservation (storage) of unfertilized eggs (§2.12), as does Spain's Law No. 35/1988 "until such time as there are sufficient guarantees as to the viability of the ovules after their freezing" (art. 11(2)). It seems clear from the wording of Spain's restriction that these prohibitions were probably due to the fact that at the time the legislation was enacted it was impossible to freeze unfertilized eggs without damaging the chromosomal matter. In this same vein, with a view to the ultimate possibility of ova storage, the Ethics Committee of the American Fertility Society viewed the "cryopreservation of oocytes as a preliminary area of investigation" and recommended that the "fertilization and/or transfer of cryopreserved oocytes be considered as a clinical experiment after the preliminary research has been completed" (p. 55S).

Fertilized eggs (zygotes/embryos) are generally stored with a view to subsequent implantation either in the woman from whom they have originated or exceptionally to a requesting couple. In addition, they may be intended for research purposes. Although, as outlined later, many countries impose or recommend a 5-year freezing period, alternative 1- and 3-year limitation periods enacted by other countries denote what each country believes is a reasonable time limit for storage given current knowledge.

The Royal Commission on New Reproductive Technologies in Canada recommended that zygotes only be stored for a 5-year period from the date of freezing or up to the death of either partner or gamete donor (rec. 171, 180). The zygote must therefore be destroyed before the end of the 5-year period if the donor dies, irrespective of any wishes the donor may have expressed in writing. Although freezing is contemplated in Bill C-47 inasmuch as its definition of zygote refers to the exclusion of "any time spent in a frozen state," it does not provide any indication as to permissible duration. France's Law No. 94-654 allows fertilized eggs to be cryopreserved for a period of 5 years. Both members of the couple must make a request for preservation in writing and must indicate every year whether they still wish to parent (art. L.152.3 Code of Public Health). The state of Victoria's Infertility Treatment Act of 1995 permits a prolongation of the general 5-year preservation period provided written consent is given for the maximum 5-year period and the authority approves of the extension (§52(4)). As such, if written consent to freezing is given for a period of less than 5 years, preservation cannot be extended beyond the time period originally provided for. Spain's 5-year cryopreservation period under Law No. 35/1988 refers to surplus preembryos resulting from IVF procedures that have not been transferred to the uterus (art. 11(3)). The law further provides that preembryos not claimed by the donors after 2 years of cryopreservation are at the disposal of the authorized banks in question (art. 11(4)). Finally, the United Kingdom's Human Fertilization and Embryology Act of 1990, under its provision for storage licenses, provides a statutory storage period for embryos not to exceed 5 years unless the embryo was created from stored gametes, in which case it could be stored for a maximum of 15 years (art. 14(4), Schedule 2, ¶2 (3)). Recently, a more flexible approach was suggested by the Dutch Health Council (1997): "[T]he normal period is still 5 years under the new regulations, but it may be extended to 10 years, if requested by the couple. Further 2-year extensions are possible after this in special cases" (p. 59).

Denmark's 1994 order stipulates that fertilized eggs may not be frozen for more than 1 year, after which they must be destroyed. However, the majority of the Danish council would forbid the cryopreservation of fertilized eggs in part because of their general opposition to IVF and in part because of the dehumanizing aspect of freezing and thawing (p. 61). Finally, Norway's 1994 act permits the freezing of embryos for a maximum period of 3 years (§2.12). The accompanying explanatory notes justify cryopreservation insofar as it will save the woman the strain of hormonal stimulation and intervention should the first therapeutic attempt fail. In addition, with a freezing period of 3 years, a second pregnancy will also be feasible without the woman having to undergo a pregnancy on the heels of the first birth.

Countries which prohibit the deliberate creation of surnumerary embryos theoretically render the issue of preservation of fertilized eggs (embryos) a moot point since only as many eggs as may be fertilized may be removed during one cycle. Although embryo conservation and storage is for all intent and purposes prohibited, many of these restrictive laws nevertheless provide exceptions, particularly in the case of illness.

Switzerland's federal constitution ["only the number of human ova that may be immediately implanted" (art. 24(2)(c))], bill relating to the federal law concerning medically assisted procreation ["only the number of fertilized eggs as are necessary to induce pregnancy during one cycle of the woman, this number not to exceed three" (art. 17(1))], and medicoethical guidelines ("in general, embryos may be kept alive only during the treatment in progress" (No. 10.3)] all forbid the conservation of surplus embryos. However, both the bill and the guidelines indicate that when a couple cannot proceed with an embryo transfer, preservation may be exceptionally allowed as long as their sole aim is ultimate implantation. No specific time limitation for preservation is outlined, although the guidelines indicate that preservation would be limited for "an appropriate period, and terminated, at the latest, on the death of one of the members of the couple" (No. 10.3). Germany's Embryo Protection Law of 1990 which, although it prohibits and penalizes any person who "attempts within one treatment cycle to transfer more than three embryos into a woman" (art. 1(1)3), also allows physicians to carry out "the preservation of a human embryo" (art. 9(3)), which suggests at least limited access to embryo conservation but not the deliberate creation of surnumerary embryos. The Netherlands' 1997 report on IVF foresees the possibility to create and preserve surnumerary embryos for the interest of parents (p. 54). Finally, in the United States, the Ethics Committee of the American Fertility Society foresees the cryopreservation of surplus embryos and believes that such preservation has a "relatively successful application." Such cultural diversity continues to be evident on the issue of anonymity.

Anonymity

The issue of anonymity with regard to third-party donors in assisted reproductive technologies concerns the debate between the paradigm of total anonymity (save perhaps in cases of medical need) and the emergence of a new "right to know one's origins" approach. Although the legislation or recommendations of the majority of countries under consideration require that anonymity be maintained and the identity of the donor safeguarded, serious medical need may override this principle.

In Canada, in the case of assisted reproduction, the Royal Commission on New Reproductive Technologies recommended that all nominative information about the donor remain strictly confidential (rec. 88, 94, 163). Moreover, like the province of Quebec, the commission endorsed the concept of nonidentifying information disclosure whereby standard nonidentifying, genetic, social, and medical information about a donor (kept by the National Reproductive Technologies Commission) would be available at any time to the parties concerned, including the child. Only exceptionally and in extraordinary circumstances of medical need as determined by a court of law would nominative information be divulged (rec. 94 (i)). Such instances would indeed be rare as the commission points out that even in the case of an inheritable medical disorder, it would usually not be necessary to release named information. The federal Bill C-47 did not raise this issue since civil rights are a provincial concern but, in the explanatory comments accompanying the bill, it foresaw the possibility of doing away with anonymity as well as creating a national registry of donors.

Similarly, Spain's Law No. 35/1988 on assisted reproduction procedures provides that donations be anonymous and that the particulars of the identity of the donor be "kept in the strictest secrecy" (§5(5)). Resultant children have the right, either personally or through their legal representatives, to obtain general nonidentifying information on the donors. In extraordinary circumstances that "entail a verified danger to the life of the child," the identity of the donor may be disclosed if "such disclosure is indispensable to avert the danger" (§5(5)).

The Ethics Committee of the American Fertility Society

also believes that anonymity should be preserved and in like manner noted that identifying information should only be released in extreme situations (p. 44S). The committee, as already mentioned, was of the opinion that although anonymity may be desirable goal, it should not preclude designated donation among friends and relatives (pp. 49S, 51S) but did not elaborate further on the impact of this policy on the best interests of the child and on family dynamics. On this subject, The Netherlands, in its 1997 report on IVF stresses the need for research about the actual consequences of anonymous versus nonanonymous donation without taking a position. Finally, although an emergency exception to the general rule of anonymity for donating and receiving couples is also provided for in France by virtue of Law No. 94-654, only nonidentifying information may be divulged (art. L.152-5 and L.152.13 Code of Public Health relating to the donation of an embryo; L.673-6 Code of Public Health relating to the donation of gametes).

Some countries do not provide for donor anonymity. In such instances, identifying information may be divulged irrespective of medical need. Under Austria's Reproductive Medicine Law of 1992, the donor does not have the right to anonymity. From the age of 14, a child born from donor insemination may ask for information on the donor's identity. Fertility clinics are under an obligation to keep records showing names of donors, place and date of birth place, nationality, address, and so on in order (§20).

In Australia, the state of Victoria's Infertility Treatment Act of 1995 provides that a person born as a result of a donor treatment procedure may, on attaining 18 years of age, apply in writing to the authority to obtain information recorded in a central register. The information may reveal both nonidentifying and identifying information about the donor (§79, 80). The act further provides that prior to releasing the information reasonable efforts must be made to advise the donor. Likewise, the donor may apply for information about the person born as a result of his donation or the parents of such person (provided appropriate age and consent requirements are satisfied), regardless of whether the information revealed is identifying or not (§76, 77).

In Switzerland, in conformity with article 24(g) of the constitution which provides that "a person shall be guaranteed access to data relating to his parentage," the bill relating to the federal law on medically assisted procreation would allow a child of 16 years of age, born from a donor treatment procedure, the right to obtain identifying information about the donor without having to demonstrate medical need. The donor would be informed of the child's request and should the donor refuse to meet the child, the child would be advised and informed of his or her rights (§27(3)). Moreover, where a justified need exists, the child, regardless of his or her age, would have a right to all information (nonidentifying as well as identifying) relating to the donor (§27(2)).

Finally, the Human Fertilisation and Embryology Act of 1990 in the United Kingdom makes limited provision for rights of access to information for those born (or think they may have been born) from assisted procreation. The act provides for the constitution of a consensual registry which allows a person, on attaining the age of 18, the right to request and obtain certain information relating either to the donor or to issues of consanguinity for the purpose of marriage (§31(4)). For the authority to divulge this information, the register must show that the applicant was, or may have been, born in consequence of treatment services and that he or she has had an opportunity to receive proper counseling about the implications of compliance with his or her request (§31(4)). Accompanying notes to the act indicate that it is the governments' stated intention that this information not reveal the identity of the donor. In extraordinary cases, however, disclosure may be permitted (§11.8 Code of Practice). Irrespective of the approach taken, all countries agree that there should be no filial link between the donor and the child.

In conclusion, it is immediately evident that although there is a certain homogeneity in the issues raised by the various countries under study, the same cannot be said with regard to the actual working out of the issues in clinical practice. Indeed, although certainty as to what may or may not be offered, to whom, under what conditions and limits is essential in an area in which we are intervening in the very "production" of life, to say nothing of the interests of the resulting child, paradoxically the detailed ethical/legal provisions are so numerous as to risk undermining the informed consent process they were meant to realize. Perhaps this is a result of our continuing unease with the new reproductive technologies, which are in constant evolution, or perhaps it is thought that the detailed enumeration will cover all real and imaginary circumstances in a way that we protect the parties concerned. It also remains to be seen whether this approach will ultimately also undermine professional responsibility and freedom of judgment. Only experience with the clinical application and the current heterogeneity of interpretations will demonstrate whether the underlying cultural differences as translated into more open or restrictive legally recognized positions are meaningful. These cultural differences are exacerbated and much more evident in the controversial subject of research.

National Positions: Research Aspects

Available Zygotes/Embryos *in Vitro*

Arguments relating to research on human embryos revolve around the difficult issue of discerning at what stage of development the status of a person should be accorded the human embryo. Those countries which legislatively or, by virtue of their guidelines and recommendations, permit (nontherapeutic) embryo research maintain that research will assist in the development of new preventive approaches to genetic disease, help improve the progress of assisted conception and preimplantation diagnosis techniques, and aid in the understanding of infertility. Other countries only allow (therapeutic) embryo research in order to improve chances of successful implantation and healthy development. As discussed here, where such research is allowed, it must take place before the 14th day of fertilization.

In Canada, the Royal Commission on New Reproductive Technologies approved of zygote/embryo research to inter alia improve existing fertility treatments and develop new technologies for assisted conception, learn about human infertility, detect and prevent genetic and chromosomal abnormalities in human beings, investigate new methods of contraception, and advance knowledge about human development and its disorders (rec. 183). The commission recommended, however, that embryos used for research not be transferred to a woman's body until such time as knowledge reaches the point at which manipulations are likely to increase the woman's chances of conceiving (rec. 189). Regulations would require that facilities conducting the research be licensed, that appropriate applications to conduct clinical trials are made, and that fully informed consent of the persons who have donated the gametes used to create the zygote be obtained (rec. 186). Except for generally prohibited practices, Bill C-47 would not distinguish between acceptable and unacceptable research. It would forbid, with criminal sanctions, embryo research except if the donors of the ovum and sperm have provided their consent.

The Human Fertilisation and Embryology Act of 1990 of the United Kingdom allows embryo research and provides that a research license may be granted provided the research protocol broadly relates to one of its existing categories of research aims (schedule 2, ¶3(2)). The aims include promoting advances in the treatment of infertility (¶3(2) (a)), increasing knowledge about the causes of miscarriages (¶3(2) (c)), developing more effective methods of contraception (¶3 (2) (d)), and developing methods for detecting the presence of gene or chromosome abnormalities in embryos before implantation (¶3(2) (e)). The act further provides that the research must "increase knowledge about the creation and development of embryos, or about disease, or enable such knowledge to be applied" (schedule 2, ¶3(3)). The licensing and supervisory roles of the act ensure that limitations of permissible research are adhered to. The Code of Practice of the Human Fertilisation and Embryology Authority provides further specifications (Part 10). It states that licenses cannot be grant for research projects involving embryo splitting with the intention of increasing the number of embryos for transfer (§10.5). Moreover, "centers should refer each research project to a properly constituted ethics committee for approval before applying for a research license" (§10.7) and "[p]roposals for research projects involving the use of embryos will be submitted for peer review to appropriate academic referees chosen by the Authority" (§10.9).

Denmark's 1992 law on the scientific ethics committee system and the examination of biomedical research project also allows embryo research provided appropriate authorization is granted and the purpose of the research is to "improve *in vitro* fertilization in order to bring about pregnancy" (§14). The law further provides that embryos that have been subject to research cannot be transferred to the woman unless there is no risk of transmitting hereditary diseases, defects, deformities, and the like (§14(4)). Spain's Law No.35/1988 on assisted reproduction procedures permits both therapeutic (§15 (1), (2)) and nontherapeutic research (§15(3)), although in the latter instance the embryo must not be viable and animal models must be shown to be insufficient for the research protocol. Appropriate authorization is required at all times.

The 1994 report of the NIH Human Embryo Research Panel in the United States was of the opinion that embryo research could be carried out provided significant scientific promise or clinical benefit was established and animal or unfertilized gametes could not be used to satisfy research goals. Valid research projects include studies aimed at improving the likelihood of a successful pregnancy, examining the fertilization process and the development of the embryo, and studying hereditary disease.

Some countries permit embryo research on surplus embryos if its goal is diagnostic (therapeutic); that is, research is allowed if it is intended to determine the development or viability of the embryo for purposes of implantation. In the Australian state of Victoria, the Infertility Treatment Act of 1995 only permits research on embryos that have not developed to syngamy and that then can be subsequently transferred to the woman (§24, 25). Therefore, the act bans all research on embryos that would harm the embryo or reduce the likelihood of a pregnancy (§24). Similarly, Austria's Reproductive Medicine Law of 1992 only permits the examination and treatment of viable cells if, taking into account the current state of medical science and experience, such action is necessary in order to achieve pregnancy (§9 (1)). Finally, by way of exception to its general prohibition against all experimentation on human embryos, France's Law No. 94-654 permits the examination of *in vitro* embryos provided there is a medical aim to this examination and the embryo is not harmed. The couple who produced the embryo must provide written consent and the National Commission of Medicine, Biology, Reproduction, and Prenatal Diagnosis must approve the study (art. L.152-8-7 Code of Public Health). A study in this context may seek to determine the embryo's chance of successful implantation or contribute to the improvement of medically assisted procreation, although more extensive research can be accomplished under the aegis of preimplantation diagnosis (Decree May 27, 1997, art. R-152-8-1). It bears mentioning that this window on research with surplus embryos is quite limited since it is very difficult, if not impossible, to show that the embryos will not be harmed.

Finally, some countries completely ban all experimentation on embryos. As such, the taking, use, and development of embryos for research purposes is completely prohibited by Germany's 1990 Embryo Protection Law. Using a human embryo for any purpose other than its preservation is considered an abuse of procreative technique punishable by fine or imprisonment (art. 1(1) (6), 2). The 1994 act relating to the application of biotechnology in medicine in Norway also precludes research on embryos without exception (§3.1), as does Switzerland's federal constitution, bill, and medicoethical guidelines. Although difficult to distinguish between embryos that become surplus and those that are specifically created, the issue of embryos deliberately created for research purposes is often considered more controversial.

Embryos Created Solely for Research Purposes

The advantage of permitting the creation of embryos for research is that by allowing fertilization to be manipulated, particular aspects of fertilization can be better understood. Several countries allow the deliberate creation of embryos for purposes of research. The Royal Commission in Canada recommended that "zygotes be created for research purposes only if gametes for this purpose are available without having to conduct additional invasive procedures" (rec. 188). In order to better frame such research, the commission set out the following conditions: The gamete donors must provide informed consent before donation, the knowledge sought must not be obtainable through studies on animal embryos or cell lines, the eggs obtained must not be specifically obtained just for research, the aim of the research must benefit human health and not be for commercial profit, and the research must be performed in a licensed facility.

In the United Kingdom, the provision in the Human Fertilisation and Embryology Act of 1990 which authorizes embryo research likewise authorizes the deliberate creation of embryos for research (schedule 2, ¶3). As mentioned previously, such research is permissible as long as it satisfies a listed aim (¶3(2)) and the requirements of the Code of Practice (United Kingdom, Part 10). Lastly, the 1994 NIH report on human embryo research in the United States also permits embryos to be created solely for research purposes, provided the research cannot otherwise be conducted and such creation is necessary for the validity of the study and "the potentially outstanding scientific and therapeutic value" expected.

Finally, the majority of recommendations and legislation not only forbid the deliberate creation of embryos for research purposes but also impose criminal prohibitions. Bill C-47, Canada's proposed Human Reproductive and Genetic Technologies Act, unlike the Royal Commissions' recommendations noted previously, would criminally sanction the creation of embryos for research. The bill stipulated that "no person shall knowingly cause the fertilization of an ovum outside the human body for purposes of research" (art. 4(1) (k)). Similarly, France's Law No. 94-654 bars the *in vitro* conception of human embryos for purposes of study, research, or experimentation. The law provides a maximum 7-year imprisonment or fine in case of violation (art. L.151.18 Penal Code). With a penalty of up to 3 years' imprisonment or fine, Germany's Embryo Protection Law of 1990 likewise penalizes any person who is responsible for the *in vitro* (extracorporeal) development of a human embryo for any purpose other than pregnancy (art. 2(2)). It remains to be seen whether research concerning the achievement of a successful pregnancy will circumvent this provision. Lastly, both Switzerland's bill and guidelines prevent the creation of embryos for research. The bill sanctions by imprisonment anyone who produces an embryo for purposes other than inducing a pregnancy (art. 29 (1)), whereas the guidelines stipulate that a doctor must abstain from collaborating in the artificial creation of embryos for purposes other than procreation (No. 12.1).

Time Limitations

In those countries that permit research, a time limitation is expressly stipulated. Despite differing language, approved research is restricted to the first 14 days of development from the time of fertilization. Beyond this period (which approximates implantation), experimentation is prohibited and any embryos used must be destroyed. For example, the Human Fertilisation and Embryology Act of 1990 of the United Kingdom (§3 (3) (a)) and the NIH report in the United States both refer to the prohibition of research after the "appearance of the primitive streak" which is no later than 14 days after the gametes are mixed. In Australia, the Infertility Treatment Act of 1995 of the state of Victoria uses slightly different language and provides that if a zygote is used, research must stop before syngamy (§26). In addition to prohibiting research conducted on embryos after the 14th day, a variety of other practices have been expressly barred by legislation.

Prohibited Practices

Generally, the countries studied enumerate experimentation practices that would not qualify as approved research for any purpose. The majority forbid germline therapy — that is, the alteration of the genetic constitution of the reproductive cells in a way that will be transmitted. Such modification then would ultimately affect the genetic patrimony of descendants [see Austria, Reproductive Medicine Law of 1992, §9 (2); Australia, Infertility Treatment Act of 1995, §39.1; Canada, Bill C-47, the proposed Human Reproductive and Genetic Technologies Act, art. 4(1); France, art. L.152-8-1 Code of Public Health; Germany, The Embryo Protection Law of 1990, art. 5(1); Norway, The Act Relating to the Application of Biotechnology in Medicine, §7(1); Switzerland, §24 of the federal constitution of May 29, 1874, and the Bill Relating to the Federal Law on Medically Assisted Procreation, art. 35; United Kingdom, the Human Fertilisation and Embryology Act of 1990, schedule 2, ¶3(4)].

Interestingly, the Human Fertilisation and Embryology Act of 1990 in the United Kingdom provides an exception to this prohibition in such circumstances (if any) as are specified in regulations (schedule 2, ¶3(4); art. 10.4 Code of Practice). Finally, the 1994 NIH report considering germline therapy to be a particularly delicate procedure recommended that "there be a presumption against funding such research for the foreseeable future." It did note, however, that the presumption could be reversed if extraordinary scientific or therapeutic merit was established.

Although technologies advanced with animals may assist in remedying infertility, all countries under study — that is, Australia (§45), Canada (Royal Commission, rec. 184; Bill C-47, art. 4(1)), Denmark (§15(4), Germany (§7(1) and (3)), Switzerland (constitution, §24), the United Kingdom (§3(2) and (3), and the United States — ban the mixing of human and animal gametes and experiments whose purpose is to enable the gestation of human embryos in the uterus of another species. However, it must be noted that Spain (Law No. 35/1988, §14(4)), the United Kingdom (Human Fertilisation and Embryology Act, schedule, 2, ¶3(5)), and the United States by virtue of the NIH allow what is referred to as the "hamster test." This test enables the chromosomal constitution of the sperm to be studied and hence contributes to the determination of the male contribution to genetic abnormalities and to infertility. This test may be extended to other animals if they are found to be a suitable source of testing.

Several countries explicitly prohibit human cloning. Cloning may be achieved either by embryo splitting or twinning (fertilized gametes) or by the recent "Dolly" technique (unfertilized gametes). The Infertility Treatment Act of 1995 of the state of Victoria in Australia bans cloning, which is defined as "to form, outside the human body, a human embryo that is genetically identical to another embryo or person" (art. 3). Similarly, the 1990 German Embryo Protection Law also prohibits "artificially caus[ing] a human embryo to develop with the same genetic information as another embryo, fetus, living person, or deceased person" (art. 6(1)). The Human Fertilisation and Embryology Act of 1990 of the United Kingdom proscribes "replacing a nucleus of a cell of an embryo with a nucleus taken from a cell of any embryo, person, or subsequent development of an embryo" (art. 3(3) (d)). Although the intent of these countries is to prohibit human cloning, it is unclear whether the latter definition includes the Dolly technique. Conversely, Denmark's Law No. 503 (§15(1)), France's Civil Code (art. 16-4), and Spain's Law No. 35/1988 (§20 (2) (B) (k)) generally prohibit research on the creation or production of "genetically identical human beings." This language may be sufficiently broad to include in its prohibition both forms of cloning since it concentrates on the result rather than the actual technique used. Furthermore, many countries, specifically Canada, Germany, Spain, Switzerland, the United Kingdom, and the United States, also bar research related to chimeras, parthenogenesis (creation of a zygote from the female gamete alone), and ectogenesis (development of a fetus to viability outside the uterus in an artificial womb), believing such manipulations to be unethical and morally reprehensive.

Other prohibited practices include the use of a mixture of sperm from different males for the purpose of medically assisted procreation (Austria and Canada) and procedures involving retrieving ovum or sperm from fetuses or cadavers with the intention of having the ovum mature outside the human body and then fertilized or implanted in a woman (Australia and Canada).

Preimplantation Diagnosis

Preimplantation diagnosis, still largely in the research stage, is in part intended to detect embryos with inherited diseases. Most countries under consideration prohibit preimplantation genetic diagnosis for sex selection. As such, procedures may not be used solely for the purpose of gender selection. However, if it is necessary for a child to be of a certain sex in order to avoid the risk of transmission of a genetic abnormality or disease, then preimplantation diagnosis of the embryos may be permissible.

In this vein, the Royal Commission on New Reproductive Technologies in Canada recommended that proposals for preimplantation diagnosis to determine the sex of an embryo for nonmedical reasons be denied (rec. 265). The commission was of the opinion that such use would "[distort] the role of preimplantation diagnosis, which is to identify the presence of severe genetic diseases at a very early stage in the development of the zygote" (art. 906). Moreover, refusing to transfer a zygote based solely on its sex was considered inconsistent with principles of respect for human life and dignity. Spain's Law No. 35/1988 similarly prohibits sex selection for nontherapeutic purposes (§20(2) (B) (n)), as does Switzerland's Bill Relating to the Federal

Law on Medically Assisted Procreation (art. 33) and its medicoethical guidelines (No. 12.3).

Germany's Embryo Protection Law of 1990 forbids sex selection except "if selection of a sperm is made by a physician in order to prevent the child from developing Duchenne-type muscular dystrophy or a sex-linked hereditary disease of similar severity"(§3). From a eugenic selection perspective, this position is interesting given that Germany prohibits embryo donation based on respect for the life of the embryo. Norway's 1994 Act Relating to the Application of Biotechnology in Medicine prohibits the examination of an embryo for the sole purpose of selecting the sex of a child (§4-3) but allows preimplantation diagnosis in the case of incurable sex-linked hereditary disease with no possibility of treatment (§4-2).

The Ethics Committee of the American Fertility Society found the use of gender selection for nondisease goals (i.e., family balancing) problematic. It noted that "it may be premature to declare that there are absolutely no circumstances under which gender selection should be used, regardless of the technology involved in achieving it" (p. 66S). Beyond preimplantation diagnosis for purposes of sex selection, the Ethics Committee of the American Fertility Society recommended the use of preimplantation diagnosis for the prevention of inherited diseases "that severely deplete the quality of human life or longevity" (p. 66S).

In this context, France's Law No. 94-654 provides that preimplantation diagnosis may be exceptionally permitted where a doctor attests to the fact that a couple has a strong possibility of giving birth to a child with a particularly grave genetic disease known to be incurable from the moment of diagnosis but prohibits it for sex selection. The anomaly or anomalies of one of the parents must be precisely identified and the couple must provide their consent to the technology in writing (art. L.162-17 Code of Public Health).

Spain's Law No. 35/1988 refers to diagnostic and therapeutic interventions on live preembryos *in vitro.* For diagnostic purposes, the law provides that it "may have as its sole objective the enhancement of its viability or the detection of hereditary diseases with a view to advising against their transfer for purposes of procreation" (§12 (1)). For therapeutic purposes, the law notes that it may have "as its sole objective the treatment or the prevention of the spread of a disease, where there are reasonable and verified prospects of success" (§13(1)).

Finally, with regard to preimplantation diagnosis, the Human Fertilisation and Embryology Act of 1990 of the United Kingdom refers to treatment and research ends. For treatment purposes, authorization can be issued for "practices designed to secure that embryos are in a suitable condition to be placed in a woman or to determine whether embryos are suitable for that purpose" (schedule 2, ¶1 (1) (d)). Relating to this subject, public consultation led the act to prohibit preimplantation diagnosis for the purpose of selecting sex and put forward this interdiction as a licensing condition of centers. For research purposes, authorization may be obtained to seek "methods for detecting the presence of gene or chromosome abnormalities in embryos before implantation" (schedule 2, ¶3(2) (e)).

Conclusion

International legislation affirms the importance of protecting both human beings and human genetic material. Despite these normative guidelines, legislative disparity at the national level underscores the difficulty of regulating the implementation and use of emerging assisted reproductive techniques. Differences undoubtedly depend on varying cultural approaches and social policies that must balance scientific advances against human dignity. Although this comparative study is illustrative of a broad spectrum of approaches from restrictive (Austria and Germany) to more liberal (United Kingdom), the nature of assisted reproduction and embryo research makes it clear that regulation, to some degree, is required. The latter area is necessarily subject to additional legislative scrutiny and oversight. The "exposed" embryo is particularly vulnerable and manipulation or selection is severely constrained if not prohibited altogether in order to ensure that accessibility does not lead to eugenic practices. Although morally reprehensible techniques such as the creation of chimeras are prohibited, preimplantation diagnosis remains an area of compromise, with the difficult choice of initiating an "at-risk" pregnancy only to face the choice of abortion at a later time in an established pregnancy. Who should support this burden—the pregnant woman or the preimplantation embryo? Different values in a given country are reflected in the broad or limited choices offered. As techniques develop and experience is gained, it is not certain whether approaches will change. What remains to be seen is whether regulatory mechanisms to ensure human dignity need to be criminally prohibitive or whether pragmatic, more flexible direction is sufficient.

Acknowledgments

We thank Sonia Le Bris for her expert advice and Me Michelle Laflamme for her invaluable assistance.

General References

INTERNATIONAL

Human Genome Organization (1996). Statement on the principled conduct of genetic research. *Genome Digest* **2.**

International Law Association (1990). Resolution on reproductive technologies and the protection of the human person, adopted at the sixty-third conference of the ILA, Warsaw,

21–27 August 1988, 41, 4. *International Digest of Health Legislation,* 723.

INTERNATIONAL PLANNED PARENTHOOD FEDERATION (1996). Charter on Sexual and Reproductive Rights, 1996, 47, 4. *International Digest of Health Legislation,* 546.

UNESCO (1996). Third session of UNESCO's International Bioethics Committee (IBC), 47, 1. *International Digest of Health Legislation,* 107.

WORLD MEDICAL ASSOCIATION (1995). Statement on *in vitro* fertilization and embryo transplantation (Madrid, October 1987). In *The World Medical Association Handbook of Declarations.* World Medical Association.

REGIONAL

CONVENTION ON HUMAN RIGHTS AND BIOMEDICINE (1996). Convention for the protection of human rights and dignity of the human being with regard to the application of biology and medicine, Directorate of Legal Affairs, Strasbourg, November 1996. *DIR/JUR* **96,** 14.

COUNCIL OF EUROPE (1989). Principles in the field of human artificial procreation, *Human Artificial Procreation,* 49, 4. *International Digest of Health Legislation,* 907.

RESOLUTION ON ARTIFICIAL INSEMINATION IN VIVO AND IN VITRO (1989). 40, 4. *International Digest of Health Legislation,* 885.

NATIONAL

AUSTRALIA Commonwealth Department of Health and Family Services, Assisted Reproductive Technology, 1996.

AUSTRALIA National Health and Medical Research Council statement on human experimentation and supplementary notes (Supplementary Note 4: *In Vitro* Fertilization and Embryo Transfer, October 1982; Supplementary Note 5: The Human Fetus and the Use of Human Fetal Tissue, October 1983).

AUSTRALIA Victoria: Infertility Treatment Act 1995 [Assented to 27 June 1995], No. 63/1995.

AUSTRIA Federal law of 1992 (serial No. 275) regulating medically assisted procreation (the reproductive medicine law) and amending the general civil code, the marriage law, and the rules of jurisdiction (1993) 44, 2. *International Digest of Health Legislation,* 247.

CANADA Civil Code of Quebec, S.Q. 1991, c.64.

CANADA (1993). Royal Commission on New Reproductive Technologies; Proceed with care: Final report, Vols. 1 and 2. Minister of Government Services, Ottawa.

CANADA (1996). Bill C-47, an act respecting human reproductive technologies and commercial transactions relating to human reproduction, 2nd Sess., 35th Parliament, 45 Elizabeth II.

DENMARK (1992). Law No. 503 of 24 June 1992 on the scientific ethics committee system and the examination of biomedical research projects, 43, 4. *International Digest of Health Legislation,* 758.

DENMARK (1992). Order No. 650 of July 1993 on the freezing and donation of human oocytes, 43, 4. *International Digest of Health Legislation,* 738.

DENMARK (1994). Order No. 852 of 28 October 1993 amending the order on the freezing and donation of human oocytes, 45, 2. *International Digest of Health Legislation,* 183.

DENMARK (1994). Order No. 392 of 17 May 1994 on the freezing and donation of human oocytes, 45, 4. *International Digest of Health Legislation,* 494.

DENMARK (1995). The Danish Council of Ethics 8th annual report on assisted reproduction.

FRANCE (1994). Law No. 94-654 of 29 July on the donation and use of elements and products of the human body, medically assisted procreation, and prenatal diagnosis, 45, 4. *International Digest of Health Legislation,* 473.

GREAT BRITAIN (1990). Human Fertilization and Embryology Act, U.K., 1990, cap. 37. In *Blackstone's Guide to the Human Fertilization and Embryology Act, 1990 Abortion and Embryo Research, the New Law* (D. Morgan and R. G. Lee, Eds.). Blackstone, London.

GREAT BRITAIN (1995). Code of Practice, Revised December 1995. Human Fertilization and Embryology Authority, London.

NORWAY (1994, June/July). The act relating to the application of biotechnology in medicine. *Bull. Med. Ethics* **8.**

SPAIN (1989). Law No. 35/1988 of 22 November 1988 on assisted procedures, 40, 1. *International Digest of Health Legislation,* 82.

SPAIN (1991). Law No. 42/1988 of 28 December 1988 on the donation and use of human embryos and fetuses or their cells, tissues, or organs, 42, 1. *International Digest of Health Legislation,* 64.

UNITED STATES (1994). NIH Human Embryo Research Panel; Major conclusions and recommendations from the final report (September 27, 1994).

UNITED STATES (1994). Ethics Committee of the American Fertility Society; Ethical considerations of assisted reproductive technologies, 62. *Fertil. Steril.* **5** (Suppl. 1), 32S.

SWEDEN (1993). Law No. 114 of 14 March 1991 on the use of certain gene technologies within the context of general medical examinations, 44, 1. *International Digest of Health Legislation,* 57.

SWEDEN (1993). Law No. 115 of 14 March 1991 concerning measures for the purposes of research or treatment in connection with fertilized human oocytes, 44, 1. *International Digest of Health Legislation,* 58.

SWITZERLAND (1990). Swiss Academy of Medical Sciences issues; Medico-ethical guidelines on medically assisted procreation, 142, 2. *International Digest of Health Legislation,* 346.

SWITZERLAND (1992). Genetics and assisted procreation, amendment of federal constitution, 13 August 1992, 43, 4. *International Digest of Health Legislation,* 745.

SWITZERLAND (1995). Avant-projet relatif a une loi fédérale concernant la procréation médicalement assistée et instituant une commission nationale d'éthique (loi sur la médecine humaine).

MATTEO ADINOLFI
JOYCE HARPER

Galton Laboratory and Department of Obstetrics and Gynaecology
University College London
London NW2 HE United Kingdom

Human Preimplantation Embryo Research

For many years, our knowledge of the biological processes that underline the early stages of human embryonic development has been based on experimental studies performed mainly in amphibia and rodents Only in recent times, with the introduction of in vitro fertilization (IVF) procedures, human preimplantation embryos have been investigated and are providing valuable information about the complex sequences of events that characterize life from fertilization to implantation.

In this chapter we summarize what has been discovered so far about the function of the oocyte maternal RNA, the early embryonic phases of development, and genetic imprinting that control the development of the human preimplantation embryo as it migrates to reach the uterine cavity and implants.

An important observation, derived from these studies, is that a large proportion of early embryos have major chromosome disorders and will not be able to survive.

The analysis of one or two blastomeres, retrieved from an 8-cell embryo, has shown it to be a safe procedure with which to perform preimplantation genetic diagnoses of a variety of inherited disorders.

We hope that this brief description of the advances in this field of study may help the readers to realize that research on human preimplantation embryos, carried out under severe, but at the same time flexible, ethical rules, has important scientific and social impact and it is essential to devise methods for the prevention or cure of severe genetic disorders.

Introduction

It would be impossible to cover the entire spectrum of past and present studies of the human preimplantation embryos in a single chapter. Consequently, our review describes only some aspects of research performed during the past few years, particularly since molecular techniques have been applied to investigate gene expression in early embryos and the origin of chromosome disorders.

Procedures developed for the treatment of infertility have provided the opportunity to gain access to a large number of human preimplantation embryos at 4 to 8 cell stages and to analyze single cells (blastomeres) for the presence of chromosomal abnormalities and the prenatal diagnoses of selected inherited disorders. However, these studies would not have been possible without the extensive body of morphological data, collected over many years, that has elucidated the origin and differentiation of the various embryonic tissues in humans and experimental animals.

Although, as far as possible, only the development of the human preimplantation embryos will be discussed, brief references to similar investigations in other species will be mentioned, particularly since the modern era of mammalian embryology is largely based on research carried out in amphibia and rodents. In fact, the most important milestones of mammalian embryonic studies can be identified in the establishment of the *in vitro* cultures of mouse embryos, the successful transfer of cultured embryos to foster mothers, the production of aggregation chimeras, and the micromanipulation of embryonic cells leading to the microinjection of tracers into progenitor cells for the analysis of their fate. The production of transgenic animals by the microinjection of exogenous DNA into the zygote pronucleus and the use of pluripotent stem cells derived from germ-cell tumors or from cultured embryos have also been two technologically powerful tools in the field of experimental mammalian embryology. So far, for technical and ethical limitations, only some of these approaches have been applied to studies of the human embryo.

The purpose of this chapter is also to emphasize how molecular and genetic investigations, performed under careful ethical supervision at early stages of human development, are justified in view of their impact on our scientific knowledge of human reproduction and, most important, on the prevention of inherited disorders.

Early Morphogenesis

Since the advent of *in vitro* fertilization (IVF) for the treatment of infertile couples, the culture of human preimplantation embryos has become commonplace and, as a result, our understanding of the early stages of human development has become clearer (Fig. 1).

Mammalian oocytes are surrounded by a layer of about 3000 cumulus cells held together by an extracellular matrix rich in hyaluronic acid. Sperm penetrate these cells, by movements of the tail, and reach the zona pellucida with the acrosome intact. Sperm first bind to the zona pellucida (primary binding); this signals the sperm to undergo the acrosome reaction and then a secondary binding occurs, followed by penetration and binding to the oolema. It is generally believed the specific structure ZP3 of the zona pellucida is the primary sperm receptor, which also facilitates the acrosome reaction. Several candidates have been put forward for the secondary receptor, such as SLIP 1 (sulfolipid immobilizing protein 1), which is found in the plasma membrane of the sperm head region and has been shown to be involved in zona pellucida binding, and PH-20, a sperm surface protein. Although many sperm are found embedded in the zona pellucida only one fertilizes the oocyte. It is thought that the region of the zona pellucida surrounding the polar body is amicrovillar (and does not bind sperm), but that the rest of the zona has many microvilli which facilitates sperm binding. Once sperm have penetrated the zona, the spindle rotates and the oocyte resumes meiosis II (MII) extruding the second polar body. Studies have shown that oocyte activation is caused by a Ca^{2+} oscillation. It was initially thought that these Ca^{2+} waves are produced by sperm binding cell surface receptors which then couple to intracellular proteins leading to the generation of a Ca^{2+} mobilizing second messenger, inositol 1, 4, 5-triphosphate. More recently it has been shown that there is a soluble sperm factor, termed oscillin, which is expressed in sperm more than any other tissue and that within minutes of entering the oocyte, is capable of causing Ca^{2+} oscillations, probably releasing Ca^{2+} via channels in the endoplasmic reticulum. Oscillin is concentrated around the equatorial segment of the sperm, which is the area that fuses with the oocyte membrane. Therefore, sperm–oocyte fusion precedes oocyte activation. Direct intracytoplasm sperm injection (ICSI) can still promote oocyte activation via the second phase of Ca^{2+} mobilization.

Within hours of oocyte penetration, arrays of microtubules grow from the sperm centrosome, the only functional centrosome in the zygote. It has been shown that these microtubules participate in the process of pronuclear association required for syngamy. Sperm centrosome defects may be the cause of some male infertility. Sperm therefore contribute at least three factors to the oocyte: the genetic material, oscillin, and the centrosome.

Pronuclei formation coincides with the onset of DNA synthesis. In humans, following insemination *in vitro,* pronuclear formation commences 3–10 hr later (Fig. 1). DNA synthesis (S phase) begins between 8 and 14 hr postinsemination and is completed in the majority of embryos by 22 hr. The male pronucleus is slightly larger than the female, but the difference is not as great as in other species, such as in mice. During this time the chromosomes present in each pronucleus consist of one chromatid (haploid, 1N); then the DNA replicates, so that each chromosome contains two chromatids, and finally the pronucleus membrane is broken down. The first mitotic spindle forms and the chromosomes line up on the spindle that pulls the two chromatids apart; finally the cytoplasm divides in half (cleavage) so that human daughter cells contain 23 pairs of chromosomes (Fig. 1).

In IVF studies, the formation of three pronuclei is observed in about 5% of cases. In 83% of these cases it is due to fertilization by more than one sperm; in the remaining cases, it is due to nonextrusion of the second polar body. It is thought that three-pronuclei oocytes form a tripolar spindle which will give rise to three daughter cells after the first mitotic division. As observed in many IVF Units, embryos produced from three-pronucleate zygotes are of good morphology (i.e., very regular and with little fragmentation). The formation of one-pronucleate oocytes is rarely observed; in IVF cycles it has been found that, in most cases, the embryos have fertilized normally, but there has been asynchronous pronuclei formation. In embryos generated by ICSI most one-pronucleate oocytes are haploid.

Chromosome analysis by fluorescent *in situ* hybridization (FISH) of multinucleated blastomeres has shown that these embryos are chromosomally abnormal.

In humans, the duration of the M phase is about 3 hr and the pronuclei break down between 28 and 30 hr postinsemination. The first cleavage division occurs around 32 hr post-

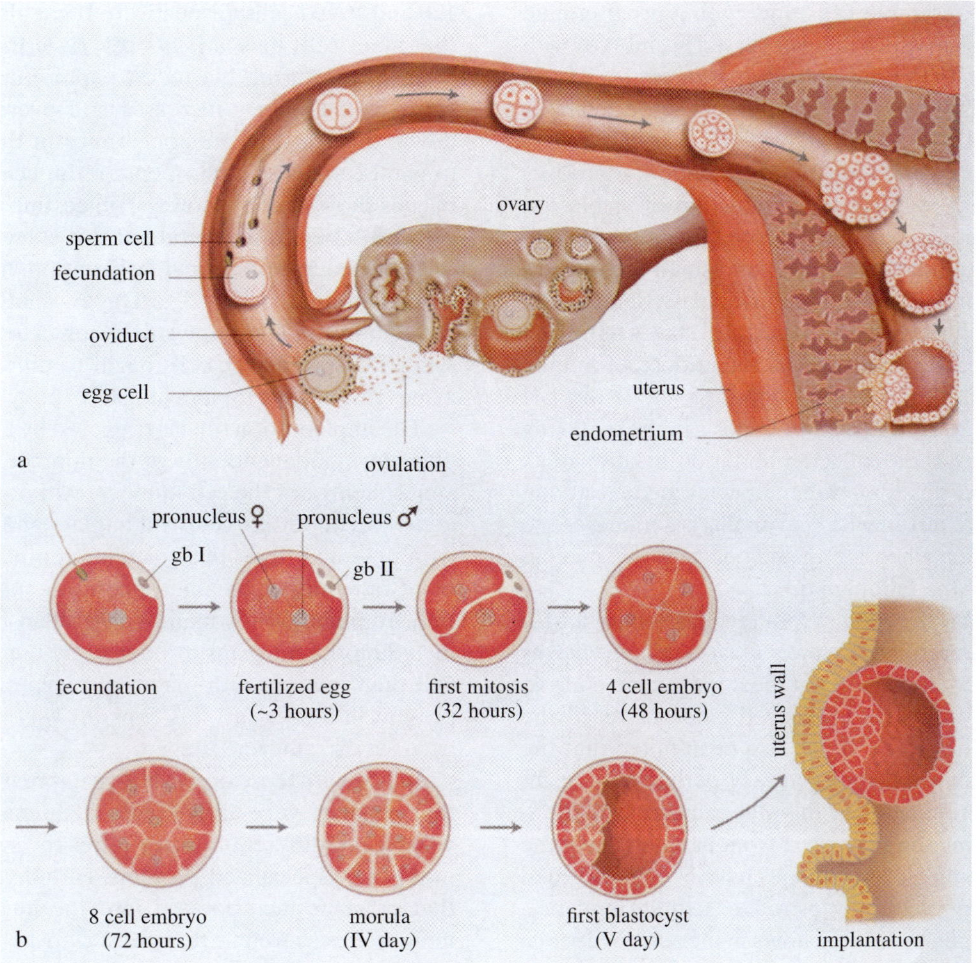

FIGURE 1 Diagrammatic representation of fertilization (a) and early development of the human embryo up to implantation (b). Following fertilization, the oocyte starts the second meiotic phase (MII) while the second polar body is expelled. Segmentation follows the fusion of the two pronuclei and formation of the zygote. Implantation occurs at the blastocyst stage (about 120 cells).

insemination, producing two daughter cells in normally fertilized embryos (Fig. 1). Embryos should have divided to 4 cells by 48 hr postinsemination, whereas the 8-cell stage is usually observed by 72 hr. In IVF procedures, embryos are usually transferred between Day 2 or Day 3 postinsemination. By Day 4 postinsemination (or ovulation), the morula stage should be reached; this is when compaction occurs and the cells flatten out over each other to maximise intercellular contacts. By Day 5 or Day 6, the cells differentiate to produce trophectoderm and inner cell mass cells. The outer, epithelial-like, polar, trophectoderm cells (giving rise to the placenta) have extensive intercellular junctions and are responsible for pumping fluid into the blastocyst. This produces the blastocoel cavity (cavitation). The inner cell mass cells are round, apolar, uniformly adhesive, and totipotent, and they give origin to the embryo proper. Lineage tracing has demonstrated that blastomeres at early cleavage-stage (2–8 cells) are not committed and contribute to both the inner cell mass and the trophectoderm. This is why one or two blastomeres can be removed from 8-cell embryos for diagnostic tests.

Up to the blastocyst stage, the size of the embryo does not increase, so that a blastocyst is approximately the same size as an oocyte (i.e., 120 μm). With each cleavage, from 2 cells to blastocyst, the cytoplasm is divided in half. Divisions are asynchronous, so that each blastomere divides independently from each other. In most mammalian species, the morphology of the blastomeres is uniform. However, in humans, irregular-sized and -shaped blastomeres as well as cytoplasmic fragmentation are a regular occurrence; embryos with extensive fragmentation have limited developmental potential. Histologically, some blastomeres of fragmented embryos resemble cells undergoing programmed cell death (apoptosis) which is not observed in embryos with good morphology. In IVF, an embryo grading system is used since embryos showing good morphology have a better chance of implanting.

During IVF procedures, an attempt is made at reproducing the events that regulate the production of viable oocytes and fertilization. For this purpose large quantities of follicle stimulating hormone (FSH) and small amounts of luteinizing hormone (LH) are administered to stimulate the production of multiple follicles. At critical stages of follicle growth and estrogen production, a single dose of human chorionic gonadotropin (hCG) is given to mimic the LH surge and to ensure resumption of meiosis in the maturing oocytes, which are then collected about 36 hr after hCG administration. At this stage, when oocytes are arrested in meiosis II, they are mixed with sperm; this is a time close to the natural situation when sperm and oocyte would meet in the fibrous end of the fallopian tube.

It is thought that the events from fertilization to blastocyst development are slightly slower *in vitro* than *in vivo,* as *in vitro* culture conditions are not ideal for human embryo growth. Indeed, only 50% of cleaved embryos reach the blastocyst stage in culture. It has also been noted that the thickness of the zona pellucida can vary, perhaps as a result of the superovulation regimes; this may cause zona hardening, thus preventing the blastocyst from hatching from the zona and implanting. Several studies have been performed to assist hatching of such embryos and to facilitate implantation, as, for example, by drilling holes in the zona pellucida.

In vivo, sperm and oocyte meet in the fibrous end of the fallopian tube; consequently, fertilization and the first cleavage stages occur as the embryos travel down the fallopian tube arriving at the uterus only at the morula stage or when the blastocyst hatches from the zona pellucida and begins to implant.

In mice, the time needed for sperm–egg fusion and second polar body emission requires 1–2 hr after insemination; pronuclei form about 6 hr after fertilization. In hamsters the interval between sperm–egg fusion and emission of the second polar body is around 2–5 hr and pronuclei form in about 4 hr. Mouse embryos undergo their first and second cleavage at intervals of approximately 20–26 hr. The events of fertilization proceed slowly in mammals, as compared to other species investigated. In *Drosophila,* for example, cleavage takes place at intervals of about 10 min and in amphibia every 30 min.

In many mammals investigated, the time required for subsequent cell cycles generally diminishes as cleavage progresses and it is also very variable so that, a few days after fertilization, cell numbers can differ greatly. For example, a rabbit embryo at 96 hr contains around 1000 cells, while mouse and rat embryos at the same age are formed by 64–128 and 16–32 cells, respectively. It has also been suggested that inner cells have a faster cell cycle than external ones.

In mice, during blastocyst expansion, asynchrony between cell divisions increases and overall division slows down; this leads to the appearance in the trophectoderm of giant cells as a result of endoreduplication. About 10% of cells show signs of apotosis, suggesting that programmed cell death begins to operate at this stage of development.

In the course of normal fertilization, human blastocysts — formed around Day 5 to Day 6 — hatch from the zona pellucida from Day 7 onward, when cytotrophoblastic and syncytiotrophoblastic cells begin to differentiate. The enzymes responsible for hatching have not yet been identified.

The implantation time is triggered by an internal embryonic clock, independently of the number of cleavages and, consequently, of the cell number. Although human blastocysts with about 120 cells will implant and grow, blastocysts with the same "age" but a smaller number of cells may implant but will not be able to continue normal differentiation. Implantation is mainly controlled by several factors including the integrins, produced by the blastocyst, which will then interact with maternal receptors not exclusively present in the uterus; this explains why implantation may occur in extrauterine sites.

In line with these observations it is reasonable to assume that, even at very early stages of human embryonic development, differentiation requires the synthesis of a large number of specialized proteins, carbohydrates, and lipids that must be incorporated into the nuclear and cellular membranes as well as the entire cytoskeleton. These macromolecular elements, essential for the biological functions of each cell, must be produced at a very early stage of development under the control of a large set of activated embryonic genes to allow the survival of the embryo.

Activation of the Zygotic Genes

Role of Maternal RNA

During early cleavage, the embryo of most — if not all — species relies on the activity of maternal RNA, mostly ribosomal and transfer RNA derived from the mature oocytes. The untranslated RNA is stored in messenger ribonucleoprotein particles, and it is bound to protective factors that will mask it from the translational machinery. Three types of translational control mechanisms, acting on maternal mRNA, have been identified: activation by cytoplasmic polyadenylation, RNA localization, and translational repression. For a brief period of time, the maternal RNA acts as a template for protein synthesis, while some of the embryonic genes start to be activated. The duration of this period of almost exclusively maternal control varies in different species. In mouse embryos, physical inoculation or treatment with transcriptional inhibitors (e.g., alpha-amantin), until the 2-cell stage, do not affect the normal sequence of morphological development; the maternal RNA decreases rapidly and it is no longer functional as a template for protein synthesis after 48 hr. In humans, data on the time sequence of these processes are still scanty, but it is assumed that the role of most of the maternal RNA ceases at around the 4-cell stage.

Transcription of Embryonic Genes

The use of genetic markers for zygote-specific products and the detection of newly synthesized messenger RNA (mRNA) have made it possible to establish when paternal genes start to be activated in mammalian embryos. In early studies, limitations of sensitive techniques and of suitable material have precluded extensive investigations of gene activation in early human embryos. Most hypotheses concerning gene expression were derived from experiments on embryos of more primitive species, such as amphibians and echinoderms, or from mice, rats, and rabbits, which could be obtained in large number.

In mouse embryos, DNA synthesis has been shown to start at early stages of development. By the electrophoretic analysis of extracts from whole mouse embryos, it was observed that many proteins begin to be produced at the 2- to 8-cell stages. Most of these proteins are constitutive, that is, they can be detected in all cells during the preimplantation period.

In early studies, the synthesis of enzymes was investigated by testing supernatants obtained from short-term cultured embryos, which were tested against labeled substrates; in this way, several glycolitic enzymes were shown to be produced at early stages of embryonic life. Molecular techniques were then introduced to investigate the expression of embryonic genes.

In early mammalian embryos, the first unequivocal evidence for the presence of enzymes synthesized under the control of paternal genes, as opposed to maternal mRNA, was provided by the detection of paternal isozymic forms of glucose phosphate isomerase (GPI-1) in extracts from mouse blastocysts. The techniques of pronuclear transplantation confirmed the paternal origin of GPI-1 synthesized by mouse embryonic cells at Day 4 of development. In mouse embryos, multiple forms of alkaline phosphatase (ALP), paternal β-glucoronidase and X-linked phosphoglycerate have been detected at the late 2-cell and early 4-cell stages, when $\beta 2$ microglobulin and hypoxanthine-guanine phosphorybosil transferase are also produced. Lactate dehydrogenase isozymes, creatine kinase isozymes, and myosin are also produced between the 2- and 8-cell stages.

The transition to embryonic transcription has also been observed in hamsters at the 2-cell stage, in rabbits between 2- and 16-cell stages and around the 8-cell stage in sheep.

TABLE I

Genes Expressed in Preimplantation Embryos

Glucose phosphate isomerase
N-acetylhexosaminidase
Alkaline phosphatase
Histamine-releasing factor
Human chorionic gonadotrophin
Collagen type IV degrading protease
Transforming growth factor-α
Epidermal growth factor
Epidermal growth factor receptor
Insulin-like growth factor II
Insulin receptors
Gap junction proteins
Pregnancy-associated $\beta 1$-glycoprotein
Plasminogen activator
Plasminogen activator inhibitor
Trombomodulin
Sex-determining region Y

In human preimplantation embryos, gene transcription starts at the 2- to 4-cell stage. A partial list of genes expressed during these early phases of development is shown in Table 1.

The presence of the secreted form of the transforming growth factor (TGF) — initially demonstrated by testing culture media of human embryos from cleavage to blastocyst stages — has been confirmed by the detection of the products of the active gene; transcripts of the epidermal growth factor (EGF) and their common receptor (EGFR) have been detected in unfertilized oocytes and human embryos between 8-cell and blastocyst stages using the reverse transcriptase polymerase chain reaction (RT-PCR) methods and specific monoclonal antibodies. The concurrent expression of both growth factors and their receptor suggests the existence of autocrine stimulation mechanisms in human preimplantation embryos.

Using the RT-PCR assay, the chromosome Y-associated genes ZFY and SRY have been shown to be expressed in 2-cell embryos and at intermediate stages of blastocyst development.

The detection of mRNA for genes not expected to be active at very early embryonic life suggests that perhaps most genes are slowly transcribed in preimplantation embryos, but later on they are switched to allow cell and tissue differentiation.

However, as mentioned, it is expected that several genes must be transcribed and translated during the very early stages of human embryonic development, since cellular division requires that many proteins, carbohydrates, and lipids should be available and that unique, complex nuclear and cytoplasmic structures should be formed. For example, studies of preimplantation mouse embryos have shown that gap junctions are essential for the maintenance of compaction, implantation, and subsequent development. In experimental animals it has been shown that abnormalities affecting these structures are among the many factors responsible for early embryonic losses.

In human preimplantation embryos, gap junction proteins and desmoplakin, a demosomal protein, have been investigated using specific antibodies. Gap junctions, first detectable at the 4-cell stage, become increasingly organized with time. In normal trophectoderm (TE), cells are linked by dense arrays of gap junctions while cells of the inner cell mass (ICM) are linked by small, punctuated gap junctions. Prior to implantation, human embryos express predominantly connexin ($C \times 43$); gap junctions containing $C \times 32$ and $C \times 26$ can be seen occasionally in TE of labeled blastocysts. Preliminary data suggests that embryos with multinucleated, apoptotic, and decompacting cells often show either increased or reduced expressions of gap junction proteins.

Much progress has also been made toward the identification of molecules involved in intercellular signaling during early embryonic development in experimental animals. It has become increasingly evident, for example, that unique patterns of specification are likely to be achieved through the production and interaction of several factors ultimately responsible for the activation or inhibition of specific genes and the establishment of distinct embryonic tissues. These processes start to occur at very early stages of embryonic development since, for example, experiments in low vertebrates have shown that the initial specification of the dorsoventral axis occurs soon after fertilization, with the reorganization of the egg cytoplasm that affects both the vegetal and animal hemisphere determining the future dorsal site.

Although in *in vitro* studies many molecules and their biological activities of inducing the formation of body pattern have been identified, as yet it has not been possible to discern their specific role *in vivo,* nor has it been feasible to conduct similar investigations using human embryos. However, many of the genes, ligands, and receptors, responsible for early patterns of differentiation in low vertebrates, are probably produced and have similar functions in human embryos. Most of these genes interact with each other and have pleiotropic effects, that is, each of them acts on different tissues and has different functions.

A group of genes, with an important role in mammalian development and growth control, was initially discovered in *Drosophila.* Their studies have provided important clues to our understanding of some human diseases. For example, mutations affecting the human *Sonic Hedgehog* gene have been shown to give rise to "holoprosencephalia," a developmental disorder characterized by anophthalmia or cyclopia and distinct facial defects. Many developmental genes interact with oncogenes and tumor suppressor genes, underlining the relationship between growth control and tumorigenesis. In fact, several oncogenes are active during embryonic life regulating, for example, placental and gonadal structure and functions.

Implantation and the success of pregnancy are also dependent upon the activation of genes and the synthesis of polypeptides that, released in small quantities from the embryonal tissues, reach specific maternal receptors located in various organs including the brain. In a short period of time they should modify the maternal hormonal system to render it suitable for the establishment of implantation and normal processes of gestation. For example, human IVF embryos over the 16-cell stage show very low survival rates when transferred to the mother. This implies that neuropeptides — capable of being released, of being transferred in the maternal circulation, and of reaching specific receptors — are produced by the human embryo at very early stages of development. Molecules controlling biological functions can be active even if produced in very small amounts and are capable of concentrating on a specific target.

So far, little is known about the nature and biological properties of many of these mediators, but pregnancy specific β1-glycoprotein, plasminogen activator, plasminogen activator inhibitor, growth hormone, growth hormone receptors, histamine releasing factor, collagenase, and trombomodulin — all produced at early stages of human embryonic development — are supposed to perform major and probably complex biological roles during implantation.

Human chorionic gonadotropin (hCG) is another hormone that is claimed to play a central role in early embryonic life. Using immunological assays, hCG has been detected in the maternal circulation at implantation and in the supernatants of *in vitro* cultures of human embryos at 4–8 cell stages of development, before differentiation between trophectoderm and inner cell mass.

Trophoblast anchoring and the extracellular matrix proteins, which mediate implantation, are also critical for the success of pregnancy. A specific type of fibrinonectin — trophouteronectin — has been implicated as one of the factors responsible for the attachment of the extravillous trophoblasts to the uterus throughout gestation and it is produced early in life.

In sheep, cows, and pigs, a protein produced by trophoblastic cells is detected early on in maternal blood and signals pregnancy. Ovine trophoblast protein-1 (oTP-1) appears to be a member of the interferon family. In the pregnant ewe, oTP-1 changes the pattern of prostaglandin synthesis in the endometrial cells that line the uterus. This induces the ovary to continue the production of progesterone, necessary to maintain the lining of the uterus during pregnancy.

Origin of Chromosomal Disorders

As a result of studies using chromosome specific polymorphic genetic markers and of investigating chromosomal disorders during gametogenesis, the majority of autosomal trisomies has been shown to originate from nondisjunction during oocyte meiosis. In trisomies affecting chromosomes 13, 15, 16, 21, and 22, the predominant stage of nondisjunction is at maternal meiosis I (MI). For chromosome 21, for example, 90% of all trisomic cases are due to maternal meiotic errors, 5% are of paternal origin, and 5% result from postzygotic defects. An excess of maternal nondisjunction at MII (70%), as compared to MI (30%), has been observed for chromosome 18 trisomies. A different picture has emerged about the origin of chromosome 8 trisomies since about half of these disorders are due to meiotic errors, while the other half are postzygotic. Trisomies 47, XYY, and 47, XXY, are, respectively, 100% and 50% of paternal origin.

Since abnormalities in the maturation of spermatocytes and oocytes are the primary source of chromosomal disorders in preimplantation embryos we will start with a short account of the detection of aneuploidies during gametogenesis.

A scheme describing the terminal stages of the maturation of oocytes and spermatocytes is reported in Fig. 2. Whereas chromosome maturation during mitosis requires that replicated copies of each chromosome should travel to the opposite poles, so that each daughter cell receives a complete set, two rounds of chromosome segregation are needed in meiosis to produce haploid gametes. The first meiotic division (MI) differs from mitosis in that homologous chromosomes first pair and then segregate; in meiosis II (MII) sister chromatids separate and travel to opposite poles. Sister chromatid cohesion must be maintained throughout the first meiotic division to delay chromatid segregation until meiosis II. It is generally believed that this cohesion establishes an inherent polarity ultimately directing sister chromatid segregation via kinetochores that capture microtubules from opposite poles. Several genes and proteins have been identified that control cohesion; their mutations may lead to meiotic nondisjunctions, gamete aneuploidies, and ultimately, embryonic chromosomal disorders.

In the majority of cases, the incidence of chromosomal anomalies in human oocytes has been investigated using unfertilized cells retrieved during IVF programs. In early studies, based on the use of conventional cytogenetic analysis of oocytes in metaphase, the incidence of chromosomal abnormalities was found to vary greatly probably as a result of the techniques employed. A study showed about 78% of human oocytes to have a normal chromosome complement, about 13% were hypohaploid, 8% were hyperhyploid, and 0.4% had structural rearrangements.

During the past few years the presence of aneuploidies in human oocytes has also been investigated using fluores-

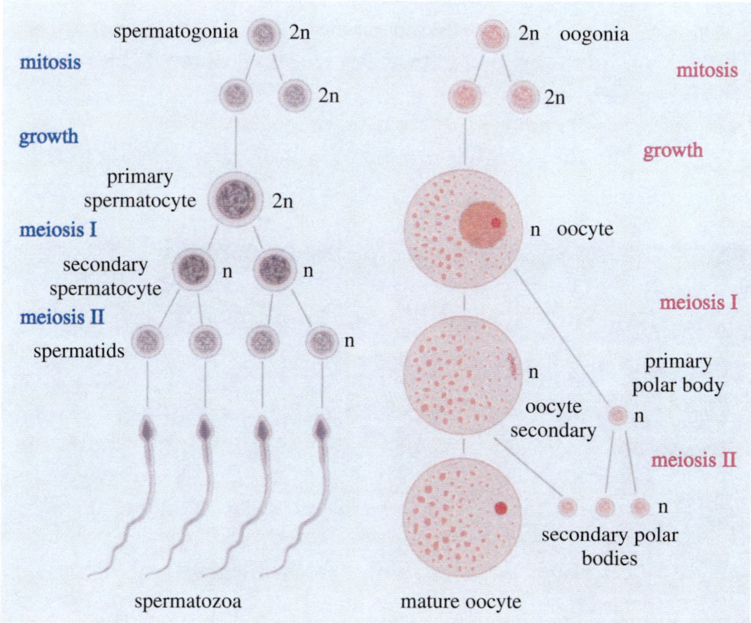

FIGURE 2 Terminal stages of the maturation of the human gametes. In spermatogenesis and oogenesis, first and second meiotic divisions result in the production of four spermatozoa and a single mature oocyte.

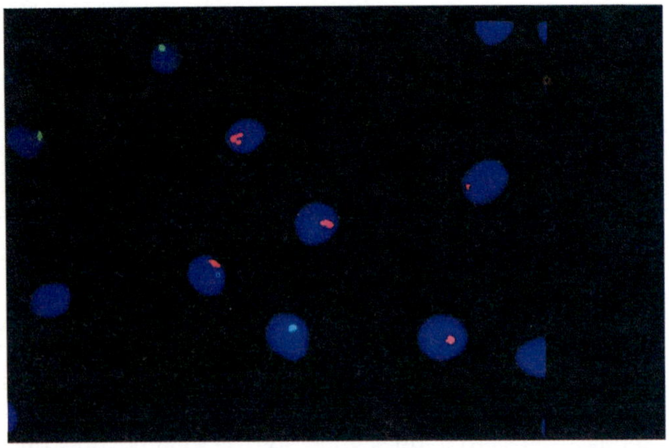

FIGURE 3 Human spermatozoa tested by FISH using probes for chromosomes X (green) and Y (red). Note all cells contain only a single X or Y chromosome.

cent *in situ* hybridization (FISH) methods with multiple DNA probes (Fig. 3). This technique is based on the use of short DNA sequences (probes) which are then hybridized with the genomic DNA present in cells in interphase or chromosomes in metaphase; the complexes are detected because either the probes are directly labeled with a fluorochrome or stained with fluorescent labeled antibodies. This method allows one to detect the correct number of specific chromosomes in a single cell and, of course, "sex" it by monitoring the presence of chromosomes X and Y.

FISH also allows the detection of chromatids of univalent chromosomes in oocytes and polar bodies and thus offers the opportunity to distinguish between nondisjunction of bivalent chromosomes and predivision of chromatids. Usually up to five probes differentially labeled are used per single cell in interphase. As a result of these studies, it is now generally accepted that the classical mechanism of nondisjunction of bivalent chromosomes and predivision of sister chromatids at MI are both responsible for the occurrence of chromosomal disorders in mature oocytes. Data also support the view that the site of the recombination influences the frequencies of nondisjunction.

A large number of investigations have also been performed to clarify the incidence of chromosomal disorders in mature spermatozoa; since they are not dividing cells, in early studies, only indirect methods could be employed to detect aneuploidies. One approach was based on the use of heterospecific fertilization of hamster oocytes. However, this approach was labor intensive, difficult, and necessarily limited to the study of only a small number of sperm capable of fertilization. In contrast, the introduction of multiple FISH analysis of sperm nuclei allows the analysis of many thousands of sperm from a single individual (Fig. 4).

As shown in Fig. 5, in spermatozoa, a disomy may arise from nondisjunction in meiosis I or II. Regarding chromosomes X and Y, a nondisjunction during meiosis I results in hyperhaploidy XY and a nondisjunction at meiosis II in a disomy XX or YY. The general consensus is that disomies and diploidies are less frequent in spermatocytes than in oocytes. The percentage of sperm without chromosome X or Y is around 0.1; the frequency of disomies for these chro-

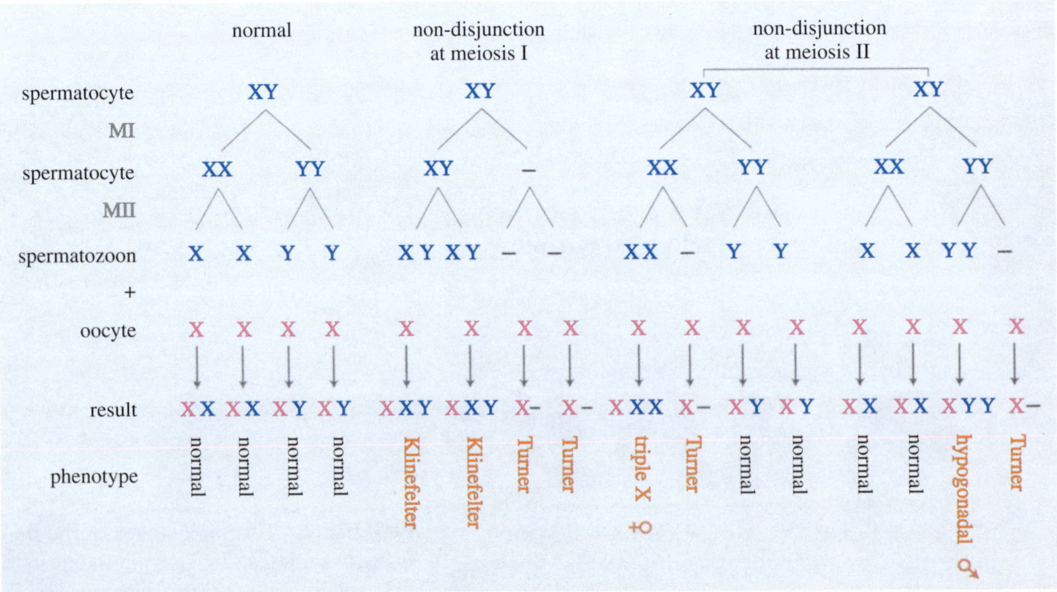

FIGURE 4 Non disjunction of chromosomes X and Y during spermatogenesis may lead to sex chromosome abnormalities such as Turner's Syndrome (45,X) or Klinefelter's Syndrome (47,XXY).

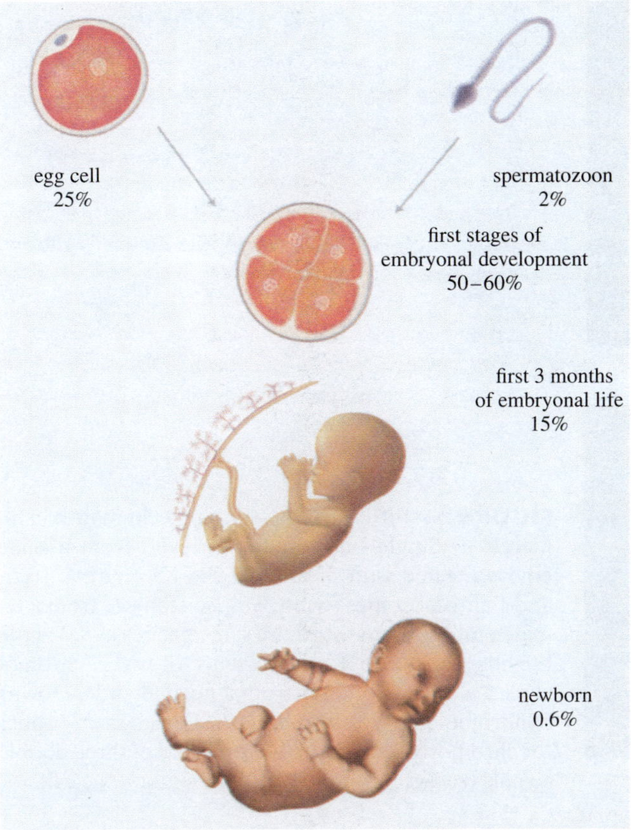

FIGURE 5 Percentages of abortions due to early embryonic or fetal abnormalities.

mosomes is near 0.4%, slightly higher than that observed for the autosomes so far investigated (0.1 to 0.34%). This suggests that the XY bivalents may be more prone to nondisjunction. In studies in which two or three fluorescent probes were used, disomy XY was found to be higher than disomy XX or YY.

Sperm with abnormal chromosomes can fertilize normal oocytes as shown by the incidence of paternally derived chromosomal disorders in embryos and neonates. While in normal individuals the ratio of X and Y bearing sperm is close to 1:1, an excess of Y bearing sperm has been observed in sperm disomic for chromosome 21. This observation is consistent with molecular studies which have identified an excess of males among paternally derived cases of trisomy 21 and suggests that a unique nondisjunctional mechanism is responsible for the preferential segregation of the extra chromosome 21 with the Y chromosome. It is generally believed that, whereas in normal reproduction diploid spermatozoa may account for some cases of triploidy, dispermy is a more frequent cause of human triploidy during conception *in vivo*. Rare cases of trispermia may occasionally occur, as inferred from some types of chromosome abnormalities seen in spontaneous abortions.

Chromosomal Disorders in Preimplantation Embryos

The successful generation of human preimplantation embryos *in vitro* has provided the possibility of examining the chromosomes of early embryos. Since biological and chromosomic studies had suggested that about 60% of human conceptions are lost at an early stage of embryonic life, it was hoped that tests on preimplantation embryos could throw some light on the causes of the high rate of early human wastage (Fig. 5).

The ideal method to study chromosome abnormalities is by karyotyping nuclei arrested in metaphase, so that chromosomes can be banded and identified. However, obtaining metaphase chromosomes from early human embryos is problematic, as only a few of these cells are available and they cannot be cultured *in vitro* for long periods of time. When metaphase chromosomes are obtained, the chromosomes are often short and banding is difficult.

In early studies, karyotype analysis of human embryos has documented the presence of mosaicism, aneuploidy, and ploidy abnormalities ranging between 16 and 30%. The incidence of aneuploidy was found to depend on the morphology of the blastomeres and the quality of the metaphase preparations. For example, only 25% of abnormalities were detected in embryos with good morphology while their frequency could reach 40–64% in fragmented or arrested embryos.

With the advent of interphase FISH, it has been possible to examine the chromosomes of embryonic single cells in interphase. Using this technique, either the whole embryo or a single blastomere can be analyzed using DNA probes that bind to specific regions of the target chromosomes (Fig. 6). A limitation of interphase FISH, compared to karyotyping, is that, due to the close proximity of chromosomes within the interphase nucleus, the number of probes that can be used on a single cell is limited. Thus the tests are usually performed employing probes specific for chromosomes X, Y, 21, 18, and 13. Using FISH tests on early embryos which are going to be transferred to the uterus, only one or two blastomeres can be analyzed; unfortunately, they may not be representative of the chromosome complement of the other embryonic cells.

As expected, FISH studies on abnormally fertilized embryos (such as polyspermic embryos) have shown that these embryos were highly abnormal, in agreement with karyotype data obtained in previous studies. Analysis of arrested embryos using three color FISH with probes for chromosomes X, Y, 21, 18, and 13 have documented numerical aberrations (mostly polyploid or mosaics) in about 60% of embryos, suggesting that if more chromosomes are examined, the frequency of abnormalities could be very high. As expected, embryos from older women were found to have

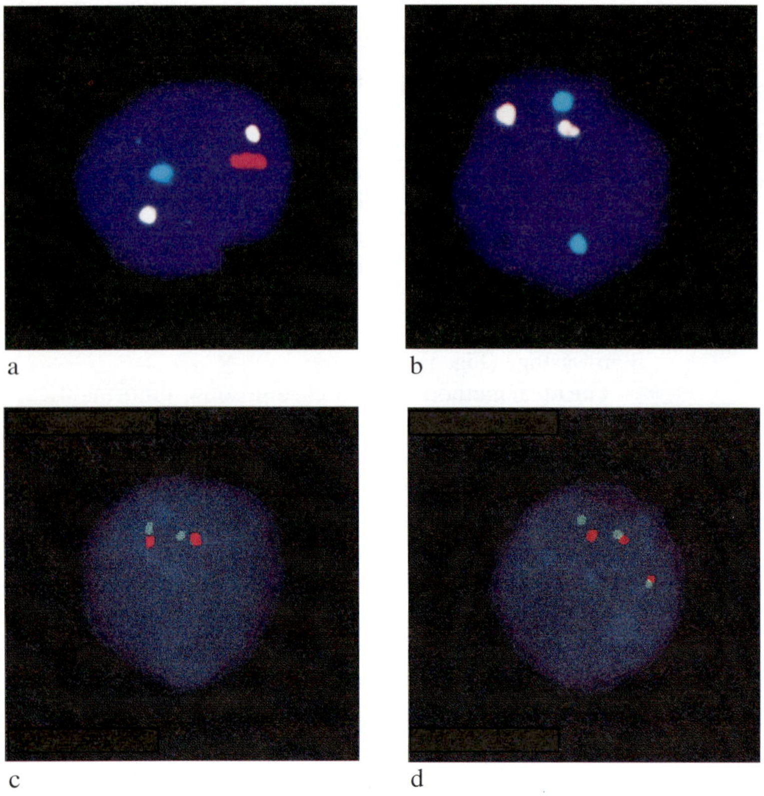

FIGURE 6 FISH analysis of single human blastomeres. a) Single blastomere retrieved form a male embryo tested with probes for the X (bleu), Y (red) and 1 chromosomes (white); b) Blastomere from a female embryo tested with the same probes as "a"; note the absence of the Y and presence of two X chromosomes; c,d) Blastomeres from a normal and a Down's Syndrome embryos tested by FISH using two probes for chromosome 21. Note the presence of three double signals (red-green) in the trisomic embryonic cell.

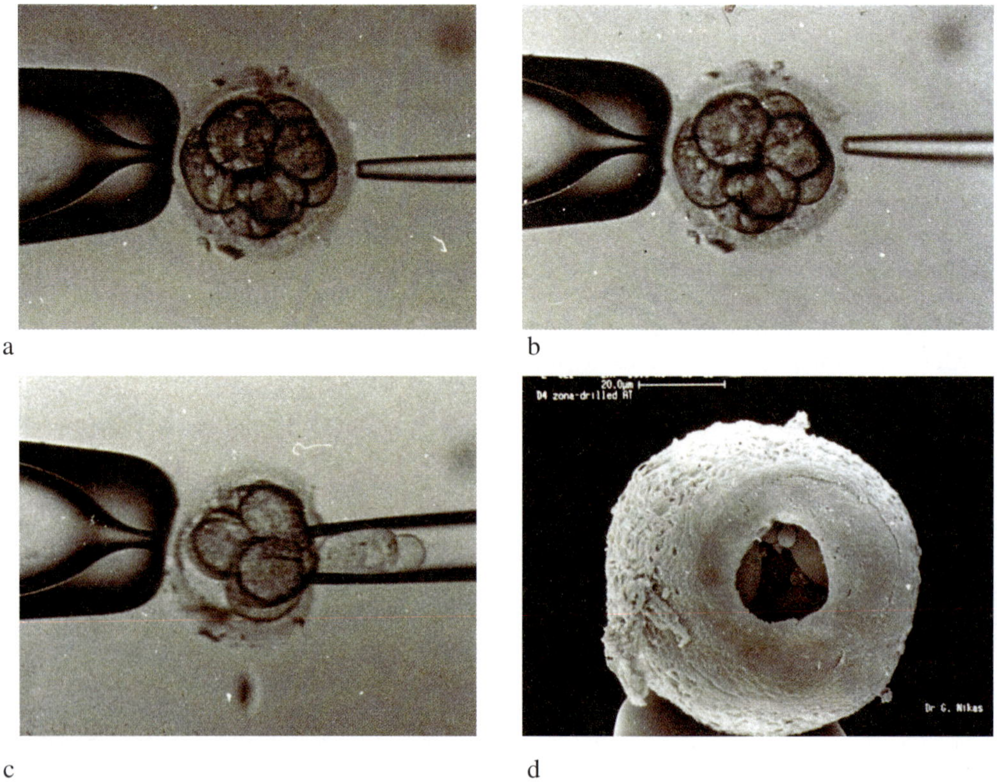

FIGURE 7 Collection of a single blastomere from a 8-cell embryo (a,b,c) and the resulting in vitro culture (d).

TABLE 2

Genetic Diagnosis of Preimplantation Embryos: Strategies Based on the PCR Technique

DISEASE	METHOD USED FOR THE GENETIC ANALYSIS OF THE EMBYRO
Cystic fibrosis	Formation of heteroduplex DNA Visualization of alleles of different sizes using fluorescent markers after PCR Digestion with restriction enzymes ARMS Quantitative fluorescent PCR
β-thalassemia	Digestion with restriction enzymes and use of an SSCP-associated marker
Sickle-cell anemia	Digestion with restriction enzymes and use of an associated marker Quantitative fluorescent PCR
Lesch–Nyham syndrome	Digestion with restriction enzymes
Tay–Sachs disease	Formation of heteroduplex DNA
Blood group Rh(D)	Allele-specific PCR
Insufficiency of acyl-CoA medium-chain fatty, acid dehydrogenase	Fluorescence SSCP and use of an associated marker
Spinal muscular atrophy	Digestion with restriction enzymes
Marfan syndrome	Use of an associated marker Digestion with restriction enzymes RT-PCR and digestion of the amplification products with restriction enzymes
Myotonic dystrophy	The defective allele is not amplifiable with PCR because it is too large
Familial adenomatous polyposis of the colon	SSCP formation of heteroduplex DNA and use of an associated marker
Retinitis pigmentosa	Site-specific mutagenesis and polymorphism of the alleles due to their different lengths
Fragile X syndrome	Use of an associated marker
Duchenne muscular dystrophy	The defective allele is not amplifiable with PCR (the primer site is missing) Use of an associated marker

higher levels of chromosome abnormalities than embryos retrieved from young subjects. Normally fertilized and developing embryos also show high levels of chromosomal abnormalities.

To try to understand the origin of these abnormalities, their pattern has been divided into four groups (Table 2). Embryos are either (a) uniformly diploid for the probes examined; (b) uniformly abnormal, such as in cases of Down's or Turner's Syndromes; (c) mosaic, where usually both diploid cells and aneuploid, haploid, or polyploid nuclei are present; and (d) chaotic, where every nucleus shows a different chromosome complement. Since mosaic and chaotic embryos are quite common and only 1 or 2 cells are analyzed from an embryo, the level of chromosome abnormalities is probably undervalued.

The chaotic group of embryos was an unexpected finding, since such embryos had not been observed in previous studies, probably because they would arrest and fail to implant at very early stages of development.

Embryos containing tetraploid cells may be part of a normal development process of the trophectoderm and represent precursors of syncytiotrophoblast cellular elements. Such cells have also been found in cattle and pigs. Indeed, from the analysis of DNA and the number of chromosomes in human blastocysts, it has been suggested that the presence of some 4N nuclei at the blastocyst stage may be a common event. The analysis of chimeras formed from diploid and tetraploid blastomeres suggests that the abnormal cells may be preferentially allocated to the trophectoderm and this would explain the origin of confined placental mosaicism (see later section).

The possible use of only polar bodies – present in the mature oocyte or in zygote – for preimplantation diagnostic tests is still the subject of controversy. When the mother is a carrier of an autosomal recessive gene, if crossing-over does not occur, the oocyte and the second polar body will have the same genotype at the disease locus. But if a crossing-over has taken place, the oocyte and the second polar body will carry different genes at the same locus. Since many chiasmata are expected to occur during meiosis, the second polar body may fail to provide reliable information about the genotype of the oocyte.

Since most of the chromosome studies performed in recent years have been carried out on preimplantation embryos generated by IVF, it has been suggested that these results may not be representative of *in vivo* development.

However, as mentioned, a high incidence of abortions (~60%) has been observed in embryos generated by natural cycles (Fig. 5) and there is no evidence that chromosome abnormalities may be induced by the *in vitro* techniques.

Preimplantation Genetic Diagnosis of Inherited Disorders

The aim of preimplantation genetic diagnosis (PGD) for the detection of inherited disease is to help couples that are at risk of transmitting the disorder to their children. The option for such couples is to conceive normally and undergo invasive prenatal diagnostic tests between 10 and 16 weeks of gestation. If the fetus is affected, the couple must decide to continue or terminate the pregnancy. For couples who have moral objections against abortion, or for those who have already undergone repeated terminations of pregnancy, preimplantation diagnosis offers a possible alternative. The couple undergo routine IVF procedures and then blastomere biopsy is performed on Day 3 postinsemination, when the embryos are around the 8-cell stage (Fig. 7). The biopsied cells are investigated to determine if they are free from the disease and unaffected embryos are transferred back to the mother so that a pregnancy starts with the parents knowing that the fetus should be normal. For a variety of single gene defects, different types of PCR assays can be used in conjunction with other preimplantation diagnostic tests. Basically the DNA sequence which contains the mutation responsible for the inherited disorder is PCR amplified and then analyzed with different technical strategies.

During the past few years PGD of Tay Sachs disease, cystic fibrosis, Lesch-Nyhan syndrome, β-thalassemia, spinal muscular atrophy, Duchenne Muscular Dystrophy (DMD), Marfan syndrome, myotonic distrophy, and several other autosomal or X-linked disorders have been performed, including the detection of the Rh(D) phenotype of the embryo when the mother is Rh(D)-negative (Table 2).

PCR assays and FISH with chromosome X and Y probes can also be used for the preimplantation diagnosis of X-linked disease. For example, for X-linked diseases such as hemophilia, it may be sufficient to select female embryos to exclude affected boys. FISH tests are also indicated for the preimplantation diagnosis of selected chromosome abnormalities.

In some cases, one of the partners may carry a balanced translocation, leading to unbalanced gametes, ultimately producing embryos with unbalanced arrangements leading to miscarriage. In some couples, either partner may have a gonadal mosaicism when, for example, in previous pregnancies they have had several embryos or pregnancies with chromosome aneuploidies. While lymphocytes have a normal karyotype, the analysis of oocytes or sperm may reveal the presence of chromosome disorders.

For accurate preimplantation diagnosis of inherited dis-

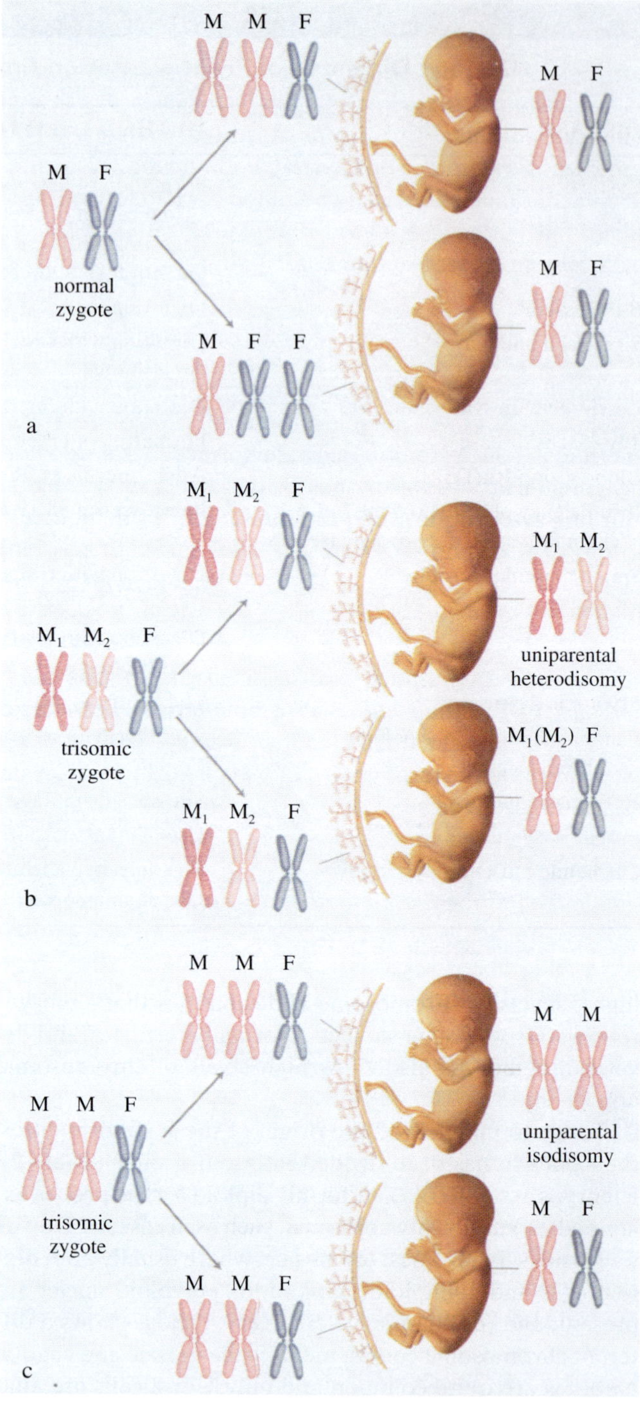

FIGURE 8 Possible origin and types of confined placental mosaicism. M and P: maternal and paternal origin of the X chromosome. (a) Mitotic non-disjunction may be more frequent in placenta than embryo, due to the origin of the embryo from only 4–6 cells out of 64 present in the blastocyst. According to a second hypothesis (b) a mitotic error occurs with the same frequency in the two types of tissues, but the extra chromosome is expelled only from the embryo. Possible origin of uniparental disomies (c).

orders a single cell from an 8-cell embryo is assumed to be representative of the whole embryo. As previously discussed, due to mosaicism and chaotic embryos, this may not be the case for the PGD of chromosome disorders. However, with regard to sexing for X-linked diseases, it appears that an XX cell is always representative of a female embryo, but XO cells may be present in either male or female embryos. Unlike single gene defects with classical Mendelian inheritance, in which the risk of affected children is high, the age-related risk of fetal aneuploidy can be too low to justify IVF and PGD. Still, there may be potential benefits in screening for the most common aneuploidies in older women who conceive by IVF to improve pregnancy rates and reduce miscarriages.

Confined Placental Mosaicism

An abnormal distribution of chromosomes in postzygotic mitotic cell divisions may result in mosaicism with cell lines showing different chromosome complements. When this event occurs in the blastocyst, the mosaicism affects both the embryonic and placental tissues. If, instead, the abnormal mitosis occurs after the differentiation of the embryonic and extraembryonic compartments, the mosaicism will be limited to either the embryo or the placenta. Confined placental mosaicism (CPM) has been noticed in early cytogenetic analysis of chorionic villus samples (CVS) when first trimester fetal karyotyping revealed discrepancies between the results of testing placental cellular elements and, later, cells obtained by amniocentesis or at birth.

When CVS is used for prenatal diagnosis, CPM is recognized as a source of both false positive and negative chromosome results. The incidence of false tests is around 1–2%, with a higher frequency in short-term *in vitro* cultures. Several mechanisms have been postulated to cause CPM (Fig. 8). First of all, it has been suggested that mitotic nondisjunctions may occur more frequently in the chorion than in the embryo, because, while the embryo is derived from only 4–6 cells of the 64 cells forming the early blastocyst,

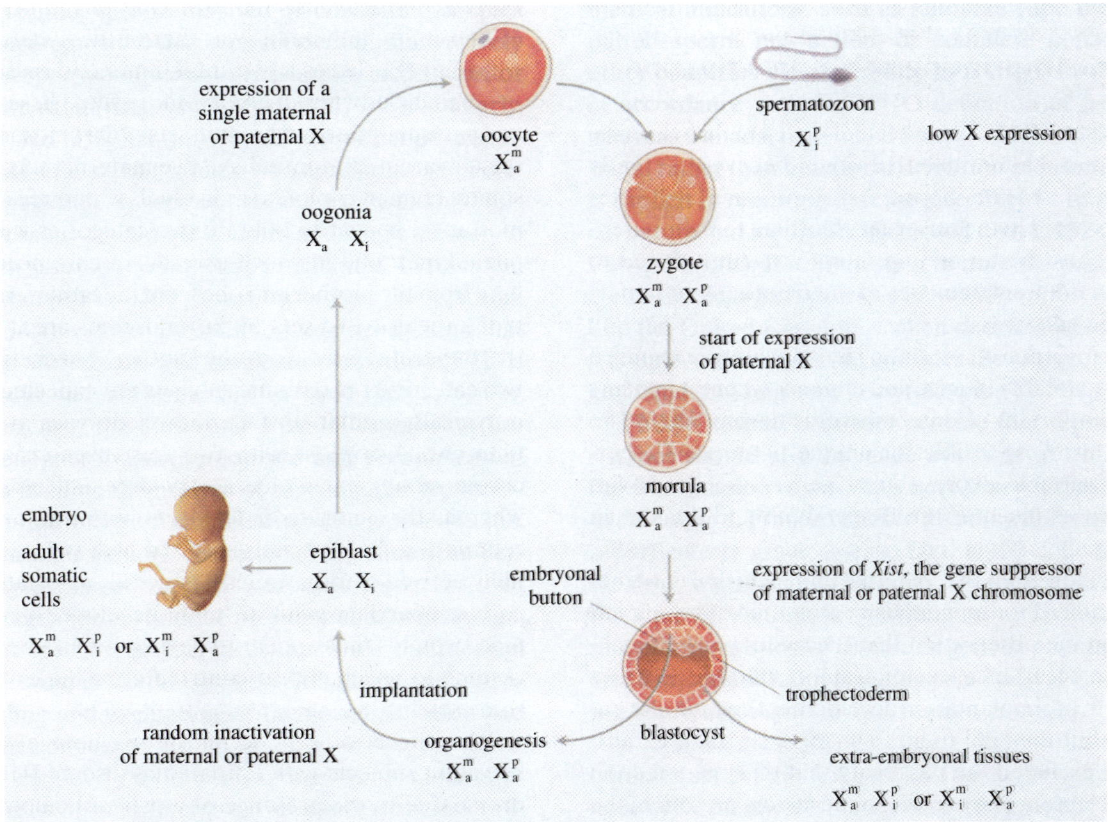

FIGURE 9 Chromosome X inactivation in different cell lineage in a female subject. XM:maternal chromosome X; XF:paternal chromosome X; Xa and Xi:active and inactive X chromosomes. In the zygote and early embryo up to the morula stage both parental X chromosomes are active (with the paternal X less active). Random inactivation in cells giving origin to the embryonic tissue starts around th blastocyst stage with the activation of the XIST gene. Stable, random inactivation of the paternal or maternal X chromosome is completed in trophectoderm and embryonic cells at the implantation stage. In rodents the paternal X chromosome is inactivated in cells giving origin to the placenta.

the remaining cells give rise to the extra embryonic tissue. Another possibility is that mitotic errors take place with the same frequency in both tissues, but that only the embryo can limit the replication of the abnormal cells. According to a third hypothesis the placenta can tolerate cells with chromosomal abnormalities and even receive positive advantages from their presence.

In testing extraembryonic tissues retrieved at 10 to 16 weeks, it has been observed that certain mosaic anomalies — such as those involving sex chromosomes, supernumerary markers, or trisomies 13 and 18 — often extend into fetal cell lineages. Others, as for instance trisomy 16, are always confined to the placenta, but can have adverse effects on the outcome of pregnancy. A third group, which includes trisomy 2, 3, 7, and 8, is usually associated with a normal course of gestation.

A positive association between chromosomal disorders in placenta and fetal growth retardation has been suggested. Although this hypothesis has been confirmed in some studies, the great majority of pregnancies with CPM proceeds uneventfully in normal live-born infants. A positive correlation has also been detected between maternal age and CPM.

Genetic Imprinting and X Inactivation

Over the past 50 years it has been discovered that chromosomes of paternal and maternal origin behave differently and the concept has developed that some genes are differentially expressed in the offspring according to their parental origin. Thus the term "imprinting" was introduced to describe the different activities of the paternally or maternally derived genomes.

The phenomenon of imprinting is now widely recognized in mammals, notably in mice and humans, where it appears to play an important role in embryonic development. Imprinting is also responsible for unique patterns of inheritance and variable expression of genes associated with several diseases (such as the Prader-Willi and Angelman Syndromes) and certain forms of cancer.

The first indication that paternal and maternal genomes are both required for mammalian embryonic development derives from the observation that successful parthenogenesis does not occur. In mice, for example, parthenogenotes can complete preimplantation development and initiate implantation, but they fail to go to term. A series of experiments have excluded the possibility that this phenomenon is due to the absence of extragenomic factors present in the sperm or to the presence of maternal lethal genes, and they have conclusively demonstrated the absolute requirement of both paternal and maternal genomes for normal embryonic development.

A key observation in these studies is the demonstration that mouse embryos with two female pronuclei develop quite normally up to Day 10 of gestation, but fail in the development of the placenta; in contrast, embryos with two male pronuclei develop more poorly but show an overgrowth of the trophoblast. From these results it was possible to conclude that epigenetic modifications during parental gametogenesis had opposite effects on the expression of selected paternal or maternal genes.

Imprinting of DNA regions and its effects have now been identified in many rodent and human chromosomes; furthermore, as mentioned, several inherited human disorders have been shown to arise from abnormal genetic imprinting. Repressed and derepressed chromatin structures constitute the initial germline-dependent imprints. *De novo* methylation and loss of methylation events start to occur during gametogenesis and will continue following fertilization and preimplantation stages. Subsequent modifications, such as DNA methylation, are influenced by these initial epigenetic effects; thus epigenetic modifications of parental alleles occur progressively and affect their potential expression.

Imprinting of some parental alleles controls cell proliferation, interaction, differentiation, specific cell lineage, and embryonic growth; it influences also penetrance and expressivity of transgenic loci. One of the first examples of imprinting in humans was that of the hydatiform mole, which is regarded as a placental tumor and is usually found in pregnancies without embryonic tissue. These moles have two paternally derived haploid sets of chromosomes. In contrast, ovarian teratomas have two maternal sets of chromosomes. Human triploids (individuals with three sets of chromosomes) appear to fall into two categories with different phenotypes: one in which two sets of chromosomes originate from the mother and one from the father, and the other type in which two sets of chromosomes are derived from the father and one from the mother. Theoretically, these two categories of trisomic subjects are expected to be phenotypically similar. In fact, the phenotypes are markedly different. A triploid with two sets of paternally derived chromosomes usually has a very large and cystic placenta, whereas the embryo, or fetus, shows intrauterine growth retardation. In contrast, a triploid with two sets of maternally derived chromosomes is generally associated as a very early miscarriage with an underdeveloped placenta. Human triploids thus appear to have different phenotypes according to which parent contributes to the extra chromosome set.

The effects of genetic imprinting have also been observed in subjects with uniparental disomy (UPD), a syndrome due to the presence of a pair of homologous chromosomes or DNA sequences derived from a single parent. UPD can result from a chromosome loss in trisomic embryos, duplication in a monosomic zygote, gamete complementation, or somatic recombination. The main effects of UPD on prenatal or postnatal development depend on the presence of imprinted genes present in the involved chromosomes. UPD has been observed in several hundred cases and shown to affect chromosome X and at least 12 autosomes. Studies of most cases are usually motivated by the detection of specific inherited diseases such as Prader-Willi Syndrome (PWS), Angelman Syndrome (AS), or Beckwith-Wiedemann Syndrome (BWS). The apparently high involvement of chromosome 15 represents a bias due to the association of this chromosome with PWS and AS.

About 70% of cases of UPD are of maternal origin and 30% are paternal; the genetic analysis can readily distinguish between isodisomies and heterodisomies in UPD cases.

UPD has also been detected in diploid cell lines retrieved from mosaic individuals. In these cases, all mosaic individuals showed phenotypic abnormalities. The presence of congenital defects in all of them has been interpreted as evidence that mosaicism per se, rather than the presence of UPD in the cell lines, could play a major role in the pathogenesis of the developmental defects.

In cases of CVS tests suggesting trisomies for chromosomes 15 or 11, the question arises about the possibility that the fetus may have UPD and further investigations are required to clarify the origin and cause of the anomalies.

A clear effect of imprinted genes is observed on the development of mouse trophoblastic cellular elements that require unique imprinting of parental genes. In this species, as well as in rats and marsupials, a preferential inactivation of the paternal X chromosome has been observed.

Chromosome X inactivation is a specialized form of imprinting that in humans and other mammals plays an important role in sex determination. The imbalance of gene expression, due to the presence of two X chromosomes in females and only one in males, is resolved by the inactivation of one X chromosome in the somatic cells of female subjects. During X inactivation, the two X chromosomes in any cell of a female mammal become differentiated early in development. One, the inactive X chromosome, assumes a compact, condensed structure, replicates its DNA asynchronously, and does not transcribe most of its genes into messenger RNA. This dose compensation ensures equal dosage of the X-linked genes in male and female subjects.

This process starts at the late blastocyst stage of embryonic development, when most genes of one of the two X chromosomes are partially down regulated in somatic cells resulting in the random expression of one of the two alleles at each locus. Consequently females have mosaic expression of paternal or maternal X-linked genes (Fig. 9). Inactivation is stable and following division each daughter cell maintains the same active X chromosomes.

The number of active X chromosomes in any cell is determined by a *counting* mechanism that maintains one active X chromosome per two sets of autosomes. In the end, only one X remains active whatever the number of X chromosomes present in an individual (i.e., 47,XXX; 47,XXY; 48,XXXX). In triploid cells (69,XXX), the ratio X to autosomes is assessed and two X chromosomes are active.

It is generally believed that many processes control X inactivation but that this phenomenon is triggered by a gene(s) mapping on the X chromosome. The concept of the existence of an X inactivation center arose from studies of patients with different types of X chromosome deletions. As a result of these investigations, an X inactivation center (XIC) was identified in the long arm of the X chromosome (Xq 13.3 region) and a gene isolated (XIST/*Xist*). Transcription of the *Xist* gene induces the inactivation of the X chromosome that bears it. However, the *Xist* gene does not seem to be needed for counting since experiments in knockout mice (experimental mice in which a specific gene has been removed) have shown that spreading of inactivation and counting are independent events. Many questions still remain to be answered about X inactivation and the role of the XIST/*Xist* gene. For example, it is not clear what is the significance of the expression of this gene in adults, since loss of the *Xist* gene in cells that have already undergone X inactivation does not cause reactivation.

The phenomenon of genetic imprinting has been the subject of particular interest in the field of assisted reproduction. In fact, many investigators have been concerned about the possible adverse effects of using immature spermatozoa for *in vitro* fertilization by the ICSI method and its variants, since immature gametes may have not yet gone through the full process of imprinting.

At present, there is no evidence suggesting that children born with the help of assisted reproductive technologies are at increased risk of diseases; however, long-term follow-ups are needed to solve the problem.

ICSI

As IVF has revolutionized the treatment of infertile patients, the use of intracytoplasmic sperm injection (ICSI) has revolutionized the treatment of infertile men. Up until 1992, donor sperm was used for the treatment of men with azo- or oligospermia or for those with sperm not capable of fertilization. Several techniques were developed during the late 1980s to try and improve the fertilizing capacity of poor sperm samples. These techniques were aimed at reducing the barrier presented by the oocyte to the sperm (i.e., the zona pellucida). Initially, holes were made in the zona pellucida, using mechanical tears (partial zona dissection, PZD). However these methods increased polyspermy and a new technique was developed based on the injection of sperm directly under the zona pellucida and into the perivitelline space (subzonal insemination, SUZI). Problems arose with this approach as, if only one sperm were used, poor fertilization rates were obtained; on the other hand, if three sperm were injected, the frequency of polyspermy was high. Despite these difficulties, several births were reported. In 1992, the first report of ICSI was published, when one sperm was injected directly into the cytoplasm. Improvements of the technique have been made either by removing the sperm tail or by ensuring that, on injection, part of the oocyte cytoplasm is gently aspirated. The ICSI procedure has now become a common treatment in most IVF units worldwide and thousands of babies have been born.

As well as being able to use sperm from men with oligospermia, the successful aspiration of sperm from the epididymis (MESA – microscopic epididymal sperm aspiration) and testis (TESA – testicular sperm aspiration) is possible and even acrosomeless sperm have been used. Reports on the use of spermatids for ICSI have been published and pregnancies obtained. Men with nonobstructive azospermia have an increased chance of carrying a chromosome abnormality (16%), so that for such patients, a karyotype should be performed to establish the risk of transmitting such an abnormality. In about 10% of cases of nonobstructive azospermia a deletion of the AZF region of the long arm of the Y chromosome is found. In these cases, fathers will transmit the deletion to their male offspring, therefore making them infertile by the same gene defect. It has also been suggested that children born as a result of ICSI have a slightly increased risk of inheriting a sex chromosome abnormality.

Fetal Gene Therapy

Over the past few years, fetal therapy has moved from a purely hypothetical concept to experimentation aimed at the treatment of selected inherited disorders. Basically, the principles of this approach are twofold. The first is based on the transplantation of immature precursors (stem cells) from an immunologically compatible donor into the fetus to replace the defective cells. A few attempts have already been made at correcting the genetic defect in fetuses known to have severe combined immunodeficiencies or thalassemia. The results are promising but still uncertain due to the limited number of cases that have been treated.

In the second approach, a normal gene is introduced by genetic molecular techniques into selected cells obtained from the affected fetus; the treated cells are then reinjected into the fetus. Currently, clinical trials are in progress in young patients with inherited single gene disorders. They include immunological diseases due to adenosine deaminase (ADA) deficiency, selected hemoglobinopathies, cystic fibrosis (CF), α1-antitrypsin deficiency, Gaucher's disease, and Hunter's Syndrome. Some of these disorders require only a low level of normal gene expression and are therefore considered to be the best candidates for gene therapy.

However, the first clinical trials in young patients with ADA deficiency have been only partially successful, highlighting the difficulty of obtaining sustained gene expression even after repeated gene deliveries. It has also become obvious that in many genetic diseases, such as lysosomal deficiencies, treatment must start at an early age to prevent irreparable damage to various tissues and organs particularly when the nervous system is also affected. For this reason, it has been suggested that gene therapy should be applied during development *in utero* when the genetically corrected stem cells may rapidly expand and a still immature fetal immune system may reduce the probability of immunological rejection. Currently, there are no plans to use gene therapy in human conceptuses, but this topic has been the object of several papers and discussions. Clearly, many complex problems need to be resolved. Furthermore, it is essential to confirm that stem cells retrieved from very early embryos are suitable for transfection and subsequent clonal expansion *in vitro*.

The ethical issues related to these new therapeutical approaches can only be solved after the evaluation of several investigations performed in experimental animals and small, carefully planned human studies.

Embryonic Mortality after Implantation

Large numbers of human conceptuses are known to die before or soon after implantation, as judged by the frequency of termination of pregnancies detected using biological, biochemical, and molecular tests. Besides the already mentioned causes of abortion due to chromosomal abnormalities, several maternal factors have been suspected of being responsible for the natural interruption of pregnancy. However, there is no general consensus of opinions on the maternal factors involved. Social, metabolic, immunological, and inherited disorders have all been held responsible for these abortions. Morphological anomalies of the uterus, the site of implantation, and placental anatomical malformations are undoubtedly causes of early abortion. Bacterial, chlamydial, and mycoplasmal infections have been implicated, as well as smoking and chronic alcoholism, as causes of abortion during the first trimester of pregnancy.

The immunological relationship between mother and fetus may lead to abortion, for example, in pregnant women with antibodies against selected phospholipids, lupus anticoagulant, and cardiolipin. Antigenic disparities between mother and fetus have long been considered to be causes of embryonic lethality, although several investigations have shown that ABO and HLA incompatibilities do not induce fetal losses. Rh incompatibility and its effects can now be readily prevented while molecular methods allow to establish the Rh(D) genotype of the father.

Lack of dissimilarities between the parents at the HLA loci has been claimed to cause chronic recurrent abortion. According to this hypothesis, sharing of HLA antigens between parents would induce an immunological maternal hyporeactivity, ultimately responsible for the absence of "blocking" antibodies, essential for the success of pregnancy. However, sharing of HLA antigens – a frequent phenomenon particularly in consanguineous marriages—is not statistically associated with increased recurrent chronic abortion. Induced immunological stimulation of the mother with her partner or third donor lymphocytes does not increase the success of pregnancy as compared to appropriated control groups. Furthermore, this approach may have deleterious effects on the fetus since a few cases of graft-versus-host reactions (GVH) have been observed in infants born to treated mothers.

Luteal deficiency has also claimed to lead to failure of implantation or abortion. Exogenous progesterone is recommended as treatment in mothers with luteal deficiencies. However, doubts have been expressed about the role of most endocrine factors in abortion and recurrent abortion.

Pregnancy losses are easier to quantify and their causes more readily identifiable when they occur later in pregnancy. More investigations are therefore needed to clarify the biological processes of implantation before the causes of its failure can be recognized and prevention of abortion at this stage of fetal life can be achieved.

Conclusion

The impact that molecular genetic techniques have made on all branches of biological science sets the scene for solving complex and challenging problems in the field of early human embryonic development. Our knowledge of the biological processes operating at the molecular level during the development of preimplantation embryos is still limited. But soon, the application of recombinant DNA technology should allow us to clarify developmental areas of fundamental scientific interest as well as to improve our understanding of the genetic mechanisms causing inherited disorders responsible for the high rate of embryonic morbidity and mortality.

The major areas of research that need to be explored to fulfill these aims are those in the field of early gene expression, pattern formation, and the development of specific cellular functions in different tissues and organs. While studies in experimental animals may provide important information, these problems can be solved only by the use of human embryonic material. It is up to the scientific and nonscientific communities to decide whether the gained results justify further carefully planned and specific experimental studies in this area of research.

References

Adinolfi, M., Sherlock, J., Cirigliano, V., and Pertl, B. (2000). Prenatal screening for aneuploidies by quantitative fluorescent polymerase chain reaction. *Comm. Genet.* **3,** 50–60.

Barrister, B., and Squire, J. (2000). Mitochondrial distribution and function in oocytes and early embryos. *Hum. Reprod.* **15** (suppl. 2), 189–198.

Carroll, J., Jones, K. T., and Wittingham, D. G. (1996). Ca^{2+} release and development of Ca^{2+} release mechanisms during oocyte maturation: A prelude to fertilization. *Rev. Reprod.* **1,** 137–143.

Cheung, A., Swann, K., and Carroll, J. (2000). The ability to generate normal Ca^{2+} transients in response to spermatozoa develops during the final stages of oocyte growth and maturation. *Hum. Reprod.* **15,** 1389–1395.

Coon, C. M., Cozzi, J., Harper, J. C., and Winston, R. M. L. (1999). Preimplantation diagnosis for couples at high risk of Down syndrome owing to parental translocation or mosaicism. *J. Med. Genet.* **36,** 45–50.

De Robertis, E. M., Larrain, J., Oelgeschlager, M., and Wessely, O. (2000). The establishment of Spemann's organizer and patterning of the vertebrate embryo. *Nature Reviews / Genetics* **1,** 171–181.

Edwards, R. G. (Ed.) (1990). *Establishing a Successful Human Pregnancy.* Raven Press, New York.

Gardner, D. K., and Schoolcraft, W. B. (1999). *In vitro* culture of human blastocyst. In *Towards Reproductive Certainty* (R. Jensen and D. Mortimer, Eds.), pp. 378–388. Parthenon Press, Conforth.

Harper, J., and Thornill, A. (2001). Embryo biopsy. In *Preimplantation Genetic Diagnosis* (J. Harper and A. Handyside, Eds.), pp. 141–164. Wiley, London.

Harper, J., and Wilton, L. (2001). FISH and embryo sexing to avoid X-linked disease. In *Preimplantation Genetic Diagnosis* (J. Harper and A. Handyside, Eds.), Wiley, London.

Larsen, W. J. (1993). *Human Embryology.* Churchill Livingstone, London.

Miller, J., and Smith, T. T. (2001). The effect of intracytoplasm sperm injection and semen parameters on blastocyst development *in vitro. Hum. Reprod.* **16,** 918–924.

Moore, T., and Reik, W. (1996). Genetic conflict in early development: Parental imprinting in normal and abnormal growth. *Rev. Reprod.* **1,** 73–77.

Ohlsson, R., Paldi, A., and Marshall Graves, J. A. (2000). Did genomic imprinting and X chromosome inactivation arise from stochastic expression? *Trend Genet.* **17,** 136–141.

Pergament, E. (2000). The application of fluorescence *in situ* hybridization to prenatal diagnosis. *Curr. Opin. Obst. Gynec.* **12,** 73–76.

Tesarik, J., and Mendoza, C. (1996). Genetic imprinting abnormalities: A new potential risk of assisted reproduction? *Mol. Hum. Reprod.* **2,** 295–298.

Tesarik, J., Souza, M., Greco E., and Mendoza, C. (1996). Spermatids as gametes: Indications and limitations. *Hum. Reprod.* **13,** 89–107.

Thorogood, P. (ed.) (1998). *Embryo, Genes, and Birth Defects.* Wiley, Chichester.

Wells, D., and Sherlock, J. (2001). Diagnosis of single gene disorders. In *Preimplantation Genetic Diagnosis* (J. C. Harper and A. H. Handyside, Eds.), pp. 165–190. Wiley, London.

Mauro Giacca
Arturo Falaschi

International Center for Genetic Engineering and Biotechnology
Trieste, Italy

Gene Therapy

The concept of gene therapy started to develop at the end of the 1980s due to our understanding of the molecular basis of several human diseases and to the development of efficient techniques for gene transfer. Gene therapy was originally aimed at the treatment of some inherited disorders with monogenic recessive inheritance through the insertion of a normal copy of the disease gene in the appropriate tissues. In addition to coding genes, gene therapy also involves the transfer of DNA and RNA fragments with catalytic or regulatory function, and it is also interested in the therapy of cancer, cardiovascular disorders, and AIDS. In most cases, transfer of therapeutic nucleic acids is obtained by means of viral vectors, the most common ones being those based on murine and avian retroviruses, adenovirus, and the adeno-associated viruses. Based on the preliminary results of the first clinical studies of gene therapy, it is still premature to conclude that some patients have already obtained clinical benefit from the application of this technology. Gene transfer in humans is feasible and safe, but several experimental goals have to be attained before some important therapeutic results are generated.

Introduction

Since the end of the 1980s, the technological and scientific revolution generated by the application of genetic engineering — that is, the possibility to specifically modify the inheritable properties of living organisms — has prompted several research groups to apply these techniques for the therapy human diseases. Predictably, the first objective has been to cure hereditary monogenic disorders with recessive inheritance. For example, should these techniques allow to specifically correct the mutations in the β chain of hemoglobin, it would be possible to cure β-thalassemia, an inherited disorder with high prevalence. Less ambitiously, since this disorder has recessive inheritance and one normal allele of the β-globin gene is sufficient to prevent the disease, the same therapeutic goal could be attained through the insertion of a normal copy of the gene in the mutated cells. Given the technical difficulties that hamper the possibility of modifying the mammalian cells' genomes at specific sites, gene therapy of inherited disorders actually developed using this latter strategy.

In the subsequent years, it was realized that this therapeutic approach could be extended to other disorders that are much more prevalent in the population compared to inherited monogenic diseases. In fact, the array of possible therapeutic genes goes beyond those with substitutive properties and includes DNA or RNA fragments with regulatory activity on cell growth or metabolism or that are capable of modulating the functions of the immune system. As a consequence, the spectrum of diseases now potentially treatable by gene therapy includes several acquired disorders, such as cancer, cardiovascular disorders, and HIV infection.

Parallel to the extension of the therapeutic possibilities of gene therapy, in the past few years several technical and conceptual problems have arisen that hamper the large-scale application of this technology. These difficulties are evident when the clinical results obtained in the more than 3000 patients so far treated with gene therapy are considered.

In this article, we describe the main strategies for gene therapy, analyze the possible therapeutic genes and the techniques used for their delivery, and highlight the results of the first clinical trials. We also discuss the technical problems that still need to be solved and the social and ethical implications of the modification of the human genome.

Gene Transfer Techniques

The success of gene therapy strictly depends on the efficiency of introduction of exogenous genetic material into the genome of the target cells. In this respect, two alternative approaches are possible. The first one is aimed at the treatment of the target tissue cells, drawn from the patients, after *in vitro* culture in the laboratory. This *ex vivo* treatment is appropriate, for example, for gene transfer into hematopoietic precursors or hepatocytes. The second possible approach is based on *in vivo* gene transfer, for example, into the muscle, the brain, or the cells of the arterial wall. A schematic representation of the different possibilities for gene therapy is shown in Fig. 1. The examples are taken from some clinical studies that will be described and discussed later. The most important available techniques for gene transfer are presented in Table 1, which indicates their advantages and disadvantages.

Naked DNA

The simple addition of naked DNA, usually as a plasmid, to the extracellular milieu is generally a very inefficient means to obtain internalization of a DNA molecule inside the cells. An exception to this rule is given by the cells of some tissues, such as skeletal and cardiac muscle, which are capable of actively endocytosing DNA fragments, which are subsequently brought to the nucleus and expressed. Relatively more efficient is the treatment of cells with short DNA fragments (oligodeoxynucleotides of 15–30 nucleotides), which can freely enter the cells if present in relatively high concentration in the extracellular medium. The nucleotide sequences of these oligonucleotides can be designed to be complementary to those of some cellular or viral mRNAs. Once inside the cells, these antisense molecules pair with their target nucleic acids and promote their degradation by cellular nucleases that specifically recognize and cleave the DNA/RNA hybrids. In the past few years, this approach has been extensively used in several laboratories to experimentally inhibit the functions of cellular genes. However, "natural" oligonucleotides, in which the different nucleotides are joined by phosphodiesteric bonds as in normal DNA, are inefficient drugs since they poorly enter the cells, are rapidly degraded by cytoplasmatic nucleases, and are prone to form unspecific interactions with nucleic acid sequences different from their specific targets. To overcome these limitations, in the past few years some chemical modifications of the backbone of these molecules have been introduced. These modifications are aimed at improving the pharmacokinetic properties of these molecules while maintaining the consecutive arrangement of the bases that confer specificity to the hybridization with the target nucleic acids (Matteucci and Wagner, 1996). Several of these chemically modified oligonucleotides have produced interesting experimental results in animal models, and some of them have been tested in clinical trials.

Chemical Methods

Although the most widely used methods for gene transfer are those based on viral vectors, at least two reasons suggest that nonviral methods will be preferred in the near future — namely, safety and ease of production. A fully synthetic gene transfer system eliminates the risk of biological contamination with a virus capable of autonomous replication or exerting toxic effects during the transduction procedure. Additionally, chemical or physical gene transfer procedures are generally much simpler than methods used for the production of viral vectors. However, these advantages of nonviral gene transfer procedures are currently largely outweighed by their relative inefficiency.

Among the chemical methods for gene transfer, the most thoroughly investigated is lipofection, which exploits the properties of cationic lipids and liposomes (Ledley, 1995). Cationic lipids exert ionic interactions with DNA and form complexes; liposomes form microscopic lamellar vesicles surrounding DNA. In both cases, through these interactions DNA is transferred efficiently through the cytoplasmatic membrane and internalized into the cellular endosomes. The particles formed by the interaction of lipids with DNA can be targeted to specific cellular receptors by the introduction of specific peptidic ligands. Among these ligands are the hemoagglutinin of Sendai virus, which increases the efficiency of fusion of liposomes with the cellular membrane, and the peptides carrying the amino acid sequence Arg-Gly-Asp (RGD), which specifically bind to integrin receptors of the cells. These procedures are still quantitatively inefficient since the transporter lipid allows the intracellular entry of several million DNA molecules, of which the majority are degraded inside endolisosomal vesicles. Only a small part escapes degradation, is brought to the nucleus, and is expressed. Of the molecules that reach the nucleus, only a few are randomly integrated into the genome.

DNA can also be transferred by forming protein/DNA complexes to exploit the property of some proteins to interact with specific receptors present on the membrane of some cell types. After the interaction is formed, the complex

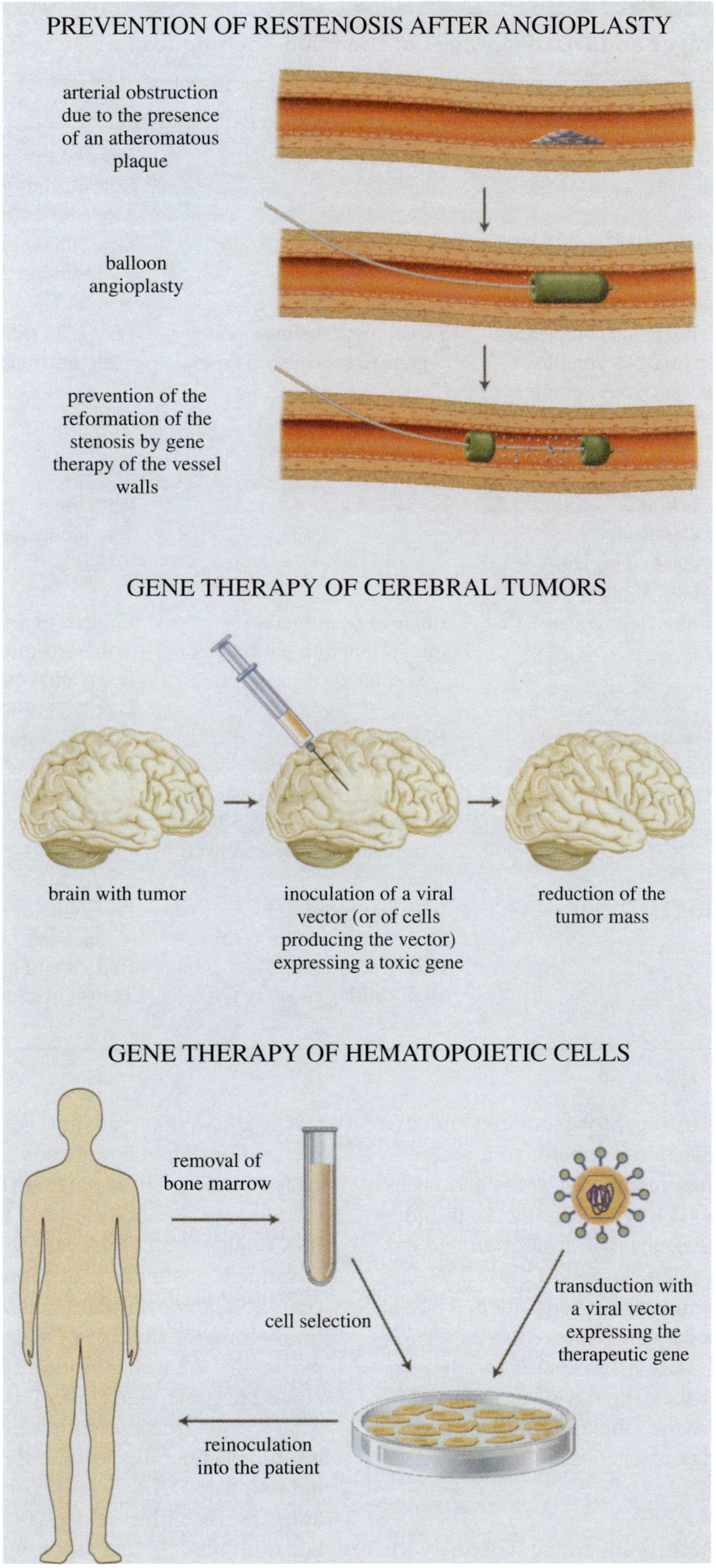

FIGURE 1 Examples of gene therapy *in vivo* (prevention of restenosis and cerebral tumor therapy) and *ex vivo* (hematopoietic cell therapy).

TABLE I

Advantages and Disadvantages of the Main Techniques for Gene Transfer

STRATEGY	METHOD	ADVANTAGES	DISADVANTAGES
Naked DNA	Plasmids and oligonucleotides	Easy production and use	Limited to some applications Low efficiency Transitory effect
Lipofection	Liposomes Cationic lipids Liposome/protein complexes	Simplicity of use Possible use *in vivo*	Low efficiency Random integration into the genome
Protein/DNA complexes	Transferrin/DNA complexes Lactoferrin/DNA complexes Poly-L-lysine/DNA complexes Asialoglycoprotein/DNA complexes Viral capsid/DNA complexes	Targeting of the therapeutic gene to specific cell types	Low gene transfer efficiency and random integration into the host cell genome
Physical methods	Microinjection Electrophoration Bombardment with DNA-coated beads (gene gun)		With the exception of the gene gun, limited to to laboratory use
Transduction with viral vectors	Retroviruses (lentiviruses)	Efficient transduction Integration into the cellular genome	Limited to actively proliferating cells (with the exception of lentiviruses) Insertional mutagenesis Low expression from the LTR in some cells
	Adeno-associated viruses	Nonpathogenic Large host tropism High titer preparations Integration in resting cell populations (muscle, liver, brain)	Limited cloning capacity (<5 kb) No packaging cell line available Requires unknown cellular function for replication
	Adenoviruses	Infect quiescent cells Can be delivered *in vivo* Does not integrate High cloning capacity (7–10 kb)	Possibility of immune and inflammatory reactions Risk of autonomous replication Transient gene transfer

is internalized into the cells by receptor-mediated endocytosis. For example, asyaloglycoprotein binds to a receptor expressed only by hepatocytes; in contrast, transferrin uses receptors expressed by most cell types. Similar to lipofection, also in this case the efficiency of transfection and expression of the transferred DNA is modest since DNA is trapped within endosomes and rapidly degraded. The simultaneous treatment of cells with endosomolytic agents (e.g., the addition of inactivated viruses, such as adenoviruses, that naturally possess the property of escaping lysosomal degradation) increases the efficiency of these procedures several thousand-fold (Curiel *et al.,* 1991).

Physical Methods

Among the physical methods, bombardment of cells with DNA-coated microscopic particles offers very interesting results (Yang *et al.,* 1990). The most popular version of this method takes advantages of a special gun for high-speed shooting of gold beads coated with the gene of interest (gene gun). These particles can cross the cell and nuclear membranes and are stopped within the nucleus, in which they release their DNA cargo. This technology, which takes advantage of similar procedures developed in the past few years to transform plant cells, offers an important opportunity for gene therapy of accessible tissues such as skin. In particular, the utilization of this procedure in animal models and in controlled clinical trials appears to be very promising for the administration of naked DNA plasmids coding for proteins that stimulate the immune system. In the dermis, this DNA is loaded by unknown mechanisms by antigen-presenting cells (APCs) and can efficiently stimulate both the cell- and antibody-mediated immune response involving T and B –lymphocytes, respectively. Given these properties, the utilization of this technology is very promising to promote DNA immunization against several pathogens by using individual cloned genes as immunogens to elicit a protective response.

Viral Vectors

The most common techniques for gene transfer are based on the utilization of viral vectors, which exploit the properties of viruses to infect their target cells at high efficiency by interacting with specific receptors on the cell membrane. Viruses belonging to different families are used for this purpose. The genomic organization of the four most common viruses used for gene transfer — animal retroviruses, adeno-associated virus, adenovirus, and lentiviruses — is schematically shown in Fig. 2. The main advantages and disadvantages of these viruses as gene therapy vectors are also summarized in Table 1.

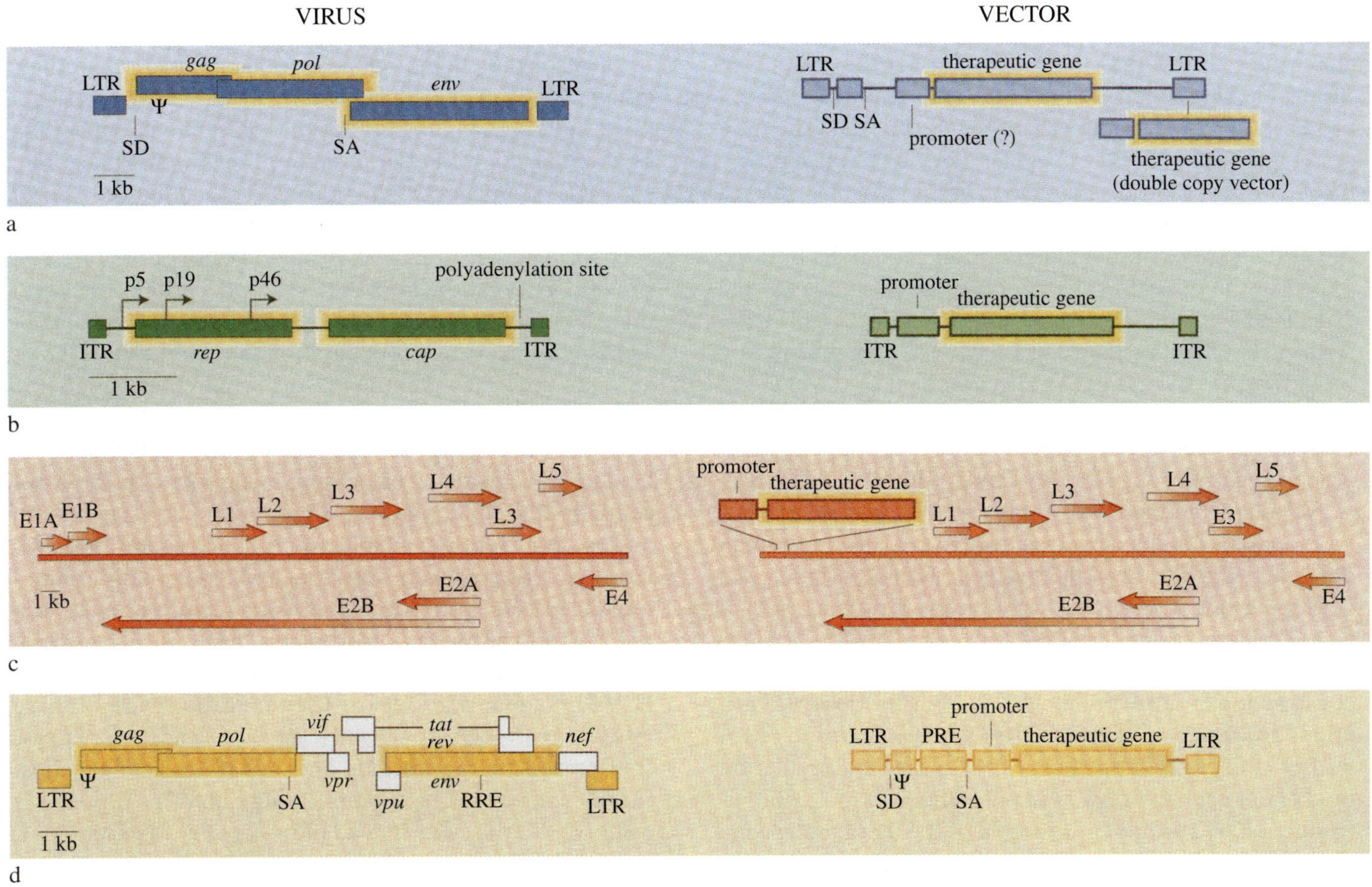

FIGURE 2 Viral vectors for gene therapy. Organization of the genomes of four different types of viruses and modifications introduced for their utilization as vectors. (a) Murine and avian retroviruses. The three viral genes (*env, gag,* and *pol*), the two LTRs flanking the proviral genome, the location of the splice donor (SD) and splice acceptor (SA) sites, and the location of the packaging signal (ψ) are indicated. In the vectors derived from these viruses, the therapeutic gene can be cloned either between the two LTRs or within the LTR. In the former case, control of therapeutic gene expression is directly exerted by the LTR or by an internal promoter. (b) Adeno-associated virus (AAV). The two viral genes (*rep* and *cap*), the two inverted terminal repeats (ITR) flanking the genome, and the location of the three viral promoters (p5, p19, and p46) and of the polyadenylation site are indicated. The viral vectors derived from this virus consist of the promoter–therapeutic gene pair flanked by the two ITRs. (c) Adenovirus. The locations of the main early (E1–E4) and late (L1–L5) transcriptional units of the virus are indicated schematically. The first-generation adenoviral vectors contain the therapeutic gene under the control of a specific promoter in the 5′ portion of the genome substituting for the E1 region. The second- and third-generation vectors (not shown) also have deletions of the E3 and E4 regions (with the exception of the E4 ORF6 transcriptional unit). (d) Lentivirus. The genetic organization of HIV-1 is represented, showing the genes common to murine retroviruses (*gag, pol,* and *env*) and those specific to this retrovirus (in white). Inside the *env* gene, the position of the Rev-responsive element (RRE) is shown. In the HIV-1-derived vectors, the packaging signal and the RRE element are retained, whereas the other viral functions are provided *in trans* by the cell lines used for the production of the recombinant virion.

Most of the current experimental and clinical applications of gene therapy take advantage of the properties of avian and murine retroviruses (Miller, 1992). These viruses have a relatively simple genetic organization, infect a wide variety of cell types at high efficiency, and their cell cycle involves stable integration into the infected cell genome in a DNA form (provirus). The genome of the integrated provirus consists of two long terminal repeat (LTR) sequences that are formed during the reverse transcription process and flank the rest of the viral genome, containing three essential genes. These genes code for the proteins of the nucleoproteic core of the virus (*gag* gene), for the proteins of the envelope (*env* gene), and for the reverse-transcriptase–protease–integrase complex (*pol* gene) (Fig. 2a). The integrated provirus acts as any cellular gene: Its replication coincides with the process of transcription, which is carried out by the cellular RNA polymerase II and is regulated by the cellular transcription factors binding to the LTR at the 5′ end of the provirus. The viral transcript initiates at the 5′ LTR and ends at the 3′ LTR of the provirus. This transcript acts either as the template mRNA for translation of viral proteins or as the viral genome that is included in a new viral particle. Viral proteins are produced directly from the genomic mRNA (for Gag and Pol products) or after splicing of this mRNA (for Env products).

Starting from the retroviral genome backbone, a retroviral vector is obtained by leaving the two LTRs and a small portion at the 5′ end of the *gag* gene (which contains the packaging signal, ψ, for inclusion in the viral particle) intact and by replacing the rest of the genome with the therapeutic gene. Several variants have been proposed for the construction of retroviral vectors. Among the possible variations are the choice of the LTRs (avian and murine retroviruses differ widely in their cell tropism and regulation of proviral gene expression in different cell types), the introduction of a promoter upstream of the therapeutic gene (e.g., specific for a certain tissue), the inclusion or the mutation of the splicing sites inside the retroviral genome, and the introduction of another gene in the vector allowing selection of transduced cells.

An alternative option is to clone the therapeutic gene directly inside the 3′ LTR of the vector, thus allowing the reverse-transcription process to copy this gene also inside the 5′ LTR in order to obtain two copies of the gene (double-copy vectors). The development of all these variants is aimed at the construction of vectors with increased efficiency of transduction and regulated expression of the therapeutic genes inside the transduced cells.

Once the retroviral vector is obtained as a plasmid DNA, this is transferred into a cell line, usually of murine derivation, which is used for the production of the recombinant virions (packaging cell line) (Miller, 1990). These cells contain the *gag, pol,* and *env* retroviral genes, which are essential for viral particle production, integrated into different locations of their genome (Fig. 3). The transferred plasmid DNA molecules then transcribe the vector RNA, which due to the ψ region is recognized by the retroviral proteins produced by the cells and is included in the virion. As a result, a complete infectious viral particle is released by the cell, which contains the recombinant vector genome. This particle is able to infect at high efficiency the cells of the target tissue, where the genomic RNA of the vector is reverse transcribed. After integration in the host cell chromosomes, the vector cannot replicate any longer and acts as a novel gene added to the genetic material of the host.

Most of the current clinical applications of gene therapy take advantage of murine or avian retroviruses, despite some disadvantages. Among these are the difficulty of obtaining high-titer viral vector preparations (maximum 1×10^7–10^9 infectious viral particles per milliliter, according to the specificity of the receptors interacting with the vector). Another drawback of these vector is the fact that proviral integration into the host cell chromosomes requires that the target cell undergo at least one cycle of division since the viral preintegration complex cannot cross an intact nuclear membrane. This requirement limits the utilization of these vectors in cells that are not actively dividing, such as neurons in the central nervous system or hematopoietic stem cells in the bone marrow. In the latter cell type, the induction of proliferation — at least *in vitro* and in the currently available experimental conditions — almost inevitably coincides with the induction of differentiation of the cells, with consequent loss of pluripotent potential. This difficulty could be overcome by the use of retroviral vectors based on the family of lentiviruses, and in particular HIV-1 (Naldini *et al.,* 1996). The replication cycle of these viruses involves a step in which the viral genome is actively transported to the nucleus and integrates into the host chromosomes even in nonreplicating cells. This property is conferred to the virus by two of its proteins, Vpr and MA, which transport the preintegration complex formed in the cytoplasm across the nuclear membrane. Lentiviral vectors are conceptually analogous to murine and avian retroviral vectors since they contain the therapeutic gene under the control of an appropriate promoter and the packaging signal flanked by the two viral LTRs (Fig. 2d). These vectors also contain the RRE region of HIV-1, which binds to the Rev protein and is required for the transport of the vector RNA from the nucleus to the cytoplasm of the cells used for the production of recombinant particles. All the viral functions required for the production of viral particles and for the regulation of this process are provided *in trans* inside the packaging cells due to the expression of two packaging plasmids which are cotransfected with the plasmid vector. Obviously, the potential utilization of HIV-1 vectors in humans raises an essential safety issue, especially related to the potential generation of infectious wild-type virus formed by accidental recombination in the packaging cells.

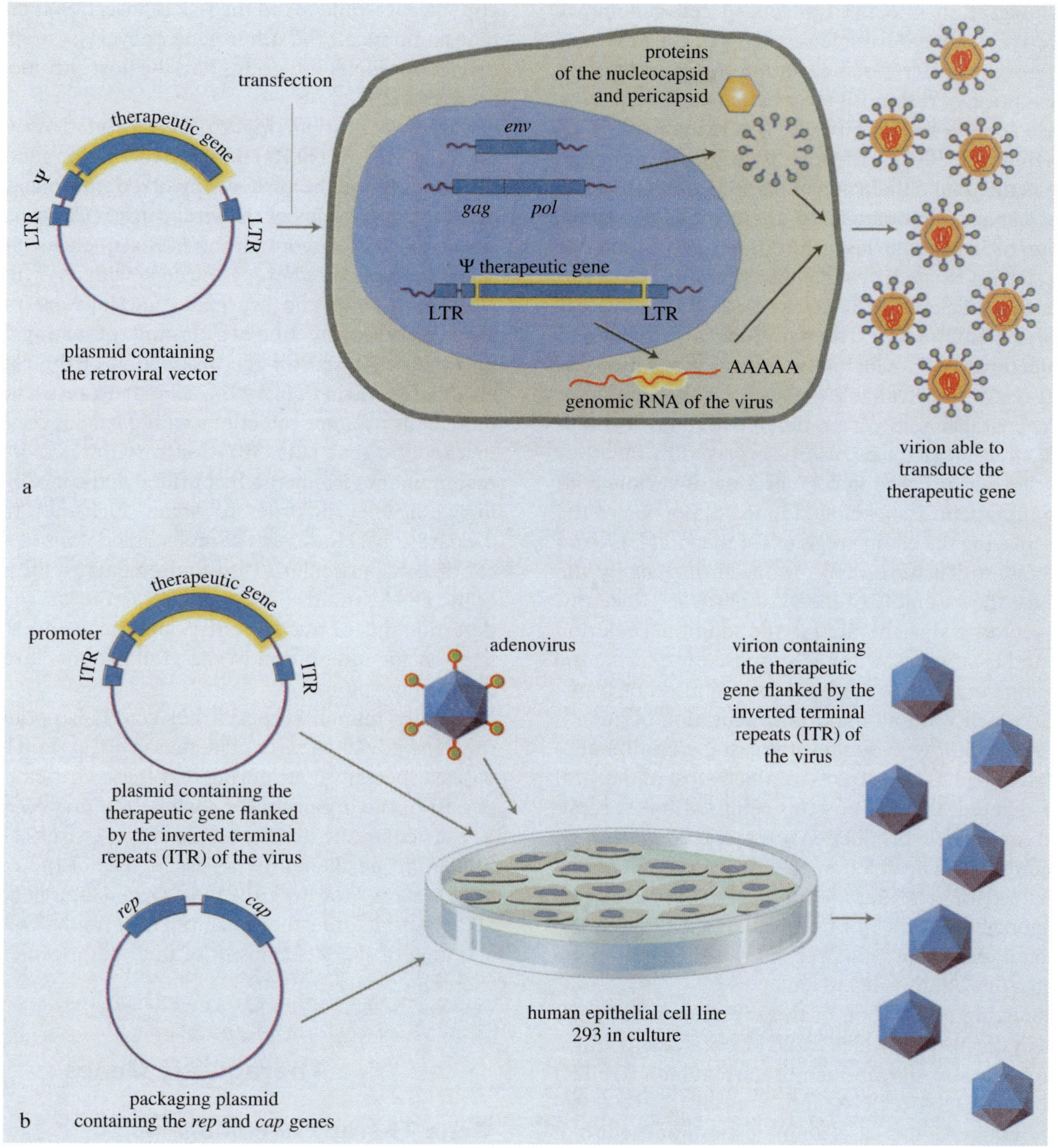

FIGURE 3 Production of recombinant murine/avian retroviral vectors and adeno-associated virus vectors. (a) Scheme of the procedure used for the production of recombinant retrovirus for gene transfer using a packaging cell line. The cells of this line contain, integrated in different locations in the genome, the genes coding for the viral proteins Env, Gag, and Pol. The viral vector, as a plasmid, is transfected into the cells and, once integrated in the genome, it starts producing a messenger RNA expressing the therapeutic gene and containing the packaging signal for the formation of the nucleocapsid (ψ). This signal is recognized by the viral proteins produced by the *gag* gene, and the RNA containing it is thus included in the viral particle that is released into the cell culture supernatant. (b) Scheme of the production of adeno-associated recombinant virus (AAV) for gene therapy. Adeno-associated vectors contain the therapeutic gene, under the control of a suitable promoter, flanked by the viral ITRs. This construct is transfected into a cultured cell line together with another plasmid expressing the viral Rep and Cap proteins. Upon adenoviral infection, which stimulates the synthesis of unknown cellular factors required for AAV replication, the recombinant viral genome is replicated and included in the viral particle. The adenovirus can be easily inactivated by heating at 56°C since AAV is resistant at this temperature.

When integration into the transduced cell genome is not required or even undesirable — such as in gene therapy of cancer, gene transfer for vaccination, or gene therapy for the prevention of restenosis after balloon angioplasty — vectors based on the adenoviruses offer very interesting advantages (Berkner, 1992). Adenoviral vectors display several features that make them appealing for gene transfer *in vivo.* They can accommodate large fragments of therapeutic DNA (up to 30 kb in the last generation vectors) and can infect a variety of tissue types (including resting cell populations) at high efficiency (often close to 100% in some experimental systems), and viral preparations can be obtained at high infectious titers. Currently, several generations of adenoviral vectors are available that differ in the amount of the wild-type viral genome that is deleted to remove potentially toxic viral functions and can be replaced by the therapeutic gene (Fig. 2c). The first generation of vectors contained the therapeutic gene cloned in the 5′ portion of the genome, replacing the coding regions for the viral E1A and E1B proteins, which were then produced *in trans* by the cells used for the assembly of the viral particles. Recently, improved vectors were generated by the additional deletion of the E3 and E4 regions, with the purpose of reducing the extent of inflammation and immune recognition of transduced cells which were an important limitation of adenoviral transduction *in vivo.* Finally, the last-generation adenoviral vectors include constructs in which most of the viral genome is deleted, and the proteins required for its replication and for particle production are provided *in trans* by the packaging cells. The deletion of viral genes is not necessarily beneficial for gene therapy applications. For example, the E3 region codes for a 19-kDa protein that protects the virus-infected cells from being recognized by the immune system and consequently destroyed.

An interesting application of these vectors is for gene therapy of cystic fibrosis, which could be obtained by transferring the cystic fibrosis transmembrane regulator (CFTR) complementary DNA (cDNA) into the cells of the respiratory epithelium. In this case, an indication for the use of adenoviral vectors is given by the natural tropism of adenovirus for the epithelial cells of the respiratory tract. This property can be exploited by direct instillation or aerosol administration of the recombinant virions.

An additional class of viral vectors for gene therapy is based on the adeno-associated viruses (AAV). These are small viruses with single-stranded DNA genomes that require auxiliary cellular functions for their replication and are characterized by their high-frequency integration into the host cell genome, even in the absence of cell proliferation. Additionally, the DNA genome of wild-type AAV preferentially integrates into a specific region of human chromosome 19 (band 19q13.3) by unknown molecular mechanisms. If also maintained by recombinant vectors, this property of AAV would be very desirable for gene therapy since it would avoid the risk of insertional mutagenesis that hampers all the other gene delivery systems in which exogenous DNA integrates into the host chromosomes in a random manner.

The AAV genome (Fig. 2b) is constituted by two inverted terminal repeats (ITRs) flanking two major genes (*rep* and *cap*) coding for the proteins involved in replication (Rep) and for the proteins of the viral capsid (Cap), respectively. AAVs exclusively contain the ITR sequences (165 nucleotides) flanking the therapeutic gene with its own promoter, whereas the *rep* and *cap* gene functions are provided *in trans* by the transfection of a plasmid containing these genes into the cells used for the production of the recombinant virions (Ferrari *et al.,* 1997). Similar to retroviral vectors, these cells produce infectious viral particles containing the therapeutic gene (Fig. 3b). Contrary to retroviruses, these recombinant virions can be purified and concentrated with high efficiency in order to obtain high-titer transducing preparations. However, the molecular details of AAV replication, and particularly the requirements for the production of the DNA strand complementary to the viral genome, the determinants of nuclear entry, and the events that lead to site-specific integration of the viral genome, are currently largely unexplored.

Finally, human herpesviruses could also potentially offer some advantages for gene therapy of certain human diseases, especially considering that these viruses can latently persist in the organism for a whole lifetime (Geller, 1993). In particular, the herpes simplex type-1 virus (HSV-1) can infect and permanently persist in cells of the central nervous system. However, the large size of their genome (150–200 kb), that limits their manipulation, and the poor understanding of the mechanisms of their pathogenicity prevent their clinical utilization.

Therapeutic Genes

Gene Therapy of Monogenic Inherited Disorders

The targeted insertion of an exogenous DNA fragment to replace a homologous sequence within a mammalian cell genome is a very unlikely event that occurs about 1000 times less frequently than random integration. As a consequence, gene therapy was originally conceived with the idea of supplying a cellular function, which was absent due to a genetic defect, by adding an extra copy of a normal version of the gene. Since in most cases human genes are too long to be easily manipulated, the transferred genetic element is usually the cDNA of the gene or, more often, only its coding region. This cDNA is added to the genome of the appropriate cells that do not have the function of the encoded protein due to mutations of their endogenous genes. In this way, gene therapy can only be useful for hereditary disor-

TABLE 2

Hereditary Monogenic Disorders Suited for Gene Therapy

DISEASE
Deficiency of α1-antitrypsin
Chronic granulomatous disease
Cystic fibrosis
Familial hypercholesterolemia
Fanconi's anemia
Gaucher's disease
Hunter's disease
Hurler's disease
Deficiency of ornithin transcarbamylase
Deficiency of purine nucloside phosphorylase
Deficiency of adenosine deaminase
X-linked immunodeficiency
Leukocyte adhesion defect
Canavan's disease
Hemophilia B

ders with autosomal or X-linked inheritance which have recessive phenotype and in which the appropriate tissue could be easily approachable. The current exclusion of disorders with dominant inheritance, in which gene therapy would require the specific correction of the mutated allele, is due to our limited understanding of the molecular mechanisms that regulate homologous recombination in mammalian cells. Table 2 provides a list of human disorders for which gene therapy has already proceeded to clinical experimentation.

Gene Therapy of Cancer

The strategies for cancer gene therapy are based on the assumption that all malignancies are due, on the one hand, to alterations of the molecular mechanisms that control cell proliferation and, on the other hand, to the inefficacy of the immune system to control this proliferation. Accordingly, the therapeutic genes can be divided into different categories. The first group consists of genes that induce cell proliferation arrest by acting on the DNA replication proteins (e.g., the proliferating cell nuclear antigen), on the mechanisms that regulate cell cycle progression and DNA damage checkpoints (e.g., cdc2 kinase, p53, and p16), or directly on oncogenes which are inappropriately activated such as c-*myc* and c-*fos* (Table 3). Alternatively, the aggression of the cells of the immune system against the tumor can be enhanced by the transfer into the tumor cells of genes coding for immunoregulatory cytokines such as interleukin-2 (IL-2), IL-12, IL-7, IL-4, and GM-CSF. Tumor immunogenicity can also be increased by transferring genes coding for exogenous histocompatibility antigens or tumor-specific antigens. For example, in B lymphocyte proliferative disorders, the specific antigen that can be used is the antibody which is expressed by the malignant clone.

An alternative strategy for gene therapy of cancer is based on the transfer of genes than can induce cell death upon the control of drug treatment (Connors, 1995). An example of this strategy is the direct intratumoral injection of viral vectors expressing the HSV-1 thymidine kinase gene. The intratumoral expression of this enzyme is harmless unless the patient is treated with the drug ganciclovir since only the viral thymidine kinase, contrary to the human enzyme, can phoshorylate and activate this drug. The result of prodrug activation is the arrest of DNA synthesis followed by cell death. This strategy is potentially very interesting since cells that are transduced with the thymidine kinase

TABLE 3

Strategies for Gene Therapy of Cancer

STRATEGY	PRINCIPLE	PROTEINS CODED BY THE THERAPEUTIC GENES
Cell cycle arrest or restoration of sensitivity to chemotherapy or both	Tumor suppressor genes	p53, Rb, BRCA1
	Inhibition of the function of some genes such as c-*myc* and c-*fos*	Oligonucleotides, antisense RNAs, ribozymes, or intracellular antibodies against oncogenes, cdc-2, cyclins, PCNA
Immunotherapy	Transfer of genes coding for immunoregulatory cytokines	IL-2, IL-12, IL-7, IL-4, GM-CSF, IFN-β and IFN-γ, IL-6, TNF
		MHC class I and II MHC alloantigens
	Transfer of genes coding for immunogenic proteins	Tumor-specific antigens
		Idiotype-specific antigens
Induction of tumor-specific cytotoxicity	Transfer of genes activating a toxic prodrug	HSV-1 thymidine kinase (plus gancyclovir)
		Cytosine deaminase (plus 5-fluorocytosine)
Increased resistance to chemotherapy	Gene transfer in hematopoietic precursors	mdr-1

gene and treated with ganciclovir release toxic metabolites that also kill nontransduced neighboring cells, thus extending the therapeutic efficacy of the treatment (*bystander effect*).

An important conceptual difference distinguishes the approaches aimed at inhibiting cell proliferation or inducing local toxicity in the tumor from those aimed at increasing tumor cell recognition by the immune system. The first ones require that all malignant cells are directly or indirectly treated by gene therapy, whereas the enhancement of the immune system would also permit recognition of nontransduced or metastatic cells. Finally, an alternative approach for gene therapy of cancer consists in the administration of genes that could increase the therapeutic index of chemotherapy (Baum *et al.,* 1996). In most cases, very high-dose chemotherapy cannot be used in cancer patients due to its toxicity for hematopoietic cells in the bone marrow. This problem could be circumvented by the transfer into the $CD34^+$ hematopoietic precursors of genes conferring increased resistance to the treatments. Among the genes that could be used for this purpose, one potentially promising candidate is *mdr-1,* which codes for a membrane transporter protein that prevents the intracellular accumulation of several anticancer drugs, such as plant alkaloids (vincristine and vinblastine), anthracycline antibiotics (adriamycin and daunorubicin), etoposide, and paclitaxel.

Vaccination

One of the most promising fields for gene therapy applications is vaccination. In this case, the gene coding for the immunogenic protein can be transiently expressed by a viral vector that does not integrate into the genome (e.g., adenovirus and poxvirus) or simply be administered as naked DNA, usually by direct intramuscular or intradermal injection. Among the important advantages of the latter approach compared to more standard vaccination procedures based on recombinant proteins are the safety of the preparation, the reduced cost (no complex purification is required), the facility of storage, and, often, the better stimulation of the immune system since the exogenous protein is directly expressed from the patient's cells within the histocompatibility antigens. Currently, naked DNA vaccination has been developed for immunization against tuberculosis, AIDS, and some cancers using tumor-specific antigens.

Noncoding DNAs and RNAs

Not just genes coding for a normal protein can be used to obtain a therapeutic effect by gene transfer. New experimental approaches are aimed at the specific inactivation of pathologic mRNAs using ribozymes, which are small RNA molecules that can specifically pair with a complementary RNA and enzymatically cleave it (Rossi, 1995). The recognition specificity of these RNA enzymes can be modified to make them reactive against specific pathologic RNAs. This strategy is promising for gene therapy of genetic disorders with autosomal dominant inheritance (in which the mRNA expressed from the disease allele should be inactivated), of some viral infections (in which the destruction of the viral mRNA is required), and in the neoplastic disorders in which new pathologic mRNAs are formed, as is the case of chronic myelogenous leukemia in which the reciprocal translocation between chromosomes 9 and 22 leads to the formation of the fusion gene *bcr-abl,* whose expression maintains the malignant phenotype.

An approach that is conceptually similar to that of ribozymes is based on the inhibition of a pathologic mRNA using antisense DNA or RNA oligonucleotides, which have a sequence complementary to that of the target mRNA. The specific pairing of the oligonucleotide with the target leads to the inactivation of the mRNA and to its degradation by cellular enzymes. Although ribozymes are usually expressed by retroviral vectors and thus produced inside the cells, antisense oligonucleotides are administered to the cells from outside. As previously discussed, this approach is limited by the kinetics of distribution and degradation of oligonucleotides *in vivo,* by the requirement for very high concentrations, and, in several cases, by their limited specificity.

Other Strategies

Another possible strategy for gene therapy is based on the specific inhibition of pathologic proteins using DNA, RNA, or proteins that bind to these targets, thus limiting their function. Some examples of this approach can be derived from the currently proposed clinical applications for gene therapy of AIDS (Gilboa and Smith, 1994). HIV-1 replication is dependent on the function of two of its proteins (Tat and Rev) that control the rate of proviral gene expression and the export of unspliced viral mRNAs from the nucleus, respectively. Both proteins act by forming specific interactions with two sequences in the viral RNA genome, named TAR (Tat-binding region) for Tat and RRE (Rev responsive element) for Rev. In the first case, the therapeutic strategies take advantage of several multimeric copies of the TAR sequence to divert the Tat protein from its physiological target on the viral RNA. In the second case, a mutated form of the protein is used (RevM10) that behaves as a transdominant negative mutant since it complexes with wild-type Rev in HIV-infected cells but then it inhibits the function of the formed complex.

Strategies for the Administration of Therapeutic Genes

As previously mentioned, it is clear that the successful development of gene therapy demands that several requirements are fulfilled, including the correct identification of the disease that could benefit from the treatment, the choice of the therapeutic gene and the signals controlling its expression, the development of a suitable gene transfer vehicle, the choice of the route and timing of administration, and the elaboration of procedures for the correct evaluation of the efficacy of treatment. In the field of hereditary disorders, currently gene therapy only considers those with recessive or X-linked inheritance. Furthermore, all the cells of the affected tissue should in principle be accessible to therapy, either directly or *ex vivo*. These considerations render the application of gene therapy problematic for inherited disorders with neurologic degeneration or affected tissues such as bone or muscle. In contrast, other tissues such as liver, bone marrow, the lymphoid system, and respiratory airways are more amenable to be treated. A specific problem is related to the use of gene therapy for disorders due to defects of secretory proteins, such as the coagulation proteins that are normally produced by the liver. In this case, it appears easier to have these proteins produced by muscle or dermis fibroblast cells rather than by hepatocytes (Jones and Tuddenham, 1995). As an alternative source of protein production, an immunologically protected neo-organ could be implanted under the skin of the patients. In these neo-organs, heterologous cells which are engineered for the production of secretory proteins can be implanted that release high amounts of these proteins in the blood of the patients.

The most difficult problem is related to the correct choice of the therapeutic gene and its delivery system. To obtain clinical benefit, gene transfer has to occur in a sufficient number of relevant cells of the target tissue. For gene therapy of inherited disorders, retroviral and adeno-associated vectors are preferred because they can permanently transfer the therapeutic gene inside the host chromosomes. In this respect, the identification of the cells of the target tissues that possess staminal properties remains a critical issue. For example, among the unanswered questions are how to identify stem cells in the respiratory epithelium to permanently deliver the cystic fibrosis gene or to purify the subtype of $CD34^+$ cells in the bone marrow that truly includes the hematopoietic stem cells.

In gene therapy laboratories, another important topic of investigation is related to the production of retroviral vectors driving expression of the therapeutic genes at levels that are sufficiently high, permanent, and, possibly, tissue specific. Especially for gene therapy of inherited disorders, it is important that the control of expression of the therapeutic gene mimics that of the endogenous gene. For example, for gene therapy of thalassemias, the expression levels of the α- and β-globin chains must be balanced inside the erythroblasts. Since the promoter sequences of most human genes are either too long or poorly characterized, it is important to identify tissue-specific promoters that are sufficiently short to be included in a vector while they still maintain sufficient strength. An example related to this issue is derived from the current gene therapy experiments aimed at expressing genes in myeloid cell lineage. Gene expression from most of the retroviral vectors that integrate into $CD34^+$ cells is progressively switched off along myeloid differentiation, unless driven by promoters that are physiologically active in mature granulocytic leukocytes and macrophages (Zentilin *et al.,* 1996).

In contrast, gene therapy of tumors has completely different requirements. When the purpose of gene transfer is the induction of a potent immune response, it is not important that all the tumor cells are treated by gene transfer or that gene therapy is permanent. For these applications, the use of adenoviral vectors is very appropriate since they can transiently express high levels of the immunogenic or immunostimulatory therapeutic proteins. Viral vectors can be directly injected inside the tumor or used to transduce tumor cells (or allogeneic tumor cell lines) cultured *in vitro,* followed by inoculation of these cells into the patients as a sort of vaccination.

Finally, HIV-1 infection is one of the disorders that will possibly benefit from gene therapy in the near future (Gilboa and Smith, 1994). In the course of HIV-1 infection, a massive replication of the virus occurs causing a rapid turnover of the $CD4^+$ cells that are infected and whose half-life is reduced to a few days. If the $CD34^+$ hematopoietic precursors or the peripheral blood $CD4^+$ cells are rendered resistant to HIV infection, this should lead to a progressive expansion of the number of resistant cells under the selective pressure of the infection.

Clinical Trials

In 1986 and 1987, the first clinical experimentation of gene therapy was conducted in a surreptitious manner in the United States on a thalassemic child, without obtaining any therapeutic benefit. At that time, the ethical behavior of the experimenters had been highly criticized by the scientific and medical communities. As a result, since that time all clinical applications of gene therapy in developed countries have to be examined and approved by local and national ethical committees. The first clinical gene therapy trial was approved by the United States Recombinant DNA Advisory Committee in October 1988, and the experimentation actually started in the first months of 1989 (Rosenberg *et al.,* 1990). This was a gene-marking trial performed in 10 patients with advanced melanoma and aimed at understanding whether lymphocytes infiltrating the tumors of these patients (tumor infiltrating lymphocytes; TILs) could be expanded *in vitro* while maintaining their property of infiltrating the tumors after readministration to the patients. During *in vitro* culture, lymphocytes were marked by transduction with a retroviral vector in order to allow their identification after reinfusion. Provirus-containing lymphocytes could be recovered in the peripheral blood of the patients during the 3 weeks following the infusion in all the patients and in the tumor biopsies for at least 2 months, thus showing both the efficacy of the *ex vivo* gene transfer procedure and the targeting properties of TILs.

In the following years, the number of gene therapy trials increased exponentially. The peak was reached at the end of 1995, due especially to the intense activity performed in the United States in which, in the second half of 1995, it was estimated that about 10 patients every week were enrolled in gene therapy trials. In subsequent years, the initial enthusiasm was progressively replaced by the perception of the general inefficacy of the proposed protocols. This led to a reduction of trials that were initiated, and more efforts started to be made on the improvement of gene transfer technologies and our understanding of the biological basis of the treated diseases. As of May 1998, more than 300 gene therapy clinical trials have been concluded or are still open, involving more than 3000 patients. Most of these trials have been performed in industrialized countries.

About 70% of these clinical studies are phase I trials, which are conducted on small numbers of patients and are mainly planned to evaluate the safety of the gene transfer procedures. Less than one-third of the trials are phase I/II or II, and these are also expected to produce a limited therapeutic benefit. Also, these studies are conducted on few patients, with the exception of a large trial conducted on more than 200 individuals infected with HIV-1. This study entails the intramuscular administration of a retroviral vector expressing the HIV-1 *env* gene as a vaccine. Only a phase III multicentric study has explicit efficacy purposes in a large number of patients. This trial involves the intracranial inoculation of a retroviral vector expressing the HSV-1 thymidine kinase gene in patients with brain tumors.

The limited number of clinical studies that are more advanced than phase I is indicative of the problems that gene therapy faces at the clinical level. Several phase I trials produced results that were not expected and were not followed by more advanced studies. For example, of the 23 clinical trials for gene therapy of cystic fibrosis, 18 started before the end of 1995 but have not reached phase II since the initial results were not encouraging.

Diseases and Therapeutic Strategies

Table 4 shows the number of clinical studies conducted in the Unites States and in other countries and is separated according to the type of disorders.

Gene Therapy of Cancer. Most of the clinical studies of gene therapy are aimed at gene therapy of cancer. The therapeutic approaches are very different and take advantage, both *in vivo* and *ex vivo,* of the transfer of genes blocking cell proliferation or coding for cytokines or antigens that stimulate the immune system, according to the considerations reported previously (Table 3). These studies are directed mainly toward gene therapy of patients with carcinomas or melanomas in advanced stage, often metastatic, where gene transfer is obtained by intratumor injection *in vivo* or by *ex vivo* transduction of the cultured tumor cells followed by reinjection. A specific application of cancer gene therapy is experimented for intrathecal tumors (glioblastomas and astrocytomas), in which the HSV-1 thymidine kinase gene is injected into the tumors by a stereotaxic device, followed by treatment of the patients with ganciclovir. This strategy leads to significant regression of experimental tumors in animal models, but much caution is necessary before evaluating the results of clinical studies in patients.

TABLE 4

Number of Gene Therapy Clinical Trials (May 1998)

DISEASE	NUMBER OF CLINICAL TRIALS: USA	NUMBER OF CLINICAL TRIALS: Other Countries
Cancer	132	60
Inherited monogenic disorders	32	16
Infectious diseases	21	6
Other diseases	7	4
Gene marking studies	30	10

Retroviral vectors are used mostly as gene transfer vehicles in cancer gene therapy trials, sometimes even by inoculating the packaging cell lines that produce the recombinant viruses. In some cases, plasmid DNA complexed with liposomes is also used. Finally, some American and European studies also take advantage of the *mdr-1* gene to confer increased resistance to bone marrow $CD34^+$ cells in patients undergoing anticancer chemotherapy. After gene transfer, these cells are eventually reinfused in the patients via autologous transplantation.

Gene Therapy of Genetic Disorders with Monogenic Inheritance. Most of the clinical applications of gene therapy for monogenic disorders (two-thirds of which are conducted in America and one-third in Europe) are phase I/II studies aimed at gene therapy of cystic fibrosis. These studies try to transfer the gene for the CFTR protein into the cells of the respiratory epithelium of the patients by adenoviral vectors of second and third generation or adeno-associated virus vectors (Boucher, 1996). The administration of the recombinant virus occurs directly *in vivo,* through bronscoscopic instillation at different levels of the respiratory tract, or by aerosol. Other monogenic diseases treated by gene therapy are the autosomal recessive form of chronic granulomatous disease (by gene transfer of the p47-*phox* cDNA in the $CD34^+$ peripheral blood stem cells), the defect of α1-antitrypsin (transfer of the gene into the respiratory tract by liposomes), the defect of adenosine deaminase (transfer of the gene in lymphocytes and $CD34^+$ cells), the group C Fanconi's anemia (transfer of the *FACC* gene in peripheral blood $CD34^+$ cells), Gaucher's disease (transfer of the glucocerebrosidase gene in $CD34^+$ cells), Hunter's disease (transfer of the iduronate-2-solphatase gene in lymphocytes), and the defect of purine nucleoside phosphorylase (transfer of the *PNP* gene in peripheral blood lymphocytes). Another disease that is considered for gene therapy is familial hypercholesterolemia, which is caused by mutations of the gene for the receptor of low-density lipoproteins (LDLs). In a clinical protocol currently under implementation, 15–20% of the patient's liver is surgically removed and the LDL receptor gene is transferred *ex vivo* into the cultured hepatocytes using a retroviral vector. At the end of the procedure, these cells are eventually reinfused into the patient's liver through the portal vein.

Gene Therapy of AIDS. Among the clinical studies of gene therapy of AIDS, some are based on the intramuscular injection of retroviral vectors coding for the Env protein, trying to elicit an immune response against the protein. Other experimentations entail the passive transfer of autologous or allogenic cytotoxic $CD8^+$ lymphocytes, activated *ex vivo* against viral antigens. This strategy is aimed at the stimulation of a vigorous adoptive response against the virus. Finally, another approach is based on the utilization of therapeutic genes rendering the transduced cell resistant to infection. These genes include ribozymes cleaving the viral genome, the transdominant negative mutant Rev M10, and a gene coding for an intracellular antibody against the viral Env protein. For gene therapy of HIV infection, efficacy of treatment can be easily monitored by measuring viral load (viremia and number of proviral DNA molecules integrated in $CD4^+$ cells) in infected patients, which is very sensitive to changes in viral replication rate.

Other Disorders. Other clinical studies also deserve special mention since their preliminary results are encouraging. Some of these studies are conducted for gene therapy of vascular disorders. They start from the observation that the proliferation of muscular cells of the arterial wall is an important response to endothelial damage and is involved in some degenerative disorders such as arteriosclerosis and restenosis after balloon angioplasty (Lafont *et al.,* 1995). This proliferation can be prevented by delivering oligonucleotides or viral vectors expressing antisense RNAs targeting the cdc-2 kinase, the B1 cyclin, the *ras* oncogene, or genes coding for cell cycle inhibitors such as *Rb* (Table 3). In the cardiovascular field, the preliminary results obtained by transferring the gene for the vascular endothelial growth factor for peripheral arteriosclerosis obliterans of the limbs to stimulate the formation of collateral blood vessels are also very promising (Isner *et al.,* 1996).

Another disease for which gene therapy is used is rheumatoid arthritis, in which the injection of genes suppressing the inflammatory process, such as the IL-1 receptor inhibitor gene, inside the affected joints appears to be beneficial (Ross *et al.,* 1996). These clinical applications of gene therapy, aimed at therapeutic end points which are less ambitious than the cure of a multisystemic disorder but certainly more realistic, are likely to be the first ones that will have success. Preliminary data obtained from the first treated patients support this optimistic view.

Gene Marking Studies. Several clinical studies of gene transfer are not aimed at obtaining therapeutic benefit for the patients but take advantage of gene transfer to gain important biological information (Brenner, 1996). These gene marking studies entail the transfer of a marker gene, usually by a retroviral vector, into specific cell populations *ex vivo* to follow the fate of these populations after their reinfusion into the patients.

Several studies in the experimental hematology field have been conducted for the comparative evaluation of different procedures for the mobilization, purification, and manipulation of hematopoietic precursors obtained from peripheral blood or bone marrow. Gene marking studies are also applied to patients undergoing autologous bone transplantation for hematopoietic malignancies. In the subset of these patient that relapse, the presence or absence of the marker gene in the tumoral cells is consistent with the source of the relapse being the transplanted cells or the residual malignant cells, respectively. In addition, gene marking of hematopoietic precursors also allows the comparison of different procedures for bone marrow purging before autologous transplantation.

Another field of application of gene marking is the study of the kinetics of the lymphocytes with antitumoral activities (Rosenberg *et al.,* 1990), namely, tumor infiltrating lymphocytes or lymphokine-activated killer cells.

Gene Transfer Procedures in Gene Therapy Trials. More than half of the clinical studies of gene therapy take advantage of retroviral vectors, usually as packaging cell supernatants or, less frequently, by direct inoculation of the packaging cells (Table 5). Among the other viral vectors, adenoviruses and adeno-associated viruses are used in about 13 and 1% of cases, respectively. DNA complexed with liposomes or with liposomes and inactivated adenovirus is used in about 20% of the current trials.

Results of Clinical Trials

Regarding the preliminary results of these trials, it is important to state that despite successful animal experimentation, no gene transfer procedure has generated therapeutic results that are long-lasting, persistent, and essential in

TABLE 5

Methods for Gene Transfer Used in Gene Therapy Clinical Studies (May 1998)

METHOD	NUMBER OF STUDIES (%)
Retrovirus	145 (45.6)
Retrovirus producing cell lines	22 (6.9)
Adenovirus	41 (12.9)
AAV	4 (1.3)
Poxvirus	18 (5.7)
Transfection	5 (1.6)
Electrophoration	2 (0.6)
Liposomes/adenovirus	1 (0.3)
Liposomes	59 (18.6)
Naked DNA	9 (2.8)
Gene gun	4 (1.3)
RNA transfer	1 (0.3)
Other	7 (2.2)

human patients. In this respect, however, it should be considered that most of the clinical trials conducted to date are phase I or I/II and thus aimed at obtaining only limited clinical results. As a result of these studies, it was observed that, with few exceptions, the current gene transfer procedures are relatively devoid of toxic effects. An important exception to this conclusion was generated by studies for gene therapy of cystic fibrosis conducted with first-generation adenoviral vectors. These vectors showed a marked tendency to elicit inflammatory responses in treated patients.

Generally, the outcome of the first clinical studies indicated that several biological problems have to be solved before gene therapy can really be successful. In most cases, these problems are related to the improvement of the gene transfer procedures, to the better understanding of the cellular and molecular mechanisms governing vector–cell interactions, and to the correct identification of the target cells. In this respect, the results obtained from the gene therapy trials of hematopoietic cells are very telling. Transduction of $CD34^+$ cells (a population of cells that is generally believed to include the subset of hematopoietic stem cells) can be achieved at high efficiency *ex vivo* using different types of vectors. However, when the transduced cells are reinfused into the patients, most of them persist for only a few weeks, having been irreversibly differentiated by the *in vitro* manipulation. With minor exceptions, only a small percentage (0.01–1%) of cells containing the transferred gene can be retrieved in the patients several years after transplantation (Barranger, 1996).

Given these observations, it is not surprising that the most important contribution of gene therapy has been obtained not from the therapeutic but from the marking studies, aimed at understanding the kinetics of several cell populations *in vivo*. For example, in the field of autologous bone marrow transplantation for hematopoietic malignancies, the gene marking studies have clearly indicated that relapse usually originates from the tumor cells that are present within the transplant.

The initial results of some cancer gene therapy studies have shown a marked reduction of the tumoral mass in several patients treated by intratumoral injection of genes stimulating the immune system, such as HLA-B7 and β2-microglobulin. These results were obtained in patients with metastatic melanoma (Nabel *et al.*, 1994) or colon cancer (Rubin *et al.*, 1994). Similarly, treatment of tumors with genes coding for cytokines, or transduction of the same genes into xenogenic cells expressing tumor antigens followed by vaccination with these cells, often results in a reduction of the tumoral mass, and sometimes also of the metastasis. In both cases, however, it is too early to make conclusions about the efficacy of these treatments in the long term. In an anecdotal case, a patient with a metastatic melanoma treated with a retrovirus expressing the interferon-γ (IFN-γ) gene showed complete remission and disappearance of the tumor for at least 7 months after treatment. Obviously, success in a single patient does not immediately imply that this treatment will have more general success. Packaging cell lines producing retroviral vectors with suicide genes have been stereotaxically injected in patients with brain glioblastomas. Intracranial injection was devoid of toxic effects. However, out of 10 patients treated in a phase II trial conducted in Germany, only 1 showed complete remission from the disease at 1 year after treatment, 4 died of tumor, and in the remaining 5 patients the disease progressed (Stockhammer *et al.*, 1997). As already reported, a large phase III trial is currently under way that involves more than 40 centers in North America and Europe and 250 patients.

Potentially interesting results were obtained by treatment of T lymphocytes specifically directed against tumor antigens. In patients undergoing allogenic bone marrow transplantation, these cells are expanded *ex vivo*, transduced with the thymidine kinase gene (which will permit their selective elimination), and reinjected into patients. In patients undergoing allogenic transplantation for malignant hematopoietic disorders, transplanted T lymphocytes play an essential role since they recognize and destroy tumoral cells (graft versus leukemia). On the other hand, they also attack the cells of the normal tissues of the patients, leading to the graft versus host disease (GvHD), which might be prevented by this strategy. Eight patients who relapsed or developed an Epstein–Barr virus-induced lymphoma after a T-cell-depleted allogenic bone marrow transplantation were treated with donor lymphocytes transduced with the thymidine kinase gene. Transduced lymphocytes survived up to 1 year and exerted antitumoral effects in five patients. Three patients developed GvHD, which was controlled by administration of ganciclovir (Bonini, 1997). Although these preliminary results appear very interesting, efficacy of this therapeutic strategy needs to be verified in a larger number of cases.

The first authorized clinical experimentation of gene therapy for hereditary disorders was conducted for the deficiency of adenosine deaminase (ADA). This is a very rare autosomal recessive disorder that appears to be amenable to treatment by gene therapy since the corrected cells naturally have a selective advantage over the cells that have the defect. The clinical trials were conducted in the United Stated by transferring the gene coding for this enzyme into peripheral blood T cells of two children using retroviral vectors (Blaese *et al.,* 1995). The results of this study, showing the persistence of genetically modified cells for years after treatment, have encouraged the realization of other trials which have been performed by gene transfer into lymphocytes or peripheral blood and cord blood $CD34^+$ hematopoietic precursors. However, it is still not possible to determine whether the patients had some benefit from these procedures since during these years they still had to continue therapy with the recombinant ADA enzyme.

The majority of the trials for genetic disorders are aimed at gene therapy of cystic fibrosis. More than 10 different protocols have been approved using adenovirus, liposomes, and, recently, AAV as delivery means. In most cases, these were phase I studies, also conducted in the nasal epithelium of patients. Generally, the results of these studies have shown that the *CFTR* gene can be transferred into the cells of the respiratory epithelium of cystic fibrosis patients, and that its expression can partially or completely correct the transepithelial voltage and ion transport defects. Commonly, this correction has shown transitory efficacy, lasting for only 1–3 weeks after treatment. It is premature to draw any conclusion about the therapeutic potential of this approach. Regarding the vectors, these clinical studies have confirmed the efficacy of adenovirus for gene transfer in this setting but have also highlighted the need for improvement of the viral backbone in order to avoid the inflammatory and immune response of the patients.

In the field of gene therapy for HIV infection, an extended clinical study is being carried out by a biotechnology company and is aimed at intramuscular immunization using a retroviral vector expressing the HIV-1 Env protein (more than 200 patients are enrolled). The preliminary data, obtained from a few tens of patients, have not shown side effects (Haubrich *et al.,* 1995), but data concerning the therapeutic efficacy are not available. In particular, this evaluation is hampered by the recent introduction of highly active antiretroviral therapies by the combination of three drugs (two inhibitors of the viral reverse transcriptase and one inhibitor of the protease), which drastically reduce viral load.

A different approach for gene therapy of HIV infection was followed by another clinical study aimed at adoptive immunotherapy by selection of $CD8^+$ cytotoxic T lymphocytes against the viral Gag proteins from the peripheral blood of patients, transfer into these cells of a fusion gene (*HyTK*) between the hygromycin-resistance gene and the thymidine kinase gene, expansion of the transduced cells *in vitro,* and reinfusion of the cells back into the patients. This treatment does not seem to cause important toxic side effects and increases the cytotoxic response against the infected cells expressing Gag. However, even in HIV-infected immune-suppressed patients, a cytotoxic response against the transduced cells was observed, targeting the fusion HyTK protein, with the consequent decline of these cells during the first 7 days after injection (Riddell *et al.,* 1996). This unexpected result again indicates how success of gene therapy will depend also on the solution of unpredictable problems, and it highlights the complexity of the biological systems in which gene transfer occurs.

Finally, for gene therapy of AIDS the most used approaches entail the delivery of genes conferring resistance to HIV infection (ribozymes, transdominant negative mutant Rev M10, and intracellular antibody against the Env protein). In addition, the recent observation that the entry of HIV into the target $CD4^+$ cells requires the presence of chemokine receptors and the finding that some individuals are resistant to HIV infection due to the lack of these receptors have prompted several groups to study the possibility of inhibiting expression of the receptors genes to induce resistance, for example, by using specific ribozymes.

Ethical Considerations

Even before clinical experimentation of gene therapy had started, the scientific community long debated about the license to modify the human genome and reached some conclusions that are almost universally accepted and, in some countries, ratified by law. These can be summarized as follows: Regarding gene therapy of somatic cells, this treatment does not modify the transmissible genetic information and no ethical or moral opposition is raised when the studies are conducted according to the common clinical practice — that is, after an in-depth evaluation of risks and benefits. Genetic modification of the germline to cure a severe inherited disorder (this modification is considered feasible today also in humans as it is in other mammals) is considered unlawful, at least by using the current technologies. In fact, as noted previously, the introduction of genetic material inside the genome of mammalian cells almost always occurs randomly, and only very seldom does the exogenous gene recombine with its chromosomal homolog. For gene therapy of somatic cells, this important limitation can be overcome by the selection of the cells in which the homologous recombination event has occurred and by allowing selection to get rid of all the other cells. This process is obviously not feasible for germline cells, in terms of cost/risk benefits, since it would imply the elimination of several embryos in which the homologous recombination event had not occurred. This selection would be much worse than the one that is currently obtained by prenatal diagnosis followed by interruption of pregnancy if indicated.

Recently, some scientists in the United States, and in particular J. Watson, have declared that the current technological advances suggest the removal of any legal constraint to gene therapy of germ cells. However, it was objected that even if genetic manipulation of germ cells is technically feasible, gene therapy of these cells offers no real improvement compared to preimplantation diagnosis consisting of *in vitro* fertilization, development of the embryo until the two- or four-cell stage, analysis of single cells for the presence of a given mutation, and reimplantation of the remaining cells if normal. Therefore, it appears that gene therapy of the germ cell line will remain forbidden under either legal or ethical considerations.

Since gene therapy of somatic cells to treat serious and otherwise hopeless diseases is ethically acceptable, and no important side effects have been observed, a paradoxical risk is that this technology will also be used for cosmetic and not exclusively medical purposes. For example, in the United States a private company recently obtained permission to start transferring the tyrosinase gene in the hair follicle cells to prevent hair loss in patients undergoing chemotherapy for neoplastic disorders. Although this application appears ethically acceptable, its success would most probably prompt the use of the same technology to treat baldness in otherwise healthy individuals. Generally, this example is very telling about the potential use of genetic engineering in humans for purposes that are not necessarily of medical priority. This risk appears to be impending, given the rapid progress in our understanding of the function of several human genes and the concomitant advancements in the development of efficient gene transfer technologies.

Conclusions

As a consequence of the unsatisfactory results of the first clinical studies, it is clear that the initial enthusiasm about gene therapy should be quenched and more careful attention paid to the solutions to the problems that arose in most fields of application. Several experimental goals still need to be attained before gene therapy can provide real therapeutic benefit to the patients. The most important questions that need to be addressed are summarized as follows:

1. More adequate methods are needed for the identification, characterization, and manipulation of the stem cells of the human tissues which are targets for gene therapy (hematopoietic system, mucous membranes, hepatocytes, and others). These studies are not strictly gene therapy studies, but their outcome is an essential requisite for successful gene therapy.
2. It is essential to develop efficient procedures for cell targeting, entailing the delivery of the therapeutic genes inside the majority of the cells in the target organ. For this purpose, new vectors targeting cell-specific receptors are required. This requisite can be fulfilled, for example, by modifying the surface protein of retroviral vectors or by inserting ligands into the viral envelope, which binds to specific proteins on the target cell surface.
3. It is important to eliminate or quench the host immune response against the engineered cells that produce exogenous proteins and against the viral proteins encoded by the vector used for gene transfer. Although immune reactivity is useful and exploitable for gene therapy of tumors, it is a major drawback for gene therapy of inherited disorders.
4. The development of *in vivo* strategies for the selection of the cells cured by gene therapy would offer a tremendous advantage in all gene therapy approaches whose outcome is limited by the relatively small number of cells that are successfully treated (e.g., gene therapy of hematopoietic stem cells).
5. Should a specific and efficient system for gene therapy become available, another essential requirement is that the therapeutic gene is expressed inside the target cells in a stable and regulated manner. In several situations, the utilization of exogenous promoters is disadvantageous since they are often inactivated inside the cells by methylation or other mechanisms. This is particularly true for some promoters of viral origin, such as the simian virus 40 promoter or the promoters of some murine and avian retroviruses. Furthermore, some important human genes, such as the insulin gene, are not continuously expressed but respond to physiological signals from the organism. In this respect, our knowledge of cell-, tissue-, and organ-specific promoters is very limited. In other cases, the genetic elements that control expression of the endogenous genes are very large. For example, the regulatory elements of the human β-globin gene are dispersed in a region as large as 100 kb, far larger that what can be currently accommodated inside a retroviral vector of the current generation.

Although the previous goals are aimed at improving the currently available technologies for gene transfer, the holy grail of gene therapy will undoubtedly be the development of strategies to insert an exogenous gene into a specific location of the genome. This will require the occurrence of targeted events of homologous recombination, similarly to bacteria, yeast, and some avian cells. The solution to this problem largely depends on advances in the understanding of the basic molecular interactions occurring inside the human cells and regulating these processes.

An alternative approach to homologous recombination is the construction of small artificial chromosomes to be added to the cellular genome. This approach will also require the solution of fundamental biological problems, such as the characterization of the elements that determine the stability of a chromosome (size and genetic elements constituting the centromere and the telomeres) and its capacity of autonomous replication (identification of the origins of DNA replication and the factors regulating their activity).

Notwithstanding the problems outlined previously, the scientific community is firmly convinced that gene therapy will play an essential role in medical practice in the near future. Since no shortcuts are allowed to expedite our understanding in this field, much experimental work will be needed to comprehend and master some of the basic processes that govern the functions of the human organism.

References Cited

Barranger, J. A. (1996). Haematopoietic stem cell gene transfer. *Gene Ther.* **3,** 379–380.

Baum, C., Margison, G. P., Eckert, H.-G., Fairbairn, L. J., Ostertag, W., and Rafferty, J. A. (1996). Gene transfer to augment the therapeutic index of anticancer chemotherapy. *Gene Ther.* **3,** 1–3.

Berkner, K. L. (1992). Expression of heterologous sequences in adenoviral vectors. *Curr. Topics Microbiol. Immunol.* **158,** 39–66.

Blaese, R. M., *et al.* (1995). T lymphocyte-directed gene therapy for ADA-SCID: Initial trial results after 4 years. *Science* **270,** 475–480.

Bonini, C. (1997). HSV-TK gene transfer into donor lymphocytes for control of allogeneic graft-versus leukemia. *Science* **276,** 1719–1724.

Boucher, R. C. (1996). Current status of CF gene therapy. *Trends Genet.* **12,** 81–84.

Brenner, M. K. (1996). Gene marking. *Gene Ther.* **3,** 278–279.

Connors, T. A. (1995). The choice of prodrugs for gene directed enzyme prodrug therapy of cancer. *Gene Ther.* **2,** 702–709.

Curiel, D. T., Agarwal, S., Wagner, E., and Cotten, M. (1991). Adenovirus enhancement of transferrin-polylysine-mediated gene delivery. *Proc. Natl. Acad. Sci. USA* **88,** 8850–8854.

Ferrari, F. K., Xiao, X., McCarty, D., and Samulski, R. J. (1997). New developments in the generation of Ad-free, high-titer rAAV gene therapy vectors. *Nat. Med.* **3,** 1295–1297.

Geller, A. I. (1993). Herpesviruses: Expression of genes in post-mitotic brain cells. *Curr. Opin. Genet. Dev.* **3,** 81–85.

Gilboa, E., and Smith, C. (1994). Gene therapy for infectious diseases: The AIDS model. *Trends Genet.* **10,** 139–144.

Haubrich, R., McCutchan, J. A., Holdredge, R., Heiner, L., Merritt, J., and Merchant, B. (1995). An open label, phase I/II clinical trial to evaluate the safety and biological activity of HIV-IT (V) (HIV-1IIIBenv/rev retroviral vector) in HIV-1-infected subjects. *Hum. Gene Ther.* **6,** 941–955.

Isner, J. M., Pieczek, A., Schainfeld, R., Blair, R., Haley, L., Asahara, T., Rosenfield, K., Razvi, S., Walsh, K., and Symes, J. F. (1996). Clinical evidence of angiogenesis after arterial gene transfer of phVEGF165 in patient with ischaemic limb. *Lancet* **348,** 370–374.

Jones, K. J., and Tuddenham, G. D. (1995). Gene therapy for the haemophilias. *Gene Ther.* **2,** 699–701.

Lafont, A., Guerot, C., and Lemarchand, P. (1995). Which gene for which restenosis? *Lancet* **346,** 1442–1443.

Ledley, F. D. (1995). Nonviral gene therapy: The promise of genes as pharmaceutical products. *Hum. Gene Ther.* **6,** 1129–1144.

Matteucci, M. D., and Wagner, R. W. (1996). In pursuit of antisense. *Nature* **384,** 20–22.

Miller, A. D. (1990). Retrovirus packaging cells. *Hum. Gene Ther.* **1,** 5–14.

Miller, A. D. (1992). Retroviral vectors. *Curr. Topics Microbiol. Immunol.* **158,** 1–24.

Nabel, G. J., Chang, A. E., Nabel, E. G., Plautz, G. E., Ensminger, W., Fox, B. A., Felgner, P., Shu, S., and Cho, K. (1994). Immunotherapy for cancer by direct gene transfer into tumors. *Hum. Gene Ther.* **5,** 57–77.

Naldini, L., Blomer, U., Gallay, P., Ory, D., Mulligan, R., Gage, F. H., Verma, I. M., and Trono, D. (1996). *In vivo* gene delivery and stable transduction of nondividing cells by a lentiviral vector. *Science* **272,** 263–267.

Riddell, S. R., *et al.* (1996). T-cell mediated rejection of gene-modified HIV-specific cytotoxic T lymphocytes in HIV-infected patients. *Nat. Med.* **2,** 216–223.

Rosenberg, S. A., *et al.* (1990). Gene transfer into humans—Immunotherapy of patients with advanced melanoma, using tumor-infiltrating lymphocytes modified by retroviral gene transduction. *N. Engl. J. Med.* **323,** 570–578.

Ross, G., Erickson, R., Knorr, D., Motulsky, A. G., Parkman, R., Samulski, J., Straus, S. E., and Smith, B. R. (1996). Gene therapy in the United States: A five-year status report. *Hum. Gene Ther.* **7,** 1781–1790.

Rossi, J. J. (1995). Controlled, targeted, intracellular expression of ribozymes: Progress and problems. *Trends Biotechnol.* **13,** 301–306.

Rubin, J., Charboneau, J. W., Reading, C., and Kovach, J. S. (1994). Phase I study of immunotherapy of hepatic metastases of colorectal carcinoma by direct gene transfer. *Hum. Gene Ther.* **5,** 1385–1399.

Stockhammer, G., *et al.* (1997). Gene therapy for glioblastoma multiform: *In vivo* tumor transduction with the herpes simplex thymidine kinase gene followed by ganciclovir. *J. Mol. Med.* **75,** 300–304.

Yang, N. S., Burkholder, J., Roberts, B., Martinell, B., and McCabe, D. (1990). *In vivo* and *in vitro* gene transfer to mammalian somatic cells by particle bombardment. *Proc. Natl. Acad. Sci. USA* **87,** 9568–9572.

Zentilin, L., Tafuro, S., Grassi, G., Garcia, R., Ventura, A., Baralle, F. E., Falaschi, A., and Giacca, M. (1996). Functional reconstitution of oxidase activity in X-linked chronic granulomatous disease by retrovirus-mediated gene transfer. *Exp. Cell Res.* **225,** 257–267.

General References

ANDERSON, W. F. (1998). Human gene therapy. *Nature* **392** (Suppl.), 25–30.

CULVER, K. W., RAM, Z., WALLBRIDGE, S., ISHII, H., OLDFIELD, E. H., and BLAESE, R. M. (1992). *In vivo* gene transfer with retroviral vector-producer cells for treatment of experimental brain tumors. *Science* **256,** 1550–1552.

KAY, M. A., LIU, D., and HOOGERBRUGGE, P. M. (1997). Gene therapy. *Proc. Natl. Acad. Sci. USA* **94,** 12,744–12,746.

MILLER, N., and WHELAN, J. (1997). Progress in transcriptionally targeted and regulatable vectors for genetic therapy. *Hum. Gene Ther.* **8,** 803–815.

VERMA, I. M., and SONIA, N. (1997). Gene therapy: Promises, problems and prospects. *Nature* **389,** 239–242.

MAURICE A. M. DE WACHTER
Consultant in Bioethics
1410 Waterloo, Belgium

The End of Life

Coping with the inevitability of death as life's end has always been a challenge to man. Western man today denies death and tries to overcome its meaninglessness by successful, healthy living and longevity. Medicine and medical technology are man's most powerful asset in conquering disease and keeping death at bay. Nevertheless, death remains inevitable, and it is therefore of interest to master the time, the condition, and the way we die. Not only the individual but also society, health care systems, and legal provisions in many Western countries seem to support this trend of death denial. It is no longer clear where the difference lies between religious and secular views of life's end. The art of dying is marked by cultural pluralism.

Life and Death

Death Is Inevitable but Not beyond Control

Death is part of life (Fig. 1). The French physiologist Claude Bernard (1878) said "La vie c'est la mort." Indeed, there is no life without death. This awareness has been for man at all times a perplexing certainty. Today's humankind tries to come to terms with this knowledge either by ignoring it or by keeping death at bay. In the twentieth century man has been successful in postponing death but not indefinitely. Life generally has become longer and its quality has improved, but its limits in time have not been overcome. In the West the traditional attitude toward death, expressed as "let nature takes its course," is now profoundly resented and resisted. The prevention of premature and untimely death has certainly been one goal that modern medicine has realized to a great extent, especially in developed countries. Sometimes, however, the price for that survival is economically and morally disproportionate and seemingly beyond reason. Worst of all, even prolonged life comes to an end. Man is destined to die.

Individual and cultural reactions to death as man's destiny have also changed considerably during the second half of the twentieth century. Philippe Ariès (1981) calls today's attitude toward death "death denied." Common attitudes characteristic of previous periods in history, though presented to us by historians, rarely play a decisive role in today's attitude of death denied. Thus, familiarity with death during the Middle Ages when "death was tamed," death of the self as the last act of a personal drama (the strict proto-

FIGURE 1 Gustav Klimt, *Death and Life;* Leopold Collection, Vienna (photo courtesy of Lessing-Magnum/ Contrasto).

type of *ars moriendi*) during the Renaissance, or death of the other as the end of relationships during the past century all may be known but do not stop us from closing our eyes in front of the undeniable certainty of death to come. The word "denial" suggests very clearly that we do not want to know, even though we are certain we shall die. The problem is not one of knowing but of denial. It is possible to identify two strategies with which man pursues his denial of death. One is the relentless medical effort to postpone death, and the other is the cultural effort to cope with the insecurity and uncertainty about what comes after death, particularly by making health and success the epitome of life. An important corollary of the first strategy is Western man's ambition to control his own dying, its time, its place, and its way.

Today, medicine disposes of an endless armamentarium of techniques and drugs that conquer disease and keep death at bay. After World War II, many developed countries invested heavily in medical technologies. Their aim was to perpetuate the momentary life-saving intervention that had resulted in a provisional success of reviving patients nearly dead. Cardiopulmonary resuscitation (Fig. 2), for example, nowadays artificially supports lives that only a few years ago would have been beyond recuperation. The machine, through mechanical ventilation, stabilizes the patient's condition and takes over. Similarly, during the 1950s the "kidney machine" was developed, saving the lives of thousands of patients suffering from kidney failure. Soon, with a first in 1954, transplantation of kidneys was to follow and provide more years to the lives of numerous patients. Heart, eye, liver, lung, and, recently, more successful xenografts have been added to the list of transplanted organs (Fig. 3). Technology sustains medicine's fight against premature death.

Eventually, however, age sets the ultimate limit. Death is inevitable. This, in turn, does not mean that man has no power over death — over how, when, and where it will occur. Neither does it mean that the social and cultural impact of dying cannot be mended. Indeed, more than in previous times, man has taken up the challenge of mastery over death. Recent developments in Western society, backed by moral and sometimes even legal arguments, show a clear tendency to recognize the "right to die." Increasingly, the personal and intimate choice of competent terminal patients in determining the time and manner of their death is being accepted and even sanctioned by society.

The inevitability of death is most burdensome for the individual human being. The species will survive, perhaps even at the price of letting individuals die. However, what about the individual's existence: Is death final for him or her? The awareness of inevitable death to come seems to create very different reactions in people. Some will be reconciled by the idea of survival in progeny, in works of art, in political or military feats, or even in the afterlife. However, for most people the so-called existential "angst" remains, despite fulfillment and success. This nearly indelible

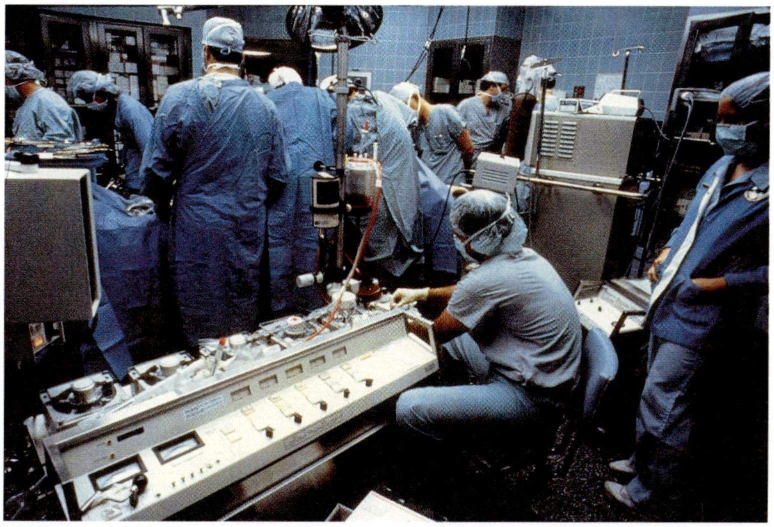

FIGURE 2 A heart–lung machine in use during a heart transplant operation (photo courtesy of Sygma/Neri).

anxiety should stimulate our moral exploration of the remedies against the fear of death (Fagot-Largeault, 1996).

Sociologists of culture seem to maintain that in most societies and cultures (Parsons, 1996)

> from such primitive peoples as the Australian Aborigines to the most sophisticated of the world religions, beliefs in the existence of an individual "soul" have persisted. . . . The persistence of that belief and the factors giving rise to it provide the framework for the problematic of death in the Western world.

From a historical viewpoint, Momeyer (1996) notes that "probably the single most persistent theme in Western philosophical reflection on death is the view that death is not the annihilation of the self but its transformation into another form of existence." Nevertheless, contemporary scientific orientations, epitomized by the biomedical technologies, have also contributed to the profound changes in people's perceptions of values and norms. Therefore, it is "not surprising that there is a great deal of bafflement, anxiety, and downright confusion in contemporary attitudes and opinions in this area of death and dying" (Parsons, 1996).

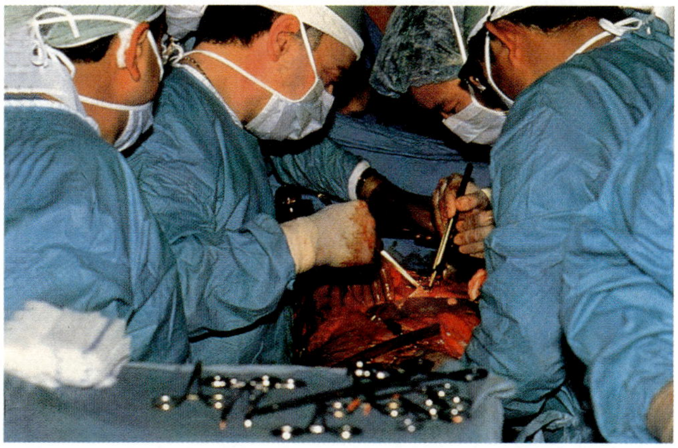

FIGURE 3 A liver transplant operation (photo courtesy of Sygma/Neri).

The way we live reflects our belief about death. The way we value life also reflects our understanding of death or our lack thereof. It is difficult to discuss the end of life without referring to the value of life. Especially in situations in which the end of life, be it by act or omission, is being considered, people have to speak out. For many, life is sacred and cannot be terminated at man's wish. The sanctity of life derives either from religion believing that God gave life to man or from the social covenant among citizens which imposes the state's duty to ensure the protection of the citizens. Whether sacred or secular in origin, the prohibition concerns all taking of life in a medical context. Life is therefore not at the disposal of any human being.

Another argument in the debate builds on the notions of the individual human being, the person. Here, again, is a value that cannot be disposed. However, personhood is defined in very different ways. For some, personhood is indestructible even by death, albeit that the person may be transformed after death. Also, it is not permissible to divide human beings into "more and less worthy" beings. For others, life has no more value than any other biological phenomenon, and personhood obtains value through experience. For instance, for utilitarians personhood exists only as long as consciousness is present. For them, the only things of value are conscious experiences. Once personhood is lost, there is no longer a normal human being.

One major change in the way we value life is reflected in the distinction between biological and biographical life as the morally relevant category. This distinction has been used to describe the transition from an objective given to a subjectively defined content (Rachels, 1975). Indeed, a biography contains all that matters for the individual: aspirations, decisions, choices, achievements, projects, and human relationships. It is said to logically follow that when an individual is without biography, killing him or her would not be the same as destroying life.

Therefore, it is fair to say that death, although inevitable, is not beyond our control. As discussed later, for some people such control is in line with heteronomous guidelines, whereas for others it is the ultimate occasion to affirm man's autonomy and his right to self-determination.

The Determination and the Definition of Death

To determine when a person is dead is a professional responsibility of physicians. To define death is another issue on which society as a whole decides. Until the 1950s, death was defined as the irreversible cessation of circulatory and respiratory functions. Since then, new technologies, such as respirators and cardiopulmonary resuscitation techniques, as well as a better understanding of neurologic functions have created the need for new criteria and a new terminology for both determining and defining death.

The single most widely known new term for the definition of death is *brain death*. Currently understood as an irreversible cessation of all functions of the entire brain, including the brain stem, the concept was first used in 1959 by the French neurologists Mollaret et Coulon under the term *coma dépassé*. During the next decade, better techniques and criteria for the determination of death were developed. In 1968, a famous Harvard Medical School committee described the neurologic criteria of irreversible coma or *whole brain death* as a state characterized by the following aspects: unreceptivity and unresponsivity, no movements or breathing, no reflexes, and a flat electroencephalogram. Another controversial term is *persistent vegetative state after brain damage,* which was first described by Jennett and Plum (1972). In the United States, in 1994 a Multi-Society Task Force on the Persistent Vegetative State suggested the use of two different categories: "persistent," indicating that the vegetative state has been present for more than 1 month, and "permanent," when the condition is considered to be irreversible, usually after 12 months.

The medical profession, in the fulfillment of its duty to determine when death has occurred, generally accepts both cardiorespiratory and brain death criteria. However, the determination of death and the definition of death may differ with regard to the moment that identifies life's demise. Here, society may face a dilemma. Will it pursue a single policy that seeks out a singular definition of death, or will it tolerate this diversity as another instance of cultural pluralism? What this may mean concretely is demonstrated by the following example: "A family with a brain-dead child may insist on care appropriate to the living until cardiopulmonary death is reached while others may insist that the child is already dead and should be treated as such" (Emanuel,1995). The implications for medical activity are equally puzzling. For those who accept brain death criteria, it would be good clinical practice to turn off the respirator and stop food and fluids. For those who accept only the cardiopulmonary criteria, this omission would be at least passive euthanasia.

Medical Practices at the End of Life

Ariès (1981) called the Western world's current attitude an attitude of denial: death denied. Such an attitude is widely influenced by medical practices that conceal, sanitize, and especially prevent death. The result of such practices can be seen in the kinds of deathbed scenes that occur in such specializing hiding places as intensive care units, oncology research facilities, and emergency rooms (Nuland, 1994). At the same time, people do not always wish that heroic measures be taken. In fact, many of our contemporaries long for the ideal of "death with dignity," not in the hospital but at home. Also, in cases in which pain cannot be controlled and death is irreversibly at the doorstep, increasing numbers of patients ask the physician's help in assisted suicide or in euthanasia. Interestingly, in The Netherlands, a much observed country because of its liberal practices of euthanasia, 65% of all deaths occur at home, and 75% of all cases of euthanasia and assisted suicide take place under the supervision of a family physician.

General Practice and/or Hospitals

Generally, today's medical doctors are not well prepared to work with seriously ill and dying patients. Medical training and practice often desensitizes them and become a handicap when they are supposed to provide sensitive care to a patient. Some countries (e.g., The Netherlands) take imitative in adjusting training of medical students and interns. In many other countries, the hospice movement has caught on, and physicians and nursing staff learn again how to accompany the dying patient.

In North America, 75% of patients die in hospitals or nursing homes. However, the trend of patients dying at home seems to be increasing. Some hospitals are aware of the disadvantages of the hospital environment in caring for the dying. They want to improve end-of-life decisions and reduce the frequency of painful, prolonged dying. A study in the United States showed that even though one-third of patients in a hospital preferred that no cardiopulmonary resuscitation be ordered, less than half of the doctors were aware of these wishes (SUPPORT, 1995). It was also noticed that pain control is not optimal, and it becomes worse in the last days before death. Nuland (1994), a surgeon and teacher of medicine, thinks that the art of medicine — that is, the ability to manage the process of death by making it as tranquil as professional kindness can — is mostly lost. It has been replaced by "the brilliance of rescue and, unfortunately, the all-too-common abandonment when rescue proves impossible."

Heroic Measures

The "acharnement thérapeutique" was quite common until the middle of the twentieth century. The nascent power of biomedical technology was such that many lost cases of previous times could be saved. Even today, heroic measures are quite often common practice. The frontiers of life-saving techniques are constantly being removed. Life, so the argument goes, is to be saved at all costs and indiscriminately. Especially in "for-profit" health care systems such as exist in the United States, the discernment about justified costs is often made unlikely. Instead, "defensive medicine" and heroic measures are often chosen regardless of the chances of success or the quality of the lives that may be extended (Lidz, 1996).

On the other hand, heroic measures have been questioned, in medical and in nonmedical circles alike, either as "acharnement" or as counterproductive and unjustified aggression of the dying patient. Today, physicians, if permitted by the legal and social system of their country, discuss the meaningfulness or the futility of certain interventions. Regarding societal reservations vis-à-vis unremitting medical measures, in the mid-1950s several religious groups introduced the distinction between ordinary and extraordinary means in medical treatment. In the 1980s, such concepts were further refined as the criterion of proportionality. Thus, the Vatican *Declaration on Euthanasia* of 1980 states that

> It is also permitted, with the patient's consent, to interrupt these means, where the results fall short of expectations. . . . It is also permissible to make do with the normal means medicine can offer. . . . When inevitable death is imminent in spite of the means used, it is permitted in conscience to take the decision to refuse forms of treatment that would only secure a precarious and burdensome prolongation of life, so long as the normal care due to the sick person in similar cases is not interrupted.

The Hospice Movement

The hospice movement originated in 1967 when Cicely Saunders opened Saint Christopher's Hospice in London. Its aim was to offer care to incurable terminal, mostly cancer patients. Often, these patients were suffering from uncontrollable pain. Adequate medication to control the pain and to restitute some comfort and quality of living has always been a primary goal of the hospice movement. Rightly, where no cure can be offered any more, the focus is on symptom treatment or palliative care. Since the hospice movement takes care of the last days of living, the setting implies that patients forego life-sustaining treatment. In fact, often no such medical facilities or equipment are in place in the hospice. This is understandable because, in principle, hospice care is not a part of hospital services. Moreover, hospice care wants to be holistic and tries to meet personal, social, and spiritual needs of patients, relatives, and friends. Because hospice care is complex and comprehensive, it has often proved impossible to organize all hospice services in hospitals and other medical facilities. Therefore, and due to the commitment of numerous volunteers, hospice care is also provided under hospice programs, focusing much more on home care.

Worldwide, the hospice movement found a strong support, from a psychological viewpoint, in the best-selling publication *On Death and Dying* by Elizabeth Kübler-Ross (1969). The main theme of her book is man's struggle to come to terms with death (Fig. 4). This struggle, according to Kübler-Ross, has five stages: denial and isolation, anger, bargaining, depression, and acceptance. For many patients and for most hospice workers, volunteers and professionals, these insights helped to serve in what Saunders called the unique period in the patient's illness when the long defeat of living can be gradually converted into a positive achievement in dying.

Certainly, the hospice movement has contributed to re-

FIGURE 4 Cover of the 1966 Italian edition of the book *On Death and Dying* by E. Kübler-Ross (photo courtesy of Rossini).

lieve not only physical pain but also the spiritual suffering of terminally ill patients. However, even the United Kingdom hospice system readily accords its benefits to fewer than 10% of those dying every year. In the United States, hospice programs care for about 15% of all deaths. It is obvious that financial constraints and the scarcity of hospice resources, inevitably leading to limited access of needy patients, are major setbacks of the hospice movement.

In the mid-1990s, the question arose again about palliative care and its relation to hospitals. A plea often heard is that even if palliative care is already part of hospital care, hospitals may still learn from the hospice movement how to improve such palliation.

Euthanasia and Physician-Assisted Suicide

In addition to resisting and postponing death, man has always faced the possibility of hastening the dying process and actively terminating one's life. The Stoics in Antiquity are but one example of several cultures that knew of self-inflicted death when the time had come. This is a moral issue which in its contemporary version of physician-assisted suicide concerns many more dying patients than ever before. Beginning in the Renaissance, the wish to have physicians not only help patients who wanted to terminate their own lives but also to perform euthanasia for them became not uncommon. Francis Bacon (1561–1626), in his book *The Advancement of Learning* (1605), was the first to use the word "euthanasia," a term which he defined as an easy, painless, and happy death. He also recommended that physicians make their art more perfect and help to alleviate the agony and the suffering of dying.

Doctors and Euthanasia

Generally, medical associations oppose both euthanasia and assisted suicide. A survey performed in 1994 (Shapiro *et al.,* 1994) shows that in most countries the medical profession still insists on the legal and moral prohibition for a physician to intentionally put an end to a patient's life. For example, the commentary to the 1996 Medical Code in France (Ordre National des Médecins, 1996) speaks of euthanasia as an irreversible intervention violating the (French) law and the European Convention for the protection of human rights and fundamental freedoms of 1950.

However, an increasing number of physicians support the right of competent, terminally ill adult patients who are suffering severely to seek a doctor's assistance in dying. The first and thus far the only country in Europe to have made euthanasia and physician-assisted suicide legally acceptable is The Netherlands. Due to a clearly defined practice of euthanasia, the medical profession was able to avoid the condemnation of doctors by the Courts of Justice. When a physician was able to prove that he or she had found himself or herself caught in a conflict of duties and had, consequently, chosen for the fiduciary relationship with a patient who was suffering unbearably and requesting euthanasia, the judges took a very lenient approach (Maas *et al.,*1991). In 1994, the duty of reporting by official notification of all cases of nonnatural death to the coroner was introduced and it was recently evaluated (Wal and Maas,1996). In other countries there have been similar, albeit less transparent, developments. Thus, 46% of British general practitioners and hospital doctors thought that they should be legally permitted to end the life of a terminally ill patient requesting euthanasia, and 44% said they should not be so permitted. Similarly, more than one-third of Danish physicians consider euthanasia and physician-assisted suicide acceptable. They base their view on the patient's right to self-determination, on the fact that a patient should not be forced to suffer, and on the patient's right to be helped to a dignified death (Polker *et al.,* 1996).

In 1996, the American Medical Association voted almost unanimously its opposition to physician-assisted suicide. Allowing physicians to participate in assisted suicide would cause more harm than good. This practice is considered fundamentally incompatible with the physician's role as a healer. A survey among American oncologists showed that 57% of them had received a request for euthanasia or assisted suicide; 14% said they had performed euthanasia or assisted suicide at the patient's request. The same survey also addressed oncology patients and the public. It was discovered that two-thirds of them found euthanasia and physician-assisted suicide acceptable for patients with unremitting pain (Emanuel *et al.,*1996). This confirms the results of repeated opinion polls showing that 63% of Americans think that physicians should be allowed to terminate the lives of incurable sick requesting euthanasia (Blendon *et al.,*1992).

Social and Legal Trends Regarding Euthanasia

In many Western countries, medical decisions about the end of life have been strongly embedded in social, moral, and legal terminology. The interaction of law and ethics in the field of medicine created the so-called new discipline of bioethics. Bioethics plays an important role at the level of the individual who wants to reflect about rights and duties and also at the level of the public debate. The intertwinement of law and ethics, especially in the field of medicine, caused a uniformization and a juridification of complex ethical issues regarding the end of life. The main actor in this field, however, is the individual human being whose position has thoroughly changed in postmodernity.

The Permissive Society: Individualization as a Cultural Phenomenon

Individualization is a prominent social trend in today's Western society. A European Values Survey, conducted in

1981 and repeated in 1990, showed that, next to de-Christianization and secularization, individualization explains why euthanasia has been widely accepted and is even believed to be in keeping with basic values and norms (Ester *et al.,*1993). Individualization represents two changes: (i) that the community has lost much of its power as a normative agent, whose norms the individual used to internalize, and (ii) that individual happiness prevails over self-sacrifice and that the individual may expect that society will also contribute to his or her happiness. Furthermore, the shift from biological life to biographical life as the morally relevant category in appreciating the value of life embodies the cultural trend of individualization. The Values Survey confirmed these trends. It revealed that attitudes toward adultery, prostitution, homosexuality, divorce, suicide, abortion, and euthanasia all referred to a common value orientation. There were in fact only two kinds of reactions, namely, strict and permissive, albeit that permissiveness could differ for various types of behavior. For instance, suicide was generally thought to be less justified than euthanasia (Moor, 1996). Also, euthanasia, understood as terminating the life of the incurably sick, was considered a morally right or neutral act. Finally, the survey repeated in 1990 showed that in most countries moral permissiveness had considerably increased during the 1980s.

The Courts and End-of-Life Decisions

Starting in the mid-1970s, a few European countries, but especially the United States, experienced increasing numbers of law cases with recourse for death and dying decisions to legal courts. Requests from patients or parents of incompetent patients were brought to court. Simultaneously, cases against doctors practicing euthanasia or assisted suicide created new jurisprudence in many countries. In some countries (e.g., The Netherlands), such cases went to the highest court. In the United States, similar court cases as well as increased legislative activity make it likely that the Supreme Court will find these issues on its docket in the near future. The Supreme Court could then either state that all citizens have a liberty interest in physician-assisted suicide or that each state may set its own policy (Capron, 1995). The point is that law lords inevitably will have to address the question of legalizing new practices at the end of life. One such example is the validity of advance directives.

Living Will and Advance Directives

Advance directives express the patient's preferences in advance of decisional incapacity due to future disease or accidents. In contrast to the "living will," which tends to be general and stated in broad terms, advance directives are concrete, specific, and practical instructions, although some insist on the importance of mentioning personal goals and values so that doctors may decide along the lines of those values. Often, such advance directives are written as legal documents. The efficacy of both the living will and advance directives remains questionable both in practice and in theory. Starting in 1991, the United States required that all patients be notified of their right to use advance directives. In the era of dialysis, a study of the role of advance directives showed that 30% of American patients had advance directives, whereas only 0.3% of German and Japanese patients had them. Interestingly, almost all physicians from these countries would stop dialysis if a family request to withdraw was also supported by an advance directive (Sehgal *et al.,*1996).

Citizens and Euthanasia

The first ever society aimed at making medical help to die legal was the British Voluntary Euthanasia Society, founded in 1935. Doctors, churchmen, and lawyers started the group's efforts. Today, this society is a member of the World Federation of Right to Die Societies. They want competent adults, suffering unbearably from an incurable illness, to receive medical help to die at their own considered and persistent request. Safeguards include a second medical opinion, psychiatric assessment, much consultation, and advance written request. It may be hoped that such measures will prevent cases of euthanasia or assisted suicide from occurring in patients who suffer from impaired judgment.

Regarding today's populations in Europe, the previously mentioned European Values Survey (Ester *et al.,* 1993) showed a direct link between the de-Christianization of societies and the permissiveness of euthanasia and assisted suicide. The Netherlands score highest with 87% of the citizens approving of euthanasia. Many other countries show similar, albeit less pronounced, trends. A group of southern European countries (Italy, Portugal, and Spain) and also Ireland score on average 35%. A group of northwestern European countries (Belgium, France, West Germany, Iceland, and the Scandinavian countries) score an average of 55% approving of euthanasia and physician-assisted suicide.

The Law, Euthanasia, and Assisted Suicide

Among European countries, many are still prohibiting euthanasia and physician-assisted suicide (de Wachter, 1996). This is the case in Belgium, Denmark, Finland, Ireland, Italy, and Spain. Other countries, such as France, Germany, and the United Kingdom, do not interdict suicide, but physician-assisted suicide remains controversial. For example, in France and Germany, assisting a patient to commit suicide would be considered as nonassisting a person in danger. Greece, The Netherlands, Portugal, and Switzerland are governed by a permissive legal regulation. In The Netherlands, in particular, physician-assisted suicide is tolerated under the same conditions as (active and voluntary) those for euthanasia. In the early 1990s in North America, several initiatives were introduced to legalize suicide and active voluntary euthanasia and to popularize methods to commit suicide especially for terminal patients with unbearable suffering.

Ethical Issues Concerning Death and Dying

The Cultural Context

The ethical debate on euthanasia and physician-assisted suicide occurs within the cultural context and the premises of Western bioethics. Anthropologists rightly point out that premises of rationalism, individualism, and naturalism still form the backdrop of the more readily apparent technical proficiency of modern medicine (Lidz, 1996). Some countries, however, resist one or several of those premises. Consequently, their attitude and ethical concerns regarding death and dying may also differ. Germany, for example, remains deeply skeptical of the so-called rational instrumentality. This type of rationality is perceived by the Germans as a manipulated justification of biomedical research even when it is destructive of environment and society. If one adds the historical burden of the Nazi atrocities, together with the breakdown of the German moral tradition, one begins to understand the bases of the German skepticism about rational instrumentality and their opposition to euthanasia (Vollmann, 1996; de Wachter, 1997). Nevertheless, on average it seems plausible to say that the premise of individualism plays a major role in the recent radical change of Western attitudes toward death and dying.

Active and Passive Euthanasia

From a moral standpoint, a distinction can be made between an act and the omission of that act. In the case of euthanasia and assisted suicide, the "active" way would be to provoke a person's death intentionally with the use of rapid and painless means, whereas the "passive" way would be to refrain from all action of curing and life saving but with the same intention and with expectations of the same result, probably over a longer period of time.

In the past, the distinction between active and passive euthanasia served, among other purposes, to separate some medical practices that could not be morally justified from other practices that could be morally justified. Sometimes the same distinction is still maintained. For example, the 1996 commentary of the French medical code of deontology mentions the doctor's duty not only to try to relieve pain and to morally support the patient but also to avoid all unreasonable obstinacy, be it in diagnostics or in therapeutics.

Given the advancement of drug treatments and the widespread use of pain-killing regimens, it is sometimes difficult to determine whether an analgesic treatment is a form of active or passive euthanasia. Also, in 1975, due to a widely debated article by J. Rachels, many authors abandoned this distinction. The main point this author made was that, from an ethical standpoint and by examining the consequences, it is impossible to say that "letting die" is not the cause of death. There is no intrinsic moral difference between "killing" and letting die. Therefore, letting die lays as much moral responsibility on doctors as killing patients (Rachels, 1975). Despite this criticism, the distinction is still extensively used by religious groups and churches, even though their moral theologians do not always support such use. Still, both maintenance and rejection of the distinction "active–passive euthanasia" have gained symbolic value. The first symbolizes a continuity with a religious past and the second a rupture with that past and the affirmation of man's autonomy.

Respect for the Patient's Autonomy

Our society places the highest possible value on a person's autonomy. Medicine has learned to respect patient autonomy, particularly by making the well-informed patient a partner in the decision making. Well-informed patients weigh benefits and burdens of various alternative therapies and are entitled to choose for no treatment if this were to better fit their personal goals and values. In the field of medicine "death and dying" is but one area, in addition to many others such as procreative freedom and the right to abortion, in which patient autonomy and informed consent are almost decisive. In particular, the adult, competent patient is entitled to speak for himself or herself. The physician, on the other hand, has an obligation to comply with an appropriate refusal. Witness the 1988 *Principles of Medical Ethics* of the American Medical Association, which states that "for humane reasons, with informed consent, a physician may do what is medically necessary to alleviate severe pain, or cease or omit treatment to permit a terminally ill patient whose death is imminent to die."

However, as asked by Tom Beauchamp (1996), must the notion of a valid authorization be continued to a valid refusal, or can a valid request authorize action in much the same way that a valid refusal authorizes inaction? The answer to this question depends on the nature of the patient–physician relationship as well as on the reasonableness of the request. Although it may be right to affirm that, in most Western countries, it is now the general consensus that "it would be both immoral and illegal for a physician not to forgo treatment in the face of a competent, authoritative refusal of any form of medical service or treatment" (Beauchamp, 1996), it is not likely that this generalization applies to all European countries.

to forgo treatment in the face of a competent, authoritative refusal of any form of medical service or treatment" (Beauchamp, 1996), it is not likely that this generalization applies to all European countries.

Food and Fluids

An important step in the debate about treatment and care for dying patients was taken during the 1980s with regard to withholding or withdrawing artificially administered food and fluids.

Traditionally, artificial nutrition and hydration were not primarily considered to be treatment. Rather, they were care, the most basic care due all dependent persons. However, the more technology and medical staff were involved in administering such support systems, the more nutrition and hydration took on the appearance of medical acts. Nevertheless, the nursing profession tends to insist on the symbolic meaning of food and fluids as the ultimate sign of our care for those in need. However, this argument is countered by the consideration that no harm should be done, for instance, by needless prolongation of the agony or by inflicting useless suffering, and/or by the argument that possibly some benefit for the patient can be gained.

The Need for Agreement on the Definition of Euthanasia

Euthanasia remains a polyvalent term in medical practice. This is, of course, not very helpful if one wants to derive normative statements on a variety of practices. Indeed, in current terminology the word euthanasia still may mean not only the active killing of the patient but also withholding or withdrawing treatment after the patient has refused to be treated, abstaining from futile treatment or even pain treatment which may hasten the patient's death. In the past, based on a classic distinction between acts and omissions, almost all of the previous forms of letting the patient die were classified as passive euthanasia. In order to break through the deadlock of messy definitions, some people have tried to better define euthanasia.

Thus, a narrower and more focused definition developed in The Netherlands during the 1970s and 1980s. Following the suggestion made by a state commission in 1985, the Dutch call euthanasia the intentional termination of the life of someone by another at the former's request. The term assisted suicide is used for exactly the same situation except with regard to the agent, who is then the patient and not the doctor, administering the lethal drug. Additional criteria used by Dutch courts of justice in pronouncing on cases of euthanasia or physician-assisted suicide are mainly the following: voluntariness — a persistent, conscious, and free request by the patient; a hopeless situation — a state of illness which both patient and physician consider to be beyond recovery and amelioration; and consultation of a medical colleague who will assess the appropriateness of the euthanasia request.

Until recently, The Netherlands was the only country in which active euthanasia was socially, medically, and legally acceptable, despite the fact that the law calling it a criminal act was not changed. In the mid-1990s, several other countries began showing similar trends. In 1996, Australia's Northern Territory implemented the first law permitting voluntary euthanasia. However, this law was immediately attacked by the federal government and remains controversial. Several states of the United States and also Japan are likely to follow the same trend. In fact, the state legislature of Oregon was the first to vote to legalize physician-assisted suicide.

Arguments in the Ethical Debate on Euthanasia

The two major arguments in favor of euthanasia and assisted suicide are respect for the patient's autonomy and the principle of nonmaleficence — that is, mercy for the patient's suffering. The first argument claims that it is the competent and free person's right to decide about death as well as about life. The second argument states that a physician may consider active euthanasia to be part of accompanying a dying patient whose suffering is such that his or her dignity is being jeopardized and destroyed.

Arguments against euthanasia and assisted suicide are numerous. First, it remains an act of killing. How can one justify that? Is it enough to refer to the free and explicit request of the patient to (be helped to) die? Second, the practice of euthanasia sets us on a slippery slope with no control to stop developments from voluntary to nonvoluntary and even involuntary termination of life. The first argument is based on the duty not to kill, and the second is based on the negative consequences that euthanasia will have for the medical profession and for society at large.

Western Religious and Secular Views of Death and Dying

Religious Views

The Judaeo-Christian tradition is based on a common source, the Bible. Although there is no single view on death and dying in the Bible, be it Old or New Testament, the value of life as such derives from life being created and given to man, for him to care and to cherish. Death is seen either as an evil power or as a natural part of life.

The bioethical implications of Jewish, Catholic, and Protestant religions (Kliever, 1996) concerning death and dying are similar. Generally, these religions seem to accept the validity of a brain-oriented definition of death. This does not always mean a radical substitute for the tradidrawing of heroic measures, and even of food and fluids, is acknowledged by many religious thinkers. Finally, it is important to mention the respect that these religious traditions have for the deceased. Mutilation by dissection and harvesting of organs for research or transplantation have long been controversial issues. Today, more liberal views are held on the matter.

With secularization pervading most Western religions, death, the moment of death, and its antecedents of dying appear in a different light. Death and dying are no longer a threshold, a transition, a new beginning, or the entrance into another life. Thus, members of religious groups find themselves wondering what the meaning of death and dying can be. They may then in their search turn to another tradition, that of the secular views.

Secular Views

The secular views about death and dying, particularly about euthanasia and assisted suicide, draw on the secular tradition as developed in the sixteenth century by Montaigne, pursued by Hume, and later by Schopenhauer and Nietzsche. These authors agree that intolerable pain, fear for a worse death, and the wish to die justify the active termination of life. This permission takes exception to the duties of the social covenant, in particular the prohibition of suicide based on the fact that the individual's life is at the disposal of society. In other words, in the secular views as well as in the religious view, positions are held that are permissive of suicide when the individual judges his or her life not to be worth living. In such cases, autodetermination prevails over duties to society. It allows the individual to terminate the covenant with society as well as with God (Fagot-Largeault, 1996).

Some seem to regret the impoverishment that occurs in today's philosophical thinking about death. Momeyer (1996) notes that

> The problem of death, understood as the struggle to make life meaningful in an increasingly secular age plagued by the temptations of nihilism, continues. The little that philosophers in the present time have had to say about death—outside of chiefly moral concerns centering on choosing death—has tended to suppose death is final, not, in any form, to be survived.

Do the different approaches taken by religious and secular views make a difference with regard to making decisions about the end of life? Some maintain that neither religion nor social class make a significant difference in patients' wishes about when and how to die (Searle and Addington-Hall, 1994). Others, surveying populations in the eastern United States, found that being Roman Catholic and being religious were strongly and independently associated with all aspects of people's attitudes to euthanasia and physician-assisted suicide (Emanuel *et al.,* 1996).

Ars Moriendi Today

Predictably, in a world in which death is kept at bay, the art of dying will not be well developed. In Western culture the "personal search to discover and affirm the ultimate meaning in one's dying is avoided or evaded as long as death is not imminent" (Bresnahan, 1996).

History offers many examples of the *ars moriendi.* However, modern man does not want to follow their lead without qualification. Also, medicine helps him to postpone death evermore. Contemporary philosophers declare the exploration of "life beyond" a meaningless enterprise. Their main concerns are man's choices of dying. Religious views do hardly better: They do not appeal to modern man. The message of eternal life does not come across for most people, even believers. Modern man, so ruthlessly self-centered at the end of the twentieth century, is facing the cultural pluralism of views of death and the diversity of appropriate ways of behavior. Perhaps man's current involvement in ignoring the meaning of death, in focusing on a successful life, prevents him from even developing interest in what may come after death. Nevertheless, the existential fear in front of the absolute certainty that this life comes to an end obliges man to make a moral evaluation of the alternatives he has at his disposal to overcome that fear.

Also, in line with the emancipatory movements of our time, new regulations about how we die have been drawn up by common consensus. Thus, advance directives, living wills, and proxy consent contribute to the art of dying as a way of remaining in control as long as possible by anticipating what may or may not be done in case of incompetency. Nevertheless, despite the ongoing expansion of the frontiers of life, death as the inevitable end of life remains. Facing this reality, Western man today has chosen for maximum control until the last moment of life's end, sometimes by taking charge of the very ending itself. What he thereby gains may be just a loss forever.

Thus, the problem of death, understood as the struggle to make life meaningful in an increasingly secular age plagued by the temptations of nihilism, continues (Momeyer, 1996).

References Cited

American Medical Association (1988). Principles of medical ethics. *J. Am. Med. Assoc.* **259,** 3.

Ariès, P. (1981). *The Hour of Our Death.* Knopf, New York.

Beauchamp. T. L. (1996). Refusals of treatment and requests for death. *The Kennedy Inst. Ethics J.* **6,** 373.

BERNARD, C. (1878). *Leçons sur les Phénomènes de la vie Communs aux Animaux et aux Végétaux.* Baillière, Paris.

BLENDON, R. J., *et al.* (1992). Should physicians aid their patients in dying? The public perspective. *J. Am. Med. Assoc.* **267**(19), 2658–2662.

BRESNAHAN, J. F. (1996). Contemporary art of dying. In *Encyclopedia of Bioethics* (W. T. Reich, Ed.). Macmillan, New York.

CAPRON, A. M. (1995). Constitutionalizing death. *Hastings Center Rep.* **25**(3), 23–24.

DE WACHTER, M. A. M. (Ed.) (1996). Euthanasia and assisted suicide in The Netherlands and in Europe. In *Proceedings of the European Conference June 1994* (Report No. EUR 16636 EN). European Commission DG XII, Brussels.

DE WACHTER, M. A. M. (1997). The European Bioethics Convention. *Hastings Center Rep.* **27**(1), 13–23.

EMANUEL, E. J., FAIRCLOUGH, D. L., DANIELS, E. R., and CLARRIDGE, B. R. (1996). Euthanasia and physician-assisted suicide: Attitudes and experiences of oncology patients, oncologists, and the public. *Lancet* **347,** 1805–1810.

EMANUEL, L. L. (1995). Reexamining death: The asymptotic model and a bounded zone definition. *Hastings Center Rep.* **25**(4), 33.

ESTER, P., HALMAN, L., and DE MOOR, R. (1993). *The Individualizing Society. Value Change in Europe and North America.* Tilburg Univ. Press, Tilburg, The Netherlands.

FAGOT-LARGEAULT, A. (1996). In *Dictionnaire d'Éthique et de Philosophie Morale* (M. Canto-Sperber, Ed.), pp. 1583–1590. PUF, Paris.

JENNET, B., and PLUM, F. (1972). Persistent vegetative state after brain damage: A syndrome in search of a name. *Lancet* **1**(7753), 734–747.

KÜBLER-ROSS, E. (1969). *On Death and Dying.* Macmillan, New York.

LIDZ, V. (1996). Death: Anthropological perspectives. In *Encyclopedia of Bioethics* (W. T. Reich, Ed.), pp. 477–487. Macmillan, New York.

MAAS, P. J. VAN DER, VAN DELDEN, J. J., and PIJNENBORG, L. (1991). Euthanasia and other medical decisions concerning the end of life. An investigation performed upon request of the commission of inquiry into the medical practice concerning euthanasia. *Health Policy* **22.**

MOMEYER, R. W. (1996). Death: Western philosophical thought. In *Encyclopedia of Bioethics* (W. T. Reich, Ed.). Macmillan, New York.

MOOR, R. (1996). Euthanasia and moral permissiveness. In *Euthanasia and Assisted Suicide in The Netherlands and in Europe* (M. A. M. de Wachter, Ed.), pp. 87–101. Brussels.

MULTI-SOCIETY TASK FORCE ON PVS (1994). Medical aspects of the persistent vegetative state. *N. Engl. J. Med.* **330,** 1499–1508, 1572–1579.

NULAND, S. B. (1994). *How We Die. Reflections on Life's Final Chapter.* Knopf, New York.

ORDRE NATIONAL DES MÉDECINS (1996). *Commentaire du Code de Déontologie Médicale.* Ordre National des Médecins, Paris.

PARSONS, T. (1996). Death in the Western world. In *Encyclopedia of Bioethics* (W. T. Reich, Ed.). Macmillan, New York.

POLKER, A. P., HOLTUG, N., JENSEN, A. B., KAPPEL, K., NIELSEN, J. K., and NORUP, M. (1996). Experiences and attitudes towards end-of-life decisions amongst Danish physicians. *Bioethics* **10,** 233–249.

RACHELS, J. (1975). Active and passive euthanasia. *N. Engl. J. Med.* **292,** 78–80.

SEARLE, C., and ADDINGTON-HALL, J. (1994). Euthanasia: Why people want to die earlier. *Social Sci. Med.* **39,** 647–654.

SEHGAL, A. R., *et al.* (1996). Advance directives and withdrawal of dialysis in the United States, Germany and Japan. *J. Am. Med. Assoc.* **276,** 1652–1656.

SHAPIRO, R. S., DERSE, A. R., GOTTLIEB, M., SCHIEDERMAYER, D., and OLSON, M. (1994). Willingness to perform euthanasia: A survey of physician attitudes. *Arch. Internal Med.* **154,** 575–584.

SUPPORT (1995). A controlled trial to improve care for seriously ill hospitalized patients. *J. Am. Med. Assoc.* **274,** 1591–1598.

VATICAN CONGREGATION FOR THE DOCTRINE OF THE FAITH (1980). *Declaration on Euthanasia,* pp. 9–10. Vatican Congregation for the Doctrine of the Faith, Rome.

VOLLMANN, J. (1996). Why does bioethics develop differently in Germany? *Bull. Eur. Soc. Philos. Med. Health Care* **4,** 13–20.

WAL, G. J. VAN DER, and MAAS, P. J. VAN DER (1996). *Euthanasie en Andere Medische Beslissingen Rondom het Levenseinde. De Praktijk en de Meldingsprocedure.* SDU Uitgevers, Den Haag.

General References

ARIÈS, P. (1974). *Western Attitudes toward Death: From the Middle Ages to the Present.* Johns Hopkins Univ. Press, Baltimore.

BATTIN, M. P. (1982). *Ethical Issues in Suicide.* Prentice Hall, Englewood Cliffs, NJ.

BECKER, E. (1973). *The Denial of Death.* Free Press, New York.

CALLAHAN, D. (1993). *The Troubled Dream of Life: Living with Mortality.* Simon & Schuster, New York.

CORR, C. A., NADE, C. C., and CORR, D. M. (1996). *Death and Dying, Life and Living,* 2nd ed. Brooks/Cole, London.

GLOVER, J. (1977). *Causing Death and Saving Lives.* Penguin, New York.

GOMEZ, C. F. (1991). *Regulating Death; Euthanasia and the Case of The Netherlands.* Free Press, New York.

HARRIS, J. (1985). *The Value of Life.* Routledge & Kegan Paul, London.

KÜBLER-ROSS, E. (1983). *On Children and Death.* Macmillan, New York.

STODDARD, S. (1992). *The Hospice Movement: A Better Way of Dying,* Rev. ed. Vintage, New York.

VOVELLE, M. (1983). *La Mort en Occident Depuis 1300 à nos Jours.* Gallimard, Paris.

YOUNGNER, S. J. (1992). Defining death: A superficial and fragile consensus. *Arch. Neurol.* **49**(5), 570–572.

DIETMAR MIETH
Interfaculty Center of Ethics in the Sciences and Humanities Tübingen, Germany

Biotechnologies, Ecology, and Ethics

In this article, I discuss many important questions pertaining to ecology and ethics in the context of biotechnologies. In the first part, I dwell on the ethical and social criteria on which the permissibility or otherwise of genetic engineering should be based, and I show that both the experimentation and the introduction on a vast scale of new biotechnologies must be preceded by a debate involving specialists of the human and the natural sciences in order to assess risks and opportunities of the new discoveries. In the second part, I provide an analysis of the concepts of nature and of creation, and I show that technological progress, which enables man to influence the destinies of living species (including his own), must be reconsidered in the light of the modern philosophy and theology of creation so as to assess not only the practical but also the ethical consequences of man's altering of the natural order.

Introduction

What Is Ethics?

Through ethics we understand the search for conduct that is good and just, given certain conditions and possibilities of action, situations (cases), individual ways of behaving, and institutions (Hartmann, 1989; Mieth, 1992; Kissling, 1993).

Ethics is different than morality. The latter can indicate the factual status (the current moral rules in a society), a religious concept (e.g., the moral doctrine of the church), or an idea of morality of the optimum. Ethics, on the contrary, is a philosophical reflection on what is good and right, open also to theological interpretations. From the scientific standpoint, ethics is autonomous — that is, independent of any external, rational, and critical influence, oriented simultaneously toward free self-obligation and institutional justice (e.g., of the legal system and of science). Moreover, it is interdisciplinary because, having to start from a context of knowledge and of experience for which it does not possess the exclusive competence, any ethics requires an interdisciplinary anchorage. Lastly, it is integrative because it seeks to connect various sorts of knowledge and to achieve the formulation of a concordant judgment.

Technological Evaluation and Ethics

Ethics and technological assessment are two different but inseparable things. Ethics must nourish neither fear of technology nor its indiscriminate exaltation. If one viewed ethics solely as a form of personal conduct, one would be expecting too little of it; on the contrary, if one took it for the embodiment of a decisional, political practice, one would be expecting too much. Among the tasks of ethics, however, is that of providing a rational guide to political action.

Ethics and technological assessment are mutually correlated. Without technological assessment (based on empirical data or on a combination of theoretical knowledge) no ethics of science and technology could exist. For its part, ethics plays an important role in technological assessment: As an integrative science, it seeks to prevent problems from being considered in isolation, making a critical comparison of various types of forecasts, facts and theories, and calculations of the cost/benefit ratio. When a comparative technological assessment is made, criteria based on ethical principles are necessary. When given criteria are introduced as anticipated, unverified judgments, ethical experience is necessary in relation to such prejudices. When the goal is to formulate correct judgments (for scientific, legal, and institutional procedures), procedures based on tried and tested sequences are necessary (study of the problem, analysis of alternatives, reciprocal correction of the various prospects, fixing of priorities according to empirical rules, verification of possible exceptions, and so on). On all these points there can be profitable cooperation among scientists, philosophers, and theologians who deal with ethical problems.

Technological assessment is especially important before a new technology or given applications are introduced to verify their possibilities, direct effects, and collateral consequences by means of a comparison and discussion of experts' opinions. An essential element in this discussion process is citizen participation (not to be confused with the right to information); another equally important task of the preventive intervention of technological assessment consists of considering the problem from the standpoint of the social desirability of the objectives, of the relation between ends and means, and of any possible alternative solutions.

There is no doubt that ethical competence should also come into play here, and that decisions should be screened on the basis of an ethical criteriology. Much of technological assessment consists of integrating data, facts, theories, and preexisting forecasts in the light of ethical considerations. In this preparatory phase, it means establishing in a given case and in a specific sector (e.g., the biotechnologies) what should be promoted, tolerated, or forbidden. Then it is necessary to assess whether the bans should be of a juridical type or whether a moral self-obligation of individuals and of the scientific, technological, and economic communities is sufficient. In addition, ethical responsibility must also endeavor to respond to questions of the institutional type relating to organization, education, transparency, and social measures, seeking a dialogue with the public institutions: the scientific community, the schools, the economic, political, and trade union world, the churches, and the media.

Social Acceptation and Ethical Good

Man's effective conduct does not constitute an ethical criterion, nor can the states of mind or the spontaneous judgments of individuals be appealed to in order to establish what is ethically right. Consequently, acceptance by society, or by significant social groups capable of orienting behavior, can undoubtedly provide indications regarding moral questions but cannot allow one to make decisions on them. If man's effective conduct were an ethical criterion, normative thought would be confused with moral sociology, and political pragmatism would be reduced to a complex of strategies of acceptation: how to allow for social acceptance or nonacceptance, how to stabilize or destabilize it, how to preserve it, how to transform it, and so forth.

Whoever faces up to the problem of ethical correctness accepts making his or her actions depend on principles, goods, values, and norms (i.e., prescriptive judgments) that correspond to the fundamental anthropological question of the capacity to discriminate between good and evil and the true and the false.

Ethical judgment in specific contexts is a mixed judgment in that it also contains descriptive elements or technical evaluations that contribute toward deciding the appropriateness or the validity of the norm, even if that does not influence the value of a norm in the abstract or in standardized situations. A change (e.g., a further development) of the technical judgment thus also influences the concrete norm, and sometimes it determines its amendment.

Ethical good must also be subject to the social discourse, in which it must not only justify itself according to the rules of a free and fair consensus but also has to do with processes of acquisition which sometimes permit only partial realization. In this sense, ethical evaluation may be viewed as an uninterrupted process which, however paradoxical it may appear, precisely by reason of its flexibility requires stable institutions able to react in concomitance.

Method of Combination of Elements of Judgment Based on Experience

Every operational proceeding on which the ethical judgment is based requires a plausible procedure (Mieth, 1989, 1993a; Christoffer, 1989). A pragmatic reduction is linked with the sequence of cognitive steps: Even one who examines a series of different prospects would like to arrive at an "operational" procedure which strives as far as possible to combine single elements of judgment. As evidenced by the considerations made previously, an inductive proceeding is indeed recommended. An inductive method verifies the validity of given ethical normative judgments starting from the concrete problem and moving more deeply into ethical reflections and not vice versa. Starting from this assumption, it is possible to outline the methodical sequence of an ethical proceeding.

The first step consists of knowing the problem or of understanding the debate of the experts; it is immediately evident, in this regard, that for one engaged in ethical evaluation, it is essential to have an adequate knowledge of scientific controversies. It is likewise indispensable that the basis for mutual confidence should be created by both parts. This can happen only through constant contact.

It is therefore necessary to seek interpretations or postulates of pertinent sense which already have a role in the experts' discussion. In this category are the criterion of safety and also the efficiency of resource management or the sense of an ecological disposal of urban or industrial wastes.

In the third instance, the alternatives have to be rationalized. Various ways of reasoning, which can integrate or oppose each other, are examined, as are their conclusive nature or their capacity to absorb the strongest counterargument. In the discussion, therefore, it is not just a question of excluding the less convincing arguments but also of asserting the alternative in which the greatest possible number of other arguments can be included. For this reason, the counterarguments must be accepted in their strongest version and not the weakest. This forms part of the practical ethic of the discourse or of a standard ethic that has to be accepted and absorbed in the scientific ethos.

Lastly, an assessment is made of the priorities. It cannot be expected that the entire process of study, analysis, and evaluation of the consequences should arrive at anything more than an excess of plausible arguments. Although there are mathematical possibilities of confuting false logical deductions, in the field of ethics a demonstrative mathematical proceeding does not exist. Also in ethics, however, false deductions are found; for instance, the motivational fallacy, which consists of inferring the legitimacy of the end pursued from the legitimacy of a motivation; the naturalistic fallacy, which consists of deducing a "having to be" from "being"; or the genetic fallacy, which consists of inferring from the appearance of an application its ethical correctness.

The procedure illustrated can be shown to be valid in all circumstances in which the search for the right conduct takes place on the basis of the assumption of communication. In fact, when individuals cannot delegate ethical evaluations and the decisions relating to their sphere of action to a specialized normative thought or to some authority (e.g. religion), they have to resort to a procedure in which it is possible for them to contribute collectively their own knowledge, their own experience, and their own reflection.

Ethical Arguments in Favor of Genetic Engineering

Genetic engineering offers numerous opportunities, which are discussed in other articles in this volume. It is our duty to thoroughly examine and sort out these opportunities. Here, I concentrate on the limits of genetic engineering because of the fact that the ethical evaluation of the opportunities this offers is clearer, simpler, and more concise to formulate. In the sector of research and that of pharmacological and immunological applications, there are well-founded hopes that not only legitimize ethically but actually impose that progress in this field be pursued, naturally with every attention and all the precautions necessary. Not just doing but also neglecting to do can be morally mistaken.

The advisability of producing less costly, simpler, and more effective drugs, of treating certain pathologies (insufficiencies of the immune system, angiopathies, tumors, and so on) with new, biologically active substances produced with the methods of genetic engineering, and of resolving ecological problems using new disposal techniques for urban and industrial wastes, their conversion into materials that can be reused, or the reduction of polluting substances cannot be deemed other than positively, although with the proviso that advantages and disadvantages are assessed in the specific context.

No judgment should ever be expressed *en bloc* either for or against genetic engineering and the biotechnologies in general; instead, it is necessary to make an assessment case by case. Moreover, this willingness to verify and discuss implies renouncing any absolute condemnation. "Slow speed ahead" is the criterion by which we should act, as noted in the first symposium on the subject of ethics in the sciences. There is a need for time for reflection not only in the field of specialized knowledge but also in that of knowledge of orientation.

Principles of economic ethics also come into play here (Wirtz, 1993): Economic good, measured on the basis of the criteria of efficiency, stability, employment, and protection of the environment, coincides with ethical good to the extent to which it adjusts to the social context, fostering its conservation and development (the social market economy). Socioethical interests cannot always be reconciled with economic imperatives (competitiveness, profit, and maintaining and creating jobs, determined exclusively by economic cycles), but the latter cannot be abolished, nor is it legitimate to violate the ethical principle of the autonomy of the economy in the name of socioethical interests.

In the following section, the general and specific considerations on the criteria relating to the ethical limits to genetic engineering must be read taking the previously mentioned points into account.

General Considerations on the Ethical Limits of Genetic Engineering

Whether one holds, as does H. Jonas, that the "worst prognosis" should have preeminence over the best one — always bearing in mind the fact that even not acting can mean risks and undesired consequences — or one agrees with the argument that a moralization of the probabilities that given results will ensue is not legitimate — when the active subjects and the passive subjects are different persons (or groups of persons) and therefore consensus is necessary — the conclusion is a strong need for careful verifications and public debate. The fact that given norms or decisions are developed between different subjects, and with consensus, does not necessarily mean that they are ethically right. Ethical principles and criteria are still indispensable for decisions.

The ethical debate on safety should be public and open to criticism. The greatest participation and the greatest possible transparency, and functionally indispensable confidence in the competent authorities and in delegated decisions, must be the rule. The debate on genetic engineering, however, cannot be reduced to a discussion on safety. It is necessary to also assess the objectives pursued and not just the means used in the genetic engineering projects, and one has to ask oneself whether they are just, appropriate, and realistic. There are ethical limits to genetic engineering regarding both goals and consequences and the means used to achieve said goals.

The aims of genetic engineering are basically medical, agricultural, ecological, economic, and military. Military aims are not justifiable ethically. The so-called "scientific" arms, whose development, possession, and application are impossible to control adequately, should be banned throughout the world. The ethical limits of the other aims should be discussed in relation to the individual contexts and to the specific means of each one of them. Thus, for example, the applications of genetic engineering in the agricultural sector (plants, animals, and microorganisms) are morally questionable if inserted in the context of an agricultural policy weighed down by problems such as overproduction, the conversion of cash crops into fodder for the production of meat with a loss of six-sevenths of the crop, the dominion of the single-crop system and the monopolizing of the varieties of seeds, little respect for the environment, the monopolization of technological know-how, and the increased dependence of the developing countries. Future applications of genetic engineering in the agricultural sector must be assessed from the economic standpoint and from the ecological standpoint.

There are also limits that concern not so much the aims as the means used and their consequences. The ethical criteria in this case cannot be entirely substituted by technological assessment since the means used, the consequences, and the risks they involve have to be assessed from the ethical standpoint. Regarding applications in the medical field the following three principles hold:

1. According to Böckle (1985), "it is not legitimate to conduct experiments on human beings, not even on virtual human beings (embryos and fetuses). In principle, forms of human life should be instrumentalized only for life itself."

2. Currently, there is no ethical justification that is sufficient for the genetic modification of germ cells. Only a few years ago, experiments in this sense were forbidden since demonstration is lacking of a controllable effect and because sufficient indications do not exist; in a 1986 document, Sperling stated, "No one wants it, no one can and no one needs it." Methods have recently been finalized that could one day enable defective genes to be replaced. Although there is not yet certainty in this regard, the ban on manipulating germ cells can be reasserted only insofar as uncontrollable effects can be excluded. However, scientists admit that in somatic experiments with retroviruses there could also be repercussions on germ cells so that the limit will be reached quickly. If gene therapies are justifiable for therapeutic purposes, the question remains open as to whether these therapies can be applied without irresponsible experiments (Mieth, 1990a, 1993b; Wimmer, 1991).

3. It is not legitimate to exert medical or social pressures on the individuals who submit themselves to consultations, examinations, experiments, or measures on their genes or who should assume responsibility therefore on behalf of minors, fetuses, or embryos.

To assess the reasonableness of the medical and ecological aims of genetic engineering, the *rule of the solution to the problem* holds, on the basis of which scientific or technological progress should not be made at the price of creating greater problems than those it resolves. This criterion imposes particularly careful observation of multifactoral effects, for example, in the case of releasing genetically modified microorganisms into the environment. The rule of the solution to the problem must be regarded only as a point of reference which in the concrete case it should be sought to approach as nearly as possible. Whether and to what extent the problems created are "greater" than those resolved can be established only by approximation. Given our cognitive and temporal limits, it is impossible to imagine all the current and future consequences, but in critical cases it is necessary to ascertain these as accurately as possible and to seek alternative paths. It is also necessary to verify that the growing need for technological safety does not have negative effects on social safety and on the freedom of citizens, just as moral responsibility to posterity as well as to the environment and to society must be introduced as a criterion whether positive or negative.

Since we have entered into new, high-risk technological dimensions, we have had a dual experience: On the one hand, overly enthusiastic expectations are deceiving, whereas on the other hand, all the results are ambivalent. The question as to whether we are on a par as human beings with the developments that we ourselves have started and their consequences crops up repeatedly. Without doubt, our responsibilities have increased in such a way as to force us to take the contingent nature of the human being into account: Whereas the products constantly "improve" (i.e., they become ever more complex and efficient), man himself does not grow better.

Even when reflection on the ethical limits of genetic engineering intervenes promptly, three difficult problems arise for ethics in the present-day world. First is the difficulty of having an overall vision in the face of the complexity of the contexts and of the specialization or the increasing differentiations in science and technology. Second is the problem of a scant consensus on ethical criteria. An example is the discussion on the safeguarding of life and of its integrity in virtual human subjects (embryos and fetuses). Furthermore, ethical–ecological and ethical–economic criteria are less uniform throughout the rest of the world than in our own latitudes. Lastly, the problem exists of enforcing ethical responsibility vis-à-vis vested interests that are oriented differently and with regard to apparently spontaneous developments. What is known can very soon be done, and what can be done is very soon done: At this point, ethics intervenes too late — that is, in the best of cases its action can only be one of compensation. Appeal is often made to the normative force of effectiveness or of the ongoing trend, using this argument as an instrument of power.

To address these problems, it would be appropriate to institutionalize ethics within the contexts of biology in relation to medicine and of technology in relation to economy. An interdisciplinary debate should be opened on the limits of the applications of genetic engineering in various sectors in an attempt to answer the following questions:

1. Do scientists make an overly optimistic evaluation of the objectives and of the possibilities of development in their own specialization?
2. How can the isolation that characterizes specialized applied research be diminished?
3. In what way should responsibility toward nature, society, and posterity be introduced?
4. How can certain problems (food, environment, and so forth) be resolved in such a way that the solutions do not create worse problems that the ones they resolve?
5. What risks can be assumed and what safety measures become necessary?
6. To what extent does the increased need for technological security entail a limitation of social security and freedom?
7. To what extent does genetic engineering increase rather than resolve problems already believed to be linked with progress and growth (e.g., the limits of the "green revolution")?
8. Is it not possible to adopt alternative techniques that will produce better results and create fewer problems?
9. Have we really entered a new dimension of man as the responsible creator of the world, and to what extent are we adequate, as human beings, to fill this role?
10. What are the possibilities of intervention for ethics in an apparently spontaneous development?

In the debate on genetic engineering, the formula "risk and opportunity evaluation" is frequently heard. However, often such an evaluation is impossible. Such is the case, for example, when the amount of the damage and the likelihood that it will happen are neither known nor quantifiable. In the terminology of the theory of games, it is thus a question of decisions under conditions of uncertainty. Nevertheless, it is possible to outline scenarios of undesired consequences.

In light of these considerations, there are numerous criteria that should be followed in the evaluation and decision-making processes: Doubts and perplexities must be expressed at the outset of the reflection so as to reduce uncertainties as much as possible; evaluations must be conducted case by case; the types of damage must be defensible at the ethical–political level — that is, the decisions must be made using a political and legal procedure oriented toward moral principles and, for instance, toward the basic questions presented previously; and a procedure must include an evaluation of the objectives, consequences, and alternatives. Moreover, the empirical technological assessment, viewed as a verification of the level of toleration, requires a normative ethical orientation as part of an ethic of technology. Lastly, in a debate between opposed arguments it is mandatory to motivate the arguments in the sense of a distribution of the onus of proof.

By reason of the close interconnection of science, technology, and economics, the problem of responsibility is posed at the origin of scientific knowledge: Anticipation of the consequences requires institutional forms of (self)-responsibility. Anticipation of the consequences, however, is not up to individual scientists; science and society are both called on to assume responsibility. To this end, connections must be established between science, technology, economics, and society, such as may be glimpsed both in the specialized press and in journals of current affairs and also in institutions and programs that interface science and society.

Applications of Genetic Engineering in the Agricultural Sector

Without wishing to deny or belittle the progress made in the growth of productivity, the balance sheet of the green revolution in the developing countries, but also the balance sheet of the mechanization, rationalization, and massive use of chemicals in agriculture, presents extremely negative aspects: the destruction of small farms, the drift from the countryside, environmental damage, scanty distributive justice, single cropping, and overproduction (Stanger, 1992). If we then consider that, as already mentioned, six-sevenths of all cash crops are lost by being converted into fodder for meat production instead of being used for human consumption, that the increase in productivity means not only tons of vegetables and butter but also serious harm caused to the soil and the ground waters, that mistaken development projects lead to soil erosion and impoverishment, that an ever-diminishing number of farms control an increasing number of products, and that the various species are diminishing while the dependence of underdeveloped countries is increasing, then one cannot help being skeptical about all the optimistic projects of those hoping to resolve the food problem by increasing productivity, without wondering what is being produced, and where, with what methods, for whom, and to whose advantage.

After mechanization, single –cropping, and chemicals, the new password is "biotechnologies," of which genetic engineering is a foremost sector. There have been important changes in animal husbandry: The cloning of embryo cells makes it possible to further accelerate progress in stock raising, and through gene transfer techniques it is possible to produce animals with completely new properties. The development of these new technologies is aimed mainly at increasing productivity (greater quantities of milk and meat), improving quality (less fat meat), obtaining more disease-resistant animal species, producing pharmacological substances from the mammary glands of animals, and making pest control more effective through genetically manipulated viruses. Great expectations can and should be placed in the macro- and microorganisms manipulated with the new methods of genetic engineering (Fig. 1), whose impetuous development has been boosted by a new wave of optimism: new products, new properties, a qualitatively superior type of growth that is less harmful to the environment, and the creation of new jobs. For a differentiation of these expectations, based on understandable intentions and aims worthy of respect, new institutions should undertake the task of posing rightful questions of bioethics, environmental ethics, social ethics, and economic ethics.

Genetic engineering in itself is not more "dangerous" than other technologies, such as information technology. However, growing urgency is forcing us to consider whether we are able to make ever-quicker progress in shorter amounts of time while time to reflect on the responsible application of the new acquisitions diminishes. In this respect, nuclear technology represents a controversial point for reflection. It is necessary to overcome the isolation of the objectives of scientific and technological research, and scientists should force themselves to look beyond their test tubes and the walls of their laboratories. Man's possibilities as responsible creator are increasing enormously, and once more this marks the entry into a new dimension. Another question that arises is whether, in the long run, man will not fall victim to the new need for security, to the growing environmental problems, and to the profound structural changes.

From the standpoint of bioethics, one should ask oneself whether the experiments on the genetic manipulation of plants and animals are not debatable also at the biological level — if the huge technological costs entailed by production, by growing transgenic plants, and by breeding transgenic animals are not greater than the single advantage that derives from them. On the other hand, the efforts of genetic engineering are directed toward obtaining ever more resistant organisms. From these efforts, considerable successes have been achieved, both with plants and with animals. The question remains, however, whether transgenic plants and animals do not prove useful only in particular conditions.

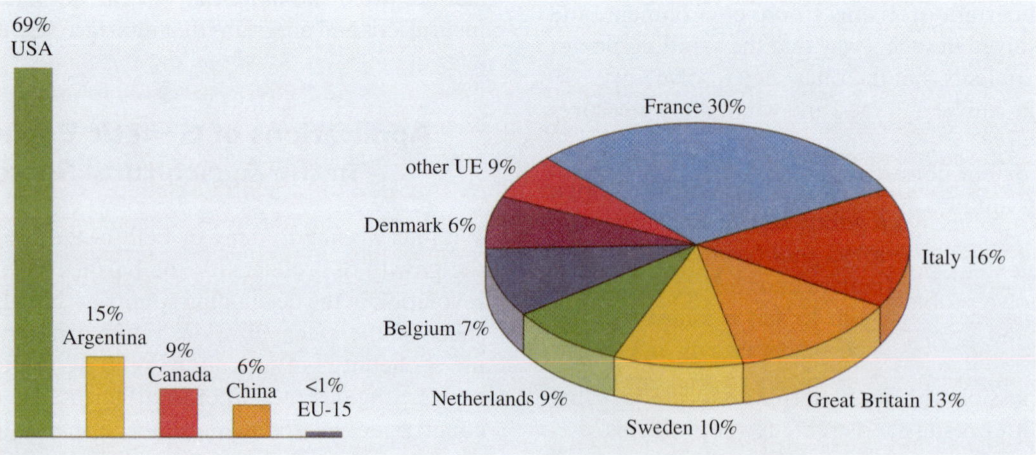

FIGURE 1 (a) The impact of biotechnologies in the agricultural sector in various geographical areas. The percentages given refer to a total of 29.6 million hectares cultivated. (b) Agrobiological experimentation in Europe. The percentages refer to a total of 1283 authorizations granted for experimentation in the period 1990–1998. Normalized data.

In a long-term perspective and in light of the changed productive requirements, a substantially positive balance of the cost/benefit ratio could change to the opposite. Releasing into the environment plants that have been genetically manipulated raises questions to which the answers are difficult and controversial: One wonders, for instance, whether transgenic plants, in view of their additional metabolic charge, have greater or lesser competitive capacities, and one also wonders whether conclusions regarding the biological integrity of these organisms can be drawn from these facts or otherwise.

From the standpoint of environmental ethics, one has to start from the fact that the biotechnologies consolidate structurally (at the lower level) a given need for chemical products. The following questions then arise: Do alternatives exist (as biological agriculture or ecological projects for sufficient autonomous production) that permit sufficient productivity? To what extent are chemical products really necessary? Does genetic engineering increase the dependence of agriculture on chemicals or could it help to reduce it? Another important question concerns the actual or potential environmental repercussions of certain experiments of genetic manipulation, particularly when new plants are provided with given selective advantages. For example, the possibility has been mooted of transferring the new properties from cultivated to self-sown plants (Wöhrmann, 1991).

From the socioethical standpoint, one should ask oneself the following: What risks can be assumed? What safety measures are necessary? Does the greater need for technological safety not risk endangering social freedom and safety? Should solutions to the food problem be accepted without doing battle against the basic causes of poverty (e.g., the production of meat preferred to that of vegetables) or imbalances in the distribution of the productive surplus? Genetic engineering promises both an improvement in the quality of products and a battle against poverty. However, how is it possible to determine the ratio between the two objectives?

From the standpoint of economic ethics, one must ask oneself whether genetic engineering does not prolong or worsen problems already present with regard to progress, growth, and dependence. Faced by the undeniable limits to growth, is it legitimate to continue to give priority to growth as opposed to distribution, to the intensity of work, and to the struggle against drug consumption? Furthermore, do alternative methods not give better results, creating less problems?

These are but some of the numerous questions that arise regarding ethical responsibility. What has been proved to be mistaken in the past cannot become right if it is still pursued with other means. That which creates new problems must not be measured only on the basis of an isolated success but also on the basis of the weight of the new problems (Stanger, 1992).

In addition, there is the question of the acceleration, differentiation, and multiplication of the single applications. Without an adequate institutionalization of ethical, political, and juridical responsibility, and without a multilateral technological assessment, the basis cannot be created for reasonable decisions. The socioethical guidance and control of the market economy must be matched by a socioethical guidance and control of technological development. In the social doctrine of the Catholic Church, the trend in recent times has been asserted of considering technological know-how and its distribution as a problem of social justice (Encyclical *Centesimus Annus,* 1991).

Regarding food produced with genetic engineering, the following measures are essential: First, it is necessary to verify the processes of producing organisms manipulated genetically and the effects of their release in the environment, as occurs with other biotechnological methods. In addition, the necessary information must be supplied to the user (product labeling) so as to safeguard the sovereignty of the consumer. Doubts relating to an insufficiency of the consumer's sovereignty do not justify placing citizens under protection, eluding the obligation of transparency. Possible discrimination of products cannot legitimize discrimination of the consumer's right to free choice. The mere indication that the products stem from genetic manipulation helps the consumer to orient his or her preferences. The indication of the type of genetic manipulation enables the consumer to reflect on his or her choices. Consequently, the label should give "neutral" information; that is, it should indicate the procedure or the material used.

Transgenic Animals

The production and raising of transgenic animals should be judged in the same way as the methods of cross-breeding and selection used in traditional stock breeding (Wimmel, 1991). It is not possible to identify an ethical limit that permits or imposes an assessment different on principle of the transgenic techniques applied to animal species. Although entry into a new dimension such as noted previously makes the ethical problem more acute, it does not change the criteria thereof. It must be demonstrated that the techniques it is intended to use possess essential advantages over alternative systems. The alternatives do not refer only to the use of different techniques but also to risk reduction strategies such as renouncing unnecessary benefits. Even more than in the case of conventional animal husbandry, in relation to transgenic techniques we have to ask ourselves what is the sense of increasing production vis-à-vis European overproduction, the need to import concentrated feed, and the level of hunger in poor countries.

To man, utility is without doubt an important element, but it is not the sole decisive criterion of assessment in relation to the treatment of animals. Equally important is the consideration of the suffering inflicted on them compared with the advantages stemming therefrom for man. Stock raising based on man's productive interests need not necessarily mean suffering and maltreatment of the animals. These are justified only if the possible advantages are significant and highly probable and if no alternative ways exist. From the standpoint of distributive justice, one must also ask oneself whether such advantages do not exclusively benefit a minority of the population.

These considerations hold for the applications of engineering in the various sectors and for the relevant processes of technical–economic implementation; that is, both for pharmaceutical applications (biomedicines) and for the qualitative improvement of the animal and plant species of productive interest.

Organisms and Microorganisms Modified Genetically

According to the experts (Mieth, 1993c; von Schell, 1993), the possibility of genetically transforming an organism represents a new technological stage. The following are the main elements: acceleration of the change with respect to conventional methods of selection; variability of the change; the possibility of obtaining exactly the change desired; overcoming the genetic barriers produced in the course of evolution; further improvement through the creation of other genetically modified microorganisms in which the desired function is enhanced; and "multiprogrammability," i.e., the possibility of introducing a plurality of activation functions that can correct each other in such a way that the action carried out is always that which is desired.

The following are the risks involved in these genetic manipulation techniques: transformation of the physiology of the organism (which can mean, for example, weakening it); vertical and horizontal interchange with other organisms (air, soil, etc.); assumption of "free" DNA, i.e., detached from the cell; multiplication or alteration of the genetically modified microorganisms through the effect of agents present in the environment; collateral effects (aside from the one desired) of the genetically modified microorganisms released into the environment; and the possibility that a possible programmed deactivation may not exclude a subsequent spontaneous activation. Total control of the activation system is impossible as occurs, for example, for the resistance developed by microorganisms toward "killer" functions or "suicide" genes.

The view of the possible risks can also be described as follows: (i) Uncontrolled dissemination of the organism and the genetic mutation can occur, (ii) unforeseeable effects of genetic manipulation can occur on the physiological properties of the organism, and (iii) there can be unexpected or unwanted effects on living beings and on the ecosystem (pathogenicity, toxic products, and repercussions on the existing species and on the biogeochemical cycles).

However one wants to classify them, there is no doubt that the individual risk factors combine and overlap. There are always elements of uncertainty in the study of the ecosystem processes or in their assessment. This means that the effects induced or explicated at the genetic level—with their direct and indirect or deferred repercussions at the level of the metabolic processes—caused by release into the environment can be observed and assessed only within the context of an ecosystem interaction. The occurrence of a certain event following given measures depends on specific circumstances that are difficult to foresee both in space and in time, and the variable effects that they produce are also difficult to foresee (von Schell, 1993). As stated by Holling (1980), one has to "expect the unexpected." It will be said that ethical responsibility regards only the foreseeable consequences. However, the very question of predictability (or unpredictability) is the object of ethical responsibility. Here, man as an ethical subject has to be placed in the center of our reflection since man is a finite, limited, and false being and cannot start from the assumption, which is at the basis of the myth of progress, of always being able to control unforeseeable consequences. Against a technological mentality which holds that, in accordance with the futurological argument of anticipating the future, man can always master the consequences of his actions, attention must be drawn to the fact that in history man has often not managed to do so and he has sometimes managed it only at the price of major sacrifices. Taking due account of man's contingent character is part of the common anthropological approach of ethical reflection.

There is a certain consensus regarding the necessity to control the release of organisms and microorganisms into the environment. In this connection, a specific criterion of safety is fixed for man, for life in general, and for the environment. The safety concerned cannot be total, but a margin of risk is accepted that is considered negligible on the basis of rigorous criteria of probability. An important role is played here by analogy with other situations in which margins of risk are normally accepted. However, there is a difference between accepting a margin of risk in a system that was introduced long before and which can hardly be replaced (e.g., the road system) and the willingness to accept in advance a risk of this type in a system not yet installed. Far greater attention to prevention is paid today than in the past, as shown by the institution of technological assessment.

Control based on the criterion of safety should be imposed by progressively phasing release into the environment, which should take place first in a closed system, then in a partly open system, and finally in a fully open system. In support of controlled experimentation for the release of genetically modified organisms and microorganisms there is a specific legal discipline. Legislation is also currently envisaged in the European Union.

The problem of experimentation and, at a later stage, of actual use stems from the fact that the single quantities have to be fixed in a relatively arbitrary fashion. This applies throughout the experimentation due to the size of the populations with which and on which these experiments are carried out, the space to be used to verify a possible effect, and the level of tolerability. We know from experience that the level of tolerability has to be fixed with a certain margin of uncertainty, which often leads to reducing it to the minimum.

A further problem consists of considering different cases to be equal on the basis of the principle of homology. Is it correct to draw from the behavior of one organism conclusions regarding the behavior of another one? It seems more reasonable to adopt, instead, a differentiated approach for each single case. Another risk is that the particular importance of the problem of release into the environment of genetically modified plants, animals, microorganisms, and viruses, and the discussions caused thereby, induces those concerned to neglect equally rigorous controls of other techniques of molecular biology.

Objectives and Criteria of Tolerability

On the basis of the overall development of the new biotechnologies, particularly experiments on the genetic manipulation of plants, animals, and microorganisms, there are legitimate interests and reasonable goals. These interests and goals cannot be questioned as motivations; however, as noted previously, we would be falling into the motivational fallacy, so to speak, if the legitimacy of the objective pursued were to be inferred from the legitimacy and the soundness of the motivations. This would be the same as shirking responsibility in the name of intentions.

The objectives pursued in the main sectors of application of genetic engineering are manifold. First, there are ecological objectives, such as the disposal of noxious substances or measures carried out on the damage caused by industrial accidents (e.g., measures performed on oil slicks). Second, there are economic objectives, which is the case for genetic engineering in the agricultural sector (Figs. 2 and 3), the goals of which are to increase productivity, make unproductive regions productive, and exercise closer control over the chemical products used. Third, there are experimental objectives, for which it means extending basic research and as far as possible verifying whether the calculations required for certain practical applications can prove to be correct in a simplified application as yet detached from any productive interests.

The ethical evaluation of interests and objectives entails a critical analysis of the goals and the relation between ends and means. In particular, one must ask oneself the following: Are they achievable (realistic) objectives? Is this the only way to achieve these objectives? Do genetically modified organisms and microorganisms create greater problems than the ones they are supposed to resolve? Can the problems be prevented rather than seeking a solution *ex post?* In this connection, consider, for example, the pre-

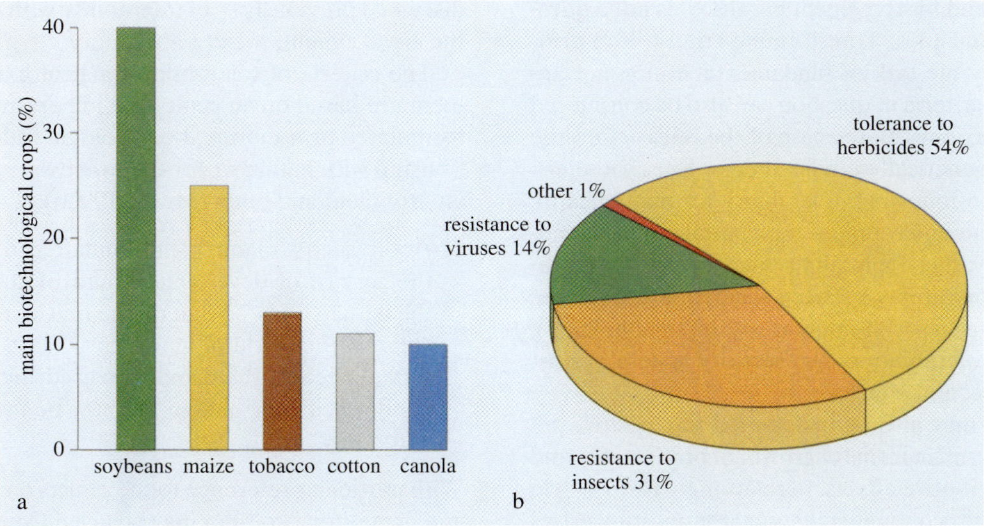

FIGURE 2 Diffusion and characteristics of the main biotechnological crops in the world. (a) Percentages of the area dedicated to transgenic crops of the total cultivated area in 1997. (b) Percentage breakdown of objectives of genetic manipulations out of total transgenic crops. Normalized data.

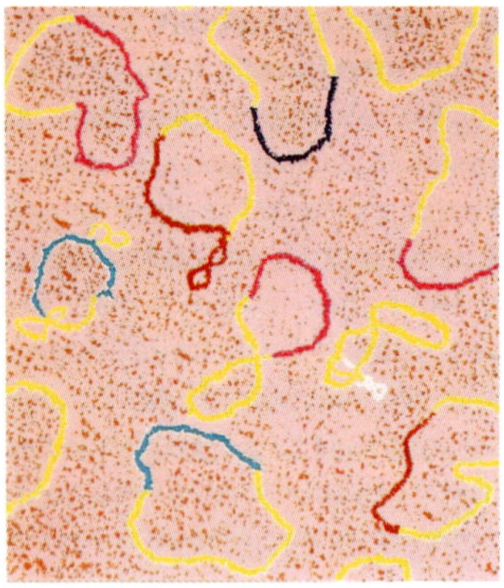

FIGURE 3 Transmission electron microscope image in false colors of DNA plasmids stemming from the bacterium *Escherichia coli,* manipulated genetically. The plasmids consist of DNA fragments capable of self-replication independently and of moving within the genome of the host cell (bacteria and yeasts). Because they are easily transferable from one cell to another, they are used as vectors for genes in genetic engineering. For example, in gene therapy plasmids with therapeutic genes are inserted in human cells to replace defective or missing genes. The DNA of plasmids are shown (in yellow) with extraneous gene segments (differently colored parts) inserted internally (photo courtesy of Murti/SPL/Neri).

vention of ecological problems or, in the economic field, the possibility of increasing productivity through a better distribution of resources.

In this case, the inductive method of ethical assessment illustrated earlier must be flanked by applied ethics, which introduces given criteria established between various subjects and with consensus and also illustrates their validity in the concrete case. These criteria are established between various subjects and by agreement because it is not a question of higher principles. Transforming criteria into principles is generally the task of fundamental ethics, not applied ethics. The criteria in question can also be considered as intermediate axioms. In the case of the release into the environment of genetically modified organisms, these criteria could be the following: tolerability for man's health; tolerability for the environment (i.e., sustainability); economic tolerability; social tolerability, understood as distributive justice, compatibility between security and freedom, safeguarding of given social categories (e.g., smallholders), and so on; and tolerability of the security system against wars, acts of force, and attacks.

Sustainability may also be understood as a positive action in the long term, in terms of growth of productivity and constructive distributive effects. In relation to Third World countries, from the economic standpoint the question arises as to whether such a costly technology in financial terms is economically "reasonable" and sustainable financially. Consider, for example, high-cost technologies, such as fixing atmospheric nitrogen (von Schell, 1993). To establish social tolerability, the question of the society's relation with technological–ecological risks must also be considered. In other words, there is a problem of participation — the democratization of decision-making processes regarding the introduction of a new technology in which society performs the function of laboratory (Krihn and Weyer, 1990).

Perhaps in this context it would be indicated to focus attention on the problem of compatibility between safety and freedom. The more the ecological and economic needs increase for safety systems, which at first seem to limit only the freedom of companies and not that of the population, the more mobility and the rights of citizens to self-determination are limited (e.g., consider nuclear technology). There are limits beyond which safety systems cannot be reconciled with democratic freedom.

The citizen's rights of freedom take on importance in many situations, such as the analysis of the genome, gene therapy, and the problem of informed consensus; participation in decisions concerning the introduction of a new technology or the placing of new products on the market; or the right of the consumer to information.

In an ethic of concrete responsibility, the inductive and the applicative methods do not exclude but rather integrate each other. The inductive method is necessary since without it the application of given criteria would be purely illustrative: To correctly guide the criteria it is necessary to develop problems starting from the concrete situation. On the other hand, any analysis of the concrete problem is blind unless it is guided by ethical criteria. Criteria without induction are empty, and induction without criteria is blind. Criteria, or intermediate axioms, require an ethical justification; for this purpose they have to be reduced to integrative principles, which can be defined in light of the criteria discussed previously — of tolerability with regard to health, the environment, society, and safety.

The criteria of tolerability for health and the environment are based on an ecological imperative, which may be formulated as a universal or "golden" rule for man's relationship with nature or for the processes of assimilation of environment and body (Mieth, 1990b):

> Act in such a manner that human institutions are in the service of development and of conservation of the human body, so that on the one hand the intrinsic value of the world of nature is as far as possible preserved, reconstituted and increased, and on the other hand specifically human life may be lived in creative autonomy.

With particular reference to the effects on the human body and on nature, two rules may be given (Mieth, 1990b): "Act in such a way that in every institution and in every measure man's contingent, transient, and frail nature is taken into due account as a reality and as human value" and "Act in such a way that the vital requirements of nature as the place of experience of man's contingent body are preserved and developed.". These two rules may be subsumed thus (Mieth, 1990b): "Act in such a way that the instruments of man's satisfactory and creative self-realization (technological and social institutions) do not endanger his physical and biological resources, but in pursuing their specific ends it is always sought to make reference to man." These universal ecological imperatives can be formulated in different ways, provided they try to achieve the maximum possible integration of various prospects.

Economic tolerability, social tolerability, and tolerability vis-à-vis safety are based on the principle of *constituent coherency,* as formulated by Gewirth (1978) and developed by Steigleder (1992): "Act constantly in harmony with the constituent rights and duties both of yourself and of the receivers of your actions." At the basis of this principle is the observation that the possibility of acting always presupposes, in one way or another, the assuming of communicative obligations. Demonstrating that original obligations and liberties are constituent elements of the action is one of the fundamental tasks of rational ethics. We observe in this connection that theological ethics must also abide by this rationality, and this means that it has to be conceived as autonomous. Theological criteria — for example, the definition of man as a creature or as the image of God — can have a sense only if their rationality is demonstrated, even before giving it a theological interpretation.

All the criteria of tolerability have to be traced formally to the rule of the solution of the problem and have to be based on it: "Act in such a way that the problems created by a solution are not greater than those resolved." The demonstration of this rule is based on the observation that it is constantly accepted in everyday life, even if it is not always respected. In the ethical discourse, however, the normative force of reality is not asserted but rather that of the convictions and principles on the basis of which we orient our existence. The fact that a rule is not respected, although it is considered just, does not alter its validity.

It seems clear (and many scientists have already drawn attention to this fact) that in the sphere of scientific, technological, and economic action the rule of the solution of the problem is not respected. In science this is due especially to isolation of the problems — an isolation that can be overcome by means of studying the contexts through technological assessment and through an interdisciplinary ethic. To this end, institutions and structures are needed that are adequate both in the field of research and in that of education. In science, there is need not only of an ethos of truth and honesty but also of an ethos of the consequences and also a social ethos, by virtue of which the social and individual standards of ethical responsibility are respected.

It would indeed be strange if the ethical principles that underlie our daily lives were not transferable to scientific, technological, and economic progress. Probably, it is precisely because of the detachment of scientific, technological, and economic action from any integrated ethic that a loss of faith in these institutions is recorded. What is being undermined is not faith in science or in the progress of knowledge, nor faith in efficiency, because science and technology enjoy the citizens' faith more than any other system. This faith, however, concerns efficiency, not ethical responsibility. Also, because in our society an impressive growth in instrumental knowledge is matched by the constant decline in a universally valid knowledge of "orientation," our sense of insecurity increases. Only through the promotion of such knowledge of orientation can this insecurity be reduced. In this regard, a new alliance is necessary between science and society (Roy *et al.,* 1991) that will guarantee, on the one hand, the full freedom of every specific sector (this, moreover, is in itself a very exacting task) and, on the other hand, the establishment of the requisite connections between the various sectors.

Spiritual Foundations and Ethical–Theological Assumptions for the Preservation of the Earth

Nature on the Horizon of Science and Technology

The premodern conception of nature was not aware of the differentiation of science as the specific sphere of knowledge nor consequently the specialization of the natural sciences. For premodern man there still existed given natural, certain, and unchangeable entities to be approached with a contemplative attitude and not with manipulative verve. Shifting the frontier represented by nature seemed over and above man's possibilities, and that confirmed his finite nature and his limitations as a creature.

The premodern conception of nature implied certainty — a certainty that also characterized cosmology, making it possible to deduce the laws of statics (i.e., the balanced order of the actualization of powers that tend toward fulfillment).

Today, we are certain neither of *teleonomy,* or finalism, nor of the cosmological order. Discussion of the case and the necessity, the image of the world delineated by quantum mechanics and by the theory of relativity, has shown instead that the confines of our scientific horizons are anything but irremovable. Only the relative certainty of experimental reproducibility has remained valid, within the context of a given theory and in a given field of reality. These "facts" and "laws," however, are no longer universal truths but conditioned truths, and they are valid insofar as they indicate with exactitude their own conditions of validity.

The appeal to nature of classical ethics was centered on

the rationality that was the key concept in which the order and the purpose of all nature were revealed. The postmodern return to nature, and in particular what van der Daele (1987) defined, with reference to ecology and biology, as "the moralization of nature," is on the contrary biocentric: "The moralization of human nature imposes that its respect be placed higher than the self-determination of human action" (van der Daele, 1987). In this development can be seen, according to van der Daele, a reversal of the trend to recognize the sole criterion of morality in man's right to freedom in respect of the freedom of others. So as not to have to decide on the integrity of the person, we take refuge in biological facts, with the risk, however, that the interpretation of these facts will be controversial.

In any case, the concept of nature becomes the concept of a limit and at the same time of a project. As a limit, it marks the confine of that which is to be condemned (e.g., the manipulation of the human genotype to create hybrids and chimeras); as a project, it attempts to grasp the "foundation of sense in man's being" (Furger, 1990) and therefore indicates in the project person the sense of nature. However, this concept of nature is philosophical, not scientific, and it is an interpretation. Can it be an adequate point of reference? Quite correctly, van der Daele (1987) draws attention to the fact that hitherto there was "an obvious natural basis for man, which had to be taken into account and which could be relied upon. . . . For this reason the idea that man is his own creator, or that artificiality is man's authentic nature, did not yet know its most extreme forms" (Fig. 4). On the contrary, today, according to van der Daele, the overcoming of some of these "limits to man's self-creation" seems to be in sight. Scientific technology is

> transforming the anthropological constants into options. Human nature is becoming contingent, able to be modified at our pleasure. This contingency is irrevocable. Whether or not to keep man's current characteristics in the future will be a decision that we will have to take with the awareness that we could even be different.

FIGURE 4 The hybrid "pregoat" chimera derived from merging sheep and goat embryos subsequently implanted, in the blastocyst stage, in a carrier mother (photo courtesy of Tomkinson/SPL/Neri).

Even if in this context the concept of contingency has a basically different significance than the theological one — instability and variability in the first case and finiteness in the other (Drewermann, 1991; Mieth, 1997) — the argument clearly emerges that the "moralization" of nature is an option that has to be justified. Furthermore, according to Furger (1990), "Insofar as the definition of human nature remains or becomes a task entrusted to science, normative recourse to nature cannot have that univocal character able to dispense us from the obligation of acting and deciding in its regard." In the gamut of possible configurations of human nature, it is necessary to decide on the basis of "different criteria than that of naturalness." Nevertheless, these other normative criteria (person, human dignity, and freedom), however much their positivity may be detached from any reference to nature and they exist in the sphere of culture and history, in my view are seeking a rational justification — an "ex negative" one, so to speak — because of the fact that they are not in contradiction with an intuitive concept of nature.

Substitutability or Reduction of the Concept of Nature

The more numerous the conceptions of nature, the more a conceptual analysis of the term implies a reasoned reduction of the possible meanings. The concept of nature may be understood as "foundation of sense" but also as a world of facts that takes on ethical relevance only in the combination of many different sectors of reality and as a particular theory within the context of a proceeding of ethical foundation of the rules of conduct. This last meaning is evident, for example, when, in the discussion of the right to protect the life of the human person, reference is made to a theory of the natural beginning of life.

The theories of needs also resort to nature as the point of reference to define operative rules of conduct. Just as in this case nature must be a nonevaluative point of reference for making evaluations, in the doctrine of the *bonum physicum* the premoral good becomes the object of evaluations, which in turn are based on natural ideas (in the sense of reasonable ones) relating to ends and values. What, then, is nature? A world of facts, a particular theory, or the foundation of sense?

In any case, we have to get used to occasionally translating every concept of nature into the equivalent intended so as to identify the reduction implicit in the use of the concept. If the foundation of sense is considered to be an adequate reduction (here "reduce" means reverting to the essential), it should be clear that the concept of nature is understood in the sense of a hermeneutic comprehension, of an interpretation of worlds, of facts, and of theoretical models. This leads to a sort of splitting of the "nature of nature," which is understood now in the sense of the essence of human nature (dignity of the person) and, as occurs in the classical philosophical models, in the sense of ultimate foundation (God, *polis,* and culture).

Despite the Kantian foundation of the ethic of free moral self-obligation in practical reason, the concept of nature as indicator of the state of things and meanings of ethical importance has maintained an important role in applied ethics (and probably also in Kantian theory). In fact, theological ethics was also used to avoid the concept of nature in fundamental ethics, even if, in continuation with classical philosophy, A. Auer speaks, for example, of "rationality of reality." On the other hand, the concept of nature returns in special ethics due to the need to give an interpretation of it in the writings of the Fathers of the Catholic Church and because the influence of St. Thomas Aquinas continues to make itself felt in moral theology and in current philosophical ethics, and also because the process of a "moralization of nature," described by van der Daele (1987) in the ecological context as a social phenomenon, represents a new challenge.

In the face of this challenge, which specifically affects applied ethics by reason of its functions of interdisciplinary integration, it is necessary to reflect once more on the cooperation of various sciences, theories, or theoretical models relating to single sectors of reality and on nature as a synonym for foundation of sense. This will be clear in the case of bioethics and technological ethics. Briefly, if the concept of nature can be substituted in its particular configurations, in its globality it remains an indicator of an indispensable task of ethics as science of integration. The philosopher O. Schwemmer (cited in Wehowski and Schwemmer, 1995) defined this task as follows:

> In my view nature has to be attributed its own autonomous existence, vis-à-vis which we are responsible. Quite apart from the fact of postulating a "right" of nature, or otherwise, with regard to nature we have an unconditional duty. This obligation stems from our awareness of having stepped out of nature, in that we have learnt to develop new possibilities of action. But it would be misunderstanding this privilege to deduce from it the right to turn nature into a mere object of manipulation. Rather, man's relationship with nature is still today as in the past a relationship of man with himself. And this is a good reason for behaving as though nature possessed rights, irrespective of how these are defined conceptually. The concept of "man's relationship with himself" is not to be interpreted in a sentimental or romantic sense. Since it expresses also the fact that nature in our life plays an absolutely autonomous role, and therefore touches us directly. This is shown by what happens when we go against nature. We fall ill, we die, without managing to recognize in some way in this dying the end of our existence, the conclusion of our project of life.

The Concept of Nature as an Indicator

What then remains after a discussion on the concept of nature in the ethical argumentation? There remains the memory, at times dangerous and at times liberating, of an attempt at philosophical–theological integration performed by scholastic philosophy, of a legacy, or of a tradition that is directed toward the future. This tradition, constantly alive in theological ethics, must be directed toward the future since in order to deal with today's ecological and biotechnological challenge of man's specific responsibility toward his world of life, an element that cannot be omitted is the dialectic between the choice of preserving integrity and legitimate manipulation.

There remains the problem of a principle able to establish a bridge between "being" and "should be," between the truth of reality, insofar as we are able to know it, and the ethical correctness of acting. The more ethics, regarded as a branch of science whose object is the logic of prescriptive knowledge, turns to the model of scientific truth, the more it gains in (scientific) respect and loses in significance. However, the more ethics moves in the sphere of a higher truth (reason and rationality) than all reality, the more it is reduced by science to the rank of practical option, of purely philosophical and religious discourse. The challenge that stems from this constitutes a commitment for the present and for the future. The term nature, we must not forget, is used in a metaphoric sense also in ordinary language. Since antiquity, the concept has been used as a metaphor for opposed things such as order and chaos. This, too, is a task of ethics — to distinguish carefully between metaphoric meaning and conceptual meaning and to motivate in detail the use of the concept.

In many aspects, the nature that surrounds us, even in our bodies, is in a certain sense an artifact. Explicating our sexuality according to the dictates of erotic culture and respecting ethics is nature and culture simultaneously, as the artist's statue is simultaneously a creation of the spirit and marble. However, since for man art also represents a surpassing of himself conditioned by nature, respect for his origin always implies the profound respect of art toward reality. This should also be the case for technology. This analogy was proposed by Schmitz (1990) in defining nature as an unyielding "residue" whose sacrifice can only mean our death.

The concept of nature links philosophical ethics of classical type with theology, not just because of its origin. The term nature is a lasting indicator of an unsteady form of thought in theology: grace and nature, creation and sin, "pure nature" as a residual analytical concept for thinking of redemption, nature and freedom, and nature and theodicy. To what extent the blend of natural and supernatural continues to animate theological reflection is testified, for instance, by the thought of K. Rahner, to whom the dimension of the "existential supernatural" in which resides man's possibility of faith is part and parcel of man's naturalness. Quite correctly, the problem of a "natural theology" is constantly reproposed; likewise, quite correctly Demmer (1990) emphasizes the need to reintegrate the empirical, cosmological, and theological dimensions of the concept of nature in the philosophical foundations of the theological doctrine of creation or to establish interdisciplinary teaching ("natural sciences for theologians").

Consequences for a Theological Ethic of Creation

To Impose a Sense or to Find a Sense? Science and technology seem to be subject to the paradigm of *imposing sense.* The fundamental passiveness that characterizes the "creatural" theological doctrine no longer finds any room here and is reduced to the contemplative component in the cognitive approach to creation. An imposed sense, however, is not yet a sense of life, the degree of filling the gap left by the removal of a "global" sense of existence. Consequently, the concept of nature is again filled with religiosity (and vice versa). A theological ethic of creation can strengthen the paradigm of the recovery of sense and with this what Schweitzer calls "reverential respect for life" (Wehowski and Schwemmer, 1995) (Fig. 5).

Instrumentalism or Meaning? The problem is that of finding a happy medium between instrumentalism (everything is available) and biologism (we are determined by nature). The solution could reside in the meaning, i.e., in the symbolic value (mediated by human language and knowledge) of all creation for man; consider, for example, the symbolic language of the poetry of nature (Mieth, 1986a). An approach of this type can have unpredictable effects for a whole series of problems. I propose a few by way of example: Is I. Illich in the right when he laments the leveling down of the sexual nature in sexual culture? Here, the symbolic value of nature could reside in the so-called bipolarity of the sexes. Should living species be safeguarded from the rational–teleological standpoint, or as symbols of the fullness of creation (Noah's ark)? How should we interpret the sufferings and fear of animals used as guinea pigs in scientific experiments? Instrumentally (i.e., from the point of view of necessity and utility) or in terms of significance (i.e., as signs of the life in us and around us), and therefore supportively? In man's reproduction is there a sense of predefined and assigned creation, or is it just a question of a mix of needs so that reproduction can take different forms according to necessity?

FIGURE 5 Albert Schweitzer shown at Lambaréné in Gabon in 1954 in the leper village founded by him (photo courtesy of E. Smith/Black Star/Contrasto).

Progress or Metamorphosis? The image of progress is that of the line (of the spiral for retarded progress), whereas the image of metamorphosis is the cycle. I give two examples for the image of metamorphosis. According to Eckhart of Hochheim (Mieth, 1986b), "there is nothing new under the sun" (*Sermons* 1, 10) because everything that is new grows old. The same idea is turned upside down by the German mystic in a "progressive" sense: "If there were nothing new, there would be nothing old!" However, the *novum* in the sense of a radical "bursting forth" occurs only in the divine sphere, as a *creatio continua.* According to Goethe, we are prisoners of the cycle of nature; in his lyrical fragment *Die Natur* (1783), he writes (cited in Marsch, 1980),

> Nature! We are surrounded and swallowed up by her, prevented from coming out and prevented from immersing ourselves in her. Uninvited and without warning, nature entangles us in her round of dancing and drags us away with her, until, wearied, we escape from her embrace. She eternally creates new forms; that which exists, that which is not yet, that which has been does not return: All is new, yet always old.

In Goethe's conception Spinoza's *Deus sive natura,* pantheism, transpires.

The image of progress, however, does not imply the idea of a balance that respects ecological requirements in science, technology, and economics. Progress, Jüngel (1969) observes, exists only as "plurality" and not as "totality." The image of metamorphosis does not exclude progress in the individual fields, and indeed it implies it. However, it refers to a balance that man has to respect and which was not created by him.

Conclusions

First, we have to revert to the concept of nature. In the natural sciences, it is occasionally the synonym of empirical facts, laws, or theories. Meanwhile, the concept has become "subjectivized," but this process is matched by a reduction of man to object. Paradoxically, the greater man's power of manipulation grows, the more he loses his sovereignty because as body he is vulnerable and as spirit he is limited and finite. When he loses his tension toward the natural world, man loses himself.

The concept of nature is an indicator of problems, not a normative concept. It calls for our responsibility but does not control it. In the relationship between nature and reason, reason is the element that has to look after nature. One could state, provocatively, that the concept of creation as a theological concept of nature can today prove useful even without religious dogmas. On a par with the concepts of "dignity of creation" (as stated in the Swiss Constitution) and of "creaturality," it points to the continuation of a residue of reverential respect: "religion bequeathed." The concept of creation is reproposed in various philosophical versions, the most interesting of which is the process interpretation: *creatio continua; liber creaturarum;* a relationship with God understood as movement, as a process. "God works, and I become," in Eckhart's words.

References Cited

BÖCKLE, F. (1985). Gentechnologie und Verantwortung. In *Gentechnologie — Chancen und Risiken* (R. Flöhl, Ed.), pp. 86–92. Schweizer, München.

CHRISTOFFER, U. (1989). *Erfahrung und Induktion. Zur Metodenlehre philosophischer und theologischer Ethik.* Freiburg i.Ue, Freiburg.

DEMMER, K. (1990). Natur und Person. In *Natur* (B. Fraling, Ed.), pp. 55–86. Freiburg i.Ue, Freiburg.

DREWERMANN, E. (1991). Endlichkeit/Transzendenz. In *Neues Handbuch theologischer Grundbegriffe,* Vol. 1, pp. 332–338. Kösel-Verlag, München.

FURGER, F. (1990). Natur und Kultur. Die sozialethische Perspektive. In *Natur im ethischen Argument* (B. Fraling, Ed.), pp. 87–104. Freiburg i. Ue-Freiburg i.Br.

GEWIRTH, A. (1978). *Reason and Morality.*

HARTMANN, K. (1989). Was ist und was will Ethik? *Philos. Jahrbuch* **93,** 1–18.

HOLLING, C. S. (Ed.) (1980). *Adoptive Environmental Assessment and Management.* Wiley, New York.

JÜNGEL, E. (1969). Der Schritt des Glaubens im Rhythmus der Welt. In *Das Problem des Fortschritts-heute* (W. von Rudolf and R. W. Meyer, Eds.), pp. 143–163. Wissenscheftliche Buchgesellschaft, Darmstadt.

KISSLING, C. (1993). *Gemeinwohl un Gerechtigkeit. Ein Vergleich von traditioneller Naturrechtsethik und kritischer Gesellschaftstheorie.* Freiburg i.Ue, Freiburg.

KREBS, H. (Ed.) (1997). *Naturethik.* Frankfurt.

KRIHN, W., and WEYER, J. (1990). Die Gesellschaft als Labor. Risikotrasformation und Risikokonstitution durch moderne Forschung. In *Risikante Entscheidungen und Katastrophepotentiale* (J. Halfmann and K. P. Japp, Eds.), pp. 89–122. Westdeutscher Verlag, Opladen.

MARSCH, E. (Ed.) (1980). *Moderne deutsche Naturlyrik.* Reclam, Stuttgart.

MIETH, D. (1986a). Das Baumgedicht. In *Bäume braucht man doch!* (H. Schweizer, Ed.), pp. 21–40. Sigmaringen, Thorbecke.

MIETH D. (Ed.) (1986b). *Meister Eckhart.* München.

MIETH, D. (1989). *Moral und Erfahrung: beitrage zur theologishethischen Hermeneutik.* Verlag Herder, Freiburg.

MIETH, D. (1990a). Genetische Testmöglichkeiten — allgemeine und grundsätzliche ethische Aspekte. In *Genetische Testmöglichkeiten. Ethische und rechtliche Fragen* (R. Baumann-Hölzle, A. Bondolfi, and H. Ruh, Eds.), pp. 75–86. Campus, Frankfurt.

MIETH, D. (1990b). Die ethische Dimension der Umwelterziehung. *Karlsruher pädagogische Beiträge* **22,** 47–72.

MIETH, D. (1992). Begründungsversuche von Ethik. In *Moraltheologie im Dienst der Kirche* (K. Demmer and K. H. Ducke, Eds.), pp. 37–51. Benno Verlag, Leipzig.

MIETH, D. (1993a). Norm und Erfahrung. *ZZE* **37,** 34–45.

MIETH, D. (1993b). Ethics of human genome analysis, European perspectives. In *Ethik Wissenschaften* **5,** 199–324.

MIETH, D. (1993c). The release of microorganisms: Ethical criteria. In *Transgenic organisms. Risk assessment of deliberate release* (K. Wörmann and K. Tomiuk, Eds.), pp. 245–256. Birkhauser Verlag, Basel.

MIETH, D. (1997). Die Krise des Fortschritts und die vergessene Endlichkeit des Menschen. In *Die widerspenstige Religion* (T. van den Hoogen, H. Küng, and J. P. Wils, Eds.), pp. 203–227. Kok Pharos, Kampen.

ROY, D. J., WYNNE, B. E., and OLD, R. W. (Eds.) (1991). *Schering Foundation Workshop 1: Bioscience Society.* Wiley, Chichester, UK.

SCHELL, T. VON (1993). Die Diskussion um die Freisetzung gentechnisch veränderter Mikroorganismen als Beispiel einer interdisziplinären Urteilsbildung (tesi di laurea). In *Ethick in den Wissenschaften.* Tübingen.

SCHMITZ, P. (1990). Natur im ökosystemischen Denken. In *Natur* (B. Fraling, Ed.).

STANGER, A. (1992). *Zur ethischen Beurteilung der Möglichkeiten und Auswirkungen der Gentechnologie in der Landwirtschaft am Beispiel der Herbizidresistenz in der Pflanzenproduktion.* Univ. Tübingen, Tübingen.

STEIGLEDER, K. (1992). *Die Begründung moralischen Sollens. Studien zur Möglichkeit einer normativen Ethik.* Attempto Verlag, Tübingen.

VAN DER DAELE, W. (1987). Die Moralisierung der menschlichen Natur und Naturbezüge in gesellschaftlichen Institutionen. *Vierteljahrsschrift Gesetzgebung Rechtswissenschaft* **2,** 351–366.

WEHOWSKI, S., and SCHWEMMER, O. (1995). *Gespräch über Ethik.* München.

WIMMER, R. (1991). Kategorische Argumente gegen die Keimbahnterapie? Eine Prüfung der Stellungnahme der Enquète-Kommission des Deutschen Bundestages. In *Ethik ohne Chance? Erkundungen im technologischen Zeitalter.* In *Ethik in den Wissenschaften* (J. P. Wils and D. Mieth, Eds.), pp. 182–209. Attempto Verlag, Tübingen.

WIRTZ, S. (1993). *Vom Mangel zum Überfluss. Die bedürfnisethische Frage in der Industriegesellschaft.* Aschendorff, Münster.

WÖHRMANN, K. (1991). In *Ethik in den Wissenschaften.* Attempto Verlag, Tübingen.

DIETMAR VON DER PFORDTEN
Chair for Social Philosophy and Philosophy of Law
University of Erfurt
99089 Erfurt, Germany

Whereas the moral treatment of humans, at least regarding the essentials (e.g., killing, harming, and violating), is subject to a wide agreement and only debated with regard to special questions, the moral treatment of non-human living beings (animals, plants, and microorganisms) is widely disputed. This article describes and discusses different opinions concerning the moral treatment of non-human beings in general and with respect to several applications, e.g., the keeping and killing of animals, experimenting with animals, the treatment of plants and microorganisms, and genetic engineering.

The Moral Consideration of Non-Human Beings

Introduction

Humans use non-human beings — animals, plants, and microorganisms — in many different ways for their purposes: for production of food and commodities, transportation, labor, medical experiments, education, entertainment (e.g., in zoos, films, and races). Although in the past, and still today within homogenous cultures, traditional moral rules concerning the treatment of non-human beings were widely accepted, the situation has changed, at least in Western countries. Reasons for this change are the intensified and more effective use of non-human beings, on the one hand (e.g., the development of large food and commodity-producing industries, the increase and refinement of animal experiments, and the new techniques of genetic engineering), and a higher awareness of ethical problems, on the other hand (e.g., by the growing exchange of different moral codes among different cultures, the higher sensibility for harm and cruelty, and a growing ethical consciousness in general). All this has led to a ceasing agreement concerning the rules for the treatment of non-human beings. The keeping and killing of animals, their use for experimentation, and the genetic manipulation of living beings are especially subject to criticism and protest.

This criticism can be ethically grounded in two fundamentally different ways. One can limit oneself to take into account human interests and consider non-human beings only as useful means to further these human interests; for example, the fear that genetic engineering can breed new types of bacteria and cause diseases in humans is an expression of human interests. One can also consider non-human beings as deserving moral consideration independent of human interests and as having moral standing independent of the usefulness for humans. The first alternative is called *anthropocentric* or, if it is weakened because of aesthetic concerns, compassion, stewardship, or the avoidance of adopting cruel habits (*anthroporelational*). The second alternative is called *nonanthropocentric.* The first alternative is only a special version of moral obligations toward humans and therefore merely an extension and application of traditional morality. The second nonanthropocentric alternative is more challenging and innovative. It can lead to changes in traditional habits and institutions (e.g., the abolishment of the use of animals for food) as well as in the prohibition of new technologies (e.g., the genetic manipulation of non-human beings).

As the title indicates, this article is mainly occupied with the second, more challenging and innovative nonanthropocentric alternative. With a few exceptions, it deals with the moral consideration of non-human beings as ends in themselves, as having moral standing. That means that the human actor has an obligation to consider in his or her decisions the indigenous or adopted strivings of the non-human beings, which are affected by his or her action.

To assign moral standing to non-human beings is not the same as to assign "rights" to them. To assign rights, either moral or legal ones, to them would require some additional conditions, which are scrutinized at the end of the article.

Historical and Current Theories

Some general moral norms concerning humans are widely accepted: You shall not kill, harm others, cause them pain, take their freedom, lie, etc. The ethical justifications of these norms differ but not so much that an agreement is impossible on central issues. The situation concerning animals and other non-human beings is different. Whereas some see no ethical problem in killing non-human beings, hurting them, or causing them pain, others do not differentiate in moral respect between non-human beings and humans so that, for them, every harming or killing of a non-human being is tantamount to harming or killing of a human being. Between these two extreme positions many gradations exist.

The variety of positions is as great in the literature as for ordinary people. The moral attitude of the majority of the people in Western countries is situated between the extreme alternatives just mentioned, but it is far more toward the first alternative. Only higher animals — mammals and perhaps vertebrates — are considered to have moral standing. Also, the main ethical problem in the treatment of higher animals is believed to be causing pain to sentient animals, which is allowed if the benefits for humans are not just minimal.

What are the ethical justifications which underlie these moral standards concerning non-human beings? What are the arguments for the different positions? In the following sections, historical ethical positions, which are held even today, and positions which were developed only recently are presented and discussed.

Theological/Christian Position

Christians believe in a teleological, hierarchical order of the universe with God on top. God has created the world. That gives him the right to rule it. Man is under God but on top of the created world. According to the book of Genesis, which is part of the Old Testament, man is entitled to rule over the world (*dominium terrae*). Non-human beings have no souls. They are not images of God as is man. They are not entitled to give names to other objects as is man. There is nothing in the Ten Commandments which explicitly limits the human treatment of non-human beings for their own sake, which gives non-human beings a moral standing, independent of the dignity of men. The 10th Commandment speaks of animals, but only as property of other men: It is forbidden to desire the property of others. The Fathers of the Church (e.g., Augustine and Thomas Aquinas) taught that non-human beings were made by God for the sake of man and that man is therefore allowed to use and kill them.

However, some citations in the Bible and, according to some interpreters, the divine revelation as a whole and the teleological "natural world order", as it arose out of Creation, dictate ethical boundaries of human behavior toward non-human beings. Unreasonable killing, harming, and cruelty are not allowed. Accordingly, animal exploitation and experimentation are restricted by limitations. However, these restrictions are obviously vague and weak. They have to be found by interpretation of the Bible and are therefore subject to doubt and discussion. Also, they have not prevented Western civilization from establishing an industrial complex, which exploits animals in an extreme and cruel way, limited only by some very weak animal protection laws.

Only recently have Christian theologians tried to interpret the biblical limitations in a stricter and narrower sense. However, the results differ. The emphasis lies on the human interference with the God-made world order and God's prohibitions, not on the acknowledgment of own strivings and a moral standing of non-human beings. Being a theological ethic, a Christian ethic is always on a first level theocentric and puts on a second level man on top of all creatures. Only on a third level can non-human beings receive some protection against man's actions by reference to the first theocentric level of the whole hierarchy (e.g., God's creation). Therefore, everything depends on how strictly these limitations of man's freedom of action are interpreted.

The new catechism of the Catholic Church (1992) understands the 7th Commandment ("Do not steal") broadly, also capturing the treatment of the non-human creation:

"Stealing" is harming God's creation. The catechism states in Sections 2415–2418 that man is allowed to use animals for his purposes, but within certain weak limits. These limits, however, are justified by reference to the dignity of men (§2418). Therefore, no clear nonanthropocentric moral standing of non-human beings is proposed. Christian philosophers have followed this line. Reasonable keeping and killing of animals is allowed, but no unnecessary cruelty is allowed (Spaemann, 1989, p. 155).

Regarding the Christian position in general and also all other theological positions, creation and teleological world order are subject to belief and not to scientific proof. Therefore, one has to accept the Christian or a similar faith in order to accept the religious or theological morality. Thus, for non-Christians and nonbelievers, the search for a secular basis of ethics and morality is necessary. However, the same holds for Christians and other believers. Because the biblical or prophetic guidelines are vague, additional rational considerations are necessary. All revelation and prophecy is received and interpreted by fallible human beings, who have limited epistemological capacities and apply inevitably the historical prejudices of their times. Therefore, the recourse to immanent rationality seems necessary to purify religious doctrines of subjective and historical elements.

Finally it should be mentioned that there existed and exist today many other religions and sects, which had and currently have a different view of the moral standing of non-human beings and which require, for example, vegetarianism. If one believes in rebirth, as do the Hindus, the non-human beings participate in the cycle of life and have a totally different standing in relation to human beings. It is impossible to give a detailed account of the doctrines of other religions here. However, one has to remember that from a world perspective the Western/Christian moral is only one proposal.

Kantian Philosophy

The Kantian ethic acknowledges only rational beings as having self-purposes (*Zwecke-an-sich*). Rational beings — that is, humans, if God and angels are omitted — are therefore the only entities which are acknowledged to have moral standing (Kant, 1911) (Fig. 1). Rationality is the necessary condition for moral standing because all nonrational beings are subject to the causal laws of nature. They are not responsible for their actions and can therefore not claim to be moral patients. Only a rational being is able to transcend the causal world and be, with its rational purposes, a member of the so-called "common kingdom of ends" (*Reich-der-Zwecke*).

The Kantian position therefore provides only an indirect, anthroporelational limitation for the treatment of higher, sensitive animals. The cruel treatment of such animals is forbidden, but only for the sake of the prevention of cruelty to humans, because it could lead to such cruelties among humans through desensitization to violence (Kant,

FIGURE 1 Frontispiece of the second edition of the paper *Grundlegung zur Metaphysik der Sitten* by Immanuel Kant, published in 1786.

1907). Kant therefore condemns torturous animal experiments on grounds of pure speculation but proposes no further limitations on such experiments, as long as they are carried out for rational human purposes (as opposed to those of other animals) and without unnecessary cruelty. The keeping and killing of animals is allowed if no unnecessary cruelty occurs. Limitations of genetic manipulations can be grounded only in human interests.

However, Kant's thesis that only rational beings with their own purposes are morally considerable seems to be unconvincing. It is easy to see that nonrational beings also have strivings and purposes of their own. There is no reason why only rational purposes should be estimated as ethically relevant. Even theorists such as Leonard Nelson (1970), who were themselves Kantians or at least stood in the Kantian tradition, criticized Kant on this position. Kant's Newtonian view of a causal determined universe is abandoned by quantum mechanics and the theory of relativity. Therefore, the factual border between nonliving entities and living entities is as great as — or as small as — that between animals and humans. The view that only rational beings transcend causal determination is outdated. A more correct view would assume that everything is subject to the laws of physics, but within this determination by the laws of physics unconscious and conscious strivings of living beings are possible and ethically decisive.

Utilitarianism

In classical utilitarianism, avoiding pain while maximizing pleasure is the decisive ethical criterion — this being not intrapersonal but interpersonal in the form of a summarizing and maximizing calculus. Higher sensitive animals participate in this calculus (Bentham, 1963). Therefore, utilitarianism is nonanthropocentric. However, the range of moral standing of non-human beings is very limited: Only higher sensitive animals are included. The question — and the difference between the various utilitarians — is only whether or not the pain of animals is analogous to that of humans. Those, such as Jeremy Bentham, who interpret pain and pleasure as being materialistic and hedonistic tend to the first position, whereas others, such as J. St. Mill, who interpret it as being intellectual and spiritual tend to the second position. The keeping and killing of animals is allowed if no pain is inflicted. Animal experimentation is justified if the consequences are positive — that is, if the gain of pleasure or utility for humans and/or animals outweighs the pain of the afflicted animals. According to this position, the testing of cosmetics or other such things, which could not cause great danger to humans or animals or gain great advantages for either, is unacceptable. The genetic manipulation of non-human beings is allowed if no humans or sentient animals are harmed or, if harm is caused, the benefits are greater than the harms.

However, avoiding pain and seeking pleasure is not the only nor even the highest human purpose. If one tries to achieve pleasure directly, one will fail (this is referred to as the *hedonistic paradox*). Human purposes are pluralistic. They include seeking acknowledgment, propagation, communication, sociality, knowledge, wisdom, and development of personality. Animals have no cultural or intellectual wishes and needs. However, even for them, self-preservation, reproduction, and free movement seem to often be more important than avoiding pain (von der Pfordten, 1995, 1996). Plants also strive for self-preservation and reproduction. Sunflowers turn their heads to the sun and many plants protect themselves against enemies with the help of certain chemical substances. Physiologically, pain is only a warning system to protect life and physical integrity from injuries and sickness. For example, experiments in which a rat crosses a hot or electric grating to reach its brood demonstrate the prioritization of the interest of procreation. Pain is only a means of the body to ensure its self-preservation against diseases and injuries. Avoiding pain is therefore not the only nor the highest aim. However, since animals have no cultural and intellectual interests, avoiding pain seems to be of higher priority for animals than for humans.

Additionally, there seems to be no reason to allow or even demand interpersonal aggregation and maximization between humans and animal entities. We do not hold this position concerning the vital interests of humans. If interpersonal aggregation would be a justified ethical criterion, we would, for example, be obliged to convict innocent people if large riots with many victims could be avoided. Robert Nozick gave the example of a lust monster which is able to draw out of everything the maximum pleasure. If maximization would be the duty, we should give everything to this monster because in this way the overall pleasure of all is maximized. Utilitarianism is based on an individualistic ground because it takes into account all sentient beings equally, but it ends up in a collectivist criterion of maximization of the total sum. If we are not willing to accept utilitarianism among humans, why should we do so concerning similar interests of animals? At this point, the argument of great scientific progress to support animal experiments fails in this abstract form for future experiments. A necessary condition would be a general allowance of the interpersonal maximizing of pleasure or utility, for which there are no grounds. The same holds for the keeping and killing of animals and genetical engineering.

Compassion and Empathy

For some theorists, compassion — or, more generally, empathy — as a feeling is decisive for ethical action (Hume, 1991; Schopenhauer, 1989; Wolf, 1990). Here, it is essential to understand that in contrast to utilitarianism, a feeling in the moral agent and not in the moral patient is decisive, which means that there is a feeling in the human actor, not in the non-human being, which is affected by the action. However, it is clear that normally the feelings of the actor occur because of some pain or harm in the entity, for which the actor feels compassion. The second difference of utilitarianism is that no interpersonal aggregation and maximization are allowed.

From this point of view, every sensitive being which feels pain and thus arouses compassion in us as ethical actors deserves ethical consideration and some action for relief. The only exception is a collective emergency of moral agent and patient or an aggression of the moral patient. The conclusions of this position concerning animal experiments can vary, but they normally lead to severe limitations or total prohibition of all animal experiments, as in the case of many authors (Wolf, 1990). Animal experiments are prohibited in much the same way that involuntary experiments with humans are prohibited. The keeping and killing of animals is restricted. However, there is no nonanthropocentric ethical limitation of genetic engineering if the manipulated non-human being will be able to develop and live without suffering. Manipulations of fetal cells are not objectionable because they are performed in a stadium of the development of the entity, in which it feels no pain.

However, compassion or empathy as a feeling within an actor do not seem to justify his or her ethical commitment. If he or she lacks this feeling, another basis is needed to command the development of the feeling. Also, if the actor feels compassion or empathy, the mere existence of the feeling within the actor is not a good reason for the prohibition or commitment of his or her actions. The feeling may be totally inappropriate toward the real situation of the other being. An ethics of compassion potentially favors criminals without feelings. It also gives the actor liberty of free action according to or against the moral duties. These difficulties result in a very unclear status of the theory. In referring to the pain of the non-human beings it is nonanthropocentric, but in referring to the empathy in the actor it is anthropocentric. There can be no guarantee that the pain of the non-human being and the empathy in the actor are congruent. Therefore, one has only two alternatives: to base the decision on the reality of the non-human being and not consider the decisiveness of the actors empathy or to base the decision on the empathy of the actor, no matter whether it matches the reality of the non-human being. The first alternative leaves an empathy ethic, and the second is not very attractive.

The historical dominance of the first three of these four ethical positions in the Western world explains why the use of non-human beings was not severely limited in past centuries. The fourth position — compassion and empathy — gained greater consent in Great Britain in the eighteenth and nineteenth centuries, in which the first animal protection law was enacted in 1822 and animal experiments were registered very early. Other countries followed — some quickly, some slowly. However, with some of the recent positions, the picture changes considerably.

Contractualism and the Discourse Theory

This is not the case, however, with the first type of the newer theories, contractualism and discourse theory. This type of view is situated in the Kantian tradition but dismisses most of the metaphysical assumptions of Kant's theory. It is the only one of the modern views which denies animals an ethical nonanthropocentric position and therefore moral standing. For this view, morals and ethics emerge out of an ideal contract (Rawls), agreement (Scanlon), or at least discourse (Habermas, 1992). The outcome of this ideal contract, agreement, or discourse is morally binding (Carruthers, 1992). Only entities which are in principle able to participate in such a contract, agreement, or discourse are possible moral patients and have moral standing. However, animals are not considered rational beings. Therefore, they lack any moral standing (Carruthers, 1992). They deserve only indirect moral consideration as within the theory of Kant. Since many people have concerns for animals and are deeply distressed at seeing an animal suffer, this may place on us an obligation not to cause suffering to animals, except for very important reasons (Carruthers, 1992). However, this holds only for public cruelties against animals. Private cruelties without the danger of becoming public are not morally wrong. Therefore, factory farming and laboratory testing are allowed (Carruthers, 1992).

There are many objections to this view. First, it violates basic moral intuitions. We do not think that it makes a difference to torture a cat privately and not in public. We also do not think that very unimportant human interests justify cruelty toward animals. Additionally, contractualism has notorious problems explaining the moral standing of children, newborn, unborn, and irreversible comatose human beings who lack rationality in a similar way as do higher animals. However, even if contractualism manages to include these types of human beings by an uncertainty or slippery slope argument, assuming that we cannot be sure when rationality within humans emerges and therefore have to consider birth as a starting point for moral protection, contractualism constructs morality and ethics in an unconvincing way. If the contract or agreement is the decisive criterion, every entity is in its moral status dependent on other entities. This contradicts the precondition of morality to acknowledge the independent moral status of other entities. Morality starts with this acknowledgment: It is not the result as contractualism assumes. Contractualism and discourse ethics wrongfully extends a problem framing and problem treatment, which is adequate for the social and political sphere of public life among humans, to the moral assessment of interactions among humans and non-human entities.

Habermas tried to extend the scope of discourse ethics beyond this anthropocentric border by constructing a so-called "moral analogs relation" between humans and animals (Habermas, 1992) (Fig. 2). If we treat animals in a sort of communicative way as alter egos, then a moral analogs situation is established. However, this idea gives animals no real moral standing. Also, it leads only to an elevated status of pets and not to a moral consideration of wild animals. There is no doubt that some sort of human treatment of animals creates duties toward them, but it seems to be questionable to construct all types of relations toward animals in this way. Non-human beings are not really consid-

FIGURE 2 Cover of the book *Erläuterungen zur Diskursethik* by Jürgen Habermas (1992).

ered and recognized as other entities, which have moral standing, if only our decision to integrate them in a symbolic way in the community of communicators provides the status of moral patients to them.

Preference-Utilitarianism

For the advanced preference-utilitarian type, "preferences" or "interests" are the basic quality to be considered of ethically relevant entities. This means a widening of the pain–pleasure criterion of classical utilitarianism. All actions which can cause harm or good to an entity are taken into account. If an entity is sensitive, its interests limit human action, but only if the action does not result in greater pleasure of other sensitive beings. Therefore, the overall calculus of interests is the sole deciding factor. Animals are subject to the calculus of the greatest total good of the greatest number as in classical utilitarianism.

Being killed is a sort of harm, but it can be outweighed by the pleasures of a happy life. This leads some more orthodox preference utilitarians to the conclusion that vegetarianism is wrong because it would cause the abandonment of animal husbandry and thus reduce the total amount of existing happiness, which can be measured in the total sum of quality-adjusted life-years of all sentient beings (Hare, 1993). The obviously absurd consequence of this view is that we have an ethical duty to bring happy animals into being, if we can do so without reducing the total sum of happiness. It should be clear that this theory is far too demanding. We are not allowed to lead our own private lives but are responsible for the rest of the world every minute, including the world of sentient animals.

Other preference-utilitarian theorists, such as P. Singer, have departed further from the roots of classical utilitarianism. They make a distinction between conscious and self-conscious rational beings (Singer, 1993). Entities which are rational and self-conscious deserve ethical recognition concerning their lives, which are, in turn, not merely subject to the overall interpersonal aggregation and maximizing calculus of all sorts of interests but can be sacrificed only if live stands against live. These rational and self-conscious entities are individuals, leading lives of their own, and cannot in any sense be regarded merely as receptacles for containing a certain quantity of happiness (Singer, 1993). These are humans.

Concerning higher animals, a clear positive physiological decision on this point is currently possible only for the great apes (chimpanzees, gorillas, and orangutans). Therefore, the lived of the great apes are not subject to the general utilitarian calculus. To be on the safe side and depending on how far we are prepared to extend the benefit of the doubt, the other mammals (whales, dolphins, monkeys, dogs, cats, cattle, pigs, etc.) should be protected in a similar way (Singer, 1993). Therefore, the keeping and killing of higher mammals for food and killing for other similar human purposes is forbidden. Thus, moral and legal rights for the great apes are demanded (Cavalieri and Singer, 1993). Animal experimentation that causes death to mammals is rejected if it is not possible to save several lives of "persons" (humans and other mammals) by an experiment that would take just one life (Singer, 1990) (Fig. 3).

FIGURE 3 Cover of the book *Animal Liberation* by Peter Singer (1990).

For nonmammals the limitations are weaker. Their lives are subject to the interpersonal aggregating and maximizing calculus as in classical utilitarianism. Therefore, their keeping and killing for the sake of human pleasures is allowed if the total sum of good is not reduced thereby. Medical or scientific experiments are justified on utilitarian grounds; that is, they must result in greater overall good.

A critique of preference-utilitarianism has first to realize that the harm or good criterion, which is decisive for the concern toward the entity in question, is still too narrow. Strivings such as that for moving, for procreation, and for sociality are not taken into account at all or not in an adequate way (authors opinions differ on this aspect) if their nonfulfillment causes no harm. Second, the aggregation principle is, as in classical utilitarianism, not convincing. We do not demand that humans sacrifice their lives, for example, by requesting living human beings to donate their organs to save several other lives. Therefore, there is no general reason to demand the same from animals. Singer's modified version tries to cope this objection. However, he provides no good argument why the natural abilities of rationality and self-consciousness should automatically generate an obligation to consider the life of the entities in question individually. Singer's view is in this respect a modern and more sophisticated version of naturalism. Singer commits the naturalistic fallacy, which means he draws a false conclusion from a fact (self-consciousness) to a prohibition (duty of preservation of the life of the being in question).

Jean-Claude Wolf developed Singer's position and departed from the utilitarian roots. For him, self-consciousness is not necessary to exclude the interpersonal utilitarian overall aggregation; consciousness alone is sufficient. A conscious being experiences not only anonymous sensations but also his or her own. The continuity of consciousness is sufficient to make death in the form of the destruction of consciousness into an ethical evil per se (Wolf, 1992). The conclusion, similar to that of U. Wolf, is that all animal experimentation that places the life or well-being of animals in danger is prohibited (Wolf, 1992).

However, this ethical justification is objectionable in a similar way to that of Singer: Consciousness is a physiological ability but cannot generate moral prohibitions without further qualifications. J.-C. Wolf also commits the naturalistic fallacy. There seems to be no good reason why the fact of consciousness should provide a reason to treat the strivings of conscious beings as ethically relevant and the strivings of nonconscious beings as ethically irrelevant or at least as subject to a maximizing calculus.

Subject of a Life

A very strong theory has been developed by Regan (1983) (Fig. 4). For him, every "subject of a life" contains inherent worth and therefore deserves respect and is a holder of a right of respect. Individuals are

> subjects of a life if they have beliefs and desires; perception, memory, and a sense of the future, including their own future; an emotional life together with feelings of pleasure and pain; preference and welfare interests; the ability to initiate action in pursuit of their desires and goals; a psychophysical identity over time; and an individual welfare etc. (p. 243)

FIGURE 4 Cover of the book *The Case for Animal Rights* by Tom Regan (1983).

Therefore, mammals which are at least 1 year old are strictly protected. This theory is mainly intuitionistic, which means it refers to our commonsense intuitions. It is concerned with duties, not with the maximization of the consequences as is utilitarianism, but its outcomes are nevertheless similar with the position of J. C. Wolf. It results in a total prohibition of animal experimentation (p. 363): "Those who accept the rights view, and who sign for animals, will not be satisfied with anything less than the total abolition of the harmful use of animals in science — in education, in toxicity testing, in basic research." The killing of animals is prohibited and the keeping severely limited.

The same objections of the view of J.-C. Wolf are valid here. There seems to be no reason why the mental abilities of an entity as merely brute facts justify moral limitations and prohibitions in favor of conscious animals and exclude the strivings of other non-human beings. The duty-oriented claims of Regan's theory provide assertions without further justification.

Equalitarianism

This type is not a real independent ethical theory in the strict sense but rather a form of argument. It accuses all higher valuing of humans or human interests which is not justified by morally relevant differences as "speciesism" (Ryder, 1991; Singer, 1990, 1993; Rollin, 1992). The claim is that there is no objective justification for inequality in the treatment of humans and higher animals (some theorists widen the circle to include only self-conscious beings, whereas others go so far as to include all conscious ones). There is no reason to think that human pain is, in general, more severe than animal pain and that human interests are worthier than those of animals: Animals lack the foresight that some pain or harm will last for a certain amount of time, which can cause some additional distress to humans, but they also lack the foresight that the pain or harm will end, which gives humans some relief.

The proponents of the equality argument seem to be right in accusing unequal treatment — that is, if the entities in question are equal in the relevant aspect. However, the first question is, Equality of what? The first decision of an ethical theory therefore must be to determine the ethically relevant criterion. The decision whether equality or inequality is appropriate can only be the second step. The discussion has made clear that pain or consciousness cannot be the only or highest relevant criterion. Therefore, the outcome of the equality argument is open until another decisive ethical criterion is established.

Protection of Life

The aforementioned positions include only sentient animals in the realm of the nonanthropocentric moral consideration. Consciousness and/or self-consciousness are seen as necessary conditions for a moral protection. However, other authors question these conditions. For them, life in all its different forms is the decisive criterion. Therefore, plants and microorganisms are also moral patients. Animal ethics has to be widened to an environmental ethics. Also, some go even further to a holistic view and also include species, ecosystems, and the biosphere. The proponents of the life-protection position only differ in the question of weighing. Some propose a hierarchy of life-forms (Attfield, 1991) and others propose an equality (Taylor, 1989).

However, these positions face a similar difficulty as the ones which view consciousness and/or self-consciousness as decisive. The mere fact of life, of the biological metabolism, cannot generate moral duties. Otherwise, one would commit an is/ought fallacy. Therefore, one must first consider the function of ethics on a more abstract level.

Ethics of the Consideration of Others

If a human being lived alone in the universe, there would exist no moral limitations. Therefore, the first condition and basis of every ethics is the existence of other beings, *moral patients,* which limit the actions of human beings as *moral agents.* This limitation does not operate in a factual way. It does not limit human actions like a big stone which blocks our path. However, it limits human actions in a normative way.

What does that mean? In what respect is a "normative limitation" different from a "factual"? Factual means "physical." The physical natural laws limit our actions. If a very heavy and immovable stone blocks our path, we cannot walk on because according to the laws of physics two objects cannot occupy the same space at the same time. Therefore, normative means "nonphysical" or "not only physical": If we are morally and normatively prohibited to lie, there exists no physical obstacle to lie. We can produce the factual phenomenon which is necessary for a lie (i.e., sound waves). The limitation is not physical but rather linguistic and mental. We experience that other persons express their wish to know the truth. Their striving for truth and the expression of this striving is the necessary condition of our moral limitation to lie. In a world in which nobody strives for truth, a prohibition of lying would be unreasonable.

Now, the decisive question for a nonanthropocentric ethics is the following: Are only linguistic and mental strivings — strivings of human beings, if God or supernatural beings are not considered — to be regarded as normative limitations? Kant and his followers made this assumption. Also, utilitarianism has only added pain and pleasure so that strivings of higher, sentient animals are also considered as normative limitations. However, what about the strivings of sentient animals to move freely and to procreate? What about their striving to socialize? What about the strivings of nonsentient animals? What about the strivings of trees for water and sunlight?

All these strivings have a physical basis. They take place within the framework of physical laws, as do human actions and utterances. An objection against the consideration of these strivings is that they are all reducible to physical forces. There exists a discussion, if biological and/or mental processes are reducible to physics. This discussion cannot be pursued here. The decisive point is that a descriptive–scientific decision in this discussion is not crucial for ethics because within a moral and ethical frame of debate, we always assume that actions can be interpreted as more than results of physical processes. We view them as more than the mere effect of physical forces. They are for us strivings of self-oriented beings, strivings for self-development and self-preservation, or, in the language of physics (which can only partially describe this process), strivings to use external energy to maintain a higher internal level of energy and to evade entropy. If we refuse a total ethical skepticism and nihilism and acknowledge moral limitations between humans, the assumption of a determination of the world by physical forces does not exclude this acknowledgment of moral limitations between humans. However, the same holds for nonhuman beings because the strivings of non-human beings are also more than the outcome of physical forces. Therefore, all non-human living beings, animals, plants, and microorganisms deserve moral standing.

However, is there also a positive argument for the moral standing of non-human beings? Why should we respect the strivings of non-human beings? A first argument against a limitation with respect to the strivings of human beings could refer to the fact that human beings acknowledge not only conscious and/or articulated strivings but also vegetative ones. We think it wrong to damage the immune system of a human being, even if the human being has never gained knowledge of his or her immune system. Therefore, consciousness or articulation of a striving is not a necessary condition for its normative force. However, if vegetative strivings of beings with consciousness and language are acknowledged as normative forceful, why should similar strivings of beings without consciousness and language lack normative forcefulness to us?

For further discussion of this crucial question, it is useful to change our factual language into a normative. What is normative in the "strivings" of other beings is not the factual status of these strivings but their normative force–their status as "interests" of others. To understand this we have to examine how the ethical realm of interchanging interests works: Decisive for ethical limitations is not the will, wish, need, or interest of the actor, but the interest of the entity which is afflicted. Its interests limit the actor's free action. The interests of the actor restrict in a second move that limitation so that an equilibrium of interests emerge, but the origin of the ethical limitation lies in the other entity concerned. The consideration of the interests of others is decisive for an ethical behavior. The following is the central ethical question: Should the interests of other entities be considered (von der Pfordten, 1995, 1996)?

My own interests are a good reason for me to act in an appropriate way. If it is in my interest to move from A to B, I shall move from A to B. There exists a noncausal, weak connection between the interests of the actor and his or her action. I am free to abstain from the action; nevertheless, I have good reason to act according my interests. However, for others, my interests are in themselves no good reason to pursue the satisfaction of this interest, at least if it would require more than only a minimal effort. If it is in my interest to move from A to B, in principle no other entity has a duty to carry me from A to B. The picture becomes clearer if one uses a space metaphor here: Every entity with interests sets up a space of interests around itself. It has weak reasons to satisfy the interests in its own interest space but no reason to interfere in the interest space of the other entities so that interference, without informed consent, is not justified. The weak interests of the other entity are good reasons to prohibit the interference. This situation requires the drawing of a borderline between the different interest spaces. This might be difficult in practice, but in most cases it is not impossible.

In the case of positive interference, the result is the same. Outside of my interest space, help is in principle not commanded. However, a model of different interest spaces is only adequate for a situation of complete strangers who have never had contact with each other, even in an indirect form. If there exists a community (it may be quite loose as in all of mankind), an agreement, or a relevant previous action, the basic model of interest spaces is superimposed by a second layer, which generates positive helping duties. However, the basic layer is never totally displaced.

With the introduction of the concept "interests" no final clarity is gained for the entity radius of that ethical justification — that is, which entities are subject to ethical consideration and which are not — because that concept is used in quite different ways. The ascription of interests is limited by some authors to humans (Frey, 1980) and by others to higher animals with sensitivity (Feinberg, 1980), but it is also extended to all living beings (Attfield, 1991) or even all homeostatic self-regulating systems such as species, ecosystems, and the biosphere. The vagueness and nondetermination of the concept interests seems to be insurmountable. Only one qualification is necessary, however, if one chooses not to follow an egoistic interpretation which would totally deny ethics and morality: The interests should be acknowledged by the actor as being the interests of the others concerned. This requires, if an ascription by God or other persons is excluded as speculative or ethically irrelevant, that the entities concerned must have self-ascribed interests. An essential part of this self-ascription of interests is the manner of self-ascription. Accordingly, man as a being with rationality, language, and sensitivity is not allowed to establish rationality, language, and/or sensitivity as necessary qualifications to ascribe interests. In that case, it is not the self-ascription of the entity whose consideration is in question but rather the ascription of the actor whose action has been called into question that is decisive. Therefore, rationality, language, and sensitivity can only be an indication of a more elaborate and intense form of the self-ascription of interests. These qualifications are not decisive for the "If" of an ascription of interests but only for the "How intense?" For the question of whether interests should be acknowledged, the only criterion is the possible perception of some self-ascription by the entity in question.

To make this notion more concrete, it could be further broken down according to the normal development of a living being into three aspects: self-origin, self-development, and self-preservation. None of these qualifications can be lacking; each one must be fulfilled, at least in a rudimentary form, if interests are to be ascribed to an entity. For the different types of entities, one can use the following distinctions and conclusions (Table 1):

Higher wild animal: Self-origin similar to humans as well as self-development and self-preservation, but lack of rationality and language or at least not as elaborate as that of humans.

Higher useful animal: Self-origin like humans, but also restricted by farming, development, and preservation.

Wild animal: Self-origin, self-development, and self-preservation, but these are not as elaborate as those of humans (e.g., lack of sensitivity).

Wild plant: Self-origin, self-development, and self-preservation, but no rationality, language, sensitivity, or movement. Thus, these are not as elaborate as those of humans or other animals.

Useful plant: Has a self-origin but is restricted, along with its development and preservation, by genetic engineering and growth.

Stone/river: Origin, development, and preservation only by external chemical and physical processes.

Species: Origin, development, and preservation completely reducible to individual (mutation) and external influences (selection).

Ecosystem and biosphere: Self-origin and development are questionable; self-preservation is completely reducible to external physical influences or behavior of the individual entities.

Computer: No self-origin (constructed by humans) but partial self-development and self-preservation in advanced models (e.g., by computerized self-control).

The conclusion of these considerations is that all non-human living individual beings, animals, plants, and micro-

TABLE I

Variability of the Fundamental Aspects in the Different Types of Entities[a]

ENTITY	SELF-ORIGIN	SELF-DEVELOP-MENT	SELF-PRESER-VATION	INTEREST
Human	xxxxx(x)	xxxxxx	xxxxxx	Yes
Higher wild animal	xxxxxx	xxxx	xxxx	Yes
Higher useful animal	xxxx(xx)	xx(xx)	xx(xx)	Yes
Wild animal	xxxxxx	xx	xx	Yes
Wild plant	xxxxxx	x	x	Yes
Useful plant	xxx(xxx)	(x)	(x)	Yes
Stone/river	—	—	—	No
Species	—	—	—	No
Ecosystem	?	?	—	No
Biosphere	?	?	—	No
Computer	—	(x)	(x)	No

[a] x's refer to the fulfillment; several x's denote a more elaborate form. This does not imply that an interpersonal utility maximization is allowed. Parentheses symbolize variabilities in occasion.

organisms must be acknowledged as moral patients with own interests and therefore have moral standing. However, all collectives, such as species, ecosystems, or the biosphere, are excluded as moral patients. Their development and existence are the result of the strivings of the constituting individuals and physical forces.

The second conclusion is that not all interests deserve an equal moral consideration. There is a decrease in the complexity of self-ascription and hence also a decrease in the moral relevance of the interests formed by self-ascription. Rational and self-conscious self-ascription is more complex and advanced and therefore deserves higher moral consideration than conscious; conscious and painful self-ascription deserves higher consideration than instinctive; instinctive deserves higher consideration than purely physiological; etc. Therefore, the strivings of plants and microorganisms deserve only a minimal consideration, the strivings of animals a small consideration, the strivings of sentient animals a higher consideration, and the strivings of humans the highest consideration.

An interpersonal aggregation and maximization of interests is not in general justified. However, it is justified if the discrepancies of the complexity of the interests are very high, such as between humans and plants and between humans and animals in certain cases, if the singular discrepancy of the interests in question is high. If a voluntarily founded community with mutually helpful duties exists, the fact that the community does exist lowers the necessary discrepancy between the value of the interests to allow interpersonal aggregation. The situation is similar between humans, in which case vital interests are not subject to interpersonal weighing. However, if a peripheral interest stands against a vital one, a weighing is allowed and necessary because living together in a society occasions basic helping – duties. If somebody is in danger of drowning, one is allowed to take a boat which is in foreign private property and use it to help. The conclusion can be summed up as follows: Wild animals in general deserve higher ethical consideration than useful ones which live in community with humans, get their care and food from them, etc. The question of whether one has to draw the categorical line between sentient and nonsentient animals or between animals and plants is thus not decisive for the ethical consideration. All interests of morally relevant entities (i.e., all interests of living beings) should be considered.

Keeping and Killing of Animals

A consequence of this is that the avoidance of pain or harm cannot be the only limiting criterion for the keeping and killing of animals. First, the avoidance of other strivings, such as injury and death, is even more urgent because they are normally superior purposes for the animal. Second, all strivings and all interests of the animal in question have to be acknowledged as moral limitations (e.g., the strivings to move, procreate, and socialize).

The use of animals is thus only allowed if the human interests are vital ones and cannot be fulfilled by other means. Because humans are able to live as vegetarians in most parts

FIGURE 5 Egg-laying hens in a battery (photo courtesy of Compassion in World Farming/Lega Antivivisezione).

of the world, vegetarianism is required in these parts. The killing of animals for food is not justified, at least not if the conditions of the land and the population size allow agriculture. The big industrial complex of rearing and using animals within farming factories which has evolved in the Western countries should be prohibited (Fig. 5). Only a small-scale milk-farming economy seems to be allowed, if the keeping conditions are good and provide enough free moving and socializing opportunities for the animals. The exposing of wild animals in zoos or amusement parks seems to be unjustified. Only wildlife reservations which provide living conditions which are similar to life in the wilderness are acceptable. With respect to humans, a killing of animals in self-defense is allowed (e.g., the swatting of a biting mosquito).

If there exists no community between humans and animals, we have no positive helping duties toward animals. Therefore, we are not responsible to help wild animals. One has to also consider that such a help would normally interfere with the ecological predator–prey cycles and result in unforeseeable damages to the ecological equilibrium and the stability of ecosystems. However, if wild animals are harmed because of human facilities, such as roads, glass windows, and circuit lines, a weak helping duty exists. If animals are kept for human purposes and live in a sort of community with humans, humans have in any case, even if damage is not caused by humans, a positive helping duty. The situation is similar to that of a child who lives with his or her parents. Therefore, if a cat returns home after a fight with other cats and has some bloody wounds, it seems clear that the owner is responsible to look after the wounds. This responsibility is the reverse of the medal of the pleasure we have from domestic animals.

Animal Experimentation

Inasmuch as animals derive ethical consideration and the interpersonal aggregation between the interests of different entities, animal experimentation to further the interests of humans (or other animals) is not allowed. However, this general consequence must be restricted for two reasons previously mentioned. First, not all interests deserve an equal moral consideration. There is a decrease in the complexity of self-ascription and hence also a decrease in the moral relevance of the interests formed by self-ascription. Rational and self-conscious self-ascription is more complex and advanced and therefore deserves higher moral consideration than conscious, conscious and painful self-ascription deserves higher consideration than instinctive, instinctive deserves higher consideration than purely physiological, etc. Second, in a voluntarily founded community with mutually helpful duties, if a high discrepancy in the urgency of interests arises, a weighing of these interests is allowed. As previously noted, wild animals in general deserve higher ethical consideration than useful ones, and wild plants deserve higher ethical consideration than useful ones.

Taken together, these two aspects suggest that the total prohibition of animal experimentation can be slightly weakened and remain well within the bounds of ethical responsibility. Experiments with wild vertebrates seem to be allowed only if the consequences are minimal, such as placing a ring on birds' feet. For animals which are reared by humans, the following hold:

1. Great apes: No experiments which cause prolonged anxiety or more than minimal injuries or pain or death.

2. Other primates, whales, dolphins, cattle, dogs, cats, sheep, pigs, etc.: Experiments which cause prolonged anxiety or more than minimal injuries or pain or death are only allowed for the purpose of testing or developing life-saving

operation methods or medicines for humans, if no other form of experimentation will yield comparable results.

3. Other mammals: Experiments which cause prolonged anxiety or more than minimal injuries or pain or death are only allowed for the purpose of testing or developing life-saving operation methods or medicines for humans, if no other form of experimentation will yield comparable results.

4. Other vertebrates: Experiments which cause prolonged anxiety or more than minimal injuries or pain or death are only allowed for the purpose of testing or developing serious medical operation methods or medicines for humans, if no other form of experimentation will yield comparable results.

5. Other animals: Experiments which cause prolonged anxiety or more than minimal injuries or pain or death are only allowed for the purpose of testing or developing medical operation methods or medicines for humans and other mammals, if no other form of experimentation will yield comparable results.

All experiments for the purpose of education, toxicity and cosmetics testing, basic research, development of military weapons, and the repetition of experiments are ethically inadmissible. In general, in all animal experimentation one has to observe the three "R"'s—*reduction, replacement,* and *refinement:* reduction of the total sum of experiments, for example, by using a registration institution to avoid multiple experimentation, replacement of experiments by using other methods (computer simulation, construction of mechanical models, and testing of plants or bacteria cultures), and refinement of the necessary experiments. The avoidance of pain does not need to be the only or highest goal in that respect, although it often is the main goal. The main goal is the avoidance of death and serious injuries. Therefore, killing of animals after experiments are performed is not allowed. This sort of killing is more objectionable than the killing of animals for food because a being will be killed that has helped humans as an object of experimentation. One could compare the situation with an employer killing one of his or her laborers after the laborer had worked all his or her life for the employer and become useless after retirement.

These findings may seem too rigid to some experimenters. However, it has to be stressed that the most urgent political task in the field of animal ethics lies not in the abolishment or severe restriction of animal experimentation but in the abolishment of the inhumane mass animal keeping practiced in the food product industry (Fig. 6). This thesis is supported by three arguments: The total number of animals used in experimentation and for food products differs enormously. For example, in Germany, 330 million animals per year are used for food products, whereas less than 2 million are used for animal experimentation, most of them mice and rats. The keeping conditions of animals used in experimentation are usually better simply because the experimenters need sane animals. Also, the justification of animal experimentation, inasmuch as it pertains to human medicine, is much more valuable than that of good taste or the preservation of traditional eating habits.

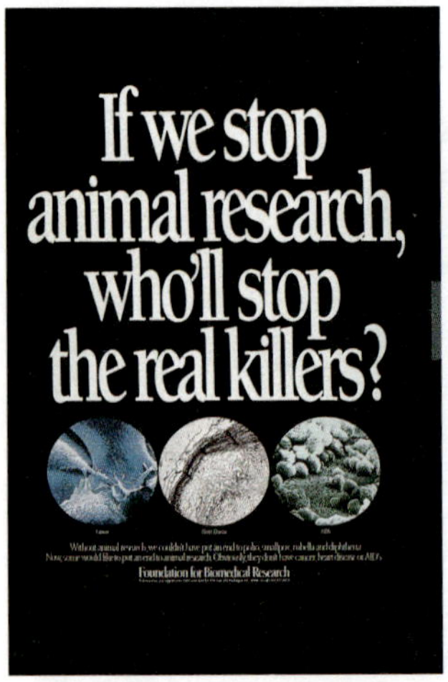

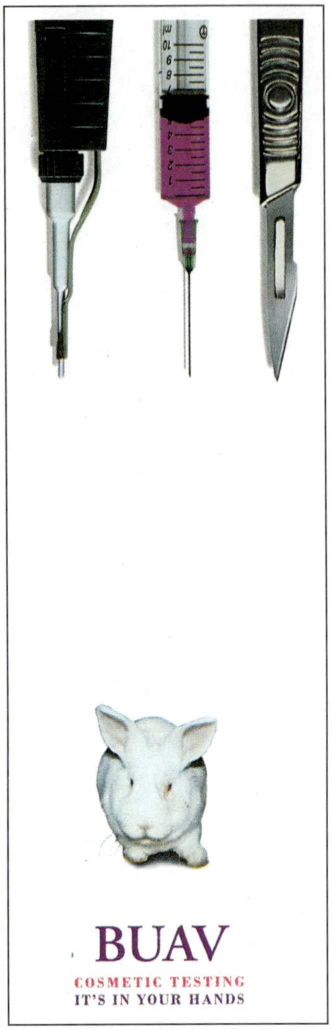

FIGURE 6 (Left) Poster in favor of the use of animals in biomedical research, produced by the Foundation for Biomedical Research. (Right) Promotion by the British Union for the Abolition of Vivisection (BUAV) against the testing of cosmetic products using animals (courtesy of Lega Antivivisezione).

Plants and Microorganisms

As already mentioned, the interests of plants and microorganisms deserve only a weak consideration. We can use them if there exists a good reason to do so. Therefore, we are allowed to cut a tree if we need firewood or wood for furniture or if the tree could fall down and damage houses or people. However, simply doing something for fun is not a good reason if other forms of amusement are possible. Therefore, we are not allowed to cut the tree just for fun or to show our friends how strong we are or to exercise. For some people, the consideration of interests of plants seems weird, but our intuitions match very well with this model. We consider that it is important, for example, that a child not trample without necessity on stones or wild flowers. As in the case of animals, there exist no positive helping duties toward wild plants. However, if our domestic plants need some water or new soil, we are required to provide these to the plants.

The moral consideration of plants has another important result: The mere addition of weak interests of plants can result in gigantic industrial projects. If a new canal or dam promises only questionable or little economic gain, the loss of billions of living beings is not outweighed. Therefore, the project seems not to be justified.

Genetic Engineering

Scientific progress has created a vast and exciting new realm of science and technique: genetics. Because of the novelty of the methods in this realm, traditional morality can provide no guidelines for an ethical evaluation. This is the case if the manipulated non-human being is healthy and suffers no pain. If an animal is created which is unhealthy or suffers from pain, this is just as bad as causing in an already existing animal illness or suffering. However, what about the moral assessment of the mere genetic manipulation? What about *in vitro* fertilization, cloning, and manipulation of singular genes if the outcome is a healthy animal without any damage or pain or a plant which is not different in most respects from natural plants? The double standard concerning animals and humans is nowhere more obvious than with regard to this question. With humans, we consider the manipulating act as bad and prohibit, for example, cloning even if it results in a similar outcome as a twin birth, whereas with animals we do not consider the act as bad.

To assess the moral features of genetic engineering, we first have to understand a basic principle of ethics, which I call the *principle of the might of the producer,* that states that the one who produces something is entitled to use it. If someone produces a car he or she is entitled to drive it, sell it, and scrap it as he or she likes, provided the using act does not interfere with the interests of others, such as if the individual drives the car dangerously or sells it fraudulently. The principle of the might of the producer can be found in many theories, e.g., in John Locke's theory of labor or in the Christian assumption of God's entitlement to rule the world because God is its creator. One result of the basic principle of the might of the producer is that if somebody has produced something, the product cannot claim a moral standing against the producer because (i) if he or she is entitled to use the product then the product cannot limit this use and (ii) he or she could have decided to construct the product in a totally different way or refrain from the construction.

The principle of the might of the producer can be seen as the reverse of the medal of the ethical position previously discussed. What gives an entity moral standing is self-ascription of interests, consisting of self-origin, self-development, and self-preservation, which means being not totally or in main parts produced by others. Therefore, if it would be possible to produce a living being in all its details, this biological machine could not claim any moral standing against its producer. Thus, genetic engineering endangers the entities which are manipulated in their moral standing.

The problem we are currently facing is not a complete production of living beings by methods of genetic engineering but a process of a step-by-step increase in manipulation: Ordinary methods of breeding, artificial insemination, *in vitro* fertilization, cloning, and modification of genes are steps on the way to creating the biological machine. For this step-by-step increase in manipulation two rules can be stated. First, the genetic manipulation is more problematic the higher the entity is situated in the hierarchy of life. Therefore, manipulating humans is more questionable than manipulating mammals, manipulating mammals is more questionable than manipulating other animals, manipulating animals is more questionable than manipulating plants, etc. Second, every further step in the process of the step-by-step increase in manipulation is more problematic than the preceding step. Therefore, the ordinary method of breeding is less problematic than artificial insemination, artificial insemination is less problematic than *in vitro* fertilization, and *in vitro* fertilization is less problematic than cloning or the direct modification of genes.

If one considers the two rules together, the result is that genetic manipulated plants seem to be acceptable only to serve important purposes (e.g., to develop medicines or to fight famines) but not to serve minor economic aims such as so-called "novel food" (Fig. 7). The genetic manipulation of animals requires even more urgent reasons. The cloning or genetic manipulation of human beings seems to be prohibited.

The Rights of Non-Human Beings

Philosophical concepts are often like new or renewed weapons. They emerge or are created to serve in a battle against certain established positions. This is very significantly the case with the ethical use of the concept of "right." Until

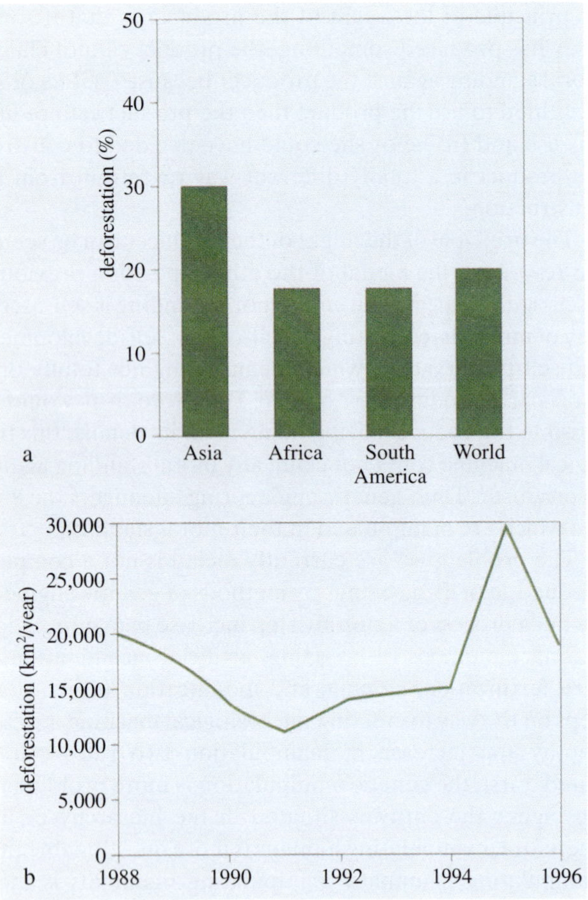

FIGURE 7 (a) Data shown as percentages of the total deforested area relative to the deforestation of tropical forests in the period between 1960 and 1990 (reproduced with permission from Bryant *et al.,* 1997). (b) Deforestation index for the Amazon region. The value for 1988 represents the mean for the period 1978–1988 (reproduced with permission from Brazilian Institute for Space Research INPE, 1998).

the late 1960s, the dominant ethical doctrine in the English-speaking world was utilitarianism, which designated the concepts "pleasure," "good," "value," "obligation," and "duty" as central concepts. Rights did not play a major role and were merely regarded as a special form or inference of obligations. However, an individualistic revolution against the collectivist, egalitarian, and maximizing assumptions of utilitarianism began and one major weapon in this battle was the concept of rights. To assign moral rights to entities, two major proposals were developed: (i) that the only source of a duty toward an entity is the entity itself and not an external "thing" such as a collective good or a common logical rule of universability as in utilitarianism, etc., and (ii) that some or all interests of this entity are not subject to interpersonal aggregation and maximization as in utilitarianism.

Even utilitarians such as P. Singer used the concept to exclude some interests of self-conscious beings from the general overall maximization. Also, nonutilitarians such as Regan used the term to assign to animals a moral position which in many respects is similar to that of humans. Therefore, many of the more radical authors and institutions in the English-speaking world, and also the international discussion which proposes the moral standing of animals, speak not only of "animal ethics" or "moral standing" but also of "animal rights" (Cavalieri and Singer, 1993; Regan, 1983).

The use of the term rights instead of ethics or interests has several purposes. A pragmatic purpose is to continue the successful "human rights movement" and expand it to include higher sentient animals, especially the great apes (Cavalieri and Singer, 1993). A second pragmatic purpose is to highlight the demand to secure the moral consideration of animals by enacting legal rights — that is, to improve the animal protection codes substantially or formulate some sort of "animal rights constitution." On a theoretical level, the use of the concept of rights excludes the derivation of animal-protecting prohibitions from human interests, which is a purely anthropocentric position. Therefore, the demand for a nonanthropocentric foundation of the moral consideration of animals is expressed. Animals are recognized as moral patients which have to be considered morally for their own sake. Finally, even if the interpersonal aggregation is allowed overall or to some degree, the concept of rights strengthens the position of the rights-holder in conflict situations.

Now, the question is whether, if in addition to the specific battle against utilitarianism in the English-speaking world, the concept of right is a useful tool. The main problem lies in the fact that its ethical, moral, and legal uses are often not carefully divided. Legal rights exist if they are established in a certain legal system. Moral rights exist if they have emerged by habit in a certain existing moral system. Ethical rights are not established and do not exist by habit, but they express the ethical position that an entity should have moral standing. However, this form of ethical consideration of others as having moral standing is the basis of ethics. Therefore, within the realm of ethics, the use of the concept of rights instead of moral standing does not add anything substantial if one is not a utilitarian and wishes to exclude interpersonal aggregation as do Singer and Regan.

Therefore, some theorists consider that the special feature of a right is that the entity in question has a valid claim against another person (the actor or a third person) to force the actor to comply with a moral rule. However, if an ethic is constructed from the basis of the individuals concerned, all duties have their basis in ethical claims.

For the realm of law, one has to consider that it is impossible to demand an automatic transformation of ethical or moral rights into legal rights. We can depict the phenomenon that even entities such as corporations, which have no moral rights, are assigned legal rights. In contrast, not all ethical or moral rights are transformed for various reasons into legal rights: The ethical right to be told the truth by others is not transformed in this general form into the penal law. Special features of the legal system, such as publicity, sociality, and the use of force, make additional reflections necessary. That does not mean that legal rights for higher animals are impossible or not desirable. However, an undifferentiated use of the term rights does not contribute to a careful discussion of the difficult ethical and legal questions.

References Cited

ATTFIELD, R. (1991). *The Ethics of Environmental Concern,* 2nd ed. Univ. of Georgia Press, Athens.

BENTHAM, J. (1963). *An Introduction to the Principles of Morals and Legislation,* 3rd ed. New York.

BRYANT, D., *et al.* (1997). *The Last Frontier Forests: Ecosystems and Economies on the Edge.* Washington, DC.

CARRUTHERS, P. (1992). *The Animal Issue. Moral Theory in Practice.* Cambridge Univ. Press, Cambridge.

CAVALIERI, P., and SINGER, P. (Eds.) (1993). *The Great Ape Project. Equality beyond Humanity.* Fourth Estate, London.

FEINBERG, J. (1980). The rights of animals and unborn generations. In *Rights, Justice and the Bounds of Liberty* (J. Feinberg, Ed.), pp. 159–184. Princeton Univ. Press, Princeton, NJ.

FREY, R. G. (1980). *Interests and Rights. The Case against Animals.* Clarendon, Oxford.

HABERMAS, J. (1992). *Erläuterungen zur Diskursethik,* 2nd ed. Suhrkamp, Frankfurt.

HARE, R. M. (1993). Why I am only a demi-vegetarian. In *Essays on Bioethics* (R. M. Hare, Ed.), pp. 219–235. Oxford Univ. Press, Oxford.

HUME, D. (1991). *An Enquiry Concerning the Principles of Morals,* 5th ed. La Salle.

KANT, I. (1907). *Die Metaphysik der Sitten, Akademieausgabe VI.* Berlin.

KANT, I. (1911). *Grundlegung zur Metaphysik der Sitten, Akademieausgabe IV.* Berlin.

NELSON, L. (1970). *System der philosophischen Ethik und Pädagogik, Gesammelte Schriften,* Vol. 9. Meiner, Hamburg.

REGAN, T. (1983). *The Case for Animal Rights.* Univ. of California Press, Berkeley.

REGAN, T., and Singer, P. (Eds.) (1989). *Animal Rights and Human Obligations,* 2nd ed. Prentice Hall, Englewood Cliffs, NJ.

ROLLIN, B. (1992). *Animal Rights and Human Morality,* 2nd ed. Prometheus, Buffalo, NY.

RYDER, R. (1991). Speciesism. In *Animal Experimentation* (R. M. Baird and S. E. Rosenbaum, Eds.), pp. 35–53. Prometheus, Buffalo, NY.

SCHOPENHAUER, A. (1989). Preisschrift über die Grundlagen der Moral. In *Die beiden Grundprobleme der Ethik.* Meiner, Hamburg.

SINGER, P. (1990). *Animal Liberation,* 2nd ed. New York.

SINGER, P. (1993). *Practical Ethics,* 2nd ed. Cambridge.

SPAEMANN, R. (1989). Glück und Wohlwollen, Stuttgart.

TAYLOR, P. W. (1989). *Respect for Nature,* 2nd ed. Princeton.

VON DER PFORDTEN, D. (1995). Der moralische und rechtliche Status von Tieren. In *Ökologische Ethik und Rechtstheorie* (J. Nida-Rümelin and D. von der Pfordten, Eds.). Baden-Baden.

VON DER PFORDTEN, D. (1996). *Ökologische Ethik.* Reinbek.

WOLF, J.-C. (1992). *Tierethik. Neue Perspektiven für Menschen und Tiere.* Freiburg/Schweiz.

WOLF, U. (1990). *Das Tier in der Moral.* Frankfurt.

General References

CLARK, S. R. L. (1977). *The Moral Status of Animals.* Oxford Univ. Press, Oxford.

LEAHY, M. P. T. (1994). *Against Liberation. Putting Animals in Perspective,* 2nd ed. Routledge, London.

MIDGLEY, M. (1983). *Animals and Why They Matter.* Univ. of Georgia Press, Athens.

NIDA-RÜMELIN, J. (1996). Tierethik I: Zu den philosophischen und ethischen Grundlagen des Tierschutzes. In *Angewandte Ethik. Die Bereichsethiken und ihre theoretische Fundierung* (J. Nida-Rümelin, Ed.), pp. 458–483. Kroner Verlag, Stuttgart.

NIDA-RÜMELIN, J., and VON DER PFORDTEN, D. (1996). Tierethik II: Zu den ethischen Grundlagen des Deutschen Tierschutzgesetzes. In *Angewandte Ethik. Die Bereichsethiken und ihre theoretische Fundierung* (J. Nida-Rümelin, Ed.), pp. 484–509. Kroner Verlag, Stuttgart.

RODD, R. (1990). *Biology, Ethics and Animals.* Oxford Univ. Press, Oxford.

ROLLIN, B. (1990). *The Unheeded Cry.* Oxford Univ. Press, Oxford.

SAPONTZIS, S. (1987). *Morals, Reason and Animals.* Temple Univ. Press, Philadelphia.

BRUNO SILVESTRINI
Institute of Pharmacology
University of Rome "La Sapienza"
Rome, Italy

The Rights of Nonhuman Living Beings: Experimentation on Animals

In this article, I discuss experimentation on animals in the context of the fundamental laws of nature as represented by self-defense, solidarity, and a gradual approach. I then trace its history and contributions to scientific and technological progress, especially in regard to the ethical–juridical principles which underlie its regulation, and I critically analyze arguments pro and con emphasizing the value of information which may be derived from the observation of natural processes. The uses of animal experimentation could be limited in part by means of a different strategy whose end is not the struggle against disease but rather the study of the latter's biological significance and the appreciation of drugs already tested in nature. The protection of the genome is discussed with reference not only to developments in molecular biology but also to the actions which man has undertaken since the start in order to channel the course of life in ways functional to his own needs. From the biological standpoint, the ethical problem of animal experimentation seems to have solutions substantially similar to those advanced by certain trends of religious and philosophical thought.

Introduction

Experimentation is a part of life, and life uses it in order to choose from among billions of possible solutions those which are best suited to the difficult path that leads it — in contrast to the degradation of the rest of the known universe — toward more organized and complex systems. As a result, man also makes use of it in his relentless search for conditions more favorable to his own survival and to his own development.

Between man's experimentation and that carried out by the rest of life commonly referred to as nature there exists a basic difference. The latter conducts the experimentation on itself, at its own risk and cost, and the information which is thus obtained is inseparable from its own tangible expression. Man, instead, has succeeded in isolating information from its original, tangible support and treating it as something separate. As a result, he can also conduct experiments on other living creatures and profit from them in the form of knowledge which may be put to uses which go well beyond the experiments whence it was derived. Many problems are born as a result, some of which will be examined in this article from the biological point of view.

The Two Fundamental Laws of Life

In the course of its development, which has been sustained by the ceaseless experimentation of every possible alternative, life has abided by two fundamental rules. They concern all living beings from viruses and unicellular organisms to the multicellular ones, which include animals and man.

The first of these laws is the all-out defense of life's own existence, which can go so far as to entail the suppression of the lives of others. For simplicity, I call it *the law of self-defense.* If life fails to abide by it, it stops, recedes, declines, and therefore becomes extinguished. Man is not exempt from this elementary obligation, both as far as his instincts are concerned and as far as culture, which in turn supports scientific and technological progress, is at issue. The former are written into his genetic code and transmitted from one generation to another through the act of conception. The latter is handed down from one person to another after they are born by means of spoken language, writing, and education.

It is not surprising, therefore, that man has since the very start made use of other living beings to satisfy his own needs. He hunted them to feed himself when he was hungry. He killed them in order to defend himself when they threatened him. He tamed them, transforming their original habitat, to use them more easily and derive foodstuffs and other useful products such as wool from them. He has used them as laborers to oversee his flocks, to work his fields, and to have himself carried along with his goods. He has selected the useful species and exterminated the useless or harmful ones by means of ever-more powerful tools, such as fertilizers, pesticides, and other poisons capable of killing every sort of living being, from bacteria and molds to wild mammals. He has also used animals for entertainment in cruel games such as races and fights to the last breath. He has gained control, finally, of the genetic code and manipulated it, thereby intervening profoundly on the very mechanisms which regulate and maintain life.

The greatest successes of modern medicine may be traced back to this basic law. Consider, for example, chemotherapeutic drugs, antibiotics, and vaccines: What are they if not terrible weapons used to counter, and in some cases to exterminate, other forms of life? In this case, the latter are pathogenic fungi, bacteria, and protozoa, but in order to defend his life man does not hesitate to also kill livestock and wild animals. He does it if he has the slightest suspicion that they may be carriers of transmittable diseases such as rabies and bovine spongiform encephalopathy, commonly known as mad cow disease. Were he not to behave in this way, on the other hand, he would once again be prey to malnutrition and to the diseases which in the past killed or maimed millions of individuals, such as polio, diphtheria, tuberculosis, and the so-called puerperal fevers. Even the stance taken by vegetarians, who in order not to kill animals feed themselves on plants, appears weak insofar as their nutrition, too, is vouchsafed by the extermination of animals and other living beings that are harmful to agricultural production.

Mors tua vita mea: This is the first law of life. There exists another one, however, which is just as fundamental and inseparable from the first. In what follows, I refer to it as the *law of solidarity.* Thanks to it, at the dawning of life, two pieces of matter, or of energy if we want to call it that, came together and in so doing produced a force which enabled them to proceed in a manner opposite to that of the surrounding world, which was slowly breaking down and becoming extinguished. From their union something more complex than a simple sum was born, something which knew how to take advantage of what had taken place up to then by memorizing it and using it, on the one hand, in order to defend its own existence and, on the other hand, to use it as a springboard toward other, even more complex and organized systems. That first expression of life duplicated itself so as to reduce the risk of being squashed by the surrounding world and of being lost. It did it at the cost of its own individual existence since duplication implies the transferal of one's self to something different, as represented by entities endowed with an individuality of their own, different from that of the parent.

Life has this extraordinary peculiarity. On the one hand, it is fully represented in every single living being, from the most elementary which appeared in the first phases of life to the most complex and recent, including man. The former have only leafed through the first pages of its project, whereas the others have read a bit further. On the other hand, it identifies itself in a global project carried forth by a myriad of living beings past, present, and future that pass it on from one to another like athletes pass the baton in a relay race.

All this would not have been possible without the contribution of both of the fundamental laws of life — self-defense and solidarity. We also find them in scientific and technological progress. However, if this is the way things stand, why is the latter often perceived as a deviation from the natural order of things? It is as if, on the one hand, there was uncontaminated nature and, on the other hand, man ravages it, thereby committing a sin he is bound to pay for dearly. This concept is not new; it has existed since the beginning. According to some, we find it in the Earthly Paradise whence man was driven away for having tasted the forbidden fruit. It reappears in the contemporary doctrine of the Catholic Church, for example, whenever natural methods of contraception are counterposed to those invented by man. It is expressed in the myth of Prometheus, who steals fire from the gods and suffers a cruel punishment for his deed. It finds many different philosophical forms, both in antiquity and in the modern era, in the naturalistic trends of thought which stand opposed to the Enlightenment. It shows up again in the ecological movements, whose members fight for the preservation of nature — which they refer to as an ecological system — also on the basis of scientific reasoning and also at the cost of putting an end to progress.

The Third Law of Life

The problem with progress is the result of the speed of the changes that it implies. Also, nature, in fact, changes all the time, but it does so in a gradual manner. For example, it has taken living beings millions of years to equip themselves with apparatuses such as eyes, ears, internal tensoceptors, and nociceptors which allow them to perceive some of the many signals coming from within and without. Likewise, it has taken them that long to pass from water to land, to take flight, and to perfect each of these capacities. The more drastic the changes, the longer it takes to adapt to them. A clear example of this is the passage from monocellular to multicellular organisms. The former, which do not have problems of cohabitation, grow rapidly, reproducing every few minutes. The latter, which are made up of cells, tissues, and various apparatuses, and which have to adapt to each other in order to survive together, have slowed down the speed of cellular duplication to once every 12–24 h or longer. Without this slowing down, the previously mentioned passage would not have been possible.

Proof of this fact is afforded by infections, which appear to the biologist — who is trained to also view things from the point of view of beings other than humans — as a classic example of an attempt at cohabitation between different biologic systems — that is, the infective agent and the infected organism. If the infection spreads rapidly, then the latter's reciprocal adaptation is impossible because one of the two prevails upon the other, either by eliminating it or by suppressing it. If the infection spreads slowly, then a reciprocal tolerance develops between infective agent and infected organism which enables both to survive. This is the case for infections that are slow or caused by eukaryotes, which duplicate physiologically every 12–24 h, and for bacteria such as those responsible for tuberculosis and syphilis, which have learned to grow slowly.

Man followed the same evolutionary track but then moved away from it in order to take the incomparably more rapid path afforded by scientific and technological development. He learned to protect himself from the cold and the elements not by means of the adaptation of his organism, which requires millions of years, but rather by means of fire, clothing, and housing, tools which he was rapidly able to put to use. To preserve and develop his experiences he made use not of his nervous system, which in the course of evolution has grown only slowly, but rather of artificial substrates foreign to his own body, such as paper and other increasingly powerful media which were developed in the span of a few generations. Once he had put together the necessary knowledge and technology, he went to on to make, in the course of a few decades, other changes — the same ones which nature had taken millions of years to accomplish. Some unexpected consequences have ensued, with respect to which the process of adaptation has shown itself to be increasingly difficult.

Consider, for example, the successes that have been achieved in the fight against infective diseases. Millions of human lives have been saved but, at the same time, demographic development has been favored, which in turn has engendered new needs in terms of food education social organization and health care long before they could possibly be satisfied. Tools for the conservation, processing, and distribution of information have become increasingly powerful; paradoxically, however, man risks being not enriched but impoverished by all these opportunities for knowledge which, because of their very abundance, end up overwhelming him.

The three laws of nature provide the general frame of reference within which animal experimentation is situated and will be discussed next.

Definitions and Generalities

Previously, I implicitly provided a definition of what I mean by experimentation. In what follows, I define it more precisely and also examine some of its more general aspects.

In a scientific context, experimentation may be defined as "a test carried out to prove a hypothesis, to study the intrinsic characteristics of something, to ascertain what possibilities exist for applying a theoretical discovery and verifying its goodness and safety." In other words, it is a particular moment in the knowing process which differs from others because of the uncertainty of results and because it entails a direct intervening of the experimenter on whatever constitutes the object of investigation. Such interventions can vary greatly and range from the measurement of a particular parameter to the reproduction of a natural phenomenon in the laboratory. Whatever the type of intervention, however, it will always imply some form of change, which may even be the sacrificing of a living being. In the absence of these two connotations — the uncertainty of results and the change necessarily implied by their pursuit — the intervention only involves a mere observation of facts. In its most common form, observation consists of the registering of a previous fact and an acquisition of knowledge which, though different, is just as important. Some consider other uses to which animals are put, such as those with a didactic end, as experimentation. In this article, however, I abide strictly by the definition that was previously provided.

In some cases, experimentation yields, in addition to new information of a general interest, a specific advantage for the subject on which it has been carried out. This occurs, for example, when an animal or human being affected by a certain disease is treated with a drug whose therapeutic value has already been ascertained but requires further definition.

In these cases, one speaks of therapeutic experimentation or of experimental therapy.

Contrary to what many people believe, in the majority of cases experimentation is not carried out on living beings but, rather, on inert materials, which may be pieces of machinery or molecules of varying complexity from inorganic salts to enzymes and DNA. This kind of experimentation also raises ethical questions of great importance, but the latter concern not the rights and protection of the experimental medium but rather the ends being pursued and the uses to which results are put. A telling example is that of mustard gas, whose experimentation had initially been forbidden because it was aimed at applications in the fields of chemical and biological weaponry, whereas subsequently it turned out to be determinant in the treatment of cancer. Experimentation on the genome has imposed the need to protect the experimental material, even when it is not endowed with a life of its own. The problem will be dealt with briefly later.

Living beings employed in experimentation include viruses, bacteria, protozoa, and other monocellular organisms and also multicellular organisms of increasing complexity, from gnats to mammals and man. Here, the problem of the protection of the experimental object stands out in all its importance since all living beings, from the most elementary to the most evolved, are carriers of "life's project," as R. Dulbecco (1989) called it. Some have fulfilled, or leafed through, only the first pages, and others have gone further, but all are equally a part of it. From this point of view, therefore, they deserve equal respect.

A substantial part of the ethical debate, however, concerns experimentation on animals, especially those endowed with a well-developed central nervous system which allows for organized thought and the conscious perception of suffering, both physical and mental, in addition to that of pleasure (D'Agostino, 1996). Unlike what takes place in nature, in the course of experimentation these beings are helpless before man, who can do what he will to them, violence included. Their condition, from this point of view, might appear somewhat similar to that of a child, of a person who is mentally ill, and of other impaired individuals, but a fundamental difference between them subsists. Humans are protected by members of their same species, who perceive it as an obligation rooted in religion, lay morality, and jurisprudence. Moreover, whoever helps a weak person knows that he or she is performing an act of solidarity, which he or she too might someday be in need of: Indirectly, therefore, he or she is also protecting himself or herself (Silvestrini, 1997). The protection of nonhumans, on the contrary, is entrusted to a different species and is a principle which the latter has never properly accepted and which has yet to be put clearly into focus.

Animal experimentation, as a result, is also a pregnant topic on the ethical level and deserves to be treated separately from the question of experimentation on humans. It is the main theme of this article, which analyzes some of its various aspects. The deontological part relating to so-called norms of proper laboratory practice, instead, has been eschewed. We are in line with the Anglo-Saxon tradition and believe, in fact, that the respect of such norms concerns research scientists' basic professional training.

A Brief Historical Account and Developments in the Set of Rules

Though ancient as a practice, experimentation on animals became established and widespread as one of the basic tools of medical progress only in the modern era. W. Harvey (1578–1657) used it to study and describe the circulation of blood (Fig. 1), demolishing the incorrect explanation that C. Galen had previously provided. F. Magendie (1783–1855) used it to examine the biological effects of many new compounds, and C. Bernard (1813–1878) used it to study the functioning of the liver, pancreas, and vasomotory nerves as well as the effects of curare and certain anesthetics. Animal experimentation was also determinant in the discovery of salvarsan, the first surely efficient chemotherapeutic substance, by P. Ehrlich (1854–1915). He also employed it to gather more information on the immune

FIGURE 1 Title page of W. Harvey's treatise *Exercitatio Anatomica de Motu Cordis et Sanguinis in Animalibus* (1628), often abbreviated *De Motu Cordis,* Paris, Biblioteque Nationale (photo courtesy of Treccani Library).

system. After that, the procedure became so widespread that it would be materially impossible even to list its most significant uses and results.

However, despite its widespread use and increasing importance, its ethical implications were at first scarcely debated. F. Bacon (1561–1626) went no further than to justify and predict its use in the name of the idea that science works with a primary aim in mind which stands out above all the others and is represented by the good of man. According to I. Kant (1724–1804), man is the only living being endowed with moral dignity. The problem of the protection of animals, on the other hand, is hardly perceived even in the monotheistic religions: In the latter, the vision of a hierarchical order of the universe with divinity at the top and man — to whom dominion over nature has been granted — immediately below prevails. In the Bible, in the Gospels, and even in the new catechism of the Catholic Church (1992), no specific mention is made of the rights of animals. In the Tenth Commandment, the latter are only considered a property of man. Therefore, the general norms governing the behavior of man are understood to apply.

The rights of animals begin to be emphatically set forth only in the second half of the past century, especially as a result of the activities of working-class movements. At the urging of the league founded by France's Power Cobbe (Fig. 2), in 1876 the government presided by B. Disraeli issued in Great Britain the Cruelty to Animals Act. After that, several countries adopt analogous norms, which prescribe punishment for acts of cruelty to animals. An example of a detailed enactment is the Animal Welfare Act (1970), which precisely establishes in the United States the norms to be followed in the confinement, transportation, and management of guinea pigs, rabbits, cats, dogs, and, in general, all animals that live in captivity, including those to be experimented on.

There are numerous collections of writings on the subject (Schiller, 1967; French, 1978), in addition to two documents by the Comitato Nazionale per la Bioetica (National Bioethics Committee (1992a, 1998) and one by the Committee on Ethics of the Consiglio Nazionale delle Ricerche (National Research Council) (1992). I will not go into details, therefore, not least because regulations and legislation are in continuous flux. In a general sense, it should be noted that the weakness of the theoretical principles governing the defense of the rights of animals is more than made up by the way animals are protected in practice, at least as far experimentation is concerned. Existing regulations and laws on the subject, in fact, are becoming increasingly stringent and detailed. Any researcher who wants to conduct an experiment is subject to a long series of obligations and outside controls in regard to stabularium conditions, the scientific premises justifying the test, the importance of expected results, the absence of alternative procedures, precautionary measures to avoid discomfort and suffering, and other aspects. The result is an objective impediment to that considerable part of scientific research which is still based on such a practice.

It should be noted that the situation as far as man is concerned is different. His fundamental rights are no longer just generically entrusted to religious, philosophical, and juridical tradition but have been solemnly upheld in documents which formally bind humanity as a whole, such as the Uni-

a b

FIGURE 2 (a) Frances Power Cobbe. (b) Anti-vivisection demonstration in Hyde Park, London, 1913 (photo courtesy of Hulton-Getty/Rochi).

versal Declaration of Human Rights (United Nations, 1948) and, in the context of the European Community, the recent Convention on Human Rights and Fundamental Freedoms (1997). Regarding experimentation, a fundamental document is the Nuremberg Code, formulated after the sentence was passed in 1947 at the end of the trial against the Nazi doctors who were responsible for experiments that had been conducted on women, children, and men. It establishes that no experimentation on man may be carried out without the agreement of the individual who lends himself or herself to it. This basic principle was then taken up in the document known as the Helsinki Declaration (1964), drawn up on the occasion of the XVIII World Medical Association Assembly, but with an important addition. Although the Nuremberg Code entrusted the experimenter with the task of obtaining the subject's consensus, in this document the need for an independent control, entrusted to an outside organization, is upheld. The same principle was again upheld during the World Medical Association Assemblies that were held in Tokyo (1975), Venice (1983), and Hong Kong (1989) and on other public occasions as well. It has thus become a sort of deontological code for researchers as well as a constant point of reference for laws and regulations enacted in defense of persons undergoing experimentation (National Bioethics Committee, 1992b; Bompiani, 1996). In brief, outside control is entrusted to public organizations such as the Food and Drug Administration in the United States and to the central and peripheral ethical committees which have played an increasingly important role in policy making.

On the practical level, however, a broad range of experiments are left to the doctor's discretion. Medicine is still a largely empirical discipline which relies on therapeutic approaches and methods, both pharmacological and surgical, that cannot be easily traced back to preestablished schemata based on solid scientific knowledge. As a result, it proceeds via a series of trials and progressive approximations, which we might define as therapeutic but which nevertheless remain a sort of experimentation carried out on patients who are not in a position to give their informed consent and who cannot enjoy the outside guarantees which, as we have seen, condition experimentation on animals and that part of experimentation on man that is carried out in strictly scientific contexts.

Pros and Cons

The supporters of animal experimentation can easily appeal to the fact that it has been determinant in the modern development of research and, therefore, of medicine as well. It has made possible the collection of precious information on the functioning of the main organs and to transform it into practical applications which have often turned out to be useful not only to man but also to domestic animals and in the protection of the environment. It has made possible the reproduction in the laboratory of many diseases, such as infections, vitamin-deficiency disorders, epilepsy, hypertension, arteriosclerosis, and rheumatic arthritis, whose study in the field would have been incomparably longer and more difficult. In this way, it made possible the systematic screening of millions of natural and synthetic compounds, many of which were found to be active and put to practical use in medicine and other fields. It has provided an opportunity to test the safety and efficacy of drugs and other substances in a manner that is more systematic and, in certain respects, more in-depth compared to what had been taking place when these tests were carried out on man, often represented by socially weak individuals such as prisoners and the dispossessed. All this implies a certain degree of damage and suffering to the animal, which is often inevitable insofar as it represents the parameter on which the evaluation is based. This negative aspect is counterbalanced by concrete advantages, not only for man but, as noted earlier, also for animals. Moreover, experimentation on animals is categorically required by health authorities for the approval of most new drugs, some of which are essential for our very survival, such as vaccines and antibiotics. There are some exceptions, which are discussed later, but they concern only particular cases. Outlawing animal experimentation, therefore, would mean putting an end to therapeutic progress, depriving it of the hope of overcoming those diseases which are even to this day without a cure, such as many forms of cancer.

Those who are against it, however, emphasize how difficult it is to transfer to man the results which have been obtained with living beings that are profoundly different on anatomical, physiological, and, consequently, pharmacological levels. This is shown by the fact that the negative effects of drugs often manifest themselves not in the course of animal experimentation but subsequently, in the course of their therapeutic application. Despite all precautions, a certain margin of uncertainty remains, so much so that everyone is agreed in considering as crucial not activity and safety tests carried out on animals but those performed on man. Furthermore, animal models are often misleading because they reproduce the symptoms and external manifestations of diseases but not their causes. As a result, they have favored the development of symptomatic instead of curative drugs, such as those used against hypertension, schizophrenia, and depression.

The debate is a live one, between apparently irreconcilable positions which, however, might find a way of coming to terms with one another were there the goodwill to do so. On the one hand, animal experimentation still appears to be indispensable for the progress of science, not only to the advantage of man but also to that of the animals. On the other hand, it can be further reduced by taking better advantage of the alternatives which exist.

The Alternatives

Even while not questioning its necessity, the knowledge and techniques which are currently available provide numerous alternatives to animal experimentation and reduce its burden on those that are subjected to it. *In vitro* tests carried out on isolated cells tissues and biological extracts often provide information which is just as valuable (Fig. 3). Molecular biology has provided, along with the new experimental tools that have accompanied its development, interesting perspectives for the detailed and not necessarily invasive study of physiological and pathological processes. Computerized models and systems, which can simulate the functioning of an organism to a considerable degree of approximation, are providing increasingly important assistance. Many instruments are available, such as the same ones that are broadly used in clinical diagnostics that allow one to probe deeply into an organism without causing it the slightest damage. For example, the study of the possible carcinogenic effects of drugs and of their metabolic derivatives, which at one time required recourse to animals, can today be partly carried out on isoated cells tissues and enzymatic systems. This explains why the debate between those in favor and those against animal experimentation, which was once especially heated, is cooling down. Today, it hinges not so much on the real utility of alternative methods as on their real importance, which is overestimated by some and underestimated by others.

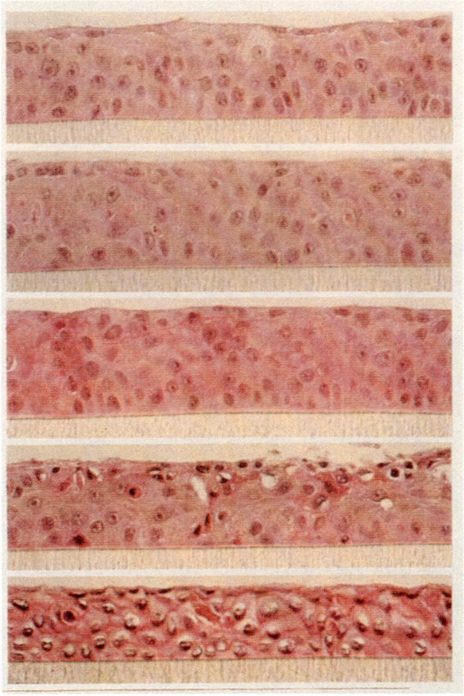

FIGURE 3 Effects of the local application of ethanol for 24 h on a reconstructed model of human epithelium (photo courtesy of Skinethic).

The problem, however, does not end here. To explain it one must remember that until recently animal experimentation has been especially useful in three sectors of biomedical research: basic research, which studies biological processes, including diseases, for the purpose of explaining their mechanism of action; applied research, which transforms knowledge into practical applications; and the toxicological testing of drugs and other substances, which is carried out before they are assayed on man and clinically employed.

Basic research does not always require, at least not in its initial phases, recourse to experimentation on animals. Many of the great discoveries, those which have affected human existence most deeply, are born from the careful and acute observation of processes which manifest themselves in nature spontaneously, without any intervention on the part of man. For example, J. Lind (1716–1794), a doctor of the British Navy, noticed a relation between scurvy — a very serious disease at the time and one which was very common among sailors — and the lack of fresh food, and he educed that the latter contained some essential principles that deteriorated with time. To prove the hypothesis, he resorted to experimentation, which he carried out directly on affected individuals. He did this without asking for either approval or authorization because such factors were not taken into consideration at the time, and yet his decision appears to have been correct from various viewpoints. First, the experiments were justified by an argument, or hypothesis, that was plausible — that there existed a causal relation between lack of fresh food and the disease. Second, there was a state of necessity represented by a disease which, in the absence of medical intervention, might have been lethal. Moreover, Lind treated his patients with substances which were not particularly risky insofar as they were foodstuffs that were being regularly used at the time. Finally, the patients were not prevented from enjoying the benefits of other therapies since none other existed at the time. In brief, he separated some sailors suffering from scurvy into homogeneous groups, to whom he administered various fresh foods. Fortunately, at the time there were commercial exchanges between Sicily and Great Britain so that he could include among the various treatments the juice of citrus fruit, which has a marked therapeutic effect.

That was probably the first great medical discovery of the modern era. Not only did it involve a medicine of proven efficacy but also it was documented by a controlled study, comparing diverse treatments with one another according to modalities which are still valid today. Furthermore, the juice of citrus fruit enjoyed some form of government registration, which enabled it to be distributed on all the ships of the British Navy. Animal experimentation was also of great importance in this instance, but only later when an experimental model of scurvy was developed which made it easier to identify the main active component in citrus fruit — vitamin C. A. Szent-Györgyi won the Nobel prize for medicine for its discovery in 1937.

Approximately one century after Lind, E. Jenner (1749–1823) observed that milkers who had contracted a slight disorder (Fig. 4), which in appearance was not unlike smallpox, then developed a resistance to the latter, which in those days manifested itself in epidemic form, killing thousands of people. He deduced that the natural defenses of an organism can be activated using stimuli which are not dangerous in themselves but capable of simulating the disease. In other words, he discovered the Hippocratic principle *similia similibus curantur,* thus opening the way to modern vaccines. He too verified the hypothesis not on animals but directly on man, according to a procedure which, in this case, raises heavy doubts of an ethical order. To demonstrate the efficacy of vaccination, in fact, in 1796 he inoculated for the first time a young man who had been previously vaccinated with some smallpox-infected pustules. He did not contract the disease and this represented a result of both enormous scientific and practical importance, butt it had been obtained at a grave risk for the subject of the experimentation. Today, the first tests would be conducted on animals and not on man.

FIGURE 4 Edward Jenner portrayed in a sculpture by Giulio Monteverde, Genoa, Civic Gallery of Modern Art (photo courtesy of Hulton-Getty/Ronchi).

Starting in 1803, a large-scale vaccination campaign was carried out in London which resulted in a drastic reduction in the number of deaths from smallpox. Later, the practice continued to spread until, in the second half of the twentieth century, it led to the total disappearance of this very serious disease.

Another interesting case is that of the discovery of insulin. In 1889, it was determined that diabetes mellitus is caused by a deficiency of the pancreas, but in the decades that followed no one succeeded in isolating the active principle that caused the disease. F. G. Banting observed on the surgical table that the obstruction of the pancreatic duct caused the degeneration of the exocrine part of the pancreas which is rich in proteolytic enzymes. Note that the latter, once they have come out of cells, have the capacity to destroy the protein compounds with which they come into contact. To avoid this phenomenon, which might have been responsible for all previous lack of success, Banting had the idea, prompted by his surgical observations, of causing the degeneration of the exocrine part of the pancreas so as to induce the disappearance of the proteolytic enzymes. Therefore, he subjected some dogs to the tying of their pancreatic duct and, after a congruous amount of time, collected and examined their pancreases. In this way, he managed to isolate the factor whose absence was responsible for diabetes. For this successful result, which had been obtained by surmounting innumerable obstacles, Banting received the Nobel prize in 1923 along with J. J. R. Macleod, who had supported and assisted him. It should be mentioned that the two, in turn, shared their prize with C. H. Best (Fig. 5) and J. B. Collip, respectively, who had helped them carry out their experiments. In this case as well, animal experimentation was critically important, but the crucial event was the interpretation of a natural event which had manifested itself spontaneously without man's intervention. The same combination can be found in many other modern discoveries, especially with regard to vitamins, vaccines, and hormones.

Once one has established the causes of diseases, it is often possible to cure them with drugs that are made out of physiological substances — ones which, if they are not present in sufficient quantities, cause the diseases, or ones which can activate a defensive mechanism on the part of the organism. The former is the case for vitamins, insulin, and other hormones, and the latter is the case for vaccines. The

FIGURE 5 Frederick G. Banting (right) and Charles Best with one of the first "patients" to be administered insulin, January, 1921 (photo courtesy of Hulton-Getty/Ronchi).

therapeutic use of these agents consists in the reintegration of the organism's composition or in the strengthening of its capabilities. Their safety, therefore, depends on the basic knowledge one has about them, which in turn allows one to avail oneself of the testing which nature has subjected them to in the course of thousands or millions of years. Their study using animals, for the purpose of verifying their safety before experimentation on man and subsequent therapeutic use, may be useful but it is not indispensable. Furthermore, in some cases these drugs are made up of proteins which differ so much between species and species that the transferal of results from animals to man ceases to be difficult and becomes simply impossible.

The study of natural phenomena is important even when it leads to the discovery of drugs that are foreign to the composition and functioning of the human organism. A classic example is antibiosis, which was first described by L. Pasteur and J. Joubert (1877). These scientists discovered that certain microorganisms secrete poisonous substances which protect from other mechanisms with which there is competition. This observation not only ushered the advent of modern antibiotics but also provided a lesson about the inconveniences that have accompanied their use. Antibiosis is a relatively primordial defensive mechanism insofar as antibiotics are weapons whose composition is fixed and which cannot be adapted to the specific characteristics of each separate aggressor. Moreover, it is easy to develop countermeasures against them, which become a form of resistance. For this reason, nature has, in the course of evolution, sought other solutions offered by the immune system, which first appears in invertebrates. The latter does not produce substances whose composition is fixed, as do antibiotics, but rather antibodies which can be modified taking into account both the specific characteristics of each aggressor and its possible counteradaptation. A close reading of natural processes would lead us to understand the inconveniences associated with antibiotics, represented for the most part by resistance, and teach us that the main path to follow in the fight against infections is not that of antibiotics or of chemotherapeutic drugs but rather that of vaccines.

The problem of experimentation on animals assumes different dimensions with the lesser known diseases. In these circumstances, animal models capable of reproducing their symptoms or external manifestations have often been used. A classic example is that of epilepsy. In 1938, Merritt and Putnam had the idea of reproducing the disease in laboratory animals by using an electric charge capable of inducing convulsions, and they used this experimental model to sift through the agents that were available at the time. This method, which is substantially empirical since the origin of epilepsy is still unknown, allowed for the identification of phenytoin, which is still used in therapy, and of many other agents.

The experimental models of diseases whose causes are known have multiplied and become of great importance in pharmacological research. From certain points of view, however, they have turned out to be misleading, as previously mentioned, and have entailed many inconveniences. Consider, for example, schizophrenia. Its causes and organic basis are unknown, as is the concrete substrate which sustains it. In animals, therefore, it is possible to reproduce its external manifestations and, with great difficulty, something which simulates its psychic symptoms such as the hallucinations that patients describe in moments of lucidity. The use of these experimental models of schizophrenia in biological research has led to the identification of symptomatic, and not curative, drugs. Moreover, attempts have been made to trace the causes of the disease through the characteristics of these drugs. It is a risky procedure which can easily lead one to confuse symptoms with causes. Had it been applied to the study of infective diseases when the pathological microorganisms had yet to be isolated but antipyretics were already available, it would have led one to conclude that infections are caused by fever.

The use of these empirical models of diseases generally leads to the discovery of drugs made of synthetic substances first synthesized by man or, in any case, lacking any relation with the composition and functioning of his organism. Before they may be employed in therapy, therefore, such substances must be carefully put to the test to verify their safety. As mentioned previously, today one can use *in vitro* assays which provide valuable information about possible toxicological risks. Unless such negative indications arise as to suggest that the substance be abandoned, the *in vitro* tests are then combined with acute, subacute, chronic, and special ones carried out on different species of animals to analyze its effects on main organs and apparatuses. As a result, one can say that animal experimentation aimed at the toxicological testing of drugs is born especially from the empirical mode of research described previously and from the characteristics of the drugs it allows one to identify.

In a more general sense, one may conclude that the real alternative measure to animal experimentation, the one which could substantially redefine its importance in biomedical research, is made up not only of *in vitro* studies, computerized models, and other methods of analogous value but also especially from a radical rethinking of the very strategy of research. On the one hand, there are laboratories in which man conducts his experimentation, both on animals and on some other materials, according to his own logic, which is that of combating rapidly and with every available means the disorders which afflict him. On the other hand, there is the great laboratory of nature, in which at every moment for many millions of years billions of experiments have been conducted on every sort of living being. Many of the problems that man struggles to resolve have already been dealt with here and have found a solution.

Unfortunately, these solutions are neither written in books nor stored in libraries or in modern information technologies, all of which are relatively easy to consult, but rather are inscribed in the very functioning of life. The reading of the latter "book" requires a very particular sensibility, which not all scientists possess. This is not the only reason why the experimentation carried out by nature is neglected. With the passage of time, an ensemble of norms that favor this form of human behavior at the expense of others have consolidated themselves about it. The granting of patents rewards those chemical substances that have been first produced by man more than their intrinsic characteristics. Without the property which is guaranteed by patents, a pharmaceutical industry which invested in research would find itself in the same position as a constructor who built a building without owning it. He or she might do it once, perhaps twice, but then he or she would go bankrupt. As a result, since the nineteenth century the pharmaceutical industry has been prompted to aim more at the synthesis of new molecules than at the therapeutic use of substances which, in themselves, are more promising but which are already known and cannot therefore be so easily patented. The industry has also insisted on the empirical screening of these new molecules rather than on basic research.

In this way, with the passage of time, the notion that empirical drugs (i.e., those which cure diseases before their causes are known) should be used only as an emergency solution — until other drugs with a solid scientific basis, which in turn must necessarily have derived an advantage from the lessons of nature, are developed — has been lost. To denounce such a situation would be a sterile operation, unless constructive proposals were not advanced at the same time. To this end, a strong cultural engagement is required in order to discuss these fundamental problems and to favor, as a result, a reconversion of industrial and academic biomedical research — a reconversion which does not present itself as a punitive exercise but rather as the result and expression of a change in the factors which have up until now conditioned the development of research, starting from the principles of patenting.

An analysis of animal experimentation which remained within the limits of its superficial aspects, without dealing with its deeper roots, would also be inadequate on the ethical level. It is to be hoped that this view of the problem may help overcome the polemics between supporters and opponents of this practice, uniting them in common initiatives capable of satisfying the aspirations of both.

The Animal Genome

The genome, the infinitesimally small particle enclosed in each of the billions upon billions of cells which compose an animal organism, is a topic of great interest on the scientific and ethical planes.

With molecular biology man has learned to manipulate the genome, erasing from it the information he deems harmful and replacing it with other, potentially useful information. In other words, he has learned how to experiment on the program, known as genome, on which life's developments depend. He does it *in vitro,* in the test-tube as one used to say, and outside of animal experimentation which raises the question of whether it is ethically correct to inflict suffering on another living being for the benefit of man. As a result, the problem is no longer one of safeguarding animals but rather of the consequences resulting from the violation of that which conserves a memory of past life and uses it, bearing external circumstances in mind, to condition its future development.

A sense of fear, widespread not only among politicians and scientists but also among the common people, has been provoked and has led to the proclamation of moratoriums and prohibition of experimentation on the genome in general, both human and animal.

However, there is nothing new in all this. Not only the experimentation nature has conducted in its vast laboratory, made up of the totality of living beings, but also that conducted by man has always rested, finally, on the genome. Man has known how to manipulate it and to deviate it from its course for many thousands of years. Such a manipulation began when man first tamed animals by preventing them from indulging in their original habits and forcing them to adopt others and when he selected those with the altered psychic and somatic characteristics which best suited his needs. This intervention entailed the conservation of particular anomalies, known as mutations, which in a logic other than the human logic represent unsuccessful attempts, which would have been immediately abandoned. Therefore, the genome is not sacred and immutable: On the contrary, it has always been changing both because of its own intrinsic force and because of the intervention of man.

What has changed is not the problem in itself but the speed of the changes that man produces in the genome. For this reason, it would make no sense to prohibit experimentation on the genome since it has been going on since the very beginning; it is instead necessary that a moment of pause be taken to reflect on these issues, which might also turn into a moratoria of manipulations allowed in molecular biology in addition to their being severely kept in check.

The careful reader will have certainly noticed that, even without having been explicitly mentioned, these considerations revolve around the three fundamental laws of life: self-defense, solidarity, and a gradual approach.

Conclusions

Every time he interfaces with nature in an attempt to dominate or to explore it, man experiences contradictory feelings of exaltation in moments of success and of confusion when these same successes turn against him, making him aware of his own weakness and inability to foresee the consequences of his own actions.

Scientists, too, are prey to the same feelings when the exaltation of having deciphered a part of the great book of nature, or of having been a protagonist in the struggle against the evils which afflict humanity, is replaced by an awareness of his own limitations. Then his vision becomes cloudy and his very victories seem to turn against him. Perhaps this feeling is more common today than in the past because never before have scientists had such powerful tools available to them, but in reality it is a problem which has always existed.

It also happens to those who try to distil an ethical rule, which is also valid for their work, from what they know. If someone is a research scientist, he or she will believe in what he or she does, not least because his or hers is not only "an abstract need for knowledge but also a concrete need to see it put to use in the best interests of man." The scientist has difficulty understanding "the rights of animals," as P. Singer defined them. Have mice rights with respect to cats, or swallows with respect to falcons? Singer recalls how this problem was already set forth by Epicurus (1960), with the usual clarity of Greek philosophical thought, more than 2000 years ago:

> "XXXI. Right according to nature is the symbol of utility so that harm be neither done nor received.
>
> XXXII. For all those animals that have been unable to define agreements so that they neither receive nor do harm to one another, there is no right or wrong, and the same is true for all those peoples who have not wanted and have not been able to define agreements so as not to receive or do harm.

Why should man be an exception to this rule?

Then the research scientist discovers that there is something true in the feelings of common people, in the lack of faith in science which moves away from nature, in the sense of solidarity with animals. Then, the scientist is also overcome with doubts and his or her certainties cease to be such.

These are moments when one goes back to basics, to the things that really count and that give meaning to all other things. The scientist discovers them in his or her own manner: the first law of life, that of self-defense, which is justifiable only to the extent that it responds to vital necessities and not to whim and the ostentation of one's own strength; the second law of life, that of solidarity, which hardships temper and make all the more indispensable the more they are imposing; and the third law, that of a gradual approach which is the law of patience and of wisdom. C. Linnaeus, inspired by his view of life which bursts forth without the intervention of man, called it *Natura non facit saltus,* but the same holds true for scientific and technological progress. If an embryo spends 9 months — which for it represents an incommensurably longer time than it does for an adult — in the mother's womb without being rejected by it, then human inventions require a period of testing which should be all the longer the more they are earth-shaking. Only in this way, perhaps, will animal experimentation cease to be a deviation from the natural order of things or, conversely, a brutal expression of the latter and become an integral part of it, illuminated by man's rationality.

This is the interpretative key offered by a biologist in all humility to the reader, in the belief that the term bioethics has, in addition to the meaning currently ascribed to it (Reich, 1994), a meaning for science of the ethical principles that result from the study of life.

References Cited

BOMPIANI, A. (1996). *Bioetica: Le Ragioni della Vita e della Scienza (Prometheus, 22)* (F. Angeli, Ed.), pp. 201–229. Milan.

D'Agostino, F. (1996). *Bioetica (Recta Ratio, 1)* (G. G. Ed.), pp. 221–247. Torino.

Dulbecco, R. (1989). *Il Progetto della Vita.* Mondadori, Milan.

Epicurus (1960). *Epicuri ratae Sententiae.* Einaudi, Torino.

French, R. D. (1978). *Animal Experimentations: Historical Aspects. Encyclopedia of Bioethics,* pp. 75–85. Free Press, New York.

National Bioethics Committee (1992a, November 17). *La Sperimentazione dei Farmaci.* Istituto Poligrafico e Zecca dello Stato.

National Bioethics Committee (1992b). Documento sulla sperimentazione animale. CNR, Ufficio Pubblicazioni e Informazioni Scientifiche.

National Bioethics Committee (1998). *La Sperimentazione sugli Animali e la Salute dei Viventi.* Istituto Poligrafico e Zecca dello Stato.

Pasteur, L., and Joubert, J. (1877). Charbonne et septicemie. *C.R. Acad. Sci.* **85,** 101–115.

Reich, W. T. (1994). Il termine "Bioetica." Nascita, provenienza, forza. *Itinerarium* **2,** 51.

Schiller, J. (1967). Claude Bernard and vivisection. *J. History Med.* **22,** 246.

Silvestrini, B. (1997). L'innovazione medica al servizio dell'uomo. *Dolentium Hominum* **34,** 77–80.

ALESSANDRO PALMIERI
ROBERTO PARDOLESI
Libera International University of Social Studies
Rome, Italy

The Patentability of Living Organisms

The finalization and perfecting of techniques for manipulating living matter enable us to glimpse many applications useful to man, but at the same time they give rise to great concern of an ethical nature. The possibility of patenting biotechnological inventions proves to be a formidable stimulating factor for scientific and technological research, attracting considerable human and economic resources to the sector. In principle, the patentability of innovations regarding living organisms is not doubted, but the problem that remains to be solved is that of the limits within which this patentability should be allowed. Whereas the United States and Japan have far more permissive legislation, which allows their enterprises to dominate a continuously expanding market, the countries of Europe have remained behind, held back essentially by motives of an ethical character. Community directives, too, while giving more space for issuing the biotechnological patent, are to a certain extent conditioned by ethical considerations—or, rather, by the persons or groups that act according to given values, dogmas, and metajuridical principles—with regard to the choices made at the political level.

Safeguarding Inventions Concerning Living Matter

The Patent and Incentives for Research

Under the influence of evocative speculations of philosophical and literary type, technological progress can as a first approximation be identified with the expansion of human capacities to modify nature, quite apart from any assessment as to the existence or otherwise of its effective improvement in qualitative terms. This is not the place for delving into the meanders of a barren discussion about the notion of progress; however, there is no doubt that this past century, characterized by a veritable boom in the scientific and the technological field, has witnessed an excessive growth of man's manipulative potentials in every possible direction. Manipulations at the macroscopic level are undoubtedly the most difficult to develop, requiring very large resources of energy. Developing the classification proposed in the 1960s by the Russian astrophysicist Kardeshev, and imagining a ladder at the top of which is the capacity to manipulate the entire universe, Barrow (1997) observed that our civilization is still on the bottom rungs since it is able to significantly alter the earth's environment – mostly with negative effects, such as the thinning of the ozone layer and global warming – and possesses the technology necessary to affect the evolution of other planets within the solar system but is brought to a halt before the hypothetical and science-fictional restructuring of galaxies and even vaster portions of the universe. There is a different situation at the microscopic level, at which significant steps forward have been taken along a path whose ultimate goal is control of the very structure of space and time. On the basis of our current technical capacities, the phase in which man was content to use and to modify only objects of his own order of magnitude, in order to cater for his needs, may be said to have been outstripped. A start has been made on the manipulation of genes and molecules, and even molecular bonds, and we are on the threshold of acting on single atoms and on the most elementary particles of matter.

This development of knowledge and of operative capabilities at the infinitely small level triggers a bitter conflict between positions that represent a great many collective and individual interests. It is here that the law, with its apparatus of rules aimed at settling conflicts and at channeling them toward more satisfactory solutions for mankind, comes — or should come — into play.

The complexity of the problems to be resolved is immediately perceptible when we consider that the acquisitions made by relatively young scientific disciplines, such as biochemistry and molecular biology, have enabled techniques to be produced for the manipulation of living matter.

All of this, while opening up prospects with great hopes for an improvement in the conditions of life of the entire planet (e.g., the discovery of therapies for serious diseases and the increased food resources made available), at the same time instills and fosters feelings of grave concern of an ethical nature, giving glimpses of disquieting scenarios in which man runs the risk of becoming the object of possible manipulations — an eventuality that becomes increasingly less remote with the advancement of the ambitious international research plan, the Human Genome Project (HGP), aimed at outlining a map of the entire alphabet of the DNA in the chromosomes of the human species.

These tensions, which are bound to grow more acute on a par with the advance of scientific knowledge, reverberate at the legal level, especially in the context of the patenting system. The patent, attributing to the inventor and to his or her heirs and assigning the exclusive right to exploit the invention for a certain period of time (at least 20 years, according to the stipulations of the most recent international agreements), is the preferred instrument (although not the only one conceptually possible) for providing incentives for scientific and technological research, to which states and the supranational bodies attribute a foremost role (e.g., Article 9 of the Italian Constitution and Title XV of the EC Treaty).

Toward an Extension of Exclusive Rights: The TRIP Agreement

Instances favorable to an ever-increasing use of the patent in relation to living matter are in perfect keeping with the trend characterizing the most recent evolutionary phase of the discipline of the rights of what are termed intangibles. We are in fact moving rapidly toward extending protection both with regard to the length of exclusive rights and from the even more disruptive standpoint of the entry into the patentability area of categories of inventions that the classical model did not take into consideration (Ghidini, 1995). These developments have been punctually reflected at competent international venues. Of particular importance, as far as the countries adhering to the General Agreement on Tariffs and Trade are concerned, was the adoption at Marrakech on April 15, 1994, at the conclusion of the Uruguay Round negotiations of the final act embodying the results of the Uruguay Round of multilateral trade negotiations, which includes the TRIP agreement (Agreement on Trade Related Aspects of Intellectual Property Rights, including trade in counterfeit goods), which fixes fairly high minimum standards of protection.

The impetus to move in the direction indicated stems essentially from the peak sectors of contemporary industry, viz. those of information technology and of biotechnology in which, as opposed to the massive investments lavished on them, the results are obtained gradually, normally in the form of incremental innovations and, less frequently, of revolutionary innovations with respect to existing technology.

Inserted in this process is the frequent search for hybrid forms of protection (e.g., the law *sui generis* foreseen regarding data banks) so as to exploit the advantages connected with the patent together with those of copyright, or the outright use of copyright, as occurs for software. Even for biotechnological innovations, the hope has been expressed, especially in America, that there will be fuller and broader copyright protection. This prospect, however, is opposed by various arguments, especially that which holds that the representation of the DNA or of a protein by means of writing or by a three-dimensional model cannot be likened to an original literary composition or a design (Stanley and Ince, 1997).

In this context, the arduous task of defining the area of protection of inventions concerning living matter, specifying the amount of protection and the respective limits, falls precisely to the discipline of patents. Also, the question of these limits seems to be the crux of the whole issue, given that the basic prejudice that biotechnological inventions, due to their peculiar nature, do not possess the requisites for obtaining protection by patent has generally gone by the board.

Protection of Biotechnological Inventions: Different Legislative Scenarios

Examining international legislation, it can be observed that, at least in the economically most developed nations, there is no absolute preclusion to recognizing the patentability of certain inventions correlative to living organisms. However, there is a clear-cut divergence between the various codes. Some countries, such as the United States and Japan, have permissive legislation, which allows their local industries to make substantial investments and to reap huge profits, dominating the world market. If one glances through the list of patent classes contained in the *Manual of Classification* of the U.S. Patent and Trademark Office (PTO), one will see, within the chemicals group or the biochemical subgroup, many categories of patents whose titles are more than eloquent regarding the extending of patent protection to the field of biotechnology: for example, class 800, titled Multicellular Living Organisms and Unmodified Parts Thereof; class 930, Peptide or Protein Sequences; or class 935, Genetic Engineering: Recombinant DNA Technology, Hybrid or Fused Cell Technology, and Manipulations of Nucleic Acids. The great success of the biotechnological patent is confirmed also by quantitative data. In the 5-year period 1990–1995, during which the increase in the total number of patent applications was approximately 2%, in the biotechnological sector there was an increase of 10%.

For their part, the European countries that signed the European Patent Convention (EPC) adopted at Munich in 1973 (which recycled an earlier convention signed at Strasbourg in 1963) are subject to a clause thereof (Article 53) in which the already limited possibility of patenting biotechnologies is an exception to the exception. Whereas, in fact, "plant varieties or animal breeds as also essentially biological procedures for the constitution of plants or of animals" are excluded from being patented, the general principle, sanctioned by Article 52 of the convention on the patentability of new inventions that imply an inventive activity and are suitable to have an industrial application, is reestablished as far as "microbiological procedures" and "products obtained through these procedures" are concerned. Despite the estimates made by the European Union, a serious policy of investments in the sector would enable industrial development to take place such as to guarantee, at the beginning of the third millennium, a turnover of 100 billion Euros, with the creation of at least 200,000 new jobs.

Europe Trying to Catch Up

Europe, left behind through a great deal of legal and extrajudicial resistance, is laboriously striving to fill the considerable gap so as to become competitive again with the North American and Asiatic giants. There is a definite and well-grounded fear that the major beneficiaries of more intense protection of biotechnological inventions would be precisely the non-European competitors since the risk of "colonization" is immanent in the entire area of industrial ownership, especially when one considers that the most conspicuous part of the applications for registration for the European Community trademark, established in accordance with EC regulation No. 94/40, comes from the United States. Nevertheless, there are hopes that the Old Continent can take up the challenge, if and when it can be sufficiently compact. The creation of financial markets of a European dimension can undoubtedly favor the subsidizing of the biotechnological industries. In the opinion of the experts in this sector, once the virtuous circle in which the investors gain profits from the companies and plow these back into them has become established, there will be nothing to prevent European industry from benefiting from the same flow of incoming capital enjoyed by its American counterpart. In any case, the European countries, in taking the steps which by now can no longer be postponed toward the legal regulation of the biotechnological patent, will be able to benefit from the specific cultural *humus* and the different degree of ethical sensibility that characterizes them, provided that they do not overly suffocate the incentives to scientific and technological research (Ricolfi, 1995).

Emblematic of the difficulties encountered in Europe is without doubt the laborious course of the Community directive on the legal protection of biotechnological inventions. The proposal, presented by the Commission on October 20, 1988, after receiving the opinions of the European Economic and Social Committee and the European Parliament, was amended in 1992 to take into account many indications concerning mainly the ethical dimension that had emerged during the parliamentary debate. The procedure foreseen by Article 189 B of the EC treaty was begun, but then the common position drawn up by the Committee of Conciliation was surprisingly rejected by Parliament in March 1995. A new directive was thus proposed, was adopted by the commission on December 13, 1995, and received the favorable vote of the Economic and Social Committee in July 1996. The competent commissions in Parliament went on to examine the proposal, which aroused both positive and negative reactions. The opinion of the Legal Affairs Commission was positive; some perplexity was expressed, regarding the patentability of human genes, by the Environment, Health and Consumer Protection Commission, which suggested that the voices of those interested in developing new therapies should not be wholly ignored; and the judgment of the Agriculture Commission was an extremely negative one, interpreting a widespread opinion that the patentability of living matter has an insurmountable ethical barrier. In the meantime, the commission's group of consultants on the subject of the ethics of biotechnology expressed its opinion in favor of the patentability of the sequences of genes for which a specific use has been determined while stressing the need to comply with moral norms. Finally, on June 18, 1997, the report of the Legal Affairs Commission was approved, with a series of amendments to the original text, and during the session of July 16, 1997, the European Parliament gave its assent, with a large majority, to the proposed directive. It was then up to the commission to vote on the measure, and on August 29, 1997, it adopted the amended proposal of the directive (published in the OGEC of October 11, 1997, C 311/12), accepting almost *en bloc* the amendments voted by Parliament.

Biotechnological Inventions in the Patenting System

The Capacities of Adaptation of the Patenting System

Substantial difficulties are encountered on the road toward the patentability of biotechnological inventions due to the necessity of framing this particular category of inventions according to the consolidated rules governing (for about a century) the system of patents (Floridia, 1997).

It is obvious that the patent system came into being as a result of mechanical inventions in an age in which these represented the archetype of things innovative. Symptomatic in this connection is the list, although it was merely by way of example, contained in Article 2585 of the Italian Civil Code of 1942, in which the object of the patent was described by means of the reference to "a machine, an instrument, a tool or a mechanical device." However, as has been keenly observed in jurisprudence, the setup of the system proved to be very flexible, managing to adapt to the stresses and strains exerted by the various typologies of inventions—first electrical and electronic ones and then chemical and pharmaceutical ones (Di Cataldo, 1985).

The truth is that it has not always and everywhere been a painless process, as demonstrated by the prolonged hostility that many national legal systems have shown to the patentability of medicines and drugs. Recall the case of Italy, in which the impossibility of obtaining a patent for "medicaments," stemming from the legislation of the nineteenth century before the unification of Italy, had been sanctioned by Royal Decree No. 1129 of June 29, 1939, and had put up a tenacious resistance to the numerous attacks based on jurisprudence and case law before capitulating with the important sentence of the Constitutional Court (No. 20) of March 20, 1978, only to perpetuate its negative effects because of questions of intertemporal law, with which the Supreme Court of Cassation has been called on to deal, as has the Council of State.

The Requisites of Patentability: Novelty, Inventive Activity, and Potential Industrial Application

The necessity of verifying the presence of the requisites to which the patentability of an invention is subject (i.e., those summarized in the trio of novelty–inventive activity–potential industrial application, as per the already mentioned Article 52 CBE and also in Article 27.1 of the TRIP agreement (to which corresponds, with a few different shades of meaning, the other trio of novelty–nonobviousness–utility foreseen in section 35 of United States Code 102)—does not appear to be a major obstacle to the recognition of the patenting of inventions pertaining to genetics and bioengineering. Naturally, certain adjustments are necessary and, as also stressed by the shrewdest North American observers, the considerable rapidity of changes in the biotechnological field means that it becomes increasingly difficult to discern clearly the elements considered to be essential, especially the elusive "nonobviousness" (Cubert, 1995).

There is no need to dwell at length on the single problem aspects concerning the notions just mentioned. However, in the sphere of definitions, it is possible to agree on some points. First, novelty means that for the invention to be patentable, it must not be included in the ensemble of knowledge made accessible to the public before the date of depositing the patent application (*the state of the technique*). Second, the verification of inventive activity serves to guarantee that the invention to be patented is not even implicitly included in the common technological patrimony. For this purpose, use is made of the criterion of the average technique—that is, it has to be established whether an expert person in the sector, having recourse to his or her ordinary capacities and in the exercise of his or her normal activity, would be able to achieve that result from the ensemble of notions that are common knowledge. Last is the so-called *potential industrial application* of the invention to be patented, which has been much discussed. According to a minority viewpoint, this requisite means also that the patentable invention must have a character of utility.

Hardly anyone doubts that biotechnological inventions, on a par with inanimate inventions, have the possibility of satisfying these requisites. From the standpoint of the potential industrial application, it has been observed that biotechnological inventions, being concerned with living matter, do not guarantee the repeatability and the reproducibility of the result with constant effects. Nevertheless, this aspect has also been found not to be preclusive for purposes of patentability. As a viable interpretation, the solution of linking the requisite of potential industrial application to a sufficient homogeneity and stability of the essential elements of the invention has been indicated (Caforio, 1995).

The conceptual difficulties relating to the potential industrial application also occur with regard to the sufficient description of the invention. In fact, another basic principle of the patents system is that the application for granting a patent must be accompanied by a sufficiently clear and complete description of the invention so that an expert of the sector may be able to actuate it. This is based on the twofold requirement of permitting a wide-ranging use of the invention once the exclusive rights have ended and of exactly identifying the invention while at the same time describing the technique used by the inventor. In the sector of microbiology, however, there are inventions that imply the use of a microorganism not accessible to the public and which do not lend themselves to being described with the clarity and precision required as a general rule. Nonetheless, this obstacle can be overcome by depositing a culture of the microorganism at a qualified collecting center (see points 28 and 28-bis of the CBE's regulations). In 1977, a treaty was signed at Budapest on the international recognition of the deposit of microorganisms, the aim of which was to avoid the repetition, in every country in which patent protection was requested, of the complex and costly operations linked with depositing. Therefore, the possibility is admitted of depositing with an equipped scientific institution, which acts as an international depositing authority.

Regarding American practice, there has emerged from the decisions of the PTO and of the lawcourts an assessment of the utility of biotechnological inventions essentially anchored to the characteristics of the concrete case. The most frequent cases of refusal to grant a patent occur when the biotechnological invention concerns a therapeutic treatment or includes numerous uses not demonstrated in the application. With regard to the debate on nonobviousness, which in the United States has reached particularly heated tones, a quick account thereof will be given when dealing with the problem of the patentability of genes, for which the way of understanding this requisite takes on decisive importance.

The Difficult Relation between Inventions and Discoveries

A further difficulty with the traditional patents discipline can occur in relation to another key rule, taken up in every system — namely, that which excludes the patentability of discoveries, as also that of scientific theories and mathematical methods. This rule presupposes a clear, well-defined line of demarcation between discovery and invention, both from the ontological point of view — the former representing the identification or the revealing of something that already exists, even if unknown, and the latter implying the creation or the elaboration of something new — and from the chronological standpoint, with the discovery occurring at an earlier moment in time than the invention.

The presumed clarity of this distinction has become blurred recently due to a complex series of considerations. In fact, emphasis has been placed on the contradiction between the choice of admitting in general the patentability of product inventions and that of excluding the patentability of given results achieved by a researcher in revealing something previously unknown since it is a question of results conceptually not unlike those obtained by one who constructs *ex novo* a product or a substance. In an attempt to solve this problem, the conventional nature of the distinction between the intangible nature of the discovery and the tangible nature of the invention has been stressed, with the hope for a unitary recomposition of the two notions, under the sign of the conceptual unity of science and technique. From another perspective, it should be observed that if discovery and invention are both made by the same researcher, this can only be regarded in a unitary fashion, with the consequence that the checking of the requisites necessary for patentability must be referred precisely at the moment of the discovery.

However, the relation outlined between discovery and invention reaches a crisis point precisely in the chemical field and even more so in the biotechnological field (Sena, 1990; Floridia, 1997). It has been stated with authority in this regard that chemical research always takes the form of inventive research. Hence, some propose that the current discipline should be interpreted in the sense that discoveries should be open to protection whenever they are taken into consideration with a view to their subsequent technical application or as a subspecies of industrially applied discoveries. Without engaging in intricate hermeneutic reconstructions, it suffices to note that in this case, too, the system seems characterized by the consideration of the dynamics of the discovery–invention relationship characteristic of the mechanical sector. It therefore seems that a review of this relationship is inevitable, even though it is not clear whether this should be an operation limited exclusively to the biotechnological sphere (with the risk of compromising the unity of the patenting system) or a real revolution of much broader scope. Most likely, under the impetus of these ferments, we are moving toward a restriction of the space reserved for nonpatentable discoveries, confined to the sole cases in which — as has already been asserted in terms of legal interpretation — the idea of solution is merely theoretical and is limited to the abstract intuition of scientific principles or to envisaging potential or latent effects.

The Intensity of Protection of Biotechnological Inventions: The Limits of the Theory of Equivalence

Mention should be made of the definition of the sphere of protection of the patent for biotechnological inventions. If in fact, in order to assess whether the reproduction with some variation of a patent belonging to others is tantamount to counterfeiting, the concept of equivalence were applied noncritically, unacceptable results would be obtained. The theory of equivalence, also developed with reference to mechanical inventions, states the following rule: For the counterfeiting of an industrial invention to exist, it is not necessary for there to be a precise reproduction and application of the inventive idea in all its elements, even accessory or secondary ones, but it suffices for there to be the essential elements and characteristics of the idea, without which that new industrial result which is achieved by the *inventio* would not be obtained.

It should be borne in mind that in biotechnology, even more than in chemistry, there is no univocal relationship between the structure and the function of a compound, and indeed modest structural modifications can lead to enormous functional differences. The use of the parameter of structural identity would lead to an excessive broadening of the protection guaranteed to the holder of the patent. For that matter, the problem has already cropped up in the chemical field regarding the so-called patents with a general formula, which are issued for compounds with a general formula which can be modified in such a way as to obtain further derived compounds whose properties are not yet known. Indiscriminate protection of patents with a general formula would frustrate research activity aimed at identifying new compounds obtainable from the basic formula or of new properties of the compound.

In the search for a satisfactory solution, it might be useful to consider the American experience and to recall the "reverse doctrine of equivalents" (Lemley, 1997). On the basis of this doctrine, protection is afforded to all who make radical improvements with respect to a patented product that has the very same function, or a similar one, in a substantially different way, although the new product is susceptible to being included in the formulation of the original patent. Arguing in this way, radical improvements are favored, to the detriment of the patent holder, and are provided with far greater incentives than would be the case with the application of the "blocking patents rule," which in practice causes a situation of stalemate (of bilateral monopoly), attributing to the holder of the original patent the power to deny to the one who has developed the invention the use of the patented technology and, conversely, to the latter the power to prevent the other from exploiting the improvements made.

One of the examples, which are indeed not numerous, in which the reverse doctrine of equivalents has been followed is in the field of biotechnology, represented by the case of *Scripps Clinic & Research Foundation v. Genentech Inc.*, discussed by the Appeals Court of the Federal Circuit in 1991. In this circumstance in which, on the one hand, there was a holder of the patent concerning a blood coagulant, the factor VIII C, purified from human blood, and, on the other hand, there was a producer of said factor obtained from bacteria through the recombinant DNA technique, the decision of the district court which had ruled that the elements of counterfeiting were present was overturned. Another possibility of application of said rules has been envisaged with the proposal that the patents possibly obtained for the sequences of segments of human genes (*express sequence tags*) should not preclude the subsequent use of the same sequences for the production of recombinant proteins. In this case, recourse to the reverse doctrine would be justified because the value of the improvements made would far outstrip those of the original invention.

The Object of the Biotechnological Patent

Premise: The Unitary Conception of Living Organisms

Generally, studies on the patentability of biotechnological inventions have an internal subdivision of the subject matter into three basic chapters: one dedicated to plants, one to microorganisms, and the third to multicellular animals.

In the contrary sense to this approach, from which in the final analysis all the unjustified disparities of treatment within the context of patentable living matter are stated to stem, it is sought to affirm a unitary vision of living matter on the basis of the acquired awareness that the DNA constitutes the common substratum of all known forms of life (Menesini, 1996).

Considering that the sectors indicated have followed clearly differentiated courses of evolution, both at the level of normative regulation and at the level of practical application, it seems appropriate to give a brief discussion of the problems peculiar to each of the three groups of inventions and then to deal specifically with the problem of the patentability of the DNA.

Forms of Protection of Plant Innovations

Historically, the agricultural sector was the one in which the typical institution of industrial ownership first came into contact with living matter. The exclusion of plant products from the list of objects susceptible to being patented was initially explained not so much by its lack of the character of "industriality" (potential industrial application)—it being clear that this notion must be regarded in such a way as not to exclude applications—in the agricultural or commercial field but rather by the considered impossibility of supplying a detailed description of the invention. Moreover, in 1930, the United States Congress passed the Plant Patent Act, which guaranteed limited protection for plant innovations, offering the possibility of obtaining a special form of patent (plant patent) for almost all plants characterized by asexual reproduction.

The response of other countries, especially those in Europe, came in 1961 in Paris with the stipulation of an international convention, which laid down uniform norms to protect new plant varieties and foresaw the setting up of a new supranational union, the Union pour la Protection des Obtentions Vegetales (UPOV), or International Union for the Protection of New Varieties of Plants, which was assigned tasks of coordination and of stimulus. The objective of the convention was to ensure protection for the inventors of plant varieties which have suitable features to distinguish them clearly from others and are sufficiently homogeneous and stable in their basic essentials. The right of these inventors was extended to the material of reproduction or propagation of the protected variety. In addition, it was sought to safeguard a series of further interests, such as those of the nonspeculative user and of the successive inventor. The protection to be granted to the new plant varieties, at the choice of the member states of the European Union, could alternatively take the form of the patent or of particular exclusive rights. Almost every country chose this last option, using certificates of "obtention." In the international scene, the choice of the Italian lawmaker is isolated because Italy preferred to make reference to the general discipline of patents, for which there are no specific provisions derogating therefrom. A few years after this act, the United States adopted the Plant Variety Protection Act, which did not interfere with the earlier legislation and was targeted exclusively to the protection of plant varieties that multiply by sexual reproduction.

Subsequently, there have been various revisions of the 1961 Paris convention which have profoundly modified the original treatment of the subject. Of particular significance is the turning point reached with the Geneva revision in 1991 allowing for the progressive abandonment of the traditional selection techniques in favor of genetic manipulation procedures. The result was that of strengthening the position of the inventors, both by extending the range of the plant material subject to protection and by enabling the inventor to enforce his or her rights at later times other than those originally foreseen. The ban on double protection was also dropped.

Following the last revision and the European Union's decision to adhere to UPOV, the EC Regulation No. 94/2100 was issued concerning Community rights for plant inventions. The discipline of these rights, defined as the sole form of Community industrial ownership on the subject, basically follows that of the convention.

Regarding the United States, it should be noted that for a long time the PTO refused to grant "ordinary" patents for plants on the assumption that they could be protected by the specific instruments of the "plant patents" and of the "plant variety certificates." This treatment was changed in 1985 with the Hibberd case, following which the way was opened to the issuance of "utility patents" for genetically modified plants (Kjeldgaard and Marsh, 1997).

From Microbiological Procedures to Microorganisms

Article 53, letter b of the Munich Convention admits the granting of European patents for microbiological procedures and for the products obtained by said procedures. In light of this explicit development, there has never been any doubt regarding the patentability of the industrial procedures foreseeing the use of microorganisms. Greater difficulty has been encountered in considering procedures aimed at the production of new microorganisms, and especially microorganisms as products obtained via microbiological procedures, as coming within the terms of the cited provision.

Not dissimilar was the tenor of the debate in the United States, until 1980 when the Supreme Court took an authoritative step with its well-known decision in the case of *Diamond v. Chakrabarty,* coming out in favor of the patentability of a genetically altered bacterium and sanctioning the important principle that living matter may form the subject of a patent when there has been human intervention in its production.

In the ensuing years, the European Patent Office (EPO) followed suit, ushering in revolutionary consequences for the problem of the patentability of living matter. Therefore, currently in the EPO's guidelines, the process by which the microorganism is isolated from the culture may be patented, as may the microorganism as such, having been obtained by a microbiological process, and also the product achieved through the metabolism of a microorganism. The national law courts also had to rule on controversies regarding industrial ownership concerning microorganisms. In Great Britain, the House of Lords, in the case of *Biogen Inc. v. Medeva PLC,* directly addressed the problem of the validity of a patent with a genetically modified microorganism as its object. In Italy, the Milan law court, with a recent order — issued in a proceeding bearing on a controversy regarding the validity of a patent concerning an immunodiagnostic kit and methods and anti-hepatitis C virus vaccine — stated that "it does not seem . . . that any real logical–legal obstacles can be identified to the patentability of the [proceeding and of the] viral product."

The possibilities of obtaining a patent for microorganisms are therefore increasing. While emphasizing that the distinction between biology and microbiology is purely conventional and the relevant confines appear susceptible to further modifications, it should be observed that for patenting purposes, it is the very concept of microorganism that is interpreted extensively. In this way, even cells and cell cultures eventually come under it, in addition to bacteria, viruses, plasmids, and DNA sequences.

The Patentability of Multicellular Animals

Even greater problems are caused by the question of the patentability of the so-called higher animals (Di Cerbo, 1992). Here, the strongest doubts gather, from a standpoint within the patenting discipline and from that of ethical compatibility. Consider the stir aroused in international public opinion by the news and the pictures of two cloned mammals, the Scottish sheep Dolly (Fig. 1), in relation to which

FIGURE 1 Dolly, the first cloned sheep (photo courtesy of Sygma/Neri).

applications have been submitted to obtain the patent and for the registration of the trademark, and the American calf Eugene.

This news, certainly not wholly unexpected among researchers in the sector, took some government authorities almost by surprise (the Italian health minister took the step of issuing an order, several times reissued, temporarily banning "any form of experimentation and measure, however carried out, aimed, even indirectly, at human or animal cloning"), mostly giving rise to negative reactions in the mass media and among opinion leaders.

As opposed to what occurs with microorganisms, Article 53, letter b of the EPC leaves very little margin for the patentability of multicellular animals. The granting of patents for animal breeds, as also for the essentially biological procedures for the constitution of animals, is in fact expressly excluded.

It seems very difficult to pass through the meshes of this provision, despite the proposals of extensive interpretation and of the adjustment thereof. However, the Appeals Commission of the EPO managed to do so. Its decision of October 3, 1990, recognized the patentability of the oncomouse, otherwise known as the Harvard mouse, which is a rodent in whose cells an oncogenic sequence had been inserted. The Appeals Commission did so after the invention had been patented in the United States, Japan, and Australia, whereas the examining division of the EPO had refused to grant the patent on the grounds of Article 53 of the EPC. In this circumstance, the Appeals Commission succeeded in stating, on the one hand, that pursuant to Article 53 not all animals as such are considered excluded from patent protection but just some categories of animals and, on the other hand, that the ban on patents for being contrary to law and order or morality is not intended to apply in the case of a request concerning an animal in which genetic mutations have been brought about if the advantages for the human species that can stem from the invention are of a prevalent character with respect to the sufferings that derive therefrom and with respect to possible environmental risks.

However, it has been sought to circumvent the ban by means of interpretation, focusing on the admissibility of patents for products obtained with microbiological procedures. The term "product" in this perspective could also refer to an animal that has been obtained by operating on its DNA, i.e., by a procedure regarded as typically microbiological.

The situation is very different in the United States, where an orientation favorable to the patenting of multicellular animals has prevailed. If patents included in the subclass Nonhuman Animals are considered, such significant titles as the following can be found: Transgenic Mice Displaying the Amyloid-Forming Pathology of Alzheimer's Disease; Method for Producing an Animal Model for Inflammatory Bowel Disease Including Ulcerative Colitis; and Virus-Resistant Transgenic Mice (Fig. 2). This orientation has

FIGURE 2 Transgenic mice obtained by Philip Leder of Harvard University Medical School (photo courtesy of J. B. Nordell/Neri).

been further strengthened by the approval of Congress of the Biotechnology Process Patent Act in 1995. The American lawmaker intends to make it easier to obtain patent protection for what is defined as the "biotechnological process" to the point of considering together both unicellular and multicellular organisms. The normative definition covers procedures aimed at modifying genetically a unicellular or multicellular organism or at inducing it in another way to express an exogenous nucleotide sequence, to inhibit, eliminate, increase, or alter the expression of an exogenous nucleotide sequence or to express a specific physiological characteristic naturally not associated with that organism; cell fusion procedures generating a cell line that expresses a specific protein; and the methods of using a product obtained with one of the procedures described previously or with a combination of these.

It should also be noted that Article 27.3 of the TRIP agreement attributes only the right, but does not make it obligatory, to exclude plants and animals from being patented, whereas it actually prevents contracting states from prohibiting the granting of patents relating to microorganisms and to nonbiological or microbiological processes for the production of plants and animals.

The amended proposal of the 1997 directive is also intended to increase patent protection for animals, even though compared with the original project a few steps backward have probably been taken. In the latest version of the normative text, the patenting of animal varieties and of essentially biological procedures for the production of animals is excluded, with these last procedures being defined as those based on cross-breeding or selection. On the other hand, whereas in the proposal for the directive drawn up in 1988 and in the amended 1992 proposal the point was made of admitting on principle the patentability of biological material including animals, a formulation was arrived at according to which inventions concerning animals can be patented if their practicability is not technically limited to a particular variety of animal.

The Patentability of DNA

The Last Frontier: Competition to Patent Genes

The last frontier of patentability, especially in the most advanced countries outside of Europe, concerns the DNA, the "spiral of life," which contains the information for the production of proteins. That is, an attempt is being made to obtain patent protection for genes (i.e., specific sequences of DNA bases that are responsible for codifying given proteins). Quite a sensation was caused by the initiative of the National Institutes of Health (NIH), an American governmental body which, a few years after the start of the PGU, submitted a patent application relating to thousands of partial DNA sequences that had been identified in the laboratory but whose specific function was unknown (Eisenberg, 1994). Against this initiative, voices of criticism were raised, both by many prestigious scientists (first and foremost James Watson, one of the discoverers of the structure of DNA, who by way of protest resigned as head of the HGP) and by the biotechnological industries, concerned about the possible establishment of a public monopoly and the creation of obstacles to the patenting of inventions constituting the development of the sequences made known. However, the PTO refused many patent applications, and the new directors of the NIH decided not to appeal and to withdraw the remaining applications.

The Debate on Nonobviousness

At a more general level, the patentability of DNA has come up against the barrier of nonobviousness. Verifying the existence of this requisite has proved to be the key for the granting of patents in the whole area of biotechnological inventions (Cubert, 1995). The frenzied development connected with biotechnology has inevitably meant a corresponding increase in the likelihood that an invention, even if new, is considered obvious. Furthermore, the difficult balancing act between maximizing the social benefit that derives from spreading knowledge of an invention and minimizing the exclusive rights attributed to the inventor goes by way of nonobviousness. The rules laid down in 1996 by the Supreme Court in the leading case of *Graham v. the John Deere Company,* in order to ascertain the obviousness of an invention, very soon proved unsatisfactory in the field of biotechnology, especially with regard to process patents, given that the procedures or processes used often depend on standard methodologies applied to new problems. On this a lively debate ensued, counterpointed by various negative decisions by the PTO based on a series of legal precedents, foremost among which was the decision of the Appeals Court of the Federal Circuit in the Durden case (Zhe Zang, 1997). It was established that although the original material and the final product might be new and nonobvious, an otherwise obvious process was in any case not patentable. These decisions have definitely had a negative impact on the biotechnology industry, which feared the affirmation of a *per se obvious rule.* It has been sought to soothe these fears by legislative means with the approval, at the end of a parliamentary debate lasting 5 years, of the Biotechnology Process Patent Act, which contains the forecast of a per se nonobvious rule, provided the process to be patented uses or produces a new and nonobvious compound and many other conditions are respected.

Although many do not consider a *per se rule* adequate, denouncing the risk of attributing too much market power to the patent holder, its adoption in the biotechnology field is justified in terms of protecting American industries against their cutthroat Japanese competitors. However, outside of the biotechnology field two recent decisions of the Federal Circuit should be mentioned, the Ochiai and Brouwer cases, which have repudiated any sort of per se rule, opting for a case-by-case type of approach (Zhe Zang, 1997).

Regarding the patentability of DNA, the problem is further complicated by the fact that its features differentiate it considerably from the majority of other chemical compounds. In general, if two polymers present a similar structure, it is to be expected that they will behave in a similar manner if used in an analogous manner. Hence, the similarity of structure of the new compound, the patenting of which is requested, means that a prima facie case of obviousness is determined, which places on the applicant the onus of proving the nonobviousness of his or her invention.

The situation of DNA is totally different since even a slight modification in the sequence can give rise to significant functional changes. Consequently, as recognized by the Federal Circuit in the Bell and Dillon cases, the doctrine of structural similarity was not appropriate with reference to the peculiar DNA-codified protein relation, in which relation lies the real value of the patent. These conclusions are contested by those who consider that in this way the balance tips too much in favor of the biotechnology industries, with no test having been drawn up to replace that of structural similarity to define the cases of prima facie obviousness (Varma and Abraham, 1996).

Patents of Genes in Light of the Rent Dissipation Theory

The criteria by which patents are granted for genes are also contested on the basis of the *rent dissipation theory,* which, in contrast to the traditional *reward theory,* justifies the patent system in terms of minimizing the total social costs (Erramouspe, 1996). According to this theory, an invention deserves the protection guaranteed by the patent when the competition to obtain it causes a lesser dispersion of resources than that foreseeable for the improvement of the invention or, in other terms, when the invention launches a strong technological signal, suggesting further improvements. Now, the patents for genes issued by the TPO, extremely remunerative for their holders, have started a veritable gene hunt, with an immense waste of resources because of the overlapping of numerous research initiatives converging on one and the same objective while knowing that only one of them is destined to be rewarded with the patent. In light of the rent dissipation theory, this wastage would be avoided by imposing stricter limits for the granting of gene patents, requiring a demonstration of the novelty and the nonobviousness of the methods for obtaining them, and also of the nonfunctionality of conventional methods, so as to avoid the dispersion of resources due to the homogeneity of the inventors. An alternative solution, even if less effective, could be that of granting patent protection only to processes for obtaining genes, excluding it for genes considered in themselves. Since the process patent provides less extensive protection than a product patent, there would be less competition in the pioneer stage, but there would be competition for the discovery of alternative procedures to identify the genes and to sequence them.

Concern of an Ethical Nature

Public Morality and Law and Order

The indication of concepts of public law and order and morality as limits to the patentability of inventions is contained in Article 53, letter a of the EPC which has been absorbed in the internal laws of the signatory countries of the convention, which in advance were not unaware of these notions. The provision cited, in fact, prohibits the granting of patents for inventions whose publication or whose implementation would be contrary to public law and order or to morality. It states, however, that to consider the implementation of an invention contrary to law and order or to public morality, the sole circumstance that it is prohibited in all the contracting states or in part of them by a legal or administrative order is not enough.

Reference to these concepts is also present, albeit in different terms, in the TRIP agreement. Article 27.2 of this is the result of an evident compromise between the position (supported by the United States, Japan, Switzerland, and the countries of north Europe) of those pressing for objections to patentability not being sanctioned, or at any rate contemplating a reduced number of them, and that (supported by the European Union and by a group of developing countries) of those intending to reaffirm the traditional limit of law and order. Once more, in fact, for the member states of the World Trade Organization, no obligation is foreseen but just the right to exclude those inventions from being patented for which a ban on commercial exploitation in their territory is necessary in order to safeguard public morality and law and order, and in particular to protect the life or the health of man, animals, or plants and to avoid a serious prejudice to the environment.

It is thus recognized that certain inventions, although in principle possessing all the technical requisites to be granted a patent, may nonetheless be excluded from any form of protection in the name of values that have nothing to do with any patenting logic properly so termed. The mentioned concepts of law and order and of morality, even if used for a function different from that which they normally perform in civil law, always maintain a feature of vagueness and indefiniteness so that they need to be filled with a concrete content when it comes to interpreting and applying them. For this purpose, the EPO's guidelines try to provide further explanatory parameters concerning the granting of patents that would be inconceivable or aberrant to the public. However, these notions also leave ample margins of discretionary powers to the interpreter. These guidelines also point out the eventuality that certain inventions might have more than one form of implementation, some legitimate and others not. The law and order limit was stressed in the case of Plant Genetic Systems/Glutamine Synthetase Inhibitors, examined by the Appeals Commission of the EPO in 1995 when, following the opposition of the Greenpeace ecological association, it was stated that inventions whose exploitation is liable to have serious negative effects on the environment (always given that the environmental threat appears substantial and current) cannot be patented, pursuant to Article 53, letter a of the EPC.

General Clauses and Prohibitions

The attention given to notions of law and order and public morality inevitably leads to addressing problems of an ethical character, which, as already mentioned, occupy a position of great importance in this matter. The fact that the legal institutions considered have a direct and immediate effect not on inanimate objects but on living organisms raises questions of importance, the answers to which are conditioned by philosophical, moral, and religious principles widely held in the various societies. A particular form of applied ethics, commonly called *bioethics,* which is currently experiencing considerable vitality in the wake of the massive use of technologies in the biological field, is attempting to find answers to these difficult questions. In this way, numerous themes that require the making of "tragic" choices are drawn into the focus of bioethics: from fertilization *in vitro* to the transplanting of organs and from euthanasia to genetic engineering. It is clear that bioethics can suggest solutions to legal operators, and in particular to legislative or government organs, which often turn to experts on the subject for advice.

In the biotechnology field, apart from the extremist positions (bordering on obscurantism) which are *a priori* set against the patentability of living organisms in their every form and manifestation, the most delicate ethical questions are posed with reference to animals and especially to human beings. Special mention should be made of plant innovations, regarding which the main problems concern aspects of environmental ethics rather than the genetic modification of plants or their patenting. The risk has in fact been observed that an uncontrolled diffusion of the techniques of genetic manipulation in agriculture could lead to a widening of the gap between the rich countries, which are the possessors of technologies, and the developing countries, whose main resource consists of possessing a great quantity of self-sown plant varieties. The subject is now regulated at the international level by the Convention on Biodiversity, approved by the Conference on the Environment and Development held in Rio de Janeiro in 1992. The contracting states undertook to foster access to genetic resources for rational uses from an ecological standpoint and to facilitate access to the technologies necessary for the conservation and the lasting use of biological diversity.

Ethical demands mostly take the form of appeals to broad-ranging principles, characterized obviously by an intrinsic — and not negligible — margin of indefiniteness. Occasionally, self-responsibility, tolerance, and respect for human dignity are evoked. With particular reference to this last mentioned fundamental principle, it is claimed that not only moral obligations but also legal ones properly so termed stem from it (Ford, 1997). These concern at least the countries belonging to the European Convention on Human Rights, if not the entire human community, since the majority of the convention's provisions may be regarded in the same way as generally accepted provisions of international law and considered as such in the interpretation of every subsequent treaty in accordance with the Vienna Convention of 1969. Parallel to the enunciation of general clauses is the imposing of specific bans, aimed at defining more precisely the limits of the legality of patents concerning biotechnologies and living organisms.

With regard to the European Community directive on the protection of biotechnologies, it should be pointed out that the most conspicuous part of the obstacles encountered along its tortuous path are connected precisely with the need to authorize exclusions relating to given processes and products obtainable with biotechnology. The list of exclusions has thus swollen to include the processes for human cloning, the methods foreseeing the use of human embryos, the procedures for the modification of the genetic identity of animals inflicting suffering on them without any substantial medical benefit, and the animals obtained with these procedures. Moreover, the patentability of the human body is prohibited, as is the mere discovery of one of its elements including the gene sequences, leaving room only for the patenting of elements isolated from the human body or otherwise produced with a technical process.

Ethics of the Invention or Ethics of the Patent?

A considerable part of the preclusions evidenced would disappear if a distinction were drawn between the ethics of the patent and the ethics of technological innovation (Spada, 1996). In this perspective, the bioethical questions relating to human intervention to modify natural processes could be proposed only at moments other than that of actually taking out the patent. The processes of technological innovation, in fact, cover various phases — research, the technical application of the results thereof, the entrepreneurial production of the goods and services that are the outcome of said results, and the production system — and each one of these lends itself to being autonomously examined from the ethical standpoint. The doubts and questions regarding the admissibility of the modification by man of biological dynamics would therefore not constitute specific questions of the biotechnological patent.

In discussing the patentability of biotechnological inventions, the real question to be asked concerns the ethical compatibility of the implementation of the invention under private law. Once this system has been accepted in relation to biotechnologies, those with concerns of an ethical character could only focus their attention either on the upstream phase (research activity) or, and perhaps mainly, on the downstream phase (marketing and use of the invention) with respect to the patent. An example of this *modus operandi* is provided by two directives, No. 90/219/EEC concerning the "confined" use (i.e., in research laboratories or industrial plants) of genetically modified microorganisms and No. 90/220/EEC concerning the deliberate emission into the environment, for both experimental and commercial purposes, of genetically modified organisms. Regarding the marketing of genetically modified products, a complex procedure has been foreseen to arrive at the granting of the necessary authorization. From the date the directive came into effect until the end of 1996, 16 applications for products to be marketed were made and seven products were authorized. Numerous, even bitter clashes were recorded among the member states and between them and the commission (e.g., the veritable battle about transgenic maize), as were other dysfunctions, to the point that a proposal was made to modify directive No. 90/220/EEC. Thus, the European Parliament adopted a resolution on July 15, 1997, in which inevitably the commission is invited to introduce ethical paradigms (as well as economic and social ones) into its decision-making procedures concerning the directive.

The Clash between Ethics and Market

Various conceptions are thus vying with each other regarding the scenario of the patentability of living organisms; to simplify, these may be grouped in two categories. On the one hand are the positions which, in the name of principles of a higher order, aim at stopping or at least limiting scientific and technological development in the field of biotechnology and which intend, therefore, to impose stringent, if not actually suffocating, limits on the use of the patenting instrument, viewing it as a formidable multiplier of capital flowing toward research, with all that stems from that for the improvement of genetic engineering techniques within a shorter time and with greater accuracy. On the opposite front is the market, with the support of a relatively small number of "pure" scientists and a few representatives of the organizations fighting against specific diseases. However, the market, alone or almost, exercises very strong pressure. Its awareness of having found a vein of gold without equal boosts its striving to overturn all barriers in the way of its full commercial exploitation of biotechnological innovations.

In addition, the pure scientists, engaged in their laboratories in increasing the patrimony of knowledge of the mechanisms that govern the biological processes, are also in need of the market for the funds they require for their costly research, especially where state funding is insufficient.

The Jurist's Role and the Conditioning Effect of Ethics

Jurists aspire to penetrate this epoch-making conflict so as to attempt the extremely arduous task of mediation between positions which, because of the strongly felt motivations underlying them, run the risk of easily becoming overly extremist (Alpa, 1993). However, once observed that the juridical discipline of the patent effectively sets no absolute preclusions (having if anything to resolve within itself many surmountable problems of coordination) and always assuming it is not wished to contest the very basis of this discipline (which frankly appears to be a countertrend compared with the extension, also of a geographical nature, made with the TRIP agreement), it is realized that the jurist's role, especially in the European context, tends to be marginalized. It is not a question of a conflict between representatives of vested interests which need to be adjusted on the basis of parameters that can be drawn from the legal system. Those speaking in the name of ethics declare themselves to be inspired by metajuridical principles and criteria, which can hardly be translated into legal rules, if not into bans. Hence, with the important exception of the United States, in which one frequently witnesses the phenomenon of "juridicizing" ethical questions debated with a wealth of arguments in the courts of every level and degree, the confrontation takes place prevalently at the level of drawing up the normative reference framework. Precisely in this phase, bioethics makes its weight felt, also through the institutionalized committees, with the result that in the current legislation in Europe "keep-out" attitudes continue to prevail over "come-in" ones. Also, when the lawmaker gives more or less substantial glimmers of hope that the patent may be granted, the position of ethics again emerges with the creation of special committees, which are specifically attributed advisory functions to the European Parliament but which most likely seem destined to play a role of surveillance — to prevent the meshes of patentability from being excessively enlarged.

The choices that have to be made at the political level are not simple: How can a proper comparison be made between propositions presented as dogmas and evaluations based on cost–benefit analyses? Whereas on the one hand it is claimed that man cannot presume to be the creator of living matter, on the other hand, it must be questioned why activities should not be encouraged which give a concrete hope of defeating diseases that are incurable with the therapeutic means currently available; for example, among the genes sequenced in the HGP are those that, if altered, are considered to be at the origin of Alzheimer's disease and diabetes.

Without doubt, certain images evoked to describe the results of bioengineering research (from headless bodies to hordes of individuals all exactly alike) are liable to arouse feelings of dismay, just as the ghosts of eugenics seem to return, possibly in the updated version of consumer eugenics (will the day come when we will be able to choose the color of the eyes or the sex of an unborn baby?). The question, however, lends itself to consideration from a different angle: If it is thought that the incentives to the manipulation of genes can lead to perfecting a method for the production of an organ usable by surgeons in otherwise impossible transplant operations, the judgment even of the average observer would tend to be one of approval rather than condemnation. Indeed, such a prospect would contribute to a decisive degree toward resolving the equally dramatic ethical problem regarding the scarcity of human organs for transplants. However, we always make use of techniques of manipulating the DNA so that establishing the line of demarcation between what is absolutely prohibited and what is allowable under certain conditions risks becoming an arbitrary operation or the outcome of impositions assimilated without discernment.

References Cited

Alpa, G. (1993). Limiti dell'intervento giuridico. In *Questioni di Bioetica* (S. Rodotà, Ed.). Laterza, Rome.

Barrow, J. D. (1997). Progresso, limiti, confini e frontiere. In *Il Patto col Diavolo* (P. Donghi, Ed.). Laterza, Rome.

Caforio, G. (1995). *Le Invenzioni Biotecnologiche nell'unità del Sistema Brevettuale.* Torino, Giappichelli.

Cubert, J. (1995). U.S. patent policy and biotechnology: Growing pains on the cutting edge. *J. Patent Trademark Office Soc.* **77,** 151.

Di Cataldo, V. (1985). Sistema brevettuale e settori della tecnica. *Riv. Dir. Comm.,* 277.

Di Cerbo, V. (1992). La brevettabilità degli animali. *Foro It.* **5,** 452.

Eisenberg, R. S. (1994). A technology policy perspective on the NIH gene patenting controversy. *Univ. Pittsburgh Law Rev.* **55,** 633.

Erramouspe, M. (1996). Staking patent claims on the human blueprint: Rewards and rent-dissipating races. *UCLA Law Rev.* **43,** 961.

Floridia, G. (1997). Profili giuridici della protezione brevettuale delle invenzioni antivirali. *Riv. Dir. Ind.,* 291.

Ford, R. (1997). The morality of biotech patents: Differing legal obligations in Europe. *EIPR* **6,** 315.

Ghidini, G. (1995). Prospettive "protezioniste" nel diritto industriale. *Riv. Dir. Ind.* **1,** 73.

Kjeldgaard, R. H., and Marsh, D. R. (1997). Recent developments in the patent protection of plant-based technology in the United States. *EIPR* **1,** 116.

Lemley, M. A. (1997). The economics of improvement in intellectual property law. *Texas Law Rev.* **75,** 989.

Menesini, V. (1996). Le invenzioni biotecnologiche fra scoperte scientifiche; Applicazioni industriali; Preoccupazioni bioetiche. *Riv. Dir. Ind.* **1,** 191.

Ricolfi, M. (1995). Bioetica valori e mercato: Il caso del brevetto biotecnologico. *Riv. Trim. Dir. Proc. Civ.* **2,** 627–640.

Sena, G. (1990). La brevettazione delle scoperte e delle invenzioni fondamentali. *Riv. Dir. Ind.* **1,** 316.

Spada, P. (1996). Etica dell'innovazione tecnologica ed etica del brevetto. *Riv. Dir. Privato,* 217.

Stanley, J., and Ince, D. C. (1997). Copyright law in biotechnology: A view from the formalist camp. *EIPR* **3,** 142.

Varma, A., and Abraham, D. (1996). DNA is different: Legal obviousness and the balance between biotech investors and the market. *Harvard J. Law Technol.* **9,** 53.

Zhe Zang, J. (1997). *In re* Ochiai, *in re* Brouwer and the Biotechnology Process Patent Act of 1995: The end of the *Durden* legacy. *IDEA J. Law Technol.* **37,** 405.

General References

Barbuto, M. (1993). Le nuove frontiere della tutela brevettuale: Dalla chimica farmacologica alla bio-tecnologia e alla botanica. *Impresa,* 2563.

Beier, F. K., and Moufang, R. (1994). Patentability of human genes and living organisms. *Riv. Dir. Ind.* **1,** 340.

Buiatti, F., Bernasconi, A., Bontempi, R., Fumagalli, N., Di Girolamo, N., Ginnelli, F., Ricolfi, M., and Violante, L. (1991). La brevettabilità delle biotecnologie: Una riflessione preliminare. *Giur. It.* **4,** 285.

Byk, C. (1997). Bioéthique (Législation, jurisprudence et avis des instances d'éthique). *JCP* **1,** 4035.

Capasso, O. (1995). DNA ricombinante e tutela brevettuale. *Dir. Ind.,* 567.

Carroll, A. E. (1995). Not always the best medicine: Biotechnology and the global impact of U.S. patent law. *Am. Univ. Law Rev.* **44,** 24–33.

Comitato Nazionale per la Bioetica (1993). *Rapporto sulla Brevettabilità degli Organismi Viventi.* Comitato Nazionale per la Bioetica, Rome.

Crespi, R. S. (1997). Recombinant DNA patents in litigation. A comparative study of some EPO and UK national court decisions. *IIC* **28,** 603.

Cripps, Y. (1997). Recombinant DNA technology—A patent case in the House of Lords. *C. Law J.* **2,** 262.

Di Cerbo, V. (1991). Il "topo di Harvard," ovvero la manipolazione genetica di animali, all'esame dell'ufficio europeo dei brevetti. *Foro It.* **4,** 178.

Ghidini, G., and Hassan, S. (1989). Biotecnologie e brevetti. *Contratto Impr.,* 1136.

Ghidini, G., and Hassan, S. (1990). *Biotecnologie, Novità Vegetali e Brevetti.* Guiffré, Milan.

Hettinger, N. (1995). Patenting life: Biotechnology, intellectual property, and environmental ethics. *B. C. Environ. Aff. Law Rev.* **22,** 267.

Karet, I. (1997). Delivering the goods? The House of Lords' decision in *Biogen v. Medeva. EIPR,* **1,** 21.

LIAO, S. A. (1995). Resolving the dilemmas between the patent law and biotechnology: An analysis of three recent biotechnology patent cases. *Santa Clara Comput. High Tech. Law J.* **11,** 229.

LUCASSEN, V. (1990). Brevetto sulla vita. *Riv. Giur. Ambiente* **1,** 55.

MARCHETTI, P., and UBERTAZZI, L. C. (1997). *Commentario breve al Diritto della Concorrenza,* 2nd ed. Cedam, Padova.

MAZZONI, C. M. (Ed.) (1998). *Una Norma Giuridica per la Bioetica.* Bologna.

MCINERNEY (1998). Biotechnology: *Biogen v. Medeva* in the House of Lords. *EIPR* **1,** 14.

MENESINI, V. (Ed.) (1996). *Il Vivente Brevettabile.* Centro Stampa dell'Università, Perugia.

RICOLFI, M. (1993). La brevettabilità della materia vivente: Fra mercato e nuovi diritti. *Giur. It.* **4,** 292.

SGARAMELLA, V. (1995). Brevettare la vita: La protezione della proprietà industriale nelle biotecnologie geniche. *Rass. Dir. Farm.,* 567.

VANZETTI, A. (Ed.) (1995). *I Nuovi Brevetti — Biotecnologie e Invenzioni Chimiche.* Giuffrè, Milan.

WAYLAND KENNET
University of Cambridge, Cambridge, United Kingdom

JOSEPH THORNTON
University of Oxford Oxford, United Kingdom

The Role of the Public Powers

Public powers, national and international, are defined as sources of law, regulation, and judgment and distinguished from professional associations, etc. The exercise of public power depends on public opinion, and a summary account is given of the latter in Western Europe. The effect of religious and philosophical traditions on bioethical law is noted, with a separate note on Islamic law. The distinctions among law, regulation, and advice are discussed. Differing types of institutions provide these in different countries. Divergences in national legal provisions give rise to problems of "medical tourism."

Medically assisted human conception is chosen as the prime exemplar to be followed though the accounts of bioethical law, etc. in different countries. Short subsidiary excursions are occasionally made into certain other areas, namely, human genetic testing, human gene therapy, cloning, patents, family planning, and population screening.

Accounts are given of five intergovernmental organizations which have been active in the bioethical field (United Nations Educational, Scientific, and Cultural Organization, World Health Organization, the European Union, the Council of Europe, and Organization for African Unity), with excerpts from relevant texts they have prepared.

The bulk of the article provides accounts of 32 countries, with mention of their bioethical legal institutions, and including excerpts from relevant legal texts.

Introduction

For the purpose of this article, the *public powers* are defined as national legislatures, national judicatures, international governmental organizations that produce conventions to which states may adhere and declarations which they may sign—when adherence or signature may either require them or persuade them to make domestic law—and international courts of justice.

Public power is exerted not only by law but also by statutory regulation (as opposed to professional regulation), advice (guidelines), and the creation, adaptation, and abolition of institutions. These sometimes intermingle.

In all countries medical and scientific professional associations make their own regulations and/or guidelines about the applications of modern biotechnology to human beings, and in quite a few this is considered to be sufficient so that no laws or government regulations have been made. This situation accounts for the nonappearance of many countries among the entries which follow. Others do not appear for reasons of space (because their arrangements are very similar to those of their neighbors) or because it has simply proved impossible to find out about their arrangements.

Public Opinion

The role of public power in the control of biotechnology as in everything else is an expression, often approximate, of public opinion or public will. Elected legislatures must punctually take account of public opinion; of course, nonelected ones must also take account of public opinion, although if they are lucky they may be able to take a little more time to do so.

Public opinion on biotechnology is exceptionally well-known in Europe because of two enquires in 1993–1995 and in 1996. The first was done for the European Commission by the Institute for Research on Intercultural Cooperation at the Universities of Brabant and Tilburg using data from a Eurobarometer study. It showed complex correlations among perceptions of biotechnology based on the *hope of utility* gradient (the perceived promise of benefit) and the *demand for public control* gradient (the perceived threat of harm) in different countries. The results are displayed in Fig. 1 and throw much light on the long-lasting geographical and linguistic communities of Western Europe. The second enquiry, while confirming that opinion varies strongly from country to country, was able to make a new generalization. Regression analysis of more than 16,000 responses to a massive questionnaire showed that in the public reaction to new developments in biotechnology (*Nature* **387,** June 26, 1997),

> First, usefulness is a precondition of support; second, people seem to accept some risk as long as there is a perception of usefulness and no moral concern; but third, and crucially, moral doubts act as a veto irrespective of people's views on use and risk.

The "risk" mentioned is that of unexpected ill effects.

Philosophical and Religious Traditions

Law, judgments, and regulations made by the authorities proper to their making naturally follow the convictions and thought forms of their makers. These differ according to their religious, philosophical, and social traditions. Thus, for instance, within monotheism there are differences between Judaism, Christianity, and Islam. Within Judaism there are many and complex differences. This is also the case within Christianity, in which there are differences between Roman Catholic, Orthodox, and Protestant. Within Western philosophy in general there are differences between, for instance, thinkers who favor rights and thinkers who favor duties and between those who favor the ethics of various calculi and those who favor the ethics of various convictions.

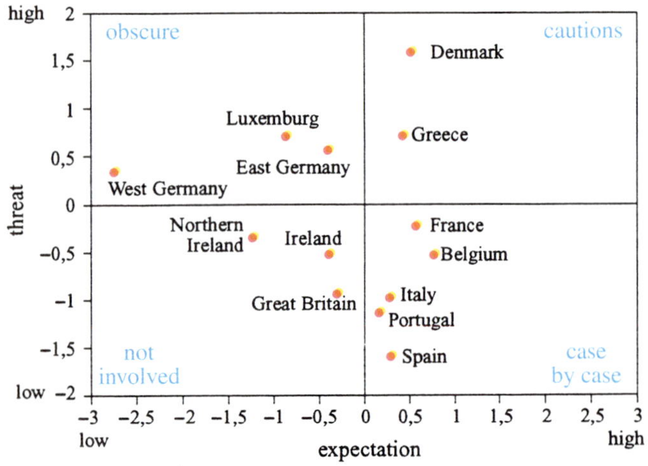

FIGURE 1 Location of each country with respect to their attitude to the factors "threat of harm" and "hope of utility."

Islam is its own world, and since it is so little known to Europe it is treated in a separate section. In Buddhist, Confucian, polytheistic, and diffusely religious societies, there are differences which are more difficult for a Western writer to perceive and express: They are perhaps more readily modified by perceived political and economic necessity (e.g., China).

In many African countries the problem of infertility is not perceived as a medical one but as a sociofamilial one. Thus, if a wife is infertile, the husband may take another wife as well, and if the husband is infertile a secret lover may be arranged for the wife. The former arrangement is of course institutionalized in Islamic polygamy.

This polymorphism of custom and belief conditions the outcome of debate on public intervention in matters which have hitherto been left to private morality, and it accounts for the majority of the differences between and among national legislatures. Such differences can be of great practical significance when a given medical treatment or research procedure is lawful in country A and unlawful in country B. The likelihood then arises that people will travel from B to A for treatment, which gives rise to moral scandal and economic resentment and may culminate in the establishment of a fully fledged branch of tourism: bioethical tourism. This poses the practical question whether it would be morally right or economically advantageous to bring national laws more closely together or, in the case of the European Union, to work toward council directives.

Law, Regulation, and Advice

Law is a collection of injunctions made directly by the legislature of a sovereign state, the failure to comply with which is punishable. A law is an item in this collection.

Regulation normally specifies in detail those actions, actors, subjects, etc. which taken together will bring a given case within a general law. Regulations may be made by a parliament direct, by a government or a single minister, or by a regulator or regulatory body appointed by a parliament or a government. The regulations may also be enforced by such a regulator or regulatory body, when it is empowered to do so. The enforcement will normally be preceded by advice, which, in more fortunate countries, is normally taken.

In some countries a single authority makes rules for and gives advice about all sorts of cases in bioethics; in others, many different authorities are created to make rules for different sorts of cases; for example, assisted reproduction, human xenografts, genetic screening, and gene therapy.

Clinical and research ethics committees, which are well developed in most countries and are of course older than the bodies we are considering here, are not public powers.

At all levels, and between them, there may be institutional provision for the obtaining and offering of advice which is separate from the regulatory power. Especially in the field of bioethics, there has been a sudden growth of such provision in the past 20 years. The advice may be given to government, parliament, the medical and paramedical professions, the research sector, the biotechnology industry, and health and welfare providers – in fact, to any party requesting it, or indeed to a people at large without any request.

Many countries have national ethical or bioethical advisory committees or councils. These are invited to comment on new questions, and some have the right to give advice on their own initiative. Their advice is usually applicable to all levels of public authority and to the exercise of the relevant professions as well.

Executives, and sometimes parliaments, often have special arrangements to keep themselves in touch with professional opinion in general and with biomedical and ethical opinion in particular. Offices of technology assessment have been created by many Western parliaments, and these naturally report to their parliaments on bioethical problems as well as on the social, economic, and ethical problems arising from other technologies. They may undertake public consultations and even direct soundings of public opinion. The two most interesting have been the United States and the Danish ones.

The Scope of This Article

The role of public power in the control of biotechnology is no different in nature, and cannot be, from its role in other spheres. However, the traditional ways of making law about the effects of new technologies have been especially challenged by biotechnology, not only because its development has been very sudden but also because it affects human beings, both present ones and future ones, in ways which go closer to the center of their physical and mental being, their humanity, than any previous development in biological science.

These ways are numerous. Some of them have been found not to require any new law: The problems that they create, if any, can be dealt with by existing law. Others, however, have given rise to social situations that are quite unfamiliar, and these are the ones which in many countries have been found to demand new law.

The subjects of new bioethical law are so numerous and each is so broad that an attempted account of the whole could only be a depthless catalog. Therefore, one subject has been chosen which, with occasional short excursions, seems most likely to throw the most light on the operation of the public powers: human medically assisted conception. This phrase is often loosely used. Here, it means conception toward which medical intervention is made at any point between the appearance of gametes in the bodies of women and men and the implantation of fertilized ova (conception), whether the gametes are those of the intending "parents" or not. This includes surrogacy. In the laws of the world, various phrases are used for these procedures, for various parts of these procedures, and for procedures which include these procedures among others. In what follows, clarity will be sought, but pedantry will be avoided.

This subject was chosen since the practice raises many ethical, social, and economic questions, all of which directly concern legislatures and judiciaries. It produces children who may or may not be permitted to know who their genetic father or mother or siblings are; it may either decrease or occasionally, by mistake, increase the number of genetically handicapped children born; its effects on the fertility of children of people who have themselves been generated by this means are still unknown; and it is expensive. For all these and many other reasons, it is universally attended to and usually legislated by national parliaments because of the creation of new and strange relationships between humans, problems of confidentiality, and also on everyday public health grounds and the grounds of general health economics.

The short excursions referred to previously are all more or less closely related to the main subject. They include genetic testing, gene therapy (both somatic and germline), human cloning, patents, family planning, and population screening.

A Note on Islamic Law

Most Muslim countries run two systems of law concurrently: *Sharia,* the revealed canonical law of Islam, for family and personal relations and for religious practice and a Western or westernoid system for administration and economic activity. The relative importance of each varies from country to country, with Sharia currently at its weightiest in Saudi Arabia, Sudan, and Afghanistan. In countries which are only partly Muslim (e.g., Indonesia), the interpenetration of different legal concepts becomes even more complicated.

Sharia is set forth in the Koran (Fig. 2), the Haddith, and the Sunnah, and it is elaborated by the analytical principles of the four or five traditional schools. According to it, every deed a human being can do is obligatory, recommended, permitted, discouraged, or forbidden. Intention is the critical factor in determining the nature of every act.

Recently, many Islamic countries have experienced a widespread demand for a return to a much wider applica-

FIGURE 2 Title page of the Maghreb Koran (*Vat. Arab.*, 210, c. 75v; courtesy of Vatican Apostolic Library, Rome).

tion of pure Sharia. However, nowhere today, with the exception of Saudi Arabia and some Gulf states, is it the dominant legal system. As the Prophet Mohammed stated, according to the Haddith, "In the beginning, if one omits a tenth of the law one will be punished, but at the end of time if one accomplishes a tenth of the law one will be saved."

The basic religious statement on the subject of human reproduction comes from the Koran:

> To Allah belongs the Kingdom of the heavens and the earth; he creates what he will; he bestows girl children upon whom he will and boy children upon whom he will; and he twins them, both boys and girls; and he makes whom he will to be barren. Surely he is omniscient and omnipotent. (XLII, p. 49)

It would be difficult to find a more comprehensively apposite statement.

Moreover, Hadith No. 2942 states, "Any woman who brings a stranger into the family can hope nothing from God, nor will he admit her to his Paradise." Hadiths, although some of them are mutually incompatible, are taken in general to be statements of the Prophet, and it is a principle of Islam that a child born out of wedlock can never inherit or claim support from the father's family. Sharia refuses to recognize adoption as creating ties of filiation. The use of donor gametes by any means would therefore create profound problems: Blood relatives would be forced to bequeath land, etc. to the incomer since in Sharia a child born 6 months or more after a marriage is conclusively presumed to be legitimate.

In dealing with bioethical issues, Islam does not speak with one voice. However, it is doubtful if it anywhere accepts artificial insemination by donor (AID) by any means: donor insemination, embryo transfer, surrogacy, etc. On the other hand, especially among the richer Muslims of the world who inhabit a moral country of their own, the law is not always closely followed. Also, in some parts of India, for instance, in comparatively wealthy cities in which a Muslim element is by no means lacking, the proportion of first babies who are boys is as high as 90%.

Medically assisted procreation as between husband and wife is of course accepted in Islam as elsewhere. The following findings of a symposium on Islam and reproduction held in Kuwait in 1982 by the Islamic Organization for Medical Sciences are very clear:

> AIH is acceptable for as long as is necessary. AID — the use of semen other than the husband's — is not permissible and if the husband is infertile beyond cure then infertility should be accepted. Although AID is not adultery, it leads to procreation outside the specific duality of a marriage contract. It is a human right to be the legitimate and real child of one's own father and the fruit of the bond of legitimate marriage. AID also risks the lie of registering the baby as the child of a man who is not the natural father, . . . increase the risk of inadvertent incest, and flout juridical inheritance laws. The crux of the issue is the switch of the concept of legitimacy from God-made to man made.

Adoption in the Western sense is not tolerated because of the lie it tells. One should cheat neither the state registrar nor the child by stating that the child is one's own. Also, surrogate motherhood is absolutely unacceptable to Islam.

The previous excerpts and statements do not constitute interventions of the public powers in the Western sense, but Islam is in many places still trying to live by God's law as Christendom did in the Middle Ages, and to Muslims God is indeed still a public power.

Intergovernmental Organizations

All intergovernmental organizations by definition affect the role of the public powers, and those which affect them in the bioethical field are quite numerous. Here, only five are discussed, two being worldwide and three regional: United Nations Educational, Scientific, and Cultural Organization (UNESCO), the World Health Organization (WHO), the European Union, the Council of Europe, and the Organization for African Unity (OAU). Of these, only the European Union constitutes a direct public power in its own right, but the others are influential enough to constitute formative elements in the operation of public power.

There are also many international and regional associa-

FIGURE 3 Detail of the face of the UNESCO building in Paris (photo courtesy of Sygma/Neri).

tions of scientific and professional people. Some of them are highly influential, but none are public powers.

UNESCO

UNESCO was set up immediately after World War II. It now includes all the states of the world except Singapore and the United States (Fig. 3).

In 1993, UNESCO set up the International Bioethics Committee, composed of relevant experts from many of its member states. In 1997, this committee finished work on a Declaration on Human Rights and the Human Genome, which is open to all countries to sign. This text places itself explicitly in the tradition of the United Nations Declaration of Human Rights of 1948. It contains sections on human dignity and the human genome, the rights of the persons concerned, research on the human genome, conditions for the exercise of scientific activity, solidarity and international cooperation, promotion of the principles set out in the declaration, and implementation of the declaration.

Like its great precursor of 1948, it is best seen as a flag to be waved by legislators, lawyers, and editors in any country of the world in which human rights are threatened in this most intimate of concerns.

European Union

The European Union (formerly European Community) is an international and in part supranational association of west European states set up in the 1950s mainly to cement the recent belligerents so closely together as to preclude any recurrence of European war.

It consists of a council of ministers (who are mandated by their national parliaments, in the sense that in every democracy the actions of the government have to satisfy the parliament), a parliament elected directly by the peoples of the member states (Fig. 4), a powerful executive commission, which executes the decisions of the council of ministers and proposes new ones to it, and a court — the Court of Justice of the European Union or European Court — which hears cases by member governments against one another and against the commission and also between the commission and private-sector companies.

The decisions of the council of ministers, having been made by member governments together, are binding on member states. The resolutions of the directly elected European Parliament are increasingly necessary in the formulation of these decisions, and the European Union is thus a public power in the full sense of the words.

Council of Ministers. The European Union is a major source of funds for scientific and other research which, to qualify, has to be done in several member states cooperatively. In December 1994, it adopted and then funded a program of research and development in biomedicine

FIGURE 4 Reunion of the European Parliament in Strasbourg (photo courtesy Reuter/Neri).

and health, of which Area 5 is devoted to human genome research:

> No research modifying, or seeking to modify, the genetic constitution of human beings by the alteration of germ cells or of any stage of embryo development which may make these alterations hereditary, nor research seeking to replace a nucleus of a cell of an embryo with a nucleus taken from a cell of any person, embryo, or subsequent development of an embryo, known as cloning, will be carried out under this program.

However, the document notes that attempts will be made to develop somatic gene therapy where conditions in Europe justify a targeted effort (e.g., cystic fibrosis). Research will also be carried out for the development of new diagnostic techniques for genetic diseases and of new therapies resulting from the most recent progress in molecular genetics and also for harmonization of genetic data banks. Research should aim to prevent the gap between diagnosis and therapy from becoming even wider.

Research on biomedical ethics will address general standards for the respect of human dignity and the protection of the individual in the context of biomedical research and its clinical applications. Topics to be considered include medically assisted procreation (MAP), including gender selection, preimplantation and prenatal diagnosis, research on embryos, use of fetal ovarian tissue, and postmenopause.

European Parliament. Among other sorts of texts, the parliament passes resolutions. Their effect can be best understood by observing the language in which they are cast. In March 1988it passed a resolution on artificial insemination *in vivo* and *in vitro* that

1. Recognized the value of life and especially the human being's right to protection, and therefore expressed its concern at the waste of embryos which *in vitro* fertilization can entail; also, it hoped that techniques and practices will be employed to eliminate this risk.
2. Called therefore for the number of egg cells fertilized *in vitro* to be limited to the number that can actually be implanted.
3. Considered that embryos should only be frozen to keep them alive where immediate transfer *in utero* is impossible for any reason arising during fertilization.
4. Was convinced that in no circumstances should it be possible to abuse *in vitro* fertilization (IVF) methods to select certain embryos, and therefore called for a prohibition of any form of genetic experiment on embryos outside the womb.
5. Considered that heterologous *in vivo* or *in vitro* insemination is not desirable — this applied to the donation of sperm or egg and frozen storage thereof.
6. Considered that, should this principle not be accepted by a member state, the following conditions should be met: ban on disavowal of paternity, informed consent, and verified sterility.
7. Considered that in general any form of surrogate motherhood should be rejected. The procuring of surrogate mothers for gain should be punishable by law.
8. Believed that the role of legislatures is to lay down the limits of science, whereas the role of ethics committees and professional bodies is purely to translate these limits into practice.

In 1989 a resolution on the ethical and legal problems of genetic engineering expressed concern about the dangers of the new technologies. The parliament

1. Declared its intention of evaluating and assessing the social, economic, ecological, health, ethical, and legal aspects of new developments in the field of genetic engineering at the research and development stage, including possible undesirable and unintentional (social) side effects.
2. Called on the commission to support its ethical and legal assessments and to submit corresponding proposals for Community legislation in the areas within its jurisdiction.
3. Called on the governments and parliaments of the member states to take corresponding action and looked to them to press for parallel initiatives in the international organizations of which their states are members.
4. Reaffirmed the principles of freedom of science and research.
5. Regarded the restraints imposed on the freedom of science and research, arising from the rights of third parties and the society they constitute, as the expression in legal terms of the responsibility assumed by society as a whole for the action of the scientist and for research.
6. Recognized that the rights from which these restraints derive are primarily from the dignity of the individual and the sum of all individuals.
7. Considered that the absolute preconditions for the use of genetic analysis are that (i) genetic analyses and counseling must be designed exclusively for the well-being of those concerned, and this includes informed consent and that the results be communicated on request. This means that a doctor shall not have the right to inform the family of those concerned without their consent; (ii) they must on no account be used for the scientifically dubious and politically unacceptable purpose of "positively improving" the population's gene pool, negatively selecting genetically undesirable characteristics, or laying down "genetic standards"; (iii) the establishing of individual gene maps may only be carried out by a doctor; the forwarding, collection, storage, and evaluation of genetic data by government or private organizations shall be prohibited; and (iv) the development of genetic strategies for the solution of social problems must not be allowed since it would undermine our ability to understand human life as a complex entity which can never be encompassed by any single scientific approach.

8. Called for a statutory ban on the selection of workers by genetic criteria.

9. Considered that insurance companies have no right to demand that genetic testing be carried out before or after the conclusion of an insurance contract or to demand to be informed of the results of past tests, and that genetic analysis should not be made a requirement for the conclusion of an insurance contract.

10. Demanded that genetic analysis in legal proceedings should be admissible only in exceptional circumstances to be determined by the judge alone and in certain limited areas; only those elements of a genome analysis may be used which are relevant to the case and do not allow conclusions to be drawn regarding the genotype as a whole.

11. Considered that gene transfers in human somatic cells are basically an acceptable form of therapy, for which informed consent must be obtained. A statutory list of the hereditary diseases suitable for this treatment was to be drawn up and reviewed periodically.

12. Called for an absolute ban on all experiments designed to reorganize the genetic makeup of humans on an arbitrary basis.

13. Called for legislation prohibiting any gene transfer to human germline cells.

14. Considered that the legal status of the human embryo must be defined to provide unequivocal protection of genetic identity.

15. Considered that the zygote also needs protection and therefore must not be subject to arbitrary experimentation; it is not enough to regulate this problem by means of guidelines within the medical profession.

16. Called for neonatal screening to be restricted to curable diseases and to be voluntary, with refusal not penalized in any way, and for the commercial use of embryos to be a criminal offense.

17. Advocated that human embryos should be frozen only for a limited period and only if they are to be implanted to produce a pregnancy in the woman from whom ova were removed for this purpose.

18. Considered that the only possible response to the possibility of producing humans by cloning and to experiments to that end must be to make them a criminal offense.

19. Called for the following to be made criminal offenses: the generation of viable hybrid embryos with various genomes and using human DNA, the fertilization of a human egg cell with animal sperm or the fertilization of an animal egg cell with human sperm to produce a viable embryo, the transfer of any of these cell combinations or embryos to a woman, and all experiments designed to generate chimeras and hybrids using human and animal genetic material.

In 1995 the parliament voted against a directive that would have allowed the patenting of animals, plants, and human genes. This is the first time that the parliament exercised the new powers given to it under the Treaty of Maastricht. In March 1997, as a result of the public excitement following the revelation by the Roslin Institute of Edinburgh that it had successfully cloned an adult sheep, the parliament, confirming its opposition to the cloning of human embryos, expressed in its resolutions of 1989 and 1993 that there is a need to ensure that the benefits of biotechnology are not lost as a result of sensationalist and alarmist information, and that adequate methods of regulating and policing developments in the field of genetics must be established. The parliament

1. Stressed that each individual has a right to his or her own genetic identity, and that human cloning is, and must continue to be, prohibited.

2. Called for the establishment of a European Union ethics committee which would assess applications of gene technology and monitor developments in this field, and also to call on the commission to submit proposals for the composition and terms of reference of the committee under the procedure set out in Article 189b of the European Community (EC) treaty, while ensuring that it is constituted with full respect for transparency and democratic principles and that all appropriate interested groups are represented.

3. Considered that, in view of the universality of the principles relating to the dignity of the human being, efforts must be made by the European Union, its member states, and the United Nations to promote worldwide agreements to ensure that such principles are applied worldwide.

4. Called on researchers and doctors engaged in research on the human genome to abstain spontaneously from participating in the cloning of human beings until the entry into force of a legally binding ban.

5. Called on the commission to prepare a recommendation concerning bioethics laying down strict limits on its research in accordance with respect for human life and to consider, if necessary, whether action at the EC level is necessary.

6. Propose EC legislation on animal cloning.

The Advisory Group. In 1991 the commission set up a Group of Advisers on the Ethical Implications of Biotechnology, with seven members. By 1987 it had published seven opinions. An opinion of December 1994, responding to a commission request, stated that somatic gene therapy is highly promising, but there is a long way to go. The ethical principles governing it should be, at the very least, as strict as those applying to good clinical practice: informed consent, review by an independent multidisciplinary body: (ethics committee), etc. It notes the tendency toward the creation of national supervisory bodies.

Germline therapy raises considerable and controversial ethical problems. The possible transmission of the modification to future generations raises specific philosophical questions. Therefore, no proposal for clinical experimentation of germline gene therapy on humans is currently even contemplated.

The ethical evaluation of somatic gene therapy protocols requires processes ensuring quality, transparency, and efficiency of this evaluation without introducing any unnecessary delays to the treatment of patients. In addition to local review systems, a national supervisory body is important to evaluate this experimental technology as thoroughly as possible. It would be helpful to standardize evaluation procedures throughout the European Union.

Somatic therapy should initially be confined to diseases for which there is currently no effective treatment. In the evaluation of risks and results, the conclusions should be published to allow public scrutiny and encourage public debate.

In February 1996 the group published an opinion on the ethical aspects of prenatal diagnosis (PND). PND, the group stated, raises ethical concerns regarding both the risk of miscarriage and the continuation or termination of pregnancy if a genetic anomaly is diagnosed. Other ethical issues are raised by the risk of misuse of the technique: eugenic abortion, abortion for the selection of sex or other characteristics for nonmedical reasons, and disclosure and misuse of the genetic information.

However, because of the rapid advance of science they did not list the problems for which PND is suitable, but preferred a "framework, based on general ethical principles"; that is, free and informed consent, equality of access, and careful and nondirective genetic counseling. Confidentiality of the test result is an ethical imperative. Genetic data should not be released except with the free and informed consent of the woman or couple. If the genetic data are relevant to the interests of other family members, the woman or couple should be strongly recommended by the genetic counselor to allow its release to them.

In September 1996 the group issued an opinion on the ethical implications of biotechnology directed toward the formulation of a directive on patenting:

1. Whatever the nature of a biotechnological invention involving elements of human origin, the directive must provide legal foundation for the refusal to grant a patent on an invention insofar as it infringes the rights of the person and the respect of human dignity. Therefore, the consideration of patentability criteria under the usual technical requirements of novelty, inventive step, and industrial application must also take these ethical principles into account.

2. The traditional distinction between discovery (not patentable) and invention (patentable) involves, in the field of biotechnology, a particular ethical dimension. It follows from this distinction that knowledge related to the human body or its elements pertains to scientific discovery and cannot be patented. It has to be clearly specified that the simple knowledge of the complete or partial structure of a gene cannot be patented. No patent can be given on the human body or on its elements. It follows that no remuneration can be made to the person from whom the samples are taken.

3. The ethical principle of the informed and free consent of the person from whom samples are taken must be respected. This principle implies that the information on this person is complete and specific, particularly regarding the potential patent application which could be made for an invention based on its use.

4. The granting of patents for inventions arising from knowledge of a human gene or part of a human gene sequence is acceptable only if, on the one hand, the identification of the function attached to the human gene or part of a human gene sequence allows new possibilities (e.g., the production of new drugs) and, on the other hand, if the intended use of the patent is sufficiently specific and identified.

The European Court of Justice. Since 1969 the European Court has held the protection of basic personal rights to be within its competence. The Single Act of 1986 expressly gave the Union competence in consumer protection, and the Treaty of the European Union of 1992 did so in health, both of which clearly include parts of the sphere of bioethics, but no decision of the court has yet explicitly been based on a bioethical principle. However, a judgment of 1994 (14.4.94. Case No. J.10/93) stated that if the applicant for a job, after receiving full information about the implications, refuses to take an AIDS test which the medical adviser of the potential employer holds necessary to judge his or her suitability for the job, the employer is not bound to employ him or her.

World Health Organization

WHO, which is part of the UN family, is an intergovernmental organization. It is the major channel through which the medical and health skills of the rich world flow to the poor world. It runs major worldwide campaigns for the extirpation of disease and the improvement of public health, holds professional conferences, and does much outreach work in ending ancient injustices, combating modern exploitation, and introducing modern standards where necessary and possible. It also produces advisory guidelines of various kinds, such as the recommendations on artificial reproduction of 1985 (Regional Office for Europe), the recommendation on recent advances in medically assisted conception and the report on the place of IVF in infertility care (both 1990), and the international technical guidelines for biomedical research involving human subjects, produced jointly with the professional Council for International Organizations of Medical Sciences in 1993. It also maintains an indispensable and easily accessible library and register of international documents and national laws.

Council of Europe

The Council of Europe was set up immediately after World War II to associate all the countries of Europe under the UN. This ambition foundered when the Cold War began. Since the end of the Cold War, though, most of the nations of Eastern Europe have joined, including Russia. Like the European Union, it has a council of ministers and a parliamentary assembly; however, the assembly consists of members of national parliaments, and the council remains solely consultative.

The most important result of the Council of Europe's work in the field of bioethics has probably been the convention for the protection of human rights and dignity of the human being with regard to the application of biology and medicine — the Convention on Human Rights and Biomedicine. It was drafted over several years in the assembly and finally passed in September 1996. It was immediately adopted by the council of ministers in November 1996, and it is now open to signature by member states. This document states the following:

1. The interests and welfare of the human being shall prevail over the sole interest of society or science (Art. 2)
2. Any from of discrimination against a person on grounds of his or her genetic heritage is prohibited (Art. 11).
3. Tests which are predictive of genetic disease or which serve either to identify the subject as a carrier of a gene responsible for a disease or to detect a genetic predisposition or susceptibility to a disease may be performed only for health purposes or for scientific research linked to health purposes and subject to appropriate genetic counseling (Art. 12).
4. An intervention seeking to modify the human genome may only be undertaken for preventive, diagnostic, or therapeutic purposes and only if its aim is not to introduce any modification in the genome of any descendants (Art. 13).
5. The use of medically assisted procreation shall not be allowed for the purpose of choosing a future child's sex, except where serious hereditary sex-related disease is to be avoided (Art. 14).
6. Freedom of scientific research (Art. 15).
7. Where the law allows research on embryos *in vitro,* it shall ensure adequate protection of the embryo (Art. 18).
8. The creation of human embryos for research purposes is prohibited (Art. 18.2).
9. Financial gain from body parts is prohibited (Art. 21).

The legal affairs committee of the European Parliament proposed that the European Parliament should give qualified support to the Council of Europe Convention, but the European Parliament rejected this proposal by 197 votes to 192.

The German government reserves its assent to a provision in the convention for a form of substituted judgment that can permit, under certain conditions, research on human beings who are incapable of giving their consent.

Organization for African Unity

In 1996 the OAU adopted a resolution on bioethics which pledges to promote in Africa the inviolability of the genetic heritage of humanity and the supervision of research and the prohibition of selective eugenic practices on embryos, especially those produced as a result of medical procedures offering assistance toward procreation.

National Laws

Argentina

Article 2 of the constitution states that the federal government supports the Roman Catholic religion. As of 1996, there were no laws or regulations governing assisted reproduction, although it has been vigorously debated and various bills have been introduced, many to protect embryos. There is a court decision that until a law is enacted, each treatment or step in treatment in new reproductive technologies requires approval from a judge. The judgment has been challenged.

In 1992 the Ministry of Health and Social Welfare created a National Bioethics Commission. Its mandate includes the establishment and oversight of research protocols. No laws explicitly require a patient's informed consent to treatment, although consent is implicit in some.

Australia

Australia is a federation. Some of the states, to which the constitutional responsibility for regulating health care practice and research belongs, and the commonwealth (federation) itself have been pioneers of legislation on human artificial reproduction.

In 1982 a Medical Research Ethics Committee was established within the commonwealth (i.e., federal) National Health and Medical Research Council, and compliance with its research guidelines became a prerequisite for federal funding. A more general National Bioethics Consultative Committee was established in1988 by a joint meeting of federal and state ministers of health and social welfare to advise ministers. Its reports have been primarily concerned with human reproduction and were intended to stimulate debate. They found little favor with churches, feminists, the disabled, right-to-lifers, and many bioethics groups, and the ministers withdrew their support.

In 1991 the Medical Research Ethics Committee and the National Bioethics Consultative Committee were consolidated into the Australian Health Ethics Committee, which has undertaken wide consultation. The Australian approach is to legislate heavily on medically assisted conception.

Australian Capital Territory. The Artificial Conception Ordinance of 1985 covers artificial insemination and the procedure of transferring into the uterus of a woman an embryo derived from an ovum fertilized outside her body.

A presumption of paternity clause states that when a married woman (or bona fide partner) has, with the consent of her husband, undergone a procedure as a result of which she has become pregnant, her husband shall, for all purposes, be conclusively presumed to be the father of any child born as a result, and if any of the semen used in the procedure was produced by a man other than the husband, that man shall, for all purposes, be conclusively presumed not to be the father of any child born as a result of the pregnancy. The consent of the man is presumed but rebuttable. There is to be a presumption of maternity, even if the ovum used was not the woman's own. The ovum producer shall not be considered to be the mother.

When a procedure is carried out in respect to a woman who is not a married woman or a married woman otherwise than with the consent of her husband, any man who produced semen used in the procedure (not being, in the case of a married woman, her husband) shall, for all purposes, be conclusively presumed not to be the father of any child born as a result of a pregnancy occurring by that procedure.

By an act of 1996, surrogacy is allowed when a woman does not have a womb or cannot sustain a pregnancy. A board of the Canberra Fertility Center must approve each case, and couples receive counseling. The prospective social mother must be less than 40 years old, and the couple must be married and have no more than one child. The legislation does not allow advertisement or payment for surrogacy arrangements.

New South Wales. The Children (Equality of Status) Amendment Act of 1984 states that should the husband or partner later withdraw consent, the donor does not assume paternity.

Under the 1984 Artificial Conception Act, the home environment and the stability of the household in which any such child would live, and also the prospective parents' physical and mental health, age, and likely emotional reaction to artificial conception, must all be taken into account.

The Human Tissue Act of 1993 states that a sperm donor must have a medical certificate stating he is free of AIDS or symptoms thereof and has engaged in no homosexual or other high-risk sexual activity since 1980. It also regulates businesses which deal in sperm.

Queensland. The 1988 Surrogate Parenthood Act states that a person shall not publish or cause to be published any advertisement, statement, notice, or other document that (i) is intended or likely to induce a person to agree to act as a surrogate parent, (ii) seeks or purports to seek a person to agree to act as a surrogate parent, or (iii) states or implies that a person is willing to act as a surrogate parent.

Any payment is illegal and contracts are null and void: There is a maximum penalty of 3 years' imprisonment. The punishment may apply whether or not the act occurred in Queensland.

South Australia. The situation is similar to the previously mentioned conditions under acts of 1984 and 1988. The Transplantation and Anatomy Act of 1985 makes it an offense to knowingly supply false or misleading information in relation to the donation of blood or semen.

The Reproductive Technology Act of 1988 established the South Australian Commission on Reproductive Technology, consisting of 11 members appointed by the governor and representing various medical bodies, a representative of the heads of churches, the law society, and 5 members appointed by the minister. Its functions are to formulate and keep under review a code of ethical practice in the use of artificial fertilization and in research involving experimentation with human reproductive material, to advise the commission on the conditions to be included in licensing artificial fertilization (AF) procedures, to carry out research on the social consequences of AF, to promote research into causes of infertility, to advise the minister on questions of reproductive technology, to promote informed public debate on ethical and social issues arising from reproductive technology, and to cooperate with other such bodies in Australia. The act also states that the welfare of any child is of paramount importance in ethical regulations for AF.

A person must not carry out an AF procedure except under a license granted by the commission. Such license will be subject to a condition preventing the application of AF procedures except for the benefit of married couples in the following circumstances: (i) The husband or wife (or both) appear to be infertile or (ii) there appears to be a risk that a genetic defect would be transmitted to a child conceived naturally. A "married couple" includes two people who are not married but who are cohabiting as husband and wife and who have cohabitated continuously as husband and wife for the immediately preceding 5 years or have, during the past 6 years, aggregated 5 years of cohabitation.

Regarding IVF, a court decision in the case *Pearce v. South Australian Health Commission* made on September 10, 1996. The plaintiff had parted from her husband in 1994 and lived alone. She applied for IVF but was turned down by the administrator of the hospital's IVF program on the ground that she did not qualify under the Reproductive Technology Act. She challenged the application of the act on the ground that it was inconsistent with the (Australian) Sex Discrimination Act, which made it unlawful for a person to discriminate in the provision of goods and services against another person on the ground of the other person's sex, marital status, or pregnancy. The Australian constitution states that when a state law is inconsistent with a law of the commonwealth, the latter shall prevail, and the court held that there was a direct conflict: The South Australian government should remove the inconsistency. South Australia should thus make reproductive technology available to single people.

Victoria. A governmental Standing Review and Advisory Committee, which includes a philosopher, two medical people, two representatives of religious bodies, a lawyer, and a teacher with an interest in community affairs, deals with ethical issues and advises the relevant minister on new techniques, etc. It may make recommendations to the minister at its own will. Surrogate motherhood is comprehensively outlawed, as it is in the states mentioned previously, with the addition that acting as a go-between is punishable by 2 years of prison.

The Family Law Act of 1975, as amended in 1987, provides that when a child is born to a woman as a result of an artificial conception procedure while the woman was married to a man, and when either (i) the procedure was carried out with their consent or(ii) under a prescribed law of the commonwealth or of a state or territory the child is a child of the woman and of the man, then whether or not the child is biologically a child of the woman and of the man, the child is their child. This applies to both the man and the woman and to a situation in which two people live as husband and wife on a bona fide domestic basis.

A law against surrogacy was passed in 1995 instituting a fine of up to $50,000 or up to 4 years in prison. It also bans destructive research on any embryo: This was previously allowed on surplus embryos from IVF procedures.

Brazil

The National Health Council was set up in 1987, and later the National Committee for Ethics in Research was established, which specifies what must be submitted to it and ensures that all research institutions have their own ethics committee to handle informed consent (in modern terms), risk–benefit equations, etc.

In 1996, this committee passed an analytical and hortatory resolution (No. 196) on research involving human subjects. This resolution begins by citing as a basis six relevant international declarations. It grounds itself on the four basic principles of bioethics — autonomy, nonmaleficence, beneficence, and justice — among others and aims at ensuring the rights and duties of the scientific community, the research subjects, and the state. These are further formulated as follows: (i) freely given and informed consent of target individuals and the protection of vulnerable groups and the legally disabled (autonomy); (ii) weighing risks and benefits, both actual and potential, individual and collective (beneficence), making a commitment to maximize benefits and minimize distress and risks; (iii) ensuring that predictable injury will be prevented (nonmaleficence); and (iv) social relevance of the research, with significant advantages to the subjects and minimization of the burden to vulnerable individuals, which guarantees equal consideration of all interests involved and preserves the sociohumanitarian purpose of research (justice and equality). These principles are elaborated in a text of (in the English translation) 18 pages.

Bulgaria

The Family Code of May 1985 states that motherhood is determined through birth, including that of children conceived with the genetic material of another woman (Art. 31).

The 1987 Ordinance of the Ministry of National Health on artificial fertilization of women states the following:

1. Spouses who desire that artificial insemination be performed shall attend a consultation center on sterility and shall establish their identity.
2. When insemination is to be performed using genetic material from a third party, the persons referred to in previously shall be required to sign a declaration.
3. Artificial insemination may not be performed using genetic material from a third party who is consanguineous within the fourth collateral degree with the woman undergoing the insemination.
4. Declarations shall be signed by the person donating the genetic material and by the woman undergoing AI, on the basis of which consanguinity can be excluded.
5. AI shall be authorized if the following conditions are fulfilled: (i) The woman is in a satisfactory state of health, in accordance with the instructions on the procedures governing the prenuptial medical examination, and (ii) the person responsible for the consultation center on sterility has established the presence of ovulation and patency of the fallopian tubes in the woman.
6. AI can be performed using the husband's genetic material if (i) a cervical secretion in the woman prevents the penetration of her husband's spermatozoa, (ii) the husband has a mild degree of reduced fertility, or (iii) one of the spouses is suffering from a disease or pathological condition that renders normal sexual relations or natural insemination impossible.
7. Third-party material can be used if the husband is infertile or has a hereditary condition.
8. Only persons of Bulgarian nationality between 18 and 40 years of age, in good physical and mental health, free from hereditary diseases, and presenting normal results in paraclinical investigations and as regards the indicators of the spermogram, may donate material free of hereditary diseases, AIDS, etc. Husbands need not undergo these tests.

Canada

Canada is a federal state, and both the federal and the provincial parliaments legislate on the use of biotechnology. The federal parliament makes criminal law, and the provincial parliaments and governments deal with health care and most family law.

In 1984 the Medical Research Council (MRC) of Canada established a Standing Committee on Ethics in experimentation, with a working group on research involving human subjects. All MRC funding goes to those observing its guidelines. The National Council on Bioethics in Human Research was established in 1989 by the MRC, National Health and Welfare Canada, and the Royal College of Physicians and Surgeons. A Royal Commission on New Reproductive Technologies, responsible to the privy council and prime minister, was later set up to review all policies, laws, etc. on forthcoming and current reproductive technologies. It reported in 1997 that a permanent National Commission on Reproductive Technologies should be set up, having, through five subcommittees, regulatory responsibilities over the whole field (including research) and a duty to further the prevention of infertility by research and education (through a sixth subcommittee). This should be "an immediate federal priority." It continued:

> A central goal of our recommendations is to enable individual Canadians to make personal decisions about their involvement with the technologies, confident in the knowledge that mechanisms are in place to assess their safety and effectiveness and to consider their ethical, legal, and social implications.

By a unanimous judgment in *Tremblay v. Daigle,* the Supreme Court of Canada found that the fetus does not possess legal status or personality under Quebec civil law.

The Children's Act of 1984 (reproducing part of Uniform Child Status Act proposed by a uniform law conference in Canada in 1980) stated that if a husband or cohabiter consents in advance to his wife's being impregnated with third-party material, then he is legally the father. A donor who is not married or cohabiting with the mother is not legally recognized as the father. Notwithstanding a married or cohabitating man's failure to consent to the insemination or consent to assume the responsibilities of parenthood, he shall be deemed to be the father of the resulting child if he has demonstrated a settled intention to treat the child as his child, unless it is proved that he did not know that the child resulted from artificial insemination.

The 1996 Human Reproductive and Genetics Technologies Act (*Lancet,* June 22, 1996, p. 1758) introduces substantial fines and/or prison terms for those engaged in practices such as buying and selling of ova, sperm, embryos, or commercial surrogacy. Prenatal sex selection for nonmedical reasons is also prohibited, as are several areas of research, including germline genetic alteration, cloning of human embryos, creation of animal–human hybrids, research involving the maturation of sperm or eggs outside the human body, and the creation of embryos for research purposes alone.

Provincial Law. The 1980 Civil Code of Quebec states that when a child has been conceived through artificial insemination, either by the father or with the consent of the spouses by a third person, no action for disavowal or contestation of paternity is admissible. The territories of Yukon, Newfoundland, and Quebec have legislation that expressly protects the interests of donors, recipients, and donor-inseminated children.

China

In the West, we try to help people who cannot have babies. In China, they do the same but then try to stop people with one child from having any more. Since the early 1980s, family-planning and population control programs have been vigorously pursued through education, discussion, and propaganda because of the obvious disadvantages of uncontrolled growth of the largest population in the world (currently 1.2 billion, or one-fifth of the whole population of Earth) living in a land of which less than half is useful and only one-tenth is cultivable. The programs are written into the constitution and are enforced by every sort of counseling and social pressure and, ultimately, by the denial of residence permits to third or later children.

Under the Maternal Health and Infant Care Law of 1995,

> If a man or woman is diagnosed at premarital checkups as having serious hereditary diseases and is considered medically as inappropriate to have children, they will be offered medical advice from the doctors, making them fully aware of the consequences to the children and the family if they do. With the consent of both the man and the woman, long-term contraceptive measures can be adopted or a sterilization operation be performed. They can then get married. However, no marriage may be contracted under Art. 6 of the marriage law of 1981 if the man or the woman is suffering from leprosy or from any other disease which is regarded by medical science as rendering a person unfit for marriage. Thus, if serious hereditary diseases in the fetus are found during prenatal checkups, doctors should responsibly inform the couples of the situation and give medical advice for prevention. They will also advise the couples to make a correct choice voluntarily and responsibly.

This quote is taken from an account of the situation kindly supplied by the Chinese ambassador in London in August 1997, which is easier to read than the official translation of the legal texts.

The general Criminal Law of 1979 was amended by the National People's Congress in March 1997. It is now so inclusive as to be virtually a criminal code. In Section 5, Crimes of Endangering Public Health, Article 336 states in part, "Whoever conducts unauthorized birth control reversal surgery, fake birth control surgery, or takes out birth control devices from the womb shall, when the circumstances are serious, be sentenced to not more than 3 years of fixed-term imprisonment."

These and other such measures have to be considered against the background of rapid and inevitable population increase caused by the arrival of modern medicine in China as elsewhere and by the unprecedented half century of internal peace and rising prosperity which, whatever its effects on individual human rights, Chinese communist governments have been able to achieve. They must also be considered in light of two other facts: (i) China is not the only country in the world with alarming population growth and (ii) it is the country which has most strikingly responded to the worldwide perception that the growth of humanity is one of the principal threats to humanity's own survival.

There is no national bioethics committee in China, but the Academy of Social Sciences, which advises the government, has an active program in bioethics. Various national and regional ad hoc advisory groups also exist. The Ministry of Health has been considering setting up a standing committee on policy issues stemming from technological innovations in health care.

Provincial Law. In 1988 the provincial government of Gansu banned childbirth among the mentally retarded, blaming illness on inherited deficiency. (Much informed Western opinion held the conditions in question were not always genetic in origin but often due to iodine deficiency and poor diet. Steps have now been taken to remedy these defects.) According to provincial officials, most of the province's 260,000 "retarded" people were sterilized. According to the *South China Morning Post* of Hong Kong, during the first year of the law's operation, more than 1000 congenitally retarded people were sterilized.

In the same letter quoted previously, the Chinese ambassador in London wrote,

> The regulation by the Gansu Province is a local one passed by the Provincial People's Congress. Should any contradictions occur after the promulgation of the Law [of 1995] on Maternal and Infant Health Care of China the latter, as a national law, should enjoy precedence over the local regulations.

So much for the law itself. However, it is administered with different strictness in different places. It is more flexible in rural than in urban areas so that in 1990 the fertility rate per woman in urban areas was 1.26 and in the countryside was 2.8. Although the laws apply to everybody, they are less strictly enforced on the national minorities than on the Han people. Between 1982 and 1990, the minorities increased at 3.27 times the rate of the Han population, and they then constituted 8% of the population. The government of China leaves policy in Tibet to the government of the Tibetan Autonomous Region, which is now following a similar course. The policies are not enforced in Tibet, in which the government of the Autonomous Region formulates its own policy.

A Court Decision. In 1988 at an intermediate court in Shanghai, a husband sought divorce on the ground of complete alienation of mutual affection. By mutual agreement he and his wife had resorted to AID to produce a child. However, once the child was born, he was unable to bear the gossip and loss of face occasioned by public knowledge of the circumstances of the birth. The court granted the divorce, gave custody of the child to the wife, and ordered the husband to support both the former wife and the child.

Human Cloning. In May 1997 the Chinese Academy of Sciences discountenanced all research into the cloning of humans, stating that sexual reproduction was fundamental for survival of the species. (Xinhua) Xu Zhihong, the vice president of the academy, said that "banning the use of cloning to copy humans is absolutely necessary to maintain ethical morality which holds together today's society."

Denmark

Denmark is distinguished among the countries of the world by the care and ingenuity which its parliament and governments have devoted to public education in bioethics and by the importance they attach to the opinion of this public. In the wake of the Helsinki II Declaration on Research on Human Subjects in 1978, the Central Scientific–Ethical Committee was created to advise and recommend on that subject.

In 1987 the parliament created a new body, the Danish Council of Ethics. It is composed of 8 members appointed by the Minister of Health and 9 by a parliamentary committee. In 1992 the Central Scientific–Ethical Committee was given statutory authority. The Danish Council of Ethics remained the primary adviser to parliament on ethical problems in the health sciences. The Danish Council of Ethics has 17 members, with a lay majority and equal numbers of men and women, appointed for 3 years, with only one reappointment. It was given a permanent secretariat and sufficient funds. The law of 1987, which set up this council, stated that it should carry out its work on the assumption that human life begins at the moment of fertilization.

Among the duties of the council were advising research ethics committees, health authorities, and other public authorities on new techniques, ethical questions, and the control of information on the hereditary diseases or characteristics of individuals or groups. It also permanently prohibited all research toward the production of genetically identical human beings, of human individuals by fusion of genetically different embryos or parts of embryos before they implant in the uterus, and of living human individuals who are hybrids with a genetic constitution including elements from other species.

In 1991 the parliament, on a proposal of the minister of labor, legislated to prohibit the use of genetic tests in job appointments and in underwriting pensions and insurance. The Danish Council of Ethics, responding, agreed on the need for legislation and with the underlying intention of the bill. It pointed out that it had been possible to analyze personal risk for a long time before the advent of genetic testing, and it urged an intensification of the confidentiality of genetic information.

The law of 1993 on the freezing and donation of human ova allows unfertilized ova to be deep frozen for future transfer to the woman who provided them, for research, and for donation. The maximum storage period for fertilized ova is 1 year. Deep freezing of fertilized human ova is permitted for return to the woman from whom they were taken and for research. The ova must be destroyed after 1 year. They must also be destroyed in the event of the death of one of the parties or of separation or divorce before 1 year has passed from the time of deep freezing. Fertilized human ova may only be thawed and used provided the couple has given written consent before each treatment cycle. Embryo donation is forbidden. The donation of unfertilized ova is allowed.

The work of the Danish Council of Ethics has been on a scale probably not attempted in any other country. There has been a large program of public education, e.g., on the ethical issues involved in the protecting human gametes. The council holds public hearings, finances local debates on TV, distributes educational material to high schools, and produces award-winning films shown on national TV and cinemas and brochures for public libraries, etc.

The Danish Council of Technology has also pioneered the form of public education-cum-consultation known as *consensus conferencing.* In this, a group of citizens is gathered, matched as near as possible to the educational, professional, age, religious, etc. profile of the population as a whole, and is invited to spend two weekends with experts on the subject under discussion to consider the ethical content of the new technology, treatment, or whatever is in question. The first stage is an account of the subject given by the experts, followed by prolonged questioning from the participants. Then, the participants go home for 2 or 3 weeks and think about it and discuss it with friends and family. Then they resume, and on the second weekend they discuss the matter among themselves but with the experts present. The experts are only allowed to intervene to correct errors of fact which may have entered into the discussion. The next phase (and this is traditionally an all-night sitting) is the drafting of a report, and here the experts are not present. A consensus may or may not emerge. The last stage is the presentation of the report to the media, which takes place immediately after the all-night session. It is not surprising, owing to the dramatic and confrontational nature of the last part of this procedure, that its success depends to some extent on the presence at all stages of an experienced chairman or moderator, who never takes sides on any question but may suggest ways of working based on his or her experience of earlier conferences.

Egypt

There is no statute law governing MAP, but two sets of guidelines exist. The first was drawn up from Sharia by a fatwa of the University of Al-Azhar in 1980, which is taken together with the teachings of the Coptic Church of Alexandria. The second was published by the International Islamic Centre for Population Studies and Research and was based on the findings of the First International Conference on Bioethics in Human Reproduction Research in the Muslim World, held in Cairo in 1991.

The Ministry of Health and the Medical Syndicate (professional association) have made regulations based on the guidelines. The guidelines state that

1. Medically assisted conception is only to be performed when there is indication for the procedure and only for married couples during the validity of the marriage contract.
2. The donation of sperm or ova is prohibited.
3. The fertilized ova may be transferred only to the uterus of the woman who produced them and only during the validity of the marriage contract.
4. Cryopreservation of the excess embryos is allowed because it decreases the risk to which the wife is exposed.
5. Selective fetal reduction in cases of multiple pregnancy is allowed with the free informed consent of the couples because it decreases maternal and fetal risk.
6. Therapeutic research on embryos is allowed, but non-therapeutic research is not.

Finland

A permanent National Research Ethics Committee was established in 1991 to advise government. The Ethical Advisory Committee of the National Agency for Welfare and Health was founded in the same year by the Ministry of Social Affairs and Health to promote the understanding of bioethics and to provide advice.

France

The French system of bioethical law is among the most thoughtful and comprehensive in the world. In 1983 the National Consultative Ethics Commission for the Life Sciences and Health Sciences (often abbreviated as National Consultative Bioethics Commission) was set up by parliamentary order. It the oldest such body in Europe, enjoys handsome funding, and is able to hold large consultations with the public, normally for a 2-day session once a year. It has 43 members, including 5 philosophers "belonging to the principal philosophic and spiritual families" who are appointed by the president of the republic, 18 members appointed by the ministers of health and of research, and 16 members appointed by the scientific and research community. It produced 34 reports in its first 10 years.

The following year, three laws were passed. They were integrated into existing codes of law — namely, the civil code, the penal code, the public health code, and the intellectual property code — and also into single laws such as the law on clean industry and the law on computerized databases. Excerpts from two of these three laws are presented here.

Law of July 1994 on Respect for the Human Body

No person may prejudice the integrity of the human species.

Any eugenic practice aimed at organizing the selection of persons is prohibited.

Without prejudice to research on the prevention and treatment of genetic diseases, no alteration may be made to genetic characters with a view to modifying a person's lineal descent.

No payment may be made to a person who participated in experimentation on his body, in the removal of elements from his body, or in the collection of products therefrom.

The human body, its elements, and its products may not be the subject of a right of inheritance.

Any agreement concerning procreation or gestation for the benefit of a third party shall be null and void.

No information may be divulged that reveals the identity of a person who has donated an element or a product of his body, or of a person who has received such an element or product.

The provisions of this chapter are a matter of public order.

The genetic study of the characteristics of a person may be undertaken only for medical purposes or for scientific research, and the prior consent is necessary.

The identification of a person by genetic fingerprinting may only be sought for an inquiry or investigation for a judicial procedure or for medical purposes or scientific research.

The only persons empowered to carry out identification by means of genetic fingerprinting shall be those who have been granted approval under the conditions established by a decree made after consulting the Council of State (now the Constitutional Council).

The law changed the Intellectual Property Code to declare that the following shall not be patentable: (i) inventions whose publication or exploitation would be contrary to public order and morality, the exploitation of such an invention not being considered as such solely because it is prohibited by a legislative or regulatory provision. In this connection, the human body and its elements and products, as well as knowledge of the total or partial structure of a human gene, may not, as such, be the subject of a patent.

The law provided for a new chapter in the penal code under Other Offenses. Under the headings Protection of the Human Species, and Protection of the Human Body, it imposed penalties for infractions of the prohibitions erected in Article 16.

Law No. 94-654 of the same year, on the donation and use of elements and products of the human body, medically assisted procreation, and prenatal diagnosis, was designed to prevent trade in human body parts. Among other things, it provided that preservation and use of tissues and cells of the human body may only be carried out by public health establishments and authorized nonprofit-making bodies. It also provided that the handling and distribution of products derived from cells intended for transplantation, somatic cell therapy, or somatic gene therapy should all be regulated by this law and by subordinate regulations.

For MAP the man and the woman making up the couple must both be living, of procreative age, married, or able to prove they have lived together for at least 2 years, and they must have given their prior consent to embryo transfer or to insemination. The embryo must be conceived with gametes that are derived from at least one of them. Taking into account the state of medical techniques, both members of the couple may decide in writing that fertilization may be attempted in a number of oocytes, which may render necessary the storage of embryos, with a view to fulfilling their wish to become parents within a period of 5 years.

As an exceptional measure, a couple who satisfy the conditions, and for whom MAP without recourse to a third-party donor is unsuccessful, may receive an embryo, subject to approval by the judicial authority and with the written consent of the couple who conceived the embryo. The judge shall investigate the familial, educational, and psychological conditions that the couple is able to offer the unborn child.

The two couples may not know each other's identity, the physician may have access to some nonidentifying information for therapeutic purposes, and there must be no payment. A human embryo may not be conceived or used for commercial or industrial purposes. The *in vitro* conception of human embryos for study, research, or experimentation is prohibited, as is all experimentation on embryos.

As an exceptional measure, the man and the woman making up a couple may agree in writing to studies being carried out on their embryos. The studies must have a medical objective, may not be harmful to the embryo, and may only be conducted after an official opinion has been obtained from a committee set up after consulting the Constitutional Council.

Before MAP, all couples are to be interviewed to discover their motivation, to counsel them about the possibility of adoption, to inform them of the likelihood of success, failure, and discomfort, and to provide them with an overview of the legislation on and regulation of medically assisted procreation. They may not proceed within 1 month of interviews. They may not proceed even then if the physician, after consultation within a multidisciplinary team, considers that they need longer reflection, in the interest of the unborn child.

A third-party donor must be a member of a couple that has procreated and must have the written consent of his or her partner. There must be no mixture of sperm, and the use of gametes from one donor may not intentionally lead to the birth of more than five children. Donation shall be unpaid.

Biological diagnosis carried out on cells removed from the embryo *in vitro* shall only be authorized on an exceptional basis and under the conditions set out hereinafter: The physician must certify that the couple, by reason of their family situation, is very likely to give birth to a child suffering from a particularly serious genetic disease recognized as incurable at the time of diagnosis, written consent of the parents must be obtained, and the diagnosis may have no purpose other than identification of this disorder and the means for its prevention and treatment.

There have also been some interesting court decisions. In 1991 the French Supreme Court decided that all surrogate motherhood contracts are null and void. However, courts seem to allow adoption to take place in these circumstances in the best interests of the child.

In 1992, the Court of Appeal of Poitiers considered the following case: A couple had entered into a plan whereby a surrogate mother would be fertilized from the husband, and the couple would then adopt the resulting child. However, the couple divorced, and the husband refused to consent to his former wife's adopting the baby. The court found in favor of the adoption because the husband had already committed to allowing it, and it also served the best interest of the child.

The same year, the Court of First Instance in Toulouse ruled that a woman whose husband had died would not be allowed to implant into herself frozen embryos created from the gametes of her and her husband while he was living since his consent only applied while he was living. It allowed the embryos to be conserved because legislation might be enacted to resolve the dispute.

Germany

Germany is among the minority of sovereign states in the world which are federations, and it is unique in the rigor with which it protects the sanctity of human life, recognizes the individuality of the fetus, and eschews all possibility of germline therapy. Both these facts stem from its experience before and during World War II. Its federal constitution was imposed by the peace of 1945 in order to discourage the resurgence of nationalism. The bioethical concern stems from the Germans' passionate postwar rejection of the Nazi ideology of racial persecution and extermination, which took, among others, the form of "eugenics."

The federal constitution guarantees freedom of scientific research (as do some others in Europe), unless that research violates the basic rights of others or other constitutional values. The Federal Minister of Health has an Advisory Ethics Committee, and in 1997 there were moves to promote this into a Federal Ethics Commission. In 1989 the Federal Constitutional Court ruled that denying persons born of donated gametes knowledge of the donor's identity is unconstitutional because it denies them knowledge of their own origin. A law of 1989 prohibited substitute motherhood.

In 1990 the Embryo Protection Law prescribed a penalty of 3 years imprisonment for transferring into a woman an unfertilized egg cell produced by another woman; artificially fertilizing an egg cell for any purpose other than bringing about a pregnancy of the woman from whom it originated; transferring, within one treatment cycle, more than three embryos into a woman; fertilizing by gamete intrafallopian transfer more than three eggs within one treatment cycle; fertilizing more egg cells from a woman than may be transferred to her within one treatment cycle; removing an embryo from a woman before completion of implantation in the uterus, with a goal of transferring it to another or of using it for another purpose that is not conducive to its preservation; fertilizing a woman who is prepared to give up her child permanently after birth (thus, a surrogate mother) or transferring a human embryo into her or artificially transferring a human sperm cell into a human egg cell without intending to bring about a pregnancy in the woman from whom the egg cell originated; or attempting to do any of these things. It also imposed penalties on anyone who transfers an artificially created human embryo, or does anything with it except for its own preservation, and for attempting to do these things.

In vitro sex selection is punishable by up to 1 year imprisonment or a fine, unless done for reasons of sex-linked hereditary disease. Up to 3 years imprisonment is prescribed for unauthorized fertilization, embryo transfer, and artificial fertilization after death; attempting to artificially fertilize an ovum without the consent of the woman and the man; attempting to transfer an embryo into a woman without her consent; and fertilizing an egg cell with the sperm of a man after his death.

Up to 5 years imprisonment is imposed for artificially altering the genetic information in a human germline cell (this does not apply when the cell is outside the body, as long as there is no possibility of its being used for fertilization) and for the artificial modification of the genetic information of germline cells that come from a different body (i.e., from a dead conceptus, a living person, or a dead person) if there is any possibility that a germline cell will develop from them or if there is any possibility that they will be transferred to an embryo, fetus, or human being.

Cloning. Any person who artificially causes a human embryo to develop with the same genetic information as another embryo, fetus, living person, or deceased person shall be punished by up to 5 years imprisonment. The same is prescribed for anyone who transfers or attempts to transfer such an embryo into a womb.

A Court Judgment. On October 25, 1991, before the Land High Court at Celle a husband contested the paternity of a child born to his wife after AID to which he had consented. The court found for the husband. German courts have reached differing opinions on this issue. This court followed the reasoning of a ruling of the federal court from 1983. In 1988, however, the Land High Court at Dusseldorf held that a husband's bringing such a suit when he has agreed to his wife's artificial insemination constitutes an abuse of the law.

Hungary

A 1989 ordinance of the Ministry of Health and Social Affairs stated that

> artificial insemination may be performed at the request of any woman living in a state of marriage who is not yet 45 years of age, is in full possession of her physical and intellectual capacities, resides permanently on Hungarian territory, and, according to medical opinion, is in all probability unable to conceive a healthy child by normal means.

In December 1991 a law on the protection of the life of the fetus was passed. There is a national bioethics committee.

India

The Prenatal Diagnostic Techniques (Regulation and Prevention of Misuse) Act of 1994 provides for

> the regulation of the use of prenatal diagnostic techniques for the purpose of detecting genetic or metabolic disorders or chromosomal abnormalities or certain congenital malformations or sex-linked disorders and for the prevention of the misuse of such techniques for the purpose of prenatal sex determination leading to female feticide, and for matters connected there with or incidental thereto.

Article 5(2) provides that "no person conducting prenatal diagnostic procedures shall communicate to the woman concerned or her relatives the sex of the fetus by words, signs, or in any other manner." Before this act, it was estimated by professionals in the field that among the well-off in Bombay about 90% of live firstborns were boys.

The act set up a Central Supervisory Board to regulate the proceedings of appropriate authorities in the union territories, each with its advisory committee.

Indonesia

A 1992 health law allows assisted reproduction as "the last effort" to help a couple have a descendant.

Ireland

In this predominantly Roman catholic country all embryo research is banned and embryos are protected by the constitution, but there is no law about medically assisted conception.

In 1987 the Medical Council and the Department of Health accepted guidelines drawn up by the Institute of Obstetricians and Gynaecologists of the Royal College of Physicians of Ireland, which state that IVF is suitable for about 5% of infertile couples: those in which the woman has a normal uterus and produces healthy ova but has damaged, diseased, or absent fallopian tubes. It is limited to married couples who have been appropriately counseled, understand the method, and give informed consent. Only sperms and eggs from the consenting couple are allowed. In 1994 about 300 women a year were receiving treatment. In no circumstances may IVF be used to create or store embryos for research.

Surrogacy is neither prohibited nor regulated. However, in common law it is clear that an agreement for surrogate motherhood is against public policy and would probably be void in court.

Israel

There is a National Council of Bioethics, responsible to the Minister of Health. Because Israel is a religious state, the task of reconciling scriptural teaching with the possibilities of modern molecular medicine falls largely to this body, the composition of which is designed with that in mind.

The Public Health Regulations (*In vitro* Fertilization) of 1987 state that

1. IVF shall be carried out only in authorized departments of accredited hospitals.
2. Ova may be extracted only for fertilization.
3. Semen may be taken only from the woman's husband or from an anonymous donor through a semen bank.
4. Both husband and wife must consent.
5. An ovum may also be implanted in an unmarried woman.
6. When a woman is widowed between fertilization and reimplantation, 1 year must elapse from the extraction of the ovum before it may be reimplanted in her.
7. No use may be made of a dead woman's ova unless she had consented to donate to another woman.
8. Informed consent must be obtained.
9. Confidentiality must be ensured.

A religious court in Haifa ruled that a Jewish child born as a result of AI can only marry a convert for fear of marrying someone closely related. Also, in 1985 the Supreme Rabbinical Court ruled that even if a husband had agreed to his wife receiving AI with donor sperm, he was exempt from paying for the maintenance of the resulting child.

In 1990 a National Ethics Committee was set up to advise the parliament: It has 44 members and is modeled on the French one. In 1991 it published a report on gene therapy.

Italy

Among the developed countries, Italy has been late in legislating to regulate the new biotechnology in its ethical aspects. In the 1990s Italian freedom from regulation became notorious, and biotourism developed, especially from the rest of Europe. The notoriety was due to a string of records. In 1992, after IVF of another woman's ovum by her own husband's sperm, a 61-year-old Italian woman gave birth to a healthy boy. In 1994 a 62-year-old woman gave birth after IVF. There was a Vatican comment on the latter from the eminent moral theologian Mgr Gino Concetti: "The means by which Mrs. Della Corte's desire was realized is in open conflict with God's plan." This pregnancy was achieved at a fertility clinic in Rome run by Dr. Severino Antinori, who justified his programs by saying that his criteria for treatment were a woman's life expectancy and her physical and mental health. He said that a 63-year-old woman with a life expectancy of between 20 and 25 years was probably the oldest woman he would treat.

Public regulation of medically assisted reproduction is limited to a 1985 circular from the Ministry of Health, which licensed public health facilities to perform only simple cases of IVF — that is, for married couples, using the husband's sperm, and leaving no extra embryos. Private centers remain unregulated.

Many bills to address medically assisted conception and other bioethical concerns have been presented in the parliament, but none has been passed. In the absence of legislation, the medical profession in 1995 adopted its own code of ethics for fertility clinics, attempting to forbid certain practices which drew complaints inside and outside Italy. The code of conduct banned "granny mums" (*mamme nonne*), the selection of sperm for the physical attributes or social status of the sperm donor, surrogate motherhood, and assisted pregnancy for lesbians and single women. Although this code does not have the force of law, doctors who break it risk expulsion or suspension from their professional bodies and would thus be unable to practice.

On the other hand, laws have been passed by the elected regional councils. For instance, in 1993 Friuli/Venezia Giulia imposed compulsory genetic counseling before a couple decided to have a child and prenatal diagnosis in order to detect and prevent diseases of the embryo and fetus due to maternal infection or genetic abnormality, especially in the case of persons, couples, or groups exposed to risk. A similar law has been passed in the Marche.

In the prevailing absence of statute law, the judgments of courts achieve a greater prominence. In November 1991 a public prosecutor appeared before the Court of Appeal of Salerno in Campania to object to the attempt by the wife of a child's father to adopt the child on the ground that it had been born by a surrogate mother. The court decided that the case should be determined on the best interests of the child, which meant that the adoption could proceed.

Japan

In Japan there is no legislation on and no public regulation of medically assisted conception, and there is no legislation on and virtually no public regulation of any other activity raising modern bioethical questions. However, in May 1997 the first step toward government standards was taken: The Ministry of Health and Welfare

> decided to set up an advisory council to discuss adopting the country's first standards of medical ethics . . . in such fields as fertilization and the examination of genes, where the use of advanced medical technology is closely related to issues of ethics and privacy.

Under the Eugenics Protection Law of 1948, abortions for the mother's health or for economic reasons are allowed, but they are not allowed for any other reason. However, it is reported that fetuses with genetic defects are in fact aborted for one or the other of these two reasons. This law reversed statutory population policy by changing the word "increase" to "stabilize and not increase." In 1996 its name was changed to Mother's Body Protection Law.

Ministry of Health and Welfare guidelines (as opposed to standards) now exist on research into gene therapy and on the safety of medicines used in such research. The first cover effectiveness and safety, quality assurance, informed consent, the protection of public health, the selection of subjects, and the information to be given to subjects and prohibit germline alterations. They also establish the grades of researchers, principal researchers, and directors of institutions; define these last; require them to have institutional review boards; require reports to the Minister of Health; prescribe the form of research protocols' demand progress reports; prescribe the way to begin a research action; specify how the director of an institution may request an opinion from the minister and the way the minister must handle such a request; prescribe how to archive results; regulate confidentiality and disclosure; and enjoin education and popularization.

The second give even more detailed injunctions on the materials, handling, processes, analyses, etc. to be used in the manufacture of drugs for gene therapy. The language here is quite directive, and the effect certainly tends toward governmental control of gene therapy.

Japan is a country in which strongly contrasting opinions coexist. Bioethical opinion is largely akin to that in the West, being influenced by the same reaction against the horrors of World War II as dominates German opinion. However, consider the following quotation from a speech by Professor Hyakudai Sakamoto of Nihon University, then the incoming President of the International Association of Bioethics, at a meeting in San Francisco:

> Sakamoto explained that in Asia the "natural" and "artificial" blend into each other, without clear dis-

tinctions. Nonnatural and artificial human activities are ultimately included in nature. Natural and artificial are not contradictory concepts. In Buddhist thought, there is no invariance or eternity: Everything is always changing. The idea of an invariant human identity that persists through all changes in a human being is somewhat foreign to the traditional Asian ethos. Therefore, Sakamoto believes that geneticists should use recombinant DNA technologies for the further evolution of humankind in the interests of human happiness and harmony. When asked whether he meant genetic enhancement, he replied that he meant positive eugenics. Although Dr. Sakamoto, an ethicist, said that most Japanese geneticists do not agree with him about artificial evolution, many of his general statements about Asian bioethics were repeated by other presenters at the conference session on Asian bioethics. Westerners could do well to listen to these voices, and to remember that Western bioethics represents only part of the world.

Kyrgyzstan

A recent law on the protection of health allows the performance of AI and embryo transplantation on a married woman with her spouse's consent and also on an unmarried woman.

Libya

By a law of 1986, AI or IVF and transfer of an embryo to the uterus are prohibited unless absolutely necessary and provided that in both situations the ovum and sperm are obtained from the married couple after obtaining their consent:

Article 18: "It is not permitted to act or intervene in any way to limit child bearing unless the married couple agree, and without jeopardizing the interest of society, or when a medical board decides that it is necessary to avoid congenital malformation or mental defect, or in consideration of a potential danger to the woman's life from pregnancy or child birth."

Article 19: "It is prohibited to carry out abortion or feticide unless it is absolutely necessary to save the mother's life."

Article 33: "A prison sentence shall be passed on any individual who intentionally prevents or attempts to prevent an individual from procreation except in the situations mentioned in Article 18."

An accompanying commentary on Articles 17 and 18 notes that IVF and AI are only permitted in the strictest sense and provided that both the sperm and the ova used are from the married couple. This is in accordance with Islam, in which a strict regard to blood relations exists. Although Islam permits child fostering, it absolutely prohibits adoption for the same reasons.

Luxembourg

In 1988 the government established a National Ethical Consultative Committee for the Life and Health Sciences with 15 members, modeled on the French one. A law of 1990 makes it a criminal offense to be a third party to a surrogacy contract.

The Netherlands

In 1989 the Minister of Health established an interim Central Committee on Medical Research Ethics, and in 1997 the parliament legislated to make this a permanent institution. In the same year, a national Animal Biotechnology Committee was established to advise the Minister for Agriculture, the Management of Nature, and Fisheries. Research on animals is illegal without his license, which he gives only following the advice of the committee. The following is an excerpt from the civil code: "An action for denying paternity shall be dismissed if the husband has consented to an act capable of leading to the conception of the child" (Art. 201/1). IVF treatment centers must be licensed by the Ministry of Welfare, Public Health, and Culture.

The Population Screening Law of 1992 provides that the authorization of the minister is required for many new screening projects: Any application for authorization shall contain a detailed description of (i) the test methods to be applied, (ii) the diseases or risk indicators to be detected, (iii) the population groups to be tested, (iv) the organization of the population screening, and (v) the measures to be taken in order to ensure the quality of the population screening. Detailed descriptions must also be submitted of possible reasons for refusal. Penalties for infraction are specified. This law implies that screening related to untreatable disorders will only be allowed in exceptional circumstances. At the insistence of the Christian Democrat Party, the government expressly stated that abortion will not be regarded as treatment, thereby virtually excluding prenatal screening which goes beyond accepted practice in clinical genetics. In 1993 all forms of commercial surrogate motherhood were banned under the penal code.

The Royal Dutch Medical Association reported in May 1996 that the age limit for women who could have IVF should be raised to 55 years. Women's life expectancy would probably make women capable of rearing the child independently. Whether IVF should be available to women beyond 55 years of age should be dependent on medical research and the balance of risks. However, these recommendations encountered reservations in the parliament.

In 1997 The Netherlands Health Council submitted wide-ranging proposals for tighter regulation of IVF, covering age limits, multiple births, and the storage of frozen embryos, which would require new legislation. These proposals state that in general IVF should be limited to women less than 40 years of age because of the declining success rates after that. For women using ova donated by younger women, where the success rate is higher, the age limit could be raised to 45 years. Over 45 years, women should only be offered IVF in the context of research since the risks are unknown. Even then, a woman should be no older than 48 years of age when she starts treatment. The proposals advise only two embryos at a time. They also urge legal time limits on the storage of embryos and clarification of the rules about who may decide what happens to them.

New Zealand

In 1985 the Department of Justice appointed an Interdepartmental Monitoring Committee on Assisted Reproductive Technologies to which some executive power was accorded. In 1994 a corresponding Ministerial Committee was formed.

The Status of Children Amendment Act of 1987 made it clear that gamete donors were not legal parents, and that for the purposes of MAP marriage was considered to include a living situation approximating marriage.

In 1993 the Minister of Justice appointed a two-person Ministerial Committee to advise on the desirability of a law or laws on bioethics. It made good use of the reports that had been made in other countries, particularly Canada. The committee established certain principles, including the Treaty of Waitangi, which incorporates the partnership between the Maori, the indigenous people of New Zealand, and the European settlers. The committee stated that while the system for accrediting medically assisted procreation clinics in New Zealand, which was run by the Australian Reproductive Technologies Accreditation Committee, was working well, legislation was needed to make it a criminal offense for clinics to operate without accreditation.

The Human Rights Act of 1993 made it unlawful to discriminate on the grounds of disability, sexual orientation, and family status, as well as the existing grounds of age and race. The Ministerial Committee concluded that this act must form the basis of the decision for medically assisted procreation. Professionals working in the field expressed concern that they would not be allowed to withhold treatment from, for example, a lesbian couple, a single person, or a 16- or 55-year-old person.

The longest chapter in the Ministerial Report is devoted to information sharing. It argued for a special code under the Privacy Act to ensure that a gamete donor will be identifiable to the offspring. This new approach to openness is partly in response to the Treaty of Waitangi and thus to the fact that the Maori sense of identity and mode of relating to others depend on the individual's *whakapapa* (lineage or ancestry). For this, individuals must of course know who their ancestors are, and the protection of *whakapapa* requires that all children born as a result of MAP should have this knowledge.

Regarding the question of whether body parts and human tissues should be bought and sold, the committee again argued that all human tissue, including embryos and gametes, has *mana* (Maori word meaning dignity or status), and that consequently commercialization should be forbidden.

Norway

The law of August 1994 on the application of biotechnology in medicine states that "the medical application of biotechnology . . . shall take place in keeping with the principles of respect for human dignity, human rights, and personal integrity and shall be . . . based on the ethical norms embodied in our Western cultural heritage."

Artificial fertilization may only be performed on a woman who is married or who is living with a man in a relationship comparable to marriage and with the informed consent of both partners. The physician shall decide whether to carry out the treatment, making the judgment on a medical and psychosocial assessment of the couple. He shall also select a suitable sperm donor, whose identity shall remain confidential. The sperm donor may not know the identity of the couple or the child. IVF may only take place with the couple's own gametes: Fertilized ova may only be used for implantation into the woman from whom they originate. Research on fertilized ova is prohibited. Embryos may not be stored for more than 3 years (a 1987 act on AF stated not more than 1 year).

Before a prenatal diagnosis takes place, permission has to be sought from the Ministry of Health. Information about the sex of a fetus may only be given in the case of sex-linked disease. Postnatal genetic testing may only be carried out with the permission of the ministry. Genetic testing shall not be carried out on children under the age of 16, unless it is to detect a condition that can be treated to prevent or reduce damage to the child's health. It is prohibited (except for health authorities) to enquire whether genetic testing has been carried out.

Human genetic makeup may only be altered by means of genetic therapy to somatic cells for the treatment or prevention of serious diseases. Biotechnology may only be used in establishments specifically approved for this purpose by the ministry. Also, the government is to appoint a biotechnology board to advise on all matters covered by the act and other questions concerning biotechnology.

Portugal

The National Council of Ethics for the Life Sciences was established in 1991. It has 20 members and is modeled on the French one. It has produced 24 reports.

Medically Assisted Conception. Under the civil code as amended in 1977, a husband who has consented to AI cannot deny his paternity. Under the criminal code of 1982, whoever performs AI on a woman without her consent shall be punished by up to 5 years imprisonment. Criminal procedure shall be initiated by personal complaint only.

A decree law of September 1986 states the conditions for the regulation by the Minister of Health of assisted reproduction, but despite recommendations from the National Council of Ethics, no regulations have yet been introduced. Authorization is not required for homologous artificial procreation using fresh sperm. All embryo research is banned.

Russia

The National Bioethics Committee forms part of the presidium of the Russian Academy of Sciences. There are two cochairmen: an academic philosopher and a vice president of the academy. There is also a national committee on medical bioethics in the Academy of Medical Sciences, headed by an academician.

The basic documents on bioethical law include the Principles of Legislation of the Russian Federation "on maintaining citizen health" of 1993, Ministry of Health Order No. 235 of 1992 on the testing of drugs on healthy volunteers, and the 1992 state law on transplants of human organs and tissue.

In 1996 a federal law came into effect on genetic engineering in other than human organisms which included "insuring ecological security." There is currently no legislation on human genetic engineering. The texts of both the 1992 and the 1996 laws were examined by international experts before adoption, respectively, through the UN Industrial Development Organization and the Organization for Economic Cooperation and Development.

An interdepartmental commission under the Minister of Science and Technology has been set up to make proposals to the government for a new legislative base in bioethics. Among other things, it is considering world experience in bioethics as concerns human rights in genetic testing, confidentiality and general ethics in gene therapy and genetic engineering as a whole, and the results of opinion polls. The Russian representation on the International Bioethics Committee informs the government of developments abroad, and current Russian policy in bioethics will be implemented on the basis of its Declaration on Human Rights and the Human Genome (personal communication, G. Koslov, First Deputy Minister of Science and Technology, 1997).

Part VII of the Principles of Legislation of the Russian Federation, adopted in July 1993 (mentioned previously), provides that every woman who has reached the age of majority and is of childbearing age is entitled to benefit from AI and embryo implantation.

Saudi Arabia

The 1997 annual message of the Saudi King and the Crown Prince to the pilgrims to Mecca included the following words:

> Dear brothers, pilgrims to the Ka'bah: In the sphere of genetic engineering, about which we hear a lot, we said on the occasion of the last pilgrimage that concerned parties in human societies hope to unite their efforts to foil any deviant efforts on the part of those active in the sphere of genetic engineering to tamper with genes, whether in humans or animals. There is also a need to limit the widespread insertion of chemical compounds into food products on account of the great dangers this presents to human health, the effects of which may only appear in the long term.

Sweden

Sweden has a National Council of Medical Ethics, created by statute. It consists of 7 members of parliament, and 11 experts. The presence of legislators is believed to be unique and reflects the Nordic conviction that since government is answerable to the people it should do what it can to respond to their views.

The Swedish law on MAP is like that of most other advanced countries, but the relaxed wording of the relevant section of the Code Relating to Parents, Guardians and Children, as amended in December 1984, is worthy of note:

> Article 6. If insemination has been undertaken on the mother with the consent of her husband or of the man with whom she cohabits under marriage-like conditions, and if in all the circumstances it is probable that the child has been conceived by the insemination, then the person who gave his consent shall be regarded as the father of the child.

Article 4 of this law broke new ground and was much discussed around the world: A child conceived as a result of AI is entitled to be told by the hospital who the donor of the sperm was "provided the child has reached sufficient maturity."

Regulations and general recommendations No. 35 of November 1989 of the National Health and Welfare Board state that in the event of the man's death, his sperm may not be used for extracorporeal fertilization, and that already fertilized ova may not be implanted in a woman.

Law No. 114 of March 1991 states that any investigation of the human genetic heritage that involves analysis of the DNA or RNA of the genes shall be subject to special authorization if it is for a general health examination. Authorization is only to be given if it is to acquire knowledge of serious pathological conditions, etc. National Health Board decisions may be appealed before a court of appeal.

Research procedures on fertilized human oocytes are subject to the consent of the oocyte and sperm donors and are only to be conducted within 14 days of fertilization. They may not be such as would cause heritable defects, and the oocytes are to be destroyed immediately after. Three fertilized oocytes may be stored frozen for 1 year.

Switzerland

The peculiarities of the Swiss system — federal and with multiple referenda — have led to many bioethical questions being regulated in the federal constitution. They appear in

Article 24 of the Federal Constitution, adopted by popular referendum on May 17, 1992. This complex article touches inter alia on genetic engineering, human genetic testing, and reproductive technologies:

1. Man and the environment shall be protected against abuse in the field of procreation and genetic engineering techniques.

2. The confederation shall issue provisions concerning the use of the germ cell and human genetic heritage. It shall thereby endeavor to ensure the protection of human dignity, personality, and the family and shall conform to the following principles: (i) Interventions affecting the genetic heritage of human gametes and embryos shall not be admissible; (ii) the germ cell and genetic heritage of nonhuman species may not be transferred to the human germ cell or fused with it; (iii) the use of methods of assisted procreation shall only be authorized when sterility or the risk of transmission of a serious disease cannot be avoided in any other manner; they shall not be authorized in order to develop certain characteristics in the child or for research purposes. The fertilization of human ova outside the woman's body shall only be authorized under the condition set by law. Only the number of human ova that may be immediately implanted may be developed as far as the embryo stage, outside the woman's body; (iv) embryo donation and all forms of surrogate motherhood shall be prohibited; and (v) the human germ cell heritage and products derived from embryos may not be the subject of commerce.

The meaning of 2(iii) has been debated ever since. In any case, since it is part of the constitution it cannot be implemented directly but stands as a general directive to parliament to formulate legislation.

As an example of cantonal differences, the legislature of the canton of St. Gallen banned AID and GIFT and was challenged before the Federal Supreme Court. The court ruled that the limitation of access to methods of procreative medicine may infringe liberty and privacy rights and thus may be unconstitutional.

The United Kingdom

The United Kingdom has no national ethics or bioethics committee. In 1978 the first test-tube baby in human history was born in northern England. To advise it on the future development of artificially assisted procreation, the British government in 1982 set up a committee under Baroness (Mary) Warnock, which reported in 1984, and a voluntary licensing system with very limited resources was set up which regulated IVF research and clinical practice for the next 6 years.

Legislation followed in the form of the Human Fertility & Embryo Act of 1990, which regulates the following:

1. Fertility treatment using donated ova or sperm or embryos created outside the body (IVF).
2. Storage of ova, sperm. and embryos.
3. Research on human embryos (only if deemed essential and only on embryos less than 14 days old: No cloning and no creation of chimeras are allowed).

Other fertility treatments (e.g., drugs) remain regulated only by the professional bodies.

The act set up a Human Fertilization and Embryology Authority (HFEA) with statutory power to license institutions while at the same time recouping their costs from them and also to regulate procedures. The authority also publishes a code of practice for treatment centers, keeps confidential information on patients, donors, and treatment (people 18 years of age can be told by the HFEA whether they were born by MAP and whether they could be related to someone they want to marry), advises those using the services, and keeps the whole field of fertility treatment and research under review, giving advice to the government as it sees fit.

The law stipulates that before treating a woman centers must also take account of the welfare of any child who may be born or who may be affected as a result of the treatment, and that centers must look for indicators of risk to the future well-being of the child, both social and medical. It also bans sex selection for nonmedical reasons and the use of fetal or cadaveric ovarian tissue for infertility treatment.

Commercial surrogacy as such is illegal, but the payment of "expenses" is permitted. The use of superovulation remains largely unregulated.

In 1991 the government, basically averse from regulation either of private enterprise or of professional freedom but acknowledging the need for a wider view of the whole field of bioethics, favored the setting up by the Nuffield Foundation, a major charitable fund, of the Nuffield Council on Bioethics, which is still active. At the same time, there began the creation of a succession of ad hoc advisory bodies set up by the government but without powers: the Gene Therapy Advisory Committee (which works with local ethical committees), the Human Genetics Advisory Committee, and, in 1997, the Advisory Committee on Genetic Testing and a Committee on Xenografts.

In 1994 articles of the HFEA came into force, which meant that a child born of MAP must be genetically related to at least one of the couple concerned, but made it acceptable for the first time to "falsify" a birth certificate in Britain, putting the name of the social father rather than the physical father on it.

In 1997 an English medical practitioner advertised his willingness to guarantee couples the sex of their child as long as they could afford to go to Italy (Naples), where he could legally do it. He has said that he was persuaded to offer this service by numerous appeals from couples, mainly from the Middle East, desperate for a son. A survey carried out by the Institute of Public Health has shown that failure to have a son is a significant factor in the high percentage of suicides among Asian women in Britain.

The United States

Several factors combine to make the processes of U.S. bioethics law and the politics surrounding them different from those of other countries. Of these, the most important is perhaps the unique history. It was American state laws, beginning in approximately 1907, which first licensed the compulsory sterilization of the "unfit," and the U.S. Supreme Court wrote in 1927 that eugenics by involuntary sterilization of the mentally retarded was constitutionally acceptable. These laws supplied a precedent for the Nazi German laws and other less famous ones, particularly in Scandinavia. However, it was also the United States, acting through nothing grander than a military tribunal, which drew up the Nuremberg Code of 1947, on which generations of national and international law-making throughout the world have been based. This is a dramatic and often unnoticed redemption story.

Another difference between the United States and the rest of the world is the longer experience of the social and ethical problems produced by the recent rapid advances in human biotechnology since most them first occurred or had been foreshadowed in the United States. Other differences include that the United States is a federal entity — hence the custom of making federal disapproval of something, typically research, known by refusing federal funding for it while not prohibiting it — and that the greater wealth of the majority of the people, which leads in this field, as in others, to more audible litigation.

The constitutional position in the United States is formally much the same as that in Canada, Germany, or any other federal union in which the public power operates at two levels, the federal and the state. In this particular federal state, there are five levels of legal compulsion:

1. Constitutional adjudication
2. Federal legislation
3. State legislation
4. Common law judgments
5. Administrative law (regulation)

Federal Structures. Since 1968 the Congress and the executive have been ceaselessly active in debating bioethical problems, legislating and attempting to legislate, assigning and reassigning responsibility, and setting up and abolishing advisory bodies with a generosity which makes it impossible to chronicle here. The tensions between finance, industry, medicine, justice, and morals have been handled with that hectic and joyous activism which defines the culture. The main long-term survivor has been the Ethical, Legal, and Social Issues programs in the Department of Energy, which has responsibilities far wider than do its European equivalents, and in the National Institutes of Health (NIH), among other sites. (Though it was the Department of Energy which also, in the 1950s and 1960s, tested the effects of nuclear radiation on unsuspecting individuals.) Research into human reproduction is among the more closely regulated sectors.

Embryo Research. In 1975 the National Commission for the Protection of Human Subjects of Biomedical and Behavioral Research examined the use of human fetuses in research, and federal guidelines were adopted. It was later discovered that they did not cover the therapeutic use of the human fetus (e.g., in the treatment of Parkinson's disease). An ad hoc panel, the NIH Human Fetal Tissue Transplantation Research Panel, was set up and in 1988 voted 17 to 14 to continue research funding using induced abortions. The finding appears to have been ignored (between 1988 and 1992 there was no federal funding for this purpose), and it was later rescinded by President Clinton. He also banned the creation of embryos for research but allowed the use of those left over from IVF treatments in accordance with state laws.

The state of Louisiana stated that a frozen embryo is a person and must be kept in perpetuity. In Illinois an attorney general stated that a woman who has given rise to a frozen embryo is to be considered pregnant. However, in most states the embryo is in legal limbo. Most states permit the use of tissue from dead fetuses for experimental transplants: Arizona, Arkansas, Illinois, Indiana, Louisiana, New Mexico, Ohio, and Oklahoma do not.

A federal law of 1993 to revise and extend the programs of the NIH stated that the decision on any abortion may not be influenced by the need for material on which to conduct research. Nevertheless, the prevailing atmosphere of freedom has given rise to some spectacular abuses. In the same year, researchers at George Washington University Hospital in Washington, DC, separated cells of blastomeres, obtained from polyspermic embryos that had no chance of implanting in the uterus and would have been discarded. The researchers obtained 48 blastomeres from the 17 polyspermic embryos or, theoretically, 48 new totipotent embryos. The Vatican newspaper termed it a step into a tunnel of madness. It was later revealed in December 1994 that they had done so without the approval of the university's institutional review board. They were disciplined and instructed to destroy their data. One resigned.

Research on human embryos, and thus on human cloning, remains unregulated at the federal level.

Medically Assisted Conception. The principal federal action has been the Uniform Parentage Act of 1984. This is a text prepared by the National Council of Commissioners on State Laws for recommendation to the states for enactment by them. It has been enacted in whole or in part in 15 states.

Section 5 states that if, with the consent of her husband, a wife is inseminated artificially with semen donated by a man not her husband, the husband is treated in law as if he were the natural father of a child thereby conceived. The husband's consent must be in writing and signed by him and his wife. The physician must certify their signatures and the date of the insemination and file the husband's consent with the state health department, where it shall be kept confidential and in a sealed file. However, the physician's failure to do so does not affect the father and child relationship. All papers and records pertaining to the insemination, whether part of the permanent record of a court or of a file held by the supervising physician or elsewhere, are subject to inspection only on an order of court for good cause shown.

Several U.S. states have prohibited the destruction of IVF embryos and/or their creation for research uses. Nevertheless, the pattern of state diversity prevails in many of the major battlegrounds of bioethics, for instance, the degree of public control over insurers' discrimination against applicants on the grounds of their genetic status, the patentability of human genetic material, and the provision of MAP treatments under private corporate insurance schemes.

State laws and public institutions to regulate and oversee bioethical problems and practices are usually single issue. Here, a listing of some of the provisions which are interesting either for their content or for their turn of phrase is presented.

Arkansas. By an act of 1985, as amended by one of 1989, a child born by means of AI to a woman who is unmarried at the time of the birth shall be, for all legal purposes, the child of the woman giving birth, except in the case of a surrogate mother, in which event the child shall be that of (i) the biological father and the woman intended to be the mother if the biological father is married, (ii) the biological father only if unmarried, or (iii) the woman intended to be the mother in cases of a surrogate mother when an anonymous donor's sperm was utilized for AI.

For birth registration purposes, in cases of surrogate mothers, the woman giving birth shall be presumed to be the natural mother and shall be listed as such on the certificate of birth, but a substituted certificate of birth may be issued on orders of a court of competent jurisdiction.

California. Under the California Probate Code, "relatives of the decedent conceived before the decedent's death but born thereafter inherit as if they had been born in the lifetime of the decedent." Read literally, frozen embryos could inherit if they are conceived before death, even though they are implanted after death.

Indiana. In the Human Reproduction Act, the general assembly declares that it is against public policy to enforce any term of a surrogate agreement that requires a surrogate to do any of the following:

1. Provide a gamete to conceive a child
2. Become pregnant
3. Consent to undergo or undergo an abortion
4. Undergo medical or psychological treatment or examination
5. Use a substance or engage in an activity only in accordance with the demands of another person
6. Waive parental rights or duties to a child
7. Terminate care, custody, or control of a child
8. Consent to a stepparent adoption

Louisiana. A husband cannot disavow paternity of a child born through AI if he consented. By an act of 1986, the use of a human ovum fertilized *in vitro* is solely for the support and contribution of the complete development of a human *in utero* implantation. No *in vitro* fertilized human ovum will be farmed or cultured solely for research purposes or any other purposes. The sale of a human ovum, fertilized human ovum, or human embryo is expressly prohibited.

An *in vitro* fertilized human ovum exists as a juridical person until such time as the *in vitro* fertilized ovum is implanted in the womb or at any other time when rights attach to an unborn child in accordance with law.

As a juridical person, the *in vitro* fertilized human ovum shall be given an identification by the medical facility for use within the medical facility which entitles such ovum to sue or be sued. The confidentiality of the *in vitro* fertilization patient shall be maintained.

A viable in vitro fertilized human ovum is a juridical person which shall not be intentionally destroyed by any natural or other juridical person or through actions any such person. An *in vitro* fertilized human ovum that fails to develop further over a 36-h period (except when the embryo is in a state of cryopreservation) is considered nonviable and is not considered a juridical person.

An *in vitro* fertilized human ovum is a juridical person who cannot be owned by the IVF patients who owe it a high duty of care and prudent administration. If the IVF patients renounce, by notorial act, their parental rights for *in utero* implantation of the *in vitro* fertilized human ovum, it shall be available for adoptive implantation in accordance with written procedures of the facility in which it is housed or stored.

Inheritance rights will not flow to the *in vitro* fertilized ovum as a juridical person, unless the *in vitro* fertilized ovum develops into an unborn child that is born in a live birth or at any other time when rights attach to an unborn child in accordance with law. As a juridical person, the embryo or child born as a result of IVF and *in vitro* fertilized ovum donation to another couple does not retain its inheritance rights from the *in vitro* fertilized patients.

New Jersey. After right-to-die cases in the 1970s and 1980s, in 1985 the New Jersey Commission on Legal and Ethical Problems in the Delivery of Health Care (NJ Bioethics Commission) was set up as a permanent legislative commission. The state supreme court decided that surrogate mother contracts can never be specifically enforceable and that women must have the right to change their minds and reassert their maternal rights to rear their children.

New York. In 1995 a New York court ruled that a woman seeking possession of five fertilized IVF embryos in a divorce settlement "has the exclusive right to determine the fate of the zygotes." (A similar case in Israel in 1995 was decided by the supreme court in the man's favor.)

Texas. Under an act of June 1987, insurance benefits for IVF procedures must be provided to the same extent as the benefits provided for other pregnancy-related procedures.

Virginia. By an act of 1991, in the case of a surrogacy that has not been approved by a court, if (i) the surrogate is married, (ii) her husband is a party to the surrogacy contract, and (iii) the surrogate exercises her right to retain custody and parental rights to the resulting child, then the surrogate and her husband are the parents. If neither of the intended parents is a genetic parent of the resulting child, the surrogate is the mother and her husband is the child's father if he is a party to the contract. The intended parents may only obtain parental rights through adoption.

After the signing and filing of the surrogate consent and report form in conformance with the requirements of the act, the intended parents are the parents of the child and the surrogate and her husband, if any, shall not be the parents of the child.

A surrogate, her husband, if any, and prospective intended parents may enter into a written agreement whereby the surrogate may relinquish all her rights and duties as parent of a child conceived through assisted conception, and the intended parents may become the parents of the child.

Ruth Chadwick
Centre for Professional Ethics
University of Central Lancashire
Preston PR1 2HE, United Kingdom

The issues concerning the allocation of health care resources include questions at different levels: whether rationing is inevitable, who should make decisions, and the rationale for consulting the public. The central ethical issues concern the contrast between individual and community and the competing conceptions of justice in allocation. Virtue ethics has also entered the debate, but apparently insoluble value conflicts remain, along with fundamental philosophical questions not only about provision of health care but also about the definition of health.

The Allocation of Resources

Introduction

The issues surrounding the allocation of resources are among the most controversial in bioethics. First, there is a debate about whether or not some form of "rationing" or "priority setting" is inevitable: Some argue that resources are sufficient if waste and ineffective interventions are eliminated, whereas others argue that this only takes us so far. As Light (1997) points out, the amount and kind of rationing affect perceptions of whether or not funding is adequate. Whatever the extent of scarcity, however, allocation decisions have to be made, and questions arise about who should make the decisions and on what criteria. There are also fundamental questions about the kind of system within which health care is delivered and what the goals of a system should be.

The Provision and Goals of Health Care

The first issue concerns national versus private insurance. One argument for national health insurance is that it is inherently more fair and equitable than private insurance, which is more likely to be available to the wealthier elements of society.

Government health policies have the overriding goals of ensuring that principles of public service replace commercial profit objectives for all publicly provided health and disability services (Lopez, 1997). In a private health insurance system, it is not only the poorer sections of society that are likely to lose out but also groups with particular needs, such as persons with a mental illness or mental handicap. Lopez proceeds to argue that the private option may be seen as a breach of solidarity—"The willingness of citizens to contribute to the welfare of others in order to improve the general level of well-being" (p. 21)—and that a universal system, in contrast, "has the capacity to place a major emphasis on health prevention and health promotion in the widest sense and to develop not just effective community participation but community control" (p. 23).

Even within a nationally funded system, however, there may be dispute about what the goals of the service should be. For example, in the United Kingdom, Culyer (1997) has argued that a principal objective of the National Health Service is maximizing the health of the whole community, and that this not only is the case but that it should be the case. Health is in general a necessary condition of flourishing, which is an ultimate good. Harris (1997), however, has criticized this position on the grounds that maximizing the health of the whole community might have the effect of systematically disadvantaging some sections of it; for example, aggregate health might be significantly improved by allowing some people to die. Culyer's position is also subject to the objection that the National Health Service has not in fact been a "health" service but an "illness" service.

The National Health Service is based on three principles: Health care should be available to all on the basis of need, it should be comprehensive, and it should be free at the point of delivery. Weale (1998) has drawn attention to the fact that it has been criticized for not being of high quality (while pointing out that this criticism tends to depend on a particular view about what is considered high quality). He examines the "inconsistent triad" of comprehensiveness, high quality, and availability to all. Although the National Health Service is criticized for sacrificing quality for comprehensiveness and availability, the United States has gone for quality at the expense of availability to all (Weale, 1998).

One solution to the inconsistent triad problem, as Weale (1998) points out, is to sacrifice comprehensiveness. It is claimed that universality in health care systems has failed (Lopez, 1997). Increased demand fueled by new technologies, an increase in life expectancy, and rising expectations have coincided with the advent of new health problems and growth in other pressures on the welfare budgets of governments. Hence, there is a realization that it is necessary to seek methods of allocating resources in the health care setting by giving priority either to some patients or some treatments. Therefore, even if it is accepted that national health insurance should be in place, in light of the problem of expanding demand many strategies have arisen to cope with this problem. One option is a two-tier system of a basic package nationally funded plus the option to buy private insurance. Other strategies include selection of programs of populations, but as Lopez points out,

> In Western industrialized countries, where the issue of access has been addressed through a variety of national programs, all of which provide nearly universal entitlement . . . recent cost containment efforts have focused on regulatory strategies. However, these strategies do not focus on restricting access through the creation of restrictive eligibility criteria. . . . Equitable access continues as the main issue in these programs. (p. 22)

Lopez suggests that the aim of a universal system is not only to provide equity in access but also to facilitate equity in health status. This, however, is precisely one of the focal points of contest in the interpretation of equity in this context. Before proceeding to discuss competing accounts of equity, however, it is necessary to examine where the relevant decisions about distribution are made.

Regarding a core or basic package, Weale (1998) suggests that one problem with this as a strategy is that the treatments suggested for exclusion tend to be on the margins anyway. Regarding the suggestion of excluding some patients, it might be argued that access to health care is a fundamental right. It could further be argued that the distinction is not entirely clear-cut because to exclude some treatments is *ipso facto* to exclude some patients.

A Right to Health Care

There are various problems with the approach of arguing for a universal right to health care or, in the context of a given society, for a right to health care for all members of that society. First, the notion of rights is very complex, and there is still dispute over whether welfare rights, as opposed to freedom rights, are genuine. There are arguments to the effect that those with a skill, such as the provision of health care, have a freedom right to sell it where they will. Second, even if it is accepted that there is a welfare right to health care, this does not resolve the question regarding what the content of this right is and of how extensive the provision should be.

At What Level Should Decisions Be Made and Who Should Make Them?

A distinction is commonly drawn between macro-, meso-, and micro levels of decision making. At the macrolevel the government sets the global budget for health service provision; at the mesolevel local planners determine what services are to be offered and how much budget each service will receive. At the microlevel individual clinicians determine what interventions to offer to which patients and which patients should be referred to consultants. Similarly, at each level there are ethical issues to be discussed.

At the macrolevel the debate frequently turns on how large central or national spending should be as opposed to private arrangements. On the one hand, there are arguments that, among the social goods, health care is so central in importance that it should correspondingly be given increased funding, if necessary at the expense of other social goods (Fig. 1). On the other hand, there is a view that extra funding does not solve the problem: A reduction of ineffective and wasteful practice is required.

At the mesolevel the debate is much more complicated. The local decision makers are managers, who by definition have to take the most effective steps to achieve the goals of the service. Making decisions at the local level between services, even if not overtly between patients, has clear implications for those who suffer from conditions that are not regarded as high priority and thus indirectly affects particular groups of patients. The considerations to be taken into account include effectiveness, cost, ethics, the particular conditions in their locality in terms of disease patterns, which may be associated with factors such as population mix and poverty. In addition, there is the matter of accountability to the local population.

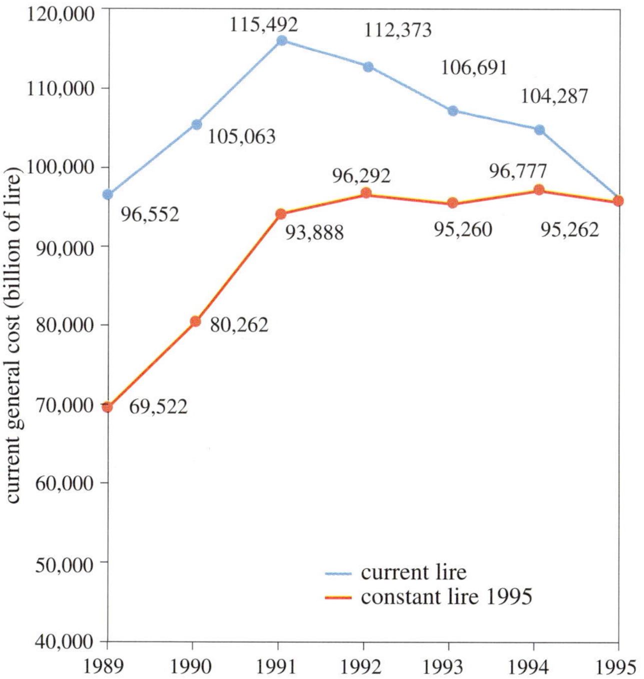

FIGURE 1 Total costs of the Italian national health service, expressed in real and constant lire. The constant values were obtained by transformation, using the ISTAT cost of living index, of the real lire values into lire values in 1995.

At the microlevel clinicians make allocation decisions about particular patients. At this level, one of the most interesting aspects of the debate turns on the purported conflict between what is held to be the traditional ethos of medicine and the necessity to allocate resources. There is a view that doctors should not be in the business of deciding between patients — that they have an obligation to do what is best for each and every patient in front of them. On the other hand, it is argued that in fact doctors have always made allocation decisions such as whether or not to refer onwards; it is simply that now the process is of necessity explicit where previously it was implicit.

There remains some discussion, however, regarding the level at which decisions should be made and whether there should be national guidelines on criteria so that clinicians, for example, are not left to shoulder the burden of priority setting. Light (1997) has put forward powerful arguments for the view that the real issues which ethics needs to deal with in this context are the political and social arrangements which force rationing decisions "downstream" to the point where doctors treat patients. That clinicians will continue to have to make allocation decisions, however, seems inevitable, although the criteria might be a matter for broader agreement, and in this connection various strategies have been proposed for consulting the public.

Why Consult the Public?

There have been well-publicized attempts to consult the public regarding allocation of health care resources, such as the much discussed experiment in Oregon. Such attempts run into the *democracy problem* (Daniels, 1993), i.e., we may not like the results. One study has shown the public to be less concerned about issues of equity, for example, than are health care professionals (Stronks *et al.,* 1997). Therefore, the question arises as to the theoretical justification for consulting the public in this way. It is not the case that the public is consulted on allocation decisions in other spheres.

As mentioned previously, one reason might be the special importance given to health. It is not just health care that has an influence on health, however. Decisions on traffic and on environmental planning can have just as much effect on people's health status. Therefore, the special importance of health cannot be the answer. Another way of approaching it might be to view the rationale for consulting the public as an educational tool. In countries in which there has been national health service provision, an expectation has grown that there will continue to be universal health coverage. Consulting the public about priorities in allocation is one way of bringing home the point that in the light of spiraling health care costs, a continuation of this is simply not feasible. However, even in countries in which universal health coverage has not been the norm, there has been a recognition of a need to contain costs, the United States being the most prominent example.

Another reason for consulting the public might be a theoretical one, based on a communitarian vision of ethics. As Zwart (1993) pointed out, this is relevant with regard to the question of setting limits to medicine. He contrasts the liberal perspective, which has tended to emphasize not only personal autonomy but also a right to choose to have health care interventions, with the communitarian perspective which, according to Zwart, should aim at developing a consensus on the goals of medicine. From this point of view, "the moral agent should not be viewed in an atomistic way, but rather as situated in a moral community from which he derives his moral identity, his substantial moral convictions, and his sense of direction" (pp. 53–54).

Therefore, from the community we derive a consensus on when, in health care, we should intervene and when it is desirable instead to accept the situation. A community consensus, however, is not the same as a majority vote, which could be prejudicial to the interests of minority groups. ten Have (1993) suggests that a communitarian approach is dependent on a "normative, deontological framework defining the meaning of community interests" (p. 45). He defines the presuppositions of the Dutch position, for example, in terms of the fundamental equality of persons, the fundamental need to protect human life, and solidarity.

It might be objected, however, that in a pluralistic, postmodern society, a communitarian consensus of this kind is not possible and that the notion of community is not available (Poole, 1991). van Willigenburg (1993) has further argued that the effectiveness of the community-oriented criterion relies on its power to exclude — to silence some voices. In this case, the question arises as to whether it is feasible to construct some kind of consensus (Moon, 1993) through ongoing dialogue between different groups. Bioethics is arguably a field in which, partly because of the special importance of health and health care, consensus is particularly desirable.

There is a problem, of course, with regard to the size of the community — whether what is envisaged is local, regional, or national. Doyal and Gough (1991) argued,

> Any local, community-based, small-scale form of need satisfaction can foster "insider" conceptions of human need and inhibit the growth of generalizable notions based on a wider collective identity. . . . The dream of a community politics which could unite different groups . . . cannot be realized in the absence of precisely such a cross-cultural and cross-group source of identity as human need. (pp. 308–309)

For the purposes of a given health care system this may be too ambitious, but there are also issues regarding the global distribution of health care resources (Attfield, 1990).

Allocation Decisions and Biomedical Ethics

The contrasting theoretical frameworks of liberalism and communitarianism will have implications for the principles of allocation that are thought acceptable and desirable and for the relationship of principles of priority setting to biomedical ethics as commonly interpreted (Fig. 2). It might be thought that medical ethics is traditionally deontological, especially insofar as it has been reflected in professional codes, which may be duty based, emphasizing the duty of the professional to protect the interests of the individual patient, thus stressing the rights of the individual patient (Fig. 3). The principle of the sanctity of life and the principle of rescue — that when something can be done acutely to avoid death, it should be done — might be thought of in this connection. This type of approach is individualistic and arguably ill suited to situations in which it is necessary to adjudicate between the competing interests of different individuals. Consequentialism might appear to offer more in this regard, but this depends on what kind of consequentialism is offered. How do we judge which consequences are desirable? This too might be interpreted in terms of promoting the benefit of the individual patient rather than a utilitarian version which adopts a maximizing approach with regard to a number of persons.

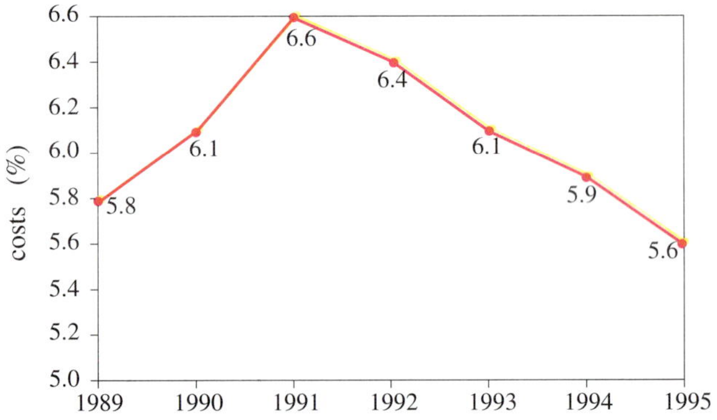

FIGURE 2 Costs of the health service in Italy as a percentage of the gross national product (source of data: Ministry of Health, Programming Department, SISTAN Statistical Office, *I° Compendio del Servizio Sanitario Nazionale Anni 1991–1995,* July 1996).

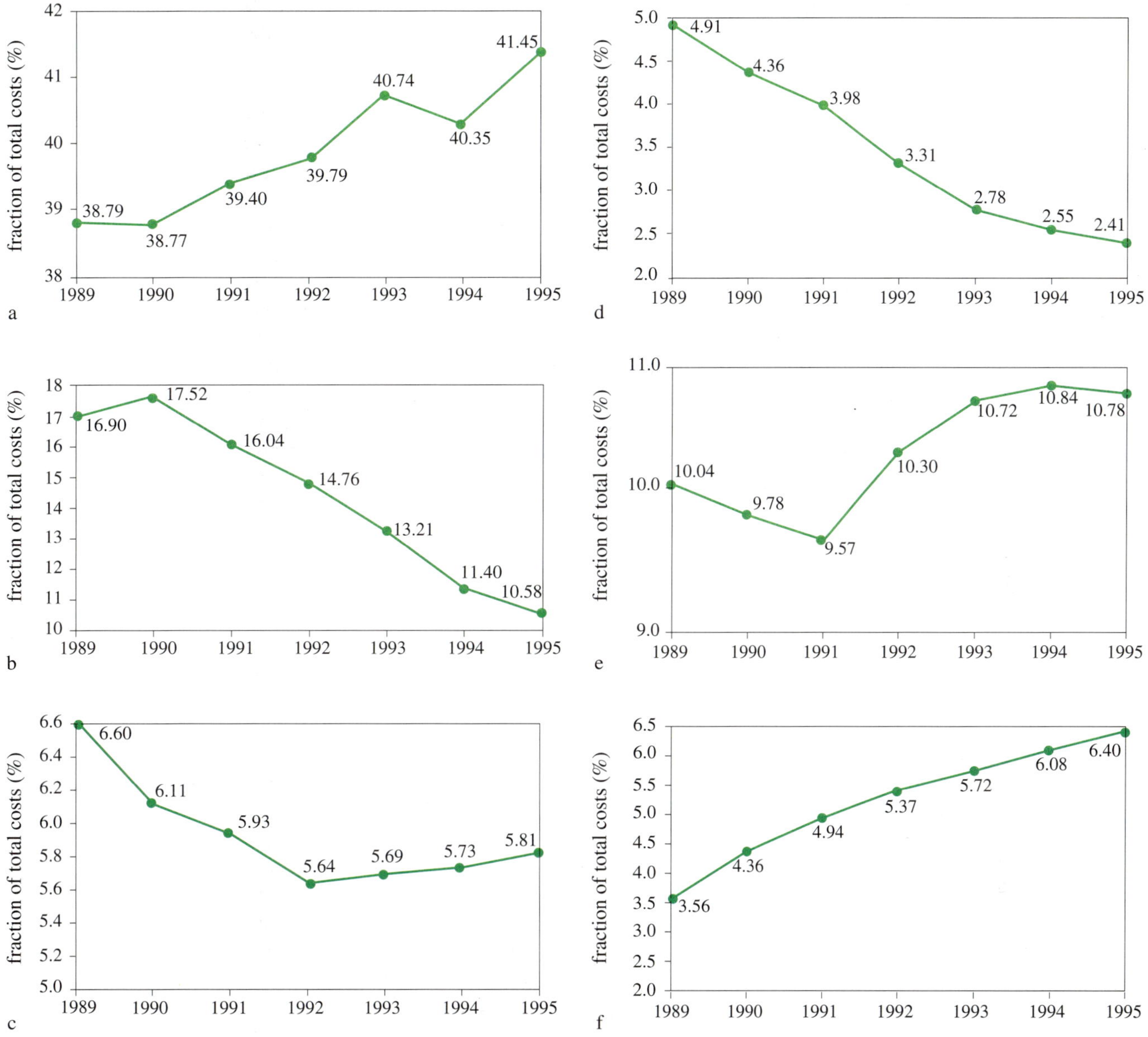

FIGURE 3 Index of some of the main cost items identified from the quarterly reports of the local units of the Italian health service during the period 1989–1995, calculated on the basis of total health service running costs. (a) Personnel. (b) Insured pharmaceutical assistance. (c) General, pediatric, and emergency medical assistance. (d) Specialistic medical assistance. (e) Insured hospital assistance. (f) Other assistance (rehabilitation, prosthetics, thermal cures, assistance to the aged and handicapped, and therapeutic communities) (source of data: Ministry of Health, Programming Department, SISTAN Statistical Office, *I° Compendio del Servizio Sanitario Nazionale Anni 1991–1995,* July 1996).

Of the four principles of biomedical ethics — autonomy, beneficence, nonmaleficence, and justice — the first three are concerned with the interests of the individual patients. Justice takes us toward a consideration of competing interests. Competing theories of social justice, however, make justice an "essentially contested concept" and thereby produce different criteria of distribution. As Daniels (1993) has argued, difficult problems arise precisely because different individuals can appeal to different though perfectly respectable principles to support their claims to health care.

What Are the Criteria to Be Observed?

Although there is a fair amount of consensus that criteria such as efficiency and effectiveness are important, it is with regard to interpretations of conceptions of justice and equity that most of the ethical discussions have taken place. Is it fair to allocate resources according to what people deserve, according to what they need, or according to what is likely to promote the best outcome?

Desert

It is sometimes said that the idea that people should get their just deserts is one of the deepest and most enduring intuitions about justice. In allocating resources, what people deserve provides a way of distinguishing between them. The idea is familiar although not without its rivals in regard to criminal justice, but whether it is appropriate in the health care setting is problematic. One aspect of it is the view that a person's or group's lifestyle may be relevant to whether or not they "deserve" priority in health care; for example, do drink heavily or smoke? If so, then perhaps they are to blame for their own health status and not deserving of priority.

This view, however, is arguably not appropriate for use in priority setting. First, the framework of praise and blame does not carry over well from the context of punishment and reward. Second, there is a problem of where to draw the line, insofar as we all take some risks with our health (e.g., in our habits with regard to exercise).

The punitive aspect of the model arises from the view of the individual as a free chooser responsible for the choices made. This is related to the particular notion of equality embodied in the desert model. People are thought to have equal opportunity to make choices about their own lives. It is showing respect for them as free choosers to evaluate these choices.

These ideas are also linked with that of individual responsibility for health. The problem is that it gives insufficient weight to the social causes of disease and the context in which choices are made. In fact, there is dispute not only regarding whether allocation decisions should be made on the basis of desert but also regarding whether particular decisions should be made on this basis. In some cases, a clinician might defend a decision to withhold a certain intervention on the grounds that the chances of success are slim, whereas this might be interpreted by onlookers as a withholding on the grounds of lifestyle, for example, when the patient has a history of drug misuse.

Need

A competing conception of just or equitable allocation holds that it is just to allocate resources according to need. Like the desert model, this has some intuitive support: Surely a just society would be one that gave individuals what they need.

The need conception is generally considered to be more egalitarian than the desert model. It is operating, however, with a different understanding of equality. This involves making a distinction between equal treatment and treatment as equals. Obviously, it is not the case that equality demands giving everyone equal treatment in the sense of the same prescription, for example, because not everyone needs a given prescription. However, people should be treated as equals in the sense that people with equal needs should be given equal consideration, whatever the individual merits of the person.

However, how is need to be defined? Health care needs have been defined by economists in terms of the capacity to benefit. This transfers the problem of defining need to the problem of what counts as benefit, and individuals, health care professionals, and representatives of the community may have different ideas about this. Individuals may feel benefit if they receive something they want, health care professionals may determine benefit in terms of restoration to or toward normal functioning, and the community may judge the issue by the facilitation of the individual's contribution to society.

If we allow individuals or groups to define their own needs, it may be feared that we will be subject to the problem known as the bottomless pit. Light (1997), however, suggested that the claim that health care needs are bottomless is an empirical question which can be examined by considering what in fact happens in a well-funded system. If health care professionals determine need, for obvious reasons they are likely to be biased in favor of their own specialisms and to perceive more unmet need in those areas. Even if we can solve the definition problem, it may still not be possible to satisfy all identified needs. Then we have to set priorities between needs. One way of doing this is to use an outcome measure.

Utility

Discussion of outcome measures naturally leads to consideration of utilitarianism. According to the utility model, priorities should be set in light of what is expected to lead to the best possible outcome. Like utilitarianism generally as a moral theory, the model as applied in the health care context has two elements: a concept of utility and a principle of maximization. In other words, it has a notion of medical success and also a principle that we should aim for as much as possible. At first, utilitarianism might seem to have little to do with justice, and it has been criticized as a tool for considering health care purchasing decisions on the grounds of its failure to accommodate the notion of the inviolability of the individual (Hunter, 1993). The notion of justice that is incorporated into utilitarianism is that in the course of the utilitarian calculation, each counts for one and no one for more than one. The focus on the outcome, however, distracts attention not only from the pattern of distribution but also from the potential significance of health state before treatment (which is emphasized in principles such as the rescue principle), where we often refer to someone as being in need of help. Nevertheless, the practical impossibility of satisfying all needs that could be identified has led to the perception of a suitable outcome measure being considered desirable.

Outcome Measures

Edgar *et al.* (1995) summarized the economic arguments for an outcome measure such as the quality adjusted life-year (QALY) as follows: Given that resources for producing health gains are scarce, that prioritizing is inevitable, and that prioritizing on the basis of explicit criteria is preferable to prioritizing without, then maximizing the health gain from available resources is a (if not the only) good criterion. The issue of whether prioritizing is inevitable was discussed previously. That explicit criteria are preferable to implicit criteria, however, is not accepted by everybody. There is an alternative view that prioritizing has always been done implicitly by health care practitioners and that this was not undesirable because, among other considerations, it caused less anxiety to patients.

Edgar *et al.* (1995) further point out that a satisfactory quality of life measure must be reliable (consistent), valid in its ability to measure what it is intended to measure, and responsive (i.e., sensitive to changes that may be small but clinically important). They argue that although QALYs have been subject to many criticisms, such as their potential to be discriminatory or that they ignore individual needs, there are ways in which these criticisms can be addressed. For example, QALY and needs approaches may be considered as complementary rather than in competition:

> The emphasis that a philosophy of need places upon normal human functioning, and thus upon health as a precondition of normal life, may provide material for determining those parameters of health that are most relevant to the delineation of quality of life. Conversely, as Williams suggests, QALYs may serve as effective instruments to measure the degree of gravity and substitutability of needs claims, and to resolve disputes between two or more claimants over a limited resource. (p. 69).

Despite such attempts to reconcile, at least partly, competing conceptions of justice, Sass (1995) has argued that "traditional forms of distributive justice will always be of limited success, in part because of human failure, but also in part because full fledged justice will require uniforming recognized needs and services."

The Relevance of Age

QUALYs have frequently been criticized on the grounds that they are ageist because distributing resources to younger people has a greater likelihood of producing a larger health gain in terms of numbers of QUALYs. The idea that age might be relevant in allocating resources has been a matter of considerable controversy. However, age might be agreed to be relevant with reference to specific conditions and interventions because some forms of intervention are simply not worth doing at certain ages. An example of this is breast cancer screening, which is said to be ineffective if not actually potentially harmful in younger women. More controversial is the debate about the provision of fertility treatment to postmenopausal women. What is mainly at issue here, however, is the suggestion that age might be relevant insofar as individuals over a certain age should not receive priority in health care resource allocation generally. The purported justifications for this will be outlined and may be aligned to a greater or lesser degree with more fundamental positions described previously.

The Fair Innings Argument

The fair innings argument has been outlined by Harris (1985) as follows:

> The fair innings argument takes the view that there is some span of years that we consider a reasonable life, a fair innings. Let's say that a fair share of life is the traditional three score and ten, 70 years. Anyone who does not reach 70 suffers, in this view, the injustice of being cut off in their prime. They have missed out on a reasonable share of life: They have been short-changed. Those, however, who do make 70 suffer no such injustice, they have not lost out but rather must consider any additional years a sort of bonus beyond that which could reasonable be hoped for. The fair innings argument requires that everyone be given an equal chance to have a fair innings, to reach the appropriate threshold but, having reached it, they have received their entitlement. The rest of their life is the sort of bonus which may be canceled when this is necessary to help others reach the threshold. (p. 91)

The fair innings argument embraces an idea of equality of opportunity. Each person should have equal opportunity to have a fair innings, but beyond that their claims do not carry equal weight.

The Natural Life Span Argument

An alternative argument for considering age to be relevant was proposed by Callahan in *Setting Limits* (1987). This is based on the notion of a natural life span:

> I will first propose a definition of a tolerable death. If we could agree on that, we would then have, in effect, the basis for a correlative idea of a natural life span; and thus, perhaps, the foundation for an appropriate goal for medicine in this approach to aging. My definition of a "tolerable death" is this: The individual event

> of death at that stage in a life span when (a) one's life possibilities have on the whole been accomplished; (b) one's moral obligations to those for whom one has responsibility have been discharged; and (c) one's death will not seem to others an offense to sense or sensibility, or tempt others to despair and rage at the finitude of human existence. The "natural life span" argument is an interpretation of the need criterion of distribution. It is an interpretation of need in terms of what is "natural" where natural is understood not biologically but biographically. The goal of a publicly funded health care system should not be to prolong life beyond this stage. This requires some rethinking on the part of the community about both the meaning of life and the tolerability of death.

Callahan's (1987) argument may be associated with the need conception of justice in priority setting. Insofar as it involves an interpretation of "need," this has to be defined at the community level rather than in terms of the individual. He says

> Need will not be a manageable idea . . . unless we forthrightly recognize that it is only in part an empirical concept. . . . It will also be a reflection of our values, what we think people require for an acceptable life. In the case of the aged, I have proposed that our ideal of old age should be achieving a life span that enables each of us to accomplish the ordinary scope of possibilities that life affords, recognizing that this may encompass a range of time rather than pointing to a precise age. On the basis of that ideal, the aged would need only those resources which would allow them a solid chance to live that long and, once they had passed that stage, to finish out their years free of pain and avoidable suffering. I will, therefore, define need in the old as primarily to achieve a natural life span and thereafter to have their suffering relieved.

In addition to the fair innings and natural life span arguments, it has already been indicated that principles of allocation based on outcomes measures such as the QALY have been accused of being ageist. The question is whether arguments of the previous kinds should be supported or whether there are antiageist considerations that justify the view that age should be irrelevant in decision making about allocation. Antiageists argue that to use age as a criterion, independently of any other considerations, is as unjust as using sex and race as criteria of allocation, and that the equal interests of people in having their life continue should be given equal weight. Harris (1997) points out, however, that although there is an important insight in the latter consideration, there may be greater injustice done to a person who is not given the chance to reach a fair innings. In the case of the application of QALYs, denial of prioritization is not in fact on the grounds of age alone but on the grounds of likely success and life expectancy.

Virtue Ethics

As previously discussed, there is a view that to examine the problem of resource allocation from the point of view of conceptions of justice is intractable. The project of finding a communitarian consensus also has its critics. Sass (1995) argued that in the old way of viewing health care systems, (communitarian) solidarity was appropriate as a principle. In the present, changing circumstances a triad of principles has arisen, at least in Europe: self-responsibility, solidarity, and subsidiarity. At the level of the basic package, solidarity with those in need and distributive justice remain appropriate. Beyond that, subsidiarity suggests that "whatever the individual can do, should not be done by the state or by social institutions or uniform services." This facilitates the notion of personal responsibility.

However, here responsibility should not be understood in terms of the desert model of justice but in terms of virtue ethics. According to Sass (1995), under the old system a patient's moral virtues were interpreted largely in terms of compliance and hope. Under the new system, the patient is to be encouraged to an attitude of acceptance and to being responsible. This "encouragement" would presumably require education and programs of public awareness and suggests the need for answers to the questions concerning what it means to live a human life, such as are examined in the work of Callahan.

This turn toward the virtuous patient need not depend on any particular view about the level at which decisions are made—whether this falls to a consensus of "local voices," national guidelines, or individual clinicians. In the latter case especially, it would arguably be desirable for a virtue ethic to be adopted for medical ethics as well. Just what the virtues in question would be requires exploration, but they are likely to include integrity and compassion.

Conclusions

Weale (1998) pointed out that value conflicts inevitable occur in this area, and this applies not only between different conceptions of justice but also between the very notion of rationing and traditional views of medical ethics, between economists' and others' definitions of "need" and "benefit," and between liberal and communitarian conceptions of political philosophy. The introduction of virtue ethics into the debate adds a further complicating factor. Our attention also needs to be drawn to the questions of politics, power, and influence that underlie these controversies and the ultimate questions that have to be addressed concerning the nature and function of a health system, which lead to further philosophical issues concerning how we should define health and its relationship to a meaningful life.

Acknowledgments

Portions of this article draw on my earlier work and on work undertaken for the National Health Service Working Party on Priority Setting (1996).

References Cited

ATTFIELD, R. (1990). The global distribution of health care resources. *J. Med. Ethics* **16,** 153–156.

CALLAHAN, D. (1987). *Setting Limits: Medical Goals in an Aging Society.* Simon & Schuster, New York.

CULYER, A. J. (1997). The rationing debate: Maximising the health of the whole community. The case for. *Br. Med. J.* **314,** 667–669.

DANIELS, N. (1993). Just rationing: Programmatic considerations. *Bioethics* **7,** 224–233.

DOYAL, L., and GOUGH, I. (1991). *A Theory of Human Need.* Guilford, New York.

EDGAR, A., *et al.* (1995). The ethical QALY: An investigation of the ethical implications of measures of quality of life applicable to a range of diseases and health states for use in the allocation of resources in prevention diagnosis and treatment. A report to the Commission of the European Communities, Centre for Applied Ethics, Cardiff, UK.

HARRIS, J. (1985). *The Value of Life.* Routledge & Kegan Paul, London.

HARRIS, J. (1997). The rationing debate: Maximising the health of the whole community. The case against: What the principal objective of the NHS should really be. *Br. Med. J.* **314,** 669–672.

HAVE, H. TEN (1993). Choosing core services in The Netherlands. *Health Care Anal.* **1,** 43–47.

HUNTER, D. J. (1993). Rationing and health gain. *Crit. Public Health* **4,** 27–33.

LIGHT, D. W. (1997). The real ethics of rationing. *Br. Med. J.* **315,** 112–115.

LOPEZ, J. (1997). Latin American health care reforms at the crossroads: An introduction. *Eurohealth* **3**(3) 21–23.

MOON, J. D. (1993). *Constructing Communitiy: Moral Pluralism and Tragic Conflicts.* Princeton Univ. Press, Princeton, NJ.

POOLE, R. (1991). *Morality and Modernity.* Routledge, London.

SASS, H. M. (1995). The new triad: Responsibility, solidarity and subsidiarity. *J. Med. Philos.* **20,** 587–594.

STRONKS, K., STRIJBIS, A. M., WENDTE, J. F., and GUNNING-SCHEPERS, L. J. (1997). Who should decide? Qualitative analysis of panel data from public, patients, healthcare professionals, and insurers on priorities in health care. *Br. Med. J.* **315,** 92–96.

WEALE, A. (1998). Rationing health care: A logical solution to an inconsistent triad. *Br. Med. J.* **316,** 410.

WILLIGENBURG, T. VAN (1993). Communitarian illusions: Or why the Dutch proposal for setting priorities in health care must fail. *Health Care Anal.* **1,** 49–52.

ZWART, H. (1993). Rationing in The Netherlands: The liberal and communitarian perspective. *Health Care Anal.* **1**(1), 53–56.

General Reference

BEAUCHAMP, T. L., and CHILDRESS, J. F. (1994). *Principles of Biomedical Ethics,* 4th ed. Oxford Univ. Press, New York.

SUBJECT INDEX

*Number in **bold** indicates volume number.*

B

E

G

H

J

N

O

P

Q

R

T

ISBN 0-12-077344-9